EDITION 7

APPLIED MATHEMATICS
FOR THE MANAGERIAL, LIFE, AND SOCIAL SCIENCES

SOO T. TAN

STONEHILL COLLEGE

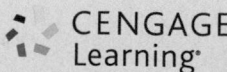
CENGAGE
Learning·

Australia · Brazil · Mexico · Singapore · United Kingdom · United States

CENGAGE
Learning®

Applied Mathematics for the Managerial, Life, and Social Sciences, **Seventh Edition**
Soo T. Tan

Vice President, General Manager: Balraj Kalsi

Product Directors: Richard Stratton and Terrence Boyle

Product Manager: Rita Lombard

Content Developer: Erin Brown

Product Assistant: Katharine Werring

Media Developer: Andrew Coppola

Marketing Manager: Julie Schuster

Content Project Manager: Cheryll Linthicum

Art Director: Vernon Boes

Manufacturing Planner: Becky Cross

Production Service: Martha Emry BookCraft

Photo and Text Researcher: Lumina Datamatics

Copy Editor: Barbara Willette

Illustrator: Jade Myers, Matrix Art Services; Graphic World, Inc.

Text Designer: Diane Beasley

Cover Designer: Irene Morris

Cover Image: Background: imagotres/iStock Vectors/Getty Images; © Ozerina Anna/ Shutterstock.com; Bacteria: CLIPAREA/ Custom Media; Crosswalk light: © cooperr/ Shutterstock.com; ATM: © Aleksandr Kurganov/Shutterstock.com

Compositor: Graphic World, Inc.

Library of Congress Control Number: 2014949639

ISBN: 978-1-305-10790-8

Cengage Learning
20 Channel Center Street
Boston, MA 02210
USA

Cengage Learning is a leading provider of customized learning solutions with office locations around the globe, including Singapore, the United Kingdom, Australia, Mexico, Brazil, and Japan. Locate your local office at **www.cengage.com/global**.

Cengage Learning products are represented in Canada by Nelson Education, Ltd.

To learn more about Cengage Learning Solutions, visit **www.cengage.com**.

Purchase any of our products at your local college store or at our preferred online store **www.cengagebrain.com**.

Printed in the United States of America
Print Number: 01   Print Year: 2014

See the complete Index of Applications at the back of the text to find out more ...

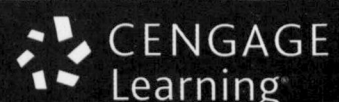

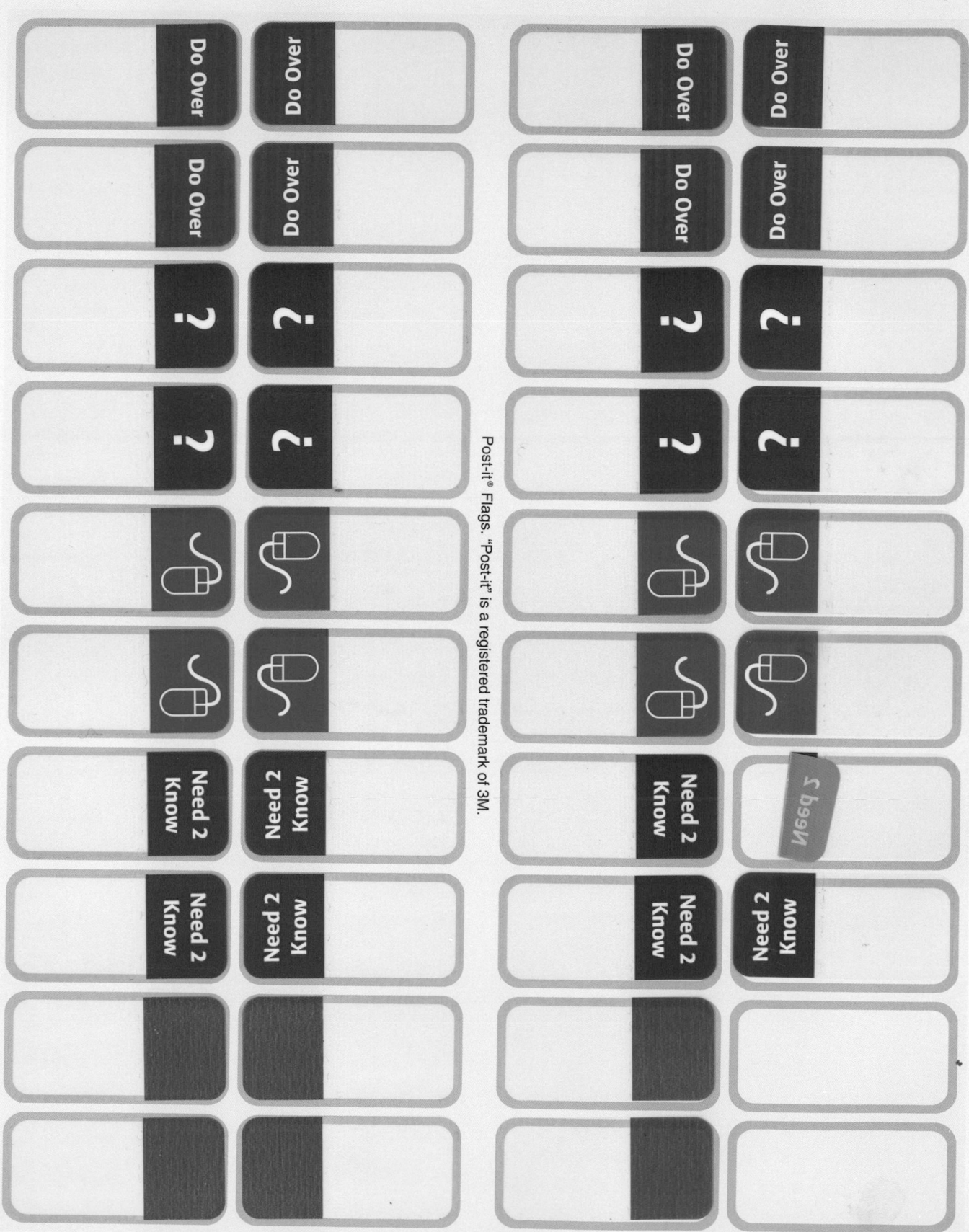

Post-it® Flags. "Post-it" is a registered trademark of 3M.

Tab it. Do it. Ace it.

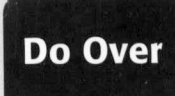

Do Over

Do you need to review something? Try again? Work it out on your own after class? Tab it.

?

Got a question for office hours? Do you need to review an example on your own to get a full understanding? Do you need to look something up before moving on? Tab it.

Do you need to see the video? Check out an online source? Complete your online homework? Tab it.

Need 2 Know

Is this going to be on the test? Need to mark a key formula? Do you need to memorize these steps? Tab it.

Do you have your own study system? Do you need to make a note? Do you want to express yourself? Tab it.

Tab it. Do it. Ace it.

ISBN 13: 978-0-495-55855-2
ISBN 10: 0-4955-5855-9

TO PAT, BILL, AND MICHAEL

About the Author

 © Cengage Learning

SOO T. TAN received his S.B. degree from Massachusetts Institute of Technology, his M.S. degree from the University of Wisconsin–Madison, and his Ph.D. from the University of California at Los Angeles. He has published numerous papers in optimal control theory, numerical analysis, and mathematics of finance. He is also the author of a series of calculus textbooks.

CONTENTS

CHAPTER 12   Calculus of Several Variables   899

PREFACE

Math plays a vital role in our increasingly complex daily life. *Applied Mathematics for the Managerial, Life, and Social Sciences* attempts to illustrate this point with its applied approach to mathematics. Students have a much greater appreciation of the material if the applications are drawn from their fields of interest and from situations that occur in the real world. This is one reason you will see so many exercises in my texts that are modeled on data gathered from newspapers, magazines, journals, and other media. In addition, many students come into this course with some degree of apprehension. For this reason, I have adopted an intuitive approach in which I try to introduce each abstract mathematical concept through an example drawn from a common life experience. Once the idea has been conveyed, I then proceed to make it precise, thereby ensuring that no mathematical rigor is lost in this intuitive treatment of the subject.

This text offers more than enough material for a two-semester or three-semester course. The following chart on chapter dependency is provided to help the instructor design a course that is most suitable for the intended audience.

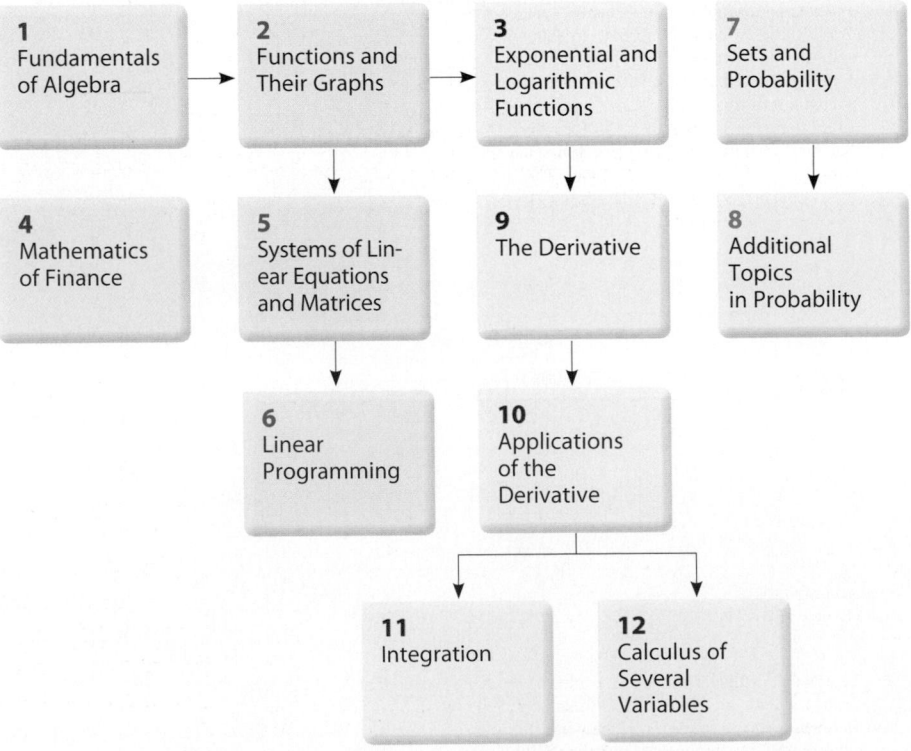

Note that in this text exponential and logarithmic functions are treated early.

x

The Approach

Presentation

Consistent with my intuitive approach, I state the results informally. However, I have taken special care to ensure that mathematical precision and accuracy are not compromised.

Problem-Solving Emphasis

Special emphasis is placed on helping students formulate, solve, and interpret the results of applied problems. Because students often have difficulty setting up and solving word problems, extra care has been taken to help them master these skills.

- Very early in the text, students are given guidelines for setting up word problems (see Section 2.7). This is followed by numerous examples and exercises to help students master this skill.
- One entire section is devoted to modeling and setting up linear programming problems (see Section 6.2).
- First, in Section 10.4, techniques of calculus are used to solve optimization problems in which the function to be optimized is given. Later, in Section 10.5, optimization problems that require the additional step of formulating the problem are treated.

Absolute Extrema on a Closed Interval

As the preceding examples show, a continuous function defined on an arbitrary interval does not always have an absolute maximum or an absolute minimum. But an important case arises often in practical applications in which both the absolute maximum and the absolute minimum of a function are guaranteed to exist. This occurs when a continuous function is defined on a *closed* interval.

Before stating this important result formally, let's look at a real-life example. The graph of the function f in Figure 60 shows the average price, $f(t)$, in dollars, of domestic airfares by days before flight. The domain of f is the closed interval $[-210, -1]$, where -210 is interpreted as 210 days before flight and -1 is interpreted as the day before flight.

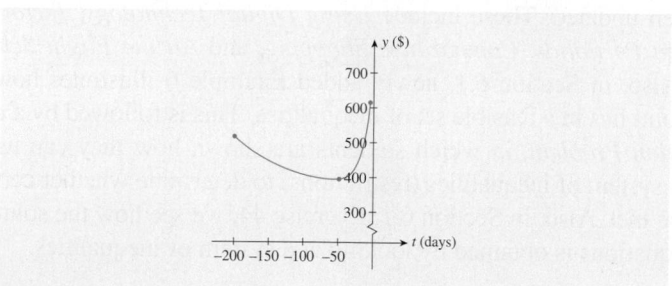

FIGURE **60**
Average price before flight
Source: Cheapair.com.

Observe that f attains the minimum value of 395 when $t = -49$ and the maximum value of 614 when $t = -1$. This result tells us that the best time to book a domestic flight is seven weeks in advance and the worst day to book a domestic flight is the day before the flight. Probably most surprising of all, booking too early can be almost as expensive as booking too late. Note that the function f is continuous on a closed interval. For such functions, we have the following theorem.

Motivation

Illustrating the practical value of mathematics in applied areas is an objective of my approach. Concepts are introduced with concrete, real-life examples wherever appropriate. These examples and other applications have been chosen from current topics and issues in the media and serve to answer a question often posed by students: "What will I ever use this for?" In this new edition, for example, the concept of finding the absolute extrema over a closed interval is introduced as shown at left.

Modeling

One important skill that every student should acquire is the ability to translate a real-life problem into a mathematical model. In Section 2.7, the modeling process is discussed, and students are asked to use models (functions) constructed from real-life data to answer questions. Additionally, students get hands-on experience constructing these models in the Using Technology sections.

New to this Edition

The focus of this revision has been the continued emphasis on illustrating the mathematical concepts in *Applied Mathematics* by using more real-life applications that are relevant to the everyday life of students and to their fields of study in the managerial, life, and social sciences. Over 300 new applications have been added in the examples and exercises. A sampling of these new applications is provided on the inside front cover pages.

Many of the exercise sets have been revamped. In particular, the exercise sets were restructured to follow more closely the order of the presentation of the material in each section and to progress more evenly from easier to more difficult problems in both the rote and applied sections of each exercise set. Additional concept questions, rote exercises, and true-or-false questions were also included.

More Specific Content Changes

Chapters 1 and 2 In Section 1.1, the introduction to real numbers was rewritten for clarity. New rote, concept, and application exercises were also added. In Section 2.2, newly added Example 8b–c illustrates how to determine whether a point lies on a line. Also, a new application, *Smokers in the United States,* has been added to the Self-Check Exercises 2.2. The U.S. federal budget deficit graphs that are used as motivation to introduce "The Algebra of Functions" in Section 2.4 have been updated to reflect the current deficit situation. In Section 2.5, Applied Example 1, *Erosion of the Middle Class,* has been added. Students are shown how the model for this application is constructed in newly added Section 2.8, *The Method of Least Squares.* New models and graphs for the *Global Warming, Social Security Trust Fund Assets,* and *Driving Costs* applications appear in Section 2.7.

Chapters 3 and 4 New application exercises were added to Chapter 3. These include *The Decline of American Idol, Renewable Energy, Bank Failures,* and *Obesity Over Time,* among others. In Chapter 4, interest-rate problems throughout the chapter were revised to reflect the current interest-rate environment. Also, in Section 4.3, two new exercises were added illustrating the new Ability-to-Repay rules for mortgages adopted by the Consumer Financial Protection Bureau in response to the financial crisis.

Chapters 5 and 6 New application exercises have been added, and many examples and exercises have been updated. These include *Using Digital Technology, Terrorism Polls, Model Investment Portfolios, Comparative Shopping,* and *Airline Flight Scheduling,* among others. Also, in Section 6.1, newly added Example 6 illustrates how to determine whether a point lies in a feasible set of inequalities. This is followed by a new application, *A Production Problem,* in which students are shown how they can use a solution set for a given system of inequalities (restrictions) to determine whether certain production goals can be met. Also, in Section 6.1, Exercise 44, we see how the solution of a system of linear equations is obtained by looking at a system of inequalities.

Chapters 7 and 8 These chapters deal with the calculations of probabilities and data analysis. The emphasis here is placed on providing data from marketing, economic, consumer, and scientific surveys that are relevant, current, and of interest to students to motivate the mathematical concepts presented. Topics in these exercises and examples include the greatest challenge on starting a new job, the number of years it will take for people to fully recover from the Great Recession, the most common cause of on-the-job distractions, the number of social media accounts that individuals have, and the financial hardship caused by gas prices.

Also, several new examples were added in these chapters. Example 13 in Section 7.1, *Cyber Privacy,* illustrates set operations. Example 5 in Section 7.2 illustrates how the solution of a system of linear equations can sometimes be used to help draw a Venn diagram. In Section 8.2, Example 8 illustrates the difference between mutually exclusive and independent events. Also, Example 12, *Predicting Travel Weather,* illustrates the calculation of the probability of independent events. Example 8, *Commuting Times,* in Section 8.5, illustrates the calculation of expected value for grouped data; and Example 4 in Section 8.6, *Married Males,* illustrates the calculation of standard deviation for grouped data. Using Technology Section 8.4 was expanded to include an example (Example 3, *Time Use of College Students*) and exercises illustrating how Excel can be used to create pie charts.

Chapters 9 and 10 A wealth of new application exercises has been added throughout these chapters. A new model and graph, *Income of American Households,* have been added as an introduction to exponential models in Section 9.7. This is followed by an analysis of the function describing the graph in the Using Technology exercises. A new subsection on relative rates of change and a new application, *Inflation,* have been added to Section 9.8. In Section 10.1, the U.S. budget deficit (surplus) graph that is used to introduce relative extrema has been updated. In the Section 10.4 exercises, the absolute extrema for this same deficit function are found in Problem 72. Also, in Section 10.4, a new application, *Average Fare Before a Flight,* has been added to introduce the concept of absolute extrema on a closed interval.

Chapters 11 and 12 Many new applications have been added to these chapters. The intuitive discussion of area and the definite integral at the beginning of Section 11.3, as illustrated by the total daily petroleum consumption of a New England state, is given a firmer mathematical footing at the end of the section by demonstrating that the petroleum consumption of the state is indeed given by the area under a curve. The 3-D art in the text and exercises in Chapter 12 has been further enhanced.

Features

Real-World Connections

Motivating Applications

Many new applied examples and exercises have been added in the Seventh Edition. Among the topics of the new applications are Cyber Monday, family insurance coverage, leveraged return, credit card debt, tax refund fraud, salaries of married women, and online video advertising.

20. ONLINE VIDEO ADVERTISING Although still a small percentage of all online advertising, online video advertising is growing. The following table gives the projected spending on Web video advertising (in billions of dollars) through 2016:

Year	2011	2012	2013	2014	2015	2016
Spending, y	2.0	3.1	4.5	6.3	7.8	9.3

a. Letting $x = 0$ denote 2011, find an equation of the least-squares line for these data.

b. Use the result of part (a) to estimate the projected rate of growth of video advertising from 2011 through 2016.

Source: eMarketer.

Portfolios

These interviews share the varied experiences of professionals who use mathematics in the workplace. Among those included are a Vice President of Wealth Management Advisor and an associate at The Mason Group.

PORTFOLIO Robert H. Mason

TITLE Vice President, Wealth Management Advisor
INSTITUTION The Mason Group

The Mason Group—a team of financial advisors at a major wire house firm—acts as an interface between clients and investment markets. To meet the needs of our private clients, it is important for us to maintain constant contact with them, adjusting their investments when the markets and allocations change and when the clients' goals change.

We often help our clients determine whether their various expected sources of income in retirement will provide them with their desired retirement lifestyle. To begin, we determine the after-tax funds needed in retirement in today's dollars in consultation with the client. Using simple arithmetic, we then look at the duration and amount of their various income flows and their current portfolio allocations (stocks, bonds, cash, etc.). Once we have made this assessment, we take into consideration future possible allocations to arrive at a range of probabilities for portfolio valuation for all years up to and including retirement.

For example, we might tell a client that, on the basis of their given investment plan, there is an 80–95% probability that the expected value of his or her portfolio will increase from $1 million to about $1.5 million by the time the client turns 90; a 50–80% chance that it will increase from $1 million to around $2 million; and a 30–50% chance that it will increase to approximately $2.8 million.

To arrive at these probabilities, we use a deterministic model that assumes that a constant annual rate of return is applied to the portfolio every year of the analysis. We also use probabilistic modeling, taking into account various factors such as economic conditions, the allocation of assets, and market volatility. By using these techniques we are able to arrive at a confidence level, without making any guarantees, that our clients will be able to attain the lifestyle in retirement they desire.

Scooter Grubb; (inset) © iStockPhoto.com/Eugene Choi

Explore and Discuss

These optional questions can be discussed in class or assigned as homework. They generally require more thought and effort than the usual exercises. They may also be used to add a writing component to the class or as team projects.

Explore and Discuss

The average price of gasoline at the pump over a 3-month period, during which there was a temporary shortage of oil, is described by the function f defined on the interval $[0, 3]$. During the first month, the price was increasing at an increasing rate. Starting with the second month, the good news was that the rate of increase was slowing down, although the price of gas was still increasing. This pattern continued until the end of the second month. The price of gas peaked at $t = 2$ and began to fall at an increasing rate until $t = 3$.

1. Describe the signs of $f'(t)$ and $f''(t)$ over each of the intervals $(0, 1)$, $(1, 2)$, and $(2, 3)$.
2. Make a sketch showing a plausible graph of f over $[0, 3]$.

Explorations and Technology

Exploring with Technology

These optional discussions appear throughout the main body of the text and serve to enhance the student's understanding of the concepts and theory presented. Often the solution of an example in the text is augmented with a graphical or numerical solution.

Refer to Example 4. Suppose Marcus wished to know how much he would have in his IRA at any time in the future, not just at the beginning of 2014, as you were asked to compute in the example.

1. Using Formula (18) and the relevant data from Example 4, show that the required amount at any time x (x measured in years, $x > 0$) is given by

$$A = f(x) = 40,000(e^{0.05x} - 1)$$

2. Use a graphing utility to plot the graph of f, using the viewing window $[0, 30] \times [0, 200,000]$.

3. Using **ZOOM** and **TRACE**, or using the function evaluation capability of your graphing utility, use the result of part 2 to verify the result obtained in Example 4. Comment on the advantage of the mathematical model found in part 1.

Using Technology

Written in the traditional example-exercise format, these optional sections show how to use the graphing calculator and Microsoft Excel 2010 as a tool to solve problems. Illustrations showing graphing calculator screens and spreadsheets are used extensively. In keeping with the theme of motivation through real-life examples, many sourced applications are included.

A *How-To Technology Index* is included at the back of the book for easy reference to Using Technology examples.

 APPLIED EXAMPLE 3 Time Use of College Students Use the data given in Table 1 to construct a pie chart.

TABLE T1	
Time Used on an Average Weekday for Full-Time University and College Students	
Time Use	**Time (in hours)**
Sleeping	8.5
Leisure and sports	3.7
Working and related activities	2.9
Educational activities	3.3
Eating and drinking	1.0
Grooming	0.7
Traveling	1.5
Other	2.4

Source: Bureau of Labor Statistics.

Solution

We begin by entering the information from Table T1 in Columns A and B on a spreadsheet. Then follow these steps:

Step 1 First, highlight the data in cells A2:A9 and B2:B9 as shown in Figure T4.

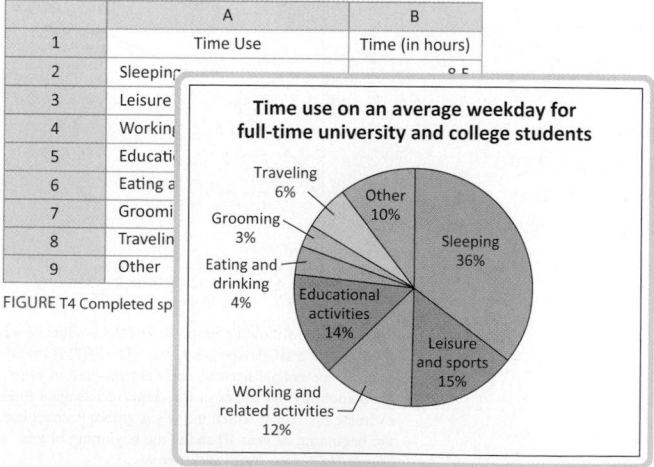

	A	B
1	Time Use	Time (in hours)
2	Sleeping	8.5
3	Leisure	
4	Working	
5	Education	
6	Eating a	
7	Grooming	
8	Traveling	
9	Other	

FIGURE T4 Completed sp

FIGURE T5
The pie chart describing the data in Table T1

Concept Building and Critical Thinking

Self-Check Exercises

Offering students immediate feedback on key concepts, these exercises begin each end-of-section exercise set and contain both rote and word problems (applications). Fully worked-out solutions can be found at the end of each exercise section. If students get stuck while solving these problems, they can get immediate help before attempting to solve the homework exercises. Applications have been included here because students often need extra practice with setting up and solving these problems.

9.3 Self-Check Exercises

1. Let $f(x) = -x^2 - 2x + 3$.
   a. Find the derivative f' of f, using the definition of the derivative.
   b. Find the slope of the tangent line to the graph of f at the point $(0, 3)$.
   c. Find the rate of change of f when $x = 0$.
   d. Find an equation of the tangent line to the graph of f at the point $(0, 3)$.
   e. Sketch the graph of f and the tangent line to the curve at the point $(0, 3)$.

2. **BANK LOSSES** The losses (in millions of dollars) due to bad loans extended chiefly in agriculture, real estate, shipping, and energy by the Franklin Bank are estimated to be
   $$A = f(t) = -t^2 + 10t + 30 \qquad (0 \le t \le 10)$$
   where t is the time in years ($t = 0$ corresponds to the beginning of 2007). How fast were the losses mounting at the beginning of 2010? At the beginning of 2012? At the beginning of 2014?

   Solutions to Self-Check Exercises 9.3 can be found on page 646.

Concept Questions

Designed to test students' understanding of the basic concepts discussed in the section, these questions encourage students to explain learned concepts in their own words.

9.3 Concept Questions

For Questions 1 and 2, refer to the following figure.

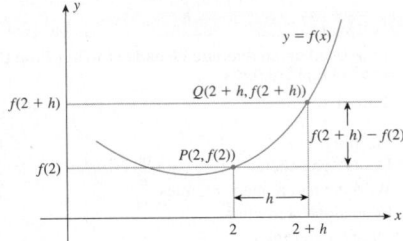

1. Let $P(2, f(2))$ and $Q(2 + h, f(2 + h))$ be points on the graph of a function f.
   a. Find an expression for the slope of the secant line passing through P and Q.
   b. Find an expression for the slope of the tangent line passing through P.

2. Refer to Question 1.
   a. Find an expression for the average rate of change of f over the interval $[2, 2 + h]$.
   b. Find an expression for the instantaneous rate of change of f at 2.
   c. Compare your answers for parts (a) and (b) with those of Question 1.

Exercises

Each section contains an ample set of exercises of a routine computational nature followed by an extensive set of modern application exercises.

9.3 Exercises

1. **AVERAGE WEIGHT OF AN INFANT** The following graph shows the weight measurements of the average infant from the time of birth ($t = 0$) through age 2 ($t = 24$). By computing the slopes of the respective tangent lines, estimate the rate of change of the average infant's weight when $t = 3$ and when $t = 18$. What is the average rate of change in the average infant's weight over the first year of life?

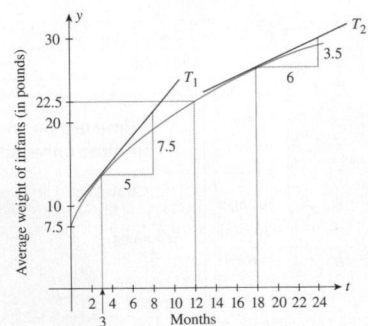

2. **FORESTRY** The following graph shows the volume of wood produced in a single-species forest. Here, $f(t)$ is measured in cubic meters per hectare, and t is measured in years. By computing the slopes of the respective tangent lines, estimate the rate at which the wood grown is changing at the beginning of year 10 and at the beginning of year 30.
   Source: The Random House Encyclopedia.

3. **TV-VIEWING PATTERNS** The following graph shows the percentage of U.S. households watching television during a 24-hr period on a weekday ($t = 0$ corresponds to 6 A.M.). By computing the slopes of the respective tangent lines, estimate the rate of change of the percent of households watching television at 4 P.M. and 11 P.M.
   Source: A. C. Nielsen Company.

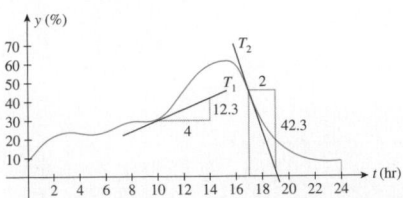

4. **CROP YIELD** Productivity and yield of cultivated crops are often reduced by insect pests. The following graph shows the relationship between the yield of a certain crop, $f(x)$, as a function of the density of aphids x. (Aphids are small insects that suck plant juices.) Here, $f(x)$ is measured in kilograms per 4000 square meters, and x is measured in hundreds of aphids per bean stem. By computing the slopes of the respective tangent lines, estimate the rate of change of the crop yield with respect to the density of aphids when that density is 200 aphids/bean stem and when it is 800 aphids/bean stem.
   Source: The Random House Encyclopedia.

Review and Study Tools

Summary of Principal Formulas and Terms

Each review section begins with the Summary, which highlights the important equations and terms, with page numbers given for quick review.

CHAPTER 2 Summary of Principal Formulas and Terms

FORMULAS

1.	Slope of a line	$m = \dfrac{y_2 - y_1}{x_2 - x_1}$
2.	Equation of a vertical line	$x = a$
3.	Equation of a horizontal line	$y = b$
4.	Point-slope form of the equation of a line	$y - y_1 = m(x - x_1)$
5.	Slope-intercept form of the equation of a line	$y = mx + b$
6.	General equation of a line	$Ax + By + C = 0$

TERMS

Cartesian coordinate system (72)	dependent variable (94)	profit function (119)
ordered pair (72)	graph of a function (95)	break-even point (121)
coordinates (72)	piecewise-defined function (97)	quadratic function (131)
parallel lines (76)	graph of an equation (98)	demand function (134)
perpendicular lines (80)	Vertical Line Test (98)	supply function (135)
function (92)	composite function (110)	market equilibrium (136)
domain (92)	linear function (116)	polynomial function (143)
range (92)	total cost function (118)	rational function (146)
independent variable (94)	revenue function (119)	power function (146)

Concept Review Questions

These questions give students a chance to check their knowledge of the basic definitions and concepts given in each chapter.

CHAPTER 2 Concept Review Questions

Fill in the blanks.

1. A point in the plane can be represented uniquely by a/an _____ pair of numbers. The first number of the pair is called the _____, and the second number of the pair is called the _____.

2. **a.** The point $P(a, 0)$ lies on the _____-axis, and the point $P(0, b)$ lies on the _____-axis.
   b. If the point $P(a, b)$ lies in the fourth quadrant, then the point $P(-a, b)$ lies in the _____ quadrant.

3. **a.** If $P_1(x_1, y_1)$ and $P_2(x_2, y_2)$ are any two distinct points on a nonvertical line L, then the slope of L is $m =$ _____.
   b. The slope of a vertical line is _____.
   c. The slope of a horizontal line is _____.
   d. The slope of a line that slants upward from left to right is _____.

4. If L_1 and L_2 are nonvertical lines with slopes m_1 and m_2, respectively, then L_1 is parallel to L_2 if and only if _____ and L_1 is perpendicular to L_2 if and only if _____.

x takes on all possible values in A is called the _____ of f. The range of f is contained in the set _____.

8. The graph of a function is the set of all points (x, y) in the xy-plane such that x is in the _____ of f and $y =$ _____. The Vertical Line Test states that a curve in the xy-plane is the graph of a function $y = f(x)$ if and only if each _____ line intersects it in at most one _____.

9. If f and g are functions with domains A and B, respectively, then (a) $(f \pm g)(x) =$ _____, (b) $(fg)(x) =$ _____, and (c) $\left(\dfrac{f}{g}\right)(x) =$ _____. The domain of $f + g$ is _____. The domain of $\dfrac{f}{g}$ is _____ with the additional condition that $g(x)$ is never _____.

10. The composition of g and f is the function with rule $(g \circ f)(x) =$ _____. Its domain is the set of all x in the domain of _____ such that _____ lies in the domain of _____.

Review Exercises

Offering a solid review of the chapter material, the Review Exercises contain routine computational exercises followed by applied problems.

CHAPTER 2 Review Exercises

In Exercises 1–6, find an equation of the line L that passes through the point $(-2, 4)$ and satisfies the given condition.

1. L is a vertical line.

2. L is a horizontal line.

3. L passes through the point $\left(3, \frac{7}{2}\right)$.

4. The x-intercept of L is 3.

5. L is parallel to the line $5x - 2y = 6$.

6. L is perpendicular to the line $4x + 3y = 6$.

7. Find an equation of the line with slope $-\frac{1}{2}$ and y-intercept -3.

8. Find the slope and y-intercept of the line with equation $3x - 5y = 6$.

9. Find an equation of the line passing through the point $(2, 3)$ and parallel to the line with equation $3x + 4y - 8 = 0$.

10. Find an equation of the line passing through the point $(-1, 3)$ and parallel to the line joining the points $(-3, 4)$ and $(2, 1)$.

11. Find an equation of the line passing through the point $(-2, -4)$ that is perpendicular to the line with equation $2x - 3y - 24 = 0$.

Before Moving On . . .

Found at the end of each chapter review, these exercises give students a chance to determine whether they have mastered the basic computational skills developed in the chapter.

CHAPTER 2 Before Moving On . . .

1. Find an equation of the line that passes through $(-1, -2)$ and $(4, 5)$.

2. Find an equation of the line that has slope $-\frac{1}{3}$ and y-intercept $\frac{4}{3}$.

3. Let

$$f(x) = \begin{cases} -2x + 1 & \text{if } -1 \le x < 0 \\ x^2 + 2 & \text{if } 0 \le x \le 2 \end{cases}$$

Find (a) $f(-1)$, (b) $f(0)$, and (c) $f\left(\frac{3}{2}\right)$.

4. Let $f(x) = \dfrac{1}{x+1}$ and $g(x) = x^2 + 1$. Find the rules for
(a) $f + g$, (b) fg, (c) $f \circ g$, and (d) $g \circ f$.

5. **POSTAL REGULATIONS** Postal regulations specify that a parcel sent by parcel post may have a combined length and girth of no more than 108 in. Suppose a rectangular package that has a square cross section of x in. $\times$ x in. is to have a combined length and girth of exactly 108 in. Find a function in terms of x giving the volume of the package.
Hint: The length plus the girth is $4x + h$ (see the accompanying figure).

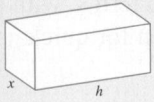

Action-Oriented Study Tabs

Convenient color-coded study tabs make it easy for students to flag pages that they want to return to later, whether for additional review, exam preparation, online exploration, or identifying a topic to be discussed with the instructor.

Instructor Resources

ENHANCED WEBASSIGN® WebAssign

Printed Access Card: 978-1-285-85758-9
Online Access Code: 978-1-285-85761-9

Exclusively from Cengage Learning, Enhanced WebAssign combines the exceptional mathematics content that you know and love with the most powerful online homework solution, WebAssign. Enhanced WebAssign engages students with immediate feedback, rich tutorial content, and interactive, fully customizable e-books (YouBook), helping students to develop a deeper conceptual understanding of their subject matter. Quick Prep and Just In Time exercises provide opportunities for students to review prerequisite skills and content, both at the start of the course and at the beginning of each section. Flexible assignment options give instructors the ability to release assignments conditionally on the basis of students' prerequisite assignment scores. Visit us at **www.cengage.com/ewa** to learn more.

COMPLETE SOLUTIONS MANUAL by Soo T. Tan

Written by the author, the Complete Solutions Manual contains solutions for all exercises in the text, including *Exploring with Technology* and *Explore and Discuss* exercises. The Complete Solutions Manual is available on the Instructor Companion Site.

CENGAGE LEARNING TESTING POWERED BY COGNERO

Cengage Learning Testing Powered by Cognero is a flexible, online system that allows you to author, edit, and manage test bank content from multiple Cengage Learning solutions; create multiple test versions in an instant; and deliver tests from your LMS, your classroom, or wherever you want. Access to Cognero is available on the Instructor Companion Site.

INSTRUCTOR COMPANION SITE

Everything you need for your course in one place! This collection of book-specific lecture and class tools is available online at **www.cengage.com/login**. Access and download PowerPoint presentations, images, solutions manual, videos, and more.

Student Resources

STUDENT SOLUTIONS MANUAL by Soo T. Tan ISBN: 978-1-305-10812-7

Giving you more in-depth explanations, this insightful resource includes fully worked-out solutions for selected exercises in the textbook, as well as problem-solving strategies, additional algebra steps, and review for selected problems.

ENHANCED WEBASSIGN® WebAssign

Printed Access Card: 978-1-285-85758-9

Online Access Code: 978-1-285-85761-9

Enhanced WebAssign (assigned by the instructor) provides you with instant feedback on homework assignments. This online homework system is easy to use and includes helpful links to textbook sections, video examples, and problem-specific tutorials.

CENGAGEBRAIN.COM

Visit **www.cengagebrain.com** to access additional course materials and companion resources. At the CengageBrain.com home page, search for the ISBN of your title (from the back cover of your book) using the search box at the top of the page. This will take you to the product page where free companion resources can be found.

Acknowledgments

I wish to express my personal appreciation to each of the following reviewers of this and previous editions, whose many suggestions have helped make a much improved book.

Paul Abraham
Kent State University—Stark

James Adair
Missouri Valley College

A.J. Alnaser
Trina University

Bill Barge
Trine University

Richard Baslaw
York College of CUNY

Denis Bell
University of North Florida

Jill Britton
Camosun College

Debra D. Bryant
Tennessee Technological University

Debra Carney
University of Denver

Michelle Dedeo
University of North Florida

Scott L. Dennison
University of Wisconsin—Oshkosh

Christine Devena
Miles Community College

Andrew Diener
Christian Brothers University

James Eby
Blinn College—Bryan Campus

Mike Everett
Santa Ana College

Kevin Ferland
Bloomsburg University

Edna Greenwood
Tarrant County College—Northwest Campus

Tao Guo
Rock Valley College

Velma Hill
York College of CUNY

Jiashi Hou
Norfolk State University

Mark Jacobson
Montana State University—Billings

Xingde Jia
Texas State University

Kristi Karber
University of Central Oklahoma

Armando Perez
Laredo Community College

Sarah Kilby
North Country Community College

Mohammed Rajah
Miracosta College

Rebecca Leefers
Michigan State University

Dennis H. Risher
Loras College

Murray Lieb
New Jersey Institute of Technology

Brian Rodas
Santa Monica College

James Liu
James Madison University

Dr. Arthur Rosenthal
Salem State College

Lia Liu
University of Illinois at Chicago

Abdelrida Saleh
Miami Dade College

Bin Lu
California State University—Sacramento

Stephanie Anne Salomone
University of Portland

Rebecca Lynn
Colorado State University

Mohammed Siddique
Virginia Union University

Theresa Manns
Salisbury University

Donald Stengel
California State University—Fresno

Mary T. McMahon
North Central College

Jennifer Strehler
Oakton Community College

Daniela Mihai
University of Pittsburgh

Beimnet Teclezghi
New Jersey City University

Kathy Nickell
College of DuPage

Ray Toland
Clarkson University

Carol Overdeep
Saint Martin's University

Justin Wyss-Gallifent
University of Maryland at College Park

Michael Paulding
Kapiolani Community College

I also wish to thank Tao Guo for the superb job he did as the accuracy checker for this text and the *Complete Solutions Manual* that accompanies the text. I also thank the editorial, production, and marketing staffs of Cengage Learning—Richard Stratton, Rita Lombard, Erin Brown, Andrew Coppola, Cheryll Linthicum, and Vernon Boes—for all of their help and support during the development and production of this edition. I also thank Martha Emry and Barbara Willette who both did an excellent job of ensuring the accuracy and readability of this edition. Simply stated, the team I have been working with is outstanding, and I truly appreciate all their hard work and efforts.

S. T. Tan

1 FUNDAMENTALS OF ALGEBRA

THIS CHAPTER CONTAINS a brief review of the algebra you will use in this course. In the process of solving many practical problems, you will need to solve algebraic equations. You will also need to simplify algebraic expressions. This chapter also contains a short review of inequalities and absolute value; their uses range from describing the domains of functions to formulating applied problems.

How much money is needed to purchase at least 100,000 shares of the Starr Communications Company? Corbyco, a giant conglomerate, wishes to purchase a minimum of 100,000 shares of the company. In Example 8, page 60, you will see how Corbyco's management determines how much money they will need for the acquisition.

1.1 Real Numbers

The Real Numbers

We use *real numbers* every day to describe various quantities such as temperature, salary, annual percentage rate, shoe size, and grade point average. Some of the symbols we use to represent real numbers are

$$3, \quad -17, \quad \sqrt{2}, \quad 0.666\ldots, \quad 113, \quad 3.9, \quad 0.12875$$

To construct the real numbers, we start with the **natural numbers** (also called counting numbers)

$$1, 2, 3, 4, \ldots$$

and adjoin other numbers to them. The **whole numbers**

$$0, 1, 2, 3, \ldots$$

are obtained by adjoining the single number 0 to the natural numbers. By adjoining the negatives of the natural numbers to the whole numbers, we obtain the **integers:**

$$\ldots, -3, -2, -1, 0, 1, 2, 3, \ldots$$

Next, we consider the **rational numbers,** numbers of the form $\frac{a}{b}$, where a and b are integers, with $b \neq 0$. Observe that every integer is a rational number, since each integer may be written in the form $\frac{a}{b}$, with $b = 1$. For example, the integer 6 may be written in the form $\frac{6}{1}$. But a rational number is not necessarily an integer, since fractions such as $\frac{1}{2}$ and $\frac{23}{25}$ are not integers.

Finally, we obtain the real numbers by adjoining the rational numbers to the **irrational numbers**—numbers that cannot be expressed in the form $\frac{a}{b}$, where a and b are integers ($b \neq 0$). Examples of irrational numbers are $\sqrt{2}$, $\sqrt{3}$, π, and so on. Thus, the **real numbers** comprise all rational numbers and irrational numbers. (See Figure 1.)

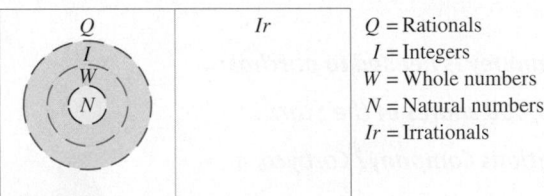

FIGURE **1**
The real numbers consist of the rational numbers and the irrational numbers.

Representing Real Numbers as Decimals

Every real number can be written as a decimal. A rational number can be represented as either a repeating decimal or a terminating decimal. For example, $\frac{2}{3}$ is represented by the repeating decimal

$$0.66666666\ldots \qquad \text{Repeating decimal; note that the integer 6 repeats.}$$

and may also be written as $0.\overline{6}$, where the bar above the 6 indicates that the 6 repeats indefinitely. The number $\frac{1}{2}$ is represented by the terminating decimal

$$0.5 \qquad \text{Terminating decimal}$$

When an irrational number is represented as a decimal, it neither terminates nor repeats. For example,

$$\sqrt{2} = 1.41421\ldots \quad \text{and} \quad \pi = 3.14159\ldots$$

Table 1 summarizes this classification of real numbers.

TABLE 1			
The Real Numbers			
Type	**Description**	**Examples**	**Decimal Representation**
Natural numbers	Counting numbers	1, 2, 3, . . .	Terminating decimals
Whole numbers	Counting numbers and 0	0, 1, 2, 3, . . .	Terminating decimals
Integers	Natural numbers, their negatives, and 0	. . . , -3, -2, -1, 0, 1, 2, 3, . . .	Terminating decimals
Rational numbers	Numbers that can be written in the form $\frac{a}{b}$, where a and b are integers and $b \neq 0$	-3, $-\frac{3}{4}$, $-0.22\overline{2}$, 0, $\frac{5}{6}$, 2, 4.3111	Terminating or repeating decimals
Irrational numbers	Numbers that cannot be written in the form $\frac{a}{b}$, where a and b are integers and $b \neq 0$	$\sqrt{2}$, $\sqrt{3}$, π, 1.414213. . . , 1.732050. . .	Nonterminating, non-repeating decimals
Real numbers	Rational and irrational numbers	All of the above	All types of decimals

Representing Real Numbers on a Number Line

Real numbers may be represented geometrically by points on a line. This *real number,* or *coordinate, line* is constructed as follows: Arbitrarily select a point on a straight line to represent the number 0. This point is called the *origin.* If the line is horizontal, then choose a point at a convenient distance to the right of the origin to represent the number 1. This determines the scale for the number line.

The point representing each positive real number x lies x units to the right of 0, and the point representing each negative real number x lies $-x$ units to the left of 0. Thus, real numbers may be represented by points on a line in such a way that corresponding to each real number there is exactly one point on a line, and vice versa (Figure 2).

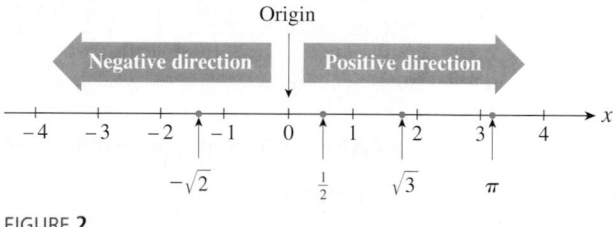

FIGURE **2**
The real number line

Operations with Real Numbers

Two real numbers may be combined to obtain a real number. The operation of addition, written $+$, enables us to combine any two numbers a and b to obtain their *sum,* denoted by $a + b$. Another operation, multiplication, written $\cdot$, enables us to combine any two real numbers a and b to form their *product,* the number $a \cdot b$ (more simply written ab). These two operations are subject to the rules of operation given in Table 2.

The operation of subtraction is defined in terms of addition. Thus,

$$a + (-b)$$

where $-b$ is the additive inverse of b, may be written in the more familiar form $a - b$, and we say that b is subtracted from a. Similarly, the operation of division is defined in terms of multiplication. Recall that the multiplicative inverse of a nonzero real number b is $\frac{1}{b}$, also written b^{-1}. Then,

$$a\left(\frac{1}{b}\right)$$

is written $\frac{a}{b}$, and we say that a is divided by b. Thus, $4\left(\frac{1}{3}\right) = \frac{4}{3}$. Remember, zero does not have a multiplicative inverse, since division by zero is not defined.

TABLE 2

Rules of Operation for Real Numbers

Rule		Example
Addition		
1. $a + b = b + a$	Commutative law of addition	$2 + 3 = 3 + 2$
2. $a + (b + c) = (a + b) + c$	Associative law of addition	$4 + (2 + 3) = (4 + 2) + 3$
3. $a + 0 = a$	Identity law of addition	$6 + 0 = 6$
4. $a + (-a) = 0$	Inverse law of addition	$5 + (-5) = 0$
Multiplication		
1. $ab = ba$	Commutative law of multiplication	$3 \cdot 2 = 2 \cdot 3$
2. $a(bc) = (ab)c$	Associative law of multiplication	$4(3 \cdot 2) = (4 \cdot 3)2$
3. $a \cdot 1 = 1 \cdot a$	Identity law of multiplication	$4 \cdot 1 = 1 \cdot 4$
4. $a\left(\dfrac{1}{a}\right) = 1 \quad (a \neq 0)$	Inverse law of multiplication	$3\left(\dfrac{1}{3}\right) = 1$
Addition and multiplication		
1. $a(b + c) = ab + ac$	Distributive law for multiplication with respect to addition	$3(4 + 5) = 3 \cdot 4 + 3 \cdot 5$

Do the operations of associativity and commutativity hold for subtraction and division? Looking first at associativity, we see that the answer is no, since

$$a - (b - c) \neq (a - b) - c \qquad 7 - (4 - 2) \neq (7 - 4) - 2, \text{ or } 5 \neq 1$$

and

$$a \div (b \div c) \neq (a \div b) \div c \qquad 8 \div (4 \div 2) \neq (8 \div 4) \div 2, \text{ or } 4 \neq 1$$

Similarly, commutativity does not hold because

$$a - b \neq b - a \qquad 7 - 4 \neq 4 - 7, \text{ or } 3 \neq -3$$

and

$$a \div b \neq b \div a \qquad 8 \div 4 \neq 4 \div 8, \text{ or } 2 \neq \frac{1}{2}$$

EXAMPLE 1 State the real number property that justifies each statement.

Statement	Property
a. $4 + (x - 2) = 4 + (-2 + x)$	Commutative law of addition
b. $(a + 2b) + c = a + (2b + c)$	Associative law of addition
c. $x(y - z + 2) = (y - z + 2)x$	Commutative law of multiplication
d. $4(xy^2) = (4x)y^2$	Associative law of multiplication
e. $x(y - 2) = xy - 2x$	Distributive law for multiplication under addition

Using the properties of real numbers listed earlier, we can derive all other algebraic properties of real numbers. Some of the more important properties are given in Tables 3–5.

TABLE 3

Properties of Negatives

Property	Example
1. $-(-a) = a$	$-(-6) = 6$
2. $(-a)b = -(ab) = a(-b)$	$(-3)4 = -(3 \cdot 4) = 3(-4)$
3. $(-a)(-b) = ab$	$(-3)(-4) = 3 \cdot 4$
4. $(-1)a = -a$	$(-1)5 = -5$

TABLE 4

Properties Involving Zero

Property

1. $a \cdot 0 = 0$
2. If $ab = 0$, then either $a = 0$ or $b = 0$, or both a and $b = 0$.

TABLE 5

Properties of Quotients

Property	Example
1. $\dfrac{a}{b} = \dfrac{c}{d}$ if $ad = bc$	$\dfrac{3}{4} = \dfrac{9}{12}$ because $3 \cdot 12 = 9 \cdot 4$
2. $\dfrac{ca}{cb} = \dfrac{a}{b}$	$\dfrac{4 \cdot 3}{4 \cdot 8} = \dfrac{3}{8}$
3. $\dfrac{a}{-b} = \dfrac{-a}{b} = -\dfrac{a}{b}$	$\dfrac{4}{-3} = \dfrac{-4}{3} = -\dfrac{4}{3}$
4. $\dfrac{a}{b} \cdot \dfrac{c}{d} = \dfrac{ac}{bd}$	$\dfrac{3}{4} \cdot \dfrac{5}{2} = \dfrac{15}{8}$
5. $\dfrac{a}{b} \div \dfrac{c}{d} = \dfrac{a}{b} \cdot \dfrac{d}{c} = \dfrac{ad}{bc}$	$\dfrac{3}{4} \div \dfrac{5}{2} = \dfrac{3}{4} \cdot \dfrac{2}{5} = \dfrac{3}{10}$
6. $\dfrac{a}{b} + \dfrac{c}{d} = \dfrac{ad + bc}{bd}$	$\dfrac{3}{4} + \dfrac{5}{2} = \dfrac{3 \cdot 2 + 4 \cdot 5}{8} = \dfrac{13}{4}$
7. $\dfrac{a}{b} - \dfrac{c}{d} = \dfrac{ad - bc}{bd}$	$\dfrac{3}{4} - \dfrac{5}{2} = \dfrac{3 \cdot 2 - 4 \cdot 5}{8} = -\dfrac{7}{4}$

Note Here and in the rest of this book, we assume that all variables are restricted so that the denominator of a quotient is not equal to zero. ∎

EXAMPLE 2 State the real number property that justifies each statement.

Statement	Property
a. $-(-4) = 4$	Property 1 of negatives
b. If $(4x - 1)(x + 3) = 0$, then $x = \dfrac{1}{4}$ or $x = -3$.	Property 2 of zero properties
c. $\dfrac{(x - 1)(x + 1)}{(x - 1)(x - 3)} = \dfrac{x + 1}{x - 3}$	Property 2 of quotients
d. $\dfrac{x - 1}{y} \div \dfrac{y + 1}{x} = \dfrac{x - 1}{y} \cdot \dfrac{x}{y + 1} = \dfrac{x(x - 1)}{y(y + 1)}$	Property 5 of quotients
e. $\dfrac{x}{y} + \dfrac{x}{y + 1} = \dfrac{x(y + 1) + xy}{y(y + 1)}$	Property 6 of quotients
$= \dfrac{xy + x + xy}{y(y + 1)} = \dfrac{2xy + x}{y(y + 1)}$	Distributive law ∎

1.1 Self-Check Exercises

State the property (or properties) that justify each statement.

1. $(3v + 2) - w = 3v + (2 - w)$

2. $(3s)(4t) = 3[s(4t)]$

3. $-(-s + t) = s - t$

4. $\dfrac{2}{-(u - v)} = -\dfrac{2}{u - v}$

Solutions to Self-Check Exercises 1.1 can be found on page 7.

1.1 Concept Questions

1. Give an example of each of the following:
   a. A natural number
   b. A whole number that is not a natural number
   c. A rational number that is not an integer
   d. An irrational number

2. Give an example of each of the following:
   a. A rational number that is represented by a terminating decimal
   b. A rational number that is represented by a repeating decimal
   c. An irrational number represented by a decimal

3. a. The associative law of addition states that $a + (b + c) = $ _____.
   b. The distributive law states that $ab + ac = $ _____.

4. a. Does the operation of associativity hold for subtraction? For division? Explain.
   b. Does zero have a multiplicative inverse? Explain.

5. What can you say about a and b if $ab \neq 0$? How about a, b, and c if $abc \neq 0$?

1.1 Exercises

In Exercises 1–10, classify the number as to type. (For example, $\frac{1}{2}$ is rational and real, whereas $\sqrt{5}$ is irrational and real.)

1. -3
2. -420
3. $\dfrac{3}{8}$
4. $-\dfrac{4}{125}$

5. $\sqrt{11}$
6. $-\sqrt{5}$
7. $\dfrac{\pi}{2}$
8. $\dfrac{2}{\pi}$

9. $2.\overline{421}$
10. $2.71828\ldots$

In Exercises 11–16, indicate whether the statement is true or false.

11. Every integer is a whole number.

12. Every integer is a rational number.

13. Every natural number is an integer.

14. Every rational number is a real number.

15. Every natural number is an irrational number.

16. Every irrational number is a real number.

In Exercises 17–36, state the real number property that justifies the statement.

17. $(2x + y) + z = z + (2x + y)$

18. $3x + (2y + z) = (3x + 2y) + z$

19. $u(3v + w) = (3v + w)u$

20. $a^2(b^2c) = (a^2b^2)c$

21. $u(2v + w) = 2uv + uw$

22. $(2u + v)w = 2uw + vw$

23. $(2x + 3y) + (x + 4y) = 2x + [3y + (x + 4y)]$

24. $(a + 2b)(a - 3b) = a(a - 3b) + 2b(a - 3b)$

25. $a - [-(c + d)] = a + (c + d)$

26. $-(2x + y)[-(3x + 2y)] = (2x + y)(3x + 2y)$

27. $0(2a + 3b) = 0$

28. If $(x - y)(x + y) = 0$, then $x = y$ or $x = -y$.

29. If $(x - 2)(2x + 5) = 0$, then $x = 2$ or $x = -\frac{5}{2}$.

30. If $x(2x - 9) = 0$, then $x = 0$ or $x = \frac{9}{2}$.

31. $\dfrac{(x + 1)(x - 3)}{(2x + 1)(x - 3)} = \dfrac{x + 1}{2x + 1}$

32. $\dfrac{(2x + 1)(x + 3)}{(2x - 1)(x + 3)} = \dfrac{2x + 1}{2x - 1}$

33. $\dfrac{a + b}{b} \div \dfrac{a - b}{ab} = \dfrac{a(a + b)}{a - b}$

34. $\dfrac{x + 2y}{3x + y} \div \dfrac{x}{6x + 2y} = \dfrac{x + 2y}{3x + y} \cdot \dfrac{2(3x + y)}{x} = \dfrac{2(x + 2y)}{x}$

35. $\dfrac{a}{b + c} + \dfrac{c}{b} = \dfrac{ab + bc + c^2}{b(b + c)}$

36. $\dfrac{x + y}{x + 1} - \dfrac{y}{x} = \dfrac{x^2 - y}{x(x + 1)}$

In Exercises 37–42, indicate whether the statement is true or false.

37. If $ab = 1$, then $a = 1$ or $b = 1$.

38. If $ab = 0$ and $a \neq 0$, then $b = 0$.

39. $a - b = b - a$

40. $a \div b = b \div a$

41. $(a - b) - c = a - (b - c)$

42. $a \div (b \div c) = (a \div b) \div c$

1.1 Solutions to Self-Check Exercises

1. Associative law of addition: $a + (b + c) = (a + b) + c$

2. Associative law of multiplication: $a(bc) = (ab)c$

3. Distributive law for multiplication: $a(b + c) = ab + ac$
Properties 1 and 4 of negatives: $-(-a) = a$; $(-1)a = -a$

4. Property 3 of quotients: $\dfrac{a}{-b} = \dfrac{-a}{b} = -\dfrac{a}{b}$

1.2 Polynomials

Exponents

Expressions such as 2^5, $(-3)^2$, and $\left(\frac{1}{4}\right)^4$ are exponential expressions. More generally, if n is a natural number and a is a real number, then a^n represents the product of the real number a and itself n times.

Exponential Notation

If a is a real number and n is a natural number, then

$$a^n = \underbrace{a \cdot a \cdot a \cdot \cdots \cdot a}_{n \text{ factors}} \qquad 3^4 = \underbrace{3 \cdot 3 \cdot 3 \cdot 3}_{4 \text{ factors}}$$

The natural number n is called the **exponent,** and the real number a is called the **base.**

EXAMPLE 1

a. $4^4 = (4)(4)(4)(4) = 256$

b. $(-5)^3 = (-5)(-5)(-5) = -125$

c. $\left(\dfrac{1}{2}\right)^3 = \left(\dfrac{1}{2}\right)\left(\dfrac{1}{2}\right)\left(\dfrac{1}{2}\right) = \dfrac{1}{8}$

d. $\left(-\dfrac{1}{3}\right)^2 = \left(-\dfrac{1}{3}\right)\left(-\dfrac{1}{3}\right) = \dfrac{1}{9}$

When we evaluate expressions such as $3^2 \cdot 3^3$, we use the following property of exponents to write the product in exponential form.

Property 1

If m and n are natural numbers and a is any real number, then

$$a^m \cdot a^n = a^{m+n} \qquad 3^2 \cdot 3^3 = 3^{2+3} = 3^5$$

To verify that Property 1 follows from the definition of an exponential expression, we note that the total number of factors in the exponential expression

$$a^m \cdot a^n = \underbrace{a \cdot a \cdot a \cdot \cdots \cdot a}_{m \text{ factors}} \cdot \underbrace{a \cdot a \cdot a \cdot \cdots \cdot a}_{n \text{ factors}}$$

is $m + n$.

EXAMPLE 2

a. $3^2 \cdot 3^3 = 3^{2+3} = 3^5 = 243$

b. $(-2)^2 \cdot (-2)^5 = (-2)^{2+5} = (-2)^7 = -128$

c. $(3x) \cdot (3x)^3 = (3x)^{1+3} = (3x)^4 = 81x^4$

Be careful to apply the exponent to the indicated base only. Note that

$$4 \cdot x^2 = 4x^2 \neq (4x)^2 = 4^2 \cdot x^2 = 16x^2$$

The exponent applies to $4x$.

The exponent applies only to the base x.

and

$$-3^2 = -9 \neq (-3)^2 = 9$$

The exponent applies to -3.

The exponent applies only to the base 3.

Polynomials

Recall that a *variable* is a letter that is used to represent any element of a given set. Unless specified otherwise, variables in this text will represent real numbers. Sometimes physical considerations impose restrictions on the values a variable may assume. For example, if the variable x denotes the number of LCD television sets sold daily in an appliance store, then x must be a nonnegative integer. At other times, restrictions must be imposed on x in order for an expression to make sense. For example, in the expression $\frac{1}{x+2}$, x cannot take on the value -2, since division by 0 is not permitted. We call the set of all real numbers that a variable is allowed to assume the *domain of the variable*.

In contrast to a variable, a *constant* is a fixed number or letter whose value remains fixed throughout a particular discussion. For example, in the expression $\frac{1}{2}gt^2$, which gives the distance in feet covered by a free-falling body near the surface of the earth, t seconds from rest, the letter g represents the constant of acceleration due to gravity (approximately 32 feet/second/second), whereas the letter t is a variable with domain consisting of nonnegative real numbers.

By combining constants and variables through addition, subtraction, multiplication, division, exponentiation, and root extraction, we obtain *algebraic expressions*. Examples of algebraic expressions are

$$3x - 4y \qquad 2x^2 - y + \frac{1}{xy} \qquad \frac{ax - b}{1 - x^2} \qquad \frac{3xy^{-2} + \pi}{x^2 + y^2 + z^2}$$

where a and b are constants and x, y, and z are variables. Intimidating as some of these expressions might be, remember that they are just real numbers. For example, if $x = 1$ and $y = 4$, then the second expression represents the number

$$2(1)^2 - 4 + \frac{1}{(1)(4)}$$

or $-\frac{7}{4}$, obtained by replacing x and y in the expression by the appropriate values.

Polynomials are an important class of algebraic expressions. The simplest polynomials are those involving *one* variable.

Polynomial in One Variable

A **polynomial** in x is an expression of the form

$$a_n x^n + a_{n-1} x^{n-1} + \cdots + a_1 x + a_0$$

where n is a nonnegative integer and $a_0, a_1, \ldots, a_n$ are real numbers, with $a_n \neq 0$.

The expressions $a_k x^k$ in the sum are called the *terms* of the polynomial. The numbers $a_0, a_1, \ldots, a_n$ are called the *coefficients* of $1, x, x^2, \ldots, x^n$, respectively. The coefficient a_n of x^n (the highest power in x) is called the *leading coefficient* of the polynomial. The nonnegative integer n gives the *degree* of the polynomial. For example, consider the polynomial

$$-2x^5 + 8x^3 - 6x^2 + 3x + 1$$

1. The terms of the polynomial are $-2x^5$, $8x^3$, $-6x^2$, $3x$, and 1.
2. The coefficients of $1, x, x^2, x^3, x^4$, and x^5 are $1, 3, -6, 8, 0$, and -2, respectively.
3. The leading coefficient of the polynomial is -2.
4. The degree of the polynomial is 5.

A polynomial that has just one term (such as $2x^3$) is called a *monomial;* a polynomial that has exactly two terms (such as $x^3 + x$) is called a *binomial;* and a polynomial that has only three terms (such as $-2x^3 + x - 8$) is called a *trinomial.* Also, a polynomial consisting of one (constant) term a_0 (such as the monomial -8) is called a *constant polynomial.* Observe that the degree of a constant polynomial a_0, with $a_0 \neq 0$, is 0 because we can write $a_0 = a_0 x^0$ and see that $n = 0$ in this situation. If all the coefficients of a polynomial are 0, it is called the *zero polynomial* and is denoted by 0. The zero polynomial is not assigned a degree.

Most of the terminology used for a polynomial in one variable carries over to the discussion of polynomials in several variables. But the *degree of a term* in a polynomial in several variables is obtained by adding the powers of all variables in the term, and the *degree of the polynomial* is given by the highest degree of all its terms. For example, the polynomial

$$2x^2 y^5 - 3xy^3 + 8xy^2 - 3y + 4$$

is a polynomial in the two variables x and y. It has five terms with degrees 7, 4, 3, 1, and 0, respectively. Accordingly, the degree of the polynomial is 7.

Adding and Subtracting Polynomials

Constant terms and terms that have the same variable and exponent are called *like* or *similar* terms. Like terms may be combined by adding or subtracting their numerical coefficients. For example,

$$3x + 7x = (3 + 7)x = 10x \qquad \text{Add like terms.}$$

and

$$\frac{1}{2}m^2 - 3m^2 = \left(\frac{1}{2} - 3\right)m^2 = -\frac{5}{2}m^2 \qquad \text{Subtract like terms.}$$

The distributive property of the real number system,

$$ab + ac = a(b + c)$$

is used to justify this procedure.

To add or subtract two or more polynomials, first remove the parentheses and then combine like terms. The resulting expression is then written in order of decreasing degree from left to right.

EXAMPLE 3

a. $(3x^3 + 2x^2 - 4x + 5) + (-2x^3 - 2x^2 - 2)$

$\quad = 3x^3 + 2x^2 - 4x + 5 - 2x^3 - 2x^2 - 2 \qquad$ Remove parentheses.

$\quad = 3x^3 - 2x^3 + 2x^2 - 2x^2 - 4x + 5 - 2 \qquad$ Group like terms together.

$\quad = x^3 - 4x + 3 \qquad\qquad\qquad\qquad\qquad$ Combine like terms.

b. $(2x^4 + 3x^3 + 4x + 6) - (3x^4 + 9x^3 + 3x^2)$

$= 2x^4 + 3x^3 + 4x + 6 - 3x^4 - 9x^3 - 3x^2$ Remove parentheses. Note that the minus sign preceding the second polynomial changes the sign of each term of that polynomial.

$= 2x^4 - 3x^4 + 3x^3 - 9x^3 - 3x^2 + 4x + 6$ Group like terms.

$= -x^4 - 6x^3 - 3x^2 + 4x + 6$ Combine like terms.   ◾

Multiplying Polynomials

To find the product of two polynomials, we again use the distributive property for real numbers. For example, to compute the product $3x(4x - 2)$, we use the distributive law to obtain

$$3x(4x - 2) = (3x)(4x) + (3x)(-2) \qquad a(b + c) = ab + ac$$
$$= 12x^2 - 6x$$

Observe that each term of one polynomial is multiplied by each term of the other. The resulting expression is then simplified by combining like terms. In general, an algebraic expression is *simplified* if none of its terms are similar.

EXAMPLE 4 Find the product of $(3x + 5)(2x - 3)$.

Solution

$$(3x + 5)(2x - 3) = 3x(2x - 3) + 5(2x - 3) \qquad \text{Distributive property}$$
$$= (3x)(2x) + (3x)(-3) \qquad \text{Distributive property}$$
$$\quad + (5)(2x) + (5)(-3)$$
$$= 6x^2 - 9x + 10x - 15 \qquad \text{Multiply terms.}$$
$$= 6x^2 + x - 15 \qquad \text{Combine like terms.} \quad ◾$$

EXAMPLE 5 Find the product of $(2t^2 - t + 3)(2t^2 - 1)$.

Solution

$$(2t^2 - t + 3)(2t^2 - 1)$$
$$= 2t^2(2t^2 - 1) - t(2t^2 - 1) + 3(2t^2 - 1) \qquad \text{Distributive property}$$
$$= (2t^2)(2t^2) + (2t^2)(-1) + (-t)(2t^2) \qquad \text{Distributive property}$$
$$\quad + (-t)(-1) + (3)(2t^2) + (3)(-1)$$
$$= 4t^4 - 2t^2 - 2t^3 + t + 6t^2 - 3 \qquad \text{Multiply terms.}$$
$$= 4t^4 - 2t^3 + 4t^2 + t - 3 \qquad \text{Combine terms.}$$

Alternative Solution   We can also find the product by arranging the polynomials vertically and multiplying:

$$
\begin{array}{r}
2t^2 - t + 3 \\
2t^2 - 1 \\
\hline
4t^4 - 2t^3 + 6t^2 \\
-2t^2 + t - 3 \\
\hline
4t^4 - 2t^3 + 4t^2 + t - 3
\end{array}
$$

   ◾

     The polynomials in Examples 4 and 5 are polynomials in one variable. The operations of addition, subtraction, and multiplication are performed on polynomials of more than one variable in the same way as they are for polynomials in one variable.

EXAMPLE 6 Multiply $(3x - y)(4x^2 - 2y)$.

Solution

$$
\begin{aligned}
(3x - y)(4x^2 - 2y) &= 3x(4x^2 - 2y) - y(4x^2 - 2y) && \text{Distributive property} \\
&= 12x^3 - 6xy - 4x^2y + 2y^2 && \text{Distributive property} \\
&= 12x^3 - 4x^2y - 6xy + 2y^2 && \text{Arrange terms in order of} \\
& && \text{descending powers of } x. \quad\blacksquare
\end{aligned}
$$

Several commonly used products of polynomials are summarized in Table 6. Since these products occur so frequently, you will find it helpful to memorize these formulas.

TABLE 6	
Special Products	
Formula	**Example**
1. $(a + b)^2 = a^2 + 2ab + b^2$	$(2x + 3y)^2 = (2x)^2 + 2(2x)(3y) + (3y)^2$ $= 4x^2 + 12xy + 9y^2$
2. $(a - b)^2 = a^2 - 2ab + b^2$	$(4x - 2y)^2 = (4x)^2 - 2(4x)(2y) + (2y)^2$ $= 16x^2 - 16xy + 4y^2$
3. $(a + b)(a - b) = a^2 - b^2$	$(2x + y)(2x - y) = (2x)^2 - (y)^2$ $= 4x^2 - y^2$

EXAMPLE 7 Use the special product formulas to compute:

a. $(2x + y)^2$ **b.** $(3a - 4b)^2$ **c.** $\left(\dfrac{1}{2}x - 1\right)\left(\dfrac{1}{2}x + 1\right)$

Solution

a. $\begin{aligned}[t] (2x + y)^2 &= (2x)^2 + 2(2x)(y) + y^2 && \text{Formula 1} \\ &= 4x^2 + 4xy + y^2 \end{aligned}$

b. $\begin{aligned}[t] (3a - 4b)^2 &= (3a)^2 - 2(3a)(4b) + (4b)^2 && \text{Formula 2} \\ &= 9a^2 - 24ab + 16b^2 \end{aligned}$

c. $\left(\dfrac{1}{2}x - 1\right)\left(\dfrac{1}{2}x + 1\right) = \left(\dfrac{1}{2}x\right)^2 - 1 = \dfrac{1}{4}x^2 - 1$ Formula 3 $\blacksquare$

Order of Operations

The common steps in Examples 1–7 have been to remove parentheses and combine like terms. If more than one grouping symbol is present, the innermost symbols are removed first. As you work through Examples 8 and 9, note the order in which the grouping symbols are removed: parentheses ( ) first, brackets [ ] second, and finally braces { }. Also, note that the operations of multiplication and division take precedence over addition and subtraction.

EXAMPLE 8 Perform the indicated operations: $2t^3 - \{t^2 - [t - (2t - 1)] + 4\}$.

Solution

$$
\begin{aligned}
2t^3 - \{t^2 &- [t - (2t - 1)] + 4\} \\
&= 2t^3 - \{t^2 - [t - 2t + 1] + 4\} && \text{Remove parentheses.} \\
&= 2t^3 - \{t^2 - [-t + 1] + 4\} && \text{Combine like terms within the brackets.} \\
&= 2t^3 - \{t^2 + t - 1 + 4\} && \text{Remove brackets.} \\
&= 2t^3 - \{t^2 + t + 3\} && \text{Combine like terms within the braces.} \\
&= 2t^3 - t^2 - t - 3 && \text{Remove braces.} \quad\blacksquare
\end{aligned}
$$

EXAMPLE 9 Simplify $2\{3 - 2[x - 2x(3 - x)]\}$.

Solution

$$
\begin{aligned}
2\{3 - 2[x - 2x(3 - x)]\} &= 2\{3 - 2[x - 6x + 2x^2]\} && \text{Remove parentheses.}\\
&= 2\{3 - 2[-5x + 2x^2]\} && \text{Combine like terms.}\\
&= 2\{3 + 10x - 4x^2\} && \text{Remove brackets.}\\
&= 6 + 20x - 8x^2 && \text{Remove braces.}\\
&= -8x^2 + 20x + 6 && \text{Write answer in order of}\\
& && \text{descending powers of } x. \quad\blacksquare
\end{aligned}
$$

1.2 Self-Check Exercises

1. Find the product of $(2x + 3y)(3x - 2y)$.

2. Simplify $3x - 2\{2x - [x - 2(x - 2)] + 1\}$.

Solutions to Self-Check Exercises 1.2 can be found on page 14.

1.2 Concept Questions

1. Which of the following are polynomial expressions?
   a. $3x^4 - 2\sqrt{x} + 6$

   b. $-5x^3 + 2x - \dfrac{1}{2}$

   c. $\dfrac{2x}{3x^2 - x + 2}$

2. a. Give an example of a polynomial of degree 4 in x.

   b. Give an example of a polynomial of degree 3 in x and y.

3. Without looking at the formulas in the text, complete the following:
   a. $(1 + b)^2 = $ _____
   b. $(a - b)^2 = $ _____
   c. $(a + b)(a - b) = $ _____

1.2 Exercises

In Exercises 1–12, evaluate the expression.

1. 3^4

2. $(-2)^5$

3. $\left(\dfrac{2}{3}\right)^3$

4. $\left(-\dfrac{3}{4}\right)^2$

5. -3^4

6. $-\left(-\dfrac{4}{5}\right)^3$

7. $-3\left(\dfrac{3}{5}\right)^3$

8. $\left(-\dfrac{2}{3}\right)^2\left(-\dfrac{3}{4}\right)^3$

9. $2^3 \cdot 2^5$

10. $(-3)^2 \cdot (-3)^3$

11. $(3y)^2(3y)^3$

12. $(-2x)^3(-2x)^2$

In Exercises 13–56, perform the indicated operations and simplify.

13. $(2x + 3) + (4x - 6)$

14. $(-3x + 2) - (4x - 3)$

15. $(7x^2 - 2x + 5) + (2x^2 + 5x - 4)$

16. $(3x^2 + 5xy + 2y) + (4 - 3xy - 2x^2)$

17. $(5y^2 - 2y + 1) - (y^2 - 4y - 8)$

18. $(2x^2 - 3x + 4) - (-x^2 + 2x - 6)$

19. $(2.4x^3 - 3x^2 + 1.7x - 6.2) - (1.2x^3 + 1.2x^2 - 0.8x + 2)$

20. $(1.4x^3 - 1.2x^2 + 3.2) - (-0.8x^3 - 2.1x - 1.8)$

21. $(3x^2)(2x^3)$

22. $(-2rs^2)(4r^2s^2)(2s)$

23. $-2x(x^2 - 2) + 4x^3$

24. $xy(2y - 3x)$

25. $2m(3m - 4) + m(m - 1)$

26. $-3x(2x^2 + 3x - 5) + 2x(x^2 - 3)$

27. $3(2a - b) - 4(b - 2a)$

28. $2(3m - 1) - 3(-4m + 2n)$

29. $(2x + 3)(3x - 2)$

30. $(3r - 1)(2r + 5)$

31. $(2x - 3y)(3x + 2y)$

32. $(5m - 2n)(5m + 3n)$

33. $(3r + 2s)(4r - 3s)$ 34. $(2m + 3n)(3m - 2n)$

35. $(0.2x + 1.2y)(0.3x - 2.1y)$

36. $(3.2m - 1.7n)(4.2m + 1.3n)$

37. $(2x - y)(3x^2 + 2y)$ 38. $(3m - 2n^2)(2m^2 + 3n)$

39. $(2x + 3y)^2$ 40. $(3m - 2n)^2$

41. $(2u - v)(2u + v)$ 42. $(3r + 4s)(3r - 4s)$

43. $(2x - 1)^2 + 3x - 2(x^2 + 1) + 3$

44. $(3m + 2)^2 - 2m(1 - m) - 4$

45. $(2x + 3y)^2 - (2y + 1)(3x - 2) + 2(x - y)$

46. $(x - 2y)(y + 3x) - 2xy + 3(x + y - 1)$

47. $(t^2 - 2t + 4)(2t^2 + 1)$ 48. $(3m^2 - 1)(2m^2 + 3m - 4)$

49. $2x - \{3x - [x - (2x - 1)]\}$

50. $3m - 2\{m - 3[2m - (m - 5)] + 4\}$

51. $x - \{2x - [-x - (1 + x)]\}$

52. $3x^2 - \{x^2 + 1 - x[x - (2x - 1)]\} + 2$

53. $(2x - 3)^2 - 3(x + 4)(x - 4) + 2(x - 4) + 1$

54. $(x - 2y)^2 + 2(x + y)(x - 3y) + x(2x + 3y + 2)$

55. $2x\{3x[2x - (3 - x)] + (x + 1)(2x - 3)\}$

56. $-3[(x + 2y)^2 - (3x - 2y)^2 + (2x - y)(2x + y)]$

57. **PROFIT OF A COMPANY** The total revenue realized in the sale of x units of the Spectra smartphone is

$$-0.04x^2 + 2000x$$

dollars/week, and the total cost incurred in manufacturing x units of the phones is

$$0.000002x^3 - 0.02x^2 + 1000x + 120,000$$

dollars/week $(0 \le x \le 50,000)$. Find an expression giving the total weekly profit of the company.
Hint: The profit is revenue minus cost.

58. **PROFIT OF A COMPANY** A manufacturer of tennis rackets finds that the total cost of manufacturing x rackets/day is given by

$$0.0001x^2 + 4x + 400$$

dollars. Each racket can be sold at a price of p dollars, where

$$p = -0.0004x + 10$$

Find an expression giving the daily profit for the manufacturer, assuming that all the rackets manufactured can be sold.
Hint: The total revenue is given by the total number of rackets sold multiplied by the price of each racket. The profit is given by revenue minus cost.

59. **REVENUE OF A COMPANY** Jake owns two gas stations in town. The projected revenue of the first gas station for the next 12 months is

$$0.2t^2 + 150t \qquad (0 \le t \le 12)$$

thousand dollars t months from now. The projected revenue of the second gas station for the next 12 months is

$$0.5t^2 + 200t \qquad (0 \le t \le 12)$$

thousand dollars t months from now. Find an expression that gives the projected total revenue realized by Jake's gas stations in month t $(0 \le t \le 12)$.

60. **REVENUE OF A COMPANY** Refer to Exercise 59. Find an expression that gives the amount by which the revenue of the second gas station will exceed that of the first gas station in month t $(0 < t \le 12)$.

61. **COST OF CABLE TELEVISION** The average price of expanded basic cable television packages in the United States from 2010 through 2012 was

$$12(0.5t^2 + 3t + 54) \qquad (0 \le t \le 2)$$

dollars/year, t years from 2010 $(t = 0)$. (Expanded basic cable is one step up from the entry-level package offered by most providers.) The average price pegged to the standard rate of inflation as defined by the Consumer Price Index (CPI) was

$$12(0.75t + 38.5) \qquad (0 \le t \le 2)$$

dollars/year for the same period. Find an expression giving the difference between what the average American cable-watching household had spent on expanded basic cable television packages in year t over what it would have spent if the price of cable had matched inflation.
Source: Consumer Reports.

62. **PRISON OVERCROWDING** The 1980s saw a trend toward old-fashioned punitive deterrence in contrast to the more liberal penal policies and community-based corrections that were popular in the 1960s and early 1970s. As a result, prisons became more crowded, and the gap between the number of people in prison and prison capacity widened. Based on figures from the U.S. Department of Justice, the number of prisoners (in thousands) in federal and state prisons is approximately

$$3.5t^2 + 26.7t + 436.2 \qquad (0 \le t \le 10)$$

and the number of inmates (in thousands) for which prisons were designed is given by

$$24.3t + 365 \qquad (0 \le t \le 10)$$

where t is measured in years, and $t = 0$ corresponds to 1984. Find an expression giving the gap between the number of prisoners and the number for which the prisons were designed at any time t.
Source: U.S. Department of Justice.

In Exercises 63–66, determine whether the statement is true or false. If it is true, explain why it is true. If it is false, give an example to show why it is false.

63. If m and n are natural numbers and a and b are real numbers, then $a^m \cdot b^n = (ab)^{m+n}$.

64. $a^{16} - b^{16} = (a^8 + b^8)(a^4 + b^4)(a^2 + b^2)(a + b)(a - b)$

65. The degree of the product of a polynomial of degree m and a polynomial of degree n is mn.

66. Suppose that p and q are polynomials of degree n. Then $p + q$ is a polynomial of degree n.

1.2 Solutions to Self-Check Exercises

1. $(2x + 3y)(3x - 2y) = 2x(3x - 2y) + 3y(3x - 2y)$
$$= 6x^2 - 4xy + 9xy - 6y^2$$
$$= 6x^2 + 5xy - 6y^2$$

2. $3x - 2\{2x - [x - 2(x - 2)] + 1\}$
$$= 3x - 2\{2x - [x - 2x + 4] + 1\}$$
$$= 3x - 2\{2x - [-x + 4] + 1\}$$
$$= 3x - 2\{2x + x - 4 + 1\}$$
$$= 3x - 2\{3x - 3\}$$
$$= 3x - 6x + 6$$
$$= -3x + 6$$

1.3 Factoring Polynomials

Factoring

Factoring a polynomial is the process of expressing it as a product of two or more polynomials. For example, by applying the distributive property, we may write

$$3x^2 - x = x(3x - 1)$$

and we say that x and $3x - 1$ are factors of $3x^2 - x$.

How do we know whether a polynomial is completely factored? Recall that an integer greater than 1 is *prime* if its only positive integer factors are itself and 1. For example, the number 3 is prime because its only factors are 3 and 1. In the same way, a polynomial is said to be prime if it cannot be expressed as a product of two or more polynomials of positive degree with integral coefficients. For example, $x^2 + 2x + 2$ is a prime polynomial whereas $x^2 - 9$ is not a prime polynomial, since $x^2 - 9 = (x + 3)(x - 3)$. Finally, a polynomial is *completely factored* if it is expressed as a product of prime polynomials.

Note   Unless otherwise mentioned, we will consider only factorization over the set of integers in this text. Hence, when the term *factor* is used, it will be understood that the factorization is to be completed over the set of integers.   ■

Common Factors

The first step in factoring a polynomial is to determine whether it contains any common factors. If it does, the common factor of highest degree is then factored out. For example, the greatest common factor of $2a^2x + 4ax + 6a$ is $2a$ because

$$2a^2x + 4ax + 6a = 2a \cdot ax + 2a \cdot 2x + 2a \cdot 3$$
$$= 2a(ax + 2x + 3)$$

EXAMPLE 1 Factor out the greatest common factor:

a. $-3t^2 + 3t$ **b.** $6a^4b^4c - 9a^2b^2$

Solution

a. Since $3t$ is a common factor of each term, we have

$$-3t^2 + 3t = 3t(-t + 1) = -3t(t - 1)$$

b. Since $3a^2b^2$ is the common factor of highest degree, we have

$$6a^4b^4c - 9a^2b^2 = 3a^2b^2(2a^2b^2c - 3)$$

Some Important Formulas

Having checked for common factors, the next step in factoring a polynomial is to express the polynomial as the product of a constant and/or one or more prime polynomials. The formulas given in Table 7 for factoring polynomials should be memorized.

TABLE 7

Factoring Formulas

Formula	Example
Difference of two squares $a^2 - b^2 = (a + b)(a - b)$	$x^2 - 36 = (x + 6)(x - 6)$ $8x^2 - 2y^2 = 2(4x^2 - y^2) = 2[(2x)^2 - y^2]$ $\qquad = 2(2x + y)(2x - y)$ $9 - a^6 = 3^2 - (a^3)^2 = (3 + a^3)(3 - a^3)$
Perfect square trinomial $a^2 + 2ab + b^2 = (a + b)^2$ $a^2 - 2ab + b^2 = (a - b)^2$	$x^2 + 8x + 16 = (x + 4)^2$ $4x^2 - 4xy + y^2 = (2x)^2 - 2(2x)(y) + y^2$ $\qquad = (2x - y)^2$
Sum of two cubes $a^3 + b^3 = (a + b)(a^2 - ab + b^2)$	$z^3 + 27 = z^3 + (3)^3$ $\qquad = (z + 3)(z^2 - 3z + 9)$
Difference of two cubes $a^3 - b^3 = (a - b)(a^2 + ab + b^2)$	$8x^3 - y^6 = (2x)^3 - (y^2)^3$ $\qquad = (2x - y^2)(4x^2 + 2xy^2 + y^4)$

Note Observe that a formula is given for factoring the sum of two cubes, but none is given for factoring the sum of two squares, since $x^2 + a^2$ cannot be factored.

EXAMPLE 2 Factor:

a. $x^2 - 9$ **b.** $16x^2 - 81y^4$ **c.** $(a - b)^2 - (a^2 + b)^2$

Solution Observe that each of the polynomials in parts (a)–(c) is the difference of two squares. Using the formulas given in Table 7, we have

a. $x^2 - 9 = x^2 - 3^2 = (x + 3)(x - 3)$

b. $16x^2 - 81y^4 = (4x)^2 - (9y^2)^2 = (4x + 9y^2)(4x - 9y^2)$

c. $(a - b)^2 - (a^2 + b)^2 = [(a - b) + (a^2 + b)][(a - b) - (a^2 + b)]$

$$= [a - b + a^2 + b][a - b - a^2 - b] \quad \text{Remove parentheses.}$$
$$= (a + a^2)(-a^2 + a - 2b) \quad \text{Combine like terms.}$$
$$= a(1 + a)(-a^2 + a - 2b)$$

EXAMPLE 3 Factor:

a. $x^2 + 4xy + 4y^2$ **b.** $4a^2 - 12ab + 9b^2$

Solution Recognizing these expressions to be perfect square trinomials, we use the appropriate formula in Table 7 to factor them. Thus,

a. $x^2 + 4xy + 4y^2 = x^2 + 2x(2y) + (2y)^2 = (x + 2y)(x + 2y) = (x + 2y)^2$

b. $4a^2 - 12ab + 9b^2 = (2a)^2 - 2(2a)(3b) + (3b)^2$
$$= (2a - 3b)(2a - 3b) = (2a - 3b)^2 \qquad ■$$

EXAMPLE 4 Factor:

a. $x^3 + 8y^3$ **b.** $27a^3 - 64b^3$

Solution

a. This polynomial is the sum of two cubes. Using the formula given in Table 7, we have

$$x^3 + 8y^3 = x^3 + (2y)^3 = (x + 2y)\left[x^2 - x(2y) + (2y)^2\right]$$
$$= (x + 2y)(x^2 - 2xy + 4y^2)$$

b. Using the formula for the difference of two cubes given in Table 7, we have

$$27a^3 - 64b^3 = \left[(3a)^3 - (4b)^3\right]$$
$$= (3a - 4b)\left[(3a)^2 + (3a)(4b) + (4b)^2\right]$$
$$= (3a - 4b)(9a^2 + 12ab + 16b^2) \qquad ■$$

Trial-and-Error Factorization

The factors of the second-degree polynomial $px^2 + qx + r$, where p, q, and r are integers, have the form

$$(ax + b)(cx + d)$$

where $ac = p$, $ad + bc = q$, and $bd = r$. Since only a limited number of choices are possible, we use a trial-and-error method to factor polynomials having this form.

For example, to factor $x^2 - 2x - 3$, we first observe that the only possible first-degree terms in each factor in the product are

$$(x \quad)(x \quad) \qquad \text{Since the coefficient of } x^2 \text{ is 1}$$

Next, we observe that the product of the constant terms is (-3). This gives us the following possible factors:

$$(x - 1)(x + 3)$$
$$(x + 1)(x - 3)$$

Looking once again at the polynomial $x^2 - 2x - 3$, we see that the coefficient of x is -2. Checking to see which set of factors yields -2 for the coefficient of x, we find that

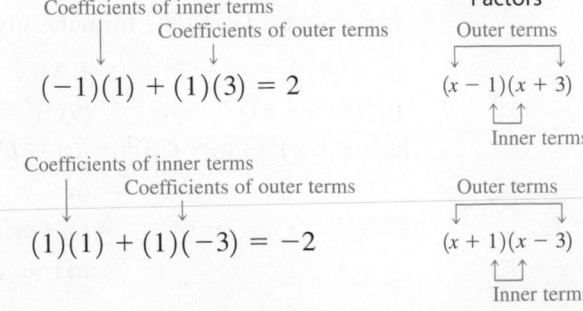

and we conclude that the correct factorization is

$$x^2 - 2x - 3 = (x + 1)(x - 3)$$

With practice, you will soon find that you can perform many of these steps mentally, and you will no longer need to write out each step.

EXAMPLE 5 Factor:

a. $3x^2 + 4x - 4$ **b.** $3x^2 - 6x - 24$

Solution

a. Using trial and error, we find that the correct factorization is

$$3x^2 + 4x - 4 = (3x - 2)(x + 2)$$

b. Since each term has the common factor 3, we have

$$3x^2 - 6x - 24 = 3(x^2 - 2x - 8)$$

Using the trial-and-error method of factorization, we find that

$$x^2 - 2x - 8 = (x - 4)(x + 2)$$

Thus, we have

$$3x^2 - 6x - 24 = 3(x - 4)(x + 2)$$

Factoring by Regrouping

A polynomial may sometimes be factored by regrouping and rearranging terms so that a common term can be factored out. This technique is illustrated in Example 6.

EXAMPLE 6 Factor:

a. $x^3 + x + x^2 + 1$ **b.** $2ax + 2ay + bx + by$

Solution

a. We begin by rearranging the terms in order of descending powers of x. Thus,

$$\begin{aligned} x^3 + x + x^2 + 1 &= x^3 + x^2 + x + 1 \\ &= x^2(x + 1) + x + 1 \qquad \text{Factor the first two terms.} \\ &= (x + 1)(x^2 + 1) \qquad \text{Factor the common term } x + 1. \end{aligned}$$

b. First, factor the common term $2a$ from the first two terms and the common term b from the last two terms. Thus,

$$2ax + 2ay + bx + by = 2a(x + y) + b(x + y)$$

Since $(x + y)$ is common to both terms of the polynomial on the right, we can factor it out. Hence,

$$2a(x + y) + b(x + y) = (x + y)(2a + b)$$

More Examples on Factoring

EXAMPLE 7 Factor:

a. $4x^6 - 4x^2$ **b.** $18x^4 - 3x^3 - 6x^2$

Solution

a. $4x^6 - 4x^2 = 4x^2(x^4 - 1)$ Common factor

$= 4x^2(x^2 - 1)(x^2 + 1)$ Difference of two squares

$= 4x^2(x - 1)(x + 1)(x^2 + 1)$ Difference of two squares

b. $18x^4 - 3x^3 - 6x^2 = 3x^2(6x^2 - x - 2)$ Common factor

$= 3x^2(3x - 2)(2x + 1)$ Trial-and-error factorization

EXAMPLE 8 Factor:

a. $3x^2y + 9x^2 - 12y - 36$ b. $(x - at)^3 - (x + at)^3$

Solution

a. $3x^2y + 9x^2 - 12y - 36 = 3(x^2y + 3x^2 - 4y - 12)$ Common factor

$= 3[x^2(y + 3) - 4(y + 3)]$ Regrouping

$= 3(y + 3)(x^2 - 4)$ Common factor

$= 3(y + 3)(x - 2)(x + 2)$ Difference of two squares

b. $(x - at)^3 - (x + at)^3$

$= [(x - at) - (x + at)]$

$\quad \cdot [(x - at)^2 + (x - at)(x + at) + (x + at)^2]$ Difference of two cubes

$= -2at(x^2 - 2atx + a^2t^2 + x^2 - a^2t^2 + x^2 + 2atx + a^2t^2)$

$= -2at(3x^2 + a^2t^2)$

Be sure you become familiar with the factorization methods discussed in this chapter because we will be using them throughout the text. As with many other algebraic techniques, you will find yourself becoming more proficient at factoring as you work through the exercises.

1.3   Self-Check Exercises

1. Factor: **a.** $4x^3 - 2x^2$ **b.** $3(a^2 + 2b^2) + 4(a^2 + 2b^2)^2$ **2.** Factor: **a.** $6x^2 - x - 12$ **b.** $4x^2 + 10x - 6$

Solutions to Self-Check Exercises 1.3 can be found on page 20.

1.3   Concept Questions

1. What is meant by the expression *factor a polynomial*? Illustrate the process with an example.

2. Without looking at the formulas in the text, complete the following formulas:

     a. $a^3 + b^3 = $ _____   **b.** $a^3 - b^3 = $ _____

1.3   Exercises

In Exercises 1–10, factor out the greatest common factor.

1. $6m^2 - 4m$

2. $4t^4 - 12t^3$

3. $9ab^2 - 6a^2b$

4. $12x^3y^5 + 16x^2y^3$

5. $10m^2n - 15mn^2 + 20mn$

6. $6x^4y - 4x^2y^2 + 2x^2y^3$

7. $3x(2x + 1) - 5(2x + 1)$

8. $2u(3v^2 + w) + 5v(3v^2 + w)$

9. $(3a + b)(2c - d) + 2a(2c - d)^2$

10. $4uv^2(2u - v) + 6u^2v(v - 2u)$

In Exercises 11–54, factor the polynomial. If the polynomial is prime, state it.

11. $2m^2 - 11m - 6$

12. $6x^2 - x - 1$

13. $x^2 - xy - 6y^2$

14. $2u^2 + 5uv - 12v^2$

15. $x^2 - 3x - 1$

16. $m^2 + 2m + 3$

17. $4a^2 - b^2$

18. $12x^2 - 3y^2$

19. $u^2v^2 - w^2$

20. $4a^2b^2 - 25c^2$

21. $z^2 + 4$

22. $u^2 + 25v^2$

23. $x^2 + 6xy + y^2$

24. $4u^2 - 12uv + 9v^2$

25. $x^2 + 3x - 4$

26. $3m^3 + 3m^2 - 18m$

27. $12x^2y - 10xy - 12y$

28. $12x^2y - 2xy - 24y$

29. $35r^2 + r - 12$

30. $6uv^2 + 9uv - 6v$

31. $9x^3y - 4xy^3$

32. $4u^4v - 9u^2v^3$

33. $x^4 - 16y^2$

34. $16u^4v - 9v^3$

35. $(a - 2b)^2 - (a + 2b)^2$

36. $2x(x + y)^2 - 8x(x + y^2)^2$

37. $8m^3 + 1$

38. $27m^3 - 8$

39. $8r^3 - 27s^3$

40. $x^3 + 64y^3$

41. $u^2v^6 - 8u^2$

42. $r^6s^6 + 8s^3$

43. $2x^3 + 6x + x^2 + 3$

44. $2u^4 - 4u^2 + 2u^2 - 4$

45. $3ax + 6ay + bx + 2by$

46. $6ux - 4uy + 3vx - 2vy$

47. $u^4 - v^4$

48. $u^4 - u^2v^2 - 6v^4$

49. $4x^3 - 9xy^2 + 4x^2y - 9y^3$

50. $4u^4 + 11u^2v^2 - 3v^4$

51. $x^4 + 3x^3 - 2x - 6$

52. $a^2 - b^2 + a + b$

53. $au^2 + (a + c)u + c$

54. $ax^2 - (1 + ab)xy + by^2$

55. SIMPLE INTEREST The accumulated amount after t years for a bank deposit of P dollars earning interest at the rate of r/year is $A = P + Prt$. Factor the expression on the right-hand side of this equation.

56. WORKER EFFICIENCY An efficiency study conducted by Elektra Electronics showed that the number of Space Commander walkie-talkies assembled by the average worker t hr after starting work at 8 A.M. is

$$-t^3 + 6t^2 + 15t \qquad (0 \le t \le 4)$$

Factor the expression.

57. REVENUE OF A COMPANY Williams Commuter Air Service realizes a monthly revenue of

$$8000x - 100x^2 \qquad (0 \le x \le 80)$$

dollars when the price charged per passenger is x dollars. Factor the expression.

58. CHEMICAL REACTION In an autocatalytic chemical reaction, the product that is formed acts as a catalyst for the reaction. If Q is the amount of the original substrate present initially and x is the amount of catalyst formed, then the rate of change of the chemical reaction with respect to the amount of catalyst present in the reaction is

$$R = kQx - kx^2 \qquad (0 \le x \le Q)$$

where k is a constant. Factor the expression on the right-hand side of the equation.

59. SPREAD OF AN EPIDEMIC The incidence (number of new cases per day) of a contagious disease spreading in a population of M people, where k is a positive constant and x denotes the number of people already infected, is given by $kMx - kx^2$. Factor this expression.

60. REVENUE OF A COMPANY The total revenue realized by the Apollo Company from the sale of x minitablets is given by $R = -0.1x^2 + 500x$ dollars. Factor the expression on the right-hand side of this equation.

61. CHARTER-FLIGHT FARE The revenue (in dollars) for a charter flight organized by the Leisure World Travel Agency is

$$R = 60,000 + 100x - x^2 \qquad (0 \le x \le 300)$$

where x denotes the number of people beyond 200 who signed up for the flight. Factor the expression on the right-hand side of the equation.

62. TEMPERATURE IN WINTER The temperature T (in degrees Fahrenheit) over a certain 24-hr period in La Crosse, Wisconsin, was determined to be

$$T = \frac{1}{2}(t^3 - 39t^2 + 360t) \qquad (0 \le t \le 24)$$

where t is the time (in hours), and $t = 0$ corresponds to 6 A.M. Factor the expression on the right-hand side of the equation.

63. CHARLES' LAW Charles' Law for gases states that if the pressure remains constant, then the volume V that a gas occupies is related to its temperature T in degrees Celsius by the equation

$$V = V_0 + \frac{V_0}{273}T$$

Factor the expression on the right-hand side of the equation.

64. REACTION TO A DRUG The strength of a human body's reaction to a dosage D of a certain drug, where k is a positive constant, is given by

$$\frac{kD^2}{2} - \frac{D^3}{3}$$

Factor this expression.

1.3 Solutions to Self-Check Exercises

1. a. The common factor is $2x^2$. Therefore,

$$4x^3 - 2x^2 = 2x^2(2x - 1)$$

b. The common factor is $a^2 + 2b^2$. Therefore,

$$3(a^2 + 2b^2) + 4(a^2 + 2b^2)^2 = (a^2 + 2b^2)[3 + 4(a^2 + 2b^2)]$$
$$= (a^2 + 2b^2)(3 + 4a^2 + 8b^2)$$

2. a. Using the trial-and-error method of factorization, we find that

$$6x^2 - x - 12 = (3x + 4)(2x - 3)$$

b. We first factor out the common factor 2. Thus,

$$4x^2 + 10x - 6 = 2(2x^2 + 5x - 3)$$

Using the trial-and-error method of factorization, we find that

$$2x^2 + 5x - 3 = (2x - 1)(x + 3)$$

and, consequently,

$$4x^2 + 10x - 6 = 2(2x - 1)(x + 3)$$

1.4 Rational Expressions

Quotients of polynomials are called **rational expressions.** Examples of rational expressions are

$$\frac{6x - 1}{2x + 3} \quad \text{and} \quad \frac{3x^2y^3 - 2xy}{4x - y}$$

Because division by zero is not allowed, the denominator of a rational expression must not be equal to zero. Thus, in the first example, $x \neq -\frac{3}{2}$, and in the second example, $y \neq 4x$.

Since rational expressions are quotients in which the variables represent real numbers, the properties of real numbers apply to rational expressions as well. For this reason, operations with rational fractions are performed in the same way as operations with arithmetic fractions.

Simplifying Rational Expressions

A rational expression is *simplified,* or reduced to lowest terms, if its numerator and denominator have no common factors other than 1 and -1. If a rational expression does contain common factors, we use the properties of the real number system to write

$$\frac{ac}{bc} = \frac{a}{b} \cdot \frac{c}{c} = \frac{a}{b} \cdot 1 = \frac{a}{b} \qquad (a, b, c \text{ are real numbers, and } bc \neq 0.)$$

This process is often called "canceling common factors." To indicate this process, we often write

$$\frac{a\cancel{c}}{b\cancel{c}} = \frac{a}{b}$$

where a slash is shown through the common factors. As another example, the rational expression

$$\frac{(x + 2)(x - 3)}{(x - 2)(x - 3)}$$

is simplified by canceling the common factors $(x - 3)$ and writing

$$\frac{(x + 2)\cancel{(x - 3)}}{(x - 2)\cancel{(x - 3)}} = \frac{x + 2}{x - 2}$$

⚠️ $\dfrac{\cancel{3} + 4x}{\cancel{3}} = 1 + 4x$ is an example of incorrect cancellation. Instead, we write

$$\frac{3 + 4x}{3} = \frac{3}{3} + \frac{4x}{3} = 1 + \frac{4x}{3}$$

EXAMPLE 1 Simplify the following expressions:

a. $\dfrac{x^2 + 2x - 3}{x^2 + 4x + 3}$ **b.** $\dfrac{3 - 4x - 4x^2}{2x - 1}$ **c.** $\dfrac{(k + 4)^2(k - 1)}{k^2 - 16}$

Solution

a. $\dfrac{x^2 + 2x - 3}{x^2 + 4x + 3} = \dfrac{(x + 3)(x - 1)}{(x + 3)(x + 1)} = \dfrac{x - 1}{x + 1}$ Factor numerator and denominator, and cancel common factors.

b. $\dfrac{3 - 4x - 4x^2}{2x - 1} = \dfrac{(1 - 2x)(3 + 2x)}{2x - 1}$

$\qquad\qquad = -\dfrac{(2x - 1)(2x + 3)}{2x - 1}$ Rewrite the term $1 - 2x$ in the equivalent form $-(2x - 1)$.

$\qquad\qquad = -(2x + 3)$ Cancel common factors.

c. $\dfrac{(k + 4)^2(k - 1)}{k^2 - 16} = \dfrac{(k + 4)^2(k - 1)}{(k + 4)(k - 4)} = \dfrac{(k + 4)(k - 1)}{k - 4}$ ■

Multiplication and Division

The operations of multiplication and division are performed with rational expressions in the same way that they are with arithmetic fractions (Table 8).

TABLE 8

Multiplication and Division of Rational Expressions

Operation	Example
If P, Q, R, and S are polynomials, then	
Multiplication	
$\dfrac{P}{Q} \cdot \dfrac{R}{S} = \dfrac{PR}{QS} \quad (Q, S \neq 0)$	$\dfrac{2x}{y} \cdot \dfrac{(x + 1)}{(y - 1)} = \dfrac{2x(x + 1)}{y(y - 1)}$
Division	
$\dfrac{P}{Q} \div \dfrac{R}{S} = \dfrac{P}{Q} \cdot \dfrac{S}{R} = \dfrac{PS}{QR} \quad (Q, R, S \neq 0)$	$\dfrac{x^2 + 3}{y} \div \dfrac{y^2 + 1}{x} = \dfrac{x^2 + 3}{y} \cdot \dfrac{x}{y^2 + 1} = \dfrac{x(x^2 + 3)}{y(y^2 + 1)}$

When the operations of multiplication and division are performed on rational expressions, the resulting expression should be simplified.

EXAMPLE 2 Perform the indicated operations and simplify:

a. $\dfrac{2x - 8}{x + 2} \cdot \dfrac{x^2 + 4x + 4}{x^2 - 16}$ **b.** $\dfrac{x^2 - 6x + 9}{3x + 12} \div \dfrac{x^2 - 9}{6x^2 + 18x}$

Solution

a. $\dfrac{2x - 8}{x + 2} \cdot \dfrac{x^2 + 4x + 4}{x^2 - 16}$

$= \dfrac{2(x - 4)}{x + 2} \cdot \dfrac{(x + 2)^2}{(x + 4)(x - 4)}$ Factor numerators and denominators.

$= \dfrac{2(x - 4)(x + 2)(x + 2)}{(x + 2)(x + 4)(x - 4)}$

$= \dfrac{2(x + 2)}{x + 4}$ Cancel the common factors $(x + 2)(x - 4)$.

b. $\dfrac{x^2 - 6x + 9}{3x + 12} \div \dfrac{x^2 - 9}{6x^2 + 18x} = \dfrac{x^2 - 6x + 9}{3x + 12} \cdot \dfrac{6x^2 + 18x}{x^2 - 9}$

$= \dfrac{(x - 3)^2}{3(x + 4)} \cdot \dfrac{6x(x + 3)}{(x + 3)(x - 3)}$

$= \dfrac{(x - 3)(x - 3)(6x)(x + 3)}{3(x + 4)(x + 3)(x - 3)}$

$= \dfrac{2x(x - 3)}{x + 4}$

Addition and Subtraction

The operations of addition and subtraction are performed on rational expressions by finding a common denominator for the fractions and then adding or subtracting the fractions. Table 9 shows the rules for fractions with common denominators.

TABLE 9		
Adding and Subtracting Fractions with Common Denominators		
Operation		**Example**
If P, Q, and R are polynomials, then		
Addition		
$\dfrac{P}{R} + \dfrac{Q}{R} = \dfrac{P + Q}{R} \quad (R \neq 0)$		$\dfrac{2x}{x + 2} + \dfrac{6x}{x + 2} = \dfrac{2x + 6x}{x + 2} = \dfrac{8x}{x + 2}$
Subtraction		
$\dfrac{P}{R} - \dfrac{Q}{R} = \dfrac{P - Q}{R} \quad (R \neq 0)$		$\dfrac{3y}{y - x} - \dfrac{y}{y - x} = \dfrac{3y - y}{y - x} = \dfrac{2y}{y - x}$

To add or subtract fractions that have different denominators, first find the least common denominator (LCD). To find the LCD of two or more rational expressions, follow these steps:

1. *Find the prime factors* of each denominator.
2. *Form the product of the different prime factors* that occur in the denominators. Use the highest power of each prime factor that appears in the denominators.

After finding the LCD, carry out the indicated operations following the procedure for adding and subtracting fractions with common denominators.

EXAMPLE 3 Perform the indicated operations and simplify:

a. $\dfrac{3x + 4}{4x} + \dfrac{4y - 2}{3y}$ **b.** $\dfrac{2x}{x^2 - 1} + \dfrac{3x + 1}{2x^2 - x - 1}$ **c.** $\dfrac{1}{x + h} - \dfrac{1}{x}$

Solution

a. $\dfrac{3x + 4}{4x} + \dfrac{4y - 2}{3y} = \dfrac{3x + 4}{4x} \cdot \dfrac{3y}{3y} + \dfrac{4y - 2}{3y} \cdot \dfrac{4x}{4x}$ LCD $= (4x)(3y) = 12xy$

$$= \dfrac{9xy + 12y}{12xy} + \dfrac{16xy - 8x}{12xy}$$

$$= \dfrac{25xy - 8x + 12y}{12xy}$$

b. $\dfrac{2x}{x^2 - 1} + \dfrac{3x + 1}{2x^2 - x - 1} = \dfrac{2x}{(x + 1)(x - 1)} + \dfrac{3x + 1}{(2x + 1)(x - 1)}$

$$= \dfrac{2x(2x + 1) + (3x + 1)(x + 1)}{(x + 1)(x - 1)(2x + 1)} \quad \begin{matrix} \text{LCD} = (2x + 1) \\ \cdot (x + 1)(x - 1) \end{matrix}$$

$$= \dfrac{4x^2 + 2x + 3x^2 + 3x + x + 1}{(x + 1)(x - 1)(2x + 1)}$$

$$= \dfrac{7x^2 + 6x + 1}{(x + 1)(x - 1)(2x + 1)}$$

c. $\dfrac{1}{x + h} - \dfrac{1}{x} = \dfrac{1}{x + h} \cdot \dfrac{x}{x} - \dfrac{1}{x} \cdot \dfrac{x + h}{x + h}$ LCD $= x(x + h)$

$$= \dfrac{x}{x(x + h)} - \dfrac{x + h}{x(x + h)}$$

$$= \dfrac{x - x - h}{x(x + h)}$$

$$= -\dfrac{h}{x(x + h)}$$

Complex Fractions

An expression that contains fractions in its numerator and/or denominator is called a **complex fraction.** The techniques used to simplify rational expressions may be used to simplify these fractions.

EXAMPLE 4 Simplify:

a. $\dfrac{1 + \dfrac{1}{x + 1}}{x - \dfrac{4}{x}}$ **b.** $\dfrac{\dfrac{1}{x} + \dfrac{1}{y}}{\dfrac{1}{x^2} - \dfrac{1}{y^2}}$

Solution

a. We first express the numerator and denominator of the given expression as a single quotient. Thus,

$$\frac{1 + \dfrac{1}{x+1}}{x - \dfrac{4}{x}} = \frac{1 \cdot \dfrac{x+1}{x+1} + \dfrac{1}{x+1}}{x \cdot \dfrac{x}{x} - \dfrac{4}{x}}$$

The LCD for the fraction in the numerator is $x + 1$, and the LCD for the fraction in the denominator is x.

$$= \frac{\dfrac{x+1+1}{x+1}}{\dfrac{x^2-4}{x}}$$

$$= \frac{\dfrac{x+2}{x+1}}{\dfrac{x^2-4}{x}}$$

We then multiply the numerator by the reciprocal of the denominator, obtaining

$$\frac{x+2}{x+1} \cdot \frac{x}{x^2-4} = \frac{x+2}{x+1} \cdot \frac{x}{(x-2)(x+2)}$$

Factor the denominator of the second fraction.

$$= \frac{x}{(x+1)(x-2)}$$

Cancel the common factors.

b. As before, we first write the numerator and denominator of the given expression as a single quotient and then simplify the resulting fraction.

$$\frac{\dfrac{1}{x} + \dfrac{1}{y}}{\dfrac{1}{x^2} - \dfrac{1}{y^2}} = \frac{\dfrac{y+x}{xy}}{\dfrac{y^2-x^2}{x^2y^2}}$$

The LCD for the fractions in the numerator is xy, and the LCD for the fractions in the denominator is x^2y^2.

$$= \frac{y+x}{xy} \cdot \frac{x^2y^2}{y^2-x^2}$$

$$\frac{a}{b} \div \frac{c}{d} = \frac{a}{b} \cdot \frac{d}{c}$$

$$= \frac{y+x}{xy} \cdot \frac{x^2y^2}{(y+x)(y-x)}$$

$$= \frac{xy}{y-x}$$

Cancel common factors.

APPLIED EXAMPLE 5 Optics The equation

$$\frac{1}{f} = \frac{1}{p} + \frac{1}{q}$$

sometimes called a **lens-maker's equation,** gives the relationship between the focal length f of a thin lens, the distance p of the object from the lens, and the distance q of its image from the lens.

a. Write the right-hand side of the equation as a single fraction.
b. Use the result of part (a) to find the focal length of the lens.

Solution

a. $\dfrac{1}{p} + \dfrac{1}{q} = \dfrac{q + p}{pq} = \dfrac{p + q}{pq}$

b. Using the result of part (a), we have

$$\frac{1}{f} = \frac{1}{p} + \frac{1}{q} = \frac{p + q}{pq}$$

from which we see that

$$f = \frac{pq}{p + q}$$

1.4 Self-Check Exercises

1. Simplify $\dfrac{3a^2b^3}{2ab^2 + 4ab} \cdot \dfrac{b^2 + 4b + 4}{6a^2b^5}$.

2. Simplify $\dfrac{\dfrac{x}{y} - \dfrac{y}{x}}{\dfrac{x^2 + 2xy + y^2}{x^2 - y^2}}$.

Solutions to Self-Check Exercises 1.4 can be found on page 27.

1.4 Concept Questions

1. **a.** What is a rational expression? Give an example.
   b. Explain why a polynomial is a rational expression but not vice versa.

2. **a.** If P, Q, R, and S are polynomials, what is $\left(\frac{P}{Q}\right)\left(\frac{R}{S}\right)$? What is $\left(\frac{P}{Q}\right) \div \left(\frac{R}{S}\right)$?
   b. If P, Q, and R are polynomials, what are $\left(\frac{P}{R}\right) + \left(\frac{Q}{R}\right)$ and $\left(\frac{P}{R}\right) - \left(\frac{Q}{R}\right)$?

1.4 Exercises

In Exercises 1–12, simplify the expression.

1. $\dfrac{28x^2}{7x^3}$

2. $\dfrac{3y^4}{18y^2}$

3. $\dfrac{4x + 12}{5x + 15}$

4. $\dfrac{12m - 6}{18m - 9}$

5. $\dfrac{6x^2 - 3x}{6x^2}$

6. $\dfrac{8y^2}{4y^3 - 4y^2 + 8y}$

7. $\dfrac{x^2 + x - 2}{x^2 + 3x + 2}$

8. $\dfrac{2y^2 - y - 3}{2y^2 + y - 1}$

9. $\dfrac{x^2 - 9}{2x^2 - 5x - 3}$

10. $\dfrac{6y^2 + 11y + 3}{4y^2 - 9}$

11. $\dfrac{x^3 + y^3}{x^2 - xy + y^2}$

12. $\dfrac{8r^3 - s^3}{2r^2 + rs - s^2}$

In Exercises 13–46, perform the indicated operations and simplify.

13. $\dfrac{6x^3}{32} \cdot \dfrac{8}{3x^2}$

14. $\dfrac{25y^4}{12y} \cdot \dfrac{3y^2}{5y^3}$

15. $\dfrac{3x^3}{8x^2} \div \dfrac{15x^4}{16x^5}$

16. $\dfrac{6x^5}{21x^2} \div \dfrac{4x}{7x^3}$

17. $\dfrac{3x}{x + 2y} \cdot \dfrac{5x + 10y}{6}$

18. $\dfrac{4y + 12}{y + 2} \cdot \dfrac{3y + 6}{2y - 1}$

19. $\dfrac{2m + 6}{3} \div \dfrac{3m + 9}{6}$

20. $\dfrac{3y - 6}{4y + 6} \div \dfrac{6y + 24}{8y + 12}$

21. $\dfrac{6r^2 - r - 2}{2r + 4} \cdot \dfrac{6r + 12}{4r + 2}$

22. $\dfrac{x^2 - x - 6}{2x^2 + 7x + 6} \cdot \dfrac{2x^2 - x - 6}{x^2 + x - 6}$

23. $\dfrac{k^2 - 2k - 3}{k^2 - k - 6} \div \dfrac{k^2 - 6k + 8}{k^2 - 2k - 8}$

24. $\dfrac{6y^2 - 5y - 6}{6y^2 + 13y + 6} \div \dfrac{6y^2 - 13y + 6}{9y^2 - 12y + 4}$

25. $\dfrac{2}{2x + 3} + \dfrac{3}{2x - 1}$ 26. $\dfrac{2x - 1}{x + 2} - \dfrac{x + 3}{x - 1}$

27. $\dfrac{3}{x^2 - x - 6} + \dfrac{2}{x^2 + x - 2}$

28. $\dfrac{4}{x^2 - 9} - \dfrac{5}{x^2 - 6x + 9}$

29. $\dfrac{2m}{2m^2 - 2m - 1} + \dfrac{3}{2m^2 - 3m + 3}$

30. $\dfrac{t}{t^2 + t - 2} - \dfrac{2t - 1}{2t^2 + 3t - 2}$

31. $\dfrac{x}{1 - x} + \dfrac{2x + 3}{x^2 - 1}$ 32. $2 + \dfrac{1}{a + 2} - \dfrac{2a}{a - 2}$

33. $x - \dfrac{x^2}{x + 2} + \dfrac{2}{x - 2}$ 34. $\dfrac{y}{y^2 - 1} + \dfrac{y - 1}{y + 1} - \dfrac{2y}{1 - y}$

35. $\dfrac{x}{x^2 + 5x + 6} + \dfrac{2}{x^2 - 4} - \dfrac{3}{x^2 + 3x + 2}$

36. $\dfrac{2x + 1}{2x^2 - x - 1} - \dfrac{x + 1}{2x^2 + 3x + 1} + \dfrac{4}{x^2 + 2x - 3}$

37. $\dfrac{x}{ax - ay} + \dfrac{y}{by - bx}$ 38. $\dfrac{ax + by}{ax - bx} + \dfrac{ay - bx}{by - ay}$

39. $\dfrac{1 + \dfrac{1}{x}}{1 - \dfrac{1}{x}}$ 40. $\dfrac{2 + \dfrac{2}{x}}{x - \dfrac{2}{x}}$

41. $\dfrac{\dfrac{1}{x} + \dfrac{1}{y}}{1 - \dfrac{1}{xy}}$ 42. $\dfrac{1 + \dfrac{x}{y}}{1 - \dfrac{x^2}{y^2}}$

43. $\dfrac{\dfrac{1}{x^2} - \dfrac{1}{y^2}}{x + y}$ 44. $\dfrac{\dfrac{1}{x^3} - \dfrac{1}{y^3}}{\dfrac{1}{x} - \dfrac{1}{y}}$

45. $\dfrac{\dfrac{1}{2(x + h)} - \dfrac{1}{2x}}{h}$ 46. $\dfrac{\dfrac{1}{(x + h)^2} - \dfrac{1}{x^2}}{h}$

47. **AVERAGE COST** The cost incurred by Herald Records in pressing x DVDs is

$$2.2 + \dfrac{2500}{x}$$

dollars per disc.

a. Write the cost per disc as a single fraction.
b. What is the total cost incurred by Herald Records in pressing x discs?

48. **AIR POLLUTION** The amount of nitrogen dioxide, a brown gas that impairs breathing, present in the atmosphere on a certain May day in the city of Long Beach is approximated by

$$A = \dfrac{136}{1 + 0.25(t - 4.5)^2} + 28 \qquad (0 \le t \le 11)$$

where A is measured in pollutant standard index (PSI) and t is measured in hours, with $t = 0$ corresponding to 7 A.M. Express A as a single fraction.
Source: Los Angeles Times.

49. **AMORTIZING A LOAN** The periodic payment R on a loan of P dollars to be amortized (paid off gradually by periodic payments of principal and interest) over n periods with interest charged at the rate of i per period is found by solving the equation

$$P = \dfrac{R}{i} - \dfrac{R}{i(1 + i)^n}$$

for R. Write the right-hand side of this equation as a single fraction.

50. **INVENTORY CONTROL** The equation

$$A = \dfrac{km}{q} + cm + \dfrac{hq}{2}$$

gives the annual cost of ordering and storing (as yet unsold) merchandise. Here, q is the size of each order, k is the cost of placing each order, c is the unit cost of the product, m is the number of units of the product sold per year, and h is the annual cost for storing each unit. Write the right-hand side of this equation as a single fraction.

51. **FOCAL LENGTH** The equation

$$\dfrac{1}{f} = \dfrac{1}{f_1} + \dfrac{1}{f_2} - \dfrac{d}{f_1 f_2}$$

gives the relationship between the combined focal length of two thin lenses with focal lengths f_1 and f_2, separated by a distance d.
a. Write the right-hand side of the equation as a single fraction.
b. Use the result of part (a) to find the combined focal length of the two lenses.

52. **CYLINDER PRESSURE** The pressure P, volume V, and temperature T of a gas in a cylinder are related by the van der Waals equation

$$P = \dfrac{kT}{V - b} + \dfrac{ab}{V^2(V - b)} - \dfrac{a}{V(V - b)}$$

where a, b, and k are constants. Write P as a single fraction.

1.4   Solutions to Self-Check Exercises

1. Factoring the numerator and denominator of each expression, we have

$$\frac{3a^2b^3}{2ab^2 + 4ab} \cdot \frac{b^2 + 4b + 4}{6a^2b^5} = \frac{3a^2b^3}{2ab(b + 2)} \cdot \frac{(b + 2)^2}{(3a^2b^3)(2b^2)}$$

$$= \frac{b + 2}{2ab(2b^2)} \quad \text{Cancel common factors.}$$

$$= \frac{b + 2}{4ab^3}$$

2. Writing the numerator of the given expression as a single quotient, we have

$$\frac{\dfrac{x}{y} - \dfrac{y}{x}}{\dfrac{x^2 + 2xy + y^2}{x^2 - y^2}} = \frac{\dfrac{x^2 - y^2}{xy}}{\dfrac{x^2 + 2xy + y^2}{x^2 - y^2}} \qquad \begin{array}{l}\text{The LCD for the fractions in the numerator is } xy.\end{array}$$

$$= \frac{\dfrac{(x + y)(x - y)}{xy}}{\dfrac{(x + y)(x + y)}{(x + y)(x - y)}} \qquad \text{Factor.}$$

$$= \frac{(x + y)(x - y)}{xy} \cdot \frac{(x + y)(x - y)}{(x + y)(x + y)} \qquad \frac{a}{b} \div \frac{c}{d} = \frac{a}{b} \cdot \frac{d}{c}$$

$$= \frac{(x - y)^2}{xy} \qquad \text{Cancel common factors.}$$

1.5   Integral Exponents

Exponents

We begin by recalling the definition of the exponential expression a^n, where a is a real number and n is a positive integer.

> **Exponential Expressions**
>
> If a is any real number and n is a natural number, then the expression a^n (read "a to the power n") is defined as the number
>
> $$a^n = \underbrace{a \cdot a \cdot a \cdots a}_{n \text{ factors}}$$
>
> Recall that the number a is the *base* and the superscript n is the *exponent*, or *power*, to which the base is raised.

Next, we extend our definition of a^n to include $n = 0$; that is, we define the expression a^0. Observe that if a is any real number and m and n are positive integers, then we have the rule

$$a^m a^n = \underbrace{(a \cdot a \cdots a)}_{m \text{ factors}}\underbrace{(a \cdot a \cdots a)}_{n \text{ factors}} = \underbrace{a \cdot a \cdots a}_{(m + n) \text{ factors}} = a^{m+n}$$

Now, if we require that this rule hold for the zero exponent as well, then we must have, upon setting $m = 0$,

$$a^0 a^n = a^{0+n} = a^n \quad \text{or} \quad a^0 a^n = a^n$$

Therefore, if $a \neq 0$, we can divide both sides of this last equation by a^n to obtain $a^0 = 1$. This motivates the following definition.

> **Zero Exponent**
>
> For any nonzero real number a,
>
> $$a^0 = 1$$
>
> The expression 0^0 is not defined.

EXAMPLE 1

a. $2^0 = 1$ **b.** $(-2)^0 = 1$ **c.** $(\pi)^0 = 1$ **d.** $\left(\dfrac{1}{3}\right)^0 = 1$ ◾

Next, we extend our definition to include expressions of the form a^n, where the exponent is a negative integer. Once again, we use the property

$$a^m a^n = a^{m+n}$$

where n is a positive integer. Now, if we require that this property hold for negative integral exponents as well, upon setting $m = -n$, we have

$$a^{-n} a^n = a^{-n+n} = a^0 = 1 \quad \text{or} \quad a^{-n} a^n = 1$$

Therefore, if $a \neq 0$, we can divide both sides of this last equation by a^n to obtain $a^{-n} = 1/a^n$. This motivates the following definition.

> **Exponential Expressions with Negative Exponents**
>
> If a is any nonzero real number and n is a positive integer, then
>
> $$a^{-n} = \frac{1}{a^n}$$

EXAMPLE 2 Write each of the following numbers without using exponents:

a. 4^{-2} **b.** 3^{-1} **c.** -2^{-3} **d.** $\left(\dfrac{2}{3}\right)^{-1}$ **e.** $\left(\dfrac{3}{2}\right)^{-3}$

Solution

a. $4^{-2} = \dfrac{1}{4^2} = \dfrac{1}{16}$ **b.** $3^{-1} = \dfrac{1}{3^1} = \dfrac{1}{3}$ **c.** $-2^{-3} = -\dfrac{1}{2^3} = -\dfrac{1}{8}$

d. $\left(\dfrac{2}{3}\right)^{-1} = \dfrac{1}{\left(\frac{2}{3}\right)^1} = \dfrac{1}{\frac{2}{3}} = \dfrac{3}{2}$ **e.** $\left(\dfrac{3}{2}\right)^{-3} = \dfrac{1}{\left(\frac{3}{2}\right)^3} = \dfrac{2^3}{3^3} = \dfrac{8}{27}$ ◾

Note In Example 2d and 2e, the intermediate steps may be omitted by observing that

$$\left(\frac{a}{b}\right)^{-n} = \left(\frac{b}{a}\right)^n \qquad \text{For example, } \left(\frac{2}{3}\right)^{-1} = \left(\frac{3}{2}\right)^1$$

since

$$\left(\frac{a}{b}\right)^{-n} = \frac{1}{\left(\frac{a}{b}\right)^n} = \frac{1}{\frac{a^n}{b^n}} = 1 \cdot \frac{b^n}{a^n} = \left(\frac{b}{a}\right)^n$$ ◾

Five basic properties of exponents are given in Table 10.

TABLE 10	
Properties of Exponents	
Property	**Example**
1. $a^m \cdot a^n = a^{m+n}$	$x^2 \cdot x^3 = x^{2+3} = x^5$
2. $\dfrac{a^m}{a^n} = a^{m-n}$	$\dfrac{x^7}{x^4} = x^{7-4} = x^3$
3. $(a^m)^n = a^{mn}$	$(x^4)^3 = x^{4 \cdot 3} = x^{12}$
4. $(ab)^n = a^n \cdot b^n$	$(2x)^4 = 2^4 \cdot x^4 = 16x^4$
5. $\left(\dfrac{a}{b}\right)^n = \dfrac{a^n}{b^n} \quad (b \neq 0)$	$\left(\dfrac{x}{2}\right)^3 = \dfrac{x^3}{2^3} = \dfrac{x^3}{8}$

It can be shown that these properties are valid for any real numbers a and b and any integers m and n.

Simplifying Exponential Expressions

The next two examples illustrate the use of the properties of exponents.

EXAMPLE 3 Simplify the expression, and write your answer using positive exponents only:

a. $(2x^3)(3x^5)$ **b.** $\dfrac{2x^5}{3x^4}$ **c.** $(x^{-2})^{-3}$ **d.** $(2u^{-1}v^3)^3$ **e.** $\left(\dfrac{2m^3n^4}{m^5n^3}\right)^{-1}$

Solution

a. $(2x^3)(3x^5) = 6x^{3+5} = 6x^8$ $\qquad$ Property 1

b. $\dfrac{2x^5}{3x^4} = \dfrac{2}{3}x^{5-4} = \dfrac{2}{3}x$ $\qquad$ Property 2

c. $(x^{-2})^{-3} = x^{(-2)(-3)} = x^6$ $\qquad$ Property 3

d. $(2u^{-1}v^3)^3 = 2^3 u^{(-1)(3)} v^{3(3)} = 8u^{-3}v^9 = \dfrac{8v^9}{u^3}$ $\qquad$ Property 4

e. $\left(\dfrac{2m^3n^4}{m^5n^3}\right)^{-1} = (2m^{3-5}n^{4-3})^{-1}$ $\qquad$ Property 2

$\qquad\qquad = (2m^{-2}n)^{-1}$ $\qquad$ Property 1

$\qquad\qquad = \dfrac{1}{2m^{-2}n} = \dfrac{m^2}{2n}$

EXAMPLE 4 Simplify the expression, and write your answer using positive exponents only:

a. $(2^2)^3 - (3^2)^2$ **b.** $(x^{-1} + y^{-1})^{-1}$ **c.** $\dfrac{2^{-4} \cdot (2^{-1})^2}{(2^0 + 1)^{-1}}$

Solution

a. $(2^2)^3 - (3^2)^2 = 2^6 - 3^4 = 64 - 81 = -17$

b. $(x^{-1} + y^{-1})^{-1} = \left(\dfrac{1}{x} + \dfrac{1}{y}\right)^{-1} = \left(\dfrac{y + x}{xy}\right)^{-1} = \dfrac{xy}{y + x}$

c. $\dfrac{2^{-4} \cdot (2^{-1})^2}{(2^0 + 1)^{-1}} = \dfrac{2^{-4} \cdot 2^{-2}}{(2)^{-1}} = 2^{-4-2+1} = 2^{-5} = \dfrac{1}{2^5} = \dfrac{1}{32}$

1.5 Self-Check Exercises

1. Simplify the expression, and write your answer using positive exponents only.

   a. $(3a^4)(4a^3)$ **b.** $\left(\dfrac{u^{-3}}{u^{-5}}\right)^{-2}$

2. Simplify the expression, and write your answer using positive exponents only.

   a. $(x^2y^3)^3(x^5y)^{-2}$ **b.** $\left(\dfrac{a^2b^{-1}c^3}{a^3b^{-2}}\right)^2$

Solutions to Self-Check Exercises 1.5 can be found on page 31.

1.5 Concept Questions

1. In the expression a^n, what restrictions, if any, are placed on a and n? What is a^0 if a is a nonzero real number? What is a^{-n} if n is a positive integer and $a \neq 0$?

2. Write all the properties of exponents, and illustrate with examples.

1.5 Exercises

In Exercises 1–20, rewrite the number without using exponents.

1. $(-2)^3$

2. $\left(-\dfrac{2}{3}\right)^4$

3. 7^{-2}

4. $\left(\dfrac{3}{4}\right)^{-2}$

5. $-\left(-\dfrac{1}{4}\right)^{-2}$

6. -4^2

7. $2^{-2} + 3^{-1}$

8. $-3^{-2} - \left(-\dfrac{2}{3}\right)^2$

9. $(0.03)^2$

10. $(-0.3)^{-2}$

11. 1996^0

12. $(18 + 25)^0$

13. $(ab^2)^0$, where $a, b \neq 0$

14. $(3x^2y^3)^0$, where $x, y \neq 0$

15. $\dfrac{2^3 \cdot 2^5}{2^4 \cdot 2^9}$

16. $\dfrac{6 \cdot 10^4}{3 \cdot 10^2}$

17. $\dfrac{2^{-3} \cdot 2^{-4}}{2^{-5} \cdot 2^{-2}}$

18. $\dfrac{4 \cdot 2^{-3}}{2 \cdot 4^{-2}}$

19. $\left(\dfrac{3^4 \cdot 3^{-3}}{3^{-2}}\right)^{-1}$

20. $\left(\dfrac{5^{-2} \cdot 5^{-2}}{5^{-5}}\right)^{-2}$

In Exercises 21–54, simplify the expression, and write your answer using positive exponents only.

21. $(2x^3)\left(\dfrac{1}{8}x^2\right)$

22. $(-2x^2)(3x^{-4})$

23. $\dfrac{3x^3}{2x^4}$

24. $\dfrac{(3x^2)(4x^3)}{2x^4}$

25. $(a^{-2})^3$

26. $(-a^2)^{-3}$

27. $(2x^{-2}y^2)^3$

28. $(3u^{-1}v^{-2})^{-3}$

29. $(4x^2y^{-3})(2x^{-3}y^2)$

30. $\left(\dfrac{1}{2}u^{-2}v^3\right)(4v^3)$

31. $(-x^2y)^3\left(\dfrac{2y^2}{x^4}\right)$

32. $\left(-\dfrac{1}{2}x^2y\right)^{-2}$

33. $\left(\dfrac{2u^2v^3}{3uv}\right)^{-1}$

34. $\left(\dfrac{a^{-2}}{2b^2}\right)^{-3}$

35. $(3x^{-2})^3(2x^2)^5$

36. $(2^{-1}r^3)^{-2}(3s^{-1})^2$

37. $\dfrac{3^0 \cdot 4x^{-2}}{16 \cdot (x^2)^3}$

38. $\dfrac{5x^2(3x^{-2})}{(4x^{-1})(x^3)^{-2}}$

39. $\dfrac{2^2u^{-2}(v^{-1})^3}{3^2(u^{-3}v)^2}$

40. $\dfrac{(3a^{-1}b^2)^{-2}}{(2a^2b^{-1})^{-3}}$

41. $(-2x)^{-2}(3y)^{-3}(4z)^{-2}$

42. $(3x^{-1})^2(4y^{-1})^3(2z)^{-2}$

43. $(a^2b^{-3})^2(a^{-2}b^2)^{-3}$

44. $(5u^2v^{-3})^{-1} \cdot 3(2u^2v^2)^{-2}$

45. $\left[\left(\dfrac{a^{-2}b^{-2}}{3a^{-1}b^2}\right)^2\right]^{-1}$

46. $\left[\left(\dfrac{x^2y^{-3}z^{-4}}{x^{-2}y^{-1}z^2}\right)^{-2}\right]^3$

47. $\left(\dfrac{3^2u^{-2}v^2}{2^2u^3v^{-3}}\right)^{-2}\left(\dfrac{3^2v^5}{4^2u}\right)^2$

48. $\left[\left(-\dfrac{2^2x^{-2}y^0}{3^2x^3y^{-2}}\right)^{-2}\right]^{-2}$

49. $\dfrac{x^{-1} - 1}{x^{-1} + 1}$

50. $\dfrac{x^{-1} - y^{-1}}{x^{-1} + y^{-1}}$

51. $\dfrac{u^{-1} - v^{-1}}{v - u}$

52. $\dfrac{(uv)^{-1}}{u^{-1} + v^{-1}}$

53. $\left(\dfrac{a^{-1} - b^{-1}}{a^{-1} + b^{-1}}\right)^{-1}$

54. $[(a^{-1} + b^{-1})(a^{-1} - b^{-1})]^{-2}$

In Exercises 55–57, determine whether the statement is true or false. If it is true, explain why it is true. If it is false, give an example to show why it is false.

55. If a and b are real numbers and m and n are natural numbers, then $a^m b^n = (ab)^{mn}$.

56. If a and b are real numbers $(b \neq 0)$ and m and n are natural numbers, then

$$\frac{a^m}{b^n} = \left(\frac{a}{b}\right)^{m-n}$$

57. If a and b are real numbers and n is a natural number, then $(a + b)^n = a^n + b^n$.

1.5 Solutions to Self-Check Exercises

1. a. $(3a^4)(4a^3) = 3 \cdot a^4 \cdot 4 \cdot a^3 = 12a^{4+3} = 12a^7$

b. $\left(\dfrac{u^{-3}}{u^{-5}}\right)^{-2} = \dfrac{u^{(-3)(-2)}}{u^{(-5)(-2)}} = \dfrac{u^6}{u^{10}} = u^{6-10} = u^{-4} = \dfrac{1}{u^4}$

2. a. $(x^2y^3)^3(x^5y)^{-2} = x^{2 \cdot 3}y^{3 \cdot 3}x^{5(-2)}y^{-2} = x^6y^9x^{-10}y^{-2}$
$$= x^{6-10}y^{9-2} = x^{-4}y^7$$
$$= \frac{y^7}{x^4}$$

b. $\left(\dfrac{a^2b^{-1}c^3}{a^3b^{-2}}\right)^2 = \dfrac{a^{2 \cdot 2}b^{(-1)(2)}c^{3 \cdot 2}}{a^{3 \cdot 2}b^{(-2)(2)}} = \dfrac{a^4b^{-2}c^6}{a^6b^{-4}}$
$$= a^{4-6}b^{-2+4}c^6 = a^{-2}b^2c^6$$
$$= \frac{b^2c^6}{a^2}$$

1.6 Solving Equations

Equations

An **equation** is a statement that two mathematical expressions are equal.

EXAMPLE 1 The following are examples of equations:

a. $2x + 3 = 7$

b. $3(2x + 3) = 4(x - 1) + 4$

c. $\dfrac{y}{y - 2} = \dfrac{3y + 1}{3y - 4}$

d. $\sqrt{z - 1} = 2$

In Example 1, the letters x, y, and z are called variables. A **variable** is a letter that represents a real number.

A **solution of an equation** involving one variable is a number that renders the equation a true statement when the number is substituted for the variable. For example, replacing the variable x in the equation $2x + 3 = 7$ by the number 2 gives

$$2(2) + 3 = 7$$
$$4 + 3 = 7$$

which is true. This shows that the number 2 is a solution of $2x + 3 = 7$. The set of all solutions of an equation is called the **solution set.** To *solve* an equation is synonymous with finding its solution set.

The standard procedure for solving an equation is to transform the given equation, using an appropriate operation, into an *equivalent* equation—that is, one having exactly the same solution(s) as the original equation. The transformations are repeated if necessary until the solution(s) are easily read off. The following properties of real numbers can be used to produce equivalent equations.

> **Equality Properties of Real Numbers**
>
> Let a, b, and c be real numbers.
>
> **1.** If $a = b$, then $a + c = b + c$ and $a - c = b - c$.   Addition and subtraction properties
>
> **2.** If $a = b$ and $c \neq 0$, then $ca = cb$ and $\dfrac{a}{c} = \dfrac{b}{c}$.   Multiplication and division properties

Thus, adding or subtracting the same number to both sides of an equation leads to an equivalent equation. Also, multiplying or dividing both sides of an equation by a *non-zero* number leads to an equivalent equation. Let's apply the procedure to the solution of some linear equations.

Linear Equations

A **linear equation** in the variable x is an equation that can be written in the form $ax + b = 0$, where a and b are constants with $a \neq 0$. A linear equation in x is also called a **first-degree equation in x** or an **equation of degree 1 in x.**

EXAMPLE 2 Solve the linear equation $8x - 3 = 2x + 9$.

Solution   We use the equality properties of real numbers to obtain the following equivalent equations, in which the aim is to isolate x.

$$8x - 3 = 2x + 9$$
$$8x - 3 - 2x = 2x + 9 - 2x \qquad \text{Subtract } 2x \text{ from both sides.}$$
$$6x - 3 = 9$$
$$6x - 3 + 3 = 9 + 3 \qquad \text{Add 3 to both sides.}$$
$$6x = 12$$
$$\frac{1}{6}(6x) = \frac{1}{6}(12) \qquad \text{Multiply both sides by } \tfrac{1}{6}.$$
$$x = 2$$

so the required solution is 2.

EXAMPLE 3 Solve the linear equation $3p + 2(p - 1) = -2p - 4$.

Solution

$$3p + 2(p - 1) = -2p - 4$$
$$3p + 2p - 2 = -2p - 4 \qquad \text{Use the distributive property.}$$
$$5p - 2 = -2p - 4 \qquad \text{Simplify.}$$
$$5p - 2 + 2p = -2p - 4 + 2p \qquad \text{Add } 2p \text{ to both sides.}$$
$$7p - 2 = -4$$
$$7p - 2 + 2 = -4 + 2 \qquad \text{Add 2 to both sides.}$$
$$7p = -2$$
$$\frac{1}{7}(7p) = \frac{1}{7}(-2) \qquad \text{Multiply both sides by } \tfrac{1}{7}.$$
$$p = -\frac{2}{7}$$

EXAMPLE 4 Solve the linear equation $\dfrac{2k + 1}{3} - \dfrac{k - 1}{4} = 1$.

Solution First multiply both sides of the given equation by 12, the LCD. Thus,

$$12\left(\frac{2k + 1}{3} - \frac{k - 1}{4}\right) = 12(1)$$

$$12 \cdot \frac{2k + 1}{3} - 12 \cdot \frac{k - 1}{4} = 12 \qquad \text{Use the distributive property.}$$

$$4(2k + 1) - 3(k - 1) = 12 \qquad \text{Simplify.}$$

$$8k + 4 - 3k + 3 = 12 \qquad \text{Use the distributive property.}$$

$$5k + 7 = 12 \qquad \text{Simplify.}$$

$$5k = 5 \qquad \text{Subtract 7 from both sides.}$$

$$k = 1 \qquad \text{Multiply both sides by } \tfrac{1}{5}. \qquad \blacksquare$$

Some Special Nonlinear Equations

The solution(s) of some nonlinear equations are found by solving a related linear equation as the following examples show.

EXAMPLE 5 Solve $\dfrac{2}{3(x + 1)} - \dfrac{x}{2(x + 1)} = \dfrac{1}{3}$.

Solution We multiply both sides of the given equation by $6(x + 1)$, the LCD. Thus,

$$6(x + 1) \cdot \frac{2}{3(x + 1)} - 6(x + 1) \cdot \frac{x}{2(x + 1)} = 6(x + 1) \cdot \frac{1}{3}$$

which, upon simplification, yields

$$4 - 3x = 2(x + 1)$$

$$4 - 3x = 2x + 2$$

$$4 - 3x - 2x = 2x + 2 - 2x \qquad \text{Subtract } 2x \text{ from both sides.}$$

$$4 - 5x = 2$$

$$4 - 5x - 4 = 2 - 4 \qquad \text{Subtract 4 from both sides.}$$

$$-5x = -2$$

$$x = \frac{2}{5} \qquad \text{Multiply both sides by } -\tfrac{1}{5}.$$

We can verify that $x = \tfrac{2}{5}$ is a solution of the original equation by substituting $\tfrac{2}{5}$ into the left-hand side of the equation. Thus,

$$\frac{2}{3\left(\frac{2}{5} + 1\right)} - \frac{\frac{2}{5}}{2\left(\frac{2}{5} + 1\right)} = \frac{2}{3\left(\frac{7}{5}\right)} - \frac{\frac{2}{5}}{2\left(\frac{7}{5}\right)}$$

$$= \frac{10}{21} - \frac{1}{7} = \frac{7}{21} = \frac{1}{3}$$

which is equal to the right-hand side. $\qquad \blacksquare$

When we solve an equation in x, we sometimes multiply both sides of the equation by an expression in x. The resulting equation may contain solution(s) that are not solution(s) of the original equation. Such a solution is called an **extraneous solution.** For example, the solution of the equation $3x = 0$ is of course 0. But multiplying both sides of this equation by the expression $(x - 2)$ leads to the equation $3x(x - 2) = 0$ whose solutions are 0 and 2. The solution 2 is not a solution of the original equation. Thus, if you multiply both sides of an equation by an expression that involves a variable, you should check whether each solution of the modified equation is indeed a solution of the original equation. This process is illustrated in Example 6.

EXAMPLE 6 Solve $\dfrac{x + 1}{x} - \dfrac{x - 1}{x + 1} = \dfrac{1}{x^2 + x}$.

Solution Multiplying both sides of the equation by the LCD, $x(x + 1)$, we obtain

$$(x + 1)^2 - x(x - 1) = 1 \qquad \text{Note: } x^2 + x = x(x + 1)$$
$$x^2 + 2x + 1 - x^2 + x = 1$$
$$3x + 1 = 1$$
$$3x = 0$$
$$x = 0$$

Since the original equation is not defined for $x = 0$ (division by 0 is not permitted), we see that 0 is an extraneous solution of the given equation and conclude, accordingly, that the given equation has no solution. ◾

Solving for a Specified Variable

Equations involving more than one variable occur frequently in practical applications. In these situations, we may be interested in solving for one of the variables in terms of the others. To obtain such a solution, we think of all the variables other than the one we are solving for as constants. This technique is illustrated in the next example.

$\$$ APPLIED EXAMPLE 7 Value of an Investment The equation $A = P + Prt$ gives the relationship between the value A of an investment of P dollars after t years when the investment earns simple interest at the rate of r percent per year. Solve the equation for (a) P, (b) t, and (c) r.

Solution

a.
$$A = P + Prt$$
$$A = P(1 + rt) \qquad \text{Factor.}$$
$$\frac{A}{1 + rt} = P \qquad \text{Multiply both sides by } \frac{1}{1 + rt}.$$

b.
$$A = P + Prt$$
$$A - P = Prt \qquad \text{Subtract } P \text{ from both sides.}$$
$$\frac{A - P}{Pr} = t \qquad \text{Multiply both sides by } \frac{1}{Pr}.$$

c.
$$A = P + Prt$$
$$A - P = Prt$$
$$\frac{A - P}{Pt} = r$$

1.6 Self-Check Exercises

1. Solve $2\left(\dfrac{x-1}{4}\right) - \dfrac{2x}{3} = \dfrac{4-3x}{12}$.

2. Solve $\dfrac{k}{2k+1} = \dfrac{3}{8}$.

Solutions to Self-Check Exercises 1.6 can be found on page 37.

1.6 Concept Questions

1. What is an equation? What is a solution of an equation? What is the solution set of an equation? Give examples.

2. Write the equality properties of real numbers. Illustrate with examples.

3. What is a linear equation in x? Give an example of one, and solve it.

1.6 Exercises

In Exercises 1–32, solve the given equation.

1. $3x = 12$

2. $2x = 0$

3. $0.3y = 2$

4. $2x + 5 = 11$

5. $3x + 4 = 2$

6. $2 - 3y = 8$

7. $-2y + 3 = -7$

8. $\dfrac{1}{3}k + 1 = \dfrac{1}{4}k - 2$

9. $\dfrac{1}{5}p - 3 = -\dfrac{1}{3}p + 5$

10. $3.1m + 2 = 3 - 0.2m$

11. $0.4 - 0.3p = 0.1(p + 4)$

12. $\dfrac{1}{3}k + 4 = -2\left(k + \dfrac{1}{3}\right)$

13. $\dfrac{3}{5}(k + 1) = \dfrac{1}{4}(2k + 4)$

14. $3\left(\dfrac{3m}{4} - 1\right) + \dfrac{m}{5} = \dfrac{42 - m}{4}$

15. $\dfrac{2x - 1}{3} + \dfrac{3x + 4}{4} = \dfrac{7(x + 3)}{10}$

16. $\dfrac{w - 1}{3} + \dfrac{w + 1}{4} = -\dfrac{w + 1}{6}$

17. $\dfrac{1}{2}[2x - 3(x - 4)] = \dfrac{2}{3}(x - 5)$

18. $\dfrac{1}{3}[2 - 3(x + 2)] = \dfrac{1}{4}\left[(-3x + 1) + \dfrac{1}{2}x\right]$

19. $(2x + 1)^2 - (3x - 2)^2 = 5x(2 - x)$

20. $x[(2x - 3)^2 + 5x^2] = 3x^2(3x - 4) + 18$

21. $\dfrac{8}{x} = 24$

22. $\dfrac{1}{x} + \dfrac{2}{x} = 6$

23. $\dfrac{2}{y - 1} = 4$

24. $\dfrac{1}{x + 3} = 0$

25. $\dfrac{2x - 3}{x + 1} = \dfrac{2}{5}$

26. $\dfrac{r}{3r - 1} = 4$

27. $\dfrac{2}{q - 1} = \dfrac{3}{q - 2}$

28. $\dfrac{y}{3} - \dfrac{2}{y + 1} = \dfrac{1}{3}(y - 3)$

29. $\dfrac{3k - 2}{4} - \dfrac{3k}{4} = \dfrac{k + 3}{k}$

30. $\dfrac{2x - 1}{3x + 2} = \dfrac{2x + 1}{3x + 1}$

31. $\dfrac{m - 2}{m} + \dfrac{2}{m} = \dfrac{m + 3}{m - 3}$

32. $\dfrac{4}{x(x - 2)} = \dfrac{2}{x - 2}$

In Exercises 33–48, solve the equation for the indicated variable.

33. $I = Prt$; r

34. $ax + by + c = 0$; y

35. $p = -3q + 1$; q

36. $w = \dfrac{kuv}{s^2}$; u

37. $R = R_0(1 + aT)$; T

38. $S = R\left[\dfrac{(1 + i)^n - 1}{i}\right]$; R

39. $S = R(1 + i)\left[\dfrac{(1 + i)^n - 1}{i}\right]$; R

40. $V = \dfrac{ax}{x + b}$; x

41. $V = C\left(1 - \dfrac{n}{N}\right)$; n

42. $r = \dfrac{2mI}{B(n+1)}$; m

43. $p = \dfrac{x+10}{x+4}$; x

44. $r = \dfrac{2mI}{B(n+1)}$; n

45. $y = 10\left(1 - \dfrac{1}{1+2x}\right)$; x

46. $\dfrac{1}{f} = \dfrac{1}{p} + \dfrac{1}{q}$; p

47. $\dfrac{1}{f} = \dfrac{1}{p} + \dfrac{1}{q} - \dfrac{d}{pq}$; q

48. $\dfrac{1}{R} = \dfrac{1}{R_1} + \dfrac{1}{R_2} + \dfrac{1}{R_3}$; R_3

49. SIMPLE INTEREST The simple interest I (in dollars) earned when P dollars is invested for a term of t years is given by $I = Prt$, where r is the (simple) interest rate per year. Solve for t in terms of I, P, and r. If Susan invests $1000 in a bank paying interest at the rate of 6%/year, how long must she leave it in the bank before it earns interest of $90?

50. TEMPERATURE CONVERSION The relationship between the temperature in degrees Fahrenheit (°F) and the temperature in degrees Celsius (°C) is $F = \frac{9}{5}C + 32$. Solve for C in terms of F. Then use the result to find the temperature in degrees Celsius corresponding to a temperature of 70°F.

51. WEISS'S LAW According to Weiss's Law of excitation of tissue, the strength S of an electric current is related to the time t the current takes to excite tissue by the formula

$$S = \frac{a}{t} + b \qquad (t > 0)$$

where a and b are positive constants. Solve this equation for t.

52. SPEED OF A CHEMICAL REACTION Certain proteins, known as enzymes, serve as catalysts for chemical reactions in living things. In 1913, Leonor Michaelis and L. M. Menten discovered the following formula:

$$V = \frac{ax}{x+b}$$

where a and b are positive constants, giving the initial rate V, in moles per liter per second, at which a reaction begins in terms of the amount of substrate x (the substance being acted upon), measured in moles per liter. Solve this equation for x.

53. LINEAR DEPRECIATION Suppose that an asset has an original value of C and is depreciated linearly over N years with a scrap value of S. Then the book value V (in dollars) of the asset at the end of t years is given by

$$V = C - \left(\frac{C-S}{N}\right)t$$

  a. Solve for C in terms of V, S, N, and t.
  b. A speed boat is being depreciated linearly over 5 years. If the scrap value of the boat is $40,000 and the book value of the boat at the end of 3 years is $70,000, what was its original value?

54. TAXABLE EQUIVALENT YIELD Taxable equivalent yield measures what you would have to earn on a taxable investment to match the yield provided by a tax-exempt municipal bond. Suppose that your total tax rate is $T\%$ and the stated interest rate on a tax-exempt bond is $r\%$/year. Then the taxable equivalent yield, $R\%$/year, is given by

$$R = \frac{r}{1-T}$$

  a. Solve for r in terms of R and T.
  b. If your taxable return is 6%/year and your total tax rate is 20%, what rate of return on a tax-exempt security do you need to match the after-tax return on a taxable security?

55. DISTRIBUTION OF INCOME The distribution of income in a certain city can be described by

$$y = (1.4 \cdot 10^{14})(x)^{-2}$$

where y is the number of families with an income of x or more dollars.
  a. How many families in this city have an income of $30,000 or more?
  b. How many families have an income of $60,000 or more?
  c. How many families have an income of $150,000 or more?

56. MOTION OF A CAR The distance s (in feet) covered by a car traveling along a straight road is related to its initial speed u (in ft/sec), its final speed v (in ft/sec), and its constant acceleration a (in ft/sec²) by the equation $v^2 = u^2 + 2as$.
  a. Solve the equation for a in terms of the other variables.
  b. A car starting from rest and accelerating at a constant rate reaches a speed of 88 ft/sec after traveling $\frac{1}{4}$ mile (1320 ft). What is its acceleration?

57. COWLING'S RULE Cowling's Rule is a method for calculating pediatric drug dosages. If a denotes the adult dosage (in milligrams) and if t is the child's age (in years), then the child's dosage (in milligrams) is given by

$$c = \left(\frac{t+1}{24}\right)a$$

  a. Solve the equation for t in terms of a and c.
  b. If the adult dose of a drug is 500 mg and a child received a dose of 125 mg, how old was the child?

58. AMOUNT OF RAINFALL The total amount of rain (in inches) after t hr during a rainfall is given by

$$T = \frac{0.8t}{t+4.1}$$

  a. Solve for t in terms of T.
  b. How long does it take for the total amount of rain to reach 0.4 in.?

1.6 Solutions to Self-Check Exercises

1. $2\left(\dfrac{x-1}{4}\right) - \dfrac{2x}{3} = \dfrac{4-3x}{12}$

$6(x-1) - 8x = 4 - 3x$ Multiply both sides by 12, the LCD.

$6x - 6 - 8x = 4 - 3x$

$-6 - 2x = 4 - 3x$

$-6 + x = 4$ Add $3x$ to both sides.

$x = 10$

2. $\dfrac{k}{2k+1} = \dfrac{3}{8}$

$8k = 3(2k+1)$ Multiply both sides by $8(2k+1)$, the LCD.

$8k = 6k + 3$

$2k = 3$

$k = \dfrac{3}{2}$

If we substitute this value of k into the original equation, we find

$$\frac{\frac{3}{2}}{2\left(\frac{3}{2}\right)+1} = \frac{\frac{3}{2}}{3+1} = \frac{3}{8}$$

which is equal to the right-hand side. So $k = \frac{3}{2}$ is the solution of the given equation.

1.7 Rational Exponents and Radicals

nth Roots of Real Numbers

Thus far, we have described the expression a^n, where a is a real number and n is an integer. We now direct our attention to a closely related topic: roots of real numbers. As we will soon see, expressions of the form a^n for fractional (rational) powers of n may be defined in terms of the roots of a.

> **nth Root of a Real Number**
>
> If n is a natural number and a and b are real numbers such that
>
> $$a^n = b$$
>
> then we say that a is the **nth root** of b.

For $n = 2$ and $n = 3$, the roots are commonly referred to as the **square roots** and **cube roots**, respectively. Some examples of roots follow:

- -2 and 2 are square roots of 4 because $(-2)^2 = 4$ and $2^2 = 4$.
- -3 and 3 are fourth roots of 81 because $(-3)^4 = 81$ and $3^4 = 81$.
- -4 is a cube root of -64 because $(-4)^3 = -64$.
- $\frac{1}{2}$ is a fifth root of $\frac{1}{32}$ because $\left(\frac{1}{2}\right)^5 = \frac{1}{32}$.

How many real roots does a real number b have?

1. *When n is even, the real nth roots of a positive real number b must come in pairs— one positive and the other negative.* For example, the real fourth roots of 81 include -3 and 3.
2. *When n is even and b is a negative real number, there are no real nth roots of b.* For example, if $b = -9$ and the real number a is a square root of b, then by definition, $a^2 = -9$. But this is a contradiction, since the square of a real number cannot be negative, and we conclude that b has no real roots in this case.
3. *When n is odd, then there is only one real nth root of b.* For example, the cube root of -64 is -4.

As you can see from the first statement, given a number b, there might be more than one real root. So to avoid ambiguity, we define the *principal* nth root of a positive real number, when n is even, to be the positive root. Thus, the principal square root of a real number, when n is even, is the positive root. For example, the principal square root of 4 is 2, and the principal fourth root of 81 is 3. Of course, the principal nth root of any real number b, when n is odd, is given by the (unique) nth root of b. For example, the principal cube root of -64 is -4, and the principal fifth root of $\frac{1}{32}$ is $\frac{1}{2}$.

A summary of the number of roots of a real number b is given in Table 11.

TABLE 11

Number of Roots of a Real Number b

Index	b	Number of Roots
n even	$b > 0$	Two real roots (one principal root)
	$b < 0$	No real roots
	$b = 0$	One real root
n odd	$b > 0$	One real root
	$b < 0$	One real root
	$b = 0$	One real root

We use the notation $\sqrt[n]{b}$, called a **radical**, to denote the principal nth root of b. The symbol $\sqrt{}$ is called a **radical sign**, and the number b within the radical sign is called the **radicand**. The positive integer n is called the **index** of the radical. For square roots ($n = 2$), we write $\sqrt{b}$ instead of $\sqrt[2]{b}$.

 A common mistake is to write $\sqrt[4]{16} = \pm 2$. This is wrong because $\sqrt[4]{16}$ denotes the principal fourth root of 16, which is the positive root 2. Of course, the negative of the fourth root of 16 is $-\sqrt[4]{16} = -(2) = -2$.

EXAMPLE 1 Determine the number of roots of each real number:

a. $\sqrt{25}$ **b.** $\sqrt[5]{0}$ **c.** $\sqrt[3]{-27}$ **d.** $\sqrt{-27}$

Solution

a. Here, $b > 0$, n is even, and there is one principal root. Thus, $\sqrt{25} = 5$.

b. Here, $b = 0$, n is odd, and there is one root. Thus, $\sqrt[5]{0} = 0$.

c. Here, $b < 0$, n is odd, and there is one root. Thus, $\sqrt[3]{-27} = -3$.

d. Here, $b < 0$, n is even, and no real root exists. Thus, $\sqrt{-27}$ is not defined.   ▪

 Note that $(-81)^{1/4}$ does not exist because n is even and $b < 0$, but $-81^{1/4} = -(81)^{1/4} = -3$. The first expression is "the fourth root of -81," whereas the second expression is "the negative of the fourth root of 81."

EXAMPLE 2 Evaluate the following radicals:

a. $\sqrt[6]{64}$ **b.** $\sqrt[5]{-32}$ **c.** $\sqrt[3]{\dfrac{8}{27}}$ **d.** $-\sqrt{\dfrac{4}{25}}$

Solution

a. $\sqrt[6]{64} = 2$ because $2^6 = 64$.

b. $\sqrt[5]{-32} = -2$ because $(-2)^5 = -32$.

c. $\sqrt[3]{\dfrac{8}{27}} = \dfrac{2}{3}$ because $\left(\dfrac{2}{3}\right)^3 = \dfrac{8}{27}$.

d. $-\sqrt{\dfrac{4}{25}} = -\dfrac{2}{5}$ because $\sqrt{\dfrac{4}{25}} = \dfrac{2}{5}$, so $-\sqrt{\dfrac{4}{25}} = -\dfrac{2}{5}$.   ▪

Rational Exponents and Radicals

In Section 1.5, we defined expressions such as 2^{-3}, $\left(\frac{1}{2}\right)^2$, and $1/\pi^3$ involving integral exponents. But how do we evaluate expressions such as $8^{1/3}$, in which the exponent is a rational number? From the definition of the nth root of a real number, we know that $\sqrt[3]{8} = 2$. Using rational exponents, we write this same result in the form $8^{1/3} = 2$. More generally, we have the following definitions.

Rational Exponents	Example
1. If n is a natural number and b is a real number, then $$b^{1/n} = \sqrt[n]{b}$$ (If $b < 0$ and n is even, $b^{1/n}$ is not defined.)	$9^{1/2} = \sqrt{9} = 3$ $(-8)^{1/3} = \sqrt[3]{-8} = -2$
2. If m/n is a rational number reduced to lowest terms (m, n natural numbers), then $$b^{m/n} = (b^{1/n})^m$$ or, equivalently, $$b^{m/n} = \sqrt[n]{b^m}$$ whenever it exists.	$(27)^{2/3} = (27^{1/3})^2 = 3^2 = 9$ $(27)^{2/3} = [(27^2)]^{1/3} = (729)^{1/3} = 9$ $(-27)^{2/3} = (-27^{1/3})^2$ $\qquad = (-3)^2 = 9$

EXAMPLE 3

a. $(64)^{1/3} = \sqrt[3]{64} = 4$

b. $(81)^{3/4} = (81^{1/4})^3 = 3^3 = 27$

c. $(-8)^{5/3} = (-8^{1/3})^5 = (-2)^5 = -32$

d. $\left(\dfrac{1}{27}\right)^{2/3} = \left[\left(\dfrac{1}{27}\right)^{1/3}\right]^2 = \left(\dfrac{1}{3}\right)^2 = \dfrac{1}{9}$

Expressions involving *negative* rational exponents are taken care of by the following definition.

Negative Exponents
$$a^{-m/n} = \frac{1}{a^{m/n}} \qquad (a \neq 0)$$

EXAMPLE 4

a. $4^{-5/2} = \dfrac{1}{4^{5/2}} = \dfrac{1}{(4^{1/2})^5} = \dfrac{1}{2^5} = \dfrac{1}{32}$

b. $(-8)^{-1/3} = \dfrac{1}{(-8)^{1/3}} = \dfrac{1}{-2} = -\dfrac{1}{2}$

All the properties of integral exponents listed in Table 10 (on page 29) hold for rational exponents. Examples 5 and 6 illustrate the use of these properties.

EXAMPLE 5

a. $\dfrac{16^{5/4}}{16^{1/2}} = 16^{5/4-1/2} = 16^{5/4-2/4} = 16^{3/4} = (16^{1/4})^3 = 2^3 = 8$ $\quad \dfrac{a^m}{a^n} = a^{m-n}$

b. $(6^{2/3})^3 = 6^{(2/3)\cdot 3} = 6^{6/3} = 6^2 = 36$ $\quad (a^m)^n = a^{mn}$

c. $\left(\dfrac{16}{81}\right)^{3/4} = \left[\left(\dfrac{16}{81}\right)^{1/4}\right]^3 = \left(\dfrac{16^{1/4}}{81^{1/4}}\right)^3 = \left(\dfrac{2}{3}\right)^3 = \dfrac{8}{27}$ $\quad \left(\dfrac{a}{b}\right)^n = \dfrac{a^n}{b^n}$ ∎

EXAMPLE 6 Evaluate each expression:

a. $2x^{1/2}(x^{2/3} - x^{1/4})$ **b.** $(x^{1/3} - y^{2/3})^2$

Solution

a. $2x^{1/2}(x^{2/3} - x^{1/4}) = 2x^{1/2}(x^{2/3}) - 2x^{1/2}(x^{1/4})$

$\qquad\qquad\qquad\qquad = 2x^{1/2+2/3} - 2x^{1/2+1/4} = 2x^{3/6+4/6} - 2x^{2/4+1/4}$

$\qquad\qquad\qquad\qquad = 2x^{7/6} - 2x^{3/4}$

b. $(x^{1/3} - y^{2/3})^2 = (x^{1/3})^2 - 2x^{1/3}y^{2/3} + (y^{2/3})^2$

$\qquad\qquad\qquad\quad = x^{2/3} - 2x^{1/3}y^{2/3} + y^{4/3}$ ∎

Simplifying Radicals

The properties of radicals given in Table 12 follow directly from the properties of exponents discussed earlier (Table 10, page 29).

TABLE 12	
Properties of Radicals	
Property	**Example**
If m and n are natural numbers and a and b are real numbers for which the indicated roots exist, then	
1. $(\sqrt[n]{a})^n = a$	$(\sqrt[3]{2})^3 = (2^{1/3})^3 = 2^1 = 2$
2. $\sqrt[n]{ab} = \sqrt[n]{a} \cdot \sqrt[n]{b}$	$\sqrt[3]{216} = \sqrt[3]{27 \cdot 8} = \sqrt[3]{27} \cdot \sqrt[3]{8} = 3 \cdot 2 = 6$
3. $\sqrt[n]{\dfrac{a}{b}} = \dfrac{\sqrt[n]{a}}{\sqrt[n]{b}}$ $(b \neq 0)$	$\sqrt[3]{\dfrac{8}{64}} = \dfrac{\sqrt[3]{8}}{\sqrt[3]{64}} = \dfrac{2}{4} = \dfrac{1}{2}$
4. $\sqrt[m]{\sqrt[n]{a}} = \sqrt[mn]{a}$	$\sqrt[3]{\sqrt{64}} = \sqrt[2\cdot 3]{64} = \sqrt[6]{64} = 2$
5. If n is even: $\sqrt[n]{a^n} = \lvert a\rvert$.	$\sqrt{(-3)^2} = \lvert -3\rvert = 3$
If n is odd: $\sqrt[n]{a^n} = a$.	$\sqrt[3]{-8} = -2$

⚠ A common error is to write $\sqrt[n]{a^n} = a$. This is not true if a is negative (see Example 5 in Table 12). Thus, unless the variable a is known to be nonnegative, the correct answer is given by Property 5.

When we work with algebraic expressions involving radicals, we usually express the radical in simplified form.

Simplifying Radicals

An expression involving radicals is simplified if the following conditions are satisfied:

1. The powers of all factors under the radical sign are less than the index of the radical.

> **2.** The index of the radical has been reduced as far as possible.
>
> **3.** No radical appears in a denominator.
>
> **4.** No fraction appears within a radical.

EXAMPLE 7 Determine whether each radical is in simplified form. If not, state which condition is violated.

a. $\dfrac{1}{\sqrt[4]{x^5}}$ **b.** $\sqrt[6]{y^2}$ **c.** $\sqrt{\dfrac{5}{4}}$

Solution   None of the three radicals are in simplified form. The radical in part (a) violates conditions 1 and 3; that is, the power of x is 5, which is greater than 4, the index of the radical, and a radical appears in the denominator. The radical in part (b) violates condition 2, since the index of the radical can be reduced; that is,

$$\sqrt[6]{y^2} = \sqrt[3 \cdot 2]{y^2} = \sqrt[3]{y}$$

The radical in part (c) violates condition 4, since there is a fraction within the radical. Rewriting the radical, we see that

$$\sqrt{\frac{5}{4}} = \frac{\sqrt{5}}{\sqrt{4}} = \frac{\sqrt{5}}{2} = \frac{1}{2}\sqrt{5}$$

EXAMPLE 8 Simplify each expression:

a. $\sqrt[3]{375}$ **b.** $\sqrt[3]{8x^3y^6z^9}$ **c.** $\sqrt[6]{81x^4y^2}$

Solution

a. $\sqrt[3]{375} = \sqrt[3]{3 \cdot 125} = \sqrt[3]{3} \cdot \sqrt[3]{5^3} = 5\sqrt[3]{3}$ $\sqrt[3]{5^3} = 5$

b. $\sqrt[3]{8x^3y^6z^9} = \sqrt[3]{2^3 \cdot (xy^2z^3)^3}$

$\qquad = \sqrt[3]{2^3} \cdot \sqrt[3]{(xy^2z^3)^3}$

$\qquad = 2xy^2z^3$ $\sqrt[3]{2^3} = 2, \quad \sqrt[3]{(xy^2z^3)^3} = xy^2z^3$

c. $\sqrt[6]{81x^4y^2} = \sqrt[6]{9^2(x^2y)^2}$

$\qquad = \sqrt[3 \cdot 2]{9^2} \cdot \sqrt[3 \cdot 2]{(x^2y)^2}$

$\qquad = \sqrt[3]{9} \cdot \sqrt[3]{x^2y}$ $\sqrt[3 \cdot 2]{9^2} = \sqrt[3]{9}, \quad \sqrt[3 \cdot 2]{(x^2y)^2} = \sqrt[3]{x^2y}$

$\qquad = \sqrt[3]{9x^2y}$

As was mentioned earlier, a simplified rational expression should not have radicals in its denominator. For example, $3/\sqrt{5}$ is *not* in simplified form, whereas $3\sqrt{5}/5$ is, since the denominator of the latter is free of radicals. How do we get rid of the radical $\sqrt{5}$ in the fraction $3/\sqrt{5}$? Obviously, multiplying by $\sqrt{5}$ does the job, since $(\sqrt{5})(\sqrt{5}) = \sqrt{25} = 5$! But we cannot multiply the denominator of a fraction by any number other than 1 without changing the fraction. So the solution is to multiply *both* the numerator and the denominator of $3/\sqrt{5}$ by $\sqrt{5}$. Equivalently, we multiply $3/\sqrt{5}$ by $\sqrt{5}/\sqrt{5}$ (which is equal to 1). Thus,

$$\frac{3}{\sqrt{5}} = \frac{3}{\sqrt{5}} \cdot \frac{\sqrt{5}}{\sqrt{5}} = \frac{3\sqrt{5}}{\sqrt{25}} = \frac{3\sqrt{5}}{5}$$

The process of eliminating a radical from the denominator of an algebraic expression is referred to as *rationalizing the denominator* and is illustrated in Examples 9 and 10.

EXAMPLE 9 Rationalize the denominator:

a. $\dfrac{1}{\sqrt{2}}$ **b.** $\dfrac{3x}{2\sqrt{x}}$ **c.** $\dfrac{x}{\sqrt[3]{y}}$

Solution

a. $\dfrac{1}{\sqrt{2}} \cdot \dfrac{\sqrt{2}}{\sqrt{2}} = \dfrac{\sqrt{2}}{2} = \dfrac{1}{2}\sqrt{2}$

b. $\dfrac{3x}{2\sqrt{x}} \cdot \dfrac{\sqrt{x}}{\sqrt{x}} = \dfrac{3x\sqrt{x}}{2x} = \dfrac{3}{2}\sqrt{x}$

c. $\dfrac{x}{\sqrt[3]{y}} \cdot \dfrac{\sqrt[3]{y^2}}{\sqrt[3]{y^2}} = \dfrac{x\sqrt[3]{y^2}}{\sqrt[3]{y^3}} = \dfrac{x\sqrt[3]{y^2}}{y}$

Note that here we have multiplied the denominator by $\sqrt[3]{y^2}$ so that we obtain $\sqrt[3]{y^3}$, which is equal to y.

In general, to rationalize a denominator involving an nth root, we multiply the numerator and the denominator by a factor that will yield a product in the denominator involving an nth power. For example,

$$\dfrac{1}{\sqrt[5]{x^3}} = \dfrac{1}{\sqrt[5]{x^3}} \cdot \dfrac{\sqrt[5]{x^2}}{\sqrt[5]{x^2}} = \dfrac{\sqrt[5]{x^2}}{\sqrt[5]{x^5}} = \dfrac{\sqrt[5]{x^2}}{x} \qquad \text{Since } \sqrt[5]{x^3} \cdot \sqrt[5]{x^2} = x^{3/5} \cdot x^{2/5} = x^{5/5} = x$$

EXAMPLE 10 Rationalize the denominator:

a. $\sqrt[3]{\dfrac{8}{3}}$ **b.** $\sqrt[3]{\dfrac{x}{y^2}}$

Solution

a. $\sqrt[3]{\dfrac{8}{3}} = \dfrac{\sqrt[3]{8}}{\sqrt[3]{3}} = \dfrac{2}{\sqrt[3]{3}} \cdot \dfrac{\sqrt[3]{3^2}}{\sqrt[3]{3^2}} = \dfrac{2\sqrt[3]{3^2}}{3} = \dfrac{2}{3}\sqrt[3]{9}$

b. $\sqrt[3]{\dfrac{x}{y^2}} = \dfrac{\sqrt[3]{x}}{\sqrt[3]{y^2}} = \dfrac{\sqrt[3]{x}}{\sqrt[3]{y^2}} \cdot \dfrac{\sqrt[3]{y}}{\sqrt[3]{y}} = \dfrac{\sqrt[3]{xy}}{\sqrt[3]{y^3}} = \dfrac{\sqrt[3]{xy}}{y}$

How do we rationalize the denominator of a fraction like $\dfrac{1}{1 - \sqrt{3}}$? Rather than multiplying by $\dfrac{\sqrt{3}}{\sqrt{3}}$ (which does not eliminate the radical in the denominator), we multiply by $\dfrac{1 + \sqrt{3}}{1 + \sqrt{3}}$, obtaining

$$\dfrac{1}{1 - \sqrt{3}} \cdot \dfrac{1 + \sqrt{3}}{1 + \sqrt{3}} = \dfrac{1 + \sqrt{3}}{1 - (\sqrt{3})^2} \qquad (a - b)(a + b) = a^2 - b^2$$

$$= \dfrac{1 + \sqrt{3}}{1 - 3} = \dfrac{1 + \sqrt{3}}{-2}$$

$$= -\dfrac{1 + \sqrt{3}}{2}$$

In general, to rationalize a denominator of the form $a + \sqrt{b}$, we multiply by $\dfrac{a - \sqrt{b}}{a - \sqrt{b}}$.

Similarly, to rationalize a denominator of the form $a - \sqrt{b}$, we multiply by $\dfrac{a + \sqrt{b}}{a + \sqrt{b}}$.
We refer to the quantities $a + \sqrt{b}$ and $a - \sqrt{b}$ as **conjugates** of each other.

EXAMPLE 11 Rationalize the denominator in the expression $\dfrac{3}{2 - \sqrt{5}}$.

Solution   The conjugate of $2 - \sqrt{5}$ is $2 + \sqrt{5}$. Therefore, we multiply the given
fraction by $\dfrac{2 + \sqrt{5}}{2 + \sqrt{5}}$, obtaining

$$
\frac{3}{2 - \sqrt{5}} \cdot \frac{2 + \sqrt{5}}{2 + \sqrt{5}} = \frac{3(2 + \sqrt{5})}{(2 - \sqrt{5})(2 + \sqrt{5})} = \frac{3(2 + \sqrt{5})}{2^2 - (\sqrt{5})^2}
$$
$$
= \frac{3(2 + \sqrt{5})}{-1}
$$
$$
= -3(2 + \sqrt{5})
$$

You may often find it easier to convert an expression containing a radical to one
involving rational exponents before evaluating the expression.

EXAMPLE 12 Evaluate each expression:

a. $\sqrt[3]{3^2} \cdot \sqrt[4]{9^3}$ **b.** $\sqrt[3]{x^2 y} \cdot \sqrt{xy}$

Solution

a. $\sqrt[3]{3^2} \cdot \sqrt[4]{9^3} = 3^{2/3} \cdot 9^{3/4} = 3^{2/3} \cdot 3^{6/4} = 3^{2/3 + 3/2} = 3^{4/6 + 9/6} = 3^{13/6}$
b. $\sqrt[3]{x^2 y} \cdot \sqrt{xy} = x^{2/3} y^{1/3} \cdot x^{1/2} y^{1/2} = x^{2/3 + 1/2} y^{1/3 + 1/2} = x^{4/6 + 3/6} y^{2/6 + 3/6} = x^{7/6} y^{5/6}$

Note how much easier it is to work with rational exponents in this case.

Solving Equations Involving Radicals

The following examples show how to solve equations involving radicals.

EXAMPLE 13 Solve $\sqrt{2x + 5} = 3$.

Solution   This equation is not a linear equation. To solve it, we square both sides
of the equation, obtaining

$$
(\sqrt{2x + 5})^2 = 3^2
$$
$$
2x + 5 = 9
$$
$$
2x = 4
$$
$$
x = 2
$$

Substituting $x = 2$ into the left-hand side of the original equation yields
$$
\sqrt{2(2) + 5} = \sqrt{9} = 3
$$
which is the same as the number on the right-hand side of the equation. Therefore,
the required solution is 3.

EXAMPLE 14 Solve $\sqrt{k^2 - 4} = k - 4$.

Solution Squaring both sides of the equation leads to

$$k^2 - 4 = (k - 4)^2$$
$$k^2 - 4 = k^2 - 8k + 16$$
$$-4 = -8k + 16$$
$$-20 = -8k$$
$$\frac{5}{2} = k$$

Substituting this value of k into the left-hand side of the original equation gives

$$\sqrt{\left(\frac{5}{2}\right)^2 - 4} = \sqrt{\frac{25}{4} - 4} = \sqrt{\frac{9}{4}} = \frac{3}{2}$$

But this is not equal to $\frac{5}{2} - 4 = -\frac{3}{2}$, which is the result obtained if the value of k is substituted into the right-hand side of the original equation. We conclude that the given equation has no solution.

1.7 Self-Check Exercises

1. Simplify each expression.

   a. $\dfrac{8^{2/3} \cdot 8^{4/3}}{8^{3/2}}$ **b.** $\left(\dfrac{3y^{-4}y^4}{z^{-2}}\right)^3$ **c.** $\sqrt{5} \cdot \sqrt{45}$

2. Rationalize the denominator in each expression.

   a. $\dfrac{1}{\sqrt[3]{xy}}$ **b.** $\dfrac{4}{3 + \sqrt{8}}$

   Solutions to Self-Check Exercises 1.7 can be found on page 46.

1.7 Concept Questions

1. What is the nth root of a real number? Give an example.

2. What is the principal nth root of a positive real number? Give an example.

3. What is meant by the statement "rationalize the denominator of an algebraic expression"? Illustrate the process with an example.

1.7 Exercises

In Exercises 1–20, rewrite the number without radicals or exponents.

1. $\sqrt{81}$

2. $\sqrt[3]{-27}$

3. $\sqrt[4]{256}$

4. $\sqrt[5]{-32}$

5. $16^{1/2}$

6. $625^{1/4}$

7. $8^{2/3}$

8. $32^{2/5}$

9. $-25^{1/2}$

10. $-16^{3/2}$

11. $(-8)^{2/3}$

12. $(-32)^{3/5}$

13. $\left(\dfrac{4}{9}\right)^{1/2}$

14. $\left(\dfrac{9}{25}\right)^{3/2}$

15. $\left(\dfrac{27}{8}\right)^{2/3}$

16. $\left(-\dfrac{8}{125}\right)^{1/3}$

17. $8^{-2/3}$

18. $81^{-1/4}$

19. $-\left(\dfrac{27}{8}\right)^{-1/3}$

20. $-\left(-\dfrac{8}{27}\right)^{-2/3}$

In Exercises 21–40, carry out the indicated operation and write your answer using positive exponents only.

21. $3^{1/3} \cdot 3^{5/3}$

22. $2^{6/5} \cdot 2^{-1/5}$

23. $\dfrac{3^{1/2}}{3^{5/2}}$

24. $\dfrac{3^{-5/4}}{3^{-1/4}}$

25. $\dfrac{2^{-1/2} \cdot 3^{2/3}}{2^{3/2} \cdot 3^{-1/3}}$

26. $\dfrac{4^{1/3} \cdot 4^{-2/5}}{4^{2/3}}$

27. $(2^{3/2})^4$

28. $[(-3)^{1/3}]^2$

29. $x^{2/5} \cdot x^{-1/5}$

30. $y^{-3/8} \cdot y^{1/4}$

31. $\dfrac{x^{3/4}}{x^{-1/4}}$

32. $\dfrac{x^{7/3}}{x^{-2}}$

33. $\left(\dfrac{x^3}{-27x^{-6}}\right)^{-2/3}$

34. $\left(\dfrac{27x^{-3}y^2}{8x^{-2}y^{-5}}\right)^{1/3}$

35. $\left(\dfrac{x^{-3}}{y^{-2}}\right)^{1/2}\left(\dfrac{y}{x}\right)^{3/2}$

36. $\left(\dfrac{r^n}{r^{5-2n}}\right)^4$

37. $x^{2/5}(x^2 - 2x^3)$

38. $s^{1/3}(2s - s^{1/4})$

39. $2p^{3/2}(2p^{1/2} - p^{-1/2})$

40. $3y^{1/3}(y^{2/3} - 1)^2$

In Exercises 41–52, write the expression in simplest radical form. (Assume that all variables are nonnegative.)

41. $\sqrt{32}$

42. $\sqrt{45}$

43. $\sqrt[3]{-54}$

44. $-\sqrt[4]{48}$

45. $\sqrt{16x^2y^3}$

46. $\sqrt{40a^3b^4}$

47. $\sqrt[3]{m^6n^3p^{12}}$

48. $\sqrt[3]{-27p^2q^3r^4}$

49. $\sqrt[3]{\sqrt{9}}$

50. $\sqrt[5]{\sqrt[3]{9}}$

51. $\sqrt[3]{\sqrt{x}}$

52. $\sqrt[3]{-\sqrt[4]{x^3}}$

In Exercises 53–68, rationalize the denominator of the expression.

53. $\dfrac{2}{\sqrt{3}}$

54. $\dfrac{3}{\sqrt{5}}$

55. $\dfrac{3}{2\sqrt{x}}$

56. $\dfrac{3}{\sqrt{xy}}$

57. $\dfrac{2y}{\sqrt{3y}}$

58. $\dfrac{5x^2}{\sqrt{3x}}$

59. $\dfrac{1}{\sqrt[3]{x}}$

60. $\sqrt{\dfrac{2x}{y}}$

61. $\dfrac{2}{1 + \sqrt{3}}$

62. $\dfrac{3}{1 - \sqrt{2}}$

63. $\dfrac{1 + \sqrt{2}}{1 - \sqrt{2}}$

64. $\dfrac{9 + \sqrt{2}}{3 - \sqrt{2}}$

65. $\dfrac{q}{\sqrt{q} - 1}$

66. $\dfrac{xy}{\sqrt{x} + \sqrt{y}}$

67. $\dfrac{y}{\sqrt[3]{x^2z}}$

68. $\dfrac{2x}{\sqrt[3]{xy^2}}$

In Exercises 69–76, simplify the expression. (Assume that all variables are positive.)

69. $\sqrt{\dfrac{16}{3}}$

70. $-\sqrt{\dfrac{8}{3}}$

71. $\sqrt[3]{\dfrac{2}{3}}$

72. $\sqrt[3]{\dfrac{81}{4}}$

73. $\sqrt{\dfrac{3}{2x^2}}$

74. $\sqrt{\dfrac{x^3y^5}{4}}$

75. $\sqrt[3]{\dfrac{2y^2}{3}}$

76. $\sqrt[3]{\dfrac{3a^3}{b^2}}$

In Exercises 77–84, simplify the expression.

77. $\dfrac{1}{\sqrt{a}} + \sqrt{a}$

78. $\dfrac{x}{\sqrt{x - y}} - \sqrt{x - y}$

79. $\dfrac{\sqrt{x}}{\sqrt{x} + \sqrt{y}} + \dfrac{\sqrt{y}}{\sqrt{x} - \sqrt{y}}$

80. $\dfrac{a}{\sqrt{a^2 - b^2}} - \dfrac{\sqrt{a^2 - b^2}}{a}$

81. $(x + 1)^{1/2} + \dfrac{1}{2} x(x + 1)^{-1/2}$

82. $\dfrac{1}{2} x^{-1/2}(x + y)^{1/3} + \dfrac{1}{3} x^{1/2}(x + y)^{-2/3}$

83. $\dfrac{\frac{2}{3}(1 + x^{1/3})x^{-1/2} - \frac{1}{3}x^{1/2} \cdot x^{-2/3}}{(1 + x^{1/3})^2}$

84. $\dfrac{\frac{1}{2} x^{-1/2}(x + y)^{1/2} - \frac{1}{2} x^{1/2}(x + y)^{-1/2}}{x + y}$

In Exercises 85–90, solve the given equation.

85. $\sqrt{3x + 1} = 2$

86. $\sqrt{2x - 3} - 3 = 0$

87. $\sqrt{k^2 - 4} = 4 - k$

88. $\sqrt{4k^2 - 3} = 2k + 1$

89. $\sqrt{k + 1} + \sqrt{k} = 3\sqrt{k}$

90. $\sqrt{x + 1} - \sqrt{x} = \sqrt{4x - 3}$

91. **DEMAND FOR TIRES** The management of Titan Tire Company has determined that x thousand Super Titan tires will be sold each week if the price per tire is p dollars, where p and x are related by the equation

$$x = \sqrt{144 - p} \qquad (0 < p \le 144)$$

Solve the equation for p in terms of x.

92. **DEMAND FOR WATCHES** The equation

$$x = 10\sqrt{\dfrac{50 - p}{p}} \qquad (0 < p \le 50)$$

gives the relationship between the number x, in thousands, of the Sicard sports watch demanded per week and the unit price p in dollars. Solve the equation for p in terms of x.

In Exercises 93–96, determine whether the statement is true or false. If it is true, explain why it is true. If it is false, explain why or give an example to show why it is false.

93. If a is a real number, then $\sqrt{a^2} = |a|$.

94. If a is a real number, then $-(a^2)^{1/4}$ is not defined.

95. If n is a natural number and a is a positive real number, then $(a^{1/n})^n = a$.

96. If a and b are positive real numbers, then $\sqrt{a^2 + b^2} = a + b$.

1.7   Solutions to Self-Check Exercises

1. a. $\dfrac{8^{2/3} \cdot 8^{4/3}}{8^{3/2}} = 8^{2/3+4/3-3/2} = 8^{6/3-3/2} = 8^{1/2} = \sqrt{8} = 2\sqrt{2}$

   b. $\left(\dfrac{3y^{-4}y^4}{z^{-2}}\right)^3 = \left(\dfrac{3y^{-4+4}}{z^{-2}}\right)^3 = \left(\dfrac{3y^0}{z^{-2}}\right)^3$

        $= (3z^2)^3 = 3^3 z^{2\cdot3} = 27z^6$

   c. $\sqrt{5} \cdot \sqrt{45} = \sqrt{5 \cdot 45} = \sqrt{225} = 15$

2. a. $\dfrac{1}{\sqrt[3]{xy}} \cdot \dfrac{\sqrt[3]{x^2y^2}}{\sqrt[3]{x^2y^2}} = \dfrac{\sqrt[3]{x^2y^2}}{\sqrt[3]{x^3y^3}} = \dfrac{\sqrt[3]{x^2y^2}}{xy}$

   b. $\dfrac{4}{3+\sqrt{8}} \cdot \dfrac{3-\sqrt{8}}{3-\sqrt{8}} = \dfrac{4(3-\sqrt{8})}{3^2 - (\sqrt{8})^2}$

        $= \dfrac{4(3-\sqrt{8})}{9-8} = 4(3-\sqrt{8})$

1.8   Quadratic Equations

The equation

$$2x^2 + 3x + 1 = 0$$

is an example of a quadratic equation. In general, a **quadratic equation** in the variable x is an equation that can be written in the form

$$ax^2 + bx + c = 0$$

where a, b, and c are constants and $a \neq 0$. We refer to this as the *standard form of a quadratic equation*. Equations such as

$$3x^2 + 4x = 1 \quad \text{and} \quad 2t^2 = t + 1$$

are quadratic equations in nonstandard form, but they can be easily transformed into standard form. For example, adding -1 to both sides of the first equation leads to $3x^2 + 4x - 1 = 0$, which is in standard form. Similarly, by subtracting $(t + 1)$ from both sides of the second equation, we obtain $2t^2 - t - 1 = 0$, which is also in standard form.

Solving by Factoring

We solve a quadratic equation in x by finding its roots. The *roots* of a quadratic equation in x are precisely the values of x that satisfy the equation. The method of solving quadratic equations by *factoring* relies on the following zero-product property of real numbers, which we restate here.

> **Zero-Product Property of Real Numbers**
>
> If a and b are real numbers and $ab = 0$, then either $a = 0$ or $b = 0$, or both a and $b = 0$.

This property says that the product of two real numbers is equal to zero if and only if one (or both) of the factors is equal to zero.

EXAMPLE 1 Solve $x^2 - 3x + 2 = 0$ by factoring.

Solution Factoring the given equation, we find that

$$x^2 - 3x + 2 = (x - 2)(x - 1) = 0$$

By the zero-product property of real numbers, we have

$$x - 2 = 0 \quad \text{or} \quad x - 1 = 0$$

from which we see that $x = 2$ or $x = 1$ are the roots of the equation.

If a quadratic equation is not in standard form, we first rewrite it in standard form and then factor the equation to find its roots.

EXAMPLE 2 Solve by factoring:

a. $2x^2 - 7x = -6$ **b.** $4x^2 = 3x$ **c.** $2x^2 = 6x - 4$

Solution

a. Rewriting the equation in standard form, we have

$$2x^2 - 7x + 6 = 0$$

Factoring this equation, we obtain

$$(2x - 3)(x - 2) = 0$$

Therefore,

$$2x - 3 = 0 \quad \text{or} \quad x - 2 = 0$$
$$x = \frac{3}{2} \qquad\qquad x = 2$$

b. We first write the equation in standard form:

$$4x^2 - 3x = 0$$

Factoring this equation, we have

$$x(4x - 3) = 0$$

Therefore,

$$x = 0 \quad \text{or} \quad 4x - 3 = 0$$
$$x = 0 \qquad\qquad x = \frac{3}{4}$$

(Observe that if we had divided $4x^2 - 3x = 0$ by x before factoring the original equation, we would have lost the solution $x = 0$.)

c. Rewriting the equation in standard form, we have

$$2x^2 - 6x + 4 = 0$$

Factoring, we have

$$2(x^2 - 3x + 2) = 0 \qquad \text{2 is a common factor.}$$
$$2(x - 2)(x - 1) = 0$$

and

$$x = 2 \quad \text{or} \quad x = 1$$

Solving by Completing the Square

The method of solution by factoring works well for equations that are easily factored. But what about equations such as $x^2 - 2x - 2 = 0$ that are not easily factored? Equations of this type may be solved by using the method of *completing the square*.

The Method of Completing the Square	Example
1. Write the equation $ax^2 + bx + c = 0$ in the form $$x^2 + \frac{b}{a}x = -\frac{c}{a}$$ Coefficient of x ⬑ ⬑ Constant term	$x^2 - 2x - 2 = 0$ $x^2 - 2x = 2$ Coefficient ⬑ ⬑ Constant of x term
where the coefficient of x^2 is 1 and the constant term is on the right side of the equation.	
2. Square half of the coefficient of x.	$\left(-\dfrac{2}{2}\right)^2 = 1$
3. Add the number obtained in Step 2 to both sides of the equation, factor, and solve for x.	$x^2 - 2x + 1 = 2 + 1$ $(x - 1)^2 = 3$ $x - 1 = \pm\sqrt{3}$ $x = 1 \pm \sqrt{3}$

EXAMPLE 3 Solve by completing the square:

a. $4x^2 - 3x - 2 = 0$

b. $6x^2 - 27 = 0$

Solution

a. Step 1   First, write

$$x^2 - \frac{3}{4}x - \frac{1}{2} = 0 \qquad \text{Divide the original equation by 4, the coefficient of } x^2.$$

$$x^2 - \frac{3}{4}x = \frac{1}{2} \qquad \text{Add } \tfrac{1}{2} \text{ to both sides of the equation so that the constant term is on the right side.}$$

Step 2   Square half of the coefficient of x, obtaining

$$\left(\frac{-\frac{3}{4}}{2}\right)^2 = \left(-\frac{3}{8}\right)^2 = \frac{9}{64}$$

Step 3   Add $\frac{9}{64}$ to both sides of the equation:

$$x^2 - \frac{3}{4}x + \frac{9}{64} = \frac{1}{2} + \frac{9}{64}$$

$$= \frac{41}{64}$$

Factoring the left side of the equation, we have

$$\left(x - \frac{3}{8}\right)^2 = \frac{41}{64}$$

$$x - \frac{3}{8} = \pm\frac{\sqrt{41}}{8}$$

$$x = \frac{3}{8} \pm \frac{\sqrt{41}}{8} = \frac{1}{8}\left(3 \pm \sqrt{41}\right)$$

b. This equation is much easier to solve because the coefficient of x is 0. As before, we write the equation in the form

$$6x^2 = 27 \quad \text{or} \quad x^2 = \frac{9}{2}$$

Taking the square root of both sides of the equation, we have

$$x = \pm\sqrt{\frac{9}{2}} = \pm\frac{3}{\sqrt{2}} = \pm\frac{3\sqrt{2}}{2}$$

Using the Quadratic Formula

By using the method of completing the square to solve the general quadratic equation,

$$ax^2 + bx + c = 0 \qquad (a \neq 0)$$

we obtain the quadratic formula (see Exercise 90, page 55). This formula can be used to solve any quadratic equation.

> The Quadratic Formula
>
> The solutions of $ax^2 + bx + c = 0$ $(a \neq 0)$ are given by
>
> $$x = \frac{-b \pm \sqrt{b^2 - 4ac}}{2a}$$

EXAMPLE 4 Use the quadratic formula to solve the following:

a. $2x^2 + 5x - 12 = 0$ **b.** $x^2 + 165x + 6624 = 0$ **c.** $x^2 = -3x + 8$

Solution

a. The equation is in standard form, with $a = 2$, $b = 5$, and $c = -12$. Using the quadratic formula, we find

$$x = \frac{-b \pm \sqrt{b^2 - 4ac}}{2a} = \frac{-5 \pm \sqrt{5^2 - 4(2)(-12)}}{2(2)}$$

$$= \frac{-5 \pm \sqrt{121}}{4} = \frac{-5 \pm 11}{4}$$

$$= -4 \quad \text{or} \quad \frac{3}{2}$$

Observe that this equation can also be solved by factoring. Thus,

$$2x^2 + 5x - 12 = (2x - 3)(x + 4) = 0$$

from which we see that the desired roots are $x = \frac{3}{2}$ or $x = -4$, as obtained earlier.

b. The equation is in standard form, with $a = 1$, $b = 165$, and $c = 6624$. Using the quadratic formula, we find

$$x = \frac{-b \pm \sqrt{b^2 - 4ac}}{2a} = \frac{-165 \pm \sqrt{165^2 - 4(1)(6624)}}{2(1)}$$

$$= \frac{-165 \pm \sqrt{729}}{2}$$

$$= \frac{-165 \pm 27}{2} = -96 \quad \text{or} \quad -69$$

In this case, using the quadratic formula is preferable to factoring the quadratic equation.

c. We first rewrite the given equation in the standard form $x^2 + 3x - 8 = 0$, from which we see that $a = 1$, $b = 3$, and $c = -8$. Using the quadratic formula, we find

$$x = \frac{-b \pm \sqrt{b^2 - 4ac}}{2a} = \frac{-3 \pm \sqrt{3^2 - 4(1)(-8)}}{2(1)} = \frac{-3 \pm \sqrt{41}}{2}$$

That is, the solutions are

$$\frac{-3 + \sqrt{41}}{2} \approx 1.7 \quad \text{or} \quad \frac{-3 - \sqrt{41}}{2} \approx -4.7$$

In this case, the quadratic formula proves quite handy!

EXAMPLE 5 Use the quadratic formula to solve $9x^2 - 12x + 4 = 0$.

Solution Using the quadratic formula with $a = 9$, $b = -12$, and $c = 4$, we find that

$$x = \frac{-b \pm \sqrt{b^2 - 4ac}}{2a} = \frac{-(-12) \pm \sqrt{(-12)^2 - 4(9)(4)}}{2(9)}$$

$$= \frac{12 \pm \sqrt{144 - 144}}{18} = \frac{12}{18} = \frac{2}{3}$$

Here, the only solution is $x = \frac{2}{3}$. We refer to $\frac{2}{3}$ as a *double root*. (This equation could also be solved by factoring. Try it!)

EXAMPLE 6 Solve the equation $\sqrt{2x - 1} - \sqrt{x + 3} + 1 = 0$.

Solution We proceed as follows:

$$\sqrt{2x - 1} = \sqrt{x + 3} - 1 \qquad \text{Add } \sqrt{x + 3} - 1 \text{ to both sides.}$$
$$2x - 1 = (\sqrt{x + 3} - 1)^2 \qquad \text{Square both sides.}$$
$$2x - 1 = x + 3 - 2\sqrt{x + 3} + 1$$
$$x - 5 = -2\sqrt{x + 3} \qquad \text{Simplify.}$$
$$(x - 5)^2 = (-2\sqrt{x + 3})^2 \qquad \text{Square both sides.}$$
$$x^2 - 10x + 25 = 4x + 12$$
$$x^2 - 14x + 13 = 0$$
$$(x - 1)(x - 13) = 0$$
$$x = 1 \quad \text{or} \quad 13$$

Next, we need to verify that these solutions of the quadratic equation are indeed the solutions of the original equation. Squaring an equation, which is the same as multiplying both sides of an equation by an expression in x, could introduce extraneous solutions. Now, substituting $x = 1$ into the original equation gives

$$\sqrt{2 - 1} - \sqrt{1 + 3} + 1 = 1 - 2 + 1 = 0$$

so $x = 1$ is a solution. On the other hand, if $x = 13$, we have

$$\sqrt{26 - 1} - \sqrt{16} + 1 = 5 - 4 + 1 = 2 \neq 0$$

so $x = 13$ is an extraneous solution. Therefore, the required solution is $x = 1$.

The following example gives an application involving quadratic equations.

APPLIED EXAMPLE 7 Book Design A production editor at a textbook publishing house decided that the pages of a book should have 1-inch margins at the top and bottom and $\frac{1}{2}$-inch margins on the sides. She further stipulated that the length of a page should be $1\frac{1}{2}$ times its width and have a printed area of exactly 51 square inches. Find the dimensions of a page of the book.

Solution Let x denote the width of a page of the book (Figure 3). Then the length of the page is $\frac{3}{2}x$. The dimensions of the printed area of the page are $\left(\frac{3}{2}x - 2\right)$ inches by $(x - 1)$ inches, so its area is $\left(\frac{3}{2}x - 2\right)(x - 1)$ square inches. Since the printed area is to be exactly 51 square inches, we must have

$$\left(\frac{3}{2}x - 2\right)(x - 1) = 51$$

Expanding the left-hand side of the equation gives

$$\frac{3}{2}x^2 - \frac{3}{2}x - 2x + 2 = 51$$

$$\frac{3}{2}x^2 - \frac{7}{2}x - 49 = 0$$

$$3x^2 - 7x - 98 = 0 \qquad \text{Multiply both sides by 2.}$$

Factoring, we have

$$(3x + 14)(x - 7) = 0$$

so $x = -\frac{14}{3}$ or $x = 7$. Since x must be positive, we reject the negative root and conclude that the required solution is $x = 7$. Therefore, the dimensions of the page are 7 inches by $\frac{3}{2}(7)$, or $10\frac{1}{2}$, inches.

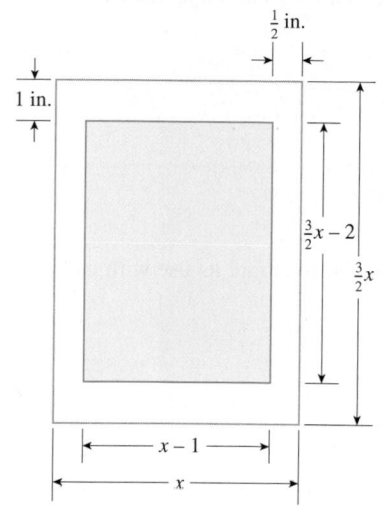

FIGURE **3**

The quantity $b^2 - 4ac$, which appears under the radical sign in the quadratic formula, is called the **discriminant.** We use the discriminant to determine the number of solutions of a quadratic equation. If the discriminant is positive, the equation has two real and distinct roots (Example 4a–c). If the discriminant is equal to 0, the equation has one double root (Example 5). Finally, if the discriminant is negative, the equation has no real roots. These results are summarized in Table 13.

TABLE 13	
Solutions of a Quadratic Equation	
Discriminant $b^2 - 4ac$	**Number of Solutions**
Positive	Two real and distinct solutions
Equal to 0	One real solution
Negative	No real solution

EXAMPLE 8 Use the discriminant to determine the number of real solutions of each equation:

a. $x^2 - 7x + 4 = 0$ **b.** $2x^2 - 3x + 4 = 0$

Solution

a. Here, $a = 1$, $b = -7$, and $c = 4$. Therefore,

$$b^2 - 4ac = (-7)^2 - 4(1)(4) = 49 - 16$$
$$= 33$$

We conclude that the equation has two real and distinct solutions.

b. Here, $a = 2$, $b = -3$, and $c = 4$. Therefore,

$$b^2 - 4ac = (-3)^2 - 4(2)(4) = 9 - 32$$
$$= -23$$

Since the discriminant is negative, we conclude that the equation has no real solution.

1.8 Self-Check Exercises

1. Solve by factoring.
   a. $x^2 - 5x + 6 = 0$
   b. $4t^2 - 4t = 3$

2. Solve the equation $2x^2 - 4x - 8 = 0$:
   a. By completing the square.
   b. By using the quadratic formula.

Solutions to Self-Check Exercises 1.8 can be found on page 55.

1.8 Concept Questions

1. What is a *quadratic equation* in x? Give an example.

2. Explain the method of completing the square. Illustrate with an example.

3. State the quadratic formula. Illustrate its use with an example.

1.8 Exercises

In Exercises 1–16, solve the equation by factoring, if required.

1. $(x + 2)(x - 3) = 0$ **2.** $(y - 3)(y - 4) = 0$

3. $x^2 - 4 = 0$ **4.** $2m^2 - 32 = 0$

5. $x^2 + x - 12 = 0$ **6.** $3x^2 - x - 4 = 0$

7. $4t^2 + 2t - 2 = 0$ **8.** $-6x^2 + x + 12 = 0$

9. $\dfrac{1}{4}x^2 - x + 1 = 0$ **10.** $\dfrac{1}{2}a^2 + a - 12 = 0$

11. $2m^2 - 7m = -6$ **12.** $6x^2 = -5x + 6$

13. $4x^2 - 9 = 0$ **14.** $8m^2 + 64m = 0$

15. $z(2z + 1) = 6$ **16.** $13m = -5 - 6m^2$

In Exercises 17–26, solve the equation by completing the square.

17. $x^2 + 2x - 8 = 0$ **18.** $x^2 - x - 6 = 0$

19. $6x^2 - 12x = 3$ **20.** $2x^2 - 6x = 20$

21. $m^2 - 3 = -m$ **22.** $p^2 - 4 = -2p$

23. $3x - 4 = -2x^2$ **24.** $10x - 5 = 4x^2$

25. $4x^2 - 13 = 0$ **26.** $7p^2 - 20 = 0$

In Exercises 27–36, solve the equation by using the quadratic formula.

27. $2x^2 - x - 6 = 0$ **28.** $6x^2 - 7x - 3 = 0$

29. $m^2 = 4m - 1$ **30.** $2x^2 = 8x - 3$

31. $8x + 3 = 8x^2$ **32.** $6p - 6 = p^2$

33. $4x = -2x^2 + 3$ **34.** $15 - 2y^2 = 7y$

35. $2.1x^2 - 4.7x - 6.2 = 0$

36. $0.2m^2 + 1.6m + 1.2 = 0$

In Exercises 37–44, solve the equation.

37. $x^4 - 5x^2 + 6 = 0$
   Hint: Let $m = x^2$. Then solve the quadratic equation in m.

38. $m^4 - 13m^2 + 36 = 0$
   Hint: Let $x = m^2$. Then solve the quadratic equation in x.

39. $y^4 - 7y^2 + 10 = 0$
   Hint: Let $x = y^2$.

40. $4x^4 - 21x^2 + 5 = 0$
   Hint: Let $y = x^2$.

41. $6(x + 2)^2 + 7(x + 2) - 3 = 0$
   Hint: Let $y = x + 2$.

42. $8(2m + 3)^2 + 14(2m + 3) - 15 = 0$
   Hint: Let $x = 2m + 3$.

43. $6w - 13\sqrt{w} + 6 = 0$
   Hint: Let $x = \sqrt{w}$.

44. $\left(\dfrac{t}{t - 1}\right)^2 - \dfrac{2t}{t - 1} - 3 = 0$
   Hint: Let $x = \dfrac{t}{t - 1}$.

In Exercises 45–64, solve the equation.

Hint: Be sure to check for extraneous solutions.

45. $\dfrac{2}{x+3} - \dfrac{4}{x} = 4$

46. $\dfrac{3y-1}{4} + \dfrac{4}{y+1} = \dfrac{5}{2}$

47. $x + 2 - \dfrac{3}{2x-1} = 0$

48. $\dfrac{x^2}{x-1} = \dfrac{3-2x}{x-1}$

49. $2 - \dfrac{7}{2y} - \dfrac{15}{y^2} = 0$

50. $6 + \dfrac{1}{k} - \dfrac{2}{k^2} = 0$

51. $\dfrac{3}{x^2-1} + \dfrac{2x}{x+1} = \dfrac{7}{3}$

52. $\dfrac{m}{m-2} - \dfrac{27}{7} = \dfrac{2}{m^2-m-2}$

53. $\dfrac{3x}{x-2} + \dfrac{4}{x+2} = \dfrac{24}{x^2-4}$

54. $\dfrac{3x}{x+1} + \dfrac{2}{x} + 5 = \dfrac{3}{x^2+x}$

55. $\dfrac{2t+1}{t-2} - \dfrac{t}{t+1} = -1$

56. $\dfrac{x}{x+1} - \dfrac{3}{x-2} + \dfrac{2}{x^2-x-2} = 0$

57. $\sqrt{u^2+u-5} = 1$

58. $\sqrt{6x^2-5x} - 2 = 0$

59. $\sqrt{2r+3} = r$

60. $\sqrt{3-4x} + 2x = 0$

61. $\sqrt{s-2} - \sqrt{s+3} + 1 = 0$

62. $\sqrt{x+1} - \sqrt{2x-5} + 1 = 0$

63. $\dfrac{1}{(x-3)^2} - \dfrac{10}{x-3} + 21 = 0$

64. $\dfrac{2}{(2x-1)^2} - \dfrac{5}{2x-1} + 3 = 0$

In Exercises 65–72, use the discriminant to determine the number of real solutions of the equation.

65. $x^2 - 6x + 5 = 0$

66. $2m^2 + 5m + 3 = 0$

67. $3y^2 - 4y + 5 = 0$

68. $2p^2 + 5p + 6 = 0$

69. $4x^2 + 12x + 9 = 0$

70. $25x^2 - 80x + 64 = 0$

71. $\dfrac{6}{k^2} + \dfrac{1}{k} - 2 = 0$

72. $(2p+1)^2 - 3(2p+1) + 4 = 0$

73. MOTION OF A BALL A person standing on the balcony of a building throws a ball directly upward. The height of the ball (in feet) as measured from the ground after t sec is given by $h = -16t^2 + 64t + 768$. When does the ball reach the ground?

74. MOTION OF A MODEL ROCKET A model rocket is launched vertically upward so that its height (measured in feet) t sec after launch is given by

$$h = -16t^2 + 384t + 4$$

a. Find the time(s) when the rocket is at a height of 1284 ft.

b. How long is the rocket in flight?

75. MOTION OF A CYCLIST A cyclist riding along a straight path has a speed of u ft/sec as she passes a tree. Accelerating at a ft/sec^2, she reaches a speed of v ft/sec t sec later, where $v = ut + at^2$. If the cyclist was traveling at 10 ft/sec and she began accelerating at a rate of 4 ft/sec^2 as she passed the tree, how long did it take her to reach a speed of 22 ft/sec?

76. PROFIT OF A VINEYARD Phillip, the proprietor of a vineyard, estimates that the profit (in dollars) from producing and selling $(x + 10,000)$ bottles of wine is $P = -0.0002x^2 + 3x + 50,000$. Find the level(s) of production that will yield a profit of $60,800.

77. DEMAND FOR SMOKE ALARMS The quantity demanded x (measured in units of a thousand) of the Sentinel smoke alarm/week is related to its unit price p (in dollars) by the equation

$$p = \dfrac{30}{0.02x^2 + 1} \qquad (0 \le x \le 10)$$

If the unit price is set at $10, what is the quantity demanded?

78. DEMAND FOR COMMODITIES The quantity demanded x (measured in units of a thousand) of a certain commodity when the unit price is set at p is given by the equation

$$p = \sqrt{-x^2 + 100}$$

If the unit price is set at $6, what is the quantity demanded?

79. SUPPLY OF SATELLITE RADIOS The quantity x of satellite radios that a manufacturer will make available in the marketplace is related to the unit price p (in dollars) by the equation

$$p = \dfrac{1}{10}\sqrt{x} + 10$$

How many satellite radios will the manufacturer make available in the marketplace if the unit price is $30?

80. OXYGEN CONTENT OF A POND When organic waste is dumped into a pond, the oxidation process that takes place reduces the pond's oxygen content. However, given time, nature will restore the oxygen content to its natural level.

Suppose that the oxygen content t days after organic waste has been dumped into the pond is given by

$$P = 100\left(\frac{t^2 + 10t + 100}{t^2 + 20t + 100}\right)$$

percent of its normal level. Find t corresponding to an oxygen content of 80%, and interpret your results.

81. **THE GOLDEN RATIO** Consider a rectangle of width x and height y (see the accompanying figure). The ratio $r = \frac{x}{y}$ satisfying the equation

$$\frac{x}{y} = \frac{x + y}{x}$$

is called the *golden ratio*. Show that

$$r = \left(\frac{1}{2}\right)(1 + \sqrt{5}) \approx 1.6$$

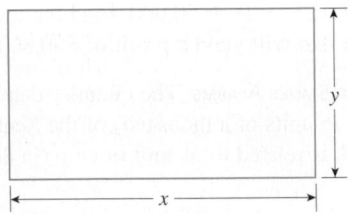

Note: A structure or a picture with a ratio of width to height equal to the golden ratio is especially pleasing to the eye. In fact, this golden ratio was used by the ancient Greeks in designing their beautiful temples and public buildings such as the Parthenon (see photo below).

Jim Winkley/Latitude/Corbis

82. **CONSTRUCTING A BOX** By cutting away identical squares from each corner of a rectangular piece of cardboard and folding up the resulting flaps, an open box may be made (see the accompanying figure). If the cardboard is 16 in.

long and 10 in. wide, find the dimensions of the resulting box if it is to have a total surface area of 144 in.^{2}.

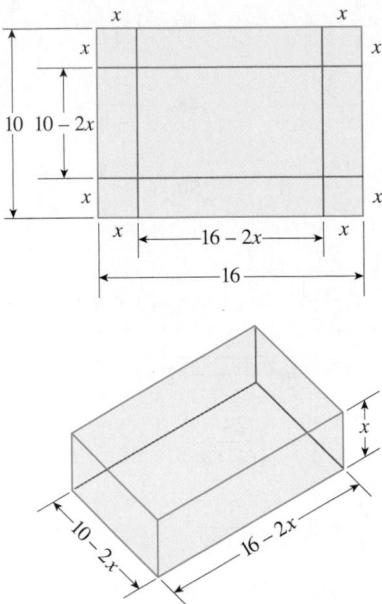

83. **ENCLOSING AN AREA** Carmen wishes to put up a fence around a proposed rectangular garden in her backyard. The length of the garden is to be twice its width, and the area of the garden is to be 200 ft^2. How many feet of fencing does she need?

84. **ENCLOSING AN AREA** George has 120 ft of fencing. He wishes to cut it into two pieces, with the purpose of enclosing two square regions. If the sum of the areas of the regions enclosed is 562.5 ft^2, how long should each piece of fencing be?

85. **WIDTH OF A SIDEWALK** A rectangular garden of length 40 ft and width 20 ft is surrounded by a path of uniform width. If the area of the walkway is 325 ft^2, what is its width?

86. **ENCLOSING AN AREA** The owner of the Rancho los Feliz has 3000 yd of fencing to enclose a rectangular piece of grazing land along the straight portion of a river. If an area of 1,125,000 yd^2 is to be enclosed, what will be the dimensions of the fenced area?

87. RADIUS OF A CYLINDRICAL CAN The surface area of a right circular cylinder is given by $S = 2\pi r^2 + 2\pi rh$, where r is the radius of the cylinder and h is its height. What is the radius of a cylinder of surface area 100 in.² and height 3 in.?

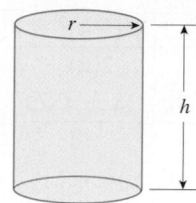

88. DESIGNING A METAL CONTAINER A metal container consists of a right circular cylinder with hemispherical ends. The surface area of the container is $S = 2\pi rl + 4\pi r^2$, where l is the length of the cylinder and r is the radius of the hemisphere. If the length of the cylinder is 4 ft and the surface area of the container is 28π ft², what is the radius of each hemisphere?

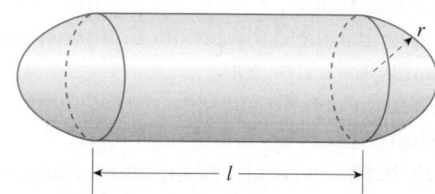

89. OIL SPILLS In calm waters, the oil spilling from the ruptured hull of a grounded oil tanker spreads in all directions. The area polluted at a certain instant of time was circular with a radius of 100 ft. A little later, the area, still circular, had increased by 4400π ft². By how much had the radius increased?

90. Derive the quadratic formula

$$x = -\frac{b}{2a} \pm \frac{\sqrt{b^2 - 4ac}}{2a}$$

for solving the equation $ax^2 + bx + c = 0$ $(a \neq 0)$ using the following steps.

a. Write the equation $ax^2 + bx + c = 0$ in the form

$$x^2 + \left(\frac{b}{a}\right)x = -\frac{c}{a}$$

b. Complete the square on the left-hand side by adding $(b/2a)^2$ to both sides of the equation in part (a) obtaining the equation

$$\left(x + \frac{b}{2a}\right)^2 = \frac{b^2 - 4ac}{4a^2}$$

c. Take the square root on both sides of the equation in part (b) to obtain the required quadratic formula.

In Exercises 91–94, determine whether the statement is true or false. If it is true, explain why it is true. If it is false, explain why or give an example to show why it is false.

91. If a and b are real numbers and $ab \neq 0$, then $a \neq 0$ or $b \neq 0$.

92. If a, b, and c are real numbers and $abc \neq 0$, then $\frac{a + b}{c}$ is a real number but $\frac{a}{b + c}$ might not be a real number.

93. If $b^2 - 4ac > 0$ and $a \neq 0$, then the roots of $ax^2 - bx + c = 0$ are the negatives of the roots of $ax^2 + bx + c = 0$.

94. If $b^2 - 4ac \neq 0$ and $a \neq 0$, then $ax^2 + bx + c = 0$ has two distinct real roots, or it has no real roots at all.

1.8 Solutions to Self-Check Exercises

1. a. Factoring the given equation, we have

$$x^2 - 5x + 6 = (x - 3)(x - 2) = 0$$

and $x = 3$ or $x = 2$.

b. Rewriting the given equation, we have

$$4t^2 - 4t - 3 = 0 \qquad \text{Add } -3 \text{ to both sides of the equation.}$$

Factoring this equation gives

$$(2t - 3)(2t + 1) = 0$$

and $t = \frac{3}{2}$ or $t = -\frac{1}{2}$.

2. a. Step 1 First write

$$x^2 - 2x - 4 = 0 \qquad \text{Divide the original equation by 2, the coefficient of } x^2.$$

$$x^2 - 2x = 4 \qquad \text{Add 4 to both sides so that the constant term is on the right side.}$$

Step 2 Square half of the coefficient of x, obtaining

$$\left(\frac{-2}{2}\right)^2 = 1$$

Step 3 Add 1 to both sides of the equation:

$$x^2 - 2x + 1 = 5$$

Factoring, we have

$$(x - 1)^2 = 5$$
$$x - 1 = \pm\sqrt{5}$$
$$x = 1 \pm \sqrt{5}$$

b. Using the quadratic formula, with $a = 2$, $b = -4$, and $c = -8$, we obtain

$$x = \frac{-b \pm \sqrt{b^2 - 4ac}}{2a}$$

$$= \frac{-(-4) \pm \sqrt{(-4)^2 - 4(2)(-8)}}{2(2)}$$

$$= \frac{4 \pm \sqrt{80}}{4} = \frac{4 \pm 4\sqrt{5}}{4} = 1 \pm \sqrt{5}$$

1.9 Inequalities and Absolute Value

Intervals

We described the system of real numbers and its properties in Section 1.1. Often, we will restrict our attention to certain subsets of the set of real numbers. For example, if x denotes the number of cars rolling off an assembly line each day in an automobile assembly plant, then x must be nonnegative; that is, $x \geq 0$. Taking this example one step further, suppose management decides that the daily production must not exceed 200 cars. Then x must satisfy the inequality $0 \leq x \leq 200$.

More generally, we will be interested in certain subsets of real numbers called finite intervals and infinite intervals. **Finite intervals** are open, closed, or half-open. The set of all real numbers that lie *strictly* between two fixed numbers a and b is called an **open interval** (a, b). It consists of all real numbers x that satisfy the inequalities $a < x < b$; it is called "open" because neither of its endpoints is included in the interval. A **closed interval** contains both of its endpoints. Thus, the set of all real numbers x that satisfy the inequalities $a \leq x \leq b$ is the closed interval $[a, b]$. Notice that brackets are used to indicate that the endpoints are included in this interval. **Half-open intervals** (also called *half-closed intervals*) contain only *one* of their endpoints. The interval $[a, b)$ is the set of all real numbers x that satisfy $a \leq x < b$, whereas the interval $(a, b]$ is described by the inequalities $a < x \leq b$. Examples of these finite intervals are illustrated in Table 14.

TABLE 14

Finite Intervals

Interval	Graph		Example	
Open: (a, b)		x	$(-2, 1)$	x
Closed: $[a, b]$		x	$[-1, 2]$	x
Half-open: $(a, b]$		x	$\left(\frac{1}{2}, 3\right]$	x
Half-open: $[a, b)$		x	$\left[-\frac{1}{2}, 3\right)$	x

Infinite intervals include the half-lines (a, ∞), $[a, \infty)$, $(-\infty, a)$, and $(-\infty, a]$, defined by the set of all real numbers that satisfy $x > a$, $x \geq a$, $x < a$, and $x \leq a$, respectively. The symbol ∞, called *infinity*, is not a real number. It is used here only for notational purposes in conjunction with the definition of infinite intervals. The notation $(-\infty, \infty)$ is used for the set of real numbers x, since, by definition, the inequalities $-\infty < x < \infty$ hold for any real number x. These infinite intervals are illustrated in Table 15.

TABLE 15

Infinite Intervals

Interval	Graph		Example	
(a, ∞)			$(2, \infty)$	
$[a, \infty)$			$[-1, \infty)$	
$(-\infty, a)$			$(-\infty, 1)$	
$(-\infty, a]$			$\left(-\infty, -\frac{1}{2}\right]$	

Inequalities

In practical applications, intervals are often found by solving one or more inequalities involving a variable. To solve these inequalities, we use the properties listed in Table 16.

TABLE 16

Properties of Inequalities

Property	Example
Let a, b, and c be any real numbers.	
1. If $a < b$ and $b < c$, then $a < c$.	$2 < 3$ and $3 < 8$, so $2 < 8$.
2. If $a < b$, then $a + c < b + c$.	$-5 < -3$, so $-5 + 2 < -3 + 2$; that is, $-3 < -1$.
3. If $a < b$ and $c > 0$, then $ac < bc$.	$-5 < -3$ and $2 > 0$, so $(-5)(2) < (-3)(2)$; that is, $-10 < -6$.
4. If $a < b$ and $c < 0$, then $ac > bc$.	$-2 < 4$ and $-3 < 0$, so $(-2)(-3) > (4)(-3)$; that is, $6 > -12$.

Similar properties hold if each inequality sign, $<$, between a and b is replaced by $\geq$, $>$, or $\leq$.

A real number is a *solution of an inequality* involving a variable if a true statement is obtained when the variable is replaced by that number. The set of all real numbers satisfying the inequality is called the *solution set*.

EXAMPLE 1 Solve $3x - 2 < 7$.

Solution Add 2 to each side of the inequality, obtaining

$$3x - 2 + 2 < 7 + 2$$
$$3x < 9$$

Next, multiply each side of the inequality by $\frac{1}{3}$, obtaining

$$\frac{1}{3}(3x) < \frac{1}{3}(9)$$

$$x < 3$$

The solution is the set of all values of x in the interval $(-\infty, 3)$.

EXAMPLE 2 Solve $-1 \leq 2x - 5 < 7$ and graph the solution set.

Solution Add 5 to each member of the double inequality, obtaining

$$4 \leq 2x < 12$$

Next multiply each member of the resulting double inequality by $\frac{1}{2}$, yielding

$$2 \leq x < 6$$

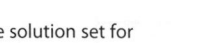

FIGURE 4
The graph of the solution set for
$-1 \leq 2x - 5 < 7$

Thus, the solution is the set of all values of x lying in the interval $[2, 6)$. The graph of the solution set is shown in Figure 4.

Solving Inequalities by Factoring

The method of factoring can be used to solve inequalities that involve polynomials of degree 2 or higher. This method relies on the principle that a polynomial changes sign only at a point where its value is 0. To find the values of x where the polynomial is equal to 0, we set the polynomial equal to 0 and then solve for x. The values obtained can then be used to help us solve the given inequality. In Examples 3 and 4, detailed steps are provided for this technique.

EXAMPLE 3 Solve $x^2 - 5x + 6 > 0$.

Solution

Step 1   *Set the polynomial in the inequality equal to 0:*

$$x^2 - 5x + 6 = 0$$

Step 2   *Factor the polynomial:*

$$(x - 3)(x - 2) = 0$$

Step 3   *Construct a sign diagram for the factors of the polynomial.* We use a $+$ to indicate that a factor is positive for a given value of x, a $-$ to indicate that it is negative, and a 0 to indicate that it is equal to 0. Now, $x - 3 < 0$ if $x < 3$, $x - 3 > 0$ if $x > 3$, and $x - 3 = 0$ if $x = 3$. Similarly, $x - 2 < 0$ if $x < 2$, $x - 2 > 0$ if $x > 2$, and $x - 2 = 0$ if $x = 2$. Using this information, we construct the sign diagram shown in Figure 5.

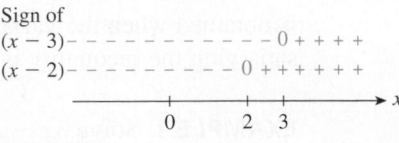

FIGURE 5

Step 4   *Determine the intervals that satisfy the given inequality.* Since $x^2 - 5x + 6 > 0$, we require that the product of the two factors be positive—that is, that both factors have the same sign. From the sign diagram, we see that the two factors have the same sign when $x < 2$ or $x > 3$. Thus, the solution set is $(-\infty, 2)$ and $(3, \infty)$.

EXAMPLE 4 Solve $x^2 + 2x - 8 < 0$.

Solution

Step 1 $x^2 + 2x - 8 = 0$

Step 2 $(x + 4)(x - 2) = 0$, so $x = -4$ or $x = 2$.

Step 3 $x + 4 > 0$ when $x > -4$, $x + 4 < 0$ when $x < -4$, and $x + 4 = 0$ when $x = -4$. Similarly, $x - 2 > 0$ when $x > 2$, $x - 2 < 0$ when $x < 2$, and $x - 2 = 0$ when $x = 2$. Using these results, we construct the sign diagram for the factors of $x^2 + 2x - 8$ (Figure 6).

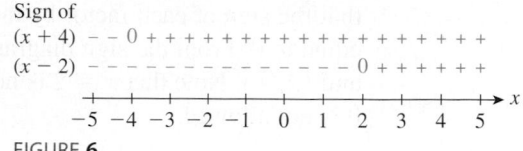

FIGURE **6**

Step 4 Since $x^2 + 2x - 8 < 0$, the product of the two factors must be negative; that is, the signs of the two factors must differ. From the sign diagram, we see that the two factors $x + 4$ and $x - 2$ have opposite signs when x lies strictly between -4 and 2. Therefore, the required solution is the interval $(-4, 2)$. ∎

Solving Inequalities Involving a Quotient

The next two examples show how an inequality involving the quotient of two algebraic expressions is solved.

EXAMPLE 5 Solve $\dfrac{x + 1}{x - 1} \geq 0$.

Solution The quotient $(x + 1)/(x - 1)$ is positive (greater than 0) when the numerator and denominator have the *same* sign. The signs of $x + 1$ and $x - 1$ are shown in Figure 7.

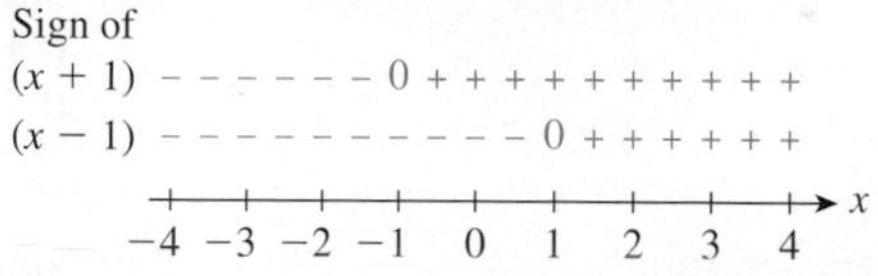

Sign of
$(x + 1)$ – – – – – – 0 + + + + + + + + + +
$(x - 1)$ – – – – – – – – – – 0 + + + + + +
$$\xrightarrow{} x$$
$-4 \quad -3 \quad -2 \quad -1 \quad 0 \quad 1 \quad 2 \quad 3 \quad 4$

FIGURE **7**

From the sign diagram, we see that $x + 1$ and $x - 1$ have the same sign when $x < -1$ or $x > 1$. The quotient $(x + 1)/(x - 1)$ is equal to 0 when $x = -1$. It is undefined at $x = 1$, since the denominator is 0 at that point. Therefore, the required solution is the set of all x in the intervals $(-\infty, -1]$ and $(1, \infty)$. ∎

EXAMPLE 6 Solve $\dfrac{2x - 1}{x - 2} \geq 1$.

Solution We rewrite the given inequality so that the right side is equal to 0:

$$\frac{2x - 1}{x - 2} - 1 \geq 0$$

$$\frac{2x - 1 - (x - 2)}{x - 2} \geq 0$$

$$\frac{2x - 1 - x + 2}{x - 2} \geq 0$$

$$\frac{x + 1}{x - 2} \geq 0$$

Next we construct the sign diagram for the factors in the numerator and the denominator (Figure 8).

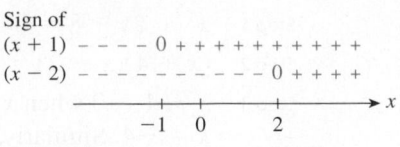

Sign of
$(x + 1)$ – – – – 0 + + + + + + + + + +
$(x - 2)$ – – – – – – – – – 0 + + + +

FIGURE **8**

Since the quotient of these two factors must be positive or equal to 0, we require that the sign of each factor be the same or that the quotient of the two factors be equal to 0. From the sign diagram, we see that the solution set is given by $(-\infty, -1]$ and $(2, \infty)$. Note that $x = 2$ is not included in the second interval, since division by 0 is not allowed.

 APPLIED EXAMPLE 7 Gross Domestic Product The gross domestic product (GDP) of a certain country is projected to be $t^2 + 2t + 50$ billion dollars t years from now. Find the time t when the GDP of the country will first equal or exceed \$58 billion.

Solution The GDP of the country will equal or exceed \$58 billion when

$$t^2 + 2t + 50 \geq 58$$

To solve this inequality for t, we first write it in the form

$$t^2 + 2t - 8 \geq 0$$
$$(t + 4)(t - 2) \geq 0$$

The sign diagram for the factors of $t^2 + 2t - 8$ is shown in Figure 9.

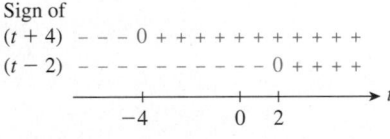

Sign of
$(t + 4)$ – – – 0 + + + + + + + + + + +
$(t - 2)$ – – – – – – – – – – – 0 + + + +

FIGURE **9**

From the sign diagram, we see that the solution set is $(-\infty, -4]$ and $[2, \infty)$. Since t must be nonnegative for the problem to be meaningful, we see that the GDP of the country is greater than or equal to \$58 billion when $t \geq 2$; that is, the GDP will first equal or exceed \$58 billion when $t = 2$, or 2 years from now.

APPLIED EXAMPLE 8 Stock Purchase The management of Corbyco, a giant conglomerate, has estimated that x thousand dollars is needed to purchase

$$100,000(-1 + \sqrt{1 + 0.001x})$$

shares of common stock of the Starr Communications Company. Determine how much money Corbyco needs in order to purchase at least 100,000 shares of Starr's stock.

Solution The amount of cash Corbyco needs to purchase at least 100,000 shares is found by solving the inequality

$$100,000(-1 + \sqrt{1 + 0.001x}) \geq 100,000$$

Proceeding, we find

$$-1 + \sqrt{1 + 0.001x} \geq 1$$
$$\sqrt{1 + 0.001x} \geq 2$$
$$1 + 0.001x \geq 4 \qquad \text{Square both sides.}$$
$$0.001x \geq 3$$
$$x \geq 3000$$

so Corbyco needs at least \$3,000,000. (Remember, x is measured in thousands of dollars.)

Absolute Value

> ### Absolute Value
>
> The **absolute value** of a number a is denoted by $|a|$ and is defined by
>
> $$|a| = \begin{cases} a & \text{if } a \geq 0 \\ -a & \text{if } a < 0 \end{cases}$$

Since $-a$ is a positive number when a is negative, it follows that the absolute value of a number is always nonnegative. For example, $|5| = 5$ and $|5| = -(-5) = 5$. Geometrically, $|a|$ is the distance between the origin and the point on the number line that represents the number a (Figure 10a and b).

The absolute value properties are given in Table 17. Property 4 is called the **triangle inequality.**

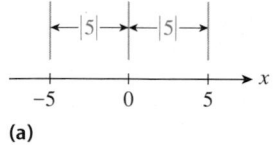

(a)

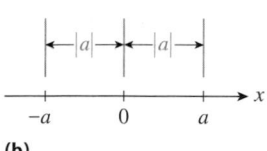

(b)

FIGURE **10**

TABLE 17	
Absolute Value Properties	
Property	**Example**
If a and b are any real numbers, then	
1. $\lvert -a \rvert = \lvert a \rvert$	$\lvert -3 \rvert = -(-3) = 3 = \lvert 3 \rvert$
2. $\lvert ab \rvert = \lvert a \rvert \lvert b \rvert$	$\lvert (2)(-3) \rvert = \lvert -6 \rvert = 6$ $= \lvert 2 \rvert \lvert -3 \rvert$
3. $\left\lvert \dfrac{a}{b} \right\rvert = \dfrac{\lvert a \rvert}{\lvert b \rvert}$	$\left\lvert \dfrac{-3}{-4} \right\rvert = \left\lvert \dfrac{3}{4} \right\rvert = \dfrac{3}{4} = \dfrac{\lvert -3 \rvert}{\lvert -4 \rvert}$
4. $\lvert a + b \rvert \leq \lvert a \rvert + \lvert b \rvert$	$\lvert 8 + (-5) \rvert = \lvert 3 \rvert = 3$ $\leq \lvert 8 \rvert + \lvert -5 \rvert = 13$

EXAMPLE 9 Evaluate each expression:

a. $\lvert \pi - 5 \rvert + 3$ **b.** $\lvert \sqrt{3} - 2 \rvert + \lvert 2 - \sqrt{3} \rvert$

Solution

a. Since $\pi - 5 < 0$, we see that $\lvert \pi - 5 \rvert = -(\pi - 5)$. Therefore,

$$\lvert \pi - 5 \rvert + 3 = -(\pi - 5) + 3 = -\pi + 5 + 3 = 8 - \pi$$

b. Since $\sqrt{3} - 2 < 0$, we see that $\lvert \sqrt{3} - 2 \rvert = -(\sqrt{3} - 2)$. Next observe that $2 - \sqrt{3} > 0$, so $\lvert 2 - \sqrt{3} \rvert = 2 - \sqrt{3}$. Therefore,

$$\lvert \sqrt{3} - 2 \rvert + \lvert 2 - \sqrt{3} \rvert = -(\sqrt{3} - 2) + (2 - \sqrt{3}) = -\sqrt{3} + 2 + 2 - \sqrt{3}$$
$$= 4 - 2\sqrt{3} = 2(2 - \sqrt{3})$$

EXAMPLE 10 Solve the inequalities $|x| \leq 5$ and $|x| \geq 5$.

Solution   We first consider the inequality $|x| \leq 5$. If $x > 0$, then $|x| = x$, so $|x| \leq 5$ implies $x \leq 5$ in this case. However, if $x < 0$, then $|x| = -x$, so $|x| \leq 5$ implies that $-x \leq 5$, or $x \geq -5$. Thus, $|x| \leq 5$ means $-5 \leq x \leq 5$ (Figure 11a). Alternatively, observe that $|x|$ is the distance from the point x to 0, so the inequality $|x| \leq 5$ implies immediately that $-5 \leq x \leq 5$.

Next, the inequality $|x| \geq 5$ states that the distance from x to 0 is greater than or equal to 5. This observation yields the result $x \geq 5$ or $x \leq -5$ (Figure 11b).

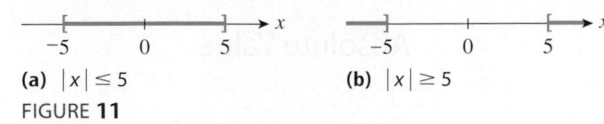

(a) $|x| \leq 5$ **(b)** $|x| \geq 5$

FIGURE **11**

The results of Example 10 may be generalized. Thus, if $k > 0$, then $|x| \leq k$ is equivalent to $-k \leq x \leq k$, and $|x| \geq k$ is equivalent to $x \geq k$ or $x \leq -k$.

EXAMPLE 11 Solve the inequality $|2x - 3| \leq 1$.

Solution   The inequality $|2x - 3| \leq 1$ is equivalent to the inequalities $-1 \leq 2x - 3 \leq 1$ (see Example 10). Then

$$2 \leq 2x \leq 4 \qquad \text{Add 3 to each member of the inequality.}$$

and

$$1 \leq x \leq 2 \qquad \text{Multiply each member of the inequality by } \tfrac{1}{2}.$$

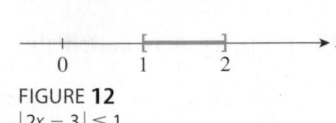

FIGURE **12**
$|2x - 3| \leq 1$

Therefore, the solution is given by the set of all x in the interval $[1, 2]$ (Figure 12).

EXAMPLE 12 Solve the inequality $|5x + 7| \geq 18$.

Solution   Referring to Example 10 once again, we see that $|5x + 7| \geq 18$ is equivalent to

$$5x + 7 \leq -18 \quad \text{or} \quad 5x + 7 \geq 18$$

That is,

$$5x \leq -25 \quad \text{or} \quad 5x \geq 11$$

$$x \leq -5 \qquad\qquad x \geq \frac{11}{5}$$

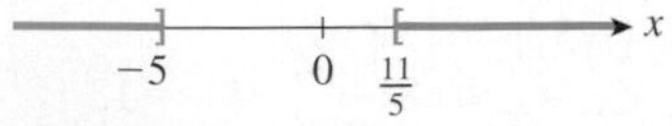

FIGURE **13**
$|5x + 7| \geq 18$

Therefore, the solution is given by the set of all x in the interval $(-\infty, -5]$ and the interval $\left[\frac{11}{5}, \infty\right)$ (Figure 13).

1.9   Self-Check Exercises

1. Solve $-1 < 2x - 1 \leq 5$, and graph the solution set.

2. Solve $6x^2 - 5x - 4 \leq 0$.

Solutions to Self-Check Exercises 1.9 can be found on page 65.

1.9 Concept Questions

1. Let a, b, c, and d be real numbers.
   a. If $a < b$ and $b < c$, write an inequality involving ad and cd, where $d < 0$. Explain.
   b. If $a < b$ and $c < d$, show that $a + c < b + d$.

2. Let a, b, and c be real numbers. If $a < b$ and $c(a - b) > 0$, what can you say about the sign of c? Explain.

3. What is the absolute value of a number a? Can $|a|$ be negative? Explain.

4. Let a and b be real numbers. Show that
   (a) $|a - b| \leq |a| + |b|$ and (b) $|a - b| = |b - a|$.

1.9 Exercises

In Exercises 1–4, determine whether the statement is true or false.

1. $-3 < -20$

2. $-5 \leq -5$

3. $\dfrac{2}{3} > \dfrac{5}{6}$

4. $-\dfrac{5}{6} < -\dfrac{11}{12}$

In Exercises 5–10, show the interval on a number line.

5. $(3, 6)$

6. $(-2, 5]$

7. $[-1, 4)$

8. $\left[-\dfrac{6}{5}, -\dfrac{1}{2}\right]$

9. $(0, \infty)$

10. $(-\infty, 5]$

In Exercises 11–16, rewrite the expression using the symbols $<$, $\leq$, $>$, or $\geq$.

11. x is less than 10.

12. a is greater than π.

13. $x + y$ is less than or equal to z.

14. $2a$ is greater than or equal to $b + 1$.

15. $3x$ is greater than -3 but less than or equal to 8.

16. $2x + 1$ is greater than or equal to a but less than b.

In Exercises 17–34, find the values of x that satisfy the inequalities.

17. $2x + 2 < 8$

18. $-6 > 4 + 5x$

19. $-4x \geq 20$

20. $-12 \leq -3x$

21. $-6 < x - 2 < 4$

22. $0 \leq x + 1 \leq 4$

23. $x + 1 > 4$ or $x + 2 < -1$

24. $x + 1 > 2$ or $x - 1 < -2$

25. $x + 3 > 1$ and $x - 2 < 1$

26. $x - 4 \leq 1$ and $x + 3 > 2$

27. $(x + 3)(x - 5) \leq 0$

28. $(2x - 4)(x + 2) \geq 0$

29. $(2x - 3)(x - 1) \leq 0$

30. $(3x - 4)(2x + 2) \leq 0$

31. $\dfrac{x + 3}{x - 2} \geq 0$

32. $\dfrac{2x - 3}{x + 1} \geq 4$

33. $\dfrac{x - 2}{x - 1} \leq 2$

34. $\dfrac{2x - 1}{x + 2} \leq 4$

In Exercises 35–44, evaluate the expression.

35. $|-6 + 2|$

36. $4 + |-4|$

37. $\dfrac{|-12 + 4|}{|16 - 12|}$

38. $\left|\dfrac{0.2 - 1.4}{1.6 - 2.4}\right|$

39. $\sqrt{3}|-2| + 3|-\sqrt{3}|$

40. $|-1| + \sqrt{2}|-2|$

41. $|\pi - 1| + 2$

42. $|\pi - 6| - 3$

43. $|\sqrt{2} - 1| + |3 - \sqrt{2}|$

44. $|2\sqrt{3} - 3| - |\sqrt{3} - 4|$

In Exercises 45–50, suppose a and b are real numbers other than 0 and $a > b$. State whether the inequality is true or false.

45. $b - a > 0$

46. $\dfrac{a}{b} > 1$

47. $a^2 > b^2$

48. $\dfrac{1}{a} > \dfrac{1}{b}$

49. $a^3 > b^3$

50. $-a < -b$

51. Write the inequality $|x - a| < b$ without using absolute values.

52. Write the inequality $|x - a| \geq b$ without using absolute values.

In Exercises 53–56, rewrite the expression using the absolute value symbol.

53. x is equal to a or $-a$, where $a > 0$.

54. $x + y$ is equal to -1 or 1.

55. a is less than or equal to 8 but greater than or equal to -8.

56. $x + 4$ is less than or equal to $a + 2b$ but greater than or equal to $-(a + 2b)$.

In Exercises 57–62, determine whether the statement is true for all real numbers a and b.

57. $|-a| = a$ **58.** $|b^2| = b^2$

59. $|a - 4| = |4 - a|$ **60.** $|a + 1| = |a| + 1$

61. $|a + b| = |a| + |b|$ **62.** $|a - b| = |a| - |b|$

In Exercises 63 and 64, state whether the inequality is true or false.

63. $x^4 + 2x^2 - 5 \geq -5$ for all x.

64. $201x^5 + 62x^3 + 26x \leq 0$ for all $x \leq 0$.

65. Find the minimum cost C (in dollars) given that

$$5(C - 25) \geq 1.75 + 2.5C$$

66. Find the maximum profit P (in dollars) given that

$$6(P - 2500) \leq 4(P + 2400)$$

67. Driving Range of a Car An advertisement for a certain car states that the EPA fuel economy is 20 mpg city and 27 mpg highway and that the car's fuel-tank capacity is 18.1 gal. Assuming ideal driving conditions, determine the driving range for the car from the foregoing data.

68. Exchanging Currency Natalie arrives in Paris and wishes to purchase at least 300 euros from a currency exchange. If the exchange rate is 0.74 € per U.S. dollar, how many U.S. dollars does she need? How many euros will she acquire in the transaction?

69. Production Levels The monthly profit realized by Auto-Time, a manufacturer of 24-hour variable timers, is $P = 6x - 48,000$ dollars if x units of the timers are manufactured and sold per month. Find the level of production of the timers if the monthly profit is between $72,000 and $75,000 per month.

70. Celsius and Fahrenheit Temperatures The relationship between Celsius (°C) and Fahrenheit (°F) temperatures is given by the formula

$$C = \frac{5}{9}(F - 32)$$

  a. If the temperature range for Montreal during the month of January is $-15° < C° < -5°$, find the range in degrees Fahrenheit in Montreal for the same period.

  b. If the temperature range for New York City during the month of June is $63° < F° < 80°$, find the range in degrees Celsius in New York City for the same period.

71. Meeting Sales Targets A salesman's monthly commission is 15% on all sales over $12,000. If his goal is to make a commission of at least $6000/month, what minimum monthly sales figures must he attain?

72. Markup on a Car The markup on a used car was at least 30% of its current wholesale price. If the car was sold for $11,200, what was the maximum wholesale price?

73. Half-Marathons Half-marathons are growing in popularity. The number of finishers, y (in thousands), in year t from 2008 through 2013 is given by

$$y = 892.2 + 239.7t \qquad (0 \leq t \leq 5)$$

where $t = 0$ corresponds to 2008. Find the year in which the number of finishers of half-marathons first equaled or exceeded 1,611,300.
Source: Competition Group and Running U.S.A.

74. Meeting Profit Goals A manufacturer of a certain commodity has estimated that her profit (in thousands of dollars) is given by the expression

$$-6x^2 + 30x - 10$$

where x (in thousands) is the number of units produced. What production range will enable the manufacturer to realize a profit of at least $14,000 on the commodity?

75. Concentration of a Drug in the Bloodstream The concentration (in milligrams per cubic centimeter) of a certain drug in a patient's bloodstream t hr after injection is given by

$$\frac{0.2t}{t^2 + 1}$$

Find the interval of time when the concentration of the drug is greater than or equal to 0.08 mg/cc.

76. Cost of Removing Toxic Pollutants A city's main well was recently found to be contaminated with trichloroethylene (a cancer-causing chemical) as a result of an abandoned chemical dump that leached chemicals into the water. A proposal submitted to the city council indicated that the cost, in millions of dollars, of removing $x\%$ of the toxic pollutants is

$$\frac{0.5x}{100 - x}$$

If the city could raise between $25 and $30 million inclusive for the purpose of removing the toxic pollutants, what is the range of pollutants that could be expected to be removed?

77. Average Speed of a Vehicle The average speed of a vehicle in miles per hour on a stretch of route 134 between 6 A.M. and 10 A.M. on a typical weekday is approximated by the expression

$$20t - 40\sqrt{t} + 50 \qquad (0 \leq t \leq 4)$$

where t is measured in hours, with $t = 0$ corresponding to 6 A.M. Over what interval of time is the average speed of a vehicle less than or equal to 35 mph?

78. Effect of Bactericide The number of bacteria in a certain culture t min after an experimental bactericide is introduced is given by

$$\frac{10{,}000}{t^2 + 1} + 2000$$

Find the time when the number of bacteria will have dropped below 4000.

79. Air Pollution Nitrogen dioxide is a brown gas that impairs breathing. The amount of nitrogen dioxide present in the atmosphere on a certain May day in the city of Long Beach measured in PSI (pollutant standard index) at time t, where t is measured in hours, and $t = 0$ corresponds to 7 A.M., is approximated by

$$\frac{136}{1 + 0.25(t - 4.5)^2} + 28 \qquad (0 \le t \le 11)$$

Find the time of the day when the amount of nitrogen dioxide is greater than or equal to 128 PSI.
Source: Los Angeles Times.

80. Motion of a Ball A ball is thrown straight up so that its height after t sec is

$$128t - 16t^2 + 4$$

ft. Determine the length of time the ball stays at or above a height of 196 ft.

81. Quality Control PAR Manufacturing Company manufactures steel rods. Suppose that the rods ordered by a customer are manufactured to a specification of 0.5 in. and are acceptable only if they are within the *tolerance limits* of 0.49 in. and 0.51 in. Letting x denote the diameter of a rod, write an inequality using absolute value to express a criterion involving x that must be satisfied in order for a rod to be acceptable.

82. Quality Control The diameter x (in inches) of a batch of ball bearings manufactured by PAR Manufacturing satisfies the inequality

$$|x - 0.1| \le 0.01$$

What is the smallest diameter a ball bearing in the batch can have? The largest diameter?

1.9   Solutions to Self-Check Exercises

1.
$$-1 < 2x - 1 \le 5$$
$$-1 + 1 < 2x - 1 + 1 \le 5 + 1 \qquad \text{Add 1 to each member of the inequality.}$$
$$0 < 2x \le 6 \qquad \text{Combine like terms.}$$
$$0 < x \le 3 \qquad \text{Multiply each member of the inequality by } \tfrac{1}{2}.$$

We conclude that the solution set is $(0, 3]$. The graph of the solution set is shown in the following figure:

2. Step 1 $6x^2 - 5x - 4 \le 0$

Step 2 $(3x - 4)(2x + 1) \le 0$

Step 3 $3x - 4 > 0$ when $x > \tfrac{4}{3}$, $3x - 4 = 0$ when $x = \tfrac{4}{3}$, and $3x - 4 < 0$ when $x < \tfrac{4}{3}$. Similarly, $2x + 1 > 0$ when $x > -\tfrac{1}{2}$, $2x + 1 = 0$ when

$x = -\tfrac{1}{2}$, and $2x + 1 < 0$ when $x < -\tfrac{1}{2}$. Using these results, we construct the following sign diagram for the factors of $6x^2 - 5x - 4$:

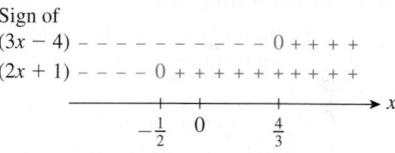

Step 4 Since $6x^2 - 5x - 4 \le 0$, the signs of the two factors must differ or be equal to 0. From the sign diagram, we see that x must lie between $-\tfrac{1}{2}$ and $\tfrac{4}{3}$, inclusive. Therefore, the required solution set is $\left[-\tfrac{1}{2}, \tfrac{4}{3}\right]$.

CHAPTER 1   Summary of Principal Formulas and Terms

FORMULAS

1. Product formula	$(a + b)^2 = a^2 + 2ab + b^2$ $(a - b)^2 = a^2 - 2ab + b^2$ $(a + b)(a - b) = a^2 - b^2$
2. Quadratic formula	$x = \dfrac{-b \pm \sqrt{b^2 - 4ac}}{2a}$

3. Difference of two squares	$a^2 - b^2 = (a + b)(a - b)$
4. Perfect square trinomial	$a^2 + 2ab + b^2 = (a + b)^2$ $a^2 - 2ab + b^2 = (a - b)^2$
5. Sum of two cubes	$a^3 + b^3 = (a + b)(a^2 - ab + b^2)$
6. Difference of two cubes	$a^3 - b^3 = (a - b)(a^2 + ab + b^2)$

TERMS

natural number (2)

whole number (2)

integer (2)

rational number (2)

irrational number (2)

real number (2)

exponent (7)

base (7)

polynomial (8)

rational expression (20)

complex fraction (23)

zero exponent (28)

equation (31)

variable (31)

solution of an equation (31)

solution set (31)

linear equation (32)

extraneous solution (33)

nth root (37)

square root (37)

cube root (37)

radical (38)

radical sign (38)

radicand (38)

index (38)

conjugate (43)

quadratic equation (46)

discriminant (51)

finite interval (56)

open interval (56)

closed interval (56)

half-open interval (56)

infinite interval (56)

absolute value (61)

triangle inequality (61)

CHAPTER 1 Concept Review Questions

Fill in the blanks.

1. a. A number of the form $\frac{a}{b}$, where a and b are integers, with $b \neq 0$ is called a/an _____ number. A rational number can be represented by either a/an _____ or _____ decimal.

b. A real number that is not rational is called _____. When such a number is represented by a decimal, it neither _____ nor _____.

2. a. Under addition, we have $a + b =$ _____, $a + (b + c) =$ _____, $a + 0 =$ _____, and $a + (-a) =$ _____.

b. Under multiplication, we have $ab =$ _____, $a(bc) =$ _____, $a \cdot 1 =$ _____, and $a\left(\frac{1}{a}\right) =$ _____ $(a \neq 0)$.

c. Under addition and multiplication, we have $a(b + c) =$ _____.

3. a. If a and b are real numbers, then $-(-a) =$ _____, $(-a)b =$ _____, $(-a)(-b) =$ _____, $(-1)a = -a$, and $a \cdot 0 = 0$.

b. If a and b are real numbers and $ab = 0$, then $a =$ _____, or $b =$ _____, or both.

4. a. An expression of the form $a_n x^n + a_{n-1}x^{n-1} + \cdots + a_0$ is called a/an _____ in _____; the nonnegative integer n is called its _____; the expression $a_k x^k$ is called the _____ of the _____; and a_k is called the _____ of x_k.

b. To add or subtract two polynomials, we add or subtract _____ terms.

5. To factor a polynomial, we express it as a/an _____ of two or more _____ polynomials. For example, $x^3 + x^2 - 2x =$ _____.

6. a. A rational expression is a quotient of _____.

b. A rational expression is simplified or reduced to lowest terms if the _____ and the _____ have no common _____ other than _____ and _____.

c. To add or subtract rational expressions, first find the least common _____ of the expressions, if necessary. Then follow the procedure for adding and subtracting _____ with common denominators.

7. A rational expression that contains fractions in its numerator or denominator is called a/an _____ fraction. An example of a complex fraction is _____.

8. a. If a is any real number and n is a natural number, then $a^n =$ _____. The number a is the _____, and the superscript n is called the _____, or _____.

b. For any nonzero real number a, $a^0 =$ _____. The expression 0^0 is _____ _____.

c. If a is any nonzero number and n is a positive integer, then $a^{-n} =$ _____.

9. a. A statement that two mathematical statements are equal is called a/an _____.

b. A variable is a letter that stands for a/an _____ belonging to a set of real numbers.

c. A linear equation in the variable x is an equation that can be written in the form _____; a linear equation in x has degree _____ in x.

10. a. If n is a natural number and a and b are real numbers, we say that a is the nth root of b if _____.

   b. If n is even, the real nth roots of a positive number b must come in _____.

   c. If n is even and b is negative, then there are _____ real roots.

   d. If n is odd, then there is only one _____ _____ of b.

11. a. If n is a natural number and b is a real number, then $\sqrt[n]{b}$ is called a/an _____; also, $\sqrt[n]{b} =$ _____.

   b. To rationalize a denominator of an algebraic expression means to eliminate a/an _____ from the denominator.

12. a. A quadratic equation is an equation in x that can be written in the form _____.

   b. A quadratic equation can be solved by _____, by _____ _____, or by using the quadratic formula. The quadratic formula is _____.

CHAPTER 1 Review Exercises

In Exercises 1–6, classify the number as to type.

1. $\dfrac{7}{8}$ **2.** $\sqrt{13}$ **3.** -2π

4. 0 **5.** $2.\overline{71}$ **6.** $3.14159\ldots$

In Exercises 7–14, evaluate the expression.

7. $\left(\dfrac{9}{4}\right)^{3/2}$ **8.** $\dfrac{5^6}{5^4}$

9. $(3 \cdot 4)^{-2}$ **10.** $(-8)^{5/3}$

11. $\left(\dfrac{16}{9}\right)^{3/2}$ **12.** $\dfrac{(3 \cdot 2^{-3})(4 \cdot 3^5)}{2 \cdot 9^3}$

13. $\sqrt[3]{\dfrac{27}{125}}$ **14.** $\dfrac{3\sqrt[3]{54}}{\sqrt[3]{18}}$

In Exercises 15–22, simplify the expression. (Assume that all variables are positive.)

15. $\dfrac{4(x^2 + y)^3}{x^2 + y}$ **16.** $\dfrac{a^6 b^{-5}}{(a^3 b^{-2})^{-3}}$

17. $\dfrac{\sqrt[4]{16x^5 yz}}{\sqrt[4]{81xyz^5}}$ **18.** $(2x^3)(-3x^{-2})\left(\dfrac{1}{6}x^{-1/2}\right)$

19. $\left(\dfrac{3xy^2}{4x^3 y}\right)^{-2}\left(\dfrac{3xy^3}{2x^2}\right)^3$ **20.** $(-3a^2 b^3)^2 (2a^{-1} b^{-2})^{-1}$

21. $\sqrt[3]{81x^5 y^{10}}\ \sqrt[3]{9xy^2}$ **22.** $\left(\dfrac{-x^{1/2} y^{2/3}}{x^{1/3} y^{3/4}}\right)^6$

In Exercises 23–30, perform the indicated operations and simplify the expression.

23. $(3x^4 + 10x^3 + 6x^2 + 10x + 3) + (2x^4 + 10x^3 + 6x^2 + 4x)$

24. $(3x - 4)(3x^2 - 2x + 3)$

25. $(2x + 3y)^2 - (3x + 1)(2x - 3)$

26. $2(3a + b) - 3[(2a + 3b) - (a + 2b)]$

27. $\dfrac{(t + 6)(60) - (60t + 180)}{(t + 6)^2}$

28. $\dfrac{6x}{2(3x^2 + 2)} + \dfrac{1}{4(x + 2)}$

29. $\dfrac{2}{3}\left(\dfrac{4x}{2x^2 - 1}\right) + 3\left(\dfrac{3}{3x - 1}\right)$

30. $\dfrac{-2x}{\sqrt{x + 1}} + 4\sqrt{x + 1}$

In Exercises 31–40, factor the expression.

31. $-2\pi^2 r^3 + 100\pi r^2$ **32.** $2v^3 w + 2vw^3 + 2u^2 vw$

33. $16 - x^2$ **34.** $12t^3 - 6t^2 - 18t$

35. $-2x^2 - 4x + 6$ **36.** $12x^2 - 92x + 120$

37. $9a^2 - 25b^2$ **38.** $8u^6 v^3 + 27u^3$

39. $6a^4 b^4 c - 3a^3 b^2 c - 9a^2 b^2$

40. $6x^2 - xy - y^2$

In Exercises 41–46, perform the indicated operations and simplify the expression.

41. $\dfrac{2x^2 + 3x - 2}{2x^2 + 5x - 3}$

42. $\dfrac{[(t^2 + 4)(2t - 4)] - (t^2 - 4t + 4)(2t)}{(t^2 + 4)^2}$

43. $\dfrac{2x - 6}{x + 3} \cdot \dfrac{x^2 + 6x + 9}{x^2 - 9}$

44. $\dfrac{3x}{x^2 + 2} + \dfrac{3x^2}{x^3 + 1}$ **45.** $\dfrac{1 + \dfrac{1}{x + 2}}{x - \dfrac{9}{x}}$

46. $\dfrac{x(3x^2 + 1)}{x - 1} \cdot \dfrac{3x^3 - 5x^2 + x}{x(x - 1)(3x^2 + 1)^{1/2}}$

In Exercises 47–54, solve the equation.

47. $8x^2 + 2x - 3 = 0$ **48.** $-6x^2 - 10x + 4 = 0$

49. $2x^2 - 3x - 4 = 0$ **50.** $x^2 + 5x + 3 = 0$

51. $2y^2 - 3y + 1 = 0$ **52.** $0.3m^2 - 2.1m - 3.2 = 0$

53. $-x^3 - 2x^2 + 3x = 0$ **54.** $2x^4 + x^2 = 1$

In Exercises 55–60, solve the equation.

55. $\frac{1}{4}x + 2 = \frac{3}{4}x - 5$ **56.** $\frac{3p + 1}{2} - \frac{2p - 1}{3} = \frac{5p}{12}$

57. $(x + 2)^2 - 3x(1 - x) = (x - 2)^2$

58. $\frac{3(2q + 1)}{4q - 3} = \frac{3q + 1}{2q + 1}$

59. $\sqrt{k - 1} = \sqrt{2k - 3}$

60. $\sqrt{x} - \sqrt{x - 1} = \sqrt{4x - 3}$

61. Solve $C = \frac{20x}{100 - x}$ for x.

62. Solve $r = \frac{2mI}{B(n + 1)}$ for I.

In Exercises 63–66, find the values of x that satisfy the inequalities.

63. $-x + 3 \le 2x + 9$ **64.** $-2 \le 3x + 1 \le 7$

65. $x - 3 > 2$ or $x + 3 < -1$

66. $2x^2 > 50$

In Exercises 67–70, evaluate the expression.

67. $|-5 + 7| + |-2|$ **68.** $\left| \dfrac{5 - 12}{-4 - 3} \right|$

69. $|2\pi - 6| - \pi$ **70.** $|\sqrt{3} - 4| + |4 - 2\sqrt{3}|$

In Exercises 71–76, find the value(s) of x that satisfy the expression.

71. $2x^2 + 3x - 2 \le 0$ **72.** $x^2 + x - 12 \le 0$

73. $\dfrac{1}{x + 2} > 2$ **74.** $|2x - 3| < 5$

75. $|3x - 4| \le 2$ **76.** $\left| \dfrac{x + 1}{x - 1} \right| = 5$

77. Rationalize the numerator:

$$\frac{\sqrt{x} - 1}{x - 1}$$

78. Rationalize the numerator:

$$\sqrt[3]{\frac{x^2}{yz^3}}$$

79. Rationalize the denominator:

$$\frac{\sqrt{x} - 1}{2\sqrt{x}}$$

80. Rationalize the denominator:

$$\frac{3}{1 + 2\sqrt{x}}$$

In Exercises 81 and 82, use the quadratic formula to solve the quadratic equation.

81. $x^2 - 2x - 5 = 0$ **82.** $2x^2 + 8x + 7 = 0$

83. Find the minimum cost C (in dollars) given that

$$2(1.5C + 80) \le 2(2.5C - 20)$$

84. Find the maximum revenue R (in dollars) given that

$$12(2R - 320) \le 4(3R + 240)$$

85. **HEALTH-CARE SPENDING** Health-care spending per person (in dollars) by the private sector includes payments by individuals, corporations, and their insurance companies and is approximated by

$$2.5t^2 + 18.5t + 509 \qquad (0 \le t \le 6)$$

where t is measured in years, and $t = 0$ corresponds to the beginning of 1994. The corresponding government spending (in dollars), including expenditures for Medicaid and other federal, state, and local government public health care, is

$$-1.1t^2 + 29.1t + 429 \qquad (0 \le t \le 6)$$

where t has the same meaning as before. Find an expression for the difference between private and government expenditures per person at any time t. What was the difference between private and government expenditures per person at the beginning of 1998? At the beginning of 2000?
Source: Health Care Financing Administration.

86. **TEMPERATURE RANGE** The relationship between the Celsius and Fahrenheit temperature scales is given by $F = \frac{9}{5}C + 32$. In a certain 24-hr period, the temperature at the Los Angeles International Airport (LAX) ranged from a low of 51.8°F to a high of 80.6°F.
  a. Write an expression involving two inequalities that gives the temperature range at LAX in degrees Fahrenheit.
  b. Find the corresponding temperature range on the Celsius scale.

87. **SUPPLY OF DESK LAMPS** The supplier of the Luminar desk lamp will make x thousand units of the lamp available in the marketplace if its unit price is p dollars, where p and x are related by the equation

$$p = 0.1x^2 + 0.5x + 15$$

If the unit price of the lamp is set at \$20, how many units will the supplier make available in the marketplace?

88. Suppose p is a polynomial of degree m and q is a polynomial of degree n, where $m > n$. What is the degree of $p - q$?

The problem-solving skills that you learn in each chapter are building blocks for the rest of the course. Therefore, it is a good idea to make sure that you have mastered these skills before moving on to the next chapter. The Before Moving On exercises that follow are designed for that purpose. After completing these exercises, you can identify the skills that you should review before starting the next chapter.

CHAPTER 1   Before Moving On . . .

1. Perform the indicated operations and simplify.
$$2(3x - 2)^2 - 3x(x + 1) + 4$$

2. Factor:
   a. $x^4 - x^3 - 6x^2$
   b. $(a - b)^2 - (a^2 + b)^2$

3. Perform the indicated operation and simplify.
$$\frac{2x}{3x^2 - 5x - 2} + \frac{x - 1}{x^2 - x - 2}$$

4. Simplify $\left(\dfrac{8x^2 y^{-3}}{9x^{-3} y^2}\right)^{-1} \left(\dfrac{2x^2}{3y^3}\right)^2$.

5. Solve $2s = \dfrac{r}{s + r}$ for r.

6. Rationalize the denominator in the expression
$$\frac{2 - \sqrt{3}}{2 + \sqrt{3}}$$

7. a. Solve $2x^2 + 5x - 12 = 0$ by factoring.
   b. Solve $m^2 - 3m - 2 = 0$.

8. Solve $\sqrt{x + 4} - \sqrt{x - 5} - 1 = 0$.

9. Find the values of x that satisfy $(3x + 2)(2x - 3) \le 0$.

10. Find the values of x that satisfy $|2x + 3| \le 1$.

2 Functions and Their Graphs

THIS CHAPTER INTRODUCES the Cartesian coordinate system, a system that allows us to represent points in the plane in terms of ordered pairs of real numbers. This in turn enables us to study geometry, using algebraic methods. Specifically, we will see how straight lines in the plane can be represented by algebraic equations. Next, we study functions, which are special relationships between two quantities. These relationships, or mathematical models, can be found in fields of study as diverse as business, economics, the social sciences, physics, and medicine. We study in detail two special classes of functions: linear functions and quadratic functions. We also look at the process of solving real-world problems using mathematics, a process called mathematical modeling. Finally, we learn how to find an algebraic representation of the straight line that "best" fits a set of data points that are scattered about a straight line.

Half of millenials (18- to 29-year-olds) "do not believe Social Security will exist" when they reach retirement age, according to a study released by the iOme Challenge organization. Are their concerns about the present system justified? In Example 2, page 145, we use a mathematical model constructed from data from the Social Security Administration to predict the year in which the assets of the current system will be depleted.

© EDHAR/Shutterstock.com

2.1 The Cartesian Coordinate System and Straight Lines

The Cartesian Coordinate System

In Section 1.1, we saw how a one-to-one correspondence between the set of real numbers and the points on a straight line leads to a coordinate system on a line (a one-dimensional space). In a similar manner, we can represent points in a plane (a two-dimensional space) by using the **Cartesian coordinate system,** which we construct as follows: Take two perpendicular lines, one of which is normally chosen to be horizontal. These lines intersect at a point O, called the **origin** (Figure 1). The horizontal line is called the ***x*-axis,** and the vertical line is called the ***y*-axis.** A number scale is set up along the x-axis, with the positive numbers lying to the right of the origin and the negative numbers lying to the left of it. Similarly, a number scale is set up along the y-axis, with the positive numbers lying above the origin and the negative numbers lying below it.

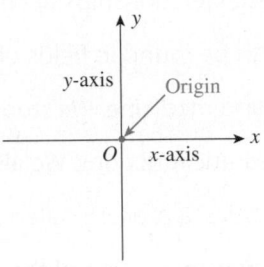

FIGURE 1
The Cartesian coordinate system

Note   The number scales on the two axes need not be the same. Indeed, in many applications, different quantities are represented by x and y. For example, x may represent the number of smartphones sold, and y may represent the total revenue resulting from the sales. In such cases, it is often desirable to choose different number scales to represent the different quantities. Note, however, that the zeros of both number scales coincide at the origin of the two-dimensional coordinate system. ◾

We can represent a point in the plane in this coordinate system by an **ordered pair** of numbers—that is, a pair (x, y) in which x is the first number and y is the second. To see this, let P be any point in the plane (Figure 2). Draw perpendicular lines from P to the x-axis and y-axis, respectively. Then the number x is precisely the number that corresponds to the point on the x-axis at which the perpendicular line through P hits the x-axis. Similarly, y is the number that corresponds to the point on the y-axis at which the perpendicular line through P crosses the y-axis.

Conversely, given an ordered pair (x, y) with x as the first number and y as the second, a point P in the plane is uniquely determined as follows: Locate the point on the x-axis represented by the number x, and draw a line through that point perpendicular to the x-axis. Next, locate the point on the y-axis represented by the number y, and draw a line through that point perpendicular to the y-axis. The point of intersection of these two lines is the point P (Figure 2).

In the ordered pair (x, y), x is called the **abscissa,** or ***x*-coordinate;** y is called the **ordinate,** or ***y*-coordinate;** and x and y together are referred to as the **coordinates** of the point P. The point P with x-coordinate equal to a and y-coordinate equal to b is often written $P(a, b)$.

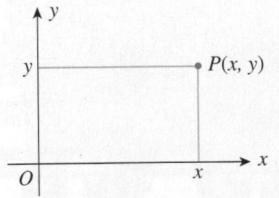

FIGURE 2
An ordered pair in the coordinate plane

The points $A(2, 3)$, $B(-2, 3)$, $C(-2, -3)$, $D(2, -3)$, $E(3, 2)$, $F(4, 0)$, and $G(0, -5)$ are plotted in Figure 3.

Note In general, $(x, y) \neq (y, x)$. This is illustrated by the points A and E in Figure 3.

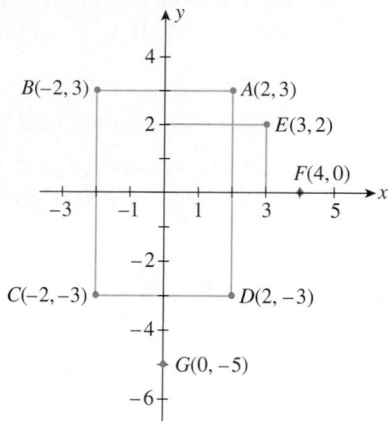

FIGURE 3
Several points in the coordinate plane

The axes divide the plane into four quadrants. Quadrant I consists of the points P with coordinates x and y, denoted by $P(x, y)$, satisfying $x > 0$ and $y > 0$; Quadrant II consists of the points $P(x, y)$ where $x < 0$ and $y > 0$; Quadrant III consists of the points $P(x, y)$ where $x < 0$ and $y < 0$; and Quadrant IV consists of the points $P(x, y)$ where $x > 0$ and $y < 0$ (Figure 4).

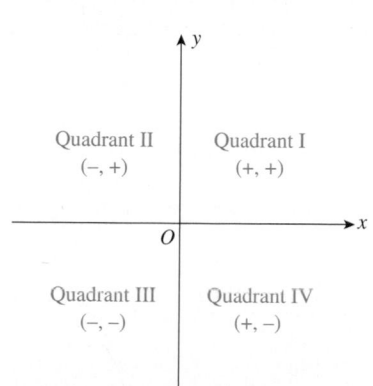

FIGURE 4
The four quadrants in the coordinate plane

Straight Lines

Businesses may depreciate certain assets such as buildings, machines, furniture, vehicles, and equipment over a period of time for income tax purposes. *Linear depreciation*, or the *straight-line method*, is often used for this purpose. The graph of the straight line shown in Figure 5 describes the book value V of a network server that has an initial value of $10,000 and that is being depreciated linearly over 5 years with a scrap value of $3000. Note that only the solid portion of the straight line is of interest here.

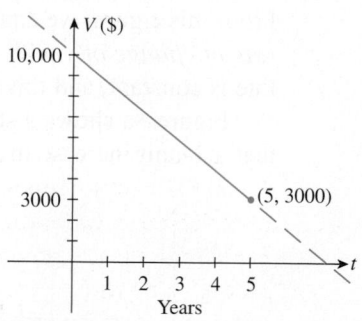

FIGURE 5
Linear depreciation of a network server

The book value of the server at the end of year t, where t lies between 0 and 5, can be read directly from the graph. But there is one shortcoming in this approach: The result depends on how accurately you draw and read the graph. A better and more accurate method is based on finding an *algebraic* representation of the depreciation line. (We continue our discussion of the linear depreciation problem in Section 2.5.)

To see how a straight line in the xy-plane may be described algebraically, we need first to recall certain properties of straight lines.

Slope of a Line

Let L denote the unique straight line that passes through the two distinct points (x_1, y_1) and (x_2, y_2). If $x_1 \neq x_2$, then we define the slope of L as follows.

> **Slope of a Nonvertical Line**
>
> If (x_1, y_1) and (x_2, y_2) are any two distinct points on a nonvertical line L, then the slope m of L is given by
>
> $$m = \frac{\Delta y}{\Delta x} = \frac{y_2 - y_1}{x_2 - x_1} \qquad (1)$$
>
> (Figure 6).
>
>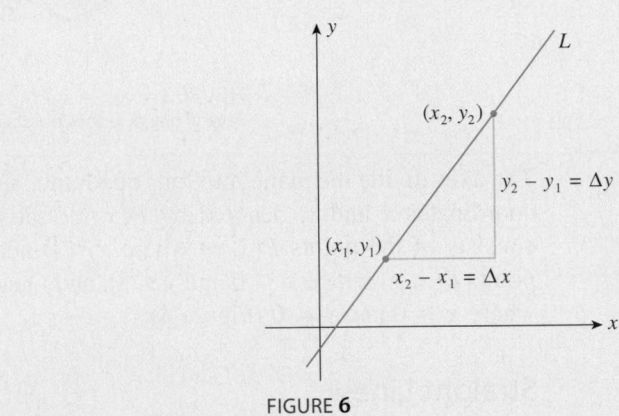
>
> FIGURE **6**

If $x_1 = x_2$, then L is a vertical line (Figure 7). Its slope is undefined, since the denominator in Equation (1) will be zero and division by zero is not allowed.

Observe that the slope of a straight line is a constant whenever it is defined. The number $\Delta y = y_2 - y_1$ (Δy is read "delta y") is a measure of the vertical change in y, and $\Delta x = x_2 - x_1$ is a measure of the horizontal change in x as shown in Figure 6. From this figure, we can see that the slope m of a straight line L is a measure of the *rate of change of y with respect to x.* Furthermore, the slope of a nonvertical straight line is constant, and this tells us that this rate of change is constant.

Figure 8a shows a straight line L_1 with slope 2. Observe that L_1 has the property that a 1-unit increase in x results in a 2-unit increase in y. To see this, let $\Delta x = 1$ in

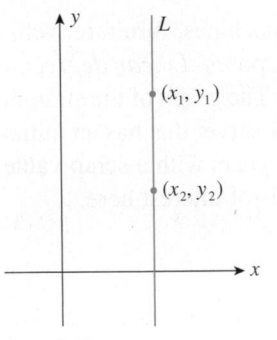

FIGURE **7**
The slope of L is undefined if $x_1 = x_2$.

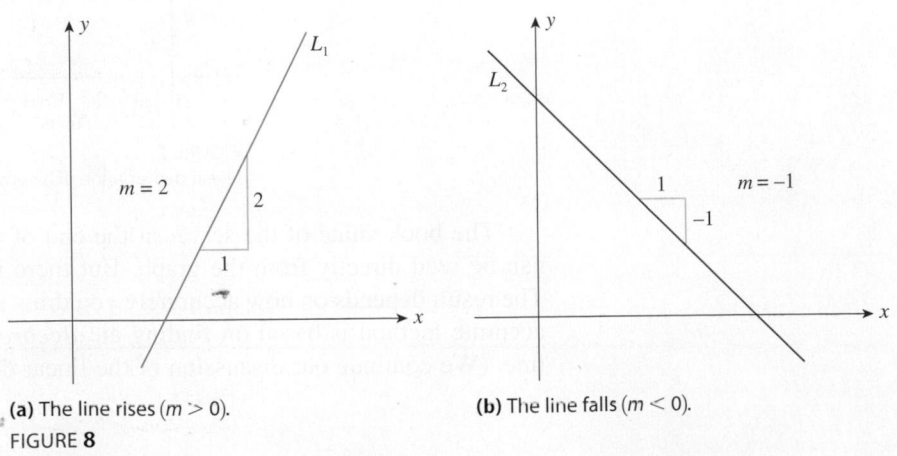

(a) The line rises ($m > 0$).   **(b)** The line falls ($m < 0$).
FIGURE **8**

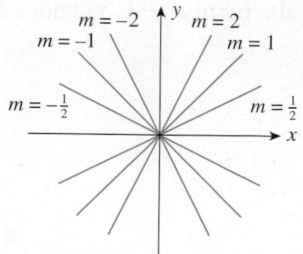

FIGURE **9**
A family of straight lines

Equation (1) so that $m = \Delta y$. Since $m = 2$, we conclude that $\Delta y = 2$. Similarly, Figure 8b shows a line L_2 with slope -1. Observe that a straight line with positive slope slants upward from left to right (y increases as x increases), whereas a line with negative slope slants downward from left to right (y decreases as x increases). Finally, Figure 9 shows a family of straight lines passing through the origin with indicated slopes.

> *Explore and Discuss*
>
> Show that the slope of a nonvertical line is independent of the two distinct points used to compute it.
> Hint: Pick any two distinct points lying on a line L. Then pick two other distinct points, $P_3(x_3, y_3)$ and $P_4(x_4, y_4)$ lying on L. Draw a picture, and use similar triangles to demonstrate that using P_3 and P_4 gives the same value as that obtained by using P_1 and P_2.

EXAMPLE 1 Sketch the straight line that passes through the point $(-2, 5)$ and has slope $-\frac{4}{3}$.

Solution First, plot the point $(-2, 5)$ (Figure 10). Next, recall that a slope of $-\frac{4}{3}$ indicates that an increase of 1 unit in the x-direction produces a *decrease* of $\frac{4}{3}$ units in the y-direction, or equivalently, a 3-unit increase in the x-direction produces a $3\left(\frac{4}{3}\right)$, or 4-unit, decrease in the y-direction. Using this information, we plot the point $(1, 1)$ and draw the line through the two points.

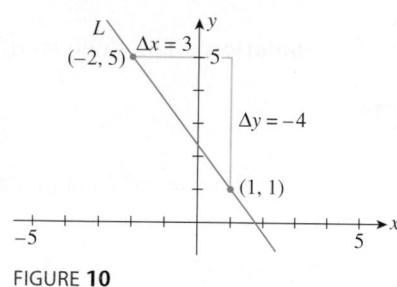

FIGURE **10**
L has slope $-\frac{4}{3}$ and passes through $(-2, 5)$.

EXAMPLE 2 Find the slope m of the line that passes through the points $(-1, 1)$ and $(5, 3)$.

Solution Choose (x_1, y_1) to be the point $(-1, 1)$ and (x_2, y_2) to be the point $(5, 3)$. Then, with $x_1 = -1$, $y_1 = 1$, $x_2 = 5$, and $y_2 = 3$, we find, using Equation (1),

$$m = \frac{y_2 - y_1}{x_2 - x_1} = \frac{3 - 1}{5 - (-1)} = \frac{2}{6} = \frac{1}{3}$$

(Figure 11). You may verify that the result obtained would be the same had we chosen the point $(-1, 1)$ to be (x_2, y_2) and the point $(5, 3)$ to be (x_1, y_1).

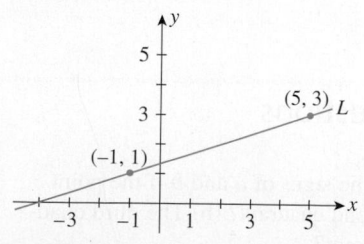

FIGURE **11**
L passes through $(5, 3)$ and $(-1, 1)$.

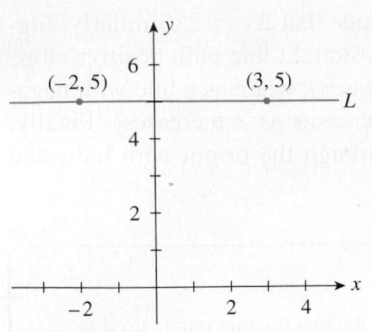

FIGURE 12
The slope of the horizontal line L is zero.

EXAMPLE 3 Find the slope of the line that passes through the points $(-2, 5)$ and $(3, 5)$.

Solution   The slope of the required line is given by

$$m = \frac{5 - 5}{3 - (-2)} = \frac{0}{5} = 0$$

(Figure 12).

Note   The slope of a horizontal line is zero.

We can use the slope of a straight line to determine whether a line is parallel to another line.

> **Parallel Lines**
> Two distinct lines are **parallel** if and only if their slopes are equal or their slopes are undefined.

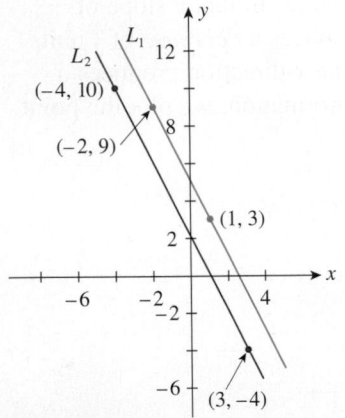

FIGURE 13
L_1 and L_2 have the same slope and hence are parallel.

EXAMPLE 4 Let L_1 be a line that passes through the points $(-2, 9)$ and $(1, 3)$, and let L_2 be the line that passes through the points $(-4, 10)$ and $(3, -4)$. Determine whether L_1 and L_2 are parallel.

Solution   The slope m_1 of L_1 is given by

$$m_1 = \frac{3 - 9}{1 - (-2)} = -2$$

The slope m_2 of L_2 is given by

$$m_2 = \frac{-4 - 10}{3 - (-4)} = -2$$

Since $m_1 = m_2$, the lines L_1 and L_2 are in fact parallel (Figure 13).

2.1 Self-Check Exercise

Determine the number a such that the line passing through the points $(a, 2)$ and $(3, 6)$ is parallel to a line with slope 4.

The solution to Self-Check Exercise 2.1 can be found on page 78.

2.1 Concept Questions

1. What can you say about the signs of a and b if the point $P(a, b)$ lies in (a) the second quadrant? (b) The third quadrant? (c) The fourth quadrant?

2. What is the slope of a nonvertical line? What can you say about the slope of a vertical line?

2.1 Exercises

In Exercises 1–6, refer to the accompanying figure and determine the coordinates of the point and the quadrant in which it is located.

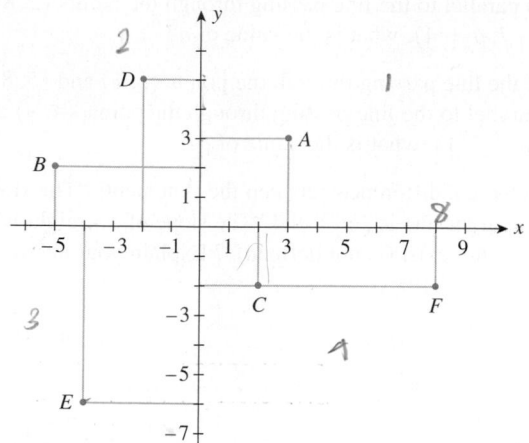

1. A   **2.** B   **3.** C

4. D   **5.** E   **6.** F

In Exercises 7–12, refer to the accompanying figure.

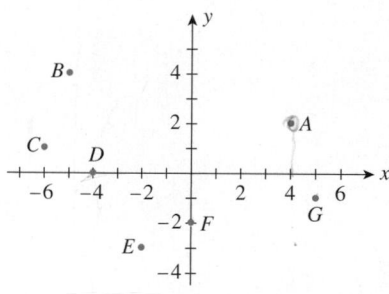

7. Which point is represented by the ordered pair $(4, 2)$?

8. What are the coordinates of point B?

9. Which points have negative y-coordinates?

10. Which point has a negative x-coordinate and a negative y-coordinate?

11. Which point has an x-coordinate that is equal to zero?

12. Which point has a y-coordinate that is equal to zero?

In Exercises 13–20, sketch a set of coordinate axes and then plot the point.

13. $(-2, 5)$ **14.** $(1, 3)$

15. $(3, -1)$ **16.** $(3, -4)$

17. $\left(8, -\frac{7}{2}\right)$ **18.** $\left(-\frac{5}{2}, \frac{3}{2}\right)$

19. $(4.5, -4.5)$ **20.** $(1.2, -3.4)$

In Exercises 21–24, find the slope of the line shown in each figure.

21.

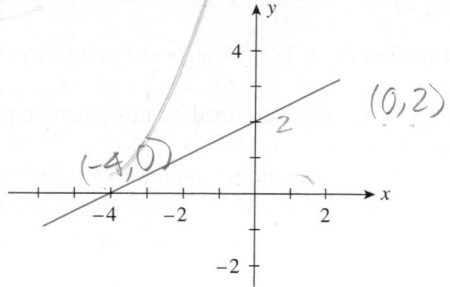

22.

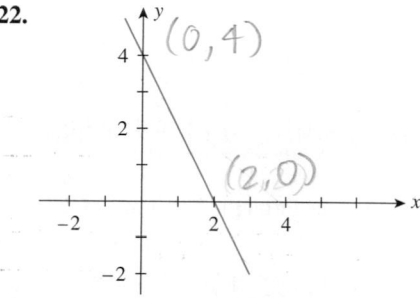

23.

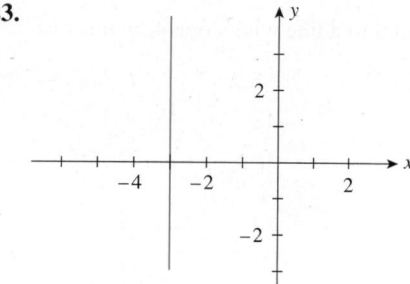

24.

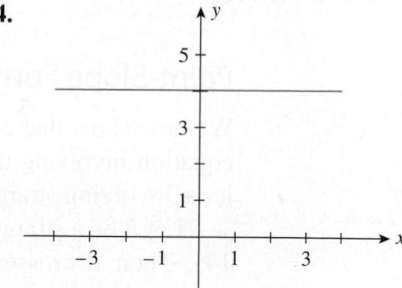

In Exercises 25–30, find the slope of the line that passes through the given pair of points.

25. $(4, 3)$ and $(5, 8)$ **26.** $(4, 5)$ and $(3, 8)$

27. $(-2, 3)$ and $(4, 8)$ **28.** $(-2, -2)$ and $(4, -4)$

29. (a, b) and (c, d)

30. $(-a + 1, b - 1)$ and $(a + 1, -b)$

31. Given the equation $y = 4x - 3$, answer the following questions:
   a. If x increases by 1 unit, what is the corresponding change in y?
   b. If x decreases by 2 units, what is the corresponding change in y?

32. Given the equation $2x + 3y = 4$, answer the following questions:
   a. Is the slope of the line described by this equation positive or negative?
   b. As x increases in value, does y increase or decrease?
   c. If x decreases by 2 units, what is the corresponding change in y?

In Exercises 33 and 34, determine whether the lines through the pairs of points are parallel.

33. $A(1, -2)$, $B(-3, -10)$ and $C(1, 5)$, $D(-1, 1)$

34. $A(2, 3)$, $B(2, -2)$ and $C(-2, 4)$, $D(-2, 5)$

35. If the line passing through the points $(1, a)$ and $(4, -2)$ is parallel to the line passing through the points $(2, 8)$ and $(-7, a + 4)$, what is the value of a?

36. If the line passing through the points $(a, 1)$ and $(5, 8)$ is parallel to the line passing through the points $(4, 9)$ and $(a + 2, 1)$, what is the value of a?

37. Is there a difference between the statements "The slope of a straight line is zero" and "The slope of a straight line does not exist (is not defined)"? Explain your answer.

2.1   Solution to Self-Check Exercise

The slope of the line that passes through the points $(a, 2)$ and $(3, 6)$ is

$$m = \frac{6 - 2}{3 - a} = \frac{4}{3 - a}$$

Since this line is parallel to a line with slope 4, m must be equal to 4; that is,

$$\frac{4}{3 - a} = 4$$

or, upon multiplying both sides of the equation by $3 - a$,

$$4 = 4(3 - a)$$
$$4 = 12 - 4a$$
$$4a = 8$$
$$a = 2$$

2.2   Equations of Lines

Point-Slope Form

We now show that every straight line lying in the xy-plane may be represented by an equation involving the variables x and y. One immediate benefit of this is that problems involving straight lines may be solved algebraically.

Let L be a straight line parallel to the y-axis (perpendicular to the x-axis) (Figure 14). Then L crosses the x-axis at some point $(a, 0)$ with x-coordinate given by $x = a$, where a is some real number. Any other point on L has the form (a, y), where y is an appropriate number. Therefore, the vertical line L is described by the sole condition

$$x = a$$

and this is accordingly an equation of L. For example, the equation $x = -2$ represents a vertical line 2 units to the left of the y-axis, and the equation $x = 3$ represents a vertical line 3 units to the right of the y-axis (Figure 15).

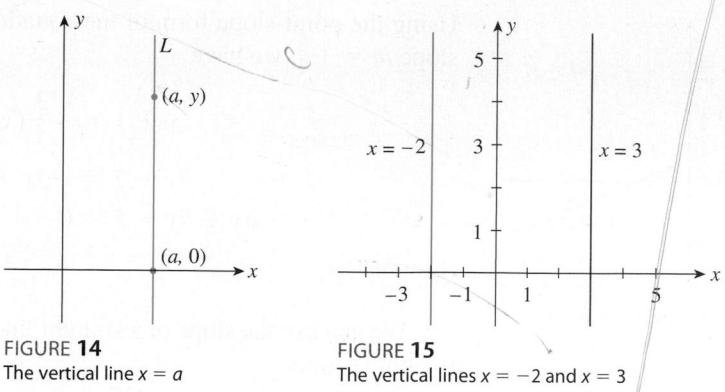

FIGURE **14**
The vertical line $x = a$

FIGURE **15**
The vertical lines $x = -2$ and $x = 3$

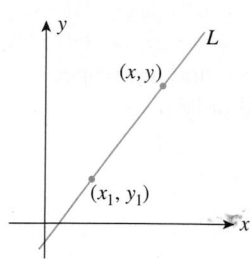

FIGURE **16**
L passes through (x_1, y_1) and has slope m.

Next, suppose L is a nonvertical line, so it has a well-defined slope m. Suppose (x_1, y_1) is a fixed point lying on L and (x, y) is a variable point on L distinct from (x_1, y_1) (Figure 16). Using Equation (1) with the point $(x_2, y_2) = (x, y)$, we find that the slope of L is given by

$$m = \frac{y - y_1}{x - x_1}$$

Upon multiplying both sides of the equation by $x - x_1$, we obtain Equation (2).

> **Point-Slope Form of an Equation of a Line**
>
> An equation of the line that has slope m and passes through the point (x_1, y_1) is given by
>
> $$y - y_1 = m(x - x_1) \tag{2}$$

Equation (2) is called the *point-slope form* of an equation of a line because it uses a given point (x_1, y_1) on a line and the slope m of the line.

EXAMPLE 1 Find an equation of the line that passes through the point $(1, 3)$ and has slope 2.

Solution Using the point-slope form of the equation of a line with the point $(1, 3)$ and $m = 2$, we obtain

$$y - 3 = 2(x - 1) \qquad y - y_1 = m(x - x_1)$$

which, when simplified, becomes

$$2x - y + 1 = 0$$

(Figure 17).

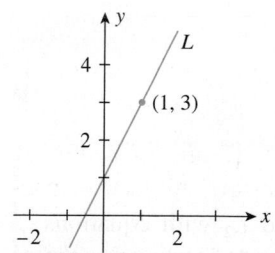

FIGURE **17**
L passes through $(1, 3)$ and has slope 2.

EXAMPLE 2 Find an equation of the line that passes through the points $(-3, 2)$ and $(4, -1)$.

Solution The slope of the line is given by

$$m = \frac{-1 - 2}{4 - (-3)} = -\frac{3}{7}$$

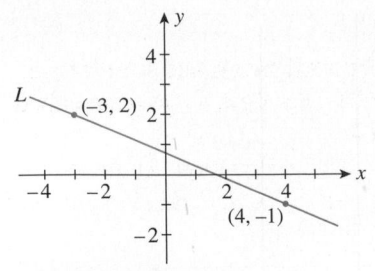

FIGURE 18
L passes through $(-3, 2)$ and $(4, -1)$.

Using the point-slope form of the equation of a line with the point $(4, -1)$ and the slope $m = -\frac{3}{7}$, we have

$$y + 1 = -\frac{3}{7}(x - 4) \qquad y - y_1 = m(x - x_1)$$

$$7y + 7 = -3x + 12$$

$$3x + 7y - 5 = 0$$

(Figure 18).

We can use the slope of a straight line to determine whether a line is perpendicular to another line.

Perpendicular Lines

If L_1 and L_2 are two distinct nonvertical lines that have slopes m_1 and m_2, respectively, then L_1 is **perpendicular** to L_2 (written $L_1 \perp L_2$) if and only if

$$m_1 = -\frac{1}{m_2}$$

If the line L_1 is vertical (so that its slope is undefined), then L_1 is perpendicular to another line, L_2, if and only if L_2 is horizontal (so that its slope is zero). For a proof of these results, see Exercise 76, page 88.

EXAMPLE 3 Find an equation of the line that passes through the point $(3, 1)$ and is perpendicular to the line of Example 1.

Solution Since the slope of the line in Example 1 is 2, it follows that the slope of the required line is given by $m = -\frac{1}{2}$, the negative reciprocal of 2. Using the point-slope form of the equation of a line, we obtain

$$y - 1 = -\frac{1}{2}(x - 3) \qquad y - y_1 = m(x - x_1)$$

$$2y - 2 = -x + 3$$

$$x + 2y - 5 = 0$$

(Figure 19).

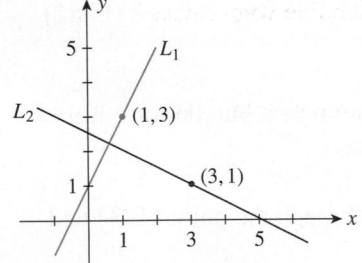

FIGURE 19
L_2 is perpendicular to L_1 and passes through $(3, 1)$.

Exploring with TECHNOLOGY

1. Use a graphing utility to plot the straight lines L_1 and L_2 with equations $2x + y - 5 = 0$ and $41x + 20y - 11 = 0$ on the same set of axes, using the standard viewing window.
   a. Can you tell whether the lines L_1 and L_2 are parallel to each other?
   b. Verify your observations by computing the slopes of L_1 and L_2 algebraically.

2. Use a graphing utility to plot the straight lines L_1 and L_2 with equations $x + 2y - 5 = 0$ and $5x - y + 5 = 0$ on the same set of axes, using the standard viewing window.
   a. Can you tell whether the lines L_1 and L_2 are perpendicular to each other?
   b. Verify your observation by computing the slopes of L_1 and L_2 algebraically.

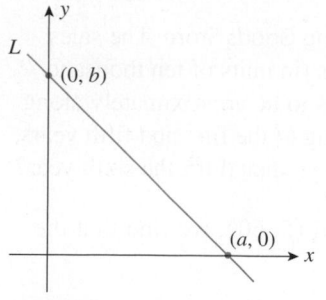

FIGURE 20
The line L has x-intercept a and y-intercept b.

Slope-Intercept Form

A straight line L that is neither horizontal nor vertical cuts the x-axis and the y-axis at, say, points $(a, 0)$ and $(0, b)$, respectively (Figure 20). The numbers a and b are called the **x-intercept** and **y-intercept**, respectively, of L.

Now, let L be a line with slope m and y-intercept b. Using Equation (2), the point-slope form of the equation of a line, with the point given by $(0, b)$ and slope m, we have

$$y - b = m(x - 0)$$
$$y = mx + b$$

> **Slope-Intercept Form of an Equation of a Line**
>
> The equation of the line that has slope m and intersects the y-axis at the point $(0, b)$ is given by
>
> $$y = mx + b \tag{3}$$

EXAMPLE 4 Find an equation of the line that has slope 3 and y-intercept -4.

Solution Using Equation (3) with $m = 3$ and $b = -4$, we obtain the required equation:

$$y = 3x - 4$$

EXAMPLE 5 Determine the slope and y-intercept of the line whose equation is $3x - 4y = 8$.

Solution Rewrite the given equation in the slope-intercept form. Thus,

$$3x - 4y = 8$$
$$-4y = -3x + 8$$
$$y = \frac{3}{4}x - 2$$

Comparing this result with Equation (3), we find $m = \frac{3}{4}$ and $b = -2$, and we conclude that the slope and y-intercept of the given line are $\frac{3}{4}$ and -2, respectively.

> **Exploring with TECHNOLOGY**
>
> 1. Use a graphing utility to plot the straight lines with equations $y = -2x + 3$, $y = -x + 3$, $y = x + 3$, and $y = 2.5x + 3$ on the same set of axes, using the standard viewing window. What effect does changing the coefficient m of x in the equation $y = mx + b$ have on its graph?
>
> 2. Use a graphing utility to plot the straight lines with equations $y = 2x - 2$, $y = 2x - 1$, $y = 2x$, $y = 2x + 1$, and $y = 2x + 4$ on the same set of axes, using the standard viewing window. What effect does changing the constant b in the equation $y = mx + b$ have on its graph?
>
> 3. Describe in words the effect of changing both m and b in the equation $y = mx + b$.

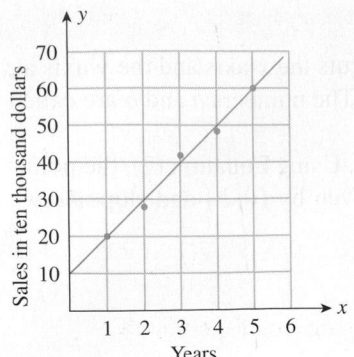

FIGURE 21
Sales of a sporting goods store

$ APPLIED EXAMPLE 6 Forecasting Sales of a Sporting Goods Store The sales manager of a local sporting goods store plotted sales (in units of ten thousand dollars) versus time for the last 5 years and found the points to lie approximately along a straight line (Figure 21). By using the points corresponding to the first and fifth years, find an equation of the *trend line*. What sales figure can be predicted for the sixth year?

Solution Using Equation (1) with the points $(1, 20)$ and $(5, 60)$, we find that the slope of the required line is given by

$$m = \frac{60 - 20}{5 - 1} = 10$$

Next, using the point-slope form of the equation of a line with the point $(1, 20)$ and $m = 10$, we obtain

$$y - 20 = 10(x - 1) \qquad y - y_1 = m(x - x_1)$$
$$y = 10x + 10$$

as the required equation.

The sales figure for the sixth year is obtained by letting $x = 6$ in the last equation, giving

$$y = 10(6) + 10 = 70$$

or $700,000.

$ APPLIED EXAMPLE 7 Appreciation in Value of a Painting Suppose a painting purchased for $50,000 is expected to appreciate in value at a constant rate of $5000 per year for the next 5 years. Use Equation (3) to write an equation predicting the value of the painting in the next several years. What will be its value 3 years from the purchase date?

Solution Let x denote the time (in years) that has elapsed since the purchase date, and let y denote the painting's value (in dollars). Then $y = 50,000$ when $x = 0$. Furthermore, the slope of the required equation is given by $m = 5000$, since each unit increase in x (1 year) implies an increase of 5000 units (dollars) in y. Using Equation (3) with $m = 5000$ and $b = 50,000$, we obtain

$$y = 5000x + 50,000 \qquad y = mx + b$$

Three years from the purchase date, the value of the painting will be given by

$$y = 5000(3) + 50,000$$

or $65,000.

Explore and Discuss

Refer to Applied Example 7. Can the equation predicting the value of the painting be used to predict long-term growth?

General Form of an Equation of a Line

We have considered several forms of the equation of a straight line in the plane. These different forms of the equation are equivalent to each other. In fact, each is a special case of the following equation.

> **General Form of a Linear Equation**
> The equation
>
> $$Ax + By + C = 0 \qquad (4)$$
>
> where A, B, and C are constants and A and B are not both zero, is called the general form of a linear equation in the variables x and y.

We now state (without proof) an important result concerning the algebraic representation of straight lines in the plane.

THEOREM 1

An equation of a straight line is a linear equation; conversely, every linear equation represents a straight line.

This result justifies the use of the adjective *linear* in describing Equation (4).

EXAMPLE 8

a. Sketch the straight line represented by the equation

$$3x - 4y - 12 = 0$$

b. Does the point $\left(2, -\frac{3}{2}\right)$ lie on L?

c. Does the point $(1, -2)$ lie on L?

Solution

a. Since every straight line is uniquely determined by two distinct points, we need to find only two points through which the line passes in order to sketch it. For convenience, let's compute the points at which the line crosses the x- and y-axes. Setting $y = 0$, we find $x = 4$, the x-intercept, so the line crosses the x-axis at the point $(4, 0)$. Setting $x = 0$ gives $y = -3$, the y-intercept, so the line crosses the y-axis at the point $(0, -3)$. A sketch of the line appears in Figure 22.

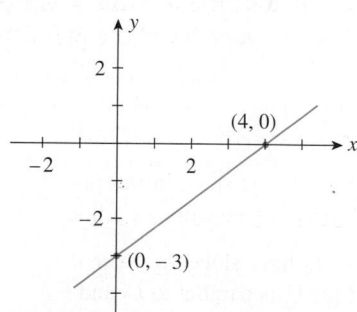

FIGURE 22
To sketch $3x - 4y - 12 = 0$, first find the x-intercept, 4, and the y-intercept, -3.

b. Substituting $x = 2$ and $y = -\frac{3}{2}$ into the left-hand side of the equation $3x - 4y - 12 = 0$ found in part (a), we obtain

$$3(2) - 4\left(-\frac{3}{2}\right) - 12 = 6 + 6 - 12 = 0$$

This shows that the equation is satisfied, and we conclude that the point $\left(2, -\frac{3}{2}\right)$ does indeed lie on L.

c. Substituting $x = 1$ and $y = -2$ into the left-hand side of the equation $3x - 4y - 12 = 0$, we obtain

$$3(1) - 4(-2) - 12 = 3 + 8 - 12 = -1$$

which is not equal to zero, the number on the right-hand side of the equation. This shows that the point $(1, -2)$ does not lie on L.

Here is a summary of the common forms of the equations of straight lines discussed in this section.

Equations of Straight Lines	
Vertical line:	$x = a$
Horizontal line:	$y = b$
Point-slope form:	$y - y_1 = m(x - x_1)$
Slope-intercept form:	$y = mx + b$
General form:	$Ax + By + C = 0$

2.2   Self-Check Exercises

1. Find an equation of the line that passes through the point $(3, -1)$ and is perpendicular to a line with slope $-\frac{1}{2}$.

2. Does the point $(3, -3)$ lie on the line with equation $2x - 3y - 12 = 0$? Sketch the graph of the line.

3. SMOKERS IN THE UNITED STATES   The following table gives the percentage of adults in the United States from 2006 through 2010 who smoked in year t. Here, $t = 0$ corresponds to the beginning of 2006.

Year, t	0	1	2	3	4
Percent, y	20.8	20.5	20.1	19.8	19.0

a. Plot the percentage of U.S. adults who smoke (y) versus the year (t) for the given years.

b. Draw the line L through the points $(0, 20.8)$ and $(4, 19.0)$, and find an equation of the line L.

c. Assuming that this trend continues, estimate the percentage of U.S. adults who smoked at the beginning of 2014.

Source: Centers for Disease Control and Prevention.

Solutions to Self-Check Exercises 2.2 can be found on page 88.

2.2   Concept Questions

1. Give (a) the point-slope form, (b) the slope-intercept form, and (c) the general form of an equation of a line.

2. Let L_1 have slope m_1 and let L_2 have slope m_2. State the conditions on m_1 and m_2 if (a) L_1 is parallel to L_2 and (b) L_1 is perpendicular to L_2.

3. Suppose a line L has equation $Ax + By + C = 0$.
   a. What is the slope of L if $B \neq 0$?
   b. What is the slope of L if $B = 0$ and $A \neq 0$?

2.2   Exercises

In Exercises 1–6, match the statement with one of the graphs (a)–(f).

1. The slope of the line is zero.

2. The slope of the line is undefined.

3. The slope of the line is positive, and its y-intercept is positive.

4. The slope of the line is positive, and its y-intercept is negative.

5. The slope of the line is negative, and its x-intercept is negative.

6. The slope of the line is negative, and its x-intercept is positive.

(a)

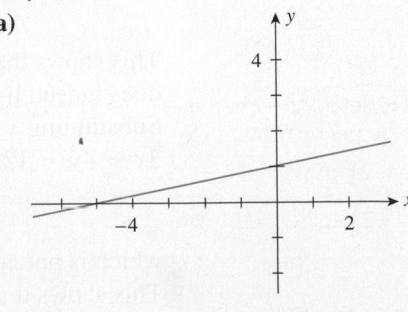

(b)

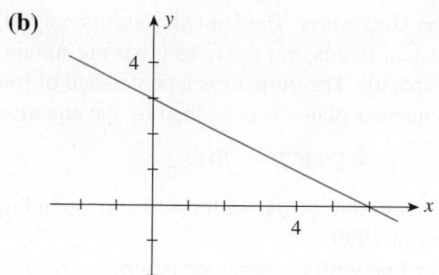

(c)

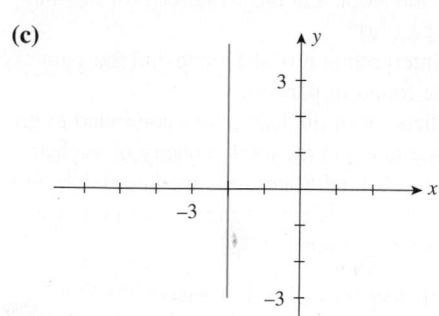

(d)

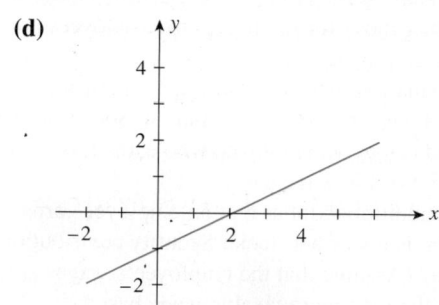

(e)

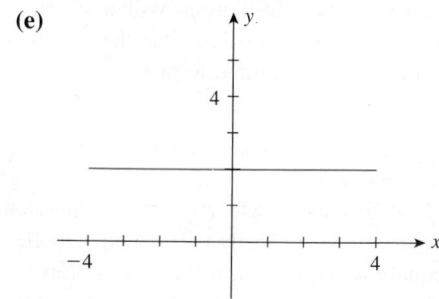

(f)

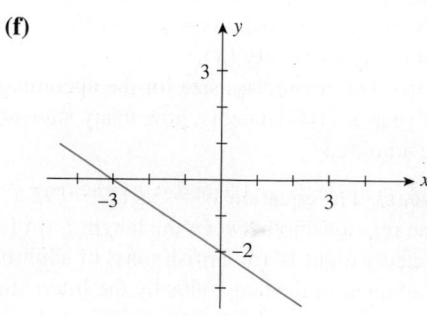

In Exercises 7 and 8, determine whether the lines through the pairs of points are perpendicular.

7. $A(-2, 5)$, $B(4, 2)$ and $C(-1, -2)$, $D(3, 6)$

8. $A(2, 0)$, $B(1, -2)$ and $C(4, 2)$, $D(-8, 4)$

9. Find an equation of the horizontal line that passes through $(-4, -3)$.

10. Find an equation of the vertical line that passes through $(0, 5)$.

In Exercises 11–14, find an equation of the line that passes through the point and has the indicated slope m.

11. $(3, -4)$; $m = 2$ **12.** $(2, 4)$; $m = -1$

13. $(-3, 2)$; $m = 0$ **14.** $(1, 2)$; $m = -\dfrac{1}{2}$

In Exercises 15–18, find an equation of the line that passes through the given points.

15. $(2, 4)$ and $(3, 7)$ **16.** $(2, 1)$ and $(2, 5)$

17. $(1, 2)$ and $(-3, -2)$ **18.** $(-1, -2)$ and $(3, -4)$

In Exercises 19–22, find an equation of the line that has slope m and y-intercept b.

19. $m = 3$; $b = 4$ **20.** $m = -2$; $b = -1$

21. $m = 0$; $b = 5$ **22.** $m = -\dfrac{1}{2}$; $b = \dfrac{3}{4}$

In Exercises 23–28, write the equation in the slope-intercept form and then find the slope and y-intercept of the corresponding line.

23. $x - 2y = 0$ **24.** $y - 2 = 0$

25. $2x - 3y - 9 = 0$ **26.** $3x - 4y + 8 = 0$

27. $2x + 4y = 14$ **28.** $5x + 8y - 24 = 0$

29. Find an equation of the line that passes through the point $(-2, 2)$ and is parallel to the line $2x - 4y - 8 = 0$.

30. Find an equation of the line that passes through the point $(-1, 3)$ and is parallel to the line passing through the points $(-2, -3)$ and $(2, 5)$.

31. Find an equation of the line that passes through the point $(2, 4)$ and is perpendicular to the line $3x + 4y - 22 = 0$.

32. Find an equation of the line that passes through the point $(1, -2)$ and is perpendicular to the line passing through the points $(-2, -1)$ and $(4, 3)$.

In Exercises 33–38, find an equation of the line that satisfies the given condition.

33. The line parallel to the x-axis and 6 units below it

34. The line passing through the origin and parallel to the line passing through the points $(2, 4)$ and $(4, 7)$

35. The line passing through the point (a, b) with slope equal to zero

36. The line passing through $(-3, 4)$ and parallel to the x-axis

37. The line passing through $(-5, -4)$ and parallel to the line passing through $(-3, 2)$ and $(6, 8)$

38. The line passing through (a, b) with undefined slope

39. Given that the point $P(-3, 5)$ lies on the line $kx + 3y + 9 = 0$, find k.

40. Given that the point $P(2, -3)$ lies on the line $-2x + ky + 10 = 0$, find k.

In Exercises 41–46, sketch the straight line defined by the linear equation by finding the x- and y-intercepts.
Hint: See Example 8.

41. $3x - 2y + 6 = 0$ **42.** $2x - 5y + 10 = 0$

43. $x + 2y - 4 = 0$ **44.** $2x + 3y - 15 = 0$

45. $y + 5 = 0$ **46.** $-2x - 8y + 24 = 0$

47. Show that an equation of a line through the points $(a, 0)$ and $(0, b)$ with $a \neq 0$ and $b \neq 0$ can be written in the form

$$\frac{x}{a} + \frac{y}{b} = 1$$

(Recall that the numbers a and b are the x- and y-intercepts, respectively, of the line. This form of an equation of a line is called the **intercept form**.)

In Exercises 48–51, use the results of Exercise 47 to find an equation of a line with the x- and y-intercepts.

48. x-intercept 3; y-intercept 4

49. x-intercept -2; y-intercept -4

50. x-intercept $-\dfrac{1}{2}$; y-intercept $\dfrac{3}{4}$

51. x-intercept 4; y-intercept $-\dfrac{1}{2}$

In Exercises 52 and 53, determine whether the points lie on a straight line.

52. $A(-1, 7)$, $B(2, -2)$, and $C(5, -9)$

53. $A(-2, 1)$, $B(1, 7)$, and $C(4, 13)$

54. John claims that the following points lie on a line: $(1.2, -9.04)$, $(2.3, -5.96)$, $(4.8, 1.04)$, and $(7.2, 7.76)$. Prove or disprove his claim.

55. Alison claims that the following points lie on a line: $(1.8, -6.44)$, $(2.4, -5.72)$, $(5.0, -2.72)$, and $(10.4, 3.88)$. Prove or disprove her claim.

56. TEMPERATURE CONVERSION The relationship between the temperature in degrees Fahrenheit (°F) and the temperature in degrees Celsius (°C) is

$$F = \frac{9}{5}C + 32$$

a. Sketch the line with the given equation.
b. What is the slope of the line? What does it represent?
c. What is the F-intercept of the line? What does it represent?

57. NUCLEAR PLANT UTILIZATION The United States is not building many nuclear plants, but the ones it has are running at nearly full capacity. The output (as a percentage of total capacity) of nuclear plants is described by the equation

$$y = 1.9467t + 70.082$$

where t is measured in years, with $t = 0$ corresponding to the beginning of 1990.
a. Sketch the line with the given equation.
b. What are the slope and the y-intercept of the line found in part (a)?
c. Give an interpretation of the slope and the y-intercept of the line found in part (a).
d. If the utilization of nuclear power continued to grow at the same rate and the total capacity of nuclear plants in the United States remained constant, by what year were the plants generating at maximum capacity?
Source: Nuclear Energy Institute.

58. SOCIAL SECURITY CONTRIBUTIONS For wages less than the maximum taxable wage base, Social Security contributions (including those for Medicare) by employees are 7.65% of the employee's wages.
a. Find an equation that expresses the relationship between the wages earned (x) and the Social Security taxes paid (y) by an employee who earns less than the maximum taxable wage base.
b. For each additional dollar that an employee earns, by how much is his or her Social Security contribution increased? (Assume that the employee's wages are less than the maximum taxable wage base.)
c. What Social Security contributions will an employee who earns $65,000 (which is less than the maximum taxable wage base) be required to make?
Source: Social Security Administration.

59. COLLEGE ADMISSIONS Using data compiled by the Admissions Office at Faber University, college admissions officers estimate that 55% of the students who are offered admission to the freshman class at the university will actually enroll.
a. Find an equation that expresses the relationship between the number of students who actually enroll (y) and the number of students who are offered admission to the university (x).
b. If the desired freshman class size for the upcoming academic year is 1100 students, how many students should be admitted?

60. WEIGHT OF WHALES The equation $W = 3.51L - 192$, expressing the relationship between the length L (in feet) and the expected weight W (in British tons) of adult blue whales, was adopted in the late 1960s by the International Whaling Commission.
a. What is the expected weight of an 80-ft blue whale?
b. Sketch the straight line that represents the equation.

61. THE NARROWING GENDER GAP Since the founding of the Equal Employment Opportunity Commission and the passage of equal-pay laws, the gulf between men's and

women's earnings has continued to close gradually. At the beginning of 1990 ($t = 0$), women's wages were 68% of men's wages, and by the beginning of 2000 ($t = 10$), women's wages were 80% of men's wages. If this gap between women's and men's wages continued to narrow *linearly,* then women's wages were what percentage of men's wages at the beginning of 2004?
Source: Journal of Economic Perspectives.

62. DECLINING NUMBER OF PAY PHONES As cell phones proliferate, the number of pay phones continues to drop. The number of pay phones from 2004 through 2009 (in millions) are shown in the following table ($x = 0$ corresponds to 2004):

Year, x	0	1	2	3	4	5
Number of Pay Phones, y	1.30	1.15	1.00	0.84	0.69	0.56

a. Plot the number of pay phones (y) versus the year (x).
b. Draw the straight line L through the points $(0, 1.30)$ and $(5, 0.56)$.
c. De
d. As
　　be
Source.

63. SPEND
　　contin
　　spend
　　The f
　　ment
　　($x = $

　　Qua
　　Perc

a. Pl
b. Dr
　　ing
c. De
d. If
　　in
　　qu
Source.

64. IDEAL
　　Club
　　table,
　　pound

　　Heig
　　Weig

a. Pl
b. Dr
　　ing
c. De
d. Us
　　de

65. COST OF A COMMODITY A manufacturer obtained the following data relating the cost y (in dollars) to the number of units (x) of a commodity produced:

Units Produced, x	0	20	40	60	80	100
Cost in Dollars, y	200	208	222	230	242	250

a. Plot the cost (y) versus the quantity produced (x).
b. Draw a straight line through the points $(0, 200)$ and $(100, 250)$.
c. Derive an equation of the straight line of part (b).
d. Taking this equation to be an approximation of the relationship between the cost and the level of production, estimate the cost of producing 54 units of the commodity.

66. CORPORATE FRAUD The number of pending corporate fraud cases stood at 545 at the beginning of 2008 ($t = 0$) and was 726 at the beginning of 2012. The growth was ‥ly linear.
　　‥n equation of the line passing through the ‥$(0, 545)$ and $B(4, 726)$.
　　‥line with the equation found in part (a).
　　‥equation found in part (a) to estimate the num-‥nding corporate fraud cases at the beginning

‥l Bureau of Investigation.

‥ Metro Department Store's annual sales (in ‥ollars) during the past 5 years were

‥s, y	5.8	6.2	7.2	8.4	9.0
	1	2	3	4	5

‥annual sales (y) versus the year (x).
‥traight line L through the points correspond-‥ first and fifth years.
‥n equation of the line L.
‥ equation found in part (c), estimate Metro's ‥les 4 years from now ($x = 9$).

　　A Nielsen survey of 3000 American ‥aged 12–74 found that 27% of them used ‥ to chat about movies in 2010. The percent-‥ in 2011 and 31% in 2012. Let $t = 0$, $t = 1$, ‥rrespond to the years 2010, 2011, and 2012,

‥why the three points $P_1(0, 27)$, $P_2(1, 29)$, and ‥) lie on a straight line L.
‥d continues, what will the percentage of ‥ers who use social media to chat about mov-‥2014?
‥quation of L. Then use this equation to find ‥cile the result obtained in part (b).
‥ survey.

In Exercises 69–74, determine whether the statement is true or false. If it is true, explain why it is true. If it is false, give an example to show why it is false.

69. Suppose the slope of a line L is $-\frac{1}{2}$ and P is a given point on L. If Q is the point on L lying 4 units to the left of P, then Q is situated 2 units above P.

70. The point $(1, k)$ lies on the line with equation $3x + 4y = 12$ if and only if $k = \frac{9}{4}$.

71. The line with equation $Ax + By + C = 0$ ($B \neq 0$) and the line with equation $ax + by + c = 0$ ($b \neq 0$) are parallel if $Ab - aB = 0$.

72. If the slope of the line L_1 is positive, then the slope of a line L_2 perpendicular to L_1 may be positive or negative.

73. The lines with equations $ax + by + c_1 = 0$ and $bx - ay + c_2 = 0$, where $a \neq 0$ and $b \neq 0$, are perpendicular to each other.

74. If L is the line with equation $Ax + By + C = 0$, where $A \neq 0$, then L crosses the x-axis at the point $(-C/A, 0)$.

75. Show that two distinct lines with equations $a_1x + b_1y + c_1 = 0$ and $a_2x + b_2y + c_2 = 0$, respectively, are parallel if and only if $a_1b_2 - b_1a_2 = 0$.
Hint: Write each equation in the slope-intercept form and compare.

76. Prove that if a line L_1 with slope m_1 is perpendicular to a line L_2 with slope m_2, then $m_1m_2 = -1$.
Hint: Refer to the accompanying figure. Show that $m_1 = b$ and $m_2 = c$. Next, apply the Pythagorean Theorem and the distance formula to the triangles OAC, OCB, and OBA to show that $1 = -bc$.

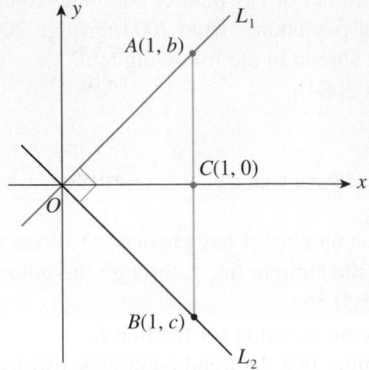

2.2 Solutions to Self-Check Exercises

1. Since the required line L is perpendicular to a line with slope $-\frac{1}{2}$, the slope of L is

$$m = -\frac{1}{-\frac{1}{2}} = 2$$

Next, using the point-slope form of the equation of a line, we have

$$y - (-1) = 2(x - 3)$$
$$y + 1 = 2x - 6$$
$$y = 2x - 7$$

2. Substituting $x = 3$ and $y = -3$ into the left-hand side of the given equation, we find

$$2(3) - 3(-3) - 12 = 3$$

which is not equal to zero (the right-hand side). Therefore, $(3, -3)$ does not lie on the line with equation $2x - 3y - 12 = 0$. (See the accompanying figure.)
Setting $x = 0$, we find $y = -4$, the y-intercept. Next, setting $y = 0$ gives $x = 6$, the x-intercept. We now draw the line passing through the points $(0, -4)$ and $(6, 0)$, as shown.

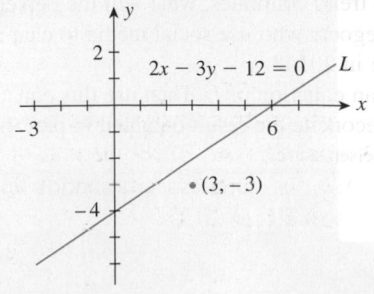

3. a. and b.

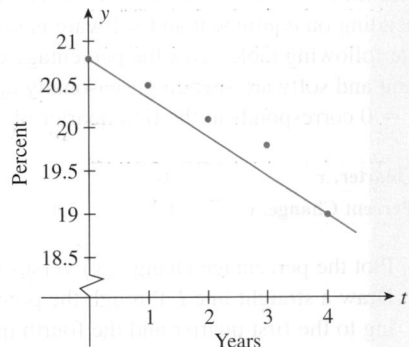

The slope of L is

$$m = \frac{19.0 - 20.8}{4 - 0} = -0.45$$

Using the point-slope form of the equation of a line with the point $(0, 20.8)$, we find

$$y - 20.8 = -0.45(t - 0) \quad \text{or} \quad y = -0.45t + 20.8$$

c. The year 2014 corresponds to $t = 8$, so the estimated percentage of U.S. adults who will be smoking is

$$y = -0.45(8) + 20.8 = 17.2$$

or 17.2%.

USING TECHNOLOGY

Graphing a Straight Line

Graphing Utility

The first step in plotting a straight line with a graphing utility is to select a suitable viewing window. We usually do this by experimenting. For example, you might first plot the straight line using the **standard viewing window** $[-10, 10] \times [-10, 10]$. If necessary, you then might adjust the viewing window by enlarging it or reducing it to obtain a sufficiently complete view of the line or at least the portion of the line that is of interest.

EXAMPLE 1 Plot the straight line $2x + 3y - 6 = 0$ in the standard viewing window.

Solution The straight line in the standard viewing window is shown in Figure T1.

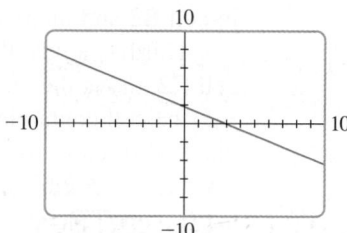

FIGURE **T1**
The straight line $2x + 3y - 6 = 0$ in the standard viewing window

EXAMPLE 2 Plot the straight line $2x + 3y - 30 = 0$ in (a) the standard viewing window and (b) the viewing window $[-5, 20] \times [-5, 20]$.

Solution

a. The straight line in the standard viewing window is shown in Figure T2a.
b. The straight line in the viewing window $[-5, 20] \times [-5, 20]$ is shown in Figure T2b. This figure certainly gives a more complete view of the straight line.

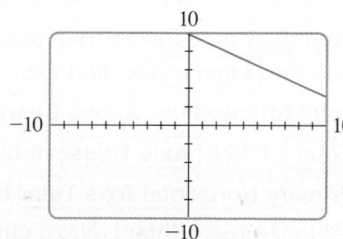

(a) The graph of $2x + 3y - 30 = 0$ in the standard viewing window

(b) The graph of $2x + 3y - 30 = 0$ in the viewing window $[-5, 20] \times [-5, 20]$

FIGURE **T2**

Excel

In the examples and exercises that follow, we assume that you are familiar with the basic features of Microsoft Excel. Please consult your Excel manual or use Excel's Help features to answer questions regarding the standard commands and operating instructions for Excel. Here, we use Microsoft Excel 2010.

EXAMPLE 3 Plot the graph of the straight line $2x + 3y - 6 = 0$ over the interval $[-10, 10]$.

Solution

1. *Write the equation in the slope-intercept form:*

$$y = -\frac{2}{3}x + 2$$

2. *Create a table of values.* First, enter the input values: Enter the values of the endpoints of the interval over which you are graphing the straight line. (Recall that we need only two distinct data points to draw the graph of a straight line. In general, we select the endpoints of the interval over which the straight line is to be drawn as our data points.) In this case, we enter -10 in cell B1 and 10 in cell C1.

   Second, enter the formula for computing the y-values: Here, we enter

   $$= -(2/3)*B1+2$$

   in cell B2 and then press ⎡Enter⎤.

   Third, evaluate the function at the other input value: To extend the formula to cell C2, move the pointer to the small black box at the lower right corner of cell B2 (the cell containing the formula). Observe that the pointer now appears as a black + (plus sign). Drag this pointer through cell C2, and then release it. The y-value, -4.66667, corresponding to the x-value in cell C1(10) will appear in cell C2 (Figure T3).

	A	B	C
1	x	-10	10
2	y	8.666667	-4.66667

FIGURE **T3**
Table of values for *x* and *y*

3. *Graph the straight line determined by these points.* First, highlight the numerical values in the table. Here, we highlight cells B1:B2 and C1:C2.

   Step 1 Click on the ⎡Insert⎤ ribbon tab, and then select ⎡Scatter⎤ from the Charts group. Select the chart subtype in the first row and second column. A chart will then appear on your worksheet.

   Step 2 From the Chart Tools group that now appears at the end of the ribbon, click the ⎡Layout⎤ tab, and then select ⎡Chart Title⎤ from the Labels group followed by ⎡Above Chart⎤. Type y =-(2/3)x + 2 and press ⎡Enter⎤. Click ⎡Axis Titles⎤ from the Labels group, and select ⎡Primary Horizontal Axis Title⎤ followed by ⎡Title Below Axis⎤. Type x and then press ⎡Enter⎤. Next, click ⎡Axis Titles⎤ again, and select ⎡Primary Vertical Axis Title⎤ followed by ⎡Vertical Title⎤. Type y and press ⎡Enter⎤.

Note: Boldfaced words/characters enclosed in a box (for example, ⎡Enter⎤) indicate that an action (click, select, or press) is required. Words/characters printed blue (for example, Chart Type) indicate words/characters that appear on the screen. Words/characters printed in a monospace font (for example, =(-2/3)*A2+2) indicate words/characters that need to be typed and entered.

Step 3 Click $\boxed{\textbf{Series1}}$, which appears on the right side of the graph, and press $\boxed{\textbf{Delete}}$.

The graph shown in Figure T4 will appear.

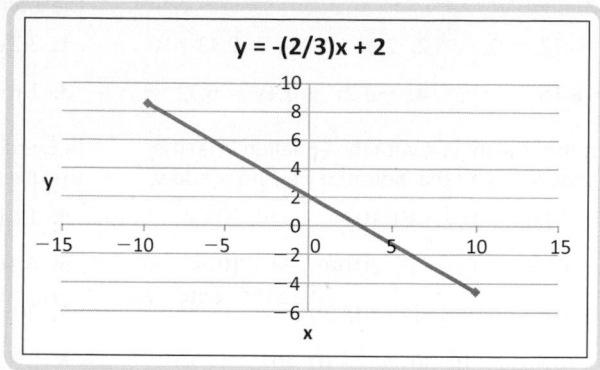

FIGURE **T4**
The graph of $y = -\frac{2}{3}x + 2$ over the interval $[-10, 10]$

If the interval over which the straight line is to be plotted is not specified, then you might have to experiment to find an appropriate interval for the *x*-values in your graph. For example, you might first plot the straight line over the interval $[-10, 10]$. If necessary, you then might adjust the interval by enlarging it or reducing it to obtain a sufficiently complete view of the line or at least the portion of the line that is of interest.

EXAMPLE 4 Plot the straight line $2x + 3y - 30 = 0$ over the intervals (a) $[-10, 10]$ and (b) $[-5, 20]$.

Solution **a.** and **b.** We first cast the equation in the slope-intercept form, obtaining $y = -\frac{2}{3}x + 10$. Following the procedure given in Example 3, we obtain the graphs shown in Figure T5.

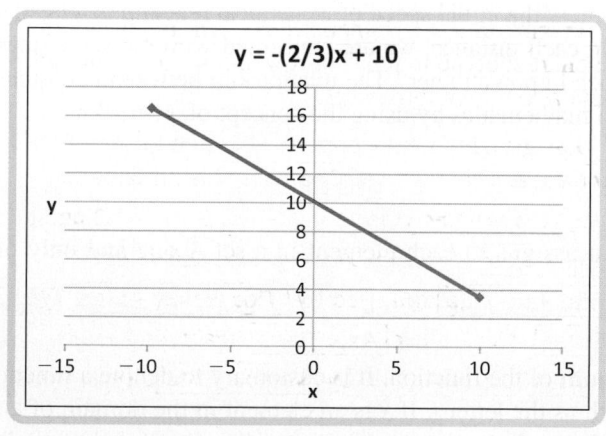

(a)

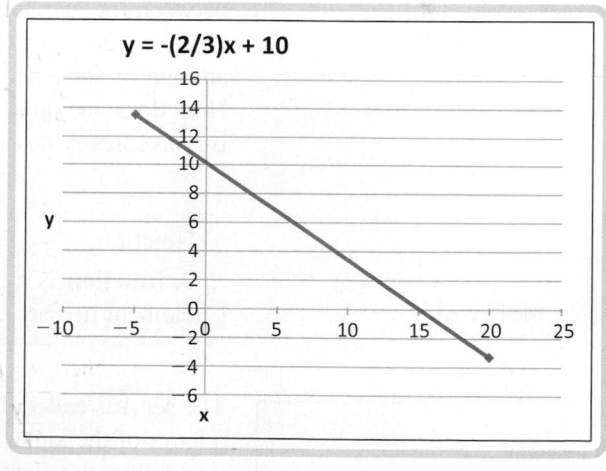

(b)

FIGURE **T5**
The graph of $y = -\frac{2}{3}x + 10$ over the intervals (a) $[-10, 10]$ and (b) $[-5, 20]$

Observe that the graph in Figure T5b includes the *x*- and *y*-intercepts. This figure certainly gives a more complete view of the straight line.

TECHNOLOGY EXERCISES

Graphing Utility

In Exercises 1–4, plot the straight line with the equation in the standard viewing window.

1. $3.2x + 2.1y - 6.72 = 0$
2. $2.3x - 4.1y - 9.43 = 0$
3. $1.6x + 5.1y = 8.16$
4. $-3.2x + 2.1y = 6.72$

In Exercises 5–8, plot the straight line with the equation in (a) the standard viewing window and (b) the indicated viewing window.

5. $12.1x + 4.1y - 49.61 = 0; [-10, 10] \times [-10, 20]$
6. $4.1x - 15.2y - 62.32 = 0; [-10, 20] \times [-10, 10]$
7. $20x + 16y = 300; [-10, 20] \times [-10, 30]$
8. $32.2x + 21y = 676.2; [-10, 30] \times [-10, 40]$

In Exercises 9–12, plot the straight line with the equation in an appropriate viewing window. (*Note:* The answer is *not* unique.)

9. $20x + 30y = 600$
10. $30x - 20y = 600$
11. $22.4x + 16.1y - 352 = 0$
12. $18.2x - 15.1y = 274.8$

Excel

In Exercises 1–4, plot the straight line with the equation over the interval $[-10, 10]$.

1. $3.2x + 2.1y - 6.72 = 0$
2. $2.3x - 4.1y - 9.43 = 0$
3. $1.6x + 5.1y = 8.16$
4. $-3.2x + 2.1y = 6.72$

In Exercises 5–8, plot the straight line with the equation over the given interval.

5. $12.1x + 4.1y - 49.61 = 0; [-10, 10]$
6. $4.1x - 15.2y - 62.32 = 0; [-10, 20]$
7. $20x + 16y = 300; [-10, 20]$
8. $32.2x + 21y = 676.2; [-10, 30]$

In Exercises 9–12, plot the straight line with the equation. (*Note:* The answer is *not* unique.)

9. $20x + 30y = 600$
10. $30x - 20y = 600$
11. $22.4x + 16.1y - 352 = 0$
12. $18.2x - 15.1y = 274.8$

2.3   Functions and Their Graphs

Functions

A manufacturer would like to know how his company's profit is related to its production level; a biologist would like to know how the size of the population of a certain culture of bacteria will change over time; a psychologist would like to know the relationship between the learning time of an individual and the length of a vocabulary list; and a chemist would like to know how the initial speed of a chemical reaction is related to the amount of substrate used. In each instance, we are concerned with the same question: How does one quantity depend upon another? The relationship between two quantities is conveniently described in mathematics by using the concept of a function.

> **Function**
>
> A **function** is a rule that assigns to each element in a set A one and only one element in a set B.

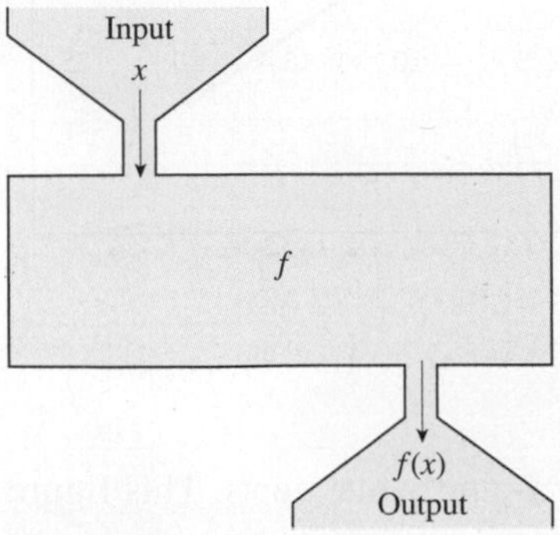

The set A is called the **domain** of the function. It is customary to denote a function by a letter of the alphabet, such as the letter f. If x is an element in the domain of a function f, then the element in B that f associates with x is written $f(x)$ (read "f of x") and is called the value of f at x. The set comprising all the values assumed by $y = f(x)$ as x takes on all possible values in its domain is called the **range** of the function f.

We can think of a function f as a machine. The domain is the set of inputs (raw material) for the machine, the rule describes how the input is to be processed, and the values of the function are the outputs of the machine (Figure 23).

FIGURE 23
A function machine

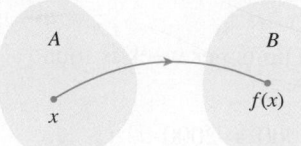

We can also think of a function f as a mapping in which an element x in the domain of f is mapped onto a unique element $f(x)$ in B (Figure 24).

Notes

1. The output $f(x)$ associated with an input x is unique. To appreciate the importance of this uniqueness property, consider a rule that associates with each item x in a department store its selling price y. Then, each x must correspond to *one and only one y*. Notice, however, that different x's may be associated with the same y. In the context of the present example, this says that different items may have the same price.
2. Although the sets A and B that appear in the definition of a function may be quite arbitrary, in this book they will denote sets of real numbers.

An example of a function may be taken from the familiar relationship between the area of a circle and its radius. Letting x and y denote the radius and area of a circle, respectively, we have, from elementary geometry,

$$y = \pi x^2 \tag{5}$$

Equation (5) defines y as a function of x, since for each admissible value of x (that is, for each nonnegative number representing the radius of a certain circle), there corresponds precisely one number $y = \pi x^2$ that gives the area of the circle. The rule defining this "area function" may be written as

$$f(x) = \pi x^2 \tag{6}$$

To compute the area of a circle of radius 5 inches, we simply replace x in Equation (6) with the number 5. Thus, the area of the circle is

$$f(5) = \pi 5^2 = 25\pi$$

or 25π square inches.

In general, to evaluate a function at a specific value of x, we replace x with that value, as illustrated in Examples 1 and 2.

EXAMPLE 1 Let the function f be defined by the rule $f(x) = 2x^2 - x + 1$. Find:

a. $f(1)$ **b.** $f(-2)$ **c.** $f(a)$ **d.** $f(a + h)$

Solution

a. $f(1) = 2(1)^2 - (1) + 1 = 2 - 1 + 1 = 2$
b. $f(-2) = 2(-2)^2 - (-2) + 1 = 8 + 2 + 1 = 11$
c. $f(a) = 2(a)^2 - (a) + 1 = 2a^2 - a + 1$
d. $f(a + h) = 2(a + h)^2 - (a + h) + 1 = 2a^2 + 4ah + 2h^2 - a - h + 1$

APPLIED EXAMPLE 2 Profit Functions ThermoMaster manufactures an indoor–outdoor thermometer at its Mexican subsidiary. Management estimates that the profit (in dollars) realizable by ThermoMaster in the manufacture and sale of x thermometers per week is

$$P(x) = -0.001x^2 + 8x - 5000$$

Find ThermoMaster's weekly profit if its level of production is (a) 1000 thermometers per week and (b) 2000 thermometers per week.

Solution

a. The weekly profit when the level of production is 1000 units per week is found by evaluating the profit function P at $x = 1000$. Thus,

$$P(1000) = -0.001(1000)^2 + 8(1000) - 5000 = 2000$$

or $2000.

b. When the level of production is 2000 units per week, the weekly profit is given by

$$P(2000) = -0.001(2000)^2 + 8(2000) - 5000 = 7000$$

or $7000.

Determining the Domain of a Function

Suppose we are given the function $y = f(x)$.* Then, the variable x is called the **independent variable**. The variable y, whose value depends on x, is called the **dependent variable**.

To determine the domain of a function, we need to find what restrictions, if any, are to be placed on the independent variable x. In general, if a function is defined by a rule relating x to $f(x)$ without specific mention of its domain, it is understood that the domain will consist of all values of x for which $f(x)$ is a real number. In this connection, you should keep in mind that (1) division by zero is not permitted and (2) the even root of a negative number is not a real number.

EXAMPLE 3 Find the domain of each function.

a. $f(x) = \sqrt{x - 1}$ b. $f(x) = \dfrac{1}{x^2 - 4}$ c. $f(x) = x^2 + 3$

Solution

a. Since the square root of a negative number is not a real number, it is necessary that $x - 1 \geq 0$. The inequality is satisfied by the set of real numbers $x \geq 1$. Thus, the domain of f is the interval $[1, \infty)$.

b. The only restriction on x is that $x^2 - 4$ be different from zero, since division by zero is not allowed. But $(x^2 - 4) = (x + 2)(x - 2) = 0$ if $x = -2$ or $x = 2$. Thus, the domain of f in this case consists of the intervals $(-\infty, -2), (-2, 2)$, and $(2, \infty)$.

c. Here, any real number satisfies the equation, so the domain of f is the set of all real numbers.

In many practical applications, the domain of a function is dictated by the nature of the problem, as illustrated in Example 4.

$\$$ APPLIED EXAMPLE 4 Packaging An open box is to be made from a rectangular piece of cardboard 16 inches long and 10 inches wide by cutting away identical squares (x inches by x inches) from each corner and folding up the

---

*It is customary to refer to a function f as $f(x)$ or by the equation $y = f(x)$ defining the function.

resulting flaps (Figure 25). Find an expression that gives the volume V of the box as a function of x. What is the domain of the function?

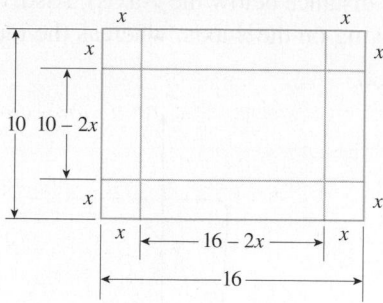 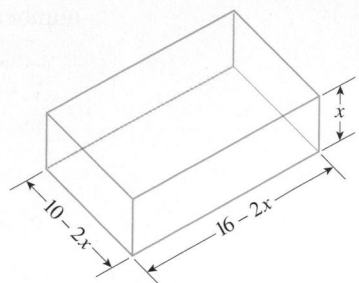

(a) The box is constructed by cutting x in. by x in. squares from each corner.

(b) The dimensions of the resulting box are $(10 - 2x)$ in. by $(16 - 2x)$ in. by x in.

FIGURE **25**

Solution The dimensions of the box are $(10 - 2x)$ inches by $(16 - 2x)$ inches by x inches, so its volume (in cubic inches) is given by

$$V = f(x) = (16 - 2x)(10 - 2x)x \qquad \text{Length · width · height}$$
$$= (160 - 52x + 4x^2)x$$
$$= 4x^3 - 52x^2 + 160x$$

Since the length of each side of the box must be greater than or equal to zero, we see that

$$16 - 2x \geq 0 \qquad 10 - 2x \geq 0 \qquad x \geq 0$$

simultaneously; that is,

$$x \leq 8 \qquad x \leq 5 \qquad x \geq 0$$

All three inequalities are satisfied simultaneously provided that $0 \leq x \leq 5$. Thus, the domain of the function f is the interval $[0, 5]$.

Graphs of Functions

If f is a function with domain A, then corresponding to each real number x in A, there is precisely one real number $f(x)$. We can also express this fact by using ordered pairs of real numbers. Write each number x in A as the first member of an ordered pair and each number $f(x)$ corresponding to x as the second member of the ordered pair. This gives exactly one ordered pair $(x, f(x))$ for each x in A.

Observe that the condition that there be one and only one number $f(x)$ corresponding to each number x in A translates into the requirement that *no two distinct ordered pairs have the same first number.*

Since ordered pairs of real numbers correspond to points in the plane, we have found a way to exhibit a function graphically.

Graph of a Function of One Variable

The **graph of a function** f is the set of all points (x, y) in the xy-plane such that x is in the domain of f and $y = f(x)$.

Figure 26 shows the graph of a function f. Observe that the y-coordinate of the point (x, y) on the graph of f gives the height of that point (the distance above the x-axis), if $f(x)$ is positive. If $f(x)$ is negative, then $-f(x)$ gives the depth of the point (x, y) (the distance below the x-axis). Also, observe that the domain of f is a set of real numbers lying on the x-axis, whereas the range of f lies on the y-axis.

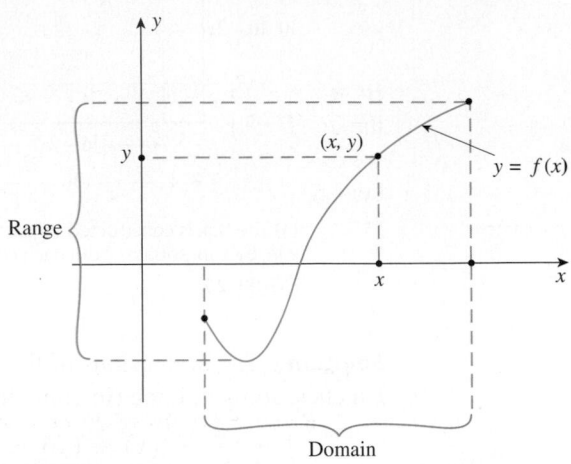

FIGURE 26
The graph of f

EXAMPLE 5 The graph of a function f is shown in Figure 27.

a. What is the value of $f(3)$? The value of $f(5)$?
b. What is the height or depth of the point $(3, f(3))$ from the x-axis? The point $(5, f(5))$ from the x-axis?
c. What is the domain of f? The range of f?

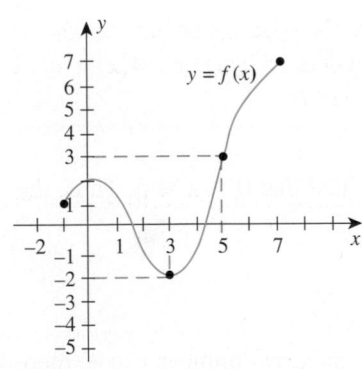

FIGURE 27
The graph of f

Solution

a. From the graph of f, we see that $y = -2$ when $x = 3$, and we conclude that $f(3) = -2$. Similarly, we see that $f(5) = 3$.
b. Since the point $(3, -2)$ lies below the x-axis, we see that the depth of the point $(3, f(3))$ is $-f(3) = -(-2) = 2$ units below the x-axis. The point $(5, f(5))$ lies above the x-axis and is located at a height of $f(5)$, or 3 units above the x-axis.
c. Observe that x may take on all values between $x = -1$ and $x = 7$, inclusive, so the domain of f is $[-1, 7]$. Next, observe that as x takes on all values in the domain of f, $f(x)$ takes on all values between -2 and 7, inclusive. (You can easily see this by running your index finger along the x-axis from $x = -1$ to $x = 7$ and observing the corresponding values assumed by the y-coordinate of each point of the graph of f.) Therefore, the range of f is $[-2, 7]$.

We can gain much information about the graph of a function by plotting a few points on its graph. Later on, we will develop more systematic and sophisticated techniques for graphing functions.

EXAMPLE 6 Sketch the graph of the function defined by the equation $y = x^2 + 1$. What is the range of f?

Solution The domain of the function is the set of all real numbers. By assigning several values to the variable x and computing the corresponding values for y, we obtain the following solutions to the equation $y = x^2 + 1$:

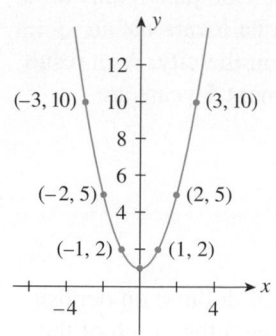

x	-3	-2	-1	0	1	2	3
y	10	5	2	1	2	5	10

FIGURE 28
The graph of $y = x^2 + 1$ is a parabola.

By plotting these points and then connecting them with a smooth curve, we obtain the graph of $y = f(x)$, which is a parabola (Figure 28). To determine the range of f, we observe that $x^2 \geq 0$ if x is any real number, and so $x^2 + 1 \geq 1$ for all real numbers x. We conclude that the range of f is $[1, \infty)$. The graph of f confirms this result visually. ∎

Exploring with TECHNOLOGY

Let $f(x) = x^2$.

1. Plot the graphs of $F(x) = x^2 + c$ on the same set of axes for $c = -2, -1, -\frac{1}{2}, 0, \frac{1}{2}, 1, 2$.

2. Plot the graphs of $G(x) = (x + c)^2$ on the same set of axes for $c = -2, -1, -\frac{1}{2}, 0, \frac{1}{2}, 1, 2$.

3. Plot the graphs of $H(x) = cx^2$ on the same set of axes for $c = -2, -1, -\frac{1}{2}, -\frac{1}{4}, 0, \frac{1}{4}, \frac{1}{2}, 1, 2$.

4. Study the family of graphs in parts 1–3, and describe the relationship between the graph of a function f and the graphs of the functions defined by (a) $y = f(x) + c$, (b) $y = f(x + c)$, and (c) $y = cf(x)$, where c is a constant.

Sometimes a function is defined by giving different formulas for different parts of its domain. Such a function is said to be a **piecewise-defined function.**

EXAMPLE 7 Sketch the graph of the function f defined by

$$f(x) = \begin{cases} -x & \text{if } x < 0 \\ \sqrt{x} & \text{if } x \geq 0 \end{cases}$$

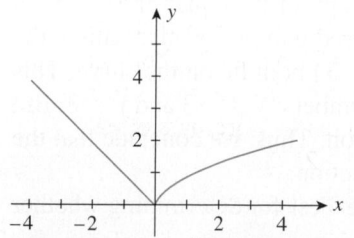

FIGURE 29
The graph of $y = f(x)$ is obtained by graphing $y = -x$ over $(-\infty, 0)$ and $y = \sqrt{x}$ over $[0, \infty)$.

Solution The function f is defined in a piecewise fashion on the set of all real numbers. In the subdomain $(-\infty, 0)$, the rule for f is given by $f(x) = -x$. The equation $y = -x$ is a linear equation in the slope-intercept form (with slope -1 and intercept 0). Therefore, the graph of f corresponding to the subdomain $(-\infty, 0)$ is the half-line shown in Figure 29. Next, in the subdomain $[0, \infty)$, the rule for f is given by $f(x) = \sqrt{x}$. The values of $f(x)$ corresponding to $x = 0, 1, 2, 3$, and 4 are shown in the following table:

x	0	1	2	3	4
$f(x)$	0	1	$\sqrt{2}$	$\sqrt{3}$	2

Using these values, we sketch the graph of the function f as shown in Figure 29. ∎

$ APPLIED EXAMPLE 8 Bank Deposits Madison Finance Company plans to open two branch offices 2 years from now in two separate locations: an industrial complex and a newly developed commercial center in the city. As a result of these expansion plans, Madison's total deposits during the next 5 years are expected to grow in accordance with the rule

$$f(x) = \begin{cases} \sqrt{2x} + 20 & \text{if } 0 \le x \le 2 \\ \dfrac{1}{2}x^2 + 20 & \text{if } 2 < x \le 5 \end{cases}$$

where $y = f(x)$ gives the total amount of money (in millions of dollars) on deposit with Madison in year x ($x = 0$ corresponds to the present). Sketch the graph of the function f.

Solution The function f is defined in a piecewise fashion on the interval $[0, 5]$. In the subdomain $[0, 2]$, the rule for f is given by $f(x) = \sqrt{2x} + 20$. The values of $f(x)$ corresponding to $x = 0, 1,$ and 2 may be tabulated as follows:

x	0	1	2
$f(x)$	20	21.4	22

Next, in the subdomain $(2, 5]$, the rule for f is given by $f(x) = \frac{1}{2}x^2 + 20$. The values of $f(x)$ corresponding to $x = 3, 4,$ and 5 are shown in the following table:

x	3	4	5
$f(x)$	24.5	28	32.5

Using the values of $f(x)$ in this table, we sketch the graph of the function f as shown in Figure 30. ■

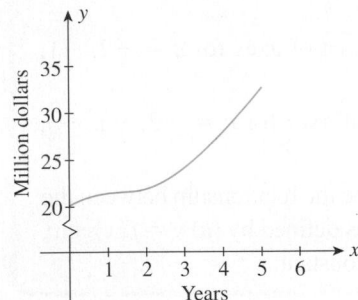

FIGURE **30**
We obtain the graph of the function $y = f(x)$ by graphing $y = \sqrt{2x} + 20$ over [0, 2] and $y = \frac{1}{2}x^2 + 20$ over (2, 5].

The Vertical Line Test

Although it is true that every function f of a variable x has a graph in the xy-plane, it is not true that every curve in the xy-plane is the graph of a function. For example, consider the curve depicted in Figure 31. This is the graph of the equation $y^2 = x$. In general, the **graph of an equation** is the set of all ordered pairs (x, y) that satisfy the given equation. Observe that the points $(9, -3)$ and $(9, 3)$ both lie on the curve. This implies that the number $x = 9$ is associated with *two* numbers: $y = -3$ and $y = 3$. But this clearly violates the uniqueness property of a function. Thus, we conclude that the curve under consideration cannot be the graph of a function.

This example suggests the following **Vertical Line Test** for determining whether a curve is the graph of a function.

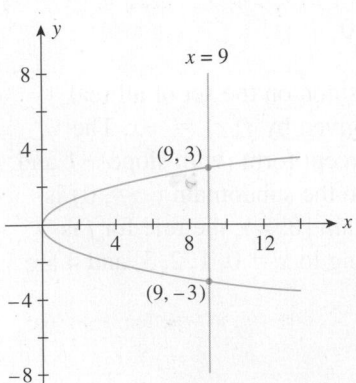

FIGURE **31**
Since a vertical line passes through the curve at more than one point, we deduce that the curve is *not* the graph of a function.

> Vertical Line Test
>
> A curve in the xy-plane is the graph of a function $y = f(x)$ if and only if each vertical line intersects it in at most one point.

EXAMPLE 9 Determine which of the curves shown in Figure 32 are the graphs of functions of x.

Solution The curves depicted in Figure 32a, c, and d are graphs of functions because each curve satisfies the requirement that each vertical line intersects the curve in at most one point. Note that the vertical line shown in Figure 32c does *not* intersect the graph because the point on the x-axis through which this line passes does not lie in the domain of the function. The curve depicted in Figure 32b is *not* the graph of a function of x because the vertical line shown there intersects the graph at three points.

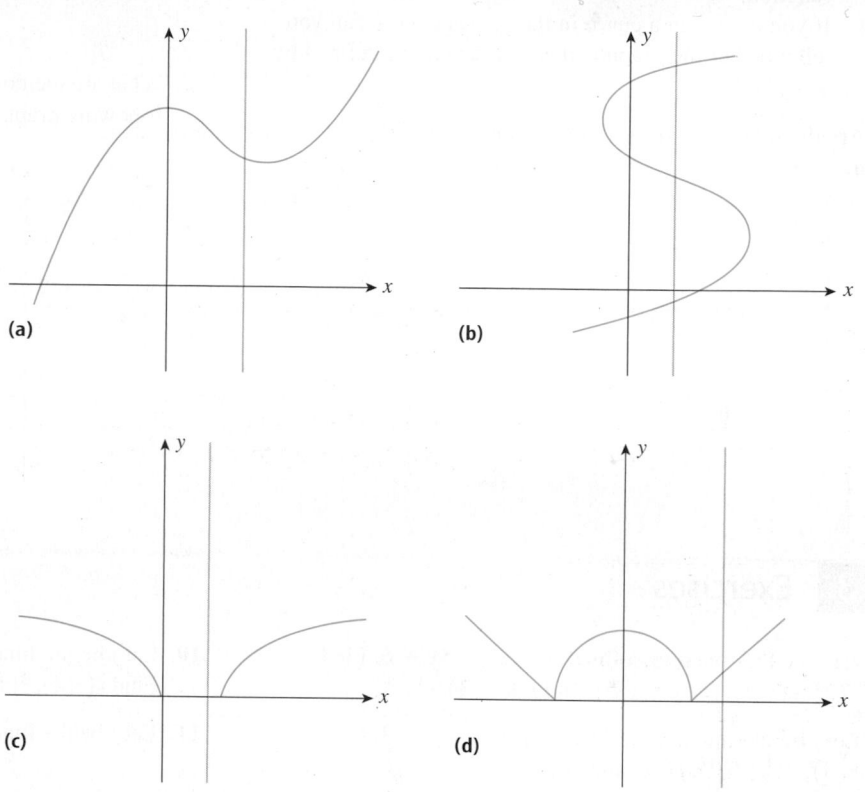

(a)

(b)

(c)

(d)

FIGURE **32**
The Vertical Line Test can be used to determine which of these curves are graphs of functions.

2.3 Self-Check Exercises

1. Let f be the function defined by

$$f(x) = \frac{\sqrt{x+1}}{x}$$

a. Find the domain of f.   **b.** Compute $f(3)$.
c. Compute $f(a + h)$.

2. Let

$$f(x) = \begin{cases} -x + 1 & \text{if } -1 \le x < 1 \\ \sqrt{x-1} & \text{if } 1 \le x \le 5 \end{cases}$$

a. Find $f(0)$ and $f(2)$.
b. Sketch the graph of f.

3. Let $f(x) = \sqrt{2x+1} + 2$. Determine whether the point $(4, 6)$ lies on the graph of f.

Solutions to Self-Check Exercises 2.3 can be found on page 104.

2.3   Concept Questions

1. a. What is a function?
  b. What is the domain of a function? The range of a function?
  c. What is an independent variable? A dependent variable?

2. a. What is the graph of a function? Use a drawing to illustrate the graph, the domain, and the range of a function.
  b. If you are given a curve in the xy-plane, how can you tell whether the graph is that of a function f defined by $y = f(x)$?

3. Are the following graphs of functions? Explain.
  a.   **b.**

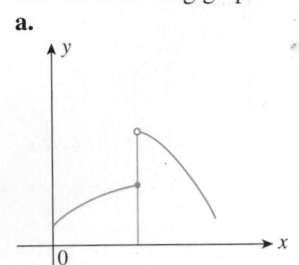 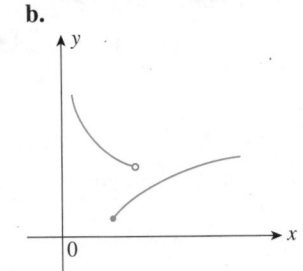

c.   **d.**

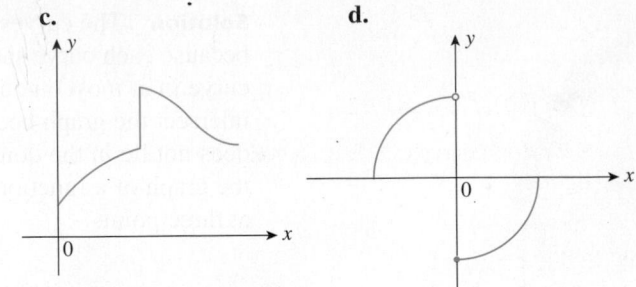

4. What are the domain and range of the function f with the following graph?

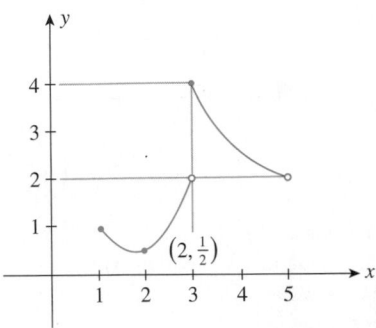

$\left(2, \frac{1}{2}\right)$

2.3   Exercises

1. Let f be the function defined by $f(x) = 5x + 6$. Find $f(3), f(-3), f(a), f(-a)$, and $f(a + 3)$.

2. Let f be the function defined by $f(x) = 4x - 3$. Find $f(4), f(\frac{1}{4}), f(0), f(a)$, and $f(a + 1)$.

3. Let g be the function defined by $g(x) = 3x^2 - 6x - 3$. Find $g(0), g(-1), g(a), g(-a)$, and $g(x + 1)$.

4. Let h be the function defined by $h(x) = x^3 - x^2 + x + 1$. Find $h(-5), h(0), h(a)$, and $h(-a)$.

5. Let f be the function defined by $f(x) = 2x + 5$. Find $f(a + h), f(-a), f(a^2), f(a - 2h)$, and $f(2a - h)$.

6. Let g be the function defined by $g(x) = -x^2 + 2x$. Find $g(a + h), g(-a), g(\sqrt{a}), a + g(a)$, and $\dfrac{1}{g(a)}$.

7. Let s be the function defined by $s(t) = \dfrac{2t}{t^2 - 1}$. Find $s(4), s(0), s(a), s(2 + a)$, and $s(t + 1)$.

8. Let g be the function defined by $g(u) = (3u - 2)^{3/2}$. Find $g(1), g(6), g(\frac{11}{3})$, and $g(u + 1)$.

9. Let f be the function defined by $f(t) = \dfrac{2t^2}{\sqrt{t - 1}}$. Find $f(2), f(a), f(x + 1)$, and $f(x - 1)$.

10. Let f be the function defined by $f(x) = 2 + 2\sqrt{5 - x}$. Find $f(-4), f(1), f(\frac{11}{4})$, and $f(x + 5)$.

11. Let f be the function defined by
$$f(x) = \begin{cases} x^2 + 1 & \text{if } x \le 0 \\ \sqrt{x} & \text{if } x > 0 \end{cases}$$
Find $f(-2), f(0)$, and $f(1)$.

12. Let g be the function defined by
$$g(x) = \begin{cases} -\dfrac{1}{2}x + 1 & \text{if } x < 2 \\ \sqrt{x - 2} & \text{if } x \ge 2 \end{cases}$$
Find $g(-2), g(0), g(2)$, and $g(4)$.

13. Let f be the function defined by
$$f(x) = \begin{cases} -\dfrac{1}{2}x^2 + 3 & \text{if } x < 1 \\ 2x^2 + 1 & \text{if } x \ge 1 \end{cases}$$
Find $f(-1), f(0), f(1)$, and $f(2)$.

3 - 13 odd

14. Let f be the function defined by

$$f(x) = \begin{cases} 2 + \sqrt{1 - x} & \text{if } x \le 1 \\ \dfrac{1}{1 - x} & \text{if } x > 1 \end{cases}$$

Find $f(0), f(1)$, and $f(2)$.

15. Refer to the graph of the function f in the following figure.

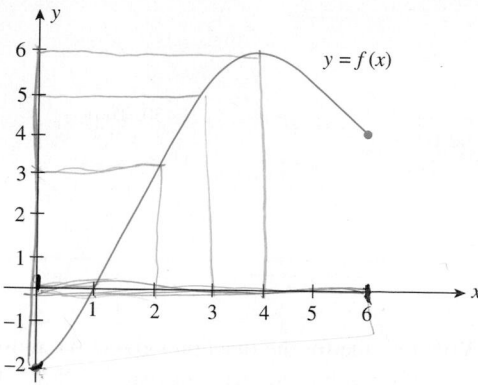

a. Find the value of $f(0)$.
b. Find the value of x for which (i) $f(x) = 3$ and (ii) $f(x) = 0$.
c. Find the domain of f.
d. Find the range of f.

16. Refer to the graph of the function f in the following figure.

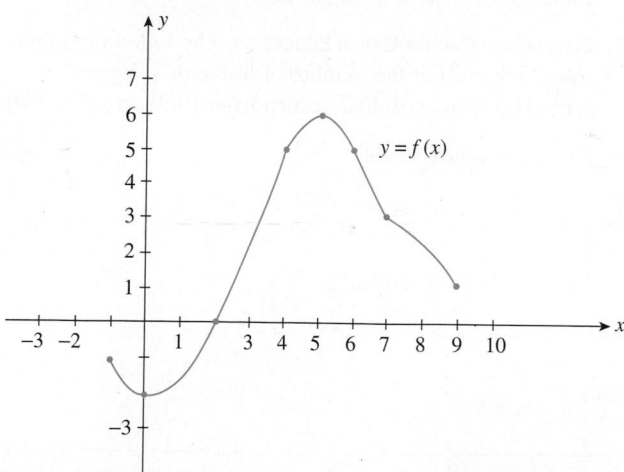

a. Find the value of $f(7)$.
b. Find the values of x corresponding to the point(s) on the graph of f located at a height of 5 units from the x-axis.
c. Find the point on the x-axis at which the graph of f crosses it. What is the value of $f(x)$ at this point?
d. Find the domain and range of f.

In Exercises 17–20, determine whether the point lies on the graph of the function.

17. $(2, \sqrt{3}); g(x) = \sqrt{x^2 - 1}$

18. $(3, 3); f(x) = \dfrac{x + 1}{\sqrt{x^2 + 7}} + 2$

19. $(-2, -3); f(t) = \dfrac{|t - 1|}{t + 1}$

20. $\left(-3, -\dfrac{1}{13}\right); h(t) = \dfrac{|t + 1|}{t^3 + 1}$

In Exercises 21 and 22, find the value of c such that the point $P(a, b)$ lies on the graph of the function f.

21. $f(x) = 2x^2 - 4x + c; P(1, 5)$

22. $f(x) = x\sqrt{9 - x^2} + c; P(2, 4)$

In Exercises 23–36, find the domain of the function.

23. $f(x) = x^2 + 3$ **24.** $f(x) = 7 - x^2$

25. $f(x) = \dfrac{3x + 1}{x^2}$ **26.** $g(x) = \dfrac{2x + 1}{x - 1}$

27. $f(x) = \sqrt{x^2 + 1}$ **28.** $f(x) = \sqrt{x - 5}$

29. $f(x) = \sqrt{5 - x}$ **30.** $g(x) = \sqrt{2x^2 + 3}$

31. $f(x) = \dfrac{x}{x^2 - 1}$ **32.** $f(x) = \dfrac{1}{x^2 + x - 2}$

33. $f(x) = (x + 3)^{3/2}$ **34.** $g(x) = 2(x - 1)^{5/2}$

35. $f(x) = \dfrac{\sqrt{1 - x}}{x^2 - 4}$ **36.** $f(x) = \dfrac{\sqrt{x - 1}}{(x + 2)(x - 3)}$

37. Let f be the function defined by the rule $f(x) = x^2 - x - 6$.
a. Find the domain of f.
b. Compute $f(x)$ for $x = -3, -2, -1, 0, \frac{1}{2}, 1, 2, 3$.
c. Use the results obtained in parts (a) and (b) to sketch the graph of f.

38. Let f be the function defined by the rule $f(x) = 2x^2 + x - 3$.
a. Find the domain of f.
b. Compute $f(x)$ for $x = -3, -2, -1, -\frac{1}{2}, 0, 1, 2, 3$.
c. Use the results obtained in parts (a) and (b) to sketch the graph of f.

In Exercises 39–50, sketch the graph of the function with the given rule. Find the domain and range of the function.

39. $f(x) = 2x^2 + 1$ **40.** $f(x) = 9 - x^2$

41. $f(x) = 2 + \sqrt{x}$ **42.** $g(x) = 4 - \sqrt{x}$

43. $f(x) = \sqrt{1 - x}$ **44.** $f(x) = \sqrt{x - 1}$

45. $f(x) = |x| - 1$ **46.** $f(x) = |x| + 1$

47. $f(x) = \begin{cases} x & \text{if } x < 0 \\ 2x + 1 & \text{if } x \ge 0 \end{cases}$

48. $f(x) = \begin{cases} 4 - x & \text{if } x < 2 \\ 2x - 2 & \text{if } x \ge 2 \end{cases}$

49. $f(x) = \begin{cases} -x + 1 & \text{if } x \le 1 \\ x^2 - 1 & \text{if } x > 1 \end{cases}$

17-49 odd

50. $f(x) = \begin{cases} -x - 1 & \text{if } x < -1 \\ 0 & \text{if } -1 \leq x \leq 1 \\ x + 1 & \text{if } x > 1 \end{cases}$

In Exercises 51–58, use the Vertical Line Test to determine whether the graph represents y as a function of x.

51.

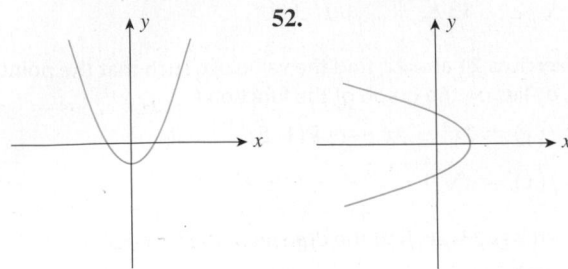

52.

53.

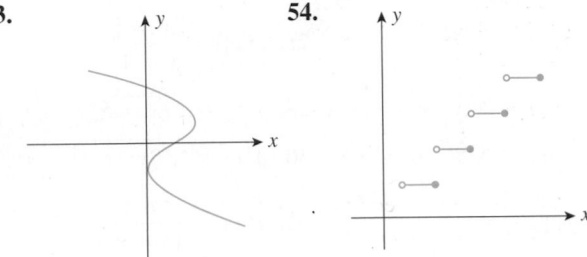

54.

55.

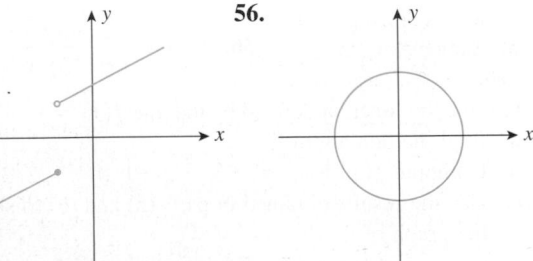

56.

57.

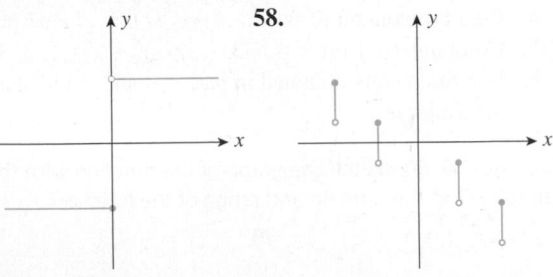

58.

59. The circumference of a circle is given by

$$C(r) = 2\pi r$$

where r is the radius of the circle. What is the circumference of a circle with a 5-in. radius?

60. The volume of a sphere of radius r is given by

$$V(r) = \frac{4}{3}\pi r^3$$

Compute $V(2.1)$ and $V(2)$. What does the quantity $V(2.1) - V(2)$ measure?

61. SURFACE AREA OF A SINGLE-CELLED ORGANISM The surface area S of a single-celled organism may be found by multiplying 4π times the square of the radius r of the cell. Express S as a function of r.

62. THE GENDER GAP The following graph shows the ratio of women's earnings to men's from 1960 through 2000.

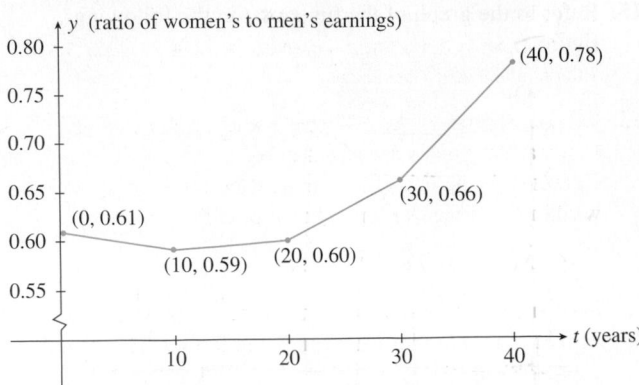

a. Write the rule for the function f giving the ratio of women's earnings to men's in year t, with $t = 0$ corresponding to 1960.
   Hint: The function f is defined piecewise and is linear over each of four subintervals.

b. In what decade(s) was the gender gap expanding? Shrinking?

c. Refer to part (b). How fast was the gender gap expanding or shrinking in each of these decades?
Source: U.S. Bureau of Labor Statistics.

63. CLOSING THE GENDER GAP IN EDUCATION The following graph shows the ratio of the number of bachelor's degrees earned by women to that of men from 1960 through 1990.

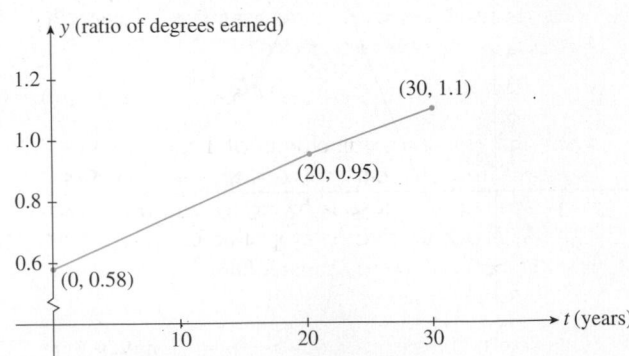

a. Write the rule for the function f giving the ratio of the number of bachelor's degrees earned by women to that of men in year t, with $t = 0$ corresponding to 1960.
   Hint: The function f is defined piecewise and is linear over each of two subintervals.

b. How fast was the ratio changing in the period from 1960 to 1980? From 1980 to 1990?

c. In what year (approximately) was the number of bachelor's degrees earned by women equal for the first time to the number earned by men?
Source: Department of Education.

64. ALZHEIMER'S DISEASE The projected number of people aged 65 and over in the U.S. population with Alzheimer's disease (in millions) is given by the function

$$P(t) = -0.0002083t^3 + 0.0157t^2 - 0.093t + 5.2$$
$$(4 \leq t \leq 40)$$

where t is measured in years, with $t = 4$ corresponding to 2014. What is the projected number of people aged 65 and over with Alzheimer's disease in 2030? In 2050?
Source: Alzheimer's Association.

65. WORKER EFFICIENCY An efficiency study conducted for Elektra Electronics showed that the number of Space Commander walkie-talkies assembled by the average worker t hr after starting work at 8 A.M. is given by

$$N(t) = -t^3 + 6t^2 + 15t \qquad (0 \leq t \leq 4)$$

How many walkie-talkies can an average worker be expected to assemble between 8 and 9 A.M.? Between 9 and 10 A.M.?

66. POLITICS Political scientists have discovered the following empirical rule, known as the "cube rule," which gives the relationship between the proportion of seats in the House of Representatives won by Democratic candidates $s(x)$ and the proportion of popular votes x received by the Democratic presidential candidate:

$$s(x) = \frac{x^3}{x^3 + (1 - x)^3} \qquad (0 \leq x \leq 1)$$

Compute $s(0.6)$ and interpret your result.

67. U.S. HEALTH-CARE INFORMATION TECHNOLOGY SPENDING As health-care costs increase, payers are turning to technology and outsourced services to keep a lid on expenses. The amount of health-care information technology (IT) spending by payer is projected to be

$$S(t) = -0.03t^3 + 0.2t^2 + 0.23t + 5.6 \qquad (0 \leq t \leq 4)$$

where $S(t)$ is measured in billions of dollars and t is measured in years, with $t = 0$ corresponding to 2004. What was the amount spent by payers on health-care IT in 2004? Assuming that the projection held true, what amount was spent by payers in 2008?
Source: U.S. Department of Commerce.

68. HOTEL RATES The average daily rate of U.S. hotels from 2006 through 2009 is approximated by the function

$$f(t) = \begin{cases} 0.88t^2 + 3.21t + 96.75 & \text{if } 0 \leq t < 2 \\ -5.58t + 117.85 & \text{if } 2 \leq t \leq 3 \end{cases}$$

where $f(t)$ is measured in dollars and $t = 0$ corresponds to 2006.
a. What was the average daily rate of U.S. hotels in 2006? In 2007? In 2008?
b. Sketch the graph of f.
Source: Smith Travel Research.

69. INVESTMENTS IN HEDGE FUNDS Investments in hedge funds have increased along with their popularity. The assets of

hedge funds (in trillions of dollars) from 2002 through 2007 are modeled by the function

$$f(t) = \begin{cases} 0.6 & \text{if } 0 \leq t < 1 \\ 0.6t^{0.43} & \text{if } 1 \leq t \leq 5 \end{cases}$$

where t is measured in years, with $t = 0$ corresponding to the beginning of 2002.
a. What were the assets in hedge funds at the beginning of 2002? At the beginning of 2003?
b. What were the assets in hedge funds at the beginning of 2005? At the beginning of 2007?
Source: Hennessee Group.

70. RISING MEDIAN AGE Increased longevity and the aging of the baby boom generation—those born between 1946 and 1965—are the primary reasons for a rising median age. The median age (in years) of the U.S. population from 1900 through 2011 is approximated by the function

$$f(t) = \begin{cases} 1.3t + 22.9 & \text{if } 0 \leq t \leq 3 \\ -0.7t^2 + 7.2t + 11.5 & \text{if } 3 < t \leq 7 \\ 2.6t + 9.4 & \text{if } 7 < t \leq 11 \end{cases}$$

where t is measured in decades, with $t = 0$ corresponding to the beginning of 1900.
a. What was the median age of the U.S. population at the beginning of 1900? At the beginning of 1950? At the beginning of 2000?
b. Sketch the graph of f.
Source: U.S. Census Bureau.

71. POSTAL REGULATIONS In 2012, the postage for parcels sent by first-class mail was raised to $1.95 for any parcel weighing less than 4 oz or fraction thereof and 17¢ for each additional ounce or fraction thereof. Any parcel not exceeding 13 oz may be sent by first-class mail. Letting x denote the weight of a parcel in ounces and $f(x)$ the postage in dollars, complete the following description of the "postage function" f:

$$f(x) = \begin{cases} \$1.95 & \text{if } 0 < x < 4 \\ \$2.12 & \text{if } 4 \leq x < 5 \\ \vdots \\ ? & \text{if } x = 13 \end{cases}$$

a. What is the domain of f?
b. Sketch the graph of f.

In Exercises 72–76, determine whether the statement is true or false. If it is true, explain why it is true. If it is false, give an example to show why it is false.

72. If $a = b$, then $f(a) = f(b)$.

73. If $f(a) = f(b)$, then $a = b$.

74. If f is a function, then $f(a + b) = f(a) + f(b)$.

75. A vertical line must intersect the graph of $y = f(x)$ at exactly one point.

76. The domain of $f(x) = \sqrt{x + 2} + \sqrt{2 - x}$ is $[-2, 2]$.

2.3 Solutions to Self-Check Exercises

1. a. The expression under the radical sign must be nonnegative, so $x + 1 \geq 0$ or $x \geq -1$. Also, $x \neq 0$ because division by zero is not permitted. Therefore, the domain of f is $[-1, 0)$ and $(0, \infty)$.

b. $f(3) = \dfrac{\sqrt{3+1}}{3} = \dfrac{\sqrt{4}}{3} = \dfrac{2}{3}$

c. $f(a + h) = \dfrac{\sqrt{(a+h)+1}}{a+h} = \dfrac{\sqrt{a+h+1}}{a+h}$

2. a. The function f is defined in a piecewise fashion. For $x = 0$, the rule is $f(x) = -x + 1$, and so $f(0) = 1$. For $x = 2$, the rule is $f(x) = \sqrt{x-1}$, and so $f(2) = \sqrt{2-1} = 1$.

b. In the subdomain $[-1, 1)$, the graph of f is the line segment $y = -x + 1$, which is a linear equation with slope -1 and y-intercept 1. In the subdomain $[1, 5]$, the graph of f is given by the rule $f(x) = \sqrt{x-1}$. From the table below,

x	1	2	3	4	5
$f(x)$	0	1	$\sqrt{2}$	$\sqrt{3}$	2

we obtain the following graph of f.

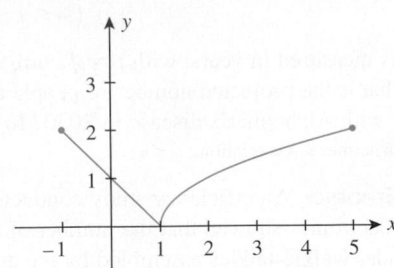

3. A point (x, y) lies on the graph of the function f if and only if the coordinates satisfy the equation $y = f(x)$. Now,

$$f(4) = \sqrt{2(4)+1} + 2 = \sqrt{9} + 2 = 5 \neq 6$$

and we conclude that the given point does *not* lie on the graph of f.

USING TECHNOLOGY Graphing a Function

Most of the graphs of functions in this book can be plotted with the help of a graphing utility. Furthermore, a graphing utility can be used to analyze the nature of a function. However, the amount and accuracy of the information obtained by using a graphing utility depend on the experience and sophistication of the user. As you progress through this book, you will see that the more knowledge of calculus you gain, the more effective the graphing utility will prove to be as a tool in problem solving.

EXAMPLE 1 Plot the graph of $f(x) = 2x^2 - 4x - 5$ in the standard viewing window.

Solution   The graph of f, shown in Figure T1a, is a parabola. From our previous work (Example 6, Section 2.3), we know that the figure does give a good view of the graph.

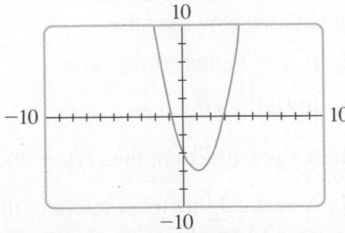

(a)

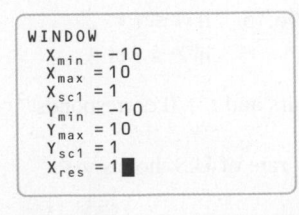

(b)

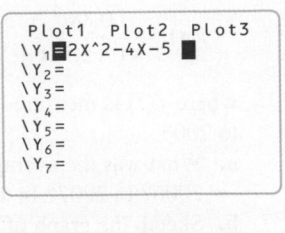

(c)

FIGURE **T1**
(a) The graph of $f(x) = 2x^2 - 4x - 5$ on $[-10, 10] \times [-10, 10]$; (b) the TI-83/84 window screen for (a); (c) the TI-83/84 equation screen

EXAMPLE 2 Let $f(x) = x^3(x - 3)^4$.

a. Plot the graph of f in the standard viewing window.
b. Plot the graph of f in the window $[-1, 5] \times [-40, 40]$.

Solution

a. The graph of f in the standard viewing window is shown in Figure T2a. Since the graph does not appear to be complete, we need to adjust the viewing window.

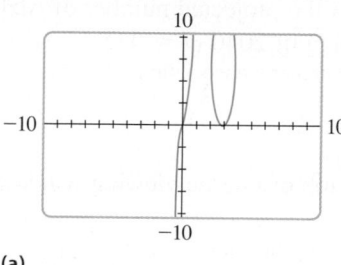
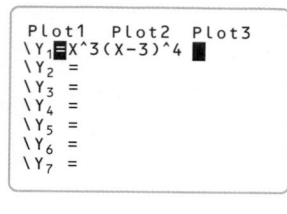

(a)   (b)

FIGURE **T2**
(a) An incomplete sketch of $f(x) = x^3(x - 3)^4$ on $[-10, 10] \times [-10, 10]$;
(b) the TI-83/84 equation screen

b. The graph of f in the window $[-1, 5] \times [-40, 40]$, shown in Figure T3a, is an improvement over the previous graph. (Later we will be able to show that the figure does in fact give a rather complete view of the graph of f.)

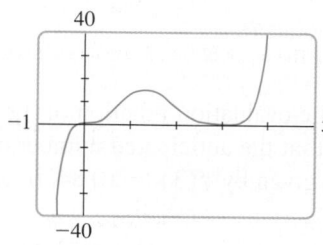
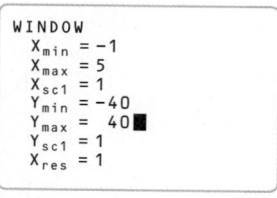

(a)   (b)

FIGURE **T3**
(a) A complete sketch of $f(x) = x^3(x - 3)^4$ is shown using the window
$[-1, 5] \times [-40, 40]$; (b) the TI-83/84 window screen

Evaluating a Function

A graphing utility can be used to find the value of a function with minimal effort, as the next example shows.

EXAMPLE 3 Let $f(x) = x^3 - 4x^2 + 4x + 2$.

a. Plot the graph of f in the standard viewing window.
b. Find $f(3)$ and verify your result by direct computation.
c. Find $f(4.215)$.

Solution

a. The graph of f is shown in Figure T4a.
b. Using the evaluation function of the graphing utility and the value 3 for x, we find $y = 5$. This result is verified by computing

$$f(3) = 3^3 - 4(3^2) + 4(3) + 2 = 27 - 36 + 12 + 2 = 5$$

c. Using the evaluation function of the graphing utility and the value 4.215 for x, we find $y = 22.679738$. Thus, $f(4.215) = 22.679738$. The efficacy of the graphing utility is clearly demonstrated here!

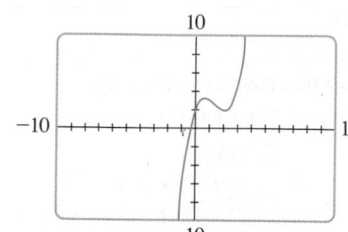

(a)

(b)

FIGURE **T4**
(a) The graph of $f(x) = x^3 - 4x^2 + 4x + 2$
in the standard viewing window;
(b) the TI-83/84 equation screen

APPLIED EXAMPLE 4 Alzheimer's Patients in the United States The number of Alzheimer's patients in the United States is approximated by

$$f(t) = -0.208t^3 + 1.571t^2 - 0.9274t + 5.1 \qquad (0 \le t \le 4)$$

where $f(t)$ is measured in millions and t is measured in decades, with $t = 0$ corresponding to the beginning of 2010.

a. Use a graphing utility to plot the graph of f in the viewing window $[0, 4] \times [0, 14]$.
b. What is the projected number of Alzheimer's patients in the United States at the beginning of 2040 $(t = 3)$?
Source: Alzheimer's Association.

Solution

a. The graph of f in the viewing window $[0, 4] \times [0, 14]$ is shown in Figure T5a.

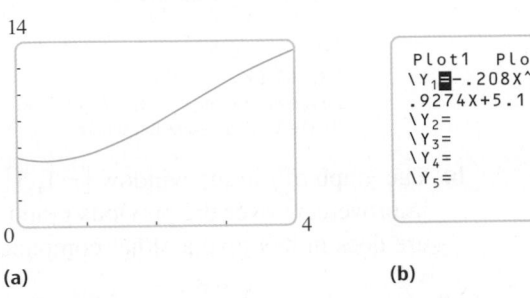

(a) (b)

FIGURE **T5**
(a) The graph of f in the viewing window [0, 4] × [0, 14]; (b) the TI-83/84 equation screen

b. Using the evaluation function of the graphing utility and the value 3 for x, we see that the anticipated number of Alzheimer's patients at the beginning of 2040 is given by $f(3) \approx 10.84$, or approximately 10.8 million.

TECHNOLOGY EXERCISES

In Exercises 1–4, plot the graph of the function f in (a) the standard viewing window and (b) the indicated window.

1. $f(x) = x^4 - 2x^2 + 8; [-2, 2] \times [6, 10]$

2. $f(x) = x^3 - 20x^2 + 8x - 10; [-20, 20] \times [-1200, 100]$

3. $f(x) = x\sqrt{4 - x^2}; [-3, 3] \times [-2, 2]$

4. $f(x) = \dfrac{4}{x^2 - 8}; [-5, 5] \times [-5, 5]$

In Exercises 5–8, plot the graph of the function f in an appropriate viewing window. (*Note:* The answer is *not* unique.)

5. $f(x) = 2x^4 - 3x^3 + 5x^2 - 20x + 40$

6. $f(x) = -2x^4 + 5x^2 - 4$

7. $f(x) = \dfrac{x^3}{x^3 + 1}$

8. $f(x) = \dfrac{2x^4 - 3x}{x^2 - 1}$

In Exercises 9–12, use the evaluation function of your graphing utility to find the value of f at the indicated value of x. Express your answer accurate to four decimal places.

9. $f(x) = 3x^3 - 2x^2 + x - 4; x = 2.145$

10. $f(x) = 5x^4 - 2x^2 + 8x - 3; x = 1.28$

11. $f(x) = \dfrac{2x^3 - 3x + 1}{3x - 2}; x = 2.41$

12. $f(x) = \sqrt{2x^2 + 1} + \sqrt{3x^2 - 1}; x = 0.62$

13. LOBBYISTS' SPENDING Lobbyists try to persuade legislators to propose, pass, or defeat legislation or to change existing laws. The amount (in billions of dollars) spent by lobbyists from 2003 through 2009, where $t = 0$ corresponds to 2003, is given by

$$f(t) = -0.0056t^3 + 0.112t^2 + 0.51t + 8 \qquad (0 \le t \le 6)$$

a. Plot the graph of f in the viewing window $[0, 6] \times [0, 15]$.

b. What amount was spent by lobbyists in the year 2005? In 2009?

Source: OpenSecrets.org.

14. SAFE DRIVERS The fatality rate in the United States (per 100 million miles traveled) by age of driver (in years) is given by the function

$$f(x) = 0.00000304x^4 - 0.0005764x^3 + 0.04105x^2$$
$$- 1.30366x + 16.579 \qquad (18 \le x \le 82)$$

a. Plot the graph of f in the viewing window $[18, 82] \times [0, 8]$.

b. What is the fatality rate for 18-year-old drivers? For 50-year-old drivers? For 80-year-old drivers?

Source: National Highway Traffic Safety Administration.

15. KEEPING WITH THE TRAFFIC FLOW By driving at a speed to match the prevailing traffic speed, you decrease the chances of an accident. According to data obtained in a university study, the number of accidents per 100 million vehicle miles, y, is related to the deviation from the mean speed, x, in miles per hour by

$$y = 1.05x^3 - 21.95x^2 + 155.9x - 327.3 \qquad (6 \le x \le 11)$$

a. Plot the graph of y in the viewing window $[6, 11] \times [20, 150]$.

b. What is the number of accidents per 100 million vehicle miles if the deviation from the mean speed is 6 mph, 8 mph, and 11 mph?

Source: University of Virginia School of Engineering and Applied Science.

<h2>2.4 The Algebra of Functions</h2>

The Sum, Difference, Product, and Quotient of Functions

Let $S(t)$ and $R(t)$ denote the federal government's spending and revenue, respectively, at any time t, measured in billions of dollars. The graphs of these functions for the period between 2006 and 2012 are shown in Figure 33.

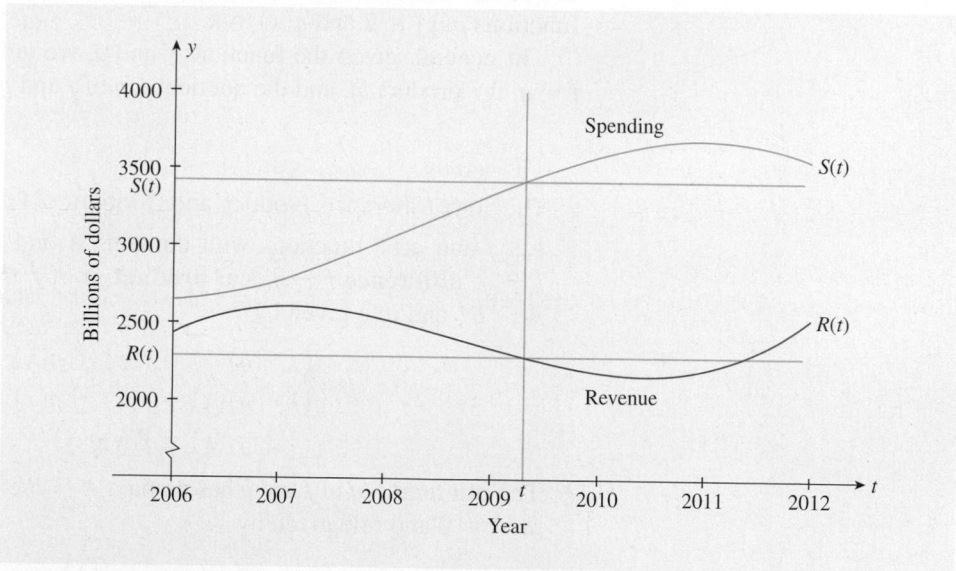

FIGURE **33**
$R(t) - S(t)$ gives the federal budget deficit (surplus) at any time t.
Source: Office of Management and Budget.

The difference $R(t) - S(t)$ gives the deficit (surplus) in billions of dollars at any time t if $R(t) - S(t)$ is negative (positive). This observation suggests that we can define a function D whose value at any time t is given by $R(t) - S(t)$. The function D, the *difference* of the two functions R and S, is written $D = R - S$ and may be called the "deficit (surplus) function," since it gives the budget deficit or surplus at any time t. It has the same domain as the functions S and R. The graph of the function D is shown in Figure 34.

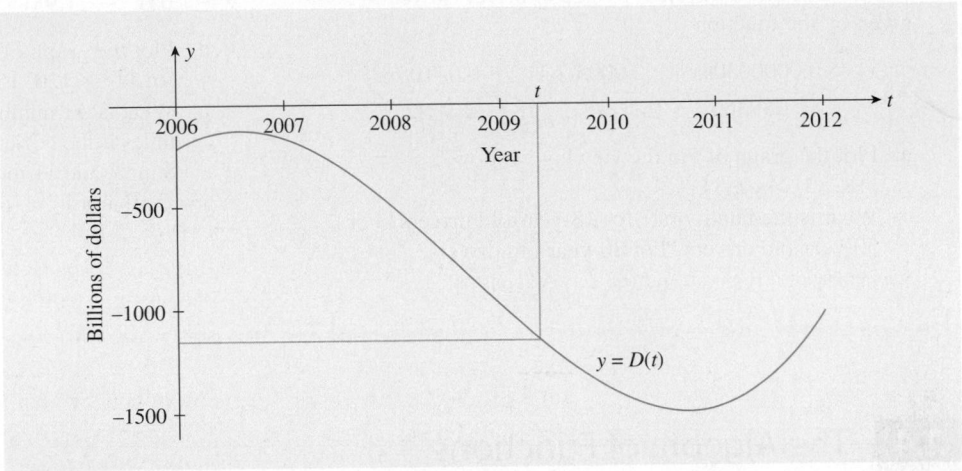

FIGURE 34
The graph of $D(t)$
Source: Office of Management and Budget.

Most functions are built up from other, generally simpler, functions. For example, we may view the function $f(x) = 2x + 4$ as the sum of the two functions $g(x) = 2x$ and $h(x) = 4$. The function $g(x) = 2x$ may in turn be viewed as the product of the functions $p(x) = 2$ and $q(x) = x$.

In general, given the functions f and g, we define the sum $f + g$, the difference $f - g$, the product fg, and the quotient f/g of f and g as follows:

The Sum, Difference, Product, and Quotient of Functions

Let f and g be functions with domains A and B, respectively. Then the **sum** $f + g$, **difference** $f - g$, and **product** fg of f and g are functions with domain $A \cap B$* and rule given by

$$(f + g)(x) = f(x) + g(x) \qquad \text{Sum}$$
$$(f - g)(x) = f(x) - g(x) \qquad \text{Difference}$$
$$(fg)(x) = f(x)g(x) \qquad \text{Product}$$

The **quotient** f/g of f and g has domain $A \cap B$ excluding all numbers x such that $g(x) = 0$ and rule given by

$$\left(\frac{f}{g}\right)(x) = \frac{f(x)}{g(x)} \qquad \text{Quotient}$$

---

*$A \cap B$ is read "A intersected with B" and denotes the set of all points common to both A and B.

EXAMPLE 1 Let $f(x) = \sqrt{x + 1}$ and $g(x) = 2x + 1$. Find the sum s, the difference d, the product p, and the quotient q of the functions f and g.

Solution Since the domain of f is $A = [-1, \infty)$ and the domain of g is $B = (-\infty, \infty)$, we see that the domain of s, d, and p is $A \cap B = [-1, \infty)$. The rules follow:

$$s(x) = (f + g)(x) = f(x) + g(x) = \sqrt{x + 1} + 2x + 1$$
$$d(x) = (f - g)(x) = f(x) - g(x) = \sqrt{x + 1} - (2x + 1) = \sqrt{x + 1} - 2x - 1$$
$$p(x) = (fg)(x) = f(x)g(x) = \sqrt{x + 1}(2x + 1) = (2x + 1)\sqrt{x + 1}$$

The rule for the quotient function q is

$$q(x) = \left(\frac{f}{g}\right)(x) = \frac{f(x)}{g(x)} = \frac{\sqrt{x + 1}}{2x + 1}$$

Its domain is $[-1, \infty)$ together with the restriction $x \neq -\frac{1}{2}$. We denote this by $[-1, -\frac{1}{2})$ and $(-\frac{1}{2}, \infty)$.

The mathematical formulation of a problem arising from a practical situation often leads to an expression that involves the combination of functions. Consider, for example, the costs incurred in operating a business. Costs that remain more or less constant regardless of the firm's level of activity are called **fixed costs.** Examples of fixed costs are rental fees and executive salaries. On the other hand, costs that vary with production or sales are called **variable costs.** Examples of variable costs are wages and costs of raw materials. The **total cost** of operating a business is thus given by the *sum* of the variable costs and the fixed costs, as illustrated in the next example.

 APPLIED EXAMPLE 2 Cost Functions Suppose Puritron, a manufacturer of water filters, has a monthly fixed cost of $10,000 and a variable cost of

$$-0.0001x^2 + 10x \qquad (0 \leq x \leq 40{,}000)$$

dollars, where x denotes the number of filters manufactured per month. Find a function C that gives the total monthly cost incurred by Puritron in the manufacture of x filters.

Solution Puritron's monthly fixed cost is always $10,000, regardless of the level of production, and it is described by the constant function $F(x) = 10{,}000$. Next, the variable cost is described by the function $V(x) = -0.0001x^2 + 10x$. Since the total cost incurred by Puritron at any level of production is the sum of the variable cost and the fixed cost, we see that the required total cost function is given by

$$C(x) = V(x) + F(x)$$
$$= -0.0001x^2 + 10x + 10{,}000 \qquad (0 \leq x \leq 40{,}000)$$

Next, the **total profit** realized by a firm in operating a business is the *difference* between the total revenue realized and the total cost incurred; that is,

$$P(x) = R(x) - C(x)$$

 APPLIED EXAMPLE 3 Profit Functions Refer to Example 2. Suppose the total revenue in dollars realized by Puritron from the sale of x water filters per month is given by the total revenue function

$$R(x) = -0.0005x^2 + 20x \qquad (0 \leq x \leq 40{,}000)$$

a. Find the total profit function—that is, the function that describes the total profit Puritron realizes in manufacturing and selling x water filters per month.
b. What is the profit when the level of production is 10,000 filters per month?

Solution

a. The total profit realized by Puritron in manufacturing and selling x water filters per month is the difference between the total revenue realized and the total cost incurred. Thus, the required total profit function is given by

$$P(x) = R(x) - C(x)$$
$$= (-0.0005x^2 + 20x) - (-0.0001x^2 + 10x + 10,000)$$
$$= -0.0004x^2 + 10x - 10,000$$

b. The profit realized by Puritron when the level of production is 10,000 filters per month is

$$P(10,000) = -0.0004(10,000)^2 + 10(10,000) - 10,000 = 50,000$$

or $50,000 per month.

Composition of Functions

Another way to build up a function from other functions is through a process known as the *composition of functions*. Consider, for example, the function h, whose rule is given by $h(x) = \sqrt{x^2 - 1}$. Let f and g be functions defined by the rules $f(x) = x^2 - 1$ and $g(x) = \sqrt{x}$. Evaluating the function g at the point $f(x)$ [remember that for each real number x in the domain of f, $f(x)$ is simply a real number], we find that

$$g(f(x)) = \sqrt{f(x)} = \sqrt{x^2 - 1}$$

which is just the rule defining the function h!

In general, the composition of a function g with a function f is defined as follows.

> **The Composition of Two Functions**
>
> Let f and g be functions. Then the composition of g and f is the function $g \circ f$ defined by
>
> $$(g \circ f)(x) = g(f(x))$$
>
> The domain of $g \circ f$ is the set of all x in the domain of f such that $f(x)$ lies in the domain of g.

The function $g \circ f$ (read "g circle f") is also called a **composite function.** The interpretation of the function $h = g \circ f$ as a machine is illustrated in Figure 35, and its interpretation as a mapping is shown in Figure 36.

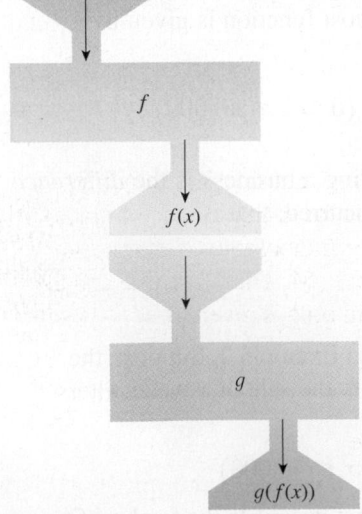

FIGURE **35**
The composite function $h = g \circ f$ viewed as a machine

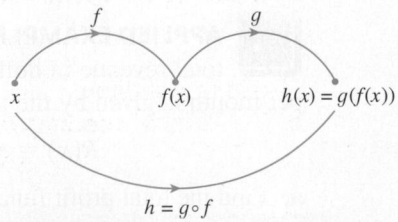

$h = g \circ f$

FIGURE **36**
The function $h = g \circ f$ viewed as a mapping

EXAMPLE 4 Let $f(x) = x^2 - 1$, and $g(x) = \sqrt{x} + 1$. Find:

a. The rule for the composite function $g \circ f$.
b. The rule for the composite function $f \circ g$.

Solution

a. To find the rule for the composite function $g \circ f$, evaluate the function g at $f(x)$. We obtain

$$(g \circ f)(x) = g(f(x)) = \sqrt{f(x)} + 1 = \sqrt{x^2 - 1} + 1$$

b. To find the rule for the composite function $f \circ g$, evaluate the function f at $g(x)$. Thus,

$$(f \circ g)(x) = f(g(x)) = (g(x))^2 - 1 = (\sqrt{x} + 1)^2 - 1$$
$$= x + 2\sqrt{x} + 1 - 1 = x + 2\sqrt{x}$$

Example 4 shows us that in general $g \circ f$ is different from $f \circ g$, so care must be taken in finding the rule for a composite function.

> ### Explore and Discuss
>
> Let $f(x) = \sqrt{x} + 1$ for $x \geq 0$, and let $g(x) = (x - 1)^2$ for $x \geq 1$.
>
> **1.** Show that $(g \circ f)(x)$ and $(f \circ g)(x) = x$. (*Note:* The function g is said to be the *inverse* of f and vice versa.)
>
> **2.** Plot the graphs of f and g together with the straight line $y = x$. Describe the relationship between the graphs of f and g.

APPLIED EXAMPLE 5 Automobile Pollution An environmental impact study conducted for the city of Oxnard indicates that under existing environmental protection laws, the level of carbon monoxide (CO) present in the air due to pollution from automobile exhaust will be $0.01x^{2/3}$ parts per million when the number of motor vehicles is x thousand. A separate study conducted by a state government agency estimates that t years from now, the number of motor vehicles in Oxnard will be $0.2t^2 + 4t + 64$ thousand.

a. Find an expression for the concentration of CO in the air due to automobile exhaust t years from now.
b. What will be the level of concentration 5 years from now?

Solution

a. The level of CO present in the air due to pollution from automobile exhaust is described by the function $g(x) = 0.01x^{2/3}$, where x is the number (in thousands) of motor vehicles. But the number of motor vehicles x (in thousands) t years from now may be estimated by the rule $f(t) = 0.2t^2 + 4t + 64$. Therefore, the concentration of CO due to automobile exhaust t years from now is given by

$$C(t) = (g \circ f)(t) = g(f(t)) = 0.01(0.2t^2 + 4t + 64)^{2/3}$$

parts per million.

b. The level of concentration 5 years from now will be

$$C(5) = 0.01[0.2(5)^2 + 4(5) + 64]^{2/3}$$
$$= (0.01)89^{2/3} \approx 0.20$$

or approximately 0.20 parts per million.

2.4 Self-Check Exercises

1. Let f and g be functions defined by the rules

$$f(x) = \sqrt{x} + 1 \quad \text{and} \quad g(x) = \frac{x}{1+x}$$

respectively. Find the rules for
   a. The sum s, the difference d, the product p, and the quotient q of f and g.
   b. The composite functions $f \circ g$ and $g \circ f$.

2. **HEALTH-CARE SPENDING** Health-care spending per person by the private sector includes payments by individuals, corporations, and their insurance companies and is approximated by the function

$$f(t) = 2.48t^2 + 18.47t + 509 \quad (0 \le t \le 6)$$

where $f(t)$ is measured in dollars and t is measured in years, with $t = 0$ corresponding to the beginning of 1994.

The corresponding government spending—including expenditures for Medicaid, Medicare, and other federal, state, and local government public health care—is

$$g(t) = -1.12t^2 + 29.09t + 429 \quad (0 \le t \le 6)$$

where t has the same meaning as before.
   a. Find a function that gives the difference between private and government health-care spending per person at any time t.
   b. What was the difference between private and government expenditures per person at the beginning of 1995? At the beginning of 2000?
   Source: Health Care Financing Administration.

Solutions to Self-Check Exercises 2.4 can be found on page 116.

2.4 Concept Questions

1. The figure below shows the graphs of a total cost function and a total revenue function. Let P, defined by $P(x) = R(x) - C(x)$, denote the total profit function.
   a. Find an expression for $P(x_1)$. Explain its significance.
   b. Find an expression for $P(x_2)$. Explain its significance.

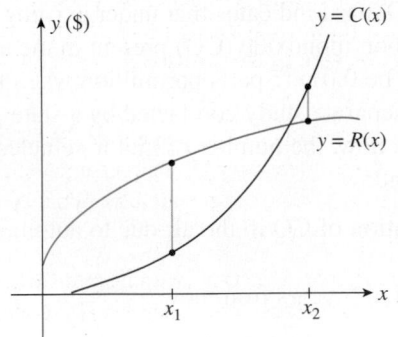

2. **a.** Explain what is meant by the sum, difference, product, and quotient of the functions f and g with domains A and B, respectively.
   b. If $f(2) = 3$ and $g(2) = -2$, what is $(f + g)(2)$? $(f - g)(2)$? $(fg)(2)$? $(f/g)(2)$?

3. Let f and g be functions, and suppose that (x, y) is a point on the graph of h. What is the value of y for $h = f + g$? $h = f - g$? $h = fg$? $h = f/g$?

4. **a.** What is the composition of the functions f and g? The functions g and f?
   b. If $f(2) = 3$ and $g(3) = 8$, what is $(g \circ f)(2)$? Can you conclude from the given information what $(f \circ g)(3)$ is? Explain.

5. Let f be a function with domain A, and let g be a function whose domain contains the range of f. If a is any number in A, must $(g \circ f)(a)$ be defined? Explain with an example.

6. The profit P (in dollars) of a one product company is given by $P = g(x)$, where x is the number of units sold. At the same time, the number of units of the product sold is given by $x = f(p)$, where p (in dollars) is the unit price of the product. Write an expression for the profit of the company in terms of the unit price it charges.

2.4 Exercises

In Exercises 1–8, let $f(x) = x^3 + 5$, $g(x) = x^2 - 2$, and $h(x) = 2x + 4$. Find the rule for each function.

1. $f + g$

2. $f - g$

3. fg

4. gf

5. $\dfrac{f}{g}$

6. $\dfrac{f - g}{h}$

7. $\dfrac{fg}{h}$

8. fgh

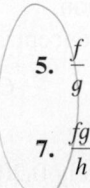

In Exercises 9–18, let $f(x) = x - 1$, $g(x) = \sqrt{x + 1}$, and $h(x) = 2x^3 - 1$. Find the rule for each function.

9. $f + g$ **10.** $g - f$ **11.** fg **12.** gf

13. $\dfrac{g}{h}$ **14.** $\dfrac{h}{g}$ **15.** $\dfrac{fg}{h}$ **16.** $\dfrac{fh}{g}$

17. $\dfrac{f - h}{g}$ **18.** $\dfrac{gh}{g - f}$

In Exercises 19–24, find the functions $f + g$, $f - g$, fg, and f/g.

19. $f(x) = x^2 + 5$; $g(x) = \sqrt{x} - 2$

20. $f(x) = \sqrt{x - 1}$; $g(x) = x^3 + 1$

21. $f(x) = \sqrt{x + 3}$; $g(x) = \dfrac{1}{x - 1}$

22. $f(x) = \dfrac{1}{x^2 + 1}$; $g(x) = \dfrac{1}{x^2 - 1}$

23. $f(x) = \dfrac{x + 1}{x - 1}$; $g(x) = \dfrac{x + 2}{x - 2}$

24. $f(x) = x^2 + 1$; $g(x) = \sqrt{x + 1}$

In Exercises 25–30, find the rules for the composite functions $f \circ g$ and $g \circ f$.

25. $f(x) = x^2 + x + 1$; $g(x) = x^2$

26. $f(x) = 3x^2 + 2x + 1$; $g(x) = x + 3$

27. $f(x) = \sqrt{x} + 1$; $g(x) = x^2 - 1$

28. $f(x) = 2\sqrt{x} + 3$; $g(x) = x^2 + 1$

29. $f(x) = \dfrac{x}{x^2 + 1}$; $g(x) = \dfrac{1}{x}$

30. $f(x) = \sqrt{x + 1}$; $g(x) = \dfrac{1}{x - 1}$

In Exercises 31–34, evaluate $h(2)$, where $h = g \circ f$.

31. $f(x) = x^2 + x + 1$; $g(x) = x^2$

32. $f(x) = \sqrt[3]{x^2 - 1}$; $g(x) = 3x^3 + 1$

33. $f(x) = \dfrac{1}{2x + 1}$; $g(x) = \sqrt{x}$

34. $f(x) = \dfrac{1}{x - 1}$; $g(x) = x^2 + 1$

In Exercises 35–42, find functions f and g such that $h = g \circ f$. (*Note:* The answer is *not* unique.)

35. $h(x) = (2x^3 + x^2 + 1)^5$ **36.** $h(x) = (3x^2 - 4)^{-3}$

37. $h(x) = \sqrt{x^2 - 1}$ **38.** $h(x) = (2x - 3)^{3/2}$

39. $h(x) = \dfrac{1}{x^2 - 1}$ **40.** $h(x) = \dfrac{1}{\sqrt{x^2 - 4}}$

41. $h(x) = \dfrac{1}{(3x^2 + 2)^{3/2}}$

42. $h(x) = \dfrac{1}{\sqrt{2x + 1}} + \sqrt{2x + 1}$

In Exercises 43–46, find $f(a + h) - f(a)$ for each function. Simplify your answer.

43. $f(x) = 3x + 4$ **44.** $f(x) = -\dfrac{1}{2}x + 3$

45. $f(x) = 4 - x^2$ **46.** $f(x) = x^2 - 2x + 1$

In Exercises 47–52, find and simplify

$$\frac{f(a + h) - f(a)}{h} \qquad (h \neq 0)$$

for each function.

47. $f(x) = x^2 + 1$ **48.** $f(x) = 2x^2 - x + 1$

49. $f(x) = x^3 - x$ **50.** $f(x) = 2x^3 - x^2 + 1$

51. $f(x) = \dfrac{1}{x}$ **52.** $f(x) = \sqrt{x}$

53. **RESTAURANT REVENUE** Nicole owns and operates two restaurants. The revenue of the first restaurant at time t is $f(t)$ dollars, and the revenue of the second restaurant at time t is $g(t)$ dollars. What does the function $F(t) = f(t) + g(t)$ represent?

54. **BIRTHRATE OF ENDANGERED SPECIES** The birthrate of an endangered species of whales in year t is $f(t)$ whales/year. This species of whales is dying at the rate of $g(t)$ whales/year in year t. What does the function $F(t) = f(t) - g(t)$ represent?

55. **VALUE OF AN INVESTMENT** The number of IBM shares that Nancy owns is given by $f(t)$. The price per share of the stock of IBM at time t is $g(t)$ dollars. What does the function $f(t)g(t)$ represent?

56. **PRODUCTION COSTS** The total cost incurred by time t in the production of a certain commodity is $f(t)$ dollars. The number of products produced by time t is $g(t)$ units. What does the function $f(t)/g(t)$ represent?

57. **CARBON MONOXIDE POLLUTION** The number of cars running in the business district of a town at time t is given by $f(t)$. Carbon monoxide pollution coming from these cars is given by $g(x)$ parts per million, where x is the number of cars being operated in the district. What does the function $g \circ f$ represent?

58. **EFFECT OF ADVERTISING ON REVENUE** The revenue of Leisure Travel is given by $f(x)$ dollars, where x is the dollar amount spent by the company on advertising. The amount spent by Leisure at time t on advertising is given by $g(t)$ dollars. What does the function $f \circ g$ represent?

59. Cost of Producing DVDs TMI, a manufacturer of blank DVDs, has a monthly fixed cost of $12,100 and a variable cost of $.60/disc. Find a function C that gives the total cost incurred by TMI in the manufacture of x discs/month.

60. Business Email The average number of email messages sent and received per corporate user per day in year t between 2011 ($t = 1$) and 2015 ($t = 5$) is projected to be

$$f(t) = 3t + 69 \qquad (1 \le t \le 5)$$

The average number of spam emails sent per corporate user per day for the period under consideration is projected to be

$$g(t) = -0.2t + 13.8 \qquad (1 \le t \le 5)$$

a. Find a function, h, giving the projected average number of legitimate (non spam) emails sent and received per corporate user per day in year t.
Hint: $h(t) = f(t) - g(t)$
b. Compute $f(5), g(5)$, and $h(5)$. Is $h(5) = f(5) - g(5)$?
Source: Radicati Group.

61. Public Transportation Budget Deficit According to the Massachusetts Bay Transportation Authority (MBTA), the projected cumulative MBTA budget deficit with a $160 million rescue package (in billions of dollars) is given by

$$D_1(t) = 0.0275t^2 + 0.081t + 0.07 \qquad (0 \le t \le 3)$$

and the budget deficit without the rescue package is given by

$$D_2(t) = 0.035t^2 + 0.21t + 0.24 \qquad (0 \le t \le 3)$$

Find the function $D = D_2 - D_1$, and interpret your result.
Source: MBTA Review.

62. Motorcycle Deaths Suppose the fatality rate (deaths per 100 million miles traveled) of motorcyclists is given by $g(x)$, where x is the percentage of motorcyclists who wear helmets. Next, suppose the percentage of motorcyclists who wear helmets at time t (t measured in years) is $f(t)$, with $t = 0$ corresponding to 2000.
a. If $f(0) = 0.64$ and $g(0.64) = 26$, find $(g \circ f)(0)$ and interpret your result.
b. If $f(6) = 0.51$ and $g(0.51) = 42$, find $(g \circ f)(6)$ and interpret your result.
c. Comment on the results of parts (a) and (b).
Source: National Highway Traffic Safety Administration.

63. Fighting Crime Suppose the reported serious crimes (crimes that include homicide, rape, robbery, aggravated assault, burglary, and car theft) that end in arrests or in the identification of suspects is $g(x)$ percent, where x denotes the total number of detectives. Next, suppose the total number of detectives in year t is $f(t)$, with $t = 0$ corresponding to 2001.

a. If $f(1) = 406$ and $g(406) = 23$, find $(g \circ f)(1)$ and interpret your result.
b. If $f(6) = 326$ and $g(326) = 18$, find $(g \circ f)(6)$ and interpret your result.
c. Comment on the results of parts (a) and (b).
Source: Boston Police Department.

64. Profit from Sale of Smartphones Apollo manufactures smartphones at a variable cost of

$$V(x) = 0.000003x^3 - 0.03x^2 + 200x$$

dollars, where x denotes the number of units manufactured per month. The monthly fixed cost attributable to the division that produces them is $100,000. The total revenue realized by Apollo from the sale of x smartphones is given by the total revenue function

$$R(x) = -0.1x^2 + 500x \qquad (0 \le x \le 5000)$$

where $R(x)$ is measured in dollars.
a. Find a function C that gives the total cost incurred by the manufacture of x smartphones.
b. Find the total profit function.
c. What is the profit when 1500 units are produced and sold each month?

65. Profit from Sale of Pagers A division of Chapman Corporation manufactures a pager. The weekly fixed cost for the division is $20,000, and the variable cost for producing x pagers/week is

$$V(x) = 0.000001x^3 - 0.01x^2 + 50x$$

dollars. The company realizes a revenue of

$$R(x) = -0.02x^2 + 150x \qquad (0 \le x \le 7500)$$

dollars from the sale of x pagers/week.
a. Find the total cost function.
b. Find the total profit function.
c. What is the profit for the company if 2000 units are produced and sold each week?

66. Federal Deficit The spending by the federal government (in trillions of dollars) in year t from 2006 ($t = 0$) through 2012 ($t = 6$) is approximately

$$S(t) = -0.015278t^3 + 0.11179t^2 + 0.02516t + 2.64 \qquad (0 \le t \le 6)$$

and the revenue realized by the federal government over the same period is approximately

$$R(t) = 0.023611t^3 - 0.19679t^2 + 0.34365t + 2.42 \qquad (0 \le t \le 6)$$

a. Find a function D that gives the approximate deficit of the federal government in year t for $0 \le t \le 6$.
b. Find the spending, revenue, and deficit for the year 2009 ($t = 3$).
c. Is $D(3) = R(3) - S(3)$?
Source: Office of Management and Budget.

67. FAMILY INSURANCE COVERAGE The average annual worker and employer contributions (in dollars) to premiums for family insurance coverage from 2005 ($t = 1$) through 2011 ($t = 7$) are approximated by the functions,

$$f(t) = 4.389t^3 - 47.833t^2 + 374.49t + 2390 \quad (1 \le t \le 7)$$

and

$$g(t) = 13.222t^3 - 132.524t^2 + 757.9t + 7481 \quad (1 \le t \le 7)$$

respectively.

  a. Find a function h giving the total premiums for family coverage.
  b. Find the average annual worker contribution to premiums in 2010 ($t = 6$), the employer's contribution to premiums in 2010, and the total contributions to premiums in 2010.
  c. Compute $h(6)$. Compare the result with the total contributions to premiums in 2010 obtained in part (b).

  Source: Kaiser/HRET Survey of Employer-Sponsored Health Benefits.

68. EFFECT OF MORTGAGE RATES ON HOUSING STARTS A study prepared for the National Association of Realtors estimated that the number of housing starts per year over the next 5 years will be

$$N(r) = \frac{7}{1 + 0.02r^2}$$

million units, where r (percent) is the mortgage rate. Suppose the mortgage rate t months from now will be

$$r(t) = \frac{5t + 75}{t + 10} \quad (0 \le t \le 24)$$

percent/year.

  a. Find an expression for the number of housing starts per year as a function of t, t months from now.
  b. Using the result from part (a), determine the number of housing starts at present, 12 months from now, and 18 months from now.

69. HOTEL OCCUPANCY RATE The occupancy rate of the all-suite Wonderland Hotel, located near an amusement park, is given by the function

$$r(t) = \frac{10}{81}t^3 - \frac{10}{3}t^2 + \frac{200}{9}t + 55 \quad (0 \le t \le 11)$$

percent, where t is measured in months and $t = 0$ corresponds to the beginning of January. Management has estimated that the monthly revenue (in thousands of dollars) is approximated by the function

$$R(r) = -\frac{3}{5000}r^3 + \frac{9}{50}r^2 \quad (0 \le r \le 100)$$

where r (percent) is the occupancy rate.

  a. What is the hotel's occupancy rate at the beginning of January? At the beginning of June?
  b. What is the hotel's monthly revenue at the beginning of January? At the beginning of June?
  Hint: Compute $R(r(0))$ and $R(r(5))$.

70. HOUSING STARTS AND CONSTRUCTION JOBS The president of a major housing construction firm reports that the number of construction jobs (in millions) created is given by

$$N(x) = 1.42x$$

where x denotes the number of housing starts. Suppose the number of housing starts in the next t months is expected to be

$$x(t) = \frac{7(t + 10)^2}{(t + 10)^2 + 2(t + 15)^2}$$

million units. Find an expression for the number of jobs created per year in the next t months. How many jobs per year will have been created 6 months and 12 months from now?

71. a. Let f, g, and h be functions. How would you define the "sum" of f, g, and h?
  b. Give a real-life example involving the sum of three functions. (*Note:* The answer is not unique.)

72. a. Let f, g, and h be functions. How would you define the "composition" of h, g, and f, in that order?
  b. Give a real-life example involving the composition of these functions. (*Note:* The answer is not unique.)

In Exercises 73–78, determine whether the statement is true or false. If it is true, explain why it is true. If it is false, give an example to show why it is false.

73. If f and g are functions with domain D, then $f + g = g + f$.

74. If $g \circ f$ is defined at $x = a$, then $f \circ g$ must also be defined at $x = a$.

75. If f and g are functions, then $f \circ g = g \circ f$.

76. If f is a function, then $(f \circ f)(x) = [f(x)]^2$.

77. If f, g, and h are functions, then $h \circ (g \circ f) = (h \circ g) \circ f$.

78. If f, g, and h are functions, then $h \circ (g + f) = h \circ g + h \circ f$.

2.4 Solutions to Self-Check Exercises

1. a. $s(x) = f(x) + g(x) = \sqrt{x} + 1 + \dfrac{x}{1+x}$

$d(x) = f(x) - g(x) = \sqrt{x} + 1 - \dfrac{x}{1+x}$

$p(x) = f(x)g(x) = (\sqrt{x} + 1) \cdot \dfrac{x}{1+x} = \dfrac{x(\sqrt{x} + 1)}{1+x}$

$q(x) = \dfrac{f(x)}{g(x)} = \dfrac{\sqrt{x} + 1}{\dfrac{x}{1+x}} = \dfrac{(\sqrt{x} + 1)(1 + x)}{x}$

b. $(f \circ g)(x) = f(g(x)) = \sqrt{\dfrac{x}{1+x} + 1}$

$(g \circ f)(x) = g(f(x)) = \dfrac{\sqrt{x} + 1}{1 + (\sqrt{x} + 1)} = \dfrac{\sqrt{x} + 1}{\sqrt{x} + 2}$

2. a. The difference between private and government health-care spending per person at any time t is given by the function d with the rule

$d(t) = f(t) - g(t) = (2.48t^2 + 18.47t + 509)$
$\qquad\qquad\quad -(-1.12t^2 + 29.09t + 429)$
$\qquad\quad = 3.6t^2 - 10.62t + 80$

b. The difference between private and government expenditures per person at the beginning of 1995 is given by

$d(1) = 3.6(1)^2 - 10.62(1) + 80$

or $72.98/person.

The difference between private and government expenditures per person at the beginning of 2000 is given by

$d(6) = 3.6(6)^2 - 10.62(6) + 80$

or $145.88/person.

2.5 Linear Functions

We now focus our attention on an important class of functions known as linear functions. Recall that a linear equation in x and y has the form $Ax + By + C = 0$, where A, B, and C are constants and A and B are not both zero. If $B \neq 0$, the equation can always be solved for y in terms of x; in fact, as we saw in Section 2.2, the equation may be cast in the slope-intercept form:

$$y = mx + b \qquad (m, b \text{ constants}) \qquad (7)$$

Equation (7) defines y as a function of x. The domain and range of this function are the set of all real numbers. Furthermore, the graph of this function, as we saw in Section 2.2, is a straight line in the plane. For this reason, the function $f(x) = mx + b$ is called a linear function.

> **Linear Function**
>
> The function f defined by
>
> $$f(x) = mx + b$$
>
> where m and b are constants, is called a **linear function.**

Linear functions play an important role in the quantitative analysis of business and economic problems. First, many problems that arise in these and other fields are linear in nature or are linear in the intervals of interest and thus can be formulated in terms of linear functions. Second, because linear functions are relatively easy to work with, assumptions involving linearity are often made in the formulation of problems. In many cases, these assumptions are justified, and acceptable mathematical models are obtained that approximate real-life situations.

The following example uses a linear function to model the percentage of middle-income adults in the United States from 1971 through 2011. In Section 2.8, we will show how this model is constructed using the *least-squares technique.* In Using Technology on pages 165–167, you will be asked to use a graphing calculator to construct other mathematical models from raw data.

APPLIED EXAMPLE 1 Erosion of the Middle Class The idea of a large, stable middle class (defined as those with annual household incomes in 2010 between $39,000 and $118,000 for a family of three) is central to America's sense of itself. The following table gives the percentage of middle-income adults in the United States from 1971 through 2011:

Year	1971	1981	1991	2001	2011
Percent, y	61	59	56	54	51

A mathematical model giving the percentage of middle-class adults in the United States for the period under consideration is given by

$$f(t) = -2.5t + 61.2 \qquad (0 \le t \le 4)$$

where t is measured in decades, with $t = 0$ corresponding to 1971.

a. Plot the data points, and sketch the graph of the function f on the same set of axes.
b. What is the rate of change of the percentage of middle-income adults in the United States over the period from 1971 through 2011?
c. Assuming that the trend continues, what will the percentage of middle-income adults in the United States be in 2021?
Source: Pew Research Center.

Solution

a. The graph of f is shown in Figure 37.

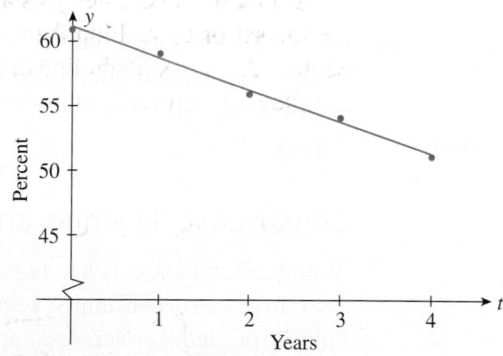

FIGURE **37**
The percentage of middle-income adults in the United States from 1971 through 2011

b. The rate of change of the percentage of middle-income adults in the United States over the period from 1971 through 2011 is -2.5% per decade, that is, a decline of 2.5% per decade.
c. The projected percentage of middle-income adults in the United States in 2021 is

$$f(5) = -2.5(5) + 61.2 = 48.7$$

or 48.7%.

In the rest of this section, we look at several applications that can be modeled by using linear functions.

Simple Depreciation

We first discussed linear depreciation in the introduction to Section 2.1 as a real-world application of straight lines. The following example illustrates how to derive an equation describing the book value of an asset that is being depreciated linearly.

APPLIED EXAMPLE 2 Linear Depreciation of a Network Server A network server has an original value of $10,000 and is to be depreciated linearly over 5 years with a $3000 scrap value. Find an expression giving the book value at the end of year t. What will be the book value of the server at the end of the second year? What is the rate of depreciation of the server?

Solution Let $V(t)$ denote the network server's book value at the end of the tth year. Since the depreciation is linear, V is a linear function of t. Equivalently, the graph of the function is a straight line. To find an equation of the straight line, observe that $V = 10,000$ when $t = 0$; this tells us that the line passes through the point $(0, 10,000)$. Similarly, the condition that $V = 3000$ when $t = 5$ says that the line also passes through the point $(5, 3000)$. The slope of the line is given by

$$m = \frac{10,000 - 3000}{0 - 5} = -\frac{7000}{5} = -1400$$

Using the point-slope form of the equation of a line with the point $(0, 10,000)$ and the slope $m = -1400$, we have

$$V - 10,000 = -1400(t - 0)$$
$$V = -1400t + 10,000$$

the required expression. The book value at the end of the second year is given by

$$V(2) = -1400(2) + 10,000 = 7200$$

or $7200. The rate of depreciation of the server is given by the negative of the slope of the depreciation line. Since the slope of the line is $m = -1400$, the rate of depreciation is $1400 per year. The graph of $V = -1400t + 10,000$ is sketched in Figure 38.

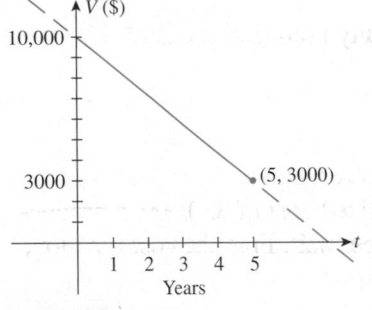

FIGURE 38
Linear depreciation of a network server

Linear Cost, Revenue, and Profit Functions

Whether a business is a sole proprietorship or a large corporation, the owner or chief executive must constantly keep track of operating costs, revenue resulting from the sale of products or services, and, perhaps most important, the profits realized. Three functions provide management with a measure of these quantities: the total cost function, the revenue function, and the profit function.

> Cost, Revenue, and Profit Functions
>
> Let x denote the number of units of a product manufactured or sold. Then the **total cost function** is
>
> $$C(x) = \text{Total cost of manufacturing } x \text{ units of the product}$$

The **revenue function** is

$$R(x) = \text{Total revenue realized from the sale of } x \text{ units of the product}$$

The **profit function** is

$$P(x) = \text{Total profit realized from manufacturing and selling } x \text{ units of the product}$$

Generally speaking, the total cost, revenue, and profit functions associated with a company will probably be nonlinear (these functions are best studied using the tools of calculus). But *linear* cost, revenue, and profit functions do arise in practice, and we will consider such functions in this section. Before deriving explicit forms of these functions, we need to recall some common terminology.

The costs that are incurred in operating a business are usually classified into two categories. Costs that remain more or less constant regardless of the firm's activity level are called **fixed costs.** Examples of fixed costs are rental fees and executive salaries. Costs that vary with production or sales are called **variable costs.** Examples of variable costs are wages and costs for raw materials.

Suppose a firm has a fixed cost of F dollars, a production cost of c dollars per unit, and a selling price of s dollars per unit. Then the *cost function $C(x)$*, the *revenue function $R(x)$*, and the *profit function $P(x)$* for the firm are given by

$$C(x) = cx + F$$
$$R(x) = sx$$
$$P(x) = R(x) - C(x) \qquad \text{Revenue} - \text{cost}$$
$$= (s - c)x - F$$

where x denotes the number of units of the commodity produced and sold. The functions C, R, and P are linear functions of x.

$ **APPLIED EXAMPLE 3** Profit Function for Puritron Water Filters Puritron, a manufacturer of water filters, has a monthly fixed cost of $20,000, a production cost of $20 per unit, and a selling price of $30 per unit. Find the cost function, the revenue function, and the profit function for Puritron.

Solution Let x denote the number of units produced and sold. Then

$$C(x) = 20x + 20,000$$
$$R(x) = 30x$$
$$P(x) = R(x) - C(x)$$
$$= 30x - (20x + 20,000)$$
$$= 10x - 20,000$$

Intersection of Straight Lines

The solution of certain practical problems involves finding the point of intersection of two straight lines. To see how such a problem may be solved algebraically, suppose we are given two straight lines L_1 and L_2 with equations

$$y = m_1 x + b_1 \quad \text{and} \quad y = m_2 x + b_2$$

(where m_1, b_1, m_2, and b_2 are constants) that intersect at the point $P(x_0, y_0)$ (Figure 39).

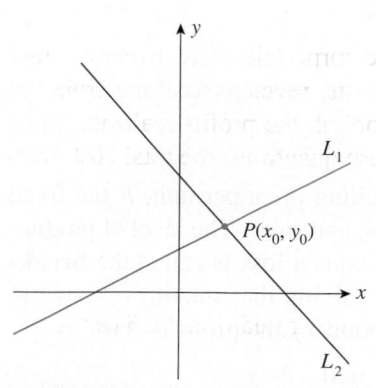

FIGURE **39**
L_1 and L_2 intersect at the point $P(x_0, y_0)$.

The point $P(x_0, y_0)$ lies on the line L_1, so it satisfies the equation $y = m_1x + b_1$. It also lies on the line L_2, so it satisfies the equation $y = m_2x + b_2$. Therefore, to find the point of intersection $P(x_0, y_0)$ of the lines L_1 and L_2, we solve the system composed of the two equations

$$y = m_1x + b_1 \quad \text{and} \quad y = m_2x + b_2$$

for x and y.

EXAMPLE 4 Find the point of intersection of the straight lines that have equations $y = x + 1$ and $y = -2x + 4$.

Solution We solve the given simultaneous equations. Substituting the value of y as given in the first equation into the second, we obtain

$$x + 1 = -2x + 4$$
$$3x = 3$$
$$x = 1$$

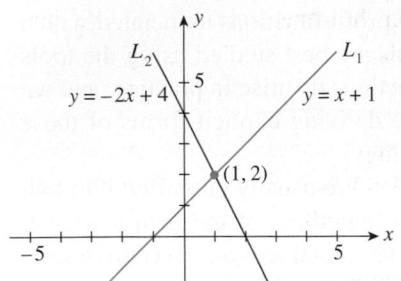

FIGURE **40**
The point of intersection of L_1 and L_2 is (1, 2).

Substituting this value of x into either one of the given equations yields $y = 2$. Therefore, the required point of intersection is (1, 2) (Figure 40).

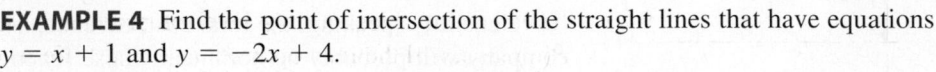

Exploring with TECHNOLOGY

1. Use a graphing utility to plot the straight lines L_1 and L_2 with equations $y = 3x - 2$ and $y = -2x + 3$, respectively, on the same set of axes in the standard viewing window. Then use **TRACE** and **ZOOM** to find the point of intersection of L_1 and L_2. Repeat using the "intersection" function of your graphing utility.

2. Find the point of intersection of L_1 and L_2 algebraically.

3. Comment on the effectiveness of each method.

We now turn to some applications involving the intersections of pairs of straight lines.

Break-Even Analysis

Consider a firm with (linear) cost function $C(x)$, revenue function $R(x)$, and profit function $P(x)$ given by

$$C(x) = cx + F$$
$$R(x) = sx$$
$$P(x) = R(x) - C(x) = (s - c)x - F$$

where c denotes the unit cost of production, s the selling price per unit, F the fixed cost incurred by the firm, and x the level of production and sales. The level of production at which the firm neither makes a profit nor sustains a loss is called the **break-even level of operation** and may be determined by solving the equations $y = C(x)$ and $y = R(x)$ simultaneously. At the level of production x_0, the profit is zero, so

$$P(x_0) = R(x_0) - C(x_0) = 0$$
$$R(x_0) = C(x_0)$$

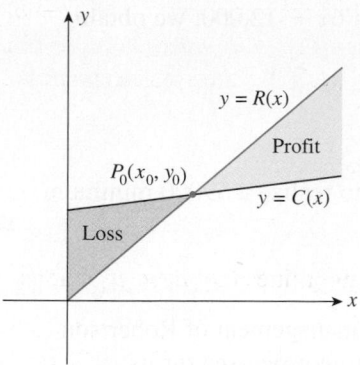

FIGURE 41
P_0 is the break-even point.

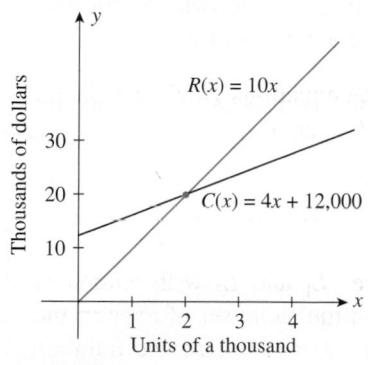

FIGURE 42
The point at which $R(x) = C(x)$ is the break-even point.

The point $P_0(x_0, y_0)$, the solution of the simultaneous equations $y = R(x)$ and $y = C(x)$, is referred to as the **break-even point;** the number x_0 and the number y_0 are called the **break-even quantity** and the **break-even revenue,** respectively.

Geometrically, the break-even point $P_0(x_0, y_0)$ is just the point of intersection of the straight lines representing the cost and revenue functions, respectively. This follows because $P_0(x_0, y_0)$, being the solution of the simultaneous equations $y = R(x)$ and $y = C(x)$, must lie on both these lines simultaneously (Figure 41).

Note that if $x < x_0$, then $R(x) < C(x)$, so $P(x) = R(x) - C(x) < 0$; thus, the firm sustains a loss at this level of production. On the other hand, if $x > x_0$, then $P(x) > 0$, and the firm operates at a profitable level.

 APPLIED EXAMPLE 5 Break-Even Level Prescott manufactures its products at a cost of $4 per unit and sells them for $10 per unit. If the firm's fixed cost is $12,000 per month, determine the firm's break-even point.

Solution The cost function C and the revenue function R are given by $C(x) = 4x + 12,000$ and $R(x) = 10x$, respectively (Figure 42).

Setting $R(x) = C(x)$, we obtain

$$10x = 4x + 12,000$$
$$6x = 12,000$$
$$x = 2000$$

Substituting this value of x into $R(x) = 10x$ gives

$$R(2000) = (10)(2000) = 20,000$$

So for a break-even operation, the firm should manufacture 2000 units of its product, resulting in a break-even revenue of $20,000 per month.

APPLIED EXAMPLE 6 Break-Even Analysis Using the data given in Example 5, answer the following questions:

a. What is the loss sustained by the firm if only 1500 units are produced and sold each month?
b. What is the profit if 3000 units are produced and sold each month?
c. How many units should the firm produce to realize a minimum monthly profit of $9000?

Solution The profit function P is given by the rule

$$P(x) = R(x) - C(x)$$
$$= 10x - (4x + 12,000)$$
$$= 6x - 12,000$$

a. If 1500 units are produced and sold each month, we have

$$P(1500) = 6(1500) - 12,000 = -3000$$

so the firm will sustain a loss of $3000 per month.
b. If 3000 units are produced and sold each month, we have

$$P(3000) = 6(3000) - 12,000 = 6000$$

or a monthly profit of $6000.

c. Substituting 9000 for $P(x)$ in the equation $P(x) = 6x - 12,000$, we obtain

$$9000 = 6x - 12,000$$
$$6x = 21,000$$
$$x = 3500$$

Thus, the firm should produce at least 3500 units to realize a $9000 minimum monthly profit.

APPLIED EXAMPLE 7 Decision Analysis The management of Robertson Controls must decide between two manufacturing processes for its model C electronic thermostat. The monthly cost of the first process is given by $C_1(x) = 20x + 10,000$ dollars, where x is the number of thermostats produced. The monthly cost of the second process is given by $C_2(x) = 10x + 30,000$ dollars. If the projected monthly sales are 800 thermostats at a unit price of $40, which process should management choose in order to maximize the company's profit?

Solution The break-even level of operation using the first process is obtained by solving the equation

$$40x = 20x + 10,000$$
$$20x = 10,000$$
$$x = 500$$

giving an output of 500 units. Next, we solve the equation

$$40x = 10x + 30,000$$
$$30x = 30,000$$
$$x = 1000$$

giving an output of 1000 units for a break-even operation using the second process. Since the projected sales are 800 units, we conclude that management should choose the first process, which will give the firm a profit.

APPLIED EXAMPLE 8 Decision Analysis Referring to Example 7, decide which process Robertson's management should choose if the projected monthly sales are (a) 1500 units and (b) 3000 units.

Solution In both cases, the production is past the break-even level. Since the revenue is the same regardless of which process is employed, the decision will be based on how much each process costs.

a. If $x = 1500$, then

$$C_1(x) = (20)(1500) + 10,000 = 40,000$$
$$C_2(x) = (10)(1500) + 30,000 = 45,000$$

Hence management should choose the first process.

b. If $x = 3000$, then

$$C_1(x) = (20)(3000) + 10,000 = 70,000$$
$$C_2(x) = (10)(3000) + 30,000 = 60,000$$

In this case, management should choose the second process.

Exploring with TECHNOLOGY

1. Use a graphing utility to plot the straight lines L_1 and L_2 with equations $y = 2x - 1$ and $y = 2.1x + 3$, respectively, on the same set of axes, using the standard viewing window. Do the lines appear to intersect?

2. Plot the straight lines L_1 and L_2, using the viewing window $[-100, 100] \times [-100, 100]$. Do the lines appear to intersect? Can you find the point of intersection using **TRACE** and **ZOOM**? Using the "intersection" function of your graphing utility?

3. Find the point of intersection of L_1 and L_2 algebraically.

4. Comment on the effectiveness of the solution methods in parts 2 and 3.

2.5 Self-Check Exercises

1. A manufacturer has a monthly fixed cost of $60,000 and a production cost of $10 for each unit produced. The product sells for $15/unit.
   a. What is the cost function?
   b. What is the revenue function?
   c. What is the profit function?
   d. Compute the profit (loss) corresponding to production levels of 10,000 and 14,000 units/month.

2. **U.S. HEALTH-CARE EXPENDITURES** Because the over-65 population will be growing more rapidly in the next few decades, health-care spending is expected to increase significantly in the coming decades. The following table gives the projected U.S. health-care expenditure (in trillions of dollars) from 2013 through 2018:

Year	2013	2014	2015	2016	2017	2018
Expenditure	2.908	3.227	3.418	3.632	3.850	4.080

A mathematical model giving the approximate U.S. health-care expenditures over the period in question is given by

$$S(t) = 0.226t + 2.954$$

where t is measured in years, with $t = 0$ corresponding to 2013.
   a. Sketch the graph of the function S and the given data on the same set of axes.
   b. Assuming that the trend continues, how much will U.S. health-care expenditures be in 2019 ($t = 6$)?
   c. What is the projected rate of increase of U.S. health-care expenditures over the period in question?

   Source: Centers for Medicare & Medicaid Services.

Solutions to Self-Check Exercises 2.5 can be found on page 127.

2.5 Concept Questions

1. a. What is a *linear function*? Give an example.
   b. What is the domain of a linear function? The range?
   c. What is the graph of a linear function?

2. What is the general form of a linear cost function? A linear revenue function? A linear profit function?

3. In the accompanying figure, $C(x)$ is the cost function and $R(x)$ is the revenue function associated with a certain product.
   a. Plot the break-even point $P(x_0, y_0)$ on the graph.
   b. Identify and mark the break-even quantity, x_0, and the break-even revenue, y_0, on the set of axes.

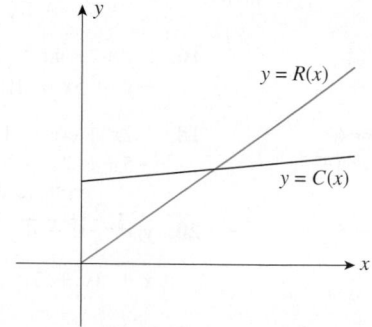

4. The value of an investment (in dollars) after t years is given by

$$V(t) = 50,000 + 4000t \qquad (t \geq 0)$$

a. What was the initial investment?
b. What is the rate of growth of the investment?

1-31 odd

2.5 Exercises

In Exercises 1–10, determine whether the equation defines y as a linear function of x. If so, write it in the form $y = mx + b$.

1. $2x + 3y = 6$

2. $-2x + 4y = 7$

3. $x = 2y - 4$

4. $2x = 3y + 8$

5. $2x - 4y + 9 = 0$

6. $3x - 6y + 7 = 0$

7. $2x^2 - 8y + 4 = 0$

8. $3\sqrt{x} + 4y = 0$

9. $2x - 3y^2 + 8 = 0$

10. $2x + \sqrt{y} - 4 = 0$

11. A manufacturer has a monthly fixed cost of $40,000 and a production cost of $8 for each unit produced. The product sells for $12/unit.
   a. What is the cost function?
   b. What is the revenue function?
   c. What is the profit function?
   d. Compute the profit (loss) corresponding to production levels of 8000 and 12,000 units.

12. A manufacturer has a monthly fixed cost of $100,000 and a production cost of $14 for each unit produced. The product sells for $20/unit.
   a. What is the cost function?
   b. What is the revenue function?
   c. What is the profit function?
   d. Compute the profit (loss) corresponding to production levels of 12,000 and 20,000 units.

13. Find the constants m and b in the linear function $f(x) = mx + b$ such that $f(0) = 2$ and $f(3) = -1$.

14. Find the constants m and b in the linear function $f(x) = mx + b$ such that $f(2) = 4$ and the straight line represented by f has slope -1.

In Exercises 15–20, find the point of intersection of each pair of straight lines.

15. $y = 3x + 4$
   $y = -2x + 14$

16. $y = -4x - 7$
   $-y = 5x + 10$

17. $2x - 3y = 6$
   $3x + 6y = 16$

18. $2x + 4y = 11$
   $-5x + 3y = 5$

19. $y = \frac{1}{4}x - 5$
   $2x - \frac{3}{2}y = 1$

20. $y = \frac{2}{3}x - 4$
   $x + 3y + 3 = 0$

In Exercises 21–24, find the break-even point for the firm whose cost function C and revenue function R are given.

21. $C(x) = 5x + 10,000; R(x) = 15x$

22. $C(x) = 15x + 12,000; R(x) = 21x$

23. $C(x) = 0.2x + 120; R(x) = 0.4x$

24. $C(x) = 150x + 20,000; R(x) = 270x$

25. **LINEAR DEPRECIATION OF AN OFFICE BUILDING** An office building worth $1 million when completed in 2008 is being depreciated linearly over 50 years. What was the book value of the building in 2013? What will it be in 2018? (Assume that the scrap value is $0.)

26. **LINEAR DEPRECIATION OF AN AUTOMOBILE** An automobile purchased for use by the manager of a firm at a price of $34,000 is to be depreciated using the straight-line method over 5 years. What will be the book value of the automobile at the end of 3 years? (Assume that the scrap value is $0.)

27. **SOCIAL SECURITY COLAs** Social Security recipients receive an automatic cost-of-living adjustment (COLA) once each year. Their monthly benefit is increased by the same percentage that consumer prices have increased during the preceding year. Suppose consumer prices have increased by 3.3% during the preceding year.
   a. Express the adjusted monthly benefit of a Social Security recipient as a function of his or her current monthly benefit.
   b. If Carlos Garcia's monthly Social Security benefit is now $1220, what will be his adjusted monthly benefit?

28. **PROFIT FROM SALE OF DIGITAL TIMERS** AutoTime, a manufacturer of electronic digital timers, has a monthly fixed cost of $48,000 and a production cost of $8 for each timer manufactured. The timers sell for $14 each.
   a. What is the cost function?
   b. What is the revenue function?
   c. What is the profit function?
   d. Compute the profit (loss) corresponding to production levels of 4000, 6000, and 10,000 timers, respectively.

29. **PROFIT FROM SALE OF LIGHT BULBS** The management of TMI finds that the monthly fixed costs attributable to the production of their 100-watt light bulbs is $12,100.00. If the cost of producing each twin-pack of light bulbs is $0.60 and each twin-pack sells for $1.15, find the company's cost function, revenue function, and profit function.

30. **LINEAR DEPRECIATION OF A TEXTILE MACHINE** In 2012, National Textile installed a new textile machine in one of its factories at a cost of $250,000. The machine is depreciated linearly over 10 years with a scrap value of $10,000.
    a. Find an expression for the textile machine's book value in the tth year of use $(0 \leq t \leq 10)$.
    b. Sketch the graph of the function of part (a).
    c. Find the machine's book value in 2016.
    d. Find the rate at which the machine is being depreciated.

31. **LINEAR DEPRECIATION OF A WORKCENTER SYSTEM** A workcenter system purchased at a cost of $60,000 in 2013 has a scrap value of $12,000 at the end of 4 years. If the straight-line method of depreciation is used,
    a. Find the rate of depreciation.
    b. Find the linear equation expressing the system's book value at the end of t years.
    c. Sketch the graph of the function of part (b).
    d. Find the system's book value at the end of the third year.

32. **LINEAR DEPRECIATION** Suppose an asset has an original value of $\$C$ and is depreciated linearly over N years with a scrap value of $\$S$. Show that the asset's book value at the end of the tth year is described by the function

$$V(t) = C - \left(\frac{C - S}{N}\right)t$$

Hint: Find an equation of the straight line passing through the points $(0, C)$ and (N, S). (Why?)

33. **LINEAR DEPRECIATION OF AN OFFICE BUILDING** Rework Exercise 25 using the formula derived in Exercise 32.

34. **LINEAR DEPRECIATION OF AN AUTOMOBILE** Rework Exercise 26 using the formula derived in Exercise 32.

35. **DRUG DOSAGES FOR CHILDREN** A method sometimes used by pediatricians to calculate the dosage of medicine for children is based on the child's surface area. If a denotes the adult dosage (in milligrams) and if S is the child's surface area (in square meters), then the child's dosage is given by

$$D(S) = \frac{Sa}{1.7}$$

    a. Show that D is a linear function of S.
       Hint: Think of D as having the form $D(S) = mS + b$. What are the slope m and the y-intercept b?
    b. If the adult dose of a drug is 500 mg, how much should a child whose surface area is 0.4 m² receive?

36. **DRUG DOSAGES FOR CHILDREN** Cowling's Rule is a method for calculating pediatric drug dosages. If a denotes the adult dosage (in milligrams) and if t is the child's age (in years), then the child's dosage is given by

$$D(t) = \left(\frac{t + 1}{24}\right)a$$

    a. Show that D is a linear function of t.
       Hint: Think of $D(t)$ as having the form $D(t) = mt + b$. What are the slope m and the y-intercept b?
    b. If the adult dose of a drug is 500 mg, how much should a 4-year-old child receive?

37. **DRINKING AND DRIVING AMONG HIGH SCHOOL STUDENTS** The percentage of high school students who drink and drive was 17.5% at the beginning of 2001 and declined linearly to 10.3% at the beginning of 2011.
    a. Find a linear function $f(t)$ giving the percentage of high school students who drink and drive in year t, where $t = 0$ corresponds to the beginning of 2001.
    b. If the trend continues, what will the percentage of high school students who drink and drive be at the beginning of 2014?
    Source: Centers for Disease Control and Prevention.

38. **GLOBAL DEFENSE SPENDING** Global defense spending stood at $1.44 trillion in 2009 and is projected to grow at the rate of $0.058 trillion per year through 2018.
    a. Find a function $f(t)$ giving the projected global defense spending in year t, where $t = 0$ corresponds to 2009.
       Hint: The graph of f lies on a straight line.
    b. What is the projected global defense spending in 2018?
    Source: Homeland Security Research.

39. **MEDIAN AGE OF EU-27 POPULATION** The European Union (EU) was established on November 1, 1993, with 12 member states. As of January 2007, through a series of enlargements, the number of members had grown to 27. Consistently low birth rates and higher life expectancy have resulted in a steady increase in the median age of the population of EU-27, as the Union is now called. The median age of the population of EU-27 (in years) is given by

$$M(t) = 0.3t + 37.9 \qquad (0 \leq t \leq 11)$$

where t is measured in years, with $t = 0$ corresponding to the year 2000.
    a. How fast was the median age of the population of EU-27 changing at any time during the period under consideration?
    b. What was the median age in 2011?
    c. What will be the median age in 2015 if the trend continues?
    Source: Eurostat.

40. **CALIFORNIA EMISSIONS CAPS** The California emissions cap is set at 400 million metric tons of carbon dioxide equivalent in 2015 and is expected to drop by 13.2 million metric tons of carbon dioxide equivalent per year through 2020.
    a. Find a linear function f giving the California emissions cap in year t, where $t = 0$ corresponds to 2015.
    b. If the same rate of decline of emissions cap is adopted through 2017, what will the emissions cap be in 2017?
    Source: California Air Resource Board.

41. EROSION OF THE MIDDLE CLASS The idea of a large, stable middle class (defined as those with annual household incomes in 2010 between $39,000 and $118,000 for a family of three), is central to America's sense of itself. But the U.S. middle class has been shrinking steadily from 61% of all adults in 1971 ($t = 0$) to 51% in 2011 ($t = 4$), where t is measured in decades. Research has shown that this decline is approximately linear. Find a linear function $f(t)$ giving the percentage of middle-income adults in decade t, where $t = 0$ corresponds to 1971.
Source: Pew Research Center.

42. U.S. AIRPLANE PASSENGER PROJECTIONS In a report issued by the U.S. Department of Transportation in 2012, it was predicted that the number of passengers boarding planes in the United States would grow steadily from the current 0.7 billion boardings/year to 1.2 billion boardings/year in 2032.
a. Find a linear function f giving the projected boardings (in billions) in year t, where $t = 0$ corresponds to 2012.
b. What is the projected annual rate of growth of boardings between 2012 and 2032?
c. How many boardings per year are projected for 2022?
Source: U.S. Department of Transportation.

43. CELSIUS AND FAHRENHEIT TEMPERATURES The relationship between temperature measured on the Celsius scale and on the Fahrenheit scale is linear. The freezing point is 0°C and 32°F, and the boiling point is 100°C and 212°F.
a. Find an equation giving the relationship between the temperature F measured on the Fahrenheit scale and the temperature C measured on the Celsius scale.
b. Find F as a function of C, and use this formula to determine the temperature in Fahrenheit corresponding to a temperature of 20°C.
c. Find C as a function of F, and use this formula to determine the temperature in Celsius corresponding to a temperature of 70°F.

44. CRICKET CHIRPING AND TEMPERATURE Entomologists have discovered that a linear relationship exists between the rate of chirping of crickets of a certain species and the air temperature. When the temperature is 70°F, the crickets chirp at the rate of 120 chirps/min, and when the temperature is 80°F, they chirp at the rate of 160 chirps/min.
a. Find an equation giving the relationship between the air temperature T and the number of chirps per minute N of the crickets.
b. Find N as a function of T, and use this function to determine the rate at which the crickets chirp when the temperature is 102°F.

45. BREAK-EVEN ANALYSIS AutoTime, a manufacturer of 24-hr variable timers, has a monthly fixed cost of $48,000 and a production cost of $8 for each timer manufactured. The units sell for $14 each.
a. Sketch the graphs of the cost function and the revenue function, and thereby find the break-even point graphically.
b. Find the break-even point algebraically.
c. Sketch the graph of the profit function.

d. At what point does the graph of the profit function cross the x-axis? Interpret your result.

46. BREAK-EVEN ANALYSIS A division of Carter Enterprises produces income tax apps for smartphones. Each income tax app sells for $8. The monthly fixed costs incurred by the division are $25,000, and the variable cost of producing each income tax app is $3.
a. Find the break-even point for the division.
b. What should be the level of sales in order for the division to realize a 15% profit over the cost of making the income tax apps?

47. BREAK-EVEN ANALYSIS A division of the Gibson Corporation manufactures bicycle pumps. Each pump sells for $9, and the variable cost of producing each unit is 40% of the selling price. The monthly fixed costs incurred by the division are $50,000. What is the break-even point for the division?

48. LEASING A TRUCK Ace Truck Leasing Company leases a certain size truck for $25/day and $.50/mi, whereas Acme Truck Leasing Company leases the same size truck for $20/day and $.60/mi.
a. Find the functions describing the daily cost of leasing from each company.
b. Sketch the graphs of the two functions on the same set of axes.
c. If a customer plans to drive at most 30 mi, from which company should he rent a truck for a single day?
d. If a customer plans to drive at least 60 mi, from which company should he rent a truck for a single day?

49. CYBER MONDAY SALES The amount (in millions of dollars) spent on Cyber Monday for the years 2009 through 2011 is given in the following table:

Year	2009	2010	2011
Sales, y	887	1028	1251

a. Plot the Cyber Monday sales (y) versus the year (t), where $t = 0$ corresponds to 2009.
b. Draw a straight line L through the points $(0, 887)$ and $(2, 1251)$.
c. Find a function f in the variable t that describes the line L found in part (b).
d. Assuming that the trend continues, use the function found in part (c) to estimate the amount consumers spent on Cyber Monday in 2014.
e. What is the rate of change in the amount consumers spent on Cyber Monday from 2009 through 2011?
Source: Comscore.

50. DECISION ANALYSIS A product may be made by using Machine I or Machine II. The manufacturer estimates that the monthly fixed costs of using Machine I are $18,000, whereas the monthly fixed costs of using Machine II are $15,000. The variable costs of manufacturing 1 unit of the product using Machine I and Machine II are $15 and $20, respectively. The product sells for $50 each.

a. Find the cost functions associated with using each machine.

b. Sketch the graphs of the cost functions of part (a) and the revenue functions on the same set of axes.

c. Which machine should management choose in order to maximize their profit if the projected sales are 450 units? 550 units? 650 units?

d. What is the profit for each case in part (c)?

51. ANNUAL SALES OF TWO PHARMACIES The annual sales of Crimson Pharmacy are expected to be given by $S = 2.3 + 0.4t$ million dollars t years from now, whereas the annual sales of Cambridge Pharmacy are expected to be given by $S = 1.2 + 0.6t$ million dollars t years from now. When will Cambridge's annual sales first surpass Crimson's annual sales?

52. LCDs VERSUS CRTs The global shipments of traditional cathode-ray tube monitors (CRTs) is approximated by the equation

$$y = -12t + 88 \qquad (0 \le t \le 3)$$

where y is measured in millions and t in years, with $t = 0$ corresponding to the beginning of 2001. The equation

$$y = 18t + 13.4 \qquad (0 \le t \le 3)$$

gives the approximate number (in millions) of liquid crystal displays (LCDs) over the same period. When did the global shipments of LCDs first overtake the global shipments of CRTs?

Source: International Data Corporation.

53. DIGITAL VERSUS FILM CAMERAS The sales of digital cameras (in millions of units) in year t is given by the function

$$f(t) = 3.05t + 6.85 \qquad (0 \le t \le 3)$$

where $t = 0$ corresponds to 2001. Over that same period, the sales of film cameras (in millions of units) is given by

$$g(t) = -1.85t + 16.58 \qquad (0 \le t \le 3)$$

a. Show that more film cameras than digital cameras were sold in 2001.

b. When did the sales of digital cameras first exceed those of film cameras?

Source: Popular Science.

54. U.S. FINANCIAL TRANSACTIONS The percentage of U.S. transactions by check between the beginning of 2001 ($t = 0$) and the beginning of 2010 ($t = 9$) is approximated by

$$f(t) = -\frac{11}{9}t + 43 \qquad (0 \le t \le 9)$$

whereas the percentage of transactions done electronically during the same period is approximated by

$$g(t) = \frac{11}{3}t + 23 \qquad (0 \le t \le 9)$$

a. Sketch the graphs of f and g on the same set of axes.

b. Find the time when transactions done electronically first exceeded those done by check.

Source: Foreign Policy.

In Exercises 55 and 56, determine whether the statement is true or false. If it is true, explain why it is true. If it is false, give an example to show why it is false.

55. Suppose $C(x) = cx + F$ and $R(x) = sx$ are the cost and revenue functions, respectively, of a certain firm. Then the firm is operating at a break-even level of production if its level of production is $F/(s - c)$.

56. If the book value V at the end of the year t of an asset being depreciated linearly is given by $V = -at + b$, where a and b are positive constants, then the rate of depreciation of the asset is a units per year.

2.5 Solutions to Self-Check Exercises

1. Let x denote the number of units produced and sold. Then

a. $C(x) = 10x + 60,000$

b. $R(x) = 15x$

c. $P(x) = R(x) - C(x) = 15x - (10x + 60,000)$
$$= 5x - 60,000$$

d. $P(10,000) = 5(10,000) - 60,000$
$$= -10,000$$
or a loss of \$10,000 per month.

$$P(14,000) = 5(14,000) - 60,000$$
$$= 10,000$$
or a profit of \$10,000 per month.

2. a. The graph of S is shown below.

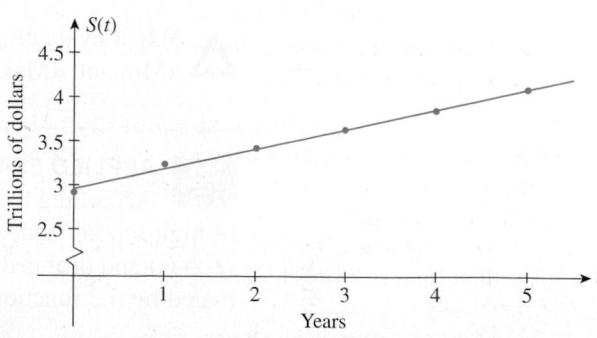

b. The projected U.S. health-care expenditure in 2019 is

$$S(6) = 0.226(6) + 2.954 = 4.31$$

or approximately \$4.31 trillion.

c. The function S is linear; hence, we see that the rate of increase of the U.S. health-care expenditures is given by the slope of the straight line represented by S, which is approximately \$0.23 trillion per year.

USING TECHNOLOGY   Linear Functions

Graphing Utility

A graphing utility can be used to find the value of a function f at a given point with minimal effort. However, to find the value of y for a given value of x in a linear equation such as $Ax + By + C = 0$, the equation must first be cast in the slope-intercept form $y = mx + b$, thus revealing the desired rule $f(x) = mx + b$ for y as a function of x.

EXAMPLE 1 Consider the equation $2x + 5y = 7$.

a. Plot the straight line with the given equation in the standard viewing window.
b. Find the value of y when $x = 2$ and verify your result by direct computation.
c. Find the value of y when $x = 1.732$.

Solution

a. The straight line with equation $2x + 5y = 7$ or, equivalently, $y = -\frac{2}{5}x + \frac{7}{5}$ in the standard viewing window is shown in Figure T1.

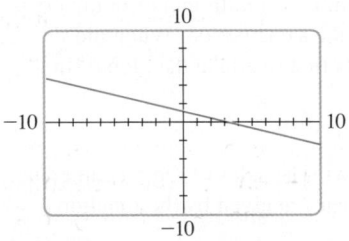

FIGURE **T1**
The straight line $2x + 5y = 7$ in the standard viewing window

b. Using the evaluation function of the graphing utility and the value of 2 for x, we find $y = 0.6$. This result is verified by computing

$$y = -\frac{2}{5}(2) + \frac{7}{5} = -\frac{4}{5} + \frac{7}{5} = \frac{3}{5} = 0.6$$

when $x = 2$.

c. Once again using the evaluation function of the graphing utility, this time with the value 1.732 for x, we find $y = 0.7072$.

⚠️ When evaluating $f(x)$ at $x = a$, remember that the number a must lie between xMin and xMax.

APPLIED EXAMPLE 2 Drinking and Driving Among High School Students
According to the Centers for Disease Control and Prevention, the percentage of high school students who drink and drive stood at 17.5% at the beginning of 2001 ($t = 0$) and dropped steadily in the following years. This percentage is approximated by the function

$$P(t) = -0.73t + 17.5 \qquad (t \geq 0)$$

a. Plot the graph of the function P in the viewing window $[0, 14] \times [0, 25]$.
b. If the trend continues, what will the percentage of high school students who drink and drive be at the beginning of 2014?
c. At what rate was the percentage of students who drink and drive dropping between 2001 and 2011?

Source: Centers for Disease Control and Prevention.

Solution

a. The graph of P is shown in Figure T2.

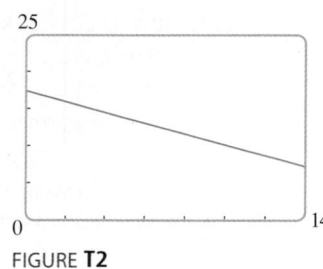

FIGURE **T2**

b. The percentage of students who drink and drive at the beginning of 2014 will be approximately

$$P(13) = 8.01$$

or 8.01%.
c. The rate at which the percentage of students who drink and drive was dropping between 2001 and 2011 is 0.73% per year.

Excel

Excel can be used to find the value of a function at a given value with minimal effort. However, to find the value of y for a given value of x in a linear equation such as $Ax + By + C = 0$, the equation must first be cast in the slope-intercept form $y = mx + b$, thus revealing the desired rule $f(x) = mx + b$ for y as a function of x.

EXAMPLE 3 Consider the equation $2x + 5y = 7$.

a. Find the value of y for $x = 0, 5$, and 10.
b. Plot the straight line with the given equation over the interval $[0, 10]$.

Solution

a. Since this is a linear equation, we first cast the equation in slope-intercept form:

$$y = -\frac{2}{5}x + \frac{7}{5}$$

Next, we create a table of values (Figure T3), following the same procedure outlined in Example 3, pages 90–91. In this case, we use the formula $= (-2/5) \ast B1 + 7/5$ for the y-values.

	A	B	C	D
1	x	0	5	10
2	y	1.4	-0.6	-2.6

FIGURE **T3**
Table of values for x and y

b. Following the procedure outlined in Example 3, we obtain the graph shown in Figure T4.

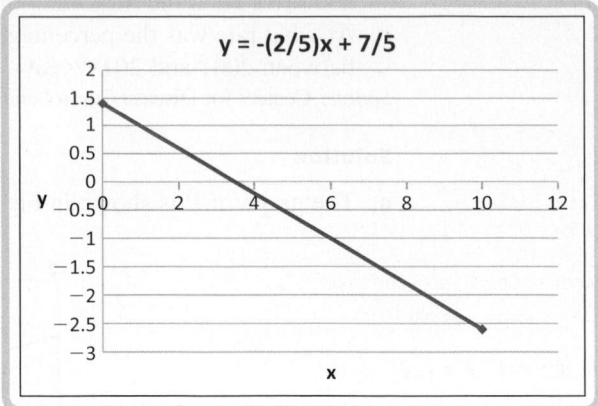

FIGURE **T4**
The graph of $y = -\frac{2}{5}x + \frac{7}{5}$ over the interval [0, 10]

APPLIED EXAMPLE 4 Drinking and Driving Among High School Students
According to the Centers for Disease Control and Prevention, the percentage of high school students who drink and drive was 17.5% at the beginning of 2001 ($t = 0$) and dropped steadily in the following years. This percentage is approximated by the function

$$P(t) = -0.73t + 17.5 \qquad (t \geq 0)$$

a. Plot the graph of the function P over the interval $[0, 14]$.
b. If the trend continues, what will the percentage of high school students who drink and drive be at the beginning of 2014?
c. At what rate was the percentage of students who drink and drive dropping between 2001 and 2011?
Source: Centers for Disease Control and Prevention.

Solution

a. Following the instructions given in Example 3, pages 90–91, we obtain the spreadsheet and graph shown in Figure T5. [*Note:* We have made the appropriate entries for the title and *x*- and *y*-axis labels. In particular, for Primary Vertical Axis Title, select **Rotated Title** and type P(t) percent].

	A	B	C
1	t	0	13
2	P(t)	17.5	8.01

(a)

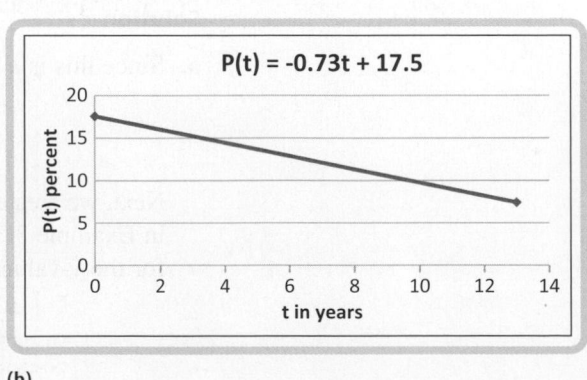

(b)

FIGURE **T5**
(a) The table of values for *t* and *P(t)* and (b) the graph showing the percentage of high school students who drink and drive.

Note: Boldfaced words/characters enclosed in a box (for example, **Enter**) indicate that an action (click, select, or press) is required. Words/characters printed blue (for example, Chart Type) indicate words/characters that appear on the screen. Words/characters printed in a monospace font (for example, = (-2/3) *A2+2) indicate words/characters that need to be typed and entered.

b. From the table of values, we see that

$$P(13) = -0.73(13) + 17.5 = 8.01$$

or 8.01%.

c. The rate at which the percentage of students who drink and drive was dropping between 2001 and 2011 is 0.73% per year.

TECHNOLOGY EXERCISES

Find the value of y corresponding to the given value of x.

1. $3.1x + 2.4y - 12 = 0$; $x = 2.1$

2. $1.2x - 3.2y + 8.2 = 0$; $x = 1.2$

3. $2.8x + 4.2y = 16.3$; $x = 1.5$

4. $-1.8x + 3.2y - 6.3 = 0$; $x = -2.1$

5. $22.1x + 18.2y - 400 = 0$; $x = 12.1$

6. $17.1x - 24.31y - 512 = 0$; $x = -8.2$

7. $2.8x = 1.41y - 2.64$; $x = 0.3$

8. $0.8x = 3.2y - 4.3$; $x = -0.4$

2.6 Quadratic Functions

Quadratic Functions

A **quadratic function** is one of the form

$$f(x) = ax^2 + bx + c$$

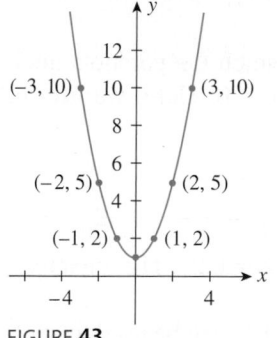

FIGURE 43
The graph of $f(x) = x^2 + 1$ is a parabola.

where a, b, and c are constants and $a \neq 0$. For example, the function $f(x) = 2x^2 - 3x + 4$ is quadratic (here, $a = 2$, $b = -3$, and $c = 4$). Also, the function $f(x) = x^2 + 1$ of Example 6, Section 2.3, is quadratic (here, $a = 1$, $b = 0$, and $c = 1$). Its graph illustrates the general shape of the graph of a quadratic function, which is called a *parabola* (Figure 43).

In general, the graph of a quadratic function is a parabola that opens upward or downward (Figure 44). Furthermore, the parabola is symmetric with respect to a vertical line called the *axis of symmetry* (shown dashed in Figure 44). This line also passes through the lowest point or the highest point of the parabola. The point of intersection of the parabola with its axis of symmetry is called the *vertex* of the parabola.

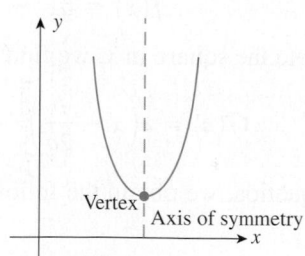

FIGURE 44
Graphs of quadratic functions are parabolas.

We can use these properties to help us sketch the graph of a quadratic function. For example, suppose we want to sketch the graph of

$$f(x) = 2x^2 - 4x + 1$$

If we complete the square in x, we obtain

$$f(x) = 2(x^2 - 2x) + 1$$

Factor out the coefficient of x^2 from the first two terms.

Adding and subtracting 2

$$= 2[x^2 - 2x + (-1)^2] + 1 - 2$$

Because of the 2 outside the brackets, we have added $2(1)$ and must therefore subtract 2.

$$= 2(x - 1)^2 - 1$$

Factor the terms within the brackets.

Observe that the first term, $2(x - 1)^2$, is nonnegative. In fact, it is equal to zero when $x = 1$ and is greater than zero if $x \neq 1$. Consequently, we see that $f(x) \geq -1$ for all values of x. This tells us that the vertex (in this case, the lowest point) of the parabola is the point $(1, -1)$. The axis of symmetry of the parabola is the vertical line $x = 1$. Finally, plotting the vertex and a few additional points on either side of the axes of symmetry of the parabola, we obtain the graph shown in Figure 45.

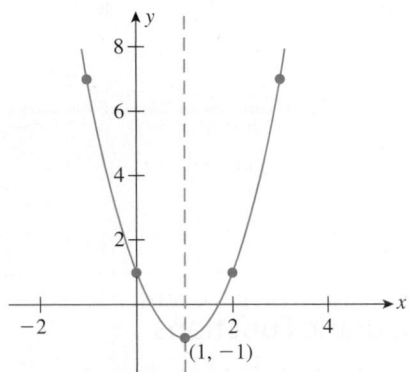

x	y
-1	7
0	1
2	1
3	7

FIGURE 45
The graph of $f(x) = 2x^2 - 4x + 1$

The x-intercepts of f, the x-coordinates of the points at which the parabola intersects the x-axis, can be found by solving the equation $f(x) = 0$. Here, we use the quadratic formula with $a = 2$, $b = -4$, and $c = 1$ to find that

$$x = \frac{-(-4) \pm \sqrt{(-4)^2 - 4(2)(1)}}{2(2)} = \frac{4 \pm \sqrt{8}}{4} = \frac{4 \pm 2\sqrt{2}}{4} = 1 \pm \frac{\sqrt{2}}{2}$$

Therefore, the x-intercepts are $1 + \sqrt{2}/2 \approx 1.71$ and $1 - \sqrt{2}/2 \approx 0.29$. The y-intercept of f (obtained by setting $x = 0$) is $f(0) = 1$.

The technique that we used to analyze $f(x) = 2x^2 - 4x + 1$ can be used to study the general quadratic function

$$f(x) = ax^2 + bx + c \qquad (a \neq 0)$$

If we complete the square in x, we find

$$f(x) = a\left(x + \frac{b}{2a}\right)^2 + \frac{4ac - b^2}{4a}$$

See Exercise 66.

From this equation, we obtain the following properties of the quadratic function f.

Properties of the Quadratic Function

$$f(x) = ax^2 + bx + c \qquad (a \neq 0)$$

1. The domain of f is the set of all real numbers, and the graph of f is a parabola.
2. If $a > 0$, the parabola opens upward, and if $a < 0$, it opens downward.

3. The vertex of the parabola is $\left(-\dfrac{b}{2a}, f\left(-\dfrac{b}{2a}\right)\right)$.

4. An equation of the axis of symmetry of the parabola is $x = -\dfrac{b}{2a}$.

5. The x-intercepts (if any) are found by solving $f(x) = 0$. The y-intercept is $f(0) = c$.

EXAMPLE 1 Given the quadratic function

$$f(x) = -2x^2 + 5x - 2$$

a. Find the vertex of the parabola.
b. Find the x-intercepts (if any) of the parabola.
c. Sketch the parabola.

Solution

a. Comparing $f(x) = -2x^2 + 5x - 2$ with the general form of the quadratic equation, we find that $a = -2$, $b = 5$, and $c = -2$. Therefore, the x-coordinate of the vertex of the parabola is

$$-\frac{b}{2a} = -\frac{5}{2(-2)} = \frac{5}{4}$$

Next, to find the y-coordinate of the vertex, we evaluate f at $x = \frac{5}{4}$, obtaining

$$f\left(\frac{5}{4}\right) = -2\left(\frac{5}{4}\right)^2 + 5\left(\frac{5}{4}\right) - 2$$

$$= -\frac{25}{8} + \frac{25}{4} - 2 \qquad -2\left(\frac{25}{16}\right) = -\frac{25}{8}$$

$$= \frac{9}{8} \qquad -\frac{25}{8} + \frac{50}{8} - \frac{16}{8} = \frac{9}{8}$$

b. To find the x-intercepts of the parabola, we solve the equation

$$-2x^2 + 5x - 2 = 0$$

using the quadratic formula with $a = -2$, $b = 5$, and $c = -2$. We find

$$x = \frac{-5 \pm \sqrt{25 - 4(-2)(-2)}}{2(-2)} \qquad x = \frac{-b \pm \sqrt{b^2 - 4ac}}{2a}$$

$$= \frac{-5 \pm \sqrt{9}}{-4}$$

$$= \frac{-5 \pm 3}{-4}$$

$$= \frac{1}{2} \quad \text{or} \quad 2$$

Thus, the x-intercepts of the parabola are $\frac{1}{2}$ and 2.

c. Since $a = -2 < 0$, the parabola opens downward. The vertex of the parabola $\left(\frac{5}{4}, \frac{9}{8}\right)$ is therefore the highest point on the curve. The parabola crosses the x-axis at the points $\left(\frac{1}{2}, 0\right)$ and $(2, 0)$. Setting $x = 0$ gives -2 as the y-intercept of the curve. Finally, using this information, we sketch the parabola shown in Figure 46.

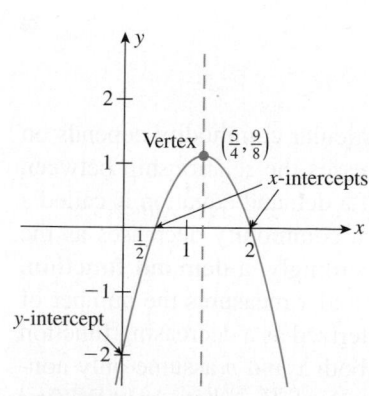

FIGURE **46**
The graph of $f(x) = -2x^2 + 5x - 2$

APPLIED EXAMPLE 2 Effect of Advertising on Profit The quarterly profit (in thousands of dollars) of Cunningham Realty is given by

$$P(x) = -\frac{1}{3}x^2 + 7x + 30 \qquad (0 \le x \le 50)$$

where x (in thousands of dollars) is the amount of money Cunningham spends on advertising per quarter. Find the amount of money Cunningham should spend on advertising to realize a maximum quarterly profit. What is the maximum quarterly profit realizable by Cunningham?

Solution The profit function P is a quadratic function, so its graph is a parabola. Furthermore, the coefficient of x^2 is $a = -\frac{1}{3} < 0$, so the parabola opens downward. The x-coordinate of the vertex of the parabola is

$$-\frac{b}{2a} = -\frac{7}{2\left(-\frac{1}{3}\right)} = \frac{21}{2} = 10.5 \qquad a = -\frac{1}{3} \text{ and } b = 7$$

The corresponding y-coordinate is

$$f\left(\frac{21}{2}\right) = -\frac{1}{3}\left(\frac{21}{2}\right)^2 + 7\left(\frac{21}{2}\right) + 30 = \frac{267}{4} = 66.75$$

Therefore, the vertex of the parabola is $\left(\frac{21}{2}, \frac{267}{4}\right)$. Since the parabola opens downward, the vertex of the parabola is the highest point on the parabola. Accordingly, the y-coordinate of the vertex gives the maximum value of P. This implies that the maximum quarterly profit of \$66,750 [remember that $P(x)$ is measured in thousands of dollars] is realized if Cunningham spends \$10,500 per quarter on advertising. The graph of P is shown in Figure 47.

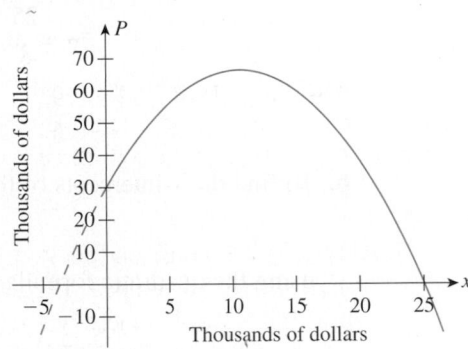

FIGURE **47**
The graph of the profit function $P(x) = -\frac{1}{3}x^2 + 7x + 30$

Demand and Supply Curves

In a free-market economy, consumer demand for a particular commodity depends on the commodity's unit price. A *demand equation* expresses the relationship between the unit price and the quantity demanded. The graph of a demand equation is called a *demand curve*. In general, the quantity demanded of a commodity decreases as the commodity's unit price increases, and vice versa. Accordingly, a **demand function,** defined by $p = f(x)$, where p measures the unit price and x measures the number of units of the commodity in question, is generally characterized as a decreasing function of x; that is, $p = f(x)$ decreases as x increases. Since both x and p assume only non-negative values, the demand curve is that part of the graph of $f(x)$ that lies in the first quadrant (Figure 48).

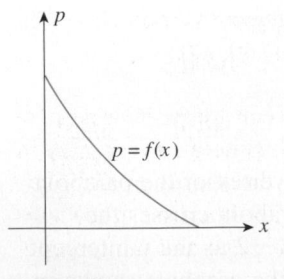

FIGURE **48**
A demand curve

 APPLIED EXAMPLE 3 Demand for Bluetooth Headsets The demand function for a certain brand of Bluetooth wireless headsets is given by

$$p = d(x) = -0.025x^2 - 0.5x + 60$$

where p is the wholesale unit price in dollars and x is the quantity demanded each month, measured in units of a thousand. Sketch the corresponding demand curve. Above what price will there be no demand? What is the maximum quantity demanded per month?

Solution The given function is quadratic, and its graph may be sketched using the methods just developed (Figure 49). The p-intercept, 60, gives the wholesale unit price above which there will be no demand. To obtain the maximum quantity demanded, set $p = 0$, which gives

$$-0.025x^2 - 0.5x + 60 = 0$$
$$x^2 + 20x - 2400 = 0 \qquad \text{Upon multiplying both sides of the equation by } -40$$
$$(x + 60)(x - 40) = 0$$

That is, $x = -60$ or $x = 40$. Since x must be nonnegative, we reject the root $x = -60$. Thus, the maximum number of headsets demanded per month is 40,000. ∎

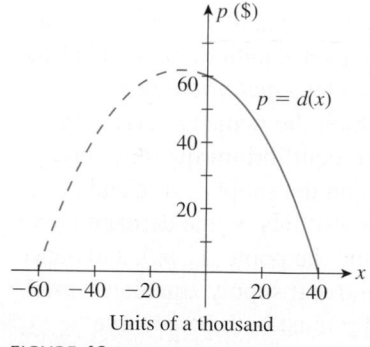

FIGURE 49
The demand curve $p = d(x)$

In a competitive market, a relationship also exists between the unit price of a commodity and the commodity's availability in the market. In general, an increase in the commodity's unit price induces the producer to increase the supply of the commodity. Conversely, a decrease in the unit price generally leads to a drop in the supply. The equation that expresses the relation between the unit price and the quantity supplied is called a *supply equation*, and its graph is called a *supply curve*. A **supply function,** defined by $p = f(x)$, is generally characterized as an increasing function

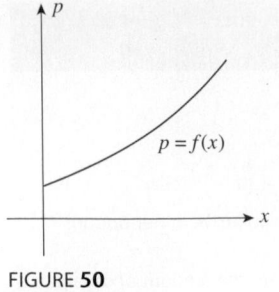

FIGURE **50**
A supply curve

of x; that is, $p = f(x)$ increases as x increases. Since both x and p assume only non-negative values, the supply curve is the part of the graph of $f(x)$ that lies in the first quadrant (Figure 50).

APPLIED EXAMPLE 4 Supply of Bluetooth Headsets The supply function for a certain brand of bluetooth wireless headsets is given by

$$p = s(x) = 0.02x^2 + 0.6x + 20$$

where p is the unit wholesale price in dollars and x stands for the quantity in units of a thousand that will be made available in the market by the supplier. Sketch the corresponding supply curve. What is the lowest price at which the supplier will make the headsets available in the market?

Solution A sketch of the supply curve appears in Figure 51. The p-intercept, 20, gives the lowest price at which the supplier will make the headsets available in the market.

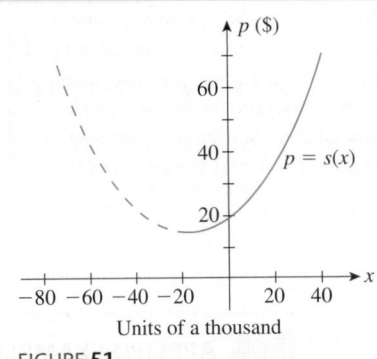

FIGURE **51**
The supply curve $p = s(x)$

Market Equilibrium

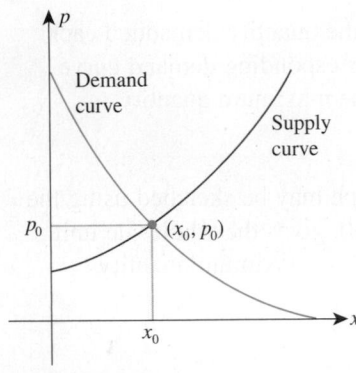

FIGURE **52**
Market equilibrium corresponds to (x_0, p_0), the point at which the supply and demand curves intersect.

Under pure competition, the price of a commodity will eventually settle at a level dictated by the following condition: The supply of the commodity will be equal to the demand for it. If the price is too high, the consumer will not buy, and if the price is too low, the supplier will not produce. **Market equilibrium** prevails when the quantity produced is equal to the quantity demanded. The quantity produced at market equilibrium is called the *equilibrium quantity,* and the corresponding price is called the *equilibrium price.*

Market equilibrium corresponds to the point at which the demand curve and the supply curve intersect. In Figure 52, x_0 represents the equilibrium quantity, and p_0 represents the equilibrium price. The point (x_0, p_0) lies on the supply curve and therefore satisfies the supply equation. At the same time, it also lies on the demand curve and therefore satisfies the demand equation. Thus, to find the point (x_0, p_0), and hence the equilibrium quantity and price, we solve the demand and supply equations simultaneously for x and p. For meaningful solutions, x and p must both be positive.

APPLIED EXAMPLE 5 Market Equilibrium Refer to Examples 3 and 4. The demand function for a certain brand of bluetooth wireless headsets is given by

$$p = d(x) = -0.025x^2 - 0.5x + 60$$

and the corresponding supply function is given by

$$p = s(x) = 0.02x^2 + 0.6x + 20$$

where p is expressed in dollars and x is measured in units of a thousand. Find the equilibrium quantity and price.

Solution We solve the following system of equations:

$$p = -0.025x^2 - 0.5x + 60$$
$$p = 0.02x^2 + 0.6x + 20$$

Substituting the first equation into the second yields

$$-0.025x^2 - 0.5x + 60 = 0.02x^2 + 0.6x + 20$$

which is equivalent to

$$0.045x^2 + 1.1x - 40 = 0$$
$$45x^2 + 1100x - 40{,}000 = 0 \qquad \text{Multiply by 1000.}$$
$$9x^2 + 220x - 8000 = 0 \qquad \text{Divide by 5.}$$
$$(9x + 400)(x - 20) = 0$$

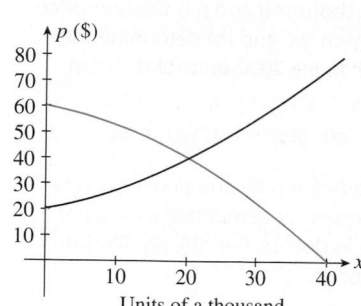

p ($)

80
70
60
50
40
30
20
10

10 20 30 40 x

Units of a thousand

FIGURE 53
The supply curve and the demand curve intersect at the point (20, 40).

Thus, $x = -\frac{400}{9}$ or $x = 20$. Since x must be nonnegative, the root $x = -\frac{400}{9}$ is rejected. Therefore, the equilibrium quantity is 20,000 headsets. The equilibrium price is given by

$$p = 0.02(20)^2 + 0.6(20) + 20 = 40$$

or $40 per headset (Figure 53).

2.6 Self-Check Exercises

Given the quadratic function

$$f(x) = 2x^2 - 3x - 3$$

1. Find the vertex of the parabola.

2. Find the x-intercepts (if any) of the parabola.

3. Sketch the parabola.

Solutions to Self-Check Exercises 2.6 can be found on page 140.

2.6 Concept Questions

1. Consider the quadratic function

$$f(x) = ax^2 + bx + c \ (a \neq 0)$$

  a. What is the domain of f?
  b. What can you say about the parabola if $a > 0$?
  c. What is the vertex of the parabola in terms of a and b?
  d. What is the axis of symmetry of the parabola?

2. a. What is a demand function? A supply function?
  b. What is market equilibrium?
  c. What are the equilibrium quantity and equilibrium price? How do you determine these quantities?

2.6 Exercises

In Exercises 1–18, find the vertex, the x-intercepts (if any), and sketch the parabola.

1. $f(x) = x^2 + x - 6$

2. $f(x) = 3x^2 - 5x - 2$

3. $f(x) = x^2 - 4x + 4$

4. $f(x) = x^2 + 6x + 9$

5. $f(x) = -x^2 + 5x - 6$

6. $f(x) = -4x^2 + 4x + 3$

7. $f(x) = 3x^2 - 5x + 1$

8. $f(x) = -2x^2 + 6x - 3$

9. $f(x) = 2x^2 - 3x + 3$

10. $f(x) = 3x^2 - 4x + 2$

11. $f(x) = x^2 - 4$

12. $f(x) = 2x^2 + 3$

13. $f(x) = 16 - x^2$

14. $f(x) = 5 - x^2$

15. $f(x) = \frac{3}{8}x^2 - 2x + 2$

16. $f(x) = \frac{3}{4}x^2 - \frac{1}{2}x + 1$

17. $f(x) = 1.2x^2 + 3.2x - 1.2$

18. $f(x) = 2.3x^2 - 4.1x + 3$

In Exercises 19–22, the parabola shown is the graph of $y = f(x) = ax^2 + bx + c$. Find the sign of (a) a, (b) b, (c) $f\left(-\dfrac{b}{2a}\right)$, and (d) $b^2 - 4ac$.

19.

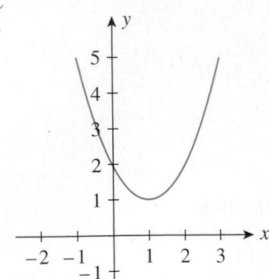

20.

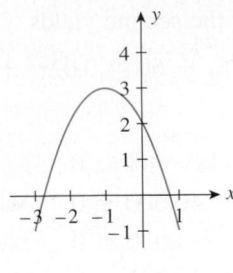

21.

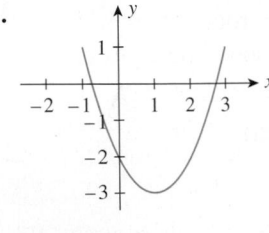

22.

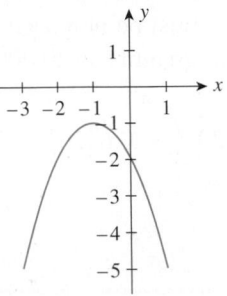

In Exercises 23–28, find the points of intersection of the graphs of the functions.

23. $f(x) = -x^2 + 4;\ g(x) = x - 2$

24. $f(x) = x^2 - 5x + 6;\ g(x) = \dfrac{1}{2}x + \dfrac{3}{2}$

25. $f(x) = -x^2 + 2x + 6;\ g(x) = x^2 - 6$

26. $f(x) = x^2 - 2x - 2;\ g(x) = -x^2 - x + 1$

27. $f(x) = 2x^2 - 5x - 8;\ g(x) = -3x^2 + x + 5$

28. $f(x) = 0.2x^2 - 1.2x - 4;\ g(x) = -0.3x^2 + 0.7x + 8.2$

29. Find all values of b for which $f(x) = -2x^2 - bx + 8$ has a maximum value of 16.

30. Find a so that $f(x) = ax^2 + 8x - 8$ has a minimum value of -24.

31. Find c so that $f(x) = -3x^2 - 4x + c$ has a maximum value of $-\dfrac{2}{3}$.

32. Find conditions on a and c so that $f(x) = ax^2 + 2x + c$ has a minimum value of 4.

33. Find c so that $f(x) = x^2 + 3x + c$ has exactly one real root.

34. Find a so that $f(x) = ax^2 + 4x + 1$ has two real roots.

35. Find all values of b so that $f(x) = 2x^2 + bx + 5$ has at least one real root.

36. Find all values of a so that $f(x) = ax^2 - 2x - 4$ has no real roots.

For the demand equations in Exercises 37 and 38, where x represents the quantity demanded in units of a thousand and p is the unit price in dollars, (a) sketch the demand curve, and (b) determine the quantity demanded when the unit price is set at $\$p$.

37. $p = -x^2 + 36;\ p = 11$ **38.** $p = -x^2 + 16;\ p = 7$

For the supply equations in Exercises 39 and 40, where x is the quantity supplied in units of a thousand and p is the unit price in dollars, (a) sketch the supply curve, and (b) determine the price at which the supplier will make 2000 units of the commodity available in the market.

39. $p = 2x^2 + 18$ **40.** $p = x^2 + 16x + 40$

In Exercises 41–44, for each pair of supply and demand equations, where x represents the quantity demanded in units of a thousand and p the unit price in dollars, find the equilibrium quantity and the equilibrium price.

41. $p = -2x^2 + 80$ and $p = 15x + 30$

42. $p = -x^2 - 2x + 100$ and $p = 8x + 25$

43. $11p + 3x - 66 = 0$ and $2p^2 + p - x = 10$

44. $p = 60 - 2x^2$ and $p = x^2 + 9x + 30$

45. CANCER SURVIVORS The number of living Americans who have had a cancer diagnosis has increased drastically since 1971. In part, this is due to more testing for cancer and better treatment for some cancers. In part, it is because the population is older, and cancer is largely a disease of the elderly. The number of cancer survivors (in millions) between 1975 ($t = 0$) and 2000 ($t = 25$) is approximately

$$N(t) = 0.0031t^2 + 0.16t + 3.6 \qquad (0 \le t \le 25)$$

   a. How many living Americans had a cancer diagnosis in 1975? In 2000?

   b. Assuming the trend continued, how many cancer survivors were there in 2005?

   Source: National Cancer Institute.

46. MOTION OF A STONE A stone is thrown straight up from the roof of an 80-ft building. The distance of the stone from the ground at any time t (in seconds) is given by

$$h(t) = -16t^2 + 64t + 80$$

   a. Sketch the graph of h.

   b. At what time does the stone reach the highest point? What is the stone's maximum height from the ground?

47. MAXIMIZING PROFIT The estimated monthly profit realizable by the Cannon Precision Instruments Corporation for manufacturing and selling x units of its model M1 cameras is

$$P(x) = -0.04x^2 + 240x - 10,000$$

dollars. Determine how many cameras Cannon should produce per month to maximize its profits.

48. MAXIMIZING PROFIT Lynbrook West, an apartment complex, has 100 two-bedroom units. The monthly profit realized from renting out x apartments is given by

$$P(x) = -10x^2 + 1760x - 50{,}000$$

dollars. How many units should be rented out to maximize the monthly rental profit? What is the maximum monthly profit realizable?

49. MAXIMIZING REVENUE The monthly revenue R (in hundreds of dollars) realized in the sale of Royal electric shavers is related to the unit price p (in dollars) by the equation

$$R(p) = -\frac{1}{2}p^2 + 30p$$

a. Sketch the graph of R.
b. At what unit price is the monthly revenue maximized?

50. EFFECT OF ADVERTISING ON PROFIT The relationship between Northwood Realty's quarterly profit, $P(x)$, and the amount of money x spent on advertising per quarter is described by the function

$$P(x) = -\frac{1}{8}x^2 + 7x + 30 \qquad (0 \le x \le 50)$$

where both $P(x)$ and x are measured in thousands of dollars.

a. Sketch the graph of P.
b. Find the amount of money the company should spend on advertising per quarter to maximize its quarterly profits.

51. BABY BOOMERS AND MEDICARE BENEFITS Aging baby boomers will put a strain on Medicare benefits unless Congress takes action. The Medicare benefits to be paid out from 2010 through 2040 are projected to be

$$B(t) = 0.09t^2 + 0.102t + 0.25 \qquad (0 \le t \le 3)$$

where $B(t)$ is measured in trillions of dollars and t is measured in decades with $t = 0$ corresponding to 2010.

a. What was the amount of Medicare benefits paid out in 2010?
b. What is the amount of Medicare benefits projected to be paid out in 2040?
Source: Social Security and Medicare Trustees' 2010 Report.

52. MEDIAN PRICE OF NEW CARS The median price of a new car in year t is given by

$$P(t) = 9.1667t^2 + 1213.3333t + 30{,}000 \qquad (0 \le t \le 8)$$

Here, $t = 0$ corresponds to the 2006 model year, and P is the manufacturer's suggested retail price for the car.

a. Show that the median price of a new car was increasing each year from 2006 through 2014.
Hint: Show that the graph of P is a parabola that opens upward, and that the t-coordinate of its vertex is negative.
b. In what year did the median price of a new vehicle first reach $35,000?
Source: Kelley Blue Book.

53. DIABETES IN THE UNITED STATES The number of adults (in millions) diagnosed with diabetes in the United States in year t is given by

$$N(t) = 0.0125t^2 + 0.475t + 20.7 \qquad (0 \le t \le 4)$$

a. Show that the number of adults diagnosed with diabetes was increasing each year from 2010 through 2014.
Hint: Show that the graph of P is a parabola that opens upward, and that the t-coordinate of its vertex is negative.
b. In what year did the number of adults diagnosed with diabetes first reach 21.7 million?
Source: Centers for Disease Control and Prevention.

54. SUPPLY FUNCTIONS The supply function for the Luminar daylight LED desk lamp is given by

$$p = 0.1x^2 + 0.5x + 15$$

where x is the quantity supplied (in thousands) and p is the unit price in dollars. Sketch the graph of the supply function. What unit price will induce the supplier to make 5000 lamps available in the marketplace?

55. MARKET EQUILIBRIUM The weekly demand and supply functions for Sportsman 5×7 tents are given by

$$p = -0.1x^2 - x + 40$$
$$p = 0.1x^2 + 2x + 20$$

respectively, where p is measured in dollars and x is measured in units of a hundred. Find the equilibrium quantity and price.

56. MARKET EQUILIBRIUM The management of the Titan Tire Company has determined that the weekly demand and supply functions for their Super Titan tires are given by

$$p = 144 - x^2$$
$$p = 48 + \frac{1}{2}x^2$$

respectively, where p is measured in dollars and x is measured in units of a thousand. Find the equilibrium quantity and price.

57. POISEUILLE'S LAW According to a law discovered by the nineteenth century physician Poiseuille, the velocity (in centimeters per second) of blood r cm from the central axis of an artery is given by

$$v(r) = k(R^2 - r^2)$$

where k is a constant and R is the radius of the artery. Suppose that for a certain artery, $k = 1000$ and $R = 0.2$ so that $v(r) = 1000(0.04 - r^2)$.

a. Sketch the graph of v.
b. For what value of r is $v(r)$ largest? Smallest? Interpret your results.

58. MOTION OF A BALL A ball is thrown straight upward from the ground and attains a height of $s(t) = -16t^2 + 128t + 4$ ft above the ground after t sec. When does the ball reach the maximum height? What is the maximum height?

59. DESIGNING A NORMAN WINDOW A Norman window has the shape of a rectangle surmounted by a semicircle (see the accompanying figure). If a Norman window is to have a perimeter of 28 ft, what should be its dimensions in order to allow the maximum amount of light through the window?

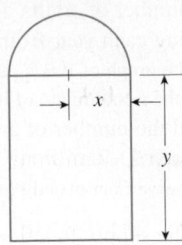

60. DISTANCE OF WATER FLOW A cylindrical tank of height h ft is filled to the top with water. If a hole is punched into the lateral side of the tank, the stream of water flowing out of the tank will reach the ground at a distance of x ft from the base of the tank where $x = 2\sqrt{y(h-y)}$ (see the accompanying figure). Find the location of the hole so that x is a maximum. What is this maximum value of x? Hint: It suffices to maximize the expression for x^2. (Why?)

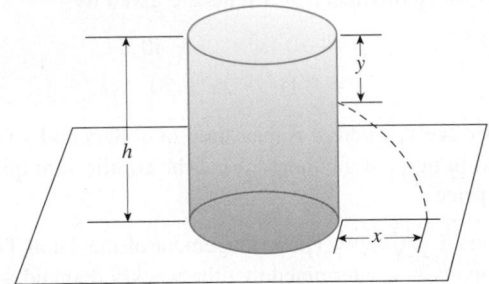

In Exercises 61–65, determine whether the statement is true or false. If it is true, explain why it is true. If it is false, explain why or give an example to show why it is false.

61. If $f(x) = ax^2 + bx + c \ (a \neq 0)$, then
$$f\left(\frac{-b + \sqrt{b^2 - 4ac}}{2a}\right) = 0$$

62. The quadratic function $f(x) = ax^2 + bx + c \ (a \neq 0)$ has no x-intercepts if $b^2 - 4ac > 0$.

63. If a and c have opposite signs, then the parabola with equation $y = ax^2 + bx + c$ intersects the x-axis at two distinct points.

64. If $b^2 = 4ac$, then the graph of the quadratic function $f(x) = ax^2 + bx + c \ (a \neq 0)$ touches the x-axis at exactly one point.

65. If a profit function is given by $P(x) = ax^2 + bx + c$, where x is the number of units produced and sold, then the level of production that yields the maximum profit is $-\dfrac{b}{2a}$ units.

66. Let $f(x) = ax^2 + bx + c \ (a \neq 0)$. By completing the square in x, show that
$$f(x) = a\left(x + \frac{b}{2a}\right)^2 + \frac{4ac - b^2}{4a}$$

2.6 Solutions to Self-Check Exercises

1. Here, $a = 2$, $b = -3$, and $c = -3$. The x-coordinate of the vertex is
$$-\frac{b}{2a} = -\frac{(-3)}{2(2)} = \frac{3}{4}$$

The corresponding y-coordinate is
$$f\left(\frac{3}{4}\right) = 2\left(\frac{3}{4}\right)^2 - 3\left(\frac{3}{4}\right) - 3 = \frac{9}{8} - \frac{9}{4} - 3 = -\frac{33}{8}$$

Therefore, the vertex of the parabola is $\left(\frac{3}{4}, -\frac{33}{8}\right)$.

2. Solving the equation $2x^2 - 3x - 3 = 0$, we find
$$x = \frac{-(-3) \pm \sqrt{(-3)^2 - 4(2)(-3)}}{2(2)} = \frac{3 \pm \sqrt{33}}{4}$$

So the x-intercepts are $\dfrac{3}{4} - \dfrac{\sqrt{33}}{4} \approx -0.7$ and $\dfrac{3}{4} + \dfrac{\sqrt{33}}{4} \approx 2.2$.

3. Since $a = 2 > 0$, the parabola opens upward. The y-intercept is -3. The graph of the parabola is shown in the accompanying figure.

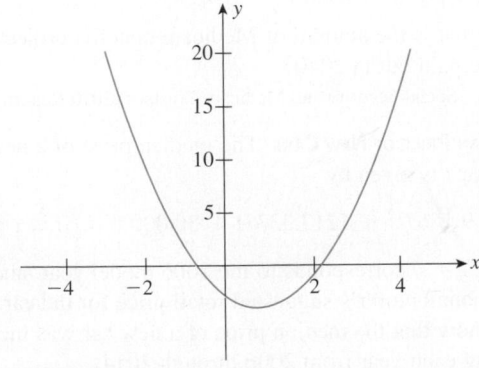

The graph of $f(x) = 2x^2 - 3x - 3$

USING TECHNOLOGY — Finding the Points of Intersection of Two Graphs

A graphing utility can be used to find the point(s) of intersection of the graphs of two functions.

EXAMPLE 1 Find the points of intersection of the graphs of

$$f(x) = 0.3x^2 - 1.4x - 3 \quad \text{and} \quad g(x) = -0.4x^2 + 0.8x + 6.4$$

Solution The graphs of both f and g in the standard viewing window are shown in Figure T1a. Using the function for finding the points of intersection of two graphs on a graphing utility, we find the point(s) of intersection, accurate to four decimal places, to be $(-2.4158, 2.1329)$ (Figure T1b) and $(5.5587, -1.5125)$ (Figure T1c). To access this function on the TI-83/84, select **5: intersect** on the Calc menu.

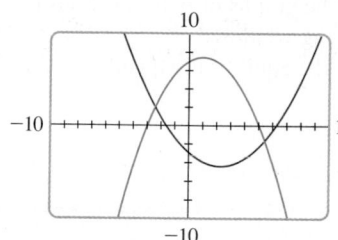
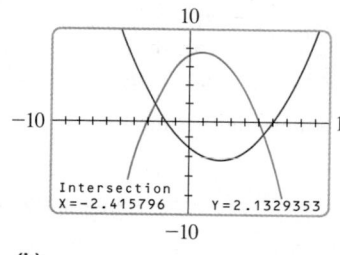
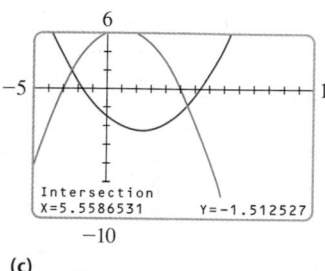

(a)   (b)   (c)

FIGURE **T1**
(a) The graphs of f and g in the standard viewing window; (b) and (c) the TI-83/84 intersection screens

EXAMPLE 2 Consider the demand and supply functions

$$p = d(x) = -0.01x^2 - 0.2x + 8 \quad \text{and} \quad p = s(x) = 0.01x^2 + 0.1x + 3$$

a. Plot the graphs of d and s in the viewing window $[0, 15] \times [0, 10]$.
b. Verify that the equilibrium point is $(10, 5)$.

Solution

a. The graphs of d and s are shown in Figure T2a.

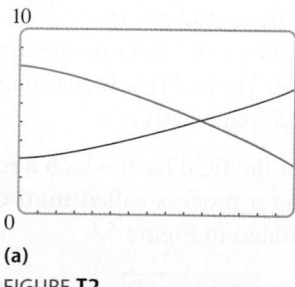
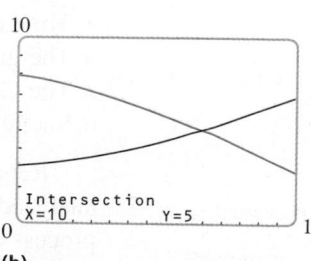

(a)   (b)

FIGURE **T2**
(a) The graphs of d and s in the window $[0, 15] \times [0, 10]$; (b) the TI-83/84 intersection screen

b. Using the function for finding the point of intersection of two graphs, we see that $x = 10$ and $y = 5$ (Figure T2b), so the equilibrium point is $(10, 5)$.

TECHNOLOGY EXERCISES

In Exercises 1–6, find the points of intersection of the graphs of the functions. Express your answer accurate to four decimal places.

1. $f(x) = 1.2x + 3.8$; $g(x) = -0.4x^2 + 1.2x + 7.5$

2. $f(x) = 0.2x^2 - 1.3x - 3$; $g(x) = -1.3x + 2.8$

3. $f(x) = 0.3x^2 - 1.7x - 3.2$; $g(x) = -0.4x^2 + 0.9x + 6.7$

4. $f(x) = -0.3x^2 + 0.6x + 3.2$; $g(x) = 0.2x^2 - 1.2x - 4.8$

5. $f(x) = -1.8x^2 + 2.1x - 2$; $g(x) = 2.1x - 4.2$

6. $f(x) = 1.2x^2 - 1.2x + 2$; $g(x) = -0.2x^2 + 0.8x + 2.1$

7. Market Equilibrium for Wall Clocks The monthly demand and supply functions for a certain brand of wall clock are given by

$$p = -0.2x^2 - 1.2x + 50$$
$$p = 0.1x^2 + 3.2x + 25$$

respectively, where p is measured in dollars and x is measured in units of a hundred.

a. Plot the graphs of both functions in an appropriate viewing window.

b. Find the equilibrium quantity and price.

8. Market Equilibrium for Digital Cameras The quantity demanded x (in units of a hundred) of Mikado digital cameras per week is related to the unit price p (in dollars) by

$$p = -0.2x^2 + 80$$

The quantity x (in units of a hundred) that the supplier is willing to make available in the market is related to the unit price p (in dollars) by

$$p = 0.1x^2 + x + 40$$

a. Plot the graphs of both functions in an appropriate viewing window.

b. Find the equilibrium quantity and price.

2.7 Functions and Mathematical Models

Mathematical Models

One of the fundamental goals in this book is to show how mathematics and, in particular, calculus can be used to solve real-world problems such as those arising from the world of business and the social, life, and physical sciences. You have already seen some of these problems earlier. Here are a few more examples of real-world phenomena that we will analyze in this and ensuing chapters.

- The erosion of the middle class (page 117)
- Global warming (page 144)
- The solvency of the U.S. Social Security trust fund (page 145)
- Time use of college students (page 556)
- The growth of tablet and smartphone users (page 656)
- The Case-Shiller Home Price Index (page 752)
- Social networks (page 820)

Regardless of the field from which a real-world problem is drawn, the problem is analyzed by using a process called **mathematical modeling.** The four steps in this process are illustrated in Figure 54.

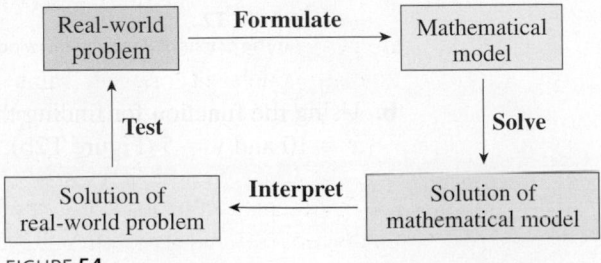

FIGURE **54**

1. **Formulate** Given a real-world problem, our first task is to formulate the problem, using the language of mathematics. The many techniques used in constructing mathematical models range from theoretical consideration of the problem on the one extreme to an interpretation of data associated with the problem on the other. For example, the mathematical model giving the accumulated amount at any time when a certain sum of money is deposited in the bank can be derived theoretically (see Chapter 4). On the other hand, many of the mathematical models in this book are constructed by studying the data associated with the problem (see Using Technology, pages 154–157). In calculus, we will be primarily concerned with how one (dependent) variable depends on one or more (independent) variables. Consequently, most of our mathematical models in calculus will involve functions of one or more variables or equations defining these functions (implicitly).

2. **Solve** Once a mathematical model has been constructed, we can use the appropriate mathematical techniques, which we will develop throughout the book, to solve the problem.

3. **Interpret** Bearing in mind that the solution obtained in Step 2 is just the solution of the mathematical model, we need to interpret these results in the context of the original real-world problem.

4. **Test** Some mathematical models of real-world applications describe the situations with complete accuracy. For example, the model describing a deposit in a bank account gives the exact accumulated amount in the account at any time. But other mathematical models give, at best, an approximate description of the real-world problem. In this case, we need to test the accuracy of the model by observing how well it describes the original real-world problem and how well it predicts past and/or future behavior. If the results are unsatisfactory, then we may have to reconsider the assumptions made in the construction of the model or, in the worst case, return to Step 1.

Many real-world phenomena, including those mentioned at the beginning of this section, are modeled by an appropriate function. In what follows, we will recall some familiar functions and give examples of real-world phenomena that are modeled by using these functions.

Polynomial Functions

A **polynomial function** of degree n is a function of the form

$$f(x) = a_n x^n + a_{n-1} x^{n-1} + \cdots + a_2 x^2 + a_1 x + a_0 \qquad (a_n \neq 0)$$

where n is a nonnegative integer and the numbers $a_0, a_1, \ldots, a_n$ are constants, called the **coefficients** of the polynomial function. For example, the functions

$$f(x) = 2x^5 - 3x^4 + \frac{1}{2}x^3 + \sqrt{2}x^2 - 6$$

and

$$g(x) = 0.001x^3 - 0.2x^2 + 10x + 200$$

are polynomial functions of degrees 5 and 3, respectively. Observe that a polynomial function is defined for every value of x and so its domain is $(-\infty, \infty)$.

A polynomial function of degree 1 ($n = 1$) has the form

$$y = f(x) = a_1 x + a_0 \qquad (a_1 \neq 0)$$

and is an equation of a straight line in the slope-intercept form with slope $m = a_1$ and y-intercept $b = a_0$ (see Section 2.2). For this reason, a polynomial function of degree 1 is called a linear function.

Linear functions are used extensively in mathematical modeling for two important reasons. First, some models are *linear* by nature. For example, the formula for converting temperature from Celsius (°C) to Fahrenheit (°F) is $F = \frac{9}{5}C + 32$, and F is a linear function of C. Second, some natural phenomena exhibit linear characteristics over a small range of values and can therefore be modeled by a linear function restricted to a small interval.

A polynomial function of degree 2 has the form

$$y = f(x) = a_2x^2 + a_1x + a_0 \qquad (a_2 \neq 0)$$

or, more simply, $y = ax^2 + bx + c$, and is called a quadratic function.

Quadratic functions serve as mathematical models for many phenomena, as Example 1 shows.

APPLIED EXAMPLE 1 Global Warming The increase in carbon dioxide (CO_2) in the atmosphere is a major cause of global warming. The Keeling curve, named after Charles David Keeling, a professor at Scripps Institution of Oceanography, gives the average amount of CO_2, measured in parts per million volume (ppmv), in the atmosphere from 1958 through 2013. Even though data were available for every year in this time interval, we'll construct the curve based only on the following randomly selected data points.

Year	1958	1970	1974	1978	1985	1991	1998	2003	2007	2010	2013
Amount CO$_2$	315	325	330	335	345	355	365	375	380	390	395

The **scatter plot** associated with these data is shown in Figure 55a. A mathematical model giving the approximate amount of CO_2 in the atmosphere during this period is given by

$$A(t) = 0.012444t^2 + 0.7485t + 313.9 \qquad (1 \leq t \leq 56)$$

where t is measured in years, with $t = 1$ corresponding to 1958. The graph of A is shown in Figure 55b.

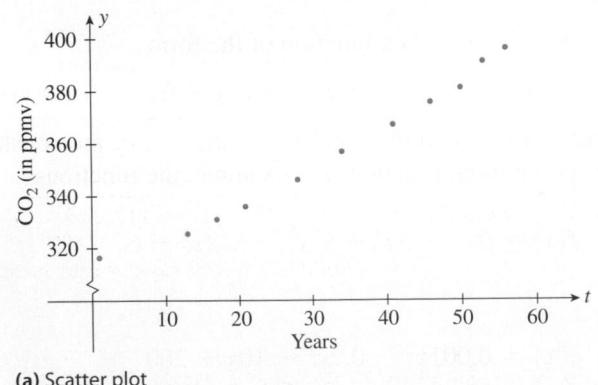

(a) Scatter plot

FIGURE 55

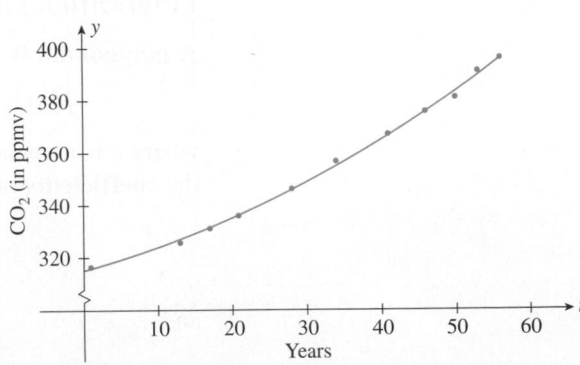

(b) The graph of A superimposed upon the scatter plot

a. Use the model to estimate the average amount of atmospheric CO_2 in 1980 ($t = 23$).

b. Assume that the trend continued, and use the model to predict the average amount of atmospheric CO_2 in 2016 ($t = 59$).

Source: Scripps Institution of Oceanography.

Solution

a. The average amount of atmospheric carbon dioxide in 1980 is given by

$$A(23) = 0.012444(23)^2 + 0.7485(23) + 313.9 \approx 337.70$$

or approximately 338 ppmv.

b. Assuming that the trend continued, the average amount of atmospheric CO_2 in 2016 will be

$$A(59) = 0.012444(59)^2 + 0.7485(59) + 313.9 \approx 401.38$$

or approximately 401 ppmv.

The next example uses a polynomial of degree 4 to help us construct a model that describes the projected assets of the Social Security trust fund.

 APPLIED EXAMPLE 2 Social Security Trust Fund Assets The projected assets of the Social Security trust fund (in trillions of dollars) from 2010 through 2033 are given in the following table.

Year	2010	2015	2020	2025	2030	2033
Assets	2.61	2.68	2.44	1.87	0.78	0

The scatter plot associated with these data are shown in Figure 56a, where $t = 0$ corresponds to 2010. A mathematical model giving the approximate value of the assets in the trust fund $A(t)$, in trillions of dollars, in year t is

$$A(t) = 0.000008140t^4 - 0.00043833t^3 - 0.0001305t^2 + 0.02202t + 2.612 \qquad (0 \le t \le 23)$$

The graph of $A(t)$ is shown in Figure 56b. (You will be asked to construct this model in Exercise 12, Using Technology Exercises 2.7.)

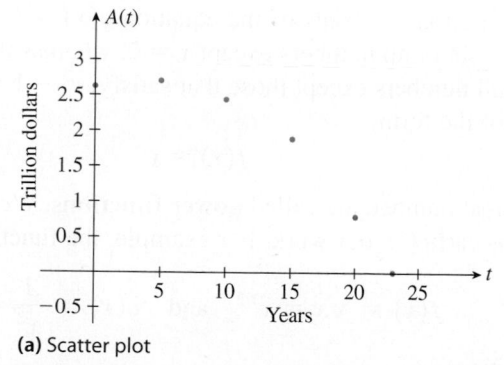

(a) Scatter plot

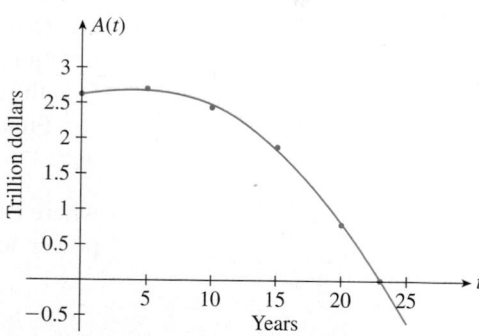

(b) The graph of A together with the scatter plot

FIGURE 56

a. The first baby boomers turned 65 in 2011. What were the assets of the Social Security system trust fund at that time? The last of the baby boomers will turn 65 in 2029. What will the assets of the trust fund be at that time?

b. Unless payroll taxes are increased significantly and/or benefits are scaled back dramatically, it is a matter of time before the assets of the current system are depleted. Use the graph of the function $A(t)$ to estimate the year in which the current Social Security system is projected to go broke.

Source: Social Security Administration.

Solution

a. The assets of the Social Security trust fund in 2011 ($t = 1$) were

$$A(1) = 0.000008140(1)^4 - 0.00043833(1)^3 - 0.0001305(1)^2$$
$$+ 0.02202(1) + 2.612 \approx 2.633$$

or approximately \$2.63 trillion. The assets of the trust fund in 2029 ($t = 19$) will be

$$A(19) = 0.000008140(19)^4 - 0.00043833(19)^3 - 0.0001305(19)^2$$
$$+ 0.02202(19) + 2.612 \approx 1.038$$

or approximately \$1.04 trillion.

b. From Figure 56b, we see that the graph of A crosses the t-axis at approximately $t = 23$. So unless the current system is changed, it is projected to go broke in 2033. (At this time, the first of the baby boomers will be 87, and the last of the baby boomers will be 69.) ∎

Rational and Power Functions

Another important class of functions is rational functions. A **rational function** is simply the quotient of two polynomials. Examples of rational functions are

$$F(x) = \frac{3x^3 + x^2 - x + 1}{x - 2}$$

$$G(x) = \frac{x^2 + 1}{x^2 - 1}$$

In general, a rational function has the form

$$R(x) = \frac{f(x)}{g(x)}$$

where $f(x)$ and $g(x)$ are polynomial functions. Since division by zero is not allowed, we conclude that the domain of a rational function is the set of all real numbers except the zeros of g—that is, the roots of the equation $g(x) = 0$. Thus, the domain of the function F is the set of all numbers except $x = 2$, whereas the domain of the function G is the set of all numbers except those that satisfy $x^2 - 1 = 0$, or $x = \pm 1$.

Functions of the form

$$f(x) = x^r$$

where r is any real number, are called **power functions.** We encountered examples of power functions earlier in our work. For example, the functions

$$f(x) = \sqrt{x} = x^{1/2} \quad \text{and} \quad g(x) = \frac{1}{x^2} = x^{-2}$$

are power functions.

Many of the functions that we encounter later will involve combinations of the functions introduced here. For example, the following functions may be viewed as combinations of such functions:

$$f(x) = \sqrt{\frac{1 - x^2}{1 + x^2}}$$

$$g(x) = \sqrt{x^2 - 3x + 4}$$

$$h(x) = (1 + 2x)^{1/2} + \frac{1}{(x^2 + 2)^{3/2}}$$

As with polynomials of degree 3 or greater, analyzing the properties of these functions is facilitated by using the tools of calculus, to be developed later.

In the next example, we use a power function to construct a model that describes the driving costs of a car.

APPLIED EXAMPLE 3 Driving Costs A study of driving costs based on a 2012 medium-sized sedan found the following average costs (car payments, gas, insurance, upkeep, and depreciation), measured in cents per mile.

Miles/year	5000	10,000	15,000	20,000
Cost/mile (¢)	161.2	78.0	61.0	52.3

A mathematical model giving the average cost in cents per mile is

$$C(x) = \frac{1910.5}{x^{1.72}} + 42.9$$

where x (in thousands) denotes the number of miles the car is driven in each year. The scatter plot associated with these data and the graph of C are shown in Figure 57. Using this model, estimate the average cost per mile of driving a 2012 medium-sized sedan 8000 miles per year and 18,000 miles per year.
Source: American Automobile Association.

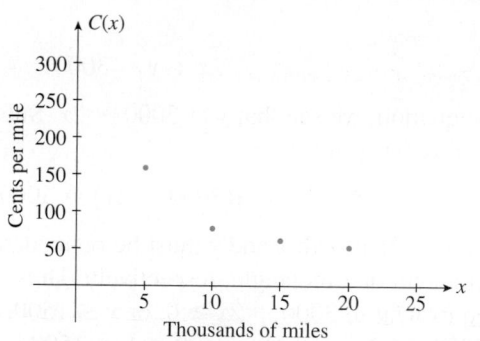
(a) Scatter plot

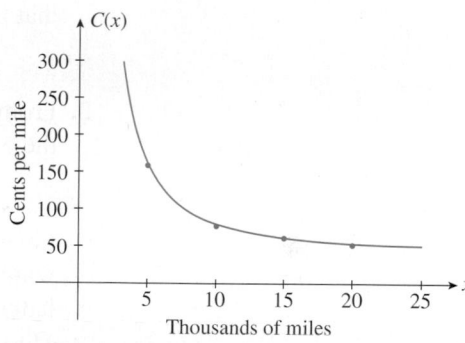
(b) The graph of the model for driving costs

FIGURE **57**

Solution The average cost per mile for driving a car 8000 miles per year is

$$C(8) = \frac{1910.5}{8^{1.72}} + 42.9 \approx 96.3$$

or approximately 96.3 cents per mile. The average cost per mile for driving it 18,000 miles per year is

$$C(18) = \frac{1910.5}{18^{1.72}} + 42.9 \approx 56.1$$

or approximately 56.1 cents per mile.

Constructing Mathematical Models

We close this section by showing how some mathematical models can be constructed by using elementary geometric and algebraic arguments.

The following guidelines can be used to construct mathematical models.

> **Guidelines for Constructing Mathematical Models**
>
> 1. Assign a letter to each variable mentioned in the problem. If appropriate, draw and label a figure.
> 2. Find an expression for the quantity sought.
> 3. Use the conditions given in the problem to write the quantity sought as a function f of one variable. Note any restrictions to be placed on the domain of f from physical considerations of the problem.

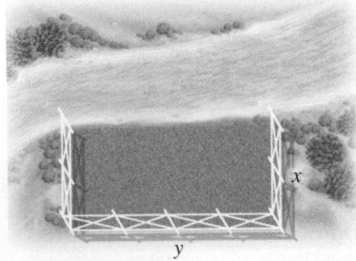

FIGURE **58**
The rectangular grazing land has width x and length y.

APPLIED EXAMPLE 4 Enclosing an Area The owner of Rancho Los Feliz has 3000 yards of fencing with which to enclose a rectangular piece of grazing land along the straight portion of a river. Fencing is not required along the river. Letting x denote the width of the rectangle, find a function f in the variable x giving the area of the grazing land if she uses all of the fencing (Figure 58).

Solution

1. This information was given.
2. The area of the rectangular grazing land is $A = xy$. Next, observe that the amount of fencing is $2x + y$ and this must be equal to 3000, since all the fencing is used; that is,

$$2x + y = 3000$$

3. From the equation, we see that $y = 3000 - 2x$. Substituting this value of y into the expression for A gives

$$A = xy = x(3000 - 2x) = 3000x - 2x^2$$

Finally, observe that both x and y must be nonnegative, since they represent the width and length of a rectangle, respectively. Thus, $x \geq 0$ and $y \geq 0$. But the latter is equivalent to $3000 - 2x \geq 0$, or $x \leq 1500$. So the required function is $f(x) = 3000x - 2x^2$ with domain $0 \leq x \leq 1500$. ■

Note Observe that if we view the function $f(x) = 3000x - 2x^2$ strictly as a mathematical entity, then its domain is the set of all real numbers. But physical considerations dictate that its domain should be restricted to the interval $[0, 1500]$. ■

APPLIED EXAMPLE 5 Charter-Flight Revenue If exactly 200 people sign up for a charter flight, Leisure World Travel Agency charges $300 per person. However, if more than 200 people sign up for the flight (assume that this is the case), then each fare is reduced by $1 for each additional person. Letting x denote the number of passengers beyond 200, find a function giving the revenue realized by the company.

Solution

1. This information was given.
2. If there are x passengers beyond 200, then the number of passengers signing up for the flight is $200 + x$. Furthermore, the fare will be $(300 - x)$ dollars per passenger.

3. The revenue will be

$$R = (200 + x)(300 - x) \qquad \text{Number of passengers} \times$$
$$= -x^2 + 100x + 60{,}000 \qquad \text{the fare per passenger}$$

Clearly, x must be nonnegative, and $300 - x \geq 0$, or $x \leq 300$. So the required function is $f(x) = -x^2 + 100x + 60{,}000$ with domain $[0, 300]$.

2.7 Self-Check Exercise

ENCLOSING A PLAYGROUND The Cunningham Day Care Center wants to enclose a playground of rectangular shape having an area of 500 ft^2 with a wooden fence. Find a function f giving the amount of fencing required in terms of the width x of the rectangular playground.

The solution to Self-Check Exercise 2.7 can be found on page 154.

2.7 Concept Questions

1. Describe mathematical modeling in your own words.

2. Define (a) a polynomial function and (b) a rational function. Give an example of each.

2.7 Exercises

In Exercises 1–6, determine whether the function is a polynomial function, a rational function, or some other function. State the degree of each polynomial function.

1. $f(x) = 3x^6 - 2x^2 + 1$

2. $f(x) = \dfrac{x^2 - 9}{x - 3}$

3. $G(x) = 2(x^2 - 3)^3$

4. $H(x) = 2x^{-3} + 5x^{-2} + 6$

5. $f(t) = 2t^2 + 3\sqrt{t}$

6. $f(r) = \dfrac{6r}{r^3 - 8}$

7. **BOUNCED-CHECK CHARGES** Overdraft fees have become an important part of a bank's total fee income. The following table gives the bank revenue from overdraft fees (in billions of dollars) from 2004 through 2009.

Year, t	0	1	2	3	4	5
Revenue, y	27.5	29	31	34	36	38

where t is measured in years, with $t = 0$ corresponding to 2004. A mathematical model giving the approximate projected bank revenue from overdraft fees over the period under consideration is given by

$$f(t) = 2.19t + 27.12 \qquad (0 \leq t \leq 5)$$

a. Plot the six data points and sketch the graph of the function f on the same set of axes.

b. Assuming that the projection held and the trend continued, what was the projected bank revenue from overdraft fees in 2010 ($t = 6$)?

c. What was the rate of increase of the bank revenue from overdraft fees over the period from 2004 through 2009?
Source: New York Times.

8. **BABY BOOMERS AND SOCIAL SECURITY BENEFITS** Aging baby boomers will put a strain on Social Security benefits unless Congress takes action. The Social Security benefits to be paid out from 2010 through 2040 are projected to be

$$S(t) = 0.1375t^2 + 0.5185t + 0.72 \qquad (0 \leq t \leq 3)$$

where $S(t)$ is measured in trillions of dollars and t is measured in decades, with $t = 0$ corresponding to 2010.

a. What was the amount of Social Security benefits paid out in 2010?

b. What is the amount of Social Security benefits projected to be paid out in 2040?
Source: Social Security and Medicare Trustees' 2010 report.

9. **MOBILE DEVICE USAGE** The average time U.S. adults spent per day on mobile devices (in minutes) for the years 2009 through 2012 is approximated by

$$f(t) = 2.25t^2 + 13.41t + 21.76 \qquad (0 \leq t \leq 3)$$

where $t = 0$ corresponds to 2009.

a. What was the average time U.S. adults spent per day on mobile devices in 2009?

b. If the trend continued through 2013, what was the average time U.S. adults spent per day on mobile devices in 2013?
Source: eMarketer.

10. U.S. GDP The gross domestic product (GDP) of the United States, in trillions of dollars, from 2011 through 2015 is approximately

$$G(t) = 0.064t^2 + 0.473t + 15.0 \qquad (0 \le t \le 4)$$

where t is measured in years, with $t = 0$ corresponding to 2011. In constructing this model, the government used actual GDP figures from 2011 and 2012 and estimates for the years 2013 through 2015.
a. What was the U.S. GDP in 2011?
b. What is the predicted U.S. GDP for 2015?
Source: World Bank.

11. Sub-Saharan African GDP The real GDP per capita of sub-Saharan Africa (in 2009 U.S. dollars) from 1990 through 2030 is projected to be

$$f(t) = 1.86251t^2 - 28.08043t + 884 \qquad (0 \le t \le 40)$$

where t is measured in years, with $t = 0$ corresponding to 1990.
a. What was the real GDP per capita of sub-Saharan Africa in 2000?
b. Assuming that the projection holds true, what will be the GDP per capita of sub-Saharan Africa in 2030?
Source: IMF.

12. U.S. Public Debt The U.S. public debt (the outstanding amount owed by the federal government of the United States from the issue of securities by the U.S. Treasury and other federal government agencies) for the years 2005 through 2011 is modeled by the function

$$f(t) = -0.03817t^3 + 0.4571t^2 - 0.1976t + 8.246$$
$$(0 \le t \le 7)$$

where $f(t)$ is measured in trillions of dollars and t is measured in years, with $t = 0$ corresponding to 2005. What was the U.S. public debt in 2005? In 2008?
Source: U.S. Department of the Treasury.

13. Workers' Expectations The percentage of workers who expect to work past age 65 has more than tripled in 30 years. The function

$$f(t) = 0.004545t^3 - 0.1113t^2 + 1.385t + 11 \quad (0 \le t \le 22)$$

gives an approximation of the percentage of workers who expect to work past age 65 in year t, where t is measured in years, with $t = 0$ corresponding to 1991. What was the percentage of workers who expected to work past age 65 in 1991? In 2013?
Source: PBS News.

14. Aging Drivers The number of fatalities due to car crashes, based on the number of miles driven, begins to climb after the driver is past age 65. Aside from declining ability as one ages, the older driver is more fragile. The number of fatalities per 100 million vehicle miles driven is approximately

$$N(x) = 0.0336x^3 - 0.118x^2 + 0.215x + 0.7 \qquad (0 \le x \le 7)$$

where x denotes the age group of drivers, with $x = 0$ corresponding to those aged 50–54, $x = 1$ corresponding to those aged 55–59, $x = 2$ corresponding to those aged 60–64, ..., and $x = 7$ corresponding to those aged 85–89. What is the fatality rate per 100 million vehicle miles driven for an average driver in the 50–54 age group? In the 85–89 age group?
Source: U.S. Department of Transportation.

15. Total Global Mobile Data Traffic In a 2009 report, equipment maker Cisco forecast the total global mobile data traffic to be

$$f(t) = 0.021t^3 + 0.015t^2 + 0.12t + 0.06 \qquad (0 \le t \le 5)$$

million terabytes/month in year t, where $t = 0$ corresponds to 2009.
a. What was the total global mobile data traffic in 2009?
b. According to Cisco, what will the total global mobile data traffic be in 2014?
Source: Cisco.

16. Leveraged Return Leanne is contemplating borrowing money from a bank to buy a bond returning 6%/year. The bank requires her to make a down payment of D% of the loan with the remaining $(100 - D)$% borrowed at an interest rate of 5%/year. Then the return on the bond using the borrowed money, called *leveraged return*, is given by

$$L = \frac{1 + 0.05D}{D}$$

a. If Leanne makes a down payment of 20% to secure the loan on the bond using borrowed money, what is the leveraged return?
b. If Leanne makes a down payment of 10% to secure the loan on the bond using borrowed money, what is the leveraged return?
Source: Scientific American.

17. Online Video Viewers As broadband Internet grows more popular, video services such as YouTube will continue to expand. The number of online video viewers (in millions) is projected to grow according to the rule

$$N(t) = 52t^{0.531} \qquad (1 \le t \le 10)$$

where $t = 1$ corresponds to 2003.
a. Sketch the graph of N.
b. How many online video viewers were there in 2012?
Source: eMarketer.com.

18. Infant Mortality Rates in Massachusetts The deaths of children younger than 1 year old per 1000 live births is modeled by the function

$$R(t) = 162.8t^{-3.025} \qquad (1 \le t \le 3)$$

where t is measured in 50-year intervals, with $t = 1$ corresponding to 1900.
a. Find $R(1)$, $R(2)$, and $R(3)$, and use your result to sketch the graph of the function R over the domain $[1, 3]$.
b. What was the infant mortality rate in 1900? In 1950? In 2000?
Source: Massachusetts Department of Public Health.

7 - 35 odd

19. OUTSOURCING OF JOBS According to a study conducted in 2003, the total number of U.S. jobs (in millions) that are projected to leave the country by year t, where $t = 0$ corresponds to 2000, is

$$N(t) = 0.0018425(t + 5)^{2.5} \qquad (0 \le t \le 15)$$

What was the projected number of outsourced jobs for 2005 ($t = 5$)? For 2013 ($t = 13$)?
Source: Forrester Research.

20. CHIP SALES The worldwide sales of flash memory chips (in billions of dollars) is approximated by

$$S(t) = 4.3(t + 2)^{0.94} \qquad (0 \le t \le 6)$$

where t is measured in years, with $t = 0$ corresponding to 2002. Flash chips are used in cell phones, digital cameras, and other products.
a. What were the worldwide flash memory chip sales in 2002?
b. What were the estimated sales for 2010?
Source: Web-Feet Research, Inc.

21. REACTION OF A FROG TO A DRUG Experiments conducted by A. J. Clark suggest that the response $R(x)$ of a frog's heart muscle to the injection of x units of acetylcholine (as a percent of the maximum possible effect of the drug) may be approximated by the rational function

$$R(x) = \frac{100x}{b + x} \qquad (x \ge 0)$$

where b is a positive constant that depends on the particular frog.
a. If a concentration of 40 units of acetylcholine produces a response of 50% for a certain frog, find the "response function" for this frog.
b. Using the model found in part (a), find the response of the frog's heart muscle when 60 units of acetylcholine are administered.

22. WALKING VERSUS RUNNING The oxygen consumption (in milliliters per pound per minute) for a person walking at x mph is approximated by the function

$$f(x) = \frac{5}{3}x^2 + \frac{5}{3}x + 10 \qquad (0 \le x \le 9)$$

whereas the oxygen consumption for a runner at x mph is approximated by the function

$$g(x) = 11x + 10 \qquad (4 \le x \le 9)$$

a. Sketch the graphs of f and g.
b. At what speed is the oxygen consumption the same for a walker as it is for a runner? What is the level of oxygen consumption at that speed?
c. What happens to the oxygen consumption of the walker and the runner at speeds beyond that found in part (b)?
Source: William McArdley, Frank Katch, and Victor Katch, *Exercise Physiology.*

23. U.S. HEALTH-CARE COSTS The U.S. health-care costs per capita (in dollars) from 2001 through 2011 can be approximated by the linear function $f(t) = at + b$, where t is measured in years, with $t = 1$ corresponding to 2001, and a and b are constants. The costs per capita in 2001 and 2011 were \$5240 and \$8680, respectively.
a. Find a and b.
b. Use the model obtained in part (a) to find the approximate per capita costs for 2005.
Source: Centers for Medicare and Medicaid Services.

24. FARMERS MARKETS Farmers markets have been growing steadily over the years. The number of such markets in the United States from 2006 through 2012 can be modeled by using a quadratic function of the form

$$f(t) = at^2 + bt + c$$

where a, b, and c are constants and t is measured in years, with $t = 0$ corresponding to 2006.
a. Find a, b, and c if $f(0) = 3173$, $f(4) = 6132$, and $f(6) = 7864$.
b. Use the model obtained in part (a) to estimate the number of farmers markets in 2014, assuming that the trend continued.
Source: U.S. Department of Agriculture.

25. SMALL BREWERIES U.S. craft-beer breweries (breweries that make fewer than 6 million barrels annually and are less than 25% owned by big breweries) have been doing booming business. The number of these small breweries from 2008 through 2012 can be modeled by using a quadratic function of the form

$$f(t) = at^2 + bt + c$$

where a, b, and c are constants and t is measured in years, with $t = 0$ corresponding to 2008.
a. Find a, b, and c if $f(0) = 1547$, $f(2) = 1802$, and $f(4) = 2403$.
b. Use the model obtained in part (a) to estimate the number of craft-beer breweries in 2014, assuming that the trend continued.
Source: Breweries Association.

26. PRICE OF IVORY According to the World Wildlife Fund, a group in the forefront of the fight against illegal ivory trade, the price of ivory (in dollars per kilogram) compiled from a variety of legal and black market sources is approximated by the function

$$f(t) = \begin{cases} 8.37t + 7.44 & \text{if } 0 \le t \le 8 \\ 2.84t + 51.68 & \text{if } 8 < t \le 30 \end{cases}$$

where t is measured in years, with $t = 0$ corresponding to the beginning of 1970.
a. Sketch the graph of the function f.
b. What was the price of ivory at the beginning of 1970? At the beginning of 1990?
Source: World Wildlife Fund.

27. **COST OF THE HEALTH-CARE BILL** The Congressional Budget Office estimates that the health-care bill passed by the Senate in November 2009, combined with a package of revisions known as the reconciliation bill, will result in a cost by year t (in billions of dollars) of

$$f(t) = \begin{cases} 5 & \text{if } 0 \le t < 2 \\ -0.5278t^3 + 3.012t^2 + 49.23t - 103.29 & \text{if } 2 \le t \le 8 \end{cases}$$

where t is measured in years, with $t = 0$ corresponding to 2010. What will be the cost of the health-care bill by 2011? By 2015?
Source: U.S. Congressional Budget Office.

28. **WORKING-AGE POPULATION** The ratio of working-age population to the elderly in the United States (including projections after 2000) is given by

$$f(t) = \begin{cases} 4.1 & \text{if } 0 \le t < 5 \\ -0.03t + 4.25 & \text{if } 5 \le t < 15 \\ -0.075t + 4.925 & \text{if } 15 \le t \le 35 \end{cases}$$

with $t = 0$ corresponding to the beginning of 1995.
   a. Sketch the graph of f.
   b. What was the ratio at the beginning of 2005? What will the ratio be at the beginning of 2020?
   c. Over what years is the ratio constant?
   d. Over what years is the decline of the ratio greatest?
Source: U.S. Census Bureau.

29. **DEMAND FOR SMOKE ALARMS** The demand function for the Sentinel smoke alarm is given by

$$p = \frac{30}{0.02x^2 + 1} \qquad (0 \le x \le 10)$$

where x (measured in units of a thousand) is the quantity demanded per week and p is the unit price in dollars.
   a. Sketch the graph of the demand function.
   b. What is the unit price that corresponds to a quantity demanded of 10,000 units?

30. **SUPPLY OF SATELLITE RADIOS** Suppliers of satellite radios will market 10,000 units when the unit price is $20 and 62,500 units when the unit price is $35. Determine the supply function if it is known to have the form

$$p = a\sqrt{x} + b \qquad (a > 0, b > 0)$$

where x is the quantity supplied and p is the unit price in dollars. Sketch the graph of the supply function. What unit price will induce the supplier to make 40,000 satellite radios available in the marketplace?

31. **DEMAND FOR COMMODITIES** Assume that the demand function for a certain commodity has the form

$$p = \sqrt{-ax^2 + b} \qquad (a \ge 0, b \ge 0)$$

where x is the quantity demanded, measured in units of a thousand and p is the unit price in dollars. Suppose the

quantity demanded is 6000 ($x = 6$) when the unit price is $8 and 8000 ($x = 8$) when the unit price is $6. Determine the demand equation. What is the quantity demanded when the unit price is set at $7.50?

32. **SUPPLY AND DEMAND EQUATIONS** Suppose the demand and supply equations for a certain commodity are given by $p = ax + b$ and $p = cx + d$, respectively, where $a < 0$, $c > 0$, and $b > d > 0$ (see the figure below).
   a. Find the equilibrium quantity and equilibrium price in terms of a, b, c, and d.
   b. Use part (a) to determine what happens to the market equilibrium if c is increased while a, b, and d remain fixed. Interpret your answer in economic terms.
   c. Use part (a) to determine what happens to the market equilibrium if b is decreased while a, c, and d remain fixed. Interpret your answer in economic terms.

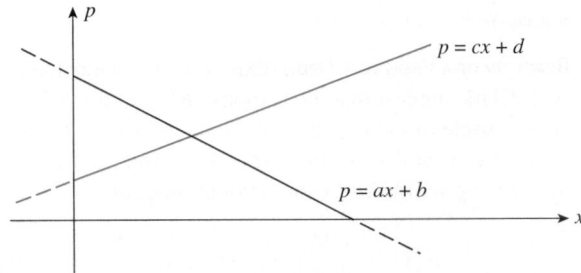

33. **ENCLOSING AN AREA** Patricia wishes to have a rectangular garden in her backyard. She has 80 ft of fencing with which to enclose her garden. Letting x denote the width of the garden, find a function f in the variable x giving the area of the garden. What is its domain?

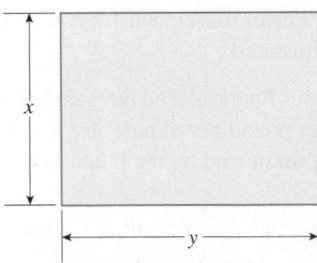

34. **ENCLOSING AN AREA** Juanita wishes to have a rectangular garden in her backyard with an area of 250 ft^2. Letting x denote the width of the garden, find a function f in the variable x giving the length of the fencing required to construct the garden. What is the domain of the function? Hint: Refer to the figure for Exercise 33. The amount of fencing required is equal to the perimeter of the rectangle, which is twice the width plus twice the length of the rectangle.

35. **PACKAGING** By cutting away identical squares from each corner of a rectangular piece of cardboard and folding up the resulting flaps, an open box can be made. If the cardboard is 15 in. long and 8 in. wide and the square

cutaways have dimensions of x in. by x in., find a function giving the volume of the resulting box.

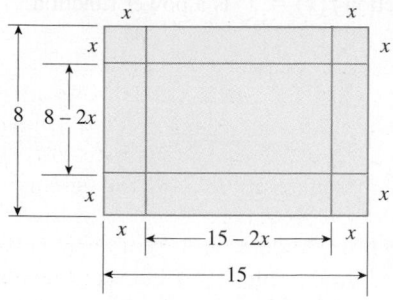

36. Packaging Costs A rectangular box is to have a square base and a volume of 20 ft³. The material for the base costs 30¢/ft², the material for the sides costs 10¢/ft², and the material for the top costs 20¢/ft². Letting x denote the length of one side of the base, find a function in the variable x giving the cost of constructing the box.

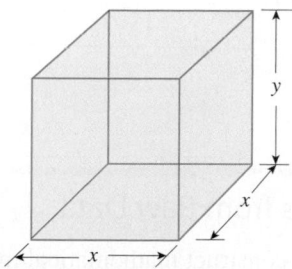

37. Area of a Norman Window A Norman window has the shape of a rectangle surmounted by a semicircle (see the accompanying figure). Suppose a Norman window is to have a perimeter of 28 ft. Find a function in the variable x giving the area of the window.

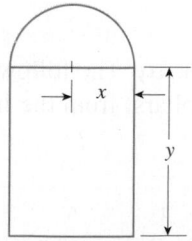

38. Yield of an Apple Orchard An apple orchard has an average yield of 36 bushels of apples per tree if tree density is 22 trees/acre. For each unit increase in tree density, the yield decreases by 2 bushels/tree. Letting x denote the number of trees beyond 22/acre, find a function in x that gives the yield of apples.

39. Book Design A book designer has decided that the pages of a book should have 1-in. margins at the top and bottom and $\frac{1}{2}$-in. margins on the sides. She further stipulated that

each page should have a total area of 50 in.². Find a function in the variable x, giving the area of the printed part of the page. What is the domain of the function?

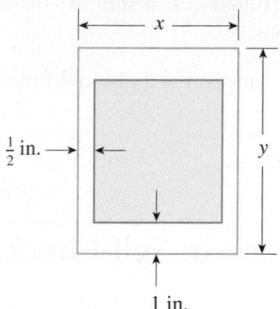

40. Profit of a Vineyard Phillip, the proprietor of a vineyard, estimates that if 10,000 bottles of wine were produced this season, then the profit would be $5/bottle. But if more than 10,000 bottles were produced, then the profit per bottle for the entire lot would drop by $0.0002 for each additional bottle sold. Assume that at least 10,000 bottles of wine are produced and sold, and let x denote the number of bottles produced and sold in excess of 10,000.
  a. Find a function P giving the profit in terms of x.
  b. What is the profit Phillip can expect from the sale of 16,000 bottles of wine from his vineyard?

41. Charter Revenue The owner of a luxury motor yacht that sails among the 4000 Greek islands charges $600/person per day if exactly 20 people sign up for the cruise. However, if more than 20 people sign up for the cruise (up to the maximum capacity of 90), the fare for all the passengers is reduced by $4/person for each additional passenger. Assume that at least 20 people sign up for the cruise, and let x denote the number of passengers above 20.
  a. Find a function R giving the revenue per day realized from the charter.
  b. What is the revenue per day if 60 people sign up for the cruise?
  c. What is the revenue per day if 80 people sign up for the cruise?

42. Oil Spills The oil spilling from the ruptured hull of a grounded tanker spreads in all directions in calm waters. Suppose the polluted area is a circle of radius r ft and the radius is increasing at the rate of 2 ft/sec.
  a. Find a function f giving the polluted area in terms of r.
  b. Find a function g giving the radius of the polluted area in terms of t.
  c. Find a function h giving the polluted area in terms of t.
  d. What is the size of the polluted area 30 sec after the hull was ruptured?

In Exercises 43–46, determine whether the statement is true or false. If it is true, explain why it is true. If it is false, give an example to show why it is false.

43. A polynomial function is a sum of constant multiples of power functions.

44. A polynomial function is a rational function, but the converse is false.

45. If $r > 0$, then the power function $f(x) = x^r$ is defined for all values of x.

46. The function $f(x) = 2^x$ is a power function.

2.7 Solution to Self-Check Exercise

Let the length of the rectangular playground be y ft (see the figure).

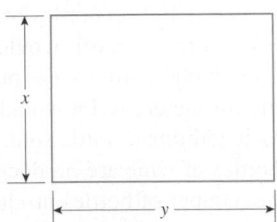

Then, the amount of fencing required is $L = 2x + 2y$. But the requirement that the area of the rectangular playground be 500 ft^2 implies that $xy = 500$, or upon solving for y, $y = 500/x$. Therefore, the amount of fencing required is

$$L = f(x) = 2x + 2\left(\frac{500}{x}\right) = 2x + \frac{1000}{x}$$

with domain $(0, \infty)$.

USING TECHNOLOGY — Constructing Mathematical Models from Raw Data

A graphing utility can sometimes be used to construct mathematical models from sets of data. For example, if the points corresponding to the given data are scattered about a straight line, then use **LinReg($ax+b$)** (linear regression) from the statistical calculations menu of the graphing utility to obtain a function (model) that approximates the data at hand. If the points seem to be scattered along a parabola (the graph of a quadratic function), then use **QuadReg** (second-degree polynomial regression), and so on. (These are functions on the TI-83/84 calculator.)

 APPLIED EXAMPLE 1 Indian Gaming Industry The following table gives the estimated gross revenues (in billions of dollars) from the Indian gaming industries from 2000 ($t = 0$) to 2008 ($t = 8$).

Year	0	1	2	3	4	5	6	7	8
Revenue	11.0	12.8	14.7	16.8	19.5	22.7	25.1	26.4	26.8

a. Use a graphing utility to find a polynomial function f of degree 4 that models the data.

b. Plot the graph of the function f, using the viewing window $[0, 8] \times [0, 30]$.

c. Use the function evaluation capability of the graphing utility to compute $f(0)$, $f(1), \ldots, f(8)$, and compare these values with the original data.

d. If the trend continued, what was the gross revenue for 2009 ($t = 9$)?

Source: National Indian Gaming Association.

Solution

a. First, enter the data using the statistical menu. Then choose **QuartReg** (fourth-degree polynomial regression) from the statistical calculations menu of a graphing utility. We find

$$f(t) = -0.00737t^4 + 0.0655t^3 - 0.008t^2 + 1.61t + 11$$

b. The graph of f is shown in Figure T1.

c. The required values, which compare favorably with the given data, follow:

t	0	1	2	3	4	5	6	7	8
$f(t)$	11.0	12.7	14.6	16.9	19.6	22.4	25.0	26.6	26.7

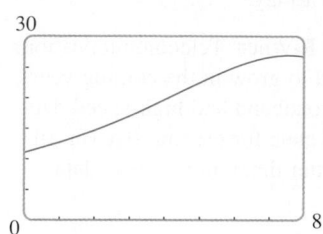

FIGURE T1
The graph of f in the viewing window
$[0, 8] \times [0, 30]$

d. The gross revenue for 2009 ($t = 9$) is given by

$$f(9) = -0.00737(9)^4 + 0.0655(9)^3 - 0.008(9)^2 + 1.61(9) + 11 \approx 24.24$$

or approximately \$24.2 billion. ∎

TECHNOLOGY EXERCISES

In Exercises 1–12, use the statistical calculations menu to construct a mathematical model associated with the given data.

1. Consumption of Bottled Water The annual per-capita consumption of bottled water (in gallons) and the scatter plot for these data follow:

Year	2001	2002	2003	2004	2005	2006
Consumption	18.8	20.9	22.4	24	26.1	28.3

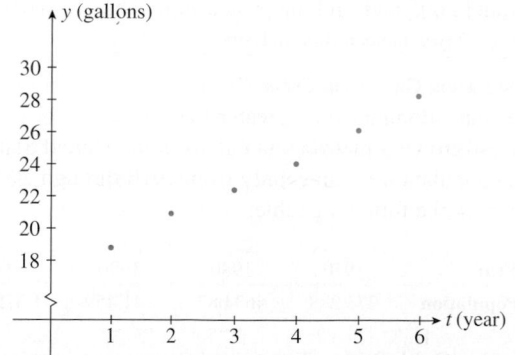

a. Use **LinReg(ax+b)** to find a first-degree (linear) polynomial regression model for the data. Let $t = 1$ correspond to 2001.

b. Plot the graph of the function f found in part (a), using the viewing window $[1, 6] \times [0, 30]$.

c. Compute the values for $t = 1, 2, 3, 4, 5$, and 6. How do your figures compare with the given data?

d. If the trend continued, what was the annual per-capita consumption of bottled water in 2008?

Source: Beverage Marketing Corporation.

2. Web Conferencing Web conferencing is a big business, and it's growing rapidly. The amount (in billions of dollars) spent on Web conferencing from the beginning of 2003 through 2010, and the scatter diagram for these data follow:

Year	2003	2004	2005	2006	2007	2008	2009	2010
Amount	0.50	0.63	0.78	0.92	1.16	1.38	1.60	1.90

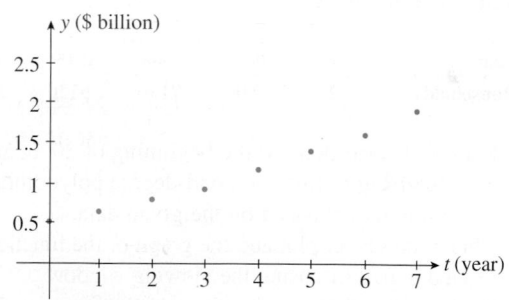

a. Let $t = 0$ correspond to the beginning of 2003 and use **QuadReg** to find a second-degree polynomial regression model based on the given data.

b. Plot the graph of the function f found in part (a) using the window $[0, 7] \times [0, 2]$.

c. Compute $f(0), f(3), f(6)$, and $f(7)$. Compare these values with the given data.

Source: Gartner Dataquest.

3. **STUDENT POPULATION** The projected total number of students in elementary schools, secondary schools, and colleges (in millions) from 1995 through 2015 is given in the following table:

Year	1995	2000	2005	2010	2015
Number	64.8	68.7	72.6	74.8	78

a. Use **QuadReg** to find a second-degree polynomial regression model for the data. Let t be measured in 5-year intervals, with $t = 0$ corresponding to the beginning of 1995.

b. Plot the graph of the function f found in part (a), using the viewing window $[0, 4] \times [0, 85]$.

c. Using the model found in part (a), what will be the projected total number of students (all categories) enrolled in 2015?

Source: U.S. National Center for Education Statistics.

4. **MOBILE DEVICE USAGE** The average time U.S. adults spent per day on mobile devices (in minutes) for the years 2009 through 2012 is shown in the following table:

Year	2009	2010	2011	2012
Average Time Spent	22	36.7	58.3	82

a. Let $t = 0$ correspond to the beginning of 2009, and use **QuadReg** to find a second-degree polynomial regression model based on the given data.

b. Obtain the scatter plot and the graph of the function f found in part (a), using the viewing window $[0, 3] \times [0, 100]$.

Source: eMarketer.

5. **TiVo OWNERS** The projected number of households (in millions) with digital video recorders that allow viewers to record shows onto a server and skip commercials are given in the following table:

Year	2006	2007	2008	2009	2010
Households	31.2	49.0	71.6	97.0	130.2

a. Let $t = 0$ correspond to the beginning of 2006, and use **QuadReg** to find a second-degree polynomial regression model based on the given data.

b. Obtain the scatter plot and the graph of the function f found in part (a), using the viewing window $[0, 4] \times [0, 140]$.

Source: Strategy Analytics.

6. **U.S. PUBLIC DEBT** The U.S. public debt (the outstanding amount owed by the federal government of the U.S. from the issue of securities by the U.S. Treasury and other federal government agencies) for the years 2005 through 2011 (in trillions of dollars) is given in the following table:

Year	2005	2006	2007	2008	2009	2010	2011
Debt	8.170	8.680	9.229	10.700	12.311	14.025	15.223

a. Let $t = 0$ correspond to the beginning of 2005 and use **CubicReg** to find a third-degree polynomial regression model based on the given data.

b. Obtain the scatter plot and the graph of the function f found in part (a), using the viewing window $[0, 7] \times [0, 18]$.

Source: U.S. Department of the Treasury.

7. **TELECOMMUNICATIONS INDUSTRY REVENUE** Telecommunications industry revenue is expected to grow in the coming years, fueled by the demand for broadband and high-speed data services. The worldwide revenue for the industry (in trillions of dollars) and the scatter diagram for these data follow:

Year	2000	2002	2004	2006	2008	2010
Revenue	1.7	2.0	2.5	3.0	3.6	4.2

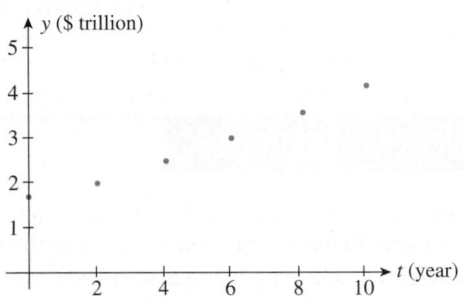

a. Let $t = 0$ correspond to the beginning of 2000 and use **CubicReg** to find a third-degree polynomial regression model based on the given data.

b. Plot the graph of the function f found in part (a), using the viewing window $[0, 10] \times [0, 5]$.

c. Find the worldwide revenue for the industry in 2001 and 2005 and find the projected revenue for 2010.

Source: Telecommunication Industry Association.

8. **POPULATION GROWTH IN CLARK COUNTY** Clark County in Nevada—dominated by greater Las Vegas—is one of the fastest-growing metropolitan areas in the United States. The population of the county from 1970 through 2000 is given in the following table:

Year	1970	1980	1990	2000
Population	273,288	463,087	741,459	1,375,765

a. Use **CubicReg** to find a third-degree polynomial regression model for the data. Let t be measured in decades, with $t = 0$ corresponding to the beginning of 1970.

b. Plot the graph of the function f found in part (a), using the viewing window $[0, 3] \times [0, 1,500,000]$.

c. Compare the values of f at $t = 0, 1, 2,$ and 3, with the given data.

Source: U.S. Census Bureau.

9. **Lobbyists' Spending** Lobbyists try to persuade legislators to propose, pass, or defeat legislation or to change existing laws. The amount (in billions of dollars) spent by lobbyists from 2003 through 2009 is shown in the following table:

Year	2003	2004	2005	2006	2007	2008	2009
Amount	8.0	8.5	9.7	10.2	11.3	12.9	13.8

   a. Use **CubicReg** to find a third-degree polynomial regression model for the data, letting $t = 0$ correspond to 2003.
   b. Plot the scatter diagram and the graph of the function f found in part (a), using the viewing window $[0, 6] \times [0, 15]$.
   c. Compare the values of f at $t = 0$, 3, and 6 with the given data.
   Source: Center for Public Integrity.

10. **Mobile Enterprise IM Accounts** The projected number of mobile enterprise instant messaging (IM) accounts (in millions) from 2006 through 2010 is given in the following table ($t = 0$ corresponds to the beginning of 2006):

Year	0	1	2	3	4
Accounts	2.3	3.6	5.8	8.7	14.9

   a. Use **CubicReg** to find a third-degree polynomial regression model based on the given data.
   b. Plot the graph of the function f found in part (a), using the viewing window $[0, 5] \times [0, 16]$.
   c. Compute $f(0), f(1), f(2), f(3)$, and $f(4)$.
   Source: The Radical Group.

11. **Nicotine Content of Cigarettes** Even as measures to discourage smoking have been growing more stringent in recent years, the nicotine content of cigarettes has been rising, making it more difficult for smokers to quit. The following table gives the average amount of nicotine in cigarette smoke (in milligrams) from 1999 through 2004:

Year	1999	2000	2001	2002	2003	2004
Yield per Cigarette	1.71	1.81	1.85	1.84	1.83	1.89

   a. Use **QuartReg** to find a fourth-degree polynomial regression model for the data. Let $t = 0$ correspond to the beginning of 1999.
   b. Plot the graph of the function f found in part (a), using the viewing window $[0, 5] \times [0, 2]$.
   c. Compute the values of $f(t)$ for $t = 0$, 1, 2, 3, 4, and 5.
   d. If the trend continued, what was the average amount of nicotine in cigarettes in 2005?
   Source: Massachusetts Tobacco Control Program.

12. **Social Security Trust Fund Assets** The projected assets of the Social Security trust fund (in trillions of dollars) from 2010 through 2033 are given in the following table:

Year	2010	2015	2020	2025	2030	2033
Assets	2.61	2.68	2.44	1.87	0.78	0

   Use **QuartReg** to find a fourth-degree polynomial regression model for the data. Let $t = 0$ correspond to 2010.
   Source: Social Security Administration.

2.8 The Method of Least Squares (Optional)

The Method of Least Squares

In Example 6, Section 2.2, we saw how a linear equation may be used to approximate the sales trend for a local sporting goods store. The *trend line*, as we saw, may be used to predict the store's future sales. Recall that we obtained the trend line in Example 6 by requiring that the line pass through two data points, the rationale being that such a line seems to *fit* the data reasonably well.

In this section, we describe a general method known as the **method of least squares** for determining a straight line that, in some sense, best fits a set of data points when the points are scattered about a straight line. To illustrate the principle behind the method of least squares, suppose, for simplicity, that we are given five data points,

$$P_1(x_1, y_1), \ P_2(x_2, y_2), \ P_3(x_3, y_3), \ P_4(x_4, y_4), \ P_5(x_5, y_5)$$

describing the relationship between the two variables x and y. By plotting these data points, we obtain a graph called a **scatter diagram** (Figure 59).

If we try to *fit* a straight line to these data points, the line will miss the first, second, third, fourth, and fifth data points by the amounts d_1, d_2, d_3, d_4, and d_5, respectively

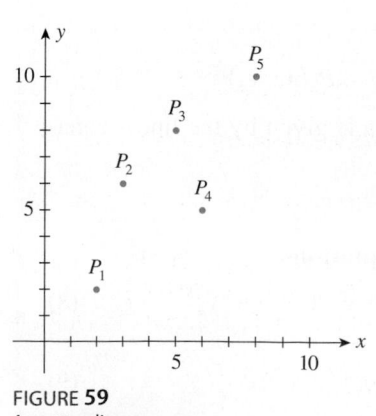

FIGURE 59
A scatter diagram

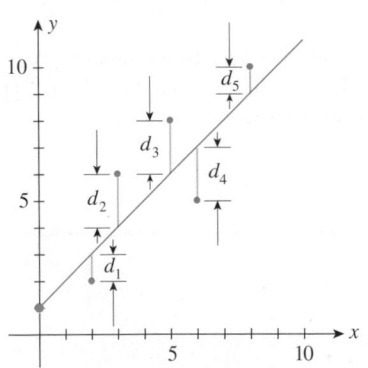

FIGURE 60
d_i is the vertical distance between the straight line and a given data point.

(Figure 60). We can think of the amounts $d_1, d_2, \ldots, d_5$ as the errors made when the values $y_1, y_2, \ldots, y_5$ are approximated by the corresponding values of y lying on the straight line L.

The **principle of least squares** states that the straight line L that fits the data points *best* is the one chosen by requiring that the sum of the squares of $d_1, d_2, \ldots, d_5$— that is,

$$d_1^2 + d_2^2 + d_3^2 + d_4^2 + d_5^2$$

be made as small as possible. In other words, the least-squares criterion calls for minimizing the sum of the squares of the errors. The line L obtained in this manner is called the **least-squares line,** or *regression line.*

The method for computing the least-squares lines that best fits a set of data points follows. (We omit the proof.)

The Method of Least Squares

Suppose we are given n data points

$$P_1(x_1, y_1), P_2(x_2, y_2), P_3(x_3, y_3), \ldots, P_n(x_n, y_n)$$

Then the least-squares (regression) line for the data is given by the linear equation (function)

$$y = f(x) = mx + b$$

where the constants m and b satisfy the **normal equations**

$$nb + (x_1 + x_2 + \cdots + x_n)m = y_1 + y_2 + \cdots + y_n \tag{8}$$

$$(x_1 + x_2 + \cdots + x_n)b + (x_1^2 + x_2^2 + \cdots + x_n^2)m$$
$$= x_1 y_1 + x_2 y_2 + \cdots + x_n y_n \tag{9}$$

simultaneously.

EXAMPLE 1 Find the least-squares line for the data

$$P_1(1, 1), P_2(2, 3), P_3(3, 4), P_4(4, 3), P_5(5, 6)$$

Solution   Here, we have $n = 5$ and

$$x_1 = 1 \qquad x_2 = 2 \qquad x_3 = 3 \qquad x_4 = 4 \qquad x_5 = 5$$
$$y_1 = 1 \qquad y_2 = 3 \qquad y_3 = 4 \qquad y_4 = 3 \qquad y_5 = 6$$

Before using Equations (8) and (9), it is convenient to summarize these data in the form of a table:

	x	y	x^2	xy
	1	1	1	1
	2	3	4	6
	3	4	9	12
	4	3	16	12
	5	6	25	30
Sum	15	17	55	61

Using this table and (8) and (9), we obtain the normal equations

$$5b + 15m = 17 \tag{10}$$
$$15b + 55m = 61 \tag{11}$$

Solving Equation (10) for b gives

$$b = -3m + \frac{17}{5} \tag{12}$$

which, upon substitution into Equation (11), gives

$$15\left(-3m + \frac{17}{5}\right) + 55m = 61$$
$$-45m + 51 + 55m = 61$$
$$10m = 10$$
$$m = 1$$

Substituting this value of m into Equation (12) gives

$$b = -3 + \frac{17}{5} = \frac{2}{5} = 0.4$$

Therefore, the required least-squares line is

$$y = x + 0.4$$

The scatter diagram and the least-squares line are shown in Figure 61.   ▪

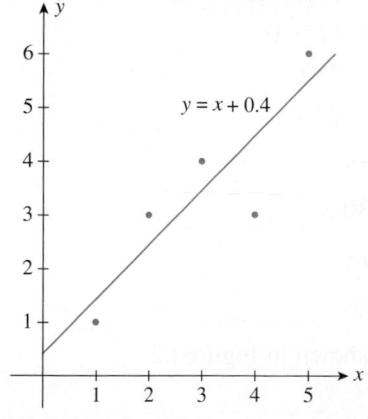

FIGURE 61
The scatter diagram and the least-squares line $y = x + 0.4$

$ **APPLIED EXAMPLE 2** Advertising and Profit   The proprietor of Leisure Travel Service compiled the following data relating the firm's annual profit to its annual advertising expenditure (both measured in thousands of dollars):

Annual Advertising Expenditure, x	12	14	17	21	26	30
Annual Profit, y	60	70	90	100	100	120

a. Determine the equation of the least-squares line for these data.

b. Draw a scatter diagram and the least-squares line for these data.

c. Use the result obtained in part (a) to predict Leisure Travel's annual profit if the annual advertising budget is $20,000.

Solution

a. The calculations required for obtaining the normal equations are summarized in the following table:

	x	y	x^2	xy
	12	60	144	720
	14	70	196	980
	17	90	289	1,530
	21	100	441	2,100
	26	100	676	2,600
	30	120	900	3,600
Sum	120	540	2646	11,530

The normal equations are

$$6b + 120m = 540 \tag{13}$$

$$120b + 2646m = 11,530 \tag{14}$$

Solving Equation (13) for b gives

$$b = -20m + 90 \tag{15}$$

which, upon substitution into Equation (14), gives

$$120(-20m + 90) + 2646m = 11,530$$
$$-2400m + 10,800 + 2646m = 11,530$$
$$246m = 730$$
$$m \approx 2.97$$

Substituting this value of m into Equation (15) gives

$$b = -20(2.97) + 90 = 30.6$$

Therefore, the required least-squares line is given by

$$y = f(x) = 2.97x + 30.6$$

b. The scatter diagram and the least-squares line are shown in Figure 62.

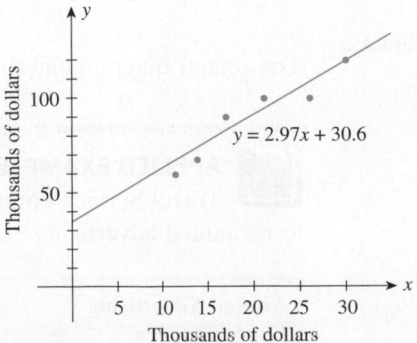

FIGURE 62
Profit versus advertising expenditure

c. Leisure Travel's predicted annual profit corresponding to an annual budget of $20,000 is given by

$$f(20) = 2.97(20) + 30.6 = 90$$

or $90,000.

■

APPLIED EXAMPLE 3 U.S. Health-Care Expenditures Because the over-65 population will be growing more rapidly in the next few decades, health-care spending is expected to increase significantly in the coming decades. The following table gives the projected U.S. health expenditures (in trillions of dollars) from 2013 through 2018, where t is measured in years, with $t = 0$ corresponding to 2013.

Year, t	0	1	2	3	4	5
Expenditure, y	2.91	3.23	3.42	3.63	3.85	4.08

Find a function giving the U.S. health-care spending between 2013 and 2018, using the least-squares technique.
Source: Centers for Medicare & Medicaid Services.

Solution The calculations required for obtaining the normal equations are summarized in the following table:

t	y	t^2	ty
0	2.91	0	0
1	3.23	1	3.23
2	3.42	4	6.84
3	3.63	9	10.89
4	3.85	16	15.40
5	4.08	25	20.40
15	21.12	55	56.76

The normal equations are

$$6b + 15m = 21.12 \qquad \textbf{(16)}$$
$$15b + 55m = 56.76 \qquad \textbf{(17)}$$

Solving Equation (16) for b gives

$$6b = -15m + 21.12$$
$$b = -2.5m + 3.52 \qquad \textbf{(18)}$$

which, upon substitution into Equation (17), gives

$$15(-2.5m + 3.52) + 55m = 56.76$$
$$-37.5m + 52.80 + 55m = 56.76$$
$$17.5m = 3.96$$
$$m \approx 0.2263$$

Substituting this value of m into Equation (18) gives

$$b \approx -2.5(0.2263) + 3.52 \approx 2.954$$

Therefore, the required function is

$$S(t) = 0.226t + 2.954$$

The scatter diagram and the least-squares lines are shown in Figure 63.

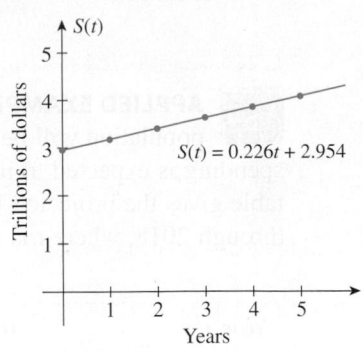

FIGURE **63**

2.8 Self-Check Exercise

Box-Office-Hit DVD Sales In a market research study for Century Communications, the following data were provided based on the projected monthly sales x (in thousands) of a DVD version of a box-office-hit adventure movie with a proposed wholesale unit price of p dollars.

x	2.2	5.4	7.0	11.5	14.6
p	38.0	36.0	34.5	30.0	28.5

Find the demand equation if the demand curve is the least-squares line for these data.

Solutions to Self-Check Exercise 2.8 can be found on page 165.

2.8 Concept Questions

1. Explain the terms (a) *scatter diagram* and (b) *least-squares line*.

2. Explain the principle of least squares in your own words.

2.8 Exercises

In Exercises 1–6, (a) find the equation of the least-squares line for the data, and (b) draw a scatter diagram for the data and graph the least-squares line.

1.

x	1	2	3	4
y	4	6	8	11

2.

x	1	3	5	7	9
y	9	8	6	3	2

3.

x	1	2	3	4	4	6
y	4.5	5	3	2	3.5	1

4.

x	1	1	2	3	4	4	5
y	2	3	3	3.5	3.5	4	5

5. $P_1(1, 3), P_2(2, 5), P_3(3, 5), P_4(4, 7), P_5(5, 8)$

6. $P_1(1, 8), P_2(2, 6), P_3(5, 6), P_4(7, 4), P_5(10, 1)$

7. **SAT VERBAL SCORES** The superintendent of schools in a large metropolitan area compiled data on the students' performance on standardized tests. The following table shows the average SAT verbal scores of high school seniors during the 5 years since the district implemented its "back to basics" program:

Year, x	1	2	3	4	5
Average Score, y	436	438	428	430	426

   a. Determine the equation of the least-squares line for these data.
   b. Draw a scatter diagram and the least-squares line for these data.
   c. Assuming that the trend continued, estimate the average SAT verbal score of high school seniors 2 years from now ($x = 7$).

8. **NET SALES** The management of Kaldor, a manufacturer of electric motors, submitted the accompanying data in its annual stockholders report. The following table shows the net sales (in millions of dollars) during the 5 years that have elapsed since the new management team took over:

Year, x	1	2	3	4	5
Net Sales, y	426	437	460	473	477

(The first year the firm operated under the new management corresponds to the time period $x = 1$, and the four subsequent years correspond to $x = 2, 3, 4,$ and 5.)

   a. Determine the equation of the least-squares line for these data.
   b. Draw a scatter diagram and the least-squares line for these data.
   c. Use the result obtained in part (a) to predict the net sales for the upcoming year.

9. **FACEBOOK USERS** End-of-year data for the number of Facebook users (in millions) from 2008 through 2011 are given in the following table:

Year	2008	2009	2010	2011
Number, y	154.5	381.8	654.5	845.0

   a. Letting $x = 0$ denote the end of 2008, find an equation of the least-squares line for these data.
   b. Use the result of part (a) to project the number of Facebook users at the end of 2015, assuming that the trend continues.
   Source: Company reports.

10. **COST OF SUMMER BLOCKBUSTERS** Hollywood is spending more and more to produce its big summer movies each year. The estimated costs of summer big-budget releases (in billions of dollars) for the years 2011 through 2013 are given in the following table:

Year	2011	2012	2013
Spending, y	2.1	2.4	2.7

   a. Letting $x = 1$ denote 2011, find an equation of the least-squares line for these data.
   b. Use the result of part (a) to estimate the amount of money Hollywood will spend in 2015 to produce its big summer movies for that year, assuming the trend continues.
   Source: Los Angeles Times.

11. **E-BOOK AUDIENCE** The number of adults (in millions) using e-book devices is expected to climb in the years ahead. The projected number of e-book readers in the United States from 2011 through 2015 is given in the following table:

Year	2011	2012	2013	2014	2015
Number, y	25.3	33.4	39.5	50.0	59.6

   a. Letting $x = 0$ denote 2011, find an equation of the least-squares line for these data.
   b. Use the result of part (a) to estimate the projected average rate of growth of the number of e-book readers between 2011 and 2015.
   Source: Forrester Research, Inc.

12. **PERCENTAGE OF THE POPULATION ENROLLED IN SCHOOL** The percentage of the population (aged 3 years or older) who were enrolled in school from 2007 through 2011 is given in the following table:

Year	2007	2008	2009	2010	2011
Percent, y	26.2	26.8	27.5	28.3	28.7

   a. Letting $x = 0$ denote 2007, find an equation of the least-squares line for these data.
   b. Use the result of part (a) to estimate the percentage of the population (aged 3 or older) who were enrolled in school in 2014, assuming that the trend continued.
   Source: U.S. Census Bureau.

13. **GLOBAL BOX-OFFICE RECEIPTS** Global ticket sales have been growing steadily over the years, reflecting the rapid growth in overseas markets, particularly in China. The sales (in billions of dollars) from 2007 through 2011 are summarized in the following table:

Year	2007	2008	2009	2010	2011
Sales, y	26.1	27.2	28.9	31.1	32.6

   a. Find an equation of the least-squares line for these data. (Let $x = 1$ represent 2007.)
   b. Use the result of part (a) to predict the global ticket sales for 2014, assuming that the trend continued.
   Source: Motion Picture Association of America.

14. FIRST-CLASS MAIL VOLUME As more and more people turn to using the Internet and phones to pay bills and to communicate, replacing letters, the first-class mail volume is expected to decline until 2020. The following table gives the volume (in billions of pieces) of first-class mail from 2007 through 2011:

Year	2007	2008	2009	2010	2011
Value, y	95.9	91.7	83.8	78.2	73.5

a. Letting $x = 1$ denote 2007, find an equation of the least-squares line for these data.
b. Use the results of part (a) to estimate the volume of first-class mail in 2014, assuming that the trend continued through that year.
Source: U.S. Postal Service.

15. GROWTH OF CREDIT UNIONS Credit union membership is on the rise. The following table gives the number (in millions) of credit union members from 2003 through 2011 in 2-year intervals:

Year	2003	2005	2007	2009	2011
Number, y	82.0	84.7	86.8	89.7	91.8

a. Letting $x = 0$ denote 2003, find an equation of the least-squares line for these data.
b. Assuming that the trend continued, estimate the number of credit union members in 2013 ($x = 5$).
Source: National Credit Union Association.

16. ONLINE VIDEO ADVERTISING Although still a small percentage of all online advertising, online video advertising is growing. The following table gives the projected spending on Web video advertising (in billions of dollars) through 2016:

Year	2011	2012	2013	2014	2015	2016
Spending, y	2.0	3.1	4.5	6.3	7.8	9.3

a. Letting $x = 0$ denote 2011, find an equation of the least-squares line for these data.
b. Use the result of part (a) to estimate the projected rate of growth of video advertising from 2011 through 2016.
Source: eMarketer.

17. U.S. OUTDOOR ADVERTISING U.S. outdoor advertising expenditure (in billions of dollars) from 2011 through 2015 is given in the following table ($x = 0$ corresponds to 2011):

Year	2011	2012	2013	2014	2015
Expenditure, y	6.4	6.8	7.1	7.4	7.6

a. Find an equation of the least-squares line for these data.
b. Use the result of part (a) to estimate the rate of change of the advertising expenditures for the period in question.
Source: Outdoor Advertising Association.

18. ONLINE SALES OF USED AUTOS The amount (in millions of dollars) of used autos sold online in the United States is expected to grow in accordance with the figures given in the following table ($x = 0$ corresponds to 2011):

Year, x	0	1	2	3	4
Sales, y	12.9	13.9	14.65	15.25	15.85

a. Find an equation of the least-squares line for these data.
b. Use the result of part (a) to estimate the sales of used autos online in 2016, assuming that the predicted trend continued.
Source: comScore Networks, Inc.

19. HOME HEALTH-CARE AND EQUIPMENT SPENDING The following table gives the projected spending on home care and durable medical equipment (in billions of dollars) from 2004 through 2016 ($x = 0$ corresponds to 2004):

Year, x	0	2	4	6	8	10	12
Spending, y	60	74	90	106	118	128	150

a. Find an equation of the least-squares line for these data.
b. Use the result of part (a) to give the approximate projected spending on home care and durable medical equipment in 2015.
c. Use the result of part (a) to estimate the projected rate of change of the spending on home care and durable medical equipment for the period from 2004 through 2016.
Source: National Association of Home Care and Hospice.

20. GLOBAL DEFENSE SPENDING The following table gives the projected global defense spending (in trillions of dollars) from the beginning of 2008 ($t = 0$) through 2015 ($t = 7$):

Year, t	0	1	2	3	4	5	6	7
Spending, y	1.38	1.44	1.49	1.56	1.61	1.67	1.74	1.78

a. Find an equation of the least-squares line for these data.
b. Use the result of part (a) to estimate the rate of change in the projected global defense spending from 2008 through 2015.
c. Assuming that the trend continues, what will the global spending on defense be in 2018?
Source: Homeland Security Research.

In Exercises 21–24, determine whether the statement is true or false. If it is true, explain why it is true. If it is false, give an example to show why it is false.

21. The least-squares line must pass through at least one data point.

22. The error incurred in approximating n data points using the least-squares linear function is zero if and only if the n data points lie on a nonvertical straight line.

23. If the data consist of two distinct points, then the least-squares line is just the line that passes through the two points.

24. A data point lies on the least-squares line if and only if the vertical distance between the point and the line is equal to zero.

2.8   Solution to Self-Check Exercise

The calculations required for obtaining the normal equations may be summarized as follows:

	x	p	x^2	xp
	2.2	38.0	4.84	83.6
	5.4	36.0	29.16	194.4
	7.0	34.5	49.00	241.5
	11.5	30.0	132.25	345.0
	14.6	28.5	213.16	416.1
Sum	40.7	167.0	428.41	1280.6

The normal equations are

$$5b + 40.7m = 167$$
$$40.7b + 428.41m = 1280.6$$

Solving this system of linear equations simultaneously, we find that

$$m \approx -0.81 \quad \text{and} \quad b \approx 40.00$$

Therefore, an equation of the least-squares line is given by

$$p = f(x) = -0.81x + 40$$

which is the required demand equation, provided that

$$0 \le x \le 49.38$$

USING TECHNOLOGY

Finding an Equation of a Least-Squares Line

Graphing Utility

A graphing utility is especially useful in calculating an equation of the least-squares line for a set of data. We simply enter the given data in the form of lists into the calculator and then use the linear regression function to obtain the coefficients of the required equation.

EXAMPLE 1 Find an equation of the least-squares line for the data

x	1.1	2.3	3.2	4.6	5.8	6.7	8.0
y	−5.8	−5.1	−4.8	−4.4	−3.7	−3.2	−2.5

Plot the scatter diagram and the least-squares line for this data.

Solution   First, we enter the data as follows:

$$x_1 = 1.1 \qquad y_1 = -5.8 \qquad x_2 = 2.3 \qquad y_2 = -5.1 \qquad x_3 = 3.2$$
$$y_3 = -4.8 \qquad x_4 = 4.6 \qquad y_4 = -4.4 \qquad x_5 = 5.8 \qquad y_5 = -3.7$$
$$x_6 = 6.7 \qquad y_6 = -3.2 \qquad x_7 = 8.0 \qquad y_7 = -2.5$$

Then, using the linear regression function from the statistics menu, we obtain the output shown in Figure T1a. Therefore, an equation of the least-squares line ($y = ax + b$) is

$$y = 0.46x - 6.3$$

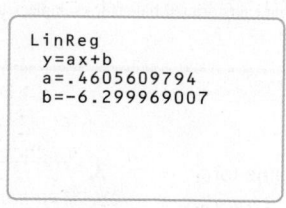

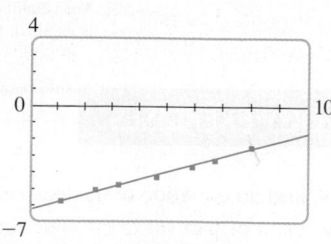

(a) The TI-83/84 linear regression screen

(b) The scatter diagram and least-squares line for the data

FIGURE **T1**

The graph of the least-squares equation and the scatter diagram for the data are shown in Figure T1b.

Excel

Excel can be used to find an equation of the least-squares line for a set of data and to plot a scatter diagram and the least-squares line for the data.

EXAMPLE 2 Find an equation of the least-squares line for the data given in the following table:

x	1.1	2.3	3.2	4.6	5.8	6.7	8.0
y	-5.8	-5.1	-4.8	-4.4	-3.7	-3.2	-2.5

Plot the scatter diagram and the least-squares line for these data.

	A	B
1	x	y
2	1.1	-5.8
3	2.3	-5.1
4	3.2	-4.8
5	4.6	-4.4
6	5.8	-3.7
7	6.7	-3.2
8	8	-2.5

FIGURE T2
Table of values for *x* and *y*

Solution

1. *Set up a table of values in two columns on a spreadsheet* (Figure T2).
2. *Plot the scatter diagram.* Highlight the numerical values in the table of values. Follow the procedure given in Example 3, Using Technology Section 2.2, page 90, selecting the first chart subtype instead of the second from the Scatter chart type. The scatter diagram will appear.
3. *Insert the least-squares line.* Select the [Layout] tab, and click on [Trendline] in the Analysis group. Next, click on [More Trendline Options...] in the same subgroup. In the Format Trendline dialog box that appears, click on [Display Equation on chart] .

$$y = 0.4606x - 6.3$$

and the least-squares line will appear on the chart (Figure T3).

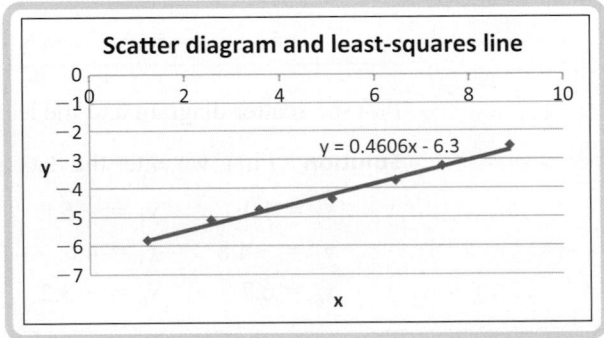

FIGURE T3
Scatter diagram and least-squares line for the given data

Note: Boldfaced words/characters enclosed in a box (for example, [Enter]) indicate that an action (click, select, or press) is required. Words/characters printed blue (for example, Chart sub-type:) indicate words/characters that appear on the screen.

TECHNOLOGY EXERCISES

In Exercises 1–4, find an equation of the least-squares line for the given data.

1.

x	2.1	3.4	4.7	5.6	6.8	7.2
y	8.8	12.1	14.8	16.9	19.8	21.1

2.

x	1.1	2.4	3.2	4.7	5.6	7.2
y	-0.5	1.2	2.4	4.4	5.7	8.1

3.

x	-2.1	-1.1	0.1	1.4	2.5	4.2	5.1
y	6.2	4.7	3.5	1.9	0.4	-1.4	-2.5

4.

x	-1.12	0.1	1.24	2.76	4.21	6.82
y	7.61	4.9	2.74	-0.47	-3.51	-8.94

5. EROSION OF THE MIDDLE CLASS The idea of a large, stable, middle class (defined as those with annual household incomes in 2010 between $39,000 and $118,000 for a family of three), is central to America's sense of itself. The following table gives the percentage of middle-income adults (y) in the United States from 1971 through 2011.

Year	1971	1981	1991	2001	2011
Percent, y	61	59	56	54	51

Let t be measured in decades with $t = 0$ corresponding to 1971.
a. Find an equation of the least-squares line for these data.
b. If this trend continues, what will the percentage of middle-income adults be in 2021?
Source: Pew Research Center.

6. MODELING WITH DATA Moody's Corporation is the holding company for Moody's Investors Service, which has a 40% share in the world credit-rating market. According to company reports, the total revenue (in billions of dollars) of the company is projected to be as follows ($x = 4$ corresponds to 2004):

Year	2004	2005	2006	2007	2008
Revenue, y	1.42	1.73	1.98	2.32	2.65

a. Find an equation of the least-squares line for these data.
b. Use the results of part (a) to estimate the rate of change of the revenue of the company for the period in question.
c. Use the result of part (a) to estimate the total revenue of the company in 2010, assuming that the trend continued.
Source: Company reports.

CHAPTER 2 **Summary of Principal Formulas and Terms**

FORMULAS

1. Slope of a line	$m = \dfrac{y_2 - y_1}{x_2 - x_1}$
2. Equation of a vertical line	$x = a$
3. Equation of a horizontal line	$y = b$
4. Point-slope form of the equation of a line	$y - y_1 = m(x - x_1)$
5. Slope-intercept form of the equation of a line	$y = mx + b$
6. General equation of a line	$Ax + By + C = 0$

TERMS

Cartesian coordinate system (72)
ordered pair (72)
coordinates (72)
parallel lines (76)
perpendicular lines (80)
function (92)
domain (92)
range (92)
independent variable (94)

dependent variable (94)
graph of a function (95)
piecewise-defined function (97)
graph of an equation (98)
Vertical Line Test (98)
composite function (110)
linear function (116)
total cost function (118)
revenue function (119)

profit function (119)
break-even point (121)
quadratic function (131)
demand function (134)
supply function (135)
market equilibrium (136)
polynomial function (143)
rational function (146)
power function (146)

CHAPTER 2 Concept Review Questions

Fill in the blanks.

1. A point in the plane can be represented uniquely by a/an _____ pair of numbers. The first number of the pair is called the _____, and the second number of the pair is called the _____.

2. **a.** The point $P(a, 0)$ lies on the _____-axis, and the point $P(0, b)$ lies on the _____-axis.
   b. If the point $P(a, b)$ lies in the fourth quadrant, then the point $P(-a, b)$ lies in the _____ quadrant.

3. **a.** If $P_1(x_1, y_1)$ and $P_2(x_2, y_2)$ are any two distinct points on a nonvertical line L, then the slope of L is $m =$ _____.
   b. The slope of a vertical line is _____.
   c. The slope of a horizontal line is _____.
   d. The slope of a line that slants upward from left to right is _____.

4. If L_1 and L_2 are nonvertical lines with slopes m_1 and m_2, respectively, then L_1 is parallel to L_2 if and only if _____ and L_1 is perpendicular to L_2 if and only if _____.

5. **a.** An equation of the line passing through the point $P(x_1, y_1)$ and having slope m is _____. This form of the equation of a line is called the _____ _____.
   b. An equation of the line that has slope m and y-intercept b is _____. It is called the _____ form of an equation of a line.

6. **a.** The general form of an equation of a line is _____.
   b. If a line has equation $ax + by + c = 0$ $(b \neq 0)$, then its slope is _____.

7. If f is a function from the set A to the set B, then A is called the _____ of f, and the set of all values of $f(x)$ as x takes on all possible values in A is called the _____ of f. The range of f is contained in the set _____.

8. The graph of a function is the set of all points (x, y) in the xy-plane such that x is in the _____ of f and $y =$ _____. The Vertical Line Test states that a curve in the xy-plane is the graph of a function $y = f(x)$ if and only if each _____ line intersects it in at most one _____.

9. If f and g are functions with domains A and B, respectively, then (a) $(f \pm g)(x) =$ _____, (b) $(fg)(x) =$ _____, and (c) $\left(\dfrac{f}{g}\right)(x) =$ _____. The domain of $f + g$ is _____. The domain of $\dfrac{f}{g}$ is _____ with the additional condition that $g(x)$ is never _____.

10. The composition of g and f is the function with rule $(g \circ f)(x) =$ _____. Its domain is the set of all x in the domain of _____ such that _____ lies in the domain of _____.

11. A quadratic function has the form $f(x) =$ _____. Its graph is a/an _____ that opens _____ if $a > 0$ and _____ if $a < 0$. Its highest point or lowest point is called its _____. The x-coordinate of its vertex is _____, and an equation of its axis of symmetry is _____.

12. **a.** A polynomial function of degree n is a function of the form _____.
   b. A polynomial function of degree 1 is called a/an _____ function; one of degree 2 is called a/an _____ function.
   c. A rational function is a/an _____ of two _____.
   d. A power function has the form $f(x) =$ _____.

CHAPTER 2 Review Exercises

In Exercises 1–6, find an equation of the line L that passes through the point $(-2, 4)$ and satisfies the given condition.

1. L is a vertical line.

2. L is a horizontal line.

3. L passes through the point $\left(3, \frac{7}{2}\right)$.

4. The x-intercept of L is 3.

5. L is parallel to the line $5x - 2y = 6$.

6. L is perpendicular to the line $4x + 3y = 6$.

7. Find an equation of the line with slope $-\frac{1}{2}$ and y-intercept -3.

8. Find the slope and y-intercept of the line with equation $3x - 5y = 6$.

9. Find an equation of the line passing through the point $(2, 3)$ and parallel to the line with equation $3x + 4y - 8 = 0$.

10. Find an equation of the line passing through the point $(-1, 3)$ and parallel to the line joining the points $(-3, 4)$ and $(2, 1)$.

11. Find an equation of the line passing through the point $(-2, -4)$ that is perpendicular to the line with equation $2x - 3y - 24 = 0$.

1-26

29-34

In Exercises 12 and 13, sketch the graph of the equation.

12. $3x - 4y = 24$ **13.** $-2x + 5y = 15$

In Exercises 14 and 15, find the domain of the function.

14. $f(x) = \sqrt{9 - x}$

15. $f(x) = \dfrac{x + 3}{2x^2 - x - 3}$

16. Let $f(x) = 3x^2 + 5x - 2$. Find:
   a. $f(-2)$
   b. $f(a + 2)$
   c. $f(2a)$
   d. $f(a + h)$

17. SALES OF PRERECORDED MUSIC The following graphs show the sales y of prerecorded music (in billions of dollars) by format as a function of time t (in years), with $t = 0$ corresponding to 1985.

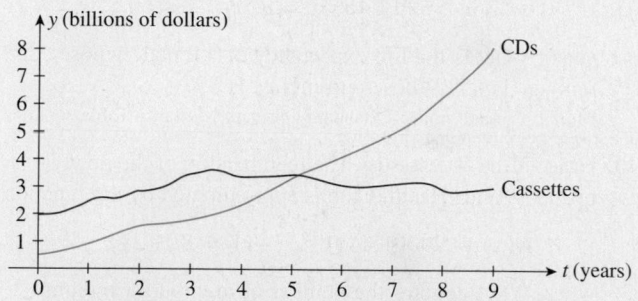

y (billions of dollars)

   a. In what years were the sales of prerecorded cassettes greater than those of prerecorded CDs?
   b. In what years were the sales of prerecorded CDs greater than those of prerecorded cassettes?
   c. In what year were the sales of prerecorded cassettes the same as those of prerecorded CDs? Estimate the level of sales in each format at that time.
   Source: Recording Industry Association of America.

18. Let $y^2 = 2x + 1$.
   a. Sketch the graph of this equation.
   b. Is y a function of x? Why?
   c. Is x a function of y? Why?

19. Sketch the graph of the function defined by

$$f(x) = \begin{cases} x + 1 & \text{if } x < 1 \\ -x^2 + 4x - 1 & \text{if } x \geq 1 \end{cases}$$

20. Let $f(x) = \frac{1}{x}$ and $g(x) = 2x + 3$. Find:
   a. $f(x)g(x)$ **b.** $f(x)/g(x)$
   c. $f(g(x))$ **d.** $g(f(x))$

In Exercises 21 and 22, find the vertex and the x-intercepts of the parabola and sketch the parabola.

21. $y = 6x^2 - 11x - 10$ **22.** $y = -4x^2 + 4x + 3$

In Exercises 23 and 24, find the point of intersection of the lines with the given equations.

23. $3x + 4y = -6$ and $2x + 5y = -11$

24. $y = \dfrac{3}{4}x + 6$ and $3x - 2y + 3 = 0$

25. Find the point of intersection of the two straight lines having the equations $7x + 9y = -11$ and $3x = 6y - 8$.

26. The cost and revenue functions for a certain firm are given by $C(x) = 12x + 20{,}000$ and $R(x) = 20x$, respectively. Find the company's break-even point.

27. DEMAND FOR CLOCK RADIOS In the accompanying figure, L_1 is the demand curve for the model A clock radios manufactured by Ace Radio, and L_2 is the demand curve for their model B clock radios. Which line has the greater slope? Interpret your results.

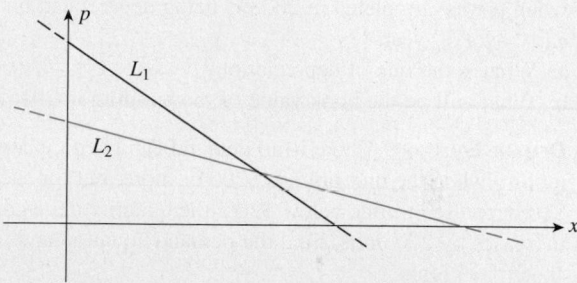

28. SUPPLY OF CLOCK RADIOS In the accompanying figure, L_1 is the supply curve for the model A clock radios manufactured by Ace Radio, and L_2 is the supply curve for their model B clock radios. Which line has the greater slope? Interpret your results.

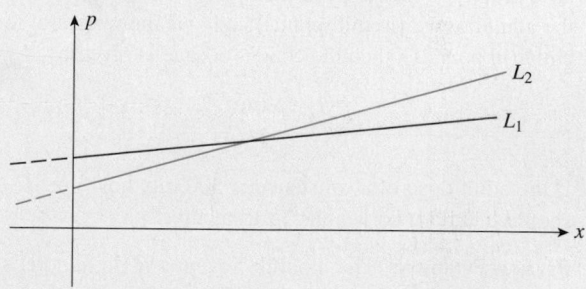

29. CONSUMPTION FUNCTION The consumption function in a certain economy is given by

$$C(y) = 0.75y + 6$$

where $C(y)$ is the personal consumption expenditure, y is the disposable personal income, and both $C(y)$ and y are measured in billions of dollars. Find $C(0)$, $C(50)$, and $C(100)$.

30. **COMPANY SALES** A company's total sales (in millions of dollars) are approximately linear as a function of time (in years). Sales in 2010 ($x = 0$) were $2.4 million, whereas sales in 2015 amounted to $7.4 million.
    a. Find an equation giving the company's sales as a function of time.
    b. What were the sales in 2013?

31. **PROFIT FUNCTIONS** A company has a fixed cost of $30,000 and a production cost of $6 for each CD it manufactures. Each CD sells for $10.
    a. What is the cost function?
    b. What is the revenue function?
    c. What is the profit function?
    d. Compute the profit (loss) corresponding to production levels of 6000, 8000, and 12,000 units, respectively.

32. **LINEAR DEPRECIATION** An office building worth $6 million when it was completed in 2008 is being depreciated linearly over 30 years.
    a. What is the rate of depreciation?
    b. What will be the book value of the building in 2018?

33. **DEMAND EQUATIONS** There is no demand for a certain commodity when the unit price is $200 or more, but for each $10 decrease in price below $200, the quantity demanded increases by 200 units. Find the demand equation and sketch its graph.

34. **SUPPLY EQUATIONS** Bicycle suppliers will make 200 bicycles available in the market per month when the unit price is $50 and 2000 bicycles available per month when the unit price is $100. Find the supply equation if it is known to be linear.

35. **CLARK'S RULE** Clark's Rule is a method for calculating pediatric drug dosages based on a child's weight. If a denotes the adult dosage (in milligrams) and w is the weight of the child (in pounds), then the child's dosage is given by

$$D(w) = \frac{aw}{150}$$

If the adult dose of a substance is 500 mg, how much should a child who weighs 35 lb receive?

36. **REVENUE FUNCTIONS** The monthly revenue R (in hundreds of dollars) realized in the sale of Royal electric shavers is related to the unit price p (in dollars) by the equation

$$R(p) = -\frac{1}{2}p^2 + 30p$$

Find the revenue when an electric shaver is priced at $30.

37. **GROWTH OF CRUISE INDUSTRY** The number of North American cruise ship passengers has been increasing over the last several decades. The number of cruise ship passengers from 1995 through 2010 is approximated by the function

$$N(t) = 0.011t^2 + 0.521t + 4.6 \qquad (0 \le t \le 15)$$

where $N(t)$ is the number of passengers (in millions) in year t, with $t = 0$ corresponding to 1995. What was the approximate number of passengers in 1995? In 2010?
Source: Cruise Lines International Association.

38. **MALE LIFE EXPECTANCY** Advances in medical science and healthier lifestyles have resulted in longer life expectancies. The life expectancy of a male whose current age is x years old is

$$f(x) = 0.0069502x^2 - 1.6357x + 93.76 \qquad (60 \le x \le 75)$$

years. What is the life expectancy of a male whose current age is 65? A male whose current age is 75?
Source: Commissioners' Standard Ordinary Mortality Table.

39. **FEMALE LIFE EXPECTANCY** Advances in medical science and healthier lifestyles have resulted in longer life expectancies. The life expectancy of a female whose current age is x years old is

$$f(x) = 0.0053694x^2 - 1.4663x + 92.74 \qquad (60 \le x \le 75)$$

years. What is the life expectancy of a female whose current age is 65? Whose current age is 75?
Source: Commissioners' Standard Ordinary Mortality Table.

40. **HEALTH CLUB MEMBERSHIP** The membership of the newly opened Venus Health Club is approximated by the function

$$N(x) = 200(4 + x)^{1/2} \qquad (1 \le x \le 24)$$

where $N(x)$ denotes the number of members x months after the club's grand opening. Find $N(0)$ and $N(12)$, and interpret your results.

41. **POPULATION GROWTH** A study prepared for a Sunbelt town's Chamber of Commerce projected that the population of the town in the next 3 years will grow according to the rule

$$P(x) = 50,000 + 30x^{3/2} + 20x$$

where $P(x)$ denotes the population x months from now. By how much will the population increase during the next 9 months? During the next 16 months?

42. **THURSTONE LEARNING CURVE** Psychologist L. L. Thurstone discovered the following model for the relationship between the learning time T and the length of a list n:

$$T = f(n) = An\sqrt{n - b}$$

where A and b are constants that depend on the person and the task. Suppose that, for a certain person and a certain task, $A = 4$ and $b = 4$. Compute $f(4)$, $f(5)$, ..., $f(12)$, and use this information to sketch the graph of the function f. Interpret your results.

43. **GLOBAL SUPPLY OF PLUTONIUM** The global stockpile of plutonium for military applications between 1990 ($t = 0$) and 2003 ($t = 13$) stood at a constant 267 tons. On the other hand, the global stockpile of plutonium for civilian use was

$$2t^2 + 46t + 733$$

tons in year t over the same period.

a. Find the function f giving the global stockpile of plutonium for military use from 1990 through 2003 and the function g giving the global stockpile of plutonium for civilian use over the same period.

b. Find the function h giving the total global stockpile of plutonium between 1990 and 2003.

c. What was the total global stockpile of plutonium in 2003?

Source: Institute for Science and International Security.

44. MARKET EQUILIBRIUM The monthly demand and supply functions for the Luminar desk lamp are given by

$$p = d(x) = -1.1x^2 + 1.5x + 40$$
$$p = s(x) = 0.1x^2 + 0.5x + 15$$

respectively, where p is measured in dollars and x in units of a thousand. Find the equilibrium quantity and price.

45. INFLATING A BALLOON A spherical balloon is being inflated at a rate of $\frac{9}{2}\pi$ ft³/min.

a. Find a function f giving the radius r of the balloon in terms of its volume V.
Hint: $V = \frac{4}{3}\pi r^3$

b. Find a function g giving the volume of the balloon in terms of time t.

c. Find a function h giving the radius of the balloon in terms of time.

d. What is the radius of the balloon after 8 min?

46. HOTEL OCCUPANCY RATE A forecast released by PricewaterhouseCoopers in June of 2004 predicted the occupancy rate of U.S. hotels between 2001 ($t = 0$) and 2005 ($t = 4$) to be

$$P(t) = \begin{cases} -0.9t + 59.8 & \text{if } 0 \le t < 1 \\ 0.3t + 58.6 & \text{if } 1 \le t < 2 \\ 56.79t^{0.06} & \text{if } 2 \le t \le 4 \end{cases}$$

percent.

a. Compute $P(0)$, $P(1)$, $P(2)$, $P(3)$, and $P(4)$.

b. Sketch the graph of P.

c. What was the predicted occupancy rate of hotels for 2004?

Source: PricewaterhouseCoopers LLP Hospitality & Leisure Research.

47. PACKAGING By cutting away identical squares from each corner of a 20-in. × 20-in. piece of cardboard and folding up the resulting flaps, an open box can be made. Denoting the length of a side of a cutaway by x, find a function of x giving the volume of the resulting box.

48. CONSTRUCTION COSTS The length of a rectangular box is to be twice its width, and its volume is to be 30 ft³. The material for the base costs 30¢/ft², the material for the sides costs 15¢/ft², and the material for the top costs 20¢/ft². Letting x denote the width of the box, find a function in the variable x giving the cost of constructing the box.

The problem-solving skills that you learn in each chapter are building blocks for the rest of the course. Therefore, it is a good idea to make sure that you have mastered these skills before moving on to the next chapter. The Before Moving On exercises that follow are designed for that purpose. After completing these exercises, you can identify the skills that you should review before starting the next chapter.

CHAPTER 2 Before Moving On ...

1. Find an equation of the line that passes through $(-1, -2)$ and $(4, 5)$.

2. Find an equation of the line that has slope $-\frac{1}{3}$ and y-intercept $\frac{4}{3}$.

3. Let

$$f(x) = \begin{cases} -2x + 1 & \text{if } -1 \le x < 0 \\ x^2 + 2 & \text{if } 0 \le x \le 2 \end{cases}$$

Find (a) $f(-1)$, (b) $f(0)$, and (c) $f\left(\frac{3}{2}\right)$.

4. Let $f(x) = \dfrac{1}{x + 1}$ and $g(x) = x^2 + 1$. Find the rules for
(a) $f + g$, (b) fg, (c) $f \circ g$, and (d) $g \circ f$.

5. POSTAL REGULATIONS Postal regulations specify that a parcel sent by parcel post may have a combined length and girth of no more than 108 in. Suppose a rectangular package that has a square cross section of x in. × x in. is to have a combined length and girth of exactly 108 in. Find a function in terms of x giving the volume of the package.
Hint: The length plus the girth is $4x + h$ (see the accompanying figure).

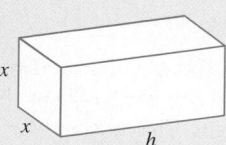

3 Exponential and Logarithmic Functions

THE EXPONENTIAL FUNCTION is without doubt the most important function in mathematics and its applications. After a brief introduction to the exponential function and its *inverse,* the logarithmic function, we explore some of the many applications involving exponential functions, such as the growth rate of a bacteria population in the laboratory, the way in which radioactive matter decays, the rate at which a factory worker learns a certain process, and the rate at which a communicable disease is spread over time. Exponential functions also play an important role in computing interest earned in a bank account, as we will see in Chapter 4.

How many cameras can a new employee at Eastman Optical assemble after completing the basic training program, and how many cameras can he assemble after being on the job for 6 months? In Example 5, page 194, you will see how to answer these questions.

3.1 Exponential Functions

Exponential Functions and Their Graphs

Suppose you deposit a sum of $1000 in an account earning interest at the rate of 10% per year *compounded continuously* (the way most financial institutions compute interest). Then, the accumulated amount at the end of t years ($0 \le t \le 20$) is described by the function f, whose graph appears in Figure 1.* This function is called an *exponential function*. Observe that the graph of f rises rather slowly at first but very rapidly as time goes by. For purposes of comparison, we have also shown the graph of the function $y = g(t) = 1000(1 + 0.10t)$, giving the accumulated amount for the same principal ($1000) but earning *simple* interest at the rate of 10% per year. The moral of the story: It is never too early to save.

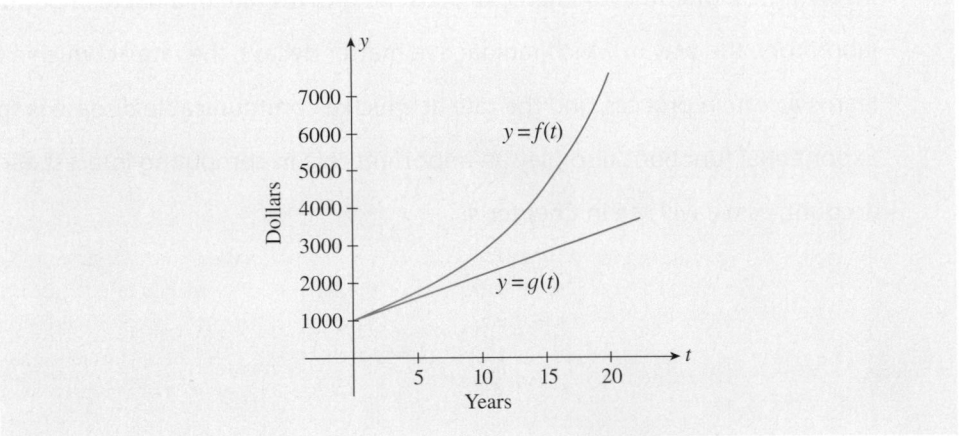

FIGURE 1
Under continuous compounding, a sum of money grows exponentially.

Exponential functions play an important role in many real-world applications, as you will see throughout this chapter.

Recall that if b is a positive number and r is any rational number, the expression b^r is a real number. It can be shown, although we will not do so here, that if x is *any* real number, then b^x is also a real number. This enables us to define an exponential function as follows:

> **Exponential Function**
> The function defined by
> $$f(x) = b^x \qquad (b > 0, b \ne 1)$$
> is called an **exponential function with base b and exponent x.** The domain of f is the set of all real numbers.

For example, the exponential function with base 2 is the function

$$f(x) = 2^x$$

---

*We will derive the rule for f in Section 4.1.

with domain $(-\infty, \infty)$. The values of $f(x)$ for selected values of x follow:

$$f(3) = 2^3 = 8 \qquad f\left(\frac{3}{2}\right) = 2^{3/2} = 2 \cdot 2^{1/2} = 2\sqrt{2} \qquad f(0) = 2^0 = 1$$

$$f(-1) = 2^{-1} = \frac{1}{2} \qquad f\left(-\frac{2}{3}\right) = 2^{-2/3} = \frac{1}{2^{2/3}} = \frac{1}{\sqrt[3]{4}}$$

Computations involving exponentials are facilitated by the properties of exponents. These properties were stated in Section 1.5, and you might want to review the material there. We use one of these properties in the following example.

EXAMPLE 1 Let $f(x) = 2^{2x-1}$. Find the value of x for which $f(x) = 16$.

Solution We want to solve the equation

$$2^{2x-1} = 16 = 2^4$$

But this equation holds if and only if

$$2x - 1 = 4 \qquad b^m = b^n \Rightarrow m = n$$

giving $x = \frac{5}{2}$.

Exponential functions play an important role in mathematical analysis. Because of their special characteristics, they are some of the most useful functions and are found in virtually every field in which mathematics is applied. To mention a few examples: Under ideal conditions, the number of bacteria present at any time t in a culture may be described by an exponential function of t; radioactive substances decay over time in accordance with an "exponential" law of decay; money left on fixed deposit and earning compound interest grows exponentially; and some of the most important distribution functions encountered in statistics are exponential.

Let's begin our investigation into the properties of exponential functions by studying their graphs.

EXAMPLE 2 Sketch the graph of the exponential function $y = 2^x$.

Solution First, as was discussed earlier, the domain of the exponential function $y = f(x) = 2^x$ is the set of real numbers. Next, putting $x = 0$ gives $y = 2^0 = 1$, the y-intercept of f. There is no x-intercept, since there is no value of x for which $y = 0$. To find the range of f, consider the following table of values:

x	-5	-4	-3	-2	-1	0	1	2	3	4	5
y	$\frac{1}{32}$	$\frac{1}{16}$	$\frac{1}{8}$	$\frac{1}{4}$	$\frac{1}{2}$	1	2	4	8	16	32

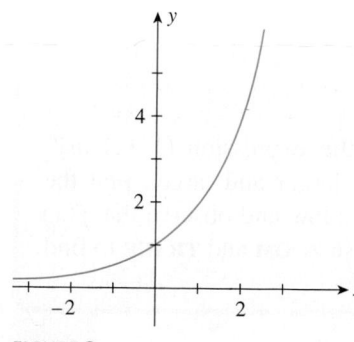

FIGURE 2
The graph of $y = 2^x$

We see from these computations that 2^x decreases and approaches zero as x decreases without bound and that 2^x increases without bound as x increases without bound. Thus, the range of f is the interval $(0, \infty)$—that is, the set of positive real numbers. Finally, we sketch the graph of $y = f(x) = 2^x$ in Figure 2.

EXAMPLE 3 Sketch the graph of the exponential function $y = (1/2)^x$.

Solution The domain of the exponential function $y = (1/2)^x$ is the set of all real numbers. The y-intercept is $(1/2)^0 = 1$; there is no x-intercept, since there is no value of x for which $y = 0$. From the following table of values

x	-5	-4	-3	-2	-1	0	1	2	3	4	5
y	32	16	8	4	2	1	$\frac{1}{2}$	$\frac{1}{4}$	$\frac{1}{8}$	$\frac{1}{16}$	$\frac{1}{32}$

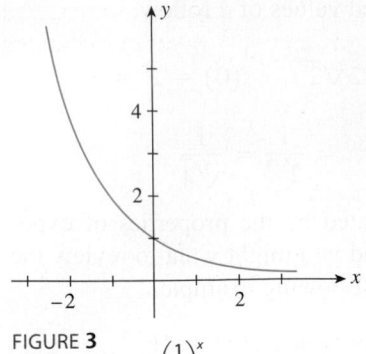

FIGURE 3
The graph of $y = \left(\dfrac{1}{2}\right)^x$

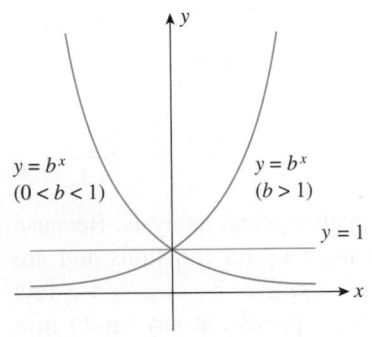

$y = b^x$
$(0 < b < 1)$

$y = b^x$
$(b > 1)$

$y = 1$

FIGURE 4
$y = b^x$ is an increasing function of x if $b > 1$, a constant function if $b = 1$, and a decreasing function if $0 < b < 1$.

TABLE 1	
m	$\left(1 + \dfrac{1}{m}\right)^m$
10	2.59374
100	2.70481
1000	2.71692
10,000	2.71815
100,000	2.71827
1,000,000	2.71828

we deduce that $(1/2)^x = 1/2^x$ increases without bound as x decreases without bound and that $(1/2)^x$ decreases and approaches zero as x increases without bound. Thus, the range of f is the interval $(0, \infty)$. The graph of $y = f(x) = (1/2)^x$ is sketched in Figure 3.

The functions $y = 2^x$ and $y = (1/2)^x$, whose graphs you studied in Examples 2 and 3, are special cases of the exponential function $y = f(x) = b^x$, obtained by setting $b = 2$ and $b = 1/2$, respectively. In general, the exponential function $y = b^x$ with $b > 1$ has a graph similar to that of $y = 2^x$, whereas the graph of $y = b^x$ for $0 < b < 1$ is similar to that of $y = (1/2)^x$ (Exercises 11 and 12 on page 177). When $b = 1$, the function $y = b^x$ reduces to the constant function $y = 1$. For comparison, the graphs of all three functions are sketched in Figure 4.

> **Properties of the Exponential Function**
>
> The exponential function $y = b^x$ $(b > 0, b \neq 1)$ has the following properties:
>
> 1. Its domain is $(-\infty, \infty)$.
> 2. Its range is $(0, \infty)$.
> 3. Its graph passes through the point $(0, 1)$.
> 4. Its graph is an unbroken curve devoid of holes or jumps.
> 5. Its graph rises from left to right if $b > 1$ and falls from left to right if $b < 1$.

The Base e

It can be shown, although we will not do so here, that as m gets larger and larger, the value of the expression

$$\left(1 + \frac{1}{m}\right)^m$$

approaches the irrational number 2.7182818. . . , which we denote by e. You may convince yourself of the plausibility of this definition of the number e by examining Table 1, which may be constructed with the help of a calculator. (Also, see the Exploring with Technology exercise that follows.)

> **Exploring with TECHNOLOGY**
>
> To obtain a visual confirmation of the fact that the expression $(1 + 1/m)^m$ approaches the number $e = 2.71828. . .$ as m gets larger and larger, plot the graph of $f(x) = (1 + 1/x)^x$ in a suitable viewing window and observe that $f(x)$ approaches 2.71828. . . as x gets larger and larger. Use **ZOOM** and **TRACE** to find the value of $f(x)$ for large values of x.

EXAMPLE 4 Sketch the graph of the function $y = e^x$.

Solution Since $e > 1$, it follows from our previous discussion that the graph of $y = e^x$ is similar to the graph of $y = 2^x$ (see Figure 2). With the aid of a calculator, we obtain the following table:

x	-3	-2	-1	0	1	2	3
y	0.05	0.14	0.37	1	2.72	7.39	20.09

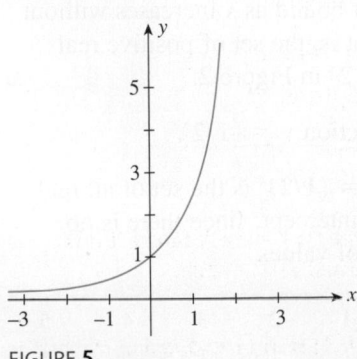

FIGURE 5
The graph of $y = e^x$

The graph of $y = e^x$ is sketched in Figure 5.

Next, we consider another exponential function to the base e that is closely related to the previous function and is particularly useful in constructing models that describe "exponential decay."

EXAMPLE 5 Sketch the graph of the function $y = e^{-x}$.

Solution Since $e > 1$, it follows that $0 < 1/e < 1$, so $f(x) = e^{-x} = 1/e^x = (1/e)^x$ is an exponential function with base less than 1. Therefore, it has a graph similar to that of the exponential function $y = (1/2)^x$. As before, we construct the following table of values of $y = e^{-x}$ for selected values of x:

x	-3	-2	-1	0	1	2	3
y	20.09	7.39	2.72	1	0.37	0.14	0.05

Using this table, we sketch the graph of $y = e^{-x}$ in Figure 6.

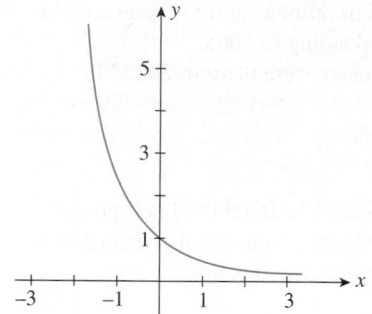

FIGURE 6
The graph of $y = e^{-x}$

3.1 Self-Check Exercises

1. Solve the equation $2^{2x+1} \cdot 2^{-3} = 2^{x-1}$.

2. Sketch the graph of $y = e^{0.4x}$.

Solutions to Self-Check Exercises 3.1 can be found on page 179.

3.1 Concept Questions

1. Define the exponential function f with base b and exponent x. What restrictions, if any, are placed on b?

2. For the exponential function $y = b^x$ $(b > 0, b \neq 1)$, state (a) its domain and range, (b) its y-intercept, (c) where its graph rises and where it falls for the case $b > 1$ and the case $b < 1$.

3.1 Exercises

In Exercises 1–10, solve the equation for x.

1. $6^{2x} = 6^6$

2. $5^{-x} = 5^3$

3. $3^{3x-4} = 3^5$

4. $10^{2x-1} = 10^{x+3}$

5. $(2.1)^{x+2} = (2.1)^5$

6. $(1.3)^{x-2} = (1.3)^{2x+1}$

7. $8^x = \left(\dfrac{1}{32}\right)^{x-2}$

8. $3^{x-x^2} = \dfrac{1}{9^x}$

9. $3^{2x} - 12 \cdot 3^x + 27 = 0$

10. $2^{2x} - 4 \cdot 2^x + 4 = 0$

In Exercises 11–20, sketch the graphs of the given functions on the same axes.

11. $y = 2^x$, $y = 3^x$, and $y = 4^x$

12. $y = \left(\dfrac{1}{2}\right)^x$, $y = \left(\dfrac{1}{3}\right)^x$, and $y = \left(\dfrac{1}{4}\right)^x$

13. $y = 2^{-x}$, $y = 3^{-x}$, and $y = 4^{-x}$

14. $y = 4^{0.5x}$ and $y = 4^{-0.5x}$

15. $y = 4^{0.5x}$, $y = 4^x$, and $y = 4^{2x}$

16. $y = e^x$, $y = 2e^x$, and $y = 3e^x$

17. $y = e^{0.5x}$, $y = e^x$, and $y = e^{1.5x}$

18. $y = e^{-0.5x}$, $y = e^{-x}$, and $y = e^{-1.5x}$

19. $y = 0.5e^{-x}$, $y = e^{-x}$, and $y = 2e^{-x}$

20. $y = 1 - e^{-x}$ and $y = 1 - e^{-0.5x}$

21. A function f has the form $f(x) = Ae^{kx}$. Find f if it is known that $f(0) = 100$ and $f(1) = 120$.
Hint: $e^{kx} = (e^k)^x$

22. If $f(x) = Axe^{-kx}$, find $f(3)$ if $f(1) = 5$ and $f(2) = 7$.
Hint: $e^{kx} = (e^k)^x$

23. If

$$f(t) = \frac{1000}{1 + Be^{-kt}}$$

find $f(5)$ given that $f(0) = 20$ and $f(2) = 30$.
Hint: $e^{kx} = (e^k)^x$

24. DECLINE OF AMERICAN IDOL After having been on the air for more than a decade, Fox's *American Idol* seemed to be suffering from viewer fatigue. The average number of viewers from the 2011 season through the 2013 season is approximated by

$$f(t) = 32.744e^{-0.252t} \qquad (1 \leq t \leq 3)$$

where $f(t)$ is measured in millions, with $t = 1$ corresponding to the 2011 season.
a. What was the average number of viewers in the 2011 season?
b. What was the average number of viewers in the 2014 season, assuming that the trend continued into that season?
Source: Nielsen Ratings.

25. RENEWABLE ENERGY Developing countries are accelerating the pace of their investment in renewable energy. According to a report by the Frankfurt School of Finance and Management, the amount of investment in renewable energy (in billions of dollars) by developing countries between 2009 ($t = 0$) and 2012 is given by

$$f(t) = 64e^{0.188t} \qquad (0 \leq t \leq 3)$$

a. Find the amount of investment in renewable energy by developing countries in each of the years 2009 through 2012 by completing the following table:

t	0	1	2	3
$f(t)$				

b. Use the information from part (a) to sketch the graph of f.
Source: Frankfurt School of Finance and Management.

26. ONLINE SHOPPERS According to a study conducted by Forrester Research, Inc., the amount of money spent by online shoppers in the United States is projected to be

$$f(t) = 105e^{0.095t} \qquad (1 \leq t \leq 6)$$

billion dollars in year t, where $t = 1$ corresponds to 2011.
a. Find the projected spending in each of the years 2011 through 2016 by completing the following table:

t	1	2	3	4	5	6
$f(t)$						

b. Use the information from part (a) to sketch the graph of f.
Source: Forrester Research, Inc.

27. INTERNET USERS IN CHINA The number of Internet users in China is approximated by

$$N(t) = 94.5e^{0.2t} \qquad (1 \leq t \leq 6)$$

where $N(t)$ is measured in millions and t is measured in years, with $t = 1$ corresponding to 2005.
a. How many Internet users were there in 2005? In 2006? In 2010?
b. Sketch the graph of N.
Source: C. E. Unterberg.

28. U.S. CELL PHONE SUBSCRIBERS The number of cell phone subscribers in the United States between the years 2000 and 2010 is approximated by the function

$$N(t) = \frac{385.474}{1 + 2.521e^{-0.214t}} \qquad (0 \leq t \leq 10)$$

where $N(t)$ is measured in millions and t is measured in years, with $t = 0$ corresponding to the year 2000. How many cell phone subscribers were there in the United States in 2000? If the trend continued, how many subscribers were there in 2012?
Source: CTIA—The Wireless Association.

29. ALTERNATIVE MINIMUM TAX The alternative minimum tax was created in 1969 to prevent the very wealthy from using creative deductions and shelters to avoid having to pay anything to the Internal Revenue Service. But it has increasingly hit the middle class. The number of taxpayers subjected to an alternative minimum tax is projected to be

$$N(t) = \frac{35.5}{1 + 6.89e^{-0.8674t}} \qquad (0 \leq t \leq 7)$$

where $N(t)$ is measured in millions and t is measured in years, with $t = 0$ corresponding to 2004. What was the projected number of taxpayers subjected to an alternative minimum tax in 2010?
Source: Brookings Institution.

30. ABSORPTION OF DRUGS The concentration of a drug in an organ at any time t (in seconds) is given by

$$x(t) = 0.08 + 0.12(1 - e^{-0.02t})$$

where $x(t)$ is measured in grams per cubic centimeter (g/cm^3).
a. What is the initial concentration of the drug in the organ?
b. What is the concentration of the drug in the organ after 20 sec?

31. ABSORPTION OF DRUGS The concentration of a drug in an organ at any time t (in seconds) is given by

$$C(t) = \begin{cases} 0.3t - 18(1 - e^{-t/60}) & \text{if } 0 \leq t \leq 20 \\ 18e^{-t/60} - 12e^{-(t-20)/60} & \text{if } t > 20 \end{cases}$$

where $C(t)$ is measured in grams per cubic centimeter (g/cm^3).

a. What is the initial concentration of the drug in the organ?

b. What is the concentration of the drug in the organ after 10 sec?

c. What is the concentration of the drug in the organ after 30 sec?

32. **ABSORPTION OF DRUGS** Jane took 100 mg of a drug in the morning and another 100 mg of the same drug at the same time the following morning. The amount of the drug (in milligrams) in her body t days after the first dose was taken is given by

$$A(t) = \begin{cases} 100e^{-1.4t} & \text{if } 0 \le t < 1 \\ 100(1 + e^{1.4})e^{-1.4t} & \text{if } t \ge 1 \end{cases}$$

What was the amount of drug in Jane's body immediately after taking the second dose? After 2 days?

In Exercises 33–36, determine whether the statement is true or false. If it is true, explain why it is true. If it is false, give an example to show why it is false.

33. If a and b are positive numbers, then $(a + b)^x = a^x + b^x$.

34. If $x < y$, then $e^x < e^y$.

35. If $0 < b < 1$ and $x < y$, then $b^x > b^y$.

36. If $e^{kx} > 1$, then $k > 0$ and $x > 0$.

3.1 Solutions to Self-Check Exercises

1.
$$2^{2x+1} \cdot 2^{-3} = 2^{x-1}$$

$$\frac{2^{2x+1}}{2^{x-1}} \cdot 2^{-3} = 1 \qquad \text{Divide both sides by } 2^{x-1}.$$

$$2^{(2x+1)-(x-1)-3} = 1$$

$$2^{x-1} = 1$$

This is true if and only if $x - 1 = 0$ or $x = 1$.

2. We first construct the following table of values:

x	-3	-2	-1	0	1	2	3	4
$y = e^{0.4x}$	0.3	0.4	0.7	1	1.5	2.2	3.3	5

Next, we plot these points and join them by a smooth curve to obtain the graph of f shown in the accompanying figure.

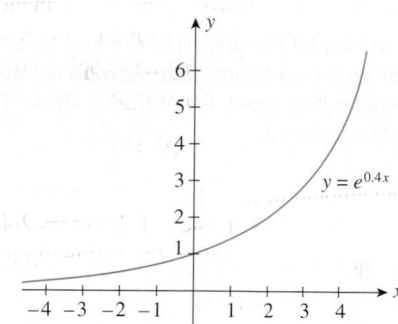

USING TECHNOLOGY

Although the proof is outside the scope of this book, it can be proved that an exponential function of the form $f(x) = b^x$, where $b > 1$, will ultimately grow faster than the power function $g(x) = x^n$ for *any* positive real number n. To give a visual demonstration of this result for the special case of the exponential function $f(x) = e^x$, we can use a graphing utility to plot the graphs of both f and g (for selected values of n) on the same set of axes in an appropriate viewing window and observe that the graph of f ultimately lies above that of g.

EXAMPLE 1 Use a graphing utility to plot the graphs of (a) $f(x) = e^x$ and $g(x) = x^3$ on the same set of axes in the viewing window $[0, 6] \times [0, 250]$ and (b) $f(x) = e^x$ and $g(x) = x^5$ in the viewing window $[0, 20] \times [0, 1{,}000{,}000]$.

Solution

a. The graphs of $f(x) = e^x$ and $g(x) = x^3$ in the viewing window $[0, 6] \times [0, 250]$ are shown in Figure T1a.

b. The graphs of $f(x) = e^x$ and $g(x) = x^5$ in the viewing window $[0, 20] \times [0, 1{,}000{,}000]$ are shown in Figure T1b.

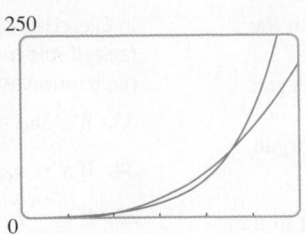

(a) The graphs of $f(x) = e^x$ and $g(x) = x^3$ in the viewing window $[0, 6] \times [0, 250]$

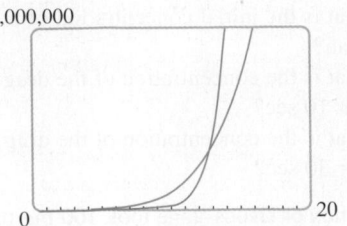

(b) The graphs of $f(x) = e^x$ and $g(x) = x^5$ in the viewing window $[0, 20] \times [0, 1,000,000]$

FIGURE **T1**

In the exercises that follow, you are asked to use a graphing utility to reveal the properties of exponential functions.

TECHNOLOGY EXERCISES

In Exercises 1 and 2, plot the graphs of the functions f and g on the same set of axes in the specified viewing window.

1. $f(x) = e^x$ and $g(x) = x^2$; $[0, 4] \times [0, 30]$

2. $f(x) = e^x$ and $g(x) = x^4$; $[0, 15] \times [0, 20{,}000]$

In Exercises 3 and 4, plot the graphs of the functions f and g on the same set of axes in an appropriate viewing window to demonstrate that f ultimately grows faster than g. (*Note:* Your answer will *not* be unique.)

3. $f(x) = 2^x$ and $g(x) = x^{2.5}$

4. $f(x) = 3^x$ and $g(x) = x^3$

5. Plot the graphs of $f(x) = 2^x$, $g(x) = 3^x$, and $h(x) = 4^x$ on the same set of axes in the viewing window $[0, 5] \times [0, 100]$. Comment on the relationship between the base b and the growth of the function $f(x) = b^x$.

6. Plot the graphs of $f(x) = (1/2)^x$, $g(x) = (1/3)^x$, and $h(x) = (1/4)^x$ on the same set of axes in the viewing window $[0, 4] \times [0, 1]$. Comment on the relationship between the base b and the growth of the function $f(x) = b^x$.

7. Plot the graphs of $f(x) = e^x$, $g(x) = 2e^x$, and $h(x) = 3e^x$ on the same set of axes in the viewing window $[-3, 3] \times [0, 10]$. Comment on the role played by the constant k in the graph of $f(x) = ke^x$.

8. Plot the graphs of $f(x) = -e^x$, $g(x) = -2e^x$, and $h(x) = -3e^x$ on the same set of axes in the viewing window $[-3, 3] \times [-10, 0]$. Comment on the role played by the constant k in the graph of $f(x) = ke^x$.

9. Plot the graphs of $f(x) = e^{0.5x}$, $g(x) = e^x$, and $h(x) = e^{1.5x}$ on the same set of axes in the viewing window $[-2, 2] \times [0, 4]$. Comment on the role played by the constant k in the graph of $f(x) = e^{kx}$.

10. Plot the graphs of $f(x) = e^{-0.5x}$, $g(x) = e^{-x}$, and $h(x) = e^{-1.5x}$ on the same set of axes in the viewing window $[-2, 2] \times [0, 4]$. Comment on the role played by the constant k in the graph of $f(x) = e^{kx}$.

11. ABSORPTION OF DRUGS The concentration of a drug in an organ at any time t (in seconds) is given by
$$x(t) = 0.08 + 0.12(1 - e^{-0.02t})$$
where $x(t)$ is measured in grams per cubic centimeter (g/cm^3).

a. Plot the graph of the function x in the viewing window $[0, 200] \times [0, 0.2]$.

b. What is the initial concentration of the drug in the organ?

c. What is the concentration of the drug in the organ after 20 sec?

12. ABSORPTION OF DRUGS Jane took 100 mg of a drug in the morning and another 100 mg of the same drug at the same time the following morning. The amount of the drug in her body t days after the first dosage was taken is given by
$$A(t) = \begin{cases} 100e^{-1.4t} & \text{if } 0 \le t < 1 \\ 100(1 + e^{1.4})e^{-1.4t} & \text{if } t \ge 1 \end{cases}$$

a. Plot the graph of the function A in the viewing window $[0, 5] \times [0, 140]$.

b. Verify the results of Exercise 32, page 179.

13. ABSORPTION OF DRUGS The concentration of a drug in an organ at any time t (in seconds) is given by
$$C(t) = \begin{cases} 0.3t - 18(1 - e^{-t/60}) & \text{if } 0 \le t \le 20 \\ 18e^{-t/60} - 12e^{-(t-20)/60} & \text{if } t > 20 \end{cases}$$
where $C(t)$ is measured in grams per cubic centimeter (g/cm^3).

a. Plot the graph of the function C in the viewing window $[0, 120] \times [0, 1]$.

b. How long after the drug is first introduced will it take for the concentration of the drug to reach a peak?

c. How long after the concentration of the drug has peaked will it take for the concentration of the drug to fall back to 0.5 g/cm³?

Hint: Plot the graphs of $y_1 = C(x)$ and $y_2 = 0.5$, and use the ISECT function of your graphing utility.

14. MODELING WITH DATA The estimated number of Internet users in China (in millions) from 2005 through 2010 are shown in the following table:

Year	2005	2006	2007	2008	2009	2010
Number	116.1	141.9	169.0	209.0	258.1	314.8

a. Use **ExpReg** to find an exponential regression model for the data. Let $t = 1$ correspond to 2005.
Hint: $a^x = e^{x \ln a}$

b. Plot the scatter diagram and the graph of the function f found in part (a).

3.2 Logarithmic Functions

Logarithms

You are already familiar with exponential equations of the form

$$b^y = x \qquad (b > 0, b \neq 1)$$

where the variable x is expressed in terms of a real number b and a variable y. But what about solving this same equation for y? You may recall from your study of algebra that the number y is called the **logarithm of x to the base b** and is denoted by $\log_b x$. It is the power to which the base b must be raised to obtain the number x.

Logarithm of x to the Base b

$$y = \log_b x \quad \text{if and only if} \quad x = b^y \qquad (b > 0, b \neq 1, \text{ and } x > 0)$$

⚠ Observe that the logarithm $\log_b x$ is defined only for positive values of x.

EXAMPLE 1

a. $\log_{10} 100 = 2$ since $100 = 10^2$

b. $\log_5 125 = 3$ since $125 = 5^3$

c. $\log_3 \dfrac{1}{27} = -3$ since $\dfrac{1}{27} = \dfrac{1}{3^3} = 3^{-3}$

d. $\log_{20} 20 = 1$ since $20 = 20^1$

EXAMPLE 2 Solve each of the following equations for x.

a. $\log_3 x = 4$ **b.** $\log_{16} 4 = x$ **c.** $\log_x 8 = 3$

Solution

a. By definition, $\log_3 x = 4$ implies $x = 3^4 = 81$.

b. $\log_{16} 4 = x$ is equivalent to $4 = 16^x = (4^2)^x = 4^{2x}$, or $4^1 = 4^{2x}$, from which we deduce that

$$2x = 1 \qquad b^m = b^n \Rightarrow m = n$$

$$x = \frac{1}{2}$$

c. Referring once again to the definition, we see that the equation $\log_x 8 = 3$ is equivalent to

$$8 = 2^3 = x^3$$
$$x = 2 \qquad a^m = b^m \Rightarrow a = b$$

The two most widely used systems of logarithms are the system of **common logarithms,** which uses the number 10 as its base, and the system of **natural logarithms,** which uses the irrational number $e = 2.71828\ldots$ as its base. Also, it is standard practice to write **log** for $\log_{10}$ and **ln** for $\log_e$.

Logarithmic Notation

$$\log x = \log_{10} x \qquad \text{Common logarithm}$$
$$\ln x = \log_e x \qquad \text{Natural logarithm}$$

The system of natural logarithms is widely used in theoretical work. Using natural logarithms rather than logarithms to other bases often leads to simpler expressions.

Laws of Logarithms

Computations involving logarithms are facilitated by the following **laws of logarithms.**

Laws of Logarithms

If m and n are positive numbers and $b > 0$, $b \neq 1$, then

1. $\log_b mn = \log_b m + \log_b n$

2. $\log_b \dfrac{m}{n} = \log_b m - \log_b n$

3. $\log_b m^n = n \log_b m$

4. $\log_b 1 = 0$

5. $\log_b b = 1$

⚠ Do not confuse the expression $\log m/n$ (Law 2) with the expression $\log m/\log n$. For example,

$$\log \frac{100}{10} = \log 100 - \log 10 = 2 - 1 = 1 \neq \frac{\log 100}{\log 10} = \frac{2}{1} = 2$$

You will be asked to prove these laws in Exercises 78–80 on page 189. Their derivations are based on the definition of a logarithm and the corresponding properties of exponents. The following examples illustrate the laws of logarithms.

EXAMPLE 3

a. $\log(2 \cdot 3) = \log 2 + \log 3$ **b.** $\ln \dfrac{5}{3} = \ln 5 - \ln 3$

c. $\log \sqrt{7} = \log 7^{1/2} = \dfrac{1}{2} \log 7$ **d.** $\log_5 1 = 0$

e. $\log_{45} 45 = 1$

EXAMPLE 4 Given that $\log 2 \approx 0.3010$, $\log 3 \approx 0.4771$, and $\log 5 \approx 0.6990$, use the laws of logarithms to find

a. $\log 15$ **b.** $\log 7.5$ **c.** $\log 81$ **d.** $\log 50$

Solution

a. Note that $15 = 3 \cdot 5$, so by Law 1 for logarithms,

$$
\begin{aligned}
\log 15 &= \log 3 \cdot 5 \\
&= \log 3 + \log 5 \\
&\approx 0.4771 + 0.6990 \\
&= 1.1761
\end{aligned}
$$

b. Observing that $7.5 = 15/2 = (3 \cdot 5)/2$, we apply Laws 1 and 2, obtaining

$$
\begin{aligned}
\log 7.5 &= \log \frac{(3)(5)}{2} \\
&= \log 3 + \log 5 - \log 2 \\
&\approx 0.4771 + 0.6990 - 0.3010 \\
&= 0.8751
\end{aligned}
$$

c. Since $81 = 3^4$, we apply Law 3 to obtain

$$
\begin{aligned}
\log 81 &= \log 3^4 \\
&= 4 \log 3 \\
&\approx 4(0.4771) \\
&= 1.9084
\end{aligned}
$$

d. We write $50 = 5 \cdot 10$ and find

$$
\begin{aligned}
\log 50 &= \log(5)(10) \\
&= \log 5 + \log 10 \\
\log 50 &\approx 0.6990 + 1 \qquad \text{Use Law 5} \\
&= 1.6990
\end{aligned}
$$

EXAMPLE 5 Expand and simplify the following expressions:

a. $\log_3 x^2 y^3$ **b.** $\log_2 \dfrac{x^2 + 1}{2^x}$ **c.** $\ln \dfrac{x^2\sqrt{x^2 - 1}}{e^x}$

Solution

a.
$$
\begin{aligned}
\log_3 x^2 y^3 &= \log_3 x^2 + \log_3 y^3 \qquad && \text{Law 1} \\
&= 2 \log_3 x + 3 \log_3 y \qquad && \text{Law 3}
\end{aligned}
$$

b.
$$
\begin{aligned}
\log_2 \frac{x^2 + 1}{2^x} &= \log_2(x^2 + 1) - \log_2 2^x \qquad && \text{Law 2} \\
&= \log_2(x^2 + 1) - x \log_2 2 \qquad && \text{Law 3} \\
&= \log_2(x^2 + 1) - x \qquad && \text{Law 5}
\end{aligned}
$$

c.
$$
\begin{aligned}
\ln \frac{x^2\sqrt{x^2 - 1}}{e^x} &= \ln \frac{x^2(x^2 - 1)^{1/2}}{e^x} \qquad && \text{Rewrite} \\
&= \ln x^2 + \ln(x^2 - 1)^{1/2} - \ln e^x \qquad && \text{Laws 1 and 2} \\
&= 2 \ln x + \frac{1}{2} \ln(x^2 - 1) - x \ln e \qquad && \text{Law 3} \\
&= 2 \ln x + \frac{1}{2} \ln(x^2 - 1) - x \qquad && \text{Law 5}
\end{aligned}
$$

Solving Equations Involving Logarithms

Examples 6 and 7 illustrate how the laws of logarithms are used to solve equations.

EXAMPLE 6 Solve $\log_3(x + 1) - \log_3(x - 1) = 1$ for x.

Solution Using the laws of logarithms, we obtain

$$\log_3(x + 1) - \log_3(x - 1) = 1$$

$$\log_3 \frac{x + 1}{x - 1} = 1 \qquad \text{Law 2}$$

$$\frac{x + 1}{x - 1} = 3^1 = 3 \qquad \text{Definition of logarithms}$$

So

$$x + 1 = 3(x - 1)$$
$$x + 1 = 3x - 3$$
$$4 = 2x$$
$$x = 2$$

EXAMPLE 7 Solve $\log x + \log(2x - 1) = \log 6$.

Solution We have

$$\log x + \log(2x - 1) = \log 6$$
$$\log x + \log(2x - 1) - \log 6 = 0$$

$$\log\left[\frac{x(2x - 1)}{6}\right] = 0 \qquad \text{Laws 1 and 2}$$

$$\frac{x(2x - 1)}{6} = 10^0 = 1 \qquad \text{Definition of logarithms}$$

So

$$x(2x - 1) = 6$$
$$2x^2 - x - 6 = 0$$
$$(2x + 3)(x - 2) = 0$$

$$x = -\frac{3}{2} \quad \text{or} \quad 2$$

Because $(2x - 1)$ is defined for $2x - 1 > 0$, or $x > \frac{1}{2}$, the domain of $\log(2x - 1)$ is the interval $\left(\frac{1}{2}, \infty\right)$. So we reject the root $-\frac{3}{2}$ of the quadratic equation and conclude that the solution of the given equation is $x = 2$.

Note Using the fact that $\log a = \log b$ if and only if $a = b$, we can also solve the equation of Example 7 as follows:

$$\log x + \log(2x - 1) = \log 6$$
$$\log x(2x - 1) = \log 6$$
$$x(2x - 1) = 6$$

The rest of the solution is the same as that in Example 7.

Logarithmic Functions and Their Graphs

The definition of a logarithm implies that if b and n are positive numbers and b is different from 1, then the expression $\log_b n$ is a real number. This enables us to define a logarithmic function as follows.

> **Logarithmic Function**
>
> The function defined by
>
> $$f(x) = \log_b x \qquad (b > 0, b \neq 1)$$
>
> is called the **logarithmic function with base b.** The domain of f is the set of all positive numbers.

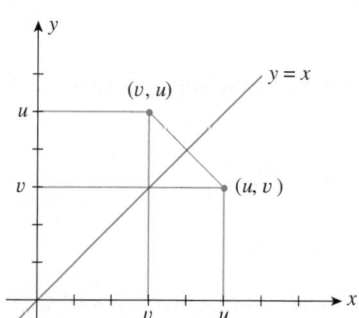

FIGURE 7
The points (u, v) and (v, u) are mirror reflections of each other.

One easy way to obtain the graph of the logarithmic function $y = \log_b x$ is to construct a table of values of the logarithm (base b). However, another method—and a more instructive one—is based on exploiting the intimate relationship between logarithmic and exponential functions.

If a point (u, v) lies on the graph of $y = \log_b x$, then

$$v = \log_b u$$

But we can also write this equation in exponential form as

$$u = b^v$$

So the point (v, u) also lies on the graph of the function $y = b^x$. Let's look at the relationship between the points (u, v) and (v, u) and the line $y = x$ (Figure 7). If we think of the line $y = x$ as a mirror, then the point (v, u) is the mirror reflection of the point (u, v). Similarly, the point (u, v) is the mirror reflection of the point (v, u). We can take advantage of this relationship to help us draw the graph of logarithmic functions. For example, if we wish to draw the graph of $y = \log_b x$, where $b > 1$, then we need only draw the mirror reflection of the graph of $y = b^x$ with respect to the line $y = x$ (Figure 8).

You may discover the following properties of the logarithmic function by taking the reflection of the graph of an appropriate exponential function (Exercises 47 and 48 on page 188).

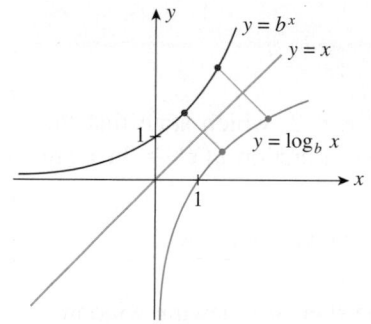

FIGURE 8
The graphs of $y = b^x$ and $y = \log_b x$ are mirror reflections of each other.

> **Properties of the Logarithmic Function**
>
> The logarithmic function $y = \log_b x$ ($b > 0$, $b \neq 1$) has the following properties:
>
> **1.** Its domain is $(0, \infty)$.
> **2.** Its range is $(-\infty, \infty)$.
> **3.** Its graph passes through the point $(1, 0)$.
> **4.** Its graph is an unbroken curve devoid of holes or jumps.
> **5.** Its graph rises from left to right if $b > 1$ and falls from left to right if $b < 1$.

EXAMPLE 8 Sketch the graph of the function $y = \ln x$.

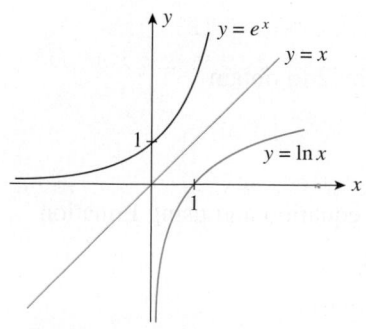

FIGURE 9
The graph of $y = \ln x$ is the mirror reflection of the graph of $y = e^x$.

Solution We first sketch the graph of $y = e^x$. Then, the required graph is obtained by tracing the mirror reflection of the graph of $y = e^x$ with respect to the line $y = x$ (Figure 9). ∎

Properties Relating the Exponential and Logarithmic Functions

We made use of the relationship that exists between the exponential function $f(x) = e^x$ and the logarithmic function $g(x) = \ln x$ when we sketched the graph of g in Example 8. This relationship is further described by the following properties, which are an immediate consequence of the definition of the logarithm of a number.

> **Properties Relating e^x and $\ln x$**
>
> $$e^{\ln x} = x \qquad \text{(for } x > 0\text{)} \qquad\qquad (1)$$
>
> $$\ln e^x = x \qquad \text{(for any real number } x\text{)} \qquad\qquad (2)$$

(Try to verify these properties.)

From Properties 1 and 2, we conclude that the composite function satisfies

$$(f \circ g)(x) = f[g(x)]$$
$$= e^{\ln x} = x \qquad \text{(for all } x > 0\text{)}$$
$$(g \circ f)(x) = g[f(x)]$$
$$= \ln e^x = x \qquad \text{(for all } x > 0\text{)}$$

Any two functions f and g that satisfy this relationship are said to be **inverses** of each other. Note that the function f undoes what the function g does, and vice versa, so the composition of the two functions in any order results in the identity function $F(x) = x$.

The relationships expressed in Equations (1) and (2) are useful in solving equations that involve exponentials and logarithms.

> **Exploring with TECHNOLOGY**
>
> You can demonstrate the validity of Properties 1 and 2, which state that the exponential function $f(x) = e^x$ and the logarithmic function $g(x) = \ln x$ are inverses of each other, as follows:
>
> 1. Sketch the graph of $(f \circ g)(x) = e^{\ln x}$, using the viewing window $[0, 10] \times [0, 10]$. Interpret the result.
> 2. Sketch the graph of $(g \circ f)(x) = \ln e^x$, using the standard viewing window. Interpret the result.

EXAMPLE 9 Solve the equation $2e^{x+2} = 5$.

Solution We first divide both sides of the equation by 2 to obtain

$$e^{x+2} = \frac{5}{2} = 2.5$$

Next, taking the natural logarithm of each side of the equation and using Equation (2), we have

$$\ln e^{x+2} = \ln 2.5$$
$$x + 2 = \ln 2.5$$
$$x = -2 + \ln 2.5$$
$$\approx -1.08$$

Explore and Discuss

Consider the equation $y = y_0 b^{kx}$, where y_0 and k are positive constants and $b > 0$, $b \neq 1$. Suppose we want to express y in the form $y = y_0 e^{px}$. Use the laws of logarithms to show that $p = k \ln b$ and hence that $y = y_0 e^{(k \ln b)x}$ is an alternative form of $y = y_0 b^{kx}$ using the base e.

EXAMPLE 10 Solve the equation $5 \ln x + 3 = 0$.

Solution Adding -3 to both sides of the equation leads to

$$5 \ln x = -3$$

$$\ln x = -\frac{3}{5} = -0.6$$

and so

$$e^{\ln x} = e^{-0.6}$$

Using Equation (1), we conclude that

$$x = e^{-0.6}$$

$$\approx 0.55$$

3.2 Self-Check Exercises

1. Sketch the graph of $y = 3^x$ and $y = \log_3 x$ on the same set of axes.

2. Solve the equation $3e^{x+1} - 2 = 4$.

Solutions to Self-Check Exercises 3.2 can be found on page 190.

3.2 Concept Questions

1. **a.** Define $y = \log_b x$.
   b. Define the logarithmic function f with base b. What restrictions, if any, are placed on b?

2. For the logarithmic function $y = \log_b x$ ($b > 0$, $b \neq 1$), state (a) its domain and range, (b) its x-intercept, (c) where its graph rises and where it falls for the case $b > 1$ and the case $b < 1$.

3. **a.** If $x > 0$, what is $e^{\ln x}$?
   b. If x is any real number, what is $\ln e^x$?

4. Let $f(x) = \ln x^2$ and $g(x) = 2 \ln x$. Are f and g identical? Hint: Look at their domains.

3.2 Exercises

In Exercises 1–10, express each equation in logarithmic form.

1. $2^6 = 64$

2. $3^5 = 243$

3. $4^{-2} = \dfrac{1}{16}$

4. $5^{-3} = \dfrac{1}{125}$

5. $\left(\dfrac{1}{3}\right)^1 = \dfrac{1}{3}$

6. $\left(\dfrac{1}{2}\right)^{-4} = 16$

7. $32^{4/5} = 16$

8. $81^{3/4} = 27$

9. $10^{-3} = 0.001$

10. $16^{-1/4} = 0.5$

In Exercises 11–16, given that $\log 3 \approx 0.4771$ and $\log 4 \approx 0.6021$, find the value of each logarithm.

11. $\log 12$

12. $\log \dfrac{3}{4}$

13. $\log 16$

14. $\log \sqrt{3}$

15. $\log 48$

16. $\log \dfrac{1}{300}$

In Exercises 17–20, write the expression as the logarithm of a single quantity.

17. $2 \ln a + 3 \ln b$

18. $\dfrac{1}{2} \ln x + 2 \ln y - 3 \ln z$

19. $\ln 3 + \dfrac{1}{2} \ln x + \ln y - \dfrac{1}{3} \ln z$

20. $\ln 2 + \dfrac{1}{2} \ln(x + 1) - 2 \ln(1 + \sqrt{x})$

In Exercises 21–28, use the laws of logarithms to expand and simplify the expression.

21. $\log x(x + 1)^4$

22. $\log x(x^2 + 1)^{-1/2}$

23. $\log \dfrac{\sqrt{x + 1}}{x^2 + 1}$

24. $\ln \dfrac{e^x}{1 + e^x}$

25. $\ln xe^{-x^2}$

26. $\ln x(x + 1)(x + 2)$

27. $\ln \dfrac{x^{1/2}}{x^2 \sqrt{1 + x^2}}$

28. $\ln \dfrac{x^2}{\sqrt{x}(1 + x)^2}$

In Exercises 29–42, use the laws of logarithms to solve the equation.

29. $\log_2 x = 3$

30. $\log_3 x = 2$

31. $\log_2 8 = x$

32. $\log_3 27 = 2x$

33. $\log_x 10^3 = 3$

34. $\log_x \dfrac{1}{16} = -2$

35. $\log_2(2x + 5) = 4$

36. $\log_4(5x - 4) = 2$

37. $\log_2 x - \log_2(x - 2) = 3$

38. $\log x - \log(x + 6) = -1$

39. $\log_5(2x + 1) - \log_5(x - 2) = 1$

40. $\log(x + 7) - \log(x - 2) = 1$

41. $\log x + \log(2x - 5) = \log 3$

42. $\log_3(x + 1) + \log_3(2x - 3) = 1$

In Exercises 43–46, sketch the graph of the equation.

43. $y = \log_3 x$

44. $y = \log_{1/3} x$

45. $y = \ln 2x$

46. $y = \ln \dfrac{1}{2} x$

In Exercises 47 and 48, sketch the graphs of the equations on the same coordinate axes.

47. $y = 2^x$ and $y = \log_2 x$

48. $y = e^{3x}$ and $y = \dfrac{1}{3} \ln x$

In Exercises 49–58, use logarithms to solve the equation for t.

49. $e^{0.4t} = 8$

50. $\dfrac{1}{3} e^{-3t} = 0.9$

51. $5e^{-2t} = 6$

52. $4e^{t-1} = 4$

53. $2e^{-0.2t} - 4 = 6$

54. $12 - e^{0.4t} = 3$

55. $\dfrac{50}{1 + 4e^{0.2t}} = 20$

56. $\dfrac{200}{1 + 3e^{-0.3t}} = 100$

57. $A = Be^{-t/2}$

58. $\dfrac{A}{1 + Be^{t/2}} = C$

59. A function f has the form $f(x) = a + b \ln x$. Find f if it is known that $f(1) = 2$ and $f(2) = 4$.

60. **AVERAGE LIFE SPAN** One reason for the increase in human life span over the years has been the advances in medical technology. The average life span for American women from 1907 through 2007 is given by

$$W(t) = 49.9 + 17.1 \ln t \qquad (1 \le t \le 6)$$

where $W(t)$ is measured in years and t is measured in 20-year intervals, with $t = 1$ corresponding to 1907.

   a. What was the average life expectancy for women in 1907?

   b. If the trend continues, what will be the average life expectancy for women in 2027 ($t = 7$)?

Source: American Association of Retired Persons (AARP).

61. **BLOOD PRESSURE** A normal child's systolic blood pressure may be approximated by the function

$$p(x) = m(\ln x) + b$$

where $p(x)$ is measured in millimeters of mercury, x is measured in pounds, and m and b are constants. Given that $m = 19.4$ and $b = 18$, determine the systolic blood pressure of a child who weighs 92 lb.

62. **MAGNITUDE OF EARTHQUAKES** On the Richter scale, the magnitude R of an earthquake is given by the formula

$$R = \log \dfrac{I}{I_0}$$

where I is the intensity of the earthquake being measured and I_0 is the standard reference intensity.

   a. Express the intensity I of an earthquake of magnitude $R = 5$ in terms of the standard intensity I_0.

   b. Express the intensity I of an earthquake of magnitude $R = 8$ in terms of the standard intensity I_0. How many times greater is the intensity of an earthquake of magnitude 8 than one of magnitude 5?

   c. In modern times, the greatest loss of life attributable to an earthquake occurred in Haiti in 2010. Known as the Haiti earthquake, it registered 7.0 on the Richter scale. How does the intensity of this earthquake compare with the intensity of an earthquake of magnitude $R = 5$?

63. **SOUND INTENSITY** The relative loudness of a sound D of intensity I is measured in decibels (db), where

$$D = 10 \log \dfrac{I}{I_0}$$

and I_0 is the standard threshold of audibility.

   a. Express the intensity I of a 30-db sound (the sound level of normal conversation) in terms of I_0.

   b. Determine how many times greater the intensity of an 80-db sound (rock music) is than that of a 30-db sound.

   c. Prolonged noise above 150 db causes permanent deafness. How does the intensity of a 150-db sound compare with that of an 80-db sound?

64. BAROMETRIC PRESSURE Halley's Law states that the barometric pressure (in inches of mercury) at an altitude of x mi above sea level is approximated by the equation

$$p(x) = 29.92e^{-0.2x} \qquad (x \geq 0)$$

If the barometric pressure as measured by a hot-air balloonist is 20 in. of mercury, what is the balloonist's altitude?

65. NEWTON'S LAW OF COOLING The temperature of a cup of coffee t min after it is poured is given by

$$T = 70 + 100e^{-0.0446t}$$

where T is measured in degrees Fahrenheit.
a. What was the temperature of the coffee when it was poured?
b. When will the coffee be cool enough to drink (say, 120°F)?

66. HEIGHT OF TREES The height (in feet) of a certain kind of tree is approximated by

$$h(t) = \frac{160}{1 + 240e^{-0.2t}}$$

where t is the age of the tree in years. Estimate the age of an 80-ft tree.

67. OBESITY OVER TIME The percentage of obese adults, ages 20 through 74 years, is projected to be

$$f(t) = \frac{46.5}{1 + 2.324e^{-0.05113t}} \qquad (0 \leq t \leq 60)$$

in year t, with $t = 0$ corresponding to 1970. Estimate the year when the percentage reached 40%.
Source: Centers for Disease Control and Prevention.

68. LENGTHS OF FISH The length (in centimeters) of a typical Pacific halibut t years old is approximately

$$f(t) = 200(1 - 0.956e^{-0.18t})$$

Suppose a Pacific halibut caught by Mike measures 140 cm. What is its approximate age?

69. ABSORPTION OF DRUGS The concentration of a drug in an organ t seconds after it has been administered is given by

$$x(t) = 0.08 + 0.12e^{-0.02t}$$

where $x(t)$ is measured in grams per cubic centimeter (g/cm^3).
a. How long would it take for the concentration of the drug in the organ to reach 0.18 g/cm^3?
b. How long would it take for the concentration of the drug in the organ to reach 0.16 g/cm^3?

70. ABSORPTION OF DRUGS The concentration of a drug in an organ t seconds after it has been administered is given by

$$x(t) = 0.08(1 - e^{-0.02t})$$

where $x(t)$ is measured in grams per cubic centimeter (g/cm^3).
a. How long would it take for the concentration of the drug in the organ to reach 0.02 g/cm^3?
b. How long would it take for the concentration of the drug in the organ to reach 0.04 g/cm^3?

71. FORENSIC SCIENCE Forensic scientists use the following law to determine the time of death of accident or murder victims. If T denotes the temperature of a body t hr after death, then

$$T = T_0 + (T_1 - T_0)(0.97)^t$$

where T_0 is the air temperature and T_1 is the body temperature at the time of death. John Doe was found murdered at midnight in his house; the room temperature was 70°F, and his body temperature was 80°F when he was found. When was he killed? Assume that the normal body temperature is 98.6°F.

In Exercises 72–76, determine whether the statement is true or false. If it is true, explain why it is true. If it is false, give an example to show why it is false.

72. $(\ln x)^3 = 3 \ln x$ for all x in $(0, \infty)$.

73. If $a > 0$ and $b > 0$, then $\ln(a + b) = \ln a + \ln b$.

74. If $b > 0$, then $e^{\ln b} = \ln e^b$.

75. $\ln a - \ln b = \ln(a - b)$ for all positive real numbers a and b.

76. $(\log_2 3)(\log_3 2) = 1$

77. a. Given that $2^x = e^{kx}$, find k.
b. Show that, in general, if b is a positive real number, then any equation of the form $y = b^x$ may be written in the form $y = e^{kx}$, for some real number k.

78. Use the definition of a logarithm to prove
a. $\log_b mn = \log_b m + \log_b n$
b. $\log_b \dfrac{m}{n} = \log_b m - \log_b n$

Hint: Let $\log_b m = p$ and $\log_b n = q$. Then, $b^p = m$ and $b^q = n$.

79. Use the definition of a logarithm to prove

$$\log_b m^n = n \log_b m$$

80. Use the definition of a logarithm to prove
a. $\log_b 1 = 0$
b. $\log_b b = 1$

3.2   Solutions to Self-Check Exercises

1. First, sketch the graph of $y = 3^x$ with the help of the following table of values:

x	-3	-2	-1	0	1	2	3
$y = 3^x$	$\frac{1}{27}$	$\frac{1}{9}$	$\frac{1}{3}$	1	3	9	27

Next, take the mirror reflection of this graph with respect to the line $y = x$ to obtain the graph of $y = \log_3 x$.

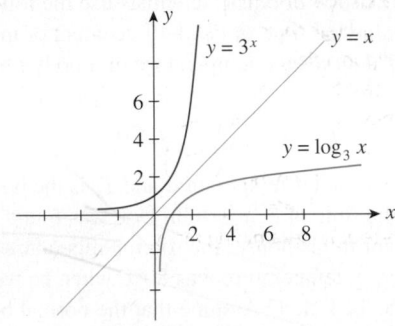

2.
$$3e^{x+1} - 2 = 4$$
$$3e^{x+1} = 6$$
$$e^{x+1} = 2$$
$$\ln e^{x+1} = \ln 2 \quad \text{Take the logarithm of both sides.}$$
$$(x + 1)\ln e = \ln 2 \quad \text{Law 3}$$
$$x + 1 = \ln 2 \quad \text{Law 5}$$
$$x = \ln 2 - 1$$
$$\approx -0.3069$$

3.3   Exponential Functions as Mathematical Models

Exponential Growth

Many problems arising from practical situations can be described mathematically in terms of exponential functions or functions closely related to the exponential function. In this section, we look at some applications involving exponential functions from the fields of the life and social sciences.

In Section 3.1, we saw that the exponential function $f(x) = b^x$ is an increasing function when $b > 1$. In particular, the function $f(x) = e^x$ has this property. Suppose that $Q(t)$ represents a quantity at time t, then one may deduce that the function $Q(t) = Q_0 e^{kt}$, where Q_0 and k are positive constants, has the following properties:

1. $Q(0) = Q_0$
2. $Q(t)$ increases "rapidly" without bound as t increases without bound (Figure 10).

Property 1 follows from the computation

$$Q(0) = Q_0 e^0 = Q_0$$

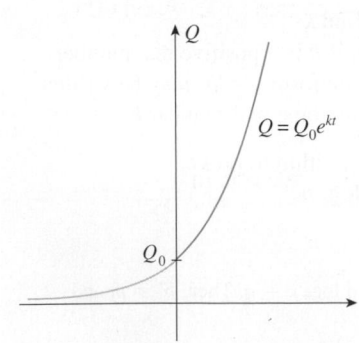

FIGURE **10**
Exponential growth

The exponential function

$$Q(t) = Q_0 e^{kt} \quad (0 \le t < \infty) \tag{3}$$

provides us with a mathematical model of a quantity $Q(t)$ that is initially present in the amount of $Q(0) = Q_0$ and whose rate of growth at any time t is directly proportional to the amount of the quantity present at time t. (See Example 5, Section 9.7.) Such a quantity is said to exhibit unrestricted **exponential growth,** and the constant k of proportionality is called the **growth constant.** Interest earned on a fixed deposit when compounded continuously exhibits exponential growth (Chapter 4). Other examples of unrestricted exponential growth follow.

PORTFOLIO Carol A. Reeb, Ph.D.

TITLE Research Associate
INSTITUTION Hopkins Marine Station, Stanford University

Historically, the world's oceans were thought to provide an unlimited source of inexpensive seafood. However, in a world in which the human population now exceeds six billion people, overfishing has pushed one third of all marine fishery stocks toward a state of collapse.

As a fishery geneticist at Hopkins Marine Station, I study commercially harvested marine populations and use exponential models in my work. The equation for determining the size of a population that grows or declines exponentially is $x_t = x_0 e^{rt}$, where x_0 is the initial population, t is time, and r is the growth or decay constant (positive for growth, negative for decay).

This equation can be used to estimate the population in the past as well as in the future. We know that the demand for seafood increased as the human population grew, eventually causing fish populations to decline. Because genetic diversity is linked to population size, the exponential function is useful to model change in fishery populations and their gene pools over time.

Interestingly, exponential functions can also be used to model the increase in the market value of seafood in the United States over the past 60 years. In general, the price of seafood has increased exponentially, although the price did stabilize briefly in 1995.

Although exponential curves are important to my work, they are not always the best fit. Exponential curves are best applied across short time frames when environments or markets are unlimited. Over longer periods, the logistic growth function is more suitable. In my research, selecting the most accurate model requires examining many possibilities.

APPLIED EXAMPLE 1 Growth of Bacteria Under ideal laboratory conditions, the number of bacteria in a culture grows in accordance with the law $Q(t) = Q_0 e^{kt}$, where Q_0 denotes the number of bacteria initially present in the culture, k is a constant determined by the strain of bacteria under consideration and other factors, and t is the elapsed time measured in hours. Suppose 10,000 bacteria are present initially in the culture and 60,000 are present 2 hours later. How many bacteria will there be in the culture at the end of 4 hours?

Solution We are given that $Q(0) = Q_0 = 10{,}000$, so $Q(t) = 10{,}000 e^{kt}$. Next, the fact that 60,000 bacteria are present 2 hours later translates into $Q(2) = 60{,}000$. Thus,

$$60{,}000 = 10{,}000 e^{2k}$$
$$e^{2k} = 6$$

Taking the natural logarithm on both sides of the equation, we obtain

$$\ln e^{2k} = \ln 6$$
$$2k = \ln 6 \qquad \text{Since } \ln e = 1$$
$$k = \frac{\ln 6}{2}$$
$$k \approx 0.8959$$

Thus, the number of bacteria present at any time t is given by

$$Q(t) \approx 10{,}000 e^{0.8959t}$$

In particular, the number of bacteria present in the culture at the end of 4 hours is given by

$$Q(4) \approx 10{,}000 e^{0.8959(4)}$$
$$\approx 360{,}000$$

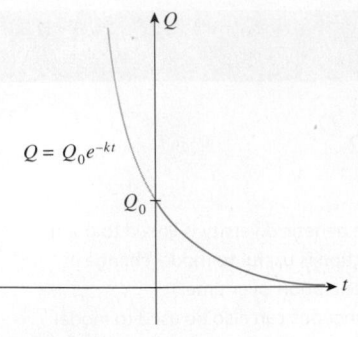

FIGURE **11**
Exponential decay

Exponential Decay

In contrast to exponential growth, a quantity exhibits **exponential decay** if it decreases at a rate that is directly proportional to its size. Such a quantity may be described by the exponential function

$$Q(t) = Q_0 e^{-kt} \qquad (0 \le t < \infty) \tag{4}$$

where the positive constant Q_0 measures the amount present initially $(t = 0)$ and k is some suitable positive number, called the **decay constant.** The choice of this number is determined by the nature of the substance under consideration and other factors. The graph of this function is sketched in Figure 11.

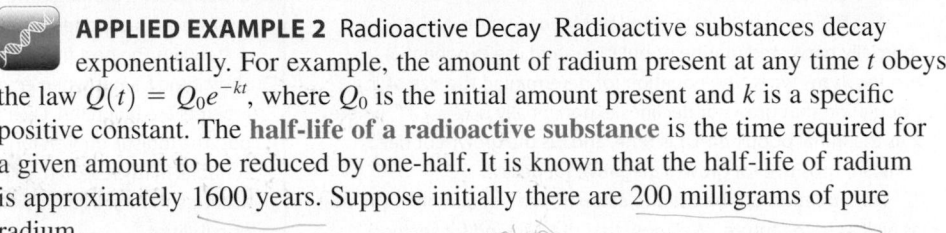

APPLIED EXAMPLE 2 Radioactive Decay Radioactive substances decay exponentially. For example, the amount of radium present at any time t obeys the law $Q(t) = Q_0 e^{-kt}$, where Q_0 is the initial amount present and k is a specific positive constant. The **half-life of a radioactive substance** is the time required for a given amount to be reduced by one-half. It is known that the half-life of radium is approximately 1600 years. Suppose initially there are 200 milligrams of pure radium.

a. Find the amount left after t years.
b. What is the amount left after 800 years?

Solution

a. The initial amount of radium present is 200 milligrams, so $Q(0) = Q_0 = 200$. Thus, $Q(t) = 200e^{-kt}$. Next, the datum concerning the half-life of radium implies that $Q(1600) = 100$, and this gives

$$100 = 200e^{-1600k}$$

$$e^{-1600k} = \frac{1}{2}$$

Taking the natural logarithm on both sides of this equation yields

$$-1600k \ln e = \ln \frac{1}{2}$$

$$-1600k = \ln \frac{1}{2} \qquad \ln e = 1$$

$$k = -\frac{1}{1600} \ln\left(\frac{1}{2}\right) \approx 0.0004332$$

Therefore, the amount of radium left after t years is

$$Q(t) = 200e^{-0.0004332t}$$

b. The amount of radium left after 800 years is

$$Q(800) = 200e^{-0.0004332(800)} \approx 141.42$$

or approximately 141 milligrams.

APPLIED EXAMPLE 3 Radioactive Decay Carbon 14, a radioactive isotope of carbon, has a half-life of 5730 years. What is its decay constant?

Solution We have $Q(t) = Q_0 e^{-kt}$. Since the half-life of the element is 5730 years, half of the substance is left at the end of that period; that is,

$$Q(5730) = Q_0 e^{-5730k} = \frac{1}{2} Q_0$$

$$e^{-5730k} = \frac{1}{2}$$

Taking the natural logarithm on both sides of this equation, we have

$$\ln e^{-5730k} = \ln \frac{1}{2}$$

$$-5730k = -0.693147$$

$$k \approx 0.000121$$

Carbon-14 dating is a well-known method used by anthropologists to establish the age of animal and plant fossils. This method assumes that the proportion of carbon 14 (C-14) present in the atmosphere has remained fairly constant over the past 50,000 years. Professor Willard Libby, recipient of the Nobel Prize in chemistry in 1960, proposed this theory.

The amount of C-14 in the tissues of a living plant or animal is fairly constant. However, when an organism dies, it stops absorbing new quantities of C-14, and the amount of C-14 in the remains diminishes because of the natural decay of the radioactive substance. Therefore, the approximate age of a plant or animal fossil can be determined by measuring the amount of C-14 present in the remains.

APPLIED EXAMPLE 4 Carbon-14 Dating A skull from an archeological site has one tenth the amount of C-14 that it originally contained. Determine the approximate age of the skull.

Solution Here,

$$Q(t) = Q_0 e^{-kt}$$
$$= Q_0 e^{-0.000121t}$$

where Q_0 is the amount of C-14 present originally and k, the decay constant, is equal to 0.000121 (see Example 3). Since $Q(t) = (1/10)Q_0$, we have

$$\frac{1}{10} Q_0 = Q_0 e^{-0.000121t}$$

$$\ln \frac{1}{10} = -0.000121t \qquad \text{Take the natural logarithm on both sides.}$$

$$t = \frac{\ln \frac{1}{10}}{-0.000121}$$

$$\approx 19,030$$

or approximately 19,030 years.

Learning Curves

The next example shows how the exponential function may be applied to describe certain types of learning processes. Consider the function

$$Q(t) = C - Ae^{-kt}$$

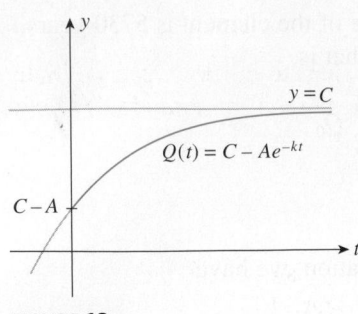

FIGURE 12
A learning curve

where C, A, and k are positive constants. The graph of the function Q is shown in Figure 12, where that part of the graph corresponding to the negative values of t is drawn with a gray line since, in practice, one normally restricts the domain of the function to the interval $[0, \infty)$. Observe that starting at $t = 0$, $Q(t)$ increases rather rapidly but then the rate of increase slows down considerably after a while. The value of $Q(t)$ never exceeds C.

This behavior of the graph of the function Q closely resembles the learning pattern experienced by workers engaged in highly repetitive work. For example, the productivity of an assembly-line worker increases very rapidly in the early stages of the training period. This productivity increase is a direct result of the worker's training and accumulated experience. But the rate of increase of productivity slows as time goes by, and the worker's productivity level approaches some fixed level due to the limitations of the worker and the machine. Because of this characteristic, the graph of the function $Q(t) = C - Ae^{-kt}$ is often called a **learning curve.**

APPLIED EXAMPLE 5 Assembly Time The Camera Division of Eastman Optical produces compact digital cameras. Eastman's training department determines that after completing the basic training program, a new, previously inexperienced employee will be able to assemble

$$Q(t) = 50 - 30e^{-0.5t}$$

model F cameras per day t months after the employee starts work on the assembly line.

a. How many model F cameras can a new employee assemble per day after basic training?
b. How many model F cameras can an employee with 1 month of experience assemble per day? An employee with 2 months of experience? An employee with 6 months of experience?
c. How many model F cameras can the average experienced employee ultimately be expected to assemble per day?

Solution

a. The number of model F cameras a new employee can assemble is given by
$$Q(0) = 50 - 30 = 20$$

b. The number of model F cameras that an employee with 1 month of experience, 2 months of experience, and 6 months of experience can assemble per day is given by

$$Q(1) = 50 - 30e^{-0.5} \approx 31.80$$
$$Q(2) = 50 - 30e^{-1} \approx 38.96$$
$$Q(6) = 50 - 30e^{-3} \approx 48.51$$

or approximately 32, 39, and 49, respectively.
c. As t gets larger and larger, $Q(t)$ approaches 50. So the average experienced employee can ultimately be expected to assemble 50 model F cameras per day.

Other applications of the learning curve are found in models that describe the dissemination of information about a product or the velocity of an object dropped into a viscous medium.

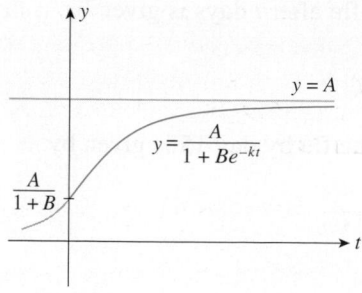

FIGURE **13**
A logistic curve

Logistic Growth Functions

Our last example of an application of exponential functions to the description of natural phenomena involves the **logistic** (also called the **S-shaped,** or **sigmoidal**) **curve,** which is the graph of the function

$$Q(t) = \frac{A}{1 + Be^{-kt}}$$

where A, B, and k are positive constants. The function Q is called a logistic growth function. The graph of the function Q is sketched in Figure 13.

Observe that $Q(t)$ increases slowly at first but more rapidly as t increases. In fact, for small positive values of t, the logistic curve resembles an exponential growth curve. However, the *rate of growth* of $Q(t)$ decreases quite rapidly as t increases and $Q(t)$ approaches the number A as t gets larger and larger, but $Q(t)$ never exceeds A.

Thus, the logistic curve exhibits both the property of rapid growth of the exponential growth curve as well as the "saturation" property of the learning curve. Because of these characteristics, the logistic curve serves as a suitable mathematical model for describing many natural phenomena. For example, if a small number of rabbits were introduced to a tiny island in the South Pacific, the rabbit population might be expected to grow very rapidly at first, but the growth rate would decrease quickly as overcrowding, scarcity of food, and other environmental factors affected it. The population would eventually stabilize at a level compatible with the life-support capacity of the environment. This level, given by A, is called the *carrying capacity* of the environment. Models describing the spread of rumors and epidemics are other examples of the application of the logistic curve.

 APPLIED EXAMPLE 6 Spread of Flu The number of soldiers at Fort MacArthur who contracted influenza after t days during a flu epidemic is approximated by the exponential model

$$Q(t) = \frac{5000}{1 + 1249e^{-kt}}$$

If 40 soldiers contracted the flu by day 7, find how many soldiers contracted the flu by day 15.

Solution The given information implies that

$$Q(7) = \frac{5000}{1 + 1249e^{-7k}} = 40$$

Thus,

$$40(1 + 1249e^{-7k}) = 5000$$

$$1 + 1249e^{-7k} = \frac{5000}{40} = 125$$

$$e^{-7k} = \frac{124}{1249}$$

$$-7k = \ln \frac{124}{1249}$$

$$k = -\frac{\ln \frac{124}{1249}}{7} \approx 0.33$$

Therefore, the number of soldiers who contracted the flu after t days is given by

$$Q(t) = \frac{5000}{1 + 1249e^{-0.33t}}$$

In particular, the number of soldiers who contracted the flu by day 15 is given by

$$Q(15) = \frac{5000}{1 + 1249e^{-15(0.33)}}$$

$$\approx 508$$

or approximately 508 soldiers.

Exploring with TECHNOLOGY

Refer to Example 6.

1. Use a graphing utility to plot the graph of the function Q, using the viewing window $[0, 40] \times [0, 5000]$.

2. How long will it take for the first 1000 soldiers to contract the flu?
   Hint: Plot the graphs of $y_1 = Q(t)$ and $y_2 = 1000$, and find the point of intersection of the two graphs.

3.3 Self-Check Exercise

IMMIGRATION INTO THE UNITED STATES Suppose the population (in millions) of a country at any time t grows in accordance with the rule

$$P = \left(P_0 + \frac{I}{k}\right)e^{kt} - \frac{I}{k}$$

where P denotes the population at any time t, k is a constant reflecting the natural growth rate of the population, I is a constant giving the (constant) rate of immigration into the

country, and P_0 is the total population of the country at time $t = 0$. The population of the United States in 1980 ($t = 0$) was 226.5 million. If the natural growth rate is 0.8% annually ($k = 0.008$) and net immigration is allowed at the rate of half a million people per year ($I = 0.5$), what is the projected population of the United States in 2015?

The solution to Self-Check Exercise 3.3 can be found on page 199.

3.3 Concept Questions

1. What is the model for unrestricted exponential growth? The model for exponential decay? What effect does the magnitude of the growth (decay) constant have on the growth (decay) of a quantity?

2. What is the half-life of a radioactive substance?

3. What is the logistic growth function? What are its characteristics?

3.3 Exercises

1. **EXPONENTIAL GROWTH** Given that a quantity $Q(t)$ is described by the exponential growth function

$$Q(t) = 300e^{0.02t}$$

where t is measured in minutes, answer the following questions:

a. What is the growth constant?
b. What quantity is present initially?
c. Complete the following table of values:

t	0	10	20	100	1000
Q					

2. **EXPONENTIAL DECAY** Given that a quantity $Q(t)$ exhibiting exponential decay is described by the function

$$Q(t) = 2000e^{-0.06t}$$

where t is measured in years, answer the following questions:
a. What is the decay constant?
b. What quantity is present initially?
c. Complete the following table of values:

t	0	5	10	20	100
Q					

3. **GROWTH OF BACTERIA** The growth rate of the bacterium *Escherichia coli*, a common bacterium found in the human intestine, is proportional to its size. Under ideal laboratory conditions, when this bacterium is grown in a nutrient broth medium, the number of cells in a culture doubles approximately every 20 min.
a. If the initial cell population is 100, determine the function $Q(t)$ that expresses the exponential growth of the number of cells of this bacterium as a function of time t (in minutes).
b. How long will it take for a colony of 100 cells to increase to a population of 1 million?
c. If the initial cell population were 1000, how would this alter our model?

4. **WORLD POPULATION** The world population at the beginning of 1990 was 5.3 billion. Assume that the population continues to grow at the rate of approximately 2%/year and find the function $Q(t)$ that expresses the world population (in billions) as a function of time t (in years), with $t = 0$ corresponding to the beginning of 1990. Using this function, complete the following table of values and sketch the graph of the function Q.

Year	1990	1995	2000	2005
World Population				

Year	2010	2015	2020	2025
World Population				

5. **WORLD POPULATION** Refer to Exercise 4.
a. If the world population continues to grow at the rate of approximately 2%/year, find the length of time t_0 required for the world population to triple in size.
b. Using the time t_0 found in part (a), what would be the world population if the growth rate were reduced to 1.8%/year?

6. **RESALE VALUE** Garland Mills purchased a certain piece of machinery 3 years ago for $500,000. Its present resale value is $320,000. Assuming that the machine's resale value decreases exponentially, what will it be 4 years from now?

7. **ATMOSPHERIC PRESSURE** If the temperature is constant, then the atmospheric pressure P (in pounds per square inch) varies with the altitude above sea level h in accordance with the law

$$P = p_0 e^{-kh}$$

where p_0 is the atmospheric pressure at sea level and k is a constant. If the atmospheric pressure is 15 lb/in.² at sea level and 12.5 lb/in.² at 4000 ft, find the atmospheric pressure at an altitude of 12,000 ft.

8. **RADIOACTIVE DECAY** The radioactive element polonium decays according to the law

$$Q(t) = Q_0 \cdot 2^{-(t/140)}$$

where Q_0 is the initial amount and the time t is measured in days. If the amount of polonium left after 280 days is 20 mg, what was the initial amount present?

9. **RADIOACTIVE DECAY** Phosphorus 32 (P-32) has a half-life of 14.2 days. If 100 g of this substance are present initially, find the amount present after t days. What amount will be left after 7.1 days?

10. **NUCLEAR FALLOUT** Strontium 90 (Sr-90), a radioactive isotope of strontium, is present in the fallout resulting from nuclear explosions. It is especially hazardous to animal life, including humans, because, upon ingestion of contaminated food, it is absorbed into the bone structure. Its half-life is 27 years. If the amount of Sr-90 in a certain area is found to be four times the "safe" level, find how much time must elapse before the safe level is reached.

11. **CARBON-14 DATING** Wood deposits recovered from an archeological site contain 20% of the C-14 they originally contained. How long ago did the tree from which the wood was obtained die?

12. **CARBON-14 DATING** The skeletal remains of the so-called Pittsburgh Man, unearthed in Pennsylvania, had lost 82% of the C-14 they originally contained. Determine the approximate age of the bones.

13. **BANK FAILURES** In the wake of the 2008 financial crisis, bank failures started spiraling upward. The number of bank failures peaked in 2010 at 157. Thereafter, the number of bank failures began to fall sharply. The number of bank failures from 2010 through 2012 is described by the function

$$f(t) = 157e^{-0.55t} \qquad (0 \le t \le 2)$$

where t is measured in years, with $t = 0$ corresponding to 2010. If the trend continued, how many bank failures were there in 2013?
Source: FDIC.

14. ONLINE SHOPPERS The number of consumers researching products or shopping online is growing steadily. According to the research firm, eMarketer, the number of online shoppers is projected to be

$$f(t) = 172.2e^{0.031t} \qquad (0 \le t \le 5)$$

million in year t, where $t = 0$ corresponds to 2010.

a. Find the projected number of online shoppers in each of the years 2010 through 2015 by completing the following table:

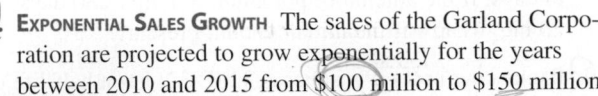

t	0	1	2	3	4	5
$f(t)$						

b. Use the information from part (a) to sketch the graph of f.

Source: eMarketer.

15. EXPONENTIAL SALES GROWTH The sales of the Garland Corporation are projected to grow exponentially for the years between 2010 and 2015 from $100 million to $150 million.

a. Find a model giving the sales of Garland Corporation in year t between 2010 ($t = 0$) and 2015 ($t = 5$).
Hint: The sales in year t are $S = S_0 e^{kt}$.

b. What were the sales of Garland Corporation in 2013?

16. LEARNING CURVES The American Court Reporting Institute finds that the average student taking Advanced Machine Shorthand, an intensive 20-week course, progresses according to the function

$$Q(t) = 120(1 - e^{-0.05t}) + 60 \qquad (0 \le t \le 20)$$

where $Q(t)$ measures the number of words (per minute) of dictation that the student can take in machine shorthand after t weeks in the course. Sketch the graph of the function Q and answer the following questions:

a. What is the beginning shorthand speed for the average student in this course?

b. What shorthand speed does the average student attain halfway through the course?

c. How many words per minute can the average student take after completing this course?

17. DRIVER'S LICENSES The percentage of teenagers and young adults with a driver's license in 2010 is approximated by

$$P(x) = 90(1 - e^{-0.37(x-15)}) \qquad (16 \le x \le 39)$$

where x is the age of the person. Sketch the graph of the function P, and answer the following questions:

a. What percentage of 16-year-olds had a driver's license in 2010?

b. What percentage of 20-year-olds had a driver's license in 2010?

c. What percentage of 39-year-olds had a driver's license in 2010?

Source: University of Michigan.

18. EFFECT OF ADVERTISING ON SALES Metro Department Store found that t weeks after the end of a sales promotion the volume of sales was given by

$$S(t) = B + Ae^{-kt} \qquad (0 \le t \le 4)$$

where $B = 50,000$ and is equal to the average weekly volume of sales before the promotion. The sales volumes at the end of the first and third weeks were $83,515 and $65,055, respectively. Assume that the sales volume is decreasing exponentially.

a. Find the decay constant k.

b. Find the sales volume at the end of the fourth week.

19. DEMAND FOR TABLET COMPUTERS Universal Instruments found that the monthly demand for its new line of Galaxy tablet computers t months after the line was placed on the market was given by

$$D(t) = 2000 - 1500e^{-0.05t} \qquad (t > 0)$$

Graph this function and answer the following questions:

a. What is the demand after 1 month? After 1 year? After 2 years? After 5 years?

b. At what level is the demand expected to stabilize?

20. RELIABILITY OF COMPUTER CHIPS The percentage of a certain brand of computer chips that will fail after t years of use is estimated to be

$$P(t) = 100(1 - e^{-0.1t})$$

What percentage of this brand of computer chips are expected to be usable after 3 years?

21. LENGTHS OF FISH The length (in centimeters) of a typical Pacific halibut t years old is approximately

$$f(t) = 200(1 - 0.956e^{-0.18t})$$

What is the length of a typical 5-year-old Pacific halibut?

22. SPREAD OF AN EPIDEMIC During a flu epidemic, the number of children in the Woodbridge Community School System who contracted influenza after t days was given by

$$Q(t) = \frac{1000}{1 + 199e^{-0.8t}}$$

a. How many children were stricken by the flu after the first day?

b. How many children had the flu after 10 days?

23. GROWTH OF A FRUIT FLY POPULATION On the basis of data collected during an experiment, a biologist found that the growth of a fruit fly (*Drosophila*) with a limited food supply could be approximated by

$$N(t) = \frac{400}{1 + 39e^{-0.16t}}$$

where t denotes the number of days since the beginning of the experiment.

a. What was the initial fruit fly population in the experiment?
b. What was the population of the fruit fly colony on the 20th day?

24. **DEMOGRAPHICS** The number of citizens aged 45–64 years is approximated by

$$P(t) = \frac{197.9}{1 + 3.274e^{-0.0361t}} \qquad (0 \le t \le 25)$$

where $P(t)$ is measured in millions and t is measured in years, with $t = 0$ corresponding to the beginning of 1990. People belonging to this age group are the targets of insurance companies that want to sell them annuities. What was the expected population of citizens aged 45–64 years in 2010? In 2015?
Source: K. G. Securities.

25. **POPULATION GROWTH IN THE TWENTY-FIRST CENTURY** The U.S. population is approximated by the function

$$P(t) = \frac{616.5}{1 + 4.02e^{-0.5t}}$$

where $P(t)$ is measured in millions of people and t is measured in 30-year intervals, with $t = 0$ corresponding to 1930. What is the expected population of the United States in 2020 $(t = 3)$?

26. **DISSEMINATION OF INFORMATION** Three hundred students attended the dedication ceremony of a new building on a college campus. The president of the traditionally female college announced a new expansion program, which included plans to make the college coeducational. The number of students who learned of the new program t hr later is given by the function

$$f(t) = \frac{3000}{1 + Be^{-kt}}$$

If 600 students on campus had heard about the new program 2 hr after the ceremony, how many students had heard about the policy after 4 hr?

27. **RADIOACTIVE DECAY** A radioactive substance decays according to the formula

$$Q(t) = Q_0e^{-kt}$$

where $Q(t)$ denotes the amount of the substance present at time t (measured in years), Q_0 denotes the amount of the substance present initially, and k (a positive constant) is the decay constant.
a. Show that half-life of the substance is $\bar{t} = (\ln 2)/k$.
b. Suppose a radioactive substance decays according to the formula

$$Q(t) = 20e^{-0.0001238t}$$

How long will it take for the substance to decay to half the original amount?

28. **LOGISTIC GROWTH FUNCTION** Consider the logistic growth function

$$Q(t) = \frac{A}{1 + Be^{-kt}}$$

Suppose the population is Q_1 when $t = t_1$ and Q_2 when $t = t_2$. Show that the value of k is

$$k = \frac{1}{t_2 - t_1} \ln\left[\frac{Q_2(A - Q_1)}{Q_1(A - Q_2)}\right]$$

29. **LOGISTIC GROWTH FUNCTION** The carrying capacity of a colony of fruit flies (*Drosophila*) is 600. The population of fruit flies after 14 days is 76, and the population after 21 days is 167. What is the value of the growth constant k?
Hint: Use the result of Exercise 28.

3.3   Solution to Self-Check Exercise

We are given that $P_0 = 226.5$, $k = 0.008$, and $I = 0.5$. So

$$P = \left(226.5 + \frac{0.5}{0.008}\right)e^{0.008t} - \frac{0.5}{0.008}$$

$$= 289e^{0.008t} - 62.5$$

Therefore, the expected population in 2015 is given by

$$P(35) = 289e^{0.28} - 62.5$$

$$\approx 319.9$$

or approximately 319.9 million.

USING TECHNOLOGY — Analyzing Mathematical Models

We can use a graphing utility to analyze the mathematical models encountered in this section.

 APPLIED EXAMPLE 1 Internet-Gaming Sales The estimated growth in global Internet-gaming revenue (in billions of dollars), as predicted by industry analysts, is given in the following table:

Year	2001	2002	2003	2004	2005	2006	2007	2008	2009	2010
Revenue	3.1	3.9	5.6	8.0	11.8	15.2	18.2	20.4	22.7	24.5

a. Use **Logistic** to find a regression model for the data. Let $t = 0$ correspond to 2001.
b. Plot the scatter diagram and the graph of the function f found in part (a) using the viewing window $[0, 9] \times [0, 35]$.

Source: Christiansen Capital/Advisors.

Solution

a. Using **Logistic,** we find

$$f(t) = \frac{27.11}{1 + 9.64e^{-0.49t}} \quad (0 \le t \le 9)$$

b. The scatter plot for the data, and the graph of f in the viewing window $[0, 9] \times [0, 35]$, are shown in Figure T1.

FIGURE T1
The graph of f in the viewing window $[0, 9] \times [0, 35]$

TECHNOLOGY EXERCISES

1. **AIR TRAVEL** Air travel has been rising dramatically in the past 30 years. In a study conducted in 2000, the FAA projected further exponential growth for air travel through 2010. The function

$$f(t) = 666e^{0.0413t} \quad (0 \le t \le 10)$$

gives the number of passengers (in millions) in year t, with $t = 0$ corresponding to 2000.
a. Plot the graph of f, using the viewing window $[0, 10] \times [0, 1000]$.
b. How many air passengers were there in 2000? What was the projected number of air passengers for 2008?

Source: Federal Aviation Administration.

2. **NEWTON'S LAW OF COOLING** The temperature of a cup of coffee t min after it is poured is given by

$$T = 70 + 100e^{-0.0446t}$$

where T is measured in degrees Fahrenheit.
a. Plot the graph of T, using the viewing window $[0, 30] \times [0, 200]$.

b. When will the coffee be cool enough to drink (say, 120°)?
   Hint: Use the **ISECT** function.

3. **COMPUTER GAME SALES** The total number of Starr Communication's newest game, Laser Beams, sold t months after its release is given by

$$N(t) = -20(t + 20)e^{-0.05t} + 400$$

thousand units. Plot the graph of N, using the viewing window $[0, 500] \times [0, 500]$.

4. **POPULATION GROWTH IN THE TWENTY-FIRST CENTURY** The U.S. population is approximated by the function

$$P(t) = \frac{616.5}{1 + 4.02e^{-0.5t}}$$

where $P(t)$ is measured in millions of people and t is measured in 30-year intervals, with $t = 0$ corresponding to 1930.
a. Plot the graph of P, using the viewing window $[0, 4] \times [0, 650]$.

b. What is the projected population of the United States in 2020 ($t = 3$)?

5. **TIME RATE OF GROWTH OF A TUMOR** The rate at which a tumor grows, with respect to time, is given by

$$R = Ax \ln \frac{B}{x} \qquad (0 < x < B)$$

where A and B are positive constants and x is the radius of the tumor in centimeters. Plot the graph of R for the case $A = B = 10$, using the viewing window $[0, 10] \times [0, 45]$.

6. **SNOWFALL ACCUMULATION** The snowfall accumulation at Logan Airport (in inches), t hr after the beginning of a 33-hr snowstorm in Boston on a certain day, follows:

Hour	0	3	6	9	12	15	18	21	24	27	30	33
Snow (in inches)	0.1	0.4	3.6	6.5	9.1	14.4	19.5	22	23.6	24.8	26.6	27

Here, $t = 0$ corresponds to noon of February 6.

a. Use **Logistic** to find a regression model for the data.
b. Plot the scatter diagram and the graph of the function f found in part (a), using the viewing window $[0, 33] \times [0, 30]$.
Source: Boston Globe.

7. **ABSORPTION OF DRUGS** The concentration of a drug in an organ at any time t (in seconds) is given by

$$C(t) = \begin{cases} 0.3t - 18(1 - e^{-t/60}) & \text{if } 0 \leq t \leq 20 \\ 18e^{-t/60} - 12e^{-(t-20)/60} & \text{if } t > 20 \end{cases}$$

where $C(t)$ is measured in grams per cubic centimeter (g/cm^3).

a. Plot the graph of C, using the viewing window $[0, 120] \times [0, 1]$.
b. What is the initial concentration of the drug in the organ?
c. What is the concentration of the drug in the organ after 10 sec?
d. What is the concentration of the drug in the organ after 30 sec?

CHAPTER 3 Summary of Principal Formulas and Terms

FORMULAS

1. Exponential function with base b	$y = b^x$, where $b > 0$ and $b \neq 1$
2. The number e	$e = 2.7182818\ldots$
3. Exponential function with base e	$y = e^x$
4. Logarithmic function with base b	$y = \log_b x \quad (x > 0)$
5. Logarithmic function with base e	$y = \ln x \quad (x > 0)$
6. Inverse properties of $\ln x$ and e^x	$\ln e^x = x \quad$ and $\quad e^{\ln x} = x$

TERMS

common logarithm (182)   growth constant (190)   half-life of a radioactive substance (192)

natural logarithm (182)   exponential decay (192)

exponential growth (190)   decay constant (192)   logistic growth function (195)

CHAPTER 3 Concept Review

Fill in the blanks.

1. The function $f(x) = x^b$ (b, a real number) is called a/an _____ function, whereas the function $g(x) = b^x$, where $b >$ _____ and $b \neq$ _____, is called a/an _____ function.

2. **a.** The domain of the function $y = 3^x$ is _____, and its range is _____.

   b. The graph of the function $y = 0.3^x$ passes through the point _____ and falls from _____ to _____.

3. **a.** If $b > 0$ and $b \neq 1$, then the logarithmic function $y = \log_b x$ has domain _____ and range _____; its graph passes through the point _____.

   b. The graph of $y = \log_b x$ _____ from left to right if $b < 1$ and _____ from left to right if $b > 1$.

4. a. If $x > 0$, then $e^{\ln x} =$ _____.
   b. If x is any real number, then $\ln e^x =$ _____.

5. a. In the unrestricted exponential growth model $Q = Q_0 e^{kt}$, Q_0 represents the quantity present _____, and k is called the _____ constant.
   b. In the exponential decay model $Q = Q_0 e^{-kt}$, k is called the _____ constant.
   c. The half-life of a radioactive substance is the _____ required for a substance to decay to _____ _____ of its original amount.

6. a. The model $Q(t) = C - Ae^{-kt}$ is called a/an _____ _____. The value of $Q(t)$ never exceeds _____.
   b. The model $Q(t) = \dfrac{A}{1 + Be^{-kt}}$ is called a/an _____ _____ _____. If the quantity $Q(t)$ is initially smaller than A, then $Q(t)$ will eventually approach _____ as t increases; the number A represents the life-support capacity of the environment and is called the _____ _____ of the environment.

CHAPTER 3   Review Exercises

In Exercises 1–4, sketch the graph of the function.

1. $f(x) = 5^x$

2. $y = \left(\dfrac{1}{5}\right)^x$

3. $f(x) = \log_4 x$

4. $y = \log_{1/4} x$

In Exercises 5–8, express each equation in logarithmic form.

5. $3^4 = 81$

6. $9^{1/2} = 3$

7. $\left(\dfrac{2}{3}\right)^{-3} = \dfrac{27}{8}$

8. $16^{-3/4} = 0.125$

In Exercises 9–12, given that $\ln 2 \approx 0.6931$, $\ln 3 \approx 1.0986$, and $\ln 5 \approx 1.6094$, find the value of the expression using the laws of logarithms.

9. $\ln 30$

10. $\ln 9$

11. $\ln 3.6$

12. $\ln 75$

In Exercises 13–15, given that $\ln 2 = x$, $\ln 3 = y$, and $\ln 5 = z$, express each of the given logarithmic values in terms of x, y, and z.

13. $\ln 30$ **14.** $\ln 3.6$ **15.** $\ln 75$

In Exercises 16–21, solve for x without using a calculator.

16. $2^{2x-3} = 8$

17. $e^{x^2+x} = e^2$

18. $3^{x-1} = 9^{x+2}$

19. $2^{x^2+x} = 4^{x^2-3}$

20. $\log_4(2x + 1) = 2$

21. $\ln(x - 1) + \ln 4 = \ln(2x + 4) - \ln 2$

In Exercises 22–35, solve for x, giving your answer accurate to four decimal places.

22. $4^x = 5$

23. $3^{-2x} = 8$

24. $3 \cdot 2^{-x} = 17$

25. $2e^{-x} = 7$

26. $0.2e^x = 3.4$

27. $e^{2x-1} = 14$

28. $5^{3x+1} = 16$

29. $2^{3x+1} = 3^{2x-3}$

30. $2^{x^2} = 12$

31. $3e^{\sqrt{x}} = 15$

32. $4e^{-0.1x} - 2 = 8$

33. $8 - e^{0.2x} = 2$

34. $\dfrac{20}{1 + 2e^{0.2x}} = 4$

35. $\dfrac{30}{1 + 2e^{-0.1x}} = 5$

36. Sketch the graph of the function $y = \log_2(x + 3)$.

37. Sketch the graph of the function $y = \log_3(x + 1)$.

38. GROWTH OF BACTERIA A culture of bacteria that initially contained 2000 bacteria has a count of 18,000 bacteria after 2 hr.
   a. Determine the function $Q(t)$ that expresses the exponential growth of the number of cells of this bacterium as a function of time t (in minutes).
   b. Find the number of bacteria present after 4 hr.

39. RADIOACTIVE DECAY The radioactive element radium has a half-life of 1600 years. What is its decay constant?

40. DEMAND FOR DVD PLAYERS VCA Television found that the monthly demand for its new line of DVD players t months after placing the players on the market is given by:

$$D(t) = 4000 - 3000e^{-0.06t} \qquad (t \geq 0)$$

Graph this function and answer the following questions:
   a. What was the demand after 1 month? After 1 year? After 2 years?
   b. At what level is the demand expected to stabilize?

41. FLU EPIDEMIC During a flu epidemic, the number of students at a certain university who contracted influenza after t days could be approximated by the exponential model

$$Q(t) = \dfrac{3000}{1 + 499e^{-kt}}$$

If 90 students contracted the flu by day 10, how many students contracted the flu by day 20?

42. U.S. INFANT MORTALITY RATE The U.S. infant mortality rate (per 1000 live births) is approximated by the function

$$N(t) = 12.5e^{-0.0294t} \qquad (0 \leq t \leq 21)$$

where t is measured in years, with $t = 0$ corresponding to 1980. What was the mortality rate in 1980? In 1990? In 2000?
   Source: U.S. Department of Health and Human Services.

43. **WORLD POPULATION** The world population is projected to grow according to the model

$$P(t) = \frac{12}{1 + 3e^{-0.2747t}} \qquad (0 \le t \le 8)$$

where $P(t)$ is measured in billions and t in decades, with $t = 0$ corresponding to 1960.
a. What was the world population in 1960?
b. What is the world population expected to be in 2040?
Source: U.S. Census Bureau.

44. **ABSORPTION OF DRUGS** The concentration of a drug in an organ at any time t (in seconds) is given by

$$x(t) = 0.08(1 - e^{-0.02t})$$

where $x(t)$ is measured in grams/cubic centimeter (g/cm^3).
a. What is the initial concentration of the drug in the organ?
b. What is the concentration of the drug in the organ after 30 sec?

The problem-solving skills that you learn in each chapter are building blocks for the rest of the course. Therefore, it is a good idea to make sure that you have mastered these skills before moving on to the next chapter. The Before Moving On exercises that follow are designed for that purpose. After completing these exercises, you can identify the skills that you should review before starting the next chapter.

CHAPTER 3 Before Moving On ...

1. Solve $e^{2x} - e^x - 6 = 0$ for x.
Hint: Let $u = e^x$.

2. Solve $\log_2(x^2 - 8x + 1) = 0$.

3. Solve the equation $\dfrac{100}{1 + 2e^{0.3t}} = 40$ for t.

4. The temperature of a cup of coffee at time t (in minutes) is

$$T(t) = 70 + ce^{-kt}$$

Initially, the temperature of the coffee was 200°F. Three minutes later, it was 180°. When will the temperature of the coffee be 150°F?

4 Mathematics of Finance

NTEREST THAT IS periodically added to the principal and thereafter itself earns interest is called *compound interest*. We begin this chapter by deriving the *compound interest formula*, which gives the amount of money accumulated when an initial amount of money is invested in an account for a fixed term and earns compound interest.

An *annuity* is a sequence of payments made at regular intervals. We derive formulas giving the *future value of an annuity* (what you end up with) and the *present value of an annuity* (the lump sum that, when invested now, will yield the same future value as that of the annuity). Then, using these formulas, we answer questions involving the amortization of certain types of installment loans and questions involving *sinking funds* (funds that are set up to be used for a specific purpose at a future date).

How much can the Jacksons afford to borrow from the bank for the purchase of a home? They have determined that after making a down payment they can afford a monthly payment of $2000. In Example 4, page 239, we learn how to determine the maximum amount they can afford to borrow if they secure a 30-year fixed mortgage at the current rate.

4.1 Compound Interest

Simple Interest

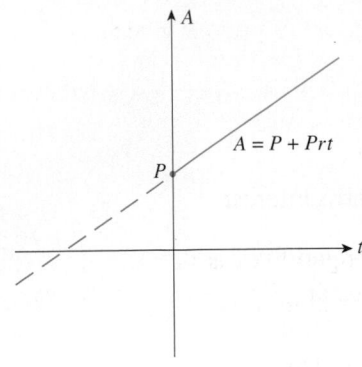

FIGURE 1
The accumulated amount is a linear function of t.

A natural application of linear functions to the business world is found in the computation of **simple interest**—interest that is computed on the original principal only. Thus, if I denotes the interest on a principal P (in dollars) at an interest rate of r per year for t years, we have

$$I = Prt$$

The **accumulated amount** A, the sum of the principal and interest after t years, is given by

$$A = P + I = P + Prt$$
$$= P(1 + rt)$$

and is a linear function of t (see Exercise 50). In business applications, we are normally interested only in the case in which t is positive, so only the part of the line that lies in Quadrant I is of interest to us (Figure 1).

> **Simple Interest Formulas**
>
> | | | |
> |---|---|---|
> | Interest: | $I = Prt$ | **(1a)** |
> | Accumulated amount: | $A = P(1 + rt)$ | **(1b)** |

EXAMPLE 1 A bank pays simple interest at the rate of 8% per year for certain deposits. If a customer deposits $1000 and makes no withdrawals for 3 years, what is the total amount on deposit at the end of 3 years? What is the interest earned in that period of time?

Solution Using Formula (1b) with $P = 1000$, $r = 0.08$, and $t = 3$, we see that the total amount on deposit at the end of 3 years is given by

$$A = P(1 + rt)$$
$$= 1000[1 + (0.08)(3)] = 1240$$

or $1240.

The interest earned over the 3-year period is given by

$$I = Prt \quad \text{Use Formula (1a).}$$
$$= 1000(0.08)(3) = 240$$

or $240.

> **Exploring with TECHNOLOGY**
>
> Refer to Example 1. Use a graphing utility to plot the graph of the function $A = 1000(1 + 0.08t)$, using the viewing window $[0, 10] \times [0, 2000]$.
>
> **1.** What is the A-intercept of the straight line, and what does it represent?
>
> **2.** What is the slope of the straight line, and what does it represent? (See Exercise 50.)

$ APPLIED EXAMPLE 2 Trust Funds An amount of $2000 is invested in a 10-year trust fund that pays 6% annual simple interest. What is the total amount of the trust fund at the end of 10 years?

Solution The total amount of the trust fund at the end of 10 years is given by

$$A = P(1 + rt)$$
$$= 2000[1 + (0.06)(10)] = 3200$$

or $3200.

A Treasury Bill (T-Bill) is a short-term debt obligation (less than or equal to 1 year) backed by the U.S. government. Rather than paying fixed interest payments, T-Bills are sold at a discount from face value. The appreciation of a T-Bill (face value − purchase price) provides the investment return to the holder.

$ APPLIED EXAMPLE 3 T-Bills Suppose that Jane buys a 26-week T-Bill with a maturity value of $10,000. If she pays $9850 for the T-Bill, what will be the rate of return on her investment?

Solution We use Formula (1b) with $A = 10,000$, $P = 9850$, and $t = \frac{26}{52} = \frac{1}{2}$. We obtain

$$A = P(1 + rt)$$

$$10,000 = 9850\left(1 + \frac{1}{2}r\right)$$
$$= 9850 + 4925r$$

Solving for r, we find

$$4925r = 150$$
$$r = \frac{150}{4925} \approx 0.0305$$

So Jane's investment will earn simple interest at the rate of approximately 3.05% per year.

Compound Interest

In contrast to simple interest, **compound interest** is earned interest that is periodically added to the principal and thereafter itself earns interest at the same rate. To find a formula for the accumulated amount, let's consider a numerical example. Suppose $1000 (the principal) is deposited in a bank for a term of 3 years, earning interest at the rate of 8% per year (called the **nominal,** or **stated, rate**) compounded annually. Then, using Formula (1b) with $P = 1000$, $r = 0.08$, and $t = 1$, we see that the accumulated amount at the end of the first year is

$$A_1 = P(1 + rt)$$
$$= 1000[1 + (0.08)(1)] = 1000(1.08) = 1080$$

or $1080.

To find the accumulated amount A_2 at the end of the second year, we use Formula (1b) once again, this time with $P = A_1$. (Remember, the principal *and* interest now earn interest over the second year.) We obtain

$$
\begin{aligned}
A_2 &= P(1 + rt) = A_1(1 + rt) \\
&= 1000[1 + 0.08(1)][1 + 0.08(1)] \\
&= 1000[1 + 0.08]^2 = 1000(1.08)^2 = 1166.40
\end{aligned}
$$

or \$1166.40.

Finally, the accumulated amount A_3 at the end of the third year is found using (1b) with $P = A_2$, giving

$$
\begin{aligned}
A_3 &= P(1 + rt) = A_2(1 + rt) \\
&= 1000[1 + 0.08(1)]^2[1 + 0.08(1)] \\
&= 1000[1 + 0.08]^3 = 1000(1.08)^3 \approx 1259.71
\end{aligned}
$$

or approximately \$1259.71.

If you reexamine our calculations, you will see that the accumulated amounts at the end of each year have the following form:

First year: $A_1 = 1000(1 + 0.08)$, or $A_1 = P(1 + r)$
Second year: $A_2 = 1000(1 + 0.08)^2$, or $A_2 = P(1 + r)^2$
Third year: $A_3 = 1000(1 + 0.08)^3$, or $A_3 = P(1 + r)^3$

These observations suggest the following general result: If P dollars is invested over a term of t years, earning interest at the rate of r per year compounded annually, then the accumulated amount is

$$
A = P(1 + r)^t \tag{2}
$$

Equation (2) was derived under the assumption that interest was compounded *annually*. In practice, however, interest is usually compounded more than once a year. The interval of time between successive interest calculations is called the **conversion period.**

If interest at a nominal rate of r per year is compounded m times a year on a principal of P dollars, then the simple interest rate per conversion period is

$$
i = \frac{r}{m} \qquad \frac{\text{Annual interest rate}}{\text{Periods per year}}
$$

For example, if the nominal interest rate is 8% per year $(r = 0.08)$ and interest is compounded quarterly $(m = 4)$, then

$$
i = \frac{r}{m} = \frac{0.08}{4} = 0.02
$$

or 2% per period.

To find a general formula for the accumulated amount when a principal of P dollars is deposited in a bank for a term of t years and earns interest at the (nominal) rate of r per year compounded m times per year, we proceed as before, using Formula (1b) repeatedly with the interest rate $i = \frac{r}{m}$. We see that the accumulated amount at the end of each period is as follows:

First period: $A_1 = P(1 + i)$
Second period: $A_2 = A_1(1 + i) = [P(1 + i)](1 + i) = P(1 + i)^2$
Third period: $A_3 = A_2(1 + i) = [P(1 + i)^2](1 + i) = P(1 + i)^3$

$\vdots$ $\qquad\qquad$ $\vdots$

nth period: $A_n = A_{n-1}(1 + i) = [P(1 + i)^{n-1}](1 + i) = P(1 + i)^n$

There are $n = mt$ periods in t years (number of conversion periods per year times the term in years). Hence the accumulated amount at the end of t years is given by

$$A = P(1 + i)^n$$

Compound Interest Formula (Accumulated Amount)

$$A = P(1 + i)^n \qquad (3)$$

where $i = \dfrac{r}{m}$, $n = mt$, and

A = Accumulated amount at the end of n conversion periods
P = Principal
r = Nominal interest rate per year
m = Number of conversion periods per year
t = Term (number of years)

Exploring with TECHNOLOGY

Let $A_1(t)$ denote the accumulated amount of $100 earning simple interest at the rate of 6% per year over t years, and let $A_2(t)$ denote the accumulated amount of $100 earning interest at the rate of 6% per year compounded monthly over t years.

1. Find expressions for $A_1(t)$ and $A_2(t)$.
2. Use a graphing utility to plot the graphs of A_1 and A_2 on the same set of axes, using the viewing window $[0, 20] \times [0, 400]$.
3. Comment on the growth of $A_1(t)$ and $A_2(t)$ by referring to the graphs of A_1 and A_2.

EXAMPLE 4 Find the accumulated amount after 3 years if $1000 is invested at 8% per year compounded (a) annually, (b) semiannually, (c) quarterly, (d) monthly, and (e) daily (assume a 365-day year).

Solution

a. Here, $P = 1000$, $r = 0.08$, and $m = 1$. Thus, $i = r = 0.08$ and $n = 3$, so Formula (3) gives

$$A = 1000(1 + 0.08)^3$$
$$\approx 1259.71$$

or $1259.71.

b. Here, $P = 1000$, $r = 0.08$, and $m = 2$. Thus, $i = \frac{0.08}{2}$ and $n = (3)(2) = 6$, so Formula (3) gives

$$A = 1000\left(1 + \frac{0.08}{2}\right)^6$$
$$\approx 1265.32$$

or $1265.32.

c. In this case, $P = 1000$, $r = 0.08$, and $m = 4$. Thus, $i = \frac{0.08}{4}$ and $n = (3)(4) = 12$, so Formula (3) gives

$$A = 1000\left(1 + \frac{0.08}{4}\right)^{12}$$
$$\approx 1268.24$$

or \$1268.24.

d. Here, $P = 1000$, $r = 0.08$, and $m = 12$. Thus, $i = \frac{0.08}{12}$ and $n = (3)(12) = 36$, so Formula (3) gives

$$A = 1000\left(1 + \frac{0.08}{12}\right)^{36}$$
$$\approx 1270.24$$

or \$1270.24.

e. Here, $P = 1000$, $r = 0.08$, $m = 365$, and $t = 3$. Thus, $i = \frac{0.08}{365}$ and $n = (3)(365) = 1095$, so Formula (3) gives

$$A = 1000\left(1 + \frac{0.08}{365}\right)^{1095}$$
$$\approx 1271.22$$

or \$1271.22. These results are summarized in Table 1.

TABLE 1				
Nominal Rate, r	Conversion Period	Interest Rate/ Conversion Period	Initial Investment	Accumulated Amount
8%	Annually ($m = 1$)	8%	\$1000	\$1259.71
8	Semiannually ($m = 2$)	4	1000	1265.32
8	Quarterly ($m = 4$)	2	1000	1268.24
8	Monthly ($m = 12$)	2/3	1000	1270.24
8	Daily ($m = 365$)	8/365	1000	1271.22

Exploring with TECHNOLOGY

Investments that are allowed to grow over time can increase in value surprisingly fast. Consider the potential growth of \$10,000 if earnings are reinvested. More specifically, suppose $A_1(t)$, $A_2(t)$, $A_3(t)$, $A_4(t)$, and $A_5(t)$ denote the accumulated values of an investment of \$10,000 over a term of t years and earning interest at the rate of 4%, 6%, 8%, 10%, and 12% per year compounded annually.

1. Find expressions for $A_1(t)$, $A_2(t)$, ..., $A_5(t)$.
2. Use a graphing utility to plot the graphs of A_1, A_2, ..., A_5 on the same set of axes, using the viewing window $[0, 20] \times [0, 100{,}000]$.
3. Use **TRACE** to find $A_1(20)$, $A_2(20)$, ..., $A_5(20)$, and then interpret your results.

Continuous Compounding of Interest

One question that arises naturally in the study of compound interest is: What happens to the accumulated amount over a fixed period of time if the interest is computed more and more frequently?

Intuition suggests that the more often interest is compounded, the larger the accumulated amount will be. This is confirmed by the results of Example 4, where we found that the accumulated amounts did in fact increase when we increased the number of conversion periods per year.

This leads us to another question: Does the accumulated amount keep growing without bound, or does it approach a fixed number when the interest is computed more and more frequently over a fixed period of time?

To answer this question, let's look again at the compound interest formula:

$$A = P(1 + i)^n = P\left(1 + \frac{r}{m}\right)^{mt} \tag{4}$$

Recall that m is the number of conversion periods per year. So to find an answer to our question, we should let m get larger and larger in Equation (4). If we let $u = \frac{m}{r}$ so that $m = ru$, then (4) becomes

$$A = P\left(1 + \frac{1}{u}\right)^{urt} \qquad \frac{r}{m} = \frac{1}{u}$$

$$= P\left[\left(1 + \frac{1}{u}\right)^u\right]^{rt} \qquad \text{Since } a^{xy} = (a^x)^y$$

Now observe that u gets larger and larger as m gets larger and larger. But from our work in Section 3.1, we know that $(1 + 1/u)^u$ approaches e as u gets larger and larger. Using this result, we can see that, as m gets larger and larger, A approaches $P(e)^{rt} = Pe^{rt}$. In this situation, we say that interest is *compounded continuously*. Let's summarize this important result.

Continuous Compound Interest Formula

$$A = Pe^{rt} \tag{5}$$

where

P = Principal

r = Nominal interest rate compounded continuously

t = Time in years

A = Accumulated amount at the end of t years

EXAMPLE 5 Find the accumulated amount after 3 years if $1000 is invested at 8% per year compounded (a) daily (assume a 365-day year) and (b) continuously.

Solution

a. Use Formula (3) with $P = 1000$, $r = 0.08$, $m = 365$, and $t = 3$. Thus, $i = \frac{0.08}{365}$ and $n = (365)(3) = 1095$, so

$$A = 1000\left(1 + \frac{0.08}{365}\right)^{(365)(3)} \approx 1271.22$$

or $1271.22.

b. Here, we use Formula (5) with $P = 1000$, $r = 0.08$, and $t = 3$, obtaining

$$A = 1000e^{(0.08)(3)}$$

$$\approx 1271.25$$

or $1271.25.

Observe that the accumulated amounts corresponding to interest compounded daily and interest compounded continuously differ by very little. The continuous compound interest formula is a very important tool in theoretical work in financial analysis.

Effective Rate of Interest

Example 4 showed that the interest actually earned on an investment depends on the frequency with which the interest is compounded. Thus, the stated, or nominal, rate of 8% per year does not reflect the actual rate at which interest is earned. This suggests that we need to find a common basis for comparing interest rates. One such way of comparing interest rates is provided by the use of the *effective rate of interest*. The **effective rate of interest** is the annual rate of interest that, when compounded annually, will yield the same accumulated amount as the nominal rate compounded m times a year (over the same term). Equivalently, the effective rate of interest is the *simple* interest rate that would produce the same accumulated amount in 1 year as the nominal rate compounded m times a year. The effective rate of interest is also called the **annual percentage yield.**

To derive a relationship between the nominal interest rate, r per year compounded m times, and its corresponding effective rate, R per year, let's assume an initial investment of P dollars. Then the accumulated amount after 1 year at a simple interest rate of R per year is

$$A = P(1 + R)$$

Also, the accumulated amount after 1 year at an interest rate of r per year compounded m times a year is

$$A = P(1 + i)^n = P\left(1 + \frac{r}{m}\right)^m \quad \text{Since } i = \frac{r}{m} \text{ and } t = 1$$

Equating the two expressions gives

$$P(1 + R) = P\left(1 + \frac{r}{m}\right)^m$$

$$1 + R = \left(1 + \frac{r}{m}\right)^m \quad \text{Divide both sides by } P.$$

If we solve the preceding equation for R, we obtain the following formula for computing the effective rate of interest.

Effective Rate of Interest Formula

$$r_{\text{eff}} = \left(1 + \frac{r}{m}\right)^m - 1 \tag{6}$$

where

r_{eff} = Effective rate of interest

r = Nominal interest rate per year

m = Number of conversion periods per year

EXAMPLE 6 Find the effective rate of interest corresponding to a nominal rate of 8% per year compounded (a) annually, (b) semiannually, (c) quarterly, (d) monthly, and (e) daily.

Solution

a. The effective rate of interest corresponding to a nominal rate of 8% per year compounded annually is, of course, given by 8% per year. This result is also confirmed by using Formula (6) with $r = 0.08$ and $m = 1$. Thus,

$$r_{\text{eff}} = (1 + 0.08) - 1 = 0.08$$

b. Let $r = 0.08$ and $m = 2$. Then Formula (6) yields

$$r_{\text{eff}} = \left(1 + \frac{0.08}{2}\right)^2 - 1$$
$$= (1.04)^2 - 1$$
$$= 0.0816$$

so the effective rate is 8.16% per year.

c. Let $r = 0.08$ and $m = 4$. Then Formula (6) yields

$$r_{\text{eff}} = \left(1 + \frac{0.08}{4}\right)^4 - 1$$
$$= (1.02)^4 - 1$$
$$\approx 0.08243$$

so the corresponding effective rate in this case is 8.243% per year.

d. Let $r = 0.08$ and $m = 12$. Then Formula (6) yields

$$r_{\text{eff}} = \left(1 + \frac{0.08}{12}\right)^{12} - 1$$
$$\approx 0.08300$$

so the corresponding effective rate in this case is 8.3% per year.

e. Let $r = 0.08$ and $m = 365$. Then Formula (6) yields

$$r_{\text{eff}} = \left(1 + \frac{0.08}{365}\right)^{365} - 1$$
$$\approx 0.08328$$

so the corresponding effective rate in this case is 8.328% per year.

If the effective rate of interest r_{eff} is known, then the accumulated amount after t years on an investment of P dollars may be more readily computed by using the formula

$$A = P(1 + r_{\text{eff}})^t$$

The 1968 Truth in Lending Act passed by Congress requires that the effective rate of interest be disclosed in all contracts involving interest charges. The passage of this act has benefited consumers because they now have a common basis for comparing the various nominal rates quoted by different financial institutions. Furthermore, knowing the effective rate enables consumers to compute the actual charges involved in a transaction. Thus, if the effective rates of interest found in Example 6 were known, then the accumulated values of Example 4 could have been readily found (see Table 2).

Explore and Discuss

Recall the effective rate of interest formula:

$$r_{\text{eff}} = \left(1 + \frac{r}{m}\right)^m - 1$$

1. Show that

$$r = m\left[(1 + r_{\text{eff}})^{1/m} - 1\right]$$

2. A certificate of deposit (CD) is known to have an effective rate of 5.3% per year. If interest is compounded monthly, find the nominal rate of interest by using the result of part 1.

TABLE 2				
Nominal Rate, r	**Frequency of Interest Payment**	**Effective Rate**	**Initial Investment**	**Accumulated Amount After 3 Years**
8%	Annually	8%	$1000	$1000(1 + 0.08)^3 \approx \$1259.71$
8	Semiannually	8.16	1000	$1000(1 + 0.0816)^3 \approx 1265.32$
8	Quarterly	8.243	1000	$1000(1 + 0.08243)^3 \approx 1268.23$
8	Monthly	8.300	1000	$1000(1 + 0.08300)^3 \approx 1270.24$
8	Daily	8.328	1000	$1000(1 + 0.08328)^3 \approx 1271.22$

Present Value

Let's return to the compound interest Formula (3), which expresses the accumulated amount at the end of n periods when interest at the rate of r is compounded m times a year. The principal P in (3) is often referred to as the **present value**, and the accumulated value A is called the **future value,** since it is realized at a future date. In certain instances, an investor might wish to determine how much money he should invest now, at a fixed rate of interest, so that he will realize a certain sum at some future date. This problem may be solved by expressing P in terms of A. Thus, from Formula (3), we find

$$P = A(1 + i)^{-n}$$

Here, as before, $i = \frac{r}{m}$, where m is the number of conversion periods per year.

> **Present Value Formula for Compound Interest**
> $$P = A(1 + i)^{-n} \tag{7}$$

EXAMPLE 7 How much money should be deposited in a bank paying interest at the rate of 6% per year compounded monthly so that at the end of 3 years, the accumulated amount will be $20,000?

Solution   Here, $r = 0.06$ and $m = 12$, so $i = \frac{0.06}{12}$ and $n = (3)(12) = 36$. Thus, the problem is to determine P given that $A = 20,000$. Using Formula (7), we obtain

$$P = 20,000\left(1 + \frac{0.06}{12}\right)^{-36}$$
$$\approx 16,713$$

or $16,713.

EXAMPLE 8 Find the present value of $49,158.60 due in 5 years at an interest rate of 10% per year compounded quarterly.

Solution   Using Formula (7) with $r = 0.1$ and $m = 4$, so that $i = \frac{0.1}{4}$, $n = (4)(5) = 20$, and $A = 49,158.6$, we obtain

$$P = 49,158.6\left(1 + \frac{0.1}{4}\right)^{-20} \approx 30,000.07$$

or approximately $30,000.

If we solve Formula (5) for P, we have

$$P = Ae^{-rt}$$

which gives the present value in terms of the future (accumulated) value for the case of continuous compounding.

> **Present Value Formula for Continuous Compound Interest**
> $$P = Ae^{-rt} \tag{8}$$

Using Logarithms to Solve Problems in Finance

The next two examples show how logarithms can be used to solve problems involving compound interest.

EXAMPLE 9 How long will it take $10,000 to grow to $15,000 if the investment earns an interest rate of 8% per year compounded quarterly?

Solution   Using Formula (3) with $A = 15,000$, $P = 10,000$, $r = 0.08$, and $m = 4$, we obtain

$$15,000 = 10,000\left(1 + \frac{0.08}{4}\right)^{4t}$$

$$(1.02)^{4t} = \frac{15,000}{10,000} = 1.5$$

Taking the logarithm on each side of the equation gives

$$\ln(1.02)^{4t} = \ln 1.5$$

$$4t \ln 1.02 = \ln 1.5 \qquad \log_b m^n = n \log_b m$$

$$4t = \frac{\ln 1.5}{\ln 1.02}$$

$$t = \frac{\ln 1.5}{4 \ln 1.02} \approx 5.12$$

So it will take approximately 5.12 years for the investment to grow from $10,000 to $15,000.

EXAMPLE 10   Find the interest rate needed for an investment of $10,000 to grow to an amount of $18,000 in 5 years if the interest is compounded monthly.

Solution   Use Formula (3) with $A = 18,000$, $P = 10,000$, $m = 12$, and $t = 5$. Thus, $i = \frac{r}{12}$ and $n = (12)(5) = 60$, so

$$18,000 = 10,000\left(1 + \frac{r}{12}\right)^{12(5)}$$

Dividing both sides of the equation by 10,000 gives

$$\frac{18,000}{10,000} = \left(1 + \frac{r}{12}\right)^{60}$$

or, upon simplification,

$$\left(1 + \frac{r}{12}\right)^{60} = 1.8$$

Now we take the logarithm on each side of the equation, obtaining

$$\ln\left(1 + \frac{r}{12}\right)^{60} = \ln 1.8$$

$$60 \ln\left(1 + \frac{r}{12}\right) = \ln 1.8$$

$$\ln\left(1 + \frac{r}{12}\right) = \frac{\ln 1.8}{60} \approx 0.009796$$

$$\left(1 + \frac{r}{12}\right) \approx e^{0.009796} \qquad \ln e^x = x$$

$$\approx 1.009844$$

and

$$\frac{r}{12} \approx 1.009844 - 1$$

$$r \approx 0.1181$$

or 11.81% per year.

APPLIED EXAMPLE 11 Real Estate Investment Blakely Investment Company owns an office building located in the commercial district of a city. As a result of the continued success of an urban renewal program, local business is enjoying a miniboom. The market value of Blakely's property is

$$V(t) = 300,000e^{\sqrt{t}/2}$$

where $V(t)$ is measured in dollars and t is the time in years from the present. If the expected rate of appreciation is 9% per year compounded continuously for the next 10 years, find an expression for the present value $P(t)$ of the market price of the property that will be valid for the next 10 years. Compute $P(7)$, $P(8)$, and $P(9)$, and then interpret your results.

Solution Using Formula (8) with $A = V(t)$ and $r = 0.09$, we find that the present value of the market price of the property t years from now is

$$\begin{aligned} P(t) &= V(t)e^{-0.09t} \\ &= 300,000e^{-0.09t + \sqrt{t}/2} \qquad (0 \le t \le 10) \end{aligned}$$

Letting $t = 7, 8$, and 9, we find

$$\begin{aligned} P(7) &= 300,000e^{-0.09(7) + \sqrt{7}/2} \approx 599,837, \text{ or } \$599,837 \\ P(8) &= 300,000e^{-0.09(8) + \sqrt{8}/2} \approx 600,640, \text{ or } \$600,640 \\ P(9) &= 300,000e^{-0.09(9) + \sqrt{9}/2} \approx 598,115, \text{ or } \$598,115 \end{aligned}$$

respectively. From the results of these computations, we see that the present value of the property's market price seems to decrease after a certain period of growth. This suggests that there is an optimal time for the owners to sell. Later, we will show that the highest present value of the property's market value is approximately $600,779 and that it occurs at time $t \approx 7.72$ years.

The returns on certain investments such as zero coupon certificates of deposit (CDs) and zero coupon bonds are compared by quoting the time it takes for each investment to triple, or even quadruple. These calculations make use of the compound interest formula.

 APPLIED EXAMPLE 12 Investment Options Jane has narrowed her investment options down to two:

1. Purchase a CD that matures in 24 years and pays interest upon maturity at the rate of 5% per year compounded daily (assume 365 days in a year).
2. Purchase a zero coupon CD that will triple her investment in the same period.

Which option will optimize Jane's investment?

Solution Let's compute the accumulated amount under option 1. Here,

$$r = 0.05 \qquad m = 365 \qquad t = 24$$

so $n = 24(365) = 8760$ and $i = \frac{0.05}{365}$. The accumulated amount at the end of 24 years (after 8760 conversion periods) is

$$A = P\left(1 + \frac{0.05}{365}\right)^{8760} \approx 3.32P$$

or $3.32P. If Jane chooses option 2, the accumulated amount of her investment after 24 years will be $3P. Therefore, she should choose option 1.

APPLIED EXAMPLE 13 IRAs Moesha has an Individual Retirement Account (IRA) with a brokerage firm. Her money is invested in a money market mutual fund that pays interest on a daily basis. Over a 2-year period in which no deposits or withdrawals were made, her account grew from $4500 to $4792.61. Find the effective rate at which Moesha's account was earning interest over that period (assume 365 days in a year).

Solution   Let r_{eff} denote the required effective rate of interest. We have

$$4792.61 = 4500(1 + r_{eff})^2$$
$$(1 + r_{eff})^2 \approx 1.06502$$
$$1 + r_{eff} \approx 1.031998 \qquad \text{Take the square root on both sides.}$$

or $r_{eff} \approx 0.031998$. Therefore, the effective rate was approximately 3.20% per year.   ∎

4.1 Self-Check Exercises

1. Find the present value of $20,000 due in 3 years at an interest rate of 5.4%/year compounded monthly.

2. **INVESTMENT INCOME** Paul is a retiree living on Social Security and the income from his investment. Currently, his $100,000 investment in a 1-year CD is yielding 4.6% interest compounded daily. If he reinvests the principal

($100,000) on the due date of the CD in another 1-year CD paying 3.2% interest compounded daily, find the net decrease in his yearly income from his investment.

Solutions to Self-Check Exercises 4.1 can be found on page 221.

4.1 Concept Questions

1. Explain the difference between simple interest and compound interest.

2. What is the difference between the accumulated amount (future value) and the present value of an investment?

3. What is the effective rate of interest?

$A = P(1 + rt)$

4.1 Exercises

1. Find the simple interest on a $500 investment made for 2 years at an interest rate of 8%/year. What is the accumulated amount?

2. Find the simple interest on a $1000 investment made for 3 years at an interest rate of 5%/year. What is the accumulated amount?

3. Find the accumulated amount at the end of 9 months on an $800 deposit in a bank paying simple interest at a rate of 6%/year.

4. Find the accumulated amount at the end of 8 months on a $1200 bank deposit paying simple interest at a rate of 7%/year.

5. If the accumulated amount is $1160 at the end of 2 years and the simple rate of interest is 8%/year, what is the principal?

6. A bank deposit paying simple interest at the rate of 5%/year grew to a sum of $3100 in 10 months. Find the principal.

7. How many days will it take for a sum of $1000 to earn $20 interest if it is deposited in a bank paying simple interest at the rate of 2.5%/year? (Use a 365-day year.)

8. How many days will it take for a sum of $1500 to earn $25 interest if it is deposited in a bank paying simple interest at the rate of 5%/year? (Use a 365-day year.)

9. A bank deposit paying simple interest grew from an initial sum of $1000 to a sum of $1075 in 9 months. Find the interest rate.

10. Determine the simple interest rate at which $1200 will grow to $1250 in 8 months.

In Exercises 11–20, find the accumulated amount A if the principal P is invested at the interest rate of r/year for t years.

11. $P = \$1000$, $r = 4\%$, $t = 8$, compounded annually

12. $P = \$1000$, $r = 5\frac{1}{2}\%$, $t = 6$, compounded annually

13. $P = \$2500$, $r = 4\%$, $t = 10$, compounded semiannually

14. $P = \$2500$, $r = 6\%$, $t = 10\frac{1}{2}$, compounded semiannually

15. $P = \$12,000$, $r = 5\%$, $t = 10\frac{1}{2}$, compounded quarterly

16. $P = \$42,000$, $r = 4\frac{3}{4}\%$, $t = 8$, compounded quarterly

17. $P = \$150,000$, $r = 4\%$, $t = 4$, compounded monthly

18. $P = \$180,000$, $r = 6\%$, $t = 6\frac{1}{4}$, compounded monthly

19. $P = \$150,000$, $r = 9\%$, $t = 3$, compounded daily

20. $P = \$200,000$, $r = 8\%$, $t = 4$, compounded daily

In Exercises 21–24, find the effective rate corresponding to the given nominal rate.

21. 6%/year compounded semiannually

22. 5%/year compounded quarterly

23. 4%/year compounded monthly

24. 4%/year compounded daily

In Exercises 25–28, find the present value of $40,000 due in 4 years at the given rate of interest.

25. 4%/year compounded semiannually

26. 4%/year compounded quarterly

27. 3%/year compounded monthly

28. 5%/year compounded daily

29. Find the accumulated amount after 4 years if $5000 is invested at 6%/year compounded continuously.

30. Find the accumulated amount after 6 years if $6500 is invested at 5%/year compounded continuously.

31. How long will it take $5000 to grow to $6500 if the investment earns interest at the rate of 6%/year compounded monthly?

32. How long will it take $12,000 to grow to $15,000 if the investment earns interest at the rate of 4%/year compounded monthly?

33. How long will it take an investment of $2000 to double if the investment earns interest at the rate of 5%/year compounded monthly?

34. How long will it take an investment of $5000 to triple if the investment earns interest at the rate of 5%/year compounded daily?

35. Find the interest rate needed for an investment of $5000 to grow to an amount of $6000 in 3 years if interest is compounded continuously.

36. Find the interest rate needed for an investment of $4000 to double in 8 years if interest is compounded continuously.

37. How long will it take an investment of $6000 to grow to $7000 if the investment earns interest at the rate of $5\frac{1}{2}\%$/year compounded continuously?

38. How long will it take an investment of $8000 to double if the investment earns interest at the rate of 8%/year compounded continuously?

39. **CONSUMER DECISIONS** Mitchell has been given the option of either paying his $300 bill now or settling it for $306 after 1 month (30 days). If he chooses to pay after 1 month, find the simple interest rate at which he would be charged.

40. **COURT JUDGMENT** Jennifer was awarded damages of $150,000 in a successful lawsuit she brought against her employer 5 years ago. Simple interest on the judgment accrues at the rate of 12%/year from the date of filing. If the case were settled today, how much would Jennifer receive in the final judgment?

41. **BRIDGE LOANS** To help finance the purchase of a new house, the Abdullahs have decided to apply for a short-term loan (a bridge loan) in the amount of $120,000 for a term of 3 months. If the bank charges simple interest at the rate of 10%/year, how much will the Abdullahs owe the bank at the end of the term?

42. **CORPORATE BONDS** David owns $20,000 worth of 10-year bonds of Ace Corporation. These bonds pay interest every 6 months at the rate of 3%/year (simple interest). How much income will David receive from this investment every 6 months? How much interest will David receive over the life of the bonds?

43. **MUNICIPAL BONDS** Maya paid $10,000 for a 7-year bond issued by a city. She received interest amounting to $3500 over the life of the bonds. What rate of (simple) interest did the bond pay?

44. **TREASURY BILLS** Isabella purchased $20,000 worth of 13-week T-Bills for $19,875. What will be the rate of return on her investment?

45. **TREASURY BILLS** Maxwell purchased $15,000 worth of 52-week T-Bills for $14,650. What will be the rate of return on his investment?

46. **COMPARING INVESTMENT RETURNS** The value of Maria's investments increased by 20% in the first year and by a further 10% in the second year. The value of Laura's

investments grew 10% in the first year, followed by a gain of 20% in the second year. Both Maria and Laura started out with a $10,000 investment. Whose investment increased more in the 2-year period? Explain.

47. **INVESTMENTS** The value of Alan's stock portfolio grew by 20% in the first year, followed by a growth of 10% in the second year. It dropped 10% and 20% in the third and fourth years, respectively. Is the value of Alan's stock portfolio after 4 years the same as that when he started out? Explain.

48. **INVESTMENTS** The value of Jack's investment portfolio fell by 20% in the first year but rebounded by 20% in the second year. Did Jack regain all of the money that he lost in the first year at the end of the second year? Explain.

49. **INVESTMENTS** The value of Arabella's stock portfolio dropped 20% in the first year. Find the annual rate of growth, compounded yearly, that she must achieve in the next 2 years to bring the value of her stock portfolio back to its initial value (its value at the beginning of the first year).

50. Write Formula (1b) in the slope-intercept form, and interpret the meaning of the slope and the A-intercept in terms of r and P.
Hint: Refer to Figure 1.

51. **HOSPITAL COSTS** If the cost of a semiprivate room in a hospital was $680/day 5 years ago and hospital costs have risen at the rate of 8%/year since that time, what rate would you expect to pay for a semiprivate room today?

52. **FAMILY FOOD EXPENDITURE** Today, a typical family of four spends $880/month for food. If inflation occurs at the rate of 3%/year over the next 6 years, how much should the typical family of four expect to spend for food 6 years from now?

53. **HOUSING APPRECIATION** The Kwans are planning to buy a house 4 years from now. Housing experts in their area have estimated that the cost of a home will increase at a rate of 5%/year during that period. If this economic prediction holds true, how much can the Kwans expect to pay for a house that currently costs $260,000?

54. **ELECTRICITY CONSUMPTION** A utility company in a western city of the United States expects the consumption of electricity to increase by 8%/year during the next decade, owing mainly to the expected increase in population. If consumption does increase at this rate, find the amount by which the utility company will have to increase its generating capacity in order to meet the needs of the area at the end of the decade.

55. **PENSION FUNDS** The managers of a pension fund have invested $1.5 million in U.S. government certificates of deposit that pay interest at the rate of 2.5%/year compounded semiannually over a period of 10 years. At the end of this period, how much will the investment be worth?

56. **RETIREMENT FUNDS** Five and a half years ago, Chris invested $10,000 in a retirement fund that grew at the rate of 6.82%/year compounded quarterly. What is his account worth today?

57. **MUTUAL FUNDS** Jodie invested $15,000 in a mutual fund 4 years ago. If the fund grew at the rate of 7.8%/year compounded monthly, what would Jodie's account be worth today?

58. **TRUST FUNDS** A young man is the beneficiary of a trust fund established for him 21 years ago at his birth. If the original amount placed in trust was $10,000, how much will he receive if the money has earned interest at the rate of 6%/year compounded annually? Compounded quarterly? Compounded monthly?

59. **INVESTMENT PLANNING** Find how much money should be deposited in a bank paying interest at the rate of 3.5%/year compounded quarterly so that at the end of 5 years, the accumulated amount will be $40,000.

60. **PROMISSORY NOTES** An individual purchased a 4-year, $10,000 promissory note with an interest rate of 5.5%/year compounded semiannually. How much did the note cost?

61. **FINANCING A COLLEGE EDUCATION** The parents of a child have just come into a large inheritance and wish to establish a trust fund for her college education. If they estimate that they will need $100,000 in 13 years, how much should they set aside in the trust now if they can invest the money at $4\frac{1}{2}$%/year compounded (a) annually, (b) semiannually, and (c) quarterly?

62. **INVESTMENTS** Anthony invested a sum of money 5 years ago in a savings account that has since paid interest at the rate of 4%/year compounded quarterly. His investment is now worth $22,289.22. How much did he originally invest?

63. **COMPARING RATES OF RETURN** In the last 5 years, Bendix Mutual Fund grew at the rate of 6.4%/year compounded quarterly. Over the same period, Acme Mutual Fund grew at the rate of 6.5%/year compounded semiannually. Which mutual fund has a better rate of return?

64. **COMPARING RATES OF RETURN** Fleet Street Savings Bank pays interest at the rate of 4.25%/year compounded weekly in a savings account, whereas Washington Bank pays interest at the rate of 4.125%/year compounded daily (assume a 365-day year). Which bank offers a better rate of interest?

65. **LOAN CONSOLIDATION** The proprietors of The Coachmen Inn secured two loans from Union Bank: one for $8000 due in 3 years and one for $15,000 due in 6 years, both at an interest rate of 8%/year compounded semiannually. The bank has agreed to allow the two loans to be consolidated into one loan payable in 5 years at the same interest rate. What amount will the proprietors of the inn be required to pay the bank at the end of 5 years?
Hint: Find the present value of the first two loans.

66. EFFECTIVE RATE OF INTEREST Find the effective rate of interest corresponding to a nominal rate of 5.5%/year compounded annually, semiannually, quarterly, and monthly.

67. ZERO COUPON BONDS Juan is contemplating buying a zero coupon bond that matures in 10 years and has a face value of $10,000. If the bond yields a return of 5.25%/year, how much should Juan pay for the bond?

68. REVENUE GROWTH OF A HOME THEATER BUSINESS Maxwell started a home theater business in 2011. The revenue of his company for that year was $240,000. The revenue grew by 20% in 2012 and by 30% in 2013. Maxwell projected that the revenue growth for his company in the next 3 years will be at least 25%/year. How much does Maxwell expect his minimum revenue to be for 2016?

69. ONLINE RETAIL SALES Online retail sales stood at $141.4 billion for the year 2004. For the next 2 years, they grew by 24.3% and 14.0% per year, respectively. For the next 3 years, online retail sales were projected to grow at 30.5%, 17.6%, and 10.5% per year, respectively. What were the projected online sales for 2009?
Source: Jupiter Research.

70. PURCHASING POWER The inflation rates in the U.S. economy for 2009 through 2012 are 2.7%, 1.5%, 3.0%, and 1.7%, respectively. What was the purchasing power of a dollar at the beginning of 2013 compared to that at the beginning of 2009?
Source: U.S. Bureau of Labor Statistics.

71. INVESTMENT OPTIONS Investment A offers an 8%/year return compounded semiannually, and Investment B offers a 7.8%/year return compounded continuously. Which investment has a higher rate of return over a 4-year period?

72. EFFECT OF INFLATION ON SALARIES Leonard's current annual salary is $65,000. Ten years from now, how much will he need to earn to retain his present purchasing power if the rate of inflation over that period is 3%/year compounded continuously?

73. SAVING FOR COLLEGE Having received a large inheritance, Jing-mei's parents wish to establish a trust for her college education. Seven years from now they need an estimated $120,000. How much should they set aside in trust now if they invest the money at 6.6%/year compounded quarterly? Continuously?

74. PENSIONS Maria, who is now 50 years old, is employed by a firm that guarantees her a pension of $40,000/year at age 65. What is the present value of her first year's pension if the inflation rate over the next 15 years is 3%/year compounded continuously? 4%/year compounded continuously? 6%/year compounded continuously?

75. REAL ESTATE INVESTMENTS An investor purchased a piece of waterfront property. Because of the development of a marina in the vicinity, the market value of the property is expected to increase according to the rule

$$V(t) = 80{,}000e^{\sqrt{t}/2}$$

where $V(t)$ is measured in dollars and t is the time (in years) from the present. If the rate of appreciation is expected to be 5%/year compounded continuously for the next 8 years, find an expression for the present value $P(t)$ of the property's market price valid for the next 8 years. What is $P(t)$ expected to be in 4 years?

76. The simple interest formula $A = P(1 + rt)$ [Formula (1b)] can be written in the form $A = Prt + P$, which is the slope-intercept form of a straight line with slope Pr and A-intercept P.
   a. Describe the family of straight lines obtained by keeping the value of r fixed and allowing the value of P to vary. Interpret your results.
   b. Describe the family of straight lines obtained by keeping the value of P fixed and allowing the value of r to vary. Interpret your results.

77. EFFECTIVE RATE OF INTEREST Suppose an initial investment of $\$P$ grows to an accumulated amount of $\$A$ in t years. Show that the effective rate (annual effective yield) is

$$r_{\text{eff}} = \left(\frac{A}{P}\right)^{1/t} - 1$$

Use the formula given in Exercise 77 to solve Exercises 78–82.

78. EFFECTIVE RATE OF INTEREST Martha invested $40,000 in a boutique 5 years ago. Her investment is worth $60,000 today. What is the effective rate (annual effective yield) of her investment?

79. HOUSING APPRECIATION Georgia purchased a house in September 2008 for $300,000. In September 2014, she sold the house and made a net profit of $66,000. Find the effective annual rate of return on her investment over the 6-year period.

80. COMMON STOCK TRANSACTION Steven purchased 1000 shares of a certain stock for $25,250 (including commissions). He sold the shares 2 years later and received $32,100 after deducting commissions. Find the effective annual rate of return on his investment over the 2-year period.

81. ZERO COUPON BONDS Nina purchased a zero coupon bond for $6724.53. The bond matures in 7 years and has a face value of $10,000. Find the effective annual rate of interest for the bond.
Hint: Assume that the purchase price of the bond is the initial investment and that the face value of the bond is the accumulated amount.

82. MONEY MARKET MUTUAL FUNDS Carlos invested $5000 in a money market mutual fund that pays interest on a daily basis. The balance in his account at the end of 8 months

(245 days) was $5070.42. Find the effective rate at which Carlos's account earned interest over this period (assume a 365-day year).

In Exercises 83–86, determine whether the statement is true or false. If it is true, explain why it is true. If it is false, give an example to show why it is false.

83. When simple interest is used, the accumulated amount is a linear function of t.

84. If interest is compounded annually, then the accumulated amount after t years is the same as the accumulated amount under simple interest over t years.

85. If interest is compounded annually, then the effective rate is the same as the nominal rate.

86. Susan's salary increased from $50,000/year to $60,000/year over a 4-year period. Therefore, Susan received annual increases of 5% over that period.

4.1 Solutions to Self-Check Exercises

1. Using Formula (7) with $A = 20,000$, $r = 0.054$, and $m = 12$ so that $i = \frac{0.054}{12} = 0.0045$ and $n = (3)(12) = 36$, we find the required present value to be

$$P = 20,000(1 + 0.0045)^{-36} \approx 17,015.01$$

or $17,015.01.

2. The accumulated amount of Paul's current investment is found by using Formula (3) with $P = 100,000$, $r = 0.046$, and $m = 365$. Thus, $i = \frac{0.046}{365}$ and $n = 365$, so the required accumulated amount is given by

$$A_1 = 100,000\left(1 + \frac{0.046}{365}\right)^{365} \approx 104,707.14$$

or $104,707.14. Next, we compute the accumulated amount of Paul's reinvestment. Using (3) with $P = 100,000$, $r = 0.032$, and $m = 365$ so that $i = \frac{0.032}{365}$ and $n = 365$, we find the required accumulated amount in this case to be

$$A_2 = 100,000\left(1 + \frac{0.032}{365}\right)^{365} \approx 103,251.61$$

or approximately $103,251.61. Therefore, Paul can expect to experience a net decrease in yearly income of approximately $104,707.14 - 103,251.61$, or $1455.53.

USING **TECHNOLOGY**

Finding the Accumulated Amount of an Investment, the Effective Rate of Interest, and the Present Value of an Investment

Graphing Utility

Some graphing utilities have built-in routines for solving problems involving the mathematics of finance. For example, the TI-83/84 **TVM SOLVER** function incorporates several functions that can be used to solve the problems that are encountered in Sections 4.1–4.3. To access the **TVM SOLVER** on the TI-83, press **2nd**, press **FINANCE**, and then select **1: TVM Solver**. To access the TVM Solver on the TI-83 plus and the TI-84, press **APPS**, press **1: Finance**, and then select **1: TVM Solver**.

EXAMPLE 1 Finding the Accumulated Amount of an Investment Find the accumulated amount after 10 years if $5000 is invested at a rate of 10% per year compounded monthly.

Note: Boldfaced words/characters enclosed in a box (for example, **Enter**) indicate that an action (click, select, or press) is required. Words/characters printed blue (for example, Chart sub-type:) indicate words/characters appearing on the screen.

Solution Use the TI-83/84 **TVM SOLVER** with the following inputs:

$$N = 120 \quad (10)(12)$$
$$I\% = 10$$
$$PV = -5000 \quad \text{Since an investment is an outflow, we enter the negative of the present value.}$$
$$PMT = 0$$
$$FV = 0$$
$$P/Y = 12 \quad \text{The number of payments each year}$$
$$C/Y = 12 \quad \text{The number of conversion periods each year}$$
$$PMT: \boxed{END} \;\; BEGIN$$

Move the cursor up to the FV line and press $\boxed{ALPHA}$ $\boxed{SOLVE}$. We obtain the display shown in Figure T1. We conclude that the required accumulated amount is $13,535.21.

```
N=120
I%=10
PV=-5000
PMT=0
■ FV=13535.20745
P/Y=12
C/Y=12
PMT:(END) BEGIN
```

FIGURE **T1**
The TI-83/84 screen showing the future value (FV) of an investment

EXAMPLE 2 Finding the Effective Rate of Interest Find the effective rate of interest corresponding to a nominal rate of 10% per year compounded quarterly.

Solution Using the TI-83/84 **TVM SOLVER Eff** function, we obtain the display shown in Figure T2. The required effective rate is approximately 10.38% per year.

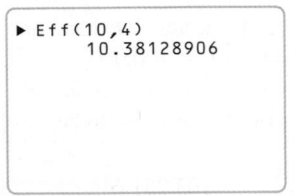

FIGURE **T2**
The TI-83/84 screen showing the effective rate of interest (Eff)

EXAMPLE 3 Finding the Present Value of an Investment Find the present value of $20,000 due in 5 years if the interest rate is 7.5% per year compounded daily.

Solution Using the TI-83/84 **TVM SOLVER** with the following inputs:

$$N = 1825 \quad (5)(365)$$
$$I\% = 7.5$$
$$PV = 0$$
$$PMT = 0$$
$$FV = 20000$$
$$P/Y = 365 \quad \text{The number of payments each year}$$
$$C/Y = 365 \quad \text{The number of conversions each year}$$
$$PMT: \boxed{END} \;\; BEGIN$$

By moving the cursor up to the PV line and pressing $\boxed{ALPHA}$ $\boxed{SOLVE}$, we obtain the display shown in Figure T3. We see that the required present value is approximately $13,746.32. Note that PV is negative because an investment is an outflow (money is paid out).

```
N=1825
I%=7.5
■ PV=-13746.3151
PMT=0
FV=20000
P/Y=365
C/Y=365
PMT:(END) BEGIN
```

FIGURE **T3**
The TI-83/84 screen showing the present value (PV) of an investment

Excel

Excel has many built-in functions for solving problems involving the mathematics of finance. Here we illustrate the use of the FV (future value), EFFECT (effective rate), and PV (present value) functions to solve problems of the type that we encountered in Section 4.1.

EXAMPLE 4 Finding the Accumulated Amount of an Investment Find the accumulated amount after 10 years if $5000 is invested at a rate of 10% per year compounded monthly.

Solution Here, we are computing the future value of a lump-sum investment, so we use the FV (future value) function. Click $\boxed{\text{Financial}}$ from the Function Library on the Formulas tab and select FV. The Function Arguments dialog box will appear (see Figure T4). In our example, the mouse cursor is in the edit box headed by Type, so a definition of that term appears near the bottom of the box. Figure T4 shows the entries for each edit box in our example.

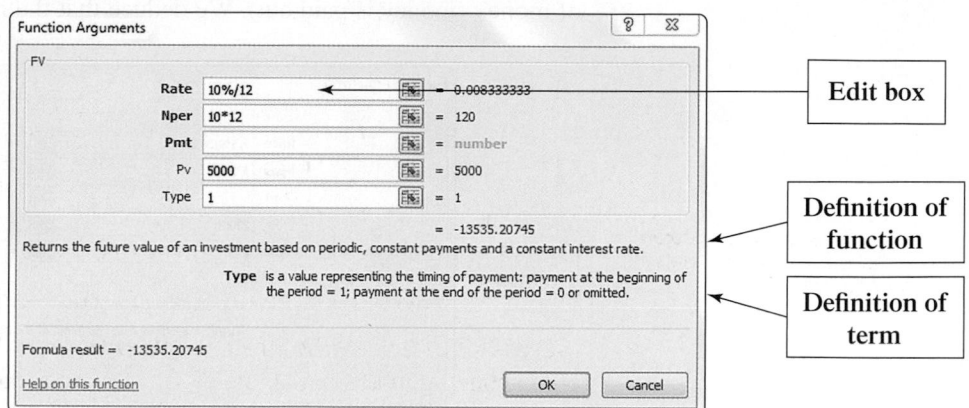

FIGURE **T4**
Excel's dialog box for computing the future value (FV) of an investment

Note that the entry for Nper is given by the total number of periods for which the investment earns interest. The Pmt box is left blank, since no money is added to the original investment. The Pv entry is 5000. The entry for Type is a 1 because the lump-sum payment is made at the beginning of the investment period. The answer, $-\$13,535.21$, is shown at the bottom of the dialog box. It is negative because an investment is considered to be an outflow of money (money is paid out). (Click OK, and the answer will also appear on your spreadsheet.)

EXAMPLE 5 Finding the Effective Rate of Interest Find the effective rate of interest corresponding to a nominal rate of 10% per year compounded quarterly.

Solution Here, we use the EFFECT function to compute the effective rate of interest. Accessing this function from the Financial function library subgroup and making the required entries, we obtain the Function Arguments dialog box shown in Figure T5. The required effective rate is approximately 10.38% per year.

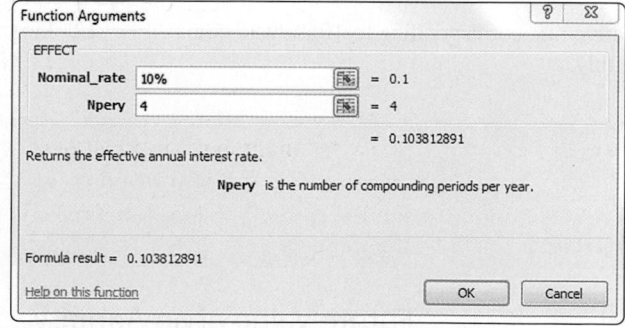

FIGURE **T5**
Excel's dialog box for the effective rate of interest function (EFFECT)

EXAMPLE 6 Finding the Present Value of an Investment Find the present value of $20,000 due in 5 years if the interest rate is 7.5% per year compounded daily.

Solution We use the PV function to compute the present value of a lump-sum investment. Accessing this function as above and making the required entries, we obtain the PV dialog box shown in Figure T6. Once again, the Pmt edit box is left blank, since no additional money is added to the original investment. The Fv entry is 20000. The answer is negative because an investment is considered to be an outflow of money (money is paid out). We deduce that the required amount is \$13,746.32.

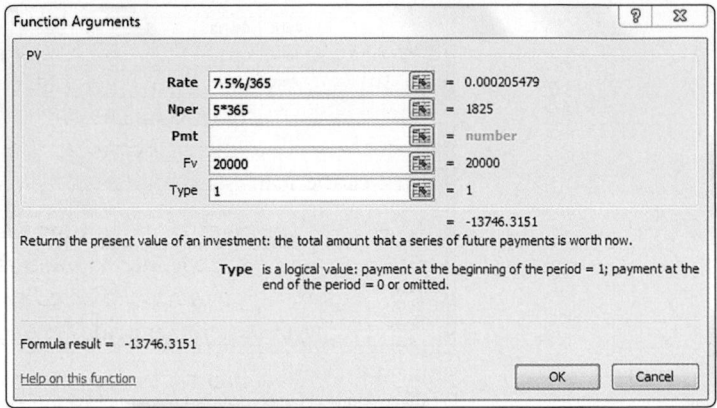

FIGURE **T6**
Excel dialog box for the present value function (PV)

TECHNOLOGY EXERCISES

1. Find the accumulated amount A if \$5000 is invested at the interest rate of $5\frac{3}{8}$%/year compounded monthly for 3 years.

2. Find the accumulated amount A if \$2850 is invested at the interest rate of $6\frac{5}{8}$%/year compounded monthly for 4 years.

3. Find the accumulated amount A if \$327.35 is invested at the interest rate of $5\frac{1}{3}$%/year compounded daily for 7 years.

4. Find the accumulated amount A if \$327.35 is invested at the interest rate of $6\frac{7}{8}$%/year compounded daily for 8 years.

5. Find the effective rate corresponding to $8\frac{2}{3}$%/year compounded quarterly.

6. Find the effective rate corresponding to $10\frac{5}{8}$%/year compounded monthly.

7. Find the effective rate corresponding to $9\frac{3}{4}$%/year compounded monthly.

8. Find the effective rate corresponding to $4\frac{3}{8}$%/year compounded quarterly.

9. Find the present value of \$38,000 due in 3 years at $8\frac{1}{4}$%/year compounded quarterly.

10. Find the present value of \$150,000 due in 5 years at $9\frac{3}{8}$%/year compounded monthly.

11. Find the present value of \$67,456 due in 3 years at $7\frac{7}{8}$%/year compounded monthly.

12. Find the present value of \$111,000 due in 5 years at $11\frac{5}{8}$%/year compounded monthly.

4.2 Annuities

Future Value of an Annuity

An **annuity** is a sequence of payments made at regular time intervals. The time period in which these payments are made is called the **term** of the annuity. Depending on whether the term is given by a *fixed time interval*, a time interval that begins at a definite date but extends indefinitely, or one that is not fixed in advance, an annuity is called an **annuity certain,** a *perpetuity*, or a *contingent annuity*, respectively. In general, the payments in an annuity need not be equal, but in many important applications they are equal. In this section, we assume that annuity payments are equal. Examples of annuities are regular deposits to a savings account, monthly home mortgage payments, and monthly insurance payments.

Annuities are also classified by payment dates. An annuity in which the payments are made at the *end* of each payment period is called an **ordinary annuity,** whereas an annuity in which the payments are made at the beginning of each period is called an *annuity due.* Furthermore, an annuity in which the payment period coincides with the interest conversion period is called a **simple annuity,** whereas an annuity in which the payment period differs from the interest conversion period is called a *complex annuity.*

In this section, we consider ordinary annuities that are certain and simple, with periodic payments that are equal in size. In other words, we study annuities that are subject to the following conditions:

1. The terms are given by fixed time intervals.
2. The periodic payments are equal in size.
3. The payments are made at the *end* of the payment periods.
4. The payment periods coincide with the interest conversion periods.

To find a formula for the accumulated amount S of an annuity, suppose a sum of $100 is paid into an account at the end of each quarter over a period of 3 years. Furthermore, suppose the account earns interest on the deposit at the rate of 8% per year, compounded quarterly. Then the first payment of $100 made at the end of the first quarter earns interest at the rate of 8% per year compounded four times a year (or $8/4 = 2\%$ per quarter) over the remaining 11 quarters and therefore, by the compound interest formula, has an accumulated amount of

$$100\left(1 + \frac{0.08}{4}\right)^{11} \quad \text{or} \quad 100(1 + 0.02)^{11}$$

dollars at the end of the term of the annuity (Figure 2).

The second payment of $100 made at the end of the second quarter earns interest at the same rate over the remaining 10 quarters and therefore has an accumulated amount of

$$100(1 + 0.02)^{10}$$

dollars at the end of the term of the annuity, and so on. The last payment earns no interest because it is due at the end of the term. The amount of the annuity is obtained by adding all the terms in Figure 2. Thus,

$$S = 100 + 100(1 + 0.02) + 100(1 + 0.02)^2 + \cdots + 100(1 + 0.02)^{11}$$

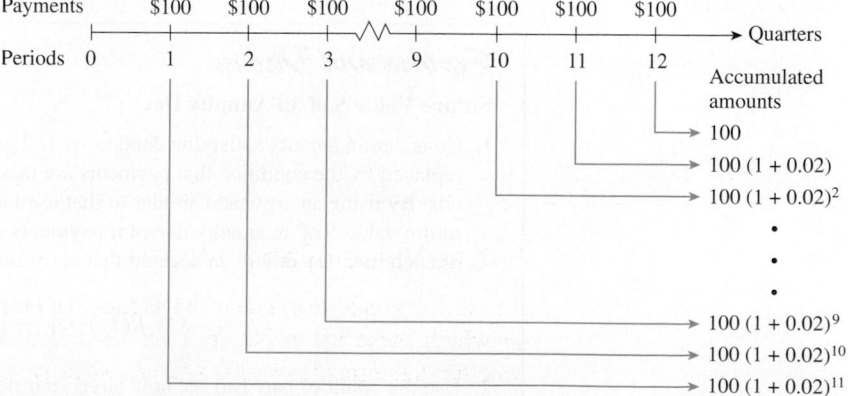

FIGURE 2
The sum of the accumulated amounts is the amount of the annuity.

The sum on the right is the sum of the first n terms of a *geometric progression* with first term 100 and common ratio $(1 + 0.02)$. We show in Section 4.4 that the sum S can be written in the more compact form

$$S = 100\left[\frac{(1 + 0.02)^{12} - 1}{0.02}\right]$$

$$\approx 1341.21$$

or approximately \$1341.21.

To find a general formula for the accumulated amount S of an annuity, suppose that a sum of \$$R$ is paid into an account at the end of each period for n periods and that the account earns interest at the rate of i per period. Then, proceeding as we did with the numerical example, we obtain

$$S = R + R(1 + i) + R(1 + i)^2 + \cdots + R(1 + i)^{n-1}$$

$$= R\left[\frac{(1 + i)^n - 1}{i}\right] \tag{9}$$

The expression inside the brackets is called the **compound-amount factor**. The quantity S in Formula (9) is realizable at some future date and is accordingly called the future value of an annuity.

Future Value of an Annuity

The **future value S of an annuity** of n payments of R dollars each, paid at the end of each investment period into an account that earns interest at the rate of i per period, is

$$S = R\left[\frac{(1 + i)^n - 1}{i}\right]$$

EXAMPLE 1 Find the amount of an ordinary annuity consisting of 12 monthly payments of \$100 that earn interest at 12% per year compounded monthly.

Solution Since i is the interest rate per *period* and since interest is compounded monthly in this case, we have $i = \frac{0.12}{12} = 0.01$. Using Formula (9) with $R = 100$, $n = 12$, and $i = 0.01$, we have

$$S = 100\left[\frac{(1.01)^{12} - 1}{0.01}\right]$$

$$\approx 1268.25 \qquad \text{Use a calculator.}$$

or \$1268.25. ◼

Explore and Discuss

Future Value S of an Annuity Due

1. Consider an annuity satisfying conditions 1, 2, and 4 on page 225 but with condition 3 replaced by the condition that payments are made at the *beginning* of the payment periods. By using an argument similar to that used to establish Formula (9), show that the future value S of an annuity due of n payments of R dollars each, paid at the beginning of each investment into an account that earns interest at the rate of i per period, is

$$S = R(1 + i)\left[\frac{(1 + i)^n - 1}{i}\right]$$

2. Use the result of part 1 to see how large your nest egg will be at age 65 if you start saving \$4000 annually at age 30, assuming a 10% average annual return; if you start saving at 35; if you start saving at 40. [Moral of the story: It is never too early to start saving!]

Refer to the preceding Explore and Discuss problem.

1. Show that if $R = 4000$ and $i = 0.1$, then $S = 44{,}000[(1.1)^n - 1]$. Using a graphing utility, plot the graph of $f(x) = 44{,}000[(1.1)^x - 1]$, using the viewing window $[0, 40] \times [0, 1{,}200{,}000]$.
2. Verify the results of part 2 by evaluating $f(35)$, $f(30)$, and $f(25)$ using the **EVAL** function.

Present Value of an Annuity

In certain instances, you may want to determine the current value P of a sequence of equal periodic payments that will be made over a certain period of time. After each payment is made, the new balance continues to earn interest at some nominal rate. The amount P is referred to as the present value of an annuity.

To derive a formula for determining the present value P of an annuity, we may argue as follows. The amount P invested now and earning interest at the rate of i per period will have an accumulated value of $P(1 + i)^n$ at the end of n periods. But this must be equal to the future value of the annuity S given by Formula (9). Therefore, equating the two expressions, we have

$$P(1 + i)^n = R\left[\frac{(1 + i)^n - 1}{i}\right]$$

Multiplying both sides of this equation by $(1 + i)^{-n}$ gives

$$P = R(1 + i)^{-n}\left[\frac{(1 + i)^n - 1}{i}\right]$$

$$= R\left[\frac{(1 + i)^n(1 + i)^{-n} - (1 + i)^{-n}}{i}\right]$$

$$= R\left[\frac{1 - (1 + i)^{-n}}{i}\right] \qquad (1 + i)^n(1 + i)^{-n} = 1$$

Present Value of an Annuity

The **present value P of an annuity** consisting of n payments of R dollars each, paid at the end of each investment period into an account that earns interest at the rate of i per period, is

$$P = R\left[\frac{1 - (1 + i)^{-n}}{i}\right] \tag{10}$$

EXAMPLE 2 Find the present value of an ordinary annuity consisting of 24 monthly payments of \$100 each and earning interest at 3% per year compounded monthly.

Solution Here, $R = 100$, $i = \frac{r}{m} = \frac{0.03}{12} = 0.0025$, and $n = 24$, so by Formula (10), we have

$$P = 100\left[\frac{1 - (1.0025)^{-24}}{0.0025}\right]$$

$$\approx 2326.60$$

or \$2326.60.

APPLIED EXAMPLE 3 Saving for a College Education As a savings program toward Alberto's college education, his parents decide to deposit $200 at the end of every month into a bank account paying interest at the rate of 6% per year compounded monthly. If the savings program began when Alberto was 6 years old, how much money would have accumulated by the time he turns 18?

Solution By the time the child turns 18, the parents would have made 144 deposits into the account. Thus, $n = 144$. Furthermore, we have $R = 200$, $r = 0.06$, and $m = 12$, so $i = \frac{0.06}{12} = 0.005$. Using Formula (9), we find that the amount of money that would have accumulated is given by

$$S = 200\left[\frac{(1.005)^{144} - 1}{0.005}\right]$$
$$\approx 42{,}030$$

or $42,030.

APPLIED EXAMPLE 4 Financing a Car After making a down payment of $6000 for an automobile, Murphy paid $600 per month for 36 months with interest charged at 6% per year compounded monthly on the unpaid balance. What was the original cost of the car? What portion of Murphy's total car payments went toward interest charges?

Solution The loan taken up by Murphy is given by the present value of the annuity

$$P = 600\left[\frac{1 - (1.005)^{-36}}{0.005}\right]$$
$$\approx 19{,}723$$

or $19,723. Therefore, the original cost of the automobile is $25,723 ($19,723 plus the $6000 down payment). The interest charges paid by Murphy are given by $(36)(600) - 19{,}723 = 1{,}877$, or $1,877.

One important application of annuities arises in the area of tax planning. During the 1980s, Congress created many tax-sheltered retirement savings plans, such as Individual Retirement Accounts (IRAs), Keogh plans, and Simplified Employee Pension (SEP) plans. These plans are examples of annuities in which the individual is allowed to make contributions (which are often tax deductible) to an investment account. The amount of the contribution is limited by congressional legislation. The taxes on the contributions and/or the interest accumulated in these accounts are deferred until the money is withdrawn—ideally during retirement, when tax brackets should be lower. In the interim period, the individual has the benefit of tax-free growth on his or her investment.

Suppose, for example, you are eligible to make a fully deductible contribution to an IRA and you are in a marginal tax bracket of 28%. Additionally, suppose you receive a year-end bonus of $2000 from your employer and have the option of depositing the $2000 into either an IRA or a regular savings account, where both accounts earn interest at an effective annual rate of 8% per year. If you choose to invest your bonus in a regular savings account, you will first have to pay taxes on the $2000, leaving $1440 to invest. At the end of 1 year, you will also have to pay taxes on the interest earned, leaving you with

Accumulated amount	−	Tax on interest	=	Net amount

$$1555.20 \quad - \quad 32.26 \quad = 1522.94$$

or $1522.94.

On the other hand, if you put the money into the IRA, the entire sum will earn interest, and at the end of 1 year, you will have $(1.08)(\$2000)$, or $2160, in your account. Of course, you will still have to pay taxes on this money when you withdraw it, but you will have gained the advantage of tax-free growth of the larger principal over the years. The disadvantage of this option is that if you withdraw the money before you reach the age of $59\frac{1}{2}$, you will be liable for taxes on both your contributions and the interest earned, *and* you will also have to pay a 10% penalty.

Note In practice, the size of the contributions an individual might make to the various retirement plans might vary from year to year. Also, he or she might make the contributions at different payment periods. To simplify our discussion, we will consider examples in which fixed payments are made at regular intervals. ■

APPLIED EXAMPLE 5 IRAs Caroline is planning to make a contribution of $2000 on January 31 of each year into a traditional IRA earning interest at an effective rate of 5% per year.

a. After she makes her 25th payment on January 31 of the year following her retirement at age 66, how much will she have in her IRA?
b. Suppose that Caroline withdraws all of her money from her traditional IRA after she makes her 25th payment in the year following her retirement at age 66 and that her investment is subjected to a tax of 28% at that time. How much money will she end up with after taxes?

Solution

a. The amount of money Caroline will have after her 25th payment into her account is found by using Formula (9) with $R = 2000$, $r = 0.05$, $m = 1$, and $t = 25$, so that $i = \frac{r}{m} = 0.05$ and $n = mt = 25$. The required amount is given by

$$S = 2000\left[\frac{(1.05)^{25} - 1}{0.05}\right]$$
$$\approx 95,454.20$$

or $95,454.20.
b. If she withdraws the entire amount from her account, she will end up with

$$(1 - 0.28)(95,454.20) \approx 68,727.02$$

that is, she will have approximately $68,727.02 after paying taxes. ■

After-tax-deferred annuities are another type of investment vehicle that allows an individual to build assets for retirement, college funds, or other future needs. The advantage gained in this type of investment is that the tax on the accumulated interest is deferred to a later date. Note that in this type of investment, the contributions themselves are not tax deductible. At first glance, the advantage thus gained may seem to be relatively inconsequential, but its true effect is illustrated by the next example.

$ **APPLIED EXAMPLE 6** Investment Analysis Both Clark and Colby are salaried individuals, 45 years of age, who are saving for their retirement 20 years from now. Both Clark and Colby are also in the 28% marginal tax bracket. Clark makes a $1000 contribution annually on December 31 into a savings account earning an effective rate of 8% per year. At the same time, Colby makes a $1000 annual payment to an insurance company for an after-tax-deferred annuity. The annuity also earns interest at an effective rate of 8% per year. (Assume that both men remain in the same tax bracket throughout this period, and disregard state income taxes.)

a. Calculate how much each man will have in his investment account at the end of 20 years.
b. Compute the interest earned on each account.
c. Show that even if the interest on Colby's investment were subjected to a tax of 28% upon withdrawal of his investment at the end of 20 years, the net accumulated amount of his investment would still be greater than that of Clark's.

Solution

a. Because Clark is in the 28% marginal tax bracket, the net yield for his investment is $(0.72)(8)$, or 5.76%, per year.
   Using Formula (9) with $R = 1000$, $r = 0.0576$, $m = 1$, and $t = 20$, so that $i = 0.0576$ and $n = mt = 20$, we see that Clark's investment will be worth

$$S = 1000\left[\frac{(1 + 0.0576)^{20} - 1}{0.0576}\right]$$
$$\approx 35,850.49$$

or $35,850.49 at his retirement.
   Colby has a tax-sheltered investment with an effective yield of 8% per year. Using Formula (9) with $R = 1000$, $r = 0.08$, $m = 1$, and $t = 20$, so that $i = 0.08$ and $n = mt = 20$, we see that Colby's investment will be worth

$$S = 1000\left[\frac{(1 + 0.08)^{20} - 1}{0.08}\right]$$
$$\approx 45,761.96$$

or $45,761.96 at his retirement.
b. Each man will have paid 20(1000), or $20,000, into his account. Therefore, the total interest earned in Clark's account will be $(35,850.49 - 20,000)$, or $15,850.49, whereas the total interest earned in Colby's account will be $(45,761.96 - 20,000)$, or $25,761.96.
c. From part (b) we see that the total interest earned in Colby's account will be $25,761.96. If it were taxed at 28%, he would still end up with $(0.72)(25,761.96)$, or $18,548.61. This amount is larger than the total interest of $15,850.49 earned by Clark. ∎

In 1997, another type of tax-sheltered retirement savings plan was created by Congress: the Roth IRA. In contrast to traditional IRAs, contributions to Roth IRAs are not tax-deferrable. However, direct contributions to Roth IRAs (but not rollovers) may be withdrawn tax-free at any time. Also, holders of a Roth IRA are not required to take minimum distributions after age $70\frac{1}{2}$.

APPLIED EXAMPLE 7 Roth IRAs Refer to Example 5. Suppose that Caroline decides to invest her money in a Roth IRA instead of a traditional IRA. Also, suppose that she is in the 28% tax bracket and remains in that bracket for the next 25 years until her retirement at age 65. If she pays taxes on $2000 and then invests the remaining $1440 into a Roth IRA earning interest at a rate of 5% per year, compounded annually, how much will she have in her Roth IRA after her 25th payment on January 31 of the year following her retirement? (Disregard state and city taxes.) How does this compare with the amount of money she would have if she had stayed with a traditional IRA and withdrawn all of her money from that account at that time? (See Example 5b.)

Solution   We use Formula (9) with $R = 1440$, $r = 0.05$, and $n = 25$, obtaining

$$S = 1440\left[\frac{(1.05)^{25} - 1}{0.05}\right]$$
$$\approx 68,727.02$$

that is, she will have approximately $68,727.02 in her account. This is the same as the amount she ended up with in her traditional IRA after paying taxes (see Example 5b).

4.2 Self-Check Exercises

1. **TRADITIONAL IRA INVESTMENT** Phyliss opened an IRA on January 31, 2000, with a contribution of $2000. She plans to make a contribution of $2000 thereafter on January 31 of each year until her retirement in the year 2019 (20 payments). If the account earns interest at the rate of 8%/year compounded yearly, how much will Phyliss have in her account when she retires?

2. **SECURING A BANK LOAN** Denver Wildcatting Company has an immediate need for a loan. In an agreement worked out with its banker, Denver assigns its royalty income of $4800/month for the next 3 years from certain oil properties to the bank, with the first payment due at the end of the first month. If the bank charges interest at the rate of 9%/year compounded monthly, what is the amount of the loan negotiated between the parties?

Solutions to Self-Check Exercises 4.2 can be found on page 233.

4.2 Concept Questions

1. Is the term of an ordinary annuity fixed or variable? Are the periodic payments all of the same size, or do they vary in size? Are the payments made at the beginning or the end of the payment period? Do the payment periods coincide with the interest conversion periods?

2. What is the difference between an ordinary annuity and an annuity due?

3. What is the future value of an annuity? Give an example.

4. What is the present value of an annuity? Give an example.

4.2 Exercises

In Exercises 1–8, find the amount (future value) of each ordinary annuity.

1. $1000/year for 10 years at 5%/year compounded annually

2. $1500/semiannual period for 8 years at 4.5%/year compounded semiannually

3. $500/semiannual period for 12 years at 6%/year compounded semiannually

4. $1800/quarter for 6 years at 4%/year compounded quarterly

5. $600/quarter for 9 years at 5%/year compounded quarterly

6. $150/month for 15 years at 6%/year compounded monthly

7. $200/month for $20\frac{1}{4}$ years at 6.5%/year compounded monthly

8. $100/week for $7\frac{1}{2}$ years at 3.5%/year compounded weekly

In Exercises 9–14, find the present value of each ordinary annuity.

9. $5000/year for 8 years at 6%/year compounded annually

10. $4000/year for 5 years at 4.5%/year compounded yearly

11. $1200/semiannual period for 6 years at 5%/year compounded semiannually

12. $3000/semiannual period for 6 years at 5.5%/year compounded semiannually

13. $800/quarter for 7 years at 6%/year compounded quarterly

14. $150/month for 10 years at 4%/year compounded monthly

15. IRAs If a merchant deposits $1500 at the end of each tax year in an IRA paying interest at the rate of 4%/year compounded annually, how much will she have in her account at the end of 25 years?

16. SAVINGS ACCOUNTS If Jackson deposits $100 at the end of each month in a savings account earning interest at the rate of 3%/year compounded monthly, how much will he have on deposit in his savings account at the end of 6 years, assuming that he makes no withdrawals during that period?

17. SAVINGS ACCOUNTS Linda has joined a Christmas Fund Club at her bank. At the end of every month, December through October inclusive, she will make a deposit of $40 in her fund. If the money earns interest at the rate of 2.5%/year compounded monthly, how much will she have in her account on December 1 of the following year?

18. KEOGH ACCOUNTS Robin, who is self-employed, contributes $5000/year into a Keogh account. How much will he have in the account after 25 years if the account earns interest at the rate of 4.5%/year compounded yearly?

19. INVESTMENT ANALYSIS Karen has been depositing $150 at the end of each month in a tax-free retirement account since she was 25. Matt, who is the same age as Karen, started depositing $250 at the end of each month in a tax-free retirement account when he was 35. Assuming that both accounts have been and will be earning interest at the rate of 4%/year compounded monthly, who will end up with the larger retirement account at the age of 65?

20. RETIREMENT PLANNING As a fringe benefit for the past 12 years, Colin's employer has contributed $100 at the end of each month into an employee retirement account for Colin that pays interest at the rate of 5%/year compounded monthly. Colin has also contributed $2000 at the end of each of the last 8 years into an IRA that pays interest at the rate of 4.5%/year compounded yearly. How much does Colin have in his retirement fund at this time?

21. INVESTMENT ANALYSIS Luis has $150,000 in his retirement account at his present company. Because he is assuming a position with another company, Luis is planning to "roll over" his assets to a new account. Luis also plans to put $3000/quarter into the new account until his retirement 20 years from now. If the new account earns interest at the rate of 4.5%/year compounded quarterly, how much will Luis have in his account at the time of his retirement? Hint: Use the compound interest formula and the annuity formula.

22. AUTO LEASING The Betzes have leased an auto for 2 years at $450/month. If money is worth 3.5%/year compounded monthly, what is the equivalent cash payment (present value) of this annuity?

23. SAVINGS ACCOUNTS The Pirerras are planning to go to Europe 3 years from now and have agreed to set aside $150/month for their trip. If they deposit this money at the end of each month into a savings account paying interest at the rate of 3%/year compounded monthly, how much money will be in their travel fund at the end of the third year?

24. INSTALLMENT PLANS Mike's Sporting Goods sells elliptical trainers under two payment plans: cash or installment. Under the installment plan, the customer pays $22/month over 3 years with interest charged on the balance at a rate of 9%/year compounded monthly. Find the cash price for an elliptical trainer if it is equivalent to the price paid by a customer using the installment plan.

25. AUTO FINANCING Lupé made a down payment of $8000 toward the purchase of a new car. To pay the balance of the purchase price, she has secured a loan from her bank at the rate of 6%/year compounded monthly. Under the terms of her finance agreement, she is required to make payments of $420/month for 36 months. What is the cash price of the car?

26. LOTTERY PAYOUTS A state lottery commission pays the winner of the Million Dollar lottery 20 installments of $50,000/year. The commission makes the first payment of $50,000 immediately and the other $n = 19$ payments at the end of each of the next 19 years. Determine how much money the commission should have in the bank initially to guarantee the payments, assuming that the balance on deposit with the bank earns interest at the rate of 4%/year compounded yearly. Hint: Find the present value of an annuity.

27. PURCHASING A HOME The Johnsons have accumulated a nest egg of $40,000 that they intend to use as a down payment toward the purchase of a new house. Because their present gross income has placed them in a relatively high tax bracket, they have decided to invest a minimum of $2400/month in monthly payments (to take advantage of the tax deduction) toward the purchase of their house. However,

into the same account at the end of that month and at the end of each subsequent month for the next 5 years. If her bank pays interest at the rate of 3%/year compounded monthly, how much will Lauren have in her account at the end of 5 years? (Assume that she makes no withdrawals during the 5-year period.)

31. **Financial Planning** Joe plans to deposit $200 at the end of each month into a bank account for a period of 2 years, after which he plans to deposit $300 at the end of each month into the same account for another 3 years. If the bank pays interest at the rate of 3.5%/year compounded monthly, how much will Joe have in his account by the end of 5 years? (Assume that no withdrawals are made during the 5-year period.)

32. **Investment Analysis** From age 25 to age 40, Jessica deposited $200 at the end of each month into a tax-free retirement account. She made no withdrawals or further contributions until age 65. Alex made deposits of $300 into his tax-free retirement account from age 40 to age

earning interest at the same rate as that of Ramos and also for a period of 30 years. Suppose that the investments of both Ramos and Vanessa are subjected to a tax of 30% at the time of their retirement and that they both wish to withdraw all of the money in their IRAs at that time.

   a. After all due taxes are paid, who will have the larger amount?

   b. How much larger will that amount be?

In Exercises 35 and 36, determine whether the stater or false. If it is true, explain why it is true. If it is false, example to show why it is false.

35. The future value of an annuity can be found by together all the payments that are paid into the a

36. If the future value of an annuity consisting of n pa of R dollars each—paid at the end of each investme period into an account that earns interest at the rate $per period—is S dollars, then

USING TECHNOLOGY

Finding the Amount of an Annuity

Graphing Utility

As was mentioned in Using Technology, Section 4.1, the TI-83/84 can facilitate the solution of problems in finance. We continue to exploit its versatility in this section.

EXAMPLE 1 Finding the Future Value of an Annuity Find the amount of an ordinary annuity of 36 quarterly payments of $220 each that earn interest at the rate of 10% per year compounded quarterly.

Solution We use the TI-83/84 **TVM SOLVER** with the following inputs:

$$N = 36$$
$$I\% = 10$$
$$PV = 0$$
$$PMT = -220 \quad \text{Recall that a payment is an outflow.}$$
$$FV = 0$$
$$P/Y = 4 \quad \text{The number of payments each year}$$
$$C/Y = 4 \quad \text{The number of conversion periods each year}$$
$$PMT: \boxed{END} \ BEGIN$$

```
N=36
I%=10
PV=0
PMT=-220
FV=12606.31078
P/Y=4
C/Y=4
PMT:(END) BEGIN
```

FIGURE T1
...84 screen showing the future ... an annuity

Move the cursor up to the FV line and press $\boxed{\text{ALPHA}}$ $\boxed{\text{SOLVE}}$. The result is displayed in Figure T1. We deduce that the desired amount is $12,606.31.

EXAMPLE 2 Finding the Present Value of an Annuity Find the present value of an ordinary annuity consisting of 48 monthly payments of $300 each and earning interest at the rate of 9% per year compounded monthly.

Solution We use the TI-83/84 **TVM SOLVER** with the following inputs:

$$N = 48$$
$$I\% = 9$$
$$PV = 0$$
$$PMT = -300 \quad \text{A payment is an outflow.}$$
$$FV = 0$$
$$P/Y = 12 \quad \text{The number of payments each year}$$
$$C/Y = 12 \quad \text{The number of conversion periods each year}$$
$$PMT: \boxed{END} \ BEGIN$$

```
N=48
I%=9
PV=12055.43457
PMT=-300
FV=0
P/Y=12
C/Y=12
PMT:(END) BEGIN
```

FIGURE T2
The TI-83/84 screen showing the present value (PV) of an ordinary annuity

By moving the cursor up to the PV line and pressing $\boxed{\text{ALPHA}}$ $\boxed{\text{SOLVE}}$, we obtain the output displayed in Figure T2. We see that the required present value of the annuity is $12,055.43.

Excel

Now we show how Excel can be used to solve financial problems involving annuities.

EXAMPLE 3 Finding the Future Value of an Annuity Find the amount of an ordinary annuity of 36 quarterly payments of $220 each that earn interest at the rate of 10% per year compounded quarterly.

Note: Boldfaced words/characters enclosed in a box (for example, $\boxed{\text{Enter}}$) indicate that an action (click, select, or press) is required. Words/characters printed blue (for example, Chart sub-type:) indicate words/characters appearing on the screen.

Solution Here, we are computing the future value of a series of equal payments, so we use the FV (future value) function. As before, we choose this function from the Financial function library to obtain the Function Arguments dialog box. After making each of the required entries, we obtain the dialog box shown in Figure T3.

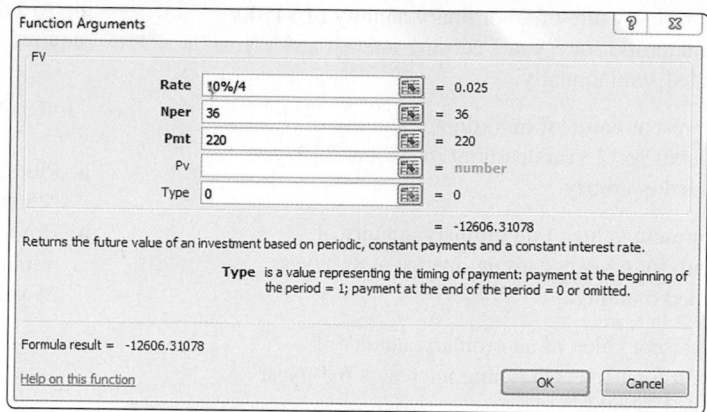

FIGURE **T3**
Excel's dialog box for the future value (FV) of an annuity

Note that a 0 is entered in the Type edit box because payments are made at the end of each payment period. Once again, the answer is negative because cash is paid out. We deduce that the desired amount is $12,606.31.

EXAMPLE 4 Finding the Present Value of an Annuity Find the present value of an ordinary annuity consisting of 48 monthly payments of $300 each and earning interest at the rate of 9% per year compounded monthly.

Solution Here, we use the PV function to compute the present value of an annuity. Accessing the PV (present value) function from the Financial function library and making the required entries, we obtain the PV dialog box shown in Figure T4. We see that the required present value of the annuity is $12,055.43.

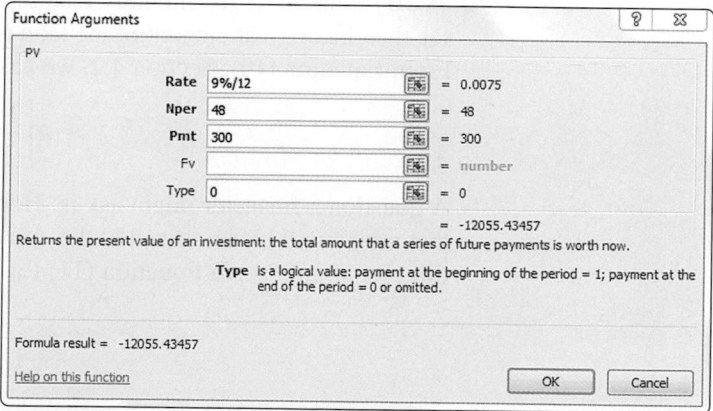

FIGURE **T4**
Excel's dialog box for computing the present value (PV) of an annuity

TECHNOLOGY EXERCISES

1. Find the amount of an ordinary annuity of 20 payments of $2500/quarter at $7\frac{1}{4}$%/year compounded quarterly.

2. Find the amount of an ordinary annuity of 24 payments of $1790/quarter at $8\frac{3}{4}$%/year compounded quarterly.

3. Find the amount of an ordinary annuity of $120/month for 5 years at $6\frac{3}{8}$%/year compounded monthly.

4. Find the amount of an ordinary annuity of $225/month for 6 years at $7\frac{5}{8}$%/year compounded monthly.

5. Find the present value of an ordinary annuity of $4500/semiannual period for 5 years earning interest at 9%/year compounded semiannually.

6. Find the present value of an ordinary annuity of $2100/quarter for 7 years earning interest at $7\frac{1}{8}$%/year compounded quarterly.

7. Find the present value of an ordinary annuity of $245/month for 6 years earning interest at $8\frac{3}{8}$%/year compounded monthly.

8. Find the present value of an ordinary annuity of $185/month for 12 years earning interest at $6\frac{5}{8}$%/year compounded monthly.

9. **ANNUITIES** At the time of retirement, Christine expects to have a sum of $500,000 in her retirement account. Assuming that the account pays interest at the rate of 5%/year compounded continuously, her accountant pointed out to her that if she made withdrawals amounting to x dollars per year $(x > 25,000)$, then the time required to deplete her savings would be T years, where

$$T = f(x) = 20 \ln\left(\frac{x}{x - 25,000}\right) \qquad (x > 25,000)$$

a. Plot the graph of f, using the viewing window $[25,000, 50,000] \times [0, 100]$.

b. How much should Christine plan to withdraw from her retirement account each year if she wants it to last for 25 years?

4.3 Amortization and Sinking Funds

Amortization of Loans

The annuity formulas derived in Section 4.2 may be used to answer questions involving the amortization of certain types of installment loans. For example, in a typical housing loan, the mortgagor makes periodic payments toward reducing his or her indebtedness to the lender, who charges interest at a fixed rate on the unpaid portion of the debt. In practice, the borrower is required to repay the lender in periodic installments, usually of the same size and over a fixed term, so that the loan (principal plus interest charges) is amortized at the end of the term.

By thinking of the monthly loan repayments R as the payments in an annuity, we see that the original amount of the loan is given by P, the present value of the annuity. From Equation (10), Section 4.2, we have

$$P = R\left[\frac{1 - (1 + i)^{-n}}{i}\right] \tag{11}$$

A question a financier might ask is: How much should the monthly installment be so that a loan will be amortized at the end of the term of the loan? To answer this question, we simply solve Equation (11) for R in terms of P, obtaining

$$R = \frac{Pi}{1 - (1 + i)^{-n}}$$

Amortization Formula

The periodic payment R on a loan of P dollars to be amortized over n periods with interest charged at the rate of i per period is

$$R = \frac{Pi}{1 - (1 + i)^{-n}} \tag{12}$$

APPLIED EXAMPLE 1 Amortization Schedule A sum of $50,000 is to be repaid over a 5-year period through equal installments made at the end of each year. If an interest rate of 8% per year is charged on the unpaid balance and interest calculations are made at the end of each year, determine the size of each installment so that the loan (principal plus interest charges) is amortized at the end of 5 years. Verify the result by displaying the amortization schedule.

Solution Substituting $P = 50,000$, $i = r = 0.08$ (here, $m = 1$), and $n = 5$ into Formula (12), we obtain

$$R = \frac{(50,000)(0.08)}{1 - (1.08)^{-5}}$$

$$\approx 12,522.82$$

giving the required yearly installment as $12,522.82.

The amortization schedule is presented in Table 3. The outstanding principal at the end of 5 years is, of course, zero. (The figure of $0.01 in Table 3 is the result of round-off errors.) Observe that initially the larger portion of the repayment goes toward payment of interest charges, but as time goes by, more and more of the payment goes toward repayment of the principal.

TABLE 3

An Amortization Schedule

End of Period	Interest Charged	Repayment Made	Payment Toward Principal	Outstanding Principal
0	—	—	—	$50,000.00
1	$4,000.00	$12,522.82	$ 8,522.82	41,477.18
2	3,318.17	12,522.82	9,204.65	32,272.53
3	2,581.80	12,522.82	9,941.02	22,331.51
4	1,786.52	12,522.82	10,736.30	11,595.21
5	927.62	12,522.82	11,595.20	0.01

Financing a Home

APPLIED EXAMPLE 2 Home Mortgage Payments The Blakelys borrowed $120,000 from a bank to help finance the purchase of a house. The bank charges interest at a rate of 5.4% per year on the unpaid balance, with interest computations made at the end of each month. The Blakelys have agreed to repay the loan in equal monthly installments over 30 years. How much should each payment be if the loan is to be amortized at the end of the term?

Solution Here, $P = 120,000$, $i = \frac{r}{m} = \frac{0.054}{12} = 0.0045$, and $n = (30)(12) = 360$. Using Formula (12), we find that the size of each monthly installment required is given by

$$R = \frac{(120,000)(0.0045)}{1 - (1.0045)^{-360}}$$

$$\approx 673.84$$

or $673.84.

APPLIED EXAMPLE 3 Home Equity Teresa and Raul purchased a house 10 years ago for $200,000. They made a down payment of 20% of the purchase price and secured a 30-year conventional home mortgage at 6% per year compounded monthly on the unpaid balance. The house is now worth $380,000. How much equity do Teresa and Raul have in their house now (after making 120 monthly payments)?

Solution Since the down payment was 20%, we know that they secured a loan of 80% of $200,000, or $160,000. Furthermore, using Formula (12) with $P = 160,000$, $i = \frac{r}{m} = \frac{0.06}{12} = 0.005$ and $n = (30)(12) = 360$, we determine that their monthly installment is

$$R = \frac{(160,000)(0.005)}{1 - (1.005)^{-360}}$$

$$\approx 959.28$$

or $959.28.

After 120 monthly payments have been made, the outstanding principal is given by the sum of the present values of the remaining installments (that is, $360 - 120 = 240$ installments). But this sum is just the present value of an annuity with $n = 240$, $R = 959.28$, and $i = 0.005$. Using Formula (10), we find

$$P = 959.28 \left[\frac{1 - (1 + 0.005)^{-240}}{0.005} \right]$$

$$\approx 133,897.04$$

or approximately $133,897. Therefore, Teresa and Raul have an equity of $380,000 - 133,897$, that is, $246,103.

Explore and Discuss and Exploring with Technology

1. Consider the amortization Formula (12):

$$R = \frac{Pi}{1 - (1 + i)^{-n}}$$

Suppose you know the values of R, P, and n and you wish to determine i. Explain why you can accomplish this task by finding the point of intersection of the graphs of the functions

$$y_1 = R \quad \text{and} \quad y_2 = \frac{Pi}{1 - (1 + i)^{-n}}$$

2. Thalia knows that her monthly repayment on her 30-year conventional home loan of $150,000 is $1100.65 per month. Help Thalia determine the interest rate for her loan by verifying or executing the following steps:

a. Plot the graphs of

$$y_1 = 1100.65 \quad \text{and} \quad y_2 = \frac{150,000x}{1 - (1 + x)^{-360}}$$

using the viewing window $[0, 0.01] \times [0, 1200]$.

b. Use the **ISECT** (intersection) function of the graphing utility to find the point of intersection of the graphs of part (a). Explain why this gives the value of i.

c. Compute r from the relationship $r = 12i$.

Explore and Discuss and Exploring with Technology

1. Suppose you secure a home mortgage loan of P with an interest rate of r per year to be amortized over t years through monthly installments of R. Show that after N installments, your outstanding principal is given by

$$B(N) = P\left[\frac{(1 + i)^n - (1 + i)^N}{(1 + i)^n - 1}\right] \quad (0 \le N \le n)$$

Hint: $B(N) = R\left[\dfrac{1 - (1 + i)^{-n+N}}{i}\right]$. To see this, study Example 3, page 238. Replace R using Formula (12).

2. Refer once again to Example 3. Using the result of part 1, show that Teresa and Raul's outstanding balance after making N payments is

$$E(N) = \frac{160{,}000(1.005^{360} - 1.005^N)}{1.005^{360} - 1} \quad (0 \le N \le 360)$$

3. Using a graphing utility, plot the graph of

$$E(x) = \frac{160{,}000(1.005^{360} - 1.005^x)}{1.005^{360} - 1}$$

using the viewing window $[0, 360] \times [0, 160{,}000]$.

4. Referring to the graph in part 3, observe that the outstanding principal drops off slowly in the early years and accelerates quickly to zero toward the end of the loan. Can you explain why?

5. How long does it take Teresa and Raul to repay half of the loan of $160,000?
   Hint: See the previous Explore and Discuss and Exploring with Technology box.

APPLIED EXAMPLE 4 Home Affordability The Jacksons have determined that after making a down payment, they could afford at most $2000 for a monthly house payment. The bank charges interest at the rate of 6% per year on the unpaid balance, with interest computations made at the end of each month. If the loan is to be amortized in equal monthly installments over 30 years, what is the maximum amount that the Jacksons can borrow from the bank?

Solution Here, $i = \frac{r}{m} = \frac{0.06}{12} = 0.005$, $n = (30)(12) = 360$, and $R = 2000$; we are required to find P. From Formula (11), we have

$$P = R\left[\frac{1 - (1 + i)^{-n}}{i}\right]$$

Substituting the numerical values for R, n, and i into this expression for P, we obtain

$$P = 2000\left[\frac{1 - (1.005)^{-360}}{0.005}\right] \approx 333{,}583.23$$

Therefore, the Jacksons can borrow at most $333,583.

An adjustable-rate mortgage (ARM) is a home loan in which the interest rate is changed periodically based on a financial index. For example, a 5/1 ARM is one that has an initial rate for the first 5 years and thereafter is adjusted every year for the remaining term of the loan. Similarly, a 7/1 ARM is one that has an initial rate for the first 7 years and thereafter is adjusted every year for the remaining term of the loan.

During the housing boom of 2001–2005, lenders aggressively promoted another type of loan—the interest-only mortgage loan—to help prospective buyers qualify for larger mortgages. With an interest-only loan, the homeowner pays only the interest on the mortgage for a fixed term, usually 5 to 7 years. At the end of that period, the borrower usually has an option to convert the loan to one that will be amortized.

APPLIED EXAMPLE 5 Home Affordability Refer to Example 4. Suppose that the bank has also offered the Jacksons (a) a 7/1 ARM with a term of 30 years and an interest rate of 5.70% per year compounded monthly for the first 7 years and (b) an interest-only loan for a term of 30 years and an interest rate of 5.94% per year for the first 7 years. If the Jacksons limit their monthly payment to $2000 per month, what is the maximum amount they can borrow with each of these mortgages?

Solution

a. Here, $i = \frac{r}{m} = \frac{0.057}{12} = 0.00475$, $n = (30)(12) = 360$, and $R = 2000$, and we want to find P. From Formula (11), we have

$$P = R\left[\frac{1 - (1 + i)^{-n}}{i}\right]$$

$$= 2000\left[\frac{1 - (1.00475)^{-360}}{0.00475}\right]$$

$$\approx 344{,}589.68$$

Therefore, if the Jacksons choose the 7/1 ARM, they can borrow at most $344,590.

b. If P denotes the maximum amount that the Jacksons can borrow, then the interest per year on their loan is $0.0594P$ dollars. But this is equal to their payments for the year, which are $(12)(2000)$ dollars. So we have

$$0.0594P = (12)(2000)$$

or

$$P = \frac{(12)(2000)}{0.0594} \approx 404{,}040.40$$

So if the Jacksons choose the 7-year interest-only loan, they can borrow at most $404,040.

APPLIED EXAMPLE 6 Adjustable Rate Mortgages Five years ago, the Campbells secured a 5/1 ARM to help finance the purchase of their home. The amount of the original loan was $350,000 for a term of 30 years, with interest at the rate of 5.76% per year, compounded monthly for the first 5 years. The Campbells' mortgage is due to reset next month, and the new interest rate will be 6.96% per year, compounded monthly.

a. What was the Campbells' monthly mortgage payment for the first 5 years?
b. What will the Campbells' new monthly mortgage payment be (after the reset)? By how much will the monthly payment increase?

Solution

a. First, we find the Campbells' monthly payment on the original loan amount. Using Formula (12) with $P = 350,000$, $i = \frac{r}{m} = \frac{0.0576}{12} = 0.0048$, and $n = mt = (12)(30) = 360$, we find that the monthly payment was

$$R = \frac{350,000(0.0048)}{1 - (1 + 0.0048)^{-360}} \approx 2044.729$$

or $2044.73 for the first 5 years.

b. To find the amount of the Campbells' new mortgage payment, we first need to find their outstanding principal. This is given by the present value of their remaining mortgage payments. Using Formula (10), with $R = 2044.729$, $i = \frac{r}{m} = \frac{0.0576}{12} = 0.0048$, and $n = mt = 360 - 5(12) = 300$, we find that their outstanding principal is

$$P = 2044.729\left[\frac{1 - (1 + 0.0048)^{-300}}{0.0048}\right] \approx 324,709.194$$

or $324,709.19.

Next, we compute the amount of their new mortgage payment for the remaining term (300 months). Using Formula (12) with $P = 324,709.194$, $i = \frac{r}{m} = \frac{0.0696}{12} = 0.0058$, and $n = mt = 300$, we find that the monthly payment is

$$R = \frac{324,709.194(0.0058)}{1 - (1 + 0.0058)^{-300}} \approx 2286.698$$

or $2286.70—an increase of $241.97.

Sinking Funds

Sinking funds are another important application of the annuity formulas. Simply stated, a **sinking fund** is an account that is set up for a specific purpose at some future date. For example, an individual might establish a sinking fund for the purpose of discharging a debt at a future date. A corporation might establish a sinking fund in order to accumulate sufficient capital to replace equipment that is expected to be obsolete at some future date.

By thinking of the amount to be accumulated by a specific date in the future as the future value of an annuity [Formula (9), Section 4.2], we can answer questions about a large class of sinking fund problems.

APPLIED EXAMPLE 7 Sinking Fund The proprietor of Carson Hardware has decided to set up a sinking fund for the purpose of purchasing a truck in 2 years' time. It is expected that the truck will cost $30,000. If the fund earns 10% interest per year compounded quarterly, determine the size of each (equal) quarterly installment the proprietor should pay into the fund. Verify the result by displaying the schedule.

Solution   The problem at hand is to find the size of each quarterly payment R of an annuity, given that its future value is $S = 30,000$, the interest earned per conversion period is $i = \frac{r}{m} = \frac{0.1}{4} = 0.025$, and the number of payments is $n = (2)(4) = 8$.

The formula for an annuity,

$$S = R\left[\frac{(1 + i)^n - 1}{i}\right]$$

when solved for R yields

$$R = \frac{iS}{(1 + i)^n - 1} \tag{13}$$

Substituting the appropriate numerical values for i, S, and n into Equation (13), we obtain the desired quarterly payment

$$R = \frac{(0.025)(30,000)}{(1.025)^8 - 1} \approx 3434.02$$

or \$3434.02. Table 4 shows the required schedule.

TABLE 4

A Sinking Fund Schedule

End of Period	Deposit Made	Interest Earned	Addition to Fund	Accumulated Amount in Fund
1	\$3,434.02	0	\$3,434.02	\$ 3,434.02
2	3,434.02	\$ 85.85	3,519.87	6,953.89
3	3,434.02	173.85	3,607.87	10,561.76
4	3,434.02	264.04	3,698.06	14,259.82
5	3,434.02	356.50	3,790.52	18,050.34
6	3,434.02	451.26	3,885.28	21,935.62
7	3,434.02	548.39	3,982.41	25,918.03
8	3,434.02	647.95	4,081.97	30,000.00

The formula derived in Example 7 is restated as follows.

Sinking Fund Payment

The periodic payment R required to accumulate a sum of S dollars over n periods with interest charged at the rate of i per period is

$$R = \frac{iS}{(1 + i)^n - 1} \tag{14}$$

$ **APPLIED EXAMPLE 8** Retirement Planning Jason is the owner of a computer consulting firm. He is currently planning to retire in 25 years and wishes to withdraw \$8000 per month from his retirement account for 25 years starting at that time. How much must he contribute each month into a retirement account earning interest at the rate of 6% per year compounded monthly to meet his retirement goal?

Solution In this case, we work backwards. We first calculate the amount of the sinking fund that Jason needs to accumulate to fund his retirement needs. Using Formula (10) with $R = 8000$, $i = \frac{r}{m} = \frac{0.06}{12} = 0.005$, and $n = 300$, we have

$$P = 8000 \left[\frac{1 - (1 + 0.005)^{-300}}{0.005} \right]$$

$$\approx 1,241,654.91$$

Next, we calculate how much Jason should deposit into his retirement account each month to accumulate $1,241,654.91 in 25 years. Here, we use Formula (14) with $S = 1{,}241{,}654.91$, $i = \frac{r}{m} = \frac{0.06}{12} = 0.005$, and $n = 300$. We have

$$R = 1{,}241{,}654.91\left[\frac{0.005}{(1 + 0.005)^{300} - 1}\right]$$

$$\approx 1791.73$$

or approximately $1791.73 per month.

Here is a summary of the formulas developed thus far in this chapter:

1. Simple and compound interest; annuities

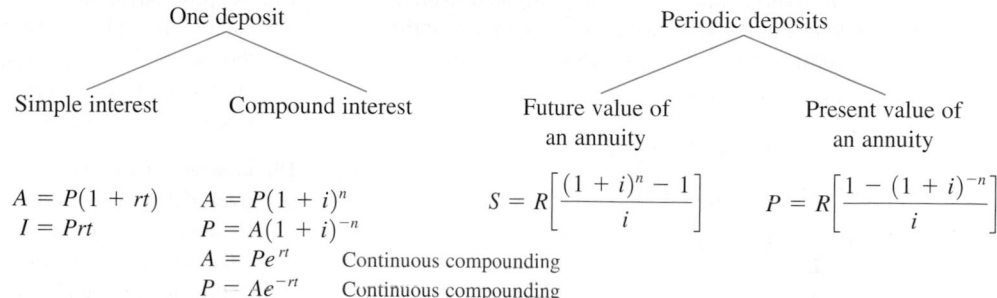

One deposit

Simple interest   Compound interest

Periodic deposits

Future value of an annuity   Present value of an annuity

$A = P(1 + rt)$
$I = Prt$

$A = P(1 + i)^n$
$P = A(1 + i)^{-n}$
$A = Pe^{rt}$ Continuous compounding
$P = Ae^{-rt}$ Continuous compounding

$S = R\left[\dfrac{(1 + i)^n - 1}{i}\right]$

$P = R\left[\dfrac{1 - (1 + i)^{-n}}{i}\right]$

2. Effective rate of interest

$$r_{\text{eff}} = \left(1 + \frac{r}{m}\right)^m - 1$$

3. Amortization

$$R = \frac{Pi}{1 - (1 + i)^{-n}} \qquad \text{Periodic payment}$$

$$P = R\left[\frac{1 - (1 + i)^{-n}}{i}\right] \qquad \text{Amount amortized}$$

4. Sinking fund

$$R = \frac{iS}{(1 + i)^n - 1} \qquad \text{Periodic payment taken out}$$

4.3 Self-Check Exercises

1. **COMPARING LOANS** The Mendozas wish to borrow $300,000 from a bank to help finance the purchase of a house. Their banker has offered the following plans for their consideration. Under Plan I, the Mendozas have 30 years to repay the loan in monthly installments with interest on the unpaid balance charged at 6.09%/year compounded monthly. Under Plan II, the loan is to be repaid in monthly installments over 15 years with interest on the unpaid balance charged at 5.76%/year compounded monthly.

   a. Find the monthly repayment for each plan.

   b. What is the difference in total payments made under each plan?

2. **PLANNING FOR RETIREMENT** Harris, a self-employed individual who is 46 years old, is setting up a defined-benefit retirement plan. If he wishes to have $250,000 in this retirement account by age 65, what is the size of each yearly installment he will be required to make into a savings account earning interest at $8\frac{1}{4}$%/year?

Solutions to Self-Check Exercises 4.3 can be found on page 248.

4.3 Concept Questions

1. Write the amortization formula.
   a. If P and i are fixed and n is allowed to increase, what will happen to R?
   b. Interpret the result of part (a).

2. Using the formula for computing a sinking fund payment, show that if the number of payments into a sinking fund increases, then the size of the periodic payment into the sinking fund decreases.

4.3 Exercises

In Exercises 1–8, find the periodic payment R required to amortize a loan of P dollars over t years with interest charged at the rate of $r\%$/year compounded m times a year.

1. $P = 100,000$, $r = 6$, $t = 10$, $m = 1$

2. $P = 40,000$, $r = 3$, $t = 15$, $m = 2$

3. $P = 5000$, $r = 4$, $t = 3$, $m = 4$

4. $P = 16,000$, $r = 5$, $t = 4$, $m = 12$

5. $P = 25,000$, $r = 3$, $t = 12$, $m = 4$

6. $P = 80,000$, $r = 5.5$, $t = 15$, $m = 12$

7. $P = 80,000$, $r = 5.5$, $t = 30$, $m = 12$

8. $P = 100,000$, $r = 5.5$, $t = 25$, $m = 12$

In Exercises 9–14, find the periodic payment R required to accumulate a sum of S dollars over t years with interest earned at the rate of $r\%$/year compounded m times a year.

9. $S = 20,000$, $r = 4$, $t = 6$, $m = 2$

10. $S = 40,000$, $r = 4$, $t = 9$, $m = 4$

11. $S = 100,000$, $r = 4.5$, $t = 20$, $m = 6$

12. $S = 120,000$, $r = 4.5$, $t = 30$, $m = 6$

13. $S = 250,000$, $r = 6.5$, $t = 25$, $m = 12$

14. $S = 350,000$, $r = 4.2$, $t = 10$, $m = 12$

15. Suppose payments were made at the end of each quarter into an ordinary annuity earning interest at the rate of 5%/year compounded quarterly. If the future value of the annuity after 5 years is $50,000, what was the size of each payment?

16. Suppose payments were made at the end of each month into an ordinary annuity earning interest at the rate of 4.5%/year compounded monthly. If the future value of the annuity after 10 years is $60,000, what was the size of each payment?

17. Suppose payments will be made for $6\frac{1}{2}$ years at the end of each semiannual period into an ordinary annuity earning interest at the rate of 3.5%/year compounded semiannually. If the present value of the annuity is $35,000, what should be the size of each payment?

18. Suppose payments will be made for $9\frac{1}{4}$ years at the end of each month into an ordinary annuity earning interest at the rate of 3.25%/year compounded monthly. If the present value of the annuity is $42,000, what should be the size of each payment?

19. **LOAN AMORTIZATION** A sum of $100,000 is to be repaid over a 10-year period through equal installments made at the end of each year. If an interest rate of 5%/year is charged on the unpaid balance and interest calculations are made at the end of each year, determine the size of each installment so that the loan (principal plus interest charges) is amortized at the end of 10 years.

20. **LOAN AMORTIZATION** What monthly payment is required to amortize a loan of $30,000 over 10 years if interest at the rate of 6%/year is charged on the unpaid balance and interest calculations are made at the end of each month?

21. **HOME MORTGAGES** Complete the following table, which shows the monthly payments on a $100,000, 30-year mortgage at the interest rates shown. Use this information to answer the following questions.

Amount of Mortgage ($)	Interest Rate (%)	Monthly Payment ($)
100,000	3	421.60
100,000	4	. . .
100,000	5	. . .
100,000	6	. . .
100,000	7	. . .
100,000	8	733.76

   a. What is the difference in monthly payments between a $100,000, 30-year mortgage secured at 7%/year and one secured at 3%/year?
   b. Use the table to calculate the monthly mortgage payments on a $150,000 mortgage at 5%/year over 30 years and a $50,000 mortgage at 5%/year over 30 years.

22. **FINANCING A HOME** The Flemings secured a bank loan of $288,000 to help finance the purchase of a house. The bank charges interest at a rate of 5%/year on the unpaid balance, and interest computations are made at the end of each month. The Flemings have agreed to repay the loan in equal monthly installments over 25 years. What should

be the size of each repayment if the loan is to be amortized at the end of the term?

23. **FINANCING A CAR** The price of a new car is $20,000. Assume that an individual makes a down payment of 25% toward the purchase of the car and secures financing for the balance at the rate of 6%/year compounded monthly.
   a. What monthly payment will she be required to make if the car is financed over a period of 36 months? Over a period of 48 months?
   b. What will the interest charges be if she elects the 36-month plan? The 48-month plan?

24. **FINANCIAL ANALYSIS** A group of private investors purchased a condominium complex for $2 million. They made an initial down payment of 10% and obtained financing for the balance. If the loan is to be amortized over 15 years at an interest rate of 6.6%/year compounded quarterly, find the required quarterly payment.

25. **FINANCING A HOME** The Taylors have purchased a $270,000 house. They made an initial down payment of $30,000 and secured a mortgage with interest charged at the rate of 6%/year on the unpaid balance. Interest computations are made at the end of each month. If the loan is to be amortized over 30 years, what monthly payment will the Taylors be required to make? What is their equity (disregarding appreciation) after 5 years? After 10 years? After 20 years?

26. **FINANCIAL PLANNING** Jessica wants to accumulate $10,000 by the end of 5 years in a special bank account, which she had opened for this purpose. To achieve this goal, Jessica plans to deposit a fixed sum of money into the account at the end of each month over the 5-year period. If the bank pays interest at the rate of 5%/year compounded monthly, how much does she have to deposit each month into her account?

27. **SINKING FUNDS** A city has $2.5 million worth of school bonds that are due in 20 years and has established a sinking fund to retire this debt. If the fund earns interest at the rate of 4%/year compounded annually, what amount must be deposited annually in this fund?

28. **TRUST FUNDS** Carl is the beneficiary of a $20,000 trust fund set up for him by his grandparents. Under the terms of the trust, he is to receive the money over a 5-year period in equal installments at the end of each year. If the fund earns interest at the rate of 5%/year compounded annually, what amount will he receive each year?

29. **SINKING FUNDS** Lowell Corporation wishes to establish a sinking fund to retire a $200,000 debt that is due in 10 years. If the investment will earn interest at the rate of 9%/year compounded quarterly, find the amount of the quarterly deposit that must be made in order to accumulate the required sum.

30. **SINKING FUNDS** The management of Gibraltar Brokerage Services anticipates a capital expenditure of $20,000 in 3 years for the purchase of new computers and has decided to set up a sinking fund to finance this purchase. If the fund earns interest at the rate of 5%/year compounded quarterly, determine the size of each (equal) quarterly installment that should be deposited in the fund.

31. **RETIREMENT ACCOUNTS** Andrea, a self-employed individual, wishes to accumulate a retirement fund of $250,000. How much should she deposit each month into her retirement account, which pays interest at the rate of 4.5%/year compounded monthly, to reach her goal upon retirement 25 years from now?

32. **STUDENT LOANS** Joe secured a loan of $12,000 3 years ago from a bank for use toward his college expenses. The bank charged interest at the rate of 4%/year compounded monthly on his loan. Now that he has graduated from college, Joe wishes to repay the loan by amortizing it through monthly payments over 10 years at the same interest rate. Find the size of the monthly payments he will be required to make.

33. **RETIREMENT ACCOUNTS** Robin wishes to accumulate a sum of $450,000 in a retirement account by the time of her retirement 30 years from now. If she wishes to do this through monthly payments into the account that earn interest at the rate of 6%/year compounded monthly, what should be the size of each payment?

34. **FINANCING COLLEGE EXPENSES** Yumi's grandparents presented her with a gift of $20,000 when she was 10 years old to be used for her college education. Over the next 7 years, until she turned 17, Yumi's parents had invested her money in a tax-free account that had yielded interest at the rate of 3.5%/year compounded monthly. Upon turning 17, Yumi now plans to withdraw her funds in equal annual installments over the next 4 years, starting at age 18. If the college fund is expected to earn interest at the rate of 4%/year, compounded annually, what will be the size of each installment?

35. **IRAS** Martin has deposited $375 in his IRA at the end of each quarter for the past 20 years. His investment has earned interest at the rate of 4%/year compounded quarterly over this period. Now, at age 60, he is considering retirement. What quarterly payment will he receive over the next 15 years? (Assume that the money is earning interest at the same rate and that payments are made at the end of each quarter.) If he continues working and makes quarterly payments of the same amount in his IRA until age 65, what quarterly payment will he receive from his fund upon retirement over the following 10 years?

36. **RETIREMENT PLANNING** Jennifer is the owner of a video game and entertainment software retail store. She is currently planning to retire in 30 years and wishes to withdraw $10,000/month for 20 years from her retirement account starting at that time. How much must she contribute each month for 30 years into a retirement account earning interest at the rate of 4%/year compounded monthly to meet her retirement goal?

37. EFFECT OF DELAYING RETIREMENT ON RETIREMENT FUNDS Refer to Example 8. Suppose that Jason delays his retirement plans and decides to continue working and contributing to his retirement fund for an additional 5 years. By delaying his retirement, he will need to withdraw only $8000/month for 20 years. In this case, how much must he contribute each month for 30 years into a retirement account earning interest at the rate of 6%/year compounded monthly to meet his retirement goal?

38. FINANCING A CAR Darla purchased a new car during a special sales promotion by the manufacturer. She secured a loan from the manufacturer in the amount of $16,000 at a rate of 4.9%/year compounded monthly. Her bank is now charging 6.5%/year compounded monthly for new car loans. Assuming that each loan would be amortized by 36 equal monthly installments, determine the amount of interest she would have paid at the end of 3 years for each loan. How much less will she have paid in interest payments over the life of the loan by borrowing from the manufacturer instead of her bank?

39. AUTO FINANCING Dan is contemplating trading in his car for a new one. He can afford a monthly payment of at most $400. If the prevailing interest rate is 4.2%/year compounded monthly for a 48-month loan, what is the most expensive car that Dan can afford, assuming that he will receive $8000 for his trade-in?

40. AUTO FINANCING Paula is considering the purchase of a new car. She has narrowed her search to two cars that are equally appealing to her. Car *A* costs $28,000, and Car *B* costs $28,200. The manufacturer of Car *A* is offering 0% financing for 48 months with zero down, while the manufacturer of Car *B* is offering a rebate of $2000 at the time of purchase plus financing at the rate of 3%/year compounded monthly over 48 months with zero down. If Paula has decided to buy the car with the lower net cost to her, which car should she purchase?

41. FINANCING A HOME Eight years ago, Kim secured a bank loan of $180,000 to help finance the purchase of a house. The mortgage was for a term of 30 years, with an interest rate of 4.5%/year compounded monthly on the unpaid balance to be amortized through monthly payments. What is the outstanding principal on Kim's house now?

42. FINANCING A HOME Sarah secured a bank loan of $200,000 for the purchase of a house. The mortgage is to be amortized through monthly payments for a term of 15 years, with an interest rate of 3%/year compounded monthly on the unpaid balance. She plans to sell her house in 5 years. How much will Sarah still owe on her house?

43. PERSONAL LOANS Two years ago, Paul borrowed $10,000 from his sister Gerri to start a business. Paul agreed to pay Gerri interest for the loan at the rate of 4%/year, compounded continuously. Paul will now begin repaying the amount he owes by amortizing the loan (plus the interest that has accrued over the past 2 years) through

monthly payments over the next 5 years at an interest rate of 3%/year compounded monthly. Find the size of the monthly payments Paul will be required to make.

44. INVESTMENT ANALYSIS Since he was 22 years old, Ben has been depositing $200 at the end of each month into a tax-free retirement account earning interest at the rate of 3.5%/year compounded monthly. Larry, who is the same age as Ben, decided to open a tax-free retirement account 5 years after Ben opened his. If Larry's account earns interest at the same rate as Ben's, determine how much Larry should deposit each month into his account so that both men will have the same amount of money in their accounts at age 65.

45. BALLOON PAYMENT MORTGAGES Emilio is securing a 7-year balloon mortgage for $280,000 to finance the purchase of his first home. The monthly payments are based on a 30-year amortization. If the prevailing interest rate is 4.5%/year compounded monthly, what will be Emilio's monthly payment? What will be his balloon payment at the end of 7 years?

46. BALLOON PAYMENT MORTGAGES Olivia plans to secure a 5-year balloon mortgage of $200,000 toward the purchase of a condominium. Her monthly payment for the 5 years is calculated on the basis of a 30-year conventional mortgage at the rate of 3%/year compounded monthly. At the end of the 5 years, Olivia is required to pay the balance owed (the "balloon" payment). What will be her monthly payment for the first 5 years, and what will be her balloon payment?

47. HOME REFINANCING Four years ago, Emily secured a bank loan of $200,000 to help finance the purchase of an apartment in Boston. The term of the mortgage is 30 years, and the interest rate is 6.5%/year compounded monthly. Because the interest rate for a conventional 30-year home mortgage has now dropped to 4.75%/year compounded monthly, Emily is thinking of refinancing her property.
a. What is Emily's current monthly mortgage payment?
b. What is Emily's current outstanding principal?
c. If Emily decides to refinance her property by securing a 30-year home mortgage loan in the amount of the current outstanding principal at the prevailing interest rate of 4.75%/year compounded monthly, what will be her monthly mortgage payment?
d. How much less would Emily's monthly mortgage payment be if she refinances?

48. HOME REFINANCING Five years ago, Diane secured a bank loan of $300,000 to help finance the purchase of a loft in the San Francisco Bay area. The term of the mortgage was 30 years, and the interest rate was 6%/year compounded monthly on the unpaid balance. Because the interest rate for a conventional 30-year home mortgage has now dropped to 4.5%/year compounded monthly, Diane is thinking of refinancing her property.
a. What is Diane's current monthly mortgage payment?
b. What is Diane's current outstanding principal?

c. If Diane decides to refinance her property by securing a 30-year home mortgage loan in the amount of the current outstanding principal at the prevailing interest rate of 4.5%/year compounded monthly, what will be her monthly mortgage payment?

d. How much less would Diane's monthly mortgage payment be if she refinances?

49. REFINANCING A HOME The Sandersons are planning to refinance their home. The outstanding principal on their original loan is $100,000 and is now to be amortized in 240 equal monthly installments at an interest rate of 5%/year compounded monthly. The new loan they expect to secure is to be amortized over the same period at an interest rate of 4.2%/year compounded monthly. How much less can they expect to pay over the life of the loan in interest payments by refinancing the loan at this time?

50. REFINANCING A HOME Josh purchased a condominium 5 years ago for $180,000. He made a down payment of 20% and financed the balance with a 30-year conventional mortgage to be amortized through monthly payments with an interest rate of 4%/year compounded monthly on the unpaid balance. The condominium is now appraised at $250,000. Josh plans to start his own business and wishes to tap into the equity that he has in the condominium. If Josh can secure a new 30-year conventional mortgage at the same rate to refinance his condominium based on a loan of 80% of the appraised value, how much cash can Josh muster for his business? (Disregard taxes.)

51. ADJUSTABLE-RATE MORTGAGES Three years ago, Samantha secured an adjustable-rate mortgage (ARM) loan to help finance the purchase of a house. The amount of the original loan was $150,000 for a term of 30 years, with interest at the rate of 5.5%/year compounded monthly. Currently, the interest rate is 4%/year compounded monthly, and Samantha's monthly payments are due to be recalculated. What will be her new monthly payment? Hint: Calculate her current outstanding principal. Then, to amortize the loan in the next 27 years, determine the monthly payment based on the current interest rate.

52. ADJUSTABLE-RATE MORTGAGES George secured an adjustable-rate mortgage (ARM) loan to help finance the purchase of his home 5 years ago. The amount of the loan was $300,000 for a term of 30 years, with interest at the rate of 6%/year compounded monthly. Currently, the interest rate for his ARM is 4.5%/year compounded monthly, and George's monthly payments are due to be reset. What will be the new monthly payment?

53. FINANCING A HOME After making a down payment of $25,000, the Meyers need to secure a loan of $280,000 to purchase a certain house. Their bank's current rate for 25-year home loans is 5.5%/year compounded monthly. The owner has offered to finance the loan at 4.9%/year compounded monthly. Assuming that both loans would be amortized over a 25-year period by 300 equal monthly installments, determine the difference in the amount of interest the Meyers would pay by choosing the seller's financing rather than their bank's.

54. REFINANCING A HOME The Martinezes are planning to refinance their home. The outstanding balance on their original loan is $150,000. Their finance company has offered them two options:

Option A: A fixed-rate mortgage at an interest rate of 4.5%/year compounded monthly, payable over a 30-year period in 360 equal monthly installments.

Option B: A fixed-rate mortgage at an interest rate of 4.25%/year compounded monthly, payable over a 15-year period in 180 equal monthly installments.

a. Find the monthly payment required to amortize each of these loans over the life of the loan.

b. How much interest would the Martinezes save if they chose the 15-year mortgage instead of the 30-year mortgage?

55. ABILITY-TO-REPAY RULE FOR MORTGAGES The Ability-to-Repay Rule, adopted by the Consumer Financial Protection Bureau in compliance with the Dodd-Frank Wall Street Reform and Consumer Protection Act, requires lenders to determine whether a consumer applying for a Qualified Mortgage can afford to repay the loan. One of the requirements is that the borrower's total monthly debt (including property taxes) cannot exceed 43% of the borrower's monthly pre-tax income. Suppose that the Foleys have applied for a $400,000 Qualified Mortgage with an interest rate of 4%/year compounded monthly and a term of 30 years. The property tax on the home they wish to purchase is $6000/year. If the Foleys' annual income is $72,000, will they qualify for the mortgage?
Source: Consumer Financial Protection Bureau.

56. ABILITY-TO-REPAY RULE FOR MORTGAGES Refer to Exercise 55. What is the maximum Qualified Mortgage with an interest rate of 3.75%/year compounded monthly and a term of 30 years the Foleys' can qualify for?

57. HOME AFFORDABILITY Suppose that the Carlsons have decided that they can afford a maximum of $3000/month for a monthly house payment. The bank has offered them (a) a 5/1 ARM for a term of 30 years with interest at the rate of 4.40%/year compounded monthly for the first 5 years and (b) an interest-only loan for a term of 30 years at the rate of 4.62%/year for the first 5 years. What is the maximum amount that they can borrow with each of these mortgages if they keep to their budget?

58. COMPARING MORTGAGES Refer to Example 5. Suppose that the Jacksons choose the 7/1 ARM and borrow the maximum amount of $344,589.68.

a. By how much will the principal of their loan be reduced at the end of 7 years?
Hint: See Example 3.

b. If they choose the 7-year interest-only mortgage, by how much will the principal of their loan be reduced after 7 years?

4.3 Solutions to Self-Check Exercises

1. a. We use Formula (12) in each instance. Under Plan I,

$$P = 300,000 \qquad i = \frac{r}{m} = \frac{0.0609}{12} = 0.005075$$

$$n = (30)(12) = 360$$

Therefore, the size of each monthly repayment under Plan I is

$$R = \frac{300,000(0.005075)}{1 - (1.005075)^{-360}}$$

$$\approx 1816.05$$

or $1816.05.

Under Plan II,

$$P = 300,000 \qquad i = \frac{r}{m} = \frac{0.0576}{12} = 0.0048$$

$$n = (15)(12) = 180$$

Therefore, the size of each monthly repayment under Plan II is

$$R = \frac{300,000(0.0048)}{1 - (1.0048)^{-180}}$$

$$\approx 2492.84$$

or $2492.84.

b. Under Plan I, the total amount of repayments will be

$$(360)(1816.05) = 653,778 \quad \text{\small Number of payments}$$
$$\text{\small × the size of each installment}$$

or $653,778. Under Plan II, the total amount of repayments will be

$$(180)(2492.84) = 448,711.20$$

or $448,711.20. Therefore, the difference in payments is

$$653,778 - 448,711.20 = 205,066.80$$

or $205,066.80.

2. We use Formula (14) with

$$S = 250,000$$
$$i = r = 0.0825 \qquad \text{\small Since } m = 1$$
$$n = 20$$

giving the required size of each installment as

$$R = \frac{(0.0825)(250,000)}{(1.0825)^{20} - 1}$$

$$\approx 5313.59$$

or $5313.59.

USING TECHNOLOGY   Amortizing a Loan

Graphing Utility

Here, we use the TI-83/84 **TVM SOLVER** function to help us solve problems involving amortization and sinking funds.

$ APPLIED EXAMPLE 1 Finding the Payment to Amortize a Loan The Wongs are considering obtaining a preapproved 30-year loan of $120,000 to help finance the purchase of a house. The mortgage company charges interest at the rate of 8% per year on the unpaid balance, with interest computations made at the end of each month. What will be the monthly installments if the loan is amortized?

Solution   We use the TI-83/84 **TVM SOLVER** with the following inputs:

$$N = 360 \qquad \text{\small (30)(12)}$$
$$I\% = 8$$
$$PV = 120000$$
$$PMT = 0$$
$$FV = 0$$
$$P/Y = 12 \qquad \text{\small The number of payments each year}$$
$$C/Y = 12 \qquad \text{\small The number of conversion periods each year}$$
$$PMT: \boxed{END} \ BEGIN$$

Note: Boldfaced words/characters enclosed in a box (for example, $\boxed{\text{Enter}}$) indicate that an action (click, select, or press) is required. Words/characters printed blue (for example, Chart sub-type:) indicate words/characters appearing on the screen.

By moving the cursor up to the PMT line and pressing ALPHA SOLVE, we obtain the output shown in Figure T1. We see that the required payment is $880.52.

```
N=360
I%=8
PV=120000
■ PMT=-880.51748...
FV=0
P/Y=12
C/Y=12
PMT:(END) BEGIN
```

FIGURE **T1**
The TI-83/84 screen showing the
monthly installment, PMT

APPLIED EXAMPLE 2 Finding the Payment in a Sinking Fund Heidi wishes to establish a retirement account that will be worth $500,000 in 20 years' time. She expects that the account will earn interest at the rate of 11% per year compounded monthly. What should be the monthly contribution into her account each month?

Solution We use the TI-83/84 **TVM SOLVER** with the following inputs:

$$N = 240 \quad (20)(12)$$
$$I\% = 11$$
$$PV = 0$$
$$PMT = 0$$
$$FV = 500000$$
$$P/Y = 12 \quad \text{The number of payments each year}$$
$$C/Y = 12 \quad \text{The number of conversion periods each year}$$
$$\text{PMT: } \boxed{\text{END}} \text{ BEGIN}$$

By moving the cursor up to the PMT line and pressing ALPHA SOLVE, we obtain the result displayed in Figure T2. We see that Heidi's monthly contribution should be $577.61. (*Note:* The display for PMT is negative because it is an outflow.)

```
N=240
I%=11
PV=0
■ PMT=-577.60862...
FV=500000
P/Y=12
C/Y=12
PMT:(END) BEGIN
```

FIGURE **T2**
The TI-83/84 screen showing the
monthly payment, PMT

Excel

Here we use Excel to help us solve problems involving amortization and sinking funds.

APPLIED EXAMPLE 3 Finding the Payment to Amortize a Loan The Wongs are considering a preapproved 30-year loan of $120,000 to help finance the purchase of a house. The mortgage company charges interest at the rate of 8% per year on the unpaid balance, with interest computations made at the end of each month. What will be the monthly installments if the loan is amortized at the end of the term?

Solution   We use the PMT function to solve this problem. Accessing this function from the Financial function library subgroup and making the required entries, we obtain the Function Arguments dialog box shown in Figure T3. We see that the desired result is $880.52. (Recall that cash you pay out is represented by a negative number.)

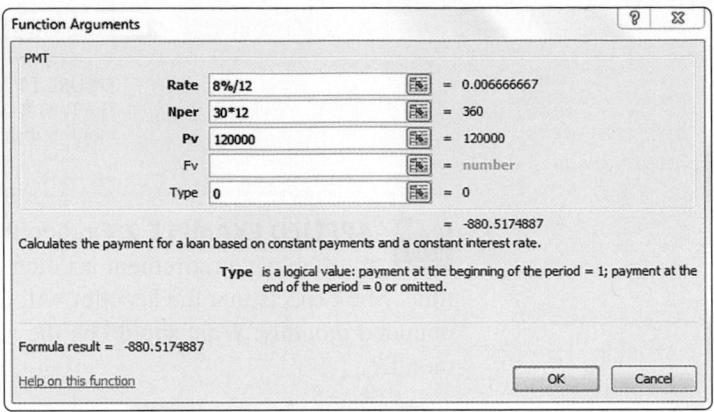

FIGURE **T3**
Excel's dialog box giving the payment function, PMT

APPLIED EXAMPLE 4   Finding the Payment in a Sinking Fund Heidi wishes to establish a retirement account that will be worth $500,000 in 20 years' time. She expects that the account will earn interest at the rate of 11% per year compounded monthly. What should be the monthly contribution into her account each month?

Solution   As in Example 3, we use the PMT function, but this time, we are given the future value of the investment. Accessing the PMT function as before and making the required entries, we obtain the Function Arguments dialog box shown in Figure T4. We see that Heidi's monthly contribution should be $577.61. (Note that the value for PMT is negative because it is an outflow.)

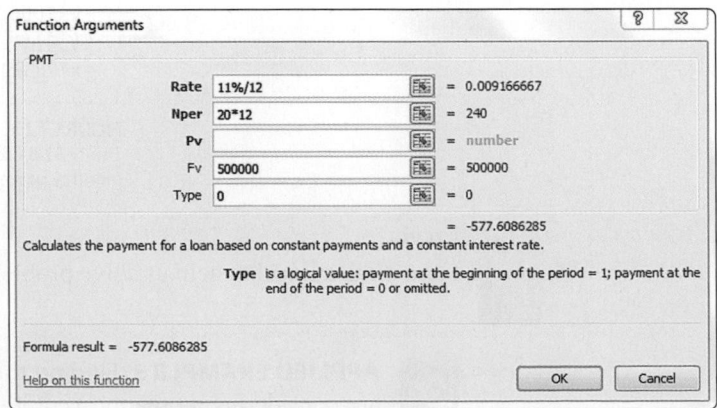

FIGURE **T4**
Excel's dialog box giving the payment function, PMT

TECHNOLOGY EXERCISES

1. Find the periodic payment required to amortize a loan of $55,000 over 120 months with interest charged at the rate of $6\frac{5}{8}$%/year compounded monthly.

2. Find the periodic payment required to amortize a loan of $178,000 over 180 months with interest charged at the rate of $5\frac{1}{8}$%/year compounded monthly.

3. Find the periodic payment required to amortize a loan of $227,000 over 360 months with interest charged at the rate of $6\frac{1}{8}$%/year compounded monthly.

4. Find the periodic payment required to amortize a loan of $150,000 over 360 months with interest charged at the rate of $4\frac{3}{8}$%/year compounded monthly.

5. Find the periodic payment required to accumulate $25,000 over 12 quarters with interest earned at the rate of $3\frac{3}{8}$%/year compounded quarterly.

6. Find the periodic payment required to accumulate $50,000 over 36 quarters with interest earned at the rate of $3\frac{7}{8}$%/year compounded quarterly.

7. Find the periodic payment required to accumulate $137,000 over 120 months with interest earned at the rate of $4\frac{3}{4}$%/year compounded monthly.

8. Find the periodic payment required to accumulate $144,000 over 120 months with interest earned at the rate of $4\frac{5}{8}$%/year compounded monthly.

9. A loan of $120,000 is to be repaid over a 10-year period through equal installments made at the end of each year. If an interest rate of 4.5%/year is charged on the unpaid balance and interest calculations are made at the end of each year, determine the size of each installment such that the loan is amortized at the end of 10 years. Verify the result by displaying the amortization schedule.

10. A loan of $265,000 is to be repaid over an 8-year period through equal installments made at the end of each year. If an interest rate of 5.4%/year is charged on the unpaid balance and interest calculations are made at the end of each year, determine the size of each installment so that the loan is amortized at the end of 8 years. Verify the result by displaying the amortization schedule.

4.4 Arithmetic and Geometric Progressions

Arithmetic Progressions

An **arithmetic progression** is a sequence of numbers in which each term after the first is obtained by adding a constant d to the preceding term. The constant d is called the **common difference**. For example, the sequence

$$2, 5, 8, 11, \ldots$$

is an arithmetic progression with common difference equal to 3.

Observe that an arithmetic progression is completely determined if the first term and the common difference are known. In fact, if

$$a_1, a_2, a_3, \ldots, a_n, \ldots$$

is an arithmetic progression with the first term given by a and common difference given by d, then by definition we have

$$a_1 = a$$
$$a_2 = a_1 + d = a + d$$
$$a_3 = a_2 + d = (a + d) + d = a + 2d$$
$$a_4 = a_3 + d = (a + 2d) + d = a + 3d$$
$$\vdots$$
$$a_n = a_{n-1} + d = a + (n - 2)d + d = a + (n - 1)d$$

Thus, we have the following formula:

---

nth Term of an Arithmetic Progression

The nth term of an arithmetic progression with first term a and common difference d is given by

$$a_n = a + (n - 1)d \qquad (15)$$

EXAMPLE 1 Find the twelfth term of the arithmetic progression

$$2, 7, 12, 17, 22, \ldots$$

Solution The first term of the arithmetic progression is $a_1 = a = 2$, and the common difference is $d = 5$; so upon setting $n = 12$ in Equation (15), we find

$$a_{12} = 2 + (12 - 1)5 = 57 \qquad \blacksquare$$

EXAMPLE 2 Write the first five terms of an arithmetic progression whose third and eleventh terms are 21 and 85, respectively.

Solution Using Equation (15), we obtain

$$a_3 = a + 2d = 21$$
$$a_{11} = a + 10d = 85$$

Subtracting the first equation from the second gives $8d = 64$, or $d = 8$. Substituting this value of d into the first equation yields $a + 16 = 21$, or $a = 5$. Thus, the required arithmetic progression is given by the sequence

$$5, 13, 21, 29, 37, \ldots \qquad \blacksquare$$

Let S_n denote the sum of the first n terms of an arithmetic progression with first term $a_1 = a$ and common difference d. Then

$$S_n = a + (a + d) + (a + 2d) + \cdots + [a + (n - 1)d] \qquad (16)$$

Rewriting the expression for S_n with the terms in reverse order gives

$$S_n = [a + (n - 1)d] + [a + (n - 2)d] + \cdots + (a + d) + a \qquad (17)$$

Adding Equations (16) and (17), we obtain

$$2S_n = [2a + (n - 1)d] + [2a + (n - 1)d]$$
$$+ \cdots + [2a + (n - 1)d]$$
$$= n[2a + (n - 1)d]$$
$$S_n = \frac{n}{2}[2a + (n - 1)d]$$

Sum of Terms in an Arithmetic Progression

The sum of the first n terms of an arithmetic progression with first term a and common difference d is given by

$$S_n = \frac{n}{2}[2a + (n - 1)d] \qquad (18)$$

EXAMPLE 3 Find the sum of the first 20 terms of the arithmetic progression of Example 1.

Solution   Letting $a = 2$, $d = 5$, and $n = 20$ in Equation (18), we obtain

$$S_{20} = \frac{20}{2}[2 \cdot 2 + 19 \cdot 5] = 990$$

APPLIED EXAMPLE 4 Company Sales Madison Electric Company had sales of $200,000 in its first year of operation. If the sales increased by $30,000 per year thereafter, find Madison's sales in the fifth year and its total sales over the first 5 years of operation.

Solution   Madison's yearly sales follow an arithmetic progression, with the first term given by $a = 200,000$ and the common difference given by $d = 30,000$. The sales in the fifth year are found by using Equation (15) with $n = 5$. Thus,

$$a_5 = 200,000 + (5 - 1)30,000 = 320,000$$

or $320,000.

Madison's total sales over the first 5 years of operation are found by using Equation (18) with $n = 5$. Thus,

$$S_5 = \frac{5}{2}[2(200,000) + (5 - 1)30,000]$$

$$= 1,300,000$$

or $1,300,000.

Geometric Progressions

A **geometric progression** is a sequence of numbers in which each term after the first is obtained by multiplying the preceding term by a constant r. The constant r is called the **common ratio.**

A geometric progression is completely determined if the first term and the common ratio are known. Thus, if

$$a_1, a_2, a_3, \ldots, a_n, \ldots$$

is a geometric progression with first term a and common ratio r, then by definition, we have

$$a_1 = a$$
$$a_2 = a_1 r = ar$$
$$a_3 = a_2 r = ar^2$$
$$a_4 = a_3 r = ar^3$$
$$\vdots$$
$$a_n = a_{n-1} r = ar^{n-1}$$

This gives the following:

> *n*th Term of a Geometric Progression
>
> The nth term of a geometric progression with first term a and common ratio r is given by
>
> $$a_n = ar^{n-1} \tag{19}$$

EXAMPLE 5 Find the eighth term of a geometric progression whose first five terms are 162, 54, 18, 6, and 2.

Solution   The common ratio is found by taking the ratio of any term other than the first to the preceding term. Taking the ratio of the fourth term to the third term, for example, gives $r = \frac{6}{18} = \frac{1}{3}$. To find the eighth term of the geometric progression, use Equation (19) with $a = 162$, $r = \frac{1}{3}$, and $n = 8$, obtaining

$$a_8 = 162\left(\frac{1}{3}\right)^7$$

$$= \frac{2}{27}$$

EXAMPLE 6 Find the tenth term of a geometric progression with positive terms and third term equal to 16 and seventh term equal to 1.

Solution   Using Equation (19) with $n = 3$ and $n = 7$, respectively, yields

$$a_3 = ar^2 = 16$$
$$a_7 = ar^6 = 1$$

Dividing a_7 by a_3 gives

$$\frac{ar^6}{ar^2} = \frac{1}{16}$$

from which we obtain $r^4 = \frac{1}{16}$, or $r = \frac{1}{2}$. Substituting this value of r into the expression for a_3, we obtain

$$a\left(\frac{1}{2}\right)^2 = 16 \quad \text{or} \quad a = 64$$

Finally, using Equation (19) once again with $a = 64$, $r = \frac{1}{2}$, and $n = 10$ gives

$$a_{10} = 64\left(\frac{1}{2}\right)^9 = \frac{1}{8}$$

To find the sum of the first n terms of a geometric progression with the first term $a_1 = a$ and common ratio r, denote the required sum by S_n. Then

$$S_n = a + ar + ar^2 + \cdots + ar^{n-2} + ar^{n-1} \tag{20}$$

Upon multiplying (20) by r, we obtain

$$rS_n = ar + ar^2 + ar^3 + \cdots + ar^{n-1} + ar^n \tag{21}$$

Subtracting Equation (21) from (20) gives

$$S_n - rS_n = a - ar^n$$
$$(1 - r)S_n = a(1 - r^n)$$

If $r \neq 1$, we may divide both sides of the last equation by $(1 - r)$, obtaining

$$S_n = \frac{a(1 - r^n)}{1 - r}$$

If $r = 1$, then (20) gives

$$S_n = a + a + a + \cdots + a \qquad n \text{ terms}$$
$$= na$$

Thus,

$$S_n = \begin{cases} \dfrac{a(1 - r^n)}{1 - r} & \text{if } r \neq 1 \\ na & \text{if } r = 1 \end{cases}$$

> **Sum of Terms in a Geometric Progression**
>
> The sum of the first n terms of a geometric progression with first term a and common ratio r is given by
>
> $$S_n = \begin{cases} \dfrac{a(1 - r^n)}{1 - r} & \text{if } r \neq 1 \\ na & \text{if } r = 1 \end{cases} \tag{22}$$

EXAMPLE 7 Find the sum of the first six terms of the following geometric progression:

$$3, 6, 12, 24, \ldots$$

Solution Here, $a = 3$, $r = \frac{6}{3} = 2$, and $n = 6$, so Equation (22) gives

$$S_6 = \frac{3(1 - 2^6)}{1 - 2} = 189$$

APPLIED EXAMPLE 8 Company Sales Michaelson Land Development Company had sales of \$1 million in its first year of operation. If sales increased by 10% per year thereafter, find Michaelson's sales in the fifth year and its total sales over the first 5 years of operation.

Solution Michaelson's yearly sales follow a geometric progression, with the first term given by $a = 1,000,000$ and the common ratio given by $r = 1.1$. The sales in the fifth year are found by using Formula (19) with $n = 5$. Thus,

$$a_5 = 1,000,000(1.1)^4 = 1,464,100$$

or \$1,464,100.

Michaelson's total sales over the first 5 years of operation are found by using Equation (22) with $n = 5$. Thus,

$$S_5 = \frac{1,000,000[1 - (1.1)^5]}{1 - 1.1}$$

$$= 6,105,100$$

or \$6,105,100.

Double Declining–Balance Method of Depreciation

In Section 2.5, we discussed the straight-line, or linear, method of depreciating an asset. Linear depreciation assumes that the asset depreciates at a constant rate. For certain assets (such as machines) whose market values drop rapidly in the early years of usage and thereafter less rapidly, another method of depreciation called the **double declining–balance method** is often used. In practice, a business firm normally employs the double declining–balance method for depreciating such assets for a certain number of years and then switches over to the linear method.

To derive an expression for the book value of an asset being depreciated by the double declining–balance method, let C (in dollars) denote the original cost of the asset and let the asset be depreciated over N years. When this method is used, the amount depreciated each year is $\frac{2}{N}$ times the value of the asset at the beginning of that year. Thus, the amount by which the asset is depreciated in its first year of use is given by $\frac{2C}{N}$, so if $V(1)$ denotes the book value of the asset at the end of the first year, then

$$V(1) = C - \frac{2C}{N} = C\left(1 - \frac{2}{N}\right)$$

Next, if $V(2)$ denotes the book value of the asset at the end of the second year, then a similar argument leads to

$$V(2) = C\left(1 - \frac{2}{N}\right) - C\left(1 - \frac{2}{N}\right)\frac{2}{N}$$

$$= C\left(1 - \frac{2}{N}\right)\left(1 - \frac{2}{N}\right)$$

$$= C\left(1 - \frac{2}{N}\right)^2$$

Continuing, we find that if $V(n)$ denotes the book value of the asset at the end of n years, then the terms $C, V(1), V(2), \ldots, V(N)$ form a geometric progression with first term C and common ratio $\left(1 - \frac{2}{N}\right)$. Consequently, the nth term, $V(n)$, is given by

$$V(n) = C\left(1 - \frac{2}{N}\right)^n \qquad (1 \le n \le N) \tag{23}$$

Also, if $D(n)$ denotes the amount by which the asset has been depreciated by the end of the nth year, then

$$D(n) = C - C\left(1 - \frac{2}{N}\right)^n$$

$$= C\left[1 - \left(1 - \frac{2}{N}\right)^n\right] \tag{24}$$

APPLIED EXAMPLE 9 Depreciation of Equipment A tractor purchased at a cost of $60,000 is to be depreciated by the double declining–balance method over 10 years. What is the book value of the tractor at the end of 5 years? By what amount has the tractor been depreciated by the end of the fifth year?

Solution We have $C = 60,000$ and $N = 10$. Thus, using Equation (23) with $n = 5$ gives the book value of the tractor at the end of 5 years as

$$V(5) = 60,000\left(1 - \frac{2}{10}\right)^5$$

$$= 60,000\left(\frac{4}{5}\right)^5 = 19,660.80$$

or $19,660.80.

The amount by which the tractor has been depreciated by the end of the fifth year is given by

$$60,000 - 19,660.80 = 40,339.20$$

or $40,339.20. You may verify the last result by using Equation (24) directly.

Exploring with **TECHNOLOGY**

A tractor purchased at a cost of $60,000 is to be depreciated over 10 years with a residual value of $0. When the double declining–balance method is used, its value at the end of n years is $V_1(n) = 60{,}000(0.8)^n$ dollars. When straight-line depreciation is used, its value at the end of n years is $V_2(n) = 60{,}000 - 6000n$. Use a graphing utility to sketch the graphs of V_1 and V_2 in the viewing window $[0, 10] \times [0, 70{,}000]$. Comment on the relative merits of each method of depreciation.

4.4 Self-Check Exercises

1. Find the sum of the first five terms of the geometric progression with first term -24 and common ratio $-\frac{1}{2}$.

2. **DEPRECIATION OF OFFICE EQUIPMENT** Office equipment purchased for $75,000 is to be depreciated by the double declining–balance method over 5 years. Find the book value at the end of 3 years.

3. Derive the formula for the future value of an annuity [Formula (9), Section 4.2].

Solutions to Self-Check Exercises 4.4 can be found on page 259.

4.4 Concept Questions

1. Suppose an arithmetic progression has first term a and common difference d.
   a. What is the formula for the nth term of this progression?
   b. What is the formula for the sum of the first n terms of this progression?

2. Suppose a geometric progression has first term a and common ratio r.
   a. What is the formula for the nth term of this progression?
   b. What is the formula for the sum of the first n terms of this progression?

4.4 Exercises

In Exercises 1–4, find the nth term of the arithmetic progression that has the given values of a, d, and n.

1. $a = 6, d = 3, n = 9$ 2. $a = -5, d = 3, n = 7$

3. $a = -15, d = \dfrac{3}{2}, n = 8$ 4. $a = 1.2, d = 0.4, n = 98$

5. Find the first five terms of the arithmetic progression whose fourth and eleventh terms are 30 and 107, respectively.

6. Find the first five terms of the arithmetic progression whose seventh and twenty-third terms are -5 and -29, respectively.

7. Find the seventh term of the arithmetic progression x, $x + y, x + 2y, \ldots$.

8. Find the eleventh term of the arithmetic progression $a + b, 2a, 3a - b, \ldots$.

9. Find the sum of the first 15 terms of the arithmetic progression 4, 11, 18, ...

10. Find the sum of the first 20 terms of the arithmetic progression 5, -1, -7, ...

11. Find the sum of the odd integers between 14 and 58.

12. Find the sum of the even integers between 21 and 99.

13. Find $f(1) + f(2) + f(3) + \cdots + f(22)$, given that $f(x) = 3x - 4$.

14. Find $g(1) + g(2) + g(3) + \cdots + g(50)$, given that $g(x) = 12 - 4x$.

15. Show that Equation (18) can be written as

$$S_n = \frac{n}{2}(a + a_n)$$

where a_n represents the last term of an arithmetic progression. Use this formula to find:
   a. The sum of the first 11 terms of the arithmetic progression whose first and eleventh terms are 3 and 47, respectively.
   b. The sum of the first 20 terms of the arithmetic progression whose first and twentieth terms are 5 and -33, respectively.

16. SALES GROWTH Moderne Furniture Company had sales of $1,500,000 during its first year of operation. If the sales increased by $160,000/year thereafter, find Moderne's sales in the fifth year and its total sales over the first 5 years of operation.

17. EXERCISE PROGRAM As part of her fitness program, Karen has taken up jogging. If she jogs 1 mi the first day and increases her daily run by $\frac{1}{4}$ mi every week, when will she reach her goal of 10 mi/day?

18. COST OF DRILLING A 100-ft oil well is to be drilled. The cost of drilling the first foot is $10.00, and the cost of drilling each additional foot is $4.50 more than that of the preceding foot. Find the cost of drilling the entire 100 ft.

19. CONSUMER DECISIONS Kunwoo wishes to go from the airport to his hotel, which is 25 mi away. The taxi rate is $2.00 for the first mile and $1.20 for each additional mile. The airport limousine also goes to his hotel and charges a flat rate of $15.00. How much money will he save by taking the airport limousine?

20. SALARY COMPARISONS Markeeta, a recent college graduate, received two job offers. Company *A* offered her an initial salary of $48,800 with guaranteed annual increases of $2000/year for the first 5 years. Company *B* offered an initial salary of $50,400 with guaranteed annual increases of $1500/year for the first 5 years.
   a. Which company is offering a higher salary for the fifth year of employment?
   b. Which company is offering more money for the first 5 years of employment?

21. SUM-OF-THE-YEARS'-DIGITS METHOD OF DEPRECIATION One of the methods that the Internal Revenue Service allows for computing depreciation of certain business property is the sum-of-the-years'-digits method. If a property valued at C dollars has an estimated useful life of N years and a salvage value of S dollars, then the amount of depreciation D_n allowed during the nth year is given by

$$D_n = (C - S)\frac{N - (n - 1)}{S_N} \qquad (0 \le n \le N)$$

where S_N is the sum of the first N positive integers. Thus,

$$S_N = 1 + 2 + \cdots + N = \frac{N(N + 1)}{2}$$

   a. Verify that the sum of the arithmetic progression $S_N = 1 + 2 + \cdots + N$ is given by

$$\frac{N(N + 1)}{2}$$

   b. If office furniture worth $6000 is to be depreciated by this method over $N = 10$ years and the salvage value of the furniture is $500, find the depreciation for the third year by computing D_3.

22. SUM-OF-THE-YEARS'-DIGITS METHOD OF DEPRECIATION The amount of depreciation allowed for a printing machine, which has an estimated useful life of 5 years and an initial value of $100,000 (with no salvage value), is $20,000/year using the straight-line method of depreciation. Determine the amount of depreciation that would be allowed for the first year if the printing machine were depreciated using the sum-of-the-years'-digits method described in Exercise 21. Which method would result in a larger depreciation of the asset in its first year of use?

In Exercises 23–28, determine which of the sequences are geometric progressions. For each geometric progression, find the seventh term and the sum of the first seven terms.

23. 4, 8, 16, 32, . . .

24. $1, -\frac{1}{2}, \frac{1}{4}, -\frac{1}{8}, \ldots$

25. $\frac{1}{2}, -\frac{3}{8}, \frac{1}{4}, -\frac{9}{64}, \ldots$

26. 0.004, 0.04, 0.4, 4, . . .

27. 243, 81, 27, 9, . . .

28. $-1, 1, 3, 5, \ldots$

29. Find the twentieth term and sum of the first 20 terms of the geometric progression $-3, 3, -3, 3, \ldots$.

30. Find the twenty-third term in a geometric progression having the first term $a = 0.1$ and ratio $r = 2$.

31. POPULATION GROWTH It has been projected that the population of a certain city in the southwest will increase by 8% during each of the next 5 years. If the current population is 200,000, what is the expected population after 5 years?

32. SALES GROWTH Metro Cable TV had sales of $2,500,000 in its first year of operation. If thereafter the sales increased by 12% of the previous year, find the sales of the company in the fifth year and the total sales over the first 5 years of operation.

33. COLAs Suppose the cost-of-living index had increased by 3% during each of the past 6 years and that a member of the EUW Union had been guaranteed an annual increase equal to 2% above the increase in the cost-of-living index over that period. What would be the present salary of a union member whose salary 6 years ago was $42,000?

34. SAVINGS PLANS The parents of a 9-year-old boy have agreed to deposit $10 in their son's bank account on his 10th birthday and to double the size of their deposit every year thereafter until his 18th birthday.
  a. How much will they have to deposit on his 18th birthday?
  b. How much will they have deposited by his 18th birthday?

35. SALARY COMPARISONS A Stenton Printing Co. employee whose current annual salary is $48,000 has the option of taking an annual raise of 8%/year for the next 4 years or a fixed annual raise of $4000/year. Which option would be more profitable to him considering his total earnings over the 4-year period?

36. BACTERIA GROWTH A culture of a certain bacteria is known to double in number every 3 hr. If the culture has an initial count of 20, what will be the population of the culture at the end of 24 hr?

37. TRUST FUNDS Sarah is the recipient of a trust fund that she will receive over a period of 6 years. Under the terms of the trust, she is to receive $10,000 the first year and each succeeding annual payment is to be increased by 15%.
  a. How much will she receive during the sixth year?
  b. What is the total amount of the six payments she will receive?

In Exercises 38–40, find the book value of office equipment purchased at a cost C at the end of the nth year if it is to be depreciated by the double declining–balance method over 10 years.

38. $C = \$20,000$, $n = 4$ **39.** $C = \$150,000$, $n = 8$

40. $C = \$80,000$, $n = 7$

41. DOUBLE DECLINING–BALANCE METHOD OF DEPRECIATION Restaurant equipment purchased at a cost of $150,000 is to be depreciated by the double declining–balance method over 10 years. What is the book value of the equipment at the end of 6 years? By what amount has the equipment been depreciated at the end of the sixth year?

42. DOUBLE DECLINING–BALANCE METHOD OF DEPRECIATION Refer to Exercise 22. Recall that a printing machine with an estimated useful life of 5 years and an initial value of $100,000 (and no salvage value) was to be depreciated. At the end of the first year, the amount of depreciation allowed was $20,000 using the straight-line method and $33,333 using the sum-of-the-years'-digits method. Determine the amount of depreciation that would be allowed for the first year if the printing machine were depreciated by the double declining–balance method. Which of these three methods would result in the largest depreciation of the printing machine at the end of its first year of use?

In Exercises 43 and 44, determine whether the statement is true or false. If it is true, explain why it is true. If it is false, give an example to show why it is false.

43. If $a_1, a_2, a_3, \ldots, a_n$ and $b_1, b_2, b_3, \ldots, b_n$ are arithmetic progressions, then $a_1 + b_1, a_2 + b_2, a_3 + b_3, \ldots, a_n + b_n$ is also an arithmetic progression.

44. If $a_1, a_2, a_3, \ldots, a_n$ and $b_1, b_2, b_3, \ldots, b_n$ are geometric progressions, then $a_1b_1, a_2b_2, a_3b_3, \ldots, a_nb_n$ is also a geometric progression.

4.4 Solutions to Self-Check Exercises

1. Use Equation (22) with $a = -24$, $n = 5$, and $r = -\frac{1}{2}$, obtaining

$$S_5 = \frac{-24\left[1 - \left(-\frac{1}{2}\right)^5\right]}{1 - \left(-\frac{1}{2}\right)}$$

$$= \frac{-24\left(1 + \frac{1}{32}\right)}{\frac{3}{2}} = -\frac{33}{2}$$

2. Use Equation (23) with $C = 75,000$, $N = 5$, and $n = 3$. The book value of the office equipment at the end of 3 years is

$$V(3) = 75,000\left(1 - \frac{2}{5}\right)^3 = 16,200$$

or $16,200.

3. We have

$$S = R + R(1 + i) + R(1 + i)^2 + \cdots + R(1 + i)^{n-1}$$

The sum on the right is easily seen to be the sum of the first n terms of a geometric progression with first term R and common ratio $(1 + i)$, so by virtue of Equation (22), we obtain

$$S = R\left[\frac{1 - (1 + i)^n}{1 - (1 + i)}\right] = R\left[\frac{(1 + i)^n - 1}{i}\right]$$

CHAPTER 4   Summary of Principal Formulas and Terms

FORMULAS

1. Simple interest (accumulated amount)	$A = P(1 + rt)$
2. Compound interest	
a. Accumulated amount	$A = P(1 + i)^n$
b. Present value	$P = A(1 + i)^{-n}$
c. Interest rate per conversion period	$i = \dfrac{r}{m}$
d. Number of conversion periods	$n = mt$
3. Continuous compound interest	
a. Accumulated amount	$A = Pe^{rt}$
b. Present value	$P = Ae^{-rt}$
4. Effective rate of interest	$r_{\text{eff}} = \left(1 + \dfrac{r}{m}\right)^m - 1$
5. Annuities	
a. Future value	$S = R\left[\dfrac{(1 + i)^n - 1}{i}\right]$
b. Present value	$P = R\left[\dfrac{1 - (1 + i)^{-n}}{i}\right]$
6. Amortization payment	$R = \dfrac{Pi}{1 - (1 + i)^{-n}}$
7. Amount amortized	$P = R\left[\dfrac{1 - (1 + i)^{-n}}{i}\right]$
8. Sinking fund payment	$R = \dfrac{iS}{(1 + i)^n - 1}$

TERMS

simple interest (206)	future value (214)	present value of an annuity (227)
accumulated amount (206)	annuity (224)	sinking fund (241)
compound interest (207)	annuity certain (224)	arithmetic progression (251)
nominal rate (stated rate) (207)	ordinary annuity (225)	common difference (251)
conversion period (208)	simple annuity (225)	geometric progression (253)
effective rate of interest (212)	future value of an annuity (226)	common ratio (253)
present value (213)		

CHAPTER 4   Concept Review Questions

Fill in the blanks.

1. a. Simple interest is computed on the _____ principal only. The formula for the accumulated amount using simple interest is $A = $ _____ .

   b. In calculations using compound interest, earned interest is periodically added to the principal and thereafter itself earns _____ . The formula for the accumulated amount using compound interest is $A = $ _____ . Solving this equation for P gives the present value formula using compound interest as $P = $ _____ .

2. The effective rate of interest is the _____ interest rate that would produce the same accumulated amount in _____ year as the _____ rate compounded _____ times a year. The formula for calculating the effective rate is $r_{\text{eff}} = $ _____.

3. A sequence of payments made at regular time intervals is called a/an _____; if the payments are made at the end of each payment period, then it is called a/an _____ _____; if the payment period coincides with the interest conversion period, then it is called a/an _____ _____.

4. The formula for the future value of an annuity is $S = $ _____. The formula for the present value of an annuity is $P = $ _____.

5. The periodic payment R on a loan of P dollars to be amortized over n periods with interest charged at the rate of i per period is $R = $ _____.

6. A sinking fund is an account that is set up for a specific purpose at some _____ date. The periodic payment R required to accumulate a sum of S dollars over n periods with interest charged at the rate of i per period is $R = $ _____.

7. An arithmetic progression is a sequence of numbers in which each term after the first is obtained by adding a/an _____ _____ to the preceding term. The nth term of an arithmetic progression is $a_n = $ _____. The sum of the first n terms of an arithmetic progression is $S_n = $ _____.

8. A geometric progression is a sequence of numbers in which each term after the first is obtained by multiplying the preceding term by a/an _____ _____. The nth term of a geometric progression is $a_n = $ _____. If $r \neq 1$, the sum of the first n terms of a geometric progression is $S_n = $ _____.

CHAPTER 4   Review Exercises

1. Find the accumulated amount after 4 years if $5000 is invested at 5%/year compounded (a) annually, (b) semiannually, (c) quarterly, and (d) monthly.

2. Find the accumulated amount after 8 years if $12,000 is invested at 3.5%/year compounded (a) annually, (b) semiannually, (c) quarterly, and (d) monthly.

3. Find the effective rate of interest corresponding to a nominal rate of 6%/year compounded (a) annually, (b) semiannually, (c) quarterly, and (d) monthly.

4. Find the effective rate of interest corresponding to a nominal rate of 5.5%/year compounded (a) annually, (b) semiannually, (c) quarterly, and (d) monthly.

5. Find the present value of $41,413 due in 5 years at an interest rate of 4.5%/year compounded quarterly.

6. Find the present value of $64,540 due in 6 years at an interest rate of 4%/year compounded monthly.

7. Find the amount (future value) of an ordinary annuity of $150/quarter for 7 years at 5%/year compounded quarterly.

8. Find the future value of an ordinary annuity of $120/month for 10 years at 4.5%/year compounded monthly.

9. Find the present value of an ordinary annuity of 36 payments of $250 each made monthly and earning interest at 4.5%/year compounded monthly.

10. Find the present value of an ordinary annuity of 60 payments of $5000 each made quarterly and earning interest at 3.5%/year compounded quarterly.

11. Find the payment R needed to amortize a loan of $22,000 at 3.5%/year compounded monthly with 36 monthly installments over a period of 3 years.

12. Find the payment R needed to amortize a loan of $10,000 at 4.6%/year compounded monthly with 36 monthly installments over a period of 3 years.

13. Find the payment R needed to accumulate $18,000 with 48 monthly installments over a period of 4 years at an interest rate of 3%/year compounded monthly.

14. Find the payment R needed to accumulate $15,000 with 60 monthly installments over a period of 5 years at an interest rate of 3.6%/year compounded monthly.

15. Find the effective rate of interest corresponding to a nominal rate of 3.6%/year compounded monthly.

16. Find the effective rate of interest corresponding to a nominal rate of 4.8%/year compounded monthly.

17. Find the present value of $119,346 due in 4 years at an interest rate of 5%/year compounded continuously.

18. **COMPANY SALES** JCN Media had sales of $1,750,000 in the first year of operation. If the sales increased by 7%/year thereafter, find the company's sales in the fourth year and the total sales over the first 4 years of operation.

19. **CDs** The manager of a money market fund has invested $4.2 million in certificates of deposit that pay interest at the rate of 5.4%/year compounded quarterly over a period of 5 years. How much will the investment be worth at the end of 5 years?

20. **SAVINGS ACCOUNTS** Emily deposited $2000 into a bank account 5 years ago. The bank paid interest at the rate of 3.2%/year compounded weekly. What is Emily's account worth today?

21. **SAVINGS ACCOUNTS** Kim invested a sum of money 4 years ago in a savings account that has since paid interest at the rate of 3.5%/year compounded monthly. Her investment is now worth $19,440.31. How much did she originally invest?

22. **SAVINGS ACCOUNTS** Andrew withdrew $5470.87 from a savings account, which he closed this morning. The account had earned interest at the rate of 3%/year compounded continuously during the 3-year period that the money was on deposit. How much did Andrew originally deposit into the account?

23. **MUTUAL FUNDS** Juan invested $24,000 in a mutual fund 5 years ago. Today his investment is worth $34,616. Find the effective annual rate of return on his investment over the 5-year period.

24. **COLLEGE SAVINGS PROGRAM** The Blakes have decided to start a monthly savings program to provide for their son's college education. How much should they deposit at the end of each month in a savings account earning interest at the rate of 3.5%/year compounded monthly so that, at the end of the tenth year, the accumulated amount will be $40,000?

25. **RETIREMENT ACCOUNTS** Mai Lee has contributed $200 at the end of each month into her company's employee retirement account for the past 10 years. Her employer has matched her contribution each month. If the account has earned interest at the rate of 5%/year compounded monthly over the 10-year period, determine how much Mai Lee now has in her retirement account.

26. **AUTOMOBILE LEASING** Maria has leased an auto for 4 years at $300/month. If money is worth 5%/year compounded monthly, what is the equivalent cash payment (present value) of this annuity? (Assume that the payments are made at the end of each month.)

27. **INSTALLMENT FINANCING** Peggy made a down payment of $400 toward the purchase of new furniture. To pay the balance of the purchase price, she has secured a loan from her bank at 6%/year compounded monthly. Under the terms of her finance agreement, she is required to make payments of $75.32 at the end of each month for 24 months. What was the purchase price of the furniture?

28. **HOME FINANCING** The Turners have purchased a house for $150,000. They made an initial down payment of $30,000

and secured a mortgage with interest charged at the rate of 4.5%/year on the unpaid balance. (Interest computations are made at the end of each month.) Assume that the loan is amortized over 30 years.
   a. What monthly payment will the Turners be required to make?
   b. What will be their total interest payment?
   c. What will be their equity (disregard depreciation) after 10 years?

29. **HOME FINANCING** Refer to Exercise 28. If the loan is amortized over 15 years:
   a. What monthly payment will the Turners be required to make?
   b. What will be their total interest payment?
   c. What will be their equity (disregard depreciation) after 10 years?

30. **SINKING FUNDS** The management of a corporation anticipates a capital expenditure of $500,000 in 5 years for the purpose of purchasing replacement machinery. To finance this purchase, a sinking fund that earns interest at the rate of 5%/year compounded quarterly will be set up. Determine the amount of each (equal) quarterly installment that should be deposited in the fund. (Assume that the payments are made at the end of each quarter.)

31. **SINKING FUNDS** The management of a condominium association anticipates a capital expenditure of $120,000 in 2 years for the purpose of painting the exterior of the condominium. To pay for this maintenance, a sinking fund will be set up that will earn interest at the rate of 5.8%/year compounded monthly. Determine the amount of each (equal) monthly installment the association will be required to deposit into the fund at the end of each month for the next 2 years.

32. **CREDIT CARD PAYMENTS** The outstanding balance on Bill's credit card account is $3200. The bank issuing the credit card is charging 9.3%/year compounded monthly. If Bill decides to pay off this balance in equal monthly installments at the end of each month for the next 18 months, how much will be his monthly payment? What is the effective rate of interest the bank is charging Bill?

33. **FINANCIAL PLANNING** Matt's parents have agreed to contribute $250/month toward the rent for his apartment in his junior year in college. The plan is for Matt's parents to deposit a lump sum in Matt's bank account on August 1 and then have Matt withdraw $250 on the first of each month starting on September 1 and ending on May 1 the following year. If the bank pays interest on the balance at the rate of 5%/year compounded monthly, how much should Matt's parents deposit into his account?

CHAPTER 4 Before Moving On . . .

1. Find the accumulated amount at the end of 3 years if $2000 is deposited in an account paying interest at the rate of 8%/year compounded monthly.

2. Find the effective rate of interest corresponding to a nominal rate of 6%/year compounded daily.

3. Find the future value of an ordinary annuity of $800/week for 10 years at 6%/year compounded weekly.

4. Find the monthly payment required to amortize a loan of $100,000 over 10 years with interest charged at the rate of 8%/year compounded monthly.

5. Find the weekly payment required to accumulate a sum of $15,000 over 6 years with interest earned at the rate of 10%/year compounded weekly.

6. a. Find the sum of the first ten terms of the arithmetic progression 3, 7, 11, 15, 19, ...

   b. Find the sum of the first eight terms of the geometric progression $\frac{1}{2}$, 1, 2, 4, 8, ...

5 Systems of Linear Equations and Matrices

THE LINEAR EQUATIONS in two variables that we studied in Chapter 2 are readily extended to cases involving more than two variables. For example, a linear equation in three variables represents a plane in three-dimensional space. In this chapter, we see how some real-world problems can be formulated in terms of systems of linear equations, and we develop two methods for solving these equations.

In addition, we see how *matrices* (rectangular arrays of numbers) can be used to write systems of linear equations in compact form. We then go on to consider some real-life applications of matrices.

Checkers Rent-A-Car is planning to expand its fleet of cars next quarter. How should the company use its budget of $18 million to meet the expected additional demand for compact and full-size cars? In Example 5, page 339, we will see how we can find the solution to this problem by solving a system of equations.

© Tom Oliveira/Shutterstock.com

5.1 Systems of Linear Equations: An Introduction

Systems of Equations

Recall that in Section 2.5, we had to solve two simultaneous linear equations to find the *break-even point*. This is an example of a real-world problem that calls for the solution of a **system of linear equations** in two or more variables. In this chapter, we take up a more systematic study of such systems.

We begin by considering a system of two linear equations in two variables. Recall that such a system may be written in the general form

$$\begin{aligned} ax + by &= h \\ cx + dy &= k \end{aligned} \tag{1}$$

where a, b, c, d, h, and k are real constants and neither a and b nor c and d are both zero.

Now let's study the nature of the **solution of a system of linear equations** in more detail. Recall that the graph of each equation in System (1) is a straight line in the plane, so geometrically, the solution to the system is the point(s) of intersection of the two straight lines L_1 and L_2, represented by the first and second equations of the system.

Given two lines L_1 and L_2, *one and only one* of the following may occur:

a. L_1 and L_2 intersect at exactly one point.
b. L_1 and L_2 are parallel and coincident.
c. L_1 and L_2 are parallel and distinct.

(See Figure 1.) In the first case, the system has a unique solution corresponding to the single point of intersection of the two lines. In the second case, the system has infinitely many solutions corresponding to the points lying on the same line. Finally, in the third case, the system has no solution because the two lines do not intersect.

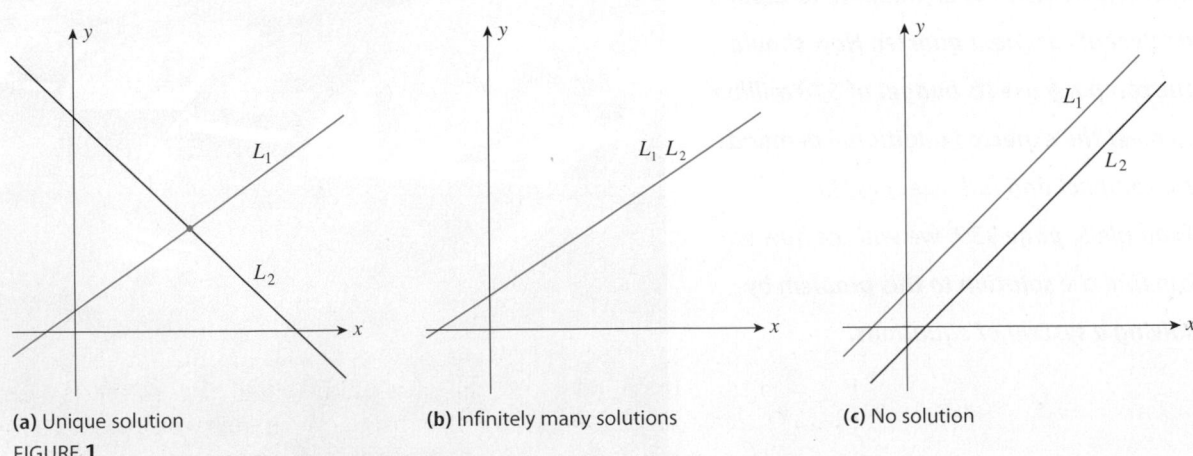

(a) Unique solution   (b) Infinitely many solutions   (c) No solution

FIGURE **1**

Explore and Discuss

Generalize the discussion on this page to the case in which there are three straight lines in the plane defined by three linear equations. What if there are n lines defined by n equations?

Let's illustrate each of these possibilities by considering some specific examples.

1. **A system of equations with exactly one solution** Consider the system

$$2x - y = 1$$
$$3x + 2y = 12$$

Solving the first equation for y in terms of x, we obtain the equation

$$y = 2x - 1$$

Substituting this expression for y into the second equation yields

$$3x + 2(2x - 1) = 12$$
$$3x + 4x - 2 = 12$$
$$7x = 14$$
$$x = 2$$

Finally, substituting this value of x into the expression for y obtained earlier gives

$$y = 2(2) - 1 = 3$$

Therefore, the unique solution of the system is given by $x = 2$ and $y = 3$. Geometrically, the two lines represented by the two linear equations that make up the system intersect at the point $(2, 3)$ (Figure 2).

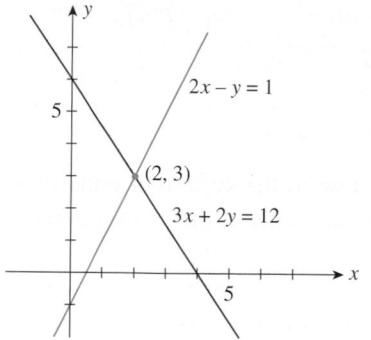

FIGURE 2
A system of equations with one solution

Note We can check our result by substituting the values $x = 2$ and $y = 3$ into the equations. Thus,

$$2(2) - (3) = 1 \quad ✓$$
$$3(2) + 2(3) = 12 \quad ✓$$

From the geometric point of view, we have just verified that the point $(2, 3)$ lies on both lines. ◾

2. **A system of equations with infinitely many solutions** Consider the system

$$2x - y = 1$$
$$6x - 3y = 3$$

Solving the first equation for y in terms of x, we obtain the equation

$$y = 2x - 1$$

Substituting this expression for y into the second equation gives

$$6x - 3(2x - 1) = 3$$
$$6x - 6x + 3 = 3$$
$$0 = 0$$

which is a true statement. This result follows from the fact that the second equation is equivalent to the first. (To see this, just multiply both sides of the first equation by 3.) Our computations have revealed that the system of two equations is equivalent to the single equation $2x - y = 1$. Thus, any ordered pair of numbers (x, y) satisfying the equation $2x - y = 1$ (or $y = 2x - 1$) constitutes a solution to the system.

In particular, by assigning the value t to x, where t is any real number, we find that $y = 2t - 1$, so the ordered pair $(t, 2t - 1)$ is a solution of the system. The variable t is called a **parameter.** For example, setting $t = 0$ gives the point $(0, -1)$ as a solution of the system, and setting $t = 1$ gives the point $(1, 1)$ as another solution. Since t represents any real number, there are infinitely many solutions of the

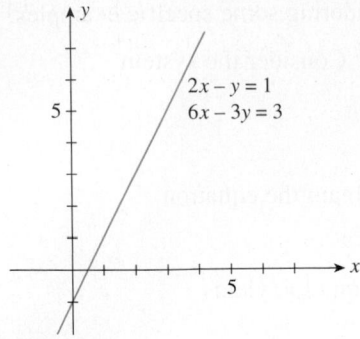

FIGURE 3
A system of equations with infinitely many solutions; each point on the line is a solution.

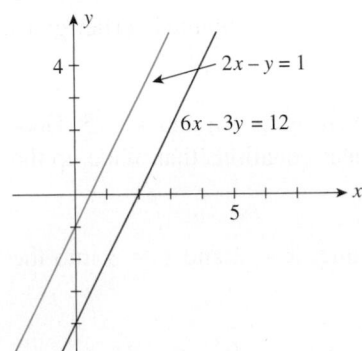

FIGURE 4
A system of equations with no solution

system. Geometrically, the two equations in the system represent the same line, and all solutions of the system are points lying on the line (Figure 3). Such a system is said to be **dependent**.

3. **A system of equations that has no solution** Consider the system

$$2x - y = 1$$
$$6x - 3y = 12$$

The first equation is equivalent to $y = 2x - 1$. Substituting this expression for y into the second equation gives

$$6x - 3(2x - 1) = 12$$
$$6x - 6x + 3 = 12$$
$$0 = 9$$

which is clearly impossible. Thus, there is no solution to the system of equations. To interpret this situation geometrically, cast both equations in the slope-intercept form, obtaining

$$y = 2x - 1$$
$$y = 2x - 4$$

We see at once that the lines represented by these equations are parallel (each has slope 2) and distinct, since the first has y-intercept -1 and the second has y-intercept -4 (Figure 4). Systems with no solutions, such as this one, are said to be **inconsistent**.

Explore and Discuss

1. Consider a system composed of two linear equations in two variables. Can the system have exactly two solutions? Exactly three solutions? Exactly a finite number of solutions?

2. Suppose at least one of the equations in a system composed of two equations in two variables is nonlinear. Can the system have no solution? Exactly one solution? Exactly two solutions? Exactly a finite number of solutions? Infinitely many solutions? Illustrate each answer with a sketch.

Note   We have used the method of substitution in solving each of these systems. If you are familiar with the method of elimination, you might want to re-solve each of these systems using this method. We will study the method of elimination in detail in Section 5.2.   ∎

In Section 2.5, we presented some real-world applications of systems involving two linear equations in two variables. Here is an example involving a system of three linear equations in three variables.

APPLIED EXAMPLE 1 Production Scheduling Ace Novelty wishes to produce three types of souvenirs: Types A, B, and C. To manufacture a Type A souvenir requires 2 minutes on Machine I, 1 minute on Machine II, and 2 minutes on Machine III. A Type B souvenir requires 1 minute on Machine I, 3 minutes on Machine II, and 1 minute on Machine III. A Type C souvenir requires 1 minute on Machine I and 2 minutes each on Machines II and III. There are 3 hours available on Machine I, 5 hours available on Machine II, and 4 hours available on Machine III for processing the order. How many souvenirs of each type should Ace Novelty make in order to use all of the available time? Formulate but do not solve the problem. (We will solve this problem in Example 7, Section 5.2.)

Solution The given information may be tabulated as follows:

	Type A	Type B	Type C	Time Available (min)
Machine I	2	1	1	180
Machine II	1	3	2	300
Machine III	2	1	2	240

We have to determine the number of each of *three* types of souvenirs to be made. So let x, y, and z denote the respective numbers of Type A, Type B, and Type C souvenirs to be made. The total amount of time that Machine I is used is given by $2x + y + z$ minutes and must equal 180 minutes. This leads to the equation

$$2x + y + z = 180 \qquad \text{Time spent on Machine I}$$

Similar considerations on the use of Machines II and III lead to the following equations:

$$x + 3y + 2z = 300 \qquad \text{Time spent on Machine II}$$
$$2x + y + 2z = 240 \qquad \text{Time spent on Machine III}$$

Since the variables x, y, and z must satisfy simultaneously the three conditions represented by the three equations, the solution to the problem is found by solving the following system of linear equations:

$$\begin{aligned} 2x + y + z &= 180 \\ x + 3y + 2z &= 300 \\ 2x + y + 2z &= 240 \end{aligned}$$

Solutions of Systems of Equations

We will complete the solution of the problem posed in Example 1 later on (page 284). For the moment, let's look at the geometric interpretation of a system of linear equations, such as the system in Example 1, to gain some insight into the nature of the solution.

A linear system composed of three linear equations in three variables x, y, and z has the general form

$$\begin{aligned} a_1x + b_1y + c_1z &= d_1 \\ a_2x + b_2y + c_2z &= d_2 \\ a_3x + b_3y + c_3z &= d_3 \end{aligned} \qquad\qquad \textbf{(2)}$$

Just as a linear equation in two variables represents a straight line in the plane, it can be shown that a linear equation $ax + by + cz = d$ (a, b, and c not all equal to zero) in three variables represents a plane in three-dimensional space. Thus, each equation in System (2) represents a *plane* in three-dimensional space, and the *solution(s) of the system* is precisely the point(s) of intersection of the three planes defined by the three linear equations that make up the system. As before, the system has one and only one solution, infinitely many solutions, or no solution, depending on whether and how the planes intersect one another. Figure 5 illustrates each of these possibilities.

In Figure 5a, the three planes intersect at a point corresponding to the situation in which System (2) has a unique solution. Figure 5b depicts a situation in which there are infinitely many solutions to the system. Here, the three planes intersect along a line, and the solutions are represented by the infinitely many points lying on this line. In Figure 5c, the three planes are parallel and distinct, so there is no point common to all three planes; System (2) has no solution in this case.

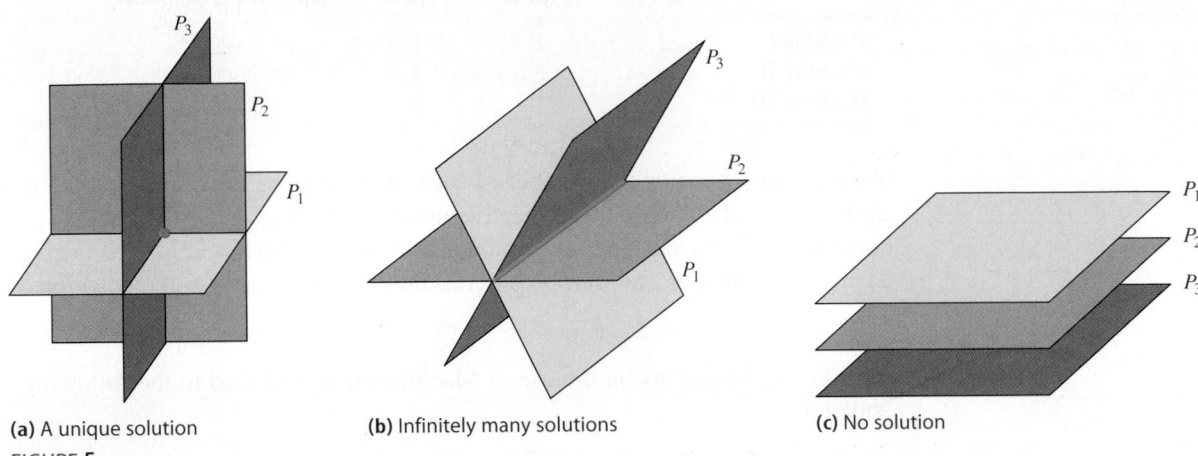

(a) A unique solution

(b) Infinitely many solutions

(c) No solution

FIGURE 5

Note The depictions in Figure 5 are by no means exhaustive. You may consider various other orientations of the three planes that would illustrate the three possible outcomes in solving a system of linear equations involving three variables. ◼

> **Linear Equations in n Variables**
>
> A linear equation in n variables, $x_1, x_2, \ldots, x_n$ is an equation of the form
>
> $$a_1 x_1 + a_2 x_2 + \cdots + a_n x_n = c$$
>
> where $a_1, a_2, \ldots, a_n$ (not all zero) and c are constants.

For example, the equation

$$3x_1 + 2x_2 - 4x_3 + 6x_4 = 8$$

is a linear equation in the four variables, x_1, x_2, x_3, and x_4.

When the number of variables involved in a linear equation exceeds three, we no longer have the geometric interpretation we had for the lower-dimensional spaces. Nevertheless, the algebraic concepts of the lower-dimensional spaces generalize to higher dimensions. For this reason, a linear equation in n variables, $a_1 x_1 + a_2 x_2 + \cdots + a_n x_n = c$, where $a_1, a_2, \ldots, a_n$ are not all zero, is referred to as an *n-dimensional hyperplane*. We may interpret the solution(s) to a system comprising a finite number of such linear equations to be the *point(s) of intersection* of the hyperplanes defined by the equations that make up the system. As in the case of systems involving two or three variables, it can be shown that only three possibilities exist regarding the nature of the solution of such a system: (1) a unique solution, (2) infinitely many solutions, or (3) no solution.

Explore and Discuss

Refer to the Note above.

Using the orientation of three planes, illustrate the outcomes in solving a system of three linear equations in three variables that result in no solution or infinitely many solutions.

5.1 Self-Check Exercises

1. Determine whether the system of linear equations

$$2x - 3y = 12$$
$$x + 2y = 6$$

has (a) a unique solution, (b) infinitely many solutions, or (c) no solution. Find all solutions whenever they exist. Make a sketch of the set of lines described by the system.

2. **CROP PLANNING** A farmer has 200 acres of land suitable for cultivating Crops A, B, and C. The cost per acre of cultivating Crops A, B, and C is \$40, \$60, and \$80,

respectively. The farmer has \$12,600 available for cultivation. Each acre of Crop A requires 20 labor-hours, each acre of Crop B requires 25 labor-hours, and each acre of Crop C requires 40 labor-hours. The farmer has a maximum of 5950 labor-hours available. If she wishes to use all of her cultivatable land, the entire budget, and all the labor available, how many acres of each crop should she plant? Formulate but do not solve the problem.

Solutions to Self-Check Exercises 5.1 can be found on page 274.

5.1 Concept Questions

1. Suppose you are given a system of two linear equations in two variables.
   a. What can you say about the solution(s) of the system of equations?
   b. Give a geometric interpretation of your answers to the question in part (a). Illustrate each answer with a sketch.

2. Suppose you are given a system of two linear equations in two variables.
   a. Explain what it means for the system to be (i) dependent and (ii) inconsistent.
   b. Illustrate each answer with a sketch.

5.1 Exercises

In Exercises 1–18, determine whether each system of linear equations has (a) one and only one solution, (b) infinitely many solutions, or (c) no solution. Find all solutions whenever they exist.

1. $x - 3y = -1$
$4x + 3y = 11$

2. $2x - 4y = -10$
$3x + 2y = 1$

3. $x + 4y = 7$
$\dfrac{1}{2}x + 2y = 5$

4. $3x - 4y = 7$
$9x - 12y = 14$

5. $x + 2y = 7$
$2x - y = 4$

6. $\dfrac{3}{2}x - 2y = 4$
$x + \dfrac{1}{3}y = 2$

7. $2x - 5y = 10$
$6x - 15y = 30$

8. $5x - 6y = 8$
$10x - 12y = 16$

9. $4x - 5y = 14$
$2x + 3y = -4$

10. $\dfrac{5}{4}x - \dfrac{2}{3}y = 3$
$\dfrac{1}{4}x + \dfrac{5}{3}y = 6$

11. $2x - 3y = 6$
$6x - 9y = 12$

12. $\dfrac{2}{3}x + y = 5$
$\dfrac{1}{2}x + \dfrac{3}{4}y = \dfrac{15}{4}$

13. $-3x + 5y = 1$
$2x - 4y = -1$

14. $-10x + 15y = -3$
$4x - 6y = -3$

15. $3x - 6y = 2$
$-\dfrac{3}{2}x + 3y = -1$

16. $\dfrac{3}{2}x - \dfrac{1}{2}y = 1$
$-x + \dfrac{1}{3}y = -\dfrac{2}{3}$

17. $0.2x + y = 1.8$
$0.4x + 0.3y = 0.2$

18. $0.3x - 0.4y = 0.2$
$-0.2x + 0.5y = 0.1$

19. Determine the value of k for which the system of linear equations

$$2x - y = 3$$
$$4x + ky = 4$$

has no solution.

20. Determine the value of k for which the system of linear equations

$$3x + 4y = 12$$
$$x + ky = 4$$

has infinitely many solutions. Then find all solutions corresponding to this value of k.

21. Determine the conditions on a and b for which the system of linear equations

$$ax - by = c$$
$$ax + by = d$$

has a unique solution. What is the solution?

22. Determine the conditions on a, b, c, and d for which the system of linear equations

$$ax + by = e$$
$$cx + dy = f$$

has a unique solution. What is the solution?

In Exercises 23–44, formulate but do not solve the problem. You will be asked to solve these problems in Section 5.2.

23. **Crop Planning** The Johnson Farm has 500 acres of land allotted for cultivating corn and wheat. The cost of cultivating corn and wheat (including seeds and labor) is $42 and $30 per acre, respectively. Jacob Johnson has $18,600 available for cultivating these crops. If he wishes to use all the allotted land and his entire budget for cultivating these two crops, how many acres of each crop should he plant?

24. **Investments** Michael Perez has a total of $2000 on deposit with two savings institutions. One pays interest at the rate of 3%/year; the other pays interest at the rate of 4%/year. If Michael earned a total of $72 in interest during a single year, how much does he have on deposit in each institution?

25. **Blended Coffee Mixtures** The Coffee Shoppe sells a gourmet coffee blend made from two coffees, one costing $8/lb and the other costing $9/lb. If the blended coffee sells for $8.60/lb, find how much of each coffee is used to obtain the desired blend. Assume that the weight of the blended coffee is 100 lb.

26. **Municipal Bonds** Kelly Fisher has a total of $30,000 invested in two municipal bonds that have yields of 4% and 5% interest per year, respectively. If the interest Kelly receives from the bonds in a year is $1320, how much does she have invested in each bond?

27. **Metro Bus Ridership** The total number of passengers riding a certain city bus during the morning shift is 1000. If the child's fare is $0.50, the adult fare is $1.50, and the total revenue from the fares in the morning shift is $1300, how many children and how many adults rode the bus during the morning shift?

28. **Apartment Complex Development** Cantwell Associates, a real estate developer, is planning to build a new apartment complex consisting of one-bedroom units and two- and three-bedroom townhouses. A total of 192 units is planned, and the number of family units (two- and three-bedroom townhouses) will equal the number of one-bedroom units. If the number of one-bedroom units will be 3 times the number of three-bedroom units, find how many units of each type will be in the complex.

29. A ball and a bat cost a total of $110. The bat costs $100 more than the ball. How much does the ball cost?

30. **Investments** Josh has invested $70,000 in two projects. The amount invested in project A exceeds that invested in project B by $20,000. How much has Josh invested in each project?

31. **Investment Planning** The annual returns on Sid Carrington's three investments amounted to $21,600: 6% on a savings account, 8% on mutual funds, and 12% on bonds. The amount of Sid's investment in bonds was twice the amount of his investment in the savings account, and the interest earned from his investment in bonds was equal to the dividends he received from his investment in mutual funds. Find how much money he placed in each type of investment.

32. **Investment Risk and Return** A private investment club has $200,000 earmarked for investment in stocks. To arrive at an acceptable overall level of risk, the stocks that management is considering have been classified into three categories: high-risk, medium-risk, and low-risk. Management estimates that high-risk stocks will have a rate of return of 15%/year; medium-risk stocks, 10%/year; and low-risk stocks, 6%/year. The members have decided that the investment in low-risk stocks should be equal to the sum of the investments in the stocks of the other two categories. Determine how much the club should invest in each type of stock if the investment goal is to have a return of $20,000/year on the total investment. (Assume that all the money available for investment is invested.)

33. **Using Digital Technology** A survey of 500 college students found that the percentage of students who went without using digital technology for up to 1 hr was 67%. The survey also determined that the percentage of students who went without using digital technology for up to 30 min exceeded the percentage of students who went without using digital technology for over 1 hr by 17%. Let x, y, and z represent the percentage of the students in the survey who went without using digital technology (a) for up to 30 min, (b) for more than 30 min but not more than 60 min, and (c) for more than 60 min, respectively. Find the values of x, y, and z.
Source: CourseSmart.

34. **Trustworthiness of Online Reviews** In a survey of 1000 adults aged 18 and older, the following question was

posed: "Are other travelers' online reviews trustworthy?" The participants were asked to answer "yes," "no," or "not sure." The survey revealed that 370 answered "no" or "not sure." It also showed that the number of those who answered "yes" exceeded the number of those who answered "no" by 340. What percentage of respondents answered (a) "yes," (b) "no," and (c) "not sure"?
Source: Alliance Global Assistance.

35. **LAWN FERTILIZERS** Lawnco produces three grades of commercial fertilizers. A 100-lb bag of grade A fertilizer contains 18 lb of nitrogen, 4 lb of phosphate, and 5 lb of potassium. A 100-lb bag of grade B fertilizer contains 20 lb of nitrogen and 4 lb each of phosphate and potassium. A 100-lb bag of grade C fertilizer contains 24 lb of nitrogen, 3 lb of phosphate, and 6 lb of potassium. How many 100-lb bags of each of the three grades of fertilizers should Lawnco produce if 26,400 lb of nitrogen, 4900 lb of phosphate, and 6200 lb of potassium are available and all the nutrients are used?

36. **BOX-OFFICE RECEIPTS** A theater has a seating capacity of 900 and charges $4 for children, $6 for students, and $8 for adults. At a certain screening with full attendance, there were half as many adults as children and students combined. The receipts totaled $5600. How many children attended the show?

37. **BUDGET ALLOCATION FOR AUTO FLEET** The management of Hartman Rent-A-Car has allocated $2.25 million to buy a fleet of new automobiles consisting of compact, intermediate-size, and full-size cars. Compacts cost $18,000 each, intermediate-size cars cost $27,000 each, and full-size cars cost $36,000 each. If Hartman purchases twice as many compacts as intermediate-size cars and the total number of cars to be purchased is 100, determine how many cars of each type will be purchased. (Assume that the entire budget will be used.)

38. **INVESTMENT RISK AND RETURN** The management of a private investment club has a fund of $200,000 earmarked for investment in stocks. To arrive at an acceptable overall level of risk, the stocks that management is considering have been classified into three categories: high-risk, medium-risk, and low-risk. Management estimates that high-risk stocks will have a rate of return of 15%/year; medium-risk stocks, 10%/year; and low-risk stocks, 6%/year. The investment in low-risk stocks is to be twice the sum of the investments in stocks of the other two categories. If the investment goal is to have an average rate of return of 9%/year on the total investment, determine how much the club should invest in each type of stock. (Assume that all the money available for investment is invested.)

39. **DIET PLANNING** A dietitian wishes to plan a meal around three foods. The percentages of the daily requirements of proteins, carbohydrates, and iron contained in each ounce of the three foods are summarized in the following table:

	Food I	Food II	Food III
Proteins (%)	10	6	8
Carbohydrates (%)	10	12	6
Iron (%)	5	4	12

Determine how many ounces of each food the dietitian should include in the meal to meet exactly the daily requirement of proteins, carbohydrates, and iron (100% of each).

40. **ASSET ALLOCATION** Mr. and Mrs. Garcia have a total of $100,000 to be invested in stocks, bonds, and a money market account. The stocks have a rate of return of 12%/year, while the bonds and the money market account pay 8%/year and 4%/year, respectively. The Garcias have stipulated that the amount invested in the money market account should be equal to the sum of 20% of the amount invested in stocks and 10% of the amount invested in bonds. How should the Garcias allocate their resources if they require an annual income of $10,000 from their investments?

41. **BOX-OFFICE RECEIPTS** For the opening night at the Opera House, a total of 1000 tickets were sold. Front orchestra seats cost $80 apiece, rear orchestra seats cost $60 apiece, and front balcony seats cost $50 apiece. The combined number of tickets sold for the front orchestra and rear orchestra exceeded twice the number of front balcony tickets sold by 400. The total receipts for the performance were $62,800. Determine how many tickets of each type were sold.

42. **PRODUCTION SCHEDULING** A manufacturer of women's blouses makes three types of blouses: sleeveless, short-sleeve, and long-sleeve. The time (in minutes) required by each department to produce a dozen blouses of each type is shown in the following table:

	Sleeveless	Short-sleeve	Long-sleeve
Cutting	9	12	15
Sewing	22	24	28
Packaging	6	8	8

The cutting, sewing, and packaging departments have available a maximum of 80, 160, and 48 labor-hours, respectively, per day. How many dozens of each type of blouse can be produced each day if the plant is operated at full capacity?

43. **BUSINESS TRAVEL EXPENSES** An executive of Trident Communications recently traveled to London, Paris, and Rome. He paid $280, $330, and $260 per night for lodging in London, Paris, and Rome, respectively, and his hotel bills totaled $4060. He spent $130, $140, and $110 per day for his meals in London, Paris, and Rome, respectively, and his expenses for meals totaled $1800. If he spent as many days in London as he did in Paris and Rome combined, how many days did he stay in each city?

44. VACATION COSTS Joan and Dick spent 2 weeks (14 nights) touring four cities on the East Coast—Boston, New York, Philadelphia, and Washington. They paid $240, $400, $160, and $200 per night for lodging in each city, respectively, and their total hotel bill came to $4040. The number of days they spent in New York was the same as the total number of days they spent in Boston and Washington, and the couple spent 3 times as many days in New York as they did in Philadelphia. How many days did Joan and Dick stay in each city?

In Exercises 45–48, determine whether the statement is true or false. If it is true, explain why it is true. If it is false, give an example to show why it is false.

45. A system composed of two linear equations must have at least one solution if the straight lines represented by these equations are nonparallel.

46. Suppose the straight lines represented by a system of three linear equations in two variables are parallel to each other. Then the system has no solution, or it has infinitely many solutions.

47. If at least two of the three lines represented by a system of three linear equations in two variables are parallel, then the system has no solution.

48. If at least two of the four lines represented by a system of four linear equations in two variables are parallel and distinct, then the system has no solution.

5.1 Solutions to Self-Check Exercises

1. Solving the first equation for y in terms of x, we obtain

$$-3y = -2x + 12$$

$$y = \frac{2}{3}x - 4$$

Next, substituting this result into the second equation of the system, we find

$$x + 2\left(\frac{2}{3}x - 4\right) = 6$$

$$x + \frac{4}{3}x - 8 = 6$$

$$\frac{7}{3}x = 14$$

$$x = 6$$

Substituting this value of x into the expression for y obtained earlier, we have

$$y = \frac{2}{3}(6) - 4 = 0$$

Therefore, the system has the unique solution $x = 6$ and $y = 0$. Both lines are shown in the accompanying figure.

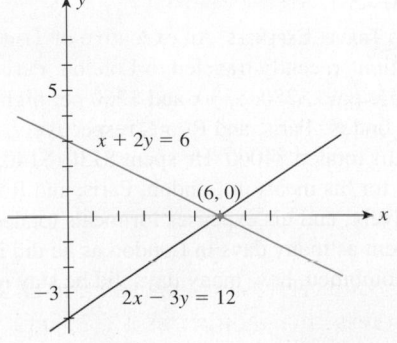

2. Let x, y, and z denote the number of acres of Crop A, Crop B, and Crop C, respectively, to be cultivated. Then the condition that all the cultivatable land be used translates into the equation

$$x + y + z = 200$$

Next, the total cost incurred in cultivating all three crops is $40x + 60y + 80z$ dollars, and since the entire budget is to be expended, we have

$$40x + 60y + 80z = 12,600$$

Finally, the amount of labor required to cultivate all three crops is $20x + 25y + 40z$ hours, and since all the available labor is to be used, we have

$$20x + 25y + 40z = 5950$$

Thus, the solution is found by solving the following system of linear equations:

$$\begin{aligned} x + y + z &= 200 \\ 40x + 60y + 80z &= 12{,}600 \\ 20x + 25y + 40z &= 5{,}950 \end{aligned}$$

5.2 Systems of Linear Equations: Unique Solutions

The Method of Elimination

The method of substitution used in Section 5.1 is well suited to solving a system of linear equations when the number of linear equations and variables is small. But for large systems, the steps involved in the procedure become difficult to manage.

The method of elimination is a suitable technique for solving systems of linear equations of any size. One advantage of this technique is its adaptability to the computer. This method involves a sequence of operations on a system of linear equations to obtain at each stage an **equivalent system**—that is, a system having the same solution as the original system. The reduction is complete when the original system has been transformed so that it is in a certain standard form from which the solution can be easily read.

The operations of the method of elimination are as follows:

1. Interchange any two equations.
2. Replace an equation by a nonzero constant multiple of itself.
3. Replace an equation by the sum of that equation and a constant multiple of any other equation.

To illustrate the method of elimination for solving systems of linear equations, let's apply it to the solution of the following system:

$$2x + 4y = 8$$
$$3x - 2y = 4$$

We begin by working with the first, or x, column. First, we transform the system into an equivalent system in which the coefficient of x in the first equation is 1:

$$2x + 4y = 8$$
$$3x - 2y = 4 \tag{3a}$$

$$x + 2y = 4$$
$$3x - 2y = 4 \tag{3b}$$

Multiply the first equation in System (3a) by $\frac{1}{2}$ (operation 2).

Next, we eliminate x from the second equation:

$$x + 2y = 4$$
$$-8y = -8 \tag{3c}$$

Replace the second equation in System (3b) by the sum of $-3 \times$ the first equation and the second equation (operation 3):

$$-3x - 6y = -12$$
$$\underline{3x - 2y = 4}$$
$$-8y = -8$$

Then we obtain the following equivalent system, in which the coefficient of y in the second equation is 1:

$$x + 2y = 4$$
$$y = 1 \tag{3d}$$

Multiply the second equation in System (3c) by $-\frac{1}{8}$ (operation 2).

Next, we eliminate y in the first equation:

$$x = 2$$
$$y = 1$$

Replace the first equation in System (3d) by the sum of $-2 \times$ the second equation and the first equation (operation 3):

$$x + 2y = 4$$
$$\underline{ - 2y = -2}$$
$$x = 2$$

This system is now in standard form, and we can read off the solution to System (3a) as $x = 2$ and $y = 1$. We can also express this solution as $(2, 1)$ and interpret it geometrically as the point of intersection of the two lines represented by the two linear equations that make up the given system of equations.

The next example involves a system of three linear equations and three variables.

EXAMPLE 1 Solve the following system of equations:

$$
\begin{aligned}
2x + 4y + 6z &= 22 \\
3x + 8y + 5z &= 27 \\
-x + y + 2z &= 2
\end{aligned}
$$

Solution First, we transform this system into an equivalent system in which the coefficient of x in the first equation is 1:

$$
\begin{aligned}
2x + 4y + 6z &= 22 \\
3x + 8y + 5z &= 27 \\
-x + y + 2z &= 2
\end{aligned}
\tag{4a}
$$

$$
\begin{aligned}
x + 2y + 3z &= 11 \\
3x + 8y + 5z &= 27 \\
-x + y + 2z &= 2
\end{aligned}
\qquad
\begin{array}{l}
\text{Multiply the first equation in} \\
\text{System (4a) by } \tfrac{1}{2}.
\end{array}
\tag{4b}
$$

Next, we eliminate the variable x from all equations except the first:

$$
\begin{aligned}
x + 2y + 3z &= 11 \\
2y - 4z &= -6 \\
-x + y + 2z &= 2
\end{aligned}
\qquad
\begin{array}{l}
\text{Replace the second equation in System (4b)} \\
\text{by the sum of } (-3) \times \text{ the first equation} \\
\text{and the second equation:} \\[4pt]
\begin{aligned}
-3x - 6y - 9z &= -33 \\
3x + 8y + 5z &= 27 \\
\hline
2y - 4z &= -6
\end{aligned}
\end{array}
\tag{4c}
$$

$$
\begin{aligned}
x + 2y + 3z &= 11 \\
2y - 4z &= -6 \\
3y + 5z &= 13
\end{aligned}
\qquad
\begin{array}{l}
\text{Replace the third equation in System (4c)} \\
\text{by the sum of the first equation and the} \\
\text{third equation:} \\[4pt]
\begin{aligned}
x + 2y + 3z &= 11 \\
-x + y + 2z &= 2 \\
\hline
3y + 5z &= 13
\end{aligned}
\end{array}
\tag{4d}
$$

Then we transform System (4d) into yet another equivalent system, in which the coefficient of y in the second equation is 1:

$$
\begin{aligned}
x + 2y + 3z &= 11 \\
y - 2z &= -3 \\
3y + 5z &= 13
\end{aligned}
\qquad
\begin{array}{l}
\text{Multiply the second equation} \\
\text{in System (4d) by } \tfrac{1}{2}.
\end{array}
\tag{4e}
$$

We now eliminate y from all equations except the second, using operation 3 of the elimination method:

$$
\begin{aligned}
x \qquad + 7z &= 17 \\
y - 2z &= -3 \\
3y + 5z &= 13
\end{aligned}
\qquad
\begin{array}{l}
\text{Replace the first equation in System (4e) by} \\
\text{the sum of the first equation and } (-2) \times \\
\text{the second equation:} \\[4pt]
\begin{aligned}
x + 2y + 3z &= 11 \\
-2y + 4z &= 6 \\
\hline
x \qquad + 7z &= 17
\end{aligned}
\end{array}
\tag{4f}
$$

$$
\begin{aligned}
x \qquad + 7z &= 17 \\
y - 2z &= -3 \\
11z &= 22
\end{aligned}
\qquad
\begin{array}{l}
\text{Replace the third equation in System (4f)} \\
\text{by the sum of } (-3) \times \text{ the second equation} \\
\text{and the third equation:} \\[4pt]
\begin{aligned}
-3y + 6z &= 9 \\
3y + 5z &= 13 \\
\hline
11z &= 22
\end{aligned}
\end{array}
\tag{4g}
$$

Multiplying the third equation by $\frac{1}{11}$ in (4g) leads to the system

$$
\begin{aligned}
x \quad\quad + 7z &= 17 \\
y - 2z &= -3 \\
z &= 2
\end{aligned}
$$

Eliminating z from all equations except the third (try it!) then leads to the system

$$
\begin{aligned}
x \quad\quad &= 3 \\
y \quad\quad &= 1 \\
z &= 2
\end{aligned}
\tag{4h}
$$

In its final form, the solution to the given system of equations can be easily read off! We have $x = 3$, $y = 1$, and $z = 2$. Geometrically, the point $(3, 1, 2)$ is the intersection of the three planes described by the three equations comprising the given system. ■

Augmented Matrices

Observe from the preceding example that the variables x, y, and z play no significant role in each step of the reduction process, except as a reminder of the position of each coefficient in the system. With the aid of **matrices,** which are rectangular arrays of numbers, we can eliminate writing the variables at each step of the reduction and thus save ourselves a great deal of work. For example, the system

$$
\begin{aligned}
2x + 4y + 6z &= 22 \\
3x + 8y + 5z &= 27 \\
-x + \ y + 2z &= 2
\end{aligned}
\tag{5}
$$

may be represented by the matrix

$$
\begin{bmatrix}
2 & 4 & 6 & 22 \\
3 & 8 & 5 & 27 \\
-1 & 1 & 2 & 2
\end{bmatrix}
\tag{6}
$$

The augmented matrix representing System (5)

The submatrix consisting of the first three columns of Matrix (6) is called the **coefficient matrix** of System (5). The matrix itself, Matrix (6), is referred to as the **augmented matrix** of System (5), since it is obtained by joining the matrix of coefficients to the column (matrix) of constants. The vertical line separates the column of constants from the matrix of coefficients.

The next example shows how much work you can save by using matrices instead of the standard representation of the systems of linear equations.

EXAMPLE 2 Write the augmented matrix corresponding to each equivalent system given in Systems (4a) through (4h).

Solution   The required sequence of augmented matrices follows.

Equivalent System	**Augmented Matrix**

a. $\begin{aligned} 2x + 4y + 6z &= 22 \\ 3x + 8y + 5z &= 27 \\ -x + \ y + 2z &= 2 \end{aligned}$ $\begin{bmatrix} 2 & 4 & 6 & 22 \\ 3 & 8 & 5 & 27 \\ -1 & 1 & 2 & 2 \end{bmatrix}$ (7a)

b. $\begin{aligned} x + 2y + 3z &= 11 \\ 3x + 8y + 5z &= 27 \\ -x + \ y + 2z &= 2 \end{aligned}$ $\begin{bmatrix} 1 & 2 & 3 & 11 \\ 3 & 8 & 5 & 27 \\ -1 & 1 & 2 & 2 \end{bmatrix}$ (7b)

c.
$$\begin{aligned} x + 2y + 3z &= 11 \\ 2y - 4z &= -6 \\ -x + y + 2z &= 2 \end{aligned}$$
$$\left[\begin{array}{ccc|c} 1 & 2 & 3 & 11 \\ 0 & 2 & -4 & -6 \\ -1 & 1 & 2 & 2 \end{array}\right]$$
(7c)

d.
$$\begin{aligned} x + 2y + 3z &= 11 \\ 2y - 4z &= -6 \\ 3y + 5z &= 13 \end{aligned}$$
$$\left[\begin{array}{ccc|c} 1 & 2 & 3 & 11 \\ 0 & 2 & -4 & -6 \\ 0 & 3 & 5 & 13 \end{array}\right]$$
(7d)

e.
$$\begin{aligned} x + 2y + 3z &= 11 \\ y - 2z &= -3 \\ 3y + 5z &= 13 \end{aligned}$$
$$\left[\begin{array}{ccc|c} 1 & 2 & 3 & 11 \\ 0 & 1 & -2 & -3 \\ 0 & 3 & 5 & 13 \end{array}\right]$$
(7e)

f.
$$\begin{aligned} x + 7z &= 17 \\ y - 2z &= -3 \\ 3y + 5z &= 13 \end{aligned}$$
$$\left[\begin{array}{ccc|c} 1 & 0 & 7 & 17 \\ 0 & 1 & -2 & -3 \\ 0 & 3 & 5 & 13 \end{array}\right]$$
(7f)

g.
$$\begin{aligned} x + 7z &= 17 \\ y - 2z &= -3 \\ 11z &= 22 \end{aligned}$$
$$\left[\begin{array}{ccc|c} 1 & 0 & 7 & 17 \\ 0 & 1 & -2 & -3 \\ 0 & 0 & 11 & 22 \end{array}\right]$$
(7g)

h.
$$\begin{aligned} x &= 3 \\ y &= 1 \\ z &= 2 \end{aligned}$$
$$\left[\begin{array}{ccc|c} 1 & 0 & 0 & 3 \\ 0 & 1 & 0 & 1 \\ 0 & 0 & 1 & 2 \end{array}\right]$$
(7h) ∎

The augmented matrix in (7h) is an example of a matrix in row-reduced form. In general, an augmented matrix with m rows and n columns (called an $m \times n$ matrix) is in **row-reduced form** if it satisfies the following conditions.

> **Row-Reduced Form of a Matrix**
> 1. Each row consisting entirely of zeros lies below all rows having nonzero entries.
> 2. The first nonzero entry in each (nonzero) row is 1 (called a **leading 1**).
> 3. In any two successive (nonzero) rows, the leading 1 in the lower row lies to the right of the leading 1 in the upper row.
> 4. If a column in the coefficient matrix contains a leading 1, then the other entries in that column are zeros.

EXAMPLE 3 Determine which of the following matrices are in row-reduced form. If a matrix is not in row-reduced form, state the condition that is violated.

a.
$$\left[\begin{array}{ccc|c} 1 & 0 & 0 & 0 \\ 0 & 1 & 0 & 0 \\ 0 & 0 & 1 & 3 \end{array}\right]$$
b.
$$\left[\begin{array}{ccc|c} 1 & 0 & 0 & 4 \\ 0 & 1 & 0 & 3 \\ 0 & 0 & 0 & 0 \end{array}\right]$$
c.
$$\left[\begin{array}{ccc|c} 1 & 2 & 0 & 0 \\ 0 & 0 & 1 & 0 \\ 0 & 0 & 0 & 1 \end{array}\right]$$

d.
$$\left[\begin{array}{ccc|c} 0 & 1 & 2 & -2 \\ 1 & 0 & 0 & 3 \\ 0 & 0 & 1 & 2 \end{array}\right]$$
e.
$$\left[\begin{array}{ccc|c} 1 & 2 & 0 & 0 \\ 0 & 0 & 1 & 3 \\ 0 & 0 & 2 & 1 \end{array}\right]$$
f.
$$\left[\begin{array}{cc|c} 1 & 0 & 4 \\ 0 & 3 & 0 \\ 0 & 0 & 0 \end{array}\right]$$

g.
$$\left[\begin{array}{ccc|c} 0 & 0 & 0 & 0 \\ 1 & 0 & 0 & 3 \\ 0 & 1 & 0 & 2 \end{array}\right]$$

Solution The matrices in parts (a)–(c) are in row-reduced form.

d. This matrix is not in row-reduced form. Conditions 3 and 4 are violated: The leading 1 in row 2 lies to the left of the leading 1 in row 1. Also, column 3 contains a leading 1 in row 3 and a nonzero element above it.

e. This matrix is not in row-reduced form. Conditions 2 and 4 are violated: The first nonzero entry in row 3 is a 2, not a 1. Also, column 3 contains a leading 1 and has a nonzero entry below it.

f. This matrix is not in row-reduced form. Condition 2 is violated: The first nonzero entry in row 2 is not a leading 1.

g. This matrix is not in row-reduced form. Condition 1 is violated: Row 1 consists of all zeros and does not lie below the nonzero rows.

The Gauss–Jordan Method

The foregoing discussion suggests the following method, called the **Gauss–Jordan elimination method,** for solving systems of linear equations using matrices. The three operations on the equations of a system (see page 275) translate into the following **row operations** on the corresponding augmented matrices.

> **Row Operations**
> **1.** Interchange any two rows.
> **2.** Replace any row by a nonzero constant multiple of itself.
> **3.** Replace any row by the sum of that row and a constant multiple of any other row.

We obtained the augmented matrices in Example 2 by using the same operations that we used on the equivalent system of equations in Example 1.

To help us describe the Gauss–Jordan elimination method using matrices, let's introduce some terminology. We begin by defining what is meant by a **unit column.**

> **Unit Column**
> A column in a coefficient matrix is called a **unit column** if one of the entries in the column is a 1 and the other entries are zeros.

For example, in the coefficient matrix of (7d) in Example 2, page 278, only the first column is in unit form; in the coefficient matrix of (7h), all three columns are in unit form. Now, the sequence of row operations that transforms the augmented matrix (7a) into the equivalent matrix (7d) in which the first column

$$\begin{matrix} 2 \\ 3 \\ -1 \end{matrix}$$

of (7a) is transformed into the unit column

$$\begin{matrix} 1 \\ 0 \\ 0 \end{matrix}$$

is called **pivoting** the matrix about the element (number) 2. Similarly, we have pivoted about the element 2 in the second column of (7d), shown circled,

$$2$$
$$②$$
$$3$$

to obtain the augmented matrix (7g), in which the second column

$$0$$
$$1$$
$$0$$

is a unit column. Finally, pivoting about the element 11 in column 3 of (7g)

$$7$$
$$-2$$
$$⑪$$

leads to the augmented matrix (7h), in which the third column

$$0$$
$$0$$
$$1$$

is a unit column. Observe that in the final augmented matrix, all three columns to the left of the vertical line are in unit form. The element about which a matrix is pivoted is called the **pivot element.**

Before looking at the next example, let's introduce the following notation for the three types of row operations.

Notation for Row Operations

Letting R_i denote the ith row of a matrix, we write:

Operation 1 $R_i \leftrightarrow R_j$ to mean: Interchange row i with row j.

Operation 2 cR_i to mean: Replace row i with c times row i.

Operation 3 $R_i + aR_j$ to mean: Replace row i with the sum of row i and a times row j.

EXAMPLE 4 Pivot the matrix about the circled element.

$$\begin{bmatrix} ③ & 5 & | & 9 \\ 2 & 3 & | & 5 \end{bmatrix}$$

Solution   We need a **1** in row 1 where the pivot element (the circled **3**) is. One way of doing this is to replace row 1 by $\frac{1}{3}$ times R_1. In other words, we use operation 2. Thus,

$$\begin{bmatrix} 3 & 5 & | & 9 \\ 2 & 3 & | & 5 \end{bmatrix} \xrightarrow{\frac{1}{3}R_1} \begin{bmatrix} 1 & \frac{5}{3} & | & 3 \\ 2 & 3 & | & 5 \end{bmatrix}$$

Next, we need to replace row 2 by a row with a **0** in the position that is currently occupied by the number **2**. This can be accomplished by replacing row 2 by the sum of row 2 and -2 times row 1. In other words, we use operation 3. Thus,

$$\begin{bmatrix} 1 & \frac{5}{3} & | & 3 \\ 2 & 3 & | & 5 \end{bmatrix} \xrightarrow{R_2 - 2R_1} \begin{bmatrix} 1 & \frac{5}{3} & | & 3 \\ 0 & -\frac{1}{3} & | & -1 \end{bmatrix}$$

Putting these two steps together, we can write the required operations as follows:

$$\begin{bmatrix} 3 & 5 & | & 9 \\ 2 & 3 & | & 5 \end{bmatrix} \xrightarrow{\frac{1}{3}R_1} \begin{bmatrix} 1 & \frac{5}{3} & | & 3 \\ 2 & 3 & | & 5 \end{bmatrix} \xrightarrow{R_2 - 2R_1} \begin{bmatrix} 1 & \frac{5}{3} & | & 3 \\ 0 & -\frac{1}{3} & | & -1 \end{bmatrix}$$

The first column, which originally contained the entry 3, is now in unit form, with a 1 where the pivot element used to be, and we are done.

Alternative Solution In the first solution, we used operation 2 to obtain a 1 where the pivot element was originally. Alternatively, we can use operation 3 as follows:

$$\begin{bmatrix} 3 & 5 & | & 9 \\ 2 & 3 & | & 5 \end{bmatrix} \xrightarrow{R_1 - R_2} \begin{bmatrix} 1 & 2 & | & 4 \\ 2 & 3 & | & 5 \end{bmatrix} \xrightarrow{R_2 - 2R_1} \begin{bmatrix} 1 & 2 & | & 4 \\ 0 & -1 & | & -3 \end{bmatrix} \quad \blacksquare$$

Note In Example 4, the two matrices

$$\begin{bmatrix} 1 & \frac{5}{3} & | & 3 \\ 0 & -\frac{1}{3} & | & -1 \end{bmatrix} \quad \text{and} \quad \begin{bmatrix} 1 & 2 & | & 4 \\ 0 & -1 & | & -3 \end{bmatrix}$$

look quite different, but they are in fact equivalent. You can verify this by observing that they represent the systems of equations

$$x + \frac{5}{3}y = 3 \qquad\qquad x + 2y = 4$$

and

$$-\frac{1}{3}y = -1 \qquad\qquad -y = -3$$

respectively, and both have the same solution: $x = -2$ and $y = 3$. Example 4 also shows that we can sometimes avoid working with fractions by using an appropriate row operation. $\blacksquare$

A summary of the Gauss–Jordan elimination method follows.

> **The Gauss–Jordan Elimination Method**
> 1. Write the augmented matrix corresponding to the linear system.
> 2. Interchange rows (operation 1), if necessary, to obtain an augmented matrix in which the first entry in the first row is nonzero. Then pivot the matrix about this entry.
> 3. Interchange the second row with any row below it, if necessary, to obtain an augmented matrix in which the second entry in the second row is nonzero. Pivot the matrix about this entry.
> 4. Continue until the final matrix is in row-reduced form.

⚠ Before writing the augmented matrix, be sure to write all equations with the variables on the left and constant terms on the right of the equal sign. Also, make sure that the variables are in the same order in all equations.

EXAMPLE 5 Solve the system of linear equations given by

$$\begin{aligned} 3x - 2y + 8z &= 9 \\ -2x + 2y + z &= 3 \\ x + 2y - 3z &= 8 \end{aligned} \tag{8}$$

Solution Using the Gauss–Jordan elimination method, we obtain the following sequence of equivalent augmented matrices:

$$\left[\begin{array}{ccc|c} ③ & -2 & 8 & 9 \\ -2 & 2 & 1 & 3 \\ 1 & 2 & -3 & 8 \end{array}\right] \xrightarrow{R_1 + R_2} \left[\begin{array}{ccc|c} 1 & 0 & 9 & 12 \\ -2 & 2 & 1 & 3 \\ 1 & 2 & -3 & 8 \end{array}\right]$$

$$\xrightarrow[R_3 - R_1]{R_2 + 2R_1} \left[\begin{array}{ccc|c} 1 & 0 & 9 & 12 \\ 0 & 2 & 19 & 27 \\ 0 & 2 & -12 & -4 \end{array}\right]$$

$$\xrightarrow{R_2 \leftrightarrow R_3} \left[\begin{array}{ccc|c} 1 & 0 & 9 & 12 \\ 0 & ② & -12 & -4 \\ 0 & 2 & 19 & 27 \end{array}\right]$$

$$\xrightarrow{\frac{1}{2}R_2} \left[\begin{array}{ccc|c} 1 & 0 & 9 & 12 \\ 0 & 1 & -6 & -2 \\ 0 & 2 & 19 & 27 \end{array}\right]$$

$$\xrightarrow{R_3 - 2R_2} \left[\begin{array}{ccc|c} 1 & 0 & 9 & 12 \\ 0 & 1 & -6 & -2 \\ 0 & 0 & ㉛ & 31 \end{array}\right]$$

$$\xrightarrow{\frac{1}{31}R_3} \left[\begin{array}{ccc|c} 1 & 0 & 9 & 12 \\ 0 & 1 & -6 & -2 \\ 0 & 0 & 1 & 1 \end{array}\right]$$

$$\xrightarrow[R_2 + 6R_3]{R_1 - 9R_3} \left[\begin{array}{ccc|c} 1 & 0 & 0 & 3 \\ 0 & 1 & 0 & 4 \\ 0 & 0 & 1 & 1 \end{array}\right]$$

The solution to System (8) is given by $x = 3$, $y = 4$, and $z = 1$. This may be verified by substitution into System (8) as follows:

$$3(3) - 2(4) + 8(1) = 9 \quad \checkmark$$
$$-2(3) + 2(4) + \quad 1 = 3 \quad \checkmark$$
$$3 \quad + 2(4) - 3(1) = 8 \quad \checkmark$$

■

When you are searching for an element to serve as a pivot, it is important to keep in mind that you may work only with the row containing the potential pivot or any row *below* it. To see what can go wrong if this caution is not heeded, consider the following augmented matrix for some linear system:

$$\left[\begin{array}{ccc|c} 1 & 1 & 2 & 3 \\ 0 & 0 & 3 & 1 \\ 0 & 2 & 1 & -2 \end{array}\right]$$

Observe that column 1 is in unit form. The next step in the Gauss–Jordan elimination procedure calls for obtaining a nonzero element in the second position of row 2. If we use row 1 (which is *above* the row under consideration) to help obtain the pivot, we might proceed as follows:

$$\left[\begin{array}{ccc|c} 1 & 1 & 2 & 3 \\ 0 & 0 & 3 & 1 \\ 0 & 2 & 1 & -2 \end{array}\right] \xrightarrow{R_2 \leftrightarrow R_1} \left[\begin{array}{ccc|c} 0 & 0 & 3 & 1 \\ 1 & 1 & 2 & 3 \\ 0 & 2 & 1 & -2 \end{array}\right]$$

As you can see, not only have we obtained a nonzero element to serve as the next pivot, but it is already a 1, thus obviating the next step. This seems like a good move. But beware—we have undone some of our earlier work: Column 1 is no longer a unit column in which a 1 appears first. The correct move in this case is to interchange row 2 with row 3 in the first augmented matrix.

> *Explore and Discuss*
>
> 1. Can the phrase "a nonzero constant multiple of itself" in a type 2 row operation be replaced by "a constant multiple of itself"? Explain.
> 2. Can a row of an augmented matrix be replaced by a row obtained by adding a constant to every element in that row without changing the solution of the system of linear equations? Explain.

The next example illustrates how to handle a situation in which the first entry in row 1 of the augmented matrix is zero.

EXAMPLE 6 Solve the system of linear equations given by

$$\begin{aligned} 2y + 3z &= 7 \\ 3x + 6y - 12z &= -3 \\ 5x - 2y + 2z &= -7 \end{aligned}$$

Solution Using the Gauss–Jordan elimination method, we obtain the following sequence of equivalent augmented matrices:

$$\begin{bmatrix} 0 & 2 & 3 & | & 7 \\ 3 & 6 & -12 & | & -3 \\ 5 & -2 & 2 & | & -7 \end{bmatrix} \xrightarrow{R_1 \leftrightarrow R_2} \begin{bmatrix} \boxed{3} & 6 & -12 & | & -3 \\ 0 & 2 & 3 & | & 7 \\ 5 & -2 & 2 & | & -7 \end{bmatrix}$$

$$\xrightarrow{\frac{1}{3}R_1} \begin{bmatrix} 1 & 2 & -4 & | & -1 \\ 0 & 2 & 3 & | & 7 \\ 5 & -2 & 2 & | & -7 \end{bmatrix}$$

$$\xrightarrow{R_3 - 5R_1} \begin{bmatrix} 1 & 2 & -4 & | & -1 \\ 0 & \boxed{2} & 3 & | & 7 \\ 0 & -12 & 22 & | & -2 \end{bmatrix}$$

$$\xrightarrow{\frac{1}{2}R_2} \begin{bmatrix} 1 & 2 & -4 & | & -1 \\ 0 & 1 & \frac{3}{2} & | & \frac{7}{2} \\ 0 & -12 & 22 & | & -2 \end{bmatrix}$$

$$\xrightarrow[R_3 + 12R_2]{R_1 - 2R_2} \begin{bmatrix} 1 & 0 & -7 & | & -8 \\ 0 & 1 & \frac{3}{2} & | & \frac{7}{2} \\ 0 & 0 & \boxed{40} & | & 40 \end{bmatrix}$$

$$\xrightarrow{\frac{1}{40}R_3} \begin{bmatrix} 1 & 0 & -7 & | & -8 \\ 0 & 1 & \frac{3}{2} & | & \frac{7}{2} \\ 0 & 0 & 1 & | & 1 \end{bmatrix}$$

$$\xrightarrow[R_2 - \frac{3}{2}R_3]{R_1 + 7R_3} \begin{bmatrix} 1 & 0 & 0 & | & -1 \\ 0 & 1 & 0 & | & 2 \\ 0 & 0 & 1 & | & 1 \end{bmatrix}$$

The solution to the system is given by $x = -1$, $y = 2$, and $z = 1$; this may be verified by substituting these values into each equation of the system.

APPLIED EXAMPLE 7 Production Scheduling Complete the solution to Example 1 in Section 5.1, page 269.

Solution To complete the solution of the problem posed in Example 1, recall that the mathematical formulation of the problem led to the following system of linear equations:

$$2x + y + z = 180$$
$$x + 3y + 2z = 300$$
$$2x + y + 2z = 240$$

where x, y, and z denote the respective numbers of Type A, Type B, and Type C souvenirs to be made.

Solving the foregoing system of linear equations by the Gauss–Jordan elimination method, we obtain the following sequence of equivalent augmented matrices:

$$\begin{bmatrix} 2 & 1 & 1 & | & 180 \\ 1 & 3 & 2 & | & 300 \\ 2 & 1 & 2 & | & 240 \end{bmatrix} \xrightarrow{R_1 \leftrightarrow R_2} \begin{bmatrix} \textcircled{1} & 3 & 2 & | & 300 \\ 2 & 1 & 1 & | & 180 \\ 2 & 1 & 2 & | & 240 \end{bmatrix}$$

$$\xrightarrow[R_3 - 2R_1]{R_2 - 2R_1} \begin{bmatrix} 1 & 3 & 2 & | & 300 \\ 0 & \textcircled{-5} & -3 & | & -420 \\ 0 & -5 & -2 & | & -360 \end{bmatrix}$$

$$\xrightarrow{-\frac{1}{5}R_2} \begin{bmatrix} 1 & 3 & 2 & | & 300 \\ 0 & 1 & \frac{3}{5} & | & 84 \\ 0 & -5 & -2 & | & -360 \end{bmatrix}$$

$$\xrightarrow[R_3 + 5R_2]{R_1 - 3R_2} \begin{bmatrix} 1 & 0 & \frac{1}{5} & | & 48 \\ 0 & 1 & \frac{3}{5} & | & 84 \\ 0 & 0 & \textcircled{1} & | & 60 \end{bmatrix}$$

$$\xrightarrow[R_2 - \frac{3}{5}R_3]{R_1 - \frac{1}{5}R_3} \begin{bmatrix} 1 & 0 & 0 & | & 36 \\ 0 & 1 & 0 & | & 48 \\ 0 & 0 & 1 & | & 60 \end{bmatrix}$$

Thus, $x = 36$, $y = 48$, and $z = 60$; that is, Ace Novelty should make 36 Type A souvenirs, 48 Type B souvenirs, and 60 Type C souvenirs in order to use all available machine time. ∎

5.2 Self-Check Exercises

1. Solve the system of linear equations

$$2x + 3y + z = 6$$
$$x - 2y + 3z = -3$$
$$3x + 2y - 4z = 12$$

using the Gauss–Jordan elimination method.

2. **CROP PLANNING** A farmer has 200 acres of land suitable for cultivating Crops A, B, and C. The cost per acre of cultivating Crop A, Crop B, and Crop C is $40, $60, and $80, respectively. The farmer has $12,600 available for land cultivation. Each acre of Crop A requires 20 labor-hours, each acre of Crop B requires 25 labor-hours, and each acre of Crop C requires 40 labor-hours. The farmer has a maximum of 5950 labor-hours available. If she wishes to use all of her cultivatable land, the entire budget, and all the labor available, how many acres of each crop should she plant?

Solutions to Self-Check Exercises 5.2 can be found on page 289.

5.2 Concept Questions

1. a. Explain what it means for two systems of linear equations to be equivalent to each other.
b. Give the meaning of the following notation used for row operations in the Gauss–Jordan elimination method:

   i. $R_i \leftrightarrow R_j$ **ii.** cR_i **iii.** $R_i + aR_j$

2. a. What is an augmented matrix? A coefficient matrix? A unit column?
b. Explain what is meant by a pivot operation.

3. Suppose that a matrix is in row-reduced form.
a. What is the position of a row consisting entirely of zeros relative to the nonzero rows?
b. What is the first nonzero entry in each row?
c. What is the position of the leading 1s in successive nonzero rows?
d. If a column contains a leading 1, then what is the value of the other entries in that column?

5.2 Exercises

In Exercises 1–4, write the augmented matrix corresponding to each system of equations.

1. $\begin{aligned} 2x - 3y &= 7 \\ 3x + y &= 4 \end{aligned}$

2. $\begin{aligned} 3x + 7y - 8z &= 5 \\ x \quad\quad + 3z &= -2 \\ 4x - 3y \quad\quad &= 7 \end{aligned}$

3. $\begin{aligned} -y + 2z &= 5 \\ 2x + 2y - 8z &= 4 \\ 3y + 4z &= 0 \end{aligned}$

4. $\begin{aligned} 3x_1 + 2x_2 \quad\quad &= 0 \\ x_1 - x_2 + 2x_3 &= 4 \\ 2x_2 - 3x_3 &= 5 \end{aligned}$

In Exercises 5–8, write the system of equations corresponding to each augmented matrix.

5. $\begin{bmatrix} 3 & 2 & | & -4 \\ 1 & -1 & | & 5 \end{bmatrix}$

6. $\begin{bmatrix} 0 & 3 & 2 & | & 4 \\ 1 & -1 & -2 & | & -3 \\ 4 & 0 & 3 & | & 2 \end{bmatrix}$

7. $\begin{bmatrix} 1 & 3 & 2 & | & 4 \\ 2 & 0 & 0 & | & 5 \\ 3 & -3 & 2 & | & 6 \end{bmatrix}$

8. $\begin{bmatrix} 2 & 3 & 1 & | & 6 \\ 4 & 3 & 2 & | & 5 \\ 0 & 0 & 0 & | & 0 \end{bmatrix}$

In Exercises 9–18, indicate whether the matrix is in row-reduced form.

9. $\begin{bmatrix} 1 & 0 & | & 3 \\ 0 & 1 & | & -2 \end{bmatrix}$

10. $\begin{bmatrix} 1 & 1 & | & 3 \\ 0 & 0 & | & 0 \end{bmatrix}$

11. $\begin{bmatrix} 0 & 1 & | & 3 \\ 1 & 0 & | & 5 \end{bmatrix}$

12. $\begin{bmatrix} 0 & 1 & | & 3 \\ 0 & 0 & | & 5 \end{bmatrix}$

13. $\begin{bmatrix} 1 & 0 & 0 & | & 3 \\ 0 & 1 & 0 & | & 4 \\ 0 & 0 & 1 & | & 5 \end{bmatrix}$

14. $\begin{bmatrix} 1 & 0 & 0 & | & -1 \\ 0 & 1 & 0 & | & -2 \\ 0 & 0 & 2 & | & -3 \end{bmatrix}$

15. $\begin{bmatrix} 1 & 0 & 1 & | & 3 \\ 0 & 1 & 0 & | & 4 \\ 0 & 0 & -1 & | & 6 \end{bmatrix}$

16. $\begin{bmatrix} 1 & 0 & | & -10 \\ 0 & 1 & | & 2 \\ 0 & 0 & | & 0 \end{bmatrix}$

17. $\begin{bmatrix} 0 & 0 & 0 & | & 0 \\ 0 & 1 & 2 & | & 4 \\ 0 & 0 & 0 & | & 0 \end{bmatrix}$

18. $\begin{bmatrix} 1 & 0 & 0 & | & 3 \\ 0 & 1 & 0 & | & 6 \\ 0 & 0 & 0 & | & 4 \\ 0 & 0 & 1 & | & 5 \end{bmatrix}$

In Exercises 19–26, pivot the system about the circled element.

19. $\begin{bmatrix} ① & 3 & | & 4 \\ 2 & 4 & | & 6 \end{bmatrix}$

20. $\begin{bmatrix} ② & 4 & | & 8 \\ 3 & 1 & | & 2 \end{bmatrix}$

21. $\begin{bmatrix} ⟨-1⟩ & 2 & | & 3 \\ 6 & 8 & | & 2 \end{bmatrix}$

22. $\begin{bmatrix} 3 & 2 & | & 6 \\ ④ & 2 & | & 5 \end{bmatrix}$

23. $\begin{bmatrix} ② & 4 & 6 & | & 12 \\ 2 & 3 & 1 & | & 5 \\ 3 & -1 & 2 & | & 4 \end{bmatrix}$

24. $\begin{bmatrix} 1 & 3 & 2 & | & 4 \\ ② & 4 & 8 & | & 6 \\ -1 & 2 & 3 & | & 4 \end{bmatrix}$

25. $\begin{bmatrix} 0 & 1 & 3 & | & 4 \\ 2 & 4 & ① & | & 3 \\ 5 & 6 & 2 & | & -4 \end{bmatrix}$

26. $\begin{bmatrix} 1 & 2 & 3 & | & 5 \\ 0 & ⟨-3⟩ & 3 & | & 2 \\ 0 & 4 & -1 & | & 3 \end{bmatrix}$

In Exercises 27–30, fill in the missing entries by performing the indicated row operations to obtain the row-reduced matrices.

27. $\begin{bmatrix} 3 & 9 & | & 6 \\ 2 & 1 & | & 4 \end{bmatrix} \xrightarrow{\frac{1}{3}R_1} \begin{bmatrix} \cdot & \cdot & | & \cdot \\ 2 & 1 & | & 4 \end{bmatrix} \xrightarrow{R_2 - 2R_1}$

$\begin{bmatrix} 1 & 3 & | & 2 \\ \cdot & \cdot & | & \cdot \end{bmatrix} \xrightarrow{-\frac{1}{5}R_2} \begin{bmatrix} 1 & 3 & | & 2 \\ \cdot & \cdot & | & \cdot \end{bmatrix} \xrightarrow{R_1 - 3R_2} \begin{bmatrix} 1 & 0 & | & 2 \\ 0 & 1 & | & 0 \end{bmatrix}$

1 — 33 odd

28. $\begin{bmatrix} 1 & 2 & | & 1 \\ 2 & 3 & | & -1 \end{bmatrix} \xrightarrow{R_2 - 2R_1} \begin{bmatrix} 1 & 2 & | & 1 \\ \cdot & \cdot & | & \cdot \end{bmatrix} \xrightarrow{-R_2}$

$\begin{bmatrix} 1 & 2 & | & 1 \\ \cdot & \cdot & | & \cdot \end{bmatrix} \xrightarrow{R_1 - 2R_2} \begin{bmatrix} 1 & 0 & | & -5 \\ 0 & 1 & | & 3 \end{bmatrix}$

29. $\begin{bmatrix} 1 & 3 & 1 & | & 3 \\ 3 & 8 & 3 & | & 7 \\ 2 & -3 & 1 & | & -10 \end{bmatrix} \xrightarrow[R_3 - 2R_1]{R_2 - 3R_1} \begin{bmatrix} 1 & 3 & 1 & | & 3 \\ \cdot & \cdot & \cdot & | & \cdot \\ \cdot & \cdot & \cdot & | & \cdot \end{bmatrix} \xrightarrow{-R_2}$

$\begin{bmatrix} 1 & 3 & 1 & | & 3 \\ \cdot & \cdot & \cdot & | & \cdot \\ 0 & -9 & -1 & | & -16 \end{bmatrix} \xrightarrow[R_3 + 9R_2]{R_1 - 3R_2}$

$\begin{bmatrix} \cdot & \cdot & \cdot & | & \cdot \\ 0 & 1 & 0 & | & 2 \\ \cdot & \cdot & \cdot & | & \cdot \end{bmatrix} \xrightarrow[-R_3]{R_1 + R_3} \begin{bmatrix} 1 & 0 & 0 & | & -1 \\ 0 & 1 & 0 & | & 2 \\ 0 & 0 & 1 & | & -2 \end{bmatrix}$

30. $\begin{bmatrix} 0 & 1 & 3 & | & -4 \\ 1 & 2 & 1 & | & 7 \\ 1 & -2 & 0 & | & 1 \end{bmatrix} \xrightarrow{R_1 \leftrightarrow R_2} \begin{bmatrix} \cdot & \cdot & \cdot & | & \cdot \\ \cdot & \cdot & \cdot & | & \cdot \\ 1 & -2 & 0 & | & 1 \end{bmatrix}$

$\xrightarrow{R_3 - R_1} \begin{bmatrix} 1 & 2 & 1 & | & 7 \\ 0 & 1 & 3 & | & -4 \\ \cdot & \cdot & \cdot & | & \cdot \end{bmatrix} \xrightarrow[R_3 + 4R_2]{R_1 - 2R_2} \begin{bmatrix} \cdot & \cdot & \cdot & | & \cdot \\ 0 & 1 & 3 & | & -4 \\ \cdot & \cdot & \cdot & | & \cdot \end{bmatrix}$

$\xrightarrow{\frac{1}{11}R_3} \begin{bmatrix} 1 & 0 & -5 & | & 15 \\ 0 & 1 & 3 & | & -4 \\ \cdot & \cdot & \cdot & | & \cdot \end{bmatrix} \xrightarrow[R_2 - 3R_3]{R_1 + 5R_3} \begin{bmatrix} 1 & 0 & 0 & | & 5 \\ 0 & 1 & 0 & | & 2 \\ 0 & 0 & 1 & | & -2 \end{bmatrix}$

31. Write a system of linear equations for the augmented matrix of Exercise 27. Using the results of Exercise 27, determine the solution of the system.

32. Repeat Exercise 31 for the augmented matrix of Exercise 28.

33. Repeat Exercise 31 for the augmented matrix of Exercise 29.

34. Repeat Exercise 31 for the augmented matrix of Exercise 30.

In Exercises 35–56, solve the system of linear equations using the Gauss–Jordan elimination method.

35. $x + y = 3$
$2x - y = 3$

36. $x - 2y = -3$
$2x + 3y = 8$

37. $x - 2y = 8$
$3x + 4y = 4$

38. $3x + y = 1$
$-7x - 2y = -1$

39. $2x - 3y = -8$
$4x + y = -2$

40. $5x + 3y = 9$
$-2x + y = -8$

41. $6x + 8y = 15$
$2x - 4y = -5$

42. $2x + 10y = 1$
$-4x + 6y = 11$

43. $3x - 2y = 1$
$2x + 4y = 2$

44. $x - \dfrac{1}{2}y = \dfrac{7}{6}$
$-\dfrac{1}{2}x + 4y = \dfrac{2}{3}$

45. $2x + y - 2z = 4$
$x + 3y - z = -3$
$3x + 4y - z = 7$

46. $x + y + z = 0$
$2x - y + z = 1$
$x + y - 2z = 2$

47. $2x + 2y + z = 9$
$x + z = 4$
$4y - 3z = 17$

48. $2x + 3y - 2z = 10$
$3x - 2y + 2z = 0$
$4x - y + 3z = -1$

49. $-x_2 + x_3 = 2$
$4x_1 - 3x_2 + 2x_3 = 16$
$3x_1 + 2x_2 + x_3 = 11$

50. $2x + 4y - 6z = 38$
$x + 2y + 3z = 7$
$3x - 4y + 4z = -19$

51. $x_1 - 2x_2 + x_3 = 6$
$2x_1 + x_2 - 3x_3 = -3$
$x_1 - 3x_2 + 3x_3 = 10$

52. $2x + 3y - 6z = -11$
$x - 2y + 3z = 9$
$3x + y = 7$

53. $2x + 3z = -1$
$3x - 2y + z = 9$
$x + y + 4z = 4$

54. $2x_1 - x_2 + 3x_3 = -4$
$x_1 - 2x_2 + x_3 = -1$
$x_1 - 5x_2 + 2x_3 = -3$

55. $x_1 - x_2 + 3x_3 = 14$
$x_1 + x_2 + x_3 = 6$
$-2x_1 - x_2 + x_3 = -4$

56. $2x_1 - x_2 - x_3 = 0$
$3x_1 + 2x_2 + x_3 = 7$
$x_1 + 2x_2 + 2x_3 = 5$

57. Determine the value(s) of k such that the following system of linear equations has a unique solution, and then find the solution in terms of k:

$$4x + 5y = 3$$
$$3x + ky = 10$$

58. Determine the value(s) of k such that the following system of linear equations has a unique solution:

$$x + 3y + z = 8$$
$$3x + 2y - 2z = 5$$
$$4x - 3y + kz = 0$$

The problems in Exercises 59–80 correspond to those in Exercises 23–44, Section 5.1. Use the results of your previous work to help you solve these problems.

59. CROP PLANNING The Johnson Farm has 500 acres of land allotted for cultivating corn and wheat. The cost of cultivating corn and wheat (including seeds and labor) is $42 and $30 per acre, respectively. Jacob Johnson has $18,600 available for cultivating these crops. If he wishes to use all the allotted land and his entire budget for cultivating these two crops, how many acres of each crop should he plant?

60. INVESTMENTS Michael Perez has a total of $2000 on deposit with two savings institutions. One pays interest at the rate of 3%/year; the other pays interest at the rate of 4%/year. If Michael earned a total of $72 in interest during a single year, how much does he have on deposit in each institution?

61. **BLENDED COFFEE MIXTURES** The Coffee Shoppe sells a gourmet coffee blend made from two coffees, one costing $8/lb and the other costing $9/lb. If the blended coffee sells for $8.60/lb, find how much of each coffee is used to obtain the desired blend. Assume that the weight of the blended coffee is 100 lb.

62. **MUNICIPAL BONDS** Kelly Fisher has a total of $30,000 invested in two municipal bonds that have yields of 4% and 5% interest per year, respectively. If the interest Kelly receives from the bonds in a year is $1320, how much does she have invested in each bond?

63. **METRO BUS RIDERSHIP** The total number of passengers riding a certain city bus during the morning shift is 1000. If the child's fare is $0.50, the adult fare is $1.50, and the total revenue from the fares in the morning shift is $1300, how many children and how many adults rode the bus during the morning shift?

64. **APARTMENT COMPLEX DEVELOPMENT** Cantwell Associates, a real estate developer, is planning to build a new apartment complex consisting of one-bedroom units and two- and three-bedroom townhouses. A total of 192 units is planned, and the number of family units (two- and three-bedroom townhouses) will equal the number of one-bedroom units. If the number of one-bedroom units will be 3 times the number of three-bedroom units, find how many units of each type will be in the complex.

65. A ball and a bat cost a total of $110. The bat costs $100 more than the ball. How much does the ball cost?

66. **INVESTMENTS** Josh has invested $70,000 in two projects. The amount invested in project A exceeds that invested in project B by $20,000. How much has Josh invested in each project?

67. **INVESTMENT PLANNING** The annual returns on Sid Carrington's three investments amounted to $21,600: 6% on a savings account, 8% on mutual funds, and 12% on bonds. The amount of Sid's investment in bonds was twice the amount of his investment in the savings account, and the interest earned from his investment in bonds was equal to the dividends he received from his investment in mutual funds. Find how much money he placed in each type of investment.

68. **INVESTMENT RISK AND RETURN** A private investment club has $200,000 earmarked for investment in stocks. To arrive at an acceptable overall level of risk, the stocks that management is considering have been classified into three categories: high-risk, medium-risk, and low-risk. Management estimates that high-risk stocks will have a rate of return of 15%/year; medium-risk stocks, 10%/year; and low-risk stocks, 6%/year. The members have decided that the investment in low-risk stocks should be equal to the sum of the investments in the stocks of the other two categories. Determine how much the club should invest in

each type of stock if the investment goal is to have a return of $20,000/year on the total investment. (Assume that all the money available for investment is invested.)

69. **USING DIGITAL TECHNOLOGY** A survey of 500 college students found that the percentage of students who went without using digital technology for up to 1 hr was 67%. The survey also determined that the percentage of students who went without using digital technology for up to 30 min exceeded the percentage of students who went without using digital technology for over 1 hr by 17%. Let x, y, and z represent the percentage of the students in the survey who went without using digital technology (a) for up to 30 min, (b) for more than 30 min but not more than 60 min, and (c) for more than 60 min, respectively. Find the values of x, y, and z.
Source: CourseSmart.

70. **TRUSTWORTHINESS OF ONLINE REVIEWS** In a survey of 1000 adults aged 18 and older, the following question was posed: "Are other travelers' online reviews trustworthy?" The participants were asked to answer "yes," "no," or "not sure." The survey revealed that 370 answered "no" or "not sure." It also showed that the number of those who answered "yes" exceeded the number of those who answered "no" by 340. What percentage of respondents answered (a) "yes," (b) "no," and (c) "not sure"?
Source: Alliance Global Assistance.

71. **LAWN FERTILIZERS** Lawnco produces three grades of commercial fertilizers. A 100-lb bag of grade A fertilizer contains 18 lb of nitrogen, 4 lb of phosphate, and 5 lb of potassium. A 100-lb bag of grade B fertilizer contains 20 lb of nitrogen and 4 lb each of phosphate and potassium. A 100-lb bag of grade C fertilizer contains 24 lb of nitrogen, 3 lb of phosphate, and 6 lb of potassium. How many 100-lb bags of each of the three grades of fertilizers should Lawnco produce if 26,400 lb of nitrogen, 4900 lb of phosphate, and 6200 lb of potassium are available and all the nutrients are used?

72. **BOX-OFFICE RECEIPTS** A theater has a seating capacity of 900 and charges $4 for children, $6 for students, and $8 for adults. At a certain screening with full attendance, there were half as many adults as children and students combined. The receipts totaled $5600. How many children attended the show?

73. **BUDGET ALLOCATION FOR AUTO FLEET** The management of Hartman Rent-A-Car has allocated $2.25 million to buy a fleet of new automobiles consisting of compact, intermediate-size, and full-size cars. Compacts cost $18,000 each, intermediate-size cars cost $27,000 each, and full-size cars cost $36,000 each. If Hartman purchases twice as many compacts as intermediate-size cars and the total number of cars to be purchased is 100, determine how many cars of each type will be purchased. (Assume that the entire budget will be used.)

74. **INVESTMENT RISK AND RETURN** The management of a private investment club has a fund of $200,000 earmarked for investment in stocks. To arrive at an acceptable overall level of risk, the stocks that management is considering have been classified into three categories: high-risk, medium-risk, and low-risk. Management estimates that high-risk stocks will have a rate of return of 15%/year; medium-risk stocks, 10%/year; and low-risk stocks, 6%/year. The investment in low-risk stocks is to be twice the sum of the investments in stocks of the other two categories. If the investment goal is to have an average rate of return of 9%/year on the total investment, determine how much the club should invest in each type of stock. (Assume that all of the money available for investment is invested.)

75. **DIET PLANNING** A dietitian wishes to plan a meal around three foods. The percentages of the daily requirements of proteins, carbohydrates, and iron contained in each ounce of the three foods are summarized in the following table:

	Food I	Food II	Food III
Proteins (%)	10	6	8
Carbohydrates (%)	10	12	6
Iron (%)	5	4	12

Determine how many ounces of each food the dietitian should include in the meal to meet exactly the daily requirement of proteins, carbohydrates, and iron (100% of each).

76. **ASSET ALLOCATION** Mr. and Mrs. Garcia have a total of $100,000 to be invested in stocks, bonds, and a money market account. The stocks have a rate of return of 12%/year, while the bonds and the money market account pay 8%/year and 4%/year, respectively. The Garcias have stipulated that the amount invested in the money market account should be equal to the sum of 20% of the amount invested in stocks and 10% of the amount invested in bonds. How should the Garcias allocate their resources if they require an annual income of $10,000 from their investments?

77. **BOX-OFFICE RECEIPTS** For the opening night at the Opera House, a total of 1000 tickets were sold. Front orchestra seats cost $80 apiece, rear orchestra seats cost $60 apiece, and front balcony seats cost $50 apiece. The combined number of tickets sold for the front orchestra and rear orchestra exceeded twice the number of front balcony tickets sold by 400. The total receipts for the performance were $62,800. Determine how many tickets of each type were sold.

78. **PRODUCTION SCHEDULING** A manufacturer of women's blouses makes three types of blouses: sleeveless, short-sleeve, and long-sleeve. The time (in minutes) required by each department to produce a dozen blouses of each type is shown in the following table:

	Sleeveless	Short-sleeve	Long-sleeve
Cutting	9	12	15
Sewing	22	24	28
Packaging	6	8	8

The cutting, sewing, and packaging departments have available a maximum of 80, 160, and 48 labor-hours, respectively, per day. How many dozens of each type of blouse can be produced each day if the plant is operated at full capacity?

79. **BUSINESS TRAVEL EXPENSES** An executive of Trident Communications recently traveled to London, Paris, and Rome. He paid $280, $330, and $260 per night for lodging in London, Paris, and Rome, respectively, and his hotel bills totaled $4060. He spent $130, $140, and $110 per day for his meals in London, Paris, and Rome, respectively, and his expenses for meals totaled $1800. If he spent as many days in London as he did in Paris and Rome combined, how many days did he stay in each city?

80. **VACATION COSTS** Joan and Dick spent 2 weeks (14 nights) touring four cities on the East Coast—Boston, New York, Philadelphia, and Washington. They paid $240, $400, $160, and $200 per night for lodging in each city, respectively, and their total hotel bill came to $4040. The number of days they spent in New York was the same as the total number of days they spent in Boston and Washington, and the couple spent 3 times as many days in New York as they did in Philadelphia. How many days did Joan and Dick stay in each city?

In Exercises 81 and 82, determine whether the statement is true or false. If it is true, explain why it is true. If it is false, give an example to show why it is false.

81. An equivalent system of linear equations can be obtained from a system of equations by replacing one of its equations by any constant multiple of itself.

82. If the augmented matrix corresponding to a system of three linear equations in three variables has a row of the form $[0 \ \ 0 \ \ 0 \ | \ a]$, where a is a nonzero number, then the system has no solution.

5.2 Solutions to Self-Check Exercises

1. We obtain the following sequence of equivalent augmented matrices:

$$\begin{bmatrix} 2 & 3 & 1 & | & 6 \\ 1 & -2 & 3 & | & -3 \\ 3 & 2 & -4 & | & 12 \end{bmatrix} \xrightarrow{R_1 \leftrightarrow R_2} \begin{bmatrix} ① & -2 & 3 & | & -3 \\ 2 & 3 & 1 & | & 6 \\ 3 & 2 & -4 & | & 12 \end{bmatrix}$$

$$\xrightarrow[R_3 - 3R_1]{R_2 - 2R_1} \begin{bmatrix} 1 & -2 & 3 & | & -3 \\ 0 & 7 & -5 & | & 12 \\ 0 & 8 & -13 & | & 21 \end{bmatrix} \xrightarrow{R_2 \leftrightarrow R_3}$$

$$\begin{bmatrix} 1 & -2 & 3 & | & -3 \\ 0 & ⑧ & -13 & | & 21 \\ 0 & 7 & -5 & | & 12 \end{bmatrix} \xrightarrow{R_2 - R_3} \begin{bmatrix} 1 & -2 & 3 & | & -3 \\ 0 & 1 & -8 & | & 9 \\ 0 & 7 & -5 & | & 12 \end{bmatrix}$$

$$\xrightarrow[R_3 - 7R_2]{R_1 + 2R_2} \begin{bmatrix} 1 & 0 & -13 & | & 15 \\ 0 & 1 & -8 & | & 9 \\ 0 & 0 & 51 & | & -51 \end{bmatrix} \xrightarrow{\frac{1}{51}R_3} \begin{bmatrix} 1 & 0 & -13 & | & 15 \\ 0 & 1 & -8 & | & 9 \\ 0 & 0 & ① & | & -1 \end{bmatrix}$$

$$\xrightarrow[R_2 + 8R_3]{R_1 + 13R_3} \begin{bmatrix} 1 & 0 & 0 & | & 2 \\ 0 & 1 & 0 & | & 1 \\ 0 & 0 & 1 & | & -1 \end{bmatrix}$$

The solution to the system is $x = 2$, $y = 1$, and $z = -1$.

2. Referring to the solution of Exercise 2, Self-Check Exercises 5.1, we see that the problem reduces to solving the following system of linear equations:

$$\begin{aligned} x + y + z &= 200 \\ 40x + 60y + 80z &= 12{,}600 \\ 20x + 25y + 40z &= 5{,}950 \end{aligned}$$

Using the Gauss–Jordan elimination method, we have

$$\begin{bmatrix} ① & 1 & 1 & | & 200 \\ 40 & 60 & 80 & | & 12{,}600 \\ 20 & 25 & 40 & | & 5{,}950 \end{bmatrix} \xrightarrow[R_3 - 20R_1]{R_2 - 40R_1} \begin{bmatrix} 1 & 1 & 1 & | & 200 \\ 0 & ⑳ & 40 & | & 4600 \\ 0 & 5 & 20 & | & 1950 \end{bmatrix}$$

$$\xrightarrow{\frac{1}{20}R_2} \begin{bmatrix} 1 & 1 & 1 & | & 200 \\ 0 & 1 & 2 & | & 230 \\ 0 & 5 & 20 & | & 1950 \end{bmatrix} \xrightarrow[R_3 - 5R_2]{R_1 - R_2} \begin{bmatrix} 1 & 0 & -1 & | & -30 \\ 0 & 1 & 2 & | & 230 \\ 0 & 0 & ⑩ & | & 800 \end{bmatrix}$$

$$\xrightarrow{\frac{1}{10}R_3} \begin{bmatrix} 1 & 0 & -1 & | & -30 \\ 0 & 1 & 2 & | & 230 \\ 0 & 0 & 1 & | & 80 \end{bmatrix} \xrightarrow[R_2 - 2R_3]{R_1 + R_3} \begin{bmatrix} 1 & 0 & 0 & | & 50 \\ 0 & 1 & 0 & | & 70 \\ 0 & 0 & 1 & | & 80 \end{bmatrix}$$

From the last augmented matrix in reduced form, we see that $x = 50$, $y = 70$, and $z = 80$. Therefore, the farmer should plant 50 acres of Crop A, 70 acres of Crop B, and 80 acres of Crop C.

USING TECHNOLOGY — Systems of Linear Equations: Unique Solutions

Solving a System of Linear Equations Using the Gauss–Jordan Method

The three matrix operations can be performed on a matrix by using a graphing utility. The commands are summarized in the following table:

	Calculator Function		
Operation	TI-83/84	TI-86	
$R_i \leftrightarrow R_j$	**rowSwap**([A], i, j)	**rSwap**(A, i, j)	or equivalent
cR_i	*row(c, [A], i)	**multR**(c, A, i)	or equivalent
$R_i + aR_j$	*row+(a, [A], j, i)	**mRAdd**(a, A, j, i)	or equivalent

When a row operation is performed on a matrix, the result is stored as an answer in the calculator. If another operation is performed on this matrix, then the matrix is erased. Should a mistake be made in the operation, the previous matrix may be lost. For this reason, you should store the results of each operation. We do this by pressing **STO**, followed by the name of a matrix, and then **ENTER**. We use this process in the following example.

EXAMPLE 1 Use a graphing utility to solve the following system of linear equations by the Gauss–Jordan method (see Example 5 in Section 5.2):

$$\begin{aligned} 3x - 2y + 8z &= 9 \\ -2x + 2y + z &= 3 \\ x + 2y - 3z &= 8 \end{aligned}$$

Solution Using the Gauss–Jordan method, we obtain the following sequence of equivalent matrices.

$$\begin{bmatrix} 3 & -2 & 8 & | & 9 \\ -2 & 2 & 1 & | & 3 \\ 1 & 2 & -3 & | & 8 \end{bmatrix} \xrightarrow{\ast\mathbf{row}+\,(1,[A],2,1)\,\blacktriangleright\,B}$$

$$\begin{bmatrix} 1 & 0 & 9 & | & 12 \\ -2 & 2 & 1 & | & 3 \\ 1 & 2 & -3 & | & 8 \end{bmatrix} \xrightarrow{\ast\mathbf{row}+\,(2,[B],1,2)\,\blacktriangleright\,C}$$

$$\begin{bmatrix} 1 & 0 & 9 & | & 12 \\ 0 & 2 & 19 & | & 27 \\ 1 & 2 & -3 & | & 8 \end{bmatrix} \xrightarrow{\ast\mathbf{row}+\,(-1,[C],1,3)\,\blacktriangleright\,B}$$

$$\begin{bmatrix} 1 & 0 & 9 & | & 12 \\ 0 & 2 & 19 & | & 27 \\ 0 & 2 & -12 & | & -4 \end{bmatrix} \xrightarrow{\ast\mathbf{row}(\frac{1}{2},[B],2)\,\blacktriangleright\,C}$$

$$\begin{bmatrix} 1 & 0 & 9 & | & 12 \\ 0 & 1 & 9.5 & | & 13.5 \\ 0 & 2 & -12 & | & -4 \end{bmatrix} \xrightarrow{\ast\mathbf{row}+\,(-2,[C],2,3)\,\blacktriangleright\,B}$$

$$\begin{bmatrix} 1 & 0 & 9 & | & 12 \\ 0 & 1 & 9.5 & | & 13.5 \\ 0 & 0 & -31 & | & -31 \end{bmatrix} \xrightarrow{\ast\mathbf{row}(-\frac{1}{31},[B],3)\,\blacktriangleright\,C}$$

$$\begin{bmatrix} 1 & 0 & 9 & | & 12 \\ 0 & 1 & 9.5 & | & 13.5 \\ 0 & 0 & 1 & | & 1 \end{bmatrix} \xrightarrow{\ast\mathbf{row}+\,(-9,[C],3,1)\,\blacktriangleright\,B}$$

$$\begin{bmatrix} 1 & 0 & 0 & | & 3 \\ 0 & 1 & 9.5 & | & 13.5 \\ 0 & 0 & 1 & | & 1 \end{bmatrix} \xrightarrow{\ast\mathbf{row}+\,(-9.5,[B],3,2)\,\blacktriangleright\,C} \begin{bmatrix} 1 & 0 & 0 & | & 3 \\ 0 & 1 & 0 & | & 4 \\ 0 & 0 & 1 & | & 1 \end{bmatrix}$$

The last matrix is in row-reduced form, and we see that the solution of the system is $x = 3$, $y = 4$, and $z = 1$. ∎

Using rref (TI-83/84 and TI-86) to Solve a System of Linear Equations

The operation **rref** (or equivalent function in your utility, if there is one) will transform an augmented matrix into one that is in row-reduced form. For example, using **rref**, we find

$$\begin{bmatrix} 3 & -2 & 8 & | & 9 \\ -2 & 2 & 1 & | & 3 \\ 1 & 2 & -3 & | & 8 \end{bmatrix} \xrightarrow{\mathbf{rref}} \begin{bmatrix} 1 & 0 & 0 & | & 3 \\ 0 & 1 & 0 & | & 4 \\ 0 & 0 & 1 & | & 1 \end{bmatrix}$$

as obtained earlier!

Using SIMULT (TI-86) to Solve a System of Equations

The operation **SIMULT** (or equivalent operation on your utility, if there is one) of a graphing utility can be used to solve a system of n linear equations in n variables, where n is an integer between 2 and 30, inclusive.

EXAMPLE 2 Use the **SIMULT** operation to solve the system of Example 1.

Solution Call for the **SIMULT** operation. Since the system under consideration has three equations in three variables, enter $n = 3$. Next, enter $a1, 1 = 3$, $a1, 2 = -2$, $a1, 3 = 8$, $b1 = 9$, $a2, 1 = -2, \ldots, b3 = 8$. Select $<$**SOLVE**$>$, and the display

$$x1 = 3$$
$$x2 = 4$$
$$x3 = 1$$

appears on the screen, giving $x = 3$, $y = 4$, and $z = 1$ as the required solution. ■

TECHNOLOGY EXERCISES

Use a graphing utility to solve the system of equations (a) by the Gauss–Jordan method, (b) using the rref operation, and (c) using SIMULT.

1.
$$\begin{aligned} x_1 - 2x_2 + 2x_3 - 3x_4 &= -7 \\ 3x_1 + 2x_2 - x_3 + 5x_4 &= 22 \\ 2x_1 - 3x_2 + 4x_3 - x_4 &= -3 \\ 3x_1 - 2x_2 - x_3 + 2x_4 &= 12 \end{aligned}$$

2.
$$\begin{aligned} 2x_1 - x_2 + 3x_3 - 2x_4 &= -2 \\ x_1 - 2x_2 + x_3 - 3x_4 &= 2 \\ x_1 - 5x_2 + 2x_3 + 3x_4 &= -6 \\ -3x_1 + 3x_2 - 4x_3 - 4x_4 &= 9 \end{aligned}$$

3.
$$\begin{aligned} 2x_1 + x_2 + 3x_3 - x_4 &= 9 \\ -x_1 - 2x_2 \quad\quad - 3x_4 &= -1 \\ x_1 \quad\quad - 3x_3 + x_4 &= 10 \\ x_1 - x_2 - x_3 - x_4 &= 8 \end{aligned}$$

4.
$$\begin{aligned} x_1 - 2x_2 - 2x_3 + x_4 &= 1 \\ 2x_1 - x_2 + 2x_3 + 3x_4 &= -2 \\ -x_1 - 5x_2 + 7x_3 - 2x_4 &= 3 \\ 3x_1 - 4x_2 + 3x_3 + 4x_4 &= -4 \end{aligned}$$

5.
$$\begin{aligned} 2x_1 - 2x_2 + 3x_3 - x_4 + 2x_5 &= 16 \\ 3x_1 + x_2 - 2x_3 + x_4 - 3x_5 &= -11 \\ x_1 + 3x_2 - 4x_3 + 3x_4 - x_5 &= -13 \\ 2x_1 - x_2 + 3x_3 - 2x_4 + 2x_5 &= 15 \\ 3x_1 + 4x_2 - 3x_3 + 5x_4 - x_5 &= -10 \end{aligned}$$

6.
$$\begin{aligned} 2.1x_1 - 3.2x_2 + 6.4x_3 + 7x_4 - 3.2x_5 &= 54.3 \\ 4.1x_1 + 2.2x_2 - 3.1x_3 - 4.2x_4 + 3.3x_5 &= -20.81 \\ 3.4x_1 - 6.2x_2 + 4.7x_3 + 2.1x_4 - 5.3x_5 &= 24.7 \\ 4.1x_1 + 7.3x_2 + 5.2x_3 + 6.1x_4 - 8.2x_5 &= 29.25 \\ 2.8x_1 + 5.2x_2 + 3.1x_3 + 5.4x_4 + 3.8x_5 &= 43.72 \end{aligned}$$

5.3 Systems of Linear Equations: Underdetermined and Overdetermined Systems

In this section, we continue our study of systems of linear equations. More specifically, we look at systems that have infinitely many solutions and those that have no solution. We also study systems of linear equations in which the number of variables is not equal to the number of equations in the system.

Solution(s) of Linear Equations

Our first two examples illustrate the situation in which a system of linear equations has infinitely many solutions.

EXAMPLE 1 A System of Equations with an Infinite Number of Solutions Solve the system of linear equations given by

$$\begin{aligned} x + 2y &= 4 \\ 3x + 6y &= 12 \end{aligned} \tag{9}$$

Solution Using the Gauss–Jordan elimination method, we obtain the following system of equivalent matrices:

$$\begin{bmatrix} ① & 2 & | & 4 \\ 3 & 6 & | & 12 \end{bmatrix} \xrightarrow{R_2 - 3R_1} \begin{bmatrix} 1 & 2 & | & 4 \\ 0 & 0 & | & 0 \end{bmatrix}$$

The last augmented matrix is in row-reduced form. Interpreting it as a system of linear equations, we see that the given System (9) is equivalent to the single equation

$$x + 2y = 4 \quad \text{or} \quad x = 4 - 2y$$

If we assign a particular value to y—say, $y = 0$—we obtain $x = 4$, giving the solution $(4, 0)$ to System (9). By setting $y = 1$, we obtain the solution $(2, 1)$. In general, if we set $y = t$, where t represents some real number (called a parameter), we obtain the solution given by $(4 - 2t, t)$. Since the parameter t may be any real number, we see that System (9) has infinitely many solutions. Geometrically, the solutions of System (9) lie on the line on the plane with equation $x + 2y = 4$. The two equations in the system have the same graph (straight line), which you can verify graphically. ∎

EXAMPLE 2 A System of Equations with an Infinite Number of Solutions Solve the system of linear equations given by

$$\begin{aligned} x + 2y - 3z &= -2 \\ 3x - y - 2z &= 1 \\ 2x + 3y - 5z &= -3 \end{aligned} \tag{10}$$

Solution Using the Gauss–Jordan elimination method, we obtain the following sequence of equivalent augmented matrices:

$$\begin{bmatrix} ① & 2 & -3 & | & -2 \\ 3 & -1 & -2 & | & 1 \\ 2 & 3 & -5 & | & -3 \end{bmatrix} \xrightarrow[R_3 - 2R_1]{R_2 - 3R_1} \begin{bmatrix} 1 & 2 & -3 & | & -2 \\ 0 & ⑦ & 7 & | & 7 \\ 0 & -1 & 1 & | & 1 \end{bmatrix} \xrightarrow{-\frac{1}{7}R_2}$$

$$\begin{bmatrix} 1 & 2 & -3 & | & -2 \\ 0 & 1 & -1 & | & -1 \\ 0 & -1 & 1 & | & 1 \end{bmatrix} \xrightarrow[R_3 + R_2]{R_1 - 2R_2} \begin{bmatrix} 1 & 0 & -1 & | & 0 \\ 0 & 1 & -1 & | & -1 \\ 0 & 0 & 0 & | & 0 \end{bmatrix}$$

The last augmented matrix is in row-reduced form. Interpreting it as a system of linear equations gives

$$\begin{aligned} x - z &= 0 \\ y - z &= -1 \end{aligned}$$

a system of two equations in the three variables x, y, and z.

Let's now single out one variable—say, z—and solve for x and y in terms of it. We obtain

$$\begin{aligned} x &= z \\ y &= z - 1 \end{aligned}$$

If we set $z = t$, where t is a parameter, then System (10) has infinitely many solutions given by $(t, t - 1, t)$. For example, letting $t = 0$ gives the solution $(0, -1, 0)$, and letting $t = 1$ gives the solution $(1, 0, 1)$. Geometrically, the solutions of System (10) lie on the straight line in three-dimensional space given by the intersection of the three planes determined by the three equations in the system. ◼

Note In Example 2, we chose the parameter to be z because it is more convenient to solve for x and y (both the x- and y-columns are in unit form) in terms of z. ◼

The next example shows what happens in the elimination procedure when the system does not have a solution.

EXAMPLE 3 A System of Equations That Has No Solution Solve the system of linear equations given by

$$\begin{aligned}
x + y + z &= 1 \\
3x - y - z &= 4 \\
x + 5y + 5z &= -1
\end{aligned} \qquad (11)$$

Solution Using the Gauss–Jordan elimination method, we obtain the following sequence of equivalent augmented matrices:

$$\left[\begin{array}{ccc|c} ① & 1 & 1 & 1 \\ 3 & -1 & -1 & 4 \\ 1 & 5 & 5 & -1 \end{array}\right] \xrightarrow[R_3 - R_1]{R_2 - 3R_1} \left[\begin{array}{ccc|c} 1 & 1 & 1 & 1 \\ 0 & -4 & -4 & 1 \\ 0 & 4 & 4 & -2 \end{array}\right]$$

$$\xrightarrow{R_3 + R_2} \left[\begin{array}{ccc|c} 1 & 1 & 1 & 1 \\ 0 & -4 & -4 & 1 \\ 0 & 0 & 0 & -1 \end{array}\right]$$

Observe that row 3 in the last matrix reads $0x + 0y + 0z = -1$—that is, $0 = -1$! We therefore conclude that System (11) is inconsistent and has no solution. Geometrically, we have a situation in which two of the planes intersect in a straight line but the third plane is parallel to this line of intersection of the two planes and does not intersect it. Consequently, there is no point of intersection of the three planes. ◼

Example 3 illustrates the following more general result of using the Gauss–Jordan elimination procedure.

Systems with No Solution

If there is a row in an augmented matrix containing all zeros to the left of the vertical line and a nonzero entry to the right of the line, then the corresponding system of equations has no solution.

It may have dawned on you that in all the previous examples, we have dealt only with systems involving exactly the same number of linear equations as there are variables. However, systems in which the number of equations differs from the number of variables also occur in practice. Indeed, we will consider such systems in Examples 4 and 5.

The following theorem provides us with some preliminary information on a system of linear equations.

THEOREM 1

a. If the number of equations is greater than or equal to the number of variables in a linear system, then one of the following is true:
  i. The system has no solution.
  ii. The system has exactly one solution.
  iii. The system has infinitely many solutions.

b. If there are fewer equations than variables in a linear system, then the system either has no solution or has infinitely many solutions.

Note   Theorem 1 may be used to tell us, before we even begin to solve a problem, what the nature of the solution may be.   ◼

Although we will not prove this theorem, you should recall that we have illustrated geometrically part (a) for the case in which there are exactly as many equations (three) as there are variables. To show the validity of part (b), let us once again consider the case in which a system has three variables. Now, if there is only one equation in the system, then it is clear that there are infinitely many solutions corresponding geometrically to all the points lying on the plane represented by the equation.

Next, if there are two equations in the system, then *only* the following possibilities exist:

1. The two planes are parallel and distinct (Figure 6a).
2. The two planes intersect in a straight line (Figure 6b).
3. The two planes are coincident (the two equations define the same plane) (Figure 6c).

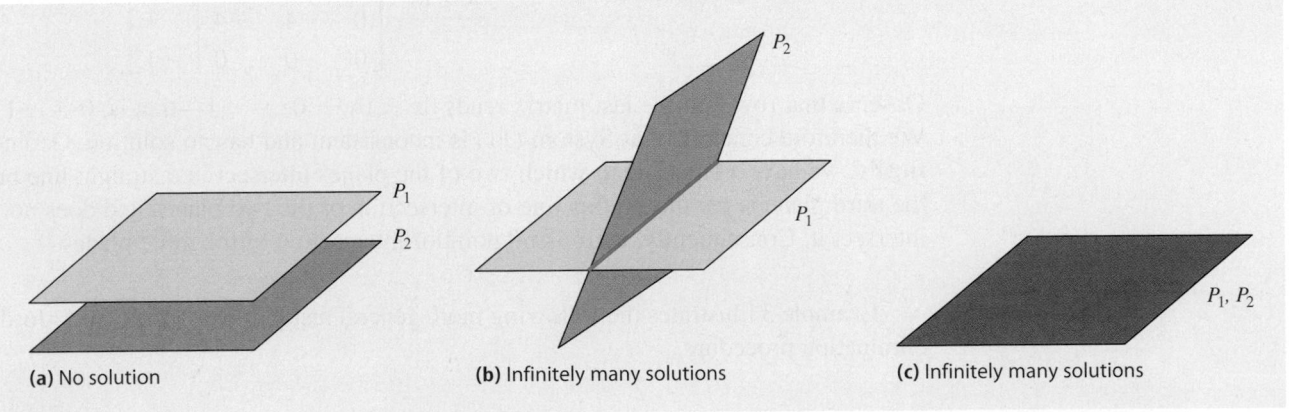

(a) No solution   **(b)** Infinitely many solutions   **(c)** Infinitely many solutions

FIGURE **6**

Thus, either there is no solution or there are infinitely many solutions corresponding to the points lying on a line of intersection of the two planes or on a single plane determined by the two equations. In the case in which two planes intersect in a straight line, the solutions will involve one parameter, and in the case in which the two planes are coincident, the solutions will involve two parameters.

Explore and Discuss

Give a geometric interpretation of Theorem 1 for a linear system composed of equations involving two variables. Specifically, illustrate what can happen if there are three linear equations in the system (the case involving two linear equations was discussed in Section 5.1). What if there are four linear equations? What if there is only one linear equation in the system?

EXAMPLE 4 A System with More Equations Than Variables Solve the following system of linear equations:

$$\begin{aligned} x + 2y &= 4 \\ x - 2y &= 0 \\ 4x + 3y &= 12 \end{aligned}$$

Solution We obtain the following sequence of equivalent augmented matrices:

$$\begin{bmatrix} \text{①} & 2 & 4 \\ 1 & -2 & 0 \\ 4 & 3 & 12 \end{bmatrix} \xrightarrow[R_3 - 4R_1]{R_2 - R_1} \begin{bmatrix} 1 & 2 & 4 \\ 0 & \text{④} & -4 \\ 0 & -5 & -4 \end{bmatrix} \xrightarrow{-\frac{1}{4}R_2}$$

$$\begin{bmatrix} 1 & 2 & 4 \\ 0 & 1 & 1 \\ 0 & -5 & -4 \end{bmatrix} \xrightarrow[R_3 + 5R_2]{R_1 - 2R_2} \begin{bmatrix} 1 & 0 & 2 \\ 0 & 1 & 1 \\ 0 & 0 & 1 \end{bmatrix}$$

The last row of the row-reduced augmented matrix implies that $0 = 1$, which is impossible, so we conclude that the given system has no solution. Geometrically, the three lines defined by the three equations in the system do not intersect at a point. (To see this for yourself, draw the graphs of these equations.) ∎

EXAMPLE 5 A System with More Variables Than Equations Solve the following system of linear equations:

$$\begin{aligned} x + 2y - 3z + w &= -2 \\ 3x - y - 2z - 4w &= 1 \\ 2x + 3y - 5z + w &= -3 \end{aligned}$$

Solution First, observe that the given system consists of three equations in four variables, so by Theorem 1b, either the system has no solution or it has infinitely many solutions. To solve it, we use the Gauss–Jordan method and obtain the following sequence of equivalent augmented matrices:

$$\begin{bmatrix} \text{①} & 2 & -3 & 1 & -2 \\ 3 & -1 & -2 & -4 & 1 \\ 2 & 3 & -5 & 1 & -3 \end{bmatrix} \xrightarrow[R_3 - 2R_1]{R_2 - 3R_1} \begin{bmatrix} 1 & 2 & -3 & 1 & -2 \\ 0 & \text{⑦} & 7 & -7 & 7 \\ 0 & -1 & 1 & -1 & 1 \end{bmatrix} \xrightarrow{-\frac{1}{7}R_2}$$

$$\begin{bmatrix} 1 & 2 & -3 & 1 & -2 \\ 0 & 1 & -1 & 1 & -1 \\ 0 & -1 & 1 & -1 & 1 \end{bmatrix} \xrightarrow[R_3 + R_2]{R_1 - 2R_2} \begin{bmatrix} 1 & 0 & -1 & -1 & 0 \\ 0 & 1 & -1 & 1 & -1 \\ 0 & 0 & 0 & 0 & 0 \end{bmatrix}$$

The last augmented matrix is in row-reduced form. Observe that the given system is equivalent to the system

$$\begin{aligned} x - z - w &= 0 \\ y - z + w &= -1 \end{aligned}$$

of two equations in four variables. Thus, we may solve for two of the variables in terms of the other two. Letting $z = s$ and $w = t$ (where s and t are any real numbers), we find that

$$\begin{aligned} x &= s + t \\ y &= s - t - 1 \\ z &= s \\ w &= t \end{aligned}$$

The solutions may be written in the form $(s + t, s - t - 1, s, t)$. Geometrically, the three equations in the system represent three hyperplanes in four-dimensional space (since there are four variables), and their "points" of intersection lie in a two-dimensional subspace of four-space (since there are two parameters). ■

Note   In Example 5, we assigned parameters to z and w rather than to x and y because x and y are readily solved in terms of z and w. ■

The following example illustrates a situation in which a system of linear equations has infinitely many solutions.

APPLIED EXAMPLE 6 Traffic Control Figure 7 shows the flow of downtown traffic in a certain city during the rush hours on a typical weekday. The arrows indicate the direction of traffic flow on each one-way road, and the average number of vehicles per hour entering and leaving each intersection appears beside each road. 5th Avenue and 6th Avenue can each handle up to 2000 vehicles per hour without causing congestion, whereas the maximum capacity of both 4th Street and 5th Street is 1000 vehicles per hour. The flow of traffic is controlled by traffic lights installed at each of the four intersections.

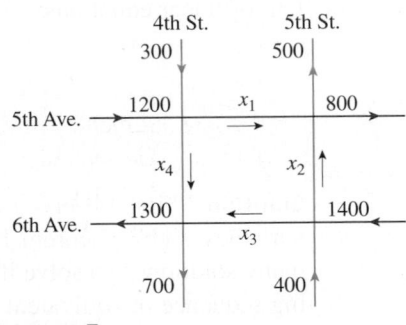

FIGURE **7**

a. Write a general expression involving the rates of flow—x_1, x_2, x_3, x_4—and suggest two possible traffic-flow patterns that will ensure no traffic congestion.
b. Suppose that the part of 4th Street between 5th Avenue and 6th Avenue is to be resurfaced and that traffic flow between the two junctions must therefore be reduced to at most 300 vehicles per hour. Find two possible traffic-flow patterns that will result in a smooth flow of traffic.

Solution

a. To avoid congestion, all traffic entering an intersection must also leave that intersection. Applying this condition to each of the four intersections in a clockwise direction beginning with the 5th Avenue and 4th Street intersection, we obtain the following equations:

$$1500 = x_1 + x_4$$
$$1300 = x_1 + x_2$$
$$1800 = x_2 + x_3$$
$$2000 = x_3 + x_4$$

This system of four linear equations in the four variables x_1, x_2, x_3, x_4 may be rewritten in the more standard form

$$
\begin{aligned}
x_1 \qquad\qquad + x_4 &= 1500 \\
x_1 + x_2 \qquad\quad &= 1300 \\
x_2 + x_3 \quad &= 1800 \\
x_3 + x_4 &= 2000
\end{aligned}
$$

Using the Gauss–Jordan elimination method to solve the system, we obtain

$$
\begin{bmatrix}
1 & 0 & 0 & 1 & \big| & 1500 \\
1 & 1 & 0 & 0 & \big| & 1300 \\
0 & 1 & 1 & 0 & \big| & 1800 \\
0 & 0 & 1 & 1 & \big| & 2000
\end{bmatrix}
\xrightarrow{R_2 - R_1}
\begin{bmatrix}
1 & 0 & 0 & 1 & \big| & 1500 \\
0 & 1 & 0 & -1 & \big| & -200 \\
0 & 1 & 1 & 0 & \big| & 1800 \\
0 & 0 & 1 & 1 & \big| & 2000
\end{bmatrix}
$$

$$
\xrightarrow{R_3 - R_2}
\begin{bmatrix}
1 & 0 & 0 & 1 & \big| & 1500 \\
0 & 1 & 0 & -1 & \big| & -200 \\
0 & 0 & 1 & 1 & \big| & 2000 \\
0 & 0 & 1 & 1 & \big| & 2000
\end{bmatrix}
$$

$$
\xrightarrow{R_4 - R_3}
\begin{bmatrix}
1 & 0 & 0 & 1 & \big| & 1500 \\
0 & 1 & 0 & -1 & \big| & -200 \\
0 & 0 & 1 & 1 & \big| & 2000 \\
0 & 0 & 0 & 0 & \big| & 0
\end{bmatrix}
$$

The last augmented matrix is in row-reduced form and is equivalent to a system of three linear equations in the four variables x_1, x_2, x_3, x_4. Thus, we may express three of the variables—say, x_1, x_2, x_3—in terms of the fourth, x_4. Setting $x_4 = t$ (t a parameter), we may write the infinitely many solutions of the system as

$$
\begin{aligned}
x_1 &= 1500 - t \\
x_2 &= -200 + t \\
x_3 &= 2000 - t \\
x_4 &= t
\end{aligned}
$$

Observe that for a meaningful solution, we must have $200 \le t \le 1000$, since x_1, x_2, x_3, and x_4 must all be nonnegative and the maximum capacity of a street is 1000. For example, picking $t = 300$ gives the flow pattern

$$
x_1 = 1200 \qquad x_2 = 100 \qquad x_3 = 1700 \qquad x_4 = 300
$$

Selecting $t = 500$ gives the flow pattern

$$
x_1 = 1000 \qquad x_2 = 300 \qquad x_3 = 1500 \qquad x_4 = 500
$$

b. In this case, x_4 must not exceed 300. Again, using the results of part (a), we find, upon setting $x_4 = t = 300$, the flow pattern

$$
x_1 = 1200 \qquad x_2 = 100 \qquad x_3 = 1700 \qquad x_4 = 300
$$

obtained earlier. Picking $t = 250$ gives the flow pattern

$$
x_1 = 1250 \qquad x_2 = 50 \qquad x_3 = 1750 \qquad x_4 = 250
$$

5.3   Self-Check Exercises

1. The following augmented matrix in row-reduced form is equivalent to the augmented matrix of a certain system of linear equations. Use this result to solve the system of equations.

$$\begin{bmatrix} 1 & 0 & -1 & | & 3 \\ 0 & 1 & 5 & | & -2 \\ 0 & 0 & 0 & | & 0 \end{bmatrix}$$

2. Solve the system of linear equations

$$\begin{aligned} 2x - 3y + z &= 6 \\ x + 2y + 4z &= -4 \\ x - 5y - 3z &= 10 \end{aligned}$$

using the Gauss–Jordan elimination method.

3. Solve the system of linear equations

$$\begin{aligned} x - 2y + 3z &= 9 \\ 2x + 3y - z &= 4 \\ x + 5y - 4z &= 2 \end{aligned}$$

using the Gauss–Jordan elimination method.

Solutions to Self-Check Exercises 5.3 can be found on page 301.

5.3   Concept Questions

1. If a system of linear equations has the same number of equations or more equations than variables, what can you say about the nature of its solution(s)?

2. If a system of linear equations has fewer equations than variables, what can you say about the nature of its solution(s)?

3. A system consists of three linear equations in four variables. Can the system have a unique solution?

5.3   Exercises

In Exercises 1–14, given that the augmented matrix in row-reduced form is equivalent to the augmented matrix of a system of linear equations, (a) determine whether the system has a solution, and (b) find the solution or solutions to the system, if they exist.

1. $\begin{bmatrix} 1 & 0 & 0 & | & 3 \\ 0 & 1 & 0 & | & -1 \\ 0 & 0 & 1 & | & 2 \end{bmatrix}$

2. $\begin{bmatrix} 1 & 0 & 0 & | & 3 \\ 0 & 1 & 0 & | & -2 \\ 0 & 0 & 1 & | & 1 \end{bmatrix}$

3. $\begin{bmatrix} 1 & 0 & | & 2 \\ 0 & 1 & | & 5 \\ 0 & 0 & | & 0 \end{bmatrix}$

4. $\begin{bmatrix} 1 & 0 & 0 & | & 3 \\ 0 & 1 & 0 & | & 1 \\ 0 & 0 & 0 & | & 0 \end{bmatrix}$

5. $\begin{bmatrix} 1 & 0 & | & 2 \\ 0 & 1 & | & 3 \\ 0 & 0 & | & -1 \end{bmatrix}$

6. $\begin{bmatrix} 1 & 0 & 0 & | & 2 \\ 0 & 0 & 0 & | & 1 \end{bmatrix}$

7. $\begin{bmatrix} 1 & 0 & 1 & | & 4 \\ 0 & 1 & 0 & | & -2 \end{bmatrix}$

8. $\begin{bmatrix} 1 & 0 & 0 & 0 & | & 3 \\ 0 & 1 & 1 & 0 & | & -1 \\ 0 & 0 & 0 & 1 & | & 2 \end{bmatrix}$

9. $\begin{bmatrix} 1 & 0 & 0 & 0 & | & 2 \\ 0 & 1 & 0 & 0 & | & 1 \\ 0 & 0 & 1 & 0 & | & 3 \\ 0 & 0 & 0 & 0 & | & 1 \end{bmatrix}$

10. $\begin{bmatrix} 1 & 0 & 0 & | & 4 \\ 0 & 1 & 0 & | & -1 \\ 0 & 0 & 1 & | & 3 \\ 0 & 0 & 0 & | & 1 \end{bmatrix}$

11. $\begin{bmatrix} 1 & 0 & 0 & 0 & | & 4 \\ 0 & 1 & 0 & 0 & | & -1 \\ 0 & 0 & 1 & 1 & | & 3 \\ 0 & 0 & 0 & 0 & | & 0 \end{bmatrix}$

12. $\begin{bmatrix} 0 & 1 & 0 & 1 & | & 3 \\ 0 & 0 & 1 & -2 & | & 4 \\ 0 & 0 & 0 & 0 & | & 0 \\ 0 & 0 & 0 & 0 & | & 0 \end{bmatrix}$

13. $\begin{bmatrix} 1 & 0 & 3 & 0 & | & 2 \\ 0 & 1 & -1 & 0 & | & 1 \\ 0 & 0 & 0 & 0 & | & 0 \\ 0 & 0 & 0 & 0 & | & 0 \end{bmatrix}$

14. $\begin{bmatrix} 1 & 0 & 3 & -1 & | & 4 \\ 0 & 1 & -2 & 3 & | & 2 \\ 0 & 0 & 0 & 0 & | & 0 \\ 0 & 0 & 0 & 0 & | & 0 \end{bmatrix}$

In Exercises 15–36, solve the system of linear equations, using the Gauss–Jordan elimination method.

15. $\begin{aligned} 2x - y &= 3 \\ x + 2y &= 4 \\ 2x + 3y &= 7 \end{aligned}$

16. $\begin{aligned} x + 2y &= 3 \\ 2x - 3y &= -8 \\ x - 4y &= -9 \end{aligned}$

17. $\begin{aligned} 3x - 2y &= -3 \\ 2x + y &= 3 \\ x - 2y &= -5 \end{aligned}$

18. $\begin{aligned} 2x + 3y &= 2 \\ x + 3y &= -2 \\ x - y &= 3 \end{aligned}$

19. $\begin{aligned} 3x - 2y &= 5 \\ -x + 3y &= -4 \\ 2x - 4y &= 6 \end{aligned}$

20. $\begin{aligned} 4x + 6y &= 8 \\ 3x - 2y &= -7 \\ x + 3y &= 5 \end{aligned}$

21. $\begin{aligned} x - 2y &= 2 \\ 7x - 14y &= 14 \\ 3x - 6y &= 6 \end{aligned}$

22. $\begin{aligned} 3x - y + 2z &= 5 \\ x - y + 2z &= 1 \\ 5x - 2y + 4z &= 12 \end{aligned}$

23. $\begin{aligned} x + 2y + z &= -2 \\ -2x - 3y - z &= 1 \\ 2x + 4y + 2z &= -4 \end{aligned}$

24. $\begin{aligned} 3y + 2z &= 4 \\ 2x - y - 3z &= 3 \\ 2x + 2y - z &= 7 \end{aligned}$

25. $\begin{aligned} 3x + 2y &= 4 \\ -\tfrac{3}{2}x - y &= -2 \\ 6x + 4y &= 8 \end{aligned}$

26. $\begin{aligned} 2x_1 - x_2 + x_3 &= -4 \\ 3x_1 - \tfrac{3}{2}x_2 + \tfrac{3}{2}x_3 &= -6 \\ -6x_1 + 3x_2 - 3x_3 &= 12 \end{aligned}$

27. $\begin{aligned} x + y - 2z &= -3 \\ 2x - y + 3z &= 7 \\ x - 2y + 5z &= 0 \end{aligned}$

28. $\begin{aligned} 2x_1 + 6x_2 - 5x_3 &= 5 \\ x_1 + 3x_2 + x_3 + 7x_4 &= -1 \\ 3x_1 + 9x_2 - x_3 + 13x_4 &= 1 \end{aligned}$

29. $\begin{aligned} x - 2y + 3z &= 4 \\ 2x + 3y - z &= 2 \\ x + 2y - 3z &= -6 \end{aligned}$

30. $\begin{aligned} x_1 - 2x_2 + x_3 &= -3 \\ 2x_1 + x_2 - 2x_3 &= 2 \\ x_1 + 3x_2 - 3x_3 &= 5 \end{aligned}$

31. $\begin{aligned} 4x + y - z &= 4 \\ 8x + 2y - 2z &= 8 \end{aligned}$

32. $\begin{aligned} x_1 + 2x_2 + 4x_3 &= 2 \\ x_1 + x_2 + 2x_3 &= 1 \end{aligned}$

33. $\begin{aligned} 2x + y - 3z &= 1 \\ x - y + 2z &= 1 \\ 5x - 2y + 3z &= 6 \end{aligned}$

34. $\begin{aligned} 3x - 9y + 6z &= -12 \\ x - 3y + 2z &= -4 \\ 2x - 6y + 4z &= 8 \end{aligned}$

35. $\begin{aligned} x + 2y - z &= -4 \\ 2x + y + z &= 7 \\ x + 3y + 2z &= 7 \\ x - 3y + z &= 9 \end{aligned}$

36. $\begin{aligned} 3x - 2y + z &= 4 \\ x + 3y - 4z &= -3 \\ 2x - 3y + 5z &= 7 \\ x - 8y + 9z &= 10 \end{aligned}$

37. **MANAGEMENT DECISIONS** The management of Hartman Rent-A-Car has allocated $1,512,000 to purchase 60 new automobiles to add to the existing fleet of rental cars. The company will choose from compact, mid-sized, and full-sized cars costing $18,000, $28,800, and $39,600 each, respectively. Find formulas giving the options available to the company. Give two specific options. (*Note*: Your answers will *not* be unique.)

38. **DIET PLANNING** A dietitian wishes to plan a meal around three foods. The meal is to include 8800 units of vitamin A, 3380 units of vitamin C, and 1020 units of calcium. The number of units of the vitamins and calcium in each ounce of the foods is summarized in the following table:

	Food I	Food II	Food III
Vitamin A	400	1200	800
Vitamin C	110	570	340
Calcium	90	30	60

Determine the amount of each food the dietitian should include in the meal to meet the vitamin and calcium requirements.

39. **DIET PLANNING** Refer to Exercise 38. In planning for another meal, the dietitian changes the requirement of vitamin C from 3380 units to 2160 units. All other requirements remain the same. Show that such a meal cannot be planned around the same foods.

40. **PRODUCTION SCHEDULING** Ace Novelty manufactures Giant Pandas, Saint Bernards, and Big Birds. Each Giant Panda requires 1.5 yd^2 of plush, 30 ft^3 of stuffing, and 5 pieces of trim; each Saint Bernard requires 2 yd^2 of plush, 35 ft^3 of stuffing, and 8 pieces of trim; and each Big Bird requires 2.5 yd^2 of plush, 25 ft^3 of stuffing, and 15 pieces of trim. If 4700 yd^2 of plush, 65,000 ft^3 of stuffing, and 23,400 pieces of trim are available, how many of each of the stuffed animals should the company manufacture if all the material is to be used? Give two specific options.

41. **ASSET ALLOCATION** Mr. and Mrs. Garcia have a total of $100,000 to be invested in stocks, bonds, and a money market account. The stocks have a rate of return of 6%/year, while the bonds and the money market account pay 4%/year and 2%/year, respectively. The Garcias have stipulated that the amount invested in stocks should be equal to the sum of the amount invested in bonds and 3 times the amount invested in the money market account. How should the Garcias allocate their resources if they require an annual income of $5,000 from their investments? Give two specific options.

42. **TRAFFIC CONTROL** The accompanying figure shows the flow of traffic near a city's Civic Center during the rush hours on a typical weekday. Each road can handle a maximum of 1000 cars/hr without causing congestion. The flow of traffic is controlled by traffic lights at each of the five intersections.

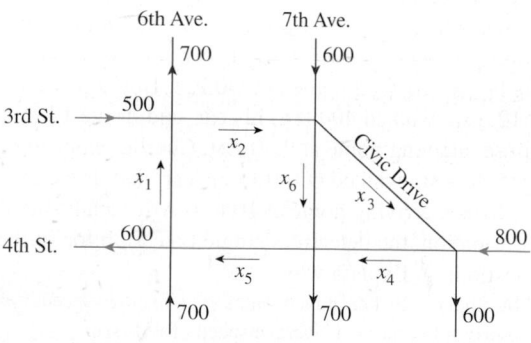

a. Set up a system of linear equations describing the traffic flow.

b. Solve the system devised in part (a), and suggest two possible traffic-flow patterns that will ensure no traffic congestion.

c. Suppose 7th Avenue between 3rd and 4th Streets is soon to be closed for road repairs. Find one possible traffic-flow pattern that will result in a smooth flow of traffic.

43. Traffic Control The accompanying figure shows the flow of downtown traffic during the rush hours on a typical weekday. Each avenue can handle up to 1500 vehicles/hr without causing congestion, whereas the maximum capacity of each street is 1000 vehicles/hr. The flow of traffic is controlled by traffic lights at each of the six intersections.

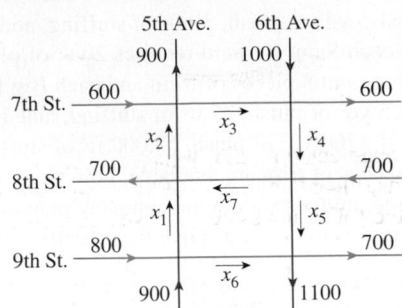

a. Set up a system of linear equations describing the traffic flow.
b. Solve the system devised in part (a), and suggest two possible traffic-flow patterns that will ensure no traffic congestion.
c. Suppose the traffic flow along 9th Street between 5th and 6th Avenues, x_6, is restricted because of sewer construction. What is the minimum permissible traffic flow along this road that will not result in traffic congestion?

44. Criminal Justice The body of Alisha was discovered in the basement of the company where she worked. The medical examiner determined that she was killed between 9 P.M. and 10 P.M. After a preliminary investigation, homicide detectives decided to question three of her coworkers: Alex, Bob, and Charlie. The detectives were told by the suspects that Alex left work at 9:18 P.M. on the day of the homicide, walked 1000 ft to his car, and drove 16 mi to his house, arriving home at 9:40 P.M. Bob left work at 9:12 P.M., walked 400 ft to his car, and drove 12 mi to his house, arriving home at 9:30 P.M. Charlie punched out at 9:35:30 P.M., walked 600 ft to his car, and drove 20 mi to his house, arriving home at 10 P.M. After analyzing this information, the detectives singled out Bob for further questioning. Explain why.

Hint: Suppose that each man walks at an average speed of v ft/sec and drives his car at an average speed of w ft/sec.

(i) Show that this leads to the equations

$$\frac{1000}{v} + \frac{84{,}480}{w} = 1320 \qquad \frac{400}{v} + \frac{63{,}360}{w} = 1080$$
$$\frac{600}{v} + \frac{105{,}600}{w} = 1470$$

(ii) Show that these equations can be expressed in the form

$$1000x + 84{,}480y = 1320 \qquad 400x + 63{,}360y = 1080$$
$$600x + 105{,}600y = 1470$$

Then solve each possible pair of equations in System (ii).

45. Determine the value of k such that the following system of linear equations has a solution, and then find the solution:

$$2x + 3y = 2$$
$$x + 4y = 6$$
$$5x + ky = 2$$

46. Determine the value of k such that the following system of linear equations has infinitely many solutions, and then find the solutions:

$$3x + 4y = 2$$
$$kx + 8y = 4$$
$$12x + 16y = 8$$

47. Determine the value of k such that the following system of linear equations has infinitely many solutions, and then find the solutions:

$$3x + 2y - z = 8$$
$$2x - 2z = 4$$
$$y + kz = 1$$

48. Determine the value of k such that the following system of linear equations has infinitely many solutions, and then find the solutions:

$$3x - 2y + 4z = 12$$
$$-9x + 6y - 12z = k$$

49. Determine the value(s) of k such that the following system of linear equations has no solution:

$$2x - 3y + 4z = 12$$
$$6x - 9y + kz = 36$$

50. Solve the system by making the appropriate substitution:

$$x^2 + 2y^2 = 9$$
$$2x^2 - 3y^2 = -10$$

51. Solve the system by making the appropriate substitution:

$$\frac{3}{x} - \frac{4}{y} = -15$$
$$\frac{5}{x} + \frac{6}{y} = 13$$

52. Solve the system by making the appropriate substitution:

$$\frac{1}{x} + \frac{1}{y} + \frac{1}{z} = -1$$
$$\frac{2}{x} + \frac{3}{y} + \frac{2}{z} = 3$$
$$\frac{2}{x} + \frac{1}{y} + \frac{2}{z} = -7$$

In Exercises 53 and 54, determine whether the statement is true or false. If it is true, explain why it is true. If it is false, give an example to show why it is false.

53. A system of linear equations having fewer equations than variables has no solution, a unique solution, or infinitely many solutions.

54. A system of linear equations having more equations than variables has no solution, a unique solution, or infinitely many solutions.

5.3 Solutions to Self-Check Exercises

1. Let x, y, and z denote the variables. Then the given row-reduced augmented matrix tells us that the system of linear equations is equivalent to the two equations

$$x \quad - z = \quad 3$$
$$y + 5z = -2$$

Letting $z = t$, where t is a parameter, we find the infinitely many solutions given by

$$x = t + 3$$
$$y = -5t - 2$$
$$z = t$$

The last augmented matrix, which is in row-reduced form, tells us that the given system of linear equations is equivalent to the following system of two equations:

$$x \quad + 2z = \quad 0$$
$$y + \quad z = -2$$

Letting $z = t$, where t is a parameter, we see that the infinitely many solutions are given by

$$x = -2t$$
$$y = -t - 2$$
$$z = t$$

2. We obtain the following sequence of equivalent augmented matrices:

$$\begin{bmatrix} 2 & -3 & 1 & | & 6 \\ 1 & 2 & 4 & | & -4 \\ 1 & -5 & -3 & | & 10 \end{bmatrix} \xrightarrow{R_1 \leftrightarrow R_2}$$

$$\begin{bmatrix} ① & 2 & 4 & | & -4 \\ 2 & -3 & 1 & | & 6 \\ 1 & -5 & -3 & | & 10 \end{bmatrix} \xrightarrow[R_3 - R_1]{R_2 - 2R_1}$$

$$\begin{bmatrix} 1 & 2 & 4 & | & -4 \\ 0 & ⑦ & -7 & | & 14 \\ 0 & -7 & -7 & | & 14 \end{bmatrix} \xrightarrow{-\frac{1}{7}R_2}$$

$$\begin{bmatrix} 1 & 2 & 4 & | & -4 \\ 0 & 1 & 1 & | & -2 \\ 0 & -7 & -7 & | & 14 \end{bmatrix} \xrightarrow[R_3 + 7R_2]{R_1 - 2R_2} \begin{bmatrix} 1 & 0 & 2 & | & 0 \\ 0 & 1 & 1 & | & -2 \\ 0 & 0 & 0 & | & 0 \end{bmatrix}$$

3. We obtain the following sequence of equivalent augmented matrices:

$$\begin{bmatrix} ① & -2 & 3 & | & 9 \\ 2 & 3 & -1 & | & 4 \\ 1 & 5 & -4 & | & 2 \end{bmatrix} \xrightarrow[R_3 - R_1]{R_2 - 2R_1}$$

$$\begin{bmatrix} 1 & -2 & 3 & | & 9 \\ 0 & 7 & -7 & | & -14 \\ 0 & 7 & -7 & | & -7 \end{bmatrix} \xrightarrow{R_3 - R_2} \begin{bmatrix} 1 & -2 & 3 & | & 9 \\ 0 & 7 & -7 & | & -14 \\ 0 & 0 & 0 & | & 7 \end{bmatrix}$$

Since the last row of the final augmented matrix is equivalent to the equation $0 = 7$, a contradiction, we conclude that the given system has no solution.

USING TECHNOLOGY

Systems of Linear Equations: Underdetermined and Overdetermined Systems

We can use the row operations of a graphing utility to solve a system of m linear equations in n unknowns by the Gauss–Jordan method, as we did in the previous technology section. We can also use the **rref** or equivalent operation to obtain the row-reduced form without going through all the steps of the Gauss–Jordan method. The **SIMULT** function, however, cannot be used to solve a system in which the number of equations and the number of variables are not the same.

EXAMPLE 1 Solve the system

$$\begin{aligned} x_1 - 2x_2 + 4x_3 &= 2 \\ 2x_1 + x_2 - 2x_3 &= -1 \\ 3x_1 - x_2 + 2x_3 &= 1 \\ 2x_1 + 6x_2 - 12x_3 &= -6 \end{aligned}$$

Solution First, we enter the augmented matrix A into the calculator as

$$A = \begin{bmatrix} 1 & -2 & 4 & 2 \\ 2 & 1 & -2 & -1 \\ 3 & -1 & 2 & 1 \\ 2 & 6 & -12 & -6 \end{bmatrix}$$

Then using the **rref** or equivalent operation, we obtain the equivalent matrix

$$\begin{bmatrix} 1 & 0 & 0 & 0 \\ 0 & 1 & -2 & -1 \\ 0 & 0 & 0 & 0 \\ 0 & 0 & 0 & 0 \end{bmatrix}$$

in reduced form. Thus, the given system is equivalent to

$$\begin{aligned} x_1 &= 0 \\ x_2 - 2x_3 &= -1 \end{aligned}$$

If we let $x_3 = t$, where t is a parameter, then we find that the solutions are $(0, 2t - 1, t)$.

TECHNOLOGY EXERCISES

Use a graphing utility to solve the system of equations using the rref or equivalent operation.

1.
$$\begin{aligned} 2x_1 - x_2 - x_3 &= 0 \\ 3x_1 - 2x_2 - x_3 &= -1 \\ -x_1 + 2x_2 - x_3 &= 3 \\ 2x_2 - 2x_3 &= 4 \end{aligned}$$

2.
$$\begin{aligned} 3x_1 + x_2 - 4x_3 &= 5 \\ 2x_1 - 3x_2 + 2x_3 &= -4 \\ -x_1 - 2x_2 + 4x_3 &= 6 \\ 4x_1 + 3x_2 - 5x_3 &= 9 \end{aligned}$$

3.
$$\begin{aligned} 2x_1 + 3x_2 + 2x_3 + x_4 &= -1 \\ x_1 - x_2 + x_3 - 2x_4 &= -8 \\ 5x_1 + 6x_2 - 2x_3 + 2x_4 &= 11 \\ x_1 + 3x_2 + 8x_3 + x_4 &= -14 \end{aligned}$$

4.
$$\begin{aligned} x_1 - x_2 + 3x_3 - 6x_4 &= 2 \\ x_1 + x_2 + x_3 - 2x_4 &= 2 \\ -2x_1 - x_2 + x_3 + 2x_4 &= 0 \end{aligned}$$

5.
$$\begin{aligned} x_1 + x_2 - x_3 - x_4 &= -1 \\ x_1 - x_2 + x_3 + 4x_4 &= -6 \\ 3x_1 + x_2 - x_3 + 2x_4 &= -4 \\ 5x_1 + x_2 - 3x_3 + x_4 &= -9 \end{aligned}$$

6.
$$\begin{aligned} 1.2x_1 - 2.3x_2 + 4.2x_3 + 5.4x_4 - 1.6x_5 &= 4.2 \\ 2.3x_1 + 1.4x_2 - 3.1x_3 + 3.3x_4 - 2.4x_5 &= 6.3 \\ 1.7x_1 + 2.6x_2 - 4.3x_3 + 7.2x_4 - 1.8x_5 &= 7.8 \\ 2.6x_1 - 4.2x_2 + 8.3x_3 - 1.6x_4 + 2.5x_5 &= 6.4 \end{aligned}$$

5.4 Matrices

Using Matrices to Represent Data

Many practical problems are solved by using arithmetic operations on the data associated with the problems. By properly organizing the data into *blocks* of numbers, we can then carry out these arithmetic operations in an orderly and efficient manner. In particular, this systematic approach enables us to use the computer to full advantage.

Let's begin by considering how the monthly output data of a manufacturer may be organized. The Acrosonic Company manufactures four different loudspeaker systems at three separate locations. The company's May output is described in Table 1.

TABLE 1	Model A	Model B	Model C	Model D
Location I	320	280	460	280
Location II	480	360	580	0
Location III	540	420	200	880

Now, if we agree to preserve the relative location of each entry in Table 1, we can summarize the set of data as follows:

$$\begin{bmatrix} 320 & 280 & 460 & 280 \\ 480 & 360 & 580 & 0 \\ 540 & 420 & 200 & 880 \end{bmatrix}$$

A matrix summarizing the data in Table 1

The array of numbers displayed here is an example of a matrix. Observe that the numbers in row 1 give the output of models A, B, C, and D of Acrosonic loudspeaker systems manufactured at Location I; similarly, the numbers in rows 2 and 3 give the respective outputs of these loudspeaker systems at Locations II and III. The numbers in each column of the matrix give the outputs of a particular model of loudspeaker system manufactured at each of the company's three manufacturing locations.

More generally, a matrix is a rectangular array of real numbers. For example, each of the following arrays is a matrix:

$$A = \begin{bmatrix} 3 & 0 & -1 \\ 2 & 1 & 4 \end{bmatrix} \qquad B = \begin{bmatrix} 3 & 2 \\ 0 & 1 \\ -1 & 4 \end{bmatrix} \qquad C = \begin{bmatrix} 1 \\ 2 \\ 4 \\ 0 \end{bmatrix} \qquad D = \begin{bmatrix} 1 & 3 & 0 & 1 \end{bmatrix}$$

The real numbers that make up the array are called the **entries,** or *elements,* of the matrix. The entries in a row in the array are referred to as a **row** of the matrix, whereas the entries in a column in the array are referred to as a **column** of the matrix. Matrix A, for example, has two rows and three columns, which may be identified as follows:

$$\begin{array}{ccc} & \text{Column 1} & \text{Column 2} & \text{Column 3} \\ \text{Row 1} & \begin{bmatrix} 3 & 0 & -1 \\ \text{Row 2} & 2 & 1 & 4 \end{bmatrix} \end{array}$$

A 2×3 matrix

The **size,** or *dimension,* **of a matrix** is described in terms of the number of rows and columns of the matrix. For example, matrix A has two rows and three columns and is said to have size 2 by 3, denoted 2×3. In general, a matrix having m rows and n columns is said to have size $m \times n$.

> **Matrix**
>
> A **matrix** is an ordered rectangular array of numbers. A matrix with m rows and n columns has size $m \times n$. The entry in the ith row and jth column of a matrix A is denoted by a_{ij}.

A matrix of size $1 \times n$—a matrix having one row and n columns—is referred to as a **row matrix,** or *row vector*, of dimension n. For example, the matrix D is a row vector of dimension 4. Similarly, a matrix having m rows and one column is referred to as a **column matrix,** or *column vector*, of dimension m. The matrix C is a column vector of dimension 4. Finally, an $n \times n$ matrix—that is, a matrix having the same number of rows as columns—is called a **square matrix.** For example, the matrix

$$\text{row} \quad \begin{bmatrix} -3 & 8 & 6 \\ 2 & \tfrac{1}{4} & 4 \\ 1 & 3 & 2 \end{bmatrix} \text{column}$$

A 3×3 square matrix

is a square matrix of size 3×3, or simply of size 3. Here, $a_{23} = 4$.

 APPLIED EXAMPLE 1 Organizing Production Data Consider the matrix

$$P = \begin{bmatrix} 320 & 280 & 460 & 280 \\ 480 & 360 & 580 & 0 \\ 540 & 420 & 200 & 880 \end{bmatrix}$$

representing the output of loudspeaker systems of the Acrosonic Company discussed earlier (see Table 1).

a. What is the size of the matrix P?

b. Find p_{24} (the entry in row 2 and column 4 of the matrix P), and give an interpretation of this number.

c. Find the sum of the entries that make up row 1 of P, and interpret the result.

d. Find the sum of the entries that make up column 4 of P, and interpret the result.

Solution

a. The matrix P has three rows and four columns and hence has size 3×4.

b. The required entry lies in row 2 and column 4 and is the number 0. This means that no model D loudspeaker system was manufactured at Location II in May.

c. The required sum is given by

$$320 + 280 + 460 + 280 = 1340$$

which gives the total number of loudspeaker systems manufactured at Location I in May as 1340 units.

d. The required sum is given by

$$280 + 0 + 880 = 1160$$

giving the output of model D loudspeaker systems at all locations of the company in May as 1160 units.

Equality of Matrices

Two matrices are said to be *equal* if they have the same size and their corresponding entries are equal. For example,

$$\begin{bmatrix} 2 & 3 & 1 \\ 4 & 6 & 2 \end{bmatrix} = \begin{bmatrix} (3-1) & 3 & 1 \\ 4 & (4+2) & 2 \end{bmatrix}$$

Also,

$$\begin{bmatrix} 1 & 3 & 5 \\ 2 & 4 & 3 \end{bmatrix} \neq \begin{bmatrix} 1 & 2 \\ 3 & 4 \\ 5 & 3 \end{bmatrix}$$

since the matrix on the left has size 2×3, whereas the matrix on the right has size 3×2, and

$$\begin{bmatrix} 2 & 3 \\ 4 & 6 \end{bmatrix} \neq \begin{bmatrix} 2 & 3 \\ 4 & 7 \end{bmatrix}$$

since the corresponding elements in row 2 and column 2 of the two matrices are not equal.

> **Equality of Matrices**
>
> Two matrices are equal if they have the same size and their corresponding entries are equal.

EXAMPLE 2 Solve the following matrix equation for x, y, and z:

$$\begin{bmatrix} 1 & x & 3 \\ 2 & y-1 & 2 \end{bmatrix} = \begin{bmatrix} 1 & 4 & z \\ 2 & 1 & 2 \end{bmatrix}$$

Solution Since the corresponding elements of the two matrices must be equal, we find that $x = 4$, $z = 3$, and $y - 1 = 1$, or $y = 2$. ∎

Addition and Subtraction

Two matrices A and B of the *same size* can be added or subtracted to produce a matrix of the same size. This is done by adding or subtracting the corresponding entries in the two matrices. For example,

$$\begin{bmatrix} 1 & 3 & 4 \\ -1 & 2 & 0 \end{bmatrix} + \begin{bmatrix} 1 & 4 & 3 \\ 6 & 1 & -2 \end{bmatrix} = \begin{bmatrix} 1+1 & 3+4 & 4+3 \\ -1+6 & 2+1 & 0+(-2) \end{bmatrix} = \begin{bmatrix} 2 & 7 & 7 \\ 5 & 3 & -2 \end{bmatrix}$$

Adding two matrices of the same size

and

$$\begin{bmatrix} 1 & 2 \\ -1 & 3 \\ 4 & 0 \end{bmatrix} - \begin{bmatrix} 2 & -1 \\ 3 & 2 \\ -1 & 0 \end{bmatrix} = \begin{bmatrix} 1-2 & 2-(-1) \\ -1-3 & 3-2 \\ 4-(-1) & 0-0 \end{bmatrix} = \begin{bmatrix} -1 & 3 \\ -4 & 1 \\ 5 & 0 \end{bmatrix}$$

Subtracting two matrices of the same size

> **Addition and Subtraction of Matrices**
>
> If A and B are two matrices of the same size, then:
>
> **1.** The *sum* $A + B$ is the matrix obtained by adding the corresponding entries in the two matrices.
>
> **2.** The *difference* $A - B$ is the matrix obtained by subtracting the corresponding entries in B from those in A.

APPLIED EXAMPLE 3 Organizing Production Data The total output of Acrosonic for June is shown in Table 2.

TABLE 2				
	Model A	**Model B**	**Model C**	**Model D**
Location I	210	180	330	180
Location II	400	300	450	40
Location III	420	280	180	740

The output for May was given earlier, in Table 1. Find the total output of the company for May and June.

Solution As we saw earlier, the production matrix for Acrosonic in May is given by

$$A = \begin{bmatrix} 320 & 280 & 460 & 280 \\ 480 & 360 & 580 & 0 \\ 540 & 420 & 200 & 880 \end{bmatrix}$$

Next, from Table 2, we see that the production matrix for June is given by

$$B = \begin{bmatrix} 210 & 180 & 330 & 180 \\ 400 & 300 & 450 & 40 \\ 420 & 280 & 180 & 740 \end{bmatrix}$$

Finally, the total output of Acrosonic for May and June is given by the matrix

$$A + B = \begin{bmatrix} 320 & 280 & 460 & 280 \\ 480 & 360 & 580 & 0 \\ 540 & 420 & 200 & 880 \end{bmatrix} + \begin{bmatrix} 210 & 180 & 330 & 180 \\ 400 & 300 & 450 & 40 \\ 420 & 280 & 180 & 740 \end{bmatrix}$$

$$= \begin{bmatrix} 530 & 460 & 790 & 460 \\ 880 & 660 & 1030 & 40 \\ 960 & 700 & 380 & 1620 \end{bmatrix}$$

The following laws hold for matrix addition.

> **Laws for Matrix Addition**
>
> If A, B, and C are matrices of the same size, then
>
> **1.** $A + B = B + A$ Commutative law
> **2.** $(A + B) + C = A + (B + C)$ Associative law

The *commutative law* for matrix addition states that the order in which matrix addition is performed is immaterial. The *associative law* states that, when adding three matrices together, we may first add A and B and then add the resulting sum to C. Equivalently, we can add A to the sum of B and C.

EXAMPLE 4 Let

$$A = \begin{bmatrix} 2 & 1 \\ 3 & -2 \\ 1 & 0 \end{bmatrix} \qquad B = \begin{bmatrix} -1 & 2 \\ 3 & 0 \\ 2 & 4 \end{bmatrix} \qquad C = \begin{bmatrix} 1 & 1 \\ 2 & 3 \\ 0 & -1 \end{bmatrix}$$

a. Show that $A + B = B + A$.
b. Show that $(A + B) + C = A + (B + C)$.

Solution

a. $A + B = \begin{bmatrix} 2 & 1 \\ 3 & -2 \\ 1 & 0 \end{bmatrix} + \begin{bmatrix} -1 & 2 \\ 3 & 0 \\ 2 & 4 \end{bmatrix} = \begin{bmatrix} 2 + (-1) & 1 + 2 \\ 3 + 3 & -2 + 0 \\ 1 + 2 & 0 + 4 \end{bmatrix} = \begin{bmatrix} 1 & 3 \\ 6 & -2 \\ 3 & 4 \end{bmatrix}$

On the other hand,

$B + A = \begin{bmatrix} -1 & 2 \\ 3 & 0 \\ 2 & 4 \end{bmatrix} + \begin{bmatrix} 2 & 1 \\ 3 & -2 \\ 1 & 0 \end{bmatrix} = \begin{bmatrix} -1 + 2 & 2 + 1 \\ 3 + 3 & 0 + (-2) \\ 2 + 1 & 4 + 0 \end{bmatrix} = \begin{bmatrix} 1 & 3 \\ 6 & -2 \\ 3 & 4 \end{bmatrix}$

so $A + B = B + A$, as was to be shown.

b. Using the results of part (a), we have

$$(A + B) + C = \begin{bmatrix} 1 & 3 \\ 6 & -2 \\ 3 & 4 \end{bmatrix} + \begin{bmatrix} 1 & 1 \\ 2 & 3 \\ 0 & -1 \end{bmatrix} = \begin{bmatrix} 2 & 4 \\ 8 & 1 \\ 3 & 3 \end{bmatrix}$$

Next,

$$B + C = \begin{bmatrix} -1 & 2 \\ 3 & 0 \\ 2 & 4 \end{bmatrix} + \begin{bmatrix} 1 & 1 \\ 2 & 3 \\ 0 & -1 \end{bmatrix} = \begin{bmatrix} 0 & 3 \\ 5 & 3 \\ 2 & 3 \end{bmatrix}$$

so

$$A + (B + C) = \begin{bmatrix} 2 & 1 \\ 3 & -2 \\ 1 & 0 \end{bmatrix} + \begin{bmatrix} 0 & 3 \\ 5 & 3 \\ 2 & 3 \end{bmatrix} = \begin{bmatrix} 2 & 4 \\ 8 & 1 \\ 3 & 3 \end{bmatrix}$$

This shows that $(A + B) + C = A + (B + C)$. ∎

A *zero matrix* is one in which all entries are zero. A zero matrix O has the property that

$$A + O = O + A = A$$

for any matrix A having the same size as that of O. For example, the zero matrix of size 3×2 is

$$O = \begin{bmatrix} 0 & 0 \\ 0 & 0 \\ 0 & 0 \end{bmatrix}$$

If A is any 3×2 matrix, then

$$A + O = \begin{bmatrix} a_{11} & a_{12} \\ a_{21} & a_{22} \\ a_{31} & a_{32} \end{bmatrix} + \begin{bmatrix} 0 & 0 \\ 0 & 0 \\ 0 & 0 \end{bmatrix} = \begin{bmatrix} a_{11} & a_{12} \\ a_{21} & a_{22} \\ a_{31} & a_{32} \end{bmatrix} = A$$

where a_{ij} denotes the entry in the ith row and jth column of the matrix A.

The matrix that is obtained by interchanging the rows and columns of a given matrix A is called the *transpose* of A and is denoted A^T. For example, if

$$A = \begin{bmatrix} 1 & 2 & 3 \\ 4 & 5 & 6 \\ 7 & 8 & 9 \end{bmatrix}$$

then

$$A^T = \begin{bmatrix} 1 & 4 & 7 \\ 2 & 5 & 8 \\ 3 & 6 & 9 \end{bmatrix}$$

> **Transpose of a Matrix**
>
> If A is an $m \times n$ matrix with elements a_{ij}, then the **transpose** of A is the $n \times m$ matrix A^T with elements a_{ji}.

Scalar Multiplication

A matrix A may be multiplied by a real number c, called a **scalar.** The scalar product, denoted by cA, is a matrix obtained by multiplying each entry of A by c. For example, the scalar product of the matrix

$$A = \begin{bmatrix} 3 & -1 & 2 \\ 0 & 1 & 4 \end{bmatrix}$$

and the scalar 3 is the matrix

$$3A = 3 \begin{bmatrix} 3 & -1 & 2 \\ 0 & 1 & 4 \end{bmatrix} = \begin{bmatrix} 9 & -3 & 6 \\ 0 & 3 & 12 \end{bmatrix}$$

> **Scalar Product**
>
> If A is a matrix and c is a real number, then the **scalar product** cA is the matrix obtained by multiplying each entry of A by c.

EXAMPLE 5 Given

$$A = \begin{bmatrix} 3 & 4 \\ -1 & 2 \end{bmatrix} \quad \text{and} \quad B = \begin{bmatrix} 3 & 2 \\ -1 & 2 \end{bmatrix}$$

find a matrix X satisfying the *matrix equation* $2X + B = 3A$.

Solution From the given equation $2X + B = 3A$, we find that

$$2X = 3A - B$$

$$= 3 \begin{bmatrix} 3 & 4 \\ -1 & 2 \end{bmatrix} - \begin{bmatrix} 3 & 2 \\ -1 & 2 \end{bmatrix}$$

$$= \begin{bmatrix} 9 & 12 \\ -3 & 6 \end{bmatrix} - \begin{bmatrix} 3 & 2 \\ -1 & 2 \end{bmatrix} = \begin{bmatrix} 6 & 10 \\ -2 & 4 \end{bmatrix}$$

$$X = \frac{1}{2} \begin{bmatrix} 6 & 10 \\ -2 & 4 \end{bmatrix} = \begin{bmatrix} 3 & 5 \\ -1 & 2 \end{bmatrix}$$

APPLIED EXAMPLE 6 Production Planning The management of Acrosonic has decided to increase its July production of loudspeaker systems by 10% (over its June output). Find a matrix giving the targeted production for July.

Solution From the results of Example 3, we see that Acrosonic's total output for June may be represented by the matrix

$$B = \begin{bmatrix} 210 & 180 & 330 & 180 \\ 400 & 300 & 450 & 40 \\ 420 & 280 & 180 & 740 \end{bmatrix}$$

The required matrix is given by

$$(1.1)B = 1.1 \begin{bmatrix} 210 & 180 & 330 & 180 \\ 400 & 300 & 450 & 40 \\ 420 & 280 & 180 & 740 \end{bmatrix}$$

$$= \begin{bmatrix} 231 & 198 & 363 & 198 \\ 440 & 330 & 495 & 44 \\ 462 & 308 & 198 & 814 \end{bmatrix}$$

and is interpreted in the usual manner.

5.4 Self-Check Exercises

1. Perform the indicated operations:

$$\begin{bmatrix} 1 & 3 & 2 \\ -1 & 4 & 7 \end{bmatrix} - 3 \begin{bmatrix} 2 & 1 & 0 \\ 1 & 3 & 4 \end{bmatrix}$$

2. Solve the following matrix equation for x, y, and z:

$$\begin{bmatrix} x & 3 \\ z & 2 \end{bmatrix} + \begin{bmatrix} 2-y & z \\ 2-z & -x \end{bmatrix} = \begin{bmatrix} 3 & 7 \\ 2 & 0 \end{bmatrix}$$

3. **GASOLINE SALES** Jack owns two gas stations, one downtown and the other in the Wilshire district. Over two consecutive days, his gas stations recorded gasoline sales represented by the following matrices:

		Regular	Regular plus	Premium
$A =$	Downtown	1200	750	650
	Wilshire	1100	850	600

and

		Regular	Regular plus	Premium
$B =$	Downtown	1250	825	550
	Wilshire	1150	750	750

Find a matrix representing the total sales of the two gas stations over the 2-day period.

Solutions to Self-Check Exercises 5.4 can be found on page 313.

5.4 Concept Questions

1. Define (a) a matrix, (b) the size of a matrix, (c) a row matrix, (d) a column matrix, and (e) a square matrix.

2. When are two matrices equal? Give an example of two matrices that are equal.

3. **a.** What condition on the size of two matrices A and B ensures that the sum of A and B exist?

   b. What condition on the size of a matrix A ensures that the scalar product of cA, where c is a real number, exists?

4. Construct a 3×3 matrix A having the property that $A = A^T$. What special characteristic does A have?

5.4 Exercises

In Exercises 1–6, refer to the following matrices:

$$A = \begin{bmatrix} 2 & -3 & 9 & -4 \\ -11 & 2 & 6 & 7 \\ 6 & 0 & 2 & 9 \\ 5 & 1 & 5 & -8 \end{bmatrix} \quad B = \begin{bmatrix} 3 & -1 & 2 \\ 0 & 1 & 4 \\ 3 & 2 & 1 \\ -1 & 0 & 8 \end{bmatrix}$$

$$C = \begin{bmatrix} 1 & 0 & 3 & 4 & 5 \end{bmatrix} \quad D = \begin{bmatrix} 1 \\ 3 \\ -2 \\ 0 \end{bmatrix}$$

1. What is the size of A? Of B? Of C? Of D?

2. Find a_{14}, a_{21}, a_{31}, and a_{43}.

3. Find b_{13}, b_{31}, and b_{43}.

4. Identify the row matrix. What is its transpose?

5. Identify the column matrix. What is its transpose?

6. Identify the square matrix. What is its transpose?

In Exercises 7–12, refer to the following matrices:

$$A = \begin{bmatrix} -1 & 2 \\ 3 & -2 \\ 4 & 0 \end{bmatrix} \quad B = \begin{bmatrix} 2 & 4 \\ 3 & 1 \\ -2 & 2 \end{bmatrix}$$

$$C = \begin{bmatrix} 3 & -1 & 0 \\ 2 & -2 & 3 \\ 4 & 6 & 2 \end{bmatrix} \quad D = \begin{bmatrix} 2 & -2 & 4 \\ 3 & 6 & 2 \\ -2 & 3 & 1 \end{bmatrix}$$

7. What is the size of A? Of B? Of C? Of D?

8. Explain why the matrix $A + C$ does *not* exist.

9. Compute $A + B$.   **10.** Compute $2A - 3B$.

11. Compute $C - D$.   **12.** Compute $4D - 2C$.

In Exercises 13–22, perform the indicated operations.

13. $\begin{bmatrix} 2 & -1 & 3 \\ 9 & 2 & 1 \end{bmatrix} + \begin{bmatrix} -1 & 2 & -1 \\ -6 & 4 & 2 \end{bmatrix}$

14. $\begin{bmatrix} 6 & 3 & 8 \\ 4 & 5 & 6 \end{bmatrix} - \begin{bmatrix} 3 & -2 & -1 \\ 0 & -5 & -7 \end{bmatrix}$

15. $\begin{bmatrix} 2 & -3 & 4 & -1 \\ 3 & 1 & 0 & 0 \end{bmatrix} + \begin{bmatrix} 4 & 3 & -2 & -4 \\ 6 & 2 & 0 & -3 \end{bmatrix}$

16. $\begin{bmatrix} 1 & 4 & -5 \\ 3 & -8 & 6 \end{bmatrix} + \begin{bmatrix} 4 & 0 & -2 \\ 3 & 6 & 5 \end{bmatrix} - \begin{bmatrix} 2 & 8 & 9 \\ -11 & 2 & -5 \end{bmatrix}$

17. $\begin{bmatrix} 1.2 & 4.5 & -4.2 \\ 8.2 & 6.3 & -3.2 \end{bmatrix} - \begin{bmatrix} 3.1 & 1.5 & -3.6 \\ 2.2 & -3.3 & -4.4 \end{bmatrix}$

18. $\begin{bmatrix} 0.06 & 0.12 \\ 0.43 & 1.11 \\ 1.55 & -0.43 \end{bmatrix} - \begin{bmatrix} 0.77 & -0.75 \\ 0.22 & -0.65 \\ 1.09 & -0.57 \end{bmatrix}$

19. $3\begin{bmatrix} 1 & 1 & -3 \\ 3 & 2 & 3 \\ 7 & -1 & 6 \end{bmatrix} + 4\begin{bmatrix} -2 & -1 & 8 \\ 4 & 2 & 2 \\ 3 & 6 & 3 \end{bmatrix}$

20. $2\begin{bmatrix} 1 & -2 \\ 2 & -1 \\ 3 & 0 \end{bmatrix} - 3\begin{bmatrix} 2 & 3 \\ 1 & -2 \\ 2 & 4 \end{bmatrix}$

21. $\dfrac{1}{2}\begin{bmatrix} 1 & 0 & 0 & -4 \\ 3 & 0 & -1 & 6 \\ -2 & 1 & -4 & 2 \end{bmatrix} + \dfrac{4}{3}\begin{bmatrix} 3 & 0 & -1 & 4 \\ -2 & 1 & -6 & 2 \\ 8 & 2 & 0 & -2 \end{bmatrix}$
$\qquad - \dfrac{1}{3}\begin{bmatrix} 3 & -9 & -1 & 0 \\ 6 & 2 & 0 & -6 \\ 0 & 1 & -3 & 1 \end{bmatrix}$

22. $0.5\begin{bmatrix} 1 & 3 & 5 \\ 5 & 2 & -1 \\ -2 & 0 & 1 \end{bmatrix} - 0.2\begin{bmatrix} 2 & 3 & 4 \\ -1 & 1 & -4 \\ 3 & 5 & -5 \end{bmatrix}$
$\qquad + 0.6\begin{bmatrix} 3 & 4 & -1 \\ 4 & 5 & 1 \\ 1 & 0 & 0 \end{bmatrix}$

In Exercises 23–26, solve for u, x, y, and z in the given matrix equation.

23. $\begin{bmatrix} 2x - 2 & 3 & 2 \\ 2 & 4 & y - 2 \\ 2z & -3 & 2 \end{bmatrix} = \begin{bmatrix} 3 & u & 2 \\ 2 & 4 & 5 \\ 4 & -3 & 2 \end{bmatrix}$

24. $\begin{bmatrix} x & -2 \\ 3 & y \end{bmatrix} + \begin{bmatrix} -2 & z \\ -1 & 2 \end{bmatrix} = \begin{bmatrix} 4 & -2 \\ 2u & 4 \end{bmatrix}$

25. $\begin{bmatrix} 1 & x \\ 2y & -3 \end{bmatrix} - 4\begin{bmatrix} 2 & -2 \\ 0 & 3 \end{bmatrix} = \begin{bmatrix} 3z & 10 \\ 4 & -u \end{bmatrix}$

26. $\begin{bmatrix} 1 & 2 \\ 3 & 4 \\ x & -1 \end{bmatrix} - 3\begin{bmatrix} y - 1 & 2 \\ 1 & 2 \\ 4 & 2z + 1 \end{bmatrix} = 2\begin{bmatrix} -4 & -u \\ 0 & -1 \\ 4 & 4 \end{bmatrix}$

In Exercises 27 and 28, let

$$A = \begin{bmatrix} -2 & 1 \\ 0 & 3 \end{bmatrix} \quad \text{and} \quad B = \begin{bmatrix} 2 & -3 \\ 1 & -2 \end{bmatrix}$$

27. Find a matrix X satisfying the matrix equation $2A + X = 3B$.

28. Find a matrix X satisfying the matrix equation $3X - A + 2B = 0$.

In Exercises 29 and 30, let

$$A = \begin{bmatrix} 2 & -4 & 3 \\ 4 & 2 & 1 \end{bmatrix} \quad B = \begin{bmatrix} 4 & 3 & 2 \\ 1 & 0 & 4 \end{bmatrix} \quad C = \begin{bmatrix} 1 & 0 & 2 \\ 3 & -2 & 1 \end{bmatrix}$$

29. Verify by direct computation the validity of the commutative law for matrix addition.

30. Verify by direct computation the validity of the associative law for matrix addition.

In Exercises 31–34, let

$$A = \begin{bmatrix} 3 & 1 \\ 2 & 4 \\ -4 & 0 \end{bmatrix} \quad \text{and} \quad B = \begin{bmatrix} 1 & 2 \\ -1 & 0 \\ 3 & 2 \end{bmatrix}$$

Verify each equation by direct computation.

31. $(3 + 5)A = 3A + 5A$ **32.** $2(4A) = (2 \cdot 4)A = 8A$

33. $4(A + B) = 4A + 4B$ **34.** $2(A - 3B) = 2A - 6B$

In Exercises 35–38, find the transpose of each matrix.

35. $\begin{bmatrix} 3 & 2 & -1 & 5 \end{bmatrix}$ **36.** $\begin{bmatrix} 4 & 2 & 0 & -1 \\ 3 & 4 & -1 & 5 \end{bmatrix}$

37. $\begin{bmatrix} 1 & -1 & 2 \\ 3 & 4 & 2 \\ 0 & 1 & 0 \end{bmatrix}$ **38.** $\begin{bmatrix} 1 & 2 & 6 & 4 \\ 2 & 3 & 2 & 5 \\ 6 & 2 & 3 & 0 \\ 4 & 5 & 0 & 2 \end{bmatrix}$

39. Cholesterol Levels Mr. Cross, Mr. Jones, and Mr. Smith all suffer from coronary heart disease. As part of their treatment, they were put on special low-cholesterol diets: Cross on Diet I, Jones on Diet II, and Smith on Diet III. Progressive records of each patient's cholesterol level were kept. At the beginning of the first, second, third, and fourth months, the cholesterol levels of the three patients were:

    Cross: 220, 215, 210, and 205
    Jones: 220, 210, 200, and 195
    Smith: 215, 205, 195, and 190

Represent this information in a 3×4 matrix.

40. Mortgage Interest Rates With interest rates low, refinancings account for a large chunk of all mortgage applications. The types of mortgages used for refinancings and home purchases in March 2012 applications are as follows:

	30-year Fixed	15-year Fixed	Fixed (Other)	Adjustable
Refinancing (%)	56.6	24.7	14.6	4.1
Home Purchase (%)	85.4	7.0	1.7	5.9

a. Write a 2×4 matrix A to represent the information.
b. Compare a_{11} with a_{12}, and interpret your result.
c. Compare a_{11} with a_{21}, and interpret your result.
Source: Mortgage Bankers Association.

41. Life Expectancy Figures for life expectancy at birth of Massachusetts residents in 2008 are 82.6, 80.5, and 91.2 years for white, black, and Hispanic women, respectively, and 78.0, 73.9, and 84.8 years for white, black, and Hispanic men, respectively. Express this information using a 2×3 matrix and a 3×2 matrix.
Source: Massachusetts Department of Public Health.

42. Terrorism Poll In a poll surveying 1508 registered California voters in August 2011, the following questions were asked: (a) Has the federal government gone too far in restricting American citizens' civil liberties to keep the country safe from terrorism? (b) Are the people responsible for airline safety in the United States going too far to protect airline passengers? The results of the poll follow.

Question (a): (I) about right (46%), (II) gone too far (33%), (III) not far enough (15%), (IV) don't know (6%).

Question (b): (I) about right (53%), (II) gone too far (24%), (III) not far enough (18%), (IV) don't know (5%).

Express this information using a 2×4 matrix.
Source: Los Angeles Times.

43. Average CD Yields The average certificate of deposit (CD) yields as of January 14, 2013, according to Bankrates' National Survey, were as follows. For the current week: 6-month CD, 0.17%; 1-year CD, 0.27%; $2\frac{1}{2}$-year CD, 0.41%; 5-year CD, 0.87%. For the previous week: 6-month CD, 0.17%; 1-year CD, 0.27%; $2\frac{1}{2}$-year CD, 0.42%; 5-year CD, 0.88%. For a year ago: 6-month CD, 0.22%; 1-year CD, 0.34%; $2\frac{1}{2}$-year CD, 0.52%; 5-year CD, 1.15%.
a. Express this information using a 3×4 matrix A.
b. What are a_{12} and a_{22}? Interpret your result.
c. What are a_{13} and a_{23}? Interpret your result.
d. What are a_{33} and a_{34}? Interpret your result.
Source: Bankrate.com.

44. Investment Portfolios The following table gives the number of shares of certain corporations held by Leslie and Tom in their respective IRA accounts at the beginning of the year:

	IBM	Facebook	Ford	Wal-Mart
Leslie	500	350	200	400
Tom	400	450	300	200

Over the year, they added more shares to their accounts, as shown in the following table:

	IBM	Facebook	Ford	Wal-Mart
Leslie	50	50	0	100
Tom	0	80	100	50

a. Write a matrix A giving the holdings of Leslie and Tom at the beginning of the year and a matrix B giving the shares they have added to their portfolios.
b. Find a matrix C giving their total holdings at the end of the year.

45. Bookstore Inventories The Campus Bookstore's inventory of books is as follows:

Hardcover: textbooks, 5280; fiction, 1680; nonfiction, 2320; reference, 1890

Paperback: fiction, 2810; nonfiction, 1490; reference, 2070; textbooks, 1940

The College Bookstore's inventory of books is as follows:

Hardcover: textbooks, 6340; fiction, 2220; nonfiction, 1790; reference, 1980

Paperback: fiction, 3100; nonfiction, 1720; reference, 2710; textbooks, 2050

a. Represent Campus's inventory as a matrix A.
b. Represent College's inventory as a matrix B.
c. Suppose that the two companies decide to merge. Write a matrix C that represents the total inventory of the newly amalgamated company.

46. Market Share of Motorcycles The market share of motorcycles in the United States in 2011 follows: Hero Moto-Corp, 44.8%; Bajaj Auto, 20.5%; TVS Motor Company, 15.0%; Honda, 13.2%; Suzuki, 2.4%; Yamaha, 2.4%; and others, 1.7%. The corresponding figures for 2012 are 45.2%, 19.1%, 14.1%, 14.9%, 2.5%, 2.6%, and 1.6%, respectively. Express this information in a 2×7 matrix. What is the sum of all the elements in the first row? In the second row? Is this expected? Which company gained the most market share between 2011 and 2012?
Source: Motorcycle Industry Council.

47. Model Investment Portfolios The following table gives Schwab's five model portfolios.

Portfolio Investment (%)	Large-cap	Small-cap	International	Bonds	Cash
Conservative	15	0	5	50	30
Moderately conservative	25	5	10	50	10
Moderate	35	10	15	35	5
Moderately aggressive	45	15	20	15	5
Aggressive	50	20	25	0	5

a. Write a matrix A representing the above data. (Use the order given in the table.)
b. What is a_{12}? Interpret this number.
c. What are $a_{13}, a_{23}, a_{33}, a_{43}$, and a_{53}? What can you conclude from this?
d. Find the sum of each row of matrix A. Are the results expected? Explain.
Source: Schwab Center for Financial Research.

48. Student Test Scores For simplicity, suppose there are five students in a class. Suppose the scores of these students on three tests are as follows:

Student	Test 1	Test 2	Test 3
A	80	84	92
B	84	86	88
C	78	76	82
D	86	82	78
E	92	94	88

a. Write a 5×3 matrix A to represent the information.
b. What are a_{11}, a_{12}, and a_{13}? What is $\frac{1}{3}(a_{11} + a_{12} + a_{13})$, and what does this quantity represent?
c. What are $a_{12}, a_{22}, a_{32}, a_{42}$, and a_{52}? What is $\frac{1}{5}(a_{12} + a_{22} + a_{32} + a_{42} + a_{52})$, and what does this quantity represent?
d. What is the average score of student E based on the three tests?
e. What is the average score of the five students for the third test?

49. Mortgage Rates The mortgage rates for the week ended February 15, 2013, in the New York region follow:

For the 30-year fixed: 3.83% in New York, 3.67% in the New York Co-ops, 3.78% in New Jersey, and 3.79% in Connecticut

For the 15-year fixed: 3.16% in New York, 2.98% in the New York Co-ops, 3.03% in New Jersey, and 3.03% in Connecticut

For the adjustable: 2.99% in New York, 2.96% in the New York Co-ops, 2.97% in New Jersey, and 2.45% in Connecticut

The mortgage rates for the week ended February 22, 2013, in the same New York region follow:

For the 30-year fixed: 3.84% in New York, 3.75% in the New York Co-ops, 3.81% in New Jersey, and 3.80% in Connecticut

For the 15-year fixed: 3.15% in New York, 2.97% in the New York Co-ops, 3.04% in New Jersey, and 3.02% in Connecticut

For the adjustable: 2.99% in New York, 2.95% in the New York Co-ops, 2.96% in New Jersey, and 2.44% in Connecticut

a. Write two 3×4 matrices, A and B, giving the mortgage rates for the three types of loans in the New York region for the week ended February 15, 2013, and for the week ended February 22, 2013, respectively.
b. What are a_{12} and b_{12}? Interpret your result.
c. What are a_{33} and b_{33}? Interpret your result.
d. What is $\frac{1}{4}(a_{11} + a_{12} + a_{13} + a_{14})$? Interpret your result.
e. What is $\frac{1}{2}(a_{34} + b_{34})$? Interpret your result.
Source: HSH.com.

50. BANKING The numbers of three types of bank accounts on January 1 at the Central Bank and its branches are represented by matrix A:

	Checking accounts	Savings accounts	Fixed-deposit accounts
Main office	2820	1470	1120
$A =$ Westside branch	1030	520	480
Eastside branch	1170	540	460

$$A = \begin{bmatrix} 2820 & 1470 & 1120 \\ 1030 & 520 & 480 \\ 1170 & 540 & 460 \end{bmatrix}$$

The number and types of accounts opened during the first quarter are represented by matrix B, and the number and types of accounts closed during the same period are represented by matrix C. Thus,

$$B = \begin{bmatrix} 260 & 120 & 110 \\ 140 & 60 & 50 \\ 120 & 70 & 50 \end{bmatrix} \quad \text{and} \quad C = \begin{bmatrix} 120 & 80 & 80 \\ 70 & 30 & 40 \\ 60 & 20 & 40 \end{bmatrix}$$

a. Find matrix D, which represents the number of each type of account at the end of the first quarter at each location.

b. Because a new manufacturing plant is opening in the immediate area, it is anticipated that there will be a 10% increase in the number of accounts at each location during the second quarter. Write a matrix $E = 1.1D$ to reflect this anticipated increase.

51. HOME SALES K & R Builders build three models of houses, M_1, M_2, and M_3, in three subdivisions I, II, and III located in three different areas of a city. The prices of the houses (in thousands of dollars) are given in matrix A:

	M_1	M_2	M_3
I	340	360	380
$A =$ II	410	430	440
III	620	660	700

K & R Builders has decided to raise the price of each house by 3% next year. Write a matrix B giving the new prices of the houses.

52. HOME SALES K & R Builders build three models of houses, M_1, M_2, and M_3, in three subdivisions I, II, and III located in three different areas of a city. The prices of the homes (in thousands of dollars) are given in matrix A:

	M_1	M_2	M_3
I	340	360	380
$A =$ II	410	430	440
III	620	660	700

The new price schedule for next year, reflecting a uniform percentage increase in each house, is given by matrix B:

	M_1	M_2	M_3
I	357	378	399
$B =$ II	430.5	451.5	462
III	651	693	735

What was the percentage increase in the prices of the houses?

Hint: Find r such that $(1 + 0.01r)A = B$.

In Exercises 53–56, determine whether the statement is true or false. If it is true, explain why it is true. If it is false, give an example to show why it is false.

53. If A and B are matrices of the same size and c is a scalar, then $c(A + B) = cA + cB$.

54. If A and B are matrices of the same size, then $A - B = A + (-1)B$.

55. If A is a matrix and c is a nonzero scalar, then $(cA)^T = (1/c)A^T$.

56. If A is a matrix, then $(A^T)^T = A$.

5.4 Solutions to Self-Check Exercises

1.
$$\begin{bmatrix} 1 & 3 & 2 \\ -1 & 4 & 7 \end{bmatrix} - 3\begin{bmatrix} 2 & 1 & 0 \\ 1 & 3 & 4 \end{bmatrix} = \begin{bmatrix} 1 & 3 & 2 \\ -1 & 4 & 7 \end{bmatrix} - \begin{bmatrix} 6 & 3 & 0 \\ 3 & 9 & 12 \end{bmatrix}$$

$$= \begin{bmatrix} -5 & 0 & 2 \\ -4 & -5 & -5 \end{bmatrix}$$

2. We are given

$$\begin{bmatrix} x & 3 \\ z & 2 \end{bmatrix} + \begin{bmatrix} 2-y & z \\ 2-z & -x \end{bmatrix} = \begin{bmatrix} 3 & 7 \\ 2 & 0 \end{bmatrix}$$

Performing the indicated operation on the left-hand side, we obtain

$$\begin{bmatrix} 2+x-y & 3+z \\ 2 & 2-x \end{bmatrix} = \begin{bmatrix} 3 & 7 \\ 2 & 0 \end{bmatrix}$$

By the equality of matrices, we have

$$2 + x - y = 3$$
$$3 + z = 7$$
$$2 - x = 0$$

from which we deduce that $x = 2$, $y = 1$, and $z = 4$.

3. The required matrix is

$$A + B = \begin{bmatrix} 1200 & 750 & 650 \\ 1100 & 850 & 600 \end{bmatrix} + \begin{bmatrix} 1250 & 825 & 550 \\ 1150 & 750 & 750 \end{bmatrix}$$

$$= \begin{bmatrix} 2450 & 1575 & 1200 \\ 2250 & 1600 & 1350 \end{bmatrix}$$

USING TECHNOLOGY | Matrix Operations

Graphing Utility

A graphing utility can be used to perform matrix addition, matrix subtraction, and scalar multiplication. It can also be used to find the transpose of a matrix.

EXAMPLE 1 Let

$$A = \begin{bmatrix} 1.2 & 3.1 \\ -2.1 & 4.2 \\ 3.1 & 4.8 \end{bmatrix} \quad \text{and} \quad B = \begin{bmatrix} 4.1 & 3.2 \\ 1.3 & 6.4 \\ 1.7 & 0.8 \end{bmatrix}$$

Find (a) $A + B$, (b) $2.1A - 3.2B$, and (c) $(2.1A + 3.2B)^T$.

Solution We first enter the matrices A and B into the calculator.

a. Using matrix operations, we enter the expression $A + B$ and obtain

$$A + B = \begin{bmatrix} 5.3 & 6.3 \\ -0.8 & 10.6 \\ 4.8 & 5.6 \end{bmatrix}$$

b. Using matrix operations, we enter the expression $2.1A - 3.2B$ and obtain

$$2.1A - 3.2B = \begin{bmatrix} -10.6 & -3.73 \\ -8.57 & -11.66 \\ 1.07 & 7.52 \end{bmatrix}$$

c. Using matrix operations, we enter the expression $(2.1A + 3.2B)^T$ and obtain

$$(2.1A + 3.2B)^T = \begin{bmatrix} 15.64 & -0.25 & 11.95 \\ 16.75 & 29.3 & 12.64 \end{bmatrix}$$

APPLIED EXAMPLE 2 Gas Station Sales John operates three gas stations at three locations, I, II, and III. Over two consecutive days, his gas stations recorded the following fuel sales (in gallons):

	Day 1			
	Regular	Regular Plus	Premium	Diesel
Location I	1400	1200	1100	200
Location II	1600	900	1200	300
Location III	1200	1500	800	500

	Day 2			
	Regular	Regular Plus	Premium	Diesel
Location I	1000	900	800	150
Location II	1800	1200	1100	250
Location III	800	1000	700	400

Find a matrix representing the total fuel sales at John's gas stations.

Solution   The fuel sales can be represented by the matrix A (day 1) and matrix B (day 2):

$$A = \begin{bmatrix} 1400 & 1200 & 1100 & 200 \\ 1600 & 900 & 1200 & 300 \\ 1200 & 1500 & 800 & 500 \end{bmatrix} \quad \text{and} \quad B = \begin{bmatrix} 1000 & 900 & 800 & 150 \\ 1800 & 1200 & 1100 & 250 \\ 800 & 1000 & 700 & 400 \end{bmatrix}$$

We enter the matrices A and B into the calculator. Using matrix operations, we enter the expression $A + B$ and obtain

$$A + B = \begin{bmatrix} 2400 & 2100 & 1900 & 350 \\ 3400 & 2100 & 2300 & 550 \\ 2000 & 2500 & 1500 & 900 \end{bmatrix}$$

Excel

First, we show how basic operations on matrices can be carried out by using Excel.

EXAMPLE 3   Given the following matrices,

$$A = \begin{bmatrix} 1.2 & 3.1 \\ -2.1 & 4.2 \\ 3.1 & 4.8 \end{bmatrix} \quad \text{and} \quad B = \begin{bmatrix} 4.1 & 3.2 \\ 1.3 & 6.4 \\ 1.7 & 0.8 \end{bmatrix}$$

a. Compute $A + B$.   **b.** Compute $2.1A - 3.2B$.

Solution

a. First, represent the matrices A and B in a spreadsheet. Enter the elements of each matrix in a block of cells as shown in Figure T1.

	A	B	C	D	E
1		A			B
2	1.2	3.1		4.1	3.2
3	-2.1	4.2		1.3	6.4
4	3.1	4.8		1.7	0.8

FIGURE **T1**
The elements of matrix A and matrix B in a spreadsheet

Second, compute the sum of matrix A and matrix B. Highlight the cells that will contain matrix $A + B$, type =, highlight the cells in matrix A, type +, highlight the cells in matrix B, and press $\boxed{\textbf{Ctrl-Shift-Enter}}$. The resulting matrix $A + B$ is shown in Figure T2.

	A	B
8		A + B
9	5.3	6.3
10	-0.8	10.6
11	4.8	5.6

FIGURE **T2**
The matrix $A + B$

Note: Boldfaced words/characters enclosed in a box (for example, $\boxed{\textbf{Enter}}$) indicate that an action (click, select, or press) is required. Words/characters printed blue (for example, Chart sub-type:) indicate words/characters that appear on the screen. Words/characters printed in a monospace font (for example, `=(-2/3)*A2+2`) indicate words/characters that need to be typed and entered.

b. Highlight the cells that will contain matrix $2.1A - 3.2B$. Type $= 2.1 *$, highlight matrix A, type $-3.2*$, highlight the cells in matrix B, and press $\boxed{\text{Ctrl-Shift-Enter}}$. The resulting matrix $2.1A - 3.2B$ is shown in Figure T3.

	A	B
		2.1A - 3.2B
13		2.1A - 3.2B
14	-10.6	-3.73
15	-8.57	-11.66
16	1.07	7.52

FIGURE T3
The matrix $2.1A - 3.2B$

APPLIED EXAMPLE 4 Gas Station Sales John operates three gas stations at three locations I, II, and III. Over two consecutive days, his gas stations recorded the following fuel sales (in gallons):

	Day 1			
	Regular	**Regular Plus**	**Premium**	**Diesel**
Location I	1400	1200	1100	200
Location II	1600	900	1200	300
Location III	1200	1500	800	500

	Day 2			
	Regular	**Regular Plus**	**Premium**	**Diesel**
Location I	1000	900	800	150
Location II	1800	1200	1100	250
Location III	800	1000	700	400

Find a matrix representing the total fuel sales at John's gas stations.

Solution The fuel sales can be represented by the matrices A (day 1) and B (day 2):

$$A = \begin{bmatrix} 1400 & 1200 & 1100 & 200 \\ 1600 & 900 & 1200 & 300 \\ 1200 & 1500 & 800 & 500 \end{bmatrix} \quad \text{and} \quad B = \begin{bmatrix} 1000 & 900 & 800 & 150 \\ 1800 & 1200 & 1100 & 250 \\ 800 & 1000 & 700 & 400 \end{bmatrix}$$

We first enter the elements of the matrices A and B onto a spreadsheet. Next, we highlight the cells that will contain the matrix $A + B$, type =, highlight A, type +, highlight B, and then press $\boxed{\text{Ctrl-Shift-Enter}}$. The resulting matrix $A + B$ is shown in Figure T4.

	A	B	C	D
23		A + B		
24	2400	2100	1900	350
25	3400	2100	2300	550
26	2000	2500	1500	900

FIGURE T4
The matrix $A + B$

Refer to the following matrices and perform the indicated operations.

$$A = \begin{bmatrix} 1.2 & 3.1 & -5.4 & 2.7 \\ 4.1 & 3.2 & 4.2 & -3.1 \\ 1.7 & 2.8 & -5.2 & 8.4 \end{bmatrix}$$

$$B = \begin{bmatrix} 6.2 & -3.2 & 1.4 & -1.2 \\ 3.1 & 2.7 & -1.2 & 1.7 \\ 1.2 & -1.4 & -1.7 & 2.8 \end{bmatrix}$$

1. $12.5A$

2. $-8.4B$

3. $A - B$

4. $B - A$

5. $1.3A + 2.4B$

6. $2.1A - 1.7B$

7. $3(A + B)$

8. $1.3(4.1A - 2.3B)$

5.5 Multiplication of Matrices

Matrix Product

In Section 5.4, we saw how matrices of the same size may be added or subtracted and how a matrix may be multiplied by a scalar (real number), an operation referred to as scalar multiplication. In this section we see how, with certain restrictions, one matrix may be multiplied by another matrix.

To define matrix multiplication, let's consider the following problem. On a certain day, Al's Service Station sold 1600 gallons of regular, 1000 gallons of regular plus, and 800 gallons of premium gasoline. If the price of gasoline on this day was \$3.79 for regular, \$3.89 for regular plus, and \$3.99 for premium gasoline, find the total revenue realized by Al's for that day.

The day's sale of gasoline may be represented by the matrix

$$A = \begin{bmatrix} 1600 & 1000 & 800 \end{bmatrix} \quad \text{Row matrix } (1 \times 3)$$

Next, we let the unit selling price of regular, regular plus, and premium gasoline be the entries in the matrix

$$B = \begin{bmatrix} 3.79 \\ 3.89 \\ 3.99 \end{bmatrix} \quad \text{Column matrix } (3 \times 1)$$

The first entry in matrix A gives the number of gallons of regular gasoline sold, and the first entry in matrix B gives the selling price for each gallon of regular gasoline, so their product $(1600)(3.79)$ gives the revenue realized from the sale of regular gasoline for the day. A similar interpretation of the second and third entries in the two matrices suggests that we multiply the corresponding entries to obtain the respective revenues realized from the sale of regular, regular plus, and premium gasoline. Finally, the total revenue realized by Al's from the sale of gasoline is given by adding these products to obtain

$$(1600)(3.79) + (1000)(3.89) + (800)(3.99) = 13,146$$

or \$13,146.

This example suggests that if we have a row matrix of size $1 \times n$,

$$A = \begin{bmatrix} a_1 & a_2 & a_3 & \cdots & a_n \end{bmatrix}$$

and a column matrix of size $n \times 1$,

$$B = \begin{bmatrix} b_1 \\ b_2 \\ b_3 \\ \vdots \\ b_n \end{bmatrix}$$

then we may define the **matrix product** of A and B, written AB, by

$$AB = \begin{bmatrix} a_1 & a_2 & a_3 \cdots a_n \end{bmatrix} \begin{bmatrix} b_1 \\ b_2 \\ b_3 \\ \vdots \\ b_n \end{bmatrix} = a_1b_1 + a_2b_2 + a_3b_3 + \cdots + a_nb_n \qquad \text{(12)}$$

EXAMPLE 1 Let

$$A = \begin{bmatrix} 1 & -2 & 3 & 5 \end{bmatrix} \quad \text{and} \quad B = \begin{bmatrix} 2 \\ 3 \\ 0 \\ -1 \end{bmatrix}$$

Then

$$AB = \begin{bmatrix} 1 & -2 & 3 & 5 \end{bmatrix} \begin{bmatrix} 2 \\ 3 \\ 0 \\ -1 \end{bmatrix} = (1)(2) + (-2)(3) + (3)(0) + (5)(-1) = -9$$

 APPLIED EXAMPLE 2 Stock Transactions Judy's stock holdings are given by the matrix

$$A = \begin{array}{ccc} \text{GM} & \text{IBM} & \text{AAPL} \\ \begin{bmatrix} 700 & 400 & 200 \end{bmatrix} \end{array}$$

At the close of trading on a certain day, the prices (in dollars per share) of these stocks are

$$B = \begin{array}{c} \text{GM} \\ \text{IBM} \\ \text{AAPL} \end{array} \begin{bmatrix} 27 \\ 205 \\ 420 \end{bmatrix}$$

What is the total value of Judy's holdings as of that day?

Solution Judy's holdings are worth

$$AB = \begin{bmatrix} 700 & 400 & 200 \end{bmatrix} \begin{bmatrix} 27 \\ 205 \\ 420 \end{bmatrix} = (700)(27) + (400)(205) + (200)(420)$$

$$= 184,900$$

or \$184,900.

Returning once again to the matrix product AB in Equation (12), observe that the number of columns of the row matrix A is *equal* to the number of rows of the column matrix B. Observe further that the product matrix AB has size 1×1 (a real number may be thought of as a 1×1 matrix). Schematically,

$$
\underset{\text{Size of } A}{1 \times n} \qquad\qquad \underset{\text{Size of } B}{n \times 1}
$$

$$
\underset{\text{Size of } AB}{\underbrace{\qquad\qquad (1 \times 1) \qquad\qquad}}
$$

More generally, if A is a matrix of size $m \times n$ and B is a matrix of size $n \times p$ (the number of columns of A equals the numbers of rows of B), then the *matrix product* of A and B, AB, is defined and is a matrix of size $m \times p$. Schematically,

$$
\underset{\text{Size of } A}{m \times n} \qquad\qquad \underset{\text{Size of } B}{n \times p}
$$

$$
\underset{\text{Size of } AB}{\underbrace{\qquad\qquad (m \times p) \qquad\qquad}}
$$

Next, let's illustrate the mechanics of matrix multiplication by computing the product of a 2×3 matrix A and a 3×4 matrix B. Suppose

$$
A = \begin{bmatrix} a_{11} & a_{12} & a_{13} \\ a_{21} & a_{22} & a_{23} \end{bmatrix}
$$

$$
B = \begin{bmatrix} b_{11} & b_{12} & b_{13} & b_{14} \\ b_{21} & b_{22} & b_{23} & b_{24} \\ b_{31} & b_{32} & b_{33} & b_{34} \end{bmatrix}
$$

From the schematic

$$
\overset{\text{Same}}{\overbrace{\qquad\qquad\qquad}}
$$
$$
\underset{\text{Size of } A}{2 \times 3} \qquad\qquad\qquad 3 \times 4 \; \underset{\text{Size of } B}{}
$$
$$
\underset{\text{Size of } AB}{\underbrace{\qquad\qquad (2 \times 4) \qquad\qquad}}
$$

we see that the matrix product $C = AB$ is defined (since the number of columns of A equals the number of rows of B) and has size 2×4. Thus,

$$
C = \begin{bmatrix} c_{11} & c_{12} & c_{13} & c_{14} \\ c_{21} & c_{22} & c_{23} & c_{24} \end{bmatrix}
$$

The entries of C are computed as follows: The entry c_{11} (the entry in the *first* row, *first* column of C) is the product of the row matrix composed of the entries from the *first* row of A and the column matrix composed of the *first* column of B. Thus,

$$
c_{11} = \begin{bmatrix} a_{11} & a_{12} & a_{13} \end{bmatrix} \begin{bmatrix} b_{11} \\ b_{21} \\ b_{31} \end{bmatrix} = a_{11}b_{11} + a_{12}b_{21} + a_{13}b_{31}
$$

The entry c_{12} (the entry in the *first* row, *second* column of C) is the product of the row matrix composed of the *first* row of A and the column matrix composed of the *second* column of B. Thus,

$$
c_{12} = \begin{bmatrix} a_{11} & a_{12} & a_{13} \end{bmatrix} \begin{bmatrix} b_{12} \\ b_{22} \\ b_{32} \end{bmatrix} = a_{11}b_{12} + a_{12}b_{22} + a_{13}b_{32}
$$

The other entries in C are computed in a similar manner.

EXAMPLE 3 Let

$$A = \begin{bmatrix} 3 & 1 & 4 \\ -1 & 2 & 3 \end{bmatrix} \quad \text{and} \quad B = \begin{bmatrix} 1 & 3 & -3 \\ 4 & -1 & 2 \\ 2 & 4 & 1 \end{bmatrix}$$

Compute AB.

Solution The size of matrix A is 2×3, and the size of matrix B is 3×3. Since the number of columns of matrix A is equal to the number of rows of matrix B, the matrix product $C = AB$ is defined. Furthermore, the size of matrix C is 2×3. Thus,

$$\begin{bmatrix} 3 & 1 & 4 \\ -1 & 2 & 3 \end{bmatrix} \begin{bmatrix} 1 & 3 & -3 \\ 4 & -1 & 2 \\ 2 & 4 & 1 \end{bmatrix} = \begin{bmatrix} c_{11} & c_{12} & c_{13} \\ c_{21} & c_{22} & c_{23} \end{bmatrix}$$

It remains now to determine the entries c_{11}, c_{12}, c_{13}, c_{21}, c_{22}, and c_{23}. We have

$$c_{11} = \begin{bmatrix} 3 & 1 & 4 \end{bmatrix} \begin{bmatrix} 1 \\ 4 \\ 2 \end{bmatrix} = (3)(1) + (1)(4) + (4)(2) = 15$$

$$c_{12} = \begin{bmatrix} 3 & 1 & 4 \end{bmatrix} \begin{bmatrix} 3 \\ -1 \\ 4 \end{bmatrix} = (3)(3) + (1)(-1) + (4)(4) = 24$$

$$c_{13} = \begin{bmatrix} 3 & 1 & 4 \end{bmatrix} \begin{bmatrix} -3 \\ 2 \\ 1 \end{bmatrix} = (3)(-3) + (1)(2) + (4)(1) = -3$$

$$c_{21} = \begin{bmatrix} -1 & 2 & 3 \end{bmatrix} \begin{bmatrix} 1 \\ 4 \\ 2 \end{bmatrix} = (-1)(1) + (2)(4) + (3)(2) = 13$$

$$c_{22} = \begin{bmatrix} -1 & 2 & 3 \end{bmatrix} \begin{bmatrix} 3 \\ -1 \\ 4 \end{bmatrix} = (-1)(3) + (2)(-1) + (3)(4) = 7$$

$$c_{23} = \begin{bmatrix} -1 & 2 & 3 \end{bmatrix} \begin{bmatrix} -3 \\ 2 \\ 1 \end{bmatrix} = (-1)(-3) + (2)(2) + (3)(1) = 10$$

so the required product AB is given by

$$AB = \begin{bmatrix} 15 & 24 & -3 \\ 13 & 7 & 10 \end{bmatrix}$$

EXAMPLE 4 Let

$$A = \begin{bmatrix} 3 & 2 & 1 \\ -1 & 2 & 3 \\ 3 & 1 & 4 \end{bmatrix} \quad \text{and} \quad B = \begin{bmatrix} 1 & 3 & 4 \\ 2 & 4 & 1 \\ -1 & 2 & 3 \end{bmatrix}$$

Then

$$AB = \begin{bmatrix} 3 \cdot 1 & + 2 \cdot 2 + 1 \cdot (-1) & 3 \cdot 3 & + 2 \cdot 4 + 1 \cdot 2 & 3 \cdot 4 & + 2 \cdot 1 + 1 \cdot 3 \\ (-1) \cdot 1 + 2 \cdot 2 + 3 \cdot (-1) & (-1) \cdot 3 + 2 \cdot 4 + 3 \cdot 2 & (-1) \cdot 4 + 2 \cdot 1 + 3 \cdot 3 \\ 3 \cdot 1 & + 1 \cdot 2 + 4 \cdot (-1) & 3 \cdot 3 & + 1 \cdot 4 + 4 \cdot 2 & 3 \cdot 4 & + 1 \cdot 1 + 4 \cdot 3 \end{bmatrix} = \begin{bmatrix} 6 & 19 & 17 \\ 0 & 11 & 7 \\ 1 & 21 & 25 \end{bmatrix}$$

$$BA = \begin{bmatrix} 1 \cdot 3 & + 3 \cdot (-1) + 4 \cdot 3 & 1 \cdot 2 & + 3 \cdot 2 + 4 \cdot 1 & 1 \cdot 1 & + 3 \cdot 3 + 4 \cdot 4 \\ 2 \cdot 3 & + 4 \cdot (-1) + 1 \cdot 3 & 2 \cdot 2 & + 4 \cdot 2 + 1 \cdot 1 & 2 \cdot 1 & + 4 \cdot 3 + 1 \cdot 4 \\ (-1) \cdot 3 + 2 \cdot (-1) + 3 \cdot 3 & (-1) \cdot 2 + 2 \cdot 2 + 3 \cdot 1 & (-1) \cdot 1 + 2 \cdot 3 + 3 \cdot 4 \end{bmatrix} = \begin{bmatrix} 12 & 12 & 26 \\ 5 & 13 & 18 \\ 4 & 5 & 17 \end{bmatrix}$$

The preceding example shows that, in general, $AB \neq BA$ for two square matrices A and B. However, the following laws are valid for matrix multiplication.

Laws for Matrix Multiplication

If the products and sums are defined for the matrices A, B, and C, then

1. $(AB)C = A(BC)$ Associative law

2. $A(B + C) = AB + AC$ Distributive law

The square matrix of size n having 1s along the main diagonal and 0s elsewhere is called the identity matrix of size n.

Identity Matrix

The **identity matrix** of size n is given by

$$I_n = \begin{bmatrix} 1 & 0 & \cdots & \cdot & 0 \\ 0 & 1 & \cdots & \cdot & 0 \\ \cdot & \cdot & \cdots & \cdot & \cdot \\ \cdot & \cdot & \cdots & \cdot & \cdot \\ \cdot & \cdot & \cdots & \cdot & \cdot \\ 0 & 0 & \cdots & \cdot & 1 \end{bmatrix} \quad n \text{ rows}$$

n columns

The identity matrix has the properties that $I_n A = A$ for every $n \times r$ matrix A and $BI_n = B$ for every $s \times n$ matrix B. In particular, if A is a square matrix of size n, then

$$I_n A = AI_n = A$$

EXAMPLE 5 Let

$$A = \begin{bmatrix} 1 & 3 & 1 \\ -4 & 3 & 2 \\ 1 & 0 & 1 \end{bmatrix}$$

Then

$$I_3 A = \begin{bmatrix} 1 & 0 & 0 \\ 0 & 1 & 0 \\ 0 & 0 & 1 \end{bmatrix} \begin{bmatrix} 1 & 3 & 1 \\ -4 & 3 & 2 \\ 1 & 0 & 1 \end{bmatrix} = \begin{bmatrix} 1 & 3 & 1 \\ -4 & 3 & 2 \\ 1 & 0 & 1 \end{bmatrix} = A$$

$$AI_3 = \begin{bmatrix} 1 & 3 & 1 \\ -4 & 3 & 2 \\ 1 & 0 & 1 \end{bmatrix} \begin{bmatrix} 1 & 0 & 0 \\ 0 & 1 & 0 \\ 0 & 0 & 1 \end{bmatrix} = \begin{bmatrix} 1 & 3 & 1 \\ -4 & 3 & 2 \\ 1 & 0 & 1 \end{bmatrix} = A$$

so $I_3 A = AI_3 = A$, confirming our result for this special case.

APPLIED EXAMPLE 6 Production Planning Ace Novelty received an order from Magic World Amusement Park for 900 Giant Pandas, 1200 Saint Bernards, and 2000 Big Birds. Ace's management decided that 500 Giant Pandas, 800 Saint Bernards, and 1300 Big Birds could be manufactured in their Los Angeles plant, and the balance of the order could be filled by their Seattle plant. Each Panda requires 1.5 square yards of plush, 30 cubic feet of stuffing, and 5 pieces of trim; each Saint Bernard requires 2 square yards of plush, 35 cubic feet of stuffing, and 8 pieces of trim; and each Big Bird requires 2.5 square yards of plush, 25 cubic feet of stuffing, and 15 pieces of trim. The plush costs \$4.50 per square yard, the stuffing costs 10 cents per cubic foot, and the trim costs 25 cents per unit.

a. How much of each type of material must be purchased for each plant?
b. What is the total cost of materials incurred by each plant and the total cost of materials incurred by Ace Novelty in filling the order?

Solution   The quantities of each type of stuffed animal to be produced at each plant location may be expressed as a 2×3 *production matrix P*. Thus,

$$P = \begin{array}{c} \text{L.A.} \\ \text{Seattle} \end{array} \begin{bmatrix} \overset{\text{Pandas}}{500} & \overset{\text{St. Bernards}}{800} & \overset{\text{Birds}}{1300} \\ 400 & 400 & 700 \end{bmatrix}$$

Similarly, we may represent the amount and type of material required to manufacture each type of animal by a 3×3 *activity matrix A*. Thus,

$$A = \begin{array}{c} \text{Pandas} \\ \text{St. Bernards} \\ \text{Birds} \end{array} \begin{bmatrix} \overset{\text{Plush}}{1.5} & \overset{\text{Stuffing}}{30} & \overset{\text{Trim}}{5} \\ 2 & 35 & 8 \\ 2.5 & 25 & 15 \end{bmatrix}$$

Finally, the unit cost for each type of material may be represented by the 3×1 *cost matrix C*.

$$C = \begin{array}{c} \text{Plush} \\ \text{Stuffing} \\ \text{Trim} \end{array} \begin{bmatrix} 4.50 \\ 0.10 \\ 0.25 \end{bmatrix}$$

a. The amount of each type of material required for each plant is given by the matrix PA. Thus,

$$PA = \begin{bmatrix} 500 & 800 & 1300 \\ 400 & 400 & 700 \end{bmatrix} \begin{bmatrix} 1.5 & 30 & 5 \\ 2 & 35 & 8 \\ 2.5 & 25 & 15 \end{bmatrix}$$

$$= \begin{array}{c} \text{L.A.} \\ \text{Seattle} \end{array} \begin{bmatrix} \overset{\text{Plush}}{5600} & \overset{\text{Stuffing}}{75,500} & \overset{\text{Trim}}{28,400} \\ 3150 & 43,500 & 15,700 \end{bmatrix}$$

b. The total cost of materials for each plant is given by the matrix PAC:

$$PAC = \begin{bmatrix} 5600 & 75,500 & 28,400 \\ 3150 & 43,500 & 15,700 \end{bmatrix} \begin{bmatrix} 4.50 \\ 0.10 \\ 0.25 \end{bmatrix}$$

$$= \begin{array}{c} \text{L.A.} \\ \text{Seattle} \end{array} \begin{bmatrix} 39,850 \\ 22,450 \end{bmatrix}$$

or \$39,850 for the L.A. plant and \$22,450 for the Seattle plant. Thus, the total cost of materials incurred by Ace Novelty is \$62,300.

Matrix Representation

Example 7 shows how a system of linear equations may be written in a compact form with the help of matrices. (We will use this matrix equation representation in Section 5.6.)

EXAMPLE 7 Write the following system of linear equations in matrix form.

$$\begin{aligned} 2x - 4y + z &= 6 \\ -3x + 6y - 5z &= -1 \\ x - 3y + 7z &= 0 \end{aligned}$$

Solution Let's write

$$A = \begin{bmatrix} 2 & -4 & 1 \\ -3 & 6 & -5 \\ 1 & -3 & 7 \end{bmatrix} \qquad X = \begin{bmatrix} x \\ y \\ z \end{bmatrix} \qquad B = \begin{bmatrix} 6 \\ -1 \\ 0 \end{bmatrix}$$

Note that A is just the 3×3 matrix of coefficients of the system, X is the 3×1 column matrix of unknowns (variables), and B is the 3×1 column matrix of constants. We now show that the required matrix representation of the system of linear equations is

$$AX = B$$

To see this, observe that

$$AX = \begin{bmatrix} 2 & -4 & 1 \\ -3 & 6 & -5 \\ 1 & -3 & 7 \end{bmatrix} \begin{bmatrix} x \\ y \\ z \end{bmatrix} = \begin{bmatrix} 2x - 4y + z \\ -3x + 6y - 5z \\ x - 3y + 7z \end{bmatrix}$$

Equating this 3×1 matrix with matrix B now gives

$$\begin{bmatrix} 2x - 4y + z \\ -3x + 6y - 5z \\ x - 3y + 7z \end{bmatrix} = \begin{bmatrix} 6 \\ -1 \\ 0 \end{bmatrix}$$

which, by matrix equality, is easily seen to be equivalent to the given system of linear equations.   ▪

5.5 Self-Check Exercises

1. Compute

$$\begin{bmatrix} 1 & 3 & 0 \\ 2 & 4 & -1 \end{bmatrix} \begin{bmatrix} 3 & 1 & 4 \\ 2 & 0 & 3 \\ 1 & 2 & -1 \end{bmatrix}$$

2. Write the following system of linear equations in matrix form:

$$\begin{aligned} y - 2z &= 1 \\ 2x - y + 3z &= 0 \\ x \quad\quad + 4z &= 7 \end{aligned}$$

3. STOCK TRANSACTIONS On June 1, the stock holdings of Ash and Joan Robinson were given by the matrix

$$A = \begin{array}{c} \\ \text{Ash} \\ \text{Joan} \end{array} \begin{array}{cccc} \text{T} & \text{TWX} & \text{IBM} & \text{GM} \\ \begin{bmatrix} 2000 & 1000 & 500 & 5000 \\ 1000 & 2500 & 1000 & 0 \end{bmatrix} \end{array}$$

and the closing prices of T, TWX, IBM, and GM were $36, $54, $205, and $27 per share, respectively. Use matrix multiplication to determine the separate values of Ash's and Joan's stock holdings as of that date.

Solutions to Self-Check Exercises 5.5 can be found on page 330.

5.5 Concept Questions

1. What is the difference between scalar multiplication and matrix multiplication? Give examples of each operation.

2. **a.** Suppose A and B are matrices whose products AB and BA are both defined. What can you say about the sizes of A and B?

   b. If A, B, and C are matrices such that $A(B + C)$ is defined, what can you say about the relationship between the number of columns of A and the number of rows of C? Explain.

5.5 Exercises

In Exercises 1–4, the sizes of matrices A and B are given. Find the size of AB and BA whenever they are defined.

1. A is of size 2×3, and B is of size 3×5.

2. A is of size 3×4, and B is of size 4×3.

3. A is of size 1×7, and B is of size 7×1.

4. A is of size 4×4, and B is of size 4×4.

5. Let A be a matrix of size $m \times n$, and let B be a matrix of size $s \times t$. Find conditions on m, n, s, and t such that both matrix products AB and BA are defined.

6. Find condition(s) on the size of a matrix A such that A^2 (that is, AA) is defined.

In Exercises 7–24, compute the indicated products.

7. $\begin{bmatrix} 1 & 2 \\ 3 & 0 \end{bmatrix} \begin{bmatrix} 1 \\ -1 \end{bmatrix}$

8. $\begin{bmatrix} -1 & 3 \\ 5 & 0 \end{bmatrix} \begin{bmatrix} 7 \\ 2 \end{bmatrix}$

9. $\begin{bmatrix} 4 & 1 & 2 \\ -1 & 2 & 4 \end{bmatrix} \begin{bmatrix} 4 \\ 1 \\ -2 \end{bmatrix}$

10. $\begin{bmatrix} 3 & 2 & -1 \\ 4 & -1 & 0 \\ -5 & 2 & 1 \end{bmatrix} \begin{bmatrix} 3 \\ -2 \\ 0 \end{bmatrix}$

11. $\begin{bmatrix} -1 & 2 \\ 3 & 1 \end{bmatrix} \begin{bmatrix} 2 & 4 \\ 3 & 1 \end{bmatrix}$

12. $\begin{bmatrix} 1 & 3 \\ -1 & 2 \end{bmatrix} \begin{bmatrix} 1 & 3 & 0 \\ 3 & 0 & 2 \end{bmatrix}$

13. $\begin{bmatrix} 2 & 1 & 2 \\ 3 & 2 & 4 \end{bmatrix} \begin{bmatrix} -1 & 2 \\ 4 & 3 \\ 0 & 1 \end{bmatrix}$

14. $\begin{bmatrix} -1 & 2 \\ 4 & 3 \\ 0 & 1 \end{bmatrix} \begin{bmatrix} 2 & 1 & 2 \\ 3 & 2 & 4 \end{bmatrix}$

15. $\begin{bmatrix} 0.1 & 0.9 \\ 0.2 & 0.8 \end{bmatrix} \begin{bmatrix} 1.2 & 0.4 \\ 0.5 & 2.1 \end{bmatrix}$

16. $\begin{bmatrix} 1.2 & 0.3 \\ 0.4 & 0.5 \end{bmatrix} \begin{bmatrix} 0.2 & 0.6 \\ 0.4 & -0.5 \end{bmatrix}$

17. $\begin{bmatrix} 6 & -3 & 0 \\ -2 & 1 & -8 \\ 4 & -4 & 9 \end{bmatrix} \begin{bmatrix} 1 & 0 & 0 \\ 0 & 1 & 0 \\ 0 & 0 & 1 \end{bmatrix}$

18. $\begin{bmatrix} 2 & 4 \\ -1 & -5 \\ 3 & -1 \end{bmatrix} \begin{bmatrix} 2 & -2 & 4 \\ 1 & 3 & -1 \end{bmatrix}$

19. $\begin{bmatrix} 3 & 0 & -2 & 1 \\ 1 & 2 & 0 & -1 \end{bmatrix} \begin{bmatrix} 2 & 1 & -2 \\ -1 & 2 & 0 \\ 0 & 0 & 1 \\ -1 & -2 & 2 \end{bmatrix}$

20. $\begin{bmatrix} 2 & 1 & -3 & 0 \\ 4 & -2 & -1 & 1 \\ -1 & 2 & 0 & 1 \end{bmatrix} \begin{bmatrix} 2 & -1 \\ 1 & 4 \\ 3 & -3 \\ 0 & -5 \end{bmatrix}$

21. $4\begin{bmatrix} 1 & -2 & 0 \\ 2 & -1 & 1 \\ 3 & 0 & -1 \end{bmatrix} \begin{bmatrix} 1 & 3 & 1 \\ 1 & 4 & 0 \\ 0 & 1 & -2 \end{bmatrix}$

22. $3\begin{bmatrix} 2 & -1 & 0 \\ 2 & 1 & 2 \\ 1 & 0 & -1 \end{bmatrix} \begin{bmatrix} 2 & 3 & 1 \\ 3 & -3 & 0 \\ 0 & 1 & -1 \end{bmatrix}$

23. $\begin{bmatrix} 1 & 0 \\ 0 & 1 \end{bmatrix} \begin{bmatrix} 4 & -3 & 2 \\ 7 & 1 & -5 \end{bmatrix} \begin{bmatrix} 1 & 0 & 0 \\ 0 & 1 & 0 \\ 0 & 0 & 1 \end{bmatrix}$

24. $2\begin{bmatrix} 3 & 2 & -1 \\ 0 & 1 & 3 \\ 2 & 0 & 3 \end{bmatrix} \begin{bmatrix} 1 & 0 & 0 \\ 0 & 1 & 0 \\ 0 & 0 & 1 \end{bmatrix} \begin{bmatrix} 1 & 2 & 0 \\ 0 & -1 & -2 \\ 1 & 3 & 1 \end{bmatrix}$

In Exercises 25 and 26, let

$$A = \begin{bmatrix} 1 & 0 & -2 \\ 1 & -3 & 2 \\ -2 & 1 & 1 \end{bmatrix} \quad B = \begin{bmatrix} 3 & 1 & 0 \\ 2 & 2 & 0 \\ 1 & -3 & -1 \end{bmatrix}$$

$$C = \begin{bmatrix} 2 & -1 & 0 \\ 1 & -1 & 2 \\ 3 & -2 & 1 \end{bmatrix}$$

25. Verify the validity of the associative law for matrix multiplication.

26. Verify the validity of the distributive law for matrix multiplication.

27. Let

$$A = \begin{bmatrix} 1 & 2 \\ 3 & 4 \end{bmatrix} \quad \text{and} \quad B = \begin{bmatrix} 2 & 1 \\ 4 & 3 \end{bmatrix}$$

Compute AB and BA, and hence deduce that matrix multiplication is, in general, not commutative.

28. Let

$$A = \begin{bmatrix} 0 & 3 & 0 \\ 1 & 0 & 1 \\ 0 & 2 & 0 \end{bmatrix} \quad B = \begin{bmatrix} 2 & 4 & 5 \\ 3 & -1 & -6 \\ 4 & 3 & 4 \end{bmatrix}$$

$$C = \begin{bmatrix} 4 & 5 & 6 \\ 3 & -1 & -6 \\ 2 & 2 & 3 \end{bmatrix}$$

a. Compute AB.
b. Compute AC.
c. Using the results of parts (a) and (b), conclude that $AB = AC$ does *not* imply that $B = C$.

29. Let

$$A = \begin{bmatrix} 3 & 0 \\ 8 & 0 \end{bmatrix} \quad \text{and} \quad B = \begin{bmatrix} 0 & 0 \\ 4 & 5 \end{bmatrix}$$

Show that $AB = 0$, thereby demonstrating that for matrix multiplication, the equation $AB = 0$ does not imply that one or both of the matrices A and B must be the zero matrix.

30. Let

$$A = \begin{bmatrix} 2 & 2 \\ -2 & -2 \end{bmatrix}$$

Show that $A^2 = 0$. Compare this with the equation $a^2 = 0$, where a is a real number.

31. Find the matrix A such that

$$A\begin{bmatrix} 1 & 0 \\ -1 & 3 \end{bmatrix} = \begin{bmatrix} -1 & -3 \\ 3 & 6 \end{bmatrix}$$

Hint: Let $A = \begin{bmatrix} a & b \\ c & d \end{bmatrix}$.

32. Find the matrix A such that

$$\begin{bmatrix} 1 & 0 \\ -1 & 3 \end{bmatrix} A = \begin{bmatrix} -1 & -3 \\ 3 & 6 \end{bmatrix}$$

Hint: Let $A = \begin{bmatrix} a & b \\ c & d \end{bmatrix}$.

33. Find a matrix B such that $AB = I$, where

$$A = \begin{bmatrix} 2 & 1 \\ -2 & 2 \end{bmatrix} \quad \text{and} \quad I = \begin{bmatrix} 1 & 0 \\ 0 & 1 \end{bmatrix}$$

Hint: See Exercises 31 and 32.

34. Find a matrix B such that $AB = I$, where

$$A = \begin{bmatrix} 1 & 2 \\ 4 & 3 \end{bmatrix} \quad \text{and} \quad I = \begin{bmatrix} 1 & 0 \\ 0 & 1 \end{bmatrix}$$

Hint: See Exercises 31 and 32.

35. A square matrix is called an *upper triangular matrix* if all its entries below the main diagonal are zero. For example, the matrix

$$A = \begin{bmatrix} a & b \\ 0 & d \end{bmatrix}$$

is a 2×2 *upper triangular matrix.*
a. Show that the sum and the product of two upper triangular matrices of size two are upper triangular matrices.
b. If A and B are two upper triangular matrices of size two, then is it true that $AB = BA$, in general?

36. Let

$$A = \begin{bmatrix} 3 & 1 \\ 0 & 2 \end{bmatrix} \quad \text{and} \quad B = \begin{bmatrix} 4 & -2 \\ 2 & 1 \end{bmatrix}$$

a. Compute $(A + B)^2$.
b. Compute $A^2 + 2AB + B^2$.
c. From the results of parts (a) and (b), show that in general, $(A + B)^2 \neq A^2 + 2AB + B^2$.

37. Let

$$A = \begin{bmatrix} 2 & 4 \\ 5 & -6 \end{bmatrix} \quad \text{and} \quad B = \begin{bmatrix} 4 & 8 \\ -7 & 3 \end{bmatrix}$$

a. Find A^T and show that $(A^T)^T = A$.
b. Show that $(A + B)^T = A^T + B^T$.
c. Show that $(AB)^T = B^T A^T$.

38. Let

$$A = \begin{bmatrix} 1 & 3 \\ -2 & -1 \end{bmatrix} \quad \text{and} \quad B = \begin{bmatrix} 3 & -4 \\ 2 & -2 \end{bmatrix}$$

a. Find A^T and show that $(A^T)^T = A$.
b. Show that $(A + B)^T = A^T + B^T$.
c. Show that $(AB)^T = B^T A^T$.

In Exercises 39–44, write the given system of linear equations in matrix form.

39. $2x - 3y = 7$
$3x - 4y = 8$

40. $2x \qquad = 7$
$3x - 2y = 12$

41. $2x - 3y + 4z = 6$
$2y - 3z = 7$
$x - y + 2z = 4$

42. $x - 2y + 3z = -1$
$3x + 4y - 2z = 1$
$2x - 3y + 7z = 6$

43. $-x_1 + x_2 + x_3 = 0$
$2x_1 - x_2 - x_3 = 2$
$-3x_1 + 2x_2 + 4x_3 = 4$

44. $3x_1 - 5x_2 + 4x_3 = 10$
$4x_1 + 2x_2 - 3x_3 = -12$
$-x_1 + x_3 = -2$

45. STOCK TRANSACTIONS Olivia's and Isabella's stock holdings are given by the matrix

$$A = \begin{array}{c} \text{Olivia} \\ \text{Isabella} \end{array} \begin{array}{cccc} \text{FB} & \text{HD} & \text{PG} & \text{SBUX} \\ \begin{bmatrix} 200 & 300 & 100 & 200 \\ 100 & 200 & 400 & 0 \end{bmatrix} \end{array}$$

At the close of trading on a certain day, the prices (in dollars per share) of the stocks are given by the matrix

$$B = \begin{array}{c} \text{FB} \\ \text{HD} \\ \text{PG} \\ \text{SBUX} \end{array} \begin{bmatrix} 27 \\ 24 \\ 63 \\ 56 \end{bmatrix}$$

a. Find AB.
b. Explain the meaning of the entries in the matrix AB.

46. AIRLINE FLIGHT SCHEDULING Pacific Airlines operates three flights between Los Angeles and Hong Kong. Matrix A gives the number of seats in each of three cabin classes on each flight.

$$A = \begin{array}{c} \text{Flight I} \\ \text{Flight II} \\ \text{Flight III} \end{array} \begin{array}{ccc} \begin{array}{c}\text{First}\\\text{class}\end{array} & \begin{array}{c}\text{Premium}\\\text{economy}\end{array} & \text{Economy} \\ \begin{bmatrix} 60 & 80 & 160 \\ 50 & 60 & 190 \\ 40 & 50 & 210 \end{bmatrix} \end{array}$$

The fares per passenger (in dollars) for each class of seats are given by matrix B:

$$B = \begin{array}{c} \text{First class} \\ \text{Premium economy} \\ \text{Economy} \end{array} \begin{bmatrix} 6000 \\ 3500 \\ 1500 \end{bmatrix}$$

The number of each type of flight operated by Pacific airlines in June is given by matrix C:

$$C = \begin{array}{c} \\ \end{array} \begin{array}{ccc} \text{Flight} & & \\ \text{I} & \text{II} & \text{III} \\ \begin{bmatrix} 8 & 6 & 6 \end{bmatrix} \end{array}$$

a. Compute AB, and explain what it represents.
b. Compute CAB, and explain its meaning.

47. RIVER CRUISE SCHEDULE PLANNING Nordic River Cruises operates four cruises between several cities in Europe. The number of each type of cruise planned for 2015 is given by matrix A:

$$A = \begin{array}{c} \\ \end{array} \begin{array}{cccc} & \text{Cruise} & & \\ \text{I} & \text{II} & \text{III} & \text{IV} \\ \begin{bmatrix} 12 & 14 & 20 & 10 \end{bmatrix} \end{array}$$

For each cruise, the classes of cabins are classified into three categories. The number of cabins in each category for each type of cruise are given by matrix B:

$$B = \begin{array}{c} \text{Cruise I} \\ \text{Cruise II} \\ \text{Cruise III} \\ \text{Cruise IV} \end{array} \begin{array}{ccc} & \text{Category} & \\ \text{A} & \text{B} & \text{C} \\ \begin{bmatrix} 20 & 30 & 40 \\ 20 & 20 & 55 \\ 15 & 35 & 45 \\ 25 & 30 & 40 \end{bmatrix} \end{array}$$

The fares per passenger (in dollars) for each category of cabins are given by matrix C:

$$C = \begin{array}{c} \text{Category A} \\ \text{Category B} \\ \text{Category C} \end{array} \begin{bmatrix} 8000 \\ 10000 \\ 7000 \end{bmatrix}$$

a. Compute AB, and explain what it represents.
b. Compute ABC, and explain its meaning.

48. FOREIGN EXCHANGE Mason has just returned to the United States from a Southeast Asian trip and wishes to exchange the various foreign currencies that he has accumulated for U.S. dollars. He has 1200 Thai bahts, 80,000 Indonesian rupiahs, 42 Malaysian ringgits, and 36 Singapore dollars. Suppose the foreign exchange rates are U.S. $0.03 for one baht, U.S. $0.0001 for one rupiah, U.S. $0.322 for one Malaysian ringgit, and U.S. $0.806 for one Singapore dollar.
a. Write a row matrix A giving the value of the various currencies that Mason holds. (*Note:* The answer is *not* unique.)
b. Write a column matrix B giving the exchange rates for the various currencies.
c. If Mason exchanges all of his foreign currencies for U.S. dollars, how many dollars will he have?

49. INVESTMENTS Ashley's stock holdings are given by matrix A:

$$A = \begin{array}{cccc} \text{Google} & \text{eBay} & \text{Priceline} & \text{Netflix} \\ \begin{bmatrix} 200 & 300 & 240 & 120 \end{bmatrix} \end{array}$$

At the close of trading on Monday, Tuesday, and Wednesday of a certain week, the prices (in dollars per share) of the stocks were given by matrix B:

$$B = \begin{array}{c} \text{Google} \\ \text{eBay} \\ \text{Priceline} \\ \text{Netflix} \end{array} \begin{array}{ccc} \text{Mon.} & \text{Tues.} & \text{Wed.} \\ \begin{bmatrix} 821.50 & 838.60 & 831.38 \\ 55.48 & 55.26 & 53.37 \\ 714.01 & 718.41 & 718.90 \\ 181.21 & 181.73 & 182.94 \end{bmatrix} \end{array}$$

Use matrix multiplication to find a matrix C giving the total value of Ashley's stock holdings on each of the three days.

50. COMPARATIVE SHOPPING Laura is planning to buy two 5-lb bags of sugar, three 5-lb bags of flour, two 1-gal cartons of milk, and three 1-dozen cartons of large eggs. The prices of these items in three neighborhood supermarkets are as follows:

	Sugar (5-lb bag)	Flour (5-lb bag)	Milk (1-gal carton)	Eggs (1-dozen carton)
Supermarket I	$3.15	$3.79	$2.99	$3.49
Supermarket II	$2.99	$2.89	$2.79	$3.29
Supermarket III	$3.74	$2.98	$2.89	$2.99

a. Write a 3×4 matrix A to represent the prices of the items in the three supermarkets.

b. Write a 4×1 matrix B to represent the quantities of the items that Laura plans to purchase in the three supermarkets.

c. Use matrix multiplication to find a matrix C that represents Laura's total outlay at each supermarket. At which supermarket should she make her purchase if she wants to minimize her cost? (Assume that she will shop at only one supermarket.)

51. **FOREIGN EXCHANGE** Ava and her friend Ella have returned to the United States from a tour of four cities: Oslo, Stockholm, Copenhagen, and Saint Petersburg. They now wish to exchange the various foreign currencies that they have accumulated for U.S. dollars. Ava has 82 Norwegian kroner, 68 Swedish kronor, 62 Danish kroner, and 1200 Russian rubles. Ella has 64 Norwegian kroner, 74 Swedish kronor, 44 Danish kroner, and 1600 Russian rubles. Suppose the exchange rates are U.S. $0.1751 for one Norwegian krone, U.S. $0.1560 for one Swedish krona, U.S. $0.1747 for one Danish krone, and U.S. $0.0325 for one Russian ruble.

a. Write a 2×4 matrix A giving the values of the various foreign currencies held by Ava and Ella. (*Note:* The answer is *not* unique.)

b. Write a column matrix B giving the exchange rate for the various currencies.

c. If both Ava and Ella exchange all their foreign currencies for U.S. dollars, how many dollars will each have?

52. **REAL ESTATE** Bond Brothers, a real estate developer, builds houses in three states. The projected number of units of each model to be built in each state is given by the matrix

$$A = \begin{array}{c} \\ NY \\ CT \\ MA \end{array} \begin{array}{cccc} \text{I} & \text{II} & \text{III} & \text{IV} \\ \left[\begin{array}{cccc} 60 & 80 & 120 & 40 \\ 20 & 30 & 60 & 10 \\ 10 & 15 & 30 & 5 \end{array}\right] \end{array}$$

The profits to be realized are $60,000, $66,000, $75,000, and $90,000, respectively, for each Model I, II, III, and IV house sold.

a. Write a column matrix B representing the profit for each type of house.

b. Find the total profit Bond Brothers expects to earn in each state if all the houses are sold.

53. **REAL ESTATE** Refer to Exercise 52. Let $B = \begin{bmatrix} 1 & 1 & 1 \end{bmatrix}$ and $C = \begin{bmatrix} 1 & 1 & 1 & 1 \end{bmatrix}$.

a. Compute BA, and explain what the entries of the matrix represent.

b. Compute AC^T, and give an interpretation of the matrix.

54. **CHARITIES** The amount of money raised by Charity I, Charity II, and Charity III (in millions of dollars) in each of the years 2013, 2014, and 2015 is represented by the matrix A:

$$A = \begin{array}{c} \\ 2013 \\ 2014 \\ 2015 \end{array} \begin{array}{ccc} & \text{Charity} & \\ \text{I} & \text{II} & \text{III} \\ \left[\begin{array}{ccc} 18.2 & 28.2 & 40.5 \\ 19.6 & 28.6 & 42.6 \\ 20.8 & 30.4 & 46.4 \end{array}\right] \end{array}$$

On average, Charity I puts 78% toward program cost, Charity II puts 88% toward program cost, and Charity III puts 80% toward program cost. Write a 3×1 matrix B reflecting the percentage put toward program cost by the charities. Then use matrix multiplication to find the total amount of money put toward program cost in each of the 3 years by the charities under consideration.

55. **BOX-OFFICE RECEIPTS** The Cinema Center consists of four theaters: Cinemas I, II, III, and IV. The admission price for one feature at the Center is $4 for children, $6 for students, and $8 for adults. The attendance for the Sunday matinee is given by the matrix

$$A = \begin{array}{c} \\ \text{Cinema I} \\ \text{Cinema II} \\ \text{Cinema III} \\ \text{Cinema IV} \end{array} \begin{array}{ccc} \text{Children} & \text{Students} & \text{Adults} \\ \left[\begin{array}{ccc} 225 & 110 & 50 \\ 75 & 180 & 225 \\ 280 & 85 & 110 \\ 0 & 250 & 225 \end{array}\right] \end{array}$$

Write a column vector B representing the admission prices. Then compute AB, the column vector showing the gross receipts for each theater. Finally, find the total revenue collected at the Cinema Center for admission that Sunday afternoon.

56. **BOX-OFFICE RECEIPTS** Refer to Exercise 55.

a. Find a 1×4 matrix B such that the entries in BA give the total number of children, the total number of students, and the total number of adults who attended the Sunday matinee. Compute BA.

b. Find a 1×3 matrix C such that the entries in AC^T give the total number of people (children, students, and adults) who attended Cinema I, Cinema II, Cinema III, and Cinema IV. Compute AC^T.

57. **VOTER AFFILIATION BY AGE** Matrix A gives the percentage of eligible voters in the city of Newton, classified according to party affiliation and age group.

$$A = \begin{array}{c} \\ \text{Under 30} \\ \text{30 to 50} \\ \text{Over 50} \end{array} \begin{array}{ccc} \text{Dem.} & \text{Rep.} & \text{Ind.} \\ \left[\begin{array}{ccc} 0.50 & 0.30 & 0.20 \\ 0.45 & 0.40 & 0.15 \\ 0.40 & 0.50 & 0.10 \end{array}\right] \end{array}$$

The population of eligible voters in the city by age group is given by the matrix B:

$$\begin{array}{ccc} \text{Under 30} & \text{30 to 50} & \text{Over 50} \end{array}$$
$$B = \begin{bmatrix} 30,000 & 40,000 & 20,000 \end{bmatrix}$$

Find a matrix giving the total number of eligible voters in the city who will vote Democratic, Republican, and Independent.

58. **401(k) RETIREMENT PLANS** Three network consultants, Alan, Maria, and Steven, each received a year-end bonus of $10,000, which they decided to invest in a 401(k) retirement plan sponsored by their employer. Under this plan, employees are allowed to place their investments in three funds: an equity index fund (I), a growth fund (II), and a global equity fund (III). The allocations of the investments (in dollars) of the three employees at the beginning of the year are summarized in the matrix

$$A = \begin{array}{c} \\ \text{Alan} \\ \text{Maria} \\ \text{Steven} \end{array} \overset{\begin{array}{ccc} & \text{Fund} & \\ \text{I} & \text{II} & \text{III} \end{array}}{\begin{bmatrix} 4000 & 3000 & 3000 \\ 2000 & 5000 & 3000 \\ 2000 & 3000 & 5000 \end{bmatrix}}$$

The returns of the three funds after 1 year are given in the matrix

$$B = \begin{array}{c} \text{Fund I} \\ \text{Fund II} \\ \text{Fund III} \end{array} \begin{bmatrix} 0.18 \\ 0.24 \\ 0.12 \end{bmatrix}$$

Which employee realized the best return on his or her investment for the year in question? The worst return?

59. **COLLEGE ADMISSIONS** A university admissions committee anticipates an enrollment of 8000 students in its freshman class next year. To satisfy admission quotas, incoming students have been categorized according to their sex and place of residence. The number of students in each category is given by the matrix

$$A = \begin{array}{c} \\ \text{In-state} \\ \text{Out-of-state} \\ \text{Foreign} \end{array} \overset{\begin{array}{cc} \text{Male} & \text{Female} \end{array}}{\begin{bmatrix} 2700 & 3000 \\ 800 & 700 \\ 500 & 300 \end{bmatrix}}$$

By using data accumulated in previous years, the admissions committee has determined that these students will elect to enter the College of Letters and Science, the College of Fine Arts, the School of Business Administration, and the School of Engineering according to the percentages that appear in the following matrix:

$$B = \begin{array}{c} \\ \text{Male} \\ \text{Female} \end{array} \overset{\begin{array}{cccc} \text{L. \& S.} & \text{Fine Arts} & \text{Bus. Ad.} & \text{Eng.} \end{array}}{\begin{bmatrix} 0.25 & 0.20 & 0.30 & 0.25 \\ 0.30 & 0.35 & 0.25 & 0.10 \end{bmatrix}}$$

Find the matrix AB that shows the number of in-state, out-of-state, and foreign students expected to enter each discipline.

60. **PRODUCTION PLANNING** Refer to Example 6 in this section. Suppose Ace Novelty received an order from another amusement park for 1200 Pink Panthers, 1800 Giant Pandas, and 1400 Big Birds. The quantity of each type of stuffed animal to be produced at each plant is shown in the following production matrix:

$$P = \begin{array}{c} \\ \text{L.A.} \\ \text{Seattle} \end{array} \overset{\begin{array}{ccc} \text{Panthers} & \text{Pandas} & \text{Birds} \end{array}}{\begin{bmatrix} 700 & 1000 & 800 \\ 500 & 800 & 600 \end{bmatrix}}$$

Each Panther requires 1.3 yd^2 of plush, 20 ft^3 of stuffing, and 12 pieces of trim. Assume that the materials required to produce the other two stuffed animals and the unit cost for each type of material are as given in Example 6.
   a. How much of each type of material must be purchased for each plant?
   b. What is the total cost of materials that will be incurred at each plant?
   c. What is the total cost of materials incurred by Ace Novelty in filling the order?

61. **COMPUTING PHONE BILLS** Cindy regularly makes long-distance phone calls to three foreign cities: London, Tokyo, and Hong Kong. The matrices A and B give the lengths (in minutes) of her calls during peak and nonpeak hours, respectively, to each of these three cities during the month of June.

$$A = \overset{\begin{array}{ccc} \text{London} & \text{Tokyo} & \text{Hong Kong} \end{array}}{\begin{bmatrix} 80 & 60 & 40 \end{bmatrix}}$$

and

$$B = \overset{\begin{array}{ccc} \text{London} & \text{Tokyo} & \text{Hong Kong} \end{array}}{\begin{bmatrix} 300 & 150 & 250 \end{bmatrix}}$$

The costs for the calls (in dollars per minute) for the peak and nonpeak periods in the month in question are given, respectively, by the matrices

$$C = \begin{array}{c} \text{London} \\ \text{Tokyo} \\ \text{Hong Kong} \end{array} \begin{bmatrix} 0.17 \\ 0.21 \\ 0.24 \end{bmatrix}$$

and

$$D = \begin{array}{c} \text{London} \\ \text{Tokyo} \\ \text{Hong Kong} \end{array} \begin{bmatrix} 0.12 \\ 0.15 \\ 0.17 \end{bmatrix}$$

Compute the matrix $AC + BD$, and explain what it represents.

62. **PRODUCTION PLANNING** The total output of loudspeaker systems of the Acrosonic Company at their three production facilities for May and June is given by the matrices A and B, respectively, where

$$A = \begin{array}{c} \\ \text{Location I} \\ \text{Location II} \\ \text{Location III} \end{array} \overset{\begin{array}{cccc} & \text{Model} & & \\ \text{A} & \text{B} & \text{C} & \text{D} \end{array}}{\begin{bmatrix} 320 & 280 & 460 & 280 \\ 480 & 360 & 580 & 0 \\ 540 & 420 & 200 & 880 \end{bmatrix}}$$

$$B = \begin{array}{c} \\ \text{Location I} \\ \text{Location II} \\ \text{Location III} \end{array} \begin{array}{cccc} & & \text{Model} & \\ \text{A} & \text{B} & \text{C} & \text{D} \\ \begin{bmatrix} 210 & 180 & 330 & 180 \\ 400 & 300 & 450 & 40 \\ 420 & 280 & 180 & 740 \end{bmatrix} \end{array}$$

The unit production costs and selling prices for these loudspeakers are given by matrices C and D, respectively, where

$$C = \begin{array}{c} \text{Model A} \\ \text{Model B} \\ \text{Model C} \\ \text{Model D} \end{array} \begin{bmatrix} 120 \\ 180 \\ 260 \\ 500 \end{bmatrix}$$

and

$$D = \begin{array}{c} \text{Model A} \\ \text{Model B} \\ \text{Model C} \\ \text{Model D} \end{array} \begin{bmatrix} 160 \\ 250 \\ 350 \\ 700 \end{bmatrix}$$

Compute the following matrices, and explain the meaning of the entries in each matrix.

a. AC **b.** AD **c.** BC **d.** BD **e.** $(A + B)C$
f. $(A + B)D$ **g.** $A(D - C)$
h. $B(D - C)$ **i.** $(A + B)(D - C)$

63. DIET PLANNING A dietitian plans a meal around three foods. The number of units of vitamin A, vitamin C, and calcium in each ounce of these foods is represented by the matrix M, where

$$M = \begin{array}{c} \text{Vitamin A} \\ \text{Vitamin C} \\ \text{Calcium} \end{array} \begin{array}{ccc} & \text{Food} & \\ \text{I} & \text{II} & \text{III} \\ \begin{bmatrix} 400 & 1200 & 800 \\ 110 & 570 & 340 \\ 90 & 30 & 60 \end{bmatrix} \end{array}$$

The matrices A and B represent the amount of each food (in ounces) consumed by a girl at two different meals, where

$$\begin{array}{ccc} & \text{Food} & \\ \text{I} & \text{II} & \text{III} \end{array}$$
$$A = \begin{bmatrix} 7 & 1 & 6 \end{bmatrix}$$

and

$$\begin{array}{ccc} & \text{Food} & \\ \text{I} & \text{II} & \text{III} \end{array}$$
$$B = \begin{bmatrix} 9 & 3 & 2 \end{bmatrix}$$

Calculate the following matrices, and explain the meaning of the entries in each matrix.

a. MA^T **b.** MB^T **c.** $M(A + B)^T$

64. SALES FORECASTING Hartman Lumber Company has two branches in the city. The sales of four of its products for the last year (in thousands of dollars) are represented by the matrix

$$A = \begin{array}{c} \text{Branch I} \\ \text{Branch II} \end{array} \begin{array}{cccc} & & \text{Product} & \\ \text{A} & \text{B} & \text{C} & \text{D} \\ \begin{bmatrix} 5 & 2 & 8 & 10 \\ 3 & 4 & 6 & 8 \end{bmatrix} \end{array}$$

For the present year, management has projected that the sales of the four products in Branch I will be 10% more than the corresponding sales for last year and the sales of the four products in Branch II will be 15% more than the corresponding sales for last year.

a. Show that the sales of the four products in the two branches for the current year are given by the matrix AB, where

$$A = \begin{bmatrix} 1.1 & 0 \\ 0 & 1.15 \end{bmatrix}$$

Compute AB.

b. Hartman has m branches nationwide. The sales of n of its products (in thousands of dollars) last year are represented by the matrix

$$B = \begin{array}{c} \text{Branch 1} \\ \text{Branch 2} \\ \vdots \\ \text{Branch } m \end{array} \begin{array}{ccccc} & & \text{Product} & & \\ 1 & 2 & 3 & \cdots & n \\ \begin{bmatrix} a_{11} & a_{12} & a_{13} & \cdots & a_{1n} \\ a_{21} & a_{22} & a_{23} & \cdots & a_{2n} \\ \vdots & \vdots & \vdots & \vdots & \vdots \\ a_{m1} & a_{m2} & a_{m3} & \cdots & a_{mn} \end{bmatrix} \end{array}$$

Also, management has projected that the sales of the n products in Branch 1, Branch 2, . . . , Branch m will be $r_1\%, r_2\%, \ldots, r_m\%$, respectively, more than the corresponding sales for last year. Write the matrix A such that AB gives the sales of the n products in the m branches for the current year.

In Exercises 65–68, determine whether the statement is true or false. If it is true, explain why it is true. If it is false, give an example to show why it is false.

65. If A and B are matrices such that AB and BA are both defined, then A and B must be square matrices of the same size.

66. If A and B are matrices such that AB is defined and if c is a scalar, then $(cA)B = A(cB) = cAB$.

67. If A, B, and C are matrices and $A(B + C)$ is defined, then B must have the same size as C, and the number of columns of A must be equal to the number of rows of B.

68. If A is a 2 × 4 matrix and B is a matrix such that ABA is defined, then the size of B must be 4 × 2.

5.5 Solutions to Self-Check Exercises

1. We compute

$$\begin{bmatrix} 1 & 3 & 0 \\ 2 & 4 & -1 \end{bmatrix} \begin{bmatrix} 3 & 1 & 4 \\ 2 & 0 & 3 \\ 1 & 2 & -1 \end{bmatrix} = \begin{bmatrix} 1(3) + 3(2) + 0(1) & 1(1) + 3(0) + 0(2) & 1(4) + 3(3) + 0(-1) \\ 2(3) + 4(2) - 1(1) & 2(1) + 4(0) - 1(2) & 2(4) + 4(3) - 1(-1) \end{bmatrix}$$

$$= \begin{bmatrix} 9 & 1 & 13 \\ 13 & 0 & 21 \end{bmatrix}$$

2. Let

$$A = \begin{bmatrix} 0 & 1 & -2 \\ 2 & -1 & 3 \\ 1 & 0 & 4 \end{bmatrix} \quad X = \begin{bmatrix} x \\ y \\ z \end{bmatrix} \quad B = \begin{bmatrix} 1 \\ 0 \\ 7 \end{bmatrix}$$

Then the given system may be written as the matrix equation

$$AX = B$$

3. Write

$$B = \begin{matrix} \text{T} \\ \text{TWX} \\ \text{IBM} \\ \text{GM} \end{matrix} \begin{bmatrix} 36 \\ 54 \\ 205 \\ 27 \end{bmatrix}$$

and compute the following:

$$AB = \begin{matrix} \text{Ash} \\ \text{Joan} \end{matrix} \begin{bmatrix} 2000 & 1000 & 500 & 5000 \\ 1000 & 2500 & 1000 & 0 \end{bmatrix} \begin{bmatrix} 36 \\ 54 \\ 205 \\ 27 \end{bmatrix}$$

$$= \begin{matrix} \text{Ash} \\ \text{Joan} \end{matrix} \begin{bmatrix} 363{,}500 \\ 376{,}000 \end{bmatrix}$$

We conclude that Ash's stock holdings were worth $363,500 and Joan's stock holdings were worth $376,000 on June 1.

USING TECHNOLOGY Matrix Multiplication

Graphing Utility

A graphing utility can be used to perform matrix multiplication.

EXAMPLE 1 Let

$$A = \begin{bmatrix} 1.2 & 3.1 & -1.4 \\ 2.7 & 4.2 & 3.4 \end{bmatrix} \quad B = \begin{bmatrix} 0.8 & 1.2 & 3.7 \\ 6.2 & -0.4 & 3.3 \end{bmatrix} \quad C = \begin{bmatrix} 1.2 & 2.1 & 1.3 \\ 4.2 & -1.2 & 0.6 \\ 1.4 & 3.2 & 0.7 \end{bmatrix}$$

Find (a) AC and (b) $(1.1A + 2.3B)C$.

Solution First, we enter the matrices A, B, and C into the calculator.

a. Using matrix operations, we enter the expression $A*C$. We obtain the matrix

$$\begin{bmatrix} 12.5 & -5.68 & 2.44 \\ 25.64 & 11.51 & 8.41 \end{bmatrix}$$

(You might need to scroll the display on the screen to obtain the complete matrix.)

b. Using matrix operations, we enter the expression $(1.1A + 2.3B)C$. We obtain the matrix

$$\begin{bmatrix} 39.464 & 21.536 & 12.689 \\ 52.078 & 67.999 & 32.55 \end{bmatrix}$$

Excel

We use the **MMULT** function in Excel to perform matrix multiplication.

EXAMPLE 2 Let

$$A = \begin{bmatrix} 1.2 & 3.1 & -1.4 \\ 2.7 & 4.2 & 3.4 \end{bmatrix} \quad B = \begin{bmatrix} 0.8 & 1.2 & 3.7 \\ 6.2 & -0.4 & 3.3 \end{bmatrix} \quad C = \begin{bmatrix} 1.2 & 2.1 & 1.3 \\ 4.2 & -1.2 & 0.6 \\ 1.4 & 3.2 & 0.7 \end{bmatrix}$$

Find (a) AC and (b) $(1.1A + 2.3B)C$.

Solution

a. First, enter the matrices A, B, and C onto a spreadsheet (Figure T1).

	A	B	C	D	E	F	G
1		A				B	
2	1.2	3.1	-1.4		0.8	1.2	3.7
3	2.7	4.2	3.4		6.2	-0.4	3.3
4							
5		C					
6	1.2	2.1	1.3				
7	4.2	-1.2	0.6				
8	1.4	3.2	0.7				

FIGURE **T1**
Spreadsheet showing the matrices *A*, *B*, and *C*

Second, compute AC. Highlight the cells that will contain the matrix product AC, which has size 2×3. Type =MMULT (, highlight the cells in matrix A, type ,, highlight the cells in matrix C, type), and press **Ctrl-Shift-Enter** . The matrix product AC shown in Figure T2 will appear on your spreadsheet.

	A	B	C
10		AC	
11	12.5	-5.68	2.44
12	25.64	11.51	8.41

FIGURE **T2**
The matrix product *AC*

b. Compute $(1.1A + 2.3B)C$. Highlight the cells that will contain the matrix product $(1.1A + 2.3B)C$. Next, type =MMULT(1.1*, highlight the cells in matrix A, type +2.3*, highlight the cells in matrix B, type ,, highlight the cells in matrix C, type), and then press **Ctrl-Shift-Enter** . The matrix product shown in Figure T3 will appear on your spreadsheet.

	A	B	C
13		(1.1A + 2.3B)C	
14	39.464	21.536	12.689
15	52.078	67.999	32.55

FIGURE **T3**
The matrix product $(1.1A + 2.3B)C$

Note: Boldfaced words/characters enclosed in a box (for example, **Enter**) indicate that an action (click, select, or press) is required. Words/characters printed blue (for example, Chart sub-type:) indicate words/characters that appear on the screen. Words/characters printed in a monospace font (for example, =(-2/3)*A2+2) indicate words/characters that need to be typed and entered.

TECHNOLOGY EXERCISES

In Exercises 1–8, refer to the following matrices, and perform the indicated operations. Round your answers to two decimal places.

$$A = \begin{bmatrix} 1.2 & 3.1 & -1.2 & 4.3 \\ 7.2 & 6.3 & 1.8 & -2.1 \\ 0.8 & 3.2 & -1.3 & 2.8 \end{bmatrix}$$

$$B = \begin{bmatrix} 0.7 & 0.3 & 1.2 & -0.8 \\ 1.2 & 1.7 & 3.5 & 4.2 \\ -3.3 & -1.2 & 4.2 & 3.2 \end{bmatrix}$$

$$C = \begin{bmatrix} 0.8 & 7.1 & 6.2 \\ 3.3 & -1.2 & 4.8 \\ 1.3 & 2.8 & -1.5 \\ 2.1 & 3.2 & -8.4 \end{bmatrix}$$

1. AC

2. CB

3. $(A + B)C$

4. $(2A + 3B)C$

5. $(2A - 3.1B)C$

6. $C(2.1A + 3.2B)$

7. $(4.1A + 2.7B)1.6C$

8. $2.5C(1.8A - 4.3B)$

In Exercises 9–12, refer to the following matrices, and perform the indicated operations. Round your answers to two decimal places.

$$A = \begin{bmatrix} 2 & 5 & -4 & 2 & 8 \\ 6 & 7 & 2 & 9 & 6 \\ 4 & 5 & 4 & 4 & 4 \\ 9 & 6 & 8 & 3 & 2 \end{bmatrix}$$

$$B = \begin{bmatrix} 2 & 6 & 7 & 5 \\ 3 & 4 & 6 & 2 \\ -5 & 8 & 4 & 3 \\ 8 & 6 & 9 & 5 \\ 4 & 7 & 8 & 8 \end{bmatrix}$$

$$C = \begin{bmatrix} 6.2 & 7.3 & -4.0 & 7.1 & 9.3 \\ 4.8 & 6.5 & 8.4 & -6.3 & 8.4 \\ 5.4 & 3.2 & 6.3 & 9.1 & -2.8 \\ 8.2 & 7.3 & 6.5 & 4.1 & 9.8 \\ 10.3 & 6.8 & 4.8 & -9.1 & 20.4 \end{bmatrix}$$

$$D = \begin{bmatrix} 4.6 & 3.9 & 8.4 & 6.1 & 9.8 \\ 2.4 & -6.8 & 7.9 & 11.4 & 2.9 \\ 7.1 & 9.4 & 6.3 & 5.7 & 4.2 \\ 3.4 & 6.1 & 5.3 & 8.4 & 6.3 \\ 7.1 & -4.2 & 3.9 & -6.4 & 7.1 \end{bmatrix}$$

9. Find AB and BA.

10. Find CD and DC. Is $CD = DC$?

11. Find $AC + AD$.

12. Find:
a. AC **b.** AD **c.** $A(C + D)$
d. Is $A(C + D) = AC + AD$?

5.6 The Inverse of a Square Matrix

The Inverse of a Square Matrix

In this section, we discuss a procedure for finding the inverse of a matrix, and we show how the inverse can be used to help us solve a system of linear equations.

Recall that if a is a nonzero real number, then there exists a unique real number a^{-1} (that is, $\frac{1}{a}$) such that

$$a^{-1}a = \left(\frac{1}{a}\right)(a) = 1$$

The use of the (multiplicative) inverse of a real number enables us to solve algebraic equations of the form

$$ax = b \tag{13}$$

Multiplying both sides of (13) by a^{-1}, we have

$$a^{-1}(ax) = a^{-1}b$$

$$\left(\frac{1}{a}\right)(ax) = \frac{1}{a}(b)$$

$$x = \frac{b}{a}$$

For example, since the inverse of 2 is $2^{-1} = \frac{1}{2}$, we can solve the equation

$$2x = 5$$

by multiplying both sides of the equation by $2^{-1} = \frac{1}{2}$, giving

$$2^{-1}(2x) = 2^{-1} \cdot 5$$

$$x = \frac{5}{2}$$

We can use a similar procedure to solve the matrix equation

$$AX = B$$

where A, X, and B are matrices of the proper sizes. To do this we need the matrix equivalent of the inverse of a real number. Such a matrix, whenever it exists, is called the **inverse of a matrix.**

> **Inverse of a Matrix**
>
> Let A be a square matrix of size n. A square matrix A^{-1} of size n such that
> $$A^{-1}A = AA^{-1} = I_n$$
> is called the inverse of A.

Let's show that the matrix

$$A = \begin{bmatrix} 1 & 2 \\ 3 & 4 \end{bmatrix}$$

has the matrix

$$A^{-1} = \begin{bmatrix} -2 & 1 \\ \frac{3}{2} & -\frac{1}{2} \end{bmatrix}$$

as its inverse. Since

$$AA^{-1} = \begin{bmatrix} 1 & 2 \\ 3 & 4 \end{bmatrix}\begin{bmatrix} -2 & 1 \\ \frac{3}{2} & -\frac{1}{2} \end{bmatrix}$$

$$= \begin{bmatrix} 1 & 0 \\ 0 & 1 \end{bmatrix} = I$$

$$A^{-1}A = \begin{bmatrix} -2 & 1 \\ \frac{3}{2} & -\frac{1}{2} \end{bmatrix}\begin{bmatrix} 1 & 2 \\ 3 & 4 \end{bmatrix}$$

$$= \begin{bmatrix} 1 & 0 \\ 0 & 1 \end{bmatrix} = I$$

we see that A^{-1} is the inverse of A, as asserted.

Explore and Discuss

In defining the inverse of a matrix A, why is it necessary to require that A be a square matrix?

Not every square matrix has an inverse. A square matrix that has an inverse is said to be **nonsingular.** A matrix that does not have an inverse is said to be **singular.** An example of a singular matrix is given by

$$B = \begin{bmatrix} 0 & 1 \\ 0 & 0 \end{bmatrix}$$

If B had an inverse given by

$$B^{-1} = \begin{bmatrix} a & b \\ c & d \end{bmatrix}$$

where a, b, c, and d are some appropriate numbers, then by the definition of an inverse, we would have $BB^{-1} = I$; that is,

$$\begin{bmatrix} 0 & 1 \\ 0 & 0 \end{bmatrix}\begin{bmatrix} a & b \\ c & d \end{bmatrix} = \begin{bmatrix} 1 & 0 \\ 0 & 1 \end{bmatrix}$$
$$\begin{bmatrix} c & d \\ 0 & 0 \end{bmatrix} = \begin{bmatrix} 1 & 0 \\ 0 & 1 \end{bmatrix}$$

which implies that $0 = 1$—an impossibility! This contradiction shows that B does not have an inverse.

A Method for Finding the Inverse of a Square Matrix

The methods of Section 5.5 can be used to find the inverse of a nonsingular matrix. To discover such an algorithm, let's find the inverse of the matrix

$$A = \begin{bmatrix} 1 & 2 \\ -1 & 3 \end{bmatrix}$$

Suppose A^{-1} exists and is given by

$$A^{-1} = \begin{bmatrix} a & b \\ c & d \end{bmatrix}$$

where a, b, c, and d are to be determined. By the definition of an inverse, we have $AA^{-1} = I$; that is,

$$\begin{bmatrix} 1 & 2 \\ -1 & 3 \end{bmatrix}\begin{bmatrix} a & b \\ c & d \end{bmatrix} = \begin{bmatrix} 1 & 0 \\ 0 & 1 \end{bmatrix}$$

which simplifies to

$$\begin{bmatrix} a + 2c & b + 2d \\ -a + 3c & -b + 3d \end{bmatrix} = \begin{bmatrix} 1 & 0 \\ 0 & 1 \end{bmatrix}$$

But this matrix equation is equivalent to the two systems of linear equations

$$\begin{cases} a + 2c = 1 \\ -a + 3c = 0 \end{cases} \text{ and } \begin{cases} b + 2d = 0 \\ -b + 3d = 1 \end{cases}$$

with augmented matrices given by

$$\left[\begin{array}{cc|c} 1 & 2 & 1 \\ -1 & 3 & 0 \end{array}\right] \text{ and } \left[\begin{array}{cc|c} 1 & 2 & 0 \\ -1 & 3 & 1 \end{array}\right]$$

Note that the matrices of coefficients of the two systems are identical. This suggests that we solve the two systems of simultaneous linear equations by writing the following augmented matrix, which we obtain by joining the coefficient matrix and the two columns of constants:

$$\left[\begin{array}{cc|cc} 1 & 2 & 1 & 0 \\ -1 & 3 & 0 & 1 \end{array}\right]$$

Using the Gauss–Jordan elimination method, we obtain the following sequence of equivalent matrices:

$$\left[\begin{array}{cc|cc} 1 & 2 & 1 & 0 \\ -1 & 3 & 0 & 1 \end{array}\right] \xrightarrow{R_2 + R_1} \left[\begin{array}{cc|cc} 1 & 2 & 1 & 0 \\ 0 & 5 & 1 & 1 \end{array}\right] \xrightarrow{-\frac{1}{5}R_2}$$

$$\left[\begin{array}{cc|cc} 1 & 2 & 1 & 0 \\ 0 & 1 & \frac{1}{5} & \frac{1}{5} \end{array}\right] \xrightarrow{R_1 - 2R_2} \left[\begin{array}{cc|cc} 1 & 0 & \frac{3}{5} & -\frac{2}{5} \\ 0 & 1 & \frac{1}{5} & \frac{1}{5} \end{array}\right]$$

Thus, $a = \frac{3}{5}$, $b = -\frac{2}{5}$, $c = \frac{1}{5}$, and $d = \frac{1}{5}$, giving

$$A^{-1} = \left[\begin{array}{cc} \frac{3}{5} & -\frac{2}{5} \\ \frac{1}{5} & \frac{1}{5} \end{array}\right]$$

The following computations verify that A^{-1} is indeed the inverse of A:

$$\left[\begin{array}{cc} 1 & 2 \\ -1 & 3 \end{array}\right]\left[\begin{array}{cc} \frac{3}{5} & -\frac{2}{5} \\ \frac{1}{5} & \frac{1}{5} \end{array}\right] = \left[\begin{array}{cc} 1 & 0 \\ 0 & 1 \end{array}\right] = \left[\begin{array}{cc} \frac{3}{5} & -\frac{2}{5} \\ \frac{1}{5} & \frac{1}{5} \end{array}\right]\left[\begin{array}{cc} 1 & 2 \\ -1 & 3 \end{array}\right]$$

The preceding example suggests a general algorithm for computing the inverse of a square matrix of size n when it exists.

Finding the Inverse of a Matrix

Given the $n \times n$ matrix A:

1. Adjoin the $n \times n$ identity matrix I to obtain the augmented matrix

$$[A \mid I]$$

2. Use a sequence of row operations to reduce $[A \mid I]$ to the form

$$[A \mid B]$$

if possible.

Then the matrix B is the inverse of A.

Note   Although matrix multiplication is not generally commutative, it is possible to prove that if A has an inverse and $AB = I$, then $BA = I$ also. Hence to verify that B is the inverse of A, it suffices to show that $AB = I$. ◼

EXAMPLE 1 Find the inverse of the matrix

$$A = \left[\begin{array}{ccc} 2 & 1 & 1 \\ 3 & 2 & 1 \\ 2 & 1 & 2 \end{array}\right]$$

Solution   We form the augmented matrix

$$\left[\begin{array}{ccc|ccc} 2 & 1 & 1 & 1 & 0 & 0 \\ 3 & 2 & 1 & 0 & 1 & 0 \\ 2 & 1 & 2 & 0 & 0 & 1 \end{array}\right]$$

and use the Gauss–Jordan elimination method to reduce it to the form $[I \mid B]$:

$$\begin{bmatrix} 2 & 1 & 1 & | & 1 & 0 & 0 \\ 3 & 2 & 1 & | & 0 & 1 & 0 \\ 2 & 1 & 2 & | & 0 & 0 & 1 \end{bmatrix} \xrightarrow{R_1 - R_2} \begin{bmatrix} -1 & -1 & 0 & | & 1 & -1 & 0 \\ 3 & 2 & 1 & | & 0 & 1 & 0 \\ 2 & 1 & 2 & | & 0 & 0 & 1 \end{bmatrix}$$

$$\xrightarrow[\substack{R_2 + 3R_1 \\ R_3 + 2R_1}]{-R_1} \begin{bmatrix} 1 & 1 & 0 & | & -1 & 1 & 0 \\ 0 & -1 & 1 & | & 3 & -2 & 0 \\ 0 & -1 & 2 & | & 2 & -2 & 1 \end{bmatrix}$$

$$\xrightarrow[\substack{-R_2 \\ R_3 - R_2}]{R_1 + R_2} \begin{bmatrix} 1 & 0 & 1 & | & 2 & -1 & 0 \\ 0 & 1 & -1 & | & -3 & 2 & 0 \\ 0 & 0 & 1 & | & -1 & 0 & 1 \end{bmatrix}$$

$$\xrightarrow[\substack{R_2 + R_3}]{R_1 - R_3} \begin{bmatrix} 1 & 0 & 0 & | & 3 & -1 & -1 \\ 0 & 1 & 0 & | & -4 & 2 & 1 \\ 0 & 0 & 1 & | & -1 & 0 & 1 \end{bmatrix}$$

The inverse of A is the matrix

$$A^{-1} = \begin{bmatrix} 3 & -1 & -1 \\ -4 & 2 & 1 \\ -1 & 0 & 1 \end{bmatrix}$$

We leave it to you to verify these results.　■

Example 2 illustrates what happens to the reduction process when a matrix A does *not* have an inverse.

EXAMPLE 2 Find the inverse of the matrix

$$A = \begin{bmatrix} 1 & 2 & 3 \\ 2 & 1 & 2 \\ 3 & 3 & 5 \end{bmatrix}$$

Solution   We form the augmented matrix

$$\begin{bmatrix} 1 & 2 & 3 & | & 1 & 0 & 0 \\ 2 & 1 & 2 & | & 0 & 1 & 0 \\ 3 & 3 & 5 & | & 0 & 0 & 1 \end{bmatrix}$$

and use the Gauss–Jordan elimination method:

$$\begin{bmatrix} 1 & 2 & 3 & | & 1 & 0 & 0 \\ 2 & 1 & 2 & | & 0 & 1 & 0 \\ 3 & 3 & 5 & | & 0 & 0 & 1 \end{bmatrix} \xrightarrow[\substack{R_3 - 3R_1}]{R_2 - 2R_1} \begin{bmatrix} 1 & 2 & 3 & | & 1 & 0 & 0 \\ 0 & -3 & -4 & | & -2 & 1 & 0 \\ 0 & -3 & -4 & | & -3 & 0 & 1 \end{bmatrix}$$

$$\xrightarrow[\substack{R_3 - R_2}]{-R_2} \begin{bmatrix} 1 & 2 & 3 & | & 1 & 0 & 0 \\ 0 & 3 & 4 & | & 2 & -1 & 0 \\ 0 & 0 & 0 & | & -1 & -1 & 1 \end{bmatrix}$$

> ### Explore and Discuss
>
> Explain in terms of solutions to systems of linear equations why the final augmented matrix in Example 2 implies that A has no inverse.
> Hint: See the discussion on page 334.

Since the entries in the last row of the 3×3 submatrix that comprises the left-hand side of the augmented matrix just obtained are all equal to zero, the latter cannot be reduced to the form $[I \mid B]$. Accordingly, we draw the conclusion that A is singular—that is, does not have an inverse.　■

More generally, we have the following criterion for determining when the inverse of a matrix does not exist.

> **Matrices That Have No Inverses**
>
> If there is a row to the left of the vertical line in the augmented matrix containing all zeros, then the matrix does not have an inverse.

A Formula for the Inverse of a 2 × 2 Matrix

Before turning to some applications, we show an alternative method that employs a formula for finding the inverse of a 2 × 2 matrix. This method will prove useful in many situations; we will see an application in Example 5. The derivation of this formula is left as an exercise (Exercise 52).

> **Formula for the Inverse of a 2 × 2 Matrix**
>
> Let
>
> $$A = \begin{bmatrix} a & b \\ c & d \end{bmatrix}$$
>
> Suppose $D = ad - bc$ is not equal to zero. Then A^{-1} exists and is given by
>
> $$A^{-1} = \frac{1}{D} \begin{bmatrix} d & -b \\ -c & a \end{bmatrix} \tag{14}$$

Note   As an aid to memorizing the formula, note that D is the product of the elements along the main diagonal minus the product of the elements along the other diagonal:

$$\begin{bmatrix} a & b \\ c & d \end{bmatrix} \qquad D = ad - bc$$

Main diagonal

Next, the matrix

$$\begin{bmatrix} d & -b \\ -c & a \end{bmatrix}$$

is obtained by interchanging a and d and reversing the signs of b and c. Finally, A^{-1} is obtained by dividing this matrix by D.

Explore and Discuss

Suppose A is a square matrix with the property that one of its rows is a nonzero constant multiple of another row. What can you say about the existence or nonexistence of A^{-1}? Explain your answer.

EXAMPLE 3   Find the inverse of

$$A = \begin{bmatrix} 1 & 2 \\ 3 & 4 \end{bmatrix}$$

Solution   We first compute $D = (1)(4) - (2)(3) = 4 - 6 = -2$. Next, we rewrite the given matrix, obtaining

$$\begin{bmatrix} 4 & -2 \\ -3 & 1 \end{bmatrix}$$

Finally, dividing this matrix by D, we obtain

$$A^{-1} = -\frac{1}{2} \begin{bmatrix} 4 & -2 \\ -3 & 1 \end{bmatrix} = \begin{bmatrix} -2 & 1 \\ \frac{3}{2} & -\frac{1}{2} \end{bmatrix}$$

Solving Systems of Equations with Inverses

We now show how the inverse of a matrix may be used to solve certain systems of linear equations in which the number of equations in the system is equal to the number of variables. For simplicity, let's illustrate the process for a system of three linear equations in three variables:

$$
\begin{aligned}
a_{11}x_1 + a_{12}x_2 + a_{13}x_3 &= b_1 \\
a_{21}x_1 + a_{22}x_2 + a_{23}x_3 &= b_2 \\
a_{31}x_1 + a_{32}x_2 + a_{33}x_3 &= b_3
\end{aligned}
\tag{15}
$$

Let's write

$$
A = \begin{bmatrix} a_{11} & a_{12} & a_{13} \\ a_{21} & a_{22} & a_{23} \\ a_{31} & a_{32} & a_{33} \end{bmatrix} \qquad X = \begin{bmatrix} x_1 \\ x_2 \\ x_3 \end{bmatrix} \qquad B = \begin{bmatrix} b_1 \\ b_2 \\ b_3 \end{bmatrix}
$$

You should verify that System (15) of linear equations may be written in the form of the matrix equation

$$
AX = B \tag{16}
$$

If A is nonsingular, then the method of this section may be used to compute A^{-1}. Next, multiplying both sides of Equation (16) by A^{-1} (on the left), we obtain

$$
A^{-1}AX = A^{-1}B \quad \text{or} \quad IX = A^{-1}B \quad \text{or} \quad X = A^{-1}B
$$

the desired solution to the problem.

In the case of a system of n equations with n unknowns, we have the following more general result.

> **Using Inverses to Solve Systems of Equations**
>
> If $AX = B$ is a linear system of n equations in n unknowns and if A^{-1} exists, then
>
> $$X = A^{-1}B$$
>
> is the unique solution of the system.

The use of inverses to solve systems of equations is particularly advantageous when we are required to solve more than one system of equations, $AX = B$, involving the same coefficient matrix, A, and different matrices of constants, B. As you will see in Examples 4 and 5, we need to compute A^{-1} just once in each case.

EXAMPLE 4 Solve the following systems of linear equations:

a.
$$
\begin{aligned}
2x + y + z &= 1 \\
3x + 2y + z &= 2 \\
2x + y + 2z &= -1
\end{aligned}
$$

b.
$$
\begin{aligned}
2x + y + z &= 2 \\
3x + 2y + z &= -3 \\
2x + y + 2z &= 1
\end{aligned}
$$

Solution We may write the given systems of equations in the form

$$
AX = B \quad \text{and} \quad AX = C
$$

respectively, where

$$
A = \begin{bmatrix} 2 & 1 & 1 \\ 3 & 2 & 1 \\ 2 & 1 & 2 \end{bmatrix} \qquad X = \begin{bmatrix} x \\ y \\ z \end{bmatrix} \qquad B = \begin{bmatrix} 1 \\ 2 \\ -1 \end{bmatrix} \qquad C = \begin{bmatrix} 2 \\ -3 \\ 1 \end{bmatrix}
$$

The inverse of the matrix A,

$$A^{-1} = \begin{bmatrix} 3 & -1 & -1 \\ -4 & 2 & 1 \\ -1 & 0 & 1 \end{bmatrix}$$

was found in Example 1. Using this result, we find that the solution of the first system (a) is

$$X = A^{-1}B = \begin{bmatrix} 3 & -1 & -1 \\ -4 & 2 & 1 \\ -1 & 0 & 1 \end{bmatrix}\begin{bmatrix} 1 \\ 2 \\ -1 \end{bmatrix}$$

$$= \begin{bmatrix} (3)(1) + (-1)(2) + (-1)(-1) \\ (-4)(1) + (2)(2) + (1)(-1) \\ (-1)(1) + (0)(2) + (1)(-1) \end{bmatrix} = \begin{bmatrix} 2 \\ -1 \\ -2 \end{bmatrix}$$

or $x = 2$, $y = -1$, and $z = -2$.

The solution of the second system (b) is

$$X = A^{-1}C = \begin{bmatrix} 3 & -1 & -1 \\ -4 & 2 & 1 \\ -1 & 0 & 1 \end{bmatrix}\begin{bmatrix} 2 \\ -3 \\ 1 \end{bmatrix} = \begin{bmatrix} 8 \\ -13 \\ -1 \end{bmatrix}$$

or $x = 8$, $y = -13$, and $z = -1$.

APPLIED EXAMPLE 5 Capital Expenditures The management of Checkers Rent-A-Car plans to expand its fleet of rental cars for the next quarter by purchasing compact and full-size cars. The average cost of a compact car is $15,000, and the average cost of a full-size car is $36,000.

a. If a total of 800 cars is to be purchased with a budget of $18 million, how many cars of each size will be acquired?

b. If the predicted demand calls for a total purchase of 1000 cars with a budget of $21 million, how many cars of each type will be acquired?

Solution Let x and y denote the number of compact and full-size cars to be purchased. Furthermore, let n denote the total number of cars to be acquired and b the amount of money budgeted for the purchase of these cars. Then

$$\begin{aligned} x + \quad\quad y &= n \\ 15,000x + 36,000y &= b \end{aligned}$$

This system of two equations in two variables may be written in the matrix form

$$AX = B$$

where

$$A = \begin{bmatrix} 1 & 1 \\ 15,000 & 36,000 \end{bmatrix} \quad X = \begin{bmatrix} x \\ y \end{bmatrix} \quad B = \begin{bmatrix} n \\ b \end{bmatrix}$$

Therefore,

$$X = A^{-1}B$$

Since A is a 2×2 matrix, its inverse may be found by using Formula (14). We find $D = (1)(36,000) - (1)(15,000) = 21,000$, so

$$A^{-1} = \frac{1}{21,000}\begin{bmatrix} 36,000 & -1 \\ -15,000 & 1 \end{bmatrix} = \begin{bmatrix} \frac{36,000}{21,000} & -\frac{1}{21,000} \\ -\frac{15,000}{21,000} & \frac{1}{21,000} \end{bmatrix}$$

Thus,

$$X = \begin{bmatrix} \frac{12}{7} & -\frac{1}{21,000} \\ -\frac{5}{7} & \frac{1}{21,000} \end{bmatrix} \begin{bmatrix} n \\ b \end{bmatrix}$$

a. Here, $n = 800$ and $b = 18,000,000$, so

$$X = A^{-1}B = \begin{bmatrix} \frac{12}{7} & -\frac{1}{21,000} \\ -\frac{5}{7} & \frac{1}{21,000} \end{bmatrix} \begin{bmatrix} 800 \\ 18,000,000 \end{bmatrix} \approx \begin{bmatrix} 514.3 \\ 285.7 \end{bmatrix}$$

Therefore, 514 compact cars and 286 full-size cars will be acquired in this case.

b. Here, $n = 1000$ and $b = 21,000,000$, so

$$X = A^{-1}B = \begin{bmatrix} \frac{12}{7} & -\frac{1}{21,000} \\ -\frac{5}{7} & \frac{1}{21,000} \end{bmatrix} \begin{bmatrix} 1000 \\ 21,000,000 \end{bmatrix} \approx \begin{bmatrix} 714.3 \\ 285.7 \end{bmatrix}$$

Therefore, 714 compact cars and 286 full-size cars will be purchased in this case.

5.6 Self-Check Exercises

1. Find the inverse of the matrix

$$A = \begin{bmatrix} 2 & 1 & -1 \\ 1 & 1 & -1 \\ -1 & -2 & 3 \end{bmatrix}$$

if it exists.

2. Solve the system of linear equations

$$\begin{aligned} 2x + y - z &= b_1 \\ x + y - z &= b_2 \\ -x - 2y + 3z &= b_3 \end{aligned}$$

where (a) $b_1 = 5$, $b_2 = 4$, $b_3 = -8$ and (b) $b_1 = 2$, $b_2 = 0$, $b_3 = 5$, by finding the inverse of the coefficient matrix.

3. **TOUR TICKETING** Grand Canyon Tours offers air and ground scenic tours of the Grand Canyon. Tickets for the $7\frac{1}{2}$-hour tour cost $169 for an adult and $129 for a child, and each tour group is limited to 19 people. On three recent fully booked tours, total receipts were $2931 for the first tour, $3011 for the second tour, and $2771 for the third tour. Determine how many adults and how many children were in each tour.

Solutions to Self-Check Exercises 5.6 can be found on page 344.

5.6 Concept Questions

1. What is the inverse of a matrix A?

2. Explain how you would find the inverse of a nonsingular matrix.

3. Give the formula for the inverse of the 2×2 matrix

$$A = \begin{bmatrix} a & b \\ c & d \end{bmatrix}$$

4. Explain how the inverse of a matrix can be used to solve a system of n linear equations in n unknowns. Does the method work for a system of m linear equations in n unknowns with $m \neq n$? Explain.

5.6 Exercises

In Exercises 1–4, show that the matrices are inverses of each other by showing that their product is the identity matrix I.

1. $\begin{bmatrix} 1 & -3 \\ 1 & -2 \end{bmatrix}$ and $\begin{bmatrix} -2 & 3 \\ -1 & 1 \end{bmatrix}$

2. $\begin{bmatrix} 4 & 5 \\ 2 & 3 \end{bmatrix}$ and $\begin{bmatrix} \frac{3}{2} & -\frac{5}{2} \\ -1 & 2 \end{bmatrix}$

3. $\begin{bmatrix} 3 & 2 & 3 \\ 2 & 2 & 1 \\ 2 & 1 & 1 \end{bmatrix}$ and $\begin{bmatrix} -\frac{1}{3} & -\frac{1}{3} & \frac{4}{3} \\ 0 & 1 & -1 \\ \frac{2}{3} & -\frac{1}{3} & -\frac{2}{3} \end{bmatrix}$

4. $\begin{bmatrix} 2 & 4 & -2 \\ -4 & -6 & 1 \\ 3 & 5 & -1 \end{bmatrix}$ and $\begin{bmatrix} \frac{1}{2} & -3 & -4 \\ -\frac{1}{2} & 2 & 3 \\ -1 & 1 & 2 \end{bmatrix}$

In Exercises 5–16, find the inverse of the matrix, if it exists. Verify your answer.

5. $\begin{bmatrix} 2 & 5 \\ 1 & 3 \end{bmatrix}$

6. $\begin{bmatrix} 2 & 3 \\ 3 & 5 \end{bmatrix}$

7. $\begin{bmatrix} 3 & -3 \\ -2 & 2 \end{bmatrix}$

8. $\begin{bmatrix} 4 & 2 \\ 6 & 3 \end{bmatrix}$

9. $\begin{bmatrix} 2 & -3 & -4 \\ 0 & 0 & -1 \\ 1 & -2 & 1 \end{bmatrix}$

10. $\begin{bmatrix} 1 & -1 & 3 \\ 2 & 1 & 2 \\ -2 & -2 & 1 \end{bmatrix}$

11. $\begin{bmatrix} 4 & 2 & 2 \\ -1 & -3 & 4 \\ 3 & -1 & 6 \end{bmatrix}$

12. $\begin{bmatrix} 1 & 2 & 0 \\ -3 & 4 & -2 \\ -5 & 0 & -2 \end{bmatrix}$

13. $\begin{bmatrix} 1 & 4 & -1 \\ 2 & 3 & -2 \\ -1 & 2 & 3 \end{bmatrix}$

14. $\begin{bmatrix} 3 & -2 & 7 \\ -2 & 1 & 4 \\ 6 & -5 & 8 \end{bmatrix}$

15. $\begin{bmatrix} 1 & 1 & -1 & 1 \\ 2 & 1 & 1 & 0 \\ 2 & 1 & 0 & 1 \\ 2 & -1 & -1 & 3 \end{bmatrix}$

16. $\begin{bmatrix} 1 & 1 & 2 & 3 \\ 2 & 3 & 0 & -1 \\ 0 & 2 & -1 & 1 \\ 1 & 2 & 1 & 1 \end{bmatrix}$

In Exercises 17–24, (a) write a matrix equation that is equivalent to the system of linear equations, and (b) solve the system using the inverses found in Exercises 5–16.

17. $2x + 5y = 3$
$x + 3y = 2$
(See Exercise 5.)

18. $2x + 3y = 5$
$3x + 5y = 8$
(See Exercise 6.)

19. $2x - 3y - 4z = 4$
$ -z = 3$
$x - 2y + z = -8$
(See Exercise 9.)

20. $x_1 - x_2 + 3x_3 = 2$
$2x_1 + x_2 + 2x_3 = 2$
$-2x_1 - 2x_2 + x_3 = 3$
(See Exercise 10.)

21. $x + 4y - z = 3$
$2x + 3y - 2z = 1$
$-x + 2y + 3z = 7$
(See Exercise 13.)

22. $3x_1 - 2x_2 + 7x_3 = 6$
$-2x_1 + x_2 + 4x_3 = 4$
$6x_1 - 5x_2 + 8x_3 = 4$
(See Exercise 14.)

23. $x_1 + x_2 - x_3 + x_4 = 6$
$2x_1 + x_2 + x_3 = 4$
$2x_1 + x_2 + x_4 = 7$
$2x_1 - x_2 - x_3 + 3x_4 = 9$
(See Exercise 15.)

24. $x_1 + x_2 + 2x_3 + 3x_4 = 4$
$2x_1 + 3x_2 - x_4 = 11$
$2x_2 - x_3 + x_4 = 7$
$x_1 + 2x_2 + x_3 + x_4 = 6$
(See Exercise 16.)

In Exercises 25–32, (a) write each system of equations as a matrix equation, and (b) solve the system of equations by using the inverse of the coefficient matrix.

25.
$x + 2y = b_1$
$2x - y = b_2$
where (i) $b_1 = 14, b_2 = 5$
and (ii) $b_1 = 4, b_2 = -1$

26.
$3x - 2y = b_1$
$4x + 3y = b_2$
where (i) $b_1 = -6, b_2 = 10$
and (ii) $b_1 = 3, b_2 = -2$

27.
$x + 2y + z = b_1$
$x + y + z = b_2$
$3x + y + z = b_3$
where (i) $b_1 = 7, b_2 = 4, b_3 = 2$
and (ii) $b_1 = 5, b_2 = -3, b_3 = -1$

28.
$x_1 + x_2 + x_3 = b_1$
$x_1 - x_2 + x_3 = b_2$
$x_1 - 2x_2 - x_3 = b_3$
where (i) $b_1 = 5, b_2 = -3, b_3 = -1$
and (ii) $b_1 = 1, b_2 = 4, b_3 = -2$

29.
$3x + 2y - z = b_1$
$2x - 3y + z = b_2$
$x - y - z = b_3$
where (i) $b_1 = 2, b_2 = -2, b_3 = 4$
and (ii) $b_1 = 8, b_2 = -3, b_3 = 6$

30.
$2x_1 + x_2 + x_3 = b_1$
$x_1 - 3x_2 + 4x_3 = b_2$
$-x_1 + x_3 = b_3$
where (i) $b_1 = 1, b_2 = 4, b_3 = -3$
and (ii) $b_1 = 2, b_2 = -5, b_3 = 0$

31.
$x_1 + x_2 + x_3 + x_4 = b_1$
$x_1 - x_2 - x_3 + x_4 = b_2$
$x_2 + 2x_3 + 2x_4 = b_3$
$x_1 + 2x_2 + x_3 - 2x_4 = b_4$
where (i) $b_1 = 1, b_2 = -1, b_3 = 4, b_4 = 0$
and (ii) $b_1 = 2, b_2 = 8, b_3 = 4, b_4 = -1$

32.
$$x_1 + x_2 + 2x_3 + x_4 = b_1$$
$$4x_1 + 5x_2 + 9x_3 + x_4 = b_2$$
$$3x_1 + 4x_2 + 7x_3 + x_4 = b_3$$
$$2x_1 + 3x_2 + 4x_3 + 2x_4 = b_4$$

where (i) $b_1 = 3$, $b_2 = 6$, $b_3 = 5$, $b_4 = 7$
and (ii) $b_1 = 1$, $b_2 = -1$, $b_3 = 0$, $b_4 = -4$

33. Let
$$A = \begin{bmatrix} 2 & 3 \\ -4 & -5 \end{bmatrix}$$

a. Find A^{-1}.
b. Show that $(A^{-1})^{-1} = A$.

34. Let
$$A = \begin{bmatrix} 6 & -4 \\ -4 & 3 \end{bmatrix}$$
and
$$B = \begin{bmatrix} 3 & -5 \\ 4 & -7 \end{bmatrix}$$

a. Find AB, A^{-1}, and B^{-1}.
b. Show that $(AB)^{-1} = B^{-1}A^{-1}$.

35. Let
$$A = \begin{bmatrix} 2 & -5 \\ 1 & -3 \end{bmatrix} \quad B = \begin{bmatrix} 4 & 3 \\ 1 & 1 \end{bmatrix} \quad C = \begin{bmatrix} 2 & 3 \\ -2 & 1 \end{bmatrix}$$

a. Find ABC, A^{-1}, B^{-1}, and C^{-1}.
b. Show that $(ABC)^{-1} = C^{-1}B^{-1}A^{-1}$.

36. Find the matrix A if
$$\begin{bmatrix} 2 & 1 \\ -1 & 3 \end{bmatrix} A = \begin{bmatrix} 3 & 2 \\ 1 & 4 \end{bmatrix}$$

37. Find the matrix A if
$$A \begin{bmatrix} 1 & 2 \\ 3 & -1 \end{bmatrix} = \begin{bmatrix} 2 & 1 \\ 3 & -2 \end{bmatrix}$$

38. **TICKET REVENUES** Rainbow Harbor Cruises charges $16/adult and $8/child for a round-trip ticket. The records show that, on a certain weekend, 1000 people took the cruise on Saturday, and 800 people took the cruise on Sunday. The total receipts for Saturday were $12,800, and the total receipts for Sunday were $9,600. Determine how many adults and children took the cruise on Saturday and on Sunday.

39. **PRICING PERSONAL PLANNERS** BelAir Publishing publishes a deluxe leather edition and a standard edition of its daily organizer. The company's marketing department estimates that x copies of the deluxe edition and y copies of the standard edition will be demanded per month when the unit prices are p dollars and q dollars, respectively,

where x, y, p, and q are related by the following system of linear equations:
$$5x + y = 1000(70 - p)$$
$$x + 3y = 1000(40 - q)$$

Find the monthly demand for the deluxe edition and the standard edition when the unit prices are set according to the following schedules:
a. $p = 50$ and $q = 25$
b. $p = 45$ and $q = 25$
c. $p = 45$ and $q = 20$

40. **DIET PLANNING** Bob, a nutritionist who works for the University Medical Center, has been asked to prepare special diets for two patients, Susan and Tom. Bob has decided that Susan's meals should contain at least 400 mg of calcium, 20 mg of iron, and 50 mg of vitamin C, whereas Tom's meals should contain at least 350 mg of calcium, 15 mg of iron, and 40 mg of vitamin C. Bob has also decided that the meals are to be prepared from three basic foods: Food A, Food B, and Food C. The special nutritional contents of these foods are summarized in the accompanying table. Find how many ounces of each type of food should be used in a meal so that the minimum requirements of calcium, iron, and vitamin C are met for each patient's meals.

	Contents (mg/oz)		
	Calcium	Iron	Vitamin C
Food A	30	1	2
Food B	25	1	5
Food C	20	2	4

41. **CROP PLANNING** Jackson Farms has allotted a certain amount of land for cultivating soybeans, corn, and wheat. Cultivating 1 acre of soybeans requires 2 labor-hours, and cultivating 1 acre of corn or wheat requires 6 labor-hours. The cost of seeds for 1 acre of soybeans is $12, the cost for 1 acre of corn is $20, and the cost for 1 acre of wheat is $8. If all resources are to be used, how many acres of each crop should be cultivated if the following hold?
a. 1000 acres of land are allotted, 4400 labor-hours are available, and $13,200 is available for seeds.
b. 1200 acres of land are allotted, 5200 labor-hours are available, and $16,400 is available for seeds.

42. **LAWN FERTILIZERS** Lawnco produces three grades of commercial fertilizers. A 100-lb bag of grade A fertilizer contains 18 lb of nitrogen, 4 lb of phosphate, and 5 lb of potassium. A 100-lb bag of grade B fertilizer contains 20 lb of nitrogen and 4 lb each of phosphate and potassium. A 100-lb bag of grade C fertilizer contains 24 lb of nitrogen, 3 lb of phosphate, and 6 lb of potassium. How many 100-lb bags of each of the three grades of fertilizers should Lawnco produce if:

a. 26,400 lb of nitrogen, 4900 lb of phosphate, and 6200 lb of potassium are available and all the nutrients are used?

b. 21,800 lb of nitrogen, 4200 lb of phosphate, and 5300 lb of potassium are available and all the nutrients are used?

43. **INVESTMENT RISK AND RETURN** A private investment club has a certain amount of money earmarked for investment in stocks. To arrive at an acceptable overall level of risk, the stocks that management is considering have been classified into three categories: high-risk, medium-risk, and low-risk. Management estimates that high-risk stocks will have a rate of return of 15%/year; medium-risk stocks, 10%/year; and low-risk stocks, 6%/year. The members have decided that the investment in low-risk stocks should be equal to the sum of the investments in the stocks of the other two categories. Determine how much the club should invest in each type of stock in each of the following scenarios. (In all cases, assume that the entire sum available for investment is invested.)

a. The club has $200,000 to invest, and the investment goal is to have a return of $20,000/year on the total investment.

b. The club has $220,000 to invest, and the investment goal is to have a return of $22,000/year on the total investment.

c. The club has $240,000 to invest, and the investment goal is to have a return of $22,000/year on the total investment.

44. **RESEARCH FUNDING** The Carver Foundation funds three nonprofit organizations engaged in alternative-energy research activities. From past data, the proportion of funds spent by each organization in research on solar energy, energy from harnessing the wind, and energy from the motion of ocean tides is given in the accompanying table.

	Proportion of Money Spent		
	Solar	Wind	Tides
Organization I	0.6	0.3	0.1
Organization II	0.4	0.3	0.3
Organization III	0.2	0.6	0.2

Find the amount awarded to each organization if the total amount spent by all three organizations on solar, wind, and tidal research is:

a. $9.2 million, $9.6 million, and $5.2 million, respectively.

b. $8.2 million, $7.2 million, and $3.6 million, respectively.

45. Find the value(s) of k such that

$$A = \begin{bmatrix} 1 & 2 \\ k & 3 \end{bmatrix}$$

has an inverse. What is the inverse of A?
Hint: Use Formula (14).

46. Find the value(s) of k such that

$$A = \begin{bmatrix} 1 & 0 & 1 \\ -2 & 1 & k \\ -1 & 2 & k^2 \end{bmatrix}$$

has an inverse.
Hint: Find the value(s) of k such that the augmented matrix $[A \mid I]$ can be reduced to the form $[I \mid B]$.

47. Find conditions on a and d such that the matrix

$$A = \begin{bmatrix} a & 0 \\ 0 & d \end{bmatrix}$$

has an inverse. A square matrix is said to be a *diagonal matrix* if all the entries not lying on the main diagonal are zero. Discuss the existence of the inverse matrix of a diagonal matrix of size $n \times n$.

48. Find conditions a, b, and d such that the 2×2 upper triangular matrix

$$A = \begin{bmatrix} a & b \\ 0 & d \end{bmatrix}$$

has an inverse. A square matrix is said to be an *upper triangular matrix* if all its entries below the main diagonal are zero. Discuss the existence of the inverse of an upper triangular matrix of size $n \times n$.

In Exercises 49–51, determine whether the statement is true or false. If it is true, explain why it is true. If it is false, give an example to show why it is false.

49. If A is a square matrix with inverse A^{-1} and c is a nonzero real number, then

$$(cA)^{-1} = \left(\frac{1}{c}\right)A^{-1}$$

50. The matrix

$$A = \begin{bmatrix} a & b \\ c & d \end{bmatrix}$$

has an inverse if and only if $ad - bc = 0$.

51. If A^{-1} does not exist, then the system $AX = B$ of n linear equations in n unknowns does not have a unique solution.

52. Let

$$A = \begin{bmatrix} a & b \\ c & d \end{bmatrix}$$

a. Find A^{-1} if it exists.
b. Find a necessary condition for A to be nonsingular.
c. Verify that $AA^{-1} = A^{-1}A = I$.

5.6 Solutions to Self-Check Exercises

1. We form the augmented matrix

$$\left[\begin{array}{ccc|ccc} 2 & 1 & -1 & 1 & 0 & 0 \\ 1 & 1 & -1 & 0 & 1 & 0 \\ -1 & -2 & 3 & 0 & 0 & 1 \end{array}\right]$$

and row-reduce as follows:

$$\left[\begin{array}{ccc|ccc} 2 & 1 & -1 & 1 & 0 & 0 \\ 1 & 1 & -1 & 0 & 1 & 0 \\ -1 & -2 & 3 & 0 & 0 & 1 \end{array}\right] \xrightarrow{R_1 \leftrightarrow R_2}$$

$$\left[\begin{array}{ccc|ccc} 1 & 1 & -1 & 0 & 1 & 0 \\ 2 & 1 & -1 & 1 & 0 & 0 \\ -1 & -2 & 3 & 0 & 0 & 1 \end{array}\right] \xrightarrow[R_3 + R_1]{R_2 - 2R_1}$$

$$\left[\begin{array}{ccc|ccc} 1 & 1 & -1 & 0 & 1 & 0 \\ 0 & -1 & 1 & 1 & -2 & 0 \\ 0 & -1 & 2 & 0 & 1 & 1 \end{array}\right] \xrightarrow[\substack{-R_2 \\ R_3 - R_2}]{R_1 + R_2}$$

$$\left[\begin{array}{ccc|ccc} 1 & 0 & 0 & 1 & -1 & 0 \\ 0 & 1 & -1 & -1 & 2 & 0 \\ 0 & 0 & 1 & -1 & 3 & 1 \end{array}\right] \xrightarrow{R_2 + R_3}$$

$$\left[\begin{array}{ccc|ccc} 1 & 0 & 0 & 1 & -1 & 0 \\ 0 & 1 & 0 & -2 & 5 & 1 \\ 0 & 0 & 1 & -1 & 3 & 1 \end{array}\right]$$

From the preceding results, we see that

$$A^{-1} = \begin{bmatrix} 1 & -1 & 0 \\ -2 & 5 & 1 \\ -1 & 3 & 1 \end{bmatrix}$$

2. a. We write the systems of linear equations in the matrix form

$$AX = B_1$$

where

$$A = \begin{bmatrix} 2 & 1 & -1 \\ 1 & 1 & -1 \\ -1 & -2 & 3 \end{bmatrix} \quad X = \begin{bmatrix} x \\ y \\ z \end{bmatrix} \quad B_1 = \begin{bmatrix} 5 \\ 4 \\ -8 \end{bmatrix}$$

Now, using the results of Exercise 1, we have

$$X = \begin{bmatrix} x \\ y \\ z \end{bmatrix} = A^{-1}B_1 = \begin{bmatrix} 1 & -1 & 0 \\ -2 & 5 & 1 \\ -1 & 3 & 1 \end{bmatrix} \begin{bmatrix} 5 \\ 4 \\ -8 \end{bmatrix} = \begin{bmatrix} 1 \\ 2 \\ -1 \end{bmatrix}$$

Therefore, $x = 1$, $y = 2$, and $z = -1$.

b. Here, A and X are as in part (a), but

$$B_2 = \begin{bmatrix} 2 \\ 0 \\ 5 \end{bmatrix}$$

Therefore,

$$X = \begin{bmatrix} x \\ y \\ z \end{bmatrix} = A^{-1}B_2 = \begin{bmatrix} 1 & -1 & 0 \\ -2 & 5 & 1 \\ -1 & 3 & 1 \end{bmatrix} \begin{bmatrix} 2 \\ 0 \\ 5 \end{bmatrix} = \begin{bmatrix} 2 \\ 1 \\ 3 \end{bmatrix}$$

or $x = 2$, $y = 1$, and $z = 3$.

3. Let x denote the number of adults, and let y denote the number of children on a tour. Since the tours are filled to capacity, we have

$$x + y = 19$$

Next, since the total receipts for the first tour were \$2931, we have

$$169x + 129y = 2931$$

Therefore, the number of adults and the number of children in the first tour are found by solving the system of linear equations

$$\begin{aligned} x + \ \ y &= \ \ 19 \\ 169x + 129y &= 2931 \end{aligned} \tag{a}$$

Similarly, we see that the number of adults and the number of children in the second and third tours are found by solving the systems

$$\begin{aligned} x + \ \ y &= \ \ 19 \\ 169x + 129y &= 3011 \end{aligned} \tag{b}$$

$$\begin{aligned} x + \ \ y &= \ \ 19 \\ 169x + 129y &= 2771 \end{aligned} \tag{c}$$

These systems may be written in the form

$$AX = B_1 \qquad AX = B_2 \qquad AX = B_3$$

where

$$A = \begin{bmatrix} 1 & 1 \\ 169 & 129 \end{bmatrix} \quad X = \begin{bmatrix} x \\ y \end{bmatrix}$$

$$B_1 = \begin{bmatrix} 19 \\ 2931 \end{bmatrix} \quad B_2 = \begin{bmatrix} 19 \\ 3011 \end{bmatrix} \quad B_3 = \begin{bmatrix} 19 \\ 2771 \end{bmatrix}$$

To solve these systems, we first find A^{-1}. Using Formula (14) with $D = (1)(129) - (1)(169) = -40$, we obtain

$$A^{-1} = -\frac{1}{40}\begin{bmatrix} 129 & -1 \\ -169 & 1 \end{bmatrix} = \begin{bmatrix} -\frac{129}{40} & \frac{1}{40} \\ \frac{169}{40} & -\frac{1}{40} \end{bmatrix}$$

Then, solving each system, we find

$$X = \begin{bmatrix} x \\ y \end{bmatrix} = A^{-1}B_1$$

$$= \begin{bmatrix} -\frac{129}{40} & \frac{1}{40} \\ \frac{169}{40} & -\frac{1}{40} \end{bmatrix} \begin{bmatrix} 19 \\ 2931 \end{bmatrix} = \begin{bmatrix} 12 \\ 7 \end{bmatrix} \tag{a}$$

$$X = \begin{bmatrix} x \\ y \end{bmatrix} = A^{-1}B_2$$

$$= \begin{bmatrix} -\frac{129}{40} & \frac{1}{40} \\ \frac{169}{40} & -\frac{1}{40} \end{bmatrix} \begin{bmatrix} 19 \\ 3011 \end{bmatrix}$$

$$= \begin{bmatrix} 14 \\ 5 \end{bmatrix} \qquad \textbf{(b)}$$

$$X = \begin{bmatrix} x \\ y \end{bmatrix} = A^{-1}B_3$$

$$= \begin{bmatrix} -\frac{129}{40} & \frac{1}{40} \\ \frac{169}{40} & -\frac{1}{40} \end{bmatrix} \begin{bmatrix} 19 \\ 2771 \end{bmatrix} = \begin{bmatrix} 8 \\ 11 \end{bmatrix} \qquad \textbf{(c)}$$

We conclude that there were:

a. 12 adults and 7 children on the first tour.
b. 14 adults and 5 children on the second tour.
c. 8 adults and 11 children on the third tour.

USING TECHNOLOGY

Finding the Inverse of a Square Matrix

Graphing Utility

A graphing utility can be used to find the inverse of a square matrix.

EXAMPLE 1 Use a graphing utility to find the inverse of

$$\begin{bmatrix} 1 & 3 & 5 \\ -2 & 2 & 4 \\ 5 & 1 & 3 \end{bmatrix}$$

Solution We first enter the given matrix as

$$A = \begin{bmatrix} 1 & 3 & 5 \\ -2 & 2 & 4 \\ 5 & 1 & 3 \end{bmatrix}$$

Then, recalling the matrix A and using the $\boxed{\mathbf{x^{-1}}}$ key, we find

$$A^{-1} = \begin{bmatrix} 0.1 & -0.2 & 0.1 \\ 1.3 & -1.1 & -0.7 \\ -0.6 & 0.7 & 0.4 \end{bmatrix}$$

EXAMPLE 2 Use a graphing utility to solve the system

$$\begin{aligned} x + 3y + 5z &= 4 \\ -2x + 2y + 4z &= 3 \\ 5x + y + 3z &= 2 \end{aligned}$$

by using the inverse of the coefficient matrix.

Solution The given system can be written in the matrix form $AX = B$, where

$$A = \begin{bmatrix} 1 & 3 & 5 \\ -2 & 2 & 4 \\ 5 & 1 & 3 \end{bmatrix} \qquad X = \begin{bmatrix} x \\ y \\ z \end{bmatrix} \qquad B = \begin{bmatrix} 4 \\ 3 \\ 2 \end{bmatrix}$$

The solution is $X = A^{-1}B$. Entering the matrices A and B in the graphing utility and using the matrix multiplication capability of the utility gives the output shown in Figure T1—that is, $x = 0$, $y = 0.5$, and $z = 0.5$.

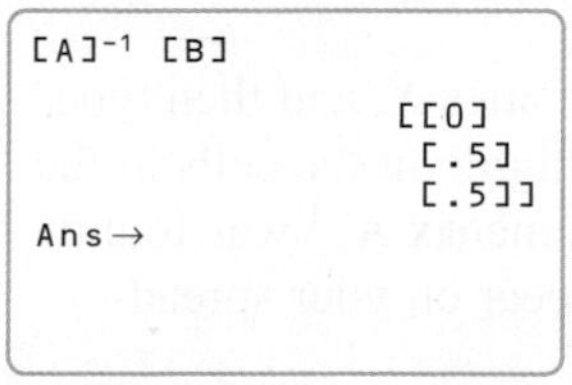

```
[A]⁻¹ [B]
            [[0]
             [.5]
             [.5]]
Ans→
```

FIGURE T1
The TI-83/84 screen showing
$A^{-1}B$

Excel

We use the function **MINVERSE** to find the inverse of a square matrix using Excel.

EXAMPLE 3 Find the inverse of

$$
A = \begin{bmatrix} 1 & 3 & 5 \\ -2 & 2 & 4 \\ 5 & 1 & 3 \end{bmatrix}
$$

Solution

1. Enter the elements of matrix A onto a spreadsheet (Figure T2).
2. Compute the inverse of the matrix A: Highlight the cells that will contain the inverse matrix A^{-1}, type = MINVERSE (, highlight the cells containing matrix A, type), and press **Ctrl-Shift-Enter**. The desired matrix will appear in your spreadsheet (Figure T2).

	A	B	C
1		Matrix A	
2	1	3	5
3	-2	2	4
4	5	1	3
5			
6		Matrix A⁻¹	
7	0.1	-0.2	0.1
8	1.3	-1.1	-0.7
9	-0.6	0.7	0.4

FIGURE **T2**
Matrix A and its inverse, matrix A^{-1}

EXAMPLE 4 Solve the system

$$
\begin{aligned}
x + 3y + 5z &= 4 \\
-2x + 2y + 4z &= 3 \\
5x + y + 3z &= 2
\end{aligned}
$$

by using the inverse of the coefficient matrix.

Solution The given system can be written in the matrix form $AX = B$, where

$$
A = \begin{bmatrix} 1 & 3 & 5 \\ -2 & 2 & 4 \\ 5 & 1 & 3 \end{bmatrix} \qquad X = \begin{bmatrix} x \\ y \\ z \end{bmatrix} \qquad B = \begin{bmatrix} 4 \\ 3 \\ 2 \end{bmatrix}
$$

The solution is $X = A^{-1}B$.

1. Enter the matrix B on a spreadsheet.
2. Compute $A^{-1}B$. Highlight the cells that will contain the matrix X, and then type =MMULT (, highlight the cells in the matrix A^{-1}, type , , highlight the cells in the matrix B, type), and press **Ctrl-Shift-Enter**. (*Note*: The matrix A^{-1} was found in Example 3.) The matrix X shown in Figure T3 will appear on your spreadsheet. Thus, $x = 0$, $y = 0.5$, and $z = 0.5$.

	A
12	Matrix X
13	5.55112E-17
14	0.5
15	0.5

FIGURE **T3**
Matrix X gives the solution to the problem.

Note: Boldfaced words/characters enclosed in a box (for example, **Enter**) indicate that an action (click, select, or press) is required. Words/characters printed blue (for example, Chart sub-type:) indicate words/characters that appear on the screen. Words/characters printed in a monospace font (for example, = (-2/3) *A2+2) indicate words/characters that need to be typed and entered.

TECHNOLOGY EXERCISES

In Exercises 1–6, find the inverse of the matrix. Round your answers to two decimal places.

1. $\begin{bmatrix} 1.2 & 3.1 & -2.1 \\ 3.4 & 2.6 & 7.3 \\ -1.2 & 3.4 & -1.3 \end{bmatrix}$
2. $\begin{bmatrix} 4.2 & 3.7 & 4.6 \\ 2.1 & -1.3 & -2.3 \\ 1.8 & 7.6 & -2.3 \end{bmatrix}$

3. $\begin{bmatrix} 1.1 & 2.3 & 3.1 & 4.2 \\ 1.6 & 3.2 & 1.8 & 2.9 \\ 4.2 & 1.6 & 1.4 & 3.2 \\ 1.6 & 2.1 & 2.8 & 7.2 \end{bmatrix}$

4. $\begin{bmatrix} 2.1 & 3.2 & -1.4 & -3.2 \\ 6.2 & 7.3 & 8.4 & 1.6 \\ 2.3 & 7.1 & 2.4 & -1.3 \\ -2.1 & 3.1 & 4.6 & 3.7 \end{bmatrix}$

5. $\begin{bmatrix} 2 & -1 & 3 & 2 & 4 \\ 3 & 2 & -1 & 4 & 1 \\ 3 & 2 & 6 & 4 & -1 \\ 2 & 1 & -1 & 4 & 2 \\ 3 & 4 & 2 & 5 & 6 \end{bmatrix}$

6. $\begin{bmatrix} 1 & 4 & 2 & 3 & 1.4 \\ 6 & 2.4 & 5 & 1.2 & 3 \\ 4 & 1 & 2 & 3 & 1.2 \\ -1 & 2 & -3 & 4 & 2 \\ 1.1 & 2.2 & 3 & 5.1 & 4 \end{bmatrix}$

In Exercises 7–10, solve the system of linear equations by first writing the system in the form $AX = B$ and then solving the resulting system by using A^{-1}. Round your answers to two decimal places.

7. $\begin{aligned} 2x - 3y + 4z &= 2.4 \\ 3x + 2y - 7z &= -8.1 \\ x + 4y - 2z &= 10.2 \end{aligned}$

8. $\begin{aligned} 3.2x - 4.7y + 3.2z &= 7.1 \\ 2.1x + 2.6y + 6.2z &= 8.2 \\ 5.1x - 3.1y - 2.6z &= -6.5 \end{aligned}$

9. $\begin{aligned} 3x_1 - 2x_2 + 4x_3 - 8x_4 &= 8 \\ 2x_1 + 3x_2 - 2x_3 + 6x_4 &= 4 \\ 3x_1 + 2x_2 - 6x_3 - 7x_4 &= -2 \\ 4x_1 - 7x_2 + 4x_3 + 6x_4 &= 22 \end{aligned}$

10. $\begin{aligned} 1.2x_1 + 2.1x_2 - 3.2x_3 + 4.6x_4 &= 6.2 \\ 3.1x_1 - 1.2x_2 + 4.1x_3 - 3.6x_4 &= -2.2 \\ 1.8x_1 + 3.1x_2 - 2.4x_3 + 8.1x_4 &= 6.2 \\ 2.6x_1 - 2.4x_2 + 3.6x_3 - 4.6x_4 &= 3.6 \end{aligned}$

CHAPTER 5 Summary of Principal Formulas and Terms

FORMULAS

1. Laws for matrix addition	
a. Commutative law	$A + B = B + A$
b. Associative law	$(A + B) + C = A + (B + C)$
2. Laws for matrix multiplication	
a. Associative law	$(AB)C = A(BC)$
b. Distributive law	$A(B + C) = AB + AC$
3. Inverse of a 2×2 matrix	If $\quad A = \begin{bmatrix} a & b \\ c & d \end{bmatrix}$ and $\quad D = ad - bc \neq 0$ then $\quad A^{-1} = \dfrac{1}{D}\begin{bmatrix} d & -b \\ -c & a \end{bmatrix}$
4. Solution of system $AX = B$ (A nonsingular)	$X = A^{-1}B$

TERMS

system of linear equations (266)	Gauss–Jordan elimination method (279)	square matrix (304)
solution of a system of linear equations (266)	row operations (279)	transpose of a matrix (308)
	unit column (279)	scalar (308)
parameter (267)	pivoting (280)	scalar product (308)
dependent system (268)	pivot element (280)	matrix product (318)
inconsistent system (268)	size of a matrix (303)	identity matrix (321)
equivalent system (275)	matrix (304)	inverse of a matrix (333)
coefficient matrix (277)	row matrix (304)	nonsingular matrix (334)
augmented matrix (277)	column matrix (304)	singular matrix (334)
row-reduced form of a matrix (278)		

CHAPTER 5 Concept Review Questions

Fill in the blanks.

1. a. Two lines in the plane can intersect at (a) exactly _____ point, (b) infinitely _____ points, or (c) _____ point.

  b. A system of two linear equations in two variables can have (a) exactly _____ solution, (b) infinitely _____ solutions, or (c) _____ solution.

2. To find the point(s) of intersection of two lines, we solve the system of _____ describing the two lines.

3. The row operations used in the Gauss–Jordan elimination method are denoted by _____, _____, and _____. The use of each of these operations does not alter the _____ of the system of linear equations.

4. a. A system of linear equations with fewer equations than variables cannot have a/an _____ solution.

  b. A system of linear equations with at least as many equations as variables may have _____ solution, _____ _____ solutions, or a/an _____ solution.

5. Two matrices are equal provided that they have the same _____ and their corresponding _____ are equal.

6. Two matrices may be added (subtracted) if they both have the same _____. To add or subtract two matrices, we add or subtract their _____ entries.

7. The transpose of a/an _____ matrix with elements a_{ij} is the matrix of size _____ with entries _____.

8. The scalar product of a matrix A by the scalar c is the matrix _____ obtained by multiplying each entry of A by _____.

9. a. For the product AB of two matrices A and B to be defined, the number of _____ of A must be equal to the number of _____ of B.

  b. If A is an $m \times n$ matrix and B is an $n \times p$ matrix, then the size of AB is _____.

10. a. If the products and sums are defined for the matrices A, B, and C, then the associative law states that $(AB)C = $ _____; the distributive law states that $A(B + C) = $ _____.

  b. If I is an identity matrix of size n, then $IA = A$ if A is any matrix of size _____.

11. A matrix A is nonsingular if there exists a matrix A^{-1} such that _____ $=$ _____ $= I$. If A^{-1} does not exist, then A is said to be _____.

12. A system of n linear equations in n variables written in the form $AX = B$ has a unique solution given by $X = $ _____ if A has an inverse.

CHAPTER 5 Review Exercises

In Exercises 1–4, perform the operations if possible.

1. $\begin{bmatrix} 1 & 2 \\ -1 & 3 \\ 2 & 1 \end{bmatrix} + \begin{bmatrix} 1 & 0 \\ 0 & 1 \\ 1 & 2 \end{bmatrix}$

2. $\begin{bmatrix} -1 & 2 \\ 3 & 4 \end{bmatrix} - \begin{bmatrix} 1 & 2 \\ 5 & -2 \end{bmatrix}$

3. $\begin{bmatrix} -3 & 2 & 1 \end{bmatrix} \begin{bmatrix} 2 & 1 \\ -1 & 0 \\ 2 & 1 \end{bmatrix}$

4. $\begin{bmatrix} 1 & 3 & 2 \\ -1 & 2 & 3 \end{bmatrix} \begin{bmatrix} 1 \\ 4 \\ 2 \end{bmatrix}$

In Exercises 5–8, find the values of the variables.

5. $\begin{bmatrix} 1 & x \\ y & 3 \end{bmatrix} = \begin{bmatrix} z & 2 \\ 3 & w \end{bmatrix}$ **6.** $\begin{bmatrix} 3 & x \\ y & 3 \end{bmatrix}\begin{bmatrix} 1 \\ 2 \end{bmatrix} = \begin{bmatrix} 7 \\ 4 \end{bmatrix}$

7. $\begin{bmatrix} 3 & a+3 \\ -1 & b \\ c+1 & d \end{bmatrix} = \begin{bmatrix} 3 & 6 \\ e+2 & 4 \\ -1 & 2 \end{bmatrix}$

8. $\begin{bmatrix} x & 3 & 1 \\ 0 & y & 2 \end{bmatrix}\begin{bmatrix} 1 & 1 \\ 3 & z \\ 4 & 2 \end{bmatrix} = \begin{bmatrix} 12 & 4 \\ 2 & 2 \end{bmatrix}$

In Exercises 9–16, compute the expressions if possible, given that

$$A = \begin{bmatrix} 1 & 3 & 1 \\ -2 & 1 & 3 \\ 4 & 0 & 2 \end{bmatrix} \quad B = \begin{bmatrix} 2 & 1 & 3 \\ -2 & -1 & -1 \\ 1 & 4 & 2 \end{bmatrix}$$

$$C = \begin{bmatrix} 3 & -1 & 2 \\ 1 & 6 & 4 \\ 2 & 1 & 3 \end{bmatrix}$$

9. $2A + 3B$

10. $3A - 2B$

11. $2(3A)$

12. $2(3A - 4B)$

13. $A(B - C)$

14. $AB + AC$

15. $A(BC)$

16. $\dfrac{1}{2}(CA - CB)$

In Exercises 17–24, solve the system of linear equations using the Gauss–Jordan elimination method.

17. $\begin{aligned} 2x - 3y &= 5 \\ 3x + 4y &= -1 \end{aligned}$ **18.** $\begin{aligned} 3x + 2y &= 3 \\ 2x - 4y &= -14 \end{aligned}$

19. $\begin{aligned} x - y + 2z &= 5 \\ 3x + 2y + z &= 10 \\ 2x - 3y - 2z &= -10 \end{aligned}$ **20.** $\begin{aligned} 3x - 2y + 4z &= 16 \\ 2x + y - 2z &= -1 \\ x + 4y - 8z &= -18 \end{aligned}$

21. $\begin{aligned} 3x - 2y + 4z &= 11 \\ 2x - 4y + 5z &= 4 \\ x + 2y - z &= 10 \end{aligned}$

22. $\begin{aligned} x - 2y + 3z + 4w &= 17 \\ 2x + y - 2z - 3w &= -9 \\ 3x - y + 2z - 4w &= 0 \\ 4x + 2y - 3z + w &= -2 \end{aligned}$

23. $\begin{aligned} 3x - 2y + z &= 4 \\ x + 3y - 4z &= -3 \\ 2x - 3y + 5z &= 7 \\ x - 8y + 9z &= 10 \end{aligned}$ **24.** $\begin{aligned} 2x - 3y + z &= 10 \\ 3x + 2y - 2z &= -2 \\ x - 3y - 4z &= -7 \\ 4x + y - z &= 4 \end{aligned}$

In Exercises 25–32, find the inverse of the matrix (if it exists).

25. $A = \begin{bmatrix} 3 & 1 \\ 1 & 2 \end{bmatrix}$ **26.** $A = \begin{bmatrix} 2 & 4 \\ 1 & 6 \end{bmatrix}$

27. $A = \begin{bmatrix} 3 & 4 \\ 2 & 2 \end{bmatrix}$ **28.** $A = \begin{bmatrix} 2 & 4 \\ 1 & -2 \end{bmatrix}$

29. $A = \begin{bmatrix} 2 & 3 & 1 \\ 1 & -1 & 2 \\ 1 & 2 & 1 \end{bmatrix}$ **30.** $A = \begin{bmatrix} 1 & 2 & 4 \\ 2 & 1 & 3 \\ -1 & 0 & 2 \end{bmatrix}$

31. $A = \begin{bmatrix} 1 & 2 & 4 \\ 3 & 1 & 2 \\ 1 & 0 & -6 \end{bmatrix}$ **32.** $A = \begin{bmatrix} 2 & 1 & -3 \\ 1 & 2 & -4 \\ 3 & 1 & -2 \end{bmatrix}$

In Exercises 33–36, compute the value of the expressions if possible, given that

$$A = \begin{bmatrix} 1 & 2 \\ -1 & 2 \end{bmatrix} \quad B = \begin{bmatrix} 3 & 1 \\ 4 & 2 \end{bmatrix} \quad C = \begin{bmatrix} 1 & 1 \\ -1 & 2 \end{bmatrix}$$

33. $(A^{-1}B)^{-1}$ **34.** $(ABC)^{-1}$

35. $(2A - C)^{-1}$ **36.** $(A + B)^{-1}$

In Exercises 37–40, write each system of linear equations in the form $AX = C$. Find A^{-1} and use the result to solve the system.

37. $\begin{aligned} 2x + 3y &= -8 \\ x - 2y &= 3 \end{aligned}$ **38.** $\begin{aligned} x - 3y &= -1 \\ 2x + 4y &= 8 \end{aligned}$

39. $\begin{aligned} x - 2y + 4z &= 13 \\ 2x + 3y - 2z &= 0 \\ x + 4y - 6z &= -15 \end{aligned}$ **40.** $\begin{aligned} 2x - 3y + 4z &= 17 \\ x + 2y - 4z &= -7 \\ 3x - y + 2z &= 14 \end{aligned}$

41. **GASOLINE SALES** Gloria Newburg operates three self-service gasoline stations in different parts of town. On a certain day, Station A sold 600 gal of premium, 800 gal of super, 1000 gal of regular gasoline, and 700 gal of diesel fuel; Station B sold 700 gal of premium, 600 gal of super, 1200 gal of regular gasoline, and 400 gal of diesel fuel; Station C sold 900 gal of premium, 700 gal of super, 1400 gal of regular gasoline, and 800 gal of diesel fuel. Assume that the price of gasoline was $3.80/gal for premium, $3.60/gal for super, and $3.40/gal for regular and that diesel fuel sold for $3.70/gal. Use matrix algebra to find the total revenue at each station.

42. **STOCK TRANSACTIONS** Jack Spaulding bought 10,000 shares of Stock X, 20,000 shares of Stock Y, and 30,000 shares of Stock Z at a unit price of $20, $30, and $50 per share, respectively. Six months later, the closing prices of Stocks X, Y, and Z were $22, $35, and $51 per share, respectively. Jack made no other stock transactions during the period in question. Compare the value of Jack's stock holdings at the time of purchase and 6 months later.

43. **INVESTMENTS** Josh's and Hannah's stock holdings are given in the following table:

	BAC	GM	IBM	ORCL
Josh	800	1200	250	1500
Hannah	600	1400	300	1200

The prices (in dollars per share) of the stocks of BAC, GM, IBM, and ORCL at the close of the stock market on a certain day are $12.57, $28.21, $214.92, and $36.34, respectively.

a. Write a 2×4 matrix A giving the stock holdings of Josh and Hannah.

b. Write a 4×1 matrix B giving the closing prices of the stocks of BAC, GM, IBM, and ORCL.

c. Use matrix multiplication to find the total value of the stock holdings of Josh and Hannah at the market close.

44. INVESTMENT PORTFOLIOS The following table gives the number of shares of certain corporations held by Jennifer and Max in their stock portfolios at the beginning of September and at the beginning of October:

	September			
	IBM	Google	Boeing	GM
Jennifer	800	500	1200	1500
Max	500	600	2000	800

	October			
	IBM	Google	Boeing	GM
Jennifer	900	600	1000	1200
Max	700	500	2100	900

a. Write matrices A and B giving the stock portfolios of Jennifer and Max at the beginning of September and at the beginning of October, respectively.

b. Find a matrix C reflecting the change in the stock portfolios of Jennifer and Max between the beginning of September and the beginning of October.

45. PRODUCTION SCHEDULING Desmond Jewelry wishes to produce three types of pendants: Type A, Type B, and Type C. To manufacture a Type A pendant requires 2 min on Machines I and II and 3 min on Machine III. A Type B pendant requires 2 min on Machine I, 3 min on Machine II, and 4 min on Machine III. A Type C pendant requires 3 min on Machine I, 4 min on Machine II, and 3 min on Machine III. There are $3\frac{1}{2}$ hr available on Machine I, $4\frac{1}{2}$ hr available on Machine II, and 5 hr available on Machine III. How many pendants of each type should Desmond make to use all the available time?

46. PETROLEUM PRODUCTION Wildcat Oil Company has two refineries, one located in Houston and the other in Tulsa. The Houston refinery ships 60% of its petroleum to a Chicago distributor and 40% of its petroleum to a Los Angeles distributor. The Tulsa refinery ships 30% of its petroleum to the Chicago distributor and 70% of its petroleum to the Los Angeles distributor. Assume that, over the year, the Chicago distributor received 240,000 gal of petroleum and the Los Angeles distributor received 460,000 gal of petroleum. Find the amount of petroleum produced at each of Wildcat's refineries.

CHAPTER 5 Before Moving On . . .

1. Solve the following system of linear equations, using the Gauss–Jordan elimination method:

$$2x + y - z = -1$$
$$x + 3y + 2z = 2$$
$$3x + 3y - 3z = -5$$

2. Find the solution(s), if it exists, of the system of linear equations whose augmented matrix in reduced form follows.

a. $\begin{bmatrix} 1 & 0 & 0 & | & 2 \\ 0 & 1 & 0 & | & -3 \\ 0 & 0 & 1 & | & 1 \end{bmatrix}$ **b.** $\begin{bmatrix} 1 & 0 & 0 & | & 3 \\ 0 & 1 & 0 & | & 0 \\ 0 & 0 & 0 & | & 1 \end{bmatrix}$

c. $\begin{bmatrix} 1 & 0 & 0 & | & 2 \\ 0 & 1 & 3 & | & 1 \\ 0 & 0 & 0 & | & 0 \end{bmatrix}$ **d.** $\begin{bmatrix} 1 & 0 & 0 & 0 & | & 0 \\ 0 & 1 & 0 & 0 & | & 0 \\ 0 & 0 & 1 & 0 & | & 0 \\ 0 & 0 & 0 & 1 & | & 0 \end{bmatrix}$

e. $\begin{bmatrix} 1 & 0 & -1 & | & 2 \\ 0 & 1 & 2 & | & 3 \end{bmatrix}$

3. Solve each system of linear equations using the Gauss–Jordan elimination method.

a. $x + 2y = 3$
$3x - y = -5$
$4x + y = -2$

b. $x - 2y + 4z = 2$
$3x + y - 2z = 1$

4. Let

$$A = \begin{bmatrix} 1 & -2 & 4 \\ 3 & 0 & 1 \end{bmatrix} \quad B = \begin{bmatrix} 1 & -1 & 2 \\ 3 & 1 & -1 \\ 2 & 1 & 0 \end{bmatrix} \quad C = \begin{bmatrix} 2 & -2 \\ 1 & 1 \\ 3 & 4 \end{bmatrix}$$

Find (a) AB, (b) $(A + C^T)B$, and (c) $C^TB - AB^T$.

5. Find A^{-1} if

$$A = \begin{bmatrix} 2 & 1 & 2 \\ 0 & -1 & 3 \\ 1 & 1 & 0 \end{bmatrix}$$

6. Solve the system

$$2x + z = 4$$
$$2x + y - z = -1$$
$$3x + y - z = 0$$

by first writing it in the matrix form $AX = B$ and then finding A^{-1}.

6 Linear Programming

MANY PRACTICAL PROBLEMS involve maximizing or minimizing a function subject to certain constraints. For example, we might wish to maximize a profit function subject to certain limitations on the amount of material and labor available. Maximization or minimization problems that can be formulated in terms of a *linear* objective function and constraints in the form of linear inequalities are called *linear programming problems*. In this chapter, we look at linear programming problems involving two variables. These problems are amenable to geometric analysis, and the method of solution introduced here will shed much light on the basic nature of a linear programming problem. Solving linear programming problems involving more than two variables requires algebraic techniques. One such technique, the *simplex method,* was developed by George Dantzig in the late 1940s and remains in wide use to this day.

How many souvenirs should Ace Novelty make in order to maximize its profit? The company produces two types of souvenirs, each of which requires a certain amount of time on two different machines. Each machine can be operated for only a certain number of hours per day. In Example 1, page 363, we show how this production problem can be formulated as a linear programming problem, and in Example 1, page 374, we solve this linear programming problem.

6.1 Graphing Systems of Linear Inequalities in Two Variables

Graphing Linear Inequalities

In Chapter 2, we saw that a linear equation in two variables x and y

$$ax + by + c = 0 \qquad \text{a, b not both equal to zero}$$

has a *solution set* that may be exhibited graphically as points on a straight line in the xy-plane. We now show that there is also a simple graphical representation for **linear inequalities** in two variables:

$$ax + by + c < 0 \qquad ax + by + c \leq 0$$
$$ax + by + c > 0 \qquad ax + by + c \geq 0$$

Before turning to a general procedure for graphing such inequalities, let's consider a specific example. Suppose we wish to graph

$$2x + 3y < 6 \tag{1}$$

We first graph the equation $2x + 3y = 6$, which is obtained by replacing the given inequality "$<$" with an equality "$=$" (Figure 1).

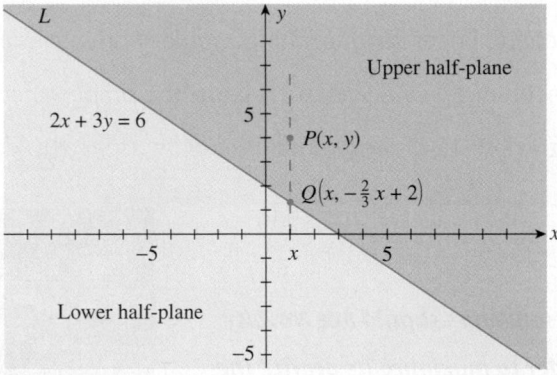

FIGURE 1
A straight line divides the xy-plane into two half-planes.

Observe that this line divides the xy-plane into two half-planes: an upper half-plane and a lower half-plane. Let's show that the upper half-plane is the graph of the linear inequality

$$2x + 3y > 6 \tag{2}$$

whereas the lower half-plane is the graph of the linear inequality

$$2x + 3y < 6 \tag{3}$$

To see this, let's write Inequalities (2) and (3) in the equivalent forms

$$y > -\frac{2}{3}x + 2 \tag{4}$$

and

$$y < -\frac{2}{3}x + 2 \tag{5}$$

The equation of the line itself is

$$y = -\frac{2}{3}x + 2 \tag{6}$$

Now pick any point $P(x, y)$ lying above the line L. Let Q be the point lying on L and directly below P (see Figure 1). Since Q lies on L, its coordinates must satisfy Equation (6); that is, the coordinates of Q are $(x, -\frac{2}{3}x + 2)$. Comparing the y-coordinates of P and Q and recalling that P lies above Q, so that its y-coordinate must be larger than that of Q, we have

$$y > -\frac{2}{3}x + 2$$

But this inequality is just Inequality (4) or, equivalently, Inequality (2). Similarly, we can show that every point lying below L must satisfy Inequality (5) and therefore Inequality (3).

This analysis shows that the lower half-plane provides a solution to our problem (Figure 2). (By convention, we draw the line as a dashed line to show that the points on L do not belong to the solution set.) Observe that the two half-planes in question are disjoint; that is, they do not have any points in common.

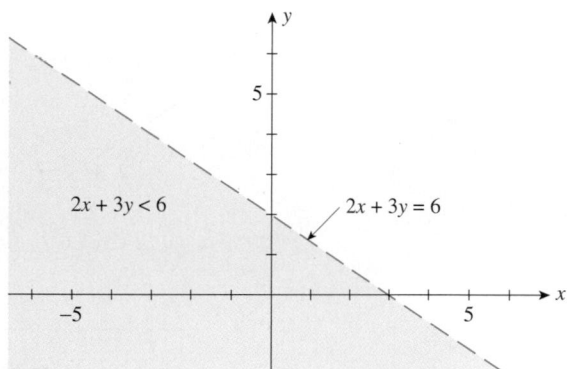

FIGURE **2**
The set of points lying below the dashed line satisfies the given inequality.

Alternatively, there is a simpler method for determining the half-plane that provides the solution to the problem. To determine the required half-plane, let's pick *any* point lying in one of the half-planes. For simplicity, pick the origin $(0, 0)$, which lies in the lower half-plane. Substituting $x = 0$ and $y = 0$ (the coordinates of this point) into the given Inequality (1), we find

$$2(0) + 3(0) < 6$$

or $0 < 6$, which is certainly true. This tells us that the required half-plane is the one containing the test point—namely, the lower half-plane.

Next, let's see what happens if we choose the point $(2, 3)$, which lies in the upper half-plane. Substituting $x = 2$ and $y = 3$ into the given inequality, we find

$$2(2) + 3(3) < 6$$

or $13 < 6$, which is false. This tells us that the upper half-plane is *not* the required half-plane, as expected. Note, too, that no point $P(x, y)$ lying on the line constitutes a solution to our problem, given the *strict* inequality $<$.

This discussion suggests the following procedure for graphing a linear inequality in two variables.

> **Procedure for Graphing Linear Inequalities**
>
> 1. Draw the graph of the equation obtained for the given inequality by replacing the inequality sign with an equal sign. Use a dashed or dotted line if the problem involves a strict inequality, $<$ or $>$. Otherwise, use a solid line to indicate that the line itself constitutes part of the solution.
> 2. Pick a test point (a, b) lying in one of the half-planes determined by the line sketched in Step 1 and substitute the numbers a and b for the values of x and y in the given inequality. For simplicity, use the origin whenever possible.
> 3. If the inequality is satisfied, the graph of the solution to the inequality is the half-plane containing the test point. Otherwise, the solution is the half-plane not containing the test point.

EXAMPLE 1 Determine the solution set for the inequality $2x + 3y \geq 6$.

Solution Replacing the inequality $\geq$ with an equality $=$, we obtain the equation $2x + 3y = 6$, whose graph is the straight line shown in Figure 3.

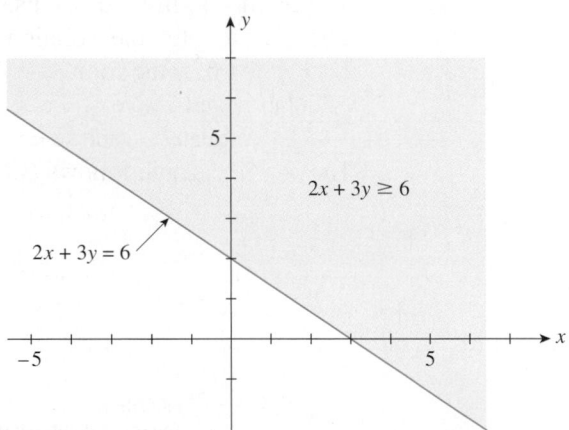

FIGURE **3**
The set of points lying on the line and in the upper half-plane satisfies the given inequality.

Instead of a dashed line as before, we use a solid line to show that all points on the line are also solutions to the inequality. Picking the origin as our test point, we find $2(0) + 3(0) \geq 6$, or $0 \geq 6$, which is false. So we conclude that the solution set is made up of the half-plane that does not contain the origin, including (in this case) the line given by $2x + 3y = 6$.

EXAMPLE 2 Graph $x \leq -1$.

Solution The graph of $x = -1$ is the vertical line shown in Figure 4. Picking the origin $(0, 0)$ as a test point, we find $0 \leq -1$, which is false. Therefore, the required solution is the *left* half-plane, which does not contain the origin.

EXAMPLE 3 Graph $x - 2y > 0$.

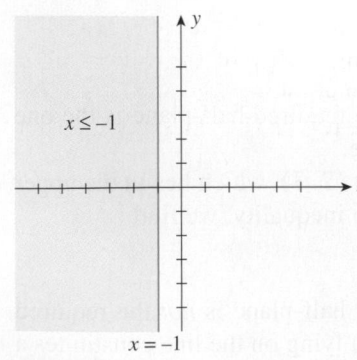

FIGURE **4**
The set of points lying on the line $x = -1$ and in the left half-plane satisfies the given inequality.

Solution We first graph the equation $x - 2y = 0$, or $y = \frac{1}{2}x$ (Figure 5). Since the origin lies on the line, we may not use it as a test point. (Why?) Let's pick $(1, 2)$

as a test point. Substituting $x = 1$ and $y = 2$ into the given inequality, we find $1 - 2(2) > 0$, or $-3 > 0$, which is false. Therefore, the required solution is the half-plane that does not contain the test point—namely, the lower half-plane.

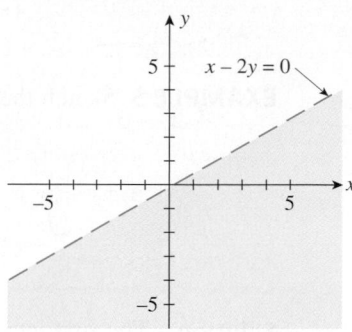

FIGURE 5
The set of points in the lower half-plane satisfies $x - 2y > 0$.

Exploring with TECHNOLOGY

A graphing utility can be used to plot the graph of a linear inequality. For example, to plot the solution set for Example 1, first rewrite the equation $2x + 3y = 6$ in the form $y = 2 - \frac{2}{3}x$. Next, enter this expression for Y_1 in the calculator, and move the cursor to the left of Y_1 (see Figure a). Then press $\boxed{\text{ENTER}}$ repeatedly, and select the icon that indicates the shading option desired. The required graph follows (see Figure b).

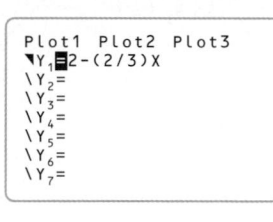

FIGURE a
TI 83/84 screen

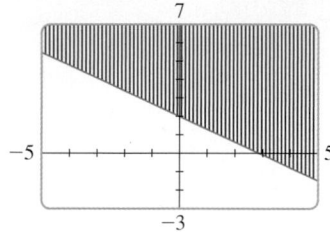

FIGURE b
Graph of the inequality $2x + 3y \geq 6$

Graphing Systems of Linear Inequalities

By the **solution set of a system of linear inequalities** in the two variables x and y, we mean the set of all points (x, y) satisfying each inequality of the system. The graphical solution of such a system may be obtained by graphing the solution set for each inequality independently and then determining the region in common with each solution set.

EXAMPLE 4 Determine the solution set for the system

$$4x + 3y \geq 12$$
$$x - y \leq 0$$

$3y \geq -4 \pm 12$
$\frac{3}{3} \qquad \frac{3}{3} \qquad \frac{3}{3}$
$y \geq -\frac{4}{3} + 4$

Solution Proceeding as in the previous examples, you should have no difficulty locating the half-planes determined by each of the linear inequalities that make up the system. These half-planes are shown in Figure 6. The intersection of the two

FIGURE 6
The set of points in the shaded area satisfies the system

$$4x + 3y \geq 12$$
$$x - y \leq 0$$

half-planes is the shaded region. A point in this region is an element of the solution set for the given system. The point $P\left(\frac{12}{7}, \frac{12}{7}\right)$, the intersection of the two straight lines determined by the equations, is found by solving the simultaneous equations

$$4x + 3y = 12$$
$$x - y = 0$$

EXAMPLE 5 Sketch the solution set for the system

$$x \geq 0$$
$$y \geq 0$$
$$x + y - 6 \leq 0$$
$$2x + y - 8 \leq 0$$

Solution The first inequality in the system defines the right half-plane—all points to the right of the y-axis plus all points lying on the y-axis itself. The second inequality in the system defines the upper half-plane, including the x-axis. The half-planes defined by the third and fourth inequalities are indicated by arrows in Figure 7. Thus, the required region—the intersection of the four half-planes defined by the four inequalities in the given system of linear inequalities—is the shaded region. The point P is found by solving the simultaneous equations $x + y - 6 = 0$ and $2x + y - 8 = 0$.

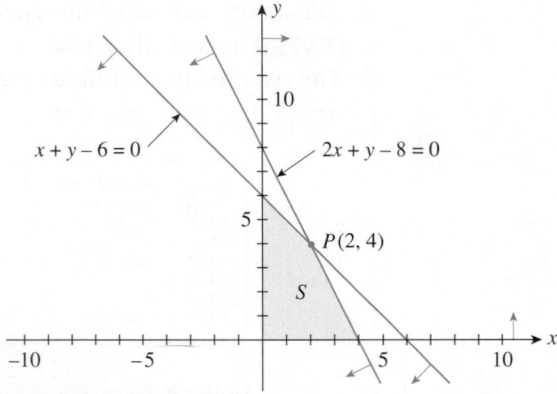

FIGURE 7
The set of points in the shaded region S, including the x- and y-axes, satisfies the given inequalities.

EXAMPLE 6 Refer to Example 5.

a. Use the graph of the solution set S of the given system of linear inequalities (Figure 7) to determine whether the point $A(1, 3)$ lies in S.
b. Repeat part (a) for the point $B(5, 3)$.

Solution

a. Referring to Figure 7, we see that $A(1, 3)$ lies in the solution set S of the given system. To prove the result algebraically, we substitute $x = 1$ and $y = 3$ into each inequality in the system. Thus,

$$1 \geq 0$$
$$3 \geq 0$$
$$1 + 3 - 6 = -2 \leq 0$$
$$2(1) + 3 - 8 = -3 \leq 0$$

Since all of these statements are true, it follows that $A(1, 3)$ does lie in S.

b. Referring to Figure 7 once again, we see that $B(5, 3)$ does not lie in S. To prove this assertion algebraically, we substitute $x = 5$ and $y = 3$ into the system of inequalities, obtaining

$$5 \geq 0$$
$$3 \geq 0$$
$$5 + 3 - 6 = 2 \leq 0$$
$$2(5) + 3 - 8 = 5 \leq 0$$

Both the third and fourth inequalities are not true. Therefore, $B(5, 3)$ does not lie in S.

APPLIED EXAMPLE 7 A Production Problem Sonoma Company manufactures a 24-bottle wooden wine rack in two versions: a standard rack and a deluxe rack. Each standard rack requires 12 minutes of fabrication time and 4 minutes of finishing time. Each deluxe rack requires 8 minutes of fabrication time and 16 minutes of finishing time. There are 6 hours of time available for fabrication and 8 hours available for finishing each day.

a. Write a system of linear inequalities that gives the restrictions placed on the number of each type of wine rack manufactured by Sonoma.
b. Graph the solution set.
c. Can Sonoma manufacture 20 each of the standard and deluxe wine racks per day? Prove your assertion.
d. Can Sonoma manufacture 15 each of the standard and 20 deluxe wine racks per day? Prove your assertion.

Solution

a. As a first step toward setting up the system of inequalities, we tabulate the given information (see Table 1).

TABLE 1			
	Standard	**Deluxe**	**Time Available**
Fabrication	12 min	8 min	360 min (6 hr)
Finishing	4 min	16 min	480 min (8 hr)

Let x denote the number of standard wine racks, and let y denote the number of deluxe wine racks to be manufactured per day. The amount of time required for fabricating these wine racks is $12x + 8y$ minutes and must not exceed 360 minutes. Thus, we have the inequality

$$12x + 8y \leq 360 \quad \text{or} \quad 3x + 2y \leq 90$$

Similarly, the total amount of time required for finishing these wine racks is $4x + 16y$ minutes and must not exceed 480 minutes. This condition leads to the inequality

$$4x + 16y \leq 480 \quad \text{or} \quad x + 4y \leq 120$$

Finally, neither x nor y can be negative, so

$$x \geq 0$$
$$y \geq 0$$

Therefore, the desired system of linear inequalities is

$$
\begin{aligned}
3x + 2y &\leq 90 \\
x + 4y &\leq 120 \\
x &\geq 0 \\
y &\geq 0
\end{aligned}
$$

b. The solution set is graphed in Figure 8.

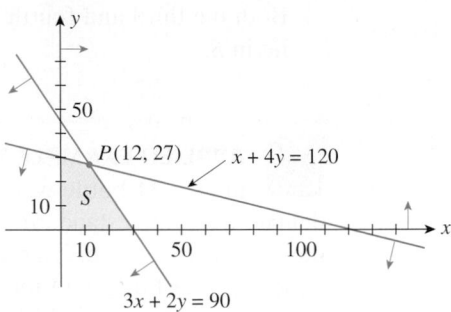

FIGURE 8

The point P is found by solving the simultaneous equations $3x + 2y = 90$ and $x + 4y = 120$.

c. A visual inspection of the solution set S suggests that this is not possible. In fact, putting $x = y = 20$ into the first inequality in the system of linear inequalities, we find

$$
3(20) + 2(20) = 100 \nleq 90
$$

This verifies the result.

d. By visual inspection of S, we see that Sonoma can manufacture 15 standard and 20 deluxe wine racks per day. To prove this, we substitute $x = 15$ and $y = 20$ into each inequality in the system, obtaining

$$
\begin{aligned}
3(15) + 2(20) &= 85 \leq 90 \\
15 + 4(20) &= 95 \leq 120 \\
15 &\geq 0 \\
20 &\geq 0
\end{aligned}
$$

We see that all of the inequalities in the system are satisfied. ∎

The solution set S, shown in Figure 8, is an example of a bounded set. Observe that the set can be enclosed by a circle. For example, if you draw a circle of radius 30 with center at the origin, you will see that the set lies entirely inside the circle. On the other hand, the solution set S, shown in Figure 6, page 355, that was found in Example 4 cannot be enclosed by a circle and is said to be unbounded.

> **Bounded and Unbounded Solution Sets**
>
> The solution set of a system of linear inequalities is **bounded** if it can be enclosed by a circle. Otherwise, it is **unbounded.**

EXAMPLE 8 Determine the graphical solution set for the following system of linear inequalities:

$$2x + y \geq 50$$
$$x + 2y \geq 40$$
$$x \geq 0$$
$$y \geq 0$$

Solution   The required solution set is the unbounded region shown in Figure 9.

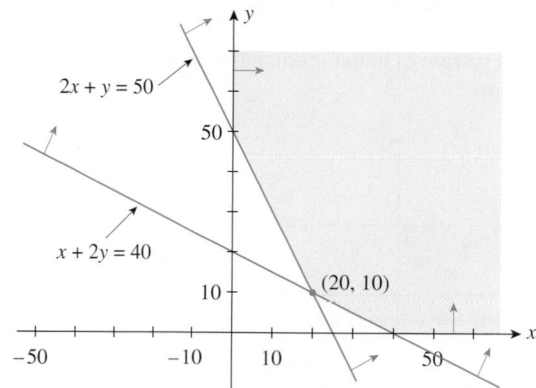

FIGURE 9
The solution set is an unbounded region.

6.1 Self-Check Exercises

1. Determine graphically the solution set for the following system of inequalities:

$$x + 2y \leq 10$$
$$5x + 3y \leq 30$$
$$x \geq 0, y \geq 0$$

2. Determine graphically the solution set for the following system of inequalities:

$$5x + 3y \geq 30$$
$$x - 3y \leq 0$$
$$x \geq 2$$

Solutions to Self-Check Exercises 6.1 can be found on page 362.

6.1 Concept Questions

1. a. What is the difference, geometrically, between the solution set of $ax + by < c$ and the solution set of $ax + by \leq c$?

   b. Describe the set that is obtained by intersecting the solution set of $ax + by \leq c$ with the solution set of $ax + by \geq c$.

2. a. What is the solution set of a system of linear inequalities?

   b. How do you find the solution of a system of linear inequalities graphically?

6.1 Exercises

In Exercises 1–10, find the graphical solution of each inequality.

1. $4x - 8 < 0$ **2.** $3y + 2 > 0$

3. $x - y \leq 0$ **4.** $3x + 4y \leq -2$

5. $x \leq -3$ **6.** $y \geq -1$

7. $2x + y \leq 4$ **8.** $-3x + 6y \geq 12$

9. $4x - 3y \leq -24$ **10.** $5x - 3y \geq 15$

In Exercises 11–18, write a system of linear inequalities that describes the shaded region.

11.

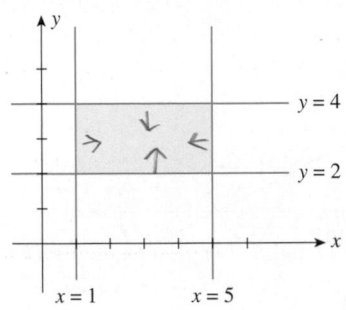

12.

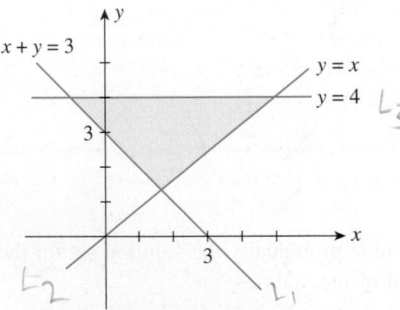

13.

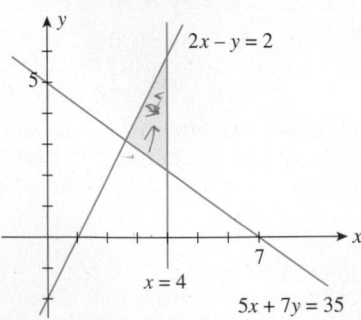

14.

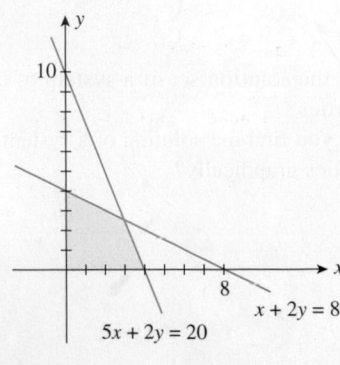

15.

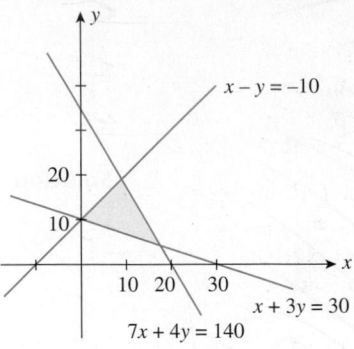

16.

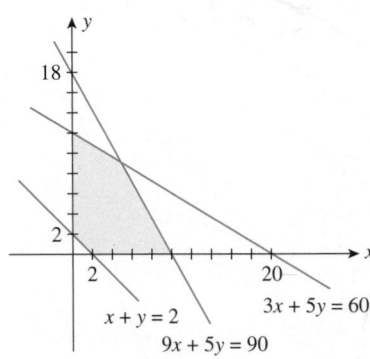

17.

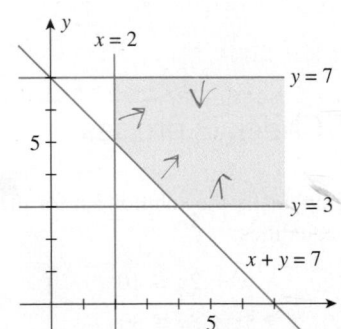

18.

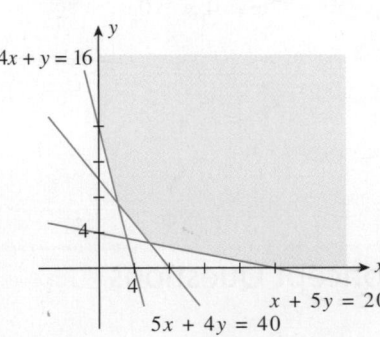

Exercises 19–22 make use of the results of Exercises 13–16.

19. Refer to the figure accompanying Exercise 13. By inspecting the figure, determine whether the point $(3, 3)$ lies in the set depicted there. Use the results of Exercise 13 to prove your assertion.

20. Refer to the figure accompanying Exercise 14. By inspecting the figure, determine whether the point $(5, 1)$ lies in the set depicted there. Use the results of Exercise 14 to prove your assertion.

21. Refer to the figure accompanying Exercise 15. By inspecting the figure, determine whether the point $(10, 10)$ lies in the set depicted there. Use the results of Exercise 15 to prove your assertion.

22. Refer to the figure accompanying Exercise 16. By inspecting the figure, determine whether the point $\left(8, \frac{18}{5}\right)$ lies in the set depicted there. Use the results of Exercise 16 to prove your assertion.

In Exercises 23–40, determine graphically the solution set for each system of inequalities and indicate whether the solution set is bounded or unbounded.

23. $2x + 4y > 16$
$-x + 3y \geq 7$

24. $3x - 2y > -13$
$-x + 2y > 5$

25. $x - y \leq 0$
$2x + 3y \geq 10$

26. $x + y \geq -2$
$3x - y \leq 6$

27. $x + 2y \geq 3$
$2x + 4y \leq -2$

28. $2x - y \geq 4$
$4x - 2y < -2$

29. $x + y \leq 6$
$0 \leq x \leq 3$
$y \geq 0$

30. $4x - 3y \leq 12$
$5x + 2y \leq 10$
$x \geq 0, y \geq 0$

31. $3x - 6y \leq 12$
$-x + 2y \leq 4$
$x \geq 0, y \geq 0$

32. $x + y \geq 20$
$x + 2y \geq 40$
$x \geq 0, y \geq 0$

33. $3x - 7y \geq -24$
$x + 3y \geq 8$
$x \geq 0, y \geq 0$

34. $3x + 4y \geq 12$
$2x - y \geq -2$
$0 \leq y \leq 3$
$x \geq 0$

35. $x + 2y \geq 3$
$5x - 4y \leq 16$
$0 \leq y \leq 2$
$x \geq 0$

36. $x + y \leq 4$
$2x + y \leq 6$
$2x - y \geq -1$
$x \geq 0, y \geq 0$

37. $6x + 5y \leq 30$
$3x + y \geq 6$
$x + y \geq 4$
$x \geq 0, y \geq 0$

38. $6x + 7y \leq 84$
$12x - 11y \leq 18$
$6x - 7y \leq 28$
$x \geq 0, y \geq 0$

39. $x - y \geq -6$
$x - 2y \leq -2$
$x + 2y \geq 6$
$x - 2y \geq -14$
$x \geq 0, y \geq 0$

40. $x - 3y \geq -18$
$3x - 2y \geq 2$
$x - 3y \leq -4$
$3x - 2y \leq 16$
$x \geq 0, y \geq 0$

41. CONCERT ATTENDANCE The Peninsula Brass Band will hold its semiannual concert in a community center that has a seating capacity of 500. The band expects that at least 200 season ticket holders will attend the concert and that at least 100 nonseason ticket holders will also attend it.
   a. Write a system of linear inequalities that gives the restrictions on the number of each type of ticket holder at the concert.
   b. Graph the solution set S for the system of linear inequalities found in part (a).
   c. Assuming that the attendance will be as expected, is it possible for 300 season ticket holders and 150 nonseason ticket holders to attend the concert? Prove your assertion.

42. MANUFACTURING FERTILIZERS Agro Products makes two types of fertilizers that are sold in 50-lb bags. A 50-lb bag of Fertilizer A contains 5 lb of nitrogen, 10 lb of phosphorus, and 20 lb of potassium. A 50-lb bag of Fertilizer B contains 6 lb of nitrogen, 4 lb of phosphorus, and 4 lb of potassium. Agro Products has 1800 lb of nitrogen, 1600 lb of phosphorus, and 2600 lb of potassium on hand.
   a. Write a system of linear inequalities that gives the restrictions to be placed on the number of bags of each type of fertilizer that Agro Products can manufacture.
   b. Graph the solution set S for the system of linear inequalities found in part (a).
   c. Is it possible for Agro Products to make 100 50-lb bags of Fertilizer A and 200 50-lb bags of Fertilizer B? Prove your assertion.

43. INVESTMENTS Louisa has earmarked at most $250,000 for investing in two companies involved in the production of renewable energy: Solaron Corporation and Windmill Corporation. She specifies that at least $50,000 must be invested in each company and that the amount invested in Solaron Corporation must not exceed 120% of that invested in Windmill Corporation.
   a. Write a system of linear inequalities that gives the restrictions placed upon Louisa's investments.
   b. Graph the solution set S for the system of linear inequalities found in part (a).
   c. Is it possible for Louisa to invest $150,000 in Solaron corporation and $100,000 in Windmill Corporation? Prove your assertion.

44. DIET PLANNING A dietitian wishes to plan a meal around four foods. The meal is to include 600 mg of phosphorus and 400 mg of magnesium. The number of units of the nutrients in each ounce of the foods (in milligrams) is summarized in the following table:

	Food A	Food B	Food C	Food D
Phosphorus	30	90	30	45
Magnesium	40	20	30	20

a. Let $x_1, x_2, x_3,$ and x_4 denote the amount of Food A, Food B, Food C, and Food D, respectively, in a meal. Write a system of linear equations in the four given variables to describe the requirements of the dietitian.

b. Use the Gauss–Jordan elimination method to solve the system of part (a) for x_1 and x_2 in terms of x_3 and x_4 (see Section 5.3).

c. Use the fact that x_1 and x_2 must be nonnegative to write a system of two linear inequalities in the two variables x_3 and x_4, and then determine graphically the solution set for the system.

d. Combine the results of parts (b) and (c) to write the solutions to the problem.

e. Suggest three possible meals that meet the nutritional requirements. Include at least one in which all four foods are used. (*Note:* Your answer is not unique.)

In Exercises 45–48, determine whether the statement is true or false. If it is true, explain why it is true. If it is false, give an example to show why it is false.

45. The solution set of a linear inequality involving two variables is either a half-plane or a straight line.

46. The solution set of the inequality $ax + by + c \le 0$ is either a left half-plane or a lower half-plane.

47. The solution set of a system of linear inequalities in two variables is bounded if it can be enclosed by a rectangle.

48. The solution set of the system

$$ax + by \le e$$
$$cx + dy \le f$$
$$x \ge 0, y \ge 0$$

where $a, b, c, d, e,$ and f are positive real numbers, is a bounded set.

6.1 Solutions to Self-Check Exercises

1. The required solution set is shown in the following figure:

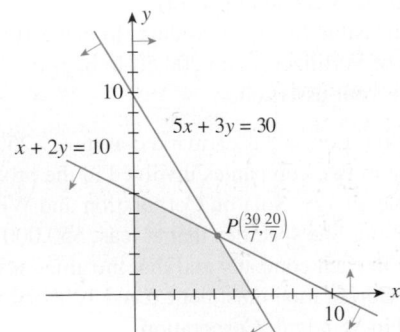

The point P is found by solving the system of equations

$$x + 2y = 10$$
$$5x + 3y = 30$$

Solving the first equation for x in terms of y gives

$$x = 10 - 2y$$

Substituting this value of x into the second equation of the system gives

$$5(10 - 2y) + 3y = 30$$
$$50 - 10y + 3y = 30$$
$$-7y = -20$$

so $y = \frac{20}{7}$. Substituting this value of y into the expression for x found earlier, we obtain

$$x = 10 - 2\left(\frac{20}{7}\right) = \frac{30}{7}$$

giving the point of intersection as $\left(\frac{30}{7}, \frac{20}{7}\right)$.

2. The required solution set is shown in the following figure:

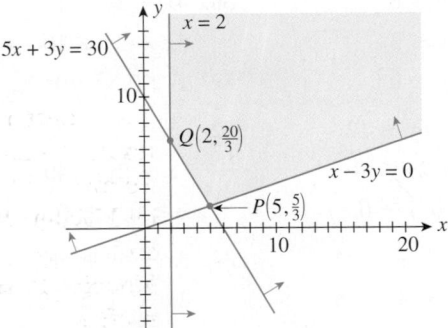

To find the coordinates of P, we solve the system

$$5x + 3y = 30$$
$$x - 3y = 0$$

Solving the second equation for x in terms of y and substituting this value of x in the first equation gives

$$5(3y) + 3y = 30$$

or $y = \frac{5}{3}$. Substituting this value of y into the second equation gives $x = 5$. Next, the coordinates of Q are found by solving the system

$$5x + 3y = 30$$
$$x = 2$$

yielding $x = 2$ and $y = \frac{20}{3}$.

6.2 Linear Programming Problems

In many business and economic problems, we are asked to optimize (maximize or minimize) a function subject to a system of equalities or inequalities. The function to be optimized is called the **objective function.** Profit functions and cost functions are examples of objective functions. The system of equalities or inequalities to which the objective function is subjected reflects the constraints (for example, limitations on resources such as materials and labor) imposed on the solution(s) to the problem. Problems of this nature are called **mathematical programming problems.** In particular, problems in which both the objective function and the constraints are expressed as linear equations or inequalities are called linear programming problems.

> **Linear Programming Problem**
>
> A **linear programming problem** consists of a linear objective function to be maximized or minimized subject to certain constraints in the form of linear equations or inequalities.

A Maximization Problem

As an example of a linear programming problem in which the objective function is to be maximized, let's consider the following simplified version of a production problem involving two variables.

APPLIED EXAMPLE 1 A Production Problem Ace Novelty wishes to produce two types of souvenirs: Type A and Type B. Each Type A souvenir will result in a profit of $1, and each Type B souvenir will result in a profit of $1.20. To manufacture a Type A souvenir requires 2 minutes on Machine I and 1 minute on Machine II. A Type B souvenir requires 1 minute on Machine I and 3 minutes on Machine II. There are 3 hours available on Machine I and 5 hours available on Machine II. How many souvenirs of each type should Ace make to maximize its profit?

Solution As a first step toward the mathematical formulation of this problem, we tabulate the given information (see Table 2).

TABLE 2	Type A	Type B	Time Available
Machine I	2 min	1 min	180 min
Machine II	1 min	3 min	300 min
Profit/Unit	$1	$1.20	

Let x be the number of Type A souvenirs, and let y be the number of Type B souvenirs to be made. Then, the total profit P (in dollars) is given by

$$P = x + 1.2y$$

which is the objective function to be maximized.

The total amount of time that Machine I is used is given by $2x + y$ minutes and must not exceed 180 minutes. Thus, we have the inequality

$$2x + y \le 180$$

Similarly, the total amount of time that Machine II is used is $x + 3y$ minutes and cannot exceed 300 minutes, so we are led to the inequality

$$x + 3y \leq 300$$

Finally, neither x nor y can be negative, so

$$x \geq 0$$
$$y \geq 0$$

To summarize, the problem here is to maximize the objective function $P = x + 1.2y$ subject to the system of inequalities

$$2x + y \leq 180$$
$$x + 3y \leq 300$$
$$x \geq 0, y \geq 0$$

The solution to this problem will be completed in Example 1, Section 6.3.   ∎

Minimization Problems

In the following linear programming problems, the objective function is to be minimized.

APPLIED EXAMPLE 2 A Nutrition Problem A nutritionist advises an individual who is suffering from iron and vitamin B deficiency to take at least 2400 milligrams (mg) of iron, 2100 mg of vitamin B_1 (thiamine), and 1500 mg of vitamin B_2 (riboflavin) over a period of time. Two vitamin pills are suitable, Brand A and Brand B. Each Brand A pill costs 6 cents and contains 40 mg of iron, 10 mg of vitamin B_1, and 5 mg of vitamin B_2. Each Brand B pill costs 8 cents and contains 10 mg of iron and 15 mg each of vitamins B_1 and B_2 (Table 3). What combination of pills should the individual purchase to meet the minimum iron and vitamin requirements at the lowest cost?

TABLE 3			
	Brand A	**Brand B**	**Minimum Requirement**
Iron	40 mg	10 mg	2400 mg
Vitamin B_1	10 mg	15 mg	2100 mg
Vitamin B_2	5 mg	15 mg	1500 mg
Cost/Pill	6¢	8¢	

Solution Let x be the number of Brand A pills, and let y be the number of Brand B pills to be purchased. The cost C (in cents) is given by

$$C = 6x + 8y$$

and is the objective function to be minimized.

The amount of iron contained in x Brand A pills and y Brand B pills is given by $40x + 10y$ mg, and this must be greater than or equal to 2400 mg. This translates into the inequality

$$40x + 10y \geq 2400$$

Similar considerations involving the minimum requirements of vitamins B_1 and B_2 lead to the inequalities

$$10x + 15y \geq 2100$$
$$5x + 15y \geq 1500$$

respectively. Thus, the problem here is to minimize $C = 6x + 8y$ subject to

$$40x + 10y \geq 2400$$
$$10x + 15y \geq 2100$$
$$5x + 15y \geq 1500$$
$$x \geq 0, y \geq 0$$

The solution to this problem will be completed in Example 2, Section 6.3. ◼

APPLIED EXAMPLE 3 A Transportation Problem Curtis-Roe Aviation Industries has two plants, I and II, that produce the Zephyr jet engines used in their light commercial airplanes. There are 100 units of the engines in Plant I and 110 units in Plant II. The engines are shipped to two of Curtis-Roe's main assembly plants, A and B. The shipping costs (in dollars) per engine from Plants I and II to the Main Assembly Plants A and B are as follows:

From	To Assembly Plant	
	A	B
Plant I	100	60
Plant II	120	70

In a certain month, Assembly Plant A needs 80 engines, whereas Assembly Plant B needs 70 engines. Find how many engines should be shipped from each plant to each main assembly plant if shipping costs are to be kept to a minimum.

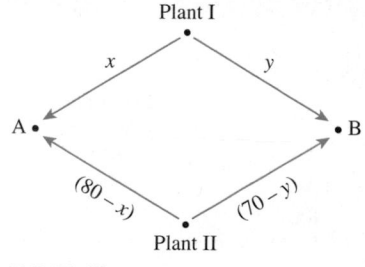

Plant I

Plant II

FIGURE **10**

Solution Let x denote the number of engines shipped from Plant I to Assembly Plant A, and let y denote the number of engines shipped from Plant I to Assembly Plant B. Since the requirements of Assembly Plants A and B are 80 and 70 engines, respectively, the number of engines shipped from Plant II to Assembly Plants A and B are $(80 - x)$ and $(70 - y)$, respectively. These numbers may be displayed in a schematic. With the aid of the accompanying schematic (Figure 10) and the shipping cost schedule, we find that the total shipping cost incurred by Curtis-Roe is given by

$$C = 100x + 60y + 120(80 - x) + 70(70 - y)$$
$$= 14{,}500 - 20x - 10y$$

Next, the production constraints on Plants I and II lead to the inequalities

$$x + y \leq 100$$
$$(80 - x) + (70 - y) \leq 110$$

The last inequality simplifies to

$$x + y \geq 40$$

Also, the requirements of the two main assembly plants lead to the inequalities

$$x \geq 0 \qquad y \geq 0 \qquad 80 - x \geq 0 \qquad 70 - y \geq 0$$

The last two may be written as $x \leq 80$ and $y \leq 70$.

Summarizing, we have the following linear programming problem: Minimize the objective (cost) function $C = 14{,}500 - 20x - 10y$ subject to the constraints

$$x + y \geq 40$$
$$x + y \leq 100$$
$$x \leq 80$$
$$y \leq 70$$

where $x \geq 0$ and $y \geq 0$.

You will be asked to complete the solution to this problem in Exercise 49, Section 6.3.

APPLIED EXAMPLE 4 A Warehouse Problem Acrosonic manufactures its Brentwood loudspeaker systems in two separate locations, Plant I and Plant II. The output at Plant I is at most 400 per month, whereas the output at Plant II is at most 600 per month. These loudspeaker systems are shipped to three warehouses that serve as distribution centers for the company. For the warehouses to meet their orders, the minimum monthly requirements of Warehouses A, B, and C are 200, 300, and 400 systems, respectively. Shipping costs from Plant I to Warehouses A, B, and C are $20, $8, and $10 per loudspeaker system, respectively, and shipping costs from Plant II to each of these warehouses are $12, $22, and $18, respectively. What should the shipping schedule be if Acrosonic wishes to meet the requirements of the distribution centers and at the same time keep its shipping costs to a minimum?

Solution The respective shipping costs (in dollars) per loudspeaker system may be tabulated as in Table 4. Letting x_1 denote the number of loudspeaker systems shipped from Plant I to Warehouse A, x_2 the number shipped from Plant I to Warehouse B, and so on leads to Table 5.

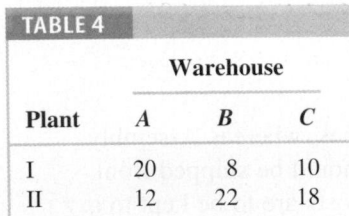

TABLE 4

Plant	Warehouse		
	A	B	C
I	20	8	10
II	12	22	18

TABLE 5

Plant	Warehouse			Max. Prod.
	A	B	C	
I	x_1	x_2	x_3	400
II	x_4	x_5	x_6	600
Min. Req.	200	300	400	

From Tables 4 and 5, we see that the cost of shipping x_1 loudspeaker systems from Plant I to Warehouse A is $20x_1$, the cost of shipping x_2 loudspeaker systems from Plant I to Warehouse B is $8x_2$, and so on. Thus, the total monthly shipping cost (in dollars) incurred by Acrosonic is given by

$$C = 20x_1 + 8x_2 + 10x_3 + 12x_4 + 22x_5 + 18x_6$$

Next, the production constraints on Plants I and II lead to the inequalities

$$x_1 + x_2 + x_3 \leq 400$$
$$x_4 + x_5 + x_6 \leq 600$$

(see Table 5). Also, the minimum requirements of each of the three warehouses lead to the three inequalities

$$x_1 + x_4 \geq 200$$
$$x_2 + x_5 \geq 300$$
$$x_3 + x_6 \geq 400$$

Summarizing, we have the following linear programming problem:

$$\text{Minimize} \quad C = 20x_1 + 8x_2 + 10x_3 + 12x_4 + 22x_5 + 18x_6$$

$$\text{subject to} \quad x_1 + x_2 + x_3 \le 400$$
$$x_4 + x_5 + x_6 \le 600$$
$$x_1 + x_4 \ge 200$$
$$x_2 + x_5 \ge 300$$
$$x_3 + x_6 \ge 400$$
$$x_1 \ge 0, x_2 \ge 0, \dots, x_6 \ge 0$$

The solution to this problem will be completed in Example 5, Section 6.5.

6.2 Self-Check Exercise

OPTIMIZING ADVERTISING EXPOSURE Gino Balduzzi, proprietor of Luigi's Pizza Palace, allocates $9000 a month for advertising in two newspapers, the *City Tribune* and the *Daily News*. The *City Tribune* charges $300 for a certain advertisement, whereas the *Daily News* charges $100 for the same ad. Gino has stipulated that the ad is to appear in at least 15 but no more than 30 editions of the *Daily News* per month. The *City Tribune* has a daily circulation of 50,000, and the *Daily News*

has a circulation of 20,000. Under these conditions, determine how many ads Gino should place in each newspaper to reach the largest number of readers. Formulate but do not solve the problem. (The solution to this problem can be found in Exercise 3 of Solutions to Self-Check Exercises 6.3.)

The solution to Self-Check Exercise 6.2 can be found on page 371.

6.2 Concept Questions

1. What is a linear programming problem?

2. Suppose you are asked to formulate a linear programming problem in two variables x and y. How would you express the fact that x and y are nonnegative? Why are these conditions often required in practical problems?

3. What is the difference between a maximization linear programming problem and a minimization linear programming problem?

6.2 Exercises

Formulate but do not solve each of the following exercises as a linear programming problem. You will be asked to solve these problems later.

1. **PRODUCTION SCHEDULING** A company manufactures two products, *A* and *B*, on two machines, I and II. It has been determined that the company will realize a profit of $3/unit of Product *A* and a profit of $4/unit of Product *B*. To manufacture a unit of Product *A* requires 6 min on Machine I and 5 min on Machine II. To manufacture a unit of Product *B* requires 9 min on Machine I and 4 min on Machine II. There are 5 hr of machine time available on Machine I and 3 hr of machine time available on Machine II in each work shift. How many units of each

product should be produced in each shift to maximize the company's profit?

2. **PRODUCTION SCHEDULING** National Business Machines manufactures two models of portable printers: A and B. Each model A costs $100 to make, and each model B costs $150. The profits are $30 for each model A and $40 for each model B portable printer. If the total number of portable printers demanded per month does not exceed 2500 and the company has earmarked no more than $600,000/month for manufacturing costs, how many units of each model should National make each month to maximize its monthly profit?

3. **PRODUCTION SCHEDULING** Kane Manufacturing has a division that produces two models of fireplace grates, model A and model B. To produce each model A grate requires 3 lb of cast iron and 6 min of labor. To produce each model B grate requires 4 lb of cast iron and 3 min of labor. The profit for each model A grate is $2.00, and the profit for each model B grate is $1.50. If 1000 lb of cast iron and 20 hr of labor are available for the production of grates per day, how many grates of each model should the division produce per day to maximize Kane's profits?

4. **PRODUCTION SCHEDULING** Refer to Exercise 3. Because of a backlog of orders for model A grates, the manager of Kane Manufacturing has decided to produce at least 150 of these grates a day. Operating under this additional constraint, how many grates of each model should Kane produce to maximize profit?

5. **PRODUCTION SCHEDULING** A division of the Winston Furniture Company manufactures dining tables and chairs. Each table requires 40 board feet of wood and 3 labor-hours. Each chair requires 16 board feet of wood and 4 labor-hours. The profit for each table is $45, and the profit for each chair is $20. In a certain week, the company has 3200 board feet of wood and 520 labor-hours available. How many tables and chairs should Winston manufacture to maximize its profits?

6. **PRODUCTION SCHEDULING** Refer to Exercise 5. If the profit for each table is $50 and the profit for each chair is $18, how many tables and chairs should Winston manufacture to maximize its profits?

7. **ALLOCATION OF FUNDS** Madison Finance has a total of $20 million earmarked for homeowner loans and auto loans. On the average, homeowner loans have a 10% annual rate of return, whereas auto loans yield a 12% annual rate of return. Management has also stipulated that the total amount of homeowner loans should be greater than or equal to 4 times the total amount of automobile loans. Determine the total amount of loans of each type Madison should extend to each category to maximize its returns.

8. **ASSET ALLOCATION** A financier plans to invest up to $500,000 in two projects. Project A yields a return of 10% on the investment, whereas Project B yields a return of 15% on the investment. Because the investment in Project B is riskier than the investment in Project A, the financier has decided that the investment in Project B should not exceed 40% of the total investment. How much should she invest in each project to maximize the return on her investment?

9. **ASSET ALLOCATION** Justin has decided to invest at most $60,000 in medium-risk and high-risk stocks. He has further decided that the medium-risk stocks should make up at least 40% of the total investment, while the high-risk stocks should make up at least 20% of the total investment. He expects that the medium-risk stocks will

appreciate by 12% and the high-risk stocks by 20% within a year. How much money should Justin invest in each type of stock to maximize the value of his investment?

10. **CROP PLANNING** A farmer plans to plant two crops, A and B. The cost of cultivating Crop A is $40/acre, whereas the cost of cultivating Crop B is $60/acre. The farmer has a maximum of $7400 available for land cultivation. Each acre of Crop A requires 20 labor-hours, and each acre of Crop B requires 25 labor-hours. The farmer has a maximum of 3300 labor-hours available. If she expects to make a profit of $150/acre on Crop A and $200/acre on Crop B, how many acres of each crop should she plant to maximize her profit?

11. **MINIMIZING MINING COSTS** Perth Mining Company operates two mines for the purpose of extracting gold and silver. The Saddle Mine costs $14,000/day to operate, and it yields 50 oz of gold and 3000 oz of silver each day. The Horseshoe Mine costs $16,000/day to operate, and it yields 75 oz of gold and 1000 oz of silver each day. Company management has set a target of at least 650 oz of gold and 18,000 oz of silver. How many days should each mine be operated so that the target can be met at a minimum cost?

12. **MINIMIZING CRUISE LINE COSTS** Deluxe River Cruises operates a fleet of river vessels. The fleet has two types of vessels: A type A vessel has 60 deluxe cabins and 160 standard cabins, whereas a type B vessel has 80 deluxe cabins and 120 standard cabins. Under a charter agreement with Odyssey Travel Agency, Deluxe River Cruises is to provide Odyssey with a minimum of 360 deluxe and 680 standard cabins for their 15-day cruise in May. It costs $44,000 to operate a type A vessel and $54,000 to operate a type B vessel for that period. How many of each type vessel should be used to keep the operating costs to a minimum?

13. **PRODUCTION SCHEDULING** Acoustical Company manufactures a DVD storage cabinet that can be bought fully assembled or as a kit. Each cabinet is processed in the fabrication department and the assembly department. If the fabrication department manufactures only fully assembled cabinets, it can produce 200 units/day; and if it manufactures only kits, it can produce 200 units/day. If the assembly department produces only fully assembled cabinets, it can produce 100 units/day; but if it produces only kits, then it can produce 300 units/day. Each fully assembled cabinet contributes $50 to the profits of the company, whereas each kit contributes $40 to its profits. How many fully assembled units and how many kits should the company produce per day to maximize its profits?

14. **FERTILIZERS** A farmer uses two types of fertilizers. A 50-lb bag of Fertilizer A contains 8 lb of nitrogen, 2 lb of phosphorus, and 4 lb of potassium. A 50-lb bag of Fertilizer B contains 5 lb each of nitrogen, phosphorus, and potassium. The minimum requirements for a field are 440 lb of nitrogen, 260 lb of phosphorus, and 360 lb of potassium. If a 50-lb bag of Fertilizer A costs $30 and a 50-lb bag of

Fertilizer B costs $20, find the amount of each type of fertilizer the farmer should use to minimize his cost while still meeting the minimum requirements.

15. **MINIMIZING CITY WATER COSTS** The water-supply manager for a Midwestern city needs to supply the city with at least 10 million gallons of potable (drinkable) water per day. The supply may be drawn from the local reservoir or from a pipeline to an adjacent town. The local reservoir has a maximum daily yield of 5 million gallons of potable water, and the pipeline has a maximum daily yield of 10 million gallons. By contract, the pipeline is required to supply a minimum of 6 million gallons/day. If the cost for 1 million gallons of reservoir water is $300 and that for pipeline water is $500, how much water should the manager get from each source to minimize daily water costs for the city?

16. **PRODUCTION SCHEDULING** Ace Novelty manufactures Giant Pandas and Saint Bernards. Each Panda requires 1.5 yd^2 of plush, 30 ft^3 of stuffing, and 5 pieces of trim; each Saint Bernard requires 2 yd^2 of plush, 35 ft^3 of stuffing, and 8 pieces of trim. The profit for each Panda is $10, and the profit for each Saint Bernard is $15. If 3600 yd^2 of plush, 66,000 ft^3 of stuffing and 13,600 pieces of trim are available, how many of each of the stuffed animals should the company manufacture to maximize profit?

17. **DIET PLANNING** A nutritionist at the Medical Center has been asked to prepare a special diet for certain patients. She has decided that the meals should contain a minimum of 400 mg of calcium, 10 mg of iron, and 40 mg of vitamin C. She has further decided that the meals are to be prepared from Foods A and B. Each ounce of Food A contains 30 mg of calcium, 1 mg of iron, 2 mg of vitamin C, and 2 mg of cholesterol. Each ounce of Food B contains 25 mg of calcium, 0.5 mg of iron, 5 mg of vitamin C, and 5 mg of cholesterol. How many ounces of each type of food should be used in a meal so that the cholesterol content is minimized and the minimum requirements of calcium, iron, and vitamin C are met?

18. **OPTIMIZING ADVERTISING EXPOSURE** Everest Deluxe World Travel has decided to advertise in the Sunday editions of two major newspapers in town. These advertisements are directed at three groups of potential customers. Each advertisement in Newspaper I is seen by 70,000 Group A customers, 40,000 Group B customers, and 20,000 Group C customers. Each advertisement in Newspaper II is seen by 10,000 Group A, 20,000 Group B, and 40,000 Group C customers. Each advertisement in Newspaper I costs $1000, and each advertisement in Newspaper II costs $800. Everest would like their advertisements to be read by at least 2 million people from Group A, 1.4 million people from Group B, and 1 million people from Group C. How many advertisements should Everest place in each newspaper to achieve its advertising goals at a minimum cost?

19. **MINIMIZING SHIPPING COSTS** TMA manufactures 37-in. high-definition LCD televisions in two separate locations: Location I and Location II. The output at Location I is at most 6000 televisions/month, whereas the output at Location II is at most 5000 televisions/month. TMA is the main supplier of televisions to Pulsar Corporation, its holding company, which has priority in having all its requirements met. In a certain month, Pulsar placed orders for 3000 and 4000 televisions to be shipped to two of its factories located in City A and City B, respectively. The shipping costs (in dollars) per television from the two TMA plants to the two Pulsar factories are as follows:

	To Pulsar Factories	
From TMA	**City A**	**City B**
Location I	$6	$4
Location II	$8	$10

Find a shipping schedule that meets the requirements of both companies while keeping costs to a minimum.

20. **SOCIAL PROGRAMS PLANNING** AntiFam, a hunger-relief organization, has earmarked between $2 million and $2.5 million (inclusive) for aid to two African countries, Country A and Country B. Country A is to receive between $1 million and $1.5 million (inclusive), and Country B is to receive at least $0.75 million. It has been estimated that each dollar spent in Country A will yield an effective return of $0.60, whereas a dollar spent in Country B will yield an effective return of $0.80. How should the aid be allocated if the money is to be utilized most effectively according to these criteria?
Hint: If x and y denote the amount of money to be given to Country A and Country B, respectively, then the objective function to be maximized is $P = 0.6x + 0.8y$.

21. **PRODUCTION SCHEDULING** A company manufactures Products A, B, and C. Each product is processed in three departments: I, II, and III. The total available labor-hours per week for Departments I, II, and III are 900, 1080, and 840, respectively. The time requirements (in hours per unit) and profit per unit for each product are as follows:

	Product A	Product B	Product C
Dept. I	2	1	2
Dept. II	3	1	2
Dept. III	2	2	1
Profit	$18	$12	$15

How many units of each product should the company produce to maximize its profit?

22. **OPTIMIZING ADVERTISING EXPOSURE** As part of a campaign to promote its annual clearance sale, the Excelsior Company decided to buy television advertising time on station KAOS. Excelsior's advertising budget is $102,000. Morning time costs $3000/minute, afternoon time

costs $1000/minute, and evening (prime) time costs $12,000/minute. Because of previous commitments, KAOS cannot offer Excelsior more than 6 min of prime time or more than a total of 25 min of advertising time over the 2 weeks in which the commercials are to be run. KAOS estimates that morning commercials are seen by 200,000 people, afternoon commercials are seen by 100,000 people, and evening commercials are seen by 600,000 people. How much morning, afternoon, and evening advertising time should Excelsior buy to maximize exposure of its commercials?

23. **PRODUCTION SCHEDULING** Custom Office Furniture Company is introducing a new line of executive desks made from a specially selected grade of walnut. Initially, three different models—A, B, and C—are to be marketed. Each model A desk requires $1\frac{1}{4}$ hr for fabrication, 1 hr for assembly, and 1 hr for finishing; each model B desk requires $1\frac{1}{2}$ hr for fabrication, 1 hr for assembly, and 1 hr for finishing; each model C desk requires $1\frac{1}{2}$ hr, $\frac{3}{4}$ hr, and $\frac{1}{2}$ hr for fabrication, assembly, and finishing, respectively. The profit for each model A desk is $26, the profit for each model B desk is $28, and the profit for each model C desk is $24. The total time available in the fabrication department, the assembly department, and the finishing department in the first month of production is 310 hr, 205 hr, and 190 hr, respectively. To maximize Custom's profit, how many desks of each model should be made in the month?

24. **ASSET ALLOCATION** A financier plans to invest up to $2 million in three projects. She estimates that Project A will yield a return of 10% on her investment, Project B will yield a return of 15% on her investment, and Project C will yield a return of 20% on her investment. Because of the risks associated with the investments, she decided to put not more than 20% of her total investment in Project C. She also decided that her investments in Projects B and C should not exceed 60% of her total investment. Finally, she decided that her investment in Project A should be at least 60% of her investments in Projects B and C. How much should the financier invest in each project if she wishes to maximize the total returns on her investments?

25. **ASSET ALLOCATION** Ashley has earmarked at most $250,000 for investment in three mutual funds: a money market fund, an international equity fund, and a growth-and-income fund. The money market fund has a rate of return of 6%/year, the international equity fund has a rate of return of 10%/year, and the growth-and-income fund has a rate of return of 15%/year. Ashley has stipulated that no more than 25% of her total portfolio should be in the growth-and-income fund and that no more than 50% of her total portfolio should be in the international equity fund. To maximize the return on her investment, how much should Ashley invest in each type of fund?

26. **OPTIMIZING PREFABRICATED HOUSING PRODUCTION** Boise Lumber has decided to enter the lucrative prefabricated housing business. Initially, it plans to offer three models: standard, deluxe, and luxury. Each house is prefabricated and partially assembled in the factory, and the final assembly is completed on site. The dollar amount of building material required, the amount of labor required in the factory for prefabrication and partial assembly, the amount of on-site labor required, and the profit per unit are as follows:

	Standard Model	Deluxe Model	Luxury Model
Material	$6,000	$8,000	$10,000
Factory Labor (hr)	240	220	200
On-site Labor (hr)	180	210	300
Profit	$3,400	$4,000	$5,000

For the first year's production, a sum of $8.2 million is budgeted for the building material; the number of labor-hours available for work in the factory (for prefabrication and partial assembly) is not to exceed 218,000 hr; and the amount of labor for on-site work is to be less than or equal to 237,000 labor-hours. Determine how many houses of each type Boise should produce (market research has confirmed that there should be no problems with sales) to maximize its profit from this new venture.

27. **MINIMIZING SHIPPING COSTS** Acrosonic of Example 4 also manufactures a model G loudspeaker system in plants I and II. The output at Plant I is at most 800 systems/month whereas the output at Plant II is at most 600/month. These loudspeaker systems are also shipped to three warehouses—A, B, and C—whose minimum monthly requirements are 500, 400, and 400, respectively. Shipping costs from Plant I to Warehouse A, Warehouse B, and Warehouse C are $16, $20, and $22 per system, respectively, and shipping costs from Plant II to each of these warehouses are $18, $16, and $14 per system, respectively. What shipping schedule will enable Acrosonic to meet the warehouses' requirements and at the same time keep its shipping costs to a minimum?

28. **OPTIMIZING PRODUCTION OF COLD FORMULAS** Beyer Pharmaceutical produces three kinds of cold formulas: Formula I, Formula II, and Formula III. It takes 2.5 hr to produce 1000 bottles of Formula I, 3 hr to produce 1000 bottles of Formula II, and 4 hr to produce 1000 bottles of Formula III. The profits for each 1000 bottles of Formula I, Formula II, and Formula III are $180, $200, and $300, respectively. For a certain production run, there are enough ingredients on hand to make at most 9000 bottles of Formula I, 12,000 bottles of Formula II, and 6000 bottles of Formula III. Furthermore, the time for the production run is limited to a maximum of 70 hr. How many bottles of each formula should be produced in this production run so that the profit is maximized?

29. **OPTIMIZING PRODUCTION OF BLENDED JUICES** CalJuice Company has decided to introduce three fruit juices made from blending two or more concentrates. These juices

will be packaged in 2-qt (64-oz) cartons. One carton of pineapple–orange juice requires 8 oz each of pineapple and orange juice concentrates. One carton of orange–banana juice requires 12 oz of orange juice concentrate and 4 oz of banana pulp concentrate. Finally, one carton of pineapple–orange–banana juice requires 4 oz of pineapple juice concentrate, 8 oz of orange juice concentrate, and 4 oz of banana pulp. The company has decided to allot 16,000 oz of pineapple juice concentrate, 24,000 oz of orange juice concentrate, and 5000 oz of banana pulp concentrate for the initial production run. The company has also stipulated that the production of pineapple–orange–banana juice should not exceed 800 cartons. Its profit on one carton of pineapple–orange juice is $1.00, its profit on one carton of orange–banana juice is $.80, and its profit on one carton of pineapple–orange–banana juice is $.90. To realize a maximum profit, how many cartons of each blend should the company produce?

30. **MINIMIZING SHIPPING COSTS** Steinwelt Piano manufactures upright and console pianos in two plants, Plant I and Plant II. The output of Plant I is at most 300/month, whereas the output of Plant II is at most 250/month. These pianos are shipped to three warehouses, which serve as distribution centers for the company. To fill current and projected future orders, Warehouse A requires at least 200 pianos/month, Warehouse B requires at least 150 pianos/month, and Warehouse C requires at least 200 pianos/month. The shipping cost of each piano from Plant I to Warehouse A, Warehouse B, and Warehouse C is $60, $60, and $80, respectively, and the shipping cost of each piano from Plant II to Warehouse A, Warehouse B, and Warehouse C is $80, $70, and $50, respectively. What shipping schedule will enable Steinwelt to meet the warehouses' requirements while keeping shipping costs to a minimum?

In Exercises 31 and 32, determine whether the statement is true or false. If it is true, explain why it is true. If it is false, give an example to show why it is false.

31. The problem

$$\begin{aligned} \text{Maximize} \quad & P = xy \\ \text{subject to} \quad & 2x + 3y \le 12 \\ & 2x + y \le 8 \\ & x \ge 0, y \ge 0 \end{aligned}$$

is a linear programming problem.

32. The problem

$$\begin{aligned} \text{Minimize} \quad & C = 2x + 3y \\ \text{subject to} \quad & 2x + 3y \le 6 \\ & x - y = 0 \\ & x \ge 0, y \ge 0 \end{aligned}$$

is a linear programming problem.

6.2 Solution to Self-Check Exercise

Let x denote the number of ads to be placed in the *City Tribune*, and let y denote the number to be placed in the *Daily News*. The total cost for placing x ads in the *City Tribune* and y ads in the *Daily News* is $300x + 100y$ dollars, and since the monthly budget is $9000, we must have

$$300x + 100y \le 9000$$

Next, the condition that the ad must appear in at least 15 but no more than 30 editions of the *Daily News* translates into the inequalities

$$y \ge 15$$
$$y \le 30$$

Finally, the objective function to be maximized is

$$P = 50{,}000x + 20{,}000y$$

To summarize, we have the following linear programming problem:

$$\begin{aligned} \text{Maximize} \quad & P = 50{,}000x + 20{,}000y \\ \text{subject to} \quad & 300x + 100y \le 9000 \\ & y \ge 15 \\ & y \le 30 \\ & x \ge 0, y \ge 0 \end{aligned}$$

6.3 Graphical Solution of Linear Programming Problems

The Graphical Method

Linear programming problems in two variables have relatively simple geometric interpretations. For example, the system of linear constraints associated with a two-dimensional linear programming problem, unless it is inconsistent, defines a planar region or a line segment whose boundary is composed of straight-line segments and/or half-lines. Such problems are therefore amenable to graphical analysis.

Consider the following two-dimensional linear programming problem:

$$\text{Maximize} \quad P = 3x + 2y$$
$$\text{subject to} \quad 2x + 3y \leq 12$$
$$2x + \ \ y \leq \ 8$$
$$x \geq 0, y \geq 0$$

(7)

The system of linear inequalities in (7) defines the planar region S shown in Figure 11. Each point in S is a candidate for the solution of the problem at hand and is referred to as a **feasible solution.** The set S itself is referred to as a **feasible set.** Our goal is to find, from among all the points in the set S, the point(s) that optimizes the objective function P. Such a feasible solution is called an **optimal solution** and constitutes the solution to the linear programming problem under consideration.

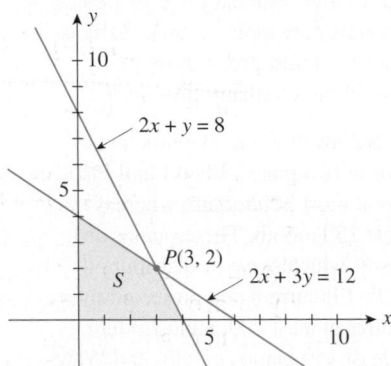

FIGURE 11
Each point in the feasible set S is a candidate
for the optimal solution.

As was noted earlier, each point $P(x, y)$ in S is a candidate for the optimal solution to the problem at hand. For example, the point $(1, 3)$ is easily seen to lie in S and is therefore in the running. The value of the objective function P at the point $(1, 3)$ is given by $P = 3(1) + 2(3) = 9$. Now, if we could compute the value of P corresponding to each point in S, then the point(s) in S that gave the largest value to P would constitute the solution set sought. Unfortunately, in most problems, the number of candidates either is too large or, as in this problem, is infinite. Therefore, this method is at best unwieldy and at worst impractical.

Let's turn the question around. Instead of asking for the value of the objective function P at a feasible point, let's assign a value to the objective function P and ask whether there are feasible points that would correspond to the given value of P. Toward this end, suppose we assign a value of 6 to P. Then the objective function P becomes $3x + 2y = 6$, a linear equation in x and y; thus, it has a graph that is a straight line L_1 in the plane. In Figure 12, we have drawn the graph of this straight line superimposed on the feasible set S.

It is clear that each point on the straight-line segment given by the intersection of the straight line L_1 and the feasible set S corresponds to the given value, 6, of P. For this reason, the line L_1 is called an **isoprofit line.** Let's repeat the process, this time assigning a value of 10 to P. We obtain the equation $3x + 2y = 10$ and the line L_2 (see Figure 12), which suggests that there are feasible points that correspond to a larger value of P. Observe that the line L_2 is parallel to the line L_1 because both lines have slope equal to $-\frac{3}{2}$, which is easily seen by casting the corresponding equations in the slope-intercept form.

In general, by assigning different values to the objective function, we obtain a family of parallel lines, each with slope equal to $-\frac{3}{2}$. Furthermore, a line corresponding to a larger value of P lies farther away from the origin than one with a smaller value of P. The implication is clear. To obtain the optimal solution(s) to the problem at hand,

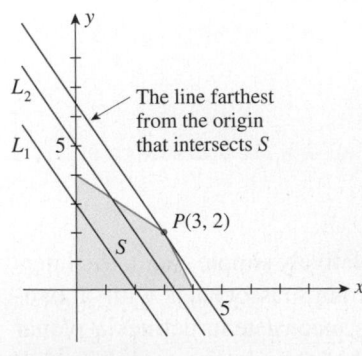

FIGURE 12
A family of parallel lines that intersect the
feasible set S

find the straight line, from this family of straight lines, that is farthest from the origin and still intersects the feasible set S. The required line is the one that passes through the point $P(3, 2)$ (see Figure 12), so the solution to the problem is given by $x = 3$, $y = 2$, resulting in a maximum value of $P = 3(3) + 2(2) = 13$.

That the optimal solution to this problem was found to occur at a vertex of the feasible set S is no accident. In fact, the result is a consequence of the following basic theorem on linear programming, which we state without proof.

THEOREM 1

Solution(s) of Linear Programming Problems

If a linear programming problem has a solution, then it must occur at a vertex, or corner point, of the feasible set S associated with the problem.

Furthermore, if the objective function P is optimized at two adjacent vertices of S, then it is optimized at every point on the line segment joining these vertices, in which case there are infinitely many solutions to the problem.

Theorem 1 tells us that our search for the solution(s) to a linear programming problem may be restricted to the examination of the set of vertices of the feasible set S associated with the problem. Since a feasible set S has finitely many vertices, the theorem suggests that the solution(s) to the linear programming problem may be found by inspecting the values of the objective function P at these vertices.

Although Theorem 1 sheds some light on the nature of the solution of a linear programming problem, it does not tell us when a linear programming problem has a solution. The following theorem states conditions that guarantee when a solution exists.

THEOREM 2

Existence of a Solution

Suppose we are given a linear programming problem with a feasible set S and an objective function $P = ax + by$.

a. If S is bounded, then P has both a maximum and a minimum value on S.

b. If S is unbounded and both a and b are nonnegative, then P has a minimum value on S provided that the constraints defining S include the inequalities $x \geq 0$ and $y \geq 0$.

c. If S is the empty set, then the linear programming problem has no solution; that is, P has neither a maximum nor a minimum value.

The **method of corners,** a simple procedure for solving linear programming problems based on Theorem 1, follows.

The Method of Corners

1. Graph the feasible set.
2. Find the coordinates of all corner points (vertices) of the feasible set.
3. Evaluate the objective function at each corner point.
4. Find the vertex that renders the objective function a maximum (minimum). If there is only one such vertex, then this vertex constitutes a unique solution to the problem. If the objective function is maximized (minimized) at two adjacent corner points of S, there are infinitely many optimal solutions given by the points on the line segment determined by these two vertices.

APPLIED EXAMPLE 1 Maximizing Profit We are now in a position to complete the solution to the production problem posed in Example 1, Section 6.2. Recall that the mathematical formulation led to the following linear programming problem:

$$\text{Maximize} \quad P = x + 1.2y$$
$$\text{subject to} \quad 2x + \ y \le 180$$
$$x + 3y \le 300$$
$$x \ge 0, y \ge 0$$

Solution The feasible set S for the problem is shown in Figure 13.

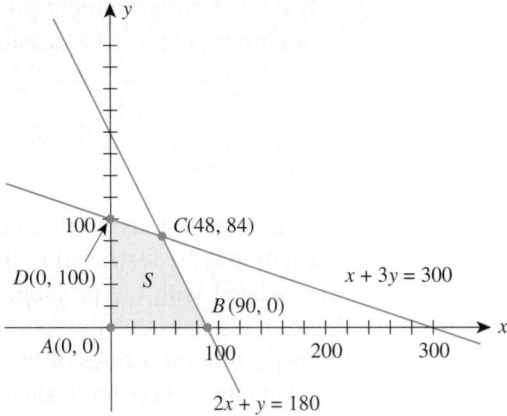

FIGURE 13
The corner point that yields the maximum profit is $C(48, 84)$.

The vertices of the feasible set are $A(0, 0)$, $B(90, 0)$, $C(48, 84)$, and $D(0, 100)$. The values of P at these vertices may be tabulated as follows:

Vertex	$P = x + 1.2y$
$A(0, 0)$	0
$B(90, 0)$	90
$C(48, 84)$	148.8
$D(0, 100)$	120

From the table, we see that the maximum of $P = x + 1.2y$ occurs at the vertex $(48, 84)$ and has a value of 148.8. Recalling what the symbols x, y, and P represent, we conclude that Ace Novelty would maximize its profit ($148.80) by producing 48 Type A souvenirs and 84 Type B souvenirs.

Explore and Discuss

Consider the linear programming problem

$$\text{Maximize} \quad P = 4x + 3y$$
$$\text{subject to} \quad 2x + \ y \le 10$$
$$2x + 3y \le 18$$
$$x \ge 0, y \ge 0$$

1. Sketch the feasible set S for the linear programming problem.
2. Draw the isoprofit lines superimposed on S corresponding to $P = 12, 16, 20,$ and 24, and show that these lines are parallel to each other.
3. Show that the solution to the linear programming problem is $x = 3$ and $y = 4$. Is this result the same as that found by using the method of corners?

APPLIED EXAMPLE 2 A Nutrition Problem Complete the solution of the nutrition problem posed in Example 2, Section 6.2.

Solution Recall that the mathematical formulation of the problem led to the following linear programming problem in two variables:

$$\text{Minimize} \quad C = 6x + 8y$$
$$\text{subject to} \quad 40x + 10y \geq 2400$$
$$10x + 15y \geq 2100$$
$$5x + 15y \geq 1500$$
$$x \geq 0, y \geq 0$$

The feasible set S defined by the system of constraints is shown in Figure 14.

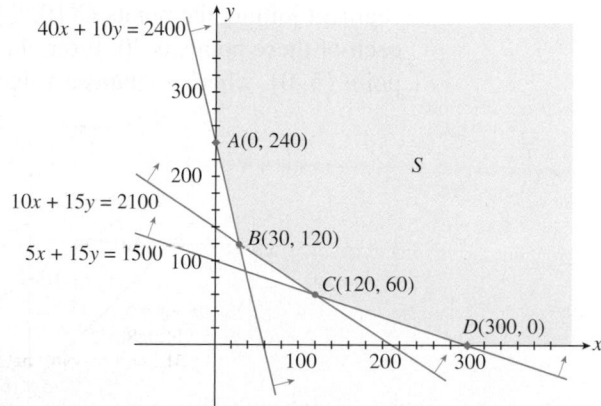

FIGURE **14**
The corner point that yields the minimum cost is $B(30, 120)$.

The vertices of the feasible set S are $A(0, 240)$, $B(30, 120)$, $C(120, 60)$, and $D(300, 0)$. The values of the objective function C at these vertices are given in the following table:

Vertex	$C = 6x + 8y$
$A(0, 240)$	1920
$B(30, 120)$	1140
$C(120, 60)$	1200
$D(300, 0)$	1800

From the table, we can see that the minimum for the objective function $C = 6x + 8y$ occurs at the vertex $B(30, 120)$ and has a value of 1140. Thus, the individual should purchase 30 Brand A pills and 120 Brand B pills at a minimum cost of $11.40. ◼

EXAMPLE 3 A Linear Programming Problem with Multiple Solutions Find the maximum and minimum of $P = 2x + 3y$ subject to the following system of linear inequalities:

$$2x + 3y \leq 30$$
$$-x + y \leq 5$$
$$x + y \geq 5$$
$$x \leq 10$$
$$x \geq 0, y \geq 0$$

Solution The feasible set S is shown in Figure 15. The vertices of the feasible set S are $A(5, 0)$, $B(10, 0)$, $C\left(10, \frac{10}{3}\right)$, $D(3, 8)$, and $E(0, 5)$. The values of the objective function P at these vertices are given in the following table:

Vertex	$P = 2x + 3y$
$A(5, 0)$	10
$B(10, 0)$	20
$C\left(10, \frac{10}{3}\right)$	30
$D(3, 8)$	30
$E(0, 5)$	15

From the table, we see that the maximum for the objective function $P = 2x + 3y$ occurs at the vertices $C\left(10, \frac{10}{3}\right)$ and $D(3, 8)$. This tells us that every point on the line segment joining the points $C\left(10, \frac{10}{3}\right)$ and $D(3, 8)$ maximizes P. The value of P at each of these points is 30. From the table, it is also clear that P is minimized at the point $(5, 0)$, where it attains a value of 10.

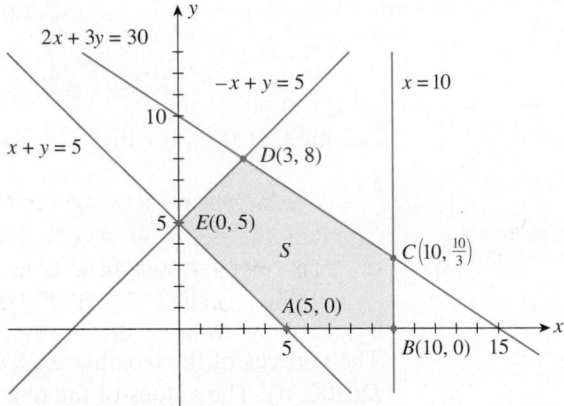

FIGURE 15
Every point lying on the line segment joining C and D maximizes P.

Explore and Discuss

Consider the linear programming problem

$$\text{Maximize} \quad P = 2x + 3y$$
$$\text{subject to} \quad 2x + \ \ y \le 10$$
$$2x + 3y \le 18$$
$$x \ge 0, y \ge 0$$

1. Sketch the feasible set S for the linear programming problem.
2. Draw the isoprofit lines superimposed on S corresponding to $P = 6, 8, 12$, and 18, and show that these lines are parallel to each other.
3. Show that there are infinitely many solutions to the problem. Is this result as predicted by the method of corners?

We close this section by examining two situations in which a linear programming problem has no solution.

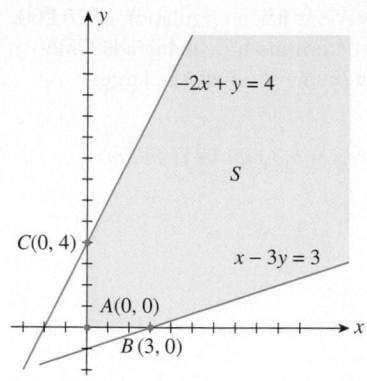

FIGURE **16**
This maximization problem has no solution because the feasible set is unbounded.

EXAMPLE 4 An Unbounded Linear Programming Problem with No Solution
Solve the following linear programming problem:

$$\text{Maximize} \quad P = x + 2y$$
$$\text{subject to} \quad -2x + y \le 4$$
$$x - 3y \le 3$$
$$x \ge 0, y \ge 0$$

Solution The feasible set S for this problem is shown in Figure 16. Since the set S is unbounded (both x and y can take on arbitrarily large positive values), we see that we can make P as large as we please by choosing x and y large enough. This problem has no solution. The problem is said to be unbounded. ∎

EXAMPLE 5 An Infeasible Linear Programming Problem Solve the following linear programming problem:

$$\text{Maximize} \quad P = x + 2y$$
$$\text{subject to} \quad x + 2y \le 4$$
$$2x + 3y \ge 12$$
$$x \ge 0, y \ge 0$$

Solution The half-planes described by the constraints (inequalities) have no points in common (Figure 17). Hence, there are no feasible points, and the problem has no solution. In this situation, we say that the problem is **infeasible,** or **inconsistent.** ∎

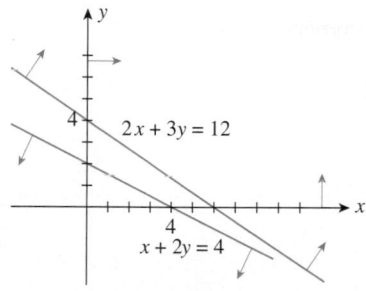

FIGURE **17**
This problem is inconsistent because there is no point that satisfies all of the given inequalities.

The situations described in Examples 4 and 5 are unlikely to occur in well-posed problems arising from practical applications of linear programming.

The method of corners is particularly effective in solving two-variable linear programming problems with a small number of constraints, as the preceding examples have amply demonstrated. However, its effectiveness decreases rapidly as the number of variables and/or constraints increases. For example, it may be shown that a linear programming problem in three variables and five constraints may have up to ten feasible corner points. The determination of the feasible corner points calls for the solution of ten 3×3 systems of linear equations and then the verification—by the substitution of each of these solutions into the system of constraints—to see whether it is, in fact, a feasible point. When the number of variables and constraints goes up to five and ten, respectively (still a very small system from the standpoint of applications in economics), the number of vertices to be found and checked for feasible corner points increases dramatically to 252, and each of these vertices is found by solving a 5×5 linear system! For this reason, the method of corners is seldom used to solve linear programming problems; its redeeming value lies in the fact that much insight is gained into the nature of the solutions of linear programming problems through its use in solving two-variable problems.

6.3 Self-Check Exercises

1. Use the method of corners to solve the following linear programming problem:

$$\text{Maximize} \quad P = 4x + 5y$$
$$\text{subject to} \quad x + 2y \le 10$$
$$5x + 3y \le 30$$
$$x \ge 0, y \ge 0$$

2. Use the method of corners to solve the following linear programming problem:

$$\text{Minimize} \quad C = 5x + 3y$$
$$\text{subject to} \quad 5x + 3y \ge 30$$
$$x - 3y \le 0$$
$$x \ge 2$$

3. OPTIMIZING ADVERTISING EXPOSURE Gino Balduzzi, proprietor of Luigi's Pizza Palace, allocates $9000 a month for advertising in two newspapers, the *City Tribune* and the *Daily News*. The *City Tribune* charges $300 for a certain advertisement, whereas the *Daily News* charges $100 for the same ad. Gino has stipulated that the ad is to appear in at least 15 but no more than 30 editions of the *Daily News* per month. The *City Tribune* has a daily circulation of 50,000, and the *Daily News* has a circulation of 20,000. Under these conditions, determine how many ads Gino should place in each newspaper to reach the largest number of readers.

Solutions to Self-Check Exercises 6.3 can be found on page 383.

6.3   Concept Questions

1. a. What is the feasible set associated with a linear programming problem?
   b. What is a feasible solution of a linear programming problem?
   c. What is an optimal solution of a linear programming problem?

2. Describe the method of corners.

6.3   Exercises

In Exercises 1–6, find the maximum and/or minimum value(s) of the objective function on the feasible set *S*.

1. $Z = 2x + 3y$

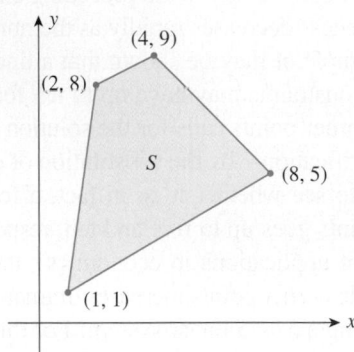

2. $Z = 3x - y$

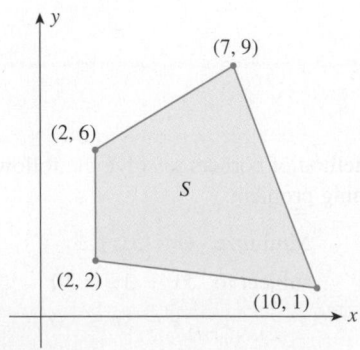

3. $Z = 2x + 3y$

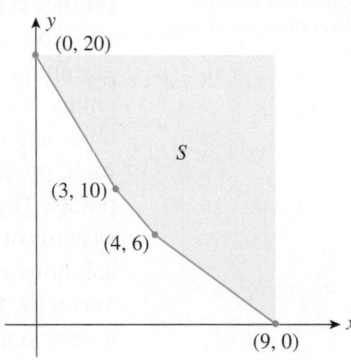

4. $Z = 7x + 9y$

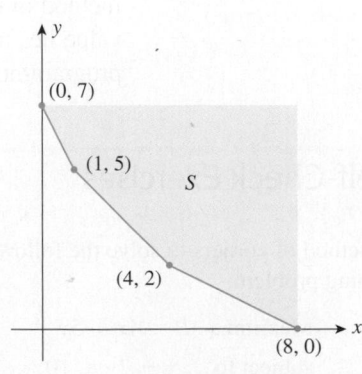

5. $Z = x + 4y$

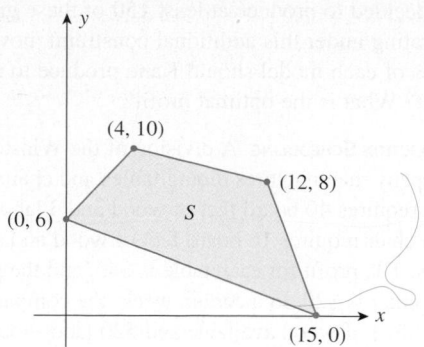

6. $Z = 3x + 2y$

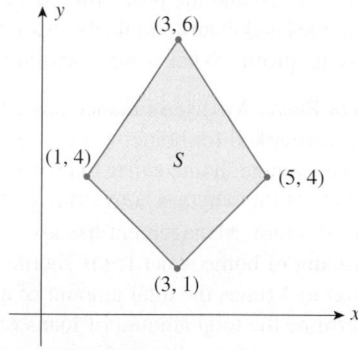

In Exercises 7–28, solve each linear programming problem by the method of corners.

7. Maximize $P = 3x + 2y$
subject to $x + y \leq 6$
$x \leq 3$
$x \geq 0, y \geq 0$

8. Maximize $P = x + 2y$
subject to $x + y \leq 4$
$2x + y \leq 5$
$x \geq 0, y \geq 0$

9. Maximize $P = 2x + y$ subject to the constraints of Exercise 8.

10. Maximize $P = 4x + 2y$
subject to $x + y \leq 8$
$2x + y \leq 10$
$x \geq 0, y \geq 0$

11. Maximize $P = x + 8y$ subject to the constraints of Exercise 10.

12. Maximize $P = 3x - 4y$
subject to $x + 3y \leq 15$
$4x + y \leq 16$
$x \geq 0, y \geq 0$

13. Maximize $P = x + 3y$
subject to $2x + y \leq 6$
$x + y \leq 4$
$x \leq 1$
$x \geq 0, y \geq 0$

14. Maximize $P = 2x + 5y$
subject to $2x + y \leq 16$
$2x + 3y \leq 24$
$y \leq 6$
$x \geq 0, y \geq 0$

15. Minimize $C = 2x + 5y$
subject to $x + y \geq 3$
$x + 2y \geq 4$
$x \geq 0, y \geq 0$

16. Minimize $C = 2x + 4y$ subject to the constraints of Exercise 15.

17. Minimize $C = 3x + 6y$
subject to $x + 2y \geq 40$
$x + y \geq 30$
$x \geq 0, y \geq 0$

18. Minimize $C = 3x + y$ subject to the constraints of Exercise 17.

19. Minimize $C = 2x + 10y$
subject to $5x + 2y \geq 40$
$x + 2y \geq 20$
$y \geq 3, x \geq 0$

20. Minimize $C = 2x + 5y$
subject to $4x + y \geq 40$
$2x + y \geq 30$
$x + 3y \geq 30$
$x \geq 0, y \geq 0$

21. Minimize $C = 10x + 15y$
subject to $x + y \leq 10$
$3x + y \geq 12$
$-2x + 3y \geq 3$
$x \geq 0, y \geq 0$

22. Maximize $P = 2x + 5y$ subject to the constraints of Exercise 21.

23. Maximize $P = 3x + 4y$
subject to $x + 2y \leq 50$
$5x + 4y \leq 145$
$2x + y \geq 25$
$y \geq 5, x \geq 0$

24. Maximize $P = 4x - 3y$ subject to the constraints of Exercise 23.

25. Maximize $P = 2x + 3y$
subject to $x + y \leq 48$
$x + 3y \geq 60$
$9x + 5y \leq 320$
$x \geq 10, y \geq 0$

26. Minimize $C = 5x + 3y$ subject to the constraints of Exercise 25.

27. Find the maximum and minimum of $P = 8x + 5y$ subject to

$$5x + 2y \geq 63$$
$$x + y \geq 18$$
$$3x + 2y \leq 51$$
$$x \geq 0, y \geq 0$$

28. Find the maximum and minimum of $P = 4x + 3y$ subject to

$$3x + 5y \geq 20$$
$$3x + y \leq 16$$
$$-2x + y \leq 1$$
$$x \geq 0, y \geq 0$$

The problems in Exercises 29–48 correspond to those in Exercises 1–20, Section 6.2. Use the results of your previous work to help you solve these problems.

29. PRODUCTION SCHEDULING A company manufactures two products, A and B, on two machines, I and II. It has been determined that the company will realize a profit of $3/unit of Product A and a profit of $4/unit of Product B. To manufacture a unit of Product A requires 6 min on Machine I and 5 min on Machine II. To manufacture a unit of Product B requires 9 min on Machine I and 4 min on Machine II. There are 5 hr of machine time available on Machine I and 3 hr of machine time available on Machine II in each work shift. How many units of each product should be produced in each shift to maximize the company's profit? What is the optimal profit?

30. PRODUCTION SCHEDULING National Business Machines manufactures two models of portable printers: A and B. Each model A costs $100 to make, and each model B costs $150. The profits are $30 for each model A and $40 for each model B portable printer. If the total number of portable printers demanded per month does not exceed 2500 and the company has earmarked no more than $600,000/month for manufacturing costs, how many units of each model should National make each month to maximize its monthly profit? What is the optimal profit?

31. PRODUCTION SCHEDULING Kane Manufacturing has a division that produces two models of fireplace grates, model A and model B. To produce each model A grate requires 3 lb of cast iron and 6 min of labor. To produce each model B grate requires 4 lb of cast iron and 3 min of labor. The profit for each model A grate is $2.00, and the profit for each model B grate is $1.50. If 1000 lb of cast iron and 20 labor-hours are available for the production of fireplace grates per day, how many grates of each model should the division produce to maximize Kane's profit? What is the optimal profit?

32. PRODUCTION SCHEDULING Refer to Exercise 31. Because of a backlog of orders for model A grates, Kane's manager had decided to produce at least 150 of these grates a day. Operating under this additional constraint, how many grates of each model should Kane produce to maximize profit? What is the optimal profit?

33. PRODUCTION SCHEDULING A division of the Winston Furniture Company manufactures dining tables and chairs. Each table requires 40 board feet of wood and 3 labor-hours. Each chair requires 16 board feet of wood and 4 labor-hours. The profit for each table is $45, and the profit for each chair is $20. In a certain week, the company has 3200 board feet of wood available and 520 labor-hours available. How many tables and chairs should Winston manufacture to maximize its profit? What is the maximum profit?

34. PRODUCTION SCHEDULING Refer to Exercise 33. If the profit for each table is $50 and the profit for each chair is $18, how many tables and chairs should Winston manufacture to maximize its profit? What is the maximum profit?

35. ALLOCATION OF FUNDS Madison Finance has a total of $20 million earmarked for homeowner loans and auto loans. On the average, homeowner loans have a 10% annual rate of return, whereas auto loans yield a 12% annual rate of return. Management has also stipulated that the total amount of homeowner loans should be greater than or equal to 4 times the total amount of automobile loans. Determine the total amount of loans of each type that Madison should extend to each category to maximize its returns. What are the optimal returns?

36. ASSET ALLOCATION A financier plans to invest up to $500,000 in two projects. Project A yields a return of 10% on the investment, whereas Project B yields a return of 15% on the investment. Because the investment in Project B is riskier than the investment in Project A, the financier has decided that the investment in Project B should not exceed 40% of the total investment. How much should she invest in each project to maximize the return on her investment? What is the maximum return?

37. ASSET ALLOCATION Justin has decided to invest at most $60,000 in medium-risk and high-risk stocks. He has further decided that the medium-risk stocks should make up at least 40% of the total investment, while the high-risk stocks should make up at least 20% of the total investment. He expects that the medium-risk stocks will appreciate by 12% and the high-risk stocks by 20% within a year. How much money should Justin invest in each type of stock to maximize the value of his investment? What is the maximum return?

38. CROP PLANNING A farmer plans to plant two crops, A and B. The cost of cultivating Crop A is $40/acre whereas the cost of cultivating Crop B is $60/acre. The farmer has a maximum of $7400 available for land cultivation. Each acre of Crop A requires 20 labor-hours, and each acre of Crop B requires 25 labor-hours. The farmer has a maximum of 3300 labor-hours available. If she expects to make a profit of $150/acre on Crop A and $200/acre on Crop B, how many acres of each crop should she plant to maximize her profit? What is the optimal profit?

39. **MINIMIZING MINING COSTS** Perth Mining Company operates two mines for the purpose of extracting gold and silver. The Saddle Mine costs $14,000/day to operate, and it yields 50 oz of gold and 3000 oz of silver each day. The Horseshoe Mine costs $16,000/day to operate, and it yields 75 oz of gold and 1000 oz of silver each day. Company management has set a target of at least 650 oz of gold and 18,000 oz of silver. How many days should each mine be operated so that the target can be met at a minimum cost? What is the minimum cost?

40. **MINIMIZING CRUISE LINE COSTS** Deluxe River Cruises operates a fleet of river vessels. The fleet has two types of vessels: A type A vessel has 60 deluxe cabins and 160 standard cabins, whereas a type B vessel has 80 deluxe cabins and 120 standard cabins. Under a charter agreement with Odyssey Travel Agency, Deluxe River Cruises is to provide Odyssey with a minimum of 360 deluxe and 680 standard cabins for their 15-day cruise in May. It costs $44,000 to operate a type A vessel and $54,000 to operate a type B vessel for that period. How many of each type vessel should be used to keep the operating costs to a minimum? What is the minimum cost?

41. **PRODUCTION SCHEDULING** Acoustical Company manufactures a DVD storage cabinet that can be bought fully assembled or as a kit. Each cabinet is processed in the fabrications department and the assembly department. If the fabrication department manufactures only fully assembled cabinets, then it can produce 200 units/day; and if it manufactures only kits, it can produce 200 units/day. If the assembly department produces only fully assembled cabinets, then it can produce 100 units/day; but if it produces only kits, then it can produce 300 units/day. Each fully assembled cabinet contributes $50 to the profits of the company whereas each kit contributes $40 to its profits. How many fully assembled units and how many kits should the company produce per day to maximize its profit? What is the optimal profit?

42. **FERTILIZERS** A farmer uses two types of fertilizers. A 50-lb bag of Fertilizer A contains 8 lb of nitrogen, 2 lb of phosphorus, and 4 lb of potassium. A 50-lb bag of Fertilizer B contains 5 lb each of nitrogen, phosphorus, and potassium. The minimum requirements for a field are 440 lb of nitrogen, 260 lb of phosphorus, and 360 lb of potassium. If a 50-lb bag of Fertilizer A costs $30 and a 50-lb bag of Fertilizer B costs $20, find the amount of each type of fertilizer the farmer should use to minimize his cost while still meeting the minimum requirements. What is the minimum cost?

43. **MINIMIZING CITY WATER COSTS** The water-supply manager for a Midwestern city needs to supply the city with at least

10 million gallons of potable (drinkable) water per day. The supply may be drawn from the local reservoir or from a pipeline to an adjacent town. The local reservoir has a maximum daily yield of 5 million gallons of potable water, and the pipeline has a maximum daily yield of 10 million gallons. By contract, the pipeline is required to supply a minimum of 6 million gallons/day. If the cost for 1 million gallons of reservoir water is $300 and that for pipeline water is $500, how much water should the manager get from each source to minimize daily water costs for the city? What is the minimum daily cost?

44. **PRODUCTION SCHEDULING** Ace Novelty manufactures Giant Pandas and Saint Bernards. Each Panda requires 1.5 yd^2 of plush, 30 ft^3 of stuffing, and 5 pieces of trim; each Saint Bernard requires 2 yd^2 of plush, 35 ft^3 of stuffing, and 8 pieces of trim. The profit for each Panda is $10, and the profit for each Saint Bernard is $15. If 3600 yd^2 of plush, 66,000 ft^3 of stuffing, and 13,600 pieces of trim are available, how many of each of the stuffed animals should the company manufacture to maximize profit? What is the maximum profit?

45. **DIET PLANNING** A nutritionist at the Medical Center has been asked to prepare a special diet for certain patients. She has decided that the meals should contain a minimum of 400 mg of calcium, 10 mg of iron, and 40 mg of vitamin C. She has further decided that the meals are to be prepared from Foods A and B. Each ounce of Food A contains 30 mg of calcium, 1 mg of iron, 2 mg of vitamin C, and 2 mg of cholesterol. Each ounce of Food B contains 25 mg of calcium, 0.5 mg of iron, 5 mg of vitamin C, and 5 mg of cholesterol. How many ounces of each type of food should be used in a meal so that the cholesterol content is minimized and the minimum requirements of calcium, iron, and vitamin C are met? What is the minimum cholesterol content?

46. **OPTIMIZING ADVERTISING EXPOSURE** Everest Deluxe World Travel has decided to advertise in the Sunday editions of two major newspapers in town. These advertisements are directed at three groups of potential customers. Each advertisement in Newspaper I is seen by 70,000 Group A customers, 40,000 Group B customers, and 20,000 Group C customers. Each advertisement in Newspaper II is seen by 10,000 Group A, 20,000 Group B, and 40,000 Group C customers. Each advertisement in Newspaper I costs $1000, and each advertisement in Newspaper II costs $800. Everest would like their advertisements to be read by at least 2 million people from Group A, 1.4 million people from Group B, and 1 million people from Group C. How many advertisements should Everest place in each newspaper to achieve its advertising goals at a minimum cost? What is the minimum cost?
Hint: Use different scales for drawing the feasible set.

47. **MINIMIZING SHIPPING COSTS** TMA manufactures 37-in. high-definition LCD televisions in two separate locations: Locations I and II. The output at Location I is at most 6000 televisions/month, whereas the output at Location II

is at most 5000 televisions/month. TMA is the main supplier of televisions to the Pulsar Corporation, its holding company, which has priority in having all its requirements met. In a certain month, Pulsar placed orders for 3000 and 4000 televisions to be shipped to two of its factories located in City A and City B, respectively. The shipping costs (in dollars) per television from the two TMA plants to the two Pulsar factories are as follows:

	To Pulsar Factories	
From TMA	**City A**	**City B**
Location I	$6	$4
Location II	$8	$10

Find a shipping schedule that meets the requirements of both companies while keeping costs to a minimum.

48. **SOCIAL PROGRAMS PLANNING** AntiFam, a hunger-relief organization, has earmarked between $2 and $2.5 million (inclusive) for aid to two African countries, Country A and Country B. Country A is to receive between $1 million and $1.5 million (inclusive), and Country B is to receive at least $0.75 million. It has been estimated that each dollar spent in Country A will yield an effective return of $0.60, whereas a dollar spent in Country B will yield an effective return of $0.80. How should the aid be allocated if the money is to be utilized most effectively according to these criteria?
Hint: If x and y denote the amount of money to be given to Country A and Country B, respectively, then the objective function to be maximized is $P = 0.6x + 0.8y$.

49. Complete the solution to Example 3, Section 6.2.

50. **VETERINARY SCIENCE** A veterinarian has been asked to prepare a diet for a group of dogs to be used in a nutrition study at the School of Animal Science. It has been stipulated that each serving should be no larger than 8 oz and must contain at least 29 units of Nutrient I and 20 units of Nutrient II. The vet has decided that the diet may be prepared from two brands of dog food: Brand A and Brand B. Each ounce of Brand A contains 3 units of Nutrient I and 4 units of Nutrient II. Each ounce of Brand B contains 5 units of Nutrient I and 2 units of Nutrient II. Brand A costs 3 cents/oz, and Brand B costs 4 cents/oz. Determine how many ounces of each brand of dog food should be used per serving to meet the given requirements at a minimum cost.

51. **MAXIMIZING INVESTMENT RETURNS** Patricia has at most $30,000 to invest in securities in the form of corporate stocks. She has narrowed her choices to two groups of stocks: growth stocks that she assumes will yield a 15% return (dividends and capital appreciation) within a year and speculative stocks that she assumes will yield a 25% return (mainly in capital appreciation) within a year. Determine how much she should invest in each group of stocks to maximize the return on her investments within a year if she has decided to invest at least 3 times as much in growth stocks as in speculative stocks. What is the maximum return?

52. **PRODUCTION SCHEDULING** Bata Aerobics manufactures two models of steppers used for aerobic exercises. Manufacturing each luxury model requires 10 lb of plastic and 10 min of labor. Manufacturing each standard model requires 16 lb of plastic and 8 min of labor. The profit for each luxury model is $40, and the profit for each standard model is $30. If 6000 lb of plastic and 60 labor-hours are available for the production of the steppers per day, how many steppers of each model should Bata produce each day to maximize its profit? What is the optimal profit?

53. **MARKET RESEARCH** Trendex, a telephone survey company, has been hired to conduct a television-viewing poll among urban and suburban families in the Los Angeles area. The client has stipulated that a maximum of 1500 families is to be interviewed. At least 500 urban families must be interviewed, and at least half of the total number of families interviewed must be from the suburban area. For this service, Trendex will be paid $6000 plus $8 for each completed interview. From previous experience, Trendex has determined that it will incur an expense of $4.40 for each successful interview with an urban family and $5 for each successful interview with a suburban family. How many urban and suburban families should Trendex interview to maximize its profit? What is the optimal profit?

In Exercises 54–57, determine whether the statement is true or false. If it is true, explain why it is true. If it is false, give an example to show why it is false.

54. An optimal solution of a linear programming problem is a feasible solution, but a feasible solution of a linear programming problem need not be an optimal solution.

55. An optimal solution of a linear programming problem can occur inside the feasible set of the problem.

56. If a maximization problem has no solution, then the feasible set associated with the linear programming problem must be unbounded.

57. Suppose you are given the following linear programming problem: Maximize $P = ax + by$ on the unbounded feasible set S shown in the accompanying figure.

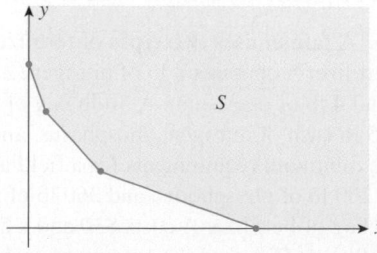

a. If $a > 0$ or $b > 0$, then the linear programming problem has no optimal solution.
b. If $a \leq 0$ and $b \leq 0$, then the linear programming problem has at least one optimal solution.

58. Suppose you are given the following linear programming problem: Maximize $P = ax + by$, where $a > 0$ and $b > 0$, on the feasible set S shown in the accompanying figure.

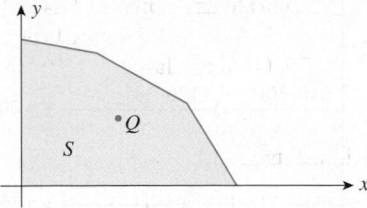

Explain, without using Theorem 1, why the optimal solution of the linear programming problem cannot occur at the point Q.

59. Suppose you are given the following linear programming problem: Maximize $P = ax + by$, where $a > 0$ and $b > 0$, on the feasible set S shown in the accompanying figure.

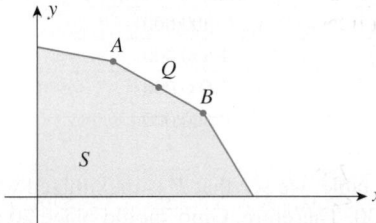

Explain, without using Theorem 1, why the optimal solution of the linear programming problem cannot occur at the point Q unless the problem has infinitely many solutions lying along the line segment joining the vertices A and B.

Hint: Let $A(x_1, y_1)$ and $B(x_2, y_2)$. Then let Q be the point $(\bar{x}, \bar{y})$, where $\bar{x} = x_1 + (x_2 - x_1)t$ and $\bar{y} = y_1 + (y_2 - y_1)t$ with $0 < t < 1$. Study the value of P at and near Q.

60. Consider the linear programming problem

$$\begin{aligned} \text{Maximize} \quad & P = 2x + 7y \\ \text{subject to} \quad & 2x + y \geq 8 \\ & x + y \geq 6 \\ & x \geq 0, y \geq 0 \end{aligned}$$

a. Sketch the feasible set S.
b. Find the corner points of S.
c. Find the values of P at the corner points of S found in part (b).
d. Show that the linear programming problem has no optimal solution. Does this contradict Theorem 2?

61. Consider the linear programming problem

$$\begin{aligned} \text{Minimize} \quad & C = -2x + 5y \\ \text{subject to} \quad & x + y \leq 3 \\ & 2x + y \leq 4 \\ & 5x + 8y \geq 40 \\ & x \geq 0, y \geq 0 \end{aligned}$$

a. Sketch the feasible set.
b. Find the solution(s) of the linear programming problem, if it exists.

62. Consider the linear programming problem

$$\begin{aligned} \text{Minimize} \quad & C = x - 4y \\ \text{subject to} \quad & x - 3y \leq -3 \\ & 2x - y \leq 4 \\ & x \geq 0, y \geq 0 \end{aligned}$$

a. Sketch the feasible set S.
b. Show that the linear programming problem has an optimal solution. Does this contradict Theorem 2? Explain.

Solutions to Self-Check Exercises

1. The feasible set S for the problem was graphed in the solution to Exercise 1, Self-Check Exercises 6.1. It is reproduced in the following figure.

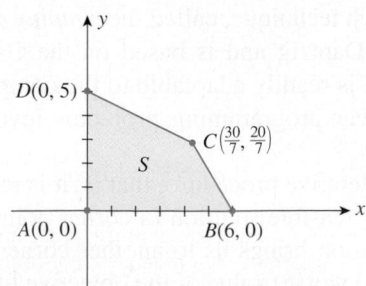

The values of the objective function P at the vertices of S are summarized in the following table:

Vertex	$P = 4x + 5y$
$A(0, 0)$	0
$B(6, 0)$	24
$C\left(\frac{30}{7}, \frac{20}{7}\right)$	$\frac{220}{7} = 31\frac{3}{7}$
$D(0, 5)$	25

From the table, we see that the maximum for the objective function P is attained at the vertex $C\left(\frac{30}{7}, \frac{20}{7}\right)$. Therefore, the solution to the problem is $x = \frac{30}{7}$, $y = \frac{20}{7}$, and $P = 31\frac{3}{7}$.

2. The feasible set S for the problem was graphed in the solution to Exercise 2, Self-Check Exercises 6.1. It is reproduced in the following figure.

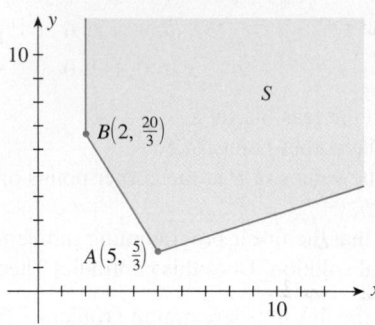

Evaluating the objective function $C = 5x + 3y$ at each corner point, we obtain the following table:

Vertex	$C = 5x + 3y$
$A\left(5, \frac{5}{3}\right)$	30
$B\left(2, \frac{20}{3}\right)$	30

We conclude that (i) the objective function is minimized at every point on the line segment joining the points $\left(5, \frac{5}{3}\right)$ and $\left(2, \frac{20}{3}\right)$, and (ii) the minimum value of C is 30.

3. Refer to Self-Check Exercise 6.2. The problem is to maximize $P = 50{,}000x + 20{,}000y$ subject to

$$300x + 100y \le 9000$$
$$y \ge 15$$
$$y \le 30$$
$$x \ge 0, y \ge 0$$

The feasible set S for the problem is shown in the following figure:

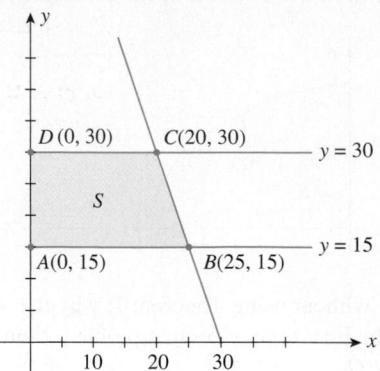

Evaluating the objective function $P = 50{,}000x + 20{,}000y$ at each vertex of S, we obtain the following table:

Vertex	$P = 50{,}000x + 20{,}000y$
$A(0, 15)$	300,000
$B(25, 15)$	1,550,000
$C(20, 30)$	1,600,000
$D(0, 30)$	600,000

From the table, we see that P is maximized when $x = 20$ and $y = 30$. Therefore, Gino should place 20 ads in the *City Tribune* and 30 in the *Daily News*.

6.4 The Simplex Method: Standard Maximization Problems

The Simplex Method

As was mentioned earlier, the method of corners is not suitable for solving linear programming problems when the number of variables or constraints is large. Its major shortcoming is that a knowledge of all the corner points of the feasible set S associated with the problem is required. What we need is a method of solution that is based on a judicious selection of the corner points of the feasible set S, thereby reducing the number of points to be inspected. One such technique, called the *simplex method*, was developed in the late 1940s by George Dantzig and is based on the Gauss–Jordan elimination method. The simplex method is readily adaptable to the computer, which makes it ideally suitable for solving linear programming problems involving large numbers of variables and constraints.

Basically, the simplex method is an iterative procedure; that is, it is repeated over and over again. Beginning at some initial feasible solution (a corner point of the feasible set S, usually the origin), each iteration brings us to another corner point of S, usually with an improved (but certainly no worse) value of the objective function. The iteration is terminated when the optimal solution is reached (if it exists).

In this section, we describe the simplex method for solving a large class of problems that are referred to as standard maximization problems.

Before stating a formal procedure for solving standard linear programming problems based on the simplex method, let's consider the following analysis of a two-variable problem. The ensuing discussion will clarify the general procedure and at the same time enhance our understanding of the simplex method by examining the motivation that led to the steps of the procedure.

A Standard Linear Programming Problem

A **standard maximization problem** is one in which

1. The objective function is to be maximized.
2. All the variables involved in the problem are nonnegative.
3. All other linear constraints may be written so that the expression involving the variables is less than or equal to a nonnegative constant.

Consider the linear programming problem presented at the beginning of Section 6.3:

$$\text{Maximize} \quad P = 3x + 2y \tag{8}$$
$$\text{subject to} \quad 2x + 3y \le 12$$
$$2x + y \le 8 \tag{9}$$
$$x \ge 0, y \ge 0$$

You can easily verify that this is a standard maximization problem. The feasible set S associated with this problem is reproduced in Figure 18, where we have labeled the four feasible corner points $A(0, 0)$, $B(4, 0)$, $C(3, 2)$, and $D(0, 4)$. Recall that the optimal solution to the problem occurs at the corner point $C(3, 2)$.

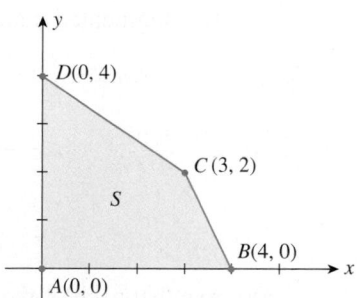

FIGURE **18**
The optimal solution occurs at $C(3, 2)$.

To solve this problem using the simplex method, we first replace the system of inequality constraints (9) with a system of equality constraints. This may be accomplished by using nonnegative variables called **slack variables.** Let's begin by considering the inequality

$$2x + 3y \le 12$$

Observe that the left-hand side of this equation is always less than or equal to the right-hand side. Therefore, by adding a nonnegative variable u to the left-hand side to compensate for this difference, we obtain the equality

$$2x + 3y + u = 12$$

For example, if $x = 1$ and $y = 1$, then $u = 7$, because

$$2(1) + 3(1) + 7 = 12$$

(You can see by referring to Figure 1 that the point $(1, 1)$ is a feasible point of S.) If $x = 2$ and $y = 1$, then $u = 5$, because

$$2(2) + 3(1) + 5 = 12$$

(So the point $(2, 1)$ is also a feasible point of S.) The variable u is a slack variable.

Similarly, the inequality $2x + y \leq 8$ is converted into the equation $2x + y + v = 8$ through the introduction of the slack variable v. System (9) of linear inequalities may now be viewed as the system of linear equations

$$
\begin{aligned}
2x + 3y + u \quad\quad &= 12 \\
2x + y \quad\quad + v &= 8
\end{aligned}
$$

where x, y, u, and v are all nonnegative.

Finally, rewriting the objective function (8) in the form $-3x - 2y + P = 0$, where the coefficient of P is $+1$, we are led to the following system of linear equations:

$$
\begin{aligned}
2x + 3y + u \quad\quad\quad &= 12 \\
2x + y \quad\; + v \quad\quad &= 8 \\
-3x - 2y \quad\quad\;\; + P &= 0
\end{aligned}
\tag{10}
$$

Since System (10) consists of three linear equations in the five variables x, y, u, v, and P, we may solve for three of the variables in terms of the other two. Thus, there are infinitely many solutions to this system expressible in terms of two parameters. Our linear programming problem is now seen to be equivalent to the following: From among all the solutions of System (10) for which x, y, u, and v are nonnegative (such solutions are called **feasible solutions**), determine the solution(s) that maximizes P.

The augmented matrix associated with System (10) is

Nonbasic variables —————— Basic variables
Column of constants

$$
\begin{array}{ccccc}
x & y & u & v & P \\
\end{array}
$$

$$
\left[
\begin{array}{ccccc|c}
2 & 3 & 1 & 0 & 0 & 12 \\
2 & 1 & 0 & 1 & 0 & 8 \\
-3 & -2 & 0 & 0 & 1 & 0
\end{array}
\right]
\tag{11}
$$

Observe that each of the u-, v-, and P-columns of the augmented matrix (11) is a unit column (see page 279). The variables associated with unit columns are called **basic variables**; all other variables are called **nonbasic variables.**

Now, the configuration of the augmented matrix (11) suggests that we solve for the basic variables u, v, and P in terms of the nonbasic variables x and y, obtaining

$$
\begin{aligned}
u &= 12 - 2x - 3y \\
v &= 8 - 2x - y \\
P &= 3x + 2y
\end{aligned}
\tag{12}
$$

Of the infinitely many feasible solutions that are obtainable by assigning arbitrary nonnegative values to the parameters x and y, a particular solution is obtained by letting $x = 0$ and $y = 0$. In fact, this solution is given by

$$x = 0 \quad\quad y = 0 \quad\quad u = 12 \quad\quad v = 8 \quad\quad P = 0$$

Such a solution, obtained by setting the nonbasic variables equal to zero, is called a **basic solution** of the system. This particular solution corresponds to the corner point $A(0, 0)$ of the feasible set associated with the linear programming problem (see Figure 18). Observe that $P = 0$ at this point.

Now, if the value of P cannot be increased, we have found the optimal solution to the problem at hand. To determine whether the value of P can in fact be improved, let's turn our attention to the objective function in Equation (8). Since the coefficients of both x and y are positive, the value of P can be improved by increasing x and/or y—that is, by moving away from the origin. Note that we arrive at the same conclusion by observing that the last row of the augmented matrix (11) contains entries that are *negative*. (Compare the original objective function, $P = 3x + 2y$, with the rewritten objective function, $-3x - 2y + P = 0$.)

Continuing our quest for an optimal solution, our next task is to determine whether it is more profitable to increase the value of x or that of y (increasing x and y simultaneously is more difficult). Since the coefficient of x is greater than that of y, a unit increase in the x-direction will result in a greater increase in the value of the objective function P than will a unit increase in the y-direction. Therefore, we should increase the value of x while holding y constant. How much can x be increased while holding $y = 0$? Upon setting $y = 0$ in the first two equations of System (12), we see that

$$
\begin{aligned}
u &= 12 - 2x \\
v &= 8 - 2x
\end{aligned}
\tag{13}
$$

Since u must be nonnegative, the first equation of System (13) implies that x cannot exceed $\frac{12}{2}$, or 6. The second equation of System (13) and the nonnegativity of v imply that x cannot exceed $\frac{8}{2}$, or 4. Thus, we conclude that x can be increased by at most 4.

Now, if we set $y = 0$ and $x = 4$ in System (12), we obtain the solution

$$
x = 4 \qquad y = 0 \qquad u = 4 \qquad v = 0 \qquad P = 12
$$

which is a basic solution to System (10), this time with y and v as nonbasic variables. (Recall that the nonbasic variables are precisely the variables that are set equal to zero.)

Let's see how this basic solution may be found by working with the augmented matrix of the system. Since x is to replace v as a basic variable, our aim is to find an augmented matrix that is equivalent to the matrix (11) and has a configuration in which the x-column is in the unit form

$$
\begin{bmatrix} 0 \\ 1 \\ 0 \end{bmatrix}
$$

replacing what is presently the form of the v-column in augmented matrix (11). This may be accomplished by pivoting about the circled number 2:

$$
\begin{array}{cccccc}
x & y & u & v & P & \text{Const.}
\end{array}
\left[
\begin{array}{ccccc|c}
2 & 3 & 1 & 0 & 0 & 12 \\
\textcircled{2} & 1 & 0 & 1 & 0 & 8 \\
-3 & -2 & 0 & 0 & 1 & 0
\end{array}
\right]
\xrightarrow{\frac{1}{2}R_2}
\begin{array}{cccccc}
x & y & u & v & P & \text{Const.}
\end{array}
\left[
\begin{array}{ccccc|c}
2 & 3 & 1 & 0 & 0 & 12 \\
\textcircled{1} & \frac{1}{2} & 0 & \frac{1}{2} & 0 & 4 \\
-3 & -2 & 0 & 0 & 1 & 0
\end{array}
\right]
\tag{14}
$$

$$\xrightarrow[R_3 + 3R_2]{R_1 - 2R_2} \begin{array}{c} \\ \begin{bmatrix} 0 & 2 & 1 & -1 & 0 & 4 \\ 1 & \frac{1}{2} & 0 & \frac{1}{2} & 0 & 4 \\ 0 & -\frac{1}{2} & 0 & \frac{3}{2} & 1 & 12 \end{bmatrix} \end{array} \qquad \textbf{(15)}$$

with column headers $x \quad y \quad u \quad v \quad P \quad$ Const.

Using System (15), we now solve for the basic variables x, u, and P in terms of the nonbasic variables y and v, obtaining

$$x = 4 - \frac{1}{2}y - \frac{1}{2}v$$

$$u = 4 - 2y + v$$

$$P = 12 + \frac{1}{2}y - \frac{3}{2}v$$

Setting the nonbasic variables y and v equal to zero gives

$$x = 4 \qquad y = 0 \qquad u = 4 \qquad v = 0 \qquad P = 12$$

as before.

We have now completed one iteration of the simplex procedure, and our search has brought us from the feasible corner point $A(0, 0)$, where $P = 0$, to the feasible corner point $B(4, 0)$, where P attained a value of 12, which is certainly an improvement! (See Figure 19.)

Before going on, let's introduce the following terminology. In what follows, refer to the augmented matrix (16), which is reproduced from the first augmented matrix in (14):

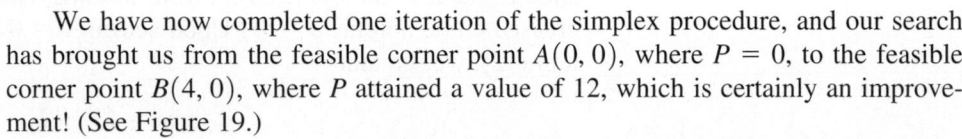

$$\begin{array}{cccccc} x & y & u & v & P & \text{Const.} \end{array}$$
$$\text{Pivot row} \rightarrow \begin{bmatrix} 2 & 3 & 1 & 0 & 0 & 12 \\ \textcircled{2} & 1 & 0 & 1 & 0 & 8 \\ -3 & -2 & 0 & 0 & 1 & 0 \end{bmatrix} \begin{array}{l} \frac{12}{2} = 6 \\ \frac{8}{2} = 4 \text{ (smallest ratio)} \\ - \end{array} \qquad \textbf{(16)}$$
$$\underset{\text{Pivot column}}{\uparrow}$$

(the negative number in the last row to the left of the vertical line with the largest absolute value)

The circled element 2 in the augmented matrix (16), which is to be converted into a 1, is called a *pivot element*. The column containing the pivot element is called the *pivot column*. The pivot column is associated with a nonbasic variable that is to be converted to a basic variable. Note that *the last entry in the pivot column is the negative number to the left of the vertical line in the last row with the largest absolute value*—precisely the criterion for choosing the direction of maximum increase in P.

The row containing the pivot element is called the *pivot row*. The pivot row can also be found by dividing each positive number in the pivot column into the corresponding number in the last column (the column of constants). *The pivot row is the one with the smallest ratio.* In augmented matrix (16), the pivot row is the second row because the ratio $\frac{8}{2}$, or 4, is less than the ratio $\frac{12}{2}$, or 6. (Compare this with the earlier analysis pertaining to the determination of the largest permissible increase in the value of x.) Then pivoting about the pivot element, we obtain the second tableau in (14).

The following is a summary of the procedure for selecting the pivot element.

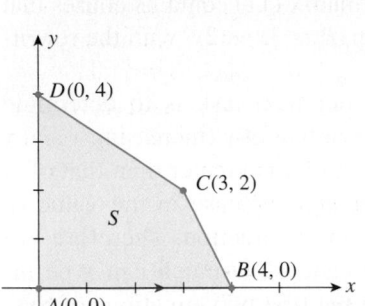

FIGURE **19**
One iteration has taken us from $A(0, 0)$, where $P = 0$, to $B(4, 0)$, where $P = 12$.

Selecting the Pivot Element

1. *Select the pivot column.* Locate the most negative entry to the left of the vertical line in the last row. The column containing this entry is the **pivot column.** (If there is more than one such column, choose any one.)

2. *Select the pivot row.* Divide each positive entry in the pivot column into its corresponding entry in the column of constants. The **pivot row** is the row corresponding to the smallest ratio thus obtained. (If there is more than one such entry, choose any one.)

3. The **pivot element** is the element common to both the pivot column and the pivot row.

Continuing with the solution to our problem, we observe that the last row of the augmented matrix (15) contains a negative number—namely, $-\frac{1}{2}$. This indicates that P is not maximized at the feasible corner point $B(4, 0)$, so another iteration is required. Without once again going into a detailed analysis, we proceed immediately to the selection of a pivot element. In accordance with the rules, we perform the necessary row operations as follows:

$$
\begin{array}{c}
\text{Pivot} \\ \text{row} \rightarrow
\end{array}
\begin{array}{ccccc}
x & y & u & v & P
\end{array}
\left[
\begin{array}{ccccc|c}
0 & ② & 1 & -1 & 0 & 4 \\
1 & \frac{1}{2} & 0 & \frac{1}{2} & 0 & 4 \\
0 & -\frac{1}{2} & 0 & \frac{3}{2} & 1 & 12
\end{array}
\right]
\begin{array}{c}
\text{Ratio} \\
\frac{4}{2} = 2 \\
\frac{4}{1/2} = 8
\end{array}
$$

$$\underset{\text{Pivot column}}{\uparrow}$$

$$
\xrightarrow{\frac{1}{2}R_1}
\begin{array}{ccccc}
x & y & u & v & P
\end{array}
\left[
\begin{array}{ccccc|c}
0 & ① & \frac{1}{2} & -\frac{1}{2} & 0 & 2 \\
1 & \frac{1}{2} & 0 & \frac{1}{2} & 0 & 4 \\
0 & -\frac{1}{2} & 0 & \frac{3}{2} & 1 & 12
\end{array}
\right]
$$

$$
\xrightarrow[R_3 + \frac{1}{2}R_1]{R_2 - \frac{1}{2}R_1}
\begin{array}{ccccc}
x & y & u & v & P
\end{array}
\left[
\begin{array}{ccccc|c}
0 & 1 & \frac{1}{2} & -\frac{1}{2} & 0 & 2 \\
1 & 0 & -\frac{1}{4} & \frac{3}{4} & 0 & 3 \\
0 & 0 & \frac{1}{4} & \frac{5}{4} & 1 & 13
\end{array}
\right]
$$

Interpreting the last augmented matrix in the usual fashion, we find the basic solution $x = 3$, $y = 2$, and $P = 13$. Since there are no negative entries in the last row, the solution is optimal, and P cannot be increased further. The optimal solution is the feasible corner point $C(3, 2)$ (Figure 20). Observe that this agrees with the solution we found using the method of corners in Section 6.3.

Having seen how the simplex method works, let's list the steps involved in the procedure. The first step is to set up the initial **simplex tableau.**

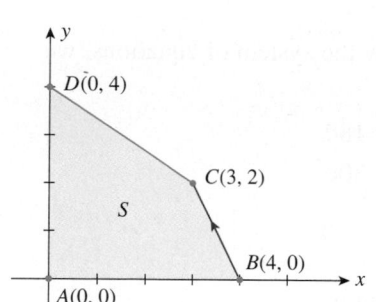

FIGURE 20
The next iteration has taken us from $B(4, 0)$, where $P = 12$, to $C(3, 2)$, where $P = 13$.

Setting Up the Initial Simplex Tableau

1. Transform the system of linear inequalities into a system of linear equations by introducing slack variables.

2. Rewrite the objective function

$$P = c_1x_1 + c_2x_2 + \cdots + c_nx_n$$

in the form

$$-c_1x_1 - c_2x_2 - \cdots - c_nx_n + P = 0$$

where all the variables are on the left and the coefficient of P is $+1$. Write this equation below the equations of Step 1.

3. Write the tableau associated with this system of linear equations.

EXAMPLE 1 Set up the initial simplex tableau for the linear programming problem posed in Example 1, Section 6.2.

Solution The problem at hand is to maximize

$$P = x + 1.2y$$

or, equivalently,

$$P = x + \frac{6}{5}y$$

subject to

$$\begin{aligned}
2x + y &\le 180 \\
x + 3y &\le 300 \\
x \ge 0, y &\ge 0
\end{aligned} \tag{17}$$

This is a standard maximization problem and may be solved by the simplex method. Since System (17) has two linear inequalities (other than $x \ge 0$, $y \ge 0$), we introduce the two slack variables u and v to convert it to a system of linear equations:

$$\begin{aligned}
2x + y + u &= 180 \\
x + 3y + v &= 300
\end{aligned}$$

Next, by rewriting the objective function in the form

$$-x - \frac{6}{5}y + P = 0$$

where the coefficient of P is $+1$, and placing it below the system of equations, we obtain the system of linear equations

$$\begin{aligned}
2x + y + u &= 180 \\
x + 3y + v &= 300 \\
-x - \frac{6}{5}y + P &= 0
\end{aligned}$$

The initial simplex tableau associated with this system is

x	y	u	v	P	Constant
2	1	1	0	0	180
1	3	0	1	0	300
-1	$-\frac{6}{5}$	0	0	1	0

Before completing the solution to the problem posed in Example 1, let's summarize the main steps of the **simplex method.**

The Simplex Method

1. *Set up the initial simplex tableau.*
2. *Determine whether the optimal solution has been reached by examining all entries in the last row to the left of the vertical line.*
   a. If all the entries are nonnegative, the optimal solution has been reached. Proceed to Step 4.
   b. If there are one or more negative entries, the optimal solution has not been reached. Proceed to Step 3.

3. *Perform the pivot operation.* Locate the pivot element and convert it to a 1 by dividing all the elements in the pivot row by the pivot element. Using row operations, convert the pivot column into a unit column by adding suitable multiples of the pivot row to each of the other rows as required. Return to Step 2.

4. *Determine the optimal solution(s).* The value of the variable heading each unit column is given by the entry lying in the column of constants in the row containing the 1. The variables heading columns not in unit form are assigned the value zero.

EXAMPLE 2 Complete the solution to the problem discussed in Example 1.

Solution The first step in our procedure, setting up the initial simplex tableau, was completed in Example 1. We continue with Step 2.

Step 2 *Determine whether the optimal solution has been reached.* First, refer to the initial simplex tableau:

x	y	u	v	P	Constant
2	1	1	0	0	180
1	3	0	1	0	300
-1	$-\frac{6}{5}$	0	0	1	0

(18)

Since there are negative entries in the last row of the initial simplex tableau, the initial solution is not optimal. We proceed to Step 3.

Step 3 *Perform the following iterations.* First, locate the pivot element:

a. Since the entry $-\frac{6}{5}$ is the most negative entry to the left of the vertical line in the last row of the initial simplex tableau, the second column in the tableau is the pivot column.

b. Divide each positive number of the pivot column into the corresponding entry in the column of constants, and compare the ratios thus obtained. We see that the ratio $\frac{300}{3}$ is less than the ratio $\frac{180}{1}$, so row 2 is the pivot row.

c. The entry 3 lying in the pivot column and the pivot row is the pivot element.

	x	y	u	v	P	Constant	Ratio
	2	1	1	0	0	180	$\frac{180}{1} = 180$
Pivot row $\rightarrow$	1	③	0	1	0	300	$\frac{300}{3} = 100$
	-1	$-\frac{6}{5}$	0	0	1	0	

$\uparrow$
Pivot column

Next, we convert this pivot element into a 1 by multiplying all the entries in the pivot row by $\frac{1}{3}$. Then, using elementary row operations, we complete the conversion of the pivot column into a unit column. The details of the iteration follow:

$$\xrightarrow{\frac{1}{3}R_2}$$

x	y	u	v	P	Constant
2	1	1	0	0	180
$\frac{1}{3}$	①	0	$\frac{1}{3}$	0	100
-1	$-\frac{6}{5}$	0	0	1	0

	x	y	u	v	P	Constant
	$\frac{5}{3}$	0	1	$-\frac{1}{3}$	0	80
$\xrightarrow[R_3 + \frac{6}{5}R_2]{R_1 - R_2}$	$\frac{1}{3}$	1	0	$\frac{1}{3}$	0	100
	$-\frac{3}{5}$	0	0	$\frac{2}{5}$	1	120

(19)

This completes one iteration. The last row of the simplex tableau contains a negative number, so an optimal solution has not been reached. Therefore, we repeat the iterative step once again, as follows:

	x	y	u	v	P	Constant		Ratio
Pivot row →	$\left(\frac{5}{3}\right)$	0	1	$-\frac{1}{3}$	0	80		$\frac{80}{5/3} = 48$
	$\frac{1}{3}$	1	0	$\frac{1}{3}$	0	100		$\frac{100}{1/3} = 300$
	$-\frac{3}{5}$	0	0	$\frac{2}{5}$	1	120		

Pivot column

	x	y	u	v	P	Constant
	$\textcircled{1}$	0	$\frac{3}{5}$	$-\frac{1}{5}$	0	48
$\xrightarrow{\frac{3}{5}R_1}$	$\frac{1}{3}$	1	0	$\frac{1}{3}$	0	100
	$-\frac{3}{5}$	0	0	$\frac{2}{5}$	1	120

	x	y	u	v	P	Constant
	1	0	$\frac{3}{5}$	$-\frac{1}{5}$	0	48
$\xrightarrow[R_3 + \frac{3}{5}R_1]{R_2 - \frac{1}{3}R_1}$	0	1	$-\frac{1}{5}$	$\frac{2}{5}$	0	84
	0	0	$\frac{9}{25}$	$\frac{7}{25}$	1	$148\frac{4}{5}$

(20)

The last row of the simplex tableau (20) contains no negative numbers, so we conclude that the optimal solution has been reached.

Step 4   *Determine the optimal solution.* Locate the basic variables in the final tableau. In this case, the basic variables (those heading unit columns) are x, y, and P. The value assigned to the basic variable x is the number 48, which is the entry lying in the column of constants and in row 1 (the row that contains the 1).

x	y	u	v	P	Constant
$\textcircled{1}$	0	$\frac{3}{5}$	$-\frac{1}{5}$	0	48
0	$\textcircled{1}$	$-\frac{1}{5}$	$\frac{2}{5}$	0	84
0	0	$\frac{9}{25}$	$\frac{7}{25}$	$\textcircled{1}$	$148\frac{4}{5}$

Similarly, we conclude that $y = 84$ and $P = 148.8$. Next, we note that the variables u and v are nonbasic and are accordingly assigned the values $u = 0$ and $v = 0$. These results agree with those obtained in Example 1, Section 6.3. ∎

EXAMPLE 3   Maximize $P = 2x + 2y + z$

           subject to $2x + y + 2z \le 14$

                     $2x + 4y + z \le 26$

                     $x + 2y + 3z \le 28$

                     $x \ge 0, y \ge 0, z \ge 0$

Solution Introducing the slack variables u, v, and w and rewriting the objective function in the standard form gives the system of linear equations

$$
\begin{aligned}
2x + \ y + 2z + u & & & = 14 \\
2x + 4y + \ z & + v & & = 26 \\
x + 2y + 3z & & + w & = 28 \\
-2x - 2y - \ z & & + P & = 0
\end{aligned}
$$

The initial simplex tableau is given by

x	y	z	u	v	w	P	Constant
2	1	2	1	0	0	0	14
2	4	1	0	1	0	0	26
1	2	3	0	0	1	0	28
−2	−2	−1	0	0	0	1	0

Since the most negative entry in the last row (-2) occurs twice, we may choose either the x- or the y-column as the pivot column. Choosing the x-column as the pivot column and proceeding with the first iteration, we obtain the following sequence of tableaus:

	x	y	z	u	v	w	P	Constant	Ratio
Pivot row →	②ᵖⁱᵛᵒᵗ	1	2	1	0	0	0	14	$\frac{14}{2} = 7$
	2	4	1	0	1	0	0	26	$\frac{26}{2} = 13$
	1	2	3	0	0	1	0	28	$\frac{28}{1} = 28$
	−2	−2	−1	0	0	0	1	0	

Pivot column ↑ (x-column)

$\xrightarrow{\frac{1}{2}R_1}$

x	y	z	u	v	w	P	Constant
①	$\frac{1}{2}$	1	$\frac{1}{2}$	0	0	0	7
2	4	1	0	1	0	0	26
1	2	3	0	0	1	0	28
−2	−2	−1	0	0	0	1	0

$\xrightarrow[\substack{R_2 - 2R_1 \\ R_3 - R_1 \\ R_4 + 2R_1}]{}$

x	y	z	u	v	w	P	Constant
1	$\frac{1}{2}$	1	$\frac{1}{2}$	0	0	0	7
0	3	−1	−1	1	0	0	12
0	$\frac{3}{2}$	2	$-\frac{1}{2}$	0	1	0	21
0	−1	1	1	0	0	1	14

Since there is a negative number in the last row of the simplex tableau, we perform another iteration, as follows:

	x	y	z	u	v	w	P	Constant	Ratio
	1	$\frac{1}{2}$	1	$\frac{1}{2}$	0	0	0	7	$\frac{7}{1/2} = 14$
Pivot row →	0	③	−1	−1	1	0	0	12	$\frac{12}{3} = 4$
	0	$\frac{3}{2}$	2	$-\frac{1}{2}$	0	1	0	21	$\frac{21}{3/2} = 14$
	0	−1	1	1	0	0	1	14	

Pivot column ↑

	x	y	z	u	v	w	P	Constant
	1	$\frac{1}{2}$	1	$\frac{1}{2}$	0	0	0	7
$\xrightarrow{\frac{1}{3}R_2}$	0	①	$-\frac{1}{3}$	$-\frac{1}{3}$	$\frac{1}{3}$	0	0	4
	0	$\frac{3}{2}$	2	$-\frac{1}{2}$	0	1	0	21
	0	-1	1	1	0	0	1	14

	x	y	z	u	v	w	P	Constant
	1	0	$\frac{7}{6}$	$\frac{2}{3}$	$-\frac{1}{6}$	0	0	5
$\xrightarrow{\substack{R_1 - \frac{1}{2}R_2 \\ R_3 - \frac{3}{2}R_2 \\ R_4 + R_2}}$	0	1	$-\frac{1}{3}$	$-\frac{1}{3}$	$\frac{1}{3}$	0	0	4
	0	0	$\frac{5}{2}$	0	$-\frac{1}{2}$	1	0	15
	0	0	$\frac{2}{3}$	$\frac{2}{3}$	$\frac{1}{3}$	0	1	18

All entries in the last row are nonnegative, so we have reached the optimal solution. We conclude that $x = 5$, $y = 4$, $z = 0$, $u = 0$, $v = 0$, $w = 15$, and $P = 18$. ∎

Explore and Discuss

Consider the linear programming problem

$$\text{Maximize} \quad P = x + 2y$$
$$\text{subject to} \quad -2x + y \leq 4$$
$$x - 3y \leq 3$$
$$x \geq 0, y \geq 0$$

1. Sketch the feasible set S for the linear programming problem and explain why the problem has an unbounded solution.
2. Use the simplex method to solve the problem as follows:
   a. Perform one iteration on the initial simplex tableau. Interpret your result. Indicate the point on S corresponding to this (nonoptimal) solution.
   b. Show that the simplex procedure breaks down when you attempt to perform another iteration by demonstrating that there is no pivot element.
   c. Describe what happens if you violate the rule for finding the pivot element by allowing the ratios to be negative and proceeding with the iteration.

The following example is constructed to illustrate the geometry associated with the simplex method when used to solve a problem in three-dimensional space. We sketch the feasible set for the problem and show the path dictated by the simplex method in arriving at the optimal solution for the problem. The use of a calculator will help in the arithmetic operations if you wish to verify the steps.

EXAMPLE 4 Geometric Illustration of Simplex Method in 3-Space

$$\text{Maximize} \quad P = 20x + 12y + 18z$$
$$\text{subject to} \quad 3x + y + 2z \leq 9$$
$$2x + 3y + z \leq 8$$
$$x + 2y + 3z \leq 7$$
$$x \geq 0, y \geq 0, z \geq 0$$

Solution Introducing the slack variables u, v, and w and rewriting the objective function in standard form, we obtain the following system of linear equations:

$$
\begin{aligned}
3x + y + 2z + u \phantom{{}+v+w+P} &= 9 \\
2x + 3y + z \phantom{{}+u} + v \phantom{{}+w+P} &= 8 \\
x + 2y + 3z \phantom{{}+u+v} + w \phantom{{}+P} &= 7 \\
-20x - 12y - 18z \phantom{{}+u+v+w} + P &= 0
\end{aligned}
$$

The initial simplex tableau is given by

x	y	z	u	v	w	P	Constant
3	1	2	1	0	0	0	9
2	3	1	0	1	0	0	8
1	2	3	0	0	1	0	7
-20	-12	-18	0	0	0	1	0

Initial Tableau starts at the point $A(0, 0, 0)$. See Figure 21, page 396.

Since the most negative entry in the last row (-20) occurs in the x-column, we choose the x-column as the pivot column. Proceeding with the first iteration, we obtain the following sequence of tableaus:

	x	y	z	u	v	w	P	Constant	Ratio
Pivot row $\rightarrow$	③	1	2	1	0	0	0	9	$\frac{9}{3} = 3$
	2	3	1	0	1	0	0	8	$\frac{8}{2} = 4$
	1	2	3	0	0	1	0	7	$\frac{7}{1} = 7$
	-20	-12	-18	0	0	0	1	0	

Pivot column

$\xrightarrow{\frac{1}{3}R_1}$

x	y	z	u	v	w	P	Constant
①	$\frac{1}{3}$	$\frac{2}{3}$	$\frac{1}{3}$	0	0	0	3
2	3	1	0	1	0	0	8
1	2	3	0	0	1	0	7
-20	-12	-18	0	0	0	1	0

First Iteration brings us to the point $B(3, 0, 0)$. See Figure 21, page 396.

$\xrightarrow[\substack{R_3 - R_1 \\ R_4 + 20R_1}]{R_2 - 2R_1}$ Pivot row

	x	y	z	u	v	w	P	Constant	Ratio
	1	$\frac{1}{3}$	$\frac{2}{3}$	$\frac{1}{3}$	0	0	0	3	9
$\rightarrow$	0	⑦⁄₃	$-\frac{1}{3}$	$-\frac{2}{3}$	1	0	0	2	$\frac{6}{7}$
	0	$\frac{5}{3}$	$\frac{7}{3}$	$-\frac{1}{3}$	0	1	0	4	$\frac{12}{5}$
	0	$-\frac{16}{3}$	$-\frac{14}{3}$	$\frac{20}{3}$	0	0	1	60	

Pivot column

Interpreting this tableau, we see that $x = 3$, $y = 0$, $z = 0$, and $P = 60$. Thus, after one iteration, we are at the point $B(3, 0, 0)$ with $P = 60$. (See Figure 21, page 396.)

Since the most negative entry in the last row is $-\frac{16}{3}$, we choose the y-column as the pivot column. Proceeding with this iteration, we obtain

$$\xrightarrow{\frac{3}{7}R_2}$$

	x	y	z	u	v	w	P	Constant
	1	$\frac{1}{3}$	$\frac{2}{3}$	$\frac{1}{3}$	0	0	0	3
	0	①	$-\frac{1}{7}$	$-\frac{2}{7}$	$\frac{3}{7}$	0	0	$\frac{6}{7}$
	0	$\frac{5}{3}$	$\frac{7}{3}$	$-\frac{1}{3}$	0	1	0	4
	0	$-\frac{16}{3}$	$-\frac{14}{3}$	$\frac{20}{3}$	0	0	1	60

Second Iteration brings us to the point $C\left(\frac{19}{7}, \frac{6}{7}, 0\right)$. See Figure 21.

$$\begin{array}{c}\xrightarrow{R_1 - \frac{1}{3}R_2} \\ R_3 - \frac{5}{3}R_2 \\ R_4 + \frac{16}{3}R_2\end{array}$$

	x	y	z	u	v	w	P	Constant		Ratio
	1	0	$\frac{5}{7}$	$\frac{3}{7}$	$-\frac{1}{7}$	0	0	$\frac{19}{7}$		$\frac{19}{5}$
	0	1	$-\frac{1}{7}$	$-\frac{2}{7}$	$\frac{3}{7}$	0	0	$\frac{6}{7}$		—
Pivot row	0	0	$\frac{18}{7}$	$\frac{1}{7}$	$-\frac{5}{7}$	1	0	$\frac{18}{7}$		1
	0	0	$-\frac{38}{7}$	$\frac{36}{7}$	$\frac{16}{7}$	0	1	$64\frac{4}{7}$		

Pivot column

Interpreting this tableau, we see that $x = \frac{19}{7}, y = \frac{6}{7}, z = 0$ and $P = 64\frac{4}{7}$. Thus, the second iteration brings us to the point $C\left(\frac{19}{7}, \frac{6}{7}, 0\right)$. (See Figure 21.)

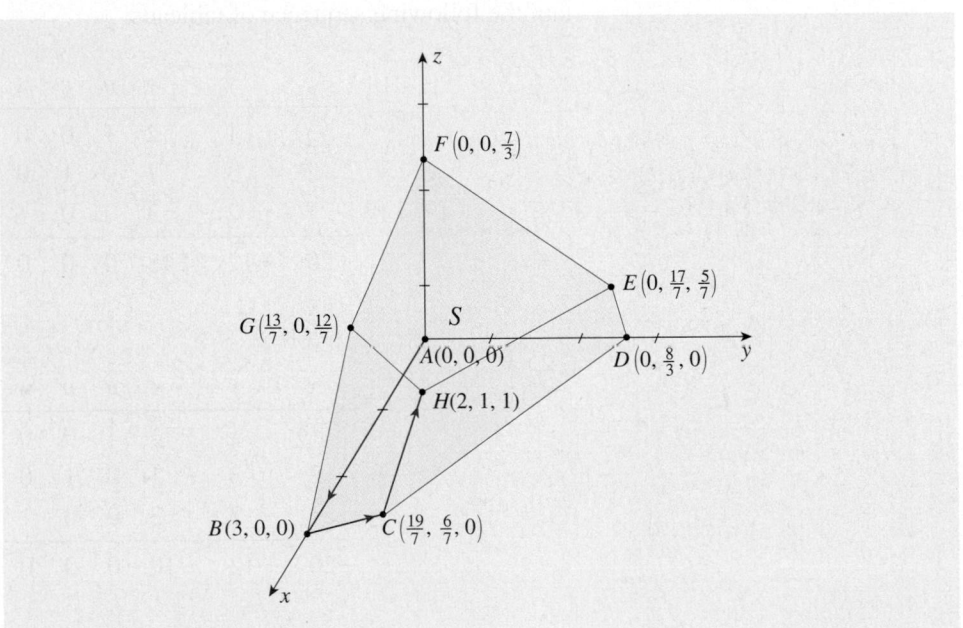

FIGURE **21**
The simplex method brings us from the point *A* to the point *H*, at which the objective function is maximized.

Since there is a negative number in the last row of the simplex tableau, we perform yet another iteration, as follows:

$$\xrightarrow{\frac{7}{18}R_3}$$

	x	y	z	u	v	w	P	Constant
	1	0	$\frac{5}{7}$	$\frac{3}{7}$	$-\frac{1}{7}$	0	0	$\frac{19}{7}$
	0	1	$-\frac{1}{7}$	$-\frac{2}{7}$	$\frac{3}{7}$	0	0	$\frac{6}{7}$
	0	0	①	$\frac{1}{18}$	$-\frac{5}{18}$	$\frac{7}{18}$	0	1
	0	0	$-\frac{38}{7}$	$\frac{36}{7}$	$\frac{16}{7}$	0	1	$64\frac{4}{7}$

<table>
<tr><td></td><td>x</td><td>y</td><td>z</td><td>u</td><td>v</td><td>w</td><td>P</td><td>Constant</td></tr>
</table>

	x	y	z	u	v	w	P	Constant
$R_1 - \frac{5}{7}R_3$	1	0	0	$\frac{7}{18}$	$\frac{1}{18}$	$-\frac{5}{18}$	0	2
$R_2 + \frac{1}{7}R_3$	0	1	0	$-\frac{5}{18}$	$\frac{7}{18}$	$\frac{1}{18}$	0	1
$R_4 + \frac{38}{7}R_3$	0	0	1	$\frac{1}{18}$	$-\frac{5}{18}$	$\frac{7}{18}$	0	1
	0	0	0	$\frac{49}{9}$	$\frac{7}{9}$	$\frac{19}{9}$	1	70

> **Third Iteration** brings us to the point $H(2, 1, 1)$. We have reached the optimal solution.

All entries in the last row are nonnegative, so we have reached the optimal solution (corresponding to the point $H(2, 1, 1)$). We conclude that $x = 2$, $y = 1$, $z = 1$, $u = 0$, $v = 0$, $w = 0$, and $P = 70$.

The feasible set S for the problem is the hexahedron shown in Figure 22. It is the intersection of the half-spaces determined by the planes P_1, P_2, and P_3 with equations $3x + y + 2z = 9$, $2x + 3y + z = 8$, $x + 2y + 3z = 7$, respectively, and the coordinate planes $x = 0$, $y = 0$, and $z = 0$. The portion of the figure showing the feasible set S is shown in Figure 21. Observe that the first iteration of the simplex method brings us from $A(0, 0, 0)$ with $P = 0$ to $B(3, 0, 0)$ with $P = 60$. The second iteration brings us from $B(3, 0, 0)$ to $C\left(\frac{19}{7}, \frac{6}{7}, 0\right)$ with $P = 64\frac{4}{7}$, and the third iteration brings us from $C\left(\frac{19}{7}, \frac{6}{7}, 0\right)$ to the point $H(2, 1, 1)$ with an optimal value of 70 for P.

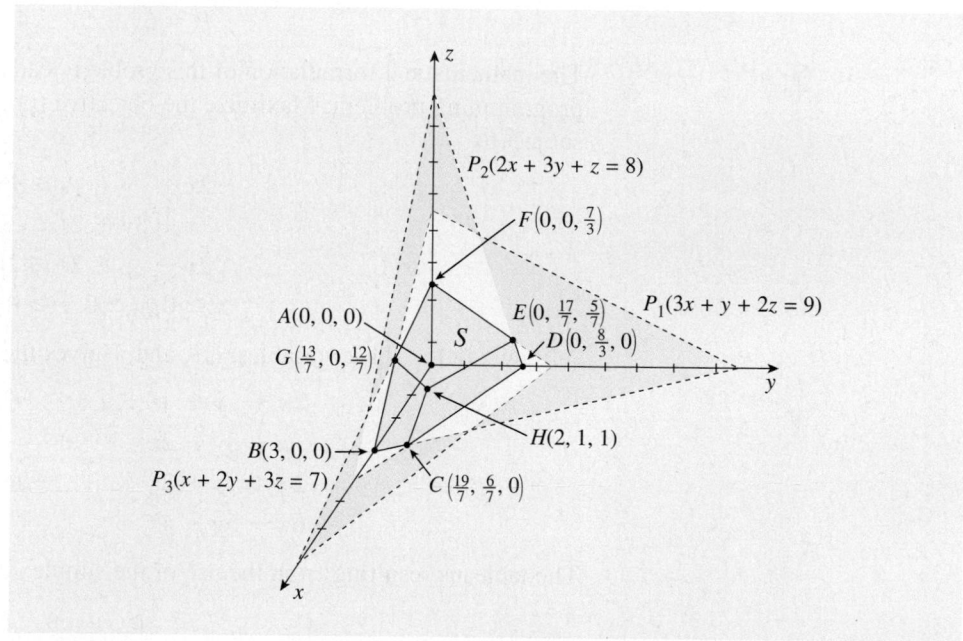

FIGURE 22
The feasible set S is obtained from the intersection of the half-spaces determined by P_1, P_2, and P_3 with the coordinate planes $x = 0$, $y = 0$, and $z = 0$.

APPLIED EXAMPLE 5 Production Planning Ace Novelty Company has determined that the profit for each Type A, Type B, and Type C souvenir that it plans to produce is $6, $5, and $4, respectively. To manufacture a Type A souvenir requires 2 minutes on Machine I, 1 minute on Machine II, and 2 minutes on Machine III. A Type B souvenir requires 1 minute on Machine I, 3 minutes on Machine II, and 1 minute on Machine III. A Type C souvenir requires 1 minute on Machine I and 2 minutes on each of Machines II and III. Each day, there are 3 hours available on Machine I, 5 hours available on Machine II, and 4 hours available on Machine III for manufacturing these souvenirs. How many souvenirs of each type should Ace Novelty make per day to maximize its profit? (Compare with Example 1, Section 5.1.)

Solution The given information is tabulated as follows:

	Type A	Type B	Type C	Time Available (min)
Machine I	2	1	1	180
Machine II	1	3	2	300
Machine III	2	1	2	240
Profit/Unit	$6	$5	$4	

Let x, y, and z denote the respective numbers of Type A, Type B, and Type C souvenirs to be made. The total amount of time that Machine I is used is given by $2x + y + z$ minutes and must not exceed 180 minutes. Thus, we have the inequality

$$2x + y + z \le 180$$

Similar considerations on the use of Machines II and III lead to the inequalities

$$x + 3y + 2z \le 300$$
$$2x + y + 2z \le 240$$

The profit resulting from the sale of the souvenirs produced is given by

$$P = 6x + 5y + 4z$$

The mathematical formulation of this problem leads to the following standard linear programming problem: Maximize the objective (profit) function $P = 6x + 5y + 4z$ subject to

$$2x + y + z \le 180$$
$$x + 3y + 2z \le 300$$
$$2x + y + 2z \le 240$$
$$x \ge 0, y \ge 0, z \ge 0$$

Introducing the slack variables u, v, and w gives the system of linear equations

$$2x + y + z + u \qquad\qquad = 180$$
$$x + 3y + 2z \qquad + v \qquad\quad = 300$$
$$2x + y + 2z \qquad\qquad + w \qquad = 240$$
$$-6x - 5y - 4z \qquad\qquad\qquad + P = 0$$

The tableaus resulting from the use of the simplex algorithm are as follows:

	x	y	z	u	v	w	P	Constant	Ratio
Pivot row →	②	1	1	1	0	0	0	180	$\frac{180}{2} = 90$
	1	3	2	0	1	0	0	300	$\frac{300}{1} = 300$
	2	1	2	0	0	1	0	240	$\frac{240}{2} = 120$
	−6	−5	−4	0	0	0	1	0	
	↑ Pivot column								

	x	y	z	u	v	w	P	Constant
$\frac{1}{2}R_1 \longrightarrow$	①	$\frac{1}{2}$	$\frac{1}{2}$	$\frac{1}{2}$	0	0	0	90
	1	3	2	0	1	0	0	300
	2	1	2	0	0	1	0	240
	−6	−5	−4	0	0	0	1	0

	x	y	z	u	v	w	P	Constant	Ratio
	1	$\frac{1}{2}$	$\frac{1}{2}$	$\frac{1}{2}$	0	0	0	90	$\frac{90}{1/2} = 180$
$\xrightarrow[R_4 + 6R_1]{\begin{subarray}{l}R_2 - R_1\\ R_3 - 2R_1\end{subarray}}$	0	$\boxed{\tfrac{5}{2}}$	$\frac{3}{2}$	$-\frac{1}{2}$	1	0	0	210	$\frac{210}{5/2} = 84$
Pivot row	0	0	1	-1	0	1	0	60	
	0	-2	-1	3	0	0	1	540	

Pivot column

	x	y	z	u	v	w	P	Constant
	1	$\frac{1}{2}$	$\frac{1}{2}$	$\frac{1}{2}$	0	0	0	90
$\xrightarrow{\frac{2}{5}R_2}$	0	①	$\frac{3}{5}$	$-\frac{1}{5}$	$\frac{2}{5}$	0	0	84
	0	0	1	-1	0	1	0	60
	0	-2	-1	3	0	0	1	540

	x	y	z	u	v	w	P	Constant
	1	0	$\frac{1}{5}$	$\frac{3}{5}$	$-\frac{1}{5}$	0	0	48
$\xrightarrow[R_4 + 2R_2]{R_1 - \frac{1}{2}R_2}$	0	1	$\frac{3}{5}$	$-\frac{1}{5}$	$\frac{2}{5}$	0	0	84
	0	0	1	-1	0	1	0	60
	0	0	$\frac{1}{5}$	$\frac{13}{5}$	$\frac{4}{5}$	0	1	708

From the final simplex tableau, we read off the solution

$$x = 48 \qquad y = 84 \qquad z = 0 \qquad u = 0 \qquad v = 0 \qquad w = 60 \qquad P = 708$$

Thus, to maximize its profit, Ace Novelty should produce 48 Type A souvenirs, 84 Type B souvenirs, and no Type C souvenirs. The resulting profit is \$708 per day. The value of the slack variable $w = 60$ tells us that 1 hour of the available time on Machine III is left unused.

Interpreting Our Results   Let's compare the results obtained here with those obtained in Example 7, Section 5.2. Recall that to use all available machine time on each of the three machines, Ace Novelty had to produce 36 Type A, 48 Type B, and 60 Type C souvenirs. This would have resulted in a profit of \$696. Example 5 shows how, through the optimal use of equipment, a company can boost its profit while reducing machine wear!

Problems with Multiple Solutions and Problems with No Solutions

As we saw in Section 6.3, a linear programming problem may have infinitely many solutions. We also saw that a linear programming problem may have no solution. How do we spot each of these phenomena when using the simplex method to solve a problem?

A linear programming problem will have infinitely many solutions if, for example, the last row to the left of the vertical line of the final simplex tableau has a zero in a column that is not a unit column, or if the final tableau contains two or more identical unit columns. Also, a linear programming problem will have *no* solution if the simplex method breaks down at some stage. For example, if at some stage there are no nonnegative ratios in our computation, then the linear programming problem has no solution (see Exercise 51).

Explore and Discuss

Consider the linear programming problem

$$\text{Maximize} \quad P = 4x + 6y$$
$$\text{subject to} \quad 2x + y \le 10$$
$$2x + 3y \le 18$$
$$x \ge 0, y \ge 0$$

1. Sketch the feasible set for the linear programming problem.
2. Use the method of corners to show that there are infinitely many optimal solutions. What are they?
3. Use the simplex method to solve the problem as follows.
   a. Perform one iteration on the initial simplex tableau and conclude that you have arrived at an optimal solution. What is the value of P, and where is it attained? Compare this result with that obtained in Step 2.
   b. Observe that the tableau obtained in part (a) indicates that there are infinitely many solutions (see the comment above on multiple solutions). Now perform another iteration on the simplex tableau using the x-column as the pivot column. Interpret the final tableau.

6.4 Self-Check Exercises

1. Solve the following linear programming problem by the simplex method:

$$\text{Maximize} \quad P = 2x + 3y + 6z$$
$$\text{subject to} \quad 2x + 3y + z \le 10$$
$$x + y + 2z \le 8$$
$$2y + 3z \le 6$$
$$x \ge 0, y \ge 0, z \ge 0$$

2. **MAXIMIZING PROFIT** The LaCrosse Iron Works makes two models of cast-iron fireplace grates: model A and model B. Producing one model A grate requires 20 lb of cast iron and 20 min of labor, whereas producing

one model B grate requires 30 lb of cast iron and 15 min of labor. The profit for a model A grate is $6, and the profit for a model B grate is $8. There are 7200 lb of cast iron and 100 labor-hours available each week. Because of a surplus from the previous week, the proprietor has decided to make no more than 150 units of model A grates this week. Determine how many of each model he should make to maximize his profit.

Solutions to Self-Check Exercises 6.4 can be found on page 405.

6.4 Concept Questions

1. Give the three characteristics of a standard maximization linear programming problem.

2. a. When the initial simplex tableau is set up, how is the system of linear inequalities transformed into a system of linear equations? How is the objective function $P = c_1x_1 + c_2x_2 + \cdots + c_nx_n$ rewritten?

   b. If you are given a simplex tableau, how do you determine whether the optimal solution has been reached?

3. In the simplex method, how is a pivot column selected? A pivot row? A pivot element?

6.4 Exercises

In Exercises 1–6, (a) write the linear programming problem as a standard maximization problem if it is not already in that form, and (b) write the initial simplex tableau.

1. Maximize $P = 2x + 4y$ subject to the constraints

$$x + 4y \le 12$$
$$x + 3y \le 10$$
$$x \ge 0, y \ge 0$$

2. Maximize $P = 3x + 5y$ subject to the constraints

$$x + 3y \le 12$$
$$-2x - 3y \ge -18$$
$$x \ge 0, y \ge 0$$

3. Maximize $P = 2x + 3y$ subject to the constraints

$$x + y \le 10$$
$$-x - 2y \ge -12$$
$$2x + y \le 12$$
$$x \ge 0, y \ge 0$$

4. Maximize $P = 2x + 5y$ subject to the constraints

$$3x + 8y \le 1$$
$$4x - 5y \le 4$$
$$2x + 7y \le 6$$
$$x \ge 0, y \ge 0$$

5. Maximize $P = x + 3y + 4z$ subject to the constraints

$$x + 2y + z \le 40$$
$$-x - y - z \ge -30$$
$$x \ge 0, y \ge 0, z \ge 0$$

6. Maximize $P = 4x + 5y + 6z$ subject to the constraints

$$2x + 3y + z \le 900$$
$$3x + y + z \le 350$$
$$4x + 2y + z \le 400$$
$$x \ge 0, y \ge 0, z \ge 0$$

In Exercises 7–16, determine whether the given simplex tableau is in final form. If so, find the solution to the associated regular linear programming problem. If not, find the pivot element to be used in the next iteration of the simplex method.

7.

x	y	u	v	P	Constant
0	1	$\frac{5}{7}$	$-\frac{1}{7}$	0	$\frac{20}{7}$
1	0	$-\frac{3}{7}$	$\frac{2}{7}$	0	$\frac{30}{7}$
0	0	$\frac{13}{7}$	$\frac{3}{7}$	1	$\frac{220}{7}$

8.

x	y	u	v	P	Constant
1	1	1	0	0	6
1	0	-1	1	0	2
3	0	5	0	1	30

9.

x	y	u	v	P	Constant
0	$\frac{1}{2}$	1	$-\frac{1}{2}$	0	2
1	$\frac{1}{2}$	0	$\frac{1}{2}$	0	4
0	$-\frac{1}{2}$	0	$\frac{3}{2}$	1	12

10.

x	y	z	u	v	P	Constant
3	0	5	1	1	0	28
2	1	3	0	1	0	16
2	0	8	0	3	1	48

11.

x	y	z	u	v	w	P	Constant
1	$-\frac{1}{3}$	0	$\frac{1}{3}$	0	$-\frac{2}{3}$	0	$\frac{1}{3}$
0	2	0	0	1	1	0	6
0	$\frac{2}{3}$	1	$\frac{1}{3}$	0	$\frac{1}{3}$	0	$\frac{13}{3}$
0	4	0	1	0	2	1	17

12.

x	y	z	u	v	w	P	Constant
$\frac{1}{2}$	0	$\frac{1}{4}$	1	$-\frac{1}{4}$	0	0	$\frac{19}{2}$
$\frac{1}{2}$	1	$\frac{3}{4}$	0	$\frac{1}{4}$	0	0	$\frac{21}{2}$
2	0	3	0	0	1	0	30
-1	0	$-\frac{1}{2}$	6	$\frac{3}{2}$	0	1	63

13.

x	y	z	s	t	u	v	P	Constant
$\frac{5}{2}$	3	0	1	0	0	-4	0	46
1	0	0	0	1	0	0	0	9
0	1	0	0	0	1	0	0	12
0	0	1	0	0	0	1	0	6
-180	-200	0	0	0	0	300	1	1800

14.

x	y	z	s	t	u	v	P	Constant
1	0	0	$\frac{2}{5}$	0	$-\frac{6}{5}$	$-\frac{8}{5}$	0	4
0	0	0	$-\frac{2}{5}$	1	$\frac{6}{5}$	$\frac{8}{5}$	0	5
0	1	0	0	0	1	0	0	12
0	0	1	0	0	0	1	0	6
0	0	0	72	0	-16	12	1	4920

15.

x	y	z	u	v	P	Constant
1	0	$\frac{3}{5}$	0	$\frac{1}{5}$	0	30
0	1	$-\frac{19}{5}$	1	$-\frac{3}{5}$	0	10
0	0	$\frac{26}{5}$	0	0	1	60

16.

x	y	z	u	v	w	P	Constant
0	$\frac{1}{2}$	0	1	$-\frac{1}{2}$	0	0	2
1	$\frac{1}{2}$	1	0	$\frac{1}{2}$	0	0	13
2	$\frac{1}{2}$	0	0	$-\frac{3}{2}$	1	0	4
-1	3	0	0	1	0	1	26

In Exercises 17–31, solve each linear programming problem by the simplex method.

17. Maximize $P = 3x + 4y$
subject to $x + y \le 4$
$2x + y \le 5$
$x \ge 0, y \ge 0$

18. Maximize $P = 5x + 3y$
subject to $x + y \le 80$
$3x \le 90$
$x \ge 0, y \ge 0$

19. Maximize $P = 10x + 12y$
subject to $x + 2y \le 12$
$3x + 2y \le 24$
$x \ge 0, y \ge 0$

20. Maximize $P = 5x + 4y$
subject to $3x + 5y \le 78$
$4x + y \le 36$
$x \ge 0, y \ge 0$

21. Maximize $P = 4x + 6y$
subject to $3x + y \le 24$
$2x + y \le 18$
$x + 3y \le 24$
$x \ge 0, y \ge 0$

22. Maximize $P = 15x + 12y$
subject to $x + y \le 12$
$3x + y \le 30$
$10x + 7y \le 70$
$x \ge 0, y \ge 0$

23. Maximize $P = 3x + 4y + z$
subject to $3x + 10y + 5z \le 120$
$5x + 2y + 8z \le 6$
$8x + 10y + 3z \le 105$
$x \ge 0, y \ge 0, z \ge 0$

24. Maximize $P = 3x + 3y + 4z$
subject to $x + y + 3z \le 15$
$4x + 4y + 3z \le 65$
$x \ge 0, y \ge 0, z \ge 0$

25. Maximize $P = 3x + 4y + 5z$
subject to $x + y + z \le 8$
$3x + 2y + 4z \le 24$
$x \ge 0, y \ge 0, z \ge 0$

26. Maximize $P = x + 4y - 2z$
subject to $3x + y - z \le 80$
$2x + y - z \le 40$
$-x + y + z \le 80$
$x \ge 0, y \ge 0, z \ge 0$

27. Maximize $P = 4x + 6y + 5z$
subject to $x + y + z \le 20$
$2x + 4y + 3z \le 42$
$2x + 3z \le 30$
$x \ge 0, y \ge 0, z \ge 0$

28. Maximize $P = x + 2y - z$
subject to $2x + y + z \le 14$
$4x + 2y + 3z \le 28$
$2x + 5y + 5z \le 30$
$x \ge 0, y \ge 0, z \ge 0$

29. Maximize $P = 12x + 10y + 5z$
subject to $2x + y + z \le 10$
$3x + 5y + z \le 45$
$2x + 5y + z \le 40$
$x \ge 0, y \ge 0, z \ge 0$

30. Maximize $P = 2x + 6y + 6z$
subject to $2x + y + 3z \le 10$
$4x + y + 2z \le 56$
$6x + 4y + 3z \le 126$
$2x + y + z \le 32$
$x \ge 0, y \ge 0, z \ge 0$

31. Maximize $P = 24x + 16y + 23z$
subject to $2x + y + 2z \le 7$
$2x + 3y + z \le 8$
$x + 2y + 3z \le 7$
$x \ge 0, y \ge 0, z \ge 0$

32. Rework Example 3 using the y-column as the pivot column in the first iteration of the simplex method.

33. Show that the linear programming problem

Maximize $P = 2x + 2y - 4z$

subject to $3x + 3y - 2z \le 100$
$5x + 5y + 3z \le 150$
$x \ge 0, y \ge 0, z \ge 0$

has infinitely many solutions and give two of them.

34. **PRODUCTION SCHEDULING** A company manufactures two products, A and B, on two machines, I and II. It has been determined that the company will realize a profit of \$3/unit on Product A and a profit of \$4/unit on Product B. To manufacture 1 unit of Product A requires 6 min on Machine I and 5 min on Machine II. To manufacture 1 unit of Product B requires 9 min on Machine I and 4 min on Machine II. There are 5 hr of machine time available on Machine I and 3 hr of machine time available on Machine II in each work shift. How many units of each product should be produced in each shift to maximize the company's profit? What is the largest profit the company can realize? Is there any time left unused on the machines?

35. **PRODUCTION SCHEDULING** National Business Machines Corporation manufactures two models of portable printers: A and B. Each model A costs $100 to make, and each model B costs $150. The profits are $30 for each model A and $40 for each model B portable printer. If the total number of portable printers demanded each month does not exceed 2500 and the company has earmarked no more than $600,000/month for manufacturing costs, find how many units of each model National should make each month to maximize its monthly profit. What is the largest monthly profit the company can make?

36. **PRODUCTION SCHEDULING** Kane Manufacturing has a division that produces two models of hibachis, model A and model B. To produce each model A hibachi requires 3 lb of cast iron and 6 min of labor. To produce each model B hibachi requires 4 lb of cast iron and 3 min of labor. The profit for each model A hibachi is $2, and the profit for each model B hibachi is $1.50. If 1000 lb of cast iron and 20 labor-hours are available for the production of hibachis each day, how many hibachis of each model should the division produce in order to maximize Kane's profit? What is the largest profit the company can realize? Is there any raw material left over?

37. **ASSET ALLOCATION** Justin has decided to invest at most $60,000 in medium-risk and high-risk stocks. He has further decided that the medium-risk stocks should make up at least 40% of the total investment, while the high-risk stocks should make up at least 20% of the total investment. He expects that the medium-risk stocks will appreciate by 12% and the high-risk stocks will appreciate by 20% within a year. How much money should Justin invest in each type of stock to maximize the value of his investment?
Hint: Write the linear programming problem in the form of a standard maximization problem.

38. **ASSET ALLOCATION** A financier plans to invest up to $500,000 in two projects. Project A yields a return of 10% on the investment, whereas Project B yields a return of 15% on the investment. Because the investment in Project B is riskier than the investment in Project A, the financier has decided that the investment in Project B should not exceed 40% of the total investment. How much should she invest in each project to maximize the return on her investment? What is the maximum return?

39. **PRODUCTION SCHEDULING** A division of the Winston Furniture Company manufactures dining tables and chairs. Each table requires 40 board feet of wood and 3 labor-hours. Each chair requires 16 board feet of wood and 4 labor-hours. The profit for each table is $45, and the profit for each chair is $20. In a certain week, the company has 3200 board feet of wood available and 520 labor-hours available. How many tables and chairs should Winston manufacture to maximize its profit? What is the maximum profit?

40. **CROP PLANNING** A farmer has 150 acres of land suitable for cultivating Crops A and B. The cost of cultivating Crop A is $40/acre, whereas the cost of cultivating Crop B is $60/acre. The farmer has a maximum of $7400 available for land cultivation. Each acre of Crop A requires 20 labor-hours, and each acre of Crop B requires 25 labor-hours. The farmer has a maximum of 3300 labor-hours available. If he expects to make a profit of $150/acre on Crop A and $200/acre on Crop B, how many acres of each crop should he plant to maximize his profit? What is the largest profit the farmer can realize? Are there any resources left over?

41. **PRODUCTION SCHEDULING** A company manufactures Products A, B, and C. Each product is processed in three departments: I, II, and III. The total available labor-hours per week for Departments I, II, and III are 900, 1080, and 840, respectively. The time requirements (in hours per unit) and profit per unit for each product are as follows:

	Product A	Product B	Product C
Dept. I	2	1	2
Dept. II	3	1	2
Dept. III	2	2	1
Profit	$18	$12	$15

How many units of each product should the company produce to maximize its profit? What is the largest profit the company can realize? Are there any resources left over?

42. **ASSET ALLOCATION** Ashley has earmarked at most $250,000 for investment in three mutual funds: a money market fund, an international equity fund, and a growth-and-income fund. The money market fund has a rate of return of 6%/year, the international equity fund has a rate of return of 10%/year, and the growth-and-income fund has a rate of return of 15%/year. Ashley has stipulated that no more than 25% of her total portfolio should be in the growth-and-income fund and that no more than 50% of her total portfolio should be in the international equity fund. To maximize the return on her investment, how much should Ashley invest in each type of fund? What is the maximum return?

43. PRODUCTION SCHEDULING Ace Novelty manufactures Giant Pandas and Saint Bernards. Each Panda requires 1.5 yd^2 of plush, 30 ft^3 of stuffing, and 5 pieces of trim; each Saint Bernard requires 2 yd^2 of plush, 35 ft^3 of stuffing, and 8 pieces of trim. The profit for each Panda is $10, and the profit for each Saint Bernard is $15. If 3600 yd^2 of plush, 66,000 ft^3 of stuffing, and 13,600 pieces of trim are available, how many of each of the stuffed animals should the company manufacture to maximize its profit? What is the maximum profit?

44. OPTIMIZING ADVERTISING EXPOSURE As part of a campaign to promote its annual clearance sale, Excelsior Company decided to buy television advertising time on station KAOS. Excelsior's television advertising budget is $102,000. Morning time costs $3000/min, afternoon time costs $1000/min, and evening (prime) time costs $12,000/min. Because of previous commitments, KAOS cannot offer Excelsior more than 6 min of prime time or more than a total of 25 min of advertising time over the 2 weeks in which the commercials are to be run. KAOS estimates that morning commercials are seen by 200,000 people, afternoon commercials are seen by 100,000 people, and evening commercials are seen by 600,000 people. How much morning, afternoon, and evening advertising time should Excelsior buy to maximize exposure of its commercials?

45. PRODUCTION SCHEDULING Custom Office Furniture is introducing a new line of executive desks made from a specially selected grade of walnut. Initially, three models—A, B, and C—are to be marketed. Each model A desk requires $1\frac{1}{4}$ hr for fabrication, 1 hr for assembly, and 1 hr for finishing; each model B desk requires $1\frac{1}{2}$ hr for fabrication, 1 hr for assembly, and 1 hr for finishing; each model C desk requires $1\frac{1}{2}$ hr, $\frac{3}{4}$ hr, and $\frac{1}{2}$ hr for fabrication, assembly, and finishing, respectively. The profit on each model A desk is $26, the profit on each model B desk is $28, and the profit on each model C desk is $24. The total time available in the fabrication department, the assembly department, and the finishing department in the first month of production is 310 hr, 205 hr, and 190 hr, respectively. To maximize Custom's profit, how many desks of each model should be made in the month? What is the largest profit the company can realize? Are there any resources left over?

46. OPTIMIZING PROFIT FOR PREFABRICATED HOUSING Boise Lumber has decided to enter the lucrative prefabricated housing business. Initially, it plans to offer three models: standard, deluxe, and luxury. Each house is prefabricated and partially assembled in the factory, and the final assembly is completed on site. The dollar amount of building material required, the amount of labor required in the factory for prefabrication and partial assembly, the amount of on-site labor required, and the profit per unit are as follows:

	Standard Model	Deluxe Model	Luxury Model
Material	$6,000	$8,000	$10,000
Factory Labor (hr)	240	220	200
On-site Labor (hr)	180	210	300
Profit	$3,400	$4,000	$5,000

For the first year's production, a sum of $8,200,000 is budgeted for the building material; the number of labor-hours available for work in the factory (for prefabrication and partial assembly) is not to exceed 218,000 hr; and the amount of labor for on-site work is to be less than or equal to 237,000 labor-hours. Determine how many houses of each type Boise should produce to maximize its profit from this new venture. (Market research has confirmed that there should be no problems with sales.)

47. ASSET ALLOCATION Sharon has a total of $200,000 to invest in three types of mutual funds: growth, balanced, and income funds. Growth funds have a rate of return of 12%/year, balanced funds have a rate of return of 10%/year, and income funds have a return of 6%/year. The growth, balanced, and income mutual funds are assigned risk factors of 0.1, 0.06, and 0.02, respectively. Sharon has decided that at least 50% of her total portfolio is to be in income funds and at least 25% in balanced funds. She has also decided that the average risk factor for her investment should not exceed 0.05. How much should Sharon invest in each type of fund to realize a maximum return on her investment? What is the maximum return?
Hint: The constraint for the average risk factor for the investment is given by $0.1x + 0.06y + 0.02z \le 0.05(x + y + z)$.

48. OPTIMIZING PROFIT FOR BLENDED JUICE DRINKS CalJuice Company has decided to introduce three fruit juices made from blending two or more concentrates. These juices will be packaged in 2-qt (64-oz) cartons. One carton of pineapple–orange juice requires 8 oz each of pineapple and orange juice concentrates. One carton of orange–banana juice requires 12 oz of orange juice concentrate and 4 oz of banana pulp concentrate. Finally, one carton of pineapple–orange–banana juice requires 4 oz of pineapple juice concentrate, 8 oz of orange juice concentrate, and 4 oz of banana pulp. The company has decided to allot 16,000 oz of pineapple juice concentrate, 24,000 oz of orange juice concentrate, and 5000 oz of banana pulp concentrate for the initial production run. The company has also stipulated that the production of pineapple–orange–banana juice should not exceed 800 cartons. Its profit on one carton of pineapple–orange juice is $1.00, its profit on one carton of orange–banana juice is $0.80, and its profit on one carton of pineapple–orange–banana juice is $0.90. To realize a maximum profit, how many cartons of each blend should the company produce? What is the largest profit it can realize? Are there any concentrates left over?

49. **OPTIMIZING PROFIT FOR COLD FORMULAS** Beyer Pharmaceutical produces three kinds of cold formulas: I, II, and III. It takes 2.5 hr to produce 1000 bottles of Formula I, 3 hr to produce 1000 bottles of Formula II, and 4 hr to produce 1000 bottles of Formula III. The profits for each 1000 bottles of Formula I, Formula II, and Formula III are $180, $200, and $300, respectively. Suppose that for a certain production run, there are enough ingredients on hand to make at most 9000 bottles of Formula I, 12,000 bottles of Formula II, and 6000 bottles of Formula III. Furthermore, suppose the time for the production run is limited to a maximum of 70 hr. How many bottles of each formula should be produced in this production run so that the profit is maximized? What is the maximum profit realizable by the company? Are there any resources left over?

50. **ASSET ALLOCATION** A financier plans to invest up to $2 million in three projects. She estimates that Project A will yield a return of 10% on her investment, Project B will yield a return of 15% on her investment, and Project C will yield a return of 20% on her investment. Because of the risks associated with the investments, she decided to put not more than 20% of her total investment in Project C. She also decided that her investments in Projects B and C should not exceed 60% of her total investment. Finally, she decided that her investment in Project A should be at least 60% of her investments in Projects B and C. How much should the financier invest in each project if she wishes to maximize the total returns on her investments? What is the maximum amount she can expect to make from her investments?

51. Consider the linear programming problem

$$\text{Maximize} \quad P = 3x + 2y$$
$$\text{subject to} \quad x - y \le 3$$
$$x \le 2$$
$$x \ge 0, y \ge 0$$

a. Sketch the feasible set for the linear programming problem.
b. Show that the linear programming problem is unbounded.
c. Solve the linear programming problem using the simplex method. How does the method break down?
d. Explain why the result in part (c) implies that no solution exists for the linear programming problem.

In Exercises 52–55, determine whether the statement is true or false. If it is true, explain why it is true. If it is false, give an example to show why it is false.

52. If at least one of the coefficients $a_1, a_2, \ldots, a_n$ of the objective function $P = a_1x_1 + a_2x_2 + \cdots + a_nx_n$ is positive, then $(0, 0, \ldots, 0)$ cannot be the optimal solution of the standard (maximization) linear programming problem.

53. Choosing the pivot row by requiring that the ratio associated with that row be the smallest ensures that the iteration will not take us from a feasible point to a nonfeasible point.

54. Choosing the pivot column by requiring that it be the column associated with the most negative entry to the left of the vertical line in the last row of the simplex tableau ensures that the iteration will result in the greatest increase or, at worse, no decrease in the objective function.

55. If, at any stage of an iteration of the simplex method, it is not possible to compute the ratios (division by zero) or the ratios are negative, then we can conclude that the standard linear programming problem may have no solution.

6.4 Solutions to Self-Check Exercises

1. Introducing the slack variables u, v, and w, we obtain the system of linear equations

$$\begin{aligned}
2x + 3y + z + u &= 10 \\
x + y + 2z + v &= 8 \\
2y + 3z + w &= 6 \\
-2x - 3y - 6z + P &= 0
\end{aligned}$$

The initial simplex tableau and the successive tableaus resulting from the use of the simplex procedure follow:

	x	y	z	u	v	w	P	Constant	Ratio
	2	3	1	1	0	0	0	10	$\frac{10}{1} = 10$
	1	1	2	0	1	0	0	8	$\frac{8}{2} = 4$
Pivot row →	0	2	③	0	0	1	0	6	$\frac{6}{3} = 2$
	−2	−3	−6	0	0	0	1	0	

Pivot column (↑), $\frac{1}{3}R_3 \rightarrow$

x	y	z	u	v	w	P	Constant
2	3	1	1	0	0	0	10
1	1	2	0	1	0	0	8
0	$\frac{2}{3}$	①	0	0	$\frac{1}{3}$	0	2
-2	-3	-6	0	0	0	1	0

$\xrightarrow[\begin{array}{c}R_1 - R_3\\ R_2 - 2R_3\\ R_4 + 6R_3\end{array}]{}$

	x	y	z	u	v	w	P	Constant	Ratio
	2	$\frac{7}{3}$	0	1	0	$-\frac{1}{3}$	0	8	$\frac{8}{2} = 4$
Pivot row →	①	$-\frac{1}{3}$	0	0	1	$-\frac{2}{3}$	0	4	$\frac{4}{1} = 4$
	0	$\frac{2}{3}$	1	0	0	$\frac{1}{3}$	0	2	—
	-2	1	0	0	0	2	1	12	

Pivot column ↑ (at x)

$\xrightarrow[\begin{array}{c}R_1 - 2R_2\\ R_4 + 2R_2\end{array}]{}$

x	y	z	u	v	w	P	Constant
0	3	0	1	-2	1	0	0
1	$-\frac{1}{3}$	0	0	1	$-\frac{2}{3}$	0	4
0	$\frac{2}{3}$	1	0	0	$\frac{1}{3}$	0	2
0	$\frac{1}{3}$	0	0	2	$\frac{2}{3}$	1	20

All entries in the last row are nonnegative, and the tableau is final. We conclude that $x = 4$, $y = 0$, $z = 2$, and $P = 20$.

2. Let x denote the number of model A grates, and let y denote the number of model B grates to be made this week. Then the profit function to be maximized is given by

$$P = 6x + 8y$$

The limitations on the availability of material and labor may be expressed by the linear inequalities

$$20x + 30y \le 7200 \quad \text{or} \quad 2x + 3y \le 720$$
$$20x + 15y \le 6000 \quad \text{or} \quad 4x + 3y \le 1200$$

Finally, the condition that no more than 150 units of model A grates be made this week may be expressed by the linear inequality

$$x \le 150$$

Thus, we are led to the following linear programming problem:

$$\begin{aligned} \text{Maximize} \quad & P = 6x + 8y \\ \text{subject to} \quad & 2x + 3y \le 720 \\ & 4x + 3y \le 1200 \\ & x \le 150 \\ & x \ge 0, \; y \ge 0 \end{aligned}$$

To solve this problem, we introduce slack variables u, v, and w and use the simplex method, obtaining the following sequence of simplex tableaus:

	x	y	u	v	w	P	Constant	Ratio
Pivot row →	2	③	1	0	0	0	720	$\frac{720}{3} = 240$
	4	3	0	1	0	0	1200	$\frac{1200}{3} = 400$
	1	0	0	0	1	0	150	—
	-6	-8	0	0	0	1	0	

Pivot column ↑ (at y)

$\xrightarrow[]{\frac{1}{3}R_1}$

x	y	u	v	w	P	Constant
$\frac{2}{3}$	①	$\frac{1}{3}$	0	0	0	240
4	3	0	1	0	0	1200
1	0	0	0	1	0	150
-6	-8	0	0	0	1	0

	x	y	u	v	w	P	Constant	Ratio
	$\frac{2}{3}$	1	$\frac{1}{3}$	0	0	0	240	$\frac{240}{2/3} = 360$
$R_2 - 3R_1$	2	0	-1	1	0	0	480	$\frac{480}{2} = 240$
$R_4 + 8R_1$ Pivot row →	①	0	0	0	1	0	150	$\frac{150}{1} = 150$
	$-\frac{2}{3}$	0	$\frac{8}{3}$	0	0	1	1920	

Pivot column ↑ (at x)

	x	y	u	v	w	P	Constant
$R_1 - \frac{2}{3}R_3$	0	1	$\frac{1}{3}$	0	$-\frac{2}{3}$	0	140
$R_2 - 2R_3$	0	0	-1	1	-2	0	180
$R_4 + \frac{2}{3}R_3$	1	0	0	0	1	0	150
	0	0	$\frac{8}{3}$	0	$\frac{2}{3}$	1	2020

The last tableau is final, and we see that $x = 150$, $y = 140$, and $P = 2020$. Therefore, LaCrosse should make 150 model A grates and 140 model B grates this week. The profit will be $2020.

USING TECHNOLOGY   The Simplex Method: Solving Maximization Problems

Graphing Utility

A graphing utility can be used to solve a linear programming problem by the simplex method, as illustrated in Example 1.

EXAMPLE 1 (Refer to Example 5, Section 6.4.) The problem reduces to the following linear programming problem:

$$\text{Maximize} \quad P = 6x + 5y + 4z$$
$$\text{subject to} \quad 2x + y + z \le 180$$
$$x + 3y + 2z \le 300$$
$$2x + y + 2z \le 240$$
$$x \ge 0, y \ge 0, z \ge 0$$

With u, v, and w as slack variables, we are led to the following sequence of simplex tableaus, where the first tableau is entered as the matrix A:

	x	y	z	u	v	w	P	Constant	Ratio
Pivot row →	②	1	1	1	0	0	0	180	$\frac{180}{2} = 90$
	1	3	2	0	1	0	0	300	$\frac{300}{1} = 300$
	2	1	2	0	0	1	0	240	$\frac{240}{2} = 120$
	−6	−5	−4	0	0	0	1	0	

Pivot column (under x)

$\xrightarrow{*\textbf{row}(\frac{1}{2}, A, 1) \blacktriangleright B}$

x	y	z	u	v	w	P	Constant
①	0.5	0.5	0.5	0	0	0	90
1	3	2	0	1	0	0	300
2	1	2	0	0	1	0	240
−6	−5	−4	0	0	0	1	0

$\xrightarrow{\begin{array}{l} *\textbf{row}+(-1, B, 1, 2) \blacktriangleright C \\ *\textbf{row}+(-2, C, 1, 3) \blacktriangleright B \\ *\textbf{row}+(6, B, 1, 4) \blacktriangleright C \end{array}}$

	x	y	z	u	v	w	P	Constant	Ratio
	1	0.5	0.5	0.5	0	0	0	90	$\frac{90}{0.5} = 180$
Pivot row →	0	②.5	1.5	−0.5	1	0	0	210	$\frac{210}{2.5} = 84$
	0	0	1	−1	0	1	0	60	
	0	−2	−1	3	0	0	1	540	

Pivot column (under y)

$\xrightarrow{*\textbf{row}(\frac{1}{2.5}, C, 2) \blacktriangleright B}$

x	y	z	u	v	w	P	Constant
1	0.5	0.5	0.5	0	0	0	90
0	①	0.6	−0.2	0.4	0	0	84
0	0	1	−1	0	1	0	60
0	−2	−1	3	0	0	1	540

$\xrightarrow{\begin{array}{l} *\textbf{row}+(-0.5, B, 2, 1) \blacktriangleright C \\ *\textbf{row}+(2, C, 2, 4) \blacktriangleright B \end{array}}$

x	y	z	u	v	w	P	Constant
1	0	0.2	0.6	−0.2	0	0	48
0	1	0.6	−0.2	0.4	0	0	84
0	0	1	−1	0	1	0	60
0	0	0.2	2.6	0.8	0	1	708

The final simplex tableau is the same as the one obtained earlier. We see that $x = 48$, $y = 84$, $z = 0$, and $P = 708$. Hence Ace Novelty should produce 48 Type A

souvenirs, 84 Type *B* souvenirs, and no Type *C* souvenirs, resulting in a profit of
$708 per day.

Excel

Solver is an Excel add-in that is used to solve linear programming problems. When
you start the Excel program, check the *Tools* menu for the *Solver* command. If it is not
there, you will need to install it. (Check your manual for installation instructions.)

EXAMPLE 2 Solve the following linear programming problem:

$$\text{Maximize} \quad P = 6x + 5y + 4z$$
$$\text{subject to} \quad 2x + y + z \le 180$$
$$x + 3y + 2z \le 300$$
$$2x + y + 2z \le 240$$
$$x \ge 0, y \ge 0, z \ge 0$$

Solution

1. Enter the data for the linear programming problem onto a spreadsheet. Enter the
   labels shown in column A and the variables with which we are working under
   Decision Variables in cells B4:B6, as shown in Figure T1. This optional step will
   help us to organize our work.

	A	B	C	D	E	F	G	H	I
1	Maximization Problem								
2							Formulas for indicated cells		
3	Decision Variables						C8: = 6*C4 + 5*C5 + 4*C6		
4		x	0				C11: = 2*C4 + C5 + C6		
5		y	0				C12: = C4 + 3*C5 + 2*C6		
6		z	0				C13: = 2*C4 + C5 + 2*C6		
7									
8	Objective Function		0						
9									
10	Constraints								
11			0	<=	180				
12			0	<=	300				
13			0	<=	240				

FIGURE **T1**
Setting up the spreadsheet for Solver

For the moment, the cells that will contain the values of the variables
(C4:C6) are left blank. In C8, type the formula for the objective function:
=6*C4+5*C5+4*C6. In C11, type the formula for the left-hand side of the
first constraint: =2*C4+C5+C6. In C12, type the formula for the left-hand side
of the second constraint: =C4+3*C5+2*C6. In C13, type the formula for the
left-hand side of the third constraint: =2*C4+C5+2*C6. Zeros will then appear
in cell C8 and cells C11:C13. In cells D11:D13, type <= to indicate that each
constraint is of the form ≤. Finally, in cells E11:E13, type the right-hand value of
each constraint—in this case, 180, 300, and 240, respectively. Note that we
need not enter the nonnegativity constraints $x \ge 0$, $y \ge 0$, and $z \ge 0$. The

resulting spreadsheet is shown in Figure T1, where the formulas that were entered for the objective function and the constraints are shown in the comment box.

2. Use Solver to solve the problem. Click the **Data** tab, and then click **Solver** in the Analysis group. The Solver Parameters dialog box will appear.

   a. The pointer will be in the Set Objective: box (refer to Figure T2). Highlight the cell on your spreadsheet containing the formula for the objective function—in this case, C8.

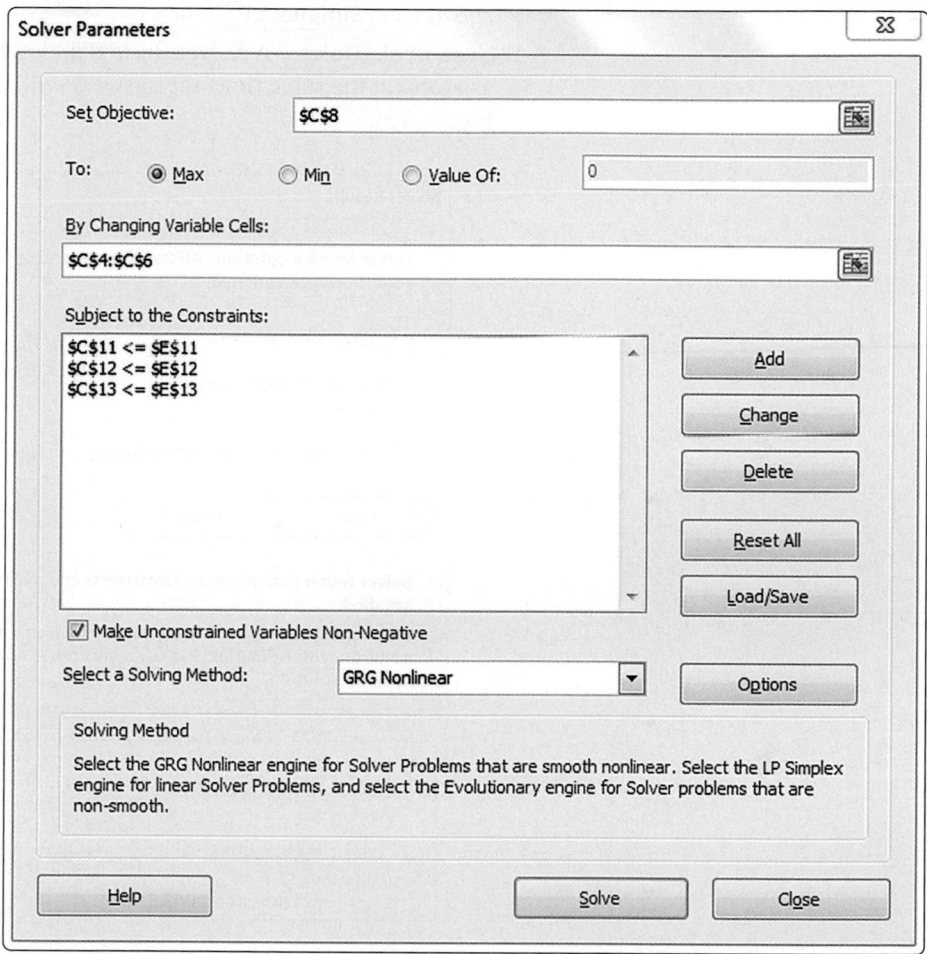

FIGURE **T2**
The completed Solver Parameters dialog box

Then, next to To:, select **Max**. Select the **By Changing Variable Cells:** box, and highlight the cells in your spreadsheet that will contain the values of the variables—in this case, C4:C6. Select the **Subject to the Constraints:** box, and then click **Add**. The Add Constraint dialog box will appear (Figure T3).

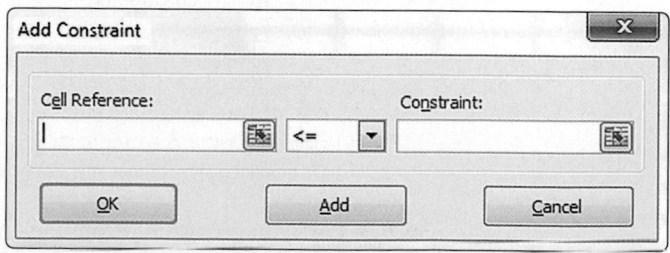

FIGURE **T3**
The Add Constraint dialog box

b. The pointer will appear in the Cell Reference: box. Highlight the cells on your spreadsheet that contain the formula for the left-hand side of the first constraint—in this case, C11. Next, select the symbol for the appropriate constraint—in this case, $\boxed{<=}$. Select the $\boxed{\textbf{Constraint:}}$ box, and highlight the value of the right-hand side of the first constraint on your spreadsheet—in this case, 180. Click $\boxed{\textbf{Add}}$, and then follow the same procedure to enter the second and third constraints. Click $\boxed{\textbf{OK}}$. The resulting Solver Parameters dialog box shown in Figure T2 will appear.

c. In the Solver Parameters dialog box, go to the Select a Solving Method: box, and select $\boxed{\textbf{Simplex LP}}$.

d. Next, click $\boxed{\textbf{Solve}}$. A Solver Results dialog box will then appear (see Figure T4), and at the same time, the answers will appear on your spreadsheet (see Figure T5).

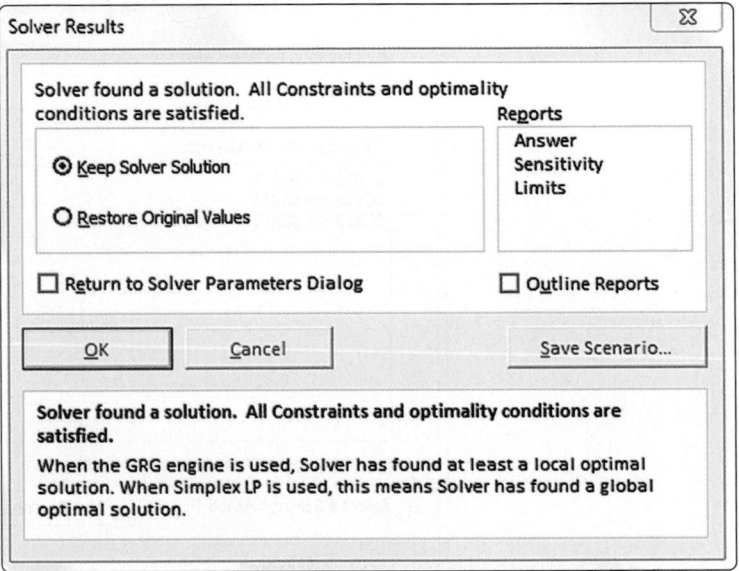

FIGURE **T4**
The Solver Results dialog box

	A	B	C	D	E
1	Maximization Problem				
2					
3	Decision Variables				
4		x	48		
5		y	84		
6		z	0		
7					
8	Objective Function		708		
9					
10	Constraints				
11			180	<=	180
12			300	<=	300
13			240	<=	240

FIGURE **T5**
Completed spreadsheet after using Solver

3. Read off your answers. From the spreadsheet, we see that the objective function attains a maximum value of 708 (cell C8) when $x = 48$, $y = 84$, and $z = 0$ (cells C4:C6).

Solve the linear programming problems.

1. Maximize $P = 2x + 3y + 4z + 2w$
subject to
$$x + 2y + 3z + 2w \leq 6$$
$$2x + 4y + z - w \leq 4$$
$$3x + 2y - 2z + 3w \leq 12$$
$$x \geq 0, y \geq 0, z \geq 0, w \geq 0$$

2. Maximize $P = 3x + 2y + 2z + w$
subject to
$$2x + y - z + 2w \leq 8$$
$$2x - y + 2z + 3w \leq 20$$
$$x + y + z + 2w \leq 8$$
$$4x - 2y + z + 3w \leq 24$$
$$x \geq 0, y \geq 0, z \geq 0, w \geq 0$$

3. Maximize $P = x + y + 2z + 3w$
subject to
$$3x + 6y + 4z + 2w \leq 12$$
$$x + 4y + 8z + 4w \leq 16$$
$$2x + y + 4z + w \leq 10$$
$$x \geq 0, y \geq 0, z \geq 0, w \geq 0$$

4. Maximize $P = 2x + 4y + 3z + 5w$
subject to
$$x - 2y + 3z + 4w \leq 8$$
$$2x + 2y + 4z + 6w \leq 12$$
$$3x + 2y + z + 5w \leq 10$$
$$2x + 8y - 2z + 6w \leq 24$$
$$x \geq 0, y \geq 0, z \geq 0, w \geq 0$$

6.5 The Simplex Method: Standard Minimization Problems

Minimization with $\leq$ Constraints

In Section 6.4, we developed a procedure, called the simplex method, for solving standard linear programming problems. Recall that a standard maximization problem satisfies three conditions:

1. The objective function is to be maximized.
2. All the variables involved are nonnegative.
3. Each linear constraint may be written so that the expression involving the variables is less than or equal to a nonnegative constant.

In this section, we see how the simplex method may be used to solve certain classes of problems that are not necessarily standard maximization problems. In particular, we see how a modified procedure may be used to solve problems involving the minimization of objective functions.

We begin by considering the class of linear programming problems that calls for the minimization of objective functions but otherwise satisfies conditions 2 and 3 for standard maximization problems. The method that is used to solve these problems is illustrated in the following example.

EXAMPLE 1

$$\text{Minimize} \quad C = -2x - 3y$$
$$\text{subject to} \quad 5x + 4y \leq 32$$
$$x + 2y \leq 10$$
$$x \geq 0, y \geq 0$$

Solution This problem involves the minimization of the objective function and is accordingly *not* a standard maximization problem. Note, however, that all other conditions for a standard maximization problem hold true. To solve a problem of this type, we observe that minimizing the objective function C is equivalent to maximizing the objective function $P = -C$. Thus, the solution to this problem may be found by solving the following associated standard maximization problem: Maximize $P = 2x + 3y$ subject to the given constraints. Using the simplex method with u and v as slack variables, we obtain the following sequence of simplex tableaus:

	x	y	u	v	P	Constant	Ratio
	5	4	1	0	0	32	$\frac{32}{4}=8$
Pivot row →	1	②	0	1	0	10	$\frac{10}{2}=5$
	−2	−3	0	0	1	0	

Pivot column (↑ under y)

$\xrightarrow{\frac{1}{2}R_2}$

	x	y	u	v	P	Constant
	5	4	1	0	0	32
	$\frac{1}{2}$	①	0	$\frac{1}{2}$	0	5
	−2	−3	0	0	1	0

Pivot row

$\xrightarrow[R_3 + 3R_2]{R_1 - 4R_2}$

	x	y	u	v	P	Constant	Ratio
	③	0	1	−2	0	12	$\frac{12}{3}=4$
	$\frac{1}{2}$	1	0	$\frac{1}{2}$	0	5	$\frac{5}{1/2}=10$
	$-\frac{1}{2}$	0	0	$\frac{3}{2}$	1	15	

Pivot column (↑ under x)

$\xrightarrow{\frac{1}{3}R_1}$

x	y	u	v	P	Constant
①	0	$\frac{1}{3}$	$-\frac{2}{3}$	0	4
$\frac{1}{2}$	1	0	$\frac{1}{2}$	0	5
$-\frac{1}{2}$	0	0	$\frac{3}{2}$	1	15

$\xrightarrow[R_3 + \frac{1}{2}R_1]{R_2 - \frac{1}{2}R_1}$

	x	y	u	v	P	Constant
	1	0	$\frac{1}{3}$	$-\frac{2}{3}$	0	4
	0	1	$-\frac{1}{6}$	$\frac{5}{6}$	0	3
	0	0	$\frac{1}{6}$	$\frac{7}{6}$	1	17

The last tableau is in final form. The solution to the standard maximization problem associated with the given linear programming problem is $x = 4$, $y = 3$, and $P = 17$, so the required solution is given by $x = 4$, $y = 3$, and $C = -17$. You may verify that the solution is correct by using the method of corners. ∎

The Dual Problem

Another special class of linear programming problems we encounter in practical applications is characterized by the following conditions:

1. The objective function is to be *minimized*.
2. All the variables involved are nonnegative.
3. All other linear constraints may be written so that the expression involving the variables is *greater than or equal to* a constant.

Such problems are called **standard minimization problems.**

A convenient method for solving this type of problem is based on the following observation. Each maximization linear programming problem is associated with a

minimization problem, and vice versa. For the purpose of identification, the given problem is called the **primal problem**; the problem related to it is called the **dual problem.** The following example illustrates the technique for constructing the dual of a given linear programming problem.

EXAMPLE 2 Write the dual problem associated with the following problem:

Minimize the objective function $C = 4x + 2y$
subject to $5x + y \geq 5$
$5x + 3y \geq 10$
$x \geq 0, y \geq 0$

} Primal problem

Solution   We first write the following tableau for the given primal problem:

x	y	Constant
5	1	5
5	3	10
4	2	

Next, we interchange the columns and rows of the foregoing tableau and head the two columns of the resulting array with the two variables u and v, obtaining the tableau:

u	v	Constant
5	5	4
1	3	2
5	10	

Interpreting the last tableau as if it were part of the initial simplex tableau for a standard maximization problem—with the exception that the signs of the coefficients pertaining to the objective function are not reversed—we construct the required dual problem as follows:

$$\left.\begin{array}{l} \text{Maximize the objective function} \quad P = 5u + 10v \\ \qquad \text{subject to} \quad 5u + 5v \le 4 \\ \qquad\qquad\qquad u + 3v \le 2 \\ \qquad\qquad\qquad u \ge 0, v \ge 0 \end{array}\right\} \begin{array}{l} \text{Dual} \\ \text{problem} \end{array}$$

Since both the primal problem and the dual problem in Example 2 involve two variables, they can be solved graphically. In fact, the feasible sets of the primal problem and the dual problem are shown in Figure 23. From the following tables of values associated with the two problems, we see that the solution to the primal problem is $x = \frac{1}{2}, y = \frac{5}{2}$, and $C = 7$, and the solution for the dual problem is $u = \frac{1}{5}, v = \frac{3}{5}$, and $P = 7$.

Table for the primal problem

Vertex	$C = 4x + 2y$
$A(2,0)$	8
$B\left(\frac{1}{2},\frac{5}{2}\right)$	7
$C(0,5)$	10

Table for the dual problem

Vertex	$P = 5u + 10v$
$A(0,0)$	0
$B\left(\frac{4}{5},0\right)$	4
$C\left(\frac{1}{5},\frac{3}{5}\right)$	7
$D\left(0,\frac{2}{3}\right)$	$\frac{20}{3}$

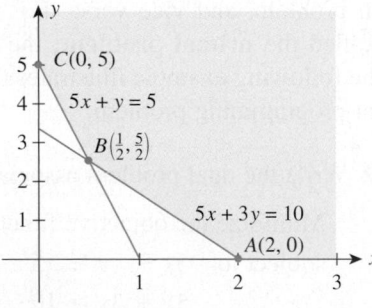

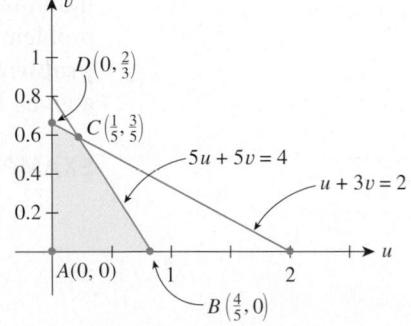

(a) Feasible set for the primal problem　　**(b)** Feasible set for the dual problem

FIGURE **23**

Observe that the *objective functions of both the primal problem and the dual problem attain the same optimal value.*

Next let's see how we can find these solutions using the simplex method. Since the dual problem is a standard maximization problem, we can use the method of Section 6.4. Proceeding, we obtain the following sequence of tableaus. Note that we use the letters x and y (the variables used in the primal problem) to denote the slack variables.

	u	v	x	y	P	Constant	Ratio	
	5	5	1	0	0	4	$\frac{4}{5}$	$\xrightarrow{\frac{1}{3}R_2}$
Pivot row $\rightarrow$	1	③	0	1	0	2	$\frac{2}{3}$	
	-5	-10	0	0	1	0		

$\qquad\qquad\qquad\qquad$ ↑
$\qquad\qquad\qquad\qquad$ Pivot
$\qquad\qquad\qquad\qquad$ column

u	v	x	y	P	Constant
5	5	1	0	0	4
$\frac{1}{3}$	1	0	$\frac{1}{3}$	0	$\frac{2}{3}$
-5	-10	0	0	1	0

$\xrightarrow[R_3 + 10R_2]{R_1 - 5R_2}$

u	v	x	y	P	Constant	Ratio	
$\boxed{\frac{10}{3}}$	0	1	$-\frac{5}{3}$	0	$\frac{2}{3}$	$\frac{1}{5}$	$\xrightarrow{\frac{3}{10}R_1}$
$\frac{1}{3}$	1	0	$\frac{1}{3}$	0	$\frac{2}{3}$	2	
$-\frac{5}{3}$	0	0	$\frac{10}{3}$	1	$\frac{20}{3}$		

u	v	x	y	P	Constant
1	0	$\frac{3}{10}$	$-\frac{1}{2}$	0	$\frac{1}{5}$
$\frac{1}{3}$	1	0	$\frac{1}{3}$	0	$\frac{2}{3}$
$-\frac{5}{3}$	0	0	$\frac{10}{3}$	1	$\frac{20}{3}$

$\xrightarrow[R_3 + \frac{5}{3}R_1]{R_2 - \frac{1}{3}R_1}$

u	v	x	y	P	Constant
1	0	$\frac{3}{10}$	$-\frac{1}{2}$	0	$\frac{1}{5}$
0	1	$-\frac{1}{10}$	$\frac{1}{2}$	0	$\frac{3}{5}$
0	0	$\frac{1}{2}$	$\frac{5}{2}$	1	7

Solution for the primal problem

Interpreting the final tableau in the usual manner, we see that the solution of the dual problem is $u = \frac{1}{5}, v = \frac{3}{5}$, and $P = 7$, as obtained earlier by using the graphical method. Next, observe that the solution of the primal problem $x = \frac{1}{2}$ and $y = \frac{5}{2}$ appears under the respective slack variables in the last row of the final tableau, as indicated. That we can always find the solution to both the optimal and dual problems in this manner is the content of the following theorem. The theorem, attributed to John von Neumann (1903–1957), is stated without proof.

THEOREM 3

The Fundamental Theorem of Duality

A primal problem has a solution if and only if the corresponding dual problem has a solution. Furthermore, if a solution exists, then:

a. The objective functions of both the primal and the dual problem attain the same optimal value.

b. The optimal solution to the primal problem appears under the slack variables in the last row of the final simplex tableau associated with the dual problem.

EXAMPLE 3 Write and solve the dual problem associated with the following problem:

$$
\left.
\begin{aligned}
\text{Minimize the objective function} \quad & C = 6x + 8y \\
\text{subject to} \quad 40x + 10y & \geq 2400 \\
10x + 15y & \geq 2100 \\
5x + 15y & \geq 1500 \\
x \geq 0, \; y & \geq 0
\end{aligned}
\right\}
\begin{aligned}
\text{Primal} \\
\text{problem}
\end{aligned}
$$

Solution   We first write down the following tableau for the given primal problem:

x	y	Constant
40	10	2400
10	15	2100
5	15	1500
6	8	

Next, we interchange the columns and rows of the foregoing tableau and head the three columns of the resulting array with the three variables u, v, and w, obtaining the tableau

	u	v	w	Constant
	40	10	5	6
	10	15	15	8
	2400	2100	1500	

Interpreting the last tableau as if it were part of the initial simplex tableau for a standard maximization problem—with the exception that the signs of the coefficients pertaining to the objective function are not reversed—we construct the required dual problem as follows:

Maximize the objective function $P = 2400u + 2100v + 1500w$

subject to $40u + 10v + 5w \leq 6$

$10u + 15v + 15w \leq 8$

$u \geq 0, v \geq 0, w \geq 0$

Dual problem

Observe that the dual problem associated with the given (primal) problem is a standard maximization problem. The solution may thus be found by using the simplex algorithm. Introducing the slack variables x and y, we obtain the system of linear equations

$$40u + 10v + 5w + x = 6$$
$$10u + 15v + 15w + y = 8$$
$$-2400u - 2100v - 1500w + P = 0$$

Continuing with the simplex algorithm, we obtain the following sequence of simplex tableaus:

	u	v	w	x	y	P	Constant	Ratio
Pivot row →	(40)	10	5	1	0	0	6	$\frac{6}{40} = \frac{3}{20}$
	10	15	15	0	1	0	8	$\frac{8}{10} = \frac{4}{5}$
	−2400	−2100	−1500	0	0	1	0	

Pivot column (under −2400)

	u	v	w	x	y	P	Constant
$\frac{1}{40}R_1$ →	(1)	$\frac{1}{4}$	$\frac{1}{8}$	$\frac{1}{40}$	0	0	$\frac{3}{20}$
	10	15	15	0	1	0	8
	−2400	−2100	−1500	0	0	1	0

	u	v	w	x	y	P	Constant	Ratio
$R_2 - 10R_1$	1	$\frac{1}{4}$	$\frac{1}{8}$	$\frac{1}{40}$	0	0	$\frac{3}{20}$	$\frac{3/20}{1/4} = \frac{3}{5}$
$R_3 + 2400R_1$ →	0	$\left(\frac{25}{2}\right)$	$\frac{55}{4}$	$-\frac{1}{4}$	1	0	$\frac{13}{2}$	$\frac{13/2}{25/2} = \frac{13}{25}$
Pivot row	0	−1500	−1200	60	0	1	360	

Pivot column (under −1500)

	u	v	w	x	y	P	Constant
$\frac{2}{25}R_2$ →	1	$\frac{1}{4}$	$\frac{1}{8}$	$\frac{1}{40}$	0	0	$\frac{3}{20}$
	0	(1)	$\frac{11}{10}$	$-\frac{1}{50}$	$\frac{2}{25}$	0	$\frac{13}{25}$
	0	−1500	−1200	60	0	1	360

		u	v	w	x	y	P	Constant
$R_1 - \frac{1}{4}R_2$		1	0	$-\frac{3}{20}$	$\frac{3}{100}$	$-\frac{1}{50}$	0	$\frac{1}{50}$
$R_3 + 1500R_2$		0	1	$\frac{11}{10}$	$-\frac{1}{50}$	$\frac{2}{25}$	0	$\frac{13}{25}$
		0	0	450	30	120	1	1140

Solution for the
primal problem

The last tableau is final. The fundamental theorem of duality tells us that the solution to the primal problem is $x = 30$ and $y = 120$ with a minimum value for C of 1140. Observe that the solution to the dual (maximization) problem may be read from the simplex tableau in the usual manner: $u = \frac{1}{50}$, $v = \frac{13}{25}$, $w = 0$, and $P = 1140$. Note that the maximum value of P is equal to the minimum value of C, as guaranteed by the fundamental theorem of duality. The solution to the primal problem agrees with the solution of the same problem that we solved in Example 2, Section 6.3, using the method of corners.

Notes

1. The dual of a standard minimization problem is always a standard maximization problem provided that the coefficients of the objective function in the primal problem are all nonnegative. Such problems can always be solved by applying the simplex method to solve the dual problem.
2. Standard minimization problems in which the coefficients of the objective function are not all nonnegative do not necessarily have a dual problem that is a standard maximization problem.

EXAMPLE 4

$$\text{Minimize} \quad C = 3x + 2y$$
$$\text{subject to} \quad 8x + y \geq 80$$
$$8x + 5y \geq 240$$
$$x + 5y \geq 100$$
$$x \geq 0, y \geq 0$$

Solution We begin by writing the dual problem associated with the given primal problem. First, we write down the following tableau for the primal problem:

x	y	Constant
8	1	80
8	5	240
1	5	100
3	2	

Next, interchanging the columns and rows of this tableau and heading the three columns of the resulting array with the three variables, u, v, and w, we obtain the tableau

u	v	w	Constant
8	8	1	3
1	5	5	2
80	240	100	

Interpreting the last tableau as if it were part of the initial simplex tableau for a standard maximization problem—with the exception that the signs of the coefficients pertaining to the objective function are not reversed—we construct the dual problem as follows: Maximize the objective function $P = 80u + 240v + 100w$ subject to the constraints

$$8u + 8v + w \le 3$$
$$u + 5v + 5w \le 2$$

where $u \ge 0$, $v \ge 0$, and $w \ge 0$. Having constructed the dual problem, which is a standard maximization problem, we now solve it using the simplex method. Introducing the slack variables x and y, we obtain the system of linear equations

$$8u + 8v + w + x \qquad = 3$$
$$u + 5v + 5w \qquad + y \qquad = 2$$
$$-80u - 240v - 100w \qquad + P = 0$$

Continuing with the simplex algorithm, we obtain the following sequence of simplex tableaus:

	u	v	w	x	y	P	Constant	Ratio
Pivot row →	8	⑧	1	1	0	0	3	$\frac{3}{8}$
	1	5	5	0	1	0	2	$\frac{2}{5}$
	−80	−240	−100	0	0	1	0	

Pivot column (under v / −240)

	u	v	w	x	y	P	Constant
$\frac{1}{8}R_1$ →	1	①	$\frac{1}{8}$	$\frac{1}{8}$	0	0	$\frac{3}{8}$
	1	5	5	0	1	0	2
	−80	−240	−100	0	0	1	0

	u	v	w	x	y	P	Constant	Ratio
$R_2 - 5R_1$	1	1	$\frac{1}{8}$	$\frac{1}{8}$	0	0	$\frac{3}{8}$	3
$R_3 + 240R_1$ → Pivot row	−4	0	㉟/8	$-\frac{5}{8}$	1	0	$\frac{1}{8}$	$\frac{1}{35}$
	160	0	−70	30	0	1	90	

Pivot column (under w / −70)

	u	v	w	x	y	P	Constant
$\frac{8}{35}R_2$ →	1	1	$\frac{1}{8}$	$\frac{1}{8}$	0	0	$\frac{3}{8}$
	$-\frac{32}{35}$	0	①	$-\frac{1}{7}$	$\frac{8}{35}$	0	$\frac{1}{35}$
	160	0	−70	30	0	1	90

	u	v	w	x	y	P	Constant
$R_1 - \frac{1}{8}R_2$	$\frac{39}{35}$	1	0	$\frac{1}{7}$	$-\frac{1}{35}$	0	$\frac{13}{35}$
$R_3 + 70R_2$ →	$-\frac{32}{35}$	0	1	$-\frac{1}{7}$	$\frac{8}{35}$	0	$\frac{1}{35}$
	96	0	0	20	16	1	92

Solution for the primal problem (under x = 20, y = 16)

The last tableau is final. The fundamental theorem of duality tells us that the solution to the primal problem is $x = 20$ and $y = 16$ with a minimum value for C of 92.

Our last example illustrates how the warehouse problem posed in Section 6.2 may be solved by duality.

 APPLIED EXAMPLE 5 A Warehouse Problem Complete the solution to the warehouse problem given in Example 4, Section 6.2 (page 366). Minimize

$$C = 20x_1 + 8x_2 + 10x_3 + 12x_4 + 22x_5 + 18x_6 \qquad\text{(21)}$$

subject to

$$
\begin{aligned}
x_1 + x_2 + x_3 & \le 400 \\
&x_4 + x_5 + x_6 \le 600 \\
x_1 + x_4 & \ge 200 \\
x_2 + x_5 & \ge 300 \\
x_3 + x_6 &\ge 400 \\
x_1 \ge 0, x_2 \ge 0, \ldots, x_6 &\ge 0
\end{aligned}
\qquad\text{(22)}
$$

Solution Upon multiplying each of the first two inequalities of (22) by -1, we obtain the following equivalent system of constraints, in which each of the expressions involving the variables is greater than or equal to a constant:

$$
\begin{aligned}
-x_1 - x_2 - x_3 & \ge -400 \\
&- x_4 - x_5 - x_6 \ge -600 \\
x_1 + x_4 & \ge 200 \\
x_2 + x_5 & \ge 300 \\
x_3 + x_6 &\ge 400 \\
x_1 \ge 0, x_2 \ge 0, \ldots, x_6 &\ge 0
\end{aligned}
$$

The problem may now be solved by duality. First, we write the tableau

x_1	x_2	x_3	x_4	x_5	x_6	Constant
-1	-1	-1	0	0	0	-400
0	0	0	-1	-1	-1	-600
1	0	0	1	0	0	200
0	1	0	0	1	0	300
0	0	1	0	0	1	400
20	8	10	12	22	18	

Interchanging the rows and columns of this tableau and heading the five columns of the resulting array of numbers by the variables $u_1, u_2, u_3, u_4,$ and u_5, we obtain the tableau

u_1	u_2	u_3	u_4	u_5	Constant
-1	0	1	0	0	20
-1	0	0	1	0	8
-1	0	0	0	1	10
0	-1	1	0	0	12
0	-1	0	1	0	22
0	-1	0	0	1	18
-400	-600	200	300	400	

from which we construct the associated dual problem:

$$\text{Maximize} \quad P = -400u_1 - 600u_2 + 200u_3 + 300u_4 + 400u_5$$

subject to

$$
\begin{aligned}
-u_1 && + u_3 && && &&\leq 20 \\
-u_1 && && + u_4 && &&\leq 8 \\
-u_1 && && && + u_5 &&\leq 10 \\
&&-u_2 + u_3 && && &&\leq 12 \\
&&-u_2 && + u_4 && &&\leq 22 \\
&&-u_2 && && + u_5 &&\leq 18 \\
\end{aligned}
$$
$$u_1 \geq 0, u_2 \geq 0, \ldots, u_5 \geq 0$$

Solving the standard maximization problem by the simplex algorithm, we obtain the following sequence of tableaus ($x_1, x_2, \ldots, x_6$ are slack variables):

	u_1	u_2	u_3	u_4	u_5	x_1	x_2	x_3	x_4	x_5	x_6	P	Constant	Ratio
	-1	0	1	0	0	1	0	0	0	0	0	0	20	—
	-1	0	0	1	0	0	1	0	0	0	0	0	8	—
Pivot row →	-1	0	0	0	①	0	0	1	0	0	0	0	10	10
	0	-1	1	0	0	0	0	0	1	0	0	0	12	—
	0	-1	0	1	0	0	0	0	0	1	0	0	22	—
	0	-1	0	0	1	0	0	0	0	0	1	0	18	18
	400	600	-200	-300	-400	0	0	0	0	0	0	1	0	

Pivot column (at -400)

	u_1	u_2	u_3	u_4	u_5	x_1	x_2	x_3	x_4	x_5	x_6	P	Constant	Ratio
	-1	0	1	0	0	1	0	0	0	0	0	0	20	—
Pivot row →	-1	0	0	①	0	0	1	0	0	0	0	0	8	8
	-1	0	0	0	1	0	0	1	0	0	0	0	10	—
	0	-1	1	0	0	0	0	0	1	0	0	0	12	—
	0	-1	0	1	0	0	0	0	0	1	0	0	22	22
	1	-1	0	0	0	0	0	-1	0	0	1	0	8	—
	0	600	-200	-300	0	0	0	400	0	0	0	1	4000	

$\xrightarrow{\begin{array}{c}R_6 - R_3 \\ R_7 + 400R_3\end{array}}$ Pivot column (at -300)

	u_1	u_2	u_3	u_4	u_5	x_1	x_2	x_3	x_4	x_5	x_6	P	Constant	Ratio
	-1	0	1	0	0	1	0	0	0	0	0	0	20	—
	-1	0	0	1	0	0	1	0	0	0	0	0	8	—
	-1	0	0	0	1	0	0	1	0	0	0	0	10	—
	0	-1	1	0	0	0	0	0	1	0	0	0	12	—
	1	-1	0	0	0	0	-1	0	0	1	0	0	14	14
Pivot row →	①	-1	0	0	0	0	0	-1	0	0	1	0	8	8
	-300	600	-200	0	0	0	300	400	0	0	0	1	6400	

$\xrightarrow{\begin{array}{c}R_5 - R_2 \\ R_7 + 300R_2\end{array}}$ Pivot column (at -300)

	u_1	u_2	u_3	u_4	u_5	x_1	x_2	x_3	x_4	x_5	x_6	P	Constant	Ratio
	0	-1	1	0	0	1	0	-1	0	0	1	0	28	28
	0	-1	0	1	0	0	1	-1	0	0	1	0	16	—
	0	-1	0	0	1	0	0	0	0	0	1	0	18	—
Pivot row →	0	-1	①	0	0	0	0	0	1	0	0	0	12	12
	0	0	0	0	0	0	-1	1	0	1	-1	0	6	—
	1	-1	0	0	0	0	0	-1	0	0	1	0	8	—
	0	300	-200	0	0	0	300	100	0	0	300	1	8800	

$\xrightarrow{\begin{array}{c}R_1 + R_6 \\ R_2 + R_6 \\ R_3 + R_6 \\ R_5 - R_6 \\ R_7 + 300R_6\end{array}}$ Pivot column (at -200)

u_1	u_2	u_3	u_4	u_5	x_1	x_2	x_3	x_4	x_5	x_6	P	Constant
0	0	0	0	0	1	0	-1	-1	0	1	0	16
0	-1	0	1	0	0	1	-1	0	0	1	0	16
0	-1	0	0	1	0	0	0	0	0	1	0	18
0	-1	1	0	0	0	0	0	1	0	0	0	12
0	0	0	0	0	0	-1	1	0	1	-1	0	6
1	-1	0	0	0	0	0	-1	0	0	1	0	8
0	100	0	0	0	0	300	100	200	0	300	1	11,200

(to the left of the fourth row: $\dfrac{R_1 - R_4}{R_7 + 200R_4}$)

The last tableau is final, and we find that

$$x_1 = 0 \qquad x_2 = 300 \qquad x_3 = 100 \qquad x_4 = 200$$
$$x_5 = 0 \qquad x_6 = 300 \qquad P = 11,200$$

Thus, to minimize shipping costs, Acrosonic should ship 300 loudspeaker systems from Plant I to Warehouse *B*, 100 systems from Plant I to Warehouse *C*, 200 systems from Plant II to Warehouse *A*, and 300 systems from Plant II to Warehouse *C*. The company's total shipping cost is $11,200.

6.5 Self-Check Exercises

1. Write the dual problem associated with the following problem:

$$\text{Minimize} \quad C = 2x + 5y$$
$$\text{subject to} \quad 4x + y \geq 40$$
$$2x + y \geq 30$$
$$x + 3y \geq 30$$
$$x \geq 0, y \geq 0$$

2. Solve the primal problem posed in Exercise 1.

Solutions to Self-Check Exercises 6.5 can be found on page 424.

6.5 Concept Questions

1. Suppose you are given the linear programming problem

$$\text{Minimize} \quad C = -3x - 5y$$
$$\text{subject to} \quad 5x + 2y \leq 30$$
$$x + 3y \leq 21$$
$$x \geq 0, y \geq 0$$

Give the associated standard maximization problem that you would use to solve this linear programming problem via the simplex method.

2. Give three characteristics of a standard minimization linear programming problem.

3. What is the primal problem associated with a standard minimization linear programming problem? The dual problem?

4. a. What does the fundamental theorem of duality tell us about the existence of a solution to a primal problem?
  b. How are the optimal values of the primal and dual problems related?
  c. Given the final simplex tableau associated with a dual problem, how would you determine the optimal solution to the associated primal problem?

6.5 Exercises

In Exercises 1–6, use the technique developed in this section to solve the minimization problem.

1. Minimize $C = -2x + y$
subject to $\quad x + 2y \le 6$
$\qquad 3x + 2y \le 12$
$\qquad x \ge 0, y \ge 0$

2. Minimize $C = -2x - 3y$
subject to $\quad 3x + 4y \le 24$
$\qquad 7x - 4y \le 16$
$\qquad x \ge 0, y \ge 0$

3. Minimize $C = -3x - 2y$ subject to the constraints of Exercise 2.

4. Minimize $C = x - 2y + z$
subject to $\quad x - 2y + 3z \le 10$
$\qquad 2x + y - 2z \le 15$
$\qquad 2x + y + 3z \le 20$
$\qquad x \ge 0, y \ge 0, z \ge 0$

5. Minimize $C = 2x - 3y - 4z$
subject to $\quad -x + 2y - z \le 8$
$\qquad x - 2y + 2z \le 10$
$\qquad 2x + 4y - 3z \le 12$
$\qquad x \ge 0, y \ge 0, z \ge 0$

6. Minimize $C = -3x - 2y - z$ subject to the constraints of Exercise 5.

In Exercises 7–10, you are given the final simplex tableau for the dual problem. Give the solution to the primal problem and the solution to the associated dual problem.

7. Problem: Minimize $C = 8x + 12y$
subject to $\quad x + 3y \ge 2$
$\qquad 2x + 2y \ge 3$
$\qquad x \ge 0, y \ge 0$

Final tableau:

u	v	x	y	P	Constant
0	1	$\frac{3}{4}$	$-\frac{1}{4}$	0	3
1	0	$-\frac{1}{2}$	$\frac{1}{2}$	0	2
0	0	$\frac{5}{4}$	$\frac{1}{4}$	1	13

8. Problem: Minimize $C = 3x + 2y$
subject to $\quad 5x + y \ge 10$
$\qquad 2x + 2y \ge 12$
$\qquad x + 4y \ge 12$
$\qquad x \ge 0, y \ge 0$

Final tableau:

u	v	w	x	y	P	Constant
1	0	$-\frac{3}{4}$	$\frac{1}{4}$	$-\frac{1}{4}$	0	$\frac{1}{4}$
0	1	$\frac{19}{8}$	$-\frac{1}{8}$	$\frac{5}{8}$	0	$\frac{7}{8}$
0	0	9	1	5	1	13

9. Problem: Minimize $C = 10x + 3y + 10z$
subject to $\quad 2x + y + 5z \ge 20$
$\qquad 4x + y + z \ge 30$
$\qquad x \ge 0, y \ge 0, z \ge 0$

Final tableau:

u	v	x	y	z	P	Constant
0	1	$\frac{1}{2}$	-1	0	0	2
1	0	$-\frac{1}{2}$	2	0	0	1
0	0	2	-9	1	0	3
0	0	5	10	0	1	80

10. Problem: Minimize $C = 2x + 3y$
subject to $\quad x + 4y \ge 8$
$\qquad x + y \ge 5$
$\qquad 2x + y \ge 7$
$\qquad x \ge 0, y \ge 0$

Final tableau:

u	v	w	x	y	P	Constant
0	1	$\frac{7}{3}$	$\frac{4}{3}$	$-\frac{1}{3}$	0	$\frac{5}{3}$
1	0	$-\frac{1}{3}$	$-\frac{1}{3}$	$\frac{1}{3}$	0	$\frac{1}{3}$
0	0	2	4	1	1	11

In Exercises 11–20, construct the dual problem associated with the primal problem. Solve the primal problem.

11. Minimize $C = 3x + 2y$
subject to $\quad 2x + 3y \ge 90$
$\qquad 3x + 2y \ge 120$
$\qquad x \ge 0, y \ge 0$

12. Minimize $C = 2x + 5y$
subject to $\quad x + 2y \ge 4$
$\qquad 3x + 2y \ge 6$
$\qquad x \ge 0, y \ge 0$

13. Minimize $C = 6x + 4y$
subject to $\quad 6x + y \ge 60$
$\qquad 2x + y \ge 40$
$\qquad x + y \ge 30$
$\qquad x \ge 0, y \ge 0$

14. Minimize $C = 10x + y$
subject to $\quad 4x + y \ge 16$
$\qquad x + 2y \ge 12$
$\qquad x \ge 2$
$\qquad x \ge 0, y \ge 0$

15. Minimize $C = 200x + 150y + 120z$
subject to $\quad 20x + 10y + z \ge 10$
$\qquad x + y + 2z \ge 20$
$\qquad x \ge 0, y \ge 0, z \ge 0$

16. Minimize $C = 40x + 30y + 11z$
subject to $\quad 2x + y + z \ge 8$
$\qquad x + y - z \ge 6$
$\qquad x \ge 0, y \ge 0, z \ge 0$

17. Minimize $C = 6x + 8y + 4z$
subject to $\quad x + 2y + 2z \ge 10$
$\qquad 2x + y + z \ge 24$
$\qquad x + y + z \ge 16$
$\qquad x \ge 0, y \ge 0, z \ge 0$

18. Minimize $\quad C = 12x + 4y + 8z$
subject to $\quad 2x + 4y + z \geq 6$
$$3x + 2y + 2z \geq 2$$
$$4x + y + z \geq 2$$
$$x \geq 0, y \geq 0, z \geq 0$$

19. Minimize $\quad C = 30x + 12y + 20z$
subject to $\quad 2x + 4y + 3z \geq 6$
$$6x + z \geq 2$$
$$6y + 2z \geq 4$$
$$x \geq 0, y \geq 0, z \geq 0$$

20. Minimize $\quad C = 8x + 6y + 4z$
subject to $\quad 2x + 3y + z \geq 6$
$$x + 2y - 2z \geq 4$$
$$x + y + 2z \geq 2$$
$$x \geq 0, y \geq 0, z \geq 0$$

21. Minimizing Cruise Line Costs Deluxe River Cruises operates a fleet of river vessels. The fleet has two types of vessels: A type A vessel has 60 deluxe cabins and 160 standard cabins, whereas a type B vessel has 80 deluxe cabins and 120 standard cabins. Under a charter agreement with Odyssey Travel Agency, Deluxe River Cruises is to provide Odyssey with a minimum of 360 deluxe and 680 standard cabins for their 15-day cruise in May. It costs $44,000 to operate a type A vessel and $54,000 to operate a type B vessel for that period. How many of each type vessel should be used to keep the operating costs to a minimum? What is the minimum cost?

22. Fertilizer Costs A farmer uses two types of fertilizers. A 50-lb bag of Fertilizer A contains 8 lb of nitrogen, 2 lb of phosphorus, and 4 lb of potassium. A 50-lb bag of Fertilizer B contains 5 lb each of nitrogen, phosphorus, and potassium. The minimum requirements for a field are 440 lb of nitrogen, 260 lb of phosphorus, and 360 lb of potassium. If a 50-lb bag of Fertilizer A costs $30 and a 50-lb bag of Fertilizer B costs $20, find the amount of each type of fertilizer the farmer should use to minimize his cost while still meeting the minimum requirements. What is the cost?

23. Diet Planning The owner of the Health Juice Bar wishes to prepare a low-calorie fruit juice with a high vitamin A and vitamin C content by blending orange juice and pink grapefruit juice. Each glass of the blended juice is to contain at least 1200 International Units (IU) of vitamin A and 200 IU of vitamin C. One ounce of orange juice contains 60 IU of vitamin A, 16 IU of vitamin C, and 14 calories; each ounce of pink grapefruit juice contains 120 IU of vitamin A, 12 IU of vitamin C, and 11 calories. How many ounces of each juice should a glass of the blend contain if it is to meet the minimum vitamin requirements while containing a minimum number of calories?

24. Optimizing Advertising Exposure Everest Deluxe World Travel has decided to advertise in the Sunday editions of two major newspapers in town. These advertisements are directed at three groups of potential customers. Each

advertisement in Newspaper I is seen by 70,000 Group A customers, 40,000 Group B customers, and 20,000 Group C customers. Each advertisement in Newspaper II is seen by 10,000 Group A, 20,000 Group B, and 40,000 Group C customers. Each advertisement in Newspaper I costs $1000, and each advertisement in Newspaper II costs $800. Everest would like their advertisements to be read by at least 2 million people from Group A, 1.4 million people from Group B, and 1 million people from Group C. How many advertisements should Everest place in each newspaper to achieve its advertising goals at a minimum cost? What is the minimum cost?

25. Minimizing Shipping Costs Acrosonic Company manufactures a model G loudspeaker system in Plants I and II. The output at Plant I is at most 800/month, and the output at Plant II is at most 600/month. Model G loudspeaker systems are also shipped to the three warehouses—A, B, and C—whose minimum monthly requirements are 500, 400, and 400 systems, respectively. Shipping costs from Plant I to Warehouse A, Warehouse B, and Warehouse C are $16, $20, and $22 per loudspeaker system, respectively, and shipping costs from Plant II to each of these warehouses are $18, $16, and $14, respectively. What shipping schedule will enable Acrosonic to meet the requirements of the warehouses while keeping its shipping costs to a minimum? What is the minimum cost?

26. Minimizing Shipping Costs Steinwelt Piano manufactures uprights and consoles in two plants, Plant I and Plant II. The output of Plant I is at most 300/month, and the output of Plant II is at most 250/month. These pianos are shipped to three warehouses that serve as distribution centers for Steinwelt. To fill current and projected future orders, Warehouse A requires a minimum of 200 pianos/month, Warehouse B requires at least 150 pianos/month, and Warehouse C requires at least 200 pianos/month. The shipping cost of each piano from Plant I to Warehouse A, Warehouse B, and Warehouse C is $60, $60, and $80, respectively, and the shipping cost of each piano from Plant II to Warehouse A, Warehouse B, and Warehouse C is $80, $70, and $50, respectively. What shipping schedule will enable Steinwelt to meet the requirements of the warehouses while keeping the shipping costs to a minimum? What is the minimum cost?

27. Minimizing Oil Refinery Costs An oil company operates two refineries in a certain city. Refinery I has an output of 200, 100, and 100 barrels of low-, medium-, and high-grade oil per day, respectively. Refinery II has an output of 100, 200, and 600 barrels of low-, medium-, and high-grade oil per day, respectively. The company wishes to produce at least 1000, 1400, and 3000 barrels of low-, medium-, and high-grade oil, respectively, to fill an order. If it costs $200/day to operate Refinery I and $300/day to operate Refinery II, determine how many days each refinery should be operated to meet the production requirements at minimum cost to the company. What is the minimum cost?

In Exercises 28 and 29, determine whether the statement is true or false. If it is true, explain why it is true. If it is false, give an example to show why it is false.

28. If a standard minimization linear programming problem has a unique solution, then so does the corresponding maximization problem with objective function $P = -C$, where $C = a_1 x_1 + a_2 x_2 + \cdots + a_n x_n$ is the objective function for the minimization problem.

29. The optimal value attained by the objective function of the primal problem may be different from that attained by the objective function of the corresponding dual problem.

6.5 Solutions to Self-Check Exercises

1. We first write down the following tableau for the given (primal) problem:

x	y	Constant
4	1	40
2	1	30
1	3	30
2	5	0

Next, we interchange the columns and rows of the tableau and head the three columns of the resulting array with the three variables, u, v, and w, obtaining the tableau

u	v	w	Constant
4	2	1	2
1	1	3	5
40	30	30	0

Interpreting the last tableau as if it were the initial tableau for a standard linear programming problem—with the exception that the signs of the coefficients pertaining to the objective function are not reversed—we construct the required dual problem as follows:

$$\text{Maximize} \quad P = 40u + 30v + 30w$$
$$\text{subject to} \quad 4u + 2v + w \le 2$$
$$u + v + 3w \le 5$$
$$u \ge 0, v \ge 0, w \ge 0$$

2. We introduce slack variables x and y to obtain the system of linear equations

$$4u + 2v + w + x = 2$$
$$u + v + 3w + y = 5$$
$$-40u - 30v - 30w + P = 0$$

Using the simplex algorithm, we obtain the sequence of simplex tableaus

	u	v	w	x	y	P	Constant	Ratio
Pivot row →	④	2	1	1	0	0	2	$\frac{2}{4} = \frac{1}{2}$
	1	1	3	0	1	0	5	$\frac{5}{1} = 5$
	−40	−30	−30	0	0	1	0	

$\xrightarrow{\frac{1}{4}R_1}$

Pivot column (under u)

u	v	w	x	y	P	Constant
①	$\frac{1}{2}$	$\frac{1}{4}$	$\frac{1}{4}$	0	0	$\frac{1}{2}$
1	1	3	0	1	0	5
−40	−30	−30	0	0	1	0

$\xrightarrow[R_3 + 40R_1]{R_2 - R_1}$

	u	v	w	x	y	P	Constant	Ratio
	1	$\frac{1}{2}$	$\frac{1}{4}$	$\frac{1}{4}$	0	0	$\frac{1}{2}$	$\frac{1/2}{1/4} = 2$
Pivot row →	0	$\frac{1}{2}$	⑪⁄₄	$-\frac{1}{4}$	1	0	$\frac{9}{2}$	$\frac{9/2}{11/4} = \frac{18}{11}$
	0	−10	−20	10	0	1	20	

$\xrightarrow{\frac{4}{11}R_2}$

Pivot column (under w)

u	v	w	x	y	P	Constant
1	$\frac{1}{2}$	$\frac{1}{4}$	$\frac{1}{4}$	0	0	$\frac{1}{2}$
0	$\frac{2}{11}$	①	$-\frac{1}{11}$	$\frac{4}{11}$	0	$\frac{18}{11}$
0	−10	−20	10	0	1	20

$\xrightarrow[R_3 + 20R_2]{R_1 - \frac{1}{4}R_2}$

	u	v	w	x	y	P	Constant	Ratio
Pivot row →	1	⑤⁄₁₁	0	$\frac{3}{11}$	$-\frac{1}{11}$	0	$\frac{1}{11}$	$\frac{1/11}{5/11} = \frac{1}{5}$
	0	$\frac{2}{11}$	1	$-\frac{1}{11}$	$\frac{4}{11}$	0	$\frac{18}{11}$	$\frac{18/11}{2/11} = 9$
	0	$-\frac{70}{11}$	0	$\frac{90}{11}$	$\frac{80}{11}$	1	$\frac{580}{11}$	

$\xrightarrow{\frac{11}{5}R_1}$

Pivot column (under v)

u	v	w	x	y	P	Constant
$\frac{11}{5}$	①	0	$\frac{3}{5}$	$-\frac{1}{5}$	0	$\frac{1}{5}$
0	$\frac{2}{11}$	1	$-\frac{1}{11}$	$\frac{4}{11}$	0	$\frac{18}{11}$
0	$-\frac{70}{11}$	0	$\frac{90}{11}$	$\frac{80}{11}$	1	$\frac{580}{11}$

$\xrightarrow[R_3 + \frac{70}{11}R_1]{R_2 - \frac{2}{11}R_1}$

u	v	w	x	y	P	Constant
$\frac{11}{5}$	1	0	$\frac{3}{5}$	$-\frac{1}{5}$	0	$\frac{1}{5}$
$-\frac{2}{5}$	0	1	$-\frac{1}{5}$	$\frac{2}{5}$	0	$\frac{8}{5}$
14	0	0	12	6	1	54

Solution for the primal problem

The last tableau is final, and the solution to the primal problem is $x = 12$ and $y = 6$ with a minimum value for C of 54.

The Simplex Method: Solving Minimization Problems

Graphing Utility

A graphing utility can be used to solve minimization problems using the simplex method.

EXAMPLE 1

$$\text{Minimize} \quad C = 2x + 3y$$
$$\text{subject to} \quad 8x + y \geq 80$$
$$3x + 2y \geq 100$$
$$x + 4y \geq 80$$
$$x \geq 0, y \geq 0$$

Solution We begin by writing the dual problem associated with the given primal problem. From the tableau for the primal problem

x	y	Constant
8	1	80
3	2	100
1	4	80
2	3	

we obtain—upon interchanging the columns and rows of this tableau and heading the three columns of the resulting array with the variables u, v, and w—the tableau

u	v	w	Constant
8	3	1	2
1	2	4	3
80	100	80	

This tells us that the dual problem is

$$\text{Maximize} \quad P = 80u + 100v + 80w$$
$$\text{subject to} \quad 8u + 3v + w \leq 2$$
$$u + 2v + 4w \leq 3$$
$$u \geq 0, v \geq 0, w \geq 0$$

To solve this standard maximization problem, we proceed as follows:

	u	v	w	x	y	P	Constant		Ratio	
Pivot row →	8	③	1	1	0	0	2		$\frac{2}{3}$	$*\mathbf{row}\left(\frac{1}{3}, A, 1\right) \blacktriangleright B$
	1	2	4	0	1	0	3		$\frac{3}{2}$	
	−80	−100	−80	0	0	1	0			

Pivot column

u	v	w	x	y	P	Constant	
2.67	①	0.33	0.33	0	0	0.67	$*\mathbf{row}+(-2, B, 1, 2) \blacktriangleright C$
1	2	4	0	1	0	3	$*\mathbf{row}+(100, C, 1, 3) \blacktriangleright B$
−80	−100	−80	0	0	1	0	

	u	v	w	x	y	P	Constant
	2.67	1	0.33	0.33	0	0	0.67
Pivot row →	−4.33	0	(3.33)	−0.67	1	0	1.67
	186.67	0	−46.67	33.33	0	1	66.67

Ratio: 2, 0.5

$$*\mathbf{row}\left(\tfrac{1}{3.33}, B, 2\right) \blacktriangleright C$$

↑ Pivot column

	u	v	w	x	y	P	Constant
	2.67	1	0.33	0.33	0	0	0.67
	−1.30	0	1	−0.2	0.3	0	0.5
	186.67	0	−46.67	33.33	0	1	66.67

$$*\mathbf{row}+(-0.33, C, 2, 1) \blacktriangleright B$$
$$*\mathbf{row}+(46.67, B, 2, 3) \blacktriangleright C$$

	u	v	w	x	y	P	Constant
	3.1	1	0	0.4	−0.1	0	0.50
	−1.3	0	1	−0.2	0.3	0	0.50
	125.93	0	0.05	23.99	14.02	1	90.03

Solution for the primal problem

From the last tableau, we see that $x = 23.99$, $y = 14.02$, and $C = 90.03$. ▪

Excel

EXAMPLE 2

Minimize $C = 2x + 3y$
subject to $8x + y \geq 80$
$3x + 2y \geq 100$
$x + 4y \geq 80$
$x \geq 0, y \geq 0$

Solution   We use Solver as outlined in Example 2, pages 408–410, to obtain the spreadsheet shown in Figure T1. (In this case, select ⎣**Min**⎦ next to To: instead of Max because this is a minimization problem. Also select ⎣**>=**⎦ in the Add Constraint dialog box because the inequalities in the problem are of the form ≥.) From the spreadsheet, we read off the solution: $x = 24$, $y = 14$, and $C = 90$.

	A	B	C	D	E	F	G	H	I	J
1	Minimization Problem									
2								Formulas for indicated cells		
3	Decision Variables							C8: = 2*C4 + 3*C5		
4		x	24					C11: = 8*C4 + C5		
5		y	14					C12: = 3*C4 + 2*C5		
6								C13: = C4 + 4*C5		
7										
8	Objective Function		90							
9										
10	Constraints									
11			206	>=	80					
12			100	>=	100					
13			80	>=	80					

FIGURE **T1**
Completed spreadsheet after using Solver

TECHNOLOGY EXERCISES

In Exercises 1–4, solve the linear programming problem by the simplex method.

1. Minimize $C = x + y + 3z$
subject to
$$2x + y + 3z \geq 6$$
$$x + 2y + 4z \geq 8$$
$$3x + y - 2z \geq 4$$
$$x \geq 0, y \geq 0, z \geq 0$$

2. Minimize $C = 2x + 4y + z$
subject to
$$x + 2y + 4z \geq 7$$
$$3x + y - z \geq 6$$
$$x + 4y + 2z \geq 24$$
$$x \geq 0, y \geq 0, z \geq 0$$

3. Minimize $C = x + 1.2y + 3.5z$
subject to
$$2x + 3y + 5z \geq 12$$
$$3x + 1.2y - 2.2z \geq 8$$
$$1.2x + 3y + 1.8z \geq 14$$
$$x \geq 0, y \geq 0, z \geq 0$$

4. Minimize $C = 2.1x + 1.2y + z$
subject to
$$x + y - z \geq 5.2$$
$$x - 2.1y + 4.2z \geq 8.4$$
$$x \geq 0, y \geq 0, z \geq 0$$

CHAPTER 6 Summary of Principal Terms

TERMS

solution set of a system of linear inequalities (355)

bounded solution set (358)

unbounded solution set (358)

objective function (363)

linear programming problem (363)

feasible solution (372)

feasible set (372)

optimal solution (372)

isoprofit line (372)

method of corners (373)

standard maximization problem (385)

slack variable (385)

basic variable (386)

nonbasic variable (386)

basic solution (387)

pivot column (388)

pivot row (389)

pivot element (389)

simplex tableau (389)

simplex method (390)

standard minimization problem (412)

primal problem (413)

dual problem (413)

CHAPTER 6 Concept Review Questions

Fill in the blanks.

1. a. The solution set of the inequality $ax + by < c$ (a, b not both zero) is a/an _____ _____ that does not include the _____ with equation $ax + by = c$.

b. If $ax + by < c$ describes the lower half-plane, then the inequality _____ describes the lower half-plane together with the line having equation _____.

2. a. The solution set of a system of linear inequalities in the two variables x and y is the set of all _____ satisfying _____ inequality of the system.

b. The solution set of a system of linear inequalities is _____ if it can be _____ by a circle.

3. A linear programming problem consists of a linear function, called a/an _____ _____ to be _____ or _____ subject to constraints in the form of _____ equations or _____.

4. a. If a linear programming problem has a solution, then it must occur at a/an _____ _____ of the feasible set.

b. If the objective function of a linear programming problem is optimized at two adjacent vertices of the feasible set, then it is optimized at every point on the _____ segment joining these vertices.

5. In a standard maximization problem: the objective function is to be _____; all the variables involved in the problem are _____; and each linear constraint may be written so that the expression involving the variables is _____ _____ or _____ _____ a nonnegative constant.

6. In setting up the initial simplex tableau, we first transform the system of linear inequalities into a system of linear _____, using _____ _____; the objective function is rewritten so that it has the form _____ and then is placed _____ the system of linear equations obtained earlier. Finally, the initial simplex tableau is the _____ matrix associated with this system of linear equations.

7. In a standard minimization problem: the objective function is to be _____; all the variables involved in the problem are _____; and each linear constraint may be written so that the expression involving the variables is _____ _____ or _____ _____ a constant.

8. The fundamental theorem of duality states that a primal problem has a solution if and only if the corresponding _____ problem has a solution. If a solution exists, then the _____ functions of both the primal and the dual problem attain the same _____ _____.

CHAPTER 6 Review Exercises

In Exercises 1 and 2, find the optimal value(s) of the objective function on the feasible set S.

1. $Z = 2x + 3y$

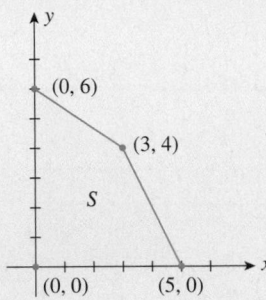

2. $Z = 4x + 3y$

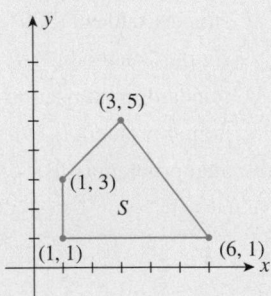

In Exercises 3–10, use the method of corners to solve the linear programming problem.

3. Maximize $P = 3x + 5y$
subject to $2x + 3y \leq 12$
$\qquad x + y \leq 5$
$\qquad x \geq 0, y \geq 0$

4. Maximize $P = 2x + 3y$
subject to $2x + y \leq 12$
$\qquad x - 2y \leq 1$
$\qquad x \geq 0, y \geq 0$

5. Minimize $C = 2x + 5y$
subject to $\quad x + 3y \geq 15$
$\qquad 4x + y \geq 16$
$\qquad x \geq 0, y \geq 0$

6. Minimize $C = 3x + 4y$
subject to $2x + y \geq 4$
$\qquad 2x + 5y \geq 10$
$\qquad x \geq 0, y \geq 0$

7. Maximize $P = 6x + 2y$
subject to $\quad x + 2y \leq 12$
$\qquad x + y \leq 8$
$\qquad 2x - 3y \geq 6$
$\qquad x \geq 0, y \geq 0$

8. Minimize $C = 2x + 7y$
subject to $\quad 3x + 5y \geq 45$
$\qquad 3x + 10y \geq 60$
$\qquad x \geq 0, y \geq 0$

9. Find the maximum and minimum values of $Q = 3x + 4y$ subject to

$$x - y \geq -10$$
$$x + 3y \geq 30$$
$$7x + 4y \leq 140$$
$$x \geq 0, y \geq 0$$

10. Find the maximum and minimum of $Q = x + y$ subject to

$$5x + 2y \geq 20$$
$$x + 2y \geq 8$$
$$x + 4y \leq 22$$
$$x \geq 0, y \geq 0$$

In Exercises 11–15, use the simplex method to solve the linear programming problem.

11. Maximize $P = 3x + 4y$
subject to $\quad x + 3y \leq 15$
$\qquad 4x + y \leq 16$
$\qquad x \geq 0, y \geq 0$

12. Maximize $P = 2x + 5y$
subject to $2x + y \leq 16$
$\qquad 2x + 3y \leq 24$
$\qquad y \leq 6$
$\qquad x \geq 0, y \geq 0$

13. Maximize $P = 3x + 2y$
subject to $\quad x + 3y \leq 18$
$\qquad 3x + 2y \leq 19$
$\qquad 3x + y \leq 15$
$\qquad x \geq 0, y \geq 0$

14. Maximize $P = 2x + 3y + 5z$
subject to $\quad x + 2y + 3z \leq 12$
$\qquad x - 3y + 2z \leq 10$
$\qquad x \geq 0, y \geq 0, z \geq 0$

15. Minimize $C = -4x - 7y$
subject to $3x + y \leq 8$
$x + 2y \leq 6$
$x \geq 0, y \geq 0$

16. Suppose that the primal problem for a linear programming problem is

Minimize $C = 2x + 5y$

subject to $3x + 2y \geq 8$

$x + 4y \geq 6$

$x \geq 0, y \geq 0$

and the final simplex tableau for the dual problem associated with the primal problem is

u	v	x	y	P	Constant
1	0	$\frac{2}{5}$	$-\frac{1}{10}$	0	$\frac{3}{10}$
0	1	$-\frac{1}{5}$	$\frac{3}{10}$	0	$\frac{11}{10}$
0	0	2	1	1	9

Find the solution to the primal problem and the solution to the dual problem.

17. Construct the dual problem associated with the following primal problem:

Minimize $C = 2x + 4y + 3z$

subject to $x + 2y + z \geq 14$

$4x + y + 2z \geq 21$

$-3x - 2y + 5z \geq 12$

$x \geq 0, y \geq 0, z \geq 0$

Do not solve.

In Exercises 18–20, use the simplex method to solve the linear programming problem.

18. Minimize $C = 3x + 2y$
subject to $2x + 3y \geq 6$
$2x + y \geq 4$
$x \geq 0, y \geq 0$

19. Minimize $C = x + 2y$
subject to $3x + y \geq 12$
$x + 4y \geq 16$
$x \geq 0, y \geq 0$

20. Minimize $C = 24x + 18y + 24z$
subject to $3x + 2y + z \geq 4$
$x + y + 3z \geq 6$
$x \geq 0, y \geq 0, z \geq 0$

In Exercises 21–23, use the method of corners to solve the linear programming problem.

21. FINANCIAL ANALYSIS An investor has decided to commit no more than $80,000 to the purchase of the common stocks of two companies, Company A and Company B. He has also estimated that there is a chance of at most a 1% capital loss on his investment in Company A and a chance of at most a 4% loss on his investment in Company B, and he has decided that together these losses should not exceed $2000. On the other hand, he expects to make a 14% profit from his investment in Company A and a 20% profit from his investment in Company B. Determine how much he should invest in the stock of each company to maximize his investment returns. What is the maximum return?

22. PRODUCTION SCHEDULING Soundex produces two models of satellite radios. Model A requires 15 min of work on Assembly Line I and 10 min of work on Assembly Line II. Model B requires 10 min of work on Assembly Line I and 12 min of work on Assembly Line II. At most, 25 labor-hours of assembly time on Line I and 22 labor-hours of assembly time on Line II are available each day. It is anticipated that Soundex will realize a profit of $12 on model A and $10 on model B. How many satellite radios of each model should be produced each day to maximize Soundex's profit? What is the maximum profit?

23. PRODUCTION SCHEDULING Kane Manufacturing has a division that produces two models of grates, model A and model B. To produce each model A grate requires 3 lb of cast iron and 6 min of labor. To produce each model B grate requires 4 lb of cast iron and 3 min of labor. The profit for each model A grate is $2.00, and the profit for each model B grate is $1.50. Available for grate production each day are 1000 lb of cast iron and 20 labor-hours. Because of a backlog of orders for model B grates, Kane's manager has decided to produce at least 180 model B grates per day. How many grates of each model should Kane produce to maximize its profit? What is the maximum profit?

In Exercises 24–26, use the simplex method to solve the linear programming problem.

24. MINIMIZING MINING COSTS Perth Mining Company operates two mines for the purpose of extracting gold and silver. The Saddle Mine costs $14,000/day to operate, and it yields 50 oz of gold and 3000 oz of silver each day. The Horseshoe Mine costs $16,000/day to operate, and it yields 75 oz of gold and 1000 oz of silver each day. Company management has set a target of at least 650 oz of gold and 18,000 oz of silver. How many days should each mine be operated at so that the target can be met at a minimum cost to the company? What is the minimum cost?

25. INVESTMENT ANALYSIS Jorge has decided to invest at most $100,000 in securities in the form of corporate stocks. He has classified his options into three groups of stocks: blue-chip stocks that he assumes will yield a 10% return (dividends and capital appreciation) within a year, growth stocks that he assumes will yield a 15% return within a year, and speculative stocks that he assumes will yield a 20% return (mainly due to capital appreciation) within a year. Because of the relative risks involved in his investment, Jorge has further decided that no more than 30% of his investment should be in growth and speculative stocks and at least 50% of his investment should be in blue-chip and speculative stocks. Determine how much Jorge should invest in each group of stocks in the hope of maximizing the return on his investments. What is the maximum return?

26. **MAXIMIZING PROFIT** A company manufactures three products, A, B, and C, on two machines, I and II. It has been determined that the company will realize a profit of $4/unit of Product A, $6/unit of Product B, and $8/unit of Product C. Manufacturing a unit of Product A requires 9 min on Machine I and 6 min on Machine II; manufacturing a unit of Product B requires 12 min on Machine I and 6 min on Machine II; manufacturing a unit of Product C requires 18 min on Machine I and 10 min on Machine II. There are 6 hr of machine time available on Machine I and 4 hr of machine time available on Machine II in each work shift. How many units of each product should be produced in each shift to maximize the company's profit? What is the maximum profit?

CHAPTER 6 Before Moving On . . .

1. Determine graphically the solution set for the following systems of inequalities.

   a. $2x + y \leq 10$
   $x + 3y \leq 15$
   $x \leq 4$
   $x \geq 0, y \geq 0$

   b. $2x + y \geq 8$
   $2x + 3y \geq 15$
   $x \geq 0$
   $y \geq 2$

2. Find the maximum and minimum values of $Z = 3x - y$ on the following feasible set.

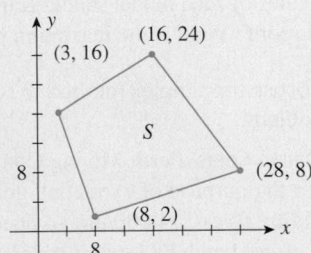

3. Use the method of corners to solve the following linear programming problem:

$$\text{Maximize} \quad P = x + 3y$$
$$\text{subject to} \quad 2x + 3y \leq 11$$
$$3x + 7y \leq 24$$
$$x \geq 0, y \geq 0$$

4. Consider the following linear programming problem:

$$\text{Maximize} \quad P = x + 2y - 3z$$
$$\text{subject to} \quad 2x + y - z \leq 3$$
$$x - 2y + 3z \leq 1$$
$$3x + 2y + 4z \leq 17$$
$$x \geq 0, y \geq 0, z \geq 0$$

Write the initial simplex tableau for the problem and identify the pivot element to be used in the first iteration of the simplex method.

5. The following simplex tableau is in final form. Find the solution to the linear programming problem associated with this tableau.

x	y	z	u	v	w	P	Constant
0	$\frac{1}{2}$	0	1	$-\frac{1}{2}$	0	0	2
0	$\frac{1}{4}$	1	0	$\frac{5}{4}$	$-\frac{1}{2}$	0	11
1	$\frac{1}{4}$	0	0	$-\frac{3}{4}$	$\frac{1}{2}$	0	2
0	$\frac{13}{4}$	0	0	$\frac{1}{4}$	$\frac{1}{2}$	1	28

6. Using the simplex method, solve the following linear programming problem:

$$\text{Maximize} \quad P = 5x + 2y$$
$$\text{subject to} \quad 4x + 3y \leq 30$$
$$2x - 3y \leq 6$$
$$x \geq 0, y \geq 0$$

7 Sets and Probability

WE OFTEN DEAL WITH well-defined collections of objects called *sets*. In this chapter, we see how sets can be combined algebraically to yield other sets. We also look at some techniques for determining the number of elements in a set and for determining the number of ways in which the elements of a set can be arranged or combined. After giving the technical meaning of the term *probability*, we see how the rules of probability are applied to many real-life situations to compute the probability of the occurrence of certain events.

In how many ways can the Futurists (a rock group) plan their concert tour to San Francisco, Los Angeles, San Diego, Denver, and Las Vegas if the three performances in California must be given consecutively? In Example 13, page 466, we will show how to determine the number of possible different itineraries.

© Pixel 4 Images/Shutterstock.com

7.1 Sets and Set Operations

Set Terminology and Notation

We often deal with collections of different kinds of objects. For example, in conducting a study of the distribution of the weights of newborn infants, we might consider the collection of all infants born at Massachusetts General Hospital during 2015. In a study of the fuel consumption of compact cars, we might be interested in the collection of hybrid cars manufactured by General Motors in the 2015 model year. Such collections are examples of sets. More specifically, a **set** is a well-defined collection of objects. Thus, a set is not just any collection of objects; a set must be well defined in the sense that if we are given an object, then we should be able to determine whether or not it belongs in the collection.

The objects of a set are called the **elements,** or *members*, **of a set** and are usually denoted by lowercase letters $a, b, c, \ldots$; the sets themselves are usually denoted by uppercase letters $A, B, C, \ldots$. The elements of a set can be displayed by listing all the elements between braces. For example, in **roster notation,** the set A consisting of the first three letters of the English alphabet is written

$$A = \{a, b, c\}$$

The set B of all letters of the alphabet can be written

$$B = \{a, b, c, \ldots, z\}$$

Another notation that is commonly used is **set-builder notation.** Here, a rule is given that describes the definite property or properties an object x must satisfy to qualify for membership in the set. In this notation, the set B is written as

$$B = \{x \mid x \text{ is a letter of the English alphabet}\}$$

and is read "B is the set of all elements x such that x is a letter of the English alphabet."

If a is an element of a set A, we write $a \in A$ and read "a belongs to A" or "a is an element of A." If the element a does not belong to the set A, however, then we write $a \notin A$ and read "a does not belong to A." For example, if $A = \{1, 2, 3, 4, 5\}$, then $3 \in A$ but $6 \notin A$.

Explore and Discuss

1. Let A denote the collection of all the days in August 2015 in which the average daily temperature at the San Francisco International Airport was approximately 75°F. Is A a set? Explain your answer.

2. Let B denote the collection of all the days in August 2015 in which the average daily temperature at the San Francisco International Airport was between 73.5°F and 81.2°F, inclusive. Is B a set? Explain your answer.

Set Equality

Two sets A and B are **equal,** written $A = B$, if and only if they have exactly the same elements.

EXAMPLE 1 Let A, B, and C be the sets

$$A = \{a, e, i, o, u\}$$
$$B = \{a, i, o, e, u\}$$
$$C = \{a, e, i, o\}$$

Then $A = B$, since they both contain exactly the same elements. Note that the order in which the elements are displayed is immaterial. Also, $A \neq C$, since $u \in A$ but $u \notin C$. Similarly, we conclude that $B \neq C$.

> **Subset**
>
> If every element of a set A is also an element of a set B, then we say that A is a **subset** of B and write $A \subseteq B$.

By this definition, two sets A and B are equal if and only if (1) $A \subseteq B$, and (2) $B \subseteq A$. You can verify this (see Exercise 70).

EXAMPLE 2 Referring to Example 1, we find that $C \subseteq B$, since every element of C is also an element of B. Also, if D is the set

$$D = \{a, e, i, o, x\}$$

then D is not a subset of A, written $D \not\subseteq A$, since $x \in D$ but $x \notin A$. Observe that $A \not\subseteq D$ as well, since $u \in A$ but $u \notin D$.

If A and B are sets such that $A \subseteq B$ but $A \neq B$, then we say that A is a **proper subset** of B. In other words, a set A is a proper subset of a set B, written $A \subset B$, if (1) $A \subseteq B$ and (2) there exists at least one element in B that is not in A. The second condition states that the set A is properly "smaller" than the set B.

EXAMPLE 3 Let $A = \{1, 2, 3, 4, 5, 6\}$ and $B = \{2, 4, 6\}$. Then B is a proper subset of A because (1) $B \subseteq A$, which is easily verified, and (2) there exists at least one element in A that is not in B—for example, the element 1.

When we refer to sets and subsets, we use the symbols $\subset$, $\subseteq$, $\supset$, and $\supseteq$ to express the idea of "containment." However, when we wish to show that an element is contained in a set, we use the symbol $\in$ to express the idea of "membership." Thus, in Example 3, we would write $1 \in A$ and *not* $\{1\} \in A$.

> **Empty Set**
>
> The set that contains no elements is called the **empty set** and is denoted by $\varnothing$.

The empty set, $\varnothing$, is a subset of every set. To see this, observe that $\varnothing$ has no elements and therefore contains no element that is not also in any set A.

Do not confuse the empty set $\varnothing = \{ \}$ with the set $\{0\}$, which is a set with one element—the number zero.

EXAMPLE 4 List all subsets of the set $A = \{a, b, c\}$.

Solution There is one subset consisting of no elements, namely, the empty set $\varnothing$. Next, observe that there are three subsets consisting of one element,

$$\{a\}, \{b\}, \{c\}$$

three subsets consisting of two elements,

$$\{a, b\}, \{a, c\}, \{b, c\}$$

and one subset consisting of three elements, the set A itself. Therefore, the subsets of A are

$$\varnothing, \{a\}, \{b\}, \{c\}, \{a, b\}, \{a, c\}, \{b, c\}, \{a, b, c\}$$ ■

In contrast with the empty set, we have, at the other extreme, the notion of a largest, or universal, set. A **universal set** is the set of all elements of interest in a particular discussion. It is the largest in the sense that all sets considered in the discussion of the problem are subsets of the universal set. Of course, different universal sets are associated with different problems, as shown in Example 5.

EXAMPLE 5

a. If the problem is to determine the ratio of female to male students in a college, then a logical choice of a universal set is the set consisting of the whole student body of the college.

b. If the problem is to determine the ratio of female to male students in the business department of the college in part (a), then the set of all students in the business department can be chosen as the universal set.   ■

We can use **Venn diagrams** to obtain a visual representation of sets. Venn diagrams are of considerable help in understanding the concepts introduced earlier as well as in solving problems involving sets. The universal set U is represented by a rectangle, and subsets of U are represented by regions lying inside the rectangle.

EXAMPLE 6 Use Venn diagrams to illustrate the following statements:

a. The sets A and B are equal.
b. The set A is a proper subset of the set B.
c. The sets A and B are not subsets of each other.

Solution   The respective Venn diagrams are shown in Figure 1a–c.

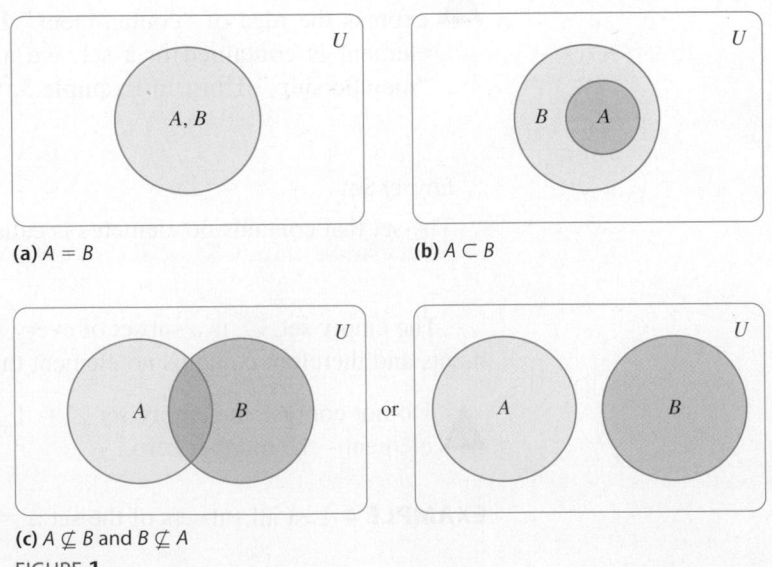

(a) $A = B$ (b) $A \subset B$

(c) $A \not\subseteq B$ and $B \not\subseteq A$

FIGURE 1   ■

Set Operations

Now that we have introduced the concept of a set, our next task is to consider operations on sets—that is, to consider ways in which sets can be combined to yield other sets. These operations enable us to combine sets in much the same way that the operations of addition and multiplication enable us to combine numbers to obtain other numbers. In what follows, all sets are assumed to be subsets of a given universal set U.

> ### Set Union
>
> Let A and B be sets. The **union** of A and B, written $A \cup B$, is the set of all elements that belong to either A or B or both.
>
> $$A \cup B = \{x \mid x \in A \text{ or } x \in B \text{ or both}\}$$

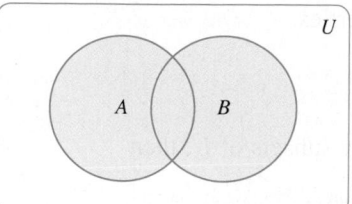

FIGURE 2
Set union $A \cup B$

The shaded portion of the Venn diagram (Figure 2) depicts the set $A \cup B$.

EXAMPLE 7 If $A = \{a, b, c\}$ and $B = \{a, c, d\}$, then $A \cup B = \{a, b, c, d\}$. ∎

> ### Set Intersection
>
> Let A and B be sets. The set of elements common to the sets A and B, written $A \cap B$, is called the **intersection** of A and B.
>
> $$A \cap B = \{x \mid x \in A \text{ and } x \in B\}$$

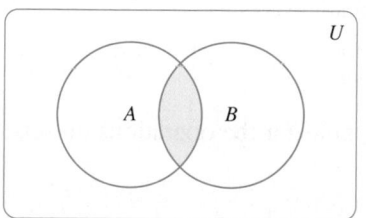

FIGURE 3
Set intersection $A \cap B$

The shaded portion of the Venn diagram (Figure 3) depicts the set $A \cap B$.

EXAMPLE 8 Let $A = \{a, b, c\}$, and let $B = \{a, c, d\}$. Then $A \cap B = \{a, c\}$. (Compare this result with Example 7.) ∎

EXAMPLE 9 Let $A = \{1, 3, 5, 7, 9\}$, and let $B = \{2, 4, 6, 8, 10\}$. Then $A \cap B = \varnothing$. ∎

The two sets of Example 9 have empty, or null, intersection. In general, the sets A and B are said to be **disjoint** if they have no elements in common—that is, if $A \cap B = \varnothing$ (see Figure 4).

EXAMPLE 10 Let U be the set of all students in the classroom. If $M = \{x \in U \mid x \text{ is male}\}$ and $F = \{x \in U \mid x \text{ is female}\}$, then $F \cap M = \varnothing$, so F and M are disjoint. ∎

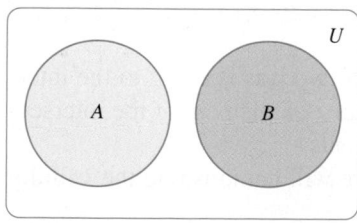

FIGURE 4
A and B are disjoint

> ### Complement of a Set
>
> If U is a universal set and A is a subset of U, then the set of all elements in U that are not in A is called the **complement** of A and is denoted A^c.
>
> $$A^c = \{x \mid x \in U \text{ and } x \notin A\}$$

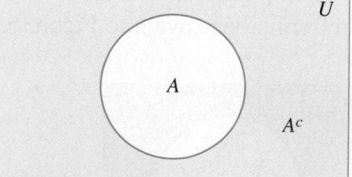

FIGURE 5
Set complementation

The shaded portion of the Venn diagram (Figure 5) shows the set A^c.

EXAMPLE 11 Let $U = \{1, 2, 3, 4, 5, 6, 7, 8, 9, 10\}$, and let $A = \{2, 4, 6, 8, 10\}$. Then $A^c = \{1, 3, 5, 7, 9\}$. ∎

> ### *Explore and Discuss*
>
> Let A, B, and C be nonempty subsets of a set U.
>
> 1. Suppose $A \cap B \neq \varnothing$, $A \cap C \neq \varnothing$, and $B \cap C \neq \varnothing$. Can you conclude that $A \cap B \cap C \neq \varnothing$? Explain your answer with an example.
> 2. Suppose $A \cap B \cap C \neq \varnothing$. Can you conclude that $A \cap B \neq \varnothing$, $A \cap C \neq \varnothing$, and $B \cap C \neq \varnothing$? Explain your answer.

The following rules hold for the operation of **complementation.** See whether you can verify them.

Set Complementation

If U is a universal set and A is a subset of U, then

a. $U^c = \varnothing$ **b.** $\varnothing^c = U$ **c.** $(A^c)^c = A$

d. $A \cup A^c = U$ **e.** $A \cap A^c = \varnothing$

The operations on sets satisfy the following properties.

Properties of Set Operations

Let U be a universal set. If A, B, and C are arbitrary subsets of U, then

$A \cup B = B \cup A$	Commutative law for union
$A \cap B = B \cap A$	Commutative law for intersection
$A \cup (B \cup C) = (A \cup B) \cup C$	Associative law for union
$A \cap (B \cap C) = (A \cap B) \cap C$	Associative law for intersection
$A \cup (B \cap C) = (A \cup B) \cap (A \cup C)$	Distributive law for union
$A \cap (B \cup C) = (A \cap B) \cup (A \cap C)$	Distributive law for intersection

---

Two additional properties, called De Morgan's Laws, hold for the operations on sets.

De Morgan's Laws

Let A and B be sets. Then

$$(A \cup B)^c = A^c \cap B^c \tag{1}$$
$$(A \cap B)^c = A^c \cup B^c \tag{2}$$

Equation (1) states that the complement of the union of two sets is equal to the intersection of their complements. Equation (2) states that the complement of the intersection of two sets is equal to the union of their complements.

We will not prove De Morgan's Laws here, but we will demonstrate the validity of Equation (2) in the following example.

EXAMPLE 12 Using Venn diagrams, show that $(A \cap B)^c = A^c \cup B^c$.

Solution $(A \cap B)^c$ is the set of elements in U but not in $A \cap B$ and is therefore the shaded region shown in Figure 6. Next, A^c and B^c are shown in Figure 7a–b. Their union, $A^c \cup B^c$, is easily seen to be equal to $(A \cap B)^c$ by referring once again to Figure 6.

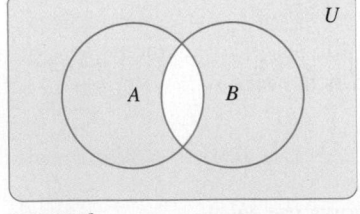

FIGURE **6**
$(A \cap B)^c$

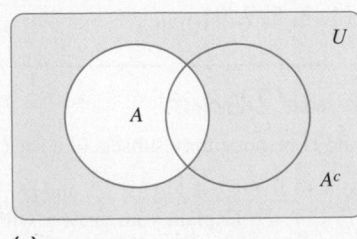

(a)

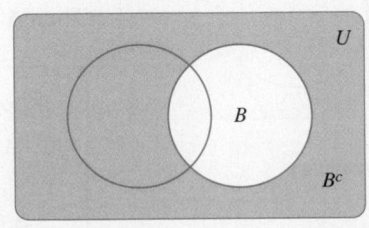

(b)

FIGURE **7**
$A^c \cup B^c$ is the set obtained by joining (a) and (b).

APPLIED EXAMPLE 13 Cyber Privacy In a poll surveying 1500 registered voters in California, the respondents were asked to rank the following companies on a scale of 0 to 10 in terms of how much they could trust these companies to keep their personal information secure, with zero meaning that they don't trust the company.

Company	Apple	Google	LinkedIn	YouTube	Facebook	Twitter
Rating	4.6	3.8	3.0	2.8	2.7	2.4

Let A denote the set of companies that have a rating higher than 2.5, let B denote the set of companies that have a rating between 2.5 and 4, and let C denote the set of companies that have a rating lower than 3. Find the following sets:

a. A, B, and C **b.** $A \cup B$ **c.** $B \cap C$ **d.** $A^c \cap B$ **e.** $A \cap B^c$

Source: Los Angeles Times.

Solution

a. $A = \{$Apple, Google, LinkedIn, YouTube, Facebook$\}$
   $B = \{$Google, LinkedIn, YouTube, Facebook$\}$
   $C = \{$YouTube, Facebook, Twitter$\}$
b. $A \cup B = \{$Apple, Google, LinkedIn, YouTube, Facebook$\} = A$
c. $B \cap C = \{$YouTube, Facebook$\}$
d. $A^c \cap B = \{$Twitter$\} \cap \{$Google, LinkedIn, YouTube, Facebook$\} = \varnothing$
e. $A \cap B^c = \{$Apple, Google, LinkedIn, YouTube, Facebook$\} \cap \{$Apple, Twitter$\}$
   $= \{$Apple$\}$

EXAMPLE 14 Let $U = \{1, 2, 3, 4, 5, 6, 7, 8, 9, 10\}$, $A = \{1, 2, 4, 8, 9\}$, and $B = \{3, 4, 5, 6, 8\}$. Verify by direct computation that $(A \cup B)^c = A^c \cap B^c$.

Solution $A \cup B = \{1, 2, 3, 4, 5, 6, 8, 9\}$, so $(A \cup B)^c = \{7, 10\}$. Moreover, $A^c = \{3, 5, 6, 7, 10\}$ and $B^c = \{1, 2, 7, 9, 10\}$, so $A^c \cap B^c = \{7, 10\}$. The required result follows.

 APPLIED EXAMPLE 15 Automobile Options Let U denote the set of all cars in a dealer's lot, and let

$$A = \{x \in U \mid x \text{ is equipped with satellite radio}\}$$
$$B = \{x \in U \mid x \text{ is equipped with a moonroof}\}$$
$$C = \{x \in U \mid x \text{ is equipped with keyless entry}\}$$

Find an expression in terms of A, B, and C for each of the following sets:

a. The set of cars with at least one of the given options
b. The set of cars with exactly one of the given options
c. The set of cars with satellite radio and keyless entry but no moonroof.

Solution

a. The set of cars with at least one of the given options is $A \cup B \cup C$ (Figure 8a).
b. The set of cars with satellite radio only is given by $A \cap B^c \cap C^c$. Similarly, we find that the set of cars with a moonroof only is given by $B \cap C^c \cap A^c$, while the set of cars with keyless entry only is given by $C \cap A^c \cap B^c$. Thus, the set of cars

with exactly one of the given options is $(A \cap B^c \cap C^c) \cup (B \cap C^c \cap A^c) \cup (C \cap A^c \cap B^c)$ (Figure 8b).

c. The set of cars with satellite radio and keyless entry but no moonroof is given by $A \cap C \cap B^c$ (Figure 8c).

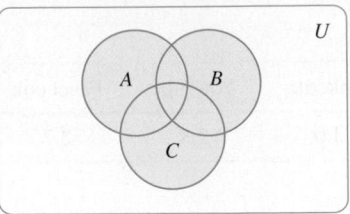

(a) The set of cars with at least one option

(b) The set of cars with exactly one option

(c) The set of cars with satellite radio and keyless entry but no moonroof

FIGURE **8**

7.1 Self-Check Exercises

1. Let $U = \{1, 2, 3, 4, 5, 6, 7\}$, $A = \{1, 2, 3\}$, $B = \{3, 4, 5, 6\}$, and $C = \{2, 3, 4\}$. Find the following sets:

a. A^c **b.** $A \cup B$ **c.** $B \cap C$
d. $(A \cup B) \cap C$ **e.** $(A \cap B) \cup C$ **f.** $A^c \cap (B \cup C)^c$

2. Politics Let U denote the set of all members of the House of Representatives. Let

$$D = \{x \in U \mid x \text{ is a Democrat}\}$$
$$R = \{x \in U \mid x \text{ is a Republican}\}$$

$$F = \{x \in U \mid x \text{ is a female}\}$$
$$L = \{x \in U \mid x \text{ is a lawyer by training}\}$$

Describe each of the following sets in words.

a. $D \cap F$ **b.** $F^c \cap R$ **c.** $D \cap F \cap L^c$

Solutions to Self-Check Exercises 7.1 can be found on page 442.

7.1 Concept Questions

1. a. What is a set? Give an example.
   b. When are two sets equal? Give an example of two equal sets.
   c. What is the empty set?

2. What can you say about two sets A and B such that
   a. $A \cup B \subseteq A$ **b.** $A \cup B = \varnothing$
   c. $A \cap B = B$ **d.** $A \cap B = \varnothing$

3. a. If $A \subset B$, what can you say about the relationship between A^c and B^c?
   b. If $A^c = \varnothing$, what can you say about A?

7.1 Exercises

In Exercises 1–4, write the set in set-builder notation.

1. The set of gold medalists in the 2014 Winter Olympic Games

2. The set of football teams in the NFL

3. $\{3, 4, 5, 6, 7\}$

4. $\{1, 3, 5, 7, 9, 11, \ldots, 39\}$

In Exercises 5–8, list the elements of the set in roster notation.

5. $\{x \mid x \text{ is a digit in the number } 352{,}646\}$

6. $\{x \mid x \text{ is a letter in the word } HIPPOPOTAMUS\}$

7. $\{x \mid 2 - x = 4 \text{ and } x \text{ is an integer}\}$

8. $\{x \mid 2 - x = 4 \text{ and } x \text{ is a fraction}\}$

In Exercises 9–14, state whether the statements are true or false.

9. a. $\{a, b, c\} = \{c, a, b\}$ **b.** $A \in A$

10. a. $\varnothing \in A$ **b.** $A \subset A$

11. a. $0 \in \varnothing$ **b.** $0 = \varnothing$

12. a. $\{\varnothing\} = \varnothing$ **b.** $\{a, b\} \in \{a, b, c\}$

13. $\{\text{Chevrolet, Cadillac, Buick}\} \subset \{x \mid x \text{ is a division of General Motors}\}$

14. $\{x \mid x \text{ is a silver medalist in the 2014 Winter Olympic Games}\} = \varnothing$

In Exercises 15 and 16, let $A = \{1, 2, 3, 4, 5\}$. Determine whether the statements are true or false.

15. a. $2 \in A$ **b.** $A \subseteq \{2, 4, 6\}$

16. a. $0 \in A$ **b.** $\{1, 3, 5\} \in A$

17. Let $A = \{1, 2, 3\}$. Which of the following sets are equal to A?
   a. $\{2, 1, 3\}$ **b.** $\{3, 2, 1\}$
   c. $\{0, 1, 2, 3\}$

18. Let $A = \{a, e, l, t, r\}$. Which of the following sets are equal to A?
   a. $\{x \mid x \text{ is a letter of the word } later\}$
   b. $\{x \mid x \text{ is a letter of the word } latter\}$
   c. $\{x \mid x \text{ is a letter of the word } relate\}$

19. List all subsets of the following sets:
   a. $\{1, 2\}$ **b.** $\{1, 2, 3\}$ **c.** $\{1, 2, 3, 4\}$

20. List all subsets of the set $A = \{\text{IBM, U.S. Steel, Union Carbide, Boeing}\}$. Which of these are proper subsets of A?

In Exercises 21–24, find the smallest possible set (i.e., the set with the least number of elements) that contains the given sets as subsets.

21. $\{1, 2\}, \{1, 3, 4\}, \{4, 6, 8, 10\}$

22. $\{1, 2, 4\}, \{a, b\}$

23. $\{\text{Jill, John, Jack}\}, \{\text{Susan, Sharon}\}$

24. $\{\text{GM, Ford, Chrysler}\}, \{\text{Daimler-Benz, Volkswagen}\}, \{\text{Toyota, Nissan}\}$

25. Use Venn diagrams to represent the following relationships:
   a. $A \subset B$ and $B \subset C$
   b. $A \subset U$ and $B \subset U$, where A and B have no elements in common
   c. The sets A, B, and C are equal.

26. Let U denote the set of all students who applied for admission to the freshman class at Faber College for the upcoming academic year, and let
   $A = \{x \in U \mid x \text{ is a successful applicant}\}$
   $B = \{x \in U \mid x \text{ is a female student who enrolled in the freshman class}\}$

$C = \{x \in U \mid x \text{ is a male student who enrolled in the freshman class}\}$

   a. Use Venn diagrams to represent the sets U, A, B, and C.
   b. Determine whether the following statements are true or false:
   i. $A \subseteq B$ **ii.** $B \subset A$ **iii.** $C \subset B$

In Exercises 27 and 28, write an expression describing the shaded portion(s) of the Venn diagram.

27. a.

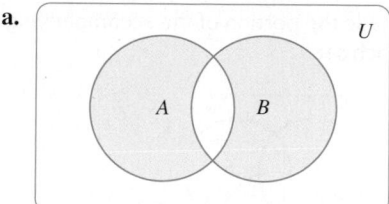

b.

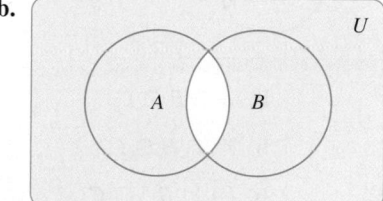

28. a.

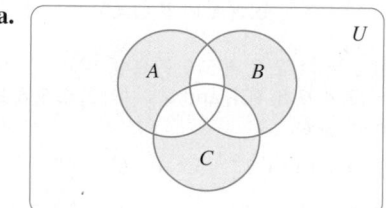

b.

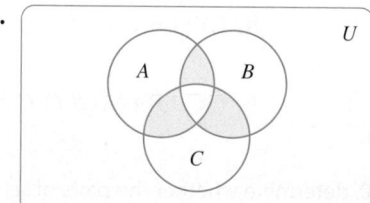

c.

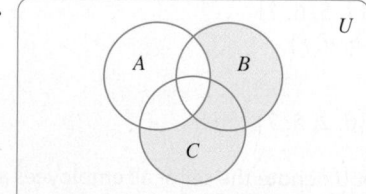

d.

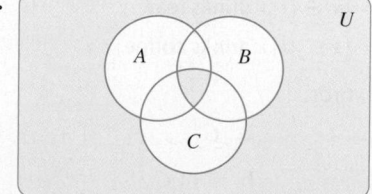

In Exercises 29 and 30, shade the portion of the accompanying figure that represents each set.

29. a. $A \cap B^c$
b. $A^c \cap B$

30. a. $A^c \cap B^c$
b. $(A \cup B)^c$

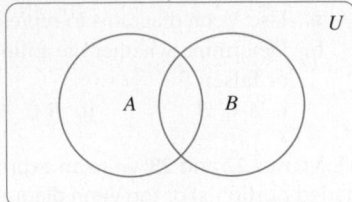

In Exercises 31–34, shade the portion of the accompanying figure that represents each set.

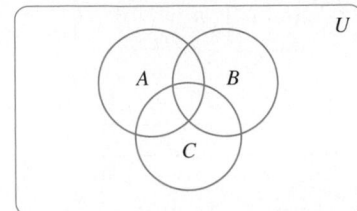

31. a. $A \cup B \cup C$ **b.** $A \cap B \cap C$

32. a. $A \cap B \cap C^c$ **b.** $A^c \cap B \cap C$

33. a. $A^c \cap B^c \cap C^c$ **b.** $(A \cup B)^c \cap C$

34. a. $A \cup (B \cap C)^c$ **b.** $(A \cup B \cup C)^c$

In Exercises 35–38, let $U = \{1, 2, 3, 4, 5, 6, 7, 8, 9, 10\}$, $A = \{1, 3, 5, 7, 9\}$, $B = \{2, 4, 6, 8, 10\}$, and $C = \{1, 2, 4, 5, 8, 9\}$. List the elements of each set.

35. a. A^c **b.** $B \cup C$ **c.** $C \cup C^c$

36. a. $C \cap C^c$ **b.** $(A \cap C)^c$ **c.** $A \cup (B \cap C)$

37. a. $(A \cap B) \cup C$ **b.** $(A \cup B \cup C)^c$
c. $(A \cap B \cap C)^c$

38. a. $A^c \cap (B \cap C^c)$ **b.** $(A \cup B^c) \cup (B \cap C^c)$
c. $(A \cup B)^c \cap C^c$

In Exercises 39 and 40, determine whether the pairs of sets are disjoint.

39. a. $\{1, 2, 3, 4\}, \{4, 5, 6, 7\}$
b. $\{a, c, e, g\}, \{b, d, f\}$

40. a. $\varnothing, \{1, 3, 5\}$
b. $\{0, 1, 3, 4\}, \{0, 2, 5, 7\}$

In Exercises 41–44, let U denote the set of all employees at Universal Life Insurance Company, and let

$$T = \{x \in U \mid x \text{ drinks tea}\}$$
$$C = \{x \in U \mid x \text{ drinks coffee}\}$$

Describe each set in words.

41. a. T^c **b.** C^c

42. a. $T \cup C$ **b.** $T \cap C$

43. a. $T \cap C^c$ **b.** $T^c \cap C$

44. a. $T^c \cap C^c$ **b.** $(T \cup C)^c$

In Exercises 45–48, let U denote the set of all employees in a hospital, and let

$$N = \{x \in U \mid x \text{ is a nurse}\}$$
$$D = \{x \in U \mid x \text{ is a doctor}\}$$
$$A = \{x \in U \mid x \text{ is an administrator}\}$$
$$M = \{x \in U \mid x \text{ is a male}\}$$
$$F = \{x \in U \mid x \text{ is a female}\}$$

Describe each set in words.

45. a. D^c **b.** N^c

46. a. $N \cup D$ **b.** $N \cap M$

47. a. $D \cap M^c$ **b.** $D \cap A$

48. a. $N \cap F$ **b.** $(D \cup N)^c$

In Exercises 49 and 50, let U denote the set of all senators in Congress, and let

$$D = \{x \in U \mid x \text{ is a Democrat}\}$$
$$R = \{x \in U \mid x \text{ is a Republican}\}$$
$$F = \{x \in U \mid x \text{ is a female}\}$$
$$L = \{x \in U \mid x \text{ is a lawyer}\}$$

Use set notation to represent each statement.

49. a. The set of all Democrats who are female
b. The set of all Republicans who are male and are not lawyers

50. a. The set of all Democrats who are female or are lawyers
b. The set of all senators who are not Democrats or are lawyers

In Exercises 51 and 52, let U denote the set of all students in the business college of a certain university. Let

$$A = \{x \in U \mid x \text{ had taken a course in accounting}\}$$
$$B = \{x \in U \mid x \text{ had taken a course in economics}\}$$
$$C = \{x \in U \mid x \text{ had taken a course in marketing}\}$$

Use set notation to represent each statement.

51. a. The set of students who have not had a course in economics
b. The set of students who have had courses in accounting and economics
c. The set of students who have had courses in accounting and economics but not marketing

52. a. The set of students who have had courses in economics but not courses in accounting or marketing
b. The set of students who have had at least one of the three courses
c. The set of students who have had all three courses

53. BEST U.S. CITY FOR ITALIAN RESTAURANTS In a reader survey conducted by *USA Today*, readers were asked to name the best U.S. city for Italian restaurants. The results follow:

City	New York	Chicago	Boston	Las Vegas	San Francisco
Respondents (%)	55	16	15	7	7

Let A denote the set of cities that were voted the best by more than 10% of the respondents, let B be the set of cities that were voted the best by between 10% and 20% of the respondents, and let C denote the set of cities that were voted the best by fewer than 10% of the respondents. Find the following sets:

a. A, B, and C **b.** $A \cup B$ **c.** $A \cap B$
d. $A^c \cap B$ **e.** $A \cap B^c$ **f.** $(A \cup B)^c$

Source: travel.usatoday.com.

54. INVENTORY LOSS The biggest cause of inventory loss, called *shrinkage*, is shoplifting, followed closely by employee theft. In a study conducted by the Center for Retail Research, the nine countries with the highest shrinkage rates, measured in the dollar amount lost for every $100 in sales, are as follows:

Country	India	Russia	Morocco	South Africa	Brazil	Mexico	Thailand	Turkey
Shrinkage Rate ($)	2.38	1.74	1.72	1.71	1.69	1.64	1.64	1.63

Let A denote the set of countries that have a shrinkage rate greater than $1.65, let B be the set of countries that have a shrinkage rate between $1.65 and $1.73, and let C be the set of countries that have a shrinkage rate less than $1.70. Find the following sets:

a. A, B, and C **b.** $A \cap B$ **c.** $A^c \cap B$
d. $A \cap B^c$ **e.** $A^c \cup B^c$

Source: Center for Retail Research Graphics.

In Exercises 55 and 56, refer to the following diagram, where U is the set of all tourists surveyed over a 1-week period in London and where

$A = \{x \in U \mid x \text{ has taken the underground [subway]}\}$
$B = \{x \in U \mid x \text{ has taken a cab}\}$
$C = \{x \in U \mid x \text{ has taken a bus}\}$

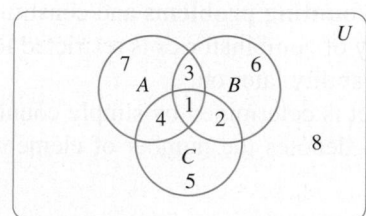

Express the indicated regions in set notation and in words.

55. a. Region 1
  b. Regions 1 and 4 together
  c. Regions 4, 5, 7, and 8 together

56. a. Region 3
  b. Regions 4 and 6 together
  c. Regions 5, 6, and 7 together

In Exercises 57–62, use Venn diagrams to illustrate each statement.

57. $A \subseteq A \cup B$; $B \subseteq A \cup B$ **58.** $A \cap B \subseteq A$; $A \cap B \subseteq B$

59. $A \cup (B \cup C) = (A \cup B) \cup C$

60. $A \cap (B \cap C) = (A \cap B) \cap C$

61. $A \cap (B \cup C) = (A \cap B) \cup (A \cap C)$

62. $(A \cup B)^c = A^c \cap B^c$

In Exercises 63 and 64, let

$$U = \{1, 2, 3, 4, 5, 6, 7, 8, 9, 10\}$$
$$A = \{1, 3, 5, 7, 9\}$$
$$B = \{1, 2, 4, 7, 8\}$$
$$C = \{2, 4, 6, 8\}$$

Verify each equation by direct computation.

63. a. $A \cup (B \cup C) = (A \cup B) \cup C$
  b. $A \cap (B \cap C) = (A \cap B) \cap C$

64. a. $A \cap (B \cup C) = (A \cap B) \cup (A \cap C)$
  b. $(A \cup B)^c = A^c \cap B^c$

In Exercises 65–68, refer to the accompanying figure, and list the points that belong to each set.

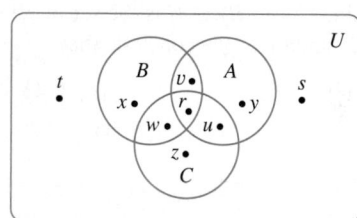

65. a. $A \cup B$ **b.** $A \cap B$

66. a. $A \cap (B \cup C)$ **b.** $(B \cap C)^c$

67. a. $(B \cup C)^c$ **b.** A^c

68. a. $(A \cap B) \cap C^c$ **b.** $(A \cup B \cup C)^c$

69. Suppose $A \subset B$ and $B \subset C$, where A and B are any two sets. What conclusion can be drawn regarding the sets A and C?

70. Verify the assertion that two sets A and B are equal if and only if (1) $A \subseteq B$ and (2) $B \subseteq A$.

In Exercises 71–80, determine whether the statement is true or false. If it is true, explain why it is true. If it is false, give an example to show why it is false.

71. A set is never a subset of itself.

72. A proper subset of a set is itself a subset of the set but not necessarily vice versa.

73. If $A \cup B = \varnothing$, then $A = \varnothing$ and $B = \varnothing$.

74. If $A \cap B = \varnothing$, then either $A = \varnothing$ or $B = \varnothing$.

75. $(A \cup A^c)^c = \varnothing$

76. If $A \subseteq B$, then $A \cap B = A$.

77. If $A \subseteq B$, then $A \cup B = B$.

78. If $A \cup B = A$, then $A \subseteq B$.

79. If $A \subset B$, then $A^c \supset B^c$.

80. $A \cap \varnothing = \varnothing$

7.1 Solutions to Self-Check Exercises

1. a. A^c is the set of all elements in U but not in A. Therefore,

$$A^c = \{4, 5, 6, 7\}$$

b. $A \cup B$ consists of all elements in A and/or B. Hence

$$A \cup B = \{1, 2, 3, 4, 5, 6\}$$

c. $B \cap C$ is the set of all elements in both B and C. Therefore,

$$B \cap C = \{3, 4\}$$

d. Using the result from part (b), we find

$$(A \cup B) \cap C = \{1, 2, 3, 4, 5, 6\} \cap \{2, 3, 4\}$$
$$= \{2, 3, 4\}$$

e. First, we compute

$$A \cap B = \{3\}$$

Next, since $(A \cap B) \cup C$ is the set of all elements in $(A \cap B)$ and/or C, we conclude that

$$(A \cap B) \cup C = \{3\} \cup \{2, 3, 4\}$$
$$= \{2, 3, 4\}$$

f. From part (a), we have $A^c = \{4, 5, 6, 7\}$. Next, we compute

$$B \cup C = \{3, 4, 5, 6\} \cup \{2, 3, 4\}$$
$$= \{2, 3, 4, 5, 6\}$$

from which we deduce that

$$(B \cup C)^c = \{1, 7\} \quad \text{The set of elements in } U \text{ but not in } B \cup C$$

Finally, using these results, we obtain

$$A^c \cap (B \cup C)^c = \{4, 5, 6, 7\} \cap \{1, 7\} = \{7\}$$

2. a. $D \cap F$ denotes the set of all elements in both D and F. Since an element in D is a Democrat and an element in F is a female representative, we see that $D \cap F$ is the set of all female Democrats in the House of Representatives.

b. Since F^c is the set of male representatives and R is the set of Republicans, it follows that $F^c \cap R$ is the set of male Republicans in the House of Representatives.

c. L^c is the set of representatives who are not lawyers by training. Therefore, $D \cap F \cap L^c$ is the set of female Democratic representatives who are not lawyers by training.

7.2 The Number of Elements in a Finite Set

Counting the Elements in a Set

The solution to some problems in mathematics calls for finding the number of elements in a set. Such problems are called **counting problems** and constitute a field of study known as **combinatorics.** Our study of combinatorics is restricted to the results that will be required for our work in probability later on.

The number of elements in a finite set is determined by simply counting the elements in the set. If A is a set, then $n(A)$ denotes the number of elements in A. For example, if

$$A = \{1, 2, 3, \ldots, 20\} \qquad B = \{a, b\} \qquad C = \{8\}$$

then $n(A) = 20$, $n(B) = 2$, and $n(C) = 1$.

The empty set has no elements in it, so $n(\varnothing) = 0$. Another result that is easily seen to be true is the following: If A and B are disjoint sets, then

$$n(A \cup B) = n(A) + n(B) \qquad \qquad (3)$$

EXAMPLE 1 If $A = \{a, c, d\}$ and $B = \{b, e, f, g\}$, then $n(A) = 3$ and $n(B) = 4$, so $n(A) + n(B) = 7$. Moreover, $A \cup B = \{a, b, c, d, e, f, g\}$ and $n(A \cup B) = 7$. Thus, Equation (3) holds true in this case. Note that $A \cap B = \varnothing$.

In the general case, A and B need not be disjoint, which leads us to the following rule.

> **Addition Rule for Sets**
>
> If A and B are finite sets, then
>
> $$n(A \cup B) = n(A) + n(B) - n(A \cap B) \qquad \textbf{(4)}$$

To see this, we observe that the set $A \cup B$ may be viewed as the union of three mutually disjoint sets with x, y, and z elements, respectively (Figure 9). This figure shows that

$$n(A \cup B) = x + y + z$$

Also,

$$n(A) = x + y \quad \text{and} \quad n(B) = y + z$$

so

$$
\begin{aligned}
n(A) + n(B) - (x + y) + (y + z) \\
= (x + y + z) + y \\
= n(A \cup B) + n(A \cap B) \qquad {\scriptstyle n(A \cap B) = y}
\end{aligned}
$$

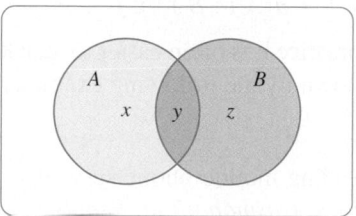

FIGURE 9
$n(A \cup B) = x + y + z$

Solving for $n(A \cup B)$, we obtain

$$n(A \cup B) = n(A) + n(B) - n(A \cap B)$$

which is the desired result.

EXAMPLE 2 Let $A = \{a, b, c, d, e\}$, and let $B = \{b, d, f, h\}$. Verify Equation (4) directly.

Solution

$$
\begin{aligned}
A \cup B = \{a, b, c, d, e, f, h\} \quad &\text{so} \quad n(A \cup B) = 7 \\
A \cap B = \{b, d\} \quad &\text{so} \quad n(A \cap B) = 2
\end{aligned}
$$

Furthermore,

$$n(A) = 5 \quad \text{and} \quad n(B) = 4$$

so

$$n(A) + n(B) - n(A \cap B) = 5 + 4 - 2 = 7 = n(A \cup B)$$

APPLIED EXAMPLE 3 Consumer Beverage Survey A survey of 100 coffee drinkers found that 70 take sugar, 60 take cream, and 50 take both sugar and cream with their coffee. How many coffee drinkers take sugar or cream with their coffee?

Solution Let U denote the set of 100 coffee drinkers surveyed, and let

$$
\begin{aligned}
A &= \{x \in U \mid x \text{ takes sugar}\} \\
B &= \{x \in U \mid x \text{ takes cream}\}
\end{aligned}
$$

Then $n(A) = 70$, $n(B) = 60$, and $n(A \cap B) = 50$. The set of coffee drinkers who take sugar or cream with their coffee is given by $A \cup B$. Using Equation (4), we find

$$n(A \cup B) = n(A) + n(B) - n(A \cap B)$$
$$= 70 + 60 - 50 = 80$$

Thus, 80 out of the 100 coffee drinkers surveyed take cream or sugar with their coffee. ∎

An equation similar to Equation (4) can be derived for the case that involves any finite number of finite sets. For example, a relationship involving the number of elements in the sets A, B, and C is given by

$$n(A \cup B \cup C) = n(A) + n(B) + n(C) - n(A \cap B) \qquad \textbf{(5)}$$
$$- n(A \cap C) - n(B \cap C) + n(A \cap B \cap C)$$

As useful as equations such as Equation (5) are, in practice it is often easier to attack a problem directly with the aid of Venn diagrams, as shown by the following example.

> *Explore and Discuss*
>
> Prove Equation (5), using an argument similar to that used to prove Equation (4). Another proof is outlined in Exercise 53 on page 451.

$ APPLIED EXAMPLE 4 Marketing Surveys A leading mobile phone service provider advertises its services in three magazines: *Cosmopolitan*, *People*, and *Time*. A survey of 500 customers by the mobile phone service provider reveals the following information:

180 learned of its services from *Cosmopolitan*.

200 learned of its services from *People*.

192 learned of its services from *Time*.

 84 learned of its services from *Cosmopolitan* and *People*.

 52 learned of its services from *Cosmopolitan* and *Time*.

 64 learned of its services from *People* and *Time*.

 38 learned of its services from all three magazines.

How many of the customers saw the manufacturer's advertisement in

a. At least one magazine?
b. Exactly one magazine?

Solution Let U denote the set of all customers surveyed, and let

$$C = \{x \in U \,|\, x \text{ learned of the services from } Cosmopolitan\}$$
$$P = \{x \in U \,|\, x \text{ learned of the services from } People\}$$
$$T = \{x \in U \,|\, x \text{ learned of the services from } Time\}$$

We begin by constructing a Venn diagram for the problem. It is best to start by filling in the number of elements in the region that is the intersection of all three sets (if any), then working with the region(s) that are the intersection of two sets, and so on. In this case, the information that 38 customers learned of the services from all three magazines translates into $n(C \cap P \cap T) = 38$ (Figure 10a). Next, the result that 64 learned of the services from *People* and *Time* translates into $n(P \cap T) = 64$. This leaves

$$64 - 38 = 26$$

who learned of the services only from *People* and *Time* (Figure 10b). Similarly, $n(C \cap T) = 52$, so

$$52 - 38 = 14$$

learned of the services only from *Cosmopolitan* and *Time*, and $n(C \cap P) = 84$, so

$$84 - 38 = 46$$

learned of the services only from *Cosmopolitan* and *People*. These numbers appear in the appropriate regions in Figure 10b.

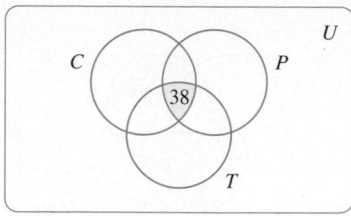

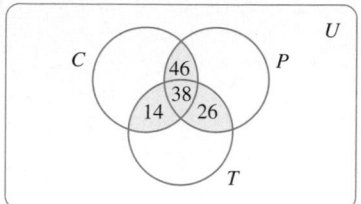

(a) All three magazines
(b) Two or more magazines
FIGURE **10**

Continuing, we have $n(T) = 192$, so the number who learned of the services only from *Time* is given by

$$192 - 14 - 38 - 26 = 114$$

(Figure 11). Similarly, $n(P) = 200$, so

$$200 - 46 - 38 - 26 = 90$$

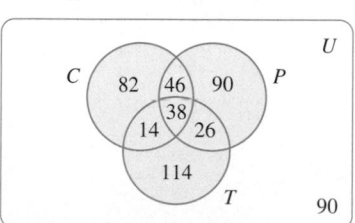

FIGURE **11**
The completed Venn diagram

learned of the services from only *People*, and $n(C) = 180$, so

$$180 - 14 - 38 - 46 = 82$$

learned of the services from only *Cosmopolitan*. Finally,

$$500 - (90 + 26 + 114 + 14 + 82 + 46 + 38) = 90$$

learned of the services from other sources.

We are now in a position to answer questions (a) and (b).

a. Referring to Figure 11, we see that the number of customers who learned of the services from at least one magazine is given by

$$n(C \cup P \cup T) = 500 - 90 = 410$$

b. The number of customers who learned of the products from exactly one magazine (Figure 12) is given by

$$n(C \cap P^c \cap T^c) + n(C^c \cap P \cap T^c) + n(C^c \cap P^c \cap T) = 82 + 90 + 114 = 286$$

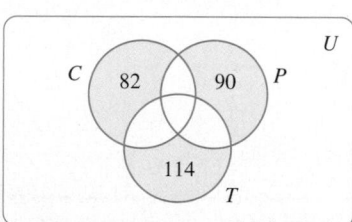

FIGURE **12**
Exactly one magazine

Our last example illustrates how we can sometimes use the solution of a system of linear equations to help us draw a Venn diagram.

EXAMPLE 5 Let A, B, and C be sets in a universal set U, and suppose $n(A \cap B \cap C^c) = 10$, $n(A \cap B^c \cap C) = 5$, $n(A^c \cap B \cap C^c) = 30$, $n(A^c \cap B^c \cap C) = 20$, $n(A^c \cap B^c \cap C^c) - 80$, $n(A) = 22$, $n(B) = 46$, and $n(U) = 156$. Use this information to complete a Venn diagram.

Solution The first five conditions lead to the Venn diagram shown in Figure 13a. Let us denote the number of elements in the three subsets that are yet to be determined by x, y, and z, as shown in Figure 13b.

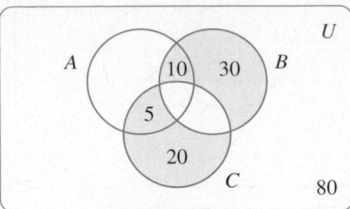

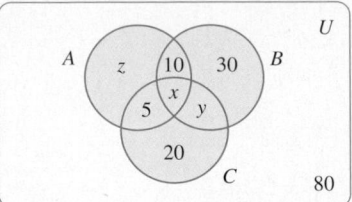

(a) The Venn diagram reflecting the first five conditions

(b) The numbers x, y, and z are to be determined.

FIGURE **13**

Then using the last three of the eight given conditions, we have

$n(A) = 22$ implies that $x + 5 + z + 10 = 22$, or $x + z = 7$

$n(B) = 46$ implies that $x + 10 + 30 + y = 46$, or $x + y = 6$

$n(U) = 156$ implies that $x + 5 + z + 10 + y + 30 + 20 + 80 = 156$, or $x + y + z = 11$

This leads to the system of equations

$$\begin{aligned} x \quad\quad + z &= 7 \\ x + y \quad\quad &= 6 \\ x + y + z &= 11 \end{aligned}$$

Subtracting the second equation from the third gives $z = 5$. Substituting this value of z into the first equation gives $x = 2$. Finally, substituting this value of x into the second equation tells us that $y = 4$. So $x = 2$, $y = 4$, and $z = 5$. This gives the completed Venn diagram shown in Figure 14.

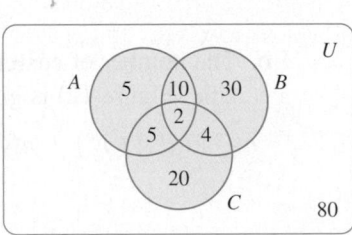

FIGURE **14**
The completed Venn diagram

7.2 Self-Check Exercises

1. Let A and B be subsets of a universal set U, and suppose that $n(U) = 100$, $n(A) = 60$, $n(B) = 40$, and $n(A \cap B) = 20$. Compute:
   a. $n(A \cup B)$ **b.** $n(A \cap B^c)$ **c.** $n(A^c \cap B)$

2. **READERSHIP SURVEY** A survey of 1000 readers of *Video Magazine* found that 166 own at least one HD player in the HD-DVD format, 161 own at least one HD player in the Blu-ray format, and 22 own HD players in both formats. How many of the readers surveyed own HD players in the HD-DVD format only? How many of the readers surveyed do not own an HD player in either format?

Solutions to Self-Check Exercises 7.2 can be found on page 451.

7.2 Concept Questions

1. a. If A and B are sets with $A \cap B = \varnothing$, what can you say about $n(A) + n(B)$? Explain.
   b. If A and B are sets satisfying $n(A \cup B) \neq n(A) + n(B)$, what can you say about $A \cap B$? Explain.

2. Let A and B be subsets of U, the universal set, and suppose that $A \cap B = \varnothing$. Is it true that $n(A) - n(B) = n(B^c) - n(A^c)$? Explain.

7.2 Exercises

In Exercises 1 and 2, verify the equation

$$n(A \cup B) = n(A) + n(B)$$

for the given disjoint sets.

1. $A = \{a, e, i, o, u\}$ and $B = \{g, h, k, l, m\}$

2. $A = \{x \mid x \text{ is a whole number between 0 and 4}\}$
   $B = \{x \mid x \text{ is a negative integer greater than } -4\}$

3. Let $A = \{2, 4, 6, 8\}$ and $B = \{6, 7, 8, 9, 10\}$. Compute:
   a. $n(A)$ **b.** $n(B)$
   c. $n(A \cup B)$ **d.** $n(A \cap B)$

4. Let $U = \{1, 2, 3, 4, 5, 6, 7, a, b, c, d, e\}$. If $A = \{1, 2, a, e\}$ and $B = \{1, 2, 3, 4, a, b, c\}$, find:
   a. $n(A^c)$ **b.** $n(A \cap B^c)$
   c. $n(A \cup B^c)$ **d.** $n(A^c \cap B^c)$

5. Verify directly that $n(A \cup B) = n(A) + n(B) - n(A \cap B)$ for the sets in Exercise 3.

6. Let $A = \{a, e, i, o, u\}$ and $B = \{b, d, e, o, u\}$. Verify by direct computation that $n(A \cup B) = n(A) + n(B) - n(A \cap B)$.

7. If $n(A) = 15$, $n(A \cap B) = 5$, and $n(A \cup B) = 30$, then what is $n(B)$?

8. If $n(A) = 10$, $n(A \cup B) = 15$, and $n(B) = 8$, then what is $n(A \cap B)$?

In Exercises 9 and 10, refer to the following Venn diagram.

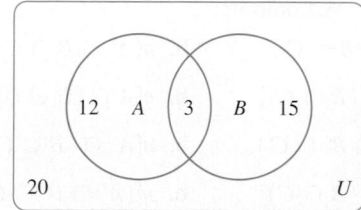

9. Find:
   a. $n(A)$ **b.** $n(A \cup B)$ **c.** $n(A^c \cap B)$
   d. $n(A \cap B^c)$ **e.** $n(U)$ **f.** $n[(A \cup B)^c]$

10. Find:
   a. $n(A \cap B)$ **b.** $n(A^c \cap B^c)$ **c.** $n[(A \cap B)^c]$
   d. $n(A^c \cup B^c)$ **e.** $n[(A \cap B^c) \cup (A^c \cap B)]$ **f.** $n(U^c)$

In Exercises 11 and 12, let A and B be subsets of a universal set U, and suppose $n(U) = 200$, $n(A) = 100$, $n(B) = 80$, and $n(A \cap B) = 40$. Compute:

11. a. $n(A \cup B)$ **b.** $n(A^c)$ **c.** $n(A \cap B^c)$

12. a. $n(A^c \cap B)$ **b.** $n(B^c)$ **c.** $n(A^c \cap B^c)$

13. Find $n(A \cup B)$ given that $n(A) = 6$, $n(B) = 10$, and $n(A \cap B) = 3$.

14. If $n(B) = 6$, $n(A \cup B) = 14$, and $n(A \cap B) = 3$, find $n(A)$.

15. If $n(A) = 4$, $n(B) = 5$, and $n(A \cup B) = 9$, find $n(A \cap B)$.

16. If $n(A) = 16$, $n(B) = 16$, $n(C) = 14$, $n(A \cap B) = 6$, $n(A \cap C) = 5$, $n(B \cap C) = 6$, and $n(A \cup B \cup C) = 31$, find $n(A \cap B \cap C)$.

17. If $n(A) = 12$, $n(B) = 12$, $n(A \cap B) = 5$, $n(A \cap C) = 5$, $n(B \cap C) = 4$, $n(A \cap B \cap C) = 2$, and $n(A \cup B \cup C) = 25$, find $n(C)$.

18. NEWSPAPER SUBSCRIBERS A survey of 1000 subscribers to the *Los Angeles Times* revealed that 900 people subscribe to the daily morning edition and 500 subscribe to both the daily morning and the Sunday editions. How many subscribe to the Sunday edition? How many subscribe to the Sunday edition only?

19. JAIL INMATES On a certain day, the Wilton County Jail held 190 prisoners accused of a crime (felony and/or misdemeanor). Of these, 130 were accused of felonies and 121 were accused of misdemeanors. How many prisoners were accused of both a felony and a misdemeanor?

20. Of 100 clock radios with digital tuners and/or CD players sold recently in a department store, 70 had digital tuners and 90 had CD players. How many radios had both digital tuners and CD players?

21. BRAND PREFERENCES OF CONSUMERS In a survey of 120 consumers conducted in a shopping mall, 80 consumers indicated that they buy Brand A of a certain product, 68 buy Brand B, and 42 buy both brands. How many consumers participating in the survey buy:
   a. At least one of these brands?
   b. Exactly one of these brands?
   c. Only Brand A?
   d. Neither of these brands?

22. **SPORTS CLUB SURVEY** In a survey of 200 members of a local sports club, 100 members indicated that they plan to attend the next Summer Olympic Games, 60 indicated that they plan to attend the next Winter Olympic Games, and 40 indicated that they plan to attend both games. How many members of the club plan to attend:
a. At least one of the two games?
b. Exactly one of the games?
c. The Summer Olympic Games only?
d. None of the games?

23. **INVESTORS' USAGE OF BROKERS** In a poll conducted among 200 active investors, it was found that 120 use discount brokers, 126 use full-service brokers, and 64 use both discount and full-service brokers. How many investors:
a. Use at least one kind of broker?
b. Use exactly one kind of broker?
c. Use only discount brokers?
d. Don't use a broker?

24. **COMMUTER TRENDS** Of 50 employees of a store located in downtown Boston, 18 people take the subway to work, 12 take the bus, and 7 take both the subway and the bus. How many employees:
a. Take the subway or the bus to work?
b. Take only the bus to work?
c. Take either the bus or the subway to work?
d. Get to work by some other means?

25. **CONSUMER SURVEY OF DESKTOP AND TABLET COMPUTER USERS** In a survey of 200 households regarding the ownership of desktop and tablet computers, the following information was obtained:

120 households own only desktop computers.

10 households own only tablet computers.

40 households own neither desktop nor tablet computers.

How many households own both desktop and tablet computers?

26. **CONSUMER SURVEY OF HDTV AND MP3 PLAYER OWNERS** In a survey of 400 households regarding the ownership of HDTVs and MP3 players, the following data were obtained:

360 households own one or more HDTVs.

170 households own one or more HDTVs and one or more MP3 players.

19 households do not own a HDTV or a MP3 player.

How many households own only one or more MP3 players?

In Exercises 27 and 28, refer to the following Venn diagram.

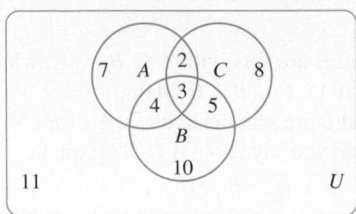

27. Find:
a. $n(A)$ **b.** $n(A \cup B)$ **c.** $n(A \cap B \cap C^c)$
d. $n[(A \cup B) \cap C^c)]$ **e.** $n[(A \cup B \cup C)^c]$

28. Find:
a. $n(A \cup B^c)$ **b.** $n[(A \cap (B \cup C)^c]$ **c.** $n(A^c)$
d. $n[(A \cap B \cap C)^c]$ **e.** $n(A^c \cup B^c \cup C^c)$

In Exercises 29–32, use the given information to draw a Venn diagram.

29. $n(A \cap B \cap C) = 3, n(A \cap B^c \cap C) = 8,$
$n(A^c \cap B^c \cap C) = 5, n(A^c \cap B \cap C^c) = 10,$
$n(A^c \cap B^c \cap C^c) = 60, n(A) = 16, n(B) = 17,$ and $n(U) = 92$

30. $n(A \cap B \cap C^c) = 2, n(A \cap B^c \cap C^c) = 10,$
$n(A \cap B^c \cap C) = 5, n(A^c \cap B \cap C) = 3,$
$n(A^c \cap B^c \cap C^c) = 9, n(B) = 10, n(C) = 12,$ and $n(A \cup B \cup C) = 27$

31. $n(A \cap B^c \cap C^c) = 4, n(A \cap B^c \cap C) = 3,$
$n(A^c \cap B^c \cap C) = 14, n(A^c \cap B \cap C^c) = 12,$
$n(A) = 15, n(B) = 22, n(C) = 24,$ and $n(A^c \cap B^c \cap C^c) = 30$

32. $n(A \cap B \cap C) = 2, n(A \cap B \cap C^c) = 5,$
$n(A^c \cap B^c \cap C) = 3, n(A^c \cap B \cap C) = 4,$
$n(A \cup B) = 20, n(B \cup C) = 19, n(A \cup C) = 21,$ and $n[(A \cup B \cup C)^c] = 20$

In Exercises 33–36, let A, B, and C be subsets of a universal set U and suppose $n(U) = 100, n(A) = 28, n(B) = 30, n(C) = 34, n(A \cap B) = 8, n(A \cap C) = 10, n(B \cap C) = 15,$ and $n(A \cap B \cap C) = 5.$ Compute:

33. a. $n(A \cup B \cup C)$ **b.** $n(A^c \cap B \cap C)$

34. a. $n[A \cap (B \cup C)]$ **b.** $n[A \cap (B \cup C)^c]$

35. a. $n(A^c \cap B^c \cap C^c)$ **b.** $n[A^c \cap (B \cup C)]$

36. a. $n[A \cup (B \cap C)]$ **b.** $n[(A^c \cap B^c \cap C^c)^c]$

37. **ON-THE-JOB DISTRACTIONS** In a survey of 500 advertising and marketing executives, the following question was posed: What is the most common cause of on-the-job distraction? The replies were as follows:

1: People stopping by your office

2: Phone calls

3: Email alerts

4: Text messages

5: Social media

6: Don't know

The number of respondents that gave each reply is given in the following table:

Reply	1	2	3	4	5	6
Number	135	130	95	50	40	50

Let A denote the set of replies that totaled more than 90, let B denote the set of replies that totaled more than 50 but less than 135, and let C denote the set of replies that totaled less than 130. Find:

a. $n(A), n(B), n(C)$ **b.** $n(A \cup B)$ **c.** $n(A \cap B^c)$

d. $n(B \cap C^c)$ **e.** $n(A \cap B^c \cap C)$

Source: The Creative Group.

38. **BRAND SWITCHING AMONG FEMALE COLLEGE STUDENTS** In a study of factors that influence brand switching by 18- to 24-year-old female college students, the following factors were identified as being significant as judged by the responses of the 439 participants in the survey:

1: Better price

2: Friend's recommendation

3: Seeing others use it

4: Interesting packaging

5: Buzz—people talking about the brand

6: An advertisement

7: Press stories

8: Entertainer/sports celebrity endorsement

The results of the survey follow:

Factor	1	2	3	4	5	6	7	8
Respondents (%)	68.8	60.4	22.8	20.8	16.8	16.8	14.9	6.9

Let A denote the set of factors that had a response greater than 20%, let B denote the set of factors that had a response between 15% and 25%, and let C denote the set of factors that had a response less than 25%. Find:

a. $n(A), n(B), n(C)$ **b.** $n(A \cap B)$ **c.** $n(A^c \cap C)$

d. $n(A \cap B^c)$ **e.** $n(A^c \cap C^c)$ **f.** $n[(A \cup B) \cap C]$

Source: Burst Media Research.

39. **BRAND SWITCHING AMONG MALE COLLEGE STUDENTS** In a study of factors that influence brand switching by 18- to 24-year-old male college students, the following factors were identified as being significant as judged by the responses of the 439 participants in the survey:

1: Better price

2: Friend's recommendation

3: Seeing others use it

4: Buzz—people talking about the brand

5: An advertisement

6: Interesting packaging

7: Press stories

8: Entertainer/sports celebrity endorsement

The results of the survey follow:

Factor	1	2	3	4	5	6	7	8
Respondents (%)	64.2	52.5	28.4	24.5	20.1	15.7	13.7	12.2

Let A denote the set of factors that had a response greater than 20%, let B denote the set of factors that had a response between 15% and 25%, and let C denote the set of factors that had a response less than 25%. Find:

a. $n(A), n(B), n(C)$ **b.** $n(A \cap B)$ **c.** $n(A^c \cap C)$

d. $n(A \cap B^c)$ **e.** $n(A^c \cap C^c)$ **f.** $n[(A \cup B) \cap C]$

Source: Burst Media Research.

40. **FEDERAL BUDGET ALLOCATION** According to the Office of Management and Budget, certain percentages of the $3.7 trillion U.S. 2012 budget were spent on the following items:

1: Social Security

2: National defense

3: Safety net programs

4: Medicare

5: Medicaid, CHIP, and other

6: Interest on the debt

7: Education

8: Transportation

9: Veterans' benefits and services

10: International affairs

11: Science and technology

12: Other

The percentages are as follows:

Item	1	2	3	4	5	6	7	8	9	10	11	12
Percent	21	18	15	13	9	6	5	4	3	2	1	3

Let A denote the set of items in which the expenditure was more than 10%, let B denote the set of items in which the expenditure was strictly between 5% and 15% (exclusive), and let C denote the set of all items in which the expenditure was less than 14%. Find:

a. $n(A), n(B), n(C)$ **b.** $n(A \cup C)$ **c.** $n(B \cap C)$

d. $n(A \cap B^c)$ **e.** $n(A^c \cap B^c)$

Source: Office of Management and Budget.

41. **SURVEY OF LEADING ECONOMISTS** A survey of the opinions of ten leading economists in a certain country showed that, because oil prices were expected to drop in that country over the next 12 months,

Seven had lowered their estimate of the consumer inflation rate.

Eight had raised their estimate of the gross national product (GNP) growth rate.

Two had lowered their estimate of the consumer inflation rate but had not raised their estimate of the GNP growth rate.

How many economists had both lowered their estimate of the consumer inflation rate and raised their estimate of the GNP growth rate for that period?

42. **STUDENT DROPOUT RATE** Data released by the Department of Education regarding the rate (percentage) of ninth-grade students who don't graduate in a certain year showed that, out of 50 states,

12 states had an increase in the dropout rate during the past 2 years.

15 states had a dropout rate of at least 30% during the past 2 years.

21 states had an increase in the dropout rate and/or a dropout rate of at least 30% during the past 2 years.

a. How many states had both a dropout rate of at least 30% and an increase in the dropout rate over the 2-year period?
b. How many states had a dropout rate that was less than 30% but that had increased over the 2-year period?

43. **STUDENT MAGAZINE PREFERENCES** A survey of 100 college students who frequent the reading lounge of a university revealed the following results:

40 read *Time*.

30 read *The New Yorker*.

25 read *Vanity Fair*.

15 read *Time* and *The New Yorker*.

12 read *Time* and *Vanity Fair*.

10 read *The New Yorker* and *Vanity Fair*.

4 read all three magazines.

How many of the students surveyed read:
a. At least one of these magazines?
b. Exactly one of these magazines?
c. Exactly two of these magazines?
d. None of these magazines?

44. **SAT SCORES** Results of a Department of Education survey of SAT test scores in 22 states showed that

10 states had an average composite SAT score of at least 1000 during the past 3 years.

15 states had an increase of at least 10 points in the average composite SAT score during the past 3 years.

8 states had both an average composite SAT score of at least 1000 and an increase in the average composite SAT score of at least 10 points during the past 3 years.

a. How many of the 22 states had composite SAT scores of less than 1000 and showed an increase of at least 10 points over the 3-year period?
b. How many of the 22 states had composite SAT scores of at least 1000 and did not show an increase of at least 10 points over the 3-year period?

45. **BRAND PREFERENCES OF CONSUMERS** The 120 consumers of Exercise 21 were also asked about their buying preferences concerning another product that is sold in the market under three labels. The results were as follows:

12 buy only those sold under Label *A*.

25 buy only those sold under Label *B*.

26 buy only those sold under Label *C*.

15 buy only those sold under Labels *A* and *B*.

10 buy only those sold under Labels *A* and *C*.

12 buy only those sold under Labels *B* and *C*.

8 buy the product sold under all three labels.

How many of the consumers surveyed buy the product sold under:
a. At least one of the three labels?
b. Labels *A* and *B* but not *C*?
c. Label *A*?
d. None of these labels?

46. **CAFETERIA STUDENT USAGE** To help plan the number of meals (breakfast, lunch, and dinner) to be prepared in a college cafeteria, a survey was conducted and the following data were obtained:

130 students ate breakfast.

180 students ate lunch.

275 students ate dinner.

68 students ate breakfast and lunch.

112 students ate breakfast and dinner.

90 students ate lunch and dinner.

58 students ate all three meals.

How many of the students ate:
a. At least one meal in the cafeteria?
b. Exactly one meal in the cafeteria?
c. Only dinner in the cafeteria?
d. Exactly two meals in the cafeteria?

47. 401(k) Investments In a survey of 200 employees of a company regarding their 401(k) investments, the following data were obtained:

141 had investments in stock funds.

91 had investments in bond funds.

60 had investments in money market funds.

47 had investments in stock funds and bond funds.

36 had investments in stock funds and money market funds.

36 had investments in bond funds and money market funds.

5 had investments only in some other vehicle.

a. How many of the employees surveyed had investments in all three types of funds?
b. How many of the employees had investments in stock funds only?

48. Newspaper Preferences of Investors In a survey of 300 individual investors regarding subscriptions to the *New York Times* (*NYT*), the *Wall Street Journal* (*WSJ*), and *USA Today* (*UST*), the following data were obtained:

122 subscribe to the *NYT*.

150 subscribe to the *WSJ*.

62 subscribe to *UST*.

38 subscribe to the *NYT* and the *WSJ*.

20 subscribe to the *NYT* and *UST*.

28 subscribe to the *WSJ* and *UST*.

36 do not subscribe to any of these newspapers.

a. How many of the individual investors surveyed subscribe to all three newspapers?
b. How many subscribe to only one of these newspapers?

In Exercises 49–52, determine whether the statement is true or false. If it is true, explain why it is true. If it is false, give an example to show why it is false.

49. If $A \cap B \neq \varnothing$, then $n(A \cup B) \neq n(A) + n(B)$.

50. If $A \subseteq B$, then $n(B) = n(A) + n(A^c \cap B)$.

51. If $n(A \cup B) = n(A) + n(B)$, then $A \cap B = \varnothing$.

52. If $n(A \cup B) = 0$ and $n(A \cap B) = 0$, then $A = \varnothing$.

53. Prove Equation (5).
   Hint: Equation (4) can be written as $n(D \cup E) = n(D) + n(E) - n(D \cap E)$. Now, put $D = A \cup B$ and $E = C$. Use Equation (4) again if necessary.

54. Find conditions on the sets A, B, and C so that
$$n(A \cup B \cup C) = n(A) + n(B) + n(C).$$

7.2 Solutions to Self-Check Exercises

1. Use the given information to construct the following Venn diagram:

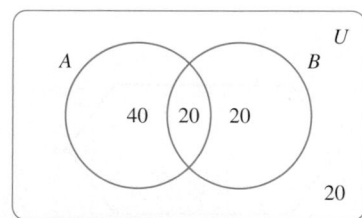

Using this diagram, we see that
a. $n(A \cup B) = 40 + 20 + 20 = 80$
b. $n(A \cap B^c) = 40$
c. $n(A^c \cap B) = 20$

2. Let U denote the set of all readers surveyed, and let

$A = \{x \in U \mid x$ owns at least one HD player in the HD-DVD format$\}$

$B = \{x \in U \mid x$ owns at least one HD player in the Blu-ray format$\}$

The fact that 22 of the readers own HD players in both formats means that $n(A \cap B) = 22$. Also, $n(A) = 166$ and $n(B) = 161$. Using this information, we obtain the following Venn diagram:

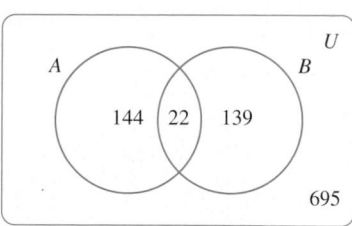

From the Venn diagram, we see that the number of readers who own HD players in only the HD format is given by

$$n(A \cap B^c) = 144$$

The number of readers who do not own an HD player in either format is given by

$$n(A^c \cap B^c) = 695$$

7.3 The Multiplication Principle

The Fundamental Principle of Counting

The solution of certain problems requires more sophisticated counting techniques than those developed in the previous section. We look at some such techniques in this and the following section. We begin by stating a fundamental principle of counting called the **multiplication principle.**

> **The Multiplication Principle**
>
> Suppose there are m ways of performing a task T_1 and n ways of performing a task T_2. Then there are mn ways of performing the task T_1 followed by the task T_2.

EXAMPLE 1 Three trunk roads connect Town A and Town B, and two trunk roads connect Town B and Town C.

a. Use the multiplication principle to find the number of ways in which a journey from Town A to Town C via Town B can be completed.

b. Verify part (a) directly by exhibiting all possible routes.

Solution

a. Since there are three ways of performing the first task (going from Town A to Town B) followed by two ways of performing the second task (going from Town B to Town C), the multiplication principle says that there are $3 \cdot 2$, or 6, ways to complete a journey from Town A to Town C via Town B.

b. Label the trunk roads connecting Town A and Town B with the Roman numerals I, II, and III, and label the trunk roads connecting Town B and Town C with the lowercase letters a and b. A schematic of this is shown in Figure 15. Then the routes from Town A to Town C via Town B can be exhibited with the aid of a **tree diagram** (Figure 16).

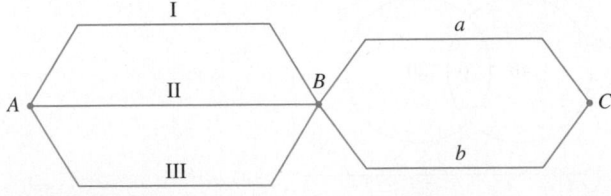

FIGURE 15
Roads from Town A to Town C

If we follow all of the branches from the initial point A to the right-hand edge of the tree, we obtain the six routes represented by six ordered pairs:

$$(\text{I}, a), (\text{I}, b), (\text{II}, a), (\text{II}, b), (\text{III}, a), (\text{III}, b)$$

where (I, a) means that the journey from Town A to Town B is made on Trunk Road I with the rest of the journey, from Town B to Town C, completed on Trunk Road a, and so forth.

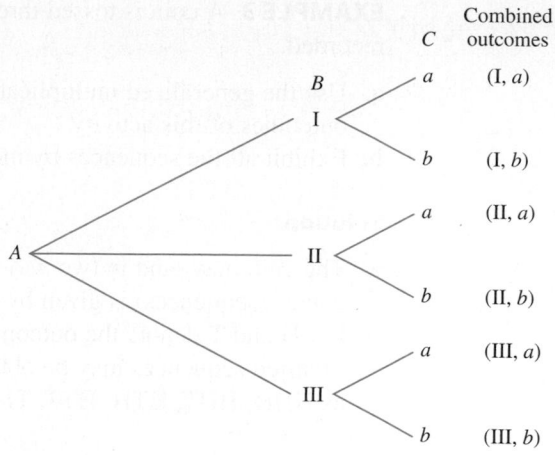

FIGURE 16
Tree diagram displaying the possible routes from Town *A* to Town *C*

Explore and Discuss

One way of gauging the performance of an airline is to track the arrival times of its flights. Suppose we denote by *E*, *O*, and *L* a flight that arrives early, on time, or late, respectively.

1. Use a tree diagram to exhibit the possible outcomes when you track two successive flights of the airline. How many outcomes are there?

2. How many outcomes are there if you track three successive flights? Justify your answer.

APPLIED EXAMPLE 2 Menu Choices Diners at Angelo's Spaghetti Bar can select their entree from 6 varieties of pasta and 28 choices of sauce. How many such combinations are there that consist of 1 variety of pasta and 1 kind of sauce?

Solution There are 6 ways of choosing a pasta followed by 28 ways of choosing a sauce, so by the multiplication principle, there are $6 \cdot 28$, or 168, combinations of this pasta dish.

The multiplication principle can be easily extended, which leads to the **generalized multiplication principle.**

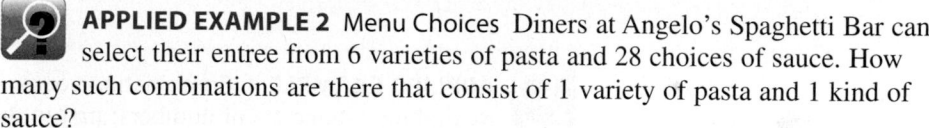

Generalized Multiplication Principle

Suppose a task T_1 can be performed in N_1 ways, a task T_2 can be performed in N_2 ways, . . . , and, finally, a task T_m can be performed in N_m ways. Then, the number of ways of performing the tasks $T_1, T_2, \ldots, T_m$ in succession is given by the product

$$N_1 N_2 \cdots N_m$$

We now illustrate the application of the generalized multiplication principle to several diverse situations.

EXAMPLE 3 A coin is tossed three times, and the sequence of heads and tails is recorded.

a. Use the generalized multiplication principle to determine the number of possible outcomes of this activity.

b. Exhibit all the sequences by means of a tree diagram.

Solution

a. The coin may land in two ways. Therefore, in three tosses, the number of outcomes (sequences) is given by $2 \cdot 2 \cdot 2$, or 8.

b. Let H and T denote the outcomes "a head" and "a tail," respectively. Then the required sequences may be obtained as shown in Figure 17, giving the sequence as HHH, HHT, HTH, HTT, THH, THT, TTH, and TTT.

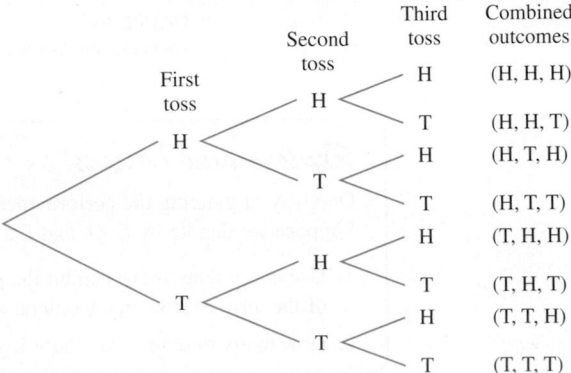

FIGURE **17**
Tree diagram displaying possible outcomes of three consecutive coin tosses

 APPLIED EXAMPLE 4 Combination Locks A combination lock is unlocked by dialing a sequence of numbers: first to the left, then to the right, and then to the left again.

a. If there are ten digits on the dial, determine the number of possible combinations.

b. How many combinations are possible if no digit is repeated?

Solution

a. There are ten choices for the first number, followed by ten for the second and ten for the third, so by the generalized multiplication principle there are $10 \cdot 10 \cdot 10$, or 1000, possible combinations.

b. Next, suppose that no number is repeated. Then, there are ten choices for the first number, followed by nine choices for the second, and eight for the third, so by the generalized multiplication principle, there are $10 \cdot 9 \cdot 8$, or 720, possible combinations.

APPLIED EXAMPLE 5 Investment Options An investor has decided to purchase shares in the stock of three companies: one engaged in aerospace activities, one involved in energy development, and one involved in electronics. After some research, the account executive of a brokerage firm has recommended that the investor consider stock from five aerospace companies, three energy development companies, and four electronics companies. In how many ways can the investor select the group of three companies from the executive's list?

Solution The investor has five choices for selecting an aerospace company, three choices for selecting an energy development company, and four choices for selecting an electronics company. Therefore, by the generalized multiplication principle, there are $5 \cdot 3 \cdot 4$, or 60, ways in which she can select a group of three companies, one from each industry group.

APPLIED EXAMPLE 6 Travel Options Tom is planning to leave for New York City from Washington, D.C., on Monday morning and has decided that he will either fly or take the train. There are five flights and two trains departing for New York City from Washington that morning. When he returns on Sunday afternoon, Tom plans to either fly or hitch a ride with a friend. There are two flights departing from New York City to Washington that afternoon. In how many ways can Tom complete this round trip?

Solution There are seven ways in which Tom can go from Washington, D.C., to New York City (five by plane and two by train). On the return trip, Tom can travel in three ways (two by plane and one by car). Therefore, by the multiplication principle, Tom can complete the round trip in $7 \cdot 3$, or 21, ways.

7.3 Self-Check Exercises

1. **SELECTING A TRAVEL PACKAGE** Encore Travel offers a "Theater Week in London" package originating from New York City. There is a choice of eight flights departing from New York City each week, a choice of five hotel accommodations, and a choice of one complimentary ticket to one of eight shows. How many such travel packages can a tourist choose from?

2. **DINNER SPECIALS** The Café Napoleon offers a dinner special on Wednesdays consisting of a choice of two entrées

(beef bourguignon and chicken basquaise); one dinner salad; one French roll; a choice of three vegetables; a choice of a carafe of burgundy, rosé, or chablis wine; a choice of coffee or tea; and a choice of six french pastries for dessert. How many combinations of dinner specials are there?

Solutions to Self-Check Exercises 7.3 can be found on page 458.

7.3 Concept Questions

1. Explain the multiplication principle, and illustrate it with a diagram.

2. Given the following tree diagram for an activity, what are the possible outcomes?

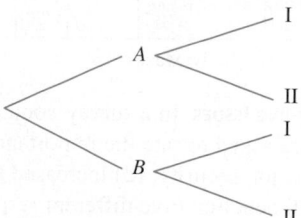

7.3 Exercises

1. **RENTAL RATES** Lynbrook West, an apartment complex financed by the State Housing Finance Agency, consists of one-, two-, three-, and four-bedroom units. The rental rate for each type of unit—low, moderate, or market—is determined by the income of the tenant. How many different rates are there?

2. **COMMUTER PASSES** Five different types of monthly commuter passes are offered by a city's local transit authority for each of three different groups of passengers: youths, adults, and senior citizens. How many different kinds of passes must be printed each month?

3. **BLACKJACK** In the game of blackjack, a two-card hand consisting of an ace and either a face card or a 10 is called a "blackjack." If a standard 52-card deck is used, determine how many blackjack hands can be dealt. (A "face card" is a jack, queen, or king.)

4. **COIN TOSSES** A coin is tossed four times, and the sequence of heads and tails is recorded.
   a. Use the generalized multiplication principle to determine the number of outcomes of this activity.
   b. Exhibit all the sequences by means of a tree diagram.

5. **WARDROBE SELECTION** A female executive selecting her wardrobe purchased two blazers, four blouses, and three skirts in coordinating colors. How many ensembles consisting of a blazer, a blouse, and a skirt can she create from this collection?

6. **COMMUTER OPTIONS** Four commuter trains and three express buses depart from City *A* to City *B* in the morning, and three commuter trains and three express buses operate on the return trip in the evening. In how many ways can a commuter from City *A* to City *B* complete a daily round trip via bus and/or train?

7. **PSYCHOLOGY EXPERIMENTS** A psychologist has constructed the following maze for use in an experiment. The maze is constructed so that a rat must pass through a series of one-way doors. How many different paths are there from start to finish?

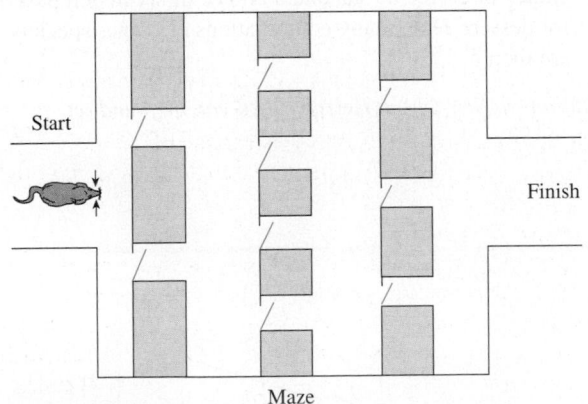

Maze

8. **UNION BARGAINING ISSUES** In a survey conducted by a union, members were asked to rate the importance of the following issues: (1) job security, (2) increased fringe benefits, and (3) health benefits. Five different responses were allowed for each issue. Among completed surveys, how many different responses to this survey were possible?

9. **HEALTH-CARE PLAN OPTIONS** A new state employee is offered a choice of ten basic health plans, three dental plans, and two vision care plans. How many different health-care plans are there to choose from if one plan is selected from each category?

10. **CODE WORDS** How many three-letter code words can be constructed from the first ten letters of the Greek alphabet if no repetitions are allowed?

11. **SOCIAL SECURITY NUMBERS** A Social Security number has nine digits. How many Social Security numbers are possible?

12. **MENU CHOICES** The New Shanghai Restaurant offers a choice of three appetizers, two choices of soups, and ten choices of entrees for its lunch special. How many different complete lunches can be ordered from the restaurant's lunch special menu?

13. **CHOOSING A PIN** Janice needs to make up a personal identification number (PIN) to be used for ATM access. She has decided that the five-digit code, to be chosen from the numbers 0 through 9, should not have a 0 as the first digit and should have an odd number as the last digit. How many such PINs are possible?

14. **MENU CHOICES** Maria's Trattoria offers mushrooms, onions, green pepper, pepperoni, Italian sausage, and anchovies as toppings for the plain cheese base of its pizzas. How many different pizzas can be made?

15. **BINARY CODES** A binary code is a method of representing text or computer processor instructions using the binary digits 0 and 1, called *bits*. How many different binary strings containing eight binary bits are possible?

16. **BUILDING A PICKUP TRUCK** The 2013 Sierra 1500 pickup truck by GMC comes with six body styles, two drive types, nine exterior colors, two interior colors, two styles of wheels, three choices of engines, and three choices of automatic transmissions. How many different models of the pickup truck can be built on the basis of these options?
    Source: GMC.

17. **MENU CHOICES** The Panini Café has a special lunch menu in which the diner chooses a soup or salad, a main course, and a dessert. There are three choices of soups, three choices of salads, five choices for the main course, and four choices of desserts. How many different three-course meals can diners order for lunch?

18. **SERIAL NUMBERS** Computers manufactured by a certain company have a serial number consisting of a letter of the alphabet followed by a four-digit number. If all the serial numbers of this type have been used, how many sets have already been manufactured?

19. **COMPUTER DATING** A computer dating service uses the results of its compatibility survey for arranging dates. The survey consists of 50 questions, each having five possible answers. How many different responses are possible if every question is answered?

20. **AUTOMOBILE COLOR AND TRIM CHOICES** The 2014 BMW 435i Coupe is offered with a choice of 13 exterior colors (10 metallic and 3 standard), 6 interior colors, and 4 trims. How many combinations involving color and trim are available for the model?
    Source: BMW.

21. **Automobile Options** The 2014 Toyota Camry comes with six grades of models, two sizes of engines, four choices of transmissions, six exterior colors, and two interior colors. How many choices of the Camry are available for a prospective buyer?
Source: Toyota.

22. **Television-Viewing Polls** An opinion poll is to be conducted among cable TV viewers. Six multiple-choice questions, each with four possible answers, will be asked. In how many different ways can a viewer complete the poll if exactly one response is given to each question?

23. **License Plate Numbers** Over the years, the state of California has used different combinations of letters of the alphabet and digits on its automobile license plates.
    a. At one time, license plates were issued that consisted of three letters followed by three digits. How many different license plates can be issued under this arrangement?
    b. Later on, license plates were issued that consisted of three digits followed by three letters. How many different license plates can be issued under this arrangement?

24. **Political Polls** An opinion poll was conducted by the Morris Polling Group. Respondents were classified according to their sex (M or F), their political affiliation (D, I, R), and the region of the country in which they reside (NW, W, C, S, E, NE).
    a. Use the generalized multiplication principle to determine the number of possible classifications.
    b. Construct a tree diagram to exhibit all possible classifications of females.

25. **License Plate Numbers** In recent years, the state of California has issued license plates using a combination of one letter of the alphabet followed by three digits, followed by another three letters of the alphabet. How many different license plates can be issued using this configuration?

26. **Exams** An exam consists of ten true-or-false questions. Assuming that every question is answered, in how many different ways can a student complete the exam? In how many ways can the exam be completed if a student can leave some questions unanswered because a penalty is assessed for each incorrect answer?

27. **Warranty Numbers** A warranty identification number for a certain product consists of a letter of the alphabet followed by a five-digit number. How many possible identification numbers are there if the first digit of the five-digit number must be nonzero?

28. **Choosing a Password** A password is to be made from a string of five characters chosen from the lowercase letters of the alphabet and the numbers 0 through 9.
    a. How many passwords are possible if there are no restrictions?
    b. How many passwords are possible if the characters must alternate between letters and numbers?

29. **Lotteries** In a state lottery, there are 15 finalists who are eligible for the Big Money Draw. In how many ways can the first, second, and third prizes be awarded if no ticket holder can win more than one prize?

30. **Combination Locks** A rolling combination four-digit padlock is unlocked by moving each of four rollers so as to produce the correct sequence. Each roller has ten digits.
    a. How many possible combinations are there?
    b. How many combinations are possible if no digit is repeated?

31. **Combination Locks** Lugano Leather Company makes an executive attaché case equipped with two rolling combination locks, each one with a provision for a three-digit number. The attaché case is unlocked by setting each lock to produce the correct combination. Each roller in each of the two locks has ten digits.
    a. How many possible combinations are there?
    b. How many combinations are possible if the lock on the left-hand side of the attaché case must end with an even number and the lock on the right-hand side must end with an odd number?

32. **Telephone Numbers**
    a. How many seven-digit telephone numbers are possible if the first digit must be nonzero?
    b. How many direct-dialing numbers for calls within the United States and Canada are possible if each number consists of a 1 plus a three-digit area code (the first digit of which must be nonzero) and a number of the type described in part (a)?

33. **Slot Machines** A "lucky dollar" is one of the nine symbols printed on each reel of a slot machine with three reels. A player receives one of various payouts whenever one or more "lucky dollars" appear in the window of the machine. Find the number of winning combinations for which the machine gives a payoff.
Hint: (a) Compute the number of ways in which the nine symbols on the first, second, and third reels can appear in the window slot and (b) compute the number of ways in which the eight symbols other than the "lucky dollar" can appear in the window slot. The difference $(a - b)$ is the number of ways in which the "lucky dollar" can appear in the window slot. Why?

34. **Staffing** Student Painters, which specializes in painting the exterior of residential buildings, has five people available to be organized into two-person and three-person teams.
    a. In how many ways can a two-person team be formed?
    b. In how many ways can a three-person team be formed?
    c. In how many ways can the company organize the available people into either two-person teams or three-person teams?

In Exercises 35 and 36, determine whether the statement is true or false. If it is true, explain why it is true. If it is false, give an example to show why it is false.

35. There are 32 three-digit odd numbers that can be formed from the digits 1, 2, 3, and 4.

36. If there are six toppings available, then the number of different pizzas that can be made is 2^5, or 32, pizzas.

7.3 Solutions to Self-Check Exercises

1. A tourist has a choice of eight flights, five hotel accommodations, and eight tickets. By the generalized multiplication principle, there are $8 \cdot 5 \cdot 8$, or 320, travel packages.

2. There is a choice of two entrées, one dinner salad, one French roll, three vegetables, three wines, two nonalco-

holic beverages, and six pastries. Therefore, by the generalized multiplication principle, there are $2 \cdot 1 \cdot 1 \cdot 3 \cdot 3 \cdot 2 \cdot 6$, or 216, combinations of dinner specials.

7.4 Permutations and Combinations

Permutations

In this section, we apply the generalized multiplication principle to the solution of two types of counting problems. Both types involve determining the number of ways in which the elements of a set can be arranged, and both play an important role in the solution of problems in probability.

We begin by considering the permutations of a set. Specifically, given a set of distinct objects, a **permutation** of the set is an arrangement of these objects in a *definite order*. To see why the order in which objects are arranged is important in certain practical situations, suppose the winning number for the first prize in a raffle is 9237. Then the number 2973, although it contains the same digits as the winning number, cannot be the first-prize winner (Figure 18). Here, the four objects—the digits 9, 2, 3, and 7—are arranged in a different order; one arrangement is associated with the winning number for the first prize, and the other is not.

FIGURE 18
The same digits appear on each ticket, but the order of the digits is different.

EXAMPLE 1 Let $A = \{a, b, c\}$.

a. Find the number of permutations of A.
b. List all the permutations of A with the aid of a tree diagram.

Solution

a. Each permutation of A consists of a sequence of the three letters a, b, c. Therefore, we may think of such a sequence as being constructed by filling in each of the three blanks

$$\underline{}\ \underline{}\ \underline{}$$

with one of the three letters. Now, there are three ways in which we can fill the first blank—we can choose a, b, or c. Having selected a letter for the first blank, we have two letters left for the second blank. Finally, there is but one way left to fill the third blank. Schematically, we have

$$\underline{\ 3\ }\ \underline{\ 2\ }\ \underline{\ 1\ }$$

Invoking the generalized multiplication principle, we conclude that there are $3 \cdot 2 \cdot 1$, or 6, permutations of the set A.

b. The tree diagram associated with this problem appears in Figure 19. The six permutations of A are abc, acb, bac, bca, cab, and cba.

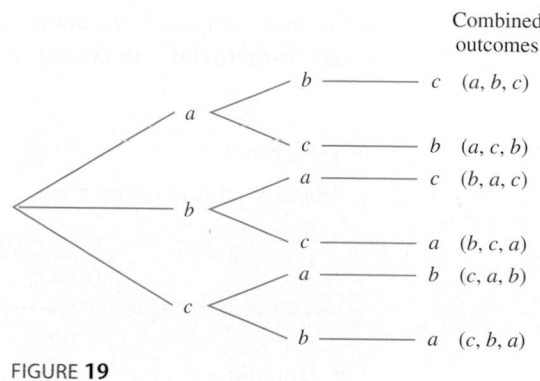

FIGURE 19
Permutations of three objects

Note Notice that when the possible outcomes are listed in the tree diagram in Example 1, order is taken into account. Thus, (a, b, c) and (a, c, b) are two different arrangements.

APPLIED EXAMPLE 2 Taking a Group Picture Find the number of ways in which a baseball team consisting of nine people can arrange themselves in a line for a group picture.

Solution We want to determine the number of permutations of the nine members of the baseball team. Each permutation in this situation consists of an arrangement of the nine team members in a line. The nine positions can be represented by nine blanks. Thus,

Position $\quad\underline{\ 1\ }\ \underline{\ 2\ }\ \underline{\ 3\ }\ \underline{\ 4\ }\ \underline{\ 5\ }\ \underline{\ 6\ }\ \underline{\ 7\ }\ \underline{\ 8\ }\ \underline{\ 9\ }$

There are nine ways to choose from among the nine players to fill the first position. When that position is filled, eight players are left, which gives us eight ways to fill the second position. Proceeding in a similar manner, we find that there are seven ways to fill the third position, and so on. Schematically, we have

Number of ways to fill each position $\quad\underline{\ 9\ }\ \underline{\ 8\ }\ \underline{\ 7\ }\ \underline{\ 6\ }\ \underline{\ 5\ }\ \underline{\ 4\ }\ \underline{\ 3\ }\ \underline{\ 2\ }\ \underline{\ 1\ }$

Invoking the generalized multiplication principle, we conclude that there are $9 \cdot 8 \cdot 7 \cdot 6 \cdot 5 \cdot 4 \cdot 3 \cdot 2 \cdot 1$, or 362,880, ways in which the baseball team can be arranged for the picture.

Whenever we are asked to determine the number of ways in which the objects of a set can be arranged in a line, order is important. For example, if we take a picture of two baseball players, A and B, then the two players can line up for the picture in two ways, AB or BA, and the two pictures will be different.

Pursuing the same line of argument used in solving the problems in the last two examples, we can derive an expression for the number of ways of permuting a set A of n distinct objects taken n at a time. In fact, each permutation may be viewed as being obtained by filling each of n blanks with one and only one element from the set. There are n ways of filling the first blank, followed by $(n - 1)$ ways of filling the second blank, and so on. Thus, by the generalized multiplication principle, there are

$$n(n - 1)(n - 2) \cdots \cdot 3 \cdot 2 \cdot 1$$

ways of permuting the elements of the set A.

Before stating this result formally, let's introduce a notation that will enable us to write in a compact form many of the expressions that follow. We use the symbol $n!$ (read "*n-factorial*") to denote the product of the first n positive integers.

> **n-Factorial**
>
> For any natural number n,
>
> $$n! = n(n - 1)(n - 2) \cdots \cdot 3 \cdot 2 \cdot 1$$
> $$0! = 1$$

For example,

$$1! = 1$$
$$2! = 2 \cdot 1 = 2$$
$$3! = 3 \cdot 2 \cdot 1 = 6$$
$$4! = 4 \cdot 3 \cdot 2 \cdot 1 = 24$$
$$5! = 5 \cdot 4 \cdot 3 \cdot 2 \cdot 1 = 120$$
$$\vdots$$
$$10! = 10 \cdot 9 \cdot 8 \cdot 7 \cdot 6 \cdot 5 \cdot 4 \cdot 3 \cdot 2 \cdot 1 = 3,628,800$$

Using this notation, we can express *the number of permutations of n distinct objects taken n at a time, denoted by $P(n, n)$, as*

$$P(n, n) = n!$$

In many situations, we are interested in determining the number of ways of permuting n distinct objects taken r at a time, where $r \leq n$. To derive a formula for computing the number of ways of permuting a set consisting of n distinct objects taken r at a time, we observe that each such permutation may be viewed as being obtained by filling each of r blanks with precisely one element from the set. Now there are n ways of filling the first blank, followed by $(n - 1)$ ways of filling the second blank, and so on. Finally, there are $(n - r + 1)$ ways of filling the rth blank. We can represent this argument schematically:

Number of ways	n	$n - 1$	$n - 2$	$\cdots$	$n - r + 1$
Position	1st	2nd	3rd		rth

Using the generalized multiplication principle, we conclude that *the number of ways of permuting n distinct objects taken r at a time, denoted by $P(n, r)$, is given by*

$$P(n, r) = \underbrace{n(n - 1)(n - 2) \cdots (n - r + 1)}_{r \text{ factors}}$$

Since

$$n(n - 1)(n - 2) \cdots (n - r + 1)$$

$$= [n(n - 1)(n - 2) \cdots (n - r + 1)] \cdot \underbrace{\frac{(n - r)(n - r - 1) \cdots \cdots 3 \cdot 2 \cdot 1}{(n - r)(n - r - 1) \cdots \cdots 3 \cdot 2 \cdot 1}}_{\text{Here we are multiplying by 1.}}$$

$$= \frac{[n(n - 1)(n - 2) \cdots (n - r + 1)][(n - r)(n - r - 1) \cdots \cdots 3 \cdot 2 \cdot 1]}{(n - r)(n - r - 1) \cdots \cdots 3 \cdot 2 \cdot 1}$$

$$= \frac{n!}{(n - r)!}$$

we have the following formula.

> **Permutations of *n* Distinct Objects**
>
> The number of *permutations* of n distinct objects taken r at a time is
>
> $$P(n, r) = \frac{n!}{(n - r)!} \tag{6}$$

Note When $r = n$, Equation (6) reduces to

$$P(n, n) = \frac{n!}{0!} = \frac{n!}{1} = n! \qquad \text{Note that } 0! = 1.$$

In other words, the number of permutations of a set of n distinct objects, taken all together, is $n!$.

EXAMPLE 3 Compute (a) $P(4, 4)$ and (b) $P(4, 2)$, and interpret your results.

Solution

a. $P(4, 4) = \dfrac{4!}{(4 - 4)!} = \dfrac{4!}{0!} = \dfrac{4!}{1} = \dfrac{4 \cdot 3 \cdot 2 \cdot 1}{1} = 24$ Note that $0! = 1.$

This gives the number of permutations of four objects taken four at a time.

b. $P(4, 2) = \dfrac{4!}{(4 - 2)!} = \dfrac{4!}{2!} = \dfrac{4 \cdot 3 \cdot 2 \cdot 1}{2 \cdot 1} = 4 \cdot 3 = 12$

This is the number of permutations of four objects taken two at a time.

EXAMPLE 4 Let $A = \{a, b, c, d\}$.

a. Use Equation (6) to compute the number of permutations of the set A taken two at a time.
b. Display the permutations of part (a) with the aid of a tree diagram.

Solution

a. Here, $n = 4$ and $r = 2$, so the required number of permutations is given by

$$P(4, 2) = \frac{4!}{(4 - 2)!} = \frac{4!}{2!} = \frac{4 \cdot 3 \cdot 2 \cdot 1}{2 \cdot 1} = 4 \cdot 3$$
$$= 12$$

b. The tree diagram associated with the problem is shown in Figure 20, and the permutations of A taken two at a time are

$$ab, \ ac, \ ad, \ ba, \ bc, \ bd, \ ca, \ cb, \ cd, \ da, \ db, \ dc$$

Combined outcomes
b (a, b)
a c (a, c)
d (a, d)
a (b, a)
b c (b, c)
d (b, d)
a (c, a)
c b (c, b)
d (c, d)
a (d, a)
d b (d, b)
c (d, c)

FIGURE 20
Permutations of four objects taken two at a time

APPLIED EXAMPLE 5 Selecting a Committee Find the number of ways in which a chairman, a vice-chairman, a secretary, and a treasurer can be chosen from a committee of eight members.

Solution   The problem is equivalent to finding the number of permutations of eight distinct objects taken four at a time. Therefore, there are

$$P(8, 4) = \frac{8!}{(8 - 4)!} = \frac{8!}{4!} = 8 \cdot 7 \cdot 6 \cdot 5 = 1680$$

ways of choosing the four officials from the committee of eight members.

The permutations considered thus far have been those involving sets of *distinct* objects. In many situations, we are interested in finding the number of permutations of a set of objects in which not all of the objects are distinct.

> **Permutations of n Objects, Not All Distinct**
>
> Given a set of n objects in which n_1 objects are alike and of one kind, n_2 objects are alike and of another kind, . . . , and n_m objects are alike and of yet another kind, so that
>
> $$n_1 + n_2 + \cdots + n_m = n$$
>
> then the number of permutations of these n objects taken n at a time is given by
>
> $$\frac{n!}{n_1! \, n_2! \cdots n_m!} \qquad (7)$$

To establish Formula (7), let's denote the number of such permutations by x. Now, if we *think* of the n_1 objects as being distinct, then they can be permuted in $n_1!$ ways. Similarly, if we *think* of the n_2 objects as being distinct, then they can be permuted in $n_2!$ ways, and so on. Therefore, if we *think* of the n objects as being distinct, then, by the generalized multiplication principle, there are $x \cdot n_1! \cdot n_2! \cdots \cdots n_m!$ permutations of these objects. But the number of permutations of a set of n distinct objects taken n at a time is just equal to $n!$. Therefore, we have

$$x(n_1! \cdot n_2! \cdots \cdots n_m!) = n!$$

from which we deduce that

$$x = \frac{n!}{n_1! \, n_2! \cdots n_m!}$$

EXAMPLE 6 Find the number of permutations that can be formed from all the letters in the word *ATLANTA*.

Solution There are seven objects (letters) involved, so $n = 7$. However, three of them are alike and of one kind (the three As), while two of them are alike and of another kind (the two Ts); hence, in this case, we have $n_1 = 3$, $n_2 = 2$, $n_3 = 1$ (the one L), and $n_4 = 1$ (the one N). Therefore, by Formula (7), there are

$$\frac{7!}{3! \, 2! \, 1! \, 1!} = \frac{7 \cdot 6 \cdot 5 \cdot 4 \cdot 3 \cdot 2 \cdot 1}{3 \cdot 2 \cdot 1 \cdot 2 \cdot 1 \cdot 1 \cdot 1} = 420$$

permutations.

APPLIED EXAMPLE 7 Management Decisions Weaver and Kline, a stock brokerage firm, has received nine inquiries regarding new accounts. In how many ways can these inquiries be directed to any three of the firm's account executives if each account executive is to handle three inquiries?

Solution If we think of the nine inquiries as being slots arranged in a row with inquiry 1 on the left and inquiry 9 on the right, then the problem can be thought of as one of filling each slot with a business card from an account executive. Then nine business cards would be used, of which three are alike and of one kind, three are alike and of another kind, and three are alike and of yet another kind. Thus, by using Formula (7) with $n = 9$ and $n_1 = n_2 = n_3 = 3$, there are

$$\frac{9!}{3! \, 3! \, 3!} = \frac{9 \cdot 8 \cdot 7 \cdot 6 \cdot 5 \cdot 4 \cdot 3 \cdot 2 \cdot 1}{3 \cdot 2 \cdot 1 \cdot 3 \cdot 2 \cdot 1 \cdot 3 \cdot 2 \cdot 1} = 1680$$

ways of assigning the inquiries.

Combinations

Until now, we have dealt with permutations of a set—that is, with arrangements of the objects of the set in which the *order* of the elements is taken into consideration. In many situations, one is interested in determining the number of ways of selecting *r* objects from a set of *n* objects without any regard to the order in which the objects are selected. Such a subset is called a **combination.**

For example, if one is interested in knowing the number of 5-card poker hands that can be dealt from a standard deck of 52 cards, then the order in which the poker hand is dealt is unimportant (Figure 21). In this situation, we are interested in determining the number of combinations of 5 cards (objects) selected from a deck (set) of 52 cards (objects). (We will solve this problem in Example 10.)

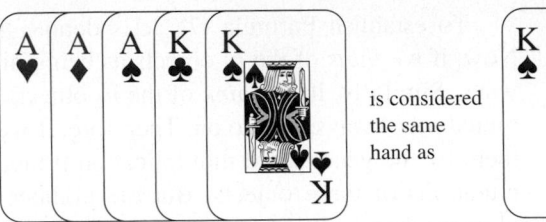

FIGURE 21

To derive a formula for determining the number of combinations of *n* objects taken *r* at a time, written

$$C(n, r) \quad \text{or} \quad \binom{n}{r}$$

we observe that each of the $C(n, r)$ combinations of *r* objects can be permuted in *r*! ways (Figure 22).

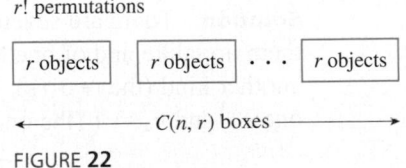

FIGURE 22

Thus, by the multiplication principle, the product $r!\, C(n, r)$ gives the number of permutations of *n* objects taken *r* at a time; that is,

$$r!\, C(n, r) = P(n, r)$$

from which we find

$$C(n, r) = \frac{P(n, r)}{r!}$$

or, using Equation (6),

$$C(n, r) = \frac{n!}{r!\,(n - r)!}$$

Combinations of *n* Objects

The number of combinations of *n* distinct objects taken *r* at a time is given by

$$C(n, r) = \frac{n!}{r!\,(n - r)!} \qquad \text{(where } r \leq n\text{)} \tag{8}$$

EXAMPLE 8 Compute and interpret the results of (a) $C(4, 4)$ and (b) $C(4, 2)$.

Solution

a. $C(4, 4) = \dfrac{4!}{4!\,(4 - 4)!} = \dfrac{4!}{4!\,0!} = 1$ Recall that $0! = 1$.

This gives 1 as the number of combinations of four distinct objects taken four at a time.

b. $C(4, 2) = \dfrac{4!}{2!\,(4 - 2)!} = \dfrac{4!}{2!\,2!} = \dfrac{4 \cdot 3 \cdot 2 \cdot 1}{2 \cdot 2} = 6$

This gives 6 as the number of combinations of four distinct objects taken two at a time.

APPLIED EXAMPLE 9 Committee Selection A Senate investigation subcommittee of four members is to be selected from a Senate committee of ten members. Determine the number of ways in which this can be done.

Solution The order in which the members of the subcommittee are selected is unimportant, so the number of ways of choosing the subcommittee is given by $C(10, 4)$, the number of combinations of ten objects taken four at a time. Hence there are

$$C(10, 4) = \frac{10!}{4!\,(10 - 4)!} = \frac{10!}{4!\,6!} = \frac{10 \cdot 9 \cdot 8 \cdot 7}{4 \cdot 3 \cdot 2 \cdot 1} = 210$$

ways of choosing such a subcommittee.

Note Remember, a combination is a selection of objects *without* regard to order. Thus, in Example 9, we used a combination formula rather than a permutation formula to solve the problem because the order of selection was not important; that is, it did not matter whether a member of the subcommittee was selected first, second, third, or fourth.

APPLIED EXAMPLE 10 Poker How many poker hands of 5 cards can be dealt from a standard deck of 52 cards?

Solution The order in which the 5 cards are dealt is not important. The number of ways of dealing a poker hand of 5 cards from a standard deck of 52 cards is given by $C(52, 5)$, the number of combinations of 52 objects taken five at a time. Thus, there are

$$\begin{aligned} C(52, 5) &= \frac{52!}{5!\,(52 - 5)!} = \frac{52!}{5!\,47!} \\ &= \frac{52 \cdot 51 \cdot 50 \cdot 49 \cdot 48}{5 \cdot 4 \cdot 3 \cdot 2 \cdot 1} \\ &= 2{,}598{,}960 \end{aligned}$$

ways of dealing such a poker hand.

The next several examples show that solving a counting problem often involves the repeated application of Equation (6) and/or (8), possibly in conjunction with the multiplication principle.

APPLIED EXAMPLE 11 Selecting Members of a Group The members of a string quartet consisting of two violinists, a violist, and a cellist are to be selected from a group of six violinists, three violists, and two cellists.

a. In how many ways can the string quartet be formed?
b. In how many ways can the string quartet be formed if one of the violinists is to be designated as the first violinist and the other is to be designated as the second violinist?

Solution

a. Since the order in which each musician is selected is not important, we use combinations. The violinists can be selected in $C(6, 2)$, or 15, ways; the violist can be selected in $C(3, 1)$, or 3, ways; and the cellist can be selected in $C(2, 1)$, or 2, ways. By the multiplication principle, there are $15 \cdot 3 \cdot 2$, or 90, ways of forming the string quartet.
b. The order in which the violinists are selected is important here. Consequently, the number of ways of selecting the violinists is given by $P(6, 2)$, or 30, ways. The number of ways of selecting the violist and the cellist remain, of course, 3 and 2, respectively. Therefore, the number of ways in which the string quartet can be formed is given by $30 \cdot 3 \cdot 2$, or 180, ways. ◼

Note The solution of Example 11 involves both a permutation and a combination. When we select two violinists from six violinists, order is not important, and we use a combination formula to solve the problem. However, when one of the violinists is designated as a first violinist, order is important, and we use a permutation formula to solve the problem. ◼

APPLIED EXAMPLE 12 Investment Options Refer to Example 5, page 454. Suppose the investor has decided to purchase shares in the stocks of two aerospace companies, two energy development companies, and two electronics companies. In how many ways can the investor select the group of six companies for the investment from the recommended list of five aerospace companies, three energy development companies, and four electronics companies?

Solution There are $C(5, 2)$ ways in which the investor can select the aerospace companies, $C(3, 2)$ ways in which she can select the companies involved in energy development, and $C(4, 2)$ ways in which she can select the electronics companies as investments. By the generalized multiplication principle, there are

$$C(5, 2)C(3, 2)C(4, 2) = \frac{5!}{2!\,3!} \cdot \frac{3!}{2!\,1!} \cdot \frac{4!}{2!\,2!}$$

$$= \frac{5 \cdot 4}{2} \cdot 3 \cdot \frac{4 \cdot 3}{2} = 180$$

ways of selecting the group of six companies for her investment. ◼

APPLIED EXAMPLE 13 Scheduling Performances The Futurists, a rock group, are planning a concert tour with performances to be given in five cities: San Francisco, Los Angeles, San Diego, Denver, and Las Vegas. In how many ways can they arrange their itinerary if:

a. There are no restrictions?
b. The three performances in California must be given consecutively?

Solution

a. The order is important here, and we see that there are

$$P(5, 5) = 5! = 120$$

ways of arranging their itinerary.

b. First, note that there are $P(3, 3)$ ways of choosing between performing in California and in the two cities outside that state. Next, there are $P(3, 3)$ ways of arranging their itinerary in the three cities in California. Therefore, by the multiplication principle, there are

$$P(3, 3)P(3, 3) = \frac{3!}{(3 - 3)!} \cdot \frac{3!}{(3 - 3)!} = 6 \cdot 6 = 36$$

ways of arranging their itinerary.

APPLIED EXAMPLE 14 U.N. Security Council Voting The United Nations Security Council consists of 5 permanent members and 10 nonpermanent members. Decisions made by the council require 9 votes for passage. However, any permanent member may veto a measure and thus block its passage. Assuming that there are no abstentions, in how many ways can a measure be passed if all 15 members of the Council vote?

Solution If a measure is to be passed, then all 5 permanent members must vote for passage of that measure. This can be done in $C(5, 5)$, or 1, way.

Next, observe that since 9 votes are required for passage of a measure, *at least* 4 of the 10 nonpermanent members must also vote for its passage. To determine the number of ways in which this can be done, notice that there are $C(10, 4)$ ways in which exactly 4 of the nonpermanent members can vote for passage of a measure, $C(10, 5)$ ways in which exactly 5 of them can vote for passage of a measure, and so on. Finally, there are $C(10, 10)$ ways in which all 10 nonpermanent members can vote for passage of a measure. Hence, there are

$$C(10, 4) + C(10, 5) + \cdots + C(10, 10)$$

ways in which at least 4 of the 10 nonpermanent members can vote for a measure. So by the multiplication principle, there are

$$C(5, 5)[C(10, 4) + C(10, 5) + \cdots + C(10, 10)]$$
$$= (1)\left[\frac{10!}{4!\, 6!} + \frac{10!}{5!\, 5!} + \cdots + \frac{10!}{10!\, 0!}\right]$$
$$= (1)(210 + 252 + 210 + 120 + 45 + 10 + 1) = 848$$

ways in which a measure can be passed.

7.4 Self-Check Exercises

1. Evaluate:
   a. 5!   **b.** $C(7, 4)$ **c.** $P(6, 2)$

2. SELECTING A SPACE SHUTTLE CREW A space shuttle crew consists of a shuttle commander, a pilot, three engineers, a scientist, and a civilian. The shuttle commander and pilot are to be chosen from 8 candidates, the three engineers from 12 candidates, the scientist from 5 candidates, and the civilian from 2 candidates. How many such space shuttle crews can be formed?

Solutions to Self-Check Exercises 7.4 can be found on page 471.

7.4 Concept Questions

1. a. What is a permutation of a set of distinct objects?
   b. How many permutations of a set of five distinct objects taken three at a time are there?

2. Given a set of ten objects in which three are alike and of one kind, three are alike and of another kind, and four are alike and of yet another kind, what is the formula for computing the permutation of these ten objects taken ten at a time?

3. a. How many combinations are there of a set of n distinct objects taken r at a time?
   b. How many combinations are there of six distinct objects taken three at a time?

7.4 Exercises

In Exercises 1–22, evaluate the given expression.

1. $3 \cdot 5!$

2. $2 \cdot 7!$

3. $\dfrac{5!}{2! \, 3!}$

4. $\dfrac{6!}{4! \, 2!}$

5. $P(5, 5)$

6. $P(6, 6)$

7. $P(5, 2)$

8. $P(5, 3)$

9. $P(n, 1)$

10. $P(k, 2)$

11. $C(6, 6)$

12. $C(8, 8)$

13. $C(7, 4)$

14. $C(9, 3)$

15. $C(5, 0)$

16. $C(6, 5)$

17. $C(9, 6)$

18. $C(10, 3)$

19. $C(n, 2)$

20. $C(7, r)$

21. $P(n, n - 2)$

22. $C(n, n - 2)$

In Exercises 23–30, classify each problem according to whether it involves a permutation or a combination.

23. In how many ways can the letters of the word *GLACIER* be arranged?

24. A 4-member executive committee is to be formed from a 12-member board of directors. In how many ways can it be formed?

25. As part of a quality-control program, 3 cell phones are selected at random for testing from 100 cell phones produced by the manufacturer. In how many ways can this test batch be chosen?

26. How many three-digit numbers can be formed by using the numerals in the set $\{3, 2, 7, 9\}$ if repetition is not allowed?

27. In how many ways can nine different books be arranged on a shelf?

28. A member of a book club wishes to purchase two books from a selection of eight books recommended for a certain month. In how many ways can she choose them?

29. How many five-card poker hands can be dealt consisting of three queens and a pair?

30. In how many ways can a six-letter security password be formed from letters of the alphabet if no letter is repeated?

31. How many four-letter permutations can be formed from the first four letters of the alphabet?

32. How many three-letter permutations can be formed from the first five letters of the alphabet?

33. In how many ways can four students be seated in a row of four seats?

34. In how many ways can five people line up at a checkout counter in a supermarket?

35. How many different batting orders can be formed for a nine-member baseball team?

36. In how many ways can the names of six candidates for political office be listed on a ballot?

37. In how many ways can a member of a hiring committee select 3 of 12 job applicants for further consideration?

38. In how many ways can an investor select four mutual funds for his investment portfolio from a recommended list of eight mutual funds?

39. Find the number of distinguishable permutations that can be formed from the letters of the word *ANTARCTICA*.

40. Find the number of distinguishable permutations that can be formed from the letters of the word *PHILIPPINES*.

41. In how many ways can the letters of the website *MySpace* be arranged if all of the letters are used and the vowels *a* and *e* must always stay in the order *ae*?

42. In how many ways can five people boarding a bus be seated if the bus has eight vacant seats?

43. How many distinct five-digit numbers can be made using the digits 1, 2, 2, 2, 7?

44. How many different signals can be made by hoisting two yellow flags, four green flags, and three red flags on a ship's mast at the same time?

45. **SUPERMARKET SITE SELECTION** In how many ways can a supermarket chain select 3 out of 12 possible sites for the construction of new supermarkets?

46. **SELECTING A READING LIST** A student is given a reading list of ten books from which he must select two for an outside reading requirement. In how many ways can he make his selections?

47. **QUALITY CONTROL** In how many ways can a quality-control engineer select a sample of 3 microprocessors for testing from a batch of 100 microprocessors?

48. **STUDY GROUP ASSIGNMENTS** A group of five students studying for a bar exam has formed a study group. Each member of the group will be responsible for preparing a study outline for one of five courses. In how many different ways can the five courses be assigned to the members of the group?

49. **TELEVISION PROGRAMMING** In how many ways can a television-programming director schedule six different commercials in the six time slots allocated to commercials during a 1-hr program?

50. **WAITING LINES** Seven people arrive at the ticket counter of Starlite Cinema at the same time. In how many ways can they line up to purchase their tickets?

51. **SELECTING A SPECIAL OCCASION CAKE** Fosselman's Ice Cream Company makes two signature cakes, a white cake and a chocolate cake. The white cake comes with one of three flavors of ice cream: burgundy-cherry, coconut-pineapple, or strawberry. The chocolate cake comes with one of four flavors: chocolate chip, chocolate raspberry, cookies-and-cream, or mint chip. Jenny is considering buying one of these cakes for her daughter's birthday. How many choices of cakes does Jenny have?
Source: Fosselman's Ice Cream Company.

52. **WEDDING CATERING** L.A. Wedding Caterers offers a wedding reception buffet. Suppose a menu is planned around four different salads, six entrees, six side dishes, and seven desserts. There are eight different choices of salads, ten different choices of entrees, eight different choices of side dishes, and ten different choices of desserts. How many menus are possible?

53. **MANAGEMENT DECISIONS** Weaver and Kline, a stock brokerage firm, has received six inquiries regarding new accounts. In how many ways can these inquiries be directed to its 12 account executives if each executive handles no more than one inquiry?

54. **CAR POOLS** A company car that has a seating capacity of six is to be used by six employees who have formed a car pool. If only four of these employees can drive, how many possible seating arrangements are there for the group?

55. **TRAVEL WARDROBE** Kaylee is planning a wardrobe for her upcoming Caribbean cruise. She has decided to bring along four blouses, four skirts, and three pairs of shorts to be selected from eight blouses, seven skirts, and six pairs of shorts. How many ensembles consisting of a blouse and either a skirt or a pair of shorts are possible?

56. **CRIMINOLOGY** The town of Carson employs 15 police officers. On a typical day, 6 of the officers are to be assigned to duty in patrol cars, 4 are assigned to the foot patrol, and the remaining 5 are assigned to duty at the station. How many different job configurations are there?

57. **BOOK EXHIBITIONS** At a college library exhibition of faculty publications, three mathematics books, four social science books, and three biology books will be displayed on a shelf. (Assume that all of the books are different.)
    a. In how many ways can the ten books be arranged on the shelf?
    b. In how many ways can the ten books be arranged on the shelf if books on the same subject matter are placed together?

58. **CONCERT SEATING** In how many ways can four married couples attending a concert be seated in a row of eight seats if:
    a. There are no restrictions?
    b. Each married couple is seated together?
    c. The members of each sex are seated together?

59. **NEWSPAPER ADS** Four items from five different departments of Metro Department Store will be featured in a one-page newspaper advertisement, as shown in the following diagram:

Advertisement

1	2	3	4
5	6	7	8
9	10	11	12
13	14	15	16
17	18	19	20

    a. In how many different ways can the 20 featured items be arranged on the page?
    b. If items from the same department must be in the same row, how many arrangements are possible?

60. **MANAGEMENT DECISIONS** C & J Realty has received 12 inquiries from prospective home buyers. In how many ways can the inquiries be directed to any four of the firm's real estate agents if each agent handles three inquiries?

61. SELECTING A BASEBALL TEAM A Little League baseball team has 12 players available for a 9-member team (no designated team positions).
   a. How many different 9-person batting orders are possible?
   b. How many different 9-member teams are possible?
   c. How many different 9-member teams and 2 alternates are possible?

62. TENNIS MATCH In the men's tennis tournament at Wimbledon, two finalists, *A* and *B*, are competing for the title, which will be awarded to the first player to win three sets. In how many different ways can the match be completed?

63. TENNIS MATCH In the women's tennis tournament at Wimbledon, two finalists, *A* and *B*, are competing for the title, which will be awarded to the first player to win two sets. In how many different ways can the match be completed?

64. JURY SELECTION In how many different ways can a panel of 12 jurors and 2 alternate jurors be chosen from a group of 30 prospective jurors?

65. U.N. VOTING Refer to Example 14. In how many ways can a measure be passed if two particular permanent and two particular nonpermanent members of the Security Council abstain from voting?

66. EXAMS A student taking an examination is required to answer exactly 10 out of 15 questions.
   a. In how many ways can the 10 questions be selected?
   b. In how many ways can the 10 questions be selected if exactly 2 of the first 3 questions must be answered?

67. TEACHING ASSISTANTSHIPS Twelve graduate students have applied for three available teaching assistantships. In how many ways can the assistantships be awarded among these applicants if:
   a. No preference is given to any student?
   b. One particular student must be awarded an assistantship?
   c. The group of applicants includes seven men and five women and it is stipulated that at least one woman must be awarded an assistantship?

68. SENATE COMMITTEES In how many ways can a subcommittee of four be chosen from a Senate committee of five Democrats and four Republicans if:
   a. All members are eligible?
   b. The subcommittee must consist of two Republicans and two Democrats?

69. CONTRACT BIDDING UBS Television Company is considering bids submitted by seven different firms for each of three different contracts. In how many ways can the contracts be awarded among these firms if no firm is to receive more than two contracts?

70. PERSONNEL SELECTION JCL Computers has five vacancies in its executive trainee program. In how many ways can the company select five trainees from a group of ten female and ten male applicants if the vacancies
   a. Can be filled by any combination of men and women?
   b. Must be filled by two men and three women?

71. COURSE SELECTION A student planning her curriculum for the upcoming year must select one of five business courses, one of three mathematics courses, two of six elective courses, and either one of four history courses or one of three social science courses. How many different curricula are available for her consideration?

72. DRIVERS' TESTS A state Motor Vehicle Department requires learners to pass a written test on the motor vehicle laws of the state. The exam consists of ten true-or-false questions, of which eight must be answered correctly to qualify for a permit. In how many different ways can a learner who answers all the questions on the exam qualify for a permit?

A list of poker hands ranked in order from the highest to the lowest is shown in the following table, along with a description and example of each hand. Use the table to answer Exercises 73–78.

Hand	Description	Example
Straight flush	5 cards in sequence in the same suit	A♥ 2♥ 3♥ 4♥ 5♥
Four of a kind	4 cards of the same rank and any other card	K♥ K♦ K♠ K♣ 2♥
Full house	3 of a kind and a pair	3♥ 3♦ 3♣ 7♥ 7♦
Flush	5 cards of the same suit that are not all in sequence	5♥ 6♥ 9♥ J♥ K♥
Straight	5 cards in sequence but not all of the same suit	10♥ J♦ Q♣ K♠ A♥
Three of a kind	3 cards of the same rank and 2 unmatched cards	K♥ K♦ K♠ 2♥ 4♦
Two pair	2 cards of the same rank and 2 cards of any other rank with an unmatched card	K♥ K♦ 2♥ 2♠ 4♣
One pair	2 cards of the same rank and 3 unmatched cards	K♥ K♦ 5♥ 2♠ 4♥

If a 5-card poker hand is dealt from a well-shuffled deck of 52 cards, how many different hands consist of the following:

73. POKER A straight flush? (Note that an ace may be played as either a high or a low card in a straight sequence—that is, A, 2, 3, 4, 5 or 10, J, Q, K, A. Hence there are ten possible sequences for a straight in one suit.)

74. POKER A straight (but not a straight flush)?

75. POKER A flush (but not a straight flush)?

76. POKER Four of a kind?

77. POKER A full house?

78. POKER Two pair?

79. BUS ROUTING The following is a schematic diagram of a city's street system between the points A and B. The City Transit Authority is in the process of selecting a route from A to B along which to provide bus service. If the company's intention is to keep the route as short as possible, how many routes must be considered?

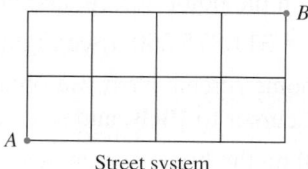

Street system

80. WORLD SERIES In the World Series, one National League team and one American League team compete for the title, which is awarded to the first team to win four games. In how many different ways can the series be completed?

81. VOTING QUORUMS A quorum (minimum) of 6 voting members is required at all meetings of the Curtis Townhomes Owners Association. If there is a total of 12 voting members in the group, find the number of ways in which this quorum can be formed.

82. CIRCULAR PERMUTATIONS Suppose n distinct objects are arranged in a circle. Show that the number of (different) circular arrangements of the n objects is $(n - 1)!$.
Hint: Consider the arrangement of the five letters A, B, C, D, and E in the accompanying figure. The permutations $ABCDE$,

$BCDEA$, $CDEAB$, $DEABC$, and $EABCD$ are not distinguishable. Generalize this observation to the case of n objects.

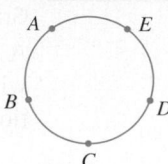

83. ROUND TABLE Refer to Exercise 82. In how many ways can five TV commentators be seated at a round table for a discussion?

84. ROUND TABLE SEATING Refer to Exercise 82. In how many ways can four men and four women be seated at a round table at a dinner party if each guest is seated between members of the opposite sex?

85. At the end of Section 6.3, we mentioned that solving a linear programming problem in three variables and five constraints by the methods of corners requires that we solve 56 3×3 systems of linear equations. Verify this assertion.

86. Refer to Exercise 85. Show that to solve a linear programming problem in five variables and ten constraints, we must solve 3003 5×5 systems of linear equations. This assertion was also made at the end of Section 6.3.

In Exercises 87–90, determine whether the statement is true or false. If it is true, explain why it is true. If it is false, give an example to show why it is false.

87. The number of permutations of n distinct objects taken all together is $n!$

88. $P(n, r) = r! \, C(n, r)$

89. The number of combinations of n objects taken $n - r$ at a time is the same as the number taken r at a time.

90. If a set of n objects consists of r elements of one kind and $n - r$ elements of another kind, then the number of permutations of the n objects taken all together is $P(n, r)$.

7.4 Solutions to Self-Check Exercises

1. a. $5! = 5 \cdot 4 \cdot 3 \cdot 2 \cdot 1 = 120$

b. $C(7, 4) = \dfrac{7!}{4! \, 3!} = \dfrac{7 \cdot 6 \cdot 5}{3 \cdot 2 \cdot 1} = 35$

c. $P(6, 2) = \dfrac{6!}{4!} = 6 \cdot 5 = 30$

2. There are $P(8, 2)$ ways of picking the shuttle commander and pilot (the order *is* important here), $C(12, 3)$ ways of picking the engineers (the order is not important here), $C(5, 1)$ ways of picking the scientist, and $C(2, 1)$ ways

of picking the civilian. By the multiplication principle, there are

$$P(8, 2) \cdot C(12, 3) \cdot C(5, 1) \cdot C(2, 1)$$

$$= \frac{8!}{6!} \cdot \frac{12!}{9! \, 3!} \cdot \frac{5!}{4! \, 1!} \cdot \frac{2!}{1! \, 1!}$$

$$= \frac{8 \cdot 7 \cdot 12 \cdot 11 \cdot 10 \cdot 5 \cdot 2}{3 \cdot 2}$$

$$= 123{,}200$$

ways in which a crew can be selected.

USING **TECHNOLOGY** Evaluating $n!$, $P(n, r)$, and $C(n, r)$

Graphing Utility

A graphing utility can be used to calculate factorials, permutations, and combinations with relative ease. Here, we use the **nPr** (permutation) and **nCr** (combination) functions of a graphing utility.

EXAMPLE 1 Use a graphing utility to find (a) 12!, (b) $P(52, 5)$, and (c) $C(38, 10)$.

a. To find 12!, first enter **12** on the home screen. Next, we obtain the **!** symbol on the TI-83/84 by pressing $\boxed{\textbf{MATH}}$, moving the cursor to **PRB,** and selecting $\boxed{\textbf{4:!}}$ by pressing $\boxed{4}$. Finally, press $\boxed{\textbf{ENTER}}$ to obtain 12! = 479,001,600. (See Figure T1.)

b. To find $P(52, 5)$, first enter **52** on the home screen. Next, we obtain the **nPr** symbol by pressing $\boxed{\textbf{MATH}}$, moving the cursor to **PRB,** and selecting $\boxed{\textbf{2: nPr}}$ by pressing $\boxed{2}$. Finally, enter **5** on the home screen, and press $\boxed{\textbf{ENTER}}$ to obtain $P(52, 5) = 52$ nPr $5 = 311{,}875{,}200$. (See Figure T1.)

c. To find $C(38, 10)$, first enter **38** on the home screen. Next, we obtain the **nCr** symbol by pressing $\boxed{\textbf{MATH}}$, moving the cursor to **PRB,** and selecting $\boxed{\textbf{3: nCr}}$ by pressing $\boxed{3}$. Finally, enter **10** on the home screen, and press $\boxed{\textbf{ENTER}}$ to obtain $C(38, 10) = 38$ nCr $10 = 472733756$. (See Figure T1.)

```
12!
        479001600
52 nPr      5
        311875200
38 nCr 10
        472733756
```

FIGURE **T1**
The TI 83/84 screen showing the entries and results for Example 1

Excel

Excel has built-in functions for calculating factorials, permutations, and combinations.

EXAMPLE 2 Use Excel to calculate (a) 12!, (b) $P(52, 5)$, and (c) $C(38, 10)$.

Solution

a. In cell A1, enter `=FACT(12)` and press $\boxed{\textbf{Shift-Enter}}$. The number 479001600 will appear.

b. In cell A2, enter `=PERMUT(52,5)` and press $\boxed{\textbf{Shift-Enter}}$. The number 311875200 will appear.

c. In cell A3, enter `=COMBIN(38,10)` and press $\boxed{\textbf{Shift-Enter}}$. The number 472733756 will appear.

Note: Boldfaced words/characters enclosed in a box (for example, $\boxed{\textbf{Enter}}$) indicate that an action (click, select, or press) is required. Words/characters printed blue (for example, Chart sub-type:) indicate words/characters that appear on the screen. Words/characters printed in a monospace font (for example, `=(-2/3)*A2+2)`) indicate words/characters that need to be typed and entered.

TECHNOLOGY EXERCISES

In Exercises 1–10, evaluate the expression.

1. 15! **2.** 20! **3.** 4(18!) **4.** $\dfrac{30!}{18!}$

5. $P(52, 7)$ **6.** $P(24, 8)$ **7.** $C(52, 7)$ **8.** $C(26, 8)$

9. $P(10, 4)C(12, 6)$ **10.** $P(20, 5)C(9, 3)C(8, 4)$

11. **Exams** A mathematics professor uses a computerized test bank to prepare her final exam. If 25 different problems are available for the first three exam questions, 40 differ-ent problems are available for the next five questions, and 30 different problems are available for the last two questions, how many different 10-question exams can she prepare? (Assume that the order of the questions within each group is not important.)

12. **Job Assignments** S & S Brokerage has received 100 inquiries from prospective clients. In how many ways can the inquiries be directed to any five of the firm's brokers if each broker handles 20 inquiries?

7.5 Experiments, Sample Spaces, and Events

Terminology

A number of specialized terms are used in the study of probability. We begin by defining the term *experiment*.

> **Experiment**
>
> An **experiment** is an activity with observable results.

The results of the experiment are called the **outcomes** of the experiment. Three examples of experiments are the following:

- Tossing a coin and observing whether it falls heads or tails
- Rolling a die and observing whether the number 1, 2, 3, 4, 5, or 6 shows up
- Testing a spark plug from a batch of 100 spark plugs and observing whether or not it is defective

In our discussion of experiments, we use the following terms:

> **Sample Point, Sample Space, and Event**
>
> **Sample point:** An outcome of an experiment
>
> **Sample space:** The set consisting of all possible sample points of an experiment
>
> **Event:** A subset of a sample space of an experiment

The sample space of an experiment is a universal set whose elements are precisely the outcomes, or the sample points, of the experiment; the events of the experiment are the subsets of the universal set. A sample space associated with an experiment that has a finite number of possible outcomes (sample points) is called a **finite sample space.**

Since the events of an experiment are subsets of a universal set (the sample space of the experiment), we may use the results for set theory given earlier to help us study probability. The event B is said to **occur** in a trial of an experiment whenever B contains the observed outcome. We begin by explaining the roles played by the empty set and a universal set when they are viewed as events associated with an experiment. The empty set, $\varnothing$, is called the *impossible event*; it cannot occur because $\varnothing$ has no elements (outcomes). Next, the universal set S is referred to as the *certain event*; it must occur because S contains all the outcomes of the experiment.

This terminology is illustrated in the next several examples.

EXAMPLE 1 Describe the sample space associated with the experiment of tossing a coin and observing whether it falls heads or tails. What are the events of this experiment?

Solution   The two outcomes are heads and tails, and the required sample space is given by $S = \{H, T\}$, where H denotes the outcome heads and T denotes the outcome tails. The events of the experiment, the subsets of S, are

$$\varnothing, \{H\}, \{T\}, S$$

Note that we have included the impossible event, $\varnothing$, and the certain event, S.

Since the events of an experiment are subsets of the sample space of the experiment, we may talk about the union and intersection of any two events; we can also consider the complement of an event with respect to the sample space.

> **Union of Two Events**
> The **union of two events** E and F is the event $E \cup F$.

Thus, the event $E \cup F$ contains the set of outcomes of E and/or F.

> **Intersection of Two Events**
> The **intersection of two events** E and F is the event $E \cap F$.

Thus, the event $E \cap F$ contains the set of outcomes common to E and F.

> **Complement of an Event**
> The **complement of event** E is the event E^c.

Thus, the event E^c is the set containing all the outcomes in the sample space S that are not in E.

Venn diagrams depicting the union, intersection, and complement of events are shown in Figure 23. These concepts are illustrated in the following example.

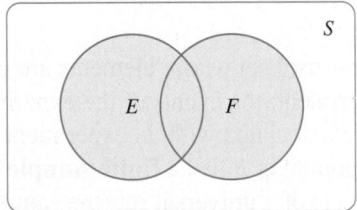

$E \cup F$

(a) The union of two events

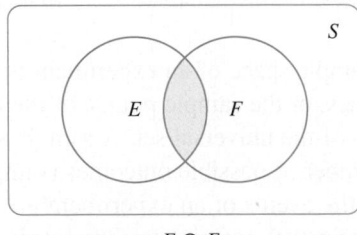

$E \cap F$

(b) The intersection of two events

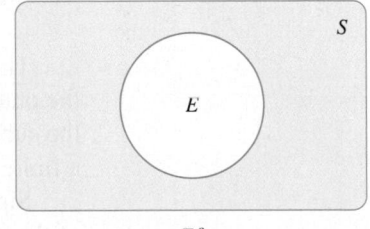

E^c

(c) The complement of the event E

FIGURE **23**

EXAMPLE 2 Consider the experiment of rolling a die and observing the number that falls uppermost. Let $S = \{1, 2, 3, 4, 5, 6\}$ denote the sample space of the experiment, and let $E = \{2, 4, 6\}$ and $F = \{1, 3\}$ be events of this experiment. List the set of outcomes of (a) $E \cup F$, (b) $E \cap F$, and (c) F^c. Interpret your results.

Solution

a. $E \cup F = \{1, 2, 3, 4, 6\}$ and is the event that the outcome of the experiment is a 1, a 2, a 3, a 4, or a 6.

b. $E \cap F = \emptyset$ is the impossible event; the number appearing uppermost when a die is rolled cannot be both even and odd at the same time.

c. $F^c = \{2, 4, 5, 6\}$ is precisely the event that the event F does not occur.

If two events cannot occur at the same time, they are said to be mutually exclusive. Using set notation, we have the following definition.

> Mutually Exclusive Events
>
> E and F are **mutually exclusive** if $E \cap F = \varnothing$.

As before, we may use Venn diagrams to illustrate these events. In this case, the two mutually exclusive events are depicted as two nonintersecting circles (Figure 24).

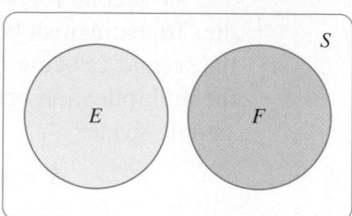

FIGURE **24**
Mutually exclusive events

EXAMPLE 3 An experiment consists of tossing a coin three times and observing the resulting sequence of heads and tails.

a. Describe the sample space S of the experiment.
b. Determine the event E that exactly two heads appear.
c. Determine the event F that at least one head appears.

Solution

a. The sample points may be obtained with the aid of a tree diagram (Figure 25).

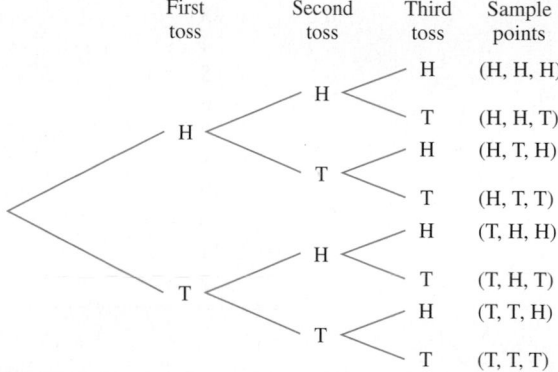

First toss	Second toss	Third toss	Sample points

FIGURE **25**

The required sample space S is given by

$$S = \{\text{HHH, HHT, HTH, HTT, THH, THT, TTH, TTT}\}$$

b. By scanning the sample space S obtained in part (a), we see that the outcomes in which exactly two heads appear are given by the event

$$E = \{\text{HHT, HTH, THH}\}$$

c. Proceeding as in part (b), we find

$$F = \{\text{HHH, HHT, HTH, HTT, THH, THT, TTH}\}$$

EXAMPLE 4 An experiment consists of rolling a pair of dice and observing the number that falls uppermost on each die.

a. Describe an appropriate sample space S for this experiment.

Explore and Discuss

1. Suppose E and F are two complementary events. Must E and F be mutually exclusive? Explain your answer.

2. Suppose E and F are mutually exclusive events. Must E and F be complementary? Explain your answer.

b. Determine the events E_2, E_3, E_4, . . . , E_{12} that the sum of the numbers falling uppermost is 2, 3, 4, . . . , 12, respectively.

Solution

a. We may represent each outcome of the experiment by an ordered pair of numbers, the first representing the number that appears uppermost on the first die and the second representing the number that appears uppermost on the second die. To distinguish between the two dice, think of the first die as being red and the second as being green. Since there are six possible outcomes for each die, the multiplication principle implies that there are 6 · 6, or 36, elements in the sample space:

$$S = \{(1, 1), (1, 2), (1, 3), (1, 4), (1, 5), (1, 6),$$
$$(2, 1), (2, 2), (2, 3), (2, 4), (2, 5), (2, 6),$$
$$(3, 1), (3, 2), (3, 3), (3, 4), (3, 5), (3, 6),$$
$$(4, 1), (4, 2), (4, 3), (4, 4), (4, 5), (4, 6),$$
$$(5, 1), (5, 2), (5, 3), (5, 4), (5, 5), (5, 6),$$
$$(6, 1), (6, 2), (6, 3), (6, 4), (6, 5), (6, 6)\}$$

b. With the aid of the results of part (a), we obtain the required list of events, shown in Table 1.

TABLE 1

Sum of Uppermost Numbers	Event
2	$E_2 = \{(1, 1)\}$
3	$E_3 = \{(1, 2), (2, 1)\}$
4	$E_4 = \{(1, 3), (2, 2), (3, 1)\}$
5	$E_5 = \{(1, 4), (2, 3), (3, 2), (4, 1)\}$
6	$E_6 = \{(1, 5), (2, 4), (3, 3), (4, 2), (5, 1)\}$
7	$E_7 = \{(1, 6), (2, 5), (3, 4), (4, 3), (5, 2), (6, 1)\}$
8	$E_8 = \{(2, 6), (3, 5), (4, 4), (5, 3), (6, 2)\}$
9	$E_9 = \{(3, 6), (4, 5), (5, 4), (6, 3)\}$
10	$E_{10} = \{(4, 6), (5, 5), (6, 4)\}$
11	$E_{11} = \{(5, 6), (6, 5)\}$
12	$E_{12} = \{(6, 6)\}$

APPLIED EXAMPLE 5 Movie Attendance The manager of a local cinema records the number of patrons attending a first-run movie at the 1 P.M. screening. The theater has a seating capacity of 500.

a. What is an appropriate sample space for this experiment?
b. Describe the event E that fewer than 50 people attend the screening.
c. Describe the event F that the theater is more than half full at the screening.

Solution

a. The number of patrons at the screening (the outcome) could run from 0 to 500. Therefore, a sample space for this experiment is

$$S = \{0, 1, 2, 3, \ldots, 500\}$$

b. $E = \{0, 1, 2, 3, \ldots, 49\}$
c. $F = \{251, 252, 253, \ldots, 500\}$

APPLIED EXAMPLE 6 Family Birth Order An experiment consists of recording, in order of their births, the sex composition of a three-child family in which the children were born at different times.

a. Describe an appropriate sample space *S* for this experiment.
b. Describe the event *E* that there are two girls and a boy in the family.
c. Describe the event *F* that the oldest child is a girl.
d. Describe the event *G* that the oldest child is a girl and the youngest child is a boy.

Solution

a. The sample points of the experiment may be obtained with the aid of the tree diagram shown in Figure 26, where *b* denotes a boy and *g* denotes a girl.

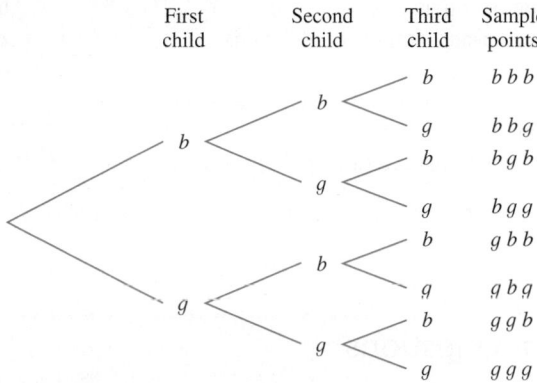

FIGURE **26**
Tree diagram for three-child families

We see from the tree diagram that the required sample space is given by

$$S = \{bbb, bbg, bgb, bgg, gbb, gbg, ggb, ggg\}$$

Using the tree diagram, we find that:

b. $E = \{bgg, gbg, ggb\}$
c. $F = \{gbb, gbg, ggb, ggg\}$
d. $G = \{gbb, ggb\}$

Sample spaces may be infinite, as illustrated in the next example.

Explore and Discuss

Think of an experiment.

1. Describe the sample point(s) and sample space of the experiment.

2. Construct two events, *E* and *F*, of the experiment.

3. Find the union and intersection of *E* and *F* and the complement of *E*.

4. Are *E* and *F* mutually exclusive? Explain your answer.

APPLIED EXAMPLE 7 Testing New Products EverBrite is developing a high-amperage, high-capacity battery as a source for powering electric cars. The battery is tested by installing it in a prototype electric car and running the car with a fully charged battery on a test track at a constant speed of 55 mph until the car runs out of power. The distance covered by the car is then observed.

a. What is the sample space for this experiment?
b. Describe the event *E* that the driving range of the prototype car under test conditions is less than 150 miles.
c. Describe the event *F* that the driving range of the prototype car is between 200 and 250 miles, inclusive.

Solution

a. Since the distance *d* covered by the car in any run may be any nonnegative number, the sample space *S* is given by

$$S = \{d \mid d \geq 0\}$$

b. The event E is given by

$$E = \{d \mid 0 \leq d < 150\}$$

c. The event F is given by

$$F = \{d \mid 200 \leq d \leq 250\}$$

7.5   Self-Check Exercises

1. SAMPLING FRUIT A sample of three apples taken from Cavallero's Fruit Stand is examined to determine whether the apples are good or rotten.
  a. What is an appropriate sample space for this experiment?
  b. Describe the event E that exactly one of the apples picked is rotten.
  c. Describe the event F that the first apple picked is rotten.

2. SAMPLING FRUIT Refer to Self-Check Exercise 1.
  a. Find $E \cup F$.
  b. Find $E \cap F$.
  c. Find F^c.
  d. Are the events E and F mutually exclusive?

Solutions to Self-Check Exercises 7.5 can be found on page 481.

7.5   Concept Questions

1. Explain what is meant by an experiment. Give an example. For the example you have chosen, describe (a) a sample point, (b) the sample space, and (c) an event of the experiment.

2. What does it mean for two events to be mutually exclusive? Give an example of two mutually exclusive events E and F. How can you prove that they are mutually exclusive?

7.5   Exercises

In Exercises 1–6, let $S = \{a, b, c, d, e, f\}$ be a sample space of an experiment, and let $E = \{a, b\}$, $F = \{a, d, f\}$, and $G = \{b, c, e\}$ be events of this experiment.

1. Find the events $E \cup F$ and $E \cap F$.

2. Find the events $F \cup G$ and $F \cap G$.

3. Find the events F^c and $E \cap G^c$.

4. Find the events E^c and $F^c \cap G$.

5. Are the events E and F mutually exclusive?

6. Are the events $E \cup F$ and $E \cap F^c$ mutually exclusive?

In Exercises 7–14, let $S = \{1, 2, 3, 4, 5, 6\}$, $E = \{2, 4, 6\}$, $F = \{1, 3, 5\}$, and $G = \{5, 6\}$.

7. Find the event $E \cup F \cup G$.

8. Find the event $E \cap F \cap G$.

9. Find the event $(E \cup F \cup G)^c$.

10. Find the event $(E \cap F \cap G)^c$.

11. Are the events E and F mutually exclusive?

12. Are the events F and G mutually exclusive?

13. Are the events E and F complementary?

14. Are the events F and G complementary?

In Exercises 15–20, let S be any sample space, and let E, F, and G be any three events associated with the experiment. Describe the events using the symbols $\cup$, $\cap$, and ^{c}.

15. The event that E and/or F occurs

16. The event that both E and F occur

17. The event that G does not occur

18. The event that E but not F occurs

19. The event that none of the events E, F, and G occurs

20. The event that E occurs but neither of the events F or G occurs

21. Consider the sample space S of Example 4, page 475.
  a. Determine the event that the number that falls uppermost on the first die is greater than the number that falls uppermost on the second die.
  b. Determine the event that the number that falls uppermost on the second die is double the number that falls uppermost on the first die.

22. Consider the sample space S of Example 4, page 475.
  a. Determine the event that the sum of the numbers falling uppermost is less than or equal to 7.
  b. Determine the event that the number falling uppermost on one die is a 4 and the number falling uppermost on the other die is greater than 4.

23. Let $S = \{a, b, c\}$ be a sample space of an experiment with outcomes a, b, and c. List all the events of this experiment.

24. Let $S = \{1, 2, 3\}$ be a sample space associated with an experiment.
  a. List all the events of this experiment.
  b. How many subsets of S contain the number 3?
  c. How many subsets of S contain either the number 2 or the number 3?

25. An experiment consists of selecting a card from a standard deck of playing cards and noting whether the card is black (B) or red (R).
  a. Describe an appropriate sample space for this experiment.
  b. What are the events of this experiment?

26. An experiment consists of selecting a letter at random from the letters in the word *MASSACHUSETTS* and observing the outcomes.
  a. What is an appropriate sample space for this experiment?
  b. Describe the event "the letter selected is a vowel."

27. An experiment consists of tossing a coin, rolling a die, and observing the outcomes.
  a. Describe an appropriate sample space for this experiment.
  b. Describe the event "a head is tossed and an even number is rolled."

28. An experiment consists of spinning the hand of the numbered disc shown in the following figure and then observing the region in which the pointer stops. (If the needle stops on a line, the result is discounted, and the needle is spun again.)

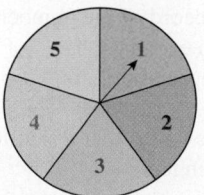

  a. What is an appropriate sample space for this experiment?

  b. Describe the event "the spinner points to the number 2."
  c. Describe the event "the spinner points to an odd number."

29. A die is rolled, and the number that falls uppermost is observed. Let E denote the event that the number shown is a 2, and let F denote the event that the number shown is an even number.
  a. Are the events E and F mutually exclusive?
  b. Are the events E and F complementary?

30. A die is rolled, and the number that falls uppermost is observed. Let E denote the event that the number shown is even, and let F denote the event that the number is an odd number.
  a. Are the events E and F mutually exclusive?
  b. Are the events E and F complementary?

31. QUALITY CONTROL A sample of three transistors taken from a local electronics store was examined to determine whether the transistors were defective (d) or nondefective (n). What is an appropriate sample space for this experiment?

32. SELECTING JOB APPLICANTS From a list of five applicants for a sales position, a, b, c, d, and e, two are selected for the next round of interviews.
  a. Describe an appropriate sample space S for this experiment.
  b. Describe the event E that the interviewees include applicant a.
  c. Describe the event F that the interviewees include applicants a and c.
  d. Describe the event G that the interviewees include applicants d and e.

33. FAMILY BIRTH ORDER An experiment consists of recording, in order of their births, the sex composition of a four-child family in which the children were born at different times.
  a. Describe an appropriate sample space S for this experiment.
  b. Describe the event E that there are three boys and a girl in the family.
  c. Describe the event F that the youngest child is a girl.
  d. Describe the event G that the oldest and the youngest children are both girls.

34. BLOOD TYPES Human blood is classified by the presence or absence of three main antigens (A, B, and Rh). When a blood specimen is typed, the presence of the A and/or B antigen is indicated by listing the letter A and/or the letter B. If neither the A nor the B antigen is present, the letter O is used. The presence or absence of the Rh antigen is indicated by the symbols $+$ or $-$, respectively. Thus, if a blood specimen is classified as AB^+, it contains the A and the B antigens as well as the Rh antigen. Similarly, O^- blood contains none of the three antigens. Using this information, determine the sample space corresponding to the different blood groups.

35. GAME SHOWS In a television game show, the winner is asked to select three prizes from five different prizes, A, B, C, D, and E.

a. Describe a sample space of possible outcomes (order is not important).

b. How many points are there in the sample space corresponding to a selection that includes A?

c. How many points are there in the sample space corresponding to a selection that includes A and B?

d. How many points are there in the sample space corresponding to a selection that includes either A or B?

36. ATMs The manager of a local bank observes how long it takes a customer to complete his transactions at the bank's automatic teller machine (ATM).

a. Describe an appropriate sample space for this experiment.

b. Describe the event that it takes a customer between 2 and 3 min to complete his transactions at the ATM.

37. PRICE CHANGES IN COMMON STOCKS Robin purchased shares of a machine tool company and shares an airline company. Let E be the event that the shares of the machine tool company increase in value over the next 6 months, and let F be the event that the shares of the airline company increase in value over the next 6 months. Using the symbols $\cup$, $\cap$, and ^{c}, describe the following events.

a. The shares in the machine tool company do not increase in value.

b. The shares in both the machine tool company and the airline company do not increase in value.

c. The shares of at least one of the two companies increase in value.

d. The shares of only one of the two companies increase in value.

38. QUALITY ASSURANCE SURVEYS The customer service department of Universal Instruments, manufacturer of the Orion tablet computer, conducted a survey among customers who had returned their purchase registration cards. Purchasers of its tablet computer were asked to report the length of time t in days before service was required.

a. Describe a sample space corresponding to this survey.

b. Describe the event E that a tablet computer required service before a period of 90 days had elapsed.

c. Describe the event F that a tablet computer did not require service before a period of 1 year had elapsed.

39. ASSEMBLY-TIME STUDIES A time study was conducted by the production manager of Vista Vision to determine the length of time in minutes required by an assembly worker to complete a certain task during the assembly of its Pulsar HDTV sets.

a. Describe a sample space corresponding to this time study.

b. Describe the event E that an assembly worker took 2 min or less to complete the task.

c. Describe the event F that an assembly worker took more than 2 min to complete the task.

40. POLITICAL POLLS An opinion poll is conducted among a state's electorate to determine the relationship between their income levels and their stands on a proposition aimed at reducing state income taxes. Voters are classified as belonging to either the low-, middle-, or upper-income group. They are asked whether they favor, oppose, or are undecided about the proposition. Let the letters L, M, and U represent the low-, middle-, and upper-income groups, respectively, and let the letters f, o, and u represent the responses—favor, oppose, and undecided, respectively.

a. Describe a sample space corresponding to this poll.

b. Describe the event E_1 that a respondent favors the proposition.

c. Describe the event E_2 that a respondent opposes the proposition and does not belong to the low-income group.

d. Describe the event E_3 that a respondent does not favor the proposition and does not belong to the upper-income group.

41. QUALITY CONTROL As part of a quality-control procedure, an inspector at Bristol Farms randomly selects ten eggs from each consignment of eggs he receives and records the number of broken eggs.

a. What is an appropriate sample space for this experiment?

b. Describe the event E that at most three eggs are broken.

c. Describe the event F that at least five eggs are broken.

42. POLITICAL POLLS In the opinion poll of Exercise 40, the voters were also asked to indicate their political affiliations: Democrat, Republican, or Independent. As before, let the letters L, M, and U represent the low-, middle-, and upper-income groups, respectively. Let the letters D, R, and I represent Democrat, Republican, and Independent, respectively.

a. Describe a sample space corresponding to this poll.

b. Describe the event E_1 that a respondent is a Democrat.

c. Describe the event E_2 that a respondent belongs to the upper-income group and is a Republican.

d. Describe the event E_3 that a respondent belongs to the middle-income group and is not a Democrat.

43. SHUTTLE BUS USAGE A certain airport hotel operates a shuttle bus service between the hotel and the airport. The maximum capacity of a bus is 20 passengers. On alternate trips of the shuttle bus over a period of 1 week, the hotel manager kept a record of the number of passengers arriving at the hotel in each bus.

a. What is an appropriate sample space for this experiment?

b. Describe the event E that a shuttle bus carried fewer than ten passengers.

c. Describe the event F that a shuttle bus arrived with a full load.

44. Sports Eight players, *A*, *B*, *C*, *D*, *E*, *F*, *G*, and *H*, are competing in a series of elimination matches of a tennis tournament in which the winner of each preliminary match will advance to the semifinals and the winners of the semifinals will advance to the finals. An outline of the scheduled matches follows. Describe a sample space listing the possible participants in the finals.

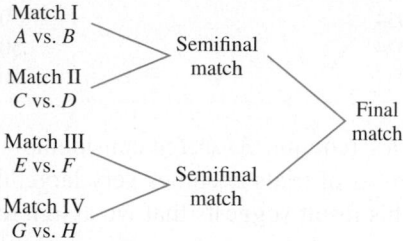

45. An experiment consists of selecting a card at random from a well-shuffled 52-card deck. Let *E* denote the event that an ace is drawn, and let *F* denote the event that a spade is drawn. Show that $n(E \cup F) = n(E) + n(F) - n(E \cap F)$.

46. Let *S* be a sample space for an experiment. Show that if *E* is any event of an experiment, then *E* and E^c are mutually exclusive.

47. Let *S* be a sample space for an experiment, and let *E* and *F* be events of this experiment. Show that the events $E \cup F$ and $E^c \cap F^c$ are mutually exclusive.
Hint: Use De Morgan's Law.

48. Let *S* be a sample space of an experiment with *n* outcomes. Determine the number of events of this experiment.

In Exercises 49 and 50, determine whether the statement is true or false. If it is true, explain why it is true. If it is false, give an example to show why it is false.

49. If *E* and *F* are mutually exclusive and *E* and *G* are mutually exclusive, then *F* and *G* are mutually exclusive.

50. The numbers 1, 2, and 3 are written separately on three pieces of paper. These slips of paper are then placed in a bowl. If you draw two slips from the bowl, one at a time and without replacement, then the sample space for this experiment consists of six elements.

7.5 Solutions to Self-Check Exercises

1. a. Let *g* denote a good apple, and let *r* denote a rotten apple. Thus, the required sample points may be obtained with the aid of a tree diagram (compare with Example 3). The required sample space is given by

$$S = \{ggg, ggr, grg, grr, rgg, rgr, rrg, rrr\}$$

b. By scanning the sample space *S* obtained in part (a), we identify the outcomes in which exactly one apple is rotten. We find

$$E = \{ggr, grg, rgg\}$$

c. Proceeding as in part (b), we find

$$F = \{rgg, rgr, rrg, rrr\}$$

2. Using the results of Self-Check Exercise 1, we find:
a. $E \cup F = \{ggr, grg, rgg, rgr, rrg, rrr\}$
b. $E \cap F = \{rgg\}$
c. F^c is the set of outcomes in *S* but not in *F*. Thus,

$$F^c = \{ggg, ggr, grg, grr\}$$

d. Since $E \cap F \neq \varnothing$, we conclude that *E* and *F* are not mutually exclusive.

7.6 Definition of Probability

Finding the Probability of an Event

Let's return to the coin-tossing experiment. The sample space of this experiment is given by $S = \{H, T\}$, where the sample points H and T correspond to the two possible outcomes, heads and tails. If the coin is *unbiased*, then there is *one chance out of two* of obtaining a head (or a tail), and we say that the *probability* of tossing a head (tail) is $\frac{1}{2}$, abbreviated

$$P(H) = \frac{1}{2} \quad \text{and} \quad P(T) = \frac{1}{2}$$

An alternative method of obtaining the values of $P(H)$ and $P(T)$ is based on continued experimentation and does not depend on the assumption that the two outcomes are equally likely. Table 2 summarizes the results of such an exercise.

TABLE 2

Tossing a Coin: As the Number of Trials Increases, the Relative Frequency Approaches .5

Number of Tosses, n	Number of Heads, m	Relative Frequency of Heads, m/n
10	4	.4000
100	58	.5800
1,000	492	.4920
10,000	5,034	.5034
20,000	10,024	.5012
40,000	20,032	.5008

Observe that the relative frequencies (column 3) differ considerably when the number of trials is small, but as the number of trials becomes very large, the relative frequency approaches the number .5. This result suggests that we assign to $P(\text{H})$ the value $\frac{1}{2}$, as before.

More generally, consider an experiment that may be repeated over and over again under independent and similar conditions. Suppose that in n trials an event E occurs m times. We call the ratio m/n the **relative frequency** of the event E after n repetitions. If this relative frequency approaches some value $P(E)$ as n becomes larger and larger, then $P(E)$ is called the **empirical probability** of E. Thus, the probability $P(E)$ of an event occurring is a measure of the proportion of the time that the event E will occur in the long run. Observe that this method of computing the probability of a head occurring is effective even when a biased coin is used in the experiment. The relative frequency distribution is often referred to as an *observed* or *empirical probability distribution.*

The **probability of an event** is a number that lies between 0 and 1, inclusive. In general, the larger the probability of an event, the more likely that the event will occur. Thus, an event with a probability of .8 is more likely to occur than an event with a probability of .6. An event with a probability of $\frac{1}{2}$, or .5, has a fifty–fifty chance of occurring.

Now suppose we are given an experiment and wish to determine the probabilities associated with certain events of the experiment. This problem could be solved by computing $P(E)$ directly for each event E of interest. In practice, however, the number of events of interest is usually quite large, so this approach is not satisfactory.

The following approach is particularly suitable when the sample space of an experiment is finite.* Let S be a finite sample space with n outcomes; that is,

$$S = \{s_1, s_2, s_3, \ldots, s_n\}$$

Then the events

$$\{s_1\}, \{s_2\}, \{s_3\}, \ldots, \{s_n\}$$

that consist of exactly one point are called the **elementary,** or **simple, events** of the experiment. They are elementary in the sense that any (nonempty) event of the experiment may be obtained by taking a finite union of suitable elementary events. The simple events of an experiment are also *mutually exclusive;* that is, given any two simple events of the experiment, only one can occur.

By assigning probabilities to each of the simple events, we obtain the results shown in Table 3. This table is called a **probability distribution** for the experiment. The function P, which assigns a probability to each of the simple events, is called a **probability function.**

The numbers $P(s_1), P(s_2), \ldots, P(s_n)$ have the following properties:

1. $0 \le P(s_i) \le 1$ $i = 1, 2, \ldots, n$
2. $P(s_1) + P(s_2) + \cdots + P(s_n) = 1$
3. $P(\{s_i\} \cup \{s_j\}) = P(s_i) + P(s_j)\,(i \ne j)$ $i = 1, 2, \ldots, n; j = 1, 2, \ldots, n$

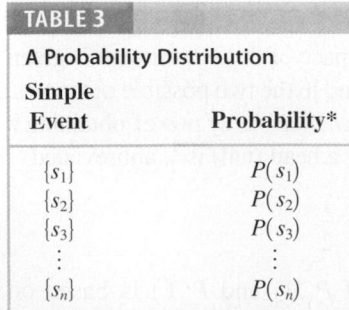

TABLE 3

A Probability Distribution

Simple Event	Probability*
$\{s_1\}$	$P(s_1)$
$\{s_2\}$	$P(s_2)$
$\{s_3\}$	$P(s_3)$
$\vdots$	$\vdots$
$\{s_n\}$	$P(s_n)$

*For simplicity, we use the notation $P(s_i)$ instead of the technically more correct $P(\{s_i\})$.

---

*For the remainder of the chapter, we assume that all sample spaces are finite.

The first property simply states that the probability of a simple event must be between 0 and 1, inclusive. The second property states that the sum of the probabilities of all simple events of the sample space is 1. This follows from the fact that the event S is certain to occur. The third property states that the probability of the union of two simple events is given by the sum of their probabilities.

Exploring with TECHNOLOGY

We can use a graphing calculator to simulate the coin-tossing experiment described earlier. Associate the outcome "a head" with the number 1 and the outcome "a tail" with the number 0. Select the function **randInt(** on the TI-83/84. (You can find this by pressing [MATH] and then moving the cursor to **PRB**.) Select **randInt(** and enter 0 , 1) and then press [ENTER] repeatedly. This generates 0s and 1s randomly, which simulates the results of tossing an unbiased coin.

As we saw earlier, there is no unique method for assigning probabilities to the simple events of an experiment. In practice, the methods that are used to determine these probabilities may range from theoretical considerations of the problem on the one extreme to the reliance on "educated guesses" on the other.

Sample spaces in which the outcomes are equally likely are called **uniform sample spaces.** Assigning probabilities to the simple events in these spaces is relatively easy.

Probability of an Event in a Uniform Sample Space
If

$$S = \{s_1, s_2, \ldots, s_n\}$$

is the sample space for an experiment in which the outcomes are equally likely, then we assign the probabilities

$$P(s_1) = P(s_2) = \cdots = P(s_n) = \frac{1}{n}$$

to each of the simple events $\{s_1\}, \{s_2\}, \ldots, \{s_n\}$.

TABLE 4	
A Probability Distribution	
Simple Event	**Probability**
$\{1\}$	$\frac{1}{6}$
$\{2\}$	$\frac{1}{6}$
$\{3\}$	$\frac{1}{6}$
$\{4\}$	$\frac{1}{6}$
$\{5\}$	$\frac{1}{6}$
$\{6\}$	$\frac{1}{6}$

EXAMPLE 1 A fair die is rolled, and the number that falls uppermost is observed. Determine the probability distribution for the experiment.

Solution The sample space for the experiment is $S = \{1, 2, 3, 4, 5, 6\}$, and the simple events are accordingly given by the sets $\{1\}, \{2\}, \{3\}, \{4\}, \{5\},$ and $\{6\}$. Since the die is assumed to be fair, the six outcomes are equally likely. We therefore assign a probability of $\frac{1}{6}$ to each of the simple events and obtain the probability distribution shown in Table 4.

Explore and Discuss
You suspect that a die is biased.

1. Describe a method you might use to show that your assertion is correct.
2. How would you assign the probability to each outcome 1 through 6 of an experiment that consists of rolling the die and observing the number that lands uppermost?

The next example shows how the *relative frequency* interpretation of probability lends itself to the computation of probabilities.

TABLE 5

Data Obtained During 200 Test Runs of an Electric Car

Distance Covered in Miles, x	Frequency of Occurrence
$0 < x \leq 50$	4
$50 < x \leq 100$	10
$100 < x \leq 150$	30
$150 < x \leq 200$	100
$200 < x \leq 250$	40
$250 < x$	16

TABLE 6

A Probability Distribution

Simple Event	Probability
$\{s_1\}$	.02
$\{s_2\}$	.05
$\{s_3\}$	.15
$\{s_4\}$	.50
$\{s_5\}$	.20
$\{s_6\}$	.08

$\boxed{\$}$ APPLIED EXAMPLE 2 Testing New Products Refer to Example 7, Section 7.5. The data shown in Table 5 were obtained in tests involving 200 test runs. Each run was made with a fully charged battery.

a. Describe an appropriate sample space for this experiment.
b. Find the empirical probability distribution for this experiment.

Solution

a. Let s_1 denote the outcome that the distance covered by the car does not exceed 50 miles; let s_2 denote the outcome that the distance covered by the car is greater than 50 miles but does not exceed 100 miles, and so on. Finally, let s_6 denote the outcome that the distance covered by the car is greater than 250 miles. Then the required sample space is given by

$$S = \{s_1, s_2, s_3, s_4, s_5, s_6\}$$

b. To compute the empirical probability distribution for the experiment, we turn to the relative frequency interpretation of probability. Accepting the inaccuracies inherent in a relatively small number of trials (200 runs), we take the probability of s_1 occurring as

$$P(s_1) = \frac{\text{Number of trials in which } s_1 \text{ occurs}}{\text{Total number of trials}}$$

$$= \frac{4}{200} = .02$$

In a similar manner, we assign probabilities to the other simple events, obtaining the probability distribution shown in Table 6. ◼

We are now in a position to give a procedure for computing the probability $P(E)$ of an arbitrary event E of an experiment.

> Finding the Probability of an Event E
>
> **1.** Determine a sample space S associated with the experiment.
> **2.** Assign probabilities to the simple events of S.
> **3.** If $E = \{s_1, s_2, s_3, \ldots, s_n\}$, where $\{s_1\}, \{s_2\}, \{s_3\}, \ldots, \{s_n\}$ are simple events, then
>
> $$P(E) = P(s_1) + P(s_2) + P(s_3) + \cdots + P(s_n)$$
>
> If E is the empty set, $\varnothing$, then $P(E) = 0$.

The principle stated in Step 3 is called the **addition principle** and is a consequence of Property 3 of the probability function (page 482). This principle allows us to find the probabilities of all other events once the probabilities of the simple events are known.

$\triangle$ The addition rule in Step 3 applies *only* to the addition of probabilities of simple events.

APPLIED EXAMPLE 3 Rolling a Pair of Dice A pair of fair dice is rolled.

a. Calculate the probability that the two dice show the same number.
b. Calculate the probability that the sum of the numbers of the two dice is 6.

Solution   From the results of Example 4, Section 7.5, page 475, we see that the sample space S of the experiment consists of 36 outcomes:

$$S = \{(1, 1), (1, 2), \ldots, (6, 5), (6, 6)\}$$

Since both dice are fair, each of the 36 outcomes is equally likely. Accordingly, we assign the probability of $\frac{1}{36}$ to each simple event.

a. The event that the two dice show the same number is given by

$$E = \{(1, 1), (2, 2), (3, 3), (4, 4), (5, 5), (6, 6)\}$$

(Figure 27). Therefore, by the addition principle, the probability that the two dice show the same number is given by

$$P(E) = P[(1, 1)] + P[(2, 2)] + \cdots + P[(6, 6)]$$

$$= \frac{1}{36} + \frac{1}{36} + \cdots + \frac{1}{36} \quad \text{Six terms}$$

$$= \frac{1}{6}$$

b. The event that the sum of the numbers of the two dice is 6 is given by

$$E_6 = \{(1, 5), (2, 4), (3, 3), (4, 2), (5, 1)\}$$

(Figure 28). Therefore, the probability that the sum of the numbers on the two dice is 6 is given by

$$P(E_6) = P[(1, 5)] + P[(2, 4)] + P[(3, 3)] + P[(4, 2)] + P[(5, 1)]$$

$$= \frac{1}{36} + \frac{1}{36} + \cdots + \frac{1}{36} \quad \text{Five terms}$$

$$= \frac{5}{36}$$

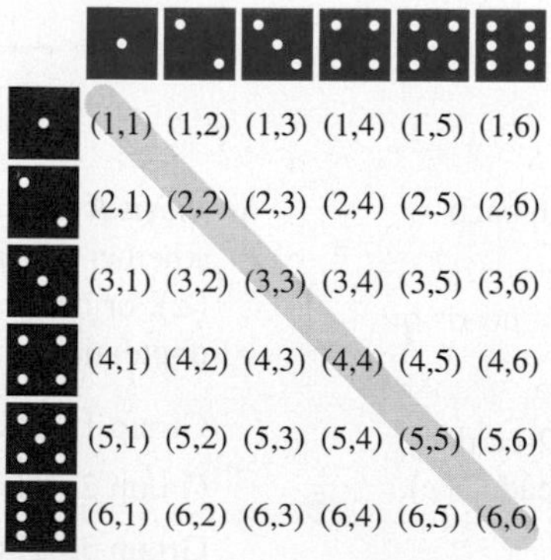

FIGURE **27**
The event that the two dice show the same number

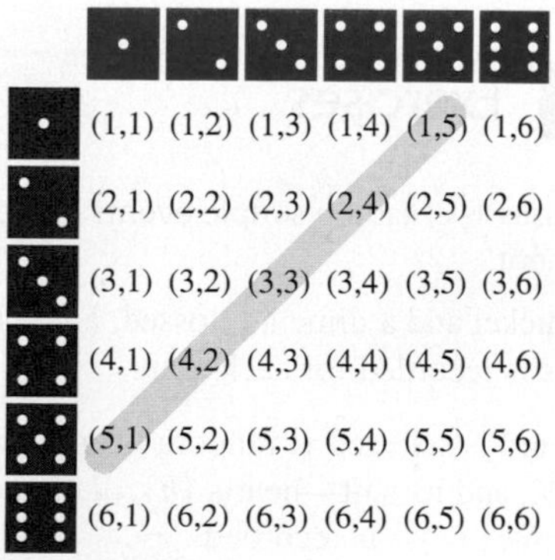

FIGURE **28**
The event that the sum of the numbers on the two dice is 6

 APPLIED EXAMPLE 4 Testing New Products Consider the experiment by EverBrite in Example 2. What is the probability that the prototype car will travel more than 150 miles on a fully charged battery?

Solution Using the results of Example 2, we see that the event that the car will travel more than 150 miles on a fully charged battery is given by $E = \{s_4, s_5, s_6\}$. Therefore, the probability that the car will travel more than 150 miles on one charge is given by

$$P(E) = P(s_4) + P(s_5) + P(s_6)$$

or, using the probability distribution for the experiment obtained in Example 2,

$$P(E) = .50 + .20 + .08 = .78$$

7.6 Self-Check Exercises

1. A biased die was rolled repeatedly, and the results of the experiment are summarized in the following table:

Outcome	1	2	3	4	5	6
Frequency of Occurrence	142	173	158	175	162	190

Using the relative frequency interpretation of probability, find the empirical probability distribution for this experiment.

2. **ACCIDENT PREVENTION** In an experiment conducted to study the effectiveness of an eye-level third brake light in the prevention of rear-end collisions, 250 of the 500 highway patrol cars of a certain state were equipped with such lights. At the end of the 1-year trial period, the records revealed that for those equipped with a third brake light there were 14 incidents of rear-end collision. There were 22 such incidents involving the cars not equipped with the accessory. On the basis of these data, what is the probability that a highway patrol car equipped with a third brake light will be rear-ended within a 1-year period? What is the probability that a car not so equipped will be rear-ended within a 1-year period?

Solutions to Self-Check Exercises 7.6 can be found on page 491.

7.6 Concept Questions

1. Define (a) a probability distribution and (b) a probability function. Give examples of each.

2. If $S = \{s_1, s_2, \ldots, s_n\}$ is the sample space for an experiment in which the outcomes are equally likely, what is the probability of each of the simple events $s_1, s_2, \ldots, s_n$? What is this type of sample space called?

3. Suppose $E = \{s_1, s_2, s_3, \ldots, s_n\}$, where E is an event of an experiment and $\{s_1\}, \{s_2\}, \{s_3\}, \ldots, \{s_n\}$ are simple events. If E is nonempty, what is $P(E)$? If E is empty, what is $P(E)$?

7.6 Exercises

In Exercises 1–8, list the simple events associated with each experiment.

1. A nickel and a dime are tossed, and the result of heads or tails is recorded for each coin.

2. A card is selected at random from a standard 52-card deck, and its suit—hearts (h), diamonds (d), spades (s), or clubs (c)—is recorded.

3. **OPINION POLLS** An opinion poll is conducted among a group of registered voters. Their political affiliation—Democrat (D), Republican (R), or Independent (I)—and their sex—male (m) or female (f)—are recorded.

4. **QUALITY CONTROL** As part of a quality-control procedure, eight circuit boards are checked, and the number of defective boards is recorded.

5. **MOVIE ATTENDANCE** In a survey conducted to determine whether movie attendance is increasing (i), decreasing (d), or holding steady (s) among various sectors of the population, participants are classified as follows:

Group 1: Those aged 10–19

Group 2: Those aged 20–29

Group 3: Those aged 30–39

Group 4: Those aged 40–49

Group 5: Those aged 50 and older

The response and age group of each participant are recorded.

6. **DURABLE GOODS ORDERS** An economist obtains data concerning durable goods orders each month. A record is

kept for a 1-year period of any increase (*i*), decrease (*d*), or unchanged movement (*u*) in the number of durable goods orders for each month as compared with the number of such orders in the same month of the previous year.

7. **BLOOD TYPES** Blood tests are given as a part of the admission procedure at the Monterey Garden Community Hospital. The blood type of each patient (A, B, AB, or O) and the presence or absence of the Rh factor in each patient's blood (Rh^+ or Rh^-) are recorded.

8. **METEOROLOGY** A meteorologist preparing a weather map classifies the expected average temperature in each of five neighboring states (MN, WI, IA, IL, MO) for the upcoming week as follows:
   a. More than 10° below average
   b. Normal to 10° below average
   c. Higher than normal to 10° above average
   d. More than 10° above average

   Using each state's abbreviation and the categories—(a), (b), (c), and (d)—the meteorologist records these data.

9. **SOCIAL MEDIA ACCOUNTS** In a survey of 1000 social media account holders, the following question was asked: How many social media accounts do you have? The results of the survey are summarized below:

Answer	1–2	3–4	5 or more
Respondents	650	200	150

   a. Determine the empirical probability distribution associated with these data.
   b. What is the probability that a participant in the survey selected at random answered that he or she had three or four accounts?
   Source: AARP.

10. **WORKPLACE** In a survey of 26,612 Parade.com visitors, the following question was asked: How do workers get ahead? The results of the survey are as follows:

Answer	Internal politics	Hard work	Initiative	Creativity
Respondents	13,572	7,185	4,790	1,065

   a. Determine the empirical probability distribution associated with these data.
   b. What is the probability that a participant in the survey selected at random answered that one gets ahead at work through hard work?
   Source: Yahoo Finance/Parade Survey.

11. **STRESS LEVEL** In a study on stress experienced by Americans, 800 adults ages 18 years and older were asked to rate their stress level as low, middle, or extreme. The results of the survey are summarized below:

Answer	Low	Middle	Extreme	No response
Respondents	272	352	160	16

   a. Determine the empirical probability distribution associated with these data.
   b. What is the probability that a participant in the survey answered that he or she had experienced an extreme stress level?
   Source: American Psychological Association.

12. **BLOOD TYPES** The percentage of the general population that has each blood type is shown in the following table. Determine the probability distribution associated with these data.

Blood Type	A	B	AB	O
Population (%)	41	12	3	44

13. **GRADE DISTRIBUTIONS** The grade distribution for a certain class is shown in the following table. Find the probability distribution associated with these data.

Grade	A	B	C	D	F
Frequency of Occurrence	4	10	18	6	2

14. **GREAT RECESSION** In a survey conducted by AlixPartners of 4980 adults 18 years old and older in June 2009, during the "Great Recession," the following question was asked: How long do you think it will take to recover your personal net worth? The results of the survey follow:

Answer (in years)	1–2	3–4	5–10	>10
Respondents	1006	1308	2113	553

   a. Determine the empirical probability distribution associated with these data.
   b. If a person who participated in the survey is selected at random, what is the probability that he or she expected that it would take 5 or more years to recover his or her personal net worth?
   Source: AlixPartners.

15. **STARTING A NEW JOB** In a survey of 420 workers, the following question was asked: What are the greatest challenges when starting a new job? The results of the survey are as follows:

Answer	New processes/ procedures	Getting to know a new boss and coworkers	New technology tools	Fitting into the corporate culture	Other
Respondents	185	84	71	50	30

   a. Determine the empirical probability distribution associated with these data.
   b. What is the probability that a participant in the survey selected at random answered that the greatest challenges when starting a new job were fitting into the corporate culture?
   Source: Accountemps.

16. SAFETY OF AMERICAN-MADE PRODUCTS The accompanying data were obtained from a survey of 1500 Americans who were asked: How safe are American-made consumer products? Determine the empirical probability distribution associated with these data.

Rating	A	B	C	D	E
Respondents	285	915	225	30	45

A: Very safe

B: Somewhat safe

C: Not too safe

D: Not safe at all

E: Don't know

17. POLITICAL VIEWS OF COLLEGE FRESHMEN In a poll conducted among 2000 college freshmen to ascertain the political views of college students, the accompanying data were obtained. Determine the empirical probability distribution associated with these data.

Political Views	A	B	C	D	E
Respondents	52	398	1140	386	24

A: Far left

B: Liberal

C: Middle of the road

D: Conservative

E: Far right

18. RED-LIGHT RUNNERS In a survey of 800 likely voters, the following question was asked: Do you support using cameras to identify red-light runners? The results of the survey follow:

Answer	Strongly support	Somewhat support	Somewhat oppose	Strongly oppose	Don't know
Respondents	360	192	88	144	16

What is the probability that a person in the survey selected at random favors using cameras to identify red-light runners?
Source: Public Opinion Strategies.

19. COOKING AT HOME In an online survey of 500 adults living with children under the age of 18 years, the participants were asked how many days per week they cook at home. The results of the survey are summarized below:

Number of Days	0	1	2	3	4	5	6	7
Respondents	25	30	45	75	55	100	85	85

Determine the empirical probability distribution associated with these data.
Source: Super Target.

20. CHECKING INTO A HOTEL ROOM In a survey of 3019 hotel guests, the following question was asked: What is the first thing you do after checking into a hotel room? The results of the survey follow:

Activity	Respondents
Adjust the thermostat	1027
Turn on the TV	755
Unpack	634
Check out the free toiletries	211
Plug in rechargeable electronics	60
Find the gym	30
Other	302

What is the probability that a person selected at random from the list of guests surveyed would, as a first activity:
a. Adjust the thermostat or plug in rechargeable electronics?
b. Unpack or find the gym?
Source: Tripadvisor.

21. TRAFFIC SURVEYS The number of cars entering a tunnel leading to an airport in a major city over a period of 200 peak hours was observed, and the following data were obtained:

Number of Cars, x	Frequency of Occurrence
$0 < x \leq 200$	15
$200 < x \leq 400$	20
$400 < x \leq 600$	35
$600 < x \leq 800$	70
$800 < x \leq 1000$	45
$x > 1000$	15

a. Describe an appropriate sample space for this experiment.
b. Find the empirical probability distribution for this experiment.

22. ARRIVAL TIMES OF COMMUTER TRAINS The arrival times of the 8 A.M. Boston-based commuter train as observed in the suburban town of Sharon over 120 weekdays is summarized below:

Arrival Time, x	Frequency of Occurrence
7:56 A.M. $< x \leq$ 7:58 A.M.	4
7:58 A.M. $< x \leq$ 8:00 A.M.	18
8:00 A.M. $< x \leq$ 8:02 A.M.	50
8:02 A.M. $< x \leq$ 8:04 A.M.	32
8:04 A.M. $< x \leq$ 8:06 A.M.	9
8:06 A.M. $< x \leq$ 8:08 A.M.	4
8:08 A.M. $< x \leq$ 8:10 A.M.	3

a. Describe an appropriate sample space for this experiment.
b. Find the empirical probability distribution for this experiment.

23. **CORRECTIVE LENS USE** According to Mediamark Research, during a certain year 84 million out of 179 million adults in the United States corrected their vision by using prescription eyeglasses, bifocals, or contact lenses. (Some respondents use more than one type.) What is the probability that an adult selected at random from the adult population uses corrective lenses?
Source: Mediamark Research.

24. **CORRECTIONAL SUPERVISION** A study conducted by the Corrections Department of a certain state revealed that 163,605 people out of a total adult population of 1,778,314 were under correctional supervision (on probation, on parole, or in jail). What is the probability that a person selected at random from the adult population in that state is under correctional supervision?

25. **LIGHTNING DEATHS** According to data obtained from the National Weather Service, 376 of the 439 people killed by lightning in the United States over a 7-year period were men. (Job and recreational habits of men make them more vulnerable to lightning.) Assuming that this trend holds in the future, what is the probability that a person killed by lightning:
   a. Is a male?   **b.** Is a female?
   Source: National Weather Service.

26. **QUALITY CONTROL** One light bulb is selected at random from a lot of 120 light bulbs, of which 5% are defective. What is the probability that the light bulb selected is defective?

27. **EFFORTS TO STOP SHOPLIFTING** According to a survey of 176 retailers, 46% of them use electronic tags as protection against shoplifting and employee theft. If one of these retailers is selected at random, what is the probability that the retailer uses electronic tags as antitheft devices?

28. If a ball is selected at random from an urn containing three red balls, two white balls, and five blue balls, what is the probability that it will be a white ball?

29. If a card is drawn at random from a standard 52-card deck, what is the probability that the card drawn is:
   a. A diamond?   **b.** A black card?
   c. An ace?

30. A pair of fair dice is rolled. What is the probability that:
   a. The sum of the numbers shown uppermost is less than 5?
   b. At least one 6 is rolled?

31. **TRAFFIC LIGHTS** What is the probability of arriving at a traffic light when it is red if the red signal is lit for 30 sec, the yellow signal for 5 sec, and the green signal for 45 sec?

32. **ROULETTE** What is the probability that a roulette ball will come to rest on an even number other than 0 or 00? (Assume that there are 38 equally likely outcomes consisting of the numbers 1–36, 0, and 00.)

In Exercises 33–35, determine whether the given experiment has a sample space with equally likely outcomes.

33. A loaded die is rolled, and the number appearing uppermost on the die is recorded.

34. Two fair dice are rolled, and the sum of the numbers appearing uppermost is recorded.

35. A ball is selected at random from an urn containing six black balls and six red balls, and the color of the ball is recorded.

36. Let $S = \{s_1, s_2, s_3, s_4, s_5, s_6\}$ be the sample space associated with an experiment having the following probability distribution:

Outcome	s_1	s_2	s_3	s_4	s_5	s_6
Probability	$\frac{1}{12}$	$\frac{1}{4}$	$\frac{1}{12}$	$\frac{1}{6}$	$\frac{1}{3}$	$\frac{1}{12}$

Find the probability of the event:
   a. $A = \{s_1, s_3\}$
   b. $B = \{s_2, s_4, s_5, s_6\}$
   c. $C = S$

37. Let $S = \{s_1, s_2, s_3, s_4, s_5\}$ be the sample space associated with an experiment having the following probability distribution:

Outcome	s_1	s_2	s_3	s_4	s_5
Probability	$\frac{1}{14}$	$\frac{3}{14}$	$\frac{6}{14}$	$\frac{2}{14}$	$\frac{2}{14}$

Find the probability of the event:
   a. $A = \{s_1, s_2, s_4\}$
   b. $B = \{s_1, s_5\}$
   c. $C = S$

38. A pair of fair dice is rolled, and the sum of the two numbers falling uppermost is observed. The probability of obtaining a sum of 2 is the same as that of obtaining a 7 since there is only one way of getting a 2—namely, by each die showing a 1; and there is only one way of obtaining a 7—namely, by one die showing a 3 and the other die showing a 4. What is wrong with this argument?

39. **SELECTING JOB APPLICANTS** Refer to Exercises 7.5, Problem 32. From a list of five applicants for a sales position, *a*, *b*, *c*, *d*, and *e*, two are selected for the next round of interviews. If the applicants are selected at random, what is the probability that the two interviewees chosen:
   a. Include applicant *a*?
   b. Include applicants *a* and *c*?
   c. Include applicants *d* and *e*?

40. **FAMILY BIRTH ORDER** Refer to Exercises 7.5, Problem 33. An experiment consists of recording the sex composition, in order of their births, of a four-child family in which the children were born at different times. Assuming that a boy is equally likely as a girl to be born into a family, what is the probability that a four-child family chosen at random will have:
   a. Three boys and a girl in the family?

b. A youngest child in the family who is a girl?

c. An oldest child and a youngest child in the family who are both girls?

41. DISPOSITION OF CRIMINAL COURT CASES Of the 98 first-degree murder cases from 2002 through the first half of 2004 in Suffolk County Superior Court, 9 cases were thrown out of the system, 62 cases were plea-bargained, and 27 cases went to trial. What is the probability that a case selected at random

a. Was settled through plea bargaining?

b. Went to trial?

Source: Boston Globe.

42. SWEEPSTAKES In a sweepstakes sponsored by Gemini Paper Products, 100,000 entries have been received. If 1 grand prize, 5 first prizes, 25 second prizes, and 500 third prizes are to be awarded, what is the probability that a person who has submitted one entry will win:

a. The grand prize?

b. A prize?

43. POLITICAL POLLS An opinion poll was conducted among a group of registered voters in a certain state concerning a proposition aimed at limiting state and local taxes. Results of the poll indicated that 35% of the voters favored the proposition, 32% were against it, and the remaining group were undecided. If the results of the poll are assumed to be representative of the opinions of the state's electorate, what is the probability that a registered voter selected at random from the electorate:

a. Favors the proposition?

b. Is undecided about the proposition?

44. SECURITY BREACHES In a survey of 106 senior information technology and data security professionals at major U.S. companies regarding their confidence that they had detected all significant security breaches in the past year, the following responses were obtained:

Answer	Very confident	Moderately confident	Not very confident	Not at all confident
Respondents	21	56	22	7

What is the probability that a respondent in the survey selected at random:

a. Had little or no confidence that he or she had detected all significant security breaches in the past year?

b. Was very confident that he or she had detected all significant security breaches in the past year?

Source: Forsythe Solutions Group.

45. GREEN COMPANIES In a survey conducted in a certain year of 1004 adults 18 years old and older, the following question was asked: How are American companies doing on protecting the environment compared with companies in other countries? The results are summarized below:

Answer	Behind	Equal	Ahead	Don't know
Respondents	382	281	251	90

If an adult in the survey is selected at random, what is the probability that he or she said that American companies are equal or ahead on protecting the environment compared with companies in other countries?

Source: GfK Roper.

46. PARENTAL INFLUENCE ON CHILDREN'S CAREER CHOICES In an online survey of 1962 executives from 64 countries conducted by Korn/Ferry International between August and October in a certain year, the executives were asked whether they would try to influence their children's career choices. Their replies: A (to a very great extent), B (to a great extent), C (to some extent), D (to a small extent), and E (not at all) are recorded below:

Answer	A	B	C	D	E
Respondents	135	404	1057	211	155

What is the probability that a randomly selected respondent's answer was D (to a small extent) or E (not at all)?

Source: Korn/Ferry International.

47. SPENDING METHODS In a survey on consumer spending methods conducted in a certain year, the following results were obtained:

Payment Method	Checks	Cash	Credit cards	Debit/ATM cards	Other
Transactions (%)	37	14	25	15	9

If a transaction tracked in this survey is selected at random, what is the probability that the transaction was paid for:

a. With a credit card or with a debit/ATM card?

b. With cash or some method other than with a check, a credit card, or a debit/ATM card?

Source: Minute/Visa USA Research Services.

48. STAYING IN TOUCH In a poll conducted by the Pew Research Center in a certain year, 2000 adults ages 18 years old and older were asked how frequently they are in touch with their parents by phone. The results of the poll are as follows:

Answer	Monthly	Weekly	Daily	Don't know	Less
Respondents (%)	11	47	32	2	8

If a person who participated in the poll is selected at random, what is the probability that the person said he or she kept in touch with his or her parents:

a. Once a week?

b. At least once a week?

Source: Pew Research Center.

49. MUSIC VENUES In a survey designed to determine where people listen to music in their home, 1000 people were asked in which room at home they were mostly likely to listen to music. The results are tabulated below:

Room	Living room	Master bedroom	Study/ home office	Kitchen	Bathroom	Other
Respondents	448	169	155	100	22	106

If a respondent is selected at random, what is the probability that he or she most likely listens to music:
a. In the living room?
b. In the study/home office or the kitchen?

Source: Phillips Electronics.

50. **RETIREMENT BENEFITS VERSUS SALARY** In a survey conducted in a certain year of 1402 workers 18 years old and older regarding their opinion on retirement benefits, the following data were obtained: 827 said that it was better to have excellent retirement benefits with a lower-than-expected salary, 477 said that it was better to have a higher-than-expected salary with poor retirement benefits, 42 said "neither," and 56 said "not sure." If a worker in the survey is selected at random, what is the probability that he or she answered that it was better to have:
a. Excellent retirement benefits with a lower-than-expected salary?
b. A higher-than-expected salary with poor retirement benefits?

Source: Transamerica Center for Retirement.

51. **AIRLINE SAFETY** In an attempt to study the leading causes of airline crashes, the following data were compiled from records of airline crashes over a 35-year period (excluding sabotage and military action):

Primary Factor	Accidents
Pilot	327
Airplane	49
Maintenance	14
Weather	22
Airport/air traffic control	19
Miscellaneous/other	15

Assume that you have just learned of an airline crash and that the data give a generally good indication of the causes of airline crashes. Give an estimate of the probability that the primary cause of the crash was due to pilot error or bad weather.

Source: National Transportation Safety Board.

52. **HOUSING APPRECIATION** In a survey conducted in fall 2006, a year before the financial crisis of 2007–2008, 800 homeowners were asked about their expectations regarding the value of their home in the next few years; the results of the survey are as follows:

Expectations	Homeowners
Decrease	48
Stay the same	152
Increase less than 5%	232
Increase 5–10%	240
Increase more than 10%	128

If a homeowner in the survey is chosen at random, what is the probability that he or she expected his or her home to:
a. Stay the same or decrease in value in the next few years?
b. Increase 5% or more in value in the next few years?

Source: S&P, RBC Capital Markets.

In Exercises 53 and 54, determine whether the statement is true or false. If it is true, explain why it is true. If it is false, give an example to show why it is false.

53. If $S = \{s_1, s_2, \ldots, s_n\}$ is a uniform sample space with n outcomes, then $0 \leq P(s_1) + P(s_2) + \cdots + P(s_n) \leq 1$.

54. Let $S = \{s_1, s_2, \ldots, s_n\}$ be a uniform sample space for an experiment. If $n \geq 5$ and $E = \{s_1, s_2, s_5\}$, then $P(E) = 3/n$.

7.6 Solutions to Self-Check Exercises

1.
$$P(1) = \frac{\text{Number of trials in which a 1 appears uppermost}}{\text{Total number of trials}}$$

$$= \frac{142}{1000}$$

$$= .142$$

Similarly, we compute $P(2), \ldots, P(6)$, obtaining the following probability distribution:

Outcome	1	2	3	4	5	6
Probability	.142	.173	.158	.175	.162	.190

2. The probability that a highway patrol car equipped with a third brake light will be rear-ended within a 1-year period is given by

$$\frac{\begin{array}{c}\text{Number of rear-end collisions involving} \\ \text{cars equipped with a third brake light}\end{array}}{\text{Total number of such cars}} = \frac{14}{250} = .056$$

The probability that a highway patrol car not equipped with a third brake light will be rear-ended within a 1-year period is given by

$$\frac{\begin{array}{c}\text{Number of rear-end collisions involving cars} \\ \text{not equipped with a third brake light}\end{array}}{\text{Total number of such cars}} = \frac{22}{250} = .088$$

7.7   Rules of Probability

Properties of the Probability Function and Their Applications

In this section, we examine some of the properties of the probability function and look at the role they play in solving certain problems. We begin by looking at the generalization of the three properties of the probability function, which were stated for simple events in Section 7.6. Let S be a sample space of an experiment, and suppose E and F are events of the experiment. We have the following properties:

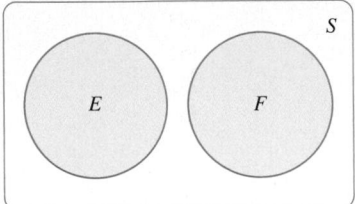

FIGURE 29
If E and F are mutually exclusive events, then $P(E \cup F) = P(E) + P(F)$.

Property 1. $P(E) \geq 0$ for any E.

Property 2. $P(S) = 1$.

Property 3. If E and F are mutually exclusive (that is, only one of them can occur or, equivalently, $E \cap F = \varnothing$), then

$$P(E \cup F) = P(E) + P(F)$$

(Figure 29).

Property 3 may be easily extended to the case involving any finite number of mutually exclusive events. Thus, if $E_1, E_2, \ldots, E_n$ are mutually exclusive events, then

$$P(E_1 \cup E_2 \cup \cdots \cup E_n) = P(E_1) + P(E_2) + \cdots + P(E_n)$$

TABLE 7	
Probability Distribution	
Score, x	**Probability**
$x > 700$	.01
$600 < x \leq 700$	.07
$500 < x \leq 600$	.19
$400 < x \leq 500$	.23
$300 < x \leq 400$	.31
$x \leq 300$	.19

APPLIED EXAMPLE 1 SAT Verbal Scores The superintendent of a metropolitan school district has estimated the probabilities associated with the SAT verbal scores of students from that district. The results are shown in Table 7. If a student is selected at random, what is the probability that his or her SAT verbal score will be:

a. More than 400?
b. Less than or equal to 500?
c. Greater than 400 but less than or equal to 600?

Solution   Let A, B, C, D, E, and F denote, respectively, the event that the score is greater than 700, greater than 600 but less than or equal to 700, greater than 500 but less than or equal to 600, and so forth. Then these events are mutually exclusive. Therefore,

a. The probability that the student's score will be more than 400 is given by

$$P(D \cup C \cup B \cup A) = P(D) + P(C) + P(B) + P(A)$$
$$= .23 + .19 + .07 + .01$$
$$= .5$$

b. The probability that the student's score will be less than or equal to 500 is given by

$$P(D \cup E \cup F) = P(D) + P(E) + P(F)$$
$$= .23 + .31 + .19 = .73$$

c. The probability that the student's score will be greater than 400 but less than or equal to 600 is given by

$$P(C \cup D) = P(C) + P(D)$$
$$= .19 + .23 = .42$$

Property 3 holds if and only if E and F are mutually exclusive. In the general case, we have the following rule:

> ### Property 4. Addition Rule
> If E and F are any two events of an experiment, then
> $$P(E \cup F) = P(E) + P(F) - P(E \cap F)$$

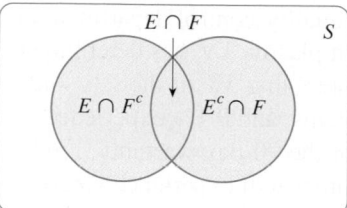

FIGURE 30
$E \cup F = (E \cap F^c) \cup (E \cap F) \cup (E^c \cap F)$

To derive this property, refer to Figure 30. Observe that we can write

$$E = (E \cap F^c) \cup (E \cap F) \quad \text{and} \quad F = (E^c \cap F) \cup (E \cap F)$$

as a union of disjoint sets. Therefore,

$$P(E) = P(E \cap F^c) + P(E \cap F) \quad \text{or} \quad P(E \cap F^c) = P(E) - P(E \cap F)$$

and

$$P(F) = P(E^c \cap F) + P(E \cap F) \quad \text{or} \quad P(E^c \cap F) = P(F) - P(E \cap F)$$

Finally, since $E \cup F = (E \cap F^c) \cup (E \cap F) \cup (E^c \cap F)$ is a union of disjoint sets, we have

$$\begin{aligned} P(E \cup F) &= P(E \cap F^c) + P(E \cap F) + P(E^c \cap F) \\ &= P(E) - P(E \cap F) + P(E \cap F) + P(F) - P(E \cap F) \quad \text{Use the earlier results.} \\ &= P(E) + P(F) - P(E \cap F) \end{aligned}$$

Note Observe that if E and F are mutually exclusive—that is, if $E \cap F = \varnothing$—then the equation of Property 4 reduces to that of Property 3. In other words, if E and F are mutually exclusive events, then $P(E \cup F) = P(E) + P(F)$. If E and F are not mutually exclusive events, then $P(E \cup F) = P(E) + P(F) - P(E \cap F)$. ◼

EXAMPLE 2 A card is drawn from a well-shuffled deck of 52 playing cards. What is the probability that it is an ace or a spade?

Solution Let E denote the event that the card drawn is an ace, and let F denote the event that the card drawn is a spade. Then

$$P(E) = \frac{4}{52} \quad \text{and} \quad P(F) = \frac{13}{52}$$

Furthermore, E and F are not mutually exclusive events. In fact, $E \cap F$ is the event that the card drawn is the ace of spades. Consequently,

$$P(E \cap F) = \frac{1}{52}$$

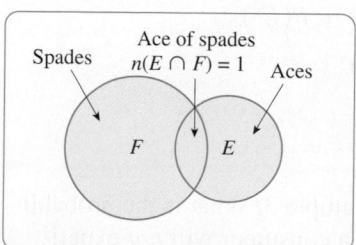

FIGURE 31
$P(E \cup F) = P(E) + P(F) - P(E \cap F)$

The event that a card drawn is an ace or a spade is $E \cup F$, with probability given by

$$\begin{aligned} P(E \cup F) &= P(E) + P(F) - P(E \cap F) \\ &= \frac{4}{52} + \frac{13}{52} - \frac{1}{52} = \frac{16}{52} = \frac{4}{13} \end{aligned}$$

(Figure 31). This result, of course, can be obtained by arguing that 16 of the 52 cards are either spades or aces of other suits. ◼

APPLIED EXAMPLE 3 Product Reliability The quality-control department of Vista Vision, manufacturer of the Pulsar 42-inch plasma TV, has determined from records obtained from the company's service centers that 3% of the sets sold experience video problems, 1% experience audio problems, and 0.1% experience both video and audio problems before the expiration of the 90-day warranty. Find the probability that a plasma TV purchased by a consumer will experience video or audio problems before the warranty expires.

Solution Let E denote the event that a plasma TV purchased will experience video problems within 90 days, and let F denote the event that a plasma TV purchased will experience audio problems within 90 days. Then

$$P(E) = .03 \qquad P(F) = .01 \qquad P(E \cap F) = .001$$

The event that a plasma TV purchased will experience video problems or audio problems before the warranty expires is $E \cup F$, and the probability of this event is given by

$$P(E \cup F) = P(E) + P(F) - P(E \cap F)$$
$$= .03 + .01 - .001$$
$$= .039$$

(Figure 32).

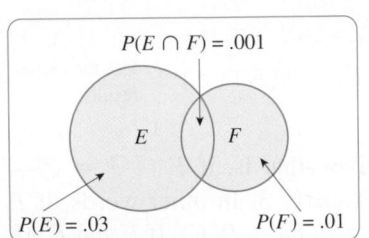

FIGURE **32**
$P(E \cup F) = P(E) + P(F) - P(E \cap F)$

Here is another property of a probability function that is of considerable aid in computing the probability of an event:

Property 5. Rule of Complements

If E is an event of an experiment and E^c denotes the complement of E, then

$$P(E^c) = 1 - P(E)$$

Property 5 is an immediate consequence of Properties 2 and 3. Indeed, we have $E \cup E^c = S$ and $E \cap E^c = \emptyset$, so

$$1 = P(S) = P(E \cup E^c) = P(E) + P(E^c)$$

and therefore,

$$P(E^c) = 1 - P(E)$$

APPLIED EXAMPLE 4 Warranties Refer to Example 3. What is the probability that a Pulsar 42-inch plasma TV bought by a consumer will *not* experience video or audio difficulties before the warranty expires?

Solution Let E denote the event that a plasma TV bought by a consumer will experience video or audio difficulties before the warranty expires. Then the event that the plasma TV will not experience either problem before the warranty expires is given by E^c, with probability

$$P(E^c) = 1 - P(E)$$
$$= 1 - .039$$
$$= .961$$

Computations Involving the Rules of Probability

We close this section by looking at two additional examples that illustrate how we use the rules of probability.

EXAMPLE 5 Let E and F be two mutually exclusive events, and suppose that $P(E) = .1$ and $P(F) = .6$. Compute:

a. $P(E \cap F)$ **b.** $P(E \cup F)$ **c.** $P(E^c)$
d. $P(E^c \cap F^c)$ **e.** $P(E^c \cup F^c)$

Solution

a. Since the events E and F are mutually exclusive—that is, $E \cap F = \varnothing$—we have $P(E \cap F) = 0$.

b. $P(E \cup F) = P(E) + P(F)$ Since E and F are mutually exclusive
$$= .1 + .6$$
$$= .7$$

c. $P(E^c) = 1 - P(E)$ Property 5
$$= 1 - .1$$
$$= .9$$

d. Observe that, by De Morgan's Law, $E^c \cap F^c = (E \cup F)^c$. Hence,

$$P(E^c \cap F^c) = P[(E \cup F)^c]$$ See Figure 33.
$$= 1 - P(E \cup F)$$ Property 5
$$= 1 - .7$$ Use the result of part (b).
$$= .3$$

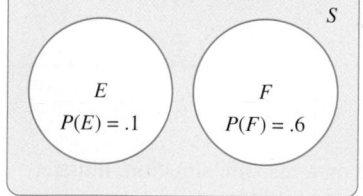

FIGURE 33
$P(E^c \cap F^c) = P[(E \cup F)^c]$

e. Again using De Morgan's Law, we find

$$P(E^c \cup F^c) = P[(E \cap F)^c]$$
$$= 1 - P(E \cap F)$$
$$= 1 - 0$$ Use the result of part (a).
$$= 1$$

EXAMPLE 6 Let E and F be two events of an experiment with sample space S. Suppose $P(E) = .2$, $P(F) = .1$, and $P(E \cap F) = .05$. Compute:

a. $P(E \cup F)$
b. $P(E^c \cap F^c)$
c. $P(E^c \cap F)$ *Hint:* Draw a Venn diagram.

Solution

a. $P(E \cup F) = P(E) + P(F) - P(E \cap F)$ Property 4
$$= .2 + .1 - .05$$
$$= .25$$

b. Using De Morgan's Law, we have
$$P(E^c \cap F^c) = P[(E \cup F)^c]$$
$$= 1 - P(E \cup F) \quad \text{Property 5}$$
$$= 1 - .25 \quad \text{Use the result of part (a).}$$
$$= .75$$

c. From the Venn diagram describing the relationship among E, F, and S (Figure 34), we have
$$P(E^c \cap F) = .05 \quad \text{The shaded subset is the event } E^c \cap F.$$

This result may also be obtained by using the relationship
$$P(E^c \cap F) = P(F) - P(E \cap F)$$
$$= .1 - .05$$
$$= .05$$

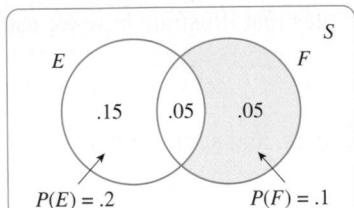

FIGURE 34
$P(E^c \cap F)$: the probability that the event F, but not the event E, will occur

---

7.7 Self-Check Exercises

1. Let E and F be events of an experiment with sample space S. Suppose $P(E) = .4$, $P(F) = .5$, and $P(E \cap F) = .1$. Compute:
  a. $P(E \cup F)$ **b.** $P(E \cap F^c)$

2. **PROBABILITY OF A SALE OR LEASE** Susan Garcia wishes to sell or lease a condominium through a realty company. The realtor estimates that the probability of finding a buyer within a month of the date the property is listed for sale or lease is .3, the probability of finding a lessee is .8, and the probability of finding both a buyer and a lessee is .1. Determine the probability that the property will be sold or leased within 1 month from the date the property is listed for sale or lease.

Solutions to Self-Check Exercises 7.7 can be found on page 502.

7.7 Concept Questions

1. Suppose that S is a sample space of an experiment, E and F are events of the experiment, and P is a probability function. Give the meaning of each of the following statements:
  a. $P(E) = 0$ **b.** $P(F) = 0.5$ **c.** $P(S) = 1$
  d. $P(E \cup F) = P(E) + P(F) - P(E \cap F)$

2. Give an example, based on a real-life situation, illustrating the property $P(E^c) = 1 - P(E)$, where E is an event and E^c is the complement of E.

7.7 Exercises

A pair of dice is rolled, and the number that appears uppermost on each die is observed. In Exercises 1–6, refer to this experiment, and find the probability of the given event.

1. The sum of the numbers is an even number.

2. The sum of the numbers is either 7 or 11.

3. A pair of 1s is thrown.

4. A double is thrown.

5. One die shows a 6, and the other is a number less than 3.

6. The sum of the numbers is at least 4.

An experiment consists of selecting a card at random from a 52-card deck. In Exercises 7–12, refer to this experiment and find the probability of the event.

7. A king of diamonds is drawn.

8. A diamond or a king is drawn.

9. A face card (i.e., a jack, queen, or king) is drawn.

10. A red face card is drawn.

11. An ace is not drawn.

12. A black face card is not drawn.

13. RAFFLES Five hundred raffle tickets were sold. What is the probability that a person holding one ticket will win the first prize? What is the probability that he or she will not win the first prize?

14. TV HOUSEHOLDS WITH REMOTE CONTROLS The results of a recent television survey of American TV households revealed that 87 out of every 100 TV households have at least one remote control. What is the probability that a randomly selected TV household does not have at least one remote control?

In Exercises 15–22, explain why the statement is incorrect.

15. The sample space associated with an experiment is given by $S = \{a, b, c\}$, where $P(a) = .3$, $P(b) = .4$, and $P(c) = .4$.

16. The probability that a bus will arrive late at the Civic Center is .35, and the probability that it will be on time or early is .60.

17. OFFICE POOLS A person participates in a weekly office pool in which he has one chance in ten of winning the purse. If he participates for 5 weeks in succession, the probability of winning at least one purse is $\frac{5}{10}$.

18. STOCK PRICES The probability that a certain stock will increase in value over a period of 1 week is .6. Therefore, the probability that the stock will decrease in value is .4.

19. A red die and a green die are tossed. The probability that a 6 will appear uppermost on the red die is $\frac{1}{6}$, and the probability that a 1 will appear uppermost on the green die is $\frac{1}{6}$. Hence the probability that the red die will show a 6 or the green die will show a 1 is $\frac{1}{6} + \frac{1}{6}$.

20. COLLEGE ADMISSIONS Joanne, a high school senior, has applied for admission to four colleges, A, B, C, and D. She has estimated that the probability that she will be accepted for admission by college A, B, C, and D is .5, .3, .1, and .08, respectively. Thus, the probability that she will be accepted for admission by at least one college is $P(A) + P(B) + P(C) + P(D) = .5 + .3 + .1 + .08 = .98$.

21. The sample space associated with an experiment is given by $S = \{a, b, c, d, e\}$. The events $E = \{a, b\}$ and $F = \{c, d\}$ are mutually exclusive. Hence the events E^c and F^c are mutually exclusive.

22. FORECASTING SALES Mark Owens, an optician, estimates that the probability that a customer coming into his store will purchase one or more pairs of glasses but not contact lenses is .40, and the probability that he will purchase one or more pairs of contact lenses but not glasses is .25. Hence Owens concludes that the probability that a customer coming into his store will purchase neither a pair of glasses nor a pair of contact lenses is .35.

23. Let E and F be two events that are mutually exclusive, and suppose $P(E) = .2$ and $P(F) = .5$. Compute:
a. $P(E \cap F)$ **b.** $P(E \cup F)$
c. $P(E^c)$ **d.** $P(E^c \cap F^c)$

24. Let E and F be two events of an experiment with sample space S. Suppose $P(E) = .6$, $P(F) = .4$, and $P(E \cap F) = .2$. Compute:
a. $P(E \cup F)$ **b.** $P(E^c)$
c. $P(F^c)$ **d.** $P(E^c \cap F)$

25. Let $S = \{s_1, s_2, s_3, s_4\}$ be the sample space associated with an experiment having the probability distribution shown in the accompanying table. If $A = \{s_1, s_2\}$ and $B = \{s_1, s_3\}$, find:
a. $P(A), P(B)$ **b.** $P(A^c), P(B^c)$
c. $P(A \cap B)$ **d.** $P(A \cup B)$
e. $P(A^c \cap B^c)$ **f.** $P(A^c \cup B^c)$

Outcome	Probability
s_1	$\frac{1}{8}$
s_2	$\frac{3}{8}$
s_3	$\frac{1}{4}$
s_4	$\frac{1}{4}$

26. Let $S = \{s_1, s_2, s_3, s_4, s_5, s_6\}$ be the sample space associated with an experiment having the probability distribution shown in the accompanying table. If $A = \{s_1, s_2\}$ and $B = \{s_1, s_5, s_6\}$, find:
a. $P(A), P(B)$ **b.** $P(A^c), P(B^c)$
c. $P(A \cap B)$ **d.** $P(A \cup B)$
e. $P(A^c \cap B^c)$ **f.** $P(A^c \cup B^c)$

Outcome	Probability
s_1	$\frac{1}{3}$
s_2	$\frac{1}{8}$
s_3	$\frac{1}{6}$
s_4	$\frac{1}{6}$
s_5	$\frac{1}{12}$
s_6	$\frac{1}{8}$

27. MAKEUP OF U.S. MOVIEGOER AUDIENCE In a survey of 3000 Americans aged 12 through 74 years, the following makeup of the moviegoer audience was obtained:

Age	12–24	25–44	45–64	65–74
Audience (%)	30	36	28	6

If a moviegoer is selected at random from the respondents of the survey, what is the probability that he or she is between:
a. 12 and 24 years of age?
b. 25 and 64 years of age?
c. 12 and 24 years of age or 65 and 74 years of age?
Source: Nielsen.

28. CLIMATE CHANGE In a survey of 1089 adults conducted by Duke University between January 16, 2013, and January 22, 2013, the following question was asked: How serious a threat is climate change? The results of the survey are summarized below:

Answer	Very serious	Somewhat serious	Not that much	Not at all
Respondents (%)	38	46	15	1

On the basis of the results of the survey, what is the probability that a person chosen at random from the survey said that climate change posed:
a. A very serious threat or a somewhat serious threat?
b. No threat?
c. A somewhat serious threat or not that much of a threat?
Source: Duke University.

29. STAY WHEN VISITING NATIONAL PARKS In a survey conducted by travel.usatoday.com, in the week of August 20, 2012, the following question was asked: Where do you stay when you visit national parks? The results of the survey follow:

Accommodation	Respondents (%)
Nearby hotel or motel	55
In-park lodge	25
In-park campground	14
Nearby campground or RV park	6

What is the probability that a randomly chosen participant in the survey said that he or she would:
a. Stay in a nearby campground or RV park?
b. Stay in an in-park lodge or an in-park campground?
c. Not stay in a nearby hotel or motel
Source: travel.usatoday.com.

30. 401(K) INVESTORS According to a study conducted in 2011 concerning the participation, by age, of 401(k) investors, the following data were obtained:

Age	20s	30s	40s	50s	60s
Percent	12	23	28	27	10

a. What is the probability that a 401(k) investor selected at random in 2011 was in his or her 20s or 60s?
b. What is the probability that a 401(k) investor selected at random in 2011 was under the age of 50?
Source: Investment Company Institute.

31. RECOVERY FROM THE GREAT RECESSION In a survey of 1140 middle-class adults who said they were worse off now than before the recession of 2008, the following question

was asked: How many years will it take you to fully recover financially? The results are as follows:

Time to Recover	Respondents (%)
Four years or less	29
More than 4 years but less than 10 years	24
Ten years or more	27
Don't know/refused	20

What is the probability that a randomly chosen participant in the survey said that he or she will:
a. Take less than 10 years to recover financially?
b. Take more than 4 years to recover financially?
c. Take 4 years or less, or 10 years or more to recover financially?
Source: Pew Research Center.

32. ELECTRICITY GENERATION Electricity in the United States is generated from many sources. The following table gives the sources as well as their shares in the production of electricity:

Source	Coal	Nuclear	Natural gas	Hydropower	Oil	Other
Share (%)	50.0	19.3	18.7	6.7	3.0	2.3

If a source for generating electricity is picked at random, what is the probability that it comes from:
a. Coal or natural gas?
b. Nonnuclear sources?
Source: Energy Information Administration.

Refer to Exercises 7.5, Problem 34, where you were asked to find the sample space corresponding to the different human blood groups. The following table gives the percent of the U.S. population having each of the eight possible blood types in the sample space. Note that the presence or absence of the Rh antigen is indicated by the symbols + or −, respectively.

Blood Types	A^+	A^-	B^+	B^-	AB^+	AB^-	O^+	O^-
Percent	35.7	6.3	8.5	1.5	3.4	0.6	37.4	6.6

Source: American Red Cross.

In Exercises 33 and 34, use the above table.

33. BLOOD TYPES
a. What is the probability that a person selected randomly from the U.S. population has a blood type that is Rh^-?
b. What is the probability that a person selected randomly from the U.S. population has a blood type that is type O?

34. BLOOD TYPES
a. What is the probability that a person selected randomly from the U.S. population has a blood type that is type O or Rh^+?
b. What is the probability that a person selected randomly from the U.S. population has a blood type that contains the A antigen?
c. What is the probability that a person selected randomly from the U.S. population has a blood type that is type AB or Rh^-?

35. **DOWNLOADING MUSIC** The following table gives the percentage of music downloaded from the United States and other countries in a certain year by U.S. users:

Country	U.S.	Germany	Canada	Italy	U.K.	France	Japan	Other
Percent	45.1	16.5	6.9	6.1	4.2	3.8	2.5	14.9

   a. Verify that the table does give a probability distribution for the experiment.
   b. What is the probability that a user who downloads music, selected at random, obtained it from either the United States or Canada?
   c. What is the probability that a U.S. user who downloads music, selected at random, does not obtain it from Italy, the United Kingdom (U.K.), or France?
   Source: Felix Oberholtzer-Gee and Koleman Strumpf.

36. **TEEN SPENDING BEHAVIOR** In a survey conducted in the spring of 2013, the following results pertaining to teen spending behavior were obtained:

Category	Share of Spending (%)
Clothing	21
Food	18
Accessories/personal care	10
Shoes	9
Car	8
Electronics	8
Music and movies	7
Video games	6
Concerts and events	6
Books	2
Furniture	2
Other	3

   a. Verify that the table does give a probability distribution for the experiment.
   b. If the expenditure for an item purchased by a teen is chosen at random, what is the probability that it was on clothing or shoes?
   Source: Piper Jeffry & Co.

37. **TIRE SAFETY** A team of automobile safety experts was asked by a television news station to conduct an experiment in which the tires of 100 randomly chosen cars of its employees were subjected to a safety inspection. It was determined that of the 100 cars inspected, 11 cars failed the tread depth test (at least one tire on the car was worn excessively), 45 cars failed the tire pressure test (at least one tire on the car was overinflated or underinflated), and 4 failed both the tread depth test and the tire pressure test. Find the probability that a car selected at random from this group of cars:
   a. Failed only the tire pressure test.
   b. Passed both the tread depth test and the tire pressure test.

38. **ALTERNATIVE FUEL VEHICLES** A survey was conducted by the local chapter of an environmental club regarding the ownership of alternative fuel vehicles (AFVs) among the members of the group. An AFV is a vehicle that runs on fuel other than petroleum fuels (petrol and diesel) or that does not involve solely petroleum fuels. It was found that of the 80 members of the club surveyed, 22 of them own at least one hybrid car, 12 of them own at least one electric car, and 4 of them own at least one hybrid and at least one electric car. If a member of the club selected at random is surveyed, what is the probability that he or she:
   a. Owns only hybrid cars?
   b. Owns no alternative fuel vehicles?

39. **FORECASTING SALES** The probability that a shopper in a certain boutique will buy a blouse is .35, that she will buy a pair of pants is .30, and that she will buy a skirt is .27. The probability that she will buy both a blouse and a skirt is .15, the probability that she will buy both a skirt and a pair of pants is .19, and the probability that she will buy both a blouse and a pair of pants is .12. Finally, the probability that she will buy all three items is .08. What is the probability that a customer selected at random will buy:
   a. Exactly one of these items?
   b. None of these items?

40. **COURSE ENROLLMENTS** Among 500 freshmen pursuing a business degree at a university, 320 are enrolled in an economics course, 225 are enrolled in a mathematics course, and 140 are enrolled in both an economics and a mathematics course. What is the probability that a freshman selected at random from this group is enrolled in:
   a. An economics and/or a mathematics course?
   b. Exactly one of these two courses?
   c. Neither an economics course nor a mathematics course?

41. **CUSTOMER SURVEYS** A leading manufacturer of kitchen appliances advertised its products in two magazines: *Good Housekeeping* and the *Ladies Home Journal*. A survey of 500 customers revealed that 140 learned of its products from *Good Housekeeping*, 130 learned of its products from the *Ladies Home Journal*, and 80 learned of its products from both magazines. What is the probability that a person selected at random from this group saw the manufacturer's advertisement in:
   a. Both magazines?
   b. At least one of the two magazines?
   c. Exactly one magazine?

42. **ASSEMBLY-TIME STUDIES** A time study was conducted by the production manager of Universal Instruments to determine how much time it took an assembly worker to complete a certain task during the assembly of its Orion tablet computers. Results of the study indicated that 20% of the workers were able to complete the task in less than 3 min, 60% of the workers were able to complete the task in 4 min or less, and 10% of the workers required more than 5 min to complete the task. If an assembly-line worker is selected at random from this group, what is the probability that:
   a. He or she will be able to complete the task in 5 min or less?
   b. He or she will not be able to complete the task within 4 min?
   c. The time taken for the worker to complete the task will be between 3 and 4 min (inclusive)?

43. USE OF LANDLINE PHONE VERSUS CELL PHONE A survey of 1000 adults aged 18 years and older conducted in July 2012 found that 910 of them had a cell phone, 670 of them had a landline phone, and 580 of them had both a cell phone and a landline phone. Find the probability that a person selected at random from this group of respondents:

a. Has only a cell phone?

b. Has only a landline phone?

Source: AARP Bulletin Poll.

44. TIME ON A DIET A survey on how long dieters stay on a diet found that 26% of them stayed on the diet for a month or less, 36% of them stayed on for more than a month but less than 6 months, 11% of them stayed on for 6 or more months but less than a year, and 27% of them stayed on for a year or more. For the survey, respondents could define "diet" any way they wanted. On the basis of this survey, what is the probability that a person selected at random from the survey said that he or she stayed on a diet for:

a. A month or less?

b. More than 1 month but less than a year?

c. Six months or more?

Source: NPD Group.

45. TEACHERS' VIEWS OF EDUCATIONAL PROBLEMS A nonprofit organization conducted a survey of 2140 metropolitan-area teachers regarding their beliefs about educational problems. The following data were obtained:

900 said that lack of parental support is a problem.

890 said that abused or neglected children are problems.

680 said that malnutrition or students in poor health is a problem.

120 said that lack of parental support and abused or neglected children are problems.

110 said that lack of parental support and malnutrition or poor health are problems.

140 said that abused or neglected children and malnutrition or poor health are problems.

40 said that lack of parental support, abuse or neglect, and malnutrition or poor health are problems.

What is the probability that a teacher selected at random from this group said that lack of parental support is the only problem hampering a student's schooling?

Hint: Draw a Venn diagram.

46. 401(K) INVESTMENTS In a survey of 200 employees of a company regarding their 401(k) investments, the following data were obtained:

141 had investments in stock funds.

91 had investments in bond funds.

60 had investments in money market funds.

47 had investments in stock funds and bond funds.

36 had investments in stock funds and money market funds.

36 had investments in bond funds and money market funds.

22 had investments in stock funds, bond funds, and money market funds.

What is the probability that an employee of the company chosen at random:

a. Had investments in exactly two kinds of investment funds?

b. Had investments in exactly one kind of investment fund?

c. Had no investment in any of the three types of funds?

47. ROLLOVER DEATHS The following table gives the number of people killed in rollover crashes in various types of vehicles in 2010:

Types of Vehicles	Cars	Light trucks	Large trucks	Buses	Other
Deaths	2748	4814	443	10	227

Find the empirical probability distribution associated with these data. If a fatality due to a rollover crash in 2010 is picked at random, what is the probability that the victim was in:

a. A car?   b. A light truck?   c. A bus or a large truck?

Source: National Highway Traffic Safety Administration.

48. SWITCHING JOBS Two hundred workers were asked: Would a better economy lead you to switch jobs? The results of the survey follow:

Answer	Very likely	Somewhat likely	Somewhat unlikely	Very unlikely	Don't know
Respondents	40	28	26	104	2

If a worker is chosen at random, what is the probability that he or she:

a. Is very unlikely to switch jobs?

b. Is somewhat likely or very likely to switch jobs?

Source: Accountemps.

49. LOSING WEIGHT In a survey, 1012 American adults were asked how many times they had tried to lose weight in their lifetime. The results of the survey follow:

Answer	Once or twice	3–10 times	More than 10 times	Never	No opinion
Respondents	253	304	81	334	40

If a person in the survey is selected at random, what is the probability that the person answered that he or she had:

a. Tried at least once to lose weight in his or her lifetime?

b. Tried between one and ten times to lose weight in his or her lifetime?

c. Never tried to lose weight in his or her lifetime?

Source: Gallup Poll.

50. WALKING ON MARS In a survey of 1000 adults conducted in 2012, the following question was asked: Will humans walk on Mars in the next 25 years? The results of the survey are summarized in the following table:

Answer	Very likely	Somewhat likely	Not very likely	Not at all likely	Not sure
Respondents	210	290	340	80	80

If a respondent in the survey is chosen at random, what is the probability that he or she answered that:
a. It is very likely or somewhat likely that humans will walk on Mars in the next 25 years?
b. It is not very likely or not at all likely that humans will walk on Mars in the next 25 years?
Source: Rasmussen Reports.

51. **RISK OF AN AIRPLANE CRASH** According to a study of Western-built commercial jets involved in crashes over a 10-year period, the percentage of airplane crashes that occur at each stage of flight are as follows:

Phase	Percent
On ground, taxiing	4
During takeoff	10
Climbing to cruise altitude	19
En route	5
Descent and approach	31
Landing	31

If one of the doomed flights in this period is picked at random, what is the probability that it crashed:
a. While taxiing on the ground or while en route?
b. During takeoff or landing?

If the study is indicative of airplane crashes in general, when is the risk of a plane crash the highest?
Source: National Transportation Safety Board.

52. **DISTRACTED DRIVING** According to a study of 100 drivers in metropolitan Washington, D.C., whose cars were equipped with cameras with sensors, the distractions and the number of incidents (crashes, near crashes, and situations that require an evasive maneuver after the driver was distracted) caused by these distractions are as follows:

Distraction	A	B	C	D	E	F	G	H	I
Driving Incidents	668	378	194	163	133	134	111	111	89

where A = Wireless device (cell phone, PDA)
 B = Passenger
 C = Something inside car
 D = Vehicle
 E = Personal hygiene
 F = Eating
 G = Something outside car
 H = Talking/singing
 I = Other

If an incident caused by a distraction is picked at random, what is the probability that it was caused by:

a. The use of a wireless device?
b. Something other than personal hygiene or eating?
Source: Virginia Tech Transportation Institute and NHTSA.

53. **PLANS TO KEEP CARS** In a survey conducted to determine how long Americans keep their cars, 2000 automobile owners were asked how long they planned to keep their present cars. The results of the survey follow:

Years Car Is Kept, x	Respondents
$0 \le x < 1$	60
$1 \le x < 3$	440
$3 \le x < 5$	360
$5 \le x < 7$	340
$7 \le x < 10$	240
$10 \le x$	560

Find the probability distribution associated with these data. What is the probability that an automobile owner selected at random from those surveyed planned to keep his or her present car:
a. Less than 5 years?
b. 3 years or more?

54. **GUN-CONTROL LAWS** A poll was conducted among 250 residents of a certain city regarding tougher gun-control laws. The results of the poll are shown in the table:

	Own Only a Handgun	Own Only a Rifle	Own a Handgun and a Rifle	Own Neither	Total
Favor Tougher Laws	0	12	0	138	150
Oppose Tougher Laws	58	5	25	0	88
No Opinion	0	0	0	12	12
Total	58	17	25	150	250

If one of the participants in this poll is selected at random, what is the probability that he or she:
a. Favors tougher gun-control laws?
b. Owns a handgun?
c. Owns a handgun but not a rifle?
d. Favors tougher gun-control laws and does not own a handgun?

55. Suppose the probability that Bill can solve a problem is p_1 and the probability that Mike can solve it is p_2. Show that the probability that Bill and Mike working independently can solve the problem is $p_1 + p_2 - p_1 p_2$.

56. Fifty raffle tickets are numbered 1 through 50, and one of them is drawn at random. What is the probability that the number is a multiple of 5 or 7? Consider the following "solution": Since 10 tickets bear numbers that are multiples of 5 and since 7 tickets bear numbers that are multiples of 7, we conclude that the required probability is

$$\frac{10}{50} + \frac{7}{50} = \frac{17}{50}$$

What is wrong with this argument? What is the correct answer?

In Exercises 57–62, determine whether the statement is true or false. If it is true, explain why it is true. If it is false, give an example to show why it is false.

57. If A is a subset of B and $P(B) = 0$, then $P(A) = 0$.

58. If A is a subset of B, then $P(A) \leq P(B)$.

59. If $E_1, E_2, \ldots, E_n$ are events of an experiment, then
$P(E_1 \cup E_2 \cup \cdots \cup E_n) = P(E_1) + P(E_2) + \cdots + P(E_n)$.

60. If E is an event of an experiment, then $P(E) + P(E^c) = 1$.

61. If $P(E) = P(F)$ and $E \subseteq F$, then $E = F$.

62. If S is a sample space of an experiment and $E \neq S$, then $P(E) < 1$.

7.7 Solutions to Self-Check Exercises

1. a. Using Property 4, we find

$$P(E \cup F) = P(E) + P(F) - P(E \cap F)$$
$$= .4 + .5 - .1$$
$$= .8$$

b. From the accompanying Venn diagram, in which the subset $E \cap F^c$ is shaded, we see that

$$P(E \cap F^c) = .3$$

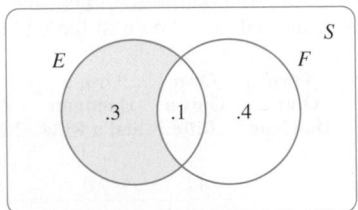

The result may also be obtained by using the relationship

$$P(E \cap F^c) = P(E) - P(E \cap F)$$
$$= .4 - .1 = .3$$

2. Let E denote the event that the realtor will find a buyer within 1 month of the date the property is listed for sale or lease, and let F denote the event that the realtor will find a lessee within the same time period. Then

$$P(E) = .3 \qquad P(F) = .8 \qquad P(E \cap F) = .1$$

The probability of the event that the realtor will find a buyer or a lessee within 1 month of the date the property is listed for sale or lease is given by

$$P(E \cup F) = P(E) + P(F) - P(E \cap F)$$
$$= .3 + .8 - .1 = 1$$

that is, a certainty.

CHAPTER 7 Summary of Principal Formulas and Terms

FORMULAS

1. Commutative laws	$A \cup B = B \cup A$ $A \cap B = B \cap A$
2. Associative laws	$A \cup (B \cup C) = (A \cup B) \cup C$ $A \cap (B \cap C) = (A \cap B) \cap C$
3. Distributive laws	$A \cup (B \cap C) = (A \cup B) \cap (A \cup C)$ $A \cap (B \cup C) = (A \cap B) \cup (A \cap C)$
4. De Morgan's laws	$(A \cup B)^c = A^c \cap B^c$ $(A \cap B)^c = A^c \cup B^c$
5. Number of elements in the union of two finite sets	$n(A \cup B) = n(A) + n(B) - n(A \cap B)$
6. Permutation of n distinct objects, taken r at a time	$P(n, r) = \dfrac{n!}{(n - r)!}$
7. Permutation of n objects, not all distinct, taken n at a time	$\dfrac{n!}{n_1! \, n_2! \cdots n_m!}$

8. Combination of n distinct objects, taken r at a time	$C(n, r) = \dfrac{n!}{r!\,(n-r)!}$
9. Probability of an event in a uniform sample space	$P(E) = \dfrac{n(E)}{n(S)}$
10. Probability of the union of two mutually exclusive events	$P(E \cup F) = P(E) + P(F)$
11. Addition rule	$P(E \cup F) = P(E) + P(F) - P(E \cap F)$
12. Rule of complements	$P(E^c) = 1 - P(E)$

TERMS

set (432)

element of a set (432)

roster notation (432)

set-builder notation (432)

set equality (432)

subset (433)

empty set (433)

universal set (434)

Venn diagram (434)

set union (435)

set intersection (435)

disjoint set (435)

complement of a set (435)

set complementation (436)

multiplication principle (452)

generalized multiplication principle (453)

permutation (458)

n-factorial (460)

combination (464)

experiment (473)

outcome (473)

sample point (473)

sample space (473)

event (473)

finite sample space (473)

union of two events (474)

intersection of two events (474)

complement of an event (474)

mutually exclusive events (475)

relative frequency (482)

empirical probability (482)

probability of an event (482)

elementary (simple) event (482)

probability distribution (482)

probability function (482)

uniform sample space (483)

addition principle (484)

CHAPTER 7 Concept Review Questions

Fill in the blanks.

1. A well-defined collection of objects is called a/an _____. These objects are called _____ of the _____.

2. Two sets having exactly the same elements are said to be _____.

3. If every element of a set A is also an element of a set B, then A is a/an _____ of B.

4. **a.** The empty set $\varnothing$ is the set containing _____ elements.
   b. The universal set is the set containing _____ elements.

5. **a.** The set of all elements in A and/or B is called the _____ of A and B.
   b. The set of all elements in both A and B is called the _____ of A and B.

6. The set of all elements in U that are not in A is called the _____ of A.

7. Applying De Morgan's Laws, we can write $(A \cup B \cup C)^c = $ _____.

8. An arrangement of a set of distinct objects in a definite order is called a/an _____; an arrangement in which the order is not important is a/an _____.

9. An activity with observable results is called a/an _____; an outcome of an experiment is called a/an _____ point, and the set consisting of all possible sample points of an experiment is called a sample _____; a subset of a sample space of an experiment is called a/an _____.

10. The events E and F are mutually exclusive if $E \cap F = $ _____.

11. A sample space in which the outcomes are equally likely is called a/an _____ sample space; if such a space contains n simple events, then the probability of each simple event is _____.

CHAPTER 7 Review Exercises

In Exercises 1–4, list the elements of each set in roster notation.

1. $\{x \mid 3x - 2 = 7 \text{ and } x \text{ is an integer}\}$

2. $\{x \mid x \text{ is a letter of the word } TALLAHASSEE\}$

3. The set whose elements are the even numbers between 3 and 11

4. $\{x \mid (x - 3)(x + 4) = 0 \text{ and } x \text{ is a negative integer}\}$

Let $A = \{a, c, e, r\}$. In Exercises 5–8, determine whether the set is equal to A.

5. $\{r, e, c, a\}$

6. $\{x \mid x \text{ is a letter of the word } career\}$

7. $\{x \mid x \text{ is a letter of the word } racer\}$

8. $\{x \mid x \text{ is a letter of the word } cares\}$

In Exercises 9–12, shade the portion of the accompanying figure that represents the given set.

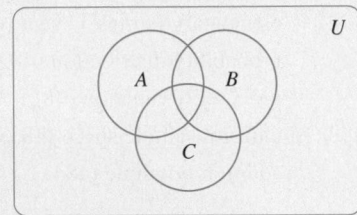

9. $A \cup (B \cap C)$

10. $(A \cap B \cap C)^c$

11. $A^c \cap B^c \cap C^c$

12. $A^c \cap (B^c \cup C^c)$

In Exercises 13–16, verify the equation by direct computation. Let $U = \{a, b, c, d, e\}$, $A = \{a, b\}$, $B = \{b, c, d\}$, and $C = \{a, d, e\}$.

13. $A \cup (B \cup C) = (A \cup B) \cup C$

14. $A \cap (B \cap C) = (A \cap B) \cap C$

15. $A \cap (B \cup C) = (A \cap B) \cup (A \cap C)$

16. $A \cup (B \cap C) = (A \cup B) \cap (A \cup C)$

For Exercises 17–20, let

$U = \{$all participants in a consumer-behavior survey conducted by a national polling group$\}$

$A = \{$consumers who avoided buying a product because it is not recyclable$\}$

$B = \{$consumers who used cloth rather than disposable diapers$\}$

$C = \{$consumers who boycotted a company's products because of its record on the environment$\}$

$D = \{$consumers who voluntarily recycled their garbage$\}$

Describe each set in words.

17. $A \cap C$ **18.** $A \cup D$

19. $B^c \cap D$ **20.** $C^c \cup D^c$

Let A and B be subsets of a universal set U and suppose $n(U) = 350$, $n(A) = 120$, $n(B) = 80$, and $n(A \cap B) = 50$. In Exercises 21–26, find the number of elements in each set.

21. $n(A \cup B)$ **22.** $n(A^c)$

23. $n(B^c)$ **24.** $n(A^c \cap B)$

25. $n(A \cap B^c)$ **26.** $n(A^c \cap B^c)$

In Exercises 27–30, evaluate each quantity.

27. $C(20, 18)$ **28.** $P(9, 7)$

29. $C(5, 3) \cdot P(4, 2)$ **30.** $4 \cdot P(5, 3) \cdot C(7, 4)$

31. Let E and F be two mutually exclusive events, and suppose $P(E) = .4$ and $P(F) = .2$. Compute:
  a. $P(E \cap F)$ **b.** $P(E \cup F)$
  c. $P(E^c)$ **d.** $P(E^c \cap F^c)$
  e. $P(E^c \cup F^c)$

32. Let E and F be two events of an experiment with sample space S. Suppose $P(E) = .3$, $P(F) = .2$, and $P(E \cap F) = .15$. Compute:
  a. $P(E \cup F)$ **b.** $P(E^c \cap F^c)$
  c. $P(E^c \cap F)$

33. Let E and F be two mutually exclusive events, and suppose $P(E) = .35$ and $P(F) = .47$. Find:
  a. $P(F^c)$ **b.** $P(E \cap F^c)$
  c. $P(E \cup F)$ **d.** $P(E^c \cap F^c)$

34. A die is loaded, and it has been determined that the probability distribution associated with the experiment of rolling the die and observing which number falls uppermost is given by the following:

Simple Event	Probability
$\{1\}$	.20
$\{2\}$	.12
$\{3\}$	.16
$\{4\}$	.18
$\{5\}$	.15
$\{6\}$	.19

a. What is the probability of the number being even?
b. What is the probability of the number being either a 1 or a 6?
c. What is the probability of the number being less than 4?

35. An urn contains six red balls, five black balls, and four green balls. If a ball is selected at random from the urn, what is the probability that a red ball or a black ball will be selected?

36. CREDIT CARD COMPARISONS A comparison of five major credit cards showed that:

3 offered cash advances.

3 offered extended payments for all goods and services purchased.

2 required an annual fee of less than $35.

2 offered both cash advances and extended payments.

1 offered extended payments and had an annual fee less than $35.

No card had an annual fee less than $35 and offered both cash advances and extended payments.

How many cards had an annual fee less than $35 and offered cash advances? (Assume that every card had at least one of the three mentioned features.)

37. STUDENT SURVEYS The Department of Foreign Languages of a liberal arts college conducted a survey of its recent graduates to determine the foreign language courses they had taken while undergraduates at the college. Of the 480 graduates:

200 had at least 1 year of Spanish.

178 had at least 1 year of French.

140 had at least 1 year of German.

33 had at least 1 year of Spanish and French.

24 had at least 1 year of Spanish and German.

18 had at least 1 year of French and German.

3 had at least 1 year of all three languages.

How many of the graduates had:
a. At least 1 year of at least one of the three languages?
b. At least 1 year of exactly one of the three languages?
c. Less than 1 year of any of the three languages?

38. In how many ways can six different DVDs be arranged on a shelf?

39. In how many ways can three pictures be selected from a group of six different pictures?

40. In how many ways can six different books, four of which are math books, be arranged on a shelf if the math books must be placed next to each other?

41. In how many ways can six people be arranged in a line for a group picture:
a. If there are no restrictions?
b. If two people in the group insist on not standing next to each other?

42. Find the number of distinguishable permutations that can be formed from the letters of each word.
a. *CINCINNATI*   **b.** *HONOLULU*

43. How many three-digit numbers can be formed from the numerals in the set $\{1, 2, 3, 4, 5\}$ if:
a. Repetition of digits is not allowed?
b. Repetition of digits is allowed?

44. AUTOMOBILE SELECTION An automobile manufacturer has three different subcompact cars in the line. Customers selecting one of these cars have a choice of three engine sizes, four body styles, and three color schemes. How many different selections can a customer make?

45. MENU SELECTIONS Two soups, five entrées, and three desserts are listed on the "Special" menu at the Neptune Restaurant. How many different selections consisting of one soup, one entrée, and one dessert can a customer choose from this menu?

46. INVESTMENTS In a survey conducted by Helena, a financial consultant, it was revealed that of her 400 clients:

300 own stocks.

180 own bonds.

160 own mutual funds.

110 own both stocks and bonds.

120 own both stocks and mutual funds.

90 own both bonds and mutual funds.

How many of Helena's clients own stocks, bonds, and mutual funds?

47. POKER From a standard 52-card deck, how many 5-card poker hands can be dealt consisting of:
a. Five clubs?   **b.** Three kings and one pair?

48. ELECTIONS In an election being held by the Associated Students Organization, there are six candidates for president, four for vice president, five for secretary, and six for treasurer. How many different possible outcomes are there for this election?

49. TEAM SELECTION There are eight seniors and six juniors in the Math Club at Jefferson High School. In how many ways can a math team consisting of four seniors and two juniors be selected from the members of the Math Club?

50. If order matters, in how many ways can two cards be drawn from a 52-card deck:
a. If the first card is replaced before the second card is drawn?
b. If the second card is drawn without replacing the first card?

51. **SEATING ARRANGEMENTS** In how many ways can seven students be assigned seats in a row containing seven desks if:
   a. There are no restrictions?
   b. Two of the students must not be seated next to each other?

52. **QUALITY CONTROL** From a shipment of 60 CPUs, 5 of which are defective, a sample of 4 CPUs is selected at random.
   a. In how many different ways can the sample be selected?
   b. How many samples contain 3 defective CPUs?
   c. How many samples do not contain any defective CPUs?

53. **RANDOM SAMPLES** A sample of 4 balls is to be selected at random from an urn containing 15 balls numbered 1 to 15. If 6 balls are green, 5 are white, and 4 are black, then:
   a. How many different samples can be selected?
   b. How many samples can be selected that contain at least 1 white ball?

54. **SEATING ARRANGEMENTS** In how many ways can three married couples be seated in a row of six seats:
   a. If there are no seating restrictions?
   b. If men and women must alternate?
   c. If each married couple must sit together?

55. **TEAM SELECTION** There are seven boys and five girls in the debate squad at Franklin High School. In how many ways can a four-member debating team be selected:
   a. If there are no restrictions?
   b. If two boys and two girls must be on the team?
   c. If at least two boys must be on the team?

56. **QUALITY CONTROL** The quality-control department of Starr Communications, a manufacturer of video-game DVDs, has determined from records that 1.5% of the DVDs sold have video defects, 0.8% have audio defects, and 0.4% have both audio and video defects. What is the probability that a DVD purchased by a customer:
   a. Will have a video or audio defect?
   b. Will not have a video or audio defect?

57. **FIGHTING INFLATION** In a survey of 2000 adults aged 18 years and older conducted in a certain year, the following question was asked: Is your family income keeping pace with the cost of living? The results of the survey follow:

Answer	Falling behind	Staying even	Increasing faster	Don't know
Respondents	800	880	240	80

Determine the empirical probability distribution associated with these data.
Source: Pew Research Center.

58. **U.S. INCOME DISTRIBUTION FOR HOUSEHOLDS** According to the U.S. Census Bureau, the income distribution for households in 2011 was as follows:

Income ($)	0–24,999	25,000–49,999	50,000–74,999
Households and Families	30,337,000	30,134,000	21,294,000

Income ($)	75,000–99,999	100,000–124,999	125,000–149,999
Households and Families	13,899,000	9,130,000	5,311,000

Income ($)	150,000–199,999	200,000–249,999	250,000 or more
Households and Families	5,875,000	2,297,000	2,808,000

Find the empirical probability distribution associated with these data.
Source: U.S. Census Bureau.

59. **WORK HABITS** In a survey of 7780 workers, the following question was asked: How often are you late for work? A summary of the results of the survey follow:

Answer	Never	At least once a week	Once a month	Once a year
Respondents	4746	1244	856	934

If a worker in the survey is chosen at random, what is the probability that he or she:
   a. Was never late for work?
   b. Was late once a year?
Source: USA Today.

60. **CONSUMER PREFERENCES** Olivia is contemplating buying a laser printer. The probability that she will buy a printer manufactured by Epson, Brother, Canon, and Hewlett-Packard is .23, .18, .31, and .28, respectively. Find the probability that she will buy a laser printer manufactured by:
   a. Epson or Canon.   b. Epson, Brother, or Canon.

61. **TRANSPORTATION FATALITIES** The following breakdown of a total of 18,598 transportation fatalities that occurred in a certain year was obtained from records compiled by the U.S. Department of Transportation (DOT).

Mode of Transportation	Car	Train	Bicycle	Plane
Number of Fatalities	16,520	845	698	535

What is the probability that a victim randomly selected from this list of transportation fatalities for that year died in:
   a. A car crash or a bicycle accident?
   b. A train or a plane accident?
Source: U.S. Department of Transportation.

62. **CREDIT CARD OWNERSHIP** A survey of 1020 adults aged 18–49 years found that 34% of them have no credit cards, 22% have one credit card, 28% have two or three credit cards, 11% have four or five credit cards, and 5% have more than five credit cards. What is the probability that a randomly chosen respondent in the survey has:
    a. No credit card?
    b. Between one and three credit cards?
    c. Four or more credit cards?
    Source: ORC International.

63. **RETIREMENT EXPECTATIONS** In a survey on retirement, participants were asked this question: Do you think that life will be better, worse, or about the same when you retire? The results of the survey follow:

Answer	Better	Worse	Same	Don't know
Respondent (%)	38	18	41	3

If a person in the survey is selected at random, what is the probability that he or she answered that life after retirement would be:
    a. The same or better?
    b. The same or worse?
    Source: Bankrate.com

64. **SALES OF PLASMA TVs** The records of Ace Electronics show that of the plasma TVs sold by the company, 26% were manufactured by Panasonic, 15.4% were manufactured by LG, 13.7% were manufactured by Samsung, 13.3% were manufactured by Philips, and 7.3% were manufactured by Hitachi. If a customer chosen at random purchases a plasma TV from Ace Electronics, what is the probability that the set was manufactured by:
    a. Panasonic, LG, Samsung, Philips, or Hitachi?
    b. A company other than those mentioned in part (a)?

65. **KEEPING UP WITH THE COST OF LIVING** In a survey of 2000 adults, 18 years old and older, conducted in 2007, the following question was asked: Is your family income keeping pace with the cost of living? The results of the survey follow:

Answer	Falling behind	Staying even	Increasing faster	Don't know
Respondents	800	880	240	80

According to the survey, what percentage of the people polled said their family income is:
    a. At least keeping pace with the cost of living?
    b. Falling behind the cost of living?
    Source: Pew Research Center.

66. **TAX PREPARATION** A survey in which people were asked how they were planning to prepare their taxes revealed the following:

Method of Preparation	Percent
Computer software	33.9
Accountant	23.6
Tax preparation service	17.4
Spouse, friend, or other relative will prepare	10.8
By hand	14.3

What is the probability that a randomly chosen participant in the survey:
    a. Was planning to use an accountant or a tax preparation service to prepare his or her taxes?
    b. Was not planning to use computer software to prepare his or her taxes and was not planning to do taxes by hand?
    Source: National Retail Federation.

67. **WOMEN'S APPAREL** In an online survey for Talbots of 1095 women ages 35 years old and older, the participants were asked what article of clothing women most want to fit perfectly. A summary of the results of the survey follows:

Article of Clothing	Respondents
Jeans	470
Black pantsuit	307
Cocktail dress	230
White shirt	22
Gown	11
Other	55

If a woman who participated in the survey is chosen at random, what is the probability that she most wants:
    a. Jeans to fit perfectly?
    b. A black pantsuit or a cocktail dress to fit perfectly?
    Source: Market Tool's Zoom Panel.

68. **SALES OF DISASTER-RECOVERY SYSTEMS** Jay sells disaster-recovery computer systems to hedge funds. He estimates the probability of Hedge Fund A purchasing a system to be .6 and that of Hedge Fund B purchasing a system to be .5. He also estimates that the probability of both hedge funds purchasing a system is .3. What is the probability that only Hedge Fund A or only Hedge Fund B will purchase a system?

CHAPTER 7 Before Moving On . . .

1. Let $U = \{a, b, c, d, e, f, g\}$, $A = \{a, d, f, g\}$, $B = \{d, f, g\}$, and $C = \{b, c, e, f\}$. Find:
   a. $A \cap (B \cup C)$
   b. $(A \cap C) \cup (B \cup C)$
   c. A^c

2. Let A, B, and C be subsets of a universal set U, and suppose that $n(U) = 120$, $n(A) = 20$, $n(A \cap B) = 10$, $n(A \cap C) = 11$, $n(B \cap C) = 9$, and $n(A \cap B \cap C) = 4$. Find $n[A \cap (B \cup C)^c]$.

3. In how many ways can four compact discs be selected from six different compact discs?

4. There are six seniors and five juniors in the Chess Club at Madison High School. In how many ways can a team consisting of three seniors and two juniors be selected from the members of the Chess Club?

5. Let $S = \{s_1, s_2, s_3, s_4, s_5, s_6\}$ be the sample space associated with an experiment having the following probability distribution:

Outcome	s_1	s_2	s_3	s_4	s_5	s_6
Probability	$\frac{1}{12}$	$\frac{2}{12}$	$\frac{3}{12}$	$\frac{2}{12}$	$\frac{3}{12}$	$\frac{1}{12}$

   Find the probability of the event $A = \{s_1, s_3, s_6\}$.

6. A card is drawn from a well-shuffled 52-card deck. What is the probability that the card drawn is a deuce or a face card?

7. Let E and F be events of an experiment with sample space S. Suppose $P(E) = .5$, $P(F) = .6$, and $P(E \cap F) = .2$. Compute:
   a. $P(E \cup F)$ b. $P(E \cap F^c)$

8 Additional Topics in Probability

IN THIS CHAPTER, we develop additional techniques for computing the probabilities of certain events, and we take a look at descriptive statistics. We begin by looking at problems involving large sample spaces. In Section 8.2, we consider the effect that the occurrence of prior events has on the probability of an event occurring, and in Section 8.3 we learn how to compute the probabilities of certain events after the occurrence of an event. Throughout our discussion, we will see how these techniques are applied to many practical problems in fields as diverse as quality control and medical research. In the rest of the chapter, we take a glimpse at statistics, the branch of mathematics concerned with the collection, analysis, and interpretation of data.

Which of two motels should a certain private equity firm purchase? In Example 4, page 561, we show how the occupancy rate and the average daily profit for each motel can be used to help us determine which motel will generate the higher daily profit.

© Minerva Studio/Shutterstock.com

8.1   Use of Counting Techniques in Probability

Further Applications of Counting Techniques

As we have seen many times before, a problem in which the underlying sample space has a small number of elements may be solved by first determining all such sample points. However, for problems involving sample spaces with a large number of sample points, this approach is neither practical nor desirable.

In this section, we see how the counting techniques studied in Chapter 7 may be employed to help us solve problems in which the associated sample spaces contain large numbers of sample points. In particular, we restrict our attention to the study of uniform sample spaces—that is, sample spaces in which the outcomes are equally likely. For such spaces, we have the following result:

> **Computing the Probability of an Event in a Uniform Sample Space**
>
> Let S be a uniform sample space, and let E be any event. Then
>
> $$P(E) = \frac{\text{Number of outcomes in } E}{\text{Number of outcomes in } S} = \frac{n(E)}{n(S)} \qquad (1)$$

EXAMPLE 1 An unbiased coin is tossed six times. What is the probability that the coin will land heads:

a. Exactly three times?
b. At most three times?
c. On the first and the last toss?

Solution

a. Each outcome of the experiment may be represented as a sequence of heads and tails. Using the generalized multiplication principle, we see that the number of outcomes of this experiment is given by 2^6, or 64. Let E denote the event that the coin lands heads exactly three times. Since there are $C(6, 3)$ ways this can occur, we see that the required probability is

$$P(E) = \frac{n(E)}{n(S)} = \frac{C(6, 3)}{64} = \frac{\dfrac{6!}{3!\,3!}}{64} \qquad \text{\small S is a sample space of the experiment.}$$

$$= \frac{\dfrac{6 \cdot 5 \cdot 4}{3 \cdot 2}}{64} = \frac{20}{64} = \frac{5}{16} = .3125$$

b. Let F denote the event that the coin lands heads at most three times. Then $n(F)$ is given by the sum of the number of ways the coin lands heads zero times (no heads!), the number of ways it lands heads exactly once, the number of ways it lands heads exactly twice, and the number of ways it lands heads exactly three times. That is,

$$n(F) = C(6, 0) + C(6, 1) + C(6, 2) + C(6, 3)$$

$$= \frac{6!}{0!\,6!} + \frac{6!}{1!\,5!} + \frac{6!}{2!\,4!} + \frac{6!}{3!\,3!}$$

$$= 1 + 6 + \frac{6 \cdot 5}{2} + \frac{6 \cdot 5 \cdot 4}{3 \cdot 2} = 42$$

Therefore, the required probability is

$$P(F) = \frac{n(F)}{n(S)} = \frac{42}{64} = \frac{21}{32} \approx .6563$$

c. Let F denote the event that the coin lands heads on the first and the last toss. Then $n(F) = 1 \cdot 2 \cdot 2 \cdot 2 \cdot 2 \cdot 1 = 2^4$, so the probability that this event occurs is

$$P(F) = \frac{2^4}{2^6}$$

$$= \frac{1}{2^2}$$

$$= \frac{1}{4}$$

EXAMPLE 2 Two cards are selected at random (without replacement) from a well-shuffled deck of 52 playing cards. What is the probability that:

a. They are both aces?   **b.** Neither of them is an ace?

Solution

a. The experiment consists of selecting 2 cards from a pack of 52 playing cards. Since the order in which the cards are selected is immaterial, the sample points are combinations of 52 cards taken 2 at a time. Now there are $C(52, 2)$ ways of selecting 52 cards taken 2 at a time, so the number of elements in the sample space S is given by $C(52, 2)$. Next, we observe that there are $C(4, 2)$ ways of selecting 2 aces from the 4 in the deck. Therefore, if E denotes the event that the cards selected are both aces, then

$$P(E) = \frac{n(E)}{n(S)}$$

$$= \frac{C(4, 2)}{C(52, 2)} = \frac{\dfrac{4!}{2!\,2!}}{\dfrac{52!}{2!\,50!}} = \frac{4 \cdot 3}{2} \cdot \frac{2}{52 \cdot 51}$$

$$= \frac{1}{221} \approx .0045$$

b. Let F denote the event that neither of the two cards selected is an ace. Since there are $C(48, 2)$ ways of selecting two cards neither of which is an ace, we find that

$$P(F) = \frac{n(F)}{n(S)} = \frac{C(48, 2)}{C(52, 2)} = \frac{\dfrac{48!}{2!\,46!}}{\dfrac{52!}{2!\,50!}} = \frac{48 \cdot 47}{2} \cdot \frac{2}{52 \cdot 51}$$

$$= \frac{188}{221} \approx .8507$$

APPLIED EXAMPLE 3 Quality Control A bin in the hi-fi department of Building 20, a bargain outlet, contains 100 blank DVDs, of which 10 are known to be defective. If a customer selects 6 of these DVDs at random, determine the probability:

a. That 2 of them are defective.
b. That at least 1 of them is defective.

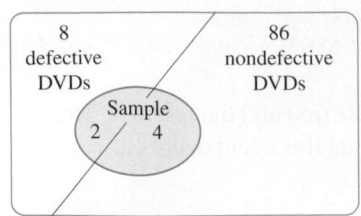

FIGURE 1
A sample of 6 DVDs selected from 90 nondefective DVDs and 10 defective DVDs

Solution

a. There are $C(100, 6)$ ways of selecting a set of 6 DVDs from the 100, and this gives $n(S)$, the number of outcomes in the sample space associated with the experiment. Next, we observe that there are $C(10, 2)$ ways of selecting a set of 2 defective DVDs from the 10 defective DVDs and $C(90, 4)$ ways of selecting a set of 4 nondefective DVDs from the 90 nondefective DVDs (Figure 1). Thus, by the multiplication principle, there are $C(10, 2) \cdot C(90, 4)$ ways of selecting 2 defective and 4 nondefective DVDs. Therefore, the probability of selecting 6 DVDs of which 2 are defective is given by

$$\frac{C(10, 2) \cdot C(90, 4)}{C(100, 6)} = \frac{\dfrac{10!}{2!\,8!}\dfrac{90!}{4!\,86!}}{\dfrac{100!}{6!\,94!}}$$

$$= \frac{10 \cdot 9}{2} \cdot \frac{90 \cdot 89 \cdot 88 \cdot 87}{4 \cdot 3 \cdot 2} \cdot \frac{6 \cdot 5 \cdot 4 \cdot 3 \cdot 2}{100 \cdot 99 \cdot 98 \cdot 97 \cdot 96 \cdot 95}$$

$$\approx .096$$

b. Let E denote the event that none of the DVDs selected is defective. Then E^c gives the event that at least 1 of the DVDs is defective. By the rule of complements,

$$P(E^c) = 1 - P(E)$$

To compute $P(E)$, we observe that there are $C(90, 6)$ ways of selecting a set of 6 DVDs that are nondefective. Therefore,

$$P(E) = \frac{C(90, 6)}{C(100, 6)}$$

$$P(E^c) = 1 - \frac{C(90, 6)}{C(100, 6)}$$

$$= 1 - \frac{\dfrac{90!}{6!\,84!}}{\dfrac{100!}{6!\,94!}}$$

$$= 1 - \frac{90 \cdot 89 \cdot 88 \cdot 87 \cdot 86 \cdot 85}{6 \cdot 5 \cdot 4 \cdot 3 \cdot 2} \cdot \frac{6 \cdot 5 \cdot 4 \cdot 3 \cdot 2}{100 \cdot 99 \cdot 98 \cdot 97 \cdot 96 \cdot 95}$$

$$\approx .478$$

The Birthday Problem

APPLIED EXAMPLE 4 The Birthday Problem A group of five people is selected at random. What is the probability that at least two of them have the same birthday?

Solution   For simplicity, we assume that none of the five people was born on February 29 of a leap year. Since the five people were selected at random, we also assume that each of them is equally likely to have any of the 365 days of a year as his or her birthday. If we let A, B, C, D, and F represent the five people, then an outcome of the experiment may be represented by (a, b, c, d, f), where the dates a, b, c, d, and f give the birthdays of A, B, C, D, and F, respectively.

We first observe that since there are 365 possibilities for each of the dates a, b, c, d, and f, the multiplication principle implies that there are

$$\underset{a}{\boxed{365}} \cdot \underset{b}{\boxed{365}} \cdot \underset{c}{\boxed{365}} \cdot \underset{d}{\boxed{365}} \cdot \underset{f}{\boxed{365}}$$

or 365^5 outcomes of the experiment. Therefore,

$$n(S) = 365^5$$

where S denotes the sample space of the experiment.

Next, let E denote the event that two or more of the five people have the same birthday. It is now necessary to compute $P(E)$. However, a direct computation of $P(E)$ is relatively difficult. It is much easier to compute $P(E^c)$, where E^c is the event that no two of the five people have the same birthday, and then use the relation

$$P(E) = 1 - P(E^c)$$

To compute $P(E^c)$, observe that there are 365 ways (corresponding to the 365 dates) on which A's birthday can occur, followed by 364 ways on which B's birthday could occur if B were not to have the same birthday as A, and so on. Therefore, by the generalized multiplication principle,

$$n(E^c) = \underset{\substack{A\text{'s}\\ \text{birthday}}}{365} \cdot \underset{\substack{B\text{'s}\\ \text{birthday}}}{364} \cdot \underset{\substack{C\text{'s}\\ \text{birthday}}}{363} \cdot \underset{\substack{D\text{'s}\\ \text{birthday}}}{362} \cdot \underset{\substack{F\text{'s}\\ \text{birthday}}}{361}$$

Thus,

$$P(E^c) = \frac{n(E^c)}{n(S)}$$
$$= \frac{365 \cdot 364 \cdot 363 \cdot 362 \cdot 361}{365^5}$$
$$P(E) = 1 - P(E^c)$$
$$= 1 - \frac{365 \cdot 364 \cdot 363 \cdot 362 \cdot 361}{365^5}$$
$$\approx .027$$

We can extend the result obtained in Example 4 to the general case involving r people. In fact, if E denotes the event that at least two of the r people have the same birthday, an argument similar to that used in Example 4 leads to the result

$$P(E) = 1 - \frac{365 \cdot 364 \cdot 363 \cdot \cdots \cdot (365 - r + 1)}{365^r}$$

By letting r take on the values 5, 10, 15, 20, . . . , 50, in turn, we obtain the probabilities that at least 2 of 5, 10, 15, 20, . . . , 50 people, respectively, have the same birthday. These results are summarized in Table 1.

The results show that in a group of 23 randomly selected people, the chances are greater than 50% that at least 2 of them will have the same birthday. In a group of 50 people, it is an excellent bet that at least 2 people in the group will have the same birthday.

TABLE 1	
Probability That at Least Two People in a Randomly Selected Group of r People Have the Same Birthday	
r	$P(E)$
5	.027
10	.117
15	.253
20	.411
22	.476
23	.507
25	.569
30	.706
40	.891
50	.970

> *Explore and Discuss*
>
> During an episode of the *Tonight Show*, a talk show host related "The Birthday Problem" to the audience—noting that in a group of 50 or more people, probabilists have calculated that the probability of at least 2 people having the same birthday is very high. To illustrate this point, he proceeded to conduct his own experiment. A person selected at random from the audience was asked to state his birthday. The host then asked whether anyone in the audience had the same birthday. The response was negative. He repeated the experiment. Once again, the response was negative. These results, observed the host, were contrary to expectations. In a later episode of the show, the host explained why this experiment had been improperly conducted. Explain why the host failed to illustrate the point he was trying to make in the earlier episode.

8.1 Self-Check Exercises

1. Four balls are selected at random without replacement from an urn containing ten white balls and eight red balls. What is the probability that all the chosen balls are white?

2. **QUALITY CONTROL** A box contains 20 microchips, of which 4 are substandard. If 2 of the chips are taken from the box, what is the probability that they are both substandard?

Solutions to Self-Check Exercises 8.1 can be found on page 517.

8.1 Concept Questions

1. What is the probability of an event E in a uniform sample space S?

2. Suppose we want to find the probability that at least two people in a group of six randomly selected people have the same birthday.

   a. If S denotes the sample space of this experiment, what is $n(S)$?
   b. If E is the event that two or more of the six people in the group have the same birthday, explain how you would use $P(E^c)$ to determine $P(E)$.

8.1 Exercises

An unbiased coin is tossed five times. In Exercises 1–4, find the probability of the given event.

1. The coin lands heads all five times.

2. The coin lands heads exactly once.

3. The coin lands heads at least once.

4. The coin lands heads more than once.

Two cards are selected at random without replacement from a well-shuffled deck of 52 playing cards. In Exercises 5–8, find the probability of the given event.

5. A pair is drawn.

6. A pair is not drawn.

7. Two black cards are drawn.

8. Two cards of the same suit are drawn.

Four balls are selected at random without replacement from an urn containing three white balls and five blue balls. In Exercises 9–12, find the probability of the given event.

9. Two of the balls are white, and two are blue.

10. All of the balls are blue.

11. Exactly three of the balls are blue.

12. Two or three of the balls are white.

Assume that the probability of a boy being born is the same as the probability of a girl being born. In Exercises 13–16, find the probability that a family with three children will have the given composition.

13. Two boys and one girl

14. At least one girl

15. No girls

16. The two oldest children are girls.

17. TAKING EXAMS An exam consists of ten true-or-false questions. If a student guesses at every answer, what is the probability that he or she will answer exactly six questions correctly?

18. PERSONNEL SELECTION Jacobs & Johnson, an accounting firm, employs 14 accountants, of whom 8 are CPAs. If a delegation of 3 accountants is randomly selected from the firm to attend a conference, what is the probability that 3 CPAs will be selected?

19. QUALITY CONTROL Two light bulbs are selected at random from a lot of 24, of which 4 are defective. What is the probability that:
  a. Both of the light bulbs are defective?
  b. At least 1 of the light bulbs is defective?

20. A customer at Cavallaro's Fruit Stand picks a sample of 3 oranges at random from a crate containing 60 oranges, of which 4 are rotten. What is the probability that the sample contains 1 or more rotten oranges?

21. QUALITY CONTROL A shelf in the Metro Department Store contains 80 colored ink cartridges for a popular ink-jet printer. Six of the cartridges are defective. If a customer selects 2 cartridges at random from the shelf, what is the probability that:
  a. Both are defective?
  b. At least 1 is defective?

22. QUALITY CONTROL Electronic baseball games manufactured by Tempco Electronics are shipped in lots of 24. Before shipping, a quality-control inspector randomly selects a sample of 8 from each lot for testing. If the sample contains any defective games, the entire lot is rejected. What is the probability that a lot containing exactly 2 defective games will still be shipped?

23. PERSONNEL SELECTION The City Transit Authority plans to hire 12 new bus drivers. From a group of 100 qualified applicants, of whom 60 are men and 40 are women, 12 names are to be selected by lot. Suppose that Mary and John Lewis are among the 100 qualified applicants.
  a. What is the probability that Mary's name will be selected? That both Mary's and John's names will be selected?

  b. If it is stipulated that an equal number of men and women are to be selected (6 men from the group of 60 men and 6 women from the group of 40 women), what is the probability that Mary's name will be selected? That Mary's and John's names will be selected?

24. SELECTION OF PUBLIC HOUSING APPLICANTS The City Housing Authority has received 50 applications from qualified applicants for eight low-income apartments. Three of the apartments are on the north side of town, and five are on the south side. If the apartments are to be assigned by means of a lottery, what is the probability that:
  a. A specific qualified applicant will be selected for one of these apartments?
  b. Two specific qualified applicants will be selected for apartments on the same side of town?

25. EXAMS A student studying for a vocabulary test knows the meanings of 12 words from a list of 20 words. If the test contains 10 words from the study list, what is the probability that at least 8 of the words on the test are words that the student knows?

26. DRIVING TESTS Four different written driving tests are administered by the Motor Vehicle Department. One of these four tests is selected at random for each applicant for a driver's license. If a group consisting of two women and three men apply for a license, what is the probability that:
  a. Exactly two of the five will take the same test?
  b. The two women will take the same test?

27. BRAND SELECTION A druggist wishes to select three brands of aspirin to sell in his store. He has five major brands to choose from: A, B, C, D, and E. If he selects the three brands at random, what is the probability that he will select:
  a. Brand B?
  b. Brands B and C?
  c. At least one of the two brands B and C?

28. BLACKJACK In the game of blackjack, a 2-card hand consisting of an ace and a face card or a 10 is called a blackjack.
  a. If a player is dealt 2 cards from a standard deck of 52 well-shuffled cards, what is the probability that the player will receive a blackjack?
  b. If a player is dealt 2 cards from 2 well-shuffled standard decks, what is the probability that the player will receive a blackjack?

29. SLOT MACHINES Refer to Exercise 33, Section 7.3, in which the "lucky dollar" slot machine was described. What is the probability that the three "lucky dollar" symbols will appear in the window of the slot machine?

30. **ROULETTE** In 1959, a world record was set for the longest run on an ungaffed (fair) roulette wheel at the El San Juan Hotel in Puerto Rico. The number 10 appeared six times in a row. What is the probability of the occurrence of this event? (Assume that there are 38 equally likely outcomes consisting of the numbers 1–36, 0, and 00.)

In the Numbers Game, a state lottery, four numbers are drawn with replacement from an urn containing balls numbered 0–9, inclusive. In Exercises 31–34, find the probability that a ticket holder has the indicated winning ticket.

31. **LOTTERIES** All four digits in exact order (the grand prize)

32. **LOTTERIES** Two specified, consecutive digits in exact order (the first two digits, the middle two digits, or the last two digits)

33. **LOTTERIES** One digit (the first, second, third, or fourth digit)

34. **LOTTERIES** Three digits in exact order

A list of poker hands, ranked in order from the highest to the lowest, is shown in the accompanying table along with a description and example of each hand. Use the table to answer Exercises 35–40.

Hand	Description	Example
Straight flush	5 cards in sequence in the same suit	A♥ 2♥ 3♥ 4♥ 5♥
Four of a kind	4 cards of the same rank and any other card	K♥ K♦ K♠ K♣ 2♥
Full house	3 of a kind and a pair	3♥ 3♦ 3♣ 7♥ 7♦
Flush	5 cards of the same suit that are not all in sequence	5♥ 6♥ 9♥ J♥ K♥
Straight	5 cards in sequence but not all of the same suit	10♥ J♦ Q♣ K♠ A♥
Three of a kind	3 cards of the same rank and 2 unmatched cards	K♥ K♦ K♠ 2♥ 4♦
Two pair	2 cards of the same rank and 2 cards of any other rank with an unmatched card	K♥ K♦ 2♥ 2♠ 4♣
One pair	2 cards of the same rank and 3 unmatched cards	K♥ K♦ 5♥ 2♠ 4♥

If a 5-card poker hand is dealt from a well-shuffled deck of 52 cards, what is the probability of being dealt the given hand?

35. **POKER** A straight flush (Note that an ace may be played as either a high or a low card in a straight sequence—that is, A, 2, 3, 4, 5 or 10, J, Q, K, A. Hence, there are ten possible sequences for a straight in one suit.)

36. **POKER** A straight (but not a straight flush)

37. **POKER** A flush (but not a straight flush)

38. **POKER** Four of a kind

39. **POKER** A full house

40. **POKER** Two pairs

41. **ZODIAC SIGNS** There are 12 signs of the Zodiac: Aries, Taurus, Gemini, Cancer, Leo, Virgo, Libra, Scorpio, Sagittarius, Capricorn, Aquarius, and Pisces. Each sign corresponds to a different calendar period of approximately 1 month. Assuming that a person is just as likely to be born under one sign as another, what is the probability that in a group of five people at least two of them:
    a. Have the same sign?
    b. Were born under the sign of Aries?

42. **BIRTHDAY PROBLEM** What is the probability that at least two of the nine justices of the U.S. Supreme Court have the same birthday?

43. **BIRTHDAY PROBLEM** Fifty people are selected at random. What is the probability that none of the people in this group have the same birthday?

44. **BIRTHDAY PROBLEM** A group of five people are selected at random. What is the probability that two of them were born on the same day of the week? (Assume that a person is equally likely to be born on any day of the week.)

45. **BIRTHDAY PROBLEM** A group of twelve people are selected at random. What is the probability that at least two of them have the same birthday?

46. **BIRTHDAY PROBLEM** There were 44 different presidents of the United States from 1789 through 2014. What is the probability that at least two of them had the same birthday? Compare your calculation with the facts by checking an almanac or some other source.

8.1 Solutions to Self-Check Exercises

1. The probability that all four balls selected are white is given by

The number of ways of selecting 4 white
balls from the 10 in the urn
$$\overline{\text{The number of ways of selecting any}}$$
4 balls from the 18 balls in the urn

$$= \frac{C(10, 4)}{C(18, 4)}$$

$$= \frac{\dfrac{10!}{4!\,6!}}{\dfrac{18!}{4!\,14!}}$$

$$= \frac{10 \cdot 9 \cdot 8 \cdot 7}{4 \cdot 3 \cdot 2} \cdot \frac{4 \cdot 3 \cdot 2}{18 \cdot 17 \cdot 16 \cdot 15}$$

$$\approx .069$$

2. The probability that both chips are substandard is given by

The number of ways of choosing any
2 of the 4 substandard chips
$$\overline{\text{The number of ways of choosing any}}$$
2 of the 20 chips

$$= \frac{C(4, 2)}{C(20, 2)}$$

$$= \frac{\dfrac{4!}{2!\,2!}}{\dfrac{20!}{2!\,18!}}$$

$$= \frac{4 \cdot 3}{2} \cdot \frac{2}{20 \cdot 19}$$

$$\approx .032$$

8.2 Conditional Probability and Independent Events

Conditional Probability

Suppose that three cities, A, B, and C, are vying to play host to the Summer Olympic Games in 2020. If each city has the same chance of winning the right to host the Games, then the probability of City A hosting the Games is $\frac{1}{3}$. Now suppose City B decides to pull out of contention because of fiscal problems. Then it would seem that City A's chances of playing host will increase. In fact, if the two remaining cities have equal chances of winning, then the probability of City A playing host to the Games is $\frac{1}{2}$.

In general, the probability of an event is affected by the occurrence of other events and/or by the knowledge of information relevant to the event. Basically, the injection of conditions into a problem modifies the underlying sample space of the original problem. This in turn leads to a change in the probability of the event.

EXAMPLE 1 Two cards are drawn without replacement from a well-shuffled deck of 52 playing cards.

a. What is the probability that the first card drawn is an ace?
b. What is the probability that the second card drawn is an ace given that the first card drawn was not an ace?
c. What is the probability that the second card drawn is an ace given that the first card drawn was an ace?

Solution

a. The sample space here consists of 52 equally likely outcomes, 4 of which are aces. Therefore, the probability that the first card drawn is an ace is $\frac{4}{52}$, or $\frac{1}{13}$.
b. The first card having been drawn, there are 51 cards left in the deck. In other words, for the second phase of the experiment, we are working in a *reduced*

sample space. If the first card drawn was not an ace, then this modified sample space of 51 points contains 4 "favorable" outcomes (the 4 aces), so the probability that the second card drawn is an ace is given by $\frac{4}{51}$.

c. If the first card drawn was an ace, then there are 3 aces left in the deck of 51 playing cards, so the probability that the second card drawn is an ace is given by $\frac{3}{51}$, or $\frac{1}{17}$.

Observe that in Example 1, the occurrence of the first event reduces the size of the original sample space. The information concerning the first card drawn also leads us to the consideration of modified sample spaces: In part (b), the deck contained four aces, and in part (c), the deck contained three aces.

The probability found in part (b) or part (c) of Example 1 is known as a **conditional probability,** since it is the probability of an event occurring given that another event has already occurred. For example, in part (b), we computed the probability of the event that the second card drawn is an ace *given that* the first card drawn was not an ace. In general, given two events A and B of an experiment, under certain circumstances one may compute the probability of the event B given that the event A has already occurred. This probability, denoted by $P(B|A)$, is called the **conditional probability of B given A.**

A formula for computing the conditional probability of B given A may be discovered with the aid of a Venn diagram. Consider an experiment with a uniform sample space S, and suppose that A and B are two events of the experiment (Figure 2).

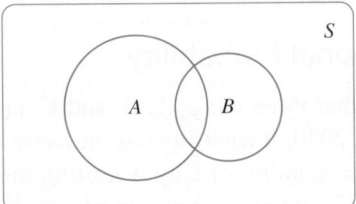
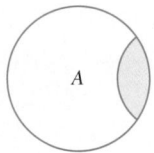

(a) Original sample space

(b) Reduced sample space A. The shaded area is $A \cap B$.

FIGURE **2**

The condition that the event A has occurred tells us that the possible outcomes of the experiment in the second phase are restricted to those outcomes (elements) in the set A. In other words, we may work with the reduced sample space A instead of the original sample space S in the experiment. Next we observe that, with respect to the reduced sample space A, the outcomes in the event B are precisely those elements in the set $A \cap B$. Consequently, the conditional probability of B given A is

$$P(B|A) = \frac{\text{Number of elements in } A \cap B}{\text{Number of elements in } A}$$

$$= \frac{n(A \cap B)}{n(A)} \qquad n(A) \neq 0$$

Dividing the numerator and the denominator by $n(S)$, the number of elements in S, we have

$$P(B|A) = \frac{\dfrac{n(A \cap B)}{n(S)}}{\dfrac{n(A)}{n(S)}}$$

which is equivalent to the following formula:

> **Conditional Probability of an Event**
>
> If A and B are events in an experiment and $P(A) \neq 0$, then the conditional probability that the event B will occur given that the event A has already occurred is
>
> $$P(B \mid A) = \frac{P(A \cap B)}{P(A)} \qquad (2)$$

EXAMPLE 2 A pair of fair dice is rolled. What is the probability that the sum of the numbers falling uppermost is 7 if it is known that one of the numbers is a 5?

Solution Let A denote the event that the sum of the numbers falling uppermost is 7, and let B denote the event that one of the numbers is a 5. From the results of Example 4, Section 7.5, we find that

$$A = \{(6, 1), (5, 2), (4, 3), (3, 4), (2, 5), (1, 6)\}$$
$$B = \{(5, 1), (5, 2), (5, 3), (5, 4), (5, 5), (5, 6),$$
$$(1, 5), (2, 5), (3, 5), (4, 5), (6, 5)\}$$

so

$$A \cap B = \{(5, 2), (2, 5)\}$$

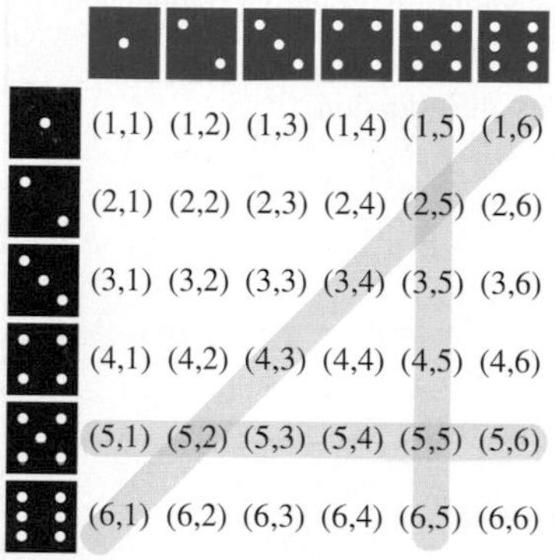

(1,1) (1,2) (1,3) (1,4) (1,5) (1,6)
(2,1) (2,2) (2,3) (2,4) (2,5) (2,6)
(3,1) (3,2) (3,3) (3,4) (3,5) (3,6)
(4,1) (4,2) (4,3) (4,4) (4,5) (4,6)
(5,1) (5,2) (5,3) (5,4) (5,5) (5,6)
(6,1) (6,2) (6,3) (6,4) (6,5) (6,6)

FIGURE 3
$A \cap B = \{(5, 2), (2, 5)\}$

(Figure 3). Since the dice are fair, each outcome of the experiment is equally likely; therefore,

$$P(A \cap B) = \frac{2}{36} \quad \text{and} \quad P(B) = \frac{11}{36} \qquad \text{Recall that } n(S) = 36.$$

Thus, the probability that the sum of the numbers falling uppermost is 7 given that one of the numbers is a 5 is, by virtue of Equation (2),

$$P(A \mid B) = \frac{\dfrac{2}{36}}{\dfrac{11}{36}} = \frac{2}{11}$$

APPLIED EXAMPLE 3 Color Blindness In a test conducted by the U.S. Army, it was found that of 1000 new recruits (600 men and 400 women), 50 of the men and 4 of the women were red-green color-blind. Given that a recruit selected at random from this group is red-green color-blind, what is the probability that the recruit is a male?

Solution Let C denote the event that a randomly selected subject is red-green color-blind, and let M denote the event that the subject is a male recruit. Since 54 out of the 1000 subjects are color-blind, we have

$$P(C) = \frac{54}{1000} = .054$$

Also, there are 50 male recruits who are red-green color-blind so

$$P(M \cap C) = \frac{50}{1000} = .05$$

Therefore, by Equation (2), the probability that a subject is male given that the subject is red-green color-blind is

$$P(M \mid C) = \frac{P(M \cap C)}{P(C)}$$

$$= \frac{.05}{.054} \approx .926$$

Explore and Discuss

Let A and B be events in an experiment, and suppose that $P(A) \neq 0$. In n trials, the event A occurs m times, the event B occurs k times, and the events A and B occur together l times.

1. Explain why it makes good sense to call the ratio l/m the conditional relative frequency of the event B given the event A.

2. Show that the relative frequencies l/m, m/n, and l/n satisfy the equation

$$\frac{l}{m} = \frac{\dfrac{l}{n}}{\dfrac{m}{n}}$$

3. Explain why the result of part 2 suggests that Equation (2),

$$P(B \mid A) = \frac{P(A \cap B)}{P(A)} \qquad P(A) \neq 0$$

is plausible.

In certain problems, the probability of an event B occurring given that A has occurred, written $P(B \mid A)$, is known, and we wish to find the probability of A *and* B occurring. The solution to such a problem is facilitated by the use of the following formula:

Product Rule

$$P(A \cap B) = P(A) \cdot P(B \mid A) \tag{3}$$

This formula is obtained from Equation (2) by multiplying both sides of the equation by $P(A)$. We illustrate the use of the Product Rule in the next several examples.

APPLIED EXAMPLE 4 Seniors with Driver's Licenses There are 300 seniors at Jefferson High School, of whom 140 are males. It is known that 80% of the males and 60% of the females have their driver's license. If a student is selected at random from this senior class, what is the probability that the student is:

a. A male and has a driver's license?
b. A female and does not have a driver's license?

Solution

a. Let M denote the event that the student is a male, and let D denote the event that the student has a driver's license. Then

$$P(M) = \frac{140}{300} \quad \text{and} \quad P(D \mid M) = .8$$

The event that the student selected at random is a male and has a driver's license is $M \cap D$, and by the Product Rule, the probability of this event occurring is given by

$$P(M \cap D) = P(M) \cdot P(D \mid M)$$
$$= \left(\frac{140}{300}\right)(.8) \approx .373$$

b. Let F denote the event that the student is a female. Then D^c is the event that the student does not have a driver's license. We have

$$P(F) = \frac{160}{300} \quad \text{and} \quad P(D^c \mid F) = 1 - .6 = .4$$

Note that we have used the rule of complements in the computation of $P(D^c \mid F)$. The event that the student selected at random is a female and does not have a driver's license is $F \cap D^c$, so by the Product Rule, the probability of this event occurring is given by

$$P(F \cap D^c) = P(F) \cdot P(D^c \mid F)$$
$$= \left(\frac{160}{300}\right)(.4) \approx .213$$

EXAMPLE 5 Two cards are drawn without replacement from a well-shuffled deck of 52 playing cards. What is the probability that the first card drawn is an ace and the second card drawn is a face card?

Solution   Let A denote the event that the first card drawn is an ace, and let F denote the event that the second card drawn is a face card. Then $P(A) = \frac{4}{52}$. After the first card is drawn, there are 51 cards left in the deck, of which 12 are face cards. Therefore, the probability of drawing a face card given that the first card drawn was an ace is given by

$$P(F \mid A) = \frac{12}{51}$$

By the Product Rule, the probability that the first card drawn is an ace and the second card drawn is a face card is given by

$$P(A \cap F) = P(A) \cdot P(F \mid A)$$
$$= \frac{4}{52} \cdot \frac{12}{51} = \frac{4}{221} \approx .018$$

Explore and Discuss

The Product Rule can be extended to the case involving three or more events. For example, if A, B, and C are three events in an experiment, then it can be shown that

$$P(A \cap B \cap C) = P(A) \cdot P(B \mid A) \cdot P(C \mid A \cap B)$$

1. Explain the formula in words.
2. Suppose 3 cards are drawn without replacement from a well-shuffled deck of 52 playing cards. Use the given formula to find the probability that the 3 cards are aces.

The Product Rule may be generalized to the case involving any finite number of events. For example, in the case involving the three events E, F, and G, it may be shown that

$$P(E \cap F \cap G) = P(E) \cdot P(F \mid E) \cdot P(G \mid E \cap F) \tag{4}$$

More on Tree Diagrams

Equation (4) and its generalizations may be used to help us solve problems that involve finite stochastic processes. A **finite stochastic process** is an experiment consisting of a finite number of stages in which the outcomes and associated probabilities of each stage depend on the outcomes and associated probabilities of the preceding stages.

We can use tree diagrams to help us solve problems involving finite stochastic processes. Consider, for example, the experiment consisting of drawing 2 cards without replacement from a well-shuffled deck of 52 playing cards. What is the probability that the second card drawn is a face card?

We may think of this experiment as a stochastic process with two stages. The events associated with the first stage are F, that the card drawn is a face card, and F^c, that the card drawn is not a face card. Since there are 12 face cards, we have

$$P(F) = \frac{12}{52} \quad \text{and} \quad P(F^c) = 1 - \frac{12}{52} = \frac{40}{52}$$

The outcomes of this trial, together with the associated probabilities, may be represented along two branches of a tree diagram as shown in Figure 4.

In the second trial, we again have two events: G, that the card drawn is a face card, and G^c, that the card drawn is not a face card. But the outcome of the second trial depends on the outcome of the first trial. For example, if the first card drawn was a face card, then the event G that the second card drawn is a face card has probability given by the *conditional probability* $P(G \mid F)$. Since the occurrence of a face card in the first draw leaves 11 face cards in a deck of 51 cards for the second draw, we see that

$$P(G \mid F) = \frac{11}{51} \qquad \text{\small The probability of drawing a face card given that a face card has already been drawn}$$

Similarly, the occurrence of a face card in the first draw leaves 40 that are other than face cards in a deck of 51 cards for the second draw. Therefore, the probability of drawing a card other than a face card in the second draw given that the first card drawn is a face card is

$$P(G^c \mid F) = \frac{40}{51}$$

Using these results, we extend the tree diagram of Figure 4 by displaying another two branches of the tree growing from its upper branch (Figure 5).

To complete the tree diagram, we compute $P(G \mid F^c)$ and $P(G^c \mid F^c)$, the conditional probabilities that the second card drawn is a face card and other than a face card, respectively, given that the first card drawn is not a face card. We find that

$$P(G \mid F^c) = \frac{12}{51} \quad \text{and} \quad P(G^c \mid F^c) = \frac{39}{51}$$

This leads to the completion of the tree diagram, shown in Figure 6, in which the branches of the tree that lead to the two outcomes of interest have been highlighted.

Having constructed the tree diagram associated with the problem, we are now in a position to answer the question posed earlier: What is the probability of the second card being a face card? Observe that Figure 6 shows the two ways in which a face card may result in the second draw—namely, the two Gs on the extreme right of the diagram.

Now, by the Product Rule, the probability that the second card drawn is a face card and the first card drawn is a face card (this is represented by the upper branch) is

$$P(G \cap F) = P(F) \cdot P(G \mid F)$$

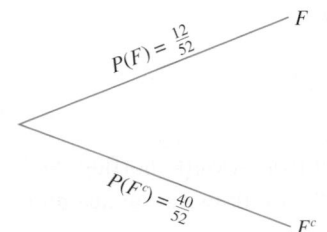

FIGURE 4
F is the event that a face card is drawn.

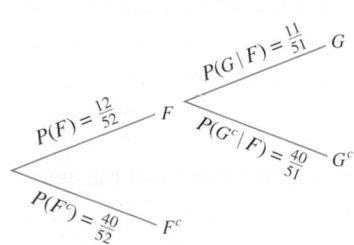

FIGURE 5
G is the event that the second card drawn is a face card.

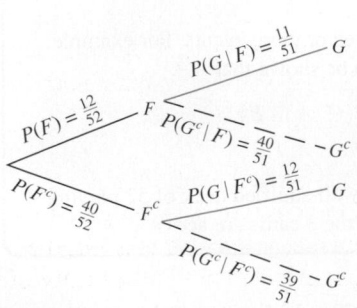

FIGURE 6
Tree diagram showing the two trials of the experiment

Similarly, the probability that the second card drawn is a face card and the first card drawn is other than a face card (this corresponds to the other branch) is

$$P(G \cap F^c) = P(F^c) \cdot P(G|F^c)$$

Observe that each of these probabilities is obtained by taking the *product of the probabilities appearing on the respective branches*. Since $G \cap F$ and $G \cap F^c$ are mutually exclusive events (why?), the probability that the second card drawn is a face card is given by

$$P(G \cap F) + P(G \cap F^c) = P(F) \cdot P(G|F) + P(F^c) \cdot P(G|F^c)$$

or, upon replacing the probabilities on the right of the expression by their numerical values,

$$P(G \cap F) + P(G \cap F^c) = \frac{12}{52} \cdot \frac{11}{51} + \frac{40}{52} \cdot \frac{12}{51}$$

$$= \frac{3}{13}$$

APPLIED EXAMPLE 6 Quality Control The panels for the Pulsar 32-inch widescreen LCD HDTVs are manufactured in three locations and then shipped to the main plant of Vista Vision for final assembly. Plants A, B, and C supply 50%, 30%, and 20%, respectively, of the panels used by the company. The quality-control department of the company has determined that 1% of the panels produced by Plant A are defective, whereas 2% of the panels produced by Plants B and C are defective. What is the probability that a randomly selected Pulsar 32-inch HDTV will have a defective panel?

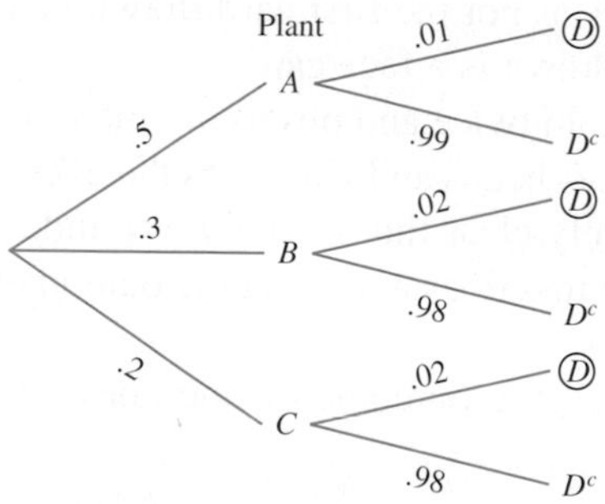

FIGURE 7
Tree diagram showing the probabilities of producing defective panels at each plant

Solution Let A, B, and C denote the events that the HDTV chosen has a panel manufactured in Plant A, Plant B, and Plant C, respectively. Also, let D denote the event that an HDTV has a defective panel. Using the given information, we draw the tree diagram shown in Figure 7. (The events that result in an HDTV with a defective panel being selected are circled.) Taking the product of the probabilities along each branch leading to such an event and then adding them, we obtain the probability that an HDTV chosen at random has a defective panel. Thus, the required probability is given by

$$(.5)(.01) + (.3)(.02) + (.2)(.02) = .005 + .006 + .004$$
$$= .015$$

APPLIED EXAMPLE 7 Quality Control A box contains eight 9-volt batteries, of which two are known to be defective. The batteries are selected one at a time without replacement and tested until a nondefective one is found. What is the probability that the number of batteries tested is (a) One? (b) Two? (c) Three?

Solution We may view this experiment as a multistage process with up to three stages. In the first stage, a battery is selected with a probability of $\frac{6}{8}$ of being nondefective and a probability of $\frac{2}{8}$ of being defective. If the battery selected is good, the experiment is terminated. Otherwise, a second battery is selected with probability of $\frac{6}{7}$ and $\frac{1}{7}$, respectively, of being nondefective and defective. If the second battery selected is good, the experiment is terminated. Otherwise, a third battery is selected with probability of 1 and 0, respectively, of its being nondefective and defective.

The tree diagram associated with this experiment is shown in Figure 8, where N denotes the event that the battery selected is nondefective and D denotes the event that the battery selected is defective.

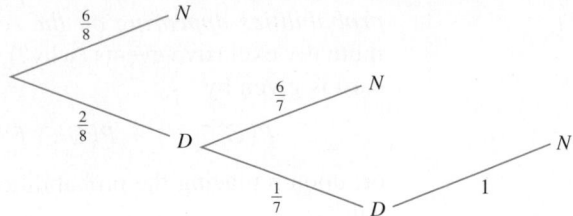

FIGURE **8**
In this experiment, batteries are selected until a nondefective one is found.

With the aid of the tree diagram, we see that (a) the probability that only one battery is selected is $\frac{6}{8} = \frac{3}{4}$, (b) the probability that two batteries are selected is $\left(\frac{2}{8}\right)\left(\frac{6}{7}\right)$, or $\frac{3}{14}$, and (c) the probability that three batteries are selected is $\left(\frac{2}{8}\right)\left(\frac{1}{7}\right)(1) = \frac{1}{28}$. ∎

Independent Events

Let's return to the experiment of drawing 2 cards in succession without replacement from a well-shuffled deck of 52 playing cards as considered in Example 5. Let E denote the event that the first card drawn is not a face card, and let F denote the event that the second card drawn is a face card. Intuitively, it is clear that the events E and F are *not* independent of each other, because whether or not the first card drawn is a face card affects the likelihood that the second card drawn is a face card.

Next, let's consider the experiment of tossing a coin twice and observing the outcomes: If H denotes the event that the first toss produces heads and T denotes the event that the second toss produces tails, then it is intuitively clear that H and T *are* independent of each other because the outcome of the first toss does not affect the outcome of the second.

In general, two events A and B are independent if the outcome of one does not affect the outcome of the other. Thus, we have

> **Independent Events**
>
> If A and B are **independent events,** then
> $$P(A \mid B) = P(A) \quad \text{and} \quad P(B \mid A) = P(B)$$

Using the Product Rule, we can find a simple test to determine the independence of two events. Suppose that A and B are independent and that $P(A) \neq 0$ and $P(B) \neq 0$. Then

$$P(B \mid A) = P(B)$$

Thus, by the Product Rule, we have

$$P(A \cap B) = P(A) \cdot P(B \mid A) = P(A) \cdot P(B)$$

Conversely, if this equation holds, then it can be seen that $P(B|A) = P(B)$; that is, A and B are independent. Accordingly, we have the following test for the independence of two events:

Test for the Independence of Two Events

Two events A and B are independent if and only if

$$P(A \cap B) = P(A) \cdot P(B) \tag{5}$$

 Do not confuse *independent* events with *mutually exclusive* events. The former pertains to how the occurrence of one event affects the occurrence of another event, whereas the latter pertains to the question of whether the events can occur at the same time.

EXAMPLE 8 Determine whether E and F are (a) mutually exclusive and (b) independent if:

a. $P(E) = .5, P(F) = .4$, and $P(E \cap F) = .2$.
b. $P(E) = .5, P(F) = .7$, and $P(E \cap F) = .3$.
c. $P(E) = .1, P(F) = .3$, and $P(E \cap F) = 0$.

Solution

a. Here, $P(E \cap F) = .2$. Since $P(E \cap F) = .2 \neq 0$, we know that $E \cap F \neq \emptyset$, and this tells us that E and F are not mutually exclusive. Next, observe that

$$P(E)P(F) = (.5)(.4) = .2$$

Since $P(E \cap F), = P(E)P(F)$, we conclude that E and F are independent.

b. Since $P(E \cap F) = .3 \neq 0$, we know that $E \cap F \neq \emptyset$, and we see that E and F are not mutually exclusive. Next, observe that

$$P(E)P(F) = (.5)(.7) = .35$$

Since $P(E \cap F) \neq P(E)P(F)$, we conclude that E and F are not independent.

c. Here, $P(E \cap F) = 0$, and this shows that E and F have no points in common. In other words, E and F are mutually exclusive. Next, observe that

$$P(E)P(F) = (.1)(.3) = .03$$

Since

$$P(E \cap F) \neq P(E)P(F)$$

we conclude that E and F are not independent. ∎

Note that part (c) of Example 8 shows that two events may be mutually exclusive but not independent and vice versa.

EXAMPLE 9 Consider the experiment consisting of tossing a fair coin twice and observing the outcomes. Show that the event of heads on the first toss and the event of tails on the second toss are independent events.

Solution The sample space of the experiment is

$$S = \{(HH), (HT), (TH), (TT)\}$$

Let A denote the event that the outcome of the first toss is a head, and let B denote the event that the outcome of the second toss is a tail. Then

$$A = \{(HH), (HT)\}$$
$$B = \{(HT), (TT)\}$$

so

$$A \cap B = \{(HT)\}$$

Next, we compute

$$P(A \cap B) = \frac{1}{4} \quad P(A) = \frac{1}{2} \quad P(B) = \frac{1}{2}$$

and observe that Equation (5) is satisfied in this case. Hence A and B are independent events, as we set out to show.

APPLIED EXAMPLE 10 Medical Surveys A survey conducted by an independent agency for the National Lung Society found that of 2000 women, 680 were heavy smokers and 50 had emphysema. Of those who had emphysema, 42 were also heavy smokers. Using the data in this survey, determine whether the events "being a heavy smoker" and "having emphysema" are independent events.

Solution Let A denote the event that a woman chosen at random in this survey is a heavy smoker, and let B denote the event that a woman chosen at random in this survey has emphysema. Then the probability that a woman is a heavy smoker and has emphysema is given by

$$P(A \cap B) = \frac{42}{2000} = .021$$

Next,

$$P(A) = \frac{680}{2000} = .34 \quad \text{and} \quad P(B) = \frac{50}{2000} = .025$$

so

$$P(A) \cdot P(B) = (.34)(.025) = .0085$$

Since $P(A \cap B) \neq P(A) \cdot P(B)$, we conclude that A and B are not independent events.

The solution of many practical problems involves more than two independent events. In such cases, we use the following result.

Explore and Discuss

Let E and F be independent events in a sample space S. Are E^c and F^c independent?

Independence of More Than Two Events

If $E_1, E_2, \ldots, E_n$ are independent events, then

$$P(E_1 \cap E_2 \cap \cdots \cap E_n) = P(E_1) \cdot P(E_2) \cdot \cdots \cdot P(E_n) \qquad (6)$$

Formula (6) states that the probability of the simultaneous occurrence of n independent events is equal to the product of the probabilities of the n events.

⚠ It is important to note that the mere requirement that the n events $E_1, E_2, \ldots, E_n$ satisfy Formula (6) is not sufficient to guarantee that the n events are indeed independent. However, a criterion does exist for determining the independence of n events; it may be found in more advanced texts on probability.

EXAMPLE 11 It is known that the three events A, B, and C are independent and that $P(A) = .2$, $P(B) = .4$, and $P(C) = .5$. Compute:

a. $P(A \cap B)$ **b.** $P(A \cap B \cap C)$

Solution   Using Formulas (5) and (6), we find

a. $P(A \cap B) = P(A) \cdot P(B)$
$$= (.2)(.4) = .08$$
b. $P(A \cap B \cap C) = P(A) \cdot P(B) \cdot P(C)$
$$= (.2)(.4)(.5) = .04$$

APPLIED EXAMPLE 12 Predicting Travel Weather Ron is planning to visit Paris on Thursday and Friday. A quick check of the weather report for Paris revealed that there is a probability of .2 that it will rain on Thursday and a probability of .4 that it will rain on Friday. Assuming that the likelihood of rain on Thursday is independent of the likelihood of rain on Friday, what is the probability that Ron will see rain on at least one of those two days?

Solution   Let A and B denote the event that it will rain on Thursday and Friday, respectively. Then the probability that it will rain on Thursday or Friday is given by $P(A \cup B)$. Using the Addition Rule, we have

$$P(A \cup B) = P(A) + P(B) - P(A \cap B)$$

Here, $P(A) = .2$ and $P(B) = .4$, and since A and B are independent events, we see that

$$P(A \cap B) = P(A)P(B) = (.2)(.4) = .08$$

Therefore,

$$P(A \cup B) = .2 + .4 - .08$$
$$= .52$$

We conclude that there is a 52% likelihood that Ron will see rain on at least one of the two days of his stay in Paris.

APPLIED EXAMPLE 13 Quality Control The Acrosonic model F loudspeaker system has four loudspeaker components: a woofer, a mid-range, a tweeter, and an electrical crossover. The quality-control manager of Acrosonic has determined that on the average, 1% of the woofers, 0.8% of the midranges, and 0.5% of the tweeters are defective, while 1.5% of the electrical crossovers are defective. Determine the probability that a loudspeaker system selected at random as it comes off the assembly line (and before final inspection) is not defective. Assume that the defects in the manufacturing of the components are unrelated.

Solution   Let A, B, C, and D denote, respectively, the events that the woofer, the midrange, the tweeter, and the electrical crossover are defective. Then

$$P(A) = .01 \qquad P(B) = .008 \qquad P(C) = .005 \qquad P(D) = .015$$

and the probabilities of the corresponding complementary events are

$$P(A^c) = .99 \qquad P(B^c) = .992 \qquad P(C^c) = .995 \qquad P(D^c) = .985$$

The event that a loudspeaker system selected at random is not defective is given by $A^c \cap B^c \cap C^c \cap D^c$. Because the events A, B, C, and D (and therefore also A^c, B^c,

C^c, and D^c) are assumed to be independent, we find that the required probability is given by

$$P(A^c \cap B^c \cap C^c \cap D^c) = P(A^c) \cdot P(B^c) \cdot P(C^c) \cdot P(D^c)$$
$$= (.99)(.992)(.995)(.985)$$
$$\approx .96$$

8.2 Self-Check Exercises

1. Let A and B be events in a sample space S such that $P(A) = .4$, $P(B) = .8$, and $P(A \cap B) = .3$. Find:
   a. $P(A|B)$ **b.** $P(B|A)$

2. **HAPPINESS WITH MARRIAGE** According to a survey cited in *Newsweek*, 29.7% of married survey respondents who married between the ages of 20 and 22 (inclusive), 26.9% of those who married between the ages of 23 and 27, and 45.1% of those who married at age 28 or older said that

"their marriage was less than 'very happy.'" Suppose that a survey respondent from each of the three age groups was selected at random. What is the probability that all three respondents said that their marriage was "less than very happy"?
Source: Marc Bain, Newsweek.

Solutions to Self-Check Exercises 8.2 can be found on page 533.

8.2 Concept Questions

1. What is conditional probability? Illustrate the concept with an example.

2. If A and B are events in an experiment and $P(A) \neq 0$, then what is the formula for computing $P(B|A)$?

3. If A and B are events in an experiment and the conditional probability $P(B|A)$ is known, give the formula that can

be used to compute the probability of the event that both A and B will occur.

4. **a.** What is the test for determining the independence of two events?
   b. What is the difference between mutually exclusive events and independent events?

8.2 Exercises

1. Let A and B be two events in a sample space S such that $P(A) = .6$, $P(B) = .5$, and $P(A \cap B) = .2$. Find:
   a. $P(A|B)$ **b.** $P(B|A)$

2. Let A and B be two events in a sample space S such that $P(A) = .4$, $P(B) = .6$, and $P(A \cap B) = .3$. Find:
   a. $P(A|B)$ **b.** $P(B|A)$

3. Let A and B be two events in a sample space S such that $P(A) = .6$ and $P(B|A) = .5$. Find $P(A \cap B)$.

4. Let A and B be the events described in Exercise 1. Find:
   a. $P(A|B^c)$ **b.** $P(B|A^c)$
   Hint: $(A \cap B^c) \cup (A \cap B) = A$

In Exercises 5–8, determine whether the events A and B are independent.

5. $P(A) = .3$, $P(B) = .6$, $P(A \cap B) = .18$

6. $P(A) = .6$, $P(B) = .8$, $P(A \cap B) = .2$

7. $P(A) = .5$, $P(B) = .7$, $P(A \cup B) = .85$

8. $P(A^c) = .3$, $P(B^c) = .4$, $P(A \cap B) = .42$

9. If A and B are independent events, $P(A) = .4$, and $P(B) = .6$, find:
   a. $P(A \cap B)$ **b.** $P(A \cup B)$
   c. $P(A|B)$ **d.** $P(A^c \cup B^c)$

10. If A and B are independent events, $P(A) = .35$, and $P(B) = .45$, find:
    a. $P(A \cap B)$ **b.** $P(A \cup B)$
    c. $P(A|B)$ **d.** $P(A^c \cup B^c)$

11. The accompanying tree diagram represents an experiment consisting of two trials:

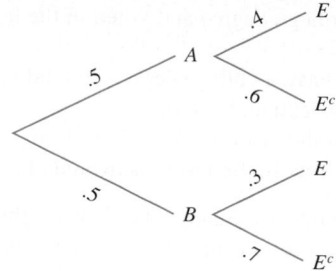

Use the diagram to find:

a. $P(A)$ **b.** $P(E\,|\,A)$

c. $P(A \cap E)$ **d.** $P(E)$

e. Does $P(A \cap E) = P(A) \cdot P(E)$?

f. Are A and E independent events?

12. The accompanying tree diagram represents an experiment consisting of two trials. Use the diagram to find:

a. $P(A)$ **b.** $P(E\,|\,A)$

c. $P(A \cap E)$ **d.** $P(E)$

e. Does $P(A \cap E) = P(A) \cdot P(E)$?

f. Are A and E independent events?

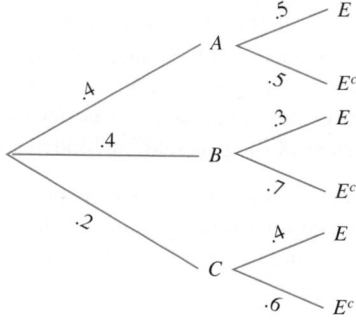

13. An experiment consists of two independent trials. The outcomes of the first trial are A and B with probabilities of occurring equal to .4 and .6. There are also two outcomes, C and D, in the second trial with probabilities of .3 and .7. Draw a tree diagram representing this experiment, and use it to find:

a. $P(A)$ **b.** $P(C\,|\,A)$

c. $P(A \cap C)$ **d.** $P(C)$

e. Does $P(A \cap C) = P(A) \cdot P(C)$?

f. Are A and C independent events?

14. An experiment consists of two independent trials. The outcomes of the first trial are A, B, and C, with probabilities of occurring equal to .2, .5, and .3, respectively. The outcomes of the second trial are E and F, with probabilities of occurring equal to .6 and .4. Draw a tree diagram representing this experiment. Use this diagram to find:

a. $P(B)$ **b.** $P(F\,|\,B)$

c. $P(B \cap F)$ **d.** $P(F)$

e. Does $P(B \cap F) = P(B) \cdot P(F)$?

f. Are B and F independent events?

15. A pair of fair dice is rolled. Let E denote the event that the number falling uppermost on the first die is 5, and let F denote the event that the sum of the numbers falling uppermost is 10.

a. Compute $P(F)$.   **b.** Compute $P(E \cap F)$.

c. Compute $P(F\,|\,E)$.   **d.** Compute $P(E)$.

e. Are E and F independent events?

16. A pair of fair dice is rolled. Let E denote the event that the number falling uppermost on the first die is 4, and let F denote the event that the sum of the numbers falling uppermost is 6.

a. Compute $P(F)$.   **b.** Compute $P(E \cap F)$.

c. Compute $P(F\,|\,E)$.   **d.** Compute $P(E)$.

e. Are E and F independent events?

17. A pair of fair dice is rolled. What is the probability that the sum of the numbers falling uppermost is less than 9, given that at least one of the numbers is a 6?

18. A pair of fair dice is rolled. What is the probability that the number landing uppermost on the first die is a 4 if it is known that the sum of the numbers landing uppermost is 7?

19. A pair of fair dice is rolled. Let E denote the event that the number landing uppermost on the first die is a 3, and let F denote the event that the sum of the numbers landing uppermost is 7. Determine whether E and F are independent events.

20. A pair of fair dice is rolled. Let E denote the event that the number landing uppermost on the first die is a 3, and let F denote the event that the sum of the numbers landing uppermost is 6. Determine whether E and F are independent events.

21. A card is drawn from a well-shuffled deck of 52 playing cards. Let E denote the event that the card drawn is black and let F denote the event that the card drawn is a spade. Determine whether E and F are independent events. Give an intuitive explanation for your answer.

22. A card is drawn from a well-shuffled deck of 52 playing cards. Let E denote the event that the card drawn is an ace and let F denote the event that the card drawn is a diamond. Determine whether E and F are independent events. Give an intuitive explanation for your answer.

23. BATTERY LIFE The probability that a battery will last 10 hr or more is .80, and the probability that it will last 15 hr or more is .15. Given that a battery has lasted 10 hr, find the probability that it will last 15 hr or more.

24. Five black balls and four white balls are placed in an urn. Two balls are then drawn in succession. What is the probability that the second ball drawn is a white ball if:

a. The second ball is drawn without replacing the first?

b. The first ball is replaced before the second is drawn?

Refer to Exercises 7.5, Problem 34, where you were asked to find the sample space corresponding to the different human blood groups. The following table gives the percent of the U.S. population having each of the eight possible blood types in the sample space. Note that the presence or absence of the Rh antigen is indicated by the symbols $+$ or $-$, respectively.

Blood Types	A^+	A^-	B^+	B^-	AB^+	AB^-	O^+	O^-
Percent	35.7	6.3	8.5	1.5	3.4	0.6	37.4	6.6

Source: American Red Cross.

In Exercises 25 and 26, use the above table.

25. **BLOOD TYPES**
   a. What is the probability that a person selected at random from the U.S. population has a blood type that is type A given that the person is Rh^-?
   b. What is the probability that a person selected at random from the U.S. population has a blood type that is Rh^+ given that the person is type B?

26. **BLOOD TYPES**
   a. What is the probability that a person selected at random from the U.S. population has a blood type that is type AB given that the person is Rh^+?
   b. What is the probability that a person selected at random from the U.S. population has a blood type that is Rh^- given that the person is type O?

27. **SELLING A CAR** Jack has decided to advertise the sale of his car by placing flyers in the student union and the dining hall of the college. He estimates that there is a probability of .3 that a potential buyer will read the advertisement and that, if it is read, the probability that the reader will buy his car will be .2. Using these estimates, find the probability that the person who reads the ad will buy Jack's car.

28. **AUDITING TAX RETURNS** A tax specialist has estimated that the probability that a tax return selected at random will be audited is .02. Furthermore, he estimates that the probability that an audited return will result in additional assessments being levied on the taxpayer is .60. What is the probability that a tax return selected at random will result in additional assessments being levied on the taxpayer?

29. **STUDENT ENROLLMENT** At a certain medical school, $\frac{1}{7}$ of the students are from a minority group. Of the students who belong to a minority group, $\frac{1}{3}$ are black.
   a. What is the probability that a student selected at random from this medical school is black?
   b. What is the probability that a student selected at random from this medical school is black if it is known that the student is a member of a minority group?

30. **EDUCATIONAL LEVEL OF VOTERS** In a survey of 1000 eligible voters selected at random, it was found that 80 had a college degree. Additionally, it was found that 80% of those who had a college degree voted in the last presidential election, whereas 55% of the people who did not have a college

degree voted in the last presidential election. Assuming that the poll is representative of all eligible voters, find the probability that an eligible voter selected at random:
   a. Had a college degree and voted in the last presidential election.
   b. Did not have a college degree and did not vote in the last presidential election.
   c. Voted in the last presidential election.
   d. Did not vote in the last presidential election.

31. The probability that Sandy takes her daughter Olivia to the supermarket on Friday is .6. If Sandy does bring Olivia to the supermarket on Friday, the probability that she buys Olivia a popsicle is .8. What is the probability that Sandy takes Olivia to the supermarket on Friday and buys her a popsicle?

32. **SOLVING PROBLEMS** The probability that Art will submit the correct solution to a certain homework problem is $\frac{1}{3}$, and the probability that Candice will submit the correct solution to it is $\frac{1}{2}$. Assuming that they work independently, what is the probability that either Art or Candice or both Art and Candice will submit the correct solution?

33. **MEDICAL SURVEY** A nationwide survey conducted by the National Cancer Society revealed the following information. Of 10,000 people surveyed, 3200 were "heavy coffee drinkers," and 160 had cancer of the pancreas. Of those who had cancer of the pancreas, 132 were heavy coffee drinkers. Using the data in this survey, determine whether the events "being a heavy coffee drinker" and "having cancer of the pancreas" are independent events.

34. **EMPLOYEE EDUCATION AND INCOME** The personnel department of Franklin National Life Insurance Company compiled the accompanying data regarding the income and education of its employees:

	Income $65,000 or Below	Income Above $65,000
Noncollege Graduate	2040	840
College Graduate	400	720

Let A be the event that a randomly chosen employee has a college degree, and let B be the event that the chosen employee's income is more than $65,000.
   a. Find each of the following probabilities: $P(A)$, $P(B)$, $P(A \cap B)$, $P(B|A)$, and $P(B|A^c)$.
   b. Are the events A and B independent events?

35. **STUDENT FINANCIAL AID** The accompanying data were obtained from the financial aid office of a certain university:

	Receiving Financial Aid	Not Receiving Financial Aid	Total
Undergraduates	4,222	3,898	8,120
Graduates	1,879	731	2,610
Total	6,101	4,629	10,730

Let A be the event that a student selected at random from this university is an undergraduate student, and let B be the event that a student selected at random is receiving financial aid.

a. Find each of the following probabilities: $P(A)$, $P(B)$, $P(A \cap B)$, $P(B \mid A)$, and $P(B \mid A^c)$.

b. Are the events A and B independent events?

36. Two cards are drawn without replacement from a well-shuffled deck of 52 playing cards.

a. What is the probability that the first card drawn is a heart?

b. What is the probability that the second card drawn is a heart if the first card drawn was not a heart?

c. What is the probability that the second card drawn is a heart if the first card drawn was a heart?

37. FAMILY COMPOSITION In a three-child family, what is the probability that all three children are girls given that at least one of the children is a girl? (Assume that the probability of a boy being born is the same as the probability of a girl being born.)

38. A coin is tossed three times. What is the probability that the coin will land heads:

a. At least twice?

b. On the second toss, given that heads were thrown on the first toss?

c. On the third toss, given that tails were thrown on the first toss?

39. TELEVISION PILOTS Max Productions has two pilots for the coming television season. The probability that the first pilot will be successful is estimated to be .9, and the probability that the second pilot will be successful is estimated to be .8. Assuming that the success of one pilot does not have a bearing on the success of the other, what is the probability that Max Productions will have at least one successful pilot for the coming television season.

40. CAR THEFT Figures obtained from a city's police department seem to indicate that of all motor vehicles reported as stolen, 64% were stolen by professionals, whereas 36% were stolen by amateurs (primarily for joy rides). Of the vehicles presumed stolen by professionals, 24% were recovered within 48 hr, 16% were recovered after 48 hr, and 60% were never recovered. Of the vehicles presumed stolen by amateurs, 38% were recovered within 48 hr, 58% were recovered after 48 hr, and 4% were never recovered.

a. Draw a tree diagram representing these data.

b. What is the probability that a vehicle stolen by a professional in this city will be recovered within 48 hr?

c. What is the probability that a vehicle stolen in this city will never be recovered?

41. SWITCHING BROADBAND SERVICE According to a survey conducted in 2010 by the Federal Communications Commission (FCC) of 3005 adults who were home broadband users, 37.5% of those surveyed had switched their service over the past 3 years. Of those who had switched service in the past 3 years, 51% were very satisfied, 39% were somewhat satisfied, and 10% were not satisfied with their service. Of the 62.5% who had not switched service in the past 3 years, 48% were very satisfied, 43% were somewhat satisfied, and 9% were not satisfied with their service.

a. What is the probability that a participant chosen at random had not switched service in the past 3 years and was very satisfied with their service?

b. What is the probability that a participant chosen at random was not satisfied with their service?

Source: FCC.

42. PROBABILITY OF TRANSPLANT REJECTION The probabilities that the three patients who are scheduled to receive kidney transplants at General Hospital will suffer rejection are $\frac{1}{2}$, $\frac{1}{3}$, and $\frac{1}{10}$. Assuming that the events (kidney rejection) are independent, find the probability that:

a. At least one patient will suffer rejection.

b. Exactly two patients will suffer rejection.

43. QUALITY CONTROL An automobile manufacturer obtains the microprocessors used to regulate fuel consumption in its automobiles from three microelectronic firms: A, B, and C. The quality-control department of the company has determined that 1% of the microprocessors produced by Firm A are defective, 2% of those produced by Firm B are defective, and 1.5% of those produced by Firm C are defective. Firms A, B, and C supply 45%, 25%, and 30%, respectively, of the microprocessors used by the company. What is the probability that a randomly selected automobile manufactured by the company will have a defective microprocessor?

44. HOUSING LOANS The chief loan officer of La Crosse Home Mortgage Company summarized the housing loans extended by the company in 2014 according to type and term of the loan. Her list shows that 70% of the loans were fixed-rate mortgages (F), 25% were adjustable-rate mortgages (A), and 5% belong to some other category (O) (mostly second trust-deed loans and home equity loans). Of the fixed-rate mortgages, 80% were 30-year loans and 20% were 15-year loans; of the adjustable-rate mortgages, 40% were 30-year loans and 60% were 15-year loans; finally, of the other loans extended, 30% were 20-year loans, 60% were 10-year loans, and 10% were for a term of 5 years or less.

a. Draw a tree diagram representing these data.

b. What is the probability that a home loan extended by La Crosse has an adjustable rate and is for a term of 15 years?

c. What is the probability that a home loan extended by La Crosse is for a term of 15 years?

45. Three cards are drawn without replacement from a well-shuffled deck of 52 playing cards. What is the probability that the third card drawn is a diamond?

46. **COLLEGE ADMISSIONS** The admissions office of a private university released the following data for the preceding academic year: From a pool of 3900 male applicants, 40% were accepted by the university, and 40% of these subsequently enrolled. Additionally, from a pool of 3600 female applicants, 45% were accepted by the university, and 40% of these subsequently enrolled. What is the probability that:
   a. A male applicant will be accepted by and subsequently will enroll in the university?
   b. A student who applies for admissions will be accepted by the university?
   c. A student who applies for admission will be accepted by the university and subsequently will enroll?

47. **QUALITY CONTROL** A box contains two defective Christmas tree lights that have been inadvertently mixed with eight nondefective lights. If the lights are selected one at a time without replacement and tested until both defective lights are found, what is the probability that both defective lights will be found after exactly three trials?

48. **NYC TOURISTS** In 2011, 21% of all tourists in New York City (NYC) were international visitors, of whom an estimated 70% visited the Empire State Building. Of the international tourists who visited the Empire State Building, approximately 40% bought at least one souvenir in the gift shop. What percentage of international tourists in NYC in 2011 visited the Empire State Building and also bought at least one souvenir from the gift shop?
   Source: nycgo.com.

49. **REAL ESTATE SALES** Mark, a real estate agent, estimates that of prospective homebuyers who see an ad for a house he has listed for sale, 24% will show up for the open house. Of those who show up for the open house, 30% will return for a second showing, and of these, 75% will make an offer to buy the house. What is the probability that a prospective home buyer who has seen Mark's ad will show up for a second viewing and ultimately make an offer to buy the house?

50. **WINNING BIDS** Brian, a landscape architect, submitted a bid on each of three home landscaping projects. He estimates that the probabilities of winning the bid on Project A, Project B, and Project C are .7, .6, and .5, respectively. Assume that the probability of winning a bid on one of the three projects is independent of winning or losing the bids on the other two projects. Find the probability that Brian will:
   a. Win all three of the bids.
   b. Win exactly two of the bids.
   c. Win exactly one bid.

51. **FAMILY PORTRAITS** The Jackson family of four is posing for a family portrait in a studio. The probability of Dad and Mom striking an acceptable pose is .9 each; the probability of daughter Janet striking an acceptable pose is .7; and the probability of baby Bob striking an acceptable pose is .3.

Assuming that the probability of each family member striking an acceptable pose is independent of the way in which the other members of the family pose, what is the probability that an acceptable photo will result after one take?

52. Two cards are drawn without replacement from a well-shuffled deck of 52 cards. Let A be the event that the first card drawn is a heart, and let B be the event that the second card drawn is a red card. Show that the events A and B are dependent events.

53. **RELIABILITY OF SECURITY SYSTEMS** Before being allowed to enter a maximum-security area at a military installation, a person must pass three independent identification tests: a voice-pattern test, a fingerprint test, and a handwriting test. If the reliability of the first test is 97%, that of the second test is 98.5%, and that of the third is 98.5%, what is the probability that this security system will allow an improperly identified person to enter the maximum-security area?

54. **QUALITY CONTROL** Copykwik has four photocopy machines: A, B, C, and D. The probability that a given machine will break down on a particular day is
$$P(A) = \frac{1}{50} \quad P(B) = \frac{1}{60} \quad P(C) = \frac{1}{75} \quad P(D) = \frac{1}{40}$$
Assuming independence, what is the probability on a particular day that:
   a. All four machines will break down?
   b. None of the machines will break down?

55. **QUALITY CONTROL** It is estimated that 0.80% of a large consignment of eggs in a certain supermarket is broken.
   a. What is the probability that a customer who randomly selects a dozen of these eggs receives at least one broken egg?
   b. What is the probability that a customer who selects these eggs at random will have to check three cartons before finding a carton without any broken eggs? (Each carton contains a dozen eggs.)

56. **RELIABILITY OF A HOME THEATER SYSTEM** In a home theater system, the probability that the video components need repair within 1 year is .01, the probability that the electronic components need repair within 1 year is .005, and the probability that the audio components need repair within 1 year is .001. Assuming that the events are independent, find the probability that:
   a. At least one of these components will need repair within 1 year.
   b. Exactly one of these components will need repair within 1 year.

57. **PRODUCT RELIABILITY** The proprietor of Cunningham's Hardware Store has decided to install floodlights on the premises as a measure against vandalism and theft. If the probability is .01 that a certain brand of floodlight will burn out within a year, find the minimum number of floodlights that must be installed to ensure that the probability that at least one of them will remain

functional for the whole year is at least .99999. (Assume that the floodlights operate independently.)

58. Suppose the probability that an event will occur in one trial is p. Show that the probability that the event will occur at least once in n independent trials is $1 - (1 - p)^n$.

59. Let E be any event in a sample space S.
   a. Are E and S mutually exclusive? Explain your answer.
   b. Are E and $\varnothing$ mutually exclusive? Explain your answer.

60. Let E and F be events such that $F \subset E$. Find $P(E \mid F)$, and interpret your result.

61. Let E and F be mutually exclusive events, and suppose $P(F) \neq 0$. Find $P(E \mid F)$, and interpret your result.

62. Let E and F be independent events; show that E and F^c are independent.

63. Suppose that A and B are mutually exclusive events and that $P(A \cup B) \neq 0$. What is $P(A \mid A \cup B)$?

64. Prove Equation (4),
$$P(A \cap B \cap C) = P(A)P(B \mid A)P(C \mid A \cap B).$$

In Exercises 65–68, determine whether the statement is true or false. If it is true, explain why it is true. If it is false, give an example to show why it is false.

65. If A and B are mutually exclusive and $P(B) \neq 0$, then $P(A \mid B) = 0$.

66. If A is an event of an experiment, then $P(A \mid A^c) \neq 0$.

67. If A and B are events of an experiment, then
$$P(A \cap B) = P(A \mid B) \cdot P(B) = P(B \mid A) \cdot P(A)$$

68. If A and B are independent events with $P(A) \neq 0$ and $P(B) \neq 0$, then $A \cap B \neq \varnothing$.

8.2 Solutions to Self-Check Exercises

1. a. $P(A \mid B) = \dfrac{P(A \cap B)}{P(B)}$
$$= \frac{.3}{.8} = \frac{3}{8}$$

b. $P(B \mid A) = \dfrac{P(A \cap B)}{P(A)}$
$$= \frac{.3}{.4} = \frac{3}{4}$$

2. Let A, B, and C denote the events that a respondent who married between the ages of 20 and 22, between the ages of 23 and 27, and at age 28 or older (respectively) said

that his or her marriage was "less than very happy." Then the probability of each of these events occurring is $P(A) = .297$, $P(B) = .269$, and $P(C) = .451$. So the probability that all three of the respondents said that his or her marriage was "less than very happy" is

$$P(A) \cdot P(B) \cdot P(C) = (.297)(.269)(.451) \approx .036$$

8.3 Bayes' Theorem

A Posteriori Probabilities

Suppose three machines, A, B, and C, produce similar engine components. Machine A produces 45% of the total components, Machine B produces 30%, and Machine C produces 25%. For the usual production schedule, 6% of the components produced by Machine A do not meet established specifications; for Machine B and Machine C, the corresponding figures are 4% and 3%, respectively. One component is selected at random from the total output and is found to be defective. What is the probability that the component selected was produced by Machine A?

The answer to this question is found by calculating the probability *after* the outcomes of the experiment have been observed. Such probabilities are called **a posteriori probabilities** in contrast to **a priori probabilities**—probabilities that give the likelihood that an event *will* occur, the subject of the last two sections.

Returning to the example under consideration, we need to determine the a posteriori probability for the event that the component selected was produced by Machine A. Toward this end, let A, B, and C denote the events that a component is produced by

Machine A, Machine B, and Machine C, respectively. We may represent this experiment with a Venn diagram (Figure 9).

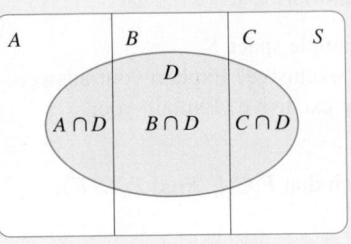

FIGURE 9
D is the event that a defective component is produced by Machine A, Machine B, or Machine C.

The three mutually exclusive events A, B, and C form a **partition** of the sample space S; that is, aside from being mutually exclusive, their union is precisely S. The event D that a component is defective is the shaded area. Again referring to Figure 9, we see that

1. The event D may be expressed as

$$D = (A \cap D) \cup (B \cap D) \cup (C \cap D)$$

2. The event that a component is defective and is produced by Machine A is given by $A \cap D$.

Thus, the a posteriori probability that a defective component selected was produced by Machine A is given by

$$P(A \mid D) = \frac{P(A \cap D)}{P(D)}$$

Upon dividing both the numerator and the denominator by $P(S)$ and observing that the events $A \cap D$, $B \cap D$, and $C \cap D$ are mutually exclusive, we obtain

$$P(A \mid D) = \frac{P(A \cap D)}{P(D)} \tag{7}$$

$$= \frac{P(A \cap D)}{P(A \cap D) + P(B \cap D) + P(C \cap D)}$$

Next, using the Product Rule, we may express

$$P(A \cap D) = P(A) \cdot P(D \mid A)$$
$$P(B \cap D) = P(B) \cdot P(D \mid B)$$
$$P(C \cap D) = P(C) \cdot P(D \mid C)$$

so Equation (7) may be expressed in the form

$$P(A \mid D) = \frac{P(A) \cdot P(D \mid A)}{P(A) \cdot P(D \mid A) + P(B) \cdot P(D \mid B) + P(C) \cdot P(D \mid C)} \tag{8}$$

which is a special case of a result known as **Bayes' Theorem.**

Observe that the expression on the right of Equation (8) involves the probabilities $P(A)$, $P(B)$, and $P(C)$ as well as the conditional probabilities $P(D \mid A)$, $P(D \mid B)$, and $P(D \mid C)$. In fact, by displaying these probabilities on a tree diagram, we obtain Figure 10. We may compute the required probability by substituting the relevant quantities into Equation (8), or we may make use of the following device:

$$P(A \mid D) = \frac{\text{Product of probabilities along the branch through } A \text{ terminating at } D}{\text{Sum of products of the probabilities along each branch terminating at } D}$$

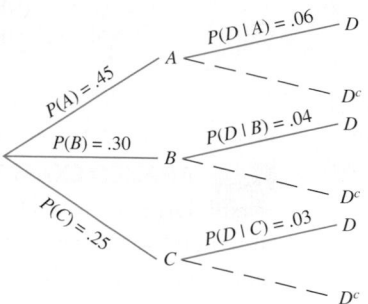

FIGURE **10**
A tree diagram displaying the probabilities that a defective component is produced by Machine A, Machine B, or Machine C

In either case, we obtain

$$P(A \mid D) = \frac{(.45)(.06)}{(.45)(.06) + (.30)(.04) + (.25)(.03)}$$
$$\approx .58$$

Before looking at any further examples, let's state the general form of Bayes' Theorem.

Bayes' Theorem

Let $A_1, A_2, \ldots, A_n$ be a partition of a sample space S, and let E be an event of the experiment such that $P(E) \neq 0$ and $P(A_i) \neq 0$ for $1 \leq i \leq n$. Then the a posteriori probability $P(A_i \mid E)$ $(1 \leq i \leq n)$ is given by

$$P(A_i \mid E) = \frac{P(A_i) \cdot P(E \mid A_i)}{P(A_1) \cdot P(E \mid A_1) + P(A_2) \cdot P(E \mid A_2) + \cdots + P(A_n) \cdot P(E \mid A_n)} \tag{9}$$

APPLIED EXAMPLE 1 Quality Control The panels for the Pulsar 32-inch widescreen LCD HDTVs are manufactured in three locations and then shipped to the main plant of Vista Vision for final assembly. Plants A, B, and C supply 50%, 30%, and 20%, respectively, of the panels used by Vista Vision. The quality-control department of the company has determined that 1% of the panels produced by Plant A are defective, whereas 2% of the panels produced by Plants B and C are defective. If a Pulsar 32-inch HDTV is selected at random and the panel is found to be defective, what is the probability that the panel was manufactured in Plant C? (Compare with Example 6, page 523.)

Solution Let A, B, and C denote the events that the set chosen has a panel manufactured in Plant A, Plant B, and Plant C, respectively. Also, let D denote the event

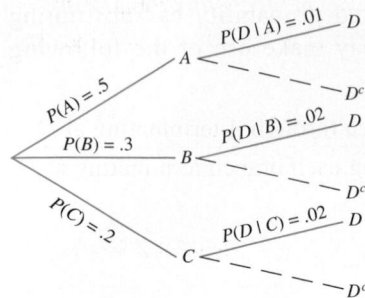

FIGURE 11

$$P(C \mid D) = \frac{\text{Product of probabilities of}}{\text{branches to } D \text{ through } C} \div \text{Sum of product of probabilities of branches leading to } D$$

that a set has a defective panel. Using the given information, we may draw the tree diagram shown in Figure 11. Next, using Formula (9), we find that the required a posteriori probability is given by

$$
\begin{aligned}
P(C \mid D) &= \frac{P(C) \cdot P(D \mid C)}{P(A) \cdot P(D \mid A) + P(B) \cdot P(D \mid B) + P(C) \cdot P(D \mid C)} \\
&= \frac{(.20)(.02)}{(.50)(.01) + (.30)(.02) + (.20)(.02)} \\
&\approx .27
\end{aligned}
$$

APPLIED EXAMPLE 2 Income Distributions A study was conducted in a large metropolitan area to determine the annual incomes of married couples in which the husbands were the sole providers and of those in which the husbands and wives were both employed. Table 2 gives the results of this study.

TABLE 2

Annual Family Income ($)	Married Couples (%)	Income Group with Both Spouses Employed (%)
150,000 and over	4	65
100,000–149,999	10	73
75,000–99,999	21	68
50,000–74,999	24	63
30,000–49,999	30	43
Under 30,000	11	28

a. What is the probability that a couple selected at random from this area has two incomes?

b. If a randomly chosen couple has two incomes, what is the probability that the annual income of this couple is $150,000 or more?

c. If a randomly chosen couple has two incomes, what is the probability that the annual income of this couple is greater than $49,999?

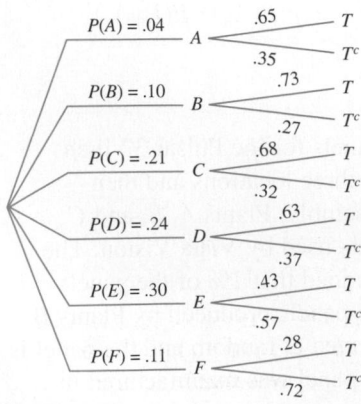

FIGURE 12

Solution Let A denote the event that the annual income of the couple is $150,000 or more; let B denote the event that the annual income is between $100,000 and $149,999; let C denote the event that the annual income is between $75,000 and $99,999; and so on. Finally, let T denote the event that both spouses are employed. The probabilities of the occurrence of these events are displayed in Figure 12.

a. The probability that a couple selected at random from this group has two incomes is given by

$$
\begin{aligned}
P(T) &= P(A) \cdot P(T \mid A) + P(B) \cdot P(T \mid B) + P(C) \cdot P(T \mid C) \\
&\quad + P(D) \cdot P(T \mid D) + P(E) \cdot P(T \mid E) + P(F) \cdot P(T \mid F) \\
&= (.04)(.65) + (.10)(.73) + (.21)(.68) + (.24)(.63) \\
&\quad + (.30)(.43) + (.11)(.28) \\
&= .5528
\end{aligned}
$$

b. Using the results of part (a) and Bayes' Theorem, we find that the probability that a randomly chosen couple has an annual income of $150,000 or more, given that both spouses are employed, is

$$P(A \mid T) = \frac{P(A) \cdot P(T \mid A)}{P(T)} = \frac{(.04)(.65)}{.5528}$$

$$\approx .047$$

c. The probability that a randomly chosen couple has an annual income greater than $49,999, given that both spouses are employed, is

$$P(A \mid T) + P(B \mid T) + P(C \mid T) + P(D \mid T)$$
$$= \frac{P(A) \cdot P(T \mid A) + P(B) \cdot P(T \mid B) + P(C) \cdot P(T \mid C) + P(D) \cdot P(T \mid D)}{P(T)}$$
$$= \frac{(.04)(.65) + (.1)(.73) + (.21)(.68) + (.24)(.63)}{.5528}$$
$$\approx .711$$

8.3 Self-Check Exercises

1. The accompanying tree diagram represents a two-stage experiment. Use the diagram to find $P(B \mid D)$.

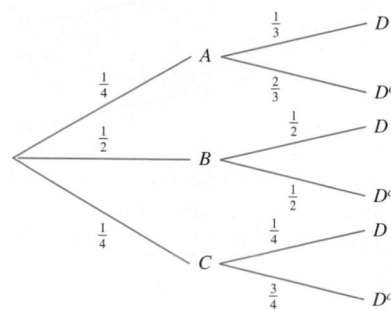

2. **POLITICS** In a recent presidential election, it was estimated that the probability that the Republican candidate would be elected was $\frac{3}{5}$ and therefore the probability that the Democratic candidate would be elected was $\frac{2}{5}$ (the two Independent candidates were given little chance of being elected). It was also estimated that if the Republican candidate were elected, then the probability that research for a new manned bomber would continue was $\frac{4}{5}$. But if the Democratic candidate were successful, then the probability that the research would continue was $\frac{3}{10}$. Research was terminated shortly after the successful presidential candidate took office. What is the probability that the Republican candidate won that election?

Solutions to Self-Check Exercises 8.3 can be found on page 544.

8.3 Concept Questions

1. What are a priori probabilities and a posteriori probabilities? Give an example of each.

2. Suppose the events A, B, and C form the partition of a sample space S, and suppose E is an event of an experiment such that $P(E) \neq 0$. Use Bayes' Theorem to write the formula for the a posteriori probability $P(A \mid E)$. (Assume that $P(A), P(B), P(C) \neq 0$.)

3. Refer to Question 2. If E is the event that a product was produced in Factory A, Factory B, or Factory C and $P(E) \neq 0$, what does $P(A \mid E)$ represent?

8.3   Exercises

In Exercises 1–3, refer to the accompanying Venn diagram. An experiment in which the three mutually exclusive events A, B, and C form a partition of the uniform sample space S is depicted in the diagram.

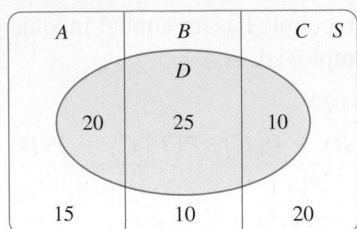

1. Using the information given in the Venn diagram, draw a tree diagram illustrating the probabilities of the events A, B, C, and D.

2. Find: **a.** $P(D)$ **b.** $P(A \mid D)$

3. Find: **a.** $P(D^c)$ **b.** $P(B \mid D^c)$

In Exercises 4–6, refer to the accompanying Venn diagram. An experiment in which the three mutually exclusive events A, B, and C form a partition of the uniform sample space S is depicted in the diagram.

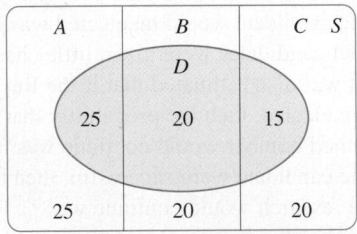

4. Using the information given in the Venn diagram, draw a tree diagram illustrating the probabilities of the events A, B, C, and D.

5. Find: **a.** $P(D)$ **b.** $P(B \mid D)$

6. Find: **a.** $P(D^c)$ **b.** $P(B \mid D^c)$

7. The accompanying tree diagram represents a two-stage experiment. Use the diagram to find:
   a. $P(A) \cdot P(D \mid A)$ **b.** $P(B) \cdot P(D \mid B)$
   c. $P(A \mid D)$

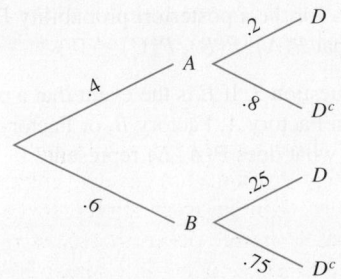

8. The accompanying tree diagram represents a two-stage experiment. Use the diagram to find:
   a. $P(A) \cdot P(D \mid A)$ **b.** $P(B) \cdot P(D \mid B)$
   c. $P(A \mid D)$

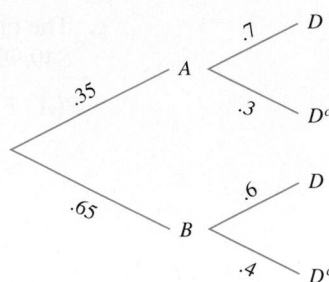

9. The accompanying tree diagram represents a two-stage experiment. Use the diagram to find:
   a. $P(A) \cdot P(D \mid A)$ **b.** $P(B) \cdot P(D \mid B)$
   c. $P(C) \cdot P(D \mid C)$ **d.** $P(A \mid D)$

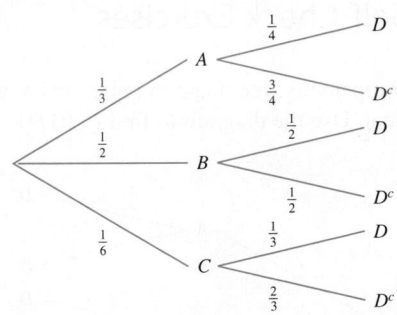

10. The accompanying tree diagram represents a two-stage experiment. Use this diagram to find:
    a. $P(A \cap D)$ **b.** $P(B \cap D)$ **c.** $P(C \cap D)$ **d.** $P(D)$
    e. Verify:

$$P(A \mid D) = \frac{P(A \cap D)}{P(D)}$$

$$= \frac{P(A) \cdot P(D \mid A)}{P(A) \cdot P(D \mid A) + P(B) \cdot P(D \mid B) + P(C) \cdot P(D \mid C)}$$

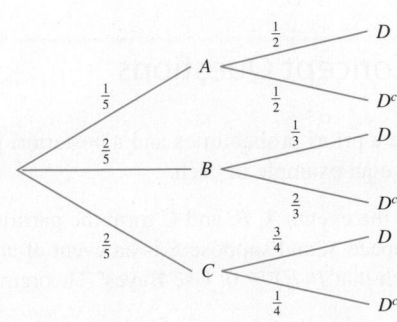

11. The accompanying diagram represents a two-stage experiment. Complete the information on the diagram, and use it to find:

  a. $P(B)$ **b.** $P(A \mid B)$
  c. $P(B^c)$ **d.** $P(A \mid B^c)$

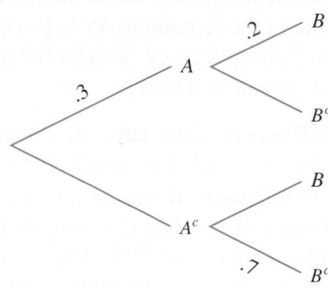

12. The accompanying diagram represents a two-stage experiment. Here, $B = (A \cup C)^c$. Complete the information on the diagram, and use it to find:

  a. $P(D)$ **b.** $P(B \mid D)$
  c. $P(D^c)$ **d.** $P(A \mid D^c)$

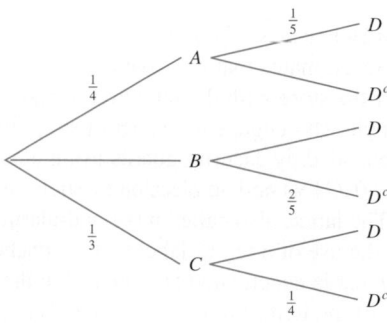

In Exercises 13–16, refer to the following experiment: Two cards are drawn in succession without replacement from a standard deck of 52 cards.

13. What is the probability that the first card is a heart given that the second card is a heart?

14. What is the probability that the first card is a heart given that the second card is a diamond?

15. What is the probability that the first card is a jack given that the second card is an ace?

16. What is the probability that the first card is a face card given that the second card is an ace?

In Exercises 17–20, refer to the following experiment: Urn *A* contains four white balls and six black balls. Urn *B* contains three white balls and five black balls. A ball is drawn from Urn *A* and then transferred to Urn *B*. A ball is then drawn from Urn *B*.

17. Represent the probabilities associated with this two-stage experiment in the form of a tree diagram.

18. What is the probability that the transferred ball was white given that the second ball drawn was white?

19. What is the probability that the transferred ball was black given that the second ball drawn was white?

20. What is the probability that the transferred ball was black given that the second ball drawn was black?

21. SEAT-BELT COMPLIANCE Data compiled by the Highway Patrol Department regarding the use of seat-belts by drivers in a certain area after the passage of a compulsory seat-belt law are shown in the accompanying table.

Drivers	Percentage of Drivers in Group	Percentage of Group Stopped for Moving Violation
Group I (using seat-belts)	.64	.002
Group II (not using seat-belts)	.36	.005

If a driver in that area is stopped for a moving violation, what is the probability that he or she:

  a. Will have a seat-belt on?
  b. Will not have a seat-belt on?

22. RETIREMENT NEEDS In a survey of 2000 adults 50 years old and older of whom 60% were retired and 40% were pre-retired, the following question was asked: Do you expect your income needs to vary from year to year in retirement? Of those who were retired, 33% answered no, and 67% answered yes. Of those who were pre-retired, 28% answered no, and 72% answered yes. If a respondent in the survey was selected at random and had answered yes to the question, what is the probability that he or she was retired?
Source: Sun Life Financial.

23. BLOOD TESTS If a certain disease is present, then a blood test will reveal it 95% of the time. But the test will also indicate the presence of the disease 2% of the time when in fact the person tested is free of that disease; that is, the test gives a false positive 2% of the time. If 0.3% of the general population actually has the disease, what is the probability that a person chosen at random from the population has the disease given that he or she tested positive?

24. OPINION POLLS In a survey to determine the opinions of Americans on health insurers, 400 baby boomers and 600 pre-boomers were asked this question: Do you believe that insurers are very responsible for high health costs? Of the baby boomers, 212 answered in the affirmative, whereas 198 of the pre-boomers answered in the affirmative. If a respondent chosen at random from those surveyed answered the question in the affirmative, what is the probability that he or she is a baby boomer? A pre-boomer?
Source: GfK Roper Consulting.

25. CRIME RATES Data compiled by the Department of Justice on the number of people arrested in a certain year for serious crimes (murder, forcible rape, robbery, etc.) revealed that 89% were male and 11% were female. Of the males, 30% were under 18, whereas 27% of the females arrested were under 18.

a. What is the probability that a person arrested for a serious crime in that year was under 18?

b. If a person arrested for a serious crime in that year is known to be under 18, what is the probability that the person is female?

Source: Department of Justice.

26. **GENDER GAP** A study of the faculty at U.S. medical schools in 2006 revealed that 32% of the faculty were women and 68% were men. Of the female faculty, 31% were full/associate professors, 47% were assistant professors, and 22% were instructors. Of the male faculty, 51% were full/associate professors, 37% were assistant professors, and 12% were instructors. If a faculty member at a U.S. medical school selected at random in 2006 held the rank of full/associate professor, what is the probability that the faculty member was female?

Source: Association of American Medical Colleges.

27. **MEDICAL RESEARCH** On the basis of data obtained from the National Institute of Dental Research, it has been determined that 42% of 12-year-olds have never had a cavity, 34% of 13-year-olds have never had a cavity, and 28% of 14-year-olds have never had a cavity. Suppose a child is selected at random from a group of 24 junior high school students that includes six 12-year-olds, eight 13-year-olds, and ten 14-year-olds. If this child does not have a cavity, what is the probability that this child is 14 years old?

Source: National Institute of Dental Research.

28. **VOTING PATTERNS** In a recent senatorial election, 50% of the voters in a certain district were registered as Democrats, 35% were registered as Republicans, and 15% were registered as Independents. The incumbent Democratic senator was reelected over her Republican and Independent opponents. Exit polls indicated that she gained 75% of the Democratic vote, 25% of the Republican vote, and 30% of the Independent vote. Assuming that the exit poll is accurate, what is the probability that a vote for the incumbent was cast by a registered Republican?

29. **MEDICAL DIAGNOSES** A study was conducted among a certain group of union members whose health insurance policies required second opinions prior to surgery. Of those members whose doctors advised them to have surgery, 20% were informed by a second doctor that no surgery was needed. Of these, 70% took the second doctor's opinion and did not go through with the surgery. Of the members who were advised to have surgery by both doctors, 95% went through with the surgery. What is the probability that a union member who had surgery was advised to do so by a second doctor?

30. **PERSONNEL SELECTION** Applicants for temporary office work at Carter Temporary Help Agency who have successfully completed an administrative assistant course are then placed in suitable positions by Nancy Dwyer and Darla Newberg. Employers who hire temporary help through the agency return a card indicating satisfaction or dissatisfaction with the work performance of those hired. From past experience it is known that 80% of the employees placed by Nancy are rated as satisfactory, and 70% of those placed by Darla are rated as satisfactory. Darla places 55% of the temporary office help at the agency, and Nancy places the remaining 45%. If a Carter office worker is rated unsatisfactory, what is the probability that he or she was placed by Darla?

31. **IMPACT OF GAS PRICES ON CONSUMERS** In a survey of 1012 adults aged 18 years and older conducted by Social Science Research Solutions, it was found that gas prices have caused financial hardship for 60% of the respondents aged 18–49 years and 55% of those who are 50 years or older. There were 742 adults aged 18–49 years and 270 adults who were 50 years or older in the survey. If a respondent in the survey selected at random reported that he or she did not experience financial hardship from gas prices, what is the probability that he or she was an adult aged 18–49 years?

Source: Social Science Research Solutions.

32. **DETECTING SHOPLIFTERS** The management of Mark's Department Store estimates that approximately 2% of all shoppers enter the store with the intention of shoplifting. Of those people who engage in the act of shoplifting, 95% are apprehended by security guards using closed-circuit television (CCTV) and an electronic article surveillance system. The latter, also called a tag-and-alarm system, involves the use of a tag or label that is attached to an item. The tag is deactivated or detached at the time of purchase of the item; failure to do so will trigger an alarm when the item is carried through the gates. Of the other 98% of the shoppers, 1% are mistakenly identified as shoplifters by security guards watching them on CCTV or have triggered the alarm because the salesperson failed to deactivate or detach the tag. If a shopper is detained by security, what is the probability that the person was in fact a shoplifter?

33. **OPINION POLLS** A survey involving 400 likely Democratic voters and 300 likely Republican voters asked the question: Do you support or oppose legislation that would require registration of all handguns? The following results were obtained:

Answer	Democrats (%)	Republicans (%)
Support	77	59
Oppose	14	31
Don't know/refused	9	10

If a randomly chosen respondent in the survey answered "oppose," what is the probability that he or she is a likely Democratic voter?

34. **OPINION POLLS** A survey involving 400 likely Democratic voters and 300 likely Republican voters asked the question: Do you support or oppose legislation that would

require trigger locks on guns, to prevent misuse by children? The following results were obtained:

Answer	Democrats (%)	Republicans (%)
Support	88	71
Oppose	7	20
Don't know/refused	5	9

If a randomly chosen respondent in the survey answered "support," what is the probability that he or she is a likely Republican voter?

35. BEVERAGE PREFERENCES In a study of scientific research on soft drinks, juices, and milk, 50 research studies were fully sponsored by the food industry, and 30 studies were conducted with no corporate ties. In those that were fully sponsored by the food industry, 14% of the participants found the products unfavorable, 23% were neutral, and 63% found the products favorable. In those that had no industry funding, 38% found the products unfavorable, 15% were neutral, and 47% found the products favorable.
a. What is the probability that a participant selected at random found the products favorable?
b. If a participant selected at random found the product favorable, what is the probability that he or she belonged to a group that participated in a corporate-sponsored study?
Source: Children's Hospital, Boston.

36. SELECTION OF SUPREME COURT JUDGES In a past presidential election, it was estimated that the probability that the Republican candidate would be elected was $\frac{3}{5}$ and therefore the probability that the Democratic candidate would be elected was $\frac{2}{5}$ (the two Independent candidates were given no chance of being elected). It was also estimated that if the Republican candidate were elected, the probability that a conservative, moderate, or liberal judge would be appointed to the Supreme Court (one retirement was expected during the presidential term) was $\frac{1}{2}$, $\frac{1}{3}$, and $\frac{1}{6}$, respectively. If the Democratic candidate were elected, the probabilities that a conservative, moderate, or liberal judge would be appointed to the Supreme Court would be $\frac{1}{8}$, $\frac{3}{8}$, and $\frac{1}{2}$, respectively. A conservative judge was appointed to the Supreme Court during the presidential term. What is the probability that the Democratic candidate was elected?

37. AGE DISTRIBUTION OF RENTERS A study conducted by the Metro Housing Agency in a Midwestern city revealed the following information concerning the age distribution of renters within the city.

Age	Adult Population (%)	Group Who Are Renters (%)
21–44	51	58
45–64	31	45
65 and over	18	60

a. What is the probability that an adult selected at random from this population is a renter?

b. If a renter is selected at random, what is the probability that he or she is in the 21–44 age bracket?
c. If a renter is selected at random, what is the probability that he or she is 45 years old or older?

38. PRODUCT RELIABILITY The estimated probability that a Brand A, a Brand B, and a Brand C plasma TV will last at least 30,000 hr is .90, .85, and .80, respectively. Of the 4500 plasma TVs that Ace TV sold in a certain year, 1000 were Brand A, 1500 were Brand B, and 2000 were Brand C. If a plasma TV set sold by Ace TV that year is selected at random and is still working after 30,000 hr of use:
a. What is the probability that it was a Brand A TV?
b. What is the probability that it was not a Brand A TV?

39. An experiment consists of randomly selecting one of three coins, tossing it, and observing the outcome—heads or tails. The first coin is a two-headed coin, the second is a biased coin such that $P(H) = .75$, and the third is a fair coin.
a. What is the probability that the coin that is tossed will show heads?
b. If the coin selected shows heads, what is the probability that this coin is the fair coin?

40. GUN OWNERS IN THE SENATE As of January 3, 2013, the U.S. Senate was made up of 53 Democrats, 45 Republicans, and 2 Independents who caucus with the Democrats. In a survey of the U.S. Senate conducted at that time, every senator was asked whether he or she owned at least one gun. Of the Democrats, 16 declared themselves gun owners; of the Republicans, 26 of them declared themselves gun owners; none of the Independents owned guns. If a senator participating in that survey was picked at random and turned out to be a gun owner, what was the probability that he or she was a Democrat?
Source: Gannett Washington Bureau.

41. RELIABILITY OF MEDICAL TESTS A medical test has been designed to detect the presence of a certain disease. Among people who have the disease, the probability that the disease will be detected by the test is .95. However, the probability that the test will erroneously indicate the presence of the disease in those who do not actually have it is .04. It is estimated that 4% of the population who take this test have the disease.
a. If the test administered to an individual is positive, what is the probability that the person actually has the disease?
b. If an individual takes the test twice and the test is positive both times, what is the probability that the person actually has the disease? (Assume that the tests are independent.)

42. RELIABILITY OF MEDICAL TESTS Refer to Exercise 41. Suppose 20% of the people who were referred to a clinic for the test did in fact have the disease. If the test administered to an individual from this group is positive, what is the probability that the person actually has the disease?

43. QUALITY CONTROL Jansen Electronics has four machines that produce identical components for use in its DVD players. The proportion of the components produced by each machine and the probability of a component produced by that machine being defective are shown in the accompanying table. What is the probability that a component selected at random:
  a. Is defective?
  b. Was produced by Machine I, given that it is defective?
  c. Was produced by Machine II, given that it is defective?

Machine	Proportion of Components Produced	Probability of Defective Component
I	.15	.04
II	.30	.02
III	.35	.02
IV	.20	.03

44. COMMUTING TIMES According to the U.S. Census Bureau, in 2011, 121,298,000 workers who do not work at home have travel times of less than 1 hr, and 10,979,000 workers who do not work at home have travel times of 1 hr or longer. Of those workers whose travel times are less than 1 hr, 91.4% drive alone or carpool, 1.6% take the subway or railroad, 2.1% take other public transportation, and 4.9% use other means. Of those workers whose travel times are 1 hr or longer, 74.0% drive alone or carpool, 11.8% take the subway or railroad, 11.2% take other public transportation, and 3.0% use other means. If a worker chosen at random drives or carpools, what is the probability that he or she has travel times of less than 1 hr?
Source: U.S. Census Bureau.

45. QUALITY CONTROL A halogen desk lamp produced by Luminar was found to be defective. The company has three factories where the lamps are manufactured. The percentage of the total number of halogen desk lamps produced by each factory and the probability that a lamp manufactured by that factory is defective are shown in the accompanying table. What is the probability that the defective lamp was manufactured in Factory III?

Factory	Total Production (%)	Probability of Defective Component
I	35	.015
II	35	.01
III	30	.02

46. OBESITY IN CHILDREN Researchers weighed 1976 3-year-olds from low-income families in 20 U.S. cities. Each child was classified by race (white, black, or Hispanic) and by weight (normal weight, overweight, or obese). The results follow:

Race	Children	Weight (%) Normal Weight	Overweight	Obese
White	406	68	18	14
Black	1081	68	15	17
Hispanic	489	56	20	24

If a child in the research study is selected at random and is found to be obese, what is the probability that the child is white? Hispanic?
Source: American Journal of Public Health.

47. AUTO-ACCIDENT RATES An insurance company has compiled the accompanying data relating the age of drivers and the accident rate (the probability of being involved in an accident during a 1-year period) for drivers within that group:

Age Group	Insured Drivers (%)	Accident Rate (%)
Under 25	16	5.5
25–44	40	2.5
45–64	30	2
65 and over	14	4

What is the probability that an insured driver selected at random:
  a. Will be involved in an accident during a particular 1-year period?
  b. Who is involved in an accident is under 25?

48. PERSONAL HABITS There were 80 male guests at a party. The number of men in each of four age categories is given in the following table. The table also gives the probability that a man in the respective age category will keep his paper money in order of denomination.

Age	Men	Keep Paper Money in Order (%)
21–34	25	90
35–44	30	61
45–54	15	80
55 and over	10	80

A man's wallet was retrieved, and the paper money in it was kept in order of denomination. What is the probability that the wallet belonged to a male guest between the ages of 35 and 44?
Source: USA Today.

49. SLEEPING WITH CELL PHONES In a survey of 920 people aged from 18 through 84 years, of whom 84 belonged to the Millenial Generation, 224 belonged to Generation X, 200 belonged to the Baby Boom Generation, and the rest belonged to the Silent Generation, the following question was asked: Who has slept with a cell phone nearby? Of those who answered in the affirmative, 83%, 68%, 50%, and 20% were from the Millenial Generation, Generation X, the Baby Boom Generation, and the Silent Generation, respectively. If a person in the survey is selected at random and has not slept with a cell phone nearby, what is the probability that the person belongs to the Millenial Generation?
Source: Pew Research Center.

50. VOTER TURNOUT BY INCOME Voter turnout drops steadily as income level declines. The following table gives the percentage of eligible voters in a certain city, categorized by income, who responded with "did not vote" in the 2000 presidential election. The table also gives the number of eligible voters in the city, categorized by income.

Income Percentile	Did Not Vote (%)	Eligible Voters
0–16	52	4,000
17–33	31	11,000
34–67	30	17,500
68–95	14	12,500
96–100	12	5,000

If an eligible voter from this city who had voted in the election is selected at random, what is the probability that this person had an income in the 17–33 percentile?
Source: The National Election Studies.

51. THE SOCIAL LADDER The following table summarizes the results of a poll conducted with 1154 adults.

Annual Household Income ($)	Respondents Within That Income Range (%)	Respondents Who Call Themselves		
		Rich (%)	Middle Class (%)	Poor (%)
Less than 15,000	11.2	0	24	76
15,000–29,999	18.6	3	60	37
30,000–49,999	24.5	0	86	14
50,000–74,999	21.9	2	90	8
75,000 and higher	23.8	5	91	4

a. What is the probability that a respondent chosen at random calls himself or herself middle class?
b. If a randomly chosen respondent calls himself or herself middle class, what is the probability that the annual household income of that individual is between $30,000 and $49,999, inclusive?
c. If a randomly chosen respondent calls himself or herself middle class, what is the probability that the individual's income is either less than or equal to $29,999 or greater than or equal to $50,000?
Source: New York Times/CBS News; Wall Street Journal Almanac.

52. SELECTION OF COLLEGE MAJORS The Office of Admissions and Records of a large western university released the accompanying information concerning the contemplated majors of its freshman class:

Major	Freshmen Choosing This Major (%)	Females Choosing This Major (%)	Males Choosing This Major (%)
Business	24	38	62
Humanities	8	60	40
Education	8	66	34
Social science	7	58	42
Natural sciences	9	52	48
Other	44	48	52

What is the probability that:
a. A student selected at random from the freshman class is a female?
b. A business student selected at random from the freshman class is a male?
c. A female student selected at random from the freshman class is majoring in business?

53. VOTER TURNOUT BY PROFESSION The following table gives the percentage of eligible voters grouped according to profession who responded with "voted" in the 2000 presidential election. The table also gives the percentage of people in a survey categorized by their profession.

Profession	Percentage Who Voted	Percentage in Each Profession
Professionals	.84	.12
White collar	.73	.24
Blue collar	.66	.32
Unskilled	.57	.10
Farmers	.68	.08
Housewives	.66	.14

If an eligible voter who participated in the survey and voted in the election is selected at random, what is the probability that this person is a housewife?
Source: The National Election Studies.

54. SMOKING AND EDUCATION According to the Centers for Disease Control and Prevention, the percentage of adults 25 years old and older who smoke, by educational level, is as follows:

Educational Level	Respondents (%)
No diploma	27
GED diploma	4.3
High school graduate	26
Some college	24
Undergraduate level	10.7
Graduate degree	8

In a group of 140 people, there were 8 with no diploma, 14 with GED diplomas, 40 high school graduates, 24 with some college, 42 with an undergraduate degree, and 12 with a graduate degree. (Assume that these categories are mutually exclusive.) If a person selected at random from this group was a smoker, what is the probability that he or she is a person with a graduate degree?
Source: Centers for Disease Control and Prevention.

8.3 Solutions to Self-Check Exercises

1. Using the probabilities given in the tree diagram and Bayes' Theorem, we have

$$P(B \mid D) = \frac{P(B) \cdot P(D \mid B)}{P(A) \cdot P(D \mid A) + P(B) \cdot P(D \mid B) + P(C) \cdot P(D \mid C)}$$

$$= \frac{\left(\frac{1}{2}\right)\left(\frac{1}{2}\right)}{\left(\frac{1}{4}\right)\left(\frac{1}{3}\right) + \left(\frac{1}{2}\right)\left(\frac{1}{2}\right) + \left(\frac{1}{4}\right)\left(\frac{1}{4}\right)} = \frac{12}{19}$$

2. Let R and D, respectively, denote the event that the Republican and the Democratic candidate won the presidential election. Then $P(R) = \frac{3}{5}$ and $P(D) = \frac{2}{5}$. Also, let C denote the event that research for the new manned bomber continued. These data may be exhibited as in the accompanying tree diagram:

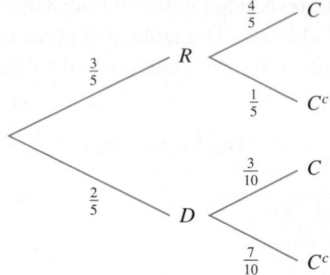

Using Bayes' Theorem, we find that the probability that the Republican candidate had won the election is given by

$$P(R \mid C^c) = \frac{P(R) \cdot P(C^c \mid R)}{P(R) \cdot P(C^c \mid R) + P(D) \cdot P(C^c \mid D)}$$

$$= \frac{\left(\frac{3}{5}\right)\left(\frac{1}{5}\right)}{\left(\frac{3}{5}\right)\left(\frac{1}{5}\right) + \left(\frac{2}{5}\right)\left(\frac{7}{10}\right)} = \frac{3}{10}$$

8.4 Distributions of Random Variables

Random Variables

In many situations, it is desirable to assign numerical values to the outcomes of an experiment. For example, if an experiment consists of rolling a die and observing the face that lands uppermost, then it is natural to assign the numbers 1, 2, 3, 4, 5, and 6, respectively, to the outcomes *one, two, three, four, five,* and *six* of the experiment. If we let X denote the outcome of the experiment, then X assumes one of these numbers. Because the values assumed by X depend on the outcomes of a chance experiment, the outcome X is referred to as a random variable.

> **Random Variable**
>
> A **random variable** is a rule that assigns a number to each outcome of a chance experiment.

More precisely, a random variable is a function with domain given by the set of outcomes of a chance experiment and range contained in the set of real numbers.

EXAMPLE 1 A coin is tossed three times. Let the random variable X denote the number of heads that occur in the three tosses.

a. List the outcomes of the experiment; that is, find the domain of the function X.
b. Find the value assigned to each outcome of the experiment by the random variable X.
c. Find the event comprising the outcomes to which a value of 2 has been assigned by X. This event is written $(X = 2)$ and is the event consisting of the outcomes in which two heads occur.

TABLE 3

Number of Heads in Three Coin Tosses

Outcome	Value of X
HHH	3
HHT	2
HTH	2
THH	2
HTT	1
THT	1
TTH	1
TTT	0

TABLE 4

Number of Coin Tosses Before Heads Appear

Outcome	Value of Y
H	1
TH	2
TTH	3
TTTH	4
TTTTH	5
⋮	⋮

Solution

a. From the results of Example 3, Section 7.5 (page 475), we see that the set of outcomes of the experiment is given by the sample space

$$S = \{\text{HHH, HHT, HTH, THH, HTT, THT, TTH, TTT}\}$$

b. The outcomes of the experiment are displayed in the first column of Table 3. The corresponding value assigned to each such outcome by the random variable X (the number of heads) appears in the second column.

c. With the aid of Table 3, we see that the event $(X = 2)$ is given by the set

$$\{\text{HHT, HTH, THH}\}$$

EXAMPLE 2 A coin is tossed repeatedly until a head occurs. Let the random variable Y denote the number of coin tosses in the experiment. What are the values of Y?

Solution   The outcomes of the experiment make up the infinite set

$$S = \{\text{H, TH, TTH, TTTH, TTTTH, \ldots}\}$$

These outcomes of the experiment are displayed in the first column of Table 4. The corresponding values assumed by the random variable Y (the number of tosses) appear in the second column.

 APPLIED EXAMPLE 3 Product Reliability A disposable flashlight is turned on and left on until its battery runs out. Let the random variable Z denote the length (in hours) of the life of the battery. What values may Z assume?

Solution   The value of Z may be any nonnegative real number; that is, the possible values of Z make up the interval $0 \le Z < \infty$.

One advantage of working with random variables—rather than working directly with the outcomes of an experiment—is that random variables are functions that may be added, subtracted, and multiplied. Because of this, results developed in the field of algebra and other areas of mathematics may be used freely to help us solve problems in probability and statistics.

A random variable is classified into three categories depending on the set of values it assumes. A random variable is called **finite discrete** if it assumes only finitely many values. For example, the random variable X of Example 1 is finite discrete because it may assume values only from the finite set of numbers $\{0, 1, 2, 3\}$. Next, a random variable is said to be **infinite discrete** if it takes on infinitely many values, which may be arranged in a sequence. For example, the random variable Y of Example 2 is infinite discrete because it assumes values from the set $\{1, 2, 3, 4, 5, \ldots\}$, which has been arranged in the form of an infinite sequence. Finally, a random variable is called **continuous** if the values it may assume comprise an interval of real numbers. For example, the random variable Z of Example 3 is continuous because the values it may assume make up the interval of nonnegative real numbers. For the remainder of this section, unless otherwise noted, *all random variables will be assumed to be finite discrete.*

Probability Distributions of Random Variables

In Section 7.6, we learned how to construct the probability distribution for an experiment. There, the probability distribution took the form of a table that gave the probabilities associated with the outcomes of an experiment. Since the random variable associated with an experiment is related to the outcomes of the experiment, it is clear

TABLE 5

Probability Distribution for the Random Variable X

x	$P(X = x)$
x_1	p_1
x_2	p_2
x_3	p_3
$\vdots$	$\vdots$
x_n	p_n

that we should be able to construct a probability distribution associated with the *random variable* rather than one associated with the outcomes of the experiment. Such a distribution is called the **probability distribution of a random variable** and may be given in the form of a formula or displayed in a table that gives the distinct (numerical) values of the random variable X and the probabilities associated with these values. Thus, if $x_1, x_2, \ldots, x_n$ are the values assumed by the random variable X with associated probabilities $P(X = x_1)$, $P(X = x_2)$, $\ldots$, $P(X = x_n)$, respectively, then the required probability distribution of the random variable X may be expressed in the form of the table shown in Table 5, where $p_i = P(X = x_i)$, $i = 1, 2, \ldots, n$.

The probability distribution of a random variable X satisfies

1. $0 \leq p_i \leq 1 \qquad i = 1, 2, \ldots, n$
2. $p_1 + p_2 + \cdots + p_n = 1$

In the next several examples, we illustrate the construction and application of probability distributions.

EXAMPLE 4

a. Find the probability distribution of the random variable associated with the experiment of Example 1.
b. What is the probability of obtaining at least two heads in the three tosses of the coin?

Solution

a. From the results of Example 1, we see that the values assumed by the random variable X are 0, 1, 2, and 3, corresponding to the events of 0, 1, 2, and 3 heads occurring, respectively. Referring to Table 3 once again, we see that the outcome associated with the event $(X = 0)$ is given by the set $\{TTT\}$. Consequently, the probability associated with the random variable X when it assumes the value 0 is given by

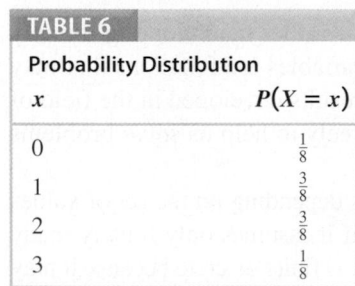

$$P(X = 0) = \frac{1}{8} \qquad \text{Note that } n(S) = 8.$$

Next, observe that the event $(X = 1)$ is given by the set $\{HTT, THT, TTH\}$, so

$$P(X = 1) = \frac{3}{8}$$

In a similar manner, we may compute $P(X = 2)$ and $P(X = 3)$, which gives the probability distribution shown in Table 6.

b. The probability of obtaining at least two heads is given by

$$P(X \geq 2) = P(X = 2) + P(X = 3)$$
$$= \frac{3}{8} + \frac{1}{8} = \frac{1}{2}$$

TABLE 6

Probability Distribution

x	$P(X = x)$
0	$\frac{1}{8}$
1	$\frac{3}{8}$
2	$\frac{3}{8}$
3	$\frac{1}{8}$

EXAMPLE 5 Let X denote the random variable that gives the sum of the faces that fall uppermost when two fair dice are rolled.

a. Find the probability distribution of X.
b. What is the probability that the sum of the faces that fall uppermost is less than or equal to 5? Between 8 and 10, inclusive?

Solution

a. The values assumed by the random variable X are 2, 3, 4, $\ldots$, 12, corresponding to the events $E_2, E_3, E_4, \ldots, E_{12}$ (see Example 4, Section 7.5). The probabilities

TABLE 7

x	$P(X = x)$
2	$\frac{1}{36}$
3	$\frac{2}{36}$
4	$\frac{3}{36}$
5	$\frac{4}{36}$
6	$\frac{5}{36}$
7	$\frac{6}{36}$
8	$\frac{5}{36}$
9	$\frac{4}{36}$
10	$\frac{3}{36}$
11	$\frac{2}{36}$
12	$\frac{1}{36}$

associated with the random variable X when X assumes the values 2, 3, 4, . . . , 12 are precisely the probabilities $P(E_2)$, $P(E_3)$, . . . , $P(E_{12})$, respectively, and may be computed in much the same way as the solution to Example 3, Section 7.6. Thus,

$$P(X = 2) = P(E_2) = \frac{1}{36}$$

$$P(X = 3) = P(E_3) = \frac{2}{36}$$

and so on. The required probability distribution of X is given in Table 7.

b. The probability that the sum of the faces that fall uppermost is less than or equal to 5 is given by

$$P(X \le 5) = P(X = 2) + P(X = 3) + P(X = 4) + P(X = 5)$$

$$= \frac{1}{36} + \frac{2}{36} + \frac{3}{36} + \frac{4}{36} = \frac{10}{36} = \frac{5}{18}$$

The probability that the sum of the faces that fall uppermost is between 8 and 10, inclusive, is given by

$$P(8 \le X \le 10) = P(X = 8) + P(X = 9) + P(X = 10)$$

$$= \frac{5}{36} + \frac{4}{36} + \frac{3}{36} = \frac{12}{36} = \frac{1}{3}$$

APPLIED EXAMPLE 6 Waiting Lines The following data give the number of cars observed waiting in line at the beginning of 2-minute intervals between 3 P.M. and 5 P.M. on a certain Friday at the drive-in teller of Westwood Savings Bank and the corresponding frequency of occurrence:

Cars	0	1	2	3	4	5	6	7	8
Frequency of Occurrence	2	9	16	12	8	6	4	2	1

a. Find the probability distribution of the random variable X, where X denotes the number of cars observed waiting in line.

b. What is the probability that the number of cars observed waiting in line in any 2-minute interval between 3 P.M. and 5 P.M. on a Friday is less than or equal to 3? Between 2 and 4, inclusive? Greater than 6?

Solution

a. The sum of the numbers in the second row of the table given above is 60. Dividing each number in the second row of the table by 60 gives the respective probabilities associated with the random variable X when X assumes the values 0, 1, 2, . . . , 8. (Here, we use the relative frequency interpretation of probability.) For example,

$$P(X = 0) = \frac{2}{60} \approx .03$$

$$P(X = 1) = \frac{9}{60} = .15$$

and so on. The resulting probability distribution is shown in Table 8.

TABLE 8

Probability Distribution

x	$P(X = x)$
0	.03
1	.15
2	.27
3	.20
4	.13
5	.10
6	.07
7	.03
8	.02

b. The probability that the number of cars observed waiting in line is less than or equal to 3 is given by

$$P(X \leq 3) = P(X = 0) + P(X = 1) + P(X = 2) + P(X = 3)$$
$$= .03 + .15 + .27 + .20 = .65$$

The probability that the number of cars observed waiting in line is between 2 and 4, inclusive, is given by

$$P(2 \leq X \leq 4) = P(2) + P(3) + P(4)$$
$$= .27 + .20 + .13 = .60$$

The probability that the number of cars observed waiting in line is greater than 6 is given by

$$P(X > 6) = P(7) + P(8)$$
$$= .03 + .02 = .05$$

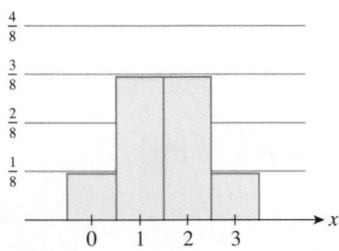

FIGURE **13**
Histogram showing the probability distribution for the number of heads occurring in three coin tosses

Histograms

A probability distribution of a random variable may be exhibited graphically by means of a **histogram.** To construct a histogram of a particular probability distribution, first locate the values of the random variable on a number line. Then, above each such number, draw a rectangle with width 1 and height equal to the probability associated with that value of the random variable. For example, the histogram of the probability distribution appearing in Table 6 is shown in Figure 13. The histograms of the probability distributions of Examples 5 and 6 are constructed in a similar manner and are displayed in Figures 14 and 15, respectively.

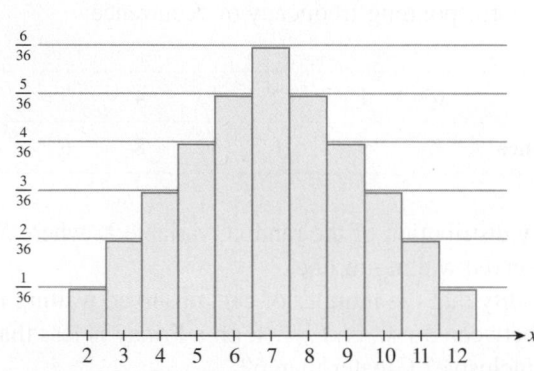

FIGURE **14**
Histogram showing the probability distribution for the sum of the uppermost faces of two dice

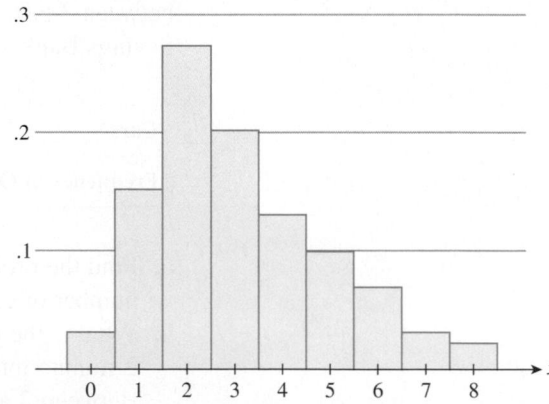

FIGURE **15**
Histogram showing the probability distribution for the number of cars waiting in line

Observe that in each histogram, the area of a rectangle associated with a value of a random variable X gives precisely the probability associated with the value of X. This follows because each such rectangle, by construction, has width 1 and height corresponding to the probability associated with the value of the random variable. Another consequence arising from the method of construction of a histogram is that *the probability associated with more than one value of the random variable X is given by the sum of the areas of the rectangles associated with those values of X.* For example, in the coin-tossing experiment of Example 1, the event of obtaining at least two heads, which corresponds to the event $(X = 2)$ or $(X = 3)$, is given by

$$P(X = 2) + P(X = 3)$$

and may be obtained from the histogram depicted in Figure 13 by adding the areas associated with the values 2 and 3 of the random variable X. We obtain

$$P(X = 2) + P(X = 3) = (1)\left(\frac{3}{8}\right) + (1)\left(\frac{1}{8}\right) = \frac{1}{2}$$

This result provides us with a method of computing the probabilities of events directly from the knowledge of a histogram of the probability distribution of the random variable associated with the experiment.

EXAMPLE 7 Suppose the probability distribution of a random variable X is represented by the histogram shown in Figure 16. Identify the part of the histogram whose area gives the probability $P(10 \le X \le 20)$.

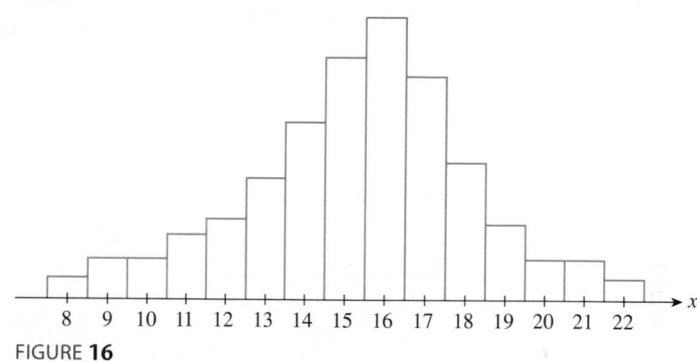

FIGURE **16**

Solution The event $(10 \le X \le 20)$ is the event consisting of outcomes related to the values $10, 11, 12, \ldots, 20$ of the random variable X. The probability of this event $P(10 \le X \le 20)$ is therefore given by the shaded area of the histogram in Figure 17.

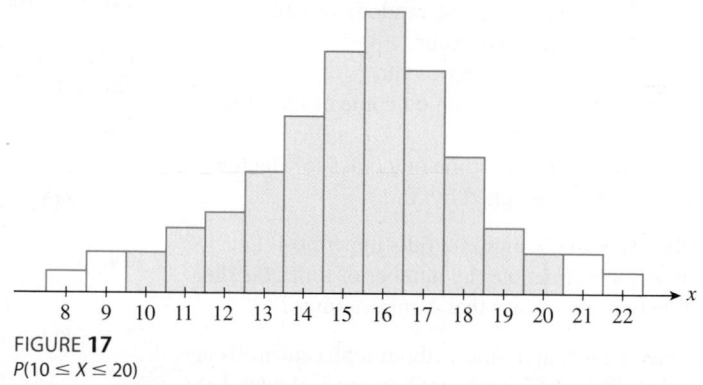

FIGURE **17**
$P(10 \le X \le 20)$

Pie charts are also widely used in business and media presentations to describe numerical data. Pie charts are often used to show the relationship of the parts to the whole. We illustrate how Excel can be used to create pie charts in Using Technology, page 556.

8.4 Self-Check Exercises

1. Three balls are selected at random without replacement from an urn containing four black balls and five white balls. Let the random variable X denote the number of black balls drawn.
   a. List the outcomes of the experiment.

b. Find the value assigned to each outcome of the experiment by the random variable X.

c. Find the event consisting of the outcomes to which a value of 2 has been assigned by X.

2. **LIBRARY USAGE** The following data, extracted from the records of Dover Public Library, give the number of books borrowed by the library's members over a 1-month period:

Books	0	1	2	3	4	5	6	7	8
Frequency of Occurrence	780	300	412	205	98	54	57	30	6

a. Find the probability distribution of the random variable X, where X denotes the number of books checked out over a 1-month period by a randomly chosen member.

b. Draw the histogram representing this probability distribution.

Solutions to Self-Check Exercises 8.4 can be found on page 553.

8.4 Concept Questions

1. What is a random variable? Give an example.

2. Give an example of (a) a finite discrete random variable, (b) an infinite discrete random variable, and (c) a continuous random variable.

3. Suppose you are given the probability distribution for a random variable X. Explain how you would construct a histogram for this probability distribution. What does the area of each rectangle in the histogram represent?

8.4 Exercises

1. Three balls are selected at random without replacement from an urn containing four green balls and six red balls. Let the random variable X denote the number of green balls drawn.
   a. List the outcomes of the experiment.
   b. Find the value assigned to each outcome of the experiment by the random variable X.
   c. Find the event consisting of the outcomes to which a value of 3 has been assigned by X.

2. A coin is tossed four times. Let the random variable X denote the number of tails that occur.
   a. List the outcomes of the experiment.
   b. Find the value assigned to each outcome of the experiment by the random variable X.
   c. Find the event consisting of the outcomes to which a value of 2 has been assigned by X.

3. A die is rolled repeatedly until a 6 falls uppermost. Let the random variable X denote the number of times the die is rolled. What are the values that X may assume?

4. Cards are selected one at a time without replacement from a well-shuffled deck of 52 cards until an ace is drawn. Let X denote the random variable that gives the number of cards drawn. What values may X assume?

5. Let X denote the random variable that gives the sum of the faces that fall uppermost when two fair dice are rolled. Find $P(X = 7)$.

6. Two cards are drawn from a well-shuffled deck of 52 playing cards. Let X denote the number of aces drawn. Find $P(X = 2)$.

In Exercises 7–12, give the range of values that the random variable X may assume and classify the random variable as finite discrete, infinite discrete, or continuous.

7. $X =$ The number of times a die is thrown until a 2 appears

8. $X =$ The number of defective iPads in a sample of eight iPads

9. $X =$ The distance in miles a commuter travels to work

10. $X =$ The number of hours a child watches television on a given day

11. $X =$ The number of times an accountant takes the CPA examination before passing

12. $X =$ The number of boys in a four-child family

In Exercises 13–16, determine whether the table gives the probability distribution of the random variable X. Explain your answer.

13.

x	-3	-2	-1	0	1	2
$P(X = x)$	0.2	0.4	0.3	-0.2	0.1	0.1

14.

x	-2	-1	0	1	2
$P(X = x)$	0.2	0.1	0.3	0.2	0.1

15.

x	1	2	3	4	5	6
$P(X = x)$	0.3	0.1	0.2	0.2	0.1	0.2

16.

x	-1	0	1	2	3
$P(X = x)$	0.3	0.1	0.2	0.2	0.2

In Exercises 17 and 18, find conditions on the numbers a and/or b such that the table gives the probability distribution of the random variable X.

17.

x	0	2	4	6	8
$P(X = x)$	0.1	0.4	a	0.1	0.2

18.

x	-1	0	1	2	4	5
$P(X = x)$	0.3	a	0.2	0.2	b	0.1

19. The probability distribution of the random variable X is shown in the following table:

x	-10	-5	0	5	10	15	20
$P(X = x)$	.20	.15	.05	.1	.25	.1	.15

Find:

a. $P(X = -10)$ **b.** $P(X \geq 5)$
c. $P(-5 \leq X \leq 5)$ **d.** $P(X \leq 20)$
e. $P(X < 5)$ **f.** $P(X = 3)$

20. The probability distribution of the random variable X is shown in the following table:

x	-5	-3	-2	0	2	3
$P(X = x)$	.17	.13	.33	.16	.11	.10

Find:

a. $P(X \leq 0)$ **b.** $P(X \leq -3)$
c. $P(-2 \leq X \leq 2)$ **d.** $P(X = -2)$
e. $P(X > 0)$ **f.** $P(X = 1)$

21. Suppose that the probability distribution of a random variable X is represented by the accompanying histogram. Shade the part of the histogram whose area gives the probability $P(17 \leq X \leq 20)$.

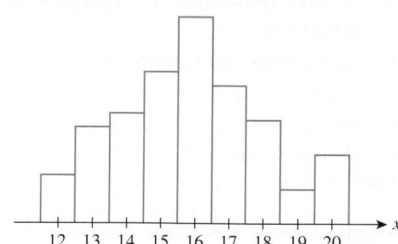

22. **Exams** An examination consisting of ten true-or-false questions was taken by a class of 100 students. The probability distribution of the random variable X, where X denotes the number of questions answered correctly by a randomly chosen student, is represented by the accompanying histogram. The rectangle with base centered on the number 8 is missing. What should be the height of this rectangle?

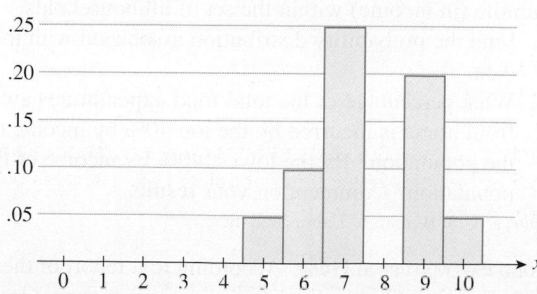

23. Two dice are rolled. Let the random variable X denote the number that falls uppermost on the first die, and let Y denote the number that falls uppermost on the second die.
a. Find the probability distributions of X and Y.
b. Find the probability distribution of $X + Y$.

24. **Emergency Fund Savings** In a survey of 1000 people, the following question was asked: How much emergency living expense savings do you have? Answers were expressed in the number of months the person could live on savings. The following results were obtained:

Group	1	2	3	4
Time (in months)	Less than 3	3–6	More than 6	No answer
Number	480	200	290	30

Let X denote the random variable that takes on the values 1, 2, 3, and 4, corresponding to the groups given in the table.
a. Find the probability distribution associated with these data.
b. If a respondent in the survey is chosen at random, what is the probability that he or she said that they had 6 or fewer months' worth of emergency living expense savings?
Source: Princeton Survey Research Associates.

25. **Money Market Rates** The interest rates paid by 30 financial institutions on a certain day for money market deposit accounts are shown in the accompanying table:

Rate (%)	2	2.25	2.55	2.56
Institutions	1	7	7	1

Rate (%)	2.58	2.60	2.65	2.85
Institutions	1	8	3	2

Let the random variable X denote the interest rate per year paid by a randomly chosen financial institution on its money market deposit accounts.
a. Find the probability distribution associated with these data.
b. Find the probability that the interest rate paid by a financial institution chosen at random is less than 2.56% per year.

26. **Distribution of Families by Size** The Public Housing Authority in a certain community conducted a survey of 1000 families to determine the distribution of families by size. The results follow:

Family Size	2	3	4	5	6	7	8
Frequency of Occurrence	350	200	245	125	66	10	4

a. Find the probability distribution of the random variable X, where X denotes the number of people in a randomly chosen family.
b. Draw the histogram corresponding to the probability distribution found in part (a).
c. Find the probability that a family chosen at random from those surveyed has more than five members.

27. WAITING LINES The accompanying data were obtained in a study conducted by the manager of SavMore Supermarket. In this study, the number of customers waiting in line at the express checkout at the beginning of each 3-min interval between 9 A.M. and 12 noon on Saturday was observed.

Customers	0	1	2	3	4
Frequency of Occurrence	1	4	2	7	14

Customers	5	6	7	8	9	10
Frequency of Occurrence	8	10	6	3	4	1

a. Find the probability distribution of the random variable X, where X denotes the number of customers observed waiting in line.
b. Draw the histogram representing the probability distribution.
c. Find the probability that the number of customers waiting in line in any 3-min interval between 9 A.M. and 12 noon is between 1 and 3, inclusive.

28. TELEVISION PILOTS After the private screening of a new television pilot, audience members were asked to rate the new show on a scale of 1 to 10 (10 being the highest rating). From a group of 140 people, the following responses were obtained:

Rating	1	2	3	4	5	6	7	8	9	10
Frequency of Occurrence	1	4	3	11	23	21	28	29	16	4

Let the random variable X denote the rating given to the show by a randomly chosen audience member.
a. Find the probability distribution associated with these data.
b. What is the probability that the new television pilot got a rating that is higher than 5?

29. SMARTPHONE OWNERSHIP BY AGE The following table gives the number of adults in the United States within each of four age groups who own smartphones.

Age (in years)	18–29	30–49	50–64	65+
Number of Owners (in thousands)	25,115	54,157	16,310	14,509

Let X denote the random variable that takes on the values 1, 2, 3, and 4, corresponding to the age groups 18–29, 30–49, 50–64, and 65+, respectively.
a. Find the probability distribution associated with these data.
b. If a smartphone owner in one of these age groups is chosen at random, what is the probability that he or she is between 18 and 49 years of age?
Source: Pew Research.

30. SMARTPHONE OWNERSHIP BY INCOME The following table gives the number of U.S. households within each of four income groups in which someone owns a smartphone.

Income ($)	<30,000	30,000–49,999	50,000–74,999	75,000+
Number of Owners (in thousands)	11,817	8,475	7,609	32,057

Let X denote the random variable that takes on the values 1, 2, 3, and 4, corresponding to the income groups <\$30,000, \$30,000–\$49,999, \$50,000–\$74,999, and \$75,000+, respectively.
a. Find the probability distribution associated with these data.
b. If a smartphone owner in one of these households is chosen at random, what is the probability that his or her household income is between \$30,000 and \$74,999?
Source: Pew Research.

31. ACADEMY MEMBERS The following table gives the number of members in different age groups of the Academy of Motion Picture Arts and Sciences (AMPAS). All academy members are allowed to vote for a winner in most categories, including Best Picture.

Group	1	2	3	4	5
Age (in years)	Under 40	40–49	50–59	60 and over	Unknown
Number	115	634	1441	3113	462

Let X be the random variable that takes on the values 1, 2, 3, 4, and 5, corresponding to the age groups in the AMPAS membership.
a. Find the probability distribution associated with these data.
b. What is the percentage of AMPAS members who are younger than 50 years old?
Source: LATIMES.com.

32. FOOD EXPENDITURE AWAY FROM HOME According to a report of the U.S. Bureau of Labor Statistics, the average annual food expenditures away from home by quintiles of household income before taxes is as follows:

Income Quintile	Lowest 20%	Second 20%	Third 20%	Fourth 20%	Highest 20%
Expenditures ($)	1038	1569	2127	3206	5151

Let the random variable X denote a randomly chosen quintile (in income) within the set of all households.
a. Find the probability distribution associated with these data.
b. What percentage of the total food expenditures away from home is incurred by the top 40% by income of the population? By the lowest 40% by income of the population? Comment on your results.
Source: U.S. Bureau of Labor Statistics.

33. FOOD EXPENDITURE AT HOME According to a report of the U.S. Bureau of Labor Statistics, the average annual food expenditures at home by quintiles of household income before taxes is as follows:

Income Quintile	Lowest 20%	Second 20%	Third 20%	Fourth 20%	Highest 20%
Expenditures ($)	2463	2999	3355	4316	5629

Let the random variable X denote a randomly chosen quintile (in income) within the set of all households.
a. Find the probability distribution associated with these data.
a. What percentage of the total food expenditures is incurred by the top 40% by income of the population? By the lowest 40% by income of the population?
Source: U.S. Bureau of Labor Statistics.

34. U.S. POPULATION BY AGE The following table gives the 2011 age distribution of the U.S. population:

Group	1	2	3	4	5	6
Age (in years)	Under 5	5–19	20–24	25–44	45–64	65 and over
Number (in thousands)	21,265	61,776	21,525	81,426	80,938	39,178

Let x denote a random variable that takes on the values 1, 2, 3, 4, 5, and 6 corresponding to the age groups given in the table.
a. Find the probability distribution associated with these data.
b. What percentage of the U.S. population is between 5 and 24 years old, inclusive?
Source: U.S. Census Bureau.

In Exercises 35 and 36, determine whether the statement is true or false. If it is true, explain why it is true. If it is false, give an example to show why it is false.

35. Suppose X is a finite discrete random variable assuming the values $x_1, x_2, \ldots, x_n$ and associated probabilities p_1, $p_2, \ldots, p_n$. Then $p_1 + p_2 + \cdots + p_n = 1$.

36. The area of a histogram associated with a probability distribution is a number between 0 and 1.

8.4 Solutions to Self-Check Exercises

1. a. Using the accompanying tree diagram, we see that the outcomes of the experiment are

$$S = \{BBB, BBW, BWB, BWW,$$
$$WBB, WBW, WWB, WWW\}$$

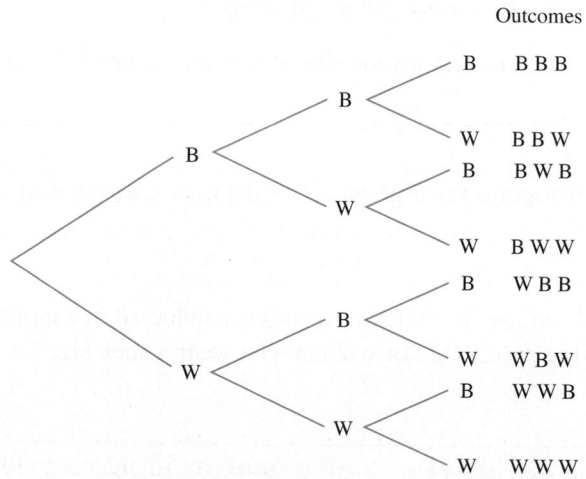

b. Using the results of part (a), we obtain the values assigned to the outcomes of the experiment as follows:

Outcome	BBB	BBW	BWB	BWW
Value	3	2	2	1

Outcome	WBB	WBW	WWB	WWW
Value	2	1	1	0

c. The required event is $\{BBW, BWB, WBB\}$.

2. a. We divide each number in the bottom row of the given table by 1942 (the sum of these numbers) to obtain the probabilities associated with the random variable X when X takes on the values 0, 1, 2, 3, 4, 5, 6, 7, and 8. For example,

$$P(X = 0) = \frac{780}{1942} \approx .402$$

$$P(X = 1) = \frac{300}{1942} \approx .154$$

The required probability distribution and histogram follow:

x	0	1	2	3	4
$P(X = x)$	.402	.154	.212	.106	.050

x	5	6	7	8
$P(X = x)$	.028	.029	.015	.003

b.

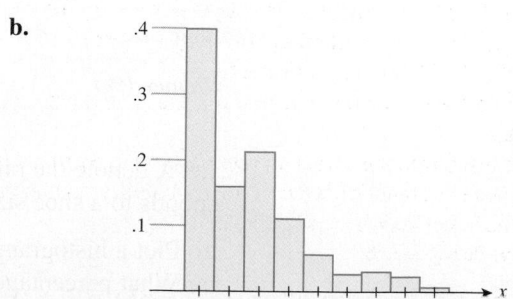

USING TECHNOLOGY

Graphing a Histogram

Graphing Utility

A graphing utility can be used to plot the histogram for a given set of data, as illustrated in the following example.

 APPLIED EXAMPLE 1 A survey of 90,000 households conducted in a certain year revealed the following percentage of women who wear a shoe size within the given ranges.

Shoe Size	<5	$5–5\frac{1}{2}$	$6–6\frac{1}{2}$	$7–7\frac{1}{2}$	$8–8\frac{1}{2}$	$9–9\frac{1}{2}$	$10–10\frac{1}{2}$	$>10\frac{1}{2}$
Women (%)	1	5	15	27	29	14	7	2

Let X denote the random variable taking on the values 1 through 8, where 1 corresponds to a shoe size less than 5, 2 corresponds to a shoe size of $5–5\frac{1}{2}$, and so on.

a. Plot a histogram for the given data.
b. What percentage of women in the survey wear a shoe size within the ranges $7–7\frac{1}{2}$ or $8–8\frac{1}{2}$?
Source: Footwear Market Insights survey.

Solution

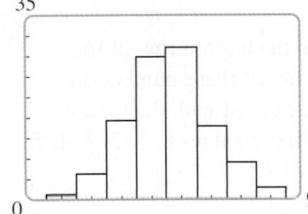

0
FIGURE **T1**
The histogram for the given data, using the viewing window [0, 9] × [0, 35]

a. Enter the values of X as $x_1 = 1, x_2 = 2, \ldots, x_8 = 8$ and the corresponding values of Y as $y_1 = 1, y_2 = 5, \ldots, y_8 = 2$. Then using the **DRAW** function from the Statistics menu, we draw the histogram shown in Figure T1.
b. The probability that a woman participating in the survey wears a shoe size within the ranges $7–7\frac{1}{2}$ or $8–8\frac{1}{2}$ is given by

$$P(X = 4) + P(X = 5) = .27 + .29 = .56$$

This tells us that 56% of the women wear a shoe size within the ranges $7–7\frac{1}{2}$ or $8–8\frac{1}{2}$.

Excel

Excel can be used to plot the histogram for a given set of data, as illustrated in the following example.

 APPLIED EXAMPLE 2 A survey of 90,000 households conducted in a certain year revealed the following percentage of women who wear a shoe size within the given ranges.

Shoe Size	<5	$5–5\frac{1}{2}$	$6–6\frac{1}{2}$	$7–7\frac{1}{2}$	$8–8\frac{1}{2}$	$9–9\frac{1}{2}$	$10–10\frac{1}{2}$	$>10\frac{1}{2}$
Women (%)	1	5	15	27	29	14	7	2

Let X denote the random variable taking on the values 1 through 8, where 1 corresponds to a shoe size less than 5, 2 corresponds to a shoe size of $5–5\frac{1}{2}$, and so on.

a. Plot a histogram for the given data.
b. What percentage of women in the survey wear a shoe size within the ranges $7–7\frac{1}{2}$ or $8–8\frac{1}{2}$?
Source: Footwear Market Insights survey.

Solution

a. Enter the given data in columns A and B onto a spreadsheet, as shown in Figure T2. Highlight the data in column B, and select $\boxed{\Sigma}$ from the Editing group under the Home tab. The sum of the numbers in this column (100) will appear in cell B10. In cell C2, type =B2/100, and then press $\boxed{\textbf{Enter}}$. To extend the formula to cell C9, select C2 and move the pointer to the small black box at the lower right corner of that cell. Drag the black $+$ that appears (at the lower right corner of cell C2) through cell C9, and then release it. The probability distribution shown in cells C2 to C9 will then appear on your spreadsheet. Then highlight the data in the Probability column, and select $\boxed{\textbf{Column}}$ from the Charts group under the Insert tab. Click on the first chart type (Clustered Column). Under the Layout tab, click on $\boxed{\textbf{Chart Title}}$ in the Labels group, and select $\boxed{\textbf{Above Chart}}$. Enter Histogram as the title. Select $\boxed{\textbf{Primary Horizontal Axis Title}}$ under Axis Titles from the Labels group, select $\boxed{\textbf{Title Below Axis}}$, and then enter X as the horizontal axis title. Next, select $\boxed{\textbf{Primary Vertical Axis Title}}$ under Axis Titles followed by $\boxed{\textbf{Rotated Title}}$ and enter Probability. Right-click a bar on the chart and select $\boxed{\textbf{Format Data Series}}$. Adjust the slider under Gap Width to 0% in the dialog box that appears, and then click $\boxed{\textbf{Close}}$. Finally, delete the Series 1 legend entry.

	A	B	C
1	X	Frequency	Probability
2	1	1	0.01
3	2	5	0.05
4	3	15	0.15
5	4	27	0.27
6	5	29	0.29
7	6	14	0.14
8	7	7	0.07
9	8	2	0.02
10		100	

FIGURE **T2**
Completed spreadsheet for Example 2

The histogram shown in Figure T3 will appear.

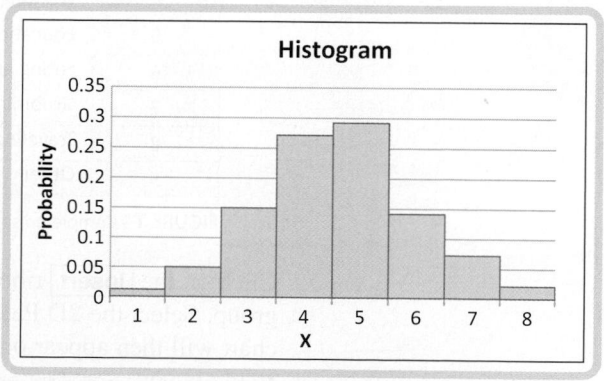

FIGURE **T3**
The histogram for the random variable *X*

Note: Boldfaced words/characters enclosed in a box (for example, $\boxed{\textbf{Enter}}$) indicate an action (click, select, or press) is required. Words/characters printed blue (for example, Chart sub-type:) indicate words/characters that appear on the screen. Words/characters printed in a monospace font (for example, =(-2/3)*A2+2) indicates words/characters that need to be typed and entered).

b. The probability that a woman participating in the survey wears a shoe size within the ranges $7-7\frac{1}{2}$ or $8-8\frac{1}{2}$ is given by

$$P(X = 4) + P(X = 5) = .27 + .29 = .56$$

This tells us that 56% of the women wear a shoe size within the ranges $7-7\frac{1}{2}$ or $8-8\frac{1}{2}$.

Excel can also be used to create pie charts, as illustrated in the following example.

 APPLIED EXAMPLE 3 Time Use of College Students Use the data given in Table T1 to construct a pie chart.

TABLE T1

Time Used on an Average Weekday for Full-Time University and College Students

Time Use	Time (in hours)
Sleeping	8.5
Leisure and sports	3.7
Working and related activities	2.9
Educational activities	3.3
Eating and drinking	1.0
Grooming	0.7
Traveling	1.5
Other	2.4

Source: Bureau of Labor Statistics.

Solution

We begin by entering the information from Table T1 in Columns A and B on a spreadsheet. Then follow these steps:

Step 1   First, highlight the data in cells A2:A9 and B2:B9 as shown in Figure T4.

	A	B
1	Time Use	Time (in hours)
2	Sleeping	8.5
3	Leisure and sports	3.7
4	Working and related activities	2.9
5	Educational activities	3.3
6	Eating and drinking	1
7	Grooming	0.7
8	Traveling	1.5
9	Other	2.4

FIGURE **T4** Completed spreadsheet

Step 2   Click on the Insert ribbon tab, and then select Pie from the Charts group. Select the 2D Pie chart subtype in the first row and first column. A chart will then appear on your worksheet.

Step 3   From the Chart Tools group that now appears on the ribbon, click the Design ribbon tab, and then select the chart appearing in the first row and first column of the Charts Layouts group. Note that this chart displays the percentage of a day (24 hours) spent on each activity. Next, select the chart appearing in row 2 and column 2 of the Chart Styles group.

Step 4 Click the [Layout] tab, and select [Chart Title] from the Labels group followed by [Above Chart]. Type the title of the chart, and click [Enter].

The pie chart shown in Figure T5 will appear.

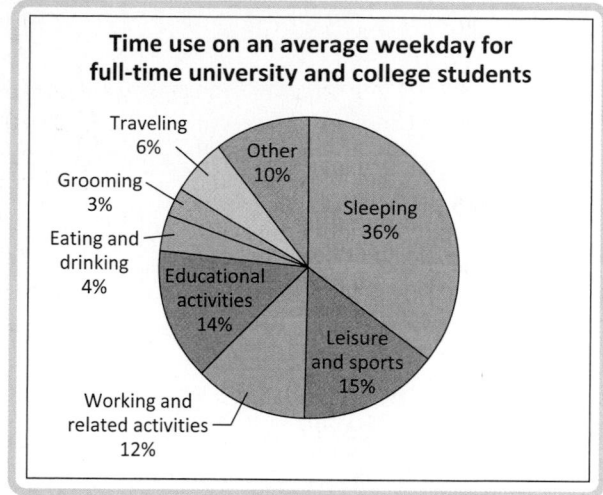

Time use on an average weekday for full-time university and college students

- Traveling 6%
- Grooming 3%
- Eating and drinking 4%
- Educational activities 14%
- Working and related activities 12%
- Other 10%
- Sleeping 36%
- Leisure and sports 15%

FIGURE **T5**
The pie chart describing the data in Table T1

TECHNOLOGY EXERCISES

1. Graph the histogram associated with the data given in Table 3, page 545. Compare your graph with that given in Figure 13, page 548.

2. Graph the histogram associated with the data given in Exercise 26, page 551.

3. Graph the histogram associated with the data given in Exercise 27, page 552.

4. Graph the histogram associated with the data given in Exercise 28, page 552.

In Exercises 5–10, use the data given in the table to construct a pie chart. (These exercises are for EXCEL only.)

5. WHO PAYS TAXES? Payroll taxes are used for Social Security and Medicare, two of the biggest items in the federal budget. The following facts regarding the people in the United States who pay taxes were reported in *Money* magazine in March 2012:

Taxpaying and Non-Taxpaying Groups	Percent
Pay income tax	53.6
Pay no income tax but do pay payroll taxes	28.3
Pay no income tax but are elderly	10.3
Pay no income tax but earn less than $20,000	6.9
Others	0.9

Source: Urban Brookings Tax Policy Center.

6. TIME SPENT PER WEEK ON THE INTERNET BY COLLEGE STUDENTS
The following table gives the time spent on the Internet by 18- to 24-year-old college students:

Time Spent	College Students (%)
20 or more hours	19.6
10 or more but fewer than 20 hours	13.4
5 or more but fewer than 10 hours	20.3
3 or more but fewer than 5 hours	22.6
Fewer than 3 hours	24.1

Source: Burst Research.

7. MAIN REASONS WHY YOUNG ADULTS SHOP ONLINE The results of an online survey by Bing among 1077 adults aged 18–34 years in November 2012 regarding the main reasons why they shopped online are summarized in the following table:

Main Reason	Percent
Better prices	37
Avoiding holiday crowds, hassles	29
Convenience	18
Better selection	13
Ships directly	3

Source: Impulse Research.

8. BRINGING SOMETHING TO A PARTY In a survey of 2008 adults conducted by American Express, the following question was asked: When invited to a party, do you contribute something even if not asked? The responses given are shown in the following table:

Response	Percent
Always	46
Sometimes	40
Rarely	6
Never	4
Not Sure	4

Source: American Express.

9. **How Money Is Spent in the U.S. for Health Care** In the United States, $2.5 trillion was spent on health care in 2009. The following table shows how this money was spent:

Where Spent	Percent
Hospitals	31
Doctors, other professionals	27
Prescription drugs (retail)	10
Nursing home care	5
Other private revenues	27

Source: "Covering Health Issues: A Sourcebook for Journalists," 6E.

10. **Percentage of Mobile Ad Revenues by Device Type** The following table gives the percentage of mobile ad revenues in the first quarter of 2013 by device type:

Device Type	Percent
iPhone	34.2
Android phone	26.2
Other	15.4
iPad	12.6
RIM	6.2
Symbian	2.5
iPod Touch	2.4
Android tablet	0.5

Source: Opera Mediaworks.

8.5 Expected Value

Mean

The average value of a set of numbers is a familiar notion to most people. For example, to compute the average of the four numbers

$$12, 16, 23, 37$$

we simply add these numbers and divide the resulting sum by 4, giving the required average as

$$\frac{12 + 16 + 23 + 37}{4} = \frac{88}{4} = 22$$

In general, we have the following definition:

Average, or Mean

The **average,** or **mean,** of the n numbers

$$x_1, x_2, \ldots, x_n$$

is $\bar{x}$ (read "x bar"), where

$$\bar{x} = \frac{x_1 + x_2 + \cdots + x_n}{n}$$

$ APPLIED EXAMPLE 1 Waiting Lines Refer to Example 6, Section 8.4. Find the average number of cars waiting in line at the bank's drive-in teller at the beginning of each 2-minute interval during the period in question.

TABLE 9	
Cars	Frequency of Occurrence
0	2
1	9
2	16
3	12
4	8
5	6
6	4
7	2
8	1

Solution The number of cars and the corresponding frequency of occurrence are reproduced in Table 9. Observe that the number 0 (of cars) occurs twice, the number 1 occurs 9 times, and so on. There are altogether

$$2 + 9 + 16 + 12 + 8 + 6 + 4 + 2 + 1 = 60$$

numbers to be averaged. Therefore, the required average is given by

$$\frac{(0 \cdot 2) + (1 \cdot 9) + (2 \cdot 16) + (3 \cdot 12) + (4 \cdot 8) + (5 \cdot 6) + (6 \cdot 4) + (7 \cdot 2) + (8 \cdot 1)}{60} \approx 3.1 \quad \textbf{(10)}$$

or approximately 3.1 cars.

Expected Value

Let's reconsider the expression on the left-hand side of Equation (10), which gives the average of the frequency distribution shown in Table 9. Dividing each term by the denominator, we may rewrite the expression in the form

$$0 \cdot \left(\frac{2}{60}\right) + 1 \cdot \left(\frac{9}{60}\right) + 2 \cdot \left(\frac{16}{60}\right) + 3 \cdot \left(\frac{12}{60}\right) + 4 \cdot \left(\frac{8}{60}\right) + 5 \cdot \left(\frac{6}{60}\right)$$
$$+ 6 \cdot \left(\frac{4}{60}\right) + 7 \cdot \left(\frac{2}{60}\right) + 8 \cdot \left(\frac{1}{60}\right)$$

Observe that each term in the sum is a product of two factors; the first factor is the value assumed by the random variable X, where X denotes the number of cars waiting in line, and the second factor is just the probability associated with that value of the random variable. This observation suggests the following general method for calculating the expected value (that is, the average or mean) of a random variable X that assumes a finite number of values from the knowledge of its probability distribution.

> **Expected Value of a Random Variable X**
>
> Let X denote a random variable that assumes the values $x_1, x_2, \ldots, x_n$ with associated probabilities $p_1, p_2, \ldots, p_n$, respectively. Then the **expected value** of X, denoted by $E(X)$, is given by
>
> $$E(X) = x_1 p_1 + x_2 p_2 + \cdots + x_n p_n \quad \textbf{(11)}$$

Note The numbers $x_1, x_2, \ldots, x_n$ may be positive, zero, or negative. For example, such a number might be positive if it represents a profit and negative if it represents a loss.

TABLE 10	
Probability Distribution	
x	$P(X = x)$
0	.03
1	.15
2	.27
3	.20
4	.13
5	.10
6	.07
7	.03
8	.02

APPLIED EXAMPLE 2 Waiting Lines Re-solve Example 1 by using the probability distribution associated with the experiment, which is reproduced in Table 10.

Solution Let X denote the number of cars waiting in line. Then the average number of cars waiting in line is given by the expected value of X—that is, by

$$E(X) = (0)(.03) + (1)(.15) + (2)(.27) + (3)(.20) + (4)(.13)$$
$$+ (5)(.10) + (6)(.07) + (7)(.03) + (8)(.02)$$
$$= 3.1 \text{ cars}$$

which agrees with the earlier result.

The expected value of a random variable X is a measure of the central tendency of the probability distribution associated with X. In repeated trials of an experiment with random variable X, the average of the observed values of X gets closer and closer to the expected value of X as the number of trials gets larger and larger. Geometrically, the expected value of a random variable X has the following simple interpretation: If a laminate is made of the histogram of a probability distribution associated with a random variable X, then the expected value of X corresponds to the point on the base of the laminate at which the laminate will balance perfectly when the point is directly over a fulcrum (Figure 18).

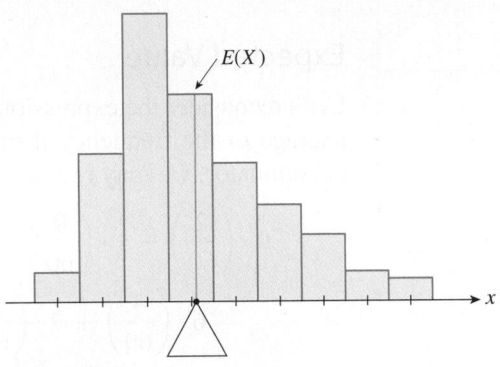

FIGURE **18**
Expected value of a random variable X

EXAMPLE 3 Let X denote the random variable that gives the sum of the faces that fall uppermost when two fair dice are rolled. Find the expected value, $E(X)$, of X.

Solution The probability distribution of X, reproduced in Table 11, was found in Example 5, Section 8.4. Using this result, we find

$$E(X) = 2\left(\frac{1}{36}\right) + 3\left(\frac{2}{36}\right) + 4\left(\frac{3}{36}\right) + 5\left(\frac{4}{36}\right) + 6\left(\frac{5}{36}\right) + 7\left(\frac{6}{36}\right)$$
$$+ 8\left(\frac{5}{36}\right) + 9\left(\frac{4}{36}\right) + 10\left(\frac{3}{36}\right) + 11\left(\frac{2}{36}\right) + 12\left(\frac{1}{36}\right)$$
$$= 7$$

Note that because of the symmetry of the histogram of the probability distribution with respect to the vertical line $x = 7$, the result could have been obtained by merely inspecting Figure 19.

TABLE 11	
Probability Distribution	
x	$P(X = x)$
2	$\frac{1}{36}$
3	$\frac{2}{36}$
4	$\frac{3}{36}$
5	$\frac{4}{36}$
6	$\frac{5}{36}$
7	$\frac{6}{36}$
8	$\frac{5}{36}$
9	$\frac{4}{36}$
10	$\frac{3}{36}$
11	$\frac{2}{36}$
12	$\frac{1}{36}$

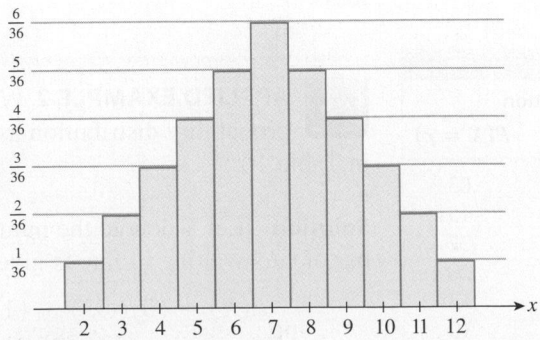

FIGURE **19**
Histogram showing the probability distribution for the sum of the uppermost faces of two dice

The next example shows how we can use the concept of expected value to help us make the best investment decision.

APPLIED EXAMPLE 4 Expected Profit A private equity group intends to purchase one of two motels currently being offered for sale in a certain city. The terms of sale of the two motels are similar, although the Regina Inn has 52 rooms and is in a slightly better location than the Merlin Motor Lodge, which has 60 rooms. Records obtained for each motel reveal that the occupancy rates, with corresponding probabilities, during the May–September tourist season are as shown in the following tables:

Regina Inn

Occupancy Rate	.80	.85	.90	.95	1.00
Probability	.19	.22	.31	.23	.05

Merlin Motor Lodge

Occupancy Rate	.75	.80	.85	.90	.95	1.00
Probability	.35	.21	.18	.15	.09	.02

The average profit per day for each occupied room at the Regina Inn is $40, whereas the average profit per day for each occupied room at the Merlin Motor Lodge is $36.

a. Find the average number of rooms occupied per day at each motel.
b. If the investors' objective is to purchase the motel that generates the higher daily profit, which motel should they purchase? (Compare the expected daily profit of the two motels.)

Solution

a. Let X denote the occupancy rate at the Regina Inn. Then the average daily occupancy rate at the Regina Inn is given by the expected value of X—that is, by

$$E(X) = (.80)(.19) + (.85)(.22) + (.90)(.31)$$
$$+ (.95)(.23) + (1.00)(.05)$$
$$= .8865$$

The average number of rooms occupied per day at the Regina Inn is

$$(.8865)(52) \approx 46.1$$

or approximately 46.1 rooms. Similarly, letting Y denote the occupancy rate at the Merlin Motor Lodge, we have

$$E(Y) = (.75)(.35) + (.80)(.21) + (.85)(.18) + (.90)(.15)$$
$$+ (.95)(.09) + (1.00)(.02)$$
$$= .8240$$

The average number of rooms occupied per day at the Merlin Motor Lodge is

$$(.8240)(60) \approx 49.4$$

or approximately 49.4 rooms.

b. The expected daily profit at the Regina Inn is given by

$$(46.1)(40) = 1844$$

or $1844. The expected daily profit at the Merlin Motor Lodge is given by

$$(49.4)(36) \approx 1778$$

or approximately $1778. From these results, we conclude that the private equity group should purchase the Regina Inn, which is expected to yield a higher daily profit.

APPLIED EXAMPLE 5 Raffles The Island Club is holding a fund-raising raffle. Ten thousand tickets have been sold for $2 each. There will be a first prize of $3000, 3 second prizes of $1000 each, 5 third prizes of $500 each, and 20 consolation prizes of $100 each. Letting X denote the net winnings (that is, winnings less the cost of the ticket) associated with a ticket, find $E(X)$. Interpret your results.

TABLE 12

Probability Distribution for a Raffle

x	$P(X = x)$
−2	.9971
98	.0020
498	.0005
998	.0003
2998	.0001

Solution The values assumed by X are $(0 − 2)$, $(100 − 2)$, $(500 − 2)$, $(1000 − 2)$, and $(3000 − 2)$—that is, $−2, 98, 498, 998,$ and 2998—which correspond, respectively, to the value of a losing ticket, a consolation prize, a third prize, and so on. The probability distribution of X may be calculated in the usual manner and appears in Table 12. Using the table, we find

$$E(X) = (−2)(.9971) + 98(.0020) + 498(.0005)$$
$$+ 998(.0003) + 2998(.0001)$$
$$= −0.95$$

This expected value gives the long-run average loss (negative gain) of a holder of one ticket; that is, if one participated in such a raffle by purchasing one ticket each time, in the long run, one may expect to lose, on the average, 95 cents per raffle.

APPLIED EXAMPLE 6 Roulette In the game of roulette as played in Las Vegas casinos, the wheel is divided into 38 compartments numbered 1 through 36, 0, and 00. One-half of the numbers 1 through 36 are red, the other half are black, and 0 and 00 are green (Figure 20). Of the many types of bets that may be placed, one type involves betting on the outcome of the color of the winning number. For example, one may place a certain sum of money on *red*. If the winning number is red, one wins an amount equal to the bet placed and the amount of the bet is returned; otherwise, one loses the amount of the bet. Find the expected value of the winnings on a $1 bet placed on *red*.

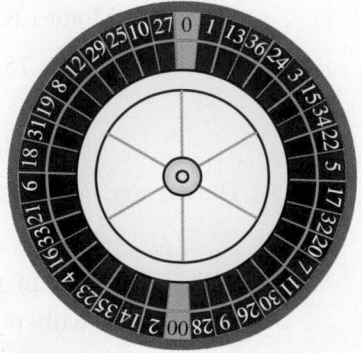

FIGURE 20
Roulette wheel

Solution   Let X be a random variable whose values are 1 and -1, which correspond to a win and a loss, respectively. The probabilities associated with the values 1 and -1 are $\frac{18}{38}$ and $\frac{20}{38}$, respectively. Therefore, the expected value is given by

$$E(X) = 1\left(\frac{18}{38}\right) + (-1)\left(\frac{20}{38}\right) = -\frac{2}{38}$$
$$\approx -0.053$$

Thus, if one places a \$1 bet on *red* over and over again, one may expect to lose, on the average, approximately 5 cents per bet in the long run.

Examples 5 and 6 illustrate games that are not "fair." Of course, most participants in such games are aware of this fact and participate in them for other reasons. In a fair game, neither party has an advantage, a condition that translates into the condition that $E(X) = 0$, where X takes on the values of a player's winnings.

APPLIED EXAMPLE 7 Fair Games Mike and Bill play a card game with a standard deck of 52 cards. Mike selects a card from a well-shuffled deck and receives A dollars from Bill if the card selected is a diamond; otherwise, Mike pays Bill a dollar. Determine the value of A if the game is to be fair.

Solution   Let X denote a random variable whose values are associated with Mike's winnings. Then X takes on the value A with probability $P(X = A) = \frac{1}{4}$ (since there are 13 diamonds in the deck) if Mike wins and takes on the value -1 with probability $P(X = -1) = \frac{3}{4}$ if Mike loses. Since the game is to be a fair one, the expected value $E(X)$ of Mike's winnings must be equal to zero; that is,

$$E(X) = A\left(\frac{1}{4}\right) + (-1)\left(\frac{3}{4}\right) = 0$$

Solving this equation for A gives $A = 3$. Thus, the card game will be fair if Bill makes a \$3 payoff when a diamond is drawn.

Expected Value for Grouped Data

The following example illustrates the technique for finding the mean for grouped data.

APPLIED EXAMPLE 8 Commuting Times The following table gives the travel time to work of workers aged 16 years and over who do not work at home:

Travel Time, x (in minutes)	$x < 10$	$10 \le x < 15$	$15 \le x < 20$	$20 \le x < 25$	$25 \le x < 30$	$30 \le x < 35$	$35 \le x < 45$	$45 \le x < 60$	$x \ge 60$
Number (in thousands)	17,725	18,915	20,503	19,577	8,069	18,122	8,466	9,921	10,979

Source: U.S. Census Bureau.

Estimate the average time it takes for a worker who does not work at home to travel to work. (Assume that "$x < 10$" means "$0 < x < 10$" and "$x \ge 60$" means "$60 \le x \le 90$.")

Solution Let X denote the random variable that gives the time a worker takes to travel to work. Next, we find the *midpoint* of each of the nine class intervals. Thus,

$$x_1 = \frac{0 + 10}{2} = 5 \qquad x_2 = \frac{10 + 15}{2} = \frac{25}{2} \qquad x_3 = \frac{15 + 20}{2} = \frac{35}{2}$$

$$x_4 = \frac{20 + 25}{2} = \frac{45}{2} \qquad x_5 = \frac{25 + 30}{2} = \frac{55}{2} \qquad x_6 = \frac{30 + 35}{2} = \frac{65}{2}$$

$$x_7 = \frac{35 + 45}{2} = \frac{80}{2} \qquad x_8 = \frac{45 + 60}{2} = \frac{105}{2} \qquad x_9 = \frac{60 + 90}{2} = \frac{150}{2}$$

Then we use these values of $x_1, x_2, \ldots, x_9$ in Equation (11), obtaining

$$\begin{aligned}
E(X) &= x_1 p_1 + x_2 p_2 + \cdots + x_9 p_9 \\
&= \left(\frac{10}{2}\right)\left(\frac{17{,}725}{132{,}277}\right) + \left(\frac{25}{2}\right)\left(\frac{18{,}915}{132{,}277}\right) + \left(\frac{35}{2}\right)\left(\frac{20{,}503}{132{,}277}\right) \\
&\quad + \left(\frac{45}{2}\right)\left(\frac{19{,}577}{132{,}277}\right) + \left(\frac{55}{2}\right)\left(\frac{8{,}069}{132{,}277}\right) + \left(\frac{65}{2}\right)\left(\frac{18{,}122}{132{,}277}\right) \\
&\quad + \left(\frac{80}{2}\right)\left(\frac{8{,}466}{132{,}277}\right) + \left(\frac{105}{2}\right)\left(\frac{9{,}921}{132{,}277}\right) + \left(\frac{150}{2}\right)\left(\frac{10{,}979}{132{,}277}\right) \\
&\approx 27.35
\end{aligned}$$

Thus, the average time it takes for a worker who does not work at home to travel to work is 27.35 minutes.

Odds

In everyday parlance, the probability of the occurrence of an event is often stated in terms of the *odds in favor of* (or *odds against*) the occurrence of the event. For example, one often hears statements such as "The odds that the Dodgers will win the World Series this season are 7 to 5" and "The odds that it will not rain tomorrow are 3 to 2." We will return to these examples later. But first, let us look at a definition that ties together these two concepts.

> **Odds In Favor Of and Odds Against**
>
> If $P(E)$ is the probability of an event E occurring, then
>
> **1.** The odds in favor of E occurring are
>
> $$\frac{P(E)}{1 - P(E)} = \frac{P(E)}{P(E^c)} \qquad P(E) \neq 1 \qquad \text{(12a)}$$
>
> **2.** The odds against E occurring are
>
> $$\frac{1 - P(E)}{P(E)} = \frac{P(E^c)}{P(E)} \qquad P(E) \neq 0 \qquad \text{(12b)}$$

Notes

1. The odds in favor of the occurrence of an event are given by the ratio of the probability of the event occurring to the probability of the event not occurring. The odds against the occurrence of an event are given by the reciprocal of the odds in favor of the occurrence of the event.

2. Whenever possible, odds are expressed as ratios of whole numbers. If the odds in favor of E are a/b, we say that the odds in favor of E are a to b. If the odds against E occurring are b/a, we say that the odds against E are b to a.

APPLIED EXAMPLE 9 Roulette Find the odds in favor of winning a bet on red in American roulette. What are the odds against winning a bet on *red*?

Solution   The probability of winning a bet here—the probability that the ball lands in a red compartment—is given by $P = \frac{18}{38}$. Therefore, using Equation (12a), we see that the odds in favor of winning a bet on *red* are

$$\frac{P(E)}{1 - P(E)} = \frac{\frac{18}{38}}{1 - \frac{18}{38}} \qquad E, \text{ event of winning a bet on } red$$

$$= \frac{\frac{18}{38}}{\frac{38 - 18}{38}}$$

$$= \frac{18}{38} \cdot \frac{38}{20}$$

$$= \frac{18}{20} = \frac{9}{10}$$

or 9 to 10. Next, using Equation (12b), we see that the odds against winning a bet on *red* are $\frac{10}{9}$, or 10 to 9.

Now suppose that the odds in favor of the occurrence of an event are a to b. Then Equation (12a) gives

$$\frac{a}{b} = \frac{P(E)}{1 - P(E)}$$

$$a[1 - P(E)] = bP(E) \qquad \text{Cross-multiply.}$$

$$a - aP(E) = bP(E)$$

$$a = (a + b)P(E)$$

$$P(E) = \frac{a}{a + b}$$

which leads us to the following result:

> **Probability of an Event (Given the Odds)**
>
> If the odds in favor of an event E occurring are a to b, then the probability of E occurring is
>
> $$P(E) = \frac{a}{a + b} \qquad (13)$$

Equation (13) is often used to determine subjective probabilities, as the next example shows.

EXAMPLE 10 Consider each of the following statements.

a. "The odds that the Dodgers will win the World Series this season are 7 to 5."
b. "The odds that it will not rain tomorrow are 3 to 2."

Express each of these odds as a probability of the event occurring.

Robert H. Mason

TITLE   Vice President, Wealth Management Advisor
INSTITUTION   The Mason Group

The Mason Group—a team of financial advisors at a major wire house firm—acts as an interface between clients and investment markets. To meet the needs of our private clients, it is important for us to maintain constant contact with them, adjusting their investments when the markets and allocations change and when the clients' goals change.

We often help our clients determine whether their various expected sources of income in retirement will provide them with their desired retirement lifestyle. To begin, we determine the after-tax funds needed in retirement in today's dollars in consultation with the client. Using simple arithmetic, we then look at the duration and amount of their various income flows and their current portfolio allocations (stocks, bonds, cash, etc.). Once we have made this assessment, we take into consideration future possible allocations to arrive at a range of probabilities for portfolio valuation for all years up to and including retirement.

For example, we might tell a client that, on the basis of the client's given investment plan, there is an 80–95% probability that the expected value of his or her portfolio will increase from $1 million to about $1.5 million by the time the client turns 90; a 50–80% chance that it will increase from $1 million to around $2 million; and a 30–50% chance that it will increase to approximately $2.8 million.

To arrive at these probabilities, we use a deterministic model that assumes that a constant annual rate of return is applied to the portfolio every year of the analysis. We also use probabilistic modeling, taking into account various factors such as economic conditions, the allocation of assets, and market volatility. By using these techniques we are able to arrive at a confidence level, without making any guarantees, that our clients will be able to attain the lifestyle in retirement they desire.

Solution

a. Using Equation (13) with $a = 7$ and $b = 5$ gives the required probability as

$$\frac{7}{7 + 5} = \frac{7}{12} \approx .5833$$

b. Here, the event is that it will not rain tomorrow. Using Equation (13) with $a = 3$ and $b = 2$, we conclude that the probability that it will not rain tomorrow is

$$\frac{3}{3 + 2} = \frac{3}{5} = .6$$

Explore and Discuss

In the movie *Casino*, the executive of the Tangiers Casino, Sam Rothstein (Robert DeNiro), fired the manager of the slot machines in the casino after three gamblers hit three "million dollar" jackpots in a span of 20 minutes. Rothstein claimed that it was a scam and that somebody had gotten into those machines to set the wheels. He was especially annoyed at the slot machine manager's assertion that there was no way to determine this. According to Rothstein, the odds of hitting a jackpot in a four-wheel machine is 1 in $1\frac{1}{2}$ million, and the probability of hitting three jackpots in a row is "in the billions." "It cannot happen! It will not happen!" To see why Rothstein was so indignant, find the odds of hitting the jackpots in three of the machines in quick succession, and comment on the likelihood of this happening.

Median and Mode

In addition to the mean, there are two other measures of central tendency of a group of numerical data: the median and the mode of a group of numbers.

> **Median**
>
> The **median** of a group of numbers arranged in increasing or decreasing order is (a) the middle number if there is an odd number of entries or (b) the mean of the two middle numbers if there is an even number of entries.

 APPLIED EXAMPLE 11 Commuting Times

   a. The times, in minutes, Susan took to go to work on nine consecutive working days were

$$46 \quad 42 \quad 49 \quad 40 \quad 52 \quad 48 \quad 45 \quad 43 \quad 50$$

What is the median of her morning commute times?

b. The times, in minutes, Susan took to return home from work on eight consecutive working days were

$$37 \quad 36 \quad 39 \quad 37 \quad 34 \quad 38 \quad 41 \quad 40$$

What is the median of her evening commute times?

Solution

a. Arranging the numbers in increasing order, we have

$$40 \quad 42 \quad 43 \quad 45 \quad 46 \quad 48 \quad 49 \quad 50 \quad 52$$

Here, we have an odd number of entries with the middle number equal to 46, and this gives the required median.

b. Arranging the numbers in increasing order, we have

$$34 \quad 36 \quad 37 \quad 37 \quad 38 \quad 39 \quad 40 \quad 41$$

Here, the number of entries is even, and the required median is

$$\frac{37 + 38}{2} = 37.5$$

> **Mode**
>
> The **mode** of a group of numbers is the number in the group that occurs most frequently.

Note

A group of numerical data may have no mode, a unique mode, or more than one mode.

EXAMPLE 12 Find the mode, if there is one, of the given group of numbers.

a. 1, 2, 3, 4, 6
b. 2, 3, 3, 4, 6, 8
c. 2, 3, 3, 3, 4, 4, 4, 8

Solution

a. The set has no mode because there isn't a number that occurs more frequently than the others.

b. The mode is 3 because it occurs more frequently than the others.

c. The modes are 3 and 4 because each number occurs three times. ■

Of the three measures of central tendency of a group of numerical data, the mean is by far the most suitable in work that requires mathematical computations.

8.5 Self-Check Exercises

1. Find the expected value of a random variable X having the following probability distribution:

x	−4	−3	−1	0	1	2
$P(X = x)$	.10	.20	.25	.10	.25	.10

2. **FORECASTED TOWNHOUSE SALES** The developer of Shoreline Condominiums has provided the following estimate of the probability that 20, 25, 30, 35, 40, 45, or 50 of the townhouses will be sold within the first month they are offered for sale.

Units	20	25	30	35	40	45	50
Probability	.05	.10	.30	.25	.15	.10	.05

How many townhouses can the developer expect to sell within the first month they are put on the market?

Solutions to Self-Check Exercises 8.5 can be found on page 573.

8.5 Concept Questions

1. What is the expected value of a random variable? Give an example.

2. What is a fair game? Is the game of roulette as played in American casinos a fair game? Why or why not?

3. a. If the probability of an event E occurring is $P(E)$, what are the odds in favor of E occurring?

b. If the odds in favor of an event occurring are a to b, what is the probability of E occurring?

8.5 Exercises

1. Find the expected value of a random variable X having the following probability distribution:

x	−5	−1	0	1	5	8
$P(X = x)$	.12	.16	.28	.22	.12	.10

2. Find the expected value of a random variable X having the following probability distribution:

x	0	1	2	3	4	5
$P(X = x)$	$\frac{1}{8}$	$\frac{1}{4}$	$\frac{3}{16}$	$\frac{1}{4}$	$\frac{1}{16}$	$\frac{1}{8}$

3. **CALCULATING GPA** During the first year at a university that uses a four-point grading system, a freshman took ten three-credit courses and received two As, three Bs, four Cs, and one D.

a. Compute this student's grade-point average.

b. Let the random variable X denote the number of points corresponding to a given letter grade. Find the probability distribution of the random variable X and compute $E(X)$, the expected value of X.

4. **FAMILY COMPOSITION** In a four-child family, what is the expected number of boys? (Assume that the probability of a boy being born is the same as the probability of a girl being born.)

5. **EXPECTED SALES** On the basis of past experience, the manager of the VideoRama Store has compiled the following table, which gives the probabilities that a customer who enters the VideoRama Store will buy 0, 1, 2, 3, or 4 DVDs. How many DVDs can a customer entering this store be expected to buy?

DVDs	0	1	2	3	4
Probability	.42	.36	.14	.05	.03

6. **CAFETERIA MILK CONSUMPTION** Records kept by the chief dietitian at the university cafeteria over a 30-week period show the following weekly consumption of milk (in gallons):

Milk	200	205	210	215	220
Weeks	3	4	6	5	4

Milk	225	230	235	240
Weeks	3	2	2	1

a. Find the average number of gallons of milk consumed per week in the cafeteria.

b. Let the random variable X denote the number of gallons of milk consumed in a week at the cafeteria. Find the probability distribution of the random variable X and compute $E(X)$, the expected value of X.

7. Expected Earnings The daily earnings X of an employee who works on a commission basis are given by the following probability distribution. Find the employee's expected earnings.

x ($)	0	25	50	75
$P(X = x)$	.07	.12	.17	.14

x ($)	100	125	150
$P(X = x)$	.28	.18	.04

8. Expected Number of Defective Products If a sample of three batteries is selected from a lot of ten, of which two are defective, what is the expected number of defective batteries?

9. Expected Number of Auto Accidents The numbers of accidents that occur at a certain intersection known as Five Corners on a Friday afternoon between the hours of 3 P.M. and 6 P.M., along with the corresponding probabilities, are shown in the following table. Find the expected number of accidents during the period in question.

Accidents	0	1	2	3	4
Probability	.935	.030	.020	.010	.005

10. Expected Demand for Magazines The owner of a newsstand in a college community estimates the weekly demand for a certain magazine as follows:

Quantity Demanded	10	11	12	13	14	15
Probability	.05	.15	.25	.30	.20	.05

Find the number of issues of the magazine that the newsstand owner can expect to sell per week.

11. Expected ATM Reliability A bank has two automatic teller machines at its main office and two at each of its three branches. The numbers of machines that break down on a given day, along with the corresponding probabilities, are shown in the following table:

Machines That Break Down	0	1	2	3	4
Probability	.43	.19	.12	.09	.04

Machines That Break Down	5	6	7	8
Probability	.03	.03	.02	.05

Find the expected number of machines that will break down on a given day.

12. Expected Sales The management of the Cambridge Company has projected the sales of its products (in millions of dollars) for the upcoming year, with the associated probabilities shown in the following table:

Sales	20	22	24	26	28	30
Probability	.05	.10	.35	.30	.15	.05

What are the expected sales for next year?

13. Prime Interest-Rate Prediction A panel of 50 economists was asked to predict the average prime interest rate for the upcoming year. The results of the survey follow:

Interest Rate (%)	2.9	3.0	3.1	3.2	3.3	3.4
Economists	3	8	12	14	8	5

On the basis of this survey, what does the panel expect the average prime interest rate to be next year?

14. Forecasted Unemployment Rates A panel of 64 economists was asked to predict the average unemployment rate for the upcoming year. The results of the survey follow:

Unemployment Rate (%)	6.5	6.6	6.7	6.8	6.9	7.0	7.1
Economists	2	4	8	20	14	12	4

On the basis of this survey, what does the panel expect the average unemployment rate to be next year?

15. Expected Value of a Lottery Ticket In a lottery, 5000 tickets are sold for $1 each. One first prize of $2000, 1 second prize of $500, 3 third prizes of $100, and 10 consolation prizes of $25 are to be awarded. What are the expected net earnings of a person who buys one ticket?

16. Life Insurance Premiums A man wishes to purchase a 5-year term-life insurance policy that will pay his beneficiary $20,000 in the event that his death occurs during the next 5 years. Using life insurance tables, he determines that the probability that he will live another 5 years is .96. What is the minimum amount that he can expect to pay for his premium?
Hint: The minimum premium occurs when the insurance company's expected profit is zero.

17. LIFE INSURANCE PREMIUMS A woman purchased a $20,000, 1-year term-life insurance policy for $260. Assuming that the probability that she will live another year is .992, find the company's expected gain.

18. LIFE INSURANCE PREMIUMS As a fringe benefit, Dennis Taylor receives a $50,000 life insurance policy from his employer. The probability that Dennis will live another year is .9935. If he purchases the same coverage for himself, what is the minimum amount that he can expect to pay for the policy? (See the Hint in Exercise 16.)

19. EXPECTED PROFIT OF A BUILDER Max built a spec house at a cost of $450,000. He estimates that he can sell the house for $580,000, $570,000, or $560,000, with probabilities .24, .40, and .36, respectively. What is Max's expected profit?

20. INVESTMENT ANALYSIS The proprietor of Midland Construction Company needs to choose one of two projects. He estimates that the first project will yield a profit of $180,000 with a probability of .7 or a profit of $150,000 with a probability of .3; the second project will yield a profit of $220,000 with a probability of .6 or a profit of $80,000 with a probability of .4. Which project should the proprietor choose if he wants to maximize his expected profit?

21. CABLE TELEVISION RIGHTS FOR A CITY The management of MultiVision, a cable TV company, intends to submit a bid for the cable television rights in one of two cities, *A* or *B*. If the company obtains the rights to City *A*, the probability of which is .2, the estimated profit over the next 10 years is $10 million; if the company obtains the rights to City *B*, the probability of which is .3, the estimated profit over the next 10 years is $7 million. The cost of submitting a bid for rights in City *A* is $250,000, and that in City *B* is $200,000. By comparing the expected profits for each venture, determine whether the company should bid for the rights in City *A* or City *B*.

22. EXPECTED AUTO SALES OF A DEALERSHIP Roger Hunt intends to purchase one of two car dealerships currently for sale in a certain city. Records obtained from each of the two dealers reveal that their weekly volume of sales, with corresponding probabilities, are as follows:

Dahl Motors

Cars Sold/Week	5	6	7	8
Probability	.05	.09	.14	.24

Cars Sold/Week	9	10	11	12
Probability	.18	.14	.11	.05

Farthington Auto Sales

Cars Sold/Week	5	6	7	8	9	10
Probability	.08	.21	.31	.24	.10	.06

The average profit per car at Dahl Motors is $543, and the average profit per car at Farthington Auto Sales is $654.

a. Find the average number of cars sold each week at each dealership.

b. If Roger's objective is to purchase the dealership that generates the higher weekly profit, which dealership should he purchase? (Compare the expected weekly profit for each dealership.)

23. EXPECTED HOME SALES OF A REALTOR Sally Leonard, a real estate broker, is relocating in a large metropolitan area where she has received job offers from Realty Company *A* and Realty Company *B*. The number of houses she expects to sell in a year at each firm and the associated probabilities are shown in the following tables:

Company *A*

Houses Sold	12	13	14	15	16
Probability	.02	.03	.05	.07	.07

Houses Sold	17	18	19	20
Probability	.16	.17	.13	.11

Houses Sold	21	22	23	24
Probability	.09	.06	.03	.01

Company *B*

Houses Sold	6	7	8	9	10
Probability	.01	.04	.07	.06	.11

Houses Sold	11	12	13	14
Probability	.12	.19	.17	.13

Houses Sold	15	16	17	18
Probability	.04	.03	.02	.01

The average price of a house in the locale of Company *A* is $308,000, whereas the average price of a house in the locale of Company *B* is $474,000. If Sally will receive a 3% commission on sales at either company, which job offer should she accept to maximize her expected yearly commission?

24. INVESTMENT ANALYSIS Bob, the proprietor of Midway Lumber, bases his projections for the annual revenues of the company on the performance of the housing market. He rates the performance of the market as very strong, strong, normal, weak, or very weak. For the next year, Bob estimates that the probabilities for these outcomes are .18, .27, .42, .10, and .03, respectively. He also thinks that the revenues corresponding to these outcomes are $20, $18.8, $16.2, $14, and $12 million, respectively. What is Bob's expected revenue for next year?

25. EXPECTED GROWTH FOR A BUSINESS Maria sees the growth of her business for the upcoming year as being tied to the gross domestic product (GDP). She believes that her business will grow (or contract) at the rate of 5%, 4.5%, 3%, 0%, or −0.5% per year if the GDP grows (or contracts) at the rate of between 2% and 2.5%, between 1.5% and 2%, between 1% and 1.5%, between 0% and 1%, and between −1% and 0%, respectively. Maria has decided to assign a probability of .12, .24, .40, .20, and .04, respectively, to these outcomes. At what rate does Maria expect her business to grow next year?

26. **WEATHER PREDICTIONS** Suppose the probability that it will rain tomorrow is .3.
   a. What are the odds that it will rain tomorrow?
   b. What are the odds that it will not rain tomorrow?

27. **EXPECTED VALUE OF A ROULETTE BET** In American roulette, as described in Example 6, a player may bet on a split (two adjacent numbers). In this case, if the player bets $1 and either number comes up, the player wins $17 and gets his $1 back. If neither comes up, he loses his $1 bet. Find the expected value of the winnings on a $1 bet placed on a split.

28. **EXPECTED VALUE OF A ROULETTE BET** If a player placed a $1 bet on *red* and a $1 bet on *black* in a single play in American roulette, what would be the expected value of her winnings?

29. **EXPECTED VALUE OF A ROULETTE BET** In European roulette, the wheel is divided into 37 compartments numbered 1 through 36 and 0. (In American roulette there are 38 compartments numbered 1 through 36, 0, and 00.) Find the expected value of the winnings on a $1 bet placed on *red* in European roulette.

30. **MALE COMMUTING TIMES** The following table gives the travel time to work of male workers in the United States aged 16 years and over who do not work at home:

Travel Time, x (in minutes)	$x < 10$	$10 \leq x < 15$	$15 \leq x < 20$	$20 \leq x < 25$	$25 \leq x < 30$
Number (in thousands)	8,734	9,362	10,341	10,201	4,192

Travel Time, x (in minutes)	$30 \leq x < 35$	$35 \leq x < 45$	$45 \leq x < 60$	$x \geq 60$
Number (in thousands)	9,991	4,681	5,729	6,638

Estimate the average time to work for male workers aged 16 years and over who do not work at home. (Assume that "$x < 10$" means "$0 < x < 10$" and "$x \geq 60$" means "$60 \leq x \leq 75$.")
Source: U.S. Census Bureau.

31. **FEMALE COMMUTING TIMES** The following table gives the travel time to work of female workers in the United States aged 16 years and over who do not work at home:

Travel Time, x (in minutes)	$x < 10$	$10 \leq x < 15$	$15 \leq x < 20$	$20 \leq x < 25$	$25 \leq x < 30$
Number (in thousands)	9,049	9,611	10,172	9,423	3,807

Travel Time, x (in minutes)	$30 \leq x < 35$	$35 \leq x < 45$	$45 \leq x < 60$	$x \geq 60$
Number (in thousands)	8,175	3,807	4,244	4,119

Estimate the average time to work for female workers aged 16 years and over who do not work at home. (Assume that "$x < 10$" means "$0 < x < 10$" and "$x \geq 60$" means "$60 \leq x \leq 75$.")
Source: U.S. Census Bureau.

32. **POPULATION BY AGE IN THE UNITED STATES** The resident population (in thousands) by age in the United States as of April 1, 2010, is summarized in the following table:

Age (in years)	Under 5	5–14	15–24	25–34	35–44	45–54
Population (in thousands)	20,201	41,026	43,626	41,064	41,071	45,007

Age (in years)	55–64	65–74	75–84	85–94	95 and over
Population (in thousands)	36,483	21,713	13,061	5,069	425

Estimate the average age of the resident population in the United States as of April 1, 2010. (Assume that "95 and over" means "95–104.")
Source: U.S. Census Bureau.

33. **POPULATION BY AGE IN CALIFORNIA** The resident population (in thousands) by age in California as of April 1, 2010, is summarized in the following table:

Age (in years)	Under 5	5–14	15–24	25–34	35–44	45–54
Population (in thousands)	2531	5097	5590	5318	5183	5252

Age (in years)	55–64	65–74	75–84	85–94	95 and over
Population (in thousands)	4036	2275	1370	555	46

Estimate the average age of the resident population in California as of April 1, 2010. (Assume that "95 and over" means "95–104.")
Source: U.S. Census Bureau.

34. The probability of an event E occurring is .8. What are the odds in favor of E occurring? What are the odds against E occurring?

35. The probability of an event E not occurring is .6. What are the odds in favor of E occurring? What are the odds against E occurring?

36. The odds in favor of an event E occurring are 9 to 7. What is the probability of E occurring?

37. The odds against an event E occurring are 2 to 3. What is the probability of E not occurring?

38. **ODDS OF MAKING A SALE** Carmen, a computer sales representative, believes that the odds are 8 to 5 that she will clinch the sale of a minicomputer to a certain company. What is the (subjective) probability that Carmen will make the sale?

39. Odds of Winning a Tennis Match Steffi believes that the odds in favor of her winning her tennis match tomorrow are 7 to 5. What is the (subjective) probability that she will win her match tomorrow?

40. Odds of Winning a Boxing Match If a sports forecaster states that the odds of a certain boxer winning a match are 4 to 3, what is the (subjective) probability that the boxer will win the match?

41. Odds of Closing a Business Deal Bob, the proprietor of Midland Lumber, believes that the odds in favor of a business deal going through are 9 to 5. What is the (subjective) probability that this deal will *not* materialize?

42. Expected Loss for Roulette Bet

a. Show that, for any number c,

$$E(cX) = cE(X)$$

b. Use this result to find the expected loss if a gambler bets \$300 on *red* in a single play in American roulette. Hint: Use the results of Example 6.

43. If X and Y are random variables and c is any constant, show that

a. $E(c) = c$
b. $E(cX) = cE(X)$
c. $E(X + Y) = E(X) + E(Y)$
d. $E(X - Y) = E(X) - E(Y)$

44. Wage Rates The frequency distribution of the hourly wage rates among blue-collar workers in a certain factory is given in the following table. Find the mean (or average) wage rate, the mode, and the median wage rate of these workers.

Wage Rate (\$)	10.70	10.80	10.90	11.00	11.10	11.20
Frequency	60	90	75	120	60	45

45. Exam Scores In an examination given to a class of 20 students, the following test scores were obtained:

40 45 50 50 55 60 60 75 75 80

80 85 85 85 85 90 90 95 95 100

a. Find the mean (or average) score, the mode, and the median score.
b. Which of these three measures of central tendency do you think is the least representative of the set of scores?

46. San Francisco Weather The normal daily minimum temperatures in degrees Fahrenheit for the months of January through December in San Francisco follow:

46.2 48.4 48.6 49.2 50.7 52.5

53.1 54.2 55.8 54.8 51.5 47.2

Find the average and the median daily minimum temperatures in San Francisco for these months.
Source: San Francisco Convention and Visitors Bureau.

47. Waiting Lines Refer to Example 6, Section 8.4. Find the median of the number of cars waiting in line at the bank's drive-in teller at the beginning of each 2-min interval during the period in question. Compare your answer to the mean obtained in Example 1, Section 8.5.

48. Boston Weather The relative humidity, in percent, in the morning for the months of January through December in Boston follows:

68 67 69 69 71 73

74 76 79 77 74 70

Find the average and the median of these humidity readings.
Source: National Weather Service Forecast Office.

49. Weight of Potato Chips The weights, in ounces, of ten packages of potato chips are as follows:

16.1 16 15.8 16 15.9 16.1 15.9 16 16 16.2

Find the average and the median of these weights.

50. Blood Types The following table gives the top ten countries in the world whose populations have the highest concentration of type O^+ blood:

Country	Saudi Arabia	Iceland	Ireland	Taiwan	Australia
Population (%)	48.0	47.6	47.0	43.9	40.0

Country	Hong Kong	Italy	Netherlands	Canada	South Africa
Population (%)	40.0	39.5	39.0	39.0	38.0

Find the mean (average), median, and mode of these concentrations of type O^+ blood in these ten countries.
Source: Bloodbook.com.

51. Blood Types The following table gives the top 12 countries in the world whose populations have the highest concentration of type A^- blood.

Country	Brazil	Spain	Norway	Australia	Netherlands	Belgium
Population (%)	8.0	8.0	7.2	7.0	7.0	7.0

Country	United Kingdom	France	Denmark	Sweden	Austria	Portugal
Population (%)	7.0	7.0	7.0	7.0	7.0	6.6

Find the mean (average), median, and mode of these concentrations of type A^- blood in these 12 countries.
Source: Bloodbook.com.

In Exercises 52 and 53, determine whether the statement is true or false. If it is true, explain why it is true. If it is false, give an example to show why it is false.

52. If the odds in favor of an event E occurring are a to b, then the probability of E^c occurring is $b/(a + b)$.

53. A game between two people is fair if the expected value to both people is zero.

8.5 Solutions to Self-Check Exercises

1. $E(X) = (-4)(.10) + (-3)(.20) + (-1)(.25)$
$\qquad + (0)(.10) + (1)(.25) + (2)(.10)$
$\qquad = -0.8$

2. Let X denote the number of townhouses that will be sold within 1 month of being put on the market. Then the number of townhouses the developer expects to sell within

1 month is given by the expected value of X—that is, by

$$E(X) = 20(.05) + 25(.10) + 30(.30) + 35(.25)$$
$$\qquad + 40(.15) + 45(.10) + 50(.05)$$
$$\qquad = 34.25$$

or 34 townhouses.

8.6 Variance and Standard Deviation

Variance

The mean, or expected value, of a random variable enables us to express an important property of the probability distribution associated with the random variable in terms of a single number. But the knowledge of the location, or central tendency, of a probability distribution alone is usually not enough to give a reasonably accurate picture of the probability distribution. Consider, for example, the two probability distributions whose histograms appear in Figure 21. Both distributions have the same expected value, or mean, of $\mu = 4$ (the Greek letter μ is read "mu"). Note that the probability distribution with the histogram shown in Figure 21a is closely concentrated about its mean μ, whereas the one with the histogram shown in Figure 21b is widely dispersed or spread about its mean.

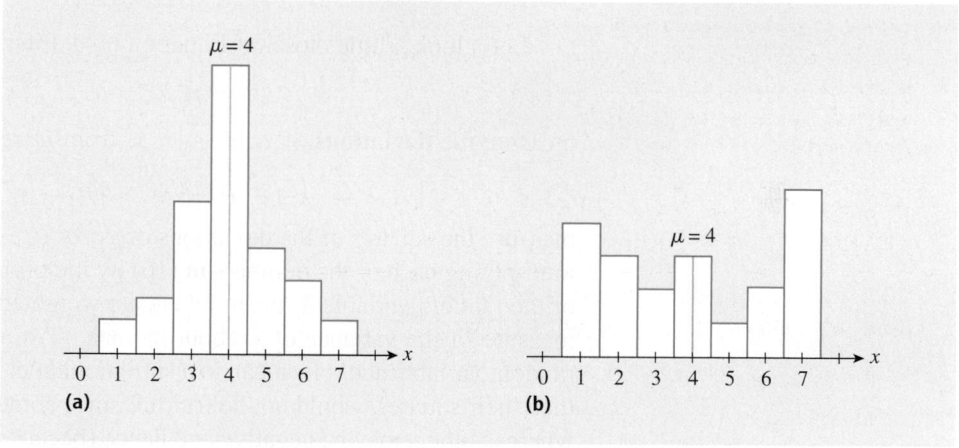

FIGURE **21**
The histograms of two probability distributions

As another example, suppose that Olivia has ten packages of Brand A potato chips and ten packages of Brand B potato chips. After carefully measuring the weights of each package, she obtains the following results:

Weight in Ounces									
Brand A	16.1	16	15.8	16	15.9	16.1	15.9	16	16	16.2
Brand B	16.3	15.7	15.8	16.2	15.9	16.1	15.7	16.2	16	16.1

In Example 3, we verify that the mean weights for each of the two brands is 16 ounces. However, a cursory examination of the data now shows that the weights of the Brand B packages exhibit much greater dispersion about the mean than do those of Brand A.

One measure of the degree of dispersion, or spread, of a probability distribution about its mean is given by the variance of the random variable associated with the probability distribution. A probability distribution with a small spread about its mean will have a small variance, whereas one with a larger spread will have a larger variance. Thus, the variance of the random variable associated with the probability distribution whose histogram appears in Figure 21a is smaller than the variance of the random variable associated with the probability distribution whose histogram is shown in Figure 21b (see Example 1). Also, as we will see in Example 3, the variance of the random variable associated with the weights of the Brand A potato chips is smaller than that of the random variable associated with the weights of the Brand B potato chips.

We now define the variance of a random variable.

Variance of a Random Variable X

Suppose a random variable has the probability distribution

x	x_1	x_2	x_3	$\cdots$	x_n
$P(X = x)$	p_1	p_2	p_3	$\cdots$	p_n

and expected value

$$E(X) = \mu$$

Then the **variance** of the random variable X is

$$\mathrm{Var}(X) = p_1(x_1 - \mu)^2 + p_2(x_2 - \mu)^2 + \cdots + p_n(x_n - \mu)^2 \qquad \textbf{(14)}$$

Let's look a little closer at Equation (14). First, note that the numbers

$$x_1 - \mu, x_2 - \mu, \ldots, x_n - \mu \qquad \textbf{(15)}$$

measure the **deviations** of $x_1, x_2, \ldots, x_n$ from μ, respectively. Thus, the numbers

$$(x_1 - \mu)^2, (x_2 - \mu)^2, \ldots, (x_n - \mu)^2 \qquad \textbf{(16)}$$

measure the squares of the deviations of $x_1, x_2, \ldots, x_n$ from μ, respectively. Next, by multiplying each of the numbers in (16) by the probability associated with each value of the random variable X, the numbers are weighted accordingly so that their sum is a measure of the variance of X about its mean. An attempt to define the variance of a random variable about its mean in a similar manner using the deviations in (15), rather than their squares, would not be fruitful, since some of the deviations may be positive whereas others may be negative and hence (because of cancellations) the sum will not give a satisfactory measure of the variance of the random variable.

EXAMPLE 1 Find the variance of the random variable X and of the random variable Y whose probability distributions are shown in the following table. These are the probability distributions associated with the histograms shown in Figure 21a–b.

x	$P(X = x)$	y	$P(Y = y)$
1	.05	1	.2
2	.075	2	.15
3	.2	3	.1
4	.375	4	.15
5	.15	5	.05
6	.1	6	.1
7	.05	7	.25

Solution The mean of the random variable X is given by

$$\mu_X = (1)(.05) + (2)(.075) + (3)(.2) + (4)(.375) + (5)(.15)$$
$$+ (6)(.1) + (7)(.05)$$
$$= 4$$

Therefore, using Equation (14) and the data from the probability distribution of X, we find that the variance of X is given by

$$Var(X) = (.05)(1 - 4)^2 + (.075)(2 - 4)^2 + (.2)(3 - 4)^2$$
$$+ (.375)(4 - 4)^2 + (.15)(5 - 4)^2$$
$$+ (.1)(6 - 4)^2 + (.05)(7 - 4)^2$$
$$= 1.95$$

Next, we find that the mean of the random variable Y is given by

$$\mu_Y = (1)(.2) + (2)(.15) + (3)(.1) + (4)(.15) + (5)(.05)$$
$$+ (6)(.1) + (7)(.25)$$
$$= 4$$

so the variance of Y is given by

$$Var(Y) = (.2)(1 - 4)^2 + (.15)(2 - 4)^2 + (.1)(3 - 4)^2$$
$$+ (.15)(4 - 4)^2 + (.05)(5 - 4)^2$$
$$+ (.1)(6 - 4)^2 + (.25)(7 - 4)^2$$
$$= 5.2$$

Note that $Var(X)$ is smaller than $Var(Y)$, which confirms our earlier observations about the spread (or dispersion) of the probability distribution of X and Y, respectively.

Standard Deviation

Because Equation (14), which gives the variance of the random variable X, involves the squares of the deviations, the unit of measurement of $Var(X)$ is the square of the unit of measurement of the values of X. For example, if the values assumed by the random variable X are measured in units of a gram, then $Var(X)$ will be measured in units involving the *square* of a gram. To remedy this situation, one normally works with the square root of $Var(X)$ rather than $Var(X)$ itself. The former is called the standard deviation of X.

Standard Deviation of a Random Variable X

The **standard deviation** of a random variable X denoted σ (pronounced "sigma"), is defined by

$$\sigma = \sqrt{Var(X)}$$
$$= \sqrt{p_1(x_1 - \mu)^2 + p_2(x_2 - \mu)^2 + \cdots + p_n(x_n - \mu)^2} \qquad (17)$$

where $x_1, x_2, \ldots, x_n$ denote the values assumed by the random variable X and $p_1 = P(X = x_1), p_2 = P(X = x_2), \ldots, p_n = P(X = x_n)$.

EXAMPLE 2 Find the standard deviations of the random variables X and Y of Example 1.

Solution From the results of Example 1, we have $\text{Var}(X) = 1.95$ and $\text{Var}(Y) = 5.2$. Taking their respective square roots, we have

$$\sigma_X = \sqrt{1.95}$$
$$\approx 1.40$$
$$\sigma_Y = \sqrt{5.2}$$
$$\approx 2.28$$

APPLIED EXAMPLE 3 Packaging Let X and Y denote the random variables whose values are the weights of the Brand A and Brand B potato chips, respectively (see page 573). Compute the means and standard deviations of X and Y and interpret your results.

Solution The probability distributions of X and Y may be computed from the given data as follows:

Brand A			Brand B		
x	Relative Frequency of Occurrence	$P(X = x)$	y	Relative Frequency of Occurrence	$P(Y = y)$
15.8	1	.1	15.7	2	.2
15.9	2	.2	15.8	1	.1
16.0	4	.4	15.9	1	.1
16.1	2	.2	16.0	1	.1
16.2	1	.1	16.1	2	.2
			16.2	2	.2
			16.3	1	.1

The means of X and Y are given by

$$\mu_X = (.1)(15.8) + (.2)(15.9) + (.4)(16.0) + (.2)(16.1)$$
$$+ (.1)(16.2)$$
$$= 16$$
$$\mu_Y = (.2)(15.7) + (.1)(15.8) + (.1)(15.9) + (.1)(16.0)$$
$$+ (.2)(16.1) + (.2)(16.2) + (.1)(16.3)$$
$$= 16$$

Therefore,

$$\text{Var}(X) = (.1)(15.8 - 16)^2 + (.2)(15.9 - 16)^2 + (.4)(16 - 16)^2$$
$$+ (.2)(16.1 - 16)^2 + (.1)(16.2 - 16)^2$$
$$= 0.012$$
$$\text{Var}(Y) = (.2)(15.7 - 16)^2 + (.1)(15.8 - 16)^2 + (.1)(15.9 - 16)^2$$
$$+ (.1)(16 - 16)^2 + (.2)(16.1 - 16)^2 + (.2)(16.2 - 16)^2$$
$$+ (.1)(16.3 - 16)^2$$
$$= 0.042$$

Explore and Discuss

A useful alternative formula for the variance is

$$\sigma^2 = E(X^2) - \mu^2$$

where $E(X^2)$ is the expected value of X^2.

1. Establish the validity of the formula.

2. Use the formula to verify the calculations in Example 3.

so the standard deviations are

$$\sigma_X = \sqrt{\text{Var}(X)}$$
$$= \sqrt{0.012}$$
$$\approx 0.11$$
$$\sigma_Y = \sqrt{\text{Var}(Y)}$$
$$= \sqrt{0.042}$$
$$\approx 0.20$$

The mean of X and that of Y are both equal to 16. Therefore, the average weight of a package of potato chips of either brand is 16 ounces. However, the standard deviation of Y is greater than that of X. This tells us that the weights of the packages of Brand B potato chips are more widely dispersed about the common mean of 16 than are those of Brand A.

Explore and Discuss

Suppose the mean weight of m packages of Brand A potato chips is μ_1 and the standard deviation from the mean of their weight distribution is σ_1. Also suppose the mean weight of n packages of Brand B potato chips is μ_2 and the standard deviation from the mean of their weight distribution is σ_2.

1. Show that the mean of the combined weights of packages of Brand A and Brand B is

$$\mu = \frac{m\mu_1 + n\mu_2}{m + n}$$

2. If $\mu_1 = \mu_2$, show that the standard deviation from the mean of the combined-weight distribution is

$$\sigma = \left(\frac{m\sigma_1^2 + n\sigma_2^2}{m + n}\right)^{1/2}$$

3. Refer to Example 3, page 576. Using the results of parts 1 and 2, find the mean and the standard deviation of the combined-weight distribution.

The following example illustrates the technique for finding the standard deviation for grouped data.

APPLIED EXAMPLE 4 Married Males The following table gives the number of married males in the United States aged 15 years and over but less than 65 years in 2011:

Age (in years)	15–19	20–34	35–44	45–54	55–64
Number (in thousands)	11,220	32,206	20,308	21,990	18,346

Source: U.S. Census Bureau.

Find the mean and the standard deviation for these data.

Solution

Let X denote the random variable that measures the number of married males. Taking X to be the midpoint of a group interval, we obtain the following probability distribution:

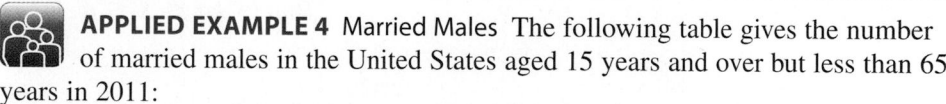

x	17	27	39.5	49.5	59.5
$P(X = x)$	$\left(\dfrac{11{,}220}{104{,}070}\right)$	$\left(\dfrac{32{,}206}{104{,}070}\right)$	$\left(\dfrac{20{,}308}{104{,}070}\right)$	$\left(\dfrac{21{,}990}{104{,}070}\right)$	$\left(\dfrac{18{,}346}{104{,}070}\right)$

The mean of X is

$$\mu = \left(\frac{11{,}220}{104{,}070}\right)(17) + \left(\frac{32{,}206}{104{,}070}\right)(27) + \left(\frac{20{,}308}{104{,}070}\right)(39.5)$$
$$+ \left(\frac{21{,}990}{104{,}070}\right)(49.5) + \left(\frac{18{,}346}{104{,}070}\right)(59.5)$$
$$\approx 38.8446$$

Next, we see that

$$\text{Var}(X) = \left(\frac{11{,}220}{104{,}070}\right)(17 - 38.8446)^2 + \left(\frac{32{,}206}{104{,}070}\right)(27 - 38.8446)^2$$
$$+ \left(\frac{20{,}308}{104{,}070}\right)(39.5 - 38.8446)^2 + \left(\frac{21{,}990}{104{,}070}\right)(49.5 - 38.8446)^2$$
$$+ \left(\frac{18{,}346}{104{,}070}\right)(59.5 - 38.8446)^2$$
$$\approx 194.1483$$

So the standard variation for these data is

$$\sigma = \sqrt{194.1483} \approx 13.93$$

Chebychev's Inequality

The standard deviation of a random variable X may be used in statistical estimations. For example, the following result, derived by the Russian mathematician P. L. Chebychev (1821–1894), gives a bound on the proportion of the values of X lying within k standard deviations of the expected value of X.

> **Chebychev's Inequality**
>
> Let X be a random variable with expected value μ and standard deviation σ. Then the probability that a randomly chosen outcome of the experiment lies between $\mu - k\sigma$ and $\mu + k\sigma$ is at least $1 - (1/k^2)$, where k is the number of standard deviations from the mean; that is,
>
> $$P(\mu - k\sigma \le X \le \mu + k\sigma) \ge 1 - \frac{1}{k^2} \qquad (18)$$

To shed some light on this result, let's take $k = 2$ in Inequality (18) and compute

$$P(\mu - 2\sigma \le X \le \mu + 2\sigma) \ge 1 - \frac{1}{2^2} = 1 - \frac{1}{4} = .75$$

This tells us that at least 75% of the outcomes of the experiment lie within 2 standard deviations of the mean (Figure 22). Taking $k = 3$ in Inequality (18), we have

$$P(\mu - 3\sigma \le X \le \mu + 3\sigma) \ge 1 - \frac{1}{3^2} = 1 - \frac{1}{9} = \frac{8}{9} \approx .89$$

This tells us that at least 89% of the outcomes of the experiment lie within 3 standard deviations of the mean (Figure 23).

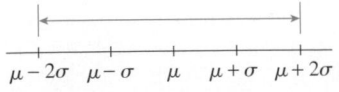

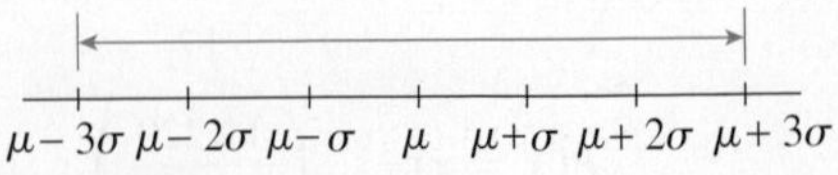

$\mu - 2\sigma \quad \mu - \sigma \quad \mu \quad \mu + \sigma \quad \mu + 2\sigma$

FIGURE **22**
At least 75% of the outcomes fall within this interval.

$\mu - 3\sigma \ \mu - 2\sigma \ \mu - \sigma \quad \mu \quad \mu + \sigma \ \mu + 2\sigma \ \mu + 3\sigma$

FIGURE **23**
At least 89% of the outcomes fall within this interval.

EXAMPLE 5 A probability distribution has a mean of 10 and a standard deviation of 1.5. Use Chebychev's inequality to find a bound on the probability that an outcome of the experiment lies between 7 and 13.

Solution   Here, $\mu = 10$ and $\sigma = 1.5$. To determine the value of k, note that $\mu - k\sigma = 7$ and $\mu + k\sigma = 13$. Substituting the appropriate values for μ and σ, we find $k = 2$. Using Chebychev's Inequality (18), we see that a bound on the probability that an outcome of the experiment lies between 7 and 13 is given by

$$P(7 \leq X \leq 13) \geq 1 - \left(\frac{1}{2^2}\right)$$

$$= \frac{3}{4}$$

that is, at least 75%.

Note   The results of Example 4 tell us that at least 75% of the outcomes of the experiment lie between $10 - 2\sigma$ and $10 + 2\sigma$—that is, between 7 and 13.

$ **APPLIED EXAMPLE 6** Industrial Accidents Great Northwest Lumber Company employs 400 workers in its mills. It has been estimated that X, the random variable measuring the number of mill workers who have industrial accidents during a 1-year period, is distributed with a mean of 40 and a standard deviation of 6. Using Chebychev's Inequality (18), find a bound on the probability that the number of workers who will have an industrial accident over a 1-year period is between 30 and 50, inclusive.

Solution   Here, $\mu = 40$ and $\sigma = 6$. We wish to estimate $P(30 \leq X \leq 50)$. To use Chebychev's Inequality (18), we first determine the value of k from the equation

$$\mu - k\sigma = 30 \quad \text{or} \quad \mu + k\sigma = 50$$

Since $\mu = 40$ and $\sigma = 6$ in this case, we see that k satisfies

$$40 - 6k = 30 \quad \text{and} \quad 40 + 6k = 50$$

from which we deduce that $k = \frac{5}{3}$. Thus, a bound on the probability that the number of mill workers who will have an industrial accident during a 1-year period is between 30 and 50 is given by

$$P(30 \leq X \leq 50) \geq 1 - \frac{1}{\left(\frac{5}{3}\right)^2}$$

$$= \frac{16}{25}$$

that is, at least 64%.

8.6 Self-Check Exercises

1. Compute the mean, variance, and standard deviation of the random variable X with probability distribution as follows:

x	-4	-3	-1	0	2	5
$P(X = x)$	.1	.1	.2	.3	.1	.2

2. **COMMUTE TIMES** James recorded the following commute times (the length of time in minutes it took him to drive to work) on ten consecutive days:

$$55 \quad 50 \quad 52 \quad 48 \quad 50 \quad 52 \quad 46 \quad 48 \quad 50 \quad 51$$

Calculate the mean and standard deviation of the random variable X associated with these data.

Solutions to Self-Check Exercises 8.6 can be found on page 585.

8.6 Concept Questions

1. **a.** What is the variance of a random variable X?
   b. What is the standard deviation of a random variable X?

2. What does Chebychev's inequality measure?

8.6 Exercises

In Exercises 1–6, the probability distribution of a random variable X is given. Compute the mean, variance, and standard deviation of X.

1.
x	1	2	3	4
$P(X = x)$	.4	.3	.2	.1

2.
x	−4	−2	0	2	4
$P(X = x)$	.1	.2	.3	.1	.3

3.
x	−2	−1	0	1	2
$P(X = x)$	1/16	4/16	6/16	4/16	1/16

4.
x	10	11	12	13	14	15
$P(X = x)$	1/8	2/8	1/8	2/8	1/8	1/8

5.
x	430	480	520	565	580
$P(X = x)$	.1	.2	.4	.2	.1

6.
x	−198	−195	−193	−188	−185
$P(X = x)$	.15	.30	.10	.25	.20

7. The following histograms represent the probability distributions of the random variables X and Y. Determine by inspection which probability distribution has the larger variance.

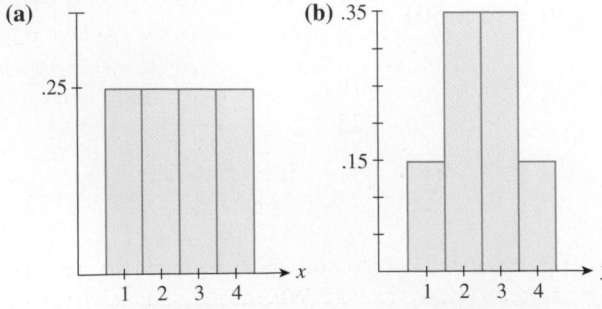

8. The following histograms represent the probability distributions of the random variables X and Y. Determine by inspection which probability distribution has the larger variance.

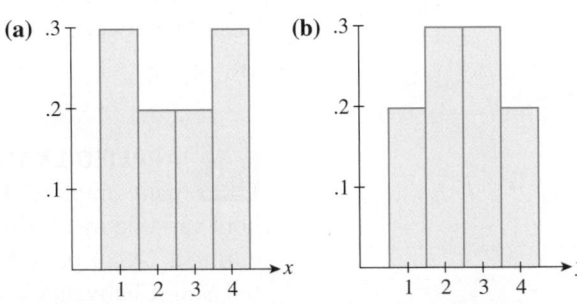

In Exercises 9 and 10, find the variance of the probability distribution for the histogram shown.

9.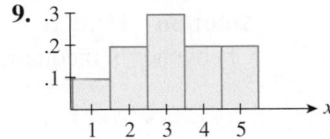

10.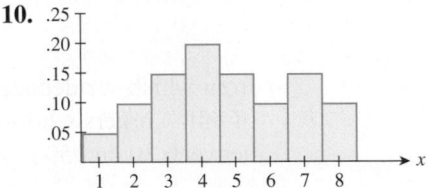

11. An experiment consists of rolling an eight-sided die (numbered 1 through 8) and observing the number that appears uppermost. Find the mean and variance of this experiment.

12. **HAPPINESS SCORE** The happiness score, by generation, conducted in April 2013 is given in the following table (percent indicates the top-two box scores on a five-point happiness scale):

Generation	Average 18+	Millennials	Generation X	Boomer	Silent Generation
Score (%)	58.6	54.1	55.7	61.0	66.7

Find the average happiness score for the five generations. What is the standard deviation for these data?
Source: Prosper Insights & Analytics.

13. **BIRTHRATES** The birthrates in the United States for the years 2003–2012 are given in the following table. (The birthrate is the number of live births/1000 population.)

Year	2003	2004	2005	2006	2007
Birthrate	14.7	14.0	14.0	14.2	14.2

Year	2008	2009	2010	2011	2012
Birthrate	14.0	13.8	13.8	13.8	13.7

a. Describe a random variable X that is associated with these data.
b. Find the probability distribution for the random variable X.
c. Compute the mean, variance, and standard deviation of X.

Source: National Center for Health Statistics.

14. **INVESTMENT ANALYSIS** Paul Hunt is considering two business ventures. The anticipated returns (in thousands of dollars) of each venture are described by the following probability distributions:

Venture A

Earnings	Probability
−20	.3
40	.4
50	.3

Venture B

Earnings	Probability
−15	.2
30	.5
40	.3

a. Compute the mean and variance for each venture.
b. Which investment would provide Paul with the higher expected return (the greater mean)?
c. In which investment would the element of risk be less (that is, which probability distribution has the smaller variance)?

15. **INVESTMENT ANALYSIS** Rosa Walters is considering investing $10,000 in two mutual funds. The anticipated returns from price appreciation and dividends (in hundreds of dollars) are described by the following probability distributions:

Mutual Fund A

Returns	Probability
−4	.2
8	.5
10	.3

Mutual Fund B

Earnings	Probability
−2	.2
6	.4
8	.4

a. Compute the mean and variance associated with the returns for each mutual fund.
b. Which investment would provide Rosa with the higher expected return (the greater mean)?
c. In which investment would the element of risk be less (that is, which probability distribution has the smaller variance)?

16. The distribution of the number of chocolate chips (x) in a cookie is shown in the following table. Find the mean and the variance of the number of chocolate chips in a cookie.

x	0	1	2	3	4
$P(X = x)$	.01	.03	.05	.11	.13

x	5	6	7	8
$P(X = x)$	.24	.22	.16	.05

17. Equation (5) can also be expressed in the form

$$\text{Var}(X) = (p_1 x_1^2 + p_2 x_2^2 + \cdots + p_n x_n^2) - \mu^2$$

Find the variance of the distribution of Exercise 1 using this equation.

18. Find the variance of the distribution of Exercise 16 using the equation

$$\text{Var}(X) = (p_1 x_1^2 + p_2 x_2^2 + \cdots + p_n x_n^2) - \mu^2$$

19. **STUCK IN TRAFFIC** The following table gives the extra travel time in hours for peak-period travelers in urban areas with more than 3 million people in a certain year:

Urban Area	Annual Hours of Delay per Traveler
Los Angeles–Long Beach–Santa Ana, CA	70
San Francisco–Oakland, CA	55
Atlanta, GA	57
Washington (DC–VA–MD)	62
Dallas–Fort Worth–Arlington, TX	53
Houston, TX	56

Find the mean of the extra travel time in that year, in hours, for peak-period travelers in urban areas with more than 3 million people. What is the standard deviation for these data?

Source: Texas Transportation Institute.

20. **COST OF TAKING TIME OFF** A survey was conducted of graduates of Harvard College 15 years after graduation. In the survey, the pay of graduates in different fields who had previously taken off 18 months, often to care for children, was compared with pay for graduates who had not taken time off. The average financial penalty for those who had taken time off is summarized in the following table:

Field	M.B.A.	J.D.	Ph.D	B.A. only	M.D.	Other, Masters only
Penalty (%)	−41	−29	−29	−25	−16	−13

Find the mean of the financial penalty for the graduates who had taken time off. What is the standard deviation for these data?

Source: Claudia Golden and Lawrence Katz, Harvard College.

21. CONVICTION RATES The following table gives the percentage of homicide cases in Suffolk County, Massachusetts, ending in pleas or verdicts of guilty from 2004 through 2009:

Year	2004	2005	2006	2007	2008	2009
Conviction Rate (%)	81	91	82	75	82	95

Find the mean of the percentage of homicide cases in Suffolk County ending in pleas or verdicts of guilty from 2004 through 2009. What is the standard deviation for these data?

Source: Suffolk County, Massachusetts, District Attorney's office.

22. NEW YORK STATE COURTS' TOTAL CASELOAD The following table gives the total caseload in the New York State courts from 2004 through 2009.

Year	2004	2005	2006	2007	2008	2009
Cases (in millions)	4.2	4.3	4.6	4.5	4.7	4.7

Find the mean of the total caseload in the New York State courts from 2004 through 2009. What is the standard deviation for these data?

Source: New York State Office of Court Administration.

23. HOURS WORKED IN SOME COUNTRIES The number of average hours worked per year per worker in the United States and five European countries in a certain year is given in the following table:

Country	U.S.	Spain	Great Britain	France	West Germany	Norway
Average Hours Worked	1815	1807	1707	1545	1428	1342

Find the average of the average hours worked per worker in that year for workers in the six countries. What is the standard deviation for these data?

Source: Office of Economic Cooperation and Development.

24. IDENTITY FRAUD The identity fraud rates in the United States for the years 2005–2011 are given in the following table:

Year	2005	2006	2007	2008	2009	2010	2011
Incidence Rate (%)	5.04	4.71	4.51	5.44	6.00	4.35	4.90

Find the average incidence rate of identity fraud for the years 2005–2011. What is the standard deviation?

Source: FBI.

25. HEALTH ISSUES IN MASSACHUSETTS CITIES A random survey of health issues, conducted by the Department of Public Health of the Commonwealth of Massachusetts, examined the results from the state's seven largest cities. These cities were selected on the basis of their diverse racial and ethnic populations. The percentage of adults reporting fair or poor health for each city in the survey is given in the following table:

City	Boston	Worcester	Springfield	Lowell
Adults Reporting Fair or Poor Health (%)	16.3	15.4	22.2	17.2

City	Fall River	Lawrence	New Bedford
Adults Reporting Fair or Poor Health (%)	23.2	30.4	26.4

Find the average percentage of adults reporting fair or poor health for the seven cities. What is the standard deviation for these data?

Source: Massachusetts Department of Public Health.

26. DIABETES IN MASSACHUSETTS CITIES A random survey of health issues, conducted by the Department of Public Health of the Commonwealth of Massachusetts, examined the results from the state's seven largest cities. These cities were selected on the basis of their diverse racial and ethnic populations. The percentage of adults with diabetes in each city in the survey is given in the following table:

City	Boston	Worcester	Springfield	Lowell
Adults with Diabetes (%)	7.2	8.2	12.1	8.7

City	Fall River	Lawrence	New Bedford
Adults with Diabetes (%)	11.1	10.9	9.3

Find the average percentage of adults with diabetes in these seven cities. What is the standard deviation for these data?

Source: Massachusetts Department of Public Health.

27. HYBRID VEHICLE MILEAGE The following table gives the mileage (in miles per gallon) of the eight 2013 model hybrid vehicles with the highest combined mileages:

Model	Toyota Prius	Ford C-Max	Ford Fusion	Lincoln MKZ	Volkswagen Jetta
Mileage (in mpg)	50	47	47	45	45

Model	Honda Insight	Lexus CT 200h	Lexus ES 300h
Mileage (in mpg)	42	42	40

Find the average mileage of the eight 2013 hybrid vehicles. What is the standard deviation for these data?

Source: U.S. Department of Energy.

28. HOUSING PRICES The market research department of the National Real Estate Company conducted a survey among 500 prospective buyers in a suburb of a large metropolitan area to determine the maximum price a prospective buyer would be willing to pay for a house. From the data collected, the distribution that follows was obtained:

Maximum Price Considered, x	$P(X = x)$
480	$\frac{10}{500}$
490	$\frac{20}{500}$
500	$\frac{75}{500}$
510	$\frac{85}{500}$
520	$\frac{70}{500}$
550	$\frac{90}{500}$
580	$\frac{90}{500}$
600	$\frac{55}{500}$
650	$\frac{5}{500}$

Compute the mean, variance, and standard deviation of the maximum price x (in thousands of dollars) that these buyers were willing to pay for a house.

29. **GOVERNMENT DEBT** The following table gives the projected debt as a percentage of the gross domestic product (GDP) of nine selected countries for 2011. The study was conducted by the Organization for Economic Co-operation and Development (OECD) in early 2010.

Country	Spain	U.S.	Germany	Portugal
GDP (%)	67	72	83	88

Country	U.K.	France	Japan	Italy	Greece
GDP (%)	89	91	113	121	127

Find the mean of the projected debt as a percentage of GDP of the nine countries under consideration. What is the standard deviation for these data?
Source: OECD.

30. **ON-TIME ARRIVALS** The following table gives the percentage of on-time arrivals in U.S. airports from 2004 through 2013:

Year	2004	2005	2006	2007	2008	2009	2010	2011	2012	2013
On-time Arrivals (%)	76.13	74.35	77.11	70.33	70.55	79.69	76.75	75.46	84.93	80.33

Find the average percentage of on-time arrivals over the 10 years from 2004 through 2013. What is the standard deviation for these data?
Source: U.S. Department of Transportation.

31. **FLIGHT CANCELLATIONS** The following table gives the percentage of flights canceled by U.S. carriers from 2004 through 2013:

Year	2004	2005	2006	2007	2008	2009	2010	2011	2012	2013
Flights Canceled (%)	2.38	3.03	1.89	3.47	3.23	1.81	3.89	4.35	1.24	1.93

Find the average percentage of flights canceled over the 10 years from 2004 through 2013. What is the standard deviation for these data?
Source: U.S. Department of Transportation.

32. **ACCESS TO CAPITAL** One of the key determinants of economic growth is access to capital. Using 54 variables to create an index of 1–7, with 7 being best possible access to capital, Milken Institute ranked the following as the top ten nations (although technically Hong Kong is not a nation) by the ability of their entrepreneurs to gain access to capital:

Country	Hong Kong	Netherlands	U.K.	Singapore	Switzerland
Index	5.70	5.59	5.57	5.56	5.55

Country	U.S.	Australia	Finland	Germany	Denmark
Index	5.55	5.31	5.24	5.23	5.22

Find the mean of the indices of the top ten nations. What is the standard deviation for these data?
Source: Milken Institute.

33. **ACCESS TO CAPITAL** Refer to Exercise 32. Milken Institute also ranked the following as the ten worst-performing nations by the ability of their entrepreneurs to gain access to capital:

Country	Peru	Mexico	Bulgaria	Brazil	Indonesia
Index	3.76	3.70	3.66	3.50	3.46

Country	Colombia	Turkey	Argentina	Venezuela	Russia
Index	3.46	3.43	3.20	2.88	2.19

Find the mean of the indices of the ten worst-performing nations. What is the standard deviation for these data?
Source: Milken Institute.

34. **LIGHTNING INJURIES** The number of injuries due to lightning in the United States from 1999 through 2008 is given in the following table:

Year	1999	2000	2001	2002	2003	2004	2005	2006	2007	2008
Number	243	371	372	256	238	279	309	245	139	207

What was the average number of injuries per year due to lightning in the United States from 1999 through 2008? What is the standard deviation for these data?
Source: National Oceanic and Atmosphere Administration.

35. **FEDERAL LIBRARIES** The federal government planned to spend more than $68.7 million in 2013 on programming, operations, and maintenance of 13 presidential libraries, including $3.5 million for the central management office. The total cost to be spent on each library is shown in the following table:

Library	Hoover	Roosevelt	Truman	Eisenhower	Kennedy	Johnson
Expenditure ($) (in millions)	2.4	4.8	4.4	3.8	7.0	5.7

Library	Nixon	Ford	Carter	Reagan	G.H.W. Bush	Clinton	G.W. Bush
Expenditure ($) (in millions)	4.9	5.2	4.7	5.7	5.4	5.3	5.9

Find the mean of the federal government expenditure for the 13 presidential libraries in 2013. What is the standard deviation for these data?
Source: National Archives and Records Administration.

36. ELECTION TURNOUT The percentage of the voting-age population who cast ballots in presidential elections from 1932 through 2008 are given in the following table:

Election Year	1932	1936	1940	1944	1948	1952	1956	1960	1964	1968
Turnout (%)	53	57	59	56	51	62	59	59	62	61

Election Year	1972	1976	1980	1984	1988	1992	1996	2000	2004	2008
Turnout (%)	55	54	53	53	50	55	49	51	55	57

Find the mean and the standard deviation of the given data.
Source: Federal Election Commission.

37. EXISTING-HOME SALES

a. The monthly supply (in millions) of single-family homes for sale in 2011 is summarized in the following table:

Month	Jan.	Feb.	Mar.	Apr.	May	June	July	Aug.	Sept.	Oct.	Nov.	Dec.
Supply	7.7	8.3	8.3	8.9	8.9	9.0	8.9	8.1	7.9	7.4	7.1	6.2

Find the mean and the standard deviation for the given data.

b. The monthly supply (in millions) of single-family homes for sale in 2012 is summarized in the following table:

Month	Jan.	Feb.	Mar.	Apr.	May	June	July	Aug.	Sept.	Oct.	Nov.	Dec.
Supply	6.1	6.1	6.1	6.4	6.4	6.5	6.3	6.0	5.5	5.3	4.8	4.4

Find the mean and the standard deviation of the given data.

c. What does a comparison of the means of the housing supplies for the two years tell you about the recovery of the Great Recession of 2009?
Source: Los Angeles Times.

38. EXAM SCORES The following table gives the scores of 30 students in a mathematics examination:

Scores	90–99	80–89	70–79	60–69	50–59
Students	4	8	12	4	2

Find the mean and the standard deviation of the distribution of the given data.
Hint: Assume that all scores lying within a group interval take the middle value of that group.

39. MARITAL STATUS OF MEN The number of married men (in thousands) between the ages of 20 and 44 in the United States in 2010 is given in the following table:

Age	20–24	25–29	30–34	35–39	40–44
Men	1182	3810	6104	7124	8195

Find the mean and the standard deviation of the given data.
Hint: See the hint for Exercise 38.
Source: U.S. Census Bureau.

40. TRAFFIC SURVEY In a survey of the distances between vehicles traveling along a stretch of Interstate Highway 5, the following data were obtained:

Distance, x (in feet)	$20 \le x < 50$	$50 \le x < 80$	$80 \le x < 110$
Frequency	232	410	143

Distance, x (in feet)	$110 \le x < 140$	$140 \le x < 170$	$170 \le x < 200$
Frequency	84	26	10

Find the mean and standard deviation for these data.

41. A probability distribution has a mean of 42 and a standard deviation of 2. Use Chebychev's inequality to find a bound on the probability that an outcome of the experiment lies between:
a. 38 and 46.   b. 32 and 52.

42. A probability distribution has a mean of 20 and a standard deviation of 3. Use Chebychev's inequality to find a bound on the probability that an outcome of the experiment lies between:
a. 15 and 25.   b. 10 and 30.

43. A probability distribution has a mean of 50 and a standard deviation of 1.4. Use Chebychev's inequality to find the value of c that guarantees the probability is at least 96% that an outcome of the experiment lies between $50 - c$ and $50 + c$.

44. Suppose X is a random variable with mean μ and standard deviation σ. If a large number of trials is observed, at least what percentage of these values is expected to lie between $\mu - 2\sigma$ and $\mu + 2\sigma$?

45. PRODUCT RELIABILITY The deluxe ionic hair dryer produced by Roland Electric has a mean expected lifetime of 24 months with a standard deviation of 3 months. Find a bound on the probability that one of these hair dryers will last between 20 and 28 months.

46. PRODUCT RELIABILITY A Christmas tree light has an expected life of 200 hr with a standard deviation of 2 hr.
a. Find a bound on the probability that one of these Christmas tree lights will last between 190 hr and 210 hr.
b. Suppose a large city uses 150,000 of these Christmas tree lights as part of its Christmas decorations. Estimate the number of lights that are likely to require replacement between 180 hr and 220 hr of use.

47. STARTING SALARIES The mean annual starting salary of a new graduate in a certain profession is $58,000 with a standard deviation of $500. Find a bound on the probability that the starting salary of a new graduate in this profession will be between $56,000 and $60,000.

48. QUALITY CONTROL Sugar packaged by a certain machine has a mean weight of 5 lb and a standard deviation of 0.02 lb. For what values of c can the manufacturer of the machinery claim that the sugar packaged by this machine has a weight between $5 - c$ and $5 + c$ lb with probability at least 96%?

In Exercises 49 and 50, determine whether the statement is true or false. If it is true, explain why it is true. If it is false, give an example to show why it is false.

49. Both the variance and the standard deviation of a random variable measure the spread of a probability distribution.

50. Chebychev's inequality is useless when $k \leq 1$.

8.6 Solutions to Self-Check Exercises

1. The mean of the random variable X is

$$\mu = (-4)(.1) + (-3)(.1) + (-1)(.2)$$
$$+ (0)(.3) + (2)(.1) + (5)(.2)$$
$$= 0.3$$

The variance of X is

$$\text{Var}(X) = (.1)(-4 - 0.3)^2 + (.1)(-3 - 0.3)^2$$
$$+ (.2)(-1 - 0.3)^2 + (.3)(0 - 0.3)^2$$
$$+ (.1)(2 - 0.3)^2 + (.2)(5 - 0.3)^2$$
$$= 8.01$$

The standard deviation of X is

$$\sigma = \sqrt{\text{Var}(X)} = \sqrt{8.01} \approx 2.83$$

2. We first compute the probability distribution of X from the given data as follows:

x	Relative Frequency of Occurrence	$P(X = x)$
46	1	.1
48	2	.2
50	3	.3
51	1	.1
52	2	.2
55	1	.1

The mean of X is

$$\mu = (.1)(46) + (.2)(48) + (.3)(50)$$
$$+ (.1)(51) + (.2)(52) + (.1)(55)$$
$$= 50.2$$

The variance of X is

$$\text{Var}(X) = (.1)(46 - 50.2)^2 + (.2)(48 - 50.2)^2$$
$$+ (.3)(50 - 50.2)^2 + (.1)(51 - 50.2)^2$$
$$+ (.2)(52 - 50.2)^2 + (.1)(55 - 50.2)^2$$
$$= 5.76$$

from which we deduce the standard deviation

$$\sigma = \sqrt{5.76}$$
$$= 2.4$$

USING TECHNOLOGY

Finding the Mean and Standard Deviation

The calculation of the mean and standard deviation of a random variable is facilitated by the use of a graphing utility.

APPLIED EXAMPLE 1 Age Distribution of Company Directors A survey conducted in a certain year of the Fortune 1000 companies revealed the following age distribution of the company directors:

Age (in years)	20–24	25–29	30–34	35–39	40–44	45–49	50–54
Directors	1	6	28	104	277	607	1142

Age (in years)	55–59	60–64	65–69	70–74	75–79	80–84	85–89
Directors	1413	1424	494	159	62	31	5

Let X denote the random variable taking on the values 1 through 14, where 1 corresponds to the age bracket 20–24, 2 corresponds to the age bracket 25–29, and so on.

a. Plot a histogram for the given data.
b. Find the mean and the standard deviation of these data. Interpret your results.
Source: Directorship.

Solution

a. Enter the values of X as $x_1 = 1$, $x_2 = 2, \ldots, x_{14} = 14$ and the corresponding values of Y as $y_1 = 1$, $y_2 = 6, \ldots, y_{14} = 5$. Then using the **DRAW** function from the Statistics menu of a graphing utility, we obtain the histogram shown in Figure T1.

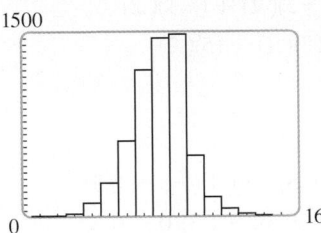

FIGURE **T1**
The histogram for the given data,
using the viewing window
$[0, 16] \times [0, 1500]$

b. Using the appropriate function from the Statistics menu, we find that $\bar{x} \approx 7.9193$ and $\sigma x \approx 1.6378$; that is, the mean of X is $\mu \approx 7.9$, and the standard deviation is $\sigma \approx 1.6$. Interpreting our results, we see that the average age of the directors is in the 55- to 60-year-old bracket.

TECHNOLOGY EXERCISES

1. a. Graph the histogram associated with the random variable X in Example 1, page 574.
   b. Find the mean and the standard deviation for these data.

2. a. Graph the histogram associated with the random variable Y in Example 1, page 574.
   b. Find the mean and the standard deviation for these data.

3. DRIVING AGE REQUIREMENTS The minimum age requirement for a regular driver's license differs from state to state. The frequency distribution for this age requirement in the 50 states is given in the following table:

Minimum Age (in years)	15	16	17	18	19	21
Frequency of Occurrence	1	15	4	28	1	1

   a. Graph the histogram associated with the random variable X associated with these data.
   b. Find the mean and the standard deviation for these data.

4. a. Graph the histogram associated with the data given in Exercise 16, page 581.
   b. Find the mean and the standard deviation for these data.

5. A sugar refiner uses a machine to pack sugar in 5-lb cartons. To check the machine's accuracy, cartons are selected at random and weighed. The results follow:

4.98	5.02	4.96	4.97	5.03
4.96	4.98	5.01	5.02	5.06
4.97	5.04	5.04	5.01	4.99
4.98	5.04	5.01	5.03	5.05
4.96	4.97	5.02	5.04	4.97
5.03	5.01	5.00	5.01	4.98

   a. Describe a random variable X that is associated with these data.
   b. Find the probability distribution for the random variable X.
   c. Compute the mean and standard deviation of X.

6. The scores of 25 students in a mathematics examination are as follows:

90	85	74	92	68	94	66
87	85	70	72	68	73	72
69	66	58	70	74	88	90
98	71	75	68			

a. Describe a random variable X that is associated with these data.

b. Find the probability distribution for the random variable X.

c. Compute the mean and standard deviation of X.

7. **HEIGHTS OF WOMEN** The following data, obtained from the records of the Westwood Health Club, give the heights (to the nearest inch) of 200 female members of the club:

Height (in inches)	62	$62\frac{1}{2}$	63	$63\frac{1}{2}$	64	$64\frac{1}{2}$	65	$65\frac{1}{2}$	66
Frequency	2	3	4	8	11	20	32	30	18

Height (in inches)	$66\frac{1}{2}$	67	$67\frac{1}{2}$	68	$68\frac{1}{2}$	69	$69\frac{1}{2}$	70	$70\frac{1}{2}$	71
Frequency	18	16	8	10	5	5	4	3	2	1

a. Plot a histogram for the given data.

b. Find the mean and the standard deviation of these data.

8. **AGE DISTRIBUTION IN A TOWN** The following table gives the distribution of the ages of the residents of the town of Monroe who are under the age of 40 years:

Age (in years)	0–3	4–7	8–11	12–15	16–19
Residents (in hundreds)	30	42	50	60	50

Age (in years)	20–23	24–27	28–31	32–35	36–39
Residents (in hundreds)	41	50	45	42	34

Let X denote the random variable taking on the values 1 through 10, where 1 corresponds to the range 0–3, . . . , and 10 corresponds to the range 36–39.

a. Plot a histogram for the given data.

b. Find the mean and the standard deviation of X.

CHAPTER 8   Summary of Principal Formulas and Terms

FORMULAS

1. Conditional probability	$P(B\mid A) = \dfrac{P(A \cap B)}{P(A)}, \ P(A) \neq 0$
2. Product rule	$P(A \cap B) = P(A) \cdot P(B\mid A)$
3. Test for independence	$P(A \cap B) = P(A) \cdot P(B)$
4. Mean of n numbers	$\bar{x} = \dfrac{x_1 + x_2 + \cdots + x_n}{n}$
5. Expected value	$E(X) = x_1 p_1 + x_2 p_2 + \cdots + x_n p_n$
6. Odds in favor of E occurring	$\dfrac{P(E)}{P(E^c)}$
7. Odds against E occurring	$\dfrac{P(E^c)}{P(E)}$
8. Probability of an event occurring given the odds	$\dfrac{a}{a + b}$
9. Variance of a random variable	$\begin{aligned} \text{Var}(X) = {} & p_1(x_1 - \mu)^2 \\ & + p_2(x_2 - \mu)^2 + \cdots \\ & + p_n(x_n - \mu)^2 \end{aligned}$
10. Standard deviation of a random variable	$\sigma = \sqrt{\text{Var}(X)}$
11. Chebychev's inequality	$P(\mu - k\sigma \leq X \leq \mu + k\sigma) \geq 1 - \dfrac{1}{k^2}$

TERMS

conditional probability (518)

finite stochastic process (522)

independent events (524)

Bayes' Theorem (534)

random variable (544)

finite discrete random variable (545)

infinite discrete random variable (545)

continuous random variable (545)

probability distribution of a random variable (546)

histogram (548)

average (mean) (558)

expected value (559)

median (567)

mode (567)

variance (574)

standard deviation (575)

CHAPTER 8 Concept Review Questions

Fill in the blanks.

1. The probability of the occurrence of event B given that the event A has already occurred is called the _____ probability of B given A.

2. If the outcome of one event does not depend on a second event, then the two events are said to be _____.

3. The probability of an event after the outcomes of an experiment have been observed is called a/an _____ _____ _____.

4. A rule that assigns a number to each outcome of a chance experiment is called a/an _____ variable.

5. If a random variable assumes only finitely many values, then it is called _____ discrete; if it takes on infinitely many values that can be arranged in a sequence, then it is called _____ discrete; if it takes on all real numbers in an interval, then it is said to be _____.

6. The expected value of a random variable X is given by the _____ of the products of the values assumed by the random variable and their associated probabilities. For example, if X assumes the values -2, 3, and 4 with associated probabilities $\frac{1}{2}$, $\frac{1}{4}$, and $\frac{1}{4}$, then its expected value is _____.

7. **a.** If the probability of an event E occurring is $P(E)$, then the odds in favor of E occurring are _____.
   b. If the odds in favor of an event E occurring are a to b, then the probability of E occurring is _____.

8. Suppose a random variable X takes on the values $x_1, x_2, \ldots, x_n$ with probabilities $p_1, p_2, \ldots, p_n$ and has a mean of μ. Then the variance of X is _____, and the standard deviation of X is _____.

CHAPTER 8 Review Exercises

1. Let E and F be two events, and suppose that $P(E) = .35$, $P(F) = .55$, and $P(E \cup F) = .70$. Find $P(E \mid F)$.

2. Suppose that $P(E) = .60$, $P(F) = .32$, and $P(E \cap F) = .22$. Are E and F independent?

3. Suppose that E and F are independent events. If $P(E) = .32$ and $P(E \cap F) = .16$, what is $P(F)$?

The accompanying tree diagram represents an experiment consisting of two trials. In Exercises 4–8, use the diagram to find the given probability.

4. $P(A \cap E)$

5. $P(B \cap E)$

6. $P(C \cap E)$

7. $P(E)$

8. $P(A \mid E)$

9. An experiment consists of tossing a fair coin three times and observing the outcomes. Let A be the event that at least one head is thrown, and let B be the event that at most two tails are thrown.
   a. Find $P(A)$.   **b.** Find $P(B)$.
   c. Are A and B independent events?

10. **QUALITY CONTROL** In a group of 20 ballpoint pens on a shelf in the stationery department of Metro Department Store, 2 are known to be defective. If a customer selects 3 of these pens, what is the probability that:
    a. At least 1 is defective?
    b. No more than 1 is defective?

11. **BIRTHDAY PROBLEM** Five people are selected at random. What is the probability that none of the people in this group were born on the same day of the week?

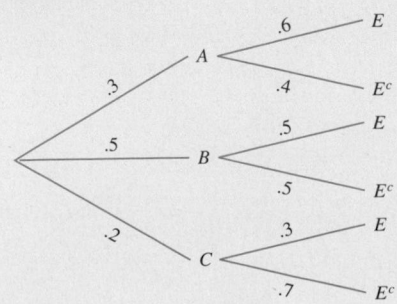

12. A pair of fair dice is rolled. What is the probability that the sum of the numbers falling uppermost is 8 if it is known that the two numbers are different?

13. A fair die is rolled three times. What is the probability that it shows an even number in the first toss, an odd number in the second toss, and a 1 on the third toss? Assume that the outcomes of the tosses are independent.

14. A fair die is rolled, a fair coin is tossed, and a card is drawn from a standard deck of 52 playing cards. Assuming these events are independent, what is the probability that the number falling uppermost on the die is a 6, the coin shows a tail, and the card drawn is a face card?

Three cards are drawn at random without replacement from a standard deck of 52 playing cards. In Exercises 15–19, find the probability of each of the given events.

15. All three cards are aces.

16. All three cards are face cards.

17. The second and third cards are red.

18. The second card is black, given that the first card was red.

19. The second card is a club, given that the first card was black.

20. Three balls are selected at random without replacement from an urn containing three white balls and four blue balls. Let the random variable X denote the number of blue balls drawn.
   a. List the outcomes of this experiment.
   b. Find the value assigned to each outcome of this experiment by the random variable X.
   c. Find the probability distribution of the random variable associated with this experiment.
   d. Draw the histogram representing this distribution.

21. LIFE INSURANCE POLICIES A man purchased a $25,000, 1-year term-life insurance policy for $375. Assuming that the probability that he will live for another year is .989, find the company's expected gain.

22. The probability distribution of a random variable X is shown in the following table:

x	$P(X = x)$
0	.1
1	.1
2	.2
3	.3
4	.2
5	.1

   a. Compute $P(1 \le X \le 4)$.
   b. Compute the mean and standard deviation of X.

23. CUSTOMER SURVEYS The sales department of Thompson Drug Company released the accompanying data concerning the sales of a certain pain reliever manufactured by the company.

Pain Reliever	Drug Sold (%)	Group Sold in Extra-Strength Dosage (%)
Group I (capsule form)	57	38
Group II (tablet form)	43	31

If a customer purchased the extra-strength dosage of this drug, what is the probability that it was in capsule form?

24. OPINION POLLS A survey involving 600 Democrats, 400 Republicans, and 200 Independents asked the question: Do you favor or oppose eliminating taxes on dividends paid to shareholders? The following results were obtained:

Answer	Democrats (%)	Republicans (%)	Independents (%)
Favor	29	66	48
Opposed	71	34	52

If a randomly chosen respondent in the survey answered "favor," what is the probability that he or she is an Independent?
Source: TechnoMetrica Market Intelligence.

25. OPINION POLLS A poll was conducted among 500 registered voters in a certain area regarding their position on a national lottery to raise revenue for the government. The results of the poll are shown in the accompanying table.

Sex	Voters Polled (%)	Favoring Lottery (%)	Not Favoring Lottery (%)	Expressing No Opinion (%)
Male	51	62	32	6
Female	49	68	28	4

What is the probability that a registered voter who:
   a. Favored a national lottery was a woman?
   b. Expressed no opinion regarding the lottery was a woman?

26. IMPACT OF GAS PRICES ON CONSUMERS In a survey of 1012 adults aged 18 years and older conducted by Social Science Research Solutions, 24% of the respondents aged 18–49 years said that they would increase their use of public transportation, whereas only 11% of those 50 years or older said they would do so. There were 742 adults aged 18–49 years and 270 adults who were 50 years or older in the survey. If a respondent in the survey selected at random reported that he or she would not increase the use of public transportation, what is the probability that he or she was 50 years or older?
Source: Social Science Research Solutions.

27. GUN OWNERS IN THE HOUSE OF REPRESENTATIVES As of January 3, 2013, the U.S. House of Representatives was made up of 200 Democrats and 232 Republicans. There were 3 seats vacant. In a survey of the House of Representatives at that time, each representative was asked whether he or she owned at least one gun. Of the Democrats, 30 declared themselves gun owners. Of the Republicans, 93 declared themselves gun owners. If a representative participating in the survey was picked at random and turned out to be a gun owner, what is the probability that he or she is a Democrat?
Source: Gannett Washington Bureau.

28. COMMUTING According to the 2011 U.S. Census, 121,298,000 workers who do not work at home have travel times of less than 1 hr, and 10,979,000 workers who do not work at home have travel times of 1 hr or longer. Of those workers whose travel times are less than 1 hr, 91.4% drive alone or carpool, 1.6% take the subway or railroad, 2.1% take other public transportation, and 4.9% use other means. Of those workers whose travel times are 1 hr or longer, 74.0% drive alone or carpool, 11.8% take the subway or railroad, 11.2% take other public transportation, and 3.0% use other means. If a worker chosen at random takes the subway or railroad, what is the probability that he or she has travel times of less than 1 hr?
Source: U.S. Census Bureau.

29. DRIVING AGE REQUIREMENTS The minimum age requirement for a regular driver's license differs from state to state. The frequency distribution for this age requirement in the 50 states is given in the following table:

Minimum Age (in years)	15	16	17	18	19	21
Frequency of Occurrence	1	15	4	28	1	1

a. Describe a random variable X that is associated with these data.
b. Find the probability distribution for the random variable X.
c. Compute the mean, variance, and standard deviation of X.

30. ANNUAL FOOD EXPENDITURE According to a report of the U.S. Bureau of Labor Statistics, the average annual food expenditures by quintiles of household income before taxes are as follows:

	Lowest 20%	Second 20%	Third 20%	Fourth 20%	Highest 20%
Income Quintile					
Expenditures ($)	3501	4568	5482	7522	10,780

Let the random variable X denote a randomly chosen quintile (in income) within the set of all households.
a. Find the probability distribution associated with these data.

b. What percentage of the total food expenditures is incurred by the top 40% by income of the population? By the lowest 40% by income of the population?
Source: U.S. Bureau of Labor Statistics.

31. TRAFFIC A traffic survey of the speeds of vehicles traveling along a stretch of Hampton Road between 4 P.M. and 6 P.M. yielded the following results:

Speed (in mph)	30–34	35–39	40–44	45–49	50–54
Probability	.07	.28	.42	.18	.05

Find the average speed of the vehicles.

32. EXPECTED PROFIT A buyer for Discount Fashions, an outlet for women's apparel, is considering buying a batch of clothing for $64,000. She estimates that the company will be able to sell it for $80,000, $75,000, or $70,000 with probabilities of .30, .60, and .10, respectively. On the basis of these estimates, what will be the company's expected gross profit?

33. HEIGHTS OF WOMEN The heights of 4000 women who participated in a recent survey were found to have a mean of 64.5 in. and a standard deviation of 2.5 in. Use Chebychev's inequality to estimate the probability that the height of a woman who participated in the survey will fall within 2 standard deviations of the mean—that is, that her height will be between 59.5 and 69.5 in.

34. NETFLIX REVENUE FROM STREAMING SUBSCRIBERS The revenue of Netflix from its streaming subscribers (in millions of dollars) for the five quarters beginning with the first quarter of 2012 are summarized in the following table:

	2012				2013
Quarter	Q1	Q2	Q3	Q4	Q1
Revenue	26.2	27.5	28.4	35.8	37.1

Find the average quarterly revenue of Netflix from its streaming subscribers for the five quarters in question. What is the standard deviation?
Source: Company reports.

35. MARITAL STATUS OF WOMEN The number of single women between the ages of 20 and 44 in the United States in 2010 is given in the following table:

Age (in years)	20–24	25–29	30–34	35–39	40–44
Women (in thousands)	8296	5026	2678	1768	1430

Find the mean and the standard deviation of the given data.
Hint: Assume that all values lying within a group interval take the middle value of that group.
Source: U.S. Census Bureau.

CHAPTER 8   Before Moving On . . .

1. Suppose A and B are independent events with $P(A) = .3$ and $P(B) = .6$. Find $P(A \cup B)$.

2. The following tree diagram represents a two-stage experiment. Use the diagram to find $P(A \mid D)$.

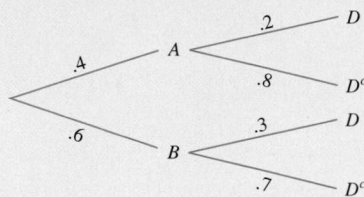

3. The values taken on by a random variable X and the frequency of their occurrence are shown in the following table. Find the probability distribution of X.

x	-3	-2	0	1	2	3
Frequency of Occurrence	4	8	20	24	16	8

4. The probability distribution of the random variable X is shown in the following table:

x	-4	-3	-1	0	1	3
$P(X = x)$	.06	.14	.32	.28	.12	.08

Find:
a. $P(X \le 0)$ b. $P(-4 \le X \le 1)$

5. Find the mean, variance, and standard deviation of a random variable X having the following probability distribution:

x	-3	-1	0	1	3	5
$P(X = x)$	.08	.24	.32	.16	.12	.08

9 The Derivative

IN THIS CHAPTER, we begin the study of differential calculus. Historically, differential calculus was developed in response to the problem of finding the tangent line to an arbitrary curve. But it quickly became apparent that solving this problem provided mathematicians with a method of solving many practical problems involving the rate of change of one quantity with respect to another. The basic tool used in differential calculus is the *derivative* of a function. The concept of the derivative is based, in turn, on a more fundamental notion: that of the *limit* of a function.

What happens to the sales of a DVD recording of a certain hit movie over a 10-year period after it is first released into the market? In Example 6, page 664, you will see how to find the rate of change of sales for the DVD over the first 10 years after its release.

9.1   Limits

Introduction to Calculus

Historically, the development of calculus by Isaac Newton (1642–1727) and Gottfried Wilhelm Leibniz (1646–1716) resulted from the investigation of the following problems:

1. Finding the tangent line to a curve at a given point on the curve (Figure 1a)
2. Finding the area of a planar region bounded by an arbitrary curve (Figure 1b)

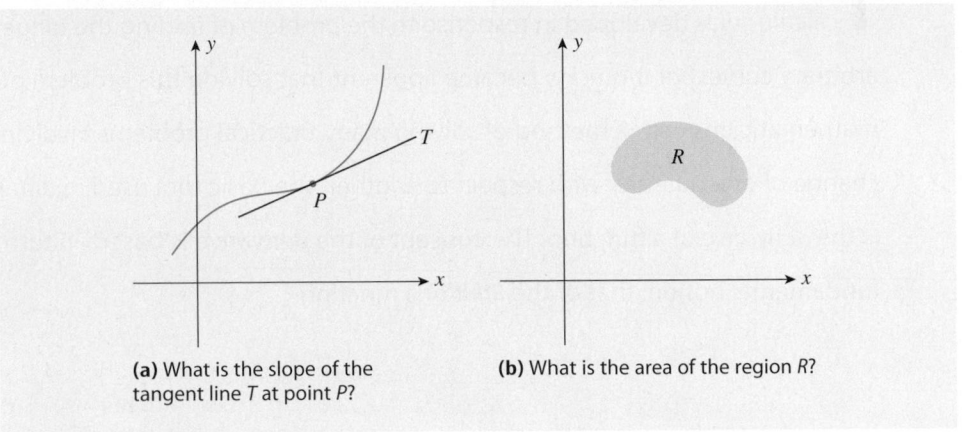

(a) What is the slope of the tangent line T at point P?

(b) What is the area of the region R?

FIGURE **1**

The tangent-line problem might appear to be unrelated to any practical applications of mathematics, but as you will see later, the problem of finding the *rate of change* of one quantity with respect to another is mathematically equivalent to the geometric problem of finding the slope of the *tangent line* to a curve at a given point on the curve. It is precisely the discovery of the relationship between these two problems that spurred the development of calculus in the seventeenth century and made it such an indispensable tool for solving practical problems. The following are a few examples of such problems:

- Finding the velocity of an object
- Finding the rate of change of a bacteria population with respect to time
- Finding the rate of change of a company's profit with respect to time
- Finding the rate of change of a travel agency's revenue with respect to the agency's expenditure for advertising

The study of the tangent-line problem led to the creation of *differential calculus*, which relies on the concept of the *derivative* of a function. The study of the area problem led to the creation of *integral calculus*, which relies on the concept of the *antiderivative*, or *integral*, of a function. (The derivative of a function and the integral of a function are intimately related, as you will see in Section 11.4.) Both the derivative of a function and the integral of a function are defined in terms of a more fundamental concept: the limit, our next topic.

A Real-Life Example

From data obtained in a test run conducted on a prototype of a maglev (magnetic levitation train), which moves along a straight monorail track, engineers have determined that the position of the maglev (in feet) from the origin at time t (in seconds) is given by

$$s = f(t) = 4t^2 \qquad (0 \le t \le 30) \tag{1}$$

where f is called the **position function** of the maglev. The position of the maglev at time $t = 0, 1, 2, 3, \ldots, 10$, measured from its initial position, is

$$f(0) = 0 \qquad f(1) = 4 \qquad f(2) = 16 \qquad f(3) = 36, \ldots \qquad f(10) = 400$$

feet (Figure 2).

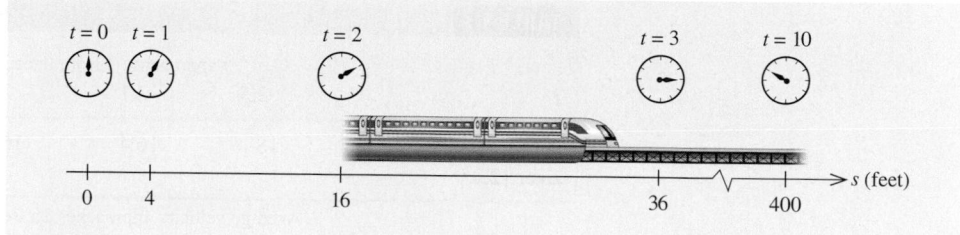

FIGURE **2**
A maglev moving along an elevated monorail track

Suppose we want to find the velocity of the maglev at $t = 2$. This is just the velocity of the maglev as shown on its speedometer at that precise instant of time. Offhand, calculating this quantity using only Equation (1) appears to be an impossible task; but consider what quantities we *can* compute using this relationship. Obviously, we can compute the position of the maglev at any time t as we did earlier for some selected values of t. Using these values, we can then compute the *average velocity* of the maglev over an interval of time. For example, the average velocity of the train over the time interval $[2, 4]$ is given by

$$\begin{aligned} \frac{\text{Distance covered}}{\text{Time elapsed}} &= \frac{f(4) - f(2)}{4 - 2} \\ &= \frac{4(4^2) - 4(2^2)}{2} \\ &= \frac{64 - 16}{2} = 24 \end{aligned}$$

or 24 feet/second.

Although this is not quite the velocity of the maglev at $t = 2$, it does provide us with an approximation of its velocity at that time.

Can we do better? Intuitively, the smaller the time interval we pick (with $t = 2$ as the left endpoint), the better the average velocity over that time interval will approximate the actual velocity of the maglev at $t = 2$.*

Now, let's describe this process in general terms. Let $t > 2$. Then, the average velocity of the maglev over the time interval $[2, t]$ is given by

$$\frac{f(t) - f(2)}{t - 2} = \frac{4t^2 - 4(2^2)}{t - 2} = \frac{4(t^2 - 4)}{t - 2} \tag{2}$$

By choosing the values of t closer and closer to 2, we obtain a sequence of numbers that give the average velocities of the maglev over smaller and smaller time intervals. As we observed earlier, this sequence of numbers should approach the *instantaneous velocity* of the train at $t = 2$.

---

*Actually, any interval containing $t = 2$ will do.

Let's try some sample calculations. Using Equation (2) and taking the sequence $t = 2.5, 2.1, 2.01, 2.001$, and 2.0001, which approaches 2, we find the following:

The average velocity over $[2, 2.5]$ is $\dfrac{4(2.5^2 - 4)}{2.5 - 2} = 18$, or 18 feet/second.

The average velocity over $[2, 2.1]$ is $\dfrac{4(2.1^2 - 4)}{2.1 - 2} = 16.4$, or 16.4 feet/second.

and so forth. These results are summarized in Table 1.

TABLE 1

		t approaches 2 from the right.			
t	2.5	2.1	2.01	2.001	2.0001
Average Velocity over $[2, t]$	18	16.4	16.04	16.004	16.0004

Average velocity approaches 16 from the right.

From Table 1, we see that the average velocity of the maglev seems to approach the number 16 as it is computed over smaller and smaller time intervals. These computations suggest that the instantaneous velocity of the train at $t = 2$ is 16 feet/second.

Note   Notice that we cannot obtain the instantaneous velocity for the maglev at $t = 2$ by substituting $t = 2$ into Equation (2) because this value of t is not in the domain of the average velocity function.   ■

Intuitive Definition of a Limit

Consider the function g defined by

$$g(t) = \frac{4(t^2 - 4)}{t - 2}$$

which gives the average velocity of the maglev [see Equation (2)]. Suppose we are required to determine the value that $g(t)$ approaches as t approaches the (fixed) number 2. If we take the sequence of values of t approaching 2 from the right-hand side, as we did earlier, we see that $g(t)$ approaches the number 16. Similarly, if we take a sequence of values of t approaching 2 from the left, such as $t = 1.5, 1.9, 1.99, 1.999$, and 1.9999, we obtain the results shown in Table 2.

TABLE 2

		t approaches 2 from the left.			
t	1.5	1.9	1.99	1.999	1.9999
$g(t)$	14	15.6	15.96	15.996	15.9996

Average velocity approaches 16 from the left.

Observe that $g(t)$ approaches the number 16 as t approaches 2—this time from the left-hand side. In other words, as t approaches 2 from *either* side of 2, $g(t)$ approaches 16. In this situation, we say that the limit of $g(t)$ as t approaches 2 is 16, written

$$\lim_{t \to 2} g(t) = \lim_{t \to 2} \frac{4(t^2 - 4)}{t - 2} = 16$$

The graph of the function g, shown in Figure 3, confirms this observation.

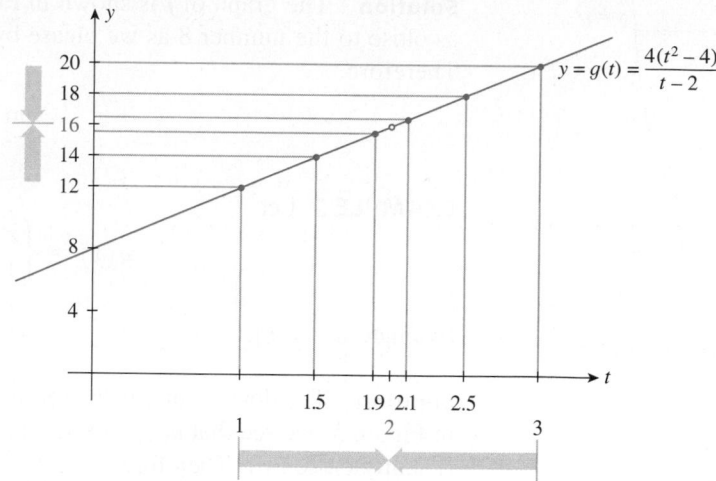

FIGURE 3
As t approaches $t = 2$ from either direction, $g(t)$ approaches $y = 16$.

Observe that the point $t = 2$ is not in the domain of the function g [for this reason, the point $(2, 16)$ is missing from the graph of g]. This, however, is inconsequential because the value, if any, of $g(t)$ at $t = 2$ plays no role in computing the limit.

This example leads to the following informal definition.

> ### Limit of a Function
> The function f has the **limit** L as x approaches a, written
> $$\lim_{x \to a} f(x) = L$$
> if the value of $f(x)$ can be made as close to the number L as we please by taking x sufficiently close to (but not equal to) a.

> ### Exploring with TECHNOLOGY
>
> **1.** Use a graphing utility to plot the graph of
> $$g(x) = \frac{4(x^2 - 4)}{x - 2}$$
> in the viewing window $[0, 3] \times [0, 20]$.
> **2.** Use **ZOOM** and **TRACE** to describe what happens to the values of $g(x)$ as x approaches 2, first from the right and then from the left.
> **3.** What happens to the y-value when you try to evaluate $g(x)$ at $x = 2$? Explain.
> **4.** Reconcile your results with those of the preceding example.

Evaluating the Limit of a Function

Let's now consider some examples involving the computation of limits.

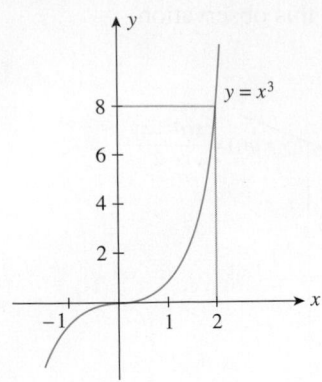

FIGURE 4
f(x) is close to 8 whenever x is close to 2.

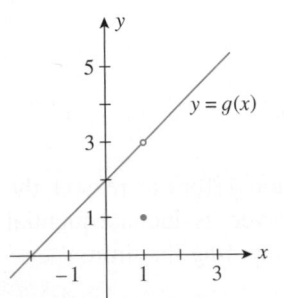

FIGURE 5
lim g(x) = 3
x→1

EXAMPLE 1 Let $f(x) = x^3$ and evaluate $\lim\limits_{x \to 2} f(x)$.

Solution   The graph of f is shown in Figure 4. You can see that $f(x)$ can be made as close to the number 8 as we please by taking x sufficiently close to 2. Therefore,

$$\lim_{x \to 2} x^3 = 8$$

EXAMPLE 2 Let

$$g(x) = \begin{cases} x + 2 & \text{if } x \neq 1 \\ 1 & \text{if } x = 1 \end{cases}$$

Evaluate $\lim\limits_{x \to 1} g(x)$.

Solution   The domain of g is the set of all real numbers. From the graph of g shown in Figure 5, we see that $g(x)$ can be made as close to 3 as we please by taking x sufficiently close to 1. Therefore,

$$\lim_{x \to 1} g(x) = 3$$

Observe that $g(1) = 1$, which is not equal to the limit of the function g as x approaches 1. [Once again, the value of $g(x)$ at $x = 1$ has no bearing on the existence or value of the limit of g as x approaches 1.]

EXAMPLE 3 Evaluate the limit of the following functions as x approaches the indicated point.

a. $f(x) = \begin{cases} -1 & \text{if } x < 0 \\ 1 & \text{if } x \geq 0 \end{cases}; x = 0$ **b.** $g(x) = \dfrac{1}{x^2}; x = 0$

Solution   The graphs of the functions f and g are shown in Figure 6.

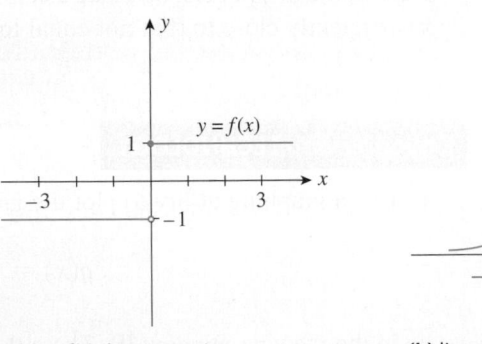

(a) $\lim\limits_{x \to 0} f(x)$ does not exist.

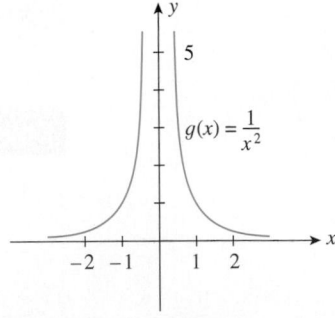

(b) $\lim\limits_{x \to 0} g(x)$ does not exist.

FIGURE 6

a. Referring to Figure 6a, we see that no matter how close x is to zero, $f(x)$ takes on the values 1 or -1, depending on whether x is positive or negative. Thus, there is no *single* real number L that $f(x)$ approaches as x approaches zero. We conclude that the limit of $f(x)$ does *not* exist as x approaches zero.

b. Referring to Figure 6b, we see that as x approaches zero (from either side), $g(x)$ increases without bound and thus does not approach any specific real number. We conclude, accordingly, that the limit of $g(x)$ does *not* exist as x approaches zero.

Explore and Discuss

Consider the graph of the function h shown in the following figure.

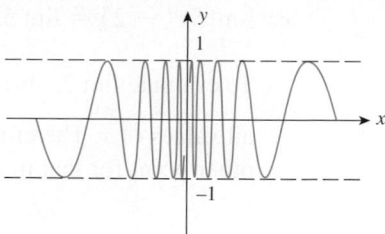

It has the property that as x approaches zero from either the right or the left, the curve oscillates more and more frequently between the lines $y = -1$ and $y = 1$.

1. Explain why $\lim\limits_{x \to 0} h(x)$ does not exist.

2. Compare this function with those in Example 3. More specifically, discuss the different ways in which the functions fail to have a limit at $x = 0$.

Until now, we have relied on knowing the actual values of a function or the graph of a function near $x = a$ to help us evaluate the limit of the function $f(x)$ as x approaches a. The following properties of limits, which we list without proof, enable us to evaluate limits of functions algebraically.

THEOREM 1

Properties of Limits

Suppose

$$\lim\limits_{x \to a} f(x) = L \quad \text{and} \quad \lim\limits_{x \to a} g(x) = M$$

Then

1. $\lim\limits_{x \to a} [f(x)]^r = \left[\lim\limits_{x \to a} f(x)\right]^r = L^r$ r, a positive constant

2. $\lim\limits_{x \to a} cf(x) = c \lim\limits_{x \to a} f(x) = cL$ c, a real number

3. $\lim\limits_{x \to a} [f(x) \pm g(x)] = \lim\limits_{x \to a} f(x) \pm \lim\limits_{x \to a} g(x) = L \pm M$

4. $\lim\limits_{x \to a} [f(x)g(x)] = \left[\lim\limits_{x \to a} f(x)\right]\left[\lim\limits_{x \to a} g(x)\right] = LM$

5. $\lim\limits_{x \to a} \dfrac{f(x)}{g(x)} = \dfrac{\lim\limits_{x \to a} f(x)}{\lim\limits_{x \to a} g(x)} = \dfrac{L}{M}$ Provided that $M \neq 0$

EXAMPLE 4 Use Theorem 1 to evaluate the following limits.

a. $\lim\limits_{x \to 2} x^3$ **b.** $\lim\limits_{x \to 4} 5x^{3/2}$ **c.** $\lim\limits_{x \to 1} (5x^4 - 2)$

d. $\lim\limits_{x \to 3} 2x^3\sqrt{x^2 + 7}$ **e.** $\lim\limits_{x \to 2} \dfrac{2x^2 + 1}{x + 1}$

Solution

a. $\lim\limits_{x \to 2} x^3 = \left[\lim\limits_{x \to 2} x\right]^3$ Property 1

$\qquad = 2^3 = 8$ $\lim\limits_{x \to 2} x = 2$

b. $\lim\limits_{x\to4} 5x^{3/2} = 5\left[\lim\limits_{x\to4} x^{3/2}\right]$ Property 2

$\qquad\qquad = 5(4)^{3/2} = 40$ Property 1

c. $\lim\limits_{x\to1}(5x^4 - 2) = \lim\limits_{x\to1} 5x^4 - \lim\limits_{x\to1} 2$ Property 3

To evaluate $\lim\limits_{x\to1} 2$, observe that the constant function $g(x) = 2$ has value 2 for all values of x. Therefore, $g(x)$ must approach the limit 2 as x approaches 1 (or any other point for that matter!). Therefore,

$$\lim\limits_{x\to1}(5x^4 - 2) = 5(1)^4 - 2 = 3$$

d. $\lim\limits_{x\to3} 2x^3 \sqrt{x^2 + 7} = 2 \lim\limits_{x\to3} x^3 \sqrt{x^2 + 7}$ Property 2

$\qquad\qquad = 2 \lim\limits_{x\to3} x^3 \lim\limits_{x\to3} \sqrt{x^2 + 7}$ Property 4

$\qquad\qquad = 2(3)^3 \sqrt{3^2 + 7}$ Properties 1 and 3

$\qquad\qquad = 2(27)\sqrt{16} = 216$

e. $\lim\limits_{x\to2} \dfrac{2x^2 + 1}{x + 1} = \dfrac{\lim\limits_{x\to2}(2x^2 + 1)}{\lim\limits_{x\to2}(x + 1)}$ Property 5

$\qquad\qquad = \dfrac{2(2)^2 + 1}{2 + 1} = \dfrac{9}{3} = 3$

Indeterminate Forms

Let's emphasize once again that Property 5 of limits is valid only when the limit of the function that appears in the denominator is not equal to zero at the number in question.

If the numerator has a limit different from zero and the denominator has a limit equal to zero, then the limit of the quotient does not exist at the number in question. This is the case with the function $g(x) = 1/x^2$ in Example 3b. Here, as x approaches zero, the numerator approaches 1 but the denominator approaches zero, so the quotient becomes arbitrarily large. Thus, as was observed earlier, the limit does not exist.

Next, consider

$$\lim\limits_{x\to2} \frac{4(x^2 - 4)}{x - 2}$$

which we evaluated earlier by looking at the values of the function for x near $x = 2$. If we attempt to evaluate this expression by applying Property 5 of limits, we see that both the numerator and denominator of the function

$$\frac{4(x^2 - 4)}{x - 2}$$

approach zero as x approaches 2; that is, we obtain an expression of the form 0/0. In this event, we say that the limit of the quotient $f(x)/g(x)$ as x approaches 2 has the **indeterminate form 0/0.**

We need to evaluate limits of this type when we discuss the derivative of a function, a fundamental concept in the study of calculus. As the name suggests, the meaningless expression 0/0 does not provide us with a solution to our problem. One strategy that can be used to solve this type of problem follows.

Strategy for Evaluating Indeterminate Forms

1. Replace the given function with an appropriate one that takes on the same values as the original function everywhere except at $x = a$.

2. Evaluate the limit of this function as x approaches a.

Examples 5 and 6 illustrate this strategy.

EXAMPLE 5 Evaluate:

$$\lim_{x \to 2} \frac{4(x^2 - 4)}{x - 2}$$

Solution Since both the numerator and the denominator of this expression approach zero as x approaches 2, we have the indeterminate form $0/0$. We rewrite

$$\frac{4(x^2 - 4)}{x - 2} = \frac{4(x - 2)(x + 2)}{(x - 2)}$$

which, upon cancellation of the common factors, is equivalent to $4(x + 2)$, provided that $x \neq 2$. Next, we replace $4(x^2 - 4)/(x - 2)$ with $4(x + 2)$ and find that

$$\lim_{x \to 2} \frac{4(x^2 - 4)}{x - 2} = \lim_{x \to 2} 4(x + 2) = 16$$

The graphs of the functions

$$f(x) = \frac{4(x^2 - 4)}{x - 2} \quad \text{and} \quad g(x) = 4(x + 2)$$

are shown in Figure 7. Observe that the graphs are identical except when $x = 2$. The function g is defined for all values of x and, in particular, its value at $x = 2$ is $g(2) = 4(2 + 2) = 16$. Thus, the point $(2, 16)$ is on the graph of g. However, the function f is not defined at $x = 2$. Since $f(x) = g(x)$ for all values of x except $x = 2$, it follows that the graph of f must look exactly like the graph of g, with the exception that the point $(2, 16)$ is missing from the graph of f. This illustrates graphically why we can evaluate the limit of f by evaluating the limit of the "equivalent" function g.

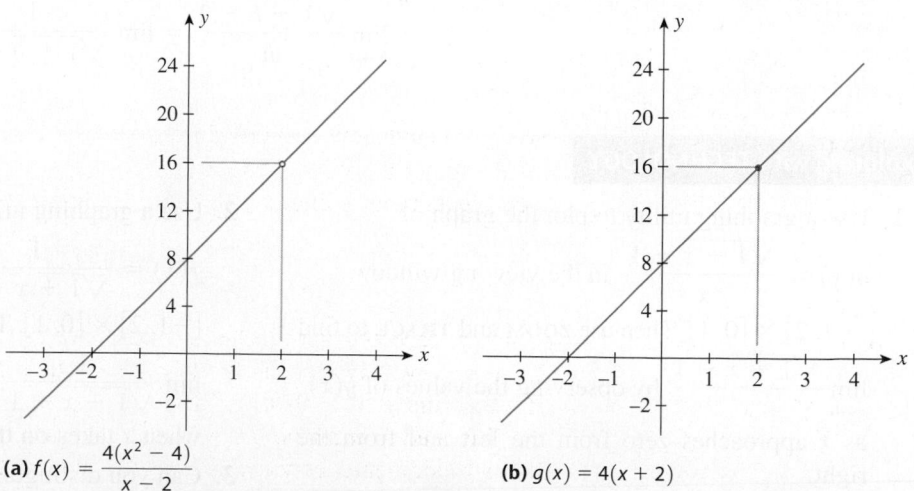

(a) $f(x) = \dfrac{4(x^2 - 4)}{x - 2}$

(b) $g(x) = 4(x + 2)$

FIGURE 7
The graphs of $f(x)$ and $g(x)$ are identical except at the point $(2, 16)$.

Note   Notice that the limit in Example 5 is the same limit that we evaluated earlier when we discussed the instantaneous velocity of a maglev at a specified time.

Exploring with TECHNOLOGY

1. Use a graphing utility to plot the graph of

$$f(x) = \frac{4(x^2 - 4)}{x - 2}$$

in the viewing window $[0, 3] \times [0, 20]$. Then use **ZOOM** and **TRACE** to find

$$\lim_{x \to 2} \frac{4(x^2 - 4)}{x - 2}$$

What happens to the y-value when you try to evaluate $f(x)$ at $x = 2$? Explain.

2. Use a graphing utility to plot the graph of $g(x) = 4(x + 2)$ in the viewing window $[0, 3] \times [0, 20]$. Then use **ZOOM** and **TRACE** to find $\lim_{x \to 2} 4(x + 2)$.

3. Can you distinguish between the graphs of f and g?

4. Reconcile your results with those of Example 5.

EXAMPLE 6 Evaluate:

$$\lim_{h \to 0} \frac{\sqrt{1 + h} - 1}{h}$$

(x^2) The algebra icon is used to indicate that the algebraic computation or problem-solving skill used in the example is reviewed on the referenced page. For instance, in Example 6, if you refer to page 43, you will find a review of the process of rationalizing an algebraic fraction. This is followed by a worked example.

Solution   Letting h approach zero, we obtain the indeterminate form 0/0. Next, we rationalize the numerator of the quotient by multiplying both the numerator and the denominator by the expression $(\sqrt{1 + h} + 1)$, obtaining

$$\frac{\sqrt{1 + h} - 1}{h} = \frac{(\sqrt{1 + h} - 1)(\sqrt{1 + h} + 1)}{h(\sqrt{1 + h} + 1)} \qquad (\sqrt{a} - \sqrt{b})(\sqrt{a} + \sqrt{b}) = a - b$$

$$= \frac{1 + h - 1}{h(\sqrt{1 + h} + 1)} \qquad\qquad (x^2)\ \text{See page 43.}$$

$$= \frac{h}{h(\sqrt{1 + h} + 1)}$$

$$= \frac{1}{\sqrt{1 + h} + 1}$$

Therefore,

$$\lim_{h \to 0} \frac{\sqrt{1 + h} - 1}{h} = \lim_{h \to 0} \frac{1}{\sqrt{1 + h} + 1} = \frac{1}{\sqrt{1} + 1} = \frac{1}{2}$$

Exploring with TECHNOLOGY

1. Use a graphing utility to plot the graph of

$$g(x) = \frac{\sqrt{1 + x} - 1}{x}$$ in the viewing window

$[-1, 2] \times [0, 1]$. Then use **ZOOM** and **TRACE** to find

$$\lim_{x \to 0} \frac{\sqrt{1 + x} - 1}{x}$$ by observing the values of $g(x)$ as x approaches zero from the left and from the right.

2. Use a graphing utility to plot the graph of

$$f(x) = \frac{1}{\sqrt{1 + x} + 1}$$ in the viewing window

$[-1, 2] \times [0, 1]$. Then use **ZOOM** and **TRACE** to find

$$\lim_{x \to 0} \frac{1}{\sqrt{1 + x} + 1}.$$ What happens to the y-value when x takes on the value zero? Explain.

3. Can you distinguish between the graphs of f and g?

4. Reconcile your results with those of Example 6.

Limits at Infinity

Up to now, we have studied the limit of a function as x approaches a (finite) number a. There are occasions, however, when we want to know whether $f(x)$ approaches a unique number as x increases without bound. Consider, for example, the function P, giving the number of fruit flies (*Drosophila*) in a container under controlled laboratory conditions, as a function of a time t. The graph of P is shown in Figure 8. You can see from the graph of P that as t increases without bound (gets larger and larger), $P(t)$ approaches the number 400. This number, called the *carrying capacity* of the environment, is determined by the amount of living space and food available, as well as other environmental factors.

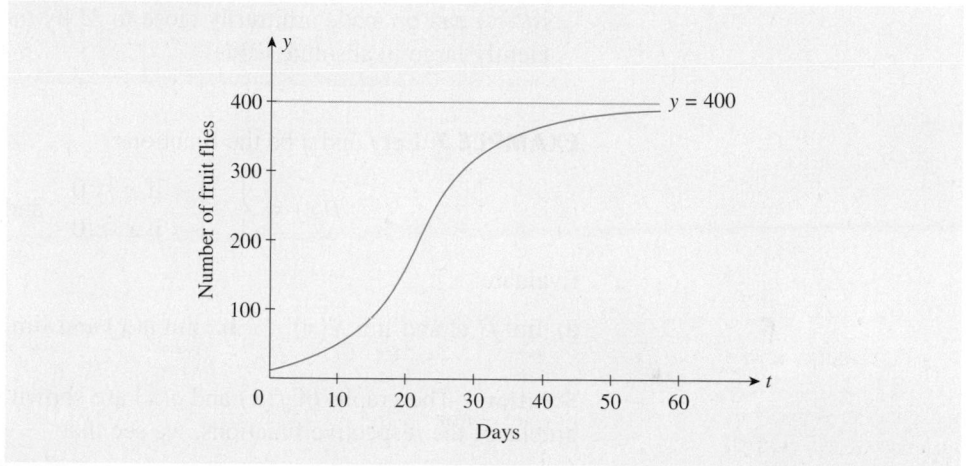

FIGURE **8**
The graph of $P(t)$ gives the population of fruit flies in a laboratory experiment.

As another example, suppose we are given the function

$$f(x) = \frac{2x^2}{1 + x^2}$$

and we want to determine what happens to $f(x)$ as x gets larger and larger. Picking the sequence of numbers 1, 2, 5, 10, 100, and 1000 and computing the corresponding values of $f(x)$, we obtain the following table of values:

x	1	2	5	10	100	1000
$f(x)$	1	1.6	1.92	1.98	1.9998	1.999998

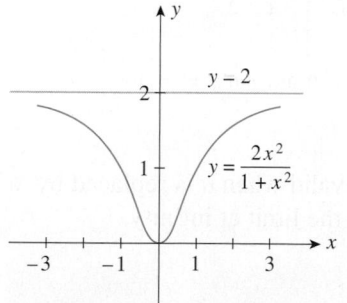

FIGURE **9**
The graph of

$$y = \frac{2x^2}{1 + x^2}$$

has a horizontal asymptote at $y = 2$.

From the table, we see that as x gets larger and larger, $f(x)$ gets closer and closer to 2. The graph of the function f shown in Figure 9 confirms this observation. We call the line $y = 2$ a **horizontal asymptote.*** In this situation, we say that the limit of the function $f(x)$ as x increases without bound is 2, written

$$\lim_{x \to \infty} \frac{2x^2}{1 + x^2} = 2$$

In the general case, the following definition for a **limit of a function at infinity** is applicable.

---

*We will discuss asymptotes in greater detail in Section 10.3.

Limit of a Function at Infinity

The function f has the limit L as x increases without bound (or as x approaches infinity), written

$$\lim_{x \to \infty} f(x) = L$$

if $f(x)$ can be made arbitrarily close to L by taking x large enough.

Similarly, the function f has the limit M as x decreases without bound (or as x approaches negative infinity), written

$$\lim_{x \to -\infty} f(x) = M$$

if $f(x)$ can be made arbitrarily close to M by taking x to be negative and sufficiently large in absolute value.

EXAMPLE 7 Let f and g be the functions

$$f(x) = \begin{cases} -1 & \text{if } x < 0 \\ 1 & \text{if } x \geq 0 \end{cases} \quad \text{and} \quad g(x) = \frac{1}{x^2}$$

Evaluate:

a. $\lim_{x \to \infty} f(x)$ and $\lim_{x \to -\infty} f(x)$ **b.** $\lim_{x \to \infty} g(x)$ and $\lim_{x \to -\infty} g(x)$

Solution The graphs of $f(x)$ and $g(x)$ are shown in Figure 10. Referring to the graphs of the respective functions, we see that

a. $\lim_{x \to \infty} f(x) = 1$ and $\lim_{x \to -\infty} f(x) = -1$ **b.** $\lim_{x \to \infty} \frac{1}{x^2} = 0$ and $\lim_{x \to -\infty} \frac{1}{x^2} = 0$

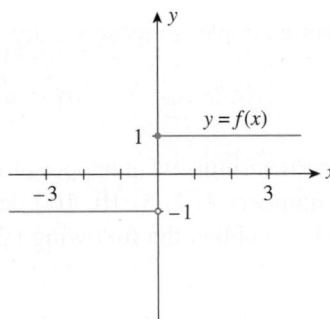

(a) $\lim_{x \to \infty} f(x) = 1$ and $\lim_{x \to -\infty} f(x) = -1$ (b) $\lim_{x \to \infty} g(x) = 0$ and $\lim_{x \to -\infty} g(x) = 0$

FIGURE **10**

All the properties of limits listed in Theorem 1 are valid when a is replaced by ∞ or $-\infty$. In addition, we have the following property for the limit at infinity.

THEOREM 2

For all $n > 0$,

$$\lim_{x \to \infty} \frac{1}{x^n} = 0 \quad \text{and} \quad \lim_{x \to -\infty} \frac{1}{x^n} = 0$$

provided that $\frac{1}{x^n}$ is defined.

1. Use a graphing utility to plot the graphs of

$$y_1 = \frac{1}{x^{0.5}} \qquad y_2 = \frac{1}{x} \qquad y_3 = \frac{1}{x^{1.5}}$$

in the viewing window $[0, 200] \times [0, 0.5]$. What can you say about $\lim\limits_{x \to \infty} \dfrac{1}{x^n}$ if $n = 0.5$, $n = 1$, and $n = 1.5$? Are these results predicted by Theorem 2?

2. Use a graphing utility to plot the graphs of

$$y_1 = \frac{1}{x} \quad \text{and} \quad y_2 = \frac{1}{x^{5/3}}$$

in the viewing window $[-50, 0] \times [-0.5, 0]$. What can you say about $\lim\limits_{x \to -\infty} \dfrac{1}{x^n}$ if $n = 1$ and $n = \dfrac{5}{3}$? Are these results predicted by Theorem 2?

Hint: To graph y_2, write it in the form $y2 = 1/(x^{\wedge}(1/3))^{\wedge}5$.

We often use the following technique to evaluate the limit at infinity of a rational function: *Divide the numerator and denominator of the expression by x^n, where n is the highest power present in the denominator of the expression.*

EXAMPLE 8 Evaluate

$$\lim_{x \to \infty} \frac{x^2 - x + 3}{2x^3 + 1}$$

Solution Since the limits of both the numerator and the denominator do not exist as x approaches infinity, the property pertaining to the limit of a quotient (Property 5) is not applicable. Let's divide the numerator and denominator of the rational expression by x^3, obtaining

$$\lim_{x \to \infty} \frac{x^2 - x + 3}{2x^3 + 1} = \lim_{x \to \infty} \frac{\dfrac{1}{x} - \dfrac{1}{x^2} + \dfrac{3}{x^3}}{2 + \dfrac{1}{x^3}}$$

$$= \frac{0 - 0 + 0}{2 + 0} = \frac{0}{2} \qquad \text{Use Theorem 2.}$$

$$= 0 \qquad\qquad\qquad\qquad\qquad \blacksquare$$

EXAMPLE 9 Let

$$f(x) = \frac{3x^2 + 8x - 4}{2x^2 + 4x - 5}$$

Compute $\lim\limits_{x \to \infty} f(x)$ if it exists.

Solution Again, we see that Property 5 is not applicable. Dividing the numerator and the denominator by x^2, we obtain

$$\lim_{x \to \infty} \frac{3x^2 + 8x - 4}{2x^2 + 4x - 5} = \lim_{x \to \infty} \frac{3 + \dfrac{8}{x} - \dfrac{4}{x^2}}{2 + \dfrac{4}{x} - \dfrac{5}{x^2}}$$

$$= \frac{\displaystyle\lim_{x \to \infty} 3 + 8 \lim_{x \to \infty} \frac{1}{x} - 4 \lim_{x \to \infty} \frac{1}{x^2}}{\displaystyle\lim_{x \to \infty} 2 + 4 \lim_{x \to \infty} \frac{1}{x} - 5 \lim_{x \to \infty} \frac{1}{x^2}}$$

$$= \frac{3 + 0 - 0}{2 + 0 - 0} \qquad \text{Use Theorem 2.}$$

$$= \frac{3}{2}$$

EXAMPLE 10 Let $f(x) = \dfrac{2x^3 - 3x^2 + 1}{x^2 + 2x + 4}$, and evaluate:

a. $\displaystyle\lim_{x \to \infty} f(x)$ **b.** $\displaystyle\lim_{x \to -\infty} f(x)$

Solution

a. Dividing the numerator and the denominator of the rational expression by x^2, we obtain

$$\lim_{x \to \infty} \frac{2x^3 - 3x^2 + 1}{x^2 + 2x + 4} = \lim_{x \to \infty} \frac{2x - 3 + \dfrac{1}{x^2}}{1 + \dfrac{2}{x} + \dfrac{4}{x^2}}$$

Since the numerator becomes arbitrarily large, whereas the denominator approaches 1 as x approaches infinity, we see that the quotient $f(x)$ gets larger and larger as x approaches infinity. In other words, the limit does not exist. In this case, we indicate this by writing

$$\lim_{x \to \infty} \frac{2x^3 - 3x^2 + 1}{x^2 + 2x + 4} = \infty$$

b. Once again, dividing both the numerator and the denominator by x^2, we obtain

$$\lim_{x \to -\infty} \frac{2x^3 - 3x^2 + 1}{x^2 + 2x + 4} = \lim_{x \to -\infty} \frac{2x - 3 + \dfrac{1}{x^2}}{1 + \dfrac{2}{x} + \dfrac{4}{x^2}}$$

In this case, the numerator becomes arbitrarily large in magnitude but negative in sign, whereas the denominator approaches 1 as x approaches negative infinity. Therefore, the quotient $f(x)$ decreases without bound, and the limit does not exist. In this case, we indicate this by writing

$$\lim_{x \to -\infty} \frac{2x^3 - 3x^2 + 1}{x^2 + 2x + 4} = -\infty$$

Example 11 gives an application of the concept of the limit of a function at infinity.

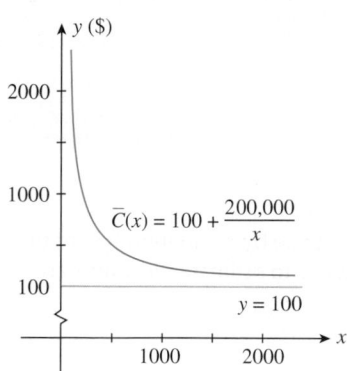

FIGURE **11**
As the level of production increases, the average cost approaches $100 per desk.

$ APPLIED EXAMPLE 11 Average Cost Functions Custom Office makes a line of executive desks. It is estimated that the total cost of making x Senior Executive Model desks is $C(x) = 100x + 200{,}000$ dollars per year, so the average cost of making x desks is given by

$$\overline{C}(x) = \frac{C(x)}{x}$$

$$= \frac{100x + 200{,}000}{x} = 100 + \frac{200{,}000}{x}$$

dollars per desk. Evaluate $\lim_{x \to \infty} \overline{C}(x)$ and interpret your result.

Solution

$$\lim_{x \to \infty} \overline{C}(x) = \lim_{x \to \infty} \left(100 + \frac{200{,}000}{x} \right)$$

$$= \lim_{x \to \infty} 100 + \lim_{x \to \infty} \frac{200{,}000}{x} = 100$$

A sketch of the graph of the function $\overline{C}(x)$ appears in Figure 11. The result we obtained is fully expected if we consider its economic implications. Note that as the level of production increases, the fixed cost per desk produced, represented by the term $(200{,}000/x)$, drops steadily. The average cost should approach a constant unit cost of production—$100 in this case.

Explore and Discuss

Consider the graph of the function f depicted in the following figure:

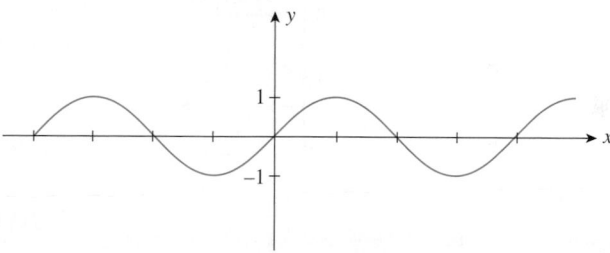

It has the property that the curve oscillates between $y = -1$ and $y = 1$ indefinitely in either direction.

1. Explain why $\lim_{x \to -\infty} f(x)$ and $\lim_{x \to \infty} f(x)$ do not exist.
2. Compare this function with those of Example 10. More specifically, discuss the different ways each function fails to have a limit at infinity or minus infinity.

9.1 Self-Check Exercises

1. Find the indicated limit if it exists.

   a. $\displaystyle \lim_{x \to 3} \frac{\sqrt{x^2 + 7} + \sqrt{3x - 5}}{x + 2}$

   b. $\displaystyle \lim_{x \to -1} \frac{x^2 - x - 2}{2x^2 - x - 3}$

2. **AVERAGE COST OF PRODUCING CDS** The average cost per compact disc (in dollars) incurred by Herald Records in

pressing x CDs is given by the average cost function

$$\overline{C}(x) = 1.8 + \frac{3000}{x}$$

Evaluate $\lim_{x \to \infty} \overline{C}(x)$ and interpret your result.

Solutions to Self-Check Exercises 9.1 can be found on page 611.

9.1 Concept Questions

1. Explain what is meant by the statement $\lim\limits_{x \to 2} f(x) = 3$.

2. **a.** If $\lim\limits_{x \to 3} f(x) = 5$, what can you say about $f(3)$? Explain.

   b. If $f(2) = 6$, what can you say about $\lim\limits_{x \to 2} f(x)$? Explain.

3. Evaluate the following, and state the property of limits that you use at each step.

   a. $\lim\limits_{x \to 4} \sqrt{x}(2x^2 + 1)$ **b.** $\lim\limits_{x \to 1} \left(\dfrac{2x^2 + x + 5}{x^4 + 1} \right)^{3/2}$

4. What is an indeterminate form? Illustrate with an example.

5. Explain in your own words the meaning of $\lim\limits_{x \to \infty} f(x) = L$ and $\lim\limits_{x \to -\infty} f(x) = M$.

9.1 Exercises

In Exercises 1–8, use the graph of the given function f to determine $\lim\limits_{x \to a} f(x)$ at the indicated value of a, if it exists.

1.
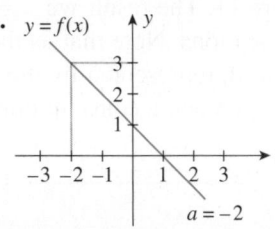
$a = -2$

2.
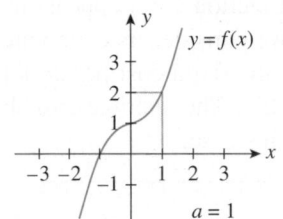
$a = 1$

3.
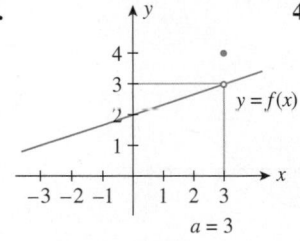
$a = 3$

4.
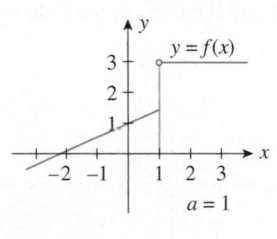
$a = 1$

5.
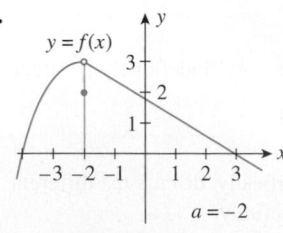
$a = -2$

6.
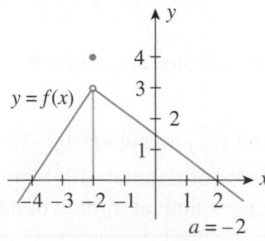
$a = -2$

7.
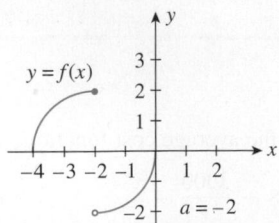
$a = -2$

8.
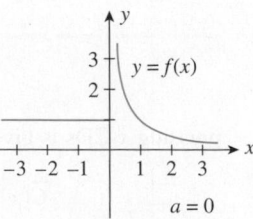
$a = 0$

In Exercises 9–16, complete the table by computing $f(x)$ at the given values of x. Use these results to estimate the indicated limit (if it exists).

9. $f(x) = x^2 + 1$; $\lim\limits_{x \to 2} f(x)$

x	1.9	1.99	1.999	2.001	2.01	2.1
$f(x)$						

10. $f(x) = 2x^2 - 1$; $\lim\limits_{x \to 1} f(x)$

x	0.9	0.99	0.999	1.001	1.01	1.1
$f(x)$						

11. $f(x) = \dfrac{|x|}{x}$; $\lim\limits_{x \to 0} f(x)$

x	-0.1	-0.01	-0.001	0.001	0.01	0.1
$f(x)$						

12. $f(x) = \dfrac{|x - 1|}{x - 1}$; $\lim\limits_{x \to 1} f(x)$

x	0.9	0.99	0.999	1.001	1.01	1.1
$f(x)$						

13. $f(x) = \dfrac{1}{(x - 1)^2}$; $\lim\limits_{x \to 1} f(x)$

x	0.9	0.99	0.999	1.001	1.01	1.1
$f(x)$						

14. $f(x) = \dfrac{1}{x - 2}$; $\lim\limits_{x \to 2} f(x)$

x	1.9	1.99	1.999	2.001	2.01	2.1
$f(x)$						

15. $f(x) = \dfrac{x^2 + x - 2}{x - 1}$; $\lim\limits_{x \to 1} f(x)$

x	0.9	0.99	0.999	1.001	1.01	1.1
$f(x)$						

16. $f(x) = \dfrac{x-1}{x-1}$; $\lim\limits_{x \to 1} f(x)$

x	0.9	0.99	0.999	1.001	1.01	1.1
$f(x)$						

In Exercises 17–22, sketch the graph of the function f, and evaluate $\lim\limits_{x \to a} f(x)$, if it exists, for the given value of a.

17. $f(x) = \begin{cases} x - 1 & \text{if } x \le 0 \\ -1 & \text{if } x > 0 \end{cases}$ $(a = 0)$

18. $f(x) = \begin{cases} x - 1 & \text{if } x \le 3 \\ -2x + 8 & \text{if } x > 3 \end{cases}$ $(a = 3)$

19. $f(x) = \begin{cases} x & \text{if } x < 1 \\ 0 & \text{if } x = 1 \\ -x + 2 & \text{if } x > 1 \end{cases}$ $(a = 1)$

20. $f(x) = \begin{cases} -2x + 4 & \text{if } x < 1 \\ 4 & \text{if } x = 1 \\ x^2 + 1 & \text{if } x > 1 \end{cases}$ $(a = 1)$

21. $f(x) = \begin{cases} |x| & \text{if } x \ne 0 \\ 1 & \text{if } x = 0 \end{cases}$ $(a = 0)$

22. $f(x) = \begin{cases} |x - 1| & \text{if } x \ne 1 \\ 0 & \text{if } x = 1 \end{cases}$ $(a = 1)$

In Exercises 23–40, find the indicated limit.

23. $\lim\limits_{x \to 2} 3$

24. $\lim\limits_{x \to -2} -3$

25. $\lim\limits_{x \to 3} x$

26. $\lim\limits_{x \to -2} -3x$

27. $\lim\limits_{x \to 1} (1 - 2x^2)$

28. $\lim\limits_{t \to 3} (4t^2 - 2t + 1)$

29. $\lim\limits_{x \to 1} (2x^3 - 3x^2 + x + 2)$

30. $\lim\limits_{x \to 0} (4x^5 - 20x^2 + 2x + 1)$

31. $\lim\limits_{s \to 0} (2s^2 - 1)(2s + 4)$ **32.** $\lim\limits_{x \to 2} (x^2 + 1)(x^2 - 4)$

33. $\lim\limits_{x \to 2} \dfrac{2x + 1}{x + 2}$ **34.** $\lim\limits_{x \to 1} \dfrac{x^3 + 1}{2x^3 + 2}$

35. $\lim\limits_{x \to 2} \sqrt{x + 2}$ **36.** $\lim\limits_{x \to -2} \sqrt[3]{5x + 2}$

37. $\lim\limits_{x \to -3} \sqrt{2x^4 + x^2}$ **38.** $\lim\limits_{x \to 2} \sqrt{\dfrac{2x^3 + 4}{x^2 + 1}}$

39. $\lim\limits_{x \to -1} \dfrac{\sqrt{x^2 + 8}}{2x + 4}$ **40.** $\lim\limits_{x \to 3} \dfrac{x\sqrt{x^2 + 7}}{2x - \sqrt{2x + 3}}$

In Exercises 41–48, find the indicated limit given that $\lim\limits_{x \to a} f(x) = 3$ and $\lim\limits_{x \to a} g(x) = 4$.

41. $\lim\limits_{x \to a} [f(x) - g(x)]$ **42.** $\lim\limits_{x \to a} 2f(x)$

43. $\lim\limits_{x \to a} [2f(x) - 3g(x)]$ **44.** $\lim\limits_{x \to a} [f(x)g(x)]$

45. $\lim\limits_{x \to a} \sqrt{g(x)}$ **46.** $\lim\limits_{x \to a} \sqrt[3]{5f(x) + 3g(x)}$

47. $\lim\limits_{x \to a} \dfrac{2f(x) - g(x)}{f(x)g(x)}$ **48.** $\lim\limits_{x \to a} \dfrac{g(x) - f(x)}{f(x) + \sqrt{g(x)}}$

In Exercises 49–62, find the indicated limit, if it exists.

49. $\lim\limits_{x \to 1} \dfrac{x^2 - 1}{x - 1}$ **50.** $\lim\limits_{x \to -2} \dfrac{x^2 - 4}{x + 2}$

51. $\lim\limits_{x \to 0} \dfrac{x^2 - x}{x}$ **52.** $\lim\limits_{x \to 0} \dfrac{2x^2 - 3x}{x}$

53. $\lim\limits_{x \to -5} \dfrac{x^2 - 25}{x + 5}$ **54.** $\lim\limits_{b \to -3} \dfrac{b + 1}{b + 3}$

55. $\lim\limits_{x \to 1} \dfrac{x}{x - 1}$ **56.** $\lim\limits_{x \to 2} \dfrac{x + 2}{x - 2}$

57. $\lim\limits_{x \to -2} \dfrac{x^2 - x - 6}{x^2 + x - 2}$ **58.** $\lim\limits_{z \to 2} \dfrac{z^3 - 8}{z - 2}$

59. $\lim\limits_{x \to 1} \dfrac{\sqrt{x} - 1}{x - 1}$ **60.** $\lim\limits_{x \to 4} \dfrac{x - 4}{\sqrt{x} - 2}$

Hint: Multiply by $\dfrac{\sqrt{x} + 1}{\sqrt{x} + 1}$.   Hint: See Exercise 59.

61. $\lim\limits_{x \to 1} \dfrac{x - 1}{x^3 + x^2 - 2x}$ **62.** $\lim\limits_{x \to -2} \dfrac{4 - x^2}{2x^2 + x^3}$

In Exercises 63–68, use the graph of the function f to determine $\lim\limits_{x \to \infty} f(x)$ and $\lim\limits_{x \to -\infty} f(x)$, if they exist.

63.
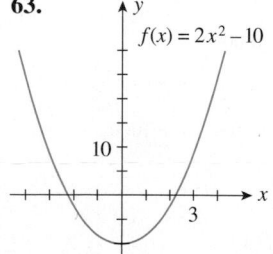
$f(x) = 2x^2 - 10$

64.
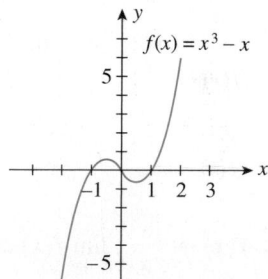
$f(x) = x^3 - x$

65.
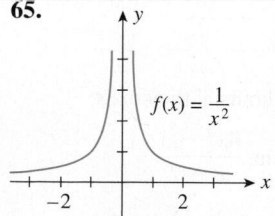
$f(x) = \dfrac{1}{x^2}$

66.
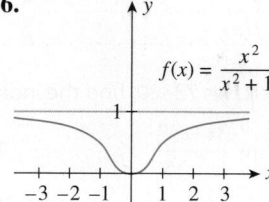
$f(x) = \dfrac{x^2}{x^2 + 1}$

67.
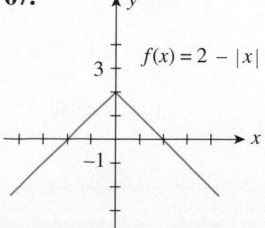
$f(x) = 2 - |x|$

68.

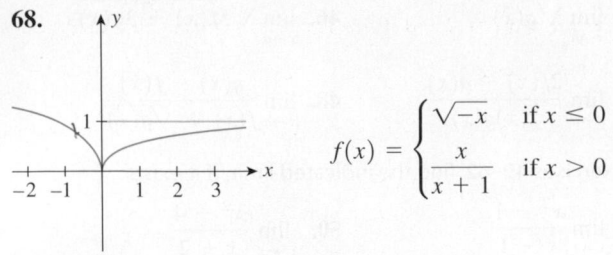

$$f(x) = \begin{cases} \sqrt{-x} & \text{if } x \leq 0 \\ \dfrac{x}{x+1} & \text{if } x > 0 \end{cases}$$

In Exercises 69–72, complete the table by computing $f(x)$ at the given values of x. Use the results to guess at the indicated limits, if they exist.

69. $f(x) = \dfrac{1}{x^2 + 1}$; $\lim\limits_{x \to \infty} f(x)$ and $\lim\limits_{x \to -\infty} f(x)$

x	1	10	100	1000
$f(x)$				

x	-1	-10	-100	-1000
$f(x)$				

70. $f(x) = \dfrac{2x}{x+1}$; $\lim\limits_{x \to \infty} f(x)$ and $\lim\limits_{x \to -\infty} f(x)$

x	1	10	100	1000
$f(x)$				

x	-5	-10	-100	-1000
$f(x)$				

71. $f(x) = 3x^3 - x^2 + 10$; $\lim\limits_{x \to \infty} f(x)$ and $\lim\limits_{x \to -\infty} f(x)$

x	1	5	10	100	1000
$f(x)$					

x	-1	-5	-10	-100	-1000
$f(x)$					

72. $f(x) = \dfrac{|x|}{x}$; $\lim\limits_{x \to \infty} f(x)$ and $\lim\limits_{x \to -\infty} f(x)$

x	1	10	100	-1	-10	-100
$f(x)$						

In Exercises 73–80, find the indicated limits, if they exist.

73. $\lim\limits_{x \to \infty} \dfrac{3x + 2}{x - 5}$

74. $\lim\limits_{x \to -\infty} \dfrac{4x^2 - 1}{x + 2}$

75. $\lim\limits_{x \to -\infty} \dfrac{3x^3 + x^2 + 1}{x^3 + 1}$

76. $\lim\limits_{x \to \infty} \dfrac{2x^2 + 3x + 1}{x^4 - x^2}$

77. $\lim\limits_{x \to -\infty} \dfrac{x^4 + 1}{x^3 - 1}$

78. $\lim\limits_{x \to \infty} \dfrac{4x^4 - 3x^2 + 1}{2x^4 + x^3 + x^2 + x + 1}$

79. $\lim\limits_{x \to \infty} \dfrac{x^5 - x^3 + x - 1}{x^6 + 2x^2 + 1}$

80. $\lim\limits_{x \to \infty} \dfrac{2x^2 - 1}{x^3 + x^2 + 1}$

81. Toxic Waste A city's main well was recently found to be contaminated with trichloroethylene, a cancer-causing chemical, as a result of an abandoned chemical dump leaching chemicals into the water. A proposal submitted to city council members indicates that the cost, measured in millions of dollars, of removing $x\%$ of the toxic pollutant is given by

$$C(x) = \frac{0.5x}{100 - x} \qquad (0 < x < 100)$$

a. Find the cost of removing 50%, 60%, 70%, 80%, 90%, and 95% of the pollutant.

b. Evaluate

$$\lim_{x \to 100} \frac{0.5x}{100 - x}$$

and interpret your result.

82. A Doomsday Situation The population of a certain breed of rabbits introduced onto an isolated island is given by

$$P(t) = \frac{72}{9 - t} \qquad (0 \leq t < 9)$$

where t is measured in months.

a. Find the number of rabbits present on the island initially (at $t = 0$).

b. Show that the population of rabbits is increasing without bound.

c. Sketch the graph of the function P.
(*Comment:* This phenomenon is referred to as a *doomsday situation.*)

83. Average Cost The average cost per disc in dollars incurred by Herald Media in pressing x DVDs is given by the average cost function

$$\overline{C}(x) = 2.2 + \frac{2500}{x}$$

Evaluate $\lim\limits_{x \to \infty} \overline{C}(x)$ and interpret your result.

84. Concentration of a Drug in the Bloodstream The concentration of a certain drug in a patient's bloodstream t hr after injection is given by

$$C(t) = \frac{0.2t}{t^2 + 1}$$

mg/cm³. Evaluate $\lim\limits_{t \to \infty} C(t)$ and interpret your result.

85. Box-Office Receipts The total worldwide box-office receipts for a long-running blockbuster movie are approximated by the function

$$T(x) = \frac{120x^2}{x^2 + 4}$$

where $T(x)$ is measured in millions of dollars and x is the number of months since the movie's release.

a. What are the total box-office receipts after the first month? The second month? The third month?

b. What will the movie gross in the long run (when x is very large)?

86. POPULATION GROWTH A major corporation is building a 4325-acre complex of homes, offices, stores, schools, and churches in the rural community of Glen Cove. As a result of this development, the planners have estimated that Glen Cove's population (in thousands) t years from now will be given by

$$P(t) = \frac{25t^2 + 125t + 200}{t^2 + 5t + 40}$$

a. What is the current population of Glen Cove?
b. What will be the population in the long run?

87. DRIVING COSTS A study of the costs of driving 2012 small-sized sedans found that the average cost per mile (car payments, gas, insurance, upkeep, and depreciation), measured in cents per mile, is approximated by the function

$$C(x) = \frac{2410}{x^{1.95}} + 32.8$$

where x denotes the number of miles (in thousands) the car is driven in a year.

a. What is the average cost per mile of driving a small-sized sedan 5000 mi/year? 10,000 mi/year? 15,000 mi/year? 20,000 mi/year? 25,000 mi/year?
b. Use part (a) to sketch the graph of the function C.
c. What happens to the average cost per mile as the number of miles driven increases without bound?
Source: American Automobile Association.

88. PHOTOSYNTHESIS The rate of production R in photosynthesis is related to the light intensity I by the function

$$R(I) = \frac{aI}{b + I^2}$$

where a and b are positive constants.

a. Taking $a = b = 1$, compute $R(I)$ for $I = 0, 1, 2, 3, 4,$ and 5.
b. Evaluate $\lim_{I \to \infty} R(I)$.
c. Use the results of parts (a) and (b) to sketch the graph of R. Interpret your results.

In Exercises 89–94, determine whether the statement is true or false. If it is true, explain why it is true. If it is false, give an example to show why it is false.

89. If $\lim_{x \to a} f(x)$ exists, then f is defined at $x = a$.

90. If $\lim_{x \to 3} g(x) = 0$ and if $\lim_{x \to 3} f(x)/g(x) = 0$ exists, then $\lim_{x \to 3} f(x) = 0$.

91. If $\lim_{x \to 2} f(x) = 3$ and $\lim_{x \to 2} g(x) = 0$, then $\lim_{x \to 2} [f(x)]/[g(x)]$ does not exist.

92. If $\lim_{x \to 3} f(x) = 0$ and $\lim_{x \to 3} g(x) = 0$, then $\lim_{x \to 3} [f(x)]/[g(x)]$ does not exist.

93. $\lim_{x \to 2} \left(\frac{x}{x + 1} + \frac{3}{x - 1} \right) = \lim_{x \to 2} \frac{x}{x + 1} + \lim_{x \to 2} \frac{3}{x - 1}$

94. $\lim_{x \to 1} \left(\frac{2x}{x - 1} - \frac{2}{x - 1} \right) = \lim_{x \to 1} \frac{2x}{x - 1} - \lim_{x \to 1} \frac{2}{x - 1}$

95. SPEED OF A CHEMICAL REACTION Certain proteins, known as enzymes, serve as catalysts for chemical reactions in living things. In 1913 Leonor Michaelis and L. M. Menten discovered the following formula giving the initial speed V (in moles per liter per second) at which the reaction begins in terms of the amount of substrate x (the substance being acted upon, measured in moles per liter) present:

$$V = \frac{ax}{x + b}$$

where a and b are positive constants. Evaluate

$$\lim_{x \to \infty} \frac{ax}{x + b}$$

and interpret your result.

96. Show by means of an example that $\lim_{x \to a} [f(x) + g(x)]$ may exist even though neither $\lim_{x \to a} f(x)$ nor $\lim_{x \to a} g(x)$ exists. Does this example contradict Theorem 1?

97. Show by means of an example that $\lim_{x \to a} [f(x)g(x)]$ may exist even though neither $\lim_{x \to a} f(x)$ nor $\lim_{x \to a} g(x)$ exists. Does this example contradict Theorem 1?

98. Show by means of an example that $\lim_{x \to a} f(x)/g(x)$ may exist even though neither $\lim_{x \to a} f(x)$ nor $\lim_{x \to a} g(x)$ exists. Does this example contradict Theorem 1?

9.1 Solutions to Self-Check Exercises

1. a. $\lim_{x \to 3} \frac{\sqrt{x^2 + 7} + \sqrt{3x - 5}}{x + 2} = \frac{\sqrt{9 + 7} + \sqrt{3(3) - 5}}{3 + 2}$

$$= \frac{\sqrt{16} + \sqrt{4}}{5}$$

$$= \frac{6}{5}$$

b. Letting x approach -1 leads to the indeterminate form $0/0$. Thus, we proceed as follows:

$$\lim_{x \to -1} \frac{x^2 - x - 2}{2x^2 - x - 3} = \lim_{x \to -1} \frac{(x + 1)(x - 2)}{(x + 1)(2x - 3)}$$

$$= \lim_{x \to -1} \frac{x - 2}{2x - 3} \quad \text{Cancel the common factors.}$$

$$= \frac{-1 - 2}{2(-1) - 3}$$

$$= \frac{3}{5}$$

2. $\displaystyle\lim_{x\to\infty} \overline{C}(x) = \lim_{x\to\infty}\left(1.8 + \frac{3000}{x}\right)$

$\qquad = \displaystyle\lim_{x\to\infty} 1.8 + \lim_{x\to\infty}\frac{3000}{x}$

$\qquad = 1.8$

Our computation reveals that as the production of CDs increases without bound, the average cost drops and approaches a unit cost of $1.80/disc.

USING TECHNOLOGY   Finding the Limit of a Function

A graphing utility can be used to help us find the limit of a function, if it exists, as illustrated in the following examples.

EXAMPLE 1   Let $f(x) = \dfrac{x^3 - 1}{x - 1}$.

a. Plot the graph of f in the viewing window $[-2, 2] \times [0, 4]$.

b. Use **ZOOM** to find $\displaystyle\lim_{x\to 1}\frac{x^3 - 1}{x - 1}$.

c. Verify your result by evaluating the limit algebraically.

Solution

a. The graph of f in the viewing window $[-2, 2] \times [0, 4]$ is shown in Figure T1a.

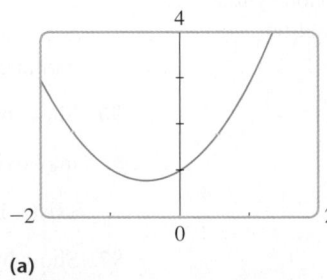
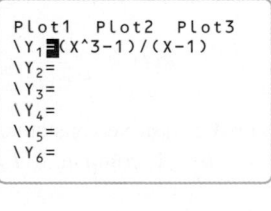

(a)   (b)

FIGURE **T1**
(a) The graph of $f(x) = (x^3 - 1)/(x - 1)$ in the viewing window $[-2, 2] \times [0, 4]$;
(b) the TI-83/84 equation screen

b. Using **ZOOM-IN** repeatedly, we see that the y-value approaches 3 as the x-value approaches 1. We conclude, accordingly, that

$$\lim_{x\to 1}\frac{x^3 - 1}{x - 1} = 3$$

c. We compute

$$\lim_{x\to 1}\frac{x^3 - 1}{x - 1} = \lim_{x\to 1}\frac{(x - 1)(x^2 + x + 1)}{x - 1}$$

$$= \lim_{x\to 1}(x^2 + x + 1) = 3$$

Note   If you attempt to find the limit in Example 1 by using the evaluation function of your graphing utility to find the value of $f(x)$ when $x = 1$, you will see that the graphing utility does not display the y-value. This happens because $x = 1$ is not in the domain of f.

EXAMPLE 2 Use **zoom** to find $\lim_{x \to 0} (1 + x)^{1/x}$.

Solution   We first plot the graph of $f(x) = (1 + x)^{1/x}$ in a suitable viewing window. Figure T2a shows a plot of f in the window $[-1, 1] \times [0, 4]$. Using **zoom-in** repeatedly, we see that $\lim_{x \to 0} (1 + x)^{1/x} \approx 2.71828$.

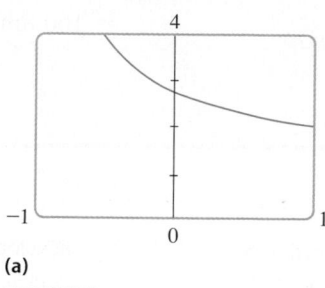
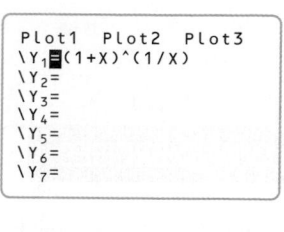

(a)   **(b)**

FIGURE **T2**
(a) The graph of $f(x) = (1 + x)^{1/x}$ in the viewing window $[-1, 1] \times [0, 4]$;
(b) the T1-83/84 equation screen

The limit of $f(x) = (1 + x)^{1/x}$ as x approaches zero, denoted by the letter e, plays a very important role in the study of mathematics and its applications (see Section 3.3). Thus,

$$\lim_{x \to 0} (1 + x)^{1/x} = e$$

where, as we have just seen, $e \approx 2.71828$.

APPLIED EXAMPLE 3 Oxygen Content of a Pond When organic waste is dumped into a pond, the oxidation process that takes place reduces the pond's oxygen content. However, given time, nature will restore the oxygen content to its natural level. Suppose the oxygen content t days after the organic waste has been dumped into the pond is given by

$$f(t) = 100 \left(\frac{t^2 + 10t + 100}{t^2 + 20t + 100} \right)$$

percent of its normal level.

a. Plot the graph of f in the viewing window $[0, 200] \times [70, 100]$.
b. What can you say about $f(t)$ when t is very large?
c. Verify your observation in part (b) by evaluating $\lim_{t \to \infty} f(t)$.

Solution

a. The graph of f is shown in Figure T3a.
b. From the graph of f, it appears that $f(t)$ approaches 100 steadily as t gets larger and larger. This observation tells us that eventually the oxygen content of the pond will be restored to its natural level.

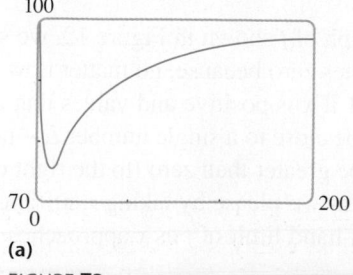
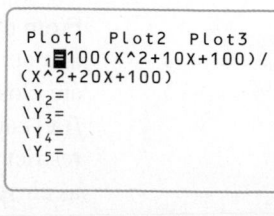

(a)   **(b)**

FIGURE **T3**
(a) The graph of f in the viewing window $[0, 200] \times [70, 100]$; (b) the TI-83/84 equation screen

c. To verify the observation made in part (b), we compute

$$\lim_{t \to \infty} f(t) = \lim_{t \to \infty} 100\left(\frac{t^2 + 10t + 100}{t^2 + 20t + 100}\right)$$

$$= 100 \lim_{t \to \infty} \left(\frac{1 + \dfrac{10}{t} + \dfrac{100}{t^2}}{1 + \dfrac{20}{t} + \dfrac{100}{t^2}}\right) = 100$$

TECHNOLOGY EXERCISES

In Exercises 1–8, find the indicated limit by first plotting the graph of the function in a suitable viewing window and then using the ZOOM-IN feature of the calculator.

1. $\displaystyle\lim_{x \to 1} \frac{2x^3 - 2x^2 + 3x - 3}{x - 1}$

2. $\displaystyle\lim_{x \to -2} \frac{2x^3 + 3x^2 - x + 2}{x + 2}$

3. $\displaystyle\lim_{x \to -1} \frac{x^3 + 1}{x + 1}$

4. $\displaystyle\lim_{x \to -1} \frac{x^4 - 1}{x - 1}$

5. $\displaystyle\lim_{x \to 1} \frac{x^3 - x^2 - x + 1}{x^3 - 3x + 2}$

6. $\displaystyle\lim_{x \to 0} \frac{\sqrt{x + 1} - 1}{x}$

7. $\displaystyle\lim_{x \to 0} (1 + 2x)^{1/x}$

8. $\displaystyle\lim_{x \to 0} \frac{2^x - 1}{x}$

9. Show that $\displaystyle\lim_{x \to 3} \frac{2}{x - 3}$ does not exist.

10. Show that $\displaystyle\lim_{x \to 2} \frac{x^3 - 2x + 1}{x - 2}$ does not exist.

11. CITY PLANNING A major developer is building a 5000-acre complex of homes, offices, stores, schools, and churches in the rural community of Marlboro. As a result of this development, the planners have estimated that Marlboro's population (in thousands) t years from now will be given by

$$P(t) = \frac{25t^2 + 125t + 200}{t^2 + 5t + 40}$$

a. Plot the graph of P in the viewing window $[0, 50] \times [0, 30]$.

b. What will be the population of Marlboro in the long run? Hint: Find $\displaystyle\lim_{t \to \infty} P(t)$.

12. AN EXTINCTION SITUATION The number of saltwater crocodiles in a certain area of northern Australia in year t is given by

$$P(t) = \frac{300e^{-0.024t}}{5e^{-0.024t} + 1}$$

a. How many crocodiles were in the population initially?

b. Show that $\displaystyle\lim_{t \to \infty} P(t) = 0$.

c. Plot the graph of P in the viewing window $[0, 200] \times [0, 70]$.

(*Comment:* This phenomenon is referred to as an *extinction situation*.)

9.2 One-Sided Limits and Continuity

One-Sided Limits

Consider the function f defined by

$$f(x) = \begin{cases} x - 1 & \text{if } x < 0 \\ x + 1 & \text{if } x \geq 0 \end{cases}$$

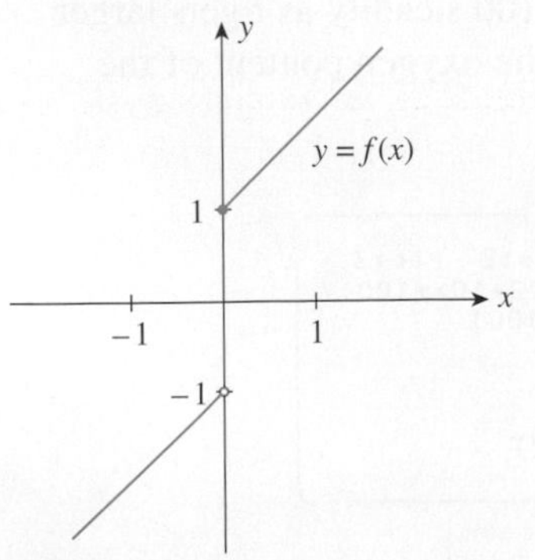

From the graph of f shown in Figure 12, we see that the function f does not have a limit as x approaches zero because, no matter how close x is to zero, $f(x)$ takes on values that are close to 1 if x is positive and values that are close to -1 if x is negative. Therefore, $f(x)$ cannot be close to a single number L—no matter how close x is to zero. Now, if we restrict x to be greater than zero (to the right of zero), then we see that $f(x)$ can be made as close to 1 as we please by taking x sufficiently close to zero. In this situation, we say that the right-hand limit of f as x approaches zero (from the right) is 1, written

$$\lim_{x \to 0^+} f(x) = 1$$

FIGURE 12
The function f does not have a limit as x approaches zero.

Similarly, we see that $f(x)$ can be made as close to -1 as we please by taking x sufficiently close to, but to the left of, zero. In this situation, we say that the left-hand limit of f as x approaches zero (from the left) is -1, written

$$\lim_{x \to 0^-} f(x) = -1$$

These limits are called **one-sided limits.** More generally, we have the following informal definitions.

One-Sided Limits

The function f has the **right-hand limit** L as x approaches a from the right, written

$$\lim_{x \to a^+} f(x) = L$$

if the values of $f(x)$ can be made as close to L as we please by taking x sufficiently close to (but not equal to) a and to the right of a.

Similarly, the function f has the **left-hand limit** M as x approaches a from the left, written

$$\lim_{x \to a^-} f(x) = M$$

if the values of $f(x)$ can be made as close to M as we please by taking x sufficiently close to (but not equal to) a and to the left of a.

The connection between one-sided limits and the two-sided limit defined earlier is given by the following theorem.

THEOREM 3

Let f be a function that is defined for all values of x close to $x = a$ with the possible exception of a itself. Then

$$\lim_{x \to a} f(x) = L \quad \text{if and only if} \quad \lim_{x \to a^+} f(x) = \lim_{x \to a^-} f(x) = L$$

Thus, the two-sided limit exists if and only if the one-sided limits exist and are equal.

EXAMPLE 1 Let

$$f(x) = \begin{cases} -x & \text{if } x \le 0 \\ \sqrt{x} & \text{if } x > 0 \end{cases} \quad \text{and} \quad g(x) = \begin{cases} -1 & \text{if } x < 0 \\ 1 & \text{if } x \ge 0 \end{cases}$$

a. Show that $\lim_{x \to 0} f(x)$ exists by studying the one-sided limits of f as x approaches $x = 0$.

b. Show that $\lim_{x \to 0} g(x)$ does not exist.

Solution

a. For $x \le 0$,

$$\lim_{x \to 0^-} f(x) = \lim_{x \to 0^-} (-x) = 0$$

and for $x > 0$, we find

$$\lim_{x \to 0^+} f(x) = \lim_{x \to 0^+} \sqrt{x} = 0$$

Thus,

$$\lim_{x \to 0} f(x) = 0$$

(Figure 13a).

b. We have

$$\lim_{x \to 0^-} g(x) = -1 \quad \text{and} \quad \lim_{x \to 0^+} g(x) = 1$$

and since these one-sided limits are not equal, we conclude that $\lim_{x \to 0} g(x)$ does not exist (Figure 13b).

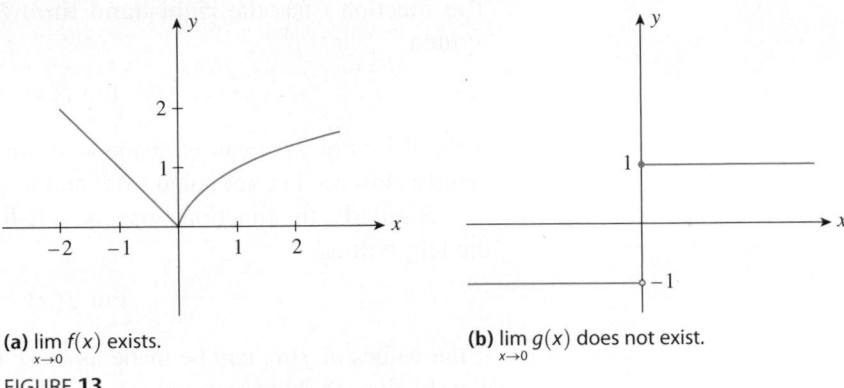

(a) $\lim_{x \to 0} f(x)$ exists. **(b)** $\lim_{x \to 0} g(x)$ does not exist.

FIGURE **13**

Continuous Functions

Continuous functions will play an important role throughout most of our study of calculus. Loosely speaking, a function is continuous at a point if the graph of the function at that point is devoid of holes, gaps, jumps, or breaks. Consider, for example, the graph of the function f depicted in Figure 14.

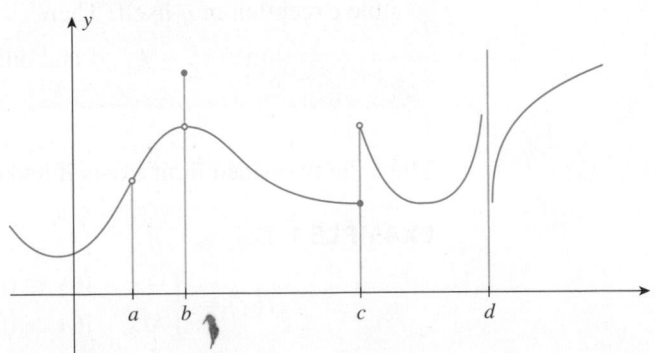

FIGURE **14**
The graph of this function is not continuous at $x = a$, $x = b$, $x = c$, and $x = d$.

Let's take a closer look at the behavior of f at or near $x = a$, $x = b$, $x = c$, and $x = d$. First, note that f is not defined at $x = a$; that is, $x = a$ is not in the domain of f, thereby resulting in a "hole" in the graph of f. Next, observe that the value of f at b, $f(b)$, is not equal to the limit of $f(x)$ as x approaches b, resulting in a "jump" in the graph of f at $x = b$. The function f does not have a limit at $x = c$ since the left-hand and right-hand limits of $f(x)$ are not equal, also resulting in a jump in the graph of f at $x = c$. Finally, the limit of f does not exist at $x = d$, resulting in a break in the graph of f. The function f is *discontinuous* at each of these numbers. It is *continuous* everywhere else.

> **Continuity of a Function at a Number**
>
> A function f is **continuous at a number** $x = a$ if the following conditions are satisfied.
>
> **1.** $f(a)$ is defined.   **2.** $\lim_{x \to a} f(x)$ exists.   **3.** $\lim_{x \to a} f(x) = f(a)$

Thus, a function f is continuous at $x = a$ if the limit of f at $x = a$ exists and has the value $f(a)$. Geometrically, f is continuous at $x = a$ if the proximity of x to a implies the proximity of $f(x)$ to $f(a)$.

If f is not continuous at $x = a$, then f is said to be **discontinuous** at $x = a$. Also, f is **continuous on an interval** if f is continuous at every number in the interval.

Figure 15 depicts the graph of a continuous function on the interval (a, b). Notice that the graph of the function over the stated interval can be sketched without lifting one's pencil from the paper.

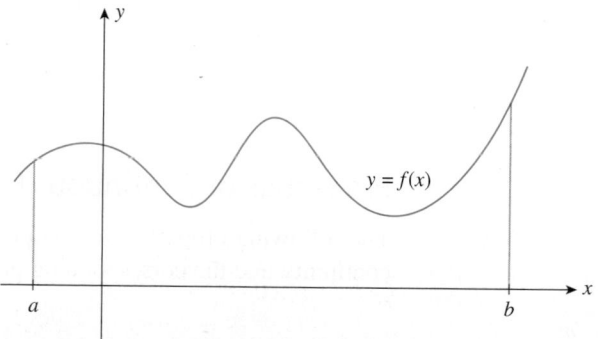

FIGURE **15**
The graph of f is continuous on the interval (a, b).

EXAMPLE 2 Find the values of x for which each function is continuous.

a. $f(x) = x + 2$ **b.** $g(x) = \dfrac{x^2 - 4}{x - 2}$ **c.** $h(x) = \begin{cases} x + 2 & \text{if } x \neq 2 \\ 1 & \text{if } x = 2 \end{cases}$

d. $F(x) = \begin{cases} -1 & \text{if } x < 0 \\ 1 & \text{if } x \geq 0 \end{cases}$ **e.** $G(x) = \begin{cases} \dfrac{1}{x} & \text{if } x > 0 \\ -1 & \text{if } x \leq 0 \end{cases}$

The graph of each function is shown in Figure 16 on the next page.

Solution

a. The function f is continuous everywhere because the three conditions for continuity are satisfied for all values of x.

b. The function g is discontinuous at $x = 2$ because g is not defined at that number. It is continuous everywhere else.

c. The function h is discontinuous at $x = 2$ because the third condition for continuity is violated; the limit of $h(x)$ as x approaches 2 exists and has the value 4, but this limit is not equal to $h(2) = 1$. It is continuous for all other values of x.

d. The function F is continuous everywhere except at $x = 0$, where the limit of $F(x)$ fails to exist as x approaches zero (see Example 3a, Section 9.1).

e. Since the limit of $G(x)$ does not exist as x approaches zero, we conclude that G fails to be continuous at $x = 0$. The function G is continuous everywhere else.

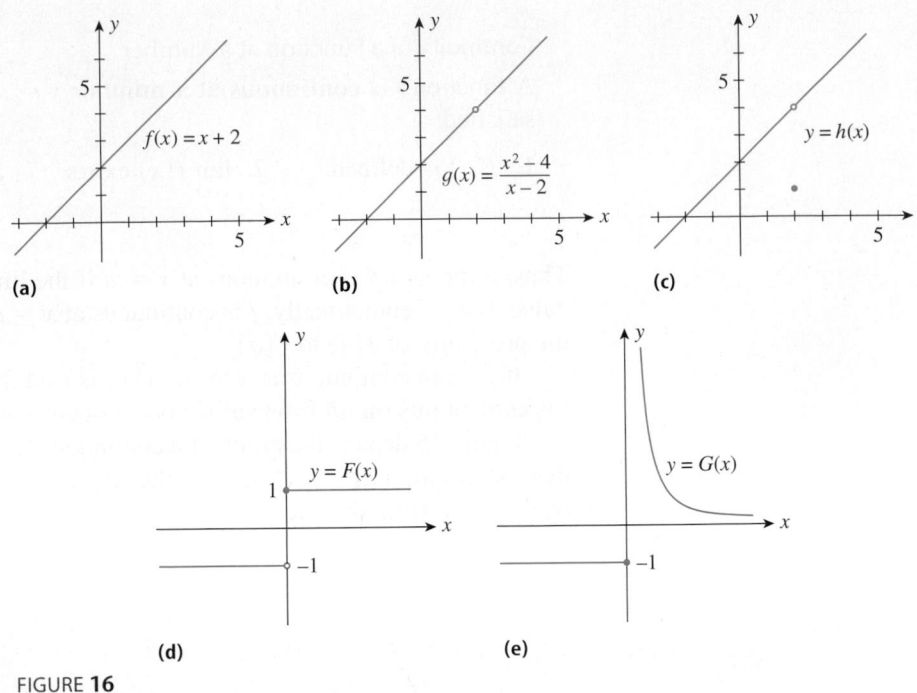

FIGURE 16

Properties of Continuous Functions

The following properties of continuous functions follow directly from the definition of continuity and the corresponding properties of limits. They are stated without proof.

> **Properties of Continuous Functions**
>
> **1.** The constant function $f(x) = c$ is continuous everywhere.
> **2.** The identity function $f(x) = x$ is continuous everywhere.
>
> *If f and g are continuous at $x = a$, then*
>
> **3.** $[f(x)]^n$, where n is a real number, is continuous at $x = a$ whenever it is defined at that number.
> **4.** $f \pm g$ is continuous at $x = a$.
> **5.** fg is continuous at $x = a$.
> **6.** f/g is continuous at $x = a$ provided that $g(a) \neq 0$.

Using these properties of continuous functions, we can prove the following results. (A proof is sketched in Exercise 97, page 627.)

> **Continuity of Polynomial and Rational Functions**
>
> **1.** A polynomial function $y = P(x)$ is continuous at every value of x.
> **2.** A rational function $R(x) = p(x)/q(x)$ is continuous at every value of x where $q(x) \neq 0$.

EXAMPLE 3 Find the values of x for which each function is continuous.

a. $f(x) = 3x^3 + 2x^2 - x + 10$ **b.** $g(x) = \dfrac{8x^{10} - 4x + 1}{x^2 + 1}$

c. $h(x) = \dfrac{4x^3 - 3x^2 + 1}{x^2 - 3x + 2}$

Solution

a. The function f is a polynomial function of degree 3, so $f(x)$ is continuous for all values of x.

b. The function g is a rational function. Observe that the denominator of g—namely, $x^2 + 1$—is never equal to zero. Therefore, we conclude that g is continuous for all values of x.

c. The function h is a rational function. In this case, however, the denominator of h is equal to zero at $x = 1$ and $x = 2$, which can be seen by factoring it. Thus,

$$x^2 - 3x + 2 = (x - 2)(x - 1)$$

We therefore conclude that h is continuous everywhere except at $x = 1$ and $x = 2$, where it is discontinuous.

Up to this point, most of the applications we have discussed involved functions that are continuous everywhere. In Example 4, we consider an application from the field of educational psychology that involves a discontinuous function.

APPLIED EXAMPLE 4 Learning Curves Figure 17 depicts the learning curve associated with a certain individual. Beginning with no knowledge of the subject being taught, the individual makes steady progress toward understanding it over the time interval $0 \le t < t_1$. In this instance, the individual's progress slows as we approach time t_1 because he fails to grasp a particularly difficult concept. All of a sudden, a breakthrough occurs at time t_1, propelling his knowledge of the subject to a higher level. The curve is discontinuous at t_1.

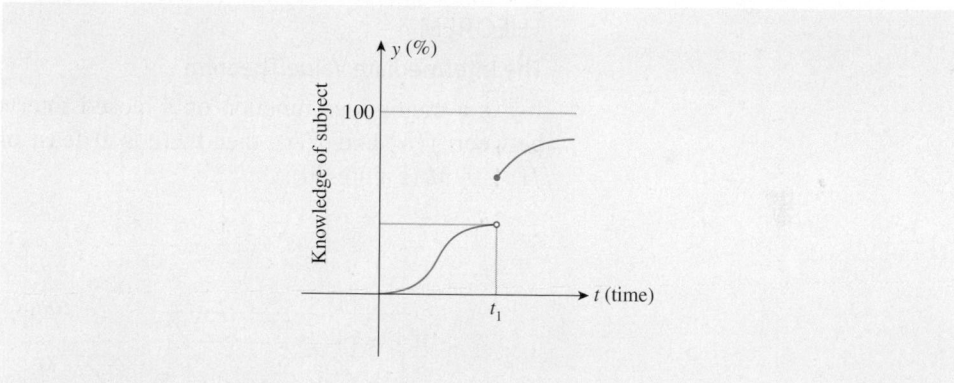

FIGURE **17**
A learning curve that is discontinuous at $t = t_1$

Intermediate Value Theorem

Let's look again at our model of the motion of the maglev on a straight stretch of track. We know that the train cannot vanish at any instant of time and it cannot skip portions of the track and reappear someplace else. To put it another way, the train cannot occupy the positions s_1 and s_2 without, at some time, occupying an intermediate position (Figure 18).

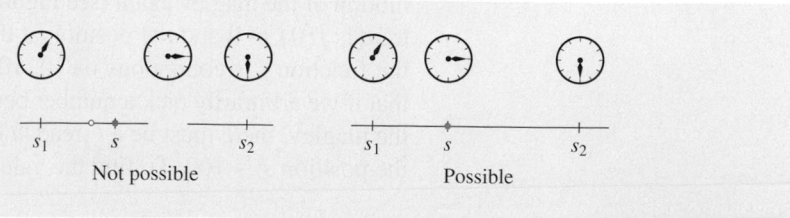

FIGURE **18**
The position of the maglev

To state this fact mathematically, recall that the position of the maglev as a function of time is described by

$$f(t) = 4t^2 \qquad (0 \le t \le 10)$$

Suppose the position of the maglev is s_1 at some time t_1 and its position is s_2 at some time t_2 (Figure 19). Then, if s_3 is any number between s_1 and s_2 giving an intermediate position of the maglev, there must be at least one t_3 between t_1 and t_2 giving the time at which the train is at s_3—that is, $f(t_3) = s_3$.

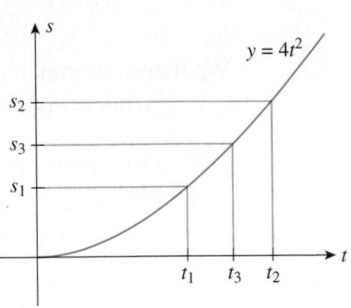

FIGURE 19
If $s_1 \le s_3 \le s_2$, then there must be at least one t_3 ($t_1 \le t_3 \le t_2$) such that $f(t_3) = s_3$.

This discussion carries the gist of the Intermediate Value Theorem. The proof of this theorem can be found in most advanced calculus texts.

THEOREM 4

The Intermediate Value Theorem

If f is a continuous function on a closed interval $[a, b]$ and M is any number between $f(a)$ and $f(b)$, then there is at least one number c in $[a, b]$ such that $f(c) = M$ (Figure 20).

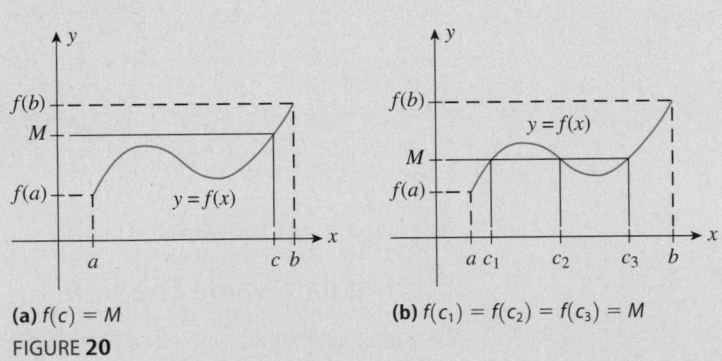

(a) $f(c) = M$

(b) $f(c_1) = f(c_2) = f(c_3) = M$

FIGURE 20

To illustrate the Intermediate Value Theorem, let's look at the example involving the motion of the maglev again (see Figure 2, page 595). Notice that the initial position of the train is $f(0) = 0$ and the position at the end of its test run is $f(10) = 400$. Furthermore, the function f is continuous on $[0, 10]$. So the Intermediate Value Theorem guarantees that if we arbitrarily pick a number between 0 and 400—say, 100—giving the position of the maglev, there must be a $\bar{t}$ (read "t bar") between 0 and 10 at which time the train is at the position $s = 100$. To find the value of $\bar{t}$, we solve the equation $f(\bar{t}) = s$, or

$$4\bar{t}^2 = 100$$

giving $\bar{t} = 5$ (t must lie between 0 and 10).

When we use Theorem 4, it is important to remember that the function f must be continuous. The conclusion of the Intermediate Value Theorem may not hold if f is not continuous (see Exercise 98, page 627).

The next theorem is an immediate consequence of the Intermediate Value Theorem. It not only tells us when a **zero of a function** f [root of the equation $f(x) = 0$] exists but also provides the basis for a method of approximating it.

THEOREM 5

Existence of Zeros of a Continuous Function

If f is a continuous function on a closed interval $[a, b]$, and if $f(a)$ and $f(b)$ have opposite signs, then there is at least one solution of the equation $f(x) = 0$ in the interval (a, b) (Figure 21).

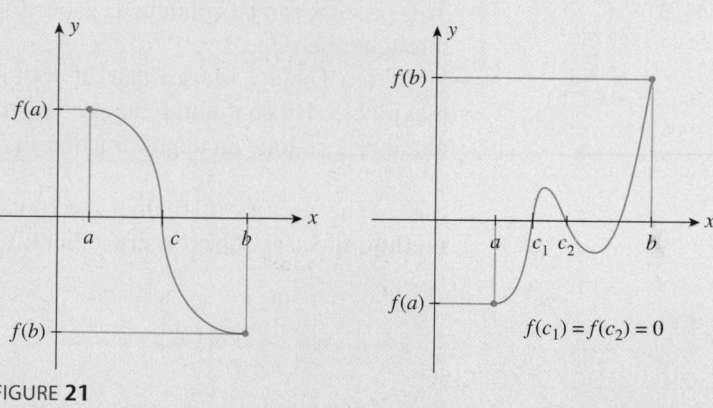

FIGURE 21
If $f(a)$ and $f(b)$ have opposite signs, there must be at least one number c $(a < c < b)$ such that $f(c) = 0$.

Geometrically, this property states that if the graph of a continuous function goes from above the x-axis to below the x-axis or vice versa, it must *cross* the x-axis. This is not necessarily true if the function is discontinuous (Figure 22).

EXAMPLE 5 Let $f(x) = x^3 + x + 1$.

a. Show that f is continuous for all values of x.
b. Compute $f(-1)$ and $f(1)$, and use the results to deduce that there must be at least one number $x = c$, where c lies in the interval $(-1, 1)$ and $f(c) = 0$.

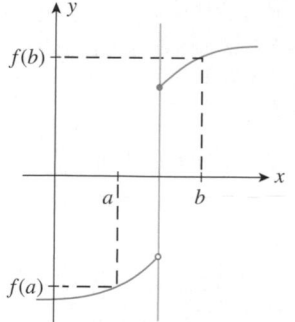

FIGURE 22
$f(a) < 0$ and $f(b) > 0$, but the graph of f does not cross the x-axis between a and b because f is discontinuous.

Solution

a. The function f is a polynomial function of degree 3 and is therefore continuous everywhere.
b. $f(-1) = (-1)^3 + (-1) + 1 = -1 \quad$ and $\quad f(1) = 1^3 + 1 + 1 = 3$

Since $f(-1)$ and $f(1)$ have opposite signs, Theorem 5 tells us that there must be at least one number $x = c$ with $-1 < c < 1$ such that $f(c) = 0$.

The next example shows how the Intermediate Value Theorem can be used to help us find a zero of a function.

EXAMPLE 6 Let $f(x) = x^3 + x - 1$. Since f is a polynomial function, it is continuous everywhere. Observe that $f(0) = -1$ and $f(1) = 1$, so Theorem 5 guarantees the existence of at least one root of the equation $f(x) = 0$ in $(0, 1)$.*

---

*It can be shown that f has precisely one zero in $(0, 1)$ (see Exercise 110, Section 10.1).

We can locate the root more precisely by using Theorem 5 once again as follows: Evaluate $f(x)$ at the midpoint of $[0, 1]$, obtaining

$$f(0.5) = -0.375$$

Because $f(0.5) < 0$ and $f(1) > 0$, Theorem 5 now tells us that the root must lie in $(0.5, 1)$.

Repeat the process: Evaluate $f(x)$ at the midpoint of $[0.5, 1]$, which is

$$\frac{0.5 + 1}{2} = 0.75$$

Thus,

$$f(0.75) \approx 0.1719$$

Because $f(0.5) < 0$ and $f(0.75) > 0$, Theorem 5 tells us that the root is in $(0.5, 0.75)$. This process can be continued. Table 3 summarizes the results of our computations through nine steps.

From Table 3, we see that the root is approximately 0.68, accurate to two decimal places. By continuing the process through a sufficient number of steps, we can obtain as accurate an approximation to the root as we please.

TABLE 3

Step	Root of $f(x) = 0$ Lies in
1	$(0, 1)$
2	$(0.5, 1)$
3	$(0.5, 0.75)$
4	$(0.625, 0.75)$
5	$(0.625, 0.6875)$
6	$(0.65625, 0.6875)$
7	$(0.671875, 0.6875)$
8	$(0.6796875, 0.6875)$
9	$(0.6796875, 0.6835937)$

Note   The process of finding the root of $f(x) = 0$ used in Example 6 is called the **method of bisection.** It is crude but effective.

9.2   Self-Check Exercises

1. Evaluate $\lim_{x \to -1^-} f(x)$ and $\lim_{x \to -1^+} f(x)$, where

$$f(x) = \begin{cases} 1 & \text{if } x < -1 \\ 1 + \sqrt{x + 1} & \text{if } x \geq -1 \end{cases}$$

Does $\lim_{x \to -1} f(x)$ exist?

2. Determine the values of x for which the given function is discontinuous. At each number where f is discontinuous, indicate which condition(s) for continuity are violated. Sketch the graph of the function.

a. $f(x) = \begin{cases} -x^2 + 1 & \text{if } x \leq 1 \\ x - 1 & \text{if } x > 1 \end{cases}$

b. $g(x) = \begin{cases} -x + 1 & \text{if } x < -1 \\ 2 & \text{if } -1 < x \leq 1 \\ -x + 3 & \text{if } x > 1 \end{cases}$

Solutions to Self-Check Exercises 9.2 can be found on page 627.

9.2   Concept Questions

1. Explain what is meant by the statement $\lim_{x \to 3^-} f(x) = 2$ and $\lim_{x \to 3^+} f(x) = 4$.

2. Suppose $\lim_{x \to 1^-} f(x) = 3$ and $\lim_{x \to 1^+} f(x) = 4$.

   a. What can you say about $\lim_{x \to 1} f(x)$? Explain.

   b. What can you say about $f(1)$? Explain.

3. Explain what it means for a function f to be continuous (a) at a number a and (b) on an interval I.

4. Suppose $\lim_{x \to a^-} f(x) = L$ and $\lim_{x \to a^+} f(x) = M$, where L and M are real numbers. What conditions on L, M, and $f(a)$ will guarantee that f is continuous at $x = a$?

5. Determine whether each function f is continuous or discontinuous. Explain your answer.

   a. $f(t)$ gives the altitude of an airplane at time t.

   b. $f(t)$ measures the total amount of rainfall at time t at the Municipal Airport.

   c. $f(s)$ measures the fare as a function of the distance s for taking a cab from Kennedy Airport to downtown Manhattan.

   d. $f(t)$ gives the interest rate charged by a financial institution at time t.

6. Explain the Intermediate Value Theorem in your own words.

9.2   Exercises

In Exercises 1–8, use the graph of the function f to find $\lim_{x \to a^-} f(x)$, $\lim_{x \to a^+} f(x)$, and $\lim_{x \to a} f(x)$ at the indicated value of a, if the limit exists.

1.
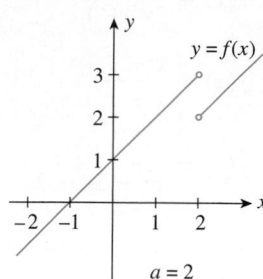
$a = 2$

2.
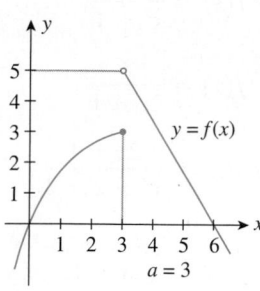
$a = 3$

3.
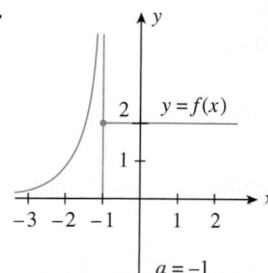
$a = -1$

4.
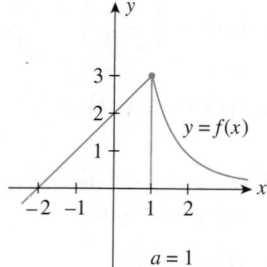
$a = 1$

5.
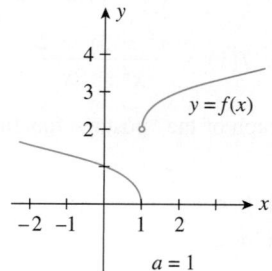
$a = 1$

6.
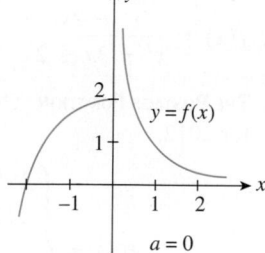
$a = 0$

7.
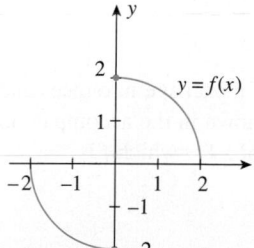
$a = 0$

8.
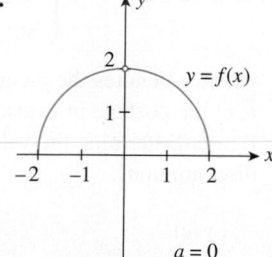
$a = 0$

In Exercises 9–14, refer to the graph of the function f and determine whether each statement is true or false.

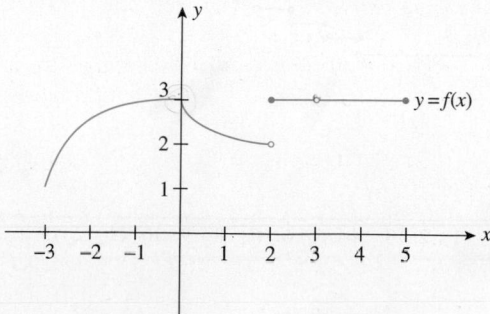

9. $\lim_{x \to -3^+} f(x) = 1$

10. $\lim_{x \to 0} f(x) = f(0)$

11. $\lim_{x \to 2^-} f(x) = 2$

12. $\lim_{x \to 2^+} f(x) = 3$

13. $\lim_{x \to 3} f(x)$ does not exist.

14. $\lim_{x \to 5^-} f(x) = 3$

In Exercises 15–20, refer to the graph of the function f and determine whether each statement is true or false.

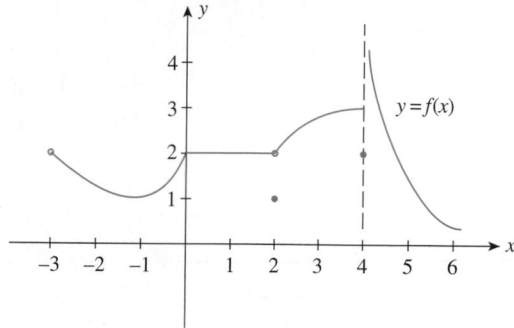

15. $\lim_{x \to -3^+} f(x) = 2$

16. $\lim_{x \to 0} f(x) = 2$

17. $\lim_{x \to 2} f(x) = 1$

18. $\lim_{x \to 4^-} f(x) = 3$

19. $\lim_{x \to 4^+} f(x)$ does not exist.

20. $\lim_{x \to 4} f(x) = 2$

In Exercises 21–38, find the indicated one-sided limit, if it exists.

21. $\lim_{x \to 1^+} (2x + 4)$

22. $\lim_{x \to 1^-} (3x - 4)$

23. $\lim_{x \to 2^-} \dfrac{x - 3}{x + 2}$

24. $\lim_{x \to 1^+} \dfrac{x + 2}{x + 1}$

25. $\lim_{x \to 0^+} \dfrac{1}{x}$

26. $\lim_{x \to 0^-} \dfrac{1}{x}$

27. $\lim_{x \to 0^+} \dfrac{x - 1}{x^2 + 1}$

28. $\lim_{x \to 2^+} \dfrac{x + 1}{x^2 - 2x + 3}$

29. $\lim_{x \to 0^+} \sqrt{x}$

30. $\lim_{x \to 2^+} 2\sqrt{x - 2}$

31. $\lim_{x \to -2^+} (2x + \sqrt{2 + x})$

32. $\lim_{x \to -5^+} x(1 + \sqrt{5 + x})$

33. $\lim_{x \to 1^-} \dfrac{1 + x}{1 - x}$

34. $\lim_{x \to 1^+} \dfrac{1 + x}{1 - x}$

35. $\lim_{x \to 2^-} \dfrac{x^2 - 4}{x - 2}$

36. $\lim_{x \to -3^+} \dfrac{\sqrt{x + 3}}{x^2 + 1}$

37. $\lim_{x \to 0^+} f(x)$ and $\lim_{x \to 0^-} f(x)$, where

$$f(x) = \begin{cases} 2x & \text{if } x < 0 \\ x^2 & \text{if } x \geq 0 \end{cases}$$

38. $\lim_{x \to 0^+} f(x)$ and $\lim_{x \to 0^-} f(x)$, where

$$f(x) = \begin{cases} -x + 1 & \text{if } x \leq 0 \\ 2x + 3 & \text{if } x > 0 \end{cases}$$

In Exercises 39–44, determine the values of x, if any, at which each function is discontinuous. At each number where f is discontinuous, state the condition(s) for continuity that are violated.

39.

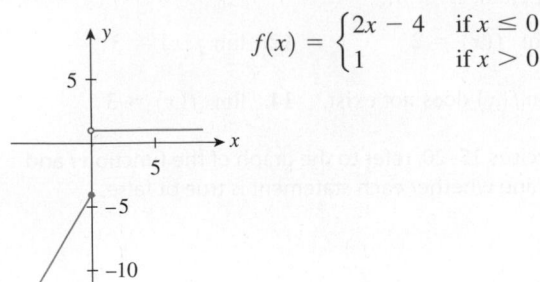

$$f(x) = \begin{cases} 2x - 4 & \text{if } x \leq 0 \\ 1 & \text{if } x > 0 \end{cases}$$

40.

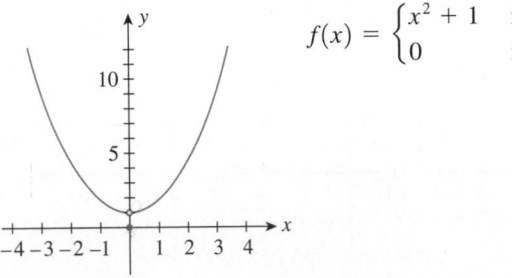

$$f(x) = \begin{cases} x^2 + 1 & \text{if } x \neq 0 \\ 0 & \text{if } x = 0 \end{cases}$$

41.

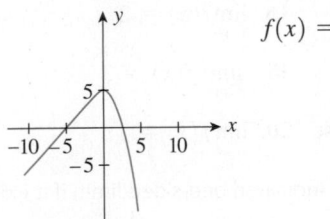

$$f(x) = \begin{cases} x + 5 & \text{if } x \leq 0 \\ -x^2 + 5 & \text{if } x > 0 \end{cases}$$

42.

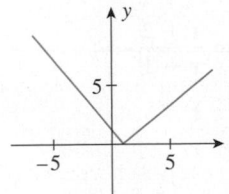

$$f(x) = |x - 1|$$

43.

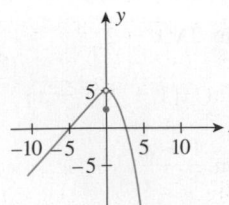

$$f(x) = \begin{cases} x + 5 & \text{if } x < 0 \\ 2 & \text{if } x = 0 \\ -x^2 + 5 & \text{if } x > 0 \end{cases}$$

44.

$$f(x) = \begin{cases} \dfrac{x^2 - 1}{x + 1} & \text{if } x \neq -1 \\ 1 & \text{if } x = -1 \end{cases}$$

In Exercises 45–56, find the values of x for which each function is continuous.

45. $f(x) = 2x^2 + x - 1$

46. $f(x) = x^3 - 2x^2 + x - 1$

47. $f(x) = \dfrac{2}{x^2 + 1}$

48. $f(x) = \dfrac{x}{2x^2 + 1}$

49. $f(x) = \dfrac{2}{2x - 1}$

50. $f(x) = \dfrac{x + 1}{x - 1}$

51. $f(x) = \dfrac{2x + 1}{x^2 + x - 2}$

52. $f(x) = \dfrac{x - 1}{x^2 + 2x - 3}$

53. $f(x) = \begin{cases} x & \text{if } x \leq 1 \\ 2x - 1 & \text{if } x > 1 \end{cases}$

54. $f(x) = \begin{cases} -2x + 1 & \text{if } x < 0 \\ x^2 + 1 & \text{if } x \geq 0 \end{cases}$

55. $f(x) = |x + 1|$

56. $f(x) = \dfrac{|x - 1|}{x - 1}$

In Exercises 57–60, determine all values of x at which the function is discontinuous.

57. $f(x) = \dfrac{2x}{x^2 - 1}$

58. $f(x) = \dfrac{1}{(x - 1)(x - 2)}$

59. $f(x) = \dfrac{x^2 - 2x}{x^2 - 3x + 2}$

60. $f(x) = \dfrac{x^2 - 3x + 2}{x^2 - 2x}$

61. THE POSTAGE FUNCTION The graph of the "postage function" for 2012,

$$f(x) = \begin{cases} 195 & \text{if } 0 < x < 4 \\ 212 & \text{if } 4 \leq x < 5 \\ \vdots \\ 348 & \text{if } 12 \leq x < 13 \\ 365 & \text{if } x = 13 \end{cases}$$

where x denotes the weight of a package in ounces and $f(x)$ the postage in cents, is shown in the accompanying figure. Determine the values of x for which f is discontinuous.

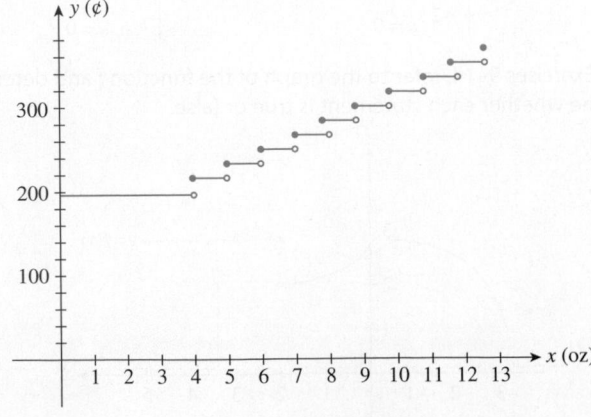

62. **INVENTORY CONTROL** As part of an optimal inventory policy, the manager of an office supply company orders 500 reams of photocopy paper every 20 days. The accompanying graph shows the *actual* inventory level of paper in an office supply store during the first 60 business days of 2015. Determine the values of t for which the "inventory function" is discontinuous, and give an interpretation of the graph.

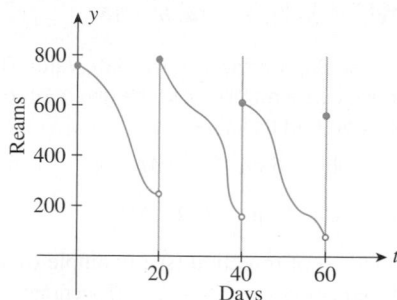

63. **LEARNING CURVES** The following graph describes the progress Michael made in solving a problem correctly during a mathematics quiz. Here, y denotes the percentage of work completed, and x is measured in minutes. Give an interpretation of the graph.

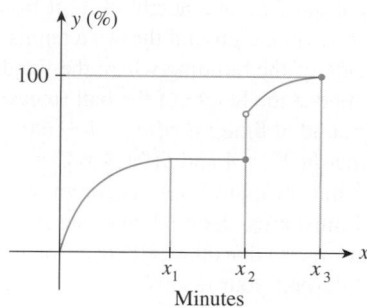

64. **AILING FINANCIAL INSTITUTIONS** Franklin Savings and Loan acquired two ailing financial institutions in 2012. One of them was acquired at time $t = T_1$, and the other was acquired at time $t = T_2$ ($t = 0$ corresponds to the beginning of 2012). The following graph shows the total amount of money on deposit with Franklin. Explain the significance of the discontinuities of the function at T_1 and T_2.

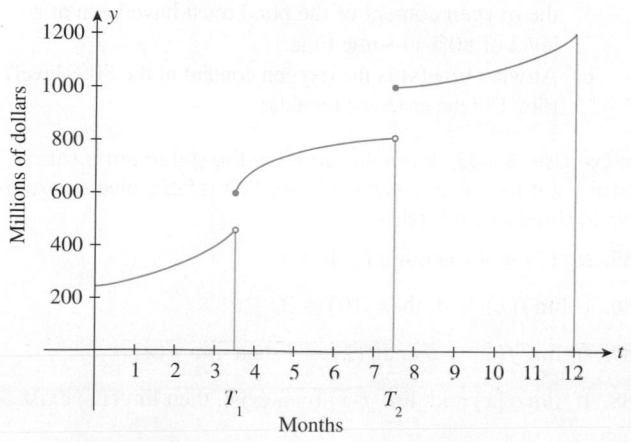

65. **ENERGY CONSUMPTION** The following graph shows the amount of home heating oil remaining in a 200-gal tank over a 120-day period ($t = 0$ corresponds to October 1). Explain why the function is discontinuous at $t = 40$, 70, 95, and 110.

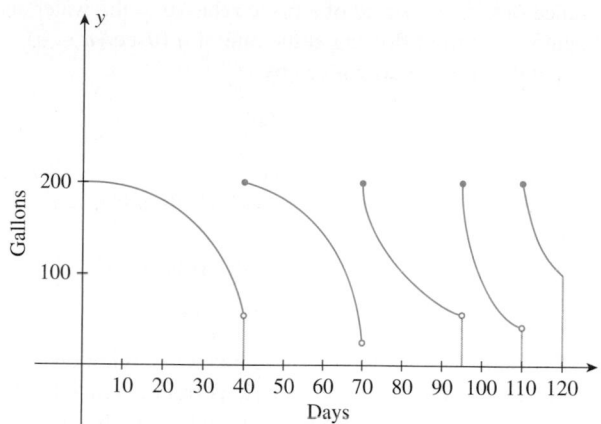

66. **PRIME INTEREST RATE** The function P, whose graph follows, gives the prime rate (the interest rate banks charge their best corporate customers) for a certain country as a function of time for the first 32 weeks in 2015. Determine the values of t for which P is discontinuous, and interpret your results.

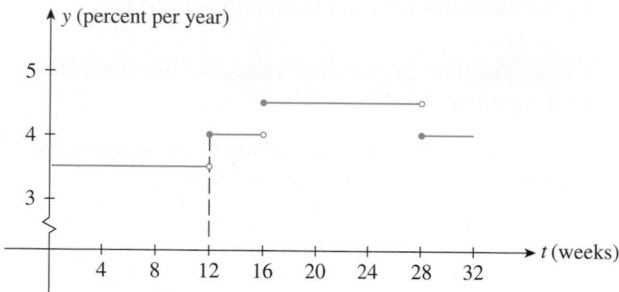

67. **PARKING FEES** The fee charged per car in a downtown parking lot is $2.00 for the first half hour and $1.00 for each additional half hour or part thereof, subject to a maximum of $10.00. Derive a function f relating the parking fee to the length of time a car is left in the lot. Sketch the graph of f and determine the values of x for which the function f is discontinuous.

68. **COMMISSIONS** The base monthly salary of a salesman working on commission is $22,000. For each $50,000 of sales beyond $100,000, he is paid a $1000 commission. Sketch a graph showing his earnings as a function of the level of his sales x. Determine the values of x for which the function f is discontinuous.

69. **COMMODITY PRICES** The function that gives the cost of a certain commodity is defined by

$$C(x) = \begin{cases} 5x & \text{if } 0 < x < 10 \\ 4x & \text{if } 10 \le x < 30 \\ 3.5x & \text{if } 30 \le x < 60 \\ 3.25x & \text{if } x \ge 60 \end{cases}$$

where x is the number of pounds of a certain commodity sold and $C(x)$ is measured in dollars. Sketch the graph of the function C, and determine the values of x for which the function C is discontinuous.

70. ENERGY EXPENDED BY A FISH Suppose a fish swimming a distance of L ft at a speed of v ft/sec relative to the water and against a current flowing at the rate of u ft/sec ($u < v$) expends a total energy given by

$$E(v) = \frac{aLv^3}{v - u}$$

where E is measured in foot-pounds (ft-lb) and a is a constant.

  a. Evaluate $\lim\limits_{v \to u^+} E(v)$, and interpret your result.

  b. Evaluate $\lim\limits_{v \to \infty} E(v)$, and interpret your result.

71. WEISS'S LAW According to Weiss's law of excitation of tissue, the strength S of an electric current is related to the time t the current takes to excite tissue by the formula

$$S(t) = \frac{a}{t} + b \qquad (t > 0)$$

where a and b are positive constants.

  a. Evaluate $\lim\limits_{t \to 0^+} S(t)$, and interpret your result.

  b. Evaluate $\lim\limits_{t \to \infty} S(t)$, and interpret your result.

(*Note:* The limit in part (b) is called the threshold strength of the current. Why?)

72. LEVERAGED RETURN The return on assets using borrowed money, called *leveraged return*, is given by

$$L = \frac{Y - (1 - D)R}{D}$$

where Y is the return of the asset, R is the cost of borrowed money, and D is the percentage of money the investor must put down to secure the loan. Suppose that both Y and R are constant.

  a. Find $\lim\limits_{D \to 0^+} L$, and interpret your result.

  b. Find $\lim\limits_{D \to 1^-} L$, and interpret your result.

Source: Scientific American.

73. Let

$$f(x) = \begin{cases} x + 2 & \text{if } x \leq 1 \\ kx^2 & \text{if } x > 1 \end{cases}$$

Find the value of k that will make f continuous on $(-\infty, \infty)$.

74. Let

$$f(x) = \begin{cases} \dfrac{x^2 - 4}{x + 2} & \text{if } x \neq -2 \\ k & \text{if } x = -2 \end{cases}$$

For what value of k will f be continuous on $(-\infty, \infty)$?

In Exercises 75–78, (a) show that the function f is continuous for all values of x in the interval $[a, b]$ and (b) prove that f must have at least one zero in the interval (a, b) by showing that $f(a)$ and $f(b)$ have opposite signs.

75. $f(x) = x^2 - 6x + 8$; $a = 1, b = 3$

76. $f(x) = 2x^3 - 3x^2 - 36x + 14$; $a = 0, b = 1$

77. $f(x) = x^3 - 2x^2 + 3x + 2$; $a = -1, b = 1$

78. $f(x) = 2x^{5/3} - 5x^{4/3}$; $a = 14, b = 16$

In Exercises 79 and 80, use the Intermediate Value Theorem to show that there exists a number c in the given interval such that $f(c) = M$. Then find its value.

79. $f(x) = x^2 - 4x + 6$ on $[0, 3]$; $M = 4$

80. $f(x) = x^2 - x + 1$ on $[-1, 4]$; $M = 7$

81. Use the method of bisection (see Example 6) to find the root of the equation $x^5 + 2x - 7 = 0$ accurate to two decimal places.

82. Use the method of bisection (see Example 6) to find the root of the equation $x^3 - x + 1 = 0$ accurate to two decimal places.

83. FALLING OBJECT Joan is looking straight out of a window of an apartment building at a height of 32 ft from the ground. A boy on the ground throws a tennis ball straight up by the side of the building where the window is located. Suppose the height of the ball (measured in feet) from the ground at time t is $h(t) = 4 + 64t - 16t^2$.

  a. Show that $h(0) = 4$ and $h(2) = 68$.

  b. Use the Intermediate Value Theorem to conclude that the ball must cross Joan's line of sight at least once.

  c. At what time(s) does the ball cross Joan's line of sight? Interpret your results.

84. OXYGEN CONTENT OF A POND The oxygen content t days after organic waste has been dumped into a pond is given by

$$f(t) = 100 \left(\frac{t^2 + 10t + 100}{t^2 + 20t + 100} \right)$$

percent of its normal level.

  a. Show that $f(0) = 100$ and $f(10) = 75$.

  b. Use the Intermediate Value Theorem to conclude that the oxygen content of the pond must have been at a level of 80% at some time.

  c. At what time(s) is the oxygen content at the 80% level? Hint: Use the quadratic formula.

In Exercises 85–93, determine whether the statement is true or false. If it is true, explain why it is true. If it is false, give an example to show why it is false.

85. If $f(2) = 4$, then $\lim\limits_{x \to 2} f(x) = 4$.

86. If $\lim\limits_{x \to 0} f(x) = 3$, then $f(0) = 3$.

87. If $\lim\limits_{x \to 2^+} f(x) = 3$ and $f(2) = 3$ then $\lim\limits_{x \to 2^-} f(x) = 3$.

88. If $\lim\limits_{x \to 3^-} f(x)$ and $\lim\limits_{x \to 3^+} f(x)$ both exist, then $\lim\limits_{x \to 3} f(x)$ exists.

89. If $f(5)$ is not defined, then $\lim\limits_{x \to 5^-} f(x)$ does not exist.

90. Suppose the function f is defined on the interval $[a, b]$. If $f(a)$ and $f(b)$ have the same sign, then f has no zero in $[a, b]$.

91. If $\lim\limits_{x \to a^-} f(x) = L$ and $\lim\limits_{x \to a^+} f(x) = L$, then $f(a) = L$.

92. If $\lim\limits_{x \to a} f(x) = L$ and $g(a) = M$, then $\lim\limits_{x \to a} f(x)g(x) = LM$.

93. If f is continuous for all $x \neq 0$ and $f(0) = 0$, then $\lim\limits_{x \to 0} f(x) = 0$.

94. a. Is the following statement true or false? Suppose f is continuous on $[a, b]$ and $f(a) < f(b)$. If M is a number that lies outside the interval $[f(a), f(b)]$, then there does not exist a number $a < c < b$ such that $f(c) = M$.
  b. Does this contradict the Intermediate Value Theorem?

95. Let $f(x) = \dfrac{x^2}{x^2 + 1}$.
  a. Show that f is continuous for all values of x.
  b. Show that $f(x)$ is nonnegative for all values of x.
  c. Show that f has a zero at $x = 0$. Does this contradict Theorem 5?

96. Let $f(x) = x - \sqrt{1 - x^2}$.
  a. Show that f is continuous for all values of x in the interval $[-1, 1]$.
  b. Show that f has at least one zero in $[-1, 1]$.
  c. Find the zeros of f in $[-1, 1]$ by solving the equation $f(x) = 0$.

97. a. Prove that a polynomial function $y = P(x)$ is continuous at every number x. Follow these steps:
  (i) Use Properties 2 and 3 of continuous functions to establish that the function $g(x) = x^n$, where n is a positive integer, is continuous everywhere.
  (ii) Use Properties 1 and 5 to show that $f(x) = cx^n$, where c is a constant and n is a positive integer, is continuous everywhere.
  (iii) Use Property 4 to complete the proof of the result.
  b. Prove that a rational function $R(x) = p(x)/q(x)$ is continuous at every point x where $q(x) \neq 0$.
  Hint: Use the result of part (a) and Property 6.

98. Show that the conclusion of the Intermediate Value Theorem does not necessarily hold if f is discontinuous on $[a, b]$.

9.2 Solutions to Self-Check Exercises

1. For $x < -1$, $f(x) = 1$, and so

$$\lim_{x \to -1^-} f(x) = \lim_{x \to -1^-} 1 = 1$$

For $x \geq -1$, $f(x) = 1 + \sqrt{x + 1}$, and so

$$\lim_{x \to -1^+} f(x) = \lim_{x \to -1^+} (1 + \sqrt{x + 1}) = 1$$

Since the left-hand and right-hand limits of f exist as x approaches -1 and both are equal to 1, we conclude that

$$\lim_{x \to -1} f(x) = 1$$

2. a. The graph of f follows:

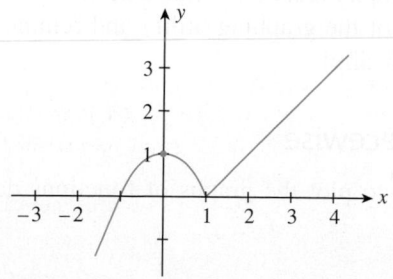

We see that f is continuous everywhere.

b. The graph of g follows:

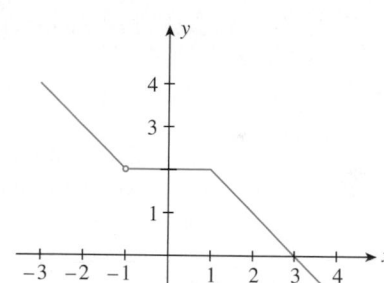

Since g is not defined at $x = -1$, it is discontinuous there. It is continuous everywhere else.

USING TECHNOLOGY   Finding the Points of Discontinuity of a Function

You can very often recognize the points of discontinuity of a function f by examining its graph. For example, Figure T1a shows the graph of $f(x) = x/(x^2 - 1)$ obtained using a graphing utility. It is evident that f is discontinuous at $x = -1$ and $x = 1$. This observation is also borne out by the fact that both these points are not in the domain of f.

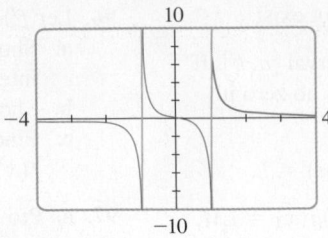
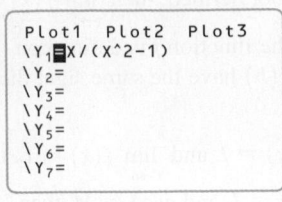

(a) **(b)**

FIGURE T1
(a) The graph of $f(x) = x/(x^2 - 1)$ in the viewing window $[-4, 4] \times [-10, 10]$;
(b) the TI-83/84 equation screen

Consider the function

$$g(x) = \frac{2x^3 + x^2 - 7x - 6}{x^2 - x - 2}$$

Using a graphing utility, we obtain the graph of g shown in Figure T2a.

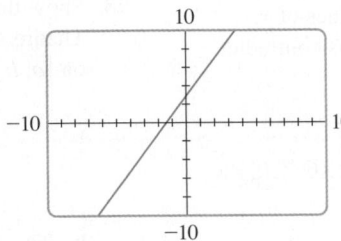

(a) **(b)**

FIGURE T2
(a) The graph of $g(x) = (2x^3 + x^2 - 7x - 6)/(x^2 - x - 2)$ in the standard
viewing window; (b) the TI-83/84 equation screen

An examination of this graph does not reveal any points of discontinuity. However, if
we factor both the numerator and the denominator of the rational expression, we see that

$$g(x) = \frac{(x + 1)(x - 2)(2x + 3)}{(x + 1)(x - 2)}$$
$$= 2x + 3$$

provided that $x \neq -1$ and $x \neq 2$, so its graph in fact looks like that shown in Figure T3.

This example shows the limitation of the graphing utility and reminds us of the
importance of studying functions analytically!

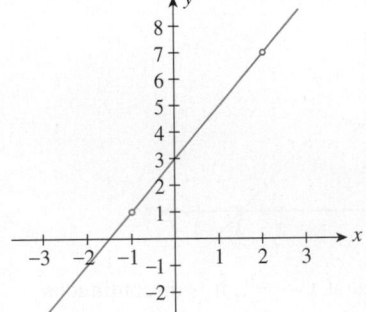

FIGURE T3
The graph of g has holes at $(-1, 1)$ and
$(2, 7)$.

Graphing Functions Defined Piecewise

The following example illustrates how to plot the graphs of functions defined in a
piecewise manner on a graphing utility.

EXAMPLE 1 Plot the graph of

$$f(x) = \begin{cases} x + 1 & \text{if } x \leq 1 \\ \dfrac{2}{x} & \text{if } x > 1 \end{cases}$$

Solution We enter the function

$$y1 = (x + 1)(x \leq 1) + (2/x)(x > 1)$$

Figure T4a shows the graph of the function in the viewing window $[-5, 5] \times [-2, 4]$.

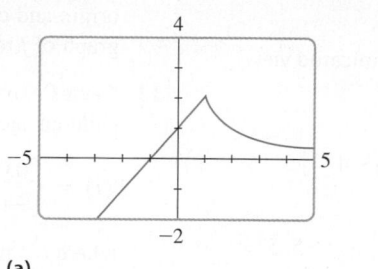

(a)

(b)

FIGURE **T4**
(a) The graph of f in the viewing window $[-5, 5] \times [-2, 4]$; (b) the TI-83/84 equation screen

APPLIED EXAMPLE 2 TV Viewing Patterns The percent of U.S. households, $P(t)$, watching television during weekdays between the hours of 4 P.M. and 4 A.M. is given by

$$P(t) = \begin{cases} 0.01354t^4 - 0.49375t^3 + 2.58333t^2 + 3.8t + 31.60704 & \text{if } 0 \le t \le 8 \\ 1.35t^2 - 33.05t + 208 & \text{if } 8 < t \le 12 \end{cases}$$

where t is measured in hours, with $t = 0$ corresponding to 4 P.M. Plot the graph of P in the viewing window $[0, 12] \times [0, 80]$.
Source: A. C. Nielsen Co.

Solution We enter the function

$$y1 = (.01354x^4 - .49375x^3 + 2.58333x^2 + 3.8x + 31.60704)(x \ge 0)(x \le 8)$$
$$+ (1.35x^2 - 33.05x + 208)(x > 8)(x \le 12)$$

Figure T5a shows the graph of P.

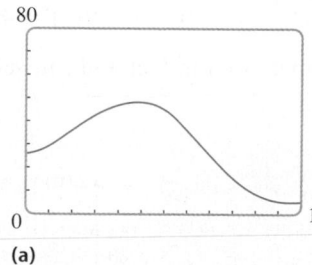

(a)

(b)

FIGURE **T5**
(a) The graph of P in the viewing window $[0, 12] \times [0, 80]$; (b) the TI-83/84 equation screen

TECHNOLOGY EXERCISES

In Exercises 1–8, plot the graph of f, and find the points of discontinuity of f. Then use analytical means to verify your observation and find all numbers where f is discontinuous.

1. $f(x) = \dfrac{2}{x^2 - x}$

2. $f(x) = \dfrac{3}{\sqrt{x}(x + 1)}$

3. $f(x) = \dfrac{6x^3 + x^2 - 2x}{2x^2 - x}$

4. $f(x) = \dfrac{2x^3 - x^2 - 13x - 6}{2x^2 - 5x - 3}$

5. $f(x) = \dfrac{2x^4 - 3x^3 - 2x^2}{2x^2 - 3x - 2}$

6. $f(x) = \dfrac{6x^4 - x^3 + 5x^2 - 1}{6x^2 - x - 1}$

7. $f(x) = \dfrac{x^3 + x^2 - 2x}{x^4 + 2x^3 - x - 2}$
Hint: $x^4 + 2x^3 - x - 2 = (x^3 - 1)(x + 2)$

8. $f(x) = \dfrac{x^3 - x}{x^{4/3} - x + x^{1/3} - 1}$

Hint: $x^{4/3} - x + x^{1/3} - 1 = (x^{1/3} - 1)(x + 1)$

In Exercises 9 and 10, plot the graph of f, in the indicated viewing window.

9. $f(x) = \begin{cases} 2 & \text{if } x \leq 0 \\ \sqrt{4 - x^2} & \text{if } x > 0; [-2, 2] \times [-4, 4] \end{cases}$

10. $f(x) = \begin{cases} -x^2 + x + 2 & \text{if } x \leq 1 \\ 2x^3 - x^2 - 4 & \text{if } x > 1; [-4, 4] \times [-5, 5] \end{cases}$

11. FLIGHT PATH OF A PLANE The function

$$f(x) = \begin{cases} 0 & \text{if } 0 \leq x < 1 \\ -0.00411523x^3 + 0.0679012x^2 \\ \quad -0.123457x + 0.0596708 & \text{if } 1 \leq x < 10 \\ 1.5 & \text{if } 10 \leq x \leq 100 \end{cases}$$

where both x and $f(x)$ are measured in units of 1000 ft, describes the flight path of a plane taking off from the origin and climbing to an altitude of 15,000 ft. Plot the graph of f to visualize the trajectory of the plane.

12. OBESE CHILDREN IN THE UNITED STATES The percentage of obese children aged 12–19 in the United States is approximately

$$P(t) = \begin{cases} 0.04t + 4.6 & \text{if } 0 \leq t < 10 \\ -0.01005t^2 + 0.945t - 3.4 & \text{if } 10 \leq t \leq 30 \end{cases}$$

where t is measured in years, with $t = 0$ corresponding to the beginning of 1970.

a. Plot the graph of $P(t)$.

b. What was the percentage of obese children aged 12–19 at the beginning of 1970? At the beginning of 1985? At the beginning of 2000?

Source: Centers for Disease Control and Prevention.

9.3 The Derivative

An Intuitive Example

We mentioned in Section 9.1 that the problem of finding the *rate of change* of one quantity with respect to another is mathematically equivalent to the problem of finding the *slope of the tangent line* to a curve at a given point on the curve. Before going on to establish this relationship, let's show its plausibility by looking at it from an intuitive point of view.

Consider the motion of the maglev discussed in Section 9.1. Recall that the position of the maglev at any time t is given by

$$s = f(t) = 4t^2 \qquad (0 \leq t \leq 30)$$

where s is measured in feet and t in seconds. The graph of the function f is sketched in Figure 23.

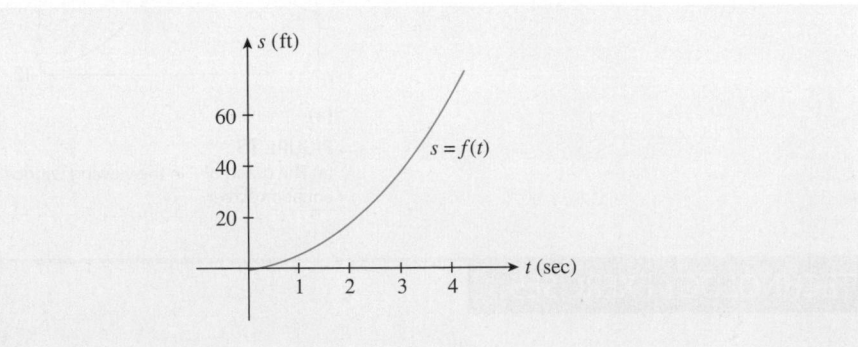

FIGURE 23
Graph showing the position s of a maglev at time t

Observe that the graph of f rises slowly at first but more rapidly as t increases, reflecting the fact that the speed of the maglev is increasing with time. This observation suggests a relationship between the speed of the maglev at any time t and the *steepness* of the curve at the point corresponding to this value of t. Thus, it would appear that we can solve the problem of finding the speed of the maglev at any time if we can find a way to measure the steepness of the curve at any point on the curve.

To discover a yardstick that will measure the steepness of a curve, consider the graph of a function f such as the one shown in Figure 24a. Think of the curve as representing a stretch of roller coaster track (Figure 24b). When the car is at the point P on the curve, a passenger sitting erect in the car and looking straight ahead will have a line of sight that is parallel to the line T, the tangent to the curve at P.

As Figure 24a suggests, the steepness of the curve—that is, the rate at which y is increasing or decreasing with respect to x—is given by the slope of the tangent line to the graph of f at the point $P(x, f(x))$. But for now we will show how this relationship can be used to estimate the rate of change of a function from its graph.

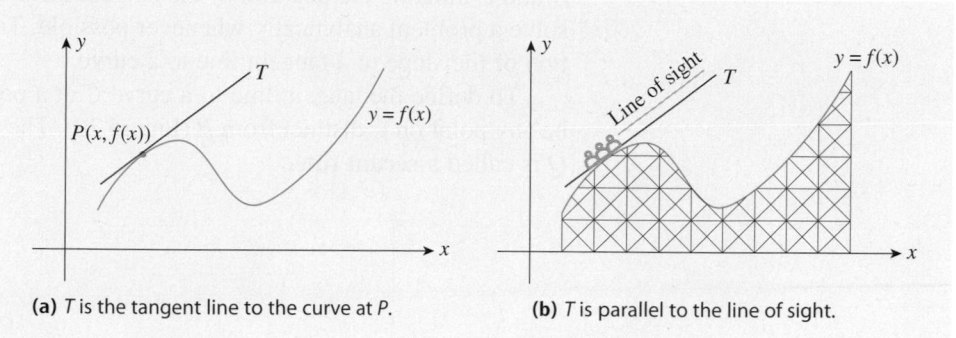

(a) T is the tangent line to the curve at P.   **(b)** T is parallel to the line of sight.

FIGURE **24**

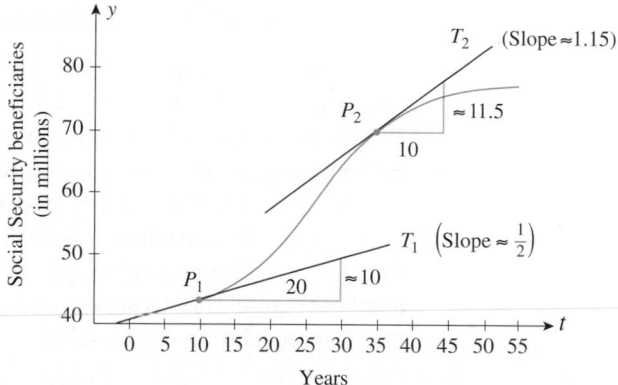

![Social Security beneficiaries icon] **APPLIED EXAMPLE 1** Increasing Number of Social Security Beneficiaries The graph of the function $y = N(t)$, shown in Figure 25, gives the number of Social Security beneficiaries from the beginning of 1990 ($t = 0$) through the year 2045 ($t = 55$).

FIGURE **25**
The number of Social Security beneficiaries from 1990 through 2045. We can use the slope of the tangent line at the indicated points to estimate the rate at which the number of Social Security beneficiaries will be changing.

Use the graph of $y = N(t)$ to estimate the rate at which the number of Social Security beneficiaries was growing at the beginning of the year 2000 ($t = 10$). How fast will the number be growing at the beginning of 2025 ($t = 35$)? [Assume that the rate of change of the function N at any value of t is given by the slope of the tangent line at the point $P(t, N(t))$.]
Source: Social Security Administration.

Solution   From the figure, we see that the slope of the tangent line T_1 to the graph of $y = N(t)$ at $P_1(10, 44.7)$ is approximately 0.5. This tells us that the quantity y is increasing at the rate of $\frac{1}{2}$ unit per unit increase in t, when $t = 10$. In other words, at the beginning of the year 2000, the number of Social Security beneficiaries was increasing at the rate of approximately 0.5 million, or 500,000, per year.

The slope of the tangent line T_2 at $P_2(35, 71.9)$ is approximately 1.15. This tells us that at the beginning of 2025, the number of Social Security beneficiaries will be growing at the rate of approximately 1.15 million, or 1,150,000, per year. ∎

Slope of a Tangent Line

In Example 1, we answered the questions raised by drawing the graph of the function N and estimating the position of the tangent lines. Ideally, however, we would like to solve a problem analytically whenever possible. To do this, we need a precise definition of the slope of a tangent line to a curve.

To define the tangent line to a curve C at a point P on the curve, fix P and let Q be any point on C distinct from P (Figure 26). The straight line passing through P and Q is called a **secant line**.

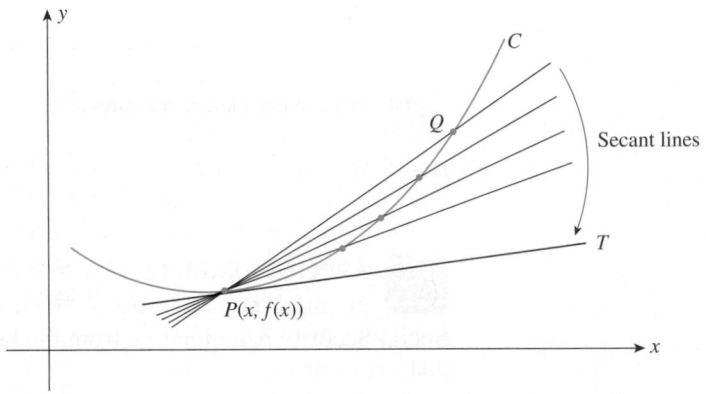

FIGURE **26**
As Q approaches P along the curve C, the secant lines approach the tangent line T.

Now, as the point Q is allowed to move toward P along the curve, the secant line through P and Q rotates about the fixed point P and approaches a fixed line through P. This fixed line, which is the limiting position of the secant lines through P and Q as Q approaches P, is the **tangent line to the graph of f at the point P.**

We can describe the process more precisely as follows. Suppose the curve C is the graph of a function f defined by $y = f(x)$. Then the point P is described by $P(x, f(x))$ and the point Q by $Q(x + h, f(x + h))$, where h is some appropriate nonzero number (Figure 27a). Observe that we can make Q approach P along the curve C by letting h approach zero (Figure 27b).

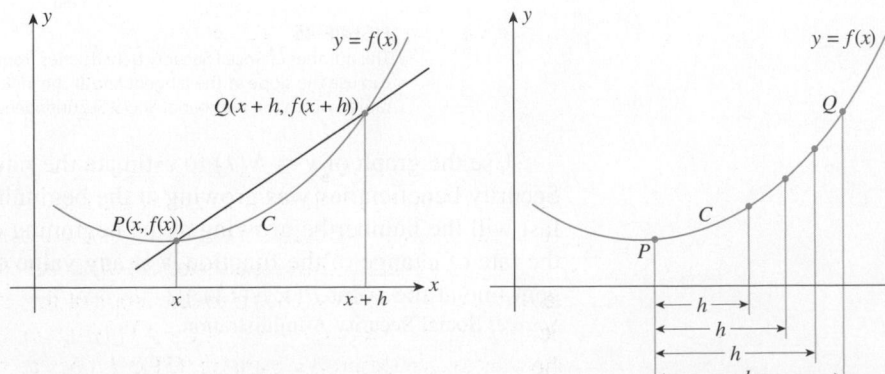

(a) The points $P(x, f(x))$ and $Q(x + h, f(x + h))$ **(b)** As h approaches zero, Q approaches P.
FIGURE **27**

Next, using the formula for the slope of a line, we can write the slope of the secant line passing through $P(x, f(x))$ and $Q(x + h, f(x + h))$ as

$$\frac{f(x + h) - f(x)}{(x + h) - x} = \frac{f(x + h) - f(x)}{h} \tag{3}$$

As we observed earlier, Q approaches P, and therefore the secant line through P and Q approaches the tangent line T as h approaches zero. Consequently, we might expect that the slope of the secant line would approach the slope of the tangent line T as h approaches zero. This leads to the following definition.

> **Slope of a Tangent Line**
>
> The slope of the tangent line to the graph of f at the point $P(x, f(x))$ is given by
>
> $$\lim_{h \to 0} \frac{f(x + h) - f(x)}{h} \tag{4}$$
>
> if it exists.

Rates of Change

We now show that the problem of finding the slope of the tangent line to the graph of a function f at the point $P(x, f(x))$ is mathematically equivalent to the problem of finding the rate of change of f at x. To see this, suppose we are given a function f that describes the relationship between the two quantities x and y—that is, $y = f(x)$. The number $f(x + h) - f(x)$ measures the change in y that corresponds to a change h in x (Figure 28).

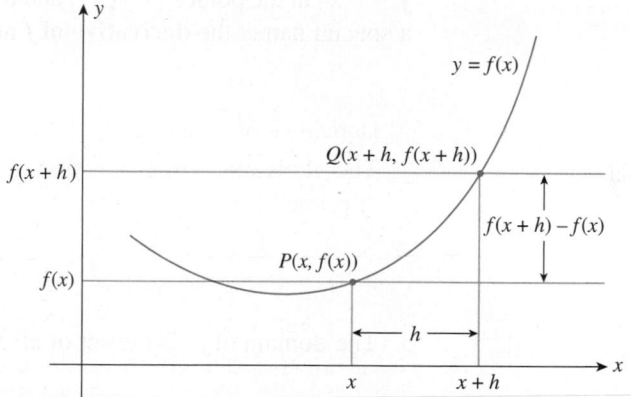

FIGURE **28**
$f(x + h) - f(x)$ is the change in y that corresponds to a change h in x.

Then, the **difference quotient**

$$\frac{f(x + h) - f(x)}{h} \tag{5}$$

measures the **average rate of change of y with respect to x** over the interval $[x, x + h]$. For example, if y measures the position of a car at time x, then quotient (5) gives the average velocity of the car over the time interval $[x, x + h]$.

Observe that the difference quotient (5) is the same as (3). We conclude that the difference quotient (5) also measures the slope of the secant line that passes through the two points $P(x, f(x))$ and $Q(x + h, f(x + h))$ lying on the graph of $y = f(x)$. Next, by taking the limit of the difference quotient (5) as h goes to zero—that is, by evaluating

$$\lim_{h \to 0} \frac{f(x + h) - f(x)}{h} \tag{6}$$

we obtain the **rate of change of f at x.** For example, if y measures the position of a car at time x, then the limit (6) gives the velocity of the car at time x. For emphasis, the rate of change of a function f at x is often called the **instantaneous rate of change of f at x.** This distinguishes it from the average rate of change of f, which is computed over an *interval* $[x, x + h]$ rather than at a *number x.*

Observe that the limit (6) is the same as (4). Therefore, the limit of the difference quotient also measures the slope of the tangent line to the graph of $y = f(x)$ at the point $(x, f(x))$. The following summarizes this discussion.

> **Average and Instantaneous Rates of Change**
>
> The **average rate of change** of f over the interval $[x, x + h]$ or **slope of the secant line** to the graph of f through the points $(x, f(x))$ and $(x + h, f(x + h))$ is
>
> $$\frac{f(x + h) - f(x)}{h} \tag{7}$$
>
> The **instantaneous rate of change** of f at x or **slope of the tangent line** to the graph of f at $(x, f(x))$ is
>
> $$\lim_{h \to 0} \frac{f(x + h) - f(x)}{h} \tag{8}$$

The Derivative

The limit (4) or (8), which measures both the slope of the tangent line to the graph of $y = f(x)$ at the point $P(x, f(x))$ and the (instantaneous) rate of change of f at x, is given a special name: the **derivative of f at x.**

> **Derivative of a Function**
>
> The derivative of a function f with respect to x is the function f' (read "f prime"),
>
> $$f'(x) = \lim_{h \to 0} \frac{f(x + h) - f(x)}{h} \tag{9}$$
>
> The domain of f' is the set of all x for which the limit exists.

Thus, the derivative of a function f is a function f' that gives the slope of the tangent line to the graph of f at *any* point $(x, f(x))$ and also the rate of change of f at x (Figure 29).

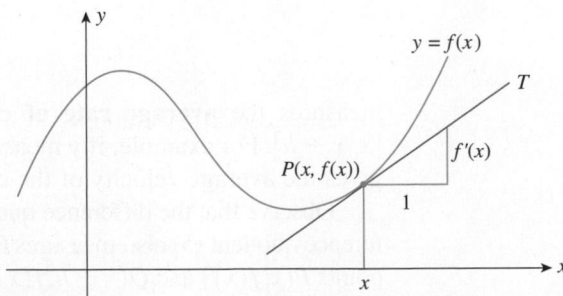

FIGURE 29
The slope of the tangent line at $P(x, f(x))$ is $f'(x)$; f changes at the rate of $f'(x)$ units per unit change in x at x.

Other notations for the derivative of f include the following:

$$D_x f(x) \qquad \text{Read "}d \text{ sub } x \text{ of } f \text{ of } x\text{"}$$

$$\frac{dy}{dx} \qquad \text{Read "}d \, y \, d \, x\text{"}$$

$$y' \qquad \text{Read "}y \text{ prime"}$$

The last two are used when the rule for f is written in the form $y = f(x)$.

The calculation of the derivative of f is facilitated by using the following four-step process.

> **Four-Step Process for Finding $f'(x)$**
>
> 1. Compute $f(x + h)$.
> 2. Form the difference $f(x + h) - f(x)$.
> 3. Form the quotient $\dfrac{f(x + h) - f(x)}{h}$.
> 4. Compute the limit $f'(x) = \displaystyle\lim_{h \to 0} \dfrac{f(x + h) - f(x)}{h}$.

EXAMPLE 2 Find the slope of the tangent line to the graph of $f(x) = 3x + 5$ at any point $(x, f(x))$.

Solution The slope of the tangent line at any point on the graph of f is given by the derivative of f at x. To find the derivative, we use the four-step process:

Step 1 $f(x + h) = 3(x + h) + 5 = 3x + 3h + 5$

Step 2 $f(x + h) - f(x) = (3x + 3h + 5) - (3x + 5) = 3h$

Step 3 $\dfrac{f(x + h) - f(x)}{h} = \dfrac{3h}{h} = 3$

Step 4 $f'(x) = \displaystyle\lim_{h \to 0} \dfrac{f(x + h) - f(x)}{h} = \lim_{h \to 0} 3 = 3$

We expect this result, since the tangent line to any point on a straight line must coincide with the line itself and therefore must have the same slope as the line. In this case, the graph of f is a straight line with slope 3.

EXAMPLE 3 Let $f(x) = x^2$.

a. Find $f'(x)$.
b. Compute $f'(2)$ and interpret your result.

Solution

a. To find $f'(x)$, we use the four-step process:

Step 1 $f(x + h) = (x + h)^2 = x^2 + 2xh + h^2$

Step 2 $f(x + h) - f(x) = x^2 + 2xh + h^2 - x^2 = 2xh + h^2 = h(2x + h)$

Step 3 $\dfrac{f(x + h) - f(x)}{h} = \dfrac{h(2x + h)}{h} = 2x + h$

Step 4 $f'(x) = \displaystyle\lim_{h \to 0} \dfrac{f(x + h) - f(x)}{h} = \lim_{h \to 0} (2x + h) = 2x$

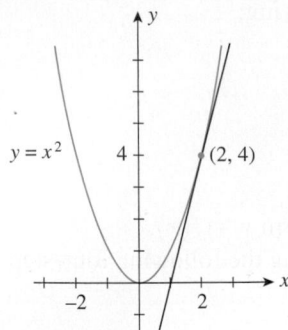

FIGURE **30**
The tangent line to the graph of
$f(x) = x^2$ at $(2, 4)$

b. $f'(2) = 2(2) = 4$. This result tells us that the slope of the tangent line to the graph of f at the point $(2, 4)$ is 4. It also tells us that the function f is changing at the rate of 4 units per unit change in x at $x = 2$. The graph of f and the tangent line at $(2, 4)$ are shown in Figure 30.

Exploring with TECHNOLOGY

1. Consider the function $f(x) = x^2$ of Example 3. Suppose we want to compute $f'(2)$, using Equation (9). Thus,

$$f'(2) = \lim_{h \to 0} \frac{f(2 + h) - f(2)}{h} = \lim_{h \to 0} \frac{(2 + h)^2 - 2^2}{h}$$

Use a graphing utility to plot the graph of

$$g(x) = \frac{(2 + x)^2 - 4}{x}$$

in the viewing window $[-3, 3] \times [-2, 6]$.

2. Use **ZOOM** and **TRACE** to find $\lim_{x \to 0} g(x)$.

3. Explain why the limit found in part 2 is $f'(2)$.

EXAMPLE 4 Let $f(x) = x^2 - 4x$.

a. Find $f'(x)$.
b. Find the point on the graph of f where the tangent line to the curve is horizontal.
c. Sketch the graph of f and the tangent line to the curve at the point found in part (b).
d. What is the rate of change of f at this point?

Solution

a. To find $f'(x)$, we use the four-step process:

Step 1 $f(x + h) = (x + h)^2 - 4(x + h) = x^2 + 2xh + h^2 - 4x - 4h$

Step 2 $f(x + h) - f(x) = x^2 + 2xh + h^2 - 4x - 4h - (x^2 - 4x)$

$$= 2xh + h^2 - 4h = h(2x + h - 4)$$

Step 3 $\dfrac{f(x + h) - f(x)}{h} = \dfrac{h(2x + h - 4)}{h} = 2x + h - 4$

Step 4 $f'(x) = \lim_{h \to 0} \dfrac{f(x + h) - f(x)}{h} = \lim_{h \to 0} (2x + h - 4) = 2x - 4$

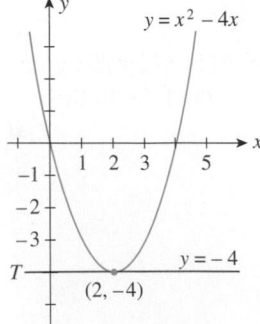

FIGURE **31**
The tangent line to the graph of
$y = x^2 - 4x$ at $(2, -4)$ is $y = -4$.

b. At a point on the graph of f where the tangent line to the curve is horizontal and hence has slope zero, the derivative f' of f is zero. Accordingly, to find such point(s), we set $f'(x) = 0$, which gives $2x - 4 = 0$, or $x = 2$. The corresponding value of y is given by $y = f(2) = -4$, and the required point is $(2, -4)$.

c. The graph of f and the tangent line are shown in Figure 31.

d. The rate of change of f at $x = 2$ is zero.

Explore and Discuss

Can the tangent line to the graph of a function intersect the graph at more than one point? Explain your answer using illustrations.

EXAMPLE 5 Let $f(x) = \dfrac{1}{x}$.

a. Find $f'(x)$.
b. Find the slope of the tangent line T to the graph of f at the point where $x = 1$.
c. Find an equation of the tangent line T in part (b).

Solution

a. To find $f'(x)$, we use the four-step process:

Step 1 $f(x + h) = \dfrac{1}{x + h}$

Step 2 $f(x + h) - f(x) = \dfrac{1}{x + h} - \dfrac{1}{x} = \dfrac{x - (x + h)}{x(x + h)} = -\dfrac{h}{x(x + h)}$

Step 3 $\dfrac{f(x + h) - f(x)}{h} = -\dfrac{h}{x(x + h)} \cdot \dfrac{1}{h} = -\dfrac{1}{x(x + h)}$ (x^2) See page 23.

Step 4 $f'(x) = \lim\limits_{h \to 0} \dfrac{f(x + h) - f(x)}{h} = \lim\limits_{h \to 0} -\dfrac{1}{x(x + h)} = -\dfrac{1}{x^2}$

b. The slope of the tangent line T to the graph of f where $x = 1$ is given by $f'(1) = -1$.

c. When $x = 1$, $y = f(1) = 1$ and T is tangent to the graph of f at the point $(1, 1)$. From part (b), we know that the slope of T is -1. Thus, an equation of T is

$$y - 1 = -1(x - 1)$$
$$y = -x + 2$$

(Figure 32).

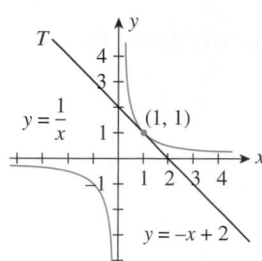

FIGURE 32
The tangent line to the graph of $f(x) = 1/x$ at $(1, 1)$

Exploring with TECHNOLOGY

1. Use the results of Example 5 to draw the graph of $f(x) = 1/x$ and its tangent line at the point $(1, 1)$ by plotting the graphs of $y_1 = 1/x$ and $y_2 = -x + 2$ in the viewing window $[-4, 4] \times [-4, 4]$.

2. Some graphing utilities draw the tangent line to the graph of a function at a given point automatically—you need only specify the function and give the x-coordinate of the point of tangency. If your graphing utility has this feature, verify the result of part 1 without finding an equation of the tangent line.

Explore and Discuss

Consider the following alternative approach to the definition of the derivative of a function: Let h be a positive number and suppose $P(x - h, f(x - h))$ and $Q(x + h, f(x + h))$ are two points on the graph of f.

1. Give a geometric and a physical interpretation of the quotient

$$\frac{f(x + h) - f(x - h)}{2h}$$

Make a sketch to illustrate your answer.

2. Give a geometric and a physical interpretation of the limit

$$\lim\limits_{h \to 0} \frac{f(x + h) - f(x - h)}{2h}$$

Make a sketch to illustrate your answer.

3. Explain why it makes sense to define

$$f'(x) = \lim\limits_{h \to 0} \frac{f(x + h) - f(x - h)}{2h}$$

4. Using the definition given in part 3, formulate a four-step process for finding $f'(x)$ similar to that given on page 635, and use it to find the derivative of $f(x) = x^2$. Compare your answer with that obtained in Example 3 on page 635.

APPLIED EXAMPLE 6 Velocity of a Car Suppose the distance (in feet) covered by a car moving along a straight road t seconds after starting from rest is given by the function $f(t) = 2t^2$ $(0 \leq t \leq 30)$.

a. Calculate the average velocity of the car over the time intervals $[22, 23]$, $[22, 22.1]$, and $[22, 22.01]$.
b. Calculate the (instantaneous) velocity of the car when $t = 22$.
c. Compare the results obtained in part (a) with that obtained in part (b).

Solution

a. We first compute the average velocity (average rate of change of f) over the interval $[t, t + h]$ using Formula (7). We find

$$\frac{f(t + h) - f(t)}{h} = \frac{2(t + h)^2 - 2t^2}{h}$$
$$= \frac{2t^2 + 4th + 2h^2 - 2t^2}{h}$$
$$= 4t + 2h$$

Next, using $t = 22$ and $h = 1$, we find that the average velocity of the car over the time interval $[22, 23]$ is

$$4(22) + 2(1) = 90$$

or 90 ft/sec. Similarly, using $t = 22$, $h = 0.1$, and $h = 0.01$, we find that its average velocities over the time intervals $[22, 22.1]$ and $[22, 22.01]$ are 88.2 and 88.02 ft/sec, respectively.

b. Using the limit (8), we see that the instantaneous velocity of the car at any time t is given by

$$\lim_{h \to 0} \frac{f(t + h) - f(t)}{h} = \lim_{h \to 0} (4t + 2h) \quad \text{Use the results from part (a).}$$
$$= 4t$$

In particular, the velocity of the car 22 seconds from rest $(t = 22)$ is given by

$$v = 4(22)$$

or 88 ft/sec.

c. The computations in part (a) show that, as the time intervals over which the average velocity of the car are computed become smaller and smaller, the average velocities over these intervals do approach 88 ft/sec, the instantaneous velocity of the car at $t = 22$.

APPLIED EXAMPLE 7 Demand for Tires The management of Titan Tire Company has determined that the weekly demand function of their Super Titan tires is given by

$$p = f(x) = 144 - x^2$$

where p, the price per tire, is measured in dollars and x is measured in units of a thousand (Figure 33).

a. Find the average rate of change in the unit price of a tire if the quantity demanded is between 5000 and 6000 tires, between 5000 and 5100 tires, and between 5000 and 5010 tires.
b. What is the instantaneous rate of change of the unit price when the quantity demanded is 5000 units?

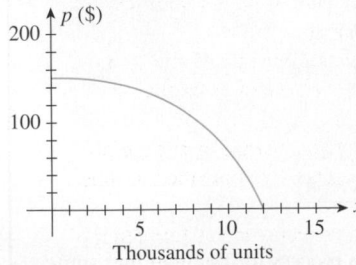

FIGURE 33
The graph of the demand function
$p = 144 - x^2$

Solution

a. The average rate of change of the unit price of a tire if the quantity demanded (in thousands) is between x and $x + h$ is

$$\frac{f(x + h) - f(x)}{h} = \frac{[144 - (x + h)^2] - (144 - x^2)}{h}$$

$$= \frac{144 - x^2 - 2xh - h^2 - 144 + x^2}{h}$$

$$= -2x - h$$

To find the average rate of change of the unit price of a tire when the quantity demanded is between 5000 and 6000 tires (that is, over the interval $[5, 6]$), we take $x = 5$ and $h = 1$, obtaining

$$-2(5) - 1 = -11$$

or $-\$11$ per 1000 tires. (Remember, x is measured in units of a thousand.) Similarly, taking $h = 0.1$ and $h = 0.01$ with $x = 5$, we find that the average rates of change of the unit price when the quantities demanded are between 5000 and 5100 and between 5000 and 5010 are $-\$10.10$ and $-\$10.01$ per 1000 tires, respectively.

b. The instantaneous rate of change of the unit price of a tire when the quantity demanded is x units is given by

$$\lim_{h \to 0} \frac{f(x + h) - f(x)}{h} = \lim_{h \to 0} (-2x - h) \quad \text{Use the results from part (a).}$$

$$= -2x$$

In particular, the instantaneous rate of change of the unit price per tire when the quantity demanded is 5000 is given by $-2(5)$, or $-\$10$ per 1000 tires. ◼

The derivative of a function provides us with a tool for measuring the rate of change of one quantity with respect to another. Table 4 lists several other applications involving this limit.

TABLE 4

Applications Involving Rate of Change

x stands for	y stands for	$\dfrac{f(a + h) - f(a)}{h}$ measures	$\displaystyle\lim_{h \to 0} \dfrac{f(a + h) - f(a)}{h}$ measures
Time	**Concentration of a drug** in the bloodstream at time x	Average rate of change in the concentration of the drug over the time interval $[a, a + h]$	Instantaneous rate of change in the concentration of the drug in the bloodstream at time $x = a$
Number of items sold	**Revenue** at a sales level of x units	Average rate of change in the revenue when the sales level is between $x = a$ and $x = a + h$	Instantaneous rate of change in the revenue when the sales level is a units
Time	**Volume of sales** at time x	Average rate of change in the volume of sales over the time interval $[a, a + h]$	Instantaneous rate of change in the volume of sales at time $x = a$
Time	**Population** of *Drosophila* (fruit flies) at time x	Average rate of growth of the fruit fly population over the time interval $[a, a + h]$	Instantaneous rate of change of the fruit fly population at time $x = a$
Temperature in a chemical reaction	**Amount of product formed in the chemical reaction** when the temperature is x degrees	Average rate of formation of chemical product over the temperature range $[a, a + h]$	Instantaneous rate of formation of chemical product when the temperature is a degrees

Differentiability and Continuity

In practical applications, one encounters continuous functions that fail to be **differentiable**—that is, do not have a derivative—at certain values in the domain of the function f. It can be shown that a continuous function f fails to be differentiable at $x = a$ when the graph of f makes an abrupt change of direction at $(a, f(a))$. We call such a point a "corner." A function also fails to be differentiable at a point where the tangent line is vertical, since the slope of a vertical line is undefined. These cases are illustrated in Figure 34.

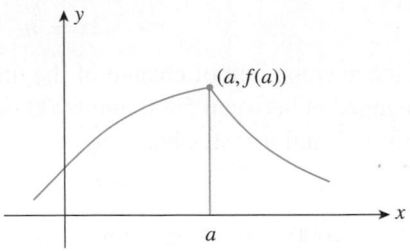 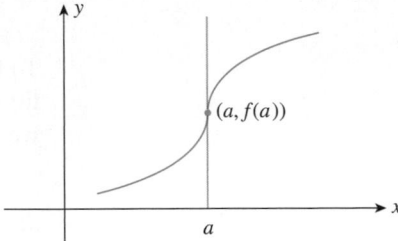

(a) The graph makes an abrupt change of direction at $x = a$.

(b) The slope at $x = a$ is undefined.

FIGURE **34**

The next example illustrates a function that is not differentiable at a point.

$ APPLIED EXAMPLE 8 Wages Mary works at the B&O department store, where, on a weekday, she is paid $10 an hour for the first 8 hours and $15 an hour for overtime. The function

$$f(x) = \begin{cases} 10x & \text{if } 0 \le x \le 8 \\ 15x - 40 & \text{if } 8 < x \end{cases}$$

gives Mary's earnings on a weekday in which she worked x hours. Sketch the graph of the function f, and explain why it is not differentiable at $x = 8$.

Solution The graph of f is shown in Figure 35. Observe that the graph of f has a corner at $x = 8$ and consequently is not differentiable at $x = 8$. ∎

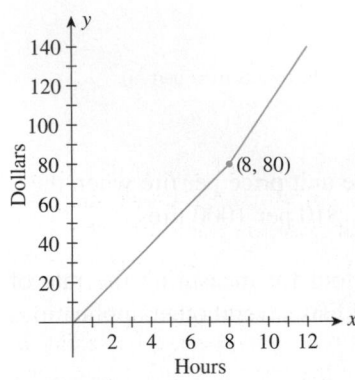

FIGURE **35**
The function f is not differentiable at $(8, 80)$.

We close this section by mentioning the connection between the continuity and the differentiability of a function at a given value $x = a$ in the domain of f. By reexamining the function of Example 8, it becomes clear that f is continuous everywhere and, in particular, when $x = 8$. This shows that in general the continuity of a function at $x = a$ does not necessarily imply the differentiability of the function at that number. The converse, however, is true: If a function f is differentiable at $x = a$, then it is continuous there.

> **Differentiability and Continuity**
> If a function is differentiable at $x = a$, then it is continuous at $x = a$.

For a proof of this result, see Exercise 62, page 646.

Explore and Discuss

Suppose a function f is differentiable at $x = a$. Can there be two tangent lines to the graphs of f at the point $(a, f(a))$? Explain your answer.

Exploring with TECHNOLOGY

1. Use a graphing utility to plot the graph of $f(x) = x^{1/3}$ in the viewing window $[-2, 2] \times [-2, 2]$.
2. Use a graphing utility to draw the tangent line to the graph of f at the point $(0, 0)$. Can you explain why the process breaks down?

EXAMPLE 9 Figure 36 depicts a portion of the graph of a function. Explain why the function fails to be differentiable at each of the numbers $x = a, b, c, d, e, f$, and g.

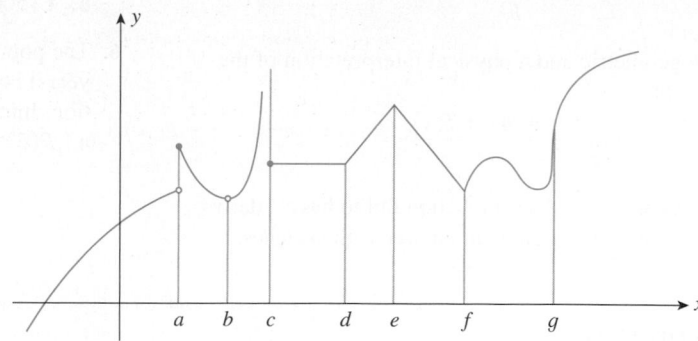

FIGURE **36**
The graph of this function is not differentiable at the numbers a–g.

Solution   The function fails to be differentiable at $x = a, b$, and c because it is discontinuous at each of these numbers. The derivative of the function does not exist at $x = d, e$, and f because it has a kink at each point on the graph corresponding to these numbers. Finally, the function is not differentiable at $x = g$ because the tangent line is vertical at the corresponding point on the graph.

9.3 Self-Check Exercises

1. Let $f(x) = -x^2 - 2x + 3$.
   a. Find the derivative f' of f, using the definition of the derivative.
   b. Find the slope of the tangent line to the graph of f at the point $(0, 3)$.
   c. Find the rate of change of f when $x = 0$.
   d. Find an equation of the tangent line to the graph of f at the point $(0, 3)$.
   e. Sketch the graph of f and the tangent line to the curve at the point $(0, 3)$.

2. **BANK LOSSES** The losses (in millions of dollars) due to bad loans extended chiefly in agriculture, real estate, shipping, and energy by the Franklin Bank are estimated to be

$$A = f(t) = -t^2 + 10t + 30 \qquad (0 \le t \le 10)$$

where t is the time in years ($t = 0$ corresponds to the beginning of 2007). How fast were the losses mounting at the beginning of 2010? At the beginning of 2012? At the beginning of 2014?

Solutions to Self-Check Exercises 9.3 can be found on page 646.

9.3 Concept Questions

For Questions 1 and 2, refer to the following figure.

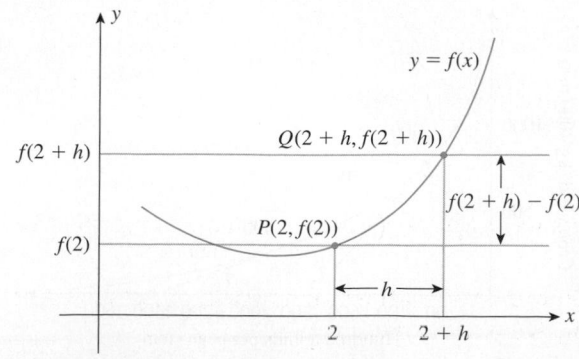

1. Let $P(2, f(2))$ and $Q(2 + h, f(2 + h))$ be points on the graph of a function f.
   a. Find an expression for the slope of the secant line passing through P and Q.
   b. Find an expression for the slope of the tangent line passing through P.

2. Refer to Question 1.
   a. Find an expression for the average rate of change of f over the interval $[2, 2 + h]$.
   b. Find an expression for the instantaneous rate of change of f at 2.
   c. Compare your answers for parts (a) and (b) with those of Question 1.

3. a. Give a geometric and a physical interpretation of the expression

$$\frac{f(x + h) - f(x)}{h}$$

b. Give a geometric and a physical interpretation of the expression

$$\lim_{h \to 0} \frac{f(x + h) - f(x)}{h}$$

4. Under what conditions does a function fail to have a derivative at a number? Illustrate your answer with sketches.

5. The total cost (in dollars) incurred in producing x units of a product is $C(x)$, where C is a differentiable function. Interpret the following:
a. $C(500)$ **b.** $C'(500)$

6. The population of a city (in thousands) at any time t (in years) is given by $P(t)$, where P is a differentiable function. Interpret the following:
a. $P(5)$ **b.** $P'(5)$

9.3 Exercises

1. AVERAGE WEIGHT OF AN INFANT The following graph shows the weight measurements of the average infant from the time of birth ($t = 0$) through age 2 ($t = 24$). By computing the slopes of the respective tangent lines, estimate the rate of change of the average infant's weight when $t = 3$ and when $t = 18$. What is the average rate of change in the average infant's weight over the first year of life?

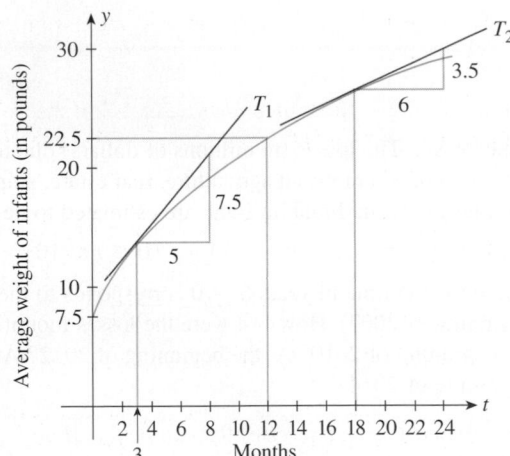

2. FORESTRY The following graph shows the volume of wood produced in a single-species forest. Here, $f(t)$ is measured in cubic meters per hectare, and t is measured in years. By computing the slopes of the respective tangent lines, estimate the rate at which the wood grown is changing at the beginning of year 10 and at the beginning of year 30.
Source: The Random House Encyclopedia.

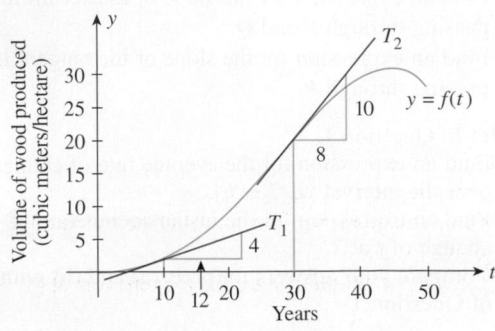

3. TV-VIEWING PATTERNS The following graph shows the percentage of U.S. households watching television during a 24-hr period on a weekday ($t = 0$ corresponds to 6 A.M.). By computing the slopes of the respective tangent lines, estimate the rate of change of the percent of households watching television at 4 P.M. and 11 P.M.
Source: A. C. Nielsen Company.

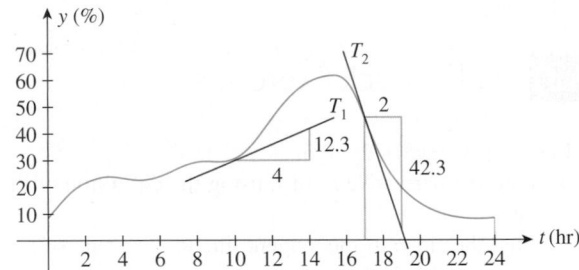

4. CROP YIELD Productivity and yield of cultivated crops are often reduced by insect pests. The following graph shows the relationship between the yield of a certain crop, $f(x)$, as a function of the density of aphids x. (Aphids are small insects that suck plant juices.) Here, $f(x)$ is measured in kilograms per 4000 square meters, and x is measured in hundreds of aphids per bean stem. By computing the slopes of the respective tangent lines, estimate the rate of change of the crop yield with respect to the density of aphids when that density is 200 aphids/bean stem and when it is 800 aphids/bean stem.
Source: The Random House Encyclopedia.

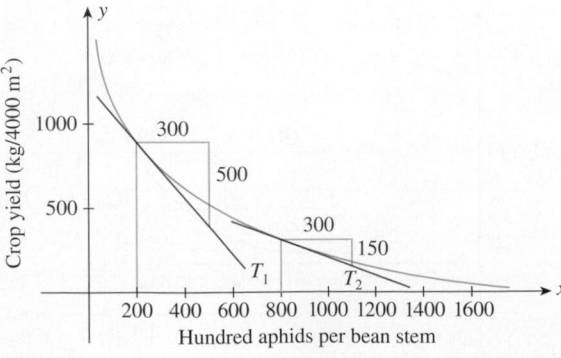

5. The positions of Car A and Car B, starting out side by side and traveling along a straight road, are given by $s = f(t)$ and $s = g(t)$, respectively, where s is measured in feet and t is measured in seconds (see the accompanying figure).

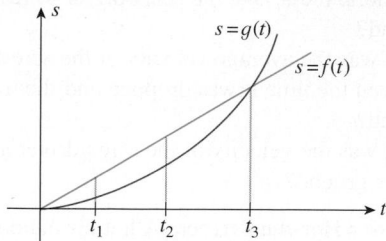

   a. Which car is traveling faster at t_1?
   b. What can you say about the speed of the cars at t_2? Hint: Compare tangent lines.
   c. Which car is traveling faster at t_3?
   d. What can you say about the positions of the cars at t_3?

6. The velocities of Car A and Car B, which start out side by side and travel along a straight road, are given by $v = f(t)$ and $v = g(t)$, respectively, where v is measured in feet per second and t is measured in seconds (see the accompanying figure).

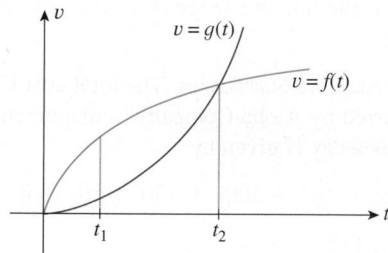

   a. What can you say about the velocity and acceleration of the two cars at t_1? (Acceleration is the rate of change of velocity.)
   b. What can you say about the velocity and acceleration of the two cars at t_2?

7. **EFFECT OF A BACTERICIDE ON BACTERIA** In the following figure, $f(t)$ gives the population P_1 of a certain bacteria culture at time t after a portion of Bactericide A was introduced into the population at $t = 0$. The graph of g gives the population P_2 of a similar bacteria culture at time t after a portion of Bactericide B was introduced into the population at $t = 0$.
   a. Which population is decreasing faster at t_1?
   b. Which population is decreasing faster at t_2?

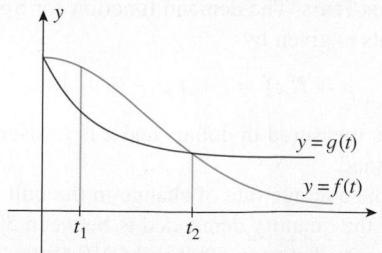

   c. Which bactericide is more effective in reducing the population of bacteria in the short run? In the long run?

8. **MARKET SHARE** The following figure shows the devastating effect the opening of a new discount department store had on an established department store in a small town. The revenue of the discount store at time t (in months) is given by $f(t)$ million dollars, whereas the revenue of the established department store at time t is given by $g(t)$ million dollars. Answer the following questions by giving the value of t at which the specified event took place.

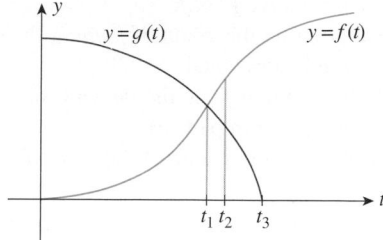

   a. The revenue of the established department store is decreasing at the slowest rate.
   b. The revenue of the established department store is decreasing at the fastest rate.
   c. The revenue of the discount store first overtakes that of the established store.
   d. The revenue of the discount store is increasing at the fastest rate.

In Exercises 9–16, use the four-step process to find the slope of the tangent line to the graph of the given function at any point.

9. $f(x) = 13$
10. $f(x) = -6$
11. $f(x) = 2x + 7$
12. $f(x) = 8 - 4x$
13. $f(x) = 3x^2$
14. $f(x) = -\dfrac{1}{2}x^2$
15. $f(x) = -x^2 + 3x$
16. $f(x) = 2x^2 + 5x$

In Exercises 17–22, find the slope of the tangent line to the graph of the function at the given point, and determine an equation of the tangent line.

17. $f(x) = 2x + 7$ at $(2, 11)$
18. $f(x) = -3x + 4$ at $(-1, 7)$
19. $f(x) = 3x^2$ at $(1, 3)$
20. $f(x) = 3x - x^2$ at $(-2, -10)$
21. $f(x) = -\dfrac{1}{x}$ at $\left(3, -\dfrac{1}{3}\right)$
22. $f(x) = \dfrac{3}{2x}$ at $\left(1, \dfrac{3}{2}\right)$
23. Let $f(x) = 2x^2 + 1$.
   a. Find the derivative f' of f.
   b. Find an equation of the tangent line to the curve at the point $(1, 3)$.
   c. Sketch the graph of f and its tangent line at $(1, 3)$.

24. Let $f(x) = x^2 + 6x$.
  a. Find the derivative f' of f.
  b. Find the point on the graph of f where the tangent line to the curve is horizontal.
  Hint: Find the value of x for which $f'(x) = 0$.
  c. Sketch the graph of f and the tangent line to the curve at the point found in part (b).

25. Let $f(x) = x^2 - 2x + 1$.
  a. Find the derivative f' of f.
  b. Find the point on the graph of f where the tangent line to the curve is horizontal.
  c. Sketch the graph of f and the tangent line to the curve at the point found in part (b).
  d. What is the rate of change of f at this point?

26. Let $f(x) = \dfrac{1}{x - 1}$.
  a. Find the derivative f' of f.
  b. Find an equation of the tangent line to the curve at the point $\left(-1, -\frac{1}{2}\right)$.
  c. Sketch the graph of f and the tangent line to the curve at $\left(-1, -\frac{1}{2}\right)$.

27. Let $y = f(x) = x^2 + x$.
  a. Find the average rate of change of y with respect to x in the interval from $x = 2$ to $x = 3$, from $x = 2$ to $x = 2.5$, and from $x = 2$ to $x = 2.1$.
  b. Find the (instantaneous) rate of change of y at $x = 2$.
  c. Compare the results obtained in part (a) with the result of part (b).

28. Let $y = f(x) = x^2 - 4x$.
  a. Find the average rate of change of y with respect to x in the interval from $x = 3$ to $x = 4$, from $x = 3$ to $x = 3.5$, and from $x = 3$ to $x = 3.1$.
  b. Find the (instantaneous) rate of change of y at $x = 3$.
  c. Compare the results obtained in part (a) with the result of part (b).

29. VELOCITY OF A CAR Suppose the distance s (in feet) covered by a car moving along a straight road after t sec is given by the function $s = f(t) = 2t^2 + 48t$.
  a. Calculate the average velocity of the car over the time intervals $[20, 21]$, $[20, 20.1]$, and $[20, 20.01]$.
  b. Calculate the (instantaneous) velocity of the car when $t = 20$.
  c. Compare the results of part (a) with the result of part (b).

30. VELOCITY OF A BALL THROWN INTO THE AIR A ball is thrown straight up with an initial velocity of 128 ft/sec, so that its height (in feet) after t sec is given by $s(t) = 128t - 16t^2$.
  a. What is the average velocity of the ball over the time intervals $[2, 3]$, $[2, 2.5]$, and $[2, 2.1]$?
  b. What is the instantaneous velocity at time $t = 2$?
  c. What is the instantaneous velocity at time $t = 5$? Is the ball rising or falling at this time?
  d. When will the ball hit the ground?

31. VELOCITY OF A FALLING OBJECT During the construction of a high-rise building, a worker accidentally dropped his portable electric screwdriver from a height of 400 ft. After t sec, the screwdriver had fallen a distance of $s = 16t^2$ ft.
  a. How long did it take the screwdriver to reach the ground?
  b. What was the average velocity of the screwdriver between the time it was dropped and the time it hit the ground?
  c. What was the velocity of the screwdriver at the time it hit the ground?

32. VELOCITY OF A HOT-AIR BALLOON A hot-air balloon rises vertically from the ground so that its height after t sec is $h = \frac{1}{2}t^2 + \frac{1}{2}t$ ft $(0 \le t \le 60)$.
  a. What is the height of the balloon at the end of 40 sec?
  b. What is the average velocity of the balloon between $t = 0$ and $t = 40$?
  c. What is the velocity of the balloon at the end of 40 sec?

33. At a temperature of 20°C, the volume V (in liters) of 1.33 g of O_2 is related to its pressure p (in atmospheres) by the formula $V = 1/p$.
  a. What is the average rate of change of V with respect to p as p increases from $p = 2$ to $p = 3$?
  b. What is the rate of change of V with respect to p when $p = 2$?

34. COST OF PRODUCING SURFBOARDS The total cost $C(x)$ (in dollars) incurred by Aloha Company in manufacturing x surfboards a day is given by
  $$C(x) = -10x^2 + 300x + 130 \qquad (0 \le x \le 15)$$
  a. Find $C'(x)$.
  b. What is the rate of change of the total cost when the level of production is ten surfboards a day?

35. EFFECT OF ADVERTISING ON PROFIT The quarterly profit (in thousands of dollars) of Cunningham Realty is given by
  $$P(x) = -\frac{1}{3}x^2 + 7x + 30 \qquad (0 \le x \le 50)$$
  where x (in thousands of dollars) is the amount of money Cunningham spends on advertising per quarter.
  a. Find $P'(x)$.
  b. What is the rate of change of Cunningham's quarterly profit if the amount it spends on advertising is $10,000/quarter $(x = 10)$ and $30,000/quarter $(x = 30)$?

36. DEMAND FOR TENTS The demand function for Sportsman 5×7 tents is given by
  $$p = f(x) = -0.1x^2 - x + 40$$
  where p is measured in dollars and x is measured in units of a thousand.
  a. Find the average rate of change in the unit price of a tent if the quantity demanded is between 5000 and 5050 tents; between 5000 and 5010 tents.
  b. What is the rate of change of the unit price if the quantity demanded is 5000?

37. A Country's GDP The gross domestic product (GDP) of a certain country is projected to be

$$N(t) = t^2 + 2t + 50 \qquad (0 \le t \le 5)$$

billion dollars t years from now. What will be the rate of change of the country's GDP 2 years and 4 years from now?

38. Growth of Bacteria Under a set of controlled laboratory conditions, the size of the population of a certain bacteria culture at time t (in minutes) is described by the function

$$P = f(t) = 3t^2 + 2t + 1$$

Find the rate of population growth at $t = 10$ min.

39. Air Temperature The air temperature at a height of h ft from the surface of the earth is $T = f(h)$ degrees Fahrenheit.
  a. Give a physical interpretation of $f'(h)$. Give units.
  b. Generally speaking, what do you expect the sign of $f'(h)$ to be?
  c. If you know that $f'(1000) = -0.05$, estimate the change in the air temperature if the altitude changes from 1000 ft to 1001 ft.

40. Revenue of a Travel Agency Suppose that the total revenue realized by the Odyssey Travel Agency is $R = f(x)$ thousand dollars if x thousand dollars are spent on advertising.
  a. What does

$$\frac{f(b) - f(a)}{b - a} \qquad (0 < a < b)$$

  measure? What are the units?
  b. What does $f'(x)$ measure? Give units.
  c. Given that $f'(20) = 3$, what is the approximate change in the revenue if Odyssey increases its advertising budget from \$20,000 to \$21,000?

In Exercises 41–46, let x and $f(x)$ represent the given quantities. Fix $x = a$ and let h be a small positive number. Give an interpretation of the quantities

$$\frac{f(a + h) - f(a)}{h} \quad \text{and} \quad \lim_{h \to 0} \frac{f(a + h) - f(a)}{h}$$

41. x denotes time, and $f(x)$ denotes the population of seals at time x.

42. x denotes time, and $f(x)$ denotes the prime interest rate at time x.

43. x denotes time, and $f(x)$ denotes a country's industrial production.

44. x denotes the level of production of a certain commodity, and $f(x)$ denotes the total cost incurred in producing x units of the commodity.

45. x denotes altitude, and $f(x)$ denotes atmospheric pressure.

46. x denotes the speed of a car in miles per hour (mph), and $f(x)$ denotes the fuel consumption of the car measured in miles per gallon (mpg).

In each of Exercises 47–52, the graph of a function is shown. For each function, state whether or not (a) $f(x)$ has a limit at $x = a$, as x approaches a, (b) $f(x)$ is continuous at $x = a$, and (c) $f(x)$ is differentiable at $x = a$. Justify your answers.

47.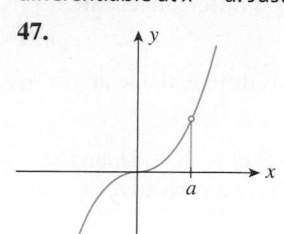

48.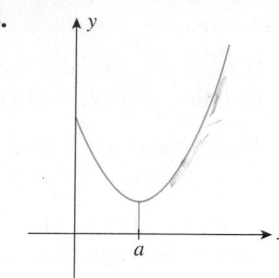

49.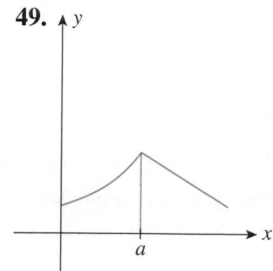

50.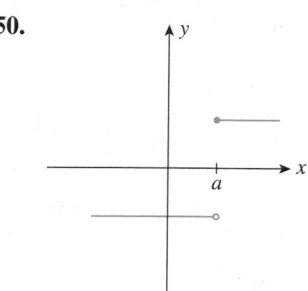

51.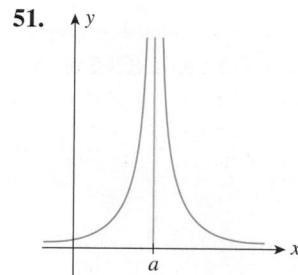

52.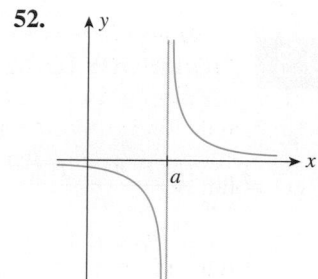

53. Velocity of a Motorcycle The distance s (in feet) covered by a motorcycle traveling in a straight line and starting from rest in t sec is given by the function

$$s(t) = -0.1t^3 + 2t^2 + 24t \quad (0 \le t \le 3)$$

Calculate the motorcycle's average velocity over the time interval $[2, 2 + h]$ for $h = 1, 0.1, 0.01, 0.001, 0.0001,$ and 0.00001, and use your results to guess at the motorcycle's instantaneous velocity at $t = 2$.

54. Rate of Change of Production Costs The daily total cost $C(x)$ incurred by Trappee and Sons for producing x cases of TexaPep hot sauce is given by

$$C(x) = 0.000002x^3 + 5x + 400$$

Calculate

$$\frac{C(100 + h) - C(100)}{h}$$

for $h = 1, 0.1, 0.01, 0.001,$ and 0.0001, and use your results to estimate the rate of change of the total cost function when the level of production is 100 cases/day.

In Exercises 55 and 56, determine whether the statement is true or false. If it is true, explain why it is true. If it is false, give an example to show why it is false.

55. If f is continuous at $x = a$, then f is differentiable at $x = a$.

56. If f is continuous at $x = a$ and g is differentiable at $x = a$, then $\lim_{x \to a} f(x)g(x) = f(a)g(a)$.

57. Sketch the graph of the function $f(x) = |x + 1|$, and show that the function does not have a derivative at $x = -1$.

58. Sketch the graph of the function $f(x) = 1/(x - 1)$, and show that the function does not have a derivative at $x = 1$.

59. Let

$$f(x) = \begin{cases} x^2 & \text{if } x \leq 1 \\ ax + b & \text{if } x > 1 \end{cases}$$

Find the values of a and b such that f is continuous and has a derivative at $x = 1$. Sketch the graph of f.

60. Sketch the graph of the function $f(x) = x^{2/3}$. Is the function continuous at $x = 0$? Does $f'(0)$ exist? Why or why not?

61. Prove that the derivative of the function $f(x) = |x|$ for $x \neq 0$ is given by

$$f'(x) = \begin{cases} 1 & \text{if } x > 0 \\ -1 & \text{if } x < 0 \end{cases}$$

Hint: Recall the definition of the absolute value of a number.

62. Show that if a function f is differentiable at $x = a$, then f must be continuous at $x = a$.
Hint: Write

$$f(x) - f(a) = \left[\frac{f(x) - f(a)}{x - a} \right] (x - a)$$

Use the Product Rule for Limits and the definition of the derivative to show that

$$\lim_{x \to a} [f(x) - f(a)] = 0$$

9.3 Solutions to Self-Check Exercises

1. a.

$$f'(x) = \lim_{h \to 0} \frac{f(x + h) - f(x)}{h}$$

$$= \lim_{h \to 0} \frac{[-(x + h)^2 - 2(x + h) + 3] - (-x^2 - 2x + 3)}{h}$$

$$= \lim_{h \to 0} \frac{-x^2 - 2xh - h^2 - 2x - 2h + 3 + x^2 + 2x - 3}{h}$$

$$= \lim_{h \to 0} \frac{h(-2x - h - 2)}{h}$$

$$= \lim_{h \to 0} (-2x - h - 2) = -2x - 2$$

b. From the result of part (a), we see that the slope of the tangent line to the graph of f at any point $(x, f(x))$ is given by

$$f'(x) = -2x - 2$$

In particular, the slope of the tangent line to the graph of f at $(0, 3)$ is

$$f'(0) = -2$$

c. The rate of change of f when $x = 0$ is given by $f'(0) = -2$, or -2 units/unit change in x.

d. Using the result from part (b), we see that an equation of the required tangent line is

$$y - 3 = -2(x - 0)$$
$$y = -2x + 3$$

e.

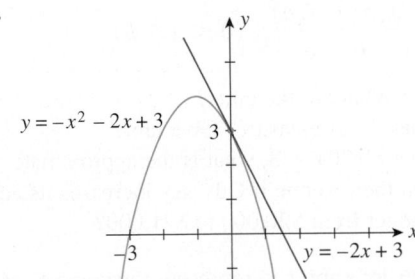

$$y = -x^2 - 2x + 3$$
$$y = -2x + 3$$

2. The rate of change of the losses at any time t is given by

$$f'(t)$$

$$= \lim_{h \to 0} \frac{f(t + h) - f(t)}{h}$$

$$= \lim_{h \to 0} \frac{[-(t + h)^2 + 10(t + h) + 30] - (-t^2 + 10t + 30)}{h}$$

$$= \lim_{h \to 0} \frac{-t^2 - 2th - h^2 + 10t + 10h + 30 + t^2 - 10t - 30}{h}$$

$$= \lim_{h \to 0} \frac{h(-2t - h + 10)}{h}$$

$$= \lim_{h \to 0} (-2t - h + 10)$$

$$= -2t + 10$$

Therefore, the rate of change of the losses suffered by the bank at the beginning of 2010 ($t = 3$) was

$$f'(3) = -2(3) + 10 = 4$$

In other words, the losses were increasing at the rate of \$4 million/year. At the beginning of 2012 ($t = 5$),

$$f'(5) = -2(5) + 10 = 0$$

and we see that the growth in losses due to bad loans was zero at this point. At the beginning of 2014 ($t = 7$),

$$f'(7) = -2(7) + 10 = -4$$

and we conclude that the losses were decreasing at the rate of \$4 million/year.

USING TECHNOLOGY Finding the Derivative of a Function for a Given Value of x

The numerical derivative operation of a graphing utility can be used to find an approximate value of the derivative of a function for a given value of x.

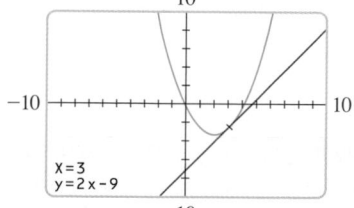

```
nDeriv(X^.5,X,2)
            .3535534017
```

FIGURE **T1**
The TI-83/84 numerical derivative screen

EXAMPLE 1 Use the numerical derivative operation of a graphing calculator to find the derivative of $f(x) = \sqrt{x}$ when $x = 2$.

Solution Access the numerical derivative operation, **nDeriv**, by pressing $\boxed{\text{MATH}}$ $\boxed{8}$. The display **nDeriv(** appears on the screen. Next, we enter the function whose derivative we want to find, the variable x, followed by the value of x at which the derivative is to be evaluated, and then press $\boxed{\text{ENTER}}$. The approximate derivative of f at $x = 2$ is shown on the screen (Figure T1).

Graphing a Tangent Line

We can use a graphing utility to plot the graph of a function f and the tangent line to the graph at a specified point on the graph.

```
          10



-10 ┼┼┼┼┼┼┼┼┼┼┼┼┼ 10


    X=3
    y=2x-9
          -10
```

FIGURE **T2**
The TI-83/84 screen showing the graph of f, the tangent line, and its equation at $(3, -3)$

EXAMPLE 2 Use a graphing utility to plot the graph of $f(x) = x^2 - 4x$ and the tangent line to the graph of f at the point $(3, -3)$.

Solution First, we plot the graph of $f(x) = x^2 - 4x$ using the standard viewing window. To draw the desired tangent line, we continue by pressing $\boxed{\text{2nd}}$ $\boxed{\text{DRAW}}$ $\boxed{5}$ to call the operation **Tangent**. Then, press $\boxed{3}$ to obtain the x-coordinate of the point of tangency. Press $\boxed{\text{ENTER}}$, and the window will display both the graph of f and the tangent line at the point $(3, -3)$ together with its equation $y = 2x - 9$ (Figure T2).

TECHNOLOGY EXERCISES

In Exercises 1–8, (a) use the numerical derivative operation to find the derivative of f for the given value of x (to two decimal places of accuracy), (b) plot the graph of f and the tangent line on the same set of axes, and (c) find an equation of the tangent line to the graph of f at the indicated point. Use a suitable viewing window.

1. $f(x) = 2x^2 + x - 3; x = 2; (2, 7)$

2. $f(x) = x + \dfrac{1}{x}; x = 1; (1, 2)$

3. $f(x) = x^{1/3}; x = 8; (8, 2)$

4. $f(x) = \dfrac{1}{\sqrt{x}}; x = 4; \left(4, \dfrac{1}{2}\right)$

5. $f(x) = x^3 + x + 1; x = 1; (1, 3)$

6. $f(x) = \dfrac{1}{x + 1}; x = 1; \left(1, \dfrac{1}{2}\right)$

7. $f(x) = x\sqrt{x^2 + 1}; x = 2; (2, 2\sqrt{5})$

8. $f(x) = \dfrac{x}{\sqrt{x^2 + 1}}; x = 1; \left(1, \dfrac{\sqrt{2}}{2}\right)$

9. ALZHEIMER'S PATIENTS IN THE UNITED STATES The number of patients with Alzheimer's disease in the United States is approximated by

$$f(t) = -0.208t^3 + 1.571t^2 - 0.9274t + 5.1 \qquad (0 \le t \le 4)$$

where $f(t)$ is measured in millions and t is measured in decades, with $t = 0$ corresponding to the beginning of 1990.

a. Plot the graph of $f(t)$ in the window $[0, 4] \times [0, 14]$.

b. How fast will the number of Alzheimer's patients in the United States be increasing at the beginning of 2020 $(t = 3)$?

Source: Alzheimer's Association.

10. MODELING WITH DATA Annual retail sales in the United States from 2000 through the year 2008 (in billions of dollars) are given in the following table:

Year	2000	2001	2002	2003	2004
Sales	2.988	3.068	3.134	3.267	3.480

Year	2005	2006	2007	2008
Sales	3.698	3.882	4.005	3.959

a. Let $t = 0$ correspond to 2000, and use **QuadReg** to find a second-degree polynomial regression model based on the given data.

b. Plot the graph of the function found in part (a) in the viewing window $[0, 8] \times [0, 5]$.

c. What were the annual retail sales in the United States in 2006 $(t = 6)$?

d. Approximately how fast were the retail sales changing in 2006 $(t = 6)$?

Source: U.S. Census Bureau.

9.4 Basic Rules of Differentiation

Four Basic Rules

The method used in Section 9.3 for computing the derivative of a function is based on a faithful interpretation of the definition of the derivative as the limit of a quotient. To find the rule for the derivative f' of a function f, we first computed the difference quotient

$$\frac{f(x + h) - f(x)}{h}$$

and then evaluated its limit as h approached zero. As you have probably observed, this method is tedious even for relatively simple functions.

The main purpose of Sections 9.4–9.7 is to derive certain rules that will simplify the process of finding the derivative of a function. We will use the notation

$$\frac{d}{dx}[f(x)] \qquad \text{Read "}d, d x \text{ of } f \text{ of } x\text{"}$$

to mean "the derivative of f with respect to x at x."

Rule 1: Derivative of a Constant

$$\frac{d}{dx}(c) = 0 \qquad (c, \text{ a constant})$$

The derivative of a constant function is equal to zero.

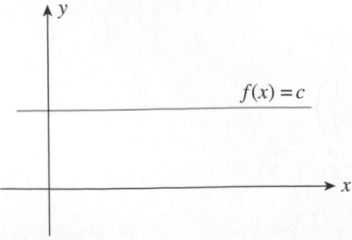

FIGURE 37
The slope of the tangent line to the graph of $f(x) = c$, where c is a constant, is zero.

We can see this from a geometric viewpoint by recalling that the graph of a constant function is a straight line parallel to the x-axis (Figure 37). Since the tangent line to a straight line at any point on the line coincides with the straight line itself, its slope [as given by the derivative of $f(x) = c$] must be zero. We can also use the definition of the derivative to prove this result by computing

$$f'(x) = \lim_{h \to 0} \frac{f(x + h) - f(x)}{h}$$

$$= \lim_{h \to 0} \frac{c - c}{h}$$

$$= \lim_{h \to 0} 0 = 0$$

EXAMPLE 1

a. If $f(x) = 28$, then

$$f'(x) = \frac{d}{dx}(28) = 0$$

b. If $f(x) = \pi^2$, then

$$f'(x) = \frac{d}{dx}(\pi^2) = 0$$

Rule 2: The Power Rule

If n is any real number, then $\dfrac{d}{dx}(x^n) = nx^{n-1}$.

Let's verify the Power Rule for the special case $n = 2$. If $f(x) = x^2$, then

$$f'(x) = \frac{d}{dx}(x^2) = \lim_{h \to 0} \frac{f(x + h) - f(x)}{h}$$

$$= \lim_{h \to 0} \frac{(x + h)^2 - x^2}{h}$$

$$= \lim_{h \to 0} \frac{x^2 + 2xh + h^2 - x^2}{h}$$

$$= \lim_{h \to 0} \frac{2xh + h^2}{h} = \lim_{h \to 0} \frac{h(2x + h)}{h}$$

$$= \lim_{h \to 0} (2x + h) = 2x$$

as we set out to show.

The Power Rule for the general case is not easy to prove, and its proof will be omitted. However, you will be asked to prove the rule for the special case $n = 3$ in Exercise 78, page 659.

EXAMPLE 2

a. If $f(x) = x$, then

$$f'(x) = \frac{d}{dx}(x) = 1 \cdot x^{1-1} = x^0 = 1$$

b. If $f(x) = x^8$, then

$$f'(x) = \frac{d}{dx}(x^8) = 8x^7$$

c. If $f(x) = x^{5/2}$, then

$$f'(x) = \frac{d}{dx}(x^{5/2}) = \frac{5}{2}x^{3/2}$$

To differentiate a function whose rule involves a radical, we first rewrite the radical using fractional powers. The resulting expression can then be differentiated by using the Power Rule.

EXAMPLE 3 Find the derivative of the following functions:

a. $f(x) = \sqrt{x}$ **b.** $g(x) = \dfrac{1}{\sqrt[3]{x}}$

Solution

a. Rewriting $\sqrt{x}$ in the form $x^{1/2}$, we obtain

(x^2) See page 39.

$$f'(x) = \frac{d}{dx}(x^{1/2})$$

$$= \frac{1}{2}x^{-1/2} = \frac{1}{2x^{1/2}} = \frac{1}{2\sqrt{x}}$$

b. Rewriting $\dfrac{1}{\sqrt[3]{x}}$ in the form $x^{-1/3}$, we obtain

$$g'(x) = \frac{d}{dx}(x^{-1/3})$$

$$= -\frac{1}{3}x^{-4/3} = -\frac{1}{3x^{4/3}}$$

In stating the remaining rules of differentiation, we assume that the functions f and g are differentiable.

Rule 3: Derivative of a Constant Multiple of a Function

$$\frac{d}{dx}[cf(x)] = c\frac{d}{dx}[f(x)] \quad (c, \text{ a constant})$$

The derivative of a constant times a differentiable function is equal to the constant times the derivative of the function.

This result follows from the following computations:

If $g(x) = cf(x)$, then

$$g'(x) = \lim_{h \to 0}\frac{g(x+h) - g(x)}{h} = \lim_{h \to 0}\frac{cf(x+h) - cf(x)}{h}$$

$$= c\lim_{h \to 0}\frac{f(x+h) - f(x)}{h}$$

$$= cf'(x)$$

EXAMPLE 4

a. If $f(x) = 5x^3$, then

$$f'(x) = \frac{d}{dx}(5x^3) = 5\frac{d}{dx}(x^3)$$

$$= 5(3x^2) = 15x^2$$

b. If $f(x) = \dfrac{3}{\sqrt{x}}$, then

$$f'(x) = \frac{d}{dx}(3x^{-1/2}) \qquad \text{Rewrite } \frac{3}{\sqrt{x}} \text{ as } \frac{3}{x^{1/2}} = 3x^{-1/2}$$

$$= 3\left(-\frac{1}{2}x^{-3/2}\right) = -\frac{3}{2x^{3/2}}$$

Rule 4: The Sum Rule

$$\frac{d}{dx}[f(x) \pm g(x)] = \frac{d}{dx}[f(x)] \pm \frac{d}{dx}[g(x)]$$

The derivative of the sum (difference) of two differentiable functions is equal to the sum (difference) of their derivatives.

This result may be extended to the sum and difference of any finite number of differentiable functions. Let's verify the rule for a sum of two functions.

If $s(x) = f(x) + g(x)$, then

$$s'(x) = \lim_{h \to 0} \frac{s(x+h) - s(x)}{h}$$

$$= \lim_{h \to 0} \frac{[f(x+h) + g(x+h)] - [f(x) + g(x)]}{h}$$

$$= \lim_{h \to 0} \frac{[f(x+h) - f(x)] + [g(x+h) - g(x)]}{h}$$

$$= \lim_{h \to 0} \frac{f(x+h) - f(x)}{h} + \lim_{h \to 0} \frac{g(x+h) - g(x)}{h}$$

$$= f'(x) + g'(x)$$

EXAMPLE 5 Find the derivatives of the following functions:

a. $f(x) = 4x^5 + 3x^4 - 8x^2 + x + 3$ **b.** $g(t) = \dfrac{t^2}{5} + \dfrac{5}{t^3}$

Solution

a. $f'(x) = \dfrac{d}{dx}(4x^5 + 3x^4 - 8x^2 + x + 3)$

$$= \frac{d}{dx}(4x^5) + \frac{d}{dx}(3x^4) - \frac{d}{dx}(8x^2) + \frac{d}{dx}(x) + \frac{d}{dx}(3)$$

$$= 20x^4 + 12x^3 - 16x + 1$$

b. Here, the independent variable is t instead of x, so we differentiate with respect to t. Thus,

$$g'(t) = \frac{d}{dt}\left(\frac{1}{5}t^2 + 5t^{-3}\right) \qquad \text{Rewrite } \frac{1}{t^3} \text{ as } t^{-3}.$$

$$= \frac{2}{5}t - 15t^{-4} = \frac{2}{5}t - \frac{15}{t^4} \qquad \text{Rewrite } t^{-4} \text{ as } \frac{1}{t^4}.$$

$$= \frac{2t^5 - 75}{5t^4}$$

EXAMPLE 6 Find the slope and an equation of the tangent line to the graph of $f(x) = 2x + 1/\sqrt{x}$ at the point $(1, 3)$.

Solution The slope of the tangent line at any point on the graph of f is given by

$$
\begin{aligned}
f'(x) &= \frac{d}{dx}\left(2x + \frac{1}{\sqrt{x}}\right) \\
&= \frac{d}{dx}\left(2x + x^{-1/2}\right) && \text{Rewrite } \frac{1}{\sqrt{x}} \text{ as } \frac{1}{x^{1/2}} = x^{-1/2}. \\
&= 2 - \frac{1}{2}x^{-3/2} && \text{Use the Sum Rule.} \\
&= 2 - \frac{1}{2x^{3/2}} && \text{Rewrite } \frac{1}{2}x^{-3/2} \text{ as } \frac{1}{2x^{3/2}}.
\end{aligned}
$$

In particular, the slope of the tangent line to the graph of f at $(1, 3)$ (where $x = 1$) is

$$
f'(1) = 2 - \frac{1}{2(1^{3/2})} = 2 - \frac{1}{2} = \frac{3}{2}
$$

Using the point-slope form of an equation of a line with slope $\frac{3}{2}$ and the point $(1, 3)$, we see that an equation of the tangent line is

$$
y - 3 = \frac{3}{2}(x - 1) \qquad y - y_1 = m(x - x_1) \qquad \text{(x²) See page 79.}
$$

or, upon simplification,

$$
y = \frac{3}{2}x + \frac{3}{2}
$$

(see Figure 38).

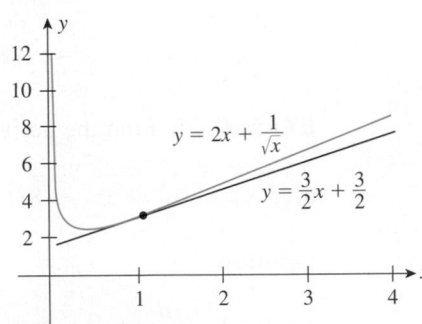

FIGURE 38
The tangent line to the graph of $f(x) = 2x + 1/\sqrt{x}$ at $(1, 3)$.

APPLIED EXAMPLE 7 Conservation of a Species A group of marine biologists at the Neptune Institute of Oceanography recommended that a series of conservation measures be carried out over the next decade to save a certain species of whale from extinction. After the conservation measures are implemented, the population of this species is expected to be

$$
N(t) = 3t^3 + 2t^2 - 10t + 600 \qquad (0 \le t \le 10)
$$

where $N(t)$ denotes the population at the end of year t. Find the rate of growth of the whale population when $t = 2$ and $t = 6$. How large will the whale population be 8 years after the conservation measures are implemented?

Solution The rate of growth of the whale population at any time t is given by

$$N'(t) = 9t^2 + 4t - 10$$

In particular, when $t = 2$ and $t = 6$, we have

$$N'(2) = 9(2)^2 + 4(2) - 10$$
$$= 34$$
$$N'(6) = 9(6)^2 + 4(6) - 10$$
$$= 338$$

Thus, the whale population's rate of growth will be 34 whales per year after 2 years and 338 per year after 6 years.

The whale population at the end of the eighth year will be

$$N(8) = 3(8)^3 + 2(8)^2 - 10(8) + 600$$
$$= 2184$$

The graph of the function N appears in Figure 39. Note the rapid growth of the population in later years, as the conservation measures begin to pay off, compared with the growth in the early years.

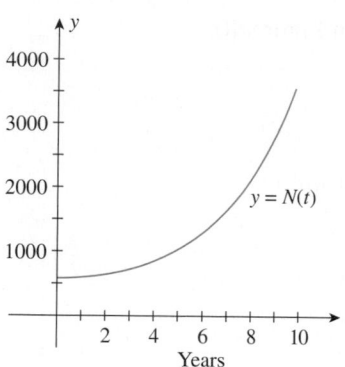

FIGURE **39**
The whale population at the end of year t is given by $N(t)$.

 APPLIED EXAMPLE 8 Altitude of a Rocket An experimental rocket lifts off vertically. Its altitude (in feet) t seconds into flight is given by

$$s = f(t) = -t^3 + 96t^2 + 5 \qquad (t \geq 0)$$

a. Find an expression v for the rocket's velocity at any time t.
b. Compute the rocket's velocity when $t = 0, 30, 50, 64,$ and 70. Interpret your results.
c. Using the results from the solution to part (b) and the observation that at the highest point in its trajectory the rocket's velocity is zero, find the maximum altitude attained by the rocket.

Solution

a. The rocket's velocity at any time t is given by

$$v = f'(t) = -3t^2 + 192t$$

b. The rocket's velocity when $t = 0, 30, 50, 64,$ and 70 is given by

$$f'(0) = -3(0)^2 + 192(0) = 0$$
$$f'(30) = -3(30)^2 + 192(30) = 3060$$
$$f'(50) = -3(50)^2 + 192(50) = 2100$$
$$f'(64) = -3(64)^2 + 192(64) = 0$$
$$f'(70) = -3(70)^2 + 192(70) = -1260$$

or 0, 3060, 2100, 0, and -1260 feet per second (ft/sec).

Thus, the rocket has an initial velocity of 0 ft/sec at $t = 0$ and accelerates to a velocity of 3060 ft/sec at $t = 30$. Fifty seconds into the flight, the rocket's velocity is 2100 ft/sec, which is less than the velocity at $t = 30$. This means that the rocket begins to decelerate after an initial period of acceleration. (Later on, we will learn how to determine the rocket's maximum velocity.)

The deceleration continues: The velocity is 0 ft/sec at $t = 64$ and -1260 ft/sec when $t = 70$. This result tells us that 70 seconds into flight, the rocket is heading back to the earth with a speed of 1260 ft/sec.

c. The results of part (b) show that the rocket's velocity is zero when $t = 64$. At this instant, the rocket's maximum altitude is

$$s = f(64) = -(64)^3 + 96(64)^2 + 5$$
$$= 131{,}077$$

or 131,077 feet. A sketch of the graph of f appears in Figure 40.

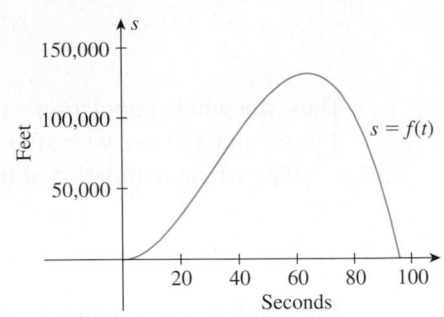

FIGURE **40**
The rocket's altitude t seconds into flight is given by $f(t)$.

Note You may have observed that the domain of the function f in Example 8 is restricted, for practical reasons, to the interval $[0, \infty)$. Since the definition of the derivative of a function f at a number a requires that f be defined in an open interval containing a, the derivative of f is not, strictly speaking, defined at 0. But notice that the function f can, in fact, be defined for all values of t, and hence it makes sense to calculate $f'(0)$. You will encounter situations such as this throughout the book, especially in exercises pertaining to real-world applications. The nature of the functions appearing in these applications obviates the necessity to consider "one-sided" derivatives.

Exploring with TECHNOLOGY

Refer to Example 8.

1. Use a graphing utility to plot the graph of the velocity function

$$v = f'(t) = -3t^2 + 192t$$

using the viewing window $[0, 120] \times [-5000, 5000]$. Then, using **ZOOM** and **TRACE** or the root-finding capability of your graphing utility, verify that $f'(64) = 0$.

2. Plot the graph of the position function of the rocket

$$s = f(t) = -t^3 + 96t^2 + 5$$

using the viewing window $[0, 120] \times [0, 150{,}000]$. Then, using **ZOOM** and **TRACE** repeatedly, verify that the maximum altitude of the rocket is 131,077 feet.

3. Use **ZOOM** and **TRACE** or the root-finding capability of your graphing utility to find when the rocket returns to the earth.

9.4 Self-Check Exercises

1. Find the derivative of each function using the rules of differentiation.
  a. $f(x) = 1.5x^2 + 2x^{1.5}$
  b. $g(x) = 2\sqrt{x} + \dfrac{3}{\sqrt{x}}$

2. Let $f(x) = 2x^3 - 3x^2 + 2x - 1$.
  a. Compute $f'(x)$.
  b. What is the slope of the tangent line to the graph of f when $x = 2$?
  c. What is the rate of change of the function f at $x = 2$?

3. GDP OF A COUNTRY A certain country's gross domestic product (GDP) (in billions of dollars) is described by the function

$$G(t) = -2t^3 + 45t^2 + 20t + 6000 \qquad (0 \le t \le 11)$$

where $t = 0$ corresponds to the beginning of 2003.
   a. At what rate was the GDP changing at the beginning of 2008? At the beginning of 2010? At the beginning of 2013?

   b. What was the average rate of growth of the GDP from the beginning of 2008 to the beginning of 2013?

Solutions to Self-Check Exercises 9.4 can be found on page 659.

9.4 Concept Questions

1. State the following rules of differentiation in your own words.
   a. The rule for differentiating a constant function
   b. The power rule
   c. The constant multiple rule
   d. The sum rule

2. If $f'(2) = 3$ and $g'(2) = -2$, find
   a. $h'(2)$ if $h(x) = 2f(x)$
   b. $F'(2)$ if $F(x) = 3f(x) - 4g(x)$

3. Suppose f and g are differentiable functions and a and b are nonzero numbers. Find $F'(x)$ if
   a. $F(x) = af(x) + bg(x)$

   b. $F(x) = \dfrac{f(x)}{a}$

4. If f is differentiable at $x = a$, is $[f'(x)](a) = \dfrac{d}{dx}[f(a)]$? Explain.

9.4 Exercises

In Exercises 1–34, find the derivative of the function f by using the rules of differentiation.

1. $f(x) = -3$

2. $f(x) = 365$

3. $f(x) = x^5$

4. $f(x) = x^7$

5. $f(x) = x^{3.1}$

6. $f(x) = x^{0.8}$

7. $f(x) = 3x^2$

8. $f(x) = -2x^3$

9. $f(r) = \pi r^2$

10. $f(r) = \dfrac{4}{3}\pi r^3$

11. $f(x) = 9x^{1/3}$

12. $f(x) = \dfrac{5}{4}x^{4/5}$

13. $f(x) = 3\sqrt{x}$

14. $f(u) = \dfrac{2}{\sqrt{u}}$

15. $f(x) = 7x^{-12}$

16. $f(x) = 0.3x^{-1.2}$

17. $f(x) = 5x^2 - 3x + 7$

18. $f(x) = x^3 - 3x^2 + 1$

19. $f(x) = -x^3 + 2x^2 - 6$

20. $f(x) = (1 + 2x^2)^2 + 2x^3$

21. $f(x) = 0.03x^2 - 0.4x + 10$

22. $f(x) = 0.002x^3 - 0.05x^2 + 0.1x - 20$

23. $f(x) = \dfrac{2x^3 - 4x^2 + 3}{x}$

24. $f(x) = \dfrac{x^3 + 2x^2 + x - 1}{x}$

25. $f(x) = 4x^4 - 3x^{5/2} + 2$

26. $f(x) = 5x^{4/3} - \dfrac{2}{3}x^{3/2} + x^2 - 3x + 1$

27. $f(x) = 5x^{-1} + 4x^{-2}$

28. $f(x) = -\dfrac{1}{3}(x^{-3} - x^6)$

29. $f(t) = \dfrac{4}{t^4} - \dfrac{3}{t^3} + \dfrac{2}{t}$

30. $f(x) = \dfrac{5}{x^3} - \dfrac{2}{x^2} - \dfrac{1}{x} + 200$

31. $f(x) = 3x - 5\sqrt{x}$

32. $f(t) = 2t^2 + \sqrt{t^3}$

33. $f(x) = \dfrac{2}{x^2} - \dfrac{3}{x^{1/3}}$

34. $f(x) = \dfrac{3}{x^3} + \dfrac{4}{\sqrt{x}} + 1$

35. Let $f(x) = 2x^3 - 4x$. Find:
   a. $f'(-2)$ **b.** $f'(0)$ **c.** $f'(2)$

36. Let $f(x) = 4x^{5/4} + 2x^{3/2} + x$. Find:
   a. $f'(4)$ **b.** $f'(16)$

In Exercises 37–40, find each limit by evaluating the derivative of a suitable function at an appropriate point.
Hint: Look at the definition of the derivative.

37. $\displaystyle\lim_{h \to 0} \dfrac{(1 + h)^3 - 1}{h}$

38. $\displaystyle\lim_{x \to 1} \dfrac{x^5 - 1}{x - 1}$
    Hint: Let $h = x - 1$.

39. $\displaystyle\lim_{h \to 0} \dfrac{3(2 + h)^2 - (2 + h) - 10}{h}$

40. $\displaystyle\lim_{t \to 0} \dfrac{1 - (1 + t)^2}{t(1 + t)^2}$

In Exercises 41–44, find the slope and an equation of the tangent line to the graph of the function f at the specified point.

41. $f(x) = 2x^2 - 3x + 4; (2, 6)$

42. $f(x) = -\dfrac{5}{3}x^2 + 2x + 2; \left(-1, -\dfrac{5}{3}\right)$

43. $f(x) = x^4 - 3x^3 + 2x^2 - x + 1; (2, -1)$

44. $f(x) = \sqrt{x} + \dfrac{1}{\sqrt{x}}; \left(4, \dfrac{5}{2}\right)$

45. Let $f(x) = x^3$.
   a. Find the point on the graph of f where the tangent line is horizontal.
   b. Sketch the graph of f, and draw the horizontal tangent line.

46. Let $f(x) = x^3 - 4x^2$. Find the points on the graph of f where the tangent line is horizontal.

47. Let $f(x) = x^3 + 1$.
   a. Find the points on the graph of f where the slope of the tangent line is equal to 12.
   b. Find the equation(s) of the tangent line(s) of part (a).
   c. Sketch the graph of f showing the tangent line(s).

48. Let $f(x) = \frac{2}{3}x^3 + x^2 - 12x + 6$. Find the values of x for which:
   a. $f'(x) = -12$ **b.** $f'(x) = 0$
   c. $f'(x) = 12$

49. Let $f(x) = \frac{1}{4}x^4 - \frac{1}{3}x^3 - x^2$. Find the points on the graph of f where the slope of the tangent line is equal to:
   a. $-2x$ **b.** 0   **c.** $10x$

50. A straight line perpendicular to and passing through the point of tangency of the tangent line is called the *normal* to the curve at that point. Find an equation of the tangent line and the normal to the curve $y = x^3 - 3x + 1$ at the point $(2, 3)$.

51. **GROWTH OF A CANCEROUS TUMOR** The volume of a spherical cancerous tumor is given by the function

$$V(r) = \frac{4}{3}\pi r^3$$

where r is the radius of the tumor in centimeters. Find the rate of change in the volume of the tumor with respect to its radius when

   a. $r = \dfrac{2}{3}$ cm   **b.** $r = \dfrac{5}{4}$ cm

52. **VELOCITY OF BLOOD IN AN ARTERY** The velocity (in centimeters per second) of blood r cm from the central axis of an artery is given by

$$v(r) = k(R^2 - r^2)$$

where k is a constant and R is the radius of the artery (see the accompanying figure). Suppose $k = 1000$ and

$R = 0.2$ cm. Find $v(0.1)$ and $v'(0.1)$, and interpret your results.

Blood vessel

53. **GROWTH OF TABLET AND SMARTPHONE USERS** The number of tablets and smartphones in use worldwide (in millions) in year t from 2010 through 2012 can be approximated by

$$f(t) = 128.1t^{1.94} \qquad (1 \le t \le 3)$$

where $t = 1$ corresponds to 2010.
   a. How many tablets and smartphones were in use in 2011?
   b. How fast was the number of tablets and smartphones changing in 2011?

Source: MIT Technology Review.

54. **U.K. MOBILE PHONE VIDEO VIEWERS** As mobile phones continue to proliferate, more and more people will watch video content on them through a mobile browser, subscription, download, or application at least once a month. In a study released in 2013, eMarketer estimated that the number of mobile phone video viewers in the United Kingdom in year t from 2011 ($t = 0$) through 2017 ($t = 6$) will be

$$P(t) = 13.86t^{0.535}$$

percent of the population.
   a. What percentage of the U.K. population is expected to watch video content on mobile phones in 2015?
   b. How fast is the percentage of mobile phone video viewers in the United Kingdom in 2015 expected to change?

Source: eMarketer.

55. **MARRIED COUPLES WITH CHILDREN** The percentage of families that were married couples with children between 1970 and 2010 is approximately

$$P(t) = \frac{49.6}{t^{0.27}} \qquad (1 \le t \le 5)$$

where t is measured in decades, with $t = 1$ corresponding to 1970.
   a. What percentage of families were married couples with children in 1970? In 1990? In 2010?
   b. How fast was the percentage of families that were married couples with children changing in 1990? In 2000?

Source: American Community Survey.

56. **EFFECT OF STOPPING ON AVERAGE SPEED** According to data from a General Motors study, the average speed of your trip A (in miles per hour) is related to the number of stops per mile you make on the trip x by the equation

$$A = \frac{26.5}{x^{0.45}}$$

Compute dA/dx for $x = 0.25$ and $x = 2$. How is the rate of change with respect to x of the average speed of your trip affected by the number of stops per mile?

Source: General Motors.

57. **DECLINE OF MIDDLE-CLASS INCOME** The share of households in the United States that are earning middle-class incomes has been in decline since the 1970s. The percentage of households with annual incomes within 50% of the median income is given by

$$P(t) = 50.3t^{-0.09} \qquad (1 \le t \le 6)$$

where t is measured in decades, with $t = 1$ corresponding to 1980.

a. What percentage of households had annual incomes within 50% of the median income in 2010?

b. How fast was the percentage of households with annual incomes within 50% of the median income decreasing in 2010?

Source: Alan Krueger.

58. **DEMAND FUNCTION FOR DESK LAMPS** The demand function for the Luminar desk lamp is given by

$$p = f(x) = -0.1x^2 - 0.4x + 35$$

where x is the quantity demanded in thousands and p is the unit price in dollars.

a. Find $f'(x)$.

b. What is the rate of change of the unit price when the quantity demanded is 10,000 units ($x = 10$)? What is the unit price at that level of demand?

59. **STOPPING DISTANCE OF A RACING CAR** During a test by the editors of an auto magazine, the distance s (in feet) traveled by the MacPherson X-2 racing car t sec after the brakes were applied conformed to the rule

$$s = f(t) = 120t - 15t^2 \qquad (t \ge 0)$$

a. Find an expression for the car's velocity v at any time t.

b. What was the car's velocity when the brakes were first applied?

c. What was the car's stopping distance for that particular test?

Hint: The stopping time is found by setting $v = 0$.

60. **MOBILE INSTANT MESSAGING ACCOUNTS** Mobile instant messaging (IM) is a small portion of total IM usage, but it is growing sharply. The function

$$P(t) = 0.257t^2 + 0.57t + 3.9 \qquad (0 \le t \le 4)$$

gives the mobile IM accounts in year t as a percentage of total enterprise IM accounts from 2006 ($t = 0$) through 2010 ($t = 4$).

a. What percentage of total enterprise IM accounts were the mobile accounts in 2008?

b. How fast was this percentage changing in 2008?

Source: The Radical Group.

61. **MEDICAL COSTS FOR A FAMILY OF FOUR** The average annual medical costs for a family of four in the United States from 2001 through 2011 is approximated by the function

$$C(t) = 22.9883t^2 + 830.358t + 7513 \qquad (1 \le t \le 11)$$

where t is measured in years, with $t = 1$ corresponding to 2001, and $C(t)$ is measured in dollars.

a. What was the approximate average annual medical costs for a family of four in 2010?

b. How fast was the approximate average medical costs for a family of four increasing in 2010?

Source: Milliman Medical Index.

62. **SPENDING ON MEDICARE** Based on the current eligibility requirement, a study conducted in 2004 showed that federal spending on entitlement programs, particularly Medicare, would grow enormously in the future. The study predicted that spending on Medicare, as a percentage of the gross domestic product (GDP), will be

$$P(t) = 0.27t^2 + 1.4t + 2.2 \qquad (0 \le t \le 5)$$

percent in year t, where t is measured in decades, with $t = 0$ corresponding to 2000.

a. How fast was the spending on Medicare, as a percentage of the GDP, growing in 2010? How fast will it be growing in 2020?

b. What was the predicted spending on Medicare in 2010? What will it be in 2020?

Source: Congressional Budget Office.

63. **CONSUMER PRICE INDEX** An economy's consumer price index (CPI) is described by the function

$$I(t) = -0.2t^3 + 3t^2 + 100 \qquad (0 \le t \le 11)$$

in year t, where $t = 0$ corresponds to 2003.

a. At what rate was the CPI changing in 2008? In 2010? In 2013?

b. What was the average rate of increase in the CPI over the period from 2008 to 2013?

64. **WORKER EFFICIENCY** An efficiency study conducted for Elektra Electronics showed that the number of Space Commander walkie-talkies assembled by the average worker during the morning shift t hr after starting work at 8 A.M. is given by

$$N(t) = -t^3 + 6t^2 + 15t \qquad (0 \le t \le 4)$$

a. Find the rate at which the average worker will be assembling walkie-talkies t hr after starting work.

b. At what rate will the average worker be assembling walkie-talkies at 10 A.M.? At 11 A.M.?

c. How many walkie-talkies will the average worker assemble between 10 A.M. and 11 A.M.?

65. **CURBING POPULATION GROWTH** Five years ago, the government of a Pacific Island country launched an extensive propaganda campaign aimed toward curbing the country's population growth. According to the Census Department,

the population (measured in thousands of people) for the following 4 years was

$$P(t) = -\frac{1}{3}t^3 + 64t + 3000$$

where t is measured in years and $t = 0$ corresponds to the start of the campaign. Find the rate of change of the population at the end of years 1, 2, 3, and 4. Was the plan working?

66. CONSERVATION OF SPECIES A certain species of turtle faces extinction because dealers collect truckloads of turtle eggs to be sold as aphrodisiacs. It is hoped that after severe conservation measures are implemented, the turtle population will grow according to the rule

$$N(t) = 2t^3 + 3t^2 - 4t + 1000 \qquad (0 \le t \le 10)$$

where $N(t)$ denotes the population at the end of year t. Find the rate of growth of the turtle population when $t = 2$ and $t = 8$. What will be the population 10 years after the conservation measures are implemented?

67. FLIGHT OF A MODEL ROCKET The altitude (in feet) of a model rocket t sec into a trial flight is given by

$$s = f(t) = -2t^3 + 12t^2 + 5 \qquad (t \ge 0)$$

  a. Find an expression v for the rocket's velocity at any time t.

  b. Compute the rocket's vertical velocity when $t = 0, 2, 4$, and 6. Interpret your results.

  c. Using the results from the solution to part (b), find the maximum altitude attained by the rocket.

    Hint: At its highest point, the velocity of the rocket is zero.

68. SUPPLY FUNCTION FOR SATELLITE RADIOS The supply function for a certain make of satellite radio is given by

$$p = f(x) = 0.0001x^{5/4} + 10$$

where x is the quantity supplied and p is the unit price in dollars.

  a. Find $f'(x)$.

  b. What is the rate of change of the unit price if the quantity supplied is 10,000 satellite radios?

69. POPULATION GROWTH FOR A RESORT TOWN A study prepared for a Sunbelt town's chamber of commerce projected that the town's population in the next 3 years will grow according to the rule

$$P(t) = 50,000 + 30t^{3/2} + 20t$$

where $P(t)$ denotes the population t months from now. How fast will the population be increasing 9 months and 16 months from now?

70. AVERAGE SPEED OF A VEHICLE ON A HIGHWAY The average speed of a vehicle on a stretch of Route 134 between 6 A.M. and 10 A.M. on a typical weekday is approximated by the function

$$f(t) = 20t - 40\sqrt{t} + 50 \qquad (0 \le t \le 4)$$

where $f(t)$ is measured in miles per hour and t is measured in hours, with $t = 0$ corresponding to 6 A.M.

  a. Compute $f'(t)$.

  b. What is the average speed of a vehicle on that stretch of Route 134 at 6 A.M.? At 7 A.M.? At 8 A.M.?

  c. How fast is the average speed of a vehicle on that stretch of Route 134 changing at 6:30 A.M.? At 7 A.M.? At 8 A.M.?

71. EFFECT OF ADVERTISING ON SALES The relationship between the amount of money x that Cannon Precision Instruments spends on advertising and the company's total sales $S(x)$ is given by the function

$$S(x) = -0.002x^3 + 0.6x^2 + x + 500 \qquad (0 \le x \le 200)$$

where x is measured in thousands of dollars. Find the rate of change of the sales with respect to the amount of money spent on advertising. Are Cannon's total sales increasing at a faster rate when the amount of money spent on advertising is (a) \$100,000 or (b) \$150,000?

72. NATIONAL HEALTH CARE EXPENDITURE The per capita health spending in the United States in year t for the years 2000 through 2011 is approximately

$$C(t) = -1.1708t^3 + 7.029t^2 + 389.69t + 4780 \quad (0 \le t \le 11)$$

dollars in year t, with $t = 0$ corresponding to 2000.

  a. What was the per capita health spending in 2010?

  b. How fast was the per capita health spending changing in 2010?

Source: NHCM Foundation.

73. OBESITY IN AMERICA The body mass index (BMI) measures body weight in relation to height. A BMI of 25 to 29.9 is considered overweight, a BMI of 30 or more is considered obese, and a BMI of 40 or more is morbidly obese. The percentage of the U.S. population that is obese is approximated by the function

$$P(t) = 0.0004t^3 + 0.0036t^2 + 0.8t + 12 \qquad (0 \le t \le 20)$$

where t is measured in years, with $t = 0$ corresponding to the beginning of 1991.

  a. What percentage of the U.S. population was deemed obese at the beginning of 1991? At the beginning of 2010?

  b. How fast was the percentage of the U.S. population that is deemed obese changing at the beginning of 1991? At the beginning of 2010?

  (*Note:* A formula for calculating the BMI of a person is given in Exercise 35, page 909.)

Source: Centers for Disease Control and Prevention.

74. AGING POPULATION The population age 65 and over (in millions) of developed countries from 2005 through 2034 is projected to be

$$f(t) = 3.567t + 175.2 \qquad (5 \le t \le 35)$$

where t is measured in years and $t = 5$ corresponds to 2005. On the other hand, the population age 65 and over of underdeveloped/emerging countries over the same period is projected to be

$$g(t) = 0.46t^2 + 0.16t + 287.8 \qquad (5 \le t \le 35)$$

a. What does the function $D = g + f$ represent?

b. Find D' and $D'(10)$, and interpret your results.

Source: U.S. Census Bureau, United Nations.

75. **SHORTAGE OF NURSES** The projected number of nurses (in millions) in year t from 2000 through 2015 is given by

$$N(t) = \begin{cases} 1.9 & \text{if } 0 \le t < 5 \\ -0.0004t^2 + 0.038t + 1.72 & \text{if } 5 \le t \le 15 \end{cases}$$

where $t = 0$ corresponds to 2000. The projected number of nursing jobs (in millions) over the same period is

$$J(t) = \begin{cases} -0.0002t^2 + 0.032t + 2 & \text{if } 0 \le t < 10 \\ -0.0016t^2 + 0.12t + 1.26 & \text{if } 10 \le t \le 15 \end{cases}$$

a. Find the rule for the function $G = J - N$ giving the gap between the supply and the demand of nurses from 2000 through 2015.

b. How fast was the gap between the supply and the demand of nurses changing in 2008? In 2012?

Source: U.S. Department of Health and Human Services.

In Exercises 76 and 77, determine whether the statement is true or false. If it is true, explain why it is true. If it is false, give an example to show why it is false.

76. If f and g are differentiable, then

$$\frac{d}{dx}[2f(x) - 5g(x)] = 2f'(x) - 5g'(x)$$

77. If $f(x) = \pi^x$, then $f'(x) = x\pi^{x-1}$.

78. Prove the Power Rule (Rule 2) for the special case $n = 3$.

Hint: Compute $\displaystyle\lim_{h \to 0} \left[\frac{(x+h)^3 - x^3}{h} \right]$.

9.4 Solutions to Self-Check Exercises

1. a. $f'(x) = \dfrac{d}{dx}(1.5x^2) + \dfrac{d}{dx}(2x^{1.5})$

$\qquad = (1.5)(2x) + (2)(1.5x^{0.5})$

$\qquad = 3x + 3x^{0.5}$

b. $g'(x) = \dfrac{d}{dx}(2x^{1/2}) + \dfrac{d}{dx}(3x^{-1/2})$

$\qquad = (2)\left(\dfrac{1}{2}x^{-1/2}\right) + (3)\left(-\dfrac{1}{2}x^{-3/2}\right)$

$\qquad = x^{-1/2} - \dfrac{3}{2}x^{-3/2} = \dfrac{1}{\sqrt{x}} - \dfrac{3}{2\sqrt{x^3}}$

2. a. $f'(x) = \dfrac{d}{dx}(2x^3) - \dfrac{d}{dx}(3x^2) + \dfrac{d}{dx}(2x) - \dfrac{d}{dx}(1)$

$\qquad = (2)(3x^2) - (3)(2x) + 2$

$\qquad = 6x^2 - 6x + 2$

b. The slope of the tangent line to the graph of f when $x = 2$ is given by

$$f'(2) = 6(2)^2 - 6(2) + 2 = 14$$

c. The rate of change of f at $x = 2$ is given by $f'(2)$. Using the results of part (b), we see that the required rate of change is 14 units/unit change in x.

3. a. The rate at which the GDP was changing at any time t $(0 < t < 11)$ is given by

$$G'(t) = -6t^2 + 90t + 20$$

In particular, the rates of change of the GDP at the beginning of the years 2008 ($t = 5$), 2010 ($t = 7$), and 2013 ($t = 10$) are given by

$$G'(5) = 320 \qquad G'(7) = 356 \qquad G'(10) = 320$$

respectively—that is, by \$320 billion/year, \$356 billion/year, and \$320 billion/year, respectively.

b. The average rate of growth of the GDP over the period from the beginning of 2008 ($t = 5$) to the beginning of 2013 ($t = 10$) is given by

$$\frac{G(10) - G(5)}{10 - 5} = \frac{[-2(10)^3 + 45(10)^2 + 20(10) + 6000]}{5}$$

$$\qquad - \frac{[-2(5)^3 + 45(5)^2 + 20(5) + 6000]}{5}$$

$$\qquad = \frac{8700 - 6975}{5}$$

or \$345 billion/year.

USING TECHNOLOGY   Finding the Rate of Change of a Function

We can use the numerical derivative operation of a graphing utility to obtain the value of the derivative at a given value of x. Since the derivative of a function f measures the rate of change of the function with respect to x, the numerical derivative operation can be used to answer questions pertaining to the rate of change of one quantity y with respect to another quantity x, where $y = f(x)$, for a specific value of x.

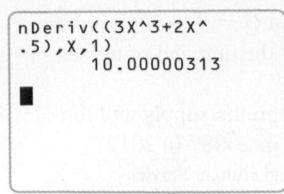

FIGURE **T1**
The TI-83/84 numerical derivative
screen for computing $f'(1)$

EXAMPLE 1 Let $y = 3t^3 + 2\sqrt{t}$.

a. Use the numerical derivative operation of a graphing utility to find how fast y is changing with respect to t when $t = 1$.
b. Verify the result of part (a), using the rules of differentiation of this section.

Solution

a. Let $f(t) = 3t^3 + 2\sqrt{t}$. Using the numerical derivative operation of a graphing utility, we find that the rate of change of y with respect to t when $t = 1$ is given by $f'(1) = 10$ (Figure T1).
b. Here, $f(t) = 3t^3 + 2t^{1/2}$, and

$$f'(t) = 9t^2 + 2\left(\frac{1}{2}t^{-1/2}\right) = 9t^2 + \frac{1}{\sqrt{t}}$$

Using this result, we see that when $t = 1$, y is changing at the rate of

$$f'(1) = 9(1^2) + \frac{1}{\sqrt{1}} = 10$$

units per unit change in t, as obtained earlier.

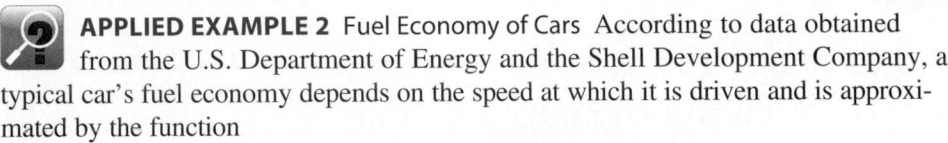

APPLIED EXAMPLE 2 Fuel Economy of Cars According to data obtained from the U.S. Department of Energy and the Shell Development Company, a typical car's fuel economy depends on the speed at which it is driven and is approximated by the function

$$f(x) = 0.00000310315x^4 - 0.000455174x^3 + 0.00287869x^2 + 1.25986x \qquad (0 \le x \le 75)$$

where x is measured in miles per hour and $f(x)$ is measured in miles per gallon (mpg).

a. Use a graphing utility to graph the function f on the interval $[0, 75]$.
b. Find the rate of change of f when $x = 20$ and when $x = 50$.
c. Interpret your results.

Source: U.S. Department of Energy and the Shell Development Company.

Solution

a. The graph is shown in Figure T2.
b. Using the numerical derivative operation of a graphing utility, we see that $f'(20) = 0.9280996$. The rate of change of f when $x = 50$ is given by $f'(50) = -0.3145009995$. (See Figure T3a and T3b.)

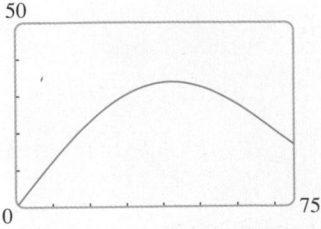

FIGURE **T2**
The graph of the function f on the
interval $[0, 75]$

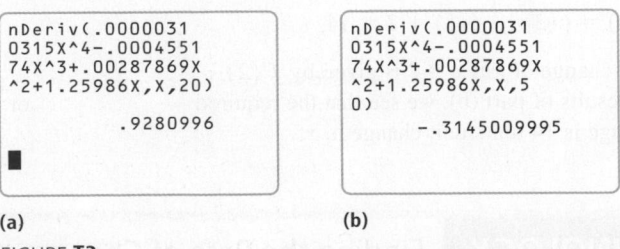

(a)   (b)

FIGURE **T3**
The TI-83/84 numerical derivative screen for computing (a) $f'(20)$ and (b) $f'(50)$

c. The results of part (b) tell us that when a typical car is being driven at 20 mph, its fuel economy increases at the rate of approximately 0.9 mpg per 1 mph increase in its speed. At a speed of 50 mph, its fuel economy decreases at the rate of approximately 0.3 mpg per 1 mph increase in its speed.

TECHNOLOGY EXERCISES

In Exercises 1–6, use the numerical derivative operation to find the rate of change of f at the given value of x. Give your answer accurate to four decimal places.

1. $f(x) = 4x^5 - 3x^3 + 2x^2 + 1$; $x = 0.5$

2. $f(x) = -x^5 + 4x^2 + 3$; $x = 0.4$

3. $f(x) = x - 2\sqrt{x}$; $x = 3$

4. $f(x) = \dfrac{\sqrt{x} - 1}{x}$; $x = 2$

5. $f(x) = x^{1/2} - x^{1/3}$; $x = 1.2$

6. $f(x) = 2x^{5/4} + x$; $x = 2$

7. CARBON MONOXIDE IN THE ATMOSPHERE The projected average global atmospheric concentration of carbon monoxide is approximated by the function

$$f(t) = 0.881443t^4 - 1.45533t^3 + 0.695876t^2 + 2.87801t + 293 \quad (0 \le t \le 4)$$

where t is measured in 40-year intervals, with $t = 0$ corresponding to the beginning of 1860, and $f(t)$ is measured in parts per million by volume.

a. Plot the graph of f in the viewing window $[0, 4] \times [280, 500]$.

b. Use a graphing utility to estimate how fast the projected average global atmospheric concentration of carbon monoxide was changing at the beginning of 1900 ($t = 1$) and at the beginning of 2010 ($t = 4$).

Source: Meadows et al., *Beyond the Limits.*

8. SPREAD OF HIV TO CHILDREN The estimated number of children newly infected with HIV through mother-to-child contact worldwide is given by

$$f(t) = -0.2083t^3 + 3.0357t^2 + 44.0476t + 200.2857 \quad (0 \le t \le 12)$$

where $f(t)$ is measured in thousands and t is measured in years, with $t = 0$ corresponding to the beginning of 1990.

a. Plot the graph of f in the viewing window $[0, 12] \times [0, 850]$.

b. How fast was the estimated number of children newly infected with HIV through mother-to-child contact worldwide increasing at the beginning of the year 2000?

Source: United Nations.

9.5 The Product and Quotient Rules; Higher-Order Derivatives

In this section, we study two more rules of differentiation: the **Product Rule** and the **Quotient Rule.** We will also see how the derivative of a function can itself be differentiated, under certain conditions, to obtain a higher-order derivative.

The Product Rule

The derivative of the product of two differentiable functions is given by the following rule:

> **Rule 5: The Product Rule**
>
> $$\frac{d}{dx}\left[f(x)g(x)\right] = f(x)g'(x) + g(x)f'(x)$$

The derivative of the product of two functions is the first function times the derivative of the second plus the second function times the derivative of the first.

The Product Rule may be extended to the case involving the product of any finite number of functions (see Exercise 89, page 674). We prove the Product Rule at the end of this section.

⚠ The derivative of the product of two functions is *not* given by the product of the derivatives of the functions; that is, in general

$$\frac{d}{dx}\left[f(x)g(x)\right] \ne f'(x)g'(x)$$

For example, if $f(x) = x$ and $g(x) = 2x^2$. Then

$$\frac{d}{dx}[f(x)g(x)] = \frac{d}{dx}[x(2x^2)] = \frac{d}{dx}(2x^3) = 6x^2$$

On the other hand, $f'(x)g'(x) = (1)(4x) = 4x$. So

$$\frac{d}{dx}[f(x)g(x)] \neq f'(x)g'(x)$$

EXAMPLE 1 Find the derivative of the function

$$f(x) = (2x^2 - 1)(x^3 + 3)$$

Solution   By the Product Rule,

$$f'(x) = (2x^2 - 1)\frac{d}{dx}(x^3 + 3) + (x^3 + 3)\frac{d}{dx}(2x^2 - 1)$$

$$= (2x^2 - 1)(3x^2) + (x^3 + 3)(4x) \qquad \text{(x^2) See page 10.}$$

$$= 6x^4 - 3x^2 + 4x^4 + 12x$$

$$= 10x^4 - 3x^2 + 12x \qquad \text{Combine like terms.}$$

$$= x(10x^3 - 3x + 12) \qquad \text{Factor out } x. \qquad ■$$

EXAMPLE 2 Differentiate (that is, find the derivative of) the function

$$f(x) = x^3(\sqrt{x} + 1)$$

Solution   First, we express the function in exponential form, obtaining

$$f(x) = x^3(x^{1/2} + 1)$$

By the Product Rule,

$$f'(x) = x^3\frac{d}{dx}(x^{1/2} + 1) + (x^{1/2} + 1)\frac{d}{dx}x^3$$

$$= x^3\left(\frac{1}{2}x^{-1/2}\right) + (x^{1/2} + 1)(3x^2)$$

$$= \frac{1}{2}x^{5/2} + 3x^{5/2} + 3x^2$$

$$= \frac{7}{2}x^{5/2} + 3x^2 \qquad ■$$

Note   We can also solve the problem by first expanding the product before differentiating f. Examples for which this is not possible will be considered in Section 9.6, where the true value of the Product Rule will be appreciated.   ▪

The Quotient Rule

The derivative of the quotient of two differentiable functions is given by the following rule:

Rule 6: The Quotient Rule

$$\frac{d}{dx}\left[\frac{f(x)}{g(x)}\right] = \frac{g(x)f'(x) - f(x)g'(x)}{[g(x)]^2} \qquad (g(x) \neq 0)$$

As an aid to remembering this expression, observe that it has the following form:

$$\frac{d}{dx}\left[\frac{f(x)}{g(x)}\right] = \frac{(\text{Denominator})\left(\begin{matrix}\text{Derivative of}\\\text{numerator}\end{matrix}\right) - (\text{Numerator})\left(\begin{matrix}\text{Derivative of}\\\text{denominator}\end{matrix}\right)}{(\text{Square of denominator})}$$

For a proof of the Quotient Rule, see Exercise 90, page 674.

The derivative of a quotient is *not* equal to the quotient of the derivatives; that is,

$$\frac{d}{dx}\left[\frac{f(x)}{g(x)}\right] \neq \frac{f'(x)}{g'(x)}$$

For example, if $f(x) = x^3$ and $g(x) = x^2$, then

$$\frac{d}{dx}\left[\frac{f(x)}{g(x)}\right] = \frac{d}{dx}\left(\frac{x^3}{x^2}\right) = \frac{d}{dx}(x) = 1$$

which is *not* equal to

$$\frac{f'(x)}{g'(x)} = \frac{\dfrac{d}{dx}(x^3)}{\dfrac{d}{dx}(x^2)} = \frac{3x^2}{2x} = \frac{3}{2}x$$

EXAMPLE 3 Find $f'(x)$ if $f(x) = \dfrac{x}{2x - 4}$.

Solution Using the Quotient Rule, we obtain

$$\begin{aligned}
f'(x) &= \frac{(2x - 4)\dfrac{d}{dx}(x) - x\dfrac{d}{dx}(2x - 4)}{(2x - 4)^2}\\[2mm]
&= \frac{(2x - 4)(1) - x(2)}{(2x - 4)^2}\\[2mm]
&= \frac{2x - 4 - 2x}{(2x - 4)^2} = -\frac{4}{(2x - 4)^2}
\end{aligned}$$

EXAMPLE 4 Find $f'(x)$ if $f(x) = \dfrac{x^2 + 1}{x^2 - 1}$.

Solution By the Quotient Rule,

$$\begin{aligned}
f'(x) &= \frac{(x^2 - 1)\dfrac{d}{dx}(x^2 + 1) - (x^2 + 1)\dfrac{d}{dx}(x^2 - 1)}{(x^2 - 1)^2}\\[2mm]
&= \frac{(x^2 - 1)(2x) - (x^2 + 1)(2x)}{(x^2 - 1)^2}\\[2mm]
&= \frac{2x^3 - 2x - 2x^3 - 2x}{(x^2 - 1)^2}\\[2mm]
&= \frac{4x}{(x^2 - 1)^2}
\end{aligned}$$

EXAMPLE 5 Find $h'(x)$ if $h(x) = \dfrac{\sqrt{x}}{x^2 + 1}$.

Solution   Rewrite $h(x)$ in the form $h(x) = \dfrac{x^{1/2}}{x^2 + 1}$. By the Quotient Rule, we find

$$
\begin{aligned}
h'(x) &= \frac{(x^2 + 1)\dfrac{d}{dx}(x^{1/2}) - x^{1/2}\dfrac{d}{dx}(x^2 + 1)}{(x^2 + 1)^2} \\[2mm]
&= \frac{(x^2 + 1)(\tfrac{1}{2}x^{-1/2}) - x^{1/2}(2x)}{(x^2 + 1)^2} \\[2mm]
&= \frac{\tfrac{1}{2}x^{-1/2}(x^2 + 1 - 4x^2)}{(x^2 + 1)^2} \qquad \text{Factor out } \tfrac{1}{2}x^{-1/2} \text{ from the numerator.} \\[2mm]
&= \frac{1 - 3x^2}{2\sqrt{x}(x^2 + 1)^2}
\end{aligned}
$$

$ **APPLIED EXAMPLE 6** Rate of Change of DVD Sales The annual sales (in millions of dollars per year) of a DVD recording of a hit movie t years from the date of release is given by

$$
S(t) = \frac{5t}{t^2 + 1}
$$

a. Find the rate at which the annual sales are changing at time t.
b. How fast are the annual sales changing at the time the DVDs are released $(t = 0)$? Two years from the date of release?

Solution

a. The rate at which the annual sales are changing at time t is given by $S'(t)$. Using the Quotient Rule, we obtain

$$
\begin{aligned}
S'(t) &= \frac{d}{dt}\left[\frac{5t}{t^2 + 1}\right] = 5\frac{d}{dt}\left[\frac{t}{t^2 + 1}\right] \\[2mm]
&= 5\left[\frac{(t^2 + 1)(1) - t(2t)}{(t^2 + 1)^2}\right] \qquad (x^2)\ \text{See page 22.} \\[2mm]
&= 5\left[\frac{t^2 + 1 - 2t^2}{(t^2 + 1)^2}\right] = \frac{5(1 - t^2)}{(t^2 + 1)^2}
\end{aligned}
$$

b. The rate at which the annual sales are changing at the time the DVDs are released is given by

$$
S'(0) = \frac{5(1 - 0)}{(0 + 1)^2} = 5
$$

That is, they are increasing at the rate of \$5 million per year per year.

Two years from the date of release, the annual sales are changing at the rate of

$$
S'(2) = \frac{5(1 - 4)}{(4 + 1)^2} = -\frac{3}{5} = -0.6
$$

That is, they are decreasing at the rate of \$600,000 per year per year.

The graph of the function S is shown in Figure 41.

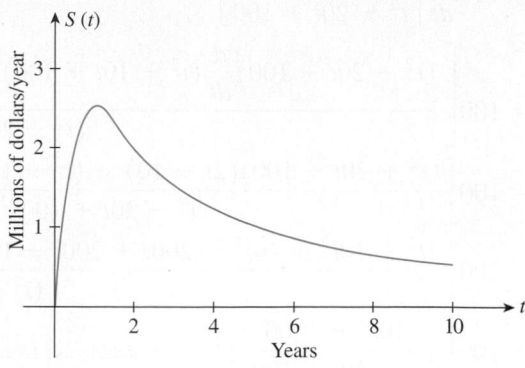

FIGURE **41**
After a spectacular rise, the annual sales begin to taper off.

Exploring with **TECHNOLOGY**

Refer to Example 6.

1. Use a graphing utility to plot the graph of the function S, using the viewing window $[0, 10] \times [0, 3]$.

2. Use **TRACE** and **ZOOM** to determine the coordinates of the highest point on the graph of S in the interval $[0, 10]$. Interpret your results.

Explore and Discuss

Suppose the revenue of a company is given by $R(x) = xp(x)$, where x is the number of units of the product sold at a unit price of $p(x)$ dollars.

1. Compute $R'(x)$ and explain, in words, the relationship between $R'(x)$ and $p(x)$ and/or its derivative.

2. What can you say about $R'(x)$ if $p(x)$ is constant? Is this expected?

APPLIED EXAMPLE 7 Oxygen-Restoration Rate in a Pond When organic waste is dumped into a pond, the oxidation process that takes place reduces the pond's oxygen content. However, given time, nature will restore the oxygen content to its natural level. Suppose the oxygen content t days after organic waste has been dumped into the pond is given by

$$f(t) = 100 \left[\frac{t^2 + 10t + 100}{t^2 + 20t + 100} \right] \qquad (0 < t < \infty)$$

percent of its normal level.

a. Derive a general expression that gives the rate of change of the pond's oxygen level at any time t.

b. How fast is the pond's oxygen content changing 1 day, 10 days, and 20 days after the organic waste has been dumped?

Solution

a. The rate of change of the pond's oxygen level at any time t is given by the derivative of the function f. Thus, the required expression is

$$f'(t) = 100 \frac{d}{dt}\left[\frac{t^2 + 10t + 100}{t^2 + 20t + 100}\right]$$

$$= 100\left[\frac{(t^2 + 20t + 100)\dfrac{d}{dt}(t^2 + 10t + 100) - (t^2 + 10t + 100)\dfrac{d}{dt}(t^2 + 20t + 100)}{(t^2 + 20t + 100)^2}\right]$$

$$= 100\left[\frac{(t^2 + 20t + 100)(2t + 10) - (t^2 + 10t + 100)(2t + 20)}{(t^2 + 20t + 100)^2}\right] \qquad (x^2) \text{ See page 22.}$$

$$= 100\left[\frac{2t^3 + 10t^2 + 40t^2 + 200t + 200t + 1000 - 2t^3 - 20t^2 - 20t^2 - 200t - 200t - 2000}{(t^2 + 20t + 100)^2}\right]$$

$$= 100\left[\frac{10t^2 - 1000}{(t^2 + 20t + 100)^2}\right] \qquad \text{Combine like terms in the numerator.}$$

b. The rate at which the pond's oxygen content is changing 1 day after the organic waste has been dumped is given by

$$f'(1) = 100\left[\frac{10 - 1000}{(1 + 20 + 100)^2}\right] \approx -6.76$$

That is, it is dropping at the rate of 6.8% per day. After 10 days, the rate is

$$f'(10) = 100\left[\frac{10(10)^2 - 1000}{(10^2 + 20(10) + 100)^2}\right] = 0$$

That is, it is neither increasing nor decreasing. After 20 days, the rate is

$$f'(20) = 100\left[\frac{10(20)^2 - 1000}{(20^2 + 20(20) + 100)^2}\right] \approx 0.37$$

That is, the oxygen content is increasing at the rate of 0.37% per day, and the restoration process has indeed begun. ◼

Higher-Order Derivatives

The derivative f' of a function f is also a function. As such, the differentiability of f' may be considered. Thus, the function f' has a derivative f'' at a point x in the domain of f' if the limit of the quotient

$$\frac{f'(x + h) - f'(x)}{h}$$

exists as h approaches zero. In other words, it is the derivative of the first derivative.

The function f'' obtained in this manner is called the **second derivative of the function** f, just as the derivative f' of f is often called the first derivative of f. Continuing in this fashion, we are led to considering the third, fourth, and higher-order derivatives of f whenever they exist. Notations for the first, second, third, and, in general, nth derivatives of a function f at a point x are

$$f'(x), f''(x), f'''(x), \ldots, f^{(n)}(x)$$

or

$$D^1 f(x), D^2 f(x), D^3 f(x), \ldots, D^n f(x)$$

If f is written in the form $y = f(x)$, then the notations for its derivatives are

$$y', y'', y''', \ldots, y^{(n)}$$

$$\frac{dy}{dx}, \frac{d^2 y}{dx^2}, \frac{d^3 y}{dx^3}, \ldots, \frac{d^n y}{dx^n}$$

or

$$D^1y, D^2y, D^3y, \ldots, D^ny$$

EXAMPLE 8 Find the derivatives of all orders of the polynomial function

$$f(x) = x^5 - 3x^4 + 4x^3 - 2x^2 + x - 8$$

Solution We have

$$f'(x) = 5x^4 - 12x^3 + 12x^2 - 4x + 1$$

$$f''(x) = \frac{d}{dx} f'(x) = 20x^3 - 36x^2 + 24x - 4$$

$$f'''(x) = \frac{d}{dx} f''(x) = 60x^2 - 72x + 24$$

$$f^{(4)}(x) = \frac{d}{dx} f'''(x) = 120x - 72$$

$$f^{(5)}(x) = \frac{d}{dx} f^{(4)}(x) = 120$$

and

$$f^{(n)}(x) = 0 \quad \text{for } n > 5$$

EXAMPLE 9 Find the third derivative of the function f defined by $y = x^{2/3}$. What is its domain?

Solution We have

$$y' = \frac{2}{3} x^{-1/3}$$

$$y'' = \left(\frac{2}{3}\right)\left(-\frac{1}{3}\right)x^{-4/3} = -\frac{2}{9} x^{-4/3}$$

so the required derivative is

$$y''' = \left(-\frac{2}{9}\right)\left(-\frac{4}{3}\right)x^{-7/3} = \frac{8}{27} x^{-7/3} = \frac{8}{27x^{7/3}}$$

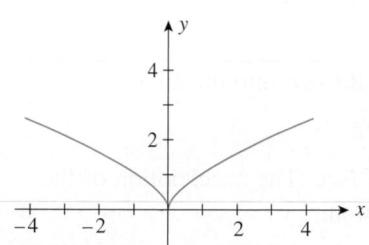

FIGURE **42**
The graph of the function $y = x^{2/3}$

The common domain of the functions f', f'', and f''' is the set of all real numbers except $x = 0$. The domain of $y = x^{2/3}$ is the set of all real numbers. The graph of the function $y = x^{2/3}$ appears in Figure 42.

Note Always simplify an expression before differentiating it to obtain the next order derivative.

Just as the derivative of a function f at a point x measures the rate of change of the function f at that point, the second derivative of f (the derivative of f') measures the rate of change of the derivative f' of the function f. The third derivative of the function f, f''', measures the rate of change of f'', and so on.

In Chapter 10, we will discuss applications involving the geometric interpretation of the second derivative of a function. The following example gives an interpretation of the second derivative in a familiar role.

APPLIED EXAMPLE 10 Acceleration of a Maglev Refer to the example on pages 594–596. The distance s (in feet) covered by a maglev moving along a straight track t seconds after starting from rest is given by the function $s = 4t^2$ ($0 \le t \le 10$). What is the maglev's acceleration at any time t?

Solution The velocity of the maglev t seconds from rest is given by

$$v = \frac{ds}{dt} = \frac{d}{dt}(4t^2) = 8t$$

The acceleration of the maglev t seconds from rest is given by the rate of change of the velocity of t—that is,

$$a = \frac{d}{dt}v = \frac{d}{dt}\left(\frac{ds}{dt}\right) = \frac{d^2s}{dt^2} = \frac{d}{dt}(8t) = 8$$

or 8 feet per second per second, normally abbreviated 8 ft/sec^2.

APPLIED EXAMPLE 11 Acceleration and Velocity of a Falling Object A ball is thrown straight up into the air from the roof of a building. The height of the ball as measured from the ground is given by

$$s = -16t^2 + 24t + 120$$

where s is measured in feet and t in seconds. Find the velocity and acceleration of the ball 3 seconds after it is thrown into the air.

Solution The velocity v and acceleration a of the ball at any time t are given by

$$v = \frac{ds}{dt} = \frac{d}{dt}(-16t^2 + 24t + 120) = -32t + 24$$

and

$$a = \frac{d^2s}{dt^2} = \frac{d}{dt}\left(\frac{ds}{dt}\right) = \frac{d}{dt}(-32t + 24) = -32$$

Therefore, the velocity of the ball 3 seconds after it is thrown into the air is

$$v = -32(3) + 24 = -72$$

That is, the ball is falling downward at a speed of 72 ft/sec. The acceleration of the ball is 32 ft/sec^2 downward at any time during the motion.

Another interpretation of the second derivative of a function—this time from the field of economics—follows. Suppose the consumer price index (CPI) of an economy between the years a and b is described by the function $I(t)$ ($a \le t \le b$) (Figure 43). Then the first derivative of I at $t = c$, $I'(c)$, where $a < c < b$, gives the rate of change of I at c. The quantity

$$\frac{I'(c)}{I(c)}$$

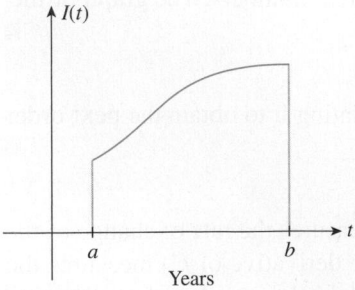

FIGURE 43
The CPI of a certain economy from year a to year b is given by $I(t)$.

measures the *inflation rate* of the economy at $t = c$. The second derivative of I at $t = c$, $I''(c)$, gives the rate of change of I' at $t = c$. Now, it is possible for $I'(t)$ to be positive and $I''(t)$ to be negative at $t = c$ (see Example 12). This tells us that at $t = c$, the economy is experiencing inflation (the CPI is increasing) but the rate at which the CPI is growing is in fact decreasing. This is precisely the situation described by an economist or a politician who claims that "inflation is slowing." One may not jump to the conclusion from the aforementioned quote that prices of goods and services are about to drop!

 APPLIED EXAMPLE 12 Inflation Rate of an Economy The function

$$I(t) = -0.2t^3 + 3t^2 + 100 \qquad (0 \le t \le 9)$$

gives the CPI of an economy, where t is measured in years, with $t = 0$ corresponding to the beginning of 2004.

a. Find the inflation rate at the beginning of 2010 ($t = 6$).
b. Show that inflation was moderating at that time.

Solution

a. We find $I'(t) = -0.6t^2 + 6t$. Next, we compute

$$I'(6) = -0.6(6)^2 + 6(6) = 14.4 \quad \text{and} \quad I(6) = -0.2(6)^3 + 3(6)^2 + 100 = 164.8$$

from which we see that the inflation rate is

$$\frac{I'(6)}{I(6)} = \frac{14.4}{164.8} \approx 0.0874$$

or approximately 8.7%.

b. We find

$$I''(t) = \frac{d}{dt}(-0.6t^2 + 6t) = -1.2t + 6$$

Since

$$I''(6) = -1.2(6) + 6 = -1.2$$

we see that I' is indeed decreasing at $t = 6$, and we conclude that inflation was moderating at that time (Figure 44).

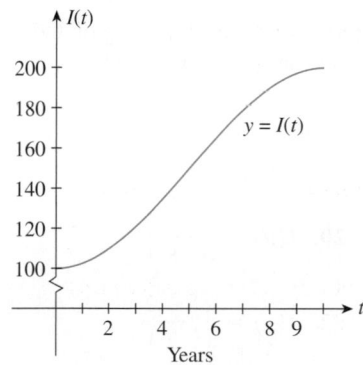

FIGURE 44
The CPI of an economy is given by $I(t)$.

Verification of the Product Rule

We will now verify the Product Rule. If $p(x) = f(x)g(x)$, then

$$p'(x) = \lim_{h \to 0} \frac{p(x + h) - p(x)}{h} = \lim_{h \to 0} \frac{f(x + h)g(x + h) - f(x)g(x)}{h}$$

By adding $-f(x + h)g(x) + f(x + h)g(x)$ (which is zero!) to the numerator and factoring, we have

$$p'(x) = \lim_{h \to 0} \frac{f(x + h)[g(x + h) - g(x)] + g(x)[f(x + h) - f(x)]}{h}$$

$$= \lim_{h \to 0} \left\{ f(x + h) \left[\frac{g(x + h) - g(x)}{h} \right] + g(x) \left[\frac{f(x + h) - f(x)}{h} \right] \right\}$$

$$= \lim_{h \to 0} f(x + h) \left[\frac{g(x + h) - g(x)}{h} \right] + \lim_{h \to 0} g(x) \left[\frac{f(x + h) - f(x)}{h} \right] \qquad \text{By Property 3 of limits}$$

$$= \lim_{h \to 0} f(x + h) \cdot \lim_{h \to 0} \frac{g(x + h) - g(x)}{h} + \lim_{h \to 0} g(x) \cdot \lim_{h \to 0} \frac{f(x + h) - f(x)}{h} \qquad \text{By Property 4 of limits}$$

$$= f(x)g'(x) + g(x)f'(x)$$

Observe that in the last link in the chain of equalities, we have used the fact that $\lim_{h \to 0} f(x + h) = f(x)$ because f is continuous at x.

9.5 Self-Check Exercises

1. Find the derivative of $f(x) = \dfrac{2x + 1}{x^2 - 1}$.

2. Find the third derivative of

$$f(x) = 2x^5 - 3x^3 + x^2 - 6x + 10$$

3. **SALES OF ADS SECURITY SYSTEMS** The total sales of ADS Security Systems in its first 2 years of operation are given by

$$S = f(t) = \frac{0.3t^3}{1 + 0.4t^2} \qquad (0 \le t \le 2)$$

where S is measured in millions of dollars and $t = 0$ corresponds to the date ADS Security Systems began operations. How fast were the sales increasing at the beginning of the company's second year of operation?

Solutions to Self-Check Exercises 9.5 can be found on page 674.

9.5 Concept Questions

1. State the rule of differentiation in your own words.
   a. Product Rule   **b.** Quotient Rule

2. If $f(1) = 3$, $g(1) = 2$, $f'(1) = -1$, and $g'(1) = 4$, find

   a. $h'(1)$ if $h(x) = f(x)g(x)$ **b.** $F'(1)$ if $F(x) = \dfrac{f(x)}{g(x)}$

3. a. What is the second derivative of a function f?
   b. How do you find the second derivative of a function f, assuming that it exists?

4. If $s = f(t)$ gives the position of an object moving on the coordinate line, what do $f'(t)$ and $f''(t)$ measure?

9.5 Exercises

In Exercises 1–30, find the derivative of each function.

1. $f(x) = 2x(x^2 + 1)$

2. $f(x) = 3x^2(x - 1)$

3. $f(t) = (t - 1)(2t + 1)$

4. $f(x) = (2x + 3)(3x - 4)$

5. $f(x) = (3x + 1)(x^2 - 2)$

6. $f(x) = (x + 1)(2x^2 - 3x + 1)$

7. $f(x) = (x^3 - 1)(x + 1)$

8. $f(x) = (x^3 - 12x)(3x^2 + 2x)$

9. $f(w) = (w^3 - w^2 + w - 1)(w^2 + 2)$

10. $f(x) = \dfrac{1}{5}x^5 + (x^2 + 1)(x^2 - x - 1) + 28$

11. $f(x) = (5x^2 + 1)(2\sqrt{x} - 1)$

12. $f(t) = (1 + \sqrt{t})(2t^2 - 3)$

13. $f(x) = (x^2 - 5x + 2)\left(x - \dfrac{2}{x}\right)$

14. $f(x) = (x^3 + 2x + 1)\left(2 + \dfrac{1}{x^2}\right)$

15. $f(x) = \dfrac{1}{x - 2}$ **16.** $g(x) = \dfrac{3}{2x + 4} + 2x^2$

17. $f(x) = \dfrac{2x - 1}{2x + 1}$ **18.** $f(t) = \dfrac{1 - 2t}{1 + 3t}$

19. $f(x) = \dfrac{1}{x^2 + x + 2}$ **20.** $f(u) = \dfrac{u}{u^2 + 1}$

21. $f(s) = \dfrac{s^2 - 4}{s + 1}$ **22.** $f(x) = \dfrac{x^3 - 2}{x^2 + 1}$

23. $f(x) = \dfrac{\sqrt{x} + 1}{x^2 + 1}$ **24.** $f(x) = \dfrac{x}{\sqrt{x} + 2}$

25. $f(x) = \dfrac{x^2 + 2}{x^2 + x + 1}$ **26.** $f(x) = \dfrac{x + 1}{2x^2 + 2x + 3}$

27. $f(x) = \dfrac{(x + 1)(x^2 + 1)}{x - 2}$

28. $f(x) = (3x^2 - 1)\left(x^2 - \dfrac{1}{x}\right)$

29. $f(x) = \dfrac{x}{x^2 - 4} - \dfrac{x - 1}{x^2 + 4}$

30. $f(x) = \dfrac{x + \sqrt{3x}}{3x - 1}$

In Exercises 31–34, suppose f and g are functions that are differentiable at $x = 1$ and that $f(1) = 2$, $f'(1) = -1$, $g(1) = -2$, and $g'(1) = 3$. Find the value of $h'(1)$.

31. $h(x) = f(x)g(x)$ **32.** $h(x) = (x^2 + 1)g(x)$

33. $h(x) = \dfrac{xf(x)}{x + g(x)}$ **34.** $h(x) = \dfrac{f(x)g(x)}{f(x) - g(x)}$

In Exercises 35–38, find the derivative of each function and evaluate $f'(x)$ at the given value of x.

35. $f(x) = (2x - 1)(x^2 + 3); x = 1$

36. $f(x) = \dfrac{2x + 1}{2x - 1}; x = 2$

37. $f(x) = \dfrac{x}{x^4 - 2x^2 - 1}; x = -1$

38. $f(x) = (\sqrt{x} + 2x)(x^{3/2} - x); x = 4$

In Exercises 39–42, find the slope and an equation of the tangent line to the graph of the function f at the specified point.

39. $f(x) = (x^3 + 1)(x^2 - 2); (2, 18)$

40. $f(x) = \dfrac{x^2}{x + 1}; \left(2, \dfrac{4}{3}\right)$ **41.** $f(x) = \dfrac{x + 1}{x^2 + 1}; (1, 1)$

42. $f(x) = \dfrac{1 + 2x^{1/2}}{1 + x^{3/2}}; \left(4, \dfrac{5}{9}\right)$

In Exercises 43–48, find the first and second derivatives of the function.

43. $f(x) = 4x^2 - 2x + 1$

44. $f(x) = -0.2x^2 + 0.3x + 4$

45. $f(x) = 2x^3 - 3x^2 + 1$

46. $g(x) = -3x^3 + 24x^2 + 6x - 64$

47. $h(t) = t^4 - 2t^3 + 6t^2 - 3t + 10$

48. $f(x) = x^5 - x^4 + x^3 - x^2 + x - 1$

In Exercises 49–52, find the third derivative of the given function.

49. $f(x) = 3x^4 - 4x^3$

50. $f(x) = 3x^5 - 6x^4 + 2x^2 - 8x + 12$

51. $f(x) = \dfrac{1}{x}$ **52.** $f(x) = \dfrac{2}{x^2}$

53. Suppose $g(x) = x^2 f(x)$ and it is known that $f(2) = 3$ and $f'(2) = -1$. Evaluate $g'(2)$.

54. Suppose $g(x) = (x^2 + 1)f(x)$ and it is known that $f(2) = 3$ and $f'(2) = -1$. Evaluate $g'(2)$.

55. Find an equation of the tangent line to the graph of the function $f(x) = (x^3 + 1)(3x^2 - 4x + 2)$ at the point $(1, 2)$.

56. Find an equation of the tangent line to the graph of the function $f(x) = \dfrac{3x}{x^2 - 2}$ at the point $(2, 3)$.

57. Let $f(x) = (x^2 + 1)(2 - x)$. Find the point(s) on the graph of f where the tangent line is horizontal.

58. Let $f(x) = \dfrac{x}{x^2 + 1}$. Find the point(s) on the graph of f where the tangent line is horizontal.

59. Find the point(s) on the graph of the function $f(x) = (x^2 + 6)(x - 5)$ where the slope of the tangent line is equal to -2.

60. Find the point(s) on the graph of the function $f(x) = \dfrac{x + 1}{x - 1}$ where the slope of the tangent line is equal to $-\dfrac{1}{2}$.

61. A straight line perpendicular to and passing through the point of tangency of the tangent line is called the *normal* to the curve at that point. Find the equation of the tangent line and the normal to the curve

$$y = \dfrac{1}{1 + x^2}$$

at the point $\left(1, \dfrac{1}{2}\right)$.

62. CONCENTRATION OF A DRUG IN THE BLOODSTREAM The concentration of a certain drug in a patient's bloodstream t hr after injection is given by

$$C(t) = \dfrac{0.2t}{t^2 + 1}$$

a. Find the rate at which the concentration of the drug is changing with respect to time.
b. How fast is the concentration changing $\frac{1}{2}$ hr, 1 hr, and 2 hr after the injection?

63. COST OF REMOVING TOXIC WASTE A city's main water reservoir was recently found to be contaminated with trichloroethylene, a cancer-causing chemical, as a result of an abandoned chemical dump leaching chemicals into the water. A proposal submitted to the city's council members indicates that the cost, measured in millions of dollars, of removing $x\%$ of the toxic pollutant is given by

$$C(x) = \dfrac{0.5x}{100 - x}$$

Find $C'(80)$, $C'(90)$, $C'(95)$, and $C'(99)$. What does your result tell you about the cost of removing *all* of the pollutant?

64. DRUG DOSAGES Thomas Young has suggested the following rule for calculating the dosage of medicine for children 1 to 12 years old. If a denotes the adult dosage (in milligrams) and if t is the child's age (in years), then the child's dosage is given by

$$D(t) = \dfrac{at}{t + 12}$$

Suppose the adult dosage of a substance is 500 mg. Find an expression that gives the rate of change of a child's dosage with respect to the child's age. What is the rate of change of a child's dosage with respect to his or her age for a 6-year-old child? A 10-year-old child?

65. EFFECT OF BACTERICIDE The number of bacteria $N(t)$ in a certain culture t min after an experimental bactericide is introduced is given by

$$N(t) = \frac{10,000}{1 + t^2} + 2000$$

Find the rate of change of the number of bacteria in the culture 1 min and 2 min after the bactericide is introduced. What is the population of the bacteria in the culture 1 min and 2 min after the bactericide is introduced?

66. DEMAND FUNCTION FOR SPORTS WATCHES The demand function for the Sicard sports watch is given by

$$d(x) = \frac{50}{0.01x^2 + 1} \qquad (0 \le x \le 20)$$

where x (measured in units of a thousand) is the quantity demanded per week and $d(x)$ is the unit price in dollars.
a. Find $d'(x)$.
b. Find $d'(5)$, $d'(10)$, and $d'(15)$, and interpret your results.

67. REVENUE FUNCTION FOR SPORTS WATCHES Refer to Exercise 66.
a. Find an expression for the revenue function R for the Sicard sports watch.
Hint: $R(x) = xd(x)$
b. Find $R'(x)$.
c. Find $R'(8)$, $R'(10)$, and $R'(12)$, and interpret your results.

68. PROFIT FUNCTION FOR SPORTS WATCHES Refer to Exercise 66. The total profit function P (in thousands of dollars) for the Sicard sports watch is given by

$$P(x) = \frac{50x}{0.01x^2 + 1} - 0.025x^3 + 0.35x^2 - 10x - 30$$

$$(0 \le x \le 20)$$

where x is measured in units of a thousand.
a. Find $P(0)$, and interpret your result.
b. Find $P'(5)$ and $P'(10)$, and interpret your results.

69. MORTGAGE RATES The average 30-year fixed mortgage rate in the United States in the first week of May in 2010 through 2012 is approximated by

$$M(t) = \frac{55.9}{t^2 - 0.31t + 11.2}$$

percent/year. Here, t is measured in years, with $t = 0$ corresponding to the first week of May in 2010.
a. What was the average 30-year fixed mortgage rate in the first week of May in 2011 ($t = 1$)?
b. How fast was the 30-year fixed mortgage rate changing in the first week of May in 2011 ($t = 1$)?
Source: Mortgage Bankers Association.

70. BOX-OFFICE RECEIPTS The total worldwide box-office receipts for a long-running movie are approximated by the function

$$T(x) = \frac{120x^2}{x^2 + 4}$$

where $T(x)$ is measured in millions of dollars and x is the number of years since the movie's release. How fast are the total receipts changing 1 year, 3 years, and 5 years after its release?

71. LEARNING CURVES From experience, Emory Secretarial School knows that the average student taking Advanced Computer Typing will progress according to the rule

$$N(t) = \frac{60t + 180}{t + 6} \qquad (t \ge 0)$$

where $N(t)$ measures the number of words per minute the student can type after t weeks in the course.
a. Find an expression for $N'(t)$.
b. Compute $N'(t)$ for $t = 1, 3, 4$, and 7, and interpret your results.
c. Sketch the graph of the function N. Does it confirm the results obtained in part (b)?
d. What will be the average student's typing speed at the end of the 12-week course?

72. UNEMPLOYMENT RATE The unemployment rate of a certain country shortly after the Great Recession was approximately

$$f(t) = \frac{5t + 300}{t^2 + 25} \qquad (0 \le t \le 4)$$

percent in year t, where $t = 0$ corresponds to the beginning of 2010. How fast was the unemployment rate of the country changing at the beginning of 2013?

73. HOME FORMALDEHYDE LEVELS A study on formaldehyde levels in 900 homes indicates that emissions of various chemicals can decrease over time. The formaldehyde level (parts per million) in an average home in the study is given by

$$f(t) = \frac{0.055t + 0.26}{t + 2} \qquad (0 \le t \le 12)$$

where t is the age of the house in years. How fast is the formaldehyde level of the average house dropping when it is new? At the beginning of its fourth year? (See Note on page 654.)
Source: Bonneville Power Administration.

74. POPULATION GROWTH A major corporation is building a 4325-acre complex of homes, offices, stores, schools, and churches in the rural community of Glen Cove. As a result of this development, the planners have estimated that Glen Cove's population (in thousands) t years from now will be given by

$$P(t) = \frac{25t^2 + 125t + 200}{t^2 + 5t + 40}$$

a. Find the rate at which Glen Cove's population is changing with respect to time.
b. What will be the population after 10 years? At what rate will the population be increasing when $t = 10$?

75. **CHANGE IN REVENUE** Suppose that the demand equation for a product is $p = D(x)$, where x is the quantity demanded. If x units of the product are sold at a price of p dollars/unit, then the revenue realized from the sales is $R(x) = xD(x)$ (number of units sold times the price per unit).
    a. Use the Product Rule to find $R'(x)$.
    b. Use the result of part (a) to find $R'(x)$ for the case of a linear demand equation $p = a - bx$, where a and b are positive constants.
    c. Verify your result by first finding an expression for $R(x)$ and then differentiating $R(x)$ directly.

76. **CHANGE IN PER-CAPITA INCOME** The gross domestic product (GDP) of a country t years from now is $g(t) = 60 + 4t$ billion dollars. The population of the country at that time is $f(t) = 3 + 0.06t$ million. How fast will the per-capita income of the country be changing 2 years from now?
    Hint: The per-capita income at time t is $g(t)/f(t)$.

77. **ACCELERATION OF A FALLING OBJECT** During the construction of an office building, a hammer is accidentally dropped from a height of 256 ft. The distance (in feet) the hammer falls in t sec is $s = 16t^2$. What is the hammer's velocity when it strikes the ground? What is its acceleration?

78. **ACCELERATION OF A CAR** The distance s (in feet) covered by a car after t sec is given by

$$s = -t^3 + 8t^2 + 20t \qquad (0 \le t \le 6)$$

Find a general expression for the car's acceleration at any time t $(0 \le t \le 6)$. Show that the car is decelerating after $2\frac{2}{3}$ sec.

79. **AUTO TRANSMISSIONS AND FUEL ECONOMY** In trying to extract maximum efficiency out of every subsystem of a vehicle, auto transmission engineers are developing transmissions with up to ten gears. This is one of the ways in which manufacturers are trying to meet stricter federal fuel economy and pollution standards. The projected percentage of new cars equipped with transmissions that have seven speeds or more is given by the function

$$P(t) = 0.38t^2 + 1.3t + 3 \qquad (0 \le t \le 20)$$

where t is measured in years, with $t = 0$ corresponding to 2010.
    a. What will the percentage of vehicles equipped with transmissions that have seven speeds or more be in 2015 according to the projection?
    b. How fast will the percentage of vehicles equipped with transmissions that have seven speeds or more be changing in 2015?
    c. What is $P''(5)$? Interpret your result.
    Source: IHS Automotive.

80. **ALZHEIMER'S DISEASE** As baby boomers enter their golden years, the number of people afflicted with Alzheimer's disease is expected to rise dramatically. In a study published in the *Journal of Neurology*, the number of people

with Alzheimer's disease in the United States age 65 years and over is projected to be

$$N(t) = 0.00525t^2 + 0.075t + 4.7 \qquad (0 \le t \le 4)$$

million in decade t, where $t = 0$ corresponds to 2010.
    a. What is the projected number of people with Alzheimer's disease in the United States age 65 years and over in 2030?
    b. How fast is the number of people with Alzheimer's disease in the United States age 65 years and over projected to grow in 2030?
    c. How fast is the rate of growth of people with Alzheimer's disease in the United States age 65 years and over projected to change in the period covered by the study?
    Source: American Academy of Neurology.

81. **CRIME RATES** The number of major crimes committed in Bronxville between 2006 and 2014 is approximated by the function

$$N(t) = -0.1t^3 + 1.5t^2 + 100 \qquad (0 \le t \le 8)$$

where $N(t)$ denotes the number of crimes committed in year t, with $t = 0$ corresponding to 2006. Enraged by the dramatic increase in the crime rate, Bronxville's citizens, with the help of the local police, organized Neighborhood Crime Watch groups in early 2010 to combat this menace.
    a. Verify that the crime rate was increasing from 2006 through 2014.
    Hint: Compute $N'(0), N'(1), \ldots, N'(8)$. (See Note on page 654.)
    b. Show that the Neighborhood Crime Watch program was working by computing $N''(4), N''(5), N''(6), N''(7)$, and $N''(8)$.

82. **GDP OF A DEVELOPING COUNTRY** A developing country's gross domestic product (GDP) from 2006 to 2015 is approximated by the function

$$G(t) = -0.2t^3 + 2.4t^2 + 60 \qquad (0 \le t \le 9)$$

where $G(t)$ is measured in billions of dollars, with $t = 0$ corresponding to 2006.
    a. Compute $G'(1), \ldots, G'(8)$.
    b. Compute $G''(1), \ldots, G''(8)$.
    c. Using the results obtained in parts (a) and (b), show that after a slow start, the GDP increases quickly and then cools off.

83. **SALES OF VINYL RECORDS** Vinyl records (LPs) almost disappeared after CDs were introduced in the early 1980s, but they have been making a comeback. Wagner's Records estimates that t months after the grand opening of the store, the sales of LPs will be

$$S(t) = 4t^3 + 2t^2 + 300t$$

units. Compute $S(6), S'(6)$, and $S''(6)$, and interpret your results.

84. MEDIAN AGE OF U.S. POPULATION The median age (in years) of the U.S. population over the decades from 1960 through 2010 is given by

$$f(t) = -0.2176t^3 + 1.962t^2 - 2.833t + 29.4 \qquad (0 \le t \le 5)$$

where t is measured in decades, with $t = 0$ corresponding to 1960.
a. What was the median age of the population in the year 2000?
b. At what rate was the median age of the population changing in the year 2000?
c. Calculate $f''(4)$, and interpret your result.
Source: U.S. Census Bureau.

85. OBESITY IN AMERICA The body mass index (BMI) measures body weight in relation to height. A BMI of 25 to 29.9 is considered overweight, a BMI of 30 or more is considered obese, and a BMI of 40 or more is morbidly obese. The percent of the U.S. population that is obese is approximated by the function

$$P(t) = 0.0004t^3 + 0.0036t^2 + 0.8t + 12 \qquad (0 \le t \le 13)$$

where t is measured in years, with $t = 0$ corresponding to the beginning of 1991. Show that the rate of the rate of change of the percent of the U.S. population that is deemed obese was positive from 1991 to 2004. What does this mean?
Source: Centers for Disease Control and Prevention.

In Exercises 86–88, determine whether the statement is true or false. If it is true, explain why it is true. If it is false, give an example to show why it is false.

86. If f and g are differentiable, then

$$\frac{d}{dx}[f(x)g(x)] = f'(x)g'(x)$$

87. If f is differentiable, then

$$\frac{d}{dx}[xf(x)] = f(x) + xf'(x)$$

88. If f is differentiable, then

$$\frac{d}{dx}\left[\frac{f(x)}{x^2}\right] = \frac{f'(x)}{2x}$$

89. Extend the Product Rule for differentiation to the following case involving the product of three differentiable functions: Let $h(x) = u(x)v(x)w(x)$, and show that $h'(x) = u(x)v(x)w'(x) + u(x)v'(x)w(x) + u'(x)v(x)w(x)$. Hint: Let $f(x) = u(x)v(x)$, $g(x) = w(x)$, and $h(x) = f(x)g(x)$, and apply the Product Rule to the function h.

90. Prove the Quotient Rule for differentiation (Rule 6).
Hint: Let $k(x) = f(x)/g(x)$, and verify the following steps:
a. $\dfrac{k(x+h) - k(x)}{h} = \dfrac{f(x+h)g(x) - f(x)g(x+h)}{hg(x+h)g(x)}$
b. By adding $[-f(x)g(x) + f(x)g(x)]$ to the numerator and simplifying, show that

$$\frac{k(x+h) - k(x)}{h} = \frac{1}{g(x+h)g(x)}$$
$$\times \left\{ \left[\frac{f(x+h) - f(x)}{h}\right] \cdot g(x) \right.$$
$$\left. - \left[\frac{g(x+h) - g(x)}{h}\right] \cdot f(x) \right\}$$

c. $k'(x) = \lim\limits_{h \to 0} \dfrac{k(x+h) - k(x)}{h}$
$= \dfrac{g(x)f'(x) - f(x)g'(x)}{[g(x)]^2}$

9.5 Solutions to Self-Check Exercises

1. We use the Quotient Rule to obtain

$$f'(x) = \frac{(x^2 - 1)\dfrac{d}{dx}(2x + 1) - (2x + 1)\dfrac{d}{dx}(x^2 - 1)}{(x^2 - 1)^2}$$

$$= \frac{(x^2 - 1)(2) - (2x + 1)(2x)}{(x^2 - 1)^2}$$

$$= \frac{2x^2 - 2 - 4x^2 - 2x}{(x^2 - 1)^2}$$

$$= \frac{-2x^2 - 2x - 2}{(x^2 - 1)^2}$$

$$= \frac{-2(x^2 + x + 1)}{(x^2 - 1)^2}$$

2. $f'(x) = \dfrac{d}{dx}(2x^5 - 3x^3 + x^2 - 6x + 10)$

$= 10x^4 - 9x^2 + 2x - 6$

$f''(x) = \dfrac{d}{dx}(10x^4 - 9x^2 + 2x - 6)$

$= 40x^3 - 18x + 2$

$f'''(x) = \dfrac{d}{dx}(40x^3 - 18x + 2) = 120x^2 - 18$

3. The rate at which the company's total sales are changing at any time t is given by

$$S'(t) = \frac{(1 + 0.4t^2)\dfrac{d}{dt}(0.3t^3) - (0.3t^3)\dfrac{d}{dt}(1 + 0.4t^2)}{(1 + 0.4t^2)^2}$$

$$= \frac{(1 + 0.4t^2)(0.9t^2) - (0.3t^3)(0.8t)}{(1 + 0.4t^2)^2}$$

Therefore, at the beginning of the second year of operation, ADS Security Systems's sales were increasing at the rate of

$$S'(1) = \frac{(1 + 0.4)(0.9) - (0.3)(0.8)}{(1 + 0.4)^2} \approx 0.52$$

or $520,000/year.

USING TECHNOLOGY The Product and Quotient Rules; Higher-Order Derivatives

EXAMPLE 1 Let $f(x) = (2\sqrt{x} + 0.5x)(0.3x^3 + 2x - \frac{0.3}{x})$. Find $f'(0.2)$.

Solution Using the numerical derivative operation of a graphing utility, we find

$$f'(0.2) = 6.4797499802$$

See Figure T1.

```
nDeriv((2X^.5+.5
X)(.3X^3+2X-.3/X),
X,.2)
        6.4797499802
■
```

FIGURE **T1**
The TI-83/84 numerical derivative screen for computing $f'(0.2)$

 APPLIED EXAMPLE 2 Importance of Time in Treating Heart Attacks
According to the American Heart Association, the treatment benefit for heart attacks depends on the time until treatment and is described by the function

$$f(t) = \frac{0.44t^4 + 700}{0.1t^4 + 7} \qquad (0 \le t \le 24)$$

where t is measured in hours and $f(t)$ is expressed as a percent.

a. Use a graphing utility to graph the function f using the viewing window $[0, 24] \times [0, 100]$.
b. Use a graphing utility to find the derivative of f when $t = 0$ and $t = 2$.
c. Interpret the results obtained in part (b).
Source: American Heart Association.

Solution

a. The graph of f is shown in Figure T2.
b. Using the numerical derivative operation of a graphing utility, we find

$$f'(0) \approx 0 \quad \text{and} \quad f'(2) \approx -28.95402429$$

(see Figure T3 on the next page).

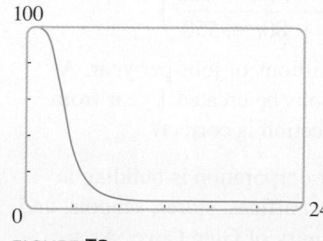

FIGURE **T2**

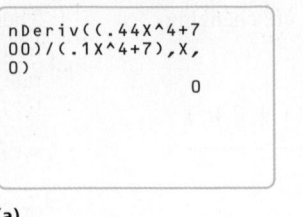

(a)   **(b)**

FIGURE **T3**
TI-83/84 numerical derivative screens (a) for computing $f'(0)$ and (b) for computing $f'(2)$

c. The results of part (b) show that there is no drop in the treatment benefit when the heart attack is treated immediately. But the treatment benefit drops off at the rate of approximately 29% per hour when the time to treatment is 2 hours. Thus, it is extremely urgent that a patient suffering a heart attack receive medical attention as soon as possible.

Some graphing utilities have the capability of numerically computing the second derivative of a function at a point. If your graphing utility has this capability, use it to work through the examples and exercises of this section.

EXAMPLE 3 Use the (second) numerical derivative operation of a graphing utility to find the second derivative of $f(x) = \sqrt{x}$ when $x = 4$.

Solution Using the (second) numerical derivative operation, we find

$$f''(4) = \text{der2}(x^{\wedge}.5, x, 4) = -.03125$$

(Figure T4).

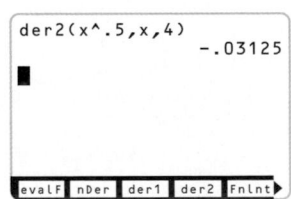

FIGURE **T4**
The TI-86 second derivative screen for computing $f''(4)$

TECHNOLOGY EXERCISES

In Exercises 1–4, use the numerical derivative operation to find the rate of change of $f(x)$ at the given value of x. Give your answer accurate to four decimal places.

1. $f(x) = (2x^2 + 1)(x^3 + 3x + 4)$; $x = -0.5$

2. $f(x) = (\sqrt{x} + 1)(2x^2 + x - 3)$; $x = 1.5$

3. $f(x) = \dfrac{\sqrt{x} - 1}{\sqrt{x} + 1}$; $x = 3$

4. $f(x) = \dfrac{\sqrt{x}(x^2 + 4)}{x^3 + 1}$; $x = 4$

5. New Construction Jobs The president of a major housing construction company claims that the number of construction jobs created in the next t months is given by

$$f(t) = 1.42 \left(\frac{7t^2 + 140t + 700}{3t^2 + 80t + 550} \right)$$

where $f(t)$ is measured in millions of jobs per year. At what rate will construction jobs be created 1 year from now, assuming that her projection is correct?

6. Population Growth A major corporation is building a 4325-acre complex of homes, offices, stores, schools, and churches in the rural community of Glen Cove. As a result of this development, the planners have estimated that Glen Cove's population (in thousands) t years from now will be given by

$$P(t) = \frac{25t^2 + 125t + 200}{t^2 + 5t + 40}$$

a. What will be the population 10 years from now?
b. At what rate will the population be increasing 10 years from now?

In Exercises 7–10, find the value of the second derivative of f at the given value of x. Express your answer correct to four decimal places.

7. $f(x) = 2x^3 - 3x^2 + 1; x = -1$

8. $f(x) = 2.5x^5 - 3x^3 + 1.5x + 4; x = 2.1$

9. $f(x) = 2.1x^{3.1} - 4.2x^{1.7} + 4.2; x = 1.4$

10. $f(x) = \dfrac{x^2 + 2x - 5}{x^3 + 1}; x = 2.1$

9.6 The Chain Rule

The population of Americans age 55 years and older as a percentage of the total population is approximated by the function

$$f(t) = 10.72(0.9t + 10)^{0.3} \qquad (0 \le t \le 20)$$

where t is measured in years with $t = 0$ corresponding to the year 2000 (Figure 45).

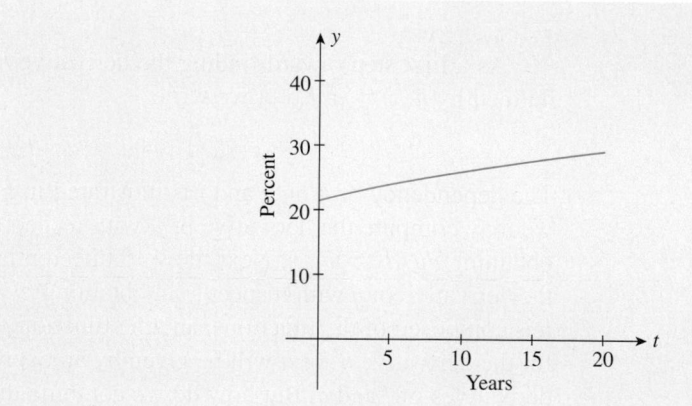

FIGURE **45**
Percentage of population of Americans age 55 years and older
Source: U.S. Census Bureau.

How fast will the population age 55 years and older be increasing at the beginning of 2015? To answer this question, we have to evaluate $f'(15)$, where f' is the derivative of f. But the rules of differentiation that we have developed up to now will not help us find the derivative of f.

In this section, we will introduce another rule of differentiation called the **Chain Rule.** When used in conjunction with the rules of differentiation developed in the last two sections, the Chain Rule enables us to greatly enlarge the class of functions that we are able to differentiate. (In Exercise 72, page 686, we will use the Chain Rule to answer the question posed in the introductory example.)

The Chain Rule

Consider the function $h(x) = (x^2 + x + 1)^2$. If we were to compute $h'(x)$ using only the rules of differentiation from the previous sections, then our approach might be to expand $h(x)$. Thus,

$$h(x) = (x^2 + x + 1)^2 = (x^2 + x + 1)(x^2 + x + 1)$$
$$= x^4 + 2x^3 + 3x^2 + 2x + 1$$

from which we find

$$h'(x) = 4x^3 + 6x^2 + 6x + 2$$

But what about the function $H(x) = (x^2 + x + 1)^{100}$? The same technique may be used to find the derivative of the function H, but the amount of work involved in this case would be prodigious! Consider, also, the function $G(x) = \sqrt{x^2 + 1}$. For each of the two functions H and G, the rules of differentiation of the previous sections cannot be applied directly to compute the derivatives H' and G'.

Observe that both H and G are **composite functions;** that is, each is composed of, or built up from, simpler functions. For example, the function H is composed of the two simpler functions $f(x) = x^2 + x + 1$ and $g(x) = x^{100}$ as follows:

$$H(x) = g[f(x)] = [f(x)]^{100}$$
$$= (x^2 + x + 1)^{100}$$

In a similar manner, we see that the function G is composed of the two simpler functions $f(x) = x^2 + 1$ and $g(x) = \sqrt{x}$. Thus,

$$G(x) = g[f(x)] = \sqrt{f(x)}$$
$$= \sqrt{x^2 + 1}$$

As a first step toward finding the derivative h' of a composite function $h = g \circ f$ defined by $h(x) = g[f(x)]$, we write

$$u = f(x) \quad \text{and} \quad y = g[f(x)] = g(u)$$

The dependency of h on g and f is illustrated in Figure 46. Since u is a function of x, we may compute the derivative of u with respect to x if f is a differentiable function, obtaining $du/dx = f'(x)$. Next, if g is a differentiable function of u, we may compute the derivative of g with respect to u, obtaining $dy/du = g'(u)$. Now, since the function h is composed of the function g and the function f, we might suspect that the rule $h'(x)$ for the derivative h' of h will be given by an expression that involves the rules for the derivatives of f and g. But how do we combine these derivatives to yield h'?

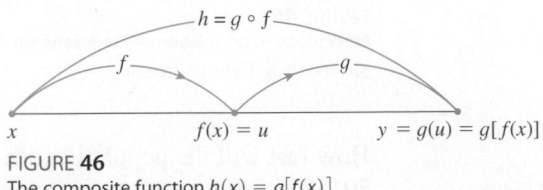

FIGURE **46**
The composite function $h(x) = g[f(x)]$

This question can be answered by interpreting the derivative of each function as the rate of change of that function. For example, suppose $u = f(x)$ changes three times as fast as x—that is,

$$f'(x) = \frac{du}{dx} = 3$$

And suppose $y = g(u)$ changes twice as fast as u—that is,

$$g'(u) = \frac{dy}{du} = 2$$

Then we would expect $y = h(x)$ to change six times as fast as x—that is,

$$h'(x) = g'(u)f'(x) = (2)(3) = 6$$

or, equivalently,

$$\frac{dy}{dx} = \frac{dy}{du} \cdot \frac{du}{dx} = (2)(3) = 6$$

This observation suggests the following result, which we state without proof.

Rule 7: The Chain Rule

If $h(x) = g[f(x)]$, then

$$h'(x) = \frac{d}{dx} g[f(x)] = g'[f(x)]f'(x) \tag{10}$$

Equivalently, if we write $y = h(x) = g(u)$, where $u = f(x)$, then

$$\frac{dy}{dx} = \frac{dy}{du} \cdot \frac{du}{dx} \tag{11}$$

Notes

1. If we label the composite function h in the following manner:

Inside function
↓
$$h(x) = g[f(x)]$$
↑
Outside function

then $h'(x)$ is just the *derivative* of the "outside function" *evaluated at* the "inside function" times the *derivative* of the "inside function."

2. Equation (11) can be remembered by observing that if we "cancel" the du's, then

$$\frac{dy}{dx} = \frac{dy}{du} \cdot \frac{du}{dx} = \frac{dy}{dx}$$

The Chain Rule for Powers of Functions

Many composite functions have the special form $h(x) = g(f(x))$, where g is defined by the rule $g(x) = x^n$ (n, a real number)—that is,

$$h(x) = [f(x)]^n$$

In other words, the function h is given by the power of a function f. The functions

$$h(x) = (x^2 + x + 1)^2 \qquad H(x) = (x^2 + x + 1)^{100} \qquad G(x) = \sqrt{x^2 + 1}$$

discussed earlier are examples of this type of composite function. By using the following corollary of the Chain Rule, the General Power Rule, we can find the derivative of this type of function much more easily than by using the Chain Rule directly.

The General Power Rule

If the function f is differentiable and $h(x) = [f(x)]^n$ (n, a real number), then

$$h'(x) = \frac{d}{dx} [f(x)]^n = n[f(x)]^{n-1} f'(x) \tag{12}$$

To see this, we observe that $h(x) = g(f(x))$ where $g(x) = x^n$, so by virtue of the Chain Rule, we have

$$h'(x) = g'[f(x)]f'(x)$$
$$= n[f(x)]^{n-1}f'(x)$$

since $g'(x) = nx^{n-1}$.

EXAMPLE 1 Let $F(x) = (3x + 1)^2$.

a. Find $F'(x)$, using the General Power Rule.
b. Verify your result without the benefit of the Chain Rule or the General Power Rule.

Solution

a. Using the General Power Rule, we obtain

$$F'(x) = 2(3x + 1)^1 \frac{d}{dx}(3x + 1)$$
$$= 2(3x + 1)(3)$$
$$= 6(3x + 1)$$

b. We first expand $F(x)$. Thus,

$$F(x) = (3x + 1)^2 = 9x^2 + 6x + 1$$

Next, differentiating, we have

$$F'(x) = \frac{d}{dx}(9x^2 + 6x + 1)$$
$$= 18x + 6$$
$$= 6(3x + 1)$$

as before.

EXAMPLE 2 Differentiate the function $G(x) = \sqrt{x^2 + 1}$.

Solution We rewrite the function $G(x)$ as

$$G(x) = (x^2 + 1)^{1/2}$$

and apply the General Power Rule, obtaining

$$G'(x) = \frac{1}{2}(x^2 + 1)^{-1/2} \frac{d}{dx}(x^2 + 1)$$

$$= \frac{1}{2}(x^2 + 1)^{-1/2} \cdot 2x = \frac{x}{\sqrt{x^2 + 1}}$$

EXAMPLE 3 Differentiate the function $f(x) = x^2(2x + 3)^5$.

Solution Applying the Product Rule followed by the General Power Rule, we obtain

$$f'(x) = x^2 \frac{d}{dx}(2x + 3)^5 + (2x + 3)^5 \frac{d}{dx}(x^2)$$

$$= (x^2)5(2x + 3)^4 \cdot \frac{d}{dx}(2x + 3) + (2x + 3)^5(2x)$$

$$= 5x^2(2x + 3)^4(2) + 2x(2x + 3)^5$$
$$= 2x(2x + 3)^4(5x + 2x + 3) = 2x(7x + 3)(2x + 3)^4$$

EXAMPLE 4 Find $f'(x)$ if $f(x) = (2x^2 + 3)^4(3x - 1)^5$.

Solution Applying the Product Rule, we have

$$f'(x) = (2x^2 + 3)^4 \frac{d}{dx}(3x - 1)^5 + (3x - 1)^5 \frac{d}{dx}(2x^2 + 3)^4$$

Next, we apply the General Power Rule to each term, obtaining

$$f'(x) = (2x^2 + 3)^4 \cdot 5(3x - 1)^4 \frac{d}{dx}(3x - 1) + (3x - 1)^5 \cdot 4(2x^2 + 3)^3 \frac{d}{dx}(2x^2 + 3)$$

$$= 5(2x^2 + 3)^4(3x - 1)^4 \cdot 3 + 4(3x - 1)^5(2x^2 + 3)^3(4x)$$

Finally, observing that $(2x^2 + 3)^3(3x - 1)^4$ is common to both terms, we can factor and simplify as follows:

$$f'(x) = (2x^2 + 3)^3(3x - 1)^4[15(2x^2 + 3) + 16x(3x - 1)]$$
$$= (2x^2 + 3)^3(3x - 1)^4(30x^2 + 45 + 48x^2 - 16x)$$
$$= (2x^2 + 3)^3(3x - 1)^4(78x^2 - 16x + 45)$$

EXAMPLE 5 Find $f'(x)$ if $f(x) = \dfrac{1}{(4x^2 - 7)^2}$.

Solution Rewriting $f(x)$ and then applying the General Power Rule, we obtain

$$f'(x) = \frac{d}{dx}\left[\frac{1}{(4x^2 - 7)^2}\right] = \frac{d}{dx}(4x^2 - 7)^{-2}$$

$$= -2(4x^2 - 7)^{-3}\frac{d}{dx}(4x^2 - 7)$$

$$= -2(4x^2 - 7)^{-3}(8x) = -\frac{16x}{(4x^2 - 7)^3}$$

EXAMPLE 6 Find the slope of the tangent line to the graph of the function

$$f(x) = \left(\frac{2x + 1}{3x + 2}\right)^3$$

at the point $(0, \frac{1}{8})$.

Solution The slope of the tangent line to the graph of f at any point x is given by $f'(x)$. To compute $f'(x)$, we use the General Power Rule followed by the Quotient Rule, obtaining

$$f'(x) = 3\left(\frac{2x + 1}{3x + 2}\right)^2 \frac{d}{dx}\left(\frac{2x + 1}{3x + 2}\right)$$

$$= 3\left(\frac{2x + 1}{3x + 2}\right)^2 \left[\frac{(3x + 2)(2) - (2x + 1)(3)}{(3x + 2)^2}\right] \qquad (x^2)\ \text{See page 22.}$$

$$= 3\left(\frac{2x + 1}{3x + 2}\right)^2 \left[\frac{6x + 4 - 6x - 3}{(3x + 2)^2}\right]$$

$$= \frac{3(2x + 1)^2}{(3x + 2)^4} \qquad \text{Combine like terms, and simplify.}$$

In particular, the slope of the tangent line to the graph of f at $(0, \frac{1}{8})$ is given by

$$f'(0) = \frac{3(0 + 1)^2}{(0 + 2)^4} = \frac{3}{16}$$

Exploring with TECHNOLOGY

Refer to Example 6.

1. Use a graphing utility to plot the graph of the function f, using the viewing window $[-2, 1] \times [-1, 2]$. Then draw the tangent line to the graph of f at the point $(0, \frac{1}{8})$.

2. For a better picture, repeat part 1 using the viewing window $[-1, 1] \times [-0.1, 0.3]$.

3. Use the numerical differentiation capability of the graphing utility to verify that the slope of the tangent line at $(0, \frac{1}{8})$ is $\frac{3}{16}$.

 APPLIED EXAMPLE 7 Growth in a Health Club's Membership The membership of The Fitness Center, which opened a few years ago, is approximated by the function

$$N(t) = 100(64 + 4t)^{2/3} \qquad (0 \le t \le 52)$$

where $N(t)$ gives the number of members at the beginning of week t.

a. Find $N'(t)$.
b. How fast was the center's membership increasing initially $(t = 0)$? (See Note on page 654.)
c. How fast was the membership increasing at the beginning of the 40th week?
d. What was the membership when the center first opened? At the beginning of the 40th week?

Solution

a. Using the General Power Rule, we obtain

$$N'(t) = \frac{d}{dt}\left[100(64 + 4t)^{2/3}\right]$$

$$= 100\frac{d}{dt}(64 + 4t)^{2/3}$$

$$= 100\left(\frac{2}{3}\right)(64 + 4t)^{-1/3}\frac{d}{dt}(64 + 4t)$$

$$= \frac{200}{3}(64 + 4t)^{-1/3}(4)$$

$$= \frac{800}{3(64 + 4t)^{1/3}}$$

b. The rate at which the membership was increasing when the center first opened is given by

$$N'(0) = \frac{800}{3(64)^{1/3}} \approx 66.7$$

or approximately 67 people per week.

c. The rate at which the membership was increasing at the beginning of the 40th week is given by

$$N'(40) = \frac{800}{3(64 + 160)^{1/3}} \approx 43.9$$

or approximately 44 people per week.

d. The membership when the center first opened is given by

$$N(0) = 100(64)^{2/3} = 100(16)$$

or approximately 1600 people. The membership at the beginning of the 40th week is given by

$$N(40) = 100(64 + 160)^{2/3} \approx 3688.3$$

or approximately 3688 people.

Explore and Discuss

The profit P of a one-product software manufacturer depends on the number of units of its product sold. The manufacturer estimates that it will sell x units of its product per week. Suppose $P = g(x)$ and $x = f(t)$, where g and f are differentiable functions and t is measured in weeks.

1. Write an expression giving the rate of change of the profit with respect to the number of units sold.
2. Write an expression giving the rate of change of the number of units sold per week.
3. Write an expression giving the rate of change of the profit per week.

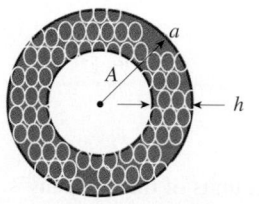

FIGURE 47
Cross section of the aorta

APPLIED EXAMPLE 8 Arteriosclerosis Arteriosclerosis begins during childhood when plaque (soft masses of fatty material) forms in the arterial walls, blocking the flow of blood through the arteries and leading to heart attacks, strokes, and gangrene. Suppose the idealized cross section of the aorta is circular with radius a cm and by year t, the thickness of the plaque (assume that it is uniform) is $h = g(t)$ cm (Figure 47). Then the area of the opening is given by $A = \pi(a - h)^2$ square centimeters (cm²).

Suppose the radius of an individual's artery is 1 cm ($a = 1$) and the thickness of the plaque in cm in year t is given by

$$h = g(t) = 1 - 0.01(10{,}000 - t^2)^{1/2} - 0.001t$$

Since the area of the arterial opening is given by

$$A = f(h) = \pi(1 - h)^2$$

the rate at which A is changing with respect to time is given by

$$\frac{dA}{dt} = \frac{dA}{dh} \cdot \frac{dh}{dt} = f'(h) \cdot g'(t) \qquad \text{By the Chain Rule}$$

$$= 2\pi(1 - h)(-1)\left[-0.01\left(\frac{1}{2}\right)(10{,}000 - t^2)^{-1/2}(-2t) - 0.001\right] \qquad \begin{array}{l}\text{Use the Chain}\\\text{Rule three}\\\text{times.}\end{array}$$

$$= -2\pi(1 - h)\left[\frac{0.01t}{(10{,}000 - t^2)^{1/2}} - 0.001\right]$$

For example, when $t = 50$,

$$h = g(50) = 1 - 0.01(10{,}000 - 2500)^{1/2} - (0.001)(50) \approx 0.08397$$

so that

$$\frac{dA}{dt} = -2\pi(1 - 0.08397)\left[\frac{(0.01)(50)}{\sqrt{10{,}000 - (50)^2}} - 0.001\right] \approx -0.027$$

That is, the area of the arterial opening is decreasing at the rate of 0.03 cm² per year.

Explore and Discuss

Suppose the population P of a certain bacteria culture is given by $P = f(T)$, where T is the temperature of the medium. Further, suppose the temperature T is a function of time t in seconds—that is, $T = g(t)$. Give an interpretation of each of the following quantities:

1. $\dfrac{dP}{dT}$ 2. $\dfrac{dT}{dt}$ 3. $\dfrac{dP}{dt}$ 4. $(f \circ g)(t)$ 5. $f'(g(t))g'(t)$

9.6 Self-Check Exercises

1. Find the derivative of

$$f(x) = -\frac{1}{\sqrt{2x^2 - 1}}$$

2. **FEMALE LIFE EXPECTANCY** Suppose the life expectancy at birth (in years) of a female in a certain country is described by the function

$$g(t) = 50.02(1 + 1.09t)^{0.1} \qquad (0 \le t \le 150)$$

where t is measured in years, with $t = 0$ corresponding to the beginning of 1900.

a. What is the life expectancy at birth of a female born at the beginning of 1980? At the beginning of 2010?
b. How fast is the life expectancy at birth of a female born at any time t changing?

Solutions to Self-Check Exercises 9.6 can be found on page 688.

9.6 Concept Questions

1. In your own words, state the Chain Rule for differentiating the composite function $h(x) = g[f(x)]$.

2. In your own words, state the General Power Rule for differentiating the function $h(x) = [f(x)]^n$, where n is a real number.

3. If $f(t)$ gives the number of units of a certain product sold by a company after t days and $g(x)$ gives the revenue (in

dollars) realized from the sale of x units of the company's products, what does $(g \circ f)'(t)$ describe?

4. Suppose $f(x)$ gives the air temperature in the gondola of a hot-air balloon when it is at an altitude of x ft from the ground and $g(t)$ gives the altitude of the balloon t min after lifting off from the ground. Find a function giving the rate of change of the air temperature in the gondola at time t.

9.6 Exercises

In Exercises 1–48, find the derivative of each function.

1. $f(x) = (2x - 1)^3$

2. $f(x) = (1 - x)^4$

3. $f(x) = (x^2 + 2)^5$

4. $f(t) = 2(t^3 - 1)^5$

5. $f(x) = (2x - x^2)^3$

6. $f(x) = 3(x^3 - x)^4$

7. $f(x) = (2x + 1)^{-2}$

8. $f(t) = \dfrac{1}{2}(2t^2 + t)^{-3}$

9. $f(x) = (x^2 - 4)^{5/2}$

10. $f(t) = (3t^2 - 2t + 1)^{3/2}$

11. $f(x) = \sqrt{3x - 2}$

12. $f(t) = \sqrt{3t^2 - t}$

13. $f(x) = \sqrt[3]{1 - x^2}$

14. $f(x) = \sqrt{2x^2 - 2x + 3}$

15. $f(x) = \dfrac{1}{(2x + 3)^3}$

16. $f(x) = \dfrac{2}{(x^2 - 1)^4}$

17. $f(t) = \dfrac{1}{\sqrt{2t - 4}}$

18. $f(x) = \dfrac{1}{\sqrt{2x^2 - 1}}$

19. $y = \dfrac{1}{(4x^4 + x)^{3/2}}$

20. $f(t) = \dfrac{4}{\sqrt[3]{2t^2 + t}}$

21. $f(x) = (3x^2 + 2x + 1)^{-2}$

22. $f(t) = (5t^3 + 2t^2 - t + 4)^{-3}$

23. $f(x) = (x^2 + 1)^3 - (x^3 + 1)^2$

24. $f(t) = (2t - 1)^4 + (2t + 1)^4$

25. $f(t) = (t^{-1} - t^{-2})^3$

26. $f(v) = (v^{-3} + 4v^{-2})^3$

27. $f(x) = \sqrt{x + 1} + \sqrt{x - 1}$

28. $f(u) = (2u + 1)^{3/2} + (u^2 - 1)^{-3/2}$

29. $f(x) = 2x^2(3 - 4x)^4$ **30.** $h(t) = t^2(3t + 4)^3$

31. $f(x) = (x - 1)^2(2x + 1)^4$

32. $g(u) = \sqrt{u + 1}(1 - 2u^2)^8$

33. $f(x) = \left(\dfrac{x + 3}{x - 2}\right)^3$ **34.** $f(x) = \left(\dfrac{x + 1}{x - 1}\right)^5$

35. $s(t) = \left(\dfrac{t}{2t + 1}\right)^{3/2}$ **36.** $g(s) = \left(s^2 + \dfrac{1}{s}\right)^{3/2}$

37. $g(u) = \sqrt{\dfrac{u + 1}{3u + 2}}$ **38.** $g(x) = \sqrt{\dfrac{2x + 1}{2x - 1}}$

39. $f(x) = \dfrac{x^2}{(x^2 - 1)^4}$ **40.** $g(u) = \dfrac{2u^2}{(u^2 + u)^3}$

41. $h(x) = \dfrac{(3x^2 + 1)^3}{(x^2 - 1)^4}$ **42.** $g(t) = \dfrac{(2t - 1)^2}{(3t + 2)^4}$

43. $f(x) = \dfrac{\sqrt{2x + 1}}{x^2 - 1}$ **44.** $f(t) = \dfrac{4t^2}{\sqrt{2t^2 + 2t - 1}}$

45. $g(t) = \dfrac{\sqrt{t + 1}}{\sqrt{t^2 + 1}}$ **46.** $f(x) = \dfrac{\sqrt{x^2 + 1}}{\sqrt{x^2 - 1}}$

47. $f(x) = (3x + 1)^4(x^2 - x + 1)^3$

48. $g(t) = (2t + 3)^2(3t^2 - 1)^{-3}$

In Exercises 49–54, find $\dfrac{dy}{du}, \dfrac{du}{dx}$, and $\dfrac{dy}{dx}$.

49. $y = u^{4/3}$ and $u = 3x^2 - 1$

50. $y = \sqrt{u}$ and $u = 7x - 2x^2$

51. $y = u^{-2/3}$ and $u = 2x^3 - x + 1$

52. $y = 2u^2 + 1$ and $u = x^2 + 1$

53. $y = \sqrt{u} + \dfrac{1}{\sqrt{u}}$ and $u = x^3 - x$

54. $y = \dfrac{1}{u}$ and $u = \sqrt{x} + 1$

55. If $g(x) = f(2x + 1)$, what is $g'(x)$?

56. If $h(x) = f(-x^3)$, what is $h'(x)$?

57. Suppose $F(x) = g(f(x))$ and $f(2) = 3, f'(2) = -3$, $g(3) = 5$, and $g'(3) = 4$. Find $F'(2)$.

58. Suppose $h = f \circ g$. Find $h'(0)$ given that $f(0) = 6$, $f'(5) = -2, g(0) = 5$, and $g'(0) = 3$.

59. Suppose $F(x) = f(x^2 + 1)$. Find $F'(1)$ if $f'(2) = 3$.

60. Let $F(x) = f(f(x))$. Does it follow that $F'(x) = [f'(x)]^2$? Hint: Let $f(x) = x^2$.

61. Suppose $h = g \circ f$. Does it follow that $h' = g' \circ f'$? Hint: Let $f(x) = x$ and $g(x) = x^2$.

62. Suppose $h = f \circ g$. Show that $h' = (f' \circ g)g'$.

In Exercises 63–66, find an equation of the tangent line to the graph of the function at the given point.

63. $f(x) = (1 - x)(x^2 - 1)^2; (2, -9)$

64. $f(x) = \left(\dfrac{x + 1}{x - 1}\right)^2; (3, 4)$

65. $f(x) = x\sqrt{2x^2 + 7}; (3, 15)$

66. $f(x) = \dfrac{8}{\sqrt{x^2 + 6x}}; (2, 2)$

67. TELEVISION VIEWERSHIP The number of viewers of a television series introduced several years ago is approximated by the function

$$N(t) = (60 + 2t)^{2/3} \qquad (1 \le t \le 26)$$

where $N(t)$ (measured in millions) denotes the number of weekly viewers of the series in the tth week. Find the rate of increase of the weekly audience at the end of week 2 and at the end of week 12. How many viewers were there in week 2? In week 24?

68. DIGITAL INFORMATION CREATION The amount of digital information created each month globally, t months after the beginning of 2008, is approximately

$$f(t) = 400\left(\dfrac{t}{12} + 1\right)^{1.09} \qquad (0 \le t \le 36)$$

billion gigabytes.
  a. How many billion gigabytes of information were created at the beginning of 2008?
  b. How fast was digital information being created at the beginning of 2010 ($t = 24$)?
Source: MIT Technology Review.

69. OUTSOURCING OF JOBS The cumulative number of jobs outsourced overseas by U.S.-based multinational companies in year t from 2005 ($t = 0$) through 2009 is approximated by

$$N(t) = -0.05(t + 1.1)^{2.2} + 0.7t + 0.9 \qquad (0 \le t \le 4)$$

where $N(t)$ is measured in millions. How fast was the number of jobs that were outsourced changing in 2008 ($t = 3$)?
Source: Forrester Research.

70. SELLING PRICE OF DVD RECORDERS The rise of digital video and the improvement to the DVD format are some of the reasons why the average selling price of stand-alone DVD recorders has dropped in recent years. The function

$$A(t) = \dfrac{699}{(t + 1)^{0.94}} \qquad (0 \le t \le 13)$$

gives the projected average selling price (in dollars) of stand-alone DVD recorders in year t, where $t = 0$ corresponds to the beginning of 2002. How fast was the average selling price of stand-alone DVD recorders falling at the beginning of 2002? How fast was it falling at the beginning of 2012?
Source: Consumer Electronics Association.

71. **BRAIN CANCER SURVIVAL RATE** Glioblastoma is the most common and most deadly of brain tumors, and it kills most patients in a little over a year. The probability of survival for patients with a glioblastoma t years after diagnosis is approximated by the function

$$P(t) = \frac{100}{(1 + 0.14t)^{9.2}} \qquad (0 \le t \le 10)$$

where P is the percent of surviving patients.
a. Compute $P(1)$ and $P(2)$, and interpret your results.
b. Compute $P'(1)$ and $P'(2)$, and interpret your results.
Source: National Cancer Institute.

72. **AGING POPULATION** The population of Americans age 55 and older as a percentage of the total population is approximated by the function

$$f(t) = 10.72(0.9t + 10)^{0.3} \qquad (0 \le t \le 20)$$

where t is measured in years, with $t = 0$ corresponding to the year 2000. At what rate was the percentage of Americans age 55 and older changing at the beginning of 2000? At what rate was the percentage of Americans age 55 and older changing in 2015? What will the percentage of the population of Americans age 55 and older be in 2015? (See Note on page 654.)
Source: U.S. Census Bureau.

73. **CONCENTRATION OF CARBON MONOXIDE (CO) IN THE AIR** According to a joint study conducted by Oxnard's Environmental Management Department and a state government agency, the concentration of CO in the air due to automobile exhaust t years from now is given by

$$C(t) = 0.01(0.2t^2 + 4t + 64)^{2/3}$$

parts per million.
a. Find the rate at which the level of CO is changing with respect to time.
b. Find the rate at which the level of CO will be changing 5 years from now.

74. **CONTINUING EDUCATION ENROLLMENT** The registrar of Kellogg University estimates that the total student enrollment in the Continuing Education division will be given by

$$N(t) = -\frac{20,000}{\sqrt{1 + 0.2t}} + 21,000$$

where $N(t)$ denotes the number of students enrolled in the division t years from now. Find an expression for $N'(t)$. How fast will the student enrollment be increasing 1 year from now? How fast will it be increasing 5 years from now?

75. **AIR POLLUTION** According to the South Coast Air Quality Management District, the level of nitrogen dioxide, a brown gas that impairs breathing, present in the atmosphere on a certain May day in downtown Los Angeles is approximated by

$$A(t) = 0.03t^3(t - 7)^4 + 60.2 \qquad (0 \le t \le 7)$$

where $A(t)$ is measured in pollutant standard index and t is measured in hours, with $t = 0$ corresponding to 7 A.M.
a. Find $A'(t)$.
b. Find $A'(1)$, $A'(3)$, and $A'(4)$, and interpret your results.
Source: Los Angeles Times.

76. **EFFECT OF LUXURY TAX ON CONSUMPTION** Government economists of a developing country determined that the purchase of imported perfume is related to a proposed "luxury tax" by the formula

$$N(x) = \sqrt{10,000 - 40x - 0.02x^2} \qquad (0 \le x \le 200)$$

where $N(x)$ measures the percentage of normal consumption of perfume when a "luxury tax" of $x\%$ is imposed on it. Find the rate of change of $N(x)$ for taxes of 10%, 100%, and 150%.

77. **PULSE RATE OF AN ATHLETE** The pulse rate (the number of heartbeats per minute) of a long-distance runner t sec after leaving the starting line is given by

$$P(t) = \frac{300\sqrt{\frac{1}{2}t^2 + 2t + 25}}{t + 25} \qquad (t \ge 0)$$

Compute $P'(t)$. How fast is the athlete's pulse rate increasing 10 sec, 60 sec, and 2 min into the run? What is her pulse rate 2 min into the run?

78. **THURSTONE LEARNING MODEL** Psychologist L. L. Thurstone suggested the following relationship between learning time T and the length of a list n:

$$T = f(n) = An\sqrt{n - b}$$

where A and b are constants that depend on the person and the task.
a. Compute dT/dn, and interpret your result.
b. For a certain person and a certain task, suppose $A = 4$ and $b = 4$. Compute $f'(13)$ and $f'(29)$, and interpret your results.

79. **OIL SPILLS** In calm waters, the oil spilling from the ruptured hull of a grounded tanker spreads in all directions. Assuming that the area polluted is a circle and that its radius is increasing at a rate of 2 ft/sec, determine how fast the area is increasing when the radius of the circle is 40 ft.

80. **ARTERIOSCLEROSIS** Refer to Example 8, page 683. Suppose the radius of an individual's artery is 1 cm and the thickness of the plaque (in centimeters) t years from now is given by

$$h = g(t) = \frac{0.5t^2}{t^2 + 10} \qquad (0 \le t \le 10)$$

How fast will the arterial opening be decreasing 5 years from now?

81. **EFFECT OF HOUSING STARTS ON JOBS** The president of a major housing construction firm claims that the number of construction jobs created is given by

$$N(x) = 1.42x$$

where x denotes the number of housing starts. Suppose the number of housing starts in the next t months is expected to be

$$x(t) = \frac{7t^2 + 140t + 700}{3t^2 + 80t + 550}$$

million units/year. Find an expression that gives the rate at which the number of construction jobs will be created t months from now. At what rate will construction jobs be created 1 year from now?

82. **HOTEL OCCUPANCY RATES** The occupancy rate of the all-suite Wonderland Hotel, located near an amusement park, is given by the function

$$r(t) = \frac{10}{81}t^3 - \frac{10}{3}t^2 + \frac{200}{9}t + 56.2 \qquad (0 \le t \le 12)$$

where t is measured in months, with $t = 0$ corresponding to the beginning of January. Management has estimated that the monthly revenue (in thousands of dollars per month) is approximated by the function

$$R(r) = -\frac{3}{5000}r^3 + \frac{9}{50}r^2 \qquad (0 \le r \le 100)$$

where r is the occupancy rate.
a. Find an expression that gives the rate of change of Wonderland's occupancy rate with respect to time.
b. Find an expression that gives the rate of change of Wonderland's monthly revenue with respect to the occupancy rate.
c. What is the rate of change of Wonderland's monthly revenue with respect to time at the beginning of January? At the beginning of July?
Hint: Use the Chain Rule to find $R'(r(0))r'(0)$ and $R'(r(6))r'(6)$.

83. **DEMAND FOR TABLET COMPUTERS** The quantity demanded per month, x, of the Zephyr tablet computer is related to the average unit price, p (in dollars), of tablet computers by the equation

$$x = f(p) = \frac{100}{9}\sqrt{810,000 - p^2}$$

It is estimated that t months from now, the average price of a tablet computer will be given by

$$p(t) = \frac{400}{1 + \frac{1}{8}\sqrt{t}} + 200 \qquad (0 \le t \le 60)$$

dollars. Find the rate at which the quantity demanded per month of the tablet computers will be changing 16 months from now.

84. **CRUISE SHIP BOOKINGS** The management of Cruise World, operators of Caribbean luxury cruises, expects that the percentage of young adults booking passage on their cruises in the years ahead will rise dramatically. They

have constructed the following model, which gives the percentage of young adult passengers in year t:

$$p = f(t) = 50\left(\frac{t^2 + 2t + 4}{t^2 + 4t + 8}\right) \qquad (0 \le t \le 5)$$

Young adults normally pick shorter cruises and generally spend less on their passage. The following model gives an approximation of the average amount of money R (in dollars) spent per passenger on a cruise when the percentage of young adults is p:

$$R(p) = 1000\left(\frac{p + 4}{p + 2}\right)$$

Find the rate at which the amount of money spent per passenger on a cruise will be changing 2 years from now.

In Exercises 85–88, determine whether the statement is true or false. If it is true, explain why it is true. If it is false, give an example to show why it is false.

85. If f and g are differentiable and $h = f \circ g$, then $h'(x) = f'[g(x)]g'(x)$.

86. If f is differentiable and c is a constant, then

$$\frac{d}{dx}[f(cx)] = cf'(cx)$$

87. If f is differentiable and $f(x) > 0$, then

$$\frac{d}{dx}\sqrt{f(x)} = \frac{f'(x)}{2\sqrt{f(x)}}$$

88. If f is differentiable, then

$$\frac{d}{dx}\left[f\left(\frac{1}{x}\right)\right] = f'\left(\frac{1}{x}\right)$$

89. In Section 9.4, we proved that

$$\frac{d}{dx}(x^n) = nx^{n-1}$$

for the special case when $n = 2$. Use the Chain Rule to show that

$$\frac{d}{dx}(x^{1/n}) = \frac{1}{n}x^{1/n - 1}$$

for any nonzero integer n, assuming that $f(x) = x^{1/n}$ is differentiable.
Hint: Let $f(x) = x^{1/n}$, so that $[f(x)]^n = x$. Differentiate both sides with respect to x.

90. With the aid of Exercise 89, prove that

$$\frac{d}{dx}(x^r) = rx^{r-1}$$

for every rational number r.
Hint: Let $r = m/n$, where m and n are integers, with $n \ne 0$, and write $x^r = (x^m)^{1/n}$.

9.6 Solutions to Self-Check Exercises

1. Rewriting, we have

$$f(x) = -(2x^2 - 1)^{-1/2}$$

Using the General Power Rule, we find

$$f'(x) = -\frac{d}{dx}(2x^2 - 1)^{-1/2}$$

$$= -\left(-\frac{1}{2}\right)(2x^2 - 1)^{-3/2}\frac{d}{dx}(2x^2 - 1)$$

$$= \frac{1}{2}(2x^2 - 1)^{-3/2}(4x)$$

$$= \frac{2x}{(2x^2 - 1)^{3/2}}$$

2. a. The life expectancy at birth of a female born at the beginning of 1980 is given by

$$g(80) = 50.02[1 + 1.09(80)]^{0.1} \approx 78.29$$

or approximately 78 years. Similarly, the life expectancy at birth of a female born at the beginning of the year 2010 is given by

$$g(110) = 50.02[1 + 1.09(110)]^{0.1} \approx 80.80$$

or approximately 81 years.

b. The rate of change of the life expectancy at birth of a female born at any time t is given by $g'(t)$. Using the General Power Rule, we have

$$g'(t) = 50.02\frac{d}{dt}(1 + 1.09t)^{0.1}$$

$$= (50.02)(0.1)(1 + 1.09t)^{-0.9}\frac{d}{dt}(1 + 1.09t)$$

$$= (50.02)(0.1)(1.09)(1 + 1.09t)^{-0.9}$$

$$= 5.45218(1 + 1.09t)^{-0.9}$$

$$= \frac{5.45218}{(1 + 1.09t)^{0.9}}$$

USING TECHNOLOGY   Finding the Derivative of a Composite Function

EXAMPLE 1 Find the rate of change of $f(x) = \sqrt{x}(1 + 0.02x^2)^{3/2}$ when $x = 2.1$.

Solution Using the numerical derivative operation of a graphing utility, we find

$$f'(2.1) = 0.5821463392$$

or approximately 0.58 unit per unit change in x. (See Figure T1.)

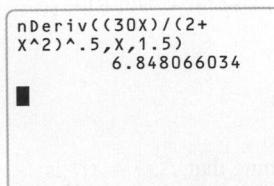

```
nDeriv(X^.5(1+.0
2X^2)^1.5,X,2.1)
    .5821463392
■
```

FIGURE T1
The TI-83/84 numerical derivative screen for computing $f'(2.1)$

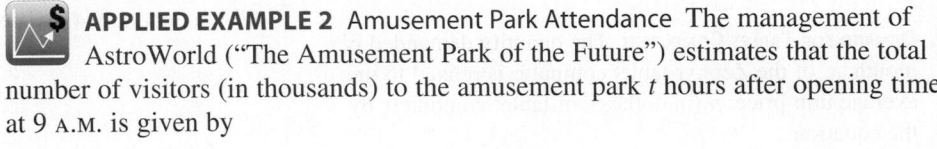

APPLIED EXAMPLE 2 Amusement Park Attendance The management of AstroWorld ("The Amusement Park of the Future") estimates that the total number of visitors (in thousands) to the amusement park t hours after opening time at 9 A.M. is given by

$$N(t) = \frac{30t}{\sqrt{2 + t^2}}$$

What is the rate at which visitors are admitted to the amusement park at 10:30 A.M.?

Solution Using the numerical derivative operation of a graphing utility, we find

$$N'(1.5) \approx 6.8481$$

or approximately 6848 visitors per hour. (See Figure T2.)

```
nDeriv((30X)/(2+
X^2)^.5,X,1.5)
        6.848066034
■
```

FIGURE T2
The TI-83/84 numerical derivative screen for computing $N'(1.5)$

TECHNOLOGY EXERCISES

In Exercises 1–6, use the numerical derivative operation to find the rate of change of f at the given value of x. Give your answer accurate to four decimal places.

1. $f(x) = \sqrt{x^2 - x^4}; x = 0.5$

2. $f(x) = x - \sqrt{1 - x^2}; x = 0.4$

3. $f(x) = x\sqrt{1 - x^2}; x = 0.2$

4. $f(x) = (x + \sqrt{x^2 + 4})^{3/2}; x = 1$

5. $f(x) = \dfrac{\sqrt{1 + x^2}}{x^3 + 2}; x = -1$

6. $f(x) = \dfrac{x^3}{1 + (1 + x^2)^{3/2}}; x = 3$

7. WATCHING TV ON SMARTPHONES The number of people watching TV on smartphones (in millions) is approximated by

$$N(t) = 11.9\sqrt{1 + 0.91t} \qquad (0 \le t \le 4)$$

where t is measured in years, with $t = 0$ corresponding to the beginning of 2007.
a. What was the rate of change of the number of people watching TV on smartphones at the beginning of 2007?
b. What was the rate of change of the number of people watching TV on smartphones expected to be at the beginning of 2011?
Source: IDC, U.S. forecast.

8. ACCUMULATION YEARS Demographic studies pertaining to investors are of particular importance to financial institutions. People from their mid-40s to their mid-50s are in the prime investing years. The function

$$N(t) = 34.4(1 + 0.32125t)^{0.15} \qquad (0 \le t \le 12)$$

gives the projected number of people in this age group in the United States (in millions) in year t, where $t = 0$ corresponds to the beginning of 1996.
a. How large was this segment of the population projected to be at the beginning of 2008?
b. How fast was this segment of the population growing at the beginning of 2008?
Source: U.S. Census Bureau.

9.7 Differentiation of Exponential and Logarithmic Functions

The Derivative of the Exponential Function

To study the effects of budget deficit-reduction plans at different income levels, it is important to know the income distribution of American households. Based on data from the U.S. Census Bureau, the graph of f shown in Figure 48 gives the percentage of American households y as a function of their annual income x (in thousands of dollars) in 2010.

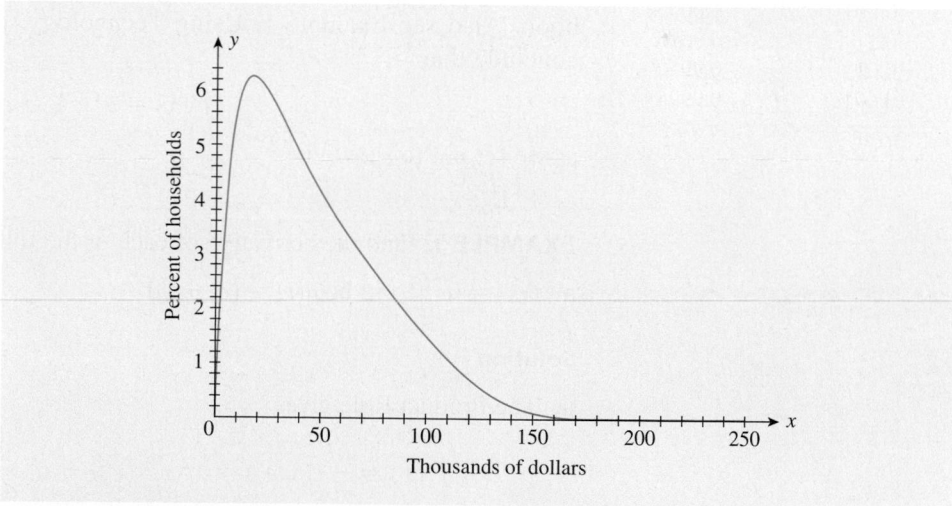

FIGURE **48**
The graph of f shows the percentage of households versus their annual income.
Source: U.S. Census Bureau.

Observe that the graph of f rises very quickly and then tapers off. From the graph of f, you can see that the bulk of American households earned less than \$150,000 per year. (We will refer to this model again in Using Technology at the end of this section.)

To analyze mathematical models involving exponential and logarithmic functions in greater detail, we need to develop rules for computing the derivative of these functions. We begin by looking at the rule for computing the derivative of the exponential function.

> **Rule 8: Derivative of the Exponential Function**
>
> $$\frac{d}{dx}(e^x) = e^x$$

Thus, the derivative of the exponential function with base e is equal to the function itself. To demonstrate the validity of this rule, we compute

$$
\begin{aligned}
f'(x) &= \lim_{h \to 0} \frac{f(x+h) - f(x)}{h} \\[2mm]
&= \lim_{h \to 0} \frac{e^{x+h} - e^x}{h} \\[2mm]
&= \lim_{h \to 0} \frac{e^x(e^h - 1)}{h} \qquad \text{Write } e^{x+h} = e^x e^h \text{ and factor.} \\[2mm]
&= e^x \lim_{h \to 0} \frac{e^h - 1}{h} \qquad \text{Why?}
\end{aligned}
$$

To evaluate

$$\lim_{h \to 0} \frac{e^h - 1}{h}$$

let's refer to Table 5, which is constructed with the aid of a calculator. From the table, we see that

$$\lim_{h \to 0} \frac{e^h - 1}{h} = 1$$

(Although a rigorous proof of this fact is possible, it is beyond the scope of this book. Also see Example 1, Using Technology, page 699.) Using this result, we conclude that

$$f'(x) = e^x \cdot 1 = e^x$$

as we set out to show.

TABLE 5	
h	$\dfrac{e^h - 1}{h}$
0.1	1.0517
0.01	1.0050
0.001	1.0005
−0.1	0.9516
−0.01	0.9950
−0.001	0.9995

EXAMPLE 1 Find the derivative of each of the following functions:

a. $f(x) = x^2 e^x$ **b.** $g(t) = (e^t + 2)^{3/2}$

Solution

a. The Product Rule gives

$$
\begin{aligned}
f'(x) &= \frac{d}{dx}(x^2 e^x) = x^2 \frac{d}{dx}(e^x) + e^x \frac{d}{dx}(x^2) \\[2mm]
&= x^2 e^x + e^x (2x) \\[2mm]
&= xe^x(x + 2) \qquad\qquad\qquad (x^2) \text{ See page 18.}
\end{aligned}
$$

b. Using the General Power Rule, we find

$$
\begin{aligned}
g'(t) &= \frac{3}{2}(e^t + 2)^{1/2} \frac{d}{dt}(e^t + 2) \\[2mm]
&= \frac{3}{2}(e^t + 2)^{1/2} e^t = \frac{3}{2} e^t (e^t + 2)^{1/2}
\end{aligned}
$$

Exploring with TECHNOLOGY

Consider the exponential function $f(x) = b^x (b > 0, b \neq 1)$.

1. Use the definition of the derivative of a function to show that

$$f'(x) = b^x \cdot \lim_{h \to 0} \frac{b^h - 1}{h}$$

2. Use the result of part 1 to show that

$$\frac{d}{dx}(2^x) = 2^x \cdot \lim_{h \to 0} \frac{2^h - 1}{h}$$

$$\frac{d}{dx}(3^x) = 3^x \cdot \lim_{h \to 0} \frac{3^h - 1}{h}$$

3. Use the technique in Using Technology, page 699, to show that (to two decimal places)

$$\lim_{h \to 0} \frac{2^h - 1}{h} = 0.69$$

and

$$\lim_{h \to 0} \frac{3^h - 1}{h} = 1.10$$

4. Conclude from the results of parts 2 and 3 that

$$\frac{d}{dx}(2^x) \approx (0.69)2^x$$

and

$$\frac{d}{dx}(3^x) \approx (1.10)3^x$$

Thus,

$$\frac{d}{dx}(b^x) = k \cdot b^x$$

where k is an appropriate constant.

5. The results of part 4 suggest that, for convenience, we pick the base b, where $2 < b < 3$, so that $k = 1$. This value of b is $e \approx 2.718281828. \ldots$ Thus,

$$\frac{d}{dx}(e^x) = e^x$$

This is why we prefer to work with the exponential function $f(x) = e^x$.

Applying the Chain Rule to Exponential Functions

To enlarge the class of exponential functions to be differentiated, we appeal to the Chain Rule to obtain the following rule for differentiating composite functions of the form $h(x) = e^{f(x)}$. An example of such a function is $h(x) = e^{x^2 - 2x}$. Here, $f(x) = x^2 - 2x$.

Rule 9: The Chain Rule for Exponential Functions

If $f(x)$ is a differentiable function, then

$$\frac{d}{dx}(e^{f(x)}) = e^{f(x)}f'(x)$$

To see this, observe that if $h(x) = g[f(x)]$, where $g(x) = e^x$, then by virtue of the Chain Rule,

$$h'(x) = g'[f(x)]f'(x) = e^{f(x)}f'(x)$$

since $g'(x) = e^x$.

As an aid to remembering the Chain Rule for Exponential Functions, observe that it has the following form:

$$\frac{d}{dx}\left(e^{f(x)}\right) = e^{f(x)} \cdot \text{derivative of the exponent}$$

$$\underset{\text{Same}}{\underbrace{\qquad}}$$

EXAMPLE 2 Find the derivative of each of the following functions:

a. $f(x) = e^{2x}$ **b.** $y = e^{-3x}$ **c.** $g(t) = e^{2t^2 + t}$

Solution

a. $f'(x) = e^{2x}\dfrac{d}{dx}(2x) = e^{2x} \cdot 2 = 2e^{2x}$

b. $\dfrac{dy}{dx} = e^{-3x}\dfrac{d}{dx}(-3x) = -3e^{-3x}$

c. $g'(t) = e^{2t^2 + t} \cdot \dfrac{d}{dt}(2t^2 + t) = (4t + 1)e^{2t^2 + t}$

EXAMPLE 3 Differentiate the function $y = xe^{-2x}$.

Solution   Using the Product Rule, followed by the Chain Rule, we find

$$\frac{dy}{dx} = x\frac{d}{dx}\left(e^{-2x}\right) + e^{-2x}\frac{d}{dx}(x)$$

$$= xe^{-2x}\frac{d}{dx}(-2x) + e^{-2x} \qquad \text{Use the Chain Rule on the first term.}$$

$$= -2xe^{-2x} + e^{-2x}$$

$$= e^{-2x}(1 - 2x)$$

EXAMPLE 4 Differentiate the function $g(t) = \dfrac{e^t}{e^t + e^{-t}}$.

Solution   Using the Quotient Rule, followed by the Chain Rule, we find

$$g'(t) = \frac{(e^t + e^{-t})\dfrac{d}{dt}(e^t) - e^t\dfrac{d}{dt}(e^t + e^{-t})}{(e^t + e^{-t})^2}$$

$$= \frac{(e^t + e^{-t})e^t - e^t(e^t - e^{-t})}{(e^t + e^{-t})^2} \qquad (x) \text{ See page 21.}$$

$$= \frac{e^{2t} + 1 - e^{2t} + 1}{(e^t + e^{-t})^2} \qquad e^0 = 1$$

$$= \frac{2}{(e^t + e^{-t})^2}$$

EXAMPLE 5 In Section 3.3, we discussed some practical applications of the exponential function

$$Q(t) = Q_0 e^{kt}$$

where Q_0 and k are positive constants and $t \in [0, \infty)$. A quantity $Q(t)$ growing according to this law experiences exponential growth. Show that for a quantity $Q(t)$ experiencing exponential growth, the rate of growth of the quantity, $Q'(t)$, at any time t is directly proportional to the amount of the quantity present.

Solution   Using the Chain Rule for Exponential Functions, we compute the derivative Q' of the function Q. Thus,

$$Q'(t) = Q_0 e^{kt} \frac{d}{dt} (kt)$$
$$= Q_0 e^{kt}(k)$$
$$= kQ_0 e^{kt}$$
$$= kQ(t) \qquad Q(t) = Q_0 e^{kt}$$

which is the desired conclusion.

The Derivative of ln x

Let's now turn our attention to the differentiation of logarithmic functions.

> **Rule 10: Derivative of ln x**
>
> $$\frac{d}{dx} \ln|x| = \frac{1}{x} \qquad (x \neq 0)$$

To derive Rule 10, suppose $x > 0$ and write $f(x) = \ln x$ in the equivalent form

$$x = e^{f(x)}$$

Differentiating both sides of the equation with respect to x, we find, using the Chain Rule,

$$1 = e^{f(x)} \cdot f'(x)$$

from which we see that

$$f'(x) = \frac{1}{e^{f(x)}}$$

or, since $e^{f(x)} = x$,

$$f'(x) = \frac{1}{x}$$

as we set out to show. You are asked to prove the rule for the case $x < 0$ in Exercise 89, page 699.

EXAMPLE 6 Find the derivative of each function:

a. $f(x) = x \ln x$ **b.** $g(x) = \dfrac{\ln x}{x}$

Solution

a. Using the Product Rule, we obtain

$$f'(x) = \frac{d}{dx} (x \ln x) = x \frac{d}{dx} (\ln x) + (\ln x) \frac{d}{dx} (x)$$

$$= x\left(\frac{1}{x}\right) + \ln x = 1 + \ln x$$

b. Using the Quotient Rule, we obtain

$$g'(x) = \frac{x \dfrac{d}{dx}(\ln x) - (\ln x)\dfrac{d}{dx}(x)}{x^2} = \frac{x\left(\dfrac{1}{x}\right) - \ln x}{x^2} = \frac{1 - \ln x}{x^2} \qquad ■$$

Explore and Discuss

You can derive the formula for the derivative of $f(x) = \ln x$ directly from the definition of the derivative, as follows.

1. Show that

$$f'(x) = \lim_{h \to 0} \frac{f(x+h) - f(x)}{h} = \lim_{h \to 0} \ln\left(1 + \frac{h}{x}\right)^{1/h}$$

2. Put $m = x/h$, and note that $m \to \infty$ as $h \to 0$. Then, $f'(x)$ can be written in the form

$$f'(x) = \lim_{m \to \infty} \ln\left(1 + \frac{1}{m}\right)^{m/x}$$

3. Finally, use both the fact that the natural logarithmic function is continuous and the definition of the number e to show that

$$f'(x) = \frac{1}{x} \ln\left[\lim_{m \to \infty}\left(1 + \frac{1}{m}\right)^{m}\right] = \frac{1}{x}$$

The Chain Rule and Logarithmic Functions

To enlarge the class of logarithmic functions to be differentiated, we appeal once again to the Chain Rule to obtain the following rule for differentiating composite functions of the form $h(x) = \ln f(x)$, where $f(x)$ is assumed to be a positive differentiable function.

> **Rule 11: Derivative of ln $f(x)$**
>
> If $f(x)$ is a differentiable function, then
>
> $$\frac{d}{dx}[\ln f(x)] = \frac{f'(x)}{f(x)} \qquad (f(x) > 0)$$

To see this, observe that $h(x) = g[f(x)]$, where $g(x) = \ln x$ $(x > 0)$. Since $g'(x) = 1/x$, we have, using the Chain Rule,

$$h'(x) = g'[f(x)]f'(x)$$
$$= \frac{1}{f(x)}f'(x) = \frac{f'(x)}{f(x)}$$

Observe that in the special case $f(x) = x$, $h(x) = \ln x$, so the derivative of h is, by Rule 10, given by $h'(x) = 1/x$.

EXAMPLE 7 Find the derivative of the function $f(x) = \ln(x^2 + 1)$.

Solution Using Rule 11, we see immediately that

$$f'(x) = \frac{\dfrac{d}{dx}(x^2 + 1)}{x^2 + 1} = \frac{2x}{x^2 + 1} \qquad ■$$

When we differentiate functions involving logarithms, the rules of logarithms may be used to advantage, as shown in Examples 8 and 9.

EXAMPLE 8 Differentiate the function $y = \ln[(x^2 + 1)(x^3 + 2)^6]$.

Solution We first rewrite the given function using the properties of logarithms:

$$
\begin{aligned}
y &= \ln[(x^2 + 1)(x^3 + 2)^6] \\
&= \ln(x^2 + 1) + \ln(x^3 + 2)^6 && \ln mn = \ln m + \ln n \\
&= \ln(x^2 + 1) + 6\ln(x^3 + 2) && \ln m^n = n \ln m
\end{aligned}
$$

Differentiating and using Rule 11, we obtain

$$
\begin{aligned}
y' &= \frac{\dfrac{d}{dx}(x^2 + 1)}{x^2 + 1} + \frac{6\dfrac{d}{dx}(x^3 + 2)}{x^3 + 2} \\
&= \frac{2x}{x^2 + 1} + \frac{6(3x^2)}{x^3 + 2} = \frac{2x}{x^2 + 1} + \frac{18x^2}{x^3 + 2}
\end{aligned}
$$

Exploring with TECHNOLOGY

Use a graphing utility to plot the graphs of $f(x) = \ln x$; its first derivative function, $f'(x) = 1/x$; and its second derivative function $f''(x) = -1/x^2$, using the same viewing window $[0, 4] \times [-3, 3]$.

1. What can you say about the graph of f'?
2. What can you say about the graph of f''?

EXAMPLE 9 Find the derivative of the function $g(t) = \ln(t^2 e^{-t^2})$.

Solution Here again, to save a lot of work, we first simplify the given expression using the properties of logarithms. We have

$$
\begin{aligned}
g(t) &= \ln(t^2 e^{-t^2}) \\
&= \ln t^2 + \ln e^{-t^2} && \ln mn = \ln m + \ln n \\
&= 2\ln t - t^2 && \ln m^n = n \ln m \quad \text{and} \quad \ln e = 1
\end{aligned}
$$

Therefore,

$$
g'(t) = \frac{2}{t} - 2t = \frac{2(1 - t^2)}{t}
$$

Example 10 involves finding the rate of change of an exponential function.

APPLIED EXAMPLE 10 Asset Depreciation An industrial asset is being depreciated at a rate so that its book value t years from now will be

$$
V(t) = 50{,}000e^{-0.4t}
$$

dollars. How fast will the book value of the asset be changing 3 years from now?

Solution The rate of change of the book value of the asset t years from now is

$$
\begin{aligned}
V'(t) &= 50{,}000\,\frac{d}{dt}e^{-0.4t} \\
&= 50{,}000(-0.4)e^{-0.4t} = -20{,}000e^{-0.4t}
\end{aligned}
$$

Therefore, 3 years from now, the book value of the asset will be changing at the rate of

$$V'(3) = -20{,}000e^{-0.4(3)} = -20{,}000e^{-1.2} \approx -6023.88$$

that is, decreasing at the rate of approximately $6024 per year.　▪

9.7　Self-Check Exercises

1. Find the first and second derivatives of $f(x) = xe^{-x}$.

2. Find an equation of the tangent line to the graph of $f(x) = x \ln(2x + 3)$ at the point $(-1, 0)$.

Solutions to Self-Check Exercises 9.7 can be found on page 699.

9.7　Concept Questions

1. State the rule for differentiating (a) $f(x) = e^x$ and (b) $g(x) = e^{f(x)}$, where f is a differentiable function.

2. Let $f(x) = e^{kx}$.
   a. Compute $f'(x)$.

   b. Use the result to deduce the sign of f' for the case $k > 0$ and the case $k < 0$.

3. State the rule for differentiating (a) $f(x) = \ln|x| (x \neq 0)$, and $g(x) = \ln f(x) [f(x) > 0]$, where f is a differentiable function.

9.7　Exercises

In Exercises 1–28, find the derivative of the function.

1. $f(x) = e^{3x}$

2. $f(x) = 3e^x$

3. $g(t) = e^{-t}$

4. $f(x) = e^{-2x}$

5. $f(x) = e^x + x^2$

6. $f(x) = 2e^x - x^2$

7. $f(x) = x^3 e^x$

8. $f(u) = u^2 e^{-u}$

9. $f(x) = \dfrac{e^x}{x}$

10. $f(x) = \dfrac{x}{e^x}$

11. $f(x) = 3(e^x + e^{-x})$

12. $f(x) = \dfrac{e^x + e^{-x}}{2}$

13. $f(w) = \dfrac{e^w + 2}{e^w}$

14. $f(x) = \dfrac{e^x}{e^x + 1}$

15. $f(x) = 2e^{3x-1}$

16. $f(t) = 4e^{3t+2}$

17. $h(x) = e^{-x^2}$

18. $f(x) = e^{x^2-1}$

19. $f(x) = 3e^{1/x}$

20. $f(x) = e^{1/(2x)}$

21. $f(x) = (e^x + 1)^{25}$

22. $f(x) = (4 - e^{-3x})^3$

23. $f(x) = e^{\sqrt{x}}$

24. $f(t) = -e^{-\sqrt{2t}}$

25. $f(x) = (x - 1)e^{3x+2}$

26. $f(s) = (s^2 + 1)e^{-s^2}$

27. $f(x) = \dfrac{e^x - 1}{e^x + 1}$

28. $g(t) = \dfrac{e^{-t}}{1 + t^2}$

In Exercises 29–32, find the second derivative of the function.

29. $f(x) = e^{-4x} + e^{3x}$

30. $f(t) = 3e^{-2t} - 5e^{-t}$

31. $f(x) = 2xe^{3x}$

32. $f(t) = t^2 e^{-2t}$

33. Find an equation of the tangent line to the graph of $y = e^{2x-3}$ at the point $\left(\frac{3}{2}, 1\right)$.

34. Find an equation of the tangent line to the graph of $y = e^{-x^2}$ at the point $\left(1, \dfrac{1}{e}\right)$.

In Exercises 35–62, find the derivative of the function.

35. $f(x) = 5 \ln x$

36. $f(x) = \ln 5x$

37. $f(x) = \ln(x + 1)$

38. $g(x) = \ln(2x + 1)$

39. $f(x) = \ln x^8$

40. $h(t) = 2 \ln t^5$

41. $f(x) = \ln \sqrt{x}$

42. $f(x) = \ln(\sqrt{x} + 1)$

43. $f(x) = \ln \dfrac{1}{x^2}$

44. $f(x) = \ln \dfrac{1}{2x^3}$

45. $f(x) = \ln(4x^2 - 5x + 3)$

46. $f(x) = \ln(3x^2 - 2x + 1)$

47. $f(x) = \ln \dfrac{2x}{x + 1}$

48. $f(x) = \ln \dfrac{x + 1}{x - 1}$

49. $f(x) = x^2 \ln x$

50. $f(x) = 3x^2 \ln 2x$

51. $f(x) = \dfrac{2 \ln x}{x}$ **52.** $f(x) = \dfrac{3 \ln x}{x^2}$

53. $f(u) = \ln(u - 2)^3$ **54.** $f(x) = \ln(x^3 - 3)^4$

55. $f(x) = \sqrt{\ln x}$ **56.** $f(x) = \sqrt{\ln x + x}$

57. $f(x) = (\ln x)^2$ **58.** $f(x) = 2(\ln x)^{3/2}$

59. $f(x) = \ln(x^3 + 1)$ **60.** $f(x) = \ln\sqrt{x^2 - 4}$

61. $f(x) = e^x \ln x$ **62.** $f(x) = e^x \ln\sqrt{x + 3}$

In Exercises 63–66, find the second derivative of the function.

63. $f(x) = \ln 2x$ **64.** $f(x) = \ln(x + 5)$

65. $f(x) = \ln(x^2 + 2)$ **66.** $f(x) = (\ln x)^2$

67. Find an equation of the tangent line to the graph of $y = x \ln x$ at the point $(1, 0)$.

68. Find an equation of the tangent line to the graph of $y = \ln x^2$ at the point $(2, \ln 4)$.

69. **ONLINE VIDEO VIEWERS** As broadband Internet grows more popular, video services such as YouTube will continue to expand. The number of online video viewers (in millions) is projected to grow from 2008 through 2013 according to the rule

$$N(t) = 135e^{0.067t} \qquad (1 \le t \le 6)$$

where $t = 1$ corresponds to 2008.
a. How many online video viewers were there in 2012?
b. How fast was the number of online video viewers changing in 2012?
Source: eMarketer.com.

70. **PHARMACEUTICAL THEFT** Pharmaceutical theft has been rising rapidly in recent years. Experts believe that pharmaceuticals are the new "street gold." The value of stolen drugs (in millions of dollars per year) from 2007 through 2009 is approximated by the function

$$f(t) = 20.5e^{0.74t} \qquad (1 \le t \le 3)$$

where t is the number of years since 2007. Find the rate of change of the value of stolen drugs at $t = 2$, and interpret your result.
Source: New York Times.

71. **OVER-100 POPULATION** Based on data obtained from the Census Bureau, the number of Americans over age 100 is expected to be

$$P(t) = 0.07e^{0.54t} \qquad (0 \le t \le 4)$$

where $P(t)$ is measured in millions and t is measured in decades, with $t = 0$ corresponding to the beginning of 2000.
a. What was the population of Americans over age 100 at the beginning of 2000? What will it be at the beginning of 2030?
b. How fast was the population of Americans over age 100 changing at the beginning of 2000? How fast will it be changing at the beginning of 2030?
Source: U.S. Census Bureau.

72. **WORLD POPULATION GROWTH** After its fastest rate of growth ever during the 1980s and 1990s, the rate of growth of world population is expected to slow dramatically in the twenty-first century. The function

$$G(t) = 1.58e^{-0.213t}$$

gives the projected annual average percent population growth per decade in the tth decade, with $t = 1$ corresponding to 2000.
a. What will the projected annual average population growth rate be in 2020 $(t = 3)$?
b. How fast will the projected annual average population growth rate be changing in 2020?
Source: U.S. Census Bureau.

73. **SALES PROMOTION** The Lady Bug, a women's clothing chain store, found that t days after the end of a sales promotion the volume of sales was given by

$$S(t) = 20{,}000(1 + e^{-0.5t}) \qquad (0 \le t \le 5)$$

dollars.
a. Find the rate of change of The Lady Bug's sales volume when $t = 1$, $t = 2$, $t = 3$, and $t = 4$.
b. In how many days will the sales volume drop below $27,400?

74. **BLOOD ALCOHOL LEVEL** The percentage of alcohol in a person's bloodstream t hr after drinking 8 fluid oz of whiskey is given by

$$A(t) = 0.23te^{-0.4t} \qquad (0 \le t \le 12)$$

a. What is the percentage of alcohol in a person's bloodstream after $\frac{1}{2}$ hr? After 8 hr?
b. How fast is the percentage of alcohol in a person's bloodstream changing after $\frac{1}{2}$ hr? After 8 hr?
Source: Encyclopedia Britannica.

75. **RATE OF BUSINESS FAILURES** The rate of business failure is highest in the first few years of the businesses' existence. According to a study of businesses that started in the second quarter of 1998, the percentage of companies still in business t years after the start of business is approximated by the function

$$f(t) = 93.1e^{-0.1626t} \qquad (1 \le t \le 7)$$

Find the rate at which the percent of businesses in existence is dropping at $t = 1, 2, 3$, and 4 years.
Source: Monthly Labor Review.

76. **AGING POPULATION** According to information obtained from the U.S. Census, the population of the United States aged 75 years and above is estimated to be

$$f(t) = \frac{72.15}{1 + 2.7975e^{-0.02145t}} \qquad (0 \le t \le 80)$$

in year t, where $t = 0$ corresponds to 2010 and $f(t)$ is measured in millions.
a. What was the population aged 75 years and over in 2010?

b. How fast is the population aged 75 years and over expected to grow in 2030?

c. What is the population aged 75 years and over expected to be in 2030?

Source: U.S. Census Bureau.

77. DEATH DUE TO STROKES Before 1950, little was known about strokes. By 1960, however, risk factors such as hypertension were identified. In recent years, CAT scans used as a diagnostic tool have helped to prevent strokes. As a result, the number of deaths due to strokes has fallen dramatically. The function

$$N(t) = 130.7e^{-0.1155t^2} + 50 \qquad (0 \le t \le 6)$$

gives the number of deaths due to stroke per 100,000 people from 1950 through 2010, where t is measured in decades, with $t = 0$ corresponding to 1950.

a. How many deaths due to strokes per 100,000 people were there in 1950?

b. How fast was the number of deaths due to strokes per 100,000 people changing in 1950? In 1960? In 1970? In 1980?

c. How many deaths due to strokes per 100,000 people were there in 2010 ($t = 6$)?

Source: American Heart Association, Centers for Disease Control and Prevention, and National Institutes of Health.

78. PRICE OF PERFUME The monthly demand for a certain brand of perfume is given by the demand equation

$$p = 100e^{-0.0002x} + 150$$

where p denotes the retail unit price (in dollars) and x denotes the quantity (in 1-oz bottles) demanded.

a. Find the rate of change of the price per bottle when $x = 1000$ and when $x = 2000$.

b. What is the price per bottle when $x = 1000$? When $x = 2000$?

79. PRICE OF WINE The monthly demand for a certain brand of table wine is given by the demand equation

$$p = 240\left(1 - \frac{3}{3 + e^{-0.0005x}}\right)$$

where p denotes the wholesale price per case (in dollars) and x denotes the number of cases demanded.

a. Find the rate of change of the price per case when $x = 1000$.

b. What is the price per case when $x = 1000$?

80. SALES OF A PRODUCT The total number of units of a new product sold t months after its introduction is given by

$$N(t) = 20,000(1 - e^{-0.05t})^2 \qquad (0 \le t \le 36)$$

a. How many units of the product were sold 24 months after its introduction?

b. How fast was the product selling 24 months after its introduction?

81. THERMOMETER READINESS A thermometer is moved from inside a house out to the deck. Its temperature t min after it has been moved is given by

$$T(t) = 30 + 40e^{-0.98t}$$

a. What is the temperature inside the house?

b. How fast is the reading on the thermometer changing 1 min after it has been taken out of the house?

c. What is the outdoor temperature?
Hint: Evaluate $\lim_{t \to \infty} T(t)$.

82. CHEMICAL REACTION Two chemicals, A and B, interact to form Chemical C. Suppose the amount (in grams) of Chemical C formed t min after the interaction begins is

$$A(t) = \frac{150(1 - e^{0.022662t})}{1 - 2.5e^{0.022662t}}$$

a. How fast is Chemical C being formed 1 min after the interaction first began?

b. How much Chemical C will there be eventually?
Hint: Evaluate $\lim_{t \to \infty} A(t)$.

83. STRAIN ON VERTEBRAE The strain (percentage of compression) on the lumbar vertebral disks in an adult human as a function of the load x (in kilograms) is given by

$$f(x) = 7.2956 \ln(0.0645012x^{0.95} + 1)$$

What is the rate of change of the strain with respect to the load when the load is 100 kg? When the load is 500 kg?
Source: Benedek and Villars, *Physics with Illustrative Examples from Medicine and Biology.*

84. STREAMING MUSIC Consumers are spending more on music from digital sources than they spend on CDs. Purchases of streamed music from services such as Pandora and Spotify are projected to grow in accordance with the model

$$f(t) = 3.7 + 0.84 \ln(t + 1) \qquad (0 \le t \le 5)$$

where $f(t)$ is measured in billions of dollars and t is measured in years, with $t = 0$ corresponding to 2012.

a. What was the amount spent by consumers on streamed music in 2012? The amount spent in 2014?

b. How fast was the amount spent by consumers on streamed music growing in 2014?
Source: Strategy Analytics.

85. WEBER–FECHNER LAW The Weber–Fechner Law

$$R = k \ln \frac{S}{S_0}$$

where k is a positive constant, describes the relationship between a stimulus S and the resulting response R. Here, S_0, a positive constant, is the threshold level.

a. Show that $R = 0$ if the stimulus is at the threshold level S_0.

b. The derivative dR/dS is the *sensitivity* corresponding to the stimulus level S and measures the capability to detect small changes in the stimulus level. Show that dR/dS is inversely proportional to S, and interpret your result.

In Exercises 86–88, determine whether the statement is true or false. If it is true, explain why it is true. If it is false, give an example to show why it is false.

86. If $f(x) = 3^x$, then $f'(x) = x \cdot 3^{x-1}$.

87. If $f(x) = e^{\pi}$, then $f'(x) = e^{\pi}$.

88. If $f(x) = \ln 5$, then $f'(x) = \dfrac{1}{5}$.

89. Prove that $\dfrac{d}{dx} \ln|x| = \dfrac{1}{x}$ $(x \neq 0)$ for the case $x < 0$.

90. Use the definition of the derivative to show that

$$\lim_{x \to 0} \frac{\ln(x+1)}{x} = 1$$

9.7 Solutions to Self-Check Exercises

1. Using the Product Rule and the Chain Rule, we obtain

$$f'(x) = x\frac{d}{dx}e^{-x} + e^{-x}\frac{d}{dx}x$$

$$= -xe^{-x} + e^{-x} = (1-x)e^{-x}$$

Using the Product Rule and the Chain Rule once again, we obtain

$$f''(x) = (1-x)\frac{d}{dx}e^{-x} + e^{-x}\frac{d}{dx}(1-x)$$

$$= (1-x)(-e^{-x}) + e^{-x}(-1)$$

$$= -e^{-x} + xe^{-x} - e^{-x} = (x-2)e^{-x}$$

2. The slope of the tangent line to the graph of f at any point $(x, f(x))$ lying on the graph of f is given by $f'(x)$. Using the Product Rule, we find

$$f'(x) = \frac{d}{dx}\left[x\ln(2x+3)\right]$$

$$= x\frac{d}{dx}\ln(2x+3) + \ln(2x+3) \cdot \frac{d}{dx}(x)$$

$$= x\left(\frac{2}{2x+3}\right) + \ln(2x+3) \cdot 1$$

$$= \frac{2x}{2x+3} + \ln(2x+3)$$

In particular, the slope of the tangent line to the graph of f at the point $(-1, 0)$ is

$$f'(-1) = \frac{-2}{-2+3} + \ln 1 = -2$$

Therefore, using the point-slope form of the equation of a line, we see that a required equation is

$$y - 0 = -2(x+1)$$
$$y = -2x - 2$$

USING TECHNOLOGY

EXAMPLE 1 At the beginning of Section 9.7, we demonstrated via a table of values of $(e^h - 1)/h$ for selected values of h the plausibility of the result

$$\lim_{h \to 0} \frac{e^h - 1}{h} = 1$$

To obtain a visual confirmation of this result, we plot the graph of

$$f(x) = \frac{e^x - 1}{x}$$

in the viewing window $[-1, 1] \times [0, 2]$ (Figure T1). From the graph of f, we see that $f(x)$ appears to approach 1 as x approaches 0. ∎

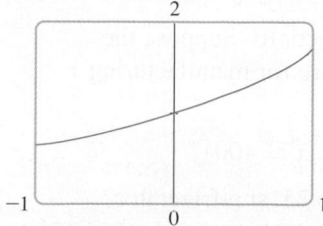

FIGURE **T1**
The graph of f in the viewing window $[-1, 1] \times [0, 2]$

The numerical derivative function of a graphing utility will yield the derivative of an exponential or logarithmic function for any value of x, just as it did for algebraic functions.

TECHNOLOGY EXERCISES

In Exercises 1–6, use the numerical derivative operation to find the rate of change of $f(x)$ at the given value of x. Give your answer accurate to four decimal places.

1. $f(x) = x^3 e^{-1/x}; \, x = -1$

2. $f(x) = (\sqrt{x} + 1)^{3/2} e^{-x}; \, x = 0.5$

3. $f(x) = x^3 \sqrt{\ln x}; \, x = 2$

4. $f(x) = \dfrac{\sqrt{x} \ln x}{x + 1}; \, x = 3.2$

5. $f(x) = e^{-x} \ln(2x + 1); \, x = 0.5$

6. $f(x) = \dfrac{e^{-\sqrt{x}}}{\ln(x^2 + 1)}; \, x = 1$

7. AN EXTINCTION SITUATION The number of saltwater crocodiles in a certain area of northern Australia in year t is given by

$$P(t) = \frac{300e^{-0.024t}}{5e^{-0.024t} + 1}$$

a. Plot the graph of P in the viewing window $[0, 200] \times [0, 70]$.

b. How many crocodiles were in the population initially?

c. What was the rate of growth initially?

(*Comment:* This phenomenon is referred to as an *extinction situation.*)

8. INCOME OF AMERICAN HOUSEHOLDS On the basis of government data, it is estimated that the percentage of American households y who earned x thousand dollars in 2010 is given by the equation

$$y = 1.168xe^{-0.00000312x^3 + 0.000659x^2 - 0.0783x} \quad (x > 0)$$

a. Plot the graph of the equation in the viewing window $[0, 150] \times [0, 10]$.

b. How fast is y changing with respect to x when $x = 10$? When $x = 50$? Interpret your results.

Source: U.S. Census Bureau.

9.8 Marginal Functions in Economics

Marginal analysis is the study of the rate of change of economic quantities. For example, an economist is not merely concerned with the value of an economy's gross domestic product (GDP) at a given time but is equally concerned with the rate at which it is growing or declining. In the same vein, a manufacturer is not only interested in the total cost corresponding to a certain level of production of a commodity but is also interested in the rate of change of the total cost with respect to the level of production, and so on. Let's begin with an example to explain the meaning of the adjective *marginal*, as used by economists.

Cost Functions

APPLIED EXAMPLE 1 Rate of Change of Cost Functions Suppose the total cost in dollars incurred each week by Polaraire for manufacturing x refrigerators is given by the total cost function

$$C(x) = 8000 + 200x - 0.2x^2 \quad (0 \le x \le 400)$$

a. What is the actual cost incurred for manufacturing the 251st refrigerator?

b. Find the rate of change of the total cost function with respect to x when $x = 250$.

c. Compare the results obtained in parts (a) and (b).

Solution

a. The actual cost incurred in producing the 251st refrigerator is the difference between the total cost incurred in producing the first 251 refrigerators and the total cost of producing the first 250 refrigerators:

$$C(251) - C(250) = [8000 + 200(251) - 0.2(251)^2]$$
$$- [8000 + 200(250) - 0.2(250)^2]$$
$$= 45{,}599.8 - 45{,}500$$
$$= 99.8$$

or $99.80.

b. The rate of change of the total cost function C with respect to x is given by the derivative of C—that is, $C'(x) = 200 - 0.4x$. Thus, when the level of production is 250 refrigerators, the rate of change of the total cost with respect to x is given by

$$C'(250) = 200 - 0.4(250)$$
$$= 100$$

or $100.

c. From the solution to part (a), we know that the actual cost for producing the 251st refrigerator is $99.80. This answer is very closely approximated by the answer to part (b), $100. To see why this is so, observe that the difference $C(251) - C(250)$ may be written in the form

$$\frac{C(251) - C(250)}{1} = \frac{C(250 + 1) - C(250)}{1} = \frac{C(250 + h) - C(250)}{h}$$

where $h = 1$. In other words, the difference $C(251) - C(250)$ is precisely the average rate of change of the total cost function C over the interval $[250, 251]$ or, equivalently, the slope of the secant line through the points $(250, 45{,}500)$ and $(251, 45{,}599.8)$. However, the number $C'(250) = 100$ is the instantaneous rate of change of the total cost function C at $x = 250$ or, equivalently, the slope of the tangent line to the graph of C at $x = 250$.

Now when h is small, the average rate of change of the function C is a good approximation to the instantaneous rate of change of the function C, or, equivalently, the slope of the secant line through the points in question is a good approximation to the slope of the tangent line through the point in question. Therefore, we may expect

$$C(251) - C(250) = \frac{C(251) - C(250)}{1} \approx \frac{C(250 + h) - C(250)}{h} \quad (h \text{ small})$$
$$\approx \lim_{h \to 0} \frac{C(250 + h) - C(250)}{h} = C'(250)$$

which is precisely the case in this example. ■

The actual cost incurred in producing an additional unit of a certain commodity given that a plant is already at a certain level of operation is called the **marginal cost.** Knowing this cost is very important to management. As we saw in Example 1, the marginal cost is approximated by the rate of change of the total cost function evaluated at the appropriate point. For this reason, economists have defined the **marginal cost function** to be the derivative of the corresponding total cost function. In other words, if C is a total cost function, then the marginal cost function is defined to be its derivative C'. Thus, the adjective *marginal* is synonymous with *derivative of.*

APPLIED EXAMPLE 2 Marginal Cost Functions A subsidiary of Elektra Electronics manufactures a portable DVD player. Management determined that the daily total cost of producing these DVD players (in dollars) is given by

$$C(x) = 0.0001x^3 - 0.08x^2 + 40x + 5000$$

where x stands for the number of DVD players produced.

a. Find the marginal cost function.
b. What is the marginal cost when $x = 200$, 300, 400, and 600?
c. Interpret your results.

Solution

a. The marginal cost function C' is given by the derivative of the total cost function C. Thus,

$$C'(x) = 0.0003x^2 - 0.16x + 40$$

b. The marginal cost when $x = 200$, 300, 400, and 600 is given by

$$C'(200) = 0.0003(200)^2 - 0.16(200) + 40 = 20$$
$$C'(300) = 0.0003(300)^2 - 0.16(300) + 40 = 19$$
$$C'(400) = 0.0003(400)^2 - 0.16(400) + 40 = 24$$
$$C'(600) = 0.0003(600)^2 - 0.16(600) + 40 = 52$$

or $20, $19, $24, and $52 per unit, respectively.

c. From the results of part (b), we see that Elektra's actual cost for producing the 201st DVD player is approximately $20. The actual cost incurred for producing one additional DVD player when the level of production is already 300 players is approximately $19, and so on. Observe that when the level of production is already 600 units, the actual cost of producing one additional unit is approximately $52. The higher cost for producing this additional unit when the level of production is 600 units may be the result of several factors, among them excessive costs incurred because of overtime or higher maintenance, production breakdown caused by greater stress and strain on the equipment, and so on. The graph of the total cost function appears in Figure 49.

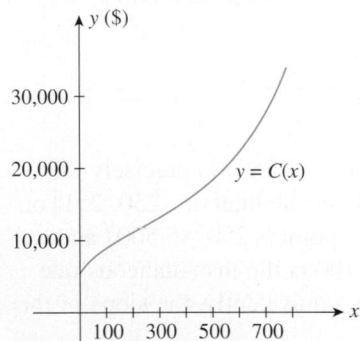

FIGURE 49
The cost of producing x DVD players is given by $C(x)$.

Average Cost Functions

Let's now introduce another marginal concept that is closely related to the marginal cost. Let $C(x)$ denote the total cost incurred in producing x units of a certain commodity. Then the **average cost** of producing x units of the commodity is obtained by dividing the total production cost by the number of units produced. This leads to the following definition:

> Average Cost Function
>
> Suppose $C(x)$ is a total cost function. Then the **average cost function,** denoted by $\bar{C}(x)$ (read "C bar of x"), is
>
> $$\frac{C(x)}{x} \tag{13}$$

The derivative $\bar{C}'(x)$ of the average cost function, called the **marginal average cost function,** measures the rate of change of the average cost function with respect to the number of units produced.

APPLIED EXAMPLE 3 Marginal Average Cost Functions The total cost of producing x units of a certain commodity is given by $C(x) = 400 + 20x$ dollars.

a. Find the average cost function $\overline{C}$.
b. Find the marginal average cost function $\overline{C}'$.
c. What are the economic implications of your results?

Solution

a. The average cost function is given by

$$\overline{C}(x) = \frac{C(x)}{x} = \frac{400 + 20x}{x}$$

$$= 20 + \frac{400}{x}$$

b. The marginal average cost function is

$$\overline{C}'(x) = -\frac{400}{x^2}$$

c. Since the marginal average cost function is negative for all admissible values of x, the rate of change of the average cost function is negative for all $x > 0$; that is, $\overline{C}(x)$ decreases as x increases. However, the graph of $\overline{C}$ always lies above the horizontal line $y = 20$, but it approaches the line, since

$$\lim_{x \to \infty} \overline{C}(x) = \lim_{x \to \infty} \left(20 + \frac{400}{x}\right) = 20$$

A sketch of the graph of the function $\overline{C}(x)$ appears in Figure 50. This result is fully expected if we consider the economic implications. Note that as the level of production increases, the fixed cost per unit of production, represented by the term $(400/x)$, drops steadily. The average cost approaches the constant unit cost of production, which is $20 in this case.

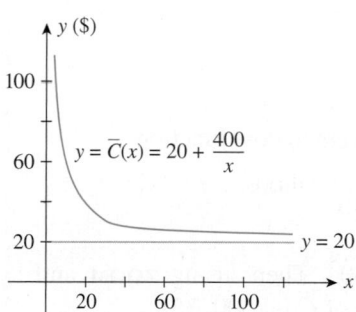

FIGURE 50
As the level of production increases, the average cost approaches $20.

 APPLIED EXAMPLE 4 Marginal Average Cost Functions Once again consider the subsidiary of Elektra Electronics. The daily total cost for producing its portable DVD players is given by

$$C(x) = 0.0001x^3 - 0.08x^2 + 40x + 5000$$

dollars, where x stands for the number of DVD players produced (see Example 2).

a. Find the average cost function $\overline{C}$.
b. Find the marginal average cost function $\overline{C}'$. Compute $\overline{C}'(500)$.
c. Sketch the graph of the function $\overline{C}$, and interpret the results obtained in parts (a) and (b).

Solution

a. The average cost function is given by

$$\overline{C}(x) = \frac{C(x)}{x} = 0.0001x^2 - 0.08x + 40 + \frac{5000}{x}$$

b. The marginal average cost function is given by

$$\overline{C}'(x) = 0.0002x - 0.08 - \frac{5000}{x^2}$$

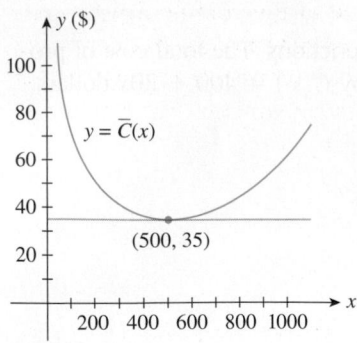

FIGURE 51
The average cost reaches a minimum
of $35 when 500 DVD players are
produced.

Also,

$$\overline{C}'(500) = 0.0002(500) - 0.08 - \frac{5000}{(500)^2} = 0$$

c. To sketch the graph of the function $\overline{C}$, observe that if x is a small positive number, then $\overline{C}(x) > 0$. Furthermore, $\overline{C}(x)$ becomes arbitrarily large as x approaches zero from the right, since the term $(5000/x)$ becomes arbitrarily large as x approaches zero. Next, the result $\overline{C}'(500) = 0$ obtained in part (b) tells us that the tangent line to the graph of the function $\overline{C}$ is horizontal at the point $(500, 35)$ on the graph. Finally, plotting the points on the graph corresponding to, say, $x = 100, 200, 300, \ldots, 900$, we obtain the sketch in Figure 51. As expected, the average cost drops as the level of production increases. But in this case, in contrast to the case in Example 3, the average cost reaches a minimum value of $35, corresponding to a production level of 500, and *increases* thereafter.

This phenomenon is typical in situations in which the marginal cost increases from some point on as production increases, as in Example 2. This situation is in contrast to that of Example 3, in which the marginal cost remains constant at any level of production. ∎

Exploring with TECHNOLOGY

Refer to Example 4.

1. Use a graphing utility to plot the graph of the average cost function

$$\overline{C}(x) = 0.0001x^2 - 0.08x + 40 + \frac{5000}{x}$$

using the viewing window $[0, 1000] \times [0, 100]$. Then, using **ZOOM** and **TRACE**, show that the lowest point on the graph of $\overline{C}$ is $(500, 35)$.

2. Draw the tangent line to the graph of $\overline{C}$ $(500, 35)$. What is its slope? Is this expected?

3. Plot the graph of the marginal average cost function

$$\overline{C}'(x) = 0.0002x - 0.08 - \frac{5000}{x^2}$$

using the viewing window $[0, 2000] \times [-1, 1]$. Then use **ZOOM** and **TRACE** to show that the zero of the function $\overline{C}'$ occurs at $x = 500$. Verify this result using the root-finding capability of your graphing utility. Is this result compatible with that obtained in part 2? Explain your answer.

Revenue Functions

Recall that a revenue function $R(x)$ gives the revenue realized by a company from the sale of x units of a certain commodity. If the company charges p dollars per unit, then

$$R(x) = px \tag{14}$$

However, the price that a company can command for the product depends on the market in which the company operates. If the company is one of many—none of which is able to dictate the price of the commodity—then in this competitive market environment, the price is determined by market equilibrium (see Section 2.6). On the other hand, if the company is the sole supplier of the product, then under this monopolistic situation, it can manipulate the price of the commodity by controlling the supply. The unit selling price p of the commodity is related to the quantity x of the commodity demanded. This relationship between p and x is called a *demand equation* (see Section 2.6). Solving the demand equation for p in terms of x, we obtain the unit price function f. Thus,

$$p = f(x)$$

and the revenue function R is given by

$$R(x) = px = xf(x)$$

The **marginal revenue** gives the actual revenue realized from the sale of an additional unit of the commodity given that sales are already at a certain level. Following an argument parallel to that applied to the cost function in Example 1, you can convince yourself that the marginal revenue is approximated by $R'(x)$. Thus, we define the **marginal revenue function** to be $R'(x)$, where R is the revenue function. The derivative R' of the function R measures the rate of change of the revenue function.

$ **APPLIED EXAMPLE 5** Marginal Revenue Functions Suppose the relationship between the unit price p in dollars and the quantity demanded x of the Acrosonic model F loudspeaker system is given by the equation

$$p = -0.02x + 400 \qquad (0 \le x \le 20{,}000)$$

a. Find the revenue function R.
b. Find the marginal revenue function R'.
c. Compute $R'(2000)$, and interpret your result.

Solution

a. The revenue function R is given by

$$\begin{aligned}
R(x) &= px \\
&= x(-0.02x + 400) \\
&= -0.02x^2 + 400x \qquad (0 \le x \le 20{,}000)
\end{aligned}$$

b. The marginal revenue function R' is given by

$$R'(x) = -0.04x + 400$$

c. $\qquad\qquad R'(2000) = -0.04(2000) + 400 = 320$

Thus, the actual revenue to be realized from the sale of the 2001st loudspeaker system is approximately \$320.

Profit Functions

Our final example of a marginal function involves the profit function. The profit function P is given by

$$P(x) = R(x) - C(x) \tag{15}$$

where R and C are the revenue and cost functions and x is the number of units of a commodity produced and sold. The **marginal profit function** $P'(x)$ measures the rate of change of the profit function P and provides us with a good approximation of the actual profit or loss realized from the sale of the $(x + 1)$st unit of the commodity (assuming that the xth unit has been sold).

APPLIED EXAMPLE 6 Marginal Profit Functions Refer to Example 5. Suppose the cost of producing x units of the Acrosonic model F loudspeaker is

$$C(x) = 100x + 200,000$$

dollars.

a. Find the profit function P.
b. Find the marginal profit function P'.
c. Compute $P'(2000)$, and interpret your result.
d. Sketch the graph of the profit function P.

Solution

a. From the solution to Example 5a, we have

$$R(x) = -0.02x^2 + 400x$$

Thus, the required profit function P is given by

$$
\begin{aligned}
P(x) &= R(x) - C(x) \\
&= (-0.02x^2 + 400x) - (100x + 200,000) \\
&= -0.02x^2 + 300x - 200,000
\end{aligned}
$$

b. The marginal profit function P' is given by

$$P'(x) = -0.04x + 300$$

c. $$P'(2000) = -0.04(2000) + 300 = 220$$

Thus, the actual profit realized from the sale of the 2001st loudspeaker system is approximately $220.

d. The graph of the profit function P appears in Figure 52.

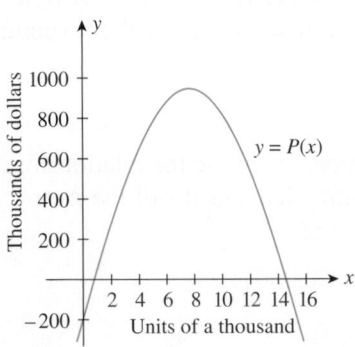

FIGURE 52
The total profit made when x loudspeakers are produced and sold is given by $P(x)$.

Relative Rate of Change

The *relative change* of the size of a quantity is defined to be

$$\frac{\text{Change in the size of the quantity}}{\text{Size of the quantity}}$$

For example, suppose that the mortgage rate increases from the current rate of 10% per year to a rate of 11% per year. Then the relative change in the mortgage rate is

$$\frac{1}{10} = 0.1 \quad \text{or} \quad 10\%$$

But if the current mortgage rate is 5% per year, then a change of 1% per year, to 6% per year, in the mortgage rate would yield a relative change in the mortgage rate of

$$\frac{1}{5} = 0.2 \quad \text{or} \quad 20\%$$

This example shows that the relative change sometimes conveys a better sense of what is going on than does a simple look at the change in the quantity itself.

Similarly, it is often more meaningful to look at the *relative rate of change* of a function at a given value of x than at the rate of change of f at x. We have the following definition.

> **Relative Rate of Change**
>
> The **relative rate of change** of a differentiable function f at x per unit change in x is
>
> $$\frac{f'(x)}{f(x)} \quad \text{or} \quad \frac{100f'(x)}{f(x)}\%$$
>
> The second expression is called the **percentage rate of change of f at x.**

 APPLIED EXAMPLE 7 Inflation An economy's consumer price index (CPI) is described by the function

$$I(t) = -0.05t^3 + 0.5t^2 + 100 \quad (0 \le t \le 4)$$

where t is measured in years, with $t = 0$ corresponding to the beginning of 2012. Find the annual percentage rate of inflation in the CPI of the country (defined as the percentage relative rate of change of I) at the beginning of 2014 ($t = 2$).

Solution The inflation rate in year t is

$$R(t) = \frac{100I'(t)}{I(t)} = 100\left[\frac{-0.15t^2 + t}{-0.05t^3 + 0.5t^2 + 100}\right]$$

So the inflation rate at the beginning of 2014 is

$$R(2) = 100\left[\frac{-0.15(2^2) + 2}{-0.05(2^3) + 0.5(2^2) + 100}\right] \approx 1.37795$$

or approximately 1.4% per year.

9.8 Self-Check Exercise

DEMAND FOR DVRs The weekly demand for Pulsar DVRs is given by the demand equation

$$p = -0.02x + 300 \quad (0 \le x \le 15{,}000)$$

where p denotes the wholesale unit price in dollars and x denotes the quantity demanded. The weekly total cost function associated with manufacturing these recorders is

$$C(x) = 0.000003x^3 - 0.04x^2 + 200x + 70{,}000$$

dollars.

a. Find the revenue function R and the profit function P.
b. Find the marginal cost function C', the marginal revenue function R', and the marginal profit function P'.
c. Find the marginal average cost function $\overline{C}'$.
d. Compute $C'(3000)$, $R'(3000)$, and $P'(3000)$, and interpret your results.

Solutions to Self-Check Exercise 9.8 can be found on page 710.

9.8 Concept Questions

1. Explain each term in your own words:
   a. Marginal cost function
   b. Average cost function
   c. Marginal average cost function
   d. Marginal revenue function
   e. Marginal profit function

2. The proprietor of a company finds that his marginal revenue and marginal cost are $3/unit and $2.80/unit, respectively, at a production level of 500 units. Should the proprietor increase production in order to increase the profit of the company? Explain.

9.8 Exercises

1. **PRODUCTION COSTS** The graph of a typical total cost function $C(x)$ associated with the manufacture of x units of a certain commodity is shown in the figure.
   a. Explain why the function C is always increasing.
   b. As the level of production x increases, the cost per unit drops so that $C(x)$ increases but at a slower pace. However, a level of production is soon reached at which the cost per unit begins to increase dramatically (owing to a shortage of raw material, overtime, breakdown of machinery due to excessive stress and strain), so $C(x)$ continues to increase at a faster pace. Use the graph of C to find the approximate level of production x_0 where this occurs.

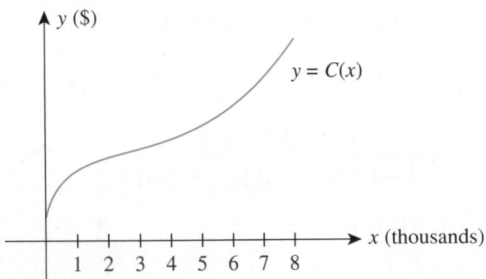

2. **PRODUCTION COSTS** The graph of a typical average cost function $A(x) = C(x)/x$, where $C(x)$ is a total cost function associated with the manufacture of x units of a certain commodity, is shown in the following figure.
   a. Explain in economic terms why $A(x)$ is large if x is small and why $A(x)$ is large if x is large.
   b. What is the significance of the numbers x_0 and y_0, the x- and y-coordinates of the lowest point on the graph of the function A?

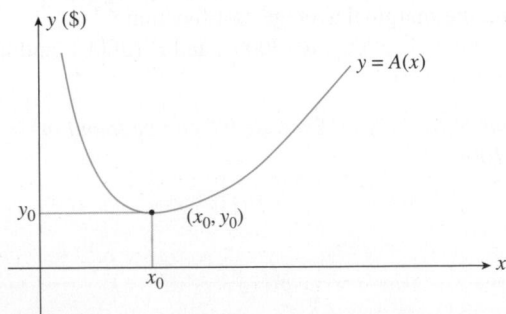

3. **MARGINAL COST** The total weekly cost (in dollars) incurred by Lincoln Records in pressing x compact discs is

   $$C(x) = 2000 + 2x - 0.0001x^2 \qquad (0 \le x \le 6000)$$

   a. What is the actual cost incurred in producing the 1001st disc and the 2001st disc?
   b. What is the marginal cost when $x = 1000$ and 2000?

4. **MARGINAL COST** A division of Ditton Industries manufactures the Futura model microwave oven. The daily cost (in dollars) of producing these microwave ovens is

   $$C(x) = 0.0002x^3 - 0.06x^2 + 120x + 5000$$

   where x stands for the number of units produced.
   a. What is the actual cost incurred in manufacturing the 101st oven? The 201st oven? The 301st oven?
   b. What is the marginal cost when $x = 100$, 200, and 300?

5. **MARGINAL AVERAGE COST FOR PRODUCING DESKS** Custom Office makes a line of executive desks. It is estimated that the total cost for making x units of their Senior Executive model is

   $$C(x) = 100x + 200{,}000$$

   dollars/year.
   a. Find the average cost function $\overline{C}$.
   b. Find the marginal average cost function $\overline{C}'$.
   c. What happens to $\overline{C}(x)$ when x is very large? Interpret your results.

6. **MARGINAL AVERAGE COST FOR PRODUCING THERMOMETERS** The management of ThermoMaster Company, whose Mexican subsidiary manufactures an indoor–outdoor thermometer, has estimated that the total weekly cost (in dollars) for producing x thermometers is

   $$C(x) = 5000 + 2x$$

   a. Find the average cost function $\overline{C}$.
   b. Find the marginal average cost function $\overline{C}'$.
   c. Interpret your results.

7. **AVERAGE COST FOR PRODUCING CDS** Find the average cost function $\overline{C}$ and the marginal average cost function $\overline{C}'$ associated with the total cost function C of Exercise 3.

8. **AVERAGE COST FOR PRODUCING MICROWAVES** Find the average cost function $\overline{C}$ and the marginal average cost function $\overline{C}'$ associated with the total cost function C of Exercise 4.

9. **MARGINAL REVENUE OF A COMMUTER AIR SERVICE** Williams Commuter Air Service realizes a monthly revenue of

   $$R(x) = 8000x - 100x^2$$

   dollars when the price charged per passenger is x dollars.
   a. Find the marginal revenue R'.
   b. Compute $R'(39)$, $R'(40)$, and $R'(41)$.
   c. Based on the results of part (b), what price should the airline charge in order to maximize their revenue?

10. **MARGINAL REVENUE FOR PRODUCING LOUDSPEAKERS** The management of Acrosonic plans to market the ElectroStat, an electrostatic speaker system. The marketing department has determined that the demand function for these speakers is

    $$p = -0.04x + 800 \qquad (0 \le x \le 20{,}000)$$

where p denotes the speaker's unit price (in dollars) and x denotes the quantity demanded.
 a. Find the revenue function R.
 b. Find the marginal revenue function R'.
 c. Compute $R'(5000)$, and interpret your results.

11. **MARGINAL PROFIT FOR PRODUCING LOUDSPEAKERS** Refer to Exercise 10. Acrosonic's production department estimates that the total cost (in dollars) incurred in manufacturing x ElectroStat speaker systems in the first year of production will be

$$C(x) = 200x + 300,000$$

 a. Find the profit function P.
 b. Find the marginal profit function P'.
 c. Compute $P'(5000)$ and $P'(8000)$.
 d. Sketch the graph of the profit function, and interpret your results.

12. **MARGINAL PROFIT FOR AN APARTMENT COMPLEX** Lynbrook West, an apartment complex, has 100 two-bedroom units. The monthly profit (in dollars) realized from renting x apartments is

$$P(x) = -10x^2 + 1760x - 50,000$$

 a. What is the actual profit realized from renting the 51st unit, assuming that 50 units have already been rented?
 b. Compute the marginal profit when $x = 50$, and compare your results with that obtained in part (a).

13. **MARGINAL COST, REVENUE, AND PROFIT FOR PRODUCING LED TVs** The weekly demand for the Pulsar 25 color LED television is

$$p = 600 - 0.05x \qquad (0 \le x \le 12,000)$$

where p denotes the wholesale unit price in dollars and x denotes the quantity demanded. The weekly total cost function associated with manufacturing the Pulsar 25 is given by

$$C(x) = 0.000002x^3 - 0.03x^2 + 400x + 80,000$$

where $C(x)$ denotes the total cost incurred in producing x sets.
 a. Find the revenue function R and the profit function P.
 b. Find the marginal cost function C', the marginal revenue function R', and the marginal profit function P'.
 c. Compute $C'(2000)$, $R'(2000)$, and $P'(2000)$, and interpret your results.
 d. Sketch the graphs of the functions C, R, and P, and interpret parts (b) and (c), using the graphs obtained.

14. **MARGINAL COST, REVENUE, AND PROFIT FOR PRODUCING LCD TVs** Pulsar manufactures a series of 20-in. flat-tube LCD televisions. The quantity x of these sets demanded each week is related to the wholesale unit price p by the equation

$$p = -0.006x + 180$$

The weekly total cost incurred by Pulsar for producing x sets is

$$C(x) = 0.000002x^3 - 0.02x^2 + 120x + 60,000$$

dollars. Answer the questions in Exercise 13 for these data.

15. **MARGINAL AVERAGE COST FOR PRODUCING LED TVs** Refer to Exercise 13.
 a. Find the average cost function $\overline{C}$ associated with the total cost function C of Exercise 13.
 b. What is the marginal average cost function $\overline{C}'$?
 c. Compute $\overline{C}'(5000)$ and $\overline{C}'(10,000)$, and interpret your results.
 d. Sketch the graph of $\overline{C}$.

16. **MARGINAL AVERAGE COST FOR PRODUCING LCD TVs** Refer to Exercise 14.
 a. Find the average cost function $\overline{C}$ associated with the total cost function C of Exercise 14.
 b. What is the marginal average cost function $\overline{C}'$?
 c. Compute $\overline{C}'(5000)$ and $\overline{C}'(10,000)$, and interpret your results.
 d. Sketch the graph of $\overline{C}$.

17. **MARGINAL REVENUE FOR GAMING MICE** The quantity of Sensitech laser gaming mice demanded each month is related to the unit price by the equation

$$p = \frac{50}{0.01x^2 + 1} \qquad (0 \le x \le 20)$$

where p is measured in thousands of dollars and x in units of a thousand.
 a. Find the revenue function R.
 b. Find the marginal revenue function R'.
 c. Compute $R'(2)$, and interpret your result.

18. **MARGINAL REVENUE** The relationship between the unit selling price p (in dollars) and the quantity demanded x (in pairs) of a certain brand of women's gloves are given by the demand equation

$$p = 100e^{-0.0001x} \qquad (0 \le x \le 20,000)$$

 a. Find the revenue function R.
    Hint: $R(x) = px$
 b. Find the marginal revenue function R'.
 c. What is the marginal revenue when $x = 10,000$?

19. **MARGINAL REVENUE** The demand function for the Viking Boat's 34-ft *Sundancer* yacht is

$$p = 200 - 0.01x \ln x$$

where x denotes the number of yachts and p is the price per yacht in hundreds of dollars.
 a. Find the revenue and the marginal revenue function for this model of yacht.
 b. Use the result of part (a) to estimate the revenue to be realized from the sale of the 500th 34-ft *Sundancer* yacht.

20. MARGINAL PROPENSITY TO CONSUME The consumption function of the U.S. economy from 1929 to 1941 is

$$C(x) = 0.712x + 95.05$$

where $C(x)$ is the personal consumption expenditure and x is the personal income, both measured in billions of dollars. Find the rate of change of consumption with respect to income, dC/dx. This quantity is called the *marginal propensity to consume.*

21. MARGINAL PROPENSITY TO CONSUME Refer to Exercise 20. Suppose a certain economy's consumption function is

$$C(x) = 0.873x^{1.1} + 20.34$$

where $C(x)$ and x are measured in billions of dollars. Find the marginal propensity to consume when $x = 10$.

22. MARGINAL PROPENSITY TO SAVE Suppose $C(x)$ measures an economy's personal consumption expenditure and x measures the personal income, both in billions of dollars. Then

$$S(x) = x - C(x) \qquad \text{Income minus consumption}$$

measures the economy's savings corresponding to an income of x billion dollars. Show that

$$\frac{dS}{dx} = 1 - \frac{dC}{dx}$$

The quantity dS/dx is called the *marginal propensity to save.*

23. MARGINAL PROPENSITY TO SAVE Refer to Exercise 22. For the consumption function of Exercise 20, find the marginal propensity to save.

24. MARGINAL PROPENSITY TO SAVE Refer to Exercise 22. For the consumption function of Exercise 21, find the marginal propensity to save when $x = 10$.

In Exercises 25–28, find the percentage rate of change of *f* at the given value of *x*.

25. $f(x) = 2x^2 + x + 1; x = 2$

26. $f(x) = \sqrt{2x^2 + 7}; x = 3$

27. $f(x) = \dfrac{x + 1}{x^3 + x + 1}; x = 2$

28. $f(x) = \left(\dfrac{x}{x^2 + 3x + 12}\right)^{3/2}; x = 1$

29. INFLATION The consumer price index (CPI) of a certain country is given by

$$I(t) = -0.02t^3 + 0.4t^2 + 120 \qquad (0 \le t \le 4)$$

where $t = 0$ corresponds to the beginning of 2013. Find the annual percentage rate of inflation in the CPI of the country at the beginning of 2014.

30. GROWTH IN INCOME The per capita income of a country t years from now is defined to be

$$C(t) = \frac{I(t)}{P(t)}$$

where $I(t)$ and $P(t)$ are the income and the population of the country in year t, respectively.
   a. Find an expression for the percentage rate of change of the per capita income of the country in year t.
   b. Use your result to find the percentage rate of change in the per capita income of the country whose income in year t is $I(t) = 10^9(300 + 12t)$ dollars and whose population in year t is $P(t) = 2 \times 10^7 e^{0.02t}$.
   c. What is the percentage rate of change of the per capita income 2 years from now?

31. PERCENTAGE RATE OF CHANGE IN REVENUE Marianne Designs operates two boutiques in two locations in the Bay Area. The revenues for Boutique I and Boutique II at time $t = a$, where t is measured in years, are $3.2 million and $2.6 million, respectively. Management estimates that the revenue for Boutique I is growing at the rate of $0.24 million/year, while that for Boutique II is growing at the rate of $0.30 million/year, at that time. What is the percentage rate of growth of the revenue of Marianne Designs at time $t = a$?
   Hint: Marianne Designs's revenue is $R = f + g$, where f and g are the revenue of Boutique I and Boutique II, respectively. Find $100R'/R$ at $t = a$.

In Exercises 32 and 33, determine whether the statement is true or false. If it is true, explain why it is true. If it is false, give an example to show why it is false.

32. If C is a differentiable total cost function, then the marginal average cost function is

$$\overline{C}'(x) = \frac{xC'(x) - C(x)}{x^2}$$

33. If the marginal profit function is positive at $x = a$, then it makes sense to decrease the level of production.

9.8 Solution to Self-Check Exercise

1. a. $R(x) = px$
$\qquad = x(-0.02x + 300)$
$\qquad = -0.02x^2 + 300x \qquad (0 \le x \le 15,000)$

$P(x) = R(x) - C(x)$
$\qquad = -0.02x^2 + 300x$
$\qquad \quad - (0.000003x^3 - 0.04x^2 + 200x + 70,000)$
$\qquad = -0.000003x^3 + 0.02x^2 + 100x - 70,000$

b. $C'(x) = 0.000009x^2 - 0.08x + 200$

$R'(x) = -0.04x + 300$

$P'(x) = -0.000009x^2 + 0.04x + 100$

c. The average cost function is

$$\overline{C}(x) = \frac{C(x)}{x}$$

$$= \frac{0.000003x^3 - 0.04x^2 + 200x + 70,000}{x}$$

$$= 0.000003x^2 - 0.04x + 200 + \frac{70,000}{x}$$

Therefore, the marginal average cost function is

$$\overline{C}'(x) = 0.000006x - 0.04 - \frac{70,000}{x^2}$$

d. Using the results from part (b), we find

$$C'(3000) = 0.000009(3000)^2 - 0.08(3000) + 200$$

$$= 41$$

That is, when the level of production is already 3000 recorders, the actual cost of producing one additional recorder is approximately $41. Next,

$$R'(3000) = -0.04(3000) + 300 = 180$$

That is, the actual revenue to be realized from selling the 3001st recorder is approximately $180. Finally,

$$P'(3000) = -0.000009(3000)^2 + 0.04(3000) + 100$$

$$= 139$$

That is, the actual profit realized from selling the 3001st DVR is approximately $139.

CHAPTER 9 Summary of Principal Formulas and Terms

FORMULAS

1. Average rate of change of f over $[x, x + h]$ or Slope of the secant line to the graph of f through $(x, f(x))$ and $(x + h, f(x + h))$ or Difference quotient	$\dfrac{f(x + h) - f(x)}{h}$	
2. Instantaneous rate of change of f at $(x, f(x))$ or Slope of tangent line to the graph of f at $(x, f(x))$ at x or Derivative of f	$\displaystyle\lim_{h \to 0} \dfrac{f(x + h) - f(x)}{h}$	
3. Derivative of a constant	$\dfrac{d}{dx}(c) = 0 \qquad (c, \text{ a constant})$	
4. Power Rule	$\dfrac{d}{dx}(x^n) = nx^{n-1}$	
5. Constant Multiple Rule	$\dfrac{d}{dx}[cf(x)] = cf'(x)$	
6. Sum Rule	$\dfrac{d}{dx}[f(x) \pm g(x)] = f'(x) \pm g'(x)$	
7. Product Rule	$\dfrac{d}{dx}[f(x)g(x)] = f(x)g'(x) + g(x)f'(x)$	

8. Quotient Rule	$\dfrac{d}{dx}\left[\dfrac{f(x)}{g(x)}\right] = \dfrac{g(x)f'(x) - f(x)g'(x)}{[g(x)]^2}$
9. Chain Rule	$\dfrac{d}{dx}\,g(f(x)) = g'(f(x))f'(x)$
10. General Power Rule	$\dfrac{d}{dx}\,[f(x)]^n = n[f(x)]^{n-1}f'(x)$
11. Derivative of the exponential function	$\dfrac{d}{dx}\,(e^x) = e^x$
12. Chain Rule for Exponential Functions	$\dfrac{d}{dx}\,(e^u) = e^u\,\dfrac{du}{dx}$
13. Derivative of the logarithmic function	$\dfrac{d}{dx}\,\ln|x| = \dfrac{1}{x}$
14. Chain Rule for Logarithmic Functions	$\dfrac{d}{dx}\,(\ln u) = \dfrac{1}{u}\,\dfrac{du}{dx}$
15. Average cost function	$\overline{C}(x) = \dfrac{C(x)}{x}$
16. Revenue function	$R(x) = px$
17. Profit function	$P(x) = R(x) - C(x)$

TERMS

limit of a function (597)	zero of a function (621)	average cost (702)
indeterminate form (600)	secant line (632)	marginal average cost function (702)
limit of a function at infinity (603)	tangent line to the graph of f (632)	marginal revenue (705)
right-hand limit of a function (615)	differentiable function (640)	marginal revenue function (705)
left-hand limit of a function (615)	second derivative of f (666)	marginal profit function (705)
continuity of a function at a number (617)	marginal cost (701)	relative rate of change 707
	marginal cost function (701)	

CHAPTER 9   Concept Review Questions

Fill in the blanks.

1. The statement $\lim\limits_{x \to a} f(x) = L$ means that the values of _____ can be made as close to _____ as we please by taking x sufficiently close to _____.

2. If $\lim\limits_{x \to a} f(x) = L$ and $\lim\limits_{x \to a} g(x) = M$, then
a. $\lim\limits_{x \to a} [f(x)]^r = $ _____, where r is a real number.
b. $\lim\limits_{x \to a} [f(x) \pm g(x)] = $ _____.
c. $\lim\limits_{x \to a} [f(x)g(x)] = $ _____.
d. $\lim\limits_{x \to a} \dfrac{f(x)}{g(x)} = $ _____ provided that _____.

3. a. The statement $\lim\limits_{x \to \infty} f(x) = L$ means that $f(x)$ can be made arbitrarily close to _____ by taking _____ large enough.

b. The statement $\lim\limits_{x \to -\infty} f(x) = M$ means that $f(x)$ can be made arbitrarily close to _____ by taking x to be _____ and sufficiently large in _____ value.

4. a. The statement $\lim\limits_{x \to a^+} f(x) = L$ is similar to the statement $\lim\limits_{x \to a} f(x) = L$, but here x is required to lie to the _____ of a.

b. The statement $\lim\limits_{x \to a^-} f(x) = L$ is similar to the statement $\lim\limits_{x \to a} f(x) = L$ but here x is required to lie to the _____ of a.

c. $\lim\limits_{x \to a} f(x) = L$ if and only if both $\lim\limits_{x \to a^-} f(x) = $ _____ and $\lim\limits_{x \to a^+} f(x) = $ _____.

5. a. If $f(a)$ is defined, $\lim\limits_{x \to a} f(x)$ exists and $\lim\limits_{x \to a} f(x) = f(a)$, then f is _____ at a.
b. If f is not continuous at a, then it is _____ at a.
c. f is continuous on an interval I if f is continuous at _____ number in the interval.

6. a. If f and g are continuous at a, then $f \pm g$ and fg are continuous at _____. Also, $\dfrac{f}{g}$ is continuous at _____, provided that _____ $\neq 0$.
b. A polynomial function is continuous _____.
c. A rational function $R = \dfrac{P}{Q}$ is continuous everywhere except at values of x where _____ $= 0$.

7. a. Suppose f is continuous on $[a, b]$ and $f(a) < M < f(b)$. Then the Intermediate Value Theorem guarantees the existence of at least one number c in _____ such that _____.
b. If f is continuous on $[a, b]$ and $f(a)f(b) < 0$, then there must be at least one solution of the equation _____ in the interval _____.

8. a. The tangent line at $P(a, f(a))$ to the graph of f is the line passing through P and having slope _____.
b. If the slope of the tangent line at $P(a, f(a))$ is m, then an equation of the tangent line at P is _____.

9. a. The slope of the secant line passing through $P(a, f(a))$ and $Q(a + h, f(a + h))$ and the average rate of change of f over the interval $[a, a + h]$ are both given by _____.
b. The slope of the tangent line at $P(a, f(a))$ and the instantaneous rate of change of f at a are both given by _____.

10. a. If c is a constant, then $\dfrac{d}{dx}(c) =$ _____.
b. The Power Rule states that if n is any real number, then $\dfrac{d}{dx}(x^n) =$ _____.
c. The Constant Multiple Rule states that if c is a constant, then $\dfrac{d}{dx}[cf(x)] =$ _____.
d. The Sum Rule states that $\dfrac{d}{dx}[f(x) \pm g(x)] =$ _____.

11. a. The Product Rule states that $\dfrac{d}{dx}[f(x)g(x)] =$ _____.
b. The Quotient Rule states that $\dfrac{d}{dx}[f(x)/g(x)] =$ _____.

12. a. The Chain Rule states that if $h(x) = g[f(x)]$, then $h'(x) =$ _____.
b. The General Power Rule states that if $h(x) = [f(x)]^n$, then $h'(x) =$ _____.

13. If C, R, P, and $\overline{C}$ denote the total cost function, the total revenue function, the profit function, and the average cost function, respectively, then C' denotes the _____ _____ function, R' denotes the _____ _____ function, P' denotes the _____ _____ function, and $\overline{C}'$ denotes the _____ _____ _____ function.

14. a. If $g(x) = e^{f(x)}$, where f is a differentiable function, then $g'(x) =$ _____.
b. If $g(x) = \ln f(x)$, where $f(x) > 0$ is differentiable, then $g'(x) =$ _____.

CHAPTER 9 Review Exercises

In Exercises 1–14, find the indicated limits, if they exist.

1. $\lim\limits_{x \to 0} (5x - 3)$

2. $\lim\limits_{x \to 1} (x^2 + 1)$

3. $\lim\limits_{x \to -1} (3x^2 + 4)(2x - 1)$

4. $\lim\limits_{x \to 3} \dfrac{x - 3}{x + 4}$

5. $\lim\limits_{x \to 2} \dfrac{x + 3}{x^2 - 9}$

6. $\lim\limits_{x \to -2} \dfrac{x^2 - 2x - 3}{x^2 + 5x + 6}$

7. $\lim\limits_{x \to 3} \sqrt{2x^3 - 5}$

8. $\lim\limits_{x \to 3} \dfrac{4x - 3}{\sqrt{x + 1}}$

9. $\lim\limits_{x \to 1^+} \dfrac{x - 1}{x(x - 1)}$

10. $\lim\limits_{x \to 1^-} \dfrac{\sqrt{x} - 1}{x - 1}$

11. $\lim\limits_{x \to \infty} \dfrac{x^2}{x^2 - 1}$

12. $\lim\limits_{x \to -\infty} \dfrac{x + 1}{x}$

13. $\lim\limits_{x \to \infty} \dfrac{3x^2 + 2x + 4}{2x^2 - 3x + 1}$

14. $\lim\limits_{x \to -\infty} \dfrac{x^2}{x + 1}$

15. Sketch the graph of the function
$$f(x) = \begin{cases} 2x - 3 & \text{if } x \leq 2 \\ -x + 3 & \text{if } x > 2 \end{cases}$$
and evaluate $\lim\limits_{x \to a^+} f(x)$, $\lim\limits_{x \to a^-} f(x)$, and $\lim\limits_{x \to a} f(x)$ at $a = 2$, if the limits exist.

16. Sketch the graph of the function
$$f(x) = \begin{cases} 4 - x & \text{if } x \leq 2 \\ x + 2 & \text{if } x > 2 \end{cases}$$
and evaluate $\lim\limits_{x \to a^+} f(x)$, $\lim\limits_{x \to a^-} f(x)$, and $\lim\limits_{x \to a} f(x)$ at $a = 2$, if the limits exist.

In Exercises 17–20, determine all values of x for which each function is discontinuous.

17. $g(x) = \begin{cases} x + 3 & \text{if } x \neq 2 \\ 0 & \text{if } x = 2 \end{cases}$

18. $f(x) = \dfrac{3x + 4}{4x^2 - 2x - 2}$

19. $f(x) = \begin{cases} \dfrac{1}{(x + 1)^2} & \text{if } x \neq -1 \\ 2 & \text{if } x = -1 \end{cases}$

20. $f(x) = \dfrac{|2x|}{x}$

21. Let $y = x^2 + 2$.
   a. Find the average rate of change of y with respect to x in the intervals $[1, 2]$, $[1, 1.5]$, and $[1, 1.1]$.
   b. Find the (instantaneous) rate of change of y at $x = 1$.

22. Use the definition of the derivative to find the slope of the tangent line to the graph of the function $f(x) = 4x + 5$ at any point $P(x, f(x))$ on the graph.

23. Use the definition of the derivative to find the slope of the tangent line to the graph of the function $f(x) = -\dfrac{1}{x}$ at any point $P(x, f(x))$ on the graph.

24. Use the definition of the derivative to find the slope of the tangent line to the graph of the function $f(x) = \frac{3}{2}x + 5$ at the point $(-2, 2)$, and determine an equation of the tangent line.

25. Use the definition of the derivative to find the slope of the tangent line to the graph of the function $f(x) = -x^2$ at the point $(2, -4)$, and determine an equation of the tangent line.

26. The graph of the function f is shown in the accompanying figure.
   a. Is f continuous at $x = a$? Why?
   b. Is f differentiable at $x = a$? Justify your answers.

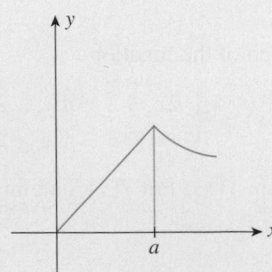

In Exercises 27–74, find the derivative of the given function.

27. $f(x) = 3x^5 - 2x^4 + 3x^2 - 2x + 1$

28. $f(x) = 4x^6 + 2x^4 + 3x^2 - 2$

29. $g(x) = -2x^{-3} + 3x^{-1} + 2$

30. $f(t) = 2t^2 - 3t^3 - t^{-1/2}$

31. $g(t) = 2t^{-1/2} + 4t^{-3/2} + 2$

32. $h(x) = x^2 + \dfrac{2}{x}$

33. $f(t) = t + \dfrac{2}{t} + \dfrac{3}{t^2}$

34. $g(s) = 2s^2 - \dfrac{4}{s} + \dfrac{2}{\sqrt{s}}$

35. $h(x) = x^2 - \dfrac{2}{x^{3/2}}$

36. $f(x) = \dfrac{x + 1}{2x - 1}$

37. $g(t) = \dfrac{t^2}{2t^2 + 1}$

38. $h(t) = \dfrac{\sqrt{t}}{\sqrt{t} + 1}$

39. $f(x) = \dfrac{\sqrt{x} - 1}{\sqrt{x} + 1}$

40. $f(t) = \dfrac{t}{2t^2 + 1}$

41. $f(x) = \dfrac{x^2(x^2 + 1)}{x^2 - 1}$

42. $f(x) = (2x^2 + x)^3$

43. $f(x) = (3x^3 - 2)^8$

44. $h(x) = (\sqrt{x} + 2)^5$

45. $f(t) = \sqrt{2t^2 + 1}$

46. $g(t) = \sqrt[3]{1 - 2t^3}$

47. $s(t) = (3t^2 - 2t + 5)^{-2}$

48. $f(x) = (2x^3 - 3x^2 + 1)^{-3/2}$

49. $f(x) = xe^{2x}$

50. $f(t) = \sqrt{t}e^t + t$

51. $g(t) = \sqrt{t}e^{-2t}$

52. $g(x) = e^x\sqrt{1 + x^2}$

53. $y = \dfrac{e^{2x}}{1 + e^{-2x}}$

54. $f(x) = e^{2x^2 - 1}$

55. $f(x) = xe^{-x^2}$

56. $g(x) = (1 + e^{2x})^{3/2}$

57. $f(x) = x^2e^x + e^x$

58. $g(t) = t \ln t$

59. $f(x) = \ln(e^{x^2} + 1)$

60. $f(x) = \dfrac{x}{\ln x}$

61. $f(x) = \dfrac{\ln x}{x + 1}$

62. $y = (x + 1)e^x$

63. $y = \ln(e^{4x} + 3)$

64. $f(r) = \dfrac{re^r}{1 + r^2}$

65. $f(x) = \dfrac{\ln x}{1 + e^x}$

66. $g(x) = \dfrac{e^{x^2}}{1 + \ln x}$

67. $h(x) = \left(x + \dfrac{1}{x}\right)^2$

68. $h(x) = \dfrac{1 + x}{(2x^2 + 1)^2}$

69. $h(t) = (t^2 + t)^4(2t^2)$

70. $f(x) = (2x + 1)^3(x^2 + x)^2$

71. $g(x) = \sqrt{x}(x^2 - 1)^3$

72. $f(x) = \dfrac{x}{\sqrt{x^3 + 2}}$

73. $h(x) = \dfrac{\sqrt{3x + 2}}{4x - 3}$

74. $f(t) = \dfrac{\sqrt{2t + 1}}{(t + 1)^3}$

In Exercises 75–84, find the second derivative of the given function.

75. $f(x) = 2x^4 - 3x^3 + 2x^2 + x + 4$

76. $g(x) = \sqrt{x} + \dfrac{1}{\sqrt{x}}$　　**77.** $h(t) = \dfrac{t}{t^2 + 4}$

78. $f(x) = xe^{-2x}$　　**79.** $h(x) = \dfrac{e^x}{1 + e^x}$

80. $f(x) = x \ln x$　　**81.** $y = \ln(3x + 1)$

82. $f(x) = (x^3 + x + 1)^2$　**83.** $f(x) = \sqrt{2x^2 + 1}$

84. $f(t) = t(t^2 + 1)^3$

85. Find $h'(0)$ if $h(x) = g(f(x))$, $g(x) = x + \dfrac{1}{x}$, and $f(x) = e^x$.

86. Find $h'(1)$ if $h(x) = g(f(x))$, $g(x) = \dfrac{x + 1}{x - 1}$, and $f(x) = \ln x$.

87. Let $f(x) = 2x^3 - 3x^2 - 16x + 3$.
   a. Find the point(s) on the graph of f at which the slope of the tangent line is equal to -4.
   b. Find the equation(s) of the tangent line(s) of part (a).

88. Let $f(x) = \frac{1}{3}x^3 + \frac{1}{2}x^2 - 4x + 1$.
   a. Find the point(s) on the graph of f at which the slope of the tangent line is equal to -2.
   b. Find the equation(s) of the tangent line(s) of part (a).

89. Find an equation of the tangent line to the graph of $y = \sqrt{4 - x^2}$ at the point $(1, \sqrt{3})$.

90. Find an equation of the tangent line to the graph of $y = x(x + 1)^5$ at the point $(1, 32)$.

91. Find an equation of the tangent line to the graph of $y = e^{-2x}$ at the point $(1, e^{-2})$.

92. Find an equation of the tangent line to the graph of $y = xe^{-x}$ at the point $(1, e^{-1})$.

93. Find the third derivative of the function

$$f(x) = \dfrac{1}{2x - 1}$$

What is the domain of f'''?

94. AVERAGE PRICE OF A COMMODITY　The average cost (in dollars) of producing x units of a certain commodity is given by

$$\overline{C}(x) = 20 + \dfrac{400}{x}$$

Evaluate $\lim\limits_{x \to \infty} \overline{C}(x)$, and interpret your results.

95. MANUFACTURING COSTS　Suppose that the total cost of manufacturing x units of a certain product is $C(x)$ dollars.
   a. What does $C'(x)$ measure? Give units.
   b. What can you say about the sign of C'?
   c. Given that $C'(1000) = 20$, estimate the additional cost to be incurred by the company in producing the 1001st unit of the product.

96. SALES OF CAMERAS　The shipments of Lica digital single-lens reflex cameras (SLRs) are projected to be

$$N(t) = 6t^2 + 200t + 4\sqrt{t} + 20,000 \qquad (0 \le t \le 4)$$

units t years from now.
   a. How many Lica SLRs will be shipped after 2 years?
   b. At what rate will the number of Lica SLRs shipped be changing after 2 years?

97. SALES OF DSPs　The annual sales of digital signal processors (DSPs) in billions of dollars is approximated by

$$S(t) = 0.14t^2 + 0.68t + 3.1 \qquad (0 \le t \le 6)$$

where t is measured in years, with $t = 0$ corresponding to the beginning of 1997.
   a. What were the sales of DSPs at the beginning of 1997? What were the sales at the beginning of 2002?
   b. How fast was the level of sales increasing at the beginning of 1997? How fast were sales increasing at the beginning of 2002?
   Source: World Semiconductor Trade Statistics.

98. U.K. DIGITAL VIDEO VIEWERS　Digital video viewing is one of the top online activities among U.K. Internet users. It is expected that between 2012 and 2017, the U.K. digital video audience will be given by

$$N(t) = 65.71t^{0.085} \qquad (2 \le t \le 7)$$

where $N(t)$ is measured in millions and t is measured in years, with $t = 2$ corresponding to 2012.
   a. How many U.K. digital video viewers will there be in 2015?
   b. How fast was the U.K. digital video audience expected to be changing in 2015?
   Source: eMarketer.

99. ADULT OBESITY　In the United States, the percentage of adults (age 20–74) classified as obese held steady through the 1960s and 1970s at around 14% but began to rise rapidly during the 1980s and 1990s. This rise in adult obesity coincided with the period when an increasing number of Americans began eating more sugar and fats. The function

$$P(t) = 0.01484t^2 + 0.446t + 15 \qquad (0 \le t \le 22)$$

gives the percentage of obese adults from 1978 $(t = 0)$ through the year 2000 $(t = 22)$.
   a. What percentage of adults were obese in 1978? In 2000?
   b. How fast was the percent of obese adults increasing in 1980 $(t = 2)$? In 1998 $(t = 20)$?
   Source: Journal of the American Medical Association.

100. CABLE TV SUBSCRIBERS　The number of subscribers to CNC Cable Television in the town of Randolph is approximated by the function

$$N(x) = 1000(1 + 2x)^{1/2} \qquad (1 \le x \le 30)$$

where $N(x)$ denotes the number of subscribers to the service in the xth week. Find the rate of increase in the number of subscribers at the end of the 12th week.

101. MALE LIFE EXPECTANCY Suppose the life expectancy of a male at birth in a certain country is described by the function

$$f(t) = 46.9(1 + 1.09t)^{0.1} \qquad (0 \le t \le 150)$$

where t is measured in years, with $t = 0$ corresponding to the beginning of 1900. How long can a male born at the beginning of 2000 in that country expect to live? What is the rate of change of the life expectancy of a male born in that country at the beginning of 2000?

102. DEMAND FOR SMARTPHONES The marketing department of Telecon has determined that the demand for their smartphones obeys the relationship

$$p = -0.02x + 600 \qquad (0 \le x \le 30{,}000)$$

where p denotes the phone's unit price (in dollars) and x denotes the quantity demanded.
a. Find the revenue function R.
b. Find the marginal revenue function R'.
c. Compute $R'(10{,}000)$, and interpret your result.

103. COST OF PRODUCING DVDs The total weekly cost in dollars incurred by Herald Media Corp. in producing x DVDs is given by the total cost function

$$C(x) = 2500 + 2.2x \qquad (0 \le x \le 8000)$$

a. What is the marginal cost when $x = 1000$ and 2000?
b. Find the average cost function $\bar{C}$ and the marginal average cost function $\bar{C}'$.

104. POPULATION GROWTH IN A SUBURB The population of a certain suburb is expected to be

$$P(t) = 30 - \frac{20}{2t + 3} \qquad (0 \le t \le 5)$$

thousand t years from now.
a. By how much will the population have grown after 3 years?
b. How fast is the population changing after 3 years?

CHAPTER 9 Before Moving On . . .

1. Find $\lim\limits_{x \to -1} \dfrac{x^2 + 4x + 3}{x^2 + 3x + 2}$.

2. Let

$$f(x) = \begin{cases} x^2 - 1 & \text{if } -2 \le x < 1 \\ x^3 & \text{if } 1 < x \le 2 \end{cases}$$

Find (a) $\lim\limits_{x \to 1^-} f(x)$ and (b) $\lim\limits_{x \to 1^+} f(x)$. Is f continuous at $x = 1$? Explain.

3. Use the definition of the derivative to find the slope of the tangent line to the graph of $x^2 - 3x + 1$ at the point $(1, -1)$. What is an equation of the tangent line?

4. Find the derivative of $f(x) = 2x^3 - 3x^{1/3} + 5x^{-2/3}$.

5. Differentiate $g(x) = x\sqrt{2x^2 - 1}$.

6. Find $\dfrac{dy}{dx}$ if $y = \dfrac{2x + 1}{x^2 + x + 1}$.

7. Find the first three derivatives of $f(x) = \dfrac{1}{\sqrt{x + 1}}$.

8. Find the slope of the tangent line to the graph of $f(x) = e^{\sqrt{x}}$.

9. Find the rate at which $y = x \ln(x^2 + 1)$ is changing at $x = 1$.

10 Applications of the Derivative

THIS CHAPTER FURTHER EXPLORES the power of the derivative as a tool to help analyze the properties of functions. We can then use this information to accurately sketch graphs of functions. We also see how the derivative is used in solving a large class of optimization problems, including finding what level of production will yield a maximum profit for a company, finding what level of production will result in minimal cost to a company, finding the maximum height attained by a rocket, finding the maximum velocity at which air is expelled when a person coughs, and a host of other problems.

How many loudspeaker systems should the Acrosonic company produce to maximize its profit? In Example 4, page 776, you will see how the techniques of calculus can be used to help answer this question.

10.1 Applications of the First Derivative

Determining the Intervals Where a Function Is Increasing or Decreasing

According to a study by the U.S. Department of Energy and the Shell Development Company, a typical car's fuel economy as a function of its speed is described by the graph shown in Figure 1. Observe that the fuel economy $f(x)$ in miles per gallon (mpg) improves as x, the vehicle's speed in miles per hour (mph), increases from 0 to 42 and then drops as the speed increases beyond 42 mph. We use the terms *increasing* and *decreasing* to describe the behavior of a function as we move from left to right along its graph.

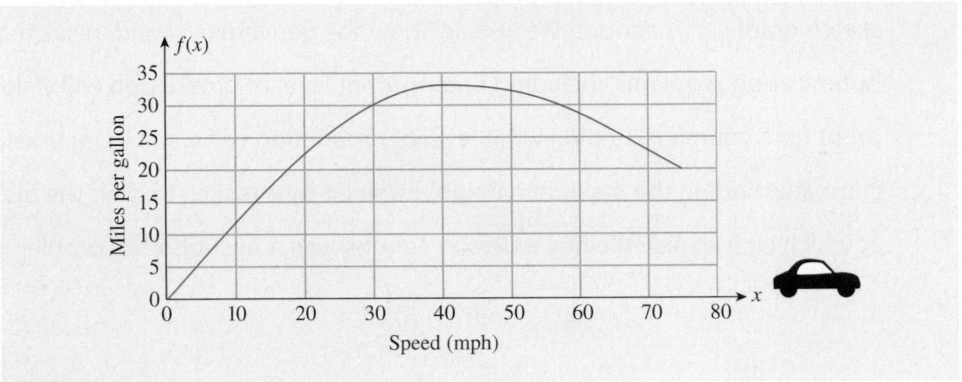

FIGURE **1**
A typical car's fuel economy improves as the speed at which it is driven increases from 0 mph to 42 mph and drops at speeds greater than 42 mph.

Sources: U.S. Department of Energy and Shell Development Co.

More precisely, we have the following definitions:

> ### Increasing and Decreasing Functions
>
> A function f is **increasing** on an interval (a, b) if for every two numbers x_1 and x_2 in (a, b), $f(x_1) < f(x_2)$ whenever $x_1 < x_2$ (Figure 2a).
>
> A function f is **decreasing** on an interval (a, b) if for every two numbers x_1 and x_2 in (a, b), $f(x_1) > f(x_2)$ whenever $x_1 < x_2$ (Figure 2b).
>
>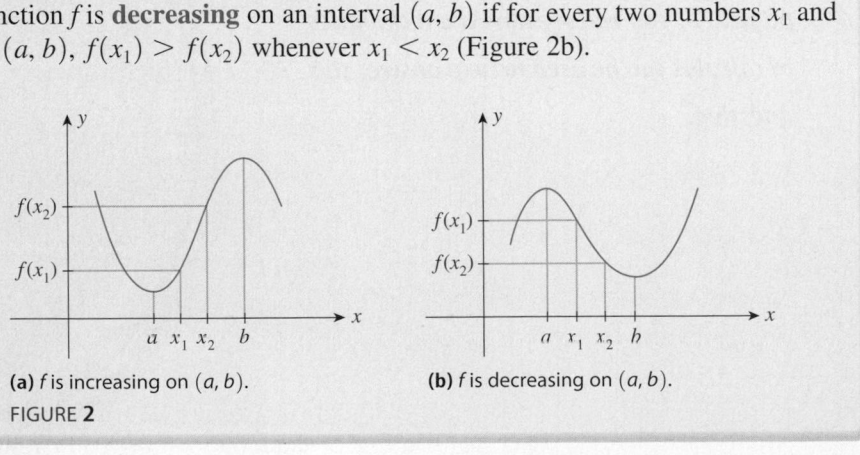
>
> **(a)** f is increasing on (a, b).   **(b)** f is decreasing on (a, b).
> FIGURE **2**

We say that f is *increasing at a number c* if there exists an interval (a, b) containing c such that f is increasing on (a, b). Similarly, we say that f is *decreasing at a number c* if there exists an interval (a, b) containing c such that f is decreasing on (a, b).

Since the rate of change of a function at $x = c$ is given by the derivative of the function at that number, the derivative lends itself naturally to being a tool for determining the intervals where a differentiable function is increasing or decreasing. Indeed, as we saw in Chapter 9, the derivative of a function at a number measures both the slope of the tangent line to the graph of the function at the point on the graph of f corresponding to that number and the rate of change of the function at that number. In fact, at a number where the derivative is positive, the slope of the tangent line to the graph is positive, and the function is increasing. At a number where the derivative is negative, the slope of the tangent line to the graph is negative, and the function is decreasing (Figure 3).

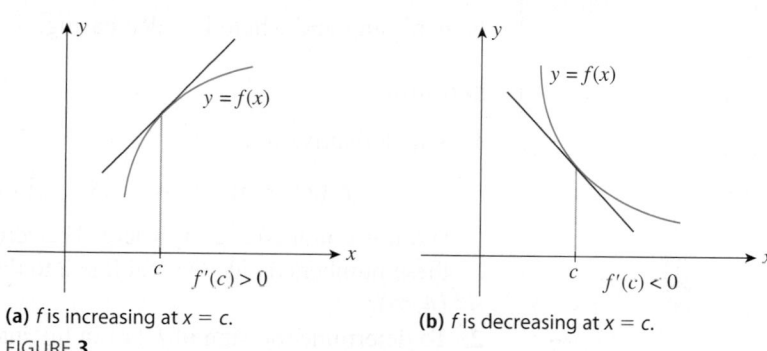

(a) f is increasing at $x = c$. **(b)** f is decreasing at $x = c$.

FIGURE **3**

These observations lead to the following important theorem, which we state without proof.

> **THEOREM 1**
>
> **a.** If $f'(x) > 0$ for every value of x in an interval (a, b), then f is increasing on (a, b).
>
> **b.** If $f'(x) < 0$ for every value of x in an interval (a, b), then f is decreasing on (a, b).
>
> **c.** If $f'(x) = 0$ for every value of x in an interval (a, b), then f is constant on (a, b).

EXAMPLE 1 Find the interval where the function $f(x) = x^2$ is increasing and the interval where it is decreasing.

Solution The derivative of $f(x) = x^2$ is $f'(x) = 2x$. Since

$$f'(x) = 2x > 0 \quad \text{if } x > 0 \qquad \text{and} \qquad f'(x) = 2x < 0 \quad \text{if } x < 0$$

f is increasing on the interval $(0, \infty)$ and decreasing on the interval $(-\infty, 0)$ (Figure 4).

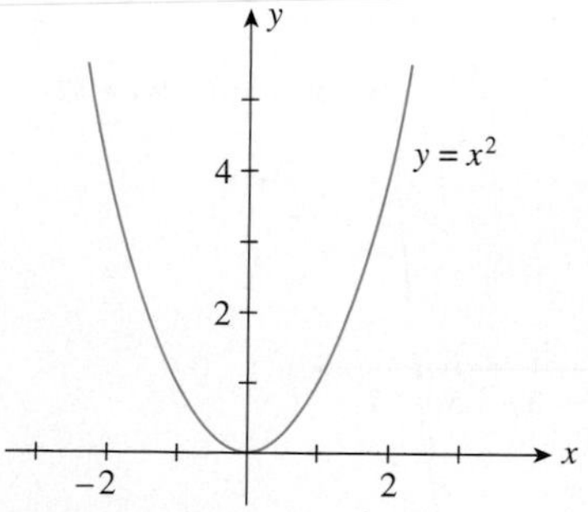

FIGURE **4**
The graph of f falls on $(-\infty, 0)$ where $f'(x) < 0$ and rises on $(0, \infty)$ where $f'(x) > 0$.

Recall that the graph of a continuous function cannot have any breaks. As a consequence, a continuous function cannot change sign unless it equals zero for some value of x. (See Theorem 5, page 621.) This observation suggests the following procedure for determining the sign of the derivative f' of a function f and hence the intervals where the function f is increasing and where it is decreasing.

> Determining the Intervals Where a Function Is Increasing or Decreasing
>
> 1. Find all values of x for which $f'(x) = 0$ or f' is discontinuous, and identify the open intervals determined by these numbers.
> 2. Select a test number c in each interval found in step 1, and determine the sign of $f'(c)$ in that interval.
>   a. If $f'(c) > 0$, f is increasing on that interval.
>   b. If $f'(c) < 0$, f is decreasing on that interval.

EXAMPLE 2 Determine the intervals where the function

$$f(x) = x^3 - 3x^2 - 24x + 32$$

is increasing and where it is decreasing.

Solution

1. The derivative of f is

$$f'(x) = 3x^2 - 6x - 24 = 3(x + 2)(x - 4) \qquad (x^2) \text{ See page 17.}$$

and it is continuous everywhere. The zeros of $f'(x)$ are $x = -2$ and $x = 4$, and these numbers divide the real line into the intervals $(-\infty, -2)$, $(-2, 4)$, and $(4, \infty)$.

2. To determine the sign of $f'(x)$ in the intervals $(-\infty, -2)$, $(-2, 4)$, and $(4, \infty)$, compute $f'(x)$ at a convenient test point in each interval. The results are shown in the following table:

Interval	Test Point c	$f'(c)$	Sign of $f'(x)$
$(-\infty, -2)$	-3	21	$+$
$(-2, 4)$	0	-24	$-$
$(4, \infty)$	5	21	$+$

(x^2) See page 58.

Using these results, we obtain the sign diagram shown in Figure 5. We conclude that f is increasing on the intervals $(-\infty, -2)$ and $(4, \infty)$ and is decreasing on the interval $(-2, 4)$. Figure 6 shows the graph of f.

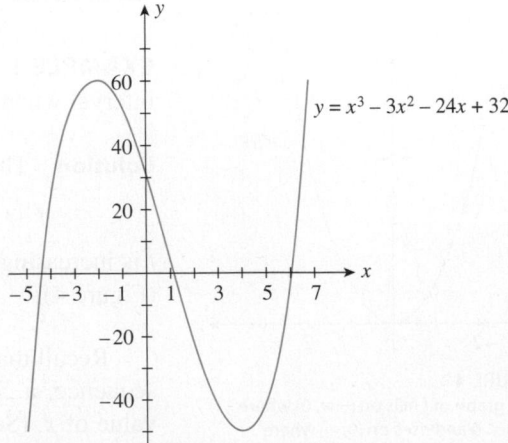

$$+ + + \, 0 \, - - - - - \, 0 \, + + + + \quad \longrightarrow x$$
$$\quad\quad -2 \quad 0 \quad\quad 4$$

FIGURE 5
Sign diagram for f'

FIGURE 6
The graph of f rises on $(-\infty, -2)$, falls on $(-2, 4)$, and rises again on $(4, \infty)$.

Note   We will learn how to sketch these graphs later. However, if you are familiar with the use of a graphing utility, you may go ahead and verify each graph.   ■

Exploring with TECHNOLOGY

Refer to Example 2.

1. Use a graphing utility to plot the graphs of $f(x) = x^3 - 3x^2 - 24x + 32$ and its derivative function $f'(x) = 3x^2 - 6x - 24$ using the viewing window $[-10, 10] \times [-50, 70]$.

2. By looking at the graph of f', determine the intervals where $f'(x) > 0$ and the intervals where $f'(x) < 0$. Next, look at the graph of f, and determine the intervals where it is increasing and the intervals where it is decreasing. Describe the relationship. Is it what you expected?

EXAMPLE 3 Find the interval where the function $f(x) = x^{2/3}$ is increasing and the interval where it is decreasing.

Solution

1. The derivative of f is

$$f'(x) = \frac{2}{3}x^{-1/3} = \frac{2}{3x^{1/3}}$$

The function f' is not defined at $x = 0$, so f' is discontinuous there. It is continuous everywhere else. Furthermore, f' is not equal to zero anywhere. The number 0 divides the real line (the domain of f) into the intervals $(-\infty, 0)$ and $(0, \infty)$.

2. Pick a test point (say, $x = -1$) in the interval $(-\infty, 0)$, and compute

$$f'(-1) = -\frac{2}{3}$$

Since $f'(-1) < 0$, we know that $f'(x) < 0$ on $(-\infty, 0)$. Next, we pick a test point (say, $x = 1$) in the interval $(0, \infty)$ and compute

$$f'(1) = \frac{2}{3}$$

Since $f'(1) > 0$, we know that $f'(x) > 0$ on $(0, \infty)$. Figure 7 shows these results in the form of a sign diagram.

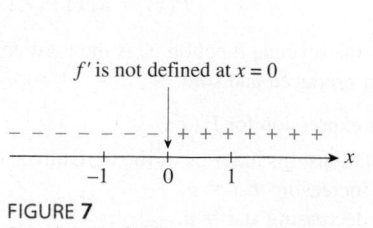

FIGURE **7**
Sign diagram for f'

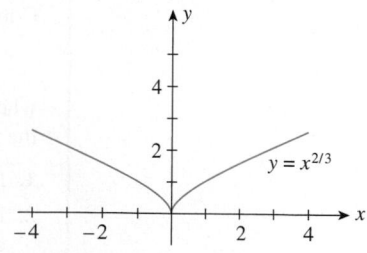

FIGURE **8**
f decreases on $(-\infty, 0)$ and increases on $(0, \infty)$.

We conclude that f is decreasing on the interval $(-\infty, 0)$ and increasing on the interval $(0, \infty)$. The graph of f, shown in Figure 8, confirms these results.   ■

EXAMPLE 4 Find the intervals where the function $f(x) = x + \dfrac{1}{x}$ is increasing and where it is decreasing.

Solution

1. The derivative of f is

$$f'(x) = 1 - \frac{1}{x^2} = \frac{x^2 - 1}{x^2} = \frac{(x + 1)(x - 1)}{x^2} \qquad \text{(x^2) See page 15.}$$

Since f' is not defined at $x = 0$, it is discontinuous there. Furthermore, $f'(x)$ is equal to zero when $(x + 1)(x - 1) = 0$ or $x = \pm 1$. Note that the value of f' is different from zero in the open intervals $(-\infty, -1)$, $(-1, 0)$, $(0, 1)$, and $(1, \infty)$.

2. To determine the sign of f' in each of these intervals, we compute $f'(x)$ at the test points $x = -2$, $-\frac{1}{2}$, $\frac{1}{2}$, and 2, respectively, obtaining $f'(-2) = \frac{3}{4}$, $f'(-\frac{1}{2}) = -3$, $f'(\frac{1}{2}) = -3$, and $f'(2) = \frac{3}{4}$. From the sign diagram for f' (Figure 9), we conclude that f is increasing on $(-\infty, -1)$ and $(1, \infty)$ and decreasing on $(-1, 0)$ and $(0, 1)$.

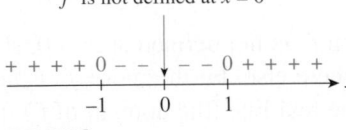

f' is not defined at $x = 0$

FIGURE 9
f' does not change sign as we move across $x = 0$.

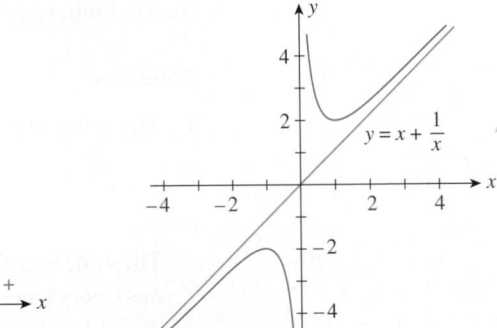

$y = x + \dfrac{1}{x}$

FIGURE 10
The graph of f rises on $(-\infty, -1)$, falls on $(-1, 0)$ and $(0, 1)$, and rises again on $(1, \infty)$.

The graph of f appears in Figure 10. Note that f' does not change sign as we move across $x = 0$. (Compare this with Example 3.)

 Example 4 reminds us that we must *not* automatically conclude that the derivative f' must change sign when we move across a number where f' is discontinuous or a zero of f'.

Explore and Discuss

Consider the profit function P associated with a certain commodity defined by

$$P(x) = R(x) - C(x) \qquad (x \geq 0)$$

where R is the revenue function, C is the total cost function, and x is the number of units of the product produced and sold.

1. Find an expression for $P'(x)$.
2. Find relationships in terms of the derivatives of R and C such that
   a. P is increasing at $x = a$.
   b. P is decreasing at $x = a$.
   c. P is neither increasing nor decreasing at $x = a$.
   Hint: Recall that the derivative of a function at $x = a$ measures the rate of change of the function at that number.
3. Explain the results of part 2 in economic terms.

> ### Exploring with TECHNOLOGY
>
> 1. Use a graphing utility to sketch the graphs of $f(x) = x^3 - ax$ for $a = -2$, -1, 0, 1, and 2, using the viewing window $[-2, 2] \times [-2, 2]$.
> 2. Use the results of part 1 to guess at the values of a for which f is increasing on $(-\infty, \infty)$.
> 3. Prove your conjecture analytically.

Relative Extrema

Besides helping us determine where the graph of a function is increasing and decreasing, the first derivative may be used to help us locate certain "high points" and "low points" on the graph of f. Knowing these points is invaluable in sketching the graphs of functions and solving optimization problems. These "high points" and "low points" correspond to the *relative (local) maxima* and *relative minima* of a function. They are so called because they are the highest or the lowest points when compared with points nearby.

The graph shown in Figure 11 gives the U.S. budget surplus (deficit) from 1996 ($t = 0$) to 2010 ($t = 14$). The relative maxima and the relative minima of the function f are indicated on the graph.

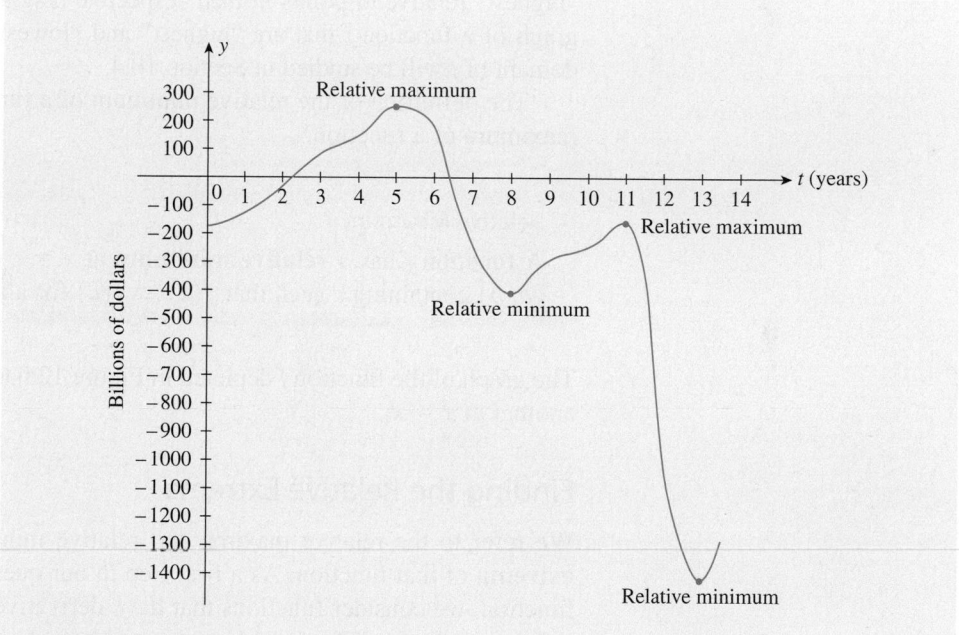

FIGURE 11
U.S. budget surplus (deficit) from 1996 to 2010
Source: Office of Management and Budget.

More generally, we have the following definition:

> ### Relative Maximum
>
> A function f has a **relative maximum** at $x = c$ if there exists an open interval (a, b) containing c such that $f(x) \leq f(c)$ for all x in (a, b).

Geometrically, this means that there is *some* interval containing $x = c$ such that no point on the graph of f with its x-coordinate in that interval can lie above the point

$(c, f(c))$; that is, $f(c)$ is the largest value of $f(x)$ in some interval around $x = c$. Figure 12 depicts the graph of a function f that has a relative maximum at $x = x_1$ and another at $x = x_3$.

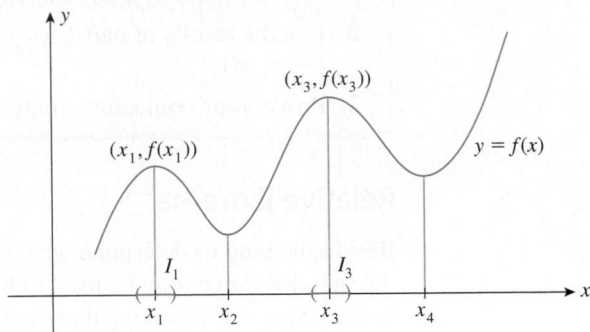

FIGURE **12**
f has a relative maximum at $x = x_1$ and at $x = x_3$.

Observe that all the points on the graph of f with x-coordinates in the interval I_1 containing x_1 (shown in blue) lie on or below the point $(x_1, f(x_1))$. This is also true for the point $(x_3, f(x_3))$ and the interval I_3. Thus, even though there are points on the graph of f that are "higher" than the points $(x_1, f(x_1))$ and $(x_3, f(x_3))$, the latter points are "highest" relative to points in their respective neighborhoods (intervals). Points on the graph of a function f that are "highest" and "lowest" with respect to *all* points in the domain of f will be studied in Section 10.4.

The definition of the relative minimum of a function parallels that of the relative maximum of a function.

> **Relative Minimum**
>
> A function f has a **relative minimum** at $x = c$ if there exists an open interval (a, b) containing c such that $f(x) \geq f(c)$ for all x in (a, b).

The graph of the function f depicted in Figure 12 has a relative minimum at $x = x_2$ and another at $x = x_4$.

Finding the Relative Extrema

We refer to the relative maxima and relative minima of a function as the **relative extrema** of that function. As a first step in our quest to find the relative extrema of a function, we consider functions that have derivatives at such points. Suppose that f is a function that is differentiable on some interval (a, b) that contains a number c and that f has a relative maximum at $x = c$ (Figure 13a).

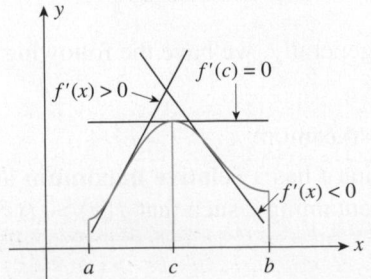

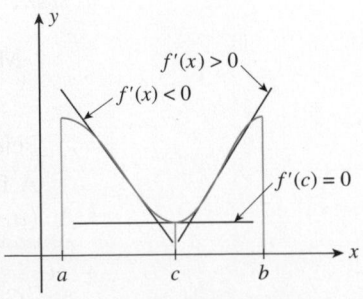

(a) f has a relative maximum at $x = c$.   **(b)** f has a relative minimum at $x = c$.

FIGURE **13**

Observe that the slope of the tangent line to the graph of f must change from positive to negative as we move across $x = c$ from left to right. Therefore, the tangent line to the graph of f at the point $(c, f(c))$ must be horizontal; that is, $f'(c) = 0$ (Figure 13a).

Using a similar argument, it may be shown that the derivative f' of a differentiable function f must also be equal to zero at $x = c$ if f has a relative minimum at $x = c$ (Figure 13b).

This analysis reveals an important characteristic of the relative extrema of a differentiable function f: *At any number c where f has a relative extremum, $f'(c) = 0$.*

 Before we develop a procedure for finding such numbers, a few words of caution are in order. First, this result tells us that if a differentiable function f has a relative extremum at a number $x = c$, then $f'(c) = 0$. The converse of this statement—if $f'(c) = 0$ at $x = c$, then f must have a relative extremum at that number—is *not* true. Consider, for example, the function $f(x) = x^3$. Here, $f'(x) = 3x^2$, so $f'(0) = 0$. However, f has neither a relative maximum nor a relative minimum at $x = 0$ (Figure 14).

Second, our result assumes that the function is differentiable and therefore has a derivative at a number that gives rise to a relative extremum. The functions $f(x) = |x|$ and $g(x) = x^{2/3}$ demonstrate that a relative extremum of a function may exist at a number at which the derivative does not exist. Both these functions fail to be differentiable at $x = 0$, but each has a relative minimum there. Figure 15 shows the graphs of these functions. Note that the slopes of the tangent lines change from negative to positive as we move across $x = 0$, just as in the case of a function that is differentiable at a value of x that gives rise to a relative minimum.

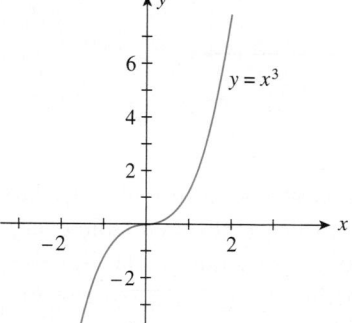

FIGURE **14**
$f'(0) = 0$, but f does not have a relative extremum at $(0, 0)$.

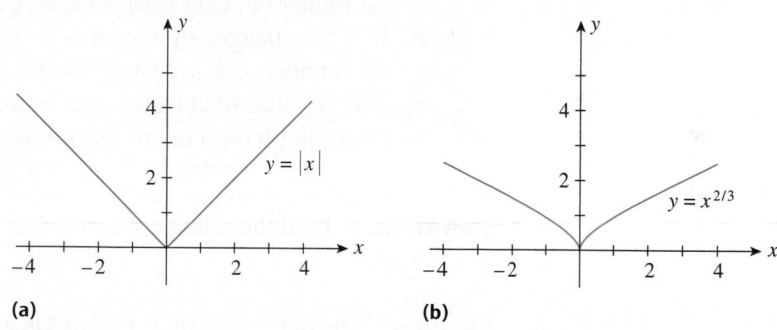

(a)   (b)

FIGURE **15**
Each of these functions has a relative extremum at $(0, 0)$, but the derivative does not exist there.

We refer to a number in the domain of f that *might* give rise to a relative extremum as a critical number.

> **Critical Number of f**
>
> A **critical number** of a function f is any number x in the domain of f such that $f'(x) = 0$ or $f'(x)$ does not exist.

Figure 16 (see next page) depicts the graph of a function that has critical numbers at $x = a, b, c, d,$ and e. Observe that $f'(x) = 0$ at $x = a, b,$ and c. Next, since there is a corner at $x = d$, $f'(x)$ does not exist there. Finally, $f'(x)$ does not exist at $x = e$ because the tangent line there is vertical. Also, observe that the critical numbers

$x = a$, b, and d give rise to relative extrema of f, whereas the critical numbers $x = c$ and $x = e$ do not.

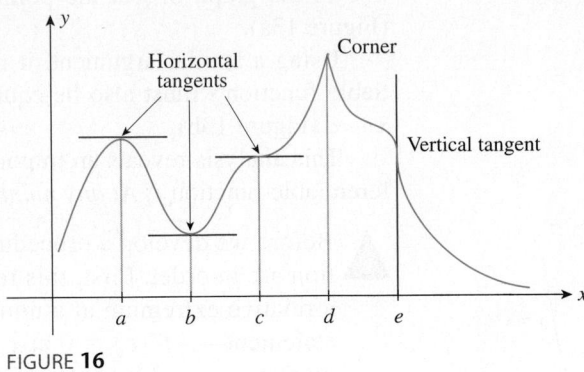

FIGURE **16**
Critical numbers of f

Having defined what a critical number is, we can now state a formal procedure for finding the relative extrema of a continuous function that is differentiable everywhere except at isolated values of x. Incorporated into the procedure is the so-called **First Derivative Test,** which helps us determine whether a number gives rise to a relative maximum or a relative minimum of the function f.

The First Derivative Test

Procedure for Finding the Relative Extrema of a Continuous Function f

1. Determine the critical numbers of f.

2. Determine the sign of $f'(x)$ to the left and right of each critical number.

   a. If $f'(x)$ changes sign from *positive* to *negative* as we move across a critical number c, then f has a relative maximum at $x = c$.

   b. If $f'(x)$ changes sign from *negative* to *positive* as we move across a critical number c, then f has a relative minimum at $x = c$.

   c. If $f'(x)$ does not change sign as we move across a critical number c, then f does not have a relative extremum at $x = c$.

EXAMPLE 5 Find the relative maxima and relative minima of the function $f(x) = x^2$.

Solution   The derivative of $f(x) = x^2$ is given by $f'(x) = 2x$. Setting $f'(x) = 0$ yields $x = 0$ as the only critical number of f. Since

$$f'(x) < 0 \quad \text{if } x < 0 \qquad \text{and} \qquad f'(x) > 0 \quad \text{if } x > 0$$

we see that $f'(x)$ changes sign from negative to positive as we move across the critical number 0. Thus, we conclude that $f(0) = 0$ is a relative minimum of f (Figure 17).

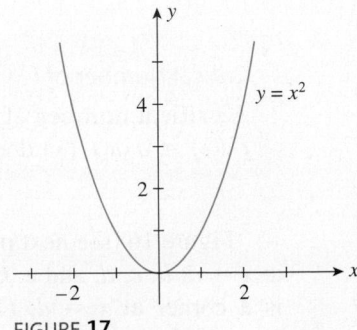

FIGURE **17**
f has a relative minimum at $x = 0$.

EXAMPLE 6 Find the relative maxima and relative minima of the function $f(x) = x^{2/3}$ (see Example 3).

Solution The derivative of f is $f'(x) = \frac{2}{3}x^{-1/3}$. As was noted in Example 3, f' is not defined at $x = 0$, is continuous everywhere else, and is not equal to zero in its domain. Thus, $x = 0$ is the only critical number of the function f.

The sign diagram obtained in Example 3 is reproduced in Figure 18. We can see that the sign of $f'(x)$ changes from negative to positive as we move across $x = 0$ from left to right. Thus, an application of the First Derivative Test tells us that $f(0) = 0$ is a relative minimum of f (Figure 19).

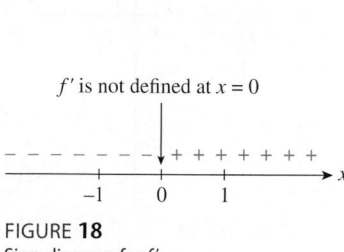

FIGURE **18**
Sign diagram for f'

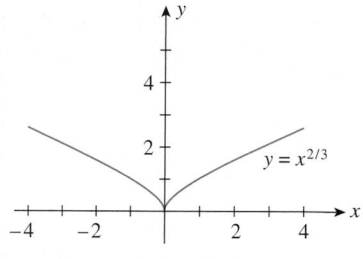

FIGURE **19**
f has a relative minimum at $x = 0$.

Explore and Discuss

Recall that the average cost function $\overline{C}$ is defined by

$$\overline{C} = \frac{C(x)}{x}$$

where $C(x)$ is the total cost function and x is the number of units of a commodity manufactured (see Section 9.8).

1. Show that

$$\overline{C}'(x) = \frac{C'(x) - \overline{C}(x)}{x} \qquad (x > 0)$$

2. Use the result of part 1 to conclude that $\overline{C}$ is decreasing for values of x at which $C'(x) < \overline{C}(x)$. Find similar conditions for which $\overline{C}$ is increasing and for which $\overline{C}$ is constant.

3. Explain the results of part 2 in economic terms.

EXAMPLE 7 Find the relative maxima and relative minima of the function

$$f(x) = x^3 - 3x^2 - 24x + 32$$

Solution The derivative of f is

$$f'(x) = 3x^2 - 6x - 24 = 3(x + 2)(x - 4) \qquad \boxed{x^2} \text{ See page 17.}$$

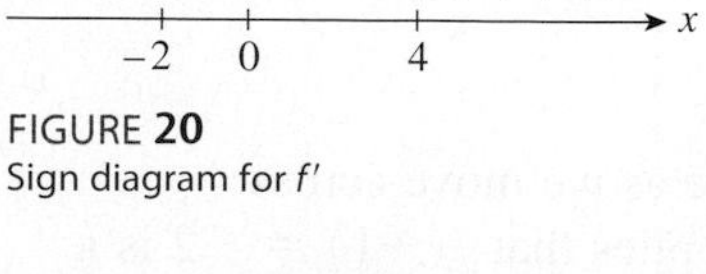

and it is continuous everywhere. The zeros of $f'(x)$, $x = -2$ and $x = 4$, are the only critical numbers of the function f. The sign diagram for f' is shown in Figure 20. Examine the two critical numbers $x = -2$ and $x = 4$ for a relative extremum using the First Derivative Test and the sign diagram for f'.

$$\overset{+\,+\,+\,0\,-\,-\,-\,-\,-\,0\,+\,+\,+\,+}{\underset{-2 \qquad 0 \qquad 4}{\vrule\quad\vrule\quad\vrule}} \longrightarrow x$$

FIGURE **20**
Sign diagram for f'

1. *The critical number* -2: Since the function $f'(x)$ changes sign from positive to negative as we move across $x = -2$ from left to right, we conclude that a relative maximum of f occurs at $x = -2$. The value of $f(x)$ when $x = -2$ is

$$f(-2) = (-2)^3 - 3(-2)^2 - 24(-2) + 32 = 60$$

2. *The critical number* 4: $f'(x)$ changes sign from negative to positive as we move across $x = 4$ from left to right, so $f(4) = -48$ is a relative minimum of f. The graph of f appears in Figure 21.

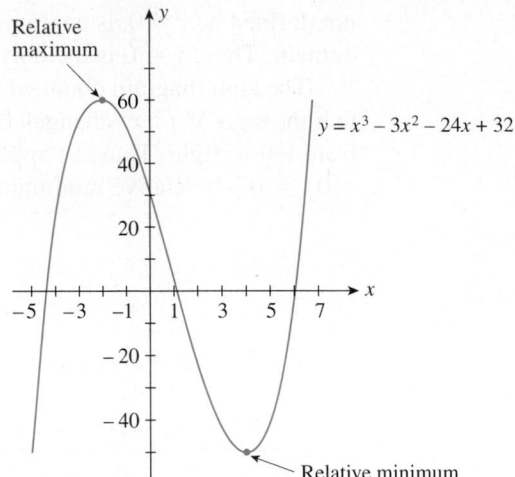

FIGURE 21
f has a relative maximum at $x = -2$ and a relative minimum at $x = 4$.

EXAMPLE 8 Find the relative maxima and the relative minima of the function

$$f(x) = x + \frac{1}{x}$$

Solution The derivative of f is

$$f'(x) = 1 - \frac{1}{x^2} = \frac{x^2 - 1}{x^2} = \frac{(x + 1)(x - 1)}{x^2}$$

Since f' is equal to zero at $x = -1$ and $x = 1$, these are critical numbers for the function f. Next, observe that f' is discontinuous at $x = 0$. However, because f is *not defined at that number,* $x = 0$ does not qualify as a critical number of f. Figure 22 shows the sign diagram for f'.

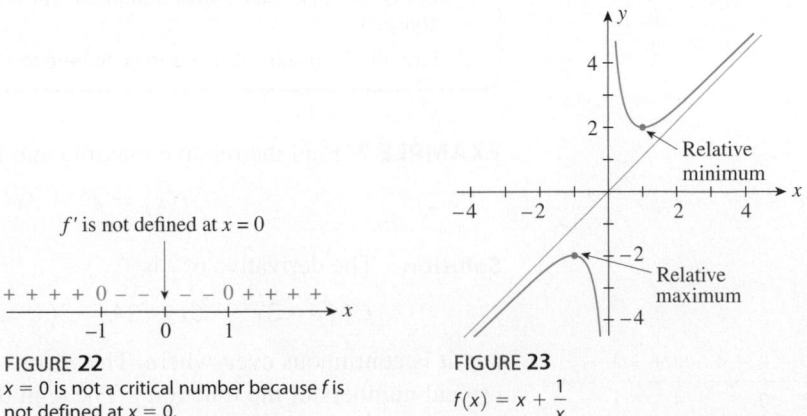

FIGURE 22
$x = 0$ is not a critical number because f is not defined at $x = 0$.

FIGURE 23
$f(x) = x + \dfrac{1}{x}$

Since $f'(x)$ changes sign from positive to negative as we move across $x = -1$ from left to right, the First Derivative Test implies that $f(-1) = -2$ is a relative maximum of the function f. Next, $f'(x)$ changes sign from negative to positive as we move across $x = 1$ from left to right, so $f(1) = 2$ is a relative minimum of the function f. The graph of f appears in Figure 23. Note that this function has a relative maximum that lies below its relative minimum.

EXAMPLE 9 Let $f(x) = xe^{-x}$, and find:

a. the intervals where f is increasing or decreasing
b. the relative extrema of f.

Solution

a. The derivative f is

$$f'(x) = \frac{d}{dx}(xe^{-x}) = x\frac{d}{dx}(e^{-x}) + e^{-x}\frac{d}{dx}(x) \qquad \text{Use the Product Rule.}$$

$$= xe^{-x}(-1) + e^{-x} = (1 - x)e^{-x}$$

Observe that f' is continuous everywhere. Next, setting $f'(x) = 0$ gives $x = 1$ (remember that $e^{-x} \neq 0$ for all values of x). Therefore, $x = 1$ is the sole critical number of f. The sign diagram of f' is shown in Figure 24. From the sign diagram, we see that f is increasing on $(-\infty, 1)$ and decreasing on $(1, \infty)$.

b. Using the sign diagram and the First Derivative Test, we see that the critical number $x = 1$ gives rise to a relative maximum of f with value $f(1) = e^{-1} = 1/e$. The graph of f shown in Figure 25 confirms our results.

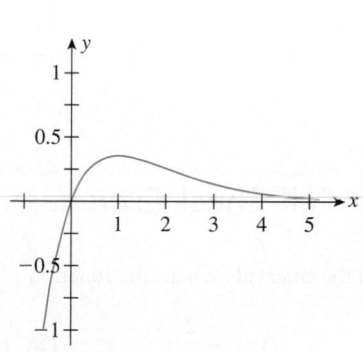

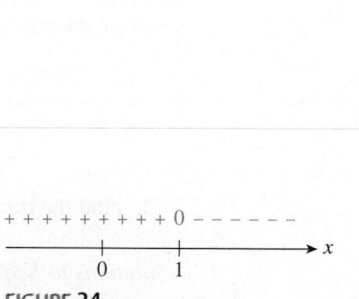

FIGURE 24
Sign diagram for f'

FIGURE 25
The graph of $f(x) = xe^{-x}$ is increasing on $(-\infty, 1)$ and decreasing on $(1, \infty)$.

APPLIED EXAMPLE 10 Profit Functions The profit function of Acrosonic Company is given by

$$P(x) = -0.02x^2 + 300x - 200,000$$

dollars, where x is the number of Acrosonic model F loudspeaker systems produced. Find where the function P is increasing and where it is decreasing.

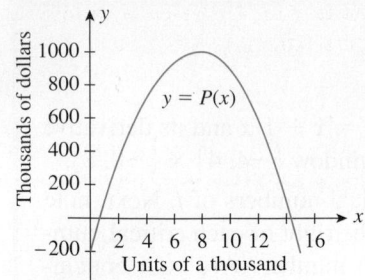

FIGURE **26**
The profit function is increasing on
(0, 7500) and decreasing on (7500, ∞).

Solution The derivative P' of the function P is

$$P'(x) = -0.04x + 300 = -0.04(x - 7500)$$

Thus, $P'(x) = 0$ when $x = 7500$. Furthermore, $P'(x) > 0$ for x in the interval $(0, 7500)$, and $P'(x) < 0$ for x in the interval $(7500, \infty)$. This means that the profit function P is increasing on $(0, 7500)$ and decreasing on $(7500, \infty)$ (Figure 26). ∎

 APPLIED EXAMPLE 11 Crime Rates The number of major crimes committed in the city of Bronxville from 2006 to 2013 is approximated by the function

$$N(t) = -0.1t^3 + 1.5t^2 + 100 \qquad (0 \le t \le 7)$$

where $N(t)$ denotes the number of crimes committed in year t, with $t = 0$ corresponding to the beginning of 2006. Find where the function N is increasing and where it is decreasing.

Solution The derivative N' of the function N is

$$N'(t) = -0.3t^2 + 3t = -0.3t(t - 10)$$

Since $N'(t) > 0$ for t in the interval $(0, 7)$, the function N is increasing throughout that interval (Figure 27).

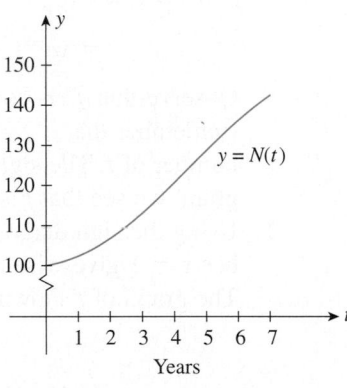

FIGURE **27**
The number of crimes, $N(t)$, is increasing over the 7-year interval. ∎

10.1 Self-Check Exercises

1. Find the intervals where the function

$$f(x) = \frac{2}{3}x^3 - x^2 - 12x + 3$$

is increasing and the intervals where it is decreasing.

2. Find the relative extrema of $f(x) = \dfrac{x^2}{1 - x^2}$.

Solutions to Self-Check Exercises 10.1 can be found on page 736.

10.1 Concept Questions

1. Explain each of the following:
   a. f is increasing on an interval I.
   b. f is decreasing on an interval I.

2. Describe a procedure for determining where a function is increasing and where it is decreasing.

3. Explain each term: (a) relative maximum and (b) relative minimum.

4. a. What is a critical number of a function f?
   b. Explain the role of critical numbers in determining the relative extrema of a function.

5. Describe the First Derivative Test, and describe a procedure for finding the relative extrema of a function.

10.1 Exercises

In Exercises 1–8, you are given the graph of a function f. Determine the intervals where f is increasing, constant, or decreasing.

1.
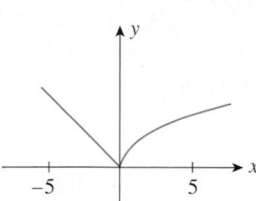

2.
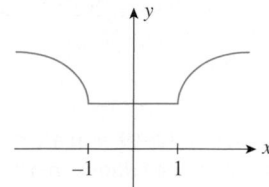

3.
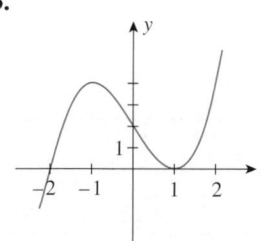

4.
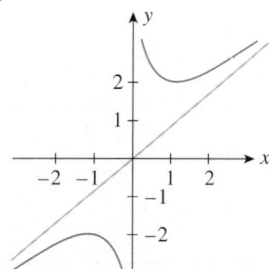

5.
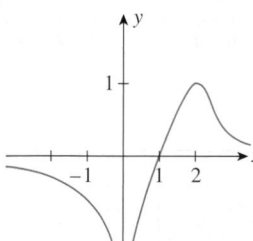

6.
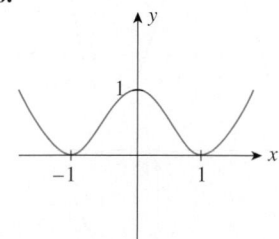

7.
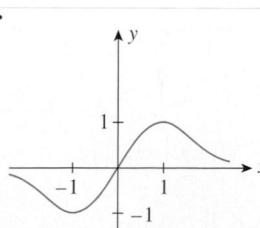

8.
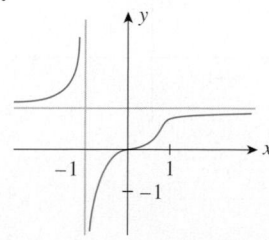

9. THE BOSTON MARATHON The graph of the function f shown in the following figure gives the elevation of the part of the Boston Marathon course that includes the notorious Heartbreak Hill. Determine the intervals (stretches of the course) where the function f is increasing (the runner is laboring), where it is constant (the runner is taking a breather), and where it is decreasing (the runner is coasting).

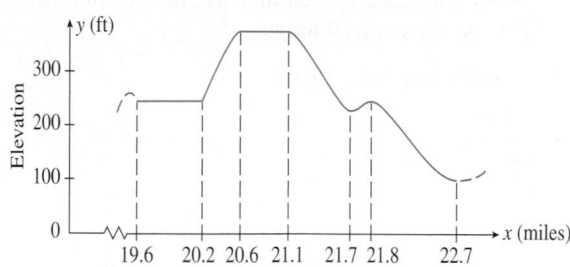

10. SOLAR PANEL POWER OUTPUT The graph of the function f shown in the accompanying figure gives the average "fixed" solar panel power output over a 15-hr period on a typical day. Determine the interval(s) where f is increasing, the interval(s) where f is constant, and the interval(s) where f is decreasing. Here, $t = 0$ corresponds to 5 A.M. Interpret your result.

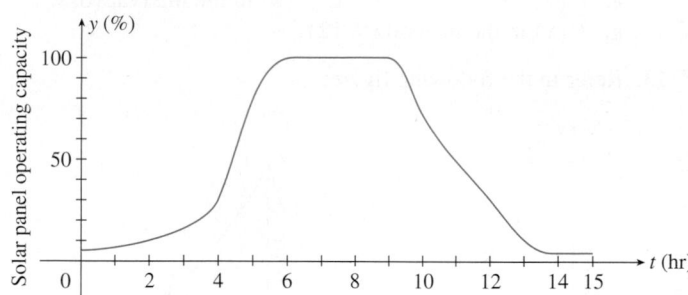

Source: Solarcity.com/California.

11. AIRCRAFT STRUCTURAL INTEGRITY Among the important factors in determining the structural integrity of an aircraft is its age. Advancing age makes planes more likely to crack. The graph of the function f, shown in the accompanying figure, is referred to as a "bathtub curve" in the airline industry. It gives the fleet damage rate (damage due to corrosion, accident, and metal fatigue) of a typical fleet of commercial aircraft as a function of the number of years of service.

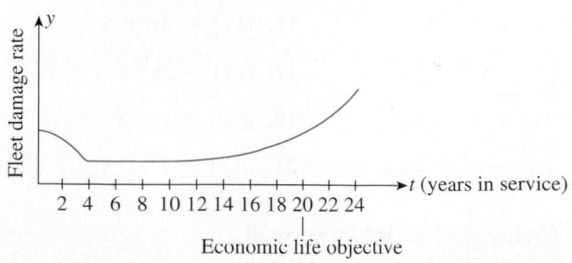

a. Determine the interval where f is decreasing. This corresponds to the time period when the fleet damage rate is dropping as problems are found and corrected during the initial "shakedown" period.

b. Determine the interval where f is constant. After the initial shakedown period, planes have few structural problems, and this is reflected by the fact that the function is constant on this interval.

c. Determine the interval where f is increasing. Beyond the time period mentioned in part (b), the function is increasing, reflecting an increase in structural defects due mainly to metal fatigue.

12. Refer to the following figure:

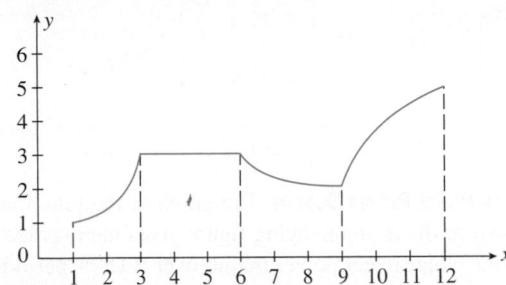

What is the sign of the following?

a. $f'(2)$ **b.** $f'(x)$ in the interval $(1, 3)$
c. $f'(4)$ **d.** $f'(x)$ in the interval $(3, 6)$
e. $f'(7)$ **f.** $f'(x)$ in the interval $(6, 9)$
g. $f'(x)$ in the interval $(9, 12)$

13. Refer to the following figure:

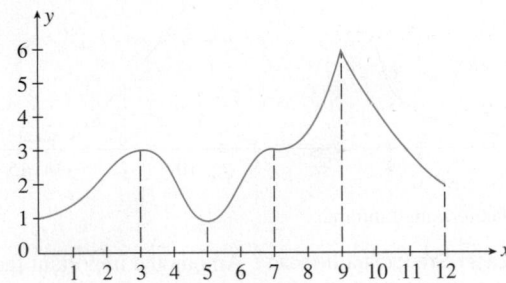

a. What are the critical numbers of f? Give reasons for your answers.
b. Draw the sign diagram for f'.
c. Find the relative extrema of f.

In Exercises 14–41, find the interval(s) where the function is increasing and the interval(s) where it is decreasing.

14. $f(x) = 4 - 5x$ **15.** $f(x) = 3x + 5$

16. $f(x) = x^2 - 3x$ **17.** $f(x) = 2x^2 + x + 1$

18. $f(x) = x^3 - 3x^2$ **19.** $g(x) = x - x^3$

20. $f(x) = x^3 - 3x + 4$ **21.** $g(x) = x^3 + 3x^2 + 1$

22. $f(x) = \dfrac{1}{3}x^3 - 3x^2 + 9x + 20$

23. $f(x) = \dfrac{2}{3}x^3 - 2x^2 - 6x - 2$

24. $g(x) = x^4 - 2x^2 + 4$ **25.** $h(x) = x^4 - 4x^3 + 10$

26. $f(x) = \dfrac{1}{x - 2}$ **27.** $h(x) = \dfrac{1}{2x + 3}$

28. $h(t) = \dfrac{t}{t - 1}$ **29.** $g(t) = \dfrac{2t}{t^2 + 1}$

30. $f(x) = x^{3/5}$ **31.** $f(x) = x^{2/3} + 5$

32. $f(x) = \sqrt{x + 1}$ **33.** $f(x) = (x - 5)^{2/3}$

34. $f(x) = \sqrt{16 - x^2}$ **35.** $g(x) = x\sqrt{x + 1}$

36. $f(x) = e^{-x^2/2}$ **37.** $f(x) = x^2 e^{-x}$

38. $f(x) = \ln x^2$ **39.** $f(x) = \dfrac{\ln x}{x}$

40. $f(x) = \dfrac{1 - x^2}{x}$ **41.** $h(x) = \dfrac{x^2}{x - 1}$

In Exercises 42–49, you are given the graph of a function f. Determine the relative maxima and relative minima, if any.

42. 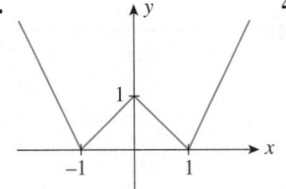   **43.**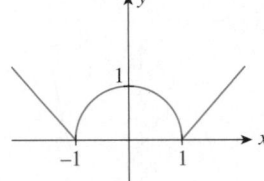

44. 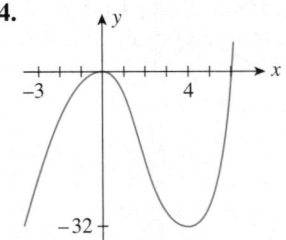   **45.**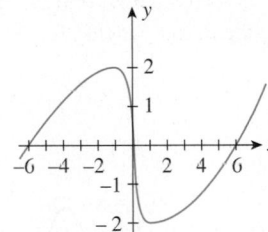

46. 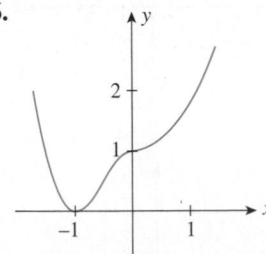   **47.**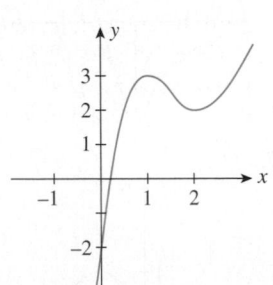

48.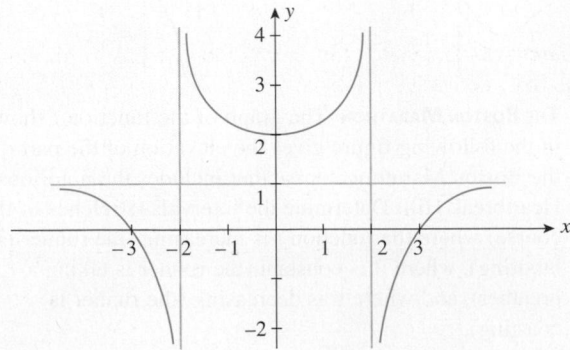

49.

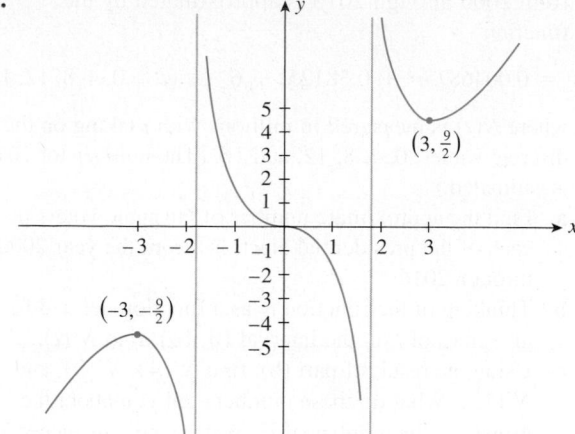

In Exercises 50 and 51, you are given the graph of the derivative, f', of a function f on an interval (a, b). Use this graph to determine where the graph of f is increasing, where f is constant, and where f is decreasing.

50.

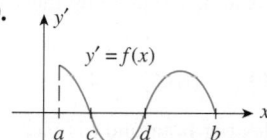

51.

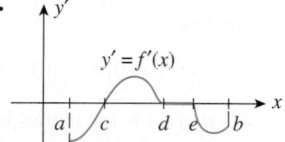

52. Motion of a Car A car travels along a straight road. Its velocity at any time t in the time interval (a, b) is shown in the accompanying figure. Describe the motion of the car.

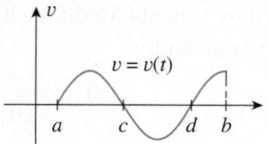

53. Profit of a Company The graph of the derivative P' of a profit function, P, is shown in the following figure. Find the levels of production, x, at which the profit of the company is increasing, stationary (neither increasing or decreasing), and decreasing.

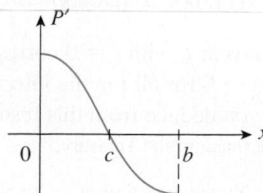

In Exercises 54–57, match the graph of the function with the graph of its derivative in (a)–(d).

(a) **(b)**

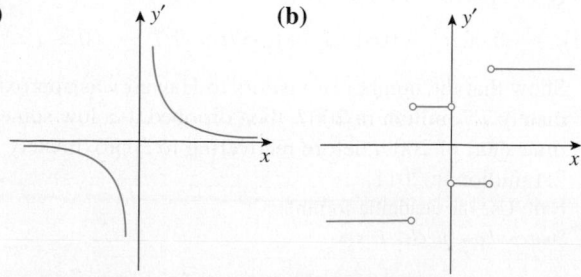

(c) **(d)**

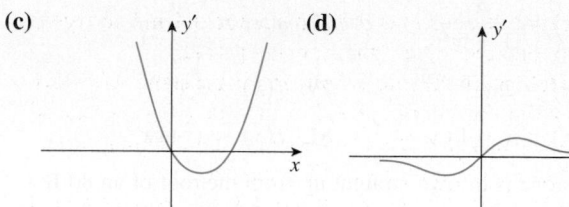

54. **55.**

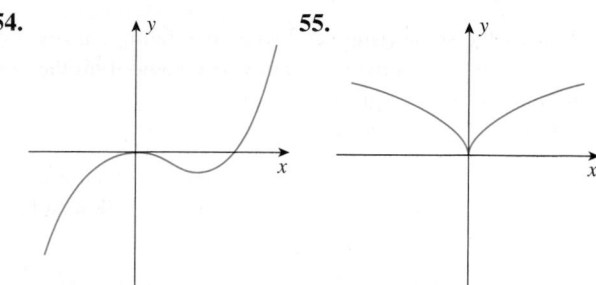

56. **57.**

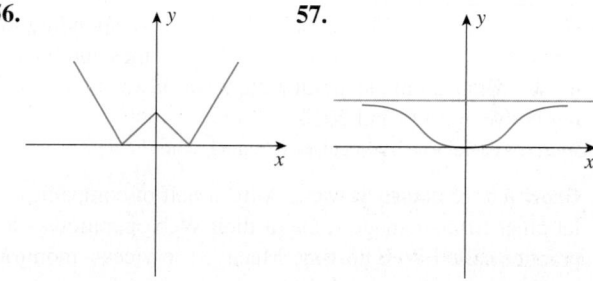

In Exercises 58–81, find the relative maxima and relative minima, if any, of each function.

58. $f(x) = x^2 - 4x$

59. $g(x) = x^2 + 3x + 8$

60. $f(x) = \frac{1}{2}x^2 - 2x + 4$

61. $h(t) = -t^2 + 6t + 6$

62. $f(x) = x^{5/3}$

63. $f(x) = x^{2/3} + 2$

64. $g(x) = x^3 - 3x^2 + 5$

65. $f(x) = x^3 - 3x + 6$

66. $F(x) = \frac{1}{3}x^3 - x^2 - 3x + 4$

67. $f(x) = \frac{1}{2}x^4 - x^2$

68. $h(x) = \frac{1}{2}x^4 - 3x^2 + 4x - 8$

69. $g(x) = x^4 - 4x^3 + 12$

70. $f(x) = 3x^4 - 2x^3 + 4$

71. $F(t) = 3t^5 - 20t^3 + 20$

72. $h(x) = \frac{x}{x+1}$

73. $g(x) = \frac{x+1}{x}$

74. $f(x) = x + \frac{9}{x} + 2$

75. $g(x) = 2x^2 + \frac{4000}{x} + 10$

76. $f(x) = \dfrac{x}{1 + x^2}$

77. $g(x) = \dfrac{x}{x^2 - 1}$

78. $f(x) = 2xe^{-x}$

79. $f(x) = x^2e^{-2x}$

80. $f(x) = x - \ln x$

81. $f(x) = x^2 \ln x$

82. A stone is thrown straight up from the roof of an 80-ft building. The distance (in feet) of the stone from the ground at any time t (in seconds) is given by

$$h(t) = -16t^2 + 64t + 80$$

When is the stone rising, and when is it falling? If the stone were to miss the building, when would it hit the ground? Sketch the graph of h.

Hint: The stone is on the ground when $h(t) = 0$.

83. **SMART METERS** The percentage of U.S. homes and businesses with digital electrical meters between 2008 and 2012 is approximated by

$$N(t) = 1.1375t^2 + 0.25t + 4.6 \qquad (0 \le t \le 4)$$

Here, t is measured in years, with $t = 0$ corresponding to 2008. Show that the percentage of U.S. homes and businesses with digital electrical meters was always increasing between 2008 and 2012.

Source: Federal Energy Regulatory Commission.

84. **GROWTH OF MANAGED SERVICES** Almost half of companies let other firms manage some of their Web operations—a practice called Web hosting. Managed services—monitoring a customer's technology services—is the fastest-growing part of Web hosting. Managed services sales are approximated by the function

$$f(t) = 0.469t^2 + 0.758t + 0.44 \qquad (0 \le t \le 10)$$

where $f(t)$ is measured in billions of dollars and t is measured in years, with $t = 0$ corresponding to 1999.

a. Find the interval where f is increasing.

b. What does your result tell you about sales in managed services from 1999 through 2009?

Source: International Data Corp.

85. **ENVIRONMENT OF FORESTS** Following the lead of the National Wildlife Federation, the Department of the Interior of a South American country began to record an index of environmental quality that measured progress and decline in the environmental quality of its forests. The index for the years 2004 through 2014 is approximated by the function

$$I(t) = \frac{1}{3}t^3 - \frac{5}{2}t^2 + 80 \qquad (0 \le t \le 10)$$

where $t = 0$ corresponds to 2004. Find the intervals where the function I is increasing and the intervals where it is decreasing. Interpret your results.

86. **VOTER TURNOUT** According to a study conducted by the Pew Research Center in 2013, the number of Hispanic voters in presidential elections in each of the years from 2000 through 2016 is approximated by the function

$$N(t) = 0.0046875t^2 + 0.38125t + 6 \qquad (t = 0, 4, 8, 12, 16)$$

where $N(t)$ is measured in millions with t taking on the discrete values, 0, 4, 8, 12, and 16. (The number for 2016 is estimated.)

a. Find the approximate number of Hispanic voters in each of the presidential elections from the year 2000 through 2016.

b. Thinking of the function N as a function defined for all values of t on the interval $(0, 16)$, find $N'(t)$.

c. Using the result of part (b), find $N'(4)$, $N'(8)$, and $N'(12)$. What do these numbers tell you about the growth in the number of Hispanic voters in succeeding elections since the year 2000?

Source: Pew Research Center.

87. **AVERAGE SPEED OF A HIGHWAY VEHICLE** The average speed of a vehicle on a stretch of Route 134 between 6 A.M. and 10 A.M. on a typical weekday is approximated by the function

$$f(t) = 20t - 40\sqrt{t} + 50 \qquad (0 \le t \le 4)$$

where $f(t)$ is measured in miles per hour and t is measured in hours, with $t = 0$ corresponding to 6 A.M. Find the interval where f is increasing and the interval where f is decreasing and interpret your results.

88. **AVERAGE COST FOR PRODUCING CDs** The average cost (in dollars) incurred by Lincoln Media each week in pressing x compact discs is given by

$$\overline{C}(x) = -0.0001x + 2 + \frac{2000}{x} \qquad (0 < x \le 6000)$$

Show that $\overline{C}(x)$ is always decreasing over the interval $(0, 6000)$.

89. **PUBLIC DEBT** The U.S. public debt outstanding from 1990 through 2011 is approximated by

$$D(t) = 0.0032t^3 - 0.0698t^2 + 0.6048t + 3.22 \qquad (0 \le t \le 21)$$

trillion dollars in year t, with $t = 0$ corresponding to 1990. Show that $D'(t) > 0$ for all t in the interval $(0, 21)$. What conclusion can you deduce from this result?

Source: U.S. Department of the Treasury.

90. **HAWAII TOURISM** Numbers of visitors to the Aloha State dropped dramatically during the recession. The number of visitors to Hawaii between 2007 ($t = 0$) and 2011 ($t = 4$) is approximately

$$N(t) = -0.062t^3 + 0.617t^2 - 1.557t + 7.7 \qquad (0 \le t \le 4)$$

Show that the number of visitors to Hawaii was approximately 7.7 million in 2007, then dropped to a low sometime short of 2009, before recovering to approximately 7.4 million in 2011.

Hint: Use the quadratic formula.

Source: Los Angeles Times.

91. DRIVING COSTS A study of driving costs based on 2012 medium-sized sedans found that the average cost (car payments, gas, insurance, upkeep, and depreciation) in cents per mile is approximately

$$C(x) = \frac{1910.5}{x^{1.72}} + 42.9 \qquad (5 \le x \le 20)$$

where x (in thousands) denotes the number of miles the car is driven each year. Show that C is a decreasing function of x on the interval $(5, 20)$. What does your result tell you about the average cost of driving a 2012 medium-sized sedan in terms of the number of miles driven?
Source: American Automobile Association.

92. FIRST-CLASS MAIL The volume of first-class mail deliveries (in billions of pieces) from 2008 through 2012 is approximated by

$$f(t) = \frac{90.7}{0.01t^2 + 0.01t + 1} \qquad (0 \le t \le 4)$$

where t is measured in years, with $t = 0$ corresponding to 2008. Show that the volume of first-class mail deliveries has been declining throughout the period under consideration.
Source: U.S. Postal Service.

93. AIR POLLUTION According to the South Coast Air Quality Management District, the level of nitrogen dioxide, a brown gas that impairs breathing, present in the atmosphere on a certain May day in downtown Los Angeles is approximated by

$$A(t) = 0.03t^3(t - 7)^4 + 60.2 \qquad (0 \le t \le 7)$$

where $A(t)$ is measured in pollutant standard index (PSI) and t is measured in hours, with $t = 0$ corresponding to 7 A.M. At what time of day is the air pollution increasing, and at what time is it decreasing?

94. DRUG CONCENTRATION IN THE BLOOD The concentration (in milligrams per cubic centimeter) of a certain drug in a patient's body t hr after injection is given by

$$C(t) = \frac{t^2}{2t^3 + 1} \qquad (0 \le t \le 4)$$

When is the concentration of the drug increasing, and when is it decreasing?

95. SMALL CAR MARKET SHARE Owing in part to an aging population and the squeeze from carbon-cutting regulations, the percentage of small and lower-midsize vehicles is expected to increase in the near future. The function

$$f(t) = \frac{5.3\sqrt{t} - 300}{\sqrt{t} - 10} \qquad (0 \le t \le 10)$$

gives the projected percentage of small and lower-midsize vehicles t years after 2005.

a. What was the percentage of small and lower-midsize vehicles in 2005? What is the projected percentage of small and lower-midsize vehicles in 2015?
b. Show that f is increasing on the interval $(0, 10)$, and interpret your results.
Source: J.D. Power Automotive.

96. AIR POLLUTION The amount of nitrogen dioxide, a brown gas that impairs breathing, present in the atmosphere on a certain May day in the city of Long Beach is approximated by

$$A(t) = \frac{136}{1 + 0.25(t - 4.5)^2} + 28 \qquad (0 \le t \le 11)$$

where $A(t)$ is measured in pollutant standard index (PSI) and t is measured in hours, with $t = 0$ corresponding to 7 A.M. Find the intervals where A is increasing and where A is decreasing, and interpret your results.
Source: Los Angeles Times.

97. ENERGY CONSUMPTION OF APPLIANCES The average energy consumption of the typical refrigerator/freezer manufactured by York Industries is approximately

$$C(t) = 1486e^{-0.073t} + 500 \qquad (0 \le t \le 20)$$

kilowatt-hours (kWh) per year, where t is measured in years, with $t = 0$ corresponding to 1972.
a. What was the average energy consumption of the York refrigerator/freezer at the beginning of 1972?
b. Prove that the average energy consumption of the York refrigerator/freezer is decreasing over the years in question.
c. All refrigerator/freezers manufactured as of January 1, 1990, must meet the 950-kWh/year maximum energy-consumption standard set by the National Appliance Conservation Act. Show that the York refrigerator/freezer satisfies this requirement.

98. POLIO IMMUNIZATION Polio, a once-feared killer, declined markedly in the United States in the 1950s after Jonas Salk developed the inactivated polio vaccine and mass immunization of children took place. The number of polio cases in the United States from the beginning of 1959 to the beginning of 1963 is approximated by the function

$$N(t) = 5.3e^{0.095t^2 - 0.85t} \qquad (0 \le t \le 4)$$

where $N(t)$ gives the number of polio cases (in thousands) and t is measured in years, with $t = 0$ corresponding to the beginning of 1959.
a. Show that the function N is decreasing over the time interval under consideration.
b. How fast was the number of polio cases decreasing at the beginning of 1959? At the beginning of 1962? (*Comment:* Since the introduction of the oral vaccine developed by Dr. Albert B. Sabin in 1963, polio in the United States has, for all practical purposes, been eliminated.
Source: Centers for Disease Control and Prevention.

99. REVENUE FUNCTIONS Suppose that the unit price of a product, p, is related to the quantity demanded, x, by the linear demand equation $p = a - bx$, where a and b are positive constants.

  a. Write an expression, R, giving the revenue realized in the sale of x units of the product.

  b. Find the levels of production at which the revenue is increasing, at which the revenue is decreasing, and at which the revenue is stationary (neither increasing nor decreasing).

  c. Find the level of production that will result in a maximum revenue. What is the maximum revenue?

100. U.S. NURSING SHORTAGE The demand for nurses between 2000 and 2015 is estimated to be

$$D(t) = 0.0007t^2 + 0.0265t + 2 \quad (0 \le t \le 15)$$

where $D(t)$ is measured in millions and $t = 0$ corresponds to the year 2000. The supply of nurses over the same time period is estimated to be

$$S(t) = -0.0014t^2 + 0.0326t + 1.9 \quad (0 \le t \le 15)$$

where $S(t)$ is also measured in millions.

  a. Find an expression $G(t)$ giving the gap between the demand and supply of nurses over the period in question.

  b. Find the interval where G is decreasing and where it is increasing. Interpret your result.

  c. Find the relative extrema of G. Interpret your result.

  Source: U.S. Department of Health and Human Services.

In Exercises 101–106, determine whether the statement is true or false. If it is true, explain why it is true. If it is false, give an example to show why it is false.

101. If f is decreasing on (a, b), then $f'(x) < 0$ for each x in (a, b).

102. If f is increasing on (a, b), then $-f$ is decreasing on (a, b).

103. If f and g are both increasing on (a, b), then $f + g$ is increasing on (a, b).

104. If $f(x)$ and $g(x)$ are positive on (a, b) and both f and g are increasing on (a, b), then fg is increasing on (a, b).

105. If f and g are decreasing on (a, b), then fg must be decreasing on (a, b).

106. If $f'(c) = 0$, then f has a relative maximum or a relative minimum at $x = c$.

107. Using Theorem 1, verify that the linear function $f(x) = mx + b$ is (a) increasing everywhere if $m > 0$, (b) decreasing everywhere if $m < 0$, and (c) constant if $m = 0$.

108. Let $f(x) = -x^2 + ax + b$. Determine the constants a and b such that f has a relative maximum at $x = 2$ and the relative maximum value is 7.

109. Let

$$f(x) = \begin{cases} -3x & \text{if } x < 0 \\ 2x + 4 & \text{if } x \ge 0 \end{cases}$$

  a. Compute $f'(x)$, and show that it changes sign from negative to positive as we move across $x = 0$.

  b. Show that f does not have a relative minimum at $x = 0$. Does this contradict the First Derivative Test? Explain your answer.

110. Refer to Example 6, page 621.

  a. Show that f is increasing on the interval $(0, 1)$.

  b. Show that $f(0) = -1$ and $f(1) = 1$, and use the result of part (a) together with the Intermediate Value Theorem to conclude that there is exactly one root of $f(x) = 0$ in $(0, 1)$.

10.1 Solutions to Self-Check Exercises

1. The derivative of f is

$$f'(x) = 2x^2 - 2x - 12 = 2(x + 2)(x - 3)$$

and it is continuous everywhere. The zeros of $f'(x)$ are $x = -2$ and $x = 3$. The sign diagram of f' is shown in the accompanying figure. We conclude that f is increasing on the intervals $(-\infty, -2)$ and $(3, \infty)$ and decreasing on the interval $(-2, 3)$.

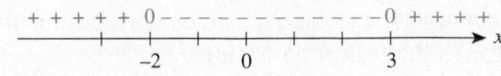

2. The derivative of f is

$$f'(x) = \frac{(1 - x^2)\dfrac{d}{dx}(x^2) - x^2 \dfrac{d}{dx}(1 - x^2)}{(1 - x^2)^2}$$

$$= \frac{(1 - x^2)(2x) - x^2(-2x)}{(1 - x^2)^2} = \frac{2x}{(1 - x^2)^2}$$

and it is continuous everywhere except at $x = \pm 1$. Since $f'(x)$ is equal to zero at $x = 0$, $x = 0$ is a critical number of f. Next, observe that $f'(x)$ is discontinuous at $x = \pm 1$, but since these numbers are not in the domain of f, they do not qualify as critical numbers of f. Finally, from the sign diagram of f' shown in the accompanying figure, we conclude that $f(0) = 0$ is a relative minimum of f.

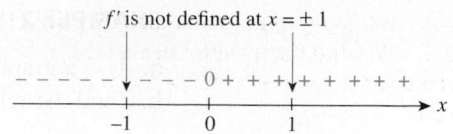

f' is not defined at $x = \pm 1$

USING TECHNOLOGY

Using the First Derivative to Analyze a Function

A graphing utility is an effective tool for analyzing the properties of functions. This is especially true when we also bring into play the power of calculus, as the following examples show.

EXAMPLE 1 Let $f(x) = 2.4x^4 - 8.2x^3 + 2.7x^2 + 4x + 1$.

a. Use a graphing utility to plot the graph of f.
b. Find the intervals where f is increasing and the intervals where f is decreasing.
c. Find the relative extrema of f.

Solution

a. The graph of f in the viewing window $[-2, 4] \times [-10, 10]$ is shown in Figure T1.
b. We compute

$$f'(x) = 9.6x^3 - 24.6x^2 + 5.4x + 4$$

and observe that f' is continuous everywhere, so the critical numbers of f occur at values of x where $f'(x) = 0$. To solve this last equation, observe that $f'(x)$ is a *polynomial function* of degree 3. The easiest way to solve the polynomial equation

$$9.6x^3 - 24.6x^2 + 5.4x + 4 = 0$$

is to use the function on a graphing utility for solving polynomial equations. (Not all graphing utilities have this function.) You can also use **TRACE** and **ZOOM**, but this will not give the same accuracy without a much greater effort.

We find

$$x_1 \approx 2.22564943249 \qquad x_2 \approx 0.63272944121 \qquad x_3 \approx -0.295878873696$$

Referring to Figure T1, we conclude that f is decreasing on $(-\infty, -0.2959)$ and $(0.6327, 2.2256)$ (correct to four decimal places) and f is increasing on $(-0.2959, 0.6327)$ and $(2.2256, \infty)$.

c. Using the evaluation function of a graphing utility, we find the value of f at each of the critical numbers found in part (b). Upon referring to Figure T1 once again, we see that $f(x_3) \approx 0.2836$ and $f(x_1) \approx -8.2366$ are relative minimum values of f and $f(x_2) \approx 2.9194$ is a relative maximum value of f.

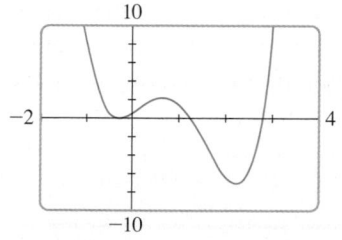

FIGURE T1
The graph of f in the viewing window $[-2, 4] \times [-10, 10]$

Note The equation $f'(x) = 0$ in Example 1 is a polynomial equation, so it is easily solved by using the function for solving polynomial equations. We could also solve the equation using the function for finding the roots of equations, but that would require much more work. For equations that are *not* polynomial equations, however, our only choice is to use the function for finding the roots of equations.

If the derivative of a function is difficult to compute or simplify and we do not require great precision in the solution, we can find the relative extrema of the function using a combination of **ZOOM** and **TRACE**. This technique, which does not require the use of the derivative of f, is illustrated in the following example.

EXAMPLE 2 Let $f(x) = x^{1/3}(x^2 + 1)^{-3/2}3^{-x}$.

a. Use a graphing utility to plot the graph of f.
b. Find the relative extrema of f.

Solution

a. The graph of f in the viewing window $[-4, 2] \times [-2, 1]$ is shown in Figure T2.
b. From the graph of f in Figure T2, we see that f has relative maxima when $x \approx -2$ and $x \approx 0.25$ and a relative minimum when $x \approx -0.75$. To obtain a better approximation of the first relative maximum, we zoom in with the cursor at approximately the point on the graph corresponding to $x \approx -2$. Then, using **TRACE**, we see that a relative maximum occurs when $x \approx -1.76$ with value $y \approx -1.01$. Similarly, we find the other relative maximum where $x \approx 0.20$ with value $y \approx 0.44$. Repeating the procedure, we find the relative minimum at $x \approx -0.86$ and $y \approx -1.07$.

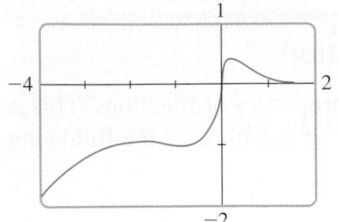

FIGURE T2
The graph of f in the viewing window
$[-4, 2] \times [-2, 1]$

You can also use the "minimum" and "maximum" functions of a graphing utility to find the relative extrema of the function.

Finally, we comment that if you have access to a computer and software such as Derive, Maple, or Mathematica, then symbolic differentiation will yield the derivative $f'(x)$ of a differentiable function. This software will also solve the equation $f'(x) = 0$ with ease. Thus, the use of a computer will simplify even more greatly the analysis of functions.

TECHNOLOGY EXERCISES

In Exercises 1–4, find (a) the intervals where f is increasing and the intervals where f is decreasing and (b) the relative extrema of f. Express your answers accurate to four decimal places.

1. $f(x) = 3.4x^4 - 6.2x^3 + 1.8x^2 + 3x - 2$

2. $f(x) = 1.8x^4 - 9.1x^3 + 5x - 4$

3. $f(x) = 2x^5 - 5x^3 + 8x^2 - 3x + 2$

4. $f(x) = 3x^5 - 4x^2 + 3x - 1$

In Exercises 5–8, use the ZOOM and TRACE features to find (a) the intervals where f is increasing and the intervals where f is decreasing and (b) the relative extrema of f. Express your answers accurate to two decimal places.

5. $f(x) = (2x + 1)^{1/3}(x^2 + 1)^{-2/3}$

6. $f(x) = [x^2(x^3 - 1)]^{1/3} + \dfrac{1}{x}$

7. $f(x) = e^{-x}\sqrt{x^2 + 1} + x^3$

8. $f(x) = \dfrac{xe^{-x^2} + x^{3/2}}{x^2 + 1}$

9. WORKERS' EXPECTATIONS The percentage of workers who expect to work past age 65 years has more than tripled in 30 years. The function

$$f(t) = 0.004545t^3 - 0.1113t^2 + 1.385t + 11 \qquad (0 \le t \le 22)$$

gives an approximation of the percentage of workers who expect to work past age 65 in year t, where t is measured in years, with $t = 0$ corresponding to 1991.
a. Plot the graph of f in the viewing window $[0, 22] \times [10, 38]$.
b. Prove that the percentage of workers who expect to work past age 65 was increasing over this interval.

10. AIR POLLUTION The amount of nitrogen dioxide, a brown gas that impairs breathing, present in the atmosphere on a certain May day in the city of Long Beach, is approximated by

$$A(t) = \dfrac{136}{1 + 0.25(t - 4.5)^2} + 28 \qquad (0 \le t \le 11)$$

where $A(t)$ is measured in pollutant standard index (PSI) and t is measured in hours, with $t = 0$ corresponding to 7 A.M. When is the PSI increasing, and when is it decreasing? At what time is the PSI highest, and what is its value at that time?

10.2 Applications of the Second Derivative

Determining the Intervals of Concavity

Consider the graphs shown in Figure 28, which give the estimated population of the world and of the United States through the year 2000. Both graphs are rising, indicating that both the U.S. population and the world population continued to increase through the year 2000. But observe that the graph in Figure 28a opens upward, whereas the graph in Figure 28b opens downward. What is the significance of this? To answer this question, let's look at the slopes of the tangent lines to various points on each graph (Figure 29).

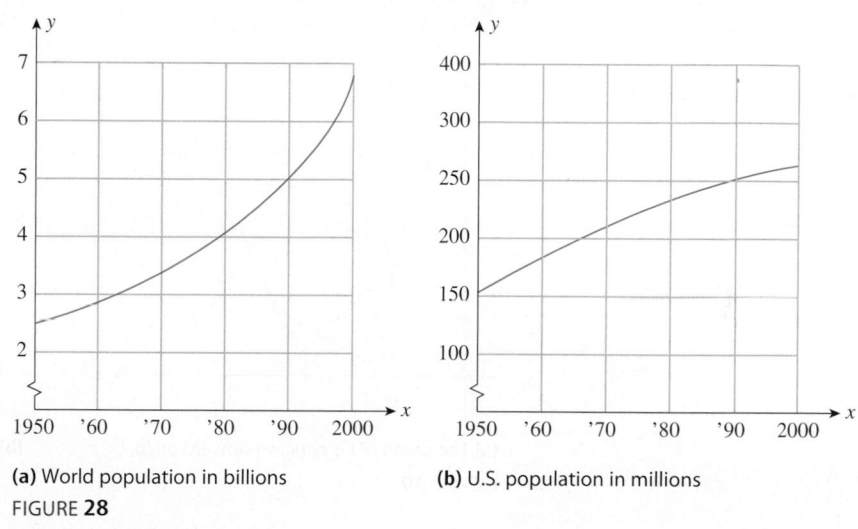

(a) World population in billions **(b)** U.S. population in millions

FIGURE **28**

Source: U.S. Department of Commerce and Worldwatch Institute.

In Figure 29a, we see that the slopes of the tangent lines to the graph are increasing as we move from left to right. Since the slope of the tangent line to the graph at a point on the graph measures the rate of change of the function at that point, we conclude that the world population not only was increasing through the year 2000 but also was increasing at an *increasing* pace. A similar analysis of Figure 29b reveals that the U.S. population was increasing, but at a *decreasing* pace.

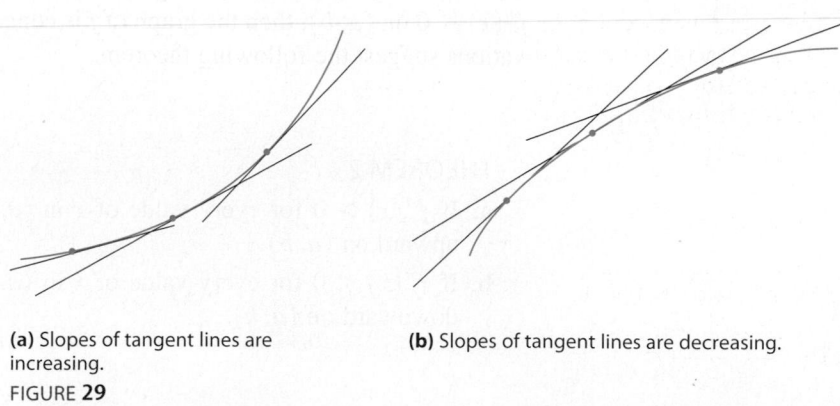

(a) Slopes of tangent lines are increasing. **(b)** Slopes of tangent lines are decreasing.

FIGURE **29**

The shape of a curve can be described by using the notion of concavity.

> **Concavity of a Function f**
>
> Let the function f be differentiable on an interval (a, b). Then,
>
> **1.** The graph of f is **concave upward** on (a, b) if f' is increasing on (a, b).
> **2.** The graph of f is **concave downward** on (a, b) if f' is decreasing on (a, b).

Geometrically, a curve is concave upward if it lies above its tangent lines (Figure 30a). Similarly, a curve is concave downward if it lies below its tangent lines (Figure 30b).

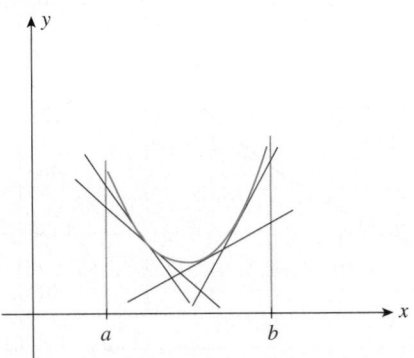
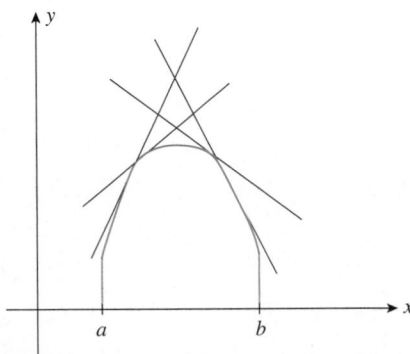

(a) The graph of f is concave upward on (a, b).   **(b)** The graph of f is concave downward on (a, b).

FIGURE **30**

We also say that the graph of f is *concave upward at a number c* if there exists an interval (a, b) containing c on which the graph of f is concave upward. Similarly, we say that the graph of f is *concave downward at a number c* if there exists an interval (a, b) containing c on which the graph of f is concave downward.

If a function f has a second derivative f'', we can use f'' to determine the intervals of concavity of the graph of the function. Recall that $f''(x)$ measures the rate of change of the slope $f'(x)$ of the tangent line to the graph of f at the point $(x, f(x))$. Thus, if $f''(x) > 0$ on an interval (a, b), then the slopes of the tangent lines to the graph of f are increasing on (a, b), so the graph of f is concave upward on (a, b). Similarly, if $f''(x) < 0$ on (a, b), then the graph of f is concave downward on (a, b). These observations suggest the following theorem.

> **THEOREM 2**
>
> **a.** If $f''(x) > 0$ for every value of x in (a, b), then the graph of f is concave upward on (a, b).
> **b.** If $f''(x) < 0$ for every value of x in (a, b), then the graph of f is concave downward on (a, b).

The following procedure, based on the conclusions of Theorem 2, may be used to determine the intervals of concavity of the graph of a function.

> Determining the Intervals of Concavity of the Graph of f
>
> **1.** Determine the values of x for which f'' is zero or where f'' is not defined, and identify the open intervals determined by these numbers.
> **2.** Determine the sign of f'' in each interval found in Step 1. To do this, compute $f''(c)$, where c is any conveniently chosen test number in the interval.
>   **a.** If $f''(c) > 0$, then the graph of f is concave upward on that interval.
>   **b.** If $f''(c) < 0$, then the graph of f is concave downward on that interval.

EXAMPLE 1 Determine where the graph of the function

$$f(x) = x^3 - 3x^2 - 24x + 32$$

is concave upward and where it is concave downward.

Solution Here,

$$f'(x) = 3x^2 - 6x - 24$$
$$f''(x) = 6x - 6 = 6(x - 1)$$

and f'' is defined everywhere. Setting $f''(x) = 0$ gives $x = 1$. The sign diagram of f'' appears in Figure 31. We conclude that the graph of f is concave downward on the interval $(-\infty, 1)$ and is concave upward on the interval $(1, \infty)$. Figure 32 shows the graph of f.

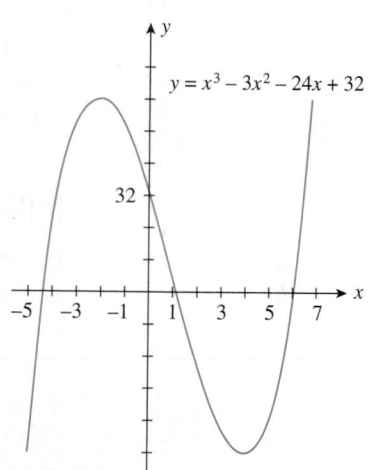

FIGURE **32**
The graph of f is concave downward on $(-\infty, 1)$ and concave upward on $(1, \infty)$.

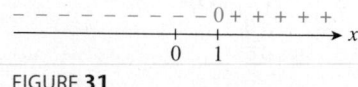

FIGURE **31**
Sign diagram for f''

Exploring with TECHNOLOGY

Refer to Example 1.

1. Use a graphing utility to plot the graph of $f(x) = x^3 - 3x^2 - 24x + 32$ and its second derivative $f''(x) = 6x - 6$ using the viewing window $[-10, 10] \times [-80, 90]$.
2. By studying the graph of f'', determine the intervals where $f''(x) > 0$ and the intervals where $f''(x) < 0$. Next, look at the graph of f, and determine the intervals where the graph of f is concave upward and the intervals where it is concave downward. Are these observations what you might have expected?

EXAMPLE 2 Determine the intervals where the graph of the function $f(x) = x + \dfrac{1}{x}$ is concave upward and where it is concave downward.

Solution   We have

$$f'(x) = 1 - \frac{1}{x^2}$$

$$f''(x) = \frac{2}{x^3}$$

We deduce from the sign diagram for f'' (Figure 33) that the graph of the function f is concave downward on the interval $(-\infty, 0)$ and concave upward on the interval $(0, \infty)$. The graph of f is sketched in Figure 34.

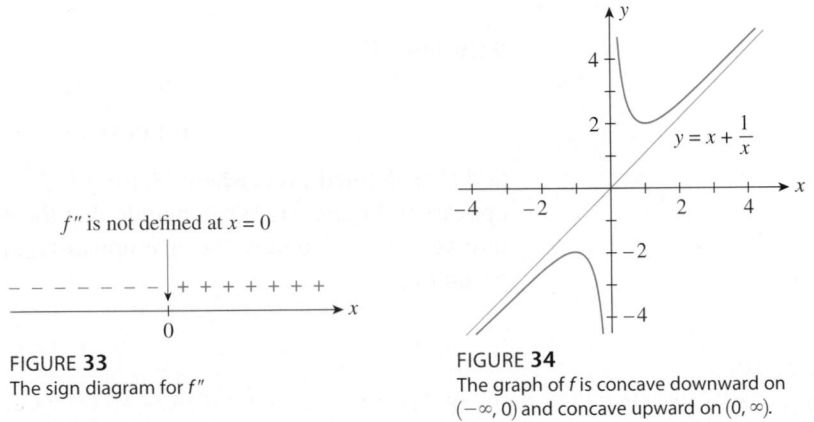

f'' is not defined at $x = 0$

FIGURE 33
The sign diagram for f''

FIGURE 34
The graph of f is concave downward on $(-\infty, 0)$ and concave upward on $(0, \infty)$.

Inflection Points

Figure 35 shows the total sales S of a manufacturer of automobile air conditioners versus the amount of money x that the company spends on advertising its product.

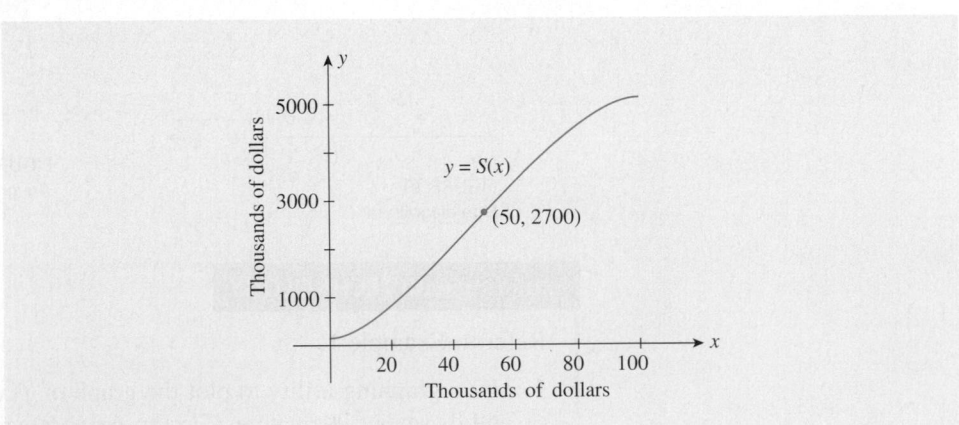

FIGURE 35
The graph of S has a point of inflection at (50, 2700).

Notice that the graph of the continuous function $y = S(x)$ changes concavity—from upward to downward—at the point (50, 2700). This point is called an *inflection point* of S. To understand the significance of this inflection point, observe that the total sales increase rather slowly at first, but as more money is spent on advertising, the total sales increase rapidly. This rapid increase reflects the effectiveness of the company's ads. However, a point is soon reached after which any additional advertising expenditure results in increased sales but at a slower rate of increase. This point, commonly known as the *point of diminishing returns,* is the point of inflection of the function S. We will return to this example later.

Let's now state formally the definition of an inflection point.

> **Inflection Point**
>
> A point on the graph of a continuous function f where the tangent line exists and where the concavity changes is called a **point of inflection** or an **inflection point.**

Observe that the graph of a function crosses its tangent line at a point of inflection (Figure 36).

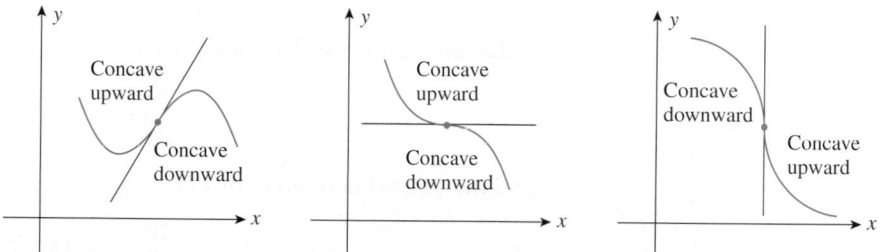

FIGURE 36
At each point of inflection, the graph of a function crosses its tangent line.

The following procedure may be used to find inflection points:

> **Finding Inflection Points**
>
> 1. Compute $f''(x)$.
> 2. Determine the numbers in the domain of f for which $f''(x) = 0$ or $f''(x)$ does not exist.
> 3. Determine the sign of $f''(x)$ to the left and right of each number c found in Step 2. If there is a change in the sign of $f''(x)$ as we move across $x = c$, then $(c, f(c))$ is an inflection point of f.

⚠ The numbers that were found in Step 2 are only *candidates* for the inflection points of f. For example, you can easily verify that $f''(0) = 0$ if $f(x) = x^4$, but a sketch of the graph of f will show that $(0, 0)$ is *not* an inflection point.

EXAMPLE 3 Find the point of inflection of the function $f(x) = x^3$.

Solution

$$f'(x) = 3x^2$$
$$f''(x) = 6x$$

Observe that f'' is continuous everywhere and is zero if $x = 0$. The sign diagram of f'' is shown in Figure 37 (see next page). From this diagram, we see that $f''(x)$ changes sign as we move across $x = 0$. Thus, the point $(0, 0)$ is an inflection point (Figure 38).

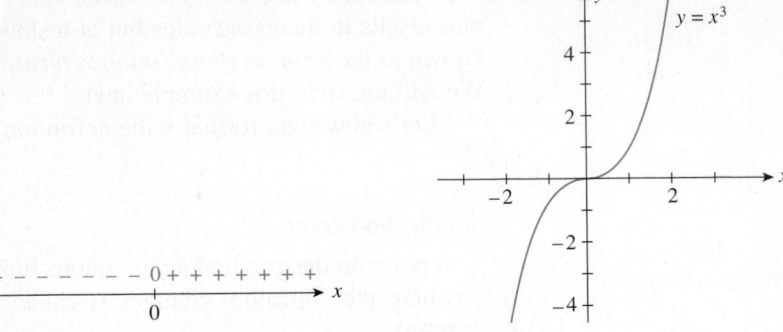

FIGURE 37
Sign diagram for f''

FIGURE 38
f has an inflection point at $(0, 0)$.

EXAMPLE 4 Determine the intervals where the graph of the function $f(x) = (x - 1)^{5/3}$ is concave upward and where it is concave downward, and find the inflection points of f.

Solution   The first derivative of f is

$$f'(x) = \frac{5}{3}(x - 1)^{2/3}$$

and the second derivative of f is

$$f''(x) = \frac{10}{9}(x - 1)^{-1/3} = \frac{10}{9(x - 1)^{1/3}}$$

We see that f'' is not defined at $x = 1$. Also, $f''(x)$ is not equal to zero anywhere. The sign diagram of f'' is shown in Figure 39. From the sign diagram, we see that the graph of f is concave downward on $(-\infty, 1)$ and concave upward on $(1, \infty)$. Next, since $x = 1$ does lie in the domain of f, we see that the point $(1, 0)$ is an inflection point (Figure 40).

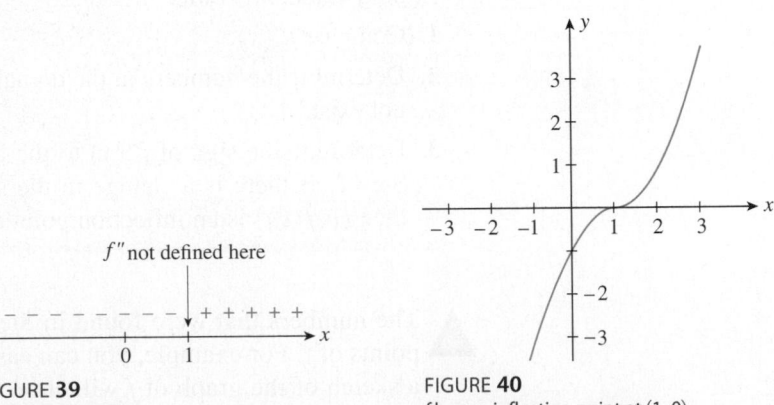

f'' not defined here

FIGURE 39
Sign diagram for f''

FIGURE 40
f has an inflection point at $(1, 0)$.

EXAMPLE 5 Find the inflection points of the function $f(x) = e^{-x^2}$.

Solution   The first derivative of f is

$$f'(x) = -2xe^{-x^2}$$

Differentiating $f'(x)$ with respect to x yields

$$f''(x) = (-2x)(-2xe^{-x^2}) - 2e^{-x^2}$$
$$= 2e^{-x^2}(2x - 1)$$

Setting $f''(x) = 0$ gives

$$2e^{-x^2}(2x^2 - 1) = 0$$

Since e^{-x^2} never equals zero for any real value of x, we see that $x = \pm 1/\sqrt{2}$ are the only candidates for inflection points of f. The sign diagram of f'', shown in Figure 41, tells us that both $x = -1/\sqrt{2}$ and $x = 1/\sqrt{2}$ give rise to inflection points of f.
Next,

$$f\left(-\frac{1}{\sqrt{2}}\right) = f\left(\frac{1}{\sqrt{2}}\right) = e^{-1/2}$$

and the inflection points of f are $(-1/\sqrt{2}, e^{-1/2})$ and $(1/\sqrt{2}, e^{-1/2})$. The graph of f appears in Figure 42.

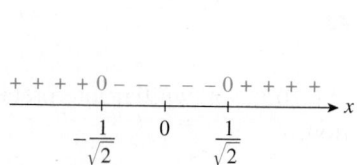

FIGURE **41**
Sign diagram for f''

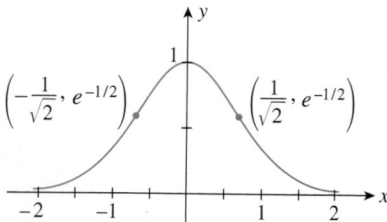

FIGURE **42**
The graph of $y = e^{-x^2}$ has two inflection points.

Explore and Discuss

1. Suppose $(c, f(c))$ is an inflection point of f. Can you conclude that f does not have a relative extremum at $x = c$? Explain your answer.

2. True or false: A polynomial function of degree 3 has exactly one inflection point.
   Hint: Study the function $f(x) = ax^3 + bx^2 + cx + d \ (a \neq 0)$.

The next example uses an interpretation of the first and second derivatives to help us sketch the graph of a function.

EXAMPLE 6 Sketch the graph of a function having the following properties:

$$f(-1) = 4$$
$$f(0) = 2$$
$$f(1) = 0$$
$$f'(-1) = 0$$
$$f'(1) = 0$$
$$f'(x) > 0 \quad \text{on } (-\infty, -1) \text{ and } (1, \infty)$$
$$f'(x) < 0 \quad \text{on } (-1, 1)$$
$$f''(x) < 0 \quad \text{on } (-\infty, 0)$$
$$f''(x) > 0 \quad \text{on } (0, \infty)$$

Solution First, we plot the points $(-1, 4)$, $(0, 2)$, and $(1, 0)$ that lie on the graph of f. Since $f'(-1) = 0$ and $f'(1) = 0$, the tangent lines at the points $(-1, 4)$ and $(1, 0)$ are horizontal. Since $f'(x) > 0$ on $(-\infty, -1)$ and $f'(x) < 0$ on $(-1, 1)$, we see that f has a relative maximum at the point $(-1, 4)$. Also, $f'(x) < 0$ on $(-1, 1)$ and $f'(x) > 0$ on $(1, \infty)$ implies that f has a relative minimum at the point $(1, 0)$ (Figure 43a).

Since $f''(x) < 0$ on $(-\infty, 0)$ and $f''(x) > 0$ on $(0, \infty)$, we see that the point $(0, 2)$ is an inflection point. Finally, we complete the graph, making use of the fact that f is increasing on $(-\infty, -1)$ and $(1, \infty)$, where it is given that $f'(x) > 0$, and f is decreasing on $(-1, 1)$, where $f'(x) < 0$. Also, make sure that the graph of f is concave downward on $(-\infty, 0)$ and concave upward on $(0, \infty)$ (Figure 43b).

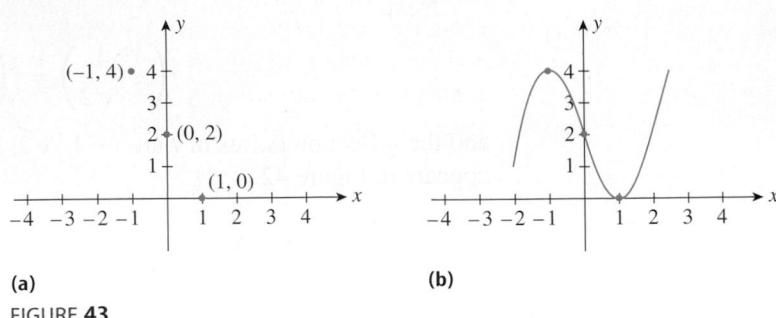

(a)

(b)

FIGURE **43**

Example 7 illustrates a familiar interpretation of the significance of an inflection point of a function.

$\$$ APPLIED EXAMPLE 7 Effect of Advertising on Sales The total sales S (in thousands of dollars) of Arctic Air Corporation, a manufacturer of automobile air conditioners, is related to the amount of money x (in thousands of dollars) the company spends on advertising its products by the formula

$$S(x) = -0.01x^3 + 1.5x^2 + 200 \qquad (0 \le x \le 100)$$

Find the inflection point of the function S.

Solution The first two derivatives of S are given by

$$S'(x) = -0.03x^2 + 3x$$
$$S''(x) = -0.06x + 3$$

Setting $S''(x) = 0$ gives $x = 50$. So $(50, S(50))$ is the only candidate for an inflection point of S. Moreover, since

$$S''(x) > 0 \quad \text{for } x < 50$$

and

$$S''(x) < 0 \quad \text{for } x > 50$$

the point $(50, 2700)$ is an inflection point of the function S. The graph of S appears in Figure 44. Notice that this is the same graph as that shown in Figure 35.

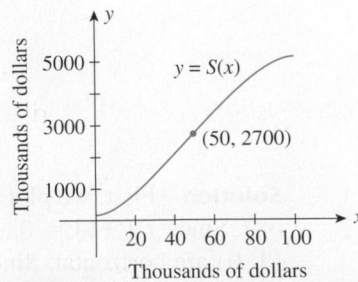

FIGURE **44**
The graph of S has a point of inflection at $(50, 2700)$.

The Second Derivative Test

We now show how the second derivative f'' of a function f can be used to help us determine whether a critical number of f gives rise to a relative extremum of f. Figure 45a shows the graph of a function that has a relative maximum at $x = c$. Observe that the graph of f is concave downward at that number. Similarly, Figure 45b shows that at a relative minimum of f, the graph is concave upward. But from our previous work, we know that the graph of f is concave downward at $x = c$ if $f''(c) < 0$ and the graph of f is concave upward at $x = c$ if $f''(c) > 0$. These observations suggest the following alternative procedure for determining whether a critical number of f gives rise to a relative extremum of f. This result is called the **Second Derivative Test** and is applicable when f'' exists.

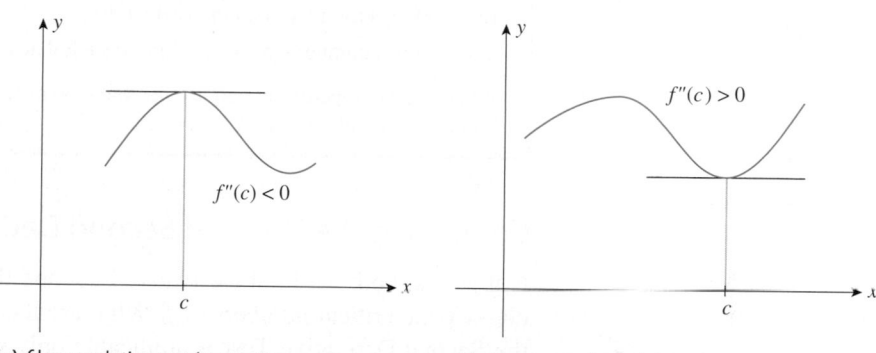

(a) f has a relative maximum at $x = c$.

(b) f has a relative minimum at $x = c$.

FIGURE **45**

The Second Derivative Test

1. Compute $f'(x)$ and $f''(x)$.

2. Find all the critical numbers of f at which $f'(x) = 0$.

3. Compute $f''(c)$ for each such critical number c.
   a. If $f''(c) < 0$, then f has a relative maximum at c.
   b. If $f''(c) > 0$, then f has a relative minimum at c.
   c. If $f''(c) = 0$ or $f''(c)$ does not exist, the test fails; that is, it is inconclusive.

Note   The Second Derivative Test does not yield a conclusion if $f''(c) = 0$ or if $f''(c)$ does not exist. In other words, $x = c$ may or may not give rise to a relative extremum (see Exercise 112, page 756). In such cases, you should revert to the First Derivative Test.

EXAMPLE 8 Determine the relative extrema of the function

$$f(x) = x^3 - 3x^2 - 24x + 32$$

using the Second Derivative Test. (See Example 7, Section 10.1.)

Solution   We have

$$f'(x) = 3x^2 - 6x - 24 = 3(x + 2)(x - 4)$$

so $f'(x) = 0$ when $x = -2$ and $x = 4$, the critical numbers of f, as in Example 7, Section 10.1. Next, we compute

$$f''(x) = 6x - 6 = 6(x - 1)$$

Since

$$f''(-2) = 6(-2 - 1) = -18 < 0$$

the Second Derivative Test implies that $f(-2) = 60$ is a relative maximum of f. Also,

$$f''(4) = 6(4 - 1) = 18 > 0$$

and the Second Derivative Test implies that $f(4) = -48$ is a relative minimum of f, which confirms the results obtained earlier. ■

Explore and Discuss

Suppose a function f has the following properties:

1. $f''(x) > 0$ for all x in an interval (a, b).
2. There is a number c between a and b such that $f'(c) = 0$.

What special property can you ascribe to the point $(c, f(c))$? Answer the question if Property 1 is replaced by the property that $f''(x) < 0$ for all x in (a, b).

Comparing the First and Second Derivative Tests

Notice that both the First Derivative Test and the Second Derivative Test are used to classify the critical numbers of f. What are the pros and cons of the two tests? Since the Second Derivative Test is applicable only when f'' exists, it is less versatile than the First Derivative Test. For example, it cannot be used to locate the relative minimum $f(0) = 0$ of the function $f(x) = x^{2/3}$.

Furthermore, the Second Derivative Test is inconclusive when f'' is equal to zero at a critical number of f, whereas the First Derivative Test always yields a conclusion; that is, it tells us whether f has a relative maximum, relative minimum, or neither. The Second Derivative Test is also inconvenient to use when f'' is difficult to compute. On the plus side, if f'' is computed easily, then we use the Second Derivative Test, since it involves just the evaluation of f'' at the critical number(s) of f. Also, the conclusions of the Second Derivative Test are important in theoretical work.

We close this section by summarizing the different roles played by the first derivative f' and the second derivative f'' of a function f in determining the properties of the graph of f. The first derivative f' tells us where f is increasing and where f is decreasing, whereas the second derivative f'' tells us where the graph of f is concave upward and where it is concave downward. These different properties of f are reflected by the signs of f' and f'' in the interval of interest. The following table shows the general characteristics of the function f for various possible combinations of the signs of f' and f'' in an interval (a, b).

Signs of f' and f''	Properties of f	General Shape of the Graph of f
$f'(x) > 0$ $f''(x) > 0$	f is increasing. The graph of f is concave upward.	
$f'(x) > 0$ $f''(x) < 0$	f is increasing. The graph of f is concave downward.	
$f'(x) < 0$ $f''(x) > 0$	f is decreasing. The graph of f is concave upward.	
$f'(x) < 0$ $f''(x) < 0$	f is decreasing. The graph of f is concave downward.	

10.2 Self-Check Exercises

1. Determine where the graph of the function
   $f(x) = 4x^3 - 3x^2 + 6$ is concave upward and
   where it is concave downward.

2. Using the Second Derivative Test, if applicable,
   find the relative extrema of the function
   $f(x) = 2x^3 - \frac{1}{2}x^2 - 12x - 10$.

3. **GDP** OF A COUNTRY A certain country's gross domestic
   product (GDP) (in millions of dollars) in year t is
   described by the function

$$G(t) = -2t^3 + 45t^2 + 20t + 6000 \qquad (0 \le t \le 11)$$

where $t = 0$ corresponds to the beginning of 2004. Find
the inflection point of the function G, and discuss its
significance.

*Solutions to Self-Check Exercises 10.2 can be found on
page 756.*

10.2 Concept Questions

1. Explain what it means for the graph of a function f to be
   (a) concave upward and (b) concave downward on an
   open interval I. Given that f has a second derivative on I
   (except at isolated numbers), how do you determine
   where the graph of f is concave upward and where it is
   concave downward?

2. What is an inflection point of a function f? How do you find
   the inflection point(s) of a function f whose rule is given?

3. State the Second Derivative Test. What are the pros and
   cons of using the First Derivative Test and the Second
   Derivative Test?

10.2 Exercises

In Exercises 1–8, you are given the graph of a function f. Determine
the intervals where the graph of f is concave upward and where it
is concave downward. Also, find all inflection points of f, if any.

1.

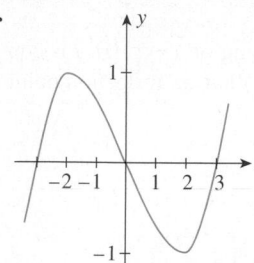

2.

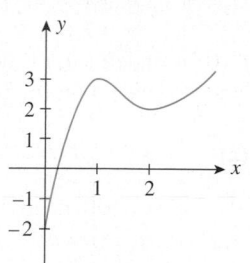

3.

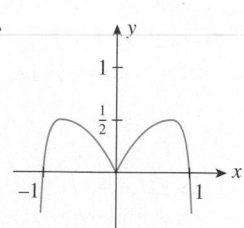

4.

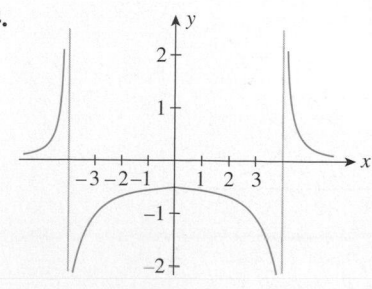

5.

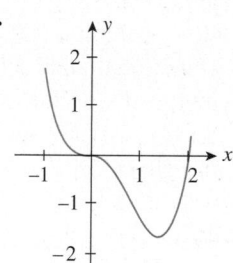

6.

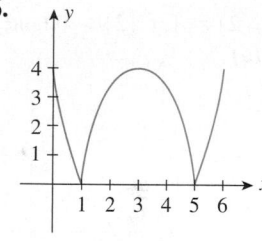

7.

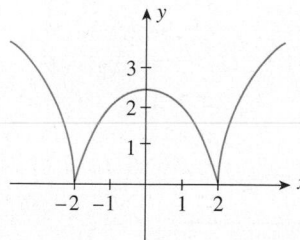

8.

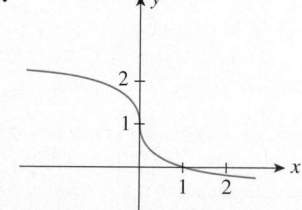

9. Refer to the graph of f shown in the following figure:

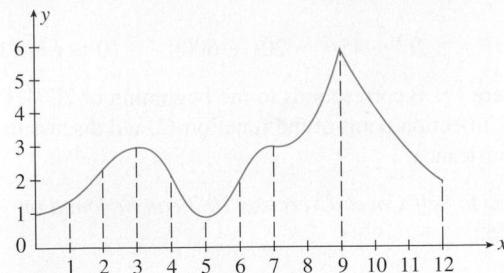

a. Find the intervals where f is concave upward and the intervals where f is concave downward.

b. Find the inflection points of f.

10. Refer to the figure for Exercise 9.

a. Explain how the Second Derivative Test can be used to show that the critical number 3 gives rise to a relative maximum of f and the critical number 5 gives rise to a relative minimum of f.

b. Explain why the Second Derivative Test cannot be used to show that the critical number 7 does not give rise to a relative extremum of f nor can it be used to show that the critical number 9 gives rise to a relative maximum of f.

In Exercises 11–14, determine which graph—(a), (b), or (c)—is the graph of the function f with the specified properties.

11. $f(2) = 1, f'(2) > 0$, and $f''(2) < 0$

(a)

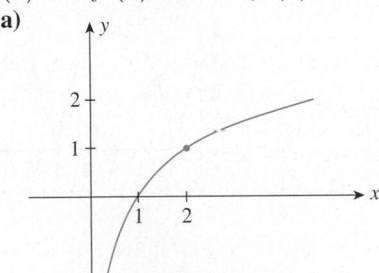

(b)

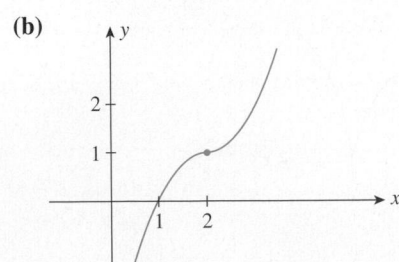

(c)

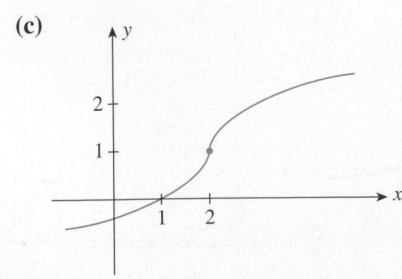

12. $f(1) = 2, f'(x) > 0$ on $(-\infty, 1)$ and $(1, \infty)$, and $f''(1) = 0$

(a)

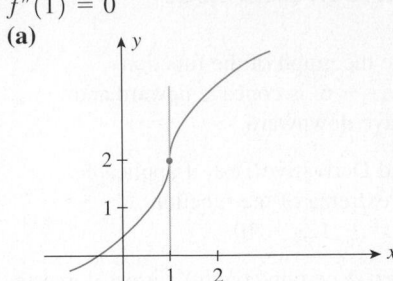

(b)

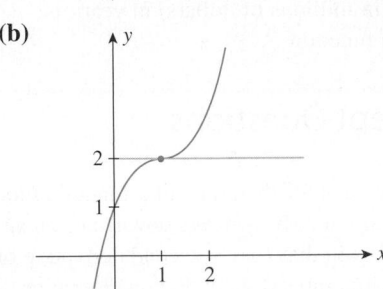

(c)

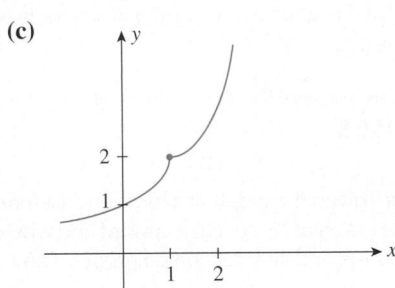

13. $f'(0)$ is undefined, f is decreasing on $(-\infty, 0)$, f is concave downward on $(0, 3)$, and f has an inflection point at $x = 3$.

(a)

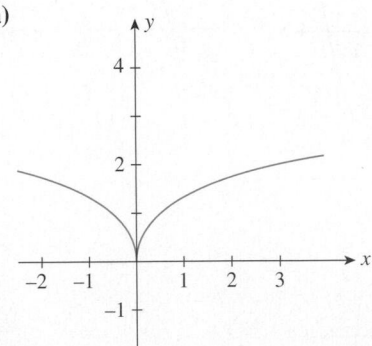

(b)

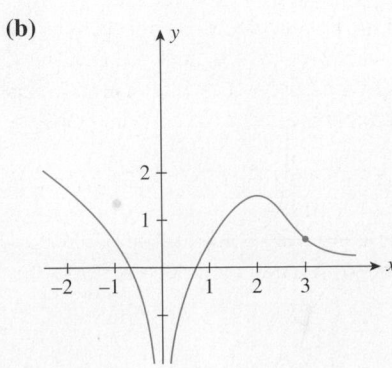

(c)

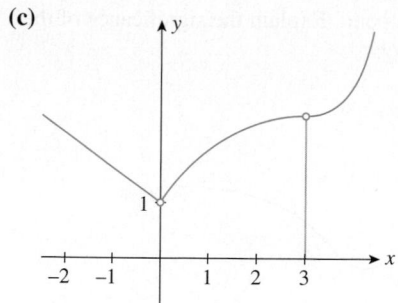

14. f is decreasing on $(-\infty, 2)$ and increasing on $(2, \infty)$, the graph of f is concave upward on $(1, \infty)$, and f has inflection points at $x = 0$ and $x = 1$.

(a)

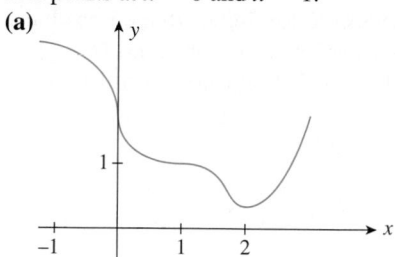

(b)

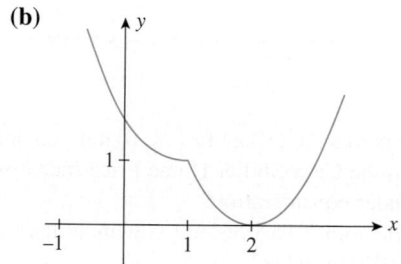

(c)

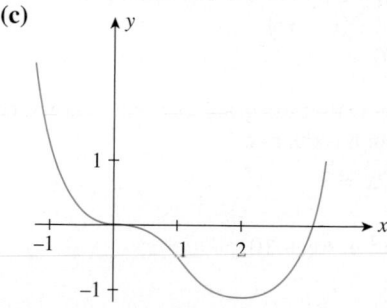

15. EFFECT OF ADVERTISING ON BANK DEPOSITS The following graphs were used by the CEO of the Madison Savings Bank to illustrate what effect a projected promotional campaign would have on its deposits over the next year. The functions D_1 and D_2 give the projected amount of money on deposit with the bank over the next 12 months with and without the proposed promotional campaign, respectively.

a. Determine the signs of $D_1'(t)$, $D_2'(t)$, $D_1''(t)$, and $D_2''(t)$ on the interval $(0, 12)$.

b. What can you conclude about the rate of change of the growth rate of the money on deposit with the

bank with and without the proposed promotional campaign?

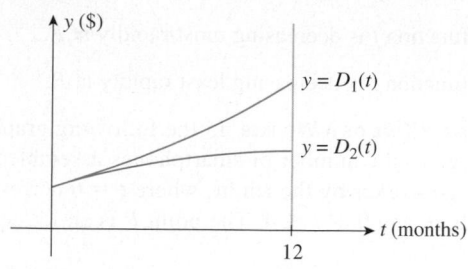

16. MOTION OF CARS Two cars start out side by side and travel along a straight road. The velocity of Car A is $f(t)$ ft/sec, and the velocity of Car B is $g(t)$ ft/sec over the interval $[0, t_2]$. Furthermore, suppose the graphs of f and g are as depicted in the accompanying figure.

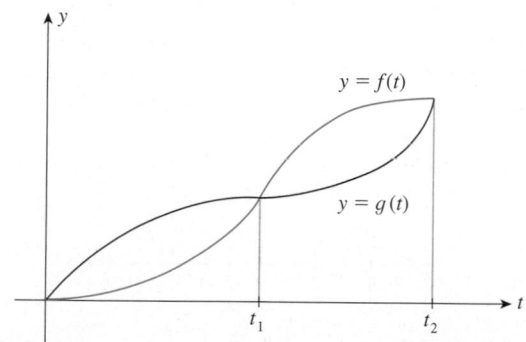

a. What can you say about the acceleration of Car A on the interval $(0, t_1)$? The acceleration of Car B on the interval $(0, t_1)$?

b. What can you say about the acceleration of Car A on the interval (t_1, t_2)? The acceleration of Car B over (t_1, t_2)?

c. What can you say about the acceleration of Car A at t_1? The acceleration of Car B at t_1?

d. At what time do both cars have the same velocity?

In Exercises 17–20, match the graphs (a), (b), (c), or (d) with the corresponding statement.

(a)   **(b)**

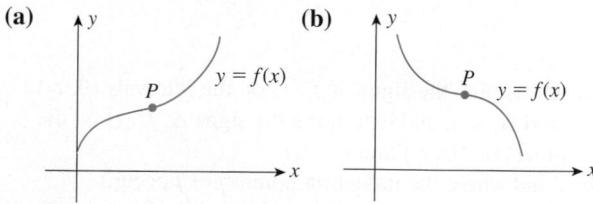

(c)   **(d)**

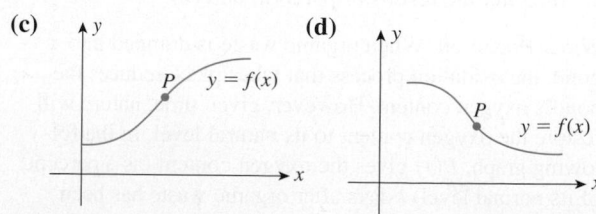

17. The function f is increasing most rapidly at P.

18. The function f is increasing least rapidly at P.

19. The function f is decreasing most rapidly at P.

20. The function f is decreasing least rapidly at P.

21. **ASSEMBLY TIME OF A WORKER** In the following graph, $N(t)$ gives the number of smartphones assembled by the average worker by the tth hr, where $t = 0$ corresponds to 8 A.M. and $0 \le t \le 4$. The point P is an inflection point.
   a. What can you say about the rate of change of the rate of change of the number of smartphones assembled by the average worker between 8 A.M. and 10 A.M.? Between 10 A.M. and 12 noon?
   b. At what time is the rate at which the smartphones are being assembled by the average worker greatest?

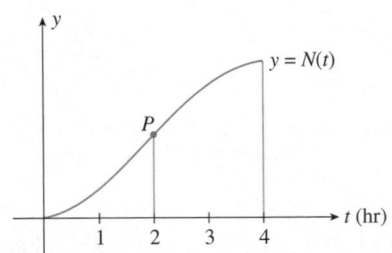

22. **RUMORS OF A RUN ON A BANK** The graph of the function f shows the total deposits with a bank t days after rumors abounded that there was a run on the bank due to heavy loan losses incurred by the bank.

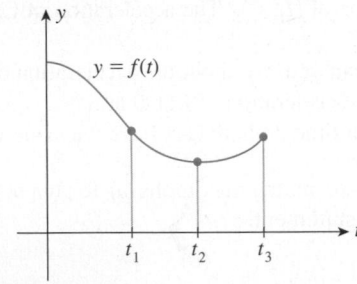

   a. Determine the signs of $f'(t)$ on the intervals $(0, t_2)$ and (t_2, t_3), and determine the signs of $f''(t)$ on the intervals $(0, t_1)$ and (t_1, t_3).
   b. Find where the inflection point(s) of f occur.
   c. Interpret the results of parts (a) and (b).

23. **WATER POLLUTION** When organic waste is dumped into a pond, the oxidation process that takes place reduces the pond's oxygen content. However, given time, nature will restore the oxygen content to its natural level. In the following graph, $P(t)$ gives the oxygen content (as a percent of its normal level) t days after organic waste has been

dumped into the pond. Explain the significance of the inflection point Q.

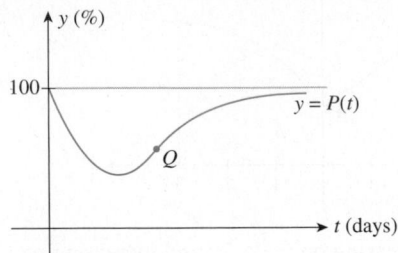

24. **CASE-SHILLER HOME PRICE INDEX** The following graph shows the change in the S&P/Case-Shiller Home Price Index based on a 20-city average from June 2001 $\left(t = \frac{1}{2}\right)$ through June 2008 $\left(t = 7\frac{1}{2}\right)$, adjusted for inflation.

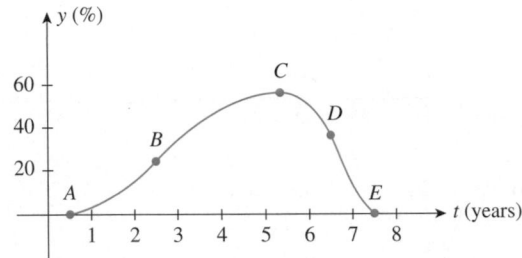

   a. What do the points $A\left(\frac{1}{2}, 0\right)$ and $E\left(7\frac{1}{2}, 0\right)$ tell you about the change in the Case-Shiller Home Price Index over the period under consideration?
   b. What does the point $C\left(5\frac{1}{3}, 56\right)$ tell you about the Case-Shiller Home Price Index?
   c. Give an interpretation of the inflection points $B\left(2\frac{1}{2}, 24\right)$ and $D\left(6\frac{1}{2}, 36\right)$.
   Source: New York Times.

In Exercises 25–28, show that the graph of the function is concave upward wherever it is defined.

25. $f(x) = 4x^2 - 12x + 7$

26. $g(x) = x^4 + \frac{1}{2}x^2 + 6x + 10$

27. $f(x) = \dfrac{1}{x^4}$ **28.** $g(x) = -\sqrt{4 - x^2}$

In Exercises 29–52, determine where the graph of the function is concave upward and where it is concave downward.

29. $f(x) = 2x^2 - 3x + 4$

30. $g(x) = -x^2 + 3x + 4$

31. $f(x) = 1 - x^3$

32. $g(x) = x^3 - x$

33. $f(x) = x^4 - 6x^3 + 2x + 8$

34. $f(x) = 3x^4 - 6x^3 + x - 8$

35. $f(x) = x^{4/7}$

36. $f(x) = \sqrt[3]{x}$

37. $f(x) = \sqrt{4 - x}$

38. $g(x) = \sqrt{x - 2}$

39. $f(x) = \dfrac{1}{x - 2}$

40. $g(x) = \dfrac{x}{x + 1}$

41. $f(x) = \dfrac{1}{2 + x^2}$

42. $g(x) = \dfrac{x}{1 + x^2}$

43. $h(t) = \dfrac{t^2}{t - 1}$

44. $f(x) = \dfrac{x + 1}{x - 1}$

45. $g(x) = x + \dfrac{1}{x^2}$

46. $h(r) = -\dfrac{1}{(r - 2)^2}$

47. $g(t) = (2t - 4)^{1/3}$

48. $f(x) = (x - 2)^{2/3}$

49. $f(x) = \dfrac{e^x - e^{-x}}{2}$

50. $f(x) = xe^x$

51. $f(x) = x^2 + \ln x^2$

52. $f(x) = \dfrac{\ln x}{x}$

In Exercises 53–66, find the inflection point(s), if any, of each function.

53. $f(x) = x^3 - 2$

54. $g(x) = x^3 - 6x$

55. $f(x) = 6x^3 - 18x^2 + 12x - 20$

56. $g(x) = 2x^3 - 3x^2 + 18x - 8$

57. $f(x) = 3x^4 - 4x^3 + 1$

58. $f(x) = x^4 - 2x^3 + 6$

59. $g(t) = \sqrt[3]{t}$

60. $f(x) = \sqrt[5]{x}$

61. $f(x) = (x - 1)^3 + 2$

62. $f(x) = (x - 2)^{4/3}$

63. $f(x) = 2e^{-x^2}$

64. $f(x) = xe^{-2x}$

65. $f(x) = x^2 \ln x$

66. $f(x) = \ln(x^2 + 1)$

In Exercises 67–84, find the relative extrema, if any, of each function. Use the Second Derivative Test if applicable.

67. $f(x) = -x^2 + 2x + 4$

68. $g(x) = 2x^2 + 3x + 7$

69. $f(x) = 2x^3 + 1$

70. $g(x) = x^3 - 6x$

71. $f(x) = \dfrac{1}{3}x^3 - 2x^2 - 5x - 5$

72. $f(x) = 2x^3 + 3x^2 - 12x - 4$

73. $g(t) = t + \dfrac{9}{t}$

74. $f(t) = 2t + \dfrac{3}{t}$

75. $f(x) = \dfrac{x}{1 - x}$

76. $f(x) = \dfrac{2x}{x^2 + 1}$

77. $f(t) = t^2 - \dfrac{16}{t}$

78. $g(x) = x^2 + \dfrac{2}{x}$

79. $g(s) = \dfrac{s}{1 + s^2}$

80. $g(x) = \dfrac{1}{1 + x^2}$

81. $g(t) = e^{t^2 - 2t}$

82. $f(x) = x^2 e^x$

83. $f(x) = \ln(x^2 + 1)$

84. $g(x) = x - \ln x$

In Exercises 85–90, sketch the graph of a function having the given properties.

85. $f(2) = 4, f'(2) = 0, f''(x) < 0$ on $(-\infty, \infty)$

86. $f(2) = 2, f'(2) = 0, f'(x) > 0$ on $(-\infty, 2), f'(x) > 0$ on $(2, \infty), f''(x) < 0$ on $(-\infty, 2), f''(x) > 0$ on $(2, \infty)$

87. $f(-2) = 4, f(3) = -2, f'(-2) = 0, f'(3) = 0,$ $f'(x) > 0$ on $(-\infty, -2)$ and $(3, \infty), f'(x) < 0$ on $(-2, 3)$, inflection point at $(1, 1)$

88. $f(0) = 0, f'(0)$ does not exist, $f''(x) < 0$ if $x \neq 0$

89. $f(0) = 1, f'(0) = 0, f(x) > 0$ on $(-\infty, \infty), f''(x) < 0$ on $(-\sqrt{2}/2, \sqrt{2}/2), f''(x) > 0$ on $(-\infty, -\sqrt{2}/2)$ and $(\sqrt{2}/2, \infty)$

90. f has domain $[-1, 1], f(-1) = -1, f(-\frac{1}{2}) = -2,$ $f'(-\frac{1}{2}) = 0, f''(x) > 0$ on $(-1, 1)$

91. DEMAND FOR RNS The following graph gives the total number of help-wanted ads for RNs (registered nurses) in 22 cities over the last 12 months as a function of time t (t measured in months).
 a. Explain why $N'(t)$ is positive on the interval $(0, 12)$.
 b. Determine the signs of $N''(t)$ on the interval $(0, 6)$ and the interval $(6, 12)$.
 c. Interpret the results of part (b).

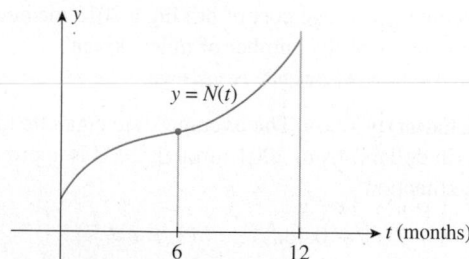

92. EFFECT OF BUDGET CUTS ON DRUG-RELATED CRIMES The graphs shown on the next page were used by a police commissioner to illustrate what effect a budget cut would have on crime in the city. The number $N_1(t)$ gives the projected number of drug-related crimes in the next 12 months. The number $N_2(t)$ gives the projected number of drug-related crimes in the same time frame if next year's budget is cut.
 a. Explain why $N_1'(t)$ and $N_2'(t)$ are both positive on the interval $(0, 12)$.
 b. What are the signs of $N_1''(t)$ and $N_2''(t)$ on the interval $(0, 12)$?

c. Interpret the results of part (b).

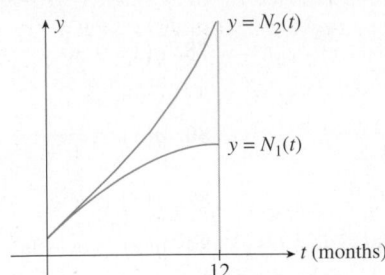

93. In the following figure, water is poured into the vase at a constant rate (in appropriate units), and the water level rises to a height of $f(t)$ units at time t as measured from the base of the vase. The graph of f follows. Explain the shape of the curve in terms of its concavity. What is the significance of the inflection point?

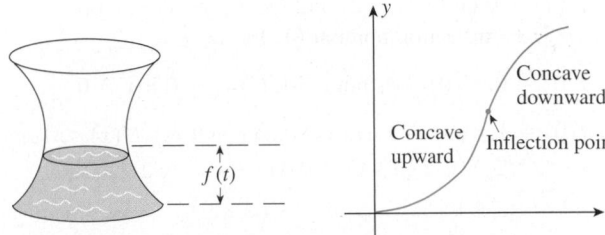

94. **DRIVING COSTS** A study of driving costs based on 2012 medium-sized sedans found that the average cost (car payments, gas, insurance, upkeep, and depreciation) in cents per mile is approximately

$$C(x) = \frac{1910.5}{x^{1.72}} + 42.9 \qquad (5 \le x \le 20)$$

where x (in thousands) denotes the number of miles the car is driven each year. Show that the graph of C is concave upward on the interval $(5, 20)$. What does your result tell you about the average cost of driving a 2012 medium-sized sedan in terms of the number of miles driven?
Source: American Automobile Association.

95. **STATE CIGARETTE TAXES** The average state cigarette tax per pack (in dollars) from 2001 through 2007 is approximated by the function

$$T(t) = 0.43t^{0.43} \qquad (1 \le t \le 7)$$

where t is measured in years, with $t = 1$ corresponding to 2001.
a. Show that the average state cigarette tax per pack was increasing throughout the period in question.
b. What can you say about the rate at which the average state cigarette tax per pack was increasing over the period in question?
Source: Campaign for Tobacco-Free Kids.

96. **GLOBAL WARMING** The increase in carbon dioxide (CO_2) in the atmosphere is a major cause of global warming. Using data obtained by Charles David Keeling, professor at Scripps Institution of Oceanography, the average amount of CO_2 in the atmosphere from 1958 through 2013 is approximated by

$$A(t) = 0.012414t^2 + 0.7485t + 313.9 \qquad (1 \le t \le 58)$$

where $A(t)$ is measured in parts per million volume (ppmv) and t in years, with $t = 1$ corresponding to 1958.
a. What can you say about the rate of change of the average amount of atmospheric CO_2 from 1958 through 2013?
b. What can you say about the rate of change of the rate of change of the average amount of atmospheric CO_2 from 1958 through 2013?
Source: Scripps Institution of Oceanography.

97. **PUBLIC DEBT** The U.S. public debt outstanding from 1990 to 2011 is approximated by

$$D(t) = 0.0032t^3 - 0.0698t^2 + 0.6048t + 3.22 \qquad (0 \le t \le 21)$$

trillion dollars in year t, where $t = 0$ is measured in years, with $t = 0$ corresponding to 1990.
a. Find the interval where the graph of D is concave downward and the interval where it is concave upward.
b. What is the inflection point of D? Interpret your result.
Source: U.S. Department of the Treasury.

98. **EFFECT OF ADVERTISING ON HOTEL REVENUE** The total annual revenue R of the Miramar Resorts Hotel is related to the amount of money x the hotel spends on advertising its services by the function

$$R(x) = -0.003x^3 + 1.35x^2 + 2x + 8000 \qquad (0 \le x \le 400)$$

where both R and x are measured in thousands of dollars.
a. Find the interval where the graph of R is concave upward and the interval where the graph of R is concave downward. What is the inflection point of the graph of R?
b. Would it be more beneficial for the hotel to increase its advertising budget slightly when the budget is $140,000 or when it is $160,000?

99. **ALZHEIMER'S DISEASE** Alzheimer's disease can occur at any age, even as young as 40 years old, but its occurrence is much more common as the years go by. The frequency of occurrence of the disease (as a percentage) is given by

$$f(t) = 0.71e^{0.7t} \qquad (1 \le t \le 5)$$

where t is measured in 5-year intervals, with $t = 1$ corresponding to an age of 70 years.
a. What is the frequency of occurrence of Alzheimer's disease for 70-year-old people? For 90-year-old people?
b. Show that f is increasing on the interval $(1, 5)$. Interpret your results.
c. Show that f is concave upward on the interval $(1, 5)$. Interpret your result.
Source: World Health Organization.

100. AVERAGE LIFE SPAN One reason for the increase in the life span over the years has been the advances in medical technology. The average life span for American women from 1907 through 2007 is given by

$$W(t) = 49.9 + 17.1 \ln t \qquad (1 \le t \le 6)$$

where $W(t)$ is measured in years and t is measured in 20-year intervals, with $t = 1$ corresponding to the beginning of 1907.
a. Show that W is increasing on $(1, 6)$.
b. What can you say about the concavity of the graph of W on the interval $(1, 6)$?

101. OFFSHORE OUTSOURCING The amount (in billions of dollars) spent by the top 15 U.S. financial institutions on IT (information technology) offshore outsourcing is approximated by

$$A(t) = 0.92(t + 1)^{0.61} \qquad (0 \le t \le 4)$$

where t is measured in years, with $t = 0$ corresponding to 2004.
a. Show that A is increasing on $(0, 4)$, and interpret your result.
b. Show that the graph of A is concave downward on $(0, 4)$. Interpret your result.
Source: Tower Group.

102. FORECASTING PROFITS As a result of increasing energy costs, the growth rate of the profit of the 4-year old Venice Glassblowing Company has begun to decline. Venice's management, after consulting with energy experts, decides to implement certain energy-conservation measures aimed at cutting energy bills. The general manager reports that, according to his calculations, the growth rate of Venice's profit should be on the increase again within 4 years. If Venice's profit (in hundreds of dollars) t years from now is given by the function

$$P(t) = t^3 - 9t^2 + 40t + 50 \qquad (0 \le t \le 8)$$

determine whether the general manager's forecast will be accurate.
Hint: Find the inflection point of the graph of P, and study the concavity of the graph of P.

103. GOOGLE'S REVENUE The revenue for Google from 2004 ($t = 0$) through 2008 ($t = 4$) is approximated by the function

$$R(t) = -0.2t^3 + 1.64t^2 + 1.31t + 3.2 \qquad (0 \le t \le 4)$$

where $R(t)$ is measured in billions of dollars.
a. Find $R'(t)$ and $R''(t)$.
b. Show that $R'(t) > 0$ for all t in the interval $(0, 4)$ and interpret your result.
Hint: Use the quadratic formula.
c. Find the inflection point of the graph of R, and interpret your result.
Source: Google company report.

104. POPULATION GROWTH IN CLARK COUNTY Clark County in Nevada—dominated by greater Las Vegas—is one of the fastest-growing metropolitan areas in the United States. The population of the county from 1970 through 2010 is approximated by the function

$$P(t) = 44{,}560t^3 - 89{,}394t^2 + 234{,}633t + 273{,}288 \qquad (0 \le t \le 4)$$

where t is measured in decades, with $t = 0$ corresponding to the beginning of 1970.
a. Show that the population of Clark County was always increasing over the time period in question.
Hint: Show that $P'(t) > 0$ for all t in the interval $(0, 4)$.
b. Show that the population of Clark County was increasing at the slowest pace some time in August 1976.
Hint: Find the inflection point of the graph of P in the interval $(0, 4)$.
Source: U.S. Census Bureau.

105. PUBLIC TRANSPORTATION BUDGET DEFICIT According to the Massachusetts Bay Transportation Authority (MBTA), the projected cumulative MBTA budget deficit with the $160 million rescue package (in billions of dollars) is given by

$$D_1(t) = 0.0275t^2 + 0.081t + 0.07 \qquad (0 \le t \le 3)$$

and the budget deficit without the rescue package is given by

$$D_2(t) = 0.035t^2 + 0.211t + 0.24 \qquad (0 \le t \le 3)$$

Here, t is measured in years, with $t = 0$ corresponding to 2011. Let $D = D_2 - D_1$.
a. Show that D is increasing on $(0, 3)$, and interpret your result.
b. Show that the graph of D is concave upward on $(0, 3)$, and interpret your result.
Source: MBTA Review.

106. WOMEN'S SOCCER Starting with the youth movement that took hold in the 1970s and buoyed by the success of the U.S. national women's team in international competition in recent years, girls and women have taken to soccer in ever-growing numbers. The function

$$N(t) = -0.9307t^3 + 74.04t^2 + 46.8667t + 3967 \qquad (0 \le t \le 16)$$

gives the number of participants in women's soccer in year t, with $t = 0$ corresponding to the beginning of 1985.
a. Verify that the number of participants per year in women's soccer had been increasing from 1985 through 2000.
Hint: Use the quadratic formula.
b. Show that the number of participants per year in women's soccer had been increasing at an increasing rate from 1985 through 2000.
Hint: Show that the sign of N'' is positive on the interval in question.
Source: NCCA News.

107. DEPENDENCY RATIO The share of the world population that is over 60 years of age compared to the rest of the working population in the world is of concern to economists. An increasing dependency ratio means that there will be fewer workers to support an aging population. The dependency ratio over the next century is forecast to be

$$R(t) = 0.00731t^4 - 0.174t^3 + 1.528t^2 + 0.48t + 19.3$$
$$(0 \le t \le 6)$$

in year t, where t is measured in decades with $t = 0$ corresponding to 2000.
a. Show that the dependency ratio will be increasing at the fastest pace around 2052.
   Hint: Use the quadratic formula.
b. What will the dependency ratio be at that time?
Source: International Institute for Applied Systems Analysis.

In Exercises 108–111, determine whether the statement is true or false. If it is true, explain why it is true. If it is false, give an example to show why it is false.

108. If the graph of f is concave upward on (a, b), then the graph of $-f$ is concave downward on (a, b).

109. If the graph of f is concave upward on (a, c) and concave downward on (c, b), where $a < c < b$, then f has an inflection point at $(c, f(c))$.

110. If a function f is defined on (a, b) and the graph of f has an inflection point at $(c, f(c))$, where $a < c < b$, then $f''(c) = 0$.

111. If c is a critical number of f where $a < c < b$ and $f''(x) < 0$ on (a, b), then f has a relative maximum at $x = c$.

112. Consider the functions $f(x) = x^3$, $g(x) = x^4$, and $h(x) = -x^4$.
a. Show that $x = 0$ is a critical number of each of the functions f, g, and h.
b. Show that the second derivative of each of the functions f, g, and h equals zero at $x = 0$.
c. Show that f has neither a relative maximum nor a relative minimum at $x = 0$, that g has a relative minimum at $x = 0$, and that h has a relative maximum at $x = 0$.

10.2 Solutions to Self-Check Exercises

1. We first compute

$$f'(x) = 12x^2 - 6x$$
$$f''(x) = 24x - 6 = 6(4x - 1)$$

Observe that f'' is continuous everywhere and has a zero at $x = \frac{1}{4}$. The sign diagram of f'' is shown in the accompanying figure.

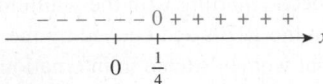

From the sign diagram for f'', we see that the graph of f is concave upward on $\left(\frac{1}{4}, \infty\right)$ and concave downward on $\left(-\infty, \frac{1}{4}\right)$.

2. First, we find the critical numbers of f by solving the equation

$$f'(x) = 6x^2 - x - 12 = 0$$

That is,

$$(3x + 4)(2x - 3) = 0$$

giving $x = -\frac{4}{3}$ and $x = \frac{3}{2}$. Next, we compute

$$f''(x) = 12x - 1$$

Since

$$f''\left(-\frac{4}{3}\right) = 12\left(-\frac{4}{3}\right) - 1 = -17 < 0$$

the Second Derivative Test implies that $f\left(-\frac{4}{3}\right) = \frac{10}{27}$ is a relative maximum of f. Also,

$$f''\left(\frac{3}{2}\right) = 12\left(\frac{3}{2}\right) - 1 = 17 > 0$$

and we see that $f\left(\frac{3}{2}\right) = -\frac{179}{8}$ is a relative minimum.

3. We compute the second derivative of G. Thus,

$$G'(t) = -6t^2 + 90t + 20$$
$$G''(t) = -12t + 90$$

Now, G'' is continuous everywhere, and $G''(t) = 0$, when $t = \frac{15}{2}$, giving $t = \frac{15}{2}$ as the only candidate for an inflection point of G. Since $G''(t) > 0$ for $t < \frac{15}{2}$ and $G''(t) < 0$ for $t > \frac{15}{2}$, we see that $\left(\frac{15}{2}, \frac{15,675}{2}\right)$ is an inflection point of G. The results of our computations tell us that the country's GDP was increasing most rapidly at the beginning of July 2011.

10.3 Curve Sketching

A Real-Life Example

As we have seen on numerous occasions, the graph of a function is a useful aid for visualizing the function's properties. From a practical point of view, the graph of a function also gives, at one glance, a complete summary of all the information captured by the function.

Consider, for example, the graph of the function giving the Dow-Jones Industrial Average (DJIA) on Black Monday, October 19, 1987 (Figure 46). Here, $t = 0$ corresponds to 9:30 A.M., when the market was open for business, and $t = 6.5$ corresponds to 4 P.M., the closing time. The following information may be gleaned from studying the graph.

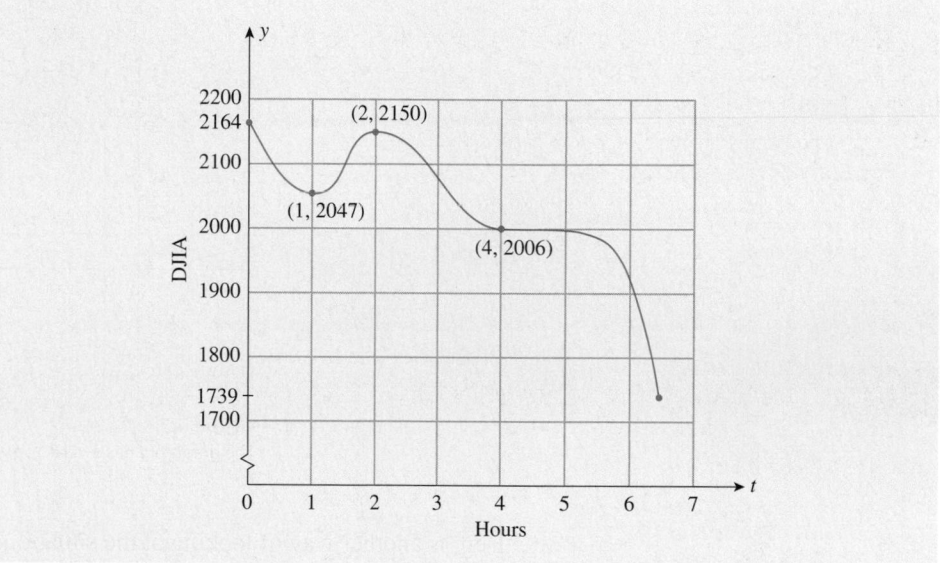

FIGURE **46**
The Dow-Jones Industrial Average on Black Monday
Source: Wall Street Journal.

The graph is *falling* rapidly from $t = 0$ to $t = 1$, reflecting the sharp drop in the index in the first hour of trading. The point $(1, 2047)$ is a *relative minimum* point of the function, and this turning point coincides with the start of an aborted recovery. The short-lived rally, represented by the portion of the graph that is *rising* on the interval $(1, 2)$, quickly fizzled out at $t = 2$ (11:30 A.M.). The *relative maximum* point $(2, 2150)$ marks the highest point of the recovery. The function is decreasing in the rest of the interval. The point $(4, 2006)$ is an *inflection point* of the function; it shows that there was a temporary respite at $t = 4$ (1:30 P.M.). However, selling pressure continued unabated, and the DJIA continued to fall until the closing bell. Finally, the graph also shows that the index opened at the high of the day [$f(0) = 2164$ is the *absolute maximum* of the function] and closed at the low of the day [$f(\frac{13}{2}) = 1739$ is the *absolute minimum* of the function], a drop of 508 points from the previous close!*

Before we turn our attention to the actual task of sketching the graph of a function, let's look at some properties of graphs that will be helpful in this connection.

---

*Absolute maxima and absolute minima of functions are covered in Section 10.4.

Vertical Asymptotes

Before going on, you might want to review the material on one-sided limits and the limit at infinity of a function (Sections 9.1 and 9.2).

Consider the graph of the function

$$f(x) = \frac{x + 1}{x - 1}$$

shown in Figure 47. Observe that $f(x)$ increases without bound (tends to infinity) as x approaches $x = 1$ from the right; that is,

$$\lim_{x \to 1^+} \frac{x + 1}{x - 1} = \infty$$

You can verify this by taking a sequence of values of x approaching $x = 1$ from the right and looking at the corresponding values of $f(x)$.

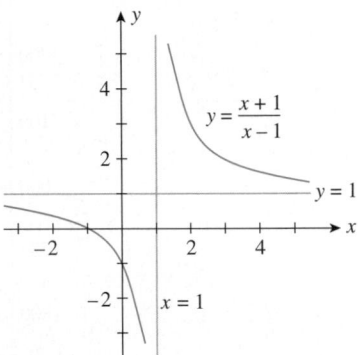

FIGURE 47
The graph of f has a vertical asymptote at $x = 1$.

Here is another way of looking at the situation: Observe that if x is a number that is a little larger than 1, then both $(x + 1)$ and $(x - 1)$ are positive, so $(x + 1)/(x - 1)$ is also positive. As x approaches $x = 1$, the numerator $(x + 1)$ approaches the number 2, but the denominator $(x - 1)$ approaches zero, so the quotient $(x + 1)/(x - 1)$ approaches infinity, as observed earlier. The line $x = 1$ is called a vertical asymptote of the graph of f.

For the function $f(x) = (x + 1)/(x - 1)$, you can show that

$$\lim_{x \to 1^-} \frac{x + 1}{x - 1} = -\infty$$

and this tells us how $f(x)$ approaches the asymptote $x = 1$ from the left.

More generally, we have the following definition:

Vertical Asymptote

The line $x = a$ is a **vertical asymptote** of the graph of a function f if

$$\lim_{x \to a^+} f(x) = \infty \quad \text{or} \quad -\infty$$

or

$$\lim_{x \to a^-} f(x) = \infty \quad \text{or} \quad -\infty$$

Note   Although a vertical asymptote of a graph is not part of the graph, it serves as a useful aid for sketching the graph.

For rational functions

$$f(x) = \frac{P(x)}{Q(x)}$$

there is a simple criterion for determining whether the graph of f has any vertical asymptotes.

> **Finding Vertical Asymptotes of Rational Functions**
>
> Suppose f is a rational function
>
> $$f(x) = \frac{P(x)}{Q(x)}$$
>
> where P and Q are polynomial functions. Then, the line $x = a$ is a vertical asymptote of the graph of f if $Q(a) = 0$ but $P(a) \neq 0$.

For the function

$$f(x) = \frac{x + 1}{x - 1}$$

considered earlier, $P(x) = x + 1$ and $Q(x) = x - 1$. Observe that $Q(1) = 0$ but $P(1) = 2 \neq 0$, so $x = 1$ is a vertical asymptote of the graph of f.

EXAMPLE 1 Find the vertical asymptotes of the graph of the function

$$f(x) = \frac{x^2}{4 - x^2}$$

Solution   The function f is a rational function with $P(x) = x^2$ and $Q(x) = 4 - x^2$. The zeros of Q are found by solving

$$4 - x^2 = 0$$

that is,

$$(2 + x)(2 - x) = 0$$

giving $x = -2$ and $x = 2$. These are candidates for the vertical asymptotes of the graph of f. Examining $x = -2$, we compute $P(-2) = (-2)^2 = 4 \neq 0$, and we see that $x = -2$ is indeed a vertical asymptote of the graph of f. Similarly, we find $P(2) = 2^2 = 4 \neq 0$, so $x = 2$ is also a vertical asymptote of the graph of f. The graph of f sketched in Figure 48 confirms these results.

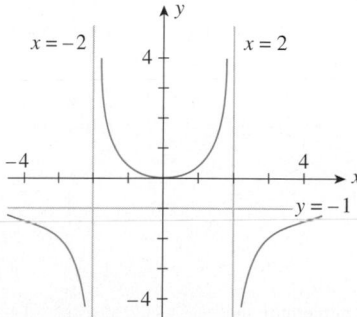

FIGURE **48**
$x = -2$ and $x = 2$ are vertical asymptotes of the graph of f.

Recall that if the line $x = a$ is a vertical asymptote of the graph of a rational function f, then *only* the denominator of $f(x)$ is equal to zero at $x = a$. If *both* $P(a)$ and $Q(a)$ are equal to zero, then $x = a$ need *not* be a vertical asymptote. For example, look at the function

$$f(x) = \frac{4(x^2 - 4)}{x - 2}$$

whose graph appears in Figure 7a in Section 9.1 (page 601).

Horizontal Asymptotes

Let's return to the function f defined by

$$f(x) = \frac{x + 1}{x - 1}$$

(Figure 49).

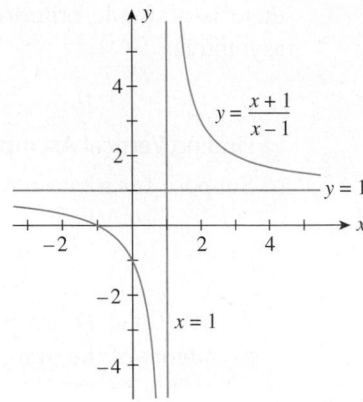

FIGURE 49
The graph of f has a horizontal asymptote
at $y = 1$.

Observe that the graph of f approaches the horizontal line $y = 1$ as x approaches infinity, and in this case, the graph of f approaches $y = 1$ as x approaches minus infinity as well. The line $y = 1$ is called a horizontal asymptote of the graph of f. More generally, we have the following definition:

> **Horizontal Asymptote**
> The line $y = b$ is a **horizontal asymptote** of the graph of a function f if either
> $$\lim_{x \to \infty} f(x) = b \quad \text{or} \quad \lim_{x \to -\infty} f(x) = b$$

For the function

$$f(x) = \frac{x + 1}{x - 1}$$

we see that

$$\lim_{x \to \infty} \frac{x + 1}{x - 1} = \lim_{x \to \infty} \frac{1 + \dfrac{1}{x}}{1 - \dfrac{1}{x}} \qquad \text{Divide numerator and denominator by } x.$$

$$= 1$$

Also,

$$\lim_{x \to -\infty} \frac{x + 1}{x - 1} = \lim_{x \to -\infty} \frac{1 + \dfrac{1}{x}}{1 - \dfrac{1}{x}}$$

$$= 1$$

In either case, we conclude that $y = 1$ is a horizontal asymptote of the graph of f, as observed earlier.

EXAMPLE 2 Find the horizontal asymptotes of the graph of the function

$$f(x) = \frac{x^2}{4 - x^2}$$

Solution We compute

$$\lim_{x \to \infty} \frac{x^2}{4 - x^2} = \lim_{x \to \infty} \frac{1}{\dfrac{4}{x^2} - 1} \qquad \text{Divide numerator and denominator by } x^2.$$

$$= -1$$

so $y = -1$ is a horizontal asymptote, as before. (Similarly, $\lim_{x \to -\infty} f(x) = -1$ as well.) The graph of f sketched in Figure 50 confirms this result.

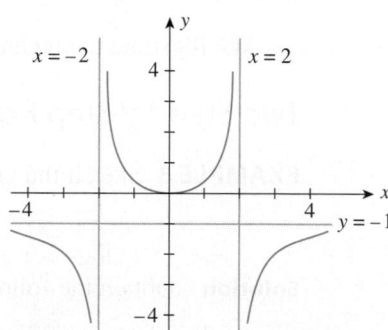

FIGURE **50**
The graph of f has a horizontal asymptote at $y = -1$.

We next state an important property of polynomial functions.

> The graph of a polynomial function has no vertical or horizontal asymptotes.

To see this, note that a polynomial function $P(x)$ can be written as a rational function with denominator equal to 1. Thus,

$$P(x) = \frac{P(x)}{1}$$

Since the denominator is never equal to zero, P has no vertical asymptotes. Next, if P is a polynomial of degree greater than or equal to 1, then

$$\lim_{x \to \infty} P(x) \quad \text{and} \quad \lim_{x \to -\infty} P(x)$$

are either infinity or minus infinity; that is, they do not exist. Therefore, P has no horizontal asymptotes.

In Sections 10.1 and 10.2, we saw how the first and second derivatives of a function are used to reveal various properties of the graph of a function f. We now show how this information can be used to help us sketch the graph of f. We begin by giving a general procedure for curve sketching.

> A Guide to Curve Sketching
>
> 1. Determine the domain of f.
> 2. Find the x- and y-intercepts of f.*
> 3. Determine the behavior of f for large absolute values of x.
> 4. Find all horizontal and vertical asymptotes of the graph of f.
> 5. Determine the intervals where f is increasing and where f is decreasing.
> 6. Find the relative extrema of f.
> 7. Determine the concavity of the graph of f.
> 8. Find the inflection points of f.
> 9. Plot a few additional points to help further identify the shape of the graph of f and sketch the graph.
>
> ————
> *The equation $f(x) = 0$ may be difficult to solve, in which case one may decide against finding the x-intercepts or to use technology, if available, for assistance.

We illustrate the techniques of curve sketching in the next two examples.

Two Step-by-Step Examples

EXAMPLE 3 Sketch the graph of the function

$$y = f(x) = x^3 - 6x^2 + 9x + 2$$

Solution   Obtain the following information on the graph of f.

Step 1   The domain of f is the interval $(-\infty, \infty)$.

Step 2   By setting $x = 0$, we find that the y-intercept is 2. The x-intercept is found by setting $y = 0$, which in this case leads to a cubic equation. Since the solution is not readily found, we will not use this information.

Step 3   Since

$$\lim_{x \to -\infty} f(x) = \lim_{x \to -\infty} (x^3 - 6x^2 + 9x + 2) = -\infty$$

$$\lim_{x \to \infty} f(x) = \lim_{x \to \infty} (x^3 - 6x^2 + 9x + 2) = \infty$$

we see that f decreases without bound as x decreases without bound and that f increases without bound as x increases without bound.

Step 4   Since f is a polynomial function, there are no asymptotes.

Step 5
$$f'(x) = 3x^2 - 12x + 9 = 3(x^2 - 4x + 3)$$
$$= 3(x - 1)(x - 3)$$

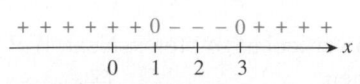

FIGURE **51**
Sign diagram for f'

Setting $f'(x) = 0$ gives $x = 1$ or $x = 3$. The sign diagram for f' shows that f is increasing on the intervals $(-\infty, 1)$ and $(3, \infty)$ and decreasing on the interval $(1, 3)$ (Figure 51).

Step 6   From the results of Step 5, we see that $x = 1$ and $x = 3$ are critical numbers of f. Furthermore, f' changes sign from positive to negative as we move across $x = 1$, so a relative maximum of f occurs at $x = 1$. Similarly, we see that a relative minimum of f occurs at $x = 3$. Now,

$$f(1) = 1 - 6 + 9 + 2 = 6$$
$$f(3) = 3^3 - 6(3)^2 + 9(3) + 2 = 2$$

so $f(1) = 6$ is a relative maximum of f and $f(3) = 2$ is a relative minimum of f.

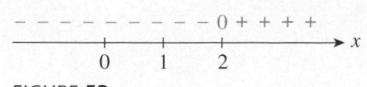

FIGURE 52
Sign diagram for f''

Step 7

$$f''(x) = 6x - 12 = 6(x - 2)$$

which is equal to zero when $x = 2$. The sign diagram of f'' shows that the graph of f is concave downward on the interval $(-\infty, 2)$ and concave upward on the interval $(2, \infty)$ (Figure 52).

Step 8 From the results of Step 7, we see that f'' changes sign as we move across $x = 2$. Next,

$$f(2) = 2^3 - 6(2)^2 + 9(2) + 2 = 4$$

so the required inflection point of f is $(2, 4)$.

Step 9 Summarizing, we have the following:

> Domain: $(-\infty, \infty)$
> Intercept: $(0, 2)$
> $\lim_{x \to -\infty} f(x); \lim_{x \to \infty} f(x): -\infty; \infty$
> Asymptotes: None
> Intervals where f is ↗ or ↘: ↗ on $(-\infty, 1)$ and $(3, \infty)$; ↘ on $(1, 3)$
> Relative extrema: Relative maximum at $(1, 6)$; relative minimum at $(3, 2)$
> Concavity: Downward on $(-\infty, 2)$; upward on $(2, \infty)$
> Point of inflection: $(2, 4)$

In general, it is a good idea to start graphing by plotting the intercept(s), relative extrema, and inflection point(s) (Figure 53). Then, using the rest of the information, we complete the graph of f, as sketched in Figure 54.

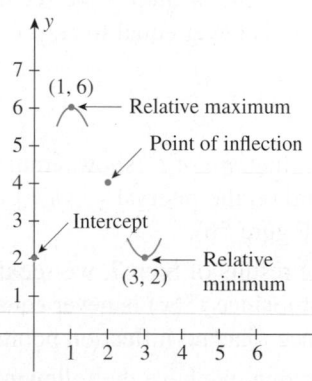

FIGURE 53
We first plot the intercept, the relative extrema, and the inflection point.

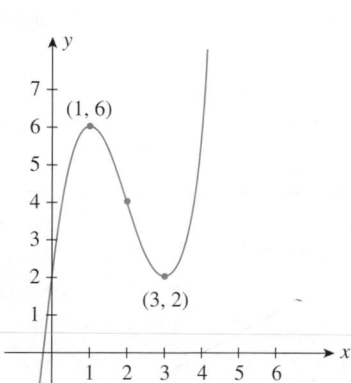

FIGURE 54
The graph of $y = x^3 - 6x^2 + 9x + 2$

Explore and Discuss

The average price of gasoline at the pump over a 3-month period, during which there was a temporary shortage of oil, is described by the function f defined on the interval $[0, 3]$. During the first month, the price was increasing at an increasing rate. Starting with the second month, the good news was that the rate of increase was slowing down, although the price of gas was still increasing. This pattern continued until the end of the second month. The price of gas peaked at $t = 2$ and began to fall at an increasing rate until $t = 3$.

1. Describe the signs of $f'(t)$ and $f''(t)$ over each of the intervals $(0, 1)$, $(1, 2)$, and $(2, 3)$.

2. Make a sketch showing a plausible graph of f over $[0, 3]$.

EXAMPLE 4 Sketch the graph of the function

$$y = f(x) = \frac{x + 1}{x - 1}$$

Solution Obtain the following information:

Step 1 f is undefined when $x = 1$, so the domain of f is the set of all real numbers other than $x = 1$.

Step 2   Setting $y = 0$ gives $x = -1$, the x-intercept of f. Next, setting $x = 0$ gives $y = -1$ as the y-intercept of f.

Step 3   Earlier, we found that

$$\lim_{x \to \infty} \frac{x + 1}{x - 1} = 1 \quad \text{and} \quad \lim_{x \to -\infty} \frac{x + 1}{x - 1} = 1$$

(see page 760). Consequently, we see that the graph of f approaches the line $y = 1$ as $|x|$ becomes arbitrarily large. For $x > 1$, $f(x) > 1$ and the graph of f approaches the line $y = 1$ from above. For $x < 1$, $f(x) < 1$, so the graph of f approaches the line $y = 1$ from below.

Step 4   The straight line $x = 1$ is a vertical asymptote of the graph of f. Also, from the results of Step 3, we conclude that $y = 1$ is a horizontal asymptote of the graph of f.

Step 5
$$f'(x) = \frac{(x - 1)(1) - (x + 1)(1)}{(x - 1)^2} = -\frac{2}{(x - 1)^2}$$

and is discontinuous at $x = 1$. The sign diagram of f' shows that $f'(x) < 0$ whenever it is defined. Thus, f is decreasing on the intervals $(-\infty, 1)$ and $(1, \infty)$ (Figure 55).

Step 6   From the results of Step 5, we see that there are no critical numbers of f, since $f'(x)$ is never equal to zero for any value of x in the domain of f.

Step 7
$$f''(x) = \frac{d}{dx}[-2(x - 1)^{-2}] = 4(x - 1)^{-3} = \frac{4}{(x - 1)^3}$$

The sign diagram of f'' shows immediately that the graph of f is concave downward on the interval $(-\infty, 1)$ and concave upward on the interval $(1, \infty)$ (Figure 56).

Step 8   From the results of Step 7, we see that there are no candidates for inflection points of f, since $f''(x)$ is never equal to zero for any value of x in the domain of f. Hence f has no inflection points.

Step 9   Summarizing, we have the following:

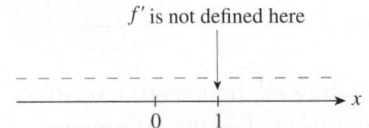

FIGURE **55**
The sign diagram for f'

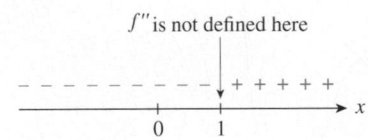

FIGURE **56**
The sign diagram for f''

> Domain: $(-\infty, 1) \cup (1, \infty)$
> Intercepts: $(-1, 0)$; $(0, -1)$
> $\lim_{x \to -\infty} f(x)$; $\lim_{x \to \infty} f(x)$: 1; 1
> Asymptotes: $x = 1$ is a vertical asymptote
> $\quad\quad\quad\quad\quad\quad y = 1$ is a horizontal asymptote
> Intervals where f is ↗ or ↘: ↘ on $(-\infty, 1)$ and $(1, \infty)$
> Relative extrema: None
> Concavity: Downward on $(-\infty, 1)$; upward on $(1, \infty)$
> Points of inflection: None

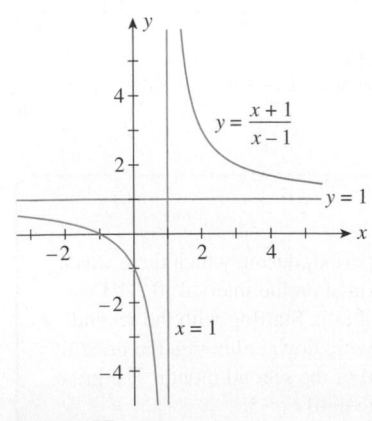

FIGURE **57**
The graph of f has a horizontal asymptote at $y = 1$ and a vertical asymptote at $x = 1$.

The graph of f is sketched in Figure 57.

10.3 Self-Check Exercises

1. Find the horizontal and vertical asymptotes of the graph of the function

$$f(x) = \frac{2x^2}{x^2 - 1}$$

2. Sketch the graph of the function

$$f(x) = \frac{2}{3}x^3 - 2x^2 - 6x + 4$$

Solutions to Self-Check Exercises 10.3 can be found on page 769.

10.3 Concept Questions

1. Explain the following terms in your own words:
   a. Vertical asymptote   b. Horizontal asymptote

2. a. How many vertical asymptotes can the graph of a function f have? Explain using graphs.
   b. How many horizontal asymptotes can the graph of a function f have? Explain using graphs.

3. How do you find the vertical asymptotes of the graph of a rational function?

4. Give a procedure for sketching the graph of a function.

10.3 Exercises

In Exercises 1–10, find the horizontal and vertical asymptotes of the graph of the function.

1.

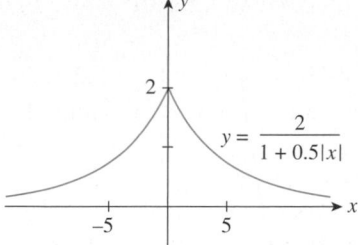

$$y = \frac{2}{1 + 0.5|x|}$$

2.

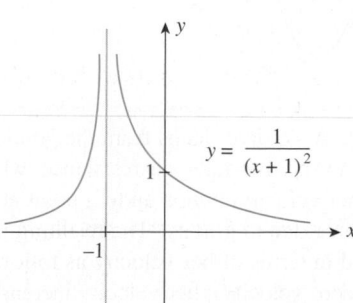

$$y = \frac{1}{(x + 1)^2}$$

3.

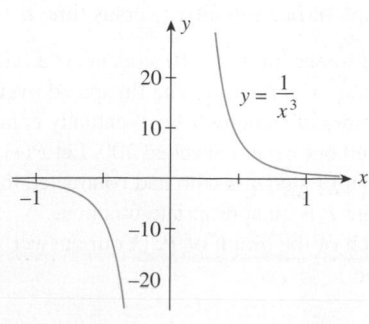

$$y = \frac{1}{x^3}$$

4.

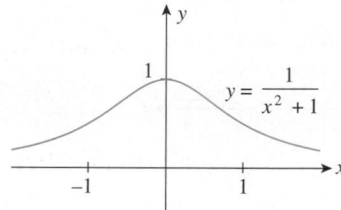

$$y = \frac{1}{x^2 + 1}$$

5.

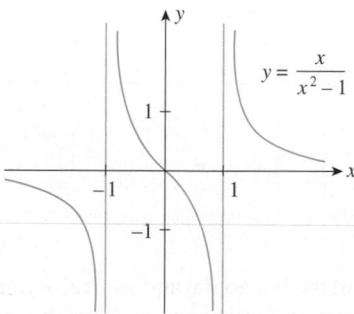

$$y = \frac{x}{x^2 - 1}$$

6.

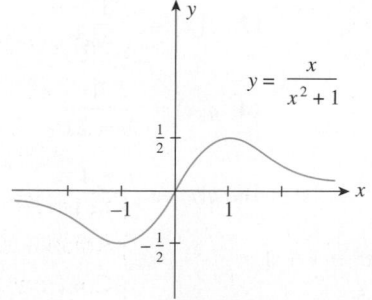

$$y = \frac{x}{x^2 + 1}$$

7.

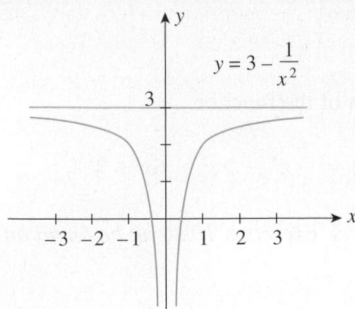

$$y = 3 - \frac{1}{x^2}$$

8.

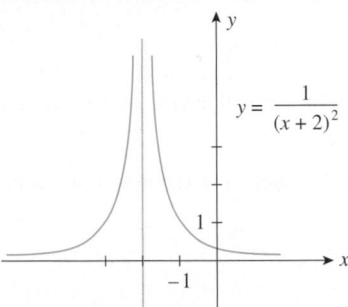

$$y = \frac{1}{(x+2)^2}$$

9.

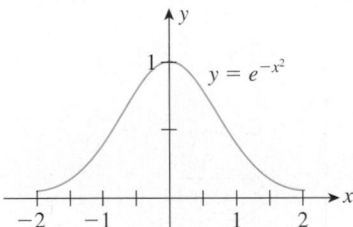

$$y = e^{-x^2}$$

10.

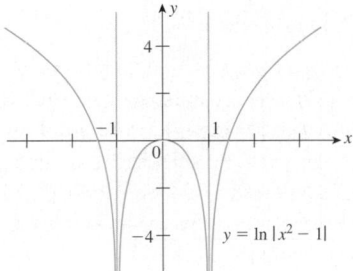

$$y = \ln|x^2 - 1|$$

In Exercises 11–28, find the horizontal and vertical asymptotes of the graph of the function. (You need not sketch the graph.)

11. $f(x) = \dfrac{1}{x}$

12. $f(x) = \dfrac{1}{x + 2}$

13. $f(x) = -\dfrac{2}{x^2}$

14. $g(x) = \dfrac{1}{1 + 2x^2}$

15. $f(x) = \dfrac{x - 2}{x + 2}$

16. $g(t) = \dfrac{t + 1}{2t - 1}$

17. $h(x) = x^3 - 3x^2 + x + 1$

18. $g(x) = 2x^3 + x^2 + 1$

19. $f(t) = \dfrac{t^2}{t^2 - 16}$

20. $g(x) = \dfrac{x^3}{x^2 - 4}$

21. $f(x) = \dfrac{3x}{x^2 - x - 6}$

22. $g(x) = \dfrac{2x}{x^2 + x - 2}$

23. $g(t) = 2 + \dfrac{5}{(t - 2)^2}$

24. $f(x) = 1 + \dfrac{2}{x - 3}$

25. $f(x) = \dfrac{x^2 - 2}{x^2 - 4}$

26. $h(x) = \dfrac{2 - x^2}{x^2 + x}$

27. $g(x) = \dfrac{x^3 - x}{x(x + 1)}$

28. $f(x) = \dfrac{x^4 - x^2}{x(x - 1)(x + 2)}$

In Exercises 29 and 30, you are given the graphs of two functions f and g. One function is the derivative function of the other. Identify each of them.

29.

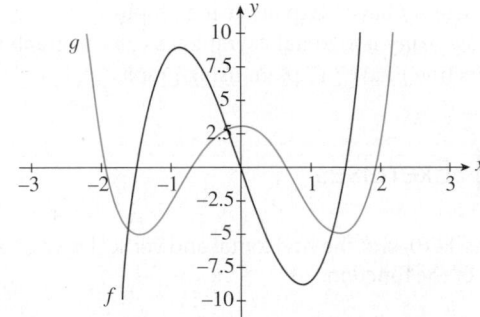

30.

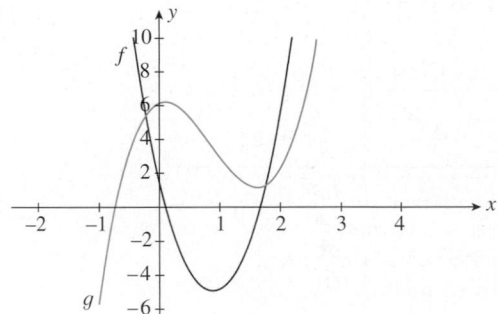

31. Terminal Velocity A skydiver leaps from the gondola of a hot-air balloon. As she free-falls, air resistance, which is proportional to her velocity, builds up to a point at which it balances the force due to gravity. The resulting motion may be described in terms of her velocity as follows: Starting at rest (zero velocity), her velocity increases and approaches a constant velocity, called the *terminal velocity*. Sketch a graph of her velocity v versus time t.

32. Spread of a Flu Epidemic Initially, 10 students at a junior high school contracted influenza. The flu spread over time, and the total number of students who eventually contracted the flu approached but never exceeded 200. Let $P(t)$ denote the number of students who had contracted the flu after t days, where P is an appropriate function.
   a. Make a sketch of the graph of P. (Your answer will *not* be unique.)

b. Where is the function increasing?
c. Does the graph of P have a horizontal asymptote? If so, what is it?
d. Discuss the concavity of the graph of P. Explain its significance.
e. Is there an inflection point on the graph of P? If so, explain its significance.

In Exercises 33–36, use the information summarized in the table to sketch the graph of f.

33. $f(x) = x^3 - 3x^2 + 1$

Domain: $(-\infty, \infty)$
Intercept: y-intercept: 1
Asymptotes: None
Intervals where f is $\nearrow$ and $\searrow$: $\nearrow$ on $(-\infty, 0)$ and $(2, \infty)$; $\searrow$ on $(0, 2)$
Relative extrema: Rel. max. at $(0, 1)$; rel. min. at $(2, -3)$
Concavity: Downward on $(-\infty, 1)$; upward on $(1, \infty)$
Point of inflection: $(1, -1)$

34. $f(x) = \dfrac{1}{9}(x^4 - 4x^3)$

Domain: $(-\infty, \infty)$
Intercepts: x-intercepts: 0, 4; y-intercept: 0
Asymptotes: None
Intervals where f is $\nearrow$ and $\searrow$: $\nearrow$ on $(3, \infty)$; $\searrow$ on $(-\infty, 3)$
Relative extrema: Rel. min. at $(3, -3)$
Concavity: Downward on $(0, 2)$; upward on $(-\infty, 0)$ and $(2, \infty)$
Points of inflection: $(0, 0)$ and $\left(2, -\dfrac{16}{9}\right)$

35. $f(x) = \dfrac{4x - 4}{x^2}$

Domain: $(-\infty, 0) \cup (0, \infty)$
Intercept: x-intercept: 1
Asymptotes: x-axis and y-axis
Intervals where f is $\nearrow$ and $\searrow$: $\nearrow$ on $(0, 2)$; $\searrow$ on $(-\infty, 0)$ and $(2, \infty)$
Relative extrema: Rel. max. at $(2, 1)$
Concavity: Downward on $(-\infty, 0)$ and $(0, 3)$; upward on $(3, \infty)$
Point of inflection: $\left(3, \dfrac{8}{9}\right)$

36. $f(x) = x - 3x^{1/3}$

Domain: $(-\infty, \infty)$
Intercepts: x-intercepts: $\pm 3\sqrt{3}$, 0; y-intercept: 0
Asymptotes: None
Intervals where f is $\nearrow$ and $\searrow$: $\nearrow$ on $(-\infty, -1)$ and $(1, \infty)$; $\searrow$ on $(-1, 1)$
Relative extrema: Rel. max. at $(-1, 2)$; rel. min. at $(1, -2)$
Concavity: Downward on $(-\infty, 0)$; upward on $(0, \infty)$
Point of inflection: $(0, 0)$

In Exercises 37–60, sketch the graph of the function, using the curve-sketching guide of this section.

37. $g(x) = 4 - 3x - 2x^3$ **38.** $f(x) = x^2 - 2x + 3$

39. $h(x) = x^3 - 3x + 1$ **40.** $f(x) = 2x^3 + 1$

41. $f(x) = -2x^3 + 3x^2 + 12x + 2$

42. $f(t) = 2t^3 - 15t^2 + 36t - 20$

43. $h(x) = \dfrac{3}{2}x^4 - 2x^3 - 6x^2 + 8$

44. $f(t) = 3t^4 + 4t^3$

45. $f(t) = \sqrt{t^2 - 4}$ **46.** $f(x) = \sqrt{x^2 + 5}$

47. $g(x) = \dfrac{1}{2}x - \sqrt{x}$ **48.** $f(x) = \sqrt[3]{x^2}$

49. $g(x) = \dfrac{2}{x - 1}$ **50.** $f(x) = \dfrac{1}{x + 1}$

51. $h(x) = \dfrac{x + 2}{x - 2}$ **52.** $g(x) = \dfrac{x}{x - 1}$

53. $f(t) = \dfrac{t^2}{1 + t^2}$ **54.** $g(x) = \dfrac{x}{x^2 - 4}$

55. $f(t) = e^t - t$ **56.** $h(x) = \dfrac{e^x + e^{-x}}{2}$

57. $f(x) = 2 - e^{-x}$ **58.** $f(x) = \dfrac{3}{1 + e^{-x}}$

59. $f(x) = \ln(x - 1)$ **60.** $f(x) = 2x - \ln x$

61. COST OF REMOVING TOXIC POLLUTANTS A city's main well was recently found to be contaminated with trichloroethylene (a cancer-causing chemical) as a result of an abandoned chemical dump that leached chemicals into the water. A proposal submitted to the city council indicated that the cost, measured in millions of dollars, of removing $x\%$ of the toxic pollutants is given by

$$C(x) = \frac{0.5x}{100 - x}$$

a. Find the vertical asymptote of the graph of C.
b. Is it possible to remove 100% of the toxic pollutant from the water?

62. AVERAGE COST OF PRODUCING DVDs The average cost per disc (in dollars) incurred by Herald Media Corporation in pressing x DVDs is given by the average cost function

$$\overline{C}(x) = 2.2 + \frac{2500}{x}$$

a. Find the horizontal asymptote of the graph of $\overline{C}$.
b. What is the limiting value of the average cost?

63. **CONCENTRATION OF A DRUG IN THE BLOODSTREAM** The concentration (in milligrams per cubic centimeter) of a certain drug in a patient's bloodstream t hr after injection is given by

$$C(t) = \frac{0.2t}{t^2 + 1}$$

a. Find the horizontal asymptote of the graph of C.
b. Interpret your result.

64. **EFFECT OF ENZYMES ON CHEMICAL REACTIONS** Certain proteins, known as enzymes, serve as catalysts for chemical reactions in living things. In 1913, Leonor Michaelis and L. M. Menten discovered the following formula giving the initial speed V (in moles per liter per second) at which the reaction begins in terms of the amount of substrate x (the substance that is being acted upon, measured in moles per liter):

$$V = \frac{ax}{x + b}$$

where a and b are positive constants.
a. Find the horizontal asymptote of the graph of V.
b. What does the result of part (a) tell you about the initial speed at which the reaction begins, if the amount of substrate is very large?

65. **GDP OF A DEVELOPING COUNTRY** A developing country's gross domestic product (GDP) from 2002 to 2010 is approximated by the function

$$G(t) = -0.2t^3 + 2.4t^2 + 60 \qquad (0 \le t \le 8)$$

where $G(t)$ is measured in billions of dollars, and t is measured in years, with $t = 0$ corresponding to 2002. Sketch the graph of the function G, and interpret your results.

66. **CRIME RATE** The number of major crimes per 100,000 committed in a city between 2006 and 2013 is approximated by the function

$$N(t) = -0.1t^3 + 1.5t^2 + 80 \qquad (0 \le t \le 7)$$

where $N(t)$ denotes the number of crimes per 100,000 committed in year t, with $t = 0$ corresponding to 2006. Enraged by the dramatic increase in the crime rate, the citizens, with the help of the local police, organized Neighborhood Crime Watch groups in early 2010 to combat this menace. Sketch the graph of the function N, and interpret your results. Is the Neighborhood Crime Watch program working?

67. **WORKER EFFICIENCY** An efficiency study showed that the total number of smartphones assembled by an average worker at Delphi Electronics t hr after starting work at 8 A.M. is given by

$$N(t) = -\frac{1}{2}t^3 + 3t^2 + 10t \qquad (0 \le t \le 4)$$

Sketch the graph of the function N, and interpret your results.

68. **CONCENTRATION OF A DRUG IN THE BLOODSTREAM** The concentration (in milligrams per cubic centimeter) of a certain drug in a patient's bloodstream t hr after injection is given by

$$C(t) = \frac{0.2t}{t^2 + 1}$$

Sketch the graph of the function C, and interpret your results.

69. **BOX-OFFICE RECEIPTS** The total worldwide box-office receipts for a long-running movie are approximated by the function

$$T(x) = \frac{120x^2}{x^2 + 4}$$

where $T(x)$ is measured in millions of dollars and x is the number of years since the movie's release. Sketch the graph of the function T, and interpret your results.

70. **SPREAD OF AN EPIDEMIC** During a flu epidemic, the total number of students on a state university campus who had contracted influenza by the xth day was given by

$$N(x) = \frac{3000}{1 + 99e^{-x}} \qquad (x \ge 0)$$

a. How many students had influenza initially?
b. Derive an expression for the rate at which the disease was being spread, and prove that the function N is increasing on the interval $(0, \infty)$.
c. Sketch the graph of N. What was the total number of students who contracted influenza during that particular epidemic?

71. **ABSORPTION OF DRUGS** A liquid carries a drug into an organ of volume V cm^3 at the rate of a cm^3/sec and leaves at the same rate. The concentration of the drug in the entering liquid is c g/cm^3. Letting $x(t)$ denote the concentration of the drug in the organ at any time t, we have $x(t) = c(1 - e^{-at/V})$.
a. Show that x is an increasing function on $(0, \infty)$.
b. Sketch the graph of x.

72. **AUTISTIC BRAIN** At birth, the autistic brain is similar in size to a healthy child's brain. Between birth and 2 years, it grows to be abnormally large, reaching its maximum size between 3 and 6 years of age. The percentage difference in size between the autistic brain and the normal brain to age 40 is approximated by

$$D(t) = 6.9te^{-0.24t} \qquad (0 \le t \le 40)$$

where t is measured in years.

a. At what ages is the difference in the size between the autistic brain and the normal brain increasing? Decreasing?

b. At what age is the difference in the size between the autistic brain and the normal brain the greatest? What is the maximum difference?

c. At what age is the difference in the size between the autistic brain and the normal brain decreasing at the fastest rate?

d. Sketch the graph of D on the interval $[0, 40]$.

Source: Newsweek.

73. CONCENTRATION OF GLUCOSE IN THE BLOODSTREAM A glucose solution is administered intravenously into the bloodstream at a constant rate of r mg/hr. As the glucose is being administered, it is converted into other substances and removed from the bloodstream. Suppose the concentration of the glucose solution at time t is given by

$$C(t) = \frac{r}{k} - \left(\frac{r}{k} - C_0\right)e^{-kt}$$

where C_0 is the concentration at time $t = 0$ and k is a positive constant. Assuming that $C_0 < r/k$, evaluate $\lim_{t \to \infty} C(t)$.

a. What does your result say about the concentration of the glucose solution in the long run?

b. Show that the function C is increasing on $(0, \infty)$.

c. Show that the graph of C is concave downward on $(0, \infty)$.

d. Sketch the graph of the function C.

In Exercises 74–76, determine whether the statement is true or false. If it is true, explain why it is true. If it is false, give an example to show why it is false.

74. If the graph of a function f has a vertical asymptote at $x = a$, then $\lim_{x \to a^-} f(x) = \infty$ or $-\infty$ and $\lim_{x \to a^+} f(x) = \infty$ or $-\infty$.

75. The graph of a function f cannot intersect its vertical asymptote.

76. The graph of a function f cannot intersect its horizontal asymptote at more than one point.

10.3 Solutions to Self-Check Exercises

1. Since

$$\lim_{x \to \infty} \frac{2x^2}{x^2 - 1} = \lim_{x \to \infty} \frac{2}{1 - \dfrac{1}{x^2}} \qquad \text{Divide the numerator and denominator by } x^2.$$

$$= 2$$

we see that $y = 2$ is a horizontal asymptote. Next, since

$$x^2 - 1 = (x + 1)(x - 1) = 0$$

implies $x = -1$ or $x = 1$, these are candidates for the vertical asymptotes of the graph of f. Since the numerator of f is not equal to zero for $x = -1$ or $x = 1$, we conclude that $x = -1$ and $x = 1$ are vertical asymptotes of the graph of f.

2. We obtain the following information on the graph of f.

(1) The domain of f is the interval $(-\infty, \infty)$.

(2) By setting $x = 0$, we find that the y-intercept is 4.

(3) Since

$$\lim_{x \to -\infty} f(x) = \lim_{x \to -\infty} \left(\frac{2}{3}x^3 - 2x^2 - 6x + 4\right) = -\infty$$

$$\lim_{x \to \infty} f(x) = \lim_{x \to \infty} \left(\frac{2}{3}x^3 - 2x^2 - 6x + 4\right) = \infty$$

we see that $f(x)$ decreases without bound as x decreases without bound and that $f(x)$ increases without bound as x increases without bound.

(4) Since f is a polynomial function, the graph of f has no asymptotes.

(5) $$f'(x) = 2x^2 - 4x - 6 = 2(x^2 - 2x - 3)$$
$$= 2(x + 1)(x - 3)$$

Setting $f'(x) = 0$ gives $x = -1$ or $x = 3$. The accompanying sign diagram for f' shows that f is increasing on the intervals $(-\infty, -1)$ and $(3, \infty)$ and decreasing on $(-1, 3)$.

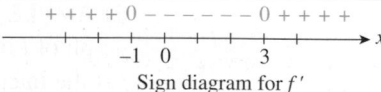

$$+ + + + 0 - - - - - - 0 + + + +$$

Sign diagram for f'

(6) From the results of Step 5, we see that $x = -1$ and $x = 3$ are critical numbers of f. Furthermore, the sign diagram of f' tells us that $x = -1$ gives rise to a relative maximum of f and $x = 3$ gives rise to a relative minimum of f. Now,

$$f(-1) = \frac{2}{3}(-1)^3 - 2(-1)^2 - 6(-1) + 4 = \frac{22}{3}$$

$$f(3) = \frac{2}{3}(3)^3 - 2(3)^2 - 6(3) + 4 = -14$$

so $f(-1) = \frac{22}{3}$ is a relative maximum of f and $f(3) = -14$ is a relative minimum of f.

(7) $f''(x) = 4x - 4 = 4(x - 1)$

which is equal to zero when $x = 1$. The accompanying sign diagram of f'' shows that the graph of f is concave downward on the interval $(-\infty, 1)$ and concave upward on the interval $(1, \infty)$.

$$- - - - - - - - 0 + + + + +$$

Sign diagram for f''

(8) From the results of Step 7, we see that $x = 1$ is the only candidate for an inflection point of f. Since $f''(x)$ changes sign as we move across the point $x = 1$ and

$$f(1) = \frac{2}{3}(1)^3 - 2(1)^2 - 6(1) + 4 = -\frac{10}{3}$$

we see that the required inflection point is $\left(1, -\frac{10}{3}\right)$.

(9) Summarizing this information, we have the following:

Domain: $(-\infty, \infty)$
Intercept: $(0, 4)$
$\lim\limits_{x \to -\infty} f(x); \lim\limits_{x \to \infty} f(x): -\infty; \infty$
Asymptotes: None
Intervals where f is ↗ or ↘: ↗ on $(-\infty, -1)$ and $(3, \infty)$;
↘ on $(-1, 3)$
Relative extrema: Rel. max. at $\left(-1, \frac{22}{3}\right)$; rel. min.
at $(3, -14)$
Concavity: Downward on $(-\infty, 1)$; upward on $(1, \infty)$
Point of inflection: $\left(1, -\frac{10}{3}\right)$

The graph of f is sketched in the accompanying figure.

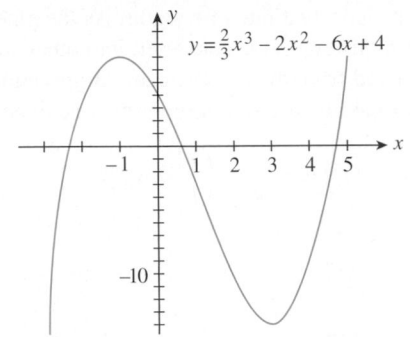

$y = \frac{2}{3}x^3 - 2x^2 - 6x + 4$

USING TECHNOLOGY Analyzing the Properties of a Function

One of the main purposes of studying Section 10.3 is to see how the many concepts of calculus come together to paint a picture of a function. The techniques of graphing also play a very practical role. For example, using the techniques of graphing developed in Section 10.3, you can tell whether the graph of a function generated by a graphing utility is reasonably complete. Furthermore, these techniques can often reveal details that are missing from a graph.

EXAMPLE 1 Consider the function $f(x) = 2x^3 - 3.5x^2 + x - 10$. A plot of the graph of f in the standard viewing window is shown in Figure T1. Since the domain of f is the interval $(-\infty, \infty)$, we see that Figure T1 does not reveal the part of the graph to the left of the y-axis. This suggests that we enlarge the viewing window accordingly. Figure T2 shows the graph of f in the viewing window $[-10, 10] \times [-20, 10]$.

The behavior of f for large values of x

$$\lim_{x \to -\infty} f(x) = -\infty \quad \text{and} \quad \lim_{x \to \infty} f(x) = \infty$$

suggests that this viewing window has captured a sufficiently complete picture of f.

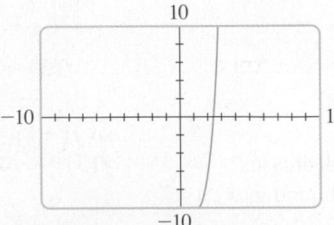

FIGURE **T1**
The graph of f in the standard viewing window

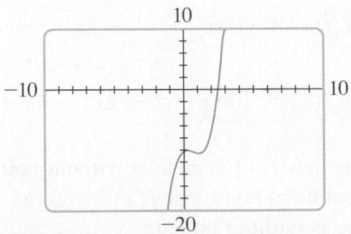

FIGURE **T2**
The graph of f in the viewing window $[-10, 10] \times [-20, 10]$

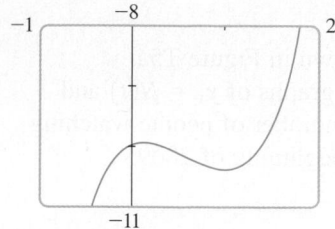

FIGURE T3
The graph of *f* in the viewing window
$[-1, 2] \times [-11, -8]$

Next, an analysis of the first derivative of *x*,

$$f'(x) = 6x^2 - 7x + 1 = (6x - 1)(x - 1)$$

reveals that *f* has critical values at $x = \frac{1}{6}$ and $x = 1$. In fact, a sign diagram of *f'* shows that *f* has a relative maximum at $x = \frac{1}{6}$ and a relative minimum at $x = 1$, details that are not revealed in the graph of *f* shown in Figure T2. To examine this portion of the graph of *f*, we use, say, the viewing window $[-1, 2] \times [-11, -8]$. The resulting graph of *f* is shown in Figure T3, which certainly reveals the hitherto missing details! Thus, through an interaction of calculus and a graphing utility, we are able to obtain a good picture of the properties of *f*.

Finding *x*-Intercepts

As was noted in Section 10.3, it is not always easy to find the *x*-intercepts of the graph of a function. But this information is very important in applications. By using the function for solving polynomial equations or the function for finding the roots of an equation, we can solve the equation $f(x) = 0$ quite easily and hence yield the *x*-intercepts of the graph of a function.

EXAMPLE 2 Let $f(x) = x^3 - 3x^2 + x + 1.5$.

a. Use the function for solving polynomial equations on a graphing utility to find the *x*-intercepts of the graph of *f*.
b. Use the function for finding the roots of an equation on a graphing utility to find the *x*-intercepts of the graph of *f*.

Solution

a. Observe that *f* is a polynomial function of degree 3, so we may use the function for solving polynomial equations to solve the equation $x^3 - 3x^2 + x + 1.5 = 0$ $[f(x) = 0]$. We find that the solutions (*x*-intercepts) are

$$x_1 \approx -0.525687120865 \qquad x_2 \approx 1.2586520225 \qquad x_3 \approx 2.26703509836$$

b. Using the graph of *f* (Figure T4), we see that $x_1 \approx -0.5$, $x_2 \approx 1$, and $x_3 \approx 2$. Using the function for finding the roots of an equation on a graphing utility and these values of *x* as initial guesses, we find

$$x_1 \approx -0.5256871209 \qquad x_2 \approx 1.2586520225 \qquad x_3 \approx 2.2670350984$$

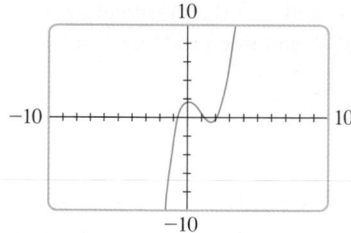

FIGURE T4
The graph of $f(x) = x^3 - 3x^2 + x + 1.5$

Note The function for solving polynomial equations on a graphing utility will solve a polynomial equation $f(x) = 0$, where *f* is a polynomial function. The function for finding the roots of an equation, however, will solve equations $f(x) = 0$ even if *f* is not a polynomial.

$ APPLIED EXAMPLE 3 TV on Smartphones The number of people watching TV on smartphones (in millions) is approximated by

$$N(t) = 11.9\sqrt{1 + 0.91t} \qquad (0 \le t \le 4)$$

where *t* is measured in years, with $t = 0$ corresponding to the beginning of 2007.

a. Use a graphing calculator to plot the graph of *N*.
b. Based on this model, when did the number of people watching TV on smartphones first exceed 20 million?

Source: IDC, U.S. forecast.

Solution

a. The graph of N in the window $[0, 4] \times [0, 30]$ is shown in Figure T5a.

b. Using the function for finding the intersection of the graphs of $y_1 = N(t)$ and $y_2 = 20$, we find $t \approx 2.005$ (see Figure T5b). So the number of people watching TV on smartphones first exceeded 20 million at the beginning of 2009.

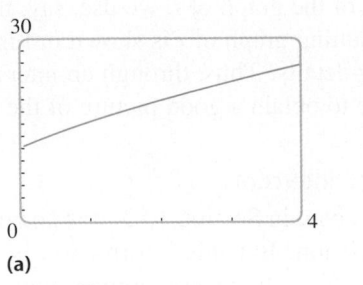

(a)   **(b)**

FIGURE **T5**
(a) The graph of N in the viewing window $[0, 4] \times [0, 30]$; (b) the graph showing the intersection of $y_1 = N(t)$ and $y_2 = 20$ on the TI-83/84.

TECHNOLOGY EXERCISES

In Exercises 1–4, use the method of Example 1 to analyze the function. (*Note:* Your answers will *not* be unique.)

1. $f(x) = 4x^3 - 4x^2 + x + 10$

2. $f(x) = x^3 + 2x^2 + x - 12$

3. $f(x) = \dfrac{1}{2}x^4 + x^3 + \dfrac{1}{2}x^2 - 10$

4. $f(x) = 2.25x^4 - 4x^3 + 2x^2 + 2$

In Exercises 5–10, find the x-intercepts of the graph of f. Give your answers accurate to four decimal places.

5. $f(x) = 0.2x^3 - 1.2x^2 + 0.8x + 2.1$

6. $f(x) = -0.2x^4 + 0.8x^3 - 2.1x + 1.2$

7. $f(x) = 2x^2 - \sqrt{x + 1} - 3$

8. $f(x) = x - \sqrt{1 - x^2}$

9. $f(x) = e^x - 2x - 2$

10. $f(x) = \ln(1 + x^2) + 2x - 3$

11. AIR POLLUTION The level of ozone, an invisible gas that irritates and impairs breathing, present in the atmosphere on a certain day in June in the city of Riverside is approximated by

$$S(t) = 1.0974t^3 - 0.0915t^4 \qquad (0 \le t \le 11)$$

where $S(t)$ is measured in pollutant standard index (PSI) and t is measured in hours, with $t = 0$ corresponding to 7 A.M. Sketch the graph of S, and interpret your results.
Source: Los Angeles Times.

12. FLIGHT PATH OF A PLANE The function

$$f(x) = \begin{cases} 0 & \text{if } 0 \le x < 1 \\ -0.0411523x^3 + 0.679012x^2 \\ \qquad -1.23457x + 0.596708 & \text{if } 1 \le x < 10 \\ 15 & \text{if } 10 \le x \le 11 \end{cases}$$

where both x and $f(x)$ are measured in units of 1000 ft, describes the flight path of a plane taking off from the origin and climbing to an altitude of 15,000 ft. Sketch the graph of f to visualize the trajectory of the plane.

10.4 Optimization I

Absolute Extrema

The graph of the function f in Figure 58 shows the average age of cars in use in the United States from the beginning of 1946 ($t = 0$) to the beginning of 2009 ($t = 63$). Observe that the highest average age of cars in use during this period is 9.3 years, whereas the lowest average age of cars in use during the same period is 5.5 years. The number 9.3, the largest value of $f(t)$ for all values of t in the interval $[0, 63]$ (the domain of f), is called the *absolute maximum value of f* on that interval. The number 5.5, the smallest value of $f(t)$ for all values of t in $[0, 63]$, is called the *absolute minimum value of f* on that interval. Notice, too, that the absolute maximum value of f is attained at the endpoint $t = 63$ of the interval, whereas the absolute minimum value of f is attained at the points $t = 12$ (corresponding to 1958) and $t = 23$ (corresponding to 1969) that lie within the interval $(0, 63)$.

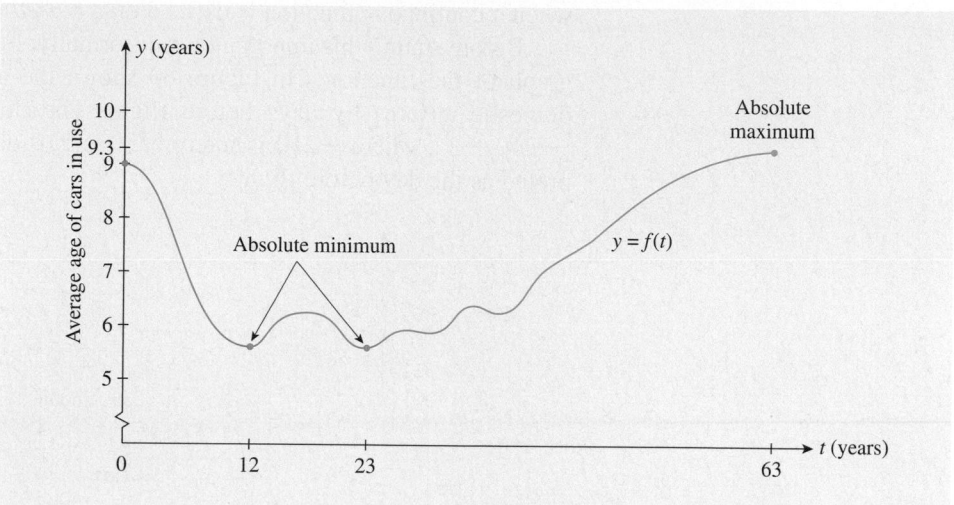

FIGURE 58
$f(t)$ gives the average age of cars in use in year t, t in $[0, 63]$.
Source: American Automobile Association.

(Incidentally, it is interesting to note that 1946 marked the first year of peace following World War II and the two years 1958 and 1969 marked the end of two periods of prosperity in recent U.S. history.)

A precise definition of the **absolute extrema** (absolute maximum or absolute minimum) of a function follows.

> **The Absolute Extrema of a Function f**
>
> If $f(x) \leq f(c)$ for all x in the domain of f, then $f(c)$ is called the **absolute maximum value** of f.
> If $f(x) \geq f(c)$ for all x in the domain of f, then $f(c)$ is called the **absolute minimum value** of f.

Figure 59 shows the graphs of several functions and gives the absolute maximum and absolute minimum of each function, if they exist.

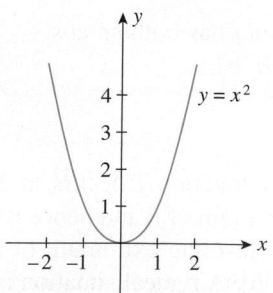

(a) $f(0) = 0$ is the absolute minimum of f; f has no absolute maximum.

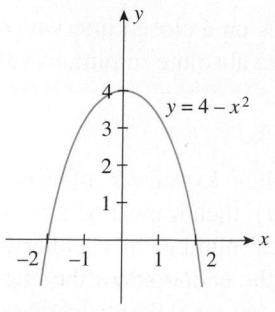

(b) $f(0) = 4$ is the absolute maximum of f; f has no absolute minimum.

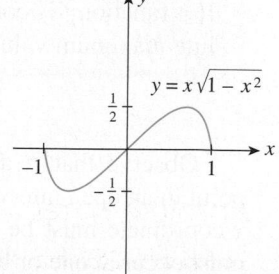

(c) $f(\sqrt{2}/2) = 1/2$ is the absolute maximum of f; $f(-\sqrt{2}/2) = -1/2$ is the absolute minimum of f.

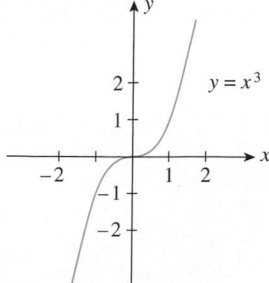

(d) f has no absolute extrema.

FIGURE 59

Absolute Extrema on a Closed Interval

As the preceding examples show, a continuous function defined on an arbitrary interval does not always have an absolute maximum or an absolute minimum. But an important case arises often in practical applications in which both the absolute maximum and the absolute minimum of a function are guaranteed to exist. This occurs when a continuous function is defined on a *closed* interval.

Before stating this important result formally, let's look at a real-life example. The graph of the function f in Figure 60 shows the average price, $f(t)$, in dollars, of domestic airfares by days before flight. The domain of f is the closed interval $[-210, -1]$, where -210 is interpreted as 210 days before flight and -1 is interpreted as the day before flight.

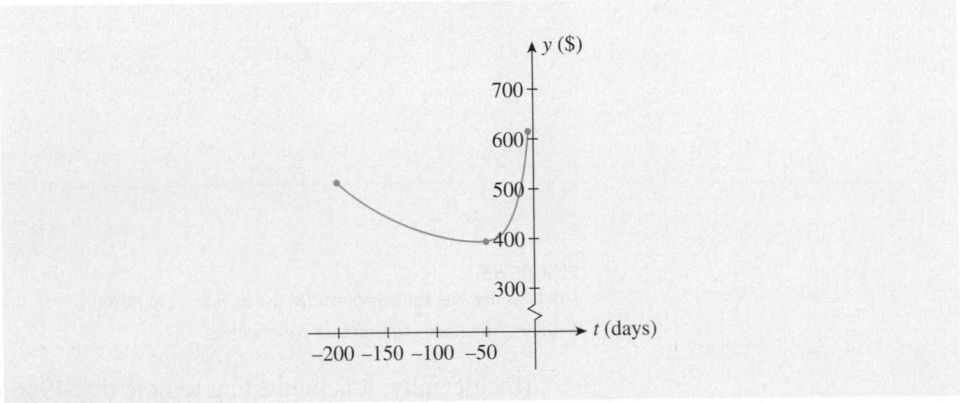

FIGURE **60**
Average price before flight
Source: Cheapair.com.

Observe that f attains the minimum value of 395 when $t = -49$ and the maximum value of 614 when $t = -1$. This result tells us that the best time to book a domestic flight is seven weeks in advance and the worst day to book a domestic flight is the day before the flight. Probably most surprising of all, booking too early can be almost as expensive as booking too late. Note that the function f is continuous on a closed interval. For such functions, we have the following theorem.

THEOREM 3

The Extreme Value Theorem

If a function f is continuous on a closed interval $[a, b]$, then f has both an absolute maximum value and an absolute minimum value on $[a, b]$.

Observe that if an absolute extremum of a continuous function f occurs at a point in an open interval (a, b), then it must be a relative extremum of f, and hence its x-coordinate must be a critical number of f. Otherwise, the absolute extremum of f must occur at one or both of the endpoints of the interval $[a, b]$. A typical situation is illustrated in Figure 61.

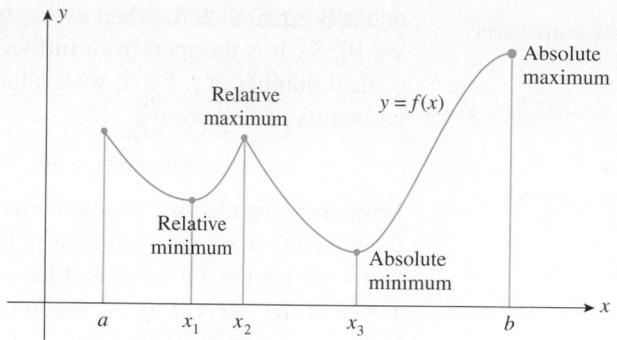

FIGURE **61**
The relative minimum of f at x_3 is the absolute minimum of f. The right endpoint b of the interval [a, b] gives rise to the absolute maximum value f(b) of f.

Here, x_1, x_2, and x_3 are critical numbers of f. The absolute minimum of f occurs at x_3, which lies in the open interval (a, b) and is a critical number of f. The absolute maximum of f occurs at b, an endpoint. This observation suggests the following procedure for finding the absolute extrema of a continuous function on a closed interval.

Finding the Absolute Extrema of f on a Closed Interval

1. Find the critical numbers of f that lie in (a, b).
2. Compute the value of f at each critical number found in Step 1 and compute $f(a)$ and $f(b)$.
3. The absolute maximum value and absolute minimum value of f will correspond to the largest and smallest numbers, respectively, found in Step 2.

EXAMPLE 1 Find the absolute extrema of the function $F(x) = x^2$ defined on the interval $[-1, 2]$.

Solution   The function F is continuous on the closed interval $[-1, 2]$ and differentiable on the open interval $(-1, 2)$. The derivative of F is

$$F'(x) = 2x$$

so 0 is the only critical number of F. Next, evaluate $F(x)$ at $x = -1$, $x = 0$, and $x = 2$. Thus,

$$F(-1) = 1 \qquad F(0) = 0 \qquad F(2) = 4$$

It follows that 0 is the absolute minimum value of F and 4 is the absolute maximum value of F. The graph of F, in Figure 62, confirms our results.   ▪

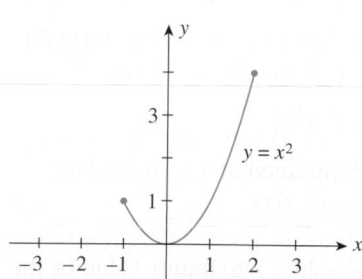

FIGURE **62**
F has an absolute minimum value of 0 and an absolute maximum value of 4.

EXAMPLE 2 Find the absolute extrema of the function

$$f(x) = x^3 - 2x^2 - 4x + 4$$

defined on the interval $[0, 3]$.

Solution   The function f is continuous on the closed interval $[0, 3]$ and differentiable on the open interval $(0, 3)$. The derivative of f is

$$f'(x) = 3x^2 - 4x - 4 = (3x + 2)(x - 2)$$

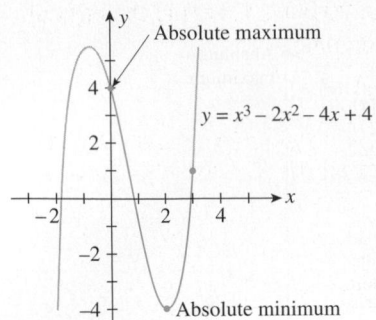

FIGURE 63
f has an absolute maximum value of 4
and an absolute minimum value of -4.

and it is equal to zero when $x = -\frac{2}{3}$ and $x = 2$. Since $x = -\frac{2}{3}$ lies outside the interval $[0, 3]$, it is dropped from further consideration, and $x = 2$ is seen to be the sole critical number of f. Next, we evaluate $f(x)$ at the critical number of f as well as the endpoints of f, obtaining

$$f(0) = 4 \qquad f(2) = -4 \qquad f(3) = 1$$

From these results, we conclude that -4 is the absolute minimum value of f and 4 is the absolute maximum value of f. The graph of f, which appears in Figure 63, confirms our results. Observe that the absolute maximum of f occurs at the endpoint $x = 0$ of the interval $[0, 3]$, while the absolute minimum of f occurs at $x = 2$, which lies in the interval $(0, 3)$.

Exploring with TECHNOLOGY

Let $f(x) = x^3 - 2x^2 - 4x + 4$. (This is the function of Example 2.)

1. Use a graphing utility to plot the graph of f, using the viewing window $[0, 3] \times [-5, 5]$. Use **ZOOM** and **TRACE** to find the absolute extrema of f on the interval $[0, 3]$ and thus verify the results obtained analytically in Example 2.

2. Plot the graph of f, using the viewing window $[-2, 1] \times [-5, 6]$. Use **ZOOM** and **TRACE** to find the absolute extrema of f on the interval $[-2, 1]$. Verify your results analytically.

EXAMPLE 3 Find the absolute maximum and absolute minimum values of the function $f(x) = x^{2/3}$ on the interval $[-1, 8]$.

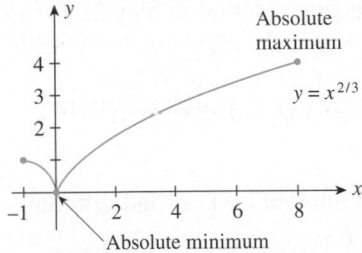

FIGURE 64
f has an absolute minimum value of
$f(0) = 0$ and an absolute maximum value
of $f(8) = 4$.

Solution The derivative of f is

$$f'(x) - \frac{2}{3}x^{-1/3} = \frac{2}{3x^{1/3}}$$

Note that f' is not defined at $x = 0$ and does not equal zero for any x. Therefore, 0 is the only critical number of f. Evaluating $f(x)$ at $x = -1, 0,$ and 8, we obtain

$$f(-1) = 1 \qquad f(0) = 0 \qquad f(8) = 4$$

We conclude that the absolute minimum value of f is 0, attained at $x = 0$, and the absolute maximum value of f is 4, attained at $x = 8$ (Figure 64).

Many real-world applications call for finding the absolute maximum value or the absolute minimum value of a given function. For example, management is interested in finding what level of production will yield the maximum profit for a company; a farmer is interested in finding the right amount of fertilizer to maximize crop yield; a doctor is interested in finding the maximum concentration of a drug in a patient's body and the time at which it occurs; and an engineer is interested in finding the dimensions of a container with a specified shape and volume that can be constructed at a minimum cost.

\$ APPLIED EXAMPLE 4 Maximizing Profits Acrosonic's total profit (in dollars) from manufacturing and selling x units of their model F loudspeaker systems is given by

$$P(x) = -0.02x^2 + 300x - 200,000 \qquad (0 \le x \le 20,000)$$

How many units of the loudspeaker system must Acrosonic produce to maximize its profits?

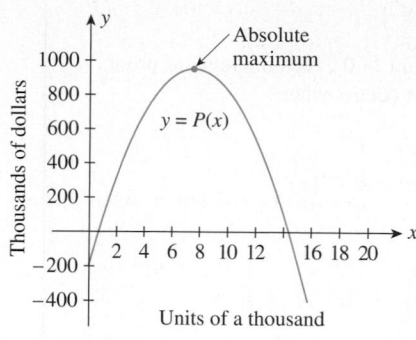

FIGURE 65
P has an absolute maximum at
(7500, 925,000).

Solution To find the absolute maximum of *P* on $[0, 20{,}000]$, first find the critical points of *P* on the interval $(0, 20{,}000)$. To do this, compute

$$P'(x) = -0.04x + 300$$

Solving the equation $P'(x) = 0$ gives $x = 7500$. Next, evaluate $P(x)$ at $x = 7500$ as well as the endpoints $x = 0$ and $x = 20{,}000$ of the interval $[0, 20{,}000]$, obtaining

$$P(0) = -200{,}000$$
$$P(7500) = 925{,}000$$
$$P(20{,}000) = -2{,}200{,}000$$

From these computations, we see that the absolute maximum value of the function *P* is 925,000. Thus, by producing 7500 units, Acrosonic will realize a maximum profit of \$925,000. The graph of *P* is sketched in Figure 65. ◼

Explore and Discuss

Recall that the total profit function *P* is defined as $P(x) = R(x) - C(x)$, where *R* is the total revenue function, *C* is the total cost function, and *x* is the number of units of a product produced and sold. (Assume that all derivatives exist.)

1. Show that at the level of production x_0 that yields the maximum profit for the company, the following two conditions are satisfied:

$$R'(x_0) = C'(x_0) \quad \text{and} \quad R''(x_0) < C''(x_0)$$

2. Interpret the two conditions in part 1 in economic terms, and explain why they make sense.

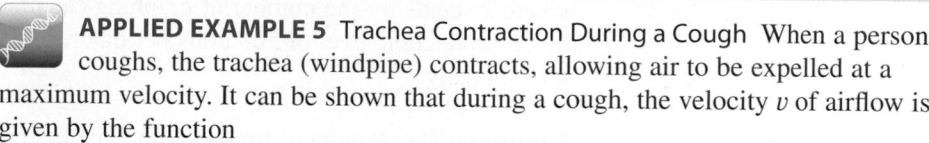

APPLIED EXAMPLE 5 Trachea Contraction During a Cough When a person coughs, the trachea (windpipe) contracts, allowing air to be expelled at a maximum velocity. It can be shown that during a cough, the velocity *v* of airflow is given by the function

$$v = f(r) = kr^2(R - r)$$

where *r* is the trachea's radius (in centimeters) during a cough, *R* is the trachea's normal radius (in centimeters), and *k* is a positive constant that depends on the length of the trachea. Find the radius *r* for which the velocity of airflow is greatest.

Solution To find the absolute maximum of *f* on $[0, R]$, first find the critical numbers of *f* on the interval $(0, R)$. We compute

$$f'(r) = 2kr(R - r) - kr^2 \qquad \text{Use the Product Rule.}$$
$$= -3kr^2 + 2kRr = kr(-3r + 2R)$$

Setting $f'(r) = 0$ gives $r = 0$ or $r = \frac{2}{3}R$, so $\frac{2}{3}R$ is the sole critical number of *f* ($r = 0$ is an endpoint). Evaluating $f(r)$ at $r = \frac{2}{3}R$, as well as at the endpoints $r = 0$ and $r = R$, we obtain

$$f(0) = 0$$
$$f\left(\frac{2}{3}R\right) = \frac{4k}{27}R^3$$
$$f(R) = 0$$

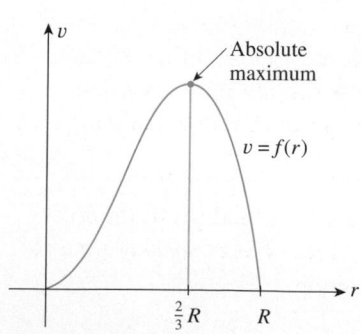

FIGURE 66
The velocity of airflow is greatest when the radius of the contracted trachea is $\frac{2}{3}R$.

from which we deduce that the velocity of airflow is greatest when the radius of the contracted trachea is $\frac{2}{3}R$—that is, when the radius is contracted by approximately 33%. The graph of the function *f* is shown in Figure 66. ◼

Explore and Discuss

Prove that if a cost function $C(x)$ is concave upward $[C''(x) > 0]$, then the level of production that will result in the smallest average production cost occurs when

$$\overline{C}(x) = C'(x)$$

that is, when the average cost $\overline{C}(x)$ is equal to the marginal cost $C'(x)$.
Hints:

1. Show that

$$\overline{C}'(x) = \frac{xC'(x) - C(x)}{x^2}$$

so the critical number of the function $\overline{C}$ occurs when

$$xC'(x) - C(x) = 0$$

2. Show that at a critical number of $\overline{C}$

$$\overline{C}''(x) = \frac{C''(x)}{x}$$

Use the Second Derivative Test to reach the desired conclusion.

 APPLIED EXAMPLE 6 Minimizing Average Cost The daily average cost function (in dollars per unit) of Elektra Electronics is given by

$$\overline{C}(x) = 0.0001x^2 - 0.08x + 40 + \frac{5000}{x} \qquad (x > 0)$$

where x stands for the number of graphing calculators that Elektra produces. Show that a production level of 500 units per day results in a minimum average cost for the company.

Solution The domain of the function $\overline{C}$ is the interval $(0, \infty)$, which is not closed. To solve the problem, we resort to the graphical method. Using the techniques of graphing from the last section, we sketch the graph of $\overline{C}$ (Figure 67).

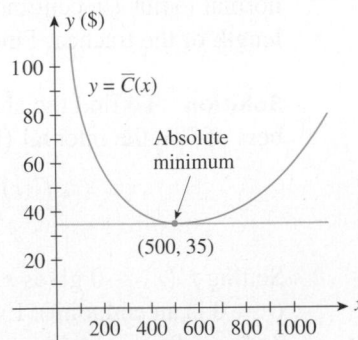

FIGURE **67**
The minimum average cost is $35 per unit.

Now,

$$\overline{C}'(x) = 0.0002x - 0.08 - \frac{5000}{x^2}$$

Substituting the given value of x, 500, into $\overline{C}'(x)$ gives $\overline{C}'(500) = 0$, so 500 is a critical number of $\overline{C}$. Next,

$$\overline{C}''(x) = 0.0002 + \frac{10,000}{x^3}$$

Thus,

$$\overline{C}''(500) = 0.0002 + \frac{10,000}{(500)^3} > 0$$

and by the Second Derivative Test, a relative minimum of the function $\overline{C}$ occurs at 500. Furthermore, $\overline{C}''(x) > 0$ for $x > 0$, which implies that the graph of $\overline{C}$ is concave upward everywhere, so the relative minimum of $\overline{C}$ must be the absolute minimum of $\overline{C}$. The minimum average cost is given by

$$\overline{C}(500) = 0.0001(500)^2 - 0.08(500) + 40 + \frac{5000}{500}$$

$$= 35$$

or $35 per unit.

Exploring with TECHNOLOGY

Refer to the preceding Explore and Discuss and Example 6.

1. Using a graphing utility, plot the graphs of

$$\overline{C}(x) = 0.0001x^2 - 0.08x + 40 + \frac{5000}{x}$$

$$C'(x) = 0.0003x^2 - 0.16x + 40$$

using the viewing window $[0, 1000] \times [0, 150]$.

Note: $C(x) = 0.0001x^3 - 0.08x^2 + 40x + 5000$. (Why?)

2. Find the point of intersection of the graphs of $\overline{C}$ and C' and thus verify the assertion in the Explore and Discuss for the special case studied in Example 6.

Our final example involves finding the absolute maximum of an exponential function.

 APPLIED EXAMPLE 7 Optimal Market Price The present value of the market price of the Blakely Office Building is given by

$$P(t) = 300,000e^{-0.09t + \sqrt{t}/2} \qquad (0 \le t \le 10)$$

Find the optimal present value of the building's market price.

Solution To find the maximum value of P over $[0, 10]$, we compute

$$P'(t) = 300,000e^{-0.09t + \sqrt{t}/2} \frac{d}{dt}\left(-0.09t + \frac{1}{2}t^{1/2}\right)$$

$$= 300,000e^{-0.09t + \sqrt{t}/2}\left(-0.09 + \frac{1}{4}t^{-1/2}\right)$$

Setting $P'(t) = 0$ gives

$$-0.09 + \frac{1}{4t^{1/2}} = 0$$

since $e^{-0.09t + \sqrt{t}/2}$ is never zero for any value of t. Solving this equation, we find

$$\frac{1}{4t^{1/2}} = 0.09$$

$$t^{1/2} = \frac{1}{4(0.09)}$$

$$= \frac{1}{0.36}$$

$$t = \left(\frac{1}{0.36}\right)^2 \approx 7.72$$

the sole critical number of the function P. Finally, evaluating $P(t)$ at the critical number as well as at the endpoints of $[0, 10]$, we have

t	0	7.72	10
$P(t)$	300,000	600,779	592,838

We conclude, accordingly, that the optimal present value of the property's market price is $600,779 and that this will occur 7.72 years from now. ■

10.4 Self-Check Exercises

1. Let $f(x) = x - 2\sqrt{x}$.
   a. Find the absolute extrema of f on the interval $[0, 9]$.
   b. Find the absolute extrema of f.

2. Find the absolute extrema of $f(x) = 3x^4 + 4x^3 + 1$ on $[-2, 1]$.

3. **FACTORY OPERATING RATE** The operating rate (expressed as a percent) of factories, mines, and utilities in a certain region of the country on the tth day of 2014 is given by the function

$$f(t) = 80 + \frac{1200t}{t^2 + 40,000} \qquad (0 \le t \le 250)$$

On which of the first 250 days of 2014 was the operating rate highest?

Solutions to Self-Check Exercises 10.4 can be found on page 786.

10.4 Concept Questions

1. Explain the following terms: (a) absolute maximum and (b) absolute minimum.

2. Describe the procedure for finding the absolute extrema of a continuous function on a closed interval.

10.4 Exercises

In Exercises 1–8, you are given the graph of a function f defined on the indicated interval. Find the absolute maximum and the absolute minimum of f, if they exist.

1.

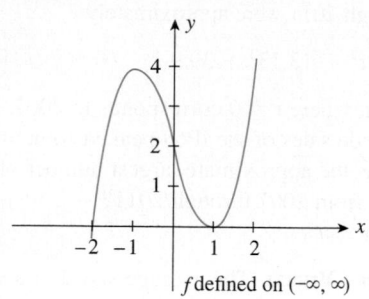

f defined on $(-\infty, \infty)$

2.

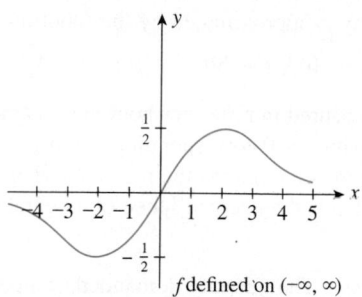

f defined on $(-\infty, \infty)$

3.

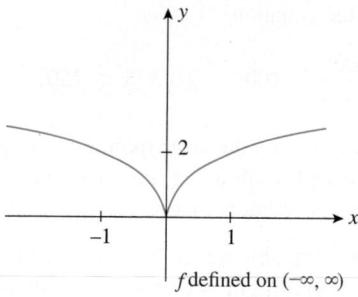

f defined on $(-\infty, \infty)$

4.

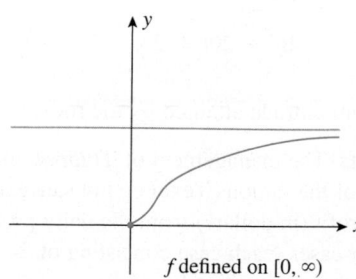

f defined on $[0, \infty)$

5.

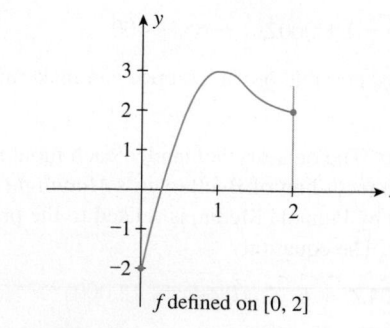

f defined on $[0, 2]$

6.

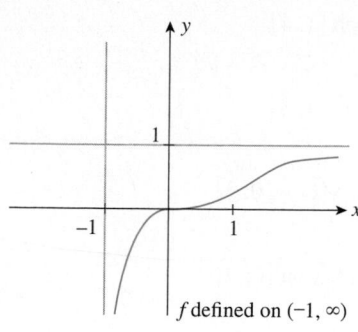

f defined on $(-1, \infty)$

7.

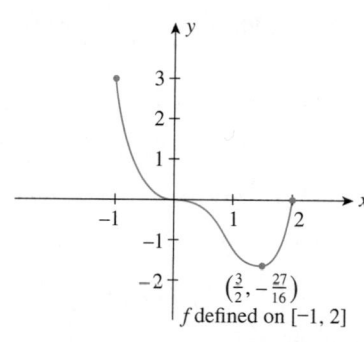

$\left(\frac{3}{2}, -\frac{27}{16}\right)$

f defined on $[-1, 2]$

8.

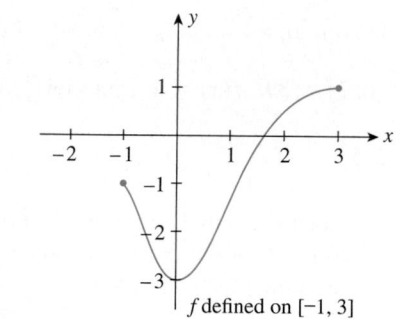

f defined on $[-1, 3]$

In Exercises 9–40, find the absolute maximum value and the absolute minimum value, if any, of each function.

9. $f(x) = 2x^2 + 3x - 4$

10. $g(x) = -x^2 + 4x + 3$

11. $h(x) = x^{1/3}$

12. $f(x) = x^{2/3}$

13. $f(x) = \dfrac{1}{1 + x^2}$

14. $f(x) = \dfrac{x}{1 + x^2}$

15. $f(x) = x^2 - 2x - 3$ on $[-2, 3]$

16. $g(x) = x^2 - 2x - 3$ on $[0, 4]$

17. $f(x) = -x^2 + 4x + 6$ on $[0, 5]$

18. $f(x) = -x^2 + 4x + 6$ on $[3, 6]$

19. $f(x) = x^3 + 3x^2 - 1$ on $[-3, 2]$

20. $g(x) = x^3 + 3x^2 - 1$ on $[-3, 1]$

21. $g(x) = 3x^4 + 4x^3$ on $[-2, 1]$

22. $f(x) = \dfrac{1}{2}x^4 - \dfrac{2}{3}x^3 - 2x^2 + 3$ on $[-2, 3]$

23. $f(x) = \dfrac{x+1}{x-1}$ on $[2, 4]$ **24.** $g(t) = \dfrac{t}{t-1}$ on $[2, 4]$

25. $f(x) = 4x + \dfrac{1}{x}$ on $[1, 4]$

26. $f(x) = 9x - \dfrac{1}{x}$ on $[1, 3]$

27. $f(x) = \dfrac{1}{2}x^2 - 2\sqrt{x}$ on $[0, 3]$

28. $g(x) = \dfrac{1}{8}x^2 - 4\sqrt{x}$ on $[0, 9]$

29. $f(x) = \dfrac{1}{x}$ on $(0, \infty)$ **30.** $g(x) = \dfrac{1}{x+1}$ on $(0, \infty)$

31. $f(x) = 3x^{2/3} - 2x$ on $[0, 3]$

32. $g(x) = x^2 + 2x^{2/3}$ on $[-2, 2]$

33. $f(x) = x^{2/3}(x^2 - 4)$ on $[-1, 2]$

34. $f(x) = x^{2/3}(x^2 - 4)$ on $[-1, 3]$

35. $f(x) = 2e^{-x^2}$ on $[-1, 1]$

36. $h(x) = e^{x^2-4}$ on $[-2, 2]$

37. $g(x) = (2x - 1)e^{-x}$ on $[0, \infty]$

38. $f(x) = xe^{-x^2}$ on $[0, 2]$ **39.** $f(x) = x - \ln x$ on $\left[\frac{1}{2}, 3\right]$

40. $g(x) = \dfrac{x}{\ln x}$ on $[2, 5]$

41. A stone is thrown straight up from the roof of an 80-ft building. The height (in feet) of the stone at any time t (in seconds), measured from the ground, is given by

$$h(t) = -16t^2 + 64t + 80$$

What is the maximum height the stone reaches?

42. MAXIMIZING PROFITS Lynbrook West, an apartment complex, has 100 two-bedroom units. The monthly profit (in dollars) realized from renting out x apartments is given by

$$P(x) = -10x^2 + 1760x - 50,000$$

To maximize the monthly rental profit, how many units should be rented out? What is the maximum monthly profit realizable?

43. STRIKE OUTS The rate at which major league players were striking out in the years 2009 through 2013 is approximately

$$f(t) = 0.136t^2 + 0.127t + 18.1 \qquad (0 \le t \le 4)$$

percent in year t, where $t = 0$ corresponds to 2009.
  a. What was the lowest rate of strikeouts over the years under consideration? When did it occur?
  b. What was the highest rate of strikeouts. When did it occur?
Source: USA Today.

44. END OF THE IPOD ERA Apple introduced the first iPod in October 2001. Sales of the portable music player grew slowly in the early years but began to grow rapidly after 2005. But the iPod era is coming to a close. Smartphones with music and video players are replacing the iPod, along with the category of device it helped to create. Sales of the iPod worldwide from 2007 through 2011 were approximately

$$N(t) = -2.65t^2 + 13.13t + 39.9 \qquad (0 \le t \le 4)$$

million in year t, where $t = 0$ corresponds to 2007. Show that the worldwide sales of the iPod peaked sometime in 2009. What was the approximate largest number of iPods sold worldwide from 2007 through 2011?
Source: Popular Mechanics.

45. AVERAGE SPEED OF A VEHICLE The average speed of a vehicle on a stretch of Route 134 between 6 A.M. and 10 A.M. on a typical weekday is approximated by the function

$$f(t) = 20t - 40\sqrt{t} + 50 \qquad (0 \le t \le 4)$$

where $f(t)$ is measured in miles per hour and t is measured in hours, with $t = 0$ corresponding to 6 A.M. At what time of the morning commute is the traffic moving at the slowest rate? What is the average speed of a vehicle at that time?

46. MAXIMIZING REVENUE The quantity demanded of a certain brand of handbags per day, x, is related to the unit price, p, in dollars, by the equation

$$p = \dfrac{100,000}{250 + x} - 100 \qquad (0 \le x \le 750)$$

Find the level of sales and the corresponding unit price of the handbags that will result in a maximum revenue per day for the company. What is that revenue?

47. FLIGHT OF A ROCKET The altitude (in feet) attained by a model rocket t sec into flight is given by the function

$$h(t) = -\dfrac{1}{3}t^3 + 4t^2 + 20t + 2 \qquad (t \ge 0)$$

Find the maximum altitude attained by the rocket.

48. MAXIMIZING PROFITS The management of Trappee and Sons, producers of the famous TexaPep hot sauce, estimate that their profit (in dollars) from the daily production and sale of x cases (each case consisting of 24 bottles) of the hot sauce is given by

$$P(x) = -0.000002x^3 + 6x - 400$$

What is the largest possible profit Trappee can make in 1 day?

49. MAXIMIZING PROFITS The quantity demanded each month of the Walter Serkin recording of Beethoven's *Moonlight Sonata*, produced by Phonola Media, is related to the price per compact disc. The equation

$$p = -0.00042x + 6 \qquad (0 \le x \le 12,000)$$

where p denotes the unit price in dollars and x is the number of discs demanded, relates the demand to the price. The total monthly cost (in dollars) for pressing and packaging x copies of this classical recording is given by

$$C(x) = 600 + 2x - 0.00002x^2 \qquad (0 \le x \le 20{,}000)$$

To maximize its profits, how many copies should Phonola produce each month?

Hint: The revenue is $R(x) = px$, and the profit is $P(x) = R(x) - C(x)$.

50. **MAXIMIZING PROFIT** A manufacturer of tennis rackets finds that the total cost $C(x)$ (in dollars) of manufacturing x rackets/day is given by $C(x) = 400 + 4x + 0.0001x^2$. Each racket can be sold at a price of p dollars, where p is related to x by the demand equation $p = 10 - 0.0004x$. If all rackets that are manufactured can be sold, find the daily level of production that will yield a maximum profit for the manufacturer.

51. **MAXIMIZING PROFIT** A division of Chapman Corporation manufactures a pager. The weekly fixed cost for the division is $20,000, and the variable cost for producing x pagers per week is

$$V(x) = 0.000001x^3 - 0.01x^2 + 50x$$

dollars. The company realizes a revenue of

$$R(x) = -0.02x^2 + 150x \qquad (0 \le x \le 7500)$$

dollars from the sale of x pagers/week. Find the level of production that will yield a maximum profit for the manufacturer.

Hint: Use the quadratic formula.

52. **MAXIMIZING PROFIT** The weekly demand for the Pulsar 40-in. high-definition television is given by the demand equation

$$p = -0.05x + 600 \qquad (0 \le x \le 12{,}000)$$

where p denotes the wholesale unit price in dollars and x denotes the quantity demanded. The weekly total cost function associated with manufacturing these sets is given by

$$C(x) = 0.000002x^3 - 0.03x^2 + 400x + 80{,}000$$

where $C(x)$ denotes the total cost incurred in producing x sets. Find the level of production that will yield a maximum profit for the manufacturer.

Hint: Use the quadratic formula.

53. **MINIMIZING AVERAGE COST** Suppose the total cost function for manufacturing a certain product is $C(x) = 0.2(0.01x^2 + 120)$ dollars, where x represents the number of units produced. Find the level of production that will minimize the average cost.

54. **MINIMIZING PRODUCTION COSTS** The total monthly cost (in dollars) incurred by Cannon Precision Instruments for manufacturing x units of the model M1 digital camera is given by the function

$$C(x) = 0.0025x^2 + 80x + 10{,}000$$

a. Find the average cost function $\overline{C}$.
b. Find the level of production that results in the smallest average production cost.
c. Find the level of production for which the average cost is equal to the marginal cost.
d. Compare the result of part (c) with that of part (b).

55. **MINIMIZING PRODUCTION COSTS** The daily total cost (in dollars) incurred by Trappee and Sons for producing x cases of TexaPep hot sauce is given by the function

$$C(x) = 0.000002x^3 + 5x + 400$$

Using this function, answer the questions posed in Exercise 54.

56. **MINIMIZING AVERAGE COST** Suppose that the total cost incurred in manufacturing x units of a certain product is given by $C(x)$, where C is a differentiable cost function. Show that the average cost is minimized at the level of production where the average cost is equal to the marginal cost.

57. **MINIMIZING PRODUCTION COSTS** Re-solve Exercise 54 using the result of Exercise 56.

58. **MAXIMIZING REVENUE** Suppose the quantity demanded per week of a certain dress is related to the unit price p by the demand equation $p = \sqrt{800 - x}$, where p is in dollars and x is the number of dresses made. To maximize the revenue, how many dresses should be made and sold each week?

Hint: $R(x) = px$

59. **MAXIMIZING REVENUE** The quantity demanded each month of the Sicard sports watch is related to the unit price by the equation

$$p = \frac{50}{0.01x^2 + 1} \qquad (0 \le x \le 20)$$

where p is measured in dollars and x is measured in units of a thousand. To yield a maximum revenue, how many watches must be sold?

60. **AIR POLLUTION** The amount of nitrogen dioxide, a brown gas that impairs breathing, present in the atmosphere on a certain May day in the city of Long Beach is approximated by

$$A(t) = \frac{136}{1 + 0.25(t - 4.5)^2} + 28 \qquad (0 \le t \le 11)$$

where $A(t)$ is measured in pollutant standard index (PSI) and t is measured in hours, with $t = 0$ corresponding to 7 A.M. Determine the time of day when the pollution is at its highest level.

61. **OXYGEN CONTENT OF A POND** When organic waste is dumped into a pond, the oxidation process that takes place reduces the pond's oxygen content. However, given time, nature will restore the oxygen content to its natural level.

Suppose the oxygen content t days after organic waste has been dumped into the pond is given by

$$f(t) = 100\left(\frac{t^2 - 4t + 4}{t^2 + 4}\right) \qquad (0 \leq t < \infty)$$

percent of its normal level.
a. When is the level of oxygen content lowest?
b. When is the rate of oxygen regeneration greatest?

62. **VELOCITY OF BLOOD** According to a law discovered by the nineteenth-century physician Jean Louis Marie Poiseuille, the velocity (in centimeters per second) of blood r cm from the central axis of an artery is given by

$$v(r) = k(R^2 - r^2)$$

where k is a constant and R is the radius of the artery. Show that the velocity of blood is greatest along the central axis.

63. **MAXIMIZING REVENUE** The average revenue is defined as the function

$$\overline{R}(x) = \frac{R(x)}{x} \qquad (x > 0)$$

Prove that if a revenue function $R(x)$ is concave downward $[R''(x) < 0]$, then the level of sales that will result in the largest average revenue occurs when $\overline{R}(x) = R'(x)$.

64. **OPTIMAL SELLING TIME** The present value of a piece of waterfront property purchased by an investor is given by the function

$$P(t) = 80{,}000e^{\sqrt{t/2} - 0.09t} \qquad (0 \leq t \leq 8)$$

Determine the optimal time (based on present value) for the investor to sell the property. What is the property's optimal present value?

65. **MAXIMUM OIL PRODUCTION** It has been estimated that the total production of oil from a certain oil well is given by

$$T(t) = -1000(t + 10)e^{-0.1t} + 10{,}000$$

thousand barrels t years after production has begun. Determine the year when the oil well will be producing at maximum capacity.

66. **BLOOD ALCOHOL LEVEL** The percentage of alcohol in a person's bloodstream t hr after drinking 8 fluid oz of whiskey is given by

$$A(t) = 0.23te^{-0.4t} \qquad (0 \leq t \leq 12)$$

At what time after drinking the alcohol is the percentage of alcohol in the person's bloodstream at its highest level? What is that level?

67. **LENGTHS OF FISH** The length (in centimeters) of a typical Pacific halibut t years old is approximately

$$f(t) = 200(1 - 0.956e^{-0.18t})$$

a. How fast is the length of a typical 5-year-old Pacific halibut increasing?

b. What is the maximum length a typical Pacific halibut can attain?
Hint: Evaluate $\lim_{x \to \infty} f(t)$.

68. **U.S. CRUDE OIL PRODUCTION** The rate of production of crude oil in the United States in year t (in millions of barrels per day) from 1900 through 2010 is approximated by the function

$$f(t) = \frac{40e^{-(t-1975)/20}}{(1 + e^{-(t-1975)/20})^2} \qquad (1900 \leq t \leq 2010)$$

What was the maximum rate of production of crude oil in the United States for the period under consideration?
Hint: Let $u = (t - 1975)/20$.
Source: U.S. Energy Information Administration.

69. **MAXIMIZING PROFIT** The manager of Seko, an information technology (IT) consulting company, estimates that the annual profit of the company, in millions of dollars, is given by

$$P(x) = 2\ln(2x + 1) + 2x - x^2 - 0.3$$

where x is the number of IT consultants (in hundreds) in its employ. Find the number of consultants the firm should hire so that its profit is maximized. What is the maximum profit?

70. **CRIME RATES** The number of major crimes committed in the city of Bronxville between 2007 and 2014 is approximated by the function

$$N(t) = -0.1t^3 + 1.5t^2 + 100 \qquad (0 \leq t \leq 7)$$

where $N(t)$ denotes the number of crimes committed in year t ($t = 0$ corresponds to 2007). Enraged by the dramatic increase in the crime rate, the citizens of Bronxville, with the help of the local police, organized Neighborhood Crime Watch groups in early 2011 to combat this menace. Show that the growth in the crime rate was maximal in 2012, giving credence to the claim that the Neighborhood Crime Watch program was working.

71. **GDP OF A DEVELOPING COUNTRY** A developing country's gross domestic product (GDP) from 2006 to 2014 is approximated by the function

$$G(t) = -0.2t^3 + 2.4t^2 + 60 \qquad (0 \leq t \leq 8)$$

where $G(t)$ is measured in billions of dollars and $t = 0$ corresponds to 2006. Show that the growth rate of the country's GDP was maximal in 2010.

72. **FEDERAL DEFICIT** The deficit of the federal government (in trillions of dollars) in year t from 2006 ($t = 0$) through 2012 ($t = 6$) is approximately

$$D(t) = -0.038898t^3 + 0.30858t^2 - 0.31849t + 0.22$$
$$(0 \leq t \leq 6)$$

What was the largest federal deficit over the period under consideration?
Hint: Use the quadratic formula.
Source: Office of Management and Budget.

73. **NEW INMATES** The number of new prison admissions (into state or federal facilities) between 2002 and 2011 is given by the function

$$N(t) = -87.244444t^3 - 2482.35t^2 + 46009.26t + 579185$$
$$(2 \le t \le 11)$$

where t is measured in years, with $t = 2$ corresponding to 2002. Show that the number of new prison admissions peaked in 2006. What was the highest number of new inmates that year?
Hint: Use the quadratic formula.
Source: Bureau of Justice Statistics.

74. **DEATH RATE FROM AIDS** The estimated number of deaths from AIDS worldwide is given by

$$f(t) = -0.0004401t^3 + 0.007t^2 + 0.112t + 0.28$$
$$(0 \le t \le 21)$$

million per year, where t is measured in years, with $t = 0$ corresponding to 1990. What was the highest rate of deaths from AIDS worldwide over the period from 1990 through 2011?
Hint: Use the quadratic formula.
Source: UNAIDS.

75. **BRAIN GROWTH AND IQs** In a study conducted at the National Institute of Mental Health, researchers followed the development of the cortex, the thinking part of the brain, in 307 children. Using repeated magnetic resonance imaging scans from childhood to the latter teens, they measured the thickness (in millimeters) of the cortex of children of age t years with the highest IQs—121 to 149. These data lead to the model

$$S(t) = 0.000989t^3 - 0.0486t^2 + 0.7116t + 1.46$$
$$(5 \le t \le 19)$$

Show that the cortex of children with superior intelligence reaches maximum thickness around age 11 years.
Hint: Use the quadratic formula.
Source: Nature.

76. **BRAIN GROWTH AND IQs** Refer to Exercise 75. The researchers at the Institute also measured the thickness (also in millimeters) of the cortex of children of age t years who were of average intelligence. These data lead to the model

$$A(t) = -0.00005t^3 - 0.000826t^2 + 0.0153t + 4.55$$
$$(5 \le t \le 19)$$

Show that the cortex of children with average intelligence reaches maximum thickness at age 6 years.
Source: Nature.

77. **WORLD POPULATION** The total world population is forecast to be

$$P(t) = 0.00074t^3 - 0.0704t^2 + 0.89t + 6.04$$
$$(0 \le t \le 10)$$

in year t, where t is measured in decades, with $t = 0$ corresponding to 2000 and $P(t)$ is measured in billions.

a. Show that the world population is forecast to peak around 2071.
Hint: Use the quadratic formula.
b. At what number will the population peak?
Source: International Institute for Applied Systems Analysis.

78. **VENTURE-CAPITAL INVESTMENT** Venture-capital investment increased dramatically in the late 1990s but came to a screeching halt after the dot-com bust. The venture-capital investment (in billions of dollars) from 1995 ($t = 0$) through 2003 ($t = 8$) is approximated by the function

$$C(t) = \begin{cases} 0.6t^2 + 2.4t + 7.6 & \text{if } 0 \le t < 3 \\ 3t^2 + 18.8t - 63.2 & \text{if } 3 \le t < 5 \\ -3.3167t^3 + 80.1t^2 & \\ \quad - 642.583t + 1730.8025 & \text{if } 5 \le t < 8 \end{cases}$$

a. In what year did venture-capital investment peak over the period under consideration? What was the amount of that investment?
b. In what year was the venture-capital investment lowest over this period? What was the amount of that investment?
Hint: Find the absolute extrema of C on each of the closed intervals $[0, 3]$, $[3, 5]$, and $[5, 8]$.
Sources: Venture One; Ernst & Young.

79. **REACTION TO A DRUG** The strength of a human body's reaction R to a dosage D of a certain drug is given by

$$R = D^2\left(\frac{k}{2} - \frac{D}{3}\right)$$

where k is a positive constant. Show that the maximum reaction is achieved if the dosage is k units.

In Exercises 80–86, determine whether the statement is true or false. If it is true, explain why it is true. If it is false, give an example to show why it is false.

80. If f is not continuous on the closed interval $[a, b]$, then f cannot have an absolute maximum value.

81. If f is defined on a closed interval $[a, b]$, then f has an absolute maximum value.

82. If $f''(x) < 0$ on (a, b) and $f'(c) = 0$, where $a < c < b$, then $f(c)$ is the absolute maximum value of f on $[a, b]$.

83. If f is continuous on $[a, b]$, f is differentiable on (a, b), and $f'(x) \neq 0$ for all x in (a, b), then the absolute maximum value of f on $[a, b]$ is $f(a)$ or $f(b)$.

84. If f is continuous on $[a, b]$, f is differentiable on (a, b), and $f'(x) > 0$ for all x in (a, b), then the absolute minimum value of f on $[a, b]$ is $f(a)$.

85. If the level of production, x, is constrained to be between a and b units, inclusive, then the maximum profit is $P(a)$, $P(b)$, or $P(c)$, where c is any number in (a, b) for which $R'(c) = C'(c)$ and $R''(c) < C''(c)$.

86. If $f''(x) > 0$ for all x in an interval I, then f must have an absolute minimum value at some number c in that interval.

87. Let f be a constant function—that is, let $f(x) = c$, where c is some real number. Show that every number a gives rise to an absolute maximum and, at the same time, an absolute minimum of f.

88. Show that a polynomial function defined on the interval $(-\infty, \infty)$ cannot have both an absolute maximum and an absolute minimum unless it is a constant function.

89. One condition that must be satisfied before Theorem 3 (page 774) is applicable is that the function f must be continuous on the closed interval $[a, b]$. Define a function f on the closed interval $[-1, 1]$ by

$$f(x) = \begin{cases} \dfrac{1}{x} & \text{if } x \in [-1, 1] \quad (x \neq 0) \\ 0 & \text{if } x = 0 \end{cases}$$

a. Show that f is not continuous at $x = 0$.
b. Show that f does not attain an absolute maximum or an absolute minimum on the interval $[-1, 1]$.
c. Confirm your results by sketching the function f.

90. One condition that must be satisfied before Theorem 3 (page 774) is applicable is that the interval on which f is defined must be a closed interval $[a, b]$. Define a function f on the *open* interval $(-1, 1)$ by $f(x) = x$. Show that f does not attain an absolute maximum or an absolute minimum on the interval $(-1, 1)$.
Hint: What happens to $f(x)$ if x is close to but not equal to $x = -1$? If x is close to but not equal to $x = 1$?

10.4 Solutions to Self-Check Exercises

1. a. The function f is continuous in its domain and differentiable on the interval $(0, 9)$. The derivative of f is

$$f'(x) = 1 - x^{-1/2} = \frac{x^{1/2} - 1}{x^{1/2}}$$

and it is equal to zero when $x = 1$. Evaluating $f(x)$ at the endpoints $x = 0$ and $x = 9$ and at the critical number 1 of f, we have

$$f(0) = 0 \qquad f(1) = -1 \qquad f(9) = 3$$

From these results, we see that -1 is the absolute minimum value of f and 3 is the absolute maximum value of f.

b. In this case, the domain of f is the interval $[0, \infty)$, which is not closed. Therefore, we resort to the graphical method. Using the techniques of graphing, we sketch the graph of f in the accompanying figure.

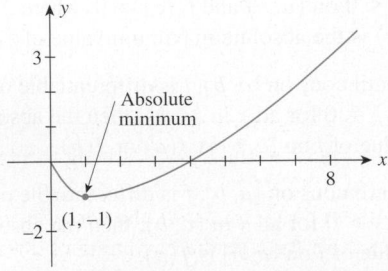

The graph of f shows that -1 is the absolute minimum value of f but f has no absolute maximum, since $f(x)$ increases without bound as x increases without bound.

2. The function f is continuous on the interval $[-2, 1]$. It is also differentiable on the open interval $(-2, 1)$. The derivative of f is

$$f'(x) = 12x^3 + 12x^2 = 12x^2(x + 1)$$

and it is continuous on $(-2, 1)$. Setting $f'(x) = 0$ gives -1 and 0 as critical numbers of f. Evaluating $f(x)$ at these critical numbers of f as well as at the endpoints of the interval $[-2, 1]$, we obtain

$$f(-2) = 17 \qquad f(-1) = 0 \qquad f(0) = 1 \qquad f(1) = 8$$

From these results, we see that 0 is the absolute minimum value of f and 17 is the absolute maximum value of f.

3. The problem is solved by finding the absolute maximum of the function f on $[0, 250]$. Differentiating $f(t)$, we obtain

$$f'(t) = \frac{(t^2 + 40,000)(1200) - 1200t(2t)}{(t^2 + 40,000)^2}$$

$$= \frac{-1200(t^2 - 40,000)}{(t^2 + 40,000)^2}$$

Upon setting $f'(t) = 0$ and solving the resulting equation, we obtain $t = -200$ or 200. Since -200 lies outside the interval $[0, 250]$, we are interested only in the critical number 200 of f. Evaluating $f(t)$ at $t = 0$, $t = 200$, and $t = 250$, we find

$$f(0) = 80 \qquad f(200) = 83 \qquad f(250) \approx 82.93$$

We conclude that the operating rate was the highest on the 200th day of 2014—that is, a little past the middle of July 2014.

USING TECHNOLOGY | Finding the Absolute Extrema of a Function

Some graphing utilities have a function for finding the absolute maximum and the absolute minimum values of a continuous function on a closed interval. If your graphing utility has this capability, use it to work through the example and exercises of this section.

EXAMPLE 1 Let $f(x) = \dfrac{2x + 4}{(x^2 + 1)^{3/2}}$.

a. Use a graphing utility to plot the graph of f in the viewing window $[-3, 3] \times [-1, 5]$.

b. Find the absolute maximum and absolute minimum values of f on the interval $[-3, 3]$. Express your answers accurate to four decimal places.

Solution

a. The graph of f is shown in Figure T1.

b. Using the function on a graphing utility for finding the absolute minimum value of a continuous function on a closed interval, we find the absolute minimum value of f to be -0.0632. Similarly, using the function for finding the absolute maximum value, we find the absolute maximum value to be 4.1593.

Note   Some graphing utilities will enable you to find the absolute minimum and absolute maximum values of a continuous function on a closed interval without having to graph the function. For example, using **fMax** on the TI-83/84 will yield the x-coordinate of the absolute maximum of f. The absolute maximum value can then be found by evaluating f at that value of x. Figure T2 shows the work involved in finding the absolute maximum of the function of Example 1.

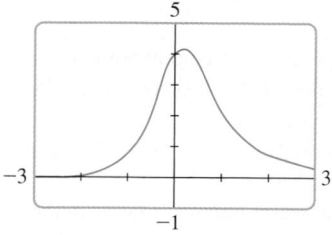

FIGURE **T1**
The graph of f in the viewing window $[-3, 3] \times [-1, 5]$

```
fMax((2X+4)/(X^2
+1)^1.5,X,-3,3)
        .1583117413
(2*.1583117413+4)/(.15
83117413^2+1)^1.5
        4.15928406
```

FIGURE **T2**
The TI-83/84 screen for Example 1

TECHNOLOGY EXERCISES

In Exercises 1–8, find the absolute maximum and the absolute minimum values of f in the given interval using the method of Example 1. Express your answers accurate to four decimal places.

1. $f(x) = 3x^4 - 4.2x^3 + 6.1x - 2; [-2, 3]$

2. $f(x) = 2.1x^4 - 3.2x^3 + 4.1x^2 + 3x - 4; [-1, 2]$

3. $f(x) = \dfrac{2x^3 - 3x^2 + 1}{x^2 + 2x - 8}; [-3, 1]$

4. $f(x) = \sqrt{x}(x^3 - 4)^2; [0.5, 1]$

5. $f(x) = \dfrac{x^3 - 1}{x^2}; [1, 3]$

6. $f(x) = \dfrac{x^3 - x^2 + 1}{x - 2}; [1, 3]$

7. $f(x) = e^{-x} \ln(x^2 + 1); [-2, 2]$

8. $f(x) = x^2 e^{-2x}; [-1, 1]$

9. BANK FAILURE The Haven Trust Bank of Duluth, Ga., founded in 2000, quickly increased its risky commercial real estate portfolio, despite many red flags from regulators. The bank failed in December 2008. The amount of construction loans of the bank as a percentage of its capital is approximated by the function

$$f(t) = -5.92t^4 + 58.89t^3 - 165.75t^2 + 56.21t + 629$$
$$(0 \le t \le 5)$$

where $t = 0$ corresponds to the beginning of 2003.

a. Plot the graph of f using the viewing window $[0, 5] \times [0, 650]$.

b. Show that at no time during the period from the beginning of 2003 through the beginning of 2008 did the amount of construction loans of the bank as a percentage of its capital fall below 415%. Note: The maximum percentage recommended by regulators in 2008 was 100%.

Source: FDIC Office of Inspector General.

10. **CONSTRUCTION LOANS** Refer to Exercise 9. The amount of construction loans of peer banks as a percentage of capital from the beginning of 2003 ($t = 0$) through the beginning of 2008 is approximated by the function

$$g(t) = -0.656t^4 + 5.693t^3 - 16.798t^2 + 36.083t^2 + 51.9$$
$$(0 \le t \le 5)$$

a. When did the amount of construction loans of peer banks as a percentage of capital first exceed the maximum of 100% as recommended by regulators in 2006?

b. What was the highest amount of construction loans as a share of capital of peer banks over the period from the beginning of 2003 through the beginning of 2008?

Source: FDIC Office of Inspector General.

11. **SICKOUTS** In a sickout by pilots of American Airlines in February 1999, the number of canceled flights from February 6 ($t = 0$) through February 14 ($t = 8$) is approximated by the function

$$N(t) = 1.2576t^4 - 26.357t^3 + 127.98t^2 + 82.3t + 43$$
$$(0 \le t \le 8)$$

where t is measured in days. The sickout ended after the union was threatened with millions of dollars in fines.

a. Show that the number of canceled flights was increasing at the fastest rate on February 8.

b. Estimate the maximum number of canceled flights in a day during the sickout.

Source: Associated Press.

12. **AVERAGE 401(K) ACCOUNT BALANCE** The following data give the average account balance (in thousands of dollars) of a 401(k) investor in year t in the United States during a certain 6-year period.

Year, t	0	1	2	3	4	5	6
Account Balance	37.5	40.8	47.3	55.5	49.4	43	40

a. Use **QuartReg** to find a fourth-degree polynomial regression model for the data. Let $t = 0$ correspond to the beginning of the first year.

b. Plot the graph of the function found in part (a), using the viewing window $[0, 6] \times [0, 60]$.

c. When was the average account balance lowest in the period under consideration? When was it highest?

d. What were the lowest average account balance and the highest average account balance during the period under consideration?

Source: Investment Company Institute.

10.5 Optimization II

In Section 10.4 we outlined a method for finding the solution to certain optimization problems in which the objective function is given. In this section, we consider problems in which we are required to first find the appropriate function to be optimized. The following guidelines will be useful for solving these problems.

> **Guidelines for Solving Optimization Problems**
>
> **1.** Assign a letter to each variable mentioned in the problem. If appropriate, draw and label a figure.
>
> **2.** Find an expression for the quantity to be optimized.
>
> **3.** Use the conditions given in the problem to write the quantity to be optimized as a function f of *one* variable. Note any restrictions to be placed on the domain of f from physical considerations of the problem.
>
> **4.** Optimize the function f over its domain using the methods of Section 10.4.

Note In carrying out Step 4, remember that if the function f to be optimized is continuous on a closed interval, then the absolute maximum and absolute minimum of f are, respectively, the largest and smallest values of $f(x)$ on the set composed of the critical numbers of f and the endpoints of the interval. If the domain of f is not a closed interval, then we resort to the graphical or some other method. ∎

Maximization Problems

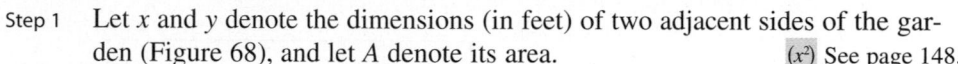

APPLIED EXAMPLE 1 Fencing a Garden A man wishes to have a rectangular-shaped garden in his backyard. He has 50 feet of fencing with which to enclose his garden. Find the dimensions for the largest garden he can have if he uses all of the fencing.

Solution

Step 1 Let x and y denote the dimensions (in feet) of two adjacent sides of the garden (Figure 68), and let A denote its area. (x^2) See page 148.

Step 2 The area of the garden

$$A = xy \tag{1}$$

is the quantity to be maximized.

Step 3 The perimeter of the rectangle, $(2x + 2y)$ feet, must equal 50 feet. Therefore, we have the equation

$$2x + 2y = 50$$

Next, solving this equation for y in terms of x yields

$$y = 25 - x \tag{2}$$

which, when substituted into Equation (1), gives

$$A = x(25 - x)$$
$$= -x^2 + 25x$$

(Remember, the function to be optimized must involve just one variable.) Since the sides of the rectangle must be nonnegative, we must have $x \geq 0$ and $y = 25 - x \geq 0$; that is, we must have $0 \leq x \leq 25$. Thus, the problem is reduced to that of finding the absolute maximum of $A = f(x) = -x^2 + 25x$ on the closed interval $[0, 25]$.

Step 4 Observe that f is continuous on $[0, 25]$, so the absolute maximum value of f must occur at the endpoint(s) of the interval or at the critical number(s) of f. The derivative of the function A is given by

$$A' = f'(x) = -2x + 25$$

Setting $A' = 0$ gives

$$-2x + 25 = 0$$

or 12.5, as the critical number of A. Next, we evaluate the function $A = f(x)$ at $x = 12.5$ and at the endpoints $x = 0$ and $x = 25$ of the interval $[0, 25]$, obtaining

$$f(0) = 0 \qquad f(12.5) = 156.25 \qquad f(25) = 0$$

We see that the absolute maximum value of the function f is 156.25. From Equation (2), we see that $y = 12.5$ when $x = 12.5$. Thus, the garden of maximum area (156.25 square feet) is a square with sides of length 12.5 feet. ■

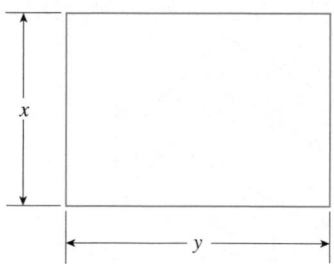

FIGURE **68**
What is the maximum rectangular area that can be enclosed with 50 feet of fencing?

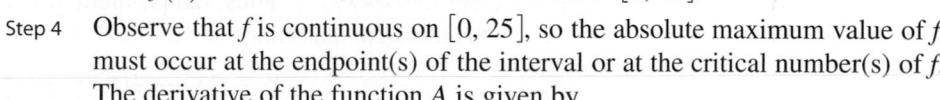

APPLIED EXAMPLE 2 Packaging By cutting away identical squares from each corner of a rectangular piece of cardboard and folding up the resulting flaps, the cardboard may be turned into an open box. If the cardboard is 16 inches long and 10 inches wide, find the dimensions of the box that will yield the maximum volume.

Solution

Step 1 Let x denote the length (in inches) of one side of each of the identical squares to be cut out of the cardboard (Figure 69), and let V denote the volume of the resulting box.

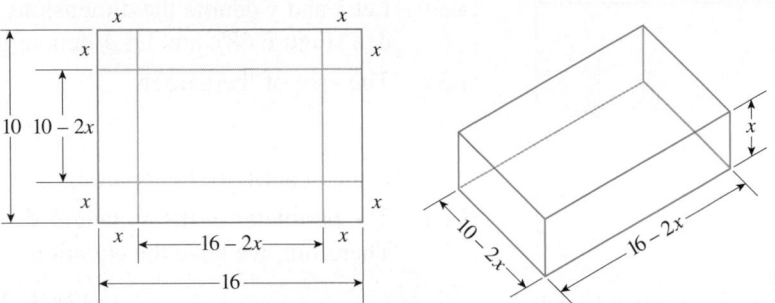

FIGURE **69**
The dimensions of the open box are $(16 - 2x)$ inches by $(10 - 2x)$ inches by x inches.

Step 2 The dimensions of the box are $(16 - 2x)$ inches by $(10 - 2x)$ inches by x inches. Therefore, its volume (in cubic inches),

$$V = (16 - 2x)(10 - 2x)x$$
$$= 4(x^3 - 13x^2 + 40x) \qquad \text{Expand the expression.}$$

is the quantity to be maximized.

Step 3 Since each side of the box must be nonnegative, x must satisfy the inequalities $x \geq 0$, $16 - 2x \geq 0$, and $10 - 2x \geq 0$. This set of inequalities is equivalent to $0 \leq x \leq 5$. Thus, the problem at hand is equivalent to that of finding the absolute maximum of

$$V = f(x) = 4(x^3 - 13x^2 + 40x)$$

on the closed interval $[0, 5]$.

Step 4 Observe that f is continuous on $[0, 5]$, so the absolute maximum value of f must be attained at the endpoint(s) or at the critical number(s) of f.
  Differentiating $f(x)$, we obtain

$$f'(x) = 4(3x^2 - 26x + 40)$$
$$= 4(3x - 20)(x - 2)$$

Upon setting $f'(x) = 0$ and solving the resulting equation for x, we obtain $x = \frac{20}{3}$ or $x = 2$. Since $\frac{20}{3}$ lies outside the interval $[0, 5]$, it is no longer considered, and we are interested only in the critical number 2 of f. Next, evaluating $f(x)$ at $x = 0$, $x = 5$ (the endpoints of the interval $[0, 5]$), and $x = 2$, we obtain

$$f(0) = 0 \qquad f(2) = 144 \qquad f(5) = 0$$

Thus, the volume of the box is maximized by taking $x = 2$. The dimensions of the box are 12 in. $\times$ 6 in. $\times$ 2 in., and the volume is 144 cubic inches.

Exploring with TECHNOLOGY

Refer to Example 2.

1. Use a graphing utility to plot the graph of

$$f(x) = 4(x^3 - 13x^2 + 40x)$$

using the viewing window $[0, 5] \times [0, 150]$. Explain what happens to $f(x)$ as x increases from $x = 0$ to $x = 5$ and give a physical interpretation.

2. Using **ZOOM** and **TRACE**, find the absolute maximum of f on the interval $[0, 5]$ and thus verify the solution for Example 2 obtained analytically.

APPLIED EXAMPLE 3 Optimal Subway Fare A city's Metropolitan Transit Authority (MTA) operates a subway line for commuters from a certain suburb to the downtown metropolitan area. Currently, an average of 6000 passengers a day take the trains, paying a fare of $3.00 per ride. The board of the MTA, contemplating raising the fare to $3.50 per ride in order to generate a larger revenue, engages the services of a consulting firm. The firm's study reveals that for each $0.50 increase in fare, the ridership will be reduced by an average of 1000 passengers a day. Therefore, the consulting firm recommends that MTA stick to the current fare of $3.00 per ride, which already yields a maximum revenue. Show that the consultants are correct.

Solution

Step 1 Let x denote the number of passengers per day, p denote the fare per ride, and R be MTA's revenue. (x^2) See page 148.

Step 2 To find a relationship between x and p, observe that the given data imply that when $x = 6000$, $p = 3$, and when $x = 5000$, $p = 3.50$. Therefore, the points $(6000, 3)$ and $(5000, 3.50)$ lie on a straight line. (Why?) To find the linear relationship between p and x, use the point-slope form of the equation of a straight line. Now, the slope of the line is

$$m = \frac{3.50 - 3}{5000 - 6000} = -0.0005$$

Therefore, the required equation is

$$p - 3 = -0.0005(x - 6000)$$
$$= -0.0005x + 3$$
$$p = -0.0005x + 6$$

Therefore, the revenue

$$R = f(x) = xp = -0.0005x^2 + 6x \qquad \text{Number of riders} \times \text{unit fare}$$

is the quantity to be maximized.

Step 3 Since both p and x must be nonnegative, we see that $0 \le x \le 12{,}000$, and the problem is that of finding the absolute maximum of the function f on the closed interval $[0, 12{,}000]$.

Step 4 Observe that f is continuous on $[0, 12{,}000]$. To find the critical number of R, we compute

$$f'(x) = -0.001x + 6$$

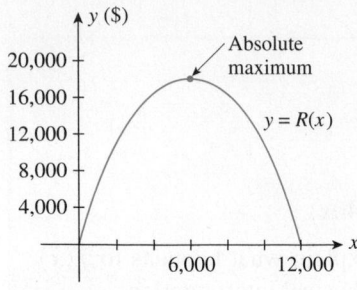

FIGURE **70**
f has an absolute maximum of 18,000 when *x* = 6000.

and set it equal to zero, giving $x = 6000$. Evaluating the function f at $x = 6000$, as well as at the endpoints $x = 0$ and $x = 12{,}000$, yields

$$f(0) = 0$$
$$f(6000) = 18{,}000$$
$$f(12{,}000) = 0$$

We conclude that a maximum revenue of \$18,000 per day is realized when the ridership is 6000 per day. The optimum price of the fare per ride is therefore \$3.00, as recommended by the consultants. The graph of the revenue function R is shown in Figure 70. ∎

Minimization Problems

 APPLIED EXAMPLE 4 Packaging Betty Moore Company requires that its corned beef hash containers have a capacity of 54 cubic inches, have the shape of a right circular cylinder, and be made of aluminum. Determine the radius and height of the container that requires the least amount of metal.

Solution

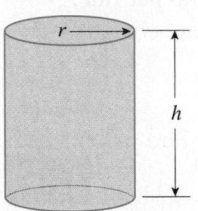

FIGURE **71**
We want to minimize the amount of material used to construct the container.

Step 1 Let the radius and height of the container be r and h inches, respectively, and let S denote the surface area of the container (Figure 71).

Step 2 The amount of aluminum used to construct the container is given by the total surface area of the cylinder. Now, the area of the base and the top of the cylinder are each πr^2 square inches, and the area of the side is $2\pi rh$ square inches. Therefore,

$$S = 2\pi r^2 + 2\pi rh \tag{3}$$

is the quantity to be minimized.

Step 3 The requirement that the volume of a container be 54 cubic inches implies that

$$\pi r^2 h = 54 \tag{4}$$

Solving Equation (4) for h, we obtain

$$h = \frac{54}{\pi r^2} \tag{5}$$

which, when substituted into Equation (3), yields

$$S = 2\pi r^2 + 2\pi r\left(\frac{54}{\pi r^2}\right)$$
$$= 2\pi r^2 + \frac{108}{r}$$

Clearly, the radius r of the container must satisfy the inequality $r > 0$. The problem now is reduced to finding the absolute minimum of the function $S = f(r)$ on the interval $(0, \infty)$.

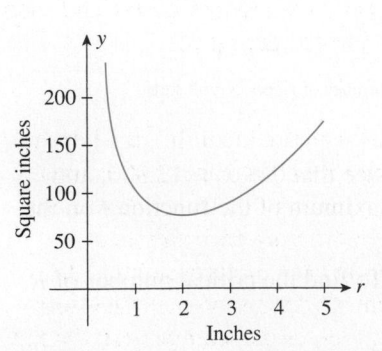

FIGURE **72**
The total surface area of the right cylindrical container is graphed as a function of *r*.

Step 4 Using the curve-sketching techniques of Section 10.3, we obtain the graph of f in Figure 72.

To find the critical number of f, we compute

$$S' = 4\pi r - \frac{108}{r^2}$$

and solve the equation $S' = 0$ for r:

$$4\pi r - \frac{108}{r^2} = 0$$

$$4\pi r^3 - 108 = 0$$

$$r^3 = \frac{27}{\pi}$$

$$r = \frac{3}{\sqrt[3]{\pi}} \approx 2 \qquad\qquad (6)$$

Next, let's show that this value of r gives rise to the absolute minimum of f. To show this, we first compute

$$S'' = 4\pi + \frac{216}{r^3}$$

Since $S'' > 0$ for $r = 3/\sqrt[3]{\pi}$, the Second Derivative Test implies that the value of r in Equation (6) gives rise to a relative minimum of f. Finally, this relative minimum of f is also the absolute minimum of f, since the graph of f is always concave upward ($S'' > 0$ for all $r > 0$). To find the height of the given container, we substitute the value of r given in Equation (6) into Equation (5). Thus,

$$h = \frac{54}{\pi r^2} = \frac{54}{\pi \left(\dfrac{3}{\pi^{1/3}}\right)^2}$$

$$= \frac{54\pi^{2/3}}{(\pi)9}$$

$$= \frac{6}{\pi^{1/3}} = \frac{6}{\sqrt[3]{\pi}}$$

$$= 2r$$

We conclude that the required container has a radius of approximately 2 inches and a height of approximately 4 inches, or twice the size of the radius.   ◼

An Inventory Problem

One problem faced by many companies is that of controlling the inventory of goods carried. Ideally, the manager must ensure that the company has sufficient stock to meet customer demand at all times. At the same time, she must make sure that this is accomplished without overstocking (incurring unnecessary storage costs) and also without having to place orders too frequently (incurring reordering costs).

APPLIED EXAMPLE 5 Inventory Control and Planning Dixie Import-Export is the sole agent for the Excalibur 250-cc motorcycle. Management estimates that the demand for these motorcycles is 10,000 per year and that they will sell at a uniform rate throughout the year. The cost incurred in ordering each shipment of motorcycles is $10,000, and the cost per year of storing each motorcycle is $200.

Dixie's management faces the following problem: Ordering too many motorcycles at one time ties up valuable storage space and increases the storage cost. On the other hand, placing orders too frequently increases the ordering costs. How large should each order be, and how often should orders be placed, to minimize ordering and storage costs?

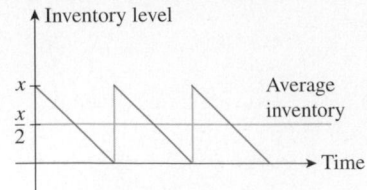

FIGURE 73
As each lot is depleted, the new lot arrives. The average inventory level is $x/2$ if x is the lot size.

Solution Let x denote the number of motorcycles in each order (the lot size). Then, assuming that each shipment arrives just as the previous shipment has been sold, the average number of motorcycles in storage during the year is $x/2$. You can see that this is the case by examining Figure 73. Thus, Dixie's storage cost for the year is given by $200(x/2)$, or $100x$ dollars.

Next, since the company requires 10,000 motorcycles for the year and since each order is for x motorcycles, the number of orders required is

$$\frac{10,000}{x}$$

This gives an ordering cost of

$$10,000\left(\frac{10,000}{x}\right) = \frac{100,000,000}{x}$$

dollars for the year. Thus, the total yearly cost incurred by Dixie, which includes the ordering and storage costs attributed to the sale of these motorcycles, is given by

$$C(x) = 100x + \frac{100,000,000}{x}$$

The problem is reduced to finding the absolute minimum of the function C on the interval $(0, 10,000]$. To accomplish this, we compute

$$C'(x) = 100 - \frac{100,000,000}{x^2}$$

Setting $C'(x) = 0$ and solving the resulting equation, we obtain $x = \pm 1000$. Since the number -1000 is outside the domain of the function C, it is rejected, leaving 1000 as the only critical number of C. Next, we find

$$C''(x) = \frac{200,000,000}{x^3}$$

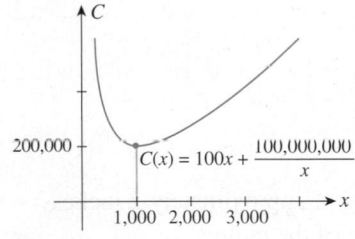

FIGURE 74
C has an absolute minimum at $(1000, 200,000)$.

Since $C''(1000) > 0$, the Second Derivative Test implies that the critical number 1000 is a relative minimum of the function C (Figure 74). Also, since $C''(x) > 0$ for all x in $(0, 10,000)$, the graph of C is concave upward everywhere, so $x = 1000$ also gives the absolute minimum of C. Thus, to minimize the ordering and storage costs, Dixie should place $10,000/1000$, or 10, orders a year, each for a shipment of 1000 motorcycles.

10.5 Self-Check Exercises

1. **MINIMIZING FENCING COSTS** A man wishes to have an enclosed vegetable garden in his backyard. If the garden is to be a rectangular area of 300 ft², find the dimensions of the garden that will minimize the amount of fencing needed.

2. **INVENTORY CONTROL AND PLANNING** The demand for Super Titan tires is 1,000,000/year. The setup cost for each production run is $4000, and the manufacturing cost is $20/tire. The cost of storing each tire over the year is $2. Assuming uniformity of demand throughout the year and instantaneous production, determine how many tires should be manufactured per production run to keep the total cost to a minimum.

Solutions to Self-Check Exercises 10.5 can be found on page 799.

10.5 Concept Questions

1. If the domain of a function f is not a closed interval, how would you find the absolute extrema of f, if they exist?

2. Refer to Example 4 (page 792). In the solution given in the example, we solved for h in terms of r, resulting in a function of r, which we then optimized with respect to r. Write S in terms of h and re-solve the problem. Which choice is better?

10.5 Exercises

1. Find the dimensions of a rectangle with a perimeter of 100 ft that has the largest possible area.

2. Find the dimensions of a rectangle of area 144 ft^2 that has the smallest possible perimeter.

3. **ENCLOSING THE LARGEST AREA** The owner of the Rancho Los Feliz has 3000 yd of fencing with which to enclose a rectangular piece of grazing land along the straight portion of a river. If fencing is not required along the river, what are the dimensions of the largest area that he can enclose? What is this area?

4. **ENCLOSING THE LARGEST AREA** Refer to Exercise 3. As an alternative plan, the owner of the Rancho Los Feliz might use the 3000 yd of fencing to enclose the rectangular piece of grazing land along the straight portion of the river and then subdivide it by means of a fence running parallel to the sides. Again, no fencing is required along the river. What are the dimensions of the largest area that can be enclosed? What is this area? (See the accompanying figure.)

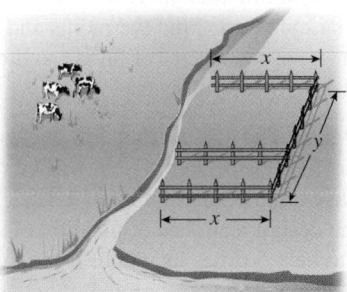

5. **MINIMIZING CONSTRUCTION COSTS** The management of the UNICO department store has decided to enclose an 800-ft^2 area outside the building for displaying potted plants and flowers. One side will be formed by the external wall of the store, two sides will be constructed of pine boards, and the fourth side will be made of galvanized steel fencing. If the pine board fencing costs $6/running foot and the steel fencing costs $3/running foot,

determine the dimensions of the enclosure that can be erected at minimum cost.

6. **PACKAGING** By cutting away identical squares from each corner of a rectangular piece of cardboard and folding up the resulting flaps, an open box may be made. If the cardboard is 15 in. long and 8 in. wide, find the dimensions of the box that will yield the maximum volume.

7. **METAL FABRICATION** If an open box is made from a tin sheet 8 in. square by cutting out identical squares from each corner and bending up the resulting flaps, determine the dimensions of the largest box that can be made.

8. **MINIMIZING PACKAGING COSTS** If an open box has a square base and a volume of 108 in.^{3} and is constructed from a tin sheet, find the dimensions of the box, assuming that a minimum amount of material is used in its construction.

9. **MINIMIZING COSTS** A pencil cup with a capacity of 36 in.^{3} is to be constructed in the shape of a rectangular box with a square base and an open top. If the material for the sides costs 15¢/in.^{2} and the material for the base costs 40¢/in.^{2}, what should the dimensions of the cup be to minimize the construction cost?

10. **MINIMIZING PACKAGING COSTS** What are the dimensions of a closed rectangular box that has a square cross section and a capacity of 128 in.^{3} and is constructed using the least amount of material?

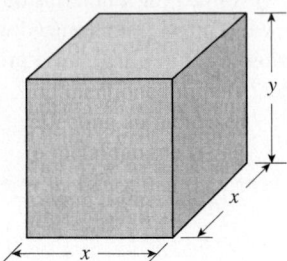

11. **MINIMIZING PACKAGING COSTS** A rectangular box is to have a square base and a volume of 20 ft^3. If the material for the base costs 30¢/ft^2, the material for the sides costs 10¢/ft^2, and the material for the top costs 20¢/ft^2, determine the dimensions of the box that can be constructed at minimum cost. (Refer to the figure for Exercise 10.)

12. **PARCEL POST REGULATIONS** Postal regulations specify that a parcel sent by priority mail may have a combined length and girth of no more than 108 in. Find the dimensions of a rectangular package that has a square cross section and the largest volume that may be sent via priority mail. What is the volume of such a package?
Hint: The length plus the girth is $4x + h$ (see the accompanying figure).

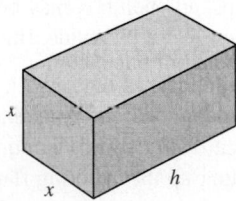

13. **BOOK DESIGN** A book designer has decided that the pages of a book should have 1-in. margins at the top and bottom and $\frac{1}{2}$-in. margins on the sides. She further stipulated that each page should have an area of 50 in.^{2} (see the accompanying figure). Determine the page dimensions that will result in the maximum printed area on the page.

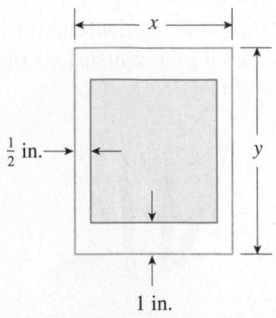

14. **PARCEL POST REGULATIONS** Postal regulations specify that a parcel sent by priority mail may have a combined length and girth of no more than 108 in. Find the dimensions of the cylindrical package of greatest volume that may be sent via priority mail. What is the volume of such a package? Compare with Exercise 12.
Hint: The length plus the girth is $2\pi r + l$.

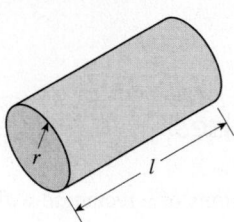

15. **MINIMIZING COSTS** A pencil cup with a capacity of 9π in.^{3} is to be constructed in the shape of a right circular cylinder with an open top. If the material for the side costs $\frac{3}{8}$ of the cost of the material for the base, what dimensions should the cup have to minimize the construction cost?

16. **MINIMIZING COSTS** For its beef stew, Betty Moore Company uses aluminum containers that have the form of right circular cylinders. Find the radius and height of a container if it has a capacity of 36 in.^{3} and is constructed using the least amount of metal.

17. **PRODUCT DESIGN** The cabinet that will enclose the Acrosonic model D loudspeaker system will be rectangular and will have an internal volume of 2.4 ft^3. For aesthetic reasons, the design team has decided that the height of the cabinet is to be 1.5 times its width. If the top, bottom, and sides of the cabinet are constructed of veneer costing 40¢/ft^2 and the front (ignore the cutouts in the baffle) and rear are constructed of particle board costing 20¢/ ft^2, what are the dimensions of the enclosure that can be constructed at a minimum cost?

18. **DESIGNING A NORMAN WINDOW** A Norman window has the shape of a rectangle surmounted by a semicircle (see the accompanying figure). If a Norman window is to have a perimeter of 28 ft, what should its dimensions be

in order to allow the maximum amount of light through the window?

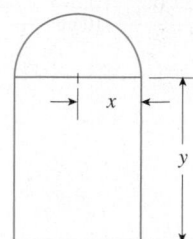

19. OPTIMAL CHARTER-FLIGHT FARE If exactly 200 people sign up for a charter flight, Leisure World Travel Agency charges $300/person. However, if more than 200 people sign up for the flight (assume that this is the case), then each fare is reduced by $1 for each additional person. Determine how many passengers will result in a maximum revenue for the travel agency. What is the maximum revenue? What would be the fare per passenger in this case?
Hint: Let x denote the number of passengers above 200. Show that the revenue function R is given by $R(x) = (200 + x)(300 - x)$.

20. MAXIMIZING YIELD An apple orchard has an average yield of 36 bushels of apples/tree if tree density is 22 trees/acre. For each unit increase in tree density, the yield decreases by 2 bushels/tree. How many trees should be planted to maximize the yield?

21. CHARTER REVENUE The owner of a luxury motor yacht that sails among the 4000 Greek islands charges $600/person/day if exactly 20 people sign up for the cruise. However, if more than 20 people sign up (up to the maximum capacity of 90) for the cruise, then every fare is reduced by $4 for each additional passenger. Assuming that at least 20 people sign up for the cruise, determine how many passengers will result in the maximum revenue for the owner of the yacht. What is the maximum revenue? What would be the fare per passenger in this case?

22. PROFIT OF A VINEYARD Phillip, the proprietor of a vineyard, estimates that the first 10,000 bottles of wine produced this season will fetch a profit of $5/bottle. But if more than 10,000 bottles are produced, then the profit per bottle for the entire lot will drop by $0.0002 for each additional bottle sold. Assuming that at least 10,000 bottles of wine are produced and sold, what is the maximum profit?

23. OPTIMAL SPEED OF A TRUCK A truck gets $600/x$ mpg when driven at a constant speed of x mph (between 50 and 70 mph). If the price of fuel is $3/gal and the driver is paid $18/hr, at what speed between 50 and 70 mph is it most economical to drive?

24. MINIMIZING COSTS Suppose the cost incurred in operating a cruise ship for one hour is $a + bv^3$ dollars, where a and b are positive constants and v is the ship's speed in miles per hour. At what speed should the ship be operated between two ports to minimize the cost?

25. STRENGTH OF A BEAM A wooden beam has a rectangular cross section of height h in. and width w in. (see the accompanying figure). The strength S of the beam is directly proportional to its width and the square of its height. What are the dimensions of the cross section of the strongest beam that can be cut from a round log of diameter 24 in.?
Hint: $S = kh^2w$, where k is a constant of proportionality.

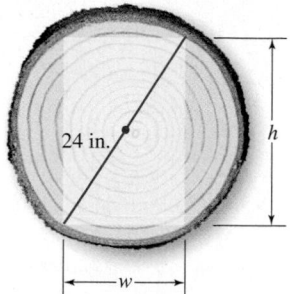

26. DESIGNING A GRAIN SILO A grain silo has the shape of a right circular cylinder surmounted by a hemisphere (see the accompanying figure). If the silo is to have a capacity of 504π ft^3, find the radius and height of the silo that requires the least amount of material to construct.
Hint: The volume of the silo is $\pi r^2h + \frac{2}{3}\pi r^3$, and the surface area (including the floor) is $\pi(3r^2 + 2rh)$.

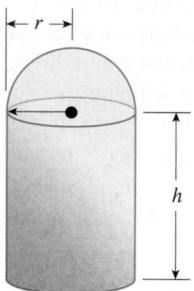

27. MINIMIZING COST OF LAYING CABLE In the following diagram, S represents the position of a power relay station located on a straight coast, and E shows the location of a marine biology experimental station on an island. A cable is to be laid connecting the relay station with the experimental station. If the cost of running the cable on land is $1.50/running foot and the cost of running the cable under water is $2.50/running foot, locate the point P that will result in a minimum cost (solve for x).

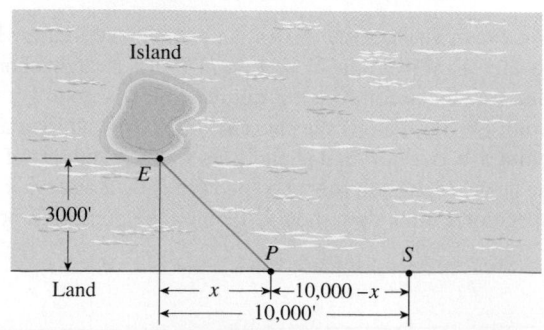

28. STORING RADIOACTIVE WASTE A cylindrical container for storing radioactive waste is to be constructed from lead and have a thickness of 6 in. (see the accompanying figure). If the volume of the outside cylinder is to be 16π ft^3, find the radius and the height of the inside cylinder that will result in a container of maximum storage capacity.

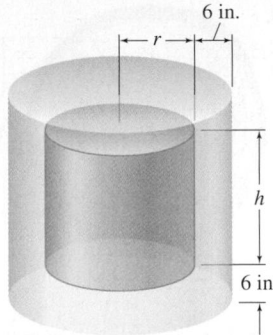

Hint: Show that the storage capacity (inside volume) is given by

$$V(r) = \pi r^2 \left[\frac{16}{(r + \frac{1}{2})^2} - 1 \right] \qquad (0 \le r \le \tfrac{7}{2})$$

29. FLIGHTS OF BIRDS During daylight hours, some birds fly more slowly over water than over land because some of their energy is expended in overcoming the downdrafts of air over open bodies of water. Suppose a bird that flies at a constant speed of 4 mph over water and 6 mph over land starts its journey at the point E on an island and ends at its nest N on the shore of the mainland, as shown in the accompanying figure. Find the location of the point P that allows the bird to complete its journey in the minimum time (solve for x).

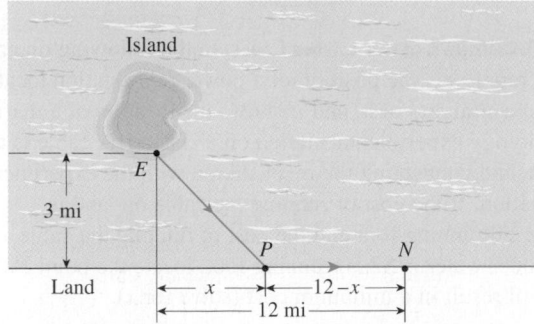

30. MINIMIZING TRAVEL TIME A woman is on a lake in a rowboat located 1 mi from the closest point P of a straight shoreline (see the accompanying figure). She wishes to get to point Q, 10 mi along the shore from P, by rowing to a point R between P and Q and then walking the rest of the distance. If she can row at a speed of 3 mph and walk at a speed of 4 mph, how should she pick the point R in order

to get to Q as quickly as possible? How much time does she require?

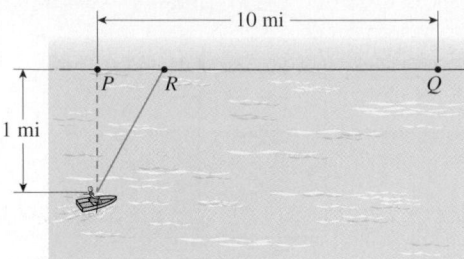

31. RACETRACK DESIGN The accompanying figure depicts a racetrack with ends that are semicircular in shape. The length of the track is 1760 ft ($\frac{1}{3}$ mi). Find l and r such that the area enclosed by the rectangular region of the racetrack is as large as possible. What is the area enclosed by the track in this case?

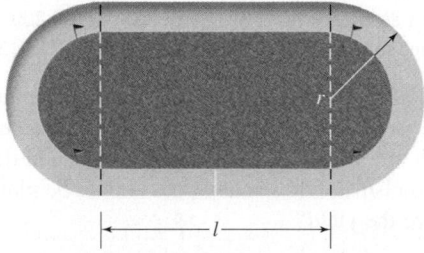

32. INVENTORY CONTROL AND PLANNING The demand for motorcycle tires imported by Dixie Import-Export is 40,000/year and may be assumed to be uniform throughout the year. The cost of ordering a shipment of tires is $400, and the cost of storing each tire for a year is $2. Determine how many tires should be in each shipment if the ordering and storage costs are to be minimized. (Assume that each shipment arrives just as the previous one has been sold.)

33. INVENTORY CONTROL AND PLANNING McDuff Preserves expects to bottle and sell 2,000,000 32-oz jars of jam at a uniform rate throughout the year. The company orders its containers from Consolidated Bottle Company. The cost of ordering a shipment of bottles is $200, and the cost of storing each empty bottle for a year is $0.40. How many orders should McDuff place per year, and how many bottles should be in each shipment if the ordering and storage costs are to be minimized? (Assume that each shipment of bottles is used up before the next shipment arrives.)

34. INVENTORY CONTROL AND PLANNING Neilsen Cookie Company sells its assorted butter cookies in containers that have a net content of 1 lb. The estimated demand for the cookies is 1,000,000 1-lb containers per year. The setup cost for each production run is $500, and the manufacturing cost is $0.50 for each container of cookies. The cost of storing each container of cookies over the year is $0.40. Assuming uniformity of demand throughout the year and instantaneous production, how many containers of cookies should Neilsen produce per production run in order to minimize the production cost?

Hint: Following the method of Example 5, show that the total production cost is given by the function

$$C(x) = \frac{500,000,000}{x} + 0.2x + 500,000$$

Then minimize the function C on the interval $(0, 1,000,000)$.

35. INVENTORY CONTROL AND PLANNING A company expects to sell D units of a certain product per year. Sales are assumed to be at a steady rate with no shortages allowed. Each time an order for the product is placed, an ordering cost of K dollars is incurred. Each item costs p dollars, and the holding cost is h dollars per item per year.

a. Show that the inventory cost (the combined ordering cost, purchasing cost, and holding cost) is

$$C(x) = \frac{KD}{x} + pD + \frac{hx}{2} \qquad (x > 0)$$

where x is the order quantity (the number of items in each order).

b. Use the result of part (a) to show that the inventory cost is minimized if

$$x = \sqrt{\frac{2KD}{h}}$$

This quantity is called the *economic order quantity* (EOQ).

36. INVENTORY CONTROL AND PLANNING Refer to Exercise 35. The Camera Store sells 960 Yamaha A35 digital cameras per year. Each time an order for cameras is placed with the manufacturer, an ordering cost of $10 is incurred. The store pays $80 for each camera, and the cost for holding a camera (mainly due to the opportunity cost incurred in tying up capital in inventory) is $12/year. Assume that the cameras sell at a uniform rate and no shortages are allowed.

a. What is the EOQ?
b. How many orders will be placed each year?
c. What is the interval between orders?

10.5 Solutions to Self-Check Exercises

1. Let x and y (measured in feet) denote the length and width of the rectangular garden.

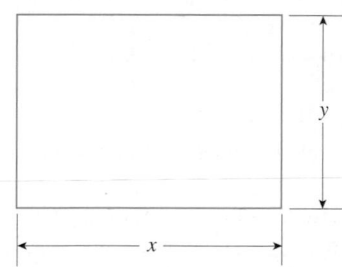

Since the area is to be 300 ft², we have

$$xy = 300$$

Next, the amount of fencing to be used is given by the perimeter, and this quantity is to be minimized. Thus, we want to minimize

$$2x + 2y$$

Since $y = 300/x$ (obtained by solving for y in the first equation), we see that the expression to be minimized is

$$f(x) = 2x + 2\left(\frac{300}{x}\right)$$

$$= 2x + \frac{600}{x}$$

for positive values of x. Now,

$$f'(x) = 2 - \frac{600}{x^2}$$

Setting $f'(x) = 0$ yields $x = -\sqrt{300}$ or $x = \sqrt{300}$. We consider only the critical number $\sqrt{300}$, since $-\sqrt{300}$ lies outside the interval $(0, \infty)$. We then compute

$$f''(x) = \frac{1200}{x^3}$$

Since

$$f''(300) > 0$$

the Second Derivative Test implies that a relative minimum of f occurs at $x = \sqrt{300}$. In fact, since $f''(x) > 0$ for all x in $(0, \infty)$, we conclude that $x = \sqrt{300}$ gives rise to the absolute minimum of f. The corresponding value of y, obtained by substituting this value of x into the equation $xy = 300$, is $y = \sqrt{300}$. Therefore, the required dimensions of the vegetable garden are approximately 17.3 ft × 17.3 ft.

2. Let x denote the number of tires in each production run. Then, the average number of tires in storage is $x/2$, so the storage cost incurred by the company is $2(x/2)$, or x dollars. Next, since the company needs to manufacture 1,000,000 tires for the year to meet the demand, the number of production runs is 1,000,000/x. This gives setup costs amounting to

$$4000\left(\frac{1,000,000}{x}\right) = \frac{4,000,000,000}{x}$$

dollars for the year. The total manufacturing cost is $20,000,000. Thus, the total yearly cost incurred by the company is given by

$$C(x) = x + \frac{4,000,000,000}{x} + 20,000,000$$

Differentiating $C(x)$, we find

$$C'(x) = 1 - \frac{4,000,000,000}{x^2}$$

Setting $C'(x) = 0$ gives 63,246 as the critical number in the interval $(0, 1,000,000)$. Next, we find

$$C''(x) = \frac{8,000,000,000}{x^3}$$

Since $C''(x) > 0$ for all $x > 0$, we see that the graph of C is concave upward for all $x > 0$. Furthermore, $C''(63,246) > 0$ implies that $x = 63,246$ gives rise to a relative minimum of C (by the Second Derivative Test). Since the graph of C is always concave upward for $x > 0$, $x = 63,246$ gives the absolute minimum of C. Therefore, the company should manufacture 63,246 tires in each production run.

CHAPTER 10 Summary of Principal Terms

TERMS

increasing function (718)	First Derivative Test (726)	vertical asymptote (758)
decreasing function (718)	concave upward (740)	horizontal asymptote (760)
relative maximum (723)	concave downward (740)	absolute extrema (773)
relative minimum (724)	point of inflection (743)	absolute maximum value (773)
relative extrema (724)	inflection point (743)	absolute minimum value (773)
critical number (725)	Second Derivative Test (747)	

CHAPTER 10 Concept Review Questions

Fill in the blanks.

1. **a.** A function f is increasing on an interval I if for any two numbers x_1 and x_2 in I, $x_1 < x_2$ implies that _____.
   b. A function f is decreasing on an interval I if for any two numbers x_1 and x_2 in I, $x_1 < x_2$ implies that _____.

2. **a.** If f is differentiable on an open interval (a, b) and $f'(x) > 0$ on (a, b), then f is _____ on (a, b).
   b. If f is differentiable on an open interval (a, b) and _____ on (a, b), then f is decreasing on (a, b).
   c. If $f'(x) = 0$ for each value of x in the interval (a, b), then f is _____ on (a, b).

3. **a.** A function f has a relative maximum at c if there exists an open interval (a, b) containing c such that _____ for all x in (a, b).
   b. A function f has a relative minimum at c if there exists an open interval (a, b) containing c such that _____ for all x in (a, b).

4. **a.** A critical number of a function f is any number in the _____ of f at which $f'(c)$ _____ or $f'(c)$ does not _____.

   b. If f has a relative extremum at c, then c must be a/an _____ _____ of f.
   c. If c is a critical number of f, then f may or may not have a/an _____ _____ at c.

5. **a.** The graph of a differentiable function f is concave upward on an interval I if _____ is increasing on I.
   b. If f has a second derivative on an open interval I and $f''(x)$ _____ on I, then the graph of f is concave upward on I.
   c. If the graph of a continuous function f has a tangent line at $P(c, f(c))$ and the graph of f changes _____ at P, then P is called an inflection point of f.
   d. Suppose f has a second derivative on an interval (a, b), containing a critical number c of f. If $f''(c) < 0$, then f has a/an _____ _____ at c. If $f''(c) = 0$, then f may or may not have a/an _____ _____ at c.

6. The line $x = a$ is a vertical asymptote of the graph f if at least one of the following is true: $\lim_{x \to a^+} f(x) = $ _____ or $\lim_{x \to a^-} f(x) = $ _____.

7. For a rational function $f(x) = \dfrac{P(x)}{Q(x)}$, the line $x = a$ is a vertical asymptote of the graph of f if $Q(a) = $ _____ but $P(a) \neq$ _____.

8. The line $y = b$ is a horizontal asymptote of the graph of a function f if either $\lim\limits_{x \to \infty} f(x) = $ _____ or $\lim\limits_{x \to -\infty} f(x) = $ _____.

9. a. A function f has an absolute maximum at c if _____ for all x in the domain D of f. The number $f(c)$ is called the _____ _____ _____ of f on D.

b. A function f has a relative minimum at c if _____ for all values of x in some _____ _____ containing c.

10. The Extreme Value Theorem states that if f is _____ on the closed interval $[a, b]$, then f has both a/an _____ maximum value and a/an _____ minimum value on $[a, b]$.

CHAPTER 10 Review Exercises

In Exercises 1–12, (a) find the intervals where the function f is increasing and where it is decreasing, (b) find the relative extrema of f, (c) find the intervals where the graph of f is concave upward and where it is concave downward, and (d) find the inflection points, if any, of f.

1. $f(x) = \dfrac{1}{3}x^3 - x^2 + x - 6$

2. $f(x) = (x - 2)^3$

3. $f(x) = x^4 - 2x^2$

4. $f(x) = x + \dfrac{4}{x}$

5. $f(x) = \dfrac{x^2}{x - 1}$

6. $f(x) = \sqrt{x - 1}$

7. $f(x) = (1 - x)^{1/3}$

8. $f(x) = x\sqrt{x - 1}$

9. $f(x) = \dfrac{2x}{x + 1}$

10. $f(x) = -\dfrac{1}{1 + x^2}$

11. $f(x) = (4 - x)e^x$

12. $f(x) = x^2 \ln x$

In Exercises 13–22, use the curve-sketching guide on page 762 to sketch the graph of the function.

13. $f(x) = x^2 - 5x + 5$

14. $f(x) = -2x^2 - x + 1$

15. $g(x) = 2x^3 - 6x^2 + 6x + 1$

16. $g(x) = \dfrac{1}{3}x^3 - x^2 + x - 3$

17. $h(x) = x\sqrt{x - 2}$

18. $h(x) = \dfrac{2x}{1 + x^2}$

19. $f(x) = \dfrac{x - 2}{x + 2}$

20. $f(x) = x - \dfrac{1}{x}$

21. $f(x) = xe^{-2x}$

22. $f(x) = x^2 - \ln x$

In Exercises 23–26, find the horizontal and vertical asymptotes of the graph of each function. Do not sketch the graph.

23. $f(x) = \dfrac{1}{2x + 3}$

24. $f(x) = \dfrac{2x}{x + 1}$

25. $f(x) = \dfrac{5x}{x^2 - 2x - 8}$

26. $f(x) = \dfrac{x^2 + x}{x(x - 1)}$

In Exercises 27–38, find the absolute maximum value and the absolute minimum value, if any, of the function.

27. $f(x) = 2x^2 + 3x - 2$

28. $g(x) = x^{2/3}$

29. $g(t) = \sqrt{25 - t^2}$

30. $f(x) = \dfrac{1}{3}x^3 - x^2 + x + 1$ on $[0, 2]$

31. $h(t) = t^3 - 6t^2$ on $[2, 5]$

32. $g(x) = \dfrac{x}{x^2 + 1}$ on $[0, 5]$

33. $f(x) = x - \dfrac{1}{x}$ on $[1, 3]$

34. $h(t) = 8t - \dfrac{1}{t^2}$ on $[1, 3]$

35. $f(t) = 3te^{-t}$ on $[-2, 2]$

36. $g(t) = \dfrac{2 \ln t}{t}$ on $[1, 2]$

37. $f(s) = s\sqrt{1 - s^2}$ on $[-1, 1]$

38. $f(x) = \dfrac{x^2}{x - 1}$ on $[-1, 3]$

39. Revenue of Competing Bookstores The graphs of R_1 and R_2 that follow show the revenue of a neighborhood bookstore and that of a branch of a national bookstore, three months after the opening of the latter $(t = 0)$ until sometime later T.

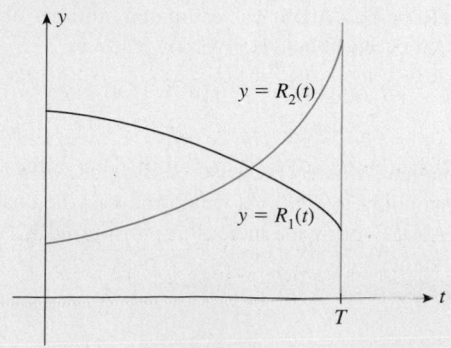

a. Find the signs of the first and second derivatives of R_1 and R_2 on the interval $(0, T)$.

b. Give an interpretation of the results obtained in part (a) in terms of the revenues of the two bookstores.

40. SPREAD OF A RUMOR Initially, a handful of students heard a rumor on campus. The rumor spread, and after t hr, the number of students who had heard it had grown to $N(t)$. The graph of the function N is shown in the figure below. Describe the spread of the rumor in terms of the speed at which it was spread. In particular, explain the significance of the inflection point P of the graph of N.

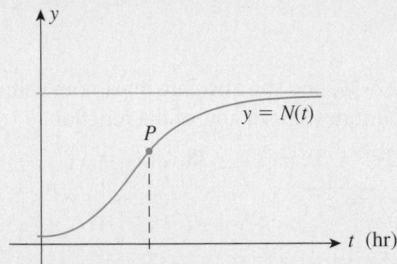

41. RELIABILITY OF COMPUTER CHIPS The percentage of a certain brand of computer chips that will fail after t years of use is estimated to be

$$P(t) = 100(1 - e^{-0.1t})$$

a. How fast will this brand of computer chips be failing after 5 years of use?

b. Evaluate $\lim_{t \to \infty} P(t)$. Did you expect this result?

42. GROWTH OF A FRUIT FLY POPULATION On the basis of data collected during an experiment, a biologist found that the growth of a fruit fly (*Drosophila*) with a limited food supply could be approximated by the exponential model

$$N(t) = \frac{400}{1 + 39e^{-0.16t}}$$

where t denotes the number of days since the beginning of the experiment.

a. What was the population of the fruit fly colony on the 20th day?

b. How fast was the population changing on the 20th day?

c. What was the maximum fruit fly population that could be expected under this laboratory condition?

43. DEATH RATE FROM AIDS The estimated number of deaths from AIDS worldwide is given by

$$f(t) = -0.0004401t^3 + 0.007t^2 + 0.112t + 0.28$$
$$(0 \leq t \leq 21)$$

million/year, where t is measured in years, with $t = 0$ corresponding to 1990. At what time was the death rate from AIDS worldwide increasing most rapidly? What was the rate?
Source: UNAIDS.

44. AGE AT FIRST MARRIAGE According to a study conducted by Rutgers University, the median age of women in the United States at first marriage is approximated by the function

$$f(t) = -0.083t^3 + 0.6t^2 + 0.18t + 20.1 \qquad (0 \leq t \leq 5)$$

where t is measured in decades and $t = 0$ corresponds to the beginning of 1960.

a. What was the median age of women at first marriage at the beginning of 1960? At the beginning of the year 2000? At the beginning of 2010?

b. When was the median age of women at first marriage changing most rapidly over the time period under consideration?
Source: The National Marriage Project, Rutgers University.

45. MAXIMIZING PROFITS Odyssey Travel Agency's monthly profit (in thousands of dollars) depends on the amount of money x (in thousands of dollars) spent on advertising each month according to the rule

$$P(x) = -x^2 + 8x + 20$$

To maximize its monthly profits, what should be Odyssey's monthly advertising budget?

46. EFFECT OF SMOKING BANS The sales (in billions of dollars) in restaurants and bars in California from 1993 ($t = 0$) through 2000 ($t = 7$) are approximated by the function

$$S(t) = 0.195t^2 + 0.32t + 23.7 \qquad (0 \leq t \leq 7)$$

a. Show that the sales in restaurants and bars continued to rise after smoking bans were implemented in restaurants in 1995 and in bars in 1998.
Hint: Show that S is increasing on the interval $(2, 7)$.

b. What can you say about the rate at which the sales were rising after smoking bans were implemented?
Source: California Board of Equalization.

47. ALTERNATIVE MINIMUM TAX Congress created the alternative minimum tax (AMT) in the late 1970s to ensure that wealthy people paid their fair share of taxes. But because of quirks in the law, even middle-income taxpayers have started to get hit with the tax. The AMT (in billions of dollars) projected to be collected by the IRS from 2001 through 2010 is

$$f(t) = 0.0117t^3 + 0.0037t^2 + 0.7563t + 4.1 \qquad (0 \leq t \leq 9)$$

where t is measured in years, with $t = 0$ corresponding to 2001.

a. Show that f is increasing on the interval $(0, 9)$. What does this result tell you about the projected amount of AMT paid over the years in question?

b. Show that f' is increasing on the interval $(0, 9)$. What conclusion can you draw from this result concerning the rate of growth of the amount of the AMT collected over the years in question?
Source: U.S. Congress Joint Economic Committee.

48. **MEASLES DEATHS** Measles is still a leading cause of vaccine-preventable death among children, but because of improvements in immunizations, measles deaths have dropped globally. The function

$$N(t) = -2.42t^3 + 24.5t^2 - 123.3t + 506 \qquad (0 \le t \le 6)$$

gives the number of measles deaths (in thousands) in sub-Saharan Africa in year t, with $t = 0$ corresponding to the beginning of 1999.
   a. How many measles deaths were there in 1999? In 2005?
   b. Show that $N'(t) < 0$ on $(0, 6)$. What does this say about the number of measles deaths from 1999 through 2005?
   c. When was the number of measles deaths decreasing most rapidly? What was the rate of decline of measles deaths at that instant of time?
   Source: Centers for Disease Control and Prevention and World Health Organization.

49. **EFFECT OF ADVERTISING ON SALES** The total sales S of Cannon Precision Instruments is related to the amount of money x that Cannon spends on advertising its products by the function

$$S(x) = -0.002x^3 + 0.6x^2 + x + 500 \qquad (0 \le x \le 200)$$

where S and x are measured in thousands of dollars. Find the inflection point of the function S, and discuss its significance.

50. **ELDERLY WORKFORCE** The percentage of men 65 years and older in the workforce from 1970 through 2000 is approximated by the function

$$P(t) = 0.00093t^3 - 0.018t^2 - 0.51t + 25 \qquad (0 \le t \le 30)$$

where t is measured in years, with $t = 0$ corresponding to the beginning of 1970.
   a. Find the interval where P is decreasing and the interval where P is increasing.
   b. Interpret the results of part (a).
   Source: U.S. Census Bureau.

51. **COST OF PRODUCING CALCULATORS** A subsidiary of Elektra Electronics manufactures graphing calculators. Management determines that the daily cost $C(x)$ (in dollars) of producing these calculators is

$$C(x) = 0.0001x^3 - 0.08x^2 + 40x + 5000$$

where x is the number of calculators produced. Find the inflection point of the function C, and interpret your result.

52. **SALES OF MOBILE PROCESSORS** The rising popularity of notebook computers is fueling the sales of mobile PC processors. In a study conducted in 2003, the sales of these chips (in billions of dollars) was projected to be

$$S(t) = 6.8(t + 1.03)^{0.49} \qquad (0 \le t \le 4)$$

where t is measured in years, with $t = 0$ corresponding to 2003.

   a. Show that S is increasing on the interval $(0, 4)$, and interpret your result.
   b. Show that the graph of S is concave downward on the interval $(0, 4)$. Interpret your result.
   Source: International Data Corp.

53. **SMALL CAR MARKET SHARE IN EUROPE** Owing in part to an aging population and the squeeze from carbon-cutting regulations, the percentage of small and lower-midsize vehicles in Europe is expected to increase in the near future. The function

$$f(t) = \frac{150\sqrt{t} + 766}{59 - \sqrt{t}} \qquad (0 \le t \le 10)$$

gives the projected percentage of small and lower-midsize vehicles t years after 2005.
   a. What was the percentage of small and lower-midsize vehicles in 2005? What is the projected percentage of small and lower-midsize vehicles in 2015?
   b. Show that f is increasing on the interval $(0, 10)$, and interpret your results.
   Source: J.D. Power Automotive.

54. **TRAFFIC FLOW ANALYSIS** The speed of traffic flow in miles per hour on a stretch of Route 123 between 6 A.M. and 10 A.M. on a typical workday is approximated by the function

$$f(t) = 20t - 40\sqrt{t} + 52 \qquad (0 \le t \le 4)$$

where t is measured in hours, with $t = 0$ corresponding to 6 A.M. Sketch the graph of f, and interpret your results.

55. **COST OF REMOVING TOXIC POLLUTANTS** The cost, measured in millions of dollars, of removing $x\%$ of a toxic pollutant is given by

$$C(x) = \frac{0.5x}{100 - x}$$

Sketch the graph of the function C, and interpret your results.

56. **INDEX OF ENVIRONMENTAL QUALITY** The Department of the Interior of an African country began to record an index of environmental quality to measure progress or decline in the environmental quality of its wildlife. The index for the years 2002 through 2012 is approximated by the function

$$I(t) = \frac{50t^2 + 600}{t^2 + 10} \qquad (0 \le t \le 10)$$

   a. Compute $I'(t)$ and show that $I(t)$ is decreasing on the interval $(0, 10)$.
   b. Compute $I''(t)$. Study the concavity of the graph of I.
   c. Sketch the graph of I.
   d. Interpret your results.

57. **MAXIMIZING PROFITS** The weekly demand for DVDs manufactured by Herald Media Corporation is given by

$$p = -0.0005x^2 + 60$$

where p denotes the unit price in dollars and x denotes the quantity demanded. The weekly total cost function associated with producing these discs is given by

$$C(x) = -0.001x^2 + 18x + 4000$$

where $C(x)$ denotes the total cost (in dollars) incurred in pressing x discs. Find the production level that will yield a maximum profit for the manufacturer.
Hint: Use the quadratic formula.

58. **MAXIMIZING PROFITS** The estimated monthly profit (in dollars) realizable by Cannon Precision Instruments for manufacturing and selling x units of its model M1 digital camera is

$$P(x) = -0.04x^2 + 240x - 10,000$$

To maximize its profits, how many cameras should Cannon produce each month?

59. **MINIMIZING AVERAGE COST** The total monthly cost (in dollars) incurred by Carlota Music in manufacturing x units of its Professional Series guitars is given by the function

$$C(x) = 0.001x^2 + 100x + 4000$$

a. Find the average cost function $\overline{C}$.
b. Determine the production level that will result in the smallest average production cost.

60. **WORKER EFFICIENCY** The average worker at Wakefield Avionics will have assembled

$$N(t) = -2t^3 + 12t^2 + 2t \qquad (0 \le t \le 4)$$

ready-to-fly radio-controlled model airplanes t hr into the 8 A.M. to 12 noon shift. At what time during this shift is the average worker performing at peak efficiency?

61. **SENIOR WORKFORCE** The percentage of women 65 years and older in the workforce from 1970 through 2000 is approximated by the function

$$P(t) = -0.0002t^3 + 0.018t^2 - 0.36t + 10 \qquad (0 \le t \le 30)$$

where t is measured in years, with $t = 0$ corresponding to the beginning of 1970.
a. Find the interval where P is decreasing and the interval where P is increasing.
b. Find the absolute minimum of P.
c. Interpret the results of parts (a) and (b).
Source: U.S. Census Bureau.

62. **AGE OF DRIVERS IN CRASH FATALITIES** The number of crash fatalities per 100,000 vehicle miles of travel in a certain year is approximated by the model

$$f(x) = \frac{15}{0.08333x^2 + 1.91667x + 1} \qquad (0 \le x \le 11)$$

where x is the age of the driver in years, with $x = 0$ corresponding to age 16. Show that f is decreasing on $(0, 11)$, and interpret your result.
Source: National Highway Traffic Safety Administration.

63. **SPREAD OF A CONTAGIOUS DISEASE** The incidence (number of new cases per day) of a contagious disease spreading in a population of M people is given by

$$R(x) = kx(M - x)$$

where k is a positive constant and x denotes the number of people already infected. Show that the incidence R is greatest when half the population is infected.

64. **MEDICAL SCHOOL APPLICANTS** According to a study from the American Medical Association, the number of medical school applicants from academic year 1997–1998 ($t = 0$) through the academic year 2008–2009 is approximated by the function

$$N(t) = \begin{cases} 0.36t^2 - 3.10t + 41.2 & \text{if } 0 \le t < 9 \\ 42.46 & \text{if } 9 \le t \le 11 \end{cases}$$

Find the years when the number of medical school applicants was increasing, when it was decreasing, and when it was approximately constant.
Source: Journal of the American Medical Association.

65. Let

$$f(x) = \begin{cases} -x^2 + 3 & \text{if } x \ne 0 \\ 2 & \text{if } x = 0 \end{cases}$$

a. Compute $f'(x)$, and show that it changes sign from positive to negative as we move across $x = 0$.
b. Show that f does not have a relative maximum at $x = 0$. Does this contradict the First Derivative Test? Explain your answer.

66. **MAXIMIZING THE VOLUME OF A BOX** A box with an open top is to be constructed from a square piece of cardboard, 10 in. wide, by cutting out a square from each of the four corners and bending up the sides. What is the maximum volume of such a box?

67. **MINIMIZING CONSTRUCTION COSTS** A man wishes to construct a cylindrical barrel with a capacity of 32π ft^3. The cost per square foot of the material for the side of the barrel is half that of the cost per square foot for the top and bottom. Help him find the dimensions of the barrel that can be constructed at a minimum cost in terms of material used.

68. **PACKAGING** You wish to construct a closed rectangular box that has a volume of 4 ft^3. The length of the base of the box will be twice as long as its width. The material for the top and bottom of the box costs 30¢/ft^2. The material for the sides of the box costs 20¢/ft^2. Find the dimensions of the least expensive box that can be constructed.

69. **INVENTORY CONTROL AND PLANNING** Lehen Vinters imports a certain brand of beer. The demand, which may be assumed to be uniform, is 800,000 cases/year. The cost of ordering a shipment of beer is $500, and the cost of storing each case of beer for a year is $2. Determine how many cases of beer should be in each shipment if the ordering and storage costs are to be kept at a minimum. (Assume that each shipment of beer arrives just as the previous one has been sold.)

70. In what interval is the quadratic function

$$f(x) = ax^2 + bx + c \qquad (a \neq 0)$$

increasing? In what interval is f decreasing?

71. Let $f(x) = x^2 + ax + b$. Determine the constants a and b such that f has a relative minimum at $x = 2$ and the relative minimum value is 7.

72. Find the values of c such that the graph of

$$f(x) = x^4 + 2x^3 + cx^2 + 2x + 2$$

is concave upward everywhere.

73. Suppose that the point $(a, f(a))$ is an inflection point of the graph of $y = f(x)$. Show that the number a gives rise to a relative extremum of the function f'.

74. Let

$$f(x) = \begin{cases} x^3 + 1 & \text{if } x \neq 0 \\ 2 & \text{if } x = 0 \end{cases}$$

a. Compute $f'(x)$, and show that it does not change sign as we move across $x = 0$.

b. Show that f has a relative maximum at $x = 0$. Does this contradict the First Derivative Test? Explain your answer.

CHAPTER 10 Before Moving On ...

1. Find the interval(s) where $f(x) = \dfrac{x^2}{1 - x}$ is increasing and where it is decreasing.

2. Find the relative maxima and inflection point of the graph of $f(x) = 4xe^{-x}$.

3. Find the intervals where the graph of

$$f(x) = \frac{1}{3}x^3 - \frac{1}{4}x^2 - \frac{1}{2}x + 1$$

is concave upward, the intervals where the graph of f is concave downward, and the inflection point(s) of f.

4. Sketch the graph of $f(x) = 2x^3 - 9x^2 + 12x - 1$.

5. Find the absolute maximum and absolute minimum values of $f(x) = 2x^3 + 3x^2 - 1$ on the interval $[-2, 3]$.

6. An open bucket in the form of a right circular cylinder is to be constructed with a capacity of 1 ft³. Find the radius and height of the cylinder if the amount of material used is minimal.

11 Integration

DIFFERENTIAL CALCULUS focuses on the problem of finding the rate of change of one quantity with respect to another. In this chapter, we begin the study of the other branch of calculus, known as integral calculus. Here, we are interested in precisely the opposite problem: If we know the rate of change of one quantity with respect to another, can we find the relationship between the two quantities? The principal tool used in the study of integral calculus is the *antiderivative* of a function, and we develop rules for antidifferentiation, or *integration*, as the process of finding the antiderivative is called. We also show that a link is established between differential and integral calculus—via the Fundamental Theorem of Calculus.

How much electricity should be produced over the next 3 years to meet the projected demand? In Example 9, page 849, you will see how the current rate of consumption can be used to answer this question.

© Songquan Deng/Shutterstock.com

11.1 Antiderivatives and the Rules of Integration

Antiderivatives

Let's return once again to the example involving the motion of the maglev (Figure 1).

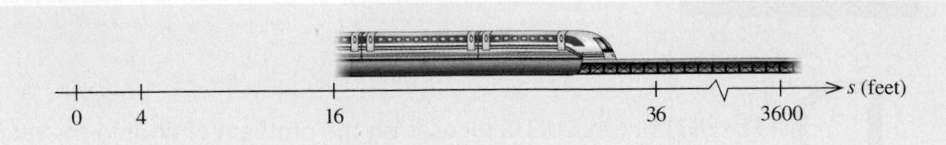

FIGURE 1
A maglev moving along an elevated monorail track

In Chapter 9, we discussed the following problem:

If we know the position of the maglev at any time t, can we find its velocity at time t?

As it turns out, if the position of the maglev is described by the position function f, then its velocity at any time t is given by $f'(t)$. Here, f'—the velocity function of the maglev—is just the derivative of f.

Now, in this chapter, we will consider precisely the opposite problem:

If we know the velocity of the maglev at any time t, can we find its position at time t?

Stated another way, if we know the velocity function f' of the maglev, can we find its position function f?

To solve this problem, we need the concept of an antiderivative of a function.

> **Antiderivative**
>
> A function F is an **antiderivative** of f on an interval I if $F'(x) = f(x)$ for all x in I.

Thus, an antiderivative of a function f is a function F whose derivative is f. For example, $F(x) = x^2$ is an antiderivative of $f(x) = 2x$ because

$$F'(x) = \frac{d}{dx}(x^2) = 2x = f(x)$$

and $F(x) = x^3 + 2x + 1$ is an antiderivative of $f(x) = 3x^2 + 2$ because

$$F'(x) = \frac{d}{dx}(x^3 + 2x + 1) = 3x^2 + 2 = f(x)$$

EXAMPLE 1 Let $F(x) = \frac{1}{3}x^3 - 2x^2 + x - 1$. Show that F is an antiderivative of $f(x) = x^2 - 4x + 1$.

Solution Differentiating the function F, we obtain

$$F'(x) = x^2 - 4x + 1 = f(x)$$

and the desired result follows. ■

EXAMPLE 2 Let $F(x) = x$, $G(x) = x + 2$, and $H(x) = x + C$, where C is a constant. Show that F, G, and H are all antiderivatives of the function f defined by $f(x) = 1$.

Solution Since

$$F'(x) = \frac{d}{dx}(x) = 1 = f(x)$$

$$G'(x) = \frac{d}{dx}(x + 2) = 1 = f(x)$$

$$H'(x) = \frac{d}{dx}(x + C) = 1 = f(x)$$

we see that F, G, and H are indeed antiderivatives of f.

Example 2 shows that once an antiderivative G of a function f is known, then another antiderivative of f may be found by adding an arbitrary constant to the function G. The following theorem states that no function other than one obtained in this manner can be an antiderivative of f. (We omit the proof.)

THEOREM 1

Let G be an antiderivative of a function f on an interval I. Then, every antiderivative F of f on I must be of the form $F(x) = G(x) + C$, where C is a constant.

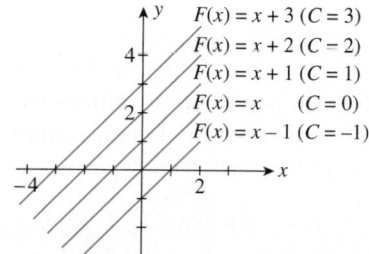

$F(x) = x + 3 \ (C = 3)$
$F(x) = x + 2 \ (C = 2)$
$F(x) = x + 1 \ (C = 1)$
$F(x) = x \quad (C = 0)$
$F(x) = x - 1 \ (C = -1)$

FIGURE 2
The graphs of some antiderivatives of
$f(x) = 1$

Returning to Example 2, we see that there are infinitely many antiderivatives of the function $f(x) = 1$. We obtain each one by specifying the constant C in the function $F(x) = x + C$. Figure 2 shows the graphs of some of these antiderivatives for selected values of C. These graphs constitute part of a family of infinitely many parallel straight lines, each having a slope equal to 1. This result is expected, since there are infinitely many curves (straight lines) with a given slope equal to 1. The antiderivatives $F(x) = x + C$ (C, a constant) are precisely the functions representing this family of straight lines.

EXAMPLE 3 Prove that the function $G(x) = x^2$ is an antiderivative of the function $f(x) = 2x$. Write a general expression for the antiderivatives of f.

Solution Since $G'(x) = 2x = f(x)$, we have shown that $G(x) = x^2$ is an antiderivative of $f(x) = 2x$. By Theorem 1, every antiderivative of the function $f(x) = 2x$ has the form $F(x) = x^2 + C$, where C is some constant. The graphs of a few of the antiderivatives of f are shown in Figure 3.

$F(x) = x^2 + 1 \ (C = 1)$
$F(x) = x^2 \quad (C = 0)$
$F(x) = x^2 - \frac{1}{2}\left(C = -\frac{1}{2}\right)$

FIGURE 3
The graphs of some antiderivatives of
$f(x) = 2x$

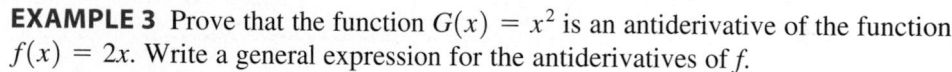

Exploring with TECHNOLOGY

Let $f(x) = x^2 - 1$.

1. Show that $F(x) = \frac{1}{3}x^3 - x + C$, where C is an arbitrary constant, is an antiderivative of f.

2. Use a graphing utility to plot the graphs of the antiderivatives of f corresponding to $C = -2$, $C = -1$, $C = 0$, $C = 1$, and $C = 2$ on the same set of axes, using the viewing window $[-4, 4] \times [-4, 4]$.

3. If your graphing utility has the capability, draw the tangent line to each of the graphs in part 2 at the point whose x-coordinate is 2. What can you say about this family of tangent lines?

4. What is the slope of a tangent line in this family? Explain how you obtained your answer.

The Indefinite Integral

The process of finding all antiderivatives of a function is called **antidifferentiation,** or **integration.** We use the symbol $\int$, called an **integral sign,** to indicate that the operation of integration is to be performed on some function f. Thus,

$$\int f(x)\,dx = F(x) + C$$

[read "the indefinite integral of f of x with respect to x equals F of x plus C"] tells us that the **indefinite integral** of f is the family of functions given by $F(x) + C$, where $F'(x) = f(x)$. The function f to be integrated is called the **integrand,** and the constant C is called a **constant of integration.** The expression dx following the integrand $f(x)$ reminds us that the operation is performed with respect to x. If the independent variable is t, we write $\int f(t)\,dt$ instead. In this sense, both t and x are "dummy variables."

Using this notation, we can write the results of Examples 2 and 3 as

$$\int 1\,dx = x + C \quad\text{and}\quad \int 2x\,dx = x^2 + K$$

where C and K are arbitrary constants.

Basic Integration Rules

Our next task is to develop some rules for finding the indefinite integral of a given function f. Because integration and differentiation are reverse operations, we discover many of the rules of integration by first making an "educated guess" at the antiderivative F of the function f to be integrated. Then this result is verified by demonstrating that $F' = f$.

> **Rule 1: The Indefinite Integral of a Constant**
>
> $$\int k\,dx = kx + C \qquad (k,\text{ a constant})$$

To prove this result, observe that

$$F'(x) = \frac{d}{dx}(kx + C) = k$$

EXAMPLE 4 Find each of the following indefinite integrals:

a. $\displaystyle\int 2\,dx$ **b.** $\displaystyle\int \pi^2\,dx$

Solution Each of the integrands has the form $f(x) = k$, where k is a constant. Applying Rule 1 in each case yields

a. $\displaystyle\int 2\,dx = 2x + C$

b. $\displaystyle\int \pi^2\,dx = \pi^2 x + C$

Next, from the rule of differentiation,

$$\frac{d}{dx}x^n = nx^{n-1}$$

we obtain the following rule of integration.

Rule 2: The Power Rule

$$\int x^n \, dx = \frac{1}{n+1} x^{n+1} + C \qquad (n \neq -1)$$

An antiderivative of a power function is another power function obtained from the integrand by increasing its power by 1 and dividing the resulting expression by the new power.

To prove this result, observe that

$$F'(x) = \frac{d}{dx}\left(\frac{1}{n+1} x^{n+1} + C\right)$$
$$= \frac{n+1}{n+1} x^n$$
$$= x^n$$
$$= f(x)$$

EXAMPLE 5 Find each of the following indefinite integrals:

a. $\displaystyle\int x^3 \, dx$ **b.** $\displaystyle\int x^{3/2} \, dx$ **c.** $\displaystyle\int \frac{1}{x^{3/2}} \, dx$

Solution Each integrand is a power function with exponent $n \neq -1$. Applying Rule 2 in each case yields the following results:

a. $\displaystyle\int x^3 \, dx = \frac{1}{4} x^4 + C$

b. $\displaystyle\int x^{3/2} \, dx = \frac{1}{\frac{5}{2}} x^{5/2} + C = \frac{2}{5} x^{5/2} + C$

c. $\displaystyle\int \frac{1}{x^{3/2}} \, dx = \int x^{-3/2} \, dx = \frac{1}{-\frac{1}{2}} x^{-1/2} + C = -2x^{-1/2} + C = -\frac{2}{x^{1/2}} + C$

These results may be verified by differentiating each of the antiderivatives and showing that the result is equal to the corresponding integrand.

The next rule tells us that a constant factor may be moved through an integral sign.

Rule 3: The Indefinite Integral of a Constant Multiple of a Function

$$\int cf(x) \, dx = c \int f(x) \, dx \qquad (c, \text{ a constant})$$

The indefinite integral of a constant multiple of a function is equal to the constant multiple of the indefinite integral of the function.

This result follows from the corresponding rule of differentiation (see Rule 3, Section 9.4).

⚠️ Only a constant can be "moved out" of an integral sign. For example, it would be incorrect to write

$$\int x^2 \, dx = x^2 \int 1 \, dx$$

In fact, $\int x^2 \, dx = \frac{1}{3} x^3 + C$, whereas $x^2 \int 1 \, dx = x^2(x + C) = x^3 + Cx^2$.

EXAMPLE 6 Find each of the following indefinite integrals:

a. $\int 2t^3\, dt$ **b.** $\int -3x^{-2}\, dx$

Solution Each integrand has the form $cf(x)$, where c is a constant. Applying Rule 3, we obtain

a. $\int 2t^3\, dt = 2 \int t^3\, dt = 2\left(\frac{1}{4}t^4 + K\right) = \frac{1}{2}t^4 + 2K = \frac{1}{2}t^4 + C$

where $C = 2K$. From now on, we will write the constant of integration as C, since any nonzero multiple of an arbitrary constant is an arbitrary constant.

b. $\int -3x^{-2}\, dx = -3 \int x^{-2}\, dx = (-3)(-1)x^{-1} + C = \frac{3}{x} + C$ ∎

Rule 4: The Sum Rule

$$\int [f(x) + g(x)]\, dx = \int f(x)\, dx + \int g(x)\, dx$$

$$\int [f(x) - g(x)]\, dx = \int f(x)\, dx - \int g(x)\, dx$$

The indefinite integral of a sum (difference) of two functions is equal to the sum (difference) of their indefinite integrals.

This result is easily extended to the case involving the sum and difference of any finite number of functions. As in Rule 3, the proof of Rule 4 follows from the corresponding rule of differentiation (see Rule 4, Section 9.4).

EXAMPLE 7 Find the indefinite integral

$$\int (3x^5 + 4x^{3/2} - 2x^{-1/2})\, dx$$

Solution Applying the extended version of Rule 4, we find that

$$\int (3x^5 + 4x^{3/2} - 2x^{-1/2})\, dx$$

$$= \int 3x^5\, dx + \int 4x^{3/2}\, dx - \int 2x^{-1/2}\, dx$$

$$= 3 \int x^5\, dx + 4 \int x^{3/2}\, dx - 2 \int x^{-1/2}\, dx \qquad \text{Rule 3}$$

$$= (3)\left(\frac{1}{6}\right)x^6 + (4)\left(\frac{2}{5}\right)x^{5/2} - (2)(2)x^{1/2} + C \qquad \text{Rule 2}$$

$$= \frac{1}{2}x^6 + \frac{8}{5}x^{5/2} - 4x^{1/2} + C \qquad \blacksquare$$

Observe that we have combined the three constants of integration, which arise from evaluating the three indefinite integrals, to obtain one constant C. After all, the sum of three arbitrary constants is also an arbitrary constant.

Rule 5: The Indefinite Integral of the Exponential Function

$$\int e^x\, dx = e^x + C$$

The indefinite integral of the exponential function with base e is equal to the function itself (except, of course, for the constant of integration).

EXAMPLE 8 Find the indefinite integral

$$\int (2e^x - x^3)\, dx$$

Solution We have

$$\int (2e^x - x^3)\, dx = \int 2e^x\, dx - \int x^3\, dx$$

$$= 2 \int e^x\, dx - \int x^3\, dx$$

$$= 2e^x - \frac{1}{4} x^4 + C$$

The last rule of integration in this section covers the integration of the function $f(x) = x^{-1}$. Remember that this function constituted the only exceptional case in the integration of the power function $f(x) = x^n$ (see Rule 2).

Rule 6: The Indefinite Integral of the Function $f(x) = x^{-1}$

$$\int x^{-1}\, dx = \int \frac{1}{x}\, dx = \ln|x| + C \qquad (x \neq 0)$$

To prove Rule 6, observe that

$$\frac{d}{dx} \ln|x| = \frac{1}{x} \qquad \text{See Rule 10, Section 9.7.}$$

EXAMPLE 9 Find the indefinite integral

$$\int \left(2x + \frac{3}{x} + \frac{4}{x^2} \right) dx$$

Solution

$$\int \left(2x + \frac{3}{x} + \frac{4}{x^2} \right) dx = \int 2x\, dx + \int \frac{3}{x}\, dx + \int \frac{4}{x^2}\, dx$$

$$= 2 \int x\, dx + 3 \int \frac{1}{x}\, dx + 4 \int x^{-2}\, dx$$

$$= 2 \left(\frac{1}{2} \right) x^2 + 3 \ln|x| + 4(-1)x^{-1} + C$$

$$= x^2 + 3 \ln|x| - \frac{4}{x} + C$$

Differential Equations

Let's return to the problem posed at the beginning of the section: *Given the derivative of a function, f', can we find the function f?* As an example, suppose we are given the function

$$f'(x) = 2x - 1 \tag{1}$$

and we wish to find $f(x)$. From what we now know, we can find f by integrating both sides of Equation (1). Thus,

$$f(x) = \int f'(x)\,dx = \int (2x - 1)\,dx = x^2 - x + C \tag{2}$$

where C is an arbitrary constant. Thus, infinitely many functions have the derivative f', each differing from the other by a constant.

Equation (1) is called a differential equation. In general, a **differential equation** is an equation that involves the derivative or differential of an unknown function. [In the case of Equation (1), the unknown function is f.] A **solution** of a differential equation is any function that satisfies the differential equation. Thus, Equation (2) gives *all* the solutions of the differential Equation (1), and it is, accordingly, called the **general solution** of the differential equation $f'(x) = 2x - 1$.

The graphs of $f(x) = x^2 - x + C$ for selected values of C are shown in Figure 4. These graphs have one property in common: For any fixed value of x, the tangent lines to these graphs have the same slope. This follows because any member of the family $f(x) = x^2 - x + C$ must have the same slope at x—namely, $2x - 1$!

Although there are infinitely many solutions to the differential equation $f'(x) = 2x - 1$, we can obtain a **particular solution** by specifying the value the function must assume at a certain value of x. For example, suppose we stipulate that the function f under consideration must satisfy the condition $f(1) = 3$ or, equivalently, the graph of f must pass through the point $(1, 3)$. Then, using the condition on the general solution $f(x) = x^2 - x + C$, we find that

$$f(1) = 1 - 1 + C = 3$$

and $C = 3$. Thus, the particular solution is $f(x) = x^2 - x + 3$ (see Figure 4).

The condition $f(1) = 3$ is an example of an initial condition. More generally, an **initial condition** is a condition imposed on the value of f at $x = a$.

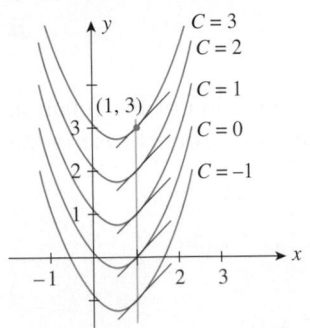

FIGURE 4
The graphs of some of the functions having the derivative $f'(x) = 2x - 1$. Observe that the slopes of the tangent lines to the graphs are the same for a fixed value of x.

Initial-Value Problems

An **initial-value problem** is one in which we are required to find a function satisfying (1) a differential equation and (2) one or more initial conditions. The following are examples of initial-value problems.

EXAMPLE 10 Find the function f if it is known that

$$f'(x) = 3x^2 - 4x + 8 \quad \text{and} \quad f(1) = 9$$

Solution We are required to solve the initial-value problem

$$\begin{cases} f'(x) = 3x^2 - 4x + 8 \\ f(1) = \qquad\quad 9 \end{cases}$$

Integrating the function f', we find

$$\begin{aligned} f(x) &= \int f'(x)\,dx \\ &= \int (3x^2 - 4x + 8)\,dx \\ &= x^3 - 2x^2 + 8x + C \end{aligned}$$

Using the condition $f(1) = 9$, we have

$$9 = f(1) = 1^3 - 2(1)^2 + 8(1) + C = 7 + C \quad \text{or} \quad C = 2$$

Therefore, the required function f is given by $f(x) = x^3 - 2x^2 + 8x + 2$. ∎

APPLIED EXAMPLE 11 Velocity of a Maglev In a test run of a maglev along a straight elevated monorail track, data obtained from reading its speedometer indicate that the velocity of the maglev at time t can be described by the velocity function

$$v(t) = 8t \qquad (0 \le t \le 30)$$

Find the position function of the maglev. Assume that initially the maglev is located at the origin of a coordinate line.

Solution Let $s(t)$ denote the position of the maglev at any time t ($0 \le t \le 30$). Then, $s'(t) = v(t)$. So we have the initial-value problem

$$\begin{cases} s'(t) = 8t \\ s(0) = 0 \end{cases}$$

Integrating both sides of the differential equation $s'(t) = 8t$, we obtain

$$s(t) = \int s'(t)\, dt = \int 8t\, dt = 4t^2 + C$$

where C is an arbitrary constant. To evaluate C, we use the initial condition $s(0) = 0$ to write

$$s(0) = 4(0)^2 + C - 0 \quad \text{or} \quad C = 0$$

Therefore, the required position function is $s(t) = 4t^2$ ($0 \le t \le 30$). ■

APPLIED EXAMPLE 12 Magazine Circulation The current circulation of the *Investor's Digest* is 3000 copies per week. The managing editor of this weekly projects a growth rate of

$$4 + 5t^{2/3}$$

copies per week, t weeks from now, for the next 3 years. On the basis of her projection, what will be the circulation of the digest 125 weeks from now?

Solution Let $S(t)$ denote the circulation of the digest t weeks from now. Then $S'(t)$ is the rate of change in the circulation in the tth week and is given by

$$S'(t) = 4 + 5t^{2/3}$$

Furthermore, the current circulation of 3000 copies per week translates into the initial condition $S(0) = 3000$. Integrating both sides of the differential equation with respect to t gives

$$S(t) = \int S'(t)\, dt = \int (4 + 5t^{2/3})\, dt$$

$$= 4t + 5\left(\frac{t^{5/3}}{\frac{5}{3}}\right) + C = 4t + 3t^{5/3} + C$$

To determine the value of C, we use the condition $S(0) = 3000$ to write

$$S(0) = 4(0) + 3(0)^{5/3} + C = 3000$$

which gives $C = 3000$. Therefore, the circulation of the digest t weeks from now will be

$$S(t) = 4t + 3t^{5/3} + 3000$$

In particular, the circulation 125 weeks from now will be

$$S(125) = 4(125) + 3(125)^{5/3} + 3000 = 12{,}875$$

copies per week.

11.1 Self-Check Exercises

1. Evaluate $\int \left(\dfrac{1}{\sqrt{x}} - \dfrac{2}{x} + 3e^x \right) dx$.

2. Find the rule for the function f given that (1) the slope of the tangent line to the graph of f at any point $P(x, f(x))$ is given by the expression $3x^2 - 6x + 3$ and (2) the graph of f passes through the point $(2, 9)$.

3. **MARKET SHARE OF AN AUTO COMPANY** Suppose United Motors' share of the new cars sold in a certain country is changing at the rate of

$$f(t) = -0.01875t^2 + 0.15t - 1.2 \qquad (0 \leq t \leq 12)$$

percent/year at year t ($t = 0$ corresponds to the beginning of 2002). The company's market share at the beginning of 2002 was 48.4%. What was United Motors' market share at the beginning of 2014?

Solutions to Self-Check Exercises 11.1 can be found on page 821.

11.1 Concept Questions

1. What is an antiderivative? Give an example.

2. If $f'(x) = g'(x)$ for all x in an interval I, what is the relationship between f and g?

3. What is the difference between an antiderivative of f and the indefinite integral of f?

4. Can the Power Rule be used to integrate $\int \dfrac{1}{x} \, dx$? Explain your answer.

11.1 Exercises

In Exercises 1–4, verify directly that F is an antiderivative of f.

1. $F(x) = \dfrac{1}{3}x^3 + 2x^2 - x + 2; f(x) = x^2 + 4x - 1$

2. $F(x) = xe^x + \pi; f(x) = e^x(1 + x)$

3. $F(x) = \sqrt{2x^2 - 1}; f(x) = \dfrac{2x}{\sqrt{2x^2 - 1}}$

4. $F(x) = x \ln x - x; f(x) = \ln x$

In Exercises 5–8, (a) verify that G is an antiderivative of f, (b) find all antiderivatives of f, and (c) sketch the graphs of a few members of the family of antiderivatives found in part (b).

5. $G(x) = 2x; f(x) = 2$
6. $G(x) = 2x^2; f(x) = 4x$

7. $G(x) = \dfrac{1}{3}x^3; f(x) = x^2$
8. $G(x) = e^x; f(x) = e^x$

In Exercises 9–50, find the indefinite integral.

9. $\int 6 \, dx$
10. $\int \sqrt{2} \, dx$

11. $\int x^3 \, dx$
12. $\int 2x^5 \, dx$

13. $\int x^{-4} \, dx$
14. $\int 3t^{-7} \, dt$

15. $\int x^{2/3} \, dx$
16. $\int 2u^{3/4} \, du$

17. $\int x^{-5/4} \, dx$
18. $\int 3x^{-2/3} \, dx$

19. $\int \dfrac{2}{x^3} \, dx$
20. $\int \dfrac{1}{3x^5} \, dx$

21. $\int \pi \sqrt{t} \, dt$
22. $\int \dfrac{3}{\sqrt{t}} \, dt$

23. $\int (3 - 4x) \, dx$
24. $\int (1 + u + u^2) \, du$

25. $\int (x^2 + x + x^{-3}) \, dx$
26. $\int (0.3t^2 + 0.02t + 2) \, dt$

27. $\int 5e^x \, dx$
28. $\int (1 + e^x) \, dx$

29. $\int (1 + x + e^x) \, dx$
30. $\int (2 + x + 2x^2 + e^x) \, dx$

31. $\int \left(4x^3 - \dfrac{2}{x^2} - 1 \right) dx$
32. $\int \left(6x^3 + \dfrac{3}{x^2} - x \right) dx$

33. $\int (x^{5/2} + 2x^{3/2} - x)\, dx$ **34.** $\int (t^{3/2} + 2t^{1/2} - 4t^{-1/2})\, dt$

35. $\int \left(\sqrt{x} + \dfrac{2}{\sqrt{x}} \right) dx$ **36.** $\int \left(\sqrt[3]{x^2} - \dfrac{1}{x^2} \right) dx$

37. $\int \left(\dfrac{u^3 + 2u^2 - u}{3u} \right) du$

Hint: $\dfrac{u^3 + 2u^2 - u}{3u} = \dfrac{1}{3}u^2 + \dfrac{2}{3}u - \dfrac{1}{3}$

38. $\int \dfrac{x^4 - 1}{x^2}\, dx$

Hint: $\dfrac{x^4 - 1}{x^2} = x^2 - x^{-2}$

39. $\int (2t + 1)(t - 2)\, dt$ **40.** $\int u^{-2}(1 - u^2 + u^4)\, du$

41. $\int \dfrac{1}{x^2}(x^4 - 2x^2 + 1)\, dx$ **42.** $\int \sqrt{t}\,(t^2 + t - 1)\, dt$

43. $\int \dfrac{ds}{(s + 1)^{-2}}$ **44.** $\int \left(\sqrt{x} + \dfrac{3}{x} - 2e^x \right) dx$

45. $\int (e^t + t^e)\, dt$ **46.** $\int \left(\dfrac{1}{x^2} - \dfrac{1}{\sqrt[3]{x^2}} + \dfrac{1}{\sqrt{x}} \right) dx$

47. $\int \dfrac{x^3 + x^2 - x + 1}{x^2}\, dx$

Hint: Simplify the integrand first.

48. $\int \dfrac{t^3 + \sqrt[3]{t}}{t^2}\, dt$

Hint: Simplify the integrand first.

49. $\int \dfrac{(\sqrt{x} - 1)^2}{x^2}\, dx$

Hint: Simplify the integrand first.

50. $\int (x + 1)^2 \left(1 - \dfrac{1}{x} \right) dx$

Hint: Simplify the integrand first.

In Exercises 51–58, find $f(x)$ by solving the initial-value problem.

51. $f'(x) = 3x + 1; f(1) = 3$

52. $f'(x) = 3x^2 - 6x; f(2) = 4$

53. $f'(x) = 3x^2 + 4x - 1; f(2) = 9$

54. $f'(x) = \dfrac{1}{\sqrt{x}}; f(4) = 2$

55. $f'(x) = 1 + \dfrac{1}{x^2}; f(1) = 3$

56. $f'(x) = e^x - 2x; f(0) = 2$

57. $f'(x) = \dfrac{x + 1}{x}; f(1) = 1$

58. $f'(x) = 1 + e^x + \dfrac{1}{x}; f(1) = 3 + e$

In Exercises 59–62, find the function f given that the slope of the tangent line to the graph of f at any point $(x, f(x))$ is $f'(x)$ and that the graph of f passes through the given point.

59. $f'(x) = \dfrac{1}{2}x^{-1/2}; (2, \sqrt{2})$

60. $f'(t) = t^2 - 2t + 3; (1, 2)$

61. $f'(x) = e^x + x; (0, 3)$ **62.** $f'(x) = \dfrac{2}{x} + 1; (1, 2)$

63. Bank Deposits Madison Finance opened two branches on September 1 ($t = 0$). Branch A is located in an established industrial park, and Branch B is located in a fast-growing new development. The net rates at which money was deposited into Branch A and Branch B in the first 180 business days are given by the graphs of f and g, respectively (see the figure). Which branch has the larger amount on deposit at the end of 180 business days? Justify your answer.

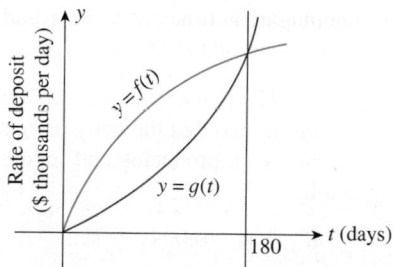

64. Velocity of a Car Two cars, side by side, start from rest and travel along a straight road. The velocity of Car A is given by $v = f(t)$, and the velocity of Car B is given by $v = g(t)$. The graphs of f and g are shown in the figure below. Are the cars still side by side after T sec? If not, which car is ahead of the other? Justify your answer.

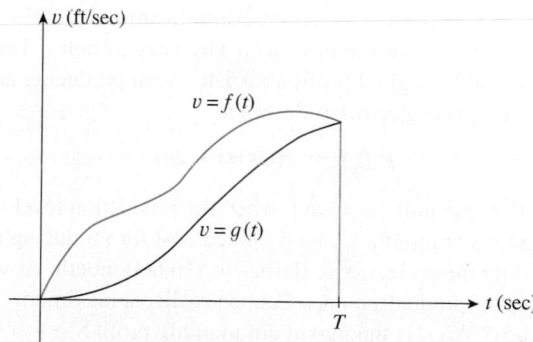

65. U.S. Smartphone Users The number of smartphone users and penetration in the United States continues to grow steadily. The number of users (in millions) from 2011 through 2015 is projected to grow at the rate of

$$R(t) = 14.3 \qquad (0 \le t \le 4)$$

million/year. The number of users in 2011 ($t = 0$) was 90.1 million. Find an expression giving the projected number of smartphone users in year t. What is the estimated number of smartphone users in 2015?
Source: eMarketer.

66. Households Owning More Than One TV Television is the most popular mass medium in the United States. The percentage of households with multiple TV sets from the year 2000 through 2010 grew at the approximate rate of

$$R(t) = 0.7 \qquad (0 \le t \le 10)$$

percent/year in year t. The percentage of American households owning multiple TV sets stood at 75% in the year 2000 ($t = 0$). Find an expression giving the approximate percentage of households owning multiple TV sets in year t. What percentage of households owned multiple TV sets in 2010?

67. Velocity of a Maglev The velocity (in feet per second) at time t of a maglev is

$$v(t) = 0.2t + 3 \qquad (0 \le t \le 120)$$

At $t = 0$, it is at the station. Find the function giving the position of the maglev at time t, assuming that the motion takes place along a straight stretch of track.

68. Revenue Functions The management of Lorimar Watch Company has determined that the daily marginal revenue function associated with producing and selling their travel clocks is given by

$$R'(x) = -0.009x + 12$$

where x denotes the number of units produced and sold and $R'(x)$ is measured in dollars per unit.
  a. Determine the revenue function $R(x)$ associated with producing and selling these clocks.
  b. What is the demand equation that relates the whole-sale unit price with the quantity of travel clocks demanded?

69. Profit Functions Cannon Precision Instruments makes an automatic electronic flash with Thyrister circuitry. The estimated marginal profit associated with producing and selling these electronic flashes is

$$P'(x) = -0.004x + 20$$

dollars per unit per month when the production level is x units per month. Cannon's fixed cost for producing and selling these electronic flashes is $16,000/month. At what level of production does Cannon realize a maximum profit? What is the maximum monthly profit?

70. Cost of Producing Guitars Carlota Music Company estimates that the marginal cost of manufacturing its Professional Series guitars is

$$C'(x) = 0.002x + 100$$

dollars/month when the level of production is x guitars/month. The fixed costs incurred by Carlota are $4000/month. Find the total monthly cost incurred by Carlota in manufacturing x guitars/month.

71. Wind Energy in China China's push to install more wind energy capacity has started paying off. As of 2012, wind energy had become the country's third largest source of energy, after coal and hydroelectric power. The amount of wind energy generated in China from 2005 through 2012 grew at the rate of

$$r(t) = 5.018t - 3.204 \qquad (0 \le t \le 7)$$

terawatt-hours/year, in year t, where $t = 0$ corresponds to 2005. The amount of wind energy generated in 2005 stood at 1.8 terawatt-hours.
  a. Find an expression giving the amount of wind energy generated in year t.
  b. How much wind energy was generated in 2012?
  c. If the trend continued into 2013, how much wind energy was generated in China in that year?
  Source: earth-policy.org.

72. Growth of National Health Costs National health expenditures are projected to grow at the rate of

$$r(t) = 0.0058t + 0.159 \qquad (0 \le t \le 13)$$

trillion dollars/year from 2002 through 2015. Here, $t = 0$ corresponds to 2002. The expenditure in 2002 was $1.60 trillion.
  a. Find a function f giving the projected national health expenditures in year t.
  b. What does your model project the national health expenditure to be in 2015?
  Source: National Health Expenditures.

73. Genetically Modified Crops The total number of acres of genetically modified crops grown in developing countries from 2006 through 2012 was changing at the rate of

$$R(t) = 150t + 14.82 \qquad (0 \le t \le 6)$$

million acres/year. The total number of acres of such crops grown in 2006 ($t = 0$) was 27.2 million acres. How many acres of genetically modified crops were grown in developing countries in 2012?
  Source: Clive James/ISAAA.

74. Velocity of a Car The velocity of a car (in feet per second) t sec after starting from rest is given by the function

$$f(t) = 2\sqrt{t} \qquad (0 \le t \le 30)$$

Find the car's position, $s(t)$, at any time t. Assume that $s(0) = 0$.

75. Ballast Dropped from a Balloon A ballast is dropped from a stationary hot-air balloon that is hovering at an altitude of 400 ft. The velocity of the ballast after t sec is $(-32t + 4)$ ft/sec.
  a. Find the height $h(t)$ of the ballast from the ground at time t.
     Hint: $h'(t) = -32t + 4$ and $h(0) = 400$.
  b. When will the ballast strike the ground?

c. Find the velocity of the ballast when it hits the ground.

Ballast

76. **POPULATION GROWTH** The development of AstroWorld ("The Amusement Park of the Future") on the outskirts of a city will increase the city's population at the rate of

$$4500\sqrt{t} + 1000$$

people/year, t years from the start of construction. The population before construction is 30,000. Determine the projected population 9 years after construction of the park has begun.

77. **CABLE TV SUBSCRIBERS** A study conducted by TeleCable estimates that the number of cable TV subscribers will grow at the rate of

$$100 + 210t^{3/4}$$

new subscribers/month, t months from the start date of the service. If 5000 subscribers signed up for the service before the starting date, how many subscribers will there be 16 months from that date?

78. **BLOOD FLOW IN AN ARTERY** Nineteenth century physician Jean Louis Marie Poiseuille discovered that the rate of change of the velocity of blood r cm from the central axis of an artery (in centimeters per second per centimeter) is given by

$$a(r) = -kr$$

where k is a constant. If the radius of an artery is R cm, find an expression for the velocity of blood as a function of r (see the accompanying figure).
Hint: $v'(r) = a(r)$ and $v(R) = 0$. (Why?)

Blood vessel

79. **FLIGHT OF A ROCKET** The velocity, in feet per second, of a rocket t sec into vertical flight is given by

$$v(t) = -3t^2 + 192t$$

Find an expression $h(t)$ that gives the rocket's altitude, in feet, t sec after liftoff. What is the altitude of the rocket 30 sec after liftoff?
Hint: $h'(t) = v(t); h(0) = 0$.

80. **QUALITY CONTROL** As part of a quality-control program, the chess sets manufactured by Jones Brothers are subjected to a final inspection before packing. The rate of increase in the number of sets checked per hour by an inspector t hr into the 8 A.M. to 12 noon shift is approximately

$$N'(t) = -3t^2 + 12t + 45 \qquad (0 \le t \le 4)$$

a. Find an expression $N(t)$ that approximates the number of sets inspected at the end of t hours.
Hint: $N(0) = 0$.

b. How many sets does the average inspector check during a morning shift?

81. **COST OF PRODUCING CLOCKS** Lorimar Watch Company manufactures travel clocks. The daily marginal cost function associated with producing these clocks is

$$C'(x) = 0.000009x^2 - 0.009x + 8$$

where $C'(x)$ is measured in dollars per unit and x denotes the number of units produced. Management has determined that the daily fixed cost incurred in producing these clocks is \$120. Find the total cost incurred by Lorimar in producing the first 500 travel clocks per day.

82. **RISK OF DOWN SYNDROME** The rate at which the risk of Down syndrome is changing is approximated by the function

$$r(x) = 0.0046411x^2 - 0.3012x + 4.9 \qquad (20 \le x \le 45)$$

where $r(x)$ is measured in percentage of all births per year and x is the maternal age at delivery.

a. Find a function f giving the risk as a percentage of all births when the maternal age at delivery is x years, given that the risk of Down syndrome at age 30 is 0.14% of all births.

b. On the basis of this model, what is the risk of Down syndrome when the maternal age at delivery is 40 years? 45 years?
Source: New England Journal of Medicine.

83. **AMOUNT OF RAINFALL** During a thunderstorm, rain was falling at the rate of

$$8\sqrt{2t} - 32t^3 \qquad (0 \le t \le 0.6)$$

in./hr.

a. Find an expression giving the total amount of rainfall after t hr.
Hint: The total amount of rainfall at $t = 0$ is zero.

b. How much rain had fallen after $\frac{1}{2}$ hr?

84. **SALARIES OF MARRIED WOMEN** The percentage of married women who earned more than their husbands over the period from 1960 ($t = 0$) through 2011 ($t = 51$) was growing at the rate of approximately

$$r(t) = -0.00025142t^2 + 0.02116t + 0.0328$$
$$(0 < t \le 51)$$

percent/year in year t. The percentage of married women earning more than their husbands in 1960 stood at 6.2%.
   a. Find an expression giving the approximate percentage of married women who earned more than their husbands in year t.
   b. Use the result of part (a) to estimate the percentage of married women who earned more than their husbands in 2013, assuming that the trend continued.
   Source: Pew Research Center.

85. **SOCIAL NETWORKS** The percentage of people age 12 years and older using social network sites and/or services "several" times a day from 2009 through 2013 grew at the rate of

$$R(t) = 5.92t^{-0.158} \qquad (1 \le t \le 5)$$

percent/year in year t, with $t = 1$ corresponding to 2009. The percentage of people age 12 years and older using social network sites and/or services in 2009 was 7%.
   a. Find an expression giving the percentage of people age 12 years and older using social network sites and/or services in year t ($1 \le t \le 5$).
   b. According to this model, what percentage of people age 12 years and older used social network sites and/or services in 2013?
   Source: Arbitron and Edison Research.

86. **COAL EXPORTS** The U.S. coal industry, under increasing pressure from tougher antipollution rules, is ratcheting up its export business. The rate of change of coal exports from 2010 through 2012 is given by

$$f(t) = 31.863t^{-0.61} \qquad (0 \le t \le 2)$$

million short tons (2000 lb) per year in year t, where t is measured in years, with $t = 0$ corresponding to 2010. U.S. coal exports in 2010 were 81.7 million short tons.
   a. Find an expression for U.S. coal exports in year t.
   b. Assuming that the trend continued through 2013, what were U.S. coal exports for that year?
   Source: U.S. Department of Energy.

87. **SURFACE AREA OF A HUMAN** Empirical data suggest that the surface area of a 180-cm-tall human body changes at the rate of

$$S'(W) = 0.131773W^{-0.575}$$

m²/kg, where W is the mass of the body in kilograms. If the surface area of a 180-cm-tall human body weighing

70 kg is 1.886277 m², what is the surface area of a human body of the same height weighing 75 kg?

88. **OZONE POLLUTION** The rate of change of the level of ozone, an invisible gas that is an irritant and impairs breathing, present in the atmosphere on a certain May day in the city of Riverside is given by

$$R(t) = 3.2922t^2 - 0.366t^3 \qquad (0 \le t \le 11)$$

(measured in pollutant standard index per hour). Here, t is measured in hours, with $t = 0$ corresponding to 7 A.M. Find the ozone level $A(t)$ at any time t, assuming that at 7 A.M. it is zero.
   Hint: $A'(t) = R(t)$ and $A(0) = 0$.
   Source: Los Angeles Times.

89. **HEIGHTS OF CHILDREN** According to the Jenss model for predicting the height of preschool children, the rate of growth of a typical infant is

$$R(t) = 1.0490t^4 - 4.2255t^3 + 12.7659t^2$$
$$- 25.7119t + 32.28 \qquad (\tfrac{1}{4} \le t \le 1)$$

cm/year, where t is measured in years. The height of a typical 3-month-old infant is 60.30 cm.
   a. What is the height of a typical infant at age t?
   b. Use the result of part (a) to estimate the height of a typical 1-year-old child.

90. **ACCELERATION OF A CAR** A car traveling along a straight road at 66 ft/sec accelerated to a speed of 88 ft/sec over a distance of 440 ft. What was the acceleration of the car, assuming that it was constant?

91. **DECELERATION OF A CAR** What constant deceleration would a car moving along a straight road have to be subjected to if it were brought to rest from a speed of 88 ft/sec in 9 sec? What would be the stopping distance?

92. **CARRIER LANDING** A pilot lands a fighter aircraft on an aircraft carrier. At the moment of touchdown, the speed of the aircraft is 160 mph. If the aircraft is brought to a complete stop in 1 sec and the deceleration is assumed to be constant, find the number of g's the pilot is subjected to during landing (1 g = 32 ft/sec²).

93. **CROSSING THE FINISH LINE** After rounding the final turn in the bell lap, two runners emerge ahead of the pack. When Runner A is 200 ft from the finish line, his speed is 22 ft/sec, a speed that he maintains until he crosses the line. At that instant of time, Runner B, who is 20 ft behind Runner A and running at a speed of 20 ft/sec, begins to sprint. Assuming that Runner B sprints with a constant acceleration, what minimum acceleration will enable him to cross the finish line ahead of Runner A?

94. **DRAINING A TANK** A tank has a constant cross-sectional area of 50 ft² and an orifice of constant cross-sectional

area of $\frac{1}{2}$ ft^2 located at the bottom of the tank (see the accompanying figure).

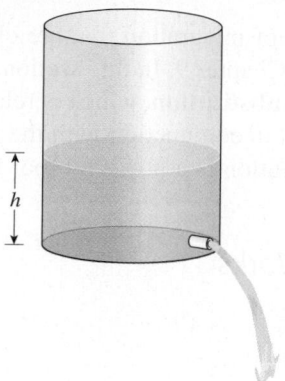

If the tank is filled with water to a height of h ft and allowed to drain, then the height of the water decreases at a rate (in feet per second) that is described by the equation

$$\frac{dh}{dt} = -\frac{1}{25}\left(\sqrt{20} - \frac{t}{50}\right) \qquad (0 \le t \le 50\sqrt{20})$$

Find an expression for the height of the water at any time t (in seconds) if its height initially is 20 ft.

95. **LAUNCHING A FIGHTER AIRCRAFT** A fighter aircraft is launched from the deck of a Nimitz-class aircraft carrier with the help of a steam catapult. If the aircraft is to attain a takeoff speed of at least 240 ft/sec after traveling 800 ft along the flight deck, find the minimum acceleration to which it must be subjected, assuming that it is constant.

In Exercises 96–100, determine whether the statement is true or false. If it is true, explain why it is true. If it is false, give an example to show why it is false.

96. If F and G are antiderivatives of f on an interval I, then $F(x) = G(x) + C$ on I.

97. If F is an antiderivative of f on an interval I, then $\int f(x)\,dx = F(x)$.

98. If f and g are integrable, then
$$\int [2f(x) - 3g(x)]\,dx = 2\int f(x)\,dx - 3\int g(x)\,dx$$

99. $\int \frac{d}{dx}[f(x)]\,dx = f(x)$

100. If f and g are integrable, then
$$\int f(x)g(x)\,dx = \left[\int f(x)\,dx\right]\left[\int g(x)\,dx\right]$$

11.1 Solutions to Self-Check Exercises

1. $\displaystyle \int\left(\frac{1}{\sqrt{x}} - \frac{2}{x} + 3e^x\right)dx = \int\left(x^{-1/2} - \frac{2}{x} + 3e^x\right)dx$

$$= \int x^{-1/2}\,dx - 2\int\frac{1}{x}\,dx + 3\int e^x\,dx$$

$$= 2x^{1/2} - 2\ln|x| + 3e^x + C$$

$$= 2\sqrt{x} - 2\ln|x| + 3e^x + C$$

2. The slope of the tangent line to the graph of the function f at any point $P(x, f(x))$ is given by the derivative f' of f. Thus, the first condition implies that

$$f'(x) = 3x^2 - 6x + 3$$

which, upon integration, yields

$$f(x) = \int (3x^2 - 6x + 3)\,dx$$
$$= x^3 - 3x^2 + 3x + k$$

where k is the constant of integration.

To evaluate k, we use the initial condition (2), which implies that $f(2) = 9$, or

$$9 = f(2) = 2^3 - 3(2)^2 + 3(2) + k$$

or $k = 7$. Hence the required function f is

$$f(x) = x^3 - 3x^2 + 3x + 7$$

3. Let $M(t)$ denote United Motors' market share at year t. Then,

$$M(t) = \int f(t)\,dt$$

$$= \int (-0.01875t^2 + 0.15t - 1.2)\,dt$$

$$= -0.00625t^3 + 0.075t^2 - 1.2t + C$$

To determine the value of C, we use the initial condition $M(0) = 48.4$, obtaining $C = 48.4$. Therefore,

$$M(t) = -0.00625t^3 + 0.075t^2 - 1.2t + 48.4$$

In particular, United Motors' market share of new cars at the beginning of 2014 is given by

$$M(12) = -0.00625(12)^3 + 0.075(12)^2$$
$$-1.2(12) + 48.4$$
$$= 34$$

or 34%.

11.2 Integration by Substitution

In Section 11.1, we developed certain rules of integration that are closely related to the corresponding rules of differentiation in Chapter 9. In this section, we introduce a method of integration called the **method of substitution,** which is related to the chain rule for differentiating functions. When used in conjunction with the rules of integration developed earlier, the method of substitution is a powerful tool for integrating a large class of functions.

How the Method of Substitution Works

Consider the indefinite integral

$$\int 2(2x + 4)^5 \, dx \tag{3}$$

One way of evaluating this integral is to expand the expression $2(2x + 4)^5$ and then integrate the resulting integrand term by term. As an alternative approach, let's see whether we can simplify the integral by making a change of variable. Write

$$u = 2x + 4$$

with differential*

$$du = 2 \, dx$$

If we formally substitute these quantities into the integral (3), we obtain

$$\int 2(2x + 4)^5 \, dx = \int (2x + 4)^5 (2 \, dx) = \int u^5 \, du$$

$$\underset{\text{Rewrite}}{\uparrow} \qquad\qquad \uparrow \begin{cases} u = 2x + 4 \\ du = 2 \, dx \end{cases}$$

Now, the last integral involves a power function and is easily evaluated by using Rule 2 of Section 11.1. Thus,

$$\int u^5 \, du = \frac{1}{6} u^6 + C$$

Therefore, using this result and replacing u by $u = 2x + 4$, we obtain

$$\int 2(2x + 4)^5 \, dx = \frac{1}{6} (2x + 4)^6 + C$$

We can verify that the foregoing result is indeed correct by computing

$$\frac{d}{dx} \left[\frac{1}{6} (2x + 4)^6 + C \right] = \frac{1}{6} \cdot 6(2x + 4)^5 (2) \qquad \text{Use the Chain Rule.}$$
$$= 2(2x + 4)^5$$

and observing that the last expression is just the integrand of the integral (3).

The Method of Integration by Substitution

To see why the approach used in evaluating the integral in (3) is successful, write

$$f(x) = x^5 \quad \text{and} \quad g(x) = 2x + 4$$

Then, $g'(x) = 2$. Furthermore, the integrand of (3) is just 2 times the composition of f and g, that is,

$$2(f \circ g)(x) = 2f(g(x))$$
$$= 2[g(x)]^5 = 2(2x + 4)^5$$

---

*If $u = f(x)$, the differential of u, written du, is $du = f'(x) \, dx$.

Therefore, the integral (3) can be written as

$$\int 2f(g(x))g'(x)\,dx \tag{4}$$

Next, let's show that an integral having the form (4) can always be written as

$$\int f(u)\,du \tag{5}$$

Suppose F is an antiderivative of f. By the Chain Rule, we have

$$\frac{d}{dx}[F(g(x))] = F'(g(x))g'(x)$$

Therefore,

$$\int F'(g(x))g'(x)\,dx = F(g(x)) + C$$

Letting $F' = f$ and making the substitution $u = g(x)$, we have

$$\int f(g(x))g'(x)\,dx = F(u) + C = \int F'(u)\,du = \int f(u)\,du$$

as we wished to show. Thus, if the transformed integral is readily evaluated, as is the case with the integral (3), then the method of substitution will prove successful.

Before we look at more examples, let's summarize the steps involved in integration by substitution.

Integration by Substitution

Step 1 Let $u = g(x)$, where $g(x)$ is part of the integrand, usually the "inside function" of a composite function $f(g(x))$.

Step 2 Find $du = g'(x)\,dx$.

Step 3 Use the substitution $u = g(x)$ and $du = g'(x)\,dx$ to convert the *entire* integral into one involving *only u*.

Step 4 Find the resulting integral.

Step 5 Replace u by $g(x)$ to obtain the final solution as a function of x.

Note Sometimes we need to consider different choices of g for the substitution $u = g(x)$ in order to carry out Step 3 and/or Step 4.

EXAMPLE 1 Find $\displaystyle\int 2x(x^2 + 3)^4\,dx$.

Solution

Step 1 Observe that the integrand involves the composite function $(x^2 + 3)^4$ with "inside function" $g(x) = x^2 + 3$. So we choose $u = x^2 + 3$.

Step 2 Find $du = 2x\,dx$.

Step 3 Use the substitution $u = x^2 + 3$ and $du = 2x\,dx$ to obtain

$$\int 2x(x^2 + 3)^4\,dx = \int \underset{\uparrow}{(x^2 + 3)^4}(2x\,dx) = \int u^4\,du$$
$$\text{Rewrite}$$

an integral involving only the variable u.

Step 4 Find the resulting integral:

$$\int u^4\,du = \frac{1}{5}u^5 + C$$

Step 5   Replacing u by $x^2 + 3$, we obtain

$$\int 2x(x^2 + 3)^4 \, dx = \frac{1}{5} (x^2 + 3)^5 + C$$

EXAMPLE 2 Find $\int 3\sqrt{3x + 1} \, dx$.

Solution

Step 1   The integrand involves the composite function $\sqrt{3x + 1}$ with "inside function" $g(x) = 3x + 1$. So let $u = 3x + 1$.

Step 2   Find $du = 3 \, dx$.

Step 3   Use the substitution $u = 3x + 1$ and $du = 3 \, dx$ to obtain

$$\int 3\sqrt{3x + 1} \, dx = \int \sqrt{3x + 1}(3 \, dx) = \int \sqrt{u} \, du$$

an integral involving only the variable u.

Step 4   Find the resulting integral:

$$\int \sqrt{u} \, du = \int u^{1/2} \, du = \frac{2}{3} u^{3/2} + C$$

Step 5   Replacing u by $3x + 1$, we obtain

$$\int 3\sqrt{3x + 1} \, dx = \frac{2}{3} (3x + 1)^{3/2} + C$$

EXAMPLE 3 Find $\int x^2(x^3 + 1)^{3/2} \, dx$.

Solution

Step 1   The integrand contains the composite function $(x^3 + 1)^{3/2}$ with "inside function" $g(x) = x^3 + 1$. So let $u = x^3 + 1$.

Step 2   Find $du = 3x^2 \, dx$.

Step 3   Use the substitution $u = x^3 + 1$ and $du = 3x^2 \, dx$, or $x^2 \, dx = \frac{1}{3} du$ to obtain

$$\int x^2(x^3 + 1)^{3/2} \, dx = \int (x^3 + 1)^{3/2}(x^2 \, dx)$$
$$= \int u^{3/2} \left(\frac{1}{3} \, du \right) = \frac{1}{3} \int u^{3/2} \, du$$

an integral involving only the variable u.

Step 4   Find the resulting integral:

$$\frac{1}{3} \int u^{3/2} \, du = \frac{1}{3} \cdot \frac{2}{5} u^{5/2} + C = \frac{2}{15} u^{5/2} + C$$

Step 5   Replacing u by $x^3 + 1$, we obtain

$$\int x^2(x^3 + 1)^{3/2} \, dx = \frac{2}{15} (x^3 + 1)^{5/2} + C$$

Explore and Discuss

Let $f(x) = x^2(x^3 + 1)^{3/2}$. Using the result of Example 3, we see that an antiderivative of f is $F(x) = \frac{2}{15} (x^3 + 1)^{5/2}$. However, in terms of u (where $u = x^3 + 1$), an antiderivative of f is $G(u) = \frac{2}{15} u^{5/2}$. Compute $F(2)$. Next, suppose we want to compute $F(2)$ using the function G instead. At what value of u should you evaluate $G(u)$ to obtain the desired result? Explain your answer.

In the remaining examples, we drop the practice of labeling the steps involved in evaluating each integral.

EXAMPLE 4 Find $\int e^{-3x}\, dx$.

Solution Let $u = -3x$, so that $du = -3\, dx$, or $dx = -\frac{1}{3}\, du$. Then,

$$\int e^{-3x}\, dx = \int e^{u}\left(-\frac{1}{3}\, du\right) = -\frac{1}{3}\int e^{u}\, du$$

$$= -\frac{1}{3} e^{u} + C = -\frac{1}{3} e^{-3x} + C$$

EXAMPLE 5 Find $\int \dfrac{x}{3x^2 + 1}\, dx$.

Solution Let $u = 3x^2 + 1$. Then, $du = 6x\, dx$, or $x\, dx = \frac{1}{6}\, du$. Making the appropriate substitutions, we have

$$\int \frac{x}{3x^2 + 1}\, dx = \int \frac{\frac{1}{6}}{u}\, du$$

$$= \frac{1}{6}\int \frac{1}{u}\, du$$

$$= \frac{1}{6}\ln|u| + C$$

$$= \frac{1}{6}\ln(3x^2 + 1) + C \qquad \text{Since } 3x^2 + 1 > 0$$

EXAMPLE 6 Find $\int \dfrac{(\ln x)^2}{2x}\, dx$.

Solution Let $u = \ln x$. Then,

$$du = \frac{d}{dx}(\ln x)\, dx = \frac{1}{x}\, dx$$

$$\int \frac{(\ln x)^2}{2x}\, dx = \frac{1}{2}\int \frac{(\ln x)^2}{x}\, dx$$

$$= \frac{1}{2}\int u^2\, du$$

$$= \frac{1}{6} u^3 + C$$

$$= \frac{1}{6}(\ln x)^3 + C$$

> *Explore and Discuss*
>
> Suppose $\int f(u)\, du = F(u) + C$.
>
> **1.** Show that $\int f(ax + b)\, dx = \dfrac{1}{a} F(ax + b) + C$.
>
> **2.** How can you use this result to facilitate the evaluation of integrals such as $\int (2x + 3)^5\, dx$ and $\int e^{3x-2}\, dx$? Explain your answer.

Examples 7 and 8 show how the method of substitution can be used in practical situations.

$ APPLIED EXAMPLE 7 Cost of Producing Solar Cell Panels In 2000, the head of the research and development department of Soloron Corporation claimed that the cost of producing solar cell panels would drop at the rate of

$$\frac{105}{2(3t + 5)^2} \qquad (0 \le t \le 15)$$

dollars per peak watt for the next t years, with $t = 0$ corresponding to the beginning of 2000. (A peak watt is the power produced at noon on a sunny day.) In 2000, the panels, which are used for photovoltaic power systems, cost \$4 per peak watt. Find an expression giving the cost per peak watt of producing solar cell panels at the beginning of year t. What was the cost at the beginning of 2012?

Solution Let $C(t)$ denote the cost per peak watt for producing solar cell panels at the beginning of year t. Then,

$$C'(t) = -\frac{105}{2(3t + 5)^2}$$

Integrating, we find that

$$C(t) = \int \frac{-105}{2(3t + 5)^2}\, dt$$

$$= -\frac{105}{2} \int (3t + 5)^{-2}\, dt$$

Let $u = 3t + 5$ so that

$$du = 3\, dt \quad \text{or} \quad dt = \frac{1}{3}\, du$$

Then,

$$C(t) = -\frac{105}{2} \left(\frac{1}{3}\right) \int u^{-2}\, du$$

$$= -\frac{35}{2} (-1)u^{-1} + k$$

$$= \frac{35}{2(3t + 5)} + k$$

where k is an arbitrary constant. To determine the value of k, note that the cost per peak watt of producing solar cell panels at the beginning of 2000 ($t = 0$) was 4, or $C(0) = 4$. This gives

$$C(0) = \frac{35}{2(5)} + k = 4$$

or $k = \frac{1}{2}$. Therefore, the required expression is given by

$$C(t) = \frac{35}{2(3t + 5)} + \frac{1}{2}$$

$$= \frac{35 + (3t + 5)}{2(3t + 5)}$$

$$= \frac{3t + 40}{2(3t + 5)}$$

The cost per peak watt for producing solar cell panels at the beginning of 2012 is given by

$$C(12) = \frac{3(12) + 40}{2[3(12) + 5]} \approx 0.93$$

or approximately $0.93 per peak watt.

Exploring with TECHNOLOGY

Refer to Example 7.

1. Use a graphing utility to plot the graph of

$$C(t) = \frac{3t + 40}{2(3t + 5)}$$

using the viewing window $[0, 15] \times [0, 5]$. Then, use the numerical differentiation capability of the graphing utility to compute $C'(12)$.

2. Plot the graph of

$$C'(t) = -\frac{105}{2(3t + 5)^2}$$

using the viewing window $[0, 15] \times [-2, 0]$. Then, use the evaluation capability of the graphing utility to find $C'(12)$. Is this value of $C'(12)$ the same as that obtained in part 1? Explain your answer.

APPLIED EXAMPLE 8 Computer Sales Projections A study prepared by the marketing department of Universal Instruments forecasts that after its new line of Galaxy Home Computers is introduced into the market, the level of sales in the tth month will be

$$2000 - 1500e^{-0.05t} \qquad (0 \le t \le 60)$$

units per month. Find an expression that gives the total number of computers that will have been sold in the first t months after they become available on the market. How many computers will Universal sell in the first year they are on the market?

Solution Let $N(t)$ denote the total number of computers that may be expected to be sold t months after their introduction in the market. Then, the rate of growth of sales is given by $N'(t)$ units per month. Thus,

$$N'(t) = 2000 - 1500e^{-0.05t}$$

so that

$$N(t) = \int (2000 - 1500e^{-0.05t})\, dt$$

$$= \int 2000\, dt - 1500 \int e^{-0.05t}\, dt$$

Upon integrating the second integral by the method of substitution, we obtain

$$N(t) = 2000t + \frac{1500}{0.05} e^{-0.05t} + C \qquad \text{Let } u = -0.05t;$$
$$\text{then } du = -0.05\, dt.$$

$$= 2000t + 30{,}000e^{-0.05t} + C$$

To determine the value of C, note that the number of computers sold at the end of month 0 is nil, so $N(0) = 0$. This gives

$$N(0) = 30,000 + C = 0 \quad \text{Since } e^0 = 1$$

or $C = -30,000$. Therefore, the required expression is given by

$$N(t) = 2000t + 30,000e^{-0.05t} - 30,000$$
$$= 2000t + 30,000(e^{-0.05t} - 1)$$

The number of computers that Universal can expect to sell in the first year is given by

$$N(12) = 2000(12) + 30,000(e^{-0.05(12)} - 1)$$
$$\approx 10,464$$

11.2   Self-Check Exercises

1. Find $\int \sqrt{2x + 5}\, dx$.

2. Find $\int \dfrac{x^2}{(2x^3 + 1)^{3/2}}\, dx$.

3. Find $\int xe^{2x^2-1}\, dx$.

4. **AUTOMOBILE POLLUTION** According to a joint study conducted by Oxnard's Environmental Management Department and a state government agency, the concentration of

carbon monoxide (CO) in the air due to automobile exhaust is increasing at the rate given by

$$f(t) = \frac{8(0.1t + 1)}{300(0.2t^2 + 4t + 64)^{1/3}}$$

parts per million (ppm) per year t. Currently, the CO concentration due to automobile exhaust is 0.16 ppm. Find an expression giving the CO concentration t years from now.

Solutions to Self-Check Exercises 11.2 can be found on page 831.

11.2   Concept Questions

1. Explain how the method of substitution works by showing the steps used to find $\int f(g(x))g'(x)\, dx$.

2. Explain why the method of substitution works for the integral $\int xe^{-x^2}\, dx$ but not for the integral $\int e^{-x^2}\, dx$.

11.2   Exercises

In Exercises 1–50, find the indefinite integral.

1. $\int 4(4x + 3)^4\, dx$

2. $\int 4x(2x^2 + 1)^7\, dx$

3. $\int (x^3 - 2x)^2(3x^2 - 2)\, dx$

4. $\int (3x^2 - 2x + 1)(x^3 - x^2 + x)^4\, dx$

5. $\int \dfrac{4x}{(2x^2 + 3)^3}\, dx$

6. $\int \dfrac{3x^2 + 2}{(x^3 + 2x)^2}\, dx$

7. $\int 3t^2\sqrt{t^3 + 2}\, dt$

8. $\int 3t^2(t^3 + 2)^{3/2}\, dt$

9. $\int 2(x^2 - 1)^9 x\, dx$

10. $\int x^2(2x^3 + 3)^4\, dx$

11. $\int \dfrac{x^4}{1 - x^5}\, dx$

12. $\int \dfrac{x^2}{\sqrt{x^3 - 1}}\, dx$

13. $\int \dfrac{2}{x - 2}\, dx$

14. $\int \dfrac{x^2}{x^3 - 3}\, dx$

15. $\int \dfrac{0.3x - 0.2}{0.3x^2 - 0.4x + 2}\, dx$

16. $\int \dfrac{2x^2 + 1}{0.2x^3 + 0.3x}\, dx$

17. $\int \dfrac{2x}{3x^2 - 1}\, dx$

18. $\int \dfrac{x^2 - 1}{x^3 - 3x + 1}\, dx$

19. $\int e^{-2x}\, dx$

20. $\int e^{-0.02x}\, dx$

21. $\int e^{2-x}\, dx$

22. $\int e^{2t+3}\, dt$

23. $\int xe^{-x^2}\, dx$

24. $\int x^2 e^{x^3-1}\, dx$

25. $\int (e^x - e^{-x})\, dx$

26. $\int (e^{2x} + e^{-3x})\, dx$

27. $\displaystyle\int \frac{2e^x}{1 + e^x}\, dx$

28. $\displaystyle\int \frac{e^{2x}}{1 + e^{2x}}\, dx$

29. $\displaystyle\int \frac{e^{\sqrt{x}}}{\sqrt{x}}\, dx$

30. $\displaystyle\int \frac{e^{-1/x}}{x^2}\, dx$

31. $\displaystyle\int \frac{e^{3x} + x^2}{(e^{3x} + x^3)^3}\, dx$

32. $\displaystyle\int \frac{e^x - e^{-x}}{(e^x + e^{-x})^{3/2}}\, dx$

33. $\displaystyle\int e^{2x}(e^{2x} + 1)^3\, dx$

34. $\displaystyle\int e^{-x}(1 + e^{-x})\, dx$

35. $\displaystyle\int \frac{\ln 5x}{x}\, dx$

36. $\displaystyle\int \frac{(\ln u)^3}{u}\, du$

37. $\displaystyle\int \frac{3}{x \ln x}\, dx$

38. $\displaystyle\int \frac{1}{x(\ln x)^2}\, dx$

39. $\displaystyle\int \frac{\sqrt{\ln x}}{x}\, dx$

40. $\displaystyle\int \frac{(\ln x)^{7/2}}{x}\, dx$

41. $\displaystyle\int \left(xe^{x^2} - \frac{x}{x^2 + 2} \right) dx$

42. $\displaystyle\int \left(xe^{-x^2} + \frac{e^x}{e^x + 3} \right) dx$

43. $\displaystyle\int \frac{x + 1}{\sqrt{x} - 1}\, dx$

44. $\displaystyle\int \frac{e^{-u} - 1}{e^{-u} + u}\, du$

Hint: Let $u = \sqrt{x} - 1$.   Hint: Let $v = e^{-u} + u$.

45. $\displaystyle\int x(x - 1)^5\, dx$

Hint: $u = x - 1$ implies $x = u + 1$.

46. $\displaystyle\int \frac{t}{t + 1}\, dt$

47. $\displaystyle\int \frac{1 - \sqrt{x}}{1 + \sqrt{x}}\, dx$

Hint: $\dfrac{t}{t + 1} = 1 - \dfrac{1}{t + 1}$.   Hint: Let $u = 1 + \sqrt{x}$.

48. $\displaystyle\int \frac{1 + \sqrt{x}}{1 - \sqrt{x}}\, dx$

49. $\displaystyle\int v^2(1 - v)^6\, dv$

Hint: Let $u = 1 - \sqrt{x}$.   Hint: Let $u = 1 - v$.

50. $\displaystyle\int x^3(x^2 + 1)^{3/2}\, dx$

Hint: Let $u = x^2 + 1$.

In Exercises 51–54, find the function f given that the slope of the tangent line to the graph of f at any point $(x, f(x))$ is $f'(x)$ and that the graph of f passes through the given point.

51. $f'(x) = 5(2x - 1)^4;\ (1, 3)$

52. $f'(x) = \dfrac{3x^2}{2\sqrt{x^3 - 1}};\ (1, 1)$

53. $f'(x) = -2xe^{-x^2 + 1};\ (1, 0)$

54. $f'(x) = 1 - \dfrac{2x}{x^2 + 1};\ (0, 2)$

55. Student Enrollment The registrar of Kellogg University estimates that the student enrollment in the Continuing Education division will grow at the rate of

$$N'(t) = 2000(1 + 0.2t)^{-3/2}$$

students/year, t years from now. If the current student enrollment is 1000, find an expression giving the enrollment t years from now. What will be the enrollment 5 years from now?

56. TV Viewers: Newsmagazine Shows The number of viewers of a weekly TV newsmagazine show, introduced in the 2009 season, has been increasing at the rate of

$$3\left(2 + \frac{1}{2}t \right)^{-1/3} \qquad (1 \le t \le 6)$$

million viewers/year in its tth year on the air, where $t = 1$ corresponds to 2009. The number of viewers of the program during its first year on the air is given by $9(5/2)^{2/3}$ million. Find how many viewers were expected in the 2014 season.

57. Credit Card Losses in the United Kingdom The mail and the Internet are major routes for fraud against merchants who sell and ship products. These merchants must often accept credit cards that are not physically present (called CNPs, credit cards not present). After peaking at 328.9 million British pounds (GBP) in 2008 ($t = 0$), CNP fraud in the United Kingdom began to decline at the rate of

$$R(t) = \frac{92.07}{t + 1} \qquad (0 \le t \le 4)$$

million GBP/year in year t through 2012. What was the amount of CNP fraud on U.K.-issued credit cards in 2012?
Source: Financial Fraud Action UK.

58. Demand: Women's Boots The rate of change of the unit price p (in dollars) of Apex women's boots is given by

$$p'(x) = \frac{-250\,x}{(16 + x^2)^{3/2}}$$

where x is the quantity demanded daily in units of a hundred. Find the demand function for these boots if the quantity demanded daily is 300 pairs ($x = 3$) when the unit price is \$50/pair.

59. Population Growth The population of a certain city is projected to grow at the rate of

$$r(t) = 400\left(1 + \frac{2t}{24 + t^2} \right) \qquad (0 \le t \le 5)$$

people/year, t years from now. The current population is 60,000. What will be the population 5 years from now?

60. Oil Spill In calm waters, the oil spilling from the ruptured hull of a grounded tanker forms an oil slick that is circular in shape. If the radius r of the circle is increasing at the rate of

$$r'(t) = \frac{30}{\sqrt{2t + 4}}$$

ft/min t min after the rupture occurs, find an expression for the radius at any time t. How large is the polluted area 16 min after the rupture occurred?
Hint: $r(0) = 0$

61. LIFE EXPECTANCY OF A FEMALE Suppose that in a certain country, the life expectancy at birth of a female is changing at the rate of

$$g'(t) = \frac{5.45218}{(1 + 1.09t)^{0.9}}$$

years/year. Here, t is measured in years, with $t = 0$ corresponding to the beginning of 1900. Find an expression $g(t)$ giving the life expectancy at birth (in years) of a female in that country if the life expectancy at the beginning of 1900 is 50.02 years. What is the life expectancy at birth of a female born in 2000 in that country?

62. LEARNING CURVES The average student enrolled in the 20-week Court Reporting I course at the American Institute of Court Reporting progresses according to the rule

$$N'(t) = 6e^{-0.05t} \qquad (0 \le t \le 20)$$

where $N'(t)$ measures the rate of change in the number of words per minute of dictation the student takes in machine shorthand after t weeks in the course. Assuming that the average student enrolled in this course begins with a dictation speed of 60 words/min, find an expression $N'(t)$ that gives the dictation speed of the student after t weeks in the course.

63. U.S. ONLINE VIDEO VIEWERS The number of online video viewers in the United States was growing at the rate of

$$r(t) = 9.045e^{0.067t} \qquad (0 \le t \le 5)$$

million viewers/year between 2008 ($t = 0$) and 2013 ($t = 5$). The number of viewers stood at 135 million in 2008.
a. Find an expression giving the number of online video viewers in year t.
b. How many viewers were there in 2012? In 2013?
Source: eMarketer.

64. VIDEO ADS In recent years, video ads have become the most popular rich media format for ad buyers as advertisers have begun to realize returns on online video advertising. According to research and forecast by Forrester, media spending on ads for the years 2012 through 2017 is projected to grow at the rate of

$$R(t) = 0.538434e^{0.234t} \qquad (0 \le t \le 5)$$

billion dollars/year in year t, with $t = 0$ corresponding to 2012. The expenditure on media spending on ads in 2012 stood at $2.9 billion.
a. Find an expression for media spending on ads in year t.
b. Assuming that the forecast was accurate, what would the media spending on ads be in 2016?
Source: Forrester.

65. U.S. ONLINE AD REVENUES Online advertisement revenues took a slight dip in 2009, but since then, they have continued to grow spectacularly. In fact, the rate of growth of online advertisement revenues from 2009 ($t = 0$) through 2012 ($t = 3$) was approximately

$$r(t) = 3.1182e^{0.163(t+1)} \qquad (0 \le t \le 3)$$

billion dollars/year in year t. Online advertisement revenues in 2009 were $22.7 billion.
a. Find an expression for the online advertisement revenue in year t.
b. What were the online advertisement revenues in 2012?
Source: IAB/PricewaterhouseCoopers.

66. AVERAGE BIRTH HEIGHT OF BOYS Using data collected at Kaiser Hospital, pediatricians estimate that the average height of male children changes at the rate of

$$h'(t) = \frac{52.8706e^{-0.3277t}}{(1 + 2.449e^{-0.3277t})^2}$$

in./year, where the child's height $h(t)$ is measured in inches and t, the child's age, is measured in years, with $t = 0$ corresponding to birth. Find an expression $h(t)$ for the average height of a boy at age t if the height at birth of an average child is 19.4 in. What is the height of an average 8-year-old boy?

67. AMOUNT OF GLUCOSE IN THE BLOODSTREAM Suppose a patient is given a continuous intravenous infusion of glucose at a constant rate of r mg/min. Then, the rate at which the amount of glucose in the bloodstream is changing at time t (in minutes) because of this infusion is given by

$$A'(t) = re^{-at}$$

mg/min, where a is a positive constant associated with the rate at which excess glucose is eliminated from the bloodstream and is dependent on the patient's metabolism rate. Derive an expression for the amount of glucose in the bloodstream at time t.
Hint: $A(0) = 0$

68. CONCENTRATION OF A DRUG IN AN ORGAN A drug is carried into an organ of volume V cm^3 by a liquid that enters the organ at the rate of a cm^3/sec and leaves it at the rate of b cm^3/sec. The concentration of the drug in the liquid entering the organ is c g/cm^3. If the concentration of the drug in the organ at time t (in seconds) is increasing at the rate of

$$x'(t) = \frac{1}{V}(ac - bx_0)e^{-bt/V}$$

g/cm^3/sec and the concentration of the drug in the organ initially is x_0 g/cm^3, show that the concentration of the drug in the organ at time t is given by

$$x(t) = \frac{ac}{b} + \left(x_0 - \frac{ac}{b}\right)e^{-bt/V}$$

In Exercises 69–72, determine whether the statement is true or false. If it is true, explain why it is true. If it is false, give an example to show why it is false.

69. If f is integrable, then $\int xf(x^2)\, dx = \frac{1}{2}\int f(x)\, dx$.

70. If f is integrable, then $\int f(ax + b)\, dx = a\int f(x)\, dx$.

71. If f is integrable, then $\int e^{kx}f(e^{kx})\, dx = \frac{1}{k}\int f(x)\, dx$. $(k \neq 0)$.

72. If f is integrable, then $\int \frac{f(\ln x)}{\ln x}\, dx = \int f(x)\, dx$.

11.2 Solutions to Self-Check Exercises

1. Let $u = 2x + 5$. Then, $du = 2\, dx$, or $dx = \frac{1}{2}\, du$. Making the appropriate substitutions, we have

$$\int \sqrt{2x + 5}\, dx = \int \sqrt{u}\left(\frac{1}{2}\, du\right) = \frac{1}{2}\int u^{1/2}\, du$$

$$= \frac{1}{2}\left(\frac{2}{3}\right)u^{3/2} + C$$

$$= \frac{1}{3}(2x + 5)^{3/2} + C$$

2. Let $u = 2x^3 + 1$, so that $du = 6x^2\, dx$, or $x^2\, dx = \frac{1}{6}\, du$. Making the appropriate substitutions, we have

$$\int \frac{x^2}{(2x^3 + 1)^{3/2}}\, dx = \int \frac{\left(\frac{1}{6}\right) du}{u^{3/2}} = \frac{1}{6}\int u^{-3/2}\, du$$

$$= \left(\frac{1}{6}\right)(-2)u^{-1/2} + C$$

$$= -\frac{1}{3}(2x^3 + 1)^{-1/2} + C$$

$$= -\frac{1}{3\sqrt{2x^3 + 1}} + C$$

3. Let $u = 2x^2 - 1$, so that $du = 4x\, dx$, or $x\, dx = \frac{1}{4}\, du$. Then,

$$\int xe^{2x^2-1}\, dx = \frac{1}{4}\int e^u\, du$$

$$= \frac{1}{4}e^u + C$$

$$= \frac{1}{4}e^{2x^2-1} + C$$

4. Let $C(t)$ denote the CO concentration in the air due to automobile exhaust t years from now. Then,

$$C'(t) = f(t) = \frac{8(0.1t + 1)}{300(0.2t^2 + 4t + 64)^{1/3}}$$

$$= \frac{8}{300}(0.1t + 1)(0.2t^2 + 4t + 64)^{-1/3}$$

Integrating, we find

$$C(t) = \int \frac{8}{300}(0.1t + 1)(0.2t^2 + 4t + 64)^{-1/3}\, dt$$

$$= \frac{8}{300}\int (0.1t + 1)(0.2t^2 + 4t + 64)^{-1/3}\, dt$$

Let $u = 0.2t^2 + 4t + 64$, so that

$$du = (0.4t + 4)\, dt = 4(0.1t + 1)\, dt$$

or

$$(0.1t + 1)\, dt = \frac{1}{4}\, du$$

Then,

$$C(t) = \frac{8}{300}\left(\frac{1}{4}\right)\int u^{-1/3}\, du$$

$$= \frac{1}{150}\left(\frac{3}{2}u^{2/3}\right) + k$$

$$= 0.01(0.2t^2 + 4t + 64)^{2/3} + k$$

where k is an arbitrary constant. To determine the value of k, we use the condition $C(0) = 0.16$, obtaining

$$C(0) = 0.16 = 0.01(64)^{2/3} + k$$

$$0.16 = 0.16 + k$$

$$k = 0$$

Therefore,

$$C(t) = 0.01(0.2t^2 + 4t + 64)^{2/3}$$

11.3 Area and the Definite Integral

An Intuitive Look

Suppose a certain state's annual rate of petroleum consumption over a 4-year period is constant and is given by the function

$$f(t) = 1.2 \qquad (0 \leq t \leq 4)$$

where t is measured in years and $f(t)$ in millions of barrels per year. Then, the state's total petroleum consumption over the period of time in question is

$$(1.2)(4 - 0) \qquad \text{Rate of consumption} \times \text{time elapsed}$$

or 4.8 million barrels. If you examine the graph of f shown in Figure 5, you will see that this total is just the area of the rectangular region bounded above by the graph of f, below by the t-axis, and to the left and right by the vertical lines $t = 0$ (the y-axis) and $t = 4$, respectively.

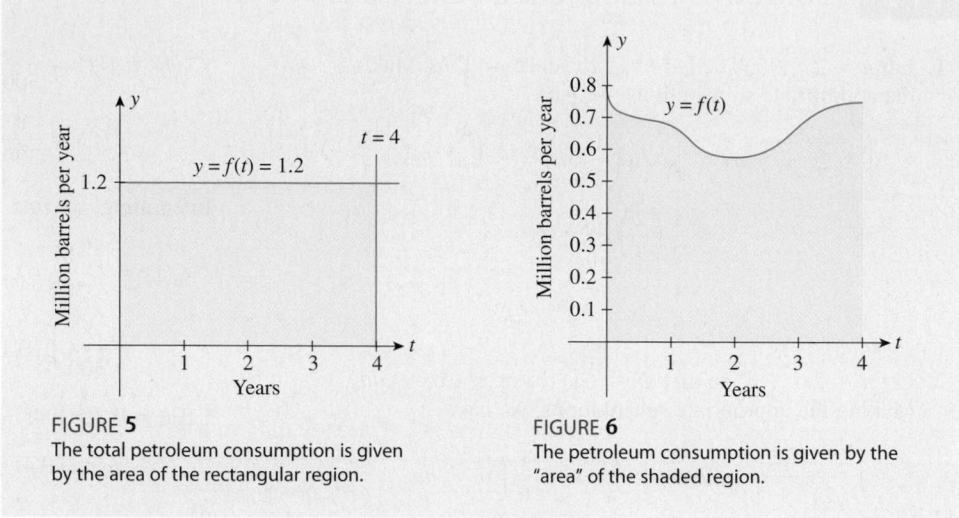

FIGURE 5
The total petroleum consumption is given by the area of the rectangular region.

FIGURE 6
The petroleum consumption is given by the "area" of the shaded region.

Figure 6 shows the actual petroleum consumption of a certain New England state over a 4-year period from 2008 ($t = 0$) to 2012 ($t = 4$). Observe that the rate of consumption is not constant; that is, the function f is not a constant function. What is the state's total petroleum consumption over this 4-year period? It seems reasonable to conjecture that it is given by the "area" of the region bounded above by the graph of f, below by the t-axis, and to the left and right by the vertical lines $t = 0$ and $t = 4$, respectively. We will show that this conjecture is justified at the end of this section.

This example raises two questions:

1. What is the "area" of the region shown in Figure 6?
2. How do we compute this area?

The Area Problem

The preceding example touches on the second fundamental problem in calculus: Calculate the area of the region bounded by the graph of a nonnegative function f, the x-axis, and the vertical lines $x = a$ and $x = b$ where $a < b$ (Figure 7). This area is called the **area of the region under the graph of f** on the interval $[a, b]$ or from a to b.

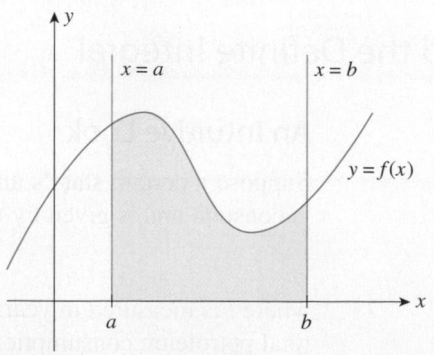

FIGURE 7
The area of the region under the graph of f on $[a, b]$

Defining Area—Two Examples

Just as we used the slopes of secant lines (quantities that we could compute) to help us define the slope of the tangent line to a point on the graph of a function, we now adopt a parallel approach and use the areas of rectangles (quantities that we can compute) to help us define the area under the graph of a function. We begin by looking at a specific example.

EXAMPLE 1 Let $f(x) = x^2$ and consider the region R under the graph of f on the interval $[0, 1]$ (Figure 8a). To obtain an approximation of the area of R, let's construct four nonoverlapping rectangles, except for their edges, as follows: Divide the interval $[0, 1]$ into four subintervals

$$\left[0, \frac{1}{4}\right], \quad \left[\frac{1}{4}, \frac{1}{2}\right], \quad \left[\frac{1}{2}, \frac{3}{4}\right], \quad \left[\frac{3}{4}, 1\right]$$

of equal length $\frac{1}{4}$. Next, construct four rectangles with these subintervals as bases and with heights given by the values of the function at the midpoints

$$\frac{1}{8}, \quad \frac{3}{8}, \quad \frac{5}{8}, \quad \frac{7}{8}$$

of each subinterval. Then, each of these rectangles has width $\frac{1}{4}$ and height

$$f\left(\frac{1}{8}\right), \quad f\left(\frac{3}{8}\right), \quad f\left(\frac{5}{8}\right), \quad f\left(\frac{7}{8}\right)$$

respectively (Figure 8b).

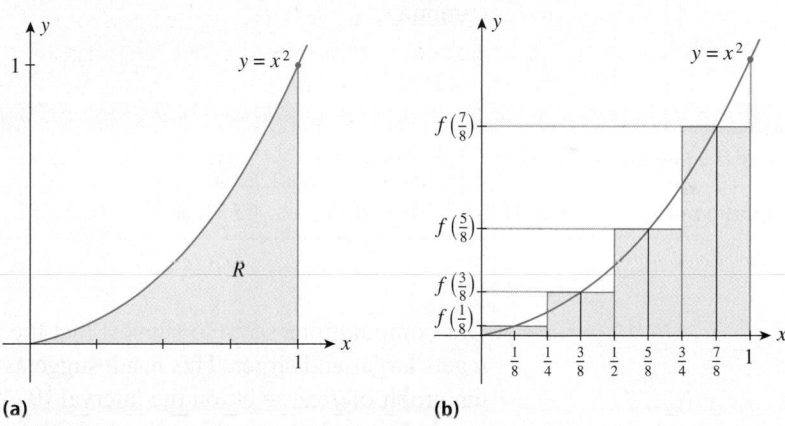

(a)   (b)

FIGURE 8
The area of the region under the graph of f on [0, 1] in (a) is approximated by the sum of the areas of the four rectangles in (b).

If we approximate the area A of R by the sum of the areas of the four rectangles, we obtain

$$A \approx \frac{1}{4} \cdot f\left(\frac{1}{8}\right) + \frac{1}{4} f\left(\frac{3}{8}\right) + \frac{1}{4} f\left(\frac{5}{8}\right) + \frac{1}{4} f\left(\frac{7}{8}\right)$$

$$= \frac{1}{4}\left[f\left(\frac{1}{8}\right) + f\left(\frac{3}{8}\right) + f\left(\frac{5}{8}\right) + f\left(\frac{7}{8}\right)\right]$$

$$= \frac{1}{4}\left[\left(\frac{1}{8}\right)^2 + \left(\frac{3}{8}\right)^2 + \left(\frac{5}{8}\right)^2 + \left(\frac{7}{8}\right)^2\right] \qquad \text{Recall that } f(x) = x^2.$$

$$= \frac{1}{4}\left(\frac{1}{64} + \frac{9}{64} + \frac{25}{64} + \frac{49}{64}\right) = \frac{21}{64}$$

or approximately 0.328125.

Following the procedure of Example 1, we can obtain approximations of the area of the region R using any number n of rectangles ($n = 4$ in Example 1). Figure 9a shows the approximation of the area A of R using 8 rectangles ($n = 8$), and Figure 9b shows the approximation of the area A of R using 16 rectangles.

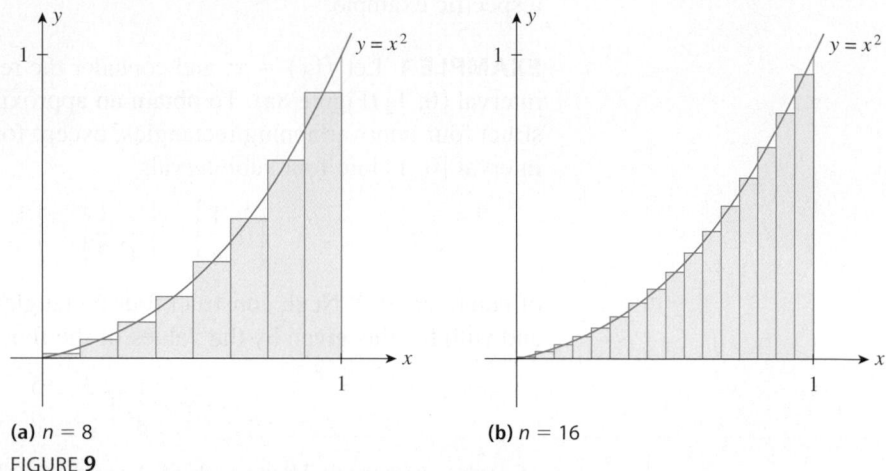

(a) $n = 8$ **(b)** $n = 16$

FIGURE **9**
As n increases, the number of rectangles increases, and the approximation improves.

These figures suggest that the approximations seem to get better as n increases. This is borne out by the results given in Table 1, which were obtained by using a computer.

TABLE 1							
Number of Rectangles, n	4	8	16	32	64	100	200
Approximation of A	0.328125	0.332031	0.333008	0.333252	0.333313	0.333325	0.333331

Our computations seem to suggest that the approximations approach the number $\frac{1}{3}$ as n gets larger and larger. This result suggests that we *define* the area of the region under the graph of $f(x) = x^2$ on the interval $[0, 1]$ to be $\frac{1}{3}$.

In Example 1, we chose the *midpoint* of each subinterval as the point at which to evaluate $f(x)$ to obtain the height of the approximating rectangle. Let's consider another example, this time choosing the *left endpoint* of each subinterval.

EXAMPLE 2 Let R be the region under the graph of $f(x) = 16 - x^2$ on the interval $[1, 3]$. Find an approximation of the area A of R using four subintervals of $[1, 3]$ of equal length and picking the left endpoint of each subinterval to evaluate $f(x)$ to obtain the height of the approximating rectangle.

Solution   The graph of f is sketched in Figure 10a. Since the length of $[1, 3]$ is 2, we see that the length of each subinterval is $\frac{2}{4}$, or $\frac{1}{2}$. Therefore, the four subintervals are

$$\left[1, \frac{3}{2}\right], \quad \left[\frac{3}{2}, 2\right], \quad \left[2, \frac{5}{2}\right], \quad \left[\frac{5}{2}, 3\right]$$

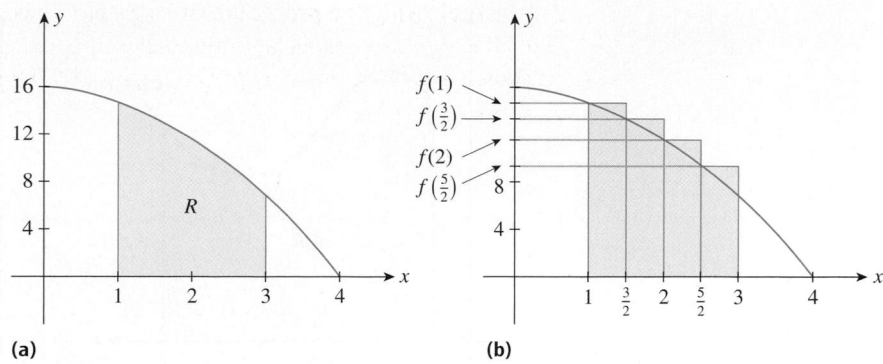

FIGURE 10
The area of R in (a) is approximated by the sum of the areas of the four rectangles in (b).

The left endpoints of these subintervals are $1, \frac{3}{2}, 2,$ and $\frac{5}{2}$, respectively, so the heights of the approximating rectangles are $f(1), f(\frac{3}{2}), f(2),$ and $f(\frac{5}{2})$, respectively (Figure 10b). Therefore, the required approximation is

$$
\begin{aligned}
A &\approx \frac{1}{2}f(1) + \frac{1}{2}f\left(\frac{3}{2}\right) + \frac{1}{2}f(2) + \frac{1}{2}f\left(\frac{5}{2}\right) \\
&= \frac{1}{2}\left[f(1) + f\left(\frac{3}{2}\right) + f(2) + f\left(\frac{5}{2}\right)\right] \\
&= \frac{1}{2}\left\{\left[16 - (1)^2\right] + \left[16 - \left(\frac{3}{2}\right)^2\right]\right. \\
&\quad \left. + \left[16 - (2)^2\right] + \left[16 - \left(\frac{5}{2}\right)^2\right]\right\} \quad \text{Recall that } f(x) = 16 - x^2. \\
&= \frac{1}{2}\left(15 + \frac{55}{4} + 12 + \frac{39}{4}\right) = \frac{101}{4}
\end{aligned}
$$

or approximately 25.25.

Table 2 shows the approximations of the area A of the region R of Example 2 when n rectangles are used for the approximation and the heights of the approximating rectangles are found by evaluating $f(x)$ at the left endpoints.

TABLE 2							
Number of Rectangles, n	4	10	100	1,000	10,000	50,000	100,000
Approximation of A	25.2500	24.1200	23.4132	23.3413	23.3341	23.3335	23.3334

Once again, we see that the approximations seem to approach a unique number as n gets larger and larger. This time, the number is $23\frac{1}{3}$. This result suggests that we *define* the area of the region under the graph of $f(x) = 16 - x^2$ on the interval $[1, 3]$ to be $23\frac{1}{3}$.

Defining Area—The General Case

Examples 1 and 2 point the way to defining the area A of the region R under the graph of an arbitrary but continuous, nonnegative function f on an interval $[a, b]$ (Figure 11a).

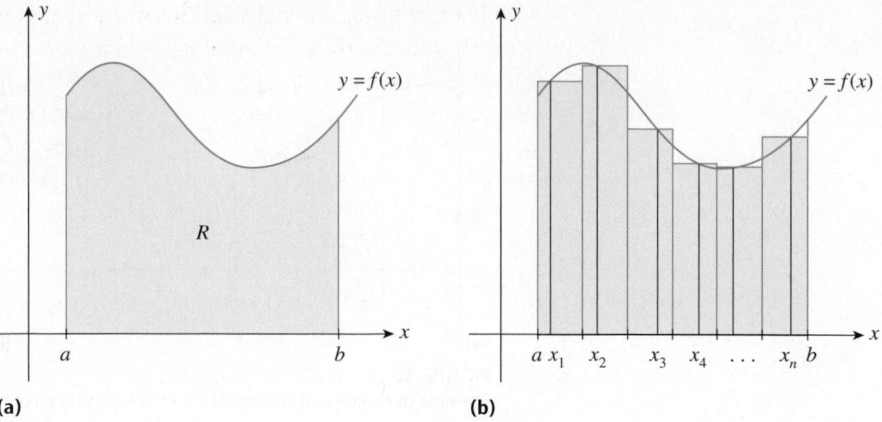

(a)   **(b)**

FIGURE **11**
The area of the region under the graph of f on $[a, b]$ in (a) is approximated by the sum of the areas of the n rectangles shown in (b).

Divide the interval $[a, b]$ into n subintervals of equal length $\Delta x = (b - a)/n$. Next, pick n arbitrary points $x_1, x_2, \ldots,$ and x_n, called *representative points*, from the first, second, $\ldots,$ and nth subintervals, respectively (Figure 11b). Then, approximating the area A of the region R by the n rectangles of width Δx and heights $f(x_1), f(x_2), \ldots, f(x_n)$, so that the areas of the rectangles are $f(x_1)\, \Delta x, f(x_2)\, \Delta x, \ldots, f(x_n)\, \Delta x$, we have

$$A \approx f(x_1)\, \Delta x + f(x_2)\, \Delta x + \cdots + f(x_n)\, \Delta x$$

The sum on the right-hand side of this expression is called a **Riemann sum** in honor of the German mathematician Bernhard Riemann (1826–1866). Now, as the earlier examples seem to suggest, the Riemann sum will approach a unique number as n becomes arbitrarily large.* We define this number to be the area A of the region R.

> **The Area Under the Graph of a Function**
>
> Let f be a nonnegative continuous function on $[a, b]$. Then, the area of the region under the graph of f is
>
> $$A = \lim_{n \to \infty} \left[f(x_1) + f(x_2) + \cdots + f(x_n) \right] \Delta x \tag{6}$$
>
> where $x_1, x_2, \ldots, x_n$ are arbitrary points in the n subintervals of $[a, b]$ of equal width $\Delta x = (b - a)/n$.

The Definite Integral

As we have just seen, the area under the graph of a continuous *nonnegative* function f on an interval $[a, b]$ is defined by the limit of the Riemann sum

$$\lim_{n \to \infty} \left[f(x_1)\, \Delta x + f(x_2)\, \Delta x + \cdots + f(x_n)\, \Delta x \right]$$

We now turn our attention to the study of limits of Riemann sums involving functions that are not necessarily nonnegative. Such limits arise in many applications of calculus.

---

*Even though we chose the representative points to be the midpoints of the subintervals in Example 1 and the left endpoints in Example 2, it can be shown that each of the respective sums will always approach the same unique number as n approaches infinity.

For example, the calculation of the distance covered by a body traveling along a straight line involves evaluating a limit of this form. The computation of the total revenue realized by a company over a certain time period, the calculation of the total amount of electricity consumed in a typical home over a 24-hour period, the average concentration of a drug in a body over a certain interval of time, and the volume of a solid—all involve limits of this type.

We begin with the following definition.

The Definite Integral

Let f be a function defined on $[a, b]$. If

$$\lim_{n \to \infty} \left[f(x_1)\,\Delta x + f(x_2)\,\Delta x + \cdots + f(x_n)\,\Delta x \right]$$

exists and is the same for all choices of representative points $x_1, x_2, \ldots, x_n$ in the n subintervals of $[a, b]$ of equal width $\Delta x = (b - a)/n$, then this limit is called the **definite integral** of f from a to b and is denoted by $\int_a^b f(x)\,dx$. Thus,

$$\int_a^b f(x)\,dx = \lim_{n \to \infty} \left[f(x_1)\,\Delta x + f(x_2)\,\Delta x + \cdots + f(x_n)\,\Delta x \right] \qquad (7)$$

The number a is called the **lower limit of integration**, and the number b is called the **upper limit of integration**.

Notes

1. If f is nonnegative, then the limit in Equation (7) is the same as the limit in (6); therefore, the definite integral gives the area under the graph of f on $[a, b]$.

2. The limit in Equation (7) is denoted by the integral sign $\int$ because, as we will see later, the definite integral and the antiderivative of a function f are related.

3. It is important to realize that the definite integral $\int_a^b f(x)\,dx$ is a *number*, whereas the indefinite integral $\int f(x)\,dx$ represents a *family of functions* (the antiderivatives of f).

4. If the limit in Equation (7) exists, we say that f is **integrable** on the interval $[a, b]$. ∎

When Is a Function Integrable?

The following theorem, which we state without proof, guarantees that a continuous function is integrable.

Integrability of a Function

Let f be continuous on $[a, b]$. Then, f is integrable on $[a, b]$; that is, the definite integral $\int_a^b f(x)\,dx$ exists.

Geometric Interpretation of the Definite Integral

If f is nonnegative and integrable on $[a, b]$, then we have the following geometric interpretation of the definite integral $\int_a^b f(x)\,dx$.

Geometric Interpretation of $\displaystyle\int_a^b f(x)\,dx$ for $f(x) \geq 0$ on $[a, b]$

If f is nonnegative and continuous on $[a, b]$, then

$$\int_a^b f(x)\,dx \tag{8}$$

is equal to the area of the region under the graph of f on $[a, b]$ (Figure 12).

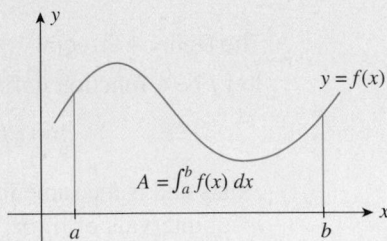

FIGURE 12
If $f(x) \geq 0$ on $[a, b]$, then $\int_a^b f(x)\,dx$ = area of the region under the graph of f on $[a, b]$.

Explore and Discuss

Suppose f is nonpositive [that is, $f(x) \leq 0$] and continuous on $[a, b]$. Explain why the area of the region below the x-axis and above the graph of f is given by $-\int_a^b f(x)\,dx$.

Next, let's extend our geometric interpretation of the definite integral to include the case where f assumes both positive and negative values on $[a, b]$. Consider a typical Riemann sum of the function f,

$$f(x_1)\,\Delta x + f(x_2)\,\Delta x + \cdots + f(x_n)\,\Delta x$$

corresponding to a partition of $[a, b]$ into n subintervals of equal width $(b - a)/n$, where $x_1, x_2, \ldots, x_n$ are representative points in the subintervals. The sum consists of n terms in which a positive term corresponds to the area of a rectangle of height $f(x_k)$ (for some positive integer k) lying above the x-axis and a negative term corresponds to the negative of the area of a rectangle of height $-f(x_k)$ lying below the x-axis. (See Figure 13, which depicts a situation with $n = 6$.)

As n gets larger and larger, the sums of the areas of the rectangles lying above the x-axis seem to give a better and better approximation of the area of the region lying above the x-axis (Figure 14). Similarly, the sums of the areas of those rectangles lying below the x-axis seem to give a better and better approximation of the area of the region lying below the x-axis.

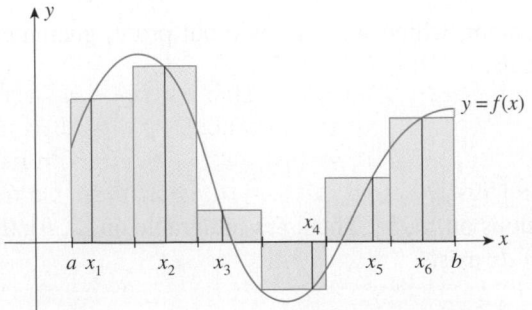

FIGURE 13
The positive terms in the Riemann sum are associated with the areas of the rectangles that lie above the x-axis, and the negative terms are associated with the areas of those that lie below the x-axis.

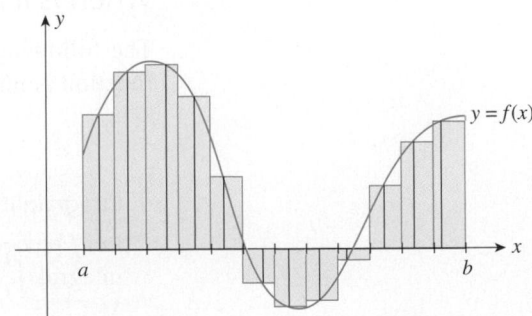

FIGURE 14
As n gets larger, the approximations get better. Here, $n = 12$, and we are approximating with twice as many rectangles as in Figure 13.

These observations suggest the following geometric interpretation of the definite integral for an arbitrary continuous function on an interval $[a, b]$.

Geometric Interpretation of $\int_a^b f(x)\,dx$ on $[a, b]$

If f is continuous on $[a, b]$, then

$$\int_a^b f(x)\,dx$$

is equal to the area of the region above $[a, b]$ minus the area of the region below $[a, b]$ (Figure 15).

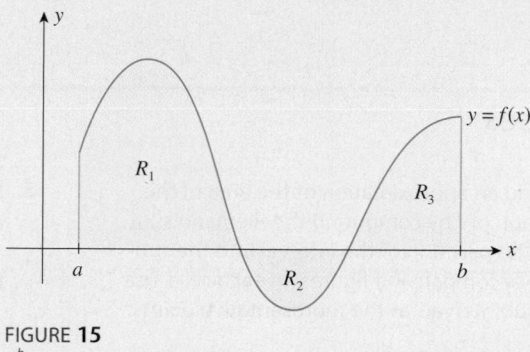

FIGURE **15**

$\int_a^b f(x)\,dx = $ Area of $R_1 - $ Area of $R_2 + $ Area of R_3

Finally, as promised at the beginning of the section, we will demonstrate that the total daily petroleum consumption of the New England state between 2008 and 2012 was given by the area under the graph of f between $t = 0$ and $t = 4$.

We begin by dividing the time interval $[0, 4]$ into n subintervals of equal length $\Delta t = 4/n$, with endpoints $t_0 = 0$, $t_1 = \Delta t$, $t_2 = 2\,\Delta t, \ldots, t_n = n\,\Delta t$. In the first sub-interval, $[t_0, t_1]$, we pick an arbitrary point t_1^*. Observe that because of the continuity of f, the values of f in $[t_0, t_1]$ do not vary appreciably from the constant value $f(t_1^*)$. So the consumption of petroleum over the period from $t = 0$ to $t = \Delta t$ is approximately $f(t_1^*)\,\Delta t$ million barrels. Similarly, we see that the petroleum consumption over the period from $t = t_1$ to $t = t_2$ is approximately $f(t_2^*)\,\Delta t$, where t_2^* is an arbitrary point in the subinterval $[t_1, t_2]$. Continuing, we see that the total petroleum consumption over the interval $[0, 4]$ is approximately

$$f(t_1^*)\,\Delta t + f(t_2^*)\,\Delta t + \cdots + f(t_n^*)\,\Delta t$$

where t_i^* is an arbitrary point in the interval $[t_{i-1}, t_i]$ $(1 \le i \le n)$. Intuitively, we see that as n gets larger and larger and the length of each subinterval gets smaller and smaller, the approximation gets better and better. Thus, as n becomes arbitrarily large, the sum approaches the limit that we take to be the total petroleum consumption on $[0, 4]$. But the sum is just the Riemann sum of the nonnegative function on $[0, 4]$, so its limit gives the area of the graph of f under $[0, 4]$.

11.3 Self-Check Exercise

Find an approximation of the area of the region R under the graph of $f(x) = 2x^2 + 1$ on the interval $[0, 3]$, using four subintervals of $[0, 3]$ of equal length and picking the midpoint of each subinterval as a representative point.

The solution to Self-Check Exercise 11.3 can be found on page 842.

11.3 Concept Questions

1. Explain how you would define the area of the region under the graph of a nonnegative continuous function f on the interval $[a, b]$.

2. Define the definite integral of a continuous function on the interval $[a, b]$. Give a geometric interpretation of

$$\int_a^b f(x)\, dx$$

for the case in which (a) f is nonnegative on $[a, b]$ and (b) f assumes both positive and negative values on $[a, b]$. Illustrate your answers graphically.

11.3 Exercises

In Exercises 1 and 2, find an approximation of the area of the region R under the graph of f by computing the Riemann sum of f corresponding to the partition of the interval into the subintervals shown in the accompanying figures. In each case, use the midpoints of the subintervals as the representative points.

1.

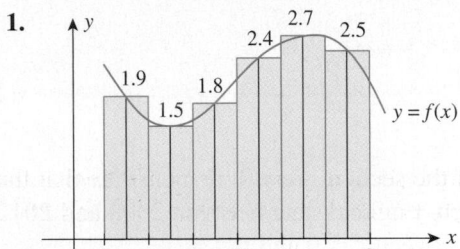

2.

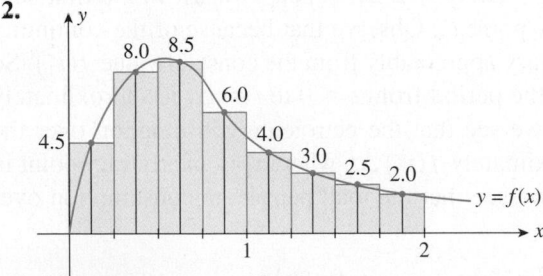

3. Let $f(x) = 3x$.
   a. Sketch the region R under the graph of f on the interval $[0, 2]$, and find its exact area using geometry.
   b. Use a Riemann sum with four subintervals of equal length ($n = 4$) to approximate the area of R. Choose the representative points to be the left endpoints of the subintervals.
   c. Repeat part (b) with eight subintervals of equal length ($n = 8$).
   d. Compare the approximations obtained in parts (b) and (c) with the exact area found in part (a). Do the approximations improve with larger n?

4. Repeat Exercise 3, choosing the representative points to be the right endpoints of the subintervals.

5. Let $f(x) = 4 - 2x$.
   a. Sketch the region R under the graph of f on the interval $[0, 2]$, and find its exact area using geometry.
   b. Use a Riemann sum with five subintervals of equal length ($n = 5$) to approximate the area of R. Choose the representative points to be the left endpoints of the subintervals.
   c. Repeat part (b) with ten subintervals of equal length ($n = 10$).
   d. Compare the approximations obtained in parts (b) and (c) with the exact area found in part (a). Do the approximations improve with larger n?

6. Repeat Exercise 5, choosing the representative points to be the right endpoints of the subintervals.

7. Let $f(x) = x^2$, and compute the Riemann sum of f over the interval $[2, 4]$, choosing the representative points to be the midpoints of the subintervals and using:
   a. Two subintervals of equal length ($n = 2$).
   b. Five subintervals of equal length ($n = 5$).
   c. Ten subintervals of equal length ($n = 10$).
   d. Can you guess at the area of the region under the graph of f on the interval $[2, 4]$?

8. Repeat Exercise 7, choosing the representative points to be the left endpoints of the subintervals.

9. Repeat Exercise 7, choosing the representative points to be the right endpoints of the subintervals.

10. Let $f(x) = x^3$, and compute the Riemann sum of f over the interval $[0, 1]$, choosing the representative points to be the midpoints of the subintervals and using:
    a. Two subintervals of equal length ($n = 2$).
    b. Five subintervals of equal length ($n = 5$).
    c. Ten subintervals of equal length ($n = 10$).
    d. Can you guess at the area of the region under the graph of f on the interval $[0, 1]$?

11. Repeat Exercise 10, choosing the representative points to be the left endpoints of the subintervals.

12. Repeat Exercise 10, choosing the representative points to be the right endpoints of the subintervals.

In Exercises 13–16, find an approximation of the area of the region R under the graph of the function f on the interval $[a, b]$. In each case, use n subintervals and choose the representative points as indicated.

13. $f(x) = x^2 + 1; [0, 2]; n = 5$; midpoints

14. $f(x) = 4 - x^2; [-1, 2]; n = 6$; left endpoints

15. $f(x) = \dfrac{1}{x}; [1, 3]; n = 4$; right endpoints

16. $f(x) = e^x; [0, 3]; n = 5$; midpoints

17. REAL ESTATE Figure (a) shows a vacant lot with a 100-ft frontage in a development. To estimate its area, we introduce a coordinate system so that the x-axis coincides with the edge of the straight road forming the lower boundary of the property, as shown in Figure (b). Then, thinking of the upper boundary of the property as the graph of a continuous function f over the interval $[0, 100]$, we see that the problem is mathematically equivalent to that of finding the area under the graph of f on $[0, 100]$. To estimate the area of the lot using a Riemann sum, we divide the interval $[0, 100]$ into five equal subintervals of length 20 ft. Then, using surveyor's equipment, we measure the distance from the midpoint of each of these subintervals to the upper boundary of the property. These measurements give the values of $f(x)$ at $x = 10, 30, 50, 70,$ and 90. What is the approximate area of the lot?

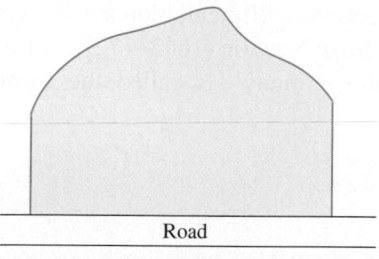

(a)

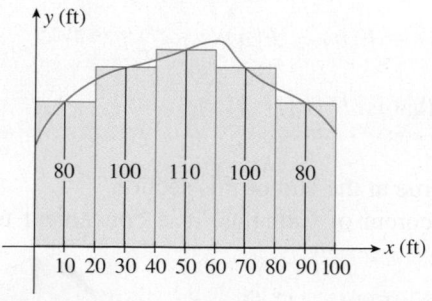

(b)

18. REAL ESTATE Use the technique of Exercise 17 to obtain an estimate of the area of the vacant lot shown in the accompanying figures.

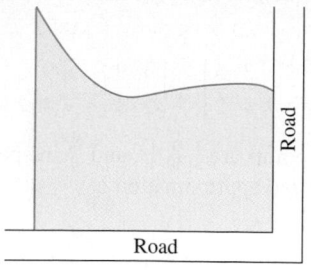

(a)

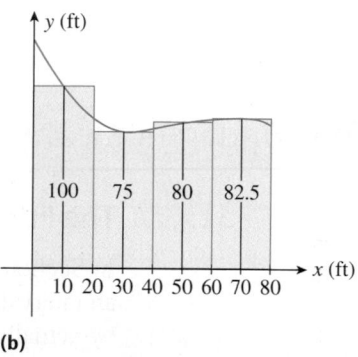

(b)

In Exercises 19 and 20, determine whether the statement is true or false. If it is true, explain why it is true. If it is false, give an example to show why it is false.

19. If f is continuous on $[a, b]$ and $\int_a^b f(x)\, dx > 0$, then $f(x) \geq 0$ for all x in $[a, b]$.

20. If f is continuous on $[a, b]$ and $f(x) \neq 0$ for all x in $[a, b]$, then $\int [f(x)]^2\, dx > 0$.

11.3 Solution to Self-Check Exercise

The length of each subinterval is $\frac{3}{4}$. Therefore, the four subintervals are

$$\left[0, \frac{3}{4}\right], \quad \left[\frac{3}{4}, \frac{3}{2}\right], \quad \left[\frac{3}{2}, \frac{9}{4}\right], \quad \left[\frac{9}{4}, 3\right]$$

The representative points are $\frac{3}{8}, \frac{9}{8}, \frac{15}{8}$, and $\frac{21}{8}$, respectively. Therefore, the required approximation is

$$
\begin{aligned}
A &= \frac{3}{4} f\left(\frac{3}{8}\right) + \frac{3}{4} f\left(\frac{9}{8}\right) + \frac{3}{4} f\left(\frac{15}{8}\right) + \frac{3}{4} f\left(\frac{21}{8}\right) \\
&= \frac{3}{4}\left[f\left(\frac{3}{8}\right) + f\left(\frac{9}{8}\right) + f\left(\frac{15}{8}\right) + f\left(\frac{21}{8}\right) \right] \\
&= \frac{3}{4}\left\{ \left[2\left(\frac{3}{8}\right)^2 + 1\right] + \left[2\left(\frac{9}{8}\right)^2 + 1\right] + \left[2\left(\frac{15}{8}\right)^2 + 1\right] \right. \\
&\quad \left. + \left[2\left(\frac{21}{8}\right)^2 + 1\right] \right\} \\
&= \frac{3}{4}\left(\frac{41}{32} + \frac{113}{32} + \frac{257}{32} + \frac{473}{32}\right) = \frac{663}{32}
\end{aligned}
$$

or approximately 20.72.

11.4 The Fundamental Theorem of Calculus

The Fundamental Theorem of Calculus

In Section 11.3, we defined the definite integral of an arbitrary continuous function on an interval $[a, b]$ as a limit of Riemann sums. Calculating the value of a definite integral by actually taking the limit of such sums is tedious and in most cases impractical. It is important to realize that the numerical results we obtained in Examples 1 and 2 of Section 11.3 were *approximations* of the respective areas of the regions in question, even though these results enabled us to *conjecture* what the actual areas might be. Fortunately, there is a much better way of finding the exact value of a definite integral.

The following theorem shows how to evaluate the definite integral of a continuous function provided that we can find an antiderivative of that function. Because of its importance in establishing the relationship between differentiation and integration, this theorem—discovered independently by Sir Isaac Newton (1642–1727) in England and Gottfied Wilhelm Leibniz (1646–1716) in Germany—is called the **Fundamental Theorem of Calculus.**

> **THEOREM 2**
>
> **The Fundamental Theorem of Calculus**
>
> Let f be continuous on $[a, b]$. Then,
>
> $$\int_a^b f(x)\, dx = F(b) - F(a) \tag{9}$$
>
> where F is any antiderivative of f; that is, $F'(x) = f(x)$.

We will explain why this theorem is true at the end of this section.

In applying the Fundamental Theorem of Calculus, it is convenient to use the notation

$$F(x)\Big|_a^b = F(b) - F(a)$$

For example, in this notation, Equation (9) is written

$$\int_a^b f(x)\, dx = F(x)\Big|_a^b = F(b) - F(a)$$

Molly H. Fisher, David C. Royster, and Diandra Leslie-Pelecky

TITLE Professors of Mathematics Education, Mathematics, and Physics
INSTITUTION University of Kentucky (Fisher and Royster); Author, *The Physics of NASCAR®* (Leslie-Pelecky)

The last thing a fan thinks about at a NASCAR Sprint Cup race is calculus. The Sprint Cup series is NASCAR's most popular and most profitable series. During a ten-month season, drivers compete in 36 races across the United States and Mexico.

Typically, a racing team is made up not only of a driver, who is the most visible, and mechanics, who are often seen at the pit stops, but also of many people who work behind the scenes, away from the track. The last group includes highly trained engineers and scientists (some of whom hold Ph.D.s in engineering, mathematics, or physics) who study how each component of a race car performs under different conditions.

According to Dr. Andrew Randolph, who is a chemical engineer and the Engine Technical Director of Earnhardt Childress Racing Engines, "a calculation we do every week is obtaining the average power over a given speed range to provide a single engine performance metric for a given track. The calculation is a simple integral of power versus speed divided by the speed delta over which the integration is performed."

Much of the data available for this purpose is in graphical form. So recognizing the relationship between the area under a curve and the corresponding definite integral is essential to the solution of the problem. Even though there are many parameters that determine how an engine in a race car performs, the problem ultimately reduces to one of finding the average value of a function as given by the integral:

$$\frac{1}{\Delta v} \int_{v_0}^{v_1} P(v) \, dv$$

This is just one of many examples of how calculus can be used in a sport that some people think is reserved for speed-hungry daredevils.

EXAMPLE 1 Let R be the region under the graph of $f(x) = x$ on the interval $[1, 3]$. Use the Fundamental Theorem of Calculus to find the area A of R, and verify your result by elementary means.

Solution The region R is shown in Figure 16a. Since f is nonnegative on $[1, 3]$, the area of R is given by the definite integral of f from 1 to 3; that is,

$$A = \int_1^3 x \, dx$$

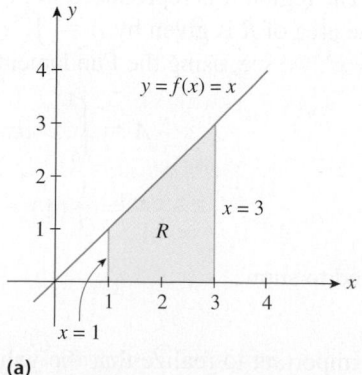

(a)

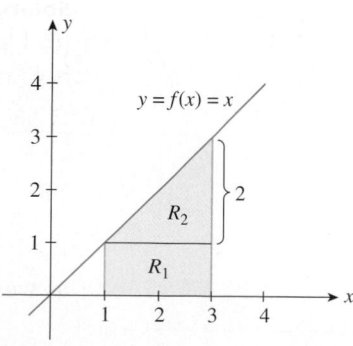

(b)

FIGURE 16
The area of R can be computed in two different ways.

To evaluate the definite integral, observe that an antiderivative of $f(x) = x$ is $F(x) = \frac{1}{2}x^2 + C$, where C is an arbitrary constant. Therefore, by the Fundamental Theorem of Calculus, we have

$$A = \int_1^3 x \, dx = \frac{1}{2}x^2 + C \Big|_1^3$$

$$= \left(\frac{9}{2} + C\right) - \left(\frac{1}{2} + C\right) = 4$$

To verify this result by elementary means, refer to Figure 16b. Observe that the area A is the area of the rectangle R_1 (width $\times$ height) plus the area of the triangle R_2 ($\frac{1}{2}$ base $\times$ height); that is,

$$2(1) + \frac{1}{2}(2)(2) = 2 + 2 = 4$$

which agrees with the result obtained earlier.

Observe that in evaluating the definite integral in Example 1, the constant of integration "dropped out." This is true in general, for if $F(x) + C$ denotes an antiderivative of some function f, then

$$F(x) + C \Big|_a^b = [F(b) + C] - [F(a) + C]$$

$$= F(b) + C - F(a) - C$$

$$= F(b) - F(a)$$

With this fact in mind, we may, in all future computations involving the evaluation of a definite integral, drop the constant of integration from our calculations.

Finding the Area of a Region Under a Curve

Having seen how effective the Fundamental Theorem of Calculus is in helping us find the area of simple regions, we now use it to find the area of more complicated regions.

EXAMPLE 2 In Section 11.3, we conjectured that the area of the region R under the graph of $f(x) = x^2$ on the interval $[0, 1]$ was $\frac{1}{3}$. Use the Fundamental Theorem of Calculus to verify this conjecture.

Solution The region R is reproduced in Figure 17. Observe that f is nonnegative on $[0, 1]$, so the area of R is given by $A = \int_0^1 x^2 \, dx$. Since an antiderivative of $f(x) = x^2$ is $F(x) = \frac{1}{3}x^3$, we see, using the Fundamental Theorem of Calculus, that

$$A = \int_0^1 x^2 \, dx = \frac{1}{3}x^3 \Big|_0^1$$

$$= \frac{1}{3}(1) - \frac{1}{3}(0) = \frac{1}{3}$$

as we wished to show.

Note It is important to realize that the value $\frac{1}{3}$ is by definition the exact value of the area of R.

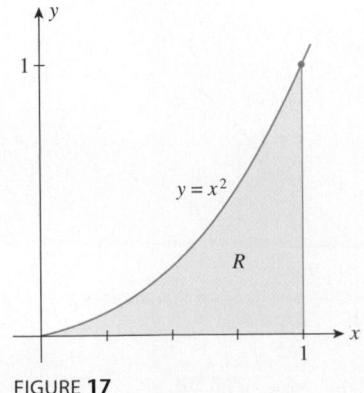

FIGURE 17
The area of R is $\int_0^1 x^2 \, dx = \frac{1}{3}$.

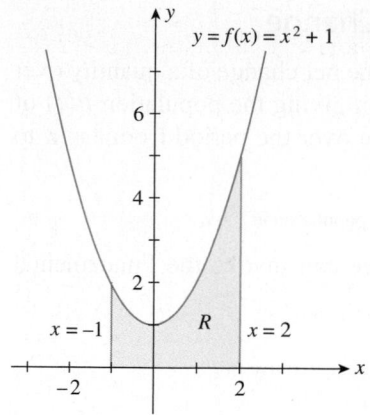

FIGURE 18
The area of R is $\int_{-1}^{2} (x^2 + 1)\, dx$.

EXAMPLE 3 Find the area of the region R under the graph of $y = x^2 + 1$ from $x = -1$ to $x = 2$.

Solution The region R under consideration is shown in Figure 18. Using the Fundamental Theorem of Calculus, we find that the required area is

$$\int_{-1}^{2} (x^2 + 1)\, dx = \left(\frac{1}{3}x^3 + x\right)\Big|_{-1}^{2}$$

$$= \left[\frac{1}{3}(2)^3 + 2\right] - \left[\frac{1}{3}(-1)^3 + (-1)\right] = 6 \qquad ◼$$

Evaluating Definite Integrals

In Examples 4 and 5, we use the rules of integration of Section 11.1 to help us evaluate the definite integrals.

EXAMPLE 4 Evaluate $\displaystyle\int_{1}^{3} (3x^2 + e^x)\, dx$.

Solution

$$\int_{1}^{3} (3x^2 + e^x)\, dx = x^3 + e^x \Big|_{1}^{3}$$

$$= (27 + e^3) - (1 + e) = 26 + e^3 - e \qquad ◼$$

EXAMPLE 5 Evaluate $\displaystyle\int_{1}^{2}\left(\frac{1}{x} - \frac{1}{x^2}\right) dx$.

Solution

$$\int_{1}^{2}\left(\frac{1}{x} - \frac{1}{x^2}\right) dx = \int_{1}^{2}\left(\frac{1}{x} - x^{-2}\right) dx$$

$$= \ln|x| + \frac{1}{x}\Big|_{1}^{2}$$

$$= \left(\ln 2 + \frac{1}{2}\right) - (\ln 1 + 1)$$

$$= \ln 2 - \frac{1}{2} \qquad \text{Recall that } \ln 1 = 0. \qquad ◼$$

Explore and Discuss

Consider the definite integral $\displaystyle\int_{-1}^{1} \frac{1}{x^2}\, dx$.

1. Show that a formal application of Equation (9) leads to

$$\int_{-1}^{1} \frac{1}{x^2}\, dx = -\frac{1}{x}\Big|_{-1}^{1} = -1 - 1 = -2$$

2. Observe that $f(x) = 1/x^2$ is positive at each value of x in $[-1, 1]$ where it is defined. So one might expect that the definite integral with integrand f has a positive value, if it exists.

3. Resolve this apparent contradiction in the result (1) and the observation (2).

The Definite Integral as a Measure of Net Change

In real-world applications, we are often interested in the net change of a quantity over a period of time. For example, suppose P is a function giving the population $P(t)$ of a city at time t. Then the *net change* in the population over the period from $t = a$ to $t = b$ is given by

$$P(b) - P(a) \qquad \text{Population at } t = b \text{ minus population at } t = a$$

If P has a continuous derivative P' in $[a, b]$, then we can invoke the Fundamental Theorem of Calculus to write

$$P(b) - P(a) = \int_a^b P'(t) \, dt \qquad \text{P is an antiderivative of P'.}$$

Thus, if we know the *rate of change* of the population at any time t, then we can calculate the net change in the population from $t = a$ to $t = b$ by evaluating an appropriate definite integral.

APPLIED EXAMPLE 6 Population Growth in Clark County As the twentieth century ended, Clark County in Nevada—dominated by Las Vegas—was one of the fastest-growing metropolitan areas in the United States. From 1970 through 2000, the population was growing at the rate of

$$R(t) = 133{,}680t^2 - 178{,}788t + 234{,}633 \qquad (0 \le t \le 3)$$

people per decade, where $t = 0$ corresponds to the beginning of 1970. What was the net change in the population over the decade from 1980 to 1990?
Source: U.S. Census Bureau.

Solution The net change in the population over the decade from 1980 $(t = 1)$ to 1990 $(t = 2)$ is given by $P(2) - P(1)$, where P denotes the population in the county at time t. But $P' = R$, so

$$
\begin{aligned}
P(2) - P(1) &= \int_1^2 P'(t) \, dt = \int_1^2 R(t) \, dt \\
&= \int_1^2 (133{,}680t^2 - 178{,}788t + 234{,}633) \, dt \\
&= 44{,}560t^3 - 89{,}394t^2 + 234{,}633t \Big|_1^2 \\
&= [44{,}560(2)^3 - 89{,}394(2)^2 + 234{,}633(2)] \\
&\quad - [44{,}560 - 89{,}394 + 234{,}633] \\
&= 278{,}371
\end{aligned}
$$

and so the net change is 278,371.

More generally, we have the following result. We assume that f has a continuous derivative, even though the integrability of f' is sufficient.

Net Change Formula

The net change in a function f over an interval $[a, b]$ is given by

$$f(b) - f(a) = \int_a^b f'(x) \, dx \qquad \text{(10)}$$

provided that f' is continuous on $[a, b]$.

Additional examples of the net change of a function follow.

APPLIED EXAMPLE 7 Production Costs The management of Staedtler Office Equipment has determined that the daily marginal cost function associated with producing battery-operated pencil sharpeners is given by

$$C'(x) = 0.000006x^2 - 0.006x + 4$$

where $C'(x)$ is measured in dollars per unit and x denotes the number of units produced. Management has also determined that the daily fixed cost incurred in producing these pencil sharpeners is $100. Find Staedtler's daily total cost for producing (a) the first 500 units and (b) the 201st through 400th units.

Solution

a. Since $C'(x)$ is the marginal cost function, its antiderivative $C(x)$ is the total cost function. The daily fixed cost incurred in producing the pencil sharpeners is $C(0)$ dollars. Since the daily fixed cost is given as $100, we have $C(0) = 100$. We are required to find $C(500)$. Let's compute $C(500) - C(0)$, the net change in the total cost function $C(x)$ over the interval $[0, 500]$. Using the Fundamental Theorem of Calculus, we find

$$C(500) - C(0) = \int_0^{500} C'(x)\, dx$$

$$= \int_0^{500} (0.000006x^2 - 0.006x + 4)\, dx$$

$$= 0.000002x^3 - 0.003x^2 + 4x \Big|_0^{500}$$

$$= [0.000002(500)^3 - 0.003(500)^2 + 4(500)]$$
$$\quad - [0.000002(0)^3 - 0.003(0)^2 + 4(0)]$$

$$= 1500$$

Therefore, $C(500) = 1500 + C(0) = 1500 + 100 = 1600$, so the total cost incurred daily by Staedtler in producing the first 500 pencil sharpeners is $1600.

b. The daily total cost incurred by Staedtler in producing the 201st through 400th units of battery-operated pencil sharpeners is given by

$$C(400) - C(200) = \int_{200}^{400} C'(x)\, dx$$

$$= \int_{200}^{400} (0.000006x^2 - 0.006x + 4)\, dx$$

$$= 0.000002x^3 - 0.003x^2 + 4x \Big|_{200}^{400}$$

$$= [0.000002(400)^3 - 0.003(400)^2 + 4(400)]$$
$$\quad - [0.000002(200)^3 - 0.003(200)^2 + 4(200)]$$

$$= 552$$

or $552.

Since $C'(x)$ is nonnegative for x in the interval $[0, \infty)$, we have the following geometric interpretation of the two definite integrals in Example 7: $\int_0^{500} C'(x)\,dx$ is the area of the region under the graph of the function C' from $x = 0$ to $x = 500$, shown in Figure 19a, and $\int_{200}^{400} C'(x)\,dx$ is the area of the region from $x = 200$ to $x = 400$, shown in Figure 19b.

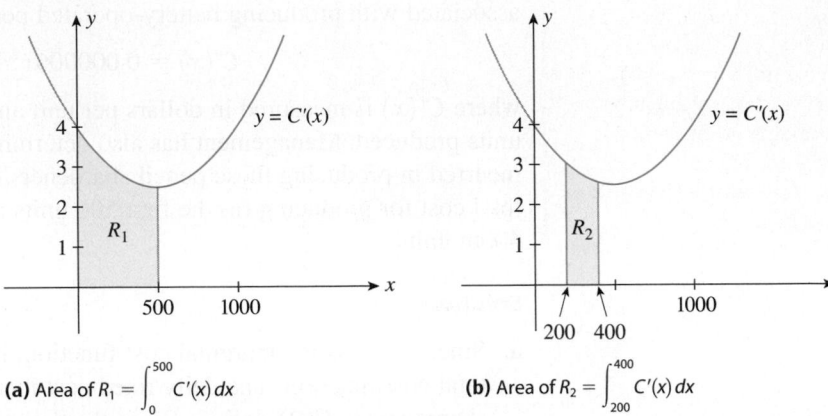

(a) Area of $R_1 = \displaystyle\int_0^{500} C'(x)\,dx$ **(b)** Area of $R_2 = \displaystyle\int_{200}^{400} C'(x)\,dx$

FIGURE **19**

💲 APPLIED EXAMPLE 8 Assembly Time of Workers An efficiency study conducted for Elektra Electronics showed that the rate at which Space Commander walkie-talkies are assembled by the average worker t hours after starting work at 8 A.M. is given by the function

$$f(t) = -3t^2 + 12t + 15 \qquad (0 \le t \le 4)$$

Determine how many walkie-talkies can be assembled by the average worker in the first hour of the morning shift.

Solution Let $N(t)$ denote the number of walkie-talkies assembled by the average worker during the first t hours after starting work in the morning shift. Then we have

$$N'(t) = f(t) = -3t^2 + 12t + 15$$

Therefore, the number of units assembled by the average worker in the first hour of the morning shift is

$$N(1) - N(0) = \int_0^1 N'(t)\,dt = \int_0^1 (-3t^2 + 12t + 15)\,dt$$

$$= -t^3 + 6t^2 + 15t \Big|_0^1 = -1 + 6 + 15$$

$$= 20$$

or 20 units. ∎

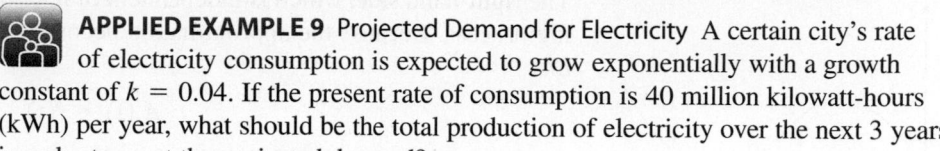

APPLIED EXAMPLE 9 Projected Demand for Electricity A certain city's rate of electricity consumption is expected to grow exponentially with a growth constant of $k = 0.04$. If the present rate of consumption is 40 million kilowatt-hours (kWh) per year, what should be the total production of electricity over the next 3 years in order to meet the projected demand?

Solution If $R(t)$ denotes the expected rate of consumption of electricity t years from now, then

$$R(t) = 40e^{0.04t}$$

million kWh per year. Next, if $C(t)$ denotes the expected total consumption of electricity over the next t years, then

$$C'(t) = R(t)$$

Therefore, the total consumption of electricity expected over the next 3 years is given by

$$
\begin{aligned}
\int_0^3 C'(t)\, dt &= \int_0^3 40e^{0.04t}\, dt \\
&= \frac{40}{0.04} e^{0.04t} \Big|_0^3 \\
&= 1000(e^{0.12} - 1) \\
&\approx 127.5
\end{aligned}
$$

or 127.5 million kWh, the amount that must be produced over the next 3 years to meet the demand. ∎

Validity of the Fundamental Theorem of Calculus

To demonstrate the plausibility of the Fundamental Theorem of Calculus for the case in which f is nonnegative on an interval $[a, b]$, let's define an "area function" A as follows. Let $A(t)$ denote the area of the region R under the graph of $y = f(x)$ from $x = a$ to $x = t$, where $a \le t \le b$ (Figure 20).

If h is a small positive number, then $A(t + h)$ is the area of the region under the graph of $y = f(x)$ from $x = a$ to $x = t + h$. Therefore, the difference

$$A(t + h) - A(t)$$

is the area of the region under the graph of $y = f(x)$ from $x = t$ to $x = t + h$ (Figure 21).

Now, the area of this last region can be approximated by the area of the rectangle of width h and height $f(t)$—that is, by the expression $h \cdot f(t)$ (see Figure 22 on the next page). Thus,

$$A(t + h) - A(t) \approx h \cdot f(t)$$

where the approximations improve as h is taken to be smaller and smaller.

Dividing both sides of the foregoing relationship by h, we obtain

$$\frac{A(t + h) - A(t)}{h} \approx f(t)$$

Taking the limit as h approaches zero, we find, by the definition of the derivative, that the left-hand side is

$$\lim_{h \to 0} \frac{A(t + h) - A(t)}{h} = A'(t)$$

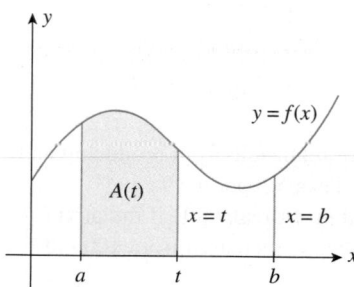

FIGURE 20
$A(t) =$ area under the graph of f from $x = a$ to $x = t$

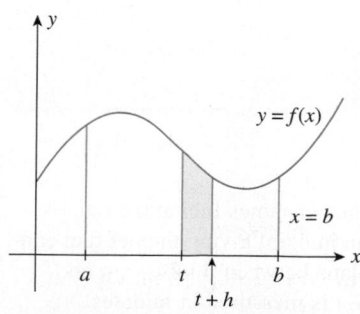

FIGURE 21
$A(t + h) - A(t) =$ area under the graph of f from $x = t$ to $x = t + h$

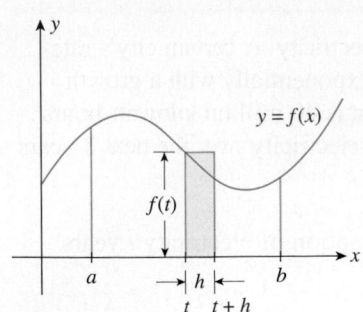

FIGURE 22
The area of the rectangle is $h \cdot f(t)$.

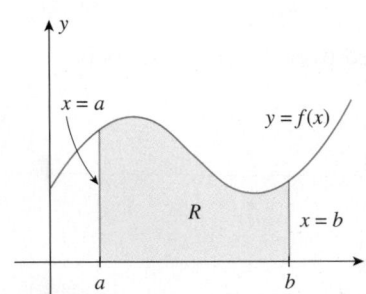

FIGURE 23
The area of R is given by $A(b)$.

The right-hand side, which is independent of h, remains constant throughout the limiting process. Because the approximation becomes exact as h approaches zero, we find that

$$A'(t) = f(t)$$

Since the foregoing equation holds for all values of t in the interval $[a, b]$, we have shown that the *area function A* is an antiderivative of the function f. By Theorem 1 of Section 11.1, we conclude that $A(x)$ must have the form

$$A(x) = F(x) + C$$

where F is any antiderivative of f and C is a constant. To determine the value of C, observe that $A(a) = 0$. This condition implies that

$$A(a) = F(a) + C = 0$$

or $C = -F(a)$. Next, since the area of the region R is $A(b)$ (Figure 23), we see that the required area is

$$A(b) = F(b) + C$$
$$= F(b) - F(a)$$

Since the area of the region R is

$$\int_a^b f(x)\, dx$$

we have

$$\int_a^b f(x)\, dx = F(b) - F(a)$$

as we set out to show.

11.4 Self-Check Exercises

1. Evaluate $\displaystyle\int_0^2 (x + e^x)\, dx$.

2. **MARGINAL PROFIT OF A COMPANY** The daily marginal profit function associated with producing and selling TexaPep hot sauce is

$$P'(x) = -0.000006x^2 + 6$$

where x denotes the number of cases (each case contains 24 bottles) produced and sold daily and $P'(x)$ is measured in dollars per unit. The fixed cost is \$400.

a. What is the total profit realizable from producing and selling 1000 cases of TexaPep per day?
b. What is the additional profit realizable if the production and sale of TexaPep is increased from 1000 to 1200 cases/day?

Solutions to Self-Check Exercises 11.4 can be found on page 853.

11.4 Concept Questions

1. State the Fundamental Theorem of Calculus.

2. State the net change formula, and use it to answer the following questions:
   a. If a company generates income at the rate of R dollars/day, explain what $\int_a^b R(t)\, dt$ measures, where a and b are measured in days with $a < b$.
   b. If a private jet airplane consumes fuel at the rate of $R(t)$ gal/min, write an integral giving the net fuel consumption by the airplane between times $t = a$ and $t = b$ $(a < b)$, where t is measured in minutes.

11.4 Exercises

In Exercises 1–4, find the area of the region under the graph of the function f on the interval $[a, b]$, using the Fundamental Theorem of Calculus. Then verify your result using geometry.

1. $f(x) = 2; [1, 4]$

2. $f(x) = 4; [-1, 2]$

3. $f(x) = 2x; [1, 3]$

4. $f(x) = -\dfrac{1}{4}x + 1; [1, 4]$

In Exercises 5–16, find the area of the region under the graph of the function f on the interval $[a, b]$.

5. $f(x) = 2x + 3; [-1, 2]$

6. $f(x) = 4x - 1; [2, 4]$

7. $f(x) = -x^2 + 4; [-1, 2]$

8. $f(x) = 4x - x^2; [0, 4]$

9. $f(x) = \dfrac{1}{x}; [1, 2]$

10. $f(x) = \dfrac{1}{x^2}; [2, 4]$

11. $f(x) = \sqrt{x}; [1, 9]$

12. $f(x) = x^3; [1, 3]$

13. $f(x) = 1 - \sqrt[3]{x}; [-8, -1]$

14. $f(x) = \dfrac{1}{\sqrt{x}}; [1, 9]$

15. $f(x) = e^x; [0, 2]$

16. $f(x) = e^x - x; [1, 2]$

In Exercises 17–40, evaluate the definite integral.

17. $\displaystyle\int_2^4 3\, dx$

18. $\displaystyle\int_{-1}^2 -2\, dx$

19. $\displaystyle\int_1^4 (2x + 3)\, dx$

20. $\displaystyle\int_{-1}^0 (4 - x)\, dx$

21. $\displaystyle\int_{-1}^3 2x^2\, dx$

22. $\displaystyle\int_0^2 8x^3\, dx$

23. $\displaystyle\int_{-2}^2 (x^2 - 1)\, dx$

24. $\displaystyle\int_1^4 \sqrt{u}\, du$

25. $\displaystyle\int_1^8 2x^{1/3}\, dx$

26. $\displaystyle\int_1^4 2x^{-3/2}\, dx$

27. $\displaystyle\int_0^1 (x^3 - 2x^2 + 1)\, dx$

28. $\displaystyle\int_1^2 (t^5 - t^3 + 1)\, dt$

29. $\displaystyle\int_1^4 \dfrac{1}{x}\, dx$

30. $\displaystyle\int_1^3 \dfrac{2}{x}\, dx$

31. $\displaystyle\int_0^4 x(x^2 - 1)\, dx$

32. $\displaystyle\int_0^2 (x - 4)(x - 1)\, dx$

33. $\displaystyle\int_1^3 (t^2 - t)^2\, dt$

34. $\displaystyle\int_{-1}^1 (x^2 - 1)^2\, dx$

35. $\displaystyle\int_{-3}^{-1} \dfrac{1}{x^2}\, dx$

36. $\displaystyle\int_1^2 \dfrac{2}{x^3}\, dx$

37. $\displaystyle\int_1^4 \left(\sqrt{x} - \dfrac{1}{\sqrt{x}}\right) dx$

38. $\displaystyle\int_0^1 \sqrt{2x}(\sqrt{x} + \sqrt{2})\, dx$

39. $\displaystyle\int_1^4 \dfrac{3x^3 - 2x^2 + 4}{x^2}\, dx$

40. $\displaystyle\int_1^2 \left(1 + \dfrac{1}{u} + \dfrac{1}{u^2}\right) du$

41. PERSONAL BANKRUPTCY The number of personal bankruptcy filings by fiscal years ending September 30 between 2010 and 2012 was declining at the rate of

$$R(t) = 0.077t + 0.0825 \qquad (0 \le t \le 2)$$

million cases/year, t years after September 30, 2010. The number of filings as of September 30, 2010, stood at approximately 1.538 million cases.

a. Estimate the change in the number of personal bankruptcy cases filed between September 30, 2010, and September 30, 2012.

b. What was the approximate number of personal bankruptcy cases filed in 2012?

Hint: If $N(t)$ denotes the number of bankruptcy filings in year t, then $N'(t) = -R(t)$.

Source: Administrative Office of the U.S. Courts.

42. HEALTH CARE COSTS According to a study conducted by the Centers for Medicare & Medicaid Services in 2010, the national spending for out-of-pocket health-care costs is projected to increase over the next several years. The amount spent annually from 2010 $(t = 0)$ is expected to grow at the rate of

$$R(t) = 1.0952t + 17.357 \qquad (0 \le t \le 6)$$

billion dollars/year in year t. The national spending in 2010 was \$317 billion. What is the projected national spending in 2016?

Source: Centers for Medicare & Medicaid Services.

43. MARGINAL COST A division of Ditton Industries manufactures a deluxe toaster oven. Management has determined that the daily marginal cost function associated with producing these toaster ovens is given by

$$C'(x) = 0.0003x^2 - 0.12x + 20$$

where $C'(x)$ is measured in dollars per unit and x denotes the number of units produced. Management has also determined that the daily fixed cost incurred in the production is \$800.

a. Find the total cost incurred by Ditton in producing the first 300 units of these toaster ovens per day.

b. What is the total cost incurred by Ditton in producing the 201st through 300th units/day?

44. MARGINAL REVENUE The management of Ditton Industries has determined that the daily marginal revenue function associated with selling x units of their deluxe toaster ovens is given by

$$R'(x) = -0.1x + 40$$

where $R'(x)$ is measured in dollars per unit.

a. Find the daily total revenue realized from the sale of 200 units of the toaster oven.

b. Find the additional revenue realized when the production (and sales) level is increased from 200 to 300 units.

45. MARGINAL PROFIT Refer to Exercise 43. The daily marginal profit function associated with the production and sales of the deluxe toaster ovens is known to be

$$P'(x) = -0.0003x^2 + 0.02x + 20$$

where x denotes the number of units manufactured and sold daily and $P'(x)$ is measured in dollars per unit.
  a. Find the total profit realizable from the manufacture and sale of 200 units of the toaster ovens per day.
    Hint: $P(200) - P(0) = \int_0^{200} P'(x)\,dx$, $P(0) = -800$.
  b. What is the additional daily profit realizable if the production and sales of the toaster ovens are increased from 200 to 220 units/day?

46. EFFICIENCY STUDIES Tempco Electronics, a division of Tempco Toys, manufactures an electronic football game. An efficiency study showed that the rate at which the games are assembled by the average worker t hr after starting work at 8 A.M. is

$$-\frac{3}{2}t^2 + 6t + 20 \qquad (0 \le t \le 4)$$

units/hr.
  a. Find the total number of games the average worker can be expected to assemble in the 4-hr morning shift.
  b. How many units can the average worker be expected to assemble in the first hour of the morning shift? In the second hour of the morning shift?

47. SPEEDBOAT RACING In a recent pretrial run for the world water speed record, the velocity of the *Sea Falcon II* t sec after firing the booster rocket was given by

$$v(t) = -t^2 + 20t + 440 \qquad (0 \le t \le 20)$$

ft/sec. Find the distance covered by the boat over the 20-sec period after the booster rocket was activated.
Hint: The distance is given by $\int_0^{20} v(t)\,dt$.

48. TABLET COMPUTERS Annual sales (in millions of units) of a certain brand of tablet computers are expected to grow in accordance with the function

$$f(t) = 0.18t^2 + 0.16t + 2.64 \qquad (0 \le t \le 6)$$

per year, where t is measured in years. How many tablet computers will be sold over the next 6 years?

49. CREDIT CARD DEBT The average U.S. household credit card debt stood at $8382 per household at the beginning of 2008 ($t = 0$) and was changing at the rate of

$$f(t) = 258t^2 - 680t - 316 \qquad (0 \le t \le 4)$$

dollars/year in year t from the beginning of 2008 until the beginning of 2012. What was the average U.S. household credit card debt at the beginning of 2012?
Source: TIM.

50. U.S. NATIONAL DEBT The U.S. national debt stood at $10.025 trillion in 2008 ($t = 0$) and was growing at the rate of

$$R(t) = 0.070251t^2 - 0.51548t + 2.1667 \qquad (0 \le t \le 4)$$

trillion dollars/year in year t between 2008 and 2012. What was the U.S. national debt in 2012?
Source: U.S. Treasury Direct.

51. AIR PURIFICATION To test air purifiers, engineers ran a purifier in a smoke-filled 10-ft $\times$ 20-ft room. While conducting a test for a certain brand of air purifier, they determined that the amount of smoke in the room was decreasing at the rate of

$$R(t) = 0.00032t^4 - 0.01872t^3 + 0.3948t^2$$
$$- 3.83t + 17.63 \qquad (0 \le t \le 20)$$

percent of the (original) amount of the smoke per minute, t min after the start of the test. How much smoke was left in the room 5 min after the start of the test? 10 min after the start of the test?
Source: Consumer Reports.

52. SOLAR PANEL PRODUCTION The manager of Sodex Corporation estimates that t months after it first began production, it was manufacturing the company's 140-watt 12-volt nominal solar panels at the rate of

$$R(t) = \frac{4t}{1 + t^2} + 3\sqrt{t}$$

hundred panels per month. What is the manager's estimate of the number of solar panels manufactured by the company during the second year?

53. CELL PHONE AD SPENDING Cell phone advertising spending is expected to grow at the rate of

$$R(t) = 0.8256t^{-0.04} \qquad (1 \le t \le 5)$$

billion dollars/year between 2007 ($t = 1$) and 2011 ($t = 5$). The cell phone ad spending in 2007 was $0.9 billion.
  a. Find an expression giving the cell phone ad spending in year t.
  b. If the trend continued, what was the cell phone ad spending in 2014?
Source: Interactive Advertising Bureau.

54. CREDIT CARD DELINQUENCY RATE The credit card delinquency rate stood at 5.7% at the beginning of 2010. It was declining at the rate of

$$R(t) = 0.150975e^{-0.0275t} \qquad (0 \le t \le 24)$$

percent t months later. What was the credit card delinquency rate at the beginning of 2012?
Source: Federal Reserve.

55. AMAZON'S GROWTH Amazon's revenue was growing at the rate of

$$R'(t) = 0.545043e^{0.291t} \qquad (0 \le t \le 6)$$

billion dollars/year in year t with $t = 0$ corresponding to 2006. The revenue of the company for 2006 was $10.71 billion.

a. By how much did the revenue of Amazon increase over the period from 2006 through 2012?

b. What was the revenue of the company in 2012?

Source: Amazon.

56. CANADIAN OIL-SANDS PRODUCTION The production of oil (in millions of barrels per day) extracted from oil sands in Canada is projected to grow according to the function

$$P(t) = \frac{4.76}{1 + 4.11e^{-0.22t}} \qquad (0 \le t \le 15)$$

where t is measured in years, with $t = 0$ corresponding to 2005. What is the expected total production of oil from oil sands over the years from 2005 until 2020 ($t = 15$)?

Hint: Multiply the integrand by $\dfrac{e^{0.22t}}{e^{0.22t}}$.

Source: Canadian Association of Petroleum Producers.

57. SENIOR CITIZENS The U.S. population aged 65 years and older (in millions) from 2000 to 2050 is projected to grow at the rate of

$$f(t) = \frac{85}{1 + 1.859e^{-0.66t}} \qquad (0 \le t \le 5)$$

where t is measured in decades, with $t = 0$ corresponding to 2000. By how much will the population aged 65 years and older increase from the beginning of 2000 until the beginning of 2030?

Hint: Multiply the integrand by $\dfrac{e^{0.66t}}{e^{0.66t}}$.

Source: U.S. Census Bureau.

58. BLOOD FLOW Consider an artery of length L cm and radius R cm. By using Poiseuille's Law (page 784), it can be shown that the rate at which blood flows through the artery (measured in cubic centimeters per second) is given by

$$V = \int_0^R \frac{k}{L} x(R^2 - x^2)\, dx$$

where k is a constant. Find an expression for V that does *not* involve an integral.

Hint: Use the substitution $u = R^2 - x^2$.

59. Find the area of the region bounded by the graph of the function $f(x) = x^4 - 2x^2 + 2$, the x-axis, and the lines $x = a$ and $x = b$, where $a < b$ and a and b are the x-coordinates of the relative maximum point and a relative minimum point of f, respectively.

60. Find the area of the region bounded by the graph of the function $f(x) = (x + 1)/\sqrt{x}$, the x-axis, and the lines $x = a$ and $x = b$ where a and b are, respectively, the x-coordinates of the relative minimum point and the inflection point of f.

In Exercises 61–64, determine whether the statement is true or false. Give a reason for your answer.

61. $\displaystyle\int_{-1}^{1} \frac{1}{x^3}\, dx = -\frac{1}{2x^2}\Big|_{-1}^{1} = -\frac{1}{2} - \left(-\frac{1}{2}\right) = 0$

62. $\displaystyle\int_{-1}^{1} \frac{1}{x}\, dx = \ln|x|\,\Big|_{-1}^{1} = \ln|1| - \ln|-1| = \ln 1 - \ln 1 = 0$

63. $\int_0^2 (1 - x)\, dx$ gives the area of the region under the graph of $f(x) = 1 - x$ on the interval $[0, 2]$.

64. The total revenue realized in selling the first 500 units of a product is given by

$$\int_0^{500} R'(x)\, dx = R(500) - R(0)$$

where R is the total revenue.

---

1. $\displaystyle\int_0^2 (x + e^x)\, dx = \frac{1}{2}x^2 + e^x\,\Big|_0^2$

$\qquad = \left[\frac{1}{2}(2)^2 + e^2\right] - \left[\frac{1}{2}(0) + e^0\right]$

$\qquad = 2 + e^2 - 1$

$\qquad = e^2 + 1$

2. a. We want $P(1000)$, but

$P(1000) - P(0) = \displaystyle\int_0^{1000} P'(x)\, dx = \int_0^{1000} (-0.000006x^2 + 6)\, dx$

$\qquad = -0.000002x^3 + 6x\,\Big|_0^{1000}$

$\qquad = -0.000002(1000)^3 + 6(1000)$

$\qquad = 4000$

So, $P(1000) = 4000 + P(0) = 4000 - 400$, or $3600/\text{day}$ [$P(0) = -$fixed cost].

b. The additional profit realizable is given by

$\displaystyle\int_{1000}^{1200} P'(x)\, dx = -0.000002x^3 + 6x\,\Big|_{1000}^{1200}$

$\qquad = [-0.000002(1200)^3 + 6(1200)]$

$\qquad\quad - [-0.000002(1000)^3 + 6(1000)]$

$\qquad = 3744 - 4000$

$\qquad = -256$

That is, the company's profit is reduced by $256/\text{day}$ if production is increased from 1000 to 1200 cases/day.

USING TECHNOLOGY　Evaluating Definite Integrals

Some graphing utilities have an operation for finding the definite integral of a function. If your graphing utility has this capability, use it to work through the example and exercises of this section.

EXAMPLE 1 Use the numerical integral operation of a graphing utility to evaluate

$$\int_{-1}^{2} \frac{2x + 4}{(x^2 + 1)^{3/2}} \, dx$$

Solution　Using the numerical integral operation of a graphing utility, we find

$$\int_{-1}^{2} \frac{2x + 4}{(x^2 + 1)^{3/2}} \, dx = \text{fnInt}((2x + 4)/(x^2 + 1)^\wedge 1.5, x, -1, 2) \approx 6.92592226 \quad \blacksquare$$

TECHNOLOGY EXERCISES

In Exercises 1–4, find the area of the region under the graph of f on the interval $[a, b]$. Express your answer to four decimal places.

1. $f(x) = 0.002x^5 + 0.032x^4 - 0.2x^2 + 2; [-1.1, 2.2]$

2. $f(x) = x\sqrt{x^3 + 1}; [1, 2]$

3. $f(x) = \sqrt{x}e^{-x}; [0, 3]$

4. $f(x) = \dfrac{\ln x}{\sqrt{1 + x^2}}; [1, 2]$

In Exercises 5–10, evaluate the definite integral.

5. $\displaystyle\int_{-1.2}^{2.3} (0.2x^4 - 0.32x^3 + 1.2x - 1) \, dx$

6. $\displaystyle\int_{1}^{3} x(x^4 - 1)^{3.2} \, dx$

7. $\displaystyle\int_{0}^{2} \frac{3x^3 + 2x^2 + 1}{2x^2 + 3} \, dx$

8. $\displaystyle\int_{1}^{2} \frac{\sqrt{x} + 1}{2x^2 + 1} \, dx$

9. $\displaystyle\int_{0}^{2} \frac{e^x}{\sqrt{x^2 + 1}} \, dx$

10. $\displaystyle\int_{1}^{3} e^{-x} \ln(x^2 + 1) \, dx$

11. Rework Exercise 51, Exercises 11.4.

12. Rework Exercise 57, Exercises 11.4.

13. MARIJUANA ARRESTS　The number of arrests for marijuana sales and possession in New York City grew at the rate of approximately

$$f(t) = 0.0125t^4 - 0.01389t^3 + 0.55417t^2$$
$$+ 0.53294t + 4.95238 \quad (0 \le t \le 5)$$

thousand/year, where t is measured in years, with $t = 0$ corresponding to the beginning of 1992. Find the approximate number of marijuana arrests in the city from the beginning of 1992 to the end of 1997.
Source: State Division of Criminal Justice Services.

14. POPULATION GROWTH　The population of a certain city is projected to grow at the rate of $9 \sqrt{t + 1} \ln \sqrt{t + 1}$ thousand people/year t years from now. If the current population is 800,000, what will be the population 45 years from now?

11.5　Evaluating Definite Integrals

This section continues our discussion of the applications of the Fundamental Theorem of Calculus.

Properties of the Definite Integral

Before going on, we list the following useful properties of the definite integral, some of which parallel the rules of integration of Section 11.1.

> **Properties of the Definite Integral**
>
> Let f and g be integrable functions; then,
>
> **1.** $\displaystyle\int_a^a f(x)\, dx = 0$
>
> **2.** $\displaystyle\int_a^b f(x)\, dx = -\int_b^a f(x)\, dx$
>
> **3.** $\displaystyle\int_a^b cf(x)\, dx = c \int_a^b f(x)\, dx$ $\quad$ (c, a constant)
>
> **4.** $\displaystyle\int_a^b \left[f(x) \pm g(x) \right] dx = \int_a^b f(x)\, dx \pm \int_a^b g(x)\, dx$
>
> **5.** $\displaystyle\int_a^b f(x)\, dx = \int_a^c f(x)\, dx + \int_c^b f(x)\, dx$ $\quad$ ($a < c < b$)

Property 5 states that if c is a number lying between a and b, so that the interval $[a, b]$ is divided into the intervals $[a, c]$ and $[c, b]$, then the integral of f over the interval $[a, b]$ may be expressed as the sum of the integral of f over the interval $[a, c]$ and the integral of f over the interval $[c, b]$.

Property 5 has the following geometric interpretation when f is nonnegative. By definition,

$$\int_a^b f(x)\, dx$$

is the area of the region under the graph of $y = f(x)$ from $x = a$ to $x = b$ (Figure 24). Similarly, we interpret the definite integrals

$$\int_a^c f(x)\, dx \quad \text{and} \quad \int_c^b f(x)\, dx$$

as the areas of the regions under the graph of $y = f(x)$ from $x = a$ to $x = c$ and from $x = c$ to $x = b$, respectively. Since the two regions do not overlap, we see that

$$\int_a^b f(x)\, dx = \int_a^c f(x)\, dx + \int_c^b f(x)\, dx$$

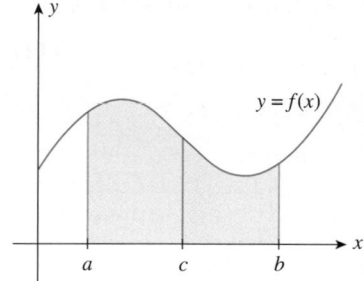

FIGURE **24**
$$\int_a^b f(x)\, dx = \int_a^c f(x)\, dx + \int_c^b f(x)\, dx$$

The Method of Substitution for Definite Integrals

Our first example shows two approaches that are generally used in evaluating a definite integral using the method of substitution.

EXAMPLE 1 Evaluate $\displaystyle\int_0^4 x\sqrt{9 + x^2}\, dx$.

Solution

Method 1 We first find the corresponding indefinite integral:

$$I = \int x\sqrt{9 + x^2}\, dx$$

Make the substitution $u = 9 + x^2$, so that

$$du = \frac{d}{dx}(9 + x^2) \, dx$$

$$= 2x \, dx$$

$$x \, dx = \frac{1}{2} du \qquad \text{Divide both sides by 2.}$$

Then

$$I = \int \frac{1}{2} \sqrt{u} \, du = \frac{1}{2} \int u^{1/2} \, du$$

$$= \frac{1}{3} u^{3/2} + C = \frac{1}{3}(9 + x^2)^{3/2} + C \qquad \begin{array}{l}\text{Substitute}\\ 9 + x^2 \text{ for } u.\end{array}$$

Using this result, we now evaluate the given definite integral:

$$\int_0^4 x\sqrt{9 + x^2} \, dx = \frac{1}{3}(9 + x^2)^{3/2} \Big|_0^4$$

$$= \frac{1}{3}[(9 + 16)^{3/2} - 9^{3/2}]$$

$$= \frac{1}{3}(125 - 27) = \frac{98}{3} = 32\tfrac{2}{3}$$

Method 2   *Changing the Limits of Integration.* As before, we make the substitution

$$u = 9 + x^2 \qquad\qquad\qquad\qquad \textbf{(11)}$$

so that

$$du = 2x \, dx$$

$$x \, dx = \frac{1}{2} du$$

Next, observe that the given definite integral is evaluated *with respect to* x with the range of integration given by the interval $[0, 4]$. If we perform the integration *with respect to* u via the substitution (11), then we must adjust the range of integration to reflect the fact that the integration is being performed with respect to the new variable u. To determine the proper range of integration, note that when $x = 0$, Equation (11) implies that

$$u = 9 + 0^2 = 9$$

which gives the required lower limit of integration with respect to u. Similarly, when $x = 4$,

$$u = 9 + 16 = 25$$

is the required upper limit of integration with respect to u. Thus, the range of integration when the integration is performed with respect to u is given by the interval $[9, 25]$. Therefore, we have

$$\int_0^4 x\sqrt{9 + x^2} \, dx = \int_9^{25} \frac{1}{2}\sqrt{u} \, du = \frac{1}{2} \int_9^{25} u^{1/2} \, du$$

$$= \frac{1}{3} u^{3/2} \Big|_9^{25} = \frac{1}{3}(25^{3/2} - 9^{3/2})$$

$$= \frac{1}{3}(125 - 27) = \frac{98}{3} = 32\tfrac{2}{3}$$

which agrees with the result obtained by using Method 1.

When you use Method 2, make sure you adjust the limits of integration to reflect integrating with respect to the new variable u.

Refer to Example 1. You can confirm the results obtained there by using a graphing utility as follows:

1. Use the numerical integration operation of the graphing utility to evaluate

$$\int_0^4 x\sqrt{9 + x^2}\, dx$$

2. Evaluate $\dfrac{1}{2}\displaystyle\int_9^{25} \sqrt{u}\, du$.

3. Conclude that $\displaystyle\int_0^4 x\sqrt{9 + x^2}\, dx = \dfrac{1}{2}\int_9^{25} \sqrt{u}\, du$.

EXAMPLE 2 Evaluate $\displaystyle\int_0^2 xe^{2x^2}\, dx$.

Solution Let $u = 2x^2$, so that $du = 4x\, dx$, or $x\, dx = \frac{1}{4} du$. When $x = 0$, $u = 0$; and when $x = 2$, $u = 8$. These give the lower and upper limits of integration with respect to u. Making the indicated substitutions, we find

$$\int_0^2 xe^{2x^2}\, dx = \int_0^8 \frac{1}{4} e^u\, du = \frac{1}{4} e^u \Big|_0^8 = \frac{1}{4}\left(e^8 - 1\right)$$

EXAMPLE 3 Evaluate $\displaystyle\int_0^1 \frac{x^2}{x^3 + 1}\, dx$.

Solution Let $u = x^3 + 1$, so that $du = 3x^2\, dx$, or $x^2\, dx = \frac{1}{3} du$. When $x = 0$, $u = 1$; and when $x = 1$, $u = 2$. These give the lower and upper limits of integration with respect to u. Making the indicated substitutions, we find

$$\int_0^1 \frac{x^2}{x^3 + 1}\, dx = \frac{1}{3}\int_1^2 \frac{du}{u} = \frac{1}{3} \ln|u| \Big|_1^2$$

$$= \frac{1}{3}\left(\ln 2 - \ln 1\right) = \frac{1}{3} \ln 2$$

Finding the Area Under a Curve

EXAMPLE 4 Find the area of the region R under the graph of $f(x) = e^{(1/2)x}$ from $x = -1$ to $x = 1$.

Solution The region R is shown in Figure 25. Its area is given by

$$A = \int_{-1}^1 e^{(1/2)x}\, dx$$

To evaluate this integral, we make the substitution

$$u = \frac{1}{2} x$$

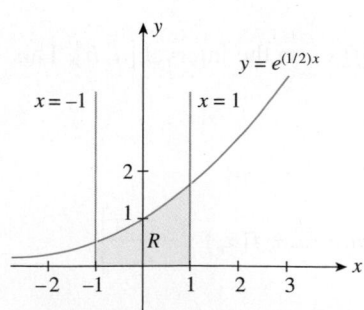

FIGURE 25

Area of $R = \displaystyle\int_{-1}^1 e^{(1/2)x}\, dx$

so that

$$du = \frac{1}{2}\, dx$$

$$dx = 2\, du$$

When $x = -1$, $u = -\frac{1}{2}$, and when $x = 1$, $u = \frac{1}{2}$. Making the indicated substitutions, we obtain

$$A = \int_{-1}^{1} e^{(1/2)x}\, dx = 2 \int_{-1/2}^{1/2} e^{u}\, du$$

$$= 2e^{u}\Big|_{-1/2}^{1/2} = 2\left(e^{1/2} - e^{-1/2}\right)$$

or approximately 2.08.

Explore and Discuss

Let f be the function defined piecewise by the rule

$$f(x) = \begin{cases} \sqrt{x} & \text{if } 0 \le x \le 1 \\ \dfrac{1}{x} & \text{if } 1 < x \le 2 \end{cases}$$

How would you use Property 5 of definite integrals to find the area of the region under the graph of f on $[0, 2]$? What is the area?

Average Value of a Function

The *average value* of a function over an interval provides us with an application of the definite integral. Recall that the average value of a set of n numbers $y_1, y_2, \ldots, y_n$ is the number

$$\frac{y_1 + y_2 + \cdots + y_n}{n}$$

Now, suppose f is a continuous function defined on $[a, b]$. Let's divide the interval $[a, b]$ into n subintervals of equal length $(b - a)/n$. Choose points $x_1, x_2, \ldots, x_n$ in the first, second, $\ldots$, nth subintervals, respectively. Then, the average value of the numbers $f(x_1), f(x_2), \ldots, f(x_n)$, given by

$$\frac{f(x_1) + f(x_2) + \cdots + f(x_n)}{n}$$

is an approximation of the average of all the values of $f(x)$ on the interval $[a, b]$. This expression can be written in the form

$$\frac{b - a}{b - a}\left[f(x_1) \cdot \frac{1}{n} + f(x_2) \cdot \frac{1}{n} + \cdots + f(x_n) \cdot \frac{1}{n} \right]$$

$$= \frac{1}{b - a}\left[f(x_1) \cdot \frac{b - a}{n} + f(x_2) \cdot \frac{b - a}{n} + \cdots + f(x_n) \cdot \frac{b - a}{n} \right]$$

$$= \frac{1}{b - a}\left[f(x_1)\,\Delta x + f(x_2)\,\Delta x + \cdots + f(x_n)\,\Delta x \right] \tag{12}$$

As n gets larger and larger, the expression (12) approximates the average value of $f(x)$ over $[a, b]$ with increasing accuracy. But the sum inside the brackets in (12) is a Riemann sum of the function f over $[a, b]$. In view of this, we have

$$\lim_{n \to \infty} \left[\frac{f(x_1) + f(x_2) + \cdots + f(x_n)}{n} \right]$$

$$= \frac{1}{b - a} \lim_{n \to \infty} [f(x_1) \Delta x + f(x_2) \Delta x + \cdots + f(x_n) \Delta x]$$

$$= \frac{1}{b - a} \int_a^b f(x) \, dx$$

This discussion motivates the following definition.

> **The Average Value of a Function**
>
> Suppose f is integrable on $[a, b]$. Then the **average value** of f over $[a, b]$ is
>
> $$\frac{1}{b - a} \int_a^b f(x) \, dx$$

EXAMPLE 5 Find the average value of the function $f(x) = \sqrt{x}$ over the interval $[0, 4]$.

Solution The required average value is given by

$$\frac{1}{4 - 0} \int_0^4 \sqrt{x} \, dx = \frac{1}{4} \int_0^4 x^{1/2} \, dx$$

$$= \frac{1}{6} x^{3/2} \Big|_0^4 = \frac{1}{6} (4^{3/2})$$

$$= \frac{4}{3}$$

APPLIED EXAMPLE 6 Automobile Financing The interest rates charged by Madison Finance on auto loans for used cars over a certain 6-month period in 2014 are approximated by the function

$$r(t) = -\frac{1}{12} t^3 + \frac{7}{8} t^2 - 3t + 12 \qquad (0 \le t \le 6)$$

where t is measured in months and $r(t)$ is the annual percentage rate. What is the average rate on auto loans extended by Madison over the 6-month period?

Solution The average rate over the 6-month period is given by

$$\frac{1}{6 - 0} \int_0^6 \left(-\frac{1}{12} t^3 + \frac{7}{8} t^2 - 3t + 12 \right) dt$$

$$= \frac{1}{6} \left(-\frac{1}{48} t^4 + \frac{7}{24} t^3 - \frac{3}{2} t^2 + 12t \right) \Big|_0^6$$

$$= \frac{1}{6} \left[-\frac{1}{48} (6^4) + \frac{7}{24} (6^3) - \frac{3}{2} (6^2) + 12(6) \right]$$

$$= 9$$

or 9% per year.

APPLIED EXAMPLE 7 Drug Concentration in a Body The amount of a certain drug in a patient's body t days after the drug has been administered is

$$C(t) = 5e^{-0.2t}$$

units. Determine the average amount of the drug present in the patient's body for the first 4 days after the drug has been administered.

Solution The average amount of the drug present in the patient's body for the first 4 days after it has been administered is given by

$$\frac{1}{4-0} \int_0^4 5e^{-0.2t}\,dt = \frac{5}{4} \int_0^4 e^{-0.2t}\,dt$$

$$= \frac{5}{4}\left[\left(-\frac{1}{0.2}\right)e^{-0.2t}\Big|_0^4\right]$$

$$= \frac{5}{4}\left(-5e^{-0.8} + 5\right)$$

$$\approx 3.44$$

or approximately 3.44 units.

We now give a geometric interpretation of the average value of a function f over an interval $[a, b]$. Suppose $f(x)$ is nonnegative, so that the definite integral

$$\int_a^b f(x)\,dx$$

gives the area under the graph of f from $x = a$ to $x = b$ (Figure 26). Observe that, in general, the "height" $f(x)$ varies from point to point. Can we replace $f(x)$ by a constant function $g(x) = k$ (which has constant height) such that the areas under each of the two functions f and g are the same? If so, since the area under the graph of g from $x = a$ to $x = b$ is $k(b - a)$, we have

$$k(b - a) = \int_a^b f(x)\,dx$$

$$k = \frac{1}{b-a} \int_a^b f(x)\,dx$$

so k is the average value of f over $[a, b]$. Thus, the average value of a function f over an interval $[a, b]$ is the height of a rectangle with base of length $(b - a)$ that has the same area as that of the region under the graph of f from $x = a$ to $x = b$.

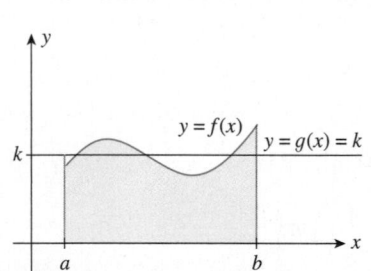

FIGURE 26
The average value of f over $[a, b]$ is k.

11.5 Self-Check Exercises

1. Evaluate $\displaystyle\int_0^2 \sqrt{2x + 5}\,dx$.

2. Find the average value of the function $f(x) = 1 - x^2$ over the interval $[-1, 2]$.

3. **Housing Prices** The median price of a house in a southwestern state between January 1, 2009, and January 1, 2014, is approximated by the function

$$f(t) = t^3 - 7t^2 + 17t + 280 \qquad (0 \le t \le 5)$$

where $f(t)$ is measured in thousands of dollars and t is expressed in years, with $t = 0$ corresponding to the beginning of 2009. Determine the average median price of a house over that time interval.

Solutions to Self-Check Exercises 11.5 can be found on page 865.

11.5 Concept Questions

1. Describe two approaches that are used to evaluate a definite integral using the method of substitution. Illustrate with the integral $\int_0^1 x^2(x^3 + 1)^2 \, dx$.

2. Define the average value of a function f over an interval $[a, b]$. Give a geometric interpretation.

11.5 Exercises

In Exercises 1–28, evaluate the definite integral.

1. $\displaystyle\int_0^2 x(x^2 - 1)^3 \, dx$

2. $\displaystyle\int_0^1 x^2(2x^3 - 1)^4 \, dx$

3. $\displaystyle\int_0^1 x\sqrt{5x^2 + 4} \, dx$

4. $\displaystyle\int_1^3 x\sqrt{3x^2 - 2} \, dx$

5. $\displaystyle\int_0^2 x^2(x^3 + 1)^{3/2} \, dx$

6. $\displaystyle\int_1^5 (2x - 1)^{5/2} \, dx$

7. $\displaystyle\int_0^1 \frac{1}{\sqrt{2x + 1}} \, dx$

8. $\displaystyle\int_0^2 \frac{x}{\sqrt{x^2 + 5}} \, dx$

9. $\displaystyle\int_1^3 (2x - 1)^4 \, dx$

10. $\displaystyle\int_1^2 (2x + 4)(x^2 + 4x - 8)^3 \, dx$

11. $\displaystyle\int_{-1}^1 x^2(x^3 + 1)^4 \, dx$

12. $\displaystyle\int_1^2 \left(x^3 + \frac{3}{4}\right)(x^4 + 3x)^{-2} \, dx$

13. $\displaystyle\int_1^5 x\sqrt{x - 1} \, dx$

14. $\displaystyle\int_1^4 x\sqrt{x + 1} \, dx$

Hint: Let $u = x + 1$.

15. $\displaystyle\int_0^2 2xe^{x^2} \, dx$

16. $\displaystyle\int_0^1 e^{-x} \, dx$

17. $\displaystyle\int_0^1 (e^{2x} + x^2 + 1) \, dx$

18. $\displaystyle\int_0^2 (e^t - e^{-t}) \, dt$

19. $\displaystyle\int_{-1}^1 xe^{x^2 + 1} \, dx$

20. $\displaystyle\int_1^4 \frac{e^{\sqrt{x}}}{\sqrt{x}} \, dx$

21. $\displaystyle\int_3^6 \frac{1}{x - 2} \, dx$

22. $\displaystyle\int_0^1 \frac{x}{1 + 2x^2} \, dx$

23. $\displaystyle\int_1^2 \frac{x^2 + 2x}{x^3 + 3x^2 - 1} \, dx$

24. $\displaystyle\int_0^1 \frac{e^x}{1 + e^x} \, dx$

25. $\displaystyle\int_1^2 \left(4e^{2u} - \frac{1}{u}\right) du$

26. $\displaystyle\int_1^2 \left(1 + \frac{1}{x} + e^x\right) dx$

27. $\displaystyle\int_1^2 \left(2e^{-4x} - \frac{1}{x^2}\right) dx$

28. $\displaystyle\int_1^2 \frac{\ln x}{x} \, dx$

In Exercises 29–34, find the area of the region under the graph of f on $[a, b]$.

29. $f(x) = x^2 - 2x + 2; [-1, 2]$

30. $f(x) = x^3 + x; [0, 1]$

31. $f(x) = \dfrac{1}{x^2}; [1, 2]$

32. $f(x) = 2 + \sqrt{x + 1}; [0, 3]$

33. $f(x) = e^{-x/2}; [-1, 2]$

34. $f(x) = \dfrac{\ln x}{4x}; [1, 2]$

In Exercises 35–44, find the average value of the function f over the indicated interval $[a, b]$.

35. $f(x) = 2x + 3; [0, 2]$

36. $f(x) = 8 - x; [1, 4]$

37. $f(x) = 2x^2 - 3; [1, 3]$

38. $f(x) = 4 - x^2; [-2, 3]$

39. $f(x) = x^2 + 2x - 3; [-1, 2]$

40. $f(x) = x^3; [-1, 1]$

41. $f(x) = \sqrt{2x + 1}; [0, 4]$

42. $f(x) = e^{-x}; [0, 4]$

43. $f(x) = xe^{x^2}; [0, 2]$

44. $f(x) = \dfrac{1}{x + 1}; [0, 2]$

45. **VELOCITY OF A CAR** A car moves along a straight road in such a way that its velocity (in feet per second) at any time t (in seconds) is given by

$$v(t) = 3t\sqrt{16 - t^2} \qquad (0 \le t \le 4)$$

Find the distance traveled by the car in the 4 sec from $t = 0$ to $t = 4$.

46. **OIL PRODUCTION** On the basis of a preliminary report by a geological survey team, it is estimated that a newly discovered oil field can be expected to produce oil at the rate of

$$R(t) = \frac{600t^2}{t^3 + 32} + 5 \qquad (0 \le t \le 20)$$

thousand barrels/year, t years after production begins. Find the amount of oil that the field can be expected to yield during the first 5 years of production, assuming that the projection holds true.

47. **NET INVESTMENT FLOW** The net investment flow (rate of capital formation) of the giant conglomerate LTF incorporated is projected to be

$$t\sqrt{\frac{1}{2}t^2 + 1}$$

million dollars/year in year t. Find the accruement on the company's capital stock in the second year.

Hint: The amount is given by

$$\int_1^2 t \sqrt{\frac{1}{2}t^2 + 1} \, dt$$

48. NEWTON'S LAW OF COOLING A bottle of white wine at room temperature (68°F) is placed in a refrigerator at 4 P.M. Its temperature after t hr is changing at the rate of

$$-18e^{-0.6t}$$

°F/hr. By how many degrees will the temperature of the wine have dropped by 7 P.M.? What will the temperature of the wine be at 7 P.M.?

49. WORLD PRODUCTION OF COAL A study proposed in 1980 by researchers from the major producers and consumers of the world's coal concluded that coal could and must play an important role in fueling global economic growth over the next 20 years. The world production of coal in 1980 was 3.5 billion metric tons. If output increased at the rate of $3.5e^{0.05t}$ billion metric tons/year in year t ($t = 0$ corresponding to 1980), determine how much coal was produced worldwide between 1980 and the end of the twentieth century.

50. DEPRECIATION: DOUBLE DECLINING–BALANCE METHOD Suppose a tractor purchased at a price of $60,000 is to be depreciated by the *double declining–balance method* over a 10-year period. It can be shown that the rate at which the book value will be decreasing is given by

$$R(t) = 13388.61e^{-0.22314t} \qquad (0 \le t \le 10)$$

dollars/year at year t. Find the amount by which the book value of the tractor will depreciate over the first 5 years of its life.

51. CELL PHONE AD SPENDING Cell phone ad spending between 2005 ($t = 1$) and 2011 ($t = 7$) is given by

$$S(t) = 0.86t^{0.96} \qquad (1 \le t \le 7)$$

where $S(t)$ is measured in billions of dollars and t is measured in years. What was the average spending per year on cell phone ads between 2005 and 2011?

Source: Interactive Advertising Bureau.

52. GLOBAL WARMING The increase in carbon dioxide (CO_2) in the atmosphere is a major cause of global warming. Using data obtained by Charles David Keeling, professor at Scripps Institution of Oceanography, the average amount of CO_2 in the atmosphere from 1958 through 2016 is approximated by

$$A(t) = 0.012414t^2 + 0.7485t + 313.9 \qquad (1 \le t \le 59)$$

where $A(t)$ is measured in parts per million volume (ppmv) and t in years, with $t = 1$ corresponding to 1958. Find the average rate of increase of the average amount of CO_2 in the atmosphere from 1958 through 2016.

Source: Scripps Institution of Oceanography.

53. PROJECTED U.S. GASOLINE USAGE The White House wants to cut gasoline usage from 140 billion gallons per year in 2007 to 128 billion gallons per year in 2017. But estimates by the Department of Energy's Energy Information Agency suggest that won't happen. In fact, the agency's projection of gasoline usage from the beginning of 2007 until the beginning of 2017 is given by

$$A(t) = 0.014t^2 + 1.93t + 140 \qquad (0 \le t \le 10)$$

where $A(t)$ is measured in billions of gallons/year and t is in years, with $t = 0$ corresponding to 2007.

a. According to the agency's projection, what will be gasoline consumption at the beginning of 2017?

b. What will be the average consumption/year over the period from the beginning of 2007 until the beginning of 2017?

Source: U.S. Department of Energy, Energy Information Agency.

54. U.S. CITIZENS 65 YEARS AND OLDER The number of U.S. citizens aged 65 years and older from 1900 through 2050 is estimated to be growing at the rate of

$$R(t) = 0.063t^2 - 0.48t + 3.87 \qquad (0 \le t \le 15)$$

million people/decade, where t is measured in decades, with $t = 0$ corresponding to 1900. Show that the average rate of growth of the number of U.S. citizens aged 65 years and older between 2000 and 2050 will be about twice the average rate between 1950 and 2000.

Source: American Heart Association.

55. BASEBALL A batted baseball is subjected to a force whose direction is constant but whose magnitude is given by

$$F(t) = -9{,}400{,}000(t^2 - 0.02t) \qquad (0 \le t \le 0.02)$$

where $F(t)$ is measured in newtons and t in seconds. The force acts only over the time interval $[0, 0.02]$. Find the impulse I exerted by the bat on the baseball where

$$I = \int_0^{0.02} F(t) \, dt$$

N-sec. What is the average force acting on the baseball?

Hint: Average force $= \dfrac{\displaystyle\int_a^b F(t) \, dt}{b - a}$

56. COMMERCIAL VEHICLE REGISTRATIONS In a report issued by the Economist Intelligence Unit in 2013, the number of commercial vehicle registrations in the United States (in thousands) by year from 2010 through 2015 is approximated by the function

$$N(t) = 1.3926t^3 - 9.2873t^2 + 74.719t + 228.3 \qquad (0 \le t \le 5)$$

where t is measured in years with $t = 0$ corresponding to 2010. If the projection holds up, what would the approximate average commercial vehicle registration per year be in the period from 2010 through 2015?

Source: Economist Intelligence Unit.

57. FLOW OF BLOOD IN AN ARTERY According to a law discovered by nineteenth century physician Jean Louis Marie Poiseuille, the velocity of blood (in centimeters per second) r cm from the central axis of an artery is given by

$$v(r) = k(R^2 - r^2)$$

where k is a constant and R is the radius of the artery. Find the average velocity of blood along a radius of the artery (see the accompanying figure).

Hint: Evaluate $\dfrac{1}{R}\displaystyle\int_0^R v(r)\, dr$.

Blood vessel

58. TOTAL KNEE REPLACEMENT PROCEDURES The number of total knee replacement procedures (in millions) performed in the United States from the beginning of 1991 through the end of 2030 is projected to be

$$f(t) = 0.0000846t^3 - 0.002116t^2 + 0.03897t + 0.16$$
$$(0 \le t \le 40)$$

where $t = 0$ corresponds to 1991. What is the projected average number of total knee replacement procedures performed in the United States in the period from 1991 through 2030?

Source: American Academy of Orthopaedic Surgeons.

59. CONCENTRATION OF A DRUG IN THE BLOODSTREAM The concentration of a certain drug in a patient's bloodstream t hr after injection is

$$C(t) = \frac{0.2t}{t^2 + 1}$$

mg/cm³. Determine the average concentration of the drug in the patient's bloodstream over the first 4 hr after the drug is injected.

60. SEAT BELT USE According to the U.S. Department of Transportation, the percentage of drivers using seat belts from 2001 through 2009 is modeled by the function

$$f(t) = 72.9(t + 1)^{0.057}$$

where t is measured in years, with $t = 0$ corresponding to the beginning of 2001. What was the average percentage use of seat belts over the period from the beginning of 2001 through the end of 2009 ($t = 9$)?

Source: U.S. Department of Transportation.

61. AVERAGE YEARLY SALES The sales of Universal Instruments in the first t years of its operation are approximated by the function

$$S(t) = t\sqrt{0.2t^2 + 4}$$

where $S(t)$ is measured in millions of dollars. What were Universal's average yearly sales over its first 5 years of operation?

62. CABLE TV SUBSCRIBERS The manager of TeleStar Cable Service estimates that the total number of subscribers to the service in a certain city t years from now will be

$$N(t) = -\frac{40,000}{\sqrt{1 + 0.2t}} + 50,000$$

Find the average number of cable television subscribers over the next 5 years if this prediction holds true.

63. CREDIT CARDS IN CHINA The total number of credit cards issued in China from 2009 through 2012 grew at the rate of

$$r(t) = 153e^{0.21t} \qquad (1 \le t \le 4)$$

million/year, where ($t = 1$) corresponds to 2009. What was the average growth in the number of credit cards issued in China for the period under consideration?

Source: Chinese Credit Card Industry Development Blue Book.

64. U.S. RETAIL E-COMMERCE SALES In a 2010 report, U.S. retail e-commerce sales (excluding travel, digital downloads, and event tickets) from 2011 through 2014 are projected to be

$$f(t) = 157.6e^{0.09t} \qquad (1 \le t \le 4)$$

billion dollars in year t, where ($t = 1$) corresponds to 2011. What is the average number of e-commerce sales in the United States over the period under consideration?

Source: eMarketer.

65. INTERNATIONAL AIR TRAFFIC International air traffic in and out of the United States is projected to grow at the rate of

$$R(t) = 6.69e^{0.0456t} \qquad (2 \le t \le 20)$$

million passengers per year between 2012 and 2030. Here, $t = 2$ corresponds to 2012. The number of passengers in 2012 was 160 million.

  a. Find an expression for the projected number of passengers in year t.

  b. What is the projected number of passengers in 2030 ($t = 20$)?

  c. What is the estimated average growth rate of international air traffic out of the United States from 2012 to 2030?

Source: Federal Aviation Administration.

66. AVERAGE PRICE OF A COMMODITY The price of a certain commodity in dollars per unit at time t (measured in weeks) is given by

$$p = 18 - 3e^{-2t} - 6e^{-t/3}$$

What is the average price of the commodity over the 5-week period from ($t = 0$) to ($t = 5$)?

67. VELOCITY OF A FALLING HAMMER During the construction of a high-rise apartment building, a construction worker accidentally drops a hammer that falls vertically a distance of h ft. The velocity of the hammer after falling a distance of x ft is $v = \sqrt{2gx}$ ft/sec ($0 \le x \le h$). Show that the average velocity of the hammer over this path is $\bar{v} = \frac{2}{3}\sqrt{2gh}$ ft/sec.

68. **Waste Disposal** When organic waste is dumped into a pond, the oxidization process that takes place reduces the pond's oxygen content. However, in time, nature will restore the oxygen content to its natural level. Suppose that the oxygen content t days after organic waste has been dumped into a pond is given by

$$f(t) = 100\left(\frac{t^2 + 10t + 100}{t^2 + 20t + 100}\right)$$

percent of its normal level. Find the average content of oxygen in the pond over the first 10 days after organic waste has been dumped into it.
Hint: Show that

$$\frac{t^2 + 10t + 100}{t^2 + 20t + 100} = 1 - \frac{10}{t + 10} + \frac{100}{(t + 10)^2}$$

69. Prove Property 1 of the definite integral.
Hint: Let F be an antiderivative of f, and use the definition of the definite integral.

70. Prove Property 2 of the definite integral.
Hint: See Exercise 69.

71. Verify by direct computation that

$$\int_1^3 x^2\, dx = -\int_3^1 x^2\, dx$$

72. Prove Property 3 of the definite integral.
Hint: See Exercise 69.

73. Verify by direct computation that

$$\int_1^9 2\sqrt{x}\, dx = 2\int_1^9 \sqrt{x}\, dx$$

74. Verify by direct computation that

$$\int_0^1 (1 + x - e^x)\, dx = \int_0^1 dx + \int_0^1 x\, dx - \int_0^1 e^x\, dx$$

What property of the definite integral is illustrated in this exercise?

75. Verify by direct computation that

$$\int_0^3 (1 + x^3)\, dx = \int_0^1 (1 + x^3)\, dx + \int_1^3 (1 + x^3)\, dx$$

What property of the definite integral is illustrated here?

76. Verify by direct computation that

$$\int_0^3 (1 + x^3)\, dx$$
$$= \int_0^1 (1 + x^3)\, dx + \int_1^2 (1 + x^3)\, dx + \int_2^3 (1 + x^3)\, dx$$

hence illustrating that Property 5 may be extended.

77. Evaluate $\displaystyle\int_3^3 (1 + \sqrt{x})e^{-x}\, dx$.

78. Evaluate $\displaystyle\int_3^0 f(x)\, dx$, given that $\displaystyle\int_0^3 f(x)\, dx = 4$.

79. Given that $\displaystyle\int_{-1}^2 f(x)\, dx = -2$ and $\displaystyle\int_{-1}^2 g(x)\, dx = 3$, evaluate

   a. $\displaystyle\int_{-1}^2 [2f(x) + g(x)]\, dx$

   b. $\displaystyle\int_{-1}^2 [g(x) - f(x)]\, dx$

   c. $\displaystyle\int_{-1}^2 [2f(x) - 3g(x)]\, dx$

80. Given that $\displaystyle\int_{-1}^2 f(x)\, dx = 2$ and $\displaystyle\int_0^2 f(x)\, dx = 3$, evaluate

   a. $\displaystyle\int_{-1}^0 f(x)\, dx$

   b. $\displaystyle\int_0^2 f(x)\, dx - \int_{-1}^0 f(x)\, dx$

In Exercises 81–86, determine whether the statement is true or false. If it is true, explain why it is true. If it is false, explain why or give an example to show why it is false.

81. $\displaystyle\int_2^2 \frac{e^x}{\sqrt{1 + x}}\, dx = 0$

82. $\displaystyle\int_1^3 \frac{dx}{x - 2} = -\int_3^1 \frac{dx}{x - 2}$

83. $\displaystyle\int_0^1 x\sqrt{x + 1}\, dx = \sqrt{x + 1}\int_0^1 x\, dx$
$$= \frac{1}{2}x^2\sqrt{x + 1}\,\Big|_0^1 = \frac{\sqrt{2}}{2}$$

84. If f' is continuous on $[0, 2]$, then
$$\int_0^2 f'(x)\, dx = f(2) - f(0)$$

85. If f and g are continuous on $[a, b]$ and k is a constant, then
$$\int_a^b [kf(x) + g(x)]\, dx = k\int_a^b f(x)\, dx + \int_a^b g(x)\, dx$$

86. If f is continuous on $[a, b]$ and $a < c < b$, then
$$\int_b^c f(x)\, dx = \int_a^c f(x)\, dx - \int_a^b f(x)\, dx$$

11.5 Solutions to Self-Check Exercises

1. Let $u = 2x + 5$. Then, $du = 2\, dx$, or $dx = \frac{1}{2}\, du$. Also, when $x = 0$, $u = 5$, and when $x = 2$, $u = 9$. Therefore,

$$\int_0^2 \sqrt{2x + 5}\, dx = \int_0^2 (2x + 5)^{1/2}\, dx$$

$$= \frac{1}{2} \int_5^9 u^{1/2}\, du$$

$$= \left(\frac{1}{2}\right)\left(\frac{2}{3} u^{3/2}\right)\Big|_5^9$$

$$= \frac{1}{3} \left[9^{3/2} - 5^{3/2}\right]$$

$$= \frac{1}{3} (27 - 5\sqrt{5})$$

2. The required average value is given by

$$\frac{1}{2 - (-1)} \int_{-1}^2 (1 - x^2)\, dx = \frac{1}{3} \int_{-1}^2 (1 - x^2)\, dx$$

$$= \frac{1}{3} \left(x - \frac{1}{3} x^3\right)\Big|_{-1}^2$$

$$= \frac{1}{3} \left[\left(2 - \frac{8}{3}\right) - \left(-1 + \frac{1}{3}\right)\right] = 0$$

3. The average median price of a house over the stated time interval is given by

$$\frac{1}{5 - 0} \int_0^5 (t^3 - 7t^2 + 17t + 280)\, dt$$

$$= \frac{1}{5} \left(\frac{1}{4} t^4 - \frac{7}{3} t^3 + \frac{17}{2} t^2 + 280t\right)\Big|_0^5$$

$$= \frac{1}{5} \left[\frac{1}{4} (5)^4 - \frac{7}{3} (5)^3 + \frac{17}{2} (5)^2 + 280(5)\right]$$

$$\approx 295.417$$

or approximately \$295,417.

USING TECHNOLOGY

Evaluating Definite Integrals for Piecewise-Defined Functions

We continue using graphing utilities to find the definite integral of a function. But here we will make use of Property 5 of the properties of the definite integral (page 855).

EXAMPLE 1 Use the numerical integral operation of a graphing utility to evaluate

$$\int_{-1}^2 f(x)\, dx$$

where

$$f(x) = \begin{cases} -x^2 & \text{if } x < 0 \\ \sqrt{x} & \text{if } x \geq 0 \end{cases}$$

Solution Using Property 5 of the definite integral, we can write

$$\int_{-1}^2 f(x)\, dx = \int_{-1}^0 -x^2\, dx + \int_0^2 x^{1/2}\, dx$$

Using a graphing utility, we find

$$\int_{-1}^2 f(x)\, dx = \text{fnInt}(-x^2, x, -1, 0) + \text{fnInt}(x^0.5, x, 0, 2)$$

$$\approx -0.333333 + 1.885618$$

$$= 1.552285$$

TECHNOLOGY EXERCISES

In Exercises 1–4, use Property 5 of the definite integral (page 855) to evaluate the definite integral accurate to five decimal places.

1. $\int_{-1}^{2} f(x)\, dx$, where

$$f(x) = \begin{cases} 2.3x^3 - 3.1x^2 + 2.7x + 3 & \text{if } x < 1 \\ -1.7x^2 + 2.3x + 4.3 & \text{if } x \geq 1 \end{cases}$$

2. $\int_{0}^{3} f(x)\, dx$, where $f(x) = \begin{cases} \dfrac{\sqrt{x}}{1 + x^2} & \text{if } 0 \leq x < 1 \\ 0.5e^{-0.1x^2} & \text{if } x \geq 1 \end{cases}$

3. $\int_{-2}^{2} f(x)\, dx$, where $f(x) = \begin{cases} x^4 - 2x^2 + 4 & \text{if } x < 0 \\ 2\ln(x + e^2) & \text{if } x \geq 0 \end{cases}$

4. $\int_{-2}^{6} f(x)\, dx$, where

$$f(x) = \begin{cases} 2x^3 - 3x^2 + x + 2 & \text{if } x < -1 \\ \sqrt{3x + 4} - 5 & \text{if } -1 \leq x \leq 4 \\ x^2 - 3x - 5 & \text{if } x > 4 \end{cases}$$

5. Crop Yield If left untreated on bean stems, aphids (small insects that suck plant juices) will multiply at an increasing rate during the summer months and reduce productivity and crop yield of cultivated crops. But if the aphids are treated in mid-June, the numbers decrease sharply to less than 100/bean stem, allowing for steep rises in crop yield. The function

$$F(t) = \begin{cases} 62e^{1.152t} & \text{if } 0 \leq t < 1.5 \\ 349e^{-1.324(t - 1.5)} & \text{if } 1.5 \leq t \leq 3 \end{cases}$$

gives the number of aphids on a typical bean stem at time t, where t is measured in months, $t = 0$ corresponding to the beginning of May. Find the average number of aphids on a typical bean stem over the period from the beginning of May to the beginning of August.

6. Absorption of Drugs Jane took 100 mg of a drug in the morning and another 100 mg of the same drug at the same time the following morning. The amount of the drug (in mg) in her body t days after the first dose was taken is given by

$$A(t) = \begin{cases} 100e^{-1.4t} & \text{if } 0 \leq t < 1 \\ 100(1 + e^{1.4})e^{-1.4t} & \text{if } t \geq 1 \end{cases}$$

Find the average amount of the drug in Jane's body over the first 2 days.

7. Absorption of Drugs The concentration of a drug in an organ t seconds after it is administered is given by

$$C(t) = \begin{cases} 0.3t - 18(1 - e^{-t/60}) & \text{if } 0 \leq t \leq 20 \\ 18e^{-t/60} - 12e^{-(t-20)/60} & \text{if } t > 20 \end{cases}$$

where $C(t)$ is measured in grams per cubic centimeter (g/cm^3). Find the average concentration of the drug in the organ over the first 30 sec after it is administered.

11.6 Area Between Two Curves

Suppose a certain country's petroleum consumption is expected to be $f(t)$ million barrels per year, t years from now, for the next 5 years. Then the country's total petroleum consumption over the period of time in question is given by the area of the region under the graph of f on the interval $[0, 5]$ (Figure 27).

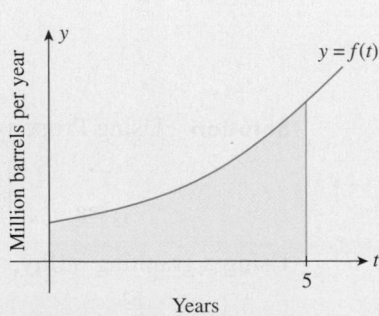

FIGURE 27
At a rate of consumption of $f(t)$ million barrels per year, the total petroleum consumption is given by the area of the region under the graph of f.

Next, suppose that because of the implementation of certain energy-conservation measures, the petroleum consumption is expected to be $g(t)$ million barrels per year instead. Then, the country's projected total petroleum consumption over the 5-year period is given by the area of the region under the graph of g on the interval $[0, 5]$ (Figure 28).

Therefore, the area of the shaded region S lying between the graphs of f and g on the interval $[0, 5]$ (Figure 29) gives the amount of petroleum that would be saved over the 5-year period because of the conservation measures.

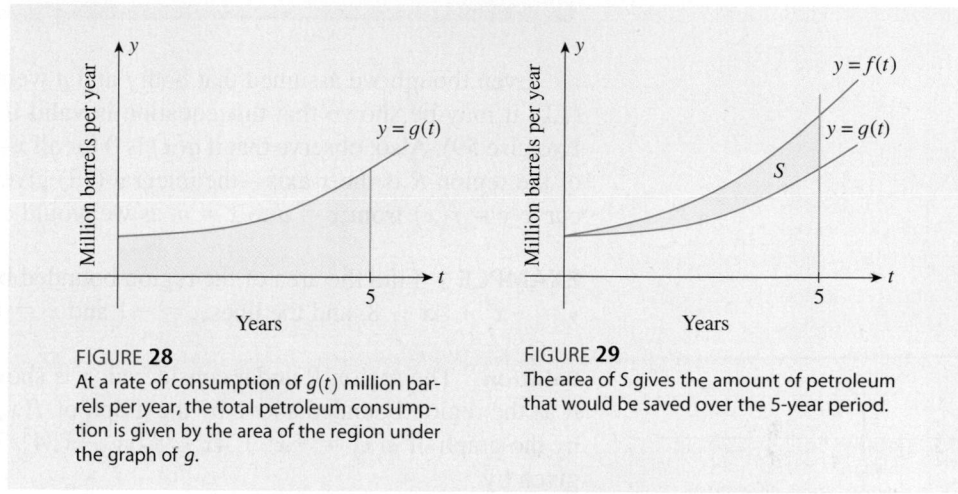

FIGURE 28
At a rate of consumption of $g(t)$ million barrels per year, the total petroleum consumption is given by the area of the region under the graph of g.

FIGURE 29
The area of S gives the amount of petroleum that would be saved over the 5-year period.

But the area of S is given by

Area of the region under the graph of f on $[a, b]$
    − Area of the region under the graph of g on $[a, b]$

$$= \int_0^5 f(t)\, dt - \int_0^5 g(t)\, dt$$

$$= \int_0^5 [f(t) - g(t)]\, dt \qquad \text{By Property 4, Section 11.5}$$

This example shows that some practical problems can be solved by finding the area of a region between two curves, which in turn can be found by evaluating an appropriate definite integral.

Finding the Area of the Region Between Two Curves

We now turn our attention to the general problem of finding the area of a plane region bounded both above and below by the graphs of functions. First, consider the situation in which the graph of one function lies above that of another. More specifically, let R be the region in the xy-plane (Figure 30) that is bounded above by the graph of a continuous function f, below by the graph of a continuous function g, where $f(x) \geq g(x) \geq 0$ on $[a, b]$, and to the left and right by the vertical lines $x = a$ and $x = b$, respectively. From the figure, we see that

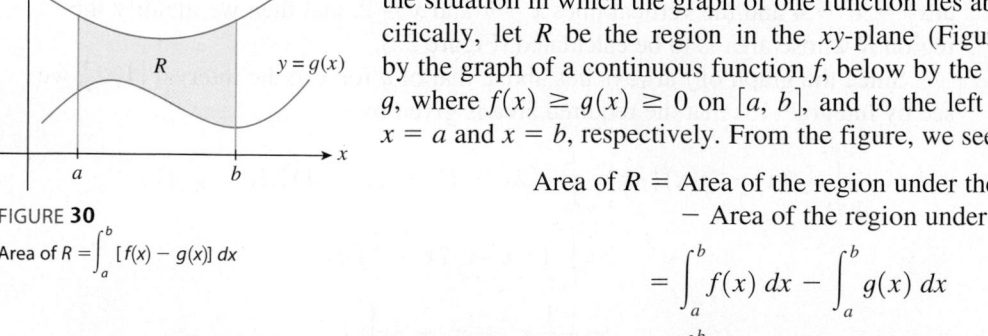

FIGURE 30
Area of $R = \int_a^b [f(x) - g(x)]\, dx$

Area of $R =$ Area of the region under the graph of $f(x)$
    − Area of the region under the graph of $g(x)$

$$= \int_a^b f(x)\, dx - \int_a^b g(x)\, dx$$

$$= \int_a^b [f(x) - g(x)]\, dx$$

upon using Property 4 of the definite integral.

The Area of the Region Between Two Curves

Let f and g be continuous functions such that $f(x) \geq g(x)$ on the interval $[a, b]$. Then the area of the region bounded above by $y = f(x)$ and below by $y = g(x)$ on $[a, b]$ is given by

$$\int_a^b \left[f(x) - g(x) \right] dx \qquad (13)$$

Even though we assumed that both f and g were nonnegative in the derivation of (13), it may be shown that this equation is valid if f and g are not nonnegative (see Exercise 59). Also, observe that if $g(x)$ is 0 for all x—that is, when the lower boundary of the region R is the x-axis—the integral (13) gives the area of the region under the curve $y = f(x)$ from $x = a$ to $x = b$, as we would expect.

EXAMPLE 1 Find the area of the region bounded by the x-axis, the graph of $y = -x^2 + 4x - 8$, and the lines $x = -1$ and $x = 4$.

Solution The region R under consideration is shown in Figure 31. We can view R as the region bounded above by the graph of $f(x) = 0$ (the x-axis) and below by the graph of $g(x) = -x^2 + 4x - 8$ on $[-1, 4]$. Therefore, the area of R is given by

$$\int_a^b \left[f(x) - g(x) \right] dx = \int_{-1}^4 \left[0 - (-x^2 + 4x - 8) \right] dx$$

$$= \int_{-1}^4 (x^2 - 4x + 8) \, dx$$

$$= \frac{1}{3}x^3 - 2x^2 + 8x \Big|_{-1}^4$$

$$= \left[\frac{1}{3}(64) - 2(16) + 8(4) \right] - \left[\frac{1}{3}(-1) - 2(1) + 8(-1) \right]$$

$$= 31\tfrac{2}{3}$$

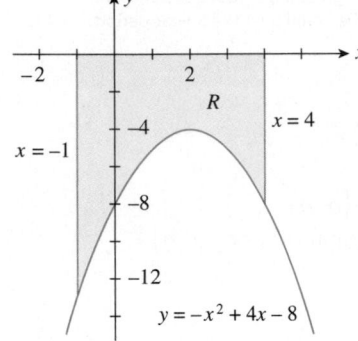

FIGURE 31

Area of $R = -\int_{-1}^4 g(x)\, dx$

EXAMPLE 2 Find the area of the region R bounded by the graphs of

$$f(x) = 2x - 1 \quad \text{and} \quad g(x) = x^2 - 4$$

and the vertical lines $x = 1$ and $x = 2$.

Solution We first sketch the graphs of the functions $f(x) = 2x - 1$ and $g(x) = x^2 - 4$ and the vertical lines $x = 1$ and $x = 2$, and then we identify the region R whose area is to be calculated (Figure 32).

Since the graph of f always lies above that of g for x in the interval $[1, 2]$, we see by integral (13) that the required area is given by

$$\int_1^2 \left[f(x) - g(x) \right] dx = \int_1^2 \left[(2x - 1) - (x^2 - 4) \right] dx$$

$$= \int_1^2 (-x^2 + 2x + 3) \, dx$$

$$= -\frac{1}{3}x^3 + x^2 + 3x \Big|_1^2$$

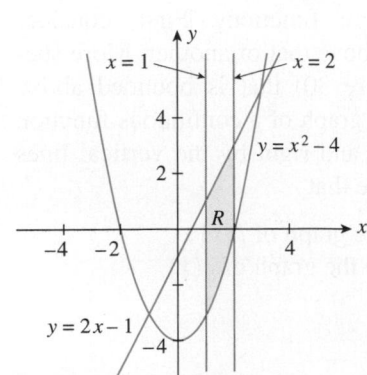

FIGURE 32

Area of $R = \int_1^2 \left[f(x) - g(x) \right] dx$

$$= \left(-\frac{8}{3} + 4 + 6 \right) - \left(-\frac{1}{3} + 1 + 3 \right) = \frac{11}{3}$$

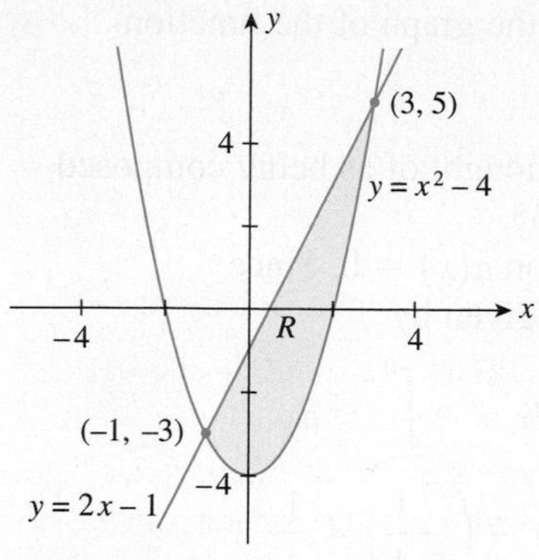

EXAMPLE 3 Find the area of the region R that is completely enclosed by the graphs of the functions

$$f(x) = 2x - 1 \quad \text{and} \quad g(x) = x^2 - 4$$

Solution The region R is shown in Figure 33. First, we find the points of intersection of the two curves. To do this, we solve the system that consists of the two equations $y = 2x - 1$ and $y = x^2 - 4$. Equating the two values of y gives

$$x^2 - 4 = 2x - 1$$
$$x^2 - 2x - 3 = 0$$
$$(x + 1)(x - 3) = 0$$

so $x = -1$ or $x = 3$. That is, the two curves intersect when $x = -1$ and $x = 3$.

Observe that we could also view the region R as the region bounded above by the graph of the function $f(x) = 2x - 1$, below by the graph of the function $g(x) = x^2 - 4$, and to the left and right by the vertical lines $x = -1$ and $x = 3$, respectively.

Next, since the graph of the function f always lies above that of the function g on $[-1, 3]$, we can use integral (13) to compute the desired area:

$$\int_a^b [f(x) - g(x)] \, dx = \int_{-1}^3 [(2x - 1) - (x^2 - 4)] \, dx$$

$$= \int_{-1}^3 (-x^2 + 2x + 3) \, dx$$

$$= -\frac{1}{3}x^3 + x^2 + 3x \Big|_{-1}^3$$

$$= (-9 + 9 + 9) - \left(\frac{1}{3} + 1 - 3\right) = \frac{32}{3}$$

$$= 10\frac{2}{3}$$

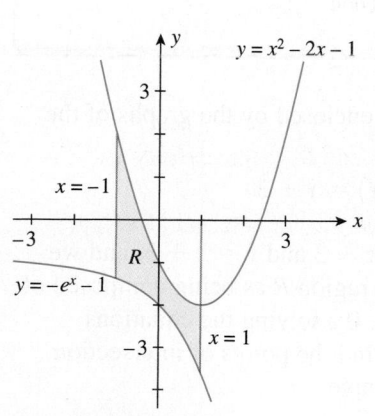

FIGURE 33

Area of $R = -\int_{-1}^3 [f(x) - g(x)] \, dx$

EXAMPLE 4 Find the area of the region R bounded by the graphs of the functions

$$f(x) = x^2 - 2x - 1 \quad \text{and} \quad g(x) = -e^x - 1$$

and the vertical lines $x = -1$ and $x = 1$.

Solution The region R is shown in Figure 34. Since the graph of the function f always lies above that of the function g, the area of the region R is given by

$$\int_a^b [f(x) - g(x)] \, dx = \int_{-1}^1 [(x^2 - 2x - 1) - (-e^x - 1)] \, dx$$

$$= \int_{-1}^1 (x^2 - 2x + e^x) \, dx$$

$$= \frac{1}{3}x^3 - x^2 + e^x \Big|_{-1}^1$$

$$= \left(\frac{1}{3} - 1 + e\right) - \left(-\frac{1}{3} - 1 + e^{-1}\right)$$

$$= \frac{2}{3} + e - \frac{1}{e} \approx 3.02$$

FIGURE 34

Area of $R = \int_{-1}^1 [f(x) - g(x)] \, dx$

Integral (13), which gives the area of the region between the curves $y = f(x)$ and $y = g(x)$ for $a \le x \le b$, is valid when the graph of the function f lies above that of the function g over the interval $[a, b]$. Example 5 shows how to use Equation (13) to find the area of a region when this condition does not hold.

EXAMPLE 5 Find the area of the region bounded by the graph of the function $f(x) = x^3$, the x-axis, and the lines $x = -1$ and $x = 1$.

Solution The region R under consideration can be thought of as being composed of the two subregions R_1 and R_2, as shown in Figure 35.

Recall that the x-axis is represented by the function $g(x) = 0$. Since $g(x) \geq f(x)$ on $[-1, 0]$, we see that the area of R_1 is given by

$$\int_a^b [g(x) - f(x)]\,dx = \int_{-1}^0 (0 - x^3)\,dx = -\int_{-1}^0 x^3\,dx$$

$$= -\frac{1}{4}x^4\Big|_{-1}^0 = 0 - \left(-\frac{1}{4}\right) = \frac{1}{4}$$

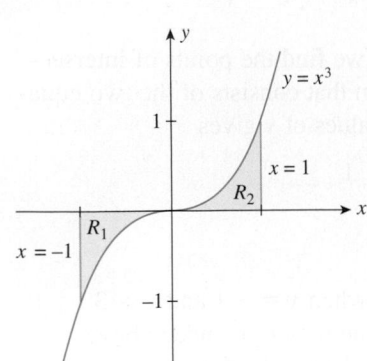

FIGURE 35
Area of R_1 = Area of R_2

To find the area of R_2, we observe that $f(x) \geq g(x)$ on $[0, 1]$, so it is given by

$$\int_a^b [f(x) - g(x)]\,dx = \int_0^1 (x^3 - 0)\,dx = \int_0^1 x^3\,dx$$

$$= \frac{1}{4}x^4\Big|_0^1 = \left(\frac{1}{4}\right) - 0 = \frac{1}{4}$$

Therefore, the area of R is $\frac{1}{4} + \frac{1}{4}$, or $\frac{1}{2}$.

By making use of symmetry, we could have obtained the same result by computing

$$-2\int_{-1}^0 x^3\,dx \quad \text{or} \quad 2\int_0^1 x^3\,dx$$

as you may verify.

Explore and Discuss

A function is *even* if it satisfies the condition $f(-x) = f(x)$, and it is *odd* if it satisfies the condition $f(-x) = -f(x)$. Show that the graph of an even function is symmetric with respect to the y-axis while the graph of an odd function is symmetric with respect to the origin. Explain why

$$\int_{-a}^a f(x)\,dx = 2\int_0^a f(x)\,dx \quad \text{if } f \text{ is even}$$

and

$$\int_{-a}^a f(x)\,dx = 0 \quad \text{if } f \text{ is odd}$$

EXAMPLE 6 Find the area of the region completely enclosed by the graphs of the functions

$$f(x) = x^3 - 3x + 3 \quad \text{and} \quad g(x) = x + 3$$

Solution First, we sketch the graphs of $y = x^3 - 3x + 3$ and $y = x + 3$, and we then identify the required region R. We can view the region R as being composed of the two subregions R_1 and R_2, as shown in Figure 36. By solving the equations $y = x + 3$ and $y = x^3 - 3x + 3$ simultaneously, we find the points of intersection of the two curves. Equating the two values of y, we have

$$x^3 - 3x + 3 = x + 3$$
$$x^3 - 4x = 0$$
$$x(x^2 - 4) = 0$$
$$x(x + 2)(x - 2) = 0$$
$$x = 0, -2, 2$$

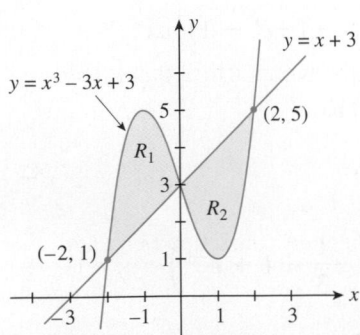

FIGURE 36
Area of R_1 + Area of R_2
$$= \int_{-2}^0 [f(x) - g(x)]\,dx$$
$$+ \int_0^2 [g(x) - f(x)]\,dx$$

Hence the points of intersection of the two curves are $(-2, 1)$, $(0, 3)$, and $(2, 5)$.

For $-2 \leq x \leq 0$, we see that the graph of the function f lies above that of the function g, so the area of the region R_1 is, by virtue of Equation (13),

$$\int_{-2}^{0} [(x^3 - 3x + 3) - (x + 3)] \, dx = \int_{-2}^{0} (x^3 - 4x) \, dx$$
$$= \frac{1}{4} x^4 - 2x^2 \Big|_{-2}^{0}$$
$$= -(4 - 8)$$
$$= 4$$

For $0 \leq x \leq 2$, the graph of the function g lies above that of the function f, and the area of R_2 is given by

$$\int_{0}^{2} [(x + 3) - (x^3 - 3x + 3)] \, dx = \int_{0}^{2} (-x^3 + 4x) \, dx$$
$$= -\frac{1}{4} x^4 + 2x^2 \Big|_{0}^{2}$$
$$= -4 + 8$$
$$= 4$$

Therefore, the required area is the sum of the areas of the two regions R_1 and R_2—that is, $4 + 4$, or 8. ■

APPLIED EXAMPLE 7 Conservation of Oil In a 2006 study for a developing country's Economic Development Board, government economists and energy experts concluded that if the Energy Conservation Bill were implemented in 2009, the country's oil consumption for the next 5 years is expected to grow in accordance with the model

$$R(t) = 20e^{0.05t}$$

where t is measured in years ($t = 0$ corresponding to the year 2009) and $R(t)$ in millions of barrels per year. Without the government-imposed conservation measures, however, the expected rate of growth of oil consumption would be given by

$$R_1(t) = 20e^{0.08t}$$

million barrels per year. Using these models, determine how much oil would have been saved from 2009 through 2014 if the bill had been implemented.

Solution Under the Energy Conservation Bill, the total amount of oil that would have been consumed between 2009 and 2014 is given by

$$\int_{0}^{5} R(t) \, dt = \int_{0}^{5} 20e^{0.05t} \, dt \qquad (14)$$

Without the bill, the total amount of oil that would have been consumed between 2009 and 2014 is given by

$$\int_{0}^{5} R_1(t) \, dt = \int_{0}^{5} 20e^{0.08t} \, dt \qquad (15)$$

The first integral in (14) may be interpreted as the area of the region under the curve $y = R(t)$ from $t = 0$ to $t = 5$. Similarly, we interpret the integrals in (15) as giving the area of the region under the curve $y = R_1(t)$ from $t = 0$ to $t = 5$. Furthermore, note that the graph of $y = R_1(t) = 20e^{0.08t}$ always lies on or above the graph of

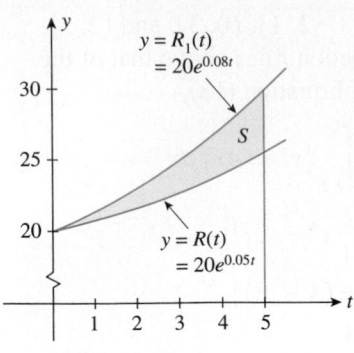

FIGURE **37**

Area of $S = \displaystyle\int_0^5 [R_1(t) - R(t)]\, dt$

$y = R(t) = 20e^{0.05t}$ $(t \geq 0)$. Thus, the area of the shaded region S in Figure 37 shows the amount of oil that would have been saved from 2009 to 2014 if the Energy Conservation Bill had been implemented. But the area of the region S is given by

$$\int_0^5 [R_1(t) - R(t)]\, dt = \int_0^5 (20e^{0.08t} - 20e^{0.05t})\, dt$$

$$= 20 \int_0^5 (e^{0.08t} - e^{0.05t})\, dt$$

$$= 20 \left(\frac{e^{0.08t}}{0.08} - \frac{e^{0.05t}}{0.05} \right) \Bigg|_0^5$$

$$= 20 \left[\left(\frac{e^{0.4}}{0.08} - \frac{e^{0.25}}{0.05} \right) - \left(\frac{1}{0.08} - \frac{1}{0.05} \right) \right]$$

$$\approx 9.3$$

Thus, the amount of oil that would have been saved is 9.3 million barrels.

Exploring with TECHNOLOGY

Refer to Example 7. Suppose we want to construct a mathematical model giving the amount of oil saved from 2009 through the year $(2009 + x)$ where $x \geq 0$. In Example 7, for instance, we take $x = 5$.

1. Show that this model is given by

$$F(x) = \int_0^x [R_1(t) - R(t)]\, dt$$
$$= 250e^{0.08x} - 400e^{0.05x} + 150$$

   Hint: You may find it helpful to use some of the results of Example 7.

2. Use a graphing utility to plot the graph of F, using the viewing window $[0, 10] \times [0, 50]$.

3. Find $F(5)$ and thus confirm the result of Example 7.

4. What is the main advantage of this model?

11.6 Self-Check Exercises

1. Find the area of the region bounded by the graphs of $f(x) = x^2 + 2$ and $g(x) = 1 - x$ and the vertical lines $x = 0$ and $x = 1$.

2. Find the area of the region completely enclosed by the graphs of $f(x) = -x^2 + 6x + 5$ and $g(x) = x^2 + 5$.

3. **PROFITS OF A HOTEL CHAIN** The management of Kane Corporation, which operates a chain of hotels, expects its profits to grow at the rate of $(1 + t^{2/3})$ million dollars/year

t years from now. However, with renovations and improvements of existing hotels and proposed acquisitions of new hotels, Kane's profits are expected to grow at the rate of $(t - 2\sqrt{t} + 4)$ million dollars/year in the next decade. What additional profits are expected over the next 10 years if the group implements the proposed plans?

Solutions to Self-Check Exercises 11.6 can be found on page 877.

11.6 Concept Questions

1. Suppose f and g are continuous functions such that $f(x) \geq g(x)$ on the interval $[a, b]$. Write an integral giving the area of the region bounded above by the graph of f, below by the graph of g, and on the left and right by the lines $x = a$ and $x = b$.

2. Write an expression in terms of definite integrals giving the area of the shaded region in the figure.

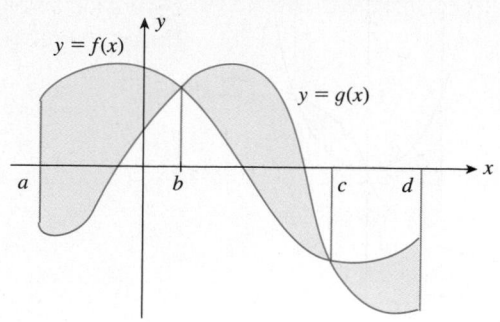

11.6 Exercises

In Exercises 1–8, find the area of the shaded region.

1.

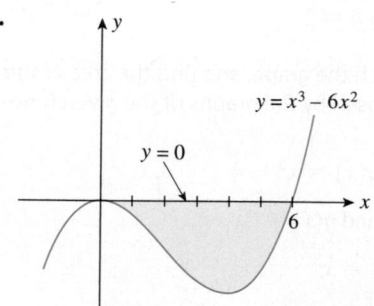

$y = x^3 - 6x^2$
$y = 0$

2.

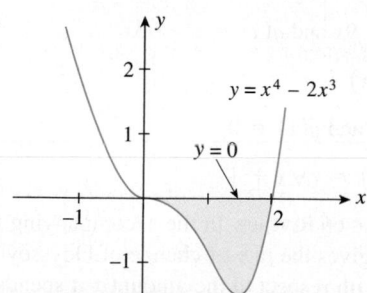

$y = x^4 - 2x^3$
$y = 0$

3.

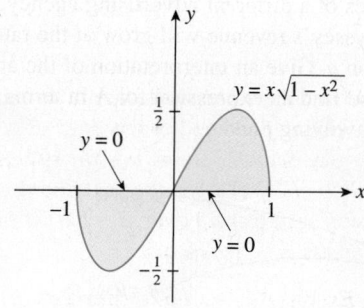

$y = x\sqrt{1 - x^2}$
$y = 0$
$y = 0$

4.

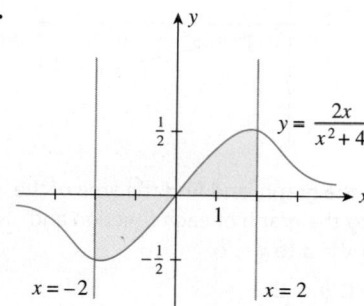

$y = \dfrac{2x}{x^2 + 4}$
$x = -2$
$x = 2$

5.

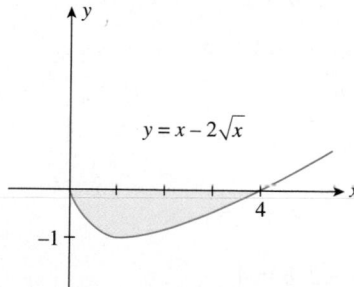

$y = x - 2\sqrt{x}$

6.

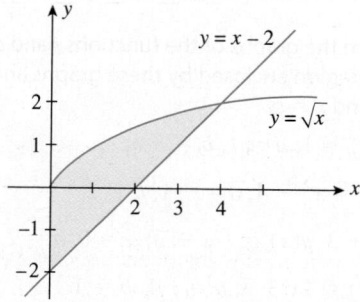

$y = x - 2$
$y = \sqrt{x}$

7.

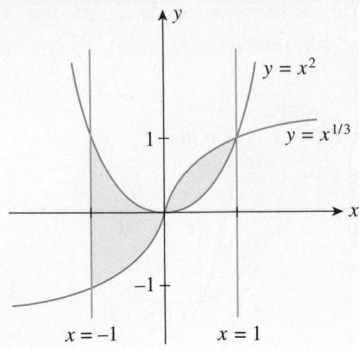

8.

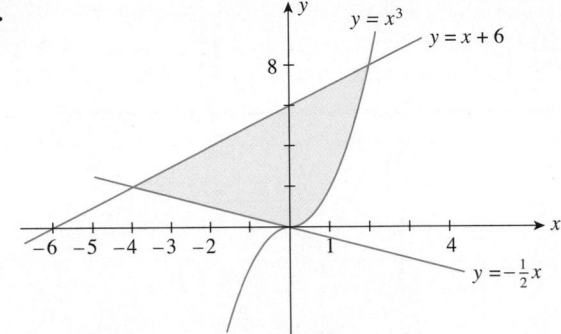

In Exercises 9–16, sketch the graph, and find the area of the region bounded below by the graph of each function and above by the x-axis from $x = a$ to $x = b$.

9. $f(x) = -x^2; a = -1, b = 2$

10. $f(x) = x^2 - 4; a = -2, b = 2$

11. $f(x) = x^2 - 5x + 4; a = 1, b = 3$

12. $f(x) = x^3; a = -1, b = 0$

13. $f(x) = -1 - \sqrt{x}; a = 0, b = 9$

14. $f(x) = \dfrac{1}{2}x - \sqrt{x}; a = 0, b = 4$

15. $f(x) = -e^{(1/2)x}; a = -2, b = 4$

16. $f(x) = -xe^{-x^2}; a = 0, b = 1$

In Exercises 17–26, sketch the graphs of the functions f and g, and find the area of the region enclosed by these graphs and the vertical lines $x = a$ and $x = b$.

17. $f(x) = x^2 + 3, g(x) = 1; a = 1, b = 3$

18. $f(x) = x + 2, g(x) = x^2 - 4; a = -1, b = 2$

19. $f(x) = -x^2 + 2x + 3, g(x) = -x + 3; a = 0, b = 2$

20. $f(x) = 9 - x^2, g(x) = 2x + 3; a = -1, b = 1$

21. $f(x) = x^2 + 1, g(x) = \dfrac{1}{3}x^3; a = -1, b = 2$

22. $f(x) = \sqrt{x}, g(x) = -\dfrac{1}{2}x - 1; a = 1, b = 4$

23. $f(x) = \dfrac{1}{x}, g(x) = 2x - 1; a = 1, b = 4$

24. $f(x) = x^2, g(x) = \dfrac{1}{x^2}; a = 1, b = 3$

25. $f(x) = e^x, g(x) = \dfrac{1}{x}; a = 1, b = 2$

26. $f(x) = x, g(x) = e^{2x}; a = 1, b = 3$

In Exercises 27–34, sketch the graph, and find the area of the region bounded by the graph of the function f and the lines $y = 0, x = a$, and $x = b$.

27. $f(x) = x; a = -1, b = 2$

28. $f(x) = x^2 - 2x; a = -1, b = 1$

29. $f(x) = -x^2 + 4x - 3; a = -1, b = 2$

30. $f(x) = x^3 - x^2; a = -1, b = 1$

31. $f(x) = x^3 - 4x^2 + 3x; a = 0, b = 2$

32. $f(x) = 4x^{1/3} + x^{4/3}; a = -1, b = 8$

33. $f(x) = e^x - 1; a = -1, b = 3$

34. $f(x) = xe^{x^2}; a = 0, b = 2$

In Exercises 35–42, sketch the graph, and find the area of the region completely enclosed by the graphs of the given functions f and g.

35. $f(x) = x + 2$ and $g(x) = x^2 - 4$

36. $f(x) = -x^2 + 4x$ and $g(x) = 2x - 3$

37. $f(x) = x^2$ and $g(x) = x^3$

38. $f(x) = x^3 + 2x^2 - 3x$ and $g(x) = 0$

39. $f(x) = x^3 - 6x^2 + 9x$ and $g(x) = x^2 - 3x$

40. $f(x) = \sqrt{x}$ and $g(x) = x^2$

41. $f(x) = x\sqrt{9 - x^2}$ and $g(x) = 0$

42. $f(x) = 2x$ and $g(x) = x\sqrt{x + 1}$

43. **EFFECT OF ADVERTISING ON REVENUE** In the accompanying figure, the function f gives the rate of change of Odyssey Travel's revenue with respect to the amount x it spends on advertising with their current advertising agency. By engaging the services of a different advertising agency, it is expected that Odyssey's revenue will grow at the rate given by the function g. Give an interpretation of the area A of the region S and find an expression for A in terms of a definite integral involving f and g.

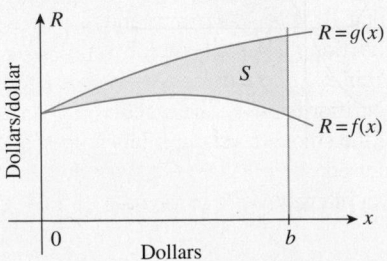

44. PULSE RATE DURING EXERCISE In the accompanying figure, the function f gives the rate of increase of an individual's pulse rate when he walked a prescribed course on a treadmill 6 months ago. The function g gives the rate of increase of his pulse rate when he recently walked the same prescribed course. Give an interpretation of the area A of the region S, and find an expression for A in terms of a definite integral involving f and g.

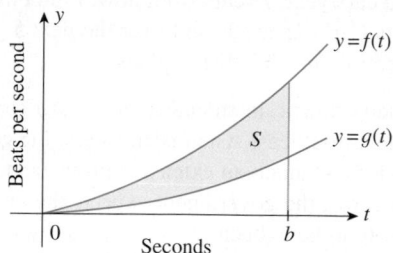

45. OIL PRODUCTION SHORTFALL Energy experts disagree about when global oil production will begin to decline. In the following figure, the function f gives the annual world oil production in billions of barrels from 1980 to 2050, according to the Department of Energy projection. The function g gives the world oil production in billions of barrels per year over the same period, according to longtime petroleum geologist Colin Campbell. Find an expression in terms of the definite integrals involving f and g giving the shortfall in the total oil production over the period in question heeding Campbell's dire warnings.
Source: U.S. Department of Energy; Colin Campbell.

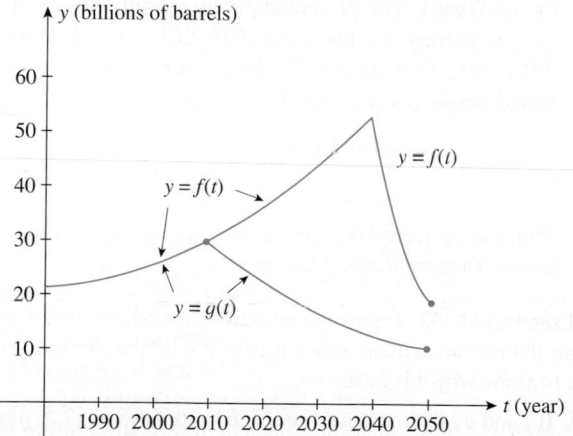

46. AIR PURIFICATION To study the effectiveness of air purifiers in removing smoke, engineers ran each purifier in a smoke-filled 10-ft × 20-ft room. In the accompanying figure, the function f gives the rate of change of the smoke level per minute, t min after the start of the test, when a Brand A purifier is used. The function g gives the rate of change of the smoke level per minute when a Brand B purifier is used.
a. Give an interpretation of the area of the region S.

b. Find an expression for the area of S in terms of a definite integral involving f and g.

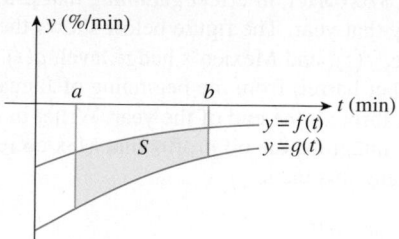

47. MOTION OF TWO CARS Two cars start out side by side and travel along a straight road. The velocity of Car 1 is $f(t)$ ft/sec over the interval $[0, T]$, the velocity of Car 2 is $g(t)$ ft/sec over the interval $[0, T]$, and $0 < T_1 < T$. Furthermore, suppose the graphs of f and g are as depicted in the accompanying figure. Let A_1 and A_2 denote the areas of the regions (shown shaded).

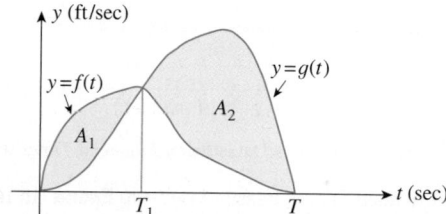

a. Write the number

$$\int_{T_1}^{T} [g(t) - f(t)]\, dt - \int_{0}^{T_1} [f(t) - g(t)]\, dt$$

in terms of A_1 and A_2.
b. What does the number obtained in part (a) represent?

48. COMPARISON OF COMPANIES' REVENUES The rate of change of the revenue of Company A over the (time) interval $[0, T]$ is $f(t)$ dollars/week, whereas the rate of change of the revenue of Company B over the same period is $g(t)$ dollars/week. The graphs of f and g are depicted in the accompanying figure. Find an expression in terms of definite integrals involving f and g giving the additional revenue that Company B will have over Company A in the period $[0, T]$.

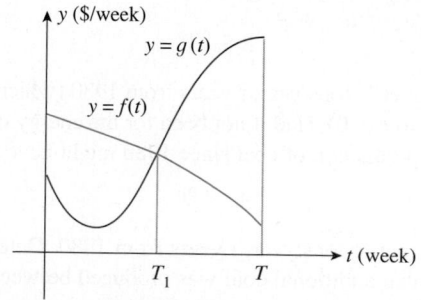

49. **MEXICO'S HEDGING TACTIC** Agustin Carstens, the Mexican finance minister, hedged all his country's oil sales for 2009 at $70/barrel, in effect gambling that prices would stay low that year. The figure below shows the U.S. crude oil price, $f(t)$, and Mexico's hedge level, $g(t) = 70$, (in dollars per barrel) from the beginning of January 2009 ($t = 0$) through the end of the year. Write, in terms of definite integrals, the oil profits that Mexico realized in 2009 using this tactic.

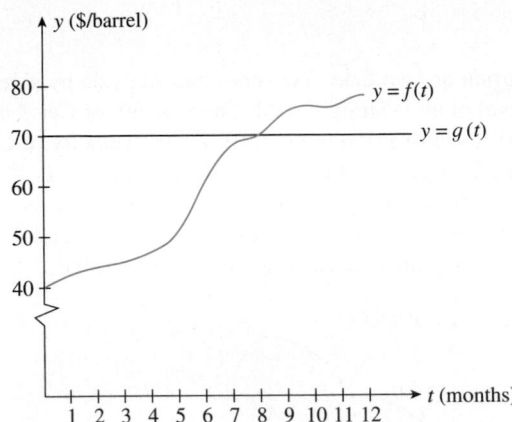

Sources: Thomson Reuters Datastream; *Financial Times* research.

50. **TURBO-CHARGED ENGINE VERSUS STANDARD ENGINE** In tests conducted by *Auto Test Magazine* on two models of the Phoenix Elite—identical except that one was equipped with a standard engine and the other with a turbo-charger—it was found that the acceleration of the former is given by

$$a = f(t) = 4 + 0.8t \quad (0 \le t \le 12)$$

ft/sec/sec, t sec after starting from rest at full throttle, whereas the acceleration of the latter is given by

$$a = g(t) = 4 + 1.2t + 0.03t^2 \quad (0 \le t \le 12)$$

ft/sec/sec. How much faster is the turbo-charged model moving than the model with the standard engine at the end of a 10-sec test run at full throttle?

51. **ALTERNATIVE ENERGY SOURCES** Because of the increasingly important role played by coal as a viable alternative energy source, the production of coal has been growing at the rate of

$$3.5e^{0.05t}$$

billion metric tons/year, t years from 1980 (which corresponds to $t = 0$). Had it not been for the energy crisis, the rate of production of coal since 1980 might have been only

$$3.5e^{0.01t}$$

billion metric tons/year, t years from 1980. Determine how much additional coal was produced between 1980 and 2000 as an alternative energy source.

52. **EFFECT OF TV ADVERTISING ON CAR SALES** Carl Williams, the proprietor of Carl Williams Auto Sales, estimates that with extensive television advertising, car sales over the next several years could be increasing at the rate of

$$5e^{0.3t}$$

thousand cars/year, t years from now, instead of at the current rate of

$$5 + 0.5t^{3/2}$$

thousand cars/year, t years from now. Find how many more cars Carl expects to sell over the next 5 years by implementing his advertising plans.

53. **POPULATION GROWTH** In an endeavor to curb population growth in a Southeast Asian island state, the government has decided to launch an extensive propaganda campaign. Without curbs, the government expects the rate of population growth to have been

$$60e^{0.02t}$$

thousand people/year, t years from now, over the next 5 years. However, successful implementation of the proposed campaign is expected to result in a population growth rate of

$$-t^2 + 60$$

thousand people/year, t years from now, over the next 5 years. Assuming that the campaign is mounted, how many fewer people will there be in that country 5 years from now than there would have been if no curbs had been imposed?

54. **BRITISH DEFICIT** The projected public spending by the British government for the years 2010–2011 ($t = 0$) through 2014–2015 ($t = 4$) is 692 billion pounds/year. The projected public revenue for the same period is

$$f(t) = \frac{292.6t + 134.4}{t + 6.9} + 500 \quad (0 \le t \le 4)$$

Find the projected deficit for the period in question.
Source: Thomson Reuters Datastream.

In Exercises 55–58, determine whether the statement is true or false. If it is true, explain why it is true. If it is false, give an example to show why it is false.

55. If f and g are continuous on $[a, b]$ and either $f(x) \ge g(x)$ for all x in $[a, b]$ or $f(x) \le g(x)$ for all x in $[a, b]$, then the area of the region bounded by the graphs of f and g and the vertical lines $x = a$ and $x = b$ is given by $\int_a^b |f(x) - g(x)| \, dx$.

56. The area of the region bounded by the graphs of $f(x) = 2 - x$ and $g(x) = 4 - x^2$ and the vertical lines $x = 0$ and $x = 2$ is given by $\int_0^2 [f(x) - g(x)] \, dx$.

57. If A denotes the area bounded by the graphs of f and g on $[a, b]$, then

$$A^2 = \int_a^b [f(x) - g(x)]^2 \, dx$$

58. If f and g are continuous on $[a, b]$ and $\int_a^b [f(t) - g(t)] \, dt > 0$, then $f(t) \geq g(t)$ for all t in $[a, b]$.

59. Show that the area of a region R bounded above by the graph of a function f and below by the graph of a function g from $x = a$ to $x = b$ is given by

$$\int_a^b [f(x) - g(x)] \, dx$$

where f and g are continuous functions.
Hint: The validity of the formula was verified earlier for the case in which both f and g were nonnegative. Now, let f and g be two functions such that $f(x) \geq g(x)$ for $a \leq x \leq b$. Then there exists some nonnegative constant c such that the translated curves

$y = f(x) + c$ and $y = g(x) + c$ lie entirely above the x-axis. The region R' has the same area as the region R (see the accompanying figures). Show that the area of R' is given by

$$\int_a^b \{[f(x) + c] - [g(x) + c]\} \, dx = \int_a^b [f(x) - g(x)] \, dx$$

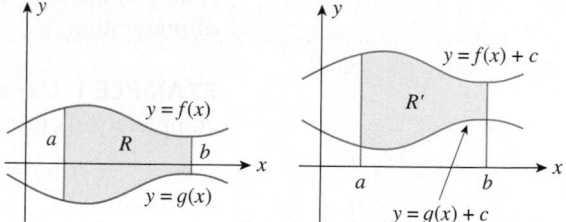

11.6   Solutions to Self-Check Exercises

1. The region in question is shown in the accompanying figure. Since the graph of the function f lies above that of the function g for $0 \leq x \leq 1$, we see that the required area is given by

$$\int_0^1 [(x^2 + 2) - (1 - x)] \, dx = \int_0^1 (x^2 + x + 1) \, dx$$

$$= \frac{1}{3}x^3 + \frac{1}{2}x^2 + x \Big|_0^1$$

$$= \frac{1}{3} + \frac{1}{2} + 1$$

$$= \frac{11}{6}$$

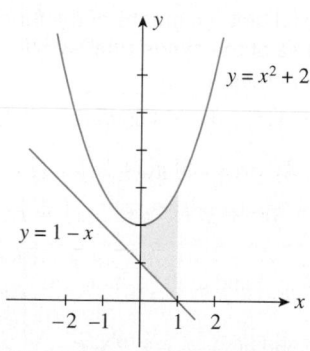

2. The region in question is shown in the accompanying figure. To find the points of intersection of the two curves, we solve the equations

$$-x^2 + 6x + 5 = x^2 + 5$$

$$2x^2 - 6x = 0$$

$$2x(x - 3) = 0$$

giving $x = 0$ or $x = 3$. Therefore, the points of intersection are $(0, 5)$ and $(3, 14)$.

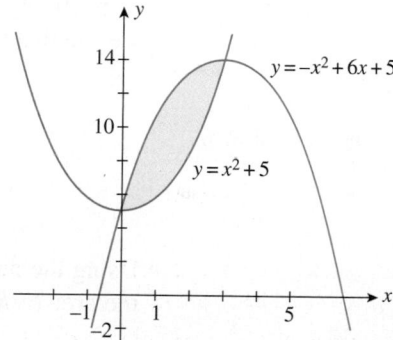

Since the graph of f always lies above that of g for $0 \leq x \leq 3$, we see that the required area is given by

$$\int_0^3 [(-x^2 + 6x + 5) - (x^2 + 5)] \, dx = \int_0^3 (-2x^2 + 6x) \, dx$$

$$= -\frac{2}{3}x^3 + 3x^2 \Big|_0^3$$

$$= -18 + 27$$

$$= 9$$

3. The additional profits realizable over the next 10 years are given by

$$\int_0^{10} [(t - 2\sqrt{t} + 4) - (1 + t^{2/3})] \, dt$$

$$= \int_0^{10} (t - 2t^{1/2} + 3 - t^{2/3}) \, dt$$

$$= \frac{1}{2}t^2 - \frac{4}{3}t^{3/2} + 3t - \frac{3}{5}t^{5/3} \Big|_0^{10}$$

$$= \frac{1}{2}(10)^2 - \frac{4}{3}(10)^{3/2} + 3(10) - \frac{3}{5}(10)^{5/3}$$

$$\approx 9.99$$

or approximately \$10 million.

USING TECHNOLOGY   Finding the Area Between Two Curves

The numerical integral operation can be used to find the area between two curves. We do this by using the numerical integral operation to evaluate an appropriate definite integral or the sum (difference) of appropriate definite integrals. In the following example, the intersection operation is also used to advantage to help us find the limits of integration.

EXAMPLE 1 Use a graphing utility to find the area of the smaller region R that is completely enclosed by the graphs of the functions

$$f(x) = 2x^3 - 8x^2 + 4x - 3 \quad \text{and} \quad g(x) = 3x^2 + 10x - 11$$

Solution   The graphs of f and t in the viewing window $[-3, 4] \times [-20, 5]$ are shown in Figure T1.

Using the intersection operation of a graphing utility, we find the x-coordinates of the points of intersection of the two graphs to be approximately -1.04 and 0.65, respectively. Since the graph of f lies above that of g on the interval $[-1.04, 0.65]$, we see that the area of R is given by

$$A \approx \int_{-1.04}^{0.65} [(2x^3 - 8x^2 + 4x - 3) - (3x^2 + 10x - 11)] \, dx$$

$$= \int_{-1.04}^{0.65} (2x^3 - 11x^2 - 6x + 8) \, dx$$

Using the numerical integration function of a graphing utility, we find $A \approx 9.87$, so the area of R is approximately 9.87.

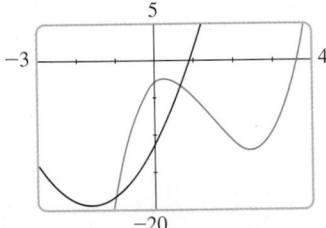

FIGURE T1
The region R is completely enclosed by the graphs of f and g.

TECHNOLOGY EXERCISES

In Exercises 1–6, (a) plot the graphs of the functions f and g, and (b) find the area of the region enclosed by these graphs and the vertical lines $x = a$ and $x = b$. Express your answers accurate to four decimal places.

1. $f(x) = x^3(x - 5)^4, g(x) = 0; a = 1, b = 3$

2. $f(x) = x - \sqrt{1 - x^2}, g(x) = 0; a = -\dfrac{1}{2}, b = \dfrac{1}{2}$

3. $f(x) = x^{1/3}(x + 1)^{1/2}, g(x) = x^{-1}; a = 1.2, b = 2$

4. $f(x) = 2, g(x) = \ln(1 + x^2); a = -1, b = 1$

5. $f(x) = \sqrt{x}, g(x) = \dfrac{x^2 - 3}{x^2 + 1}; a = 0, b = 3$

6. $f(x) = \dfrac{4}{x^2 + 1}, g(x) = x^4; a = -1, b = 1$

In Exercises 7–12, (a) plot the graphs of the functions f and g, and (b) find the area of the region totally enclosed by the graphs of these functions.

7. $f(x) = 2x^3 - 8x^2 + 4x - 3$ and $g(x) = -3x^2 + 10x - 10$

8. $f(x) = x^4 - 2x^2 + 2$ and $g(x) = 4 - 2x^2$

9. $f(x) = 2x^3 - 3x^2 + x + 5$ and $g(x) = e^{2x} - 3$

10. $f(x) = \dfrac{1}{2}x^2 - 3$ and $g(x) = \ln x$

11. $f(x) = xe^{-x}$ and $g(x) = x - 2\sqrt{x}$

12. $f(x) = e^{-x^2}$ and $g(x) = x^4$

13. Refer to Example 1. Find the area of the larger region that is completely enclosed by the graphs of the functions f and g.

11.7   Applications of the Definite Integral to Business and Economics

In this section, we consider several applications of the definite integral in the fields of business and economics.

Consumers' and Producers' Surplus

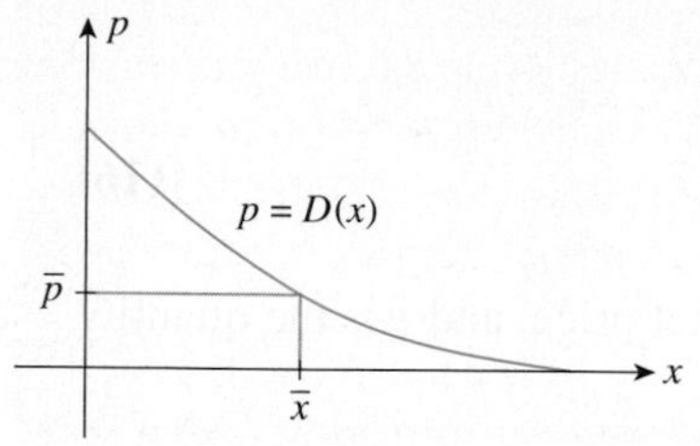

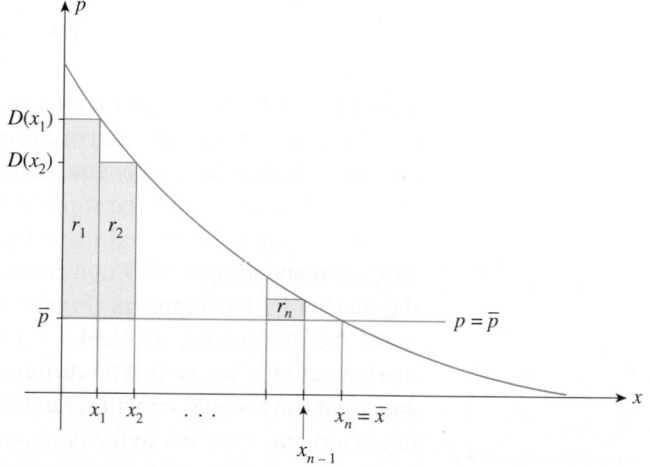

FIGURE **38**
$D(x)$ is a demand function.

We begin by deriving a formula for computing the consumers' surplus. Suppose $p = D(x)$ is the demand function that relates the unit price p of a commodity to the quantity x demanded of it. Furthermore, suppose a fixed unit market price $\bar{p}$ has been established for the commodity and corresponding to this unit price the quantity demanded is $\bar{x}$ units (Figure 38). Then those consumers who would be willing to pay a unit price higher than $\bar{p}$ for the commodity would in effect experience a savings. This difference between what the consumers *would* be willing to pay for $\bar{x}$ units of the commodity and what they *actually* pay for them is called the **consumers' surplus.**

To derive a formula for computing the consumers' surplus, divide the interval $[0, \bar{x}]$ into n subintervals, each of length $\Delta x = \bar{x}/n$, and denote the right endpoints of these subintervals by $x_1, x_2, \ldots, x_n = \bar{x}$ (Figure 39).

FIGURE **39**
Approximating consumers' surplus by the sum of the areas of the rectangles $r_1, r_2, \ldots, r_n$

From Figure 39, we see that there are consumers who would pay a unit price of at least $D(x_1)$ dollars for the first Δx units of the commodity instead of the market price of $\bar{p}$ dollars per unit. The savings to these consumers is approximated by

$$D(x_1)\,\Delta x - \bar{p}\,\Delta x = [D(x_1) - \bar{p}]\,\Delta x$$

which is the area of the rectangle r_1. Pursuing the same line of reasoning, we find that the savings to the consumers who would be willing to pay a unit price of at least $D(x_2)$ dollars for the next Δx units (from x_1 through x_2) of the commodity, instead of the market price of $\bar{p}$ dollars per unit, is approximated by

$$D(x_2)\,\Delta x - \bar{p}\,\Delta x = [D(x_2) - \bar{p}]\,\Delta x$$

Continuing, we approximate the total savings to the consumers in purchasing $\bar{x}$ units of the commodity by the sum

$$[D(x_1) - \bar{p}]\,\Delta x + [D(x_2) - \bar{p}]\,\Delta x + \cdots + [D(x_n) - \bar{p}]\,\Delta x$$
$$= [D(x_1) + D(x_2) + \cdots + D(x_n)]\,\Delta x - \underbrace{[\bar{p}\,\Delta x + \bar{p}\,\Delta x + \cdots + \bar{p}\,\Delta x]}_{n \text{ terms}}$$

$$= [D(x_1) + D(x_2) + \cdots + D(x_n)]\,\Delta x - n\bar{p}\,\Delta x$$
$$= [D(x_1) + D(x_2) + \cdots + D(x_n)]\,\Delta x - \bar{p}\,\bar{x}$$

Now, the first term in the last expression is the Riemann sum of the demand function $p = D(x)$ over the interval $[0, \bar{x}]$ with representative points $x_1, x_2, \ldots, x_n$. Letting n approach infinity, we obtain the following formula for the consumers' surplus CS.

Consumers' Surplus

The consumers' surplus is given by

$$CS = \int_0^{\bar{x}} D(x)\, dx - \bar{p}\,\bar{x} \tag{16}$$

where D is the demand function, $\bar{p}$ is the unit market price, and $\bar{x}$ is the quantity sold.

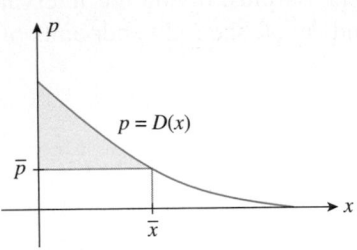

FIGURE **40**
Consumers' surplus

The consumer's surplus is given by the area of the region bounded above by the demand curve $p = D(x)$ and below by the horizontal line $p = \bar{p}$ from $x = 0$ to $x = \bar{x}$ (Figure 40). We can also see this if we rewrite Equation (16) in the form

$$CS = \int_0^{\bar{x}} [D(x) - \bar{p}]\, dx$$

and interpret the result geometrically.

Analogously, we can derive a formula for computing the producers' surplus. Suppose $p = S(x)$ is the supply equation that relates the unit price p of a certain commodity to the quantity x that the supplier will make available in the market at that price.

Again, suppose a fixed market price $\bar{p}$ has been established for the commodity, and, corresponding to this unit price, a quantity of $\bar{x}$ units will be made available in the market by the suppliers (Figure 41). Then the suppliers who would be willing to make the commodity available at a lower price stand to gain from the fact that the market price is set at $\bar{p}$. The difference between what the suppliers actually receive and what they would be willing to receive is called the **producers' surplus.** Proceeding in a manner similar to the derivation of the equation for computing the consumers' surplus, we find that the producers' surplus PS is defined as follows:

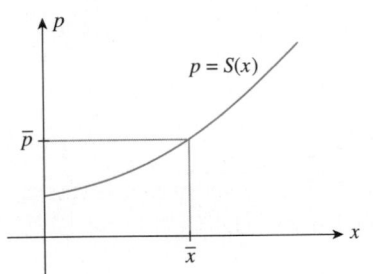

FIGURE **41**
$S(x)$ is a supply function.

Producers' Surplus

The producers' surplus is given by

$$PS = \bar{p}\,\bar{x} - \int_0^{\bar{x}} S(x)\, dx \tag{17}$$

where $S(x)$ is the supply function, $\bar{p}$ is the unit market price, and $\bar{x}$ is the quantity supplied.

Geometrically, the producers' surplus is given by the area of the region bounded above by the horizontal line $p = \bar{p}$ and below by the supply curve $p = S(x)$ from $x = 0$ to $x = \bar{x}$ (Figure 42).

We can also show that the last statement is true by converting Equation (17) to the form

$$PS = \int_0^{\bar{x}} [\bar{p} - S(x)]\, dx$$

and interpreting the definite integral geometrically.

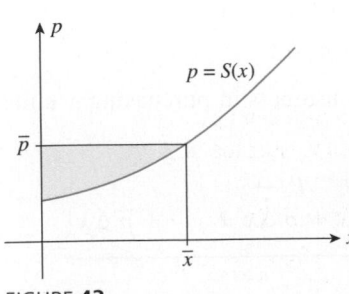

FIGURE **42**
Producers' surplus

EXAMPLE 1 The demand function for a certain make of ten-speed bicycle is given by

$$p = D(x) = -0.001x^2 + 250$$

where p is the unit price in dollars and x is the quantity demanded in thousands of units. The supply function for these bicycles is given by

$$p = S(x) = 0.0006x^2 + 0.02x + 100$$

where p stands for the unit price in dollars and x stands for the number of bicycles that the supplier will put on the market, in units of a thousand. Determine the consumers' surplus and the producers' surplus if the market price of a bicycle is set at the equilibrium price.

Solution   Recall that the equilibrium price is the unit price of the commodity when market equilibrium occurs. We determine the equilibrium price by solving for the point of intersection of the demand curve and the supply curve (Figure 43). To solve the system of equations

$$p = -0.001x^2 \qquad\qquad + 250$$
$$p = 0.0006x^2 + 0.02x + 100$$

we simply substitute the second equation into the first, obtaining

$$0.0006x^2 + 0.02x + 100 = -0.001x^2 + 250$$
$$0.0016x^2 + 0.02x - 150 = 0$$
$$16x^2 + 200x - 1,500,000 = 0$$
$$2x^2 + 25x - 187,500 = 0$$

Factoring this last equation, we obtain

$$(2x + 625)(x - 300) = 0$$

Thus, $x = -625/2$ or $x = 300$. The first number lies outside the interval of interest, so we are left with the solution $x = 300$, with a corresponding value of

$$p = -0.001(300)^2 + 250 = 160$$

Thus, the equilibrium point is $(300, 160)$; that is, the equilibrium quantity is 300,000, and the equilibrium price is \$160. Setting the market price at \$160 per unit and using Formula (16) with $\bar{p} = 160$ and $\bar{x} = 300$, we find that the consumers' surplus is given by

$$CS = \int_0^{300} (-0.001x^2 + 250)\, dx - (160)(300)$$

$$= \left(-\frac{1}{3000}x^3 + 250x \right)\Bigg|_0^{300} - 48,000$$

$$= -\frac{300^3}{3000} + (250)(300) - 48,000$$

$$= 18,000$$

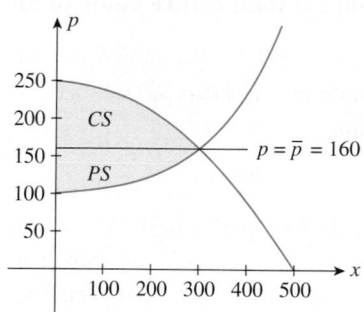

FIGURE 43
Consumers' surplus and producers' surplus when market price = equilibrium price

or \$18,000,000. (Recall that x is measured in units of a thousand.)
   Next, using Equation (17), we find that the producers' surplus is given by

$$PS = (160)(300) - \int_0^{300} (0.0006x^2 + 0.02x + 100)\, dx$$

$$= 48,000 - (0.0002x^3 + 0.01x^2 + 100x)\Bigg|_0^{300}$$

$$= 48,000 - [(0.0002)(300)^3 + (0.01)(300)^2 + 100(300)]$$

$$= 11,700$$

or \$11,700,000.

The Future and Present Value of an Income Stream

Suppose a firm generates a stream of income over a period of time—for example, the revenue generated by a large chain of retail stores over a 5-year period. As the income is realized, it is reinvested and earns interest at a fixed rate. The **accumulated future income stream** over the 5-year period is the amount of money the firm ends up with at the end of that period.

The definite integral can be used to determine this accumulated, or total, future income stream over a period of time. The total future value of an income stream gives us a way to measure the value of such a stream. To find the **total future value of an income stream,** suppose that

$$R(t) = \text{Rate of income generation at any time } t \quad \text{In dollars per year}$$

$$r = \text{Interest rate compounded continuously}$$

$$T = \text{Term} \quad \text{In years}$$

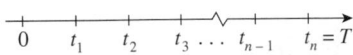

FIGURE 44
The time interval [0, T] is partitioned into n subintervals.

Let's divide the time interval $[0, T]$ into n subintervals of equal length $\Delta t = T/n$ and denote the right endpoints of these intervals by $t_1, t_2, \ldots, t_n = T$, as shown in Figure 44.

If R is a continuous function on $[0, T]$, then $R(t)$ will not differ by much from $R(t_1)$ in the subinterval $[0, t_1]$ provided that the subinterval is small (which is true if n is large). Therefore, the income generated over the time interval $[0, t_1]$ is approximately

$$R(t_1)\, \Delta t \quad \text{Constant rate of income} \times \text{length of time}$$

dollars. The future value of this amount, T years from now, calculated as if it were earned at time t_1, is

$$[R(t_1)\, \Delta t] e^{r(T-t_1)} \quad \text{Equation (5), Section 4.1}$$

dollars. Similarly, the income generated over the time interval $[t_1, t_2]$ is approximately $R(t_2)\, \Delta t$ dollars and has a future value, T years from now, of approximately

$$[R(t_2)\, \Delta t] e^{r(T-t_2)}$$

dollars. Therefore, the sum of the future values of the income stream generated over the time interval $[0, T]$ is approximately

$$R(t_1)e^{r(T-t_1)}\, \Delta t + R(t_2)e^{r(T-t_2)}\, \Delta t + \cdots + R(t_n)e^{r(T-t_n)}\, \Delta t$$
$$= e^{rT}[R(t_1)e^{-rt_1}\, \Delta t + R(t_2)e^{-rt_2}\, \Delta t + \cdots + R(t_n)e^{-rt_n}\, \Delta t]$$

dollars. But this sum is just the Riemann sum of the function $e^{rT}R(t)e^{-rt}$ over the interval $[0, T]$ with representative points $t_1, t_2, \ldots, t_n$. Letting n approach infinity, we obtain the following result.

> ### Accumulated or Total Future Value of an Income Stream
>
> The accumulated, or total, future value after T years of an income stream of $R(t)$ dollars per year, earning interest at the rate of r per year compounded continuously, is given by
>
> $$A = e^{rT} \int_0^T R(t)e^{-rt}\, dt \tag{18}$$

APPLIED EXAMPLE 2 Income Stream Crystal Car Wash recently bought an automatic car-washing machine that is expected to generate $40,000 in revenue per year for the next 5 years. If the income is reinvested in a business earning interest at the rate of 12% per year compounded continuously, find the total accumulated value of this income stream at the end of 5 years.

Solution We are required to find the total future value of the given income stream after 5 years. Using Equation (18) with $R(t) = 40,000$, $r = 0.12$, and $T = 5$, we see that the required value is given by

$$e^{0.12(5)} \int_0^5 40,000 e^{-0.12t}\, dt$$

$$= e^{0.6} \left(-\frac{40,000}{0.12} e^{-0.12t} \right)\Big|_0^5 \qquad \text{Integrate using the substitution } u = -0.12t.$$

$$= -\frac{40,000 e^{0.6}}{0.12} \left(e^{-0.6} - 1 \right) \approx 274,039.60$$

or approximately $274,040.

Another way of measuring the value of an income stream is by considering its present value. The **present value of an income stream** of $R(t)$ dollars per year over a term of T years, earning interest at the rate of r per year compounded continuously, is the principal P that will yield the same accumulated value as the income stream itself when P is invested today for a period of T years at the same rate of interest. In other words,

$$Pe^{rT} = e^{rT} \int_0^T R(t) e^{-rt}\, dt$$

Dividing both sides of the equation by e^{rT} gives the following result.

Present Value of an Income Stream

The present value of an income stream of $R(t)$ dollars per year, earning interest at the rate of r per year compounded continuously, is given by

$$PV = \int_0^T R(t) e^{-rt}\, dt \qquad\qquad (19)$$

APPLIED EXAMPLE 3 Investment Analysis The owner of a local cinema is considering two alternative plans for renovating and improving the theater. Plan A calls for an immediate cash outlay of $250,000, whereas Plan B requires an immediate cash outlay of $180,000. It has been estimated that adopting Plan A would result in a net income stream generated at the rate of

$$f(t) = 630,000$$

dollars per year, whereas adopting Plan B would result in a net income stream generated at the rate of

$$g(t) = 580,000$$

dollars per year for the next 3 years. If the prevailing interest rate for the next 3 years is 10% per year, which plan will generate the higher net income by the end of 3 years?

Solution Since the initial outlay is $250,000, we find—using Equation (19) with $R(t) = 630,000$, $r = 0.1$, and $T = 3$—that the present value of the net income under Plan A is given by

$$\int_0^3 630,000e^{-0.1t}\,dt - 250,000$$

$$= \frac{630,000}{-0.1}e^{-0.1t}\Big|_0^3 - 250,000 \qquad \begin{array}{l}\text{Integrate using the}\\\text{substitution } u = -0.1t.\end{array}$$

$$= -6,300,000e^{-0.3} + 6,300,000 - 250,000$$

$$\approx 1,382,845$$

or approximately $1,382,845.

To find the present value of the net income under Plan B, we use Equation (19) with $R(t) = 580,000$, $r = 0.1$, and $T = 3$, obtaining

$$\int_0^3 580,000e^{-0.1t}\,dt - 180,000$$

dollars. Proceeding as in the previous computation, we see that the required value is $1,323,254 (see Exercise 12, page 889).

Comparing the present value of each plan, we conclude that Plan A would generate the higher total net income over the 3 years. ◼

The Amount and Present Value of an Annuity

An annuity is a sequence of payments made at regular time intervals. The time period in which these payments are made is called the *term* of the annuity. Although the payments need not be equal in size, they are equal in many important applications, and we will assume that they are equal in our discussion. Examples of annuities are regular deposits to a savings account, monthly home mortgage payments, and monthly insurance payments.

The **amount of an annuity** is the sum of the payments plus the interest earned. A formula for computing the amount of an annuity A can be derived with the help of Equation (18). Let

$$P = \text{Size of each payment in the annuity}$$

$$r = \text{Interest rate compounded continuously}$$

$$T = \text{Term of the annuity (in years)}$$

$$m = \text{Number of payments per year}$$

Assume that the payments into the annuity constitute a constant income stream of $R(t) = mP$ dollars per year. With this value of $R(t)$, Equation (18) yields

$$A = e^{rT}\int_0^T R(t)e^{-rt}\,dt$$

$$= e^{rT}\int_0^T mPe^{-rt}\,dt$$

$$= mPe^{rT}\left[-\frac{e^{-rt}}{r}\right]\Big|_0^T$$

$$= mPe^{rT}\left[-\frac{e^{-rT}}{r} + \frac{1}{r}\right]$$

$$= \frac{mP}{r}(e^{rT} - 1) \qquad \text{Since } e^{rT}\cdot e^{-rT} = 1$$

This leads us to the following formula.

Amount of an Annuity

The amount of an annuity is

$$A = \frac{mP}{r}(e^{rT} - 1) \tag{20}$$

where P, r, T, and m are as defined earlier.

$ APPLIED EXAMPLE 4 IRAs On January 1, 1998, Marcus Chapman deposited $2000 into an Individual Retirement Account (IRA) paying interest at the rate of 5% per year compounded continuously. Assuming that he deposited $2000 annually into the account, how much did he have in his IRA at the beginning of 2014?

Solution We use Equation (20), with $P = 2000$, $r = 0.05$, $T = 16$, and $m = 1$, obtaining

$$A = \frac{2000}{0.05}(e^{0.8} - 1)$$

$$\approx 49,021.64$$

Thus, Marcus had approximately $49,022 in his account at the beginning of 2014. ◾

Exploring with TECHNOLOGY

Refer to Example 4. Suppose Marcus wished to know how much he would have in his IRA at any time in the future, not just at the beginning of 2014, as you were asked to compute in the example.

1. Using Formula (18) and the relevant data from Example 4, show that the required amount at any time x (x measured in years, $x > 0$) is given by

$$A = f(x) = 40,000(e^{0.05x} - 1)$$

2. Use a graphing utility to plot the graph of f, using the viewing window $[0, 30] \times [0, 200,000]$.

3. Using **ZOOM** and **TRACE**, or using the function evaluation capability of your graphing utility, use the result of part 2 to verify the result obtained in Example 4. Comment on the advantage of the mathematical model found in part 1.

Using Equation (19), we can derive the following formula for the present value of an annuity.

Present Value of an Annuity

The present value of an annuity is given by

$$PV = \frac{mP}{r}(1 - e^{-rT}) \tag{21}$$

where P, r, T, and m are as defined earlier.

APPLIED EXAMPLE 5 Sinking Funds Tomas Perez, the proprietor of a hardware store, wants to establish a fund from which he will withdraw $1000 per month for the next 10 years. If the fund earns interest at the rate of 6% per year compounded continuously, how much money does he need to establish the fund?

Solution We want to find the present value of an annuity with $P = 1000$, $r = 0.06$, $T = 10$, and $m = 12$. Using Equation (21), we find

$$PV = \frac{12{,}000}{0.06}\left(1 - e^{-(0.06)(10)}\right)$$

$$\approx 90{,}237.70$$

Therefore, Tomas needs approximately $90,238 to establish the fund.

Lorenz Curves and Income Distributions

One method used by economists to study the distribution of income in a society is based on the **Lorenz curve,** named after American statistician M.O. Lorenz. To describe the Lorenz curve, let $f(x)$ denote the proportion of the total income received by the poorest $100x\%$ of the population for $0 \le x \le 1$. Using this terminology, $f(0.3) = 0.1$ simply states that the lowest 30% of the income recipients receive 10% of the total income.

The function f has the following properties:

1. The domain of f is $[0, 1]$.
2. The range of f is $[0, 1]$.
3. $f(0) = 0$ and $f(1) = 1$.
4. $f(x) \le x$ for every x in $[0, 1]$.
5. f is increasing on $[0, 1]$.

The first two properties follow from the fact that both x and $f(x)$ are fractions of a whole. Property 3 is a statement that 0% of the income recipients receive 0% of the total income and 100% of the income recipients receive 100% of the total income. Property 4 follows from the fact that the lowest $100x\%$ of the income recipients cannot receive more than $100x\%$ of the total income. A typical Lorenz curve is shown in Figure 45.

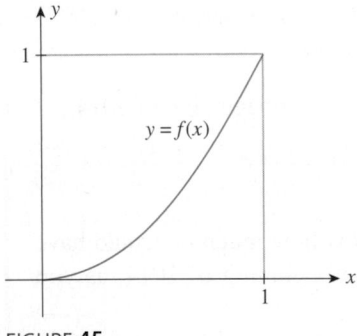

FIGURE 45
A Lorenz curve

APPLIED EXAMPLE 6 Lorenz Curves A developing country's income distribution is described by the function

$$f(x) = \frac{19}{20}x^2 + \frac{1}{20}x$$

a. Sketch the Lorenz curve for the given function.
b. Compute $f(0.2)$ and $f(0.8)$, and interpret your results.

Solution

a. The Lorenz curve is shown in Figure 46.

b.
$$f(0.2) = \frac{19}{20}(0.2)^2 + \frac{1}{20}(0.2) = 0.048$$

Thus, the lowest 20% of the people receive 4.8% of the total income.

$$f(0.8) = \frac{19}{20}(0.8)^2 + \frac{1}{20}(0.8) = 0.648$$

Thus, the lowest 80% of the people receive 64.8% of the total income.

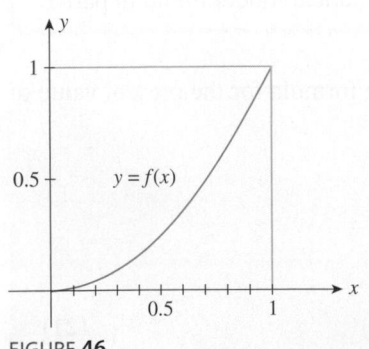

FIGURE 46
The Lorenz curve $f(x) = \dfrac{19}{20}x^2 + \dfrac{1}{20}x$

Next, let's consider the Lorenz curve described by the function $y = f(x) = x$. Since exactly $100x\%$ of the total income is received by the lowest $100x\%$ of income

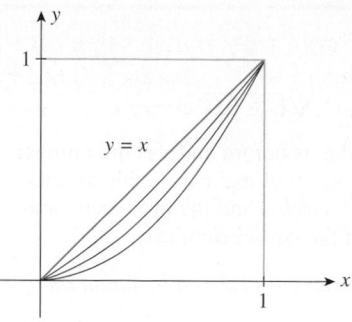

recipients, the line $y = x$ is called the **line of complete equality.** For example, 10% of the total income is received by the lowest 10% of income recipients, 20% of the total income is received by the lowest 20% of income recipients, and so on. Now, it is evident that the closer a Lorenz curve is to this line, the more equitable the income distribution is among the income recipients. But the proximity of a Lorenz curve to the line of complete equality is reflected by the area between the Lorenz curve and the line $y = x$ (Figure 47). The closer the curve is to the line, the smaller the enclosed area.

This observation suggests that we may define a number, called the coefficient of inequality of a Lorenz curve, as the ratio of the area between the line of complete equality and the Lorenz curve to the area under the line of complete equality. Since the area under the line of complete equality is $\frac{1}{2}$, we see that the coefficient of inequality is given by the following formula.

> **Coefficient of Inequality of a Lorenz Curve**
> The coefficient of inequality, or **Gini Index,** of a Lorenz curve is
>
> $$L = 2 \int_0^1 [x - f(x)]\, dx \qquad (22)$$

The coefficient of inequality is a number between 0 and 1. For example, a coefficient of zero implies that the income distribution is perfectly uniform.

APPLIED EXAMPLE 7 Income Distributions In a study conducted by a certain country's Economic Development Board with regard to the income distribution of certain segments of the country's workforce, it was found that the Lorenz curves for the distributions of incomes of medical doctors and of movie actors are described by the functions

$$f(x) = \frac{14}{15} x^2 + \frac{1}{15} x \quad \text{and} \quad g(x) = \frac{5}{8} x^4 + \frac{3}{8} x$$

respectively. Compute the coefficient of inequality for each Lorenz curve. Which profession has the more equitable income distribution?

Solution The required coefficients of inequality are, respectively,

$$L_1 = 2 \int_0^1 \left[x - \left(\frac{14}{15} x^2 + \frac{1}{15} x \right) \right] dx = 2 \int_0^1 \left(\frac{14}{15} x - \frac{14}{15} x^2 \right) dx$$

$$= \frac{28}{15} \int_0^1 (x - x^2)\, dx = \frac{28}{15} \left(\frac{1}{2} x^2 - \frac{1}{3} x^3 \right) \Big|_0^1$$

$$= \frac{14}{45} \approx 0.311$$

$$L_2 = 2 \int_0^1 \left[x - \left(\frac{5}{8} x^4 + \frac{3}{8} x \right) \right] dx = 2 \int_0^1 \left(\frac{5}{8} x - \frac{5}{8} x^4 \right) dx$$

$$= \frac{5}{4} \int_0^1 (x - x^4)\, dx = \frac{5}{4} \left(\frac{1}{2} x^2 - \frac{1}{5} x^5 \right) \Big|_0^1$$

$$= \frac{15}{40} = 0.375$$

We conclude that in this country, the incomes of medical doctors are more evenly distributed than the incomes of movie actors.

11.7 Self-Check Exercise

CONSUMERS' AND PRODUCERS' SURPLUS The demand function for a certain make of exercise bicycle that is sold exclusively through cable television is

$$p = d(x) = \sqrt{9 - 0.02x}$$

where p is the unit price in hundreds of dollars and x is the quantity demanded per week. The corresponding supply function is given by

$$p = s(x) = \sqrt{1 + 0.02x}$$

where p has the same meaning as before and x is the number of exercise bicycles the supplier will make available at price p. Determine the consumers' surplus and the producers' surplus if the unit price is set at the equilibrium price.

The solution to Self-Check Exercise 11.7 can be found on page 891.

11.7 Concept Questions

1. **a.** Define consumers' surplus. Give a formula for computing it.
   b. Define producers' surplus. Give a formula for computing it.

2. **a.** Define the accumulated (future) value of an income stream. Give a formula for computing it.
   b. Define the present value of an income stream. Give a formula for computing it.

3. Define the amount of an annuity. Give a formula for computing it.

4. Explain the following terms: (a) Lorenz curve and (b) coefficient of inequality of a Lorenz curve.

11.7 Exercises

1. **CONSUMERS' SURPLUS** The demand function for a certain make of replacement cartridges for a water purifier is given by

   $$p = -0.01x^2 - 0.1x + 6$$

   where p is the unit price in dollars and x is the quantity demanded each week, measured in units of a thousand. Determine the consumers' surplus if the market price is set at \$4/cartridge.

2. **CONSUMERS' SURPLUS** The demand function for a certain brand of CD is given by

   $$p = -0.01x^2 - 0.2x + 8$$

   where p is the unit price in dollars and x is the quantity demanded each week, measured in units of a thousand. Determine the consumers' surplus if the market price is set at \$5/disc.

3. **CONSUMERS' SURPLUS** It is known that the quantity demanded of a certain make of portable hair dryer is x hundred units/week and the corresponding unit price is

   $$p = \sqrt{225 - 5x}$$

   dollars. Determine the consumers' surplus if the market price is set at \$10/unit.

4. **PRODUCERS' SURPLUS** The supplier of the portable hair dryers in Exercise 3 will make x hundred units of hair dryers available in the market when the unit price is

   $$p = \sqrt{36 + 1.8x}$$

   dollars. Determine the producers' surplus if the market price is set at \$9/unit.

5. **PRODUCERS' SURPLUS** The supply function for the CDs of Exercise 2 is given by

   $$p = 0.01x^2 + 0.1x + 3$$

   where p is the unit price in dollars and x stands for the quantity that will be made available in the market by the supplier, measured in units of a thousand. Determine the producers' surplus if the market price is set at the equilibrium price.

6. **CONSUMERS' SURPLUS** The demand function for a certain brand of mattress is given by

   $$p = 600e^{-0.04x}$$

   where p is the unit price in dollars and x (in units of a hundred) is the quantity demanded per month.
   a. Find the number of mattresses demanded per month if the unit price is set at \$400/mattress.
   b. Use the results of part (a) to find the consumers' surplus if the selling price is set at \$400 per mattress.

7. **PRODUCERS' SURPLUS** The manufacturer of the brand of mattresses of Exercise 6 will make x hundred units available in the market when the unit price is

$$p = 100 + 80e^{0.05x}$$

dollars.
   a. Find the number of mattresses the manufacturer will make available in the market place if the unit price is set at \$250/mattress.
   b. Use the result of part (a) to find the producers' surplus if the unit price is set at \$250/mattress.

8. **CONSUMERS' SURPLUS** The demand function for a certain model of Blu-ray player is given by

$$p = \frac{600}{0.5x + 2}$$

where p is the unit price in dollars and x (in units of a thousand) is the quantity demanded per week. What is the consumers' surplus if the selling price is set at \$200/unit?

9. **PRODUCERS' SURPLUS** The manufacturer of the Blu-ray players in Exercise 8 will make x thousand units available in the market per week when the unit price is

$$p = 100\left(0.5x + \frac{0.4}{1 + x}\right)$$

dollars. What is the producers' surplus if the selling price is set at \$160/unit?

10. **CONSUMERS' AND PRODUCERS' SURPLUS** The management of the Titan Tire Company has determined that the quantity demanded x of their Super Titan tires per week is related to the unit price p by the relation

$$p = 144 - x^2$$

where p is measured in dollars and x is measured in units of a thousand. Titan will make x units of the tires available in the market if the unit price is

$$p = 48 + \frac{1}{2}x^2$$

dollars. Determine the consumers' surplus and the producers' surplus when the market unit price is set at the equilibrium price.

11. **CONSUMERS' AND PRODUCERS' SURPLUS** The quantity demanded x (in units of a hundred) of the Mikado miniature cameras per week is related to the unit price p (in dollars) by

$$p = -0.2x^2 + 80$$

and the quantity x (in units of a hundred) that the supplier is willing to make available in the market is related to the unit price p (in dollars) by

$$p = 0.1x^2 + x + 40$$

If the market price is set at the equilibrium price, find the consumers' surplus and the producers' surplus.

12. Refer to Example 3, page 883. Verify that

$$\int_0^3 580,000e^{-0.1t}\, dt - 180,000 \approx 1,323,254$$

13. **FUTURE VALUE OF AN INVESTMENT** The newly opened Mario's Trattoria is expected to produce a continuous income stream at the rate of

$$R(t) = 120,000$$

dollars/year for the next 4 years. If the prevailing interest rate is 3.5%/year compounded continuously, find the future value of this income stream.

14. **FUTURE VALUE OF AN INVESTMENT** An investment is projected to generate a continuous revenue stream at the rate of

$$R(t) = 30,000e^{0.03t}$$

dollars/year for the next 3 years. If the income stream is invested in a bank that pays interest at the rate of 4.5%/year compounded continuously, find the total accumulated value of this income stream at the end of 3 years.

15. **PRESENT VALUE OF AN INVESTMENT** Suppose an investment is expected to generate income at the rate of

$$R(t) = 200,000$$

dollars/year for the next 5 years. Find the present value of this investment if the prevailing interest rate is 8%/year compounded continuously.

16. **FRANCHISES** Camille purchased a 15-year franchise for a fitness club that is expected to generate income at the rate of

$$R(t) = 400,000$$

dollars/year. If the prevailing interest rate is 8%/year compounded continuously, find the present value of the franchise.

17. **THE AMOUNT OF AN ANNUITY** Find the amount of an annuity if \$250/month is paid into it for a period of 20 years, earning interest at the rate of 4%/year compounded continuously.

18. **THE AMOUNT OF AN ANNUITY** Find the amount of an annuity if \$400/month is paid into it for a period of 20 years, earning interest at the rate of 5%/year compounded continuously.

19. **VALUE OF AN INVESTMENT** An investment generates a continuous income stream of

$$R(t) = 20e^{0.08t}$$

thousand dollars/year at time t for the next 5 years. The prevailing rate of interest is 3%/year compounded continuously for this 5-year period.
   a. What is the future value of the investment at the end of 5 years?
   b. What is the present value of the investment over the term of 5 years?

20. Business Decisions Sharon, the owner of the Brentwood Motel, is planning to renovate all the rooms in her motel. There are two plans before her. Plan A calls for an immediate cash outlay of $600,000, whereas Plan B calls for an immediate outlay of $400,000. Sharon estimates that adopting Plan A would yield an income stream of

$$f(t) = 3{,}050{,}000e^{0.02t}$$

dollars/year for the next 5 years, whereas adopting Plan B would yield an income stream of

$$g(t) = 3{,}200{,}000$$

dollars/year for the next 5 years. If the prevailing rate of interest is 3%/year compounded continuously, which plan will yield the higher net income at the end of 5 years?

21. The Amount of an Annuity Aiso deposits $150/month in a savings account paying 5%/year compounded continuously. Estimate the amount that will be in his account after 15 years.

22. Custodial Accounts The Armstrongs wish to establish a custodial account to finance their children's education. If they deposit $200 monthly for 10 years in a savings account paying 4%/year compounded continuously, how much will their savings account be worth at the end of this period?

23. IRA Accounts Refer to Example 4, page 885. Suppose Marcus made his IRA payment on April 1, 1998, and annually thereafter. If interest is paid at the same initial rate, approximately how much did Marcus have in his account at the beginning of 2014?

24. Present Value of an Annuity Estimate the present value of an annuity if payments are $800 monthly for 12 years and the account earns interest at the rate of 5%/year compounded continuously.

25. Present Value of an Annuity Estimate the present value of an annuity if payments are $1200 monthly for 15 years and the account earns interest at the rate of 6%/year compounded continuously.

26. Lottery Payments A state lottery commission pays the winner of the "Million Dollar" lottery 20 annual installments of $50,000 each. If the prevailing interest rate is 6%/year compounded continuously, find the present value of the winning ticket.

27. Reverse Annuity Mortgages Sinclair wishes to supplement his retirement income by $300/month for the next 10 years. He plans to obtain a reverse annuity mortgage (RAM) on his home to meet this need. Estimate the amount of the mortgage he will require if the prevailing interest rate is 5%/year compounded continuously.

28. Reverse Annuity Mortgage Refer to Exercise 27. Leah wishes to supplement her retirement income by $400/month for the next 15 years by obtaining a RAM. Estimate the amount of the mortgage she will require if the prevailing interest rate is 6%/year compounded continuously.

29. Gini Index The income distribution of a certain country is described by the function

$$f(x) = 0.3x^{1.5} + 0.7x^{2.5}$$

Find the Gini Index for the country.

30. Gini Index A certain country's income distribution is described by the function

$$f(x) = \frac{e^{0.1x} - 1}{e^{0.1} - 1}$$

Find the Gini Index for the country.

31. Lorenz Curves A certain country's income distribution is described by the function

$$f(x) = \frac{15}{16}x^2 + \frac{1}{16}x$$

a. Sketch the Lorenz curve for this function.
b. Compute $f(0.4)$ and $f(0.9)$ and interpret your results.

32. Lorenz Curves A certain country's income distribution is described by the function

$$f(x) = \frac{14}{15}x^2 + \frac{1}{15}x$$

a. Sketch the Lorenz curve for this function.
b. Compute $f(0.3)$ and $f(0.7)$.

33. Lorenz Curves In a study conducted by a certain country's Economic Development Board, it was found that the Lorenz curve for the distribution of income of college teachers was described by the function

$$f(x) = \frac{13}{14}x^2 + \frac{1}{14}x$$

and that of lawyers by the function

$$g(x) = \frac{9}{11}x^4 + \frac{2}{11}x$$

a. Compute the coefficient of inequality for each Lorenz curve.
b. Which profession has a more equitable income distribution?

34. Lorenz Curves In a study conducted by a certain country's Economic Development Board, it was found that the Lorenz curve for the distribution of income of stockbrokers was described by the function

$$f(x) = \frac{11}{12}x^2 + \frac{1}{12}x$$

and that of high school teachers by the function

$$g(x) = \frac{5}{6}x^2 + \frac{1}{6}x$$

a. Compute the coefficient of inequality for each Lorenz curve.

b. Which profession has a more equitable income distribution?

11.7 Solution to Self-Check Exercise

We find the equilibrium price and equilibrium quantity by solving the system of equations

$$p = \sqrt{9 - 0.02x}$$
$$p = \sqrt{1 + 0.02x}$$

simultaneously. Substituting the first equation into the second, we have

$$\sqrt{9 - 0.02x} = \sqrt{1 + 0.02x}$$

Squaring both sides of the equation then leads to

$$9 - 0.02x = 1 + 0.02x$$
$$x = 200$$

Therefore,

$$p = \sqrt{9 - 0.02(200)}$$
$$= \sqrt{5} \approx 2.24$$

The equilibrium price is $2.24, and the equilibrium quantity is 200. The consumers' surplus is given by

$$CS \approx \int_0^{200} \sqrt{9 - 0.02x}\, dx - (2.24)(200)$$

$$= \int_0^{200} (9 - 0.02x)^{1/2}\, dx - 448$$

$$= -\frac{1}{0.02}\left(\frac{2}{3}\right)(9 - 0.02x)^{3/2}\Big|_0^{200} - 448 \qquad \text{Integrate by substitution.}$$

$$= -\frac{1}{0.03}(5^{3/2} - 9^{3/2}) - 448$$

$$\approx 79.32$$

or approximately $7932.

Next, the producers' surplus is given by

$$PS = (2.24)(200) - \int_0^{200} \sqrt{1 + 0.02x}\, dx$$

$$= 448 - \int_0^{200} (1 + 0.02x)^{1/2}\, dx$$

$$= 448 - \frac{1}{0.02}\left(\frac{2}{3}\right)(1 + 0.02x)^{3/2}\Big|_0^{200}$$

$$= 448 - \frac{1}{0.03}(5^{3/2} - 1)$$

$$\approx 108.66$$

or approximately $10,866.

USING TECHNOLOGY   Business and Economic Applications: Technology Exercises

1. Re-solve Example 1, Section 11.7, using a graphing utility.
   Hint: Use the intersection operation to find the equilibrium quantity and the equilibrium price. Use the numerical integral operation to evaluate the definite integral.

2. Re-solve Exercise 11, Section 11.7, using a graphing utility.
   Hint: See Exercise 1.

3. Consumers' and Producers' Surplus The demand function for a certain brand of travel alarm clocks is given by

$$p = -0.01x^2 - 0.3x + 10$$

where p is the unit price in dollars and x is the quantity demanded each month, measured in units of a thousand. The supply function for this brand of clocks is given by

$$p = -0.01x^2 + 0.2x + 4$$

where p has the same meaning as before and x is the quantity, in thousands, the supplier will make available in the marketplace per month. Determine the consumers' surplus and the producers' surplus when the market unit price is set at the equilibrium price.

4. **CONSUMERS' AND PRODUCERS' SURPLUS** The quantity demanded of a certain make of compact disc organizer is x thousand units per week when the corresponding unit price is

$$p = \sqrt{400 - 8x}$$

dollars. The supplier of the organizers will make x thousand units available in the market when the unit price is

$$p = 0.02x^2 + 0.04x + 5$$

dollars. Determine the consumers' surplus and the producers' surplus when the market unit price is set at the equilibrium price.

5. **RETURN ON INVESTMENTS** Investment A is expected to generate income at the rate of

$$R_1(t) = 50{,}000 + 10{,}000\sqrt{t}$$

dollars/year over the 5 years, and Investment B is expected to generate income at the rate of

$$R_2(t) = 50{,}000 + 6000t$$

dollars/year over the same period of time. If the prevailing interest rate for the next 5 years is 10%/year, which investment will generate the higher net income over of 5 years?

6. **RETURN ON INVESTMENTS** Investment A is expected to generate income at the rate of

$$R_1(t) = 40{,}000 + 5000t + 100t^2$$

dollars/year for the next 10 years, and Investment B is expected to generate income at the rate of

$$R_2(t) = 60{,}000 + 2000t$$

dollars/year over the same period of time. If the prevailing interest rate for the next 10 years is 8%/year, which investment will generate the higher net income over the 10 years?

CHAPTER 11   Summary of Principal Formulas and Terms

FORMULAS

1. Indefinite integral of a constant	$\displaystyle\int k\,du = ku + C$
2. Power Rule	$\displaystyle\int u^n\,du = \frac{u^{n+1}}{n+1} + C \quad (n \neq -1)$
3. Constant Multiple Rule	$\displaystyle\int k f(u)\,du = k\int f(u)\,du$ $(k,\ \text{a constant})$

4. Sum Rule	$\int [f(u) \pm g(u)]\, du$ $= \int f(u)\, du \pm \int g(u)\, du$
5. Indefinite integral of the exponential function	$\int e^u\, du = e^u + C$
6. Indefinite integral of $f(u) = \dfrac{1}{u}$	$\int \dfrac{du}{u} = \ln|u| + C$
7. Method of substitution	$\int F'(g(x))g'(x)\, dx = \int F'(u)\, du$
8. Definite integral as the limit of a sum	$\int_a^b f(x)\, dx = \lim_{n \to \infty} S_n,$ where S_n is a Riemann sum
9. Fundamental Theorem of Calculus	$\int_a^b f(x)\, dx = F(b) - F(a),$ where $F'(x) = f(x)$
10. Average value of f over $[a, b]$	$\dfrac{1}{b - a} \int_a^b f(x)\, dx$
11. Area between two curves	$\int_a^b [f(x) - g(x)]\, dx,$ where $f(x) \geq g(x)$
12. Consumers' surplus	$CS = \int_0^{\bar{x}} D(x)\, dx - \bar{p}\,\bar{x}$
13. Producers' surplus	$PS = \bar{p}\,\bar{x} - \int_0^{\bar{x}} S(x)\, dx$
14. Accumulated (future) value of an income stream	$A = e^{rT} \int_0^T R(t)e^{-rt}\, dt$
15. Present value of an income stream	$PV = \int_0^T R(t)e^{-rt}\, dt$
16. Amount of an annuity	$A = \dfrac{mP}{r}(e^{rT} - 1)$
17. Present value of an annuity	$PV = \dfrac{mP}{r}(1 - e^{-rT})$
18. Coefficient of inequality of a Lorenz curve	$L = 2 \int_0^1 [x - f(x)]\, dx$

TERMS

antiderivative (808)

antidifferentiation (810)

integration (810)

indefinite integral (810)

integrand (810)

constant of integration (810)

differential equation (814)

initial-value problem (814)

Riemann sum (836)

definite integral (837)

lower limit of integration (837)

upper limit of integration (837)

Lorenz curve (886)

line of complete equality (887)

CHAPTER 11   Concept Review Questions

Fill in the blanks.

1. a. A function F is an antiderivative of f on an interval I if _____ for all x in I.
   b. If F is an antiderivative of f on an interval I, then every antiderivative of f on I has the form _____.

2. a. $\int cf(x)\, dx =$ _____
   b. $\int [f(x) \pm g(x)]\, dx =$ _____

3. a. A differential equation is an equation that involves the derivative or differential of a/an _____ function.
   b. A solution of a differential equation on an interval I is any _____ that satisfies the differential equation.

4. If we let $u = g(x)$, then $du =$ _____, and the substitution transforms the integral $\int f(g(x))g'(x)\, dx$ into the integral _____ involving only u.

5. a. If f is continuous and nonnegative on an interval $[a, b]$, then the area of the region under the graph of f on $[a, b]$ is given by _____.
   b. If f is continuous on an interval $[a, b]$, then $\int_a^b f(x)\, dx$ is equal to the areas of the regions lying above the x-axis and bounded by the graph of f on $[a, b]$ _____ the areas of the regions lying below the x-axis and bounded by the graph of f on $[a, b]$.

6. a. The Fundamental Theorem of Calculus states that if f is continuous on $[a, b]$, then $\int_a^b f(x)\, dx =$ _____, where F is a/an _____ of f.

b. The net change in a function f over an interval $[a, b]$ is given by $f(b) - f(a) =$ _____, provided that f' is continuous on $[a, b]$.

7. a. If f is continuous on $[a, b]$, then the average value of f over $[a, b]$ is the number _____.
   b. If f is a continuous and nonnegative function on $[a, b]$, then the average value of f over $[a, b]$ may be thought of as the _____ of the rectangle with base lying on the interval $[a, b]$ and having the same _____ as the region under the graph of f on $[a, b]$.

8. If f and g are continuous on $[a, b]$ and $f(x) \geq g(x)$ for all x in $[a, b]$, then the area of the region between the graphs of f and g and the vertical lines $x = a$ and $x = b$ is $A =$ _____.

9. a. The consumers' surplus is given by $CS =$ _____.
   b. The producers' surplus is given by $PS =$ _____.

10. a. The accumulated value after T years of an income stream of $R(t)$ dollars/year, earning interest of r/year compounded continuously, is given by $A =$ _____.
   b. The present value of an income stream is given by $PV =$ _____.

11. The amount of an annuity is $A =$ _____.

12. The coefficient of inequality of a Lorenz curve is $L =$ _____.

CHAPTER 11   Review Exercises

In Exercises 1–20, find each indefinite integral.

1. $\displaystyle\int (x^3 + 2x^2 - x)\, dx$ **2.** $\displaystyle\int \left(\frac{1}{3}x^3 - 2x^2 + 8\right) dx$

3. $\displaystyle\int \left(x^4 - 2x^3 + \frac{1}{x^2}\right) dx$ **4.** $\displaystyle\int (x^{1/3} - \sqrt{x} + 4)\, dx$

5. $\displaystyle\int x(2x^2 + x^{1/2})\, dx$ **6.** $\displaystyle\int (x^2 + 1)(\sqrt{x} - 1)\, dx$

7. $\displaystyle\int \left(x^2 - x + \frac{2}{x} + 5\right) dx$ **8.** $\displaystyle\int \sqrt{2x + 1}\, dx$

9. $\displaystyle\int (3x - 1)(3x^2 - 2x + 1)^{1/3}\, dx$

10. $\displaystyle\int x^2(x^3 + 2)^{10}\, dx$ **11.** $\displaystyle\int \frac{x - 1}{x^2 - 2x + 5}\, dx$

12. $\displaystyle\int 2e^{-2x}\, dx$ **13.** $\displaystyle\int \left(x + \frac{1}{2}\right) e^{x^2 + x + 1}\, dx$

14. $\displaystyle\int \frac{e^{-x} - 1}{(e^{-x} + x)^2}\, dx$ **15.** $\displaystyle\int \frac{(\ln x)^5}{x}\, dx$

16. $\displaystyle\int \frac{\ln x^2}{x}\, dx$ **17.** $\displaystyle\int x^3(x^2 + 1)^{10}\, dx$

18. $\displaystyle\int x\sqrt{x + 1}\, dx$ **19.** $\displaystyle\int \frac{x}{\sqrt{x - 2}}\, dx$

20. $\displaystyle\int \frac{3x}{\sqrt{x + 1}}\, dx$

In Exercises 21–32, evaluate each definite integral.

21. $\displaystyle\int_0^1 (2x^3 - 3x^2 + 1)\, dx$

22. $\displaystyle\int_0^2 (4x^3 - 9x^2 + 2x - 1)\, dx$

23. $\displaystyle\int_1^4 (\sqrt{x} + x^{-3/2})\, dx$ **24.** $\displaystyle\int_0^1 20x(2x^2 + 1)^4\, dx$

25. $\displaystyle\int_{-1}^0 12(x^2 - 2x)(x^3 - 3x^2 + 1)^3\, dx$

26. $\displaystyle\int_4^7 x\sqrt{x - 3}\, dx$ **27.** $\displaystyle\int_0^2 \frac{x}{x^2 + 1}\, dx$

28. $\displaystyle\int_0^1 \frac{dx}{(5-2x)^2}$

29. $\displaystyle\int_0^2 \frac{4x}{\sqrt{1+2x^2}}\,dx$

30. $\displaystyle\int_0^2 xe^{(-1/2)x^2}\,dx$

31. $\displaystyle\int_{-1}^0 \frac{e^{-x}}{(1+e^{-x})^2}\,dx$

32. $\displaystyle\int_1^e \frac{\ln x}{x}\,dx$

In Exercises 33–36, find the function f given that the slope of the tangent line to the graph at any point $(x, f(x))$ is $f'(x)$ and that the graph of f passes through the given point.

33. $f'(x) = 3x^2 - 4x + 1;\, (1, 1)$

34. $f'(x) = \dfrac{x}{\sqrt{x^2+1}};\, (0, 1)$

35. $f'(x) = 1 - e^{-x};\, (0, 2)$

36. $f'(x) = \dfrac{\ln x}{x};\, (1, -2)$

In Exercises 37 and 38, refer to the following figure.

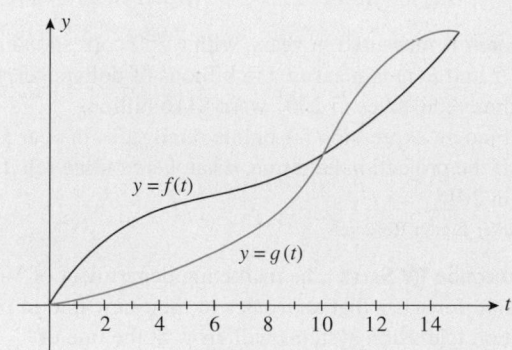

37. Motion of Two Cars Suppose $f(t)$ and $g(t)$ (both measured in feet per second) give the velocities of Car A and Car B, respectively, t seconds after starting side by side on a straight road.
   a. Give a physical interpretation of the integral

   $$\int_0^T [f(t) - g(t)]\,dt \qquad (0 < t < 14)$$

   b. At what time is the distance between the two cars greatest? Write an integral giving this distance.

38. Revenues of Companies Suppose $f(t)$ and $g(t)$ give the rate of change of the revenues (measured in dollars per day) of two new branches, A and B, of an office supply company t weeks after their simultaneous grand opening.
   a. Give an interpretation of the integral

   $$\int_0^T [f(t) - g(t)]\,dt \qquad (0 < t < 14)$$

   b. At what time over the first two weeks since opening is the difference between the revenues of the two branches greatest? Write an integral giving this difference.

39. Let $f(x) = -2x^2 + 1$, and compute the Riemann sum of f over the interval $[1, 2]$ by partitioning the interval into five subintervals of the same length $(n = 5)$, where the points p_i $(1 \le i \le 5)$ are taken to be the *right* endpoints of the respective subintervals.

40. Growth Rate of Smartphone Users The rate of growth of smartphone users, as a percentage of mobile phone users, is projected to be

$$R(t) = 10.8 \qquad (0 \le t \le 2)$$

percent/year. The percentage in October 2011 was 38.5. Find an expression giving the percentage of mobile phone users that were smartphone users in year t. What was the projected percentage of mobile phone users that were smartphone users in October 2013?
Source: eMarketer.

41. Computer Resale Value Franklin National Life Insurance Company purchased new computers for $200,000. If the rate at which the computers' resale value changes is given by the function

$$V'(t) = 3800(t - 10)$$

where t is the length of time since the purchase date and $V'(t)$ is measured in dollars per year, find an expression $V(t)$ that gives the resale value of the computers after t years. How much would the computers sell for after 6 years?

42. Marginal Cost Functions The management of National Electric has determined that the daily marginal cost function associated with producing their automatic drip coffeemakers is given by

$$C'(x) = 0.00003x^2 - 0.03x + 20$$

where $C'(x)$ is measured in dollars per unit and x denotes the number of units produced. Management has also determined that the daily fixed cost incurred in producing these coffeemakers is $500. What is the total cost incurred by National in producing the first 400 coffeemakers/day?

43. Marginal Revenue Functions Refer to Exercise 42. Management has also determined that the daily marginal revenue function associated with producing and selling their coffeemakers is given by

$$R'(x) = -0.03x + 60$$

where x denotes the number of units produced and sold and $R'(x)$ is measured in dollars per unit.
   a. Determine the revenue function $R(x)$ associated with producing and selling these coffeemakers.
   b. What is the demand equation relating the unit price to the quantity of coffeemakers demanded?

44. Measuring Temperature The temperature on a certain day as measured at the airport of a city is changing at the rate of

$$T'(t) = 0.15t^2 - 3.6t + 14.4 \qquad (0 \le t \le 4)$$

°F/hr, where t is measured in hours, with $t = 0$ corresponding to 6 A.M. The temperature at 6 A.M. was 24°F.
  a. Find an expression giving the temperature T at the airport at any time between 6 A.M. and 10 A.M.
  b. What was the temperature at 10 A.M.?

45. **DVD Sales** The total number of DVDs sold to U.S. dealers for rental and sale from 1999 through 2003 grew at the rate of approximately

$$R(t) = -0.03t^2 + 0.218t - 0.032 \qquad (0 \le t \le 4)$$

billion discs/year, where t is measured in years, with $t = 0$ corresponding to 1999. The total number of DVDs sold as of 1999 was 0.1 billion.
  a. Find an expression giving the total number of DVDs sold by year t.
  b. How many DVDs were sold in 2003?
  Source: Adams Media.

46. **Air Pollution** On an average summer day, the level of carbon monoxide (CO) in a city's air is 2 parts per million (ppm). An environmental protection agency's study predicts that, unless more stringent measures are taken to protect the city's atmosphere, the CO concentration present in the air will increase at the rate of

$$0.003t^2 + 0.06t + 0.1$$

ppm/year, t years from now. If no further pollution-control efforts are made, what will be the CO concentration on an average summer day 5 years from now?

47. **Marginal Cost Functions** The management of a division of Ditton Industries has determined that the daily marginal cost function associated with producing their hot-air corn poppers is given by

$$C'(x) = 0.00003x^2 - 0.03x + 10$$

where $C'(x)$ is measured in dollars/unit and x denotes the number of units manufactured. Management has also determined that the daily fixed cost incurred in producing these corn poppers is $600. Find the total cost incurred by Ditton in producing the first 500 corn poppers.

48. **U.S. Census** The number of Americans aged 45–54 years (which stood at 25 million at the beginning of 1990) grew at the rate of

$$R(t) = 0.00933t^3 + 0.019t^2 - 0.10833t + 1.3467$$

million people/year, t years from the beginning of 1990. How many Americans aged 45 to 54 were added to the population between 1990 and 2000?
  Source: U.S. Census Bureau.

49. **Commuter Trends** Because of the increasing cost of fuel, the manager of the City Transit Authority estimates that the number of commuters using the city subway system will increase at the rate of

$$3000(1 + 0.4t)^{-1/2} \qquad (0 \le t \le 36)$$

commuters per month, t months from now. If 100,000 commuters are currently using the system, find an

expression giving the total number of commuters who will be using the subway t months from now. How many commuters will be using the subway 6 months from now?

50. **Supply: Women's Boots** The rate of change of the unit price p (in dollars) of Apex women's boots is given by

$$p'(x) = \frac{240}{(5 - x)^2}$$

where x is the number of pairs in units of a hundred that the supplier will make available in the market daily when the unit price is p/pair. Find the supply equation for these boots if the quantity the supplier is willing to make available is 200 pairs daily $(x = 2)$ when the unit price is $50/pair.

51. **Online Retail Sales** Since the inception of the Web, online commerce has enjoyed phenomenal growth. But growth, led by such major sectors as books, tickets, and office supplies, is expected to slow in the coming years. The projected growth of online retail sales is given by

$$R(t) = 15.82e^{-0.176t} \qquad (0 \le t \le 4)$$

where t is measured in years, with $t = 0$ corresponding to 2007 and $R(t)$ is measured in billions of dollars per year. Online retail sales in 2007 were $116 billion.
  a. Find an expression for online retail sales in year t.
  b. If the projection held true, what were online retail sales in 2011?
  Source: Jupiter Research.

52. **Projection TV Sales** The marketing department of Vista Vision forecasts that total sales of their new line of projection television systems will grow at the rate of

$$3000 - 2000e^{-0.04t} \qquad (0 \le t \le 24)$$

systems/month once they are introduced into the market. Find an expression giving the total number of the projection television systems that Vista may expect to sell in the t months after they are put on the market. How many of these systems can Vista expect to sell during the first year?

53. **Sales: Loudspeakers** Sales of the Acrosonic model F loudspeaker systems have been growing at the rate of

$$f'(t) = 2000(3 - 2e^{-t})$$

systems/year, where t denotes the number of years these loudspeaker systems have been on the market. Determine the number of loudspeaker systems that were sold in the first 5 years after they appeared on the market.

54. Find the area of the region under the curve $y = 3x^2 + 2x + 1$ from $x = -1$ to $x = 2$.

55. Find the area of the region under the curve $y = e^{2x}$ from $x = 0$ to $x = 2$.

56. Find the area of the region bounded by the graph of the function $y = 1/x^2$, the x-axis, and the lines $x = 1$ and $x = 3$.

57. Find the area of the region bounded by the curve $y = -x^2 - x + 2$ and the x-axis.

58. Find the area of the region bounded by the graphs of the functions $f(x) = e^x$ and $g(x) = x$ and the vertical lines $x = 0$ and $x = 2$.

59. Find the area of the region that is completely enclosed by the graphs of $f(x) = x^4$ and $g(x) = x$.

60. Find the area of the region between the curve $y = x(x - 1)(x - 2)$ and the x-axis.

61. OIL PRODUCTION On the basis of current production techniques, the rate of oil production from a certain oil well t years from now is estimated to be

$$R_1(t) = 100e^{0.05t}$$

thousand barrels/year. On the basis of a new production technique, however, it is estimated that the rate of oil production from that oil well t years from now will be

$$R_2(t) = 100e^{0.08t}$$

thousand barrels/year. Determine how much additional oil will be produced over the next 10 years if the new technique is adopted.

62. Find the average value of the function

$$f(x) = \frac{x}{\sqrt{x^2 + 16}}$$

over the interval $[0, 3]$.

63. AVERAGE TEMPERATURE The temperature (in degrees Fahrenheit) in Boston over a 12-hr period on a certain December day was given by

$$T = -0.05t^3 + 0.4t^2 + 3.8t + 5.6 \qquad (0 \le t \le 12)$$

where t is measured in hours, with $t = 0$ corresponding to 6 A.M. Determine the average temperature on that day over the 12-hr period from 6 A.M. to 6 P.M.

64. AVERAGE VELOCITY OF A TRUCK A truck traveling along a straight road has a velocity (in feet per second) at time t (in seconds) given by

$$v(t) = \frac{1}{12}t^2 + 2t + 44 \qquad (0 \le t \le 5)$$

What is the average velocity of the truck over the time interval from $t = 0$ to $t = 5$?

65. MEMBERSHIP IN CREDIT UNIONS The membership in Massachusetts credit unions grew at the rate of

$$R(t) = -0.0039t^2 + 0.0374t + 0.0046 \qquad (0 \le t \le 9)$$

million members/year in year t between 1994 ($t = 0$) and 2003 ($t = 9$). Find the average rate of growth of membership in Massachusetts credit unions over the period in question.

Source: Massachusetts Credit Union League.

66. DEMAND FOR DIGITAL CAMCORDER TAPES The demand function for a brand of blank digital camcorder tapes is given by

$$p = -0.01x^2 - 0.2x + 23$$

where p is the unit price in dollars and x is the quantity demanded each week, measured in units of a thousand. Determine the consumers' surplus if the unit price is $8/tape.

67. CONSUMERS' AND PRODUCERS' SURPLUS The quantity demanded x (in units of a hundred) of the Sportsman 5×7 tents, per week, is related to the unit price p (in dollars) by the relation

$$p = -0.1x^2 - x + 40$$

The quantity x (in units of a hundred) that the supplier is willing to make available in the market is related to the unit price by the relation

$$p = 0.1x^2 + 2x + 20$$

If the market price is set at the equilibrium price, find the consumers' surplus and the producers' surplus.

68. RETIREMENT ACCOUNT SAVINGS Chi-Tai plans to deposit $4000/year in his defined benefit retirement account. If interest is compounded continuously at the rate of 8%/year, how much will he have in his retirement account after 20 years?

69. INSTALLMENT CONTRACTS Glenda sold her house under an installment contract whereby the buyer gave her a down payment of $20,000 and agreed to make monthly payments of $925/month for 30 years. If the prevailing interest rate is 6%/year compounded continuously, find the present value of the purchase price of the house.

70. PRESENT VALUE OF A FRANCHISE Alicia purchased a 10-year franchise for a health spa that is expected to generate income at the rate of

$$R(t) = 80,000$$

dollars/year. If the prevailing interest rate is 10%/year compounded continuously, find the present value of the franchise.

71. INCOME DISTRIBUTION OF A COUNTRY A certain country's income distribution is described by the function

$$f(x) = \frac{17}{18}x^2 + \frac{1}{18}x$$

   a. Sketch the Lorenz curve for this function.
   b. Compute $f(0.3)$ and $f(0.6)$, and interpret your results.
   c. Compute the coefficient of inequality for this Lorenz curve.

72. POPULATION GROWTH The population of a certain Sunbelt city, currently 80,000, is expected to grow exponentially in the next 5 years with a growth constant of 0.05. If the prediction comes true, what will be the average population of the city over the next 5 years?

CHAPTER 11 | Before Moving On . . .

1. Find $\int \left(2x^3 + \sqrt{x} + \dfrac{2}{x} - \dfrac{2}{\sqrt{x}} \right) dx$.

2. Find f if $f'(x) = e^x + x$ and $f(0) = 2$.

3. Find $\int \dfrac{x}{\sqrt{x^2 + 1}}\, dx$.

4. Evaluate $\displaystyle\int_0^1 x\sqrt{2 - x^2}\, dx$.

5. Find the area of the region completely enclosed by the graphs of $y = x^2 - 1$ and $y = 1 - x$.

12 Calculus of Several Variables

UP TO NOW, we have dealt with functions involving one variable. However, many situations involve functions of two or more variables. For example, the Consumer Price Index (CPI) compiled by the Bureau of Labor Statistics depends on the price of more than 95,000 consumer items. To study such relationships, we need the notion of a function of several variables, the first topic in this chapter. Next, generalizing the concept of the derivative of a function of one variable, we study the *partial derivatives* of a function of two or more variables. Using partial derivatives, we study the rate of change of a function with respect to one variable while holding all other variables constant. We then learn how to find the extremum values of a function of several variables. For example, we will learn how a manufacturer can maximize her profits by producing the optimal quantity of her products.

Suppose the government of a certain country wishes to increase the country's productivity at a certain time. Should the government encourage capital investment or an increase in the expenditure on labor at that time? In Applied Example 4, page 918, you will see how we can use the production function for the country to answer this question.

Peter Parks/AFP/Getty Images

12.1 Functions of Several Variables

Up to now, our study of calculus has been restricted to functions of one variable. In many practical situations, however, the formulation of a problem results in a mathematical model that involves a function of two or more variables. For example, suppose Ace Novelty determines that the profits are $6, $5, and $4 for three types of souvenirs it produces. Let x, y, and z denote the number of Type A, Type B, and Type C souvenirs to be made; then the company's profit is given by

$$P = 6x + 5y + 4z$$

and P is a function of the three variables, x, y, and z.

Functions of Two Variables

Although this chapter deals with real-valued functions of several variables, most of our definitions and results are stated in terms of a function of two variables. One reason for adopting this approach, as you will soon see, is that there is a geometric interpretation for this special case, which serves as an important visual aid. We can then draw upon the experience gained from studying the two-variable case to help us understand the concepts and results connected with the more general case, which, by and large, is just a simple extension of the lower-dimensional case.

A Function of Two Variables

A real-valued **function of two variables** f consists of

1. A set A of ordered pairs of real numbers (x, y) called the **domain** of the function.
2. A rule that associates with each ordered pair in the domain of f one and only one real number, denoted by $z = f(x, y)$.

The variables x and y are called **independent variables,** and the variable z, which is dependent on the values of x and y, is referred to as a **dependent variable.**

As in the case of a real-valued function of one real variable, the number $z = f(x, y)$ is called the **value of f** at the point (x, y). And, unless specified, the domain of the function f will be taken to be the largest possible set for which the rule defining f is meaningful.

EXAMPLE 1 Let f be the function defined by

$$f(x, y) = x + xy + y^2 + 2$$

Compute $f(0, 0)$, $f(1, 2)$, and $f(2, 1)$.

Solution We have

$$f(0, 0) = 0 + (0)(0) + 0^2 + 2 = 2$$
$$f(1, 2) = 1 + (1)(2) + 2^2 + 2 = 9$$
$$f(2, 1) = 2 + (2)(1) + 1^2 + 2 = 7$$

The domain of a function of two variables $f(x, y)$ is a set of ordered pairs of real numbers and may therefore be viewed as a subset of the xy-plane.

EXAMPLE 2 Find the domain of each function.

a. $f(x, y) = x^2 + y^2$ **b.** $g(x, y) = \dfrac{2}{x - y}$ **c.** $h(x, y) = \sqrt{1 - x^2 - y^2}$

Solution

a. $f(x, y)$ is defined for all real values of x and y, so the domain of the function f is the set of all points (x, y) in the xy-plane.

b. $g(x, y)$ is defined for all $x \neq y$, so the domain of the function g is the set of all points in the xy-plane except those lying on the line $y = x$ (Figure 1a).

c. We require that $1 - x^2 - y^2 \geq 0$ or $x^2 + y^2 \leq 1$, which is just the set of all points (x, y) lying on and inside the circle of radius 1 with center at the origin (Figure 1b).

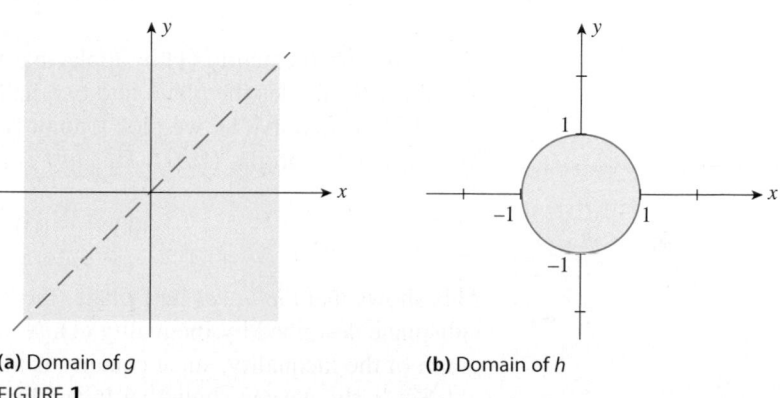

(a) Domain of g (b) Domain of h

FIGURE **1**

$ APPLIED EXAMPLE 3 Revenue Functions Acrosonic manufactures a bookshelf loudspeaker system that may be bought fully assembled or in a kit. The demand equations that relate the unit prices, p and q, to the quantities demanded weekly, x and y, of the assembled and kit versions of the loudspeaker systems are given by

$$p = 300 - \frac{1}{4}x - \frac{1}{8}y \quad \text{and} \quad q = 240 - \frac{1}{8}x - \frac{3}{8}y$$

a. What is the weekly total revenue function $R(x, y)$?

b. What is the domain of the function R?

Solution

a. The weekly revenue realizable from the sale of x units of the assembled speaker systems at p dollars per unit is given by xp dollars. Similarly, the weekly revenue realizable from the sale of y units of the kits at q dollars per unit is given by yq dollars. Therefore, the weekly total revenue function R is given by

$$R(x, y) = xp + yq$$

$$= x\left(300 - \frac{1}{4}x - \frac{1}{8}y\right) + y\left(240 - \frac{1}{8}x - \frac{3}{8}y\right) \quad \text{(x^2) See page 10.}$$

$$= -\frac{1}{4}x^2 - \frac{3}{8}y^2 - \frac{1}{4}xy + 300x + 240y$$

b. To find the domain of the function R, let's observe that the quantities x, y, p, and q must be nonnegative. This observation leads to the following system of linear inequalities:

$$300 - \frac{1}{4}x - \frac{1}{8}y \geq 0 \tag{1}$$

$$240 - \frac{1}{8}x - \frac{3}{8}y \geq 0 \tag{2}$$

$$x \geq 0 \tag{3}$$

$$y \geq 0 \tag{4}$$

This system of linear inequalities defines a region D in the xy-plane that is the domain of R. To sketch D, we first draw the line defined by the equation

$$300 - \frac{1}{4}x - \frac{1}{8}y = 0$$

obtained from Inequality (1) by replacing it by an equality (see Figure 2). Observe that this line divides the plane into two half-planes. To find the half-plane determined by Inequality (1), we pick a point lying in one of the half-planes. For simplicity, we pick the origin, $(0, 0)$. This *test point* satisfies Inequality (1), since

$$300 - \frac{1}{4}(0) - \frac{1}{8}(0) \geq 0$$

This shows that the *lower* half-plane (the half-plane containing the test point) is the half-plane described by Inequality (1). Note that the line itself is included in the graph of the inequality, since equality is also allowed in Inequality (1).

Similarly, we can show that Inequality (2) defines the lower half-plane determined by the equation

$$240 - \frac{1}{8}x - \frac{3}{8}y = 0$$

obtained from replacing Inequality (2) by an equation. Once again, the line itself is included in the graph.

Finally, Inequalities (3) and (4) together define the first quadrant with the positive x- and y-axis. Therefore, the region D is obtained by taking the intersection of the two half-planes in the first quadrant.

The domain D of the function R is sketched in Figure 2. ∎

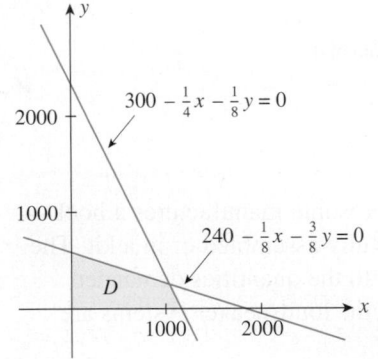

FIGURE 2
The domain of $R(x, y)$

Explore and Discuss

Suppose the total profit of a two-product company is given by $P(x, y)$, where x denotes the number of units of the first product produced and sold and y denotes the number of units of the second product produced and sold. Fix $x = a$, where a is a positive number such that (a, y) is in the domain of P. Describe and give an economic interpretation of the function $f(y) = P(a, y)$. Next, fix $y = b$, where b is a positive number such that (x, b) is in the domain of P. Describe and give an economic interpretation of the function $g(x) = P(x, b)$.

$\$$ APPLIED EXAMPLE 4 Home Mortgage Payments The monthly payment that amortizes a loan of A dollars in t years when the interest rate is r per year compounded monthly is given by

$$P = f(A, r, t) = \frac{Ar}{12\left[1 - \left(1 + \frac{r}{12}\right)^{-12t}\right]}$$

Find the monthly payment for a home mortgage of \$270,000 to be amortized over 30 years when the interest rate is 6% per year, compounded monthly.

Solution   Letting $A = 270{,}000$, $r = 0.06$, and $t = 30$, we find the required monthly payment to be

$$P = f(270{,}000, 0.06, 30) = \frac{270{,}000(0.06)}{12\left[1 - \left(1 + \frac{0.06}{12}\right)^{-360}\right]}$$
$$\approx 1618.79$$

or approximately \$1618.79.   ∎

Graphs of Functions of Two Variables

To graph a function of two variables, we need a three-dimensional coordinate system. This is readily constructed by adding a third axis to the plane Cartesian coordinate system in such a way that the three resulting axes are mutually perpendicular and intersect at O. Observe that, by construction, the zeros of the three number scales coincide at the origin of the **three-dimensional Cartesian coordinate system** (Figure 3).

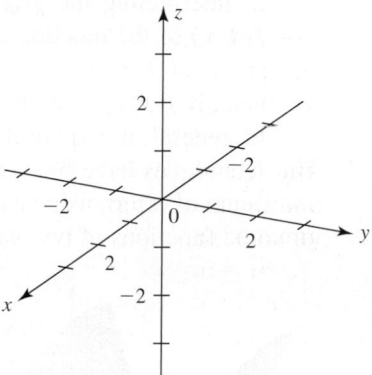

FIGURE **3**
The three-dimensional Cartesian coordinate system

A point in three-dimensional space can now be represented uniquely in this coordinate system by an **ordered triple** of numbers (x, y, z), and, conversely, every ordered triple of real numbers (x, y, z) represents a point in three-dimensional space (Figure 4a). For example, the points $A(2, 3, 4)$, $B(1, -2, -2)$, $C(2, 4, 0)$, and $D(0, 0, 4)$ are shown in Figure 4b.

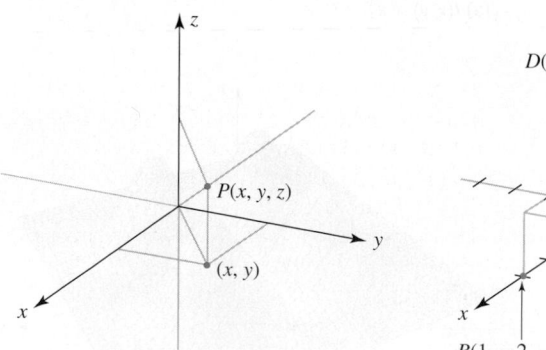

(a) A point in three-dimensional space

(b) Some sample points in three-dimensional space

FIGURE **4**

Now, if $f(x, y)$ is a function of two variables x and y, the domain of f is a subset of the xy-plane. If we denote $f(x, y)$ by z, then the totality of all points (x, y, z), that is, $(x, y, f(x, y))$, makes up the **graph** of the function f and is, except for certain degenerate cases, a surface in three-dimensional space (Figure 5).

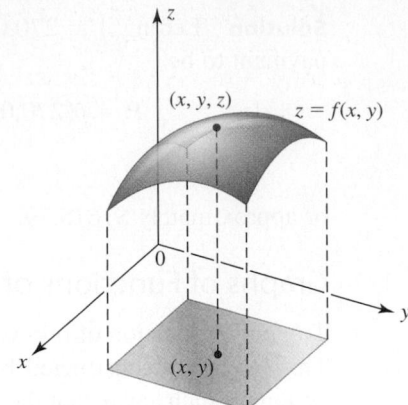

FIGURE **5**
The graph of a function in three-dimensional space

In interpreting the graph of a function $f(x, y)$, one often thinks of the value $z = f(x, y)$ of the function at the point (x, y) as the "height" of the point (x, y, z) on the graph of f. If $f(x, y) > 0$, then the point (x, y, z) is $f(x, y)$ units above the xy-plane; if $f(x, y) < 0$, then the point (x, y, z) is $|f(x, y)|$ units below the xy-plane.

In general, it is quite difficult to draw the graph of a function of two variables. But techniques have been developed that enable us to generate such graphs with a minimum of effort, using a computer. Figure 6 shows the computer-generated graphs of some functions of two variables.

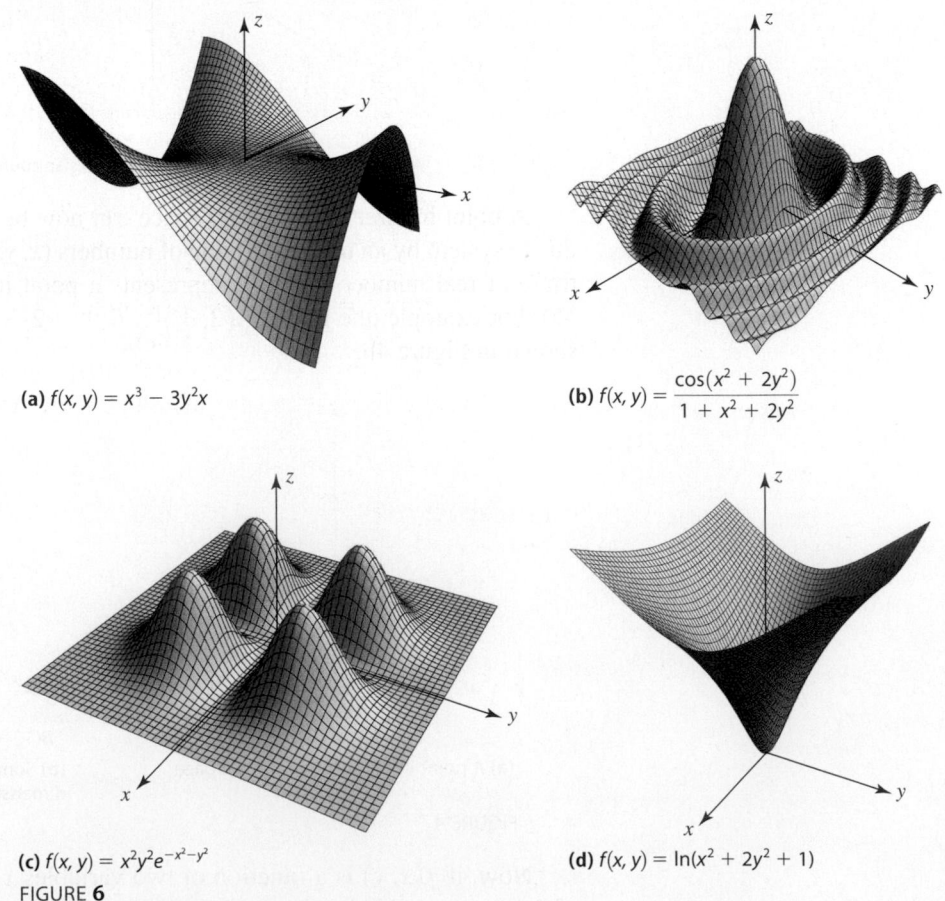

(a) $f(x, y) = x^3 - 3y^2x$

(b) $f(x, y) = \dfrac{\cos(x^2 + 2y^2)}{1 + x^2 + 2y^2}$

(c) $f(x, y) = x^2y^2e^{-x^2-y^2}$

(d) $f(x, y) = \ln(x^2 + 2y^2 + 1)$

FIGURE **6**
Four computer-generated graphs of functions of two variables

Level Curves

We can visualize the graph of a function of two variables by using *level curves*. To define the level curve of a function f of two variables, let $z = f(x, y)$ and consider the trace of f in the plane $z = k$ (k, a constant), as shown in Figure 7a. If we project this trace onto the xy-plane, we obtain a curve C with equation $f(x, y) = k$, called a *level curve* of f (Figure 7b).

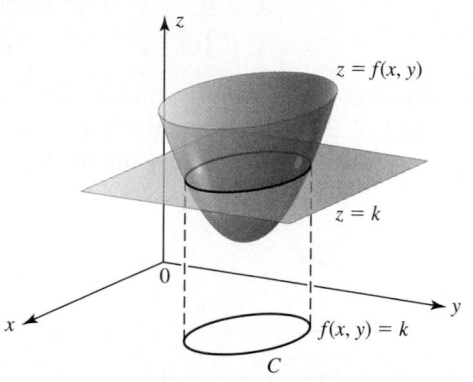

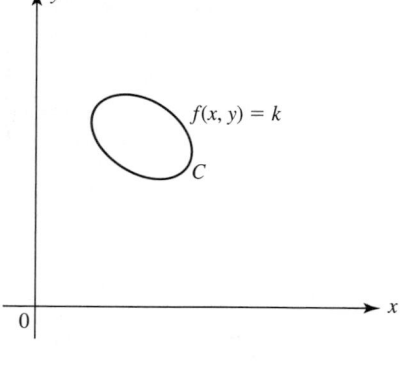

(a) The level curve C with equation $f(x, y) = k$ is the projection of the trace of f in the plane $z = k$ onto the xy-plane.

(b) The level curve C

FIGURE **7**

> ### Level Curves
>
> The **level curves** of a function f of two variables are the curves with equations $f(x, y) = k$, where k is a constant in the range of f.

Notice that the level curve with equation $f(x, y) = k$ is the set of all points in the domain of f corresponding to the points on the surface $z = f(x, y)$ having the same height or depth k. By drawing the level curves corresponding to several admissible values of k, we obtain a *contour map*. The map enables us to visualize the surface represented by the graph of $z = f(x, y)$: We simply lift or depress the level curve to see the "cross sections" of the surface. Figure 8a shows a hill, and Figure 8b shows a contour map associated with that hill.

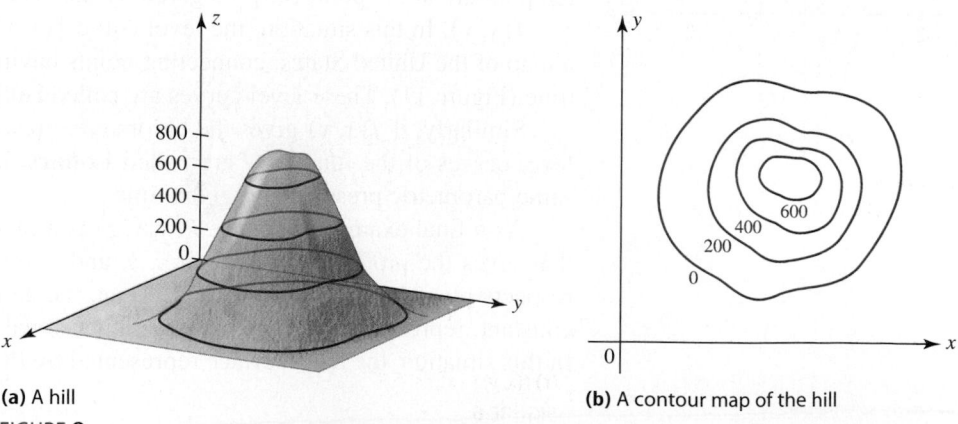

(a) A hill

(b) A contour map of the hill

FIGURE **8**

EXAMPLE 5 Sketch a contour map for the function $f(x, y) = x^2 + y^2$.

Solution The level curves are the graphs of the equation $x^2 + y^2 = k$ for nonnegative numbers k. Taking $k = 0, 1, 4, 9$, and 16, for example, we obtain

$$
\begin{aligned}
k = 0: & \quad x^2 + y^2 = 0 \\
k = 1: & \quad x^2 + y^2 = 1 \\
k = 4: & \quad x^2 + y^2 = 4 = 2^2 \\
k = 9: & \quad x^2 + y^2 = 9 = 3^2 \\
k = 16: & \quad x^2 + y^2 = 16 = 4^2
\end{aligned}
$$

The five level curves are concentric circles with center at the origin and radius given by $r = 0, 1, 2, 3$, and 4, respectively (Figure 9a). A sketch of the graph of $f(x, y) = x^2 + y^2$ is included for your reference in Figure 9b.

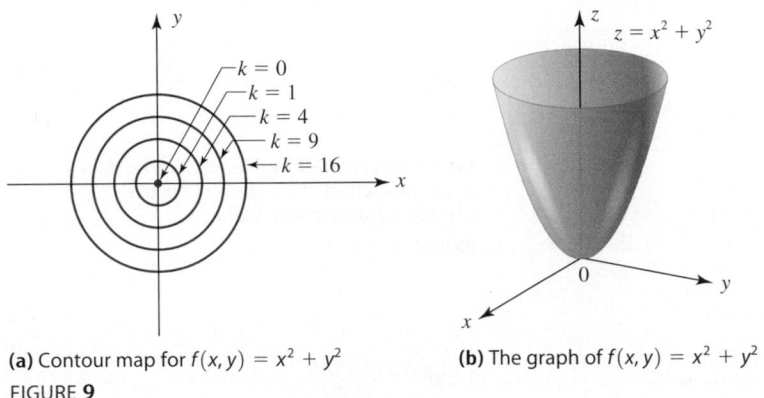

(a) Contour map for $f(x, y) = x^2 + y^2$

(b) The graph of $f(x, y) = x^2 + y^2$

FIGURE **9**

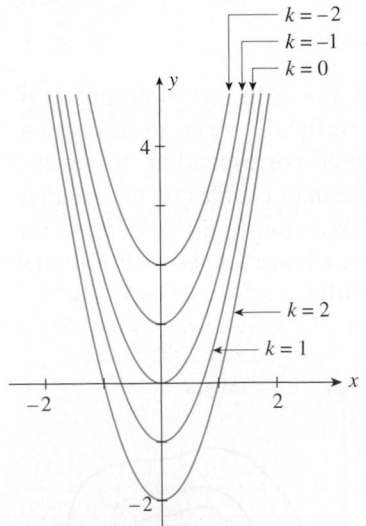

FIGURE **10**
Level curves for $f(x, y) = 2x^2 - y$

EXAMPLE 6 Sketch the level curves for the function $f(x, y) = 2x^2 - y$ corresponding to $z = -2, -1, 0, 1$, and 2.

Solution The level curves are the graphs of the equation $2x^2 - y = k$ or $y = 2x^2 - k$ for $k = -2, -1, 0, 1$, and 2. The required level curves are shown in Figure 10.

Level curves of functions of two variables are found in many practical applications. For example, if $f(x, y)$ denotes the temperature at a location within the continental United States with longitude x and latitude y at a certain point in time, then the temperature at the point (x, y) is given by the "height" of the surface, represented by $z = f(x, y)$. In this situation, the level curve $f(x, y) = k$ is a curve superimposed on a map of the United States, connecting points having the same temperature at a given time (Figure 11). These level curves are called **isotherms.**

Similarly, if $f(x, y)$ gives the barometric pressure at the location (x, y), then the level curves of the function f are called **isobars,** lines connecting points having the same barometric pressure at a given time.

As a final example, suppose $P(x, y, z)$ is a function of three variables x, y, and z that gives the profit realized when x, y, and z units of three products, A, B, and C, respectively, are produced and sold. Then, the equation $P(x, y, z) = k$, where k is a constant, represents a surface in three-dimensional space called a **level surface** of P. In this situation, the level surface represented by $P(x, y, z) = k$ represents the product mix that results in a profit of exactly k dollars. Such a level surface is called an **iso-profit surface.**

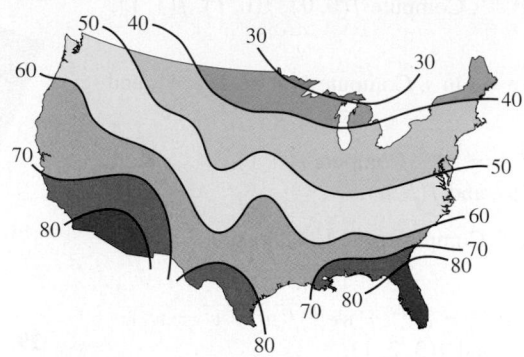

FIGURE **11**
Isotherms: curves connecting points that have the same temperature

---

12.1 Self-Check Exercises

1. Let $f(x, y) = x^2 - 3xy + \sqrt{x + y}$. Compute $f(1, 3)$ and $f(-1, 1)$. Is the point $(-1, 0)$ in the domain of f?

2. Find the domain of $f(x, y) = \dfrac{1}{x} + \dfrac{1}{x - y} - e^{x+y}$.

3. **EFFECT OF ADVERTISING ON REVENUE** Odyssey Travel Agency has a monthly advertising budget of $20,000. Odyssey's management estimates that if they spend x dollars on newspaper advertising and y dollars on television adver-

tising, then the monthly revenue will be

$$f(x, y) = 30x^{1/4}y^{3/4}$$

dollars. What will be the monthly revenue if Odyssey spends $5000/month on newspaper ads and $15,000/month on television ads? If Odyssey spends $4000/month on newspaper ads and $16,000/month on television ads?

Solutions to Self-Check Exercises 12.1 can be found on page 912.

12.1 Concept Questions

1. What is a function of two variables? Give an example of a function of two variables, and state its rule of definition and domain.

2. If f is a function of two variables and (a, b) is in the domain of f with $c = a$ and $d = b$, what can you say about the relationship between $f(a, b)$ and $f(c, d)$?

3. Define (a) the graph of $f(x, y)$ and (b) a level curve of f.

4. Suppose f is a function of two variables, and let $P = f(x, y)$ denote the profit realized when x units of Product A and y units of Product B are produced and sold. Give an interpretation of the level curve of f, defined by the equation $f(x, y) = k$, where k is a positive constant.

12.1 Exercises

1. Let $f(x, y) = 2x + 3y - 4$. Compute $f(0, 0), f(1, 0)$, $f(0, 1), f(1, 2)$, and $f(2, -1)$.

2. Let $g(x, y) = 2x^2 - y^2$. Compute $g(1, 2), g(2, 1)$, $g(1, 1), g(-1, 1)$, and $g(2, -1)$.

3. Let $f(x, y) = x^2 + 2xy - x + 3$. Compute $f(1, 2)$, $f(2, 1), f(-1, 2)$, and $f(2, -1)$.

4. Let $h(x, y) = (x + y)/(x - y)$. Compute $h(0, 1)$, $h(-1, 1), h(2, 1)$, and $h(\pi, -\pi)$.

5. Let $g(s, t) = 3s\sqrt{t} + t\sqrt{s} + 2$. Compute $g(1, 2)$, $g(2, 1)$, $g(0, 4)$, and $g(4, 9)$.

6. Let $f(x, y) = xye^{x^2+y^2}$. Compute $f(0, 0)$, $f(0, 1)$, $f(1, 1)$, and $f(-1, -1)$.

7. Let $h(s, t) = s \ln t - t \ln s$. Compute $h(1, e)$, $h(e, 1)$, and $h(e, e)$.

8. Let $f(u, v) = (u^2 + v^2)e^{uv^2}$. Compute $f(0, 1)$, $f(-1, -1)$, $f(a, b)$, and $f(b, a)$.

9. Let $g(r, s, t) = re^{s/t}$. Compute $g(1, 1, 1)$, $g(1, 0, 1)$, and $g(-1, -1, -1)$.

10. Let $g(u, v, w) = (ue^{vw} + ve^{uw} + we^{uv})/(u^2 + v^2 + w^2)$. Compute $g(1, 2, 3)$ and $g(3, 2, 1)$.

In Exercises 11–18, find the domain of the function.

11. $f(x, y) = 2x + 3y$

12. $g(x, y, z) = x^2 + y^2 + z^2$

13. $h(u, v) = \dfrac{uv}{u - v}$

14. $f(s, t) = \sqrt{s^2 + t^2}$

15. $g(r, s) = \sqrt{rs}$

16. $f(x, y) = e^{-xy}$

17. $h(x, y) = \ln(x + y - 5)$

18. $h(u, v) = \sqrt{4 - u^2 - v^2}$

In Exercises 19–24, sketch the level curves of the function corresponding to each value of z.

19. $f(x, y) = 2x + 3y$; $z = -2, -1, 0, 1, 2$

20. $f(x, y) = -x^2 + y$; $z = -2, -1, 0, 1, 2$

21. $f(x, y) = 2x^2 + y$; $z = -2, -1, 0, 1, 2$

22. $f(x, y) = xy$; $z = -4, -2, 2, 4$

23. $f(x, y) = \sqrt{16 - x^2 - y^2}$; $z = 0, 1, 2, 3, 4$

24. $f(x, y) = e^x - y$; $z = -2, -1, 0, 1, 2$

25. Find an equation of the level curve of $f(x, y) = \sqrt{x^2 + y^2}$ that contains the point $(3, 4)$.

26. Find an equation of the level surface of $f(x, y, z) = 2x^2 + 3y^2 - z$ that contains the point $(-1, 2, -3)$.

In Exercises 27 and 28, match the graph of the surface with one of the contour maps labeled (a) and (b).

(a)

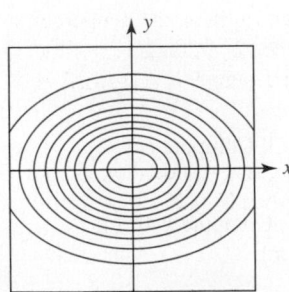

(b)

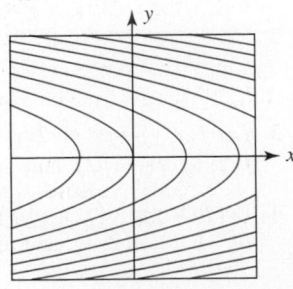

27. $f(x, y) = x + y^2$

28. $f(x, y) = e^{1-2x^2-4y^2}$

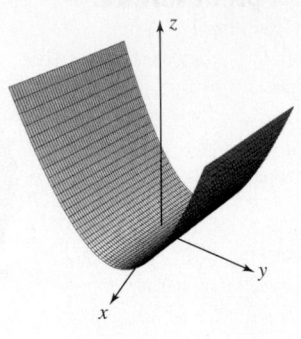

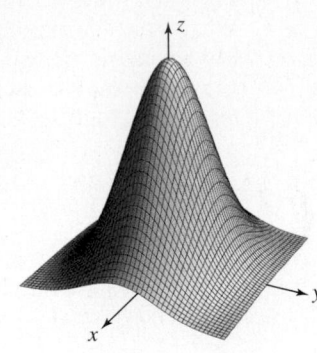

29. Can two level curves of a function f of two variables x and y intersect? Explain.

30. A *level set* of f is the set $S = \{(x, y) \mid f(x, y) = k\}$, where k is in the range of f. Let

$$f(x, y) = \begin{cases} 0 & \text{if } x^2 + y^2 < 1 \\ x^2 + y^2 - 1 & \text{if } x^2 + y^2 \geq 1 \end{cases}$$

Sketch the level set of f for $k = 0$ and 3.

31. The volume of a cylindrical tank of radius r and height h is given by

$$V = f(r, h) = \pi r^2 h$$

Find the volume of a cylindrical tank of radius 1.5 ft and height 4 ft.

32. **IQs** The IQ (intelligence quotient) of a person whose mental age is m years and whose chronological age is c years is defined as

$$f(m, c) = \frac{100m}{c}$$

What is the IQ of a 9-year-old child who has a mental age of 13.5 years?

33. **PRICE-TO-EARNINGS RATIO** The current P/E ratio (price-to-earnings ratio) of a stock is defined as

$$R = \frac{P}{E}$$

where P is the current market price of a single share of the stock and E is the earnings per share for the most recent 12-month period.
a. What is the domain of the function R?
b. The earnings per share of IBM Corporation for 2011 were $13.09, and its price per share on March 19, 2012 was $205.56. What was the P/E ratio of IBM at that time?
Source: IBM Corporate Annual Report.

34. **CURRENT DIVIDEND YIELD** The current dividend yield for common stock is calculated by using the formula

$$Y = \frac{D}{P}$$

where D is the most recent full-year dividend and P is the current share price (both measured in dollars).
a. What is the domain of the function Y?
b. The annualized dividend of IBM Corporation for the year 2011 was $3, and its price per share was $205.56 on March 19, 2012. What was the current dividend yield for the common stock of IBM at that time?
Source: IBM Corporation Annual Report.

35. BODY MASS INDEX The body mass index (BMI) is used to identify, evaluate, and treat overweight and obese adults. The BMI value for an adult of weight w (in kilograms) and height h (in meters) is defined to be

$$M = f(w, h) = \frac{w}{h^2}$$

According to federal guidelines, an adult is overweight if he or she has a BMI value greater than 25 but less than 30 and is "obese" if the value is greater than or equal to 30.
a. What is the BMI of an adult who weighs in at 80 kg and stands 1.8 m tall?
b. What is the maximum weight for an adult of height 1.8 m, who is not classified as overweight or obese?

36. POISEUILLE'S LAW Poiseuille's Law states that the resistance R of blood flowing in a blood vessel of length l and radius r is given by

$$R = f(l, r) = \frac{kl}{r^4}$$

where k is a constant that depends on the viscosity of blood. What is the resistance, in terms of k, of blood flowing through an arteriole 4 cm long and of radius 0.1 cm?

37. COST FUNCTION FOR A LOUDSPEAKER SYSTEM Acrosonic manufactures a bookshelf loudspeaker system that may be bought fully assembled or in a kit. The costs in labor and material incurred in manufacturing a fully assembled system and a kit are $200 and $120, respectively. In addition, Acrosonic has fixed costs of $20,000/month.
a. Write the monthly cost function C for Acrosonic in terms of the number of fully assembled systems x and the number of kits y manufactured.
b. What is the domain of C?
c. What is the total cost incurred in manufacturing 1000 fully assembled systems and 200 kits in a month?

38. REVENUE FUNCTION FOR ELECTRIC CARS Bell Motors manufactures two models of electric cars. The demand equations giving the relationship between the unit price, p and q, and the number of cars demanded per year, x and y, of Models S1 and S2 are

$$p = 60{,}000 - 4x - 2y \quad \text{and} \quad q = 50{,}000 - 2x - 4y$$

respectively.
a. What is the total yearly revenue function $R(x, y)$?
b. What is the domain of R? Sketch the domain R.

c. Is the point (3000, 2000) in the domain of R? Interpret your result.
Hint: Show that $x = 3000$ and $y = 2000$ satisfy the system of inequalities obtained in part (b).
d. What is the total revenue realized by Bell Motors if it sells 3000 Model S1s and 2000 Model S2s?

39. REVENUE FUNCTIONS FOR DESKS Country Workshop manufactures both finished and unfinished furniture for the home. The estimated quantities demanded each week of its roll-top desks in the finished and unfinished versions are x and y units when the corresponding unit prices are

$$p = 200 - \frac{1}{5}x - \frac{1}{10}y$$

$$q = 160 - \frac{1}{10}x - \frac{1}{4}y$$

dollars, respectively.
a. What is the weekly total revenue function $R(x, y)$?
b. Find the domain of the function R.

40. REVENUE FUNCTION FOR DESKS For the total revenue function $R(x, y)$ of Exercise 39, compute $R(100, 60)$ and $R(60, 100)$. Interpret your results.

41. REVENUE FUNCTIONS FOR A BOOK PUBLISHER Weston Publishing publishes a deluxe edition and a standard edition of its Spanish–English dictionary. Weston's management estimates that the number of deluxe editions demanded is x copies/day and the number of standard editions demanded is y copies/day when the unit prices are

$$p = 20 - 0.005x - 0.001y$$

$$q = 15 - 0.001x - 0.003y$$

dollars, respectively.
a. Find the daily total revenue function $R(x, y)$.
b. Find the domain of the function R.

42. REVENUE FUNCTION FOR A BOOK PUBLISHER For the total revenue function $R(x, y)$ of Exercise 41, compute $R(300, 200)$ and $R(200, 300)$. Interpret your results.

43. VOLUME OF A GAS The volume of a certain mass of gas is related to its pressure and temperature by the formula

$$V = \frac{30.9T}{P}$$

where the volume V is measured in liters, the temperature T is measured in kelvins (obtained by adding 273 to the Celsius temperature), and the pressure P is measured in millimeters of mercury pressure.
a. Find the domain of the function V.
b. Calculate the volume of the gas at standard temperature and pressure—that is, when $T = 273$ K and $P = 760$ mm of mercury.

44. **SURFACE AREA OF A HUMAN BODY** An empirical formula by E. F. Dubois relates the surface area S of a human body (in square meters) to its weight W (in kilograms) and its height H (in centimeters). The formula, given by

$$S = 0.007184W^{0.425}H^{0.725}$$

is used by physiologists in metabolism studies.
   a. Find the domain of the function S.
   b. What is the surface area of a human body that weighs 70 kg and has a height of 178 cm?

45. **ESTIMATING THE WEIGHT OF A TROUT** A formula for estimating the weight of a trout (from measurements) is

$$W = \frac{LG^2}{800}$$

where L is its length and G is its girth (the distance around the body of the fish at its largest point), both measured in inches. The weight of the fish W is in pounds.
   a. What is the domain of the function W?
   b. Sue caught a trout and measured its length to be 20 in. and its girth to be 12 in. What is its approximate weight?

46. **ESTIMATING THE WEIGHT OF A FISH** A formula for estimating the weight of a trout (from measurements) is

$$W = \frac{LG^2}{800}$$

where L is its length and G is its girth (the distance around the body of the fish at its largest point), both measured in inches. The weight of the fish W is in pounds. A trout caught by Ashley is 20% longer and has a girth that is 10% shorter than the one caught by Jane. Whose catch is heavier? By how much does the weight of the trout caught by Ashley differ from the weight of the one caught by Jane?

47. **PRODUCTION FUNCTION FOR A COUNTRY** Suppose the output of a certain country is given by

$$f(x, y) = 100x^{3/5}y^{2/5}$$

billion dollars if x billion dollars are spent on labor and y billion dollars are spent on capital. Find the output if the country spent $32 billion on labor and $243 billion on capital.

48. **PRODUCTION FUNCTION** Economists have found that the output of a finished product, $f(x, y)$, is sometimes described by the function

$$f(x, y) = ax^b y^{1-b}$$

where x stands for the amount of money expended on labor, y stands for the amount expended on capital, and a and b are positive constants with $0 < b < 1$.
   a. If p is a positive number, show that

$$f(px, py) = pf(x, y)$$

   b. Use the result of part (a) to show that if the amount of money expended for labor and capital are both increased by $r\%$, then the output is also increased by $r\%$.

49. **ARSON FOR PROFIT** A study of arson for profit was conducted by a team of paid civilian experts and police detectives appointed by the mayor of a large city. It was found that the number of suspicious fires in that city in 2014 was very closely related to the concentration of tenants in the city's public housing and to the level of reinvestment in the area in conventional mortgages by the ten largest banks. In fact, the number of fires was closely approximated by the formula

$$N(x, y) = \frac{100(1000 + 0.03x^2y)^{1/2}}{(5 + 0.2y)^2}$$

$$(0 \le x \le 150; 5 \le y \le 35)$$

where x denotes the number of persons per census tract and y denotes the level of reinvestment in the area in cents per dollar deposited. Using this formula, estimate the total number of suspicious fires in the districts of the city where the concentration of public housing tenants was 100/census tract and the level of reinvestment was 20 cents/dollar deposited.

50. **CONTINUOUSLY COMPOUNDED INTEREST** If a principal of P dollars is deposited in an account earning interest at the rate of r/year compounded continuously, then the accumulated amount at the end of t years is given by

$$A = f(P, r, t) = Pe^{rt}$$

dollars. Find the accumulated amount at the end of 3 years if a sum of $10,000 is deposited in an account earning interest at the rate of 6%/year.

51. **HOME MORTGAGES** The monthly payment that amortizes a loan of A dollars in t years when the interest rate is r/year, compounded monthly, is given by

$$P = f(A, r, t) = \frac{Ar}{12\left[1 - \left(1 + \frac{r}{12}\right)^{-12t}\right]}$$

   a. What is the monthly payment for a home mortgage of $300,000 that will be amortized over 30 years with an interest rate of 4%/year? An interest rate of 6%/year?
   b. Find the monthly payment for a home mortgage of $300,000 that will be amortized over 20 years with an interest rate of 6%/year.

52. **HOME MORTGAGES** Suppose a home buyer secures a bank loan of A dollars to purchase a house. If the interest rate charged is r/year compounded monthly and the loan is to be amortized in t years, then the principal repayment at the end of i months is given by

$$B = f(A, r, t, i)$$
$$= A\left[\frac{\left(1 + \frac{r}{12}\right)^i - 1}{\left(1 + \frac{r}{12}\right)^{12t} - 1}\right] \qquad (0 \le i \le 12t)$$

Suppose the Blakelys borrow a sum of $280,000 from a bank to help finance the purchase of a house and the bank charges interest at a rate of 6%/year. If the Blakelys agree to repay the loan in equal installments over 30 years, how much will they owe the bank after the 60th payment (5 years)? The 240th payment (20 years)?

53. **WILSON LOT-SIZE FORMULA** The Wilson lot-size formula in economics states that the optimal quantity Q of goods for a store to order is given by

$$Q = f(C, N, h) = \sqrt{\frac{2\,CN}{h}}$$

where C is the cost of placing an order, N is the number of items the store sells per week, and h is the weekly holding cost for each item. Find the most economical quantity of ten-speed bicycles to order if it costs the store $20 to place an order, $5 to hold a bicycle for a week, and the store expects to sell 40 bicycles/week.

54. **WIND POWER** The power output (in watts) of a certain brand of wind turbine generators is estimated to be

$$P = f(R, V) = 0.772R^2V^3$$

where R is the radius (in meters) of a rotor blade and V is the wind speed (in meters per second). Estimate the power output of a model of these generators if its radius is 30 m and the wind speed is 16 m/s.

55. **INTERNATIONAL AMERICA'S CUP CLASS** Drafted by an international committee in 1989, the rules for the new International America's Cup Class (IACC) include a formula that governs the basic yacht dimensions. The formula

$$f(L, S, D) \leq 42$$

where

$$f(L, S, D) = \frac{L + 1.25S^{1/2} - 9.80D^{1/3}}{0.388}$$

balances the rated length L (in meters), the rated sail area S (in square meters), and the displacement D (in cubic meters). All changes in the basic dimensions are trade-offs. For example, if you want to pick up speed by increasing the sail area, you must pay for it by decreasing the length or increasing the displacement, both of which slow down the boat. Show that Yacht A of rated length 20.95 m, rated sail area 277.3 m^2, and displacement 17.56 m^3 and the longer and heavier Yacht B with $L = 21.87$, $S = 311.78$, and $D = 22.48$ both satisfy the formula.

Source: americascup.com.

56. **FORCE GENERATED BY A CENTRIFUGE** A centrifuge is a machine designed for the specific purpose of subjecting materials to a sustained centrifugal force. The actual amount of centrifugal force, F, expressed in dynes (1 gram of force = 980 dynes) is given by

$$F = f(M, S, R) = \frac{\pi^2 S^2 MR}{900}$$

where S is in revolutions per minute (rpm), M is in grams, and R is in centimeters. Show that an object revolving at the rate of 600 rpm in a circle with radius of 10 cm generates a centrifugal force that is approximately 40 times gravity.

57. **IDEAL GAS LAW** According to the *ideal gas law*, the volume V of an ideal gas is related to its pressure P and temperature T by the formula

$$V = \frac{kT}{P}$$

where k is a positive constant. Describe the level curves of V and give a physical interpretation of your result.

58. **ISOQUANTS** Let $f(x, y)$ denote the output of a country if x units of its resources are spent on labor and y units are spent on capital. Then the level curve of f with equation $f(x, y) = k$, where k denotes a constant output, is called an *isoquant*. Each point (x, y) on the isoquant corresponds to a level of investment in labor, x, and investment in capital, y, that results in the same level of output k. Suppose the output of a certain country is

$$f(x, y) = 100x^{3/4}y^{1/4}$$

billion dollars if x billion dollars are spent on labor and y billion dollars are spent on capital.

a. Find the output if the country spends $81 billion on labor and $16 billion on capital.

b. Find an equation giving the relationship between x and y if the output remains constant at the level found in part (a).

c. Complete the following table for the equation found in part (b).

x	50	60	70	80	90
y					

Hint: Solve the equation for y in terms of x.

d. Use the results of part (c) to sketch the isoquant of f corresponding to the constant output found in part (b).

In Exercises 59–64, determine whether the statement is true or false. If it is true, explain why it is true. If it is false, explain why, or give an example to show why it is false.

59. If h is a function of x and y, then there are functions f and g of one variable such that

$$h(x, y) = f(x) + g(y)$$

60. If f is a function of x and y and a is a real number, then

$$f(ax, ay) = af(x, y)$$

61. The domain of $f(x, y) = 1/(x^2 - y^2)$ is $\{(x, y)\,|\,y \neq x\}$.

62. Every point on the level curve $f(x, y) = c$ corresponds to a point on the graph of f that is c units above the xy-plane if $c > 0$ and $|c|$ units below the xy-plane if $c < 0$.

63. f is a function of x and y if and only if for any two points $P_1(x_1, y_1)$ and $P_2(x_2, y_2)$ in the domain of f, $f(x_1, y_1) = f(x_2, y_2)$ implies that $P_1(x_1, y_1) = P_2(x_2, y_2)$.

64. If f is a function of x and y, and k is a real number in the range of f, then there exists at most one point $P(a, b)$ in the domain of f such that $f(a, b) = k$.

12.1 Solutions to Self-Check Exercises

1. $f(1, 3) = 1^2 - 3(1)(3) + \sqrt{1 + 3} = -6$

$f(-1, 1) = (-1)^2 - 3(-1)(1) + \sqrt{-1 + 1} = 4$

The point $(-1, 0)$ is not in the domain of f because the term $\sqrt{x + y}$ is not defined when $x = -1$ and $y = 0$. In fact, the domain of f consists of all real values of x and y that satisfy the inequality $x + y \geq 0$, the shaded half-plane shown in the accompanying figure.

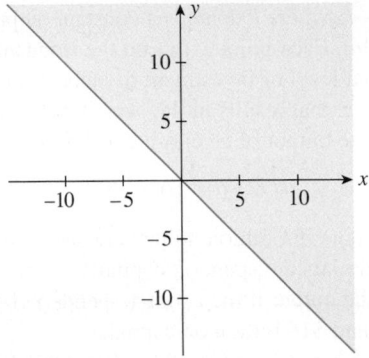

2. Since division by zero is not permitted, we see that $x \neq 0$ and $x - y \neq 0$. Therefore, the domain of f is the set of all points in the xy-plane except for the y-axis $(x = 0)$ and the straight line $x = y$.

3. If Odyssey spends \$5000/month on newspaper ads $(x = 5000)$ and \$15,000/month on television ads $(y = 15,000)$, then its monthly revenue will be given by

$$f(5000, 15,000) = 30(5000)^{1/4}(15,000)^{3/4}$$
$$\approx 341,926.06$$

or approximately \$341,926. If the agency spends \$4000/month on newspaper ads and \$16,000/month on television ads, then its monthly revenue will be given by

$$f(4000, 16,000) = 30(4000)^{1/4}(16,000)^{3/4}$$
$$\approx 339,411.26$$

or approximately \$339,411.

12.2 Partial Derivatives

Partial Derivatives

For a function $f(x)$ of one variable x, there is no ambiguity when we speak about the rate of change of $f(x)$ with respect to x, since x must be constrained to move along the x-axis. The situation becomes more complicated, however, when we study the rate of change of a function of two or more variables. For example, the domain D of a function of two variables $f(x, y)$ is a subset of the plane, so if (a, b) is any point in the domain of f, there are infinitely many directions from which one can approach the point (a, b) (Figure 12). We may therefore ask for the rate of change of f at (a, b) along any of these directions.

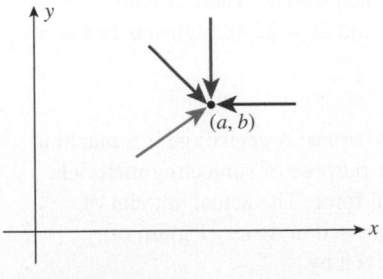

FIGURE 12
We can approach a point in the plane from infinitely many directions.

However, we will not deal with this general problem. Instead, we will restrict ourselves to studying the rate of change of the function $f(x, y)$ at a point (a, b) in each of two *preferred directions*—namely, the direction parallel to the x-axis and the direction parallel to the y-axis. Let $y = b$, where b is a constant, so that $f(x, b)$ is a function of the one variable x. Since the equation $z = f(x, y)$ is the equation of a surface, the

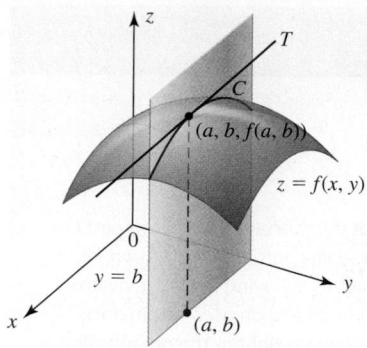

FIGURE 13
The curve *C* is formed by the intersection of the plane *y = b* with the surface *z = f(x, y)*.

equation $z = f(x, b)$ is the equation of the curve *C* on the surface formed by the intersection of the surface and the plane $y = b$ (Figure 13).

Because $f(x, b)$ is a function of one variable x, we may compute the derivative of f with respect to x at $x = a$. This derivative, obtained by keeping the variable y fixed at b and differentiating the resulting function $f(x, b)$ with respect to x, is called the **first partial derivative of f with respect to x** at (a, b), written

$$\frac{\partial z}{\partial x}(a, b) \quad \text{or} \quad \frac{\partial f}{\partial x}(a, b) \quad \text{or} \quad f_x(a, b)$$

Thus,

$$\frac{\partial z}{\partial x}(a, b) = \frac{\partial f}{\partial x}(a, b) = f_x(a, b) = \lim_{h \to 0} \frac{f(a + h, b) - f(a, b)}{h}$$

provided that the limit exists. The first partial derivative of f with respect to x at (a, b) measures both the slope of the tangent line T to the curve C and the rate of change of the function f in the x-direction when $x = a$ and $y = b$. We also write

$$\frac{\partial f}{\partial x}\bigg|_{(a, b)} = f_x(a, b)$$

Similarly, we define the **first partial derivative of f with respect to y** at (a, b), written

$$\frac{\partial z}{\partial y}(a, b) \quad \text{or} \quad \frac{\partial f}{\partial y}(a, b) \quad \text{or} \quad f_y(a, b)$$

as the derivative obtained by keeping the variable x fixed at a and differentiating the resulting function $f(a, y)$ with respect to y. That is,

$$\frac{\partial z}{\partial y}(a, b) = \frac{\partial f}{\partial y}(a, b) = f_y(a, b)$$

$$= \lim_{k \to 0} \frac{f(a, b + k) - f(a, b)}{k}$$

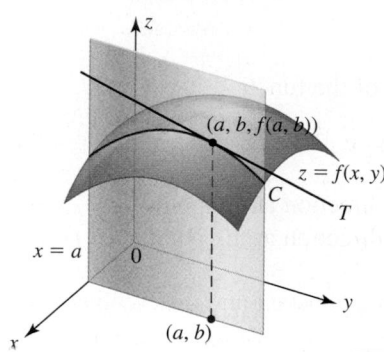

FIGURE 14
The first partial derivative of *f* with respect to *y* at *(a, b)* measures the slope of the tangent line *T* to the curve *C* with *x* held constant.

if the limit exists. The first partial derivative of f with respect to y at (a, b) measures both the slope of the tangent line T to the curve C, obtained by holding x constant (Figure 14), and the rate of change of the function f in the y-direction when $x = a$ and $y = b$. We write

$$\frac{\partial f}{\partial y}\bigg|_{(a, b)} = f_y(a, b)$$

Before looking at some examples, let's summarize these definitions.

First Partial Derivatives of $f(x, y)$

Suppose $f(x, y)$ is a function of the two variables x and y. Then the **first partial derivative of f** with respect to x at the point (x, y) is

$$\frac{\partial f}{\partial x} = \lim_{h \to 0} \frac{f(x + h, y) - f(x, y)}{h}$$

provided that the limit exists. The first partial derivative of f with respect to y at the point (x, y) is

$$\frac{\partial f}{\partial y} = \lim_{k \to 0} \frac{f(x, y + k) - f(x, y)}{k}$$

provided that the limit exists.

TITLE   Principal Software Engineer
INSTITUTION   Iron Mountain

Iron Mountain is the world's largest provider of document management solutions. Although the company has its roots in paper document storage, it has become the leader in digital document solutions as well. In addition to providing online backup and digital archiving services, Iron Mountain offers various digital document management tools.

As a principal software engineer at Iron Mountain, I help to develop eDiscovery products and services designed specifically for lawyers. When lawyers search for information relating to a corporate lawsuit, for example, it can be very difficult for them to find the information they need if they are searching through terabytes of unorganized data. To address this challenge, Iron Mountain has developed a variety of techniques to organize data automatically based on the statistical analysis of words and phrases contained within documents and email messages.

Although we use statistical analysis to initiate the organization of data, we apply calculus to refine our products and services.

When word sets are very large, it is important to focus on particular words and phrases that convey vital information. At Iron Mountain, we use optimization techniques involving partial derivatives to detect the most informative words and phrases within documents and allot them an apportioned weight in the organization process.

The methods outlined above allow us to organize data into folders—or clusters—of related documents. Ultimately, this enables us to improve upon traditional search capabilities by grouping search results into concept folders, minimizing cost and reducing the time needed for lawyers to find relevant documents.

With the aid of mathematical analysis, we provide robust eDiscovery products and services that radically improve legal workflow. As the amount of electronically stored information grows, our application of statistics and calculus allows us to remain at the forefront of our field.

EXAMPLE 1 Find the partial derivatives $\dfrac{\partial f}{\partial x}$ and $\dfrac{\partial f}{\partial y}$ of the function

$$f(x, y) = x^2 - xy^2 + y^3$$

What is the rate of change of the function f in the x-direction at the point $(1, 2)$? What is the rate of change of the function f in the y-direction at the point $(1, 2)$?

Solution To compute $\dfrac{\partial f}{\partial x}$, think of the variable y as a constant and differentiate the resulting function of x with respect to x. Let's write

$$f(x, y) = x^2 - xy^2 + y^3$$

where the variable y to be treated as a constant is shown in color. Then,

$$\frac{\partial f}{\partial x} = 2x - y^2$$

To compute $\dfrac{\partial f}{\partial y}$, think of the variable x as being fixed—that is, as a constant—and differentiate the resulting function of y with respect to y. In this case,

$$f(x, y) = x^2 - xy^2 + y^3$$

so

$$\frac{\partial f}{\partial y} = -2xy + 3y^2$$

The rate of change of the function f in the x-direction at the point $(1, 2)$ is given by

$$f_x(1, 2) = \frac{\partial f}{\partial x}\bigg|_{(1, 2)} = 2(1) - 2^2 = -2$$

That is, f decreases 2 units for each unit increase in the x-direction, y being kept constant ($y = 2$). The rate of change of the function f in the y-direction at the point $(1, 2)$ is given by

$$f_y(1, 2) = \frac{\partial f}{\partial y}\bigg|_{(1, 2)} = -2(1)(2) + 3(2)^2 = 8$$

That is, f increases 8 units for each unit increase in the y-direction, x being kept constant ($x = 1$). ∎

Explore and Discuss

Refer to the Explore and Discuss on page 902. Suppose management has decided that the projected sales of the first product is a units. Describe how you might help management decide how many units of the second product the company should produce and sell in order to maximize the company's total profit. Justify your method to management. Suppose, however, that management feels that b units of the second product should be manufactured and sold. How would you help management decide how many units of the first product to manufacture in order to maximize the company's total profit?

EXAMPLE 2 Compute the first partial derivatives of each function.

a. $f(x, y) = \dfrac{xy}{x^2 + y^2}$ **b.** $g(s, t) = (s^2 - st + t^2)^5$

c. $h(u, v) = e^{u^2 - v^2}$ **d.** $f(x, y) = \ln(x^2 + 2y^2)$

Solution

a. To compute $\dfrac{\partial f}{\partial x}$, think of the variable y as a constant. Thus,

$$f(x, y) = \frac{xy}{x^2 + y^2}$$

Then using the Quotient Rule, we have

$$\frac{\partial f}{\partial x} = \frac{(x^2 + y^2)y - xy(2x)}{(x^2 + y^2)^2} = \frac{x^2 y + y^3 - 2x^2 y}{(x^2 + y^2)^2}$$

$$= \frac{y(y^2 - x^2)}{(x^2 + y^2)^2}$$

upon simplification and factorization. To compute $\dfrac{\partial f}{\partial y}$, think of the variable x as a constant. Thus,

$$f(x, y) = \frac{xy}{x^2 + y^2}$$

Then using the Quotient Rule once again, we obtain

$$\frac{\partial f}{\partial y} = \frac{(x^2 + y^2)x - xy(2y)}{(x^2 + y^2)^2} = \frac{x^3 + xy^2 - 2xy^2}{(x^2 + y^2)^2}$$

$$= \frac{x(x^2 - y^2)}{(x^2 + y^2)^2}$$

b. To compute $\dfrac{\partial g}{\partial s}$, we treat the variable t as if it were a constant. Thus,

$$g(s, t) = (s^2 - st + t^2)^5$$

Using the General Power Rule, we find

$$\frac{\partial g}{\partial s} = 5(s^2 - st + t^2)^4 \cdot (2s - t)$$
$$= 5(2s - t)(s^2 - st + t^2)^4$$

To compute $\dfrac{\partial g}{\partial t}$, we treat the variable s as if it were a constant. Thus,

$$g(s, t) = (s^2 - st + t^2)^5$$
$$\frac{\partial g}{\partial t} = 5(s^2 - st + t^2)^4 (-s + 2t)$$
$$= 5(2t - s)(s^2 - st + t^2)^4$$

c. To compute $\dfrac{\partial h}{\partial u}$, think of the variable v as a constant. Thus,

$$h(u, v) = e^{u^2 - v^2}$$

Using the Chain Rule for Exponential Functions, we have

$$\frac{\partial h}{\partial u} = e^{u^2 - v^2} \cdot 2u = 2u e^{u^2 - v^2}$$

Next, we treat the variable u as if it were a constant,

$$h(u, v) = e^{u^2 - v^2}$$

and we obtain

$$\frac{\partial h}{\partial v} = e^{u^2 - v^2} \cdot (-2v) = -2v e^{u^2 - v^2}$$

d. To compute $\dfrac{\partial f}{\partial x}$, think of the variable y as a constant. Thus,

$$f(x, y) = \ln(x^2 + 2y^2)$$

so the Chain Rule for Logarithmic Functions gives

$$\frac{\partial f}{\partial x} = \frac{2x}{x^2 + 2y^2}$$

Next, treating the variable x as if it were a constant, we find

$$f(x, y) = \ln(x^2 + 2y^2)$$
$$\frac{\partial f}{\partial y} = \frac{4y}{x^2 + 2y^2}$$

To compute the partial derivative of a function of several variables with respect to one variable—say, x—we think of the other variables as if they were constants and differentiate the resulting function with respect to x.

Explore and Discuss

1. Let (a, b) be a point in the domain of $f(x, y)$. Put $g(x) = f(x, b)$, and suppose that g is differentiable at $x = a$. Explain why you can find $f_x(a, b)$ by computing $g'(a)$. How would you go about calculating $f_y(a, b)$ using a similar technique? Give a geometric interpretation of these processes.

2. Let $f(x, y) = x^2 y^3 - 3x^2 y + 2$. Use the method of Problem 1 to find $f_x(1, 2)$ and $f_y(1, 2)$.

EXAMPLE 3 Compute the first partial derivatives of the function

$$w = f(x, y, z) = xyz - xe^{yz} + x \ln y$$

Solution Here, we have a function of three variables, x, y, and z, and we are required to compute

$$\frac{\partial f}{\partial x}, \frac{\partial f}{\partial y}, \frac{\partial f}{\partial z}$$

To compute f_x, we think of the other two variables, y and z, as fixed, and we differentiate the resulting function of x with respect to x, thereby obtaining

$$f_x = yz - e^{yz} + \ln y$$

To compute f_y, we think of the other two variables, x and z, as constants, and we differentiate the resulting function of y with respect to y. We then obtain

$$f_y = xz - xze^{yz} + \frac{x}{y}$$

Finally, to compute f_z, we treat the variables x and y as constants and differentiate the function f with respect to z, obtaining

$$f_z = xy - xye^{yz}$$

Exploring with TECHNOLOGY

Refer to the Explore and Discuss on page 916. Let

$$f(x, y) = \frac{e^{\sqrt{xy}}}{(1 + xy^2)^{3/2}}$$

1. Compute $g(x) = f(x, 1)$, and use a graphing utility to plot the graph of g in the viewing window $[0, 2] \times [0, 2]$.
2. Use the differentiation operation of your graphing utility to find $g'(1)$ and hence $f_x(1, 1)$.
3. Compute $h(y) = f(1, y)$, and use a graphing utility to plot the graph of h in the viewing window $[0, 2] \times [0, 2]$.
4. Use the differentiation operation of your graphing utility to find $h'(1)$ and hence $f_y(1, 1)$.

The Cobb–Douglas Production Function

For an economic interpretation of the first partial derivatives of a function of two variables, let's turn our attention to the function

$$f(x, y) = ax^b y^{1-b} \tag{5}$$

where a and b are positive constants with $0 < b < 1$. This function is called the **Cobb–Douglas production function.** Here, x stands for the amount of money expended for labor, y stands for the cost of capital equipment (buildings, machinery, and other tools of production), and the function f measures the output of the finished product (in suitable units) and is called, accordingly, the production function.

The partial derivative f_x is called the **marginal productivity of labor.** It measures the rate of change of production with respect to the amount of money expended for labor, with the level of capital expenditure held constant. Similarly, the partial derivative f_y, called the **marginal productivity of capital,** measures the rate of change of production with respect to the amount expended on capital, with the level of labor expenditure held fixed.

 APPLIED EXAMPLE 4 Marginal Productivity A certain country's production in the early years following World War II is described by the function

$$f(x, y) = 30x^{2/3}y^{1/3}$$

units, when x units of labor and y units of capital were used.

a. Compute f_x and f_y.

b. What are the marginal productivity of labor and the marginal productivity of capital when the amounts expended on labor and capital are 125 units and 27 units, respectively?

c. Should the government have encouraged capital investment rather than increasing expenditure on labor to increase the country's productivity?

Solution

a. $f_x = 30 \cdot \dfrac{2}{3} x^{-1/3} y^{1/3} = 20 \left(\dfrac{y}{x} \right)^{1/3}$

$f_y = 30x^{2/3} \cdot \dfrac{1}{3} y^{-2/3} = 10 \left(\dfrac{x}{y} \right)^{2/3}$

b. The required marginal productivity of labor is given by

$$f_x(125, 27) = 20 \left(\frac{27}{125} \right)^{1/3} = 20 \left(\frac{3}{5} \right)$$

or 12 units per unit increase in labor expenditure (capital expenditure is held constant at 27 units). The required marginal productivity of capital is given by

$$f_y(125, 27) = 10 \left(\frac{125}{27} \right)^{2/3} = 10 \left(\frac{25}{9} \right)$$

or $27\frac{7}{9}$ units per unit increase in capital expenditure (labor outlay is held constant at 125 units).

c. From the results of part (b), we see that a unit increase in capital expenditure resulted in a much faster increase in productivity than a unit increase in labor expenditure would have. Therefore, the government should have encouraged increased spending on capital rather than on labor during the early years of reconstruction. ▪

Substitute and Complementary Commodities

For another application of the first partial derivatives of a function of two variables in the field of economics, let's consider the relative demands of two commodities. We say that the two commodities are **substitute** (competitive) **commodities** if a decrease in the demand for one results in an increase in the demand for the other. Examples of substitute commodities are coffee and tea. Conversely, two commodities are referred to as **complementary commodities** if a decrease in the demand for one results in a decrease in the demand for the other as well. Examples of complementary commodities are automobiles and automobile tires.

We now derive a criterion for determining whether two commodities A and B are substitute or complementary. Suppose the demand equations that relate the quantities demanded, x and y, to the unit prices, p and q, of the two commodities are given by

$$x = f(p, q) \quad \text{and} \quad y = g(p, q)$$

Let's consider the partial derivative $\partial f/\partial p$. Since f is the demand function for Commodity A, we see that for fixed q, f is typically a decreasing function of p—that is, $\partial f/\partial p < 0$. Now, if the two commodities were substitute commodities, then the quantity demanded of Commodity B would increase with respect to p—that is, $\partial g/\partial p > 0$. A similar argument with p fixed shows that if A and B are substitute commodities, then $\partial f/\partial q > 0$. Thus, the two commodities A and B are substitute commodities if

$$\frac{\partial f}{\partial q} > 0 \quad \text{and} \quad \frac{\partial g}{\partial p} > 0$$

Similarly, A and B are complementary commodities if

$$\frac{\partial f}{\partial q} < 0 \quad \text{and} \quad \frac{\partial g}{\partial p} < 0$$

Substitute and Complementary Commodities

Two commodities A and B are substitute commodities if

$$\frac{\partial f}{\partial q} > 0 \quad \text{and} \quad \frac{\partial g}{\partial p} > 0 \tag{6}$$

Two commodities A and B are complementary commodities if

$$\frac{\partial f}{\partial q} < 0 \quad \text{and} \quad \frac{\partial g}{\partial p} < 0 \tag{7}$$

 APPLIED EXAMPLE 5 Substitute and Complementary Commodities
Suppose that the daily demand for butter is given by

$$x = f(p, q) = \frac{3q}{1 + p^2}$$

and the daily demand for margarine is given by

$$y = g(p, q) = \frac{2p}{1 + \sqrt{q}} \qquad (p > 0, q > 0)$$

where p and q denote the prices per pound (in dollars) of butter and margarine, respectively, and x and y are measured in millions of pounds. Determine whether these two commodities are substitute, complementary, or neither.

Solution We compute

$$\frac{\partial f}{\partial q} = \frac{3}{1 + p^2} \quad \text{and} \quad \frac{\partial g}{\partial p} = \frac{2}{1 + \sqrt{q}}$$

Since

$$\frac{\partial f}{\partial q} > 0 \quad \text{and} \quad \frac{\partial g}{\partial p} > 0$$

for all values of $p > 0$ and $q > 0$, we conclude that butter and margarine are substitute commodities.

Second-Order Partial Derivatives

The first partial derivatives $f_x(x, y)$ and $f_y(x, y)$ of a function $f(x, y)$ of the two variables x and y are also functions of x and y. As such, we may differentiate each of the functions f_x and f_y to obtain the **second-order partial derivatives of f** (Figure 15). Thus, differentiating the function f_x with respect to x leads to the second partial derivative

$$f_{xx} = \frac{\partial^2 f}{\partial x^2} = \frac{\partial}{\partial x}(f_x)$$

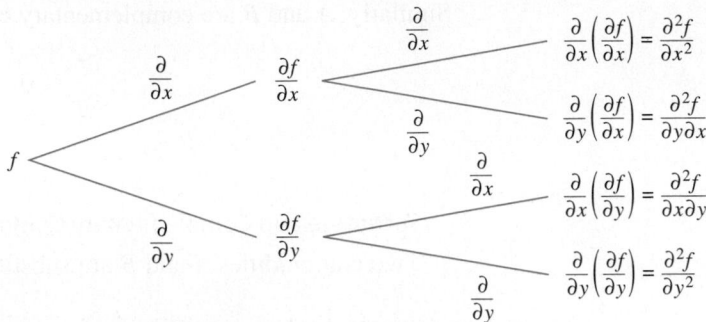

FIGURE **15**
A schematic showing the four second-order partial derivatives of f

However, differentiation of f_x with respect to y leads to the second partial derivative

$$f_{xy} = \frac{\partial^2 f}{\partial y \partial x} = \frac{\partial}{\partial y}(f_x)$$

Similarly, differentiation of the function f_y with respect to x and with respect to y leads to

$$f_{yx} = \frac{\partial^2 f}{\partial x \partial y} = \frac{\partial}{\partial x}(f_y)$$

$$f_{yy} = \frac{\partial^2 f}{\partial y^2} = \frac{\partial}{\partial y}(f_y)$$

respectively. Although it is not always true that $f_{xy} = f_{yx}$, they are equal, however, if both f_{xy} and f_{yx} are continuous. We might add that this is the case in most practical applications.

EXAMPLE 6 Find the second-order partial derivatives of the function

$$f(x, y) = x^3 - 3x^2y + 3xy^2 + y^2$$

Solution The first partial derivatives of f are

$$f_x = \frac{\partial}{\partial x}(x^3 - 3x^2y + 3xy^2 + y^2)$$
$$= 3x^2 - 6xy + 3y^2$$
$$f_y = \frac{\partial}{\partial y}(x^3 - 3x^2y + 3xy^2 + y^2)$$
$$= -3x^2 + 6xy + 2y$$

Therefore,

$$f_{xx} = \frac{\partial}{\partial x}(f_x) = \frac{\partial}{\partial x}(3x^2 - 6xy + 3y^2)$$
$$= 6x - 6y = 6(x - y)$$

$$f_{xy} = \frac{\partial}{\partial y}(f_x) = \frac{\partial}{\partial y}(3x^2 - 6xy + 3y^2)$$
$$= -6x + 6y = 6(y - x)$$

$$f_{yx} = \frac{\partial}{\partial x}(f_y) = \frac{\partial}{\partial x}(-3x^2 + 6xy + 2y)$$
$$= -6x + 6y = 6(y - x)$$

$$f_{yy} = \frac{\partial}{\partial y}(f_y) = \frac{\partial}{\partial y}(-3x^2 + 6xy + 2y)$$
$$= 6x + 2$$

Note that $f_{xy} = f_{yx}$ everywhere.

EXAMPLE 7 Find the second-order partial derivatives of the function

$$f(x, y) = e^{xy^2}$$

Solution   We have

$$f_x = \frac{\partial}{\partial x}(e^{xy^2})$$
$$= y^2 e^{xy^2}$$

$$f_y = \frac{\partial}{\partial y}(e^{xy^2})$$
$$= 2xye^{xy^2}$$

so the required second-order partial derivatives of f are

$$f_{xx} - \frac{\partial}{\partial x}(f_x) = \frac{\partial}{\partial x}(y^2 e^{xy^2})$$
$$= y^4 e^{xy^2}$$

$$f_{xy} = \frac{\partial}{\partial y}(f_x) = \frac{\partial}{\partial y}(y^2 e^{xy^2})$$
$$= 2ye^{xy^2} + 2xy^3 e^{xy^2} \qquad (x^2) \text{ See page 16.}$$
$$= 2ye^{xy^2}(1 + xy^2)$$

$$f_{yx} = \frac{\partial}{\partial x}(f_y) = \frac{\partial}{\partial x}(2xye^{xy^2})$$
$$= 2ye^{xy^2} + 2xy^3 e^{xy^2}$$
$$= 2ye^{xy^2}(1 + xy^2)$$

$$f_{yy} = \frac{\partial}{\partial y}(f_y) = \frac{\partial}{\partial y}(2xye^{xy^2})$$
$$= 2xe^{xy^2} + (2xy)(2xy)e^{xy^2}$$
$$= 2xe^{xy^2}(1 + 2xy^2)$$

Note that $f_{xy} = f_{yx}$ everywhere.

12.2 Self-Check Exercises

1. Compute the first partial derivatives of
   $f(x, y) = x^3 - 2xy^2 + y^2 - 8$.

2. Find the first partial derivatives of
   $f(x, y) = x \ln y + ye^x - x^2$ at $(0, 1)$, and
   interpret your results.

3. Find the second-order partial derivatives of the
   function of Self-Check Exercise 1.

4. MARGINAL PRODUCTIVITY A certain country's production is described by the function

   $$f(x, y) = 60x^{1/3}y^{2/3}$$

   when x units of labor and y units of capital are used.

   a. What are the marginal productivity of labor and the
   marginal productivity of capital when the amounts
   expended on labor and capital are 125 units and
   8 units, respectively?
   b. Should the government encourage capital investment
   rather than increased expenditure on labor at this time
   to increase the country's productivity?

   *Solutions to Self-Check Exercises 12.2 can be found on
   page 925.*

12.2 Concept Questions

1. **a.** What is the partial derivative of $f(x, y)$ with respect to
   x at (a, b)?
   b. Give a geometric interpretation of $f_x(a, b)$ and a practical interpretation of $f_x(a, b)$.

2. **a.** What are substitute commodities and complementary
   commodities? Give an example of each.
   b. Suppose $x = f(p, q)$ and $y = g(p, q)$ are demand
   functions for two commodities A and B, respectively.

   Give conditions for determining whether A and B are
   substitute or complementary commodities.

3. List all second-order partial derivatives of a function of
   two variables.

4. How many second-order partial derivatives are there for a
   function of three variables f, assuming all such derivatives
   exist? List all of them.

12.2 Exercises

1. Let $f(x, y) = x^2 + 2y^2$.
   a. Find $f_x(2, 1)$ and $f_y(2, 1)$.
   b. Interpret the numbers in part (a) as slopes.
   c. Interpret the numbers in part (a) as rates of change.

2. Let $f(x, y) = 9 - x^2 + xy - 2y^2$.
   a. Find $f_x(1, 2)$ and $f_y(1, 2)$.
   b. Interpret the numbers in part (a) as slopes.
   c. Interpret the numbers in part (a) as rates of change.

In Exercises 3–24, find the first partial derivatives of the function.

3. $f(x, y) = 2x + 3y + 5$ 4. $f(x, y) = 2xy$

5. $g(x, y) = 2x^2 + 4y + 1$ 6. $f(x, y) = 1 + x^2 + y^2$

7. $f(x, y) = \dfrac{2y}{x^2}$ 8. $f(x, y) = \dfrac{x}{1 + y}$

9. $g(u, v) = \dfrac{u - v}{u + v}$ 10. $f(x, y) = \dfrac{x^2 - y^2}{x^2 + y^2}$

11. $f(s, t) = (s^2 - st + t^2)^3$ 12. $g(s, t) = s^2t + st^{-3}$

13. $f(x, y) = (x^2 + y^2)^{2/3}$ 14. $f(x, y) = x\sqrt{1 + y^2}$

15. $f(x, y) = e^{xy+1}$ 16. $f(x, y) = (e^x + e^y)^5$

17. $f(x, y) = x \ln y + y \ln x$ 18. $f(x, y) = x^2e^{y^2}$

19. $g(u, v) = e^u \ln v$ 20. $f(x, y) = \dfrac{e^{xy}}{x + y}$

21. $f(x, y, z) = xyz + xy^2 + yz^2 + zx^2$

22. $g(u, v, w) = \dfrac{2uvw}{u^2 + v^2 + w^2}$

23. $h(r, s, t) = e^{rst}$ 24. $f(x, y, z) = xe^{y/z}$

In Exercises 25–34, evaluate the first partial derivatives of the
function at the given point.

25. $f(x, y) = x^2y + xy^2; (1, 2)$

26. $f(x, y) = x^2 + xy + y^2 + 2x - y; (-1, 2)$

27. $f(x, y) = x\sqrt{y} + y^2; (2, 1)$

28. $g(x, y) = \sqrt{x^2 + y^2}; (3, 4)$

29. $f(x, y) = \dfrac{x}{y}; (1, 2)$ 30. $f(x, y) = \dfrac{x + y}{x - y}; (1, -2)$

31. $f(x, y) = e^{xy}; (1, 1)$ 32. $f(x, y) = e^x \ln y; (0, e)$

33. $f(x, y, z) = x^2yz^3; (1, 0, 2)$

34. $f(x, y, z) = x^2y^2 + z^2; (1, 1, 2)$

In Exercises 35–42, find the second-order partial derivatives of the function. In each case, show that the mixed partial derivatives f_{xy} and f_{yx} are equal.

35. $f(x, y) = x^2y + xy^3$

36. $f(x, y) = x^3 + x^2y + x + 4$

37. $f(x, y) = x^2 - 2xy + 2y^2 + x - 2y$

38. $f(x, y) = x^3 + x^2y^2 + y^3 + x + y$

39. $f(x, y) = \sqrt{x^2 + y^2}$

40. $f(x, y) = x\sqrt{y} + y\sqrt{x}$

41. $f(x, y) = e^{-x/y}$

42. $f(x, y) = \ln(1 + x^2y^2)$

43. PRODUCTIVITY OF A COUNTRY The productivity of a South American country is given by the function

$$f(x, y) = 20x^{3/4}y^{1/4}$$

when x units of labor and y units of capital are used.

a. What are the marginal productivity of labor and the marginal productivity of capital when the amounts expended on labor and capital are 256 units and 16 units, respectively?

b. Should the government encourage capital investment rather than increased expenditure on labor at this time to increase the country's productivity?

44. PRODUCTIVITY OF A COUNTRY The productivity of a country in Western Europe is given by the function

$$f(x, y) = 40x^{4/5}y^{1/5}$$

when x units of labor and y units of capital are used.

a. What are the marginal productivity of labor and the marginal productivity of capital when the amounts expended on labor and capital are 32 units and 243 units, respectively?

b. Should the government encourage capital investment rather than increased expenditure on labor at this time to increase the country's productivity?

45. PRODUCTION FUNCTION FOR CORDLESS LED CANDLES Luminar Corporation manufactures cordless LED window candles. The number of candles it can manufacture per day, P, depends on the amount of labor utilized, x (in work-hours per day) and the expenditure on capital investment, y (in dollars per day) according to the production function

$$P(x, y) = x^2 + 5x + 2xy + 3y^2 + 2y$$

Find the marginal productivities if $x = 400$ and $y = 300$, and interpret your results.

46. PRODUCTIVITY OF A COMPANY The number of souvenir coffee mugs (in hundreds) that Ace Novelty can produce monthly is given by the production function

$$P(x, y) = 0.2x^2 + 2x + 3xy + 0.4y^2 + 3y$$

where x denotes the amount of labor utilized (measured in thousands of work-hours per month) and y denotes the

expenditure on capital investment (in thousands of dollars per month). Find the marginal productivities if 10,000 work-hours per month are utilized and a capital investment of \$5000/month is made. Interpret your results.

47. LAND PRICES The rectangular region R shown in the following figure represents a city's financial district. The price of land within the district is approximated by the function

$$p(x, y) = 200 - 10\left(x - \frac{1}{2}\right)^2 - 15(y - 1)^2$$

where $p(x, y)$ is the price of land at the point (x, y) in dollars per square foot and x and y are measured in miles. Compute

$$\frac{\partial p}{\partial x}(0, 1) \quad \text{and} \quad \frac{\partial p}{\partial y}(0, 1)$$

and interpret your results.

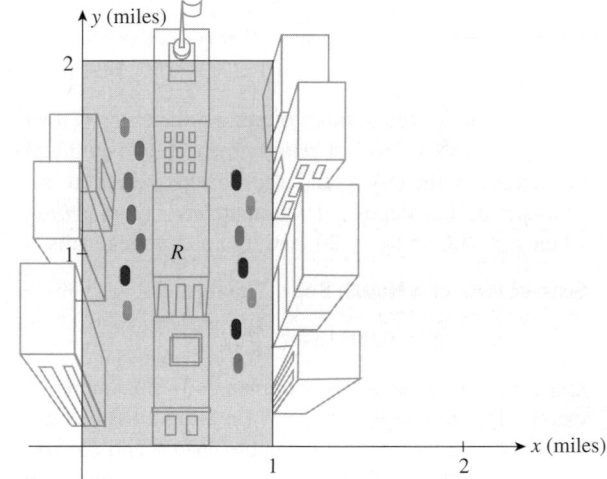

48. REVENUE FUNCTIONS The total weekly revenue (in dollars) of Country Workshop associated with manufacturing and selling their rolltop desks is given by the function

$$R(x, y) = -0.2x^2 - 0.25y^2 - 0.2xy + 200x + 160y$$

where x denotes the number of finished units and y denotes the number of unfinished units manufactured and sold each week. Compute $\partial R/\partial x$ and $\partial R/\partial y$ when $x = 300$ and $y = 250$. Interpret your results.

49. PROFIT FUNCTIONS The monthly profit (in dollars) of Bond and Barker Department Store depends on the level of inventory x (in thousands of dollars) and the floor space y (in thousands of square feet) available for display of the merchandise, as given by the equation

$$P(x, y) = -0.02x^2 - 15y^2 + xy$$
$$+ 39x + 25y - 20,000$$

Compute $\partial P/\partial x$ and $\partial P/\partial y$ when $x = 4000$ and $y = 150$. Interpret your results. Repeat with $x = 5000$ and $y = 150$.

50. Hotel Room Demand The number of rooms in demand at the hotels of the Goodwood Hotel Group is given by

$$D(x, y) = 500e^{-0.02(x-120)}(1 - 0.7e^{-0.0001y})$$
$$(120 \le x \le 280, 0 \le y \le 30{,}000)$$

where x is the daily room rate in dollars and y is the weekly spending on advertisement, also in dollars.

a. Show that if y is held fast, then D is a decreasing function of x; and if x is held fast, then D is an increasing function of y. Interpret your results.

b. If the spending on advertisement is held fast at $25,000/week, find the approximate change in the demand for the number of rooms if the daily room rate is cut by $1 from the current $200/night.

51. Arson Study A study of arson for profit conducted for a certain city found that the number of suspicious fires is approximated by the formula

$$N(x, y) = \frac{120\sqrt{1000 + 0.03x^2y}}{(5 + 0.2y)^2}$$
$$(0 \le x \le 150, 5 \le y \le 35)$$

where x denotes the number of persons per census tract and y denotes the level of reinvestment in conventional mortgages by the city's ten largest banks measured in cents per dollars deposited. Compute $\partial N/\partial x$ and $\partial N/\partial y$ when $x = 100$ and $y = 20$, and interpret your results.

52. Surface Area of a Human Body The formula

$$S = 0.007184W^{0.425}H^{0.725}$$

gives the surface area S of a human body (in square meters) in terms of its weight W (in kilograms) and its height H (in centimeters). Compute $\partial S/\partial W$ and $\partial S/\partial H$ when $W = 70$ kg and $H = 180$ cm. Interpret your results.

53. Wind Chill Factor The wind chill temperature is the temperature that you would feel in still air that is the same as the actual temperature when the presence of wind is taken into consideration. The following table gives the wind chill temperature $T = f(t, s)$ in degrees Fahrenheit in terms of the actual air temperature t in degrees Fahrenheit and the wind speed s in miles per hour (mph).

		Wind speed (mph)					
	s $\backslash$ t	10	15	20	25	30	35	40
Actual air temperature (°F)	30	21.2	19.0	17.4	16.0	14.9	13.9	13.0
	32	23.7	21.6	20.0	18.7	17.6	16.6	15.8
	34	26.2	24.2	22.6	21.4	20.3	19.4	18.6
	36	28.7	26.7	25.2	24.0	23.0	22.2	21.4
	38	31.2	29.3	27.9	26.7	25.7	24.9	24.2
	40	33.6	31.8	30.5	29.4	28.5	27.7	26.9

a. Estimate the rate of change of the wind chill temperature T with respect to the actual air temperature when the wind speed is constant at 25 mph and the actual air temperature is 34°F.

Hint: Show that it is given by

$$\frac{\partial T}{\partial t}(34, 25) \approx \frac{f(36, 25) - f(34, 25)}{2}$$

b. Estimate the rate of change of the wind chill temperature T with respect to the wind speed when the actual air temperature is constant at 34°F and the wind speed is 25 mph.

Source: National Weather Service.

54. Wind Chill Factor A formula used by meteorologists to calculate the wind chill temperature (the temperature that you feel in still air that is the same as the actual temperature when the presence of wind is taken into consideration) is

$$T = f(t, s) = 35.74 + 0.6215t - 35.75s^{0.16} + 0.4275ts^{0.16}$$
$$(s \ge 1)$$

where t is the actual air temperature in degrees Fahrenheit and s is the wind speed in miles per hour.

a. What is the wind chill temperature when the actual air temperature is 32°F and the wind speed is 20 mph?

b. If the temperature is 32°F, by how much approximately will the wind chill temperature change if the wind speed increases from 20 mph to 21 mph?

55. Engine Efficiency The efficiency of an internal combustion engine is given by

$$E = \left(1 - \frac{v}{V}\right)^{0.4}$$

where V and v are the respective maximum and minimum volumes of air in each cylinder.

a. Show that $\partial E/\partial V > 0$, and interpret your result.

b. Show that $\partial E/\partial v < 0$, and interpret your result.

56. Volume of a Gas The volume V (in liters) of a certain mass of gas is related to its pressure P (in millimeters of mercury) and its temperature T (in kelvins) by the law

$$V = \frac{30.9T}{P}$$

Compute $\partial V/\partial T$ and $\partial V/\partial P$ when $T = 300$ and $P = 800$. Interpret your results.

57. Complementary and Substitute Commodities In a survey conducted by a video magazine, it was determined that the demand equation for DVD players is given by

$$x = f(p, q) = 10{,}000 - 10p + 0.2q^2$$

and the demand equation for Blu-ray players is given by

$$y = g(p, q) = 5000 + 0.8p^2 - 20q$$

where p and q denote the unit prices (in dollars) for the DVD and Blu-ray players, respectively, and x and y denote the number of DVD and Blu-ray players demanded per week. Determine whether these two products are substitute, complementary, or neither.

58. COMPLEMENTARY AND SUBSTITUTE COMMODITIES In a survey, it was determined that the demand equation for DVD players is given by

$$x = f(p, q) = 10,000 - 10p - e^{0.5q}$$

and the demand equation for blank DVD discs is given by

$$y = g(p, q) = 50,000 - 4000q - 10p$$

where p and q denote the unit prices, respectively, and x and y denote the number of DVD players and the number of blank DVD discs demanded each week. Determine whether these two products are substitute, complementary, or neither.

59. COMPLEMENTARY AND SUBSTITUTE COMMODITIES Refer to Exercise 39, Exercises 12.1. Show that the finished and unfinished home furniture manufactured by Country Workshop are substitute commodities.
Hint: Solve the system of equations for x and y in terms of p and q.

60. According to the *ideal gas law*, the volume V (in liters) of an ideal gas is related to its pressure P (in pascals) and temperature T (in kelvins) by the formula

$$V = \frac{kT}{P}$$

where k is a constant. Show that

$$\frac{\partial V}{\partial T} \cdot \frac{\partial T}{\partial P} \cdot \frac{\partial P}{\partial V} = -1$$

61. COBB–DOUGLAS PRODUCTION FUNCTION Show that the Cobb–Douglas production function $P = kx^{\alpha}y^{1-\alpha}$, where $0 < \alpha < 1$, satisfies the equation

$$x\frac{\partial P}{\partial x} + y\frac{\partial P}{\partial y} = P$$

This equation is called Euler's equation.

62. Let $f(x, y, z) = x^2y + xy^2 + yz^3 + xye^{2z}$. Find f_{xz}, f_{yz}, and f_{zz}.

In Exercises 63–68, determine whether the statement is true or false. If it is true, explain why it is true. If it is false, give an example to show why it is false.

63. If $f_x(x, y)$ is defined at (a, b), then $f_y(x, y)$ must also be defined at (a, b).

64. If $f(x, y)$ is continuous at (a, b), then both $f_x(a, b)$ and $f_y(a, b)$ exist.

65. If $f_x(x, y) = 0$ and $f_y(x, y) = 0$ for all x and y, then f must be a constant function.

66. If $f_x(a, b) < 0$, then f is decreasing with respect to x near (a, b).

67. If $f_{xy}(x, y)$ and $f_{yx}(x, y)$ are both continuous for all values of x and y, then $f_{xy} = f_{yx}$ for all values of x and y.

68. If both f_{xy} and f_{yx} are defined at (a, b), then f_{xx} and f_{yy} must be defined at (a, b).

12.2 Solutions to Self-Check Exercises

1. $f_x = \dfrac{\partial f}{\partial x} = 3x^2 - 2y^2$

$f_y = \dfrac{\partial f}{\partial y} = -2x(2y) + 2y$

$\quad = 2y(1 - 2x)$

2. $f_x = \ln y + ye^x - 2x; f_y = \dfrac{x}{y} + e^x$

In particular,

$$f_x(0, 1) = \ln 1 + 1e^0 - 2(0) = 1$$

$$f_y(0, 1) = \frac{0}{1} + e^0 = 1$$

The results tell us that at the point $(0, 1)$, $f(x, y)$ increases 1 unit for each unit increase in the x-direction, y being kept constant; $f(x, y)$ also increases 1 unit for each unit increase in the y-direction, x being kept constant.

3. From the results of Self-Check Exercise 1,

$$f_x = 3x^2 - 2y^2$$

Therefore,

$$f_{xx} = \frac{\partial}{\partial x}(3x^2 - 2y^2) = 6x$$

$$f_{xy} = \frac{\partial}{\partial y}(3x^2 - 2y^2) = -4y$$

Also, from the results of Self-Check Exercise 1,

$$f_y = 2y(1 - 2x)$$

Thus,

$$f_{yx} = \frac{\partial}{\partial x}[2y(1 - 2x)] = -4y$$

$$f_{yy} = \frac{\partial}{\partial y}[2y(1 - 2x)] = 2(1 - 2x)$$

4. a. The marginal productivity of labor when the amounts expended on labor and capital are x and y units, respectively, is given by

$$f_x(x, y) = 60\left(\frac{1}{3}x^{-2/3}\right)y^{2/3} = 20\left(\frac{y}{x}\right)^{2/3}$$

In particular, the required marginal productivity of labor is given by

$$f_x(125, 8) = 20\left(\frac{8}{125}\right)^{2/3} = 20\left(\frac{4}{25}\right)$$

or 3.2 units/unit increase in labor expenditure, capital expenditure being held constant at 8 units. Next, we compute

$$f_y(x, y) = 60x^{1/3}\left(\frac{2}{3}y^{-1/3}\right) = 40\left(\frac{x}{y}\right)^{1/3}$$

and deduce that the required marginal productivity of capital is given by

$$f_y(125, 8) = 40\left(\frac{125}{8}\right)^{1/3} = 40\left(\frac{5}{2}\right)$$

or 100 units/unit increase in capital expenditure, labor expenditure being held constant at 125 units.

b. The results of part (a) tell us that the government should encourage increased spending on capital rather than on labor.

USING TECHNOLOGY Finding Partial Derivatives at a Given Point

Suppose $f(x, y)$ is a function of two variables and we wish to compute

$$f_x(a, b) = \left.\frac{\partial f}{\partial x}\right|_{(a, b)}$$

Recall that in computing $\partial f/\partial x$, we think of y as being fixed. But in this situation, we are evaluating $\partial f/\partial x$ at (a, b). Therefore, we set y equal to b. Doing this leads to the function g of one variable, x, defined by

$$g(x) = f(x, b)$$

It follows from the definition of the partial derivative that

$$f_x(a, b) = g'(a)$$

Thus, the value of the partial derivative $\partial f/\partial x$ at a given point (a, b) can be found by evaluating the derivative of a function of one variable. In particular, the latter can be found by using the numerical derivative operation of a graphing utility. We find $f_y(a, b)$ in a similar manner.

EXAMPLE 1 Let $f(x, y) = (1 + xy^2)^{3/2}e^{x^2y}$. Find (a) $f_x(1, 2)$ and (b) $f_y(1, 2)$.

Solution

a. Define $g(x) = f(x, 2) = (1 + 4x)^{3/2}e^{2x^2}$. Using the numerical derivative operation to find $g'(1)$, we obtain

$$f_x(1, 2) = g'(1) \approx 429.583225$$

b. Define $h(y) = f(1, y) = (1 + y^2)^{3/2}e^y$. Using the numerical derivative operation to find $h'(2)$, we obtain

$$f_y(1, 2) = h'(2) \approx 181.7468642$$

TECHNOLOGY EXERCISES

For each of the functions in Exercises 1–6, compute

$$\frac{\partial f}{\partial x} \quad \text{and} \quad \frac{\partial f}{\partial y}$$

at the given point.

1. $f(x, y) = \sqrt{x}(2 + xy^2)^{1/3}$; $(1, 2)$

2. $f(x, y) = \sqrt{xy}(1 + 2xy)^{2/3}$; $(1, 4)$

3. $f(x, y) = \dfrac{x + y^2}{1 + x^2y}$; $(1, 2)$

4. $f(x, y) = \dfrac{xy^2}{(\sqrt{x} + \sqrt{y})^2}$; $(4, 1)$

5. $f(x, y) = e^{-xy^2}(x + y)^{1/3}$; $(1, 1)$

6. $f(x, y) = \dfrac{\ln(\sqrt{x} + y^2)}{x^2 + y^2}$; $(4, 1)$

12.3 Maxima and Minima of Functions of Several Variables

Maxima and Minima

In Chapter 10, we saw that the solution of a problem often reduces to finding the extreme values of a function of one variable. In practice, however, situations also arise in which a problem is solved by finding the absolute maximum or absolute minimum value of a function of two or more variables.

For example, suppose Scandi Company manufactures computer desks in both assembled and unassembled versions. Its profit P is therefore a function of the number of assembled units, x, and the number of unassembled units, y, manufactured and sold per week; that is, $P = f(x, y)$. A question of paramount importance to the manufacturer is: How many assembled and unassembled desks should the company manufacture per week to maximize its weekly profit? Mathematically, the problem is solved by finding the values of x and y that will make $f(x, y)$ a maximum.

In this section, we will focus our attention on finding the extrema of a function of two variables. As in the case of a function of one variable, we distinguish between the relative (or local) extrema and the absolute extrema of a function of two variables.

Relative Extrema of a Function of Two Variables

Let f be a function defined on a region R containing the point (a, b). Then f has a **relative maximum** at (a, b) if $f(x, y) \leq f(a, b)$ for all points (x, y) that are sufficiently close to (a, b). The number $f(a, b)$ is called a **relative maximum value.** Similarly, f has a **relative minimum** at (a, b), with **relative minimum value** $f(a, b)$, if $f(x, y) \geq f(a, b)$ for all points (x, y) that are sufficiently close to (a, b).

Loosely speaking, f has a relative maximum at (a, b) if the point $(a\,b, f(a, b))$ is the highest point on the graph of f when compared with all nearby points. A similar interpretation holds for a relative minimum.

If the inequalities in this last definition hold for *all* points (x, y) in the domain of f, then f has an **absolute maximum** (or **absolute minimum**) at (a, b) with **absolute maximum value** (or **absolute minimum value**) $f(a, b)$. Figure 16 shows the graph of a function with relative maxima at (a, b) and (e, g) and a relative minimum at (c, d). The absolute maximum of f occurs at (e, g), and the absolute minimum of f occurs at (h, i).

Observe that in the case of a function of one variable, a relative extremum (relative maximum or relative minimum) may or may not be an absolute extremum.

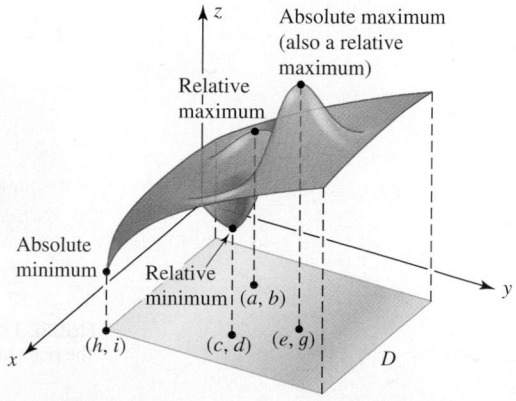

FIGURE 16
The relative and absolute extrema of the function f over the domain D

Now let's turn our attention to the study of relative extrema of a function. Suppose that a differentiable function $f(x, y)$ of two variables has a relative maximum (relative minimum) at a point (a, b) in the domain of f. From Figure 17, it is clear that at the point (a, b) the slopes of the "tangent lines" to the surface in any direction must be zero. In particular, this implies that both

$$\frac{\partial f}{\partial x}(a, b) \quad \text{and} \quad \frac{\partial f}{\partial y}(a, b)$$

must be zero.

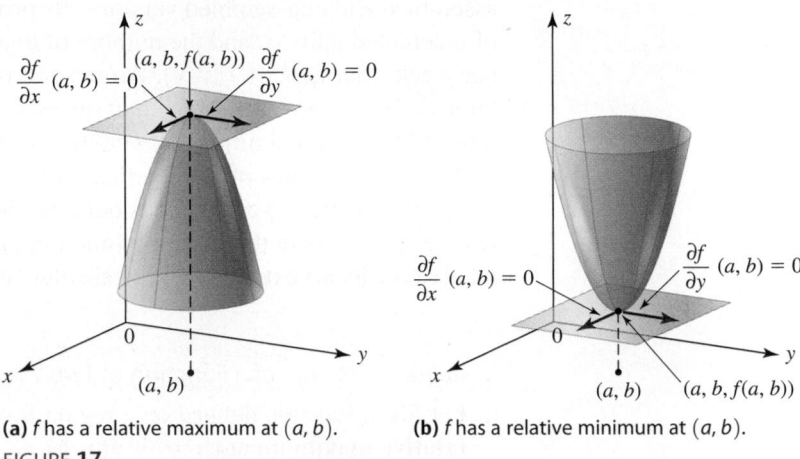

(a) f has a relative maximum at (a, b).

(b) f has a relative minimum at (a, b).

FIGURE **17**

Lest we be tempted to jump to the conclusion that a differentiable function f satisfying both the conditions

$$\frac{\partial f}{\partial x}(a, b) = 0 \quad \text{and} \quad \frac{\partial f}{\partial y}(a, b) = 0$$

at a point (a, b) must have a relative extremum at the point (a, b), let's examine the graph of the function f depicted in Figure 18. Here, both

$$\frac{\partial f}{\partial x}(a, b) = 0 \quad \text{and} \quad \frac{\partial f}{\partial y}(a, b) = 0$$

but f has neither a relative maximum nor a relative minimum at the point (a, b) because some nearby points are higher and some are lower than the point $(a, b, f(a, b))$. The point $(a, b, f(a, b))$ is called a **saddle point.**

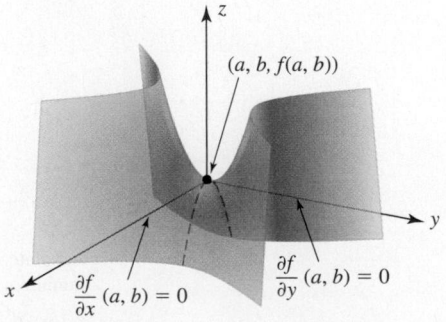

FIGURE **18**
The point $(a, b, f(a, b))$ is called a saddle point.

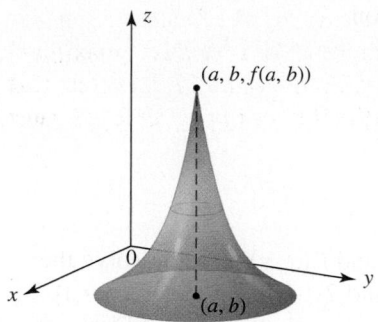

FIGURE 19
f has a relative maximum at (a, b), but neither $\partial f/\partial x$ nor $\partial f/\partial y$ exists at (a, b).

Finally, an examination of the graph of the function *f* depicted in Figure 19 should convince you that *f* has a relative maximum at the point (a, b). But both $\partial f/\partial x$ and $\partial f/\partial y$ fail to be defined at (a, b).

To summarize, a function *f* of two variables can have a relative extremum only at a point (a, b) in its domain where $\partial f/\partial x$ and $\partial f/\partial y$ both exist and are equal to zero at (a, b) or at least one of the partial derivatives does not exist. As in the case of one variable, we refer to a point in the domain of *f* that *may* give rise to a relative extremum as a critical point. The precise definition follows.

Critical Point of *f*

A **critical point** of *f* is a point (a, b) in the domain of *f* such that both

$$\frac{\partial f}{\partial x}(a, b) = 0 \quad \text{and} \quad \frac{\partial f}{\partial y}(a, b) = 0$$

or at least one of the partial derivatives does not exist.

To determine the nature of a critical point of a function $f(x, y)$ of two variables, we use the second partial derivatives of *f*. The resulting test, which helps us to classify these points, is called the **Second Derivative Test** and is incorporated in the following procedure for finding and classifying the relative extrema of *f*.

Determining Relative Extrema

1. Find the critical points of $f(x, y)$ by solving the system of simultaneous equations

$$f_x(x, y) = 0$$
$$f_y(x, y) = 0$$

2. The Second Derivative Test: Let

$$D(x, y) = f_{xx}(x, y)f_{yy}(x, y) - f_{xy}^2(x, y)$$

Then

a. $D(a, b) > 0$ and $f_{xx}(a, b) < 0$ implies that $f(x, y)$ has a **relative maximum** at the point (a, b).

b. $D(a, b) > 0$ and $f_{xx}(a, b) > 0$ implies that $f(x, y)$ has a **relative minimum** at the point (a, b).

c. $D(a, b) < 0$ implies that $f(x, y)$ has neither a relative maximum nor a relative minimum at the point (a, b). The point $(a, b, f(a, b))$ is called a saddle point.

d. $D(a, b) = 0$ implies that the test is inconclusive, so some other technique must be used to solve the problem.

Note We can replace $f_{xx}(a, b)$ by $f_{yy}(a, b)$ in parts 2a and 2b because $D(a, b) > 0$ implies that $f_{xx}(a, b)$ and $f_{yy}(a, b)$ must have the same sign. ◼

EXAMPLE 1 Find the relative extrema of the function

$$f(x, y) = x^2 + y^2$$

Solution We have

$$f_x(x, y) = 2x$$
$$f_y(x, y) = 2y$$

To find the critical point(s) of f, we set $f_x(x, y) = 0$ and $f_y(x, y) = 0$ and solve the resulting system of simultaneous equations $2x = 0$ and $2y = 0$. We obtain $x = 0$, $y = 0$, or $(0, 0)$, as the sole critical point of f. Next, we apply the Second Derivative Test to determine the nature of the critical point $(0, 0)$. We compute

$$f_{xx}(x, y) = 2 \qquad f_{xy}(x, y) = 0 \qquad f_{yy}(x, y) = 2$$

and

$$D(x, y) = f_{xx}(x, y)f_{yy}(x, y) - f_{xy}^2(x, y) = (2)(2) - 0 = 4$$

In particular, $D(0, 0) = 4$. Since $D(0, 0) > 0$ and $f_{xx}(0, 0) = 2 > 0$, we conclude that $f(x, y)$ has a relative minimum at the point $(0, 0)$. The relative minimum value, 0, also happens to be the absolute minimum of f. The graph of the function f, shown in Figure 20, confirms these results.

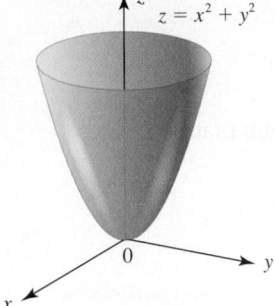

FIGURE 20
The graph of $f(x, y) = x^2 + y^2$

EXAMPLE 2 Find the relative extrema of the function

$$f(x, y) = 3x^2 - 4xy + 4y^2 - 4x + 8y + 4$$

Solution We have

$$f_x(x, y) = 6x - 4y - 4$$
$$f_y(x, y) = -4x + 8y + 8$$

To find the critical points of f, we set $f_x(x, y) = 0$ and $f_y(x, y) = 0$ and solve the resulting system of simultaneous equations

$$6x - 4y = 4$$
$$-4x + 8y = -8$$

Multiplying the first equation by 2 and the second equation by 3, we obtain the equivalent system

$$12x - 8y = 8$$
$$-12x + 24y = -24$$

Adding the two equations gives $16y = -16$, or $y = -1$. We substitute this value for y into either equation in the system to get $x = 0$. Thus, the only critical point of f is the point $(0, -1)$. Next, we apply the Second Derivative Test to determine whether the point $(0, -1)$ gives rise to a relative extremum of f. We compute

$$f_{xx}(x, y) = 6 \qquad f_{xy}(x, y) = -4 \qquad f_{yy}(x, y) = 8$$

and

$$D(x, y) = f_{xx}(x, y)f_{yy}(x, y) - f_{xy}^2(x, y) = (6)(8) - (-4)^2 = 32$$

Since $D(0, -1) = 32 > 0$ and $f_{xx}(0, -1) = 6 > 0$, we conclude that $f(x, y)$ has a relative minimum at the point $(0, -1)$. The value of $f(x, y)$ at the point $(0, -1)$ is given by

$$f(0, -1) = 3(0)^2 - 4(0)(-1) + 4(-1)^2 - 4(0) + 8(-1) + 4 = 0$$

Explore and Discuss

Suppose $f(x, y)$ has a relative extremum (relative maximum or relative minimum) at a point (a, b). Let $g(x) = f(x, b)$ and $h(y) = f(a, y)$. Assuming that f and g are differentiable, explain why $g'(a) = 0$ and $h'(b) = 0$. Explain why these results are equivalent to the conditions $f_x(a, b) = 0$ and $f_y(a, b) = 0$.

EXAMPLE 3 Find the relative extrema of the function

$$f(x, y) = 4y^3 + x^2 - 12y^2 - 36y + 2$$

Solution To find the critical points of f, we set $f_x = 0$ and $f_y = 0$ simultaneously, obtaining

$$f_x(x, y) = 2x = 0$$
$$f_y(x, y) = 12y^2 - 24y - 36 = 0$$

The first equation implies that $x = 0$. The second equation implies that

$$y^2 - 2y - 3 = 0$$
$$(y + 1)(y - 3) = 0$$

that is, $y = -1$ or 3. Therefore, there are two critical points of the function f: $(0, -1)$ and $(0, 3)$.

Next, we apply the Second Derivative Test to determine the nature of each of the two critical points. We compute

$$f_{xx}(x, y) = 2 \qquad f_{xy}(x, y) = 0 \qquad f_{yy}(x, y) = 24y - 24 = 24(y - 1)$$

Therefore,

$$D(x, y) = f_{xx}(x, y)f_{yy}(x, y) - f_{xy}^2(x, y) = 48(y - 1)$$

For the point $(0, -1)$,

$$D(0, -1) = 48(-1 - 1) = -96 < 0$$

Since $D(0, -1) < 0$, we conclude that the point $(0, -1)$ gives a saddle point of f. For the point $(0, 3)$,

$$D(0, 3) = 48(3 - 1) = 96 > 0$$

Since $D(0, 3) > 0$ and $f_{xx}(0, 3) > 0$, we conclude that the function f has a relative minimum at the point $(0, 3)$. Furthermore, since

$$f(0, 3) = 4(3)^3 + (0)^2 - 12(3)^2 - 36(3) + 2$$
$$= -106$$

we see that the relative minimum value of f is -106.

Explore and Discuss

1. Refer to the Second Derivative Test. Can the condition $f_{xx}(a, b) < 0$ in part 2a be replaced by the condition $f_{yy}(a, b) < 0$? Explain your answer. How about the condition $f_{xx}(a, b) > 0$ in part 2b?
2. Let $f(x, y) = x^4 + y^4$.
   a. Show that $(0, 0)$ is a critical point of f and that $D(0, 0) = 0$.
   b. Explain why f has a relative (in fact, an absolute) minimum at $(0, 0)$. Does this contradict the Second Derivative Test? Explain your answer.

As in the case of a practical optimization problem involving a function of one variable, the solution to an optimization problem involving a function of several variables calls for finding the *absolute* extremum of the function. Determining the absolute extremum of a function of several variables is more difficult than merely finding the relative extrema of the function. However, in many situations, the absolute extremum of a function actually coincides with the largest relative extremum of the function that occurs in the interior of its domain. We assume that the problems considered here belong to this category. Furthermore, the existence of the absolute extremum (solution) of a practical problem is often deduced from the geometric or practical nature of the problem.

$ APPLIED EXAMPLE 4 Maximizing Profits The total weekly revenue (in dollars) that Acrosonic realizes in producing and selling its bookshelf loudspeaker systems is given by

$$R(x, y) = -\frac{1}{4}x^2 - \frac{3}{8}y^2 - \frac{1}{4}xy + 300x + 240y$$

where x denotes the number of fully assembled units and y denotes the number of kits produced and sold each week. The total weekly cost attributable to the production of these loudspeakers is

$$C(x, y) = 180x + 140y + 5000$$

dollars, where x and y have the same meaning as before. Determine how many assembled units and how many kits Acrosonic should produce per week to maximize its profit. What is the maximum profit?

Solution The contribution to Acrosonic's weekly profit stemming from the production and sale of the bookshelf loudspeaker systems is given by

$$\begin{aligned}
P(x, y) &= R(x, y) - C(x, y) \\
&= \left(-\frac{1}{4}x^2 - \frac{3}{8}y^2 - \frac{1}{4}xy + 300x + 240y\right) - (180x + 140y + 5000) \\
&= -\frac{1}{4}x^2 - \frac{3}{8}y^2 - \frac{1}{4}xy + 120x + 100y - 5000
\end{aligned}$$

To find the relative maximum of the profit function $P(x, y)$, we first locate the critical point(s) of P. Setting $P_x(x, y)$ and $P_y(x, y)$ equal to zero, we obtain

$$P_x = -\frac{1}{2}x - \frac{1}{4}y + 120 = 0$$

$$P_y = -\frac{3}{4}y - \frac{1}{4}x + 100 = 0$$

Solving the first of these equations for y yields

$$y = -2x + 480$$

which, upon substitution into the second equation, yields

$$\begin{aligned}
-\frac{3}{4}(-2x + 480) - \frac{1}{4}x + 100 &= 0 \\
6x - 1440 - x + 400 &= 0 \\
x &= 208
\end{aligned}$$

We substitute this value of x into the equation $y = -2x + 480$ to get

$$y = 64$$

Therefore, the function P has the sole critical point $(208, 64)$. To show that the point $(208, 64)$ is a solution to our problem, we use the Second Derivative Test. We compute

$$P_{xx} = -\frac{1}{2} \qquad P_{xy} = -\frac{1}{4} \qquad P_{yy} = -\frac{3}{4}$$

So

$$D(x, y) = \left(-\frac{1}{2}\right)\left(-\frac{3}{4}\right) - \left(-\frac{1}{4}\right)^2 = \frac{3}{8} - \frac{1}{16} = \frac{5}{16}$$

Since $D(208, 64) > 0$ and $P_{xx}(208, 64) < 0$, the point $(208, 64)$ yields a relative maximum of P. It can be shown that this relative maximum is also the absolute maximum of P. We conclude that Acrosonic can maximize its weekly profit by manufacturing 208 assembled units and 64 kits of their bookshelf loudspeaker systems. The maximum weekly profit realizable from the production and sale of these loudspeaker systems is given by

$$\begin{aligned} P(208, 64) &= -\frac{1}{4}(208)^2 - \frac{3}{8}(64)^2 - \frac{1}{4}(208)(64) \\ &\quad + 120(208) + 100(64) - 5000 \\ &= 10{,}680 \end{aligned}$$

or $10,680.

■

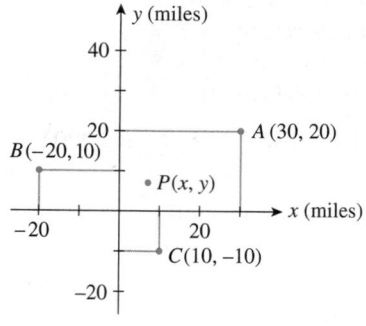

y (miles)

$B(-20, 10)$

$A(30, 20)$

$P(x, y)$

x (miles)

-20 20

$C(10, -10)$

-20

FIGURE 21
Locating a site for a television relay station

APPLIED EXAMPLE 5 Locating a Television Relay Station Site A television relay station will serve Towns A, B, and C, whose relative locations are shown in Figure 21. Determine a site for the location of the station if the sum of the squares of the distances from each town to the site is minimized.

Solution Suppose the required site is located at the point $P(x, y)$. With the aid of the distance formula, we find that the square of the distance from Town A to the site is

$$(x - 30)^2 + (y - 20)^2 \qquad d^2 = (x_2 - x_1)^2 + (y_2 - y_1)^2$$

The respective distances from Towns B and C to the site are found in a similar manner, so the sum of the squares of the distances from each town to the site is given by

$$\begin{aligned} f(x, y) &= (x - 30)^2 + (y - 20)^2 + (x + 20)^2 \\ &\quad + (y - 10)^2 + (x - 10)^2 + (y + 10)^2 \end{aligned}$$

To find the relative minimum of $f(x, y)$, we first find the critical point(s) of f. Using the Chain Rule to find $f_x(x, y)$ and $f_y(x, y)$ and setting each equal to zero, we obtain

$$\begin{aligned} f_x &= 2(x - 30) + 2(x + 20) + 2(x - 10) = 6x - 40 = 0 \\ f_y &= 2(y - 20) + 2(y - 10) + 2(y + 10) = 6y - 40 = 0 \end{aligned}$$

from which we deduce that $\left(\frac{20}{3}, \frac{20}{3}\right)$ is the sole critical point of f. Since

$$f_{xx} = 6 \qquad f_{xy} = 0 \qquad f_{yy} = 6$$

we have

$$D(x, y) = f_{xx}f_{yy} - f_{xy}^2 = (6)(6) - 0 = 36$$

Since $D\left(\frac{20}{3}, \frac{20}{3}\right) > 0$ and $f_{xx}\left(\frac{20}{3}, \frac{20}{3}\right) > 0$, we conclude that the point $\left(\frac{20}{3}, \frac{20}{3}\right)$ yields a relative minimum of f. Thus, the required site has coordinates $x = \frac{20}{3}$ and $y = \frac{20}{3}$.

■

12.3   Self-Check Exercises

1. Let $f(x, y) = 2x^2 + 3y^2 - 4xy + 4x - 2y + 3$.
   a. Find the critical point of f.
   b. Use the Second Derivative Test to classify the nature of the critical point.
   c. Find the relative extremum of f, if it exists.

2. **MAXIMIZING PROFIT** Robertson Controls manufactures two basic models of setback thermostats: a standard mechanical thermostat and a deluxe electronic thermostat. Robertson's monthly revenue (in hundreds of dollars) is

$$R(x, y) = -\frac{1}{8}x^2 - \frac{1}{2}y^2 - \frac{1}{4}xy + 20x + 60y$$

where x (in units of a hundred) denotes the number of mechanical thermostats manufactured each month and y (in units of a hundred) denotes the number of electronic thermostats manufactured each month. The total monthly cost incurred in producing these thermostats is

$$C(x, y) = 7x + 20y + 280$$

hundred dollars. Find how many thermostats of each model Robertson should manufacture each month to maximize its profits. What is the maximum profit?

Solutions to Self-Check Exercises 12.3 can be found on page 937.

12.3   Concept Questions

1. Explain the terms (a) relative maximum of a function $f(x, y)$ and (b) absolute maximum of a function $f(x, y)$.

2. a. What is a critical point of a function $f(x, y)$?
   b. Explain the role of a critical point in determining the relative extrema of a function of two variables.

3. Explain how the Second Derivative Test is used to determine the relative extrema of a function of two variables.

4. In (a)–(d), suppose that (a, b) is a critical point of f. Are the given conditions sufficient for you to determine whether f has a relative maximum, a relative minimum, or a saddle point at (a, b)? Explain.
   a. $f_{xx}(a, b) = -2, f_{xy}(a, b) = 3, f_{yy}(a, b) = -5$
   b. $f_{xx}(a, b) = 3, f_{xy}(a, b) = 3, f_{yy}(a, b) = 2$
   c. $f_{xx}(a, b) = 1, f_{xy}(a, b) = 2, f_{yy}(a, b) = 4$
   d. $f_{xx}(a, b) = 2, f_{xy}(a, b) = 2, f_{yy}(a, b) = 4$

12.3   Exercises

In Exercises 1–20, find the critical point(s) of the function. Then use the Second Derivative Test to classify the nature of each point, if possible. Finally, determine the relative extrema of the function.

1. $f(x, y) = 1 - 2x^2 - 3y^2$

2. $f(x, y) = x^2 - xy + y^2 + 1$

3. $f(x, y) = x^2 - y^2 - 2x + 4y + 1$

4. $f(x, y) = 2x^2 + y^2 - 4x + 6y + 3$

5. $f(x, y) = x^2 + 2xy + 2y^2 - 4x + 8y - 1$

6. $f(x, y) = x^2 - 4xy + 2y^2 + 4x + 8y - 1$

7. $f(x, y) = 2x^3 + y^2 - 9x^2 - 4y + 12x - 2$

8. $f(x, y) = 2x^3 + y^2 - 6x^2 - 4y + 12x - 2$

9. $f(x, y) = x^3 + y^2 - 2xy + 7x - 8y + 4$

10. $f(x, y) = 2y^3 - 3y^2 - 12y + 2x^2 - 6x + 2$

11. $f(x, y) = x^3 - 3xy + y^3 - 2$

12. $f(x, y) = x^3 - 2xy + y^2 + 5$

13. $f(x, y) = xy + \frac{4}{x} + \frac{2}{y}$

14. $f(x, y) = \frac{x}{y^2} + xy$

15. $f(x, y) = x^2 - e^{y^2}$

16. $f(x, y) = e^{x^2 - y^2}$

17. $f(x, y) = e^{x^2 + y^2}$

18. $f(x, y) = e^{xy}$

19. $f(x, y) = \ln(1 + x^2 + y^2)$

20. $f(x, y) = xy + \ln x + 2y^2$

21. **MAXIMIZING PROFIT** The total weekly revenue (in dollars) of the Country Workshop realized in manufacturing and selling its rolltop desks is given by

$$R(x, y) = -0.2x^2 - 0.25y^2 - 0.2xy + 200x + 160y$$

where x denotes the number of finished units and y denotes the number of unfinished units manufactured and sold each week. The total weekly cost attributable to the manufacture of these desks is given by

$$C(x, y) = 100x + 70y + 4000$$

dollars. Determine how many finished units and how many unfinished units the company should manufacture each week to maximize its profit. What is the maximum profit realizable?

22. MAXIMIZING PROFIT The total daily revenue (in dollars) that Weston Publishing realizes in publishing and selling its English-language dictionaries is given by

$$R(x, y) = -0.005x^2 - 0.003y^2 - 0.002xy$$
$$+ 20x + 15y$$

where x denotes the number of deluxe copies and y denotes the number of standard copies published and sold daily. The total daily cost of publishing these dictionaries is given by

$$C(x, y) = 6x + 3y + 200$$

dollars. Determine how many deluxe copies and how many standard copies Weston should publish each day to maximize its profits. What is the maximum profit realizable?

23. MAXIMUM PRICE OF LAND The rectangular region R shown in the accompanying figure represents the financial district of a city. The price of land within the district is approximated by the function

$$p(x, y) = 200 - 10\left(x - \frac{1}{2}\right)^2 - 15(y - 1)^2$$

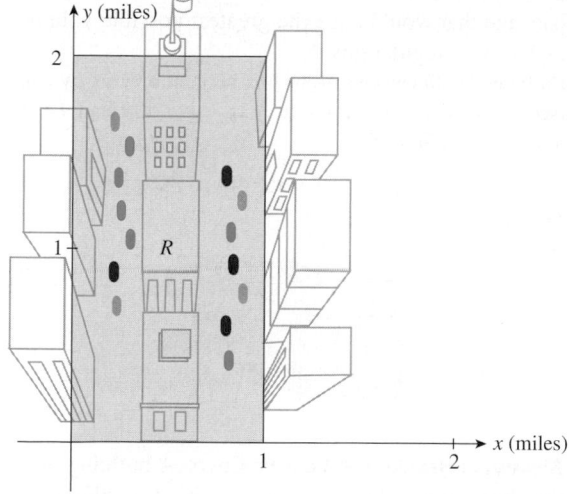

where $p(x, y)$ is the price of land at the point (x, y) in dollars per square foot and x and y are measured in miles. At what point within the financial district is the price of land highest?

24. MAXIMIZING PROFIT C&G Imports imports two brands of white wine, one from Germany and the other from Italy. The German wine costs \$4/bottle, and the Italian wine costs \$3/bottle. It has been estimated that if the German wine sells for p dollars/bottle and the Italian wine sells for for q dollars/bottle, then

$$2000 - 150p + 100q$$

bottles of the German wine and

$$1000 + 80p - 120q$$

bottles of the Italian wine will be sold each week. Determine the unit price for each brand that will allow C&G to realize the largest possible weekly profit.

25. MAXIMIZING REVENUE The management of Cal Supermarkets has determined that the quantity demanded per week of their 90% lean ground sirloin, x, and the quantity demanded per week of their 80% ground beef, y (both measured in pounds), are related to their unit prices p and q (in dollars), respectively, by the equations

$$x = 6400 - 400p - 200q \quad \text{and} \quad y = 5600 - 200p - 400q$$

a. What is the total revenue function $R(p, q)$?
   Hint: $R(p, q) = xp + yq$
b. What price should Cal Supermarkets charge for each product to maximize its weekly revenue? How many pounds of each product will then be sold? What is the maximum revenue?

26. MAXIMIZING PROFIT Johnson's Household Products has a division that produces two sizes of bar soap. The demand equations that relate the prices p and q (in dollars per hundred bars), to the quantities demanded, x and y (in units of a hundred), of the 3.5-oz size bar soap and the 5-oz bath size bar soap are given by

$$p = 80 - 0.01x - 0.005y \quad \text{and} \quad q = 60 - 0.005x - 0.015y$$

The fixed cost attributed to the division is \$10,000/week, and the cost for producing 100 3.5-oz size bars and 100 5-oz bath size bars is \$8 and \$12, respectively.
a. What is the weekly profit function $P(x, y)$?
b. How many of the 3.5-oz size bars and how many of the 5-oz bath size bars should the division produce per week to maximize its profit? What is the maximum weekly profit?

27. MAXIMIZING PROFIT Johnson's Household Products has a division that produces two types of toothpaste: a regular toothpaste and a whitening tooth paste. The demand equations that relate the prices, p and q (in dollars per thousand units), to the quantities demanded weekly, x and y (in units of a thousand), of the regular toothpaste and the whitening toothpaste are given by

$$p = 3000 - 20x - 10y \quad \text{and} \quad q = 4000 - 10x - 30y$$

respectively. The fixed cost attributed to the division is \$20,000/week, and the cost for producing 1000 tubes of regular and 1000 tubes of whitening toothpaste is \$400 and \$500, respectively.
a. What is the weekly total revenue function $R(x, y)$?
b. What is the weekly total cost function $C(x, y)$?
c. What is the weekly profit function $P(x, y)$?
d. How many tubes of regular and whitening toothpaste should be produced weekly to maximize the division's profit? What is the maximum weekly profit?

28. **DETERMINING THE OPTIMAL SITE FOR A POWER STATION** An auxiliary electric power station will serve three communities, A, B, and C, whose relative locations are shown in the accompanying figure. Determine where the power station should be located if the sum of the squares of the distances from each community to the site is minimized.

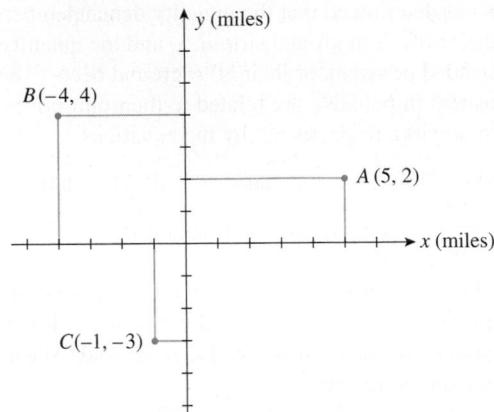

29. **LOCATING A RADIO STATION** The following figure shows the locations of three neighboring communities. The operators of a newly proposed radio station have decided that the site $P(x, y)$ for the station should be chosen so that the sum of the squares of the distances from the site to each community is minimized. Find the location of the proposed radio station.

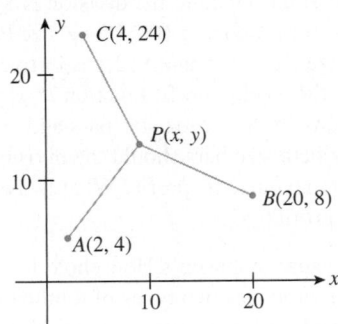

30. **PARCEL POST REGULATIONS** Postal regulations specify that a parcel sent by parcel post may have a combined length and girth of no more than 130 in. Find the dimensions of a cylindrical package of greatest volume that can be sent through the mail. What is the volume of such a package?
Hint: The length plus the girth is $2\pi r + l$.

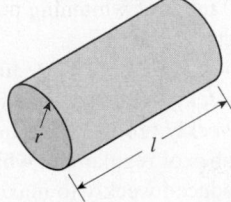

31. **PACKAGING** An open rectangular box having a volume of 108 in.^{3} is to be constructed from a tin sheet. Find the dimensions of such a box if the amount of material used in its construction is to be minimal.
Hint: Let the dimensions of the box be x in. by y in. by z in. Then $xyz = 108$, and the amount of material used is given by $S = xy + 2yz + 2xz$. Show that

$$S = f(x, y) = xy + \frac{216}{x} + \frac{216}{y}$$

Minimize $f(x, y)$.

32. **PACKAGING** An open rectangular box having a surface area of 300 in.^{2} is to be constructed from a tin sheet. Find the dimensions of the box if the volume of the box is to be as large as possible. What is the maximum volume?
Hint: Let the dimensions of the box be $x \times y \times z$ (see the figure that follows). Then the surface area is $xy + 2xz + 2yz$, and its volume is xyz.

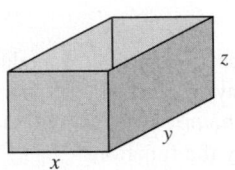

33. **PACKAGING** Postal regulations specify that the combined length and girth of a parcel sent by parcel post may not exceed 130 in. Find the dimensions of the rectangular package that would have the greatest possible volume under these regulations.
Hint: Let the dimensions of the box be x in. by y in. by z in. (see the figure below). Then $2x + 2z + y = 130$, and the volume $V = xyz$. Show that

$$V = f(x, z) = 130xz - 2x^2z - 2xz^2$$

Maximize $f(x, z)$.

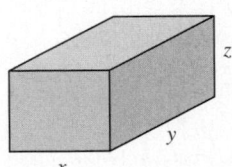

34. **MINIMIZING HEATING AND COOLING COSTS** A building in the shape of a rectangular box is to have a volume of 12,000 ft^3 (see the figure). It is estimated that the annual heating and cooling costs will be \$2/ft^2 for the top, \$4/ft^2 for the front and back, and \$3/ft^2 for the sides. Find the dimensions of the building that will result in a minimal annual heating and cooling cost. What is the minimal annual heating and cooling cost?

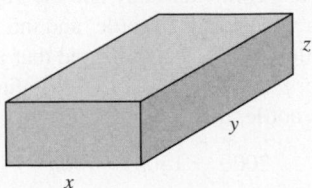

35. PACKAGING An open box having a volume of 48 in.^{3} is to be constructed. If the box is to include a partition that is parallel to a side of the box, as shown in the figure, and the amount of material used is to be minimal, what should be the dimensions of the box?

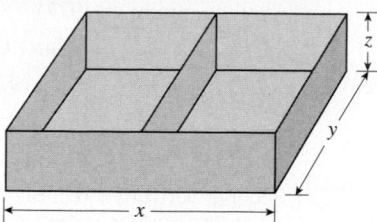

In Exercises 36–42, determine whether the statement is true or false. If it is true, explain why it is true. If it is false, give an example to show why it is false.

36. If $f_x(a, b) = 0$ and $f_y(a, b) = 0$, then f must have a relative extremum at (a, b).

37. If (a, b) is a critical point of f and both the conditions $f_{xx}(a, b) < 0$ and $f_{yy}(a, b) < 0$ hold, then f has a relative maximum at (a, b).

38. If $f(x, y)$ has a relative maximum at (a, b), then $f_x(a, b) = 0$ and $f_y(a, b) = 0$.

39. Let $h(x, y) = f(x) + g(y)$. If $f(x) > 0$ and $g(y) < 0$, then h cannot have a relative maximum or a relative minimum at any point.

40. If $f(x, y)$ satisfies $f_{xx}(a, b) \neq 0, f_{xy}(a, b) = 0, f_{yy}(a, b) \neq 0$, and $f_{xx}(a, b) + f_{yy}(a, b) = 0$ at the critical point (a, b) of f, then f cannot have a relative extremum at (a, b).

41. Suppose $h(x, y) = f(x) + g(y)$, where f and g have continuous second derivatives near a and b, respectively. If a is a critical number of f, b is a critical number of g, and $f''(a)g''(b) > 0$, then h has a relative extremum at (a, b).

42. If f_{xx} and f_{yy} have opposite signs at a critical point (a, b) of f, then f has a saddle point at (a, b).

12.3 Solutions to Self-Check Exercises

1. a. To find the critical point(s) of f, we solve the system of equations

$$f_x = 4x - 4y + 4 = 0$$
$$f_y = -4x + 6y - 2 = 0$$

obtaining $x = -2$ and $y = -1$. Thus, the only critical point of f is the point $(-2, -1)$.

b. We have $f_{xx} = 4, f_{xy} = -4$, and $f_{yy} = 6$, so

$$D(x, y) = f_{xx}f_{yy} - f_{xy}^2$$
$$= (4)(6) - (-4)^2 = 8$$

Since $D(-2, -1) > 0$ and $f_{xx}(-2, -1) > 0$, we conclude that f has a relative minimum at the point $(-2, -1)$.

c. The relative minimum value of $f(x, y)$ at the point $(-2, -1)$ is

$$f(-2, -1) = 2(-2)^2 + 3(-1)^2 - 4(-2)(-1)$$
$$+ 4(-2) - 2(-1) + 3$$
$$= 0$$

2. Robertson's monthly profit is

$$P(x, y) = R(x, y) - C(x, y)$$
$$= \left(-\frac{1}{8}x^2 - \frac{1}{2}y^2 - \frac{1}{4}xy + 20x + 60y\right) - (7x + 20y + 280)$$
$$= -\frac{1}{8}x^2 - \frac{1}{2}y^2 - \frac{1}{4}xy + 13x + 40y - 280$$

The critical point of P is found by solving the system

$$P_x = -\frac{1}{4}x - \frac{1}{4}y + 13 = 0$$

$$P_y = -\frac{1}{4}x - y + 40 = 0$$

giving $x = 16$ and $y = 36$. Thus, $(16, 36)$ is the critical point of P. Next,

$$P_{xx} = -\frac{1}{4} \qquad P_{xy} = -\frac{1}{4} \qquad P_{yy} = -1$$

and

$$D(x, y) = P_{xx}P_{yy} - P_{xy}^2$$
$$= \left(-\frac{1}{4}\right)(-1) - \left(-\frac{1}{4}\right)^2 = \frac{3}{16}$$

Since $D(16, 36) > 0$ and $P_{xx}(16, 36) < 0$, the point $(16, 36)$ yields a relative maximum of P. We conclude that the monthly profit is maximized by manufacturing 1600 mechanical and 3600 electronic setback thermostats each month. The maximum monthly profit realizable is

$$P(16, 36) = -\frac{1}{8}(16)^2 - \frac{1}{2}(36)^2 - \frac{1}{4}(16)(36)$$
$$+ 13(16) + 40(36) - 280$$
$$= 544$$

or \$54,400.

CHAPTER 12 Summary of Principal Terms

TERMS

function of two variables (900)

domain (900)

three-dimensional Cartesian coordinate system (903)

level curve (905)

first partial derivatives of f (913)

Cobb–Douglas production function (917)

marginal productivity of labor (917)

marginal productivity of capital (917)

substitute commodity (918)

complementary commodity (918)

second-order partial derivatives of f (920)

relative maximum (927)

relative maximum value (927)

relative minimum (927)

relative minimum value (927)

absolute maximum (927)

absolute minimum (927)

absolute maximum value (927)

absolute minimum value (927)

saddle point (928)

critical point (929)

Second Derivative Test (929)

CHAPTER 12 Concept Review Questions

Fill in the blanks.

1. The domain of a function f of two variables is a subset of the _____-plane. The rule of f associates with each _____ _____ in the domain of f one and only one _____ _____, denoted by $z =$ _____.

2. If the function f has rule $z = f(x, y)$, then x and y are called _____ variables, and z is a/an _____ variable. The number z is also called the _____ of f at (x, y).

3. The graph of a function f of two variables is the set of all points (x, y, z), where _____ and (x, y) is the domain of _____. The graph of a function of two variables is a/an _____ in three-dimensional space.

4. The trace of the graph of $f(x, y)$ in the plane $z = k$ is the curve with equation _____ lying in the plane $z = k$. The projection of the trace of f in the plane $z = k$ onto the xy-plane is called the _____ _____ of f. The contour map associated with f is obtained by drawing the _____ _____ of f corresponding to several admissible values of _____.

5. The partial derivative $\partial f/\partial x$ of f at (x, y) can be found by thinking of y as a/an _____ in the expression for f and differentiating this expression with respect to _____ as if it were a function of x alone.

6. The number $f_x(a, b)$ measures the _____ of the tangent line to the curve C obtained by intersecting the graph of f with the plane $y = b$ at the point _____. It also measures the rate of change of f with respect to _____ at the point (a, b) with y held fixed with value _____.

7. A function $f(x, y)$ has a relative maximum at (a, b) if $f(x, y)$ _____ $f(a, b)$ for all points (x, y) that are sufficiently close to _____. The absolute maximum value of $f(x, y)$ is the number $f(a, b)$ such that $f(x, y)$ _____ $f(a, b)$ for all (x, y) in the _____ of f.

8. A critical point of $f(x, y)$ is a point (a, b) in the _____ of f such that _____ _____ _____ or at least one of the partial derivatives of f does not _____. A critical point of f is a/an _____ for a relative extremum of f.

CHAPTER 12 Review Exercises

1. Let $f(x, y) = \dfrac{xy}{x^2 + y^2}$. Compute $f(0, 1)$, $f(1, 0)$, and $f(1, 1)$. Does $f(0, 0)$ exist?

2. Let $f(x, y) = \dfrac{xe^y}{1 + \ln xy}$. Compute $f(1, 1)$, $f(1, 2)$, and $f(2, 1)$. Does $f(1, 0)$ exist?

3. Let $h(x, y, z) = xye^z + \dfrac{x}{y}$. Compute $h(1, 1, 0)$, $h(-1, 1, 1)$, and $h(1, -1, 1)$.

4. Find the domain of the function $f(u, v) = \dfrac{\sqrt{u}}{u - v}$.

5. Find the domain of the function $f(x, y) = \dfrac{x - y}{x + y}$.

6. Find the domain of the function
$$f(x, y) = x\sqrt{y} + y\sqrt{1 - x}$$

7. Find the domain of the function
$$f(x, y, z) = \dfrac{xy\sqrt{z}}{(1 - x)(1 - y)(1 - z)}$$

In Exercises 8–11, sketch the level curves of the function corresponding to each value of z.

8. $z = f(x, y) = 2x + 3y$; $z = -2, -1, 0, 1, 2$

9. $z = f(x, y) = y - x^2$; $z = -2, -1, 0, 1, 2$

10. $z = f(x, y) = \sqrt{x^2 + y^2}$; $z = 0, 1, 2, 3, 4$

11. $z = f(x, y) = e^{xy}$; $z = 1, 2, 3$

In Exercises 12–21, compute the first partial derivatives of the function.

12. $f(x, y) = x^2y^3 + 3xy^2 + \dfrac{x}{y}$

13. $f(x, y) = x\sqrt{y} + y\sqrt{x}$ **14.** $f(u, v) = \sqrt{uv^2 - 2u}$

15. $f(x, y) = \dfrac{x - y}{y + 2x}$ **16.** $g(x, y) = \dfrac{xy}{x^2 + y^2}$

17. $h(x, y) = (2xy + 3y^2)^5$ **18.** $f(x, y) = (xe^y + 1)^{1/2}$

19. $f(x, y) = (x^2 + y^2)e^{x^2+y^2}$

20. $f(x, y) = \ln(1 + 2x^2 + 4y^4)$

21. $f(x, y) = \ln\left(1 + \dfrac{x^2}{y^2}\right)$

In Exercises 22–27, compute the second-order partial derivatives of the function.

22. $f(x, y) = x^3 - 2x^2y + y^2 + x - 2y$

23. $f(x, y) = x^4 + 2x^2y^2 - y^4$

24. $f(x, y) = (2x^2 + 3y^2)^3$ **25.** $g(x, y) = \dfrac{x}{x + y^2}$

26. $g(x, y) = e^{x^2+y^2}$ **27.** $h(s, t) = \ln\left(\dfrac{s}{t}\right)$

28. Let $f(x, y, z) = x^3y^2z + xy^2z + 3xy - 4z$. Compute $f_x(1, 1, 0)$, $f_y(1, 1, 0)$, and $f_z(1, 1, 0)$, and interpret your results.

In Exercises 29–34, find the critical point(s) of the functions. Then use the Second Derivative Test to classify the nature of each of these points, if possible. Finally, determine the relative extrema of each function.

29. $f(x, y) = 2x^2 + y^2 - 8x - 6y + 4$

30. $f(x, y) = x^2 + 3xy + y^2 - 10x - 20y + 12$

31. $f(x, y) = x^3 - 3xy + y^2$

32. $f(x, y) = x^3 + y^2 - 4xy + 17x - 10y + 8$

33. $f(x, y) = e^{2x^2+y^2}$

34. $f(x, y) = \ln(x^2 + y^2 - 2x - 2y + 4)$

35. IQs The IQ (intelligence quotient) of a person whose chronological age is c and whose mental age is m is defined as

$$I(c, m) = \frac{100m}{c}$$

Describe the level curves of I. Sketch the level curves corresponding to $I = 90, 100, 120, 180$. Interpret your results.

36. REVENUE FUNCTIONS A division of Ditton Industries makes a 16-speed and a 10-speed electric blender. The company's management estimates that x units of the 16-speed model and y units of the 10-speed model are demanded daily when the unit prices are

$$p = 80 - 0.02x - 0.1y$$
$$q = 60 - 0.1x - 0.05y$$

dollars, respectively.
a. Find the daily total revenue function $R(x, y)$.
b. Find the domain of the function R.
c. Compute $R(100, 300)$ and interpret your result.

37. DEMAND FOR CD PLAYERS In a survey conducted by *Home Entertainment* magazine, it was determined that the demand equation for CD players is given by

$$x = f(p, q) = 900 - 9p - e^{0.4q}$$

whereas the demand equation for audio CDs is given by

$$y = g(p, q) = 20{,}000 - 3000q - 4p$$

where p and q denote the unit prices (in dollars) for the CD players and audio CDs, respectively, and x and y denote the number of CD players and audio CDs demanded per week. Determine whether these two products are substitute, complementary, or neither.

38. MAXIMIZING REVENUE Odyssey Travel Agency's monthly revenue depends on the amount of money x (in thousands of dollars) spent on advertising per month and the number of agents y in its employ in accordance with the rule

$$R(x, y) = -x^2 - 0.5y^2 + xy + 8x + 3y + 20$$

Determine the amount of money the agency should spend per month and the number of agents it should employ to maximize its monthly revenue.

39. MINIMIZING FENCING COSTS The owner of the Rancho Grande wants to enclose a rectangular piece of grazing land along the straight portion of a river and then subdivide it using a fence running parallel to the sides that are perpendicular to the river. No fencing is required along the river. If the material for the sides costs \$3/running yard and the material for the divider costs \$2/running yard, what will be the dimensions of a 303,750-yd^2 pasture if the cost of fencing is kept to a minimum?

40. MAXIMIZING THE WEIGHTS OF FISH A pond is stacked with x bass and y trout (in hundreds). The average weights of bass and trout after 1 year are $\left(8 - x - \frac{1}{2}y\right)$ lb and $\left(11 - \frac{1}{2}x - 2y\right)$ lb, respectively. Find the number of each species of fish in the pond that will make the total weight of fish a maximum.

CHAPTER 12 Before Moving On . . .

1. Find the domain of

$$f(x, y) = \frac{\sqrt{x} + \sqrt{y}}{(1 - x)(2 - y)}$$

2. Sketch the level curves of the function $f(x, y) = x + 2y^2$ corresponding to $z = -3, -2, -1, 0, 1, 2, 3$.

3. Find $f_x(1, 2)$ and $f_y(1, 2)$ if $f(x, y) = x^3y - 2x^2y^2 + 3xy^3$ and interpret your results.

4. Find the first and second partial derivatives of

$$f(x, y) = x^2y + e^{xy}$$

5. Find the relative extrema, if any, of

$$f(x, y) = 2x^3 + 2y^3 - 6xy - 5$$

Answers

CHAPTER 1

Exercises 1.1, page 6

1. Integer, rational, and real   3. Rational and real

5. Irrational and real   7. Irrational and real

9. Rational and real   11. False   13. True   15. False

17. Commutative law of addition

19. Commutative law of multiplication   21. Distributive law

23. Associative law of addition   25. Property 1 of negatives

27. Property 1 of zero properties   29. Property 2 of zero properties

31. Property 2 of quotients   33. Properties 2 and 5 of quotients

35. Property 6 of quotients and distributive law   37. False

39. False   41. False

Exercises 1.2, page 12

1. 81   3. $\frac{8}{27}$ 5. -81 7. $-\frac{81}{125}$ 9. 256

11. $243y^5$ 13. $6x - 3$ 15. $9x^2 + 3x + 1$

17. $4y^2 + 2y + 9$ 19. $1.2x^3 - 4.2x^2 + 2.5x - 8.2$

21. $6x^5$ 23. $2x^3 + 4x$ 25. $7m^2 - 9m$

27. $14a - 7b$ 29. $6x^2 + 5x - 6$

31. $6x^2 - 5xy - 6y^2$ 33. $12r^2 - rs - 6s^2$

35. $0.06x^2 - 0.06xy - 2.52y^2$ 37. $6x^3 - 3x^2y + 4xy - 2y^2$

39. $4x^2 + 12xy + 9y^2$ 41. $4u^2 - v^2$

43. $2x^2 - x + 2$ 45. $4x^2 + 6xy + 9y^2 - x + 2y + 2$

47. $2t^4 - 4t^3 + 9t^2 - 2t + 4$ 49. $-2x + 1$ 51. $-3x - 1$

53. $x^2 - 10x + 50$ 55. $22x^3 - 20x^2 - 6x$

57. $-0.000002x^3 - 0.02x^2 + 1000x - 120,000$

59. $0.7t^2 + 350t$ 61. $6t^2 + 27t + 186$

63. False   65. False

Exercises 1.3, page 18

1. $2m(3m - 2)$ 3. $3ab(3b - 2a)$ 5. $5mn(2m - 3n + 4)$

7. $(3x - 5)(2x + 1)$ 9. $(2c - d)(3a + b + 4ac - 2ad)$

11. $(2m + 1)(m - 6)$ 13. $(x - 3y)(x + 2y)$ 15. Prime

17. $(2a - b)(2a + b)$ 19. $(uv - w)(uv + w)$ 21. Prime

23. Prime   25. $(x + 4)(x - 1)$ 27. $2y(3x + 2)(2x - 3)$

29. $(7r - 4)(5r + 3)$ 31. $xy(3x - 2y)(3x + 2y)$

33. $(x^2 - 4y)(x^2 + 4y)$ 35. $-8ab$

37. $(2m + 1)(4m^2 - 2m + 1)$ 39. $(2r - 3s)(4r^2 + 6rs + 9s^2)$

41. $u^2(v^2 - 2)(v^4 + 2v^2 + 4)$ 43. $(2x + 1)(x^2 + 3)$

45. $(3a + b)(x + 2y)$ 47. $(u - v)(u + v)(u^2 + v^2)$

49. $(x + y)(2x - 3y)(2x + 3y)$ 51. $(x^3 - 2)(x + 3)$

53. $(au + c)(u + 1)$ 55. $P(1 + rt)$ 57. $100x(80 - x)$

59. $kx(M - x)$ 61. $(200 + x)(300 - x)$ 63. $\frac{V_0}{273}(273 + T)$

Exercises 1.4, page 25

1. $\frac{4}{x}$ 3. $\frac{4}{5}$ 5. $\frac{2x - 1}{2x}$ 7. $\frac{x - 1}{x + 1}$ 9. $\frac{x + 3}{2x + 1}$

11. $x + y$ 13. $\frac{1}{2}x$ 15. $\frac{2}{5}x^2$ 17. $\frac{5x}{2}$ 19. $\frac{4}{3}$

21. $\frac{3(3r - 2)}{2}$ 23. $\frac{k + 1}{k - 2}$ 25. $\frac{10x + 7}{(2x + 3)(2x - 1)}$

27. $\frac{5x - 9}{(x - 3)(x + 2)(x - 1)}$ 29. $\frac{4m^3 - 3}{(2m^2 - 2m - 1)(2m^2 - 3m + 3)}$

31. $-\frac{x^2 - x - 3}{(x + 1)(x - 1)}$ 33. $\frac{2(x^2 - x + 2)}{(x + 2)(x - 2)}$

35. $\frac{x^3 - 2x^2 + 3x + 24}{(x + 3)(x + 2)(x - 2)(x + 1)}$ 37. $\frac{bx - ay}{ab(x - y)}$

39. $\frac{x + 1}{x - 1}$ 41. $\frac{x + y}{xy - 1}$ 43. $\frac{y - x}{x^2y^2}$ 45. $-\frac{1}{2x(x + h)}$

47. **a.** $\frac{2.2x + 2500}{x}$ **b.** $2.2x + 2500$ 49. $\frac{R[(1 + i)^n - 1]}{i(1 + i)^n}$

51. **a.** $\frac{f_1 + f_2 - d}{f_1 f_2}$ **b.** $f = \frac{f_1 f_2}{f_1 + f_2 - d}$

Exercises 1.5, page 30

1. -8 3. $\frac{1}{49}$ 5. -16 7. $\frac{7}{12}$ 9. 0.0009   11. 1

13. 1   15. $\frac{1}{32}$ 17. 1   19. $\frac{1}{27}$ 21. $\frac{1}{4}x^5$ 23. $\frac{3}{2x}$

25. $\frac{1}{a^6}$ 27. $\frac{8y^6}{x^6}$ 29. $\frac{8}{xy}$ 31. $-2x^2y^5$ 33. $\frac{3}{2uv^2}$

35. $864x^4$ 37. $\frac{1}{4x^8}$ 39. $\frac{4u^4}{9v^5}$ 41. $\frac{1}{1728x^2y^3z^2}$

43. $\frac{a^{10}}{b^{12}}$ 45. $9a^2b^8$ 47. $\frac{u^8}{16}$ 49. $\frac{1 - x}{1 + x}$

51. $\frac{1}{uv}$ 53. $\frac{b + a}{b - a}$ 55. False   57. False

Exercises 1.6, page 35

1. $x = 4$ **3.** $y = \frac{20}{3}$ **5.** $x = -\frac{2}{3}$ **7.** $y = 5$

9. $p = 15$ **11.** $p = 0$ **13.** $k = 4$ **15.** $x = 2$ **17.** $x = 8$

19. $x = \frac{1}{2}$ **21.** $x = \frac{1}{3}$ **23.** $y = \frac{3}{2}$ **25.** $x = \frac{17}{8}$

27. $q = -1$ **29.** $k = -2$ **31.** No solution

33. $r = \frac{I}{Pt}$ **35.** $q = -\frac{p}{3} + \frac{1}{3}$ **37.** $T = \frac{R - R_0}{aR_0}$

39. $R = \dfrac{iS}{(1 + i)[(1 + i)^n - 1]}$ **41.** $n = \dfrac{N}{C}(C - V)$

43. $x = \dfrac{2(5 - 2p)}{p - 1}$ **45.** $x = \dfrac{y}{2(10 - y)}$ **47.** $q = \dfrac{f(p - d)}{p - f}$

49. $t = \dfrac{I}{Pr}; \dfrac{3}{2}$ years **51.** $t = \dfrac{a}{S - b}$

53. a. $C = \dfrac{NV - St}{N - t}$ **b.** \$115,000

55. a. 155,556 **b.** 38,889 **c.** 6222

57. a. $t = \dfrac{24c - a}{a}$ **b.** 5 years

Exercises 1.7, page 44

1. 9 **3.** 4 **5.** 4 **7.** 4 **9.** -5 **11.** 4

13. $\frac{2}{3}$ **15.** $\frac{9}{4}$ **17.** $\frac{1}{4}$ **19.** $-\frac{2}{3}$ **21.** 9 **23.** $\frac{1}{9}$

25. $\frac{3}{4}$ **27.** 64 **29.** $x^{1/5}$ **31.** x **33.** $\dfrac{9}{x^6}$ **35.** $\dfrac{y^{5/2}}{x^3}$

37. $x^{12/5} - 2x^{17/5}$ **39.** $4p^2 - 2p$ **41.** $4\sqrt{2}$ **43.** $-3\sqrt[3]{2}$

45. $4xy\sqrt{y}$ **47.** m^2np^4 **49.** $\sqrt[3]{3}$ **51.** $\sqrt[6]{x}$ **53.** $\dfrac{2\sqrt{3}}{3}$

55. $\dfrac{3\sqrt{x}}{2x}$ **57.** $\dfrac{2\sqrt{3y}}{3}$ **59.** $\dfrac{\sqrt[3]{x^2}}{x}$ **61.** $\sqrt{3} - 1$

63. $-(1 + \sqrt{2})^2$ **65.** $\dfrac{q(\sqrt{q} + 1)}{q - 1}$ **67.** $\dfrac{y\sqrt[3]{xz^2}}{xz}$ **69.** $\dfrac{4\sqrt{3}}{3}$

71. $\dfrac{\sqrt[3]{18}}{3}$ **73.** $\dfrac{\sqrt{6}}{2x}$ **75.** $\dfrac{\sqrt[3]{18y^2}}{3}$ **77.** $\dfrac{\sqrt{a}(1 + a)}{a}$

79. $\dfrac{x + y}{x - y}$ **81.** $\dfrac{\sqrt{x + 1}(3x + 2)}{2(x + 1)}$ **83.** $\dfrac{3 + x^{1/3}}{6x^{1/2}(1 + x^{1/3})^2}$

85. $x = 1$ **87.** $k = \frac{5}{2}$ **89.** $k = \frac{1}{3}$ **91.** $p = 144 - x^2$

93. True **95.** True

Exercises 1.8, page 52

1. $x = -2, 3$ **3.** $x = -2, 2$ **5.** $x = -4, 3$

7. $t = -1, \frac{1}{2}$ **9.** $x = 2$ **11.** $m = 2, \frac{3}{2}$ **13.** $x = -\frac{3}{2}, \frac{3}{2}$

15. $z = -2, \frac{3}{2}$ **17.** $x = -4, 2$ **19.** $x = 1 - \dfrac{\sqrt{6}}{2}, 1 + \dfrac{\sqrt{6}}{2}$

21. $m = -\frac{1}{2} - \frac{1}{2}\sqrt{13}, -\frac{1}{2} + \frac{1}{2}\sqrt{13}$

23. $x = -\dfrac{3}{4} - \dfrac{\sqrt{41}}{4}, -\dfrac{3}{4} + \dfrac{\sqrt{41}}{4}$ **25.** $x = \pm\dfrac{\sqrt{13}}{2}$

27. $x = -\frac{3}{2}, 2$ **29.** $m = 2 \pm \sqrt{3}$ **31.** $x = \frac{1}{2} \pm \frac{1}{4}\sqrt{10}$

33. $x = -1 \pm \frac{1}{2}\sqrt{10}$ **35.** $x = -0.93, 3.17$

37. $x = \pm\sqrt{2}, \pm\sqrt{3}$ **39.** $y = \pm\sqrt{2}, \pm\sqrt{5}$

41. $x = -\frac{7}{2}, -\frac{5}{3}$ **43.** $w = \frac{4}{9}, \frac{9}{4}$ **45.** $x = -2, -\frac{3}{2}$

47. $x = -\frac{5}{2}, 1$ **49.** $y = -2, \frac{15}{4}$ **51.** $x = -8, 2$

53. $x = -\frac{16}{3}$ **55.** $t = -1 \pm \frac{1}{2}\sqrt{6}$ **57.** $u = -3, 2$

59. $r = 3$ **61.** $s = 6$ **63.** $x = \frac{22}{7}, \frac{10}{3}$

65. Two real solutions **67.** No real solutions

69. One real solution **71.** Two real solutions

73. 9.2 sec **75.** 1.41 sec after passing the tree **77.** 10,000

79. 40,000 **81.** $r = 1.62$ **83.** 60 ft **85.** 2.5 ft

87. 2.76 in. **89.** 20 ft **91.** False **93.** True

Exercises 1.9, page 63

1. False **3.** False

5.

7.

9.

11. $x < 10$ **13.** $x + y \le z$ **15.** $-3 < 3x \le 8$

17. $(-\infty, 3)$ **19.** $(-\infty, -5]$ **21.** $(-4, 6)$

23. $(-\infty, -3)$ and $(3, \infty)$ **25.** $(-2, 3)$ **27.** $[-3, 5]$

29. $\left[1, \frac{3}{2}\right]$ **31.** $(-\infty, -3]$ and $(2, \infty)$ **33.** $(-\infty, 0]$ and $(1, \infty)$

35. 4 **37.** 2 **39.** $5\sqrt{3}$ **41.** $\pi + 1$ **43.** 2 **45.** False

47. False **49.** True **51.** $a - b < x < a + b$ **53.** $|x| = a$

55. $|a| \le 8$ **57.** False **59.** True **61.** False

63. True **65.** \$50.70 **67.** $[362, 488.7]$

69. Between 20,000 and 20,500 **71.** \$52,000 **73.** 2011

75. Between $\frac{1}{2}$ hr and 2 hr after the injection

77. Between 6:15 A.M. and 8:15 A.M.

79. Between 10:18 A.M. and 12:42 P.M.

81. $|x - 0.5| \le 0.01$

Chapter 1 Concept Review Questions, page 66

1. a. rational; repeating; terminating
b. irrational; terminates; repeats

2. a. $b + a$; $(a + b) + c$; a; 0 **b.** ba; $(ab)c$; $1 \cdot a = a$; 1
c. $ab + ac$

3. a. a; $-(ab) = a(-b)$; ab **b.** 0; 0

4. a. polynomial; x; degree; term; polynomial; coefficient　**b.** like

5. product; prime; $x(x + 2)(x - 1)$

6. a. polynomials　**b.** numerator; denominator; factors; 1; -1
c. denominator; fractions

7. complex; $\dfrac{1 + \dfrac{1}{x}}{1 - \dfrac{1}{y}}$

8. a. $\underbrace{a \cdot a \cdot a \cdots\cdot a}_{n \text{ factors}}$; base; exponent; power　**b.** 1; not defined

c. $\dfrac{1}{a^n}$

9. a. equation　**b.** number　**c.** $ax + b = 0$; 1

10. a. $a^n = b$　**b.** pairs　**c.** no　**d.** real root

11. a. radical; $b^{1/n}$　**b.** radical

12. a. $ax^2 + bx + c = 0$

b. factoring; completing the square; $x = \dfrac{-b \pm \sqrt{b^2 - 4ac}}{2a}$

Chapter 1 Review Exercises, page 67

1. Rational and real　　**2.** Irrational and real

3. Irrational and real　　**4.** Whole, integer, rational, and real

5. Rational and real　　**6.** Irrational and real　　**7.** $\frac{27}{8}$

8. 25　　**9.** $\frac{1}{144}$　　**10.** -32　　**11.** $\frac{64}{27}$　　**12.** $\frac{1}{4}$　　**13.** $\frac{3}{5}$

14. $3\sqrt[3]{3}$　　**15.** $4(x^2 + y)^2$　　**16.** $\dfrac{a^{15}}{b^{11}}$　　**17.** $\dfrac{2x}{3z}$　　**18.** $-x^{1/2}$

19. $6xy^7$　　**20.** $\frac{9}{2}a^5b^8$　　**21.** $9x^2y^4$　　**22.** $\dfrac{x}{y^{1/2}}$

23. $5x^4 + 20x^3 + 12x^2 + 14x + 3$　　**24.** $9x^3 - 18x^2 + 17x - 12$

25. $-2x^2 + 9y^2 + 12xy + 7x + 3$　　**26.** $3a - b$

27. $\dfrac{180}{(t + 6)^2}$　　**28.** $\dfrac{15x^2 + 24x + 2}{4(3x^2 + 2)(x + 2)}$

29. $\dfrac{78x^2 - 8x - 27}{3(2x^2 - 1)(3x - 1)}$　　**30.** $\dfrac{2\sqrt{x + 1}(x + 2)}{x + 1}$

31. $-2\pi r^2(\pi r - 50)$　　**32.** $2vw(v^2 + w^2 + u^2)$　　**33.** $(4 - x)(4 + x)$

34. $6t(2t - 3)(t + 1)$　　**35.** $-2(x + 3)(x - 1)$

36. $4(3x - 5)(x - 6)$　　**37.** $(3a - 5b)(3a + 5b)$

38. $u^3(2uv + 3)(4u^2v^2 - 6uv + 9)$　　**39.** $3a^2b^2(2a^2b^2c - ac - 3)$

40. $(2x - y)(3x + y)$　　**41.** $\dfrac{x + 2}{x + 3}$　　**42.** $\dfrac{4(t^2 - 4)}{(t^2 + 4)^2}$　　**43.** 2

44. $\dfrac{3x(2x^3 + 2x + 1)}{(x^2 + 2)(x^3 + 1)}$　　**45.** $\dfrac{x}{(x + 2)(x - 3)}$

46. $\dfrac{x\sqrt{1 + 3x^2}(3x^2 - 5x + 1)}{(x - 1)^2}$　　**47.** $x = -\frac{3}{4}, \frac{1}{2}$

48. $x = -2, \frac{1}{3}$　　**49.** $x = \dfrac{3 \pm \sqrt{41}}{4}$　　**50.** $x = \dfrac{-5 \pm \sqrt{13}}{2}$

51. $y = \frac{1}{2}, 1$　　**52.** $m = -1.2871, 8.2871$　　**53.** $x = 0, -3, 1$

54. $x = \dfrac{\pm\sqrt{2}}{2}$　　**55.** $x = 14$　　**56.** $p = -2$　　**57.** $x = -\frac{5}{3}, 0$

58. $q = -\frac{6}{17}$　　**59.** $k = 2$　　**60.** $x = 1$　　**61.** $x = \dfrac{100C}{20 + C}$

62. $I = \dfrac{rB(n + 1)}{2m}$　　**63.** $[-2, \infty)$　　**64.** $[-1, 2]$

65. $(-\infty, -4)$ and $(5, \infty)$　　**66.** $(-\infty, -5)$ and $(5, \infty)$　　**67.** 4

68. 1　　**69.** $\pi - 6$　　**70.** $8 - 3\sqrt{3}$　　**71.** $[-2, \frac{1}{2}]$　　**72.** $[-4, 3]$

73. $(-2, -\frac{3}{2})$　　**74.** $(-1, 4)$　　**75.** $[\frac{2}{3}, 2]$　　**76.** $\frac{2}{3}; \frac{3}{2}$

77. $\dfrac{1}{\sqrt{x} + 1}$　　**78.** $\dfrac{x}{z\sqrt[3]{xy}}$　　**79.** $\dfrac{x - \sqrt{x}}{2x}$　　**80.** $\dfrac{3(1 - 2\sqrt{x})}{1 - 4x}$

81. $x = 1 \pm \sqrt{6}$　　**82.** $x = -2 \pm \dfrac{\sqrt{2}}{2}$　　**83.** $\$100$

84. $\$400$　　**85.** $3.6t^2 - 10.6t + 80$; $\$95.20$; $\$146$

86. a. $51.8° \le F \le 80.6°$　　**b.** $11° \le C \le 27°$

87. 5000　　**88.** m

Chapter 1 Before Moving On, page 69

1. $3(5x^2 - 9x + 4)$

2. a. $x^2(x - 3)(x + 2)$　　**b.** $a(a + 1)(a - 2b - a^2)$

3. $\dfrac{5x^2 - 1}{(3x + 1)(x - 2)(x + 1)}$　　**4.** $\dfrac{1}{2xy}$　　**5.** $\dfrac{2s^2}{1 - 2s}$

6. $7 - 4\sqrt{3}$　　**7. a.** -4 or $\frac{3}{2}$　　**b.** $\frac{1}{2}(3 + \sqrt{17})$ or $\frac{1}{2}(3 - \sqrt{17})$

8. 21　　**9.** $[-\frac{2}{3}, \frac{3}{2}]$　　**10.** $[-2, -1]$

CHAPTER 2

Exercises 2.1, page 77

1. $(3, 3)$; Quadrant I　　**3.** $(2, -2)$; Quadrant IV

5. $(-4, -6)$; Quadrant III　　**7.** A　　**9.** E, F, and G　　**11.** F

13–19. See the following figure.

13. $(-2, 5)$　**15.** $(3, -1)$　**17.** $\left(8, -\frac{7}{2}\right)$　**19.** $(4.5, -4.5)$

21. $\frac{1}{2}$　　**23.** Not defined　　**25.** 5

27. $\frac{5}{6}$　　**29.** $\dfrac{d - b}{c - a}$　$(a \ne c)$

31. a. 4   **b.** −8   **33.** Parallel

35. $a = -5$ **37.** Yes

Exercises 2.2, page 84

1. (e)   **3.** (a)   **5.** (f)   **7.** Perpendicular

9. $y = -3$ **11.** $y = 2x - 10$ **13.** $y = 2$

15. $y = 3x - 2$ **17.** $y = x + 1$ **19.** $y = 3x + 4$

21. $y = 5$ **23.** $y = \frac{1}{2}x; m = \frac{1}{2}; b = 0$

25. $y = \frac{2}{3}x - 3; m = \frac{2}{3}; b = -3$

27. $y = -\frac{1}{2}x + \frac{7}{2}; m = -\frac{1}{2}; b = \frac{7}{2}$

29. $y = \frac{1}{2}x + 3$ **31.** $y = \frac{4}{3}x + \frac{4}{3}$ **33.** $y = -6$

35. $y = b$ **37.** $y = \frac{2}{3}x - \frac{2}{3}$ **39.** $k = 8$

41. 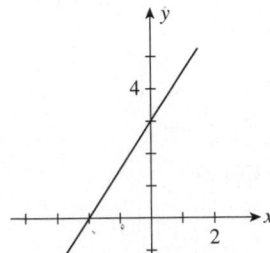   **43.**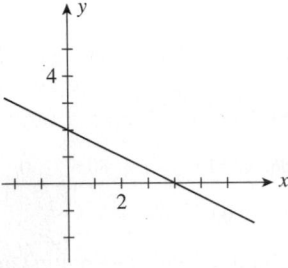

45.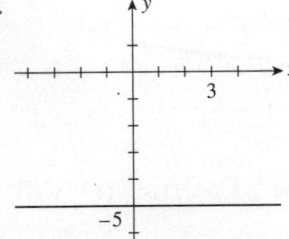

49. $y = -2x - 4$ **51.** $y = \frac{1}{8}x - \frac{1}{2}$ **53.** Yes

55. The points do not lie on a straight line.

57. a.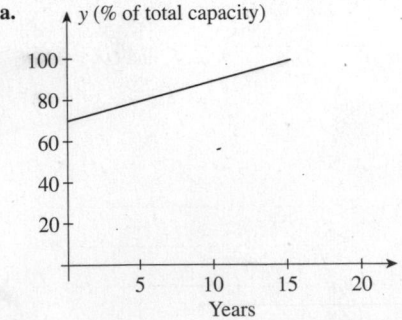

b. 1.9467; 70.082
c. The capacity utilization has been increasing by 1.9467% each year since 1990 when it stood at 70.082%.
d. In the first half of 2005

59. a. $y = 0.55x$ **b.** 2000   **61.** 84.8%

63. a. and b.

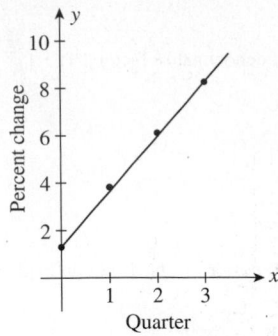

c. $y = 2.3x + 1.3$
d. 10.5%

65. a. and b.

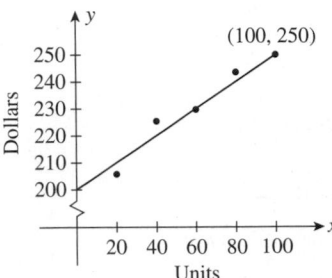

c. $y = \frac{1}{2}x + 200$ **d.** $227

67. a. and b.

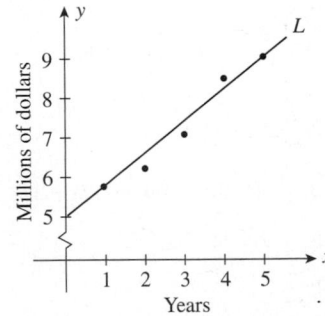

c. $y = 0.8x + 5$ **d.** $12.2 million

69. True   **71.** True   **73.** True

Using Technology Exercises 2.2, page 92

Graphing Utility

1.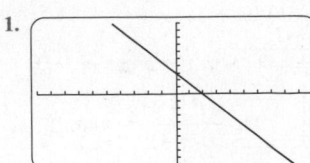

3.

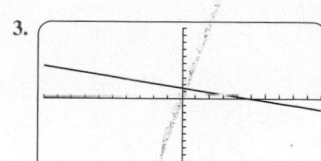

5. a.

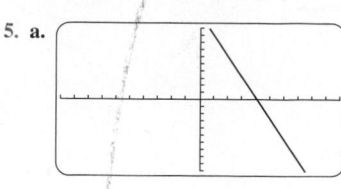

b.

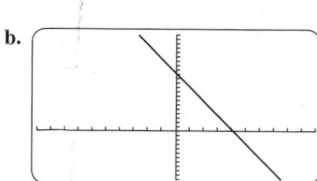

7. a.

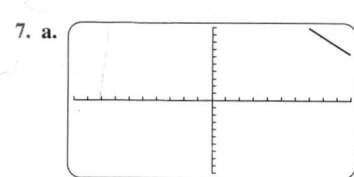

b.

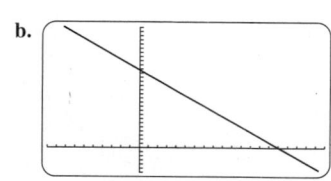

9.

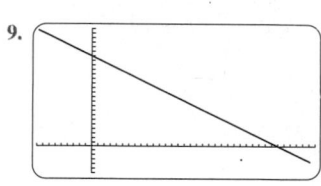

11.

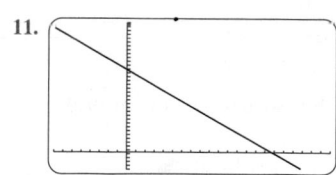

Excel

1.   3.2x + 2.1y - 6.72 = 0

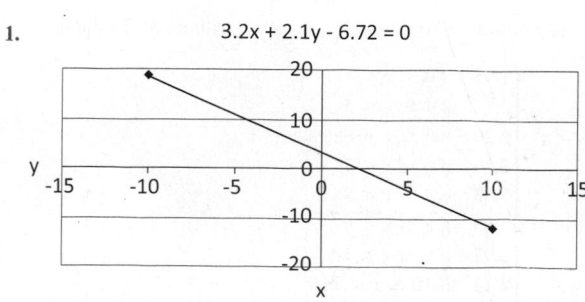

3.   1.6x + 5.1y = 8.16

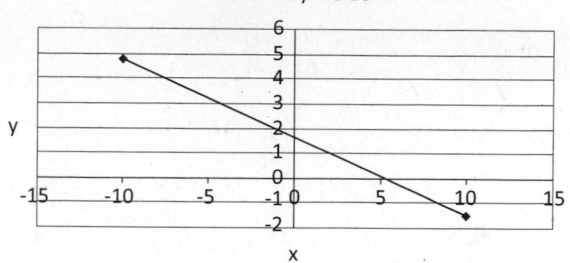

5.   12.1x + 4.1y = 49.61

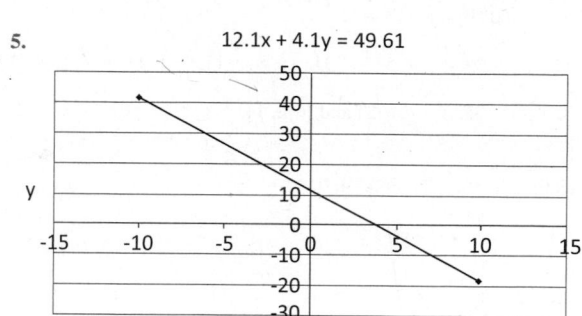

7.   20x + 16y = 300

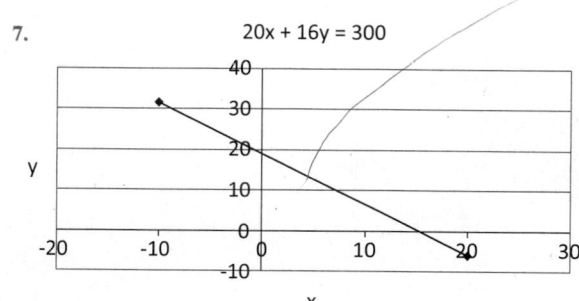

9.   20x + 30y = 600

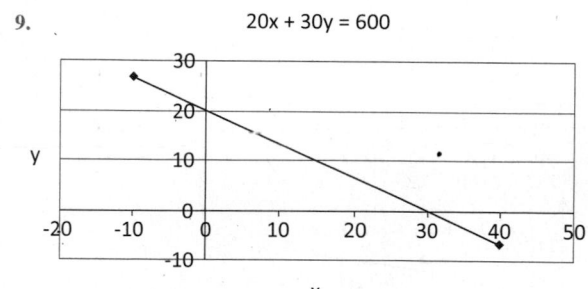

11.   22.4x + 16.1y - 352 = 0

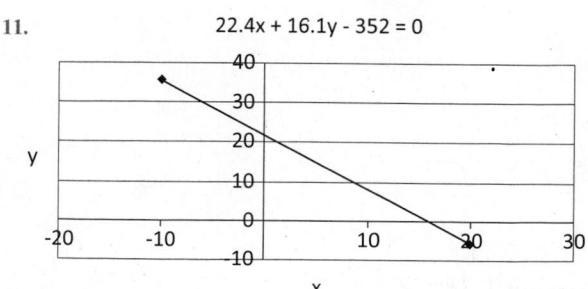

Exercises 2.3, page 100

1. $21, -9, 5a + 6, -5a + 6, 5a + 21$

3. $-3, 6, 3a^2 - 6a - 3, 3a^2 + 6a - 3, 3x^2 - 6$

5. $2a + 2h + 5$, $-2a + 5$, $2a^2 + 5$, $2a - 4h + 5$, $4a - 2h + 5$

7. $\dfrac{8}{15}$, 0, $\dfrac{2a}{a^2 - 1}$, $\dfrac{2(2 + a)}{a^2 + 4a + 3}$, $\dfrac{2(t + 1)}{t(t + 2)}$

9. 8, $\dfrac{2a^2}{\sqrt{a - 1}}$, $\dfrac{2(x + 1)^2}{\sqrt{x}}$, $\dfrac{2(x - 1)^2}{\sqrt{x - 2}}$ **11.** $5, 1, 1$ **13.** $\frac{5}{2}, 3, 3, 9$

15. a. -2 **b.** (i) $x = 2$; (ii) $x = 1$ **c.** $[0, 6]$ **d.** $[-2, 6]$

17. Yes   **19.** Yes   **21.** 7 **23.** $(-\infty, \infty)$

25. $(-\infty, 0)$ and $(0, \infty)$

27. $(-\infty, \infty)$ **29.** $(-\infty, 5]$ **31.** $(-\infty, -1)$, $(-1, 1)$, and $(1, \infty)$

33. $[-3, \infty)$ **35.** $(-\infty, -2)$ and $(-2, 1]$

37. a. $(-\infty, \infty)$
b. $6, 0, -4, -6, -\frac{25}{4}, -6, -4, 0$
c.

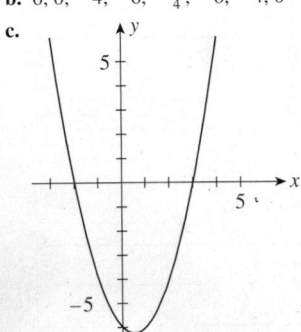

39.

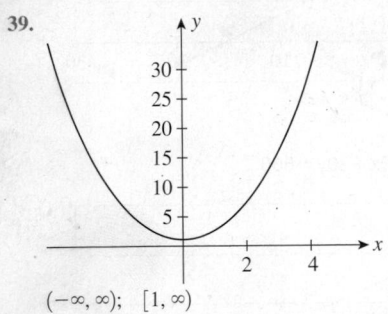

$(-\infty, \infty)$; $[1, \infty)$

41.

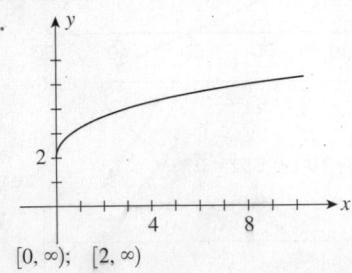

$[0, \infty)$; $[2, \infty)$

43.

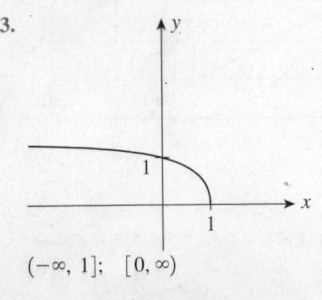

$(-\infty, 1]$; $[0, \infty)$

45.

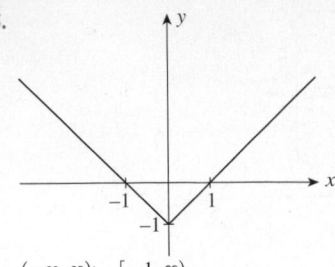

$(-\infty, \infty)$; $[-1, \infty)$

47.

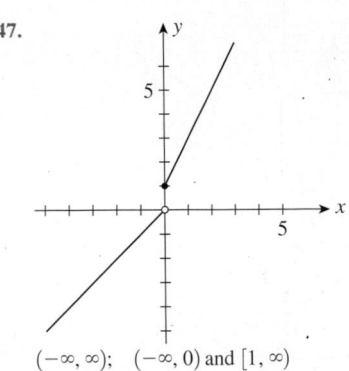

$(-\infty, \infty)$; $(-\infty, 0)$ and $[1, \infty)$

49.

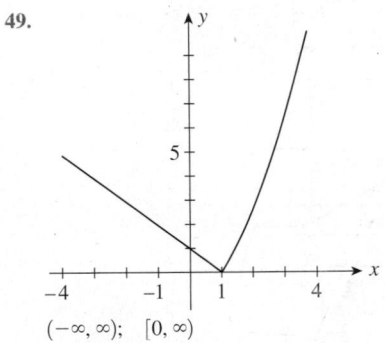

$(-\infty, \infty)$; $[0, \infty)$

51. Yes   **53.** No   **55.** Yes   **57.** Yes

59. 10π in.   **61.** $S(r) = 4\pi r^2$

63. a. $f(t) = \begin{cases} 0.0185t + 0.58 & \text{if } 0 \le t \le 20 \\ 0.015t + 0.65 & \text{if } 20 < t \le 30 \end{cases}$

b. 0.0185/year from 1960 through 1980; 0.015/year from 1980 through 1990

c. 1983

65. 20; 26   **67.** $5.6 billion; $7.8 billion

69. a. $0.6 trillion; $0.6 trillion   **b.** $0.96 trillion; $1.2 trillion

71. $f(x) = \begin{cases} 1.95 & \text{if } 0 < x < 4 \\ 2.12 & \text{if } 4 \le x < 5 \\ 2.29 & \text{if } 5 \le x < 6 \\ 2.46 & \text{if } 6 \le x < 7 \\ 2.63 & \text{if } 7 \le x < 8 \\ 2.80 & \text{if } 8 \le x < 9 \\ 2.97 & \text{if } 9 \le x < 10 \\ 3.14 & \text{if } 10 \le x < 11 \\ 3.31 & \text{if } 11 \le x < 12 \\ 3.48 & \text{if } 12 \le x < 13 \\ 3.65 & \text{if } x = 13 \end{cases}$

a. $(0, 13]$

b.

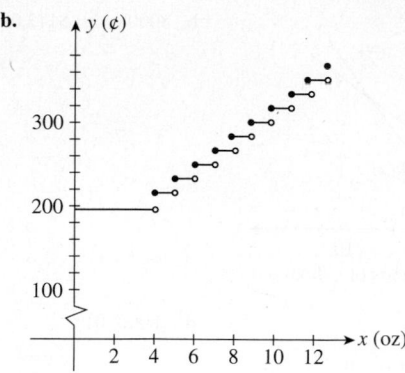

73. False   **75.** False

Using Technology Exercises 2.3, page 106

1. a.

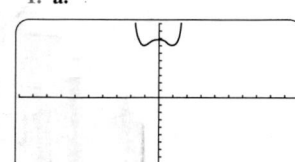

b.

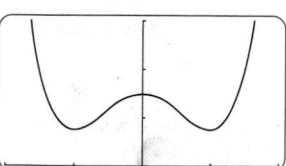

3. a.

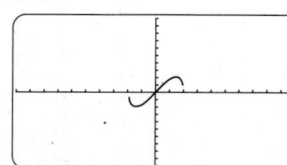

b.

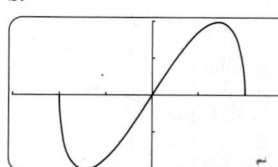

5.

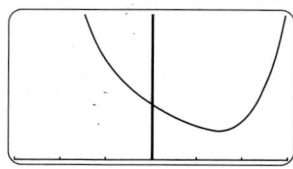

7.

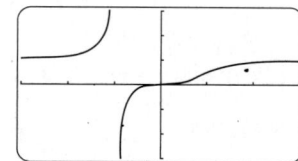

9. 18.5505   **11.** 4.1616

13. a.

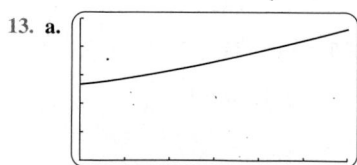

b. \$9.4 billion, \$13.9 billion

15. a.

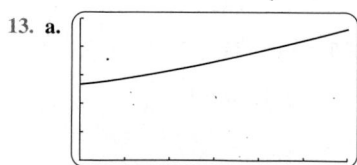

b. 44.7; 52.7; 129.2

Exercises 2.4, page 112

1. $f(x) + g(x) = x^3 + x^2 + 3$

3. $f(x)g(x) = x^5 - 2x^3 + 5x^2 - 10$

5. $\dfrac{f(x)}{g(x)} = \dfrac{x^3 + 5}{x^2 - 2}$

7. $\dfrac{f(x)g(x)}{h(x)} = \dfrac{x^5 - 2x^3 + 5x^2 - 10}{2x + 4}$

9. $f(x) + g(x) = x - 1 + \sqrt{x + 1}$

11. $f(x)g(x) = (x - 1)\sqrt{x + 1}$

13. $\dfrac{g(x)}{h(x)} = \dfrac{\sqrt{x + 1}}{2x^3 - 1}$

15. $\dfrac{f(x)g(x)}{h(x)} = \dfrac{(x - 1)\sqrt{x + 1}}{2x^3 - 1}$

17. $\dfrac{f(x) - h(x)}{g(x)} = \dfrac{x - 2x^3}{\sqrt{x + 1}}$

19. $f(x) + g(x) = x^2 + \sqrt{x} + 3;$

$f(x) - g(x) = x^2 - \sqrt{x} + 7;$

$f(x)g(x) = (x^2 + 5)(\sqrt{x} - 2); \dfrac{f(x)}{g(x)} = \dfrac{x^2 + 5}{\sqrt{x} - 2}$

21. $f(x) + g(x) = \dfrac{(x - 1)\sqrt{x + 3} + 1}{x - 1};$

$f(x) - g(x) = \dfrac{(x - 1)\sqrt{x + 3} - 1}{x - 1};$

$f(x)g(x) = \dfrac{\sqrt{x + 3}}{x - 1}; \dfrac{f(x)}{g(x)} = (x - 1)\sqrt{x + 3}$

23. $f(x) + g(x) = \dfrac{2(x^2 - 2)}{(x - 1)(x - 2)};$

$f(x) - g(x) = \dfrac{-2x}{(x - 1)(x - 2)};$

$f(x)g(x) = \dfrac{(x + 1)(x + 2)}{(x - 1)(x - 2)}; \dfrac{f(x)}{g(x)} = \dfrac{(x + 1)(x - 2)}{(x - 1)(x + 2)}$

25. $f(g(x)) = x^4 + x^2 + 1; g(f(x)) = (x^2 + x + 1)^2$

27. $f(g(x)) = \sqrt{x^2 - 1} + 1; g(f(x)) = x + 2\sqrt{x}$

29. $f(g(x)) = \dfrac{x}{x^2 + 1}; g(f(x)) = \dfrac{x^2 + 1}{x}$ **31.** 49

33. $\dfrac{\sqrt{5}}{5}$ **35.** $f(x) = 2x^3 + x^2 + 1$ and $g(x) = x^5$

37. $f(x) = x^2 - 1$ and $g(x) = \sqrt{x}$

39. $f(x) = x^2 - 1$ and $g(x) = \dfrac{1}{x}$

41. $f(x) = 3x^2 + 2$ and $g(x) = \dfrac{1}{x^{3/2}}$ **43.** $3h$

45. $-h(2a + h)$ **47.** $2a + h$

49. $3a^2 + 3ah + h^2 - 1$.   **51.** $-\dfrac{1}{a(a + h)}$

53. The total revenue in dollars from both restaurants at time t

55. The value in dollars of Nancy's shares of IBM at time t

57. The carbon monoxide pollution from cars in parts per million at time t

59. $C(x) = 0.6x + 12{,}100$

61. $0.0075t^2 + 0.13t + 0.17; D$ gives the difference in year t between the deficit without the rescue package and the deficit with the rescue package.

63. a. 23; In 2002, 23% of reported serious crimes ended in the arrests or in the identification of the suspects.
   b. 18; In 2007, 18% of reported serious crimes ended in the arrests or in the identification of the suspects.

65. a. $C(x) = 0.000001x^3 - 0.01x^2 + 50x + 20{,}000$
   b. $P(x) = -0.000001x^3 - 0.01x^2 + 100x - 20{,}000$
   c. \$132,000

67. a. $17.6111t^3 - 180.357t^2 + 1132.39t + 9871$
   b. \$3862.98; \$10,113.49; \$13,976.46   **c.** \$13,976.46

69. a. 55%; 98.2%   **b.** \$444,700; \$1,167,600

71. a. $s(x) = f(x) + g(x) + h(x)$

73. True   **75.** False   **77.** True

Exercises 2.5, page 124

1. Yes; $y = -\frac{2}{3}x + 2$ **3.** Yes; $y = \frac{1}{2}x + 2$

5. Yes; $y = \frac{1}{2}x + \frac{9}{4}$ **7.** No   **9.** No

11. a. $C(x) = 8x + 40{,}000$ **b.** $R(x) = 12x$
   c. $P(x) = 4x - 40{,}000$
   d. Loss of \$8000; profit of \$8000

13. $m = -1; b = 2$ **15.** $(2, 10)$

17. $\left(4, \frac{2}{3}\right)$ **19.** $(-4, -6)$

21. 1000 units; \$15,000   **23.** 600 units; \$240

25. \$900,000; \$800,000

27. a. $y = 1.033x$ **b.** \$1260.26

29. $C(x) = 0.6x + 12{,}100; R(x) = 1.15x;$
   $P(x) = 0.55x - 12{,}100$

31. a. \$12,000/year   **b.** $V = 60{,}000 - 12{,}000t$

c.

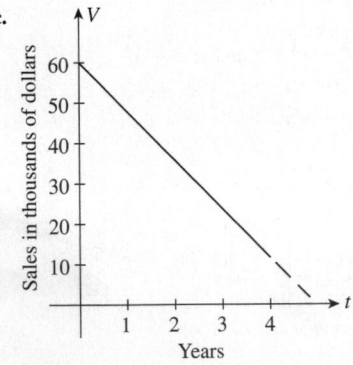

   d. \$24,000

33. \$900,000; \$800,000

35. a. $m = a/1.7; b = 0$ **b.** 117.65 mg

37. a. $f(t) = -0.72t + 17.5$ **b.** 8.14%

39. a. 0.3 year/year   **b.** 41.2 years   **c.** 42.4 years

41. $f(t) = -2.5t + 61$

43. a. $F = \frac{9}{5}C + 32$ **b.** 68°F   **c.** 21.1°C

45. a.

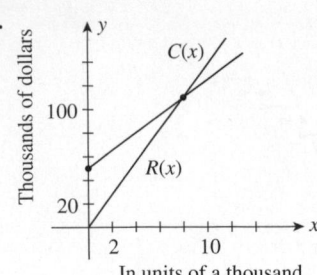

   In units of a thousand

   b. 8000 units; \$112,000

c.

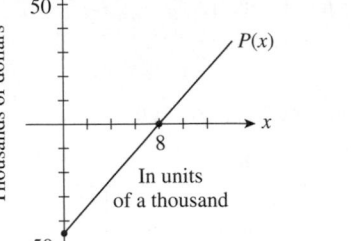

   In units of a thousand

   d. $(8000, 0)$

47. 9259 units; \$83,331

49. a and b.

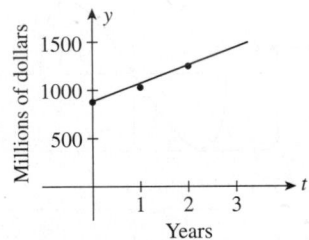

   c. $f(t) = 182t + 887$ **d.** \$1797 million   **e.** \$182 million/year

51. In $5\frac{1}{2}$ years

53. b. End of 2003

55. True

Using Technology Exercises 2.5, page 131

1. 2.2875   **3.** 2.880952381   **5.** 7.2851648352

7. 2.4680851064

Exercises 2.6, page 137

1. Vertex: $\left(-\frac{1}{2}, -\frac{25}{4}\right)$; x-intercepts: $-3, 2$

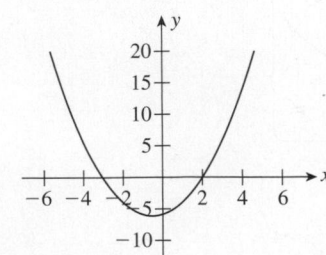

3. Vertex: $(2, 0)$; x-intercept: 2

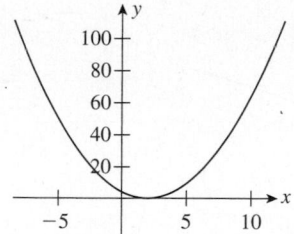

5. Vertex: $\left(\frac{5}{2}, \frac{1}{4}\right)$; x-intercepts: 2, 3

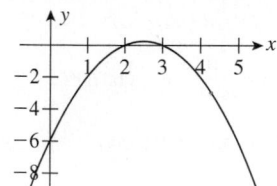

7. Vertex: $\left(\frac{5}{6}, -\frac{13}{12}\right)$; x-intercepts: 0.2324, 1.4343

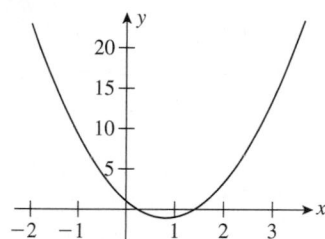

9. Vertex: $\left(\frac{3}{4}, \frac{15}{8}\right)$; no x-intercepts

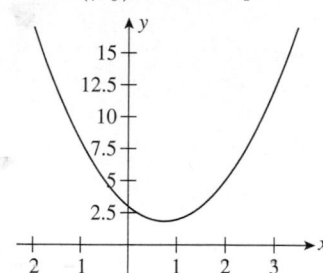

11. Vertex: $(0, -4)$; x-intercepts: ± 2

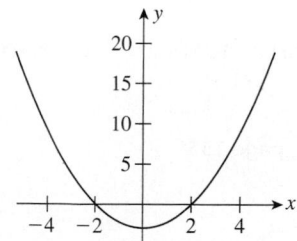

13. Vertex: $(0, 16)$; x-intercepts: ± 4

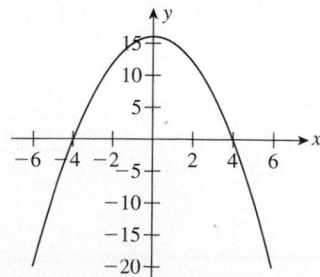

15. Vertex: $\left(\frac{8}{3}, -\frac{2}{3}\right)$; x-intercepts: $\frac{4}{3}$, 4

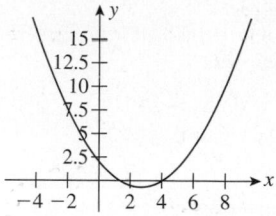

17. Vertex: $\left(-\frac{4}{3}, -\frac{10}{3}\right)$; x-intercepts: $\frac{1}{3}$, -3

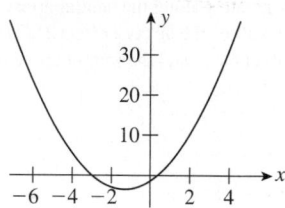

19. a. positive   **b.** positive   **c.** positive   **d.** negative

21. a. positive   **b.** negative   **c.** negative   **d.** positive

23. $(2, 0)$; $(-3, -5)$ **25.** $(-2, -2)$; $(3, 3)$

27. $(-1.1205, 0.1133)$, $(2.3205, -8.8332)$

29. ± 8 **31.** -2 **33.** $\frac{9}{4}$ **35.** $b \le -2\sqrt{10}$ or $b \ge 2\sqrt{10}$

37. a.

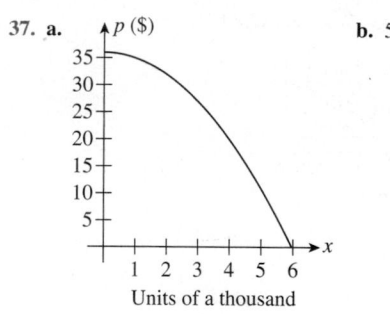

Units of a thousand

b. 5000

39. a.

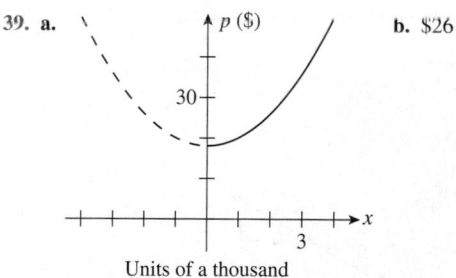

Units of a thousand

b. $26

41. 2500; $67.50   **43.** 11,000; $3

45. a. 3.6 million; 9.5 million   **b.** 11.2 million   **47.** 3000

49. a.

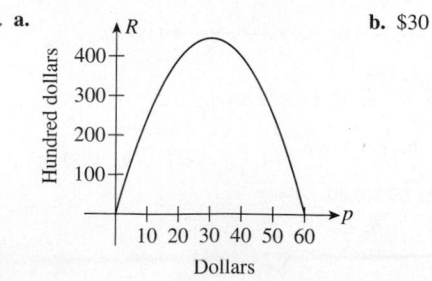

Dollars

b. $30

51. a. $250 billion **b.** $1.366 trillion

53. b. 2012 **55.** 500; $32.50

57. a.

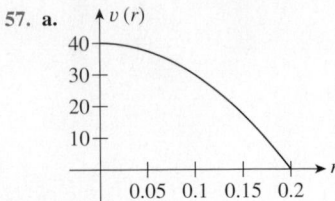

b. 0, 0.2. The velocity of blood is greatest along the central artery ($r = 0$) and smallest along the wall of the artery ($r = 0.2$). The maximum velocity is $v(0) = 40$ cm/sec, and the minimum velocity is $v(0.2) = 0$ cm/sec.

59. $\dfrac{28}{\pi + 4}$ ft by $\dfrac{28}{\pi + 4}$ ft **61.** True

63. True **65.** True

Using Technology Exercises 2.6, page 142

1. $(-3.0414, 0.1503); (3.0414, 7.4497)$

3. $(-2.3371, 2.4117); (6.0514, -2.5015)$

5. $(-1.1055, -6.5216); (1.1055, -1.8784)$

7. a.

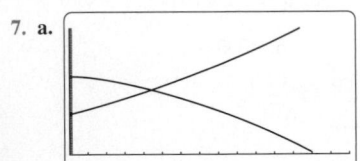

b. 438 wall clocks; $40.92

Exercises 2.7, page 149

1. Polynomial function; degree 6

3. Polynomial function; degree 6

5. Some other function

7. a.

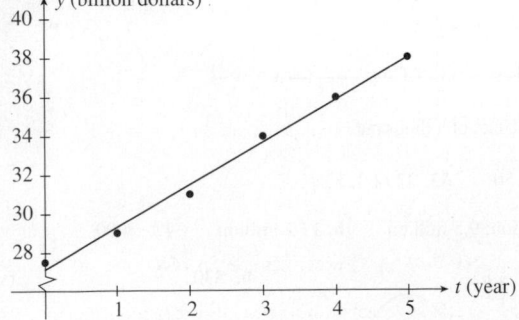

b. $40.26 billion **c.** $2.19 billion/year

9. a. 21.76 min **b.** 111.4 min

11. a. $789.45 **b.** $2740.80

13. 11%; 36%

15. a. 0.06 million terabytes **b.** 3.66 million terabytes

17. a.

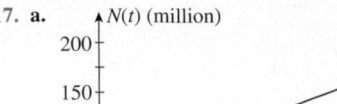
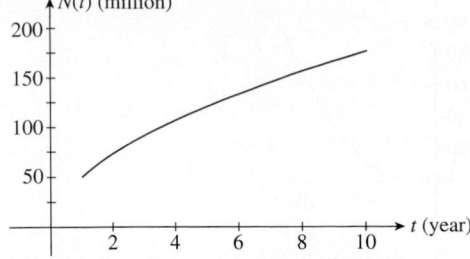

b. 177,000,000

19. 582,650; 2,532,700 **21. a.** $R(x) = \dfrac{100x}{40 + x}$ **b.** 60%

23. a. 344; 4896 **b.** $6616 **25. a.** 43.25; 41; 1547 **b.** 3350

27. $5 billion; $152 billion

29. a.

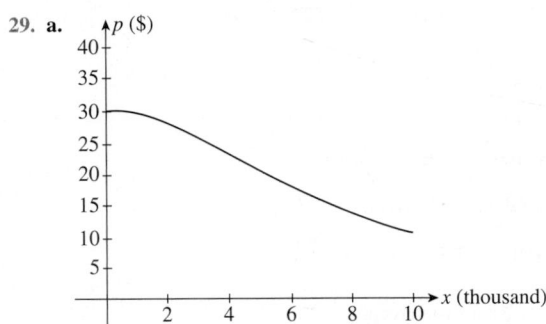

b. $10

31. $p = \sqrt{-x^2 + 100}$; 6614

33. $f(x) = 40x - x^2$; $[0, 40]$

35. $f(x) = (15 - 2x)(8 - 2x)x$

37. $f(x) = 28x - \left(\dfrac{\pi}{2} + 2\right)x^2$

39. $f(x) = -2x + 52 - \dfrac{50}{x}$; $\left[1, \frac{25}{2}\right]$

41. a. $R(x) = -4x^2 + 520x + 12,000$ **b.** $26,400 **c.** $28,800

43. False **45.** False

Using Technology Exercises 2.7, page 155

1. a. $f(t) = 1.85t + 16.9$

b.

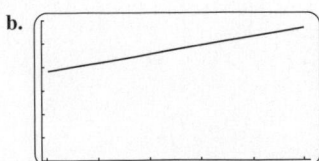

c.

t	1	2	3	4	5	6
y	18.8	20.6	22.5	24.3	26.2	28.0

d. 31.7 gal

3. a. $f(t) = -0.221t^2 + 4.14t + 64.8$

b.

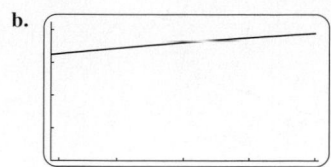

c. 77.8 million

5. a. $f(t) = 2.4t^2 + 15t + 31.4$

b.

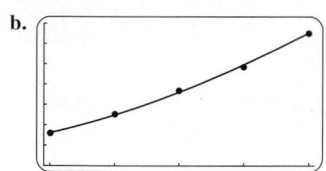

7. a. $f(t) = -0.00081t^3 + 0.0206t^2 + 0.125t + 1.69$

b.

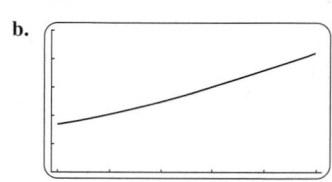

c. $1.8 trillion; $2.7 trillion; $4.2 trillion

9. a. $-0.0056t^3 + 0.112t^2 + 0.51t + 8$

b.

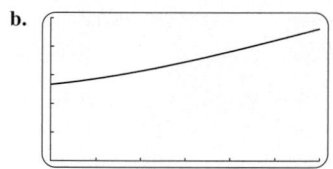

c. $8 billion, $10.4 billion, $13.9 billion

11. a. $f(t) = 0.00125t^4 - 0.0051t^3 - 0.0243t^2 + 0.129t + 1.71$

b.

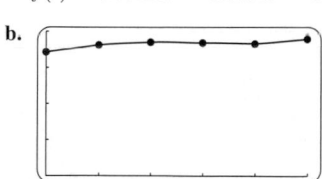

c. 1.71 mg; 1.81 mg; 1.85 mg; 1.84 mg; 1.83 mg; 1.89 mg
d. 2.13 mg/cigarette

Exercises 2.8, page 162

1. a. $y = 2.3x + 1.5$
b.

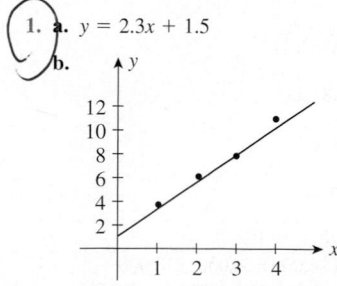

3. a. $y = -0.77x + 5.74$

b.

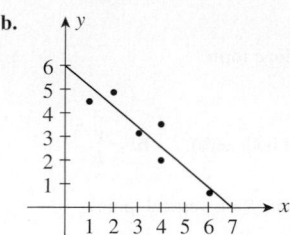

5. a. $y = 1.2x + 2$

b.

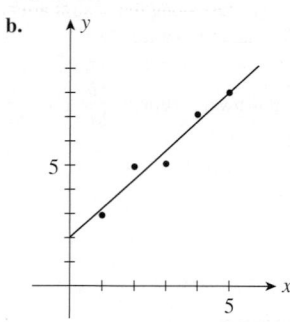

7. a. $y = -2.8x + 440$

b.   **c.** 420

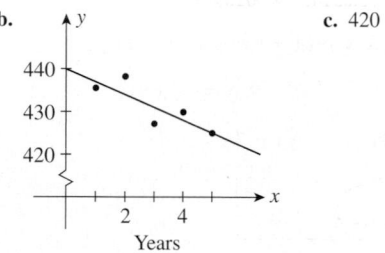

9. a. $y = 234.4x + 157.3$ **b.** 1798.1 million

11. a. $y = 8.52x + 24.52$ **b.** 8.52 million/year

13. a. $y = 1.69x + 24.11$ **b.** $37.6 billion

15. a. $y = 2.46x + 82.1$ **b.** 94.4 million

17. a. $y = 0.3x + 6.46$ **b.** $0.3 billion/year

19. a. $y = 7.25x + 60.21$ **b.** $139.96 billion   **c.** $7.25 billion/year

21. False   **23.** True

Using Technology Exercises 2.8, page 166

1. $y = 2.3596x + 3.8639$

3. $y = -1.1948x + 3.5525$

5. a. $y = -2.5t + 61.2$ **b.** 48.7%

Chapter 2 Concept Review Questions, page 168

1. ordered; abscissa (x-coordinate); ordinate (y-coordinate)

2. a. x; y **b.** third

3. a. $\dfrac{y_2 - y_1}{x_2 - x_1}$ **b.** undefined   **c.** 0   **d.** positive

4. $m_1 = m_2$; $m_1 = -\dfrac{1}{m_2}$

5. a. $y - y_1 = m(x - x_1)$; point-slope form
b. $y = mx + b$; slope-intercept

6. a. $Ax + By + C = 0$ $(A, B,$ not both zero) **b.** $-\dfrac{a}{b}$

7. domain; range; B **8.** domain, $f(x)$; vertical, point

9. $f(x) \pm g(x)$; $f(x)g(x)$; $\dfrac{f(x)}{g(x)}$; $A \cap B$; $A \cap B$; 0

10. $g[f(x)]$; f; $f(x)$; g

11. $ax^2 + bx + c$; parabola; upward; downward; vertex; $\dfrac{-b}{2a}$; $x = \dfrac{-b}{2a}$

12. a. $P(x) = a_n x^n + a_{n-1} x^{n-1} + \cdots + a_1 x + a_0$
    $(a_n \neq 0$; n, a positive integer$)$
b. linear; quadratic
c. quotient; polynomials
d. x^r $(r,$ a real number$)$

Chapter 2 Review Exercises, page 168

1. $x = -2$ **2.** $y = 4$ **3.** $y = -\frac{1}{10}x + \frac{19}{5}$

4. $y = -\frac{4}{5}x + \frac{12}{5}$ **5.** $y = \frac{5}{2}x + 9$ **6.** $y = \frac{3}{4}x + \frac{11}{2}$

7. $y = -\frac{1}{2}x - 3$ **8.** $\frac{3}{5}$; $-\frac{6}{5}$ **9.** $y = -\frac{3}{4}x + \frac{9}{2}$

10. $y = -\frac{3}{5}x + \frac{12}{5}$ **11.** $y = -\frac{3}{2}x - 7$

12.

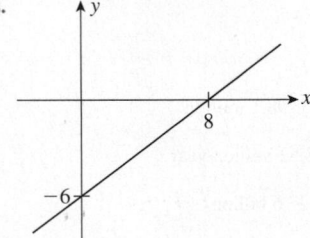

13.

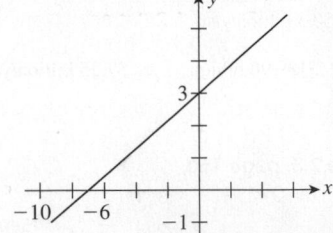

14. $(-\infty, 9]$ **15.** $(-\infty, -1)$ and $\left(-1, \frac{3}{2}\right)$ and $\left(\frac{3}{2}, \infty\right)$

16. a. 0   **b.** $3a^2 + 17a + 20$ **c.** $12a^2 + 10a - 2$
d. $3a^2 + 6ah + 3h^2 + 5a + 5h - 2$

17. a. From 1985 to 1990
b. From 1990 on
c. 1990; $3.5 billion

18. a.

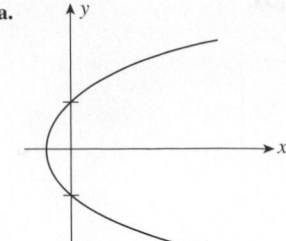

b. No   **c.** Yes

19.

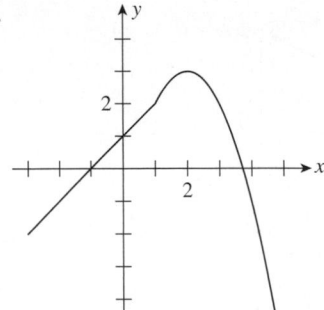

20. a. $\dfrac{2x + 3}{x}$ **b.** $\dfrac{1}{x(2x + 3)}$

c. $\dfrac{1}{2x + 3}$ **d.** $\dfrac{2}{x} + 3$

21. Vertex: $\left(\frac{11}{12}, -\frac{361}{24}\right)$; x-intercepts: $-\frac{2}{3}, \frac{5}{2}$

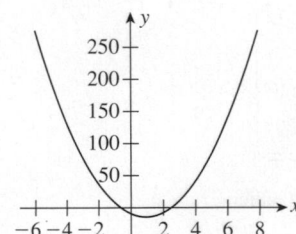

22. Vertex: $\left(\frac{1}{2}, 4\right)$; x-intercepts: $-\frac{1}{2}, \frac{3}{2}$

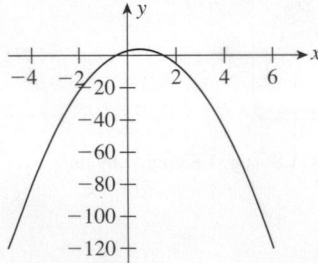

23. $(2, -3)$ **24.** $\left(6, \frac{21}{2}\right)$ **25.** $\left(-2, \frac{1}{3}\right)$

26. $(2500, 50{,}000)$ **27.** L_2 **28.** L_2

29. $6 billion; $43.5 billion; $81 billion

30. a. $f(x) = x + 2.4$ **b.** $5.4 million

31. a. $C(x) = 6x + 30{,}000$ **b.** $R(x) = 10x$
c. $P(x) = 4x - 30{,}000$ **d.** $(\$6{,}000)$; $2000; $18,000

32. a. $200,000/year   **b.** $4,000,000

33. $p = -0.05x + 200$

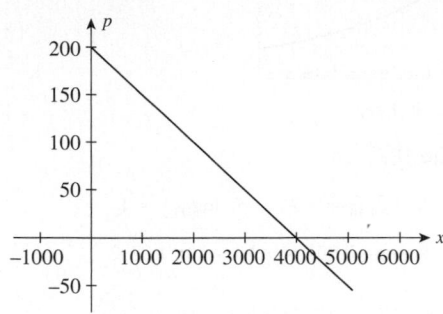

34. $p = \frac{1}{36}x + \frac{400}{9}$ **35.** 117 mg   **36.** $45,000

37. 4.6 million; 14.89 million   **38.** 16.8 years; 10.18 years

39. 20.1 years; 13 years   **40.** 400; 800   **41.** 990; 2240

42.

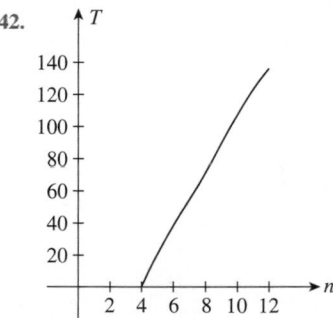

As the length of the list increases, the time taken to learn the list increases by a very large amount.

43. a. $f(t) = 267; g(t) = 2t^2 + 46t + 733$

b. $f(t) + g(t) = 2t^2 + 46t + 1000$ **c.** 1936 tons

44. 5000; $20

45. a. $r = f(V) = \sqrt[3]{(3V)/(4\pi)}$ **b.** $g(t) = \frac{9}{2}\pi t$

c. $h(t) = \frac{3}{2}\sqrt[3]{t}$ **d.** 3 ft

46. a. 59.8%; 58.9%; 59.2%; 60.7%, 61.7%

b.

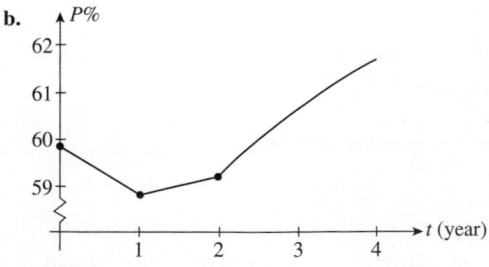

c. 60.7%

47. $x(20 - 2x)^2$

48. $100x^2 + \dfrac{1350}{x}$

Chapter 2 Before Moving On, page 171

1. $y = \frac{7}{5}x - \frac{3}{5}$

2. $y = -\frac{1}{3}x + \frac{4}{3}$

3. a. 3   **b.** 2   **c.** $\frac{17}{4}$

4. a. $\dfrac{1}{x + 1} + x^2 + 1$

b. $\dfrac{x^2 + 1}{x + 1}$ **c.** $\dfrac{1}{x^2 + 2}$

d. $\dfrac{1}{(x + 1)^2} + 1$

5. $V(x) = 108x^2 - 4x^3$

CHAPTER 3

Exercises 3.1, page 177

1. 3   **3.** 3   **5.** 3   **7.** $\frac{5}{4}$ **9.** 1 or 2

11.

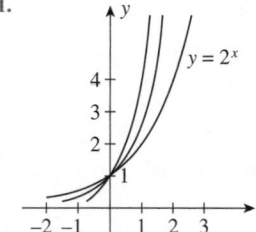

13.

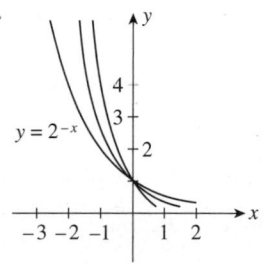

15.

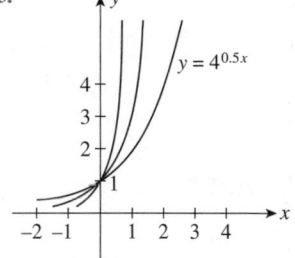

17.

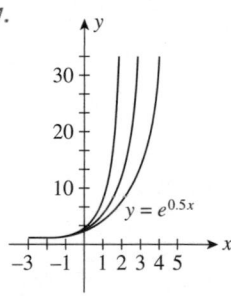

19.

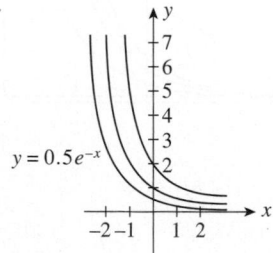

21. $f(x) = 100\left(\frac{6}{5}\right)^x$ **23.** 54.6

25. a.

t	0	1	2	3
$f(t)$	64.0	77.2	93.2	112.5

b.

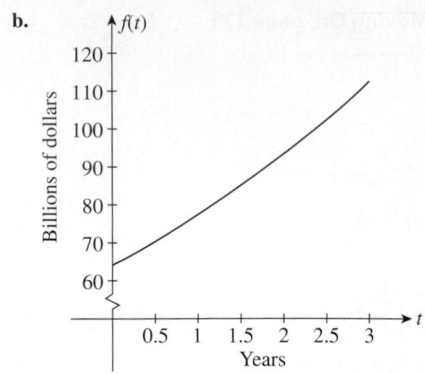

27. a. 115,423,000; 140,977,00; 313,751,000

b.

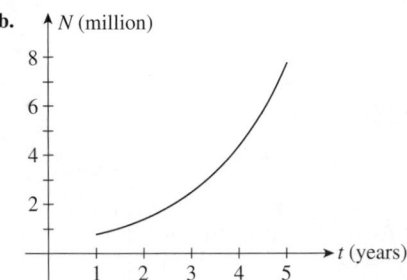

29. 34,210,000

31. a. 0 g/cm^3 **b.** 0.2367 g/cm^3 **c.** 0.7598 g/cm^3

33. False **35.** True

Using Technology Exercises 3.1, page 180

1.

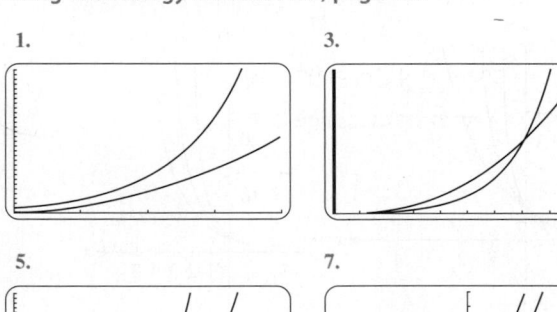

3.

5.

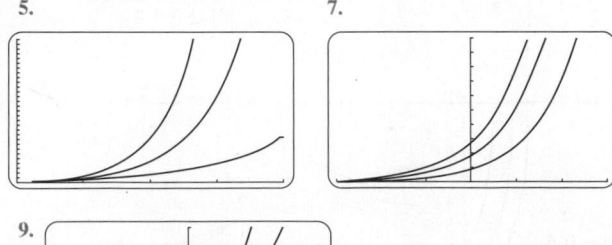

7.

9.

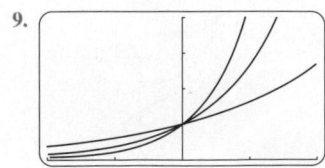

11. a.

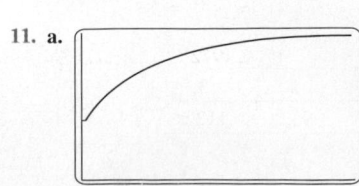

b. 0.08 g/cm^3 **c.** 0.12 g/cm^3

13. a.

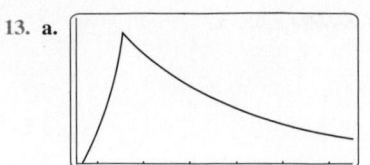

b. 20 sec **c.** 35.1 sec

Exercises 3.2, page 187

1. $\log_2 64 = 6$ **3.** $\log_4 \frac{1}{16} = -2$ **5.** $\log_{1/3} \frac{1}{3} = 1$

7. $\log_{32} 16 = \frac{4}{5}$ **9.** $\log_{10} 0.001 = -3$ **11.** 1.0792

13. 1.2042 **15.** 1.6813 **17.** $\ln a^2 b^3$ **19.** $\ln \frac{3\sqrt{xy}}{\sqrt[3]{z}}$

21. $\log x + 4\log(x + 1)$ **23.** $\frac{1}{2}\log(x + 1) - \log(x^2 + 1)$

25. $\ln x - x^2$ **27.** $-\frac{3}{2}\ln x - \frac{1}{2}\ln(1 + x^2)$

29. $x = 8$ **31.** $x = 3$ **33.** $x = 10$ **35.** $x = \frac{11}{2}$

37. $x = \frac{16}{7}$ **39.** $x = \frac{11}{3}$ **41.** $x = 3$

43.

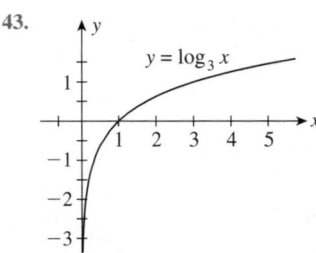

45.

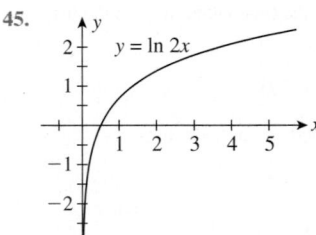

47.

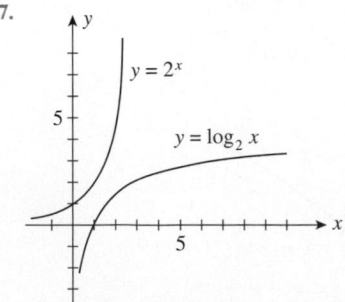

49. 5.1986 **51.** −0.0912 **53.** −8.0472 **55.** −4.9041

57. $-2\ln\left(\dfrac{A}{B}\right)$ **59.** $f(x) = 2 + 2.8854\ln x$ **61.** 106 mm

63. a. $10^3 I_0$ **b.** 100,000 times greater
c. 10,000,000 times greater

65. a. 170°F **b.** 15.54 min **67.** 2022

69. a. 9.1 sec   **b.** 20.3 sec   **71.** 1:30 P.M.

73. False   **75.** False   **77. a.** ln 2

Exercises 3.3, page 196

1. a. 0.02   **b.** 300

c.

t	0	10	20	100	1000
Q	300	366	448	2217	1.46×10^{11}

3. a. $Q(t) = 100e^{0.035t}$ **b.** 266 min   **c.** $Q(t) = 1000e^{0.035t}$

5. a. 54.93 years   **b.** 14.25 billion

7. 8.7 lb/in.^{2} **9.** $Q(t) = 100e^{-0.049t}$; 70.7 g

11. 13,412 years ago   **13.** 30

15. a. $S(t) = 100e^{0.081t}$ **b.** $127.5 million

17.

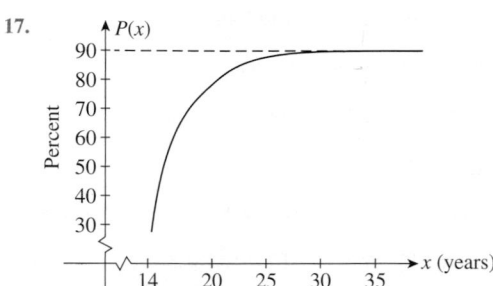

a. 27.8%   **b.** 75.8%   **c.** 90.0%

19.

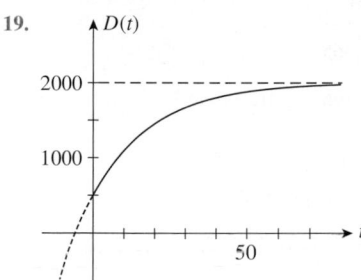

a. 573 computers/month; 1177 computers/month; 1548 computers/month; 1925 computers/month
b. 2000 computers/month

21. 122 cm   **23. a.** 10   **b.** 154

25. 325 million   **27. b.** 5599 years

29. 0.14

Using Technology Exercises 3.3, page 200

1. a.

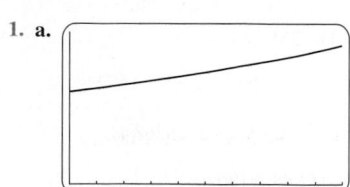

b. 666 million, 926.8 million

3.

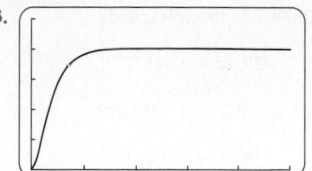

5.

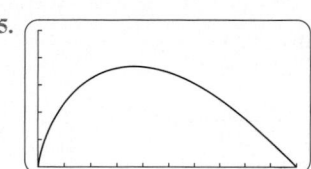

7. a.

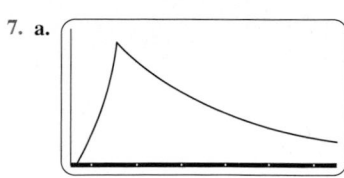

b. 0   **c.** 0.237 g/cm^3
d. 0.760 g/cm^3

Chapter 3 Concept Review, page 201

1. power; 0; 1; exponential

2. a. $(-\infty, \infty)$; $(0, \infty)$ **b.** $(0, 1)$; left; right

3. a. $(0, \infty)$; $(-\infty, \infty)$; $(1, 0)$ **b.** falls; rises

4. a. x **b.** x

5. a. initially; growth   **b.** decay   **c.** time; one half

6. a. learning curve; C
b. logistic growth model; A, carrying capacity

Chapter 3 Review Exercises, page 202

1.

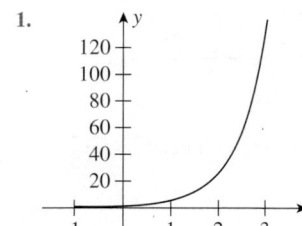

2.

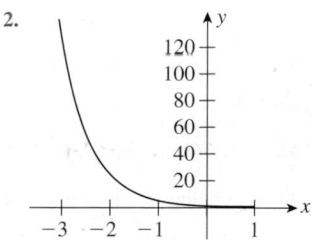

3.

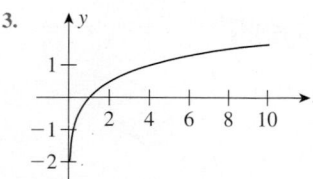

4.

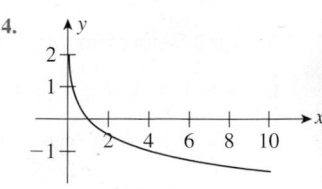

5. $\log_3 81 = 4$ **6.** $\log_9 3 = \frac{1}{2}$ **7.** $\log_{2/3} \frac{27}{8} = -3$

8. $\log_{16} 0.125 = -\frac{3}{4}$ **9.** 3.4011   **10.** 2.1972   **11.** 1.2809

12. 4.3174   **13.** $x + y + z$ **14.** $x + 2y - z$ **15.** $y + 2z$

16. $x = 3$ **17.** $x = -2, 1$ **18.** $x = -5$ **19.** $x = -2, 3$

20. $x = \frac{15}{2}$ **21.** $x = 2$ **22.** $x \approx 1.1610$ **23.** $x \approx -0.9464$

24. $x \approx -2.5025$ **25.** $x \approx -1.2528$ **26.** $x \approx 2.8332$

27. $x \approx 1.8195$ **28.** $x \approx 0.2409$ **29.** $x \approx 33.8672$

30. $x \approx \pm 1.8934$ **31.** $x \approx 2.5903$ **32.** $x = -9.1629$

33. $x \approx 8.9588$ **34.** $x \approx 3.4657$ **35.** $x \approx -9.1629$

36.

$y = \log_2(x + 3)$

37.

$y = \log_3(x + 1)$

38. a. $Q(t) = 2000e^{0.01831t}$ **b.** 162,000 **39.** $k \approx 0.0004$

40.

$D(t)$ **a.** 1175; 2540; 3289 **b.** 4000

41. 970

42. 12.5/1000 live births; 9.3/1000 live births; 6.9/1000 live births

43. a. 3 billion **b.** 9 billion

44. a. 0 g/cm³ **b.** 0.0361 g/cm³

Chapter 3 Before Moving On, page 203

1. $x = \ln 3$ **2.** $x = 0$ or 8

3. -0.9589 **4.** 8.7 min

CHAPTER 4

Exercises 4.1, page 217

1. \$80; \$580 **3.** \$836 **5.** \$1000 **7.** 292 days

9. 10%/year **11.** \$1368.57 **13.** \$3714.87

15. \$20,219.60 **17.** \$175,979.80 **19.** \$196,488.13

21. 6.09%/year **23.** 4.07%/year **25.** \$34,139.61

27. \$35,482.13 **29.** \$6356.25 **31.** 4.4 years **33.** 13.9 years

35. 6.08%/year **37.** 2.8 years **39.** 24%/year

41. \$123,000 **43.** 5%/year **45.** 2.39%/year

47. No. It is less than its original value. **49.** 11.8%/year

51. \$999.14/day **53.** \$316,032 **55.** \$1.92 million

57. \$20,471.64 **59.** \$33,603.85

61. a. \$56,427.16 **b.** \$56,073 **c.** \$55,892.84

63. Acme Mutual Fund **65.** \$23,227.22 **67.** \$5994.86

69. \$339.79 billion **71.** Investment *A*

73. \$75,888.25 \$75,602.68 **75.** $80,000e^{(\sqrt{t}/2 - 0.05t)}$; \$178,043.27

79. 3.37% **81.** 5.83% **83.** True **85.** True

Using Technology Exercises 4.1, page 224

1. \$5872.78 **3.** \$475.49 **5.** 8.95%/year

7. 10.20%/year **9.** \$29,743.30 **11.** \$53,303.25

Exercises 4.2, page 231

1. \$12,577.89 **3.** \$17,213.24 **5.** \$27,069.30 **7.** \$100,289.96

9. \$31,048.97 **11.** \$12,309.32 **13.** \$18,181.37

15. \$62,468.86 **17.** \$445.54 **19.** Karen **21.** \$753,031.24

23. \$5643.08 **25.** \$21,805.83

27. Between \$462,692 and \$568,365

29. Between \$343,493 and \$419,366 **31.** \$16,883

33. \$61,464.19 **35.** False

Using Technology Exercises 4.2, page 235

1. \$59,622.15 **3.** \$8453.59 **5.** \$35,607.23 **7.** \$13,828.60

9. a.

b. \$35,038.78/year

Exercises 4.3, page 244

1. \$13,586.80 **3.** \$444.24 **5.** \$622.13

7. \$454.23 **9.** \$1491.19 **11.** \$516.76

13. \$333.85 **15.** \$2216.02 **17.** \$3033.55

19. \$12,950.46 **21. a.** \$243.70 **b.** \$805.23; \$268.41

23. a. \$456.33; \$352.28 **b.** \$1427.88; \$1909.44

25. \$1438.92; \$46,669.74; \$69,154.44; \$140,391.51 **27.** \$83,954.38

29. \$3135.48 **31.** \$452.08 **33.** \$447.98

35. $1014.94; $1947.04   **37.** $1,111.63   **39.** $25,645.51

41. $152,670.69   **43.** $194.65   **45.** $1418.72; $243,673.79

47. a. $1264.14   **b.** $190,119.14   **c.** $991.75   **d.** $272.39

49. $10,413.60   **51.** $725.43   **53.** $29,658   **55.** Yes

57. a. $599,088.30   **b.** $779,220.78

Using Technology Exercises 4.3, page 251

1. $628.02   **3.** $1379.28   **5.** $1988.41

7. $894.12   **9.** $15,165.46

Exercises 4.4, page 257

1. 30   **3.** $-\frac{9}{2}$ **5.** $-3, 8, 19, 30, 41$ **7.** $x + 6y$

9. 795   **11.** 792   **13.** 671

15. a. 275   **b.** -280 **17.** At the beginning of the 37th week

19. $15.80   **21. b.** $800   **23.** GP; 256; 508   **25.** Not a GP

27. GP; $\frac{1}{3}$; $364\frac{1}{3}$ **29.** 3; 0   **31.** 293,866

33. $56,284   **35.** Annual raise of 8%/year

37. a. $20,113.57   **b.** $87,537.38   **39.** $25,165.82

41. $39,321.60; $110,678.40   **43.** True

Chapter 4 Concept Review Questions, page 260

1. a. original; $P(1 + rt)$ **b.** interest; $P(1 + i)^n$; $A(1 + i)^{-n}$

2. simple; one; nominal; m; $\left(1 + \dfrac{r}{m}\right)^m - 1$

3. annuity; ordinary annuity; simple annuity.

4. a. $R\left[\dfrac{(1 + i)^n - 1}{i}\right]$ **b.** $R\left[\dfrac{1 - (1 + i)^{-n}}{i}\right]$

5. $\dfrac{Pi}{1 - (1 + i)^{-n}}$ **6.** future; $\dfrac{iS}{(1 + i)^n - 1}$

7. constant d; $a + (n - 1)d$; $\dfrac{n}{2}[2a + (n - 1)d]$

8. constant r; ar^{n-1}; $\dfrac{a(1 - r^n)}{1 - r}$

Chapter 4 Review Exercises, page 261

1. a. $6077.53   **b.** $6092.01   **c.** $6099.45   **d.** $6104.48

2. a. $15,801.71   **b.** $15,839.15   **c.** $15,858.23   **d.** $15,871.09

3. a. 6%   **b.** 6.09%   **c.** 6.1363%   **d.** 6.168%

4. a. 5.5%   **b.** 5.576%   **c.** 5.6145%   **d.** 5.6408%

5. $33,110.52   **6.** $50,789.23   **7.** $4991.91

8. $18,143.77   **9.** $8404.23   **10.** $232,624.14   **11.** $644.65

12. $297.92   **13.** $353.42   **14.** $228.55   **15.** 3.660%

16. 4.907%   **17.** $97,712.24   **18.** $2,143,825.25; $7,769,900.25

19. $5,491,922   **20.** $2346.91   **21.** $16,904.04   **22.** $5000

23. 7.6%   **24.** $278.88   **25.** $62,112.91   **26.** $13,026.89

27. $2099.44   **28. a.** $608.02   **b.** $98,887.20   **c.** $53,893.05

29. a. $917.99   **b.** $45,238.20   **c.** $100,760   **30.** $22,160.19

31. $4727.67   **32.** $191.15; 9.707%/year   **33.** $2203.83

Chapter 4 Before Moving On, page 263

1. $2540.47   **2.** 6.18%/year   **3.** $569,565.47   **4.** $1213.28

5. $35.13   **6. a.** 210   **b.** 127.5

CHAPTER 5

Exercises 5.1, page 271

1. Unique solution; $(2, 1)$ **3.** No solution

5. Unique solution; $(3, 2)$

7. Infinitely many solutions; $\left(t, \frac{2}{5}t - 2\right)$; t, a parameter

9. Unique solution; $(1, -2)$

11. No solution   **13.** Unique solution; $\left(\frac{1}{2}, \frac{1}{2}\right)$

15. Infinitely many solutions; $\left(2t + \frac{2}{3}, t\right)$; t, a parameter

17. $(-1, 2)$ **19.** $k = -2$

21. $a \neq 0$ and $b \neq 0$; $\left(\dfrac{c + d}{2a}, \dfrac{d - c}{2b}\right)$

23. $\begin{aligned} x + \quad y &= 500 \\ 42x + 30y &= 18{,}600 \end{aligned}$ **25.** $\begin{aligned} x + y &= 100 \\ 8x + 9y &= 860 \end{aligned}$

27. $\begin{aligned} x + \quad y &= 1000 \\ 0.5x + 1.5y &= 1300 \end{aligned}$ **29.** $\begin{aligned} x + y &= 110 \\ y - x &= 100 \end{aligned}$

31. $\begin{aligned} 0.06x + 0.08y + 0.12z - 21{,}600 \\ z = 2x \\ 0.12z = 0.08y \end{aligned}$

33. $\begin{aligned} x + y + z &= 100 \\ x + y \quad\;\; &= 67 \\ x \quad\;\; - z &= 17 \end{aligned}$

35. $\begin{aligned} 18x + 20y + 24z &= 26{,}400 \\ 4x + \;4y + \;3z &= 4{,}900 \\ 5x + \;4y + \;6z &= 6{,}200 \end{aligned}$

37. $\begin{aligned} 18{,}000x + 27{,}000y + 36{,}000z &= 2{,}250{,}000 \\ x \quad\quad\quad\quad\quad &= 2y \\ x + \quad\; y + \quad\; z &= 100 \end{aligned}$

39. $\begin{aligned} 10x + \;6y + \;8z &= 100 \\ 10x + 12y + \;6z &= 100 \\ 5x + \;4y + 12z &= 100 \end{aligned}$

41. $\begin{aligned} x + \quad y + \quad z &= 1{,}000 \\ 80x + 60y + 50z &= 62{,}800 \\ x + \quad y - 2z &= 400 \end{aligned}$ **43.** $\begin{aligned} 280x + 330y + 260z &= 4060 \\ 130x + 140y + 110z &= 1800 \\ x - \quad y - \quad z &= 0 \end{aligned}$

45. True   **47.** False

Exercises 5.2, page 285

1. $\begin{bmatrix} 2 & -3 & | & 7 \\ 3 & 1 & | & 4 \end{bmatrix}$

3. $\begin{bmatrix} 0 & -1 & 2 & | & 5 \\ 2 & 2 & -8 & | & 4 \\ 0 & 3 & 4 & | & 0 \end{bmatrix}$

5. $\begin{aligned} 3x + 2y &= -4 \\ x - y &= 5 \end{aligned}$ 7. $\begin{aligned} x + 3y + 2z &= 4 \\ 2x &= 5 \\ 3x - 3y + 2z &= 6 \end{aligned}$

9. Yes   11. No   13. Yes   15. No   17. No

19. $\begin{bmatrix} 1 & 3 & | & 4 \\ 0 & -2 & | & -2 \end{bmatrix}$ 21. $\begin{bmatrix} 1 & -2 & | & -3 \\ 0 & 20 & | & 20 \end{bmatrix}$

23. $\begin{bmatrix} 1 & 2 & 3 & | & 6 \\ 0 & -1 & -5 & | & -7 \\ 0 & -7 & -7 & | & -14 \end{bmatrix}$

25. $\begin{bmatrix} -6 & -11 & 0 & | & -5 \\ 2 & 4 & 1 & | & 3 \\ 1 & -2 & 0 & | & -10 \end{bmatrix}$

27. $\begin{bmatrix} 3 & 9 & | & 6 \\ 2 & 1 & | & 4 \end{bmatrix} \xrightarrow{\frac{1}{3}R_1} \begin{bmatrix} 1 & 3 & | & 2 \\ 2 & 1 & | & 4 \end{bmatrix}$

$\xrightarrow{R_2 - 2R_1} \begin{bmatrix} 1 & 3 & | & 2 \\ 0 & -5 & | & 0 \end{bmatrix} \xrightarrow{-\frac{1}{5}R_2}$

$\begin{bmatrix} 1 & 3 & | & 2 \\ 0 & 1 & | & 0 \end{bmatrix} \xrightarrow{R_1 - 3R_2} \begin{bmatrix} 1 & 0 & | & 2 \\ 0 & 1 & | & 0 \end{bmatrix}$

29. $\begin{bmatrix} 1 & 3 & 1 & | & 3 \\ 3 & 8 & 3 & | & 7 \\ 2 & -3 & 1 & | & -10 \end{bmatrix} \begin{array}{l} \xrightarrow{R_2 - 3R_1} \\ \xrightarrow{R_3 - 2R_1} \end{array}$

$\begin{bmatrix} 1 & 3 & 1 & | & 3 \\ 0 & -1 & 0 & | & -2 \\ 0 & -9 & -1 & | & -16 \end{bmatrix} \xrightarrow{-R_2}$

$\begin{bmatrix} 1 & 3 & 1 & | & 3 \\ 0 & 1 & 0 & | & 2 \\ 0 & -9 & -1 & | & -16 \end{bmatrix} \begin{array}{l} \xrightarrow{R_1 - 3R_2} \\ \xrightarrow{R_3 + 9R_2} \end{array}$

$\begin{bmatrix} 1 & 0 & 1 & | & -3 \\ 0 & 1 & 0 & | & 2 \\ 0 & 0 & -1 & | & 2 \end{bmatrix} \begin{array}{l} \xrightarrow{R_1 + R_3} \\ \xrightarrow{-R_3} \end{array}$

$\begin{bmatrix} 1 & 0 & 0 & | & -1 \\ 0 & 1 & 0 & | & 2 \\ 0 & 0 & 1 & | & -2 \end{bmatrix}$

31. $(2, 0)$ 33. $(-1, 2, -2)$ 35. $(2, 1)$

37. $(4, -2)$ 39. $(-1, 2)$ 41. $\left(\frac{1}{2}, \frac{3}{2}\right)$

43. $\left(\frac{1}{2}, \frac{1}{4}\right)$ 45. $(6, -2, 3)$ 47. $(19, -7, -15)$

49. $(3, 0, 2)$ 51. $(1, -2, 1)$

53. $(-20, -28, 13)$ 55. $(4, -1, 3)$

57. $k \neq \frac{15}{4}; x = \dfrac{3k - 50}{4k - 15}, y = \dfrac{31}{4k - 15}$

59. 300 acres of corn, 200 acres of wheat

61. In 100 lb of blended coffee, use 40 lb of the \$8/lb coffee and 60 lb of the \$9/lb coffee.

63. 200 children and 800 adults

65. The bat costs \$105, and the ball costs \$5.

67. \$40,000 in a savings account, \$120,000 in mutual funds, \$80,000 in bonds

69. $x = 50, y = 17,$ and $z = 33$

71. 400 bags of grade A fertilizer; 600 bags of grade B fertilizer; 300 bags of grade C fertilizer

73. 60 compact, 30 intermediate-size, and 10 full-size cars

75. 4 oz of Food I, 2 oz of Food II, 6 oz of Food III

77. 240 front orchestra seats, 560 rear orchestra seats, 200 front balcony seats

79. 7 days in London, 4 days in Paris, and 3 days in Rome

81. False

Using Technology Exercises 5.2, page 291

1. $x_1 = 3; x_2 = 1; x_3 = -1; x_4 = 2$

3. $x_1 = 5; x_2 = 4; x_3 = -3; x_4 = -4$

5. $x_1 = 1; x_2 = -1; x_3 = 2; x_4 = 0; x_5 = 3$

Exercises 5.3, page 298

1. **a.** One solution   **b.** $(3, -1, 2)$

3. **a.** One solution   **b.** $(2, 5)$ 5. No solution

7. **a.** Infinitely many solutions
   b. $(4 - t, -2, t); t,$ a parameter

9. **a.** No solution

11. **a.** Infinitely many solutions
    b. $(4, -1, 3 - t, t); t,$ a parameter

13. **a.** Infinitely many solutions
    b. $(2 - 3s, 1 + s, s, t); s, t,$ parameters

15. $(2, 1)$ 17. No solution   19. $(1, -1)$

21. $(2 + 2t, t); t,$ a parameter   23. $(4 + t, -3 - t, t)$

25. $\left(\frac{4}{3} - \frac{2}{3}t, t\right); t,$ a parameter   27. No solution

29. $\left(-1, \frac{17}{7}, \frac{23}{7}\right)$

31. $\left(1 - \frac{1}{4}s + \frac{1}{4}t, s, t\right); s, t,$ parameters

33. No solution   35. $(2, -1, 4)$

37. $x = 20 + z, y = 40 - 2z;$ 25 compact cars, 30 mid-sized cars, and 5 full-sized cars; 30 compact cars, 20 mid-sized cars, and 10 full-sized cars

41. \$60,000 in stocks, \$30,000 in bonds, and \$10,000 in money-market account; \$70,000 in stocks, \$10,000 in bonds, and \$20,000 in money-market account

43. a.

$$
\begin{array}{rcl}
x_1 \qquad\qquad\qquad\qquad + x_6 \qquad &=& 1700 \\
x_1 - x_2 \qquad\qquad\qquad\quad + x_7 &=& 700 \\
x_2 - x_3 \qquad\qquad\qquad &=& 300 \\
- x_3 + x_4 \qquad\qquad &=& 400 \\
- x_4 + x_5 \quad + x_7 &=& 700 \\
x_5 + x_6 \qquad &=& 1800
\end{array}
$$

b. $(1700 - s, 1000 - s + t, 700 - s + t, 1100 - s + t, 1800 - s,$ $s, t)$; $(900, 1000, 700, 1100, 1000, 800, 800)$; $(1000, 1100, 800,$ $1200, 1100, 700, 800)$

c. x_6 must have at least 300 cars/hr.

45. $k = 6$ **47.** $k = 1$; $(2 + a, 1 - a, a)$; a, a parameter

49. $k = 12$ **51.** $x = -1$ and $y = \frac{1}{3}$ **53.** False

Using Technology Exercises 5.3, page 302

1. $(1 + t, 2 + t, t)$; t, a parameter

3. $\left(-\frac{17}{7} + \frac{6}{7}t, 3 - t, -\frac{18}{7} + \frac{1}{7}t, t\right)$; t, a parameter

5. No solution

Exercises 5.4, page 310

1. 4×4; 4×3; 1×5; 4×1 **3.** 2; 3; 8

5. D; $D^T = \begin{bmatrix} 1 & 3 & -2 & 0 \end{bmatrix}$ **7.** 3×2; 3×2; 3×3; 3×3

9. $\begin{bmatrix} 1 & 6 \\ 6 & -1 \\ 2 & 2 \end{bmatrix}$ **11.** $\begin{bmatrix} 1 & 1 & -4 \\ -1 & -8 & 1 \\ 6 & 3 & 1 \end{bmatrix}$ **13.** $\begin{bmatrix} 1 & 1 & 2 \\ 3 & 6 & 3 \end{bmatrix}$

15. $\begin{bmatrix} 6 & 0 & 2 & -5 \\ 9 & 3 & 0 & -3 \end{bmatrix}$ **17.** $\begin{bmatrix} -1.9 & 3.0 & -0.6 \\ 6.0 & 9.6 & 1.2 \end{bmatrix}$

19. $\begin{bmatrix} -5 & -1 & 23 \\ 25 & 14 & 17 \\ 33 & 21 & 30 \end{bmatrix}$ **21.** $\begin{bmatrix} \frac{7}{2} & 3 & -1 & \frac{10}{3} \\ -\frac{19}{6} & \frac{2}{3} & -\frac{17}{2} & \frac{23}{3} \\ \frac{29}{3} & \frac{17}{6} & -1 & -2 \end{bmatrix}$

23. $u = 3$, $x = \frac{5}{2}$, $y = 7$, and $z = 2$

25. $x = 2$, $y = 2$, $z = -\frac{7}{3}$, and $u = 15$ **27.** $\begin{bmatrix} 10 & -11 \\ 3 & -12 \end{bmatrix}$

35. $\begin{bmatrix} 3 \\ 2 \\ -1 \\ 5 \end{bmatrix}$ **37.** $\begin{bmatrix} 1 & 3 & 0 \\ -1 & 4 & 1 \\ 2 & 2 & 0 \end{bmatrix}$

39. $\begin{bmatrix} 220 & 215 & 210 & 205 \\ 220 & 210 & 200 & 195 \\ 215 & 205 & 195 & 190 \end{bmatrix}$

41.

	White	Black	Hispanic
Women	82.6	80.5	91.2
Men	78.0	73.9	84.8

	Women	Men
White	82.6	78.0
Black	80.5	73.9
Hispanic	91.2	84.8

43. a.

	6-month	1-year	$2\frac{1}{2}$-year	5-year
Current week	0.17	0.27	0.41	0.87
Previous week	0.17	0.27	0.42	0.88
1 year ago	0.22	0.34	0.52	1.15

b. $a_{12} = 0.27$, $a_{22} = 0.27$ **c.** $a_{13} = 0.41$, $a_{23} = 0.42$
d. $a_{33} = 0.52$, $a_{34} = 1.15$

45. a. $A =$

	Textbooks	Fiction	Nonfiction	Reference
Hardcover	5280	1680	2320	1890
Paperback	1940	2810	1490	2070

b. $B =$

	Textbooks	Fiction	Nonfiction	Reference
Hardcover	6340	2220	1790	1980
Paperback	2050	3100	1720	2710

c. $C =$

	Textbooks	Fiction	Nonfiction	Reference
Hardcover	11,620	3900	4110	3870
Paperback	3,990	5910	3210	4780

47. a.

$A =$

	Large-cap	Small-cap	International	Bonds	Cash
Conservative	15	0	5	50	30
Moderately conservative	25	5	10	50	10
Moderate	35	10	15	35	5
Moderately aggressive	45	15	20	15	5
Aggressive	50	20	25	0	5

b. $a_{12} = 0$ **c.** $a_{13} = 5$, $a_{23} = 10$, $a_{33} = 15$, $a_{43} = 20$, and $a_{53} = 25$
d. 100

49. a. $A =$

	NY	NY Co-ops	NJ	CT
30-year fixed	3.83	3.67	3.78	3.79
15-year fixed	3.16	2.98	3.03	3.03
Adjustable	2.99	2.96	2.97	2.45

$B =$

	NY	NY Co-ops	NJ	CT
30-year fixed	3.84	3.75	3.81	3.80
15-year fixed	3.15	2.97	3.04	3.02
Adjustable	2.99	2.95	2.96	2.44

b. $a_{12} = 3.67$, $b_{12} = 3.75$ **c.** $a_{33} = 2.97$, $b_{33} = 2.96$
d. 3.7675 **e.** 2.445

51. $B = \begin{bmatrix} 350.2 & 370.8 & 391.4 \\ 422.3 & 442.9 & 453.2 \\ 638.6 & 679.8 & 721 \end{bmatrix}$ **53.** True   **55.** False

Using Technology Exercises 5.4, page 317

1. $\begin{bmatrix} 15 & 38.75 & -67.5 & 33.75 \\ 51.25 & 40 & 52.5 & -38.75 \\ 21.25 & 35 & -65 & 105 \end{bmatrix}$

3. $\begin{bmatrix} -5 & 6.3 & -6.8 & 3.9 \\ 1 & 0.5 & 5.4 & -4.8 \\ 0.5 & 4.2 & -3.5 & 5.6 \end{bmatrix}$

5. $\begin{bmatrix} 16.44 & -3.65 & -3.66 & 0.63 \\ 12.77 & 10.64 & 2.58 & 0.05 \\ 5.09 & 0.28 & -10.84 & 17.64 \end{bmatrix}$

7. $\begin{bmatrix} 22.2 & -0.3 & -12 & 4.5 \\ 21.6 & 17.7 & 9 & -4.2 \\ 8.7 & 4.2 & -20.7 & 33.6 \end{bmatrix}$

Exercises 5.5, page 324

1. 2×5; not defined   **3.** 1×1; 7×7

5. $n = s$; $m = t$ **7.** $\begin{bmatrix} -1 \\ 3 \end{bmatrix}$

9. $\begin{bmatrix} 13 \\ -10 \end{bmatrix}$ **11.** $\begin{bmatrix} 4 & -2 \\ 9 & 13 \end{bmatrix}$

13. $\begin{bmatrix} 2 & 9 \\ 5 & 16 \end{bmatrix}$ **15.** $\begin{bmatrix} 0.57 & 1.93 \\ 0.64 & 1.76 \end{bmatrix}$

17. $\begin{bmatrix} 6 & -3 & 0 \\ -2 & 1 & -8 \\ 4 & -4 & 9 \end{bmatrix}$ **19.** $\begin{bmatrix} 5 & 1 & -6 \\ 1 & 7 & -4 \end{bmatrix}$

21. $\begin{bmatrix} -4 & -20 & 4 \\ 4 & 12 & 0 \\ 12 & 32 & 20 \end{bmatrix}$ **23.** $\begin{bmatrix} 4 & -3 & 2 \\ 7 & 1 & -5 \end{bmatrix}$

27. $AB = \begin{bmatrix} 10 & 7 \\ 22 & 15 \end{bmatrix}$; $BA = \begin{bmatrix} 5 & 8 \\ 13 & 20 \end{bmatrix}$ **31.** $A = \begin{bmatrix} -2 & -1 \\ 5 & 2 \end{bmatrix}$

33. $B = \begin{bmatrix} \frac{1}{3} & -\frac{1}{6} \\ \frac{1}{3} & \frac{1}{3} \end{bmatrix}$ **35. b.** No **37. a.** $A^T = \begin{bmatrix} 2 & 5 \\ 4 & -6 \end{bmatrix}$

39. $AX = B$, where $A = \begin{bmatrix} 2 & -3 \\ 3 & -4 \end{bmatrix}$, $X = \begin{bmatrix} x \\ y \end{bmatrix}$, and $B = \begin{bmatrix} 7 \\ 8 \end{bmatrix}$

41. $AX = B$, where $A = \begin{bmatrix} 2 & -3 & 4 \\ 0 & 2 & -3 \\ 1 & -1 & 2 \end{bmatrix}$, $X = \begin{bmatrix} x \\ y \\ z \end{bmatrix}$,

and $B = \begin{bmatrix} 6 \\ 7 \\ 4 \end{bmatrix}$

43. $AX = B$, where $A = \begin{bmatrix} -1 & 1 & 1 \\ 2 & -1 & -1 \\ -3 & 2 & 4 \end{bmatrix}$, $X = \begin{bmatrix} x_1 \\ x_2 \\ x_3 \end{bmatrix}$,

and $B = \begin{bmatrix} 0 \\ 2 \\ 4 \end{bmatrix}$

45. a. $AB = \begin{bmatrix} 30,100 \\ 32,700 \end{bmatrix}$

b. The first entry shows that the value of Olivia's total stock holdings are \$30,100; the second shows that the value of Isabella's stock holdings are \$32,700.

47. a. $AB = \begin{bmatrix} 1070 & 1640 & 2550 \end{bmatrix}$ **b.** $ABC = \begin{bmatrix} 42,810,000 \end{bmatrix}$

49. $C = \begin{bmatrix} 374,051.60 & 378,524.00 & 376,775.80 \end{bmatrix}$

51. a. $A = \begin{array}{c} \text{Ava} \\ \text{Ella} \end{array} \begin{bmatrix} \overset{\text{N}}{\underset{\text{kroner}}{}} & \overset{\text{S}}{\underset{\text{kronor}}{}} & \overset{\text{D}}{\underset{\text{kroner}}{}} & \overset{\text{R}}{\underset{\text{rubles}}{}} \\ 82 & 68 & 62 & 1200 \\ 64 & 74 & 44 & 1600 \end{bmatrix}$

b. $B = \begin{array}{c} \text{N} \\ \text{S} \\ \text{D} \\ \text{R} \end{array} \begin{bmatrix} 0.1751 \\ 0.1560 \\ 0.1747 \\ 0.0325 \end{bmatrix}$ **c.** Ava: \$74.80; Ella: \$82.44

53. a. $\begin{bmatrix} 90 & 125 & 210 & 55 \end{bmatrix}$; the entries give the respective total number of Model I, II, III, and IV houses built in the three states.

b. $\begin{bmatrix} 300 \\ 120 \\ 60 \end{bmatrix}$; the entries give the respective total number of Model I, II, III, and IV houses built in all three states.

55. $B = \begin{bmatrix} 4 \\ 6 \\ 8 \end{bmatrix}$; $AB = \begin{bmatrix} 1960 \\ 3180 \\ 2510 \\ 3300 \end{bmatrix}$; \$10,950

57. $BA = \begin{array}{ccc} \text{Dem.} & \text{Rep.} & \text{Ind.} \\ [41,000 & 35,000 & 14,000] \end{array}$

59. $AB = \begin{bmatrix} 1575 & 1590 & 1560 & 975 \\ 410 & 405 & 415 & 270 \\ 215 & 205 & 225 & 155 \end{bmatrix}$

61. $[136.80]$; it represents Cindy's long-distance bill for phone calls to London, Tokyo, and Hong Kong.

63. a. $\begin{bmatrix} 8800 \\ 3380 \\ 1020 \end{bmatrix}$ **b.** $\begin{bmatrix} 8800 \\ 3380 \\ 1020 \end{bmatrix}$ **c.** $\begin{bmatrix} 17,600 \\ 6,760 \\ 2,040 \end{bmatrix}$

65. False **67.** True

Using Technology Exercises 5.5, page 332

1. $\begin{bmatrix} 18.66 & 15.2 & -12 \\ 24.48 & 41.88 & 89.82 \\ 15.39 & 7.16 & -1.25 \end{bmatrix}$

3. $\begin{bmatrix} 20.09 & 20.61 & -1.3 \\ 44.42 & 71.6 & 64.89 \\ 20.97 & 7.17 & -60.65 \end{bmatrix}$

5. $\begin{bmatrix} 32.89 & 13.63 & -57.17 \\ -12.85 & -8.37 & 256.92 \\ 13.48 & 14.29 & 181.64 \end{bmatrix}$

7. $\begin{bmatrix} 128.59 & 123.08 & -32.50 \\ 246.73 & 403.12 & 481.52 \\ 125.06 & 47.01 & -264.81 \end{bmatrix}$

9. $\begin{bmatrix} 87 & 68 & 110 & 82 \\ 119 & 176 & 221 & 143 \\ 51 & 128 & 142 & 94 \\ 28 & 174 & 174 & 112 \end{bmatrix}$

$\begin{bmatrix} 113 & 117 & 72 & 101 & 90 \\ 72 & 85 & 36 & 72 & 76 \\ 81 & 69 & 76 & 87 & 30 \\ 133 & 157 & 56 & 121 & 146 \\ 154 & 157 & 94 & 127 & 122 \end{bmatrix}$

11. $\begin{bmatrix} 170 & 18.1 & 133.1 & -106.3 & 341.3 \\ 349 & 226.5 & 324.1 & 164 & 506.4 \\ 245.2 & 157.7 & 231.5 & 125.5 & 312.9 \\ 310 & 245.2 & 291 & 274.3 & 354.2 \end{bmatrix}$

Exercises 5.6, page 340

5. $\begin{bmatrix} 3 & -5 \\ -1 & 2 \end{bmatrix}$ **7.** Does not exist

9. $\begin{bmatrix} 2 & -11 & -3 \\ 1 & -6 & -2 \\ 0 & -1 & 0 \end{bmatrix}$ **11.** Does not exist

13. $\begin{bmatrix} -\frac{13}{10} & \frac{7}{5} & \frac{1}{2} \\ \frac{2}{5} & -\frac{1}{5} & 0 \\ -\frac{7}{10} & \frac{3}{5} & \frac{1}{2} \end{bmatrix}$

15. $\begin{bmatrix} 3 & 4 & -6 & 1 \\ -2 & -3 & 5 & -1 \\ -4 & -4 & 7 & -1 \\ -4 & -5 & 8 & -1 \end{bmatrix}$

17. a. $AX = B$, where $A = \begin{bmatrix} 2 & 5 \\ 1 & 3 \end{bmatrix}$; $X = \begin{bmatrix} x \\ y \end{bmatrix}$; $B = \begin{bmatrix} 3 \\ 2 \end{bmatrix}$

 b. $x = -1$; $y = 1$

19. a. $AX = B$, where $A = \begin{bmatrix} 2 & -3 & -4 \\ 0 & 0 & -1 \\ 1 & -2 & 1 \end{bmatrix}$; $X = \begin{bmatrix} x \\ y \\ z \end{bmatrix}$; $B = \begin{bmatrix} 4 \\ 3 \\ -8 \end{bmatrix}$

 b. $x = -1$; $y = 2$; $z = -3$

21. a. $AX = B$, where $A = \begin{bmatrix} 1 & 4 & -1 \\ 2 & 3 & -2 \\ -1 & 2 & 3 \end{bmatrix}$; $X = \begin{bmatrix} x \\ y \\ z \end{bmatrix}$; $B = \begin{bmatrix} 3 \\ 1 \\ 7 \end{bmatrix}$

 b. $x = 1$; $y = 1$; $z = 2$

23. a. $AX = B$, where $A = \begin{bmatrix} 1 & 1 & -1 & 1 \\ 2 & 1 & 1 & 0 \\ 2 & 1 & 0 & 1 \\ 2 & -1 & -1 & 3 \end{bmatrix}$; $X = \begin{bmatrix} x_1 \\ x_2 \\ x_3 \\ x_4 \end{bmatrix}$; $B = \begin{bmatrix} 6 \\ 4 \\ 7 \\ 9 \end{bmatrix}$

 b. $x_1 = 1$; $x_2 = 2$; $x_3 = 0$; $x_4 = 3$

25. b. (i) $x = 4.8$ and $y = 4.6$
  (ii) $x = 0.4$ and $y = 1.8$

27. b. (i) $x = -1$; $y = 3$; $z = 2$
  (ii) $x = 1$; $y = 8$; $z = -12$

29. b. (i) $x = -\frac{2}{17}$; $y = -\frac{10}{17}$; $z = -\frac{60}{17}$
  (ii) $x = 1$; $y = 0$; $z = -5$

31. b. (i) $x_1 = 1$; $x_2 = -4$; $x_3 = 5$; $x_4 = -1$
  (ii) $x_1 = 12$; $x_2 = -24$; $x_3 = 21$; $x_4 = -7$

33. a. $A^{-1} = \begin{bmatrix} -\frac{5}{2} & -\frac{3}{2} \\ 2 & 1 \end{bmatrix}$

35. a. $ABC = \begin{bmatrix} 4 & 10 \\ 2 & 3 \end{bmatrix}$; $A^{-1} = \begin{bmatrix} 3 & -5 \\ 1 & -2 \end{bmatrix}$;

 $B^{-1} = \begin{bmatrix} 1 & -3 \\ -1 & 4 \end{bmatrix}$; $C^{-1} = \begin{bmatrix} \frac{1}{8} & -\frac{3}{8} \\ \frac{1}{4} & \frac{1}{4} \end{bmatrix}$

37. $\begin{bmatrix} \frac{5}{7} & \frac{3}{7} \\ -\frac{3}{7} & \frac{8}{7} \end{bmatrix}$

39. a. 3214; 3929 **b.** 4286; 3571 **c.** 3929; 5357

41. a. 400 acres of soybeans; 300 acres of corn; 300 acres of wheat
 b. 500 acres of soybeans; 400 acres of corn; 300 acres of wheat

43. a. $80,000 in high-risk stocks; $20,000 in medium-risk stocks; $100,000 in low-risk stocks
 b. $88,000 in high-risk stocks; $22,000 in medium-risk stocks; $110,000 in low-risk stocks
 c. $56,000 in high-risk stocks; $64,000 in medium-risk stocks; $120,000 in low-risk stocks

45. All values of k except $k = \frac{3}{2}$; $\dfrac{1}{3 - 2k} \begin{bmatrix} 3 & -2 \\ -k & 1 \end{bmatrix}$

47. A^{-1} exists provided that $ad \neq 0$; every entry along the main diagonal is not equal to zero.

49. True **51.** True

Using Technology Exercises 5.6, page 347

1. $\begin{bmatrix} 0.36 & 0.04 & -0.36 \\ 0.06 & 0.05 & 0.20 \\ -0.19 & 0.10 & 0.09 \end{bmatrix}$

3. $\begin{bmatrix} 0.01 & -0.09 & 0.31 & -0.11 \\ -0.25 & 0.58 & -0.15 & -0.02 \\ 0.86 & -0.42 & 0.07 & -0.37 \\ -0.27 & 0.01 & -0.05 & 0.31 \end{bmatrix}$

5. $\begin{bmatrix} 0.30 & 0.85 & -0.10 & -0.77 & -0.11 \\ -0.21 & 0.10 & 0.01 & -0.26 & 0.21 \\ 0.03 & -0.16 & 0.12 & -0.01 & 0.03 \\ -0.14 & -0.46 & 0.13 & 0.71 & -0.05 \\ 0.10 & -0.05 & -0.10 & -0.03 & 0.11 \end{bmatrix}$

7. $x = 1.2$; $y = 3.6$; $z = 2.7$

9. $x_1 = 2.50$; $x_2 = -0.88$; $x_3 = 0.70$; $x_4 = 0.51$

Chapter 5 Concept Review Questions, page 348

1. a. one, many, no **b.** one, many, no **2.** equations

3. $R_i \leftrightarrow R_j$, cR_i; $R_i + aR_j$, solution

4. a. unique **b.** no, infinitely many, unique

5. size, entries **6.** size; corresponding

7. $m \times n$, $n \times m$; a_{ji} **8.** cA, c

9. a. columns, rows **b.** $m \times p$

10. a. $A(BC)$, $AB + AC$ **b.** $n \times r$

11. $A^{-1}A$, AA^{-1}; singular **12.** $A^{-1}B$

Chapter 5 Review Exercises, page 349

1. $\begin{bmatrix} 2 & 2 \\ -1 & 4 \\ 3 & 3 \end{bmatrix}$ **2.** $\begin{bmatrix} -2 & 0 \\ -2 & 6 \end{bmatrix}$ **3.** $[-6 \quad -2]$ **4.** $\begin{bmatrix} 17 \\ 13 \end{bmatrix}$

5. $x = 2$; $y = 3$; $z = 1$; $w = 3$ **6.** $x = 2$; $y = -2$

7. $a = 3$; $b = 4$; $c = -2$; $d = 2$; $e = -3$

8. $x = -1$; $y = -2$; $z = 1$

9. $\begin{bmatrix} 8 & 9 & 11 \\ -10 & -1 & 3 \\ 11 & 12 & 10 \end{bmatrix}$ **10.** $\begin{bmatrix} -1 & 7 & -3 \\ -2 & 5 & 11 \\ 10 & -8 & 2 \end{bmatrix}$

11. $\begin{bmatrix} 6 & 18 & 6 \\ -12 & 6 & 18 \\ 24 & 0 & 12 \end{bmatrix}$ **12.** $\begin{bmatrix} -10 & 10 & -18 \\ 4 & 14 & 26 \\ 16 & -32 & -4 \end{bmatrix}$

13. $\begin{bmatrix} -11 & -16 & -15 \\ -4 & -2 & -10 \\ -6 & 14 & 2 \end{bmatrix}$ **14.** $\begin{bmatrix} 5 & 20 & 19 \\ -2 & 20 & 8 \\ 26 & 10 & 30 \end{bmatrix}$

15. $\begin{bmatrix} -3 & 17 & 8 \\ -2 & 56 & 27 \\ 74 & 78 & 116 \end{bmatrix}$ **16.** $\begin{bmatrix} \frac{3}{2} & -2 & -5 \\ \frac{11}{2} & -1 & 11 \\ \frac{7}{2} & -3 & 0 \end{bmatrix}$

17. $x = 1; y = -1$ **18.** $x = -1; y = 3$

19. $x = 1; y = 2; z = 3$

20. $(2, 2t - 5, t); t$, a parameter **21.** No solution

22. $x = 1; y = -1; z = 2; w = 2$

23. $x = 1; y = 0; z = 1$ **24.** $x = 2; y = -1; z = 3$

25. $\begin{bmatrix} \frac{2}{5} & -\frac{1}{5} \\ -\frac{1}{5} & \frac{3}{5} \end{bmatrix}$ **26.** $\begin{bmatrix} \frac{3}{4} & -\frac{1}{2} \\ -\frac{1}{8} & \frac{1}{4} \end{bmatrix}$

27. $\begin{bmatrix} -1 & 2 \\ 1 & -\frac{3}{2} \end{bmatrix}$ **28.** $\begin{bmatrix} \frac{1}{4} & \frac{1}{2} \\ \frac{1}{8} & -\frac{1}{4} \end{bmatrix}$

29. $\begin{bmatrix} \frac{5}{4} & \frac{1}{4} & -\frac{7}{4} \\ -\frac{1}{4} & -\frac{1}{4} & \frac{3}{4} \\ -\frac{3}{4} & \frac{1}{4} & \frac{5}{4} \end{bmatrix}$ **30.** $\begin{bmatrix} -\frac{1}{4} & \frac{1}{2} & -\frac{1}{4} \\ \frac{7}{8} & -\frac{3}{4} & -\frac{5}{8} \\ -\frac{1}{8} & \frac{1}{4} & \frac{3}{8} \end{bmatrix}$

31. $\begin{bmatrix} -\frac{1}{5} & \frac{2}{5} & 0 \\ \frac{2}{3} & -\frac{1}{3} & \frac{1}{3} \\ -\frac{1}{30} & \frac{1}{15} & -\frac{1}{6} \end{bmatrix}$ **32.** $\begin{bmatrix} 0 & -\frac{1}{5} & \frac{2}{5} \\ -2 & 1 & 1 \\ -1 & \frac{1}{5} & \frac{3}{5} \end{bmatrix}$

33. $\begin{bmatrix} \frac{3}{2} & 1 \\ -\frac{7}{2} & -1 \end{bmatrix}$ **34.** $\begin{bmatrix} \frac{11}{24} & -\frac{7}{8} \\ -\frac{1}{12} & \frac{1}{4} \end{bmatrix}$

35. $\begin{bmatrix} \frac{2}{5} & -\frac{3}{5} \\ \frac{1}{5} & \frac{1}{5} \end{bmatrix}$ **36.** $\begin{bmatrix} \frac{4}{7} & -\frac{3}{7} \\ -\frac{3}{7} & \frac{4}{7} \end{bmatrix}$

37. $A^{-1} = \begin{bmatrix} \frac{2}{7} & \frac{3}{7} \\ \frac{1}{7} & -\frac{2}{7} \end{bmatrix}; x = -1; y = -2$

38. $A^{-1} = \begin{bmatrix} \frac{2}{5} & \frac{3}{10} \\ -\frac{1}{5} & \frac{1}{10} \end{bmatrix}; x = 2; y = 1$

39. $A^{-1} = \begin{bmatrix} 1 & -\frac{2}{5} & \frac{4}{5} \\ -1 & 1 & -1 \\ -\frac{1}{2} & \frac{3}{5} & -\frac{7}{10} \end{bmatrix}; x = 1; y = 2; z = 4$

40. $A^{-1} = \begin{bmatrix} 0 & \frac{1}{7} & \frac{2}{7} \\ -1 & -\frac{4}{7} & \frac{6}{7} \\ -\frac{1}{2} & -\frac{1}{2} & \frac{1}{2} \end{bmatrix}; x = 3; y = -1; z = 2$

41. Station A: \$11,150; Station B: \$10,380; Station C: \$13,660

42. \$2,300,000; \$2,450,000; an increase of \$150,000

43. a. $A = \begin{bmatrix} 800 & 1200 & 250 & 1500 \\ 600 & 1400 & 300 & 1200 \end{bmatrix}$ **b.** $B = \begin{bmatrix} 12.57 \\ 28.21 \\ 214.92 \\ 36.34 \end{bmatrix}$

c. Josh: \$152,148; Hannah: \$155,120

44. a. $A = \begin{matrix} & \text{IBM} & \text{Google} & \text{Boeing} & \text{GM} \\ \text{Jennifer} & 800 & 500 & 1200 & 1500 \\ \text{Max} & 500 & 600 & 2000 & 800 \end{matrix}$;

$B = \begin{matrix} & \text{IBM} & \text{Google} & \text{Boeing} & \text{GM} \\ \text{Jennifer} & 900 & 600 & 1000 & 1200 \\ \text{Max} & 700 & 500 & 2100 & 900 \end{matrix}$

b. $C = \begin{matrix} & \text{IBM} & \text{Google} & \text{Boeing} & \text{GM} \\ \text{Jennifer} & 100 & 100 & -200 & -300 \\ \text{Max} & 200 & -100 & 100 & 100 \end{matrix}$

45. 30 of each type

46. Houston: 100,000 gal; Tulsa: 600,000 gal

Chapter 5 Before Moving On, page 350

1. $\left(\frac{2}{3}, -\frac{2}{3}, \frac{5}{3}\right)$

2. a. $(2, -3, 1)$ **b.** No solution **c.** $(2, 1 - 3t, t), t$, a parameter
d. $(0, 0, 0, 0)$ **e.** $(2 + t, 3 - 2t, t), t$, a parameter

3. a. $(-1, 2)$ **b.** $\left(\frac{4}{7}, -\frac{5}{7} + 2t, t\right), t$, a parameter

4. a. $\begin{bmatrix} 3 & 1 & 4 \\ 5 & -2 & 6 \end{bmatrix}$ **b.** $\begin{bmatrix} 14 & 3 & 7 \\ 14 & 5 & 1 \end{bmatrix}$ **c.** $\begin{bmatrix} 0 & 5 & 3 \\ 4 & -1 & -11 \end{bmatrix}$

5. $\begin{bmatrix} 3 & -2 & -5 \\ -3 & 2 & 6 \\ -1 & 1 & 2 \end{bmatrix}$ **6.** $(1, -1, 2)$

CHAPTER 6

Exercises 6.1, page 360

1. **3.**

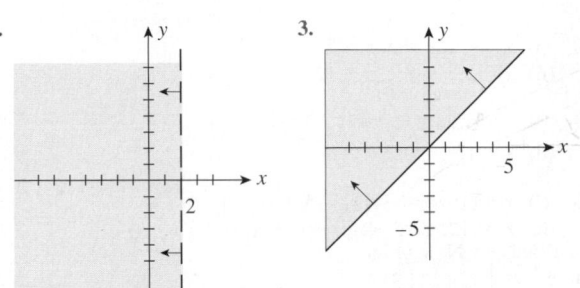

5. **7.**

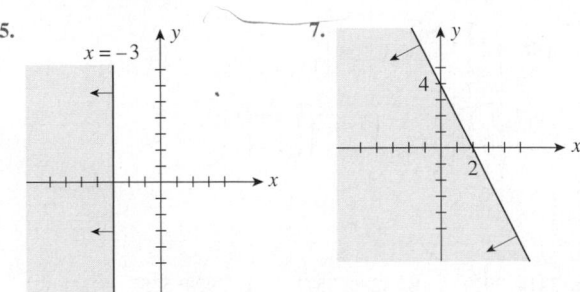

9.

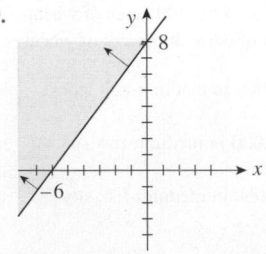

11. $x \geq 1; x \leq 5; y \geq 2; y \leq 4$

13. $2x - y \geq 2; 5x + 7y \geq 35; x \leq 4$

15. $x - y \geq -10; 7x + 4y \leq 140; x + 3y \geq 30$

17. $x + y \geq 7; x \geq 2; y \geq 3; y \leq 7$

19. $(3, 3)$ lies in S. **21.** $(10, 10)$ lies in S.

23.

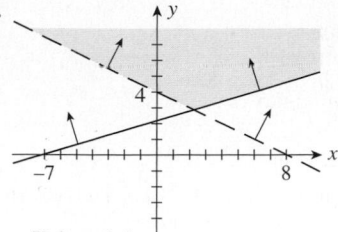

Unbounded

25.

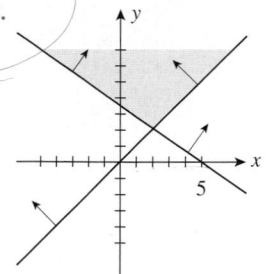

Unbounded

27.

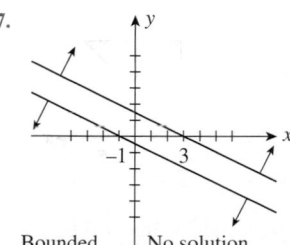

Bounded   No solution

29.

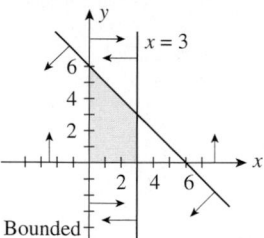

Bounded

31.

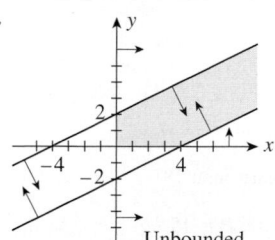

Unbounded

33.

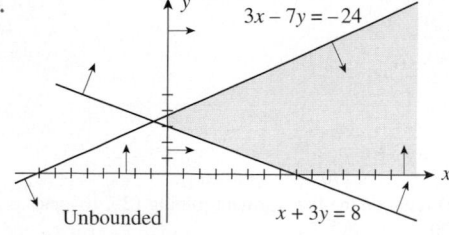

Unbounded $x + 3y = 8$

35.

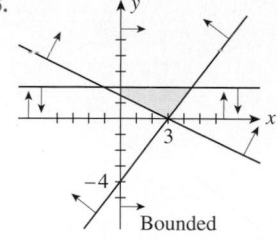

Bounded

37.

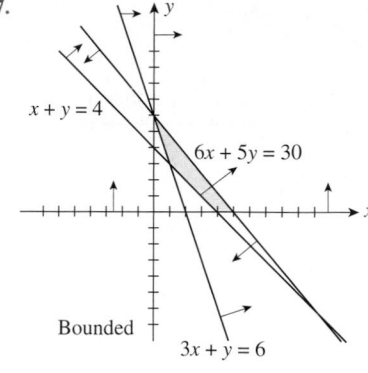

Bounded

39.

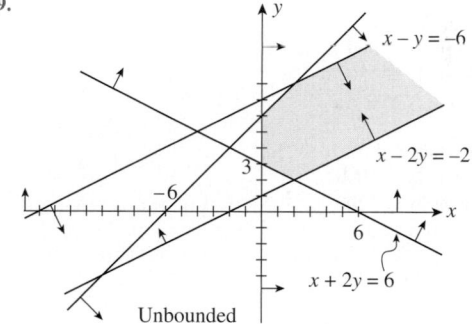

Unbounded

41. a. $x + y \leq 500$
    $x \geq 200$
    $y \geq 100$

b.

$x = 200$

$y = 100$

$x + y = 500$

c. Yes

43. a.
$$x + y \leq 250{,}000$$
$$x \geq 50{,}000$$
$$y \geq 50{,}000$$
$$5x - 6y \leq 0$$

b.

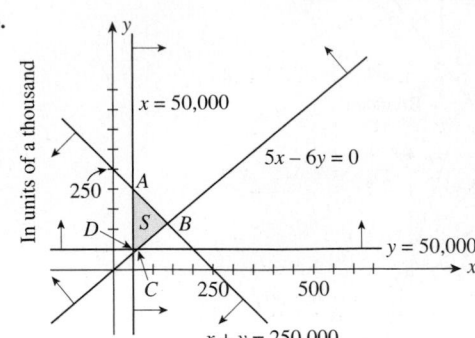

In units of a thousand

$x = 50{,}000$

$5x - 6y = 0$

250 · A

D · S · B

$y = 50{,}000$

C · 250 · 500

$x + y = 250{,}000$

In units of a thousand

c. No

45. False   **47.** True

Exercises 6.2, page 367

1. Maximize $P = 3x + 4y$
   subject to $6x + 9y \leq 300$
            $5x + 4y \leq 180$
            $x \geq 0, y \geq 0$

3. Maximize $P = 2x + 1.5y$
   subject to $3x + 4y \leq 1000$
            $6x + 3y \leq 1200$
            $x \geq 0, y \geq 0$

5. Maximize $P = 45x + 20y$
   subject to $40x + 16y \leq 3200$
            $3x + 4y \leq 520$
            $x \geq 0, y \geq 0$

7. Maximize $P = 0.1x + 0.12y$
   subject to $x + y \leq 20$
            $x - 4y \geq 0$
            $x \geq 0, y \geq 0$

9. Maximize $P = 0.12x + 0.20y$
   subject to $x + y \leq 60{,}000$
            $3x - 2y \geq 0$
            $x - 4y \leq 0$
            $x \geq 0, y \geq 0$

11. Minimize $C = 14{,}000x + 16{,}000y$
   subject to $50x + 75y \geq 650$
            $3000x + 1000y \geq 18{,}000$
            $x \geq 0, y \geq 0$

13. Maximize $P = 50x + 40y$
   subject to $\frac{1}{200}x + \frac{1}{200}y \leq 1$
            $\frac{1}{100}x + \frac{1}{300}y \leq 1$
            $x \geq 0, y \geq 0$

15. Minimize $C = 300x + 500y$
   subject to $x + y \geq 10$
            $x \leq 5$
            $y \leq 10$
            $y \geq 6$
            $x \geq 0$

17. Minimize $C = 2x + 5y$
   subject to $30x + 25y \geq 400$
            $x + 0.5y \geq 10$
            $2x + 5y \geq 40$
            $x \geq 0, y \geq 0$

19. Minimize $C = 64{,}000 - 2x - 6y$
   subject to $x + y \leq 6000$
            $x + y \geq 2000$
            $x \leq 3000$
            $y \leq 4000$
            $x \geq 0, y \geq 0$

21. Maximize $P = 18x + 12y + 15z$
   subject to $2x + y + 2z \leq 900$
            $3x + y + 2z \leq 1080$
            $2x + 2y + z \leq 840$
            $x \geq 0, y \geq 0, z \geq 0$

23. Maximize $P = 26x + 28y + 24z$
   subject to $\frac{5}{4}x + \frac{3}{2}y + \frac{3}{2}z \leq 310$
            $x + y + \frac{3}{4}z \leq 205$
            $x + y + \frac{1}{2}z \leq 190$
            $x \geq 0, y \geq 0, z \geq 0$

25. Maximize $P = 0.06x + 0.1y + 0.15z$
   subject to $x + y + z \leq 250{,}000$
            $-x - y + 3z \leq 0$
            $-x + y - z \leq 0$
            $x \geq 0, y \geq 0, z \geq 0$

27. Minimize $C = 16x_1 + 20x_2 + 22x_3 + 18x_4 + 16x_5 + 14x_6$
   subject to $x_1 + x_2 + x_3 \leq 800$
            $x_4 + x_5 + x_6 \leq 600$
            $x_1 + x_4 \geq 500$
            $x_2 + x_5 \geq 400$
            $x_3 + x_6 \geq 400$
            $x_1, x_2, \ldots, x_6 \geq 0$

29. Maximize $P = x + 0.8y + 0.9z$
   subject to $8x + 4z \leq 16{,}000$
            $8x + 12y + 8z \leq 24{,}000$
            $4y + 4z \leq 5000$
            $z \leq 800$
            $x \geq 0, y \geq 0, z \geq 0$

31. False

Exercises 6.3, page 378

1. Max: 35; min: 5   **3.** No max. value; min: 18

5. Max: 44; min: 15   **7.** $x = 3; y = 3; P = 15$

9. Any point (x, y) lying on the line segment joining $\left(\frac{5}{2}, 0\right)$ and $(1, 3)$; $P = 5$

11. $x = 0; y = 8; P = 64$

13. $x = 0; y = 4; P = 12$

15. $x = 4; y = 0; C = 8$

17. Any point (x, y) lying on the line segment joining $(20, 10)$ and $(40, 0)$; $C = 120$

19. $x = 14; y = 3; C = 58$

21. $x = 3; y = 3; C = 75$

23. $x = 15$; $y = 17.5$; $P = 115$

25. $x = 10$; $y = 38$; $P = 134$

27. Min: $x = 9$; $y = 9$; $P = 117$
    Max: $x = 15$; $y = 3$; $P = 135$

29. 20 Product A, 20 Product B; $140

31. 120 model A, 160 model B; $480

33. 40 tables; 100 chairs; $3800

35. $16 million in homeowner loans, $4 million in auto loans;
    $2.08 million

37. $24,000 in medium-risk stocks; $36,000 in high-risk stocks; $10,080

39. Saddle Mine: 4 days; Horseshoe Mine: 6 days; $152,000

41. 50 fully assembled units, 150 kits; $8500

43. Reservoir: 4 million gallons; pipeline: 6 million gallons; $4200

45. Infinitely many solutions; 10 oz of Food A and 4 oz of Food B or
    20 oz of Food A and 0 oz of Food B, etc., with a minimum value of
    40 mg of cholesterol

47. 2000 televisions from Location I to City A and 4000 televisions from
    Location I to City B; 1000 televisions from Location II to City A and
    0 televisions from Location II to City B; $36,000

49. 80 from I to A, 20 from I to B, 0 from II to A, 50 from II to B;
    $12,700

51. $22,500 in growth stocks and $7500 in speculative stocks; maximum
    return; $5250

53. 750 urban, 750 suburban; $10,950

55. False   57. **a.** True   **b.** True

61. **a.**

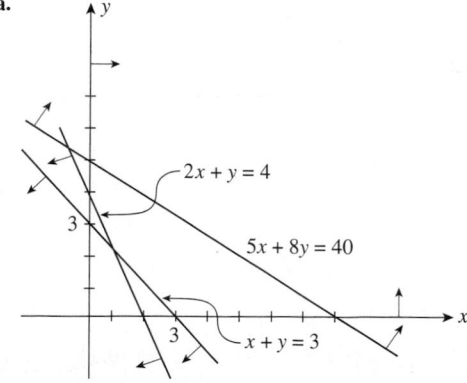

b. No solution

Exercises 6.4, page 401

1. **a.** It is already in standard form.
   b.

x	y	u	v	P	Constant
1	4	1	0	0	12
1	3	0	1	0	10
-2	-4	0	0	1	0

3. **a.** Maximize $P = 2x + 3y$
   subject to $x + \ y \le 10$
   $\qquad\qquad\ x + 2y \le 12$
   $\qquad\quad\ 2x + \ y \le 12$
   $\qquad\qquad\ x \ge 0, y \ge 0$

   b.

x	y	u	v	w	P	Constant
1	1	1	0	0	0	10
1	2	0	1	0	0	12
2	1	0	0	1	0	12
-2	-3	0	0	0	1	0

5. **a.** Maximize $P = x + 3y + 4z$
   subject to $x + 2y + z \le 40$
   $\qquad\qquad\ x + \ y + z \le 30$
   $\qquad\quad\ x \ge 0, y \ge 0, z \ge 0$

   b.

x	y	z	u	v	P	Constant
1	2	1	1	0	0	40
1	1	1	0	1	0	30
-1	-3	-4	0	0	1	0

7. In final form; $x = \frac{30}{7}$, $y = \frac{20}{7}$, $u = 0$, $v = 0$; $P = \frac{220}{7}$

9. Not in final form; pivot element is $\frac{1}{2}$, lying in the first row, second
   column.

11. In final form; $x = \frac{1}{3}$, $y = 0$, $z = \frac{13}{3}$, $u = 0$, $v = 6$, $w = 0$;
    $P = 17$

13. Not in final form; pivot element is 1, lying in the third row, second
    column.

15. In final form; $x = 30$, $y = 10$, $z = 0$, $u = 0$, $v = 0$; $P = 60$;
    $x = 30$, $y = 0$, $z = 0$, $u = 10$, $v = 0$; $P = 60$; among others

17. $x = 0$, $y = 4$, $u = 0$, $v = 1$; $P = 16$

19. $x = 6$, $y = 3$, $u = 0$, $v = 0$; $P = 96$

21. $x = 6$, $y - 6$, $u = 0$, $v = 0$, $w = 0$; $P = 60$

23. $x = 0$, $y = 3$, $z = 0$, $u = 90$, $v = 0$, $w = 75$; $P = 12$

25. $x = 0$, $y = 4$, $z = 4$, $u = 0$, $v = 0$; $P = 36$

27. $x = 15$, $y = 3$, $z = 0$, $u = 2$, $v = 0$, $w = 0$; $P = 78$

29. $x = \frac{5}{4}$, $y = \frac{15}{2}$, $z = 0$, $u = 0$, $v = \frac{15}{4}$, $w = 0$; $P = 90$

31. $x = 2$, $y = 1$, $z = 1$, $u = 0$, $v = 0$, $w = 0$; $P = 87$

33. $x = 30$, $y = 0$, $z = 0$, $P = 60$, and $x = 0$, $y = 30$, $z = 0$, $P = 60$,
    among others

35. No model A, 2500 model B; $100,000

37. Medium-risk stocks: $24,000; high-risk stocks: $36,000; $10,080

39. 40 tables; 100 chairs; $3800

41. 180 units of Product A, 140 units of Product B, and 200 units of
    Product C; $7920; no

43. 800 Giant Pandas; 1200 Saint Bernards; $26,000

45. 80 units of model A, 80 units of model B, and 60 units of model C; maximum profit: $5760; no

47. Growth funds: $50,000; balanced funds: $50,000; income funds: $100,000; $17,000

49. 9000 bottles of Formula I, 7833 bottles of Formula II, 6000 bottles of Formula III; maximum profit: $4986.60; yes, ingredients for 4167 bottles of Formula II

51. a.

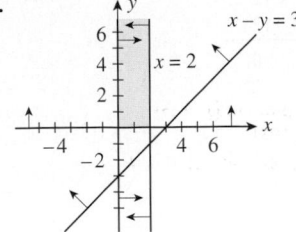

c. The ratios cannot be computed.

53. True **55.** True

Using Technology Exercises 6.4, page 411

1. $x = 1.2, y = 0, z = 1.6, w = 0; P = 8.8$

3. $x = 1.6, y = 0, z = 0, w = 3.6; P = 12.4$

Exercises 6.5, page 422

1. $x = 4, y = 0; C = -8$

3. $x = 4, y = 3; C = -18$

5. $x = 0, y = 13, z = 18, w = 14; C = -111$

7. $x = \frac{5}{4}, y = \frac{1}{4}, u = 2, v = 3; \ C = P = 13$

9. $x = 5, y = 10, z = 0, u = 1, v = 2; C = P = 80$

11. Maximize $P = 90u + 120v$
subject to $2u + 3v \le 3$
$3u + 2v \le 2; \quad x = 0, y = 60; C = 120$
$u \ge 0, v \ge 0$

13. Maximize $P = 60u + 40v + 30w$
subject to $6u + 2v + w \le 6$
$u + v + w \le 4; \quad x = 10, y = 20; C = 140$
$u \ge 0, v \ge 0, w \ge 0$

15. Maximize $P = 10u + 20v$
subject to $20u + v \le 200$
$10u + v \le 150; \quad x = 0, y = 0, z = 10; C = 1200$
$u + 2v \le 120$
$u \ge 0, v \ge 0$

17. Maximize $P = 10u + 24v + 16w$
subject to $u + 2v + w \le 6$
$2u + v + w \le 8; \quad x = 8, y = 0, z = 8; C = 80$
$2u + v + w \le 4$
$u \ge 0, v \ge 0, w \ge 0$

19. Maximize $P = 6u + 2v + 4w$
subject to $2u + 6v \le 30$
$4u + 6w \le 12; \quad x = \frac{1}{3}, y = \frac{4}{3}, z = 0; C = 26$
$3u + v + 2w \le 20$
$u \ge 0, v \ge 0, w \ge 0$

21. 2 type A vessels; 3 type B vessels; $250,000

23. 8 oz of orange juice; 6 oz of pink grapefruit juice; 178 calories

25. Plant I: 500 to Warehouse A and 200 to Warehouse B; Plant II: 200 to Warehouse B and 400 to Warehouse C; $20,800

27. Operate Refinery I for 2 days, and operate Refinery II for 6 days; $2200

29. False

Using Technology Exercises 6.5, page 427

1. $x = 1.333333, y = 3.333333, z = 0;$ and $C = 4.66667$

3. $x = 0.9524, y = 4.2857, z = 0; C = 6.0952$

Chapter 6 Concept Review Questions, page 427

1. a. half-plane; line **b.** $ax + by \le c; ax + by = c$

2. a. points; each **b.** bounded; enclosed

3. objective function; maximized; minimized; linear; inequalities

4. a. corner point **b.** line

5. maximized; nonnegative; less than; equal to

6. equations; slack variables; $-c_1x_1 - c_2x_2 - \cdots - c_nx_n + P = 0$; below; augmented

7. minimized; nonnegative; greater than; equal to

8. dual; objective; optimal value

Chapter 6 Review Exercises, page 428

1. Max: 18—any point (x, y) lying on the line segment joining $(0, 6)$ and $(3, 4)$; min: 0

2. Max: 27—any point (x, y) lying on the line segment joining $(3, 5)$ and $(6, 1)$; min: 7

3. $x = 0, y = 4; P = 20$

4. $x = 0, y = 12; P = 36$

5. $x = 3, y = 4; C = 26$

6. $x = 1.25, y = 1.5; C = 9.75$

7. $x = 8, y = 0; P = 48$

8. $x = 20, y = 0; C = 40$

9. Max: $x = \frac{100}{11}, y = \frac{210}{11}; Q = \frac{1140}{11}$; min: $x = 0, y = 10; Q = 40$

10. Max: $x = 22, y = 0; Q = 22$; min: $x = 3, y = \frac{5}{2}; Q = \frac{11}{2}$

11. $x = 3, y = 4, u = 0, v = 0; P = 25$

12. $x = 3, y = 6, u = 4, v = 0, w = 0; P = 36$

13. $x = \frac{11}{3}, y = 4, u = \frac{7}{3}, v = 0, w = 0; P = 19$

14. $x = \frac{56}{5}, y = \frac{2}{5}, z = 0, u = 0, v = 0; P = \frac{118}{5}$

15. $x = 2, y = 2, u = 0, v = 0; C = -22$

16. Primal problem: $x = 2$, $y = 1$; $C = 9$; dual problem: $u = \frac{3}{10}$, $v = \frac{11}{10}$; $P = 9$

17. Maximize $P = 14u + 21v + 12w$
subject to $u + 4v - 3w \le 2$
$2u + v - 2w \le 4$
$u + 2v + 5w \le 3$
$u \ge 0, v \ge 0, w \ge 0$

18. $x = \frac{3}{2}$, $y = 1$; $C = \frac{13}{2}$

19. $x = \frac{32}{11}$, $y = \frac{36}{11}$; $C = \frac{104}{11}$

20. $x = \frac{3}{4}$, $y = 0$, $z = \frac{7}{4}$; $C = 60$

21. $40,000 in each company; $13,600

22. 60 model A satellite radios; 60 model B satellite radios; $1320

23. 93 model A, 180 model B; $456

24. Saddle Mine: 4 days; Horseshoe Mine: 6 days; $152,000

25. $70,000 in blue-chip stocks; $0 in growth stocks; $30,000 in speculative stocks; maximum return: $13,000

26. 0 unit of Product A, 30 units of Product B, 0 unit of Product C; $P = \$180$

Chapter 6 Before Moving On, page 430

1. a.

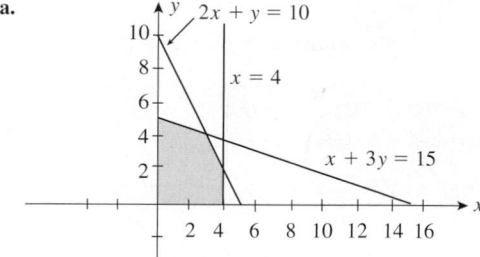

b.

2. Min: -7; max: 76

3. Max: $x = 0$, $y = \frac{24}{7}$; $P = \frac{72}{7}$

4.

x	y	z	u	v	w	P	Constant
2	①	-1	1	0	0	0	3
1	-2	3	0	1	0	0	1
3	2	4	0	0	1	0	17
-1	-2	3	0	0	0	1	0

5. $x = 2$, $y = 0$, $z = 11$, $u = 2$, $v = 0$, $w = 0$; $P = 28$

6. $x = 6$, $y = 2$; $u = 0$, $v = 0$; $P = 34$

CHAPTER 7

Exercises 7.1, page 438

1. $\{x \mid x$ is a gold medalist in the 2014 Winter Olympic Games$\}$

3. $\{x \mid x$ is an integer greater than 2 and less than 8$\}$

5. $\{2, 3, 4, 5, 6\}$ **7.** $\{-2\}$

9. a. True   **b.** False   **11. a.** False   **b.** False

13. True   **15. a.** True   **b.** False   **17. a.** and **b.**

19. a. $\varnothing, \{1\}, \{2\}, \{1, 2\}$
b. $\varnothing, \{1\}, \{2\}, \{3\}, \{1, 2\}, \{1, 3\}, \{2, 3\}, \{1, 2, 3\}$
c. $\varnothing, \{1\}, \{2\}, \{3\}, \{4\}, \{1, 2\}, \{1, 3\}, \{1, 4\}, \{2, 3\}, \{2, 4\}, \{3, 4\}, \{1, 2, 3\}, \{1, 2, 4\}, \{1, 3, 4\}, \{2, 3, 4\}, \{1, 2, 3, 4\}$

21. $\{1, 2, 3, 4, 6, 8, 10\}$

23. $\{$Jill, John, Jack, Susan, Sharon$\}$

25. a.

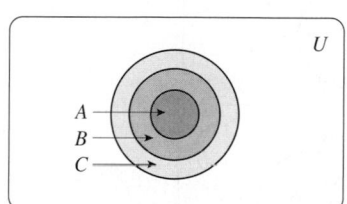

b.

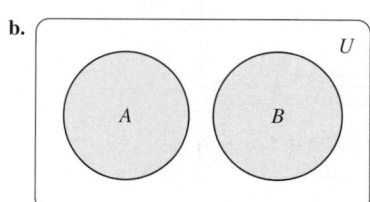

c.

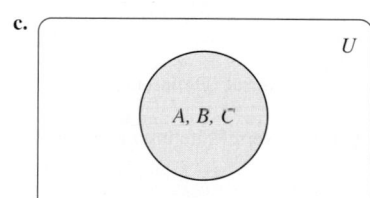

27. a. $(A \cap B^c) \cup (A^c \cap B)$ **b.** $(A \cap B)^c$ or $A^c \cup B^c$

29. a.

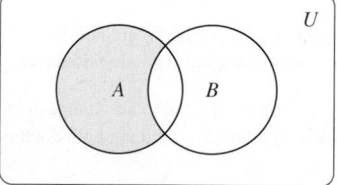

b.

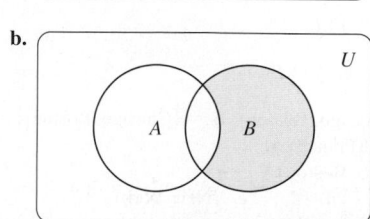

31. a.

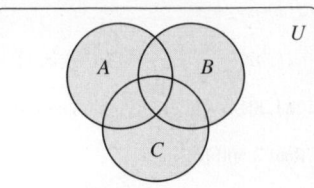

b.

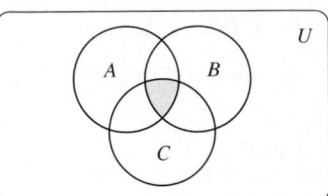

33. a.

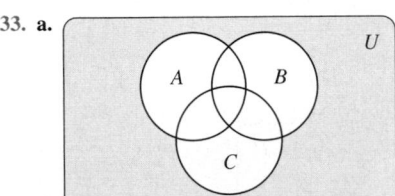

b.

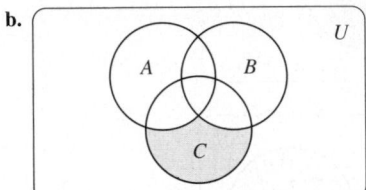

35. a. {2, 4, 6, 8, 10}   **b.** {1, 2, 4, 5, 6, 8, 9, 10}
c. {1, 2, 3, 4, 5, 6, 7, 8, 9, 10}

37. a. C = {1, 2, 4, 5, 8, 9}   **b.** $\varnothing$ **c.** {1, 2, 3, 4, 5, 6, 7, 8, 9, 10}

39. a. Not disjoint   **b.** Disjoint

41. a. The set of all employees at Universal Life Insurance who do not drink tea
b. The set of all employees at Universal Life Insurance who do not drink coffee

43. a. The set of all employees at Universal Life Insurance who drink tea but not coffee
b. The set of all employees at Universal Life Insurance who drink coffee but not tea

45. a. The set of all employees in a hospital who are not doctors
b. The set of all employees in a hospital who are not nurses

47. a. The set of all employees in a hospital who are female doctors
b. The set of all employees in a hospital who are both doctors and administrators

49. a. $D \cap F$ **b.** $R \cap F^c \cap L^c$

51. a. B^c **b.** $A \cap B$ **c.** $A \cap B \cap C^c$

53. a. A = {New York, Chicago, Boston}; B = {Chicago, Boston};
C = {Las Vegas, San Francisco}
b. {New York, Chicago, Boston}
c. {Chicago, Boston}   **d.** $\varnothing$ **e.** {New York}
f. {Las Vegas, San Francisco}

55. a. $A \cap B \cap C$; the set of tourists who have taken the underground, a cab, and a bus over a 1-week period in London
b. $A \cap C$; the set of tourists who have taken the underground and a bus over a 1-week period in London
c. B^c; the set of tourists who have not taken a cab over a 1-week period in London

57.

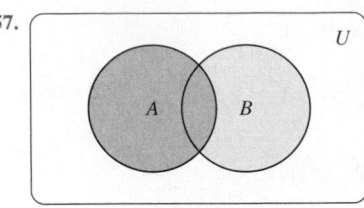

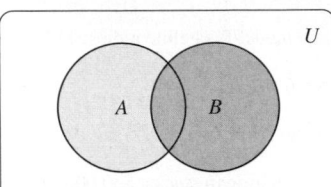

59.

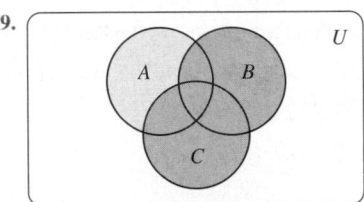

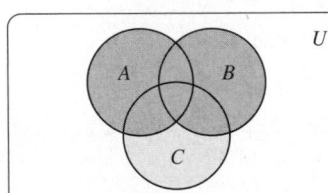

61.

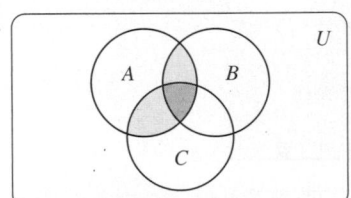

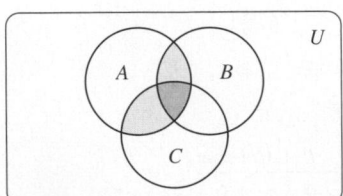

65. a. x, y, v, r, w, u **b.** v, r

67. a. s, t, y **b.** t, z, w, x, s **69.** $A \subset C$

71. False   **73.** True   **75.** True

77. True   **79.** True

Exercises 7.2, page 447

3. a. 4   **b.** 5   **c.** 7   **d.** 2   **7.** 20

9. a. 15   **b.** 30   **c.** 15   **d.** 12   **e.** 50   **f.** 20

11. a. 140   **b.** 100   **c.** 60

13. 13   **15.** 0   **17.** 13   **19.** 61

21. a. 106   **b.** 64   **c.** 38   **d.** 14

23. a. 182   **b.** 118   **c.** 56   **d.** 18   **25.** 30

27. a. 16   **b.** 31   **c.** 4   **d.** 21   **e.** 11

29.

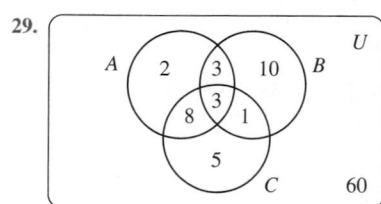

31.

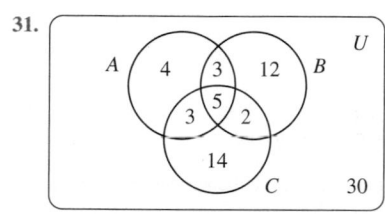

33. a. 64   **b.** 10   **35. a.** 36   **b.** 36

37. a. 3, 2, 4   **b.** 3   **c.** 1   **d.** 1   **e.** 0

39. a. 5, 3, 5   **b.** 2   **c.** 3   **d.** 3   **e.** 0   **f.** 3

41. 5   **43. a.** 62   **b.** 33   **c.** 25   **d.** 38

45. a. 108   **b.** 15   **c.** 45   **d.** 12

47. a. 22   **b.** 80

49. True   **51.** True

Exercises 7.3, page 455

1. 12   **3.** 64   **5.** 24   **7.** 24   **9.** 60

11. 1 billion   **13.** 45,000   **15.** $2^8 = 256$ **17.** 120

19. 5^{50} **21.** 576   **23. a.** 17,576,000   **b.** 17,576,000

25. 456,976,000   **27.** 2,340,000   **29.** 2730

31. a. $10^6 = 1,000,000$ **b.** 250,000   **33.** 217   **35.** True

Exercises 7.4, page 468

1. 360   **3.** 10   **5.** 120   **7.** 20   **9.** n **11.** 1

13. 35   **15.** 1   **17.** 84   **19.** $\dfrac{n(n-1)}{2}$ **21.** $\dfrac{n!}{2}$

23. Permutation   **25.** Combination

27. Permutation   **29.** Combination

31. $P(4, 4) = 24$ **33.** $P(4, 4) = 24$ **35.** $P(9, 9) = 362,880$

37. $C(12, 3) = 220$ **39.** 151,200   **41.** 2520   **43.** 20

45. $C(12, 3) = 220$ **47.** $C(100, 3) = 161,700$

49. $P(6, 6) = 720$ **51.** $C(3, 1) + C(4, 1) = 7$

53. $P(12, 6) = 665,280$ **55.** $C(8, 4)[C(7, 4) + C(6, 3)] = 3850$

57. a. $P(10, 10) = 3,628,800$
b. $P(3, 3)P(4, 4)P(3, 3)P(3, 3) = 5184$

59. a. $P(20, 20) = 20!$
b. $P(5, 5)P[(4, 4)]^5 = 5!(4!)^5 = 955,514,880$

61. a. $P(12, 9) = 79,833,600$
b. $C(12, 9) = 220$ **c.** $C(12, 9) \cdot C(3, 2) = 660$

63. $2\{C(2, 2) + [C(3, 2) - C(2, 2)]\} = 6$

65. $C(3, 3)[C(8, 6) + C(8, 7) + C(8, 8)] = 37$

67. a. $C(12, 3) = 220$ **b.** $C(11, 2) = 55$
c. $C(5, 1)C(7, 2) + C(5, 2)C(7, 1) + C(5, 3) = 185$

69. $P(7, 3) + C(7, 2)P(3, 2) = 336$

71. $C(5, 1)C(3, 1)C(6, 2)[C(4, 1) + C(3, 1)] = 1575$

73. $10C(4, 1) = 40$ **75.** $C(4, 1)C(13, 5) - 40 = 5108$

77. $13C(4, 3) \cdot 12C(4, 2) = 3744$ **79.** $C(6, 2) = 15$

81. $C(12, 6) + C(12, 7) + C(12, 8) + C(12, 9) +$
$C(12, 10) + C(12, 11) + C(12, 12) = 2510$

83. $4! = 24$ **87.** True   **89.** True

Using Technology Exercises 7.4, page 472

1. $1.307674368 \times 10^{12}$ **3.** $2.56094948229 \times 10^{16}$

5. 674,274,182,400   **7.** 133,784,560

9. 4,656,960   **11.** 658,337,004,000

Exercises 7.5, page 478

1. $\{a, b, d, f\}$; $\{a\}$ **3.** $\{b, c, e\}$; $\{a\}$ **5.** No   **7.** S

9. $\varnothing$ **11.** Yes   **13.** Yes   **15.** $E \cup F$ **17.** G^c

19. $(E \cup F \cup G)^c$

21. a. $\{(2, 1), (3, 1), (4, 1), (5, 1), (6, 1), (3, 2), (4, 2), (5, 2), (6, 2),$
$(4, 3), (5, 3), (6, 3), (5, 4), (6, 4), (6, 5)\}$
b. $\{(1, 2), (2, 4), (3, 6)\}$

23. $\varnothing, \{a\}, \{b\}, \{c\}, \{a, b\}, \{a, c\}, \{b, c\}, S$

25. a. $S = \{B, R\}$ **b.** $\varnothing, \{B\}, \{R\}, S$

27. a. $S = \{(H, 1), (H, 2), (H, 3), (H, 4), (H, 5), (H, 6), (T, 1),$
$(T, 2), (T, 3), (T, 4), (T, 5), (T, 6)\}$
b. $\{(H, 2), (H, 4), (H, 6)\}$

29. a. No   **b.** No

31. $S = \{ddd, ddn, dnd, ndd, dnn, ndn, nnd, nnn\}$

33. a. $S = \{bbbb, bbbg, bbgb, bbgg, bgbb, bgbg, bggb, bggg, gbbb,$
$\quad gbbg, gbgb, gbgg, ggbb, ggbg, gggb, gggg\}$
b. $E = \{bbbg, bbgb, bgbb, gbbb\}$
c. $F = \{bbbg, bbgg, bgbg, bggg, gbbg, gbgg, ggbg, gggg\}$
d. $G = \{gbbg, gbgg, ggbg, gggg\}$

35. a. $\{ABC, ABD, ABE, ACD, ACE, ADE, BCD, BCE, BDE, CDE\}$
b. 6   **c.** 3   **d.** 6

37. a. E^c **b.** $E^c \cap F^c$ **c.** $E \cup F$
d. $(E \cap F^c) \cup (E^c \cap F)$

39. a. $\{t \mid t > 0\}$ **b.** $\{t \mid 0 < t \le 2\}$ **c.** $\{t \mid t > 2\}$

41. a. $S = \{0, 1, 2, 3, \ldots, 10\}$ **b.** $E = \{0, 1, 2, 3\}$
c. $F = \{5, 6, 7, 8, 9, 10\}$

43. a. $S = \{0, 1, 2, \ldots, 20\}$
b. $E = \{0, 1, 2, \ldots, 9\}$ **c.** $F = \{20\}$

49. False

Exercises 7.6, page 486

1. $\{(H, H)\}, \{(H, T)\}, \{(T, H)\}, \{(T, T)\}$

3. $\{(D, m)\}, \{(D, f)\}, \{(R, m)\}, \{(R, f)\}, \{(I, m)\}, \{(I, f)\}$

5. $\{(1, i)\}, \{(1, d)\}, \{(1, s)\}, \{(2, i)\}, \{(2, d)\}, \{(2, s)\}, \ldots,$
$\{(5, i)\}, \{(5, d)\}, \{(5, s)\}$

7. $\{(A, Rh^+)\}, \{(A, Rh^-)\}, \{(B, Rh^+)\}, \{(B, Rh^-)\},$
$\{(AB, Rh^+)\}, (AB, Rh^-)\}, \{(O, Rh^+)\}, \{(O, Rh^-)\}$

9. a.

Answer	1–2	3–4	5 or more
Probability	.65	.20	.15

b. .20

11. a.

Answer	Low	Middle	Extreme	No Response
Probability	.34	.44	.20	.02

b. .20

13.

Grade	A	B	C	D	F
Probability	.10	.25	.45	.15	.05

15. a.

Answer	New processes/ procedures	Getting to know a new boss and coworkers	New technology tools	Fitting into the corporate culture	Other
Probability	.44	.20	.17	.12	.07

b. .12

17.

Event	A	B	C	D	E
Probability	.026	.199	.570	.193	.012

19.

Number of Days	0	1	2	3	4	5	6	7
Probability	.05	.06	.09	.15	.11	.20	.17	.17

21. a. $S = \{(0 < x \le 200), (200 < x \le 400),$
$\quad (400 < x \le 600), (600 < x \le 800),$
$\quad (800 < x \le 1000), (x > 1000)\}$

b.

Cars, x	Probability
$0 < x \le 200$	.075
$200 < x \le 400$	.1
$400 < x \le 600$	.175
$600 < x \le 800$	.35
$800 < x \le 1000$	.225
$x > 1000$	.075

23. .469   **25. a.** .856   **b.** .144   **27.** .46

29. a. $\frac{1}{4}$ **b.** $\frac{1}{2}$ **c.** $\frac{1}{13}$ **31.** $\frac{3}{8}$ **33.** No   **35.** Yes

37. a. $\frac{3}{7}$ **b.** $\frac{3}{14}$ **c.** 1   **39. a.** .4   **b.** .1   **c.** .1

41. a. .633   **b.** .276   **43. a.** .35   **b.** .33

45. .530   **47. a.** .4   **b.** .23

49. a. .448   **b.** .255   **51.** .783   **53.** True

Exercises 7.7, page 496

1. $\frac{1}{2}$ **3.** $\frac{1}{36}$ **5.** $\frac{1}{9}$ **7.** $\frac{1}{52}$

9. $\frac{3}{13}$ **11.** $\frac{12}{13}$ **13.** .002; .998

15. $P(a) + P(b) + P(c) \ne 1$

17. Since the five events are not mutually exclusive, Property 3 cannot be used; that is, he could win more than one purse.

19. The two events are not mutually exclusive; hence, the probability of the given event is $\frac{1}{6} + \frac{1}{6} - \frac{1}{36} = \frac{11}{36}$.

21. $E^c \cap F^c = \{e\} \ne \emptyset$

23. a. 0   **b.** .7   **c.** .8   **d.** .3

25. a. $\frac{1}{2}, \frac{3}{8}$ **b.** $\frac{1}{2}, \frac{5}{8}$ **c.** $\frac{1}{8}$ **d.** $\frac{3}{4}$ **e.** $\frac{1}{4}$ **f.** $\frac{7}{8}$

27. a. .30   **b.** .64   **c.** .36   **29. a.** .06   **b.** .39   **c.** .45

31. a. .53   **b.** .51   **c.** .56   **33. a.** .15   **b.** .44

35. b. .52   **c.** .859   **37. a.** .41   **b.** .48

39. a. .24   **b.** .46   **41. a.** .16   **b.** .38   **c.** .22

43. a. .33   **b.** .09   **45.** .332

47. a. .333   **b.** .584   **c.** .055   **49. a.** .63   **b.** .55   **c.** .33

51. a. .09   **b.** .41; during descent and approach and landing

53. a. .43   **b.** .75   **57.** True   **59.** False   **61.** True

Chapter 7 Concept Review Questions, page 503

1. set; elements; set   **2.** equal   **3.** subset

4. a. no   **b.** all   **5. a.** union   **b.** intersection

6. complement   **7.** $A^c \cap B^c \cap C^c$ **8.** permutation; combination

9. experiment; sample; space; event   **10.** $\emptyset$ **11.** uniform; $\frac{1}{n}$

Chapter 7 Review Exercises, page 504

1. $\{3\}$ **2.** $\{A, E, H, L, S, T\}$

3. $\{4, 6, 8, 10\}$ **4.** $\{-4\}$ **5.** Yes

6. Yes   **7.** Yes   **8.** No

9.

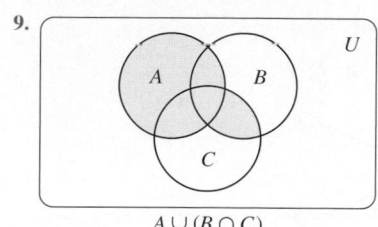

$A \cup (B \cap C)$

10.

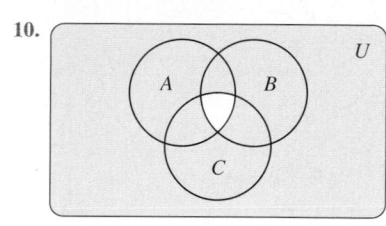

$(A \cap B \cap C)^c$

11.

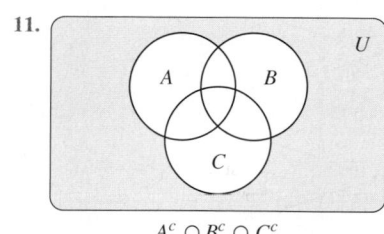

$A^c \cap B^c \cap C^c$

12.

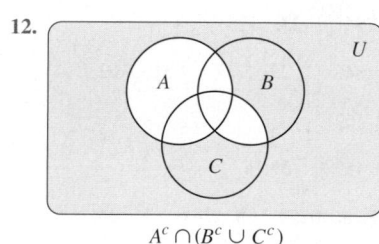

$A^c \cap (B^c \cup C^c)$

17. The set of all participants in a consumer-behavior survey who both avoided buying a product because it is not recyclable and boycotted a company's products because of its record on the environment.

18. The set of all participants in a consumer-behavior survey who avoided buying a product because it is not recyclable and/or voluntarily recycled their garbage.

19. The set of all participants in a consumer-behavior survey who both did not use cloth diapers rather than disposable diapers and voluntarily recycled their garbage.

20. The set of all participants in a consumer-behavior survey who did not boycott a company's products because of its record on the environment and/or did not voluntarily recycle their garbage.

21. 150   **22.** 230   **23.** 270   **24.** 30   **25.** 70   **26.** 200

27. 190   **28.** 181,440   **29.** 120   **30.** 8400

31. a. 0   **b.** .6   **c.** .6   **d.** .4   **e.** 1

32. a. .35   **b.** .65   **c.** .05

33. a. .53   **b.** .35   **c.** .82   **d.** .18

34. a. .49   **b.** .39   **c.** .48

35. $\frac{11}{15}$ **36.** None   **37. a.** 446   **b.** 377   **c.** 34   **38.** 720

39. 20   **40.** 144   **41. a.** 720   **b.** 480

42. a. 50,400   **b.** 5040   **43. a.** 60   **b.** 125   **44.** 108

45. 30   **46.** 80   **47. a.** 1287   **b.** 288   **48.** 720

49. 1050   **50. a.** 2704   **b.** 2652   **51. a.** 5040   **b.** 3600

52. a. 487,635   **b.** 550   **c.** 341,055   **53. a.** 1365   **b.** 1155

54. a. 720   **b.** 72   **c.** 48   **55. a.** 495   **b.** 210   **c.** 420

56. a. .019   **b.** .981

57.

Answer	Falling behind	Staying even	Increasing faster	Don't know
Probability	.40	.44	.12	.04

58.

Income ($)	0–24,999	25,000–49,999	50,000–74,999
Probability	.251	.249	.176

Income ($)	75,000–99,999	100,000–124,999	125,000–149,999
Probability	.115	.075	.044

Income ($)	150,000–199,999	200,000–249,999	250,000 or more
Probability	.049	.019	.023

59. a. .61   **b.** .12   **60. a.** .54   **b.** .72

61. a. .926   **b.** .074   **62. a.** .34   **b.** .50   **c.** .16

63. a. .79   **b.** .59   **64. a.** .757   **b.** .243

65. a. .56   **b.** .4   **66. a.** .41   **b.** .518

67. a. .429   **b.** .490   **68.** .5

Chapter 7 Before Moving On, page 508

1. a. $\{d, f, g\}$ **b.** $\{b, c, d, e, f, g\}$ **c.** $\{b, c, e\}$ **2.** 3

3. 15   **4.** 200   **5.** $\frac{5}{12}$ **6.** $\frac{4}{13}$ **7. a.** .9   **b.** .3

CHAPTER 8

Exercises 8.1, page 514

1. $\frac{1}{32}$ **3.** $\frac{31}{32}$

5. $P(E) = 13C(4, 2)/C(52, 2) \approx .059$

7. $C(26, 2)/C(52, 2) \approx .245$

9. $[C(3, 2)C(5, 2)]/C(8, 4) = 3/7$

11. $[C(5, 3)C(3, 1)]/C(8, 4) = 3/7$ **13.** $C(3, 2)/8 = 3/8$

15. 1/8   **17.** $C(10, 6)/2^{10} \approx .205$

19. a. $C(4, 2)/C(24, 2) \approx .022$
   b. $1 - C(20, 2)/C(24, 2) \approx .312$

21. a. $C(6, 2)/C(80, 2) \approx .005$
   b. $1 - C(74, 2)/C(80, 2) \approx .145$

23. a. .12; $C(98, 10)/C(100, 12) \approx .013$
    b. .15; .015

25. $[C(12, 8)C(8, 2) + C(12, 9)C(8, 1) + C(12, 10)]/C(20, 10) \approx .085$

27. a. $\frac{3}{5}$ **b.** $C(3, 1)/C(5, 3) = .3$ **c.** $1 - C(3, 3)/C(5, 3) = .9$

29. $\frac{1}{729}$ **31.** .0001   **33.** .1   **35.** $40/C(52, 5) \approx .0000154$

37. $[4C(13, 5) - 40]/C(52, 5) \approx .00197$

39. $[13C(4, 3) \cdot 12C(4, 2)]/C(52, 5) \approx .00144$

41. a. .618   **b.** .059   **43.** .030   **45.** .167

Exercises 8.2, page 528

1. a. .4   **b.** .33   **3.** .3   **5.** Independent

7. Independent   **9. a.** .24   **b.** .76   **c.** .4   **d.** .76

11. a. .5   **b.** .4   **c.** .2   **d.** .35   **e.** No   **f.** No

13. a. .4   **b.** .3   **c.** .12   **d.** .30   **e.** Yes   **f.** Yes

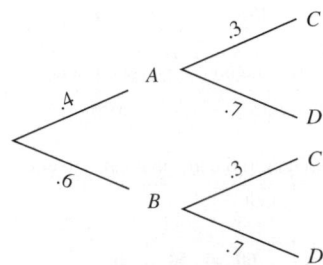

15. a. $\frac{1}{12}$ **b.** $\frac{1}{36}$ **c.** $\frac{1}{6}$ **d.** $\frac{1}{6}$ **e.** No

17. $\frac{4}{11}$ **19.** Independent   **21.** Not independent   **23.** .1875

25. a. .42   **b.** .85   **27.** .06   **29. a.** $\frac{1}{21}$ **b.** $\frac{1}{3}$

31. .48   **33.** Not independent

35. a. .757; .569; .393; .520; .720   **b.** Not independent

37. $\frac{1}{7}$ **39.** .98   **41. a.** .3   **b.** .09375   **43.** .014

45. .25   **47.** $\frac{1}{15}$ **49.** .054   **51.** .1701   **53.** .0000068

55. a. .092   **b.** .008   **57.** 3   **59. a.** No   **b.** Yes

61. 0   **63.** $\dfrac{P(A)}{P(A) + P(B)}$ **65.** True   **67.** True

Exercises 8.3, page 538

1.

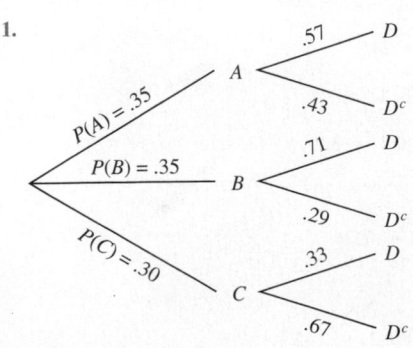

3. a. .45   **b.** .22   **5. a.** .48   **b.** .33

7. a. .08   **b.** .15   **c.** .348

9. a. $\frac{1}{12}$ **b.** $\frac{1}{4}$ **c.** $\frac{1}{18}$ **d.** $\frac{3}{14}$

11.

```
                              .2
                                    B
                      A
              .3              .8
                                    B^c
                              .3
                                    B
              .7      A^c
                              .7
                                    B^c
```

    a. .27   **b.** .22   **c.** .73   **d.** .33

13. $\frac{4}{17}$ **15.** $\frac{4}{51}$

17.

```
                      2/5        4/9  W
                            W
                                 5/9  B
                      3/5        1/3  W
                            B
                                 2/3  B
```

19. $\frac{9}{17}$ **21. a.** .416   **b.** .584   **23.** .125

25. a. .297   **b.** .10   **27.** .348   **29.** .927   **31.** .710

33. .3758   **35. a.** .57   **b.** .691

37. a. .543   **b.** .545   **c.** .455   **39. a.** $\frac{3}{4}$ **b.** $\frac{2}{9}$

41. a. .497   **b.** .959   **43. a.** .025   **b.** .24   **c.** .24

45. .407   **47. a.** .03   **b.** .29   **49.** .028

51. a. .763   **b.** .276   **c.** .724   **53.** .1337

Exercises 8.4, page 550

1. a. See part (b).
   b.

Outcome	GGG	GGR	GRG	RGG
Value	3	2	2	2

Outcome	GRR	RGR	RRG	RRR
Value	1	1	1	0

   c. {GGG}

3. Any positive integer   **5.** $\frac{1}{6}$

7. Any positive integer; infinite discrete

9. $x \geq 0$; continuous

11. Any positive integer; infinite discrete

13. No. The probability assigned to a value of the random variable X cannot be negative.

15. No. The sum of the probabilities exceeds 1.

17. $a = .2$

19. a. .20 **b.** .60 **c.** .30 **d.** 1 **e.** .40 **f.** 0

21.

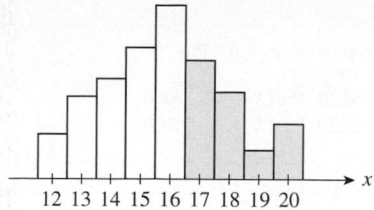

23. a.

x	1	2	3	4	5	6
$P(X = x)$	$\frac{1}{6}$	$\frac{1}{6}$	$\frac{1}{6}$	$\frac{1}{6}$	$\frac{1}{6}$	$\frac{1}{6}$

y	1	2	3	4	5	6
$P(Y = y)$	$\frac{1}{6}$	$\frac{1}{6}$	$\frac{1}{6}$	$\frac{1}{6}$	$\frac{1}{6}$	$\frac{1}{6}$

b.

$x + y$	2	3	4	5	6	7
$P(X + Y = x + y)$	$\frac{1}{36}$	$\frac{2}{36}$	$\frac{3}{36}$	$\frac{4}{36}$	$\frac{5}{36}$	$\frac{6}{36}$

$x + y$	8	9	10	11	12
$P(X + Y = x + y)$	$\frac{5}{36}$	$\frac{4}{36}$	$\frac{3}{36}$	$\frac{2}{36}$	$\frac{1}{36}$

25. a.

x	2	2.25	2.55	2.56	2.58	2.6	2.65	2.85
$P(X = x)$	$\frac{1}{30}$	$\frac{7}{30}$	$\frac{7}{30}$	$\frac{1}{30}$	$\frac{1}{30}$	$\frac{8}{30}$	$\frac{3}{30}$	$\frac{2}{30}$

b. $\frac{1}{2}$

27. a.

x	0	1	2	3	4
$P(X = x)$	.017	.067	.033	.117	.233

x	5	6	7	8	9	10
$P(X = x)$	.133	.167	.100	.050	.067	.017

b.

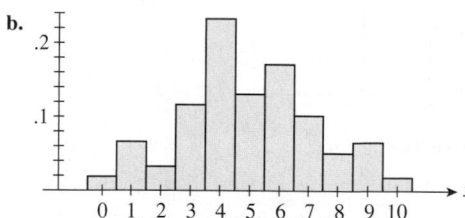

c. .217

29. a.

x	1	2	3	4
$P(X = x)$	.228	.492	.148	.132

b. .72

31. a.

x	1	2	3	4	5
$P(X = x)$	.020	.110	.250	.540	.080

b. .13

33. a.

x	1	2	3	4	5
$P(X = x)$	.131	.160	.179	.230	.300

b. .530; .291

35. True

Using Technology Exercises 8.4, page 557

Graphing Utility

1.

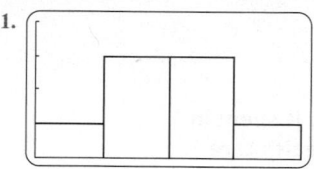

3.

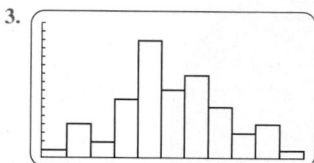

Excel

1.

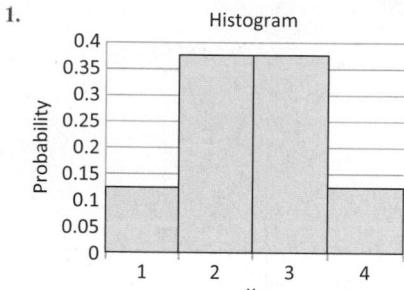

3.

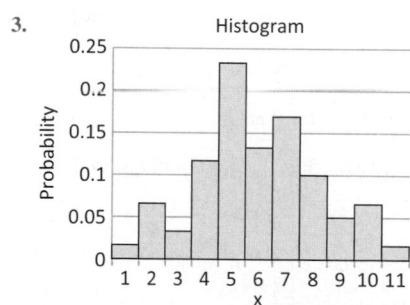

5.

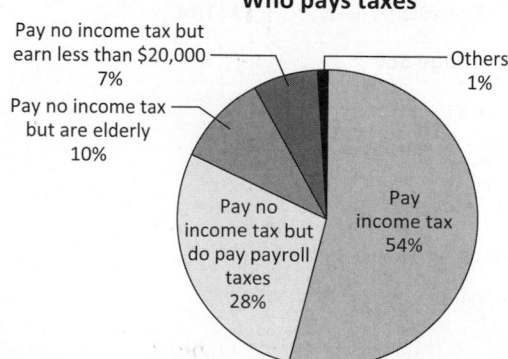

7.

Why young adults shop online

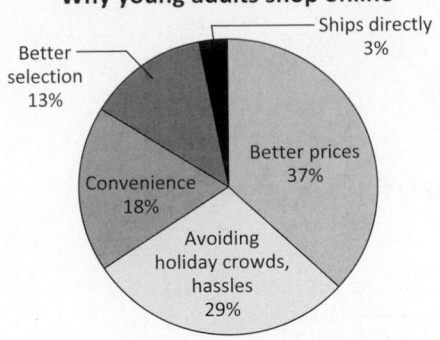

- Ships directly 3%
- Better selection 13%
- Better prices 37%
- Convenience 18%
- Avoiding holiday crowds, hassles 29%

9.

How money is spent in U.S. for health care

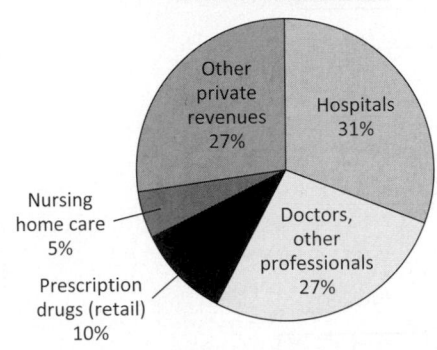

- Other private revenues 27%
- Hospitals 31%
- Nursing home care 5%
- Doctors, other professionals 27%
- Prescription drugs (retail) 10%

Exercises 8.5, page 568

1. 0.86

3. a. 2.6

b.

x	0	1	2	3	4	
$P(X = x)$	0	.1	.4	.3	.2	; 2.6

5. 0.91   **7.** $78.50   **9.** 0.12   **11.** 1.73   **13.** 3.16%

15. −39¢   **17.** $100   **19.** $118,800

21. City B **23.** Company B **25.** 2.86%

27. −5.3¢   **29.** −2.7¢   **31.** 25.3 min   **33.** 36.2 years

35. 2 to 3; 3 to 2   **37.** .4   **39.** $\frac{7}{12}$ **41.** $\frac{5}{14}$

45. a. Mean: 74; mode: 85; median: 80   **b.** Mode

47. 3; close   **49.** 16; 16; 16

51. Mean: 7.15; median: 7; mode: 7   **53.** True

Exercises 8.6, page 580

1. $\mu = 2$, Var$(X) = 1$, $\sigma = 1$

3. $\mu = 0$, Var$(X) = 1$, $\sigma = 1$

5. $\mu = 518$, Var$(X) = 1891$, $\sigma \approx 43.5$

7. Figure (a)   **9.** 1.56

11. $\mu = 4.5$, Var$(X) = 5.25$

13. a. Let X = the annual birthrate during the years 2003–2012.

b.

x	13.7	13.8	14.0	14.2	14.7
$P(X = x)$	.1	.3	.3	.2	.1

c. $\mu = 14.02$, Var$(X) = 0.0776$, $\sigma \approx 0.2786$

15. a. Mutual Fund A: $\mu = \$620$, Var$(X) = 267,600$;
Mutual Fund B: $\mu = \$520$, Var$(X) = 137,600$

b. Mutual Fund A

c. Mutual Fund B

17. 1   **19.** 58.833 hr; 5.70 hr   **21.** 84.33%; 6.67%

23. 1607 hr; 182 hr   **25.** 21.59%; 5.20%   **27.** 44.75 mi; 3.07 mi

29. 94.56%; 19.94%   **31.** 2.722%; 0.969%   **33.** 3.324; 0.4497

35. $5.0154 million; $1.07 million

37. a. 8.0583 million; 0.82 million   **b.** 5.825 million; 0.65 million

c. The average monthly supply of single-family homes for sale dropped from 2011 to 2012 as the economy recovered from the Great Recession.

39. 35.28 thousand; 5.92 thousand

41. a. At least .75
b. At least .96

43. 7   **45.** At least $\frac{7}{16}$

47. At least $\frac{15}{16}$ **49.** True

Using Technology Exercises 8.6, page 586

1. a.

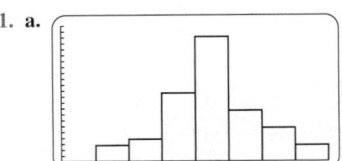

b. $\mu = 4$, $\sigma \approx 1.40$

3. a.

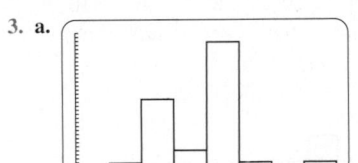

b. $\mu = 17.34$, $\sigma \approx 1.11$

5. a. Let X denote the random variable that gives the weight of a carton of sugar.

b.

x	4.96	4.97	4.98	4.99	5.00	5.01
$P(X = x)$	$\frac{3}{30}$	$\frac{4}{30}$	$\frac{4}{30}$	$\frac{1}{30}$	$\frac{1}{30}$	$\frac{5}{30}$

x	5.02	5.03	5.04	5.05	5.06
$P(X = x)$	$\frac{3}{30}$	$\frac{3}{30}$	$\frac{4}{30}$	$\frac{1}{30}$	$\frac{1}{30}$

c. $\mu \approx 5.00$; $\sigma \approx 0.03$

7. a.

b. 65.875; 1.73

Chapter 8 Concept Review Questions, page 588

1. conditional   **2.** independent   **3.** a posteriori probability

4. random   **5.** finite; infinite; continuous   **6.** sum; .75

7. a. $\dfrac{P(E)}{P(E^c)}$ **b.** $\dfrac{a}{a+b}$

8. $p_1(x_1 - \mu)^2 + p_2(x_2 - \mu)^2 + \cdots + p_n(x_n - \mu)^2$; $\sqrt{\text{Var}(X)}$

Chapter 8 Review Exercises, page 588

1. .364   **2.** No   **3.** .5   **4.** .18   **5.** .25   **6.** .06

7. .49   **8.** .37   **9. a.** $\frac{7}{8}$ **b.** $\frac{7}{8}$ **c.** No

10. a. .284   **b.** .984   **11.** .150   **12.** $\frac{2}{15}$ **13.** $\frac{1}{24}$ **14.** $\frac{1}{52}$

15. .00018   **16.** .00995   **17.** .245   **18.** .510   **19.** .245

20. a. {WWW, BWW, WBW, WWB, BBW, BWB, WBB, BBB}

b.

Outcome	WWW	BWW	WBW	WWB
Value of X	0	1	1	1

Outcome	BBW	BWB	WBB	BBB
Value of X	2	2	2	3

c.

x	0	1	2	3
$P(X = x)$	$\frac{1}{35}$	$\frac{12}{35}$	$\frac{18}{35}$	$\frac{4}{35}$

d.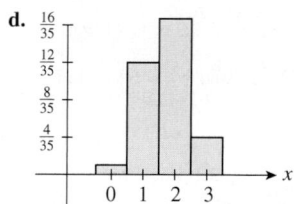

21. $100   **22. a.** .8   **b.** $\mu = 2.7$; $\sigma \approx 1.42$

23. .619   **24.** .180   **25. a.** .513   **b.** .390

26. .30   **27.** .244   **28.** .60

29. a. X gives the minimum age requirement for a regular driver's license.

b.

x	15	16	17	18	19	21
$P(X = x)$	.02	.30	.08	.56	.02	.02

c. 17.34; 1.2244; 1.11

30. a.

x	1	2	3	4	5
$P(X = x)$	.110	.143	.172	.236	.338

b. 57.4%; 25.3%

31. 41.3 mph   **32.** $12,000   **33.** At least .75

34. $31.0 million; $4.523 million   **35.** $\mu = 27.58$; $\sigma = 6.32$

Chapter 8 Before Moving On, page 591

1. .72   **2.** .308

3.

x	−3	−2	0	1	2	3
$P(X = x)$	.05	.1	.25	.3	.2	.1

4. a. .8   **b.** .92   **5.** 0.44; 4.0064; 2.0016

CHAPTER 9

Exercises 9.1, page 608

1. $\lim\limits_{x \to -2} f(x) = 3$ **3.** $\lim\limits_{x \to 3} f(x) = 3$ **5.** $\lim\limits_{x \to -2} f(x) = 3$

7. The limit does not exist.

9.

x	1.9	1.99	1.999
$f(x)$	4.61	4.9601	4.9960

x	2.001	2.01	2.1
$f(x)$	5.004	5.0401	5.41

$\lim\limits_{x \to 2}(x^2 + 1) = 5$

11.

x	−0.1	−0.01	−0.001
$f(x)$	−1	−1	−1

x	0.001	0.01	0.1
$f(x)$	1	1	1

The limit does not exist.

13.

x	0.9	0.99	0.999
$f(x)$	100	10,000	1,000,000

x	1.001	1.01	1.1
$f(x)$	1,000,000	10,000	100

The limit does not exist.

15.

x	0.9	0.99	0.999	1.001	1.01	1.1
$f(x)$	2.9	2.99	2.999	3.001	3.01	3.1

$\lim\limits_{x \to 1} \dfrac{x^2 + x - 2}{x - 1} = 3$

17.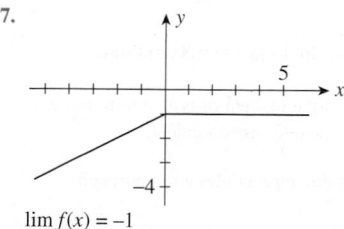

$\lim\limits_{x \to 0} f(x) = -1$

19.

$\lim\limits_{x \to 1} f(x) = 1$

21.

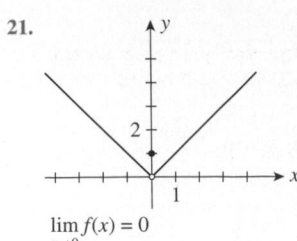

$$\lim_{x \to 0} f(x) = 0$$

23. 3 **25.** 3 **27.** −1 **29.** 2

31. −4 **33.** $\frac{5}{4}$ **35.** 2 **37.** $\sqrt{171} = 3\sqrt{19}$

39. $\frac{3}{2}$ **41.** −1 **43.** −6 **45.** 2

47. $\frac{1}{6}$ **49.** 2 **51.** −1 **53.** −10

55. The limit does not exist. **57.** $\frac{5}{3}$ **59.** $\frac{1}{2}$ **61.** $\frac{1}{3}$

63. $\lim_{x \to \infty} f(x) = \infty$; $\lim_{x \to -\infty} f(x) = \infty$ **65.** 0; 0

67. $\lim_{x \to \infty} f(x) = -\infty$; $\lim_{x \to -\infty} f(x) = -\infty$

69.

x	1	10	100	1000
$f(x)$	0.5	0.009901	0.0001	0.000001

x	−1	−10	−100	−1000
$f(x)$	0.5	0.009901	0.0001	0.000001

$\lim_{x \to \infty} f(x) = 0$ and $\lim_{x \to -\infty} f(x) = 0$

71.

x	1	5	10	100
$f(x)$	12	360	2910	2.99×10^6

x	1000	−1	−5
$f(x)$	2.999×10^9	6	−390

x	−10	−100	−1000
$f(x)$	−3090	-3.01×10^6	-3.0×10^9

$\lim_{x \to \infty} f(x) = \infty$ and $\lim_{x \to -\infty} f(x) = -\infty$

73. 3 **75.** 3 **77.** $\lim_{x \to -\infty} f(x) = -\infty$ **79.** 0

81. a. $0.5 million; $0.75 million; $1.17 million; $2 million; $4.5 million; $9.5 million
  b. The limit does not exist; as the percent of pollutant to be removed approaches 100, the cost becomes astronomical.

83. $2.20; the average cost of producing x DVDs will approach $2.20/disc in the long run.

85. a. $24 million; $60 million; $83.1 million **b.** $120 million

87. a. 137.3¢/mi; 59.8¢/mi; 45.1¢/mi; 39.8¢/mi; 37.3¢/mi

  b.

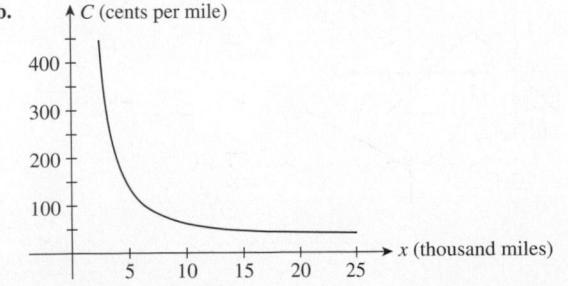

  c. It approaches 32.8¢/mi.

89. False **91.** True **93.** True

95. a moles/L/sec **97.** No

Using Technology Exercises 9.1, page 614

1. 5 **3.** 3 **5.** $\frac{2}{3}$ **7.** $e^2 \approx 7.38906$

11. a.

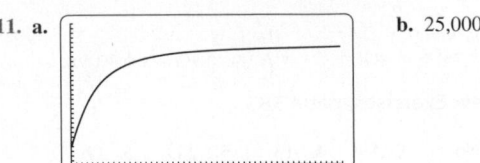

  b. 25,000

Exercises 9.2, page 623

1. 3; 2; the limit does not exist.

3. The limit does not exist; 2; the limit does not exist.

5. 0; 2; the limit does not exist.

7. −2; 2; the limit does not exist. **9.** True **11.** True

13. False **15.** True **17.** False **19.** True **21.** 6

23. $-\frac{1}{4}$ **25.** The limit does not exist. **27.** −1 **29.** 0

31. −4 **33.** The limit does not exist. **35.** 4 **37.** 0; 0

39. $x = 0$; conditions 2 and 3 **41.** Continuous everywhere

43. $x = 0$; condition 3 **45.** $(-\infty, \infty)$ **47.** $(-\infty, \infty)$

49. $\left(-\infty, \frac{1}{2}\right) \cup \left(\frac{1}{2}, \infty\right)$ **51.** $(-\infty, -2) \cup (-2, 1) \cup (1, \infty)$

53. $(-\infty, \infty)$ **55.** $(-\infty, \infty)$ **57.** −1 and 1 **59.** 1 and 2

61. f is discontinuous at $x = 4, 5, \ldots, 13$.

63. Michael makes progress toward solving the problem until $x = x_1$. Between $x = x_1$ and $x = x_2$, he makes no further progress. But at $x = x_2$ he suddenly achieves a breakthrough, and at $x = x_3$ he proceeds to complete the problem.

65. Conditions 2 and 3 are not satisfied at each of these points.

67.

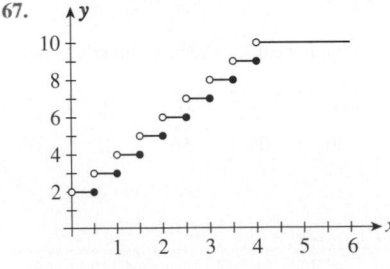

f is discontinuous at $x = \frac{1}{2}, 1, 1\frac{1}{2}, \ldots, 4$.

69.

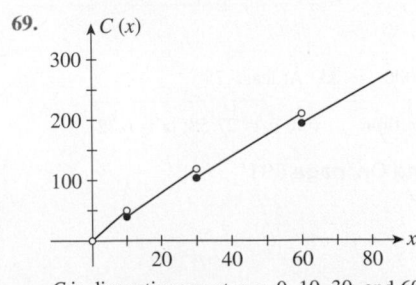

C is discontinuous at $x = 0, 10, 30,$ and 60.

71. a. ∞; as the time taken to excite the tissue is made shorter and shorter, the strength of the electric current gets stronger and stronger.
b. b; as the time taken to excite the tissue is made longer and longer, the strength of the electric current gets weaker and weaker and approaches b.

73. 3

75. a. f is a polynomial of degree 2. **b.** $f(1) = 3$ and $f(3) = -1$

77. a. f is a polynomial of degree 3. **b.** $f(-1) = -4$ and $f(1) = 4$

79. 0.59 **81.** 1.34

83. c. $\frac{1}{2}$; $\frac{7}{2}$; Joan sees the ball on its way up $\frac{1}{2}$ sec after it was thrown and again 3 sec later.

85. False **87.** False **89.** False **91.** False **93.** False

Using Technology Exercises 9.2, page 629

1. $x = 0, 1$ **3.** $x = 0, \frac{1}{2}$ **5.** $x = -\frac{1}{2}, 2$ **7.** $x = -2, 1$

9.
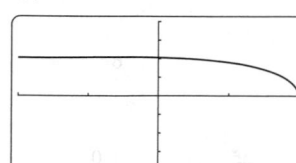

11.
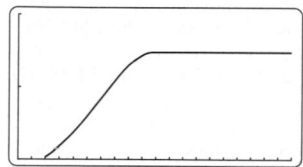

Exercises 9.3, page 642

1. 1.5 lb/month; 0.58 lb/month; 1.25 lb/month

3. 3.1%/hr; −21.2%/hr

5. a. Car A **b.** They are traveling at the same speed.
c. Car B **d.** Both cars covered the same distance; they are again side-by-side.

7. a. P_2 **b.** P_1 **c.** Bactericide B; Bactericide A

9. 0 **11.** 2 **13.** $6x$ **15.** $-2x + 3$ **17.** 2; $y = 2x + 7$

19. 6; $y = 6x - 3$ **21.** $\frac{1}{9}$; $y = \frac{1}{9}x - \frac{2}{3}$

23. a. $4x$ **b.** $y = 4x - 1$

c.
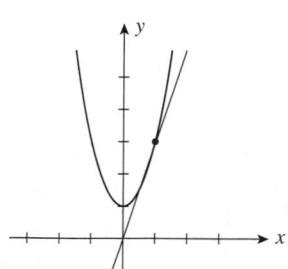

25. a. $2x - 2$ **b.** $(1, 0)$

c.
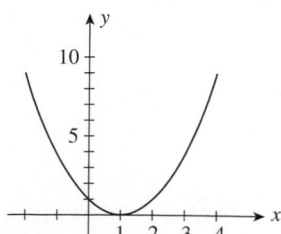

d. 0

27. a. 6; 5.5; 5.1 **b.** 5
c. The computations in part (a) illustrate that as h approaches zero, the average rate of change approaches the instantaneous rate of change.

29. a. 130 ft/sec; 128.2 ft/sec; 128.02 ft/sec **b.** 128 ft/sec
c. The computations in part (a) illustrate that as the time intervals over which the average velocity are computed become smaller and smaller, the average velocity approaches the instantaneous velocity of the car at $t = 20$.

31. a. 5 sec **b.** 80 ft/sec **c.** 160 ft/sec

33. a. $-\frac{1}{6}$ L/atm **b.** $-\frac{1}{4}$ L/atm

35. a. $-\frac{2}{3}x + 7$
b. $333 per $1000 spent on advertising; $-$13,000 per $1000 spent on advertising

37. $6 billion/year; $10 billion/year

39. a. $f'(h)$ gives the instantaneous rate of change of the temperature with respect to height at a given height h, in °F per foot.
b. Negative **c.** ≈ -0.05°F

41. Average rate of change of the seal population over $[a, a + h]$; instantaneous rate of change of the seal population at $x = a$

43. Average rate of change of the country's industrial production over $[a, a + h]$; instantaneous rate of change of the country's industrial production at $x = a$

45. Average rate of change of atmospheric pressure with respect to altitude over $[a, a + h]$; instantaneous rate of change of atmospheric pressure with respect to altitude at $x = a$

47. a. Yes **b.** No **c.** No **49. a.** Yes **b.** Yes **c.** No

51. a. No **b.** No **c.** No

53. 32.1, 30.939, 30.814, 30.8014, 30.8001, 30.8000; 30.8 ft/sec

55. False

57.
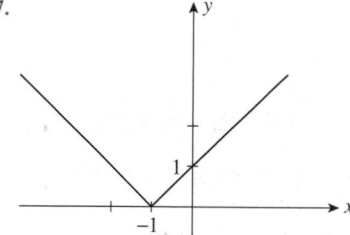

59. $a = 2, b = -1$
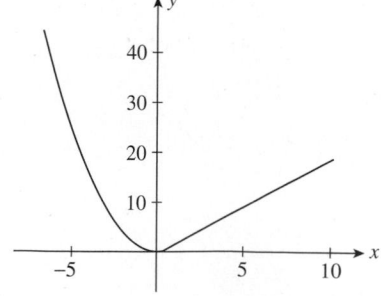

Using Technology Exercises 9.3, page 647

1. a. 9

b.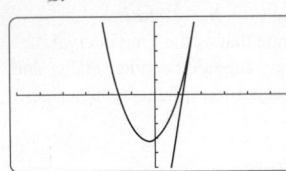

c. $y = 9x - 11$

3. a. $\frac{1}{12}$

b.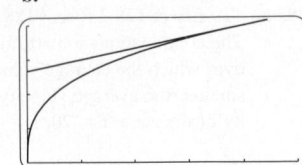

c. $y = \frac{1}{12}x + \frac{4}{3}$

5. a. 4

b.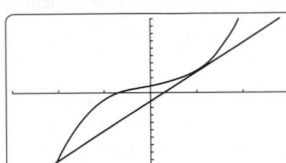

c. $y = 4x - 1$

7. a. 4.02

b.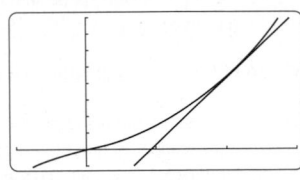

c. $y = 4.02x - 3.57$

9. a.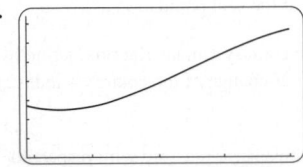

b. 2.88 million/decade

Exercises 9.4, page 655

1. 0 **3.** $5x^4$ **5.** $3.1x^{2.1}$ **7.** $6x$

9. $2\pi r$ **11.** $\dfrac{3}{x^{2/3}}$ **13.** $\dfrac{3}{2\sqrt{x}}$ **15.** $-84x^{-13}$

17. $10x - 3$ **19.** $-3x^2 + 4x$ **21.** $0.06x - 0.4$

23. $4x - 4 - \dfrac{3}{x^2}$ **25.** $16x^3 - \dfrac{15}{2}x^{3/2}$ **27.** $-\dfrac{5}{x^2} - \dfrac{8}{x^3}$

29. $-\dfrac{16}{t^5} + \dfrac{9}{t^4} - \dfrac{2}{t^2}$ **31.** $3 - \dfrac{5}{2\sqrt{x}}$ **33.** $-\dfrac{4}{x^3} + \dfrac{1}{x^{4/3}}$

35. a. 20 **b.** -4 **c.** 20 **37.** 3 **39.** 11

41. $m = 5; y = 5x - 4$ **43.** $m = 3; y = 3x - 7$

45. a. $(0, 0)$

b.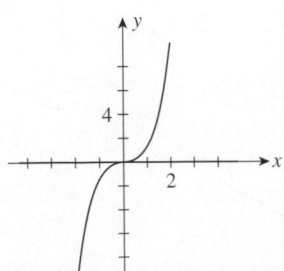

47. a. $(-2, -7), (2, 9)$
b. $y = 12x + 17$ and $y = 12x - 15$

c.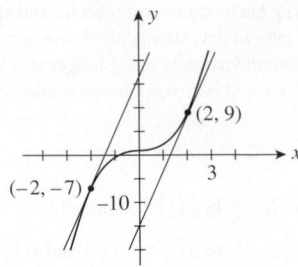

49. a. $(0, 0); \left(1, -\frac{13}{12}\right)$ **b.** $(0, 0); \left(2, -\frac{8}{3}\right); \left(-1, -\frac{5}{12}\right)$
c. $(0, 0); \left(4, \frac{80}{3}\right); \left(-3, \frac{81}{4}\right)$

51. a. $\dfrac{16\pi}{9}$ cm³/cm **b.** $\dfrac{25\pi}{4}$ cm³/cm

53. a. 491.5 million **b.** 476.8 million/year

55. a. 49.6%; 36.9%; 32.1% **b.** -3.3%/decade; -2.3%/decade

57. a. 44.4% **b.** 1%/decade

59. a. $120 - 30t$ **b.** 120 ft/sec **c.** 240 ft

61. a. \$18,115.41 **b.** \$1290.12/year

63. a. 15 points/year; 12.6 points/year; 0 points/year
b. 10 points/year

65. 63,000 people/year; 60,000 people/year; 55,000 people/year; 48,000 people/year; yes

67. a. $-6t^2 + 24t$ **b.** 0 ft/sec; 24 ft/sec; 0 ft/sec; -72 ft/sec
c. 69 ft

69. 155 people/month; 200 people/month

71. $-0.006x^2 + 1.2x + 1$; (a)

73. a. 12%; 31.2% **b.** 0.8%/year; 1.4%/year

75. a. $G(t) = \begin{cases} -0.0002t^2 + 0.032t + 0.1 & \text{if } 0 \le t < 5 \\ 0.0002t^2 - 0.006t + 0.28 & \text{if } 5 \le t < 10 \\ -0.0012t^2 + 0.082t - 0.46 & \text{if } 10 \le t < 15 \end{cases}$
b. 2800 jobs/year; 53,200 jobs/year

77. False

Using Technology Exercises 9.4, page 661

1. 1 **3.** 0.4226 **5.** 0.1613

7. a.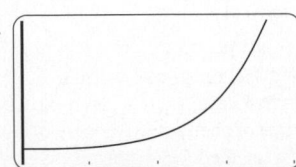

b. 3.4295 ppm/40 years; 164.239 ppm/40 years

Exercises 9.5, page 670

1. $2x(2x) + (x^2 + 1)(2)$, or $6x^2 + 2$

3. $(t - 1)(2) + (2t + 1)(1)$, or $4t - 1$

5. $(3x + 1)(2x) + (x^2 - 2)(3)$, or $9x^2 + 2x - 6$

7. $(x^3 - 1)(1) + (x + 1)(3x^2)$, or $4x^3 + 3x^2 - 1$

9. $(w^3 - w^2 + w - 1)(2w) + (w^2 + 2)(3w^2 - 2w + 1)$, or
$5w^4 - 4w^3 + 9w^2 - 6w + 2$

11. $(5x^2 + 1)(x^{-1/2}) + (2x^{1/2} - 1)(10x)$, or $\dfrac{25x^2 - 10x\sqrt{x} + 1}{\sqrt{x}}$

13. $\dfrac{(x^2 - 5x + 2)(x^2 + 2)}{x^2} + \dfrac{(x^2 - 2)(2x - 5)}{x}$, or $\dfrac{3x^4 - 10x^3 + 4}{x^2}$

15. $\dfrac{-1}{(x - 2)^2}$ **17.** $\dfrac{2(2x + 1) - (2x - 1)(2)}{(2x + 1)^2}$, or $\dfrac{4}{(2x + 1)^2}$

19. $-\dfrac{2x + 1}{(x^2 + x + 2)^2}$ **21.** $\dfrac{s^2 + 2s + 4}{(s + 1)^2}$

23. $\dfrac{(\frac{1}{2}x^{-1/2})[x^2 + 1 - 4x^{3/2}(x^{1/2} + 1)]}{(x^2 + 1)^2}$, or $\dfrac{-3x^2 - 4x^{3/2} + 1}{2\sqrt{x}(x^2 + 1)^2}$

25. $\dfrac{2x^3 + 2x^2 + 2x - 2x^3 - x^2 - 4x - 2}{(x^2 + x + 1)^2}$, or $\dfrac{x^2 - 2x - 2}{(x^2 + x + 1)^2}$

27. $\dfrac{(x - 2)(3x^2 + 2x + 1) - (x^3 + x^2 + x + 1)}{(x - 2)^2}$, or
$\dfrac{2x^3 - 5x^2 - 4x - 3}{(x - 2)^2}$

29. $\dfrac{(x^2 - 4)(x^2 + 4)(2x + 8) - (x^2 + 8x - 4)(4x^3)}{(x^2 - 4)^2(x^2 + 4)^2}$, or
$\dfrac{-2x^5 - 24x^4 + 16x^3 - 32x - 128}{(x^2 - 4)^2(x^2 + 4)^2}$

31. 8   **33.** −9   **35.** $2(3x^2 - x + 3)$; 10

37. $\dfrac{-3x^4 + 2x^2 - 1}{(x^4 - 2x^2 - 1)^2}$; $-\dfrac{1}{2}$ **39.** 60; $y = 60x - 102$

41. $-\frac{1}{2}$; $y = -\frac{1}{2}x + \frac{3}{2}$ **43.** $8x - 2$; 8   **45.** $6x^2 - 6x$; $6(2x - 1)$

47. $4t^3 - 6t^2 + 12t - 3$; $12(t^2 - t + 1)$ **49.** $72x - 24$ **51.** $-\dfrac{6}{x^4}$

53. 8   **55.** $y = 7x - 5$ **57.** $(\frac{1}{3}, \frac{50}{27})$; $(1, 2)$

59. $(\frac{4}{3}, -\frac{770}{27})$; $(2, -30)$ **61.** $y = -\frac{1}{2}x + 1$; $y = 2x - \frac{3}{2}$

63. 0.125, 0.5, 2, 50; the cost of removing (essentially) all of the pollutant is prohibitively high.

65. −5000/min; −1600/min; 7000; 4000

67. a. $\dfrac{50x}{0.01x^2 + 1}$ **b.** $\dfrac{50(1 - 0.01x^2)}{(0.01x^2 + 1)^2}$
c. 6.69, 0, −3.70; the revenue is increasing at the rate of approximately $6700/thousand watches/week when the level of sales is 8000 watches/week; the rate of change of the revenue is $0/thousand watches/week when the level of sales is 10,000 watches/week, and the revenue is decreasing at the rate of approximately $3700/thousand watches/week when the sales are 12,000 watches/week.

69. a. 4.7%/year   **b.** Dropping at the rate of 0.67%/year/year

71. a. $\dfrac{180}{(t + 6)^2}$ **b.** 3.7; 2.2; 1.8; 1.1
c.

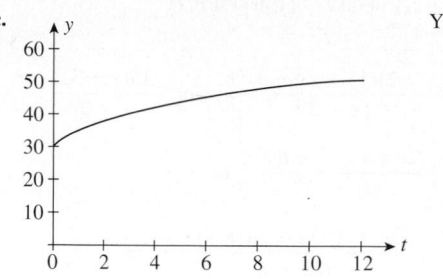

Yes

d. 50 words/min

73. Dropping at the rate of 0.0375 ppm/year; dropping at the rate of 0.006 ppm/year

75. a. $xD'(x) + D(x)$ **b.** $a - 2bx$ **77.** 128 ft/sec; 32 ft/sec^2

79. a. 19%   **b.** 5.1%/year
c. 0.76%/year/year; the percentage of vehicles equipped with transmissions that have seven speeds or more is increasing at the rate of 0.76%/year/year.

81. a. and b.

t	0	1	2	3	4	5	6	7
$N'(t)$	0	2.7	4.8	6.3	7.2	7.5	7.2	6.3
$N''(t)$				0.6	0	−0.6	−1.2	

83. 2736; 756 records/month; 148 records/month/month

85. The proportion of the U.S. population that was obese was increasing at an increasing rate from 1991 through 2004.

87. True

Using Technology Exercises 9.5, page 676

1. 0.8750   **3.** 0.0774   **5.** 31,312/year   **7.** −18   **9.** 15.2762

Exercises 9.6, page 684

1. $6(2x - 1)^2$ **3.** $10x(x^2 + 2)^4$

5. $3(2x - x^2)^2(2 - 2x)$, or $6x^2(1 - x)(2 - x)^2$ **7.** $\dfrac{-4}{(2x + 1)^3}$

9. $5x(x^2 - 4)^{3/2}$ **11.** $\dfrac{3}{2\sqrt{3x - 2}}$ **13.** $\dfrac{-2x}{3(1 - x^2)^{2/3}}$

15. $-\dfrac{6}{(2x + 3)^4}$ **17.** $\dfrac{-1}{(2t - 4)^{3/2}}$ **19.** $-\dfrac{3(16x^3 + 1)}{2(4x^4 + x)^{5/2}}$

21. $-2(3x^2 + 2x + 1)^{-3}(6x + 2)$ or $-4(3x + 1)(3x^2 + 2x + 1)^{-3}$

23. $3(x^2 + 1)^2(2x) - 2(x^3 + 1)(3x^2)$, or $6x(2x^2 - x + 1)$

25. $3(t^{-1} - t^{-2})^2(-t^{-2} + 2t^{-3})$

27. $\dfrac{1}{2\sqrt{x - 1}} + \dfrac{1}{2\sqrt{x + 1}}$

29. $2x^2(4)(3 - 4x)^3(-4) + (3 - 4x)^4(4x)$, or $(-12x)(4x - 1)(3 - 4x)^3$

31. $8(x - 1)^2(2x + 1)^3 + 2(x - 1)(2x + 1)^4$, or
$6(x - 1)(2x - 1)(2x + 1)^3$

33. $3\left(\dfrac{x + 3}{x - 2}\right)^2\left[\dfrac{(x - 2)(1) - (x + 3)(1)}{(x - 2)^2}\right]$, or $-\dfrac{15(x + 3)^2}{(x - 2)^4}$

35. $\dfrac{3}{2}\left(\dfrac{t}{2t + 1}\right)^{1/2}\left[\dfrac{(2t + 1)(1) - t(2)}{(2t + 1)^2}\right]$, or $\dfrac{3t^{1/2}}{2(2t + 1)^{5/2}}$

37. $\dfrac{1}{2}\left(\dfrac{u + 1}{3u + 2}\right)^{-1/2}\left[\dfrac{(3u + 2)(1) - (u + 1)(3)}{(3u + 2)^2}\right]$, or
$-\dfrac{1}{2\sqrt{u + 1}\,(3u + 2)^{3/2}}$

39. $\dfrac{(x^2 - 1)^4(2x) - x^2(4)(x^2 - 1)^3(2x)}{(x^2 - 1)^8}$, or $\dfrac{(-2x)(3x^2 + 1)}{(x^2 - 1)^5}$

41. $\dfrac{2x(x^2 - 1)^3(3x^2 + 1)^2[9(x^2 - 1) - 4(3x^2 + 1)]}{(x^2 - 1)^8}$, or
$-\dfrac{2x(3x^2 + 13)(3x^2 + 1)^2}{(x^2 - 1)^5}$

43. $\dfrac{(2x + 1)^{-1/2}[(x^2 - 1) - (2x + 1)(2x)]}{(x^2 - 1)^2}$, or
$-\dfrac{3x^2 + 2x + 1}{\sqrt{2x + 1}(x^2 - 1)^2}$

45. $\dfrac{(t^2 + 1)^{1/2}(\frac{1}{2})(t + 1)^{-1/2}(1) - (t + 1)^{1/2}(\frac{1}{2})(t^2 + 1)^{-1/2}(2t)}{t^2 + 1}$, or
$-\dfrac{t^2 + 2t - 1}{2\sqrt{t + 1}(t^2 + 1)^{3/2}}$

47. $4(3x + 1)^3(3)(x^2 - x + 1)^3 + (3x + 1)^4(3)(x^2 - x + 1)^2(2x - 1)$,
or $3(3x + 1)^3(x^2 - x + 1)^2(10x^2 - 5x + 3)$

49. $\frac{4}{3}u^{1/3}$; $6x$; $8x(3x^2 - 1)^{1/3}$

51. $-\dfrac{2}{3u^{5/3}}$; $6x^2 - 1$; $-\dfrac{2(6x^2 - 1)}{3(2x^3 - x + 1)^{5/3}}$

53. $\frac{1}{2}u^{-1/2} - \frac{1}{2}u^{-3/2}$; $3x^2 - 1$; $\dfrac{(3x^2 - 1)(x^3 - x - 1)}{2(x^3 - x)^{3/2}}$

55. $2f'(2x + 1)$ 57. -12 59. 6   61. No

63. $y = -33x + 57$ 65. $y = \frac{43}{5}x - \frac{54}{5}$

67. 0.333 million/week; 0.305 million/week; 16 million; 22.7 million

69. 102,000/year

71. **a.** 30%; 10.3%; the probability of survival 1 year after diagnosis is
approximately 30%, and after 2 years is approximately 10.3%.
b. -33.8%/year; -10.4%/year; after 1 year, the probability of survival is dropping at the rate of approximately 34%/year, and after
2 years is dropping at the rate of approximately 10.4%/year.

73. **a.** $0.0267(0.2t^2 + 4t + 64)^{-1/3}(0.1t + 1)$ **b.** 0.0090 ppm/year

75. **a.** $0.03[3t^2(t - 7)^4 + t^3(4)(t - 7)^3]$, or $0.21t^2(t - 3)(t - 7)^3$
b. 90.72; 0; -90.72; at 8 A.M. the level of nitrogen dioxide is
increasing; at 10 A.M. the level stops increasing; at 11 A.M. the
level is decreasing.

77.
$300\left[\dfrac{(t + 25)\frac{1}{2}(\frac{1}{2}t^2 + 2t + 25)^{-1/2}(t + 2) - (\frac{1}{2}t^2 + 2t + 25)^{1/2}(1)}{(t + 25)^2}\right]$,
or $\dfrac{3450t}{(t + 25)^2\sqrt{\frac{1}{2}t^2 + 2t + 25}}$; 2.9 beats/min/sec, 0.7 beats/min/sec,
0.2 beats/min/sec, 179 beats/min

79. 160π ft²/sec

81.
$(1.42)\left[\dfrac{(3t^2 + 80t + 550)(14t + 140) - (7t^2 + 140t + 700)(6t + 80)}{(3t^2 + 80t + 550)^2}\right]$,
or $\dfrac{1.42(140t^2 + 3500t + 21,000)}{(3t^2 + 80t + 550)^2}$; 31,312 jobs/year/month

83. $\dfrac{2500p}{9\sqrt{t}\sqrt{810,000 - p^2}(1 + \frac{1}{8}\sqrt{t})^2}$; 19 tablet computers/month

85. True   87. True

Using Technology Exercises 9.6, page 688

1. 0.5774   3. 0.9390   5. -4.9498

7. 5,414,500 people/year; 2,513,600 people/year

Exercises 9.7, page 696

1. $3e^{3x}$ 3. $-e^{-t}$ 5. $e^x + 2x$ 7. $x^2e^x(x + 3)$

9. $\dfrac{e^x(x - 1)}{x^2}$ 11. $3(e^x - e^{-x})$ 13. $-\dfrac{2}{e^w}$

15. $6e^{3x-1}$ 17. $-2xe^{-x^2}$ 19. $-\dfrac{3e^{1/x}}{x^2}$

21. $25e^x(e^x + 1)^{24}$ 23. $\dfrac{e^{\sqrt{x}}}{2\sqrt{x}}$ 25. $e^{3x+2}(3x - 2)$

27. $\dfrac{2e^x}{(e^x + 1)^2}$ 29. $16e^{-4x} + 9e^{3x}$ 31. $6e^{3x}(3x + 2)$

33. $y = 2x - 2$ 35. $\dfrac{5}{x}$ 37. $\dfrac{1}{x + 1}$ 39. $\dfrac{8}{x}$

41. $\dfrac{1}{2x}$ 43. $-\dfrac{2}{x}$ 45. $\dfrac{8x - 5}{4x^2 - 5x + 3}$

47. $\dfrac{1}{x(x + 1)}$ 49. $x(1 + 2\ln x)$ 51. $\dfrac{2(1 - \ln x)}{x^2}$

53. $\dfrac{3}{u - 2}$ 55. $\dfrac{1}{2x\sqrt{\ln x}}$ 57. $\dfrac{2\ln x}{x}$

59. $\dfrac{3x^2}{x^3 + 1}$ 61. $\dfrac{(x\ln x + 1)e^x}{x}$ 63. $-\dfrac{1}{x^2}$

65. $\dfrac{2(2 - x^2)}{(x^2 + 2)^2}$ 67. $y = x - 1$

69. **a.** 188.7 million   **b.** 12.6 million viewers/year

71. **a.** 70,000; 353,700   **b.** 37,800/decade; 191,000/decade

73. a. −$6065/day/day; −$3679/day/day; −$2231/day/day;
    −$1353/day/day
   b. 2 days

75. 12.9%/year; 10.9%/year; 9.3%/year; and 7.9%/year

77. a. 181
   b. 0/decade; −27/decade; −38/decade; −32/decade
   c. 52/decade

79. a. −1.68¢/case/case   **b.** $40.36/case

81. a. 70°F   **b.** −14.7°F/min   **c.** 30°F

83. 0.0580%/kg; 0.0133%/kg

87. False

Using Technology Exercises 9.7, page 700

1. 5.4366   **3.** 12.3929   **5.** 0.1861

7. a.

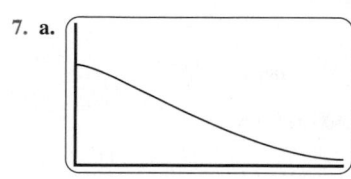

b. 50   **c.** −0.2/year

Exercises 9.8, page 708

1. a. $C(x)$ is always increasing because as the number of units x
      produced increases, the amount of money that must be spent
      on production also increases.
   b. 4000

3. a. $1.80; $1.60   **b.** $1.80; $1.60

5. a. $100 + \dfrac{200{,}000}{x}$ **b.** $-\dfrac{200{,}000}{x^2}$
   c. $\overline{C}(x)$ approaches $100 if the production level is very high.

7. $\dfrac{2000}{x} + 2 - 0.0001x;\ -\dfrac{2000}{x^2} - 0.0001$

9. a. $8000 - 200x$ **b.** 200, 0, −200   **c.** $40

11. a. $-0.04x^2 + 600x - 300{,}000$
   b. $-0.08x + 600$ **c.** 200; −40
   d. The profit increases as production increases, peaking at 7500
      units; beyond this level, profit falls.

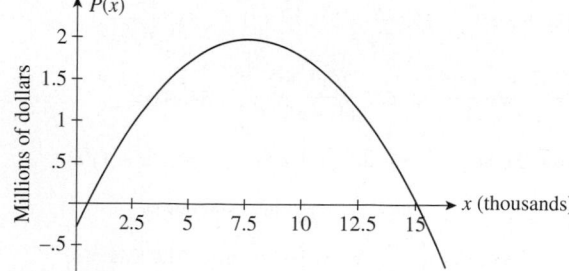

13. a. $600x - 0.05x^2;\ -0.000002x^3 - 0.02x^2 + 200x - 80{,}000$
   b. $0.000006x^2 - 0.06x + 400;\ 600 - 0.1x;$
      $-0.000006x^2 - 0.04x + 200$
   c. 304; 400; 96

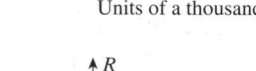

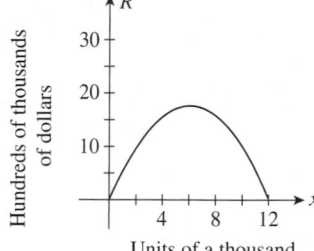

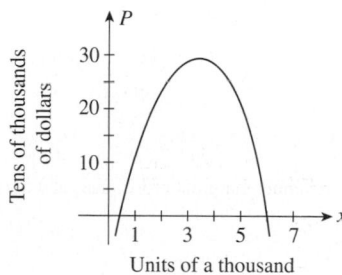

15. a. $0.000002x^2 - 0.03x + 400 + \dfrac{80{,}000}{x}$
   b. $0.000004x - 0.03 - \dfrac{80{,}000}{x^2}$
   c. −0.0132; 0.0092; the marginal average cost is negative (average
      cost is decreasing) when 5000 units are produced and positive
      (average cost is increasing) when 10,000 units are produced.
   d.

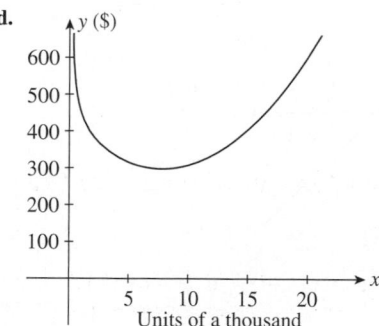

17. a. $\dfrac{50x}{0.01x^2 + 1}$ **b.** $\dfrac{50 - 0.5x^2}{(0.01x^2 + 1)^2}$
   c. $44,380; when the level of production is 2000 units, the reve-
      nue increases at the rate of $44,380 per additional 1000 units
      produced.

19. a. $200 - 0.01x - 0.02x \ln x$ **b.** $13,301

21. 1.21   **23.** 0.288

25. 81.82%/unit change in x **27.** -84.85%/unit change in x

29. 0.615%   **31.** 9.31%/year   **33.** False

Chapter 9 Concept Review Questions, page 712

1. $f(x); L; a$

2. a. L^r **b.** $L \pm M$ **c.** LM **d.** $\dfrac{L}{M}; M \neq 0$

3. a. $L; x$ **b.** M; negative; absolute

4. a. right   **b.** left   **c.** $L; L$

5. a. continuous   **b.** discontinuous   **c.** every

6. a. $a; a; g(a)$ **b.** everywhere   **c.** $Q(x)$

7. a. $[a, b]; f(c) = M$ **b.** $f(x) = 0; (a, b)$

8. a. $f'(a)$ **b.** $y = f(a) + m(x - a)$

9. a. $\dfrac{f(a+h) - f(a)}{h}$ **b.** $\lim\limits_{h \to 0} \dfrac{f(a+h) - f(a)}{h}$

10. a. 0   **b.** nx^{n-1} **c.** $cf'(x)$ **d.** $f'(x) \pm g'(x)$

11. a. $f(x)g'(x) + g(x)f'(x)$ **b.** $\dfrac{g(x)f'(x) - f(x)g'(x)}{[g(x)]^2}$

12. a. $g'[f(x)]f'(x)$ **b.** $n[f(x)]^{n-1}f'(x)$

13. marginal cost; marginal revenue; marginal profit; marginal average cost

14. a. $e^{f(x)}f'(x)$ **b.** $\dfrac{f'(x)}{f(x)}$

Chapter 9 Review Exercises, page 713

1. -3 **2.** 2   **3.** -21 **4.** 0   **5.** -1

6. The limit does not exist.   **7.** 7   **8.** $\frac{9}{2}$ **9.** 1   **10.** $\frac{1}{2}$

11. 1   **12.** 1   **13.** $\frac{3}{2}$ **14.** The limit does not exist.

15.

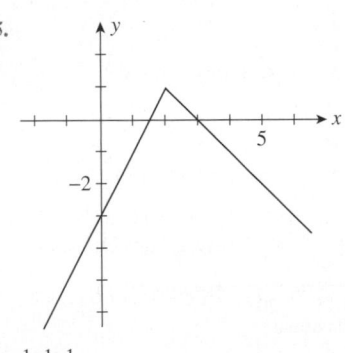

1; 1; 1

16.

4; 2; the limit does not exist.

17. $x = 2$ **18.** $x = -\frac{1}{2}, 1$ **19.** $x = -1$ **20.** $x = 0$

21. a. 3; 2.5; 2.1   **b.** 2   **22.** 4

23. $\dfrac{1}{x^2}$ **24.** $\frac{3}{2}; y = \frac{3}{2}x + 5$

25. $-4; y = -4x + 4$ **26. a.** Yes   **b.** No

27. $15x^4 - 8x^3 + 6x - 2$ **28.** $24x^5 + 8x^3 + 6x$

29. $\dfrac{6}{x^4} - \dfrac{3}{x^2}$ **30.** $4t - 9t^2 + \frac{1}{2}t^{-3/2}$ **31.** $-\dfrac{1}{t^{3/2}} - \dfrac{6}{t^{5/2}}$

32. $2x - \dfrac{2}{x^2}$ **33.** $1 - \dfrac{2}{t^2} - \dfrac{6}{t^3}$ **34.** $4s + \dfrac{4}{s^2} - \dfrac{1}{s^{3/2}}$

35. $2x + \dfrac{3}{x^{5/2}}$ **36.** $\dfrac{(2x-1)(1) - (x+1)(2)}{(2x-1)^2}$, or $-\dfrac{3}{(2x-1)^2}$

37. $\dfrac{(2t^2+1)(2t) - t^2(4t)}{(2t^2+1)^2}$, or $\dfrac{2t}{(2t^2+1)^2}$

38. $\dfrac{(t^{1/2}+1)\frac{1}{2}t^{-1/2} - t^{1/2}(\frac{1}{2}t^{-1/2})}{(t^{1/2}+1)^2}$, or $\dfrac{1}{2\sqrt{t}(\sqrt{t}+1)^2}$

39. $\dfrac{(x^{1/2}+1)(\frac{1}{2}x^{-1/2}) - (x^{1/2}-1)(\frac{1}{2}x^{-1/2})}{(x^{1/2}+1)^2}$, or $\dfrac{1}{\sqrt{x}(\sqrt{x}+1)^2}$

40. $\dfrac{(2t^2+1)(1) - t(4t)}{(2t^2+1)^2}$, or $\dfrac{1 - 2t^2}{(2t^2+1)^2}$

41. $\dfrac{(x^2-1)(4x^3+2x) - (x^4+x^2)(2x)}{(x^2-1)^2}$, or $\dfrac{2x(x^4 - 2x^2 - 1)}{(x^2-1)^2}$

42. $3(4x+1)(2x^2+x)^2$

43. $8(3x^3-2)^7(9x^2)$, or $72x^2(3x^3-2)^7$

44. $5(x^{1/2}+2)^4 \cdot \dfrac{1}{2}x^{-1/2}$, or $\dfrac{5(\sqrt{x}+2)^4}{2\sqrt{x}}$

45. $\dfrac{1}{2}(2t^2+1)^{-1/2}(4t)$, or $\dfrac{2t}{\sqrt{2t^2+1}}$

46. $-2t^2(1-2t^3)^{-2/3}$, or $-\dfrac{2t^2}{(1-2t^3)^{2/3}}$

47. $-4(3t^2-2t+5)^{-3}(3t-1)$, or $-\dfrac{4(3t-1)}{(3t^2-2t+5)^3}$

48. $-\frac{3}{2}(2x^3-3x^2+1)^{-5/2}(6x^2-6x)$, or $-9x(x-1)(2x^3-3x^2+1)^{-5/2}$

49. $(2x+1)e^{2x}$ **50.** $\dfrac{e^t}{2\sqrt{t}} + \sqrt{t}e^t + 1$ **51.** $\dfrac{1-4t}{2\sqrt{t}e^{2t}}$

52. $\dfrac{e^x(x^2+x+1)}{\sqrt{1+x^2}}$ **53.** $\dfrac{2(e^{2x}+2)}{(1+e^{-2x})^2}$ **54.** $4xe^{2x^2-1}$

55. $(1-2x^2)e^{-x^2}$ **56.** $3e^{2x}(1+e^{2x})^{1/2}$ **57.** $(x+1)^2e^x$

58. $\ln t + 1$ **59.** $\dfrac{2xe^{x^2}}{e^{x^2}+1}$ **60.** $\dfrac{\ln x - 1}{(\ln x)^2}$

61. $\dfrac{x - x\ln x + 1}{x(x+1)^2}$ **62.** $(x+2)e^x$ **63.** $\dfrac{4e^{4x}}{e^{4x}+3}$

64. $\dfrac{(r^3 - r^2 + r + 1)e^r}{(1+r^2)^2}$ **65.** $\dfrac{1 + e^x(1 - x\ln x)}{x(1+e^x)^2}$

66. $\dfrac{(2x^2 + 2x^2 \cdot \ln x - 1)e^{x^2}}{x(1 + \ln x)^2}$

67. $2\left(x + \dfrac{1}{x}\right)\left(1 - \dfrac{1}{x^2}\right)$, or $\dfrac{2(x^2 + 1)(x^2 - 1)}{x^3}$

68. $\dfrac{(2x^2 + 1)^2(1) - (1 + x)2(2x^2 + 1)(4x)}{(2x^2 + 1)^4}$, or $-\dfrac{6x^2 + 8x - 1}{(2x^2 + 1)^3}$

69. $(t^2 + t)^4(4t) + 2t^2 \cdot 4(t^2 + t)^3(2t + 1)$, or $4t^2(5t + 3)(t^2 + t)^3$

70. $(2x + 1)^3 \cdot 2(2x^2 + x)(2x + 1) + (x^2 + x)^2 3(2x + 1)^2(2)$, or $2(2x + 1)^2(x^2 + x)(7x^2 + 7x + 1)$

71. $x^{1/2} \cdot 3(x^2 - 1)^2(2x) + (x^2 - 1)^3 \cdot \dfrac{1}{2}x^{-1/2}$, or

$\dfrac{(13x^2 - 1)(x^2 - 1)^2}{2\sqrt{x}}$

72. $\dfrac{(x^3 + 2)^{1/2}(1) - x \cdot \frac{1}{2}(x^3 + 2)^{-1/2} \cdot 3x^2}{x^3 + 2}$, or $\dfrac{4 - x^3}{2(x^3 + 2)^{3/2}}$

73. $\dfrac{(4x - 3)\frac{1}{2}(3x + 2)^{-1/2}(3) - (3x + 2)^{1/2}(4)}{(4x - 3)^2}$, or

$-\dfrac{12x + 25}{2\sqrt{3x + 2}(4x - 3)^2}$

74. $\dfrac{(t + 1)^3\frac{1}{2}(2t + 1)^{-1/2}(2) - (2t + 1)^{1/2} \cdot 3(t + 1)^2(1)}{(t + 1)^6}$, or

$-\dfrac{5t + 2}{\sqrt{2t + 1}(t + 1)^4}$

75. $2(12x^2 - 9x + 2)$ **76.** $-\dfrac{1}{4x^{3/2}} + \dfrac{3}{4x^{5/2}}$

77. $\dfrac{(t^2 + 4)^2(-2t) - (4 - t^2)(2)(t^2 + 4)(2t)}{(t^2 + 4)^4}$, or $\dfrac{2t(t^2 - 12)}{(t^2 + 4)^3}$

78. $4e^{-2x}(x - 1)$ **79.** $\dfrac{e^x(1 - e^x)}{(1 + e^x)^3}$ **80.** $\dfrac{1}{x}$ **81.** $-\dfrac{9}{(3x + 1)^2}$

82. $2(15x^4 + 12x^2 + 6x + 1)$

83. $2(2x^2 + 1)^{-1/2} + 2x\left(-\dfrac{1}{2}\right)(2x^2 + 1)^{-3/2}(4x)$, or $\dfrac{2}{(2x^2 + 1)^{3/2}}$

84. $(t^2 + 1)^2(14t) + (7t^2 + 1)(2)(t^2 + 1)(2t)$, or $6t(t^2 + 1)(7t^2 + 3)$

85. 0 **86.** -2

87. a. $(2, -25)$ and $(-1, 14)$
b. $y = -4x - 17$; $y = -4x + 10$

88. a. $\left(-2, \frac{25}{3}\right)$ and $\left(1, -\frac{13}{6}\right)$
b. $y = -2x + \frac{13}{3}$; $y = -2x - \frac{1}{6}$

89. $y = -\dfrac{\sqrt{3}}{3}x + \dfrac{4}{3}\sqrt{3}$ **90.** $y = 112x - 80$

91. $y = -\dfrac{1}{e^2}(2x - 3)$ **92.** $y = \dfrac{1}{e}$

93. $-\dfrac{48}{(2x - 1)^4}$; $\left(-\infty; \dfrac{1}{2}\right) \cup \left(\dfrac{1}{2}, \infty\right)$ **94.** 20

95. a. $C'(x)$ gives the instantaneous rate of change of the total manufacturing cost C in dollars with respect to the quantity produced
when x units of the product are produced.
b. Positive
c. $20

96. a. 20,430   **b.** 225 cameras/year

97. a. $3.1 billion; $10 billion
b. $0.68 billion/year; $2.08 billion/year

98. a. 75.3 million   **b.** 1.3 million viewers/year

99. a. 15%; 31.99%   **b.** 0.51%/year; 1.04%/year

100. 200 subscribers/week

101. 75 years; 0.07 year/year

102. a. $-0.02x^2 + 600x$ **b.** $-0.04x + 600$
c. 200; the sale of the 10,001st phone will bring a revenue of $200.

103. a. $2.20; $2.20   **b.** $\dfrac{2500}{x} + 2.2$; $-\dfrac{2500}{x^2}$

104. a. 4445 people   **b.** 494 people/year

Chapter 9 Before Moving On, page 716

1. 2   **2. a.** 0   **b.** 1; no

3. -1; $y = -x$ **4.** $6x^2 - \dfrac{1}{x^{2/3}} - \dfrac{10}{3x^{5/3}}$ **5.** $\dfrac{4x^2 - 1}{\sqrt{2x^2 - 1}}$

6. $-\dfrac{2x^2 + 2x - 1}{(x^2 + x + 1)^2}$ **7.** $-\dfrac{1}{2(x + 1)^{3/2}}$; $\dfrac{3}{4(x + 1)^{5/2}}$; $-\dfrac{15}{8(x + 1)^{7/2}}$

8. $\dfrac{e^{\sqrt{x}}}{2\sqrt{x}}$ **9.** $1 + \ln 2$

CHAPTER 10

Exercises 10.1, page 731

1. Decreasing on $(-\infty, 0)$ and increasing on $(0, \infty)$

3. Increasing on $(-\infty, -1)$ and $(1, \infty)$ and decreasing on $(-1, 1)$

5. Decreasing on $(-\infty, 0)$ and $(2, \infty)$ and increasing on $(0, 2)$

7. Decreasing on $(-\infty, -1)$ and $(1, \infty)$ and increasing on $(-1, 1)$

9. Increasing on $(20.2, 20.6)$ and $(21.7, 21.8)$, constant on $(19.6, 20.2)$ and $(20.6, 21.1)$, and decreasing on $(21.1, 21.7)$ and $(21.8, 22.7)$

11. a. f is decreasing on $(0, 4)$ **b.** f is constant on $(4, 12)$
c. f is increasing on $(12, 24)$

13. a. 3, 5, 7, and 9
b.

c. Relative maxima at $(3, 3)$ and $(9, 6)$; relative minimum at $(5, 1)$

15. Increasing on $(-\infty, \infty)$

17. Decreasing on $\left(-\infty, -\frac{1}{4}\right)$ and increasing on $\left(-\frac{1}{4}, \infty\right)$

19. Decreasing on $\left(-\infty, -\dfrac{\sqrt{3}}{3}\right)$ and $\left(\dfrac{\sqrt{3}}{3}, \infty\right)$ and increasing on $\left(-\dfrac{\sqrt{3}}{3}, \dfrac{\sqrt{3}}{3}\right)$

21. Increasing on $(-\infty, -2)$ and $(0, \infty)$ and decreasing on $(-2, 0)$

23. Increasing on $(-\infty, -1)$ and $(3, \infty)$ and decreasing on $(-1, 3)$

25. Decreasing on $(-\infty, 3)$ and increasing on $(3, \infty)$

27. Decreasing on $\left(-\infty, -\dfrac{3}{2}\right)$ and increasing on $\left(-\dfrac{3}{2}, \infty\right)$

29. Increasing on $(-1, 1)$ and decreasing on $(-\infty, -1)$ and $(1, \infty)$

31. Decreasing on $(-\infty, 0)$ and increasing on $(0, \infty)$

33. Decreasing on $(-\infty, 5)$ and increasing on $(5, \infty)$

35. Decreasing on $\left(-1, -\dfrac{2}{3}\right)$ and increasing on $\left(-\dfrac{2}{3}, \infty\right)$

37. Increasing on $(0, 2)$; decreasing on $(-\infty, 0)$ and $(2, \infty)$

39. Increasing on $(0, e)$; decreasing on (e, ∞)

41. Increasing on $(-\infty, 0)$ and $(2, \infty)$; decreasing on $(0, 1)$ and $(1, 2)$

43. Relative maximum: $f(0) = 1$; relative minima: $f(-1) = 0$ and $f(1) = 0$

45. Relative maximum: $f(-1) = 2$; relative minimum: $f(1) = -2$

47. Relative maximum: $f(1) = 3$; relative minimum: $f(2) = 2$

49. Relative maximum: $f(-3) = -\dfrac{9}{2}$; relative minimum: $f(3) = \dfrac{9}{2}$

51. Decreasing on (a, c), where $f'(x) < 0$; increasing on (c, d), where $f'(x) > 0$; constant on (d, e), where $f'(x) = 0$; decreasing on (e, b), where $f'(x) < 0$

53. Increasing on $(0, c)$; neither increasing nor decreasing at c; decreasing on (c, b)

55. (a)   57. (d)   59. Relative minimum: $g\left(-\dfrac{3}{2}\right) = \dfrac{23}{4}$

61. Relative maximum: $h(3) = 15$

63. Relative minimum: $f(0) = 2$

65. Relative maximum: $f(-1) = 8$; relative minimum: $f(1) = 4$

67. Relative maximum: $f(0) = 0$; relative minima: $f(-1) = -\dfrac{1}{2}$ and $f(1) = -\dfrac{1}{2}$

69. Relative minimum: $g(3) = -15$

71. Relative minimum: $F(-2) = 84$; relative maximum: $F(2) = -44$

73. None   75. Relative minimum: $g(10) = 610$ 77. None

79. Relative maximum: $f(1) = e^{-2}$; relative minimum: $f(0) = 0$

81. Relative minimum: $f(e^{-1/2}) = -\dfrac{1}{2e}$

85. Declining from the beginning of 2004 until the end 2009; increasing after 2009

87. f is decreasing on $(0, 1)$ and increasing on $(1, 4)$. The average speed decreases from 6 A.M. to 7 A.M. and then picks up from 7 A.M. to 10 A.M.

89. The U.S. public debt outstanding was increasing throughout the period under consideration.

91. The average cost of driving a 2012 medium-sized sedan decreases as the number of miles driven increases.

93. The pollution is increasing from 7 A.M. to 10 A.M. and decreasing from 10 A.M. to 2 P.M.

95. **a.** 30%; 41.4%
    b. The percentage of small and lower-midsize vehicles was increasing from 2005 through 2015.

97. **a.** 1986 kWh/year

99. **a.** $R(x) = ax - bx^2$
    b. Increasing on $\left(0, \dfrac{a}{2b}\right)$; stationary at $x = \dfrac{a}{2b}$; and decreasing on $\left(\dfrac{a}{2b}, \dfrac{a}{b}\right)$
    c. $\dfrac{a}{2b}$; $\dfrac{a^2}{4b}$

101. False   103. True   105. False

109. **a.** -3 if $x < 0$, 2 if $x \geq 0$ **b.** No

Using Technology Exercises 10.1, page 738

1. **a.** f is decreasing on $(-\infty, -0.2934)$ and increasing on $(-0.2934, \infty)$.
   b. Relative minimum: $f(-0.2934) = -2.5435$

3. **a.** f is increasing on $(-\infty, -1.6144)$ and $(0.2390, \infty)$ and decreasing on $(-1.6144, 0.2390)$.
   b. Relative maximum: $f(-1.6144) = 26.7991$; relative minimum: $f(0.2390) = 1.6733$

5. **a.** f is decreasing on $(-\infty, -1)$ and $(0.33, \infty)$ and increasing on $(-1, 0.33)$.
   b. Relative maximum: $f(0.33) = 1.11$; relative minimum: $f(-1) = -0.63$

7. **a.** Decreasing on $(-\infty, 0.40)$ and increasing on $(0.40, \infty)$
   b. Relative maximum: $(0.40, 0.79)$

9. **a.**

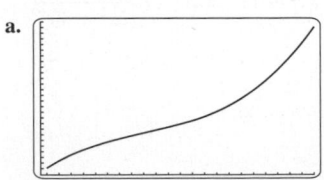

Exercises 10.2, page 749

1. Concave downward on $(-\infty, 0)$ and concave upward on $(0, \infty)$; inflection point: $(0, 0)$

3. Concave downward on $(-\infty, 0)$ and $(0, \infty)$

5. Concave upward on $(-\infty, 0)$ and $(1, \infty)$ and concave downward on $(0, 1)$; inflection points: $(0, 0)$ and $(1, -1)$

7. Concave downward on $(-\infty, -2), (-2, 2)$, and $(2, \infty)$

9. **a.** Concave upward on $(0, 2), (4, 6), (7, 9)$, and $(9, 12)$ and concave downward on $(2, 4)$ and $(6, 7)$
   b. $\left(2, \dfrac{5}{2}\right), (4, 2), (6, 2)$, and $(7, 3)$

11. (a)   **13.** (b)

15. a. $D_1'(t) > 0$, $D_2'(t) > 0$, $D_1''(t) > 0$, and $D_2''(t) < 0$ on $(0, 12)$
   b. With or without the proposed promotional campaign, the deposits will increase; with the promotion, the deposits will increase at an increasing rate; without the promotion, the deposits will increase at a decreasing rate.

17. (c)   **19.** (d)

21. a. Between 8 A.M. and 10 A.M., the rate of change of the rate of change of the number of smartphones assembled is increasing; between 10 A.M. and 12 noon, it is decreasing.
   b. At 10 A.M.

23. At the time t_0, corresponding to its t-coordinate, the restoration process is working at its peak.

29. Concave upward on $(-\infty, \infty)$

31. Concave upward on $(-\infty, 0)$; concave downward on $(0, \infty)$

33. Concave upward on $(-\infty, 0)$ and $(3, \infty)$; concave downward on $(0, 3)$

35. Concave downward on $(-\infty, 0)$ and $(0, \infty)$

37. Concave downward on $(-\infty, 4)$

39. Concave downward on $(-\infty, 2)$; concave upward on $(2, \infty)$

41. Concave upward on $\left(-\infty, -\frac{\sqrt{6}}{3}\right)$ and $\left(\frac{\sqrt{6}}{3}, \infty\right)$; concave downward on $\left(-\frac{\sqrt{6}}{3}, \frac{\sqrt{6}}{3}\right)$

43. Concave downward on $(-\infty, 1)$; concave upward on $(1, \infty)$

45. Concave upward on $(-\infty, 0)$ and $(0, \infty)$

47. Concave upward on $(-\infty, 2)$; concave downward on $(2, \infty)$

49. Concave upward on $(0, \infty)$; concave downward on $(-\infty, 0)$

51. Concave upward on $(-\infty, -1)$ and $(1, \infty)$; concave downward on $(-1, 0)$ and $(0, 1)$

53. $(0, -2)$ **55.** $(1, -20)$ **57.** $(0, 1)$ and $\left(\frac{2}{3}, \frac{11}{27}\right)$

59. $(0, 0)$ **61.** $(1, 2)$ **63.** $\left(-\frac{\sqrt{2}}{2}, 2e^{-1/2}\right)$; $\left(\frac{\sqrt{2}}{2}, 2e^{-1/2}\right)$

65. $\left(e^{-3/2}, -\frac{3}{2}e^{-3}\right)$ **67.** Relative maximum: $f(1) = 5$ **69.** None

71. Relative maximum: $f(-1) = -\frac{7}{3}$, relative minimum: $f(5) = -\frac{115}{3}$

73. Relative maximum: $g(-3) = -6$; relative minimum: $g(3) = 6$

75. None   **77.** Relative minimum: $f(-2) = 12$

79. Relative maximum: $g(1) = \frac{1}{2}$; relative minimum: $g(-1) = -\frac{1}{2}$

81. Relative minimum: $f(1) = \frac{1}{e}$ **83.** Relative minimum: $f(0) = 0$

85.
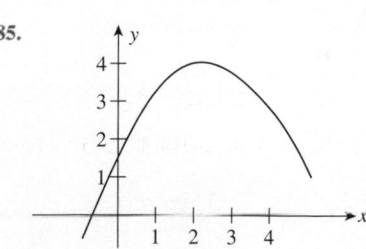

87.
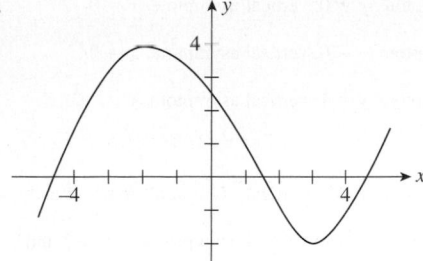

89.
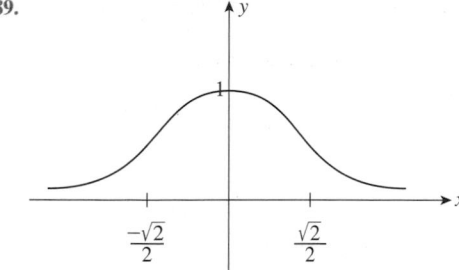

91. a. N is increasing on $(0, 12)$.
   b. $N''(t) < 0$ on $(0, 6)$ and $N''(t) > 0$ on $(6, 12)$
   c. The rate of growth of the number of help-wanted advertisements was decreasing over the first 6 months of the year and increasing over the last 6 months.

93. $f(t)$ increases at an increasing rate until the water level reaches the middle of the vase at which time (corresponding to the inflection point) $f(t)$ is increasing at the fastest rate. The water rises faster when the vase is narrower. After that, $f(t)$ increases at a decreasing rate until the vase is filled.

95. b. The rate of increase of the average state cigarette tax was decreasing from 2001 to 2007.

97. a. Concave downward on $(0, 7.27)$ and concave upward on $(7.27, 21)$
   b. $(7.27, 5.16)$; the U.S. public debt was increasing at a decreasing rate from 1990 through the end of the first quarter in 1997 and then continued to increase but at an accelerated pace from that point on.

99. a. 1.43%; 23.51%
   b. The frequency of Alzheimer's disease increases with age in the age range under consideration.
   c. The frequency of Alzheimer's disease increases at an increasing rate with age in the age range under consideration.

103. a. $-0.6t^2 + 3.28t + 1.31$; $-1.2t + 3.28$
   c. $(2.73, 14.93)$; Google's revenue was increasing at the fastest rate in late August 2006.

107. b. 44   **109.** False   **111.** True

Exercises 10.3, page 765

1. Horizontal asymptote: $y = 0$

3. Horizontal asymptote: $y = 0$; vertical asymptote: $x = 0$

5. Horizontal asymptote: $y = 0$; vertical asymptotes: $x = -1$ and $x = 1$

7. Horizontal asymptote: $y = 3$; vertical asymptote: $x = 0$

9. Horizontal asymptote: $y = 0$

11. Horizontal asymptote: $y = 0$; vertical asymptote: $x = 0$

13. Horizontal asymptote: $y = 0$; vertical asymptote: $x = 0$

15. Horizontal asymptote: $y = 1$; vertical asymptote: $x = -2$

17. None

19. Horizontal asymptote: $y = 1$; vertical asymptotes: $t = -4$ and $t = 4$

21. Horizontal asymptote: $y = 0$; vertical asymptotes: $x = -2$ and $x = 3$

23. Horizontal asymptote: $y = 2$; vertical asymptote: $t = 2$

25. Horizontal asymptote: $y = 1$; vertical asymptotes: $x = -2$ and $x = 2$

27. None   **29.** f is the derivative function of the function g.

31.

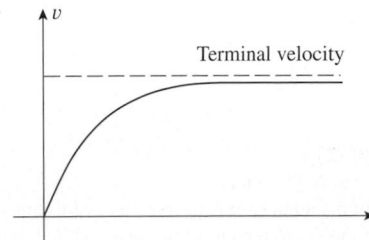

33.

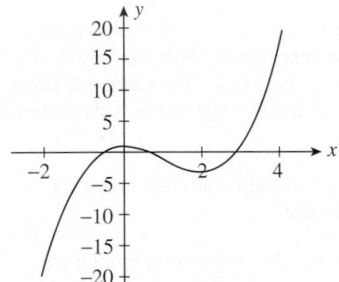

35.

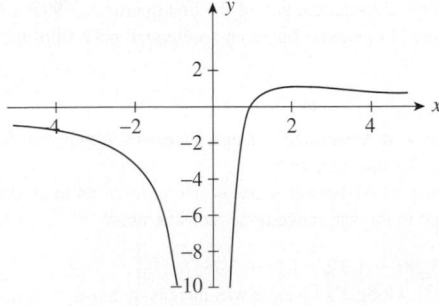

37.

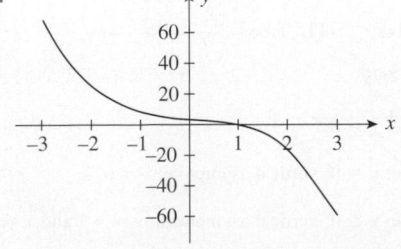

39.

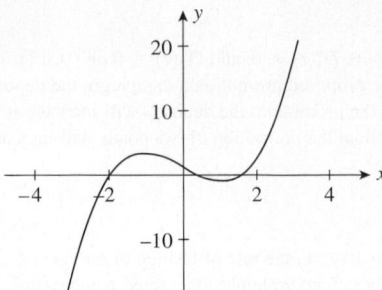

41.

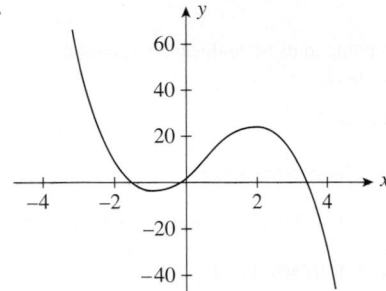

43.

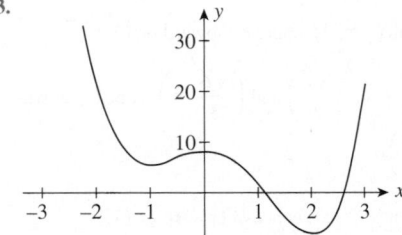

45.

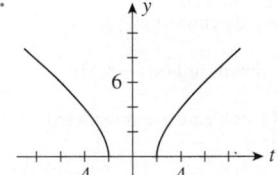

47.

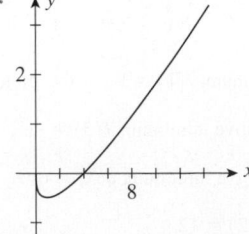

49.

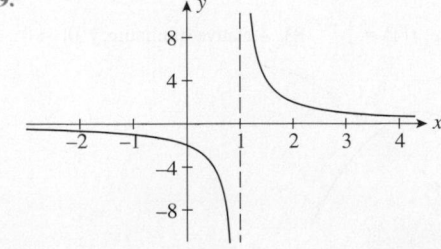

51.
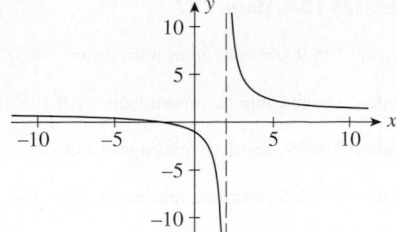

53.
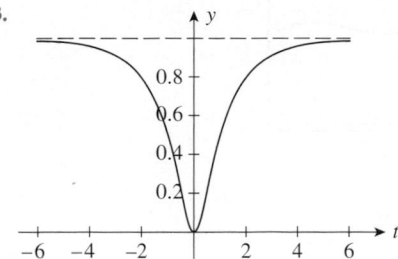

55.
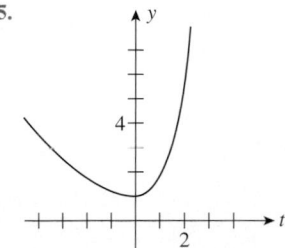

57.
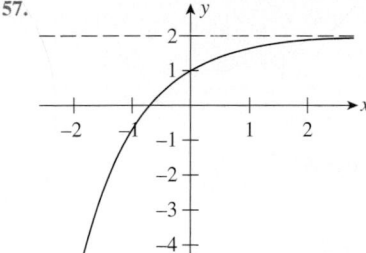

59.
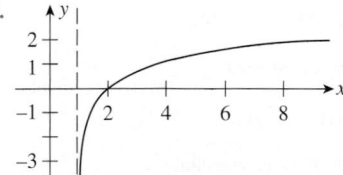

61. a. $x = 100$ **b.** No

63. a. $y = 0$
  b. As time passes, the concentration of the drug decreases and approaches zero.

65.
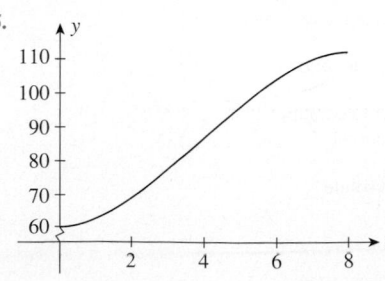

67.
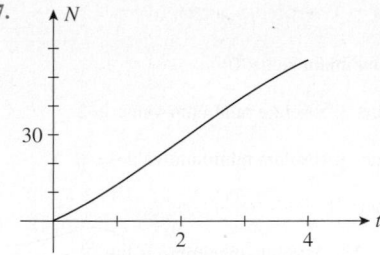

69.
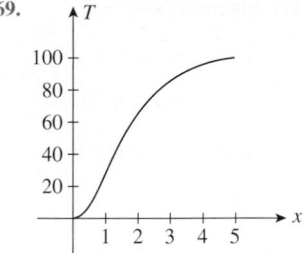

71. b.
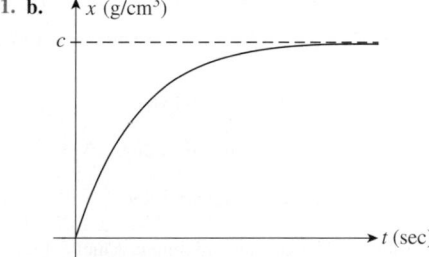

73. a. $\dfrac{r}{k}$

  d.
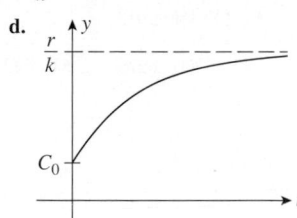

75. False

Using Technology Exercises 10.3, page 772

1.
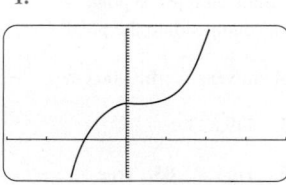

3.
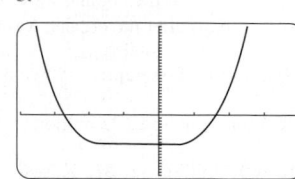

5. -0.9733; 2.3165, 4.6569 **7.** 1.5142 **9.** -0.7680; 1.6873

11.
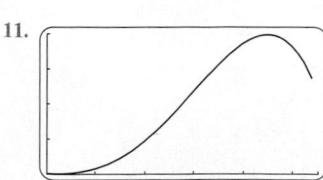

Exercises 10.4, page 781

1. None **3.** Absolute minimum value: 0

5. Absolute maximum value: 3; absolute minimum value: -2

7. Absolute maximum value: 3; absolute minimum value: $-\frac{27}{16}$

9. Absolute minimum value: $-\frac{41}{8}$

11. No absolute extrema **13.** Absolute maximum value: 1

15. Absolute maximum value: 5; absolute minimum value: -4

17. Absolute maximum value: 10; absolute minimum value: 1

19. Absolute maximum value: 19; absolute minimum value: -1

21. Absolute maximum value: 16; absolute minimum value: -1

23. Absolute maximum value: 3; absolute minimum value: $\frac{5}{3}$

25. Absolute maximum value: $\frac{65}{4}$; absolute minimum value: 5

27. Absolute maximum value ≈ 1.04; absolute minimum value: -1.5

29. No absolute extrema

31. Absolute maximum value: 1; absolute minimum value: 0

33. Absolute maximum value: 0; absolute minimum value: -3

35. Absolute maximum value: 2; absolute minimum value: $\frac{2}{e}$

37. Absolute maximum value: $2e^{-3/2}$; absolute minimum value: -1

39. Absolute maximum value: $3 - \ln 3$; absolute minimum value: 1

41. 144 ft **43. a.** 18.1%; 2009 **b.** 20.81%; 2013

45. 7 A.M.; 30 mph **47.** 268.7 ft **49.** 5000 copies **51.** 3333

53. 110

55. a. $0.000002x^2 + 5 + \dfrac{400}{x}$ **b.** 464 cases/day

   c. 464 cases/day **d.** Same

57. a. $0.0025x + 80 + \dfrac{10,000}{x}$ **b.** 2000 **c.** 2000 **d.** Same

59. 10,000 watches

61. a. 2 days after the organic waste was dumped into the pond
   b. 3.5 days after the organic waste was dumped into the pond

65. Tenth year of operation **67. a.** 14 cm/year **b.** 200 cm

69. 150 consultants; $3.22 million **73.** 749,833

77. b. 9.075 billion **81.** False **83.** True **85.** True

89. c.

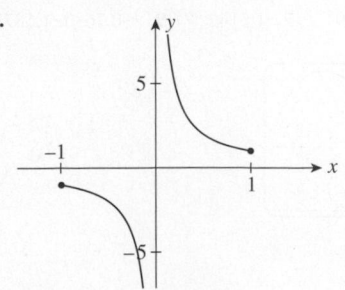

Using Technology Exercises 10.4, page 787

1. Absolute maximum value: 145.9; absolute minimum value: -4.3834

3. Absolute maximum value: 16; absolute minimum value: -0.1257

5. Absolute maximum value: 2.8889; absolute minimum value: 0

7. Absolute maximum value: 11.8922; absolute minimum value: 0

9. a.

11. b. approximately 1145

Exercises 10.5, page 795

1. 25 ft $\times$ 25 ft

3. 750 yd $\times$ 1500 yd; 1,125,000 yd^2

5. $10\sqrt{2}$ ft $\times$ $40\sqrt{2}$ ft

7. $\frac{16}{3}$ in. $\times$ $\frac{16}{3}$ in. $\times$ $\frac{4}{3}$ in.

9. 3 in. $\times$ 3 in. $\times$ 4 in.

11. 2 ft $\times$ 2 ft $\times$ 5 ft

13. 5 in. $\times$ 10 in.

15. $r = 1.5$ in., $h = 4$ in.

17. $\frac{2}{3}\sqrt[3]{9}$ ft $\times$ $\sqrt[3]{9}$ ft $\times$ $\frac{2}{5}\sqrt[3]{9}$ ft

19. 250; $62,500; $250

21. 85; $28,900; $340 **23.** 60 mph

25. $w \approx 13.86$ in.; $h \approx 19.60$ in.

27. $x = 2250$ ft **29.** $x \approx 2.68$

31. 440 ft; 140 ft; 184,874 ft^2 **33.** 45; 44,445

Chapter 10 Concept Review, page 800

1. a. $f(x_1) < f(x_2)$ **b.** $f(x_1) > f(x_2)$

2. a. increasing **b.** $f'(x) < 0$ **c.** constant

3. a. $f(x) \le f(c)$ **b.** $f(x) \ge f(c)$

4. a. domain; $= 0$; exist **b.** critical number
   c. relative extremum

5. a. $f'(x)$ **b.** > 0 **c.** concavity
   d. relative maximum; relative extremum

6. $\pm\infty; \pm\infty$ **7.** 0; 0 **8.** $b; b$

9. a. $f(x) \le f(c)$; absolute maximum value
   b. $f(x) \ge f(c)$; open interval

10. continuous; absolute; absolute

Chapter 10 Review Exercises, page 801

1. **a.** f is increasing on $(-\infty, \infty)$ **b.** No relative extrema
   c. Concave down on $(-\infty, 1)$; concave up on $(1, \infty)$
   d. $\left(1, -\frac{17}{3}\right)$

2. **a.** f is increasing on $(-\infty, \infty)$ **b.** No relative extrema
   c. Concave down on $(-\infty, 2)$; concave up on $(2, \infty)$
   d. $(2, 0)$

3. **a.** f is increasing on $(-1, 0)$ and $(1, \infty)$ and decreasing on $(-\infty, -1)$ and $(0, 1)$
   b. Relative maximum value: 0; relative minimum value: -1
   c. Concave up on $\left(-\infty, -\frac{\sqrt{3}}{3}\right)$ and $\left(\frac{\sqrt{3}}{3}, \infty\right)$; concave down on $\left(-\frac{\sqrt{3}}{3}, \frac{\sqrt{3}}{3}\right)$
   d. $\left(-\frac{\sqrt{3}}{3}, -\frac{5}{9}\right); \left(\frac{\sqrt{3}}{3}, -\frac{5}{9}\right)$

4. **a.** f is increasing on $(-\infty, -2)$ and $(2, \infty)$ and decreasing on $(-2, 0)$ and $(0, 2)$
   b. Relative maximum value: -4; relative minimum value: 4
   c. Concave down on $(-\infty, 0)$; concave up on $(0, \infty)$
   d. None

5. **a.** f is increasing on $(-\infty, 0)$ and $(2, \infty)$; decreasing on $(0, 1)$ and $(1, 2)$
   b. Relative maximum value: 0; relative minimum value: 4
   c. Concave up on $(1, \infty)$; concave down on $(-\infty, 1)$
   d. None

6. **a.** f is increasing on $(1, \infty)$ **b.** No relative extrema
   c. Concave down on $(1, \infty)$ **d.** None

7. **a.** f is decreasing on $(-\infty, \infty)$ **b.** No relative extrema
   c. Concave down on $(-\infty, 1)$; concave up on $(1, \infty)$
   d. $(1, 0)$

8. **a.** f is increasing on $(1, \infty)$ **b.** No relative extrema
   c. Concave down on $\left(1, \frac{4}{3}\right)$; concave up on $\left(\frac{4}{3}, \infty\right)$
   d. $\left(\frac{4}{3}, \frac{4\sqrt{3}}{9}\right)$

9. **a.** f is increasing on $(-\infty, -1)$ and $(-1, \infty)$
   b. No relative extrema
   c. Concave down on $(-1, \infty)$; concave up on $(-\infty, -1)$
   d. None

10. **a.** f is decreasing on $(-\infty, 0)$ and increasing on $(0, \infty)$
    b. Relative minimum value: -1
    c. Concave down on $\left(-\infty, -\frac{\sqrt{3}}{3}\right)$ and $\left(\frac{\sqrt{3}}{3}, \infty\right)$; concave up on $\left(-\frac{\sqrt{3}}{3}, \frac{\sqrt{3}}{3}\right)$
    d. $\left(-\frac{\sqrt{3}}{3}, -\frac{3}{4}\right); \left(\frac{\sqrt{3}}{3}, -\frac{3}{4}\right)$

11. **a.** f is increasing on $(-\infty, 3)$ and decreasing on $(3, \infty)$
    b. Relative maximum value: e^3
    c. Concave down on $(-\infty, 2)$; concave down on $(2, \infty)$
    d. $(2, 2e^2)$

12. **a.** f is decreasing on $(0, e^{-1/2})$ and increasing on $(e^{-1/2}, \infty)$
    b. Relative minimum value; $-\frac{1}{2}e^{-1}$
    c. Concave down on $(0, e^{-3/2})$; concave up on $(e^{-3/2}, \infty)$
    d. $(e^{-3/2}, -\frac{3}{2}e^{-3})$

13.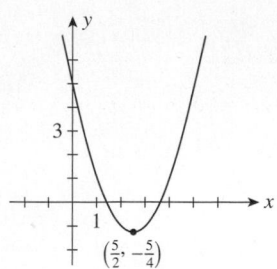

14.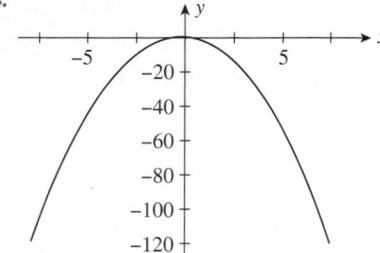

15.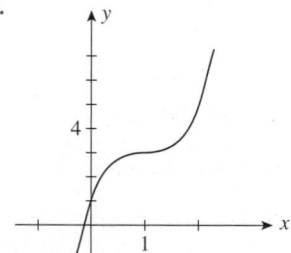

16.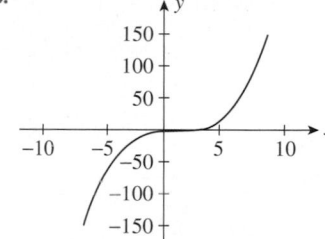

17.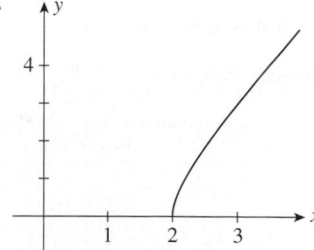

18.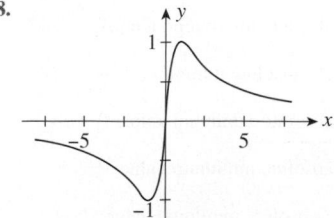

19.

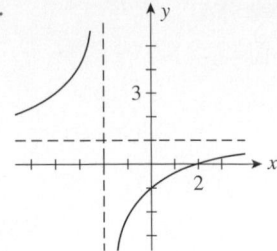

20.

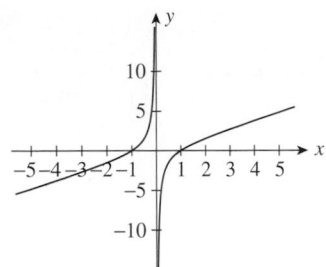

21.

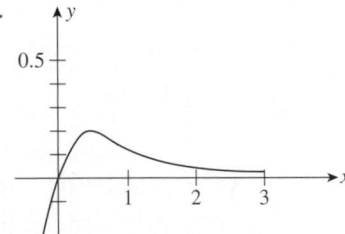

22.

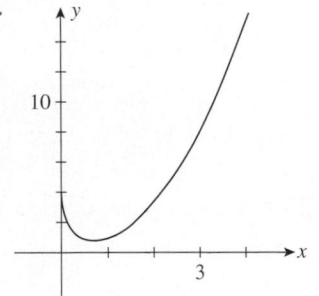

23. Horizontal asymptote: $y = 0$; vertical asymptote: $x = -\frac{3}{2}$

24. Horizontal asymptote: $y = 2$; vertical asymptote: $x = -1$

25. Horizontal asymptote: $y = 0$; vertical asymptotes: $x = -2, x = 4$

26. Horizontal asymptote: $y = 1$; vertical asymptote: $x = 1$

27. Absolute minimum value: $-\frac{25}{8}$

28. Absolute minimum value: 0

29. Absolute maximum value: 5; absolute minimum value: 0

30. Absolute maximum value: $\frac{5}{3}$; absolute minimum value: 1

31. Absolute maximum value: -16; absolute minimum value: -32

32. Absolute maximum value: $\frac{1}{2}$; absolute minimum value: 0

33. Absolute maximum value: $\frac{8}{3}$; absolute minimum value: 0

34. Absolute maximum value: $\frac{215}{9}$; absolute minimum value: 7

35. Absolute maximum value: $\frac{3}{e}$; absolute minimum value: $-6e^2$

36. Absolute maximum value: ln 2; absolute minimum value: 0

37. Absolute maximum value: $\frac{1}{2}$; absolute minimum value: $-\frac{1}{2}$

38. No absolute extrema

39. a. The sign of $R_1'(t)$ is negative; the sign of $R_2'(t)$ is positive. The sign of $R_1''(t)$ is negative and the sign of $R_2''(t)$ is positive.
   b. The revenue of the neighborhood bookstore is decreasing at an increasing rate while the revenue of the new branch of the national bookstore is increasing at an increasing rate.

40. The rumor spread initially with increasing speed. The rate at which the rumor is spread reaches a maximum at the time corresponding to the t-coordinate of the point P on the curve. Thereafter, the speed at which the rumor is spread decreases.

41. a. 6.1%/year   **b.** 100%

42. a. 154   **b.** 15/day   **c.** 400

43. March 1995; 1 million deaths/year

44. a. 20.1; 25.1; 25.63   **b.** Early 1984

45. $4000

46. b. The sales continued to accelerate through the years under consideration.

47. a. The amount of AMT paid was increasing over those years.
   b. The amount of AMT paid accelerated over those years.

48. a. 506,000; 125,480
   b. The number of measles deaths was dropping from 1999 through 2005.
   c. April 2002; approximately -41 thousand deaths/year/year

49. (100, 4600); sales increase at an increasing rate until $100,000 is spent on advertising; after that, any additional expenditure results in increased sales but at a slower rate of increase.

50. a. Decreasing on (0, 21.4); increasing on (21.4, 30)
   b. The percentage of men 65 years and older in the workforce was decreasing from 1970 until mid-1991 and increasing from mid-1991 through 2000.

51. (267, 11,874); the rate of increase is lowest when 267 calculators are produced.

53. a. 13.0%, 22.2%

54.

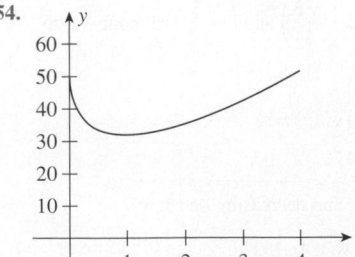

55.

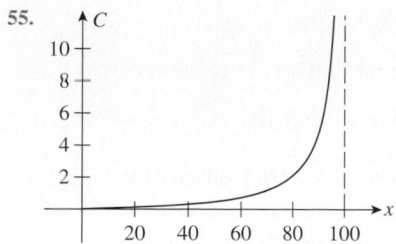

56. a. $I'(t) = -\dfrac{200t}{(t^2 + 10)^2}$

b. $I''(t) = \dfrac{-200(10 - 3t^2)}{(t^2 + 10)^3}$; concave up on $\left(\dfrac{\sqrt{10}}{3}, \infty\right)$; concave

down on $\left(0, \dfrac{\sqrt{10}}{3}\right)$

c.

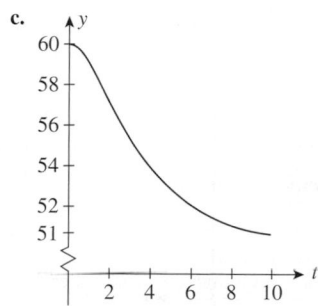

d. The rate of decline in the environmental quality of the wildlife was increasing the first 1.8 years. After that time, the rate of decline decreased.

57. 168 discs **58.** 3000 cameras

59. a. $0.001x + 100 + \dfrac{4000}{x}$ **b.** 2000 guitars **60.** 10 A.M.

61. a. Decreasing on $(0, 12.7)$; increasing on $(12.7, 30)$
b. 7.9
c. The percent of women 65 years and older in the workforce was decreasing from 1970 to Sept. 1982 and increasing from Sept. 1982 to 2000. It reached a minimum value of 7.9% in Sept. 1982.

62. b. As the age of the driver increases from 16 to 27 years, the predicted number of crash fatalities drops.

64. The number of applicants was decreasing from 1997–1998 to 2001–2002, increasing from 2001–2002 to 2006–2007, and constant from 2006–2007 to 2008–2009.

65. a. $f'(x) = \begin{cases} -2x & \text{if } x \neq 0 \\ 0 & \text{if } x = 0 \end{cases}$ **b.** No

66. 74.07 in.^{3} **67.** Radius: 2 ft; height: 8 ft

68. 1 ft $\times$ 2 ft $\times$ 2 ft **69.** 20,000 cases

70. If $a > 0$, f is decreasing on $\left(-\infty, -\dfrac{b}{2a}\right)$ and increasing on

$\left(-\dfrac{b}{2a}, \infty\right)$; if $a < 0$, f is increasing on $\left(-\infty, -\dfrac{b}{2a}\right)$ and decreasing

on $\left(-\dfrac{b}{2a}, \infty\right)$.

71. $a = -4$; $b = 11$ **72.** $c \geq \dfrac{3}{2}$

74. a. $f'(x) = 3x^2$ if $x \neq 0$ **b.** No

Chapter 10 Before Moving On, page 805

1. Decreasing on $(-\infty, 0)$ and $(2, \infty)$; increasing on $(0, 1)$ and $(1, 2)$

2. Relative maximum: $(1, 4e^{-1})$; inflection point: $(2, 8e^{-2})$

3. Concave downward on $\left(-\infty, \dfrac{1}{4}\right)$; concave upward on $\left(\dfrac{1}{4}, \infty\right)$; $\left(\dfrac{1}{4}, \dfrac{83}{96}\right)$

4.

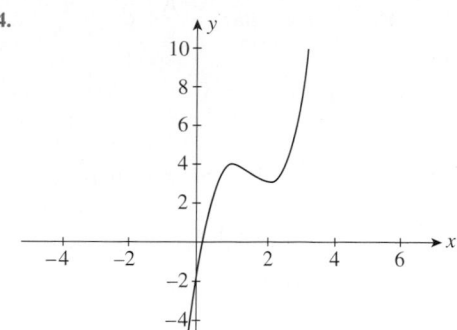

5. Absolute minimum value: -5; absolute maximum value: 80

6. $r = h = \dfrac{1}{\sqrt[3]{\pi}}$ (ft)

CHAPTER 11

Exercises 11.1, page 816

5. b. $y = 2x + C$

c.

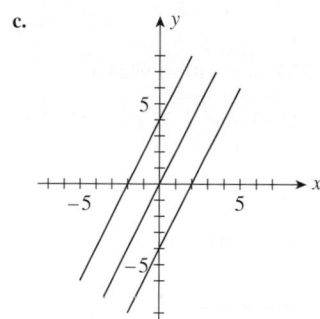

7. b. $y = \dfrac{1}{3}x^3 + C$

c.

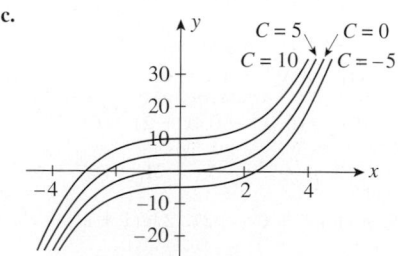

9. $6x + C$ **11.** $\dfrac{1}{4}x^4 + C$ **13.** $-\dfrac{1}{3x^3} + C$

15. $\dfrac{3}{5}x^{5/3} + C$ **17.** $-\dfrac{4}{x^{1/4}} + C$ **19.** $-\dfrac{1}{x^2} + C$

21. $\dfrac{2}{3}\pi t^{3/2} + C$ **23.** $3x - 2x^2 + C$

25. $\dfrac{1}{3}x^3 + \dfrac{1}{2}x^2 - \dfrac{1}{2x^2} + C$ **27.** $5e^x + C$

29. $x + \dfrac{1}{2}x^2 + e^x + C$ **31.** $x^4 + \dfrac{2}{x} - x + C$

33. $\frac{2}{7}x^{7/2} + \frac{4}{5}x^{5/2} - \frac{1}{2}x^2 + C$ **35.** $\frac{2}{3}x^{3/2} + 4\sqrt{x} + C$

37. $\frac{1}{9}u^3 + \frac{1}{3}u^2 - \frac{1}{3}u + C$ **39.** $\frac{2}{3}t^3 - \frac{3}{2}t^2 - 2t + C$

41. $\frac{1}{3}x^3 - 2x - \frac{1}{x} + C$ **43.** $\frac{1}{3}s^3 + s^2 + s + C$

45. $e^t + \frac{t^{e+1}}{e+1} + C$ **47.** $\frac{1}{2}x^2 + x - \ln|x| - \frac{1}{x} + C$

49. $\ln|x| + \frac{4}{\sqrt{x}} - \frac{1}{x} + C$ **51.** $\frac{3}{2}x^2 + x + \frac{1}{2}$

53. $x^3 + 2x^2 - x - 5$ **55.** $x - \frac{1}{x} + 3$ **57.** $x + \ln|x|$

59. $\sqrt{x}$ **61.** $e^x + \frac{1}{2}x^2 + 2$ **63.** Branch A

65. $14.3t + 90.1$; 147.3 million **67.** $0.1t^2 + 3t$

69. 5000 units; $34,000

71. a. $2.509t^2 - 3.204t + 1.8$ **b.** 102.313 terawatt-hours
c. 136.744 terawatt-hours

73. 2816.1 acres

75. a. $-16t^2 + 4t + 400$ **b.** $t = 5.13$ sec
c. 160.16 ft/sec downward

77. 21,960 **79.** $-t^3 + 96t^2$; 59,400 ft

81. $3370 **83. a.** $\frac{16\sqrt{2}}{3}t^{3/2} - 8t^4$ **b.** 2.2 in.

85. a. $7.031t^{0.842} - 0.031$ **b.** 27.23% **87.** 1.9424 m^2

89. a. $-1.056375t^4 + 4.2553t^3 - 12.85595t^2 + 32.28t + 52.9710$
b. 75.59 cm

91. $9\frac{7}{9}$ ft/sec^2; 396 ft **93.** 0.924 ft/sec^2

95. a. At least 36 ft/sec^2 **97.** False **99.** False

Exercises 11.2, page 828

1. $\frac{1}{5}(4x + 3)^5 + C$ **3.** $\frac{1}{3}(x^3 - 2x)^3 + C$

5. $-\frac{1}{2(2x^2 + 3)^2} + C$ **7.** $\frac{2}{3}(t^3 + 2)^{3/2} + C$

9. $\frac{1}{10}(x^2 - 1)^{10} + C$ **11.** $-\frac{1}{5}\ln|1 - x^5| + C$

13. $2\ln|x - 2| + C$ **15.** $\frac{1}{2}\ln(0.3x^2 - 0.4x + 2) + C$

17. $\frac{1}{3}\ln|3x^2 - 1| + C$ **19.** $-\frac{1}{2}e^{-2x} + C$ **21.** $-e^{2-x} + C$

23. $-\frac{1}{2}e^{-x^2} + C$ **25.** $e^x + e^{-x} + C$ **27.** $2\ln(1 + e^x) + C$

29. $2e^{\sqrt{x}} + C$ **31.** $-\frac{1}{6(e^{3x} + x^3)^2} + C$ **33.** $\frac{1}{8}(e^{2x} + 1)^4 + C$

35. $\frac{1}{2}(\ln 5x)^2 + C$ **37.** $3\ln|\ln x| + C$ **39.** $\frac{2}{3}(\ln x)^{3/2} + C$

41. $\frac{1}{2}e^{x^2} - \frac{1}{2}\ln(x^2 + 2) + C$

43. $\frac{2}{3}(\sqrt{x} - 1)^3 + 3(\sqrt{x} - 1)^2 + 8(\sqrt{x} - 1) + 4\ln|\sqrt{x} - 1| + C$

45. $\frac{(6x + 1)(x - 1)^6}{42} + C$

47. $4\sqrt{x} - x - 4\ln(1 + \sqrt{x}) + C$

49. $-\frac{1}{252}(1 - v)^7(28v^2 + 7v + 1) + C$

51. $\frac{1}{2}[(2x - 1)^5 + 5]$ **53.** $e^{-x^2+1} - 1$

55. $21,000 - \frac{20,000}{\sqrt{1 + 0.2t}}$; 6858 **57.** 180.7 million GBP

59. 62,286 **61.** $50.02(1 + 1.09t)^{0.1}$; 80

63. a. $135e^{0.067t}$ **b.** 188.722 million

65. a. $19.1301e^{0.163(t+1)} + 3.5699$ **b.** $40.3 billion

67. $\frac{r}{a}(1 - e^{-at})$ **69.** True **71.** True

Exercises 11.3, page 840

1. 4.27

3. a. 6

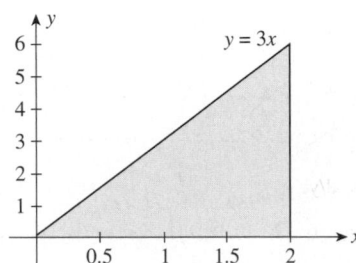

b. 4.5 **c.** 5.25 **d.** Yes

5. a. 4

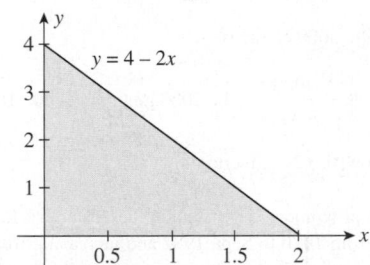

b. 4.8 **c.** 4.4 **d.** Yes

7. a. 18.5 **b.** 18.64 **c.** 18.66 **d.** $\approx 18\frac{2}{3}$

9. a. 25 **b.** 21.12 **c.** 19.88 **d.** ≈ 19.9

11. a. 0.0625 **b.** 0.16 **c.** 0.2025 **d.** ≈ 0.2

13. 4.64 **15.** 0.95 **17.** 9400 ft^2 **19.** False

Exercises 11.4, page 851

1. 6 **3.** 8 **5.** 12 **7.** 9 **9.** $\ln 2$ **11.** $17\frac{1}{3}$ **13.** $18\frac{1}{4}$

15. $e^2 - 1$ **17.** 6 **19.** 24 **21.** $\frac{56}{3}$ **23.** $\frac{4}{3}$ **25.** $\frac{45}{2}$

27. $\frac{7}{12}$ **29.** $\ln 4$ **31.** 56 **33.** $\frac{256}{15}$ **35.** $\frac{2}{3}$

37. $\frac{8}{3}$ **39.** $\frac{39}{2}$

41. a. A decrease of 0.319 million cases **b.** 1.219 million cases

43. a. $4100 **b.** $900 **45. a.** $2800 **b.** $219.20

47. $10,133\frac{1}{3}$ ft **49.** $7182 **51.** 46%; 24%

53. a. $0.86t^{0.96} + 0.04$ **b.** $6.37 billion

55. a. $8.86 billion   **b.** $19.57 billion   **57.** 149.14 million

59. $\frac{23}{15}$ **61.** False   **63.** False

Using Technology Exercises 11.4, page 854

1. 6.1787   **3.** 0.7873   **5.** -0.5888 **7.** 2.7044

9. 3.9973   **11.** 46%; 24%   **13.** 60,156

Exercises 11.5, page 861

1. 10   **3.** $\frac{19}{15}$ **5.** $\frac{484}{15}$ **7.** $\sqrt{3} - 1$ **9.** $\frac{1562}{5}$

11. $\frac{32}{15}$ **13.** $\frac{272}{15}$ **15.** $e^4 - 1$ **17.** $\frac{1}{2}e^2 + \frac{5}{6}$ **19.** 0

21. $\ln 4$ **23.** $\frac{1}{3}(\ln 19 - \ln 3)$ **25.** $2e^4 - 2e^2 - \ln 2$

27. $\frac{1}{2}(e^{-4} - e^{-8} - 1)$ **29.** 6   **31.** $\frac{1}{2}$ **33.** $2(\sqrt{e} - \frac{1}{e})$

35. 5   **37.** $\frac{17}{3}$ **39.** -1 **41.** $\frac{13}{6}$ **43.** $\frac{1}{4}(e^4 - 1)$

45. 64 ft   **47.** $\approx$$2.24 million

49. 120.3 billion metric tons   **51.** $3.24 billion/year

53. a. 160.7 billion gal/year   **b.** 150.1 billion gal/year/year

55. 12.53 N-sec; 626.5 N   **57.** $\frac{2}{3}kR^2$ cm/sec   **59.** 0.071 mg/cm^3

61. $6\frac{1}{3}$ million   **63.** 262.9 million cards/year

65. a. $146.711e^{0.0456t} - 0.72$ **b.** 364.5 million
    c. 11.4 million/year

75. Property 5   **77.** 0   **79. a.** -1 **b.** 5   **c.** -13

81. True   **83.** False   **85.** True

Using Technology Exercises 11.5, page 866

1. 7.71667   **3.** 17.56487   **5.** 159/bean stem

7. 0.48 g/cm^3/day

Exercises 11.6, page 873

1. 108   **3.** $\frac{2}{3}$ **5.** $2\frac{2}{3}$ **7.** $1\frac{1}{2}$ **9.** 3   **11.** $3\frac{1}{3}$

13. 27   **15.** $2(e^2 - e^{-1})$ **17.** $12\frac{2}{3}$ **19.** $3\frac{1}{3}$ **21.** $4\frac{3}{4}$

23. $12 - \ln 4$ **25.** $e^2 - e - \ln 2$ **27.** $2\frac{1}{2}$ **29.** $7\frac{1}{3}$ **31.** $\frac{3}{2}$

33. $e^3 - 4 + \dfrac{1}{e}$ **35.** $\frac{125}{6}$ **37.** $\frac{1}{12}$ **39.** $\frac{71}{6}$ **41.** 18

43. S is the additional revenue that Odyssey Travel could realize by switching to the new agency; $S = \int_0^b [g(x) - f(x)]\, dx$

45. Shortfall $= \int_{2010}^{2050} [f(t) - g(t)]\, dt$

47. a. $A_2 - A_1$ **b.** The distance Car 2 is ahead of Car 1 after T sec

49. $840 - \int_0^{12} f(t)\, dt$ **51.** 42.8 billion metric tons   **53.** 57,179

55. True   **57.** False

Using Technology Exercises 11.6, page 878

1. a.

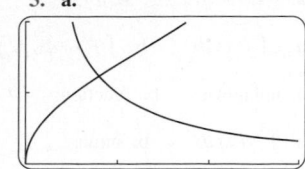

b. 1074.2857

3. a.

b. 0.9961

5. a.

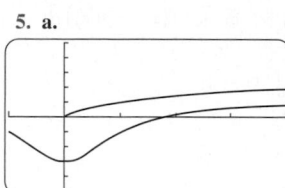

b. 5.4603

7. a.

b. 25.8549

9. a.

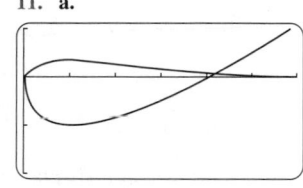

b. 10.5144

11. a.

b. 3.5799

13. 207.43

Exercises 11.7, page 888

1. $11,667   **3.** $6667   **5.** $11,667

7. a. 1257/month   **b.** $48,583   **9.** $199,548

11. Consumers' surplus: $13,333; producers' surplus: $11,667

13. $515,224.45   **15.** $824,200   **17.** $91,916

19. a. $131,996   **b.** $113,610   **21.** $40,212

23. $47,916   **25.** $142,423   **27.** $28,330   **29.** 0.360

31. a.

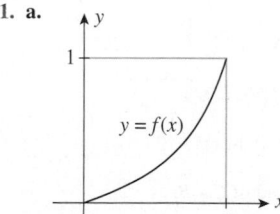

b. 0.175; 0.816

33. a. 0.31; 0.49   **b.** College teachers

Using Technology Exercises 11.7, page 891

1. Consumers' surplus: $18,000,000; producers' surplus: $11,700,000

3. Consumers' surplus: $33,120; producers' surplus: $2880

5. Investment A

Chapter 11 Concept Review, page 894

1. a. $F'(x) = f(x)$ **b.** $F(x) + C$

2. a. $c\int f(x)\,dx$ **b.** $\int f(x)\,dx \pm \int g(x)\,dx$

3. a. unknown **b.** function **4.** $g'(x)\,dx;\ \int f(u)\,du$

5. a. $\int_a^b f(x)\,dx$ **b.** minus

6. a. $F(b) - F(a)$; antiderivative **b.** $\int_a^b f'(x)\,dx$

7. a. $\dfrac{1}{b-a}\int_a^b f(x)\,dx$ **b.** area; area **8.** $\int_a^b [f(x) - g(x)]\,dx$

9. a. $\int_0^{\bar{x}} D(x)\,dx - \bar{p}\,\bar{x}$ **b.** $\bar{p}\,\bar{x} - \int_0^{\bar{x}} S(x)\,dx$

10. a. $e^{rT}\int_0^T R(t)e^{-rt}\,dt$ **b.** $\int_0^T R(t)e^{-rt}\,dt$

11. $\dfrac{mP}{r}(e^{rT} - 1)$ **12.** $2\int_0^1 [x - f(x)]\,dx$

Chapter 11 Review Exercises, page 894

1. $\frac{1}{4}x^4 + \frac{2}{3}x^3 - \frac{1}{2}x^2 + C$ **2.** $\frac{1}{12}x^4 - \frac{2}{3}x^3 + 8x + C$

3. $\frac{1}{5}x^5 - \frac{1}{2}x^4 - \frac{1}{x} + C$ **4.** $\frac{3}{4}x^{4/3} - \frac{2}{3}x^{3/2} + 4x + C$

5. $\frac{1}{2}x^4 + \frac{2}{5}x^{5/2} + C$ **6.** $\frac{2}{7}x^{7/2} - \frac{1}{3}x^3 + \frac{2}{3}x^{3/2} - x + C$

7. $\frac{1}{3}x^3 - \frac{1}{2}x^2 + 2\ln|x| + 5x + C$ **8.** $\frac{1}{3}(2x + 1)^{3/2} + C$

9. $\frac{3}{8}(3x^2 - 2x + 1)^{4/3} + C$ **10.** $\dfrac{(x^3 + 2)^{11}}{33} + C$

11. $\frac{1}{2}\ln(x^2 - 2x + 5) + C$ **12.** $-e^{-2x} + C$

13. $\frac{1}{2}e^{x^2 + x + 1} + C$ **14.** $\dfrac{1}{e^{-x} + x} + C$ **15.** $\frac{1}{6}(\ln x)^6 + C$

16. $(\ln x)^2 + C$ **17.** $\dfrac{(11x^2 - 1)(x^2 + 1)^{11}}{264} + C$

18. $\frac{2}{15}(3x - 2)(x + 1)^{3/2} + C$ **19.** $\frac{2}{3}(x + 4)\sqrt{x - 2} + C$

20. $2(x - 2)\sqrt{x + 1} + C$ **21.** $\frac{1}{2}$ **22.** -6 **23.** $\frac{17}{3}$

24. 242 **25.** -80 **26.** $\frac{132}{5}$ **27.** $\frac{1}{2}\ln 5$ **28.** $\frac{1}{15}$

29. 4 **30.** $1 - \dfrac{1}{e^2}$ **31.** $\dfrac{e - 1}{2(1 + e)}$ **32.** $\frac{1}{2}$

33. $f(x) = x^3 - 2x^2 + x + 1$ **34.** $f(x) = \sqrt{x^2 + 1}$

35. $f(x) = x + e^{-x} + 1$ **36.** $f(x) = \frac{1}{2}(\ln x)^2 - 2$

37. a. It gives the distance Car A is ahead of Car B.
b. $t = 10,\ \int_0^{10} [f(t) - g(t)]\,dt$

38. a. It gives the amount by which the revenue of Branch A exceeds that of Branch B.
b. $t = 10,\ \int_0^{10} [f(t) - g(t)]\,dt$

39. -4.28 **40.** 60.1%

41. $V(t) = 1900(t - 10)^2 + 10{,}000$; \$40,400 **42.** \$6740

43. a. $-0.015x^2 + 60x$ **b.** $p = -0.015x + 60$

44. a. $0.05t^3 - 1.8t^2 + 14.4t + 24$ **b.** 56°F

45. a. $-0.01t^3 + 0.109t^2 - 0.032t + 0.1$ **b.** 1.076 billion

46. 3.375 ppm **47.** \$3100 **48.** 37.7 million

49. $N(t) = 15{,}000\sqrt{1 + 0.4t} + 85{,}000$; 112,659

50. $p = \dfrac{240}{5 - x} - 30$

51. a. $S(t) = 205.89 - 89.89e^{-0.176t}$ **b.** \$161.43 billion

52. $3000t - 50{,}000(1 - e^{-0.04t})$; 16,939

53. 26,027 **54.** 15 **55.** $\frac{1}{2}(e^4 - 1)$

56. $\frac{2}{3}$ **57.** $\frac{9}{2}$ **58.** $e^2 - 3$ **59.** $\frac{3}{10}$ **60.** $\frac{1}{2}$

61. 234,500 barrels **62.** $\frac{1}{3}$ **63.** 26°F

64. 49.7 ft/sec **65.** 67,600/year **66.** \$270,000

67. Consumers' surplus: \$2083; producers' surplus: \$3333

68. \$197,652 **69.** \$174,420 **70.** \$505,696

71. a.

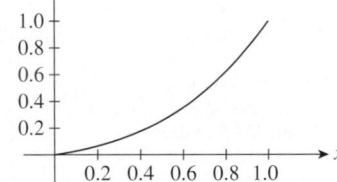

b. 0.1017; 0.3733 **c.** 0.315

72. 90,888

Chapter 11 Before Moving On, page 898

1. $\frac{1}{2}x^4 + \frac{2}{3}x^{3/2} + 2\ln|x| - 4\sqrt{x} + C$ **2.** $e^x + \frac{1}{2}x^2 + 1$

3. $\sqrt{x^2 + 1} + C$ **4.** $\frac{1}{3}(2\sqrt{2} - 1)$ **5.** $\frac{9}{2}$

CHAPTER 12

Exercises 12.1, page 907

1. $f(0, 0) = -4; f(1, 0) = -2; f(0, 1) = -1; f(1, 2) = 4;$ $f(2, -1) = -3$

3. $f(1, 2) = 7; f(2, 1) = 9; f(-1, 2) = 1; f(2, -1) = 1$

5. $g(1, 2) = 4 + 3\sqrt{2}; g(2, 1) = 8 + \sqrt{2}; g(0, 4) = 2; g(4, 9) = 56$

7. $h(1, e) = 1; h(e, 1) = -1; h(e, e) = 0$

9. $g(1, 1, 1) = e; g(1, 0, 1) = 1; g(-1, -1, -1) = -e$

11. All real values of x and y

13. All real values of u and v except those satisfying the equation $u = v$

15. All real values of r and s satisfying $rs \geq 0$

17. All real values of x and y satisfying $x + y > 5$

19.

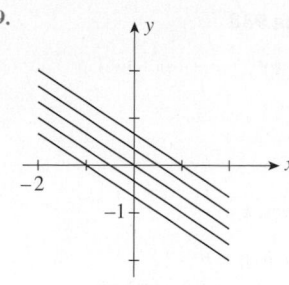

21.

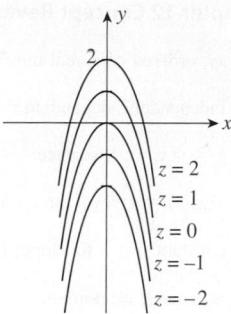

23.

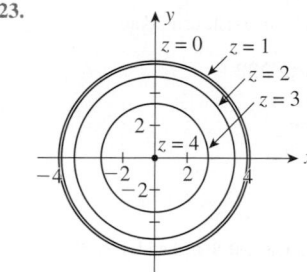

25. $\sqrt{x^2 + y^2} = 5$ **27.** (b)   **29.** No   **31.** 9π ft^3

33. a. P and E are real numbers, with $E \neq 0$.   **b.** 15.7

35. a. 24.69   **b.** 81 kg

37. a. $200x + 120y + 20,000$
b. The set of all x and y such that $x \geq 0$ and $y \geq 0$
c. $244,000

39. a. $-\frac{1}{5}x^2 - \frac{1}{4}y^2 - \frac{1}{5}xy + 200x + 160y$
b. The set of all points (x, y) satisfying $200 - \frac{1}{5}x - \frac{1}{10}y \geq 0$,
$160 - \frac{1}{10}x - \frac{1}{4}y \geq 0, x \geq 0, y \geq 0$

41. a. $-0.005x^2 - 0.003y^2 - 0.002xy + 20x + 15y$
b. The set of all ordered pairs (x, y) for which
$20 - 0.005x - 0.001y \geq 0$
$15 - 0.001x - 0.003y \geq 0, x \geq 0, y \geq 0$

43. a. The set of all ordered pairs (P, T), where P and T are positive numbers
b. 11.10 L

45. a. The domain of W is the set of all ordered pairs (L, G), where $L > 0$ and $G > 0$.
b. 3.6 lb

47. $7200 billion   **49.** 103

51. a. $1432.25; $1798.65
b. $2149.29

53. 18

57. The level curves of V have equation $\dfrac{kT}{P} = C$ (C, a positive constant).

The level curves are a family of straight lines $T = \left(\dfrac{C}{k}\right)P$ lying in

the first quadrant, since k, T, and P are positive. Every point on the level curve $V = C$ gives the same volume C.

59. False   **61.** False   **63.** False

Exercises 12.2, page 922

1. a. 4; 4
b. $f_x(2, 1) = 4$ says that the slope of the tangent line to the curve of intersection of the surface $z = x^2 + 2y^2$ and the plane $y = 1$ at the point $(2, 1, 6)$ is 4. $f_y(2, 1) = 4$ says that the slope of the tangent line to the curve of intersection of the surface $z = x^2 + 2y^2$ and the plane $x = 2$ at the point $(2, 1, 6)$ is 4.
c. $f_x(2, 1) = 4$ says that the rate of change of $f(x, y)$ with respect to x with y held fixed with a value of 1 is 4 units/unit change in x. $f_y(2, 1) = 4$ says that the rate of change of $f(x, y)$ with respect to y with x held fixed with a value of 2 is 4 units/unit change in y.

3. $f_x = 2; f_y = 3$ **5.** $g_x = 4x; g_y = 4$ **7.** $f_x = -\dfrac{4y}{x^3}; f_y = \dfrac{2}{x^2}$

9. $g_u = \dfrac{2v}{(u + v)^2}; g_v = -\dfrac{2u}{(u + v)^2}$

11. $f_s = 3(2s - t)(s^2 - st + t^2)^2; f_t = 3(2t - s)(s^2 - st + t^2)^2$

13. $f_x = \dfrac{4x}{3(x^2 + y^2)^{1/3}}; f_y = \dfrac{4y}{3(x^2 + y^2)^{1/3}}$

15. $f_x = ye^{xy+1}; f_y = xe^{xy+1}$

17. $f_x = \ln y + \dfrac{y}{x}; f_y = \dfrac{x}{y} + \ln x$ **19.** $g_u = e^u \ln v; g_v = \dfrac{e^u}{v}$

21. $f_x = yz + y^2 + 2xz; f_y = xz + 2xy + z^2; f_z = xy + 2yz + x^2$

23. $h_r = ste^{rst}; h_s = rte^{rst}; h_t = rse^{rst}$ **25.** $f_x(1, 2) = 8; f_y(1, 2) = 5$

27. $f_x(2, 1) = 1; f_y(2, 1) = 3$ **29.** $f_x(1, 2) = \frac{1}{2}; f_y(1, 2) = -\frac{1}{4}$

31. $f_x(1, 1) = e; f_y(1, 1) = e$

33. $f_x(1, 0, 2) = 0; f_y(1, 0, 2) = 8; f_z(1, 0, 2) = 0$

35. $f_{xx} = 2y; f_{xy} = 2x + 3y^2 = f_{yx}; f_{yy} = 6xy$

37. $f_{xx} = 2; f_{xy} = f_{yx} = -2; f_{yy} = 4$

39. $f_{xx} = \dfrac{y^2}{(x^2 + y^2)^{3/2}}; f_{xy} = f_{yx} = -\dfrac{xy}{(x^2 + y^2)^{3/2}};$
$f_{yy} = \dfrac{x^2}{(x^2 + y^2)^{3/2}}$

41. $f_{xx} = \dfrac{1}{y^2}e^{-x/y}; f_{xy} = \dfrac{y - x}{y^3}e^{-x/y} = f_{yx};$
$f_{yy} = \dfrac{x}{y^3}\left(\dfrac{x}{y} - 2\right)e^{-x/y}$

43. a. 7.5; 40   **b.** Yes

45. 1405; 2602; this says that an increase of 1 work-hour/day with capital expenditures held fixed at $300/day will result in an increase of approximately 1405 candles/day. An increase of $1/day in capital expenditures with work-hours/day held fixed at 400 will result in an increase of approximately 2602 candles/day.

47. $p_x = 10$—at $(0, 1)$, the price of land is changing at the rate of $10/ft^2/mile change to the right; $p_y = 0$—at $(0, 1)$, the price of land is constant/mile change upward.

49. 29; -475; this says that when the floor space is held fixed at 150,000 ft^2, the monthly profit increases at the rate of \$29 per thousand-dollar increase in the inventory when the inventory is \$4,000,000, and when the inventory is held fixed at \$4,000,000, the monthly profit decreases at the rate of \$475 per 1000 ft^2 increase in floor space when the floor space is 150,000 ft^2; -11; 525

51. 1.06; -2.85; If the level of reinvestment is held constant at 20 cents per dollar deposited, the number of fires will grow at the rate of approximately 1 fire per increase of 1 person per census tract when the number of people per census tract is 100. If the number of people per census tract is held constant at 100 per tract, the number of fires will decrease at a rate of approximately 2.9 per increase of 1 cent per dollar deposited for reinvestment when the level of reinvestment is 20 cents per dollar deposited.

53. a. 1.3°F/°F   **b.** -0.22°F/mi/hr

57. Substitute commodities

63. False   **65.** True   **67.** True

Using Technology Exercises 12.2, page 926

1. 1.3124; 0.4038   **3.** -1.8889; 0.7778

5. -0.3863; -0.8497

Exercises 12.3, page 934

1. $(0, 0)$; relative maximum value: $f(0, 0) = 1$

3. $(1, 2)$; saddle point: $f(1, 2) = 4$

5. $(8, -6)$; relative minimum value: $f(8, -6) = -41$

7. $(1, 2)$ and $(2, 2)$; saddle point: $f(1, 2) = -1$; relative minimum value: $f(2, 2) = -2$

9. $\left(-\frac{1}{3}, \frac{11}{3}\right)$ and $(1, 5)$; saddle point: $f\left(-\frac{1}{3}, \frac{11}{3}\right) = -\frac{319}{27}$; relative minimum value: $f(1, 5) = -13$

11. $(0, 0)$ and $(1, 1)$; saddle point: $f(0, 0) = -2$; relative minimum value: $f(1, 1) = -3$

13. $(2, 1)$; relative minimum value: $f(2, 1) = 6$

15. $(0, 0)$; saddle point: $f(0, 0) = -1$

17. $(0, 0)$; relative minimum value: $f(0, 0) = 1$

19. $(0, 0)$; relative minimum value: $f(0, 0) = 0$

21. 200 finished units and 100 unfinished units; \$10,500

23. Price of land (\$200/ft^2) is highest at $\left(\frac{1}{2}, 1\right)$.

25. a. $400(-p^2 - q^2 - pq + 16p + 14q)$
b. \$6/lb for ground sirloin, \$4/lb for ground beef; 3200 lb ground sirloin, 2800 lb of ground beef; \$30,400

27. a. $-20x^2 - 30y^2 - 20xy + 3000x + 4000y$
b. $400x + 500y + 20,000$
c. $-20x^2 - 30y^2 - 20xy + 2600x + 3500y - 20,000$
d. 43,000 regular tubes; 44,000 whitening tubes; \$112,900

29. At $\left(\frac{26}{3}, 12\right)$ **31.** 6 in. $\times$ 6 in. $\times$ 3 in.

33. $21\frac{2}{3}$ in. $\times$ $43\frac{1}{3}$ in. $\times$ $21\frac{2}{3}$ in.   **35.** 6 in. $\times$ 4 in. $\times$ 2 in.

37. False   **39.** False   **41.** True

Chapter 12 Concept Review, page 938

1. xy; ordered pair; real number; $f(x, y)$

2. independent; dependent; value

3. $z = f(x, y)$; f; surface

4. $f(x, y) = k$; level curve; level curves; k

5. constant; x **6.** slope; $(a, b, f(a, b))$; x; b

7. $\leq$; (a, b); $\leq$; domain

8. domain; $f_x(a, b) = 0$ and $f_y(a, b) = 0$; exist; candidate

Chapter 12 Review Exercises, page 938

1. $0, 0, \frac{1}{2}$; no   **2.** $e, \dfrac{e^2}{1 + \ln 2}, \dfrac{2e}{1 + \ln 2}$; no

3. $2, -(e + 1), -(e + 1)$

4. The set of all ordered pairs (u, v) such that $u \neq v$ and $u \geq 0$

5. The set of all ordered pairs (x, y) such that $y \neq -x$

6. The set of all ordered pairs (x, y) such that $x \leq 1$ and $y \geq 0$

7. The set of all ordered triplets (x, y, z) such that $z \geq 0$ and $x \neq 1$, $y \neq 1$, and $z \neq 1$

8. $2x + 3y = z$

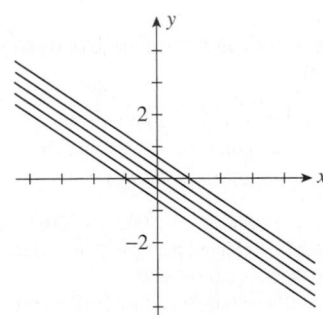

9. $z = y - x^2$

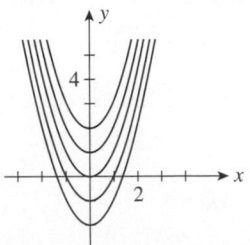

10. $z = \sqrt{x^2 + y^2}$

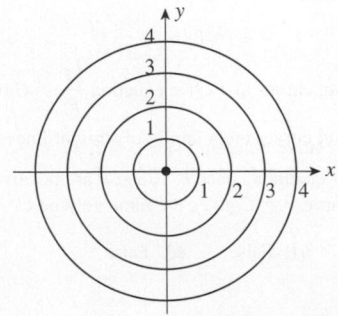

11. $z = e^{xy}$

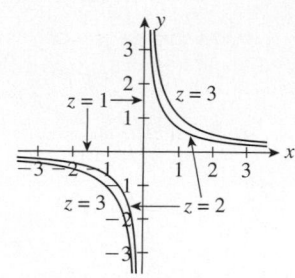

12. $f_x = 2xy^3 + 3y^2 + \dfrac{1}{y}$; $f_y = 3x^2y^2 + 6xy - \dfrac{x}{y^2}$

13. $f_x = \sqrt{y} + \dfrac{y}{2\sqrt{x}}$; $f_y = \dfrac{x}{2\sqrt{y}} + \sqrt{x}$

14. $f_u = \dfrac{v^2 - 2}{2\sqrt{uv^2 - 2u}}$; $f_v = \dfrac{uv}{\sqrt{uv^2 - 2u}}$

15. $f_x = \dfrac{3y}{(y + 2x)^2}$; $f_y = -\dfrac{3x}{(y + 2x)^2}$

16. $g_x = \dfrac{y(y^2 - x^2)}{(x^2 + y^2)^2}$; $g_y = \dfrac{x(x^2 - y^2)}{(x^2 + y^2)^2}$

17. $h_x = 10y(2xy + 3y^2)^4$; $h_y = 10(x + 3y)(2xy + 3y^2)^4$

18. $f_x = \dfrac{e^y}{2(xe^y + 1)^{1/2}}$; $f_y = \dfrac{xe^y}{2(xe^y + 1)^{1/2}}$

19. $f_x = 2x(1 + x^2 + y^2)e^{x^2+y^2}$; $f_y = 2y(1 + x^2 + y^2)e^{x^2+y^2}$

20. $f_x = \dfrac{4x}{1 + 2x^2 + 4y^4}$; $f_y = \dfrac{16y^3}{1 + 2x^2 + 4y^4}$

21. $f_x = \dfrac{2x}{x^2 + y^2}$; $f_y = -\dfrac{2x^2}{y(x^2 + y^2)}$

22. $f_{xx} = 6x - 4y$; $f_{xy} = -4x = f_{yx}$; $f_{yy} = 2$

23. $f_{xx} = 12x^2 + 4y^2$; $f_{xy} = 8xy = f_{yx}$; $f_{yy} = 4x^2 - 12y^2$

24. $f_{xx} = 12(2x^2 + 3y^2)(10x^2 + 3y^2)$;
$f_{xy} = 144xy(2x^2 + 3y^2) = f_{yx}$;
$f_{yy} = 18(2x^2 + 3y^2)(2x^2 + 15y^2)$

25. $g_{xx} = \dfrac{-2y^2}{(x + y^2)^3}$; $g_{xy} = \dfrac{2y(x - y^2)}{(x + y^2)^3} = g_{yx}$;
$g_{yy} = \dfrac{2x(3y^2 - x)}{(x + y^2)^3}$

26. $g_{xx} = 2(1 + 2x^2)e^{x^2+y^2}$; $g_{xy} = 4xye^{x^2+y^2} = g_{yx}$;
$g_{yy} = 2(1 + 2y^2)e^{x^2+y^2}$

27. $h_{ss} = -\dfrac{1}{s^2}$; $h_{st} = h_{ts} = 0$; $h_{tt} = \dfrac{1}{t^2}$

28. 3; 3; −2

29. (2, 3); relative minimum value: $f(2, 3) = -13$

30. (8, −2); saddle point at $f(8, -2) = -8$

31. (0, 0) and $\left(\frac{3}{2}, \frac{9}{4}\right)$; saddle point at $f(0, 0) = 0$; relative minimum value: $f\left(\frac{3}{2}, \frac{9}{4}\right) = -\frac{27}{16}$

32. $\left(-\frac{1}{3}, \frac{13}{3}\right)$, (3, 11); saddle point at $f\left(-\frac{1}{3}, \frac{13}{3}\right) = -\frac{445}{27}$; relative minimum value: $f(3, 11) = -35$

33. (0, 0); relative minimum value: $f(0, 0) = 1$

34. (1, 1); relative minimum value: $f(1, 1) = \ln 2$

35. $k = \dfrac{100\, m}{c}$

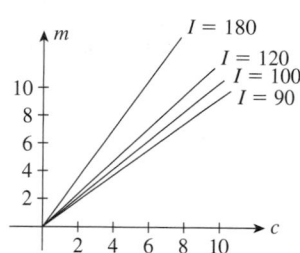

36. a. $R(x, y) = -0.02x^2 - 0.2xy - 0.05y^2 + 80x + 60y$
b. The set of all points satisfying $0.02x + 0.1y \le 80$, $0.1x + 0.05y \le 60$, $x \ge 0$, $y \ge 0$
c. 15,300; the revenue realized from the sale of 100 16-speed and 300 10-speed electric blenders is $15,300.

37. Complementary

38. The company should spend $11,000 on advertising and employ 14 agents to maximize its revenue.

39. 337.5 yd × 900 yd   **40.** 300 bass, 200 trout

Chapter 12 Before Moving On, page 940

1. All real values of x and y satisfying $x \ge 0$, $x \ne 1$, $y \ge 0$, $y \ne 2$

2.

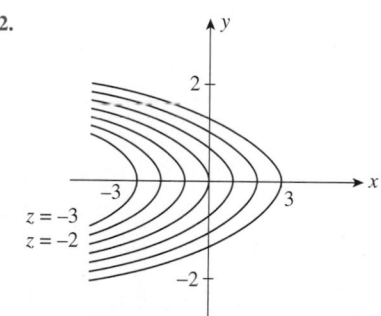

3. 14; 29; at the point (1, 2), $f(x, y)$ increases at the rate of 14 units for each unit increase in x with y held constant at a value of 2; $f(x, y)$ increases at the rate of 29 units per unit increase in y with x held fixed at 1.

4. $f_x = 2xy + ye^{xy}$; $f_{xx} = 2y + y^2e^{xy}$; $f_{xy} = 2x + (xy + 1)e^{xy} = f_{yx}$;
$f_y = x^2 + xe^{xy}$; $f_{yy} = x^2e^{xy}$

5. Relative minimum value: $f(1, 1) = -7$

INDEX

How-To Technology Index

Basic Rules of Differentiation

1. $\dfrac{d}{dx}(c) = 0$ $(c,\text{ a constant})$

2. $\dfrac{d}{dx}(u^n) = nu^{n-1}\dfrac{du}{dx}$

3. $\dfrac{d}{dx}(u \pm v) = \dfrac{du}{dx} \pm \dfrac{dv}{dx}$

4. $\dfrac{d}{dx}(cu) = c\dfrac{du}{dx}$ $(c,\text{ a constant})$

5. $\dfrac{d}{dx}(uv) = u\dfrac{dv}{dx} + v\dfrac{du}{dx}$

6. $\dfrac{d}{dx}\left(\dfrac{u}{v}\right) = \dfrac{v\dfrac{du}{dx} - u\dfrac{dv}{dx}}{v^2}$

7. $\dfrac{d}{dx}(e^u) = e^u\dfrac{du}{dx}$

8. $\dfrac{d}{dx}(\ln u) = \dfrac{1}{u}\cdot\dfrac{du}{dx}$

Basic Rules of Integration

1. $\displaystyle\int du = u + C$

2. $\displaystyle\int kf(u)\,du = k\int f(u)\,du$ $(k,\text{ a constant})$

3. $\displaystyle\int [f(u) \pm g(u)]\,du = \int f(u)\,du \pm \int g(u)\,du$

4. $\displaystyle\int u^n\,du = \dfrac{u^{n+1}}{n+1} + C$ $(n \neq -1)$

5. $\displaystyle\int e^u\,du = e^u + C$

6. $\displaystyle\int \dfrac{du}{u} = \ln|u| + C$

Formulas

Equation of a Straight Line

a. point-slope form: $y - y_1 = m(x - x_1)$
b. slope-intercept form: $y = mx + b$
c. general form: $Ax + By + C = 0$

Compound Interest

$$A = P(1 + i)^n \qquad (i = r/m, n = mt)$$

where A is the accumulated amount at the end of n conversion periods, P is the principal, r is the interest rate per year, m is the number of conversion periods per year, and t is the number of years.

Effective Rate of Interest

$$r_{\text{eff}} = \left(1 + \frac{r}{m}\right)^m - 1$$

where r_{eff} is the effective rate of interest, r is the nominal interest rate per year, and m is the number of conversion periods per year.

Future Value of an Annuity

$$S = R\left[\frac{(1 + i)^n - 1}{i}\right]$$

Present Value of an Annuity

$$P = R\left[\frac{1 - (1 + i)^{-n}}{i}\right]$$

Amortization Formula

$$R = \frac{Pi}{1 - (1 + i)^{-n}}$$

Sinking Fund Payment

$$R = \frac{iS}{(1 + i)^n - 1}$$

The Number of Permutations of n Distinct Objects Taken r at a Time

$$P(n, r) = \frac{n!}{(n - r)!}$$

The Number of Permutations of n Objects, Not All Distinct

$$\frac{n!}{n_1!\, n_2! \cdots n_m!}, \quad \text{where } n_1 + n_2 + \cdots + n_m = n$$

The Number of Combinations of n Distinct Objects Taken r at a Time

$$C(n, r) = \frac{n!}{r!\,(n - r)!}$$

The Product Rule for Probability

$$P(A \cap B) = P(A) \cdot P(B \mid A)$$

Bayes' Formula

$$P(A_i \mid E) = \frac{P(A_i) \cdot P(E \mid A_i)}{P(A_1) \cdot P(E \mid A_1) + P(A_2) \cdot P(E \mid A_2) + \cdots + P(A_n) \cdot P(E \mid A_n)}$$

Expected Value of a Random Variable

$$E(X) = x_1 p_1 + x_2 p_2 + \cdots + x_n p_n$$

List of Applications

 BUSINESS AND ECONOMICS

(continued)

Minimizing shipping costs, 369, 370, 371, 381, 423
Minimizing travel time, 798
Mobile device usage, 149, 156
Mobile instant messaging accounts, 157, 657
Model investment portfolios, 312
Money market mutual funds, 220
Money market rates, 551
Mortgage interest rates, 311, 312, 672
Movie attendance, 476, 486
Municipal bonds, 218, 272, 287
Music venues, 490
Mutual funds, 219, 262
National health-care expenditures, 658, 818
Net investment flow, 861
Net sales, 163
Netflix revenue from streaming subscribers, 590
New construction jobs, 676
Newspaper preferences of investors, 451
Newspaper subscriptions, 447
Nuclear plant utilization, 86
NYC tourists, 532
Occupancy rate, 561, 687
Oil production, 861, 897
Oil production shortfall, 873
Oil spills, 55, 686, 829
Online retail sales, 220, 896
Online sales of used autos, 164
Online video advertising, 164
Online video viewers, 150, 697
Optimal charter flight fare, 797
Optimal market price, 779
Optimal selling time, 784
Optimal speed of a truck, 797
Optimal subway fare, 791
Optimizing advertising exposure, 367, 369, 378, 381, 404
Optimizing production schedules, 370
Organizing production data, 304, 306
Out-of-pocket health care costs, 851
Outsourcing of jobs, 151, 685, 755
Packaging, 54, 94, 152, 153, 171, 576, 792, 804, 936, 937
Pension funds, 219, 220
Percentage of mobile ad revenue by device type, 558
Percentage rate of change in revenue, 710
Personal bankruptcy, 851
Personal consumption expenditure, 710
Personal loans, 245
Personnel selection, 515, 540
Petroleum consumption, 831, 867
Petroleum production, 350
Plans to keep cars, 501
Point of diminishing returns, 742
Predicting travel weather, 527
Prefab housing, 370, 404
Price-to-earnings ratio, 908
Pricing, 342, 698
Prime interest rate, 569, 625
Producers' surplus, 880, 888, 889, 891, 892, 897
Product reliability, 494, 532, 541, 545, 584
Product safety surveys, 455
Production costs, 708, 847
Production planning, 308, 322, 328, 357, 363, 370, 397
Production scheduling, 268, 273, 284, 288, 299, 350, 367, 368, 369, 380, 402, 429
Productivity of a company, 910
Productivity of a country, 910, 923
Profit of a vineyard, 53, 153, 797
Projected U.S. gasoline usage, 862
Projection TV sales, 896
Promissory notes, 291
Public debt, 734, 754
Public transportation budget deficit, 114, 755
Purchasing a home, 232, 233
Purchasing power, 220
Quality assurance surveys, 480
Quality control, 65, 479, 489, 506, 511, 515, 523, 527, 531, 535, 542, 584, 588
Rate of business failures, 697
Rate of change of cost functions, 700
Rate of change of production costs, 645
Rate comparisons, 218, 219
Real estate, 216, 220, 327, 532, 841
Recovery from the great recession, 498
Recycling, 504
Reliability of computer chips, 198, 802
Reliability of a home theater system, 532
Reliability of security systems, 532
Renewable energy, 178
Resale value, 197
Retirement benefits versus salary, 491
Retirement planning, 219, 233, 242, 243, 245, 246, 262, 507, 539, 897
Revenue of a charter yacht, 797
Revenue of competing bookstores, 801
Revenue of Moody's, 167
Revenue of a travel agency, 645
Reverse annuity mortgage, 890

Ridership, 272, 287
Risk of an airplane crash, 501
Roth IRAs, 231, 233
Safety of American-made products, 488
Salaries of married women, 820
Salary comparisons, 258, 259
Sales of disaster-recovery systems, 507
Sales forecasts, 329
Sales growth and decay, 258
Sales promotions, 697
Sales of vinyl records, 673
Satellite TV subscribers, 62
Saving for college, 220, 228
Security breaches, 490
Security system sales, 670
Selling price of DVD recorders, 685
Shoplifting, 489, 540
Shortage of nurses, 659
Shuttle bus usage, 480
Sickouts, 788
Sinking funds, 241, 249, 262, 886
Small breweries, 151
Small car market share, 735, 803
Smart meters, 734
Smartphone ownership, 552
Social media, 87, 487
Social networks, 820
Social Security beneficiaries, 631
Social Security benefits, 124
Social Security contributions, 86
Solar panel power output, 731
Solar panel production, 852
Solvency of the Social Security system, 145, 731
Spending on Medicare, 657
Spending methods, 490
Staffing, 457
Starting salaries, 584
State cigarette taxes, 754
Stock transactions, 60, 220, 318, 323, 326, 349, 480, 487
Streaming music, 698
Student loans, 245
Sub-Saharan African GDP, 150
Substitute commodities, 919, 924, 925
Sum-of-the-years' digits method of depreciation, 258
Supply and demand, 53, 68, 139, 658
Switching broadband service, 531
Switching jobs, 500
Tax planning, 229
Tax preparation, 507
Taxable equivalent yield, 36
Teen spending behavior, 499
Telecommunications industry revenue, 156
Television pilots, 531, 552
Testing new products, 477, 484, 485
TIVO owners, 156
Tour revenue, 340, 342
Transportation fatalities, 506
Transportation problem, 365
Treasury bills, 207, 218
Trust funds, 207, 219, 259
U.K. digital video viewers, 715
U.K. mobile phone video viewers, 656
U.S. airplane passenger projections, 126
U.S. crude oil production, 784
U.S. financial transactions, 127
U.S. GDP, 150
U.S. health-care expenditures, 161
U.S. health-care information technology spending, 103
U.S. income distribution for households, 506
U.S. national debt, 852
U.S. online viewers, 830
U.S. outdoor advertising, 164
U.S. public debt, 150, 156
U.S. retail e-commerce sales, 863
U.S. smartphone users, 817
Unemployment rate, 569, 672
Use of landline phones versus cell phones, 500
Using digital technology, 272, 287
Venture-capital investments, 785
Video ads, 830
Wages, 572, 640
Waiting lines, 469, 547, 552, 558, 559, 572
Walking on Mars, 500
Warehouse problem, 366, 419
Warranties, 494
Waste disposal, 864
Watching TV on smartphones, 689
Web conferencing, 155
Who pays taxes, 557
Wilson lot size formula, 910
Wind energy in China, 818
Wind power, 911
Winning bids, 532
Work habits, 506
Worker efficiency, 19, 103, 657, 768, 804, 852
Workers' expectations, 150

Workplace, 487
World production of coal, 862
Yield of an apple orchard, 153, 797
Zero coupon bonds, 220, 221

 SOCIAL SCIENCES

Academy members, 552
Accident prevention, 486
Age distribution of company directors, 585
Age distribution of renters, 541
Age distribution in a town, 587
Age of drivers in crash fatalities, 804
Age at first marriage, 802
Aging drivers, 150
Aging population, 658, 686, 697, 862
Air pollution, 686, 735, 738, 772, 783, 828, 896
Arrival times, 488
Arson for profit, 910, 924
Auto-accident rates, 542
Automobile pollution, 111
Birthrates, 581
Bringing something to the party, 557
Bus routing, 471
Car theft, 531
Checking into a hotel room, 488
Closing the gender gap in education, 102
College admissions, 86, 328, 497, 532
College savings program, 262
Committee selection, 462, 465
Commuter trends, 448
Commuting times, 542, 563, 567, 571, 590
Compliance with seat belt laws, 539
Conservation of oil, 871
Consumer surveys, 443, 448
Continuing education enrollment, 686
Conviction rates, 582
Cooking at home, 488
Correctional supervision, 489
Cost of removing toxic waste, 64, 767
Course enrollments, 499
Course selection, 470
Court judgment, 210
Crime, 300, 469, 539, 673, 730, 768, 784
Cube Rule, 103
Curbing population growth, 657
Decline of *American Idol*, 178
Decline of middle-class income, 657
Demographics, 199
Disposition of criminal cases, 490
Dissemination of information, 199
Distracted driving, 501
Distribution of families by size, 551
Distribution of incomes, 36, 536, 887, 897
Downloading music, 499
Drinking and driving among high school students, 125, 128, 130
Drivers' tests, 515
Driving age requirements, 586, 590
Education and income, 530
Educational level of voters, 530
Effect of budget cuts on crime rate, 753
Effect of smoking bans, 802
Efforts to stop shoplifting, 489
Elderly workforce, 803
Election turnout, 584
Elections, 505, 530, 537, 540, 543, 584, 734
Electricity generation, 498
Emergency fund savings, 551
Employee education and income, 530
Erosion of the middle class, 117, 126, 127
Exams, 457, 515, 551, 584
Family composition, 477, 479, 489, 531, 568
Federal libraries, 583
Fighting crime, 114
Financing a college education, 228, 262
Flight cancellations, 583
Food expenditures away from home, 552
Gender gap, 102, 540
Grade distributions, 487
Growth in income, 710
Gun owners in Congress, 541, 590
Gun-control laws, 501
Happiness with marriage, 528
Happiness score, 580
Health-care spending, 68, 112, 123, 161
Identity fraud, 582
Immigration, 196
Index of environmental quality, 801
Jail inmates, 447, 785
Jury selection, 470
Learning curves, 198, 619, 625, 672, 830
Library usage, 550